CONTENTS

Drugs: Synonyms and Properties

Second Edition

Edited by

G W A Milne

Ashgate

Aldershot • Burlington USA • Sydney

Published by
Ashgate Publishing Limited
Gower House
Croft Road
Aldershot
Hampshire GU11 3HR
England

Ashgate Publishing Company
131 Main Street
Burlington, VT 05401-5600 USA

First published 2000
Second edition 2002

British Library Cataloguing in Publication Data
Drugs: synonyms and properties. - 2nd ed.
1. Drugs - Encyclopedias 2. Drugs - Handbooks, manuals, etc.
3. Synonyms
I. Milne, G.W.A. (George William Anthony), 1937-
615.1'03

Library of Congress Control Number: 2001094289

ISBN 0 566 08491 0

Printed and bound in Great Britain by MPG Books Ltd, Bodmin, Cornwall

PREFACE

The United States Pharmacopeia contains over 14,000 drugs, each of which is a specific chemical compound with a biological activity that is medically useful. In the year since the first edition of *Drugs: Synonyms and Properties* was published, some 50 new drugs have appeared and a variety of new uses for some of the established drugs have been identified. These data are included in this new edition.

Most of these drugs have been developed commercially by the international pharmaceutical industry and, as a consequence, are associated with one or more trade names. Several names are frequently assigned to drugs, mainly for marketing purposes. Drugs are often manufactured and sold under license agreements with the original developer and acquire new names at this stage. In addition to all these non-systematic names which may be associated with a drug, it will have several systematic, chemical names, developed according to the rules of nomenclature, which allow precise identification of the chemical involved. As a result of these factors, drugs usually have a number of aliases; an average of 5-6 is typical and very commonly-used drugs such as aspirin can have well over 100 different names or synonyms. Further, when drugs are sold in combination an entirely new family of names is developed.

The abundance of names has resulted in confusion. The consumer seeking information often has only one trivial name, for example one used in an advertisement, and this name contains no information on the makeup or the use of the drug. A drug's chemical name, which defines precisely its molecular structure, is essential to the scientist; however, with only the chemical name and not the trade name, the investigator has no information on a drug's ownership, licensing status, or the practical considerations regarding its use.

Drugs: Synonyms and Properties has been compiled to bring clarity to the situation. Some 10,000 representative drugs in common use are described here. Every drug has been given an entry name and, within each drug category, entries are arranged alphabetically by that name. The entry name is the most common name for the drug, and is usually the US Adopted Name, that is, the name under which the drug is listed in the USP.

Of the 14,000-plus drugs contained in the USP, some 5,000 are in routine use. All of these, along with an additional 5,000 or so, are described in Part I. Compounds whose therapeutic activity is either trivial or obscure have been omitted from the database.

Drugs: Synonyms and Properties is organized by therapeutic category, with each category presented in alphabetical order. Drugs that have distinct activities, for example an antiarrhythmic with coronary vasodilator properties, are found in each of the appropriate sections (*Antiarrhythmics* and *Vasodilators, Coronary*). There is thus some duplication throughout Part I. Most authorities identify between 200 and 300 therapeutic categories. In this edition, some new categories such as *Ganglionic Blockers* and *Substance Abuse Inhibitors* have been added and, as in the first edition, some trivial categories have been omitted. As a result, Part I now has 204 categories. This organization allows the reader, at a glance, to determine the number of drugs available as well as the different chemical families which may be present in a specific category. It is interesting, for example, to see how the database, with 275 penicillin-based antibiotics, 367 antihypertensives and 155 tranquillizers, but only 75 drugs used to manage Parkinson's disease, tends to reflect First World society and its perceived needs. The increasing sophistication of pharmaceutical research may also be discerned in mechanistically specific categories such as ACE Inhibitors (51 drugs) or Histamine H2 Receptor Antagonists (17 drugs).

Within each record, the drug described is associated with its Chemical Abstracts Service Registry Number (CAS RN). (See Part II, CAS RN Index.) The RN is assigned by the CAS with great care to prevent association of the same number to more than one compound and has proved to be a useful unique identifier. Different chemicals can have the same molecular formula or may share a trivial name synonym, and the CAS RN is the only datum which uniquely identifies a chemical to the exclusion of all other chemicals. Each record also provides, as available, two other numeric identifiers for the chemical. These are the monograph number from the Twelfth Edition of the *Merck Index* and the European Inventory of Existing Commercial Chemical Substances (EINECS) number. (See Part II, EINECS Number Index.) The chemical name, molecular formula, and a list of trade names and synonyms are provided (see Part II, Name and Synonym Index); the physical properties of each compound are described; and the known biological activity and acute toxicity are recorded. Finally, the manufacturers and suppliers of the drug are listed. (See Part III, Manufacturer and Supplier Directory.)

Established drugs are assets which pharmaceutical companies may trade with one another and also license to third parties, exclusively or otherwise. In addition, corporate mergers, which have become increasingly common in recent years, may obscure a drug manufacturer's actual identity. For these reasons, both the owner and the supplier(s) of any specific drug are often difficult to determine with precision and, in this book, an attempt has been made to identify suppliers. The owner, if distinct from the supplier, is also identified but in a number of cases, where this is unclear, the original patent holder is listed.

HOW TO USE THIS BOOK

Drugs: Synonyms and Properties is divided into three parts. A brief description of each part is given below.

PART I: Main Entries

The main entries in this part are classified according to therapeutic class. Each category lists the most commonly used name of the drug, usually the US Adopted Name, in alphabetical order along with synonyms and other important data. Each entry is identical in structure, enabling the reader to select specific information efficiently. A unique entry number has been assigned to every entry. The three indexes in Part II allow quick cross-referencing according to the entry number in Part I by CAS Number, EINECS number, or synonym. The Manufacturer and Supplier Directory in Part III provides convenient access to information on where and how to obtain the drug of interest.

Entry Structure

A typical entry is shown on the following page. The first line contains, in boldface, the entry number (**9647**) and the name of the material (**Fluoxetine Hydrochloride**). The second line gives the Chemical Abstracts Service (CAS) Registry Number for the compound (59333-67-4), the corresponding monograph number from the Twelfth Edition of the *Merck Index* (4222) and the European Inventory of Existing Commercial Chemical Substances (EINECS) number (260-101-2). These numbers always appear in the same position (left, center, or right) enabling the reader to determine which source they belong to. Whenever CAS Registry Numbers are used in the text, they are enclosed in brackets, for example [59333-67-4]. The molecular formula and systematic chemical name of the compound are provided. A list of synonyms follows, including proprietary and trivial names.

A description of the material and its known therapeutic uses then follows. Whenever possible, physical properties are presented. These include melting point, boiling point, and optical rotation, as well as density or specific gravity, uv absorption, solubility and acute toxicity, usually limited to oral dosage in rodents. Finally, the companies who supply the product are listed.

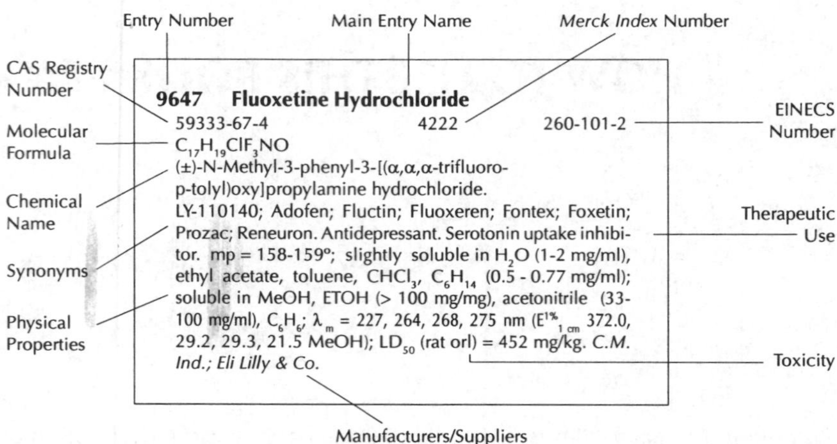

PART II: Indexes

This part contains three indexes. The purpose of each is described below:

- CAS Registry Number Index
  This index enables the reader to locate the entry number and thereby find the main entry for a drug based on its CAS Registry Number.

- EINECS Number Index
  This index enables the reader to locate the entry number and thereby find the main entry for a drug based on its EINECS number.

- Name and Synonym Index
  This is the master index containing all chemical and proprietary names found in Part I. It is the most convenient place for the reader to start if a name or synonym for a drug is known. This index enables the reader to locate the entry number in Part I which relates to the main entry for that drug.

PART III: Manufacturer and Supplier Directory

This part contains a listing of pharmaceutical manufacturers and suppliers. Arranged alphabetically by company name, this directory provides information to help the reader contact the organization directly.

GLOSSARY OF UNITS

Name	**Description**
acute toxicity	Wherever possible the units of toxicity are LD_{50}, the dose which is lethal to 50% of the test animals. In most cases, acute toxicity is measured with the rat, orally administered, and the result is reported as LD_{50} (rat orl) = 50 mg/kg. Other species (for example, mus = mouse; rbt = rabbit; pgn = pigeon; gpg = guinea pig) and occasionally the animal's sex (m = male; f = female) may be cited as might other administration routes (sc = subcutaneous; ihl = inhalation; im = intramuscular; ip = intraperitoneal; iv = intravenous). Chronic toxicity data are not given.
boiling point	When measured at atmospheric pressure, boiling points are cited with no pressure, e.g. bp = 167°. At other pressures, the pressure is also cited, e.g. $bp_{0.01}$ = 167°.
density	The density (d) of a material is dependent on its temperature. The measurement temperature is given as a superscript; thus a density of 1.123 measured at 25° will appear as d^{25} = 1.123. If the measurement is explicitly referenced to the density of water at 4°, the citation will carry both a superscript and a subscript, as in d^{25}_{4} = 1.123. Specific gravities are denoted by the abbreviation sg.

mass	Unless otherwise specified, mass is expressed in a multiple of grams (g), such as micrograms (µg; 10^{-6} g), milligrams (mg; 10^{-3} g), kilograms (kg; 10^{3} g), etc.
melting point	Melting points are cited in degrees Celsius (°C) unless otherwise specified. When melting is accompanied by decomposition, the notation (dec) is added.
optical rotation	Optical rotations (α) are cited with the measurement temperature superscripted, and the measurement wavelength (often the sodium D line at 549 nm) subscripted, as in $[\alpha]^{25}_{D}$ = 105°. When mutarotation can occur, the rotation given is an equilibrium value, measured after some time interval, which is cited, as in $[\alpha]^{25}_{D}$ = 105° (14 hr).
refractive index	Denoted by the letter n, the refractive index is usually determined at a temperature which is cited as a superscript, as in n^{25} = 1.5432. The wavelength of the light used in the measurement is cited as a subscript, as in n^{25}_{546} = 1.5432. Most commonly, the sodium D line (wavelength 549 nm) is used, and in such cases the subscript is a D, as in n^{25}_{D} = 1.5432.
temperature	When no units are cited, the temperature given is in degrees Celsius (°C).
uv absorption	The ultraviolet absorption maxima given by the material are cited in nanometers (nm = 10^{-9} m) and the absorptivities (E, A, ε, or log ε, all of which are unitless) may also be given.
volume	Volume is expressed in liters (l) or milliliters (ml) unless otherwise specified.

ABBREVIATIONS AND SYMBOLS

Terms	**Definition**
abs	absolute
abs config	absolute configuration
Ac–	acetyl (CH_3CO–)
ACE	angiotensin-converting enzyme
ACTH	adrenocorticotrophic hormone
AIDS	acquired immunodeficiency syndrome
alc	alcohol, alcoholic
amp.(s)	ampule(s)
AMP	adenosine 5′-monophosphate
aq	aqueous
atm	atmosphere, atmospheric
Bn–	benzyl (C_7H_7–)
bp	boiling point
BPH	benign prostatic hypertrophy
Bu–	butyl (C_2H_5–)
Bz–	benzoyl (C_6H_5CO–)
c	concentration (usually in g/100 ml)
C	Celsius (temperature scale)
cAMP	cyclic AMP
cc	cubic centimeters (mililiters)
CCK	cholecystokinin
CCL	*Candida* cylindrical lipase

CCl_4	carbon tetrachloride
$CHCl_3$	chloroform
CH_2Cl_2	methylene chloride
CH_3CN	acetonitrile
C_5H_5N	pyridine
C_6H_6	benzene
C_7H_8	toluene
cm	centimeter
CNS	central nervous system
CoA	coenzyme A
COD	cyclooctadiene
COMT	catechol-O-methyltransferase
d	density
d-	dextro(rotatory)
dck	duck
dec	decompose, decomposition
DIPT	diisopropyltartrate
dl-	racemic
DL-	racemic
DMA	dimethylacetamide
DMF	dimethylformamide
DMSO	dimethylsulfoxide
DNA	deoxyribonucleic acid
DOPA	dihydroxyphenylalanine
(E)-	*entgegen* (opposite)
EC	Enzyme Commission
ED	effective dose
EDTA	ethylenediamine tetraacetic acid
ee	enantiomeric equivalent
e.g.	for example
EINECS	European Inventory of Existing Commercial Chemical Substances
endo-	stereochemical descriptor
Et–	ethyl (C_2H_5–)
Et_2O	diethyl ether
EtOAc	ethyl acetate
EtOH	ethanol
exo-	stereochemical descriptor
F	Fahrenheit (temperature scale)
FMOC	fluoromethoxycarbonyl ($C_2F_3O_2$–)
g	gram(s)
g/l	grams/liter
gal	gallon(s)
GI	gastrointestinal
GLC	gas–liquid chromatography

gpg	guinea pig
gvg	gavage
HCl	hydrochloric acid
HIV	human immunodeficiency virus
H_2O	water
H_2SO_4	sulfuric acid
HMG-CoA	3-hydroxy-3-methylglutaryl coenzyme A
hmtr	hamster
hr	hour
HT	hydroxytryptamine (serotonin)
ihl	inhalation
im	intramuscular
inj	injection
ip	intraperitoneal
iPr–	isopropyl ($(CH_3)_2CH$–)
IR	infrared
iv	intravenous
kcal	kilocalories
l	liter
l-	levo(rotatory)
λ (lambda)	wavelength
LC	lethal concentration
LC_{50}	median lethal concentration
LD	lethal dose
LD_{50}	median lethal dose
log	common logarithm
LSR	lanthnide shift reagent
MAO	monoamine oxidase
max	maximum, maxima
Me–	methyl (CH_3–)
Me_2CO	acetone
MeOH	methanol
mg	milligram
min	minimum, minima, minute
MLD	minimum lethal dose
mp	melting point
μg	microgram
mμ	millimicron (nanometer)
Ms–	mesyl (CH_3O_2S–)
mus	mouse
N	normal, normality
NBD	norbornadiene
nm	nanometer (10^{-9} m)
NMO	N-methylmorpholine N-oxide
NMR	nuclear magnetic resonance

NSAID	nonsteroidal antiinflammatory drug
NSC	National Service Center (of the National Cancer Institute)
NTP	normal temperature, pressure
o-	ortho
OD	optical density
orl	oral
p-	para
pgn	pigeon
Ph–	phenyl (C_6H_5–)
pH	acid–base scale (log of reciprocal hydrogen ion concentration)
pK	log of the reciprocal of the dissociation constant
PLE	pig liver esterase
pOH	acid-base scale (log of reciprocal hydroxyl ion concentration)
ppb	parts-per-billion
PPL	porcine pancreatic lipase
ppm	parts-per-million
Pr–	propyl (C_3H_7–)
(R)	rectus (stereochemical descriptor)
rbt	rabbit
RNA	ribonucleic acid
(S)	sinister (stereochemical descriptor)
S-	symmetical
sc	subcutaneous
SDM	site directed mutagenesis
sec	second
sec-	secondary
SG, sg	specific gravity
sp.	species (specific)
spp.	species (plural)
STP	standard temperature, pressure
tabl.	tablet
TBHP	tert-butyl hydroperoxide
temp	temperature
tert-	tertiary
THF	tetrahydrofuran
THP	tetrahydropyranyl (C_5H_7O–)
Ts–	tosyl ($C_7H_7O_7S$–)
TSCA	Toxic Substances Control Act
UK	United Kingdom
USA	United States of America
USAN	United States Adopted Names
USP	United States Pharmacopeia
uv	ultraviolet
v/v	volume in volume

VIS	visible
viz.	namely
w/v	weight in volume
w/w	weight in weight
wt	weight
(Z)-	*zusammen* (on the same side)
>	greater than
<	less than
~	approximately
Å	Angstrom units (10^{-8} cm)

TABLE OF THERAPEUTIC CATEGORIES

PART I

MAIN ENTRIES

Abortifacients

1 Epostane
80471-63-2 3664
$C_{22}H_{31}NO_3$
4α,5-Epoxy-3,17β-dihydroxy-4,17-dimethyl-5α-androst-2-ene-2-carbonitrile.
Win-32729. Abortifacient. mp = 191-194°; $[\alpha]_D^{25}$ = 67.4° (c = 1 C_5H_5N). *Sterling Winthrop Inc.*

2 Gemeprost
64318-79-2 4393 264-775-9
$C_{23}H_{38}O_5$
Methyl (E)-7-[(1R,2R,3R)-3-hydroxy-2-[(E)-(3R)-3-hydroxy-4,4-dimethyl-1-octenyl]-5-oxocyclopentyl]-2-heptenoate.
16,16-dimethyl-trans-Δ^2-PGE_1 methyl ester; ONO-802; SC-37681; Cergem; Cervagem; Cervageme; Preglandin. Abortifacient; oxytocic; prostaglandin. Analog of prostaglandin E_1. Used in termination of first trimester pregnancy. *Ono Pharm.; Searle G.D. & Co.*

3 Lilopristone
97747-88-1
$C_{22}H_{36}O_5$
11β-[p-(Dimethylamino)phenyl]-17β-hydroxy-17-[(Z)-3-hydroxypropenyl]estra-4,9-diene-3-one.
13β-configured (type II) progestin antagonist. Antiprogestin; antigestagin; abortifacient.

4 Mifepristone
84371-65-3 6273
$C_{29}H_{35}NO_2$
11β-[p-(Dimethylamino)phenyl]-17β-hydroxy-17-(1-propynyl)estra-4,9-dien-3-one.
RU-38486; RU-486; Mifegyne. Abortifacient. Used as a post-coital contraceptive. mp = 150°; $[\alpha]_D^{20}$ = 138.5° (c = 0.5 $CHCl_3$). *Roussel-UCLAF.*

5 Prostaglandin E_2
363-24-6 8064 206-656-6
$C_{20}H_{32}O_5$
(5Z,11α,13E,15S)-11,15-Dihydroxy-9-oxoprosta-5,13-dien-1-oic acid.
dinoprostone; PGE_2; U-12062; Cerviprost; Dinoprost; Enzaprost F; Glandin; Minprositin E_2; Prepidil; Propess; Prostarmon-E; Prostin E_2. Abortifacient; oxytocic. Prostaglandin. mp = 66-68°; $[\alpha]_D^{26}$ = -61° (c = 1 THF). *Pharmacia & Upjohn.*

6 Prostaglandin $F_{2\alpha_a}$
551-11-1 8065
$C_{20}H_{34}O_5$
(E,Z)-(1R,2R,3R,5S)-7-[3,5-Dihydroxy-2-[(3S)-(3-hydroxy-1-octenyl)]cyclopentyl]-5-heptenoic acid.
dinoprost; $PGF_{2\alpha}$; U-14583; Enzaprost F; Glandin; Prostarmon F. Abortifacient; oxytocic. Prostaglandin. mp = 25-35°; $[\alpha]_D^{25}$ = 23.5° (c = 1 THF); freely soluble in MeOH, EtOH, EtOAc, $CHCl_3$; slightly soluble in H_2O; LD_{50} (rbt iv) = 2.5-5.0 mg/kg, (rbt im) = 2.5-5.0 mg/kg. *Pharmacia & Upjohn.*

7 Prostaglandin F_{2a} Tromethamine Salt
38562-01-5 8065 254-002-3
$C_{24}H_{45}NO_8$
(5Z,9α,11α,13E,15S)-9,11,15-Trihydroxyprosta-5,13-dien-1-oic acid salt with 2-amino-2-hydroxymethyl-1,3-propanediol.
$PGF_{2\alpha}$ THAM; U-14583E; Lutalyse; Prostin F_2 Alpha. Abortifacient; oxytocic. Prostaglandin. mp = 100-101°; soluble in H_2O (> 20 g/100 ml). *Pharmacia & Upjohn.*

8 Sulprostone
60325-46-4 9165 262-173-0
$C_{23}H_{31}NO_7S$
(Z)-7-[(1R,2R,3R)-3-Hydroxy-2-[(E)-(3R)-3-hydroxy-4-phenoxy-1-butenyl)]-5-oxocyclopentyl]-N-(methylsulfonyl)-5-heptenamide.
CP-34089; SHB-286; ZK-57671; Nalador. Prostaglandin. Abortifacient. Analog of prostaglandin E_2. Uterine stimulant. Colorless oil. *Pfizer Intl.; Schering AG.*

9 Trichosanthin
60318-52-7 9780
GLQ-223.
Tian Hua Fen. Abortifacient. A basic polypeptide of 224 units. The active component of the traditional Chinese medicine, Tian Hua Fen (*Radix trichosanthis*). Crystalline solid.

ACE Inhibitors

10 Alacepril
74258-86-9 202
$C_{20}H_{26}N_2O_5S$
N-[1-[(S)-3-Mercapto-2-methylpropionyl]-L-prolyl]-3-phenyl-L-alanine acetate (ester).
DU-1219; Cetapril. Angiotensin-converting enzyme inhibitor. Used to treat hypertension. mp= 155-156°; $[\alpha]_D^{25}$ = -81.3° (c = 1.02 EtOH); LD_{50} (rat orl) > 5000 mg/kg, (rat sc) > 3000 mg/kg, (rat ip) ≅ 2000 mg/kg, (mus orl) > 5000 mg/kg, (mus sc) > 3000 mg/kg, (mus ip) ≅ 3000 mg/kg. *Dainippon Pharm.*

11 Benazepril
86541-75-5 1058
$C_{24}H_{28}N_2O_5$
(3S)-3-[[(1S)-1-Carboxy-3-phenylpropyl]amino]-2,3,4,5-tetrahydro-2-oxo-1H-1-benzazepine-1-acetic acid 3-ethyl ester.
CGS-14824A; Briem; Cibacen; Cibacène; Lotensin. Angiotensin-converting enzyme inhibitor. Used to treat hypertension. Orally active peptidyldipeptide hydrolase inhibitor. mp= 148-149°; $[\alpha]_D$ = -159° (c = 1.2 EtOH). *Ciba-Geigy Corp.*

12 Benazepril Hydrochloride
86541-74-4 1058
$C_{24}H_{29}ClN_2O_5$
(3S)-3-[[(1S)-1-Carboxy-3-phenylpropyl]amino]-2,3,4,5-tetrahydro-2-oxo-1H-1-benzazepine-1-acetic acid 3-ethyl ester hydrochloride.
Lotensin; CGS-14824A HCl; component of: Lotensin-HCT, Lotrel, Lotrel capsules. Angiotensin-converting enzyme inhibitor. Used to treat hypertension.

Orally active peptidyldipeptide hydrolase inhibitor. mp = 188-190°; $[\alpha]_D$ = -141° (c = 0.9 EtOH). *Ciba-Geigy Corp.*

13 Benazeprilate
86541-78-8 1058
$C_{22}H_{24}N_2O_5$
(3S)-3-[[(1S)-1-Carboxy-3-phenylpropyl]amino]-2,3,4,5-tetrahydro-2-oxo-1H-1-benzazepine-1-acetic acid.
CGS-14831. Angiotensin-converting enzyme inhibitor. Used to treat hypertension. mp = 270-272°; $[\alpha]_D$ = -200.5° (c = 1 in 3% aq. NaOH). *Ciba-Geigy Corp.*

14 Captopril
62571-86-2 1817 263-607-1
$C_9H_{15}NO_3S$
1-[(2S)-3-Mercapto-2-methylpropionyl]-L-proline.
Capoten; SQ-14225; Acediur; Acepril; Aceplus; Alopresin; Acepress; Capoten; Captolane; Captoril; Cesplon; Dilabar; Garranil; Hipertil; Lopirin; Lopril; Tensobon; Tensoprel; component of: Capozide, Acezide, Captea, Ecazide. Angiotensin-converting enzyme inhibitor. Used to treat hypertension. Orally active peptidyldipeptide hydrolase inhibitor. mp = 103-104°, 86°, 87-88°, 104-105°; $[\alpha]_D^{22}$ = -131.0° (c = 1.7 EtOH); freely soluble in H_2O, EtOH, $CHCl_3$, CH_2Cl_2; LD_{50} (mus iv) = 1040 mg/kg, (mus orl) = 6000 mg/kg. *Apothecon; Bristol-Myers Squibb Co; Squibb E.R. & Sons.*

15 Ceronapril
111223-26-8 2038
$C_{21}H_{33}N_2O_6P$
1-[(2S)-6-Amino-2-hydroxyhexanoyl]-L-proline hydrogen (4-phenylbutyl)phosphonate (ester).
SQ-29852; Ceranapril. Angiotensin-converting enzyme inhibitor. Used to treat hypertension. mp = 190-195°; $[\alpha]_D$ = -47.5° (c = 1 MeOH). *Bristol-Myers Squibb Co.*

16 Cilazapril
92077-78-6 2332
$C_{22}H_{31}N_3O_5.H_2O$
(1S,9S)-9-[[(S)-1-Carboxy-3-phenylpropyl]amino]-octahydro-10-oxo-6H-pyridazino[1,2-a][1,2]diazepine-1-carboxylic acid 9-ethyl ester monohydrate.
Inhibace; Ro-31/2848/006; Dynorm; Initiss; Justor; Vascase. Angiotensin-converting enzyme inhibitor. Used to treat hypertension. mp = 95-97°; $[\alpha]_D^{20}$= -62.51° (c = 1 EtOH). *Hoffmann-LaRoche Inc.*

17 Cilazaprilat
90139-06-3 2332
$C_{20}H_{27}N_{35}$
N-(1S,9S)-1-Carboxy-10-oxoperhydropyridazino[1,2-α][1,2]diazepine-9-yl-4-phenyl-L-homoalanine.
Ro-31-3113. Angiotensin-converting enzyme inhibitor. Used to treat hypertension. mp = 242°; $[\alpha]_D^{20}$= -74.7° (c = 0.5 0.1M NaOH). *Hoechst Roussel Pharm. Inc.*

18 Delapril
83435-66-9 2928
$C_{26}H_{32}N_2O_5$
Ethyl(S)-2-[[(S)-1-[(carboxymethyl)-2-indanylcarbamoyl] ethyl]amino]-4-phenylbutyrate.
Alindapril; Indalapril. Angiotensin-converting enzyme inhibitor. Used to treat hypertension. *Takeda Chem. Ind. Ltd.*

19 Delapril Hydrochloride
83435-67-0 2928
$C_{26}H_{33}ClN_2O_5$
Ethyl (S)-2-[[(S)-1-[(carboxymethyl)-2-indanylcarbamoyl]-ethyl]amino]-4-phenylbutyrate monohydrochloride.
REV-6000A; CV-3317; Adecut; Cupressin. Angiotensin-converting enzyme inhibitor. Used to treat hypertension. mp = 166-170°; $[\alpha]_D^{22}$ = 18.5° (c = 1 MeOH). *Takeda Chem. Ind. Ltd.*

20 Enalapril
75847-73-3 3605
$C_{20}H_{28}N_2O_5$
1-[N-[(S)-1-Carboxy-3-phenylpropyl]-L-alanyl]-L-proline 1'-ethyl ester.
Angiotensin-converting enzyme inhibitor. Used to treat hypertension. Orally active peptidyldipeptide hydrolase inhibitor. *Merck & Co.Inc.*

21 Enalapril Maleate
76095-16-4 3605 278-375-7
$C_{24}H_{32}N_2O_9$
1-[N-[(S)-1-Carboxy-3-phenylpropyl]-L-alanyl]-L-proline 1'-ethyl ester maleate (1:1).
Enacard; Renitec; Vasotec; MK-421; Amprace; Bitensil; Cardiovet; Enaloc; Enapren; Glioten; Hipoartel; Innovace; Lotrial; Olivin; Pres; Reniten; Renivace; Xanef; component of: Vaseretic, Acesistem, Co-Renitec, Innozide, Renacor, Xynertec. Angiotensin-converting enzyme inhibitor. Used to treat hypertension. mp = 143-144.5°; $[\alpha]_D^{25}$= -42.2° (c = 1 MeOH); soluble in H_2O (2.5 g/100 ml), EtOH (8 g/100 ml), MeOH (20 g/100 ml). *Merck & Co.Inc.*

22 Enalaprilat
84680-54-6 3606 278-459-3
$C_{18}H_{24}N_2O_5.2H_2O$
1-[N-[(S)-1-Carboxy-3-phenylpropyl]-L-alanyl]-L-proline dihydrate.
enalaprilic acid; MK-422; Vasotec Injection; Vasotec IV. Angiotensin-converting enzyme inhibitor. Used to treat hypertension. mp = 148-151°; $[\alpha]_D$ = -67.0° (0.1M HCl). *Merck & Co.Inc.*

23 Enalaprilat [anhydrous]
76420-72-9 3606
$C_{18}H_{24}N_2O_5$
1-[N-[(S)-1-Carboxy-3-phenylpropyl]-L-alanyl]-L-proline.
enalaprilic acid. Angiotensin-converting enzyme inhibitor. Active metabolite of enalipril. Used to treat hypertension. *Merck & Co.Inc.*

24 Fasidotril
135038-57-2
$C_{23}H_{25}NO_6S$
N-[(S)-α-(Mercaptomethyl)-3,4-(methylenedioxy)hydrocinnamoyl]-L-alanine benzyl ester acetate (ester).
An ACE inhibitor.

25 Fosinopril
98048-97-6 4282
$C_{30}H_{46}NO_7P$
(4S)-4-Cyclohexyl-1-[[(R)-[(S)-1-hydroxy-2-methylpropoxy](4-phenylbutyl)phosphinyl]acetyl-L-proline propionate (ester).
fosenopril. Angiotensin-converting enzyme inhibitor. Used to treat hypertension. [diacid (SQ-27519)]: mp = 149-153°; $[\alpha]_D$= -24° (c = 1 MeOH). *Squibb E.R. & Sons.*

26 Fosinopril Sodium
88889-14-9 4282
$C_{30}H_{45}NNaO_7P$
(4S)-4-Cyclohexyl-1-[[(R)-[(S)-1-hydroxy-2-methylpropoxy](4-phenylbutyl)phosphinyl]acetyl-L-proline propionate (ester) sodium salt.
SQ-28555; Monopril; Acecor; Secorvas; Staril. Angiotensin-converting enzyme inhibitor. Used to treat hypertension. *Mead Johnson Labs.; Mead Johnson Pharmaceuticals.*

27 Goralatide
120081-14-3
$C_{20}H_{33}N_5O_9$
1-[N^2-[N-(N-acetyl-L-seryl)-L-α-aspartyl]-L-lysyl]-L-proline.
Angiotensin-converting enzyme inhibitor.

28 Idrapril
127420-24-0
$C_{11}H_{18}N_2O_5$
(1S,2R)-2-[[(Hydroxycarbamoyl)methyl]methylcarbamoyl]-cyclohexanecarboxylic acid.
Angiotensin-converting enzyme inhibitor.

29 Imidapril
89371-37-9 4947
$C_{20}H_{27}N_3O_6$
(S)-3-(N-[(S)-1-Ethoxycarbonyl-3-phenylpropyl]-L-alanyl)-1-methyl-2-oxoimidazoline-4-carboxylic acid.
Angiotensin-converting enzyme inhibitor. Used to treat hypertension. mp = 139-140°; $[\alpha]_D^{20}$ = -71.7° (c = 0.5 EtOH). *Tanabe Seiyaku Co. Ltd.*

30 Imidapril Hydrochloride
89396-94-1 4947
$C_{20}H_{28}ClN_3O_6$
(S)-3-(N-[(S)-1-Ethoxycarbonyl-3-phenylpropyl]-L-alanyl)-1-methyl-2-oxoimidazoline-4-carboxylic acid monohydrochloride.
TA-6366; Novaloc; Tanapril. Angiotensin-converting enzyme inhibitor. Used to treat hypertension. mp = 214-216° (dec); $[\alpha]_D^{20}$= -64.1° (c = 0.5 EtOH). *Tanabe Seiyaku Co. Ltd.*

31 Imidaprilat
89371-44-8 4947
$C_{18}H_{23}N_3O_6$
(S)-3-(N-[(S)-1-Carboxyl-3-phenylpropyl]-L-alanyl)-1-methyl-2-oxoimidazoline-4-carboxylic acid.
imidaprilate. Angiotensin-converting enzyme inhibitor. Used to treat hypertension. mp = 239-241°; $[\alpha]_D^{19}$= -88.4° (c = 1 5% $NaHCO_3$). *Tanabe Seiyaku Co. Ltd.*

32 Libenzapril
109214-55-3
$C_{18}H_{25}N_3O_5$
N-[(3S)-1-(Carboxymethyl)-2,3,4,5-tetrahydro-2-oxo-1H-1-benzazepin-3-yl]-L-lysine.
CGS-16617. Angiotensin-converting enzyme inhibitor. Used to treat hypertension. *Ciba-Geigy Corp.*

33 Lisinopril
83915-83-7 5540
$C_{21}H_{31}N_3O_5.2H_2O$
1-[N^2[(S)-1-Carboxy-3-phenylpropyl]-L-lysyl]-L-proline dihydrate.
MK-521; RS-10029; Acerbon; Alapril; Carace; Coric; Novatec; Prinil; Prinivil; Tensopril; Vivatec; Zestril; component of: Prinzide. Antihypertensive. Angiotensin-converting enzyme inhibitor. Orally active peptidyldipeptide hydrolase inhibitor. λ_m = 246, 254, 258, 261, 267 nm ($A_{1\ cm}^{1\%}$ 4.0, 4.5, 5.1, 5.1, 3.7 0.1N NaOH), 246 253, 258, 264, 267 nm ($A_{1\ cm}^{1\%}$ 3.2, 3.9, 4.5, 3.0, 2.8 0.1NHCl); $[\alpha]_{405}^{25}$ = -120° (c = 1 0.25M $Zn(OAc)_2$ pH 6.4). *Merck & Co.Inc.*

34 Moexipril
103775-10-6
$C_{27}H_{34}N_2O_7$
(3S)-2-[(2S)-N-[(1S)-1-Carboxy-3-phenylpropyl]-L-alanyl]-1,2,3,4-tetrahydro-6,7-dimethoxy-3-isoquinolinecarboxylic acid 2-ethyl ester.
RS-10029. Angiotensin-converting enzyme inhibitor. Used to treat hypertension. *Schwarz Pharma Kremers Urban Co.*

35 Moexipril Hydrochloride
82586-52-5
$C_{27}H_{35}ClN_2O_7$
(3S)-2-[(2S)-N-[(1S)-1-Carboxy-3-phenylpropyl]-L-alanyl]-1,2,3,4-tetrahydro-6,7-dimethoxy-3-isoquinolinecarboxylic acid 2-ethyl ester hydrochloride.
Univasc; SPM-925; CI 925; RS-10085-197. Angiotensin-converting enzyme inhibitor. Used to treat hypertension. *Schwarz Pharma Kremers Urban Co.*

36 Moexiprilat
103775-14-0
$C_{25}H_{30}N_2O_7$
(3S)-2-[(2S)-N-[(1S)-1-Carboxy-3-phenylpropyl]-alanyl]-1,2,3,4-tetrahydro-6,7-dimethoxy-3-isoquinolinecarboxylic acid.
Angiotensin-converting enzyme inhibitor. Used to treat hypertension. *Schwarz Pharma Kremers Urban Co.*

37 Moveltipril
85856-54-8 6368
$C_{19}H_{30}N_2O_5S$
(-)-1-[(2S)-3-Mercapto-2-methylpropionyl]-L-proline ester with N-(cyclohexyl-carbonyl)thio-D-alanine.
altiopril. Angiotensin-converting enzyme inhibitor. Used to treat hypertension. mp = 113-116°; $[\alpha]_D$ = 14.2° (c = 1.05 MeOH). *Chugai Pharm. Co. Ltd.*

38 Moveltipril Calcium
85921-53-5 6368
$C_{38}H_{58}CaN_4O_{10}S_2$
(-)-1-[(2S)-3-Mercapto-2-methylpropionyl]-L-proline ester with N-(cyclohexylcarbonyl)thio-D-alanine calcium salt (2:1).
MC-838; Lowpres. Angiotensin-converting enzyme inhibitor. Used to treat hypertension. mp ≅ 190°; $[\alpha]_D^{20}$= -48° to -52° (c = 1 MeOH); very soluble in H_2O, MeOH; soluble in EtOH, $CHCl_3$; insoluble in Me_2CO, EtOAc; LD_{50} (mmus orl) > 10.0 g/kg, (mmus ip) = 2.1 g/kg, (mmus sc) = 3.0 g/kg, (fmus orl) > 10 g/kg, (fmus ip) = 2.3 g/kg, (fmus sc) = 3.8 g/kg, (mrat orl) > 10.0 g/kg, (mrat ip) = 1.3 g/kg, (mrat sc) = 3.4 g/kg, (frat orl) > 10.0 g/kg, (frat ip) = 1.3 g/kg, (frat sc) = 3.9 g/kg, (mdog orl) > 6.0 g/kg, (fdog orl) > 6.0 g/kg. *Chugai Pharm. Co. Ltd.*

39 Pentopril
82924-03-6
$C_{18}H_{23}NO_5$
Ethyl (αR,σR,2S)-2-carboxy-α,σ-dimethyl-δ-oxo-1-indoline valerate.
CGS-13945. Angiotensin-converting enzyme inhibitor. Used to treat hypertension. *Ciba-Geigy Corp.*

40 Perindopril
82834-16-0
$C_{19}H_{32}N_2O_5$
(2S,3aS,7aS)-1-[(S)-N-[(S)-1-Carboxybutyl]alanyl]-hexahydro-2-indoline carboxylic acid 1-ethyl ester.
S-9490; McN-A-2833; Coversyl. Ace inhibitor. *Ortho-McNeill; Solvay Pharm. Inc.*

41 Perindopril tert-Butylamine
107133-36-8 7311
$C_{23}H_{43}N_3O_5$
(2S,3aS,7aS)-1-[(S)-N-[(S)-1-Carboxybutyl]alanyl]-hexahydro-2-indolinecarboxylic acid tert-butylamine salt.
perindopril ebumine; perinodpril erbimune; S-9490-3; McN-A-2833-109; Aceon; Coversum; Coversyl; Procaptan. Angiotensin-converting enzyme inhibitor. Used to treat hypertension. *McNeil Pharm.*

42 Quinapril
85441-61-8 8233
$C_{25}H_{30}N_2O_5$
(S)-2-[(S)-N-[(S)-1-Carboxy-3-phenylpropyl]alanyl]-1,2,3,4-tetrahydro-3-isoquinolinecarboxylic acid 1-ethyl ester.
component of: Accuretic; Acequide; Koretic. Antihypertensive. Angiotensin-converting enzyme inhibitor. Orally active peptidyldipeptide hydrolase inhibitor. *Parke-Davis.*

43 Quinapril Hydrochloride
82586-55-8 8233
$C_{25}H_{31}ClN_2O_5$
(S)-2-[(S)-N-[(S)-1-Carboxy-3-phenylpropyl]alanyl]-1,2,3,4-tetrahydro-3-isoquinolinecarboxylic acid 1-ethyl ester monohydrochloride.
CI-906; PD-109452-2; Accupril; Accuprin; Accupro; Acequin; Acuitel; Korec; Quinazil; component of: Accuretic; Acequide; Korectic. Angiotensin-converting enzyme inhibitor. Used to treat hypertension. mp = 120-130°, 119-121.5°; $[\alpha]_D^{23}$ = 14.5° (c = 1.2 EtOH), $[\alpha]_D^{25}$ = 15.4° (c = 2.0 MeOH); LD_{50} (mmus orl) = 1739 mg/kg, (mmus iv) = 504 mg/kg, (fmus orl) = 1840 mg/kg, (fmus iv) = 523 mg/kg, (mrat orl) = 4280 mg/kg, (mrat iv) = 158 mg/kg, (frat orl) = 3541 mg/kg, (frat iv) = 107 mg/kg. *Parke-Davis.*

44 Quinaprilat
82768-85-2 8233
$C_{23}H_{26}N_2O_5$
(S)-2-[(S)-N-[(S)-1-Carboxy-3-phenylpropyl]alanyl]-1,2,3,4-tetrahydro-3-isoquinolinecarboxylic acid .
CI-928. Angiotensin-converting enzyme inhibitor. Used to treat hypertension. mp = 166-168°; $[\alpha]_D^{23}$= 20.9° (c = 1 MeOH). *Parke-Davis.*

45 Ramipril
87333-19-5 8283
$C_{23}H_{32}N_2O_5$
(2S,3aS,6aS)-1-[(S)-N-[(S)-1-Carboxy-3-phenylpropyl]alanyl]octahydrocyclopenta-[b]-pyrrole-2-carboxylic acid.
Altace; HOE-498; Cardace; Delix; Pramace; Quark; Ramace; Triatec; Tritace; Unipril; Vesdil. Angiotensin-converting enzyme inhibitor. Used to treat hypertension. Orally active peptidyldipeptide hydrolase inhibitor. mp = 109°; $[\alpha]_D^{24}$= 33.2° (c= 1 0.1N ethanolic HCl); LD_{50} (mmus iv) = 1194 mg/kg, (mmus orl) = 10933 mg/kg, (fmus iv) = 1158 mg/kg, (fmus orl) = 10048 mg/kg, (mrat iv) = 687 mg/kg, (mrat orl) > 10000 mg/kg, (frat iv) = 608 mg/kg, (frat orl) > 10000 mg/kg. *Hoechst Roussel Pharm. Inc.*

46 Sampatrilat
129981-36-8
$C_{26}H_{40}N_4O_9S$
N-[[1-[(S)-3-[(S)-6-Amino-2-methanesulfonamidohexanamido]-2-carboxypropyl]cyclopentyl]carbonyl]-L-tyrosine.
A novel dual inhibitor of both angiotensin-converting enzyme (ACE) and neutral endopeptidase (NEP). Antihypertensive.

47 Spirapril
83647-97-6 8905
$C_{22}H_{30}N_2O_5S_2.1/2H_2O$
(8S)-7-[(S)-N-[(S)-1-Carboxy-3-phenylpropyl]-alanyl]-1,4-dithia-7-azaspiro[4.4]nonane-8-carboxylic acid 1-ethyl ester hemihydrate.
Angiotensin-converting enzyme inhibitor. Used to treat hypertension. $[\alpha]_D^{26}$ = -29.5° (c = 0.2 EtOH). *Schering-Plough HealthCare Products.*

48 Spirapril Hydrochloride
94841-17-5 8905
$C_{22}H_{31}ClN_2O_5S_2$
(8S)-7-[(S)-N-[(S)-1-Carboxy-3-phenylpropyl]-alanyl]-1,4-dithia-7-azaspiro[4.4]nonane-8-carboxylic acid 1-ethyl ester monohydrochloride.
Sch-33844; TI-211-950; Renormax; Renpress; Sandopril. Angiotensin-converting enzyme inhibitor. Used to treat hypertension. mp = 192-194° (dec); $[\alpha]_D^{26}$= -11.2° (c = 0.4 EtOH). *Schering-Plough HealthCare Products.*

49 **Spiraprilat**
83602-05-5 8905
$C_{20}H_{26}N_2O_5S_2$
(8S)-7-[(S)-N-[(S)-1-Carboxy-3-phenylpropyl]alanyl]-1,4-dithia-7-azaspiro[4.4]nonane-8-carboxylic acid.
Sch-33861; spiraprilic acid. Angiotensin-converting enzyme inhibitor. Used to treat hypertension. mp = 163-165° (dec); $[\alpha]_D^{26}$ = 4.1° (c = 0.4 EtOH). *Schering-Plough HealthCare Products.*

50 **Temocapril**
111902-57-9 9287
$C_{23}H_{28}N_2O_5S_2$
(+)-(2S,6R)-6-[[(1S)-1-Carboxy-3-phenylpropyl]amino]tetrahydro-5-oxo-2-(2-thienyl)-1,4-thiazepine-4(5H)-acetic acid 6-ethyl ester.
Angiotensin-converting enzyme inhibitor. Used to treat hypertension. mp =168°; $[\alpha]^{23}$ = 40° (c = 1.1 DMF). *Sankyo Co. Ltd.*

51 **Temocapril Hydrochloride**
110221-44-8 9287
$C_{23}H_{29}ClN_2O_5S_2$
(+)-(2S,6R)-6-[[(1S)-1-Carboxy-3-phenylpropyl]amino]tetrahydro-5-oxo-2-(2-thienyl)-1,4-thiazepine-4(5H)-acetic acid 6-ethyl ester monohydrochloride.
CS-622. Angiotensin-converting enzyme inhibitor. Used to treat hypertension. mp= 187° (dec); $[\alpha]_D^{25}$ = 47.7° (c = 1 DMF); LD_{50} (mus orl) > 5000 mg/kg, (rat orl) > 5000 mg/kg, (dog orl) > 800 mg/kg. *Sankyo Co. Ltd.*

52 **Temocaprilate**
110221-53-9 9287
$C_{21}H_{24}N_2O_5S_2$
(+)-(2S,6R)-6-[[(1S)-1-carboxy-3-phenylpropyl]amino]tetrahydro-5-oxo-2-(2-thienyl)-1,4-thiazepine-4(5H)-acetic acid.
temocaprilat; RS-5139. Angiotensin-converting enzyme inhibitor. Used to treat hypertension. mp = 246° (dec); $[\alpha]_D^{25}$ = 63.4° (c = 1 DMF). *Sankyo Co. Ltd.*

53 **Teprotide**
35115-60-7
$C_{53}H_{76}N_{14}O_{12}$
5-Oxo-L-prolyl-L-tryptophyl-L-prolyl-L-arginyl-L-prolyl-L-glutaminyl-L-isoleucyl-L-prolyl-L-proline.
SQ-20881. Angiotensin-converting enzyme inhibitor. Used to treat hypertension. *Bristol-Myers Squibb Co.*

54 **Trandolapril**
87679-37-6 9703
$C_{24}H_{34}N_{2}O_5$
(2S,3aR,7aS)-1-[(S)-N-[(S)-1-Carboxy-3-phenylpropyl]alanyl]hexahydro-2-indolinecarboxylic acid 1-ethyl ester.
RU-44570; Odrik; Gopten. Angiotensin-converting enzyme inhibitor. Used to treat hypertension. *Roussel-UCLAF.*

55 **Trandolaprilate**
87679-71-8 9703
$C_{22}H_{30}N_{2}O_5$
(2S,3aR,7aS)-1-[(S)-N-[(S)-1-Carboxy-3-phenylpropyl]-alanyl]hexahydro 2-indolinecarboxylic acid.
RU-44403; trandolaprilat. Angiotensin-converting enzyme inhibitor. Used to treat hypertension. *Roussel-UCLAF.*

56 **Utibapril**
109683-61-6
$C_{22}H_{31}N_3O_5S$
(S)-2-tert-Butyl-4-[(S)-N-[(S)-1-Carboxy-3-phenylpropyl]alanyl]-Δ^2-1,3,4-thiadiazoline-5-carboxylic acid 4-ethyl ester.
FPL-63547. Angiotensin-converting enzyme (ACE) inhibitor with a proposed tissue-specific inhibitory profile.

57 **Utibaprilat**
109683-79-6
$C_{20}H_{27}N_3O_5S$
(S)-2-tert-Butyl-4-[(S)-N-[(S)-1-carboxy-3-phenylpropyl]alanyl]-Δ^2-1,3,4-thiadiazoline-5-carboxylic acid.
Angiotensin-converting enzyme (ACE) inhibitor with a proposed tissue-specific inhibitory profile.

58 **Zabicipril**
83059-56-7
$C_{23}H_{32}N_2O_5$
(3S)-2-[(2S)-N-[(1S)-1-Carboxy-2-phenylpropyl]-alanyl]-2-azabicyclo[2.2.2]octane-3-carboxylic acid 1-ethyl ester.
S-9650. Prodrug of zabiciprilat. Angiotensin I-converting enzyme inhibitor.

59 **Zabiciprilat**
90103-92-7
$C_{21}H_{28}N_2O_5$
(S)-2-[(S)-N-[(S)-1-Carboxy-2-phenylpropyl]-alanyl]-2-azabicyclo[2.2.2]octane-3-carboxylic acid.
S-10211. Active metabolite of zabicipril. Angiotensin I-converting enzyme inhibitor.

60 **Zofenopril Calcium**
81938-43-4
$C_{44}H_{44}CaN_2O_8S_4$
(4S)-N-[(s)-3-Mercapto-2-methylpropionyl]-4-(phenylthio)-L-proline benzoate (ester) calcium salt.
SQ-26991; Zoprace. Angiotensin-converting enzyme inhibitor. Used to treat hypertension. *Bristol-Myers Squibb Co.*

Adrenocortical Suppressants

61 **Aminoglutethimide**
125-84-8 460 204-756-4
$C_{13}H_{16}N_2O_2$
2-(p-Aminophenyl)-2-ethylglutarimide.
p-Aminoglutethimide; Ba-16038; NSC-330915; Cytadren; Elipten; Orimeten. Adrenocortical suppressant; anticonvulsant; antineoplastic agent. Aromatase inhibitor. Has been used in treatment of Cushing's syndrome and other adrenal hormone disorders and in palliative treatment of breast cancer. mp = 149-150°; insoluble in H_2O; poorly soluble in EtOAc, EtOH; readily soluble in other organic solvents; [hydrochloride]: mp = 223-225°, soluble in H_2O. *Ciba-Geigy Corp.*

62 Trilostane
13647-35-3 9827 237-133-0
$C_{20}H_{27}NO_3$
4α,5α-Epoxy-3,17β-dihydroxyandrost-2-ene-2-carebonitrile.
Win-24540; Desopan; Modrastane; Modrenal. Adrenocortical suppressan; used in the treatment of breast cancer. In clinical trials prior to 1983. mp = 257.8-270°; $[\alpha]_D^{25}$ = 137.4° (c = 1 C_5H_5N); λ_m = 252 nm (ε 8300 EtOH). *Sterling Winthrop Inc.*

Adrenocorticotropic Hormones

63 ACTH
9002-60-2 136 232-659-7
Corticotropin.
adrenocorticotropin; adrenocorticotrophin; Acethropan; Acortan; Acorto; Acthar; Acton; Actonar; Adrenomone; Alfatrofin; Cibacthen; Corstiline; Cortiphyson; Cortrophimn; Isactid; Reacthin; Solacthyl; Tubex; Ser-Tyr-Ser-Met-Glu-His-Phe-Arg-Trp-Gly-Lys-Pro-Val-Gly-Lys-Lys-Arg-Arg-Pro-Val-Lys-Val-Tyr-Pro-NH_2. Single chain peptide with 39 amino acid residues; isolated from the pituitary gland. Adrenocorticotropic hormone, used to stimulate glucocorticoid production. Freely soluble in H_2O; appreciably soluble in 70% aqueous alcohol or Me_2OH. *CIBA plc; Parke-Davis; Pharmacia & Upjohn.*

64 Cosyntropin
16960-16-0 2617 241-031-1
$C_{136}H_{210}N_{40}O_{31}S$
α^{1-24}-Corticotropin.
tetracosactide; β^{1-24}-corticotropin; tetracosactrin; Actholain; Cortrosinta; Cortrosyn; Synacthen. Analog of ACTH. Used to stimulate glucocorticoid production. [hexaacetate tetradecahydrate]: $[\alpha]_D^{22}$ = -88° ±2° (c = 0.511 1% AcOH). *CIBA plc.*

65 Ebiratide
105250-86-0
$C_{48}H_{73}N_{11}O_{10}S$
L-Methionyl-L-glutamyl-L-histidyl-L-phenylalanyl-D-lysyl-N-(8-aminooctyl-L-phenylalaninamide S,S-dioxide.
An adrenocorticotropic hormone. An ACTH analog.

66 Giractide
24870-04-0 4435
$C_{100}H_{156}N_{34}O_{22}S$
1-Glycine-18-L-argininamide-$\alpha^{(1-18)}$-corticotropin.
Gly-Tyr-Ser-Met-Glu-His-Phe-Arg-Trp-Gly-Lys-Pro-Val-Gly-Lys-Lys-Arg-Arg-NH_2. Polypeptide corresponding to the first 18 residues of corticotropin with the 1-serine replaced by glycine. Used to stimulate glucocorticoid production. $[\alpha]_D^{23.5}$ = -51.4° ± 1.9° (c = 0.472 0.1N AcOH); λ_m = 281, 288 nm (ε 6750, 6490 0.1N NaOH). *Shionogi & Co. Ltd.*

67 Giractide Hexaacetate Salt
29365-11-5 4435
$C_{112}H_{180}N_{34}O_{34}S$
1-Glycine-18-L-argininamide-$\alpha^{(1-18)}$-corticotropin hexaacetate salt.
S-50022; Acthormon. Polypeptide corresponding to the first 18 residues of corticotropin with the 1-serine replaced by glycine. Used to stimulate glucocorticoid production. *Shionogi & Co. Ltd.*

Aldose Reductase Inhibitors

68 Epalrestat
82159-09-9 3640
$C_{15}H_{13}NO_3S_2$
(E,E)-5-(2-Methyl-3-phenyl-2-propenylidene)-4-oxo-2-thioxo-3-thiazolidineacetic acid.
ONO-2235; Kinedak; Sorbistat. Aldose reductase inhibitor, affects sorbitol levels in diabetic rats and is used in treatment of diabetes-related neuropathy. mp = 210-217°; [N-methyl-D-glucamine salt]: mp = 163-165°. *Ono Pharm.*

69 Imirestat
89391-50-4
$C_{15}H_8F_2N_2O_2$
2,7-Difluorospiro[fluorene-9,4'-imidazolidine]-2',5'-dione.
Aldose reductase inhibitor.

70 Sorbinil
68367-52-2 8871
$C_{11}H_9FN_2O_3$
(S)-6-Fluoro-2,3-dihydrospiro[4H-1-benzopyran-4,4'-imidazolidine]-2',5'-dione.
CP-45634. Spirohydantoin aldose reductase inhibitor. Used in treatment of diabetic neuropathy. mp = 241-243°; $[\alpha]_D^{25}$ = 54.0° (c = 1 in MeOH). *Pfizer Inc.*

71 Tolrestat
82964-04-3 9663
$C_{16}H_{14}F_3NO_3S$
N-[[6-Methoxy-5-(trifluoromethyl)-1-naphthalenyl]thioxomethyl]-N-methylglycine.
tolrestatin; AY-27773; Alredase; Lorestat. Orally active aldose reductase inhibitor. Used in treatment of diabetic neuropathy. mp = 164-165°; [methyl ester]: mp = 109-110°. *Wyeth-Ayerst Labs.*

72 Zenarestat
112733-06-9
$C_{17}H_{11}BrClFN_2O_4$
3-(4-Bromo-2-fluorobenzyl)-7-chloro-3,4-dihydro-2,4-dioxo-1(2H)-quinazoline acetic acid.
FR-74366; FK-366. Aldose reductase inhibitor.

73 Zopolrestat
110703-94-1 10325
$C_{19}H_{12}F_3N_3O_3S$
3,4-Dihydro-4-oxo-3-[[5-(trifluoromethyl)-2-benzothiazolyl]methyl]-1-phthalazine acetic acid.
CP-73850. Antidiabetic agent. Aldose reductase inhibitor used in treatment of diabetic neuropathy. mp = 197-198°. *Pfizer Inc.*

Aldosterone Antagonists

74 Canrenoate Potassium
2181-04-6 1795 218-554-9
$C_{22}H_{29}KO_4$
Potassium 17-hydroxy-3-oxo-17α-pregna-4,6-diene-21-carboxylate.
SC-14266; Kanrenol; Soldactone; Venactone. Diuretic. Aldosterone antagonist. *Searle G.D. & Co.*

75 Canrenoic Acid
4138-96-9 223-963-0
$C_{22}H_{30}O_4$
17-Hydroxy-3-oxo-17α-pregna-4,6-diene-21-carboxylic acid.
Diuretic. Aldosterone antagonist. *Searle G.D. & Co.*

76 Canrenone
976-71-6 1795 213-554-5
$C_{22}H_{28}O_3$
17-Hydroxy-3-oxo-17α-pregna-4,6-diene-21-carboxylic acid σ lactone.
SC-9376; Phanurane. Diuretic. Aldosterone antagonist. mp = 149-151°; $[\alpha]_D$ = 24.5° ($CHCl_3$); λ_m = 283 nm (ε 26700). *Searle G.D. & Co.*

77 Dicirenone
41020-79-5
$C_{26}H_{36}O_5$
17-Hydroxy-3-oxo-17α-pregn-4-ene-7,21-dicarboxylic acid σ-lactone isopropyl ester.
SC-26304. Aldosterone antagonist, used as a hypotensive agent. *Searle G.D. & Co.*

78 Eplerenone
107724-20-9
$C_{24}H_{30}O_6$
9,11α-Epoxy-17-hydroxy-3-oxo-17α-pregn-4-ene-7α,21-dicarboxylic acid σ-lactone methyl ester.
SC-66110. Aldosterone antagonist, used as an antihypertensive agent. *Searle G.D. & Co.*

79 Mespirenone
87952-98-5
$C_{25}H_{30}O_4S$
15α,16α-Dihydro-17-hydroxy-7α-mercapto-3-oxo-3'H-cyclopropa[15,16]-17α-pregna-1,4,15-triene-21-carboxylic acid γ-lactone acetate.
Mineralocorticoid receptor blocker; specific inhibitor of adrenocortical mineralocorticoid synthesis. Aldosterone antogonist.

80 Mexrenoate Potassium
43169-54-6
$C_{24}H_{33}KO_6.2H_2O$
7-Methyl-21-potassium-17-hydroxy-3-oxo-17α-pregn-4-ene-7α,21-dicarboxylate dihydrate.
SC-26714. Aldosterone antagonist, used as an antihypertensive agent. *Searle G.D. & Co.*

81 Mexrenoic Acid
41020-68-2
$C_{24}H_{35}O_6$
7-Methyl-17-hydroxy-3-oxo-17α-pregn-4-ene-7α,21-dicarboxylic acid.
Aldosterone antagonist, used as an antihypertensive agent. *Searle G.D. & Co.*

82 Prorenoate Potassium
49847-97-4
$C_{23}H_{31}KO_4$
Potassium 6,7-dihydro-17-hydroxy-3-oxo-3'H-cyclopropa[6,7]-17α-pregna-4,6-diene-21-carboxylate.
SC-23992. Aldosterone antagonist, used as an antihypertensive agent. *Searle G.D. & Co.*

83 Spironolactone
52-01-7 8917 200-133-6
$C_{24}H_{32}O_4S$
17-Hydroxy-7α-mercapto-3-oxo-17α-pregn-4-ene-21-carboxylic acid σ-lactone acetate.
Abbolactone; Aldactone; Aldactazide; SC-9420; Aldace; Aldopur; Almatol; Altex; Aquareduct; Deverol; Diatensec; Dira; Duraspiron; Euteberol; Lacalmin; Lacdene; Laractone; Nefurofan; Osiren; Osyrol; Sagisal; Sincomen; Spiretic; Spiroctan; Spiroderm; Spirolone; Spiro-Tablinen; Supra-Puren; Suracton; Urusonin; Verospiron; Xenalon. Diuretic. Aldosterone antagonist. Potassium sparing. Used for edema in cirrhosis of the liver, nephrotic syndrome, congestive heart failure, potentiation of thiazide and loop diuretics, hypertension and Conn's syndrome. mp = 134-135°, 201-202°; $[\alpha]_D^{20}$ = -33.5° ($CHCl_3$); λ_m = 238 nm (ε 20200); insoluble in H_2O, soluble in most organic solvents. *Abbott Labs Inc.; Parke-Davis; Searle G.D. & Co.*

84 Spirorenone
74220-07-8 277-770-1
$C_{24}H_{28}O_3$
(6R,7R,8R,9S,10R,13S,14R,15S,16S,17S)-3',4',6,7,8,9,11,12,13,14,15,16,20,21-Tetradecahydro-10,13-dimethylspiro[17H-dicyclopropa[6,7:15,16]cyclopenta[a]phenanthrene-17,2'(5'H)-furan]-3(10H),5'-dione.
Shows affinity for mineralocorticoid receptors. Aldosterone antagonist.

α-Adrenergic Agonists

85 Adrafinil
63547-13-7 168 264-303-1
$C_{15}H_{15}NO_3S$
2-[(Diphenylmethyl)sulfinyl]acetohydroxamic acid.
CRL-40028; Olmifon. Antidepressant. An α-adrenergic agonist. mp = 159-160°; soluble in H_2O (< 1g/l); LD_{50} (mus ip) < 2048 mg/kg, (mus orl) = 1950 mg/kg. *C.M. Ind.; Lab. Lafon France.*

86 Adrenalone
99-45-6 170 202-756-9
$C_9H_{11}NO_3$
3',4'-Dihydroxy-2-(methylamino)acetophenone.
Kephrine; Stryphnone; Stypnone. Adrenergic (ophthalmic). Hemostatic. An α-adrenergic agonist. mp = 235-236° (dec); sparingly soluble in H_2O, EtOH, Et_2O. *Bayer Corp. Pharm. Div.; Hoechst AG.*

87 Adrenalone Hydrochloride
62-13-5 170 200-525-7
$C_9H_{12}ClNO_3$
3',4'-Dihydroxy-2-(methylamino)acetophenone hydrochloride.
Styphnonasal. Adrenergic (ophthalmic); hemostatic. An α-adrenergic agonist. mp = 243°; soluble in H_2O, EtOH; insoluble in Et_2O. *Bayer Corp. Pharm. Div.; Hoechst AG.*

88 Amidephrine
3354-67-4 418
$C_{10}H_{16}N_2O_3S$
3'-[1-Hydroxy-2-(methylamino)ethyl] methanesulfonanilide.
amidefrine; MJ-1996. Vasoconstrictor; decongestant (nasal); adrenergic. α-adrenergic receptor agonist. mp = 159-161°. *Mead Johnson Labs.*

89 Amidephrine Mesylate
1421-68-7 418
$C_{11}H_{20}N_2O_6S_2$
3'-[1-Hydroxy-2-(methylamino)ethyl] methanesulfonanilide monomethanesulfonate (salt).
amidephrine monomethanesulfonate; amidefrine mesilate; MJ-5190; Dircol; Fentrinol; Nalde. Vasoconstrictor; decongestant (nasal); adrenergic. α-adrenergic receptor agonist. mp = 207-209°; LD_{50} (frat orl) = 13-36 mg/kg. *Mead Johnson Labs.*

90 Apraclonidine
66711-21-5 791
$C_9H_{10}Cl_2N_4$
2-[(4-Amino-2,6-dichlorophenyl)imino]imidazolidine.
p-aminoclonidine; aplonidine; NC-14. An α_2-adrenergic agonist. Used in management of post-surgical elevated intraocular pressure. mp > 230°. *Alcon Labs.*

91 Apraclonidine Dihydrochloride
73217-88-6 791
$C_9H_{12}Cl_4N_4$
2-[(4-Amino-2,6-dichlorophenyl)imino]imidazolidine dihydrochloride.
An α_2-adrenergic agonist. Used in management of post-surgical elevated intraocular pressure. λ_m = 254, 304 nm (ε 1800, 2500 EtOH). *Alcon Labs.*

92 Apraclonidine Hydrochloride
73218-79-8 791
$C_9H_{11}Cl_3N_4$
2-[(4-Amino-2,6-dichlorophenyl)imino]imidazolidine monohydrochloride.
AL02145; lopidine. An α_2-adrenergic agonist. Used in management of post-surgical elevated intraocular pressure. *Alcon Labs.*

93 Budralazine
36798-79-5 1492
$C_{14}H_{16}N_4$
4-Methyl-3-penten-2-one (1-phthalazinyl)-hydrazone.
DJ-1461; Buterazine. An α-adrenergic agonist. Used as an antihypertensive agent. Direct-acting vasodilator with central sympathoinhibitory activity. A derivative of hydralazine. mp = 132-133°; λ_m = 208, 240, 289, 357 nm (ε = 27000, 89000, 20000, 15000 MeOH); LD_{50} (mus orl) = 1820 mg/kg, (mus ip) = 4020 mg/kg, (rat orl) = 620 mg/kg, (rat ip) = 3570 mg/kg. *Daiichi Seiyaku.*

94 Clonidine
4205-90-7 2450 224-119-4
$C_9H_9Cl_2N_3$
2-[(2,6-Dichlorophenyl)imino]imidazolidine.
Catapres-TTS; ST-155-BS. An α_2-adrenergic agonist used as an antihypertensive and antidyskinetic. mp = 130°. *Boehringer Ingelheim GmbH.*

95 Clonidine Hydrochloride
4205-91-8 2450 224-121-5
$C_9H_{10}Cl_3N_3$
2-[(2,6-Dichlorophenyl)imino]imidazolidine hydrochloride.
Catapres; ST-155; component of: Combipres. An α_2-adrenergic agonist used as an antihypertensive and antidyskinetic. mp = 305°; soluble in H_2O (7.7 g/100 ml at 20°, 16.6 g/100 ml at 60°), MeOH (17.25 g/100 ml), EtOH (4 g/100 ml), $CHCl_3$ (0.02 g/100 ml); λ_m = 213, 271, 302 nm (ε 8290, 713, 340 H_2O); LD_{50} (mus orl) = 328 mg/kg, (mus iv) = 18 mg/kg, (rat orl) = 270 mg/kg, (rat iv) = 29 mg/kg. *Boehringer Ingelheim GmbH; Parke-Davis.*

96 Cyclopentamine
102-45-4 2808
$C_9H_{19}N$
N,α-Dimethylcyclopentaneethylamine .
cyclopentadrine; Clopane; Cyclonarol; Cyklosal; Sinos. An α-adrenergic agonist. Has vasoconstrictor properties and is used as a nasal decongestant. bp_{30} = 83-86°; n_D^{25} = 1.4500. *Eli Lilly & Co.*

97 Cyclopentamine Hydrochloride
538-02-3 2808 208-681-8
$C_9H_{20}ClN$
N,α-Dimethylcyclopentaneethylamine hydrochloride.
Clopane hydrochloride; component of: Aerolone Solution. An α-adrenergic agonist. Has vasoconstrictor properties and is used as a nasal decongestant. mp = 113-115°; freely soluble in H_2O. *Eli Lilly & Co.*

98 Detomidine
76631-46-4 2981
$C_{12}H_{14}N_2$
4-(2,3-Dimethylbenzyl)imidazole.
MPV-253 AII. An α_2-adrenergic agonist with sedative and analgesic properties. mp = 114-116°; LD_{50} (mus iv) = 35 mg/kg; [hydrochloride (Domosedan)]: mp = 160°. *Farmos Group Ltd.*

99 Dimetofrine
22950-29-4 3314 245-348-6
$C_{11}H_{17}NO_4$
4-Hydroxy-3,5-dimethoxy-α-[(methylamino)methyl]benzyl alcohol.
dimethophrine; dimetrophine. An α-adrenergic agonist. Used as an antihypotensive agent. mp = 178° (dec). *Zambeletti.*

100 Dimetofrine Hydrochloride
22775-12-8 3314 245-212-6
$C_{11}H_{18}ClNO_4$
4-Hydroxy-3,5-dimethoxy-α-[(methylamino)methyl]benzyl alcohol hydrochloride.
Pressamina. An α-adrenergic agonist. Used as an antihypotensive agent. mp = 171-173°. *Zambeletti.*

101 Dipivefrin
52365-63-6 3400
$C_{19}H_{29}NO_5$
(±)-3,4-Dihydroxy-α-[(methylamino)methyl]benzyl alcohol 3,4-dipivalate.
dipivalyl epinephrine; DPE. An α-adrenergic (ophthalmic) agonist. Used as an antiglaucoma agent. mp = 146-147°. *Alcon Labs; Allergan Inc.*

102 Dipivefrin Hydrochloride
64019-93-8 3400 264-609-5
$C_{19}H_{30}ClNO_5$
(±)-3,4-Dihydroxy-α-[(methylamino)methyl]benzyl alcohol 3,4-dipivalate hydrochloride.
Diopine; d Epifrin; Diphemin; Pivalephrine; Propine. An α-adrenergic (ophthalmic) agonist. Used as an antiglaucoma agent. mp = 158-159°; soluble in H_2O, EtOH. *Alcon Labs; Allergan Inc.*

103 Dopamine
51-61-6 3479 200-110-0
$C_8H_{11}NO_2$
4-(2-Aminoethyl)pyrocatechol.
3-hydroxytyramine. Adrenergic; used as an antihypotensive agent and cardiotonic. Endogenous catecholamine with α- and β-adrenergic activity. *Astra USA Inc.; Elkins-Sinn; Parke-Davis.*

104 Dopamine Hydrochloride
62-31-7 3479 200-527-8
$C_8H_{12}ClNO_2$
4-(2-Aminoethyl)pyrocatechol hydrochloride.
Dopastat; ASL-279; Cardiosteril; Dynatra; Inovan; Inotropin. Adrenergic; used as an antihypotensive and cardiotonic. Dec 241°; freely soluble in H_2O; soluble in EtOH, MeOH; insoluble in Et_2O, $CHCl_3$, C_6H_6, C_7H_8, petroleum ether; [hydrobromide]: mp = 210-214° (dec). *Astra USA Inc.; Elkins-Sinn; Parke-Davis.*

105 Ephedrine Hydrochloride
134-71-4 3645 205-153-9
$C_{10}H_{16}ClNO$
(±)-α-[1-(Methylamino)ethyl]benzenemethanol hydrochloride.
racephedrine hydrochloride [dl-form]; Ephetonin [dl-form]. α,β-Adrenergic agonist (bronchodilator). Found in Ma Huang and other Ephedra species. [(dl)-form]: mp = 187-188°; soluble in H_2O (25 g/100 ml); somewhat soluble in EtOH; nearly insoluble in Et_2O; pH 6. *Bristol Laboratories; Eli Lilly & Co.; Glaxo Labs.; Parke-Davis; Poythress; Sterling Winthrop Inc.; Whitehall Labs. Inc.*

106 dl-Ephedrine
90-81-3 3645 202-017-0
$C_{10}H_{15}NO$
(±)-α-[1-(Methylamino)ethyl]benzenemethanol.
1-phenyl-2-methylaminopropanol; racemic ephedrine; racephedrine. Adrenergic (bronchodilator). α- and β-adrenergic agonist. Occurs in Ma Huang and other Ephedra species. The d-isomer is used as a decongestant. mp = 79°; soluble in H_2O, EtOH, Et_2O, $CHCl_3$, oils.

107 l-Ephedrine
299-42-3 3645 206-080-5
$C_{10}H_{15}NO$
l-α-[1-(Methylamino)ethyl]benzenemethanol.
l-ephedrine; (-)-ephedrine; component of: Bena-Fedrin, Bronkotabs, Mudrane GG Elixir, Primatene, Quadrinal, Tedral. Bronchodilator. An α-adrenergic agonist. Waxy solid; may contain up to 1/2 mole H_2O; mp = 34°; bp = 255°; somewhat soluble in H_2O (5 g/100 ml); more soluble in EtOH (500 g/100 ml), $CHCl_3$, Et_2O, oils. *Eli Lilly & Co.; Knoll Pharm. Co.; Parke-Davis; Whitehall-Robins.*

108 l-Ephedrine Hydrochloride
50-98-6 3645 200-074-6
$C_{10}H_{16}ClNO$
l-α-[1-(Methylamino)ethyl]benzenemethanol hydrochloride.
Ephedral; Sanedrine; component of: Bena-Fedrin, Bronkotabs, Mudrane GG Elixir, Primatene, Quadrinal, Tedral. An α-adrenergic agonist. Bronchodilator. mp = 216-220°; $[\alpha]_D^{25}$ = -33° to -35.5° (c = 5); soluble in H_2O (33.3 g/100 ml), EtOH (7.1 g/100 ml); nearly insoluble in Et_2O, $CHCl_3$. *Chemoterapico; Eli Lilly & Co.; Hybridon Inc.; Knoll Pharm. Co.; Parke-Davis; Whitehall-Robins.*

109 l-Ephedrine Sulfate
134-72-5 3645 205-154-4
$C_{20}H_{32}N_2O_6$
l-α-[1-(Methylamino)ethyl]benzenemethanol sulfate salt (2:1).
Isofedrol; component of: Bronkaid, Pazo Ointment, Pazo Suppository. Bronchodilator. Adrenergic agonist. mp = 245° (dec); $[\alpha]_D^{25}$ = -29.5° to -32.0° (c = 5); soluble in H_2O (83 g/100 ml), EtOH (1.05 g/100 ml); freely soluble in hot EtOH. *Boehringer Mannheim GmbH; Bristol-Myers Squibb HIV Products; Eli Lilly & Co.; Sterling Health U.S.A.*

110 Epinephrine
51-43-4 3656 200-098-7
$C_9H_{13}NO_3$
(-)-3,4-Dihydroxy-α-[(methylamino)methyl]benzyl alcohol.
l-methylaminoethanolcatechol; adrenalin; levorenen; Bronkaid Mist; Epifrin; Epiglaufrin; Eppy; Glaucon; Glauposine; Primatene Mist; Simplene; Sus-phrine; Suprarenaline; component of: Citanest Forte. Bronchodilator; cardiostimulant; mydriatic; antiglaucoma. Endogenous catecholamine with combined α- and β-agonist activity. Principal sympathomimetic hormone produced by the adrenal medulla. mp = 211-212°; dec 215°; $[\alpha]_D^{25}$ = -50.0° to -53.5° (in 0.6N HCl); slightly soluble in H_2O, EtOH; soluble in aqueous solutions of mineral acids; insoluble in aqueous solutions of ammonia and alkali carbonates; insoluble in Et_2O, Me_2CO, oils; LD_{50} (mus ip) = 4 mg/kg. *Alcon Labs; Allergan Inc.; Astra Sweden; Bristol-Myers Squibb Co; CIBA Vision AG; Elkins-Sinn; Evans Medical Ltd.; Parke-Davis; Sterling Health U.S.A.; Whitehall-Robins; Wyeth-Ayerst Labs.*

111 Epinephrine d-Bitartrate
51-42-3 3656 200-097-1
$C_{13}H_{19}NO_9$
(-)-3,4-Dihydroxy-α-[(methylamino)methyl]benzyl alcohol (+)-tartrate (1:1) salt.
Asmatane Mist; Asthmahaler Epitrate; Bronitin Mist; Bronkaid Mist Suspension; Epitrate; Medihaler-Epi; Primatene Mist Suspension; Suprarenin; component of: Asthmahaler, E-Pilo. Bronchodilator; cardiostimulant; mydriatic; antiglaucoma. Adrenergic (opthalmic). mp = 147-154° (some decomposition); soluble in H_2O (30 g/100 ml); slightly soluble in EtOH. *3M Pharm.; CIBA Vision Corp.; Menley & James Labs Inc.; Sterling Health U.S.A.; Whitehall-Robins; Wyeth-Ayerst Labs.*

112 Epinephrine Hydrochloride
55-31-2 3656
$C_9H_{14}ClNO_3$
(-)-3,4-Dihydroxy-α-[(methylamino)methyl]benzyl alcohol hydrochloride.
Adrenalin; Epifrin; Glaucon; Suprarenin. Bronchodilator; cardiostimulant; mydriatic; antiglaucoma. An α-adrenergic agonist. *Alcon Labs; Allergan Inc.; Astra Chem. Ltd.; Bristol-Myers Squibb Co; CIBA Vision AG; Elkins-Sinn; Forest Pharm. Inc.; Parke-Davis; Sterling Health U.S.A.; Whitehall Labs. Inc.; Wyeth-Ayerst Labs.*

113 dl-Epinephrine
329-65-7 3656 206-347-6
$C_{20}H_{32}N_2O_6$
(±)-α-[1-(Methylamino)ethyl]benzenemethanol sulfate salt (2:1).
racepinefrine; racepinephrine. An α-adrenergic agonist. Used as a bronchodilator, cardiostimulant, mydriatic; antiglaucoma agent. Slightly soluble in H_2O, EtOH. *Alcon Labs; Allergan Inc.; Astra Chem. Ltd.; Bristol-Myers Squibb Co; CIBA Vision AG; Elkins-Sinn; Forest Pharm. Inc.; Parke-Davis; Sterling Health U.S.A.; Whitehall Labs. Inc.; Wyeth-Ayerst Labs.*

114 dl-Epinephrine Hydrochloride
329-63-5 3656 206-346-0
$C_{20}H_{33}ClN_2O_6$
(±)-α-[1-(Methylamino)ethyl]benzenemethanol hydrochloride salt.
Asthmanefrin; Vaponefrin. An α-adrenergic agonist. Used as a bronchodilator, cardiostimulant, mydriatic; antiglaucoma agent. mp = 157°, soluble in H_2O; sparingly soluble in EtOH. *Alcon Labs; Allergan Inc.; Astra Chem. Ltd.; Bristol-Myers Squibb Co; CIBA Vision AG; Elkins-Sinn; Forest Pharm. Inc.; Parke-Davis; Sterling Health U.S.A.; Whitehall Labs. Inc.; Wyeth-Ayerst Labs.*

115 Fenoxazoline
4846-91-7 4025 225-437-6
$C_{13}H_{18}N_2O$
4,5-Dihydro-2-[[2-(1-methylethyl)phenoxy]methyl]-1H-imidazole.
phenoxazoline. CNS Stimulant. An α-adrenergic agonist with sympathomimetic properties. *Lab. Dausse.*

116 Fenoxazoline Hydrochloride
21370-21-8 4025
$C_{13}H_{19}ClN_2O$
4,5-Dihydro-2-[[2-(1-methylethyl)phenoxy]methyl]-1H-imidazole monohydrochloride.
Aturgyl; Snup. An α-adrenergic agonist with sympathomimetic properties. CNS Stimulant. mp = 174°; soluble in H_2O, EtOH. *Lab. Dausse.*

117 Guanabenz
5051-62-7 4585 225-750-8
$C_8H_8Cl_2N_4$
[(2,6-Dichlorobenzylidene)amino]guanidine.
Wy-8678; NSC-68982. An α_2-adrenergic agonist used as an antihypertensive agent. mp = 227-229° (dec). *Sandoz Pharm. Corp.*

118 Guanabenz Monoacetate
23256-50-0 4585 245-534-7
$C_{10}H_{12}Cl_2N_4O_2$
[(2,6-Dichlorobenzylidene)amino]guanidine monoacetate.
Wytensin; Wy-8678 acetate; Rexitene; Tenelid. A centrally acting α_2-adrenergic agonist used as an antihypertensive agent. mp = 192.5° (dec); soluble in H_2O (1100 mg/100 ml), EtOH (5000 mg/ml), propylene glycol (10000 mg/100 ml), $CHCl_3$ (60 mg/100 ml), EtOAc (100 mg/100 ml). *Sandoz Pharm. Corp.; Wyeth Labs.*

119 Guanfacine
29110-47-2 4590 249-442-8
$C_9H_9Cl_2N_3O$
N-Amidino-2-(2,6-dichlorophenyl)acetamide.
Centrally active α_2-adrenergic agonist used as an antihypertensive. mp = 225-227°. *Robins, A. H. Co.; Sandoz Pharm. Corp.*

120 Guanfacine Hydrochloride
29110-48-3 4590 249-443-3
$C_9H_{10}Cl_3N_3O$
N-Amidino-2-(2,6-dichlorophenyl)acetamide hydrochloride.
BS-100-141; LON-798; Estulic; Tenex. Centrally active α_2-adrenergic agonist used as an antihypertensive. mp = 213-216°; LD_{50} (mus orl) = 165 mg/kg. *Robins, A. H. Co.; Sandoz Pharm. Corp.*

121 Hydroxyamphetamine
1518-86-1 4855
$C_9H_{13}NO$
(±)-p-(2-Aminopropyl)phenol.
p-hydroxyphenylisopropylamine; α-methyltyramine; Paredrine; Paredrinex; Pulsoton. α-Adrenergic (ophthalmic) agonist. Used as a mydriatic agent. mp = 125-126°; soluble in H_2O, EtOH, $CHCl_3$, EtOAc; [iodide]: mp=155°, soluble in H_2O, EtOH, Me_2CO; [hydrochloride]: mp=171-172°; soluble in H_2O, EtOH; insoluble in Et_2O. *Allergan Inc.; SmithKline Beecham Pharm.*

122 Hydroxyamphetamine Hydrobromide
306-21-8 4855 206-181-4
$C_9H_{14}BrNO$
(±)-p-(2-Aminopropyl)phenol hydrobromide.
Paredrine hydrobromide; component of: Paremyd. An α-adrenergic (ophthalmic) agonist. Used as a mydriatic agent. mp = 171-172°; soluble in H_2O, EtOH; insoluble in Et_2O. *Allergan Inc.; SmithKline Beecham Pharm.*

123 Ibopamine
66195-31-1 4921 266-229-5
$C_{17}H_{25}NO_4$
4-[2-(Methylamino)ethyl]-o-phenylene diisobutyrate.
SB-7505; 3,4-di-o-isobutyryl epinine. Cardiotonic. Inotropic agent with dopaminergic and adrenergic agonist activities. *Simes S.p.A.; SmithKline Beecham Pharm.*

124 Ibopamine Hydrochloride
75011-65-3 4921 278-056-2
$C_{17}H_{26}Cl5NO_4$
4-[2-(Methylamino)ethyl]-o-phenylene diisobutyrate hydrochloride.
SB-7505; Inopamil; Scandine. Cardiotonic. Inotropic with dopaminergic and adrenergic agonist activities. mp = 132°. *Simes S.p.A.; SmithKline Beecham Pharm.*

125 Indanazoline
40507-78-6 4967
$C_{12}H_{15}N_3$
2-(4-Indanylamino)-2-imidazoline.
An α-adrenergic agonist. Used as a decongestant and vasoconstrictor. mp = 109-113°. *Nordmark.*

126 Indanazoline Hydrochloride
40507-80-0 4967 254-945-0
$C_{12}H_{16}ClN_3$
2-(4-Indanylamino)-2-imidazoline hydrochloride.
EV-A-16; Farial. An α-adrenergic agonist. Used as a decongestant and vasoconstrictor. mp = 182-184°; LD_{50} (mmus orl) = 179 mg/kg, (mmus iv) = 22.3 mg/kg, (fmus orl) = 233 mg/kg, (fmus iv) = 26.9 mg/kg, (mrat orl) = 481 mg/kg, (mrat iv) = 16.3 mg/kg, (frat orl) = 542 mg/kg, (frat iv) = 17.6 mg/kg. *Nordmark.*

127 Indanidine
85392-79-6
$C_{11}H_{13}N_5$
4-(2-Imidazolin-2-ylamino)-2-methyl-2H-indazole.
An α-adrenergic agonist. A partial α1-adrenoceptor agonist; acts as a vasopressor.

128 Isometheptene
503-01-5 5202 207-959-6
$C_9H_{19}N$
6-Methylamino-2-methylheptene.
Octin; Octon; Octanil. An α-adrenergic agonist. Used as a sympathomimetic and, in veterinary medicine, as an antispasmodic. d = 0.795; bp = 176-178°, bp_7 = 58-59°; insoluble in H_2O; soluble in EtOH, Et_2O, Me_2CO, $CHCl_3$; [bitartrate]: mp = 78-80°; soluble in H_2O, EtOH; [mucate ($C_{24}H_{48}N_2O_8$)]: mp = 152°; soluble in H_2O, EtOH; insoluble in Et_2O, $CHCl_3$; LD_{50} (dog orl) = 148 mg/kg. *Knoll Pharm. Co.*

129 Isometheptene Hydrochloride
6168-86-1 5202 228-211-5
$C_9H_{20}ClN$
6-Methylamino-2-methylheptene hydrochloride.
Octin. An α-adrenergic agonist. Used as a sympathomimetic and, in veterinary medicine, as an antispasmodic. mp = 68-69°; soluble in H_2O, EtOH; LD_{50} (mus iv) = 17.5 mg/kg, (mus sc) = 171 mg/kg. *Knoll Pharm. Co.*

130 Levlofexidine
81447-78-1
$C_{11}H_{12}Cl_2N_2O$
(-)-2-[1-(2,6-Dichlorophenoxy)ethyl]-2-imidazoline.
(-)-lofexidine. A stereoselective α_2-adrenoceptor agonist. Imidazoline derivative. Used in the treatment of hypertension.

131 Levopropylhexedrine
6192-97-8 8045 228-245-0
$C_{10}H_{21}N$
(-)-N,α-Dimethylcyclohexaneethylamine.
l-propylhexedrine. Adrenergic (vasoconstrictor); decongestant. *See* propylhexedrine. Oily liquid; bp_9 = 80-81°; n_D^{20} = 1.4590. *SmithKline Beecham Pharm.*

132 Mephentermine
100-92-5 5897 202-901-6
$C_{11}H_{17}N$
N,α,α-Trimethylphenethylamine.
An α-adrenergic agonist. Used as an antihypotensive. Soluble in EtOH, Et_2O; insoluble in H_2O; LD_{50} (mus ip) = 110 mg/kg. *Wyeth-Ayerst Labs.*

133 Mephentermine Sulfate Dihydrate
6190-60-9 5897
$C_{22}H_{36}N_2O_4S$
N,α,α-Trimethylphenethylamine sulfate dihydrate.
Wyamine sulfate; Wyamine; Mephine. An α-adrenergic agonist. Used as an antihypotensive. Soluble in H_2O (5 g/100 ml), insoluble in $CHCl_3$. *Wyeth-Ayerst Labs.*

134 Metaraminol
54-49-9 5993
$C_9H_{13}NO_2$
(-)-α-(1-Aminoethyl)-m-hydroxybenzyl alcohol.
metaradrine; Pressonex. An α-adrenergic agonist; used as an antihypotensive agent. [oxalate dihydrate ($C_{11}H_{15}NO_6$)]: mp = 190°; $[\alpha]_D^{20}$ = -21.66°; soluble in H_2O. *Merck & Co.Inc.; Sterling Winthrop Inc.*

135 Metaraminol Bitartrate
33402-03-8 5993 251-502-3
$C_{13}H_{19}NO_8$
(-)-α-(1-Aminoethyl)-m-hydroxybenzyl alcohol tartrate (1:1) salt.
Aramine; Icoral B; Pressorol. An α-adrenergic agonist; used as an antihypotensive agent. mp = 176-177°; freely soluble in H_2O. *Merck & Co.Inc.; Sterling Winthrop Inc.*

136 Metaraminol Hydrochloride
5967-52-2 5993
$C_9H_{14}ClNO_2$
(-)-α-(1-Aminoethyl)-m-hydroxybenzyl alcohol hydrochloride.
Aramine; Pressonex Bitartrate. An α-adrenergic agonist; used as an antihypotensive agent. $[\alpha]_D^{20}$ = -19.75°; soluble in H_2O; LD_{50} (mus ip) = 440 mg/kg. *Merck & Co.Inc.; Sterling Winthrop Inc.*

137 Methoxamine
390-28-3 6067 206-867-3
$C_{11}H_{17}NO_3$
(±)-α-(1-Aminoethyl)-2,5-dimethoxybenzyl alcohol.
2,5-dimethoxynorephedrine. α-Adrenergic agonist; vasoconstrictor. Used as an antihypotensive. *Burroughs Wellcome Inc.; Glaxo Wellcome Inc.*

138 Methoxamine Hydrochloride
61-16-5 6067 200-499-7
$C_{11}H_{18}ClNO_3$
(±)-α-(1-Aminoethyl)-2,5-dimethoxybenzyl alcohol hydrochloride.
Vasoxine; Vasoxyl; Vasylox. α-Adrenergic agonist (vasoconstrictor). Used as an antihypotensive. mp = 212-216°; soluble in H_2O (40 g/100 ml), EtOH (8.4 g/100 ml); insoluble in Et_2O, C_6H_6; $CHCl_3$. *Burroughs Wellcome Inc.; Glaxo Wellcome Inc.*

139 Methyldopa
555-30-6 6132 209-089-2
$C_{10}H_{13}NO_4$
3-Hydroxy-α-methyl-L-tyrosine.
AMD; α-methyldopa; MK-351; Aldomet; Aldometil; Aldomine; Dopamet; Dopegyt; Elanpres; Equibar; Lederdopa; Medomet; Medopa; Medopren; Methoplain; Sembrina; Presinol. Antihypertensive agent. Activates central α_2-adrenergic receptors. [l-form sesquihydrate]: mp = 300° (dec); $[\alpha]_D^{23}$ = -4.0° ± 0.5° (c = 1 0.1N HCl); λ_m = 281 nm (ε 2780); soluble in H_2O (1 g/100 ml at 25°), insoluble in common organic solvents; [d-form]: soluble in H_2O (1.8 g/100 ml). *Merck & Co.Inc.*

140 Methyldopa Ethyl Ester Hydrochloride
2508-79-4 6132 219-720-3
$C_{12}H_{18}ClNO_4$
3-Hydroxy-α-methyl-L-tyrosine ethyl ester hydrochloride.
methyldopate hydrochloride. Antihypertensive agent. Activates central α_2-adrenergic receptors. Soluble in H_2O (1-30 g/100 ml at 25°). *Merck & Co.Inc.*

141 Methylhexaneamine
105-41-9 6159 203-296-1
$C_7H_{17}N$
4-Methyl-2-hexanamine.
1,3-dimethylamylamine; Forthane. An α-adrenergic agonist. d = 0.7620 - 0.7655; bp = 130-135°; slightly soluble in H_2O; soluble in EtOH, $CHCl_3$, Et_2O; LD_{50} (mus ip) = 185 mg/kg. *Eli Lilly & Co.*

142 Metizoline
17692-22-7 6223
$C_{13}H_{14}N_2S$
2-[(2-Methylbenzo[b]thien-3-yl)methyl]-2-imidazoline.
benazoline. An α-adrenergic agonist. Used as a vasoconstrictor and nasal decongestant. mp = 156-157°. *E. Merck; Marion Merrell Dow Inc.*

143 Metizoline Hydrochloride
5090-37-9 6223 225-811-9
$C_{13}H_{15}ClN_2S$
2-[(2-Methylbenzo[b]thien-3-yl)methyl]-2-imidazoline monohydrochloride.
EX-10-781; RMI-10482A; Ellsyl; Elsyl; Eunasin. An α-adrenergic agonist. Used as a vasoconstrictor and nasal decongestant. mp = 244-246°; slightly soluble in H_2O; soluble in alcohols; nearly insoluble in $CHCl_3$; LD_{50} (mus ip) = 49 mg/kg; (mus ip) = 9.1 mg/kg; (mus orl) = 155 mg/kg; (rat orl) = 74 mg/kg. *E. Merck; Marion Merrell Dow Inc.; Merck & Co.Inc.*

144 Midodrine
42794-76-3 6272 255-945-3
$C_{12}H_{18}N_2O_4$
(±)-2-Amino-N-(β-hydroxy-2,5-dimethoxyphenethyl)acetamide.
St-1085. α-Adrenergic; vasoconstrictor. Used as an antihypotensive. *OSSW; Roberts Pharm. Corp.*

145 Midodrine Hydrochloride
3092-17-9 6272
$C_{12}H_{19}ClN_2O_4$
(±)-2-Amino-N-(β-hydroxy-2,5-dimethoxyphenethyl)-acetamide monohydrochloride.
Pro-Amatine; A-4020 Linz; St. Peter 224; ST-1085; Alphamine; Amatine; Gutron; Hipertan; Metligene; Midamine. An α-adrenergic agonist. Used as a vasoconstrictor and antihypotensive. mp = 192-193°. *OSSW; Roberts Pharm. Corp.*

146 Mivazerol
125472-02-8
$C_{11}H_{11}N_3O_2$
α-Imidazol-4-yl-2,3-cresotamide.
Selective imidazole α_2-adrenergic agonist.

147 Modafinil
68693-11-8 6311
$C_{15}H_{15}NO_2S$
2-[(Diphenylmethyl)sulfinyl]acetamide.
CRC-40476; CEP-1538; 2-(benzhydrylsulfinyl)acetamide; Provigil; DEP 1538; CRC 40476. CNS stimulant and α_1-adrenergic agonist. Antinarcoleptic. Used for treatment of narcoplesy and hypersomnia. mp = 164-166°. *Celphalone; Lab. Lafon France.*

148 Moxonidine
75438-57-2 6375
$C_9H_{12}ClN_5O$
4-Chloro-5-(2-imidazolin-2-ylamino)-6-methoxy-2-methylpyrimidine.
BDF-5895; Cynt; Physiotens. Vasoconstrictor; antihypertensive. An α-adrenergic agonist. mp = 217-219° (dec). *Beiersdorf AG.*

149 Moxonidine Hydrochloride
75438-58-3 6375
$C_9H_{13}Cl_2N_5O$
4-Chloro-5-(2-imidazolin-2-ylamino)-6-methoxy-2-methylpyrimidine hydrochloride.
Vasoconstrictor; antihypertensive. An α-adrenergic agonist. mp = 189°. *Beiersdorf AG.*

150 Naphazoline
835-31-4 6455 212-641-5
$C_{14}H_{14}N_2$
2-(1-Naphthylmethyl)-2-imidazoline.
Privine. Used as a vasoconstrictor and nasal decongestant. α-Adrenergic agonist (vasoconstrictor).

Alcon Labs; Bristol-Myers Squibb HIV Products; CIBA Vision AG; Ciba-Geigy Corp.; Pilkington Barnes Hind; Ross Products.

151 Naphazoline Hydrochloride
550-99-2 6455 208-989-2
$C_{14}H_{15}ClN_2$
2-(1-Naphthylmethyl)-2-imidazoline monohydrochloride.
Ak-Con; Albalon; Allerest Eye Drops; Clera; Coldan; Comfort Eye Drops; Degest-2; Iridina Due; Naphcon; Niazol; Opcon; Privine Hydrochloride; Rhinatin; Rhinoperd; Sanorin; Sanorin-Spofa; Strictylon; Vasoclear; Vasocon; component of: Abalon-A Liquifilm, Clear Eyes, Naphcon-A, VasoClear A, Vasocon-A. Used as a vasoconstrictor and nasal decongestant. α-Adrenergic agonist (vasoconstrictor). mp = 255-260°; λ_m = 223, 270, 280, 287, 291 nm ($E^{1\%}_{1cm}$ 3622, 239, 286, 196, 198 EtOH); pKa (25°C) = 10.35; soluble in H_2O (40 g/100 ml), EtOH; slightly soluble in $CHCl_3$; insoluble in C_6H_6, Et_2O; LD_{50} (rat sc) = 385 mg/kg. *Alcon Labs; Allergan Inc.; Bristol-Myers Squibb HIV Products; CIBA Vision AG; Ciba-Geigy Corp.; Fisons plc; Iolab; Lovens Komiske Fabrik AS; Patchem AG; Ross Products.*

152 Norepinephrine
51-41-2 6788 200-096-6
$C_8H_{11}NO_3$
(-)-α-(Aminomethyl)-3,4-dihydroxybenzyl alcohol.
Noradrenaline; levarterenol; Adrenor; Levophed. An α-adrenergic agonist. Used as a vasopressor and anti-hypotensive agent. mp = 216.5-218° (dec); $[\alpha]^{25}_D$= -37.3° (c = 5 H_2O). *Sterling Winthrop Inc.*

153 Norepinephrine d-Bitartrate
69815-49-2 6788
$C_{12}H_{17}NO_9$
(-)-α-(Aminomethyl)-3,4-dihydroxybenzyl alcohol tartrate (1:1) salt monohydrate.
Levophed; Levarterenol bitartrate; Aktamin; Binodrenal. Vasopressor used as an antihypotensive agent. An α-adrenergic agonist. mp = 102-104°; $[\alpha]^{25}_D$= -10.7° (c = 1.6 H_2O); freely soluble in H_2O. *Sterling Winthrop Inc.*

154 Norepinephrine Hydrochloride
329-56-6 6788 206-345-5
$C_8H_{12}ClNO_3$
(-)-α-(Aminomethyl)-3,4-dihydroxybenzyl alcohol hydrochloride.
Aterenol. An α-adrenergic agonist. Used as a vasopressor and antihypotensive agent. mp = 145.2-146.4°; $[\alpha]^{25}_D$ = -40° (c = 6); soluble in H_2O. *Sterling Winthrop Inc.*

155 Norfenefrine
536-21-0 6792 208-626-8
$C_8H_{11}NO_2$
α-(Aminomethyl)-m-hydroxybenzyl alcohol.
norphenylephrine. An α-adrenergic agonist. *CIBA plc.*

156 dl-Norfenefrine Hydrochloride
15308-34-6 6792 239-351-1
$C_8H_{12}ClNO_2$
(±)-α-(Aminomethyl)-m-hydroxybenzyl alcohol hydrochloride.
WV-569; Coritat; Depot-Novadral; Energona; Esbuphon; Molycor-R; Novadral; Stagural; Tonolift; Vingsal; Zondel. An α-adrenergic agonist. mp = 159-160°; λ_m = 274 nm ($E^{1\%}_{1cm}$ = 91.21); soluble in H_2O. *CIBA plc.*

157 Octodrine
543-82-8 6854 208-851-1
$C_8H_{19}N$
1,5-Dimethylhexylamine.
SK&F-51; Vaporpac. Used as a nasal decongestant; local anesthetic. An α-adrenergic (vasoconstrictor) agonist. Viscous liquid; [dl-form]: bp = 154-156°; [hydrochloride]: soluble in H_2O; LD_{50} (mus ip) = 59 mg/kg, (rat ip) = 41.5 mg/kg. *SmithKline Beecham Pharm.*

158 Octopamine
104-14-3 6856 203-179-5
$C_8H_{11}NO_2$
(±)-α-(Aminomethyl)-p-hydroxybenzyl alcohol.
ND50; norsympatol; norsynephrine; p-hydroxy-phenylethanolamine. An α-adrenergic agonist. The phenol analog of noradrenaline formed by the hydroxylation of tyramine by the enzyme dopamine β-hydroxylase.

159 D-(-)-Octopamine
876-04-0 6856
$C_8H_{11}NO_2$
D-α-(Aminomethyl)-p-hydroxybenzyl alcohol.
An α-adrenergic agonist. mp = 160°, > 250° (dec); $[\alpha]^{25}_D$= -56.0° (0.1N HCl), -37.4° (H_2O).

160 DL-Octopamine Hydrochloride
770-05-8 6856 212-216-4
$C_8H_{12}ClNO_2$
(±)-α-(Aminomethyl)-p-hydroxybenzyl alcohol hydrochloride.
Epirenor; Norden; Norfen; Norphen (ampules). An α-adrenergic agonist. mp = 170° (dec); freely soluble in H_2O.

161 Oxymetazoline
1491-59-4 7100 216-079-1
$C_{16}H_{24}N_2O$
6-tert-Butyl-3-(2-imidazolin-2-ylmethyl)-2,4-dimethylphenol.
H-990; Hazol; Navasin; Nezeril; Rhinofrenol; Rhinolitan; Sinerol; component of: Drixin. An α-adrenergic agonist (vasoconstrictor). Used as a nasal decongestant. mp = 181-183°. *Boehringer Ingelheim GmbH; Bristol-Myers Squibb HIV Products; E. Merck; Menley & James Labs Inc.; Schering-Plough Pharm.; Sterling Health U.S.A.; Whitehall-Robins.*

162 Oxymetazoline Hydrochloride
2315-02-8 7100 219-015-0
$C_{16}H_{25}ClN_2O$
6-tert-Butyl-3-(2-imidazolin-2-ylmethyl)-2,4-dimethylphenol monohydrochloride.
Sch-9384; Afrazine; Afrin; Iliadin; Nafrine; Nasivin; Oxilin; Sinex; Allerest 12 Hour Nasal Spray; 4-Way Nasal 12 Hour Spray; Dristan Long-Lasting Nasal Mist; Duration; Neo-Synephrine 12 Hour; Neo-Synephrine 12 Hour NTZ; Nostrilla; Ocuclear; Sch-9384; Afrazine; Iliadin; Nafrine; Nasivin; Oxilin; Sinex; component of: Benzedrex Nasal Spray 12 Hour, Drixin. An α-adrenergic

agonist. Used as a nasal decongestant. mp = 300-303° (dec); soluble in H_2O, EtOH; insoluble in Et_2O, $CHCl_3$, C_6H_6; LD_{50} (mus orl) = 10 mg/kg. *Boehringer Ingelheim GmbH; Bristol-Myers Squibb HIV Products; E. Merck; Fisons plc; Menley & James Labs Inc.; Schering-Plough Pharm.; Sterling Health U.S.A.; Whitehall-Robins.*

163 Phenylephrine

59-42-7 7440 200-424-8

$C_9H_{13}NO_2$

(R)-3-Hydroxy-α-[(methylamino)methyl]-benzenemethanol.

Mydriatic; decongestant. α-Adrenergic agonist. mp = 169-172°.

164 Phenylephrine Hydrochloride

61-76-7 7440 200-517-3

$C_9H_{14}ClNO_2$

(R)-m-Hydroxy-α[(methylamino)methyl]benzyl alcohol hydrochloride.

metaoxedrin; Adrianol; Ak-Dilate; Ak-Nefrin; Alcon Efrin; Biomydrin; Isophrin; m-Sympatol; Mezaton; Mydfrin; Neophryn; Neo-Synephrine; Nostril; Pyracort D; Prefrin; component of: Afrin 4-Way Nasal Spray Regular and Menthol, Anaplex HD, Benzedrex Nasal Spray Regular, Cerose-DM, Cyclomydril, Dristan Cold Multi-Symptom, Dristan Nasal Spray, Entex Capsules and Liquid, Histalet Forte, Hycomine Compound, Isopto Frin, Naldecon, Novahistine, Op-Isophrin-Z, Phenergan VC, Prefrin Liquifilm, Preparation H Cream, Preparation H Ointment, PV Tussin Syrup, Relief, RU-Tuss, Tympagesic, Vasosulf, Zincfrin. Mydriatic; decongestant. α-Adrenergic agonist. mp = 140-145°; $[\alpha]_D^{25}$ = -46.2° to -47.2°; soluble in H_2O, EtOH; LD_{50} (rat ip) = 17 ± 1.1 mg/kg, (rat sc) = 33.0 ± 2 mg/kg. *Alcon Labs; Boehringer Ingelheim Ltd.; Boots Pharmaceuticals Inc.; Bristol-Myers Squibb Co; CIBA Vision AG; DuPont-Merck Pharm.; Fisons plc; Marion Merrell Dow Inc.; Norwich Eaton; Parke-Davis; Robins, A. H. Co.; Schering AG; Schering-Plough Pharm.; Solvay Animal Health Inc.; Sterling Winthrop Inc.; Whitehall Labs. Inc.; Wyeth-Ayerst Labs.*

165 (±)-Phenylpropanolamine

14838-15-4 7461 238-900-2

$C_9H_{13}NO$

(±)-α-(1-Aminoethyl)benzenemethanol.

An α-adrenergic agonist; sympathomimetic amine. Used as an anorexic and nasal decongestant. mp = 101-101.5°. *Apothecon; Bristol-Myers Squibb HIV Products; DuPont-Merck Pharm.; Fisons plc; Lemmon Co.; Mead Johnson Nutritionals; Menley & James Labs Inc.; Parke-Davis; Procter & Gamble Pharm. Inc.; Sandoz Pharm. Corp.; Schering-Plough Pharm.; SmithKline Beecham Pharm.; Solvay Pharm. Inc.; Whitehall-Robins.*

166 Phenylpropanolamine Hydrochloride

154-41-6 7461 205-826-7

$C_9H_{14}ClNO$

(±)-α-(1-Aminoethyl)benzenemethanol hydrochloride.

1-phenyl-2-amino-1-propanol hydrochloride; mydriatin; Kontexin; Monydrin; Obestat; Propadrine; component of: A.R.M., Chlor-Trimeton Allergy/Sinus/Headache, Comtrex Liquigels, Comtrex Non-Drowsy Liquigels, Contac, Contac 12 Hour Caplets, Contac 12 Hour Capsules, Contac Severe Cold Maximum Strength, Coricidin D, Coricidin Sinus, Corsym Capsules, Demazin, Demilets, Diet Gard, Dimetapp Allergy Sinus, Dimetapp Cold and Allergy, Dimetapp Cold and Cough; Dimetapp DM, Dimetapp Elixir, Dimetapp Extentabs, Dimetapp Liquigels, Dimetapp 4 Hour Tablets, Entex Capsules and Liquid, Histalet Forte, Hycomine, Hycomine Pediatric, Naldecon, Naldecon CX, Ornade, Rhinex D-Lay Tablets, Robitussin CF, Sine-Off Sinus Medicine Tablets Aspirin Formula, Sinubid, Teldrin Timed Release Allergy Capsules, Triaminic, Trind, Trind-DM, Tuss-Ornade. Anorexic and decongestant. Psychomotor stimulant. Used in veterinary medicine as a bronchodilator and nasal decongestant. Sympathomimetic amine. Related to norpseudoephedrine. mp = 190-194°; soluble in H_2O, EtOH; insoluble in Et_2O, $CHCl_3$, C_6H_6; pKa = 9.44 ± 0.04; LD_{50} (rat orl) = 1490 mg/kg; [(+)-form hydrochloride]: mp = 171-172°; $[\alpha]_D^{25}$ = 32° (H_2O); [(±)-form base]: mp = 101-101.5. *Apothecon; Boots Pharmaceuticals Inc.; Bristol-Myers Squibb HIV Products; DuPont-Merck Pharm.; Fisons plc; Lemmon Co.; Mead Johnson Nutritionals; Menley & James Labs Inc.; Parke-Davis; Procter & Gamble Pharm. Inc.; Robins, A. H. Co.; Sandoz Pharm. Corp.; Schering-Plough Pharm.; SmithKline Beecham Pharm.; Solvay Pharm. Inc.; Taiho; Whitehall-Robins.*

167 Phenylpropylmethylamine

93-88-9 7462

$C_{10}H_{15}N$

N-β-Dimethylbenzeneethanamine.

Vonedrine. An α-adrenergic agonist. bp = 205-210°, bp_{15} = 95-96°; d_4^{25} = 0.915 to 0.925; soluble in H_2O (1.2 g/100 ml); soluble in EtOH, Et_2O, C_6H_6; [hydrochloride]: mp = 144-148°.

168 Pholedrine

370-14-9 7485 206-725-0

$C_{10}H_{15}NO$

p-(2-Methylaminopropyl)phenol.

Knoll H_{75}. An α-adrenergic agonist. Used as an antihypotensive. mp = 162-163°; slightly soluble in H_2O; soluble in EtOH, Et_2O. *Knoll Pharm. Co.*

169 Pholedrine Sulfate

6114-26-7 7485 228-083-0

$C_{20}H_{32}N_2O_8S$

p-(2-Methylaminopropyl)phenol sulfate (2:1) salt.

Paredrinol; Pulsotyl; Veritol. An α-adrenergic agonist. Used as an antihypotensive. mp = 320-323° (dec); soluble in H_2O; LD_{50} (rat sc) = 500 mg/kg. *Knoll Pharm. Co.*

170 Propylhexedrine

101-40-6 8045 202-939-3

$C_{10}H_{21}N$

(±)-N,α-Dimethylcyclohexaneethylamine.

Dristan Inhaler; component of: Benzedrex. An α-adrenergic agonist. Used as a mydriatic and nasal decongestant. bp = 205°, bp_{20} = 92-93°; d_4^{25} = 0.8501; slightly soluble in H_2O; soluble in EtOH, Et_2O, $CHCl_3$. *Menley & James Labs Inc.; Whitehall-Robins.*

171 d-Propylhexedrine

6556-29-2 8045

$C_{10}H_{21}N$

d-N,α-Dimethylcyclohexaneethylamine.

An α-adrenergic agonist. Used as a mydriatic and nasal

decongestant. bp_{10} = 82-83°. *Menley & James Labs Inc.; Whitehall-Robins.*

172 d-Propylhexedrine Hydrochloride

6192-96-7 8045

$C_{10}H_{22}ClN$

d-N,α-Dimethylcyclohexaneethylamine hydrochloride.

An α-adrenergic agonist used as a mydriatic and nasal decongestant. mp = 138-139° (dec); $[\alpha]_D^{26}$ = 14.73°; soluble in H_2O. *Menley & James Labs Inc.; Whitehall-Robins.*

173 dl-Propylhexedrine

3595-11-7 8045 222-741-0

$C_{10}H_{21}N$

(±)-N,α-Dimethylcyclohexaneethylamine.

An α-adrenergic agonist. Used as a mydriatic and nasal decongestant. bp = 205°, bp_{20} = 92-93°; d_4^{25} = 0.8501; slightly soluble in H_2O; soluble in EtOH, Et_2O, $CHCl_3$. *Menley & James Labs Inc.; Whitehall-Robins.*

174 dl-Propylhexedrine Hydrochloride

6192-95-6 8045

$C_{10}H_{22}ClN$

(±)-N,α-Dimethylcyclohexaneethylamine hydrochloride.

An α-adrenergic agonist. Used as a mydriatic and nasal decongestant. mp = 127-128° (dec); soluble in H_2O. *Menley & James Labs Inc.; Whitehall-Robins.*

175 l-Propylhexedrine

6192-97-8 8045 228-245-0

$C_{10}H_{21}N$

l-N,α-Dimethylcyclohexaneethylamine.

An α-adrenergic agonist. Used as a mydriatic and nasal decongestant. bp_9 = 80-81°. *Menley & James Labs Inc.; Whitehall-Robins.*

176 l-Propylhexedrine Ethylphenylbarbiturate

4388-82-3 8045 224-504-7

$C_{22}H_{33}N_3O_3$

l-N,α-Dimethylcyclohexaneethylamine ethylphenylbarbiturate.

barbexaclone; Maliasin. An α-adrenergic agonist (vasoconstrictor). Used as a mydriatic and nasal decongestant. *See* propylhexedrine. *Menley & James Labs Inc.; SmithKline Beecham Pharm.; Whitehall-Robins.*

177 l-Propylhexedrine Hydrochloride

6192-98-9 8045 228-246-6

$C_{10}H_{22}ClN$

l-N,α-Dimethylcyclohexaneethylamine hydrochloride.

An α-adrenergic agonist used as a mydriatic and nasal decongestant. mp = 138-139° (dec); $[\alpha]_D^{26}$ = -14.74°; soluble in H_2O. *Menley & James Labs Inc.; Whitehall-Robins.*

178 d-Pseudoephedrine

90-82-4 3645 202-018-6

$C_{10}H_{15}NO$

[S-(R*,R*)]-α-[1-(methylamino)ethyl]benzenemethanol.

d-isoephedrine; d-+ly-ephedrine. Decongestant. α-adrenergic agonist. mp = 119°; $[\alpha]_D^{20}$ = 51° (c = 0.6 in EtOH); pH (0.5% aqueous solution) = 10.8; sparingly soluble in H_2O; soluble in EtOH, Et_2O.

179 d-Pseudoephedrine Hydrochloride

345-78-8 3645 206-462-1

$C_{10}H_{16}ClNO$

d-α-[1-(Methylamino)ethyl]benzenemethanol hydrochloride.

component of: Novafed, PediaCare Decongestant Drops, Sudafed, Sudafed Liquid Children's, Actifed, Actifed Plus, Actifed Sinus Daytime, Actifed Sinus Nighttime, Actifed with Codeine Cough Syrup, Advil Cold & Sinus, Allent, Allergy/Sinus Comtrex Tablets, Bayer Select Allergy-Sinus; Bayer Select Flu Relief, Bayer Select Head & Chest Cold, Bayer Select Head Cold, Bayer Select Nighttime Cold, Bayer Select Sinus Pain Relief, Brexin EX, Brexin L.A., Children's Tylenol Flu Liquid, Comtrex Caplets, Comtrex Liquid, Comtrex Non-Drowsy Caplets, Comtrex Tablets, Congestac; Contac Day & Night Allergy Day Caplets, Contac Day & Night Allergy Night Caplets, Contac Day & Night Cold/Flu Day Caplets, Contac Day & Night Cold/Flu Night Caplets, Contac Severe Cold Formula Non-Drowsy, Co-Pyronil 2, Dimacol, Dimetapp Children's Cold & Fever, Dimetapp Decongestant Non-Drowsy, Dimetapp Pediatric Decongestant Drops, Dorcol, Dristan Allergy, Dristan Cold & Cough, Dristan Cold Maximum Strength Multi-Symptom; Dristan Cold No Drowsiness Maximum Strength, Dristan Sinus, Drixoral Cough & Congestion, Excedrin Sinus Caplets, Excedrin Sinus Tablets, Histalet Syrup, Histalet X Tablets, Intensin, Lodrane LD, Lodrane Liquid, Naldegesic, Nasatab LA, Nucofed, Ornex/Maximum Strength Ornex, Ornex Severe Cold Formula, PediaCare Cold Formula, PediaCare Cough/Cold Formula, PediaCare Decongestant Drops + Cough, Phenergan-D, Robitussin CF; Robitussin Cold & Cough, Robitussin Cold Cough & Flu, Robitussin Maximum Strength Cough & Cold, Robitussin Night Relief, Robitussin Nightime Cold Formula, Robitussin PE, Robitussin Pediatric Cough & Cold, Robitussin Pediatric Drops, Robitussin Pediatric Night Relief, Robitussin Severe Congestion, Rondec, Rondec DM, RU-Tuss DE, Semprex-D, Sine-Aid IB Caplets, Sine-Aid Maximum Strength, Sine-Off Maximum Strength No Drowsiness Formula Caplets, Sine-Off Sinus Medicine Caplets, Sudafed Cold & Cough, Sudafed Plus, Sudafed Cold Formula, Sudafed Sinus, TheraFlu, Tussar-DM, Tylenol Allergy Sinus, Tylenol Allergy Sinus Nighttime Caplets, Tylenol Children's Cold Tablets, Tylenol Children's Cough Cold Chewable Tablets, Tylenol Cold & Flu Multi-Symptom, Tylenol Cold & Flu No Drowsiness, Tylenol Cold Medication Caplets Liquid & Tablets, Tylenol Cold No Drowsiness, Tylenol Cold Severe Congestion Caplets, Tylenol Cough + Decongestant Liquid, Tylenol Flu Nighttime Gelcaps, Tylenol Flu Nighttime Powder, Tylenol Flu No Drowsiness Gelcaps, Tylenol Sinus Medication Maximum Strength, Viro-Med. The d-form is a nasal decongestant. An α-adrenergic agonist. Sympathomimetic. Used in veterinary medicine as an antihypotensive, mydriatic, antiallergic, and CNS stimulant. mp = 181-182°; $[\alpha]_D^{20}$ = 62° (c = 0.8); λ_m = 208, 251, 257, 264 nm (ε 8300, 161, 201, 161 EtOH); soluble in H_2O, EtOH, $CHCl_3$; LD_{50} (mus ip) = 200 mg/kg. *Ascher B.F. & Co.; Bristol-Myers Squibb HIV Products; ECR Pharm.; Eli Lilly & Co.; Glaxo Wellcome Inc.; McNeil Consumer Products Co.; Menley & James Labs Inc.; Merrell Pharm. Inc.; Rhône-Poulenc Rorer Pharm. Inc.; Roberts Pharm. Corp.; Ross Products; Sandoz Pharm. Corp.; Savage Labs; SmithKline Beecham Pharm.; Solvay Pharm. Inc.; Sterling Health U.S.A.; Whitehall-Robins.*

180 d-Pseudoephedrine Sulfate
7460-12-0 3645 231-243-2
$C_{20}H_{32}N_2O_6S$
d-α-[1-(Methylamino)ethyl]benzenemethanol sulfate (2:1).
Sch-4855; Afrin Tablets; Afrinol; Chlor-Trimeton ND; Drixoral ND; Duration Tablets; component of: Chlor-Trimeton Decongestant, Claritin D, Disophrol; Drixoral Allergy/Sinus, Drixoral Cold & Allergy, Drixoral Cold & Flu, Polaramine Expectorant, Trinalin. The d-form is a nasal decongestant. An α-adrenergic agonist. Sympathomimetic. Used in veterinary medicine as an antihypotensive, mydriatic, antiallergic, and CNS stimulant. *ICI ; Schering-Plough HealthCare Products; Schering-Plough Pharm.*

181 Rilmenidene
54187-04-1 8388 259-021-0
$C_{10}H_{16}N_2O$
2-[(Dicyclopropylmethyl)amino]-2-oxazoline.
oxaminozoline; S-3341; [phosphate]: Hyperium; S-3341-3. An α_2-adrenoceptor agonist used as an anti-hypertensive agent. mp= 106-107°; [phosphate]: soluble in H_2O (19 g/100 ml), MeOH (7 g/100 ml), $CHCl_3$, EtOH (0.7 g/100 ml); LD_{50} (mus orl) = 375 mg/kg, (rat orl) = 295 mg/kg; [fumarate]: mp = 170°. *Sci. Union et Cie France.*

182 Romifidine
65896-16-4
$C_9H_9BrFN_3$
2-(2-Bromo-6-fluoroanilino-2-imidazoline.
STH-2130. An α-adrenergic agonist. Sedative (veterinary). *Boehringer Ingelheim Pharm. Inc.*

183 Romifidine Hydrochloride
65896-14-2
$C_9H_{10}BrClFN_3$
2-(2-Bromo-6-fluoroanilino-2-imidazoline hydrochloride.
STH-2130-Cl; Sedivet. An α-adrenergic agonist. Sedative (veterinary). *Boehringer Ingelheim Pharm. Inc.*

184 Synephrine
94-07-5 9189 202-300-9
$C_9H_{13}NO_2$
(RS)-1-(4-Hydroxyphenyl)-2-(methylamino)ethanol.
Oxedrine; Analeptin; Ethaphene; Parasympatol; Simpalon; Synephrin; Synthenate. Vasopressor; antihypotensive agent. An α-adrenergic agonist. mp = 184-185°. *Boehringer Ingelheim GmbH.*

185 Synephrine Hydrochloride
5985-28-4 9189 227-804-6
$C_9H_{14}ClNO_2$
(RS)-1-(4-Hydroxyphenyl)-2-(methylamino)ethanol hydrochloride.
Oxedrine hydrochloride. Vasopressor; antihypotensive agent. An α-adrenergic agonist. mp = 151-152°; soluble in H_2O. *Boehringer Ingelheim GmbH.*

186 Synephrine Tartaric Acid Monoester
6414-49-9 9189
$C_{13}H_{17}NO_7$
(RS)-1-(4-Hydroxyphenyl)-2-(methylamino)ethanol tartrate (1:1) ester.
Neupentedrin; Pentedrin. Used as a vasopressor and antihypotensive agent. An α-adrenergic agonist. *Boehringer Ingelheim GmbH.*

187 Synephrine Tartrate
16589-24-5 9189 240-647-8
$C_{22}H_{32}N_2O_{10}$
(RS)-1-(4-Hydroxyphenyl)-2-(methylamino)ethanol tartrate (2:1) salt.
Corvasymton; Simpadren; Sympathol. An α-adrenergic agonist. Used as a vasopressor and antihypotensive agent. mp = 188-190° (dec); soluble in H_2O, EtOH. *Boehringer Ingelheim GmbH.*

188 Talipexole
101626-70-4 9209
$C_{10}H_{15}N_3S$
6-Allyl-2-amino-5,6,7,8-tetrahydro-4H-thiazolo[4,5-d]azepine.
An α_2-adrenergic and dopamine D_2 receptor agonist. Used as an antiparkinsonian. *Boehringer Ingelheim GmbH.*

189 Talipexole Dihydrochloride
36085-73-1 9209
$C_{10}H_{17}Cl_2N_3S$
6-Allyl-2-amino-5,6,7,8-tetrahydro-4H-thiazolo[4,5-d]azepine dihydrochloride.
B-HT-920; Domin. An α_2-adrenergic and dopamine D_2 receptor agonist. Used as an antiparkinsonian. mp = 245° (dec). *Boehringer Ingelheim GmbH.*

190 Tetrahydrozoline
84-22-0 9358 201-522-3
$C_{13}H_{16}N_2$
2-(1,2,3,4-Tetrahydro-1-naphthyl)-2-imidazoline.
tetryzoline. Adrenergic (vasoconstrictor); decongestant. An α-adrenergic agonist. *Alcon Labs; Marion Merrell Dow Inc.; Pfizer Inc.; Ross Products; Wyeth-Ayerst Labs.*

191 Tetrahydrozoline Hydrochloride
522-48-5 9358 208-329-3
$C_{13}H_{17}ClN_2$
2-(1,2,3,4-Tetrahydro-1-naphthyl)-2-imidazoline hydrochloride.
Murine Plus; Rhinopront; Soothe; Tinarhinin; Tyzanol; Tyzine; Visine; Yxin; component of: Collyrium Fresh-Eye Drops. An α-adrenergic agonist (vasoconstrictor). Used as a nasal decongestant. mp = 256-257° (dec); λ_m = 264.5, 271.5 nm ($A^{1\%}_{1\,cm}$ 17.5, 15.5); freely soluble in H_2O, EtOH; slightly soluble in $CHCl_3$; insoluble in Et_2O. *Alcon Labs; Key Pharm.; Marion Merrell Dow Inc.; Pfizer Inc.; Ross Products; Wyeth-Ayerst Labs.*

192 Tiamenidine
31428-61-2 9558
$C_8H_{10}ClN_3S$
2-[(2-Chloro-4-methyl-3-thienyl)amino]-2-imidazoline.
HOE-440; Symcor Base TTS; Thiamendidine. An α-adrenoceptor agonist related to clonidine and used as an antihypertensive agent. mp = 152°. *Hoechst Roussel Pharm. Inc.*

193 Tiamenidine Hydrochloride
51274-83-0 9558 257-100-4
$C_8H_{11}Cl_2N_3S$
2-[(2-Chloro-4-methyl-3-thienyl)amino]-2-imidazoline hydrochloride.
HOE-42-440; Sundralen; Symcor; component of: Symcorad. An α-adrenoceptor agonist related to clonidine. Used as an antihypertensive agent. mp = 228-229°; LD_{50} (rat iv) = 40 mg/kg, (mus iv) = 45 mg/kg, (mus sc) = 170 mg/kg, (mus orl) = 400 mg/kg. *Hoechst Roussel Pharm. Inc.*

194 Tramazoline
1082-57-1 9702 214-105-6
$C_{13}H_{17}N_3$
2-[(5,6,7,8-Tetrahydro-1-naphthyl)amino]-2-imidazoline.
An α-adrenergic agonist. Used as a nasal decongestant. mp = 142-143°. *Boehringer Ingelheim GmbH; Thomae GmbH Dr. Karl.*

195 Tramazoline Hydrochloride Monohydrate
3715-90-0 9702 223-064-3
$C_{13}H_{20}ClN_3O$
2-[(5,6,7,8-Tetrahydro-1-naphthyl)amino]-2-imidazoline monohydrochloride monohydrate.
KB-227; Biciron; Ellatun; Rhinaspray; Rhinogutt; Rhinospray; Rinogutt; Towk. An α-adrenergic agonist (vasoconstrictor). Used as a nasal decongestant. mp = 172-174°; soluble in H_2O; LD_{50} (mus orl) = 195 mg/kg. *Boehringer Ingelheim GmbH; Thomae GmbH Dr. Karl.*

196 Tuaminoheptane
123-82-0 9934 204-655-5
$C_7H_{17}N$
1-Methylhexylamine.
2-heptanamine; 2-aminoheptane; Tuamine. Decongestant (nasal); adrenergic (vasoconstrictor). α-Adrenergic agonist. Topical vasoconstrictor. Volatile liquid; bp_{760} = 142-144°; d_D^{25} = 0.7600-0.7660; n_D^{25} = 1.4150-1.4200; pH (1% aqueous solution) = 11.45; slightly soluble in H_2O; soluble in EtOH, Et_2O, petroleum ether, $CHCl_3$, C_6H_6. *Eli Lilly & Co.*

197 Tuaminoheptane Sulfate
6411-75-2 9934 229-113-5
$C_{14}H_{36}N_2O_4S$
Bis[(1-methylhexyl)ammonium] sulfate.
2-heptanamine sulfate (2:1); Heptedrine. Decongestant (nasal); adrenergic (vasoconstrictor). α-Adrenergic agonist. Topical vasoconstrictor. Soluble in H_2O; pH (1% aqueous solution) = 5.4. *Eli Lilly & Co.*

198 Tymazoline
24243-97-8 9965
$C_{14}H_{20}N_2O$
2-(2-Isopropyl-5-methylphenoxymethyl)-2-imidazoline.
Pernazene. An α-adrenergic agonist. Used as a nasal decongestant. [hydrochloride]: mp = 215-217°, 223.5-225°. *CIBA plc.*

199 Tyramine
51-67-2 9966 200-115-8
$C_8H_{11}NO$
4-(2-Aminoethyl)phenol.
An α-adrenergic agonist. mp = 164-165°; bp_{25} = 205-207°, bp_2 = 166°; soluble in H_2O (1.05 g/100 ml at 15°), EtOH (10 g/100 ml at 76°); sparingly soluble in C_6H_6, xylene.

200 Tyramine Hydrochloride
60-19-5 9966 200-462-5
$C_8H_{12}ClNO$
4-(2-Aminoethyl)phenol hydrochloride.
Mydrial; Uteramin. An α-adrenergic agonist. mp = 269°; soluble in H_2O.

201 Xylometazoline
526-36-3 10219 208-390-6
$C_{16}H_{24}N_2$
2-[[4-(1,1-Dimethylethyl)-2,6-dimethylphenyl]methyl]-4,5-dihydro-1H-imidazole.
Otrivin. Decongestant; adrenergic (vasoconstrictor). α-Adrenergic agonist. Used as a nasal decongestant. mp = 131-133°. *Ciba-Geigy Corp.; Sterling Winthrop Inc.*

202 Xylometazoline Hydrochloride
1218-35-5 10219 214-936-4
$C_{16}H_{25}ClN_2$
2-[[4-(1,1-Dimethylethyl)-2,6-dimethylphenyl]methyl]-4,5-dihydro-1H-imidazole monohydrochloride.
Neo-Rinoleina; Neo-Synephrine II; Novorin; Olynth; Otriven; Otrivin hydrochloride; Otrix; Therapin; Xymelin. An α-adrenergic agonist. Used as a nasal decongestant. Soluble in H_2O (< 3 g/100 ml), MeOH, EtOH; nearly insoluble in C_6H_6, Et_2O. *CIBA plc; Ciba-Geigy Corp.; Sterling Winthrop Inc.*

α-Adrenergic Antagonists

203 Abanoquil
90402-40-7
$C_{22}H_{25}N_3O_4$
4-Amino-2-(3,4-dihydro-6,7-dimethoxy-2(1H)-isoquinolyl)-6,7-dimethoxy quinoline.
An α-adrenergic blocker and antiarrhythmic.

204 Adimolol
78459-19-5
$C_{25}H_{29}N_3O_3$
(±)-1-[3-[[2-Hydroxy-3-(1-naphthyloxy)propyl]amino]-3-methylbutyl]-2-benzimidazolinone.
A long acting β-adrenoceptor blocker.

205 Amosulalol
85320-68-9 614
$C_{18}H_{24}N_2O_5S$
(±)-5-[1-Hydroxy-2-[[2-(o-methoxyphenoxy)ethyl]-amino]ethyl]-o-toluenesulfonamide.
An α_1-adrenergic blocking agent with antihypertensive activity. *Yamanouchi U.S.A. Inc.*

206 Amosulalol Hydrochloride
93633-92-2 614
$C_{18}H_{25}ClN_2O_5S$
(±)-5-[1-Hydroxy-2-[[2-(o-methoxyphenoxy)-ethyl]amino]ethyl]-o-toluenesulfonamide hydrochloride.
YM-09538; Lowgan. An α_1-adrenergic blocking agent

with antihypertensive activity. mp = 158-160°; [R(-)-form]: mp = 158°; $[\alpha]_D^{20}$ = -30.4° (c = 1 MeOH); [S(+)-form]: mp = 158°; $[\alpha]_D^{20}$ = 30.7° (c = 1 MeOH). *Yamanouchi U.S.A. Inc.*

207 Arotinolol
68377-92-4 827
$C_{15}H_{21}N_3O_2S_3$
(±)-5-[2-[[3-(tert-Butylamino)-2-hydroxypropyl]thio]-4-thiazolyl]-2-thiophenecarboxamide.
Antianginal agent. Also antihypertensive and antiarrhythmic. Possesses both α- and β-adrenergic blocking activity. A propanolamine derivative. mp = 148-149°. *Sumitomo Pharm. Co. Ltd.*

208 Arotinolol Hydrochloride
68377-91-3 827
$C_{15}H_{22}ClN_3O_2S_3$
(±)-5-[2-[[3-(tert-Butylamino)-2-hydroxypropyl]thio]-4-thiazolyl]-2-thiophenecarboxamide hydrochloride.
S-596; ARL; Almarl. Antianginal agent. Also antihypertensive and antiarrhythmic. Possesses both α- and β-adrenergic blocking activity. A propanolamine derivative. mp = 234-235.5°; LD_{50} (mus iv) = 86 mg/kg, (mus ip) = 360 mg/kg, (mus orl) > 5000 mg/kg. *Sumitomo Pharm. Co. Ltd.*

209 Aspidosperma
1398-11-4
Plant derived substances with adrenergic blocking activities for a variety of urogenital tissues.

210 Dapiprazole
72822-12-9 2884
$C_{19}H_{27}N_5$
5,6,7,8-Tetrahydro-3-[2-(4-o-tolyl-1-piperazinyl)-ethyl]-s-triazolo[4,3-a]pyridine.
An α_1-adrenergic blocking agent with antiglaucoma and miotic activity. mp = 158-160°. *Angelini Francesco.*

211 Dapiprazole Hydrochloride
72822-13-0 2884
$C_{19}H_{28}ClN_5$
5,6,7,8-Tetrahydro-3-[2-(4-o-tolyl-1-piperazinyl)-ethyl]-s-triazolo[4,3-a]pyridine hydrochloride.
AF-2139; Glamidolo; Reversil; Rev-Eyes. An α_1-adrenergic blocking agent with antiglaucoma and miotic activity. mp = 206-207°; LD_{50} (mus ip) = 260 mg/kg. *Angelini Francesco.*

212 Dexefaroxan
143249-88-1
$C_{13}H_{16}N_2O$
(+)-(R)-2-(2-Ethyl-2,3-dihydro-2-benzofuranyl)-2-imidazoline.
Selective α_2 antagonist.

213 Dibenamine
55-43-6 200-234-5
$C_{16}H_{19}Cl_2N$
Dibenzyl(2-chloroethyl)ammonium chloride.
An α-adrenergic blocking agent.

214 Doxazosin
74191-85-8 189
$C_{23}H_{25}N_5O_5$
1-(4-Amino-6,7-dimethoxy-2-quinazolinyl)-4-(1,4-benzodioxan-2-ylcarbonyl)piperazine.
UK-33274. Antihypertensive. Also used in the treatment of benign prostatic hyperplasia. Selective α-adrenergic blocker related to Prazosin. [monohydrochloride]: mp = 289-290°. *Pfizer Intl.; Roerig Div. Pfizer Pharm.*

215 Doxazosin Monomethanesulfonate
77883-43-3 3489
$C_{24}H_{29}N_5O_8S$
1-(4-Amino-6,7-dimethoxy-2-quinazolinyl)-4-(1,4-benzodioxan-2-ylcarbonyl)piperazine monomethanesulfonate.
Cardura; UK-33274-27; doxazosin mesylate; Alfadil; Cardenalin; Cardular; Cardura; Cardran; Diblocin; Normothen; Supressin. Antihypertensive. Also used in the treatment of benign prostatic hyperplasia. Selective α-adrenergic blocker related to Prazosin. *Pfizer Intl.; Roerig Div. Pfizer Pharm.*

216 Efaroxan
89197-32-0
$C_{13}H_{16}N_2O$
(±)-2-(2-Ethyl-2,3-dihydro-2-benzofuranyl)-2-imidazoline.
An α_2-adrenoceptor antagonist.

217 Ergoloid Mesylates
8067-24-1 3692
$C_{36}H_{45}N_5O_8S$
Dihydroergotoxine monomethanesulfonate (salt).
co-dercrine mesylate; CCK-179; Circanol; Coristin; Dacoren; DCCK; Deapril-ST; Decril; Dulcion; Ergodesit; Ergohydrin; Ergoplus; Hydergine; Lysergin; Novofluen; Orphol; Pérénan; Progeril; Redergin; Sponsin; Trigot. A mixture of hydrogenated ergot alkaloids, particularly dihydroergocornine mesylate, dihydroergocristine mesylate, and α- and β-dihydroergocryptine mesylates. An α_1-adrenergic blocking agent used for treatment of impaired mental function.

218 Fenspiride
5053-06-5 4041 225-751-3
$C_{15}H_{20}N_2O_2$
8-Phenethyl-1-oxa-3,8-diazaspiro[4.5]decan-2-one.
decaspiride; DESP. An α_1-adrenergic blocking agent used as a bronchodilator. An α_1-adrenergic blocking agent used as a bronchodilator. bp_2 = 126-127°. *Marion Merrell Dow Inc.*

219 Fenspiride Hydrochloride
5053-08-7 4041 225-752-9
$C_{15}H_{21}ClN_2O_2$
8-Phenethyl-1-oxa-3,8-diazaspiro[4.5]decan-2-one monohydrochloride.
NAT 333; NDR-5998A; Decaspir; Fluiden; Pneumorel; Respiride; Tegencia; Viarespan. An α_1-adrenergic blocking agent used as a bronchodilator. Dec 232-233° (dec); soluble in H_2O; LD_{50} (mus iv) = 106 mg/kg, (rat orl) = 437 mg/kg. *Marion Merrell Dow Inc.*

220 Indoramin

26844-12-2 5000 248-041-5

$C_{22}H_{25}N_3O$

N-[1-(2-Indol-3-ylethyl)-4-piperidyl]-benzamide.

Wy-21901. An α_1-adrenergic blocking agent with antihypertensive and bronchodilating activity. mp = 208-210°. *Wyeth-Ayerst Labs.*

221 Indoramin Hydrochloride

38821-52-2 5000 254-136-2

$C_{22}H_{26}ClN_3O$

N-[1-(2-Indol-3-ylethyl)-4-piperidyl]benzamide hydrochloride.

Wy-21901 HCl; Baratol; Doralese; Vidora; Wydora; Wypres; Wypresin. An α_1-adrenergic blocking agent with antihypertensive and bronchodilating activity. mp = 230-232°, 258-260°. *Wyeth-Ayerst Labs.*

222 Labetalol

36894-69-6 5341 253-258-3

$C_{19}H_{24}N_2O_3$

5-[1-Hydroxy-2-[(1-methyl-3-phenylpropyl)-amino]ethyl]salicylamide.

ibidomide; Dilevalol [as (R,R)-isomer]. Competitive α- and β-adrenergic receptor antagonist. Used as an antihypertensive. *Glaxo Wellcome Inc.; Key Pharm.*

223 Labetalol Hydrochloride

32780-64-6 5341 251-211-1

$C_{19}H_{25}ClN_2O_3$

5-[1-Hydroxy-2-[(1-methyl-3-phenylpropyl)-amino]ethyl]salicylamide hydrochloride.

Normodyne; Trandate; Sch-15719W; AH-5158A; Amipress; Ipolab; Labelol; Labracol; Presdate; Pressalolo; Vescal. Competitive α- and β-adrenergic receptor antagonist. Used as an antihypertensive. mp = 187-189°; soluble in H_2O, EtOH; insoluble in Et_2O, $CHCl_3$; LD_{50} (mmus ip) = 114 mg/kg, (mmus iv) = 47 mg/kg, (mmus orl) = 1450 mg/kg, (fmus ip) = 120 mg/kg, (fmus iv) = 54 mg/kg, (fmus orl) = 1800 mg/kg, (mrat ip) = 113 mg/kg, (mrat iv) = 60 mg/kg, (mrat orl) = 4550 mg/kg, (frat ip) = 107 mg/kg, (frat iv) = 53 mg/kg, (frat ol) = 4000 mg/kg. *C.M. Ind.; Glaxo Wellcome Inc.; Key Pharm.*

224 Midaglizole

66529-17-7

$C_{16}H_{17}N_3$

(±)-2-[α-(2-Imidazolin-2-ylmethyl)benzyl]pyridine.

Selective α2-adrenergic antagonist.

225 Naftopidil

57149-07-2 6443

$C_{24}H_{28}N_2O_3$

4-(o-Methoxyphenyl)-α-[(1-naphthyloxy)methyl]-1-piperazineethanol.

KT-611; Avishot; Flivas. An α_1 adrenergic blocker and serotonin $5HT_{1A}$ receptor agonist. Used as an antihypertensive agent. Also used in treatment of BPH. mp = 125-126°, 125-129°; insoluble in H_2O; LD_{50} (mus orl) = 1300 mg/kg, (rat orl) = 6400 mg/kg. *Boehringer Ingelheim Ltd.; Boehringer Mannheim GmbH; C.M. Ind.*

226 (±)-Naftopidil

132295-16-0 6443

$C_{24}H_{28}N_2O_3$

(±)-4-(o-Methoxyphenyl)-α-[(1-naphthyloxy)methyl]-1-piperazineethanol.

KT-611; Avishot; Flivas. An α_1 adrenergic blocker and serotonin $5HT_{1A}$ receptor agonist. Used as an antihypertensive agent. Also used in treatment of BPH. mp = 125-126°, 125-129°; insoluble in H_2O; LD_{50} (mus orl) = 1300 mg/kg, (rat orl) = 6400 mg/kg. *Boehringer Mannheim GmbH.*

227 Naftopidil Dihydrochloride

57149-08-3 6443

$C_{24}H_{30}Cl_2N_2O_3$

4-(o-Methoxyphenyl)-α-[(1-naphthyloxy)methyl]-1-piperazineethanol dihydrochloride.

α_1-Adrenergic blocker and serotonin $5HT_{1A}$ receptor agonist used as an antihypertensive. Also used to treat BPH. mp = 212-213°. *Boehringer Mannheim GmbH.*

228 Nicergoline

27848-84-6 6580 248-694-6

$C_{24}H_{26}BrN_3O_3$

10-Methoxy-1,6-dimethylergoline-8β-methanol 5-bromonicotinate (ester).

nicotergoline; nimergoline; MNE; FI-6714; Cergodum; Circo-Maren; Dilasenil; Duracebrol; Ergotop; Ergobel; Memoq; Nicergolent; Sermion; Vasospan. An α_1-adrenergic blocking agent with vasodilating properties. mp = 136-138°; LD_{50} (mmus orl) = 860 mg/kg, (mmus iv) = 46 mg/kg, (mrat orl) = 2800 mg/kg, (mrat iv) = 43 mg/kg. *Farmitalia Carlo Erba Ltd.*

229 Phenoxybenzamine

59-96-1 7409 200-446-8

$C_{18}H_{22}ClNO$

N-(2-Chloroethyl)-N-(1-methyl-2-phenoxyethyl)benzylamine.

bensylyt; 688-A. Antihypertensive. Antipheochromocytoma. α-Adrenergic blocker. CAUTION: The hydrochloride may be a carcinogen. mp = 38-40°; soluble in C_6H_6. *SmithKline Beecham Pharm.*

230 Phenoxybenzamine Hydrochloride

63-92-3 7409 200-569-7

$C_{18}H_{23}Cl_2NO$

N-(2-Chloroethyl)-N-(1-methyl-2-phenoxyethyl)benzylamine hydrochloride.

Dibenzyline; Dibenzylin; Dibenyline; Dibenzyran. Antihypertensive. Antipheochromocytoma. α-Adrenergic blocker. CAUTION: The hydrochloride may be a carcinogen. mp = 137.5-140°; soluble in EtOH, propylene glycol; sparingly soluble in H_2O. *SmithKline Beecham Pharm.*

231 Phentolamine

50-60-2 7417 200-053-1

$C_{17}H_{19}N_3O$

m-[N-(2-Imidazolin-2-ylmethyl)-p-toluidino]phenol.

Regitine; C-7337. An α-adrenergic blocker used as an antihypertensive agent. Used in the diagnosis and treatment of pheochromocytoma. mp = 174-175°. *Ciba-Geigy Corp.*

232 Phentolamine Hydrochloride
73-05-2 7417 200-793-5
$C_{17}H_{20}ClN_3O$
m-[N-(2-Imidazolin-2-ylmethyl)-p-toluidino]phenol hydrochloride.
Regitine hydrochloride. An α-adrenergic blocker with antihypertensive properties. Used in the diagnosis and treatment of pheochromocytoma. mp = 239-240°; soluble in H_2O (2 g/100 ml), EtOH (1.43 g/100 ml); slightly soluble in $CHCl_3$; insoluble in Me_2CO, EtOAc; LD_{50} (rat iv) = 75 mg/kg, (rat sc) = 275 mg/kg, (rat orl) = 1250 mg/kg. *Ciba-Geigy Corp.*

233 Phentolamine Methanesulfonate
65-28-1 7417 200-604-6
$C_{18}H_{23}N_3O_4S$
m-[N-(2-Imidazolin-2-ylmethyl)-p-toluidino]phenol monomethanesulfonate.
phentolamine mesylate; Regitine; Rogitine. An α-adrenergic blocker used as an antihypertensive agent. Also used in the diagnosis and treatment of pheochromocytoma. mp = 177-181°; soluble in H_2O (2 g/100 ml), EtOH (4.35 g/100 ml), $CHCl_3$ (0.15 g/100 ml). *Ciba-Geigy Corp.*

234 Prazosin
19216-56-9 7897 242-885-8
$C_{19}H_{21}N_5O_4$
1-(4-Amino-6,7-dimethoxy-2-quinazolinyl)-4-(2-furoyl)piperazine.
furazosin. An α_1-adrenergic blocking agent used as an anti-hypertensive agent and also in treatment of BPH. mp = 278-280°. *Brocades-Stheeman & Pharmacia; Pfizer Intl.*

235 Prazosin Hydrochloride
19237-84-4 7897 242-903-4
$C_{19}H_{22}ClN_5O_4$
1-(4-Amino-6,7-dimethoxy-2-quinazolinyl)-4-(2-furoyl)piperazine hydrochloride.
Minipress; CP-12299-1; Alpress LP; Duramipress; Eurex; Hypovase; Peripress; Sinetens. An α_1-adrenergic blocking agent used as an antihypertensive agent and also in treatment of BPH. Soluble in Me_2CO (0.72 mg/100 ml), MeOH (640 mg/100 ml), EtOH (84 mg/100 ml), DMF (130 mg/100 ml), dimethylacetamide (120 mg/100 ml), H_2O (140 mg/100 ml at pH 3.5), $CHCl_3$ (4.1 mg/100 ml); λ_m = 246, 329 nm (a_M = 137 ± 3, 27.6 ± 0.3 MeOH/1% HCl). *Brocades-Stheeman & Pharmacia; Pfizer Inc.*

236 Tamsulosin
106133-20-4 9217
$C_{20}H_{28}N_2O_5S$
(-)-(R)-5-[2-[[2-(o-Ethoxyphenoxy)ethyl]amino]propyl]-2-methoxybenzenesulfonamide.
amsulosin. Specific α_1-adrenoceptor antagonist. Used in treatment of benign prostatic hypertrophy. *Boehringer Ingelheim Pharm. Inc.; Yamanouchi U.S.A. Inc.*

237 dl-Tamsulosin Hydrochloride
80223-99-0 9217
$C_{20}H_{29}ClN_2O_5S$
(±)-(R)-5-[2-[[2-(o-Ethoxyphenoxy)ethyl]amino]propyl]-2-methoxybenzenesulfonamide hydrochloride.
LY-253351; YM-617; Amsulosin. Used in treatment of benign prostatic hypertrophy. Specific α_1-adrenoceptor antagonist. mp = 254-256°. *Boehringer Ingelheim Pharm. Inc.; Yamanouchi U.S.A. Inc.*

238 (R)-Tamsulosin Hydrochloride
106463-17-6 9217
$C_{20}H_{29}ClN_2O_5S$
(-)-(R)-5-[2-[[2-(o-Ethoxyphenoxy)ethyl]amino]propyl]-2-methoxybenzenesulfonamide hydrochloride.
LY-253351; R-(-)-YM-12617; YM-12617-1; YM-617; Harnal. Specific α_1-adrenoceptor antagonist. Used in treatment of benign prostatic hypertrophy. mp = 228-230°; $[\alpha]_D^{24}$ = -4.0° (c = 0.35 MeOH). *Boehringer Ingelheim Pharm. Inc.; Yamanouchi U.S.A. Inc.*

239 (S)-Tamsulosin Hydrochloride
106463-19-8 9217
$C_{20}H_{29}ClN_2O_5S$
(+)-(S)-5-[2-[[2-(o-Ethoxyphenoxy)ethyl]amino]propyl]-2-methoxybenzenesulfonamide hydrochloride.
YM-12617-2. Specific α_1-adrenoceptor antagonist. Used in treatment of benign prostatic hypertrophy. mp = 228-230°; $[\alpha]_D^{24}$ = +4.2° (c = 0.36 MeOH). *Boehringer Ingelheim Pharm. Inc.; Yamanouchi U.S.A. Inc.*

240 Terazosin [anhydrous]
63074-08-8 9297
$C_{19}H_{25}N_5O_4$
1-(4-Amino-6,7-dimethoxy-2-quinazolinyl)-4-(tetrahydro-2-furoyl)piperazine.
An α_1-adrenergic blocker related to prazosin. Used as an antihypertensive agent and in treatment of BPH. mp = 272.6-274°; λ_m = 212, 245, 330 nm (a 65.7, 127.5, 24.0 H_2O); soluble in MeOH (3.37 g/100 ml), H_2O (2.97 g/100 ml), EtOH (0.41 g/100 ml), $CHCl_3$ (0.12 g/100 ml), Me_2CO (.1 mg/100 ml); insoluble in C_6H_{14}. *Abbott Labs Inc.*

241 Terazosin Hydrochloride Dihydrate
70024-40-7 9297
$C_{19}H_{26}ClN_5O_4$
1-(4-Amino-6,7-dimethoxy-2-quinazolinyl)-4-(tetrahydro-2-furoyl)piperazine hydrochloride dihydrate.
Abbott-45975; Heitrin; Hytracin; Hytrin; Hytrinex; Itrin; Urodie; Vasocard; Vasomet; Vicard. Used to treat hypertension and benign antiprostatic hypertrophy. An α_1-adrenergic blocker related to prazosin. mp = 271-274°; soluble in H_2O (2.42 g/100 ml); LD_{50} (mrat iv) = 277 mg/kg, (frat iv) = 293 mg/kg; [anhydrous hydrochloride]: mp = 278-279°; soluble in H_2O (76.12 g/100 ml); LD_{50} (mus iv) = 259.3 mg/kg. *Abbott Labs Inc.*

242 Tibalosin
63996-84-9
$C_{21}H_{27}NOS$
(±)-erythro-2,3-Dihydro-α-[1-[(4-phenylbutyl)-amino]ethyl]benzo[b]thiophene-5-methanol.
CP-804-S. Heterocyclic aminoalcohol related to ifenprodil. Antagonist acting at the polyamine site of the N-methyl-D-aspartate (NMDA) subtype of glutamate receptor. Antihypertensive; alpha 1-adrenoceptor antagonist.

243 Tolazoline
59-98-3 9645 200-448-9
$C_{10}H_{12}N_2$
2-Benzyl-2-imidazoline.
phenylmethylimidazoline. Vasodilator (peripheral). An imidazoline α-adrenergic blocker. *CIBA plc.*

244 Tolazoline Hydrochloride
59-97-2 9645 200-447-3
$C_{10}H_{13}ClN_2$
2-Benzyl-2-imidazoline hydrochloride.
priscoline hydrochloride; Lambral; Priscol; Priscoline; Vaso-Dilatan. Vasodilator (peripheral). An imidazoline α-adrenergic blocker. mp = 174°; soluble in H_2O, EtOH, $CHCl_3$; slightly soluble in Et_2O, EtOAc; pH (2.5%) = 4.9-5.3. *CIBA plc.*

245 Trimazosin
35795-16-5 9828 252-732-7
$C_{20}H_{29}N_5O_6$
2-Hydroxy-2-methylpropyl 4-(4-amino-6,7,8-trimethoxy-2-quinazolinyl)-1-piperazine carboxylate.
An α_1-adrenergic receptor antagonist with antihypertensive activity. mp = 158-159°. *Pfizer Inc.*

246 Trimazosin Hydrochloride Monohydrate
53746-46-6 9828
$C_{20}H_{30}ClN_5O_6.H_2O$
2-Hydroxy-2-methylpropyl 4-(4-amino-6,7,8-trimethoxy-2-quinazolinyl)-1-piperazine carboxylate monohydrochloride monohydrate.
CP-19106-1; Cardovar; Supres. An α_1-adrenergic receptor antagonist with antihypertensive activity. mp = 166-169° (dec). *Pfizer Inc.*

247 Tropodifene
15790-02-0
$C_{25}H_{29}NO_4$
Tropine 3-(p-hydroxyphenyl-2-phenylpropionate (ester) acetate (ester).
An α_1-adrenergic blocker.

248 Yohimbine
146-48-5 10236 205-672-0
$C_{21}H_{26}N_2O_3$
17α-Hydroxyyohimban-16α-carboxylic acid.
quebrachine; corynine; aphrodine. Indole alkaloid with α_2-adrenergic blocking activity. Used as a mydriatic. mp = 234°, 235-237°; $[\alpha]_D^{20}$ = 50.9 - 62.2° (EtOH), 108° (C_5H_5N); λ_m = 226, 280, 291 nm (log ε 4.56, 3.88, 3.80 MeOH); sparingly soluble in H_2O; soluble in EtOH, $CHCl_3$, hot C_6H_6; moderately soluble in Et_2O.

α-Glucosidase Inhibitors

249 Acarbose
56180-94-0 15 260-030-7
$C_{25}H_{43}NO_{18}$
O-4,6-Dideoxy-4-[[(1S,4R,5R,6S)-4,5,6-trihydroxy-3-(hydroxymethyl)-2-cyclohexen-1-yl]amino]-α-D-glucopyranosyl-(1→4)-O-α-D-glucopyrranosyl-(1→4)-D-glucose.
Precose; Bay g 5421; Glucobay. An α-glucosidase inhibitor with antidiabetic activity. $[\alpha]_D^{18}$ = 165° (c = 0.4 H_2O). *Bayer AG.*

250 Miglitol
72432-03-2 6274 276-661-6
$C_8H_{17}NO_5$
(2R,3R,4R,5S)-1-(2-Hydroxyethyl)-2-(hydroxymethyl)-3,4,5-piperidinetriol.
Bay m 1099. The hydroxyethyl derivative of 1-deoxynojirimycin. An α-glucosidase inhibitor with antidiabetic activity. mp = 114°. *Bayer AG.*

251 Voglibose
83480-29-9
$C_{10}H_{21}NO_7$
3,4-Dideoxy-4-[[2-hydroxy-1-(hydroxymethyl)ethyl]-amino]-2-C-(hydroxymethyl)-D-epiinositol.
AO-128. An α-Glucosidase inhibitor.

Anabolics

252 Androisoxazole
360-66-7 674
$C_{21}H_{31}NO_2$
17β-Hydroxy-17α-methylandrostano[3,2-c]isoxazole.
Neo-Ponden. An anabolic agent. mp = 169-170°; $[\alpha]_D$ = 19°; λ_m = 226 nm (log ε = 3.71 EtOH).

253 Androstenediol
521-17-5 677 208-306-8
$C_{19}H_{30}O_2$
Androst-5-ene-3β,17β-diol.
An anabolic agent. mp = 184°; $[\alpha]_D^{18}$ = -55.5° (c = 0.4 iPrOH); insoluble in H_2O. *Nopco.*

254 Androstenediol 17-Acetate
5937-72-4 677
$C_{21}H_{32}O_3$
Androst-5-ene-3β,17β-diol 17-acetate.
An anabolic agent. mp= 146.5-148.5°; $[\alpha]_D^{18}$= -62.4° (EtOH). *Nopco.*

255 Androstenediol 3-Acetate
1639-43-6 677 216-681-4
$C_{21}H_{32}O_3$
Androst-5-ene-3β,17β-diol 3-acetate.
An anabolic agent. mp= 147-148°. *Nopco.*

256 Androstenediol 3-Acetate-17-benzoate
5953-63-9 677 227-717-3
$C_{28}H_{36}O_4$
Androst-5-ene-3β,17β-diol 3-acetate-17-benzoate.
An anabolic agent. mp = 180-182°. *Nopco.*

257 Androstenediol Diacetate
2099-26-5 677 218-264-2
$C_{23}H_{34}O_4$
Androst-5-ene-3β,17β-diacetate.
An anabolic agent. mp = 165-166°; $[\alpha]_D^{18}$= -56.5° (EtOH). *Nopco.*

258 Androstenediol Dipropionate
2297-30-5 677 218-943-3
$C_{25}H_{38}O_4$
Androst-5-ene-3β,17β-dipropionate.
Bisexovis; Stenandiol. An anabolic agent. *Nopco.*

259 Bolandiol
19793-20-5 1352
$C_{18}H_{28}O_2$
Estr-4-ene-3β,17β-diol.
An anabolic agent. mp = 169-172°. *Searle G.D. & Co.*

260 Bolandiol Dipropionate
1986-53-4 1352
$C_{24}H_{36}O_4$
Estr-4-ene-3β,17β-diol dipropionate.
SC-7525; norpropandrolate; Anabiol; Storinol. An anabolic agent. *Searle G.D. & Co.*

261 Bolasterone
1605-89-6 1353 216-519-2
$C_{21}H_{32}O_2$
17β-Hydroxy-7α,17-dimethylandrost-4-en-3-one.
7α,17-dimethyltestosterone; U-19763; NSC-66233. An anabolic agent. mp = 163-165°. *Pharmacia & Upjohn.*

262 Bolmantalate
1491-81-2
$C_{29}H_{40}O_3$
17β-Hydroxyestr-4-en-3-one 1-adamantanecarboxylate.
38851. An anabolic agent. *Eli Lilly & Co.*

263 Clostebol
1093-58-9 2475 214-133-9
$C_{19}H_{27}ClO_2$
4-Chloro-17β-hydroxyandrost-4-en-3-one.
4-chlorotestosterone. An anabolic agent. mp = 188-190°; $[\alpha]_D^{20}$ = 148° ($CHCl_3$); λ_m = 256 nm (log ε 4.13 EtOH). *Farmitalia Carlo Erba Ltd.; Julian.*

264 Clostebol Acetate
855-19-6 2475 212-720-4
$C_{21}H_{29}ClO_3$
4-Chloro-17β-acetoxyandrost-4-en-3-one.
Macrobin; Steranabol. An anabolic agent. mp = 228-230°; $[\alpha]_D$= 118° ±4° ($CHCl_3$); λ_m = 255 nm (ε 13300); soluble in EtOH. *Farmitalia Carlo Erba Ltd.; Julian.*

265 Ethylestrenol
965-90-2 3851 213-523-6
$C_{20}H_{32}O$
19-Norpregn-4-en-17β-ol.
Orabolin; Durabolin-O; Orgaboral; Maxibolin; Orgabolin. An anabolic agent. mp = 209-212°; $[\alpha]_D^{25}$ = -105° ($CHCl_3$); soluble in H_2O; LD_{50} (rat ip) = 104 mg/kg, (rat sc) = 270 mg/kg, (rat orl) > 1000 mg/kg, (mus ip) = 187 mg/kg, (mus sc) = 293 mg/kg. *Organon Inc.*

266 Formebolone
2454-11-7 4266 219-523-2
$C_{21}H_{28}O_4$
11α,17β-Dihydroxy-17-methyl-3-oxoandrosta-1,4-diene-2-carboxaldehyde.
formyldienolone; Esiclene. An anabolic agent. mp = 209-212°; $[\alpha]_D^{25}$= -105° ($CHCl_3$); soluble in H_2O; LD_{50} (rat ip) = 104 mg/kg, (rat sc) = 270 mg/kg, (rat orl) > 1000 mg/kg, (mus ip) = 187 mg/kg, (mus sc) = 293 mg/kg. *Lab. Prod. Biol. Braglia.*

267 Methandriol
521-10-8 6017 208-301-0
$C_{20}H_{32}O_2$
17α-Methylandrost-5-ene-3β,17β-diol.
Methostan; methylandrostenediol; MAD; mestenediol; Masdiol; Metocryst; Metildiolo; Androdiol; Metidione; Nabadial; Neosteron; Diolandrone; Stenediol; Protandren; Neostene; Crestabolic; Diolostene; Metendiol; Metandiol; Methandiol; Methanabol; Methostan; Neutrormone; Neutrosteron; Androteston-M; Megabion; Notandron. An anabolic agent. mp = 205.5-206.5°; $[\alpha]_D^{20}$ = -73° (EtOH); slightly soluble in organic solvents, insoluble in H_2O. *Schering-Plough HealthCare Products.*

268 Methandriol Diacetate
2061-86-1 6017 218-167-5
$C_{24}H_{36}O_4$
17α-Methylandrost-5-ene-3β,17β-diacetate.
An anabolic agent. mp= 145-146°; $[\alpha]_D^{21}$= -59° (c = 0.984 EtOH). *Schering-Plough HealthCare Products.*

269 Methandriol Dipropionate
3593-85-9 6017 222-735-8
$C_{26}H_{40}O_4$
17α-Methylandrost-5-ene-3β,17β-dipropionate.
Probolin. An anabolic agent. *Schering-Plough HealthCare Products.*

270 Methenolone
153-00-4 6044 205-812-0
$C_{20}H_{30}O_2$
17β-Hydroxy-1-methyl-5α-androst-1-en-3-one.
mêtênolone. An anabolic agent. mp = 149.5-152°, 160-161°; $[\alpha]_D$ = 58.9°. *Schering AG.*

271 Methenolone 17-Acetate
434-05-9 6044 207-097-0
$C_{22}H_{32}O_3$
17β-Hydroxy-1-methyl-5α-androst-1-en-3-one 17-acetate.
SH-567; SQ-16496; NSC-74226; Primobolan Tablets; Primonabol. An anabolic agent. mp = 138-139°; λ_m = 240 nm (ε 13300 MeOH); soluble in MeOH, Et_2O, $CHCl_3$. *Schering AG.*

272 Methenolone 17-Enanthate
303-42-4 6044 206-141-6
$C_{27}H_{42}O_3$
17β-Hydroxy-1-methyl-5α-androst-1-en-3-one 17-heptanoate.
SH-601; SQ-16374; NSC-64967; methenolone enanthate; Primobolan-Depot; Primonabol Depot. An anabolic agent. *Schering AG.*

273 Methyltrienolone
965-93-5 6211
$C_{19}H_{24}O_2$
17β-Hydroxy-17α-methylestra-4,9,11-trien-3-one.

metribolone; R-1881. An anabolic agent. mp = 170°; $[\alpha]_D^{20}$= -58.7° (c = 0.5 EtOH). *Roussel-UCLAF.*

274 Mibolerone
3704-09-4 6262 223-046-5
$C_{20}H_{30}O_2$
17β-Hydroxy-7α,17-dimethylestr-4-en-3-one.
U-10997; Cheque; Matenon. An anabolic steroid with androgenic properties. Soluble in H_2O (4.54 mg/100 ml at 37°). *Pharmacia & Upjohn.*

275 Nandrolone Cyclohexanecarboxylate
18470-94-5 6452 242-351-4
$C_{25}H_{36}O_3$
17β-Hydroxyestr-4-en-3-one cyclohexanecarboxylate.
19-nortestosterone hexahydrobenzoate; Nolongandron; Nor-Durandron. An anabolic steroid with androgenic properties. mp = 88-89°; $[\alpha]_D^{20}$ = 50° (c = 0.5 $CHCl_3$). *Forest Pharm. Inc.; Organon Inc.*

276 Nandrolone Cyclohexanepropionate
912-57-2 6452 213-013-3
$C_{27}H_{40}O_3$
17β-Hydroxyestr-4-en-3-one cyclohexanepropionate.
19-nortestosterone cyclohexylpropionate; Sanabolicum. An anabolic steroid with androgenic properties. *Forest Pharm. Inc.; Organon Inc.*

277 Nandrolone Decanoate
360-70-3 6452 206-639-3
$C_{28}H_{44}O_3$
17β-Hydroxyestr-4-en-3-one decanoate.
19-nortestosterone decanoate; Deca-Durabol; Deca-Hybolin; Hybolin decanoate; Deca-Durabolin; Nandrobolic L.A.; Retabolil. An anabolic steroid with androgenic properties. mp= 32-35°; soluble in EtOH, Et_2O, Me_2CO, $CHCl_3$; insoluble in H_2O. *Forest Pharm. Inc.; Organon Inc.*

278 Nandrolone Dodecanoate
26490-31-3 6452 247-739-7
$C_{30}H_{48}O_3$
17β-Hydroxyestr-4-en-3-one dodecanoate.
nandrolone laurate; Laurabolin. An anabolic steroid with androgenic properties. *Forest Pharm. Inc.; Organon Inc.*

279 Nandrolone Furylpropionate
7642-64-0 6452 231-580-5
$C_{25}H_{32}O_4$
17β-Hydroxyestr-4-en-3-one furylpropionate.
19-nortestosterone furylpropionate; NFP; Demelon. An anabolic steroid with androgenic properties. *Forest Pharm. Inc.; Organon Inc.*

280 Nandrolone Phenpropionate
62-90-8 6452 200-551-9
$C_{27}H_{34}O_3$
17β-Hydroxyestr-4-en-3-one hydrocinnamate.
Durabolin; Nandrobolic; NSC-23162; Activin; Durabol; Strabolene; Superanabolon; Nandrolin. An anabolic steroid with androgenic properties. mp = 95-96°; $[\alpha]_D$= 58° ($CHCl_3$). *Forest Pharm. Inc.; Organon Inc.*

281 Nandrolone p-Hexyloxyphenylpropionate
52279-57-9 6452 257-810-4
$C_{33}H_{46}O_4$
17β-Hydroxyestr-4-en-3-one p-Hexyloxyphenyl-propionate.
Anador; Anadur; 19-nortestosterone-3-(p-hexyloxyphenyl)propionate. An anabolic steroid with androgenic properties. mp = 53-55°; $[\alpha]_D$ = 45° (c = 1 dioxane). *Forest Pharm. Inc.; Organon Inc.*

282 Nandrolone Propionate
7207-92-3 6452 230-587-0
$C_{21}H_{30}O_3$
17β-Hydroxyestr-4-en-3-one propionate.
19-nortestosterone propionate. An anabolic steroid with androgenic properties. mp = 55-60°; $[\alpha]_D^{23.5}$= 58° ($CHCl_3$); λ_m = 240 nm (ε 16650). *Forest Pharm. Inc.; Organon Inc.*

283 Norbolethone
1235-15-0 6779
$C_{21}H_{32}O_2$
(±)-13-Ethyl-17β-hydroxy-18,19-dinorpregn-4-en-3-one.
Wy-3475; Genabol. An anabolic agent. mp = 144-145°; λ_m = 241 nm (ε 16500); LD_{50} (mus orl) > 5010 mg/kg; [d-form]: mp = 175-176°; $[\alpha]_D$= 20.7° ($CHCl_3$); [l-form]: mp = 172-175.5°; $[\alpha]_D$= -18.1° ($CHCl_3$). *Wyeth-Ayerst Labs.*

284 Oxabolone
4721-69-1 7034 225-212-2
$C_{18}H_{26}O_3$
4,17β-Dihydroxy-estr-4-en-3-one.
19-norandrost-4-ene-4,17β-diol-3-one. An anabolic agent. mp = 188-190°; λ_m = 278 nm (ε 11600). *Farmitalia Carlo Erba Ltd.*

285 Oxabolone 17-Cyclopentanepropionate
1254-35-9 7034 215-011-8
$C_{26}H_{38}O_4$
4,17β-Dihydroxy-estr-4-en-3-one 17-cyclopentanepropionate.
FI-5852; oxabolone cypionate; Steranabol Long-Acting; Steranabol ritardo. An anabolic agent. mp = 158-160°; $[\alpha]_D^{20}$= 30° (c = 1 $CHCl_3$); λ_m = 276 nm ($E^{1\%}_{1\,cm}$ = 315EtOH); soluble in $CHCl_3$, dioxane, C_6H_6; less soluble in MeOH; insoluble in H_2O, C_6H_{16}. *Farmitalia Carlo Erba Ltd.*

286 Oxymesterone
145-12-0 7099 205-646-9
$C_{20}H_{30}O_3$
4,17β-Dihydroxy-17α-methylandrost-4-en-3-one.
oxymestrone; Anamidol; Oranabol; Theranabol. An anabolic steroid with androgenic properties. Federally controlled substance (anabolic steroid). mp = 169-171°; $[\alpha]_D^{20}$= 69° (EtOH); λ_m = 278 nm ($E^{1\%}_{1\,cm}$ 406 EtOH); soluble in $CHCl_3$, Me_2CO, EtOH; insoluble in H_2O. *Farmitalia Carlo Erba Ltd.*

287 Pizotyline Hydrochloride
73391-87-4 7671
$C_{19}H_{22}ClNS$
4-(9,10-Dihydro-4H-benzo[4,5]cyclohepta[1,2-b]thien-4-ylidene)-1-methylpiperidine hydrochloride.

Anabolic; antimigraine; appetite stimulant; antidepressant. Tricyclic serotonin antagonist, specific to migraine. mp=261-263° (dec). *Sandoz Pharm. Corp.*

288 Pizotyline Malate
5189-11-7 7671 225-970-4
$C_{23}H_{27}NSO_5$
4-(9,10-Dihydro-4H-benzo[4,5]cyclohepta[1,2-b]thien-4-ylidene)-1-methylpiperidine compound with malic acid (1:1).
Sandomigran; Sanmigran; Sanomigran; Mosegor. Antimigraine; appetite stimulant; antidepressant. Tricyclic serotonin antagonist. mp = 185-186° (dec). *Sandoz Pharm. Corp.*

289 Pizotyline Malate
5189-1-7 7671
$C_{23}H_{27}NO_5S$
4-(9,10-Dihydro-4H-benzo[4,5]cyclohepta[1,2-b]thien-4-ylidene)-1-methylpiperidine malate.
Sandomigran; Sanmigran; Sanomigran; Mosegor. Anabolic; antidepressant; serotonin inhibitor. Used to alleviate migraine. mp = 185-186° (dec). *Sandoz Pharm. Corp.*

290 Quinbolone
2487-63-0 8236
$C_{24}H_{32}O_2$
17β-(1-Cyclopenten-1-yloxy)androsta-1,4-dien-3-one.
Anabolicum Vister. An anabolic agent. mp = 133-135°; $[\alpha]_D^{20}$= 61° (dioxane); λ_m = 244-245 nm ($e_{1\ cm}^{1\%}$ 430-450 dioxane); soluble in organic solvents; sparingly soluble in EtOH, C_6H_{14}; soluble in sesame oil (4.0-4.5 g/100 ml), insoluble in H_2O.

291 Stenbolone
5197-58-0 8962
$C_{20}H_{30}O_2$
17β-Hydroxy-2-methyl-5α-androst-1-en-3-one.
An anabolic agent. The acetate is used as an anabolic. mp = 155-158°; $[\alpha]_D$ = 52° ($CHCl_3$), $[\alpha]_D^{26}$= 47° ($CHCl_3$); λ_m = 241 nm (log ε = 3.99 EtOH). *Schering AG; Syntex Intl. Ltd.*

292 Stenbolone Acetate
1242-56-4 8962
$C_{22}H_{32}O_3$
17β-Hydroxy-2-methyl-5α-androst-1-en-3-one acetate.
Anatrofin. An anabolic agent. The acetate is used as an anabolic. mp= 155-158°; $[\alpha]_D$ = 32° ($CHCl_3$), $[\alpha]_D^{26}$= 60° ($CHCl_3$); λ_m = 241 nm (log ε 4.03). *Schering AG; Syntex Intl. Ltd.*

293 Trenbolone
10161-33-8 9716
$C_{18}H_{22}O_2$
17β-Hydroxyestra-4,9,11-trien-3-one.
An anabolic agent. mp = 186°, 183-186°; $[\alpha]_D^{20}$= 19° (c = 0.45 EtOH); λ_m = 239, 340.5 nm (ε = 5260, 28000). *Roussel-UCLAF.*

294 Trenbolone Acetate
10161-34-9 9716 233-432-5
$C_{20}H_{24}O_3$
17β-Hydroxyestra-4,9,11-trien-3-one acetate.
Finaplix; RU-1697. An anabolic agent. mp = 96-97°; $[\alpha]_D^{20}$= 36.8° (c = 0.37 MeOH). *Roussel-UCLAF.*

295 Trenbolone Cyclohexylmethyl Carbonate
23454-33-3 9716 245-669-1
$C_{26}H_{34}O_4$
17β-17-[(Cyclohexylmethoxy)carbonyl]oxyestra-4,9,11-trien-3-one.
Parabolan. An anabolic agent. mp = 90-95°; $[\alpha]_D^{20}$= 41.6° (c = 0.5 etOH). *Roussel-UCLAF.*

296 Zeranol
26538-44-3 247-769-0
$C_{18}H_{26}O_5$
(3S,7X)-3,4,5,6,7,8,9,10,11,12-Decahydro-7,14,16-trihydroxy-3-methyl1H-2-benzoxacyclotetradecin-1-one.
Ralabol; Ralgro; THFES(HM); P-1496; MK-188. An anabolic agent. *Pitman-Moore Inc.*

Analgesics, Dental

297 Chlorbutanol
57-15-8 2180 200-317-6
$C_4H_7Cl_3O$
1,1,1-Trichloro-2-methyl-2-propanol.
trichlorobutanol; β,β,β-Trichloro-tert-butyl alcohol; ace-tone chloroform; chlorbutol; Chloretone; Coliquifilm; Methaform; Sedaform. Analgesic, dental. Federally controlled substance. CAUTION: may be habit forming. [anhydrous]: mp = 97°; [hemihydrate]: mp = 78°; bp_{760} = 167°; bp_{246} = 135°; soluble in hot H_2O, alcohol, glycerol, $CHCl_3$, Et_2O, Me_2CO, petroleum ether, glacial acetic acid, oils; MLD (rbt orl) = 213 mg/kg. *Dynamit Nobel GmbH.*

298 Clove
84961-50-2 2479 284-638-7
caryophyllus. Analgesic, dental. Dried flower buds of *Eugenia caryophyllata Thunb. (Caryophyllus aromaticus l.), Myrtaceae.* Made up of 15-18% eugenol, caryophyllin, tannin, gum, resin. Pharmaceutic aid (flavor); used in manufacture of oil of clove, eugenol; used in baking, confections.

299 Eugenol
97-53-0 3944 202-589-1
$C_{10}H_{12}O_2$
4-Allyl-2-methoxyphenol.
2-Methoxy-4-(2-propenyl)phenyl; allylguaiacol; eugenic acid; caryophyllic acid. Analgesic, dental. Liquid; bp = 255°; mp ≅ -9°; d_4^{20} = 1.0664; n_D^{20} = 1.5410; nearly insoluble in H_2O; miscible with alcohol, $CHCl_3$, Et_2O, oils; LD_{50} (rat orl) = 2680 mg/kg.

300 Oil of Clove
8000-34-8 6893
Local anesthetic, dental; counterirritant; carminative. Volatile oil from dried flower buds of *Eugenia caryophyllata Thunb. (Caryophyllus aromaticus l.), Myrtaceae.* Made up of 82-87% eugenol, acetyleugenol, caryophyllene, furfural, vanillin, methyl amyl ketone. Used in confectionery, tooth-powders, microscopy. Liquid; bp = 250°; d_D^{25}= 1.038-1.060; α_D^{25} = -1°10'; n_D^{20} = 1.530; insoluble in H_2O; soluble in alcohol, Et_2O, glacial acetic acid.

Analgesics, Narcotic

301 Alfentanil
71195-58-9 236 71195-58-9
$C_{21}H_{32}N_6O_3$
N-[1-[2-(4-Ethyl-5-oxo-2-tetrazolin-1-yl)-ethyl]-4-(methoxymethyl)-4-piperidyl]-propionanilide.
Analgesic, narcotic. Tetrazole derivative of fentanyl; a federally controlled substance (opiate). *Janssen Pharm. Inc.*

302 Alfentanil Hydrochloride
70879-28-6 236 70879-28-6
$C_{21}H_{35}ClN_6O_4$
N-[1-[2-(4-Ethyl-5-oxo-2-tetrazolin-1-yl)ethyl]-4-(methoxymethyl)-4-piperidyl]propionanilide monohydrochloride monohydrate.
R-39209; Alfenta; Limifen; Rapifen. Analgesic, narcotic. Tetrazole derivative of fentanyl; a federally controlled substance (opiate). mp = 140.8°; soluble in H_2O; LD_{50} (rat iv) = 47.5 mg/kg, (dog iv) = 20 mg/kg. *Janssen Pharm. Inc.*

303 Alfentanil Hydrochloride [anhydrous]
69049-06-5 236 273-846-3
$C_{21}H_{33}ClN_6O_3$
N-[1-[2-(4-Ethyl-5-oxo-2-tetrazolin-1-yl)ethyl]-4-(methoxymethyl)-4-piperidyl]propionanilide monohydrochloride.
Analgesic, narcotic. Tetrazole derivative of fentanyl; a federally controlled substance (opiate). *Janssen Pharm. Inc.*

304 Alletorphine
23758-80-7
$C_{27}H_{35}NO_4$
17-Allyl-17-demethyl-7α-((R)-1-hydroxy-1-methylbutyl)-6,14-endo-ethenotetrahydrooripavine.
R&S-218-M. Analgesic, narcotic. A semisynthetic morphine analgesic.

305 Allylprodine
25384-17-2 304
$C_{18}H_{25}NO_2$
1-Methyl-4-phenyl-3-(2-propenyl)-4-piperidinol propanoate.
3-allyl-1-methyl-4-phenyl-4-propionyloxypiperidine; NIH-7440; Ro-2-7113; Alperidine. Analgesic, narcotic. Federally controlled substance (opiate).

306 Alphameprodine
468-51-9 207-409-5
$C_{17}H_{25}NO_2$
cis-3-Ethyl-1-methyl-4-phenyl-4-propionyloxypiperidine.
Analgesic, narcotic.

307 Alphaprodine
14405-05-1 317 201-011-5
$C_{16}H_{23}NO_2$
1,3-Dimethyl-4-phenyl-4-piperidinol propanoate.
α-prodine. Analgesic, narcotic. Federally controlled substance (opiate). *Hoffmann-LaRoche Inc.*

308 dl-Alphaprodine Hydrochloride
561-78-4 317
$C_{16}H_{24}ClNO_2$
1,3-Dimethyl-4-phenyl-4-piperidinol propanoate monohydrochloride.
Nu-1196; Nisentil; Nisintel; Prisilidene. Analgesic, narcotic. Federally controlled substance (opiate). *Hoffmann-LaRoche Inc.*

309 Anileridine
144-14-9 695
$C_{22}H_{28}N_2O_2$
1-[2-(4-Aminophenyl)ethyl]-4-phenyl-4-piperidinecarboxlic acid ethyl ester.
Alidine; Apidol; Leritine; Nipecotan. Analgesic, narcotic. Federally controlled substance (opiate). mp = 83°. *Merck & Co.Inc.*

310 Anileridine Dihydrochloride
126-12-5 695 204-770-0
$C_{22}H_{30}Cl_2N_2O_2$
1-[2-(4-Aminophenyl)ethyl]-4-phenyl-4-piperidinecarboxlic acid ethyl ester dihydrochloride.
N-β-(p-aminophenyl)ethylnormeperidine. Analgesic, narcotic. Dihydrochloride of anileridine. Federally controlled substance. mp = 280-287° (dec); freely soluble in H_2O, MeOH; soluble in EtOH (8 mg/ml); λ_m = 235, 289 nm ($A^{1\%}_{1\,cm}$ 293, 34.5 90% MeOH); pH of aqueous solution = 2.0-2.5. *Merck & Co.Inc.*

311 Anpirtoline
98330-05-3
$C_{10}H_{13}ClN_2S$
4-[(6-Chloro-2-pyridyl)thio]piperidine.
Narcotic analgesic. *Carter-Wallace.*

312 Asimadoline
153205-46-0
$C_{27}H_{30}N_2O_2$
N-[(αS)-α[[(3S)-3-hydroxy-1-pyrrolidinyl]methyl]-benzyl]-N-methyl-2,2-diphenylacetamide.
A non-narcotic analgesic.

313 Benzylmorphine
14297-87-1 1176 238-230-0
$C_{24}H_{25}NO_3$
7,8-Didehydro-4,5-epoxy-17-methyl-3-(phenylmethoxy)morphinan-6-ol.
Analgesic, narcotic. Federally controlled substance (opiate). mp = 132°; slightly soluble in H_2O; freely soluble in 50% alc, C_6H_6. *Merck & Co.Inc.*

314 Betameprodine
468-50-8 207-407-4
$C_{17}H_{25}NO_2$
β-3-Ethyl-1-methyl-4-phenyl-4-propionyloxypiperidine.
trans-3-Ethyl-1-methyl-4-phenyl-4-piperidinol propanoate (ester); Nu-1932. Analgesic, narcotic.

315 Betaprodine
468-59-7 317 207-411-6
$C_{16}H_{24}ClNO_2$
β-1,3-Dimethyl-4-phenyl-4-piperidinol propanoate.

Nu-1779. Analgesic, narcotic. Federally controlled substance (opiate). *Hoffmann-LaRoche Inc.*

316 Bezitramide
15301-48-1 1241 239-335-4
$C_{31}H_{32}N_4O_2$
1-(3-Cyano-3,3-diphenylpropyl)-4-(2-oxo-3-propionyl-1-benzimidazolinyl)piperidine.
1-[1-(3-cyano-3,3-diphenylpropyl)-4-piperidinyl]-1,3-dihydro-3-(1-oxopropyl)-2H-benzimidazol-2-one; R-4845; bezitramide; Burgodin. Analgesic, narcotic. Federally controlled substance (opiate). mp = 145-149°; soluble (> 1 g/100 ml) in EtOAc, Me_2CO, C_6H_6, $CHCl_3$; almost insoluble in H_2O, dilute acids; LD_{50} (mus orl) = 2101 mg/kg, (rat orl) = 141 mg/kg. *Janssen Pharm. Inc.*

317 Bremazocine
71990-00-6
$C_{20}H_{29}O_2$
6-ethyl-1,2,3,4,5,6-hexahydro-3-[(1-hydroxycyclopropyl)methyl]-11,11-dimethyl-2,6-methano-3-benzazocin-8-ol.
A narcotic analgesic.

318 Brifentanil
101345-71-5
$C_{20}H_{29}FN_6O_3$
(±)-cis-N-[1-[2-(4-Ethyl-5-oxo-2-tetrazolin-1-yl)ethyl]-3-methyl-4-piperidyl]-2'-fluoro-2-methoxyacetanilide.
Analgesic, narcotic. *Anaquest.*

319 Brifentanil Hydrochloride
117268-95-8
$C_{20}H_{30}ClFN_6O_3$
(±)-cis-N-[1-[2-(4-Ethyl-5-oxo-2-tetrazolin-1-yl)ethyl]-3-methyl-4-piperidyl]-2'-fluoro-2-methoxyacetanilide monohydrochloride.
A3331. Analgesic, narcotic. *Anaquest.*

320 Buprenorphine
52485-79-7 1522 257-950-6
$C_{29}H_{41}NO_4$
[5α,7α(S)]-17-(Cyclopropylmethyl)-α-(1,1'-dimethylethyl)-4,5-epoxy-18,19-dihydro-3-hydroxy-6-methoxy-α-methyl-6,14-ethenomorphinan-7-methanol.
RX-6029-M. Analgesic, narcotic. An analgesic that demonstrates narcotic agonist-antagonist properties. Federally controlled substance (narcotic). mp = 209°. *Lederle Labs.; Norwich Eaton.*

321 Buprenorphine Hydrochloride
53152-21-9 1522
$C_{29}H_{42}ClNO_4$
[5α,7α(S)]-17-(Cyclopropylmethyl)-α-(1,1-dimethylethyl)-4,5-epoxy-18,19-dihydro-3-hydroxy-6-methoxy-α-methyl-6,14-ethenomorphinan-7-methanol hydrochloride.
CL-112302; NIH-8805; UM-952; Buprenex; Lepetan; Temgesic. Analgesic, narcotic. Federally controlled substance (narcotic). *Lederle Labs.; Norwich Eaton.*

322 Butorphanol
42408-82-2 1565 255-808-8
$C_{21}H_{29}NO_2$
(-)-17-(Cyclobutylmethyl)morphinan-3,14-diol.
levo-BC-2627. Analgesic, narcotic; antitussive (veterinary). Mixed opioid agonist-antagonist. mp = 215-217°; $[\alpha]_D$ = -70.0° (c = 0.1 in MeOH). *Bristol-Myers Nutritional Group.*

323 Butorphanol Tartrate
58786-99-5 1565
$C_{27}H_{35}NO_8$
(-)-17-(Cyclobutylmethyl)morphinan-3,14-diol D-(-)-tartrate (1:1).
levo-BC-2627 tartrate; Stadol; Torate; Torbugesic; Torutrol (veterinary). Analgesic, narcotic; antitussive (veterinary). Mixed opioid agonist-antagonist. mp = 217-219°; $[\alpha]_D^{22}$ = -64.0° (c = 0.4 in MeOH); LD_{50} (mus iv) = 40-57 mg/kg, (mus orl) = 395-527 mg/kg. *Bristol-Myers Nutritional Group.*

324 Carfentanil
59708-52-0
$C_{24}H_{30}N_2O_3$
Methyl 1-phenethyl-4-(N-phenylpropionamido)isonipecotate.
Analgesic, narcotic. *Janssen Pharm. Ltd.*

325 Carfentanil Citrate
61380-27-6
$C_{30}H_{38}N_2O_{10}$
Methyl 1-phenethyl-4-(N-phenylpropionamido)isonipecotate citrate (1:1).
R-33799. Analgesic, narcotic. *Janssen Pharm. Ltd.*

326 Clonitazene
3861-76-5 2451
$C_{20}H_{23}ClN_4O_2$
2-(p-Chlorobenzyl)-1-(2-diethylaminoethyl)-5-nitrobenzimidazole.
Analgesic, narcotic. Federally controlled substance (opiate). mp = 75-76°. *CIBA plc.*

327 Codeine
6059-47-8 2525
$C_{18}H_{23}NO_4$
7,8-Didehydro-4,5α-epoxy-3-methoxy-17-methyl morphinan-6α-ol monohydrate.
Analgesic, narcotic; antitussive. Federally controlled substance (opiate). mp = 154 - 156°; d_4^{20} = 1.32; $[\alpha]_D^{15}$ = -136° (c= 2, EtOH); pK = 6.05; soluble in H_2O (0.8 g/100 ml); more soluble in organic solvents.

328 Codeine [anhydrous]
76-57-3 2525 200-969-1
$C_{18}H_{21}NO_3$
7,8-Didehydro-4,5α-epoxy-3-methoxy-17-methyl morphinan-6α-ol.
Methylmorphine; morphine monomethyl ether; morphine 3-methyl ether; Codicept. Analgesic, narcotic; antitussive. Federally controlled substance (opiate).

329 Codeine Hydrochloride
1422-07-7 2525 215-829-5
$C_{18}H_{22}ClNO_3$
7,8-Dehydro-4,5α-epoxy-3-methoxy-17-methyl morphinan-6α-ol monohydrochloride.
Analgesic, narcotic; antitussive. Federally controlled substance (opiate). mp = 280° (dec); $[\alpha]_D^{22}$ = -108°; soluble

in H_2O (5 g/100 ml), less soluble in organic solvents; LD_{50} (mus sc) = 300 mg/kg.

330 Codeine Methyl Bromide
125-27-9 2526
$C_{19}H_{24}BrNO_3$
7,8-Didehydro-4,5α-epoxy-3-methoxy-17-methyl morphinan-6α-ol methyl bromide.
Eucodin. Analgesic, narcotic; antitussive. Federally controlled substance (opiate). CAUTION: May be habit forming. mp ≅ 260°; soluble in H_2O, hot MeOH; sparingly soluble in alcohol; insoluble in $CHCl_3$, Et_2O.

331 Codeine Phosphate
41444-62-6 2528
$C_{18}H_{24}NO_7P.0.5H_2O$
7,8-Dehydro-4,5α-epoxy-3-methoxy-17-methyl morphinan-6α-ol phosphate (1:1) (salt) hemihydrate.
Colrex Compound; Ambenyl Cough Syrup; component of: Actifed with Codeine Cough Syrup, ASA and Codeine Compound, Codis, Empracet Codeine Phosphate, Empirin with Codeine Tablets, Isoclor Expectorant C, Naldecon-CX, Nucofed, Percogesic with Codeine, Tussar-2, Tussar-SF, Tussi-Organidin, Tylenol. Analgesic, narcotic; antitussive. Federally controlled substance (opiate). Soluble in H_2O; slightly soluble in alcohol. *Apothecon; Eli Lilly & Co.; Elkins-Sinn; Fisons Pharm. Div.; Glaxo Wellcome Inc.; Lemmon Co.; McNeil Pharm.; Parke-Davis; Rhône-Poulenc Rorer Pharm. Inc.; Roberts Pharm. Corp.; Solvay Pharm. Inc.; Wallace Labs; Wyeth-Ayerst Labs.*

332 Codeine Phosphate [anhydrous]
52-28-8 2528 200-137-8
$C_{18}H_{24}NO_7P$
7,8-Dehydro-4,5α-epoxy-3-methoxy-17-methyl morphinan-6α-ol phosphate (1:1) (salt).
Galcodine. Analgesic, narcotic; antitussive. Federally controlled substance (opiate). *Elkins-Sinn; Parke-Davis; Solvay Pharm. Inc.; Wyeth-Ayerst Labs.*

333 Codeine Sulfate
6854-40-6 2529
$C_{36}H_{48}N_2O_{10}S$
7,8-Dehydro-4,5α-epoxy-3-methoxy-17-methyl morphinan-6α-ol sulfate (2:1) (salt) trihydrate.
component of: Copavin. Analgesic, narcotic; antitussive. Federally controlled substance (opiate). Soluble in H_2O (3.3 g/100 ml); less soluble in organic solvents; pH = 5.0; protect from light. *Wyeth-Ayerst Labs.*

334 Codeine Sulfate [anhydrous]
1420-53-7 2529 215-818-5
$C_{36}H_{42}N_2O_{10}S.2H_2O$
7,8-Dehydro-4,5α-epoxy-3-methoxy-17-methyl morphinan-6α-ol sulfate (2:1) (salt).
Analgesic, narcotic; antitussive. Federally controlled substance (opiate). *Wyeth-Ayerst Labs.*

335 Cogazocine
57653-29-9
$C_{21}H_{31}NO$
3-(Cyclobutylmethyl)-6-ethyl-1,2,3,4,5,6-hexahydro-11,11-dimethyl-2,6-methano-3-benzazocin-8-ol.
Narcotic analgesic.

336 Conorphone
72060-05-0
$C_{23}H_{29}NO_3$
17-(Cyclopropylmethyl)-4,5α-epoxy-8β-ethyl-3-methoxymorphinan-6-one.
Codorphone. Analgesic, narcotic.

337 Conorphone Hydrochloride
70865-14-4
$C_{23}H_{30}ClNO_3$
17-(Cyclopropylmethyl)-4,5α-epoxy-8β-ethyl-3-methoxymorphinan-6-one hydrochloride.
TR-5109. Analgesic, narcotic.

338 Cyprenorphine
4406-22-8 2838
$C_{26}H_{33}NO_4$
6,7,8,14-N-(Cyclopropylmethyl)tetrahydro-7α-(1-hydroxy-1-methylethyl)-6,14-endo-ethenonororipavine.
Analgesic; etorphine antagonist (veterinary). Narcotic antagonist closely related to diprenorphine. Federally controlled substance (opium derivative). mp = 234° (7α- or 7β-linkage unspecified). *Reckitt & Colman Inc.*

339 Cyprenorphine Hydrochloride
16550-22-4 2838
$C_{26}H_{34}ClNO_4$
6,7,8,14-N-(Cyclopropylmethyl)tetrahydro-7α-(1-hydroxy-1-methylethyl)-6,14-endo-ethenonororipavine hydrochloride.
M-285. Analgesic; etorphine antagonist (veterinary). Narcotic antagonist closely related to diprenorphine. Federally controlled substance (opium derivative). mp = 248°. *Reckitt & Colman Inc.*

340 Desomorphine
427-00-9 2972 207-045-7
$C_{17}H_{21}NO_2$
4,5-Epoxy-3-hydroxy-N-methylmorphinan.
dihydroepoxymorphine. Analgesic, narcotic. Federally controlled substance (opium derivative). mp = 189°; sublimes in high vacuum ≅ 140-170°; $[\alpha]_D^{28}$ = -77° (c = 1.6 in MeOH); soluble in Me_2CO, EtOAc. *Hoffmann-LaRoche Inc.*

341 Dextromoramide
357-56-2 2997 206-613-1
$C_{25}H_{32}N_2O_2$
(+)-4-[2-Methyl-4-oxo-3,3-diphenyl-4-(1-pyrrolidinyl)-butyl]morphine.
pyrrolamidol; R-875; SKF-5137; Jetrium; Dimorlin; d-Moramid(e); Palfium; Palphium. Analgesic, narcotic. mp = 180-184°; $[\alpha]_D^{20}$ = 25.5° (c = 5 C_6H_6); λ_m = 254, 260, 264 nm (0.01N iPrOH-HCl);soluble in 0.1N HCl, EtOH (50 g/100 ml); MeOH (40 g/100 ml); Me_2CO (50 g/100 ml); EtOAc (40 g/100 ml); C_6H_6 (5 g/100 ml); $CHCl_3$ (5 g/100 ml); soluble in Et_2O. *Janssen Pharm. Inc.*

342 Dextromoramide Tartrate
2922-44-3 2997 220-870-7
$C_{29}H_{38}N_2O_8$
(+)-4-[2-Methyl-4-oxo-3,3-diphenyl-4-(1-pyrrolidinyl)-butyl]morpholine bitartrate.
dextromoramide bitartrate; Dimorlin Tartrate. Analgesic, narcotic. Neurotropic. Used as an antipsychotic.

Federally controlled substance (opiate). Dec 189-192°; solubility (w/v) at 25°: in H_2O 20%, $CHCl_3$ 30%, MeOH 40%, EtOH 100%, Me_2CO 100%. *Janssen Pharm. Inc.; SmithKline Beecham Pharm.*

343 Dextrorphan
125-73-5 5496 204-754-3
$C_{17}H_{23}NO$
(+)-cis-1,3,4,9,10,10a-Hexahydro-11-methyl-2H-10,4a-iminoethanophenanthren-6-ol.
d-form of levorphanol; Ro-1-6794. Analgesic, narcotic. Federally controlled substance (opiate). *See* levorphanol. mp = 198-199°; $[\alpha]_D^{20}$ = +34.6° (c = 3 in H_2O); soluble in H_2O. *Hoffmann-LaRoche Inc.*

344 Dextrorphan Hydrochloride
69376-27-8 5496
17-Methyl-9α,13α,14α-morphinan-3-ol hydrochloride.
Ro-1-6794/706. Analgesic, narcotic. Vasospastic therapy adjunct. Federally controlled substance (opiate). *Hoffmann-LaRoche Inc.*

345 Dezocine
53648-55-8 2998
$C_{16}H_{23}NO$
(-)-13β-Amino-5,6,7,8,9,10,11α,12-octahydro-5α-methyl-5,11-methanobenzocyclodecen-3-ol.
WY-16255, Dalgan. Analgesic, narcotic. Synthetic opiate agonist-antagonist. *Am. Home Products.*

346 Diacetylmorphine
561-27-3 3012 209-217-7
$C_{21}H_{23}NO_5$
(5α,6α)-7,8-Didehydro-4,5-epoxy-17-methylmorphinan-3,6-diol diacetate (ester).
heroin; diamorphine; acetomorphine; China white. Narcotic analgesic prepared from morphine and acetyl chloride. Federally controlled substance (opium derivative). CAUTION: May be habit forming. mp = 173°; bp_{12} = 272-274°; $[\alpha]_D^{25}$ = -166° (c = 1.49 in MeOH); soluble in $CHCl_3$ (66 g/100 ml), alc (3 g/100 ml), Et_2O (1 g/100 ml), H_2O (0.1 g/100 ml); slightly soluble in ammonia, sodium carbonate solution; LD_{50} (mus iv) = 59 μmol/kg.

347 Diacetylmorphine Hydrochloride
1502-95-0 3012 216-124-5
$C_{21}H_{26}ClNO_6$
(5α,6α)-7,8-Didehydro-4,5-epoxy-17-methylmorphinan-3,6-diol diacetate (ester) hydrochloride monohydrate.
Narcotic analgesic prepared from morphine and acetyl chloride. Federally controlled substance (opium derivative). CAUTION: May be habit forming. mp = 243-244°; $[\alpha]_D^{25}$ = -156° (c = 1.044); soluble in 2 parts H_2O, 11 parts alcohol; insoluble in Et_2O.

348 Diacetylnalorphine
2748-74-5
$C_{23}H_{27}NO_5$
(-)-(5R,6S)-9α-allyl-4,5-epoxymorphin-7-en-3,6-diyldiacetate.
Narcotic analgesic.

349 Diampromide
552-25-0 3032
$C_{21}H_{28}N_2O$
N-[2-(Methylphenethylamino)propyl]propionanilide.
Analgesic, narcotic. Federally controlled substance (opiate). Liquid; $bp_{0.5}$ = 174-175°; n_D^{26} = 1.546. *Am. Cyanamid.*

350 Diethylthiambutene
86-14-6 9429
$C_{16}H_{21}NS_2$
3-Diethylamino-1,1-di-(2'-thienyl)-1-butene.
thiambutene; 191C49; NIH-4185. Analgesic, narcotic. Federally controlled substance (opiate). $bp_{0.03}$ = 122-128°; LD_{50} (mus ip) = 90 mg/kg. *Burroughs Wellcome Inc.*

351 Diethylthiambutene Hydrochloride
132-19-4 9429
$C_{16}H_{22}ClNS_2$
3-Diethylamino-1,1-di-(2'-thienyl)-1-butene hydrochloride.
diethylthiambutene hydrochloride; Themalon. Analgesic, narcotic. Federally controlled substance (opiate). mp = 152-153°. *Burroughs Wellcome Inc.*

352 Dihydrocodeine
125-28-0 3214 204-732-3
$C_{18}H_{23}NO_3$
4,5α-Epoxy-3-methoxy-17-methylmorphinan-6α-ol.
dihydroneopine; drocode; Codhydrine; Dehacodin; DF-118; DH-codeine; Didrate; Dihydrin; Hydrocodin; Nadeine; Novicodin; Paracodin; Parzone; Rapacodin. Analgesic, narcotic; antitussive. Federally controlled substance (opiate). mp = 112-113°; bp_{15} = 248°.

353 Dihydrocodeine Bitartrate
5965-13-9 3214 227-747-7
$C_{22}H_{29}NO_9$
4,5α-Epoxy-3-methoxy-17-methylmorphinan-6α-ol (+)-tartrate (salt).
dihydrocodeinone bitartrate; DF-118; Fortuss; DHC; Dico; component of: Synalgos-DC, DHCplus. Analgesic, narcotic; antitussive. Federally controlled substance (opiate). *See* hydrocodone bitartrate. Anhydrous crystals from MeOH (66.8% dihydrocodeine): mp = 192-193°; medicinal grade: mp = 186-190°; $[\alpha]_D^{25} \cong$ -73° (c = 1 in H_2O); soluble in H_2O (1 g/4.5 ml); sparingly soluble in alcohol; insoluble in Et_2O. *Purdue Pharma L.P.; Wyeth-Ayerst Labs.*

354 Dihydromorphine
509-60-4 3220 208-100-8
$C_{17}H_{21}NO_3$
(5α,6α)-4,5-epoxy-17-methylmorphinan-3,6-diol.
7,8-dihydromorphine. Analgesic, narcotic. Federally controlled substance (opiate). CAUTION: may be habit forming. Monohydrate: mp = 157°; Insoluble in H_2O; soluble in Me_2CO, alc, $CHCl_3$, dilute acids.

355 Dihydromorphine Hydrochloride
1421-28-9 3220
$C_{17}H_{22}ClNO_3$
(5α,6α)-4,5-epoxy-17-methylmorphinan-3,6-diol hydrochloride.
paramorphan; paramorfan. Analgesic, narcotic. Federally

controlled substance (opiate). CAUTION: may be habit forming. mp > 280°; $[\alpha]_D^{25}$ = -112° (c = 1.6); soluble in H_2O; sparingly soluble in absolute alcohol.

356 Dimenoxadol
509-78-4 3253
$C_{20}H_{25}NO_3$
2-Dimethylaminoethyl 1-ethoxy-1,1-diphenylacetate.
Lokarin. Analgesic, narcotic. Federally controlled substance (opiate).

357 Dimepheptanol
545-90-4 3254
$C_{21}H_{29}NO$
6-Dimethylamino-4,4-diphenyl-3-heptanol.
NIH-2933; bimethadol; Methadol; Pangerin. Analgesic, narcotic. Federally controlled substance (opiate). [β-dl-form]: mp = 127-128°; [β-d-form]: mp = 106-107°; $[\alpha]_D^{20}$ = +178° (c = .063 in alcohol); [β-l-form]: mp = 105-107°; $[\alpha]_D^{20}$ = -178° (c = 1.04 in alcohol). *Bristol Laboratories; Merck & Co.Inc.*

358 Dimethylthiambutene
524-84-5 3309
$C_{14}H_{17}NS_2$
3-Dimethylamino-1,1-di-(2'-thienyl)-1-butene.
338C48; NIH-4542; Ohton; Aminobutene; Dimethibutin; Kobaton; Takaton. Analgesic, narcotic. Federally controlled substance (opiate). Oil; $bp_{0.05}$ = 157-158°; soluble in $CHCl_3$, Et_2O; LD_{50} (mus sc) = 1.21 mg/10 kg; (dog iv) = 20.3 mg/10 kg. *Burroughs Wellcome Inc.*

359 Dioxaphetyl Butyrate
467-86-7 3354 207-402-7
$C_{22}H_{27}NO_3$
Ethyl 4-morpholino-2,2-diphenylbutyrate.
Amidalgon; Spasmoxal; Spasmoxale. Antispasmodic; narcotic analgesic. Federally controlled substance (opiate). [hydrochloride ($C_{22}H_{28}ClNO_3$)]: mp = 168-169°. *Sterling Winthrop Inc.; Winthrop.*

360 Dipipanone
467-83-4 3398 207-399-2
$C_{24}H_{31}NO$
4,4-Diphenyl-6-piperidino-3-heptanone.
378C48; Hoechst 10805; Piperidyl Amidone; Pamedon(e); Fenpidon; Pipadone. Analgesic, narcotic. Federally controlled substance (opiate). *Burroughs Wellcome Inc.*

361 Dipipanone Hydrochloride
856-87-1 3398 278-307-6
$C_{24}H_{32}ClNO$
4,4-Diphenyl-6-piperidino-3-heptanone hydrochloride.
Analgesic, narcotic. Federally controlled substance (opiate). mp = 123-126° [also reported as 126-127°]. *Burroughs Wellcome Inc.*

362 Eptazocine
72522-13-5 3673
$C_{15}H_{21}NO$
(-)-(1S,6S)-2,3,4,5,6,7-Hexahydro-1,4-dimethyl-1,6-methano-1H-4-benzazonin-10-ol.
ST-2121; Sedapain. Analgesic, narcotic. Opioid agonist-antagonist analgesic, related to pentazocine.

363 Ethylmethylthiambutene
441-61-2 3875 207-128-8
$C_{15}H_{19}NS_2$
3-Ethylmethylamino-1,1-di-(2'-thienyl)-1-butene.
ethylmethiambutene; 1C50; NIH-5145; Emethibutin. Analgesic, narcotic. Federally controlled substance (opiate). $bp_{0.01}$ = 110-113°; bp_{18} = 75-76°. *Burroughs Wellcome Inc.*

364 Ethylmorphine
125-30-4 3876 200-970-7
$C_{19}H_{23}NO_3$
7,8-Didehydro-4,5-epoxy-3-ethoxy-17-methyl-morphinan-6-ol.
Analgesic, narcotic; antitussive. Federally controlled substance (opiate). mp = 199-201°.

365 Ethylmorphine Hydrochloride
76-58-4 3876 204-734-4
$C_{19}H_{28}ClNO_5$
7,8-Didehydro-4,5-epoxy-3-ethoxy-17-methyl-morphinan-6-ol hydrochloride dihydrate.
Codethyline; Dionin. Analgesic, narcotic; antitussive. Federally controlled substance (opiate). mp ≅ 123°; soluble in H_2O (10 g/100 ml), alc (4 g/100 ml); slightly soluble in $CHCl_3$, Et_2O; LD_{50} (mus sc) = 264.6 mg/kg; (mus orl) = 77.1 mg/kg; [anhydrous form]: mp ≅ 170°.

366 Ethylmorphine Methiodide
6696-59-9 3876
$C_{20}H_{26}INO_3$
7,8-Didehydro-4,5-epoxy-3-ethoxy-17-methyl-morphinan-6-ol methiodode.
ethyl-N-methylmorphinium iodide; Trachyl. Analgesic, narcotic; antitussive. Federally controlled substance (opiate).

367 Ethylnarceine Hydrochloride
12246-80-9 6508 235-480-2
$C_{25}H_{32}ClNO_8$
Ethyl-6-[[6-[2-(dimethylamino)ethyl]-4-methoxy-1,3-benzodioxol-5-yl]acetyl]-2,3-dimethoxybenzoic acid hydrochloride.
Narcyl. Analgesic, narcotic; antitussive. *See* narceine. mp = 208-210°; soluble in H_2O, EtOH, $CHCl_3$; insoluble in Et_2O.

368 Etonitazene
911-65-9 3929 213-009-1
$C_{22}H_{28}N_4O_3$
1-(2-Diethylaminoethyl)-2-(p-ethoxybenzyl)-5-nitrobenzimidazole.
Ba-20684; NIH-7607. Analgesic, narcotic. Binds selectively to mu-opioid receptors. Federally controlled substance (opiate). *CIBA plc.*

369 Etorphine
14521-96-1 3932 238-535-9
$C_{25}H_{33}NO_4$
[5α,7α(R)]-4,5-Epoxy-3-hydroxy-6-methoxy-α,17-dimethyl-α-propyl-6,14-ethenomorphinan-7-methanol.
19-propylorvinol. Narcotic analgesic. Used to immobilize

large animals (veterinary). Federally controlled substance (opiate derivative). CAUTION: very small quantities may cause respiratory paralysis and death. mp = 214-217°; mp [7β isomer] = 280°. *Lemmon Co.*

370 Etorphine Hydrochloride
13764-49-3 3932

$C_{23}H_{34}ClNO_4$

[5α,7α(R)]-4,5-Epoxy-3-hydroxy-6-methoxy-α,17-dimethyl-α-propyl-6,14-ethenomorphinan-7-methanol hydrochloride.

M-99; component (with acepromazine) of Immobilon. Narcotic analgesic. Used to immobilize large animals (veterinary). Federally controlled substance (opiate derivative). CAUTION: very small quantities may cause respiratory paralysis and death. mp = 266-267°; [7β isomer hydrochloride]: dec 290°. *Lemmon Co.*

371 Fentanyl
437-38-7 4043 207-113-6

$C_{22}H_{28}N_2O$

N-(1-Phenylethyl-4-piperidinyl)propionanilide.

phentanyl; R-4263. Analgesic, narcotic; tranquilizer (veterinary). Federally controlled substance (opiate). mp = 83-84°. *Abbott Laboratories Inc.*

372 Fentanyl Citrate
990-73-8 4043 213-588-0

$C_{28}H_{36}N_2O_8$

N-(1-Phenylethyl-4-piperidinyl)propionanilide citrate (1:1).

Fentanest; Leptanol; Pentanyl; Sublimaze. Analgesic, narcotic; tranquilizer (veterinary). Federally controlled substance (opiate). mp = 149-151°; soluble in H_2O (1 g/40 ml), less soluble in organic solvents; LD_{50} (mus iv) = 11.2 mg/kg, (msu sc) = 62 mg/kg. *Abbott Laboratories Inc.*

373 Hydrocodone
125-29-1 4826 204-733-9

$C_{18}H_{21}NO_3$

4,5α-Epoxy-3-methoxy-17-methylmorphinan-6-one.

dihydrocodeinone; Bekadid; Dicodid. Analgesic, narcotic; antitussive. Federally controlled substance (opiate). CAUTION: may be habit forming. mp = 198°; soluble in alcohol, dilute acids; insoluble in H_2O; λ_m = 280 nm (ε 1310); LD_{50} (mus sc) = 85.7 mg/kg. *Merck & Co.Inc.*

374 Hydrocodone Bitartrate
34195-34-1 4826

$C_{22}H_{27}NO_9.21/2H_2O$

4,5α-Epoxy-3-methoxy-17-methylmorphinan-6-one tartrate (1:1) hydrate (2:5).

Calmodid; Codinovo; Dicodid; Duodin; Hydrokon; Kolikodal; Orthoxycol; Mercodinone; Norgan; Synkonin; component of: Anexsia, Anodynos-DHC, Hycodan, Hycomine, Hycomine Compound, Hycomine Pediatric, Hycotuss, Hy-Phen, Kwelcof, PV Tussin Syrup, PV Tussin Tablet, Vicodin, Zydone. Analgesic, narcotic; antitussive. Federally controlled substance (opiate). CAUTION: may be habit forming. mp = 118-128°; soluble in H_2O (6/25 g/100 ml), 95% EtOH (0.67 g/100 g); nearly insoluble in Et_2O, $CHCl_3$; pH (2% aqueous solution) = 3.6. *Ascher B.F. & Co.; DuPont-Merck Pharm.; Knoll Pharm. Co.; Marion Merrell Dow Inc.; Merck & Co.Inc.; SmithKline Beecham Pharm.; Solvay Pharm. Inc.*

375 Hydrocodone Bitartrate [anhydrous]
143-71-5 4826 205-608-1

$C_{22}H_{27}NO_9$

4,5α-Epoxy-3-methoxy-17-methylmorphinan-6-one tartrate (1:1).

Analgesic, narcotic; antitussive. Federally controlled substance (opiate). CAUTION: may be habit forming. *See* dihydrocodeine bitartrate. *Merck & Co.Inc.*

376 Hydromorphone
466-99-9 4847 207-383-5

$C_{17}H_{19}NO_3$

4,5α-Epoxy-3-hydroxy-17-methylmorphinan-6-one.

dihydromorphinone; Dimorphone; Novolaudon. Analgesic, narcotic. Federally controlled substance (opiate). CAUTION: may be habit forming. mp = 266-267°; $[\alpha]_D^{25} \cong -194°$ (c = 0.98 in dioxane). *Mallinckrodt Inc.*

377 Hydromorphone Hydrochloride
71-68-1 4847 200-762-6

$C_{17}H_{20}ClNO_3$

4,5α-Epoxy-3-hydroxy-17-methylmorphinan-6-one hydrochloride.

dihydromorphinone hydrochloride; Dilaudid; Laudicon; Hymorphan. Analgesic, narcotic. Federally controlled substance (opiate). CAUTION: may be habit forming. Dec 301-315°; $[\alpha]_D^{25} \cong -133°$; soluble in 3 parts H_2O; sparingly soluble in alcohol; LD_{50} (mus iv) = 61-96 mg/kg. *Knoll Pharm. Co.; Mallinckrodt Inc.; Wyeth-Ayerst Labs.*

378 Hydroxypethidine
468-56-4 4883 207-410-0

$C_{15}H_{21}NO_3$

4-(m-Hydroxyphenyl)-1-methylpiperidine-4-carboxylic acid ethyl ester.

Win-771; Hoechst 10446; demidone; oxipethidine; oxypetidin; emidone; Biphenal. Analgesic, narcotic. Federally controlled substance (opiate). mp = 110°. *Hoechst Marion Roussel Inc.; Sterling Winthrop Inc.*

379 Isomethadone
466-40-0 5201

$C_{21}H_{27}NO$

6-Dimethylamino-5-methyl-4,4-diphenyl-3-hexanone.

Win-1783; isoamidone; Isadanone; Liden (l-form). Analgesic, narcotic. Federally controlled substance (opiate). [dl-form]: bp_{12} = 215°; [d-form]: oil; $[\alpha]_D^{25}$ = +20.8;[l-form]: oily liquid; $bp_{0.6}$ = 162-165°; $[\alpha]_D^{25}$ = -20° (c = 1.5 in 95% EtOH). *Sterling Health U.S.A.*

380 Ketobemidone
469-79-4 5312 207-421-0

$C_{15}H_{21}NO_2$

4-(m-Hydroxyphenyl)-1-methyl-4-propionylpiperidine.

Analgesic, narcotic. Federally controlled substance (opiate). mp = 156-157°. *CIBA plc; Sterling Winthrop Inc.*

381 Ketobemidone Hydrochloride
5965-49-1 5312 227-749-8

$C_{15}H_{22}ClNO_2$

4-(m-Hydroxyphenyl)-1-methyl-4-propionylpiperidine hydrochloride.

Hoechst-10720; Win-1539; Cliradon; Cymidon; Ketogan; Ketogin. An analgesic, narcotic. Federally controlled substance (opiate). mp = 201-202°; soluble in H_2O;

slightly soluble in alcohol. *CIBA plc; Sterling Winthrop Inc.*

382 Ketorfanol
79798-39-3
$C_{20}H_{25}NO_2$
17-(Cyclopropylmethyl)-4-hydroxymorphinan-6-one.
SBW-22. Analgesic.

383 Levomethadone
125-58-6
$C_{21}H_{27}NO$
(-)-(R)-6-(Dimethylamino)-4,4-diphenyl-3-heptanone.
Analgesic, narcotic.

384 Levomethadyl
1477-40-3 5492
$C_{21}H_{27}N$
(-)-6-(Dimethylamino)-4,4-diphenyl-3-heptanol.

385 Levomethadyl Acetate
34433-66-4 5492
$C_{23}H_{31}NO_2$
(-)-6-(Dimethylamino)-4,4-diphenyl-3-heptanol acetate (ester).
levo-α-acetylmethadol; α-l-acetylmethadol; LAAM. Used in treatment of narcotic addiction. Longest acting enantiomer of methadyl acetate.

386 Levomethadyl Acetate Hydrochloride
43033-72-3 5492
$C_{23}H_{32}ClNO_2$
(-)-6-(Dimethylamino)-4,4-diphenyl-3-heptanol acetate (ester) hydrochloride.
ORLAAM. Used in treatment of narcotic addiction. Longest acting enantiomer of methadyl acetate. mp = 215°; $[\alpha]_D^{25}$ = -60° (c = 0.2); soluble in H_2O; LD_{50} (mus sc) = 110.0 mg/kg, (mus orl) = 172.8 mg/kg.

387 Levomoramide
5666-11-5 227-123-4
$C_{25}H_{32}N_2O_2$
(-)-4-[2-Methyl-4-oxo-3,3-diphenyl-4-(1-pyrrolidinyl)butyl]morpholine.
Inactive isomer of moramide, a narcotic analgesic. Opiate.

388 Levorphanol
77-07-6 5496
$C_{17}H_{23}NO$
17-Methylmorphinan-3-ol.
Ro-1-5431; levorphan; lemoran. Analgesic, narcotic. Orally active synthetic morphine analog. Federally controlled substance (opiate) mp = 198-199°; $[\alpha]_D^{20}$ = -56° (c = 3 in absolute alcohol); [dl-form]: mp = 215-253°; [d-form]: mp = 198-199°; $[\alpha]_D^{20}$ = +56.3° (c = 3 in absolute alcohol). *Hoffmann-LaRoche Inc.*

389 Levorphanol Tartrate [anhydrous]
125-72-4 5496
$C_{21}H_{29}NO_7$
17-Methylmorphinan-3-ol tartrate (1:1) (salt).
Analgesic, narcotic. Orally active synthetic morphine analog. Federally controlled substance (opiate). mp = 206-208°. *Hoffmann-LaRoche Inc.*

390 Levorphanol Tartrate Dihydrate
5985-38-6 5496
$C_{21}H_{33}NO_9$
17-Methylmorphinan-3-ol tartrate (1:1) (salt) dihydrate.
Ro-1-5431/7; Ro-1-6794 (d-form); dextrorphan (d-form); racemorphan (dl-form); methorphinan (dl-form); Dromoran; Levo-Dromoran. Analgesic, narcotic. Orally active synthetic morphine analog. Federally controlled substance (opiate). mp = 113-115°; $[\alpha]_D^{20}$ = -14° (c = 3 in H_2O); pH (0.2% aqueous solution) = 3.4-4.0; soluble in H_2O (2 g/100 ml), alcohol, Et_2O. *Hoffmann-LaRoche Inc.*

391 Lofentanil
61380-40-3 5586
$C_{25}H_{32}N_2O_3$
(-)-Methyl cis-3-methyl-1-phenethyl-4-(N-phenylpopionamido)isonipecotate.
Analgesic, narcotic. *Abbott Laboratories Inc.*

392 Lofentanil Oxalate
61380-41-4 5586 262-750-7
$C_{27}H_{34}N_2O_7$
(-)-Methyl cis-3-methyl-1-phenethyl-4-(N-phenylpopionamido)isonipecotate oxalate (1:1).
R-34995. Analgesic, narcotic. mp = 177°. *Abbott Laboratories Inc.*

393 Meperidine
57-42-1 5894 200-329-1
$C_{15}H_{21}NO_2$
Ethyl 1-methyl-4-phenylisonipecotate.
isonipecaine; pethidine. Analgesic, narcotic. Federally controlled substance (opiate). *Astra Sweden; Marion Merrell Dow Inc.; Sterling Winthrop Inc.*

394 Meperidine Hydrochloride
50-13-5 5894 200-013-3
$C_{15}H_{22}ClNO_2$
Ethyl 1-methyl-4-phenylisonipecotate hydrochloride.
Algil; Alodan; Centralgin; Demerol hydrochloride; Dispadol; Dolantin; Dolestine; Dolosal; Mefedina. Analgesic, narcotic. Federally controlled substance (opiate). mp = 186-189°; soluble in H_2O, Me_2CO, EtOAc; slightly soluble in alcohol; insoluble in C_6H_6, Et_2O; LD_{50} (rat orl) = 170 mg/kg. *Astra Sweden; Marion Merrell Dow Inc.; Sterling Winthrop Inc.*

395 Meptazinol
54340-58-8 5910 259-109-9
$C_{15}H_{23}NO$
m-(3-Ethylhexahydro-1-methyl-1H-azepin-3-yl)-phenol.
Analgesic, narcotic. Mixed opioid agonist-antagonist. mp = 127-133°. *Wyeth Labs.*

396 Meptazinol Hydrochloride
59263-76-2 5910
$C_{15}H_{24}ClNO$
m-(3-Ethylhexahydro-1-methyl-1H-azepin-3-yl)-phenol hydrochloride.
WY-22811 HCl; IL-22811-HCl; Meptid. Analgesic, narcotic. Mixed opioid agonist-antagonist. *Wyeth Labs.*

397 Metazocine
3734-52-9 5995 223-097-3
$C_{15}H_{21}NO$
2'-Hydroxy-2,5,9-trimethyl-6,7-benzomorphan.
methobenzmorphan. Analgesic, narcotic. Federally controlled substance (opiate). mp = 232-235°.

398 Methadone
76-99-3 6008 200-996-9
$C_{21}H_{27}NO$
6-(Dimethylamino)-4,4-diphenyl-3-heptanone.
Analgesic, narcotic. Federally controlled substance (opiate). [dl-form]: mp = 235°; soluble in H_2O (12 g/100 ml), alc (8 g/100 ml), isopropanol (2.4 g/100 ml); LD_{50} (rat orl) = 95 mg/kg;l-form: mp = 241°; $[\alpha]_D^{20}$ = -169° (c = 2.5); LD_{50} (rat sc) = 44 mg/kg. *Abbott Labs Inc.; Eli Lilly & Co.; Mallinckrodt Inc.; Merck & Co.Inc.; Sterling Winthrop Inc.*

399 Methadone Hydrochloride
1095-90-5 6008 214-140-7
$C_{21}H_{28}ClNO$
6-(Dimethylamino)-4,4-diphenyl-3-heptanone hydrochloride.
dl-form: AN-148; Hoechst-10820; Adanon; Algidon; Algolysin; Amidon; Butalgin; Depridol; Dolophine; Fenadone; Heptadon; Heptanon; Ketalgin; Mecodin; Mephenon; Methadose; Miadone; Moheptan; Phenadone; Tussol; Levadone; l-form: Levothyl; L-Polamidon. Analgesic, narcotic. Federally controlled substance (opiate). mp = 235°; uv max = 292 nm; λ_m = 292 nm; soluble in H_2O (12 g/100 ml), less soluble in organic solvents; LD_{50} (rat orl) = 95 mg/kg. *Abbott Labs Inc.; Eli Lilly & Co.; Mallinckrodt Inc.; Merck & Co.Inc.; Sterling Winthrop Inc.*

400 Methadyl Acetate
509-74-0 6009
$C_{23}H_{31}NO_2$
6-(Dimethylamino)-4,4-diphenyl-3-heptanol acetate (ester).
acetylmethadol; acemethadone; amidolacetatel race-acetylmetadol; alphacetylmethadol (αd-form); levomethadyl acetate (α-l-form); betacetylmethadol (β-l-form). Analgesic, narcotic. Congener of methadone. The α-dl-form is more active and less toxic than the β-dl-form. The α-l-form is less active but longer acting than the α-d-form. Federally controlled substance (opiate). *See* levomethadyl acetate. [αd-form]: mp = 215°; $[\alpha]_D^{25}$ = +61.2° (c = 0.2); soluble in H_2O; LS_{50} (mus orl) = 130 mg/kg, (mus sc) =72 mg/kg. *Bristol Laboratories; Merck & Co.Inc.*

401 Methyldesorphine
16008-36-9 240-139-6
$C_{18}H_{21}NO_2$
6-Methyl-Δ^6-deoxymorphine.
MK-57. Analgesic, narcotic.

402 Methyldihydromorphine
509-56-8
$C_{18}H_{23}NO_3$
3,6-Dihydroxy-6,N-dimethyl-4,5-epoxymorphinan.
7,8-dihydro-6-methylmorphine. Analgesic, narcotic.

403 Metopon
143-52-2 6234
$C_{18}H_{21}NO_3$
(5α)-4,5-Epoxy-3-hydroxy-5,17-dimethylmorphinan-3-one.
methyldihydromorphinone. Analgesic, narcotic. Federally controlled substance (opiate). Sinters at 235°; mp = 243-245°; $[\alpha]_D^{24}$ = -141° (c = 1 in alc); slightly soluble in organic solvents.

404 Metopon Hydrochloride
124-92-5 6234
$C_{18}H_{22}ClNO_3$
(5α)-4,5-Epoxy-3-hydroxy-5,17-dimethylmorphinan-3-one hydrochloride.
methyldihydromorphinone hydrochloride. Analgesic, narcotic. Federally controlled substance (opiate). Dec 315-318°; $[\alpha]_D^{24}$ = -105° (c = 1); freely soluble in H_2O; sparingly soluble in alc; insoluble in C_6H_6; pH (1% aqueous solution) ≅ 5.0.

405 Morpheridine
469-81-8 6357
$C_{20}H_{30}N_2O_3$
1-(2-Morpholinoethyl)-4-phenylpiperidine-4-carboxylic acid ethyl ester.
morpholinoethyl morpethidine. Narcotic analgesic. Federally controlled substance (opiate). Liquid; $bp_{0.5}$ = 188-192°; n_D^{18} = 1.5276.

406 Morpheridine Dihydrochloride
6357
$C_{20}H_{32}Cl_2N_2O_3$
1-(2-Morpholinoethyl)-4-phenylpiperidine-4-carboxylic acid ethyl ester dihydrochloride.
TA-1. Narcotic analgesic. Federally controlled substance (opiate). mp = 264-266° (dec); LD_{50} (mus iv) = 45 mg/kg, (mus ip) = 118 mg/kg.

407 Morphine
57-27-2 6359 200-320-2
$C_{17}H_{19}NO_3$
(5α,6α)-Didehydro-4,5-epoxy-17-methylmorphinan-3,6-diol.
morphium; morphia; Dulcontin; Duromorph; Morphina; Nepenthe. Analgesic, narcotic. Principle alkaloid of opium, which contains 9-14% anhydrous morphine. Occurs naturally as the (-)-form. Also found in normal brain, blood and liver. Sometimes administered as the tartrate. Federally controlled substance (opiate). CAUTION: may be habit forming. Dec 254°; [monohydrate]: d = 1.32; λ_m = 285 nm (acid); $[\alpha]_D^{25}$ = -132° (mEtOH); soluble in H_2O (20 mg/100 ml); soluble in organic solvents.

408 Morphine Hydrochloride
52-26-6 6360 200-136-2
$C_{17}H_{20}ClNO_3$
(5α,6α)-Didehydro-4,5-epoxy-17-methylmorphinan-3,6-diol hydrochloride.
Epimore. Analgesic, narcotic. Federally controlled substance (opiate). CAUTION: may be habit forming. mp = 200° (dec); $[\alpha]_D^{25}$ = -113.5° (c= 2.2 H_2O); soluble in

H_2O (57 mg/ml), EtOH; less soluble in organic solvents; LD_{50} (mus iv) = 226-318 mg/kg.

409 Morphine Sulfate
64-31-3 6361 200-582-8
$C_{34}H_{40}N_2O_{10}S$
7,8-Didehydro-4,5α-epoxy-17-methylmorphinan-3,6α-diol sulfate (2:1) (salt).
Kapanol; Moscontin; MS Contin; MST Continus; Oblioser; Oramorph; Relipain; Roxanol. Analgesic, narcotic. Federally controlled substance (opiate). CAUTION: may be habit forming. *Abbott Labs Inc.; Elkins-Sinn; Purdue Pharma L.P.; Wyeth-Ayerst Labs.*

410 Morphine Sulfate Pentahydrate
6211-15-0 6361
$C_{34}H_{50}N_2O_{15}S$
7,8-Didehydro-4,5α-epoxy-17-methylmorphinan-3,6α-diol sulfate (2:1) (salt) pentahydrate.
MST 10 Mundipharma; MST 30 Mundipharma. Analgesic, narcotic. Federally controlled substance (opiate). CAUTION: may be habit forming. mp = 250° (dec); $[\alpha]_D^{25}$ = -109° (c = 4 H_2O); soluble in H_2O (65 g/100 ml); less soluble in organic solvents.

411 Myrophine
467-18-5 6420
$C_{38}H_{51}NO_4$
3-Benzyloxy-N-methyl-6-myristoyloxy-4,5-epoxymorphin-7-en.
myristate; myrocodine; myrophinium; Leucodiniene; myricodine; Peronine. Analgesic, narcotic. Federally controlled substance (opiate derivative). mp = 41°.

412 Nalbuphine
20594-83-6 6444 243-901-6
$C_{21}H_{27}NO_4$
17-(Cyclobutylmethyl)-4,5α-epoxymorphinan-3,6α,14-triol.
N-cyclobutylmethyl-14-hydroxydihydronormorphine.
Analgesic; narcotic. Mixed opioid agonist-antagonist. mp = 230.5°. *Abbott Labs Inc.; DuPont-Merck Pharm.*

413 Nalbuphine Hydrochloride
23277-43-2 6444 245-549-9
$C_{21}H_{28}ClNO_4$
17-(Cyclobutylmethyl)-4,5α-epoxymorphinan-3,6α,14-triol.
EN-2234A; Nubain. Analgesic; narcotic. Mixed opioid agonist-antagonist. *Abbott Labs Inc.; DuPont-Merck Pharm.*

414 Nalmexone
17767-26-9
$C_{21}H_{25}NO_4$
4,5α-Epoxy-3,14-dihydroxy-17-(3-methyl-2-butenyl)morphinan-6-one.
Analgesic; antagonist to narcotics.

415 Nalmexone Hydrochloride
16676-27-0
$C_{21}H_{26}ClNO_4$
4,5α-Epoxy-3,14-dihydroxy-17-(3-methyl-2-butenyl)morphinan-6-one hydrochloride.
EN-1620A. Analgesic; antagonist to narcotics.

416 Narceine
131-28-2 6508 205-021-0
$C_{23}H_{27}NO_3$
6-[[6-[2-(Dimethylamino)ethyl]-4-methoxy-1,3-benzodioxol-5-yl]acetyl]-2,3-dimethoxybenzoic acid.
Analgesic, narcotic; antitussive Occurs in opium (0.1-0.5%). Hygroscopic; mp = 138°; λ_m = 270 nm (log ε 398 EtOH); trihydrate: mp = 176°; soluble in H_2O (1.3 g/100 ml); insoluble in C_6H_6, $CHCl_3$, Et_2O, petroleum ether; soluble in dilute mineral acids; moderately soluble in hot alcohol.

417 Nicocodine
3688-66-2 222-990-5
$C_{24}H_{24}N_2O_4$
6-Nicotinoyl morphine.
Analgesic (narcotic).

418 Nicodicodine
808-24-2 212-365-5
$C_{24}H_{26}N_2O_4$
6-Nicotinoyl dihydrocodeine.
Analgesic (narcotic).

419 Nicomorphine
639-48-5 6607 211-357-9
$C_{29}H_{25}N_3O_5$
(5α,6α)-7,8-Didehydro-4,5-epoxy-17-methylmorphinan-3,6-diol di-3-pyridinecarboxylate (ester).
morphine dinicotinate ester; morphine ester with nicotinic acid; morphine bis(nicotinate); morphine bis(pyridine-3-carboxylate); nicotinic acid morphine ester; morphine dinicotinate; Gewalan; Vilan. Analgesic, narcotic. Federally controlled substance (opiate derivative). mp = 178-178.5°; nearly insoluble in H_2O; soluble in EtOH.

420 Noracymethadol
1477-39-0 6778 216-027-8
$C_{22}H_{29}NO_2$
6-(Methylamino)-4,4-diphenyl-3-heptanol acetate (ester).
Analgesic, narcotic. Metabolite of methadyl acetate. Federally controlled substance (opiate). *Eli Lilly & Co.*

421 Noracymethadol Hydrochloride
5633-25-0 6778
$C_{22}H_{30}ClNO_2$
6-(Methylamino)-4,4-diphenyl-3-heptanol acetate (ester) hydrochloride.
NIH-7667; 30109. Analgesic, narcotic. Federally controlled substance (opiate). mp = 216-217°. *Eli Lilly & Co.*

422 Norcodeine
467-15-2 6783 207-388-2
$C_{17}H_{19}NO_3$
(5α,6α)-7,8-Didehydro-4,5-epoxy-3-methoxymorphinan-6-ol.
N-desmethylcodeine; normorphine 3-methyl ether. Analgesic, narcotic. mp = 185°; sparingly soluble in H_2O, Et_2O; moderately soluble in Me_2CO; freely soluble in hot alcohol.

423 Norlevorphanol
1531-12-0 6801 216-236-4
$C_{16}H_{21}NO$
(-)-3-Hydroxymorphinan.
morphinan-3-ol; NIH-7539. Analgesic, narcotic. Federally controlled substance (opiate). mp = 270-272°; $[\alpha]_D^{21}$ = -42° (c = 1 in MeOH). *Hoffmann-LaRoche Inc.*

424 Normethadone
467-85-6 6804 207-401-1
$C_{20}H_{25}NO$
6-Dimethylamino-4-diphenyl-3-hexanone.
isoamidone I; desmethylmethadone; phenyldimazone; Hoechst 10582. Antitussive; analgesic, narcotic. Federally controlled substance (opiate). Oily liquid; bp_3 = 164-167°. *Hoechst Marion Roussel Inc.*

425 Normethadone Hydrochloride
847-84-7 6804
$C_{20}H_{26}ClNO$
6-Dimethylamino-4-diphenyl-3-hexanone hydrochloride.
Ticarda. Antitussive; analgesic, narcotic. Federally controlled substance (opiate). mp = 174-175°; soluble in H_2O, alc; pH (1% aqueous solution) ≅ 5; LD_{50} (mus sc) = 90 mg/kg. *Hoechst Marion Roussel Inc.*

426 Normorphine
466-97-7 6806 207-381-4
$C_{16}H_{17}NO_3$
4,5-Epoxy-3,6-dihydroxymorphin-7-ene.
desmethylmorphine. Analgesic, narcotic. Federally controlled substance (opiate derivative). mp = 276-277°; sesquahydrate: mp = 273°; sparingly soluble in H_2O; Insoluble in $CHCl_3$.

427 Norpipanone
561-48-8 6810 209-220-3
$C_{23}H_{29}NO$
4,4-Diphenyl-6-(1-piperidyl)-3-hexanone.
Hoechst 10495; Hexalgon. Analgesic, narcotic. Federally controlled substance (opiate). *Hoechst Marion Roussel Inc.*

428 Norpipanone Hydrochloride
6033-41-6 6810
$C_{23}H_{30}ClNO$
4,4-Diphenyl-6-(1-piperidyl)-3-hexanone hydrochloride.
Orfenso. Analgesic, narcotic. Federally controlled substance (opiate). mp = 181-182°; soluble in H_2O, alcohol. *Hoechst Marion Roussel Inc.*

429 Ocfentanil
101343-69-5
$C_{22}H_{27}FN_2O_2$
2'-Fluoro-2-methoxy-N-[(1-phenethyl)-4-piperidyl]acetanilide monohydrochloride.
Analgesic, narcotic. *Anaquest.*

430 Ocfentanil Hydrochloride
112964-97-3
$C_{22}H_{28}ClFN_2O_2$
2'-Fluoro-2-methyoxy-N-[(1-phenethyl)-4-piperidyl]acetanilide.
A-3217. Analgesic, narcotic. *Anaquest.*

431 Opium
8008-60-4 6986 232-368-5
gum opium; crude opium. Analgesic, narcotic; hypnotic. Made up of about 20 alkaloids, which consititute about 25% of the opium [including morphine, 10%; noscapine 4-8%; codeine 0.8-2.5%; papaverine 0.5-2/5%; thebaine (paramorphine) 0.5-2%], along with meconic acid, lactic and sulfuric acids, sugar, and some resinous and waxy compounds. Used largely in the manufacture of morphine, codeine and other opium alkaloids. Federally controlled substance (opiate). CAUTION: may be habit forming. *See* morphine; noscapine; codeine; papaverine.

432 Oxycodone
76-42-6 7093 200-960-2
$C_{18}H_{21}NO_4$
4,5α-Epoxy-14-hydroxy-3-methoxy-17-methylmorphinan-6-one.
dihydrohydroxycodeinone; NSC-19043; Dihydrone; Prodalone [as pectinate]. Analgesic, narcotic. Federally controlled substance (opiate). mp = 218-220°; insoluble in H_2O, Et_2O, base; soluble in alcohol, $CHCl_3$; [tautomeric form]: mp = 219-220°; tautomeric form is more soluble in alcohol than other form. *DuPont-Merck Pharm.*

433 Oxycodone Hydrochloride
124-90-3 7093 204-717-1
$C_{18}H_{22}ClNO_4$
4,5α-Epoxy-14-hydroxy-3-methoxy-17-methylmorphinan-6-one hydrochloride.
Dinarkon; Eubine; Eucodal; Eukodal; Eutagen; Oxikon; Oxycon; Pancodine; Tecodin; Tekodin; Thecodine; Thekodin; component of: Percodan. Analgesic, narcotic. Federally controlled substance (opiate). Dec 270-272°; $[\alpha]_D^{20}$ = -125° (c = 2.5); soluble in H_2O (10 g/100 ml); slightly soluble in alcohol. *DuPont-Merck Pharm.*

434 Oxycodone Terephthalate
64336-55-6
$C_{44}H_{48}N_2O_{12}$
4,5α-Epoxy-14-hydroxy-3-methoxy-17-methylmorphinan-6-one 1,4-benzenedicarboxylate (2:1) (salt).
Analgesic, narcotic. *DuPont-Merck Pharm.*

435 Oxymorphone
76-41-5 7103 200-959-7
$C_{17}H_{19}NO_4$
4,5α-Epoxy-3,14-dihydroxy-17-methylmorphinan-6-one.
dihydroxymorphinone; dihydro-14-hydroxymorphinone; 14-hydroxydihydromorphinone. Analgesic, narcotic. Federally controlled substance (opiate). mp = 248-249°; levarotory; soluble in boiling Me_2CO and $CHCl_3$; readily soluble in aqueous alkalies; moderately soluble in boiling EtOH; sparingly soluble in C_6H_6. *DuPont-Merck Pharm.*

436 Oxymorphone Hydrochloride
357-07-3 7103 206-610-5
$C_{17}H_{20}ClNO_4$
4,5α-Epoxy-3,14-dihydroxy-17-methylmorphinan-6-one hydrochloride.
Numorphan. Analgesic, narcotic. Federally controlled substance (opiate). *DuPont-Merck Pharm.*

437 Papaveretum
8002-76-4 7150
concentrated opium; Omnopon; Pantopon. Analgesic, narcotic. A mixture of the hydrochlorides of the opium alkaloids in their approximate natural proportions: approx 50% morphine, 3% codeine, 20% noscapine, 5% papaverine. Exhibits biological action of morphine and other alkaloids present in opium. Federally controlled substance (opiate). Yellowish-gray crystal powder; freely soluble in H_2O.

438 Pentamorphone
68616-83-1
$C_{22}H_{28}N_2O_3$
7,8-Didehydro-4,5α-epoxy-3-hydroxy-17-methyl-14-(pentylamino)morphinan-6-one.
RX-77989; A-4492. Analgesic, narcotic. *Anaquest.*

439 Pentazocine
359-83-1 7261 206-634-6
$C_{19}H_{27}NO$
(2R*,6R*,11R*)-1,2,3,4,5,6-Hexahydro-6,11-dimethyl-3-(3-methyl-2-butenyl)-2,6-methano-3-benzazocin-8-ol.
Win-20228; NSC-107430; Fortral; Pentagin; Soseton. Analgesic, narcotic. Mixed opioid agonist-antagonist. Federally controlled substance. mp = 145-147°; LD_{50} (rat sc) = 175 mg/kg. *Sterling Health U.S.A.*

440 Pentazocine Hydrochloride
64024-15-3 7261
$C_{19}H_{28}ClNO$
(2R*,6R*,11R*)-1,2,3,4,5,6-Hexahydro-6,11-dimethyl-3-(3-methyl-2-butenyl)-2,6-methano-3-benzazocin-8-ol hydrochloride.
Fortralin Tablets; Fortralgesic Tablets. Analgesic, narcotic. Mixed opioid agonist-antagonist. Federally controlled substance. *Sterling Winthrop Inc.*

441 Pentazocine Lactate
146-95-1 7261
$C_{22}H_{33}NO_4$
(2R*,6R*,11R*)-1,2,3,4,5,6-Hexahydro-6,11-dimethyl-3-(3-methyl-2-butenyl)-2,6-methano-3-benzazocin-8-ol lactate (salt).
Liticon; Pentalgine; Talwin Injection. Analgesic, narcotic. Mixed opioid agonist-antagonist. Federally controlled substance. *Sterling Winthrop Inc.*

442 Phenadoxone
467-84-5 7348 207-400-6
$C_{23}H_{29}NO_2$
6-Morpholino-4,4-diphenyl-3-heptanone.
heptazone; morphodone. Analgesic, narcotic. Federally controlled substance (opiate).

443 Phenadoxone Hydrochloride
545-91-5 7348 208-895-1
$C_{23}H_{30}ClNO_2$
6-Morpholino-4,4-diphenyl-3-heptanone hydrochloride.
Hoechst 10600; CB-11; Hepagin; Heptalgin; Heptalin; Heptone; Supralgin. Analgesic, narcotic. Federally controlled substance (opiate). mp = 224-225° (some dec); λ_m = 260, 295 nm; soluble in H_2O, alcohol, EtOH, $CHCl_3$; slightly soluble in Me_2CO; almost insoluble in C_6H_6, EtOAc; LD_{50} (mus iv) = 47.5 mg/kg.

444 Phenampromide
129-83-9 7353 204-967-1
$C_{17}H_{26}N_2O$
N-(1-Methyl-2-piperidinoethyl)propionanilide.
phenampromid. Federally controlled substance (opiate). Liquid; $bp_{0.2}$ = 124-128°; n_D^{28} = 1.518. *Am. Cyanamid.*

445 Phenazocine
127-35-5 7360 204-835-3
$C_{22}H_{27}NO$
1,2,3,4,5,6-Hexahydro-6,11-dimethyl-3-(2-phenylethyl)-2,6-methano-3-benzazocin-8-ol.
phenethylazocine; phenobenzorphan. Analgesic, narcotic. Federally controlled substance (opiate). [(±)-form]: mp = 181-182°; [(-)-form]: mp = 159-159.5°; $[\alpha]_D^{20}$ = -122° (c = 0.74 95% EtOH). *SmithKline Beecham Pharm.*

446 (±)-Phenazocine Hydrobromide
1239-04-9 7360
$C_{22}H_{28}BrNO$
1,2,3,4,5,6-Hexahydro-6,11-dimethyl-3-(2-phenylethyl)-2,6-methano-3-benzazocin-8-ol Hydrobromide.
NIH-7519; SKF-6574; Narphen; Prinadol; Xenagol. Analgesic, narcotic. Federally controlled substance (opiate). mp = 166-170°; LD_{50} (mus sc) = 332 mg/kg. *SmithKline Beecham Pharm.*

447 (-)-Phenazocine Hydrobromide
7360
$C_{22}H_{28}BrNO$
(-)-1,2,3,4,5,6-Hexahydro-6,11-dimethyl-3-(2-phenylethyl)-2,6-methano-3-benzazocin-8-ol Hydrobromide.
(-)-phenazocine hydrobromide. Analgesic, narcotic. Federally controlled substance (opiate). mp = 284-287°; $[\alpha]_D^{20}$ = -84.1° (c = 1.12, 95% EtOH); LD_{50} (mus sc) = 147 mg/kg. *SmithKline Beecham Pharm.*

448 Phenomorphan
468-07-5 7399 207-403-2
$C_{24}H_{29}NO$
3-Hydroxy-N-phenethylmorphinan.
NIH-7274. Federally controlled substance (opiate). mp = 243-245°; [hydrobromide]: mp = 300-301°; $[\alpha]_D^{20}$ = -63.12° (c = 3.27 alc); [methyl bromide]: mp = 239-240°; $[\alpha]_D^{20}$ = -42.81° (c = 1.555 MeOH); [D-tartrate monohydrate]: mp = 125-126°; $[\alpha]_D^{20}$ = -42.75° (c = 0.983 H_2O).

449 Phenoperidine
562-26-5 7400 209-229-2
$C_{23}H_{29}NO_3$
1-(3-Hydroxy-3-phenylpropyl)-4-phenylpiperidine-4-carboxylic acid ethyl ester.
phenoperidin; R-1406. Analgesic, narcotic. Federally controlled substance (opiate). *Eli Lilly & Co.; Merck & Co.Inc.*

450 Phenoperidine Hydrochloride
3627-49-4 7400 222-846-1
$C_{23}H_{30}ClNO_3$
1-(3-Hydroxy-3-phenylpropyl)-4-phenylpiperidine-4-carboxylic acid ethyl ester hydrochloride.
Lealgin; Operidine. Analgesic, narcotic. Federally controlled substance (opiate). mp = 200-202°; soluble in H_2O. *Eli Lilly & Co.; Merck & Co.Inc.*

451 Piminodine
13495-09-5 7587 236-817-6
$C_{23}H_{30}N_2O_2$
4-Phenyl-1-[3-(phenylamino)propyl]-4-piperidinecarboxylic acid ethyl ester.
Analgesic, narcotic. Federally controlled substance (opiate). *Sterling Winthrop Inc.*

452 Piminodine Esylate
7081-52-9 7587
$C_{25}H_{35}N_2O_5S$
4-Phenyl-1-[3-(phenylamino)propyl]-4-piperidinecarboxylic acid ethyl ester monoethanesulfonate.
piminodine esilate; piminodine ethanesulfonate; NIH-7590; Win-14098; Alvodine; Pimadin. Analgesic, narcotic. Federally controlled substance (opiate). *Sterling Winthrop Inc.*

453 Piritramide
302-41-0 7653
$C_{27}H_{34}N_4O$
1'-(3-Cyano-3,3-diphenylpropyl)-[1,4'-bipiperidine]-4'-carboxamide.
pirinitramide; A-65; R-3365; Dipidolor; Pirdolan. Analgesic, narcotic. Federally controlled substance (opiate). mp = 149-150°. *Abbott Laboratories Inc.*

454 Proheptazine
77-14-5 7959 201-006-8
$C_{17}H_{25}NO_2$
1,3-Dimethyl-4-phenyl-4-propionyloxyaza-cycloheptane.
dimepheprimine; WY-757. Analgesic, narcotic. Federally controlled substance (opiate). Liquid; $bp_{0.03}$ = 126°; n_D^{21} = 1.5215. *Am. Home Products.*

455 Promedol
64-39-1 7968 200-583-3
$C_{17}H_{25}NO_2$
1,2,5-Trimethyl-4-phenyl-4-propionyloxypiperidine.
dimethylmeperidine; α-promedol [α isomer]; isopromedol [β isomer hydrochloride]; trimeperidine [γ isomer hydrochloride]; γ-promedol [γ isomer]. Analgesic, narcotic; antitussive. Federally controlled substance (opiate). [hydrochloride, α isomer]: mp = 153-154°; [hydrochloride, β isomer]: mp = 183-184°; [hydrochloride, γ isomer]: mp = 222-223°.

456 Properidine
561-76-2 209-222-4
$C_{16}H_{23}NO_2$
1-Methyl-4-phenylpiperidine-4-carboxylic acid isopropyl ester.
Analgesic, narcotic.

457 Propiram
15686-91-6 8017 239-775-7
$C_{16}H_{25}N_3O$
N-(1-Methyl-2-piperidinoethyl)-N-2-pyridinylpropionamide.
Analgesic, narcotic. Federally controlled substance (opiate). bp = 162-163°. *Bayer AG; Miles Inc.*

458 Propiram Fumarate
13717-04-9 8017
$C_{20}H_{29}N_3O_5$
N-(1-Methyl-2-piperidinoethyl)-N-2-pyridinylpropionamide fumarate (1:1).
Bay-4503; FBA-4503; Algeril; Dirame. Analgesic, narcotic. Federally controlled substance (opiate). LD_{50} (mus orl) = 874 mg/kg, (rat orl) = 1657 mg/kg. *Bayer AG; Miles Inc.*

459 Propoxyphene
469-62-5 8024 207-420-5
$C_{22}H_{29}NO_2$
(2S,3R)-(+)-4-(Dimethylamino)-3-methyl-1,2-diphenyl-2-butanol propionate (ester).
dextropropoxyphene; α-d-propoxyphene; racemic propoxypene (α-dl-form); diméprotane (α-dl-form); levopropoxyphene (α-l-form). Analgesic, narcotic. Bulk dextropropoxyphene (non-dosage form) is a federally controlled substantce (opiate); dextropropoxyphene is a controlled substance (narcotic). *See* levopropoxyphene (α-l-form). mp = 75-76°; $[\alpha]_D^{25}$ = +67.3° (c = 0.06 in $CHCl_3$); [β-dl-form]: mp = 187-188°; more soluble than α-form. *Eli Lilly & Co.*

460 Propoxyphene Hydrochloride
1639-60-7 8024 216-683-5
$C_{22}H_{30}ClNO_2$
(2S,3R)-(+)-4-(Dimethylamino)-3-methyl-1,2-diphenyl-2-butanol propionate (ester) hydrochloride.
SK-65; Antalvic; Darvon; Deprancol; Develin; Dolene; Dolocap; Erantin; Femadol; Harmar; Propoxychel; Proxagesic; component of: Darvon with ASA, SK-65 Apap, SK-65 Compound, Wygesic. Analgesic, narcotic. Federally controlled substance (opiate). mp = 163-165°; $[\alpha]_D^{25}$ = + 59.8° (c = 0.06 in H_2O); soluble in H_2O, alcohol, $CHCl_3$, Me_2CO; practically insoluble in C_6H_6, Et_2O; LD_{50} (mus iv) = 28 mg/kg, (mus ip) 111 mg/kg, (mus sc) = 211 mg/kg, (mus orl) = 282 mg/kg. *Eli Lilly & Co.*

461 Propoxyphene Napsylate
26570-10-5 8024
$C_{32}H_{39}NO_6S$
(αS,3R)-α-[2-(Dimethylamino)-1-methyl]-α-phenylphenethyl propionate compound with 2-napthalenesulfonic acid (1:1) monohydrate.
Darvon-N; Doloxene; component of: Darvocet-N, Propacet. Analgesic, narcotic. Federally controlled substance (opiate). LD_{50} (fmus orl) = 990 mg/kg. *Eli Lilly & Co.*

462 Propoxyphene Napsylate [anhydrous]
17140-78-2 8024 241-205-7
$C_{32}H_{37}NO_5S$
(αS,3R)-α-[2-(Dimethylamino)-1-methyl]-α-phenylphenethyl propionate compound with 2-napthalenesulfonic acid (1:1).
Analgesic, narcotic. Federally controlled substance (opiate). *Eli Lilly & Co.*

463 Racemoramide
545-59-5 2997 208-893-0
$C_{25}H_{32}N_2O_2$
(±)-4-[2-Methyl-4-oxo-3,3-diphenyl-4-(1-pyrrolidinyl)-butyl]morpholine.

R-610. Analgesic, narcotic. *See* dextromoramide. *Janssen Pharm. Inc.*

464 **Racemorphan**
297-90-5 5496 206-048-0
$C_{17}H_{23}NO$
(±)-3-Hydroxy-N-methylmorphinan.
dl-form of levorphanol; methorphinan. Analgesic, narcotic. Federally controlled substance (opiate). *See* levorphanol. mp = 251-253°. *Hoffmann-LaRoche Inc.*

465 **Remifentanil**
132875-61-7 8300
$C_{20}H_{28}N_2O_5$
4-Carboxy-4-(N-phenylpropionamido)-1-piperidinepropionic acid dimethyl ester.
remifentanyl. Analgesic, narcotic. *Glaxo Labs.*

466 **Remifentanil Hydrochloride**
132539-07-2 8300
$C_{20}H_{29}ClN_2O_5$
4-Carboxy-4-(N-phenylpropionamido)-1-piperidinepropionic acid dimethyl ester monohydrochloride.
GI-87084B. Analgesic, narcotic. *Glaxo Labs.*

467 **Sufentanil**
56030-54-7 9056
$C_{22}H_{30}N_2O_2S$
N-[4-(Methoxymethyl)-1-[2-(2-thienyl)ethyl]-4-piperidinyl]propionalnalide.
R-30730; sufentanyl. Analgesic, narcotic. Federally controlled substance (opiate). mp = 96.6°; LD_{50} (mus iv) = 18.7 mg/kg. *Abbott Laboratories Inc.*

468 **Sufentanil Citrate**
60561-17-3 9056
$C_{28}H_{38}N_2O_9S$
N-[4-(Methoxymethyl)-1-[2-(2-thienyl)ethyl]-4-piperidinyl]propionalnalide citrate (1:1).
R-33800; Sufenta. Analgesic, narcotic. Federally controlled substance (opiate). *Abbott Laboratories Inc.*

469 **Thebacon**
466-90-0 3215 207-377-2
$C_{20}H_{23}NO_4$
3-Methoxy-N-methyl-4,5-epoxymorphin-6-en-6-yl acetate.
dihydrocodeinone enol acetate; acetyldemethyldihydrothebaine; acetyldihydrochodeinone; demethyldihydrothebaine acetate (ester). Analgesic, narcotic; antitussive. Federally controlled substance (opiate). mp = 154°; soluble in most organic solvents; practically insoluble in H_2O.

470 **Thebacon Hydrochloride**
20236-82-2 3215
$C_{20}H_{24}ClNO_4$
3-Methoxy-N-methyl-4,5-epoxymorphin-6-en-6-yl acetate hydrochloride.
dihydrocodeinone enol acetate hydrochloride; Acedicon. Analgesic, narcotic; antitussive. Federally controlled substance (opiate). mp = 132-135° (dec); soluble in H_2O; stable in boiling H_2O.

471 **Tifluadom**
81656-30-6
$C_{22}H_{20}FN_3OS$
(±)-N-[[5-(o-Fluorophenyl)-2,3-dihydro-1-methyl-1H-1,4-benzodiazepin-2-yl]methyl]-3-thiophenecarboxamide.
KC-5103. A benzodiazepine. Human kappa opoiod receptor to the agonist. Analgesic; diuretic.

472 **Tilidine dl-trans**
20380-58-9 9578 243-774-7
$C_{17}H_{23}NO_2$
(±)-Ethyl trans-2-(dimethylamino)-1-phenyl-3-cyclohexene-1-carboxylate.
Analgesic, narcotic. Federally controlled substance (opiate). $bp_{0.01}$ = 96°°; [cis-(±)-form]: $bp_{0.01}$ ± 98°. *Parke-Davis; Warner-Lambert.*

473 **Tilidine Hydrochloride (+)-trans**
24357-97-9 9578
$C_{17}H_{24}ClNO_2$
(+)-Ethyl trans-2-(dimethylamino)-1-phenyl-3-cyclohexene-1-carboxylate.
Analgesic, narcotic; antitussive. Federally controlled substance (opiate). *Parke-Davis; Warner-Lambert.*

474 **Tilidine Hydrochloride dl-trans**
27107-79-5 9578 248-226-0
$C_{17}H_{24}ClNO_2$
(±)-Ethyl trans-2-(dimethylamino)-1-phenyl-3-cyclohexene-1-carboxylate hydrochloride.
Gö-1261C (hemihydrate); W-5759A (hemihydrate); Lucayan (hemihydrate); Valaron (hemihydrate). Analgesic, narcotic; antitussive. Federally controlled substance (opiate). mp = 159°; soluble in H_2O; LD_{50} (mus ig) = 437 mg/kg, (mus sv) = 490 mg/kg, (mus iv) = 52 mg/kg; [hydrochloride hemihydrate]: mp = 125°. *Parke-Davis; Warner-Lambert.*

475 **Xorphanol**
77287-89-9
$C_{23}H_{31}NO$
17-(Cyclobutylmethyl)-8β-methyl-6-methylene-morphinan-3ol.
Analgesic, narcotic; antitussive.

476 **Xorphanol Mesylate**
77287-90-2
$C_{24}H_{35}NO_4S$
17-(Cyclobutylmethyl)-8β-methyl-6-methylene-morphinan-3ol methane sulfonate (salt).
TR-5379M. Analgesic, narcotic; antitussive.

Analgesics, Non-narcotic

477 **Aceclofenac**
89796-99-6 19
$C_{16}H_{13}Cl_2NO_4$
2-[(2,6-Dichlorophenyl)amino]benzeneacetic acid carboxymethyl ester.
PR-82/3; Airtal; Biofenac; Tresquim. Anti-inflammatory; analgesic. Arylacetic acid derivative. mp = 149-150°; λ_m = 275 nm (log ε 4.14 EtOH). *Ajinomoto Co. Inc.*

478 Acemetacin
53164-05-9 27 258-403-4
$C_{21}H_{18}ClNO_6$
1-(4-Chlorobenzoyl)-5-methoxy-2-methyl-1H-indole-3-acetic acid carboxymethyl ester.
TV-1322; Acemix; Emflex; Rantudil; Rheumibis; Solart. Anti-inflammatory. Derivative of indomethacin. mp = 150-153°; LD_{50} (mmus orl) = 55.5 mg/kg, (mmus iv) = 34.1 mg/kg. *Troponwerke Dinklage.*

479 Acetaminophen
103-90-2 45 203-157-5
$C_8H_9NO_2$
4'-Hydroxyacetanilide.
p-hydroxyacetanilide; p-acetamidophenol; p-aceamino-phenol; N-acetyl-p-aminophenol; paracetamol; Abensanil; Acamol; Acetalgin; Alpiny; Amadil; Anaflon; Anhiba; Apamide; APAP; Banesin; Ben-u-ron; Bickie-mol; Calpol; Captin; Claratal; Cetadol; Dafalgan; Datril; Dirox; Disprol; Doliprane; Dolprone; Dymadon; Enelfa; Eneril; Eu-Med; Exdol; Febrilex; Finimal; Gelocatil; Hedex; Homoolan; Korum; Momentum; Naprinol; Nebs; Nobedon; Ortensan; Pacemol; Paldesic; Panadol; Panaleve; Panasorb; Panets; Panodil; Parelan; Paraspen; Parmol; Pasolind N; Phenaphen; Salzone; Tabalgin; Tapar; Tempra; Tralgon; Tylenol; Valadol; component of: Actifed Plus, Allerest Sinus Pain Formula, Anexsia, Aspirin-Free Anacin, Children's Tylenol Cold Tablets, Contac Cough & Sore Throat Formula, Contac Jr Non-drowsy Formula, Contac Nighttime Cold Medicine, Contac Severe Cold Formula, Coricidin, Darvoset-N, Dristan Cold (Multisymptom), Dristan Cold (No Drowsiness), Empracet, Endecon, Gemnisyn, Headache Strength Allerest, Hycomine Compound, Hy-Phen, Intensin, Liquiprin, Maximum Strength Sine-Aid, Midol, Midol Maximum Strength, Midol PMS, Naldegesic, Naldetuss, Ornex, Percocet, Percogesic with Codeine, Propacet, Quiet World, Rhinex D-Lay Tablets, Sinarest, Sine-Off Maxiumum Strength Allergy/Sinus, Sine-Off Maximum Strength No Drowsiness Formula Caplets, Sinubid, St Joseph's Cold Tablets for Children, Sudafed Sinus, Supac, Teen Midol, TheraFlu, Tylenol Allergy Sinus, Tylenol Cold and Flu Multi-Symptom, Tylenol Cold Medication Caplets, Liquid and Tablets, Tylenol Cold Night TIme Liquid, Tylenol Cold No Drowsiness, Tylenol PM Tablets and Caplets, Tylenol with Codeine, Tylox, Vanquish, Vicodin, Wygesic, Zydone. Analgesic; antipyretic; anti-inflammatory. mp = 169-170.5°; d_4^{21} = 1.293; λ_m = 250 nm (ε 13800 EtOH); slightly soluble in cold H_2O, Et_2O; more soluble in hot H_2O; soluble in alcohol, dimethylformamide, ethylene dichloride, Me_2CO, EtOAc; nearly insoluble in petroleum ether, pentane, C_6H_6; LD_{50} (mus orl) = 338 mg/kg, (mus ip) = 500 mg/kg. *Bristol-Myers Squibb Co; Forest Pharm. Inc.; McNeil Pharm.; Mead Johnson Nutritionals; Novo Nordisk Pharm. Inc.; Parke-Davis; Robins, A. H. Co.;Sterling Health U.S.A.; Warner-Lambert.*

480 Acetaminosalol
118-57-0 46 204-261-3
$C_{15}H_{13}NO_4$
4'-Hydroxyacetanilide salicylate.
p-acetamidophenyl salicylate; p-acetylaminophenol salicylic acid ester; phenestal; Phenosal; Salophen. An analgesic with antipyretic and anti-inflammatory properties. mp = 187°; nearly insoluble in petroleum ether, cold H_2O; more soluble in warm H_2O; soluble in alcohol, Et_2O, C_6H_6; incompatible with alkalies and alkaline solutions.

481 Acetanilide
103-84-4 47 203-150-7
C_8H_9NO
N-Phenylacetamide.
antifebrin; acetylaniline; acetylaminobenzene. Antipyretic; analgesic. mp = 113-115°; bp = 304-305°; d_4^{15}= 1.219; pK (28°) = 13.0; soluble in H_2O, alcohol, $CHCl_3$, Me_2CO, glycerol, dioxane, Et_2O, C_6H_6; nearly insoluble in petroleum ether; LD_{50} (rat ig) = 800 mg/kg.

482 Acetylsalicylic Acid
50-78-2 102 200-064-1
$C_9H_8O_4$
2-Acetoxybenzoic acid.
salycylacetylsalicylic acid; Aspirin; A.S.A; Acetyonyl; Bayer; Bufferin; Diplosal Acetate; Alka-Seltzer; Anacin; Ascriptin; Coricidin; Coricidin D; Darvon Compound; Easprin; Ecotrin; Empirin; Excedrin; Gelprin; Robaxisal; Vanquish; Ascoden-30; Measurin; Norgesic; Persistin; St Joseph's Aspirin for Adults; Supac; Triaminicin; ZORprin; Acidum Acetylsalicylicum; Acetilum Acidulatum; A.s.a.; Acenterine; Acetosal; Acetosalic Acid; Acetosalin; Acetylin; Acetylsal; Acylpyrin; Aspro; Asteric; Caprin; Colfarit; Duramax; Ecm; Endydol; Entrophen; Enterosarine; Helicon; Xaxa; Rhodine; Salacetin; Salcetogen; Saletin; Acesal; Acetisal; Acetonyl; Acimetten; Acisal; A.s.a. Empirin; Asagran; Asatard; Aspalon; Aspergum; Aspirdrops; AC 5230; Benaspir; Entericin; Extren; Bialpirinia; Contrheuma Retard; Crystar; Delgesic; Dolean Ph 8; Enterophen; Globoid; Idragin; Neuronika; Novid; Polopiryna; Solpyron; Claradin; Nu-seals; Solprin. Analgesic; anti-inflammatory. mp = 159°; nearly insoluble in H_2O, C_6H_6; soluble in alkaline solutions. *Boehringer Ingelheim GmbH.*

483 Alclofenac
22131-79-9 220 244-795-4
$C_{11}H_{11}ClO_3$
3-Chloro-4-(2-propenyloxy)benzeneacetic acid.
W-7320; Allopydin; Epinal; Medifenac; Mervan; Neoston; Prinalgin; Reufenac; Zumaril. An analgesic with antipyretic and anti-inflammatory properties. Arylacetic acid derivative. mp = 92-93°; LD_{50} (rat orl) = 1050 mg/kg, (rat sc) = 600 mg/kg, (rat ip) = 555 mg/kg. *Continental Pharma Inc.*

484 Alminoprofen
39718-89-3 308 254-604-6
$C_{13}H_{17}NO_2$
p-[(2-Methylalyly)amino]hydratropic acid.
EB-382; Minalfene. Anti-inflammatory; analgesic. mp = 107°; LD_{50} (mus orl) = 2400 mg/kg.

485 Aloxiprin
9014-67-9 316
2-(Acetyloxy)benzoic acid polymer with aluminum oxide.
polyoxyaluminum acetylsalicylate; Alaprin; Lyman; Palaprin; Rumatral; Superpyrin. Analgesic. Polymeric condensation product of aluminum oxide and aspirin. *Ajinomoto Co. Inc.*

486 Aluminum Bis(acetylsalicylate)
23413-80-1 338
$C_{18}H_{15}AlO_9$
Bis[2-(acetyloxy)benzoato-O']hydroxyaluminum.
aluminum diacetylsalicylate; acetylsalicylate aluminum; aluminum diaspirin; aluminum aspirin; monohydroxyaluminum diacetylsalicylate; monohydroxyaluminum bis(acetylsalicylate); Rumasal. Analgesic; antipyretic. Nearly insoluble in H_2O, alcohol, Et_2O; decomposes in dilute acids, dilute alkalies, alkali carbonates.

487 Aminobenzoate Potassium
138-84-1 7766 205-338-4
$C_7H_6KNO_2$
p-Aminobenzoic acid potassium salt.
potassium p-aminobenzoate; KPABA; Potoba. Antifibrotic; analgesic. Used in idiopathic pulmonary fibrosis. Soluble in H_2O; nearly insoluble in Et_2O. *Glenwood Inc.*

488 Aminobenzoate Sodium
555-06-6 209-080-3
$C_7H_6NaNO_2$
p-Aminobenzoic acid monosodium salt.
sodium p-aminobenzoate. Analgesic.

489 Aminochlorthenoxazin
3567-76-8 455
$C_{10}H_{11}ClN_2O_2$
6-Amino-2-(2-chloroethyl)-2,3-dihydro-4H-1,3-benzoxazin-4-one.
component of: Dereuma. Antipyretic; analgesic. mp = 164°; nearly insoluble in H_2O; LD_{50} (mus ip) ≅ 1950 mg/kg, (mus orl) ≅ 10000 mg/kg.

490 Aminochlorthenoxazin Hydrochloride
3443-15-0 455
$C_{12}H_{12}Cl_2N_2O_2$
6-Amino-2-(2-chloroethyl)-2,3-dihydro-4H-1,3-benzoxazin-4-one monohydrochloride.
A-350; ICI-350. Antipyretic; analgesic. mp = 209-210°; soluble in H_2O; LD_{50} (mus ip) ≅ 920 mg/kg, (mus orl) ≅ 2250 mg/kg, (mus iv) ≅ 290.

491 Aminophenazone Cyclamate
747-30-8 495
$C_{20}H_{30}N_4O_4S$
4-Dimethylamino-2,3-dimethyl-1-phenyl-3-pyrazolin-5-one cyclohexylsulfamate.
aminopyrine cyclohexylsulfamate. Analgesic; antipyretic. *See* aminopyrine.

492 Aminopicoline
695-34-1 486 211-780-9
$C_6H_8N_2$
4-Methyl-2-pyridinamine.
2-amino-4-picoline; component of: Ascensil, Askensil. Analgesic; cardiac stimulant. mp = 100-100.5°; bp_{11} = 115-117°; soluble in H_2O, alcohols, dimethylformamide, coal tar bases; slightly soluble in petroleum ether, aliphatic hydrocarbons. *Raschig GmbH.*

493 Aminopicoline Camphorsulfonate
12261-97-1 486
$C_{16}H_{24}N_2O_4S$
4-Methyl-2-pyridinamine camphorsulfonate.
2-amino-4-picoline camphorsulfonate; Piricardio; Varunax. Analgesic; cardiac stimulant. *Raschig GmbH.*

494 Aminopropylon
3690-04-8 492 222-999-4
$C_{16}H_{22}N_4O_2$
N-(2,3-Dihydro-1,5-dimethyl-3-oxo-2-phenyl-1H-pyrazol-4-yl)-2-(dimethylamino)propanamide.
Amipylo. Analgesic. mp = 181°; soluble in H_2O.

495 Aminopyrine
58-15-1 495 200-365-8
$C_{13}H_{17}N_3O$
4-Dimethylamino-2,3-dimethyl-1-phenyl-3-pyrazolin-5-one.
amidopyrine; aminophenazone; dimethylaminophenyldimethylpyrazone; Amidofebrin; Amidopyrazoline; Anafebrina; Brufaneuxol; Dimapyrin; Dipirin; Febrinina; Itamidone; Mamallet-A; Netsusarin; Novamidon; Piridol; Polinalin; Pyradone; Pyramidon. Analgesic; antipyretic. CAUTION: may cause agranulocytosis. mp = 107-109°; soluble in alcohol (1 g/1.5 ml), C_6H_6 (1 g/12 ml), $CHCl_3$ (1 g/1 ml), Et_2O (1g/13 ml), H_2O (1g/18 ml); LD_{50} (rat orl) = 1700 mg/kg.

496 Aminopyrine Bicamphorate
94442-12-3 495
$C_{33}H_{49}N_3O_9$
4-Dimethylamino-2,3-dimethyl-1-phenyl-3-pyrazolin-5-one bicamphorate.
Pyramidon bicamphorate. Analgesic; antipyretic. mp = 94°; soluble in H_2O with gradual decomposition; soluble in alcohol.

497 Aminopyrine Salicylate
603-57-6 495 210-049-1
$C_{20}H_{23}N_3O_4$
4-Dimethylamino-2,3-dimethyl-1-phenyl-3-pyrazolin-5-one salicylate.
Pyramidon salicylate. Analgesic; antipyretic. mp = 70°; soluble in H_2O (6.25 g/100 ml), alc (16.7 g/100 ml).

498 Ammonium Salicylate
528-94-9 584 208-444-9
$C_7H_9NO_3$
2-Hydroxybenzoic acid monoammonium salt.
salicylic acid monoammonium salt; Salicyl-Vasogen. Analgesic. Used topically to loosen psoriatic scales. Soluble in H_2O (100 g/100 ml), EtOH (33.3 g/100 ml).

499 Amtolmetin Guacil
87344-06-7 637
$C_{24}H_{24}N_2O_5$
N-[(1-Methyl-5-p-toluoylpyrrol-2-yl)acetyl]glycine o-methoxyphenyl ester.
ST-679; MED-15; amtolmethin guacil. Analgesic; anti-inflammatory. Ester prodrug of tolmetin. Arylacetic acid derivative. mp = 117-120°; soluble in common organic solvents; LD_{50} (mmus ip) = 1370 mg/kg, (mrat ip) = 1100 mg/kg, (mmus orl) > 1500 mg/kg, (mrat orl) = 1450 mg/kg. *Sigma-Tau Pharm. Inc.*

500 Anidoxime
34297-34-2
$C_{21}H_{27}N_3O_3$
3-(Diethylamino)propiophenone, O-[(p-methoxyphenyl)-carbamoyl]oxime. Analgesic.

501 Anilopam
53716-46-4
$C_{20}H_{26}N_2O$
(-)-3-(p-Aminophenethyl)-2,3,4,5-tetrahydro-8-methoxy-2-methyl-1H-3-benzazepine dihydrochloride.
Analgesic. *Fisons plc.*

502 Anilopam Hydrochloride
53716-45-3
$C_{20}H_{28}Cl_2N_2O$
(-)-3-(p-Aminophenethyl)-2,3,4,5-tetrahydro-8-methoxy-2-methyl-1H-3-benzazepine.
786-723. Analgesic. *Fisons plc.*

503 Anirolac
66635-85-6
$C_{16}H_{15}NO_4$
(±)-5-p-Anisoyl-2,3-dihydro-1H-pyrrolizine-1-carboxylic acid.
RS-37326. Anti-inflammatory; analgesic. *Syntex Labs. Inc.*

504 Antipyrine
60-80-0 757 200-486-6
$C_{11}H_{12}N_2O$
2,3-Dimethyl-1-phenyl-3-pyrazolin-5-one.
phenyldimethyl isopropyl pyrazolone; dimethyloxy-chinizin; dimethyloxyquinazine; oxydimethylquinizine; phenazone; Analgesine; Anodynine; Parodyne; Phenyl-one; Sedatine; component of: Auralgan. Analgesic; also as as an analgesic and antipyretic in veterinary medicine as well in treatment of laminitis in horses. Used as an indicator in hepatic drug metabolism. mp = 119-113°; soluble in H_2O (<1 g/ml), alcohol (1 g/1.3 ml), $CHCl_3$ (1 g/1 ml), Et_2O (1 g/43 ml); LD_{50} (rat orl) = 1.8 g/kg.

505 Antipyrine Acetylsalicylate
569-84-6 757
$C_{20}H_{20}N_2O_5$
2,3-Dimethyl-1-phenyl-3-pyrazolin-5-one acetylsalicylate.
Acetopyrine; Acopyrine; Acetasol. Analgesic. Has been used as an antipyretic and analgesic in veterinary medicine. mp = 63-65°; soluble in hot H_2O, $CHCl_3$, alcohol; sparingly soluble in Et_2O, cold H_2O (1 g/400 ml).

506 Antipyrine Methylethylglycolate
5794-16-1 757
$C_{19}H_{20}N_2O_4$
2,3-Dimethyl-1-phenyl-3-pyrazolin-5-one methylethylglycolate.
antipyrine 2-hydroxy-2-methylbutyrate; Astrolin. Analgesic. Has been used as an antipyretic and analgesic in veterinary medicine. mp = 64-65.5°; soluble in H_2O, alcohol.

507 Antipyrine Salicylacetate
520-07-0 757
$C_{20}H_{20}N_2O_6$
2,3-Dimethyl-1-phenyl-3-pyrazolin-5-one salicylacetate.
alpha-carboxy-o-anisic acid compound with antipyrine; Pyrosal. Analgesic. Has been used as an antipyretic and analgesic in veterinary medicine. mp = 149-150°; sparingly soluble in H_2O; soluble in alcohol.

508 Antrafenine
55300-29-3 763
$C_{30}H_{26}F_6N_4O_2$
2-[4-(α,α,α-Trifluoro-m-tolyl)-1-piperazinyl]ethyl N-[7-(trifluoromethyl)-4-quinolyl]anthranilate.
SL-73.033; Stakane. Analgesic with minimal anti-inflammatory and antipyretic activity, related structurally to floctafenine. mp = 88°; LD_{50} (mus orl) = 4000 mg/kg. *Synthelabo Pharmacie.*

509 Apazone
13539-59-8 768 236-913-8
$C_{16}H_{20}N_4O_2$
5-(Dimethylamino)-9-methyl-2-propyl-1H-pyrazolo[1,2-a][1,2,4]benzotriazine-1,3(2H)-dione.
azapropazone; AHR-3018; MI-85; NSC-102824; Cinnamin; Sinnamin. Anti-inflammatory; analgesic. A pyrazolone. mp = 228°. *Siegfried AG.*

510 Apazone Dihydrate
22304-30-9 768
$C_{16}H_{24}N_4O_4$
5-(Dimethylamino)-9-methyl-2-propyl-1H-pyrazolo[1,2-a][1,2,4]benzotriazine-1,3(2H)-dione dihydrate.
MI-85Di; Azapren; Prolixan; Rheumox; Tolyprin. Anti-inflammatory; analgesic. A pyrazolone. mp = 247-248°. *Siegfried AG.*

511 Aspirin
50-78-2 886 200-064-1
$C_9H_8O_4$
Acetylsalicylic acid.
o-carboxyphenyl acetate; 2-(acetyloxy)benzoic acid; acetate salicylic acid; salicylic acid acetate; Acenterline; Aceticyl; Acetosal; Acetosalic Acid; Acetosalin; Acetylin; Acetyl-SAL; Acimetten; Acylpyrin; Arthrisin, A.S.A; Asatard; Aspro; Asteric; Caprin; Claradin; Colfarit; Contrheuma retard; Duramax; ECM; Ecotrin; Empirin; Encaprin; Endydol; Entrophen; Enterosarine; Helicon; Levius; Longasa; Measurin; Neuronika; Platet; Rhodine; Salacetin; Salcetogen; Saletin; Solrin; Solpyron; Xaxa; Alka-seltzer; Anacin; Ascriptin; Bufferin; Coricidin D; Darvon compound; Excedrin; Gelprin; Robaxisal; Vanquish; Ascoden-30; Coricidin; Norgesic; Persistin; Supac; Triaminicin; acetophen; acidum acetylsalicylicum; acetilum acidulatum; Acetonyl; Adiro; acenterine; acetosal; acetosalic acid; acetosalin; acetylin; acetylsal; acylpyrin;asteric; caprin; colfarit; entrophen; enterosarine; rhodine; salacetin; salcetogen; saletin; acesal; acetilsalicilico; acetisal; acetonyl; asagran; asatard; aspalon; aspergum; aspirdrops; AC 5230; benaspir; entericin; extren; bialpirinia; contrheuma retard; Crystar; Delgesic; Dolean ph 8; enterophen; globoid; idragin. Analgesic; antipyretic; anti-inflammatory. Also has platelet aggregation inhibiting, antithrombotic, and antirheumatic properties. mp = 135; d = 1.40; λ_m = 229, 277 nm ($E^{1\%}_{1cm}$ 484, 68 0.1N H_2SO_4); pH (25°) = 3.49; slightly soluble in H_2O (1 g/300 ml at 25°, 1 g/100 ml at 37°); soluble in alc (1 g/5 ml), $CHCl_3$ (1 g/17 ml), Et_2O (≅1 g/10 ml); LD_{50} (mus orl) = 1100 mg/kg. *Boots Pharmaceuticals Inc.; Bristol-Myers Squibb Co; Eli Lilly &*

Co.; *Parke-Davis; Schering-Plough Pharm.; SmithKline Beecham Pharm.; Sterling Health U.S.A.; Sterling Winthrop Inc.; Upjohn Ltd.*

512 Benorylate
5003-48-5 1074 225-674-5
$C_{17}H_{15}NO_5$
4-Acetamidophenyl salicylate acetate.
benorilate; fenasprate; WIN-11450; Benoral; Benortan; Quinexin; Salipran. Analgesic; antipyretic; anti-inflammatory. Derivative of salicylic acid. mp = 175-176°; LD_{50} (mus orl) = 2000 mg/kg, (rat orl) = 10000 mg/kg, (mus ip) = 1255 mg/kg, (rat ip) = 1830 mg/kg. *Sterling Winthrop Inc.*

513 Benoxaprofen
51234-28-7 1075 257-069-7
$C_{16}H_{12}ClNO_3$
2-(4-Chlorophenyl)-α-methyl-5-benzoxazoleacetic acid.
Compound 90459; Coxigon; Opren; Oraflex; Uniprofen. Anti-inflammatory; analgesic. Arylpropionic acid derivative. mp = 189-190°; LD_{50} (mus orl) = 800 mg/kg. *Eli Lilly & Co.*

514 Benzpiperylon
53-89-4 1152 200-187-0
$C_{22}H_{25}N_3O$
4-Benzyl-2-(1-methyl-4-piperidyl)-5-phenyl-4-pyrazolin-3-one.
benzpiperilone; benzpiperylone; KB-95; Benzometan; Humedil; Reublonil; Telon. Analgesic; anti-inflammatory. A pyrazolone. Dec 181-183°; pk_1 = 6.73, pK_2 = 9.13; LD_{50} (mus iv) = 160 mg/kg, (mus orl) = 1880 mg/kg. *Sandoz Pharm. Corp.*

515 Benzydamine
642-72-8 1157 211-388-8
$C_{19}H_{23}N_3O$
1-Benzyl-3-[3-(dimethylamino)propoxy]-1H-indazole.
benzindamine. Analgesic; antipyretic; anti-inflammatory. Anticholinergic. $bp_{0.05}$ = 160°. *3M Pharm.; Angelini Pharm. Inc.*

516 Benzydamine Hydrochloride
132-69-4 1157 205-076-0
$C_{19}H_{24}ClN_3O$
1-Benzyl-3-[3-(dimethylamino)propoxy]-1H-indazole monohydrochloride.
Difflam Oral Rinse; Difflam Cream; AP-1288; BRL-1288; AF-864; Afloben; Andolex; Benalgin; Benzyrin; Difflam; Dorinamin; Enzamin; Imotryl; Ririlim; Riripen; Salyzoron; Saniflor; Tamas; Tantum; Verax. Analgesic; antipyretic; anti-inflammatory. Anticholinergic. mp = 160°; λ_m = 306 nm ($E^{1\%}_{1cm}$ 160); soluble in H_2O, EtOH, $CHCl_3$, n-butanol; LD_{50} (mus ip) = 110 mg/kg, (rat ip) = 100 mg/kg, (mus orl) = 515 mg/kg, (rat orl) = 1050 mg/kg. *Angelini Pharm. Inc.*

517 Bermoprofen
72619-34-2 1199
$C_{18}H_{16}O_4$
(±)-10,11-Dihydro-α,8-dimethyl-11-oxodibenz[b,f]oxepin-2-acetic acid.
AD-1590; AJ-1590; Dibenon. Anti-inflammatory; analgesic; antipyretic. Prostaglandin synthetase inhibitor; a derivative of arylpropionic acid derivative. mp = 128-129°; LD_{50} (mus orl) = 500 mg/kg, (rat orl) = 147 mg/kg. *Dainippon Pharm.*

518 Bicifadine
71195-57-8
$C_{12}H_{15}N$
(±)-1-p-Tolyl-3-azabicylco[3.1.0]hexane.
Analgesic. *Parke-Davis.*

519 Bicifadine Hydrochloride
66504-75-4
$C_{12}H_{16}ClN$
(±)-1-p-Tolyl-3-azabicylco[3.1.0]hexane hydrochloride.
CL-220075. Analgesic. *Parke-Davis.*

520 Bromadoline
67579-24-2
$C_{15}H_{21}BrN_2O$
trans-p-Bromo-N-[2-(dimethylamino)-cyclohexyl]benzamide.
Analgesic. *Upjohn Ltd.*

521 Bromadoline Maleate
81447-81-6
$C_{19}H_{25}BrN_2O_5$
trans-p-Bromo-N-[2-(dimethylamino)cyclohexyl]benzamide maleate (1:1).
U-47931E. Analgesic. *Upjohn Ltd.*

522 Bromfenac
91714-94-2 1411
$C_{15}H_{12}BrNO_3$
2-Amino-3-(p-bromobenzoyl)phenyl]acetic acid.
AHR-10282. Analgesic; anti-inflammatory. Prostaglandin sythetase inhibitor; analog of amfenac, an arylacetic acid derivative. *Robins, A. H. Co.*

523 Bromfenac Sodium
120638-55-3 1411
$C_{15}H_{11}BrNNaO_3.1/2H_2O$
Sodium [2-amino-3-(p-bromobenzoyl)phenyl]acetate sesquihydrate.
AHR-10282B. Analgesic; anti-inflammatory. Arylacetic acid derivative. mp = 284-286° (dec); pKa = 4.29; soluble in H_2O, MeOH, dilute base; insoluble in $CHCl_3$, dilute acid. *Robins, A. H. Co.*

524 Bromoacetanilide
103-88-8 1420 203-154-9
C_8H_8BrNO
N-(4-Bromophenyl)acetamide.
p-bromoacetanalide; monobromoacetanalide; bromo-analide; Antisepsin; Aespsin; Bromoantifebrin. Analgesic; antipyretic. mp = 168°; d = 172; nearly insoluble in cold H_2O; slightly soluble in hot H_2O; soluble in C_6H_6; $CHCl_3$, EtOAc; moderately soluble in alcohol. *Monsanto Co.*

525 5-Bromosalicylic Acid Acetate
1053-53-3 1459 203-154-9
$C_9H_7BrO_4$
2-(Acetoxy)-5-bromobenzoic acid.
bromo-aspirin. Analgesic. mp = 168-169°; slightly soluble in H_2O (1 g/1500 ml); more soluble in alcohol (1 g/9 ml), Et_2O (1 g/21 ml).

526 Bucetin
1083-57-4 1479 214-109-8
$C_{15}H_{17}NO_3$
3-Hydroxy-p-butyrophenetidide.
Betadid. Analgesic. Phenetidine derivative with analgesic and antipyretic activity. mp = 160°; sparingly soluble in H_2O; LD_{50} (mus ip) = 790 mg/kg, (mus orl) = 2800 mg/kg. *Hoechst Marion Roussel Inc.*

527 Bufexamac
2438-72-4 1497 219-451-1
$C_{12}H_{17}NO_3$
2-(p-Butoxyphenyl)acetohydroxamic acid.
CP-1044-J3; Droxarol; Droxaryl; Feximac; Malipuran; Mofenar; Norfemac; Parfenac; Parfenal. Anti-inflammatory; analgesic; antipyretic. Arylacetic acid derivative. mp = 153-155°; nearly insoluble in H_2O; LD_{50} (mus orl) > 8 g/mg. *C.M. Ind.; Pfizer Inc.*

528 Bumadizone
3583-64-0 1507 222-710-1
$C_{19}H_{22}N_2O_3$
Butylmalonic acid mono(1,2-diphenylhydrazide).
bumadizon. Analgesic; antipyretic; anti-inflammatory. Main product of hydrolysis of phenylbutazone. mp = 116-117°, 77-79°; λ_m = 234,264 nm (ε 16200, 3700 0.1 N NaOH). *Ciba-Geigy Corp.*

529 Bumadizone Calcium Salt Hemihydrate
69365-73-7 1507
$C_{38}H_{42}CaN_4O_6 \cdot 1/2H_2O$
Butylmalonic acid mono(1,2-diphenylhydrazide) calcium salt (2:1) hemihydrate.
Bumaflex; Eumotol; Rheumatol. Analgesic; antipyretic; anti-inflammatory. Arylbutyric acid derivative. Dec 154°; soluble in $CHCl_3$; EtOH, Et_2O; slightly soluble in H_2O; LD_{50} (mus orl) = 2500 mg/kg, (rat orl) = 1250 mg/kg, (mus iv) = 258 mg/kg, (rat iv) = 263 mg/kg. *Ciba-Geigy Corp.*

530 Butacetin
2109-73-1 1532
$C_{12}H_{17}NO_2$
4'-tert-Butoxyacetanilide.
BW-63-90; NSC-106564; Tromal. Analgesic; antidepressant. mp = 130°. *Burroughs Wellcome Inc.*

531 Butixirate
19992-80-4 10205 243-454-7
$C_{28}H_{33}NO_2$
(±)-α-Ethyl-4-biphenylacetic acid compound with trans-4-phenylcyclohexylamine (1:1).
xenbucin phenylcyclohexylamine; MG-5771; Flectar. Analgesic; antirheumatic; anti-inflammatory. mp = 224-227°; LD_{50} (mus ip) = 183 μM/kg. *Maggioni Farmaceutici S.p.A.*

532 Calcium Acetylsalicylate
69-46-5 1684 200-707-6
$C_{18}H_{14}CaO_8$
2-(Acetyloxy)benzoic acid calcium salt.
salicylic acid acetate calcium salt; acetylsalicylic acid calcium salt; calcium aspirin; soluble aspirin; Ascal; Cal-Aspirin; Dispril; Disprin; Kalmopyrin; Kalsetal; Solaspin; Tylcalsin. Analgesic; antipyretic; anti-inflammatory. Derivative of salicylic acid. Soluble in H_2O (16.6 g/100 ml), EtOH (1.25 g/100 ml). *Lee Labs.*

533 Calcium Acetylsalicylate Carbamide
5749-67-7 1684
$C_{19}H_{18}CaN_2O_9$
2-(Acetyloxy)benzoic acid calcium salt complex with urea.
carbaspirin calcium; carbasalate calcium; acetylsalicylic acid calcium salt complex with urea; urea calcium acetylsalicylate; carbaspirin calcium; Alcacyl; Calurin; Solupsan. Analgesic; antipyretic; anti-inflammatory. Derivative of salicylic acid. Dec 243-245°; soluble in H_2O (23.1 g/100 ml at 37°), pH = 4.8. *Lee Labs.*

534 Capsaicin
404-86-4 1811 206-969-8
$C_{18}H_{27}NO_3$
(E)-N-[(4-Hydroxy-3-methoxyphenyl)-methyl]-8-methyl-6-nonenamide.
trans-8-methyl-N-vanillyl-6-nonenamide; N-(4-hydroxy-3-methoxybenzyl)-8-methylnon-trans-6-enamide; Axsain; Mioton; Zostrix. Topical analgesic. Also used in neurobiological research. Pepper principle in (e.g.) capsicum. mp = 65°; $bp_{0.01}$ = 210-220°; λ_m = 227, 281 nm (ε = 7000, 2500); insoluble in H_2O; soluble in organic solvents.

535 Capsicum
84603-55-4 1813 283-256-8
Cayenne Pepper.
Topical counterirritant (veterinary); carminative; stomachic (veterinary). Dried fruit of *Capsicum frutescens L. Solanaceae or Capsicum Annuum.*

536 Carbamepazine
298-46-4 1826 206-062-7
$C_{15}H_{12}N_2O$
5H-Dibenz[b,f]azepine-5-carboxamine.
G-32883; Biston; Calepsin; Carbelan; Epitol; Finlepsin; Sirtal; Stazepine; Tegretal; Tegretol; Telesmin; Timonil. Analgesic; anticonvulsant. mp = 190-193°; Soluble in alcohol, Me_2CO, propylene glycol; nearly insoluble in H_2O; LD_{50} (mus orl) = 3750 mg/kg, (rat orl) = 4025 mg/kg. *Ciba-Geigy Corp.; Lemmon Co.*

537 Carbiphene
15687-16-8 1846
$C_{28}H_{34}N_2O_2$
2-Ethoxy-N-methyl-N-[2-(methylphenethylamino)ethyl]-2,2-diphenylacetamide.
etomide; etymide. Analgesic. *Bristol-Myers Squibb Co.*

538 Carbiphene Hydrochloride
467-22-1 1846
$C_{28}H_{35}ClN_2O_2$
2-Ethoxy-N-methyl-N-[2-(methylphenethylamino)ethyl]-2,2-diphenylacetamide monohydrochloride.
SQ-10269; NSC-106959; Bandol; Jubalon. Analgesic. mp = 163-165°. *Bristol-Myers Squibb Co.*

539 Carsalam
2037-95-8 1915
$C_9H_8NO_3$
2H-1,3-Benzoxazine-2,4(3H)-dione.

caronylsalicylamide; CSA; Beaprine. Analgesic. mp = 229-230°.

540 Chlorthenoxazine
132-89-8 2247 205-082-3
$C_{10}H_{10}ClNO_2$
2-(2-Chloroethyl)-2,3-dihydro-4H-1,3-benzoxazin-4-one.
chlorthenoxazin; AP-67; Apirazin; Ossapirina; Ossazone; Piroxina; Reumagrip; Valmorin; Valtorin; component of: Trigatan, Fiobrol. Antipyretic; analgesic. mp = 146-147° (dec); λ_m = 297.5 nm; soluble in $CHCl_3$.

541 Choline Salicylate
2016-36-6 2265 217-948-8
$C_{12}H_{19}NO_4$
2-(Hydroxyethyl)trimethylammonium salicylate.
choline salicylic acid salt; salicylic acid choline salt; Actasal; Arthropan; Artrobione; Audax; Mundisal. Analgesic; antipyretic. mp = 49-50°; soluble in H_2O, polar organic solvents, insoluble in non-polar organic solvents. *Omnium Chim.*

542 Cinchophen
132-60-5 2347 205-067-1
$C_{16}H_{11}NO_2$
2-Phenylquinoline-4-carboxylic acid.
2-phenylcinchoninic acid; Agotan; Alutyo; Artam; Atocin; Atophan; Cinconal; Mylofanol; Phenoquin; Polyphlogin; Quinofen; Quinophan; Rheumatan; Rheumin; Tophol; Tophosan; Vantyl; Viophan. Quinoline derivative formerly used in treatment of chronic gout. Used experimentally to induce ulcers. Formerly used as an analgesic. mp = 213-216°; nearly insoluble in H_2O; more soluble in $CHCl_3$ (0.25 g/100 ml), Et_2O (1 g/100 ml), alcohol (0.83 g/100 ml). *Parke-Davis.*

543 Ciprefadol
59889-36-0
$C_{19}H_{27}NO$
(-)-m-[2-(Cyclopropylmethyl)-1,3,4,5,6,7,8,8aα-octahydro-4aβ(2H)-isoquinolyl]phenol.
Analgesic. *Eli Lilly & Co.*

544 Ciprefadol Succinate
60719-85-9
$C_{23}H_{33}NO_5$
(-)-m-[2-(Cyclopropylmethyl)-1,3,4,5,6,7,8,8aα-octahydro-4aβ(2H)-isoquinolyl]phenol succinate (1:1).
Compound 112878. Analgesic. *Eli Lilly & Co.*

545 Ciramadol
63269-31-8 2375
$C_{15}H_{23}NO_2$
(-)-(1R,2R)-2-[(R)-α-(Dimethylamino)-m-hydroxybenzyl]cyclohexanol.
WY-15705. Analgesic. Has narcotic agonist-antagonist properties. mp = 191-193°; $[\alpha]_D^{25}$ = -46.92° (c = 1.061 MeOH). *Am. Home Products; Wyeth Labs.*

546 Ciramadol Hydrochloride
63323-46-6 2375
$C_{15}H_{24}ClNO_2$
(-)-(1R,2R)-2-[(R)-α-(Dimethylamino)-m-hydroxybenzyl]cyclohexanol hydrochloride.
WY-15705 HCl; Ciradol. Analgesic. mp = 255-257°; $[\alpha]_D^{25}$ = -15.31° (c = 1.067 MeOH). *Am. Home Products; Wyeth Labs.*

547 Clometacin
25803-14-9 2443 247-271-3
$C_{19}H_{16}ClNO_4$
3-(p-Chlorobenzoyl)-6-methoxy-2-methylindole-1-acetic acid.
clometazin; mindolic acid; R-3959; C-1656; Dupéran. Analgesic. mp = 242°. *Roussel-UCLAF.*

548 Clonixeril
21829-22-1
$C_{16}H_{17}ClN_2O_4$
2,3-Dihydroxypropyl 2-(3-chloro-o-toluidino)-nicotinate.
Sch-12707. Analgesic. *Schering-Plough HealthCare Products.*

549 Clonixin
17737-65-4 2453 241-730-1
$C_{13}H_{11}ClN_2O_2$
4-Chloro-N-(2,6-dimethylpiperidino)-3-sulfamoylbenzamide.
clonixic acid; CBA-93626; Sch-10304. Analgesic. mp = 233-235°; LD_{50} (mus orl) = 415 mg/kg, (mus ip) = 198 mg/kg, (mus sc) = 296 mg/kg. *Schering-Plough HealthCare Products.*

550 Clonixin Lysine Salt
55837-30-4 2453
$C_{19}H_{25}ClN_4O_4$
4-Chloro-N-(2,6-dimethylpiperidino)-3-sulfamoylbenzamide lysine salt.
L-104; Clonix; Deltar; Dolagial; Dorixina. Analgesic. *Schering-Plough HealthCare Products.*

551 Cloracetadol
15687-05-5 239-776-2
$C_{10}H_{10}Cl_3NO_3$
N-[4-(2,2,2-Trichloro-1-hydroxyethoxy)phenyl]-acetamide.
Analgesic; antipyretic.

552 Cropropamide
633-47-6 2659 211-193-8
$C_{13}H_{24}N_2O_2$
N-[1-(Dimethylcarbonylpropyl)-N-propylcrotonamide.
component of: prethcamide, Micoren, Respirot. Analgesic. $bp_{0.25}$ = 128-130°; soluble in H_2O, Et_2O. *Ciba-Geigy Corp.*

553 Cropropamide [E]
3544-46-5
$C_{13}H_{24}N_2O_2$
N-[1-(Dimethylcarbamoyl)propyl]-N-propylcrotonamide [E].
Analgesic. *Ciba-Geigy Corp.*

554 Crotetamide
6168-76-9 2662 228-208-9
$C_{12}H_{22}N_2O_2$
N-(1-Dimethylcarbamoylpropyl)-N-ethylcrotonamide.
crotethamid; crotethamide; component of: prethcamide,

Micoren, Respirot. Analgesic. $bp_{0.03}$ = 132-134°; soluble in H_2O, Et_2O. *Ciba-Geigy Corp.*

555 Cyclazocine
3572-80-3 2773 222-689-9
$C_{18}H_{25}NO$
3-(Cyclopropylmethyl)-1,2,3,4,5,6-hexahydro-6,11-dimethyl-2,6-methano-3-benzazocin-8-ol.
NSC-107429; Win-20740. Analgesic with mixed narcotic agonist-antagonist properties. mp = 210-204°. *Sterling Winthrop Inc.*

556 Delmetacin
16401-80-2
$C_{18}H_{15}NO_3$
1-Benzoyl-2-methylindole-3-acetic acid.
Anti-inflammatory; antipyretic; analgesic.

557 Denaverine
3579-62-2
$C_{24}H_{33}NO_3$
2-(Dimethylamino)ethyl (2-ethylbutoxy)diphenylacetate.
Analgesic.

558 Dexibuprofen
51146-56-6 4925
$C_{13}H_{18}O_2$
(+)-(S)-p-Isobutylhydratropic acid.
Analgesic; anti-inflammatory. Cyclooxygenase inhibitor. *Merck & Co.Inc.*

559 Dexibuprofen Lysine
141505-32-0 4925
$C_{19}H_{34}N_2O_5$
(+)-(S)-p-Isobutylhydratropic acid
L-lysine salt monohydrate.
Ibuprofen S-form L-lysine salt; ML-223; MK-223; Doctrin. Analgesic; anti-inflammatory. S-Form of ibuprofen L-lysine salt. Cyclooxygenase inhibitor. *Merck & Co.Inc.*

560 Dexibuprofen Lysine [anhydrous]
113403-10-4 4925
$C_{19}H_{3N2}O_4$
(+)-(S)-p-Isobutylhydratropic acid
L-lysine salt.
L-669445. Analgesic; anti-inflammatory. S-Form of ibuprofen L-lysine salt. Cyclooxygenase inhibitor. *Merck & Co.Inc.*

561 Dexindoprofen
53086-13-8 4999 258-351-2
$C_{17}H_{15}NO_3$
(+)-(S)-p-(1-Oxo-2-isoindolinyl)hydratropic acid.
(+)-form indoprofen; Nedius. Analgesic; anti-inflammatory. mp = 205-207°; $[\alpha]_D^{20}$ = +48° (c = 0.05 DMSO); LD_{50} (rat iv) = 31.98 mg/kg, (rat orl) = 33.75 mg/kg. *Chugai Pharm. Co. Ltd.*

562 Dexketoprofen
22161-81-5
$C_{16}H_{14}O_3$
(S)-2-(3-Benzoylphenyl)propionic acid.
Analgesic.

563 Dexoxadrol
4741-41-7
$C_{20}H_{23}NO_2$
(+)-2-(2,2-Diphenyl-1,3-dioxolan-4-yl)piperidine.
d-dioxadrol. CNS stimulant; analgesic. *Upjohn Ltd.*

564 Dexoxadrol Hydrochloride
631-06-1
$C_{20}H_{24}ClNO_2$
(+)-2-(2,2-Diphenyl-1,3-dioxolan-4-yl)piperidine hydrochloride.
CL-911C; U-22559A; NSC-526062; d-dioxadrol hydrochloride. CNS stimulant; analgesic. *Upjohn Ltd.*

565 Dexpemedolac
114030-44-3
$C_{22}H_{23}NO_3$
(1S,4R)-4-Benzyl-1-ethyl-1,3,4,9-tetrahydropyrano-[3,4-b]indole-1-acetic acid.
WAY-PEM-420. Analgesic. *Wyeth-Ayerst Labs.*

566 Diacetamate
2623-33-8
$C_{10}H_{11}NO_3$
4-Acetamidophenyl acetate.
Analgesic.

567 Difenamizole
20170-20-1 3181
$C_{20}H_{22}N_4O$
2-(Dimethylamino)-N-(1,3-diphenylpyrazol-5-yl)-propionamide.
AP-14; Pasalin. Analgesic; anti-inflammatory. A pyrazole. mp = 123-128°, 120-122°; soluble in Me_2CO, $CHCl_3$, C_6H_6; nearly insoluble in H_2O; LD_{50} (mus iv) = 103 mg/kg, (mus ip) = 186 mg/kg, (mus sc) = 525 mg/kg, (mus orl) = 560 mg/kg. *Takeda Chem. Ind. Ltd.*

568 Diflunisal
22494-42-4 3190 245-034-9
$C_{13}H_8F_2O_3$
2',4'-Difluoro-4-hydroxy-3-biphenyl-carboxylic acid.
MK-647; Adomal; Difludol; Dolisal; Dolobid; Dolobis; Flovacil; Fluniget; Fluodonil; Flustar. Analgesic; anti-inflammatory. Derivative of salicylic acid. mp = 210-211°; sparingly soluble in H_2O; LD_{50} (fmus orl) = 439 mg/kg. *E. Merck; N.V. Philips.*

569 Dihydroxyaluminum Acetylsalicylate
53230-06-1 3226
$C_9H_9AlO_6$
Dihydroxy(acetylsalicylato)aluminum.
dihydroxyaluminum aspirin. Analgesic; antipyretic. May contain one or more molecules of $Al(OH)_3$. Stable in aqueous suspension at neutral pH; decomposes below pH 4. *Keystone Chemurgic.*

570 Dimefadane
5581-40-8
$C_{17}H_{19}N$
N,N-Dimethyl-3-phenyl-1-indanamine.
SKF-1340. Analgesic. *SmithKline Beecham Pharm.*

571 Dipyrocetyl
486-79-3 3413 207-641-7
$C_{11}H_{10}O_6$
2,3-Bis(acetyloxy)benzoic acid.
2,3-diacetoxybenzoic acid; Artromialgina; Movirene. Analgesic; antipyretic. mp = 148-150°, 146-170°; soluble in organic solvents; nearly insoluble in H_2O. *Hoechst Roussel Pharm. Inc.*

572 Dipyrone
5907-38-0 3414
$C_{13}H_{18}N_3NaO_5S$
Sodium (antipyrinylmethylamino)methanesulfonate monohydrate.
sodium methylaminoantipyrine methanesulfonate; methylmelubrin; methampyrone; metamizol; analgin; sulpyrin; Alginodia; Algocalmin; Bonpyrin; Conmel; Divarine; Dolazon; D-Pron; Dya-Tron; Espyre; Farmolisina; Feverall; Fevonil; Keypyrone; Metilon; Minalgin; Narone; Nartate; Nevralgina; Nolotil; Novacid. Analgesic; antipyretic. Soluble in H_2O; less soluble in EtOH; nearly insoluble in Et_2O, Me_2CO, C_6H_6, $CHCl_3$. *Farmitalia Societa Farmaceutici; Hoechst Roussel Pharm. Inc.; Sterling Winthrop Inc.*

573 Dipyrone [anhydrous]
68-89-3 3414
$C_{13}H_{16}N_3NaO_4S$
Sodium (antipyrinylmethylamino)methanesulfonate.
Analgesic; antipyretic. *Farmitalia Societa Farmaceutici; Hoechst Roussel Pharm. Inc.; Sterling Winthrop Inc.*

574 Dipyrone Magnesium Salt
6150-97-6 3414
$C_{26}H_{32}MgN_6NaO_8S_2$
Sodium (antipyrinylmethylamino)methanesulfonate magnesium salt (2:1).
Magnopyrol. Analgesic; antipyretic. *Farmitalia Societa Farmaceutici; Hoechst Roussel Pharm. Inc.; Sterling Winthrop Inc.*

575 Doxpicomine
62904-71-6
$C_{12}H_{18}N_2O_2$
(-)-3-[(Dimethylamino)-m-dioxan-5-ylmethyl]pyridine.
doxpicodin. Analgesic.

576 Doxpicomine Hydrochloride
69494-04-8
$C_{12}H_{19}ClN_2O_2$
(-)-3-[(Dimethylamino)-m-dioxan-5-ylmethyl]pyridine monohydrochloride.
LY-108380; doxpicodin hydrochloride. Analgesic. *Eli Lilly & Co.*

577 Drinidene
53394-92-6
$C_{10}H_9NO$
2-(Aminomethylene)-1-indanone.
CP-224877. Analgesic. *Pfizer Inc.*

578 Emorfazone
38957-41-4 3603 254-220-9
$C_{11}H_{17}N_3O_3$
4-Ethoxy-2-methyl-5-(4-morpholinyl)-3(2H)-pyridazinone.
M-73101; Nandron; Pentoyl. Anti-inflammatory; analgesic. mp = 89-91°; LD_{50} (mus ip) = 700 mg/kg. *Robert et Carriere.*

579 Enadoline
124378-77-4
$C_{24}H_{32}N_2O_3$
N-Methyl-N-[(5R,7S,8S)-7-(1-pyrrolidinyl-1-oxa-spiro-[4.5]-dec-8yl]-4-benzofuranacetamide.
Analgesic. *Parke-Davis.*

580 Enadoline Hydrochloride
124439-07-2
$C_{24}H_{33}ClN_2O_3$
N-Methyl-N-[(5R,7S,8S)-7-(1-pyrrolidinyl-1-oxa-spiro-[4.5]-dec-8yl]-4-benzofuranacetamide hydrochloride.
CI-977. Analgesic. *Parke-Davis.*

581 Enfenamic Acid
23049-93-6 3620
$C_{15}H_{15}NO_2$
N-Phenethylanthranilic acid.
RH-8; Tromaril. Anti-inflammatory; analgesic. Amino-arylcarboxylic acid derivative. mp = 116-117°; insoluble in H_2O; LD_{50} (mus ip) = 575 mg/kg, (mus orl) > 2000 mg/kg. *CSIR New Delhi.*

582 Epirizole
18694-40-1 3659 242-507-1
$C_{11}H_{14}N_4O_2$
4-Methoxy-2-(5-methoxy-3-methylpyrazol-1-yl)-6-methylpyrimidine.
mepirizole; DA-398; Mebron. Analgesic; anti-inflammatory; antipyretic. mp = 90-92°; slightly soluble in H_2O; soluble in dilute acids, EtOH, C_6H_6, dichloroethane, Et_2O, Me_2CO; LD_{50} (mus orl) = 820 mg/kg. *Daiichi Pharm. Corp.*

583 Ergotamine
113-15-5 3703 204-023-9
$C_{33}H_{35}N_5O_5$
12'-Hydroxy-2'-methyl-5'α-(phenylmethyl)-ergotaman-3',6',18-trione.
Ergoton-A [as succinate]. Analgesic, specific to migraine. Antimigraine. An alkaloid from ergot, a dark colored fungus, which attacks damp rye and other grasses, and which, through flour, causes ergotism. Has been used in medicine as a vasoconstrictor. Dec 212-214°; $[\alpha]_D^{20}$ = -160° ($CHCl_3$); insoluble in H_2O, petroleum ether; soluble in MeOH (1.4 g/100 ml), EtOH (0.33 g/100 ml), Me_2CO (0.67 g/100 ml); freely soluble in $CHCl_3$, C_5H_5N, AcOH; soluble in EtOAc; slightly soluble in. *3M Pharm.; Fisons Pharm. Div.; Organon Inc.; Parke-Davis; Sandoz Pharm. Corp.*

584 Ergotamine Succinate
3703
$C_{70}H_{76}N_{10}O_{14}$
12'-Hydroxy-2'-methyl-5'α-(phenylmethyl) ergotaman-3',6',18-trione succinate.
Ergoton-A. Has been used as an antimigraine. *3M Pharm.; Fisons Pharm. Div.; Organon Inc.; Parke-Davis; Sandoz Pharm. Corp.*

585 Ergotamine Tartrate
379-79-3 3703 206-835-9
$C_{70}H_{76}N_{10}O_{16}$
12'-Hydroxy-2'-methyl-5'α-(phenylmethyl)-ergotaman-3',6',18-trione [R-(R*,R*)]-2,3-dihydroxy-butanedioate (2:1) salt.
Ergate; Ergomar; Ergostat; Ergotartrat; Exmigra; Femergin; Gynergin; Lingraine; Lingran; Medihaler Ergotamine; component of: Cafergot, Wigraine. Analgesic, specific to migraine. Antimigraine. Has also been used as an oxytocic in veterinary medicine. Serotonin receptor agonist. mp = 203° (dec); $[\alpha]_D^{25}$ = -125 to -155° (c = 0.4 $CHCl_3$); soluble in H_2O (2 mg/ml), EtOH; LD_{50} (rat iv) = 80 mg/kg. *3M Pharm.; Fisons Pharm. Div.; Organon Inc.; Parke-Davis; Sandoz Pharm. Corp.*

586 Etersalate
62992-61-4 3759 263-780-3
$C_{19}H_{19}NO_6$
2-(Acetyloxy)benzoic acid
2-[4-(acetylamino)phenoxy]ethyl ester.
eterilate; etherylate; eterylate; Daital. Analgesic; anti-inflammatory; antipyretic. Derivative of aspirin. mp = 139-141°; LD_{10} (rat orl) = 7000 mg/kg.

587 Ethenzamide
938-73-8 3776 213-346-4
$C_9H_{11}NO_2$
2-Ethoxybenzamide.
ethbenzamide; salicylamide; o-ethyl ether; Lucamide; Protopyrin; Trancalgyl. Analgesic. mp = 132-134°; LD_{50} (mus orl) = 1160 mg/kg.

588 Ethoheptazine
77-15-6 3789 201-007-3
$C_{16}H_{23}NO_2$
Ethyl hexahydro-1-methyl-4-phenyl-1H-azepine-4-carboxylic acid.
ethyl heptazine; WY-401. Analgesic. Liquid; bp_1 = 133-134°; d_D^{26} = 1.038; n_D^{26} = 1.5220; [hydrochloride]: mp = 151-153°; [methobromide]: mp = 215-217°. *Am. Home Products.*

589 Ethoheptazine Citrate
2085-42-9 3789
$C_{22}H_{31}NO_9$
Ethyl hexahydro-1-methyl-4-phenyl-1H-azepine-4-carboxylate
citrate (1:1).
Zactane. Analgesic. *Am. Home Products.*

590 Ethoxazene
94-10-0 3796
$C_{14}H_{16}N_4O$
4-[(p-Ethoxyphenyl)azo]-m-phenylenediamine.
p-ethoxychrisoidine; SN-612; Carmurit; Cystural. Analgesic. *Bristol-Myers Squibb Co.*

591 Ethoxazene Hydrochloride
2313-87-3 3796
$C_{14}H_{17}ClN_4O$
4-[(p-Ethoxyphenyl)azo]-m-phenylenediamine chloride.
SQ-2128; NSC-7214; Serenium. Analgesic. Insoluble in H_2O. *Bristol-Myers Squibb Co.*

592 Etodolac
41340-25-4 3920
$C_{17}H_{21}NO_3$
1,8-Diethyl-1,3,4,9-tetrahydropyranol[3,4-b]indole-1-acetic acid.
etodolic acid; AY-24236; Edolan; Lodine; Ramodar; Tedolan; Ultradol; Zedolac. Anti-inflammatory; analgesic. Arylacetic acid derivative. mp = 145-148°. *Am. Home Products; Wyeth-Ayerst Labs.*

593 Etofenamate
30544-47-9 3922 250-231-8
$C_{18}H_{18}F_3NO_4$
2-(2-Hydroxyethoxy)ethyl-N-(α,α,α-trifluoro-m-tolyl)anthranilate.
B-577; Bay-d-1107; TV-485; WHR-5020; Bayrogel; Glasel; Rheumon gel; Traumon Gel. Anti-inflammatory; analgesic. Percutaneously active antiphlogistic agent. Aminoarylcarboxylic acid derivative. Viscous oil; thermolabile at 180°; $bp_{0.001}$ = 130-135°; λ_m = 286 nm ($E_{1cm}^{1\%}$ 423 in MeOH); soluble in lower alcohols, organic solvents; barely soluble in H_2O (0.16 mg/100 ml at 22°); LD_{50} (mrat orl) = 292 mg/kg, (mrat iv) = 140 mg/kg, (mrat ip) = 373 mg/kg, (mrat sc) = 643 mg/kg, (frat orl) = 470 mg/kg, (frat iv) = 226 mg/kg, (frat ip) = 397 mg/kg, (frat sc) = 568 mg/kg. *Farbenfabriken Bayer AG; Troponwerke Dinklage.*

594 Fampridine
504-24-5 3974 207-987-9
$C_5H_6N_2$
4-Aminopyridine.
4-AP; EL-970; 4-pyridylamine; 4-pyridinamine; γ-aminopyridine. Analgesic; antimigraine. Used in treatment of multiple sclerosis. mp = 158-159°; soluble in H_2O, EtOH, less soluble in non-polar organic solvents.

595 Famprofazone
22881-35-2
$C_{24}H_{31}N_3O$
3-(N,α-Dimethylphenethylaminomethyl)-4-isopropyl-2-methyl-1-phenyl-3-pyrazolin-5-one.
component of: Gewodin. Analgesic.

596 Felbinac
5728-52-9 3989 227-233-2
$C_{14}H_{12}O_2$
4-Biphenylacetic acid.
CL-83544; L-141; LJC-10141; BPAA; Napageln; Traxam. Anti-inflammatory; analgesic. Metabolite of fenbufen. Arylacetic acid derivative. mp = 164-165°; LD_{50} = 164 mg/kg. *Lederle Labs.*

597 Felbinac Ethyl Ester
14062-23-8 3989
$C_{16}H_{16}O_2$
4-Biphenylacetic acid ethyl ester.
LM-001; Diatec. Anti-inflammatory; analgesic. Arylacetic acid derivative. *Lederle Labs.*

598 Fenoprofen
31879-05-7 4021 250-850-3
$C_{15}H_{14}O_3$
(±)-m-Phenoxyhydratropic acid.
Lilly-53858. Anti-inflammatory; analgesic. An aryl-

propionic acid derivative. Oil; $bp_{0.11}$ = 168-171°; n_D^{25} = 1.5742; pKa = 7.3. *Eli Lilly & Co.*

599 Fenoprofen Calcium
53746-45-5 4021
$C_{30}H_{30}CaO_8$
Calcium (±)-m-phenoxyhydratropate dihydrate.
Lilly-69323; Fenopron; Fepron; Feprona; Nalfon; Nalgesic; Progesic. Anti-inflammatory; analgesic. Arylpropionic acid derivative. Soluble in H_2O (2.5 mg/ml), more soluble in EtOH, less soluble in organic solvents; pKa = 4.5; aqueous solutions sensitive to uv light; LD_{50} (mus orl) = 800 mg/kg. *Eli Lilly & Co.*

600 Filenadol
78168-92-0
$C_{14}H_{19}NO_4$
(±)-erythro-α-Methyl-β-[3,4-(methylenedioxy)phenyl]-4-morpholineethanol.
Analgesic.

601 Floctafenine
23779-99-9 4140 245-881-4
$C_{20}H_{17}F_3N_2O_4$
2,3-Dihydroxypropyl N-[8-(trifluoromethyl)-4-quinolyl]anthranilate.
R-4138; RU-15750; Diralgan; Idalon; Idarac; Novodolan. Analgesic. mp = 179-180°; soluble in alcohol, Me_2CO; slightly soluble in Et_2O, $CHCl_3$; insoluble in H_2O; LD_{50} (mmus orl) = 3400 mg/kg, (mrat orl) = 960 mg/kg, (mmus iv) = 180 mg/kg, (mrat iv) = 160 mg/kg.

602 Flufenamic Acid
530-78-9 4167 208-494-1
$C_{14}H_{10}F_3NO_2$
N-(α,α,α-Trifluoro-m-tolyl)anthranilate.
CI-440; CN-27544; INF-1837; NSC-82699; Achless; Ansatin; Arlef; Fullsafe; Meralen; Paraflu; Parlef; Ristogen; Sastridex; Surika; Tecramine. Anti-inflammatory; analgesic. Aminoarylcarboxylic acid derivative. mp = 125°; LD_{50} (mus orl) = 715 mg/kg. *Parke-Davis.*

603 Flufenamic Acid Aluminum Salt
61891-34-7 4167
$C_{42}H_{27}AlF_9N_3O_6$
N-(α,α,α-Trifluoro-m-tolyl)anthranilate aluminum salt (3:1).
aluminum flufenamate; Alfenamin; Opyrin. Anti-inflammatory; analgesic. Aminoarylcarboxylic acid derivative. *Parke-Davis.*

604 Flufenamic Acid Butyl Ester
67330-25-0 4167
$C_{18}H_{18}F_3NO_2$
N-(α,α,α-Trifluoro-m-tolyl)anthranilate butyl ester.
ufenamate; Fenazol; Combec. Anti-inflammatory; analgesic. Aminoarylcarboxylic acid derivative. *Parke-Davis.*

605 Flufenisal
22494-27-5
$C_{15}H_{11}FO_4$
4'-Fluoro-4-hydroxy-3-biphenylcarboxylic acid acetate.
Analgesic. *Merck & Co.Inc.*

606 Flunixin
38677-85-9 4182
$C_{14}H_{11}F_3N_2O_2$
2-(2-Methyl-3-trifluoromethylanilino)-nicotinic acid.
Sch-14714. Analgesic; anti-inflammatory; antipyretic. Cyclooxygenase inhibitor. mp = 226-228°; pKa' = 5.28. *Schering-Plough HealthCare Products.*

607 Flunixin Meglumine
42461-84-7 4182 255-836-0
$C_{21}H_{28}F_3N_3O_7$
2-(2-Methyl-3-trifluoromethylanilino)-nicotinic acid compound with 1-deoxy-1-(methylamino)-D-glucitol (1:1).
Banamine; Finadyne. Analgesic; anti-inflammatory; antipyretic. Cyclooxygenase inhibitor. mp = 135-139°; soluble in H_2O. *Schering-Plough HealthCare Products.*

608 Fluoresone
2924-67-6 4197 220-889-0
$C_8H_9FO_2S$
Ethyl p-fluorophenyl sulfone.
Bripadon; Caducid. Anticonvulsant; analgesic; anxiolytic. A tricyclic benzodiazepine derivative. mp = 41°; LD_{50} (mus orl) = 2500 mg/kg, 850 mg/kg, 542 mg/kg.

609 Flupirtine
56995-20-1 4227 260-503-8
$C_{15}H_{17}FN_4O_2$
[2-Amino-6-[[(4-fluorophenyl)methyl]-amino]-3-pyridinyl]carbamic acid ethyl ester.
D-9998. Analgesic (CNS). Substituted pyridine with central analgesic properties. mp = 115-116°; LD_{50} (mus orl) = 617 mg/kg, (rat orl) = 1660 mg/kg. *Carter-Wallace; Degussa-Hüls Corp.*

610 Flupirtine Maleate
75507-68-5 4227
$C_{19}H_{21}FN_4O_6$
[2-Amino-6-[[(4-fluorophenyl)methyl]-amino]-3-pyridinyl]carbamate maleate (1:1).
W-2964M; Katadolon. Analgesic (CNS). mp = 175.5-176°. *Carter-Wallace; Degussa-Hüls Corp.*

611 Fluproquazone
40507-23-1 4230
$C_{18}H_{17}FN_2O$
4-(p-Fluorophenyl)-1-isopropyl-7-methyl-2(1H)-quinazolinone.
RF-46-790; Tormosyl. Analgesic. Fluoro analog of proquazone. mp = 172-174°; LD_{50} (mus orl) > 5000 mg/kg, (rat orl) > 3000 mg/kg, (rbt orl) = 742 mg/kg. *Sandoz Pharm. Corp.*

612 Fluradoline
71316-84-2
$C_{17}H_{16}FNOS$
2-[(-8-Fluorodibenz[b,f]oxepin-10-yl)thio]-N-methylethylamine.
Analgesic. *Hoechst Roussel Pharm. Inc.*

613 Fluradoline Hydrochloride
77590-97-7
$C_{17}H_{17}ClFNOS$.
2-[(-8-Fluorodibenz[b,f]oxepin-10-yl)thio]-N-methylethylamine hydrochloride.
HP-494; P-76-2494A. Analgesic. *Hoechst Roussel Pharm.*

614 Flurbiprofen
5104-49-4 4234 225-827-6
$C_{15}H_{13}FO_2$
2-Fluoro-α-methyl[1,1'-biphenyl]-4-acetic acid.
BTS-18322; U-27182; Adfeed; Ansaid; Antadys; Cebutid; Froben; Flurofen; Ocufen; Stayban; Zepolas. Anti-inflammatory; analgesic. Arylpropionic acid derivative. mp = 110-111°. *Boots Pharmaceuticals Inc.; C.M. Ind.; Upjohn Ltd.*

615 Fosfosal
6064-83-1 4281 227-993-5
$C_7H_7O_6P$
Salicylic acid dihydrogen phosphate.
salicyl phosphate; UR-1521; Disdolen. Analgesic. Salicylic acid derivative. mp = 168-170°; soluble in H_2O, EtOH, Me_2CO; insoluble in nonpolar organic solvents; LD_{50} (mmus iv) = 94 mg/kg, (mmus ip) = 352 mg/kg, (mmus iv) = 117 mg/kg, (mmus orl) = 1702 mg/kg.

616 Gentisic Acid
490-79-9 4404 207-718-5
$C_7H_6O_4$
2,5-Dihydroxybenzoic acid.
5-hydroxysalicylic acid. Analgesic; anti-inflammatory. Occurs in gentian root. Derivative of salicylic acid. mp = 199-200°; pK (25°) = 2.93; soluble in H_2O (1 part in 200 at 5°), alcohol, Et_2O; nearly insoluble in carbon disulfide, $CHCl_3$, C_6H_6. *Monsanto Co.; Schering AG.*

617 Gentisic Acid Sodium Salt
4955-90-2 4404 207-718-5
$C_7H_5NaO_4$
2,5-Dihydroxybenzoic acid sodium salt.
sodium gentisate; Gentinatre; Gentisod; Legential; Gentisine UCB. Analgesic; anti-inflammatory. Soluble in H_2O. *Monsanto Co.; Schering AG.*

618 Glafenine
3820-67-5 4443 223-315-7
$C_{19}H_{17}ClN_2O_4$
2,3-Dihydroxypropyl N-(7-chloro-4-quinolinyl)-anthranilate.
glycerylaminophenaquine; glaphenine; R-1707; Glifan; Glifanan; Privadol. Analgesic. mp = 165-170°; nearly insoluble in H_2O; slightly soluble in absolute alcohol; Me_2CO, Et_2O, C_6H_6, $CHCl_3$; soluble in dilute alkalies and acids; pKa (20°) = 7.2; λ_m = 356, 225, 255 (MeOH); LD_{50} (mus orl) > 2000 mg/kg. *Roussel-UCLAF.*

619 Guacetisal
55482-89-8 886 259-663-1
$C_{16}H_{14}O_5$
2-(Acetyloxy)benzoic acid 2-methoxyphenyl ester.
Broncaspin; Guaiaspir. Analgesic; antipyretic; anti-inflammatory.

620 Ibufenac
1553-60-2 4924 216-302-2
$C_{12}H_{16}O_2$
4-(2-Methylpropyl)benzeneacetic acid.
RD-11654; Dytransin; Ibunac. Analgesic; anti-inflammatory. Arylacetic acid derivative. mp = 85-87°; slightly soluble in H_2O; soluble in organic solvents; LD_{50} (mus orl) = 1800 mg/kg. *Boots Pharmaceuticals Inc.*

621 Imidazole Salicylate
36364-49-5 4949
$C_{10}H_{10}N_2O_3$
Salicylic acid compound with imidazole (1:1).
salizolo; IFT-182; Flogozen; Selezen. Anti-inflammatory; analgesic; antipyretic. Derivative of salicylic acid. Also used in heat-sensitive copying materials. mp = 123-124°; λ_m = 300 mn ($E^{1\%}_{1cm}$ 182.5); soluble in H_2O (>100 mg/ml); LD_{50} (mmus sc) = 763 mg/kg, (mmus iv) = 422 mg/kg, (mmus orl) = 1121 mg/kg. *Italfarmaco S.p.A.*

622 Indomethacin
53-86-1 4998 200-186-5
$C_{19}H_{16}ClNO_4$
1-(p-Chlorobenzoyl)-5-methoxy-2-methylindol-3-acetic acid.
Amuno; Argun; Artracin; Artrinovo; Bonidin; Catlep; Chibro-Amuno; Chrono-Indicid; Confortid; Dolcidium; Durametacin; Elmetacin; Idomethine; Imbrilon; Inacid; Indacin; Indocid; Indocin; Indomed; Indomee; Indomethine; Indomod; Indo-Phlogont; Indoptic; Indoptol; Indorektal; Indo-Tablinen; Indoxen; Inflazon; Infrocin; Inteban SP; Lausit; Mezolin; Mikametan; Mobilan; Rheumacin LA; Tannex; Vonum; Liometacen [meglumine salt]. Anti-inflammatory; analgesic; antipyretic. Arylacetic acid derivative which blocks prostaglandin biosynthesis. mp = 155-162°; λ_m = 230, 260, 319 mn (ε 10800, 16200, 6290 EtOH); pKa = 4.5; soluble in EtOH, Et_2O, Me_2CO, castor oil; nearly insoluble in H_2O; decomposes in strong alkalai; LD_{50} (rat ip) = 13 mg/kg. *Gruppo Lepetit S.p.A.; Lemmon Co.; Merck & Co.Inc.; Sumitomo Pharm. Co. Ltd.*

623 Indomethacin Sodium Trihydrate
74252-25-8 4998
$C_{19}H_{21}ClNNaO_7$
Sodium 1-(p-chlorobenzoyl)-5-methoxy-2-methylindol-3-acetic acid trihydrate.
Indocin IV; Osmosin. Anti-inflammatory; analgesic; antipyretic. Arylacetic acid derivative which blocks prostaglandin biosynthesis. pH (1% aqueous solution) = 8.4; soluble in MeOH, H_2O; slightly soluble in $CHCl_3$, Me_2CO. *Merck & Co.Inc.*

624 Indoprofen
31842-01-0 4999 250-833-0
$C_{17}H_{15}NO_3$
p-(1-Oxo-2-isoindolinyl)-hydratropic acid.
IPP; K-4277; Bor-ind; Flosin; Flosint; Isindone; Praxis; Reumofene. Analgesic; anti-inflammatory. *See* indoprofen. mp = 213-214°; LD_{50} (rat orl) = 61 mg/kg. *Farmitalia Carlo Erba Ltd.*

625 Ipsalazide
80573-03-1
$C_{16}H_{13}N_3O_6$
(E)-p-[(3-Carboxy-4-hydroxyphenyl)azo]hippuric acid.
Analgesic, used to treat inflammatory bowel disease.

626 Isofezolac
50270-33-2 5188 256-512-1
$C_{23}H_{18}N_2O_2$
1,3,4-Triphenylpyrazole-5-acetic acid.
LM-22102; Sofenac. Anti-inflammatory; analgesic; antipyretic. Prostaglandin sythetase inhibitor. Arylacetic acid derivative. mp = 200°; LD_{50} (mus orl) = 215 mg/kg, (rat orl) = 13 mg/kg.

627 Isoladol
530-34-7 5194
$C_{16}H_{19}NO_3$
2-Amino-1,2-bis(p-methoxyphenyl)ethanol.
isodianidylethanolamine; di-p-methoxyphenylhydroxyethylamine; Evadol. Analgesic. [dl-form]: mp = 135.5°; [l-form]: mp = 111-112°; soluble in EtOAc; $[\alpha]_D^{20}$ = -150° (c = 0.952 absolute alcohol).

628 Isoladol Hydrochloride
5934-19-0 5194
$C_{16}H_{20}ClNO_3$
2-Amino-1,2-bis(p-methoxyphenyl)ethanol hydrochloride.
LC-2 (as dl-form); Evaclin (as dl-form). Analgesic. [dl-form]: dec ≅200°; [l-form]: dec 188-190°; soluble in H_2O; $[\alpha]_D^{20}$ = -99.0° (c = 0.5805); soluble in H_2O.

629 Isonixin
57021-61-1 5209 260-521-6
$C_{14}H_{14}N_2O_2$
2-Hydroxy-2',6'-nicotinoxylidide.
Nixyn. Analgesic; anti-inflammatory. Aminoarylcarboxylic acid derivative. mp = 266-267°; soluble in $CHCl_3$, strong alkali; nearly insoluble in acids, H_2O; LD_{50} (mmus orl) >7000 mg/kg, (mus, rat ip) > 2000 mg/kg. *Hermes (GB) Ltd.*

630 Isoprazone
56463-68-4
$C_{15}H_{18}N_2O$
1-(4-Amino-2-methyl-5-phenylpyrrol-3-yl)-2-methyl-1-propanone.
Analgesic; antipyretic.

631 Ketazocine
36292-69-0
$C_{18}H_{23}NO_2$
(2R*,6S*,11S*)-3-(Cyclopropylmethyl)-3,4,5,6-tetrahydro-8-hydroxy-6,11-dimethyl-2,6-methano-3-benzazocin-1(2H)-one.
Win-34276. Analgesic. *Sterling Winthrop Inc.*

632 Ketoprofen
22071-15-4 5316 244-759-8
$C_{16}H_{14}O_3$
m-Benzoylhydratropic acid.
RP-19583; Alrheumat; Alrheumun; Capisten; Dexal; Epatec; Fastum; Iso-K; Kefenid; Ketopron; Lertus; Menamin; Meprofen; Orudis; Orugesic; Oruvail; Oscorel; Profenid; Toprec; Toprek. Anti-inflammatory; analgesic. mp = 94°; λ_m = 255 nm (log ε 4.33 MeOH); soluble in Et_2O, alc, Me_2CO, $CHCl_3$, DMF, EtOAc; slightly soluble in H_2O; LD_{50} (rat orl) = 101 mg/kg. *Rhône-Poulenc Rorer Pharm. Inc.; Wyeth-Ayerst Labs.*

633 Ketoprofen Lysine Salt
57469-78-0 5316
$C_{22}H_{28}O_5$
m-Benzoylhydratropic acid lysine salt.
Artrosilene. Anti-inflammatory; analgesic. Arylpropionic acid derivative. *Rhône-Poulenc Rorer Pharm. Inc.; Wyeth-Ayerst Labs.*

634 Ketorolac
74103-06-3 5318
$C_{15}H_{13}NO_3$
(±)-5-Benzoyl-2,3-dihydro-1H-pyrrolizine-1-carboxylic acid.
RS-37619. Analgesic; anti-inflammatory. Prostaglandin biosynthesis inhibitor. Arylcarboxylic acid derivative. mp = 160-161°; λ_m = 245, 312 nm (ε 7080, 17400 MeOH); pKa = 3.5; LD_{50} (mus orl) ≅ 200 mg/kg. *Merck & Co.Inc.; Syntex Labs. Inc.*

635 Ketorolac Tromethamine
74103-07-4 5318
$C_{19}H_{24}N_2O_6$
(±)-5-Benzoyl-2,3-dihydro-1H-pyrrolizine-1-carboxylic acid compound with 2-amino-2-(hydroxymethyl)-1,3-propanediol (1:1).
Acular; Dolac; Lixidol; Tarazyn; Toradol; Toratex. Analgesic; anti-inflammatory. Arylcarboxylic acid derivative. *Syntex Labs. Inc.*

636 Lefetamine
7262-75-1 5455
$C_{16}H_{19}N$
(-)-N,N-Dimethyl-1,2-diphenylethylamine.
Analgesic. Centrally acting analgesic with seterochemical resemblace to morphine. bp_6 = 142-147°; $[\alpha]_D^{20}$(EtOH) = -124.2°.

637 Lefetamine Hydrochloride
14148-99-3 5455
$C_{16}H_{20}ClN$
(-)-N,N-Dimethyl-1,2-diphenylethylamine hydrochloride.
Santenol, SPA. Analgesic. mp = 218-219°; $[\alpha]_D^{20}(H_2O)$ = -91.7°.

638 Letimide
26513-90-6
$C_{14}H_{18}N_2O_3$
2-[3-(Diethylamino)ethyl]-2H-1,3-benzoxazine-2,4(3H)-dione.
Analgesic. *Miles Inc.*

639 Letimide Hydrochloride
21791-39-9
$C_{14}H_{19}ClN_2O_3$
2-[3-(Diethylamino)ethyl]-2H-1,3-benzoxazine-2,4(3H)-dione monohydrochloride.
MA-1443. Analgesic. *Miles Inc.*

640 Levonantradol
71048-87-8
$C_{27}H_{35}NO_4$
(-)-(6S,6aR,9R,10aR)-5,6,6a,7,8,9,10,10a-Octahydro-6-methyl-3-[(R)-1-methyl-4-phenylbutoxy]-1,9-phenanthridinediol 1-acetate.
Analgesic. *Pfizer Inc.*

641 Levonantradol Hydrochloride
70222-86-5
$C_{27}H_{36}ClNO_4$
(-)-(6S,6aR,9R,10aR)-5,6,6a,7,8,9,10,10a-Octahydro-6-methyl-3-[(R)-1-methyl-4-phenylbutoxy]-1,9-phenanthridinediol 1-acetate hydrochloride.
CP-50556-1. Analgesic. *Pfizer Inc.*

642 Lobuprofen
96128-90-4
$C_{25}H_{33}ClN_2O_2$
2-[4-(m-Chlorophenyl)-1-piperazinyl]ethyl (±)-p-isobutylhydratropate.
Analgesic.

643 Lofemizole
65571-68-8 265-818-4
$C_{10}H_9ClN_2$
4-(p-Chlorophenyl)-5-methylimidazole.
Anti-inflammatory; analgesic; antipyretic. *Farmatis S.R.L.*.

644 Lofemizole Hydrochloride
70169-80-1
$C_{10}H_{10}Cl_2N_2$
4-(p-Chlorophenyl)-5-methylimidazole monohydrochloride.
Anti-inflammatory; analgesic; antipyretic. *Farmatis S.R.L.*.

645 Lonaprofen
41791-49-5 255-555-3
$C_{14}H_{13}ClO_3$
Methyl-2-[(1-chloro-2-naphthyl)oxy]propionate.
Analgesic; anti-inflammatory.

646 Lorcinadol
104719-71-3
$C_{17}H_{19}ClN_4$
(E)-3-Chloro-6-(4-cinnamyl-1-piperazinyl)pyridazine.
R-62818. Analgesic. *Janssen Pharm. Inc.*

647 Lornoxicam
70374-39-9 5612
$C_{13}H_{10}ClN_3O_4S_2$
6-Chloro-4-hydroxy-2-methyl-N-2-pyridyl-2H-thieno[2,3-e]-1,2-thiazine-3-carboxamide 1,1-dioxide.
chlortenoxicam; Ro-13-9297; TS-110. Anti-inflammatory; analgesic. Cyclooxygenase inhibitor; structurally similar to tenoxicam. A thiazinecarboxamide. mp = 225-230° (dec); λ_m = 371 nm; pKa_2 = 4.7; LD_{50} (mus, rat, dog, rbt, mky orl) > 10 mg/kg. *Hoffmann-LaRoche Inc.*

648 Losmiprofen
74168-08-4
$C_{17}H_{15}ClO_4$
(±)-2-[[3-(p-Chlorobenzoyl)-o-tolyl]oxy]propionic acid.
Analgesic; anti-inflammatory.

649 Loxoprofen
68767-14-6 5619
$C_{15}H_{18}O_3$
(±)-p-[(2-Oxocyclopentyl)methyl]-hydratropic acid.
Nonsteroidal anti-inflammatory prodrug; analgesic. Arylpropionic acid derivative. Active metabolite is the trans-cycloohydroxypentane. Oil; $bp_{0.3}$ = 190-195°; mp = 108.5-111°. *Sankyo Co. Ltd.*

650 Loxoprofen Sodium
80382-23-6 5619
$C_{15}H_{17}NaO_3$
(±)-p-[(2-Oxocyclopentyl)methyl]hydratropate sodium salt.
CS-600; Loxonin. Nonsteroidal anti-inflammatory prodrug; analgesic. Arylpropionic acid derivative. Active metabolite is the trans-cycloohydroxypentane. *Sankyo Co. Ltd.*

651 Lysine Acetylsalicylate
62952-06-1 5668 263-769-3
$C_{15}H_{22}N_2O_6$
DL-Lysine mono[2-(acetyloxy)benzoate].
aspirin DL-lysine; aspirin lysine salt; lysine monosalicylate acetate; LAS; Aspidol; Delgesic; Flectadol; Lysal; Quinvet; Venopirin; Vetalgina. Analgesic; antipyretic; anti-inflammatory. Water soluble, injectable aspirin derivative. mp = 154-156°; soluble in H_2O; slightly soluble in EtOH; insoluble in organic solvents.

652 Mabuprofen
82821-47-4 280-048-9
$C_{15}H_{23}NO_2$
(±)-N-(2-Hydroxyethyl)-p-isobutylhydratropamide.
Analgesic; anti-inflammatory. Analgesic; anti-inflammatory.

653 Magnesium Acetylsalicylate
132-49-0 5689 205-062-4
$C_{18}H_{14}MgO_8$
2-(Acetyloxy)benzoic acid magnesium salt.
magnesium aspirin; Apyron; Fyracyl; Magisal; Magnespirin; Novacetyl. Analgesic; anti-inflammatory. Soluble in H_2O; less soluble in organic solvents.

654 Medetomidine
86347-14-0 5830
$C_{13}H_{16}N_2$
(±)-4-(α,2,3-Triethylbenzyl)imidazole.
MPV-785; Domitor. Sedative (veterinary); analgesic (veterinary). Alpha-adrenergic agonist. *Farmos Group Ltd.*

655 Medetomidine Hydrochloride
86347-15-1 5830
$C_{13}H_{17}ClN_2$
(±)-4-(α,2,3-Triethylbenzyl)imidazole monohydrochloride.
CI-473; CN-35355; INF-3355; Bafhameritin-M; Bonabol; Coslan; Lysalgo; Mefenacid; Namphen; Parkemed; Ponalar; Ponstan; Ponstel; Ponstil; Ponstyl; Pontal; Tanston; Vialidon. Sedative (veterinary); analgesic (veterinary). *Farmos Group Ltd.*

656 Mefanamic Acid
61-68-7 5842 200-513-1
$C_{15}H_{15}NO_2$
2-[(2,3-Dimethylphenyl)amino]benzoic acid.
in-M; Bonabol; Coslan; Lysalgo; Mefenacid; Namphen; Parkemed; Ponalar; Ponstan; Ponstel; Ponstil; Ponstyl; Pontal; Tanston; Vialidon. Anti-inflammatory; analgesic. mp = 230-231° (effervesces); pKa = 4.2; λ_m = 285, 340 mn (0.1N NaOH); soluble in H_2O with pH = 7.1; soluble in alkali hydroxide solutions; sparingly soluble in Et_2O, $CHCl_3$; slightly soluble in EtOH; LD_{50} (rat orl) = 790 mg/kg. *Parke-Davis.*

657 Menabitan
83784-21-8
$C_{37}H_{56}N_2O_3$
(±)-8-(1,2-Dimethylheptyl)-1,3,4,5-tetrahydro-5,5-dimethyl-2-(2-propynyl)-2H-[1]benzopyrano[4,3-c]pyridin-10-yl α,2-dimethyl-1-piperidinebutyrate.
Analgesic.

658 Menabitan Hydrochloride
58019-50-4
$C_{37}H_{58}Cl_2N_2O_3$
(±)-8-(1,2-Dimethylheptyl)-1,3,4,5-tetrahydro-5,5-dimethyl-2-(2-propynyl)-2H-[1]benzopyrano[4,3-c]pyridin-10-yl α,2-dimethyl-1-piperidinebutyrate dihydrochloride.
SP-204. Analgesic.

659 Metacetamol
621-42-1 210-687-0
$C_8H_9NO_2$
3'-Hydroxyacetanilide.
BS-749. A nonhepatotoxic positional isomer of acetaminophen. Used as an analgesic and antipyretic. LD_{50} (mus) = 1025 mg/kg.

660 Metergotamine
22336-84-1
$C_{34}H_{37}N_5O_5$
1-Methylergotamine.
MY-25 [as bitartrate]. Analgesic; antimigraine.

661 Methopholine
2154-04-1 6229
$C_{20}H_{24}ClNO_2$
1-(Chlorophenethyl)-1,2,3,4-tetrahydro-6,7-dimethoxy-2-methylisoquinoline.
metofoline; ARC I-K-1; NIH-7672; Ro-4-1778/1. Analgesic. mp = 110-111°; LD_{50} (mus sc) = 180 mg/kg. *Hoffmann-LaRoche Inc.*

662 Methotrimeprazine
60-99-1 6066 200-495-5
$C_{19}H_{24}N_2OS$
(-)-3-(2-Methoxyphenothiazin-10-yl)-2-methylpropyldimethylamine.
2-methoxytrimeprazine; levomepromazine; levomeprazine; CL-36467; CL-39743; RP-7044; SKF-5116; Levoprome; Neozine; Nirvan; Nozinan; Sinogan-Debil; Tisercin. Analgesic. $[\alpha]_D^{20}$ = -17° (c = 5 in $CHCl_3$). *Rhône-Poulenc Rorer Pharm. Inc.*

663 Methotrimeprazine Maleate
7104-38-3 6066 230-412-8
$C_{23}H_{28}N_2O_5S$
(-)-3-(2-Methoxyphenothiazin-10-yl)-2-methylpropyldimethylamine maleate.
levomepromazine maleate; Minozinan; Milezin; Neuractil; Neruocil; Sofmin; Veractil. Analgesic. Dec 190°; slightly soluble in H_2O (0.3 g/100 ml), EtOH. *Rhône-Poulenc Rorer Pharm. Inc.*

664 Metkephamid
66960-34-7
$C_{29}H_{40}N_6O_6S$
L-Tyrosyl-D-alanylglycyl-L-phenylalanyl-N^2-methyl-L-methioninamide.
metkefamide. Analgesic. *Eli Lilly & Co.*

665 Metkephamid Acetate
66960-35-8
$C_{31}H_{44}N_6O_8S$
L-Tyrosyl-D-alanylglycyl-L-phenylalanyl-N^2-methyl-L-methioninamide monoacetate (salt).
metkefamide acetate; LY-127623. Analgesic. *Eli Lilly & Co.*

666 Mexoprofen
37529-08-1
$C_{16}H_{22}O_2$
p-(trans-2-Methylcyclohexyl)hydratropic acid.
Analgesic.

667 Mimbane
3277-59-6
$C_{20}H_{26}N_2$
1-Methylyohimbane.
Analgesic. *Parke-Davis.*

668 Mimbane Hydrochloride
5560-73-6
$C_{20}H_{27}ClN_2$
1-Methylyohimbane monohydrochloride.
W-2291A. Analgesic. *Parke-Davis.*

669 Mirfentanil
117523-47-4
$C_{22}H_{24}N_4O_2$
N-(1-Phenethyl-4-piperidyl)-N-pyranizyl-2-furamide.
A-3508. Analgesic. *Anaquest.*

670 Mirfentanil Hydrochloride
119413-53-5
$C_{22}H_{25}ClN_4O_2$
N-(1-Phenethyl-4-piperidyl)-N-pyranizyl-2-furamide monohydrochloride.
A-3508.HCl. Analgesic. *Anaquest.*

671 Miroprofen
55843-86-2
$C_{16}H_{14}N_2O_2$
p-Imidazo[1,2-a]pyridin-2-ylhydratropic acid.
Y-0213. Analgesic.

672 Mofezolac
78967-07-4 6314
$C_{19}H_{17}NO_5$
3,4-Bis(p-methoxyphenyl)-5-isoxazoleacetic acid.
N-22. Anti-inflammatory; analgesic. Prostaglandin biosynthase inhibitor. Derivative of arylacetic acid. mp = 147.5°; λ_m = 236 mn (ε 18300 MeOH); soluble in organic solvents; slightly soluble in H_2O; LD_{50} (mmus orl) = 1528 mg/kg, (mmus ip) = 275 mg/kg, (mmus sc) = 612 mg/kg.

673 Molinazone
5581-46-4
$C_{11}H_{12}N_4O_2$
3-Morpholino-1,2,3-benzotriazin-4(3H)-one.
Analgesic. *Farbenfabriken Bayer AG.*

674 Morazone
6536-18-1 6349 229-447-1
$C_{23}H_{27}N_3O_2$
4-[(3-Methyl-2-phenyl-4-morpholino)methyl]antipyrine.
R-445; Tarugan. Analgesic; anti-inflammatory; antipyretic. A pyrazolone. mp = 149-150°; soluble in $CHCl_3$, MeOH, Me_2CO; slightly soluble in Et_2O. *Ravensberg.*

675 Morazone Hydrochloride
50321-35-2 6349 256-540-4
$C_{23}H_{28}ClN_3O_2$
4-[(3-Methyl-2-phenyl-4-morpholino)methyl]antipyrine monohydrochloride.
Analgesic; anti-inflammatory; antipyretic. A pyrazolone. mp = 171-172° (dec); soluble in H_2O. *Ravensberg.*

676 Morpholine
110-91-8 6362 203-815-1
C_4H_9NO
Tetrahydro-2H-1,4-oxazine.
diethylene oximide; diethylene imidoxide. Analgesic; antipyretic; anti-inflammatory. Also used as solvent for resins, waxes, casein, dyes. CAUTION: Overexposure may cause visual disturbance, nose irritation, coughing, respiratory irritation, eye and skin irritation, liver and kidney damage. Liquid; mp = -4.9°; bp_{760} = 128.9°; bp_6 = 20.0°; d_D^{20} = 1.007; n_D^{20} = 1.4540; miscible with H_2O, many organic solvents, oils; LD_{50} (frat orl) = 1050 mg/kg.

677 Morpholine Salicylate
147-90-0 6362 205-703-8
$C_{11}H_{15}NO_4$
Tetrahydro-2H-1,4-oxazine salicylate.
Retarcyl; Deposal. Analgesic; antipyretic; anti-inflammatory. mp = 110-111°; soluble in H_2O, alcohols, EtOAc, Me_2CO, C_6H_6, $CHCl_3$; nearly insoluble in C_7H_8, xylene, petroleum ether, Et_2O, CCl_4.

678 Moxazocine
58239-89-7
$C_{18}H_{25}NO_2$
(-)-(2R,6S,11R)-3-(Cyclopropylmethyl)-1,2,3,4,5,6-hexahydro-11-methoxy-6-methyl-2,6-methano-3-benzazocin-8-ol.
levo-BL-4566. Analgesic; antitussive. *Bristol Myers Squibb Pharm. Ltd.*

679 Myfadol
4575-34-2
$C_{21}H_{25}NO_2$
2-[3-(m-Hydroxyphenyl)-2,3-dimethylpiperidino]-acetophenone.
Analgesic.

680 Nabitan
66556-74-9
$C_{35}H_{52}N_2O_3$
8-(1,2-Dimethylheptyl)-1,3,4,5-tetrahydro-5,6-dimethyl-2-(2-propynyl)-2H-[1]benzopyrano[4,3-c]pyridin-10-yl 1-piperidinebutyrate.
nabutan. Analgesic.

681 Nabitan Hydrochloride
49637-08-3
$C_{35}H_{53}ClN_2O_3$
8-(1,2-Dimethylheptyl)-1,3,4,5-tetrahydro-5,6-dimethyl-2-(2-propynyl)-2H-[1]benzopyrano[4,3-c]pyridin-10-yl 1-piperidinebutyrate monohydrochloride.
nabutan hydrochloride; SP-106; NIB. Analgesic.

682 Nabumetone
42924-53-8 6428
$C_{15}H_{16}O_2$
4-(6-Methoxy-2-naphthyl)-2-butanone.
BRL-14777; Arthaxan; Balmox; Consolan; Nabuser; Relafen; Relifen; Relifex. Anti-inflammatory; analgesic. Nonacidic, lipophillic prodrug, metabolized *in vivo* to 6-methoxy-2-napthylacetic acid, a prostaglandin sythesis inhibitor. mp = 80°. *Beecham Res. Labs. UK; C.M. Ind.*

683 Namoxyrate
1234-71-5
$C_{20}H_{27}NO_3$
α-Ethyl-4-biphenylacetic acid compound with 2-dimethylaminoethanol (1:1).
W-1769-A. Analgesic. *Parke-Davis.*

684 Nantradol
65511-41-3
$C_{27}H_{35}NO_4$
(±)-5,6,6aβ,7,8,9α,10,10aα-Octahydro-6β-methyl-3-(1-methyl-4-phenylbutoxy)-1,9-phenanthridinediol 1-acetate.
Analgesic. *Pfizer Inc.*

685 Nantradol Hydrochloride
56511-42-4
$C_{27}H_{36}ClNO_4$
(±)-5,6,6aβ,7,8,9α,10,10aα-Octahydro-6β-methyl-3-(1-methyl-4-phenylbutoxy)-1,9-phenanthridinediol 1-acetate hydrochloride.
CP-44001-1. Analgesic. *Pfizer Inc.*

686 Naproxen
22204-53-1 6504 244-838-7
$C_{14}H_{14}O_3$
(+)-6-Methoxy-α-methyl-2-napthaleneacetic acid.
MNPA; RS-3540; Bonyl; Diocodal; Dysmenalgit; Equiproxen; Floginax; Laraflex; Laser; Naixan; Napren; Naprium; Naprius; Naprosyn; Naprosyne; Naprux; Naxen; Nycopren; Panoxen; Prexan; Proxen; Proxine; Reuxen; Veradol; Xenar. Anti-inflammatory; analgesic; antipyretic. Arylpropionic acid derivative. mp = 152-

154°; $[\alpha]_D$ = +66° (c = 1 in $CHCl_3$); nearly insoluble in H_2O; soluble in organic solvents; LD_{50} (mus orl) = 1234 mg/kg, (rat orl) = 534 mg/kg, (rat ip) = 575 mg/kg. *Syntex Labs. Inc.*

687 Naproxen Piperazine
70981-66-7 6504
$C_{32}H_{38}N_2O_6$
(+)-6-Methoxy-α-methyl-2-napthaleneacetate piperazine salt (2:1).
piproxen; Numidan. Anti-inflammatory; analgesic; antipyretic. Arylpropionic acid derivative. *Syntex Labs. Inc.*

688 Naproxen Sodium
26159-34-2 6504 247-486-2
$C_{14}H_{13}NaO_3$
(-)-Sodium 6-methoxy-α-methyl-2-napthaleneacetate.
RS-3560; Aleve; Anaprox; Apranax; Axer; Alfa; Flanax; Gynestrel; Miranax; Primeral; Synflex. Anti-inflammatory; analgesic; antipyretic. Arylpropionic acid derivative. mp = 244-246°; $[\alpha]_D$ = -11° (in MeOH). *Syntex Labs. Inc.*

689 Naproxol
26159-36-4
$C_{14}H_{16}O_2$
(-)-6-Methoxy-β-methyl-2-napthaleneethanol.
RS-4034. Anti-inflammatory; analgesic; antipyretic. *Syntex Labs. Inc.*

690 Nexeridine
53716-48-6
$C_{19}H_{29}NO_2$
1-[2-Dimethylamino)-1-methylethyl]-2-phenylcyclohexanol acetate (ester).
Analgesic. *Fisons Pharm. Div.*

691 Nexeridine Hydrochloride
53716-47-5
$C_{19}H_{30}ClNO_2$
1-[2-Dimethylamino)-1-methylethyl]-2-phenylcyclohexanol acetate (ester) hydrochloride.
673-082. Analgesic. *Fisons Pharm. Div.*

692 Nicafenine
64039-88-9
$C_{24}H_{19}ClN_4O_3$
N-(7-Chloro-4-quinolyl)anthranilic acid ester with N-(2-hydroxyethyl)nicotinamide.
Analgesic.

693 Nicoboxil
13912-80-6 237-684-7
$C_{12}H_{17}NO_3$
2-Butoxyethyl nicotinate.
butoxyethyl nicotinate. Analgesic (topical). A capsaicinoid.

694 Nifenazone
2139-47-1 6619 218-387-1
$C_{17}H_{16}N_4O_2$
N-Antipyrinylnicotinamide.
N-nicotinoylaminoantipyrine; Dolongan; Nicopyron; Nikofezon; Nipazine; Phenicazone; Reupiron; Thylin. Analgesic; antipyretic. mp = 252-253°, 256-258°; slightly soluble in H_2O, Me_2CO, ethylacetate, Et_2O; soluble in hot H_2O, alcohol, $CHCl_3$, dilute acids.

695 5'-Nitro-2'-propoxyacetanilide
553-20-8 6726
$C_{11}H_{14}N_2O_4$
N-(5-Nitro-2-propoxyphenyl)acetamide.
Falimint. Antipyretic; analgesic. mp = 102.5-103.5°. *Schwartz's Essencefabriken.*

696 Nitrous Oxide
10024-97-2 6751 233-032-0
N_2O
Nitrogen oxide.
dinitrogen monoxide; dinitrogen oxide; laughing gas; hyponitrous acid anhydride; factitious air. Anasthetic (inhaled); analgesic. CAUTION: asphyxiant, narcotic in high concentrations. Less irritating than other oxides of nitrogen. Gas; dissociates > 300°; mp = -90.91°; bp_{760} = -88.46°; d^{-89}(liquid) = 1.226; d(gas) = 1.53; soluble in sulfuric acid, alcohol, Et_2O, oils. *DuPont-Merck Pharm.*

697 Nonivamide
2444-46-4 219-484-1
$C_{17}H_{27}NO_3$
N-Vanillylnonamide.
nonylic acid vanillylamide; nonanoyl vanillylamide. Topical analgesic. Antinociceptive. Capsaicin derivative. Acts by depleting stores of substance P from sensory neurons.

698 Octazamide
56391-55-0
$C_{13}H_{15}NO_2$
5-Benzoylhexahydro-1H-furo[3,4-c]pyrrole.
ICI-US-457. Analgesic. *ICI.*

699 Odalprofen
137460-88-9
$C_{20}H_{20}N_2O_2$
Methyl (±)-m-(α-imidazol-1-ylbenzyl)-hydratropate.
Analgesic.

700 Oil of Partridge Berry
Analgesic.

701 Oil of Wintergreen
Analgesic.

702 Olvanil
58493-49-5
$C_{26}H_{43}NO_3$
N-[(4-Hydroxy-3-methoxyphenyl)methyl]-9-octadecenamide.
N-vanillyloleamide; NE-19550. Analgesic. *Norwich Eaton.*

703 Oxapadol
56969-22-3 260-484-6
$C_{17}H_{14}N_2O_2$
4,5-Dihydro-1-phenyl-1,4-epxoy-1H,3H-[1,4]oxazepino[4,3-a]benzimidazole.
MD-720111. Analgesic.

704 Oxetorone
26020-55-3 7068 247-411-3
$C_{21}H_{21}NO_2$
3-Benzofuro[3,2-c][1]benzoxepin-6(12H)-ylidene-N,N-dimethyl-1-propanamine.
Analgesic (specific to migraine). Serotonin and histamine antagonist with antimigraine activity. *Labaz S.A.*

705 Oxetorone Fumarate
34522-46-8 7068
$C_{25}H_{25}NO_6$
3-Benzofuro[3,2-c][1]benzoxepin-6(12H)-ylidene-N,N-dimethyl-1-propanamine fumarate.
L-6257; Nocertone; Oxedix. Analgesic (specific to migraine). Serotonin and histamine antagonist with antimigraine activity. mp = 160°. *Labaz S.A.*

706 Parsalmide
30653-83-9 7177 250-274-2
$C_{14}H_{18}N_2O_2$
5-Amino-N-butyl-2-(2-propynyloxy)benzamide.
MY-41-6; Sinovial; Parsal. Anti-inflammatory; analgesic. Derivative of salicylic acid. mp = 83-85°; λ_m = 220, 327 nm (absolute alcohol), 284 nm (log ε 3.39 H_2O, pH 2), 313 nm (log ε 3.40 H_2O, pH 8.5); soluble in many organic solvents; slightly soluble in H_2O; nearly insoluble in petroleum ether, cyclohexane.

707 Pelubiprofen
69956-77-0
$C_{16}H_{18}O_3$
(±)-p-[[(E)-2-Oxocyclohexylidene]methyl]hydratropic acid.
Analgesic.

708 Pemedolac
114716-16-4
$C_{22}H_{23}NO_3$
(±)-cis-4-Benzyl-1-ethyl-1,3,4,9-tetrahydropyrano[3,4-b]indole-1-acetic acid.
AY-30715. Analgesic. *Wyeth-Ayerst Labs.*

709 Perisoxal
2055-44-9 7318
$C_{16}H_{20}N_2O_2$
3-(2-Piperidino-1-hydroxyethyl)-5-phenylisoxazole.
Anti-inflammatory; analgesic. mp = 107-108°.

710 Perisoxal Citrate
2139-25-5 7318
$C_{38}H_{48}N_4O_9$
3-(2-Piperidino-1-hydroxyethyl)-5-phenylisoxazole citrate (2:1).
Isoxal. Anti-inflammatory; analgesic. mp = 135-145°; LD_{50} (mus sc) = 416 mg/kg.

711 Phenacetin
62-44-2 7344 200-533-0
$C_{10}H_{13}NO_2$
N-(4-Ethoxyphenyl)acetamide.
p-acetophenetitide; p-ethoxyacetanilide; para-acetphenetidin; acetophenetidin; acetphenetidin; Phenin; component of: P-A-C compound, APC Tablets. Analgesic; antipyretic. CAUTION: may be a carcinogen. mp = 134-135°; soluble in H_2O (1 g in 1310 ml cold, 82 ml boiling), EtOH (1 g in 15 ml cold, 2.8 ml boiling), Et_2O (1 g/90 ml), $CHCl_3$ (1 g/14 ml); LD_{50} (rat orl) = 1650 mg/kg. *Marion Merrell Dow Inc.; Monsanto Co.; Upjohn Ltd.*

712 Phenazopyridine
94-78-0 7175 202-363-2
$C_{11}H_{11}N_5$
2,6-Diamino-3-(phenylazo)pyridine.
Phenazopyridine. Analgesic (urinary tract). *Boehringer Mannheim GmbH.*

713 Phenazopyridine Hydrochloride
136-40-3 7361 205-243-8
$C_{11}H_{12}ClN_5$
3-(Phenylazo)-2,6-pyridinediamine monohydrochloride.
Bisteril; Mallophene; Phenazodine; Pyridacil; Pyridiate; Pyridium; Sedural; Uridinal; Urodine. Analgesic (urinary tract). [free base]: mp = 139°; soluble in ethylene and propylene glycols, AcOH; slightly soluble in alcohol, lanolin, glycerol; insoluble in Me_2CO, C_6H_6, $CHCl_3$, Et_2O, C_7H_8; LD_{50} (rat orl) = 403 mg/kg. *Boehringer Mannheim GmbH.*

714 Phenocoll
103-97-9 7388 203-163-8
$C_{10}H_{14}N_2O_2$
2-Amino-4'-ethoxyacetanilide.
phenokoll; Phenamine. Antipyretic; analgesic. mp = 100.5°. *Schering-Plough HealthCare Products.*

715 Phenocoll Salicylate
140-47-6 7388
$C_{17}H_{20}N_2O_5$
2-Amino-4'-ethoxyacetanilide salicylate.
Salocoll. Antipyretic; analgesic. Soluble in H_2O (0.5% cold, 5% hot). *Schering-Plough HealthCare Products.*

716 Phenopyrazone
3426-01-5 7401 222-324-3
$C_{15}H_{12}N_2O_2$
1,4-Diphenylpyrazolidine-3,5-dione.
Analgesic; antipyretic. mp = 233-234°. *Knoll Pharm. Co.*

717 Phenyl Acetylsalicylate
134-55-4 7424 205-147-6
$C_{15}H_{12}O_4$
2-(Acetyloxy)benzoic acid phenyl ester.
acetylphenylsalicylate; Acetylsalol; Spiroform; Vesipyrin. Analgesic; anti-inflammatory; antipyretic. mp = 97°; bp_{11} = 198°; insoluble in H_2O; soluble in alcohol, Et_2O.

718 Phenyl Salicylate
118-55-8 7464 204-259-2
$C_{13}H_{10}O_3$
2-Hydroxybenzoic acid phenyl ester.
Salol. Analgesic; anti-inflammatory; antipyretic. Also used in the manufacture of plastics, laquers, adhesives, waxes, polishes; used in suntan oils and creams. Has some light absorbing and plasticizer properties. mp = 41-43°; bp_{12} = 173°; d = 1.25; slightly soluble in H_2O (1 g/6670 ml); more soluble in alcohol (17 g/100 ml), C_6H_6 (67 g/100 ml), amyl alcohol (20 g/100 ml), liquid paraffin (10 g/100 ml), almond oil (25 g/100 ml); soluble in Me_2CO .

719 Phenyramidol
553-69-5 7474 209-044-7
$C_{13}H_{14}N_2O$
α-[(2-Pyridylamino)methyl]benzyl alcohol.
fenyramidol. Analgesic; muscle relaxant (skeletal). mp = 82-85°; λ_m = 243, 303 nm (log ε = 4.24, 3.63 95% EtOH); [methiodide ($C_{14}H_{17}IN_2O$)]: mp = 164-166°.

720 Phenyramidol Hydrochloride
326-43-2 7474 206-308-3
$C_{13}H_{15}ClN_2O$
α-[(2-Pyridylamino)methyl]benzyl alcohol hydrochloride.
IN-511; MJ-505; NSC-17777; Abbolexin; Anabloc; Analexin; Cabral; Miodar. Skeletal muscle relaxant; analgesic. mp = 140-142°; soluble in H_2O.

721 Picenadol
79201-85-7
$C_{16}H_{25}NO$
(±)-trans-m-(1,3-Dimethyl-4-propyl-4-piperidyl)-phenol.
Analgesic. *Eli Lilly & Co.*

722 Picenadol Hydrochloride
74685-16-8
$C_{16}H_{26}ClNO$
(±)-trans-m-(1,3-Dimethyl-4-propyl-4-piperidyl)-phenol hydrochloride.
LY-150720. Analgesic. *Eli Lilly & Co.*

723 Piketoprofen
60576-13-8 7576
$C_{22}H_{20}N_2O_2$
m-Benzoyl-N-(4-methyl-2-pyridyl)hydratropamide.
Calmatel (aerosol). Anti-inflammatory (topical). Derivative of arylpropionic acid; derivative of ketoprofen. Oil; soluble in CH_2Cl_2, EtOH; insoluble in H_2O.

724 Piketoprofen Hydrochloride
59512-37-7 7576
$C_{22}H_{21}ClN_2O_2$
m-Benzoyl-N-(4-methyl-2-pyridyl)hydratropamide monohydrochloride.
Calmatel (cream). Analgesic; anti-inflammatory (topical). Derivative of arylpropionic acid; derivative of ketoprofen. mp = 180-182°.

725 Pinadoline
38955-22-5
$C_{19}H_{19}Cl_2N_3O_3$
1-[8-Chlorodibenz[b,f][1,4]oxazepin-10(11H-yl)carbonyl]-2-(5-chlorovaleryl)-hydrazine.
SC-25469. Analgesic. *Searle G.D. & Co.*

726 Pipebuzone
27315-91-9 7610 248-398-7
$C_{25}H_{32}N_4O_2$
4-Butyl-4-[(4-methyl-1-piperazinyl)methyl]-1,2-diphenyl-3,5-pyrazolidinedione.
LD-4644; Elarzone. Anti-inflammatory; analgesic; antipyretic. Derivative of phenylbutazone, an anti-inflammatory agent. A pyrazolone. mp = 129°.

727 Piperylone
2531-04-6 7632 219-788-4
$C_{17}H_{22}N_3O$
4-Ethyl-1,2-dihydro-2-(1-methyl-4-piperidinyl)-5-phenyl-3H-pyrazol-3-one.
PR-66; component of: Palerol, Pelerol. Analgesic. Dec 162-163°; pK_1 = 5.90; pK_2 = 9.15; sparingly soluble in H_2O; soluble in EtOH, isopropanol, Me_2CO, EtOAc. *Sandoz Pharm. Corp.*

728 Pirfenidone
53179-13-8
$C_{12}H_{11}NO$
5-Methyl-1-phenyl-2(1H)pyridone.
AMR-69. Analgesic; anti-inflammatory; antipyretic.

729 Piroxicam
36322-90-4 7661 252-974-3
$C_{15}H_{13}N_3O_4S$
4-Hydroxy-2-methyl-N-2-pyridinyl-2H-1,2-benzothiazine-3-carboxamide 1,1-dioxide.
CP-16171; Artroxicam; Baxo; Bruxicam; Caliment; Erazon; Feldene; Flogobene; Geldene; Impronta!; Larapam; Pirkam; Piroflex; Reudene; Riacen; Roxicam; Roxiden; Sasulen; Solocalm; Zunden. Anti-inflammatory. A thiazinecarboxamide. mp = 198-200°; pKa = 6.3 (2:1 dioxane:H_2O); LD_{50} (mus orl) = 360 mg/kg. *Pfizer Inc.*

730 Piroxicam Cinnamate
87234-24-0 7661
$C_{24}H_{19}N_3O_5S$
4-Hydroxy-2-methyl-N-2-pyridinyl-2H-1,2-benzothiazine-3-carboxamide 1,1-dioxide cinnamic acid ester.
cinnoxicam; SPA-S-510; Sinartol; Zelis; Zen. Anti-inflammatory. A thiazinecarboxamide. *Pfizer Inc.*

731 Piroxicam Compound with β-Cyclodextrin
121696-62-6 7661
$C_{57}H_{83}N_3O_{39}S$
4-Hydroxy-2-methyl-N-2-pyridinyl-2H-1,2-benzothiazine-3-carboxamide 1,1-dioxide compound with β-cyclodextrin.
Brexin; Cicladol; Cycladol. Anti-inflammatory; analgesic; antirheumatic. A thiazinecarboxamide. *Pfizer Inc.*

732 Piroxicam Olamine
85056-47-9 7661
$C_{17}H_{20}N_4O_5S$
4-Hydroxy-2-methyl-N-2-pyridinyl-2H-1,2-benzothiazine-3-carboxamide 1,1-dioxide compound with 2-aminoethanol.
CP-16171-85. Anti-inflammatory; analgesic. A thiazine-carboxamide. *Pfizer Inc.*

733 p-Lactophenetide
539-08-2 5355 208-708-3
$C_{11}H_{15}NO_3$
N-(4-Ethoxyphenyl)-2-hydroxypropanamide.
4'-ethoxylactanilide; Fenolactine; Lactophenin; Phenolactine. Analgesic; antipyretic mp = 117-118°; somewhat soluble in H_2O (1 g/330 ml cold, 1 g/55 ml hot); more soluble in alcohol (1 g/8.5 ml); slightly soluble in Et_2O, petroleum ether.

734 Populin
99-17-2 7753 202-737-5
$C_{20}H_{22}O_8$
2-(Hydroxymethyl)phenyl-β-D-glucopyranoside 6-benzoate.
populoside; salicin benzoate. Found in bark and leaves of various *Populus* species. mp = 179°; $[\alpha]_D^{20}$ = -2.0 (c = 5 C_5H_5N); $[\alpha]_D^{25}$ = -29.7 (c = 5 Me_2CO); soluble in alcohol; slightly soluble in H_2O; nearly insoluble in Et_2O.

735 Pravadoline
92623-83-1
$C_{23}H_{26}N_2O_3$
p-Methoxyphenyl 2-methyl-1-(2-morpholinoethyl)indol-2-yl ketone.
Analgesic. *Sterling Winthrop Inc.*

736 Pravadoline Maleate
92623-84-2
$C_{27}H_{30}N_2O_7$
p-Methoxyphenyl 2-methyl-1-(2-morpholinoethyl)indol-2-yl ketone maleate.
Win-48098-6. Analgesic. *Sterling Winthrop Inc.*

737 Prethcamide
8015-51-8 2659, 2662
N-[1-(Dimethylcarbamoyl)propyl]-N-propylcrotonamide mixture with N-[1-(dimethylcarbamoyl)propyl]-N-ethylcrotonamide.
Micoren, Respirot. Respiratory stimulant; analgesic. Soluble in H_2O. *Ciba-Geigy Corp.*

738 Prodilidine
3734-17-6
$C_{15}H_{21}NO_2$
1,2-Dimethyl-3-phenyl-3-pyrrolidinol propionate (ester).
Analgesic.

739 Prodilidine Hydrochloride
3734-16-5
$C_{15}H_{22}ClNO_2$
1,2-Dimethyl-3-phenyl-3-pyrrolidinol propionate (ester) hydrochloride.
A-1981-12; CI-427. Analgesic.

740 Prodolic Acid
36505-82-5
$C_{16}H_{19}NO_3$
1,3,4,9-Tetrahydro-1-propylpyrano[3,4-b]indole-1-acetic acid.
AY-23289. Anti-inflammatory.

741 Profadol
428-37-5
$C_{14}H_{21}NO$
m-(1-Methyl-3-propyl-3-pyrrolidinyl)phenol.
Analgesic. *Parke-Davis.*

742 Profadol Hydrochloride
2324-94-9
$C_{14}H_{22}ClNO$
m-(1-Methyl-3-propyl-3-pyrrolidinyl)phenol hydrochloride.
CI-572; A-2205; Centrac. Analgesic. *Parke-Davis.*

743 Propacetamol
66532-85-2 7976 266-390-1
$C_{14}H_{20}N_2O_3$
N,N-Diethylglycine 4-(acetylamino)phenyl ester.
4-acetamidophenyl (diethylamino)acetate; N,N-diethylglycine ester with 4'-hydroxyacetanilide. Analgesic; antipyretic. Injectable prodrug of acetaminophen. Oil. *Hexachemie.*

744 Propacetamol Hydrochloride
66532-86-3 7976
$C_{14}H_{22}ClN_2O_3$
N,N-Diethylglycine 4-(acetylamino)phenyl ester hydrochloride.
UP-34101; Pro-Dafalgan. Analgesic; antipyretic. Injectable prodrug of acetaminophen. mp = 228°; soluble in H_2O. *Hexachemie.*

745 Propyphenazone
479-92-5 8056 207-539-2
$C_{14}H_{18}N_2O$
4-Isopropyl-2,3-dimethyl-1-phenyl-3-pyrazolin-5-one.
4-isopropylphenazone; isopropylphenazone; Budirol; Causyth; Cibalgina; Eufibron; Isopropchin. Analgesic; antipyretic; anti-inflammatory. A pyrazolone. mp = 103°; soluble in H_2O (0.24 g/100 g at 16.5°), organic solvents. *Hoffmann-LaRoche Inc.*

746 Proxorphan
69815-38-9
$C_{19}H_{25}NO_2$
(-)-(4aR,5R,10bS)-13-(Cyclopropylmethyl)-4,4a,5,6-tetrahydro-3H-5,10b-(iminoethano)-1H-naptho[1,2-c]-pyran-9-ol.
Analgesic; antitussive. *Bristol-Myers Squibb Pharm. R&D.*

747 Proxorphan Tartrate
69815-39-0
$C_{42}H_{56}N_2O_{10}$
(-)-(4aR,5R,10bS)-13-(Cyclopropylmethyl)-4,4a,5,6-tetrahydro-3-H-5,10b-(iminoethano)-1H-naptho[1,2-c]-pyran-9-ol D-(-)-tratrate (2:1).
BL-5772M. Analgesic; antitussive. *Bristol-Myers Squibb Pharm. R&D.*

748 Pyrroliphene
15686-97-2
$C_{23}H_{29}NO_2$
(+)-α-Benzylβ-methyl-α-phenyl-1-pyrrolidinepropanol acetate (ester).
pyrrolifene. Analgesic. *Eli Lilly & Co.*

749 Pyrroliphene Hydrochloride
5591-44-6
$C_{23}H_{30}ClNO_2$
(+)-α-Benzylβ-methyl-α-phenyl-1-pyrrolidinepropanol acetate (ester) hydrochloride.
pyrrolifene hydrochloride; Lilly-31518. Analgesic. *Eli Lilly & Co.*

750 Ramifenazone
3615-24-5 8282 222-791-3
$C_{14}H_{19}N_3O$
4-Isopropylamino-2,3-dimethyl-1-phenyl-3-pyrazolin-5-one.
4-isopropylaminoantipyrine; isopropylaminophenazone;

isopyrin; component of: Tomanol (with phenylbutazine). Analgesic; anti-inflammatory; antipyretic. Has been evaluated as an antimigraine in combination with phenylbutazone. A pyrazolone. mp = 80°; LD_{50} (mus ip) = 843 mg/kg, (mus orl) = 1070 mg/kg.

751 Rimazolium
35615-72-6 8391
$C_{13}H_{19}N_2O_5^+$
3-(Ethoxycarbonyl)-6,7,8,9-tetrahydro-1,6-dimethyl-4-oxo-4H-pyrido[1,2-a]pyrimidinium ion.
Analgesic.

752 Rimazolium Metilsulfate
28610-84-6 8391 249-103-4
$C_{14}H_{22}N_2O_7S$
3-(Ethoxycarbonyl)-6,7,8,9-tetrahydro-1,6-dimethyl-4-oxo-4H-pyrido[1,2-a]pyrimidinium methyl sulfate.
MZ-144; MZ-0780; Ro-11-780; Dolcuran; Probon; Probonal; Rimagina. Analgesic. mp = 165-166°; λ_m = 336, 258 nm (ε 3630, 27500); soluble in H_2O; LD_{50} (rat iv) = 220 mg/kg, (rat ip) = 720 mg/kg, (rat orl) = 1600 mg/kg.

753 Rofecoxib
162011-90-7
$C_{17}H_{14}O_4S$
4-[4-(Methylsulfonyl)phenyl]-3-phenyl-2(5H)-furanone.
MK-966; Vioxx. Analgesic; anti-inflammatory. A COX-2 specific inhibitor used in the treatment of osteoarthritis. Sparingly soluble in Me_2CO; slightly soluble in MeOH, isopropyl acetate; very slightly soluble in EtOH; practically insoluble in octanol; insoluble in H_2O. *Merck & Co.Inc.*

754 Salacetamide
487-48-9 8471 207-656-9
$C_9H_9NO_3$
N-Acetyl-2-hydroxybenzamide.
N-acetylsalicylamide; acetsalicylamide; Actylamide. Analgesic; anti-inflammatory; antipyretic. Derivative of salicylic acid. mp = 148°.

755 Salcolex
28038-04-2
$C_{24}H_{46}MgN_2O_{12}S{\cdot}4H_2O$
Choline salicylate (salt) compound with magnesium sulfate (2:1) tetrahydrate.
Analgesic; anti-inflammatory; antipyretic. *Wallace Labs.*

756 Salcolex [anhydrous]
54194-00-2
$C_{24}H_{38}MgN_2O_{12}S$
Choline salicylate (salt) compound with magnesium sulfate (2:1).
Analgesic; anti-inflammatory; antipyretic. *Wallace Labs.*

757 Salethamide
46803-81-0
$C_{13}H_{20}N_2O_2$
N-[2-(Diethylamino)ethyl]salicylamide.
saletamide. Analgesic. *Miles Inc.*

758 Salethamide Maleate
24381-55-3
$C_{17}H_{24}N_2O_6$
N-[2-(Diethylamino)ethyl]salicylamide maleate (1:1) salt.
saletamide maleate; MA-593. Analgesic. *Miles Inc.*

759 Salicin
138-52-3 8476 205-331-6
$C_{13}H_{18}O_7$
2-(Hydroxymethyl)phenyl-β-D-glucopyranoside.
saligenin-β-D-glucopyranoside; salicoside; salicyl alcohol glucoside. Analgesic. Obtained from bark of poplar (*Populus) and willow (Salix*), as well as leaves and flowers of willow. mp = 199-202°; $[\alpha]_D^{25}$ = -62 to -67 (c = 3); $[\alpha]_D^{20}$ = -45.6 (c = 0.6 absolute alcohol); soluble in H_2O, alcohol; more soluble in alkalies, C_5H_5N, glacial AcOH; nearly insoluble in Et_2O, $CHCl_3$.

760 Salicylamide
65-45-2 8480 200-609-3
$C_7H_7NO_2$
2-Hydroxybenzamide.
Cidal; Salamid; Samid; Salizell; Salymid; Urtosal. Analgesic. Derivative of salicylic acid. mp = 140°; soluble in H_2O (20 mg/ml), soluble in organic solvents; LD_{50} (mus orl) = 1.4 g/kg. *Lemmon Co.*

761 Salicylamide O-Acetic Acid
25359-22-6 8481
$C_9H_9NO_4$
[2-(Aminocarbonyl)phenoxy]acetic acid.
o-(carbamylphenoxy)acetic acid. Analgesic; anti-inflammatory; antipyretic. Derivative of salicylic acid. mp = 221°; soluble in aqueous alkali. *Yoshitomi.*

762 Salicylamide O-Acetic Acid Sodium Salt
3785-82-8 8481
$C_9H_8NNaO_4$
[2-(Aminocarbonyl)phenoxy]acetate sodium salt (1:1).
Salizell ampules. Analgesic; anti-inflammatory; antipyretic. Derivative of salicylic acid. mp = 212-215°. *Yoshitomi.*

763 Salicylate Meglumine
23277-50-1
$C_{14}H_{23}NO_8$
Salicylic acid compound with 1-deoxy-1-(methylamino)-D-glucitol.
PFA-186. Antirheumatic; analgesic. Derivative of salicylic acid. *Purdue Pharma L.P.*

764 Salicylsulfuric Acid
89-45-2 8486
$C_7H_6O_6S$
2-(Sulfoxy)benzoic acid.
salicylic acid, acid sulfate; salicylic acid sulfuric acid ester. Analgesic; anti-inflammatory. Derivative of salicylic acid.

765 Salicylsulfuric Acid Sodium Salt
6155-64-2 8486
$C_7H_5NaO_6S$
2-(Sulfoxy)benzoate monosodium salt.
sodium salicylsulfate; Salcyl; Salcylix. Analgesic; anti-

inflammatory. Derivative of salicylic acid. Soluble in H_2O; insoluble in organic solvents.

766 Salsalate
552-94-3 8491 209-027-4
$C_{14}H_{10}O_5$
2-Hydroxybenzoic acid 2-carboxyphenyl ester.
disalicylic acid; salicyloxysalicylic acid; salicylsalicylic acid; NSC-49171; Disalcid; Disalgesic; Mono-Gesic; Salflex. Analgesic; anti-inflammatory. Nonacetylated aspirin analog. mp = 148-149°; nearly insoluble in H_2O; sparingly soluble in organic solvents. *3M Pharm.; Boehringer Mannheim GmbH; Fisons Pharm. Div.*

767 Salverine
6376-26-7 8494 228-944-0
$C_{19}H_{24}N_2O_2$
2-[2-(Diethylamino)-ethoxy]benzanilide.
M-811; component: of Montamed. Analgesic. mp = 173°.

768 Simetride
154-82-5 8682
$C_{28}H_{38}N_2O_6$
1,4-Bis[(2-methoxy-4-propylphenoxy)acetyl]piperazine.
AP2; Kyorin AP2. Analgesic. mp = 128-130°, 138.5-140°; nearly insoluble in H_2O, EtOH; soluble in $CHCl_3$; LD_{50} (mus ip) = 15000 mg/kg. *Kyorin Pharm. Co. Ltd.*

769 Sodium Salicylate
54-21-7 8819 200-198-0
$C_7H_5NaO_3$
2-Hydroxybenzoic acid monosodium salt.
Alysine; Idocyl; Enterosalicyl; Enterosalil. Analgesic; antipyretic. Used as a preservative. Soluble in H_2O (1.1 g/ml), less soluble in organic solvents; incompatible with ferric salts, lime water, spirit nitrous Et_2O, mineral acids, iodine, lead acetate, silver nitrate, sodium phosphate; LD_{50} (rat ip)= 780 mg/kg. *Marion Merrell Dow Inc.*

770 Spiradoline
87151-85-7
$C_{22}H_{30}Cl_2N_2O_2$
(±)-2-(3,4-Dichlorophenyl)-N-methyl-N-[(5R*,7S*,8S*)-7-(1-pyrrolidinyl)-1-oxaspiro[4.5]dec-8-yl]acetamide.
Analgesic. *Upjohn Ltd.*

771 Spiradoline Mesylate
87173-97-5
$C_{23}H_{34}Cl_2N_2O_5S$
(±)-2-(3,4-Dichlorophenyl)-N-methyl-N-[(5R*,7S*,8S*)-7-(1-pyrrolidinyl)-1-oxaspiro[4.5]dec-8-yl]acetamide monomethanesulfonate.
U-62066E. Analgesic. *Upjohn Ltd.*

772 Suprofen
40828-46-4 9180 255-096-9
$C_{14}H_{12}O_3S$
2-Thenoylhydratropic acid.
sutoprofen; R-25061; Masterfen; Profenol; Srendam; Supranol; Suprol; Sulproltin; Suprocil; Topalgic. Anti-inflammatory; analgesic. Prostaglindin biosynthesis inhibitor. A derivative of arylpropionic acid. mp = 124.3°; pKa = 3.91; λ_m = 266, 292 nm (ε 15700, 15600 0.01N HCL-90% iPrOH); soluble in alcohols, Et_2O, $CHCl_3$, Me_2CO, polyethylene glycol, alkalies; slightly soluble in H_2O; LD_{50} (mus orl) = 590 mg/kg, (dog orl) = 160 mg/kg, (rat orl) = 353 mg/kg, (gpg orl) = 280 mg/kg. *Abbott Laboratories Inc.; Alcon Labs; McNeil Pharm.*

773 Talmetacin
67489-39-8
$C_{27}H_{20}ClNO_6$
(±)-1-(p-Chlorobenzoyl)-5-methoxy-2-methylindole-3-acetic acid ester with 3-hydroxyphthalide.
BA-7605-06. Analgesic; anti-inflammatory; antipyretic. *Lab. Bago S.A.; Resfar S.R.L.*

774 Talniflumate
66898-62-2 9214
$C_{21}H_{13}F_3N_2O_4$
Phthalidyl 2-(α,α,α-trifluoro-m-toluidino)nicotinate.
BA-7602-06; Somalgen. Analgesic; anti-inflammatory. Derivative of niflumic acid, an anti-inflammatory agent. An aminoarylcarboxylic acid derivative. mp = 165-166°; λ_m = 287, 357 nm (ε 25600, 7800 $CHCl_3$); LD_{50} (rat orl) = 1200 mg/kg. *Lab. Bago S.A.*

775 Talosalate
66898-60-0
$C_{17}H_{12}O_6$
1,3-Dihydro-3-oxo-1-isobenzofuranyl ester 2-(acetoxy)benzoic acid.
BA-7604-02. Analgesic; anti-inflammatory. *Lab. Bago S.A.*

776 Tazadolene
87936-75-2
$C_{16}H_{21}N$
(±)-1-[(E)-2-Benzylidenecyclohexyl]azetidine.
Analgesic. *Upjohn Ltd.*

777 Tazadolene Succinate
87936-82-1
$C_{20}H_{26}NO_4$
(±)-1-[(E)-2-Benzylidenecyclohexyl]azetidine succinate (1:1).
U-53996H. Analgesic. *Upjohn Ltd.*

778 Tebufelone
112018-00-5
$C_{20}H_{28}O_2$
3',5'-Di-tert-butyl-4'-hydroxy-5-hexynophenone.
NE-11740. Analgesic; anti-inflammatory. *Norwich Eaton.*

779 Tenoxicam
59804-37-4 9293
$C_{13}H_{11}N_3O_4S_2$
4-Hydroxy-2-methyl-N-2-pyridyl-2H-thieno[2,3-e]-1,2-thiazene-3-carboxamide 1,1-dioxide.
Ro-12-0068; Alganex; Dolmen; Liman; Mobiflex; Rexalgan; Tilatil; Tilcotil. Anti-inflammatory; analgesic. A thiazinecarboxamide. mp = 209-213° (dec). *Hoffmann-LaRoche Inc.*

780 Terofenamate
29098-15-5 9312 249-434-4
$C_{17}H_{17}Cl_2NO_3$
Ethoxymethyl N-(2,6-dichloro-m-tolyl)anthranilate.
etoclofene; Etofen. Anti-inflammatory; analgesic.

Derivative of anthranilic acid. mp = 73-73°; LD_{50} (mus orl) = 918 mg/kg, (rat orl) = 307 mg/kg, (mus ip) = 300 mg/kg, (rat ip) = 274 mg/kg. *Lusofarmico.*

781 Tetrandrine
518-34-3 9369
$C_{38}H_{42}N_2O_6$
(1β)-6,6',7,12-Tetramethoxy-2,2'-dimethylberbaman.
Phaenthine [l-form]. Analgesic; antipyretic. Found in root of *Staphania tetandra s. Moore, Menispermaceae.* Present in Chinese drug han-fang-chi. mp = 217-218°; [l-form]: mp = 210°; $[\alpha]_D^{26}$ = +252.4° ($CHCl_3$); $[\alpha]_D^{26}$ = -278° ($CHCl_3$); nearly insoluble in H_2O, petroleum ether; soluble in Et_2O and some other organic solvents.

782 Tetriprofen
28168-10-7
$C_{15}H_{18}O_2$
p-1-Cyclohexen-1-ylhydratropic acid.
47-210 [as sodium]. Analgesic; anti-inflammatory.

783 Tetrydamine
17289-49-5
$C_9H_{15}N_3$
4,5,6,7-Tetrahydro-2-methyl-3-(methylamino)-2H-indazol.
POLI-67. Analgesic; anti-inflammatory. *Polichimica Sap.*

784 Tifurac
97483-17-5
$C_{18}H_{14}O_4S$
7-[p-(Methylthio)benzoyl]-5-benzofuranacetic acid.
Analgesic. *Syntex Labs. Inc.*

785 Tifurac Sodium
102488-97-1
$C_{18}H_{13}NaO_4S.xH_2O$
Sodium 7-[p-(methylthio)benzoyl]-5-benzofuranacetate hydrate.
RS-82917-030. Analgesic. *Syntex Labs. Inc.*

786 Tinoridine
24237-54-5 9590 246-102-0
$C_{17}H_{20}N_2O_2S$
Ethyl 2-amino-6-benzyl-4,5,6,7-tetrahydrothieno-[2,3-c]pyridine-3-carboxylate.
Y-3642. Analgesic; anti-inflammatory; antipyretic. mp = 112-113°; slightly soluble in H_2O; LD_{50} (mus orl) = 5400 mg/kg, (rat orl) > 10200 mg/kg, (mus ip) = 1600 mg/kg, (rat ip) = 1250 mg/kg. *Yoshitomi.*

787 Tinoridine Hydrochloride
25913-34-2 9590
$C_{17}H_{21}ClN_2O_2S$
Ethyl 2-amino-6-benzyl-4,5,6,7-tetrahydrothieno-[2,3-c]pyridine-3-carboxylate hydrochloride.
Dimaten, Noflamin. Analgesic; anti-inflammatory; antipyretic. mp = 234-235° (dec); slightly soluble in MeOH; less soluble in H_2O, Et_2O, Me_2CO, C_6H_6; LD_{50} (mus orl) = 1601 mg/kg. *Yoshitomi.*

788 Tiopinac
61220-69-7
$C_{16}H_{12}O_3S$
6,11-Dihydro-11-oxodibenzo[b,e]thiepin-3-acetic acid.
RS-40974. Analgesic; anti-inflammatory; antipyretic. *Syntex Labs. Inc.*

789 Tolfenamic Acid
13710-19-5 9650 237-264-3
$C_{14}H_{12}ClNO_2$
N-(3-Chloro-o-tolyl)anthranilic acid.
GEA-6414; Clotam; Tolfedine; Tolfine. Anti-inflammatory; analgesic. Prostaglandin biosynthesis inhibitor. Derivative of anthranilic acid, related structurally to mefanamic and flufenamic acids, anti-inflammatory agents. mp = 207-207.5°. *Gea A/S; Parke-Davis.*

790 Tonazocine
71461-18-2
$C_{23}H_{35}NO_2$
(±)-1-[(2R*,6S*,11S*)-(1,2,3,4,5,6-Hexahydro-8-hydroxy-3,6,11-trimethyl-2,6-methano-3-benzazocin-11-yl)-3-octanone.
Analgesic. *Sterling Winthrop Inc.*

791 Tonazocine Mesylate
73789-00-1
$C_{24}H_{39}NO_5S$
(±)-1-[(2R*,6S*,11S*)-(1,2,3,4,5,6-Hexahydro-8-hydroxy-3,6,11-trimethyl-2,6-methano-3-benzazocin-11-yl)-3-octanone methanesulfonate.
Win-42156-2. Analgesic. *Sterling Winthrop Inc.*

792 Tramadol
27203-92-5 9701 248-319-6
$C_{16}H_{25}NO_2$
(±)-2-[(Dimethylamino)methyl]-1-(3-methoxyphenyl)cyclohexanol.
CG-315E; E-265; U-26255A. Analgesic. *Grunenthal.*

793 Tramadol Hydrochloride
22204-88-2 9701
$C_{16}H_{26}ClNO_2$
trans-(±)-2-[(Dimethylamino)methyl]-1-(3-methoxyphenyl)cyclohexanol hydrochloride.
Crispin; Melanate; Tramal; Ultram; Zydol. Analgesic. mp = 180-181°; soluble in H_2O; LD_{50} (rat orl) = 228 mg/kg. *Grunenthal; McNeil Pharm.; Upjohn Ltd.*

794 Trefentanil
120656-74-8
$C_{25}H_{31}FN_6O_2$
N-[1-[2-(4-Ethyl-4,5-dihydro-5-oxo-1H-tetrazol-1-yl)-ethyl]-4-phenyl-4-piperidinyl]-N-(2-fluorophenyl)propanamide.
A-2665. Analgesic. *Anaquest.*

795 Trefentanil Hydrochloride
120656-93-1
$C_{25}H_{32}ClFN_6O_2$
N-[1-[2-(4-Ethyl-4,5-dihydro-5-oxo-1H-tetrazol-1-yl)-ethyl]-4-phenyl-4-piperidinyl]-N-(2-fluorophenyl)-propanamide monohydrochloride.
A-2665-HCl. Analgesic. *Anaquest.*

796 Trichlorourethan
107-69-7 9778 203-511-9
$C_3H_4Cl_3NO_3$
2,2,2-Trichloroethanol carbamate.
carbamic acid trichloroethyl ester. Sedative/hypnotic.

mp = 64-65°; soluble in H_2O (1 g/100 ml), very soluble in $CHCl_3$, EtOH, Et_2O; soluble in C_6H_6; sparingly soluble in petroleum ether.

797 Trolamine
102-71-6 9798 203-049-8
$C_6H_{15}NO_3$
2,2',2''-Nitrilotriethanol.
triethanolamine. Analgesic. Viscous liquid; mp = 21.57°; bp_{760} = 335.4°; d_D^{20} = 1.1242; d_D^{60} = 1.0985; n_D^{20} = 1.4285; pH (0.1N aqueous solution) = 10.5; miscible with H_2O; soluble in C_6H_6 (4.2% at 25°); less soluble in Et_2O (1.6%), CCl_4 (0.4%); n-heptane (<0.1%). *Union Carbide Corp.*

798 Trolamine Salicylate
2174-16-5 9798
$C_{13}H_{21}NO_6$
2,2',2''-Nitrilotriethanol salicylic acid.
Mobisyl; Myoflex. Analgesic. *Union Carbide Corp.*

799 Tropesin
65189-78-8 9909 265-607-7
$C_{28}H_{24}ClNO_6$
(±)-1-(4-Chlorobenzoyl)-5-methoxy-2-methyl-1H-indole-3-acetic acid 2-carboxy-2-phenyl ester.
Repanidal. Anti-Inflammatory; analgesic. Tropic acid ester of indomethacin, an arylacetic acid derivative. mp = 130-132°, 127-129°, 128-130°; LD_{50} (frat orl) = 140 mg/kg. *SPOFA.*

800 Veradoline
79201-80-2
$C_{20}H_{26}N_2O_2$
(±)-2-(p-Aminophenethyl)-1,2,3,4-tetrahydro-6,7-dimethoxy-1-methylisoquinoline.
Analgesic. *Fisons Pharm. Div.*

801 Veradoline Hydrochloride
76448-47-0
$C_{20}H_{28}Cl_2N_2O_2$
(±)-2-(p-Aminophenethyl)-1,2,3,4-tetrahydro-6,7-dimethoxy-1-methylisoquinoline dihydrochloride.
PR-870-714A. Analgesic. *Fisons Pharm. Div.*

802 Verilopam
68318-20-7
$C_{20}H_{26}N_2O_2$
3-(p-Aminophenethyl)-2,3,4,5-tetrahydro-7,8-dimethoxy-1H-3-benzazepine.
Analgesic. *Fisons Pharm. Div.*

803 Verilopam Hydrochloride
67394-31-4
$C_{20}H_{28}Cl_2N_2O_2$
3-(p-Aminophenethyl)-2,3,4,5-tetrahydro-7,8-dimethoxy-1H-3-benzazepine dihydrochloride.
PR-0818-516A. Analgesic. *Fisons Pharm. Div.*

804 Viminol
21363-18-8 10117 244-347-8
$C_{21}H_{31}ClN_2O$
1-(o-Chlorobenzyl)-α-[(di-sec-butylamino)methyl]pyrrole-2-methanol.
diviminol; Z-424. Analgesic. Composed of six stereoisomers. $bp_{0.1\ mm}$ = 160-165°. *Whitefin Holding.*

805 Viminol Hydroxybenzoate
23784-10-3 10117
$C_{28}H_{37}ClN_2O_4$
1-(o-Chlorobenzyl)-α-[(di-sec-butylamino)methyl]-pyrrole-2-methanol p-hydroxybenzoate.
Dividol. Analgesic. mp = 128-130° (dec); LD_{50} (rat ip) = 205.9 mg/kg. *Whitefin Holding.*

806 Volazocine
15686-68-7
$C_{18}H_{25}N$
3-(Cyclopropylmethyl)-1,2,3,4,5,6-hexahydro-cis-6,11-dimethyl-2,6-methano-3-benzazocine.
Win-23200. Analgesic. *Sterling Winthrop Inc.*

807 Xylazine
7361-61-7 10213 230-902-1
$C_{12}H_{16}N_2S$
5,6-Dihydro-2-(2,6-xylidino)-4H-1,3-thiazine.
Bay 1470; Bay Va 1470; Wh-7286. Sedative (veterinary); analgesic (veterinary); muscle relaxant (veterinary). mp = 140-142°, 136-139°; soluble in dilute acids, C_6H_6, Me_2CO, $CHCl_3$; slightly soluble in petr Et_2O; nearly insoluble in H_2O, alkalies; LD_{50} (mus iv) = 42 mg/kg, (mus orl) = 240 mg/kg, (rat orl) = 130 mg/kg. *Bayer Corp. Pharm. Div.*

808 Xylazine Hydrochloride
23076-35-9 10213
$C_{12}H_{17}ClN_2S$
5,6-Dihydro-2-(2,6-xylidino)-4H-1,3-thiazine hydrochloride.
Narcosyl; Rompun; Xylapan; Xylasol. Sedative (veterinary); analgesic (veterinary); muscle relaxant (veterinary). *Bayer AG.*

809 Zenazocine Mesylate
74559-85-6
$C_{24}H_{39}NO_5S$
(±)-1-[(2R*,6S*,11S*)-1,2,3,4,5,6-Hexahydro-8-hydroxy-3,6,11-trimethyl-2,6-methano-3-benzazocin-11-yl]-6-methyl-3-heptanone methanesulfonate (salt).
Win-42964-4. Analgesic. *Sterling Winthrop Inc.*

810 Zomepirac
33369-31-2 10322 251-474-2
$C_{15}H_{14}ClNO_3$
5-(4-Chlorobenzoyl)-1,4-dimethyl-1H-pyrrole-2-acetic acid.
Analgesic; anti-inflammatory. Arylacetic acid derivative. mp = 178-179°. *McNeil Pharm.*

811 Zomepirac Sodium [anhydrous]
64092-48-4 10322 264-669-2
$C_{15}H_{13}ClNNaO_3$
5-(4-Chlorobenzoyl)-1,4-dimethyl-1H-pyrrole-2-acetic acid sodium salt.
Analgesic; anti-inflammatory. Arylacetic acid derivative. *McNeil Pharm.*

812 Zomepirac Sodium Dihydrate
64092-49-5 10322
$C_{15}H_{17}ClNNaO_5$
5-(4-Chlorobenzoyl)-1,4-dimethyl-1H-pyrrole-2-acetic acid sodium salt dihydrate.
McN-2783-21-98; Zomax; Zomaxin; Zopirac. Analgesic; anti-inflammatory. Arylacetic acid derivative. mp = 295-296°. *McNeil Pharm.*

813 Zucapsaicin
25775-90-0
$C_{18}H_{27}NO_3$
(Z)-8-Methyl-N-vanillyl-6-nonenamide.
Topical analgesic. *GenDerm.*

Androgens

814 Boldenone
846-48-0 1354 212-686-0
$C_{19}H_{26}O_2$
17β-Androsta-1,4-dien-3-one.
dehydrotestosterone. Androgen. Has properties of an anabolic steroid. mp = 164-166°; $[\alpha]_D^{25}$ = 25° ($CHCl_3$); [acetate]: mp = 151-153°. *Olin Mathieson; Schering-Plough HealthCare Products.*

815 Boldenone 10-Undecanoate
13103-34-9 1354 236-024-5
$C_{30}H_{40}O_3$
17β-Androsta-1,4-dien-3-one 10-undecenoate.
Equipoise; Ba-9038; Parenabol. Androgen. Has properties of an anabolic steroid. *Solvay Animal Health Inc.*

816 Cloxotestosterone
53608-96-1 2482
$C_{21}H_{29}Cl_3O_3$
17β-(2,2,2-Trichloro-1-hydroxyethoxy)androst-4-en-3-one.
testosterone 17-chloral hemiacetal. Androgen. mp = 200-201°; λ_m = 241 nm (ε = 16300 EtOH). *Lovens Komiske Fabrik AS.*

817 Cloxotestosterone Acetate
13867-82-8 2482
$C_{23}H_{31}Cl_3O_4$
17β-(2,2,2-Trichloro-1-acetoxyethoxy)androst-4-en-3-one.
Caprosem. Androgen. mp = 192-193°; λ_m = 241 nm (ε = 16400 EtOH). *Lovens Komiske Fabrik AS.*

818 Fluoxymesterone
76-43-7 4223 200-961-8
$C_{20}H_{29}FO_3$
9-Fluoro-11β,17β-dihydroxy-17-methylandrost-4-en-3-one.
Halotestin; Ora-Testryl; NSC-12165; Androsterolo; Oratestin; Testoral; Ultandren. Androgen. mp = 270° (dec); $[\alpha]_D$ = 109° (EtOH); λ_m = 240 nm (ε 16700 EtOH); soluble in C_5H_5N; moderately soluble in Me_2CO, $CHCl_3$; sparingly soluble in MeOH; insoluble in H_2O, Et_2O, C_6H_6, C_6H_{12}. *Bristol-Myers Squibb Co; Pharmacia & Upjohn.*

819 Mestanolone
521-11-9 5973 208-302-6
$C_{20}H_{32}O_2$
17β-Hydroxy-17α-methyl-5α-androstan-3-one.
Androstalone. Androgen. mp = 192-193°; insoluble in H_2O; soluble in Me_2CO, EtOH, Et_2O, EtOAc. *Ciba-Geigy Corp.*

820 Mesterolone
1424-00-6 5974 215-836-3
$C_{20}H_{32}O_2$
17β-Hydroxy-1α-methyl-5α-androstan-3-one.
SH-723; NSC-75054. Androgen. mp = 203.5-205.0°; $[\alpha]_D^{20}$ = 17.6° (c = 0.875 $CHCl_3$); [acetate]: mp = 169-170°; $[\alpha]_D^{25}$ = 16.5° (c = 0.88 $CHCl_3$). *Schering AG.*

821 Methandrostenolone
72-63-9 6018 200-787-2
$C_{20}H_{28}O_2$
17β-Hydroxy-17α-methylandrosta-1,4-dien-3-one.
NSC-42722; Danabol; Nerobol; Nabolin; Sterolon; Dianabol. Androgen. mp = 163-164°; $[\alpha]_D^{26}$ = 0° (c = 1.15 $CHCl_3$)λ_m = 245 nm (ε = 15600). *Ciba-Geigy Corp.*

822 17-Methyltestosterone
58-18-4 6206 200-366-3
$C_{20}H_{30}O_2$
17β-Hydroxy-17α-methylandrost-4-en-3-one.
Android; Metandren; Oraviron; Oreton Methyl; Synandrets; Synandrotabs; Testred; Glosso-Stérandryl; Neohombreol M; Orchisterone-M; Perandren; component of: Estan, Estratest, Premarin with Methyltestosterone. Androgen. mp = 161-166°; $[\alpha]_D^{25}$ = 69° to 75° (dioxane); soluble in EtOH, MeOH, Et_2O, most organic solvents; insoluble in H_2O. *Ciba-Geigy Corp.; Eli Lilly & Co.; ICN Pharm. Inc.; Pfizer Inc.; Schering-Plough HealthCare Products; Solvay Pharm. Inc.; Wyeth-Ayerst Labs.*

823 17α-Methyltestosterone 3-cyclopentyl enol ether
67-81-2 6207 200-670-6
$C_{25}H_{38}O_2$
(17β)-3-(Cyclopentyloxy)-17-methylandrosta- 3,5-dien-17-ol.
RP-12222; Pandrocine; Penmestrol. Androgen. mp = 148-152°; $[\alpha]_D$ = -150° (dioxane).

824 Mibolerone
3704-09-4 6262 223-046-5
$C_{20}H_{30}O_2$
17β-Hydroxy-7α,17-dimethylestr-4-en-3-one.
U-10997; Cheque; Matenon. Anabolic steroid with androgenic properties. Soluble in H_2O (4.54 mg/100 ml at 37°). *Pharmacia & Upjohn.*

825 Nandrolone
434-22-0 6452 207-101-0
$C_{18}H_{26}O_2$
17β-Hydroxyestr-4-en-3-one.
19-nortestosterone. Anabolic steroid with androgenic properties. mp = 112°, 124°; $[\alpha]_D^{22}$ = 55° (c = 0.93 $CHCl_3$); λ_m = 241 nm (ε 17000 EtOH); soluble in EtOH, Et_2O, $CHCl_3$; [benzoate]:mp = 174-175°; $[\alpha]_D^{20}$ = 104.5° (EtOH).

826 Nandrolone Cyclohexanecarboxylate
18470-94-5 6452 242-351-4
$C_{25}H_{36}O_3$
17β-Hydroxyestr-4-en-3-one cyclohexane-carboxylate.
19-nortestosterone hexahydrobenzoate; Nolongandron; Nor-Durandron. Anabolic steroid with androgenic properties. mp = 88-89°; $[\alpha]_D^{20}$ = 50° (c = 0.5 $CHCl_3$). *Forest Pharm. Inc.; Organon Inc.*

827 Nandrolone Cyclohexanepropionate
912-57-2 6452 213-013-3
$C_{27}H_{40}O_3$
17β-Hydroxyestr-4-en-3-one cyclohexane-propionate.
19-nortestosterone cyclohexylpropionate; Sanabolicum. An anabolic steroid with androgenic properties. *Forest Pharm. Inc.; Organon Inc.*

828 Nandrolone Decanoate
360-70-3 6452 206-639-3
$C_{28}H_{44}O_3$
17β-Hydroxyestr-4-en-3-one decanoate.
19-nortestosterone decanoate; Deca-Durabol; Deca-Hybolin; Hybolin decanoate; Deca-Durabolin; Nandrobolic L.A.; Retabolil. Anabolic steroid with androgenic properties. mp= 32-35°; soluble in EtOH, Et_2O, Me_2CO, $CHCl_3$; insoluble in H_2O. *Forest Pharm. Inc.; Organon Inc.*

829 Nandrolone Dodecanoate
26490-31-3 6452 247-739-7
$C_{30}H_{48}O_3$
17β-Hydroxyestr-4-en-3-one dodecanoate.
nandrolone laurate; Laurabolin. Anabolic steroid with androgenic properties. *Forest Pharm. Inc.; Organon Inc.*

830 Nandrolone Furylpropionate
7642-64-0 6452 231-580-5
$C_{25}H_{32}O_4$
17β-Hydroxyestr-4-en-3-one furylpropionate.
19-nortestosterone furylpropionate; NFP; Demelon. Anabolic steroid with androgenic properties. *Forest Pharm. Inc.; Organon Inc.*

831 Nandrolone Phenpropionate
62-90-8 6452 200-551-9
$C_{27}H_{34}O_3$
17β-Hydroxyestr-4-en-3-one hydrocinnamate.
Durabolin; Nandrobolic; NSC-23162; Activin; Durabol; Strabolene; Superanabolon; Nandrolin. Anabolic steroid with androgenic properties. mp = 95-96°; $[\alpha]_D$= 58° ($CHCl_3$). *Forest Pharm. Inc.; Organon Inc.*

832 Nandrolone p-Hexyloxyphenylpropionate
52279-57-9 6452 257-810-4
$C_{33}H_{46}O_4$
17β-Hydroxyestr-4-en-3-one p-hexyloxyphenyl-propionate.
Anador; Anadur; 19-nortestosterone-3-(p-hexyloxy-phenyl)propionate. Anabolic steroid with androgenic properties. mp = 53-55°; $[\alpha]_D$ = 45° (c = 1 dioxane). *Forest Pharm. Inc.; Organon Inc.*

833 Nandrolone Propionate
7207-92-3 6452 230-587-0
$C_{21}H_{30}O_3$
17β-Hydroxyestr-4-en-3-one propionate.
19-nortestosterone propionate. Anabolic steroid with androgenic properties. mp = 55-60°; $[\alpha]_D^{23.5}$ = 58° ($CHCl_3$); λ_m = 240 nm (ε 16650). *Forest Pharm. Inc.; Organon Inc.*

834 Nisterime acetate
51354-31-5
$C_{27}H_{35}ClN_2O_5$
2α-Chloro-17β-hydroxy-5α-androstan-3-one O-(p-nitrophenyl)oxime acetate (ester).
ORF-9326. Androgen. *Ortho Pharm. Corp.*

835 Norethandrolone
52-78-8 6789 200-153-5
$C_{20}H3_{00}2$
17α-Ethyl-17β-hydroxyestr-4-en-3-one.
Nilevar; Solevar. Androgen. mp = 140-141°; λ_m = 240 nm (ε 16500); soluble in EtOH, C_6H_6, Et_2O, EtOAc; insoluble in H_2O. *Searle G.D. & Co.*

836 Normethandrone
514-61-4 6805 208-183-0
$C_{19}H_{28}O_2$
17α-Methyl-17β-hydroxyestr-4-en-3-one.
methylnortestosterone; normethandrolone; normetandrone; Orgasteron; Metalutin; Methalutin. Androgen. mp = 156-158°; $[\alpha]_D$ = 33°; λ_M = 240 nm (log ε 4.23 EtOH). *Syntex Intl. Ltd.*

837 Oxandrolone
53-39-4 7054 200-172-9
$C_{19}H_{30}O_3$
17β-Hydroxy-17α-methyl-2-oxa-5α-androstan-3-one.
Anavar; SC-11585; NSC-67068; Lonavar; Provitar; Vasorome. Androgen. mp = 235-238°; $[\alpha]_D^{25}$= -23°, λ_m = 287, 579 nm. *Searle G.D. & Co.*

838 Oxymesterone
145-12-0 7099 205-646-9
$C_{20}H_{30}O_3$
4,17β-Dihydroxy-17α-methylandrost-4-en-3-one.
oxymestrone; Anamidol; Oranabol; Theranabol. An anabolic steroid with androgenic properties. Federally controlled substance (anabolic steroid). mp = 169-171°; $[\alpha]_D^{20}$= 69° (EtOH); λ_m = 278 nm ($E_{1\,cm}^{1\%}$ 406 EtOH); soluble in $CHCl_3$, Me_2CO, EtOH; insoluble in H_2O. *Farmitalia Carlo Erba Ltd.*

839 Oxymetholone
434-07-1 7101 207-098-6
$C_{21}H_{32}O_3$
17β-Hydroxy-2-(hydroxymethylene)-17-methyl-5α-androstan-3-one.
Adroyd; Anadrol; anasterone; Anapolon; Pardroyd; Plenastril; Protanabol; Nastenon; Synasteron. Androgen. mp = 178-180°; $[\alpha]_D$ = 38°; λ_m = 285 nm (log ε 3.99); [enol acetate]: mp = 144-148°; $[\alpha]_D$ = 27° (EtOH); λ_m= 255 nm (log ε 4.09); [enol propionate]: mp = 135°; $[\alpha]_D$ = 26° (EtOH); λ_m= 257 nm (logε 4.11); [enol benzoate]:

mp = 188-190°; $[\alpha]_D$ = 0° (EtOH); λ_m= 230 nm (logε 4.19). *Parke-Davis; Syntex Intl. Ltd.*

840 Prasterone
53-43-0 7892 200-175-5
$C_{19}H_{28}O_2$
3β-Hydroxyandrost-5-en-17-one.
DHEA; Astenile; Deandros; Diandrone; Psicosterone. Androgen. mp = 140-141°, 152-153°; $[\alpha]_D^{18}$ = 10.9° (c = 0.4 EtOH); soluble in C_6H_6, EtOH, Et_2O; sparingly soluble in $CHCl_3$, petroleum ether.

841 Prasterone Sodium Sulfate
1099-87-2 7892 214-152-2
$C_{19}H_{27}NaO_5S$
3β-Hydroxyandrost-5-en-17-one sodium sulfate.
sodium dehydroepiandrosterone sulfate; DHA-S; Mylis. Androgen. mp = 154° (dec); soluble in MeOH; slightly soluble in EtOH, H_2O; insoluble in Me_2CO, $CHCl_3$, C_6H_6; LD_{50} (mmus orl) > 10000 mg/kg, (mmus sc) = 899 mg/kg, (mmus ip) = 460 mg/kg, (mmus iv) = 274 mg/kg, (mrat orl) > 10000 mg/kg, (mrat sc) = 1005 mg/kg, (mrat ip) = 559 mg/kg, (mrat iv) = 468 mg/kg.

842 Silandrone
5055-42-5 225-755-5
$C_{22}H_{36}O_2Si$
17β-(Trimethylsiloxy)androst-4-en-3-one.
SC-16148; NSC-95147. Androgen. *Searle G.D. & Co.*

843 Stanolone
521-18-6 8950 208-307-3
$C_{19}H_{30}O_2$
17β-Hydroxy-5α-androstan-3-one.
Neodrol; androstanolone; Anabolex; Andractim; Androlone. Androgen. mp = 181°; $[\alpha]_D^{20}$= 32.4° (EtOH); soluble in EtOH, Me_2CO, Et_2O, EtOAc; insoluble in H_2O. *Pfizer Inc.; Schering AG.*

844 Stanozolol
10418-03-8 8951 233-894-8
$C_{21}H_{32}N_2O$
17-Methyl-2'H-5α-androst-2-eno- [3,2-c]pyrazol-17β-ol.
Winstrol; Win-14833; NSC-43193; androstanazolestanazol; Stromba; Strombaject. Androgen. mp = 229.8-242°; $[\alpha]_D$ = 35.7° ($CHCl_3$), 48.6° (EtOH); λ_m = 223 nm (ε 4740). *Sterling Winthrop Inc.*

845 Testosterone
58-22-0 9322 200-370-5
$C_{19}H_{28}O_2$
17β-Hydroxyandrost-4-en-3-one.
Andro 100; Mertestate; Oreton; Synandrol F; Testoderm; Testolin; Testro AQ; Virosterone. Androgen. mp = 155°; $[\alpha]_D^{24}$= 109° (c = 4 EtOH); λ_m = 238 nm; soluble in EtOH, Et_2O, other organic solvents; insoluble in H_2O. *Forest Pharm. Inc.; Pfizer Intl.; Schering AG; Sterling Winthrop Inc.*

846 Testosterone Acetate
1045-69-8 9322 213-876-6
$C_{21}H_{30}O_3$
17-[(1-Oxoacetyl)oxy]androst-4-en-3-one.
Androgen. mp = 140-141°. *Forest Pharm. Inc.; Mead Johnson Pharmaceuticals; Savage Labs; Schering AG.*

847 Testosterone Cypionate
58-20-8 9322 200-368-4
$C_{27}H_{40}O_3$
17-(Cyclopentyl-1-oxopropoxy)androst-4-en-3-one.
depAndro 100; depAndro 200; Depo; Depovirin; Depotest; Depo-Testosterone; Pertestis; Virilon; component of: Depo-Testadiol. Androgen. mp = 101-102°; $[\alpha]_D^{25}$= 87° ($CHCl_3$); soluble in oils. *Forest Pharm. Inc.; Hoechst Roussel Pharm. Inc.; Pharmacia & Upjohn.*

848 Testosterone Enanthate
315-37-7 9322 206-253-5
$C_{26}H_{40}O_3$
17-[(1-Oxoheptyl)oxy]androst-4-en-3-one.
Andro L.A. 200; Delatestryl; NSC-17591; Androtardyl; Everone; Primoteston; Testinon; Testo-Enant; component of: Androgyn L.A., Ditate. Androgen. mp = 36-37.5°. *Forest Pharm. Inc.; Mead Johnson Pharmaceuticals; Savage Labs; Schering AG.*

849 Testosterone Ketolaurate
5874-98-6 9322 227-540-1
$C_{31}H_{48}O_4$
17-[(1,3-Dioxododecyl)oxy]androst-4-en-3-one.
Androgen. *Forest Pharm. Inc.; Organon Inc.*

850 Testosterone Phenylacetate
5704-03-0 9322
$C_{27}H_{34}O_3$
17-[(Phenylacetyl)oxy]androst-4-en-3-one.
Androgen. *Forest Pharm. Inc.; Organon Inc.*

851 Testosterone Propionate
57-85-2 9322 200-351-1
$C_{22}H_{32}O_3$
17-[(1-Oxopropanoyl)oxy]androst-4-en-3-one.
Anertan; Enarmon; Neo-Hombreol; Orchisterone; Perandren; Synandrol; Synerone; Testex; Testoviron; Virormone. Androgen. mp = 118-122°; $[\alpha]_D^{25}$= 83° to 90° (c = 1 dioxane); soluble in EtOH, Et_2O, C_5H_5N, other organic solvents; insoluble in H_2O. *Forest Pharm. Inc.; Mead Johnson Pharmaceuticals; Savage Labs; Schering AG.*

852 Tiomesterone
2205-73-4 9596 218-614-4
$C_{24}H_{34}O_4S_2$
1α,7α-Diacetylthio-17β-hydroxy-17α-methylandrost-4-en-3-one.
STA-307; Thiomesterone; Emdabol; Emdabolin; Protabol. Androgen. mp = 205-206°; $[\alpha]_D$ = -66° (dioxane); λ_m = 237.5 nm (ε 19800). *E. Merck.*

853 Trestolone
3764-87-2
$C_{19}H_{28}O_2$
17β-Hydroxy-7α-methylestr-4-en-3-one.
Androgen.

854 Trestolone Acetate
6157-87-5
$C_{21}H_{30}O_3$
17β-Hydroxy-7α-methylestr-4-en-3-one acetate.
U-15,614; NSC-69948. Androgen.

Anesthetics, Inhalation

855 Aliflurane
56689-41-9
$C_4H_3ClF_4O$
2-Chloro-1,2,3,3-tetrafluorocyclopropyl methyl ether.
26P. Inhalation anesthetic.

856 Chloroform
67-66-3 2193 200-663-8
$CHCl_3$
Trichloromethane.
Once used as an inhalation asesthetic. Has properties similar to halothane. Used as a solvent. No longer used clinically. CAUTION: Overexposure can cause dizziness, disorientation, nausea, headache, anasthesia. Hepatotoxic. May cause cardiac arrhythmias. Possible carcinogen. Volatile liquid; mp = - 63.5°; bp = 61-62°; d_{20}^{20} = 1.484; n_D^{20} = 1.4476; soluble in H_2O (0.5 ml/100 ml); miscible with alcohol, Et_2O, petroleum ether, CCl_4, CS_2; LD_{50} (14-day rat orl) = 2.18 ml/kg.

857 Cyclopropane
75-19-4 2817 200-847-8
C_3H_6
Cyclopropnane.
trimethylene. Inhalation anesthetic. mp = -127°; bp = -33°; 1 l. at NTP weighs 1.879 g.; soluble in H_2O (1 vol/2.7 vols); freely soluble in EtOH, Et_2O.

858 Desflurane
57041-67-5 2965
$C_3H_2F_6O$
(±)-2-Difluoromethyl 1,2,2,2-tetrafluoroethyl ether.
Suprane; I-653. Inhalation anesthetic. bp = 23.5°; d = 1.44; vapor pressure = 700 mm at 22°. *Anaquest.*

859 Enflurane
13838-16-9 3621 237-553-4
$C_3H_2ClF_5O$
2-Chloro-1,1,2-trifluoroethyl difluoromethyl ether.
Ethrane; Anesthetic Compound No. 347; NSC-115944; methylflurether; Alyrane; Efrane. Inhalation anesthetic. bp = 56.5°; d_{25}^{25} = 1.5167; soluble in organic solvents.

860 Ethyl Ether
60-29-7 3852 200-467-2
$C_4H_{10}O$
1,1'-Oxybisethane.
diethyl ether; diethyl oxide; ethyl oxide; sulfuric ether; anesthetic ether. Inhalation anesthetic. mp = -123.3°; bp = 34.6°, bp_{400} = 17.9°, bp_{200} = 2.2°, bp_{100} = -11.5°; bp_{10} = -48.1°, bp_1 = -74.3°; d_4^{10} = 0.7249, d_4^{20} = 0.7134, d_4^{30} = 0.7019; slightly soluble in H_2O, soluble in most organic solvents.

861 Ethylene
74-85-1 3837 200-815-3
C_2H_4
Ethene.
elayl; olefiant gas. Inhalation anesthetic. mp = -169.4°; bp_{700} = -102.4°; 1 liter; NTP weighs 1.260 g; soluble in H_2O (0.25 vol/vol at 0°, 0.11 vol/vol at 25°), EtOH (2 vols/vol), Et_2O, Me_2CO, C_6H_6; LC (mus ihl) = 950000 ppm.

862 Fluroxene
406-90-6 4237 206-977-1
$C_4H_5F_3O$
2,2,2-Trifluoroethyl vinyl ether.
Fluoromar. Inhalation anesthetic. bp_{751} = 42.5°; d^{20} = 1.135. *Anaquest.*

863 Halopropane
679-84-5 4632
$C_3H_3BrF_4$
3-Bromo-1,1,2,2-tetrafluoropropane.
FHD-3; Tebron. Inhalation anesthetic. bp = 74°; d_{20} = 1.81. *E. I. DuPont de Nemours Inc.*

864 Halothane
151-67-7 4634 205-796-5
$C_2HBrClF_3$
2-Bromo-2-chloro-1,1,1-trifluoroethane.
Fluothane; Rhodialothan. Inhalation anesthetic. bp = 50.2°, bp_{243} = 20°; slighlty soluble in H_2O (0.345 g/100 ml); soluble in organics. *Hoechst AG; Wyeth-Ayerst Labs.*

865 Isoflurane
26675-46-7 5191 247-897-7
$C_3H_2ClF_5O$
1-Chloro-2,2,2-trifluoroethyl difluoromethyl ether.
Aerrane; Forane; Compound 469; Forene. Inhalation anesthetic. bp = 48.5°; sg = 1.45; non-flammable; soluble in most organic solvents. *Anaquest.*

866 Methoxyflurane
76-38-0 6072 200-956-0
$C_3H_4Cl_2F_2O$
2,2-Dichloro-1,1-difluoroethyl methyl ether.
Penthrane; NSC-110432; DA-759; Metofane; Pentrane. Inhalation anesthetic. bp = 105°, bp_{100} = 51°; mp = -35°; d_4^{20} = 1.4262, 1.4226. *Abbott Labs Inc.*

867 Methyl Propyl Ether
557-17-5 6192 209-158-7
$C_4H_{10}O$
Methyl n-propyl ether.
1-methoxypropane; Neothyl. Inhalation anesthetic. bp_{761} = 38.8°; d^0_4 = 0.7494, d_4^{13} = 0.7356; soluble in H_2O (3.65 g/100 ml), EtOH, Et_2O.

868 Nitrous Oxide
10024-97-2 6751 233-032-0
N_2O
Nitrogen oxide.
dinitrogen monoxide; laughing gas; factitious air. Inhalation anesthetic. mp = -91°; bp = -88.5°.

869 Norflurane
811-97-2 212-377-0
$C_2H_2F_4$
1,1,1,2-Tetrafluoroethane.
HFA-134a. Inhalation anesthetic. *Marion Merrell Dow Inc.*

870 Roflurane
679-90-3
$C_3H_4BrF_3O$
2-Bromo-1,1,2-trifluoroethyl methyl ether.
DA-893. Inhalation anesthetic. *Marion Merrell Dow Inc.*

871 Sevoflurane
28523-86-6 8621
$C_4H_3F_7O$
Fluoromethyl-2,2,2-trifluoro-1-(trifluoromethyl)ethyl ether.
Ultane; MR6S4; Sevofrane. Inhalation anesthetic. bp = 58.5°, bp_{751} = 58.1-58.2°; d_4^{23}= 1.505. *Abbott Labs Inc.*

872 Teflurane
124-72-1 9265
C_2HBrF_4
2-Bromo-1,1,1,2-tetrafluoroethane.
Abbott-16900; DA-708; Terflurane. Inhalation anesthetic. bp = 8°. *Abbott Labs Inc.; Marion Merrell Dow Inc.*

873 Trichloroethylene
79-01-6 9769 201-167-4
C_2HCl_3
1,1,2-Trichloroethene.
Chlorylen; Ethinyl trichloride; Tri-Clene; Trilene; Trichloren; Algylen; Trimar; Triline; Trethylene; Westrosol; Chlorylen; Gemalgene; Germalgene. Inhalation anesthetic. d^4_4 = 1.4904, d_4^{15}= 1.4695, d_4^{20}= 1.4642, d_4^{25}= 1.4559; bp = 86.9°; bp_{400} = 67.0°, bp_{200} = 48.0°, bp_{100} = 31.4°, bp_{60} = 20.0°, bp_{20} = -1.0°, bp_{10} = -12.4°, bp_5 = -22.8°, bp_1 = -43.8°; soluble in H_2O (0.11 g/100 ml), EtOH, Et_2O, $CHCl_3$; LD_{50} (rat orl) = 7.16 mg/kg, LC (rat ihl 4 hr) = 8000 ppm. *Schering-Plough HealthCare Products.*

874 Vinyl Ether
109-93-3 10133 203-720-5
C_4H_6O
1,1'-Oxybisethene.
Inhalation anesthetic. bp = 28.4°; d_{20}^{20} = 0.774, d_4^{20}= 0.773; soluble in H_2O (0.55 g/100 ml), soluble in organic solvents; LC (mus ihl) = 51233 ppm. *Merck & Co.Inc.*

875 Xenon
7440-63-3 10206 231-172-7
Xe
Inhalation anesthetic. Soluble in H_2O (10.8 g/100 g); bp = -108°.

Anesthetics, Intravenous

876 Acetamidoeugenol
305-13-5 44
$C_{16}H_{23}NO_3$
N,N-Diethyl-2-[2-methoxy-4-(2-propenyl)phenoxy]acetamide.
2-M-4-A; G-29505; Detrovel; Estil. Intravenous anesthetic. $bp_{0.001}$ = 143-146°; soluble in H_2O. *Ciba-Geigy Corp.*

877 Alfadolone Acetate
23930-37-2 233 245-942-5
$C_{21}H_{34}O_4$
3α,21-Dihydroxy-5α-pregnane-11,20-dione.
GR 2/1574; alphadolone acetate. Intravenous anesthetic. Often used in combination with alfaxalone. mp = 175-177°; $[\alpha]_D^{26}$= 97° (c = 1.02 $CHCl_3$). *Glaxo Labs.*

878 Alfaxalone
23930-19-0 235
$C_{21}H_{32}O_3$
3α-Hydroxy-5α-pregnane-11,20-dione.
GR 2/234; alphaxalone; combination with alfadolone acetate: alphadione; alfadione; CT-1341; Alfatésin; Althesin; Saffan. Intravenous anesthetic. Often used in combination with alfadalone. mp = 172-174°; $[\alpha]_D^{26}$ = 113.4° (c = 1.2 $CHCl_3$); [combination with alfadolone acetate]: LD_{50} (mus iv) = 54.7 mg/kg. *Glaxo Labs.*

879 Buthalital
468-65-5 1553
$C_{11}H_{16}N_2O_2S$
5-Allyl-5-isobutyl-2-thiobarbituric acid.
Intravenous anesthetic. Intravenous anesthetic. mp = 147°.

880 Buthalital Sodium
510-90-7 1553
$C_{11}H_{15}N_2NaO_2S$
Sodium 5-allyl-5-isobutyl-2-thiobarbiturate.
Baytinal; Buthalitone Sodium; Thialisobumalnatrium; Transithal; Ulbreval. Intravenous anesthetic. Intravenous anesthetic. Freely soluble in H_2O; partly soluble in EtOH; insoluble in Et_2O, C_6H_6.

881 Climazolam
59467-77-5
$C_{18}H_{13}Cl_2N_3$
8-Chloro-6-(o-chlorophenyl)-1-methyl-H-imidazo-[1,5-a][1,4]-benzodiazepine.
Intravenous anesthetic.

882 Diazepam
439-14-5 3042 207-122-5
$C_{16}H_{13}ClN_2O$
7-Chloro-1,3-dihydro-1-methyl-5-phenyl-2H-1,4-benzodiazepin-2-one.
Alupram; Valium; Valrelease; LA-111; Ro- 5-2807; Wy-3467; NSC-77518; Apaurin; Atensine; Atilen; Bialzepam; Calmpose; Ceregulart; Dialar; Diazemuls; Dipam; Eridan; Eurosan; Evacalm; Faustan; Gewacalm; Horizon; Lamra; Lembrol; Levium; Mandrozep; Neurolytril; Noan; Novazam; Paceum; Pacitran; Paxate; Paxel; Pro-Pam; Q-Pam; Relanium; Sedapam; Seduxen; Servizepam; Setonil; Solis; Stesolid; Tranquase; Tranquo-Puren; Tranquo-Tablinen; Unisedil; Valaxona; Valiquid; Valium; Valium Injectable; Valrelease; Vival; Vivol. Anxiolytic; skeletal muscle relaxant; sedative/hypnotic. Also used as an intravenous anesthetic. Pharmaceutical preparation for the treatment of depression. A benzodiazepine. mp = 125-126°; soluble in DMF, $CHCl_3$, C_6H_6; Me_2CO, EtOH;

slightly soluble in H_2O; LD_{50} (rat orl) = 710 mg/kg. *Berk Pharm. Ltd.; Hoffmann-LaRoche Inc.*

883 Etomidate
33125-97-2 3927 251-385-9
$C_{14}H_{16}N_2O_2$
(+)-Ethyl-1-(α-methylbenzyl)imidazole-5-carboxylate.
Amidate; R-16659; Hypnomidate. Sedative/hypnotic. Carboxilated derivative of imidazole. Also used as an intravenous anesthetic and as a supplement to maintain anesthesia. mp = 67°; $[\alpha]_D^{20}$ = 66° (c = 1 EtOH); λ_m = 240 nm (ε 12200 iPrOH); soluble in H_2O (0.0045 mg/100 ml), $CHCl_3$, MeOH, EtOH, Me_2CO, propylene glycol; LD_{50} (mus iv) = 29.5 mg/kg, (rat iv) = 14.8 - 24.3 mg/kg. *Abbott Laboratories Inc.; Abbott Labs Inc.*

884 Etoxadrol
28189-85-7 3933
$C_{16}H_{23}N_O2$
(+)-2-(2-Ethyl-2-phenyl-1,3-dioxolan-4-yl)piperidine.
Intravenous anesthetic. *Cutter Labs.*

885 Etoxadrol (+)-Hydrochloride
23239-37-4 3933
$C_{16}H_{24}ClN_O2$
(+)-2-(2-Ethyl-2-phenyl-1,3-dioxolan-4-yl)piperidine hydrochloride.
Thoxan; CL-1848C. Intravenous anesthetic. mp = 221.5-222°; $[\alpha]_D^{25}$ = 16.63°. *Cutter Labs.*

886 Hexobarbital
56-29-1 4742 200-264-9
$C_{12}H_{16}N_2O_3$
5-(1-Cyclohexen-1-yl)-1,5-dimethylbarbituric acid.
methylhexabarbital; methexenyl; enhexymal; hexobarbitone; Citodon; Citopan; Cyclonal; Dorico; Evipal; Evipan; Hexanastab Oral; Noctivane; Sombucaps; Sombulex; Somnalert. Sedative/ hypnotic. Intravenous anesthetic. mp = 145-147°; poorly soluble in H_2O (0.033 g/100 ml); soluble in EtOH, MeOH, Et_2O, $CHCl_3$, Me_2CO, C_6H_6. *3M Pharm.; Sterling Winthrop Inc.*

887 Hexobarbital Sodium
50-09-9 4742 200-009-1
$C_{12}H_{15}N_2NaO_3$
5-(1-Cyclohexen-1-yl)-1,5-dimethyl-2,4,6(1H,3H,5H)-pyrimidinetrione sodium salt.
hexobarbital soluble; Cyclonal Sodium; Dorico Soluble; Evipal Sodium; Evipan Sodium; Hexanastab; Hexenal; Methexenyl Sodium; Noctivane Sodium; Narcosan Soluble; Privenal. Sedative/hypnotic. Intravenous anesthetic. Federally controlled substance. CAUTION: may be habit forming. Soluble in H_2O, EtOH, MeOH, Me_2CO; insoluble in $CHCl_3$, Et_2O, C_6H_6. *3M Pharm.; Sterling Winthrop Inc.*

888 Hydroxydione
53-10-1 4868
$C_{21}H_{32}O_3$
21-Hydroxy-5β-pregnane-3,20-dione.
Intravenous anesthetic. mp = 195-197°; $[\alpha]_D^{20}$ = 95° ($CHCl_3$). *Pfizer Inc.*

889 Hydroxydione Sodium Succinate
303-01-5 4868
$C_{25}H_{35}NaO_6$
21-Hydroxy-5β-pregnane-3,20-dione sodium succinate.
Viadril; Pressuren. IV anesthetic. Dec 193-203°; λ_m = 280 nm (ε 93.2); soluble in H_2O, Me_2CO, $CHCl_3$. *Pfizer Inc.*

890 Ketamine
6740-88-1 5306 229-804-1
$C_{13}H_{16}ClNO$
(±)-2-(o-Chlorophenyl)-2-(methylamino)cyclohexanone.
Intravenous anesthetic. Ion channel blocker. A cyclohexanone derivative mp = 92-93°; λ_m = 301, 276, 268, 261 nm ($A_{1\,cm}^{1\%}$ 5.0, 7.0, 9.8, 10.5 0.01N NaOH/95% MeOH). *Bristol-Myers Squibb Pharm. R&D; Fermenta Animal Health Co.; Parke-Davis.*

891 Ketamine Hydrochloride
1867-66-9 5306 217-484-6
$C_{13}H_{17}Cl_2NO$
(±)-2-(o-Chlorophenyl)-2-(methylamino)cyclohexanone hydrochloride.
Ketaject; Ketalar; Ketaset; Ketavet; Ketanarkon; Ketanest; Vetalar; CI-581; CL369; CN-52,372-2; component of: Ketaset Plus. Intravenous anesthetic. Ion channel blocker. A cyclohexanone derivative. mp = 262-263°; soluble in H_2O (20 g/100 ml); LD_{50} (rat ip) = 229 ± 5 mg/kg, (mus ip) = 224 ± 4 mg/kg. *Bristol-Myers Squibb Pharm. R&D; Fermenta Animal Health Co.; Parke-Davis.*

892 Methohexital
18652-93-2 6061
$C_{14}H_{18}N_2O_3$
(±)-5-Allyl-1-methyl-5-(1-methyl-2-pentynyl)barbituric acid.
A short-actiong intravenous anesthetic. Used to prepare injectable methohexital sodium. *Eli Lilly & Co.*

893 Methohexital Sodium
22151-68-4 6061 261-774-5
$C_{14}H_{17}N_2NaO_3$
Sodium 5-allyl-1-methyl-5-(1-methyl-2-pentynyl)barbiturate.
methohexitone sodium; Brevital; Brevital Sodium; Brevimytal Sodium; Brietal Sodium. Intravenous anesthetic. Ultra-short-acting barbiturate. Soluble in H_2O. *Eli Lilly & Co.*

894 Midazolam
59467-70-8 6270
$C_{18}H_{13}ClFN_3$
8-Chloro-6-(o-fluorophenyl)-1-methyl-4H-imidazo-[1,5-a][1,4]-benzodiazepine.
Intravenous anesthetic. Benzodiazepine derivative. mp = 158-160°; λ_m = 220 nm (ε 30000 iPrOH); soluble in H_2O. *Hoffmann-LaRoche Inc.*

895 Midazolam Hydrochloride
59467-96-8 6270 261-776-6
$C_{18}H_{14}Cl_2FN_3$
8-Chloro-6-(o-fluorophenyl)-1-methyl-4H-imidazo-[1,5-a][1,4]-benzodiazepine hydrochloride.
Hypnovel; Rocam; Versed; Ro-21-3981/003. Intravenous anesthetic. *Hoffmann-LaRoche Inc.*

896 Midazolam Maleate
59467-94-6 6270 261-775-0
$C_{22}H_{17}ClFN_3O_4$
8-Chloro-6-(o-fluorophenyl)-1-methyl-4H-imidazo-[1,5-a][1,4]-benzodiazepine maleate.
Ro-21-3981/001; Dormicum; Flormidal. Intravenous anesthetic. mp = 114-117°; LD_{50} (mus orl) = 760 mg/kg, (mus iv) = 86 mg/kg. *Hoffmann-LaRoche Inc.*

897 Phencyclidine
77-10-1 7364
$C_{17}H_{25}N$
1-(1-Phenylcyclohexyl)piperidine.
HOG; PCP; CI-395; angel dust. Intravenous anesthetic. Ion channel blocker. mp = 46-46.5°; $bp_{1.0}$ = 135-137°; λ_m = 252, 257.5, 262, 268.5 ($E^{1\%}_{1\ cm}$ 7.9, 11.2, 13.0, 9.7 0.1N HCl); [hydrobromide]: mp = 214-218°. *Parke-Davis.*

898 Phencyclidine Hydrochloride
956-90-1 7364
$C_{17}H_{26}ClN$
1-(1-Phenylcyclohexyl)piperidine hydrochloride.
Sernyl; Sernylan; CI-395; CN-25,253-2; GP-121; NSC-40902. Intravenous anesthetic. Ion channel blocker. mp = 233-235°; λ_m = 254, 258, 262.5, 269 nm ($E^{1\%}_{1\ cm}$ 7.9, 10.8, 12.7, 10.0 EtOH); LD_{50} (mus orl) = 76.5 mg/kg. *Parke-Davis.*

899 Propanidid
1421-14-3 7986 215-822-7
$C_{18}H_{27}NO5$
Propyl [4-[(diethylcarbamoyl)-methoxy]-3-methoxyphenyl]acetate.
Epontol; Sombrevin; FBA 1420; Bayer 1420; WH 5668; TH-2180. Intravenous anesthetic. $bp_{0.7}$ = 210-212°; insoluble in H_2O; soluble in EtOH, $CHCl_3$; LD_{50} (rat orl) > 10000 mg/kg. *3M Pharm.; Bayer Corp. Pharm. Div.*

900 Propofol
2078-54-8 8020 218-206-6
$C_{12}H_{18}O$
2,6-Diisopropylphenol.
Diprivan; Disoprivan; Rapinovet; ICI-35868. Rapidly acting intravenous anesthetic. An alkylphenol. bp_{30} = 136°, bp_{17} = 126°; mp = 19°; d_{20} 0.955. *Ethyl Corp.; Zeneca Pharm.*

901 Sodium Oxybate
502-85-2 4860 207-953-3
$C_4H_7NaO_3$
Sodium 4-hydroxybutyrate.
Somsanit; Gamma-OH; σ-OH; sodium σ-oxybutyrate; Somatomax PM; Wy-3478; NSC-84223. Intravenous anesthetic. Used as an adjunct in anesthesia. LD_{50} (mrat ip) = 2000 mg/kg, (frat ip) = 1650 mg/kg. *Wyeth-Ayerst Labs.*

902 Thialbarbital
467-36-7 9428 207-390-3
$C_{13}H_{16}N_2O_2S$
5-Allyl-5-(2-cyclohexen-1-yl)-2-thiobarbituric acid.
thialbarbitone; Kemithal. Intravenous anesthetic. mp = 148-150°; [sodium salt ($C_{13}H_{15}N_2NaO_2S$)]: mp = 130-132°; soluble in H_2O, EtOH; insoluble in C_6H_6, Et_2O. *Abbott Labs Inc.*

903 Thiamylal
77-27-0 9437 201-018-3
$C_{12}H_{18}N_2O_2S$
5-Allyl-5-(1-methylbutyl)-2-thiobarbituric acid.
thioseconal. Intravenous anesthetic. mp = 132-133°. *Abbott Labs Inc.; Parke-Davis; Yoshitomi.*

904 Thiamylal Sodium
337-47-3 9437 206-415-5
$C_{12}H_{17}N_2NaO_2S$
Sodium 5-allyl-5-(1-methylbutyl)-2-thiobarbiturate.
Surital. Intravenous anesthetic. Ultra-short-acting barbiturate. *Abbott Labs Inc.; Parke-Davis; Yoshitomi.*

905 Thiobutabarbital
2095-57-0 9457 218-260-0
$C_{10}H_{16}N_2O_2S$
5-sec-Butyl-5-ethyl-2-thiobarbituric acid.
thibutabarbital. Intravenous anesthetic. mp = 163-165°. *Abbott Labs Inc.*

906 Thiobutabarbital Sodium Salt
947-08-0 9457 213-425-3
$C_{10}H_{15}N_2NaO_2S$
Sodium 5-sec-butyl-5-ethyl-2-thiobarbiturate.
Inactin; Brevinarcon. Intravenous anesthetic. Readily soluble in H_2O. *Abbott Labs Inc.*

907 Thiopental Sodium
71-73-8 9487 200-763-1
$C_{11}H_{17}N_2NaO_2S$
Sodium 5-ethyl-5-(1-methylbutyl)-2-thiobarbiturate.
Pentothal; thiomebumal sodium; penthiobarbital sodium; thiopentone sodium; thionembutal; Intraval Sodium; Nesdonal Sodium; Pentothal Sodium; Trapanal. Intravenous anesthetic; anticonvulsant. Ultra-short-acting barbiturate. Soluble in H_2O, EtOH; insoluble in C_6H_6, Et_2O, petroleum ether; LD_{50} (mus ip) = 149 mg/kg, (mus iv) = 78 mg/kg. *Abbott Labs Inc.*

Anesthetics, Local

908 Ambucaine
119-29-9 402
$C_{17}H_{28}N_2O_3$
2-Diethylaminoethyl 4-amino-2-butoxybenzoate.
ambutoxate. Local anesthetic. *Sterling Res. Labs.*

909 Ambucaine Hydrochloride
5870-25-7 402
$C_{17}H_{29}ClN_2O_3$
2-Diethylaminoethyl 4-amino-2-butoxybenzoate hydrochloride.
Win-3706; Sympocaine. Local anesthetic. mp = 127°; [dihydrochloride]: mp = 156.8-159° (dec). *Sterling Res. Labs.*

910 Amolanone
76-65-3 610
$C_{20}H_{23}NO_2$
3-[2-(Diethylamino)ethyl]-3-phenyl-2(3H)-benzofuranone.
AP-43; Amethone. Local anesthetic. mp = 43-44°; $bp_{2.0}$ = 192-194°. *Abbott Labs Inc.*

911 Amolanone Hydrochloride
6009-67-2 610
$C_{20}H_{24}ClNO_2$
3-[2-(Diethylamino)ethyl]-3-phenyl-2(3H)-benzofuranone hydrochloride.
Amethone hydrochloride. Local anesthetic. mp = 152-153°; soluble in H_2O. *Abbott Labs Inc.*

912 Amylocaine
644-26-8
$C_{14}H_{21}NO_2$
1-(Dimethylaminomethyl)-1-methylpropyl benzoate.
Local anesthetic.

913 Amylocaine Hydrochloride
532-59-2 656 208-541-6
$C_{14}H_{22}ClNO_2$
1-(Dimethylaminomethyl)-1-methylpropyl benzoate hydrochloride.
Stovaine; amyleine hydrochloride. Local anesthetic. Dec 177-179°; soluble in H_2O (50 g/100 ml), EtOH (33 g/100 ml); insoluble in Et_2O.

914 Aptocaine
19281-29-9 242-932-2
$C_{14}H_{20}N_2O$
2-Methyl-1-pyrrolidineaceto-o-toluidide.
Used as a local anesthetic.

915 Benoxinate
99-43-4 1076
$C_{17}H_{28}N_2O_3$
2-(Diethylamino)ethyl 4-amino-3-butoxybenzoate.
oxibuprokain; oxybuprocaine. Local anesthetic. bp_2 = 215-218°. *Pilkington Barnes Hind; Sandoz Pharm. Corp.*

916 Benoxinate Hydrochloride
5987-82-6 1076 227-808-8
$C_{17}H_{29}ClN_2O_3$
2-(Diethylamino)ethyl 4-amino-3-butoxybenzoate hydrochloride.
Dorsacaine; Conjuncain; Cebesine; Novesine; Benoxil; Lacrimin; component of: Fluress. Local anesthetic. mp ≅ 155°, 157-160°; very soluble in H_2O, $CHCl_3$; soluble in EtOH; insoluble in Et_2O. *Pilkington Barnes Hind; Sandoz Pharm. Corp.*

917 Benzocaine
94-09-7 1116 202-303-5
$C_9H_{11}NO_2$
Ethyl p-aminobenzoate.
Americaine; Baby Anbesol; Aethoform; Anesthesin; Anesthone; Orthesin; Parathesin; component of: Anbesol, Anbesol Maximum Strength, Auralgan, Cough-X, Dermoplast, Outgro, Solarcaine Aerosol, Tympagesic. Local anesthetic. mp = 88-90°; soluble in H_2O (0.04 g/100 ml), EtOH (20 g/100 ml), $CHCl_3$ (50 g/100 ml), Et_2O (25 g/100 ml). *Ascher B.F. & Co.; Fisons Pharm. Div.; Savage Labs; Schering-Plough Pharm.; Whitehall Labs. Inc.; Wyeth-Ayerst Labs.*

918 Betoxycaine
3818-62-0 1235
$C_{19}H_{32}N_2O_4$
2-[2-(Diethylamino)ethoxy]ethyl-3-amino-4-butoxybenzoate.
Local anesthetic. *Corbière.*

919 Betoxycaine Hydrochloride
5003-47-4 1235
$C_{19}H_{33}ClN_2O_4$
2-[2-(Diethylamino)ethoxy]ethyl-3-amino-4-butoxybenzoate hydrochloride.
Millicaine. Local anesthetic. mp = 117°; soluble in MeOH. *Corbière.*

920 Biphenamine
3572-52-9 1275 222-686-2
$C_{19}H_{23}NO_3$
2-(Diethylamino)ethyl-2-hydroxy-3-biphenylcarboxylate.
xenysalate. Local anesthetic. Antibacterial and antifungal. Oily liquid; soluble in H_2O. *Wallace Labs.*

921 Biphenamine Hydrochloride
5560-62-3 1275 226-930-9
$C_{19}H_{24}ClNO_3$
2-(Diethylamino)ethyl-2-hydroxy-3-biphenylcarboxylate hydrochloride.
Sebaclen; Sebaklen; component of: Alvinine Shampoo. Antibacterial; antifungal; anesthetic (topical). *Wallace Labs.*

922 Bumecaine
30103-44-7
$C_{18}H_{28}N_2O$
1-Butyl-2',4',6'-trimethyl-2-pyrrolidinecarboxanilide.
Used as a local anesthetic.

923 Bupivacaine
2180-92-9 1520 218-553-3
$C_{18}H_{28}N_2O$
(±)-1-Butyl-2',6'-pipecoloxylidide.
Anekain; Marcaine. Local anesthetic. mp = 107-108°. *Astra Chem. Ltd.; Sterling Winthrop Inc.*

924 Bupivacaine Hydrochloride
14252-80-3 1520
$C_{18}H_{29}ClN_2O.H_2O$
(±)-1-Butyl-2',6'-pipecoloxylidide hydrochloride monohydrate.
Marcaine; Marcaina; AH-2250; Carbostesin; Sensorcaine; Win-11,318; LAC-43. Local anesthetic. mp = 255-256°; soluble in H_2O (4.0 g/100 ml), EtOH (12.5 g/100 ml); slightly soluble in Me_2CO, $CHCl_3$, Et_2O; LD_{50} (mus iv) = 7.8 mg/kg, (mis sc) = 82 mg/kg. *Astra Chem. Ltd.; Sterling Winthrop Inc.*

925 Butacaine
149-16-6 1531 205-734-7
$C_{18}H_{30}N_2O_2$
3-(Dibutylamino)-1-propanol-p-aminobenzoate (ester).

Butelline. Local anesthetic. $bp_{0.11}$ = 178-182°; [hydrochloride ($C_{18}H_{31}ClN_2O_2$)]: mp = 157-158.5°. *Abbott Labs Inc.*

926 Butacaine Sulfate
149-15-5 1531 205-733-1
$C_{36}H_{62}N_4O_8$
3-(Dibutylamino)-1-propanol-p-aminobenzoate (ester) sulfate (2:1).
Butyn Sulfate. Local anesthetic. mp = 138.5-139.5°, 100-103°; soluble in H_2O (>100 g/100 ml); soluble in EtOH, Me_2CO; slightly soluble in $CHCl_3$; insoluble in Et_2O; LD_{50} (mus iv) = 12.4 mg/kg. *Abbott Labs Inc.*

927 Butamben
94-25-7 1538 202-317-1
$C_{11}H_{15}NO_2$
Butyl p-aminobenzoate.
Butsein; Butoform; Planoform; Scuroform. Local anesthetic. mp = 57-59°; bp_8 = 174°; soluble in H_2O (0.0143 g/100 ml); soluble in EtOH, Et_2O, $CHCl_3$. *Abbott Labs Inc.*

928 Butamben Picrate
577-48-0 1538
$C_{38}H_{33}N_5O_{11}$
Butyl p-aminobenzoate picrate (2:1).
Abbott-34842. Local anesthetic. *Abbott Labs Inc.*

929 Butanilicaine
3785-21-5 1542
$C_{13}H_{19}ClN_2O$
2-(Butylamino)-6'-chloro-o-acetotoluidide.
Local anesthetic. mp = 45-46°; $bp_{0.001}$ = 145°, $bp_{0.5}$ = 166-167°; [hydrochloride ($C_{13}H_{20}Cl_2N_2O$)]: mp = 232°, 236-239°; soluble in H_2O; LD_{50} (mus sc) = 700 mg/kg. *Cilag-Chemie Ltd.; Hoechst AG.*

930 Butethamine
2090-89-3 1552
$C_{13}H_{20}N_2O_2$
2-[(2-Methylpropyl)amino]ethanol 4-aminobenzoate.
Local anesthetic. [formate ($C_{14}H_{22}N_2O_4$)]: mp = 136-139°; freely soluble in H_2O, EtOH; soluble in $CHCl_3$, Et_2O; slightly soluble in C_6H_6. *Novocol Chem.*

931 Butethamine Hydrochloride
553-68-4 1552
$C_{13}H_{21}ClN_2O_2$
2-[(2-Methylpropyl)amino]ethanol
4-aminobenzoate
hydrochloride.
Ibylcaine; Monocaine. Local anesthetic. mp = 192-196°; soluble in H_2O; slightly soluble in EtOH, $CHCl_3$, C_6H_6; insoluble in Et_2O. *Novocol Chem.*

932 Butoxycaine
3772-43-8 1566
$C_{17}H_{27}NO_3$
4-Butoxybenzoic acid
2-(diethylamino)ethyl ester.
Local anesthetic. *Squibb E.R. & Sons.*

933 Butoxycaine Hydrochloride
2350-32-5 1566
$C_{17}H_{28}ClNO_3$
4-Butoxybenzoic acid
2-(diethylamino)ethyl ester
hydrochloride.
Stadacain. Local anesthetic. mp = 146°. *Squibb E.R. & Sons.*

934 Carcainium Chloride
1042-42-8
$C_{18}H_{22}ClN_3O_2$
Dimethylbis[(phenylcarbamoyl)methyl]ammonium chloride.
Has properties of a local anesthetic.

935 Carticaine
23964-58-1 1920
$C_{13}H_{20}N_2O_3S$
4-Methyl-3-[2-(propylamino)propionamido]-2-thiophenecarboxylic acid methyl ester.
40045; HOE-45; Carticaine. Local anesthetic. $bp_{0.3}$ = 162-167°. *Hoechst AG.*

936 Carticaine Hydrochloride
23964-57-0 1920 245-957-7
$C_{13}H_{21}ClN_2O_3S$
4-Methyl-3-[2-(propylamino)propionamido]-
2-thiophenecarboxylic acid methyl ester
hydrochloride.
HOE-045; HOE-40045; Ultacain. Local anesthetic. mp = 177-178°; LD_{50} (mus iv) = 37 mg/kg. *Hoechst AG.*

937 Chloroprocaine
133-16-4 2210
$C_{13}H_{19}ClN_2O_2$
2-(Diethylamino)ethyl 4-amino-2-chlorobenzoate.
Local anesthetic. *Astra USA Inc.*

938 Chloroprocaine Hydrochloride
3858-89-7 2210 223-371-2
$C_{13}H_{20}Cl_2N_2O_2$
2-(Diethylamino)ethyl 4-amino-
2-chlorobenzoate
hydrochloride.
Nesacaine. Local anesthetic. mp = 176-178°; soluble in H_2O (4.5 g/100 ml), EtOH (1 g/100 ml). *Astra USA Inc.*

939 Clodacaine
5626-25-5
$C_{16}H_{26}ClN_3O$
2'-Chloro-2-[[2-(diethylamino)ethyl]-
ethylamino]acetanilide.
Local anesthetic.

940 Cocaethylene
529-38-4 2516
$C_{18}H_{23}NO_4$
[1R-(exo,exo)]-3-(Benzoyloxy)-
8-methyl-8-azabicyclo[3.2.1]octane-
2-carboxylate acid.
ethylbenzoylecgonine; Homocaine. Local anesthetic. mp = 109°; insoluble in H_2O, soluble in EtOH, Et_2O. *Astra USA Inc.*

941 Cocaine
50-36-2 2517 200-032-7
$C_{17}H_{21}NO_4$
2-Methyl-3β-hydroxy-1αH,5αH-tropane-2β-carboxylate benzoate (ester).
benzoylethylecgonine. Local anesthetic. A federally controlled substance. mp = 98°; $bp_{0.1}$ = 187-188°; $[\alpha]_D^{18}$ = -35° (50% EtOH), $[\alpha]_D^{20}$ = -16° (c = 4 $CHCl_3$); soluble in H_2O (0.16 g/100 ml at 20°, 0.37 g/100 ml at 80°), EtOH (0.15 g/100 ml), $CHCl_3$ (1.43 g/100 ml), Et_2O (0.29 g/100 ml); soluble in Me_2CO, EtOAc, CS_2; LD_{50} (rat iv) = 17.5 mg/kg. *Astra USA Inc.*

942 Cocaine Hydrochloride
53-21-4 2517 200-167-1
$C_{17}H_{22}ClNO_4$
2-Methyl-3β-hydroxy-1αH,5αH-tropane-2β-carboxylate benzoate (ester) hydrochloride.
cocaine muriate. Local anesthetic. A federally controlled substance. mp ≅ 195°; $[\alpha]_D$ = -72° (c = 2 H_2O, pH 4.5); soluble in H_2O (250 g/100 ml), EtOH (31 g/100 ml at 20°, 50 g/100 ml at 75°)$CHCl_3$ (8 g/100 ml), glycerol; insoluble in Et_2O. *Astra USA Inc.*

943 Cyclomethycaine
139-62-8 2804
$C_{22}H_{33}NO_3$
3-(2-Methylpiperidino)propyl p-(cyclohexyloxy)benzoate.
Surfacaine; Surfathesin; Topocaine. Local anesthetic. *Eli Lilly & Co.*

944 Cyclomethycaine Hydrochloride
537-61-1 2804
$C_{22}H_{34}ClNO_3$
3-(2-Methylpiperidino)propyl p-(cyclohexyloxy)benzoate hydrochloride.
Surfacaine. Local anesthetic. Dec 178-180°; soluble in H_2O (> 1 g/100 ml). *Eli Lilly & Co.*

945 Diamocaine Cyclamate
23469-05-8
$C_{37}H_{63}N_5O_7S_2$
1-(2-Anilinoethyl)-4-[2-(diethylamino)ethoxy]-4-phenylpiperidine bis(cyclohexanesulfamate).
R-10,948. Local anesthetic. *Abbott Laboratories Inc.*

946 Dibucaine
85-79-0 3081 201-632-1
$C_{20}H_{29}N_3O_2$
2-Butoxy-n-[2-(diethylamino)ethyl]-cinchoninamide.
Local anesthetic. mp = 64°. *Ciba-Geigy Corp.*

947 Dibucaine Hydrochloride
61-12-1 3081 200-498-1
$C_{20}H_{30}ClN_3O_2$
2-Butoxy-N-[2-(diethylamino)ethyl]-cinchoninamide hydrochloride.
Nupercaine hydrochloride; Nupercainal; benzolin; Percaine; Cincaine; Sovcaine; Cinchocaine. Local anesthetic. Dec 90-98°; λ_m = 247, 320 nm (ε 24700, 8810 1N HCl); soluble in H_2O (200 g/100 ml), EtOH, Me_2CO, $CHCl_3$; slightly soluble in C_6H_6, C_7H_8, EtOAc; insoluble in Et_2O. *Ciba-Geigy Corp.*

948 Dimethisoquin
86-80-6 3267 201-700-0
$C_{17}H_{24}N_2O$
3-Butyl-1-[2-(dimethylamino)ethoxy]isoquinoline.
quinisocaine. Local anesthetic. bp_3 = 155-157°; n_D^{20} = 1.5486. *SmithKline Beecham Pharm.*

949 Dimethisoquin Hydrochloride
2773-92-4 3267 220-468-1
$C_{17}H_{25}ClN_2O$
3-Butyl-1-[2-(dimethylamino)ethoxy]isoquinoline hydrochloride.
Isochinol; Pruralgan; Pruralgin; Quotane. Local anesthetic. Soluble in H_2O, EtOH, $CHCl_3$; insoluble in Et_2O; LD_{50} (rat ip) = 45-50 mg/kg. *SmithKline Beecham Pharm.*

950 Dimethocaine
94-15-5 3270
$C_{16}H_{26}N_2O_2$
3-(Dimethylamino)-2,2-dimethyl-1-propanol p-aminobenzoate.
Larocaine. Local anesthetic. [hydrochloride ($C_{16}H_{27}ClN_2O_2$)]: mp = 196-197°; soluble in H_2O (33 g/100 ml), EtOH (10 g/100 ml at 25°, 20 g/100 ml at 75°); insoluble in Et_2O; LD_{50} (mus sc) = 300 mg/kg.

951 Diperodon
51552-99-9
$C_{22}H_{27}N_3O_4$
3-Piperidino-1,2-propanediol dicarbanilate (ester) monohydrate.
Local anesthetic.

952 Diperodon Hydrochloride
537-12-2 3360 208-659-8
$C_{22}H_{28}ClN_3O_4$
3-Piperidino-1,2-propanediol dicarbanilate (ester) hydrochloride.
Diothane Hydrochloride; Proctodon. Local anesthetic. Dec 195-200°; slightly soluble in H_2O (< 1 g/100 ml); soluble in EtOH, Me_2CO, EtOAc; insoluble in C_6H_6, Et_2O. *Marion Merrell Dow Inc.*

953 Dyclonine
586-60-7 3523
$C_{18}H_{27}NO_2$
4'-Butoxy-3-piperidinopropiophenone.
Local anesthetic. *Astra Chem. Ltd.*

954 Dyclonine Hydrochloride
536-43-6 3523 208-633-6
$C_{18}H_{28}ClNO_2$
4'-Butoxy-3-piperidinopropiophenone hydrochloride.
Dyclone; Tanaclone. Local anesthetic. mp = 175-176°; soluble in H_2O, EtOH, Me_2CO. *Astra Chem. Ltd.*

955 Ecgonidine
484-93-5 3540 207-610-8
$C_9H_{13}NO_2$
(1R)-8-Methyl-8-azabicyclo[3.2.1]oct-2-ene-2-carboxylic acid.

anhydroecgonine. Local anesthetic. [dl-form]: dec 235-236°; soluble in H_2O, sparingly soluble in EtOH; [l-form]: mp = 235°; $[\alpha]_D^{14}$ = -84.6° (c = 1.7). *Olin Res. Ctr.*

956 Ecgonine
481-37-8 3541 207-565-4
$C_9H_{15}NO_3$
[1R*-(exo,exo)]-3-Hydroxy-8-methyl-8-azabicyclo[3.2.1]octane-2-carboxylic acid.
Local anesthetic. [dl-form trihydrate]: mp = 93-118°.

957 Ecgonine Hydrochloride
5796-31-6 3541
$C_9H_{16}ClNO_3$
[1R-(exo,exo)]-3-Hydroxy-8-methyl-8-azabicyclo[3.2.1]octane-2-carboxylic acid hydrochloride.
Local anesthetic. mp = 246°; $[\alpha]_D^{15}$ = -59° (c = 10), soluble in H_2O, slightly soluble in EtOH.

958 l-Ecgonine Monohydrate
5796-30-5 3541
$C_9H_{15}NO_3$
l-[1R-(exo,exo)]-3-Hydroxy-8-methyl-8-azabicyclo[3.2.1]octane-2-carboxylic acid.
Local anesthetic. mp = 198°; $[\alpha]_D^{15}$ = -45° (c = 5); soluble in H_2O (20 g/100 ml), EtOH (1.5 g/100 ml), MeOH (5 g/100 ml); EtOAc (1.3 g/100 ml); sparingly soluble in Me_2CO, Et_2O, C_6H_6, petroleum ether.

959 Embutramide
15687-14-6 239-780-4
$C_{17}H_{27}NO_3$
N-(β,β-Dimethyl-m-methoxyphenethyl)-4-hydroxybutyramide.
Embutane; T-61; HOE-18 680. Local anesthetic. *Hoechst Marion Roussel Inc.*

960 Ethyl Chloride
75-00-3 3829 200-830-5
C_2H_5Cl
Chloroethane.
monochlorethane; chlorethyl; aethylis chloridum; ether chloratus; ether hydrochloric; ether muriatic; Kelene; Chelen; Anodynon; Narcotile; Chloryle Anesthetic. Local anesthetic. mp = -138.7°; bp = 12.3°; d_4^0 = 0.9214; soluble in H_2O (0.574 g/100 ml), EtOH (48.3 g/100 ml); soluble in Et_2O.

961 Etidocaine
36637-18-0 3907 253-143-8
$C_{17}H_{28}N_2O$
(±)-2-(Ethylpropylamino-2',6'-butyroxylidide.
Local anesthetic. *Astra Chem. Ltd.*

962 Etidocaine Hydrochloride
36637-19-1 3907 253-144-3
$C_{17}H_{29}ClN_2O$
(±)-2-(Ethylpropylamino-2',6'-butyroxylidide hydrochloride.
Duranest; W-19053. Local anesthetic. mp = 203-203.5°; LD_{50} (fmus iv) = 6.7 mg/kg, (fmus sc) = 99 mg/kg. *Astra Chem. Ltd.*

963 β-Eucaine
500-34-5 3939
$C_{15}H_{21}NO_2$
2,2,6-Trimethyl-4-piperidinol benzoate (ester).
eucaine-B; betacaine. Local anesthetic. mp = 70-71°; [hydrochloride ($C_{15}H_{22}ClNO_2$)]: mp = 277-279°; soluble in H_2O (3.3 g/100 ml), EtOH (2.86 g/100 ml), $CHCl_3$ (16.67 g/100 ml); MLD (rat iv) = 15-25 mg/kg; [d-form]: mp = 57-58°; [d-form hydrochloride]: $[\alpha]_D$ = 11.5° (H_2O); [l-form]: mp = 57-58°; [l-form hydrochloride]: mp = 244-245°; $[\alpha]_D$ = -11.3° (H_2O).

964 Euprocin
1301-42-4 3950
$C_{24}H_{34}N_2O_2$
(8α,9R)-10,11-Dihydro-6'-(3-methylbutoxy)-cinchonan-9-ol.
hydrocupreine isopentyl ether; isoamylhydrocupreine; isopentylhydrocupreine; $O^{6'}$-isopentylhydrocupreine; eucupreine; Eucupin. Antiseptic; local anesthetic. A quinoline. mp = 152°; soluble in EtOH, Et_2O, $CHCl_3$; insoluble in H_2O; LD_{50} (mus sc) = 300 mg/kg, (rbt iv) = 13 mg/kg. *Schering-Plough HealthCare Products.*

965 Euprocin Hydrochloride
18984-80-0 3950
$C_{24}H_{38}Cl_2N_2O_3$
(8α,9R)-10,11-Dihydro-6'-(3-methylbutoxy)cinchonan-9-ol dihydrochloride monohydrate.
isoamylhydrocupreine hydrochloride; isopentylhydrocupreine dihydrochloride; WI-287; component of: Otodyne. Antiseptic; local anesthetic. A quinoline. Freely soluble in EtOH, soluble in H_2O (6.6 g/100 ml). *Schering AG.*

966 Febuverine
7077-33-0
$C_{28}H_{38}N_2O_4$
1,4-Piperazinediethanol di(2-phenylbutyrate ester).
Local anesthetic.

967 Fenalcomine
34616-39-2 3996
$C_{20}H_{27}NO_2$
α-Ethyl-p-[2-[(α-methylphenethyl)amino]ethoxy]benzyl alcohol.
Cardiotonic. Local anesthetic. *LaRoche-Navarron.*

968 Fenalcomine Hydrochloride
34535-83-6 3996 252-075-6
$C_{20}H_{28}ClNO_2$
α-Ethyl-p-[2-[(α-methylphenethyl)amino]ethoxy]benzyl alcohol hydrochloride.
Cordoxene. Cardiotonic. Local anesthetic. *LaRoche-Navarron.*

969 Fomocaine
17692-39-6 4259
$C_{20}H_{25}NO_2$
4-[3-(α-Phenoxy-p-tolyl)propyl]morpholine.
P-652; Erbocain. Local anesthetic. mp = 52-53°; $bp_{1.1}$ = 238-240°; λ_m = 220, 269 nm (ε 15820, 1373 EtOH); LD_{50} (mus iv) = 175 mg/kg. *Promonta.*

970 Hexylcaine
532-77-4
$C_{16}H_{23}NO_2$
1-(Cyclohexylamino)-2-propanol benzoate (ester).
Local anesthetic. *Merck & Co.Inc.*

971 Hexylcaine Hydrochloride
532-76-3 4746 208-544-2
$C_{16}H_{24}ClNO_2$
1-(Cyclohexylamino)-2-propanol benzoate (ester) hydrochloride.
D-109; Cyclaine. Local anesthetic. mp = 177-178.5°; soluble in H_2O (< 12 g/100 ml). *Merck & Co.Inc.*

972 Hydroxytetracaine
490-98-2 4894
$C_{15}H_{24}N_2O_3$
2-Dimethylaminoethyl-4-butylaminosalicylate.
Rhenocain; Salicain. Local anesthetic. [hydrochloride ($C_{15}H_{25}ClN_2O_3$)]: mp = 157°; soluble in H_2O (≅ 4 g/100 ml at 20°). *Rhinepreussen AG.*

973 Isobutamben
94-14-4 202-308-2
$C_{11}H_{15}NO_2$
Isobutyl p-aminobenzoate.
Local anesthetic.

974 Leucinocaine Mesylate
135-44-4 5476 205-191-6
$C_{18}H_{32}N_2O_5S$
2-(Diethylamino)-4-methyl-1-pentanol p-aminobenzoate (ester).
Panthesin. Local anesthetic. mp = 171°; soluble in H_2O (33 g/100 ml), EtOH (33 g/100 ml); insoluble in Et_2O, $CHCl_3$, EtOAc, Me_2CO, C_6H_6; MLD (rbt iv) = 20 mg/kg, (rbt isp) = 10.8 mg/kg.

975 Levobupivacaine
27262-47-1
$C_{18}H_{28}N_2O$
(-)-1-Butyl-2',6'-pipecoloxylidide.
(-)-bupivacaine; ropivacaine; Naropin (hydrochloride). Anesthetic (local). S-(-)-enantiomer of bupivacaine. Used in obstetric anesthesia. *Astra USA Inc.*

976 Levoxadrol
4792-18-1 3352
$C_{20}H_{23}NO_2$
(-)-2-(2,2-Diphenyl-1,3-dioxolan-4-yl)piperidine.
Smooth muscle relaxant; local anesthetic. *Cutter Labs.*

977 Levoxadrol Hydrochloride
23257-58-1 3352
$C_{20}H_{24}ClNO_2$
(-)-2-(2,2-Diphenyl-1,3-dioxolan-4-yl)piperidine hydrochloride.
CL-912C; NSC-526063; Levoxan. Local anesthetic. mp = 248-254°; $[\alpha]_D^{20}$ = -34.5° (c = 2 MeOH); LD_{50} (mus orl) = 230 mg/kg.

978 Levoxadrol Hydrochloride
23257-58-1 3352
$C_{20}H_{24}ClNO_2$
(-)-2-(2,2-Diphenyl-1,3-dioxolan-4-yl)piperidine hydrochloride.
CL-912C; NSC-526063; Levoxan. Smooth muscle relaxant; local anesthetic. mp = 248-254°; $[\alpha]_D^{20}$ = -34.5° (c = 2 MeOH); LD_{50} (mus orl) = 230 mg/kg. *Cutter Labs.*

979 Lidocaine
137-58-6 5505 205-302-8
$C_{14}H_{22}N_2O$
2-(Diethylamino)-2',6'-acetoxylidide.
Cuivasil; Dalcaine; Duncaine; Leostesin; Lida-Mantle; Lidothesin; Lignocaine; Rucaina; Solarcaine; Xylocaine; Xylocitin; Xylotox; component of: Cracked Heel Relief Cream, Emla Cream, Lidaform HC, Lidamantle HC, Neosporin Plus. Local anesthetic; antiarrhythmic (class IB). Long-acting membrane stabilizing agent. mp = 68-69°; bp_4 = 180-182°, bp_2 = 159-160°; soluble in EtOH, Et_2O, C_6H_6, $CHCl_3$; insoluble in H_2O. *Astra USA Inc.; Bayer Corp.; Glaxo Wellcome Inc.; Schering-Plough Animal Health; Schering-Plough Pharm.*

980 Lidocaine Hydrochloride
6108-05-0 5505 200-803-8
$C_{14}H_{23}ClN_2O.H_2O$
2-(Diethylamino)-2',6'-acetoxylidide hydrochloride monohydrate.
Dalcaine; Lidesthesin; Lignavet; Odontalg; Sedagul; Xylocaine; Xylocard; Xyloneural; component of: Solarcaine. Class IB antiarrhythmic. Local anesthetic. Long-acting membrane stabilizing agent. mp = 77-78°; [anhydrous form]: mp = 127-129°; very soluble in H_2O, EtOH; less soluble in $CHCl_3$; insoluble in Et_2O; LD_{50} (mus orl) = 292 mg/kg, (mus ip) = 105 mg/kg, (mus iv) = 19.5 mg/kg. *Astra Chem. Ltd.; Astra Pharm. Ltd.; Bayer Corp.; Bristol Myers Squibb Pharm. Ltd.; Carrington Labs Inc.; Elkins-Sinn; Forest Pharm. Inc.; Glaxo Wellcome plc; Schering-Plough HealthCare Products; Sterling Health U.S.A.*

981 Lotucaine
52304-85-5
$C_{18}H_{29}NO_2$
2,2,5,5-Tetramethyl-α-[(o-tolyloxy)methyl]-1-pyrrolidineethanol.
MY-33-7 [as hydrochloride]. Local anesthetic.

982 Mepivacaine
96-88-8 5905 202-543-0
$C_{15}H_{22}N_2O$
1-Methyl-2',6'-pipecoloxylidide.
Local anesthetic. mp = 150-151°. *Astra Chem. Ltd.; Sterling Winthrop Inc.*

983 Mepivacaine Hydrochloride
1722-62-9 5905 217-023-9
$C_{15}H_{23}ClN_2O$
1-Methyl-2',6'-pipecoloxylidide hydrochloride.
Carbocaine; Polocaine; Carbocaina; Chlorocain; Meaverin; Mepicaton; Mepident; Mepivastesin; Optocain; Scandicain. Local anesthetic. mp = 262-264°; soluble in H_2O; LD_{50} (mus sc) = 280 mg/kg, (rat sc) = 500 mg/kg. *Astra Chem. Ltd.; Sterling Winthrop Inc.*

984 Meprylcaine
495-70-5 5909
$C_{14}H_{21}NO_2$
2-Methyl-2-(propylamino)-1-propanol benzoate (ester).
Oracaine. Local anesthetic. mp = 205°; $[\alpha]_D$ = 200°. *Mizzy.*

985 Meprylcaine Hydrochloride
956-03-6 5909 213-475-6
$C_{14}H_{22}ClNO_2$
2-Methyl-2-(propylamino)-1-propanol benzoate (ester) hydrochloride.
Local anesthetic. mp = 150-151°; soluble in H_2O, EtOH. *Mizzy.*

986 Metabutoxycaine
3624-87-1 5978
$C_{17}H_{28}N_2O_3$
3-Amino-2-butoxybenzoic acid 2-(diethylamino)ethyl ester.
Local anesthetic. *Novocol Chem.*

987 Metabutoxycaine Hydrochloride
550-01-6 5978
$C_{17}H_{29}ClN_2O_3$
3-Amino-2-butoxybenzoic acid 2-(diethylamino)ethyl ester hydrochloride.
Primacaine. Local anesthetic. mp = 117-119°; λ_m = 313 nm ($A^{1\%}_{1\ cm}$ 6.70 H_2O); soluble in H_2O, EtOH; moderately soluble in Me_2CO, $CHCl_3$. *Novocol Chem.*

988 Methyl Chloride
74-87-3 6121 200-817-4
CH_3Cl
Chloromethane.
Local anesthetic. mp = -97°; bp = -23.7°; soluble in H_2O (303 ml/100 ml), C_6H_6 (4723 ml/100 ml), CCl_4 (3756 ml/100 ml), EtOH (3740 ml/100 ml); LC (mus ihl) = 3146 ppm.

989 Myrtecaine
7712-50-7 6422 231-735-7
$C_{17}H_{31}NO$
2-[2-(6,6-Dimethyl-2-norpinen-2-yl)ethoxy]triethylamine.
Nopoxamine. Local anesthetic. bp_{2-3} = 135-140°.

990 Naepaine
2188-67-2 6435
$C_{14}H_{22}N_2O_2$
2-(Pentylamino)ethanol p-aminobenzoate (ester).
Amylsine. Local anesthetic. mp = 66°; [hydrochloride ($C_{14}H_{24}Cl_2N_2O_2$)]: mp = 153.5°, 176°; soluble in H_2O; sparingly soluble in EtOH; insoluble in Et_2O, $CHCl_3$, C_6H_6.

991 Octacaine
13912-77-1 6839
$C_{14}H_{22}N_2O$
3-Diethylaminobutyranilide.
Local anesthetic. mp = 46-47°; bp_1 = 200°; soluble in HCl, Et_2O, EtOH, C_6H_6. *E. Geistlich Sohne.*

992 Octacaine Hydrochloride
59727-70-7 6839
$C_{14}H_{23}ClN_2O$
3-Diethylaminobutyranilide hydrochloride.
Amplicain. Local anesthetic. mp = 132-134°; soluble in H_2O. *E. Geistlich Sohne.*

993 Octodrine
543-82-8 6854 208-851-1
$C_8H_{19}N$
1,5-Dimethylhexylamine.
SK&F-51; Vaporpac. Used as a nasal decongestant, local anesthetic. An α-adrenergic (vasoconstrictor) agonist. Viscous liquid; [dl-form]: bp = 154-156°; [hydrochloride]: soluble in H_2O; LD_{50} (mus ip) = 59 mg/kg, (rat ip) = 41.5 mg/kg. *SmithKline Beecham Pharm.*

994 Orthocaine
536-25-4 7011 208-627-3
$C_8H_9NO_3$
Methyl 3-amino-4-hydroxybenzoate.
Orthoform; Orthoform New. Local anesthetic. mp = 111°, 143°; insoluble in cold H_2O; moderately soluble in hot H_2O; soluble in EtOH (16.67 g/100 ml), Et_2O (2 g/100 ml).

995 Oxethazaine
126-27-2 7067 204-780-5
$C_{28}H_{41}N_3O_3$
2,2'-[(2-Hydroxyethyl)imino]bis[N-(α,α-dimethylphenethyl)-N-methylacetamide].
Mucaine; Muthesa; Oxaine M; Tepilta; Wy-806; Storocain; Topicain; component of: Betalgil. Local anesthetic. Used in combination with aluminum and magnesium hydroxides. mp = 104-104.5°; insoluble in H_2O. *Lederle Labs.*

996 Oxethazaine Hydrochloride
13930-31-9 7067 237-698-3
$C_{28}H_{42}ClN_3O_3$
2,2'-[(2-Hydroxyethyl)imino]bis[N-(α,α-dimethylphenethyl)-N-methylacetamide] hydrochloride.
Emoren. Local anesthetic. Used in combination with aluminum and magnesium hydroxides. mp = 146-147°; soluble in H_2O; LD_{50} (mus orl) = 370.5 mg/gk, (mus im) = 229.3 mg/kg, (mus iv) = 3.34 mg/kg, (rat orl) = 580.6 mg/kg, (mus im) = 465.9 mg/kg, (mus iv) = 1.2 mg/kg, (rbt iv) = 0.50 mg/kg. *Lederle Labs.*

997 Parethoxycaine
94-23-5 7171
$C_{15}H_{23}NO_3$
4-Ethoxybenzoic acid 2-(diethylamino)ethyl ester.
diethoxin. Local anesthetic. *Squibb E.R. & Sons.*

998 Parethoxycaine Hydrochloride
136-46-9 7171 205-246-4
$C_{15}H_{24}ClNO_3$
4-Ethoxybenzoic acid 2-(diethylamino)ethyl ester hydrochloride.
Maxicaine. Local anesthetic. mp = 172.5-173.5°. *Squibb E.R. & Sons.*

999 Phenacaine
101-93-9 7342
$C_{18}H_{22}N_2O_2$
N,N'-Bis(p-ethoxyphenyl)acetamidine.
Holocaine. Local anesthetic (topical, opthalmic). Used in veterinary medicine as an ocular anesthetic. *Abbott Labs Inc.*

1000 Phenacaine Hydrochloride
620-99-5 7342 210-662-4
$C_{18}H_{23}ClN_2O_2$
N,N'-Bis(p-ethoxyphenyl)acetamidine hydrochloride.
Local anesthetic. *Abbott Labs Inc.*

1001 Phenacaine Hydrochloride Monohydrate
6153-19-1 7342
$C_{18}H_{23}ClN_2O_2.H_2O$
N,N'-Bis(p-ethoxyphenyl)acetamidine hydrochloride monohydrate.
Holocaine Hydrochloride. Local anesthetic. mp = 190-192°; soluble in H_2O (2 g/100 ml); freely soluble in boiling H_2O, EtOH, $CHCl_3$; insoluble in Et_2O. *Abbott Labs Inc.*

1002 Phenol
108-95-2 7390 203-632-7
C_6H_6O
Carbolic acid.
phenic acid; phenylic acid; phenyl hydroxide; hydroxybenzene; oxybenzene; component of: Anbesol, Campho-Phenique Cold Sore Gel, Campho-Phenique Gel, Campho-Phenique Liquid. Local anesthetic. Antipruritic. CAUTION: Systemic toxicity can arise from cutaneous absorption, ingestion, or inhalation. mp = 40.85°; bp = 812°; d = 1.071; soluble in H_2O (6.6 g/100 ml), C_6H_6 (8.3 g/100 ml); very soluble in EtOH, $CHCl_3$, Et_2O, CS_2; insoluble in petroleum ether; LD_{50} (rat orl) = 530 mg/kg. *Robins, A. H. Co.,; Sterling Health U.S.A.; Whitehall Labs. Inc.*

1003 Pinolcaine
28240-18-8
$C_{23}H_{29}NO_2$
D-(+)-1-Methyl-1-(1-methyl-2-piperidyl)ethyl diphenylacetate.
Local anesthetic.

1004 Piperocaine
136-82-3 7627 205-262-1
$C_{16}H_{23}NO_2$
3-(2-Methylpiperidino)propyl benzoate.
Neothesin; Metycaine. Local anesthetic. *Eli Lilly & Co.*

1005 Piperocaine Hydrochloride
533-28-8 7627
$C_{16}H_{24}ClNO_2$
3-(2-Methylpiperidino)propyl benzoate hydrochloride.
Metycaine hydrochloride. Local anesthetic. mp = 172-175°; soluble in H_2O (100 g/100 ml); soluble in EtOH, $CHCl_3$; insoluble in Et_2O; LD_{50} (rat sc) = 1300 mg/kg, (rat iv) = 20 mg/kg. *Eli Lilly & Co.*

1006 Piridocaine
87-21-8 7649
$C_{14}H_{20}N_2O_2$
2-Piperidineethanol 2-aminobenzoate (ester).
PD-14; Lucaine. Local anesthetic. [hydrochloride ($C_{14}H_{21}ClN_2O_2$)]: mp = 209-211°. *Maltbie Chem.*

1007 Polidocanol
3055-99-0 7717 221-284-4
$C_{12}H_{35}(OCH_2CH_2)_nOH$ (average polymer, n = 9)
α-Dodecyl-ω-hydroxypoly(oxy-1,2-ethanediyl).
polyethylene glycol (9) monodecyl ether; dodecyl alcohol polyoxyethylene ether; hydroxypolyethoxydodecane; laureth 9; polyoxyethylene lauryl ether; polyethylene glycol monododecyl ether; Aethoxysklerol; Aetoxisclerol; Atlas G-4829; Hetoxol L-9; Lipal 9LA; Thesit. Sclerosing agent; antipruritic; anesthetic (topical). Also used as a solvent, non-ionic emulsifier, pharmaceutic aid (surfactant), spermatacide. Soluble in H_2O, EtOH, C_7H_8; miscible with hot mineral, natural and synthetic oils; miscible with fats and fatty alcohols; LD_{50} (mus orl) = 1170 mg/kg, (mus iv) 125 mg/kg. *Kreussler Chemische-Fabrik.*

1008 Pramoxine
140-65-8 7888 205-425-7
$C_{17}H_{27}NO_3$
4-[3-(p-Butoxyphenoxy)propyl]morpholine.
pramocaine; proxazocain. Local anesthetic. bp_6 = 196°; $bp_{2.8}$ = 183-184°.

1009 Pramoxine Hydrochloride
637-58-1 7888 211-293-1
$C_{17}H_{28}ClNO_3$
4-[3-(p-Butoxyphenoxy)propyl]morpholine hydrochloride.
Proctofoam; Tronolaine; Tronothane; component of: Epifoam, Itch-X, Proctocream-HC, ProctoFoam-HC, ProctoFoam-NS. Local anesthetic. mp = 181-183°; soluble in H_2O; LD_{50} (mus iv) = 79.5 mg/kg. *Abbott Labs Inc.; Ross Products; Schwarz Pharma Kremers Urban Co.*

1010 Pribecaine
55837-22-4
$C_{16}H_{23}NO_3$
3-Piperidinopropyl m-anistate.
Local anesthetic.

1011 Prilocaine
721-50-6 7924 211-957-0
$C_{13}H_{20}N_2O$
2-(Propylamino)-o-propionotoluidide.
propitocaine. Local anesthetic. mp = 37-38°; $bp_{0.1}$ = 159-162°. *Astra USA Inc.*

1012 Prilocaine Hydrochloride
1786-81-8 7924 217-244-0
$C_{13}H_{21}ClN_2O$
2-(Propylamino)-o-propionotoluidide hydrochloride.
Citanest Plain; Xylonest; L-67; component of: Citanest Forte. Local anesthetic. mp = 167-168°; readily soluble in H_2O. *Astra USA Inc.*

1013 Procaine
59-46-1 7937 200-426-9
$C_{13}H_{20}N_2O_2$
2-(Diethylamino)ethyl p-aminobenzoate.
Local anesthetic. mp = 61°; soluble in H_2O (0.5 g/100 ml); soluble in EtOH, Et_2O, C_6H_6, $CHCl_3$; LD_{50} (mus ip) = 195 mg/kg, (mus iv) = 45 mg/kg.

1014 Procaine Butyrate
136-55-0 7937
$C_{17}H_{28}N_2O_4$
2-(Diethylamino)ethyl p-aminobenzoate butyrate.
Probutylin. Local anesthetic. Soluble in H_2O, EtOH, oils.

1015 Procaine Dihydrate
6192-89-8 7937
$C_{13}H_{21}ClN_2O_2.2H_2O$
2-(Diethylamino)ethyl p-aminobenzoate hydrochloride dihydrate.
Local anesthetic. mp = 51°.

1016 Procaine Hydrochloride
51-05-8 7937 200-077-2
$C_{13}H_{21}ClN_2O_2$
2-(Diethylamino)ethyl p-aminobenzoate hydrochloride.
Novocain; Anestil; Enrpo; Gero; Jenacaine; Medaject; Naucaine; Neocaine; Omnicain; Planocaine; Rocain; Syntocain; component of: Flo-Cillin. Local anesthetic. mp = 153-156°; soluble in H_2O (100 g/100 ml), EtOH (3.33 g/100 ml); slightly soluble in $CHCl_3$; insoluble in Et_2O; LD_{50} (mus sc) = 660 ± 60 mg/kg. *Bristol-Myers Squibb Co; Sterling Winthrop Inc.*

1017 Procaine Nitrate
6192-92-3 7937
$C_{13}H_{21}N_3O_5$
2-(Diethylamino)ethyl p-aminobenzoate nitrate.
Local anesthetic. mp = 100-102°; soluble in H_2O, EtOH.

1018 Propanocaine
493-76-5 7988 207-778-2
$C_{20}H_{25}NO_2$
α-(2-Diethylaminoethyl)benzyl benzoate.
Local anesthetic. Yellow oil.

1019 Proparacaine
499-67-2 7991 207-884-9
$C_{16}H_{26}N_2O_3$
2-(Diethylamino)ethyl 3-amino-4-propoxy benzoate.
proxymetacaine. Local anesthetic. Used in ophthalmology.

1020 Proparacaine Hydrochloride
5875-06-9 7991 227-541-7
$C_{16}H_{27}ClN_2O_3$
2-(Diethylamino)ethyl 3-amino-4-propoxy benzoate hydrochloride.
Alcaine; Ophthaine; Ophthetic; Ak-Taine. Local anesthetic. Used in ophthalmology. mp = 182-183.3°; λ_m = 225, 270, 300 nm (MeOH); soluble in H_2O, EtOH, MeOH; insoluble in Et_2O, C_6H_6. *Alcon Labs; Allergan Inc.; Apothecon; Solvay Pharm. Inc.*

1021 Propipocaine
3670-68-6 8016
$C_{17}H_{25}NO_2$
3-Piperidino-4'-propoxypropiophenone.
Local anesthetic.

1022 Propipocaine Hydrochloride
1155-49-3 8016
$C_{17}H_{26}ClNO_2$
3-Piperidino-4'-propoxypropiophenone hydrochloride.
falicain; Urocomb. Local anesthetic. mp = 166°; soluble in H_2O, EtOH, Me_2CO.

1023 Propoxycaine
86-43-1 8023 201-670-9
$C_{16}H_{26}N_2O_3$
2-(Diethylamino)ethyl 4-amino-2-propoxybenzoate.
Local anesthetic. *Cook-Waite Labs. Inc.; Sterling Res. Labs.*

1024 Propoxycaine Hydrochloride
550-83-4 8023 208-988-7
$C_{16}H_{27}ClN_2O_3$
2-(Diethylamino)ethyl 4-amino-2-propoxybenzoate hydrochloride.
Blockain hydrochloride; Ravocaine hydrochloride. Local anesthetic. mp = 148-150°; freely soluble in H_2O; soluble in EtOH, $CHCl_3$; sparingly soluble in Et_2O; insoluble in Me_2CO, $CHCl_3$. *Cook-Waite Labs. Inc.; Sterling Res. Labs.*

1025 Pseudococaine
478-73-9 8097
$C_{17}H_{21}NO_4$
[1R-(2-endo-3-exo)]-3-(Benzolyloxy)-8-methyl-8-azabicyclo[3.2.1]octane-2-carboxylic acid methyl ester.
depsococaine; dextrocaine; isococaine; Delcaine. Local anesthetic. mp = 47°; $[\alpha]_D^{20}$ = 42° (c = 5 $CHCl_3$); slightly soluble in H_2O; freely soluble in Et_2O, C_6H_6, $CHCl_3$, petroleum ether.

1026 Pseudococaine Hydrochloride
6363-57-1 8097
$C_{17}H_{22}ClNO_4$
[1R-(2-endo-3-exo)]-3-(Benzolyloxy)-8-methyl-8-azabicyclo[3.2.1]octane-2-carboxylic acid methyl ester hydrochloride.
Local anesthetic. mp = 210°; $[\alpha]_D^{20}$ = 41° (c = 5); soluble in H_2O.

1027 Pseudococaine n-Propyl Ester Analog
55608-72-5 8097
$C_{19}H_{25}NO_4$
[1R-(2-endo-3-exo)]-3-(Benzolyloxy)-8-methyl-8-azabicyclo[3.2.1]octane-2-carboxylic acid n-propyl ester.
Neopsicaine. Local anesthetic.

1028 Pseudococaine Tartrate
1176-03-0 8097 214-643-1
$C_{21}H_{27}NO_{10}$
[1R-(2-endo-3-exo)]-3-(Benzolyloxy)-8-methyl-8-azabicyclo[3.2.1]octane-2-carboxylic acid tartrate.
Psicaine. Local anesthetic. mp = 139°; $[\alpha]_D^{20}$ = 43° (c = 5 H_2O); soluble in H_2O (25 g/100 ml), EtOH.

1029 Pyrrocaine
2210-77-7 8197
$C_{14}H_{20}N_2O$
1-Pyrrolidinoaceto-2',6'-xylidide.
EN-1010; NSC-52644; Endocaine; Dynacaine. Local anesthetic. mp = 83°. *Endo Pharm. Inc.*

1030 Pyrrocaine Hydrochloride
2210-64-2 8197 218-642-7
$C_{14}H_{21}ClN_2O$
1-Pyrrolidinoaceto-2',6'-xylidide hydrochloride.
Local anesthetic. mp = 205°; soluble in H_2O, EtOH, iPrOH; insoluble in $CHCl_3$, Et_2O. *Endo Pharm. Inc.*

1031 Quatacaine
17692-45-4
$C_{14}H_{22}N_2O$
2-Methyl-2-(propylamino)-o-propionotoluidide.
LA-012 [as hydrochloride]. Local anesthetic.

1032 Risocaine
94-12-2 202-306-1
$C_{10}H_{13}NO_2$
Propyl p-aminobenzoate.
NSC-23516. Local anesthetic.

1033 Rodocaine
38821-80-6
$C_{18}H_{25}ClN_2O$
trans-6'-Chloro-2,3,4,4a,5,6,7,7a-octahydro-1H-1-pyridine-1-propiono-o-toluidide.
R 19,317; R 22,700 (hydrochloride). Local anesthetic. *Abbott Laboratories Inc.*

1034 Ropivacaine
84057-95-4 8417
$C_{17}H_{26}N_2O$
(-)-1-Propyl-2',6'-pipecoloxylidide.
LEA-103. Local anesthetic. mp = 144-146°; $[\alpha]_D^{25}$ = -82.0° (c = 2 MeOH). *Apothekernes.*

1035 Ropivacaine Hydrochloride
98717-15-8 8417
$C_{17}H_{27}ClN_2O$
(-)-1-Propyl-2',6'-pipecoloxylidide hydrochloride.
Local anesthetic. mp = 260-262°; $[\alpha]_D^{25}$ = -6.6° (c = 2 H_2O). *Apothekernes.*

1036 Ropivacaine Hydrochloride Monohydrate
132112-35-7 8417
$C_{17}H_{27}ClN_2O.H_2O$
(-)-1-Propyl-2',6'-pipecoloxylidide hydrochloride monohydrate.
Local anesthetic. mp = 269.5-270.6°; $[\alpha]_D^{20}$ = -7.28° (c = 2 H_2O). *Apothekernes.*

1037 Salicyl Alcohol
90-01-7 8477 201-960-5
$C_7H_8O_2$
o-Hydroxybenzyl alcohol.
Saligenin, Saligenol; Salicain. Local anesthetic. mp = 86-87°; d = 1.16; soluble in H_2O 6.6 g/100 ml); very soluble in EtOH, Et_2O; soluble in C_6H_6.

1038 Sameridine
143257-97-0
$C_{21}H_{34}N_2O$
N-Ethyl-1-hexyl-N-methyl-4-phenylisonipecotamide.
Local anesthetic with opioid properties.

1039 Tetracaine
94-24-6 9330 202-316-6
$C_{15}H_{24}N_2O_2$
2-(Dimethylamino)ethyl p-(butylamino)benzoate.
Medihaler-Tetracaine; Metraspray; Pontocaine. Local anesthetic. *3M Pharm.; Abbott Labs Inc.; Sterling Winthrop Inc.*

1040 Tetracaine Hydrochloride
136-47-0 9330 205-248-5
$C_{15}H_{25}ClN_2O_2$
2-(Dimethylamino)ethyl p-(butylamino)benzoate hydrochloride.
Anethaine; Decicain, Pantocaine; Tonexol; Pontocaine Hydrochloride; component of: Fulvidex. Local anesthetic. mp = 147-150°; λ_m 225, 310 nm (ε 14108, 26352 H_2O), 229, 281, 312 nm ($E_{1\,cm}^{1\%}$ 509, 55, 76 0.1N H_2SO_4), 226, 310 nm (ε 7586, 29512 MeOH), 308 nm (ε 27542 $CHCl_3$); soluble in H_2O (13.3 g/100 ml), EtOH (2.5 g/100 ml), $CHCl_3$ (3.3 g/100 ml); insoluble in Et_2O, C_6H_6, Me_2CO; LD_{50} (mus ip) = 70 mg/kg, (fmus iv) = 13 mg/kg, (fmus sc) = 35 mg/kg. *Abbott Labs Inc.; Alcon Labs; CIBA Vision AG; Schering-Plough Animal Health; Sterling Winthrop Inc.*

1041 Tolycaine
3686-58-6 9679 222-976-9
$C_{15}H_{22}N_2O_3$
Methyl 2-[2-(diethylamino)acetamido]-m-toluate.
Local anesthetic. bp_5 = 190-192°. *Bayer Corp. Pharm. Div.; Schenley.*

1042 Tolycaine Hydrochloride
7210-92-6 9679 230-590-7
$C_{15}H_{23}ClN_2O_3$
Methyl 2-[2-(diethylamino)acetamido]-m-toluate hydrochloride.
Baycain. Local anesthetic. mp = 139-140.5°. *Bayer Corp. Pharm. Div.; Schenley.*

1043 Trapencaine
104485-01-0
$C_{22}H_{34}N_2O_3$
(±)-trans-2-(1-Pyrrolidinyl)cyclohexyl m-(pentyloxy)carbanilate.
pentacaine. Local anesthetic; antiulcerative; gastroprotective.

1044 Trimecaine
616-68-2 9830 210-487-3
$C_{15}H_{24}N_2O$
2-Diethylamino-2',4',6'-trimethylacetanilide.
Mesocaine; Mesidicaine; Mesokain. Local anesthetic. mp = 4°; bp_6 = 187°, $bp_{0.6}$ = 154-155°; [hydrochloride ($C_{15}H_{25}ClN_2O$)]: mp = 140°; LD_{50} (mus sc) = 295 mg/kg. *Astra USA Inc.*

1045 Vadocaine
72005-58-4
$C_{18}H_{28}N_2O_2$
(±)-6'-Methoxy-2-methyl-1-piperidinepropiono-2',4'-xylidide.
OR-K-242. Local anesthetic; antitussive. Anilide derivative.

1046 Zolamine
553-13-9 10319
$C_{15}H_{21}N_3OS$
2-[[2-(Dimethylamino)ethyl]-(p-methoxybenzyl)-amino]thiazole.
Local anesthetic. Antihistaminic. Isostere of pyrilamine. bp_7 = 217-219°. *Schering-Plough HealthCare Products.*

1047 Zolamine Hydrochloride
1155-03-9 10319
$C_{15}H_{22}ClN_3OS$
2-[[2-(Dimethylamino)ethyl]-(p-methoxybenzyl)-amino]thiazole hydrochloride.
194-B; WI 291; component of: Otodyne. Antihistaminic. Isostere of pyrilamine with local anesthetic properties. mp = 167.5-167.8°; soluble in H_2O. *Schering-Plough HealthCare Products.*

Angiotensin II Antagonists

1048 Candesartan
139481-59-7 1788
$C_{24}H_{20}N_6O_3$
2-Ethoxy-1-[p-(o-1H-tetrazol-5-ylphenyl)benzyl]-7-benz-imidazolecarboxylic acid.
CV-11974. Non-peptidic angiotensin II type-1 receptor antagonist. Used as an antihypertensive. mp = 183-185°. *Takeda Chem. Ind. Ltd.*

1049 Candesartan Cilixetil
145040-37-5 1788
$C_{33}H_{34}N_6O_6$
(±)-1-Hydroxyethyl 2-ethoxy-1-[p-(o-1H-tetrazol-5-ylphenyl)benzyl]-7-benzimidazolecarboxylate cyclohexylcarbonate (ester).
TCY-116. Nonpeptidic angiotensin II type-1 receptor antagonist. Used as an antihypertensive. Ester prodrug of Candesartan. mp = 163° (dec). *Takeda Chem. Ind. Ltd.*

1050 Eprosartan
133040-01-4 3669
$C_{23}H_{24}N_2O_4S$
(E)-2-Butyl-1-(p-carboxybenzyl)-α-2-thenylimidazole-5-acrylic acid.
SK&F-108566. Non-peptidic angiotensin II receptor antagonist. Used as an antihypertensive. mp = 260-261°. *SmithKline Beecham Pharm.*

1051 Eprosartan Mesylate
144143-96-4 3669
$C_{24}H_{28}N_2O_7S_2$
(E)-2-Butyl-1-(p-carboxybenzyl)-α-2-thenylimidazole-5-acrylic acid monomethanesulfonate.
SK&F-108566J; Eprosartan methanesulfonate. Non-peptidic angiotensin II receptor antagonist. Used as an antihypertensive. *SmithKline Beecham Pharm.*

1052 Irbesartan
138402-11-6 5097
$C_{25}H_{28}N_6O$
2-Butyl-3-[p-(o-1H-tetrazol-5-ylphenyl)benzyl]-1,3-diazaspiro[4.4]non-1-en-4-one.
BMS-186295; SR-47436. Non-peptidic angiotensin II type-1 receptor antagonist. Used as an antihypertensive. mp = 180-181°. *Bristol-Myers Squibb Pharm. R&D; Sanofi Winthrop.*

1053 Losartan
114798-26-4 5613
$C_{22}H_{23}ClN_6O$
2-Butyl-4-chloro-1-[p-(o-1H-tetrazol-5-ylphenyl)benzyl]-imidazole-5-methanol.
Non-peptide angiotensin II receptor antagonist used as an antihypertensive agent. mp = 183.5-184.5°. *DuPont-Merck Pharm.; Merck & Co.Inc.*

1054 Losartan Monpotassium Salt
124750-99-8 5613
$C_{22}H_{22}ClKN_6O$
2-Butyl-4-chloro-1-[p-(o-1H-tetrazol-5-ylphenyl)benzyl]-imidazole-5-methanol monopotassium salt.
Cozaar; DuP 753; Du Pont 753; DUP-753; MK-954; component of: Hyzaar. Non-peptide angiotensin II receptor antagonist. Used as an antihypertensive. *DuPont-Merck Pharm.; Merck & Co.Inc.*

1055 Saralasin
34273-10-4 8518
$C_{42}H_{65}N_{13}O_{10}$
N-[1-[N-[N-[N-[N-[N^2-(N-Methylglycyl)-L-arginyl]-L-valyl]-L-tyrosyl]-L-valyl]-L-histidyl]-L-prolyl]-L-alanine.
Sar-Arg-Val-Tyr-Val-His-Pro-Ala; 1-sar-8-ala-angiotensin II. Antihypertensive. Peptide angiotensin receptor blocking agent; specific antagonist of angiotensin II. Proposed as a diagnostic aid in identifying angiotensin-dependent hypertension. *Norwich Eaton.*

1056 Saralasin Acetate
39698-78-7 8518
$C_{42}H_{65}N_{13}O_{10}.xC_2H_4O_2.xH_2O$
N-[1-[N-[N-[N-[N-[N^2-(N-Methylglycyl)-L-arginyl]-L-valyl]-L-tyrosyl]-L-valyl]-L-histidyl]-L-prolyl]-L-alanine acetate (salt) hydrate.
P-113; Sarenin. Antihypertensive. Peptide angiotensin receptor blocking agent. Proposed as a diagnostic aid in identifying angiotensin-dependent hypertension. mp = 256°; soluble in H_2O, 5% aqueous dextrose, 90-95% alcohol; LD_{50} (mmus iv) = 1171 mg/kg. *Norwich Eaton.*

1057 Telmisartan
144701-48-4
$C_{33}H_{30}N_4O_2$
4'-[[4-Methyl-6-(1-methyl-2-benzimidazolyl)-2-propyl-1-benzimidazolyl]methyl]-2-biphenylcarboxylic acid.
BIBR 277 SE. Non-peptidic angiotensin II receptor antagonist. Used as an antihypertensive. *Boehringer Ingelheim GmbH.*

1058 Valsartan
137862-53-4 10051
$C_{24}H_{29}N_5O_3$
N-[p-(o-1H-Tetrazol-5-ylphenyl)benzyl]-N-valeryl-L-valine.

CGP-48933. A non-peptide angiotensin II AT_1-receptor antagonist. Used as an antihypertensive agent. mp = 116-117°. *Ciba-Geigy Corp.*

1059 Zolasartan
145781-32-4
$C_{20}H_{24}BrClN_6O_3$
1-[[3-Bromo-2-(o-1H-tetrazol-5-ylphenyl)-5-benzofuranyl]methyl]-2-butyl-4-chloroimidazole-5-carboxylic acid.
GR-117289. Selective, potent, orally active, long-acting nonpeptide angiotensin II type 1 receptor antagonist. Analog of losartan. Antihypertensive.

Anorexics

1060 Aminorex
2207-50-3 497
$C_9H_{10}N_2O$
2-Amino-5-phenyl-2-oxazoline.
McN-742; NSC-66592; aminoxafen; aminoxaphen; Menocil [fumarate]; Apiquel. Anorexic. Appetite suppressant. Serotonin (5-HT) transporter. Apiquel is the fumarate. mp = 136-138°. *McNeil Pharm.*

1061 Amphecloral
5581-35-1 620
$C_{11}H_{12}Cl_3N$
α-Methyl-N-(2,2,2-trichloroethylidene)-phenethylamine.
Acutran; Amfecloral. Anorexic. [dl-form]: $bp_{0.5}$ = 95°; [d-form]: $[\alpha]_D$ = 49.9° (c = 5 dioxane).

1062 Amphetamine
300-62-9 623 206-096-2
$C_9H_{13}N$
(±)-α-Methylphenethylamine.
Actedron; Allodene; Adipan; Sympatedrine; Psychedrine; Isomyn; Isoamyne; Mecodrin; Norephedrane; Novydrine; Elastonon; Ortedrine; Phenedrine; Profamina; Propisamine; Sympamine; Simpatedrin. Anorexic; CNS stimulant. Psychomotor stimulant. d_4^{25} = 0.913; bp = 200-203°, bp_{13} = 82-85°; soluble in H_2O, EtOH, Et_2O; LD_{50} (rat sc) = 180 mg/kg. *Interco Fribourg.*

1063 Benzphetamine
156-08-1 1151
$C_{17}H_{21}N$
N-Benzyl-N,α-dimethylphenethylamine.
Anorexic; CNS stimulant. $bp_{0.02}$ = 127°; insoluble in H_2O; soluble in MeOH, EtOH, Et_2O, C_6H_6, $CHCl_3$, Me_2CO. *Pharmacia & Upjohn.*

1064 Benzphetamine Hydrochloride
5411-22-3 1151 226-489-2
$C_{17}H_{22}ClN$
N-Benzyl-N,α-dimethylphenethylamine hydrochloride.
Didrex; Inapetyl. Anorexic; CNS stimulant. mp = 129-130°; soluble in H_2O, EtOH; dextrorotatory. *Pharmacia & Upjohn.*

1065 Chlorphentermine
461-78-9 2235 207-314-9
$C_{10}H_{14}ClN$
4-Chloro-α,α-dimethylbenzeneethanamine.
clorfentermina; Lucofen; Teramine. Anorexic; CNS stimulant. Anticholinergic. Federally controlled substance (stimulant). bp_2 = 100-102°. *Parke-Davis.*

1066 Chlorphentermine Hydrochloride
151-06-4 2235 205-782-9
$C_{10}H_{15}Cl_2N$
4-Chloro-α,α-dimethylbenzeneethanamine hydrochloride.
Pre-Sate; S-62; W 2426; NSC-76098. Anorexic; CNS stimulant. mp = 234°; soluble in H_2O (> 20 g/100 ml); LD_{50} (mus orl) = 270 mg/kg, (mus sc) = 267 mg/kg. *Parke-Davis.*

1067 Clobenzorex
13364-32-4 2421
$C_{16}H_{18}ClN$
(+)-N-(o-Chlorobenzyl)-α-methylphenethylamine.
SD 271-12. Anorexic. $bp_{0.1}$ = 132-134°. *Soc. Ind. Fabric. Antiboit.*

1068 Clobenzorex Hydrochloride
5843-53-8 2421
$C_{16}H_{19}Cl_2N$
(+)-N-(o-Chlorobenzyl)-α-methylphenethylamine hydrochloride.
Ba-7205; SD-271-12; Dinintel; Rexigen. Anorexic. mp = 182-183°; $[\alpha]_D^{20}$ = 26.3 (H_2O); soluble in H_2O, EtOH; less soluble in MeOH, $CHCl_3$; LD_{50} (mus ip) = 103 mg/kg. *Soc. Ind. Fabric. Antiboit.*

1069 Cloforex
14261-75-7 2440
$C_{13}H_{18}ClNO_2$
Ethyl (p-chloro-α,α-dimethylphenethyl)carbamate.
D-237; Avicol SL; Frenapyl; Oberex. Anorexic. mp = 52.5-53°; $bp_{0.005}$ = 88-90°. *Troponwerke Dinklage.*

1070 Clominorex
3876-10-6
$C_9H_9ClN_2O$
2-Amino-5-(p-chlorophenyl)-2-oxazoline.
McN-1107. Anorexic. *McNeil Pharm.*

1071 Clortermine
10389-73-8 2472
$C_{10}H_{14}ClN$
o-Chloro-α,α-dimethylphenethylamine.
Anorexic. Federally controlled substance (stimulant). bp_{16} = 116-118°. *Ciba-Geigy Corp.*

1072 Clortermine Hydrochloride
10389-72-7 2472
$C_{10}H_{15}Cl_2N$
o-Chloro-α,α-dimethylphenethylamine hydrochloride.
Su-10568; Voranil. Anorexic. Federally controlled substance (stimulant). mp = 245-246°; LD_{50} (rat orl) = 332 ± 23 mg/kg. *Ciba-Geigy Corp.*

1073 Cyclexedrine
532-52-5 2776
$C_{10}H_{21}N$
N,β-Dimethylcyclohexaneethanamine.
ethylhexedrine; isopropylhexedrine. Anorexic. Sympathomimetic. *Knoll Pharm. Co.*

1074 Cyclexedrine Hydrochloride
64011-61-6 2776
$C_{10}H_{22}ClN$
N,β-Dimethylcyclohexaneethanamine hydrochloride.
Eventin. Anorexic. Sympathomimetic. mp = 138-140°; soluble in H_2O, EtOH, $CHCl_3$; slightly soluble in Et_2O; LD_{50} (mus orl) = 304 mg/kg. *Knoll Pharm. Co.*

1075 Dexfenfluramine
3239-44-9 4015
$C_{12}H_{16}F_3N$
(+)-(S)-N-Ethyl-α-methyl-m-(trifluoromethyl)-phenethylamine.
Redux; S 5614; fenfluramine; S-768. Anorexic. Psychomotor stimulant. Systemic appetite suppressant. Federally controlled substance (stimulant). bp_{12} = 108-112°; $[\alpha]_D^{25}$ = 9.5° (c = 8 EtOH); LD_{50} (mus ip) = 144 mg/kg, (rat orl) = 114.6 mg/kg; [l-form]: $[\alpha]_D^{25}$ = -9.6° (c = 8 EtOH); LD_{50} (rat orl) = 195 mg/kg; [dl-form]: bp_{12} = 108-112°; LD_{50} (ip mus) = 114 mg/kg. *Interneuron Pharm. Inc.; Wyeth Labs.*

1076 Dexfenfluramine Hydrochloride
3239-45-0 4015 221-806-0
$C_{12}H_{17}ClF_3N$
(+)-(S)-N-Ethyl-α-methyl-m-(trifluoromethyl)-phenethylamine hydrochloride.
S 5614 HCl; fenfluramine hydrochloride; Acino; Adipomin; Obedrex; Pesos; Ponderal; Ponderax; Ponderex; Pondimin; Rotondin; Adifax; Glypolix; Isomeride. Anorexic. Psychomotor stimulant. Systemic appetite suppressant. Federally controlled substance (stimulant). mp = 160-161°; [l-form]: mp = 160-161°; [dl-form]: mp = 166°. *Interneuron Pharm. Inc.; Wyeth Labs.*

1077 Dextroamphetamine Sulfate
51-63-8 2996 200-111-6
$C_{18}H_{28}N_2O_4S$
(+)-α-Methylphenethylamine sulfate.
Dexedrine; Dexamyl; Eskatrol; Dexampex; Dexedrine sulfate; Afatin; Dexamphetamine; d-Amfetasul; Domafate; Obesedrin; Dexten; Maxiton; Sympamin; Simpamina-D; Albemap; Dadex; Ardex; Dexalone; Amsustain; Betafedrina; d-Betafedrine; Diocurb; component of: Carboxyphen, Bontril. CNS stimulant; anorexic. Veterinary medicine: sympathomimetic. Federally controlled substance (stimulant). mp > 300°; $[\alpha]_D^{20}$ = 21.8° (c = 2); soluble in H_2O (10 g/100 ml), EtOH (0.2 g/100 ml); LD_{50} (mus orl) = 10 mg/kg. *SmithKline Beecham Pharm.*

1078 d-Furfenorex
3776-93-0 4327 223-233-1
$C_{15}H_{19}NO$
(+)-N-Methyl-N-(α-methylphenethyl)furfurylamine.
d-furfurylmethylamphetamine. Anorexic. $bp_{0.1}$ = 103-106°. *S.I.F.A.*

1079 d-Furfenorex Cyclamate
3776-92-9 4327 223-232-6
$C_{21}H_{32}N_2O_4S$
(+)-N-Methyl-N-(α-methylphenethyl)furfurylamine cyclohexylsulfamate.
d-furfurylmethylamphetamine cyclohexylsulfamate. Anorexic. mp = 90-92°; $[\alpha]_D^{20}$ = 12.7° (c = 1 H_2O). *S.I.F.A.*

1080 Diethylpropion
90-84-6 3175 202-019-1
$C_{13}H_{19}NO$
2-(Diethylamino)propiophenone.
α-benzoyltriethylamine; amfepramone. Anorexic; CNS stimulant. *3M Pharm.; Marion Merrell Dow Inc.*

1081 Diethylpropion Hydrochloride
134-80-5 3175 205-281-5
$C_{13}H_{20}ClNO$
2-(Diethylamino)propiophenone hydrochloride.
Tenuate; Tepanil; Anfamon; Anorex; Danylen; Dobesin; Frekentine; Keramik; Keramin; Magrene; Modulor; Moderatan; Parabolin; Prefamone; Regenon; Tenuate Dospan; Tylinal. Anorexic; CNS stimulant. mp = 168° (dec). *3M Pharm.; Marion Merrell Dow Inc.*

1082 Difemetorex
13862-07-2 3362
$C_{20}H_{25}NO$
2-(Diphenylmethyl)-1-piperidineethanol.
Ba-28289; Cleofil; Diphemethoxidine. Anorexic; anticholinergic. mp = 106-107°; $bp_{0.1}$ = 180-181°; [hydrochloride]: mp = 166-167°. *Ciba-Geigy Corp.*

1083 dl-Furfenorex
14817-79-9 4327
$C_{15}H_{19}NO$
(±)-N-Methyl-N-(α-methylphenethyl)furfurylamine.
furfurylmethylamphetamine. Anorexic. $bp_{0.1}$ = 103-106°. *S.I.F.A.*

1084 dl-Furfenorex Cyclamate
14611-84-8 4327
$C_{21}H_{42}N_2O_4S$
(±)-N-Methyl-N-(α-methylphenethyl)furfurylamine cyclohexylsulfamate.
furfurylmethylamphetamine cyclohexylsulfamate; Frugalan; SD-27115; E-106-E. Anorexic. mp = 113-114°. *S.I.F.A.*

1085 Etilamfetamine
457-87-4 3809
$C_{11}H_{17}N$
N-Ethyl-α-methylphenethylamine.
N-Ethylamphetamine; Adiparthrol; Apetinil. Anorexic; CNS stimulant. Federally controlled substance (stimulant). bp_{14} = 104.5-106°; [hydrochloride, d-form]: mp = 154-156°; $[\alpha]_D^{15}$ = 17.2° (c = 2 H_2O); [hydrochloride, l-form]: mp = 155-156°; $[\alpha]_D^{25}$ = -17.3° (c = 2 H_2O). *Sterling Winthrop Inc.*

1086 Fenbutrazate
4378-36-3 4005 224-480-8
$C_{23}H_{29}NO_3$
2-(3-Methyl-2-phenylmorpholino)ethyl 2-phenylbutyrate.

phenbutrazate. Anorexic. Viscous oil; $bp_{0.005}$ = 235-240°; soluble in MeOH. *Ravensberg.*

1087 Fenbutrazate Hydrochloride
6474-85-7 4005 229-330-5
$C_{23}H_{30}ClNO_3$
2-(3-Methyl-2-phenylmorpholino)ethyl 2-phenylbutyrate hydrochloride.
R-381; component of: Cafilon, Filon (with phenmetrazine theoclate). Anorexic. mp = 154°; soluble in EtOH, Me_2CO; slightly soluble in H_2O; insoluble in Et_2O, C_6H_6; LD_{50} (mus orl) = 3200 mg/kg, (mus sc) = 2800 mg/kg. *Ravensberg.*

1088 Fenfluramine
404-82-0 4015 206-968-2
$C_{12}H_{16}F_3N$
N-Ethyl-α-methyl-m-(trifluoromethyl)-phenethylamine.
S-768. Anorexic. Psychomotor stimulant. bp_{12} = 108-112°; LD_{50} (mus ip) = 144 mg/kg; [l-form]: $[\alpha]_D^{25}$ = -9.6° (c = 8 EtOH); LD_{50} (rat orl) = 195 mg/kg; [d-form]: $[\alpha]_D^{25}$ = 9.5° (c = 8 EtOH); LD_{50} (rat orl) = 114.6 mg/kg. *C.M. Ind.; Taiho.*

1089 Fenfluramine Hydrochloride
458-24-2 4015 207-276-3
$C_{12}H_{17}ClF_3N$
N-Ethyl-α-methyl-m-(trifluoromethyl)-phenethylamine hydrochloride.
Ponderax PA; Pondimin; Acino; Adipomin; Obedrex; Pesos; Ponderex; Pondimin; Rotondin; Adifax [as d-form]; Glypolix [as d-form]; Isomeride [as d-form]. Anorexic. Psychomotor stimulant. mp = 166°; [d-form]: mp = 160-161°; [l-form]: mp = 160-161°. *C.M. Ind.; Taiho.*

1090 Fenisorex
34887-52-0
$C_{16}H_{16}FNO$
cis-7-Fluoro-1-phenyl-3-isochromanmethylamine.
Anorexic. *3M Pharm.*

1091 Fenproporex
15686-61-0 4036 239-772-0
$C_{12}H_{16}N_2$
(±)-3-[(α-Methylphenethyl)amino]propionitrile.
(±)-N-2-cyanoethylamphetamine. Anorexic. bp_2 = 126-127°. *Bottu.*

1092 Fenproporex Hydrochloride
18305-29-8 4036 242-191-5
$C_{12}H_{17}ClN_2$
(±)-3-[(α-Methylphenethyl)amino]propionitrile hydrochloride.
Gacilin; Solvolip. Anorexic. mp = 146°; soluble in H_2O, EtOH. *Bottu.*

1093 Fludorex
15221-81-5
$C_{11}H_{14}F_3NO$
β-Methoxy-N-methyl-m-(trifluoromethyl)-phenethylamine.
Win-11464. Anorexic; anti-emetic. *Sterling Winthrop Inc.*

1094 Fluminorex
720-76-3
$C_{10}H_9F_3N_2O$
2-Amino-5-(α,α,α-trifluoro-p-tolyl)-2-oxazoline.
McN-1231. Anorexic. *McNeil Pharm.*

1095 Levofenfluramine
37577-24-5 4015
$C_{12}H_{16}F_3N$
(-)-(R)-N-Ethyl-α-methyl-m-(trifluoromethyl)phenethylamine.
Anorexic. $[\alpha]_D^{25}$ = -9.6° (c = 8 EtOH); LD_{50} (rat orl) = 195 mg/kg. *Interneuron Pharm. Inc.*

1096 Levophacetoperane
24558-01-8 5493
$C_{14}H_{19}NO_2$
1-Phenyl-1-(2'-piperidyl)-1-acetoxymethane.
RP-8228; phacetoperane. Anorexic. *Rhône-Poulenc.*

1097 Levophacetoperane Hydrochloride
23257-56-9 5493 245-536-8
$C_{14}H_{20}ClNO_2$
1-Phenyl-1-(2'-piperidyl)-1-acetoxymethane hydrochloride.
Lidepran. Anorexic. mp = 229-230°; levorotatory. *Rhône-Poulenc.*

1098 Mazindol
22232-71-9 5801 244-857-0
$C_{16}H_{13}ClN_2O$
5-(p-Chlorophenyl)-2,5-dihydro-3H-imidazo[2,1-a]isoindol-5-ol.
SaH-42548; 42548; Magrilon; Mazanor; Mazildene; Sanorex; Terenac; Teronac. Anorexic; CNS stimulant. Federally controlled substance (stimulant). mp = 215-217°, 198-199°; λ_m = 223, 268.5, 272 nm (ε 19000, 4400, 4400 95% EtOH); insoluble in H_2O, soluble in EtOH. *Am. Home Products; Sandoz Pharm. Corp.; Wyeth-Ayerst Labs.*

1099 Mefenorex
17243-57-1 5843 241-279-0
$C_{12}H_{18}ClN$
N-(3-Chloropropyl)-α-methylphenethylamine.
Anorexic. *Hoffmann-LaRoche Inc.*

1100 Mefenorex Hydrochloride
5586-87-8 5843 226-985-9
$C_{12}H_{19}Cl_2N$
N-(3-Chloropropyl)-α-methylphenethylamine hydrochloride.
Ro-4-5282; Incital; Pondinil; Pondinol; Rondimen. Anorexic. mp = 128-130°. *Hoffmann-LaRoche Inc.*

1101 Metamfepramone
15351-09-4 5984 239-384-1
$C_{11}H_{15}NO$
2-(Dimethylamino)propiophenone.
N-methylephedrone; MG-559; Effilone [as hydrochloride]. Anorexic. [dl-form]: bp_{13} = 126°; [l-form]: $[\alpha]_d^{26}$ = -60°; [dl-form hydrochloride; Effilone]: mp = 201-202°, 202-204° (dec); [l-form hydrochloride]: mp = 197-199°; $[\alpha]_D^{26}$ = -52.5°.

1102 Methamphetamine
537-46-2 6015 208-668-7
$C_{10}H_{15}N$
(+)-(S)-N-α-Dimethylphenethylamine.
methyl-β-phenylisopropylamine; Norodin. Anorexic. Psychomotor stimulant. Federally controlled substance (stimulant). *Abbott Labs Inc.; Sterling Winthrop Inc.*

1103 Methamphetamine Hydrochloride
51-57-0 6015 200-106-9
$C_{10}H_{16}ClN$
(+)-(S)-N-α-Dimethylphenethylamine hydrochloride.
Desoxyn; Desyphed; Speed; Meth; Ice; Amphedroxyn; Desfedrin; Desoxyfed; Destim; Doxephrin; Drinalfa; Efroxine; Gerobit; Hiropon; Isophen; Madrine; Methampex; Methedrine; Methylisomin; Pervitin; Semoxydrine; Soxysympamine; Syndrox; Tonedron. Anorexic. Psychomotor stimulant. Federally controlled substance (stimulant). mp = 170-175°; $[\alpha]_D^{25}$ = 14 - 20°; soluble in H_2O, EtOH, $CHCl_3$; insoluble in other organic solvents; LD_{50} (mus ip) = 70 mg/kg. *Abbott Labs Inc.; Sterling Winthrop Inc.*

1104 Norpseudoephedrine
36393-56-3 6811 253-014-6
$C_9H_{13}NO$
(R*,R*)-α-(1-Aminoethyl)benzenemethanol.
pseudonoephedrine; Katine; Cathine. Anorexic. [d-form]: mp = 77.5-78°; $[\alpha]_{546}^{20}$ = 37.9° (c = 3 MeOH); soluble in EtOH,$CHCl_3$, Et_2O.

1105 Norpseudoephedrine Hydrochloride
2153-98-2 6811 218-446-1
$C_9H_{14}ClNO$
(R*,R*)-α-(1-Aminoethyl)benzenemethanol hydrochloride.
Amorphan; Adiposetten; Exponcit N; Fasupond; Fugoa; Minusin. Anorexic. [d-form]: mp = 180-181°; $[\alpha]_D^{20}$ = 43.2°; LD_{50} (mus sc) = 275 mg/kg; [dl-form]: mp = 169-171°.

1106 Pentorex
434-43-5 7275 207-102-6
$C_{11}H_{17}N$
2-Phenyl-2-methyl-3-butylamine.
α,α,β-trimethylmehethylamine; phenpentermine. Anorexic. *Nordmark.*

1107 Pentorex (dl)
46029-44-1 7275
$C_{11}H_{17}N$
DL-2-Phenyl-2-methyl-3-butylamine.
DL-phenpentermine. Anorexic. bp_{20} = 109-111°. *Nordmark.*

1108 Pentorex Hydrochloride
5585-52-4 7275
$C_{11}H_{18}ClN$
2-Phenyl-2-methyl-3-butylamine hydrochloride.
α,α,β-trimethylmehethylamine hydrochloride. Anorexic. mp = 164-166°. *Nordmark.*

1109 Pentorex Hydrogen Tartrate
22232-55-9 7275
$C_{15}H_{23}NO_6$
2-Phenyl-2-methyl-3-butylamine hydrogen D-tartrate.
Liprodene; Modatrop. Anorexic. mp = 160-162°; $[\alpha]_D^{20}$ = 13.4° (c = 0.8 H_2O); [d-form]: mp = 167-169°; $[\alpha]_D^{20}$ = 17.9°; [l-form]: mp = 164-166°; $[\alpha]_D^{20}$ = +3.45°. *Nordmark.*

1110 Phendimetrazine
634-03-7 7365 211-204-6
$C_{12}H_{17}NO$
(2S,3S)-3,4-Dimethyl-2-phenylmorpholine.
3-Phenyl-2-methylmorpholine; A-66. Anorexic. bp_{12} = 138-140°; bp_1 = 104°. *Boehringer Ingelheim GmbH; Forest Pharm. Inc.; Wyeth-Ayerst Labs.*

1111 Phendimetrazine Bitartrate
50-58-8 7365 200-051-0
$C_{16}H_{23}NO_7$
(2S,3S)-3,4-Dimethyl-2-phenylmorpholine L-(+)-tartrate (1:1).
Metra; Plegine; Prelu-2; Statobex; Marsin; NeoZine; Preludin. Anorexic. mp = 182°; soluble in H_2O (2.5 g/ml), EtOH, $CHCl_3$ (0.5 g/ml). *Boehringer Ingelheim GmbH; Forest Pharm. Inc.; Wyeth-Ayerst Labs.*

1112 Phenmetrazine
134-49-6 7385 205-143-4
$C_{11}H_{15}NO$
3-Methyl-2-phenylmorpholine.
A-66. Anorexic; CNS stimulant. Federally controlled substance (stimulant). bp_{12} = 138-140°, bp_1 = 104°. *Boehringer Ingelheim GmbH.*

1113 Phenmetrazine Hydrochloride
1707-14-8 7385 216-950-6
$C_{11}H_{16}ClNO$
3-Methyl-2-phenylmorpholine hydrochloride.
Preludin; Marsin; Neo-Zine. Anorexic; CNS stimulant. Federally controlled substance (stimulant). mp = 182°; soluble in H_2O (250 g/100 ml), 95% EtOH (50 g/100 ml), $CHCl_3$ (50 g/100 ml); poorly soluble in Et_2O. *Boehringer Ingelheim GmbH.*

1114 Phentermine
122-09-8 7415 204-522-1
$C_{10}H_{15}N$
α,α-Dimethylphenethylamine.
Inoamin. Anorexic; CNS stimulant. Systemic appetite suppressant. Psychomotor stimulant. bp_{750} = 205°; bp_{21} = 100°. *Fisons plc; Lemmon Co.; Marion Merrell Dow Inc.; SmithKline Beecham Pharm.; Wyeth-Ayerst Labs.*

1115 Phentermine Hydrochloride
1197-21-3 7415 214-821-9
$C_{10}H_{16}ClN$
α,α-Dimethylphenethylamine hydrochloride.
Adipex-P; Fastin; Wilpo; Obermine Black & Yellow. Anorexic; CNS stimulant. Systemic appetite suppressant. Psychomotor stimulant. mp = 198°. *Lemmon Co.; SmithKline Beecham Pharm.*

1116 Phenylpropanolamine Hydrochloride
154-41-6 7461 205-826-7
$C_9H_{14}ClNO$
(±)-α-(1-Aminoethyl)benzenemethanol hydrochloride.
1-phenyl-2-amino-1-propanol hydrochloride; mydriatin; Kontexin; Monydrin; Obestat; Propadrine; component of: A.R.M., Chlor-Trimeton Allergy/Sinus/Headache, Comtrex Liquigels, Comtrex Non-Drowsy Liquigels, Contac, Contac 12 Hour Caplets, Contac 12 Hour Capsules, Contac Severe Cold Maximum Strength, Coricidin D, Coricidin Sinus, Corsym Capsules, Demazin, Demilets, Diet Gard, Dimetapp Allergy Sinus, Dimetapp Cold and Allergy, Dimetapp Cold and Cough; Dimetapp DM, Dimetapp Elixir, Dimetapp Extentabs, Dimetapp Liquigels, Dimetapp 4 Hour Tablets, Entex Capsules and Liquid, Histalet Forte, Hycomine, Hycomine Pediatric, Naldecon, Naldecon CX, Ornade, Rhinex D-Lay Tablets, Robitussin CF, Sine-Off Sinus Medicine Tablets Aspirin Formula, Sinubid, Teldrin Timed Release Allergy Capsules, Triaminic, Trind, Trind-DM, Tuss-Ornade. Anorexic and decongestant. Psychomotor stimulant. Used in veterinary medicine as a bronchodilator and nasal decongestant. Sympathomimetic amine. Related to norpseudoephedrine. mp = 190-194°; soluble in H_2O, EtOH; insoluble in Et_2O, $CHCl_3$, C_6H_6; pKa = 9.44 ± 0.04; LD_{50} (rat orl) = 1490 mg/kg; [(+)-form hydrochloride]: mp = 171-172°; $[\alpha]_D^{25}$ = 32° (H_2O); [(±)-form base]: mp = 101-101.5. *Apothecon; Boots Pharmaceuticals Inc.; Bristol-Myers Squibb HIV Products; DuPont-Merck Pharm.; Fisons plc; Lemmon Co.; Mead Johnson Nutritionals; Menley & James Labs Inc.; Parke-Davis; Procter & Gamble Pharm. Inc.; Robins, A. H. Co.; Sandoz Pharm. Corp.; Schering-Plough Pharm.; SmithKline Beecham Pharm.; Solvay Pharm. Inc.; Taiho; Whitehall-Robins.*

1117 Picilorex
62510-56-9 7551
$C_{14}H_{18}ClN$
3-(p-Chlorophenyl)-5-cyclopropyl-2-methylpyrrolidine.
UP-507-04. Anorexic. *Hexachemie.*

1118 Picilorex Hydrochloride
56109-02-5 7551
$C_{14}H_{19}Cl_2N$
3-(p-Chlorophenyl)-5-cyclopropyl-2-methylpyrrolidine hydrochloride.
Roxenan. Anorexic. mp = 191°. *Hexachemie.*

1119 Sibutramine
106650-56-0 8629
$C_{17}H_{26}ClN$
(±)-1-(p-Chlorophenyl)-α-isobutyl-N,N-dimethylcyclobutanemethylamine.
Anorexic; antidepressant. Monoamine reuptake inhibitor. *Boots Co.*

1120 Sibutramine Hydrochloride [anhydrous]
84485-00-7 8629
$C_{17}H_{27}Cl_2N$
(±)-1-(p-Chlorophenyl)-α-isobutyl-N,N-dimethylcyclobutanemethylamine hydrochloride.
Anorexic; antidepressant. Monoamine reuptake inhibitor. *Boots Co.; Boots Pharmaceuticals Inc.*

1121 Sibutramine Hydrochloride Monohydrate
125494-59-9 8629
$C_{17}H_{27}Cl_2N.H_2O$
(±)-1-(p-Chlorophenyl)-α-isobutyl-N,N-dimethylcyclobutanemethylamine hydrochloride monohydrate.
BTS-54524; Reductil; Meridia. Anorexic; antidepressant. Monoamine reuptake inhibitor. mp = 193-195.5°. *Boots Co.; Boots Pharmaceuticals Inc.*

1122 Tiflorex
53993-67-2
$C_{12}H_{16}F_3NS$
(+)-N-Ethyl-α-methyl-m-[(trifluoromethyl)thio]-phenethylamine.
Enhances central serotoninergic transmission. Anorexic.

Antacids

1123 Alexitol Sodium
66813-51-2 232
Sodium polyhydroxyaluminum monocarbonate hexitol complex.
aluminum sodium carbonate hexitol complex; Actal. Antacid. Tasteless, odorless powder; insoluble in H_2O, soluble in mineral acids. *Sterling Health U.S.A.*

1124 Algeldrate
1330-44-5
$AlH_3O_3.xH_2O$
Aluminum hydroxide hydrate.
W-4600. Antacid. *Parke-Davis.*

1125 Almadrate Sulfate
12125-11-0
$Al_4H_6Mg_2O_{14}S.xH_2O$
Aluminum magnesium hydroxide oxide sulfate.
W-4425. Antacid. *Parke-Davis.*

1126 Almagate
66827-12-1 307
$C_2H_{14}Mg_6O_{20}.4H_2O$
[Carbonato(2)]heptahydroxy(aluminum)trimagnesium dihydrate.
Almax. Antacid. Loses H_2O at 240°. *Anphar.*

1127 Aluminum Hydroxide
21645-51-2 355 244-492-7
H_3AlO_3
Aluminum hydroxide.
Amphojel; Dialume; Simeco; component of: Arthritis Pain Formula Maximum Strength, Calcitrel, Camalox, Gelusil, Kestomatin, Kudrox, Maalox, Maalox HRF, Maalox Plus, Simeco Suspension, Tricreamalate, Trsiogel, Wingel. Antacid with antihyperphosphatemic properties. White bulky amorphous powder; insoluble in H_2O, soluble in alkaline or acid solutions. *Wyeth Labs.*

1128 Aluminum Magnesium Silicate
12511-31-8 362 235-682-0
$Al_2MgO_8Si_2$
Magnesium aluminum silicate.
Ervasil [as hydrate]; Gelusil [as hydrate]; Ultin [as

hydrate]. Antacid. Also used as a suspending and thickening agent. *Fuji Chem. Ind.*

1129 Aluminum Phosphate
7784-30-7 371 232-056-9
AlO_4P
Aluminum orthophosphate.
Aluphos; Fosfalugel; Ulcocid; Phosphalujel; Phosphalugel; Phosphalutab. Antacid. mp > 1460°; d^{23} = 2.56; insoluble in H_2O, AcOH, slighly soluble in concentrated HCl, HNO_3.

1130 Azulene
275-51-4 956 205-993-6
$C_{10}H_8$
Bicyclo[5.3.0]deca-2,4,6,8,10-pentaene.
cyclopentacycloheptene; Azusalen [as sodium sulfonate]. The sodium sulfonate derivative is used as an antacid. Blue crystals; mp = 98.5-99°; insoluble in H_2O, soluble in organic solvents.

1131 Basic Aluminum Carbonate Gel
1339-92-0 1033 215-670-1
An aluminum hydroxide - aluminum carbonate gel.
Basaljel. Phosphorus-binding agent; antacid.

1132 Bismuth Aluminate
12284-76-3 1298 235-552-3
$Al_6Bi_2O_{12}$
Aluminum bismuth oxide.
almuth; Bisminate. Antacid. [decahydrate]: Light powder, insoluble in H_2O.

1133 Bismuth Phosphate
10049-01-1 1316 233-161-2
BiO_4P
Phosphoric acid bismuth salt (1:1).
Bismugel. Antacid; topical protectant. d^{15} = 6.323; slightly soluble in H_2O, dilute acids; insoluble in EtOH, AcOH.

1134 Bismuth Subgallate
22650-86-8 1325
$C_7H_5BiO_6$
Basic bismuth gallate.
gallic acid bismuth basic salt; B.S.G.; Dermatol. Topical protectant; astringent; antacid. Topical protectant, astringent and antacid. Insoluble in H_2O, EtOH, $CHCl_3$, Et_2O; soluble in dilute alkaline solutions, hot mineral acids.

1135 Bismuth Subnitrate
1304-85-4 1326 215-136-8
Bismuth hydroxide nitrate oxide.
bismuth oxynitrate; bismuth subnitricum; bismuthyl nitrate; bismuth white; magistery of bismuth; novismuth; paint white; Spanish white. Antacid. Insoluble in H_2O, EtOH; soluble in dilute HCl and HNO_3.

1136 Bismuth Subsalicylate
14882-18-9 1327 238-953-1
$C_7H_5BiO_4$
2-Hydroxybenzoic acid bismuth (3+) salt.
basic bismuth salicylate; oxo(salicylato)bismuth; Bismogenol Tosse Inj.; Stabisol. Antidiarrheal; antacid; antiulcerative. Also used as a lupus erythematosus suppressant. Insoluble in H_2O, EtOH. *Mobay.*

1137 Calcium Carbonate
471-34-1 1697 207-439-9
$CaCO_3$
Carbonic acid calcium salt (1:1).
Cal-Sup; Children's Mylanta Upset Stomach Relief; Chooz; Mylanta Soothing Lozenges; Calcit; Calcichew; Calcidia; Citrical; component of: Bufferin Arthritis Strength, Bufferin Extra Strength, Bufferin Regular, Calcitrel, Camalox, Di-Gel Tablets, Mylanta Gelcaps, Mylanta Tablets, Titralac, Tylenol Headache Plus. Calcium replenisher; antacid. mp = 825°, 1339° (102.5 atm); $d_{25.2}$ = 2.711; insoluble in H_2O. *3M Pharm.; Bristol-Myers Squibb HIV Products; Johnson & Johnson-Merck Consumer Pharm. ; McNeil Consumer Products Co.; Rhône-Poulenc Rorer Pharm. Inc.; Schering-Plough Pharm.; Sterling Winthrop Inc.*

1138 Dihydroxyaluminum Aminoacetate
13682-92-3 3227 237-193-8
$C_2H_6AlNO_4$
(Glycinato-N,O)dihydroxyaluminum.
aluminum glycinate; Ada; Alamine; Alminate; Alubasine; Alzinox; Aspogen; Dimothyn; Doraxamin; Elcosal; Robalate. Antacid. Insoluble in H_2O; forms stable suspensions in H_2O readily; pH of H_2O suspension = 7.4.

1139 Dihydroxyaluminum Sodium Carbonate
539-68-4 3228
CH_2AlNaO_5
[Carbonato(1-)-O]dihydroxyaluminum monosodium salt.
aluminum sodium carbonate hydroxide; Kompensan; Minicid. Antacid. d = 2.144; pH of H_2O suspension = 9.7.

1140 Ebimar
9013-42-7 3535
Sulfate polysaccharide extracted from *Chondrus crispus*.
Polysaccharide C 16; degraded carageenan. Antacid. $[\alpha]_D$ = 34°; freely soluble in H_2O, acids. *Evans Medical Ltd.*

1141 Magaldrate
1317-26-6 5685
Aluminum magnesium hydroxide monohydrate.
monalium hydrate; AY-5710; Dynese; Riopan; Ripon. Antacid. White, odorless, crystalline powder. *Byk Gulden Lomberg GmbH.*

1142 Magnesium Carbonate [anhydrous]
546-93-0 5696 208-915-9
$(MgCO_3)_4.Mg(OH)_2$
Carbonic acid magnesium salt (1:1) mixture with magnesium hydroxide hydrate.
Antacid. Slightly soluble in H_2O, insoluble in EtOH.

1143 Magnesium Carbonate Dihydrate
23389-33-5 5696
$(MgCO_3)_4.Mg(OH)_2.2H_2O$
Carbonic acid magnesium salt (1:1) mixture with magnesium hydroxide hydrate, dihydrate.
Marinco C; component of: Bufferin Arthritis Styrength, Bufferin Extra Strength, Bufferin Regular, Maalox HRF.

Antacid. Laxative-cathartic. Slightly soluble in H_2O (0.03 g/100 ml), insoluble in EtOH. *Bristol-Myers Squibb Pharm. R&D; Rhône-Poulenc Rorer Pharm. Inc.*

1144 Magnesium Carbonate Hydroxide
39409-82-0 5696
$(MgCO_3)_4.Mg(OH)_2.5H_2O$
Carbonic acid magnesium salt (1:1) mixture with magnesium hydroxide hydrate.
Marinco C; component of: Bufferin Arthritis Strength, Bufferin Extra Strength, Bufferin Regular, Maalox HRF. Antacid. Laxative-cathartic. Slightly soluble in H_2O (0.03 g/100 ml), insoluble in EtOH. *Bristol-Myers Squibb Pharm. R&D; Rhône-Poulenc Rorer Pharm. Inc.*

1145 Magnesium Hydroxide
1309-42-8 5706 215-170-3
H_2MgO_2
Magnesium hydrate.
Mint-O-Mag; Marinco H; Betalgil; Lederscon; Mucaine; Muthesa; Oxaine M; Phillips Magnesia Tablets; Phillips Milk of Magnesia Liquid; Simeco; Tepilta; component of: Arthritis Pain Formula Maximum Strength, Ascriptin, Calcitrel, Camalox, Di-Gel Liquid, Di-Gel Tablets, Gelusil, Haley's MO, Kestyomatin, Kudrox, Maalox, Maalox Plus, Mylanta Gelcaps, Mylanta Liquid, Mylanta Tablets, Simeco Suspension, Wingel. Antacid; cathartic; laxative (veterinary). Insoluble in H_2O, soluble in dilute acids. *Bristol-Myers Squibb Pharm. R&D; Johnson & Johnson-Merck Consumer Pharm. ; Lederle Labs.; Parke-Davis; Rhône-Poulenc Rorer Pharm. Inc.; Schering-Plough Pharm.; Schwarz Pharma Kremers Urban Co.; Sterling Health U.S.A.; Sterling Winthrop Inc.; Whitehall-Robins.*

1146 Magnesium Oxide
1309-48-4 5713 215-171-9
MgO
Magnesium oxide.
magnesia; calcined magnesia; magnesia usta; Magcal; Maglite. Antacid. Used in veterinary medicine as an antacid, laxative, and in treatment of hypomagnesemia. mp = 2800°; slightly soluble in H_2O, soluble in dilute acids, insoluble in EtOH; pH of saturated solution = 10.5.

1147 Magnesium Peroxide
14452-57-4 5717 238-438-1
MgO_2
Magnesium dioxide.
magnesium perhydrol; magnesium superoxol. Antiseptic; antacid. Peroxide. The commercial product contains 15-25% MgO_2, the balance being $Mg(OH)_2$. Insoluble in H_2O; gradually dec in H_2O with liberation of O_2; soluble in dilute acids, forming H_2O_2.

1148 Magnesium Phosphate, Tribasic
7757-87-1 5720 231-824-0
$Mg_3O_8P_2$
Trimagnesium phosphate.
tertiary magnesium phosphate. Antacid. [pentahydrate]: Insoluble in H_2O, soluble in dilute mineral acids.

1149 Magnesium Silicates
5727
$Mg_2Si_3O_8$
Magnesium trisilicate.
Magnosil; Petimin; Trisomin. Antacid. Insoluble in H_2O. *Merck & Co.Inc.*

1150 Polyethadine
9003-23-0
$(C_4H_6O_2)_m.(C_2H_5N)_n$
2,2'-Bioxirane polymer with aziridine.
1,2:3,4-diepoxybutane polymer with ethyleneimine. Antacid. *Eli Lilly & Co.*

1151 Potassium Citrate
866-84-2 7785 212-755-5
$C_6H_5K_3O_7$
Citric acid potassium salt.
Urocit-K. Antacid. Soluble in H_2O (153.8 g/100 ml), glycerol (40 g/100 ml); insoluble in EtOH; aqueous solution has a pH of 6.5.

1152 Potassium Citrate Monobasic
866-83-1 7785
$C_6H_5K_3O_7.H_2O$
Citric acid potassium salt monohydrate.
Urocit-K. Antacid. Soluble in H_2O (153.8 g/100 ml), glycerol (40 g/100 ml); insoluble in EtOH; pH (aqueous solution) = 6.5.

1153 Potassium Glucaldrate
23835-15-6
$C_6H_{16}AlKO_{11}$
Potassium diaqua[gluconato(2-)]-dihydroxyaluminate(1-).
Aciquel; McN-R-1162-22. Antacid. *McNeil Pharm.*

1154 Silodrate
12408-47-8
$Al_2Mg_2O_{11}Si_3.xH_2O$
Magnesium aluminometasilicate.
MP-1051; Simaldrate; aluminosilicic acid magnesium salt (1:2) hydrate. Antacid.

1155 Sodium Bicarbonate
144-55-8 8726 205-633-8
$CHNaO_3$
Sodium hydrogen carbonate.
sodium acid carbonate; baking soda. Antacid. Soluble in H_2O (10 g/100 ml at 25°, 8.3 g/100 ml at 18°), insoluble in EtOH; pH (aqueous solution) = 8.3.

Anthelmintics

1156 Alantolactone
546-43-0 208 208-899-3
$C_{15}H_{20}O_2$
[3aR-(3aα,5β,8aβ,9aα)]-3a,5,6,7,8,8a,9,9a-Octahydro-5,8a-dimethyl-3-methylene-naphtho-[2,3-b]furan-2(3H)-one.
helenin; alant camphor; elecampane camphor; inula camphor; Eupatal. Anthelmintic. Targets nematodes. mp = 78-79°; bp = 275°; $[\alpha]_D$ = 175° ($CHCl_3$); λ_m = 212 nm (ε 9500 EtOH); freely soluble in EtOH, $CHCl_3$, C_6H_6, Et_2O, oils; insoluble in H_2O.

1157 Albendazole Oxide
54029-12-8
$C_{12}H_{15}N_3O_3S$
Methyl 5-(propylsulfinyl)-2-benzimidazolecarbamate.
Anthelmintic. *SmithKline Beecham Animal Health; SmithKline Beecham Pharm.*

1158 Amidantel
49745-00-8
$C_{13}H_{19}N_3O_2$
4'-[[1-(Dimethylamino)ethylidene]amino]-2-methoxyacetanilide.
Anthelmintic. Used against hookworms, ascarids in dogs.

1159 Amocarzine
36590-19-9 608
$C_{18}H_{21}N_5O_2S$
4-Methyl-4'-(p-nitroanilino)thio-1-piperazinecarboxanilide.
CGP-6140. Derivative of amoscanate. Anthelmintic. Targets nematodes. mp = 191-196°; soluble in CH_3CN. *Ciba-Geigy Corp.*

1160 Amoscanate
26328-53-0 613
$C_{13}H_9N_3O_2S$
4-Isothiocyanato-N-(4-nitrophenyl)benzeneamine.
nithiocyamine; C-9333-Go; CGP-4540. Anthelmintic. Targets schistosoma. mp = 196-198°.

1161 Amphotalide
1673-06-9 626 216-809-9
$C_{19}H_{20}N_2O_3$
2-[5-(4-Aminophenoxy)pentyl]-1H-isoindole-1,3(2H)-dione.
RP-6171; Schistomide. Anthelmintic. Targets schistosoma. mp = 113-114°. *May & Baker Ltd.*

1162 Anthiolimine
305-97-5 720 206-173-0
$C_{12}H_9Li_6O_{12}S_3Sb$
Mercaptobutanedioic acid antimony(3+) lithium salt (3:1:6).
lithium antimony thiomalate; Anthiomaline. Trematodes. Very soluble in H_2O, slightly soluble in organic solvents. *Rhône-Poulenc.*

1163 Antimony Potassium Tartrate
28300-74-5 741
$C_8H_4K_2O_{12}Sb_2.3H_2O$
Dipotassium bis[μ-[2,3-dihydroxybutanedioato(4-)-01,02:03,04]]diantimonate(2-) trihydrate steroisomer.
tartar emetic; tartrated antimony; tartarized antimony; potassium antimonyl tartrate. Used as a mordant in the textiles and leather industries and as an anthelmintic. Targets schistosoma. d = 2.6; $[\alpha]_D^{20}$ = 140.69° (c = 2 H_2O), 139.25° (c = 2 in glycerol); soluble in H_2O (8.3 g/100 ml at 25°, 33.3 g/100 ml at 100°), glycerol (6.7 g/100 ml); insoluble in EtOH; LD_{50} (mus sc) = 55 mg/kg, (mus iv) = 65 mg/kg. *Stauffer Chem. Co.*

1164 Antimony Sodium Tartrate
34521-09-0 743 252-070-9
$C_4H_4NaO_7Sb$
Antimony sodium oxide L-(+)-tartrate.
sodium antimonyl tartrate; Emeto-Na; Stibunal. Anthelmintic. Targets schistosoma. Soluble in H_2O (66.6 g/100 ml); LD_{50} (mus iv) = 25 mg/kg.

1165 Antimony Sodium Thioglycollate
539-54-8 744
$C_4H_4NaO_4S_2Sb$
[(5-Oxo-1,3,2-oxathiostibolan-2-yl)thio]acetic acid sodium salt.
antimony sodium thioacetate. Anthelmintic. Targets schistosoma. Freely soluble in H_2O, unstable in alkali.

1166 Antimony Thioglycollamide
6533-78-4 746
$C_6H_{12}N_3O_3S_3Sb$
Thioantimonic acid tris(2-amino-2-oxoethyl) ester.
mercaptoacetamide antimony derivative; antimony thioglycollic acid triamide. Anthelmintic. Targets schistosoma. mp = 140°; soluble in H_2O (0.5 g/100 ml), insoluble in Et_2O.

1167 Arecoline
63-75-2 815 200-565-5
$C_8H_{13}NO_2$
1,2,5,6-Tetrahydro-1-methyl-3-pyridinecarboxylic acid methyl ester.
arecaline; arecholine; methylarecaidin. Cholinergic alkaloid from the seeds of the betel nut palm. Anthelmintic. Targets Cestodes. bp = 209°, bp_7 = 92-93°, bp_{12} = 105°; d^{20} = 1.0495; miscible with H_2O, EtOH, Et_2O; soluble in $CHCl_3$; LD_{50} (mus sc) = 100 mg/kg, (dog sc) = 5 mg/kg. *Nopco.*

1168 Arecoline Hydrobromide
300-08-3 815 206-087-3
$C_8H_{14}BrNO_2$
1,2,5,6-Tetrahydro-1-methyl-3-pyridinecarboxylic acid methyl ester hydrobromide.
Derivative of the cholinergic alkaloid from the seeds of the betel nut palm. Anthelmintic. Targets Cestodes. mp = 169-171°; soluble in H_2O (100 g/100 ml), EtOH (10 g/100 ml at 25°, 50 g/100 ml at 76°); slightly soluble in $CHCl_3$, Et_2O. *Nopco.*

1169 Arecoline p-Stibonobenzoic Acid
17162-36-6 815
$C_{15}H_{20}NO_7Sb$
1,2,5,6-Tetrahydro-1-methyl-3-pyridinecarboxylic acid methyl ester compound with stibonobenzoic acid.
arecoline p-stibonobenzoic acid; Anthelin. Derivative of the cholinergic alkaloid from the seeds of the betel nut palm. Anthelmintic. Targets Cestodes. *Nopco.*

1170 Ascaridole
512-85-6 864 208-147-4
$C_{10}H_{16}O_2$
1-Methyl-4-(1-methylethyl)-2,3-dioxabicyclo-[2.2.2]oct-5-ene.
1,4-peroxido-p-menthene-2; Ascarisin. Anthelmintic. Targets nematodes. d_4^{20} = 1.0103, d_{20}^{20} = 1.0113; mp = 3.3°; $bp_{0.2}$ = 39-40°; $[\alpha]_D^{20}$ = ±0.0°; soluble in C_6H_{14}, C_5H_{12}, EtOH, C_7H_8, C_6H_6, castor oil.

1171 Aspidin
584-28-1 881
$C_{25}H_{32}O_8$
2-[[2,6-Dihydroxy-4-methoxy-3-methyl-5-(1-oxobutyl)phenyl]methyl]-3,5-dihydroxy-4,4-dimethyl-6-(1-oxobutyl)-2,5-cyclohexadien-1-one.
polystichin. Active principle of fern root. Anthelmintic. Targets Cestodes. mp = 124-125°; λ_m = 230, 290 nm (ε 25500, 21300, cyclohexane); soluble in Et_2O, C_6H_6, $CHCl_3$; sparingly soluble in MeOH, EtOH, Me_2CO.

1172 Aspidinol
519-40-4 882
$C_{12}H_{16}O_4$
1-(2,6-Dihydroxy-4-methoxy-3-methylphenyl)-1-butanone.
Found in extracts of male fern. Anthelmintic. Targets Cestodes. mp = 156-161°; soluble in EtOH, Et_2O, $CHCl_3$; sparingly soluble in H_2O, C_6H_6.

1173 Becanthone
15351-04-9 1044
$C_{22}H_{28}N_2O_2S$
1-[[2-[Ethyl(2-hydroxy-2-methylpropyl)-amino]ethyl]amino]-4-methyl-9H-thio-xanthen-9-one.
becantone. Anthelmintic. Targets schistosoma. *Sterling Winthrop Inc.*

1174 Becanthone Hydrochloride
5591-22-0 1044
$C_{22}H_{29}ClN_2O_2S$
1-[[2-[Ethyl(2-hydroxy-2-methylpropyl)-amino]ethyl]amino]-4-methyl-9H-thio-xanthen-9-one hydrochloride.
Win-13820; Loranil. Anthelmintic. Targets schistosoma. mp = 157.6-160.4°. *Sterling Winthrop Inc.*

1175 Bephenium
7181-73-9 1187 230-546-7
$C_{17}H_{22}NO^+$
Benzylmethyl(2-phenoxyethyl)amine.
Anthelmintic. Targets nematodes. [chloride]: mp = 135-136°; [bromide]: mp = 144.5-146°; [iodide]: mp = 146-147°; [Pamoate (biphenium embonate; Frantin)]: mp = 144-146°. *Glaxo Wellcome Inc.*

1176 Bephenium Hydroxynaphthoate
3818-50-6 1187 223-306-8
$C_{28}H_{29}NO_4$
Benzyldimethyl(2-phenoxyethyl)ammonium 3-hydroxy-2-naphthoate.
Alcopar; Alcopara; Befeniol; Lecibis; Nemex. Anthelmintic. Targets nematodes. mp = 170-171°. *Glaxo Wellcome Inc.*

1177 Bithionoloxide
844-26-8
$C_{12}H_6Cl_4O_3S$
6,6'-Sulfinylbis(2,4-dichlorophenol).
Used in veterinary medicine as an anthelmintic targeting trematodes.

1178 Bitoscanate
4044-65-9 1345 223-741-3
$C_8H_4N_2S_2$
1,4-Diisothiocyanatobenzene.
Jonit. Anthelmintic. Targets nematodes. mp = 132°. *Hoechst AG.*

1179 Carbon Tetrachloride
56-23-5 1864 200-262-8
CCl_4
Tetrachloromethane.
tetrachloromethane; perchloromethane; Necatorina; Benzinoform. Anthelmintic. Targets nematodes. d_{25}^{25} = 1.589; mp = -23°; bp = 76.7°; soluble in H_2O (50 mg/100 ml); miscible with EtOH, $CHCl_3$, C_6H_6, Et_2O, CS_2, petroleum ether, oils; LC_{50} (mus ihl) = 9528 ppm; LD_{55} (rat orl) = 2920 mg/kg, (mus orl) = 12100-14400 mg/kg, (dog orl) = 2300 mg/kg, (mus ip) = 4100 mg/kg, (mus sc) = 30400 mg/kg.

1180 Carvacrol
499-75-2 1923 207-889-6
$C_{10}H_{14}O$
2-Methyl-5-(1-methylethyl)phenol.
2-hydroxy-p-cymene; isopropyl o-cresol; isothymol. Anthelmintic. Targets nematodes. Used as a general disinfectant. Phenol found in the oil of origanum, thyme, marjoram, and summer savory. An essential oil. Liquid; mp 0°; bp_{760} = 237-238°; bp_{18} = 118-122°; bp_3 = 93°; d_4^{20} = 0.976; d_{25}^{25} = 0.9751; n_D^{20} = 1.52295; λ_m = 277.5 (log ε 3.262 in EtOH); volatile with steam; nearly insoluble in H_2O; soluble in alcohol, Et_2O; LD_{50} (rbt orl) = 100 mg/kg.

1181 Cyclobendazole
31431-43-3 2781 250-637-5
$C_{13}H_{13}N_3O_3$
Methyl 5-(cyclopropylcarbonyl)-2-benzimidazolecarbamate.
R-17147; CC-2481; Haptocil. Anthelmintic. Targets nematodes. mp = 250.5°. *Janssen Pharm. Ltd.*

1182 Diammonium Embelate
3595
$C_{17}H_{34}N_2O_4$
2,5-Dihydroxy-3-undecyl-2,5-cyclohexadiene-1,4-dione diammonium salt.
ammonium embelate; embelin diammonium salt. Anthelmintic (cestodes). Mucous membrane irritant. *See* Embelin. Soluble in H_2O, dilute alcohol.

1183 Dichlorophen
97-23-4 3120 202-567-1
$C_{13}H_{10}Cl_2O_2$
2,2'-Methylenebis(4-chlorophenol).
dichlorophene; G-4. Anthelmintic. Targets Cestodes.

1184 Diethylcarbamazine
90-89-1 3165 202-023-3
$C_{10}H_{21}N_3O$
N,N-Diethyl-4-methyl-1-piperazinecarboxamide.
carbamazine; 84L; RP-3799; Carbilazine; Caricide; Cypip; Ethodryl; Notézine; Spatonin; [phosphate]:

Ditrazin. Anthelmintic. Targets nematodes. mp = 47-49°; bp_3 = 108.5-111°; [hydrochloride]: mp = 156.5-157°.

1185 Diethylcarbamazine Citrate
1642-54-2 3165 216-696-6
$C_{16}H_{29}N_3O_8$
N,N-Diethyl-4-methyl-1-piperazinecarboxamide citrate.
Banocide; Dec; Dirocide; Filaribits; Filazine; Franocide; Hetrazan; Loxuran; Longicid. Anthelmintic. Targets nematodes. mp = 141-143°; soluble in H_2O (> 75 g/100 ml); soluble in EtOH; insoluble in C_6H_6, Me_2CO, Et_2O, $CHCl_3$; LD_{50} (rat orl) = 1.38 g/kg.

1186 Difetarsone
3639-19-8
$C_{14}H_{18}As_2N_2O_6$
N,N-Ethylenediarsanilic acid.
Bemarsal Anthelmintic, used for the treatment of Trichuris trichiura (whipworm) infestation.

1187 Diphenane
101-71-3 3365
$C_{14}H_{13}NO_2$
α-Phenyl-p-cresolcarbamate.
Anthelmintic. Targets nematodes. mp = 147-150°; insoluble in H_2O; soluble EtOH, MeOH, $CHCl_3$, Et_2O, C_6H_6.

1188 Dithiazanine Iodide
514-73-8 3434 208-186-7
$C_{23}H_{23}IN_2S_2$
3-Ethyl-2-[5-(3-ethyl-2-benzothiazolinylidene)-1,3-pentadienyl]benzothiazolium iodide.
Developed as a photographic sensitizer. Anthelmintic. Targets nematodes. mp = 248° (dec); insoluble in H_2O.

1189 Doramectin
117704-25-3 3483
$C_{50}H_{74}O_{14}$
25-cyclohexyl-5-O-demethyl-25-de-(1-methylpropyl)avermectin A_{1a}.
25-cyclohexylavermectin B_1; UK-67994; Dectomax; Endectocide, used to treat Sheep scabies (veterinary). mp= 116-119°. *Pfizer Intl.*

1190 Dymanthine
124-28-7 3525 204-694-8
$C_{20}H_{43}N$
N,N-Dimethyloctadecylamine.
N,N-Dimethylstearylamine; N,N-Dimethyloctadecylamine; Armeen DM 18D; N,N-Dimethyl-1-octadecanamine; 18. Anthelmintic. Targets nematodes. mp = 23°. *Pfizer Inc.*

1191 Dymanthine Hydrochloride
1613-17-8 3525 216-559-0
$C_{20}H_{44}ClN$
N,N-Dimethyloctadecylamine hydrochloride.
Dimantine; N-n-Octadecyl-N,N-dimethyl amine; GS-1339; NSC-5547; Thelmesan. Anthelmintic. Targets nematodes. *Pfizer Inc.*

1192 Embelin
550-24-3 3595 208-979-8
$C_{17}H_{26}O_4$
2,5-Dihydroxy-3-undecyl-2,5-cyclohexadiene-1,4-dione.
embelic acid. Anthelmintic. Targets Cestodes.

1193 Eprinomectin
133305-89-2 3667
Mixture of eprinomectin B_{1a} (90%) and eprinomectin B_{1b} (10%).
Anthelmintic targeting nematodes.

1194 Fenfluthrin
75867-00-4
$C_{15}H_{11}Cl_2F_5O_2$
2,3,4,5,6-Pentafluorobenzyl (1R,3S)-3-(2,2-dichlorovinyl)-2,2-dimethylcyclopropanecarboxylate.
Synthetic pyrethroid used as an insecticide and anthelmintic.

1195 Fexinidazole
59729-37-2
$C_{12}H_{13}N_3O_3S$
1-Methyl-2-[[p-(methylthio)phenoxy]methyl]-5-nitroimidazole.
Anthelmintic.

1196 Gentian Violet
548-62-9 4401 208-953-6
$C_{25}H_{30}ClN_3$
[4-[Bis[4-(dimethylamino)phenyl]methylene]-2,5-cyclohexadien-1-ylidene]dimethylammonium chloride.
C.I. Basic Violet 3; hexamethylpararosaniline chloride; aniline violet; crystal violet; methylrosaniline chloride; C.I. 42555; Adergon; Axuris; Badil; Gentiaverm; Meroxylan; Meroxyl; Pyoktanin; Vianin; Viocid. Anthelmintic. Targets nematodes. Insoluble in Et_2O; soluble in H_2O, $CHCl_3$, EtOH (10 g/100 ml), glycerin (6.7 g/100 ml); LD_{50} (mus orl) = 1200 mg/kg, (rat orl) = 1000 mg/kg.

1197 4-Hexylresorcinol
136-77-6 4750 205-257-4
$C_{12}H_{18}O_2$
4-Hexyl-1,3-benzenediol.
ST-37; Ascaryl; Caprokol; Crystoids; Gelovermin; Sucrets; Worm-Agen. Used as a topical antiseptic. Anthelmintic. Targets nematodes. mp = 67.5-69°; bp = 333-335°, bp_{6-7} = 1768-180°, bp_{13-14} = 198-200°; soluble in H_2O (50 mg/100 ml), Et_2O, $CHGCl_3$, Me_2CO, EtOH; LD_{50} (rat orl) = 550 mg/kg. *Merck & Co.Inc.*

1198 Hycanthone
3105-97-3 4795 221-463-7
$C_{20}H_{24}N_2O_2S$
1-[[2-(Diethylamino)ethyl]amino]-4-(hydroxymethyl)-9H-thioxanthen-9-one.
Etrenol [as mesylate]. Metabolite of lucanthone. Anthelmintic. Targets schistosoma. mp = 100.6-102.8°; λ_m = 233, 258, 329, 438 nm (ε 19400, 37000, 9700, 6600 EtOH); sensitive to acid; [hydrochloride]: mp = 173-176° (dec). *Sterling Health U.S.A.*

1199 Ivermectin

70288-86-7 5264 274-536-0

$C_{48}H_{74}O_{14}$

(2aE,4E,8E)-(5'S,6S,6'R,7S,11R,13R,15S,17aR,20R,20aR,-20bS)-6'-(S)-sec-Butyl-3',4',5',6,6',7,10,11,14,15,17a,20,-20a,20b)-tetradecahydro-20,20b-dihydroxy-5',6,8,19-tetramethyl-17-oxospiro[11,15-methano-2H,13H,17H]-furo[4,3,2-pq][2,6]benzodioxacyclooctadecin-13,2'-[2H]pyran]-7-yl 2,6-dideoxy-4-O-(2,6-dideoxy-3-O-methyl-α-L-arabino-hexopyranosyl)-3-O-methyl-α-L-arabino-hexopyranoside.

22,23-dihydro C-076B; MK-933; Cardomec; Cardotek 30; Eqvalan; Heartgard 30; Ivomec; Mectizan; Zimecterin. Mixture of avermectins, primarily avermectin 22,23-dihydroavermectin B_{1a}. Anthelmintic. Targets onchocerca (filarial worms). $[\alpha]_D = 71.5° \pm 3°$; λ_m = 238, 245 nm (ε 27100, 30100 MeOH); slightly soluble in H_2O (0.4 mg/100 ml); insoluble in hydrocarbons; very soluble in MEK, prolylene glycol, polyethylene glycol.

1200 Kainic Acid

487-79-6 5289

$C_{10}H_{15}NO_4$

[2S-(2α,3β,4β)]-2-Carboxy-4-(1-methylethenyl)-3-pyrrolideneacetic acid.

digenic acid; α-kainic acid; L_s-xylo-kainic acid; Digenin; Helminal. Anthelmintic. Targets nematodes. mp = 251° (dec); $[\alpha]_D^{24}$ = -14.8° (c = 1.01); soluble in H_2O, insoluble in Et_2O.

1201 α-Kosin

568-50-3 5333

$C_{25}H_{32}O_8$

5,5'-Methylenebis[4,6-dihydroxy-2-methoxy-3-methylisobutyrophenone].

From flowers of Hagenia abyssinica J. J. Gmel. Co-occurs with β-kosin. Anthelmintic. Targets Cestodes. mp= 160-160.5°;λ_m = 227, 290 nm (ε 30800, 24400); soluble in EtOH, C_6H_6, $CHCl_3$, Et_2P, AcOH; [tetraacetate]; mp = 124°.

1202 β-Kosin

5333

$C_{25}H_{32}O_8$

5,5'-Methylenebis[2,4,6-trihydroxy-3-methylisobutyrophenone]

4'-methyl ether.

From flowers of Hagenia abyssinica J. J. Gmel. Co-occurs with α-kosin. Anthelmintic. Targets Cestodes. mp = 120°; λ_m = 228, 292 nm (ε 30300, 21260).

1203 Levamisole

14769-73-4 5486 238-836-5

$C_{11}H_{12}N_2S$

(-)-2,3,5,6-Tetrahydro-6-phenylimidazol-[2,1-b]thiazole.

Levovermax; Totalon; tetramisole [as DL-form]; tetramizole [as DL-form]; dexamisole [as D(+)-form]. Immunomodulator with anthelmintic activity (targets nematodes). mp = 60-61.5°; $[\alpha]_D^{25}$ = -85.1° (c = 10 $CHCl_3$); [DL-form]: mp = 87-89°; [D-(+)-form]: mp = 60-61.5°; $[\alpha]_D^{25}$ = 85.1° (c = 10 $CHCl_3$). *Am. Cyanamid; Bayer Corp. Pharm. Div.; ICI ; Janssen Pharm. Inc.; McNeil Pharm.*

1204 Levamisole Hydrochloride

16595-80-5 5486 240-654-6

$C_{11}H_{13}Cl2N_2S$

(-)-2,3,5,6-Tetrahydro-6-phenylimidazol-[2,1-b]thiazole hydrochloride.

R-12654; Ascaridil; Decaris; Ergamisol; Ergamisole; Levacide; Levadin; Levasole; Meglum; Nemicide; Nilverm; Ripercol; Solaskil; Spartakon; Tramisol; Bayer 9051 [as DL-form]; McN-JR-8299 [as DL-form]; R-8299 [as DL-form]; R-12563 [as D-(+)-form]. Immunomodulator with anthelmintic activity (targets nematodes). mp = 227-229°; $[\alpha]_D^{20}$ = -124° ± 2° (c = 0.9 H_2O); soluble in H_2O; [DL-form]: mp = 264-265°; soluble in H_2O (21 g/100 ml), MeOH, propylene glycol; sparingly soluble in EtOH, $CHCl_3$, C_6H_{14}, Me_2CO; LD_{50} (mus iv) = 22 mg/kg, (mus sc) = 84 mg/kg, (mus orl) = 210 mg/kg, (rat iv) = 24 mg/kg, (rat sc) = 130 mg/kg, (rat orl) = 480 mg/kg; [D-(+)-form]: mp = 227-227.5°; $[\alpha]_D^{20}$ = 125° (c = 0.7 H_2O). *Am. Cyanamid; Bayer Corp. Pharm. Div.; ICI ; Janssen Pharm. Inc.; McNeil Pharm.*

1205 Lucanthone Hydrochloride

548-57-2 5620 208-951-5

$C_{20}H_{25}ClN_2OS$

1-[[2-(Diethylamino)ethyl]amino]-4-methyl-9H-thioxanthen-9-one hydrochloride.

MS-752; RP-3735; Miracil D; Nilodin; Miracol; Tixantone. Anthelmintic. Targets schistosoma. mp = 195-196°; soluble in H_2O, slightly soluble in EtOH; [lucanthone]: mp = 64-65°, soluble in most organic solvents.

1206 Mebendazole

31431-39-7 5807 250-635-4

$C_{16}H_{13}N_3O_3$

Methyl 5-benzoyl-2-benzimidazolecarbamate.

Vermox; R-17635; Bantenol; Equivurm Plus; Lomper; Mebenvet; Noverme; Ovitelmin; Pantelmin; Telmin; Vermicidin; Vermirax. Anthelmintic. Targets nematodes. mp = 288.5°; insoluble in H_2O, EtOH, Et_2O, $CHCl_3$; soluble in formic acid; LD_{50} (sheep orl) > 80 mg/kg, (mus, rat, chk) > 40 mg/kg. *Janssen Pharm. Ltd.*

1207 Naphthalene

91-20-3 6457 202-049-5

$C_{10}H_8$

Naphthalene.

naphthalin; naphthene; tar camphor. Anthelmintic. Targets Cestodes. mp = 80.2°; bp = 217.9°; d_4^{20} = 1.162; d_4^{100} = 0.9628; insoluble in H_2O; soluble in EtOH or MeOH (7.7 g/100 ml), C_6H_6 or C_7H_8 (28.6 g/100 ml), $CHCl_3$ or CCl_4 (50 g/100 ml), CS_2 (83.3 g/100 ml).

1208 2-Naphthol

135-19-3 6471 205-182-7

$C_{10}H_8O$

2-Naphthalenol.

beta-naphthol; β-naphthol; β-hydroxynaphthalene; isonaphthol; C.I. Azoic Coupling Component 1; C.I. Developer 5; C.I. 37500. Formerly used as an anthelmintic. Targets nematodes. mp = 121-123°; bp = 285-286°; d = 1.22; λ_m = 226, 265, 275, 286, 320, 331 (ε

91194, 3911, 4559, 3301, 1861, 2163 EtOH); soluble in H_2O (0.1 g/100 ml at 35°, 1.25 g/100 ml at 100°), EtOH 125 g/100 ml), $CHCl_3$ (5.9 g/100 ml), Et_2O (76.9 g/100 ml), glycerol, olive oil.

1209 Niclosamide
50-65-7 6602 200-056-8
$C_{13}H_8Cl_2N_2O_4$
2',5-Dichloro-4'-nitrosalicyanilide.
Niclocide; Yomesan; BAY-2353; Cestocide; Niclocide; Ruby; Trédémine. Anthelmintic. Targets Cestodes. mp = 225-230°; insoluble in H_2O; sparingly soluble in EtOH, $CHCl_3$, Et_2O. *Bayer AG.*

1210 Niclosamide Ethanolamine Salt
1420-04-8 6602 215-811-7
$C_{15}H_{15}Cl_2N_3O_5$
2',5-Dichloro-4'-nitrosalicyanilide ethanolamine salt.
clonitrilide; Bayluscid. Anthelmintic. Targets Cestodes. mp = 204°. *Bayer AG.*

1211 Niridazole
61-57-4 6656 200-512-6
$C_6H_6N_4O_3S$
1-(5-Nitro-2-thiazolyl)-2-imidazolidinone.
nitrothiamidazol; Ba-32644; Ciba 32644-Ba; Ambilhar. Anthelmintic. Targets schistosoma. mp = 260-262°. *CIBA plc.*

1212 Nitroclofene
39224-48-1
$C_{13}H_8Cl_2N_2O_6$
4,6'-Dichloro-4',6-dinitro-2,2'-methylenediphenol.
Anthelmintic.

1213 Oltipraz
64224-21-1 264-736-6
$C_8H_6N_2S_3$
4-Methyl-5-(pyrazinyl)-3H-1,2-dithiole-3-thione.
RP-35972. An antischistosomal drug with chemoprotective properties. Anticarcinogen.

1214 Oxamniquine
21738-42-1 7051 244-556-4
$C_{14}H_{21}N_3O_3$
1,2,3,4-Tetrahydro-2-[[(1-methylethyl)amino]methyl]-7-nitro-6-quinolinemethanol.
UK-4271; Mansil; Vansil. Anthelmintic. Targets schistosoma. mp = 147-149°; soluble in Me_2CO, $CHCl_3$, MeOH, H_2O (0.03 g/100 ml); λ_m = 205.5, 249.5, 389.5 ($A^{1\%}_{1\ ccm}$ 486, 695, 62.5 MeOH); LD_{50} (mus im) > 2000 mg/kg, (mus orl) = 1300 mg/kg, (rbt im) > 1000 mg/kg, (rbt orl) = 800 mg/kg. *Pfizer Inc.*

1215 Oxantel
36531-26-7 7055
$C_{13}H_{16}N_2O$
(E)-m-[2-(1,4,5,6-Tetrahydro-1-methyl-2-pyrimidinyl)vinyl]phenol.
CP-14445. Anthelmintic. Targets nematodes. [hydrochloride]: mp = 207-208°; λ_m = 231, 274 nm (ε 12700, 20100, H_2O). *Pfizer Inc.*

1216 Oxantel Pamoate
68813-55-8 7055 272-332-6
$C_{36}H_{32}N_2O_7$
(E)-m-[2-(1,4,5,6-Tetrahydro-1-methyl-2-pyrimidinyl)-vinyl]phenol 4,4'-methylenebis[3-hydroxy-2-naphthoate] (1:1) (salt).
Telopar; CP-14445-16; oxantel ebonate. Anthelmintic. Targets nematodes. *Pfizer Inc.*

1217 Papain
9001-73-4 7148 232-627-2
vegetable pepsin; Arbuz; Caroid; Nematolyt; Papayotin; Summetrin; Tromasin; Velardon; Vermizym; component of: Panafil. Proteolytic enzyme. Digestive aid. A proteolytic enzyme isolated from the fruit and leaves of *Carica papaya*. Used as a digestive aid, also as a debridant, an anthelmintic targeting Nematodes and as an agent which can prevent adhesions. λ_m = 278 nm ($A^{1\%}_{1\ cm}$ 25.0); insoluble in most organic solvents. *Rystan Co. Inc.; Sterling Winthrop Inc.*

1218 Pelletierine
4396-01-4 7200 224-523-0
$C_8H_{15}NO$
1-(2-Piperidinyl)-2-propanone.
2-acetonylpiperidine; punicine; isopelletierine, (±)-pelletierine. Anthelmintic. Targets Cestodes. bp = 195°; bp_{11} = 102-107°; d_4^{20} = 0.988; soluble in H_2O (5 g/100 ml), EtOH, Et_2O, $CHCl_3$.

1219 Pelletierine Hydrochloride
5984-61-2 7200
$C_8H_{16}ClNO$
1-(2-Piperidinyl)-2-propanone hydrochloride.
Anthelmintic. Targets Cestodes. mp = 145°; soluble in H_2O, EtOH.

1220 Piperazine
110-85-0 7617 203-808-3
$C_4H_{10}N_2$
Piperazine.
hexahydropyrazine; piperazidine; diethylenediamine. Anthelmintic. Targets nematodes. mp = 106°; bp = 146°; freely soluble in H_2O, glycerol, glycols, EtOH (50 g/100 ml); insoluble in Et_2O. *Union Carbide Corp.*

1221 Piperazine Adipate
142-88-1 7617 205-569-0
$C_{10}H_{20}N_2O_4$
Piperazine compound with hexanedioic acid (1:1).
Entacyl; Oxyzin; Vermicompren; Nometan; Oxypaat; Pipadox; Oxurasin. Anthelmintic. Targets nematodes. mp = 256-257°; soluble in H_2O (5.53 g/100 ml at 20°, 6.61 g/100 ml at 30°, 10.14 g/100 ml at 56.3°), MeOH (0.02 g/100 ml at 25°); insoluble in EtOH, iPrOH, dioxane; LD_{50} (mus orl) = 115400 g/kg, (rat orl) = 7900 mg/kg. *BDH Laboratory Supplies.*

1222 Piperazine Citrate
144-29-6 7617 205-622-8
$C_{24}H_{46}N_6O_{14}$
Tripiperazine dicitrate.
Helmezine; Oxucide; Patazine; Pinozan; Pipizan Citrate; Pipracid (syrup); Rhomex; Ta-Verm; Worm Away.

Anthelmintic. Targets nematodes. mp = 182-187° (dec); insoluble in EtOH, Et_2O, $CHCl_3$. *Sterling Winthrop Inc.*

1223 Praziquantel
55268-74-1 7896 259-559-6
$C_{19}H_{24}N_2O_2$
2-(Cyclohexylcarbonyl)-1,2,3,6,7,11b-hexahydro-4H-pyrazino[2,1-a]isoquinolin-4-one.
EMBAY 8440; Biltricide; Cesol; Droncit. Anthelmintic. Targets schistosoma. mp = 136-138°; soluble in H_2O (0.04 g/100 ml), EtOH (9.7 g/100 ml), $CHCl_3$ (56.7 g/100 ml); LD_{50} (mus orl) = 2000-3000 mg/kg, (mus sc) > 3000 mg/kg, (rat orl) = 2000-3000 mg/kg, (rat sc) > 3000 mg/kg. *E. Merck.*

1224 Pyrantel
15686-83-6 8139 239-774-1
$C_{11}H_{14}N_2S$
(E)-1,4,5,6-Tetrahydro-1-methyl-2-[2-(2-thienyl)vinyl]pyrimidine.
Anthelmintic. Targets nematodes. mp = 178-179°. *Pfizer Inc.*

1225 Pyrantel Pamoate
22204-24-6 8139 244-837-1
$C_{34}H_{30}N_2O_6S$
(E)-1,4,5,6-Tetrahydro-1-methyl-2-[2-(2-thienyl)vinyl]pyrimidine compound with 4,4'-methylenebis[3-hydroxy-2-naphthoate].
Antiminth; Combantrin; Cobantril; Early Bird; Helmex; Helmintox; Piranver; CP-10423-16; component of: Drontal, Drontal Plus, HeartGard Plus. Anthelmintic. Targets nematodes. Insoluble in H_2O. *Bayer Corp.; Merck & Co.Inc.; Roerig Div. Pfizer Pharm.*

1226 Pyrantel Tartrate
33401-94-4 8139 251-501-8
$C_{15}H_{20}N_2O_6S$
(E)-1,4,5,6-Tetrahydro-1-methyl-2-[2-(2-thienyl)vinyl]pyrimidine tartrate (1:1).
Banminth; Strongid; CP-10423-18. Anthelmintic. Targets nematodes. mp = 148-150°; λ_m = 312 nm (log ε 4.27 H_2O). *Pfizer Inc.*

1227 Pyrvinium Pamoate
3546-41-6 8206 222-596-3
$C_{75}H_{70}N_6O_6$
6-(Dimethylamino)-2-[2-(2,5-dimethyl-1-phenyl-pyrrol-3-yl)vinyl]-1-methylquinolinium 4,4'-methylenebis[3-hydroxy-2-naphthoate] (2:1).
Povan; pyrvinium embonate; viprynium embonate; Molevac; Neo-Oxypaat; Pamovin; Poquil; Povanil; Pyrcon; Tru; Vanquin; Vermitibier. Anthelmintic. Targets nematodes. mp = 210-215°; λ_m = 236, 356, 503 nm; insoluble in H_2O, Et_2O; slightly soluble in EtOH, $CHCl_3$, methoxyethanol. *Parke-Davis.*

1228 Quinacrine
83-89-6 8225 201-508-7
$C_{23}H_{30}ClN_3O$
6-Chloro-9-[[4-(diethylamino)-1-methylbutyl]amino]-2-methoxyacridine.
Mepacrine; Atabrine. Anthelmintic; antimalarial. Targets cestodes. *Sterling Winthrop Inc.*

1229 Quinacrine Dihydrochloride Dihydrate
6151-30-0 8225
$C_{23}H_{31}Cl_2N_3O.2H_2O$
6-Chloro-9-[[4-(diethylamino)-1-methylbutyl]amino]-2-methoxyacridine dihydrochloride dihydrate.
Atabrine hydrochloride; RP-866; SN-390. Anthelmintic; antimalarial. Targets cestodes. Dec 248-250°; soluble in H_2O (2.8 g/100 ml); slightly soluble in EtOH, MeOH; insoluble in C_6H_6, Et_2O, Me_2CO. *Sterling Winthrop Inc.*

1230 Quinacrine Methanesulfonate Monohydrate
6598-46-5 8225
$C_{25}H_{38}ClN_3O_7S_2.H_2O$
6-Chloro-9-[[4-(diethylamino)-1-methylbutyl]amino]-2-methoxyacridine monomethanesulfonate monohydrate.
Anthelmintic; antimalarial. Targets cestodes. Soluble in H_2O (33 g/100 ml at 15°), EtOH (2.8 g/100 ml). *Sterling Winthrop Inc.*

1231 Quintiofos
1776-83-6 217-208-4
$C_{17}H_{16}NO_2PS$
O-Ethyl O-(8-quinolyl)phenylphosphonothioate.
Bayer 9037. Ixodicide. *Bayer AG.*

1232 α-Santonin
481-06-1 8509 207-560-7
$C_{15}H_{18}O_3$
[3S-(3α,3aα,5aβ,9bβ)]-3a,5,5a,9b-Tetrahydro-3,5a,9-trimethylnaphtho[1,2-b]furan-2,8(3H,4H)dione.
l-Santonin. Anthelmintic. Targets nematodes. [(-)-form]: mp = 170-173°; $[\alpha]_D^{25}$ = -170° to -175° (c = 2 EtOH); d = 1.187; soluble in H_2O (0.02 g/100 ml at 25°, 0.4 g/100 ml at 100°), 50% EtOH (0.36 g/100 ml at 25°, 10 g/100 ml at 76°), 90% EtOH (2.3 g/100 ml at 25°, 33.3 g/100 ml at 76°), Et_2O (0.8 g/100 ml at 25°), 1.4 g/100 ml at 34.6°), $CHCl_3$ (23.2 g/100 ml at 25°); [±-form]: mp = 181°; λ_m = 241 nm (log ε 4.10 EtOH); [(+)-form]: mp = 172°; $[\alpha]_D^{20}$ = 165.9° (C = 1.92 EtOH).

1233 Sodium Antimonylgluconate
12550-17-3 742 235-699-3
$C_6H_8NaO_7Sb$
Triostam. Anthelmintic. Targets schistosoma. Trivalent antimony complex with sodium gluconate. Soluble in H_2O, insoluble in organic solvents. *Burroughs Wellcome Inc.*

1234 Stibocaptate
27279-76-1 8966
$C_{12}H_6Na_6O_{12}S_6Sb_2$
2,2'-[(1,2-Dicarboxy-1,2-ethanediyl)bis(thio)]bis-1,3,2-dithiastibolane-4,5-dicarboxylic acid hexasodium salt.
TWSb; Ro-4-1544/6; SB-58; Astiban. Anthelmintic. Targets schistosoma. Soluble in H_2O, LD_{50} (mus sc) = 500 mg Sb/kg. *Hoffmann-LaRoche Inc.*

1235 Stibophen
15489-16-4 8967
$C_{12}H_4Na_5O_{16}S_4Sb.7H_2O$
(T-4)-Bis[4,5-dihydroxy-1,3-benzenedisulfonato(4-)-O^4,O^5-]-antimonate(5-) pentasodium heptahydrate.
Sdt-91; Fuadin; Fouadin; Fantorin; Neoantimosan;

Repodral. Anthelmintic. Targets schistosoma. Soluble in cold H_2O; insoluble in EtOH, Et_2O, $CHCl_3$; Me_2CO, petroleum ether; LD_{50} (rbt iv) ≅ 90 mg/kg. *Heyden Chem.; I.G. Farben; Sterling Winthrop Inc.*

1236 Stilbazium Iodide
3784-99-4 8971 223-247-8
$C_{31}H_{36}IN_3$
1-Ethyl-2,6-bis[2-[4-(1-pyrrolidinyl)phenyl]ethenyl]-pyridinium iodide.
BW-61-32; Monopar. Anthelmintic. Targets nematodes. mp = 282-283°; insoluble in hot MeOH; LD_{50} (mus ip) = 7 mg/kg, (mus orl) = 1360 mg/kg. *Burroughs Wellcome Inc.*

1237 Suramin Sodium
129-46-4 9181 204-949-3
$C_{51}H_{34}N_6Na_6O_{23}S_6$
8,8'-[Carbonylbis[imino-3,1-phenylene-carbonylimino-(4-methyl-3,1-phenylene)-carbonylimino]]bis-1,3,5-naphthalene-trisulfonic acid hexasodium salt.
suramin hexasodium; Bayer 205; Fourneau 309; Antrypol; Germanin; Moranyl; Naganol; Naphuride. Antineoplastic; antiviral. Anthelmintic against nematodes and antiprotozoal against Trypanosoma. Freely soluble in H_2O; poorly soluble in EtOH; insoluble in Et_2O, $CHCl_3$, petroleum ether; LD_{50} (mus iv) ≅ 620 mg/kg. *Bayer AG; Parke-Davis.*

1238 Tetrachloroethylene
127-18-4 9332 204-825-9
C_2Cl_4
Tetrachloroethene.
perchloroethylene; ethylene tetrachloride; Nema; Tetracap; Tetropil; Perclene; Ankilostin; Didakene. Anthelmintic. Targets nematodes and trematodes. d_4^{15} = 1.6311; d_4^{20} = 1.6230; bp = 121°; mp ≅ -22°; almost insoluble in H_2O, miscible with most organic solvents; LD_{50} (mus orl) = 8850 mg/kg, LC_{50} (mus ihl) = 5925 ppm.

1239 Thiabendazole
148-79-8 9426 205-725-8
$C_{10}H_7N_3S$
2-(4-Thiazolyl)-1H-benzimidazole.
MK-360; Omnizole; Thiaben; Thibenzole; Bovizole; Eprofil; Equizole; Mintezol; Top Form Wormer; Mertect; Lombristop; Minzolum; Nemapan; Polival; TBZ; Tecto. Anthelmintic. Targets nematodes. mp = 304-305°; λ_m = 298 nm (ε 23330 MeOH); slightly soluble in H_2O (3.84 g/100 ml at pH 2.2), soluble in DMF, DMSO; slightly soluble in alcohols, esters, chlorocarbons; LD_{50} (mus orl) = 3600 mg/kg, (rat orl) = 3100 mg/kg, (rbt orl) > 3800 mg/kg. *Merck & Co.Inc.*

1240 Thiabendazole Hypophosphite
28558-32-9 9426
$C_{10}H_8ClN_3S$
2-(4-Thiazolyl)-1H-benzimidazole hydrochloride.
Anthelmintic. Targets nematodes. Liquid; d^{25} = 1.103. *Merck & Co.Inc.*

1241 Thymol
89-83-8 9540 201-944-8
$C_{10}H_{14}O$
5-Methyl-2-(1-methylethyl)phenol.
3-p-cymenol; thyme camphor; m-thymol. Anthelmintic. Targets nematodes. bp ≅ 233°; mp = 51.5°; d_4^{25} = 0.9699; soluble in H_2O (0.1 g/100 ml), EtOH (100 g/100 ml), $CHCl_3$ (143 g/100 ml); Et_2O (66.7 g/100 ml), olive oil (142.8 g/100ml at 25°); LD_{50} (rat orl) = 980 mg/kg.

1242 Thymol N-Isoamylcarbamate
578-20-1 9550
$C_{16}H_{25}NO_2$
Isoamylcarbamic acid thymyl ester.
Egressin. Anthelmintic. Targets nematodes. mp = 57°; insoluble in H_2O. *E. Merck.*

1243 Triclofenol Piperazine
5714-82-9 9787
$C_{16}H_{16}Cl_6N_2O_2$
2,4,5-Trichlorophenol compound with piperazine (2:1).
Cl-416; Ranestol. Anthelmintic. Targets nematodes. mp = 109-110°. *Parke-Davis.*

1244 Urea Stibamine
1340-35-8 10010
MF Unknown
Carbostibamide.
Anthelmintic, targeting nematodes and Schistosoma; antiprotozoal against Leishmania. Chemical composition uncertain. Active principle thought to be a substituted urea: sym-diphenylcarbamido-4,4-distibinic acid. Soluble in H_2O, partly soluble in EtOH, Et_2O. *Bristol-Myers Squibb Co.*

Antiacne Agents

1245 Adapalene
106685-40-9
$C_{28}H_{28}O_3$
6-[3-(1-Adamantyl)-4-methoxyphenyl]-2-naphthoic acid.
Differin Gel; CD-271. Used to treat acne. *Galderma Labs Inc.*

1246 Algestone Acetophenide
24356-94-3 239 246-195-8
$C_{29}H_{36}O_4$
(R)-16α,17-Dihydroxypregn-4-ene-3,20-dione cyclic acetal with acetophenone.
Neolutin Depositum; P-DHP; SQ-15,101; Deladroxone; Droxone. Progestogen. Used to treat acne. mp = 150-151°; $[\alpha]_D^{23}$ = 51° ($CHCl_3$). *Bristol-Myers Squibb Co.*

1247 Azelaic Acid
123-99-9 938 204-669-1
$C_9H_{16}O_4$
1,7-Heptanedicarboxylic acid.
ZK-62498; lepargylic acid; anchoic acid; Skinoren; nonanedioic acid. Used to treat acne. mp = 106.5°; bp_{100} = 286.5°, bp_{50} = 265°; bp_{15} = 237°; bp_{10} = 225°; $d_4^{10.6}$ = 1.0291; soluble in H_2O (0.1 g/100 ml at 1°, 0.24 g/100 ml at 20°, 0.82 g/100 ml at 50°; 2.2 g/100 ml at 65°);

freely soluble in boiling H_2O or EtOH; soluble in Et_2O (1.88 g/100 ml at 11°, 2.68 g/100 ml at 15°); [dimethyl ester ($C_{11}H_{20}O_4$)]: d_4^{20}= 1.0026; mp = -3.9°; bp_8 = 140°. *Schering AG.*

1248 Benzoyl Peroxide
94-36-0 1149 202-327-6
$C_{14}H_{10}O_4$
Dibenzoyl peroxide.
Acne-Aid Cream; Benoxyl; Benzac; Benzac W; Benzagel; Benzamycin; Brevoxyl; Clear By Design; Desquam-X; Dry and Clear; Epi-Clear; Fostex BPO Bar, Gel and Wash; Loroxide; Panoxyl; Persa-Gel; pHisoAc BP; Sulfoxyl; Vanoxide; benzoyl superoxide; Acetoxyl; Acnegel; Akneroxide L; Benoxyl; Benzagel 10; Benzaknen; Debroxide; Desanden; Lucidol; Nericur; Oxy-5; Oxy-L; PanOxyl; Peroxyderm; Preoxydex; Persadox; Sanoxit; Theraderm; Xerac BP 5; Xerac BP 10. Used to treat acne. Keratolytic. mp = 103-106°; sparingly soluble in H_2O, EtOH; soluble in C_6H_6, $CHCl_3$, Et_2O, CS_2 (2.5 g/100 ml), olive oil (2 g/100 ml). *Bristol-Myers Squibb Co; Dermik Labs. Inc.; Galderma Labs Inc.; Hyland Div. Baxter ; Ortho Pharm. Corp.; SmithKline Beecham Pharm.; Sterling Winthrop Inc.; Taiho; Westwood-Squibb Pharm. Inc.*

1249 Cioteronel
89672-11-7 2372
$C_{16}H_{28}O_2$
(±)-Hexahydro-4-(5-methoxyheptyl)-2(1H)-pentalenone.
X-Andron; CPC-10997; Cyoctol; Exandron. Antiandrogen with antialopecia properties. Used to treat acne. Clear, colorless oil. *CBD Corp.*

1250 Cyproterone
2098-66-0 2844
$C_{22}H_{27}ClO_3$
6-Chloro-1β,2β-dihydro-17-hydroxy-3'H-cyclopropa[1,2]-pregna-1,4,6-triene-3,20-dione.
SH-881; SH-80881. Antiandrogen with antialopecia properties. Used in combination with estrogen to treat acne. mp = 237.5-240°. *Schering AG.*

1251 Cyproterone Acetate
427-51-0 2844 207-048-3
$C_{24}H_{29}ClO_4$
6-Chloro-1β,2β-dihydro-17-hydroxy-3'H-cyclopropa[1,2]-pregna-1,4,6-triene-3,20-dione acetate.
SH-714; NSC-81430; CPA; Androcur; Cyprostat. Antiandrogen with antialopecia properties. Used in combination with estrogen to treat acne. mp = 200-201°; λ_m = 281 nm (ε 17280 MeOH). *Schering AG.*

1252 Erythromycin Salnacedin
149908-23-6
$C_{49}H_{80}N_2O_{18}S.2H_2O$
Erythromycin compound with N-acetyl-L-cysteine salicylate (ester) (1:1) dihydrate.
G-101; SCY-Er. Used to treat acne. *Genta Inc.*

1253 Inocoterone
83646-97-3
$C_{16}H_{24}O_2$
17β-Hydroxy-2,5-seco-A-dinorestr-9-en-5-one.
Used to treat acne. *Hoechst Roussel Pharm. Inc.*

1254 Inocoterone Acetate
83646-86-0
$C_{18}H_{26}O_3$
17β-Hydroxy-2,5-seco-A-dinorestr-9-en-5-one acetate.
RU-882; RU-38882. Used to treat acne. *Hoechst Roussel Pharm. Inc.*

1255 Isotretinoin Anisatil
127471-94-7
$C_{29}H_{36}O_4$
p-Methoxyphenacyl 13-cis-retinoate.
GR-116526X. Used to treat acne. *Glaxo Wellcome plc.*

1256 Motretinide
56281-36-8 6367 260-094-6
$C_{23}H_{31}NO_2$
all-trans-N-Ethyl-9-(4-methoxy-2,3,6-trimethylphenyl)-3,7-dimethyl-2,4,6,8-nonatetraenamide.
Tasmaderm; Ro-11-1430. Keratolytic. Used to treat acne. mp = 179-180°. *Hoffmann-LaRoche Inc.*

1257 Resorcinol
108-46-3 8323 203-585-2
$C_6H_6O_2$
1,3-Benzenediol.
Resorcin; m-dihydroxybenzene; component of: Acnomel, Rezamid, Sulforcin. Keratolytic. Used to treat acne. mp = 109-111°; bp = 280°; d = 1.272; soluble in H_2O (111.1 g/100 ml at 20°, 500 g/100 ml at 80°), EtOH (111.1 g/100 ml); freely soluble in Et_2O, glycerol; slightly soluble in $CHCl_3$. *Dermik Labs. Inc.; Galderma Labs Inc.; Menley & James Labs Inc.*

1258 Retinoic Acid
302-79-4 8333 206-129-0
$C_{20}H_{28}O_2$
[all-E]-3,7-Dimethyl-9-(2,6,6-trimethyl-1-cyclohexen-1-yl)-2,4,6,8-nonatrienoic acid.
all-trans-Retinoic acid; vitamin A acid; tretinoin; NSC-122578; Aberel; Airol; Aknoten; Cordes Vas; Dermairol; Epi-Aberel; Eudyna; Retin-A; Vesanoid; Vesnaroid. Antineoplastic agent. Also a keratolytic, used to treat acne. mp = 180-182°; λ_m = 351 nm (ε 45000 MeOH); LD_{50} (10 day) (mus ip) = 790 mg/kg, (mus orl) = 2200 mg/kg, (rat ip) = 790 mg/kg, (rat orl) = 2000 mg/kg. *Degussa Ltd.; Hoffmann-LaRoche Inc.*

1259 13-cis-Retinoic Acid
4759-48-2 8333 225-296-0
$C_{20}H_{28}O_2$
(13-cis)-3,7-Dimethyl-9-(2,6,6-trimethyl-1-cyclohexen-1-yl)-2,4,6,8-nonatrienoic acid.
13-RA; Ro-4-3780; retinoic acid, 9Z form; 13-cis-Vitamin A acid; 13-cis-retinoic acid; cis-retinoic acid; neovitamin a acid; isotretinoin; Ro-4-3780; Accutane; Isotrex; Roaccutane; Tasmar; Isotretinoin; Accure; IsotrexGel; Roaccutane; 193. Keratolytic. Used to treat acne. mp = 174-175°; λ_m = 354 nm (ε 39800); LD_{50} (20 day) (mus ip) = 904 mg/kg, (mus orl) = 3389 mg/kg, (rat ip) = 901 mg/kg, (rat orl) > 4000 mg/kg. *Hoffmann-LaRoche Inc.*

1260 Tazarotene
118292-40-3 9249
$C_{21}H_{21}NO_2S$
Ethyl 6-[(4,4dimethylthiochroman-6-yl)ethynyl]nicotinate.

Zorac; AGN-190168. Keratolytic. Used to treat acne and psoriasis. White solid. *Allergan Inc.*

1261 Temarotene
75078-91-0
$C_{23}H_{28}$
1,2,3,4-Tetrahydro-1,1,4,4-tetramethyl-6-[(E)-α-methylstyryl]naphthalene.
Ro-15-0778. A synthetic retinoid chemopreventive and chemotherapeutic agent. Immunomodulator; antiacne.

1262 Tetroquinone
319-89-1 9385 206-275-5
$C_6H_4O_6$
2,3,5,6-Tetrahydroxy-2,5-cyclohexadien-1,4-dione.
Tetrahydroxy-p-benzoquinone; HPEK-1; THQ; NSC-112931; Kelox. Systemic keratolytic. Used to treat acne. Slightly soluble in cold H_2O, freely soluble in hot H_2O or EtOH, poorly soluble in Et_2O; behaves like a dibasic acid.

1263 Tioxolone
4991-65-5 9600 225-653-0
$C_7H_4O_3S$
6-Hydroxy-1,3-benzoxathiol-2-one.
thioxolone; Camyna; Stepin; component of: Psoil [with hydrocortisone]. Antiseborrheic. Used to treat acne. mp = 160°; nearly insoluble in H_2O; soluble in EtOH, iPrOH, propylene glycol, Et_2O, C_6H_6, C_7H_8; hydrolyzed by alkali. *Boehringer Ingelheim GmbH; Sterling Winthrop Inc.; Tillots Pharma; Winthrop.*

Antiallergy Agents

1264 Acitazanolast
114607-46-4
$C_9H_7N_5O_3$
3'-(1H-Tetrazol-5-yl)oxanilic acid.
Antiallergic; antihistaminic.

1265 Amlexanox
68302-57-8 515
$C_{16}H_{14}N_2O_4$
2-Amino-7-isopropyl-5-oxo-5H-[1]benzopyrano-[2,3-b]pyridine-3-carboxylic acid.
AA-673; CHX-3673; Elics; Solfa; amoxanox. Antiallergic; antiasthmatic. An orally active lipo-oxygenase inhibitor. mp > 300°. *Chemex Pharm.; Takeda Chem. Ind. Ltd.*

1266 Astemizole
68844-77-9 891 272-441-9
$C_{28}H_{31}FN_4O$
1-(p-Fluorobenzyl)-2-[[1-(p-methoxyphenethyl)-4-piperidyl]amino]benzimidazole.
Hismanal; R-43512; Astemisan; Histamen; Histaminos; Histazol; Kelp; Laridal; Metodik; Novo-Nastizol A; Paralergin; Retolen; Waruzol. Antiallergic; antihistaminic. Non-sedating type histamine H_1-receptor antagonist. mp = 149.1°; soluble in organic solvents, insoluble in H_2O; λ_m = 219, 249, 286 nm (ε 27250, 6480, 8634 EtOH), 209, 277 nm (ε 57889, 18073, 0.1N HCl). *Janssen Pharm. Ltd.*

1267 Azelastine
58581-89-8 939
$C_{22}H_{24}ClN_3O$
4-(p-Chlorobenzyl)-2-(hexahydro-1-methyl-1H-azepin-4-yl)-1(2H)-phthalazinone.
Antiallergic; antiasthmatic; antihistaminic. Orally active histamine H_1 receptor antagonist. Oil, soluble in CH_2Cl_2; gives a crystalline monohydrate. *Asta-Werke AG; Carter-Wallace.*

1268 Azelastine Hydrochloride
79307-93-0 939
$C_{22}H_{25}Cl_2N_3O$
4-(p-Chlorobenzyl)-2-(hexahydro-1-methyl-1H-azepin-4-yl)-1(2H)-phthalazinone monohydrochloride.
A-5610; W-2979M; E-0659; Allergodil; Astelin; Azeptin; Rhinolast. Antiallergic; antiasthmatic; antihistaminic. Orally active H_2-histamine receptor antagonist. mp = 225-229°; LD_{50} (mmus iv) = 36.5 mg/kg, (mmus ip) = 56.4 mg/kg, (mmus sc) = 63.0 mg/kg, (mmus orl) = 124 mg/kg, (fmus iv) = 35.5 mg/kg, (fmus ip) = 42.8 mg/kg, (fmus sc) = 54.2 mg/kg, (fmus orl) = 139 mg/kg, (mrat iv) = 26.9 mg/kg, (mrat ip) = 4. *Asta-Werke AG; Carter-Wallace.*

1269 Bufrolin
54867-56-0
$C_{18}H_{16}N_2O_6$
6-Butyl-1,4,7,10-tetrahydro-4,10-dioxo-1,7-phenanthraline-2,8-dicarboxylic acid.
An antiallergic and antihistaminic drug.

1270 Carebastine
90729-42-3
$C_{32}H_{37}NO_4$
p-[4-[4-(Diphenylmethoxy)piperidino]butyryl]-α-methylhydratropic acid.
Antiallergic; antihistaminic.

1271 Cloxacepride
65569-29-1
$C_{22}H_{27}Cl_2N_3O_4$
5-Chloro-4[2-(p-chlorophenoxy)acetamido]-N-[2-(diethylamino)ethyl]-o-anisamide.
Orally active antiallergic agent.

1272 Cromolyn
16110-51-3 2658 240-279-8
$C_{23}H_{12}O_{11}$
5,5'-[(2-Hydroxytrimethylene)dioxy]bis[4-oxo-4H-1-benzopyran-2-carboxylic acid].
cromoglycic acid; Duracroman. Antiallergic; prophylactic antiasthmatic. mp = 241-242° (dec); [monohydrate]: mp = 216-217°. *Bausch & Lomb Pharm. Inc.; Fisons plc.*

1273 Cromolyn Disodium Salt
15826-37-6 2658 239-926-7
$C_{23}H_{14}Na_2O_{11}$
Sodium 5,5'-[(2-hydroxytrimethylene)dioxy]bis[4-oxo-4H-1-benzopyran-2-carboxylate].
cromolyn sodium; disodium cromoglycate; DSCG; FPL-670; Aarane; Alercrom; Alerion; Allergocrom; Colimune; Crolom; Cromovet; Fivent; Gastrocom; Gastrofrenal; Inostral; Intal; Introl; Irtan; Lomudal; Lomupren; Lomusol; Lomuspray; Nalcrom; Nalcron; Nasalcrom; Nasmil; Opticrom; Rynacrom; Sofro; Vividrin. Antiallergic;

prophylactic antiasthmatic. Soluble in H_2O, insoluble in organic solvents; LD_{50} (rat orl) > 8000 mg/kg. *Bausch & Lomb Pharm. Inc.; Fisons plc.*

1274 Dioxamate
3567-40-6
$C_{15}H_{29}NO_4$
(2-Methyl-2-nonyl-1,3-dioxolan-4-yl)methyl carbamate.
Antiallergic.

1275 Doqualast
64019-03-0
$C_{13}H_8N_2O_3$
11-Oxo-11H-pyrido[2,1-b]quinazoline-2-carboxylic acid.
Antiallergic.

1276 Eclazolast
80263-73-6
$C_{12}H_{12}ClNO_4$
2-Ethoxyethyl 5-chloro-2-benzoxazolecarboxylate.
RHC-2871. Antiallergic. Mediator release inhibitor. *C.M. Ind.*

1277 Emedastine Difumarate
87233-62-3 3597
$C_{25}H_{34}N_4O_9$
1-(2-Ethoxyethyl)-2-(hexahydro-4-methyl-1H-1,4-diazepin-1-yl)benzimidazole fumarate (1:2).
AL-3432A; KB-2413; LY-188695. Antiallergic; antihistaminic; prophylactic antiasthmatic. Histamine H_1-receptor antagonist. mp = 148-151°; LD_{50} (gpg orl) = 744 mg/kg. *Kanebo Pharm. Ltd.*

1278 Fenpiprane
3540-95-2 4031
$C_{20}H_{25}N$
1-(3,3-Diphenylpropyl)piperidine.
component of: Aspasan. Antiallergic; antispasmodic. mp = 41-42.5°; bp_8 = 210-220°. *Hoechst Roussel Pharm. Inc.; Sterling Winthrop Inc.; Winthrop-Stearns.*

1279 Fenpiprane Hydrochloride
3329-14-4 4031 222-049-9
$C_{20}H_{26}ClN$
1-(3,3-Diphenylpropyl)piperidine hydrochloride.
component of: Efosin. Antiallergic; antispasmodic. mp = 216-217°. *Hoechst Roussel Pharm. Inc.; Sterling Winthrop Inc.; Winthrop-Stearns.*

1280 Flezelastine
135381-77-0
$C_{29}H_{30}FN_3O$
(±)-4-(p-Fluorobenzyl)-2-(hexahydro-1-phenethyl-1H-azepin-4-yl)-1(2H)-phthalazinone.
Antiasthmatic; antiallergic.

1281 Ibudilast
50847-11-5 4923
$C_{14}H_{18}N_2O$
1-(2-Isopropylpyrazolo[1,5-a]pyridin-3-yl)-2-methyl-1-propanone.
KC-404; Ketas. Antiallergic; antiasthmatic; vasodilator (cerebral). Phosphodiesterase inhibitor. Leukotriene D_4 antagonist. mp = 53.5-54°; slightly soluble in H_2O, freely soluble in organic solvents; LD_{50} (mus iv) = 260 mg/kg. *Kyorin Pharm. Co. Ltd.*

1282 Lodoxamide
53882-12-5 5585
$C_{11}H_6ClN_3O_6$
N,N'-(2-Chloro-5-cyano-m-phenylene)dioxamic acid.
Antiallergic; antiasthmatic. Used topically to treat allergic conjunctivitis. mp = 212° (dec); λ_m = 239.5 nm (ε 23800 0.1N NaOH). *Alcon Labs; Pharmacia & Upjohn.*

1283 Lodoxamide Diethyl Ester
53882-13-6 5585
$C_{15}H_{14}ClN_3O_6$
Diethyl N,N'-(2-chloro-5-cyano-m-phenylene)dioxamate.
U-42718. Antiallergic; antiasthmatic. Used topically to treat allergic conjunctivitis. mp = 177-179°. *Alcon Labs; Pharmacia & Upjohn.*

1284 Lodoxamide Tromethamine Salt
63610-09-3 5585
$C_{19}H_{28}N_5O_{12}$
N,N'-(2-Chloro-5-cyano-m-phenylene)dioxamic acid compound with 2-amino-2-(hydroxymethyl)-1,3-propanediol.
Alomide; U-42585E; lodoxamide trometamol. Antiallergic; antiasthmatic. Used topically to treat allergic conjunctivitis. *Alcon Labs; Pharmacia & Upjohn.*

1285 Mequitamium Iodide
101396-42-3
$C_{21}H_{25}IN_2S$
(±)-1-Methyl-3-(phenothiazin-10-ylmethyl)-quinuclidinium iodide.
LG-30435. The (+)-(S)-enantiomer 1β is 10x more potent than (-)-(R)-enantiomer 1α as a histamine antagonist; the two enantiomers show the same antimuscarinic activity in vitro. Antihistaminic; antiallergic; antimuscarinic.

1286 Minocromil
85118-44-1
$C_{18}H_{16}N_2O_6$
6-(Methylamino)-4-oxo-10-propyl-4H-pyrano[3,2-g]quinoline-2,8-dicarboxylic acid.
FPL-59360. Antiallergic. Prophylactic antiasthmatic. *Fisons plc.*

1287 Nedocromil
69049-73-6 6524
$C_{19}H_{17}NO_7$
9-Ethyl-6,9-dihydro-4,6-dioxo-10-propyl-4H-pyrano[3,2-g]quinoline-2,8-dicarboxylic acid.
FPL-59002. Antiallergic. Prophylactic antiasthmatic. mp = 298-300° (dec). *Fisons plc.*

1288 Nedocromil Calcium
101626-68-0 6524
$C_{19}H_{15}CaNO_7$
Calcium 9-ethyl-6,9-dihydro-4,6-dioxo-10-propyl-4H-pyrano[3,2-g]quinoline-2,8-dicarboxylate.
FPL-59002KC. Antiallergic; antiasthmatic. *Fisons plc.*

1289 Nedocromil Sodium
69049-74-7 6524
$C_{19}H_{15}NNa_2O_7$
Sodium 9-ethyl-6,9-dihydro-4,6-dioxo-10-propyl-4H-pyrano[3,2-g]quinoline-2,8-dicarboxylate.
FPL-59002KP; Rapitil; Tilade; Tilarin. Antiallergic. Prophylactic antiasthmatic. *Fisons plc.*

1290 Nivimedone
49561-92-4
$C_{11}H_8NO_4$
5,6-Dimethyl-2-aci-nitro-1,3-indandione.
Antiallergic.

1291 Nivimedone Sodium Salt
62077-09-2
$C_{11}H_8NNaO_4.H_2O$
5,6-Dimethyl-2-aci-nitro-1,3-indandione sodium salt monohydrate.
Antiallergic.

1292 Olopatadine
113806-05-6
$C_{21}H_{23}NO_3$
11-[(Z)-3-(Dimethylamino)propylidene]-6,11-sihydrodibenz[b,e]oxepin-2-acetic acid.
Antiallergic. *Kyowa Hakko Kogyo Co. Ltd.*

1293 Olopatadine Hydrochloride
140462-76-6
$C_{21}H_{24}ClNO_3$
11-[(Z)-3-(Dimethylamino)propylidene]-6,11-sihydrodibenz[b,e]oxepin-2-acetic acid hydrochloride.
KW-4679; ALO-4943A. Antiallergic. *Kyowa Hakko Kogyo Co. Ltd.*

1294 Oxatomide
60607-34-3 7058 262-320-9
$C_{27}H_{30}N_4O$
1-[3-[4-(Diphenylmethyl)-1-piperazinyl]propyl]-2-benzimidazolinone.
R-35443; Celtect; Cobiona; Dasten; Tinset. Antihistaminic; antiallergic; antiasthmatic. mp = 153.6°; LD_{50} (rat orl) > 2560 mg/kg, (rat iv) = 30 mg/kg, (gpg orl) = 320 mg/kg, (gpg iv) = 27 mg/kg, (mus orl) >2560 mg/kg, (mus iv) = 27 mg/kg. *Abbott Laboratories Inc.*

1295 Pemirolast
69372-19-6 7205
$C_{10}H_8N_6O$
9-Methyl-3-(1H-tetrazol-5-yl)-4H-pyrido[1,2-a]-pyrimidin-4-one.
Antiallergic. mp = 310-311° (dec). *Bristol-Myers Squibb Co.*

1296 Pemirolast Potassium Salt
100299-08-9 7205
$C_{10}H_7KN_6O$
9-Methyl-3-(1H-tetrazol-5-yl)-4H-pyrido[1,2-a]-pyrimidin-4-one potassium salt.
Pemilaston; BMY-26517; TBX. Antiallergic. Mediator release inhibitor. Very soluble in H_2O. *Bristol-Myers Squibb Co.*

1297 Pentigetide
62087-72-3 7270
$C_{22}H_{36}N_8O_{11}$
N^2-[1-[N-(N-L-α-Aspartyl-L-seryl)-L-α-aspartyl]-L-prolyl]-L-arginine.
Pentyde; HEPP; Pentapeptide DSDPR; DSDPR; Human IgE Pentapeptide; H-Asp-Ser-Asp-Pro-Arg-OH. Antiallergic. $[\alpha]_D^{20}$ = -78.6° (c = 1 H_2O). *Immunetech Pharm.*

1298 Picumast
39577-19-0 7573
$C_{25}H_{29}ClN_2O_3$
7-[3-[4-(p-Chlorobenzyl)-1-piperazinyl]propoxy]-3,4-dimethylcoumarin.
Antiallergic. Histamine antagonist. mp = 112-114°, 115-117°. *Boehringer Mannheim GmbH.*

1299 Picumast Dihydrochloride
39577-20-3 7573
$C_{25}H_{31}Cl_3N_2O_3$
7-[3-[4-(p-Chlorobenzyl)-1-piperazinyl]propoxy]-3,4-dimethylcoumarin dihydrochloride.
BM-15100; Auteral. Antiallergic. Histamine antagonist. mp = 266-268°. *Boehringer Mannheim GmbH.*

1300 Pirquinozol
65950-99-4
$C_{11}H_9N_3O_2$
2-(Hydroxymethyl)pyrazolo[1,5-c]-quinazolin-5(6H0-one.
SQ-13847. Antiallergic. *Bristol-Myers Squibb Co.*

1301 Probicromil Calcium
71144-97-3
$C_{17}H_{10}CaO_8$
Calcium 4,6-dioxo-10-propyl-4H,6H-benzo-[1,2-b:5,4-b']dipyran-2,8-dicarboxylate (1:1).
FPL-58668KC. Antiallergic, used prophylactically. *Fisons plc.*

1302 Proxicromil
60400-92-2
$C_{17}H_{18}O_5$
6,7,8,9-Tetrahydro-5-hydroxy-4-oxo-10-propyl-4H-naphtho[2.3-b]pyran-2-carboxylic acid.
Antiallergic. *Fisons plc.*

1303 Quinotolast
101193-40-2
$C_{17}H_{12}N_6O_3$
4-Oxo-1-phenoxy-N-1H-tetrazoyl-5-yl-4H-quinolizine-3-carboxamide.
FK-021. Antiallergic.

1304 Repirinast
73080-51-0 8305
$C_{20}H_{21}NO_5$
Isopentyl-5,6-dihydro-7,8-dimethyl-4,5-dioxo-4H-pyrano[3.2-c]quinoline-2-carboxylate.
MY-5116; Romet. Antiallergic; antiasthmatic. mp = 236-241°; LD_{50} (rat orl) > 5000 mg/kg, (rat sc) > 5000 mg/kg, (mus orl) > 5000 mg/kg, (mus sc) > 5000 mg/kg. *Mitsubishi Kasei.*

1305 Suplatast Tosylate
94055-76-2 9178
$C_{23}H_{33}NO_7S_2$
(±)-[2-[[p-(3-Ethoxy-2-hydroxypropoxy)phenyl]-carbamoyl]ethyl] dimethylsulfonium p-toluenesulfonate.
IPD-1151T. Antiasthmatic; antiallergic. Inhibits interleukin-4 gene expression. mp = 70-73°; LD_{50} (mmus iv) = 81 mg/kg, (mrat iv) = 96 mg/kg, (frat iv) = 93 mg/kg, (mus orl) > 12500 mg/kg, (rat orl) > 10000 mg/kg. *Taiho.*

1306 Tazanolast
82989-25-1 9248
$C_{13}H_{15}N_5O_3$
Butyl 3'-(1H-tetrazol-5-yl)oxanilate.
TO-188; WP-833; Tazalest; Tazanol. Antiallergic. *Wakamoto Pharm. Co. Ltd.*

1307 Tetrazolast
95104-27-1
$C_{10}H_6N_8$
4-(1H-Tetrazol-5-yl)tetrazolo[1,5-a]quinoline.
Antiallergic; antiasthmatic. *Marion Merrell Dow Inc.*

1308 Tetrazolast Meglumine
133008-33-0
$C_{17}H_{23}N_9O_5.H_2O$
4-(1H-Tetrazol-5-yl)tetrazolo[1,5-a]quinoline compound with 1-deoxy-1-(methylamino)-D-glucitol (1:1).
MDL-26024G0. Antiallergic; antiasthmatic. *Marion Merrell Dow Inc.; Merrell Dow Pharm. Inc.*

1309 Thiazinamium Chloride
4320-13-2
$C_{18}H_{23}ClN_2S$
Trimethyl(1-methyl-2-phenothiazin-10-ylethyl)-ammonium chloride.
WY-460E. Antiallergic. Antihistaminic.

1310 Tiacrilast
78299-53-3
$C_{12}H_{10}N_2O_3S$
(E)-6-(Methylthio)-4-oxo-3(4H)-quiazolineacrylic acid.
Ro-22-3747/000. Antiallergic. *Hoffmann-LaRoche Inc.*

1311 Tiacrilast Sodium
11868-63-4
$C_{12}H_9N_2NaO_3S.H_2O$
Sodium (E)-6-(methylthio)-4-oxo-3(4H)-quiazolineacrylate.
Ro-22-3747/007. Antiallergic. *Hoffmann-LaRoche Inc.*

1312 Tioxamast
74531-88-7
$C_{14}H_{14}N_2O_4S$
Ethyl [4-(p-methoxyphenyl)-2-thiazolyl]oxamate.
F-1865. Inhibits release and synthesis of certain mediators of allergy. Antiallergic.

1313 Tiprinast Meglumine
83198-90-7
$C_{19}H_{31}N_3O_8S$
3,4-Dihydro-6-isobutyl-5-methyl-4-oxothieno[2,3-d]-pyrimidine-2-carboxylic acid compound with 1-deoxy-1-(methylamino)-D-glucitol (1:1).
MJ-12175-170. Antiallergic.

1314 Tixanox
40691-50-7
$C_{15}H_{10}O_5S$
7-(Methylsulfinyl)-9-oxoxanthene-2-carboxylic acid.
Antiallergic. *Syntex Labs. Inc.*

1315 Tranilast
53902-12-8 9705
$C_{18}H_{17}NO_5$
N-(3,4-Dimethoxycinnamoyl)anthranilic acid.
MK-341; N-5'; Rizaben. Antiallergic; antiasthmatic. mp = 211-213°; LD_{50} (mmus orl) = 780 mg/kg, (mmus ip) = 410 mg/kg, (mmus sc) = 2630 mg/kg, (fmus orl) = 680 mg/kg, (fmus ip) = 385 mg/kg, (fmus sc) = 2820 mg/kg, (mrat orl) = 1600 mg/kg, (mrat ip) = 405 mg/kg, (mrat sc) = 3630 mg/kg, (frat orl) = 11. *C.M. Ind.; Kissei; Merck & Co.Inc.*

1316 Traxanox
58712-69-9 9711
$C_{13}H_6ClN_5O_2$
9-Chloro-7-(1H-tetrazol-5-yl)-5H-[1]benzopyrano[2,3-b]pyridine-5-one.
Y-12141[as sodium salt pentahydrate]; Clearnal. Antiallergic; antiasthmatic. mp > 300°. *Yoshitomi.*

Antialopecial Agents

1317 Cioteronel
89672-11-7 2372
$C_{16}H_{28}O_2$
(±)-Hexahydro-4-(5-methoxyheptyl)-2(1H)-pentalenone.
X-Andron; CPC-10997; Cyoctol; Exandron. Antiandrogen with antialopecia properties. Used to treat acne. Clear, colorless oil. *CBD Corp.*

1318 Cyproterone
2098-66-0 2844
$C_{22}H_{27}ClO_3$
6-Chloro-1β,2β-dihydro-17-hydroxy-3'H-cyclopropa[1,2]-pregna-1,4,6-triene-3,20-dione.
SH-881; SH-80881. Antiandrogen with antialopecia properties. Used in combination with estrogen to treat acne. mp = 237.5-240°. *Schering AG.*

1319 Cyproterone Acetate
427-51-0 2844 207-048-3
$C_{24}H_{29}ClO_4$
6-Chloro-1β,2β-dihydro-17-hydroxy-3'H-cyclopropa[1,2]-pregna-1,4,6-triene-3,20-dione acetate.
SH-714; NSC-81430; CPA; Androcur; Cyprostat. Antiandrogen with antialopecia properties. Used in combination with estrogen to treat acne. mp = 200-201°; λ_m = 281 nm (ε 17280 MeOH). *Schering AG.*

1320 Diphencyprone
886-38-4 3366 212-948-4
$C_{15}H_{10}O$
2,3-Diphenyl-2-cyclopropen-1-one.
DPC. Contact allergen, used as an antialopecia agent. mp = 87-90°; d = 1.202; λ_m = 297, 282, 226, 220 (log ε 4.3, 4.25, 4.13, 4.16 CH_3CN); [anhydrous form]: mp = 118-120°.

1321 Finasteride
98319-26-7 4125
$C_{23}H_{36}N_2O_2$
N-tert-Butyl-3-oxo-4-aza-5α-androst-1-ene 17β-carboxamide.
Propecia; Proscar; MK-906; Chibro-Proscar; Finastid; Prostide. A 5α-reductase inhibitor (testosterone → dihydrotestosterone converting enzyme). Formerly used as a treatment for benign prostatic hypertrophy. Also reported to have antialopecia properties and is now used to treat alopecia. mp = 257°, 252-254°; $[\alpha]_D$ = -59° (c = 1 MeOH); freely soluble in $CHCl_3$, DMSO, MeOH, EtOH, n-PrOH; sparingly soluble in propylene glycol, polyethylene glycol 400; very slightly soluble in H_2O, acids, bases. *Merck & Co.Inc.*

1322 Minoxidil
38304-91-5 6290 253-874-2
$C_9H_{15}N_5O$
2,4-Diamino-6-piperidinopyrimidine 3-oxide.
PDP; U-10858; Alopexil; Alostil; Loniten; Lonolox; Minoximen; Normoxidil; Pierminox; Prexidil; Regaine; Rogaine; Tricoxidil. Antialopecia agent with antihypertensive properties. Orally active. A piperidinopyrimidine derivative that activates potassium channels, producing vascular smooth muscle hyperpolarization and relaxation. mp = 248°, 259-261° (dec); λ_m = 230, 261, 285 nm (ε 35210, 11210, 11790 EtOH); 232, 280 nm (ε 26350, 28350 0.1N H_2SO_4), 231, 261.5, 285 nm (ε 36100, 11400, 12040 0.1N KOH); soluble in propylene glycol (7.5 g/100 ml), MeOH (4.4 g/100 ml), EtOH (2.9 g/100 ml), iPrOH (0.67 g/100 ml), DMSO (0.65 g/100 ml), H_2O (0.22 g/100 ml), $CHCl_3$ (0.05 g/100 ml); slightly soluble in Me_2CO, EtOAc, Et_2O, C_6H_6, CH_3CN; LD_{50} (rat iv) = 49 mg/kg, (mus iv) = 51 mg/kg. *Pharmacia & Upjohn.*

Antiamebic Agents

1323 Arsthinol
119-96-0 852 204-361-7
$C_{11}H_{14}AsNO_3S_2$
3-Hydroxypropylene ester of 3-acetamido-4-hydroxydithiobenzenearsonous acid.
Mercaptoarsenol; Balarsen. Antiamebic. mp = 163-166°; slightly soluble in H_2O, Et_2O; soluble in EtOH (2.9 g/ml).

1324 Berythromycin
527-75-3
$C_{37}H_{67}NO_{12}$
12-Deoxyerythromycin.
Abbott-24091. Antiamebic macrolide antibacterial. *Abbott Labs Inc.*

1325 Bialamicol
493-75-4 1242
$C_{28}H_{40}N_2O_2$
5,5'-Diallyl-α,α'-bis(diethylamino)-m,m'-bitolyl-4,4'-diol.
Camoform; Biallylamicol; SN-6771; PAA-701. Antiamebic. *Parke-Davis.*

1326 Bialamicol Dihydrochloride
3624-96-2 1242
$C_{28}H_{42}Cl_2N_2O_2$
5,5'-Diallyl-α,α'-bis(diethylamino)-m,m'-bitolyl-4,4'-diol dihydrochloride.
Camoform hydrochloride; CAM-807; CI-301; PAA-701; NSC-6386. Antiamebic. mp = 209-210°; soluble in H_2O. *Parke-Davis.*

1327 Carbarsone
121-59-5 1830 204-484-6
$C_7H_9AsN_2O_4$
N-Carbamoylarsanilic acid.
p-arsonophenylurea; p-ureidobenzene-arsonic acid; N-carbamylarsanilic acid; Amabevan; Ameban; Amibiarson; Arsambide; Carb-O-Sep; Histocarb; Fenarsone; Leucarsone; Aminarsone; Amebarsone. Antiamebic. Antihistomonad in turkeys. mp = 174°; slightly soluble in H_2O, EtOH; insoluble in Et_2O, $CHCl_3$; LD_{50} (rat orl) = 510 mg/kg. *Sankyo Co. Ltd.*

1328 Cephaeline
483-17-0 2020 207-591-6
$C_{28}H_{38}N_2O_4$
7',10,11-Trimethoxyemetan-6'-ol.
dihydropsychotrine; desmethylemetine. Antiamebic; emetic. An alkaloid of ipecac. mp= 115-116°; $[\alpha]_D^{20}$= -43.4° (c = 2 $CHCl_3$); insoluble in H_2O; soluble in MeOH, EtOH, Me_2CO, $CHCl_3$; less soluble in Et_2O, petroleum ether; [dihydrochloride heptahydrate]: mp = 270°; $[\alpha]_D^{20}$= +25.0° (c = 2); soluble in H_2O; moderately soluble in Me_2CO, $CHCl_3$; nearly insoluble in C_6H_6; [Dihydrobromide heptahydrate]: mp = 293°; soluble in H_2O; moderately soluble in alcohol; Me_2CO; nearly insoluble in C_6H_6.

1329 Chlorbetamide
97-27-8 2126
$C_{11}H_{11}Cl_4NO_2$
2,2-Dichloro-N-2,4-dichlorobenzyl-N-2-hydroxyethylacetamide.
Mantomide; Win-5047; Pontalin. Antiamebic. mp = 112.4-113.4°; slightly soluble in H_2O, more soluble in EtOH (< 5 g/100 ml). *Sterling Winthrop Inc.*

1330 Chloroquine
54-05-7 2215 200-191-2
$C_{18}H_{26}ClN_3$
7-Chloro-4-[[4-(diethylamino)-1-methylbutyl]-amino]quinoline.
Aralen; SN-7618; RP-3377; Artrichin; Bemaphate; Capquin; Nivaquine B; Resoquine; Reumachlor; Sanoquin; nivaquine [as sulfate]. Antiamebic; antimalarial; antirheumatic. Also used as a lupus erythematosus suppressant. mp = 87°. *Sterling Winthrop Inc.*

1331 Chloroquine Dihydrochloride
3545-67-3 2215 222-592-1
$C_{18}H_{28}Cl_3N_3$
7-Chloro-4-[[4-(diethylamino)-1-methylbutyl]-amino]quinoline dihydrochloride.
Aralen hydrochloride. Used as an antiamebic and

antimalarial. Also acts as an antirheumatic agent. *Sterling Winthrop Inc.*

1332 Chloroquine Diphosphate
50-63-5 2215 200-055-2
$C_{18}H_{32}ClN_3P_2O_8$
7-Chloro-4-[[4-(diethylamino)-1-methylbutyl]amino]quinoline diphosphate.
Aralen phosphate; Arechin; Avloclor; Imagon; Malaquin; Resochin; Tresochin. Antiamebic; antimalarial; antirheumatic. Also has activity as a lupus erythematosus suppressant. mp = 193-195°, 215-218°; soluble in H_2O, insoluble in organic solvents. *Sterling Winthrop Inc.*

1333 Chlorphenoxamide
3576-64-5 2233 222-694-6
$C_{17}H_{16}Cl_2N_2O_5$
2,2-Dichloro-N-(2-hydroxyethyl)-n-[[4-(4-nitrophenoxy)phenyl]-methyl]acetamide.
clefamide; chlorophenoxamide; Mebinol. Antiamebic. mp = 136-137°; insoluble in H_2O; soluble in EtOH, Me_2CO, dioxane; LD_{50} (mus orl) > 5000 mg/kg, (mus ip) = 2000 mg/kg. *Farmitalia Carlo Erba Ltd.*

1334 Chlortetracycline
57-62-5 2245 200-341-7
$C_{22}H_{23}ClN_2O_8$
7-Chloro-4-(dimethylamino)-1,4,4a,5,5a,6,11,12a-octahydro-3,6,10,12,12a-pentahydroxy-6-methyl-1,11-dioxo-2-naphthacenecarboxamide.
7-chlorotetracycline; Acronize; Aureocina; Aureomycin; Biomitsin; Centraureo; Chrusomykine; Orospray. Tetracycline antibiotic; antiamebic; antiprotozoal. mp = 168-169°; $[\alpha]_D^{23}$= -275.0° (MeOH); λ_m = 230, 262.5 367.5 nm (0.1NHCl), 255 285 345 nm (0.1N NaOH); soluble in H_2O (0.05-0.06 g/100 ml); very soluble in aqueous solutions at pH > 8.5; freely soluble in cellosolves, dioxane, carbitol; soluble in MeOH, EtOH, BuOH, Me_2CO, EtOAc, C_6H_6; insoluble in Et_2O, petroleum ether. *Am. Cyanamid; Fermenta Animal Health Co.; Lederle Labs.*

1335 Chlortetracycline Hydrochloride
64-72-2 2245 200-591-7
$C_{22}H_{24}Cl_2N_2O_8$
7-Chloro-4-(dimethylamino)-1,4,4a,5,5a,6,11,12a-octahydro-3,6,10,12,12a-pentahydroxy-6-methyl-1,11-dioxo-2-naphthacenecarboxamide monohydrochloride.
Aureomycin; Fermycin Soluble; Aureociclina; Isphamycin. Tetracycline antibiotic; antiamebic; antiprotozoal. Dec > 210°; $[\alpha]_D^{23}$= -240°; soluble in H_2O (0.86 g/100 ml at 28°), MeOH (1.74 g/100 ml at 28°), EtOH (0.17 g/100 ml at 28°); insoluble in Me_2CO, Et_2O, $CHCl_3$, dioxane; LD_{50} (rat orl) = 10300 mg/kg. *Fermenta Animal Health Co.; Lederle Labs.*

1336 Clamoxyquin
2545-39-3
$C_{17}H_{24}ClN_3O$
5-Chloro-7-[[[3-(diethylamino)propyl]amino]methyl]-8-quinolinol.
Antiamebic. *Parke-Davis.*

1337 Clamoxyquin Hydrochloride
4724-59-8
$C_{17}H_{26}Cl_3N_3O$
5-Chloro-7-[[[3-(diethylamino)propyl]amino]methyl]-8-quinolinol dihydrochloride.
Clamoxyl; CI-433; CN-17900-2B; PAA-3854; NSC-20246. Antiamebic. *Parke-Davis.*

1338 Dehydroemetine
4914-30-1 2924 225-542-7
$C_{29}H_{38}N_2O_4$
3-Ethyl-9,10-dimethoxy-1,6,7,11b-tetrahydro-2-[(1,2,3,4-tetrahydro-6,7-dimethoxy-1-isoquinolyl)methyl]-4H-benzo[a]quinolizine.
Ro-1-9334/19; 2,3-dehydroemetine; 2-dehydroemetine; Damatin [as (±)-form dihydrochloride]; Mebadin [as (±)-form dihydrochloride]. Antiamebic. The (-) form is therapeutically active. Analog of emetine. mp = 94-96°; $[\alpha]_D$ = -183°; [(±)-form dihydrochloride]: mp = 235°. *Hoffmann-LaRoche Inc.*

1339 Dibromopropamidine
496-00-4 3073
$C_{17}H_{18}Br_2N_4O_2$
4,4'-(Trimethylenedioxy)bis(3-bromobenzamidine).
dibrompropamidine. Antiseptic; antiamebic. Also used as a preservative in cosmetics. *May & Baker Ltd.*

1340 Dibromopropamidine Isethionate
614-87-9 3073 210-399-5
$C_{21}H_{30}Br_2N_4O_{10}S_2$
4,4'-(Trimethylenedioxy)bis(3-bromobenzamidine) di(2-hydroxyethanesulfonate) (ester).
dibrompropamidine isethionate; Brolene Ointment; Brulidine. Antiseptic; antiamebic. Also used as a preservative in cosmetics. mp = 226°; soluble in H_2O (0.5 g/ml), EtOH (1.6 g/100 ml), glycerol; insoluble in Et_2O, $CHCl_3$, petroleum ether; incompatible with chlorides, sulfates, many organic anions (forms sparingly soluble salts). *May & Baker Ltd.*

1341 Difetarsone
515-76-4 3394 208-209-0
$C_{14}H_{16}As_2N_2Na_2O_6$
N,N-Ethylenediarsanilic acid disodium salt.
RP-4763; diphetarsone; Amebarsin; Bemarsal; Rodameb. Antiamebic. Soluble in H_2O; less soluble in EtOH; insoluble in Me_2CO, $CHCl_3$.

1342 Diloxanide
579-38-4 3246 209-439-4
$C_9H_9Cl_2NO_2$
2,2-Dichloro-4'-hydroxy-N-methylacetanilide.
Entamide; Ame-Boots; [2-furoic acid ester] furamide; Histomibal; Miforon. Antiamebic. mp= 175°.

1343 Emetine
483-18-1 3600 207-592-1
$C_{29}H_{40}N_2O_4$
6',7',10,11-Tetramethoxyemetan.
cephaeline methyl ether. Antiamebic. mp = 74°; $[\alpha]_D^{20}$= -50° (c = 2 $CHCl_3$); soluble in MeOH, EtOH, $CHCl_3$,

Me_2CO, EtOAc, $CHCXl_3$; less soluble in H_2O, petroleum ether; LD_{50} (rat ip) = 12.1 mg/kg.

1344 Emetine Dihydrochloride
316-42-7 3600 206-259-8
$C_{29}H_{42}Cl_2N_2O_4$
6',7',10,11-Tetramethoxyemetan dihydrochloride.
Hemometina. Antiamebic. mp = 235-255°; $[\alpha]_D$ = 11° (c = 1) to 21° (c = 8); soluble in H_2O (143 mg/ml), EtOH; LD_{50} (mus sc) = 32 mg/kg, (mus orl) = 30 mg/kg (calc. as free base).

1345 Etofamide
25287-60-9
$C_{19}H_{20}Cl_2N_2O_5$
2,2-Dichloro-N-(2-ethoxyethyl)-n-[(p-nitrophenoxy)-benzyl]acetamide.
Antiamebic.

1346 Fumagillin
23110-15-8 4308 245-433-8
$C_{26}H_{34}O_7$
[3R-[3α,4α(2R*,3R*),5β,6β(all E)]]-2,4,6,8-Decatetraenedioic acid mono[5-methoxy-4-[2-methyl-3-(3-methyl-2-butenyl)oxiranyl]-1-oxapsiro[2.5]oct-6-yl] ester.
Amebacilin; Fugillin; Fumadil B; Fumidil. Antiamebic; antiprotozoal. mp = 194-195°; $[\alpha]_D^{25}$ = -26.6° (c = 1 in 95% EtOH); λ_m = 335, 351 nm (A 156.0, 1465.); insoluble in H_2O, soluble in most organic solvents; LD_{50} (mus sc) = 800 mg/kg.

1347 Glaucarubin
1448-23-3 4444
$C_{25}H_{36}O_{10}$
[1β,2α,11β,12α,15β(S)]-11,20-Epoxy-1,2,11,12-tetrahydroxy-15-(2-hydroxy-2-methyl-1-oxobutoxy)picras-3-en-16-one.
Glaumeba; α-Kirondrin. Antiamebic. mp = 250-255° (dec)$[\alpha]_D^{25}$ = 45° (c = 1.7 C_5H_5N), 69° (c = 0.6 MeOH); soluble in NaOH solutions, insoluble in NaHCO3 solutions, slightly soluble in H_2O (< 1.8 mg/ml). *Marion Merrell Dow Inc.*

1348 Glycobiarsol
116-49-4 4503 204-143-1
$C_8H_9AsBiNO_6$
(Hydrogen N-glycololoylarsanilato)oxobismuth.
Dysentulin; Milibis; Viasept; Wintodon. Antiamebic. Slightly soluble in H_2O, EtOH; insoluble in Et_2O, $CHCl_3$, C_6H_6.

1349 8-Hydroxy-7-iodo-5-quinolinesulfonic Acid
547-91-1 4872 208-938-4
$C_9H_6INO_4S$
7-Iodo-8-hydroxyquinoline-5-sulfonic acid.
Ferron; Loretin. Antiamebic; antiseptic. mp = 260-270°; soluble in H_2O (2 g/l at 25°, 5.9 g/l at 100°), slightly soluble in EtOH, insoluble in other organic solvents.

1350 Iodochlorhydroxyquin
130-26-7 5052 204-984-4
C_9H_5ClINO
5-Chloro-7-iodo-8-quinolinol.
clioquinol; chloroiodoquin; iodochlorohydroxyquinoline; iodochloroxyquinoline; Amebil; Alchloquin; Amoenol; Bactol; Barquinol; Budoform; Chinoform; Clioquinol; Cliquinol; Cort-Quin; Eczecidin; Enteroquinol; Entero-Septol; Entero-Vioform; Enterozol; Entrokin;Hi-Eneterol; Iodoenterol; Nioform; Nystaform; Quinambicide; Quin-O-Crème; Rheaform Boluses; Rometin; Vioform; Vioformio; component of: Domeform-HC, Formtone-HC, Lidaform-HC, Nystaform, Nystaform-HC, Racet, Vioform-Hydrocortisone. Antiamebic; topical anti-infective. Also used as an intestinal anti-infective (veterinary). Quinoline. CAUTION: Has been linked with occurrence of subacute myelo-optic neuropathe. Dec 178-179°; λ_m = 266 nm ($A^{1\%}_{1\ cm}$ 1120 in 0.1 N MeOHic NaOH), 269 nm ($A^{1\%}_{1\ cm}$ 1120 (MeOH/KOH), 255 nm ($A^{1\%}_{1\ cm}$ 1570 EtOH); slightly soluble in $CHCl_3$, AcOH; nearly insoluble in H_2O, cold alcohol, Et_2O; LD_{50} (cat orl) = 400 mg/kg. *Bayer AG; Ciba-Geigy Corp.; Dermik Labs. Inc.; Lemmon Co.; Marion Merrell Dow Inc.*

1351 Iodoquinol
83-73-8 5063 201-497-9
$C_9H_5I_2NO$
5,7-Diiodo-8-quinolinol.
SS-578; Diodoquin; Disoquin; Floraquin; Dyodin; Dinoleine; Searlewuin; Diodoxylin; Rafamebin; Ioquin; Direxiode; Stanquinate; Yodoxin; Zoaquin; Enterosept; Embequin. Antiamebic. mp = 200-215° (dec); insoluble in H_2O; sparingly soluble in EtOH, Et_2O, Me_2CO; soluble in hot C_5H_5N, dioxane. *Searle G.D. & Co.*

1352 Liroldine
105102-20-3
$C_{20}H_{20}F_2N_4$
2,2'-[(3,3'-Difluoro-4,4'-biphenylylene)dinitrolo]-dipyridoline.
HL-707. Antiamebic. LD_{50} (mus ip) = 940 mg/kg.

1353 Panidazole
13752-33-5 237-334-3
$C_{11}H_{12}N_4O_2$
4-[2-(2-Methyl-5-nitroimidazole-1-yl)ethyl]pyridine.
Ameobicide.

1354 Paromomycin
7542-37-2 7173 231-423-0
$C_{23}H_{45}N_5O_{14}$
O-2,6-diamino-2,6-dideoxy-α-L-idopyranosyl-(1→3)-O-β-D-ribofuranosyl-(1→5)-O-[2-amino-2-deoxy-α-D-glucopyranosyl-(1→4)-2-deoxystreptamine.
paromomycin I; amminosidin; catenulin; crestomycin; estomycin; hydroxymycin; monomycin A; neomycin E; paucimycin; R-400. Aminoglycoside antibiotic; antiamebic. Oligosaccaride antibiotic found in *Streptomyces*. $[\alpha]_D^{25}$ = 65° ±3°; soluble in H_2O; less soluble in EtOH, MeOH; LD_{50} (rat orl) = 1625 mg/kg, (rat sc) > 650 mg/kg, (rat iv) = 156 mg/kg, (mus orl) > 2275 mg/kg, (mus sc) = 423 mg/kg, (mus iiv) = 90 mg/kg. *Parke-Davis; Pfizer Inc.*

1355 Paromomycin Sulfate
1263-89-4 7173 215-031-7
$C_{23}H_{45}N_5O_{14}.xH_2SO_4$
O-2,6-diamino-2,6-dideoxy-α-L-idopyranosyl-(1→3)-O-β-D-ribofuranosyl-(1→5)-O-[2-amino-2-deoxy-α-D-glucopyranosyl-(1→4)-2-deoxystreptamine sulfate.

Humatin; 1600 Antibiotic; FI-5853; Aminoxidin; Aminosidine; Farmiglucin; Farminosidin; Gabbromicina; Gabbromycin; Gabbroral; Humagel; Pargonyl; Paramicina; Paricina; Sinosid. Aminoglycoside antibiotic; antiamebic. $[\alpha]_D^{20}$ = 50.5° (c = 1.5 H_2O, pH 6); LD_{50} (mus orl) > 15000 mg/kg, (mus sc) = 700 mg/kg, (mus iv) = 110 mg/kg. *Parke-Davis; Pfizer Inc.*

1356 Phanquone
84-12-8 7337 201-516-0
$C_{12}H_6N_2O_2$
4,7-Phenanthroline-5,6-quinone.
Phanquinone; Phanchinone; phanquone; Ciba 11925; Entobex. Antiamebic. mp = 295° (dec); λ_m = 261 nm (ε 10000).

1357 Polybenzarsol
54531-52-1 7723
Polymer of formaldehyde with 4-hydroxybenzene-arsonic acid.
Benzocal; Benzodol. Antiamebic. Soluble in H_2O, alcoholic NaOH; LD_{50} (mmus ip) = 235 mg/kg. *Marion Merrell Dow Inc.*

1358 Propamidine
104-32-5 7981 203-195-2
$C_{17}H_{20}N_4O_2$
4,4'-(Trimethylenedioxy)dibenzamidine.
4,4'-diamidino-α,ω-diphenoxypropane. Antiprotozoal (Trypanosoma); antiamebic; anti-infective (topical, veterinary). *May & Baker Ltd.*

1359 Propamidine Isethionate
140-63-6 7981 205-423-6
$C_{21}H_{32}N_4O_{10}S_2$
4,4'-(Trimethylenedioxy)dibenzamidine ethanesulfonic acid.
M&B-782; Brolene Drops. Topical anti-infective; antiamebic. Used as an antiprotozoal against Trypanosoma and Babesia. mp = 235°; soluble in H_2O (20 g/100 ml), EtOH (3 g/100 ml), glycerol; insoluble in Et_2O, $CHCl_3$, petroleum ether. *May & Baker Ltd.*

1360 Quinfamide
62265-68-3 8241 263-478-1
$C_{16}H_{13}Cl_2NO_4$
2-Furoic acid ester with 1-(dichloroacetyl)-1,2,3,4-tetrahydro-6-quinolinol.
Amenide; Amenox; Win-40014. Antiamebic. mp = 150.5-151°; soluble in Me_2CO, EtOH. *Sterling Winthrop.*

1361 Satranidazole
56302-13-7
$C_8H_{11}N_5O_5S$
1-(1-Methyl-5-nitroimidazol-2-yl)-3-(methylsulfonyl)-2-imidazolidinone.
A nitroimidazole with high selective toxicity for anaerobic prokaryotes and eukaryotes. Antiamebic; antimicrobial; radiosensitizer.

1362 Secnidazole
3366-95-8 8562 222-134-0
$C_7H_{11}N_3O_3$
α,2-Dimethyl-5-nitroimidazole-1-ethanol.
PM-185184; RP-14539; Flagentyl. Antiamebic; antiprotozoal against Trichomonas. mp = 76°. *Rhône-Poulenc Rorer Pharm. Inc.*

1363 Sulfarside
1134-98-1 9110
$C_6H_9AsN_2O_5S$
[2-Amino-4-(aminosulfonyl)phenyl]arsinous acid.
RP-4482; Bemarside. Antiamebic. The sodium salt is used as an antiamebic. *Rhône-Poulenc Rorer Pharm. Inc.*

1364 Symetine
15599-45-8
$C_{30}H_{48}N_2O_2$
4,4'-(Ethylenedioxy)bis[N-hexyl-N-methylbenzylamine].
Antiamebic. *Eli Lilly & Co.*

1365 Symetine Hydrochloride
5585-62-6
$C_{30}H_{50}Cl_2N_2O_2$
4,4'-(Ethylenedioxy)bis[N-hexyl-N-methylbenzylamine] dihydrochloride.
Antiamebic. *Eli Lilly & Co.*

1366 Teclozan
5560-78-1 9262 226-934-0
$C_{20}H_{28}Cl_4N_2O_4$
N,N'-(p-Phenylenedimethylene)bis[2,2-dichloro-N-(2-ethoxyethyl)acetamide.
Falmonox; Win-13146; Win-AM-13146; NSC-107433; teclosan; teclosine; teclozine. Antiamebic. mp = 137.6-143.9°; LD_{50} (mus orl) > 8000 mg/kg. *Sterling Winthrop Inc.*

1367 Tetracycline
60-54-8 9337 200-481-9
$C_{22}H_{24}N_2O_8$
(4S,4aS,5aS,6S,12aS)-4-(Dimethylamino)-1,4,4a,5,5a,6,11,12a-octahydro-3,6,10,12,12a-pentahydroxy-6-methyl-1,11-dioxo-2-naphthacenecarboxamide.
tsiklomitsin; deschlorobiomycin; Abricycline; Ambramycin; Bio-Tetra; Cyclomycin; Dumocyclin; Liquamycin; Mysteclin-F; Talsutin; Tetradecin; component of: Mysteclin-F. Antiamebic; antibacterial; antirickettsial. Tetracycline antibiotic produced in *Streptomyces* species. [trihydrate]: Dec 170-175°; $[\alpha]_D^{25}$ = -257.9° (0.1 N HCl), -239° (MeOH); λ_m = 220, 268, 355 nm (ε 13000, 18040, 13320 0.1N HCl); soluble in H_2O (1.7 mg/ml), MeOH (> 20 mg/ml); LD_{50} (rat orl) = 707mg/kg, (mus orl) = 808 mg/kg. *Bristol-Myers Squibb Co; Pfizer Inc.*

1368 Tetracycline Hydrochloride
64-75-5 9337 200-593-8
$C_{22}H_{25}ClN_2O_8$
(4S,4aS,5aS,6S,12aS)-4-(Dimethylamino)-1,4,4a,5,5a,6,11,12a-octahydro-3,6,10,12,12a-pentahydroxy-6-methyl-1,11-dioxo-2-naphthacenecarboxamide monohydrochloride.
Achro; Achromycin; Ala-Tet; Ambracyn; Ambramicina; Bristaciclina; Cefracycline; Criseociclina; Cyclopar; Diocyclin; Helvecyclin; Hostacyclin; Imex; Mediletten, Mephacyclin; Panmycin; Partrex; Polycycline; Purocyclina; Quadracyclin; Remicyclin; Riocyclin; Robitet; Ro-Cycline; Sanclomycine; Steclin; Sumycin; Supramycin; Sustamycin; Tefilin; Tetrabakat; Tetrabid; Tetrablet; tetrabon; Tetrachel; Tetracompren; Tetracyn;

Tetrakap; Tetralution; Tetramavan; Tetramycin; Tetrosol; TetraSURE; Topicycline; Totomycin; Triphacyclin; Unicin; Vetquamycin-324. Tetracycline antibiotic; antiamebic; antiprotozoal. Dec 214°; $[\alpha]_D^{25}$= -257.9° (c = 0.5 0.1N HCl); soluble in H_2O, MeOH, EtOH; insoluble in Et_2O, petroleum ether; LD_{50} (rat orl) = 6443 mg/kg. *Apothecon; Bristol-Myers Squibb Pharm. R&D; Fermenta Animal Health Co.; Lederle Labs.; Parke-Davis; Pharmacia & Upjohn; Robins, A. H. Co.*

1369 Thiocarbamizine
91-71-4 9458
$C_{21}H_{17}AsN_2O_5S_2$
2,2'-[[[4-[(Aminocarbonyl)amino]phenyl]-arsinidene]bis(thio)]bis[benzoic acid].
thiocarbamisin. Antiamebic. Sparingly soluble in H_2O, EtOH; insoluble in acids; soluble in alkali. *Eli Lilly & Co.*

1370 Thiocarbarsone
120-02-5 9459
$C_{11}H_{13}AsN_2O_5S_2$
2,2'-[[[4-[(Aminocarbonyl)amino]phenyl]-arsinidene]bis(thio)]bis[acetic acid].
Antiamebic. Sparingly soluble in H_2O, EtOH; insoluble in acids; soluble in alkali. *Eli Lilly & Co.*

1371 Tinidazole
19387-91-8 9588 243-014-4
$C_8H_{13}N_3O_4S$
1-[2-(Ethylsulfonyl)ethyl]-2-methyl-5-nitroimidazole.
CP-12574; Fasigin; Fasigyn; Pletil; Simplotan; Sorquetan; Tricolam; Trimonase. Antiamebic; antifungal; antiprotozoal against Giardia, Trichomonas. mp = 127-128°; LD_{50} (mus orl) > 3600 mg/kg, (mus ip) > 2000 mg/kg. *Pfizer Inc.*

Antianginals

1372 Abanoquil
90402-40-7
$C_{22}H_{25}N_3O_4$
4-Amino-2-(3,4-dihydro-6,7-dimethoxy-2(1H)-isoquinolyl)-6,7-dimethoxy quinoline.
An alpha-adrenergic blocker and antiarrhythmic.

1373 Acebutolol
37517-30-9 16 253-539-0
$C_{18}H_{28}N_2O_4$
(±)-3'-Acetyl-4'-[2-hydroxy-3-(1-methyl-ethylamino)propoxy]butyranilide.
Monitan; Sectral; Prent. Antianginal agent. Class II antiarrhythmic agent. Cardioselective β-adrenergic blocking agent. Has antiarrhythmic and antihypertensive properties. mp = 119-123°. *Wyeth Labs.*

1374 Acebutolol Hydrochloride
34381-68-5 16 251-980-3
$C_{18}H_{29}ClN_2O_4$
(±)-3'-Acetyl-4'-[2-hydroxy-3-(1-methyl-ethylamino)propoxy]butyranilide hydrochloride.
IL-17803A; M&B-17803A; Acetanol; Neptall; Sectral. Antianginal agent. Class II antiarrhythmic agent. Cardioselective β-adrenergic blocking agent. Has antiarrhythmic and antihypertensive properties. mp = 141-143°; soluble in H_2O (200 mg/ml), EtOH (70 mg/ml). *Wyeth Labs.*

1375 Alprenolol
13655-52-2 321 237-140-9
$C_{15}H_{23}NO_2$
1-(o-Allylphenoxy)-3-(isopropylamino)-2-propanol.
H 56/28. Antianginal; antiarrhythmic. Cardioselective β-adrenergic blocking agent. Has antiarrhythmic and antihypertensive properties. *C.M. Ind.; ICI.*

1376 Alprenolol Hydrochloride
13707-88-5 321 237-244-4
$C_{15}H_{24}ClNO_2$
1-(o-Allylphenoxy)-3-(isopropylamino)-2-propanol hydrochloride.
Applobal; Aprobal; Aptine; Aptol Duriles; Gubernal; Regletin; Yobir. Antianginal; class II antiarrhythmic. Cardioselective β-adrenergic blocking agent. Has antiarrhythmic and antihypertensive properties. mp = 107-109°; LD_{50} (mus orl) = 278.0 mg/kg, (rat orl) = 597.0 mg/kg, (rbt orl) = 337.3 mg/kg. *C.M. Ind.; ICI.*

1377 Amiodarone
1951-25-3 504 217-772-1
$C_{25}H_{29}I_2NO_3$
2-Butyl-3-benzofuranyl 4-[2-(diethylamino)ethoxy]-3,5-diiodophenyl ketone.
Cordarone; L-3428; SKF-33134-A. Antianginal; class III antiarrhythmic. Blocks both α- and β-receptors. Ventricular antiarrhythmic agent. A benzofuran derivative. *Wyeth-Ayerst Labs.*

1378 Amiodarone Hydrochloride
19774-82-4 504 243-293-2
$C_{25}H_{30}ClI_2NO_3$
2-Butyl-3-benzofuranyl 4-[2-(diethylamino)ethoxy]-3,5-diiodophenyl ketone hydrochloride.
L-3428; Amiodar; Ancoron; Angiodarona; Atlansil; Cordarex; Cordarone; Cordarone X; Miocard; Miodaron; Ortacrone; Ritmocardyl; Rythmarone; Trangorex. Antianginal; class III antiarrhythmic. Blocks both α- and β-receptors. Ventricular antiarrhythmic agent. A benzofuran derivative. mp = 156°, 159 ± 2°; λ_m = 208, 242 nm ($E_{1\,cm}^{1\%}$ 662 ± 8, 623 ± 10 MeOH); soluble in EtOH (1.28 g/100 ml), MeOH (9.98 g/100 ml), $CHCl_3$ (44.51 g/100 ml), n-PrOH (0.13 g/100 ml), Et_2O (0.17 g/100 ml), THF (0.60 g/100 ml), C_6H_6 (0.65 g/100 ml), CH_2Cl_2 (19.20 g/100 ml), CH_3CN (0.32 g/100 ml), 1-octanol (0.30 g/100 ml), H_2O (0.07 g/100 ml), C_6H_{14} (0.03 g/100 ml), petroleum ether (0.001 g/100 ml). *Wyeth-Ayerst Labs.*

1379 Amlodipine
88150-42-9 516
$C_{20}H_{25}ClN_2O_5$
3-Ethyl-5-methyl (±)-2-[(2-aminoethoxy)methyl]-4-(o-chlorophenyl)-1,4-dihydro-6-methyl-3,5-pyridinedicarboxylate.
Antianginal agent. Dihydropyridine calcium channel blocker. Has antihypertensive properties. *Ciba-Geigy Corp.; Pfizer Inc.*

1380 Amlodipine Besylate
111470-99-6 516
$C_{26}H_{31}ClN_2O_8S$
3-Ethyl-5-methyl (±)-2-[(2-aminoethoxy)methyl]-4-(o-chlorophenyl)-1,4-dihydro-6-methyl-3,5-pyridinedicarboxylate monobenzenesulfonate.
Norvasc; UK-48340-26; Antacal; Istin; Monopina; component of: Lotrel. Antianginal agent. Dihydropyridine calcium channel blocker. Has antihypertensive properties. mp = 178-179°. *Ciba-Geigy Corp.; Pfizer Inc.*

1381 Amlodipine Maleate
88150-47-4 516
$C_{24}H_{29}ClN_2O_9$
3-Ethyl-5-methyl (±)-2-[(2-aminoethoxy)methyl]-4-(o-chlorophenyl)-1,4-dihydro-6-methyl-3,5-pyridinedicarboxylate maleate.
UK-48340-11. Antianginal agent. Dihydropyridine calcium channel blocker. Has antihypertensive properties. *Pfizer Inc.*

1382 Amyl Nitrite
110-46-3 5137 203-770-8
$C_5H_{11}NO_2$
Isopentyl nitrite.
isoamyl nitrite; pentyl nitrite; Amyl Nitrite, Vaporole. Vasodilator; antianginal. Organic nitrate. Unstable, flammable liquid; decomposes on exposure to air, light; bp = 97-99° (volatilizes at lower temperatures); d_{25}^{25} = 0.875; n_D^{21} = 1.3781; slightly soluble in H_2O; miscible with EtOH, $CHCl_3$, Et_2O; incompatible with alcohol, antipyrine, caustic alkalies, alkaline carbonates, potassium iodide, bromides, ferrous salts. *Burroughs Wellcome Inc.*

1383 Anipamil
83200-10-6 280-213-5
$C_{34}H_{52}N_2O_2$
2-[3-[(m-Methoxyphenethyl)methylamino]propyl]-2-(m-methoxyphenyl)tetradecanenitrile.
Antianginal; class IV antiarrhythmic agent. Analog of verapamil.

1384 Arotinolol
68377-92-4 827
$C_{15}H_{21}N_3O_2S_3$
(±)-5-[2-[[3-(tert-Butylamino)-2-hydroxypropyl]thio]-4-thiazolyl]-2-thiophenecarboxamide.
Antianginal agent. Also antihypertensive and antiarrhythmic. Possesses both α- and β-adrenergic blocking activity. A propanolamine derivative. mp = 148-149°. *Sumitomo Pharm. Co. Ltd.*

1385 Arotinolol Hydrochloride
68377-91-3 827
$C_{15}H_{22}ClN_3O_2S_3$
(±)-5-[2-[[3-(tert-Butylamino)-2-hydroxypropyl]thio]-4-thiazolyl]-2-thiophenecarboxamide hydrochloride.
S-596; ARL; Almarl. Antianginal agent. Also antihypertensive and antiarrhythmic. Possesses both α- and β-adrenergic blocking activity. A propanolamine derivative. mp = 234-235.5°; LD_{50} (mus iv) = 86 mg/kg, (mus ip) = 360 mg/kg, (mus orl) > 5000 mg/kg. *Sumitomo Pharm. Co. Ltd.*

1386 Atenolol
29122-68-7 892 249-451-7
$C_{14}H_{22}N_2O_3$
2-[p-[2-Hydroxy-3-(isopropylamino)propoxy]-phenyl]acetamide.
ICI-66082; AteHexal; Atenol; Cuxanorm; Ibinolo; Myocord; Prenormine; Seles Beta; Selobloc; Teno-basan; Tenoblock; Tenormin; Uniloc. Antianginal; class II antiarrhythmic agent. Cardioselective β-adrenergic blocking agent. Has antiarrhythmic and antihypertensive properties. mp = 146-148°, 150-152°; λ_m = 225, 275, 283 nm (MeOH); very soluble in MeOH; soluble in AcOH, DMSO; less soluble in Me_2CO, dioxane; insoluble in CH_3CN, EtOAc, $CHCl_3$; LD_{50} (mus orl) = 2000 mg/kg, (mus iv) = 98.7 mg/kg, (rat orl) = 3000 mg/kg, (rat iv) = 59.24 mg/kg. *Apothecon; C.M. Ind.; ICI ; Lemmon Co.; Zeneca Pharm.*

1387 Barnidipine
104713-75-9 1031
$C_{27}H_{29}N_3O_6$
(+)-(3'S,4S)-1-Benzyl-3-pyrrolidinyl methyl 1,4-dihydro-2,6-dimethyl-4-(m-nitrophenyl)-3,5-pyridinedicarboxylate.
Mepirodipine. Antianginal agent. Dihydropyridine calcium channel blocker. Has antihypertensive properties. mp = 137-139°; $[\alpha]_D^{20}$ = 64.8 (c = 1 MeOH). *Yamanouchi U.S.A. Inc.*

1388 Barnidipine Hydrochloride
104757-53-1 1031
$C_{27}H_{30}ClN_3O_6$
(+)-(3'S,4S)-1-Benzyl-3-pyrrolidinyl methyl 1,4-dihydro-2,6-dimethyl-4-(m-nitrophenyl)-3,5-pyridinedicarboxylate hydrochloride.
YM-09730-5; Hypoca. Antianginal agent. Dihydropyridine calcium channel blocker. Also has antihypertensive properties. mp = 226-228°; $[\alpha]_D^{20}$ = 116.4° (c = 1 MeOH); insoluble in H_2O; LD_{50} (mrat orl) = 105 mg/kg, (frat orl) = 113 mg/kg. *Yamanouchi U.S.A. Inc.*

1389 Bepridil
64706-54-3 1188
$C_{24}H_{34}N_2O$
1-[2-(N-Benzylanilino)-1-(isobutoxymethyl)ethyl]-pyrrolidine.
Antianginal. A pyrrolidine calcium channel blocker with antianginal and antiarrhythmic (class IV) properties. $bp_{0.1}$ = 184°; $bp_{0.5}$ = 192°; n_D^{20} = 1.5538. *McNeil Pharm.; Wallace Labs.*

1390 Bepridil Hydrochloride
74764-40-2 1188
$C_{24}H_{37}ClN_2O_3$
1-[2-(N-Benzylanilino)-1-(isobutoxymethyl)ethyl]-pyrrolidine monohydrochloride monohydrate.
CERM-1978; Angopril; Bepadin; Cordium; Vascor. Antianginal. Calcium channel blocker with antianginal and antiarrhythmic (class IV) properties. mp 91°; LD_{50} (mus orl) = 1955 mg/kg, (mus iv) = 23.5 mg/kg. *C.M. Ind.; McNeil Pharm.; Wallace Labs.*

1391 Betaxolol
63659-18-7 1229
$C_{18}H_{29}NO_3$
(±)-1-[p-[2-(Cyclopropylmethoxy)ethyl]phenoxy]-3-(isopropylamino)-2-propanol.
Antianginal agent with antihypertensive and antiglaucoma properties. Cardioselective β_1 adrenergic blocker. mp = 70-72°. *Alcon Labs; Synthelabo Pharmacie.*

1392 Betaxolol Hydrochloride
63659-19-8 1229 264-384-3
$C_{18}H_{30}ClNO_3$
(±)-1-[p-[2-(Cyclopropylmethoxy)ethyl]phenoxy]-3-(isopropylamino)-2-propanol hydrochloride.
SLD-212; SL-75212; Betoptic; Betoptima; Kerlone. Antianginal agent with antihypertensive and antiglaucoma properties. Cardioselective β_1 adrenergic blocker. mp = 116°; LD_{50} (mus orl) = 94 mg/kg, (mus iv) = 37 mg/kg. *Alcon Labs; Synthelabo Pharmacie.*

1393 Bevantolol
59170-23-9 1238
$C_{20}H_{27}NO_4$
(±)-1-[(3,4-Dimethoxyphenethyl)amino]-3-(m-toloxy)-2-propanol.
Antianginal agent. Cardioselective β-adrenergic blocking agent. Has antiarrhythmic and antihypertensive properties. *Parke-Davis.*

1394 Bevantolol Hydrochloride
42864-78-8 1238
$C_{20}H_{28}ClNO_4$
(±)-1-[(3,4-Dimethoxyphenethyl)amino]-3-(m-toloxy)-2-propanol hydrochloride.
Vantol; CI-775; Ranestol; Sentiloc. Antianginal agent. Cardioselective β-adrenergic blocking agent. Has antiarrhythmic and antihypertensive properties. mp = 137-138°. *Parke-Davis.*

1395 Bucumolol
58409-59-9 1489
$C_{17}H_{23}NO_4$
8-[(3-tert-Butylamino)-2-hydroxypropoxy]-5-methylcoumarin.
Antianginal agent. A β-adrenergic blocker with antianginal and antiarrhythmic properties. *Sankyo Co. Ltd.*

1396 Bucumolol Hydrochloride
30073-40-6 1489
$C_{17}H_{24}ClNO_4$
8-[(3-tert-Butylamino)-2-hydroxypropoxy]-5-methylcoumarin hydrochloride.
CS-359; Bucumarol. Antianginal agent. A β-adrenergic blocker with antianginal and antiarrhythmic properties. mp = 226-228°; LD_{50} (mmus orl) = 676 mg/kg, (mmus iv) = 33.1 mg/kg, (fmus orl) = 692 mg/kg, (fmus iv) = 31.6 mg/kg. *Sankyo Co. Ltd.*

1397 Bufetolol
53684-49-4 1496
$C_{18}H_{29}NO_4$
1-(tert-Butylamino)-3-[o-[(tetrahydrofurfuryl)-oxy]phenoxy]-2-propanol.
Antianginal agent. A β-adrenergic blocker with antianginal and antiarrhythmic properties. $bp_{0.07}$ = 180-186°. *Yoshitomi.*

1398 Bufetolol Hydrochloride
35108-88-4 1496 252-369-4
$C_{18}H_{30}ClNO_4$
1-(tert-Butylamino)-3-[o-[(tetrahydrofurfuryl)-oxy]phenoxy]-2-propanol hydrochloride.
Y-6124; Adobiol. Antianginal agent. A β-adrenergic blocker with antianginal and antiarrhythmic properties. mp = 153.5-157°, 151-154°; soluble in H_2O, MeOH, AcOH; slightly soluble in C_6H_6; insoluble in Et_2O; LD_{50} (mus orl) = 409 mg/kg, (mus sc) = 507 mg/kg, (rat orl) = 1142 mg/kg, (rat sc) = 1904 mg/kg. *Yoshitomi.*

1399 Bufuralol
54340-62-4 1504 259-112-5
$C_{16}H_{23}NO_2$
α-[(tert-Butylamino)methyl]-7-ethyl-2-benzofuranmethanol.
Antianginal; antihypertensive agent. A β-adrenergic blocker with peripheral vasodilating properties. *Hoffmann-LaRoche Inc.*

1400 Bufuralol Hydrochloride
59652-29-8 1504
$C_{16}H_{24}ClNO_2$
α-[(tert-Butylamino)methyl]-7-ethyl-2-benzofuranmethanol hydrochloride.
Ro-3-4787; Angium. Antianginal; antihypertensive agent. A β-adrenergic blocker with peripheral vasodilating properties. mp = 146°; LD_{50} (mus iv) = 29.7 mg/kg, (mus ip) = 88.0 mg/kg, (mus orl) = 177 mg/kg, (rat sc) = 1400 mg/kg, (rat orl) = 750 mg/kg; [(+) form]: mp = 122-123°; $[\alpha]_{365}^{20}$ = 135.0° (c = 1.0 EtOH); [(-) form]: mp = 122-123°; $[\alpha]^{20}$ = -136.0 (c = 1.0 EtOH). *Hoffmann-LaRoche Inc.*

1401 Bunitrolol
34915-68-9 1514
$C_{14}H_{20}N_2O_2$
o-[3-(tert-Butylamino)-2-hydroxypropoxy]benzonitrile.
Ko-1366. Antianginal agent with antiarrhythmic and antihypertensive properties. A β-adrenergic blocker. *Boehringer Ingelheim GmbH.*

1402 Bunitrolol Hydrochloride
23093-74-5 1514 245-427-5
$C_{14}H_{21}ClN_2O_2$
o-[3-(tert-Butylamino)-2-hydroxypropoxy]benzonitrile hydrochloride.
Betriol; Stresson. Antianginal agent with antiarrhythmic and antihypertensive properties. A β-adrenergic blocker. mp = 163-165°; LD_{50} (mus orl) = 1344-1440 mg/kg, (mus ip) = 264-265 mg/kg, (rat orl) = 639-649 mg/kg, (rat ip) = 222-225 mg/kg. *Boehringer Ingelheim GmbH.*

1403 Bupranolol
14556-46-8 1521
$C_{14}H_{22}ClNO_2$
1-(tert-Butylamino)-3-[(6-chloro-m-tolyl)-oxy]-2-propanol.
bupranol; Ophtorenin. Antianginal agent with antiarrhythmic, antihypertensive and antiglaucoma properties. A β-adrenergic blocker.

1404 Bupranolol Hydrochloride
15148-80-8 1521 239-208-3
$C_{14}H_{23}Cl_2NO_2$
1-(tert-Butylamino)-3-[(6-chloro-m-tolyl)-oxy]-2-propanol hydrochloride.
B-1312; KL-255; Betadran; Betadrenol; Looser; Panimit. Antianginal agent with antiarrhythmic, antihypertensive and antiglaucoma properties. A β-adrenergic blocker. mp = 220-222°.

1405 Butoprozine
62228-20-0
$C_{18}H_{38}N_2O_2$
p-[3-(Dibutylamino)propoxy]phenyl 2-ethyl-3-indolizinyl ketone .
Antiarrhythmic; antianginal. Calcium channel blocker. *Labaz S.A.*

1406 Butoprozine Hydrochloride
62134-34-3 263-427-3
$C_{18}H_{39}ClN_2O_2$
p-[3-(Dibutylamino)propoxy]phenyl 2-ethyl-3-indolizinyl ketone monohydrochloride.
L-9394. Antiarrhythmic; antianginal. Calcium channel blocker. *Labaz S.A.*

1407 Carazolol
57775-29-8 1822 260-945-1
$C_{18}H_{22}N_2O_2$
1-(Carbazol-4-yloxy)-3-isopropylamino)-2-propanol.
BM-51052; Conducton; Suacron. Antianginal agent with antiarrhythmic and antihypertensive properties. Used for treatment of stress in pigs (veterinary). A β-adrenergic blocker. [hydrochloride]: mp = 234-235°. *Boehringer Mannheim GmbH.*

1408 Carteolol
51781-06-7 1917
$C_{16}H_{24}N_2O_3$
5-[3-[(1,1-Dimethylethyl)amino]-2-hydroxypropyl]-3,4-dihydro-2(1H)-quinolinone .
Antianginal agent with antiarrhythmic, antihypertensive and antiglaucoma properties. A β-adrenergic blocker. *Abbott Labs Inc.; Otsuka Am. Pharm.*

1409 Carteolol Hydrochloride
51781-21-6 1917 257-415-7
$C_{16}H_{25}ClN_2O_3$
5-[3-[(1,1-Dimethylethyl)amino]-2-hydroxypropyl]-3,4-dihydro-2(1H)-quinolinone hydrochloride.
Carteolol hydrochloride, Abbott 43326; OPC-1085; Arteoptic; Caltidren; Carteol; Cartrol; Endak; Mikelan; Optipress; Tenalet; Tenalin; Teoptic. Antianginal agent with antiarrhythmic, antihypertensive and antiglaucoma properties. A β-adrenergic blocker. mp = 278°; LD_{50} (mrat orl) = 1380 mg/kg, (mrat iv) = 158 mg/kg, (mrat ip) = 380 mg/kg, (mmus orl) = 810 mg/kg, (mmus iv) = 54.5 mg/kg, (mmus ip) = 380 mg/kg. *Abbott Labs Inc.; Otsuka Am. Pharm.*

1410 Carvediol
72956-09-3 1924
$C_{24}H_{26}N_2O_4$
(±)-1-(Carbazol-4-yloxy)-3-[[2-(o-methoxyphenoxy)ethyl]amino]-2-propanol.
BM-14190; DQ-2466; Coreg; Dilatrend; Dimitone; Eucardic; Kredex; Querto. Antianginal; antihypertensive. Non-selective β-adrenergic blocker with vasodilating activity. mp = 114-115°. *Boehringer Mannheim GmbH; SmithKline Beecham Pharm.*

1411 Celiprolol
56980-93-9 2007 260-497-7
$C_{20}H_{33}N_3O_4$
3-[3-Acetyl-4-[3-(tert-butylamino)-2-hydroxypropoxy]phenyl]-1,1-diethylurea.
ST-1396. Antianginal; antihypertensive agent. Cardioselective $β_1$ adrenergic blocker. mp = 110-112°. *Hoechst Marion Roussel Inc.*

1412 Celiprolol Hydrochloride
57470-78-7 2007 260-752-2
$C_{20}H_{34}ClN_3O_4$
3-[3-Acetyl-4-[3-(tert-butylamino)-2-hydroxypropoxy]phenyl]-1,1-diethylurea monohydrochloride.
Celectol; Corliprol; Selecor; Selectol. Antianginal; antihypertensive agent. Cardioselective $β_1$ adrenergic blocker. mp = 197-200° (dec); soluble in H_2O (15.1 g/100 ml), MeOH (18.2 g/100 ml), EtOH (1.61 g/100 ml), $CHCl_3$ (0.42 g/100 ml); $λ_m$ = 221, 324 nm ($E_{1\%}$ 652, 57 H_2O), 231, 324 nm ($E_{1\%}$ 660, 60 0.01N HCl), 231, 324 nm ($E_{1\%}$ 640, 60 0.01N NaOH), 232, 329 nm ($E_{1\%}$ 775, 58 MeOH); LD_{50} (mmus iv) = 56.2 mg/kg, (mmus orl) = 1834 mg/kg, (mrat iv) = 68.3 mg/kg, (mrat orl) = 3826 mg/kg. *Hoechst Marion Roussel Inc.*

1413 Cinepazet
23887-41-4 2349 245-927-3
$C_{20}H_{28}N_2O_6$
Ethyl-4-(3,4,5-trimethoxycinnamoyl)-1-piperazineacetate.
ethyl cinepazate. Antianginal agent. mp = 130°. *Delandale Labs. Ltd.*

1414 Cinepazet Maleate
50679-07-7 2349 256-709-2
$C_{24}H_{32}N_2O_{10}$
Ethyl-4-(3,4,5-trimethoxycinnamoyl)-1-piperazineacetate maleate.
MD-6753; Vascoril. Antianginal agent. mp = 96°; [hydrochloride]: mp = 200° (dec); LD_{50} (mus iv) = 300 mg/kg, (mus orl) = 1300 mg/kg. *Delandale Labs. Ltd.*

1415 Desrazoxane
24584-09-6 8295
$C_{11}H_{16}N_4O_4$
(+)-(S)-4,4'-Propylenedi-2,6-piperazinedione.
(+)-razoxane; dexrazoxane; (+)-4,4'-propylenedi-2,6-piperazinedione; (+)-(3,5,3',5'-tetraoxo)-1,2-dipiperazinopropane; ICI-59118; ICRF-159; NSC-129943; ADR-529; ICRF-187; NSC-169780; Zinecard; Cardioxane [hydrochloride]; Eucardion [hydrochloride]. Antianginal agent. Cardioprotectant. The racemate is used as an antineoplastic. mp = 193°; $[α]_D$ = 11.35° (c = 5, DMF); soluble in H_2O (10 mg/ml), 0.1N HCl (35-43 mg/ml), NaOH (6.7-10 mg/ml), EtOH (1 mg/ml), MeOH (7.1-10 mg/ml). *Pharmacia & Upjohn.*

1416 Devapamil
92302-55-1
$C_{26}H_{36}N_2O_3$
2-(3,4-Dimethoxyphenyl)-2-isopropyl-5-[(m-methoxyphenethyl)methylamino]valeronitrile.
Antianginal; antiarrhythmic. Calcium channel blocker with coronary vasodilating properties.

1417 Diltiazem
42399-41-7 3247 255-796-4
$C_{22}H_{26}N_2O_4S$
(+)-5-[2-(Dimethylamino)ethyl]-cis-2,3-dihydro-3-hydroxy-2-(p-methoxyphenyl)-1,5-benzothiazepin-4(5H)-one acetate (ester).
Antianginal; antihypertensive; antiarrhythmic (class IV). Calcium channel blocker with coronary vasodilating properties. *Bristol Myers Squibb Pharm. Ltd.; Forest Pharm. Inc.; Hoechst Marion Roussel Inc.; Lemmon Co.; Rhône-Poulenc Rorer Pharm. Inc.; Shionogi & Co. Ltd.; Tanabe Seiyaku Co. Ltd.*

1418 d-cis-Diltiazem Hydrochloride
33286-22-5 3247 251-443-3
$C_{22}H_{27}ClN_2O_4S$
(+)-5-[2-(Dimethylamino)ethyl]-cis-2,3-dihydro-3-hydroxy-2-(p-methoxyphenyl)-1,5-benzothiazepin-4(5H)-one acetate (ester) monohydrochloride.
CRD-401; RG-83606; Adizem; Altiazem; Anginyl; Angizem; Britiazim; Bruzem; Calcicard; Cardizem; Citizem; Cormax; Deltazen; Diladel; Dilpral; Dilrene; Dilzem; Dilzene; Herbesser; Masdil; Tildiem. Antianginal; antihypertensive; class IV antiarrhythmic. Calcium channel blocker with coronary vasodilating properties. mp = 207.5-212°; $[\alpha]_D^{24}$ = +98.3 ± 1.4° (c = 1.002 in MeOH); soluble in H_2O, MeOH, $CHCl_3$; slightly soluble in absolute EtOH; practically insoluble in C_6H_6; LD_{50} (mmus iv) = 61 mg/kg, (mmus sc) = 260 mg/kg, (mmus orl) = 740 mg/kg, (fmus iv) = 58 mg/kg, (fmus sc) = 280 mg/kg, (fmus orl) = 640 mg/kg, (mrat iv) = 38 mg/kg, (mrat sc) = 520 mg/kg, (mrat orl) = 560 mg/kg, (frat iv) = 39 mg/kg, (frat sc) = 550 mg/kg, (frat orl) = 610 mg/kg. *Bristol Myers Squibb Pharm. Ltd.; Forest Pharm. Inc.; Hoechst Marion Roussel Inc.; Lemmon Co.; Rhône-Poulenc Rorer Pharm. Inc.; Tanabe Seiyaku Co. Ltd.*

1419 Diltiazem Malate
144604-00-2 3247
$C_{26}H_{32}N_2O_9S$
(+)-5-[2-(Dimethylamino)ethyl]-cis-2,3-dihydro-3-hydroxy-2-(p-methoxyphenyl)-1,5-benzothiazepin-4(5H)-one acetate (ester) (S)-malate (1:1) monohydrochloride.
MK-793. Antianginal; antihypertensive; class IV antiarrhythmic. Calcium channel blocker with coronary vasodilating activity. Class IV antiarrhythmic. *Bristol Myers Squibb Pharm. Ltd.; Forest Pharm. Inc.; Hoechst Marion Roussel Inc.; Lemmon Co.; Rhône-Poulenc Rorer Pharm. Inc.*

1420 Dioxadilol
80743-08-4
$C_{16}H_{25}NO_4$
(±)-1-(1,4-Benzodioxan-2-ylmethoxy)-3-(tert-butylamino)-2-propanol.
Has beta adrenolytic activity; antihypertensive, antianginal and antiarrhythmic, but less potent than propranolol.

1421 Dopropidil
79700-61-1
$C_{20}H_{35}NO_2$
1-[1-(Isobutoxymethyl)-2-[[1-(1-propynyl)-cyclohexyl]oxy]ethyl]pyrrolidine.
ORG-30701. Antianginal. *Organon Inc.*

1422 Elgodipine
119413-55-7 3587
$C_{29}H_{33}FN_2O_6$
2-[(p-Fluorobenzyl)methylamino]ethyl isopropyl (±)-1,4-dihydro-2,6-dimethyl-4-[2,3-(methylenedioxy)phenyl]-3,5-pyridinedicarboxylate.
Antianginal agent. A dihydropyridine calcium channel blocker. *Inst. Invest. Desarr.; Quimicobiol.*

1423 Elgodipine Hydrochloride
121489-04-1 3587
$C_{29}H_{34}ClFN_2O_6$
2-[(p-Fluorobenzyl)methylamino]ethyl isopropyl (±)-1,4-dihydro-2,6-dimethyl-4-[2,3-(methylenedioxy)phenyl]-3,5-pyridinedicarboxylate hydrochloride.
IQB-875. Antianginal agent. A dihydropyridine calcium channel blocker. mp = 194-195°; LD_{50} (mus ip) = 30-40 mg/kg. *Inst. Invest. Desarr.; Quimicobiol.*

1424 Epanolol
86880-51-5 3641
$C_{20}H_{23}N_3O_4$
(±)-N-[2-[[3-(o-Cyanophenoxy)-2-hydroxypropyl]amino]ethyl]-2-(p-hydroxyphenyl)acetamide.
ICI-141292; Visacor. Antianginal; antihypertensive agent. Cardioselective β_1-adrenergic blocker with sympathomimetic activity. mp = 118-120°. *ICI.*

1425 Erythrityl Tetranitrate
7297-25-8 3716 230-734-9
$C_4H_6N_4O_{12}$
(R*,S*)-1,2,3,4-Butanetetroltetranitrate.
erythritol tetranitrate; erythrol tetranitrate; eritrityl tetranitrate; tetranitrol; tetranitrin; nitroerythrite; NSC-106566; Cardilate; Cardiloid. Coronary vasodilator used as an antianginal. Oral/sublingual/buccal tablets; for treatment of angina pectoris. An organic nitrate. CAUTION: Explosive. Sold in tablet form only; tablets are nonexplosive. mp = 61°; soluble in EtOH, Et_2O, glycerol; insoluble in H_2O; explodes on percussion. *Burroughs Wellcome Inc.; Glaxo Wellcome plc.*

1426 Felodipine
86189-69-7 3991
$C_{18}H_{19}Cl_2NO_4$
Ethyl methyl 4-(2,3-dichlorophenyl)-1,4-dihydro-2,6-dimethyl-3,5-pyridinedicarboxylate.
Plendil; H-154/82; Agon; Feloday; Flodil; Hydac; Munobal; Prevex; Splendil. Antianginal; antihypertensive agent. Dihydropyridine calcium channel blocker sold as the racemate. mp= 145°. *Astra USA Inc.; Merck & Co.Inc.*

1427 dl-Felodipine
72509-76-3 3991
$C_{18}H_{19}Cl_2NO_4$
(±) Ethyl methyl 4-(2,3-dichlorophenyl)-1,4-dihydro-2,6-dimethyl-3,5-pyridinedicarboxylate.
Plendil; H-154/82; Agon; Feloday; Flodil; Hydac;

Munobal; Prevex; Splendil. Antianginal; antihypertensive agent. Dihydropyridine calcium channel blocker sold as the racemate. mp= 145°. *Astra USA Inc.; Merck & Co.Inc.*

1428 Flusoxolol
84057-96-5
$C_{22}H_{30}FNO_4$
(S)-1-[p-[2-[(p-Fluorophenethyl)oxy]ethoxy]phenoxy]-3-(isopropylamino)-2-propanol.
Antihypertensive; antianginal; antiarrhythmic. A selective β_1-adrenoceptor agonist.

1429 Gallopamil
16662-47-8 4369
$C_{28}H_{40}N_2O_5$
5-[(3,4-Dimethoxyphenethyl)methylamino]-2-isopropyl-2-(3,4,5-trimethoxyphenyl)valeronitrile.
methoxyverapamil; D-600. Antianginal agent. Calcium channel blocking agent, related to verapamil. Pale yellow viscous oil. *Knoll Pharm. Co.*

1430 Gallopamil Hydrochloride
16662-46-7 4369 240-704-7
$C_{28}H_{40}N_2O_5$
5-[(3,4-Dimethoxyphenethyl)methylamino]-2-isopropyl-2-(3,4,5-trimethoxyphenyl)valeronitrile hydrochloride.
Algoclor; Procorum. Antianginal agent. Calcium channel blocking agent, related to verapamil. mp = 145-148°. *Knoll Pharm. Co.*

1431 d-Gallopamil Hydrochloride
38176-09-9 4369
$C_{28}H_{40}N_2O_5$
d-5-[(3,4-Dimethoxyphenethyl)methylamino]-2-isopropyl-2-(3,4,5-trimethoxyphenyl)valeronitrile hydrochloride.
Algoclor; Procorum. Antianginal agent. Calcium channel blocking agent, related to verapamil. mp = 160.5=161.5°; $[\alpha]_D^{25}$= 11.7° (c = 5.02 EtOH). *Knoll Pharm. Co.*

1432 l-Gallopamil Hydrochloride
36222-39-6 4369
$C_{28}H_{40}N_2O_5$
l-5-[(3,4-Dimethoxyphenethyl)methylamino]-2-isopropyl-2-(3,4,5-trimethoxyphenyl)valeronitrile hydrochloride.
Algoclor; Procorum. Antianginal agent. Calcium channel blocking agent, related to verapamil. mp = 160.5-161.5°; $[\alpha]_D^{25}$= -11.7° (c = 5.04 EtOH). *Knoll Pharm. Co.*

1433 Imolamine
318-23-0 4959 206-267-1
$C_{14}H_{20}N_4O$
4-[2-(Diethylamino)ethyl]-5-imino-3-phenyl-Δ^2-1,2,4-oxadiazoline.
Antianginal agent. $bp_{0.2}$ = 165°.

1434 Imolamine Hydrochloride
15823-89-9 4959 239-920-4
$C_{14}H_{21}ClN_4O$
4-[2-(Diethylamino)ethyl]-5-imino-3-phenyl-Δ^2-1,2,4-oxadiazoline hydrochloride.
LA-1211; Angolon; Irrigor. Antianginal agent. mp = 154-155°.

1435 Indenolol
60607-68-3 4974 262-323-5
$C_{15}H_{21}NO_2$
1-[1H-Inden-4 (or -7)-yloxy]-3-[(1-methylethyl)amino]-2-propanol.
YB-2; Sch-28316Z. Antihypertensive; antiarrhythmic; antianginal agent. A β-adrenergic blocker. A 1:2 tautomeric mixture of the 4- and 7- indenyloxy isomers. mp = 88-89°. *Schering AG; Yamanouchi U.S.A. Inc.*

1436 Indenolol Hydrochloride
81789-85-7 4974
$C_{15}H_{22}ClNO_2$
1-[1H-Inden-4 (or -7)-yloxy]-3-[(1-methylethyl)amino]-2-propanol hydrochloride.
Pulsan; Securpres. Antihypertensive; antiarrhythmic; antianginal. Non-selective β-adrenergic blocker. mp = 147-148°; LD_{50} (mus iv) = 26 mg/kg. *Schering AG; Yamanouchi U.S.A. Inc.*

1437 Isosorbide Dinitrate
87-33-2 5245 201-740-9
$C_6H_8N_2O_8$
1,4:3,6-Dianhydro-D-glucitol dinitrate.
Astridine; Cardio 10; Cardis; Carvanil; Carvasin; Cedocard; Corovliss; Dignionitrat; Dilatrate; Diniket; Disorlon; Duranitrat; EureCor; Flindix; Frandol; Glentonin; IBD; Imtack; Isdin; Iso-Bid; Isocard; Isoket; Iso-Mack; Iso-Puren; Isorbid; Isordil; Isordil Tembids; Isostenase; Isotrate; Langoran; Laserdil; Maycor; Myorexon; Nitorol; Nitrol; Nitrosorbonl Nosim; Rifloc Retard; Rigedal; Risordan; Soni-Slo; Sorbangil; Sorbichew; Sorbid SA; Sorbitrate; Sorquad; Vascardin; Vasorbate; Vasotrate; SDM-25; SDM-40; component of: BiDil, Dilatrate-SR. Coronary vasodilator. An organic nitrate used to treat angina pectoris. mp = 70°; $[\alpha]_D^{20}$ = 135° (EtOH); soluble in H_2O (1.1 mg/ml), more soluble in organic solvents. *ICI ; Medco Res. Inc.; Reed & Carnrick; Schwarz Pharma Kremers Urban Co.; Tillots Pharma; Wyeth Labs; Wyeth-Ayerst Labs; Zeneca Pharm.*

1438 Isosorbide Mononitrate
16051-77-7 5245 240-197-2
$C_6H_9NO_6$
1,4:3,6-Dianhydro-D-glucitol-5-mononitrate.
isosorbide-5-mononitrate; Corangin; Elan; Elantan; Imdur; ISMO; Isomonat; Monicor; Monit; Mono-Cedocard; Monoclair; Monoket; Momo Mack; Monosorb; Olicard; Pentacard; BM-22.145; IS 5-MN; AHR-4698. Coronary vasodilator used as an antianginal agent. An organic nitrate used to treat angina pectoris. A metabolite of isosorbide dinitrate. mp = 88-91°. *Boehringer Mannheim GmbH; Key Pharm.; Schwarz Pharma Kremers Urban Co.; Wyeth-Ayerst Labs.*

1439 Isradipine
75695-93-1 5260
$C_{19}H_{21}N_3O_5$
Isopropyl methyl (±)-4-(4-benzoxofurazanyl)-1,4-dihydro-2,6-dimethyl-3,5-pyridinedicarboxylate.
isrodipine; DynaCirc; PN-200-110; Clivoten; Dynacrine; Esradin; Lomir; Prescal; Rebriden. Antianginal; antihypertensive agent. Dihydropyridine calcium channel blocker. mp = 168-170°; [S(+) form (PN-205-

033)]: mp = 142°; $[\alpha]_D^{2-}$ = 6.7° (c = 1.5 EtOH); [R(-) form (PN-205-034)]: mp = 140°; $[\alpha]_D^{2-}$ = -6.7° (c = 1.67 EtOH). *Sandoz Pharm. Corp.*

1440 Lemildipine
125729-29-5
$C_{20}H_{22}Cl_2N_2O_6$
3-Isopropyl 5-methyl (±)-4-(2,3-dichlorophenyl)-1,4-dihydro-2-(hydroxymethyl)-6-methyl-3,5-pyridinedicarboxylate carbamate (ester).
Dihydropyridine calcium channel blocker.

1441 Lidoflazine
3416-26-0 5507 222-312-8
$C_{30}H_{35}F_2N_3O$
4-[4,4-Bis(p-fluorophenyl)butyl]-1-piperazineaceto-2',6'-xylidide.
McN-JR-7094; R-7904; Angex; Clinium; Klinium; Ordiflazine; Corflazine. Coronary vasodilator; antianginal agent. Used for treatment of angina pectoris. mp = 159-161°; soluble in $CHCl_3$, less soluble in other organic solvents, insoluble in H_2O. *Abbott Laboratories Inc.; Janssen Pharm. Inc.; McNeil Pharm.*

1442 Limaprost
88852-12-4 5514
$C_{22}H_{36}O_5$
(E)-7-[(1R,2R,3R)-3-Hydroxy-2-2[(E)(3S,5S)-3-hydroxy-5-methyl-1-nonenyl]-5-oxocyclopentyl]-2-heptenoic acid.
limaprost α-cyclodextrin clathrate; ONO-1206; OP-1206; Opalmon; Prorenal. Antianginal agent. Derivative of Prostaglandin E_1. mp = 97-100°. *Ono Pharm.; Warner-Lambert.*

1443 Linsidomine
33876-97-0
$C_6H_{10}N_4O_2$
3-Morpholinosydnone imine.
SIN-1; 3-morpholinosydnonimine. A spontaneous donor of nitric oxide and active metabolite of the antianginal drug molsidomine. Nitric oxide donor.

1444 Mepindolol
23694-81-7 5901 245-831-1
$C_{15}H_{22}N_2O_2$
1-[Isopropylamino]-3-[(2-methyl-indol-4-yl)oxy]-2-propanol.
SH-E-222 [as sulfate salt]; Betagon [as sulfate salt]; Corindolan [as sulfate salt]; Mepicor [as sulfate salt]. Antianginal; antihypertensive agent. A β-adrenergic blocker. mp = 100-102°, 95-97°. *Sandoz Pharm. Corp.*

1445 Metoprolol
37350-58-6 6235 253-483-7
$C_{15}H_{25}NO_3$
1-(Isopropylamino)-3-[p-(2-methoxyethyl)phenoxy]-2-propanol.
CGP-2175; H-93/26. Class II antiarrhythmic agent. Also has antihypertensive and antianginal activity. A β-adrenergic blocker which lacks intrinsic sympathomimetic activity. *Apothecon; Astra Chem. Ltd.; Ciba-Geigy Corp.; Lemmon Co.*

1446 Metoprolol Succinate
98418-47-4 6235
$C_{34}H_{56}N_2O_{10}$
1-(Isopropylamino)-3-[p-(2-methoxyethyl)phenoxy]-2-propanol succinate (2:1) (salt).
Toprol XL; H-93/26 succinate. Antianginal; class II antiarrhythmic; antihypertensive. A β-adrenergic blocker. Lacks inherent sympathomimetic activity. *Apothecon; Astra Chem. Ltd.; Astra Sweden; Ciba-Geigy Corp.; Lemmon Co.*

1447 Metoprolol Tartrate
56392-17-7 6235 260-148-9
$C_{34}H_{56}N_2O_{12}$
1-(Isopropylamino)-3-[p-(2-methoxyethyl)phenoxy]-2-propanol (2:1) dextro-tartrate salt.
CGP-2175E; HCTCGP 2175E; Beloc; Betaloc; Lopressor; Lopresor; Prelis; Seloken; Selopral; Selo-Zok; component of: Lopressor HCT. Antianginal; class II antiarrhythmic; antihypertensive. A β-adrenergic blocker. Lacks inherent sympathomimetic activity. Soluble in MeOH (50 g/100 ml), H_2O (> 100 g/100 ml), $CHCl_3$ (49.6 g/100 ml), Me_2CO (0.11 g/100 ml), CH_3CN (0.089 g/100 ml); insoluble in C_6H_{14}; LD_{50} (fmus iv) = 118 mg/kg, (fmus orl) = 3090 mg/kg, (mrat iv) ≅ 90 mg/kg, (mrat orl) = 3090 mg/kg, (mrat iv) ≅ 90 mg/kg, (mrat orl) = 3090 mg/kg. *Apothecon; Astra Chem. Ltd.; Ciba-Geigy Corp.; Lemmon Co.; Sandoz Nutrition.*

1448 Molsidomine
25717-80-0 6316 247-207-4
$C_9H_{14}N_4O_4$
N-Carboxy-3-morpholinosynonimine ethyl ester.
morsydomine; SIN-10; CAS-276; Corvaton; Corvasal; Molsidolat; Morial; Motazomin. Coronary vasodilator; antianginal agent. Non-benzene aromatic, heterocyclic and mesoionic compound. mp = 140-141°; soluble in $CHCl_3$, dilute HCl, EtOH, EtOAc, MeOH; sparingly soluble in H_2O, Me_2CO, EtOH; slightly soluble in Et_2O, petroleum Et_2O; pK (100°) = 3.0 ± 0.1; λ_m = 326 nm; LD_{50} (mmus sc) = 780 mg/kg, (mmus iv) = 860 mg, (mmus ip) = 700 mg/kg, (mmus orl) = 830 mg/kg, (fmus sc) = 750 mg/kg, (fmus iv) = 800 mg/kg, (fmus ip) = 760 mg/kg, (fmus orl) = 840 mg/kg, (mrat sc) = 1380 mg/kg, (mrat iv) = 830 mg/kg, (mrat ip) = 1250 mg/kg, (mrat orl) = 1050 mg/kg, (frat sc) = 1350 mg/kg, (frat iv) = 760 mg/kg, (frat ip) = 1250 mg/kg, (frat orl) = 1200 mg/kg. *Hoechst Roussel Pharm. Inc.; Takeda Chem. Ind. Ltd.*

1449 Monatepil
132019-54-6
$C_{28}H_{30}FN_3OS$
(±)-N-(6,11-Dihydrodibenzo[b,e]thiepin-11-yl)-4-(p-fluorophenyl)-1-piperazinebutyramide.
Antiarrhythmic agent with antianginal and antihypertensive properties. *Dainippon Pharm.*

1450 Monatepil Maleate
132046-06-1
$C_{32}H_{34}FN_3O_5S$
(±)-N-(6,11-Dihydrodibenzo[b,e]thiepin-11-yl)-4-(p-fluorophenyl)-1-piperazinebutyramide maleate (1:1).
AJ-2615. Antiarrhythmic agent with antianginal and antihypertensive properties. *Dainippon Pharm.*

1451 Nadolol
42200-33-9 6431 255-706-3
$C_{17}H_{27}NO_4$
1-(tert-Butylamino)-3-[(5,6,7,8-tetrahydro-cis-6,7-dihydroxy-1-naphthyl)oxy]-2-propanol.
Corgard; SQ-11725; Anabet; Solgol. Antihypertensive; antianginal agent. Class II antiarrhythmic. A β-adrenergic blocker. mp = 124-136°; λ_m = 270, 278 nm ($E^{1\%}_{1\,cm}$ 37.5, 39.1, MeOH); pKa = 9.67; poorly soluble in organic solvents; insoluble in Me_2CO, C_6H_6, Et_2O, C_6H_{14}; LD_{50} (rat orl) = 5300 mg/kg, (mus orl) = 4500 mg/kg. *Apothecon; Bristol-Myers Squibb Co.*

1452 Nicardipine
55985-32-5 6579 259-932-3
$C_{26}H_{29}N_3O_6$
2-(Benzylmethylamino) ethyl methyl 1,4-dihydro-2,6-dimethyl-4-(m-nitrophenyl)-3,5-pyridinedicarboxylate.
Antianginal; antihypertensive; vasodilator. Dihydropyridine calcium channel blocker. Has antihypertensive properties. *Syntex Labs. Inc.; Yamanouchi U.S.A. Inc.*

1453 Nicardipine Hydrochloride
54527-84-3 6579 259-198-4
$C_{26}H_{30}Cl9N_3O_6$
2-(Benzylmethylamino) ethyl methyl 1,4-dihydro-2,6-dimethyl-4-(m-nitrophenyl)-3,5-pyridinedicarboxylate monohydrochloride.
YC-93; RS-69216; RS-69216-XX-07-0; Barizin; Bionicard; Cardene; Dacarel; Lecibral; Lescodil; Loxen; Nerdipina; Nicant; Nicardal; Nicarpin; Nicapress; Nicodel; Nimicor; Perdipina; Perdipine; Ranvil; Ridene; Rycarden; Rydene; Vasodin; Vasonase. Antianginal; antihypertensive agent. Dihydropyridine calcium channel blocker. Has antihypertensive properties. [α form]: mp = 179-181°; [β form]: mp = 168-170°; LD_{50} (mrat orl) = 634 mg/kg), (mrat iv) = 18.1 mg/kg, (frat orl) = 557 mg/kg, (frat iv) = 25.0 mg/kg, (mmus orl) = 634 mg/kg, (mmus iv) = 20.7 mg/kg, (fmus orl) = 650 mg/kg, (fmus iv) = 19.9 mg. *Syntex Labs. Inc.; Yamanouchi U.S.A. Inc.*

1454 Nicorandil
65141-46-0 6608 265-514-1
$C_8H_9N_3O_4$
N-(2-Hydroxyethyl)nicotinamide nitrate (ester).
SG-75; Ikorel; Perisalol; Sigmart. Coronary vasodilator used as an antianginal agent. Nicotinamide derivative with dual mechanism as both nitrovasodilator and potassium channel activator. mp = 92-93°; LD_{50} (rat orl) = 1200-1300 mg/kg, (rat iv) = 800-1000 mg/kg. *Chugai Pharm. Co. Ltd.; Upjohn Ltd.*

1455 Nifedipine
21829-25-4 6617 244-598-3
$C_{17}H_{18}N_2O_6$
Dimethyl 1,4-dihydro-2,6-dimethyl-4-(o-nitrophenyl)-3,5-pyridinedicarboxylate.
Bay a 1040; Adalat; Adalate; Adapress; Aldipin; Alfadat; Anifed; Aprical; Bonacid; Camont; Chronadalate; Citilat; Coracten; Cordicant; Cordilan; Corotrend; Duranifin; Ecodipi; Hexadilat; Introcar; Kordafen; Nifedicor; Nifedin; Nifelan; Nifelat; Nifensar XL; Orix; Oxcord; Pidilat; Procardia; Sepamit; Tibricol; Zenusin. Coronary vasodilator used as an antianginal agent. Dihydropyridine calcium channel blocker. mp = 172-174°; λ_m = 340, 235 nm (ε 5010, 21590 MeOH), 338, 238 nm (ε 5740, 20600 0.1N HCl), 340, 238 nm (5740, 20510 0.1N NaOH); soluble in Me_2CO (25.0 g/100 ml), CH_2Cl_2 (16 g/100ml), $CHCl_3$ (14 g/100 ml), EtOAc (5 g/100 ml), MeOH (2.6 g/100 ml), EtOH (1.7 g/100 ml); LD_{50} (mus orl) = 494 mg/kg, (mus iv) = 4.2 mg/kg, (rat orl) = 1022 mg/kg, (rat iv) = 15.5 mg/kg. *Bayer AG; Miles Inc.; Pfizer Inc.; Pratt Pharm.*

1456 Nifenalol
7413-36-7 6618 231-023-6
$C_{11}H_{16}N_2O_3$
α-[(Isopropylamino)methyl]-p-nitrobenzyl alcohol.
isophenethanol; INPEA. Antiarrhythmic agent with antianginal properties. A β-adrenergic blocker. mp = 98°. *Selvi.*

1457 Nifenalol Hydrochloride
5704-60-9 6618 227-194-1
$C_{11}H_{17}ClN_2O_3$
(±)-α-[[(1-Methylethyl)amino]methyl]-4-nitrobenzenemethanol hydrochloride.
Inpea. Antianginal; antiarrhythmic agent. A β-adrenergic blocker. mp = 181°. *Selvi.*

1458 Niguldipine
113165-32-5
$C_{36}H_{39}N_3O_6$
(+)-(S)-3-(4,4-Diphenylpiperidino)propyl methyl 1,4-dihydro-2,6-dimethyl-4-(n-nitrophenyl)-3,5-pyridinecarboxylate.
B-844-39 [as hydrochloride]. Selective T-type calcium channel blocker. A dihydropyridine derivative.

1459 Nilvadipine
75530-68-6 6637
$C_{19}H_{19}N_3O_6$
5-Isopropyl 3-methyl 2-cyano-1,4-dihydro-6-methyl-4-(m-nitrophenyl)-3,5-pyridinedicarboxylate.
nivadipine; nivaldipine; CL-287389; FK-235; FR-34235; SK&F-102362; Escor; Nivadil. Antianginal; antiarrhythmic agent. Dihydropyridine calcium channel blocker. Has antihypertensive properties. mp = 148-150°; [(+) form]: $[\alpha]_D^{20}$ = 222.42° (c = 1 MeOH); [(-) form]: $[\alpha]_D^{20}$ = -219.62° (c= 1 MeOH). *Fujisawa Pharm. USA Inc.; SmithKline Beecham Pharm.*

1460 Nipradilol
81486-22-8 6655
$C_{15}H_{22}N_2O_6$
8-[2-Hydroxy-3-(isopropylamino)propoxy]-3-chromanol.
Nipradolol; K-351; Hypadil. Antianginal; antihypertensive agent. A β-adrenergic blocker with vasodilating activity. mp = 107-116°, 110-122°; LD_{50} (mus iv) = 74.0 mg/kg, (rat iv) = 73.0 mg/kg, (mus orl) = 540 mg/kg. *Kowa Chem. Ind. Co. Ltd.*

1461 Nisoldipine
63675-72-9 6658 264-407-7
$C_{20}H_{24}N_2O_6$
1,4-Dihydro-2,6-dimethyl-4-(2-nitrophenyl)-3,5-pyridinedicarboxylic acid methyl 2-methylpropyl ester.
Bay k 5552; Baymycard; Nisocor; Norvasc; Sular; Syscor; Zadipina. Antianginal; antihypertensive agent. Dihydropyridine calcium channel blocker. mp = 151-152°. *Bayer AG; Miles Inc.; Zeneca Pharm.*

1462 Nitroglycerin
55-63-0 6704 200-240-8
$C_3H_5N_3O_9$
1,2,3-Propanetriol trinitrate.
glyceryl trinitrate; glycerol nitric acid triester; nitroglycerol; trinitroglycerol; glonoin; trinitrin; blasting gelatin; blasting oil; SNG; SDM-7; SDM-17; SDM-20; manufacture of dynamite. CAUTION: Accute poisoning can cause nausea, vomiting abdominal cramps, headache, mental confusion, delirium, bradypnea, bradycardia, paralysis, convulsions, methemoglobinemia, cyanosis, circulatory collapse, death. Chronic poisoning can cause severe headaches, hallucinations, skin rashes. Alcohol aggravates symptoms. Toxic effects may occur by ingestion, inhalation, absorption. [labile form]:mp = 2.8°; [stable form]: mp = 13.5°; begins to decompose at ≅ 50°; d_{15}^{15} = 1.5991; n_D^{15} = 1.474; heat of combustion = 1580 cal/g; slightly soluble in H_2O (0.125 g/100 ml), EtOH (0.5 g/100 ml); more soluble in MeOH (2.25 g/100 ml), CS_2 (15 g/100 ml); miscible with Et_2O, Me_2CO, glacial AcOH, EtOAc, C_6H_6, nitrobenzene, C_5H_5N, $CHCl_3$, ethylene bromide, dichloroethylene; sparingly soluble in petroleum Et_2O, liquid petrolatum, glycerol. *3M Pharm.; Ciba-Geigy Corp.; Hoechst Marion Roussel Inc.; ICI Americas Inc.; Key Pharm.; KV Pharm.; Marion Merrell Dow Inc.; Parke-Davis; Rhône-Poulenc Rorer Pharm. Inc.; Schwarz Pharma Kremers Urban Co.; Searle G.D. & Co.; U.S. Ethicals Inc.; Zeneca Pharm.*

1463 Oxprenolol
6452-71-7 7086 229-257-9
$C_{15}H_{23}NO_3$
1-(o-Allyloxyphenoxy)-3-isopropylamino-2-propanol .
Antianginal; antihypertensive; class IV antiarrhythmic. A β-adrenergic blocker. mp = 78-80°. *Bayer AG; Ciba-Geigy Corp.*

1464 Oxprenolol Hydrochloride
6452-73-9 7086 229-260-5
$C_{15}H_{24}ClNO_3$
1-(o-Allyloxyphenoxy)-3-isopropylamino-2-propanol hydrochloride.
Ba-39089; Coretal; Laracor; Paritane; Slow-Pren; Trasicor; Trasacor. Antianginal; antihypertensive; class IV antiarrhythmic. A β-adrenergic blocker. mp = 107-109°. *Bayer AG; Ciba-Geigy Corp.*

1465 Oxyfedrine
15687-41-9 7096
$C_{19}H_{23}NO_3$
3-[[(αS,βR)-β-Hydroxy-α-methylphenethyl]amino]-3'-methoxypropiophenone.
oxyphedrine. Cardiotonic; antianginal. Used to treat cardiac insufficiency. Partial β-adrenergic agonist with coronary vasodilating and positive inotropic effects. *Degussa-Hüls Corp.*

1466 L-Oxyfedrine Hydrochloride
16777-42-7 7096 240-828-1
$C_{19}H_{24}ClNO_3$
3-[[(αS,βR)-β-Hydroxy-α-methylphenethyl]amino]-3'-methoxypropiophenone hydrochloride.
Ildamen; Modacor. Antianginal agent. Used in treatment of coronary insufficiency. [l-form]: mp = 192-194°; LD_{50} (mus iv) = 29 mg/kg; [dl-form]: mp = 173-175°; LD_{50} (mus iv) = 34 mg/kg. *Degussa-Hüls Corp.*

1467 Ozagrel
82571-53-7 7115
$C_{13}H_{12}N_2O_2$
(E)-3-[4-(1H-Imidazol-1-ylmethyl)cinnamic acid.
OKY-046. Antianginal; antithrombotic. Thromboxane synthetase inhibitor. mp = 223-224°. *Kissei; Ono Pharm.*

1468 Ozagrel Hydrochloride
78712-43-3 7115
$C_{13}H_{13}ClN_2O_2$
(E)-3-[4-(1H-Imidazol-1-ylmethyl)cinnamic acid hydrochloride.
Antianginal; antithrombotic. Thromboxane synthetase inhibitor. mp = 214-217°. *Kissei; Ono Pharm.*

1469 Ozagrel Sodium
7115
$C_{13}H_{11}N_2NaO_2$
Sodium (E)-3-[4-(1H-imidazol-1-ylmethyl)cinnamate.
Cataclot; Xanbon. Antianginal; antithrombotic. Thromboxane synthetase inhibitor. LD_{50} (mmus iv) = 1940 mg/kg, (mmus orl) = 3800 mg/kg, (mmus sc) = 2450 mg/kg, (fmus iv) = 1580 mg/kg, (fmus orl) = 3600 mg/kg, (fmus sc) = 2100 mg/kg, (mrat iv) = 1150 mg/kg, (mrat orl) = 5900 mg/kg, (mrat sc) = 2300 mg/kg, (frat iv) = 1300 mg/kg, (frat orl) = 5700 mg/kg, (frat sc) = 2250 mg/kg. *Kissei; Ono Pharm.*

1470 Penbutolol
38363-40-5 7209
$C_{18}H_{29}NO_2$
(S)-1-(tert-Butylamino)-3-(o-cyclopentylphenoxy)-2-propanol.
Antianginal; antihypertensive; antiarrhythmic. A β-adrenergic calcium channel blocker with antiarrhythmic activity. mp = 68-72°; $[\alpha]_D^{20}$ = -11.5° (c = 1 MeOH); soluble in MeOH, EtOH, $CHCl_3$. *Hoechst AG.*

1471 Penbutolol Sulfate
38363-32-5 7209 253-906-5
$C_{36}H_{40}N_2O_8S$
(S)-1-(tert-Butylamino)-3-(o-cyclopentylphenoxy)-2-propanol sulfate (salt) (2:1).
HOE-893d; HOE-39-893d; Betapressin; Levatol; Paginol. Antianginal; antihypertensive; antiarrhythmic. A β-adrenergic calcium channel blocker with antiarrhythmic activity. mp = 216-218° (dec); $[\alpha]_D^{20}$ = -24.6° (c = 1 MeOH). *Hoechst AG.*

1472 Pentaerythritol Tetranitrate
78-11-5 7249 201-084-3
$C_5H_8N_4O_{12}$
2,2-Bis(hydroxymethyl)-1,3-propanediol tetranitrate.
PETN; nitropentaerythritol; penthrit; niperyt; Angitet; Cardiacap; Dilcoran-80; Lentrat; Hasethrol; Metranil; Mycardol, Neo-Corovas; Nitropenta; Nitropenton; Pentral 80; Pentafilin; Pentafin; Pentitrate; Penthrit; Pentrite; Pentritol; Pentanitrine; Pentryate; Pergitral; Peritrate; Perityl; Prevangor; Quintrate; Subicard; Terpate; Vasodiatol; component of: Miltrate, SDM No. 23, SDM No. 35. Coronary vasodilator. Used as an antianginal. An organic nitrate. CAUTION: Explodes on percussion. More sensitive to shock than TNT. Dilution with an inert

ingredient helps to prevent accidental explosions. mp = 140°; d_D^{20} = 1.773; soluble in Me_2CO; sparingly soluble in EtOH, Et_2O; nearly insoluble in H_2O. *ICI Americas Inc.; Parke-Davis; Rhône-Poulenc Rorer Pharm. Inc.; Wallace Labs.*

1473 Pindolol
13523-86-9 7597 236-867-9
$C_{14}H_{20}N_2O_2$
1-(Indol-4-yloxy)-3-(isopropylamino)-2-propanol.
LB-46; Visken; prinodolol; Betapindol; Blocklin L; Calvisken; Decreten; Durapindol; Glauco-Visken; Pectobloc; Pinbetol; Pynastin. Antianginal agent with antiarrhythmic, antihypertensive and antiglaucoma properties. A β-adrenergic blocker. mp = 171-163°. *Sandoz Pharm. Corp.*

1474 Primidolol
67227-55-8
$C_{17}H_{23}N_3O_4$
1-[2-[[2-Hydroxy-3-(o-toloxy)propyl]amino]-ethyl]thymine.
UK-11443. Antihypertensive; antianginal; antiarrhythmic (cardiac depressant). *Pfizer Intl.*

1475 Pronetalol
54-80-8 7974
$C_{15}H_{19}NO$
2-Isopropyl-1-(naphth-2-yl)ethanol.
pronethalol; nethalide. Antianginal; antihypertensive; antiarrhythmic. A β-adrenergic blocker. mp = 108°. *ICI ; Wyeth-Ayerst Labs.*

1476 Pronetalol Hydrochloride
51-02-5 7974
$C_{15}H_{20}ClNO$
2-Isopropylamino-1-(naphth-2-yl)ethanol hydrochloride.
pronethalol hydrochloride; ICI-38174; AY-6204; Alderlin. Antianginal; antihypertensive; antiarrhythmic. A β-adrenergic blocker. mp = 184°; LD_{50} (mus orl) = 512 mg/kg, (mus iv) = 46 mg/kg. *ICI ; Wyeth-Ayerst Labs.*

1477 Propranolol
525-66-6 8025 208-378-0
$C_{16}H_{21}NO_2$
1-(Isopropylamino)-3-(1-naphthyloxy)-2-propanol .
Avlocardyl; Euprovasin. Antianginal; antihypertensive; antiarrhythmic (class II) agent. A β-adrenergic blocker. mp = 96°. *ICI ; Parke-Davis; Quimicobiol; Wyeth-Ayerst Labs.*

1478 Propranolol Hydrochloride
318-98-9 8025 206-268-7
$C_{16}H_{22}ClNO_2$
1-(Isopropylamino)-3-(1-naphthyloxy)-2-propanol hydrochloride.
AY-64043; ICI-45520; NSC-91523; Angilol; Apsolol; Bedranol; Beprane; Berkolol; Beta-Neg; Beta-Tablinen; Beta-Timelets; Cardinol; Caridolol; Deralin; Dociton; Duranol; Efektolol; Elbol; Frekven; Inderal; Indobloc; Intermigran; Kemi S; Oposim; Prano-Puren; Prophylux; Propranur; Pylapron; Rapynogen; Sagittol; Servanolol; Sloprolol; Sumial; Tesnol. Antianginal; antihypertensive; antiarrhythmic (class II) agent. β-Adrenergic blocker. mp = 163-164°; soluble in H_2O, alcohol; insoluble in Et_2O, C_6H_6, EtOAc; LD_{50} (mus orl) = 565 mg/kg, (mus iv) = 22 mg/kg, (mus ip) = 107 mg/kg. *ICI ; Parke-Davis; Quimicobiol; Wyeth Labs.*

1479 Ranolazine
95635-55-5 8287
$C_{24}H_{33}N_3O_4$
(±)-4-[2-Hydroxy-3-(o-methoxyphenoxy)propyl]-1-piperazineaceto-2',6'-xylidide.
Antianginal agent. anti-ischemic agent which modulates myocardial metabolism. *Syntex Labs. Inc.*

1480 Ranolazine Hydrochloride
95635-56-6 8287
$C_{24}H_{35}Cl_2N_3O_4$
(±)-4-[2-Hydroxy-3-(o-methoxyphenoxy)propyl]-1-piperazineaceto-2',6'-xylidide dihydrochloride.
RS-43285. Antianginal agent. Anti-ischemic agent which modulates myocardial metabolism. mp = 164-166°; soluble in H_2O. *Syntex Labs. Inc.*

1481 Semotiadil
116476-13-2 8590
$C_{29}H_{32}N_2O_6S$
(R)-2-[2-[3-[[2-(1,3-Benzodioxol-5-yloxy)-ethyl]methylamino]propoxy]-5-methoxyphenyl]-4-methyl-2H-1,4-benzothiazin-3(4H)-one.
sesamodil; DS-4823. Antianginal; antihypertensive; antiarrhythmic. Benzothiazine calcium antagonist. 8590.

1482 Semotiadil Fumarate
116476-14-3 8590
$C_{33}H_{36}N_2O_{10}S$
(R)-2-[2-[3-[[2-(1,3-Benzodioxol-5-yloxy)-ethyl]methylamino]propoxy]-5-methoxyphenyl]-4-methyl-2H-1,4-benzothiazin-3(4H)-one fumarate.
SD-3211. Antianginal; antihypertensive; antiarrhythmic. Benzothiazine calcium antagonist. mp = 134-135°; $[\alpha]_D^{25}$ = 195° (DMSO). 8590.

1483 Sotalol
3930-20-9 8876
$C_{12}H_{20}N_2O_3S$
4'-[1-Hydroxy-2-(isopropylamino)ethyl]-methanesulfonanilide.
Antianginal; class II and III antiarrhythmic; antihypertensive. A β-adrenergic blocker. λ_m = 242.2, 275.2 nm ($CHCl_3$). *Berlex Labs Inc.; Bristol-Myers Squibb Co.*

1484 Sotalol Hydrochloride
959-24-0 8876 213-496-0
$C_{12}H_{21}ClN_2O_3S$
4'-[1-Hydroxy-2-(isopropylamino)ethyl]methanesulfonanilide monohydrochloride.
MJ-1999; Beta-Cardone; Betapace; Darob; Sotacor; Sotalex. Antianginal; class II and III antiarrhythmic; antihypertensive. A β-adrenergic blocker. mp = 206.5-207°, 218-219°; soluble in H_2O, less soluble in $CHCl_3$; LD_{50} (mmus orl) = 2600 mg/kg, (mmus ip) = 670 mg/kg, (mrat orl) = 3450 mg/kg, (mrat ip) = 680 mg/kg, (rbt orl) = 1000 mg/kg, (dog ip) = 330 mg/kg. *Berlex Labs Inc.; Bristol-Myers Squibb Co.*

1485 Terodiline
15793-40-5 9311
$C_{20}H_{27}N$
N-tert-Butyl-1-methyl-3,3-diphenylpropylamine.
Coronary vasodilator used as an antianginal agent. Used also in treatment of urinary incontinence. Calcium antagonist with anticholinergic and vasodilatory activity. Liquid; $bp_{1.0}$ = 130-132°. *Marion Merrell Dow Inc.*

1486 Terodiline Hydrochloride
7082-21-5 9311
$C_{20}H_{28}ClN$
N-tert-Butyl-1-methyl-3,3-diphenylpropylamine hydrochloride.
Bicor; Mictrol; Micturin; Micturol. Coronary vasodilator used as an antianginal agent. Used also in treatment of urinary incontinence. Calcium antagonist with anticholinergic and vasodilatory activity. mp = 178-180°; soluble in EtOH, slightly soluble in Et_2O. *Marion Merrell Dow Inc.*

1487 Timolol
91524-16-2 9585
$C_{13}H_{24}N_4O_3S.1/2H_2O$
(S)-1-(tert-Butylamino)-3-[(4-morpholino-1,2,5-thiadiazol-3-yl)oxy]-2-propanol hemihydrate.
Antianginal agent with antiarrhythmic (class II), anti-hypertensive and antiglaucoma properties. A β-adrenergic blocker. [(±) form]: mp = 71.5-72.5°. *Merck & Co.Inc.*

1488 Timolol Maleate
26921-17-5 9585 248-111-5
$C_{17}H_{28}N_4O_7S$
(S)-1-(tert-Butylamino)-3-[(4-morpholino-1,2,5-thiadiazol-3-yl)oxy]-2-propanol maleate (1:1).
Blocadren; Timoptic; Timoptol; MK-950; Aquanil; Betim; Proflax; Temserin; Tenopt; Timacar; Timacor; component of: Cosopt, Timolide. Antianginal agent with antiarrhythmic (class II), antihypertensive and antiglaucoma properties. A β-adrenergic blocker. mp = 201.5-202.5°; $[\alpha]_{405}^{24}$ = -12.0° (c = 5 1N HCl), $[\alpha]_D^{25}$ = -4.2°; λ_m = 294 nm ($A_{1\ cm}^{1\%}$ 200 0.1N HCl); soluble in EtOH, MeOH; poorly soluble in $CHCl_3$, cyclohexane; insoluble in isooctane, Et_2O. *Merck & Co.Inc.*

1489 Tolamolol
38103-61-6 253-783-8
$C_{19}H_{24}N_2O_4$
4-[2-[[2-Hydroxy-3-(2-methylphenoxy)propyl]-amino]ethoxy]benzamide.
Vasodilator (coronary); cardiac depressant (antiarrhythmic); antiadrenergic (β-receptor). *CIBA plc; Pfizer Inc.*

1490 Toliprolol
2933-94-0 9653 220-905-6
$C_{13}H_{21}NO_2$
1-(Isopropylamino)-3-(m-tolyloxy)-2-propanol.
MHIP. Antianginal; antihypertensive agent. A β-adrenergic blocker. mp = 75-76°, 79°. *Boehringer Ingelheim GmbH; ICI.*

1491 Toliprolol Hydrochloride
306-11-6 9653 206-177-2
$C_{13}H_{22}ClNO_2$
1-(Isopropylamino)-3-(m-tolyloxy)-2-propanol hydrochloride.
ICI-45763; Ko-592; Doberol; Sinorytmal. Antianginal; antihypertensive agent. A β-adrenergic blocker. mp = 120-121°; λ_m = 270 nm ($E_{1\ cm}^{1\%}$ 498 H_2O). *Boehringer Ingelheim GmbH; ICI.*

1492 Tosifen
32295-18-4 250-983-7
$C_{17}H_{20}N_2O_3S$
(S)-1-(α-Methylphenethyl)-3-(p-tolylsulfonyl)urea.
Sch-11973. Antianginal agent. *Schering AG.*

1493 Trimetazidine
5011-34-7 9835 225-690-2
$C_{14}H_{22}N_2O_3$
1-(2,3,4-Trimethoxybenzyl)piperazine.
40045. Antianginal agent. Coronary vasodilator. $bp_{2.0}$ = 200-205°. *Sci. Union et Cie France.*

1494 Trimetazidine Dihydrochloride
13171-25-0 9835 236-117-0
$C_{14}H_{24}Cl_2N_2O_3$
1-(2,3,4-Trimethoxybenzyl)piperazine dihydrochloride.
Kyurinett; Vastarel F; Yoshimilon. Antianginal agent. Coronary vasodilator. mp = 225-228°; LD_{50} (mmus iv = 91 mg/kg, (mmus ip) = 264 mg/kg, (mmusorl) = 528 mg/kg, (fmus iv) = 107 mg/kg, (fmus ip) = 245 mg/kg, (fmus orl) = 608 mg/kg, (mrat iv) = 124 mg/kg, (mrat ip) = 327 mg/kg, (mrat orl) = 1147 mg/kg, (frat iv) = 124 mg/kg, (frat ip) = 288 mg/kg, (frat orl) = 987 mg/kg. *Sci. Union et Cie France.*

1495 Trolnitrate
7077-34-1 9900 230-376-3
$C_6H_{12}N_4O_9$
2,2',2''-Nitrilotrisethanol trinitrate (ester).
Antianginal agent. *Bristol-Myers Squibb Co; Schering-Plough HealthCare Products.*

1496 Trolnitrate Phosphate
588-42-1 9900 209-617-1
$C_6H_{18}N_4O_{17}P_2$
2,2',2''-Nitrilotrisethanol trinitrate (ester) phosphate (1:2) salt.
triethanolamine trinitrate biphosphate; trinitro-triethanolamine diphosphate; Angitrit; Bentonyl; Duronitrin; Metamed; Nitretamin; Nitroduran; Ortin; Praentiron; Vasomed. Antianginal. mp = 107-109°. *Bristol-Myers Squibb Co; Schering-Plough HealthCare Products.*

1497 Verapamil
52-53-9 10083 200-145-1
$C_{27}H_{38}N_2O_4$
5-[(3,4-Dimethoxyphenethyl)methylamino]-2-(3,4-dimethoxyphenyl)-2-isopropylvaleronitrile.
D-365; CP-16533-1; iproveratril. Antianginal; class IV antiarrhythmic agent. Coronary vasodilator with calcium

channel blocking activity. $bp_{0.001}$ = 243-246°; insoluble in H_2O; slightly soluble in C_6H_6, C_6H_{16}, Et_2O; soluble in EtOH, MeOH, Me_2CO, EtOAc, $CHCl_3$. *Bristol-Myers Squibb Co.*

1498 Verapamil Hydrochloride
152-11-4 10083 205-800-5
$C_{27}H_{39}ClN_2O_4$
5-[(3,4-Dimethoxyphenethyl)methylamino]-2-(3,4-dimethoxyphenyl)-2-isopropylvaleronitrile hydrochloride.
Arapamyl; Berkatens; Calan; Cardiagutt; Cardibeltin; Cordilox; Dignover; Drosteakard; Geangin; Isoptin; Quasar; Securon; Univer; Vasolan; Veracim; Veramex; Veraptin; Verelan; Verexamil. Antianginal; class IV antiarrhythmic agent. Coronary vasodilator with calcium channel blocking activity. mp = 138.5-140.5°; soluble in H_2O (7 g/100 g, 83 mg/ml), EtOH (26 mg/ml), propylene glycol (93 mg/ml), MeOH (> 100 mg/ml), iPrOH (4.6 mg/ml), EtOAc (1.0 mg/ml), DMF (> 100 mg/ml), CH_2Cl_2 (> 100 mg/ml), C_6H_{14} (0.001 mg/ml); LD_{50} (rat iv) = 16 mg/kg, (mus iv) = 8 mg/kg. *Knoll Pharm. Co.; Lederle Labs.; Parke-Davis; Searle G.D. & Co.*

1499 Xemilofiban
149820-74-6
$C_{18}H_{22}N_4O_4$
Ethyl (3S)-3-[3-[(p-amidinophenyl)carbamoyl]-propionamido]-4-pentynoate.
Used in treatment of unstable angina. Prevents post-recanalization reocclusion of coronary vessels. *Searle G.D. & Co.*

1500 Xemilofiban Hydrochloride
156586-91-3
$C_{18}H_{23}ClN_4O_4$
Ethyl (3S)-3-[3-[(p-amidinophenyl)carbamoyl]-propionamido]-4-pentynoate hydrochloride.
SC-54684A. Used in treatment of unstable angina. Prevents post-recanalization reocclusion of coronary vessels. *Searle G.D. & Co.*

1501 Zatebradine
85175-67-3 10245
$C_{26}H_{36}N_2O_5$
3-[3-[(3,4-Dimethoxyphenethyl)methylamino]-propyl]-1,3,4,5-tetrahydro-7,8-dimethoxy-2H-3-benzazepin-2-one.
Antianginal agent. Specific bradycardic agent; sinus node inhibitor. *Boehringer Ingelheim GmbH.*

1502 Zatebradine Hydrochloride
91940-87-3 10245
$C_{26}H_{37}ClN_2O_5$
3-[3-[(3,4-Dimethoxyphenethyl)methylamino]-propyl]-1,3,4,5-tetrahydro-7,8-dimethoxy-2H-3-benzazepin-2-one hydrochloride.
UL-FS-49. Antianginal agent. Specific bradycardic agent; sinus node inhibitor. mp = 168°, 185° (2 crystalline modifications); soluble in H_20. *Boehringer Ingelheim GmbH.*

Antiarrhythmics

1503 Abanoquil
90402-40-7
$C_{22}H_{25}N_3O_4$
4-Amino-2-(3,4-dihydro-6,7-dimethoxy-2(1H)-isoquinolyl)-6,7-dimethoxy quinoline.
An alpha-adrenergic blocker and antiarrhythmic.

1504 Acebutolol
37517-30-9 16 253-539-0
$C_{18}H_{28}N_2O_4$
(±)-3'-Acetyl-4'-[2-hydroxy-3-(1-methylethylamino)-propoxy]butyranilide.
Monitan; Sectral; Prent. Antianginal agent. Class II antiarrhythmic agent. Cardioselective β-adrenergic blocking agent. Has antiarrhythmic and antihypertensive properties. mp = 119-123°. *Wyeth Labs.*

1505 Acebutolol Hydrochloride
34381-68-5 16 251-980-3
$C_{18}H_{29}ClN_2O_4$
(±)-3'-Acetyl-4'-[2-hydroxy-3-(1-methylethylamino)-propoxy]butyranilide hydrochloride.
IL-17803A; M&B-17803A; Acetanol; Neptall; Sectral. Antianginal agent. Class II antiarrhythmic agent. Cardioselective β-adrenergic blocking agent. Has antiarrhythmic and antihypertensive properties. mp = 141-143°; soluble in H_2O (2 g/100 ml), EtOH (7.0 g/100 ml). *Wyeth Labs.*

1506 Acecainide
32795-44-1 17
$C_{15}H_{23}N_3O_2$
4'-[[2-(Diethylamino)ethyl]carbamoyl]acetanilide.
N-acetylprocainamide. Antiarrhythmic. Cardiac depressant. *Bristol-Myers Squibb Co.*

1507 Acecainide Hydrochloride
34118-92-8 17 251-831-2
$C_{15}H_{24}ClN_3O_2$
4'-[[2-(Diethylamino)ethyl]carbamoyl]acetanilide monohydrochloride.
ASL-601; NAPA. Antiarrhythmic. Cardiac depressant. mp = 190-193°. *Bristol-Myers Squibb Co; C.M. Ind.*

1508 Actisomide
96914-39-5
$C_{23}H_{35}N_3O$
(±)-cis-4-[2-(Diisopropylamino)ethyl]-4,4a,5,6,7,8-hexahydro-1-methyl-4-phenyl-3H-pyrido-[1,2-c]pyrimidin-3-one.
SC-36602. Antiarrhythmic. Cardiac depressant. *Searle G.D. & Co.*

1509 Adenosine
58-61-7 152 200-389-9
$C_{10}H_{13}N_5O_4$
9-β-D-Ribofuranosyl-9H-purin-6-amine.
adenine riboside; Adenocard; Adenocor; Adenoscan. Antiarrhythmic. A nucleo-side found widely in natural sources. mp = 234-235°; $[\alpha]_D^{11}$ = -61.7° (c = 0.706 H_2O); λ_m = 260 nm (ε 15100); insoluble in EtOH.

1510 Ajmaline
4360-12-7 194 224-439-4
$C_{20}H_{26}N_2O_2$
Ajmalan-17,20-diol.
rauwolfine; Gilurytmal; Cardiorythmine; Ritmos; Tachmalin. Antiarrhythmic agent with antihypertensive properties. [MeOH solvate]: mp = 158-160°; $[\alpha]_D^{18}$ = 131° (c = 0.4 $CHCl_3$); [anhydrous form]: mp = 205-207°; $[\alpha]_D^{20}$ = 144° (c = 0.8 $CHCl_3$); λ_m = 247, 295 nm (log ε 3.94, 3.49 EtOH); soluble in EtOH, MeOH, Et_2O, $CHCl_3$; poorly soluble in H_2O.

1511 Alinidine
33178-86-8 244
$C_{12}H_{13}Cl_2N_3$
2-(N-Allyl-2,6-dichloroanilino)-2-imidazoline.
ST-567. Antiarrhythmic. Specific bradycardiac agent. Analog of clonidine. mp = 127-129°, 130-131°; [hydrobromide ($C_{12}H_{14}BrCl_2N_3$)]: mp = 193-194°. *Boehringer Ingelheim Pharm. Inc.*

1512 Almokalant
123955-10-2
$C_{18}H_{28}N_2O_3S$
(±)-p-[3-[Ethyl[3-(propylsulfinyl)propyl]amino]-2-hydroxypropoxy]benzonitrile.
A class III antiarrhythmic drug.

1513 Alprafenone
124316-02-5
$C_{25}H_{35}NO_4$
(±)-3[3-[2-Hydroxy-3-(tert-pentylamino)-propoxy]-4-methoxyphenyl]-4'-methylpropiophenone.
A new class I antiarrhythmic agent.

1514 Alprenolol
13655-52-2 321 237-140-9
$C_{15}H_{23}NO_2$
1-(o-Allylphenoxy)-3-(isopropylamino)-2-propanol.
H-56/28. Antianginal agent. antiarrhythmic agent. Cardioselective β-adrenergic blocking agent. Has antiarrhythmic and antihypertensive properties. *C.M. Ind.; ICI.*

1515 Alprenolol Hydrochloride
13707-88-5 321 237-244-4
$C_{15}H_{24}ClNO_2$
1-(o-Allylphenoxy)-3-(isopropylamino)-2-propanol hydrochloride.
Applobal; Aprobal; Aptine; Aptol Duriles; Gubernal; Regletin; Yobir. Antianginal agent. Class II antiarrhythmic agent. Cardioselective β-adrenergic blocking agent. Has antiarrhythmic and antihypertensive properties. mp = 107-109°; LD_{50} (mus orl) = 278.0 mg/kg, (rat orl) = 597.0 mg/kg, (rbt orl) = 337.3 mg/kg. *C.M. Ind.; ICI.*

1516 Amafolone
50588-47-1
$C_{19}H_{31}NO_2$
3α-Amino-2β-hydroxy-5α-androstan-17-one.
SC-35135. A new aminosteroidal antiarrhythmic agent.

1517 Ambasilide
83991-25-7
$C_{21}H_{25}N_3O$
3-(p-Aminobenzoyl)-7-benzyl-3,7-diazabicyclo[3.3.1]nonane.
A new class III antiarrhythmic.

1518 Amiodarone
1951-25-3 504 217-772-1
$C_{25}H_{29}I_2NO_3$
2-Butyl-3-benzofuranyl 4-[2-(diethylamino)ethoxy]-3,5-diiodophenyl ketone.
Cordarone; L-3428; SKF-33134-A. Antianginal; class III antiarrhythmic. Blocks both α- and β-receptors. Ventricular antiarrhythmic agent. A benzofuran derivative. *Wyeth-Ayerst Labs.*

1519 Amiodarone Hydrochloride
19774-82-4 504 243-293-2
$C_{25}H_{30}ClI_2NO_3$
2-Butyl-3-benzofuranyl 4-[2-(diethylamino)ethoxy]-3,5-diiodophenyl ketone hydrochloride.
L-3428; Amiodar; Ancoron; Angiodarona; Atlansil; Cordarex; Cordarone; Cordarone X; Miocard; Miodaron; Ortacrone; Ritmocardyl; Rythmarone; Trangorex. Antianginal; class III antiarrhythmic. Blocks both α- and β-receptors. Ventricular antiarrhythmic agent. A benzofuran derivative. mp = 156°, 159 ± 2°; λ_m = 208, 242 nm ($E_{1\,cm}^{1\%}$ 662 ± 8, 623 ± 10 MeOH); soluble in EtOH (1.28 g/100 ml), MeOH (9.98 g/100 ml), $CHCl_3$ (44.51 g/100 ml), n-PrOH (0.13 g/100 ml), Et_2O (0.17 g/100 ml), THF (0.60 g/100 ml), C_6H_6 (0.65 g/100 ml), CH_2Cl_2 (19.20 g/100 ml), CH_3CN (0.32 g/100 ml), 1-octanol (0.30 g/100 ml), H_2O (0.07 g/100 ml), C_6H_{14} (0.03 g/100 ml), petroleum ether (0.001 g/100 ml). *Wyeth-Ayerst Labs.*

1520 Amoproxan
22661-76-3 611
$C_{22}H_{35}NO_7$
3,4,5-Trimethoxybenzoic acid 1-[(isopentyloxy)methyl]-2-morpholino ethyl ester.
Antiarrhythmic agent. *C.E.R.M.*

1521 Amoproxan Hydrochloride
22661-96-7 611
$C_{22}H_{36}ClNO_7$
3,4,5-Trimethoxybenzoic acid 1-[(isopentyloxy)methyl]-2-morpholino ethyl ester hydrochloride.
CERM-730; Mederel. Antiarrhythmic agent. mp = 145°; soluble in H_2O, EtOH; slightly soluble in EtOAc. *C.E.R.M.*

1522 Anipamil
83200-10-6 280-213-5
$C_{34}H_{52}N_2O_2$
2-[3-[(m-Methoxyphenethyl)methylamino]-propyl]-2-(m-methoxyphenyl)-tetradecanenitrile.
Antianginal; class IV antiarrhythmic agent. Analog of verapamil.

1523 Aprindine
37640-71-4 793
$C_{22}H_{30}N_2$
N-(2,3-Dihydro-1H-inden-2-yl)-N',N'-diethyl-N-phenyl-1,3-propanediamine.
compd 99170; AC-1802; Lilly 99170. Class I antiarrhythmic agent. *Christiaens S.A.; Eli Lilly & Co.*

1524 Aprindine Hydrochloride
33237-74-0 793 251-418-7
$C_{22}H_{31}ClN_2$
N-(2,3-Dihydro-1H-inden-2-yl)-N',N'-diethyl-N-phenyl-1,3-propanediamine hydrochloride.
compd 83846; Amidonal; Aspenon; Fibocil; Fiboran; Ritmusin. Class I antiarrhythmic agent. mp = 120-121°. *Christiaens S.A.; Eli Lilly & Co.*

1525 Arotinolol
68377-92-4 827
$C_{15}H_{21}N_3O_2S_3$
(±)-5-[2-[[3-(tert-Butylamino)-2-hydroxypropyl]thio]-4-thiazolyl]-2-thiophenecarboxamide.
Antianginal; antihypertensive; antiarrhythmic. Has both α- and β-adrenergic blocking activity. A propanolamine derivative. mp = 148-149°. *Sumitomo Pharm. Co. Ltd.*

1526 Arotinolol Hydrochloride
68377-91-3 827
$C_{15}H_{22}ClN_3O_2S_3$
(±)-5-[2-[[3-(tert-Butylamino)-2-hydroxypropyl]thio]-4-thiazolyl]-2-thiophenecarboxamide hydrochloride.
S-596; ARL; Almarl. Antianginal agent. Also antihypertensive; antiarrhythmic. Possesses both α- and β-adrenergic blocking activity. A propanolamine derivative. mp = 234-235.5°; LD_{50} (mus iv) = 86 mg/kg, (mus ip) = 360 mg/kg, (mus orl) > 5000 mg/kg. *Sumitomo Pharm. Co. Ltd.*

1527 Artilide
133267-19-3
$C_{19}H_{34}N_2O_3S$
(+)-4'-[(R)-4-(Dibutylamino)-1-hydroxybutyl]-methanesulfonanilide.
Antiarrhythmic. Cardiac depressant. *Pharmacia & Upjohn.*

1528 Artilide Fumarate
133267-20-6
$C_{42}H_{72}N_4O_{10}S_2$
(+)-4'-[(R)-4-(Dibutylamino)-1-hydroxybutyl]-methanesulfonanilide fumarate.
U-88943E. Antiarrhythmic. Cardiac depressant. *Pharmacia & Upjohn.*

1529 Asocainol
77400-65-8
$C_{27}H_{31}NO_3$
(±)-6,7,8,9-Tetrahydro-2,12-dimethoxy-7-methyl-6-phenethyl-5H-dibenz[d,f]azonin-1-ol.
A class I antiarrhythmic agent.

1530 Atenolol
29122-68-7 892 249-451-7
$C_{14}H_{22}N_2O_3$
2-[p-[2-Hydroxy-3-(isopropylamino)propoxy]phenyl]-acetamide.
ICI-66082; AteHexal; Atenol; Cuxanorm; Ibinolo; Myocord; Prenormine; Seles Beta; Selobloc; Teno-basan; Tenoblock; Tenormin; Uniloc. Antianginal; class II antiarrhythmic agent. Cardioselective β-adrenergic blocking agent. Has antiarrhythmic and antihypertensive properties. mp = 146-148°, 150-152°; λ_m = 225, 275, 283 nm (MeOH); very soluble in MeOH; soluble in AcOH, DMSO; less soluble in Me_2CO, dioxane; insoluble in CH_3CN, EtOAc, $CHCl_3$; LD_{50} (mus orl) = 2000 mg/kg, (mus iv) = 98.7 mg/kg, (rat orl) = 3000 mg/kg, (rat iv) = 59.24 mg/kg. *Apothecon; C.M. Ind.; ICI ; Lemmon Co.; Zeneca Pharm.*

1531 Azimilide
149908-53-2 943
$C_{23}H_{28}ClN_5O_3$
1-[[5-(p-Chlorophenyl)furfurylidene]amino]-3-[4-(4-methyl-1-piperazinyl)butyl]-hydantoin.
Class III antiarrhythmic agent. Potassium channel blocker. *Procter & Gamble Pharm. Inc.*

1532 Azimilide Dihydrochloride
149888-94-8 943
$C_{23}H_{30}Cl_3N_5O_3$
1-[[5-(p-Chlorophenyl)furfurylidene]amino]-3-[4-(4-methyl-1-piperazinyl)butyl]-hydantoin dihydrochloride.
NE-10064. Class III antiarrhythmic agent. Potassium channel blocker. *Procter & Gamble Pharm. Inc.*

1533 Barucainide
79784-22-8
$C_{22}H_{30}N_2O_2$
4-Benzyl-1,3-dihydro-7-[4-isopropylamino)butoxy]-6-methylfuro[3,4-c]pyridine.
A class IB antiarrhythmic agent.

1534 Bepridil
64706-54-3 1188
$C_{24}H_{34}N_2O$
1-[2-(N-Benzylanilino)-1-(isobutoxymethyl)-ethyl]pyrrolidine.
A calcium channel blocker with antianginal and antiarrhythmic (class IV) properties. $bp_{0.1}$ = 184°; $bp_{0.5}$ = 192°; n_D^{20} = 1.5538. *McNeil Pharm.; Wallace Labs.*

1535 Bepridil Hydrochloride
74764-40-2 1188
$C_{24}H_{37}ClN_2O_3$
1-[2-(N-Benzylanilino)-1-(isobutoxymethyl)-ethyl]pyrrolidine monohydrochloride monohydrate.
CERM-1978; Angopril; Bepadin; Cordium; Vascor. Antianginal. Calcium channel blocker with antianginal and antiarrhythmic (class IV) properties. mp 91°; LD_{50} (mus orl) = 1955 mg/kg, (mus iv) = 23.5 mg/kg. *C.M. Ind.; McNeil Pharm.; Wallace Labs.*

1536 Berlafenone
18965-97-4
$C_{19}H_{25}NO_2$
(±)-1-(2-Biphenylyloxy)-3-(tert-butylamino)-2-propanol.
A class I antiarrhythmic agent.

1537 Bertosamil
126825-36-3
$C_{19}H_{36}N_2$
3'-Isobutyl-7'-isopropylspiro-[cyclohexane-1,9'-[3,7]diazabicyclo-[3.3.1]nonane.
An antiarrhythmic agent. Potassium channel blocker.

1538 Bevantolol
59170-23-9 1238
$C_{20}H_{27}NO_4$
(±)-1-[(3,4-Dimethoxyphenethyl)amino]-3-(m-toloxy)-2-propanol.
Antianginal agent. Cardioselective β-adrenergic blocking agent. Has antiarrhythmic and antihypertensive properties. *Parke-Davis.*

1539 Bevantolol Hydrochloride
42864-78-8 1238
$C_{20}H_{28}ClNO_4$
(±)-1-[(3,4-Dimethoxyphenethyl)amino]-3-(m-toloxy)-2-propanol hydrochloride.
Vantol; CI-775; Ranestol; Sentiloc. Antianginal agent. Cardioselective β-adrenergic blocking agent. Has antiarrhythmic and antihypertensive properties. mp = 137-138°. *Parke-Davis.*

1540 Bidisomide
103810-45-3 1251
$C_{22}H_{34}ClN_3O_2$
(±)-α-(o-Chlorophenyl)-α-[2-(N-isopropyl-acetamido)ethyl]-1-piperidine-butyramide.
SC-40230. Class I antiarrhythmic agent. Sodium channel blocker. mp = 140-141°. *Searle G.D. & Co.*

1541 Bometolol
65008-93-7
$C_{25}H_{32}N_2O_7$
(±)-8-(Acetonyloxy)-5-[3-[(3,4-dimethoxy-phenethyl)amino]-2-hydroxypropoxy]-3,4-dihydrocarbostyril.
A calcium channel blocker with antiarrhythmic properties.

1542 Bretylium
59-41-6 1395
$C_{11}H_{17}BrN^+$
(o-Bromobenzyl)ethyldimethylammonium.
Adrenergic; class III antiarrhythmic agent. *Astra Chem. Ltd.; Burroughs Wellcome Inc.; Elkins-Sinn.*

1543 Bretylium Tosylate
61-75-6 1395 200-516-8
$C_{18}H_{24}BrNO_3S$
(o-Bromobenzyl)ethyldimethylammonium p-toluenesulfonate.
Bretylate; Bretylan; Bretylol; Darenthin; Ornid; ASL-603. Adrenergic; class III antiarrhythmic agent. Used in the treatment of cardiac arrhythmias. mp = 97-99°; λ_m = 278, 271, 264 nm; soluble in H_2O, organic solvents; LD_{50} (mus orl) = 400 mg/kg, (mus im) = 250 mg/kg. *Astra Chem. Ltd.; Burroughs Wellcome Inc.; Elkins-Sinn.*

1544 Brinazarone
89622-90-2
$C_{25}H_{32}N_2O_2$
p-[3-(tert-Butylamino)propoxy]phenyl 2-isopropyl-3-indolizinyl ketone.
A calcium channel blocker and potential antiarrhythmic agent.

1545 Bucainide
51481-62-0
$C_{21}H_{35}N_3$
1-Hexyl-4-(N-isobutylbenzimidoyl)piperazine.
Cardiac depressant; antiarrhythmic agent.

1546 Bucainide Maleate
51481-63-1
$C_{29}H_{43}N_3O_8$
1-Hexyl-4-(N-isobutylbenzimidoyl)piperazine maleate.
Cardiac depressant; antiarrhythmic agent.

1547 Bucromarone
78371-66-1
$C_{29}H_{37}NO_4$
2-[4-[3-(Dibutylamino)propoxyl]-3,5-dimethyl-benzoyl]chromone.
Cardiac depressant; antiarrhythmic agent. *Bristol-Myers Squibb Co.*

1548 Bucumolol
58409-59-9 1489
$C_{17}H_{23}NO_4$
8-[(3-tert-Butylamino)-2-hydroxypropoxy]-5-methylcoumarin.
Antianginal agent. A β-adrenergic blocker with antianginal and antiarrhythmic properties. *Sankyo Co. Ltd.*

1549 Bucumolol Hydrochloride
30073-40-6 1489
$C_{17}H_{24}ClNO_4$
8-[(3-tert-Butylamino)-2-hydroxypropoxy]-5-methylcoumarin hydrochloride.
CS-359; Bucumarol. Antianginal agent. A β-adrenergic blocker with antianginal and antiarrhythmic properties. mp = 226-228°; LD_{50} (mmus orl) = 676 mg/kg, (mmus iv) = 33.1 mg/kg, (fmus orl) = 692 mg/kg, (fmus iv) = 31.6 mg/kg. *Sankyo Co. Ltd.*

1550 Bufetolol
53684-49-4 1496
$C_{18}H_{29}NO_4$
1-(tert-Butylamino)-3-[o-[(tetrahydrofurfuryl)-oxy]phenoxy]-2-propanol.
Antianginal agent. A β-adrenergic blocker with antianginal and antiarrhythmic properties. $bp_{0.07}$ = 180-186°. *Yoshitomi.*

1551 Bufetolol Hydrochloride
35108-88-4 1496 252-369-4
$C_{18}H_{30}ClNO_4$
1-(tert-Butylamino)-3-[o-[(tetrahydrofurfuryl)-oxy]phenoxy]-2-propanol hydrochloride.
Y-6124; Adobiol. Antianginal agent. A β-adrenergic blocker with antianginal and antiarrhythmic properties. mp = 153.5-157°, 151-154°; soluble in H_2O, MeOH,

AcOH; slightly soluble in C_6H_6; insoluble in Et_2O; LD_{50} (mus orl) = 409 mg/kg, (mus sc) = 507 mg/kg, (rat orl) = 1142 mg/kg, (rat sc) = 1904 mg/kg. *Yoshitomi.*

1552 Bunaftine
32421-46-8 1509 251-027-1
$C_{21}H_{30}N_2O$
N-Butyl-N-[2-(diethylamino)ethyl]-1-naphthamide.
bunaphtide; bunaphtine; Meregon [as citrate]. Class III antiarrhythmic agent. $bp_{0.1}$ = 178°; LD_{50} (mus ip) = 122°.

1553 Bunitrolol
34915-68-9 1514
$C_{14}H_{20}N_2O_2$
o-[3-(tert-Butylamino)-2-hydroxypropoxy]-benzonitrile.
Ko-1366. Antianginal agent with antiarrhythmic and antihypertensive properties. A β-adrenergic blocker. *Boehringer Ingelheim GmbH.*

1554 Bunitrolol Hydrochloride
23093-74-5 1514 245-427-5
$C_{14}H_{21}ClN_2O_2$
o-[3-(tert-Butylamino)-2-hydroxypropoxy]benzonitrile hydrochloride.
Betriol; Stresson. Antianginal agent with antiarrhythmic and antihypertensive properties. A β-adrenergic blocker. mp = 163-165°; LD_{50} (mus orl) = 1344-1440 mg/kg, (mus ip) = 264-265 mg/kg, (rat orl) = 639-649 mg/kg, (rat ip) = 222-225 mg/kg. *Boehringer Ingelheim GmbH.*

1555 Bupranolol
14556-46-8 1521
$C_{14}H_{22}ClNO_2$
1-(tert-Butylamino)-3-[(6-chloro-m-tolyl)-oxy]-2-propanol.
bupranol; Ophtorenin. Antianginal agent with antiarrhythmic, antihypertensive and antiglaucoma properties. A β-adrenergic blocker.

1556 Bupranolol Hydrochloride
15148-80-8 1521 239-208-3
$C_{14}H_{23}Cl_2NO_2$
1-(tert-Butylamino)-3-[(6-chloro-m-tolyl)-oxy]-2-propanol hydrochloride.
B-1312; KL-255; Betadran; Betadrenol; Looser; Panimit. Antianginal agent with antiarrhythmic, antihypertensive and antiglaucoma properties. A β-adrenergic blocker. mp = 220-222°.

1557 Butidrine
55837-18-8 1558 259-849-2
$C_{16}H_{25}NO$
5,6,7,8-Tetrahydro-α-[[(1-isopropyl)amino]methyl]-2-naphthalenemethanol.
Antiarrhythmic agent. A β-adrenergic blocker. *Holding Ceresia.*

1558 Butidrine Hydrochloride
1506-12-3 1558
$C_{16}H_{26}ClNO$
5,6,7,8-Tetrahydro-α-[[(1-isopropyl)amino]methyl]-2-naphthalenemethanol hydrochloride.
butydrine hydrochloride; Betabloc; Recetan. Antiarrhythmic agent. A β-adrenergic blocker. mp = 129-130°; LD_{50} (mus iv) = 20.2 mg/kg, (mus orl) = 235 mg/kg. *Holding Ceresia.*

1559 Butobendine
55769-65-8 1560
$C_{32}H_{48}N_2O_{10}$
(+)-(S,S)-Ethylenebis[(methylimino)(2-ethylethylene)]-bis(3,4,5-trimethoxybenzoate).
Antiarrhythmic. Increases cardiac blood flow. mp = 60-62°, 64-65°; $[\alpha]_D^{20}$ = 2.4° (c = 5 EtOH). *Polfa.*

1560 Butobendine Dihydrochloride
55769-64-7 1560
$C_{32}H_{50}Cl_2N_2O_{10}$
(+)-(S,S)-Ethylenebis[(methylimino)(2-ethylethylene)]-bis(3,4,5-trimethoxybenzoate) dihydrochloride.
M-71; Craviten. Antiarrhythmic. Increases cardiac blood flow. mp = 83-113°, 81-83°; $[\alpha]_D^{20}$ = -6.4° (c = 2.5 etOH), -7.5° (c = 5 H_2O), -5.5° (c = 5 C_5H_5N); soluble in H_2O, $CHCl_3$, EtOH; slightly soluble in MeOH; insoluble in C_6H_6, Et_2O, CCl_4; LD_{50} (rat ip)= 142 mg/kg, (rat iv) = 15.8 mg/kg, (mus ip) = 550 mg/kg, (rbt iv) = 5.1 mg/kg. *Polfa.*

1561 Butoprozine
62228-20-0
$C_{18}H_{38}N_2O_2$
p-[3-(Dibutylamino)propoxy]phenyl 2-ethyl-3-indolizinyl ketone.
Antiarrhythmic; antianginal agent. Calcium channel blocker. *Labaz S.A.*

1562 Butoprozine Hydrochloride
62134-34-3 263-427-3
$C_{18}H_{39}ClN_2O_2$
p-[3-(Dibutylamino)propoxy]phenyl 2-ethyl-3-indolizinyl ketone monohydrochloride.
L-9394. Antiarrhythmic; antianginal agent. Calcium channel blocker. *Labaz S.A.*

1563 Capobenate Sodium
27276-25-1 1799 248-381-4
$C_{16}H_{22}NNaO_6$
Sodium 6-(3,4,5-trimethoxybenzamido)hexanoate.
C-3; sodium capobenate; Capben. Cardiac depressant; antiarrhythmic agent. *Inst. Chemioter.*

1564 Capobenic Acid
21434-91-3 1799 244-387-6
$C_{16}H_{23}NO_6$
6-(3,4,5-Trimethoxybenzamido)hexanoic acid.
C-3; ATBAC; TB-ACA. Cardiac depressant; antiarrhythmic agent. mp = 121-123°; λ_m = 214, 259 nm (EtOH); soluble in EtOH, Me_2CO, $CHCl_3$, alkaline solutions; insoluble in H_2O, Et_2O, Cl_4; LD_{50} (rat ip) = 2500 mg/kg. *Inst. Chemioter.*

1565 Carazolol
57775-29-8 1822 260-945-1
$C_{18}H_{22}N_2O_2$
1-(Carbazol-4-yloxy)-3-isopropylamino)-2-propanol.
BM-51052; Conducton; Suacron. Antianginal agent with antiarrhythmic and antihypertensive properties. Used for treatment of stress in pigs (veterinary). A β-adrenergic

blocker. [hydrochloride]: mp = 234-235°. *Boehringer Mannheim GmbH.*

1566 Carocainide
66203-00-7 266-233-7
$C_{18}H_{25}N_3O_5$
1-[4,7-Dimethoxy-6-[2-(1-pyrrolidinyl)ethoxy]-5-benzofuranyl]-3-methylurea.
An antiarrhythmic agent. A benzofuran derivative with sodium channel blocking activity.

1567 Carteolol
51781-06-7 1917
$C_{16}H_{24}N_2O_3$
5-[3-[(1,1-Dimethylethyl)amino]-2-hydroxypropyl]-3,4-dihydro-2(1H)-quinolinone.
Antianginal agent with antiarrhythmic, antihypertensive and antiglaucoma properties. A β-adrenergic blocker. *Abbott Labs Inc.; Otsuka Am. Pharm.*

1568 Carteolol Hydrochloride
51781-21-6 1917 257-415-7
$C_{16}H_{25}ClN_2O_3$
5-[3-[(1,1-Dimethylethyl)amino]-2-hydroxypropyl]-3,4-dihydro-2(1H)-quinolinone hydrochloride.
Carteolol hydrochloride, Abbott 43326; OPC-1085; Arteoptic; Caltidren; Carteol; Cartrol; Endak; Mikelan; Optipress; Tenalet; Tenalin; Teoptic. Antianginal agent with antiarrhythmic, antihypertensive and antiglaucoma properties. A β-adrenergic blocker. mp = 278°; LD_{50} (mrat orl) = 1380 mg/kg (mrat iv) = 158 mg/kg, (mrat ip) = 380 mg/kg, (mmus orl) = 810 mg/kg, (mmus iv) = 54.5 mg/kg, (mmus ip) = 380 mg/kg. *Abbott Labs Inc.; Otsuka Am. Pharm.*

1569 Cifenline
53267-01-9 2330 258-453-7
$C_{18}H_{18}N_2$
(±)-2-(2,2-Diphenylcyclopropyl)-4,5-dihydro-1H-imidazole.
Cibenzoline; (±)-2-(2,2-diphenylcyclopropyl)-2-imidazoline; 1-(2-Δ^2-imidazolinyl)-2,2-diphenylcyclopropane; cibenzoline; Ro-22-7796; UP-33-901. Antiarrhythmic agent. Posesses potent sodium channel blocking action and moderate calcium channel blocking action. mp = 103-104°; LD_{100} (rat iv) = 64 mg/kg. *Hexachemie; Hoffmann-LaRoche Inc.*

1570 Cifenline Succinate
100678-32-8 2330
$C_{22}H_{24}N_2O_4$
(±)-2-(2,2-Diphenylcyclopropyl)-4,5-dihydro-1H-imidazole succinate.
Ro-22-7796/001; Cibenol; Cipralan; Exacor. Antiarrhythmic agent. Posesses potent sodium channel blocking action and moderate calcium channel blocking action. mp = 165°. *Hexachemie; Hoffmann-LaRoche Inc.*

1571 Clofilium Phosphate
68379-03-3
$C_{21}H_{39}ClNO_4P$
[4-(p-Chlorophenyl)butyl]diethylheptylammonium phosphate.
LY-150378. Antiarrhythmic agent; cardiac depressant. A potassium channel blocker. *Eli Lilly & Co.*

1572 Cloranolol
39563-28-5 2464
$C_{13}H_{19}Cl_2NO_2$
1-(tert-Butylamino)-3-(2,5-dichlorophenoxy)-2-propanol.
A β-adrenergic blocker used as an antiarrhythmic agent. mp = 82-83°. *Res. Inst. Pharm. Chem.*

1573 Cloranolol Hydrochloride
54247-25-5 2464 259-044-6
$C_{13}H_{20}Cl_3NO_2$
1-(tert-Butylamino)-3-(2,5-dichlorophenoxy)-2-propanol hydrochloride.
GYKI-40199; Tobanum. A β-adrenergic blocker used as an antiarrhythmic agent. mp = 210-212°. *Res. Inst. Pharm. Chem.*

1574 Dazolicine
61477-97-2
$C_{17}H_{24}ClN_3S$
8-Chloro-3,4,5,6-tetrahydro-6-[(1-isopropyl-2-imidazolin-2-yl)methyl]-2H-1,6-benzothaizocine.
dazolicin; ucb B 192. Antiarrhythmic with direct membrane action.

1575 Devapamil
92302-55-1
$C_{26}H_{36}N_2O_3$
2-(3,4-Dimethoxyphenyl)-2-isopropyl-5-[(m-methoxyphenethyl)methylamino]valeronitrile.
Antianginal; antiarrhythmic. Calcium channel blocker.

1576 Dexpropranolol
5051-22-9 225-749-2
$C_{16}H_{21}NO_2$
(+)-1-(Isopropylamino)-3-(1-naphthyloxy)-2-propanol.
A β-adrenergic blocker used as an antiarrhythmic agent. *ICI ; Wyeth-Ayerst Labs.*

1577 Dexpropranolol Hydrochloride
13071-11-9 235-961-7
$C_{16}H_{22}ClNO_2$
(+)-1-(Isopropylamino)-3-(1-naphthyloxy)-2-propanol hydrochloride.
AY-20,694; ICI-47319. A β-adrenergic blocker used as an antiarrhythmic agent. *ICI ; Wyeth-Ayerst Labs.*

1578 Dexsotalol
30236-32-9
$C_{12}H_{20}N_2O_3S$
(+)-(S)-4'-[1-hydroxy-2-(isopropylamino)-ethyl]methanesulfonanilide.
Antiarrhythmic agent; cardiac depressant. *Bristol-Myers Squibb Co.*

1579 Dexsotalol Hydrochloride
4549-94-4
$C_{12}H_{21}ClN_2O_3S$
(+)-(S)-4'-[1-Hydroxy-2-(isopropylamino)-ethyl]methanesulfonanilide monohydrochloride.
BMY-05763-1-D. Antiarrhythmic agent; cardiac depressant. *Bristol-Myers Squibb Co.*

1580 Diltiazem
42399-41-7 3247 255-796-4
$C_{22}H_{26}N_2O_4S$
(+)-5-[2-(Dimethylamino)ethyl]-cis-2,3-dihydro-3-hydroxy-2-(p-methoxyphenyl)-1,5-benzothiazepin-4(5H)-one acetate (ester).
Antianginal; antihypertensive; antiarrhythmic (class IV). Calcium channel blocker with coronary vasodilating properties. *Bristol Myers Squibb Pharm. Ltd.; Forest Pharm. Inc.; Hoechst Marion Roussel Inc.; Lemmon Co.; Rhône-Poulenc Rorer Pharm. Inc.; Shionogi & Co. Ltd.; Tanabe Seiyaku Co. Ltd.*

1581 d-cis-Diltiazem Hydrochloride
33286-22-5 3247 251-443-3
$C_{22}H_{27}ClN_2O_4S$
(+)-5-[2-(Dimethylamino)ethyl]-cis-2,3-dihydro-3-hydroxy-2-(p-methoxyphenyl)-1,5-benzothiazepin-4(5H)-one acetate (ester) monohydrochloride.
CRD-401; RG-83606; Adizem; Altiazem; Anginyl; Angizem; Britiazim; Bruzem; Calcicard; Cardizem; Citizem; Cormax; Deltazen; Diladel; Dilpral; Dilrene; Dilzem; Dilzene; Herbesser; Masdil; Tildiem. Antianginal; antihypertensive; class IV antiarrhythmic. Calcium channel blocker with coronary vasodilating properties. mp = 207.5-212°; $[\alpha]_D^{24}$ = +98.3 ± 1.4° (c = 1.002 in MeOH); soluble in H_2O, MeOH, $CHCl_3$; slightly soluble in absolute EtOH; practically insoluble in C_6H_6; LD_{50} (mmus iv) = 61 mg/kg, (mmus sc) = 260 mg/kg, (mmus orl) = 740 mg/kg, (fmus iv) = 58 mg/kg, (fmus sc) = 280 mg/kg, (fmus orl) = 640 mg/kg, (mrat iv) = 38 mg/kg, (mrat sc) = 520 mg/kg, (mrat orl) = 560 mg/kg, (frat iv) = 39 mg/kg, (frat sc) = 550 mg/kg, (frat orl) = 610 mg/kg. *Bristol Myers Squibb Pharm. Ltd.; Forest Pharm. Inc.; Hoechst Marion Roussel Inc.; Lemmon Co.; Rhône-Poulenc Rorer Pharm. Inc.; Tanabe Seiyaku Co. Ltd.*

1582 Diltiazem Malate
144604-00-2 3247
$C_{26}H_{32}N_2O_9S$
(+)-5-[2-(Dimethylamino)ethyl]-cis-2,3-dihydro-3-hydroxy-2-(p-methoxyphenyl)-1,5-benzothiazepin-4(5H)-one acetate (ester) (S)-malate (1:1) monohydrochloride.
MK-793. Antianginal; antihypertensive; class IV antiarrhythmic. Calcium channel blocker with coronary vasodilating activity. Class IV antiarrhythmic. *Bristol Myers Squibb Pharm. Ltd.; Forest Pharm. Inc.; Hoechst Marion Roussel Inc.; Lemmon Co.; Rhône-Poulenc Rorer Pharm. Inc.*

1583 Dioxadilol
80743-08-4
$C_{16}H_{25}NO_4$
(±)-1-(1,4-Benzodioxan-2-ylmethoxy)-3-(tert-butylamino)-2-propanol.
Has beta adrenolytic activity; antihypertensive, antianginal and antiarrhythmic, but less potent than propranolol.

1584 Diprafenone
81447-80-5
$C_{23}H_{31}NO_3$
(±)-2'-[2-Hydroxy-3-(tert-pentylamino)propoxy]-3-phenylpropiophenone.
Class I antiarrhythmic.

1585 Disobutamide
68284-69-5
$C_{23}H_{38}ClN_3O$
α-(o-Chlorophenyl)-α-[2-(diisopropylamino)ethyl]-1-piperidinebutyramide.
SC-31828. Antiarrhythmic agent; cardiac depressant. *Searle G.D. & Co.*

1586 Disopyramide
3737-09-5 3424 223-110-2
$C_{21}H_{29}N_3O$
α-[2-(Diisopropylamino)ethyl]-α-phenyl-2-pyridineacetamide.
SC-7031; H-3292; Dicorantil; Isorythm; Lispine; Ritmodan; Rythmodan. Class IA antiarrhythmic agent; cardiac depressant. mp = 94.5-95°; LD_{50} (mus ip) = 175 mg/kg. *Searle G.D. & Co.*

1587 Disopyramide Phosphate
22059-60-5 3424 244-756-1
$C_{21}H_{32}N_3O_5P$
α-[2-(Diisopropylamino)ethyl]-α-phenyl-2-pyridineacetamide phosphate.
SC-13957; Norpace; Dirythmin SA; Diso-Duriles; Rythmodul. Class IA antiarrhythmic agent; cardiac depressant. *KV Pharm.; Searle G.D. & Co.*

1588 Dofetilide
115256-11-6 3469
$C_{19}H_{27}N_3O_5S_2$
β-[(p-Methanesulfonamidophenethyl)-methylamino]methanesulfono-p-phenetidide.
UK-68798. Class III antiarhythmic agent. Potassium channel blocker. mp = 147-149°, 151-152°, 161°. *Pfizer Inc.*

1589 Drobuline
58473-73-7
$C_{19}H_{25}NO$
(±)-1-(Isopropylamino)-4,4-diphenyl-2-butanol.
Compd. 122587. Antiarrhythmic agent; cardiac depressant. *Eli Lilly & Co.*

1590 Dronedarone
141626-36-0
$C_{31}H_{44}N_2O_5S$
N-[2-Butyl-3-[p-[3-[(dibutylamino)propoxy]benzoyl]-5-benzofuranyl]methanesulfonamide.
SR-33589. Class III antiarrhythmic. A noniodinated benzofuran derivative.

1591 Droxicainide
78421-12-2
$C_{16}H_{24}N_2O_2$
(±)-1-(2-Hydroxyethyl)-2',6'-pipecoloxylidide.
AL-S-1249. Class I antiarrhythmic.

1592 Edifolone
90733-40-7
$C_{24}H_{37}NO_4$
10-(2-Aminoethyl)estr-5-ene-3,17-dione cyclic bis(ethylene acetal).
Antiarrhythmic agent; cardiac depressant. *Searle G.D. & Co.*

1593 Edifolone Acetate
90733-42-9
$C_{26}H_{41}NO_6$
10-(2-Aminoethyl)estr-5-ene-3,17-dione cyclic bis(ethylene acetal) acetate.
SC-35135. Antiarrhythmic agent; cardiac depressant. *Searle G.D. & Co.*

1594 Emilium Tosylate
30716-01-9
$C_{19}H_{27}NO_4S$
Ethyl (m-methoxybenzyl)dimethylammonium p-toluenesulfonate.
emilium tosilate. Antiarrhythmic agent; cardiac depressant.

1595 Encainide
37612-13-8 3609
$C_{22}H_{28}N_2O_2$
(±)-4-Methoxy-N-[2-[2-(1-methyl-2-piperidinyl)-ethyl]phenyl]benzamide.
Class IC antiarrhythmic agent; cardiac depressant. *Bristol Laboratories.*

1596 Encainide Hydrochloride
66794-74-9 3609
$C_{22}H_{29}ClN_2O_2$
(±)-4-Methoxy-N-[2-[2-(1-methyl-2-piperidinyl)-ethyl]phenyl]benzamide hydrochloride.
Enkade; Enkaid; Encainide hydrochloride; MJ-9067. Class IC antiarrhythmic agent; cardiac depressant. mp = 131.5-132.5°; freely soluble in H_2O, EtOH; insoluble in heptane; LD_{50} (mus orl) = 86 mg/kg, (mus iv) = 16 mg/kg, (dog orl) = 43 mg/kg, (dog iv) = 17 mg/kg. *Bristol Laboratories.*

1597 Epicainide
66304-03-8
$C_{21}H_{26}N_2O_2$
N-[(1-Ethyl-2-pyrrolidinyl)methyl]benzilamide.
Antiarrhythmic.

1598 Eproxindine
83200-08-2
$C_{23}H_{29}N_3O_3$
(±)-N-[3-(Diethylamino)-2-hydroxypropyl]-3-methoxy-1-phenylindole-2-carboxamide.
Antiarrhythmic.

1599 Ersentilide
125279-79-0
$C_{21}H_{26}N_4O_5S$
4'-[(2S)-2-Hydroxy-3-[[2-(p-imidazol-1-ylphenoxy)-ethyl]amino]propoxy]methanesulfonanilide.
Antiarrhythmic.

1600 Esmolol
103598-03-4 3741
$C_{16}H_{25}NO_4$
(±)-Methyl p-[2-hydroxy-3-(isopropylamino)-propoxy]hydrocinnamate.
Class II antiarrhythmic. Ultra-short-acting β-adrenergic blocker. mp = 48-50°. *Am. Hospital Supply.*

1601 Esmolol Hydrochloride
81161-17-3 3741
$C_{16}H_{26}ClNO_4$
(±)-Methyl p-[2-hydroxy-3-(isopropylamino)propoxy]-hydrocinnamate hydrochloride.
ASL-8052; Brevibloc. Class II antiarrhythmic. Ultra-short-acting β-adrenergic blocker. mp = 85-86°. *Am. Hospital Supply.*

1602 Falipamil
77862-92-1
$C_{24}H_{32}N_2O_5$
2-[3-(3,4-Dimethoxyphenethyl)methylamino]-propyl]-5,6-dimethoxy-phthalimidine.
Calcium channel blocker used as an antiarrhythmic agent.

1603 Flecainide
54143-55-4 4136
$C_{17}H_{20}F_6N_2O_3$
N-(2-Piperidylmethyl)-2,5-bis(2,2,2-trifluoroethoxy)-benzamide.
Class IC antiarrhythmic agent; cardiac depressant. λ_m = 205, 230, 300 nm ($E^{1\%}_{1\,cm}$ 521, 219, 59 EtOH). *3M Pharm.*

1604 Flecainide Acetate
54143-56-5 4136 258-997-5
$C_{19}H_{24}F_6N_2O_5$
N-(2-Piperidylmethyl)-2,5-bis(2,2,2-trifluoroethoxy)-benzamide monoacetate.
Tambocor; R-818; Almarytm; Apocard; Ecrinal. Class IC antiarrhythmic agent; cardiac depressant. mp = 145-147°; soluble in H_2O (4.8 g/100 ml), EtOH (30.0 g/100 ml). *3M Pharm.*

1605 Flusoxolol
84057-96-5
$C_{22}H_{30}FNO_4$
(S)-1-[p-[2-[(p-Fluorophenethyl)oxy]ethoxy]phenoxy]-3-(isopropylamino)-2-propanol.
Antihypertensive; antianginal; antiarrhythmic. A selective β_1-adrenoceptor agonist.

1606 Hydroquinidine
1435-55-8 4851 215-862-5
$C_{20}H_{26}N_2O_2$
(9S)-10,11-Dihydro-6'-methoxycinchonan-9-ol.
hydroconchinine. Class 1A antiarrhythmic agent. mp = 169°; $[\alpha]_D^{20}$ = 231° (c = 2.02 EtOH), 299° (c = 0.82 0.1N HCl); soluble in EtOH; less soluble in H_2O, Et_2O. *Hoffmann-LaRoche Inc.*

1607 Hydroquinidine Hydrochloride
1476-98-8 4851 216-024-1
$C_{20}H_{27}ClN_2O_2$
(9S)-10,11-Dihydro-6'-methoxycinchonan-9-ol hydrochloride.
Serecor. Class 1A antiarrhythmic agent. mp = 273-274°; $[\alpha]_D^{26}$ = 184° (c = 1.3); freely soluble in MeOH, $CHCl_3$; less soluble in H_2O; insoluble in Me_2CO. *Hoffmann-LaRoche Inc.*

1608 Ibutilide
122647-31-8 4927

$C_{20}H_{36}N_2O_3S$
(±)-4'-[4-(Ethylheptylamino)-1-hydroxybutyl]methanesulfonanilide.
Class III antiarhythmic agent. *Pharmacia & Upjohn.*

1609 Ibutilide Fumarate
122647-32-9 4927

$C_{44}H_{76}N_4O_{10}S_2$
(±)-4'-[4-(Ethylheptylamino)-1-hydroxybutyl]methanesulfonanilide fumarate (2:1) (salt).
U-70226-E. Class III antiarhythmic agent. mp = 117-119°; λ_m = 228, 267 nm (ε 16670, 894 EtOH). *Pharmacia & Upjohn.*

1610 Indecainide
74517-78-5 4971

$C_{20}H_{24}N_2O$
9-[3-(Isopropylamino)propyl]fluorene-9-carboxamide.
ricainide. Antiarrhythmic agent; cardiac depressant. mp = 94-95°. *Eli Lilly & Co.*

1611 Indecainide Hydrochloride
73681-12-6 4971

$C_{20}H_{25}ClN_2O$
9-[3-(Isopropylamino)propyl]fluorene-9-carboxamide monohydrochloride.
Decabid; LY-135837. Antiarrhythmic agent; cardiac depressant. mp = 216.5-217°, 203-204°; LD_{50} (mmus orl) = 100 mg/kg, (fmus orl) = 96 mg/kg, (mrat orl) = 103 mg/kg, (frat orl) = 82 mg/kg. *Eli Lilly & Co.*

1612 Indenolol
60607-68-3 4974 262-323-5

$C_{15}H_{21}NO_2$
1-[1H-Inden-4 (or -7)-yloxy]-3-(isopropylamino)-2-propanol.
Sch-28316Z; YB-2. Antiarrhythmic; antihypertensive; antianginal agent. Non-selective β-adrenergic blocker. A 1:2 tautomeric mixture of the 4- and 7- indenyloxy isomers. mp = 88-89°. *Schering-Plough HealthCare Products; Yamanouchi U.S.A. Inc.*

1613 Indenolol Hydrochloride
81789-85-7 4974

$C_{15}H_{22}ClNO_2$
1-[1H-Inden-4 (or -7)-yloxy]-3-[(1-methylethyl)amino]-2-propanol hydrochloride.
Pulsan; Securpres. Antihypertensive; antiarrhythmic; antianginal. Non-selective β-adrenergic blocker. mp = 147-148°; LD_{50} (mus iv) = 26 mg/kg. *Schering AG; Yamanouchi U.S.A. Inc.*

1614 Ipazilide
115436-73-2

$C_{24}H_{30}N_4O$
N-[3-(Diethylamino)propyl]-4,5-diphenylpyrazole-1-acetamide.
Antiarrhythmic agent; cardiac depressant. Has class I and class III actions. *Sterling Winthrop Inc.*

1615 Ipazilide Fumarate
115436-74-3

$C_{28}H_{34}N_4O_5$
N-[3-(Diethylamino)propyl]-4,5-diphenylpyrazole-1-acetamide fumarate.
Antiarrhythmic agent; cardiac depressant. Has class I and class III actions. *Sterling Winthrop Inc.*

1616 Ipratropium Bromide
66985-17-9 5089

$C_{20}H_{30}BrNO_3.H_2O$
(8r)-3α-Hydroxy-8-isopropyl-1αH,5αH-tropanium bromide monohydrate.
Sch-1000-Br-monohydrate; Atem; Atrovent; Bitrop; Itrop; Narilet; Rinatec; component of: Combivent. An anticholinergic bronchodilator and antiarrhythmic. Quaternary ammonium compound. mp = 230-232°; soluble in H_2O, MeOH, EtOH; insoluble in Et_2O, $CHCl_3$, fluorohydrocarbons; LD_{50} (mmus orl) = 1001 mg/kg, (mmus iv) = 12.29 mg/kg, (mmus sc) = 300 mg/kg, (fmus orl) = 1083 mg/kg, (fmus iv) = 14.97 mg/kg, (fmus sc) = 340 mg/kg, (mrat orl) = 1663 mg/kg, (mrat iv) = 15.89 mg/kg, (frat orl) = 1779 mg/kg, (frat iv) = 15.70 mg/kg. *Boehringer Ingelheim Ltd.; Schering AG.*

1617 Ipratropium Bromide [anhydrous]
22254-24-6 5089 244-873-8

$C_{20}H_{30}BrNO_3$
(8r)-3α-Hydroxy-8-isopropyl-1αH,5αH-tropanium bromide monohydrate.
N-isopropylnoratropinium bromomethylate; Sch-100. An anticholinergic bronchodilator and antiarrhythmic. Quaternary ammonium compound. *Boehringer Ingelheim Ltd.*

1618 Lidocaine
137-58-6 5505 205-302-8

$C_{14}H_{22}N_2O$
2-(Diethylamino)-2',6'-acetoxylidide.
Cuivasil; Dalcaine; Duncaine; Leostesin; Lida-Mantle; Lidothesin; Lignocaine; Rucaina; Solarcaine; Xylocaine; Xylocitin; Xylotox; component of: Cracked Heel Relief Cream, Emla Cream, Lidaform HC, Lidamantle HC, Neosporin Plus. Local anesthetic; antiarrhythmic (class IB). A long-acting membrane stabilizing agent against ventricular arrhythmia. mp = 68-69°; bp_4 = 180-182°, bp_2 = 159-160°; soluble in EtOH, Et_2O, C_6H_6, $CHCl_3$; insoluble in H_2O. *Astra USA Inc.; Bayer Corp.; Glaxo Wellcome Inc.; Schering-Plough Animal Health; Schering-Plough Pharm.*

1619 Lidocaine Hydrochloride
6108-05-0 5505 200-803-8

$C_{14}H_{23}ClN_2O.H_2O$
2-(Diethylamino)-2',6'-acetoxylidide hydrochloride monohydrate.
Dalcaine; Lidesthesin; Lignavet; Odontalg; Sedagul; Xylocaine; Xylocard; Xyloneural; component of: Solarcaine. Class IB antiarrhythmic agent. Local anesthetic. Long-acting membrane stabilizing agent against ventricular arrhythmia. mp = 77-78°; [anhydrous form]: mp = 127-129°; very soluble in H_2O, EtOH; less soluble in $CHCl_3$; insoluble in Et_2O; LD_{50} (mus orl) = 292 mg/kg, (mus ip) = 105 mg/kg, (mus iv) = 19.5 mg/kg. *Astra Chem. Ltd.; Astra Pharm. Ltd.; Bayer Corp.; Bristol*

Myers Squibb Pharm. Ltd.; Carrington Labs Inc.; Elkins-Sinn; Forest Pharm. Inc.; Glaxo Wellcome plc; Schering-Plough HealthCare Products; Sterling Health U.S.A.*

1620 Lorajmine
47562-08-3 5607 256-322-9
$C_{22}H_{27}ClN_2O_3$
Ajmaline 17-(chloroacetate).
MCAA; 17-chloroacetylajmaline. Antiarrhythmic agent; cardiac depressant. mp = 232-238°; $[\alpha]_D^{20}$ = 27.5° (c = 1 $CHCl_3$). *Sterling Winthrop Inc.*

1621 Lorajmine Hydrochloride
40819-93-0 5607
$C_{22}H_{28}Cl_2N_2O_3$
Ajmaline 17-(chloroacetate) monohydrochloride.
Win-11831; Nevergor; Ritmos Elle. Antiarrhythmic agent; cardiac depressant. mp = 243-246°; $[\alpha]_D^{20}$ = 29° (c = 1 EtOH); LD_{50} (mus ip) = 176 mg/kg, (mus orl) = 370 mg/kg, (rat ip) = 139 mg/kg, (rat orl) = 480 mg/kg. *Sterling Winthrop Inc.*

1622 Lorcainide
59729-31-6 5610
$C_{22}H_{27}ClN_2O$
4'-Chloro-N-(1-isopropyl-4-piperidinyl)-2-phenylacetanilide.
Class IC antiarrhythmic agent. Also has cardiac depressant properties. LD_{50} (mmus iv) = 18.8 mg/kg, (mmus orl) = 483 mg/kg, (mrat iv) = 19.3 mg/kg, (mrat orl) = 395 mg/kg, (frat iv) = 18.6 mg/kg, (frat orl) = 435 mg/kg. *Abbott Laboratories Inc.*

1623 Lorcainide Hydrochloride
58934-46-6 5610 261-504-6
$C_{22}H_{28}Cl_2N_2O$
4'-Chloro-N-(1-isopropyl-4-piperidinyl)-2-phenylacetanilide monohydrochloride.
R-15889; Ro-13-1042; Lopantrol; Lorivox; Remivox. Class IC antiarrhythmic agent. Also has cardiac depressant properties. mp = 263°. *Abbott Laboratories Inc.*

1624 Magnesium Sulfate [anhydrous]
7487-88-9 5731 231-298-2
$MgSO_4$
Sulfuric acid magnesium salt (1:1).
Kieserite [as monohydrate]. Magnesium replenisher. The heptahydrate is an anticonvulsant and cathartic. Also effective in termination of refractory ventricular tachyarrhythmias. Loss of deep tendon reflex is a sign of overdose. *Astra USA Inc.; Dow Chem. U.S.A.; McGaw Inc.*

1625 Magnesium Sulfate Heptahydrate
10034-99-8 5731
$MgSO_4.7H_2O$
Sulfuric acid magnesium salt (1:1) heptahydrate.
bitter salts; Epsom salts. Magnesium replenisher. The heptahydrate is an anticonvulsant and cathartic. Also effective in termination of refractory ventricular tachyarrhythmias. Loss of deep tendon reflex is a sign of overdose. d = 1.67; soluble in H_2O (71 g/100 ml at 20°, 91 g/100 ml at 40°), slightly soluble in EtOH. *Astra USA Inc.; Dow Chem. U.S.A.; McGaw Inc.*

1626 Meobentine
46464-11-3 5889
$C_{11}H_{17}N_3O$
1-(p-Methoxybenzyl)-2,3-dimethylguanidine.
Class III antiarrhythmic agent. *Burroughs Wellcome Inc.*

1627 Meobentine Sulfate
58503-79-0 5889
$C_{22}H_{36}N_6O_6S$
1-(p-Methoxybenzyl)-2,3-dimethylguanidine sulfate.
Rythmatine. Class III antiarrhythmic agent. mp = 273-274°. *Burroughs Wellcome Inc.*

1628 Metoprolol
37350-58-6 6235 253-483-7
$C_{15}H_{25}NO_3$
1-(Isopropylamino)-3-[p-(2-methoxyethyl)phenoxy]-2-propanol.
CGP-2175; H-93/26. Class II antiarrhythmic agent. Also has antihypertensive and antianginal activity. A β-adrenergic blocker which lacks intrinsic sympathomimetic activity. *Apothecon; Astra Chem. Ltd.; Ciba-Geigy Corp.; Lemmon Co.*

1629 Metoprolol Fumarate
119637-66-0 6235
$C_{34}H_{54}N_2O_{10}$
1-(Isopropylamino)-3-[p-(2-methoxyethyl)phenoxy]-2-propanol fumarate (2:1) (salt).
Lopressor OROS; CGP-2175C. Antianginal; class II antiarrhythmic; antihypertensive. A β-adrenergic blocker. Lacks inherent sympathomimetic activity. Insoluble in EtOAc, Me_2CO, Et_2O, heptane. *Ciba-Geigy Corp.*

1630 Metoprolol Succinate
98418-47-4 6235
$C_{34}H_{56}N_2O_{10}$
1-(Isopropylamino)-3-[p-(2-methoxyethyl)-phenoxy]-2-propanol
succinate (2:1) (salt).
Toprol XL; H-93/26 succinate. Antianginal; class II antiarrhythmic; antihypertensive. A β-adrenergic blocker. Lacks inherent sympathomimetic activity. *Apothecon; Astra Chem. Ltd.; Astra Sweden; Ciba-Geigy Corp.; Lemmon Co.*

1631 Metoprolol Tartrate
56392-17-7 6235 260-148-9
$C_{34}H_{56}N_2O_{12}$
1-(Isopropylamino)-3-[p-(2-methoxyethyl)-phenoxy]-2-propanol (2:1)
dextro-tartrate salt.
CGP-2175E; HCTCGP 2175E; Beloc; Betaloc; Lopressor; Lopresor; Prelis; Seloken; Selopral; Selo-Zok; component of: Lopressor HCT. Antianginal; class II antiarrhythmic; antihypertensive. A β-adrenergic blocker. Lacks inherent sympathomimetic activity. Soluble in MeOH (50 g/100 ml), H_2O (> 100 g/100 ml), $CHCl_3$ (49.6 g/100 ml), Me_2CO (0.11 g/100 ml), CH_3CN (0.089 g/100 ml); insoluble in C_6H_{14}; LD_{50} (fmus iv) = 118 mg/kg, (fmus orl) = 3090 mg/kg, (mrat iv) ≅ 90 mg/kg, (mrat orl) = 3090 mg/kg, (mrat iv) ≅ 90 mg/kg, (mrat orl) = 3090 mg/kg. *Apothecon; Astra Chem. Ltd.; Ciba-Geigy Corp.; Lemmon Co.; Sandoz Nutrition.*

1632 Mexiletine
31828-71-4 6257 250-825-7
$C_{11}H_{17}NO$
1-Methyl-2-(2,6-xylyloxy)ethylamine.
Class IB antiarrhythmic. *Boehringer Ingelheim GmbH.*

1633 Mexiletine Hydrochloride
5370-01-4 6257 226-362-1
$C_{11}H_{18}ClNO$
1-Methyl-2-(2,6-xylyloxy)ethylamine hydrochloride.
Mexitil; Ko-1173Cl; Katen; ritalmex. Class IB antiarrhythmic agent. mp = 203-205°; LD_{50} (mrat orl) = 350 mg/kg, (mrat iv) = 27 mg/kg, (frat orl) = 400 mg/kg, (frat iv) = 30 mg/kg, (mmus orl) = 310 mg/kg, (mmus iv) = 43 mg/kg), (fmus orl) = 400 mg/kg, (fmus iv) = 50 mg/kg, (mrbt orl) = 180 mg/kg, (frbt orl) = 160 mg/kg. *Boehringer Ingelheim GmbH.*

1634 Modecainide
82522-70-1
$C_{22}H_{28}N_2O_3$
(±)-2'-[2-(1-Methyl-2-piperidyl)ethyl]vanillanilide.
BMY-40327. Antiarrhythmic agent; cardiac depressant. *Bristol-Myers Squibb Co.*

1635 Monatepil
132019-54-6
$C_{28}H_{30}FN_3OS$
(±)-N-(6,11-Dihydrodibenzo[b,e]thiepin-11-yl)-4-(p-fluorophenyl)-1-piperazinebutyramide.
Antiarrhythmic agent with antianginal and antihypertensive properties. *Dainippon Pharm.*

1636 Monatepil Maleate
132046-06-1
$C_{32}H_{34}FN_3O_5S$
(±)-N-(6,11-Dihydrodibenzo[b,e]thiepin-11-yl)-4-(p-fluorophenyl)-1-piperazinebutyramide maleate (1:1).
AJ-2615. Antiarrhythmic agent with antianginal and antihypertensive properties. *Dainippon Pharm.*

1637 Moricizine
31883-05-3 6351 250-854-5
$C_{22}H_{25}N_3O_4S$
Ethyl 10-(3-morpholinopropionyl)phenothiazine-2-carbamate.
Ethmozine; EN-313; ethmosine; moracizine. Class IB antiarrhythmic. mp = 156-157°. *Roberts Pharm. Corp.*

1638 Moricizine Hydrochloride
29560-58-8 6351
$C_{22}H_{26}ClN_3O_4S$
Ethyl 10-(3-morpholinopropionyl)phenothiazine-2-carbamate hydrochloride.
Ethmozine. Class IB antiarrhythmic agent. mp = 189° (dec); soluble in H_2O, EtOH; LD_{50} (mus iv) = 36 mg/kg, (mus ip) = 131 mg/kg, (rat iv) = 12 mg/kg. *Roberts Pharm. Corp.*

1639 Moxaprindine
53076-26-9 258-347-0
$C_{23}H_{32}N_2O$
N,N-Diethyl-N''-(1-methyoxy-2-indanyl)-n''-phenyl-1,3-propanediamine.
Derivative of aprindine. Antiarrhythmic.

1640 Nadolol
42200-33-9 6431 255-706-3
$C_{17}H_{27}NO_4$
1-(tert-Butylamino)-3-[(5,6,7,8-tetrahydro-cis-6,7-dihydroxy-1-naphthyl)oxy]-2-propanol.
Corgard; SQ-11725; Anabet; Solgol. Antihypertensive and antianginal agent. Class II antiarrhythmic. A β-adrenergic blocker. mp = 124-136°; λ_m = 270, 278 nm ($E^{1\%}_{1\ cm}$ 37.5, 39.1, MeOH); pKa = 9.67; poorly soluble in organic solvents; insoluble in Me_2CO, C_6H_6, Et_2O, C_6H_{14}; LD_{50} (rat orl) = 5300 mg/kg, (mus orl) = 4500 mg/kg. *Apothecon; Bristol-Myers Squibb Co.*

1641 Nadoxolol
54063-51-3 6432
$C_{14}H_{16}N_2O_3$
3-Hydroxy-4-(1-naphthyloxy)butyramidoxime.
Antiarrhythmic agent. Antihypertensive, antianginal. Used for the treatment of cardiovascular disorders. A β-adrenergic blocker. *Orsymonde.*

1642 Nadoxolol Hydrochloride
35991-93-6 6432 252-825-2
$C_{14}H_{17}ClN_2O_3$
3-Hydroxy-4-(1-naphthyloxy)butyramidoxime hydrochloride.
LL-1530; Bradyl. Antihypertensive; antianginal. Used for the treatment of cardiovascular disorders. Antiarrhythmic agent. A β-adrenergic blocker. mp = 188°; soluble in H_2O, EtOH, MeOH; insoluble in Et_2O; LD_{50} (mus iv) = 180 mg/kg, (mus orl) = 1000 mg/kg. *Orsymonde.*

1643 Nicainoprol
76252-06-7 278-403-8
$C_{21}H_{27}N_3O_3$
(±)-1,2,3,4-Tetrahydro-8-[2-hydroxy-3-(isopropylamino)propoxy]-1-nicotinoylquinoline.
Antiarrhythmic. A β-adrenergic blocker.

1644 Nifenalol
7413-36-7 6618 231-023-6
$C_{11}H_{16}N_2O_3$
α-[(Isopropylamino)methyl]-p-nitrobenzyl alcohol.
isophenethanol; INPEA. Antiarrhythmic agent with antianginal properties. A β-adrenergic blocker. mp = 98°. *Selvi.*

1645 Nifenalol Hydrochloride
5704-60-9 6618 227-194-1
$C_{11}H_{17}ClN_2O_3$
α-[(Isopropylamino)methyl]-p-nitrobenzyl alcohol hydrochloride.
Inpea. Antiarrhythmic agent with antianginal properties. A β-adrenergic blocker. mp = 181°. *Selvi.*

1646 Oxiramide
13958-40-2
$C_{25}H_{34}N_2O_2$
N-[4-(2,6-Dimethylpiperidino)butyl]-2-phenoxy-2-phenylacetamide.
CI-661. Antiarrhythmic agent; cardiac depressant. *Parke-Davis.*

1647 Oxprenolol
6452-71-7 7086 229-257-9
$C_{15}H_{23}NO_3$
1-(o-Allyloxyphenoxy)-3-isopropylamino-2-propanol.
Antianginal; antihypertensive; class IV antiarrhythmic. A β-adrenergic blocker. mp = 78-80°. *Bayer AG; Ciba-Geigy Corp.*

1648 Oxprenolol Hydrochloride
6452-73-9 7086 229-260-5
$C_{15}H_{24}ClNO_3$
1-(o-Allyloxyphenoxy)-3-isopropylamino-2-propanol hydrochloride.
Ba-39089; Coretal; Laracor; Paritane; Slow-Pren; Trasicor; Trasacor. Antianginal; antihypertensive; class IV antiarrhythmic. A β-adrenergic blocker. mp = 107-109°. *Bayer AG; Ciba-Geigy Corp.*

1649 Penbutolol
38363-40-5 7209
$C_{18}H_{29}NO_2$
(S)-1-(tert-Butylamino)-3-(o-cyclopentylphenoxy)-2-propanol.
Antianginal; antihypertensive; antiarrhythmic. A β-adrenergic calcium channel blocker with antiarrhythmic activity. mp = 68-72°; $[\alpha]_D^{20}$ = -11.5° (c = 1 MeOH); soluble in MeOH, EtOH, $CHCl_3$. *Hoechst AG.*

1650 Penbutolol Sulfate
38363-32-5 7209 253-906-5
$C_{36}H_{40}N_2O_8S$
(S)-1-(tert-Butylamino)-3-(o-cyclopentylphenoxy)-2-propanol sulfate (salt) (2:1).
HOE-893d; HOE-39-893d; Betapressin; Levatol; Paginol. Antianginal; antihypertensive; antiarrhythmic. A β-adrenergic calcium channel blocker with antiarrhythmic activity. mp = 216-218° (dec); $[\alpha]_D^{20}$ = -24.6° (c = 1 MeOH). *Hoechst AG.*

1651 Pentisomide
96513-83-6 7271
$C_{19}H_{33}N_3O$
(±)-α-[2-(Diisopropylamino)ethyl]-α-isobutyl-2-pyridineacetamide.
CM-7857; ME-3202; penticainide; propisomide. Class I antiarrhythmic agent. mp = 108-109°; soluble in H_2O. *C.M. Ind.*

1652 Phenytoin
57-41-0 7475 200-328-6
$C_{15}H_{12}N_2O_2$
5,5-Diphenylhydantoin.
diphenylhydantoin; Dilantin; Difhydan; Dihycon; Di-Hydan; Ekko; Hydantin; Hydantol; Lehydan; Lepitoin; Phenhydan; Zentropil; component of: Mebroin. Anticonvulsant; class IB antiarrhythmic. mp = 295-298°; insoluble in H_2O; soluble in EtOH (1.67 g/100 ml), Me_2CO (3.32 g/100 ml); LD_{50} (mus iv) = 92 mg/kg, (mus sc) = 110 mg/kg. *Parke-Davis; Sterling Winthrop Inc.*

1653 Phenytoin Sodium
630-93-3 7475 211-148-2
$C_{15}H_{11}N_2NaO_2$
5,5-Diphenylhydantoin sodium salt.
phenytoin soluble; Antisacer; Danten; Diphantoine; Diphenin; Diphenylan sodium; Epanutin; Minetoin; Tacosal; Solantyl; component of: Beuthanasia-D. Anticonvulsant; class IB antiarrhythmic. Soluble in EtOH (9.5 g/100 ml), H_2O (1.5 g/100 ml); insoluble in Et_2O, $CHCl_3$; LD_{50} (mus orl) = 490 mg/kg. *Elkins-Sinn; Schering-Plough Animal Health.*

1654 Pilsicainide
88069-67-4 7581
$C_{17}H_{24}N_2O$
Tetrahydro-1H-pyrrolizine-7a(5H)-aceto-2',6'-xylidide.
Antiarrhythmic agent; cardiac depressant. Structural analog of lidocaine. *Suntory Ltd.*

1655 Pilsicainide Hydrochloride
88069-49-2 7581
$C_{17}H_{25}ClN_2O.1/2H_2O$
Tetrahydro-1H-pyrrolizine-7a(5H)-aceto-2',6'-xylidide hydrochloride.
SUN-1165; Sunrythm. Antiarrhythmic agent; cardiac depressant. Structural analog of lidocaine. mp = 212-214°; LD_{50} (mus sc) = 410 mg/kg. *Suntory Ltd.*

1656 Pincainide
83471-41-4
$C_{16}H_{24}N_2O$
2,3,4,5,6,7-Hexahydro-1H-azepine-1-aceto-2',6'-xylidide.
IQB-M-81. A β-amino anilide. Antiarrhythmic.

1657 Pindolol
13523-86-9 7597 236-867-9
$C_{14}H_{20}N_2O_2$
1-(Indol-4-yloxy)-3-(isopropylamino)-2-propanol.
LB-46; Visken; prinodolol; Betapindol; Blocklin L; Calvisken; Decreten; Durapindol; Glauco-Visken; Pectobloc; Pinbetol; Pynastin. Antianginal; antiarrhythmic (class II); antihypertensive; antiglaucoma properties. A β-adrenergic blocker. mp = 171-163°. *Sandoz Pharm. Corp.*

1658 Pirmenol
68252-19-7 7656
$C_{22}H_{30}N_2O$
(±)-cis-2,6-Dimethyl-α-phenyl-α-2-pyridyl-1-piperidinebutanol.
Class IA antiarrhythmic agent. mp = 70-71°. *Parke-Davis.*

1659 Pirmenol Hydrochloride
61477-94-9 7656
$C_{22}H_{31}ClN_2O$
(±)-cis-2,6-Dimethyl-α-phenyl-α-2-pyridyl-1-piperidinebutanol monohydrochloride.
Pirmavar; CI-845. Class IA antiarrhythmic agent. mp = 171-172°; LD_{50} (mus iv) = 20.8 mg/kg, (mus orl) = 215.5 mg/kg, (rat iv) = 23.6 mg/kg, (rat orl) = 359.9 mg/kg, (dog iv) > 7.0 mg/kg, (dog orl) > 40.0 mg/kg. *Parke-Davis.*

1660 Pirolazamide
39186-49-7
$C_{23}H_{29}N_3O$
Hexahydro-α,α-diphenylpyrolo[1,2-a]pyrazine-2(1H)-butyramide.
SC-26438. Antiarrhythmic; cardiac depressant. *Searle G.D. & Co.*

1661 Practolol
6673-35-4 7882 229-712-1
$C_{14}H_{22}N_2O_3$
4'-[2-Hydroxy-3-(isopropylamino)propoxy]acetanilide.
AY-21011; ICI-50172; Dalzic; Eraldin. A β-adrenergic blocker. Antiarrhythmic. mp = 134-136°; soluble in i-PrOH; [hydrochloride monhydrate]: mp = 140-142°; [R(+) form]: mp = 130-131.5°; $[\alpha]^{25}_{365}$ = 4.3° (EtOH), $[\alpha]^{25}_{578}$ = 3.5° (EtOH); [R(+) hydrochloride]: $[\alpha]^{25}_{436}$ = 26.0° (EtOH), $[\alpha]^{25}_{578}$ = 14.0° (EtOH). *ICI ; Wyeth-Ayerst Labs.*

1662 Prajmaline
35080-11-6 7883
$[C_{23}H_{33}2O_2]^+$
N-Propylajmaline.
N4-propylajalmalinium; prajmalium. Antiarrhythmic agent. *Boehringer Ingelheim GmbH.*

1663 Prajmaline Bitartrate
2589-47-1 7883 219-975-0
$C_{27}H_{38}N_2O_8$
N-Propylajmalinium tartrate.
NPAB; GT-1012; Neo-Gilurytmal. Antiarrhythmic agent. mp = 149-152°; LD_{50} (mus orl) = 43 mg/kg, (mus iv) = 1.7 mg/kg. *Boehringer Ingelheim GmbH.*

1664 Pranolium Chloride
42879-47-0
$C_{18}H_{26}ClNO_2$
[2-Hydroxy-3-(1-naphthyloxy)propyl]isopropyldimethylammonium chloride.
SC-27761; dimethylpropranolol [pranolium]; UM-272 [pranolium]. Antiarrhythmic agent. A quaternary analog of propranolol. *Searle G.D. & Co.*

1665 Prifuroline
70833-07-7
$C_{14}H_{16}N_2O$
4-(2-Benzofuranyl)-2-(dimethylamino)-1-pyrroline.
Antiarrhythmic.

1666 Primidolol
67227-55-8
$C_{17}H_{23}N_3O_4$
1-[2-[[2-Hydroxy-3-(o-toloxy)propyl]amino]ethyl]thymine.
UK-11443. Antihypertensive; antianginal; antiarrhythmic (cardiac depressant). *Pfizer Intl.*

1667 Procainamide
51-06-9 7936 200-078-8
$C_{13}H_{21}N_3O$
p-Amino-N-[2-(diethylamino)ethyl]benzamide.
Class IA antiarrhythmic agent. *Apothecon; Elkins-Sinn; Parke-Davis.*

1668 Procainamide Hydrochloride
614-39-1 7936 210-381-7
$C_{13}H_{22}ClN_3O$
p-Amino-N-[2-(diethylamino)ethyl]benzamide monohydrochloride.
Procanbid; Pronestyl; Amisalin; Novocamid; Procamide; Procan-SR; Procapan; Pronestyl. Class IA antiarrhythmic agent. mp = 165-169°; λ_m = 278 nm; freely soluble in H_2O, less soluble in organic solvents; LD_{50} (rat orl) > 2 g/kg. *Apothecon; Elkins-Sinn; Parke-Davis.*

1669 Pronetalol
54-80-8 7974
$C_{15}H_{19}NO$
2-Isopropyl-1-(naphth-2-yl)ethanol.
pronethalol; nethalide. Antianginal; antihypertensive; antiarrhythmic. A β-adrenergic blocker. mp = 108°. *ICI ; Wyeth-Ayerst Labs.*

1670 Pronetalol Hydrochloride
51-02-5 7974
$C_{15}H_{20}ClNO$
2-Isopropylamino-1-(naphth-2-yl)ethanol hydrochloride.
pronethalol hydrochloride; ICI-38174; AY-6204; Alderlin. Antianginal; antihypertensive; antiarrhythmic. A β-adrenergic blocker. mp = 184°; LD_{50} (mus orl) = 512 mg/kg, (mus iv) = 46 mg/kg. *ICI ; Wyeth-Ayerst Labs.*

1671 Propafenone
54063-53-5 7978 258-955-6
$C_{21}H_{27}NO_3$
2'-[2-Hydroxy-3-(propylamino)propoxy]-3-phenylpropiophenone.
SA-79. Class IC antiarrhythmic agent. *Helopharm; Knoll Pharm. Co.*

1672 Propafenone Hydrochloride
34183-22-7 7978 251-867-9
$C_{21}H_{28}ClNO_3$
2'-[2-Hydroxy-3-(propylamino)propoxy]-3-phenylpropiophenone hydrochloride.
Rythmol; Arythmol; Pronon; Rytmonorm. Class IC antiarrhythmic agent. Soluble in EtOH, CCl_4, hot H_2O; less soluble in cold H_2O; insoluble in Et_2O; LD_{50} (rat iv) = 18.8 mg/kg, (rat orl) = 700 mg/kg. *C.M. Ind.; Helopharm; Knoll Pharm. Co.*

1673 Propranolol
525-66-6 8025 208-378-0
$C_{16}H_{21}NO_2$
1-(Isopropylamino)-3-(1-naphthyloxy)-2-propanol.
Avlocardyl; Euprovasin. Antianginal; antihypertensive; antiarrhythmic (class II). A β-adrenergic blocker. mp = 96°. *ICI ; Parke-Davis; Quimicobiol; Wyeth-Ayerst Labs.*

1674 Propranolol Hydrochloride
318-98-9 8025 206-268-7
$C_{16}H_{22}ClNO_2$
1-(Isopropylamino)-3-(1-naphthyloxy)-2-propanol hydrochloride.
AY-64043; ICI-45520; NSC-91523; Angilol; Apsolol; Bedranol; Beprane; Berkolol; Beta-Neg; Beta-Tablinen; Beta-Timelets; Cardinol; Caridolol; Deralin; Dociton; Duranol; Efektolol; Elbol; Frekven; Inderal; Indobloc; Intermigran; Kemi S; Oposim; Prano-Puren; Prophylux; Propranur; Pylapron; Rapynogen; Sagittol; Servanolol; Sloprolol; Sumial; Tesnol. Antianginal; antihypertensive; antiarrhythmic (class II) agent. A β-adrenergic blocker. mp = 163-164°; soluble in H_2O, alcohol; insoluble in Et_2O, C_6H_6, EtOAc; LD_{50} (mus orl) = 565 mg/kg, (mus iv) = 22 mg/kg, (mus ip) = 107 mg/kg. *ICI ; Parke-Davis; Quimicobiol; Wyeth Labs.*

1675 Pyrinoline
1740-22-3 8173
$C_{27}H_{20}N_4O$
3-(Di-2-pyridylmethylene)-α,α-di-2-pyridyl-1,4-cyclopentadiene-1-methanol.
McN-1210; Surexin. Antiarrhythmic agent; cardiac depressant. mp = 146-148°. *McNeil Pharm.*

1676 Quinacainol
86024-64-8
$C_{21}H_{30}N_2O$
(±)-2-tert-Butyl-α-[2-(4-piperidyl)ethyl]-4-quinolinemethanol.
PK-10139. Class IA antiarrhythmic.

1677 Quindonium Bromide
130-81-4
$C_{16}H_{20}BrNO$
2,3,3a,5,6,11,12,12a-Octahydro-8-hydroxy-1H-benzo[a]cyclopenta[f]quinolizinium bromide.
W3366A. Antiarrhythmic agent; cardiac depressant. *Parke-Davis.*

1678 Quinidine
56-54-2 8244 200-279-0
$C_{20}H_{24}N_2O_2$
(8R,9S)-6'-Methoxycinchonan-9-ol.
Pitayin; conquinine, β-quinine; Coccinine; Conchinine; Cin-quin; Quinicardine; Quinidex; Cardioquin; Quindine; Quinaglute; conquinine; pitayine; (+)-quinidine; β-quinidine; Kinidin; (+)-Quindine. Class 1A antiarrhythmic agent. Antimalarial. A dextrorotatory stereoisomer of quinine. Found in cinchona bark. mp = 174-175°; $[\alpha]_D^{15}$ = 230° (c = 1.8 $CHCl_3$), $[\alpha]_D^{17}$ = 258° (EtOH), $[\alpha]_D^{17}$ = 322° (c = 1.6 2M HCl); soluble in H_2O (0.05 g/100 ml at 25°, 0.125 g/100 ml at 100°), EtOH (2.78 g/100 ml), Et_2O (1.78. *Eli Lilly & Co.; Merrell Pharm. Inc.; Parke-Davis.*

1679 Quinidine Gluconate
7054-25-3 8244 230-333-9
$C_{26}H_{36}N_2O_9$
(8R,9S)-6'-Methoxycinchonan-9-ol gluconate.
gluconic acid quinidine salt; Duraquin; Quinaglute. Class IA antiarhythmic; antimalarial. mp = 175-176.5°; soluble in H_2O (11.1 g/100 ml), EtOH (1.67 g/100 ml). *Berlex Labs Inc.; Parke-Davis.*

1680 Quinidine Sulfate
6591-63-5 8244
$(C_{20}H_{24}N_2O)_2.H_2SO_4.2H_2O$
6'-Methoxy-(9S)-cinchonan-9-ol sulfate (2:1) (salt) dihydrate.
Cin-Quin; Quinidex; Quinora; Quinicardine; Quinidex Extentabs. Class IA antiarrhythmic; antimalarial. Cardiac depressant. $[\alpha]_D^{25}$ +212° (95% EtOH), +260° (dilute HCl); pKa 4.2, 8.8; pH (1% aqueous solution) = 6.0-6.8; slightly soluble in H_2O (0.01 mg/ml), boiling H_2O (0.07 mg/ml), alcohol (0.1 mg/ml), MeOH (0.33 mg/ml), $CHCl_3$ (0.08 mg/ml); insolubl. *Key Pharm.; Parke-Davis; Robins, A. H. Co.; Solvay Pharm. Inc.; Whitehall-Robins.*

1681 Quinidine Sulfate
6591-63-5 8244
$(C_{20}H_{24}N_2O)_2.H_2SO_4.2H_2O$
6'-Methoxy-(9S)-cinchonan-9-ol sulfate (2:1) (salt) dihydrate.
Cin-Quin; Quinidex; Quinora; Quinicardine; Quinidex Extentabs. Class IA antiarrhythmic; antimalarial. Cardiac depressant. $[\alpha]_D^{25}$ +212° (95% EtOH), +260° (dilute HCl); pKa 4.2, 8.8; pH (1% aqueous solution) = 6.0-6.8; slightly soluble in H_2O (0.01 mg/ml), boiling H_2O (0.07 mg/ml), alcohol (0.1 mg/ml), MeOH (0.33 mg/ml), $CHCl_3$ (0.08 mg/ml); insolubl. *Key Pharm.; Parke-Davis; Robins, A. H. Co.; Solvay Pharm. Inc.; Whitehall-Robins.*

1682 Quinidine Sulfate [anhydrous]
50-54-4 8244 200-046-3
$(C_{20}H_{24}N_2O)_2.H_2SO_4$
6'-Methoxy-(9S)-cinchonan-9-ol sulfate (2:1) (salt).
Class IA antiarrhythmic; antimalarial. Cardiac depressant.

1683 Quinidine Sulfate [anhydrous]
50-54-4 8244 200-046-3
$(C_{20}H_{24}N_2O)_2.H_2SO_4$.
6'-Methoxy-(9S)-cinchonan-9-ol sulfate (2:1) (salt).
Class IA antiarrhythmic; antimalarial. Cardiac depressant.

1684 Recainam
74738-24-2
$C_{15}H_{25}N_3O$
1-[3-(Isopropylamino)propyl]-3-(2,6-xylyl)urea.
Wy-42362. Class IC antiarrhythmic agent. *Wyeth-Ayerst Labs.*

1685 Recainam Hydrochloride
74752-07-1
$C_{15}H_{26}ClN_3O$
1-[3-(Isopropylamino)propyl]-3-(2,6-xylyl)urea hydrochloride.
Vanorm; WY-42362 HCl. Class IC antiarrhythmic agent. *Wyeth-Ayerst Labs.*

1686 Recainam Tosylate
74752-08-2
$C_{22}H_{33}N_3O_4S$
1-[3-(Isopropylamino)propyl]-3-(2,6-xylyl) urea mono-p-toluenesulfonate.
Wy-42362 tosylate. Class IC antiarrhythmic agent. *Wyeth-Ayerst Labs.*

1687 Risotilide
120688-08-6
$C_{15}H_{27}N_3O_4S_2$
4'-[Isopropyl[2-(isopropylamino)ethyl]sulfamoyl]methanesulfanilamide.
Antiarrhythmic agent. Potassium channel blocker. *Wyeth-Ayerst Labs.*

1688 Risotilide Hydrochloride
116907-13-2
$C_{15}H_{28}ClN_3O_4S_2$
4'-[Isopropyl[2-(isopropylamino)ethyl]sulfamoyl]methanesulfanilamide hydrochloride.
Wy-48986. Antiarrhythmic agent. Potassium channel blocker. *Wyeth-Ayerst Labs.*

1689 Ropitoin
56079-81-3
$C_{30}H_{33}N_3O_3$
5-(p-Methoxyphenyl)-5-phenyl-3-[3-(4-phenylpiperidino)propyl]hydantoin.
Antiarrhythmic; cardiac depressant. *Bayer AG.*

1690 Ropitoin Hydrochloride
56079-80-2
$C_{30}H_{34}ClN_3O_3$
5-(p-Methoxyphenyl)-5-phenyl-3-[3-(4-phenylpiperidino)propyl]hydantoin hydrochloride.
TR-2985. Antiarrhythmic; cardiac depressant. *Bayer AG.*

1691 Sematilide
101526-83-4 8586
$C_{14}H_{23}N_3O_3S$
N-[2-(Diethylamino)ethyl]-p-menthanesulfonamidobenzenesulfonamide.
CK-1752. Class III antiarrhythmic agent. *Schering AG.*

1692 Sematilide Hydrochloride
101526-62-9 8586
$C_{14}H_{24}ClN_3O_3S$
N-[2-(Diethylamino)ethyl]-p-menthanesulfonamidobenzenesulfonamide monohydrochloride.
CK-1752A. Class III antiarrhythmic agent. mp = 141-142°, 137°, 142°; λ_m = 289 nm (ε 19100 0.1N NaOH), 254 nm (ε 17800 0.1N HCl); LD_{50} (mus ip) = 250-300 mg/kg, (mus iv) = 96 mg/kg, (mus orl) = 1800 mg/kg, (rat iv) = 92 mg/kg, (rat orl) = 3200 mg/kg, (dog iv) = 143-175 mg/kg, (dog orl) = 500-1000 mg/kg. *Schering AG.*

1693 Sotalol
3930-20-9 8876
$C_{12}H_{20}N_2O_3S$
4'-[1-Hydroxy-2-(isopropylamino)ethyl]-methanesulfonanilide.
Antianginal; class II and III antiarrhythmic; antihypertensive. A β-adrenergic blocker. Has anti-arrhythmic, antihypertensive and antianginal properties. λ:sm = 242.2, 275.2 nm ($CHCl_3$). *Berlex Labs Inc.; Bristol-Myers Squibb Co.*

1694 Sotalol Hydrochloride
959-24-0 8876 213-496-0
$C_{12}H_{21}ClN_2O_3S$
4'-[1-Hydroxy-2-(isopropylamino)ethyl]methanesulfonanilide monohydrochloride.
MJ-1999; Beta-Cardone; Betapace; Darob; Sotacor; Sotalex. Antianginal; class II and III antiarrhythmic; antihypertensive. A β-adrenergic blocker. mp = 206.5-207°, 218-219°; soluble in H_2O, less soluble in $CHCl_3$; LD_{50} (mmus orl) = 2600 mg/kg, (mmus ip) = 670 mg/kg, (mrat orl) = 3450 mg/kg, (mrat ip) = 680 mg/kg, (rbt orl) = 1000 mg/kg, (dog ip) = 330 mg/kg. *Berlex Labs Inc.; Bristol-Myers Squibb Co.*

1695 Stirocainide
78372-27-7
$C_{22}H_{34}N_2O$
(E)-2-Benzylidenecycloheptonane(E)-O-[2-(diisopropylamino)ethyl]oxime.
Th-494. Antiarrhythmic.

1696 Suricainide
85053-46-9
$C_{18}H_{31}N_3O_3S$
3-[2-(Diethylamino)ethyl]-1-isopropyl-1-[2-(phenylsulfonyl)ethyl]urea.
Antiarrhythmic agent with cardiodepressant properties. *Whitehall-Robins.*

1697 Suricainide Maleate
85053-47-0
$C_{22}H_{35}N_3O_7S$
3-[2-(Diethylamino)ethyl]-1-isopropyl-1-[2-(phenylsulfonyl)ethyl]urea maleate (1:1).
AHR-10718. Antiarrhythmic agent with cardiodepressant properties. *Whitehall-Robins.*

1698 Talinolol
57460-41-0 9208
$C_{20}H_{33}N_3O_3$
(±)-1-[p-[3-(tert-Butylamino)-2-hydroxypropoxy]phenyl]-3-cyclohexylurea.
02-115; Cordanum. Antiarrhythmic agent with antihypertensive properties. A β-adrenergic blocker related to practolol. mp = 142-144°; LD_{50} (rat orl) = 1180 mg/kg, (rat ip) = 54.3 mg/kg, (rat iv) = 29.7 mg/kg, (mus orl) = 593 mg/kg, (mus ip) = 74.7 mg/kg, (mus iv) = 25.0 mg/kg. *Ciba-Geigy Corp.*

1699 Tedisamil
90961-53-8
$C_{19}H_{32}N_2$
3',7'-Bis(cyclopropylmethyl)spiro[cyclopentane-1,9'-[3,7]diazabicyclo[3.3.1]nonane].
KC-8857. Sinus node inhibitor. Antiarrhythmic (class III).

1700 Terikalant
121277-96-1
$C_{24}H_{31}NO_3$
(-)-(S)-1-[2-(4-Chromanyl)ethyl]-4-(3,4-dimethoxyphenyl)piperidine.
RP-62719. Blocker of inwardly rectifying K+ currents. Antiarrhythmic (class III).

1701 Tilisolol
85136-71-6 9579
$C_{17}H_{24}N_2O_3$
(±)-4-[3-(tert-Butylamino)-2-hydroxypropoxy]-2-methylisocarbostyril.
Antiarrhythmic; antihypertensive. A β-adrenergic blocker. *Nisshin Denka K.K.*

1702 Tilisolol Hydrochloride
62774-96-3 9579
$C_{17}H_{25}ClN_2O_3$
(±)-4-[3-(tert-Butylamino)-2-hydroxypropoxy]-2-methylisocarbostyril hydrochloride.
N-696; Selecal. Antiarrhythmic; antihypertensive. A β-adrenergic blocker. mp = 203-205°; LD_{50} (mmus orl) = 1393 mg/kg, (mmus sc) = 1219 mg/kg, (mmus ip) = 578 mg/kg, (mmus iv) = 74.3 mg/kg, (fmus orl) = 1290 mg/kg, (fmus sc) = 1245 mg/kg, (fmus ip) = 557 mg/kg, (fmus iv) = 104.7 mg/kg, (mrat orl) = 145 mg/kg, (mrat sc) = . *Nisshin Denka K.K.*

1703 Timolol
91524-16-2 9585
$C_{13}H_{24}N_4O_3S.1/2H_2O$
(S)-1-(tert-Butylamino)-3-[(4-morpholino-1,2,5-thiadiazol-3-yl)oxy]-2-propanol hemihydrate.
Antianginal agent with antiarrhythmic (class II), antihypertensive and antiglaucoma properties. A β-adrenergic blocker. [(±) form]: mp = 71.5-72.5°. *Merck & Co.Inc.*

1704 Timolol Maleate
26921-17-5 9585 248-111-5
$C_{17}H_{28}N_4O_7S$
(S)-1-(tert-Butylamino)-3-[(4-morpholino-1,2,5-thiadiazol-3-yl)oxy]-2-propanol maleate (1:1).
Blocadren; Timoptic; Timoptol; MK-950; Aquanil; Betim; Proflax; Temserin; Tenopt; Timacar; Timacor; component of: Cosopt, Timolide. Antianginal agent with antiarrhythmic (class II), antihypertensive and antiglaucoma properties. A β-adrenergic blocker. mp = 201.5-202.5°; $[\alpha]^{24}_{405}$ = -12.0° (c = 5 1N HCl), $[\alpha]^{25}_{D}$ = -4.2°; λ_m = 294 nm ($A^{1\%}_{1\ cm}$ 200 0.1N HCl); soluble in EtOH, MeOH; poorly soluble in $CHCl_3$, cyclohexane; insoluble in isooctane, Et_2O. *Merck & Co.Inc.*

1705 Tiracizine
83275-56-3
$C_{21}H_{25}N_3O_3$
Ethyl 5-(N,N-dimethylglycyl)-10,11-dihydro-5H-dibenz[b,f]azepine-3-carbamate.
AWD-19-166 [as hydrochloride]; GS 015 [as hydrochloride]; Bonnecor [as hydrochloride]. Antiarrhythmic (class I).

1706 Tocainide
41708-72-9 9629 255-505-0
$C_{11}H_{16}N_2O$
2-Amino-2',6'-propionoxylidide.
W-36095. Class IB antiarrhythmic agent with cardiodepressant properties. *Merck & Co.Inc.*

1707 Tocainide Hydrochloride
71395-14-7 9629 275-361-2
$C_{11}H_{17}ClN_2O$
2-Amino-2',6'-propionoxylidide hydrochloride.
Tonocard; Taquidil; Xylotocan. Class IB antiarrhythmic agent with cardiodepressant properties. mp = 246-247°. *Merck & Co.Inc.*

1708 R-(-)-Tocainide Hydrochloride
53984-74-0 9629
$C_{11}H_{17}ClN_2O$
(R)-(-)-2-Amino-2',6'-propionoxylidide.
Class IB antiarrhythmic agent with cardiodepressant properties. mp = 265-266°; $[\alpha]_D$ = -42.16° (c = 2.63 MeOH). *Merck & Co.Inc.*

1709 S-(+)-Tocainide Hydrochloride
53984-76-2 9629
$C_{11}H_{17}ClN_2O$
(S)-(+)-2-Amino-2',6'-propionoxylidide.
Class IB antiarrhythmic agent with cardiodepressant properties. mp = 264.5°; $[\alpha]_D$ = +42.35° (c = 2.63 MeOH). *Merck & Co.Inc.*

1710 Tolamolol
38103-61-6 253-783-8
$C_{19}H_{24}N_2O_4$
4-[2-[[2-Hydroxy-3-(2-methylphenoxy)propyl]amino]ethoxy]benzamide.
Vasodilator (coronary); cardiac depressant (antiarrhythmic); antiadrenergic (β-receptor). *CIBA plc; Pfizer Inc.*

1711 Transcainide
88296-62-2
$C_{22}H_{35}N_3O_2$
(±)-trans-4-(Dimethylamino)-1-(2-hydroxycyclohexyl)-2',6'-isonipectoxylidide.
R-54718. Antiarrhythmic agent; cardiac depressant. *Abbott Laboratories Inc.*

1712 Verapamil
52-53-9 10083 200-145-1
$C_{27}H_{38}N_2O_4$
5-[(3,4-Dimethoxyphenethyl)methylamino]-2-(3,4-dimethoxyphenyl)-2-isopropylvaleronitrile.
D-365; CP-16533-1; iproveratril. Antianginal; class IV antiarrhythmic; coronary vasodilator;calcium channel blocker. $bp_{0.001}$ = 243-246°; insoluble in H_2O; slightly soluble in C_6H_6, C_6H_{16}, Et_2O; soluble in EtOH, MeOH, Me_2CO, EtOAc, $CHCl_3$. *Bristol-Myers Squibb Co.*

1713 Verapamil Hydrochloride
152-11-4 10083 205-800-5
$C_{27}H_{39}ClN_2O_4$
5-[(3,4-Dimethoxyphenethyl)methylamino]-2-(3,4-dimethoxyphenyl)-2-isopropylvaleronitrile hydrochloride.
Arapamyl; Berkatens; Calan; Cardiagutt; Cardibeltin; Cordilox; Dignover; Drosteakard; Geangin; Isoptin; Quasar; Securon; Univer; Vasolan; Veracim; Veramex; Veraptin; Verelan; Verexamil. Antianginal; class IV antiarrhythmic agent. Coronary vasodilator with calcium channel blocking activity. mp = 138.5-140.5°; soluble in H_2O (7 g/100 ml), EtOH (2.6 g/100 ml), propylene glycol (93 g/100 ml), MeOH (> 10.0 g/100 ml), iPrOH (0.46 g/100 ml), EtOAc (0.1 g/100 ml), DMF (> 10.0 g/100 ml), CH_2Cl_2 (> 10.0 g/100 ml), C_6H_{14} (0.1 g/100 ml); LD_{50} (rat iv) = 16 mg/kg, (mus iv) = 8 mg/kg. *Knoll Pharm. Co.; Lederle Labs.; Parke-Davis; Searle G.D. & Co.*

1714 Viquidil
84-55-9 10141 201-540-1
$C_{20}H_{24}N_2O_2$
1-(6-Methoxy-4-quinolyl)-3-(3-vinyl-4-piperidyl)-1-propanone.
LM-192; chinicine; mequiverine; quinotoxine; quinotoxol. Vasodilator with antiarrhythmic properties. $[\alpha]_D$= 43°; yellow oil; slightly soluble in H_2O; freely soluble in EtOH, $CHCl_3$, Et_2O. *Polaroid.*

1715 Viquidil Hydrochloride
52211-63-9 10141 257-739-9
$C_{20}H_{25}ClN_2O_2$
1-(6-Methoxy-4-quinolyl)-3-(3-vinyl-4-piperidyl)-1-propanone hydrochloride.
Desclidium; Permiran. Vasodilator; antiarrhythmic. mp = 184 ± 4°; λ_m = 246, 355 nm; soluble in EtOH, poorly soluble in H_2O, insoluble in Me_2CO. *Polaroid.*

1716 Xibenolol
81584-06-7 10209
$C_{15}H_{25}NO_2$
(±)-1-(tert-Butylamino)-3-(2,3-xylyloxy)-2-propanol.
Antiarrhythmic; non-selective β-adrenergic blocker. mp = 57°; $bp_{0.7}$ = 134-136°; λ_m = 271.2, 274, 279.3 nm (ε 10800, 10700, 11100 EtOH). *Teikoku Hormone Mfg. Co. Ltd.*

1717 Xibenolol Hydrochloride
59708-57-5 10209
$C_{15}H_{26}ClNO_2$
(±)-1-(tert-Butylamino)-3-(2,3-xylyloxy)-2-propanol hydrochloride.
D-32; Selapin; Rhythminal; Rythminal. Antiarrhythmic agent. Non-selective β-adrenergic blocker. mp = 135-137°. *Teikoku Hormone Mfg. Co. Ltd.*

Antiarthritics

1718 N-Acetylglucosamine
7512-17-6 4466
$C_8H_{15}NO_6$
2-Acetylamino-2-deoxy-β-D-glucopyranose.
Antiarthritic agent. Also used as a pharmaceutic aid. mp = 205°; $[\alpha]_D^{18}$ = 64° → 40.9°.

1719 Allocupreide Sodium
5965-40-2 265
$C_{10}H_{11}CuN_2NaO_2S$
Sodium 3-(3-allyl-S-cupropseudoyhioureido)benzoate.
Ebesal; cupralyl-natrium; Cuprelon; Cupralene; Cuprion. Antirheumatic agent. Soluble in H_2O, insoluble in organic solvents. *Hoechst Roussel Pharm. Inc.*

1720 Atiprimod
123018-47-3
$C_{22}H_{44}N_2$
2-[3-(Diethylamino)propyl]-8,8-dipropyl-2-azaspiro[4,5]decane.
Antirheumatic agent with anti-inflammatory properties. *SmithKline Beecham Pharm.*

1721 Atiprimod Dihydrochloride
130065-61-1
$C_{22}H_{46}Cl_2N_2$
2-[3-(Diethylamino)propyl]-8,8-dipropyl-2-azaspiro[4,5]decane dihydrochloride.
SK&F-106615-A2. Antirheumatic agent with anti-inflammatory properties. *SmithKline Beecham Pharm.*

1722 Atiprimod Dimaleate
183063-72-1
$C_{30}H_{52}N_2O_8$
2-[3-(Diethylamino)propyl]-8,8-dipropyl-2-azaspiro[4,5]decane maleate (1:2).
SKF-106615-A12. Antirheumatic agent with anti-inflammatory properties. *SmithKline Beecham Pharm.*

1723 Auranofin
34031-32-8 911 251-801-9
$C_{20}H_{34}AuO_9PS$
(1-Thio-β-D-glucopyranosato)(triethylphosphine)gold 2,3,4,6-tetraacetate.
Ridaura; Ridauran; SKF-39162; Aktil; Crisinor; Crisofin. Antirheumatic agent. mp = 110-111°; LD_{50} (rat orl) = 265 mg/kg, (mus orl) = 310 mg/kg. *SmithKline Beecham Pharm.*

1724 Aurothioglucose
12192-57-3 917 235-365-7
$C_6H_{11}AuO_5S$
(1-Thio-D-glucopyranosato)gold.
Solganal; Solganal B; Aureotan; Aurumine; Oronol. Antirheumatic agent. Has been used to produce obesity in animals. Soluble in H_2O, slightly soluble in propylene glycol, insoluble in other organic solvents. *Schering AG.*

1725 Aurothioglycanide
16925-51-2 918
C_8H_8AuNOS
S-Gold derivative of 2-mercapto-acetanilide.
aurothioglycollic acid anilide; Lauron. Antirheumatic agent. mp = 238-241°; insoluble in H_2O, organic solvents; administered as a suspension in sesame oil.

1726 Azathioprine
446-86-6 935 207-175-4
$C_9H_7N_7O_2S$
6-[(1-Methyl-4-nitroimidazol-5-yl)thio]-purine.
BW-57-322; NSC-39084; Azamune; Azanin; Azoran; Imuran; Imurel. Immunosuppressant and antirheumatic agent. Used in combination with cyclophosphamide and hydroxychloroquine in treatment of rheumatoid arthritis. Listed as a known carcinogen. Dec 243-244°; λ_m 276 nm (ε 18200 MeOH), 280 nm (ε 17300 0.1N HCl), 285 nm (ε 15500 0.1N NaOH); slightly soluble in H_2O, $CHCl_3$, EtOH. *Glaxo Wellcome Inc.*

1727 Azathioprine Sodium
55774-33-9 935
$C_9H_6N_7NaO_2S$
6-[(1-Methyl-4-nitroimidazol-5-yl)thio]purine sodium salt.
Imuran Injection; Imurek. Immunosuppressant and antirheumatic agent. Has been used in treatment of rheumatoid arthritis. Soluble in H_2O. *Glaxo Wellcome Inc.*

1728 Bucillamine
65002-17-7 1481
$C_7H_{13}NO_3S_2$
N-(2-Mercapto-2-methylpropionyl)-L-cysteine.
tiobutarit; thiobutarit; DE-019; SA-96; Rimatil. Immunomodulator; antirheumatic. mp = 139-140°; $[\alpha]_D^{25}$ = 32.3° (c = 1.0 EtOH); LD_{50} (mus ip) = 2285 mg/kg, (mus iv) = 989.6 mg/kg. *Santen Pharm. Co. Ltd.*

1729 Calcium 3-Aurothio-2-Propanol-1-Sulfonate
5743-29-3 1689
$C_6H_{12}Au_2CaO_8S_4$
[2-Hydroxy-3-mercapto-1-propanesulfonato(2-)-O^2,S^3-]aurate(1-) calcium (2:1).
Chrisanol; Chrysanol; Krizanol; component of: Oleochrysine. Antirheumatic agent.

1730 Celecoxib
169590-42-5
$C_{17}H_{14}F_3N_3O_2S$
4-[5-(4-Methylphenyl)-3-(trifluoromethyl)-1H-pyrazol-1-yl] benzenesulfonamide.
A COX-2 inhibitor. Anti-inflammatory, used as an NSAID, especially in pain control in arthritis. *Searle G.D. & Co.*

1731 Chloroquine
54-05-7 2215 200-191-2
$C_{18}H_{26}ClN_3$
7-Chloro-4-[[4-(diethylamino)-1-methylbutyl]amino]-quinoline.
Aralen; SN-7618; RP-3377; Artrichin; Bemaphate; Capquin; Nivaquine B; Resoquine; Reumachlor; Sanoquin; nivaquine [as sulfate]. Antiamebic; antimalarial; antirheumatic. Also used as a lupus erythematosus suppressant. mp = 87°. *Sterling Winthrop Inc.*

1732 Chloroquine Dihydrochloride
3545-67-3 2215 222-592-1
$C_{18}H_{28}Cl_3N_3$
7-Chloro-4-[[4-(diethylamino)-1-methylbutyl]amino]-quinoline dihydrochloride.
Aralen hydrochloride. Antiamebic; antimalarial. Also an antirheumatic agent. *Sterling Winthrop Inc.*

1733 Chloroquine Diphosphate
50-63-5 2215 200-055-2
$C_{18}H_{32}ClN_3P_2O_8$
7-Chloro-4-[[4-(diethylamino)-1-methylbutyl]amino]quinoline diphosphate.
Aralen phosphate; Arechin; Avloclor; Imagon; Malaquin; Resochin; Tresochin. Antirheumatic agent. Also antiamebic, antimalarial. Also has activity as a lupus erythematosus suppressant. mp = 193-195°, 215-218°; soluble in H_2O, insoluble in organic solvents. *Sterling Winthrop Inc.*

1734 Clobuzarit
22494-47-9 2426 245-035-4
$C_{17}H_{17}ClO_3$
2-[(4'-Chloro-4-biphenylyl)methoxy]-2-methylpropionic acid.
ICI-55897; Clozic. Antirheumatic agent. mp = 154-155°. *ICI.*

1735 Cuproxoline
13007-93-7 2740
$C_{34}H_{56}CuN_6O_{14}S_4$
Bis(dihydrogen 8-hydroxy-5,7-quinolinedisulfonato)-copper compound with diethylamine (1:4).
copper DOS; Dicuprene; Cujec; Cuprimyl. Antirheumatic agent. Used in veterinary medicine as a copper supplement. Soluble in H_2O, giving a dark green solution; LD_{50} (rat im) = 126 mg/kg.

1736 Diacerein
13739-02-1 3003 237-310-2
$C_{19}H_{12}O_8$
9,10-Dihydro-4,5-dihydroxy-9,10-dioxo-2-anthroic acid diacetate.
diacerhein; diacetylrhein; DAR; SF-277; Artrodar; Fisiodar. Antiarthritic agent. mp = 217-218°. *Proter.*

1737 Enbrel
185243-69-0
1-235-Tumor necrosis factor receptor (human) fusion protein with 236-467-immunoglobulin G1 (human σ1-chain Fc fragment).
Etanercept; Enbrel; rhu TNFR:Fc. Antiarthritis agent.

1738 Etanercept
185243-69-0
A dimeric fusion protein consisting of the extracellular ligad-binding portion of the human 75 kilodalton (p75) tumor necrosis factor receptor linked to the Fc portion of human IgG1.
An immunomodulator, used in treatment of arthritis. *Immunex Corp.*

1739 Glucosamine
3416-24-8 4466 222-311-2
$C_6H_{13}NO_5$
2-Amino-2-deoxy-β-D-glucopyranose.
Antiarthritic agent. Also used as a pharmaceutic aid.

1740 Glucosamine Sulfate salt
29031-19-4 4466 249-379-6
$C_6H_{13}NO_5.xH_2SO_4$
2-Amino-2-deoxy-β-D-glucopyranose sulfate (salt).
Dona. Antiarthritic agent. Also used as a pharmaceutic aid.

1741 α-Glucosamine
28905-11-5 4466
$C_6H_{13}NO_5$
2-Amino-2-deoxy-α-D-glucopyranose.
Antiarthritic agent. Also used as a pharmaceutic aid. mp = 88°; $[\alpha]_D^{20}$ = 100° → 47.5° after 30 minutes.

1742 β-Glucosamine
28905-10-4 4466
$C_6H_{13}NO_5$
2-Amino-2-deoxy-β-D-glucopyranose.
Antiarthritic agent. Also used as a pharmaceutic aid. mp = 110° (dec); $[\alpha]_D^{20}$ = 28° → 47.5° after 3 minutes; very soluble in H_2O; soluble in boiling MeOH 2.63 g/100 ml; sparingly soluble in cold MeOH or EtOH; insoluble in Et_2O, $CHCl_3$.

1743 Gold Sodium Thiomalate
12244-57-4 4536 235-479-7
$C_4H_3AuNa_2O_4S + C_4H_4AuNaO_4S$
Mercaptosuccinic acid monogold(1+) sodium salt.
Myochrysine; Kidon; Myocrisin; Shiosol; Tauredon. Antirheumatic agent. Very soluble in H_2O; insoluble in EtOH, Et_2O. *Merck & Co.Inc.*

1744 Gold Sodium Thiosulfate
10233-88-2 4537 233-563-8
$AuNa_3O_6S_4.2H_2O$
Sodium aurotiosulfate.
sodium dithiosulfatoaurate. Antirheumatic agent. d = 3.09; soluble in H_2O (50 g/100 ml), insoluble in most organic solvents; LD_{50} (rbt iv) = 100 mg/kg.

1745 Hydroxychloroquine
118-42-3 4863 204-249-8
$C_{18}H_{26}ClN_3O$
2-[[4-[(7-Chloro-4-quinolyl)amino]pentyl]-ethylamino]ethanol.
oxychloroquine; oxichloroquine. Antimalarial and lupus erythematosus suppressant. Antirheumatic agent. Also an antiamebic, antimalarial. Used in combination with cyclophosphamide and azathioprine in treatment of rheumatoid arthritis. mp = 89-91°. *Apothecon; Sterling Winthrop Inc.*

1746 Hydroxychloroquine Sulfate
747-36-4 4863 212-019-3
$C_{18}H_{28}ClN_3O_5S$
2-[[4-[(7-Chloro-4-quinolyl)amino]pentyl]-ethylamino]ethanol sulfate (1:1) (salt).
Plaquenil Sulfate; Ercoquin; Quensyl. Antirheumatic agent. Also antiamebic, antimalarial. Also has activity as a lupus erythematosus suppressant. Two forms, mp = 198° and 240°; freely soluble in H_2O, insoluble in most organic solvents. *Apothecon; Sterling Winthrop Inc.*

1747 Kebuzone
853-34-9 5300 212-715-7
$C_{19}H_{18}N_2O_3$
4-(3-Oxobutyl)-1,2-diphenyl-3,5-pyrolizidinedione.
KPB; Chebutan; Chepirol; Chetazolidin; Chetil; Copirene; Ketason; Ketazone; Pecnon; Phloguron; Recheton. Antirheumatic agent. mp = 115.5-116.5° or 127.5-128.5°. *Ciba-Geigy Corp.*

1748 Leflunomide
75706-12-6
5-methyl-N-(4-(trifluoromethyl)phenyl)-4-isoxazolecarboxamide.
Arava; HWA 486; Leflunomida; Leflunomide; Leflunomidum; 5-Methyl-N-(4-(trifluoromethyl)phenyl)-4-isoxazolecarboxamide; SU 101 (pharmaceutical) 5-Methylisoxazole-4-carboxylic acid (4-trifluoromethyl)-anilide; N-(4'-Trifluoromethylphenyl)-5-methylisoxazole-4-carboxamide. Antiarthritis agent.

1749 Lobenzarit
63329-53-3 5581
$C_{14}H_{10}ClNO_4$
4-Chloro-2,2'-iminodibenzoic acid.
CCA. Antirheumatic agent. mp > 306°; LD_{50} (mrat orl) = 2100 mg/kg, (frat orl) = 2600 mg/kg. *CIBA plc.*

1750 Lobenzarit Disodium Salt
64808-48-6 5581
$C_{14}H_8ClNNa_2O_4$
Disodium 4-chloro-2,2'-iminodibenzoate.
Carfenil. Antirheumatic agent. mp = 388° (dec). *Chugai Pharm. Co. Ltd.*

1751 Lodelaben
111149-90-7
$C_{25}H_{41}ClO_3$
(±)-2-Chloro-4-(1-hydroxyoctadecyl)-benzoic acid.
Declaben; SC-39026. Antiarthritic agent. Also used as an adjunct in emphysema therapy. *Searle G.D. & Co.*

1752 Melittin
20449-79-0 5867
$C_{131}H_{229}N_{39}O_{31}$
Meilittin I.
Forapin; Gly-Ile-Gly-Ala-Val-Leu-Lys-Val-Leu-Thr-Thr-Gly-Leu-Prol-Ala-Leu-Ile-Ser-Trp-Ile-Lys-Arg-Lys-Arg-Gln-Gln-NH_2. Polypeptide (26-mer) derived from the venom of the honeybee. Used as an antirheumatic agent. $[\alpha]_D^{21}$ = -89.52° (c = 0.409).

1753 Methotrexate
59-05-2 6065 200-413-8
$C_{20}H_{22}N_8O_5$
L-(+)-N-[[(2,4-Diamino-6-pteridinyl)methyl]-methylamino]benzoyl]glutamic acid.
4-amino-10-methylfolic acid; amethopterin; methyl-aminopterin; CL-14377; EMT-25299; NSC-740; MTX; Emtexate; Ledertrexate; Metatrexan; A-Methopterin; Methopterin; Mexate; Rheumatrex. Used as both an antineoplastic and antirheumatic. Folic acid antagonist. [monohydrate]: mp = 185-204°; λ_m = 244, 307 nm (0.1N HCl); soluble in aqueous solutions with some decomposition; LD_{50} (rat orl) = 135 mg/kg. *Bristol-Myers Oncology; Lederle Labs.*

1754 Methotrexate Disodium Salt
7413-34-5 6065 231-022-0
$C_{20}H_{20}N_8Na_2O_5$
L-(+)-N-[p-[[(2,4-Diamino-6-pteridinyl)methyl]-methylamino]benzoyl]glutamic acid disodium salt.
Folex; Mexate. Antineoplastic; antirheumatic agent. *Bristol-Myers Oncology; Lederle Labs.*

1755 Methotrexate Monohydrate
6745-93-3 6065
$C_{20}H_{22}N_8O_5.H_2O$
L-(+)-N-[p-[[(2,4-Diamino-6-pteridinyl)methyl]-methylamino]benzoyl]glutamic acid monohydrate.
Antineoplastic; antirheumatic agent. mp = 185-204° (dec); λ_m = 244, 307 nm (0.1N HCl), 257, 302, 370 nm (0.1N NaOH); LD_{50} (rat iv) = 14 mg/kg. *Bristol-Myers Oncology; Lederle Labs.*

1756 Myoral
645-74-9 6411 211-454-6
$C_4H_4Au_2CaO_4S_2$
Calcium [mercaptoacetato(2-)-O,aurate(1-) 91:2).
[(carboxymethyl)thio]gold calcium salt; calcium aurothioglycollate; mercaptoacetic acid calcium salt gold derivative. Antirheumatic agent. Also used as a copper chelator in treatment of Wilson's disease. Insoluble in H_2O, used as a suspension in oil.

1757 Penicillamine
52-67-5 7214 200-148-8
$C_5H_{11}NO_2S$
D-3-Mercaptovaline.
Depen; Cuprimine; β-Thiovaline; Cuprenil; Depamine; Mercaptyl; Pendramine; Perdolat; Sufirtan; Trolovol. Antirheumatic agent. Also used as a copper chelator in treatment of Wilson's disease. mp = 202-206°; $[\alpha]_D^{25}$ = -63° (c = 1 C_5H_5N); LD_{50} (rat orl) > 10000 mg/kg, (rat ip) > 660 mg/kg. *Merck & Co.Inc.; Wallace Labs.*

1758 Penicillamine Hydrochloride
2219-30-9 7214 218-727-9
$C_5H_{12}ClNO_2S$
3-Mercaptovaline hydrochloride.
Distamine; Metalcaptase. Antirheumatic agent. Also used as a copper chelator in treatment of Wilson's disease. mp = 177.5° (dec); $[\alpha]_D^{25}$ = -63° (1N NaOH); soluble in H_2O, EtOH; LD_{50} (mus iv) = 2289 mg/kg.

1759 DL-Penicillamine
52-66-4 7214 200-147-2
$C_5H_{11}NO_2S$
DL-3-Mercaptovaline.
Antirheumatic agent. Also used as a copper chelator in treatment of Wilson's disease. mp = 201° (dec); LD_{50} (rat orl) = 365 mg/kg.

1760 DL-Penicillamine Hydrochloride
22572-05-0 7214 245-093-0
$C_5H_{12}ClNO_2S$
DL-3-Mercaptovaline hydrochloride.
Antirheumatic agent. Also used as a copper chelator in treatment of Wilson's disease. mp = 145-148°.

1761 L-Penicillamine
1113-41-3 7214 214-203-9
$C_5H_{11}NO_2S$
L-3-Mercaptovaline.
Antirheumatic agent. Also used as a copper chelator in treatment of Wilson's disease. mp = 190-194°; $[\alpha]_D^{25}$ = 63° (1N NaOH); LD_{50} (rat ip) = 350 mg/kg.

1762 Remicade
Inflixmab. Antiarthritis agent.

1763 Tiflamizole
62894-89-7
$C_{17}H_{10}F_6N_2O_2S$
4,5-Bis(p-fluorophenyl)-2-[1,1,2,2-tetrafluoro-ethyl)-sulfonyl]imidazole.
Thioimidazole derivative. Cyclooxygenase inhibitor in prostaglandin synthesis. Anti-inflammatory; antiarthritic.

Antiasthmatics

1764 Amlexanox
68302-57-8 515
$C_{16}H_{14}N_2O_4$
2-Amino-7-isopropyl-5-oxo-5H-[1]-benzopyrano[2,3-b]pyridine-3-carboxylic acid.
amoxanox; AA-673; CHX-3673; Elics; Solfa. Antiallergic; antiasthmatic. An orally active lipo-oxygenase inhibitor. mp > 300°. *Chemex Pharm.; Takeda Chem. Ind. Ltd.*

1765 Andolast
132640-22-3
$C_{15}H_{11}N_9O$
4,4'-Di-1H-tetrazol-5-ylbenzanilide.
CR-2039. A potential antiasthmatic drug.

1766 Apaflurane
431-89-0 207-079-2
C_3HF_7
1,1,1,2,3,3,3-Heptafluoropropane.
Used as a propellant to deliver antiasthmatic drugs such as beclomethasone propionate.

1767 Azelastine
58581-89-8 939
$C_{22}H_{24}ClN_3O$
4-(p-Chlorobenzyl)-2-(hexahydro-1-methyl-1H-azepin-4-yl)-1(2H)-phthalazinone.
Antiallergic; antiasthmatic; antihistaminic. Orally active histamine H_1 receptor antagonist. Oil, soluble in CH_2Cl_2; gives a crystalline monohydrate. *Asta-Werke AG; Carter-Wallace.*

1768 Azelastine Hydrochloride
79307-93-0 939
$C_{22}H_{25}Cl_2N_3O$
4-(p-Chlorobenzyl)-2-(hexahydro-1-methyl-1H-azepin-4-yl)-1(2H)-phthalazinone monohydrochloride.
A-5610; W-2979M; E-0659; Allergodil; Astelin; Azeptin; Rhinolast. Antiallergic; antiasthmatic; antihistaminic. Orally active H_2-histamine receptor antagonist. mp = 225-229°; LD_{50} (mmus iv) = 36.5 mg/kg, (mmus ip) = 56.4 mg/kg, (mmus sc) = 63.0 mg/kg, (mmus orl) = 124 mg/kg, (fmus iv) = 35.5 mg/kg, (fmus ip) = 42.8 mg/kg, (fmus sc) = 54.2 mg/kg, (fmus orl) = 139 mg/kg, (mrat iv) = 26.9 mg/kg, (mrat ip) = 4. *Asta-Werke AG; Carter-Wallace.*

1769 Beclomethasone
4419-39-0 1047 224-585-9
$C_{22}H_{29}ClO_5$
9-Chloro-11β,17,21-trihydroxy-16β-methylpregna-1,4-diene-3,20-dione.
9α-chloro-16β-methylprednisolone; beclometasone. Inhalant form used as an antiasthmatic. Also has antiallergic and topical anti-inflammatory properties. Glucocorticoid. *Glaxo Wellcome Inc.; Key Pharm.; Merck & Co.Inc.; Schering AG; Schevico.*

1770 Beclomethasone Dipropionate
5534-09-8 1047 226-886-0
$C_{28}H_{37}ClO_7$
9-Chloro-11β,17,21-trihydroxy-16β-methyl-pregna-1,4-diene-3,20-dione 17,21-dipropionate.
Sch-8020W; Aerobec; Aldecin; Anceron; Andion; Beclovent Inhaler; Beclacin; Belcoforte; Belcomet; Belchlorhinol; Becloval; Becodisks; Beconase; Becotide; Clenil-A; Entyderma; Inalone O; Inalone R; Korbutone; Propaderm; Rino-Clenil; Sanasthmax; Sanasthymyl; Vancenase; Vanceril; Viarex; Viarox. Antiallergic; antiasthmatic (inhalant); anti-inflammatory. Glucocorticoid. mp = 117-120°; $[\alpha]_D$ = 98° (c = 1 dioxane); λ_m = 238 nm (ε 15990 EtOH). *Glaxo Labs.; Key Pharm.; Merck & Co.Inc.; Schering-Plough HealthCare Products.*

1771 Budesonide
51333-22-3 1490 257-139-7
$C_{25}H_{34}O_6$
(R,S)-11β,16α,17,21-Tetrahydroxypregna-1,4-diene-3,20-dione cyclic 16,17-acetal with butyraldehyde.
S-1320; Bidien; Budeson; Cortivent; Micronyl; Preferid; Pulmicort; Rhinocort; Spirocort. Anti-inflammatory;

antiasthmatic. Non-halogenated glucocorticoid related to triamcinolone hexacetonide with a high ratio of topical to systemic activity. A mixture of two isomers in which the S-isomer varies between 40-51%. mp = 221-232° (dec); $[\alpha]_D^{25}$= 98.9 (c = 0.28 CH_2Cl_2). *Astra Chem. Ltd.; Bofors.*

1772 Bunaprolast
99107-52-5
$C_{17}H_{20}O_3$
2-Butyl-4-methoxy-1-naphthol acetate.
U-66858. Antiasthmatic. *Pharmacia.*

1773 Butixocort
120815-74-9
$C_{25}H_{36}O_5S$
11β,17-Dihydroxy-21-mercaptopregn-4-ene-3,20-dione 17-butyrate.
Antiasthmatic; similar to beclomethasone.

1774 Cromitrile Sodium
53736-52-0
$C_{20}H_{14}N_5NaO_5$
(±)-p-[2-Hydroxy-3-[[4-oxo-2-(1H-tetrazol-5-yl)-4H-1-benzopyran-5-yl]oxy]propoxy]benzonitrile monosodium salt.
TR-2855. Antiasthmatic. *Bayer Corp.*

1775 Cromoglicate Lisetil
110816-79-0
$C_{33}H_{36}N_2O_{12}$
Diethyl-5,5'-[(2-hydroxytrimethylene)dioxy]bis[40oxo-4H-benzopyran-2-carboxylate] ester with L-lysine.
Bronchodilator, used as an antiasthmatic.

1776 Cromolyn
16110-51-3 2658 240-279-8
$C_{23}H_{12}O_{11}$
5,5'-[(2-Hydroxytrimethylene)dioxy]-bis[4-oxo-4H-1-benzopyran-2-carboxylic acid].
cromoglycic acid; Duracroman. Antiallergic; prophylactic antiasthmatic. mp = 241-242° (dec); [monohydrate]: mp = 216-217°. *Bausch & Lomb Pharm. Inc.; Fisons plc.*

1777 Cromolyn Disodium Salt
15826-37-6 2658 239-926-7
$C_{23}H_{14}Na_2O_{11}$
Sodium 5,5'-[(2-hydroxytrimethylene)dioxy]-bis[4-oxo-4H-1-benzopyran-2-carboxylate].
cromolyn sodium; disodium cromoglycate; DSCG; FPL-670; Aarane; Alercrom; Alerion; Allergocrom; Colimune; Crolom; Cromovet; Fivent; Gastrocom; Gastrofrenal; Inostral; Intal; Introl; Irtan; Lomudal; Lomupren; Lomusol; Lomuspray; Nalcrom; Nalcron; Nasalcrom; Nasmil; Opticrom; Rynacrom; Sofro; Vividrin. Antiallergic; prophylactic antiasthmatic. Soluble in H_2O, insoluble in organic solvents; LD_{50} (rat orl) > 8000 mg/kg. *Bausch & Lomb Pharm. Inc.; Fisons plc.*

1778 Dexamethasone
50-02-2 2986 200-003-9
$C_{22}H_{29}FO_5$
9-Fluoro-11β,17,21-trihydroxy-16α-methylpregna-1,4-diene-3,20-dione.
hexadecadrol; Aeroseb-Dex; Corson; Cortisumman; Decacort; Decaderm; Decadron; Decalix; Decasone; Dekacort; Deltafluorene; Deronil; Deseronil; Dexacortal; Dexacortin; Dexafarma; Dexa-Mamallet; Dexameth; Dexamonozon; Dexapos; Dexa-sine; Dexasone; Dexin-oral; Dinormon; Fluormone; Isopto-Dex; Lokalison F; Loverine; Luxazone; Maxidex; Millicorten; Pet-Derm III; component of: Azimycin, Azium, Deronil, Dexacidin, Fulvidex, Maxitrol, Naquasone, Tobradex, Tresaderm. Antiasthmatic glucocorticoid. Used in diagnosis of Cushing's syndrome, depression. mp = 262-264°, 278-281°; $[\alpha]_D^{25}$= 77.5° (dioxane); slightly soluble in H_2O (0.010 g/100 ml), soluble in Me_2CO, EtOH, $CHCl_3$. *Alcon Labs; Allergan Inc.; CIBA Vision AG; Merck & Co.Inc.; Organon Inc.; Pharmacia & Upjohn; Schering-Plough Animal Health; Schering-Plough HealthCare Products; Solvay Pharm. Inc.*

1779 Dexamethasone 21-Acetate
1177-87-3 2986 214-646-8
$C_{24}H_{31}FO_6$
9-Fluoro-11β,17,21-trihydroxy-16α-methylpregna-1,4-diene-3,20-dione 21-acetate.
Dalalone DP; Dalalone LA; Decadronal; Decadron-LA; Dectancyl; Dexacortisyl. Anti-inflammatory; glucocorticoid; diagnostic aid (Cushing's Syndrome, depression). mp = 215-221°, 229-231°, 238-240°; $[\alpha]_D^{25}$ = 73° ($CHCl_3$); $[\alpha]_D$ = 77.6°; λ_m = 239 nm (ε 14900). *Forest Pharm. Inc.; Merck & Co.Inc.*

1780 Dexamethasone 21-Phosphate Disodium Salt
2392-39-4 2986 219-243-0
$C_{22}H_{28}FNa_2O_8P$
9-Fluoro-11β,17,21-trihydroxy-16α-methylpregna-1,4-diene-3,20-dione 21-phosphate disodium salt.
dexamethasone sodium; dexamethasone 21-phosphate disodium salt; dexamethasone 21-(dihydrogen phosphate) disodium salt; Ak-Dex; Baldex; Colvasone; Dalalone; Decadron; Dexabene; Dezone; Fortecortin; Hexadrol; Maxidex Ointment; Oradexon; Orgadrone; Solu-Deca; component of: NeoDecadron. Anti-inflammatory; glucocorticoid (antiasthmatic); diagnostic aid (Cushing's Syndrome, depression). Used in diagnosis of Cushing's syndrome and depression. mp = 233-235°; $[\alpha]_D$ = 257° (H_2O); $[\alpha]_D^{25}$= 74° ± 4° (c = 1); λ_m = 238-239 nm (ε 14000); soluble in H_2O. *Alcon Labs; Elkins-Sinn; Forest Pharm. Inc.; Merck & Co.Inc.; Organon Inc.*

1781 Enofelast
127035-60-3
$C_{16}H_{15}FO$
(E)-4'-Fluoro-3,5-dimethyl-4-stilbenol.
BI-L-239. Antiasthmatic. *Boehringer Ingelheim GmbH.*

1782 Filaminast
141184-34-1
$C_{15}H_{20}N_2O_4$
3'-(Cyclopentyloxy)-4'-methoxyacetophenone (E)-O-carbamoyloxime.
WAY-PDA-641. A selective phosphodiesterase IV inhibitor. Used as an antiasthmatic. *Wyeth-Ayerst Labs.*

1783 Flezelastine
135381-77-0
$C_{29}H_{30}FN_3O$
(±)-4-(p-Fluorobenzyl)-2-(hexahydro-1-phenethyl-1H-azepin-4-yl)-1(2H)-phthalazinone.
Antiasthmatic; antiallergic.

1784 Flunisolide
3385-03-3 4180 222-193-2
$C_{24}H_{31}FO_6$
6α-Fluoro-11β,16α,17,21-tetrahydroxypregna-1,4-diene-3,20-dione cyclic 16,17 acetal with acetone.
RS-3999; Aerobid; Bronalide; Lunis; Nasalide; Rhinalar; Synaclyn; Syntaris. Glucocoriticoid; antiasthmatic. Synthetic fluorinated corticosteroid related to prednisolone. *Syntex Labs. Inc.*; 21;169.

1785 Flunisolide 21-Acetate
4533-89-5 4180 224-871-3
$C_{26}H_{33}FO_7$
6α-Fluoro-11β,16α,17,21-tetrahydroxypregna-1,4-diene-3,20-dione cyclic 16,17 acetal with acetone 21-acetate.
RS-1320. An antiasthmatic glucocorticoid; anti-inflammatory. *Syntex Labs. Inc.*

1786 Ibudilast
50847-11-5 4923
$C_{14}H_{18}N_2O$
1-(2-Isopropylpyrazolo[1,5-a]pyridin-3-yl)-2-methyl-1-propanone.
KC-404; Ketas. Antiallergic; antiasthmatic; vasodilator (cerebral). Phosphodiesterase inhibitor. Leukotriene D_4 antagonist. mp = 53.5-54°; slightly soluble in H_2O, freely soluble in organic solvents; LD_{50} (mus iv) = 260 mg/kg. *Kyorin Pharm. Co. Ltd.*

1787 Iralukast
151581-24-7
$C_{38}H_{37}F_3O_8S$
7-[[(1S,2E,4Z)-9-(4-Acetyl-3-hydroxy-2-propylphenoxy)-1-[(αR)-α-hydroxy-m-(trifluoromethyl)benzyl]-2,4-nonadienyl]thio-4-oxo-4H-1-benzopyran-2-carboxylic acid.
A cysteinyl-leukotriene antagonist, used to treat asthma.

1788 Isamoxole
57067-46-6
$C_{12}H_{20}N_2O_2$
N-Butyl-2-methyl-N-(4-methyl-2-isoxazolyl)propionamide.
compound 90606. An antiasthmatic. *Eli Lilly & Co.*

1789 Ketotifen
34580-13-7 5319 252-099-7
$C_{19}H_{19}NOS$
4,9-Dihydro-4-(1-methyl-4-piperidylidene)-10H-benzo[4,5]cyclohepta[1,2-b]thiophen-10-one.
An antiasthmatic. mp = 152-153°. *Sandoz Pharm. Corp.*

1790 Ketotifen Fumarate
34580-14-8 5319 252-100-0
$C_{23}H_{23}NO_5S$
4,9-Dihydro-4-(1-methyl-4-piperidylidene)-10H-benzo[4,5]cyclohepta[1,2-b]thiophen-10-one fumarate.
Zaditen; HC-20511; HC-20511 fumarate; Allerkif; Totifen; Zasten. An antiasthmatic. mp = 192° (dec). *Sandoz Pharm. Corp.*

1791 Levcromakalim
94535-50-9 5487
$C_{16}H_{18}N_2O_3$
(3S,4R)-3-Hydroxy-2,2-dimethyl-4-(2-oxo-1-pyrrolidinyl)-6-chromancarbonitrile.
BRL-38227; levkromakalim; lemakalim; (-)-cromakalim; BRL-34915 [(±)-form]; cromakalim [(±)-form]. Antihypertensive; antiasthmatic. Potassium channel activator. mp = 242-244°; $[\alpha]_D^{26}$ = -52.2° (c = 1 $CHCl_3$); [(+)-form]: mp = 243-245°; $[\alpha]_D^{26}$ = 53.5° (c = 1 $CHCl_3$); [(±)-form]: mp = 230-231°. *SmithKline Beecham Pharm.*

1792 Levcromakalim
94535-50-9 5487
$C_{16}H_{18}N_2O_3$
(3S,4R)-3-Hydroxy-2,2-dimethyl-4-(2-oxo-1-pyrrolidinyl)-6-chromancarbonitrile.
BRL-38227; levkromakalim; lemakalim; (-)-cromakalim; BRL-34915 [(±)-form]; cromakalim [(±)-form]. Antihypertensive; antiasthmatic. Potassium channel activator. mp = 242-244°; $[\alpha]_D^{26}$ = -52.2° (c = 1 $CHCl_3$); [(+)-form]: mp = 243-245°; $[\alpha]_D^{26}$ = 53.5° (c = 1 $CHCl_3$); [(±)-form]: mp = 230-231°. *SmithKline Beecham Pharm.*

1793 Lodoxamide
53882-12-5 5585
$C_{11}H_6ClN_3O_6$
N,N'-(2-Chloro-5-cyano-m-phenylene)-dioxamic acid.
Antiallergic; antiasthmatic. Used topically to treat allergic conjunctivitis. mp = 212° (dec); λ_m = 239.5 nm (ε 23800 0.1N NaOH). *Alcon Labs; Pharmacia & Upjohn.*

1794 Lodoxamide Diethyl Ester
53882-13-6 5585
$C_{15}H_{14}ClN_3O_6$
Diethyl N,N'-(2-chloro-5-cyano-m-phenylene)-dioxamate.
U-42718. Antiallergic; antiasthmatic. Used topically to treat allergic conjunctivitis. mp = 177-179°. *Alcon Labs; Pharmacia & Upjohn.*

1795 Lodoxamide Tromethamine Salt
63610-09-3 5585
$C_{19}H_{28}N_5O_{12}$
N,N'-(2-Chloro-5-cyano-m-phenylene)dioxamic acid compound with 2-amino-2-(hydroxymethyl)-1,3-propanediol.
Alomide; U-42585E; lodoxamide trometamol. Antiallergic; antiasthmatic. Used topically to treat allergic conjunctivitis. *Alcon Labs; Pharmacia & Upjohn.*

1796 Montelukast
158966-92-8 6340
$C_{35}H_{36}ClNO_3S$
1-[[[(R)-m-[(E)-2-(7-Chloro-2-quinolyl)vinyl]-α-[o-(1-hydroxy-1-methylethyl)phenethyl]benzyl]-thio]methyl]cyclopropane acetic acid.
Selective leukotriene D_4 receptor antagonist. Used as an antiasthmatic. *Merck & Co.Inc.*

1797 Montelukast Monosodium Salt
151767-02-1 6340
$C_{35}H_{35}NNaO_3S$
Sodium 1-[[[(R)-m-[(E)-2-(7-chloro-2-quinolyl)vinyl]-α-[o-(1-hydroxy-1-methylethyl)phenethyl]benzyl]-thio]methyl]cyclopropane acetate.
Singulair; MK-476. Selective leukotriene D_4 receptor antagonist. Used as an antiasthmatic. *Merck & Co.Inc.*

1798 Nafagrel
97901-21-8
$C_{15}H_{16}N_2O_2$
(±)-5,6,7,8-Tetrahydro-6-(imidazol-1-ylmethyl)-2-naphthoic acid.
DP-1904. Selective thromboxane synthetase inhibitor. Antiasthmatic.

1799 Nedocromil
69049-73-6 6524
$C_{19}H_{17}NO_7$
9-Ethyl-6,9-dihydro-4,6-dioxo-10-propyl-4H-pyrano[3,2-g]quinoline-2,8-dicarboxylic acid.
FPL-59002. Antiallergic. Prophylactic antiasthmatic. mp = 298-300° (dec). *Fisons plc.*

1800 Nedocromil Calcium
101626-68-0 6524
$C_{19}H_{15}CaNO_7$
Calcium 9-ethyl-6,9-dihydro-4,6-dioxo-10-propyl-4H-pyrano[3,2-g]quinoline-2,8-dicarboxylate.
FPL-59002KC. Antiallergic; antiasthmatic. *Fisons plc.*

1801 Nedocromil Sodium
69049-74-7 6524
$C_{19}H_{15}NNa_2O_7$
Sodium 9-ethyl-6,9-dihydro-4,6-dioxo-10-propyl-4H-pyrano[3,2-g]quinoline-2,8-dicarboxylate.
FPL-59002KP; Rapitil; Tilade; Tilarin. Antiallergic. Prophylactic antiasthmatic. *Fisons plc.*

1802 Osanetant
160492-56-8
SR-142801. A tachykinin NK3 receptor antagonist. Reduces histamine-induced airway hyperresponsiveness.

1803 Oxarbazole
35578-20-2
$C_{21}H_{19}NO_4$
9-Benzoyl-1,2,3,4-tetrahydro-6-methoxycarbazole-3-carboxylic acid.
Win-34284. An antiasthmatic. *Sterling Res. Labs.*

1804 Oxatomide
60607-34-3 7058 262-320-9
$C_{27}H_{30}N_4O$
1-[3-[4-(Diphenylmethyl)-1-piperazinyl]propyl]-2-benzimidazolinone.
R-35443; Celtect; Cobiona; Dasten; Tinset. Antihistaminic; antiallergic; antiasthmatic. mp = 153.6°; LD_{50} (rat orl) > 2560 mg/kg, (rat iv) = 30 mg/kg, (gpg orl) = 320 mg/kg, (gpg iv) = 27 mg/kg, (mus orl) >2560 mg/kg, (mus iv) = 27 mg/kg. *Abbott Laboratories Inc.*

1805 Piclamilast
144035-83-6
$C_{18}H_{18}Cl_2N_2O_3$
3-(Cyclopentyloxy)-N-(3,5-dichloro-4-pyridyl)-p-anisamide.
RP-73401. A selective phosphodiesterase IV inhibitor. Used as an antiasthmatic. *Rhône-Poulenc Rorer Pharm. Inc.*

1806 Pipoxizine
55837-21-3
$C_{24}H_{31}NO_3$
2-[2-[2-[4-(Diphenylmethylene)piperidino]-ethoxy]ethoxy]ethanol.
Antiasthmatic.

1807 Piriprost
79672-88-1
$C_{26}H_{35}NO_4$
(4R,5R)-1,4,5,6-Tetrahydro-5-hydroxy-4-[(E)(3S)-3-hydroxy-1-octenyl]-1-phenylcyclopenta[b]pyrrole-2-valeric acid.
U-60257. An antiasthmatic. *Pharmacia.*

1808 Piriprost Potassium
88851-62-1
$C_{26}H_{34}KNO_4$
Potassium (4R,5R)-1,4,5,6-tetrahydro-5-hydroxy-4-[(E)(3S)-3-hydroxy-1-octenyl]-1-phenyl-cyclopenta[b]pyrrole-2-valerate.
U-60257B. An antiasthmatic. *Pharmacia.*

1809 Pirolate
55149-05-8
$C_{16}H_{15}N_3O_5$
Ethyl 1,4-dihydro-7,8-dimethoxy-4-oxopyrimido-[4,5-b]quinoline-2-carboxylate.
CP-32287. An antiasthmatic. *Pfizer Inc.*

1810 Pranlukast
103177-37-3 7889
$C_{27}H_{23}N_5O_4$
N-[4-Oxo-2-(1H-tetrazol-5-yl)-4H-1-benzopyran-8-yl]-p-(4-phenylbutoxy)benzamide.
ONO-1078; ONO-RS-411; Onon; SB-205312. A leukotriene antagonist used as an antiasthmatic. [hemihydrate]: mp = 244-245°; LD_{50} (mmus iv) > 1000 mg/kg. *Ono Pharm.*

1811 Quazolast
86048-40-0
$C_{12}H_7ClN_2O_3$
Methyl 5-chlorooxazolo[4,5-h]quionline-2-carboxylate.
RHC-3988. A mediator release inhibitor used as an antiasthmatic. *Rhône-Poulenc Rorer Pharm. Inc.*

1812 Quiflapon
136668-42-3
$C_{34}H_{35}ClN_2O_3S$
3-(tert-Butylthio)-1-(p-chlorobenzyl)-α,α-dimethyl-5-(2-quinolylmethoxy)indole-2-propionic acid.
An antiasthmatic also used to suppress inflammatory bowel disease. *Merck & Co.Inc.*

1813 Quiflapon Sodium
147030-01-1
$C_{34}H_{34}ClN_2NaO_3S$
Sodium 3-(tert-butylthio)-1-(p-chlorobenzyl)-α,α-dimethyl-5-(2-quinolylmethoxy)indole-2-propionate.
MK-591. An antiasthmatic also used to suppress inflammatory bowel disease. *Merck & Co.Inc.*

1814 Ramatroban
116649-85-5
$C_{21}H_{21}FN_2O_4S$
(3R)-3-(4-Fluorophenylsulfonamido)-1,2,3,4-tetrahydro-9-carbazo lepropanoic acid.
BAY-u3405. A thromboxane A2 receptor antagonist. Antithrombotic antiasthmatic. *Bayer AG.*

1815 Repirinast
73080-51-0 8305
$C_{20}H_{21}NO_5$
Isopentyl-5,6-dihydro-7,8-dimethyl-4,5-dioxo-4H-pyrano[3.2-c]quinoline-2-carboxylate.
MY-5116; Romet. Antiallergic; antiasthmatic. mp = 236-241°; LD_{50} (rat orl) > 5000 mg/kg, (rat sc) > 5000 mg/kg, (mus orl) > 5000 mg/kg, (mus sc) > 5000 mg/kg. *Mitsubishi Kasei.*

1816 Seratodrast
112665-43-7 8603
$C_{22}H_{26}O_4$
(±)-2,4,5-Trimethyl-3,6-dioxo-ς-phenyl-1,4-cyclophexadiene-1-heptanoic acid.
AA-2414; A-73001; Abbott-73001; ABT-001. A thromboxane A_2-receptor antagonist. Used as an antiasthmatic. mp = 128-129°. *Abbott Labs Inc.*

1817 Suplatast Tosylate
94055-76-2 9178
$C_{23}H_{33}NO_7S_2$
(±)-[2-[[p-(3-Ethoxy-2-hydroxypropoxy)phenyl]-carbamoyl]ethyl] dimethylsulfonium p-toluenesulfonate.
IPD-1151T. Antiasthmatic; antiallergic. Inhibits interleukin-4 gene expression. mp = 70-73°; LD_{50} (mmus iv) = 81 mg/kg, (fmus iv) = 96 mg/kg, (mrat iv) = 96 mg/kg, (frat iv) = 93 mg/kg, (mus orl) > 12500 mg/kg, (rat orl) > 10000 mg/kg. *Taiho.*

1818 Tetrazolast Meglumine
133008-33-0
$C_{17}H_{23}N_9O_5.H_2O$
4-(1H-Tetrazol-5-yl)tetrazolo[1,5-a]quinoline compound with 1-deoxy-1-(methylamino)-D-glucitol (1:1).
MDL-26024G0. Antiallergic; antiasthmatic. *Marion Merrell Dow Inc.; Merrell Dow Pharm. Inc.*

1819 Tiaramide
32527-55-2 9564 251-083-7
$C_{15}H_{18}ClN_3O_3S$
4-[(5-Chloro-2-oxo-3-benzothiazolyl)acetyl]-1-piperazineethanol.
tialamide. Antiasthmatic; antiallergic; anti-inflammatory Arylacetic acid derivative. pKa = 6.2. *Fujisawa Pharm. USA Inc.; Rhône-Poulenc Rorer Pharm. Inc.*

1820 Tiaramide Hydrochloride
35941-71-0 9564 252-802-7
$C_{15}H_{18}Cl_2N_3O_3S$
4-[(5-Chloro-2-oxo-3-benzothiazolyl)acetyl]-1-piperazineethanol monohydrochloride.
NTA-194; FK-1160; Solantal. Antiasthmatic; antiallergic; anti-inflammatory Arylacetic acid derivative. mp = 159-161°; pH (10% aqueous solution) = 3.4-3.7; soluble in H_2O; slightly soluble in organic solvents; LD_{50} (mmus iv) = 178 mg/kg, (mmus ip) = 298 mg/kg, (mmus orl) = 564 mg/kg, (mrat iv) = 203 mg/kg, (mrat ip) = 540 mg/kg, (mrat orl) = 3600 . *Fujisawa Pharm. USA Inc.; Rhône-Poulenc Rorer Pharm. Inc.*

1821 Tibenelast Sodium
105102-18-9
$C_{13}H_{13}NaO_4S$
Sodium 5,6-diethoxybenzo[b]thiophene-2-carboxylate.
LY-186655. A bronchodilator; antiasthmatic. Phosphodiesterase inhibitor. *Eli Lilly & Co.*

1822 Tranilast
53902-12-8 9705
$C_{18}H_{17}NO_5$
N-(3,4-Dimethoxycinnamoyl)anthranilic acid.
MK-341; N-5'; Rizaben. Antiallergic; antiasthmatic. mp = 211-213°; LD_{50} (mmus orl) = 780 mg/kg, (mmus ip) = 410 mg/kg, (mmus sc) = 2630 mg/kg, (fmus orl) = 680 mg/kg, (fmus ip) = 385 mg/kg, (fmus sc) = 2820 mg/kg, (mrat orl) = 1600 mg/kg, (mrat ip) = 405 mg/kg, (mrat sc) = 3630 mg/kg, (frat orl) = 11. *C.M. Ind.; Kissei; Merck & Co.Inc.*

1823 Traxanox
58712-69-9 9711
$C_{13}H_6ClN_5O_2$
9-Chloro-7-(1H-tetrazol-5-yl)-5H-[1]benzopyrano-[2,3-b]pyridine-5-one.
Y-12141[as sodium salt pentahydrate]; Clearnal. Antiallergic; antiasthmatic. mp > 300°. *Yoshitomi.*

1824 Triamcinolone Acetonide
76-25-5 9728 200-948-7
$C_{24}H_{31}FO_6$
(11β,16α)-9-Fluoro-11,21-dihydroxy-16,17-[(1-methyl-ethylidene)bis(oxy)]-pregna-1,4-diene-3,20-dione.
9α-fluoro-16α-hydroxyprednisolone; Adcortyl; Azmacort; Delphicort; Extracort; Ftorocort; Kenacort-A; Kenalog; Ledercort Cream; Nasacort; Respicort; Rineton; Solodelf; TAC-3; TAC-40; Tramacin; Triacet; Triam; Triamonide 40; Tricinolon; Trymex; Vetalog; Volon A; Volonimat; component of: Mycolog II, Myco-Triacet II, Mytrex, Panolog. Glucocorticoid; antiasthmatic (inhalant); antiallergic (nasal). mp = 292-294°; $[\alpha]_D^{23}$ = +109° (c = 0.75 $CHCl_3$); λ_m = 238nm (ε 14600 EtOH); sparingly soluble in MeOH, Me_2OH, EtOAc. 3261; *Am. Cyanamid; Bristol-Myers Squibb Co; Forest Pharm. Inc.; Herbert; Johnson & Johnson Med. Inc.; Lemmon Co.; Olin Mathieson; Rhône-Poulenc Rorer Pharm. Inc.; Savage Labs.*

1825 Verofylline
66172-75-6
$C_{12}H_{18}N_4O_2$
(±)-1,8-Dimethyl-3-(2-methylbutyl)xanthine.
CK-0383. Bronchodilator; antiasthmatic. *Berlex Labs Inc.*

1826 Zafirlukast
107753-78-6 10241
$C_{31}H_{33}N_3O_6S$
Cyclopentyl-3-[2-methoxy-4-[(o-tolylsulfonyl)-carbamoyl]benzyl]-1-methylindole-5-carbamate.
Accolate; ICI-204219. A leukotriene D_4 antagonist. Used as an antiasthmatic. mp = 138-140°. *ICI ; Zeneca Pharm.*

1827 Zardaverine
101975-10-4
$C_{12}H_{10}F_2N_2O_3$
6-[4-(Difluoromethoxy-3-methoxyphenyl]-3(2H)-pyridazinone.
Dual phosphodiesterase 3/4 inhibitor. Inhibits the proliferation of human peripheral blood mononuclear cells. Bronchodilator; antiasthmatic.

1828 Zileuton
111406-87-2 10253
$C_{11}H_{12}N_2O_2S$
(±)-1-(1-Benzo[b]thien-2-ylethyl)-1-hydroxyurea.
A-64077; Abbott-64077; Leutrol; Zyflo. Antiasthmatic; anti-inflammatory. Inhibitor of 5-lipoxygenase, the initial enzyme in the biosynthesis of leukotrienes from arachidonic acid. mp = 157-158°. *Abbott Labs Inc.*

2,4-Diaminopyrimidine Antibiotics

1829 Aditoprim
56066-63-8
$C_{15}H_{21}N_5O_2$
2,4-Diamino-5-[4-(dimethylamino)-3,5-dimethoxybenzyl]pyrimidine.
Antibacterial.

1830 Baquiloprim
102280-35-3
$C_{17}H_{20}N_6$
5-[(2,4-Diamino-5-pyrimidinyl)methyl]-8-(dimethylamino)-7-methylquinoline.
Diaminopyrimidine antibacterial.

1831 Brodimoprim
56518-41-3 1401 260-238-8
$C_{14}H_{16}BrNO_2$
2,4-Diamino-5-(4-bromo-3,5-dimethoxybenzyl)-pyrimidine.
Ro-105970; Hyprim; Unitrim. Dihydrofolate reductase inhibitor. A 2,4-diaminopyrimidine antibiotic. mp = 225-228°. *Hoffmann-LaRoche Inc.*

1832 Ciadox
65884-46-0
$C_{12}H_9N_5O_3$
Cyanoacetic acid (2-quinoxalinylmethylene)hydrazide N^1,N^4-dioxide.
Antibacterial.

1833 Diaveridine
5355-16-8 226-333-3
$C_{13}H_{16}N_4O_2$
5-[(3,4-Dimethoxyphenyl)methyl]-2,4-pyrimidinediamine.
BW-49-210; NSC-408735. A 2,4-diaminopyrimidine antibiotic. *Glaxo Wellcome Inc.*

1834 Epiroprim
73090-70-7
$C_{19}H_{23}N_5O_2$
2,4-Diamino-5-(3,5-diethoxy-4-pyrrol-1-ylbenzyl)-pyrimidine.
Antibacterial.

1835 Metioprim
68902-57-8
$C_{14}H_{18}N_4O_2S$
5-[[3,5-Dimethoxy-4-(methylthio)phenyl]methyl]-2,4-pyrimidinediamine.
A 2,4-diaminopyrimidine antibiotic. *Heumann Pharma GmbH.*

1836 Ormetoprim
6981-18-6 230-246-6
$C_{14}H_{18}N_4O_2$
5-[(4,5-Dimethoxy-2-methylphenyl)methyl]-2,4-pyrimidinediamine.
Ro-5-9754; NSC-95072. A 2,4-diaminopyrimidine antibiotic. *Hoffmann-LaRoche Inc.*

1837 Tetroxoprim
53808-87-0 9386 258-789-4
$C_{16}H_{22}N_4O_4$
5-[[3,5-Dimethoxy-4-(2-methoxyethoxy)phenyl]methyl]-2,4-pyrimidinediamine.
HE-781. A 2,4-diaminopyrimidine antibiotic. mp = 153-156°, 160.1°; soluble in H_2O (0.265 g/100 ml at 28°), $CHCl_3$ (6.90 g/100 ml), n-octanol (0.161 g/100 ml); LD_{50} (rat orl) = 1357 mg/kg. *Heumann Pharma GmbH.*

1838 Tetroxoprim Mixture with Sulfadiazine
73173-12-3 9386
co-tetroxazine; Biroxin; Sterinor; Tibirox; Troximin. Antibiotic. *Heumann Pharma GmbH.*

1839 Trimethoprim
738-70-5 9840 212-006-2
$C_{14}H_{18}N_4O_3$
5-[(3,4,5-Trimethoxyphenyl)methyl]-2,4-pyrimidinediamine.
Instalac; Monotrim; Proloprim; Syraprim; Tiempe; Trimanyl; Trimogal; Trimopan; Trimpex; Uretrim; Wellcoprim. A 2,4-diaminopyrimidine antibiotic. mp = 199-203°; soluble in dimethylacetamide (13.86 g/100 ml), benzyl alcohol (7.29 g/100 ml), propylene glycol (2.57 g/100 ml), $CHCl_3$ (1.82 g/100 ml), MeOH (1.21 g/100 ml), H_2O (0.04 g/100 ml), Et_2O (0.003 g/100 ml), C_6H_6 (0.002 g/100 ml); LD_{50} (mus orl) = 7000 mg/kg. *Glaxo Wellcome Inc.; Hoffmann-LaRoche Inc.*

Aminoglycoside Antibiotics

1840 Amikacin
37517-28-5 425 253-538-5
$C_{22}H_{43}N_5O_{13}$
(S)-O-3-Amino-3-deoxy-α-D-glucopyranosyl-(1→6)-O-[6-amino-6-deoxy-α-D-glucopyranosyl-(1→4)]-N^1-(4-amino-2-hydroxy-1-oxobutyl)-2-deoxy-D-streptamine.
Lukadin. Aminoglycoside antibiotic. Antibacterial. Amikacin is used in combination with other drugs to treat *mycobacterium avium* complex (MAC). Recent Public Health Service recommendations suggest either clarithromycin or azithromycin as the first line treatment for MAC, along with amikacin. [sesquihydrate]: mp = 203-204°; $[\alpha]_D^{23}$ = 99° (c = 1 H_2O); LD_{50} (mus iv pH 6.6) = 7.4 mg/kg, (mus iv pH 7.4) = 560 mg/kg. *Bristol-Myers Squibb Pharm. R&D.*

1841 Amikacin Sulfate
39831-55-5 425 254-648-6
$C_{22}H_{43}N_5O_{13}.2H_2SO_4$
(S)-O-3-Amino-3-deoxy-α-D-glucopyranosyl-(1→6)-O-[6-amino-6-deoxy-α-D-glucopyranosyl-(1→4)]-N[1]-(4-amino-2-hydroxy-1-oxobutyl)-2-deoxy-D-streptamine sulfate.
Amikin; Amiklin; BB-K8; Biklin; Fabianol; Kaminax; Mikavir; Novamin; Pierami; Amiglyde-V; Amikavet. Aminoglycoside antibiotic. Antibacterial. Amikacin is used in combination with other drugs to treat *mycobacterium avium* complex (MAC). Recent Public Health Service recommendations suggest either clarithromycin or azithromycin as the first line treatment for MAC, along with amikacin. Amorphous solid; dec = 220-230°; $[\alpha]_D^{22}$ = + 74.75° (H_2O). *Bristol-Myers Squibb Pharm. R&D.*

1842 Apramycin
37321-09-8 792 253-460-1
$C_{21}H_{41}N_5O_{11}$
O-4-Amino-4-deoxy-α-D-glucopyranosyl-(1→8)-O-(8R)-2-amino-2,3,7-trideoxy-7-(methylamino)-D-glycero-α-allo-octodialdo-1,5:8,4-dipyranosyl-(1→4)-2-deoxy-D-streptamine.
nebramycin factor 2; EL-857; EL-857/820; 47657; Ambylan; Apralan. Aminoglycoside antibiotic. [monohydrate]: mp = 245-247°; pKa (H_2O): 8.5, 7.8, 6.2, 5.4; very soluble in H_2O, slightly soluble in lower alcohols. *Eli Lilly & Co.*

1843 Arbekacin
51025-85-5 809
$C_{22}H_{44}N_6O_{10}$
(S)-O-3-Amino-3-deoxy-α-D-glucopyranosyl-(1→6)-O-[2,6-diamino-2,3,4,6-tetradeoxy-α-D-erythro-hexopyranosyl-(1→4)]-N[1]-(4-amino-2-hydroxy-1-oxobutyl)-2-deoxy-D-streptamine.
AHB-DKB; HABA-DKB; HBK; 1665-RB; Habekacin. Aminoglycoside antibiotic. [dicarbonate ($C_{24}H_{48}N_6O_{16}$)]: mp = 178° (dec); $[\alpha]_D^{24}$ = 86.8° (c = 0.77 H_2O); LD_{50} (mus iv) > 150 mg/kg. *Microbiochem. Res. Found.*

1844 Astromicin Sulfate
72275-67-3
$C_{17}H_{35}N_5O_6 \pm H_2SO_4$
4-Amino-1-[(aminoacetyl)methylamino]-1,4-dideoxy-3-O-(2,6-diamino-2,3,4,6,7-pentadeoxy-β-L-lyxo-heptopyranosyl)-6-O-methyl-L-chiro-inositol sulfate.
fortimicin A sulfate. Aminoglycoside antibiotic. *Abbott Labs Inc.; Kyowa Hakko Kogyo Co. Ltd.*

1845 Bambermycins
11015-37-5 979 234-246-7
Bambermycins.
Moenomycin; flavophospholipol; Flavomycin. Aminoglycoside antibiotic. Antibiotic complex containing moenomycins A, B_1, B_2 and C. Dec ≅200°; λ_m 258 nm ($E^{1\%}_{1cm}$ 60 H_2O pH 7); soluble in H_2O, MeOH, DMF; less soluble in EtOH, PrOH; slightly soluble in Et_2O, EtOAc; insoluble in C_6H_6, $CHCl_3$; LD_{50} (mus orl, sc, ip) > 2000mg/kg, (mus iv) = 1400 mg/kg. *Hoechst AG.*

1846 Betamicin
36889-15-3
$C_{19}H_{38}N_4O_{10}$
O-6-Amino-6-deoxy-α-D-glucopyranosyl-(1→4)-O-[3-deoxy-4-C-methyl-3-(methylamino)-β-L-arabinopyranosyl-(1→6)-2-deoxy-D-streptamine.
Aminoglycoside antibiotic. *Schering-Plough HealthCare Products.*

1847 Betamicin Sulfate
43169-50-2
$C_{19}H_{38}N_4O_{10}.xH_2SO_4$
O-6-Amino-6-deoxy-α-D-glucopyranosyl-(1→4)-O-[3-deoxy-4-C-methyl-3-(methylamino)-β-L-arabinopyranosyl-(1→6)-2-deoxy-D-streptamine sulfate.
Sch-14342. Aminoglycoside antibiotic. *Schering-Plough HealthCare Products.*

1848 Butikacin
59733-86-7
$C_{22}H_{45}N_5O_{12}$
O-3-Amino-3-deoxy-α-D-glucopyranosyl-(1→6)-O-[6-amino-6-deoxy-α-D-glucopyranosyl-(1→4)-N[1]-[(S)-4-amino-2-hydroxybutyl]-2-deoxy-D-streptamine.
UK-18892. Aminoglycoside antibiotic. *Pfizer Inc.*

1849 Butirosin
12772-35-9 1559
$C_{21}H_{41}N_5O_{12}$
O-2,6-Diamino-2,6-dideoxy-α-D-glucopyranosyl-(1→4)-O-[β-D-xylofuranosyl-(1→5)]-N[1]-(4-amino-2-hydroxy-1-oxobutyl)-2-deoxy-D-streptamine.
Ambutyrosin. Aminoglycoside antibiotic. Antibiotic complex obtained from *Bacillus circulans* and containing two major components, Butirosin A (80-85%) and Butirosin B (15-20%). *Parke-Davis.*

1850 Butirosin A
34291-02-6 1559
$C_{21}H_{41}N_5O_{12}$
(S)-O-2,6-Diamino-2,6-dideoxy-α-D-glucopyranosyl-(1→4)-O-[β-D-xylofuranosyl-(1→5)]-N[1]-(4-amino-2-hydroxy-1-oxobutyl)-2-deoxy-D-streptamine.
Aminoglycoside antibiotic. Melts over a range from 149°; $[\alpha]_D^{25}$ = 26° (c = 1.46 H_2O). *Parke-Davis.*

1851 Butirosin B
34291-03-7 1559
$C_{21}H_{41}N_5O_{12}$
1-N-[(S)-4-Amino-2-(hydroxybutyryl))]-ribostamycin.
Aminoglycoside antibiotic. Melts over a range from 146°; $[\alpha]_D^{25}$ = 33° (c = 1.50 H_2O). *Parke-Davis.*

1852 Butirosin Sulfate Dihydrate
51022-98-1 1559
$C_{21}H_{45}N_5O_{20}.2H_2O$
O-2,6-Diamino-2,6-dideoxy-α-D-glucopyranosyl-(1→4)-O-[β-D-xylofuranosyl-(1→5)]-N[1]-(4-amino-2-hydroxy-1-oxobutyl)-2-deoxy-D-streptamine sulfate dihydrate.
CI-642. Aminoglycoside antibiotic. No sharp mp, dec ≅225°; $[\alpha]_D^{25}$ = +29° (c = 2 in H_2O); pKa.g (H_2O) = 5.5, 7.2,

8.5, 9.4; very soluble in H_2O, moderately soluble in MeOH, slightly soluble in EtOH; LD_{50} (mus iv) = 450-500 mg/kg. *Parke-Davis.*

1853 Dibekacin
34493-98-6 3052 252-064-6
$C_{18}H_{37}N_5O_8$
O-3-Amino-3-deoxy-α-D-glucopyranosyl-(1→6)-O-[2,6-diamino-2,3,4,6-tetradeoxy-α-D-erythro-hexopyranosyl-(1→4)]-2-deoxy-D-streptamine.
DKB; 3', 4'-dideoxykanamycin B; debecacin. Aminoglycoside antibiotic. Semisynthetic analog of kanamycin. Effective against kanamycin-resistant bacteria. $[\alpha]_D^{20}$ = +132° (c= 0.65); LD_{50} (mus iv) = 61.0-68.0 mg/kg, (mus ip) = 373.0-380.0 mg/kg. *Allen & Hanbury; Microbiochem. Res. Found.*

1854 Dibekacin Sulfate
58580-55-5 3052 261-341-0
$C_{18}H_{37}N_5O_{12}S$
O-3-Amino-3-deoxy-α-D-glucopyranosyl-(1→6)-O-[2,6-diamino-2,3,4,6-tetradeoxy-α-D-erythro-hexopyranosyl-(1→4)]-2-deoxy-D-streptamine sulfate.
Débékacyl; Icacine; Kappabi; Orbicin; Panamicin; Panimycin; Tokocin. Aminoglycoside antibiotic. Soluble in H_2O, insoluble in organic solvents. *Allen & Hanbury; Microbiochem. Res. Found.*

1855 Dihydrostreptomycin
128-46-1 3222 204-888-2
$C_{21}H_{41}N_7O_{12}$
O-2-Deoxy-2-(methylamino)-α-L-glucopyranosyl-(1→2)-O-5-deoxy-3-C-(hydroxymethyl)-α-L-lyxofuranosyl-(1→4)-N,N'-bis(aminoiminomethyl)-D-streptamine.
DHSM; DST; Abiocine; Vibriomycin. Aminoglycoside antibiotic. Tuberculostatic. mp > 300°. *Am. Cyanamid; Bristol-Myers Squibb Co; Heyden Chem.; Merck & Co.Inc.; Olin Res. Ctr.; Pfizer Inc.; Schenley; Takeda Chem. Ind. Ltd.*

1856 Dihydrostreptomycin Pantothenate
3563-84-6 3222 222-637-5
$C_{30}H_{58}N_8O_{17}$
O-2-Deoxy-2-(methylamino)-α-L-glucopyranosyl-(1→2)-O-5-deoxy-3-C-(hydroxymethyl)-α-L-lyxofuranosyl-(1→4)-N,N'-bis(aminoiminomethyl)-D-streptamine pantothenate.
Didrothenat; Pantostrep. Aminoglycoside antibiotic. Tuberculostatic. *Am. Cyanamid; Bristol-Myers Squibb Pharm. R&D; Heyden Chem.; Merck & Co.Inc.; Olin Res. Ctr.; Pfizer Inc.; Schenley; Takeda Chem. Ind. Ltd.*

1857 Dihydrostreptomycin Sesquisulfate
5490-27-7 3222 226-823-7
$C_{41}H_{88}N_{14}O_{36}$
O-2-Deoxy-2-(methylamino)-α-L-glucopyranosyl-(1→2)-O-5-deoxy-3-C-(hydroxymethyl)-α-L-lyxofuranosyl-(1→4)-N,N'-bis(aminoiminomethyl)-D-streptamine sulfate (2:3) (salt).
Didromycine; Double-mycin; Sol-Mycin; Streptomagna. Aminoglycoside antibiotic. Tuberculostatic. Dec 255-265°, 250°; $[\alpha]_D^{25}$ = -88.5° (c = 1); very soluble in H_2O; soluble in 50% MeOH/H_2O (crystals: 0.08 g/100 ml; powder: 10 g/100 ml); at 28° soluble in H_2O (> 2 g/100 ml), MeOH (0.035 g/100 ml), EtOH (0.010 g/100 ml). *Am. Cyanamid; Bristol-Myers Squibb Pharm. R&D; Heyden Chem.; Merck & Co.Inc.; Olin Res. Ctr.; Pfizer Inc.; Schenley; Takeda Chem. Ind. Ltd.*

1858 Dihydrostreptomycin Trihydrochloride
6533-54-6 3222
$C_{21}H_{44}Cl_3N_7O_{12}$
O-2-Deoxy-2-(methylamino)-α-L-glucopyranosyl-(1→2)-O-5-deoxy-3-C-(hydroxymethyl)-α-L-lyxofuranosyl-(1→4)-N,N'-bis(aminoiminomethyl)-D-streptamine trihydrochloride.
Aminoglycoside antibiotic. Dec 190-195°; soluble in MeOH (> 100 g/100 ml). *Am. Cyanamid; Bristol-Myers Squibb Pharm. R&D; Heyden Chem.; Merck & Co.Inc.; Olin Res. Ctr.; Pfizer Inc.; Schenley; Takeda Chem. Ind. Ltd.*

1859 Fortimicin A
55779-06-1 4276
$C_{17}H_{35}N_5O_6$
4-Amino-1-[(aminoacetyl)methylamino]-1,4-dideoxy-3-O-(2,6-diamino-2,3,4,6,7-pentadeoxy-β-L-lyxo-heptopyranosyl)-6-O-methyl-L-chiro-inositol.
Astromicin; Abbott 44747. Aminoglycoside antibiotic. Aminoglycoside antibiotic complex produced by *Micromonospora olivoasterospora.* mp > 200° (dec); $[\alpha]_D^{25}$ = 87.5° (c = 0.1 H_2O); soluble in H_2O, MeOH, EtOH; insoluble in organic solvents; [sulfate salt]: LD_{50} (mus iv) = 380 mg/kg, (mus sc) = 400 mg/kg. *Abbott Labs Inc.; Kyowa Hakko Kogyo Co. Ltd.*

1860 Fortimicin B
54783-95-8 4276
$C_{15}H_{32}N_4O_5$
4-Amino-1,4-dideoxy-3-O-(2,6-diamino-2,3,4,6,7-pentadeoxy-β-L-lyxo-heptopyranosyl)-6-O-methyl-1-(methylamino)-L-chiro-inositol.
Aminoglycoside antibiotic. mp = 101-03°; $[\alpha]_D^{25}$ = 22.2° (c = 0.1 H_2O); soluble in H_2O, MeOH, EtOH; insoluble in organic solvents. *Abbott Labs Inc.; Kyowa Hakko Kogyo Co. Ltd.*

1861 Gentamicin
1403-66-3 4398 215-765-8
gentamycin. Aminoglycoside antibiotic. Antibiotic complex produced by *Micromonospora purpurea.* Contains Gentamicn C_1, C_2 and C_{1a} and also Gentamicin A. *Schering-Plough HealthCare Products.*

1862 Gentamicin A
13291-74-2 4398
$C_{18}H_{36}N_4O_{10}$
O-2-Amino-2-deoxy-α-D-glucopyranosyl-(1→4)-O-[3-deoxy-3-(methylamino)-α-D-xylopyranosyl-(1→6)-2-deoxy-D-streptamine.
Aminoglycoside antibiotic. *Schering-Plough HealthCare Products.*

1863 Gentamicin C Complex Sulfate
1405-41-0 4398 215-778-9
Alcomicin; Bristagen; Cidomycin; Duragentam; Garamycin; Garasol; Genoptic; Gentacin; Gentak; Gentalline; Gentalyn; Gentibioptal; Genticin; Gentocin; Gentogram; Gent-Ophtal; Gentrasul; Lugacin; Nichogencin; Ophtagram; Pangram; Refobacin; Septopal; Sulmycin; U-gencin. Aminoglycoside antibiotic. mp =

218-237°; $[\alpha]_D^{25}$ = 102°; soluble in ethylene glycol, formamide; LD_{50} (mus ip) = 430 mg/kg, (mus sc) = 485 mg/kg, (mus orl) > 9050 mg/kg. *Schering-Plough HealthCare Products.*

1864 Gentamicin C_1
25876-10-2 4398
$C_{21}H_{43}N_5O_7$
Aminoglycoside antibiotic. mp = 94-100°; $[\alpha]_D^{25}$ = 158°. *Schering-Plough HealthCare Products.*

1865 Gentamicin C_{1a}
26098-04-4 4398
$C_{20}H_{41}N_5O_7$
Aminoglycoside antibiotic. mp = 107-124°; $[\alpha]_D^{25}$ = 160°. *Schering-Plough HealthCare Products.*

1866 Gentamicin C_2
25876-11-3 4398
$C_{19}H_{39}N_5O_7$
O-3-Deoxy-4-C-methyl-3-(methylamino)-β-L-arabinopyranosyl-(1→6)-O-[2,6-diamino-2,3,4,6-tetradeoxy-α-D-erythro-hexopyranosyl-(1→4)-2-deoxy-D-streptamine.
Antibiotic. *Schering-Plough HealthCare Products.*

1867 Heliomycin
11029-70-2
$C_{23}H_{18}O_6$
resistomycin; Kanamycin A sulfate. Antibacterial.

1868 Isepamicin
58152-03-7 5121 261-143-4
$C_{22}H_{43}N_5O_{12}$
(S)-O-6-Amino-6-deoxy-α-D-glucopyranosyl-(1→4)-O-[3-deoxy-4-C-methyl-3-(mehtylamino)-β-L-arabino-pyranosyl-(1→6)]-N[1]-(3-amino-2-hydroxy-1-oxo-propyl)-2-deoxy-D-streptamine.
HAPA-B; Sch-21420. Aminoglycoside antibiotic. *Schering-Plough HealthCare Products.*

1869 Isepamicin Sulfate
67814-76-0 5121
$C_{22}H_{45}N_5O_{16}S$
(S)-O-6-Amino-6-deoxy-α-D-glucopyranosyl-(1→4)-O-[3-deoxy-4-C-methyl-3-(mehtylamino)-β-L-arabinopyranosyl-(1→6)]-N[1]-(3-amino-2-hydroxy-1-oxo-propyl)-2-deoxy-D-streptamine sulfate.
Exacin; Isepacin. Aminoglycoside antibiotic. $[\alpha]_D^{26}$ = 110.9° (c = 1 H_2O); LD_{50} (mmus iv) = 234 mg/kg, (mmus sc) = 3312 mg/kg, (mmus orl) > 5000 mg/kg, (fmus iv) = 236 mg/kg, (fmus sc) = 3320 mg/kg, (fmus orl) > 5000 mg/kg, (mrat iv) = 489 mg/kg, (mrat sc) = 3451, (mrat orl) > 5000 mg/kg, (frat iv) = 476 mg/kg, (frat sc) = 3392 mg/kg, (frat orl) > 5000 mg/kg. *Schering-Plough HealthCare Products.*

1870 Kanamycin
8063-07-8 5293 232-512-7
Antibiotic complex produced by *Streptomyces kanamyceticus*. Consists of three main components, kanamycin A, B and C. *Apothecon; Bristol-Myers Squibb Pharm. R&D; Merck & Co.Inc.; SmithKline Beecham Pharm.*

1871 Kanamycin A
59-01-8 5293 200-411-7
$C_{18}H_{36}N_4O_{11}$
O-3-Amino-3-deoxy-α-D-glucopyranosyl-(1→6)-O-[6-amino-6-deoxy-α-D-glucopyranosyl-(1→4)]-2-deoxy-D-streptamine.
Antibiotic complex produced by *Streptomyces kanamyceticus*. $[\alpha]_D^{24}$ = +146° (0.1 N H_2SO_4); LD_{50} (mus iv) = 583 mg/kg. *Apothecon; Bristol-Myers Squibb Pharm. R&D; Merck & Co.Inc.; SmithKline Beecham Pharm.*

1872 Kanamycin A Sulfate
25389-94-0 5293 246-933-9
$C_{18}H_{38}N_4O_{15}S$
O-3-Amino-3-deoxy-α-D-glucopyranosyl-(1→6)-O-[6-amino-6-deoxy-α-D-glucopyranosyl-(1→4)]-2-deoxy-D-streptamine sulfate.
Cantrex; Cristalomicina; Enterokanacin; Kamycin; Kasmynex; Kanabristol; Kanacedin; Kanamytrex; Kanasig; Kanatrol; Kanicin; Kannasyn; Kantrex; Kantrox; Klebcil; Otokalixin; Resistomycin; Ophtalmokalixan; Kantrexil; Kano; Kanescin; Kanaqua; component of: Amforol. Aminoglycoside antibiotic. From *Streptomyces kanamyceticus*. Dec > 250°; freely soluble in H_2O; insoluble in most organic solvents; LD_{50} (mus orl) = 20700 mg/kg, (mus ip) = 1450 mg/kg. *Apothecon; Bristol-Myers Squibb Pharm. R&D; Merck & Co.Inc.; SmithKline Beecham Pharm.*

1873 Kanamycin B
4696-76-8 5293 225-170-5
$C_{18}H_{37}N_5O_{10}$
Aminodeoxykanamycin.
bekanamycin; NK-1006. Aminoglycoside antibiotic. From *Streptomyces kanamyceticus*. mp = 178-182° (dec); $[\alpha]_D^{18}$ = +130° (c = 0.5 H_2O); $[\alpha]_D^{21}$ = +114° (c = 0.98 in H_2O); soluble in H_2O, formamide; slightly soluble in $CHCl_3$, iPrOH; insoluble in common alcohols, nonpolar solvents; LD_{50} (mus iv) = 136 mg/kg. *Apothecon; Bristol-Myers Squibb Pharm. R&D; Merck & Co.Inc.; SmithKline Beecham Pharm.*

1874 Kanamycin B Sulfate
5293
$C_{18}H_{37}N_5O_{10} \cdot H_2SO_4$
Coltrecin; Kanendomycin; Kanendos. Aminoglycoside antibiotic. From *Streptomyces kanamyceticus*.

1875 Kanamycin C
2280-32-2 5293
$C_{18}H_{36}N_4O_{11}$
Aminoglycoside antibiotic. From *Streptomyces kanamyceticus*. Dec > 270°; $[\alpha]_D^{20}$ = +126° (H_2O); soluble in H_2O; slightly soluble in formamide; nearly insoluble in alcohols and nonpolar solvents.

1876 Micronomicin
52093-21-7 6269
$C_{20}H_{41}N_5O_7$
O-2-Amino-2,3,4,6-tetradeoxy-6-(methylamino)-α-D-erythro-hexopyranosyl-(1→4)-O-[3-deoxy-4-C-methyl-3-(methylamino)-β-L-arabinopyranosyl-(1→6)]-2-deoxy-D-streptamine.
gentamicin C_{2b}; antibiotic KW-1062; KW-1062; XK-62-2;

[sulfate ($C_{40}H_{92}N_{10}O_{34}S$)]: 6'-N-methylbentamicin C_{1a} hemipentasulfate; Sagamicin; Santemycin. Aminoglycoside antibiotic. mp = 260° (dec); $[\alpha]_D^{20}$ = 116° (c = 1 H_2O); soluble in H_2O, MeOH; insoluble in $CHCl_3$, EtOAc, C_6H_6, petroleum ether; LD_{50} (mus iv) = 93 mg/kg. *Schering-Plough HealthCare Products.*

1877 Neomycin
1404-04-2 6542 215-766-3
Mycifradin; Fradiomycin; Neomin; Neolate; Neomas; Pimavecort; Vonamycin Powder V. Aminoglycoside antibiotic. Antibiotic produced by *Streptomyces fradiae.* Consists of Neomycins A, B and C. Has a hypolipidemic effect when administered orally, probably through the formation of insoluble complexes with bile acids in the intestine. Insoluble in organic solvents; slightly soluble in H_2O (< 25 g/100 ml). *Merck & Co.Inc.; Pharmacia & Upjohn.*

1878 Neomycin A
3947-65-7 6517
$C_{12}H_{26}N_4O_6$
2-Deoxy-4-O-(2,6-diamino-2,6-dideoxy-α-D-glucopyranosyl)-D-streptamine.
Neamine. Aminoglycoside antibiotic. Dec 225-226°; $[\alpha]_D^{25}$ = 112.8° (c = 1); [tetrahydrochloride ($C_{12}H_{30}Cl_4N_4O_6$)]: dec 250-260°; $[\alpha]_D^{25}$ = 83° (c = 1); [N-acetyl derivative ($C_{20}H_{38}N_4O_{10}$)]: mp = 334-336°; $[\alpha]_D^{25}$ = 87° (c = 1). *Merck & Co.Inc.; Pharmacia & Upjohn.*

1879 Neomycin B
119-04-0 6542 204-292-2
$C_{23}H_{46}N_6O_{13}$
Neomycin B.
antibiotique EF 185; Enterfram; Framygen; Framycetin; Soframycin; Actilin. Aminoglycoside antibiotic. *Merck & Co.Inc.; Pharmacia & Upjohn.*

1880 Neomycin B Hydrochloride
25389-99-5 6542
$C_{23}H_{47}ClN_6O_{13}$
Neomycin B hydrochloride.
Aminoglycoside antibiotic. $[\alpha]_D^{25}$ = 57° (H_2O); soluble in H_2O (1.5 g/100 ml), MeOH (0.57 g/100 ml), EtOH (0.065 g/100 ml), iPrOH (0.005 g/100 ml), cyclohexane (0.006 g/100 ml), C_6H_6 (0.003 g/100 ml); insoluble in Me_2CO, Et_2O. *Merck & Co.Inc.; Pharmacia & Upjohn.*

1881 Neomycin B Sulfate
1405-10-3 6542
$C_{23}H_{48}N_6O_{17}S$
Neomycin B sulfate.
Biosol; Bykomycin; Endomixin; Fraquinol; Myacine; Neosulf; Neomix; Neobrettin; Nivemycin; Tuttomycin. Aminoglycoside antibiotic. $[\alpha]_D^{20}$ = 54° (c = 2 H_2O); soluble in H_2O (0.63 g/100 ml), MeOH (0.0225 g/100 ml), EtOH (0.95 g/100 ml), iPrOH (0.008 g/100 ml), isoamyl alcohol (0.0247 g/100 ml), cyclohexane (0.008 g/100 ml), C_6H_6 (0.005 g/100 ml); insoluble in Et_2O, Me_2CO, $CHCl_3$. *Merck & Co.Inc.; Pharmacia & Upjohn.*

1882 Neomycin C
66-86-4 6542
$C_{23}H_{46}N_6O_{13}$
Neomycin C.
Aminoglycoside antibiotic. On hydrolysis, gives neomycin A and neobiosamine. *Merck & Co.Inc.; Pharmacia & Upjohn.*

1883 Neomycin Undecylenate
1406-04-8 6543 215-793-0
Neodecyllin; component of: Otalgine. Aminoglycoside antibiotic; antifungal. *Penick; Purdue Pharma L.P.*

1884 Netilmicin
56391-56-1 6563 260-146-8
$C_{21}H_{41}N_5O_7$
O-3-Deoxy-4-C-methyl-3-(methylamino)-β-L-arabinopyranosyl-(1→6)-O-[2,6-diamino-2,3,4,6-tetradeoxy-α-D-glycero-hex-4-enopyranosyl-(1→4)]-2-deoxy-N[1]-ethyl-D-streptamine.
1-N-ethylsisomicin; Sch-20569. Aminoglycoside antibiotic. *Schering-Plough HealthCare Products.*

1885 Netilmicin Sulfate
56391-57-2 6563 260-147-3
$(C_{21}H_{41}N_5O_7).5H_2SO_4$
O-3-Deoxy-4-C-methyl-3-(methylamino)-β-L-arabinopyranosyl-(1→6)-O-[2,6-diamino-2,3,4,6-tetradeoxy-α-D-glycero-hex-4-enopyranosyl-(1→4)]-2-deoxy-n[1]-ethyl-D-streptamine sulfate.
Certomycin; Netillin; Netilyn; Netromicine; Netromycin; Nettacin; Vectacin; Zetamicin. Aminoglycoside antibiotic. $[\alpha]_D^{26}$ = 164° (c = 3 H_2O); LD_{50} (mus iv) = 40 mg/kg, (mus ip) = 125 mg/kg, (mus sc) = 175 mg/kg. *Schering-Plough HealthCare Products.*

1886 Paromomycin
7542-37-2 7173 231-423-0
$C_{23}H_{45}N_5O_{14}$
O-2,6-diamino-2,6-dideoxy-α-L-idopyranosyl-(1→3)-O-β-D-ribofuranosyl-(1→5)-O-[2-amino-2-deoxy-α-D-glucopyranosyl-(1→4)-2-deoxystreptamine.
paromomycin I; amminosidin; catenulin; crestomycin; estomycin; hydroxymycin; monomycin A; neomycin E; paucimycin; R-400. Antibiotic; antiamebic. Oligosaccaride antibiotic found in *Streptomyces.* $[\alpha]_D^{25}$ = 65° ±3°; soluble in H_2O; less soluble in EtOH, MeOH; LD_{50} (rat orl) = 1625 mg/kg, (rat sc) > 650 mg/kg, (rat iv) = 156 mg/kg, (mus orl) > 2275 mg/kg, (mus sc) = 423 mg/kg, (mus iiv) = 90 mg/kg. *Parke-Davis; Pfizer Inc.*

1887 Paromomycin Sulfate
1263-89-4 7173 215-031-7
$C_{23}H_{45}N_5O_{14}.xH_2SO_4$
O-2,6-diamino-2,6-dideoxy-α-L-idopyranosyl-(1→3)-O-β-D-ribofuranosyl-(1→5)-O-[2-amino-2-deoxy-α-D-glucopyranosyl-(1→4)-2-deoxystreptamine sulfate.
Humatin; 1600 Antibiotic; FI-5853; Aminoxidin; Aminosidine; Farmiglucin; Farminosidin; Gabbromicina; Gabbromycin; Gabbroral; Humagel; Pargonyl; Paramicina; Paricina; Sinosid. Antibiotic; antiamebic. $[\alpha]_D^{20}$ = 50.5° (c = 1.5 H_2O, pH 6); LD_{50} (mus orl) > 15000 mg/kg, (mus sc) = 700 mg/kg, (mus iv) = 110 mg/kg. *Parke-Davis; Pfizer Inc.*

1888 Ribostamycin
25546-65-0 8373 247-091-5
$C_{17}H_{34}N_4O_{10}$
O-2,6-Diamino-2,6-dideoxy-α-D-glucopyranosyl-(1→4)-O-[β-D-ribofuranosyl-(1→5)]-2-deoxy-D-streptamine.
SF-733 antibiotic. Aminoglycoside antibiotic. mp = 192-195°, 178-180° (dec); $[\alpha]_D^{23}$ = 42°; soluble in H_2O; slightly soluble in MeOH; insoluble in Me_2CO, n-BuOH, EtOAc, C_6H_6, C_6H_{14}, Et_2O. *Meiji Seika Kaisha Ltd.*

1889 Ribostamycin Sulfate
53797-35-6 8373 258-783-1
$C_{17}H_{36}N_4O_{14}S$
O-2,6-Diamino-2,6-dideoxy-α-D-glucopyranosyl-(1→4)-O-[β-D-ribofuranosyl-(1→5)]-2-deoxy-D-streptamine sulfate.
Ibistacin; Mandamycine; Ribostamin; Ribomycine; Vistamycin. Aminoglycoside antibiotic. $[\alpha]_D^{20}$ = 39° (c = 1); LD_{50} (mus iv) = 225 mg/kg. *Meiji Seika Kaisha Ltd.*

1890 Sisomicin
32385-11-8 8695 251-018-2
$C_{19}H_{37}N_5O_7$
O-3-Deoxy-4-C-methyl-3-(methylamino)-β-L-arabinopyranosyl-(1→6)-O-[2,6-diamino-2,3,4,6-tetradeoxy-α-D-glycero-hex-4-enopyranosyl-(1→4)]-2-deoxy-D-streptamine.
Extramycin®; rickamicin; antibiotic 6640; Sch-13475. Aminoglycoside antibiotic. A gentamicin-like aminoglycoside antibiotic produced by *Micromonospora inyoesis*. [monohydrate]: mp = 198-201°; $[\alpha]_D^{26}$ = 189° (c = 0.3); [penta-N-acetate ($C_{29}H_{47}N_5O_{12}$)]: mp = 188-198° (dec); $[\alpha]_{DS}^{26}$ = 200° (c = 0.3). *Schering-Plough HealthCare Products.*

1891 Sisomicin Sulfate
53179-09-2 8695 258-414-4
$C_{19}H_{39}N_5O_{11}S$
O-3-Deoxy-4-C-methyl-3-(methylamino)-β-L-arabinopyranosyl-(1→6)-O-[2,6-diamino-2,3,4,6-tetradeoxy-α-D-glycero-hex-4-enopyranosyl-(1→4)]-2-deoxy-D-streptamine sulfate.
Baymicin; Extramycin; Mensiso; Siseptin; Sisobiotic; Sisolline; Sisomin. Aminoglycoside antibiotic. LD_{50} (mus iv) = 34 mg/kg, (mus ip) = 221 mg/kg, (mus sc) = 288 mg/kg. *Schering-Plough HealthCare Products.*

1892 Streptomycin
57-92-1 8983 200-355-3
$C_{21}H_{39}N_7O_{12}$
O-2-Deoxy-2-(methylamino)-α-L-glucopyranosyl-(1→2)-O-5-deoxy-3-C-formyl-α-L-lyxofuranosyl-(1→4)-N,N'-bis(aminoiminomethyl)-D-streptamine.
streptomycin A. Aminoglycoside antibiotic (tuberculostatic). [trihydrochloride ($C_{21}H_{42}Cl_3N_7O_{12}$)]: $[\alpha]_D^{26}$ = -84°; soluble in H_2O (> 2g/100 ml); MeOH (> 2 g/100 ml), EtOH (0.09 g/100 ml), iPrOH (0.012 g/100 ml), isoamyl alcohol (0.012 g/100 ml), petroleum ether (0.002 g/100 ml), CCl_4 (0.004 g/100 ml), Et_2O (0.001 g/100 ml). *Olin Res. Ctr.*

1893 Streptomycin B
128-45-0 8984 204-887-7
$C_{27}H_{49}N_7O_{17}$
α-L-Mannopyranosyl-(1→4)-O-2-deoxy-2-(methylamino)-α-L-glucopyranosyl-(1→2)-O-5-deoxy-3-C-formyl-α-L-lyxofuranosyl-(1→4)-N,N'-bis(aminoiminomethyl)-D-streptamine.
mannosidostreptomycin; mannosylstreptomycin. Aminoglycoside antibiotic (tuberculostatic). [trihydrochloride monohydrate ($C_{27}H_{54}Cl_3N_7O_{20}$)]: mp = 190-200° (dec); $[\alpha]_D^{25}$ = -47° (c = 11.35 H_2O). *Squibb E.R. & Sons.*

1894 Streptomycin Sesquisulfate
3810-74-0 8983 223-286-0
$C_{42}H_{84}N_{14}O_{36}S_3$
O-2-Deoxy-2-(methylamino)-α-L-glucopyranosyl-(1→2)-O-5-deoxy-3-C-formyl-α-L-lyxofuranosyl-(1→4)-N,N'-bis(aminoiminomethyl)-D-streptamine sulfate (2:3).
streptomycin sulfate; AgriStrep; Streptobrettin; Strycin; Vetstrep; component of: Intromycin. Aminoglycoside antibiotic (tuberculostatic). Soluble in H_2O (> 2 g/100 ml), MeOH (0.085 g/100 ml), EtOH (0.030 g/100 ml), iPrOH (0.001 g/100 ml), petroleum ether (0.0015 g/100 ml), CCl_4 (0.0035 g/100 ml), Et_2O (0.0035 g/100 ml). *Bristol-Myers Squibb Pharm. R&D; Enzon Inc.; Roerig Div. Pfizer Pharm.*

1895 Streptonicozid
5667-71-0 8985 227-128-1
$C_{54}H_{94}N_{20}O_{36}S_3$
4-Pyridinecarboxylic acid hydrazide hydrazone with O-2-deoxy-2-(methylamino)-α-L-glucopyranosyl-(1→2)-O-5-deoxy-3-C-formyl-α-L-lyxofuranosyl-(1→4)-N,N'-bis(aminoiminomethyl)-D-streptamine sulfate (2:3) (salt).
streptomyclidine isonicotinyl hydrazine sulfate; stretoniazide; Strazide; Streptohydrazid. Aminoglycoside antibiotic with tuberculostatic properties. Dec 230°; soluble in H_2O (> 2 g/100 ml), EtOH (0.0115 g/100 ml), CCl_4 (0.0025 g/100 ml), Et_2O (0.031 g/100 ml). *Pfizer Intl.*

1896 Tobramycin
32986-56-4 9628 251-322-5
$C_{18}H_{37}N_5O_9$
O-3-Amino-3-deoxy-α-D-glucopyranosyl-(1→6)-O-[2,6-diamino-2,3,6-trideoxy-α-D-ribo-hexopyranosyl-(1→4)-]-2-deoxy-D-streptamine.
nebramycin factor 6; NF 6; Gernebcin; Tobracin; Tobradistin; Tobralex; Tobramaxin; Tobrex. Aminoglycoside antibiotic. Soluble in H_2O; $[\alpha]_D$ = 128°; LD_{50} (mus sc) = 441 mg/kg, (rat sc) = 969 mg/kg. *Alcon Labs; Eli Lilly & Co.*

1897 Tobramycin Sulfate
79645-27-5 9628
$C_{36}H_{84}N_{10}O_{38}S_5$
O-3-Amino-3-deoxy-α-D-glucopyranosyl-(1→6)-O-[2,6-diamino-2,3,6-trideoxy-α-D-ribo-hexopyranosyl-(1→4)-]-2-deoxy-D-streptamine slufate.
Nebcin; Obracin; Tobra. Aminoglycoside antibiotic. *Apothecon; Bristol-Myers Squibb Pharm. R&D; Eli Lilly & Co.; Elkins-Sinn.*

Amphenicol Antibiotics

1898 Azidamfenicol
13838-08-9 941 237-552-9
$C_{11}H_{13}N_5O_5$
2-Azido-N-[2-hydroxy-1-(hydroxymethyl)-2-(4-nitrophenyl)ethyl]acetamide.
azidoamphenicol; Leukomycin N. Amphenicol antibiotic. mp = 107°; $[\alpha]_D^{20}$ = -20° (c = 1.6 EtOAc); soluble in H_2O (< 2 g/100 ml). *Bayer Corp. Pharm. Div.*

1899 Chloramphenicol
56-75-7 2120 200-287-4
$C_{11}H_{12}Cl_2N_2O_5$
[R-(R*,R*)]-2,2-Dichloro-N-[2-hydroxy-1-(hydroxymethyl)-2-(4-nitrophenyl)ethyl]acetamide.
Ak-Chlor; Amphicol; Anacetin; Aquamycetin; Chemicetina; Chloramex; Chlorasol; Chloricol; Chlorocid; Chloromycetin; Chloroptic; Cloramfen; Clorocyn; Enicol; Farmicetina; Fenicol; Globenicol; Intramycetin; Kemicetine; Leukomycin; Micoclorina; Mychel; Mycinol; Novomycetin; Ophthochlor; Pantovernil; Paraxin; Quemicetina; Ronphenil; Sintomicetina; Sno Phenicol; Synthomycetin; Tevcocin; Tifomycine; Veticol; Viceton. Antibacterial; antirickettsial. A broad spectrum antibiotic obtained from cultures of the soil bacterium *streptomyces venezuelae*. mp = 150.5-151.5°; $[\alpha]_D^{27}$ = 18.6° (c = 4.86 EtOH), $[\alpha]_D^{25}$ = -25.5° (EtOAc); λ_m 278 nm ($E_D^{1\%}$298); soluble in H_2O (0.25 g/100 ml at 25°), propylene glycol (15.1 g/100 ml at 25°); very soluble in MeOH, EtOH, BuOH, EtOAc; Me_2CO; fairly soluble in Et_2O; insoluble in C_6H_6, petroleum ether. *Allergan Inc.; Chinoin; Fermenta Animal Health Co.; Fujisawa Pharm. USA Inc.; Parke-Davis; Schering-Plough Animal Health; TechAmerica.*

1900 Chloramphenicol Monosuccinate Arginine Salt
34327-18-9 2120
$C_{21}H_{30}Cl_2N_6O_{10}$
D-Threo-N-dichloroacetyl-1-p-nitrophenyl-2-amino-1,3-propanediol monosuccinate arginine salt.
chloramphenicol arginine succinate; Paraxin Succinate A. Antibacterial; antirickettsial. mp = 135-145° (dec).

1901 Chloramphenicol Monosuccinate Sodium Salt
982-57-0 2120 213-568-1
$C_{15}H_{15}Cl_2N_2NaO_8$
[R-(R*,R*)]-2,2-Dichloro-N-[2-hydroxy-1-(hydroxymethyl)-2-(4-nitrophenyl)ethyl]acetamide monosuccinate sodium.
chloramphenicol sodium succinate; protophenicol. Antibacterial; antirickettsial. A broad spectrum antibiotic obtained from cultures of the soil bacterium *streptomyces venezuelae*. Freely soluble in H_2O (50 g/100 ml). *Chinoin.*

1902 Chloramphenicol Palmitate
530-43-8 2120 208-477-9
$C_{27}H_{42}Cl_2N_2O_6$
D-Threo-(-)-2,2-dichloro-N-[β-hydroxy-α-(hydroxymethyl)-p-nitrophenyl]acetamide α-palmitate.
Chlorambon; Chloropal; Clorolifarina. Antibacterial; antirickettsial. A broad spectrum antibiotic obtained from cultures of the soil bacterium *streptomyces venezuelae*. mp = 90°; $[\alpha]_D^{26}$ = 24.6° (c = 5 EtOH); λ_m 271 nm ($E_{1\ cm}^{1\%}$ 179 EtOH); slightly soluble in H_2O (0.105 g/100 ml at 28°), petroleum ether (0.0225 g/100 ml); freely soluble in MeOH, EtOH, $CHCl_3$, Et_2O, C_6H_6. *Chinoin; Parke-Davis.*

1903 Chloramphenicol Pantothenate Complex
31342-36-6 2120
$C_{62}H_{80}CaCl_8N_{10}O_{30}$
N-(2,4-Dihydroxy-3,3-dimethyl-1-oxobutyl)-β-alanine calcium salt (2:1) with [R-(R*,R*)]-2,2-Dichloro-N-[2-hydroxy-1-(hydroxymethyl)-2-(4-nitrophenyl)ethyl]-acetamide (1:4).
chloramphenicol pantothenate calcium complex (4:1); Pantothenic acid calcium salt (2:1) compound with D-threo-(-)-2,2-dichloro-N[β-hydroxy-α-(hydroxymethyl)-p-nitrophenyl]acetamide (1:4); Pantofenicol. Antibacterial; antirickettsial. A broad spectrum antibiotic obtained from cultures of the soil bacterium *streptomyces venezuelae*. *Chinoin; Eprova AG; Pluriquimica, Portugal.*

1904 Florfenicol
73231-34-2 4145
$C_{12}H_{14}Cl_2FNO_4S$
2,2-Dichloro-N-[1-fluoromethyl)-2-hydroxy-2-[4-(methylsulfonyl)phenyl]ethyl]acetamide.
Nuflor; Sch-25298; Aquafen. Antibacterial; antirickettsial. mp = 153-154°; soluble in H_2O. *Schering-Plough HealthCare Products.*

1905 Thiamphenicol
15318-45-3 . 9436 239-355-3
$C_{12}H_{15}Cl_2NO_5S$
2,2-Dichloro-N-[2-hydroxy-1-(hydroxymethyl)-2-[4-(methylsulfonyl)phenyl]ethyl]acetamide.
Thiocymetin; Win-5063-2. Antibacterial; antirickettsial. The D-isomer is an antibiotic; the DL-form is used to control fowl cholera. mp = 164.3-166.3°; $[\alpha]_D^{25}$ = 12.9° (EtOH); λ_m 224, 266, 274 nm (ε 13700 800 700 EtOH); soluble in H_2O, EtOH. *Sterling Winthrop Inc.*

1906 DL-Thiamphenicol
847-25-6 9436
$C_{12}H_{15}Cl_2NO_5S$
(DL)-2,2-Dichloro-N-[2-hydroxy-1-(hydroxymethyl)-2-[4-(methylsulfonyl)phenyl]ethyl]acetamide.
Thiocymetin; Win-5063-2; raceophenidol; racephenicol; Dexawin. Antibacterial; antirickettsial. The D-isomer is an antibiotic; the DL-form is used to control fowl cholera. *Sterling Winthrop Inc.*

Ansamycin Antibiotics

1907 Ciclotropium Bromide
85166-20-7
$C_{24}H_{36}BrNO_2$
(8r)-3α-Hydroxy-8-isopropyl-1αH,5αH-tropanium bromide α-phenylcyclopentaneacetate.
Antibacterial (ansamycin). Has vagolytic and antianginal properties.

1908 Rifamide
2750-76-7 8381 220-390-8
$C_{43}H_{58}N_2O_{13}$
N,N-Diethyl-2-[(1,2-dihydro-5,6,7,19,21-pentahydroxy-23-methoxy2,412,16,18,20,22-heptamethyl-1,11-dioxo-2,7-(epoxypentadeca[1,11,13]trienimino)naphtho[2,1-b]furan-9-yl)oxy]acetamide 21-acetate.
Rifamycin M-14; NSC-133099. Antibacterial (ansamycin). mp = 40-170° (dec); $[\alpha]_D^{20}$ = -48.7° (c = 0.4 MeOH); λ_m 222, 302, 421 nm (ε 42820, 20770, 16200 pH 7.38); LD_{50} (mus orl) = 2450 mg/kg, (mus sc) = 640 mg/kg, (mus ip) = 320 mg/kg, (mus iv) = 315 mg/kg, (rat orl) > 4000 mg/kg, (rat sc) = 2500 mg/kg, (rat ip) = 5354 mg/kg, (rat iv) = 380 mg/kg. *Marion Merrell Dow Inc.*

1909 Rifampin
13292-46-1 8382 236-312-0
$C_{43}H_{58}N_4O_{12}$
5,6,9,17,19,21-Hexahydroxy-23-methoxy-2,4,12,16,18,20,22-heptamethyl-8-[N-(4-methyl-1-piperazinyl)formimidoyl]-2,7-(epoxypentadeca[1,11,13]trienimino)naphtho[2,1-b]furan-1,11(2H)dione 21-acetate.
refampicin; rifaldazine; rifamycin AMP; R/AMP; Abrifam; Eremfat; Rifa; Rifadin; Rifadine; Rifaprodin; Rifoldin; Rimactan; Rimactane; L-5103 Lepetit; Ba 41166/E; NSC-113926. Antibacterial (ansamycin). Tuberculostatic. Dec 183-188°; λ_m 237, 255,334, 475 nm (ε 33200, 32100, 27000, 15400 pH 7.38); freely soluble in CH_3Cl, DMSO; soluble in EtOAc, MeOH, THF; slightly soluble in H_2O, Me_2CO, CCl_4; LD_{50} (mus orl) = 885 mg/kg, (mus iv) = 260 mg/kg, (mus . *Ciba-Geigy Corp.; Marion Merrell Dow Inc.; Merrell Pharm. Inc.*

1910 Rifampin SV
6998-60-3 8384
$C_{37}H_{47}NO_{12}$
5,6,9,17,19,21-Hexahydroxy-23-methoxy-2,4,12,16,18,20,22-heptamethyl-2,7-(epoxypentadeca[1,11,13]trienimino)naphtho[2,1-b]furan-1,11(2H)dione 21-acetate.
rifomycin SV; rifamicine SV. Antibacterial (ansamycin). mp 300° (dec > 140°); $[\alpha]_D^{20}$ = -4° (MeOH); λ_m = 223, 314, 445 nm ($E_{1\,cm}^{1\%}$ 586, 322, 204 phosphate buffer pH 7.3); slightly soluble in H_2O, petroleum ether; soluble in MeOH, EtOH, Me_2CO, EtOAc; LD_{50} (mus iv) = 550 mg/gk, (mus ip) = 625 mg/kg, (mus orl) = 2120 mg/kg.

1911 Rifapentine
61379-65-5 8385 262-743-9
$C_{47}H_{64}N_4O_{12}$
3-[N-(4-Cyclopentyl-1-piperazinyl)formimidoyl]rifamycin. MDL-473. Ansamycin antibacterial; tuberculostatic. mp = 179-180°; λ_m 475, 334 nm (ε 15200, 26700); LD_{50} (mus orl) > 2000 mg/kg, 3300 mg/kg, (mus ip) = 750 mg/kg, 710 mg/kg. *Marion Merrell Dow Inc.*

1912 Rifaximin
80621-81-4 8386
$C_{43}H_{51}N_3O_{11}$
(2S,16Z,18E,20S,21S,22R,23R,24R,25S,26S,27S,28E)-5,6,21,23,25-Pentahydroxy-27-methoxy-2,4,11,16,20,22,24,26-oxtamethyl-2,7-(epoxypentadeca[1,11,13]trienimino)benzofuro[4,5-e]pyrido[1,2-a]benzimidazole-1,15(2H)-dione 25-acetate.
Antibacterial (ansamycin). mp = 200-205° (dec); λ_m 232, 260, 320, 370, 450 nm ($E_{1\,cm}^{1\%}$ 489, 339, 295, 216, 119, 159); soluble in alcohols, EtOAc, $CHCl_3$, C_7H_8; LD_{50} (rat orl) > 2000 mg/kg. *Alfa Wassermann S.p.A.*

β-Lactam Antibiotics

1913 Amantocillin
10004-67-8
$C_{19}H_{27}N_3O_4S$
6-(3-Amino-1-adamantanecarboxamido)-3,3-dimethyl-7-oxo-4-thia-1-azabicyclo[3.2.0]heptane-2-carboxylic acid. Antibacterial.

1914 Amdinocillin
32887-01-7 251-277-1
$C_{15}H_{23}N_3O_3S$
[2S-(2α,5α,6β)]-6-[[(Hexahydro-1H-azepin-1-yl)methylene]amino]-3,3-dimethyl-7-oxa-4-thia-1-azabicyclo[3.2.0]heptane-2-carboxylic acid .
Coactin. Antibacterial. *Hoffmann-LaRoche Inc.*

1915 Amdinocillin Pivoxil
32886-97-8 409 251-276-6
$C_{21}H_{33}N_3O_5S$
[2S-(2α,5α,6β)]-6-[[(Hexahydro-1H-azepin-1-yl)methylene]amino]-3,3-dimethyl-7-oxa-4-thia-1-azabicyclo[3.2.0]heptane-2-carboxylic acid (2,2-dimethyl-1-oxopropoxy)methyl ester.
pivamdinocillin; pivmecillinam; FL-1039; Selexid (susp.); component of: Augmentin. Antibacterial. mp = 118.5-119.5°; $[\alpha]_D^{20}$ = +231° (c=1 in 96% EtOH); LD_{50} (mus iv) = 475-480 mg/kg, (mus sc) = 1736-1930 mg/kg, (mus orl) = 3020 mg/kg, (rat iv) = 465 mg/kg, (rat sc) = 1935-2100 mg/kg, (rat orl) = 9500-10000 mg/kg. *Hoffmann-LaRoche Inc.*

1916 Amoxicillin [anhydrous]
26787-78-0 617 248-003-8
$C_{16}H_{19}N_3O_5S$
[2S-[2α,5α,6β(S*)]]-6-[[Amino(4-hydroxyphenyl)acetyl]amino]-3,3-dimethyl-7-oxo-4-thia-1-azabicyclo[3.2.0]heptane-2-carboxylic acid.
amoxycillin; AMPC; Amocilline; Amolin; Amopenixin; Amoram; Amoxipen; Anemolin; Aspenil; Betamox; Bristamox; Cabermox; Delacillin; Efpenix; Grinsil; Helvamox; Optium; Ospamox; Pasetocin; Penamox; Penimox; Piramox; Sawacillin; Simoxil; Sumox; Widecillin; Wymox; Amoxicillin Chewable Tablets; Amoxil; Larotid; Polymox; Trimox; Utimox. Antibacterial. *Apothecon; Hoffmann-LaRoche Inc.; Lemmon Co.; Parke-Davis; SmithKline Beecham Pharm.; Wyeth-Ayerst Labs.*

1917 Amoxicillin Trihydrate
61336-70-7 617
$C_{16}H_{19}N_3O_5S.3H_2O$
[2S-[2α,5α,6β(S*)]]-6-[[Amino(4-hydroxyphenyl)acetyl]amino]-3,3-dimethyl-7-oxo-4-thia-1-azabicyclo[3.2.0]heptane-2-carboxylic acid trihydrate.
BRL-2333; Agram; Alfamox; Almodan; Amodex; Amoxi; Amoxidal; Amoxidin; Amoxil; Amoxillat; Amoxi-Wolff; Amoxypen; Ardine; AX 250; Clamoxyl; Cuxacillin; Dura AX; Flemoxin; Hiconcil; Ibiamox; Larocin (obsolete);

Larotid; Moxal; Moxaline; Neamoxyl; Polymox; Raylina; Robamox; Sigamopen; Silamox; Trimox; Uro-Clamoxyl; Utimox; Zamocillin. Antibacterial. $[\alpha]_D^{20} = +246°$ (c = 0.1); λ_m 230, 274 nm (ε 10850, 1400 EtOH); λ_m 229, 272 nm (ε 9500, 1080 0.1N HCl); λ_m = 248, 291 nm (ε 2200, 3000 KOH); soluble in H_2O (0.4 g/100 ml), MeOH (0.75 g/100 ml), EtOH (0.34 g/100 ml); insoluble in C_6H_6, EtOAc, CH_3CN, C_6H_{14}.

1918 Ampicillin
69-53-4 628 200-709-7
$C_{16}H_{19}N_3O_4S$
[2S-[2α,5α,6β(S*)]]-6-[(Aminophenylacetyl)amino]-3,3-dimethyl-7-oxo-4-thia-1-azabicyclo[3.2.0]heptane-2-carboxylic acid.
AY-6108; BRL-1341; P-50; Albipen; Amfipen; Amipenix; Ampipenin; Ampitab; Bonapicillin; Britacil; Doktacillin; Domicillin; Dumopen; Grampenil; Nuvapen; Omnipen; Pénicline; Tokiocillin. Antibacterial. *Apothecon; Bristol-Myers Squibb Pharm. R&D; Parke-Davis; Wyeth-Ayerst Labs.*

1919 Ampicillin Sodium
69-52-3 628 200-708-1
$C_{16}H_{18}NaN_3O_4S$
[2S-[2α,5α,6β(S*)]]-6-[(Aminophenylacetyl)amino]-3,3-dimethyl-7-oxo-4-thia-1-azabicyclo[3.2.0]heptane-2-carboxylic acid sodium salt.
Alpen-N; Amcill-S; Ampicin; Cilleral; Omnipen-N; Pen A/N; Penbritin-S; Pentrex; Polycillin-N; Principen/N; Synpenin; Viccillin. Antibiotic. *Apothecon; Roerig Div. Pfizer Pharm.; SmithKline Beecham Pharm.; Wyeth-Ayerst Labs.*

1920 Ampicillin Trihydrate
7177-48-2 628
$C_{16}H_{19}N_3O_4S.3H_2O$
[2S-[2α,5α,6β(S*)]]-6-[(Aminophenylacetyl)amino]-3,3-dimethyl-7-oxo-4-thia-1-azabicyclo[3.2.0]heptane-2-carboxylic acid trihydrate.
Alpen; Amblosin; Amcill; Ampilag; Ampilar; Ampi-Tablinen; Amplital; Austrapen; Binotal; Cetampin; Cymbi; Pen A; Penbristol; Penbritin; Penbrock; Pensyn; Pentrexyl; Polycillin; Princillin; Principen; Rosampline; Totacillin; Totalciclina; Totapen; Ukapen; Ultrabion; Vidopen. Antibacterial. *Apothecon; Bristol-Myers Squibb Pharm. R&D; Parke-Davis; Wyeth-Ayerst Labs.*

1921 Apalcillin
63469-19-2 766
$C_{25}H_{23}N_5O_6S$
[2S-[2α,5α,6β(S*)]]-6-[[[[(4-Hydroxy-1,5-naphthyridin-3-yl)carbonyl]amino]phenylacetyl]amino]-3,3-dimethyl-7-oxo-4-thia-1-azabicyclo[3.2.0]heptane-2-carboxylic acid.
Antibacterial. *Sumitomo Pharm. Co. Ltd.*

1922 Apalcillin Sodium
58795-03-2 766 261-446-1
$C_{25}H_{22}N_5NaO_6S$
[2S-[2α,5α,6β(S*)]]-6-[[[[(4-Hydroxy-1,5-naphthyridin-3-yl)carbonyl]amino]phenylacetyl]amino]-3,3-dimethyl-7-oxo-4-thia-1-azabicyclo[3.2.0]heptane-2-carboxylic acid sodium salt.
PC-904; Lumota; Palcin. Antibacterial. Soluble in H_2O. *Sumitomo Pharm. Co. Ltd.*

1923 Aspoxicillin
63358-49-6 887
$C_{21}H_{27}N_5O_7S$
[2S-(2α,5α,6β)]-N-Methyl-D-asparaginyl-N-(2-carboxy-3,3-dimethyl-7-oxo-4-thia-1-azabicyclo[3.2.0]hept-6-yl)-D-2-(4-hydroxyphenyl)glycinamide.
ASPC; TA-058; Doyle. Antibacterial. mp = 195-198° (dec). *Tanabe Seiyaku Co. Ltd.*

1924 Azidocillin
17243-38-8 942 241-278-5
$C_{16}H_{17}N_5O_4S$
[2S-[2α,5α,6β(S*)]]-6-[(Azidophenylacetyl)amino]-3,3-dimethyl-7-oxo-4-thia-1-azabicyclo[3.2.0]heptane-2-carboxylic acid.
SPC-97D; BRL-2351. Antibacterial. *SmithKline Beecham Pharm.*

1925 Azidocillin Potassium Salt
22647-32-1 942
$C_{16}H_{16}KN_5O_4S$
[2S-[2α,5α,6β(S*)]]-6-[(Azidophenylacetyl)amino]-3,3-dimethyl-7-oxo-4-thia-1-azabicyclo[3.2.0]heptane-2-carboxylic acid potassium salt.
Nalpen. Antibacterial. mp = 194° (dec). *SmithKline Beecham Pharm.*

1926 Azidocillin Sodium Salt
35334-12-4 942
$C_{16}H_{16}N_5NaO_4S$
[2S-[2α,5α,6β(S*)]]-6-[(Azidophenylacetyl)amino]-3,3-dimethyl-7-oxo-4-thia-1-azabicyclo[3.2.0]heptane-2-carboxylic acid sodium salt.
Globacillin; Longatren. Antibacterial. *SmithKline Beecham Pharm.*

1927 Azlocillin
37091-66-0 947 253-348-2
$C_{20}H_{23}N_5O_6S$
[2S-[2α,5α,6β(S*)]]-3,3-Dimethyl-7-oxo-6-[[[[(2-oxo-1-imidazolidinyl)carbonyl]amino]phenyl]acetyl]amino]-4-thia-1-azabicyclo[3.2.0]heptane-2-carboxylic acid.
Bay e 6905. Antibacterial. *Bayer Corp. Pharm. Div.*

1928 Azlocillin Sodium
37091-65-9 947 253-347-7
$C_{20}H_{22}N_5NaO_6S$
[2S-[2α,5α,6β(S*)]]-3,3-Dimethyl-7-oxo-6-[[[[(2-oxo-1-imidazolidinyl)carbonyl]amino]phenyl]avetyl]amino]-4-thia-1-azabicyclo[3.2.0]heptane-2-carboxylic acid sodium salt.
Azlin; Securopen. Antibacterial. Soluble in H_2O, MeOH, DMF. *Bayer Corp. Pharm. Div.*

1929 Aztreonam
78110-38-0 955 278-839-9
$C_{13}H_{17}N_5O_8S_2$
[2S-[2α,3β(Z)]]-2-[[[1-(2-Amino-4-thiazolyl)-2-[(2-methyl-4-oxo-1-sulfo-3-azetidinyl)amino]-2-oxoethylidene]amino]oxy]-2-methylpropanoic acid.
azthreonam; SQ-26776; Azactam; Azonam; Aztreon; Nebactam; Primbactam. Antibacterial. A totally synthetic monocyclic monolactam antibiotic with a high resistance to β-lactamases. Dec 227°; very slightly soluble in EtOH; slightly soluble in MeOH; soluble in DMF, DMSO;

practically insoluble in C_7H_8, $CHCl_3$, EtOAc. *Squibb E.R. & Sons.*

1930 Bacampicillin
50972-17-3 961
$C_{21}H_{27}N_3O_7S$
[2S-[2α,5α,6β(S*)]]-6-[(Aminophenylacetyl)amino]-3,3-dimethyl-7-oxo-4-thia-1-azabicyclo[3.2.0]heptane-2-carboxylic acid 1-[(ethoxycarbonyl)oxy]ethyl ester .
Antibacterial. *Roerig Div. Pfizer Pharm.*

1931 Bacampicillin Hydrochloride
37661-08-8 961 253-580-4
$C_{21}H_{28}ClN_3O_7S$
[2S-[2α,5α,6β(S*)]]-6-[(Aminophenylacetyl)amino]-3,3-dimethyl-7-oxo-4-thia-1-azabicyclo[3.2.0]heptane-2-carboxylic acid 1-[(ethoxycarbonyl)oxy]ethyl ester hydrochloride.
Ambacamp; Ambaxin; Bacacil; Bacampicine; Penglobe; Spectrobid. Antibacterial. mp = 171-176° (dec); $[\alpha]_D^{20}$ = 161°; soluble in H_2O; LD_{50} (mus orl) = 8529 mg/kg, (mus ip) = 176 mg/kg, (mus sc) = 9475 mg/kg, (mus iv) = 184 mg/kg. . *Roerig Div. Pfizer Pharm.*

1932 Bacmecillinam
50846-45-2
$C_{20}H_{31}N_3O_6S$
(2S,5R,6R)-[[(Hexahydro-1H-azepin-1-yl)methylene]-amino]-3,3-dimethyl-7-oxo-4-thia-1-azabicyclo-[3.2.0]heptane-2-carboxylic acid ester with ethyl 1-hydroxyethyl carbonate.
Penicillin antibacterial.

1933 Benzylpenicillin Sodium
69-57-8 1178 200-710-2
$C_{16}H_{17}N_2NaO_4S$
[2S-(2α,5α,6β)]-3,3-Dimethyl-7-oxo-6-[(phenylacetyl)amino]-4-thia-1-azabicyclo[3.2.0]heptane-2-carboxylic acid monosodium salt.
Crystapen; sodium penicillin G; penicillin G sodium; sodium penicillin II; sodium benzylpenicillinate; benzylpenicillinic acid sodium salt; penicillin; American penicillin; Monocillin; Nalpen G; Novocillin; Penilaryn; Pen-A-Brasive; Veticillin. Antibacterial. d = 1.41; $[\alpha]_D^{24.8}$ = +301° (c = 2.0); λ_m 252, 258.6, 264.4 nm (E_M about 300, 240, 180, H_2O); very soluble in H_2O, glycerol, primary alcohols; practically insoluble in Et_2O, Me_2CO, $CHCl_3$, ethyl and amyl acetate, fixed oils, liquid petrolatum. *Merck & Co.Inc.*

1934 Benzylpenicillinic Acid
61-33-6 1177 200-506-3
$C_{16}H_{18}N_2O_4S$
[2S-(2α,5α,6β)]-3,3-Dimethyl-7-oxo-6-[(phenylacetyl)amino]-4-thia-1-azabicyclo-[3.2.0]heptane-2-carboxylic acid.
free benzylpenicillin; free penicillin G; free penicillin II; Pfizerpen; Benzylpenicillin. Antibacterial. $[\alpha]_D^{20}$ = +269° (MeOH solution 50 ml, prepared from 350 mg benzylpenicillin sodium); sparingly soluble in H_2O; soluble in MeOH, EtOH, Et_2O, EtOAc, C_6H_6, $CHCl_3$, Me_2CO, insoluble in petroleum ether. *Pfizer Inc.*

1935 Biapenern
120410-24-4 1243
$C_{15}H_{18}N_4O_4S$
[4R-[4α,5β,6β(R*)]]-6-[[2-Carboxy-6-(1-hydroxyethyl)-4-methyl-7-oxo-1-azabicyclo[3.2.0]hept-2-en-3-yl]thio]-6,7-dihydro-5H-pyrazolo[1,2a][1,2,4]triazol-4-ium hydroxide inner salt.
Biapenem; LJC-10627; L-627; CL-186-815. Antibacterial. $[\alpha]_D^{20}$ = -32.9° (c = 0.5). *Lederle Labs.*

1936 Carbenicillin
4697-36-3 1838 225-171-0
$C_{17}H_{18}N_2O_6S$
[2S-(2α,5α,6β)]-6-[(Carboxyphenylacetyl)amino]-3,3-dimethyl-7-oxo-4-thia-1-azabicyclo[3.2.0]heptane-2-carboxylic acid.
Antibacterial. *Pfizer Inc.; SmithKline Beecham Pharm.*

1937 Carbenicillin Disodium
4800-94-6 1838 225-360-8
$C_{17}H_{16}N_2Na_2O_6S$
[2S-(2α,5α,6β)]-6-[(Carboxyphenylacetyl)amino]-3,3-dimethyl-7-oxo-4-thia-1-azabicyclo[3.2.0]heptane-2-carboxylic acid disodium salt.
BRL-2064; CP-15639-2; Anabactyl; Carbapen; Carbecin; Geocillin; Geopen; Hyoper; Microcillin; Pyocianil; Pyopen. Antibacterial. *Pfizer Inc.; SmithKline Beecham Pharm.*

1938 Carbenicillin Indanyl Sodium
26605-69-6 1838 247-845-3
$C_{26}H_{25}N_2NaO_6S$
[2S-(2α,5α,6β)]-6-[[3-[(2,3-Dihydro-1H-inden-5-yl)oxy]-1,3-dioxo-2-phenylproply]amino]-3,3-dimethyl-7-oxa-4-thia-1-azabicyclo[3.2.0]heptane-2-carboxylic acid sodium salt.
Carindacillin Sodium; CP-15464-2; Carindapen; Geocillin; G.U.-Pen. Antibacterial. *Pfizer Inc.*

1939 Carbenicillin Phenyl
27025-49-6 1838 248-171-2
$C_{23}H_{22}N_2O_6S$
[2S-(2α,5α,6β)]-6-[(Carboxyphenylacetyl)amino]-3,3-dimethyl-7-oxo-4-thia-1azabicyclo[3.2.0]heptane-2-carboxylic acid phenyl.
Carfecillin. Antibacterial. *SmithKline Beecham Pharm.*

1940 Carbenicillin Phenyl Sodium
21649-57-0 1838 244-496-9
$C_{23}H_{21}N_2NaO_6S$
[2S-(2α,5α,6β)]-6-[(Carboxyphenylacetyl)amino]-3,3-dimethyl-7-oxo-4-thia-1-azabicyclo[3.2.0]heptane-2-carboxylic acid phenyl sodium.
carfecillin sodium; BRL-3475; Gripenin-O; Urocarf; Uticillin; carbenicillin phenyl sodium. Antibacterial. $[\alpha]_D^{20}$ = 216.2° (H_2O). *SmithKline Beecham Pharm.*

1941 Carindacillin
35531-88-5 1888
$C_{26}H_{26}N_2O_6S$
[2S-(2α,5α,6β)]-6-[[3-[(2,3-Dihydro-1H-inden-5-yl)oxy]-1,3-dioxo-2-phenylpropyl]amino]-3,3-dimethyl-7-oxa-4-thia-1-azabicyclo[3.2.0]heptane-2-carboxylic acid .
N-(2-Carboxy-3,3-dimethyl-7-oxo-4-thia-1-azabicyclo-[3.2.0]hept-6-yl)-2-phenylmalonamic acid 1-(5-indanyl) ester; α-(5-indanyloxycarbonyl)-benzyl-penicillin;

carbenicillin indanyl ester; CP-15464. Antibacterial. *Pfizer Inc.*

1942 Carumonam
87638-04-8 1922
$C_{12}H_{14}N_6O_{10}S_2$
[2S-[2α,3α(Z)]]-[[[2-[[2-[[Aminocarbonyl)oxy]methyl]-4-oxo-1-sulfo-3-azetidinyl]amino]-1-(2-amino-4-thiazolyl)-2-oxoethylidene]amino]oxy]acetic acid.
Antibacterial. $[\alpha]_D^{26}$ = -45° (c = 1 in DMSO). *Hoffmann-LaRoche Inc.*

1943 Carumonam Sodium
86832-68-0 1922
$C_{12}H_{12}N_6Na_2O_{10}S_2$
[2S-[2α,3α(Z)]]-[[[2-[[2-[[Aminocarbonyl)oxy]methyl]-4-oxo-1-sulfo-3-azetidinyl]amino]-1-(2-amino-4-thiazolyl)-2-oxoethylidene]amino]oxy]acetic acid disodium salt.
AMA-1080; Ro-17-2301; Amasulin; Mobactam. Antibacterial. *Hoffmann-LaRoche Inc.*

1944 Cefaclor
70356-03-5 1962
$C_{15}H_{14}ClN_3O_4S.H_2O$
[6R-[6α,7β(R*)]]-7-[(Aminophenylacetyl)amino]-3-chloro-8-oxo-5-thia-1-azabicyclo[4.2.0]oct-2-ene-2-carboxylic acid monohydrate.
compd 99638; Alfacet; Alfatil; Ceclor; Distaclor; Panacef; Panoral. Antibacterial. λ_m = 265 nm (ε = 6800); soluble in H_2O, insoluble in organic solvents. *Eli Lilly & Co.*

1945 Cefaclor [anhydrous]
53994-73-3 1962 258-909-5
$C_{15}H_{14}ClN_3O_4S$
[6R-[6α,7β(R*)]]-7-[(Aminophenylacetyl)amino]-3-chloro-8-oxo-5-thia-1-azabicyclo[4.2.0]oct-2-ene-2-carboxylic acid.
Antibacterial. *Eli Lilly & Co.*

1946 Cefadroxil
66592-87-8 1963 256-555-6
$C_{16}H_{17}N_3O_5S.H_2O$
[6R-[6α,7β(R*)]]-7-[[Amino-(4-hydroxyphenyl)acetyl]amino]-3-methyl-8-oxo-5-thia-1-azabicyclo[4.2.0]oct-2-ene 2-carboxylic acid monohydrate.
BL-S578; MJF-11567-3; Baxan; Bidocef; Cefa-Drops; Cefamox; Ceforal; Cephos; Duracef; Duricef; Kefroxil; Oracéfal; Sedral; Ultracef. Antibacterial. mp = 197° (dec). *Bristol-Myers Squibb Pharm. R&D.*

1947 Cefadroxil Hemihydrate
119922-85-9
$C_{16}H_{17}N_3O_5S.1/2H_2O$
[6R-[6α,7β(R*)]]-7-[[Amino-(4-hydroxyphenyl)acetyl]amino]-3-methyl-8-oxo-5-thia-1-azabicyclo[4.2.0]oct-2-ene 2-carboxylic acid hemihydrate.
Antibacterial. *Bristol-Myers Squibb Pharm. R&D.*

1948 Cefalonium
5575-21-3 226-948-7
$C_{20}H_{18}N_4O_5S_2$
3-(4-Carbamoylpyridylmethyl)-8-oxo-7-(phenylacetamido)-5-thia-1-azabicyclo[4.2.0]oct-2-ene-2-carboxylic acid.
Cepravin D.C. Cephalosporin antibiotic; used to treat mastitis in cows.

1949 Cefaloram
859-07-4 212-725-1
$C_{18}H_{18}N_2O_6S$
(Acetoxymethyl)-8-oxo-7-(phenylacetamido)-5-thia-1-azabicyclo[4.2.0]oct-2-ene-2-carboxylic acid.
Antibacterial.

1950 Cefamandole
34444-01-4 1964 252-030-0
$C_{18}H_{18}N_6O_5S_2$
[6R-[6α,7β(R*)]]-7-[(Hydroxyphenylacetyl)amino]-3-[[(1-methyl-1H-tetrazol-5-yl)thio]methyl]-8-oxo-5-thia-1-azabicyclo[4.2.0]oct-2-ene 2-carboxylic acid.
CMT; compd 83405. Antibacterial. *Eli Lilly & Co.*

1951 Cefamandole Nafate
42540-40-9 1964 255-877-4
$C_{19}H_{17}N_6NaO_6S_2$
[6R-[6α,7β(R*)]]-7-[(Hydroxyphenylacetyl)amino]-3-[[(1-methyl-1H-tetrazol-5-yl)thio]methyl]-8-oxo-5-thia-1-azabicyclo[4.2.0]oct-2-ene 2-carboxylic acid salt with sodium sulfate.
Bergacef; Cedol; Cefam; Cefiran; Cemado; Cemandil; Fado; Kefadol; Kefandol; Lampomandol; Mandokef; Mandol; Mandolsan; Neocefal; Pavecef. Antibacterial. mp = 190° (dec); λ_m = 269 nm (ε 10800 H_2O); soluble in H_2O, EtOH; insoluble in organic solvents. *Eli Lilly & Co.*

1952 Cefamandole Sodium
30034-03-8 250-009-0
$C_{18}H_{17}N_6NaO_5S_2$
[6R-[6α,7β(R*)]]-7-[(Hydroxyphenylacetyl)amino]-3-[[(1-methyl-1H-tetrazol-5-yl)thio]methyl]-8-oxo-5-thia-1-azabicyclo[4.2.0]oct-2-ene 2-carboxylic acid sodium salt.
Antibacterial. *Eli Lilly & Co.*

1953 Cefaparole
51627-20-4 257-325-8
$C_{19}H_{19}N_5O_5S_3$
(6R,7R)-7-[(R)-2-Amino-2-(p-hydroxyphenyl)acetamido]-3-[[(5-methyl-1,3,4-thiadiazol-2-yl)thio]methyl]-8-oxo-5-thia-1-azabicyclo[4.2.0]oct-2-ene-2-carboxylic acid.
110264. Antibacterial. *Eli Lilly & Co.*

1954 Cefatrizine
51627-14-6 1965 257-324-2
$C_{18}H_{11}8N_6O_5S_2$
[6R-[6α,7β(R*)]]-7-[[Amino(4-hydroxyphenyl)acetyl]-amino]-8-oxo-3-[(1H-1,2,3-triazol-4-ylthio)methyl]-5-thia-1-azabicyclo[4.2.0]oct-2-ene-2-carboxylic acid.
BL-S640; SKF-60771; S-640P; Cefaperos. Antibacterial. LD_{50} (mmus ip) = 6880 mg/kg, (fmus ip) = 6410 mg/kg, (mrat ip) = 4325 mg/kg, (frat ip) = 4325 mg/kg. *Bristol-Myers Squibb Pharm. R&D.*

1955 Cefazaflur
58665-96-6
$C_{13}H_{13}F_3N_6O_4S_3$
(6R,7R)-3-[[(1-Methyl-1H-tetrazol-5-yl)thio]methyl]-8-oxo-7-[2-[(trifluoromethyl)thio]acetamido]-5-thia-1-azabicyclo[4.2.0]oct-2-ene-2-carboxylic acid.
Antibacterial. *SmithKline Beecham Pharm.*

1956 Cefazaflur Sodium
52123-49-6
$C_{13}H_{12}F_3N_6NaO_4S_3$
Sodium (6R,7R)-3-[[(1-methyl-1H-tetrazol-5-yl)thio]-methyl]-8-oxo-7-[2-[(trifluoromethyl)thio]acetamido]-5-thia-1-azabicyclo[4.2.0]oct-2-ene-2-carboxylate.
SK&F-59962. Antibacterial. *SmithKline Beecham Pharm.*

1957 Cefazedone
56187-47-4 1966
$C_{18}H_{15}Cl_2N_5O_5S_3$
(6R,7R)-7-[[(3,5-Dichloro-4-oxo-1(4H)-pyridinyl)acetyl]amino]-3-[[(5-methyl-1,3,4-thiadiazol-2-yl)thio]methyl]-8-5-thia-1-azabicyclo[4.2.0]oct-2-ene-2-carboxylic acid.
EMD-30087. Antibacterial. *E. Merck.*

1958 Cefazedone Sodium
63521-15-3 1966 264-293-9
$C_{18}H_{14}Cl_2N_5NaO_5S_3$
(6R-trans)-7-[[(3,5-Dichloro-4-oxo-1(4H)-pyridinyl)acetyl]amino]-3-[[(5-methyl-1,3,4-thiadiazol-2-yl)thio]methyl]-8-5-thia-1-azabicyclo[4.2.0]oct-2-ene-2-carboxylic acid sodium salt.
Refosporin. Antibacterial. LD_{50} (mus iv) = 6800 mg/kg, (rat iv) = 4225 mg/kg, (rbt iv) = 3200 mg/kg, (dog iv) = 3000 mg/kg. *Bayer Corp. Pharm. Div.*

1959 Cefazolin
25953-19-9 1967 247-362-8
$C_{14}H_{14}N_8O_4S_3$
(6R-trans)-3-[[(5-Methyl-1,3,4-thiadiazol-2-yl)thio]methyl]-8-oxo-7-[(1H-tetrazol-1-ylacetyl)amino]-5-thia-1-azabicyclo[4.2.0]oct-2-ene 2-carboxylic acid.
CEZ. Antibacterial. mp = 198-200° (dec); λ_m 272 nm (ε 13150 pH 6.4); easily soluble in DMF, C_5H_5N; soluble in Me_2CO, aqueous dioxane, aqueous EtOH, slightly soluble in MeOH; practically insoluble in $CHCl_3$, C_6H_6, Et_2O. *C.M. Ind.; Fujisawa Pharm. USA Inc.*

1960 Cefazolin Sodium
27164-46-1 1967 248-278-4
$C_{14}H_{13}N_8NaO_4S_3$
(6R-trans)-3[[(5-Methyl-1,3,4-thiadiazol-2-yl)thio]methyl]-8-oxo-7-[(1H-tetrazol-1-ylacetyl)amino]-5-thia-1-azabicyclo[4.2.0]oct-2-ene-2-carboxylic acid sodium salt.
sodium CEZ; SKF-41558; Acef; Ancef; Atirin; Biazolina; Bor-Cefazol; Cefacidal; Cefamedin; Cefamezin; Cefazil; Cefazina; Elzogram; Firmacef; Gramaxin; Kefzol; Lampocef; Liviclina; Totacef; Zolicef. Antibacterial. Crystallizes in three forms, readily soluble in H_2O, less soluble in organic solvents; LD_{50} (rat ip) = 7.4 mg/kg. *Fujisawa Pharm. USA Inc.*

1961 Cefbuperazone
76610-84-9 1968
$C_{22}H_{29}N_9O_9S_2$
[6R-[6α,7α,7(2R*)]]-7-[[2-[[(4-Ethyl-2,3-dioxo-1-piperazinyl)carbonyl]amino]-3-hydroxy-1-oxobutyl]amino]-7-methoxy-3-[[(1-methyl-1H-tetrazol-5-yl)thio]methyl]-8-oxo-5-thia-1-azabicyclo[4.2.0]oct-2-ene-2-carboxylic acid.
Antibacterial. mp = 118-120°. *Toyama Chem. Co. Ltd.*

1962 Cefcanel
41952-52-7
$C_{19}H_{18}N_4O_5S_3$
(6R,7R)-7-[(R)-Mandelamido]-3-[[(5-methyl-1,3,4-thiadiazol-2-yl)thio]methyl]-8-oxo-5-thia-1-azabicyclo[4.2.0]oct-2-ene-2-carboxylic acid.
Cephalosporin antibiotic.

1963 Cefcanel Daloxate
97275-40-6
$C_{27}H_{27}N_5O_9S_3$
(6R,7R)-7-[(R)-Mandelamido]-3-[[(5-methyl-1,3,4-thiadiazol-2-yl)thio]methyl]-8-oxo-5-thia-1-azabicyclo[4.2.0]oct-2-ene-2-carboxylic acid cyclic 2,3-carbonate ester with L-alanine.
Cephalosporin antibiotic.

1964 Cefcapene
135889-00-8 1969
$C_{17}H_{19}N_5O_6S_2$
[6R-[6α,7β(Z)]]-3-[[(Aminocarbonyl)oxy]methyl]-7-[[2-(2-amino-4-thiazolyl)-1-oxo-2-pentenyl]amino]-8-oxo-5-thia-1-azabicyclo[4.2.0]oct-2-ene-2-carboxylic acid .
Cefcapene pivoxil free acid; S-1006. Antibacterial. *Shionogi & Co. Ltd.*

1965 Cefcapene Pivoxil
105889-45-0
$C_{23}H_{29}N_5O_8S_2$
[6R-[6α,7β(Z)]]-3-[[(Aminocarbonyl)oxy]methyl]-7-[[2-(2-amino-4-thiazolyl)-1-oxo-2-pentenyl]amino]-8-oxo-5-thia-1-azabicyclo[4.2.0]oct-2-ene-2-carboxylic acid (2,2-dimethyl-1-oxopropoxy)methyl ester.
S-1108; Flumax. Antibacterial. *Shionogi & Co. Ltd.*

1966 Cefclidin
105239-91-6 1970
$C_{21}H_{26}N_8O_6S_2$
[6R-[6α,7β(Z)]]-4-(Aminocarbonyl)-1-[[7-[[(5-amino-1,2,4-thiadiazol-3-yl)(methoxyimino)acetyl]-amino]-2-carboxy-8-oxo-5-thia-1-azabicyclo[4.2.0]oct-2-ene-3-yl]methyl-1-azoniabicyclo[2.2.2]octane inner salt.
E-1040. Antibacterial. LD_{50} (mmus iv) > 10000 mg/kg, (mmus orl) > 10000 mg/kg, (mmus sc) > 10000 mg/kg, (mmus im) > 5000 mg/kg (fmus iv) > 10000 mg/kg, (fmus orl) > 10000 mg/kg, (fmus sc) > 10000 mg/kg, (fmus im) > 5000 mg/kg, (mrat iv) = 2236 mg/kg, (mrat orl) > 10000 mg/kg, (mrat sc) > 10000 mg/kg, (mrat im) > 5000 mg/kg, (frat iv) = 2147 mg/kg, (frat orl) > 10000 mg/kg, (frat sc) > 10000 mg/kg, (frat im) > 5000 mg/kg. *Eisai Co. Ltd.*

1967 Cefdaloxime
80195-36-4
$C_{14}H_{15}N_5O_6S_2$
(+)-(6R,7R)-7-[2-(2-Amino-4-thiazolyl)glyoxylamido]-3-(methoxymethyl)-8-oxo-5-thia-1-azabicyclo[4.2.0]oct-2-ene-2-carboxylic acid 7^2-(Z)-oxime.
Antibacterial.

1968 Cefdinir
91832-40-5 1971
$C_{14}H_{13}N_5O_5S_2$
[6R-[6α,7β(Z)]]-7-[[(2-Amino-4-thiazolyl)(hydroxyimino)-acetyl]amino]-3-ethenyl-8-oxo-5-thia-1-azabicyclo[4.2.0]oct-2-ene-2-carboxylic acid.
FK-482; BMY-28488. Antibacterial. mp = 170° (dec); λ_m

223, 286 (ε 17400, 19700 pH 7 phosphate buffer). *Fujisawa Pharm. USA Inc.*

1969 Cefditoren
104145-95-1 1972
$C_{19}H_{18}N_6O_5S_3$
[6R-[3(Z),6α,7β(Z)]]-7-[[(2-Amino-4-thiazolyl)(methoxyimino)acetyl]amino]-3-[2-(4-methyl-5-thiazolyl)ethenyl]-8-oxo-5-thia-1-azabicyclo[4.2.0]oct-2-ene-2-carboxylic acid.
ME-1206. Antibacterial. *Meiji Seika Kaisha Ltd.*

1970 Cefditoren Pivoxil
117467-28-4 1972
$C_{25}H_{28}N_6O_7S_3$
[6R-[3(Z),6α,7β(Z)]]-7-[[(2-Amino-4-thiazolyl)(methoxyimino)acetyl]amino]-3-[2-(4-methyl-5-thiazolyl)ethenyl]-8-oxo-5-thia-1-azabicyclo[4.2.0]oct-2-ene-2-carboxylic acid pivaloyloxymethyl ester.
ME-1207; Meiact. Antibacterial. mp = 127-129°; $[\alpha]_D^{20}$ = -48.5° (c = 0.5 in MeOH). *Meiji Seika Kaisha Ltd.*

1971 Cefepime
88040-23-7 1973
$C_{19}H_{24}ClN_6O_5S_2$
[6R-[6α,7β(Z)]]-1-[[7-[[2-Amino-4-thiazolyl)(methoxyimino)acetyl]amino]-2-carboxy-8-oxo-5-thia-1-azabicyclo[4.2.0]oct-2-en-3-yl]methyl]-1-methylpyrrolidinium inner salt.
BMY-28142. Antibacterial. mp = 150° (dec); λ_m 235, 257 nm (ε16700, 16100 pH 7 phosphate buffer). *Bristol-Myers Squibb Pharm. R&D.*

1972 Cefepime Hydrochloride
123171-59-5 1973
$C_{19}H_{25}ClN_6O_5S_2.HCl.H_2O$
[6R-[6α,7β(Z)]]-1-[[7-[[2-Amino-4-thiazolyl)(methoxyimino)acetyl]amino]-2-carboxy-8-oxo-5-thia-1-azabicyclo[4.2.0]oct-2-en-3-yl]methyl]-1-methylpyrrolidinium monohydrochloride monohydrate.
Axepim. Antibacterial. *Bristol-Myers Squibb Pharm. R&D.*

1973 Cefetamet
65052-63-3 1974
$C_{14}H_{15}N_5O_5S_2$
[6R-[6α,7β(Z)]]-7-[[(2-Amino-4-thiazolyl)(methoxyimino)-acetyl]amino]-8-oxo-5-thia-1-azabicyclo[4.2.0]oct-2-ene-2-carboxylic acid.
deacetoxycefotaxime. Antibacterial. *Roussel-UCLAF.*

1974 Cefetecol
127182-67-6
$C_{20}H_{17}N_5O_9S_2$
(6R,7R)-7-[2-(2-Amino-4-thiazolyl)glyoxylamido]-8-oxo-5-thia-1-azabicyclo[4.2.0]-oct-2-ene-2-carboxylic acid.
GR-69153X. Antibacterial. *Glaxo Labs.*

1975 Cefetrizole
65307-12-2
$C_{16}H_{15}N_5O_4S_3$
(6R,7R)-8-Oxo-7-[2-(2-thienyl)acetamido]-3-[(s-triazolo-3-ylthio)methyl]-5-thia-1-azabicyclo[4.2.0]oct-2-ene-2-carboxylic acid.
Cephalosporin antibiotic.

1976 Cefivitril
66474-36-0
$C_{15}H_{15}N_7O_4S_3$
(6R,7R)-7-[2-[[(Z)-2-Cyanovinyl]thio]acetamido]-3-[[(1-methyl-1H-tetrazol-5-yl)thio]methyl]-8-oxo-5-thia-1-azabicyclo[4.2.0]oct-2-ene-2-carboxylic acid.
Cephalosporin antibiotic.

1977 Cefixime
79350-37-1 1975
$C_{16}H_{15}N_5O_7S_2$
[6R-[6α,7β(Z)]]-7-[[(2-Amino-4-thiazolyl)[(carboxymethoxy)imino]acetyl]amino]-3-ethyenyl-8-oxo-5-thia-1-azabicyclo[4.2.0]oct-2-ene-2-carboxylic acid.
FK-027; FR-17027; CL-284635; Cefixoral; Cefspan; Cephoral; Oroken; Suprax; Unixime. Antibacterial. *Fujisawa Pharm. USA Inc.; Lederle Labs.*

1978 Cefluprenam
116853-25-9
$C_{20}H_{25}FN_8O_6S_2$
(-)-[(E)-3-[(6R,7R)-7-[2-(5-Amino-1,2,4-thiadiazol-3-yl)glyoxylamido]-2-carboxy-8-oxo-5-thia-1-azabicyclo-[4.2.0]oct-2-en-3-yl]allyl](carbamoylmethyl)ethyl-methylammonium hydroxide inner salt 7^2-(Z)-[O-(fluoromethyl)oxime].
Cephalosporin antibiotic.

1979 Cefmenoxime
65085-01-0 1976
$C_{16}H_{17}N_9O_5S_3$
[6R-[6α,7β(Z)]]-7-[[(2-amino-4-thiazolyl)-(methoxyimino)acetyl]amino]-3-[[(1-methyl-1H-tetrazol-5-yl)thio]methyl]-8-oxo-5-thia-1-azabicyclo[4.2.0]oct-2-ene-2-carboxylic acid.
SCE-1365. Antibacterial. *Abbott Labs Inc.; Roussel-UCLAF; Takeda Chem. Ind. Ltd.*

1980 Cefmenoxime Hydrochloride
75738-58-8 1976 278-299-4
$C_{32}H_{35}ClN_{18}O_{10}S_6$
[6R-[6α,7β(Z)]]-7-[[(2-Amino-4-thiazolyl)-(methoxyimino)acetyl]amino]-3-[[(1-methyl-1H-tetrazol-5-yl)thio]methyl]-8-oxo-5-thia-1-azabicyclo[4.2.0]-oct-2-ene-2-carboxylic acid hydrochloride (syn isomer).
(6R,7R)-7-[2-(2-amino-4-thiazolyl)glyoxylamido]-3-[[(1-methyl-1H-tetrazol-5-yl)thio]-methyl]-8-oxo-5-thia-1-aza-bicyclo[4.2.0]oct-2-ene-2-carboxylic acid 7^2-(Z)-(O-methyloxime hydrochloride; Cefmax; SCE 1365; Abbott 50192; Bestcall; Cefmax; Cemix; Tacef. Antibacterial. *Abbott Labs Inc.; Roussel-UCLAF; Takeda Chem. Ind. Ltd.*

1981 Cefmetazole
56796-20-4 1977 260-384-2
$C_{15}H_{17}N_7O_5S_3$
(6R-cis)-7-[[[(Cyanomethyl)thio]acetyl]amino]-7-methoxy-3-[[(1-methyl-1H-tetrazol-5-yl)thio]-methyl]-8-oxo-5-thia-1-azabicyclo[4.2.0]oct-2-ene-2-carboxylic acid.
CS-1170; SKF-83088. Antibacterial. Very soluble in H_2O, MeOH; soluble in Me_2CO; LD_{50} (rat iv) > 5000 mg/kg. *Squibb E.R. & Sons.*

1982 Cefmetazole Sodium
56796-39-5 1977
$C_{15}H_{16}N_7NaO_5S_3$
(6R-cis)-7-[[[(Cyanomethyl)thio]acetyl]amino]-7-methoxy-3-[[(1-methyl-1H-tetrazol-5-yl)thio]methyl]-8-oxo-5-thia-1-azabicyclo[4.2.0]oct-2-ene-2-carboxylic acid sodium salt.
Cefmetazon; Metafar; Metazol; Zefazone. Antibacterial. Very soluble in H_2O, MeOH; soluble in Me_2CO; LD_{50} (rat iv) > 5000 mg/kg. *Sankyo Co. Ltd.*

1983 Cefminox
84305-41-9 1978
$C_{16}H_{21}N_7O_7S_3$
[6R-[6α,7α,7(S*)]]-7-[[[(2-Amino-2-carboxyethyl)-thio]acetyl]amino]-7-methoxy-3-[[(1-methyl-1H-tetrazol-5-yl)thio]methyl]-8-oxo-5-thia-1-azabicyclo[4.2.0]oct-2-ene-2-carboxylic acid.
Cephamycin antibacterial. Semisynthetic, broad-spectrum antibiotic. *Meiji Seika Kaisha Ltd.*

1984 Cefminox
75481-73-1
$C_{16}H_{21}N_7O_7S_3$
(6R,7S)-7-[2-[[(S)-2-Amino-2-carboxyethyl]-thio]acetamido]-7-methoxy-3-[[(1-methyl-1H-tetrazol-5-yl)thio]methyl]-8-oxo-5-thia-1-azabicyclo[4.2.0]oct-2-ene-2-carboxylic acid.
Cephalosporin antibiotic.

1985 Cefminox Sodium Salt
75498-96-3 1978
$C_{16}H_{20}N_7NaO_7S_3.7H_2O$
[6R-[6α,7α,7(S*)]]-7-[[[(2-Amino-2-carboxyethyl)-thio]acetyl]amino]-7-methoxy-3-[[(1-methyl-1H-tetrazol-5-yl)thio]methyl]-8-oxo-5-thia-1-azabicyclo[4.2.0]oct-2-ene-2-carboxylic acid sodium salt heptahydrate.
MT-141; Meicelin. Antibacterial. mp = 90-91°; LD_{50} (mmus iv) = 6100 mg/kg, (fmus iv) = 5200 mg/kg, (mrat iv) = 6600 mg/kg, (mrat ip) = 8600 mg/kg, (mrat orl) > 15000 mg/kg, (frat iv) = 5700 mg/kg, (frat ip) = 8550 mg/kg, (frat orl) > 15000 mg/kg. *Meiji Seika Kaisha Ltd.*

1986 Cefodizime
69739-16-8 1979
$C_{20}H_{20}N_6O_7S_4$
[6R-(6α,7β)]-7-[[(2-Amino-4-thiazolyl)methoxyimino)-acetyl]amino]-3-[[[5-(carboxymethyl)-4-methyl-2-thiazolyl]thio]methyl]-8-oxo-5-thia-1-azabicyclo[4.2.0]-oct-2-ene-2-carboxylic acid.
Antibacterial. $[\alpha]_D^{25}$ = -55.9°; λ_m 228,260, 288 nm (log ε 4.25, 4,25, 4,20, H_2O). *Hoechst AG.*

1987 Cefodizime Disodium Salt
86329-79-5 1979
$C_{20}H_{18}N_6Na_2O_7S_4$
[6R-(6α,7β)]-7-[[(2-Amino-4-thiazolyl)methoxyimino)-acetyl]amino]-3-[[[5-(carboxymethyl)-4-methyl-2-thiazolyl]thio]methyl]-8-oxo-5-thia-1-azabicyclo[4.2.0]-oct-2-ene-2-carboxylic acid disodium salt.
HR-221; THR-221; Kenicef; Timecef. Antibacterial. Soluble in H_2O (≅27.0 g/100 ml); LD_{50} (mus iv) = 4000-8000 mg/kg, (rbt iv) = 4000-8000 mg/kg, (rat iv) = 4000-8000 mg/kg, (rat sc) = 15000-17500 mg/kg, (rat ip) = 8000-22000 mg/kg. *Hoechst AG.*

1988 Cefonicid
61270-58-4 1980
$C_{18}H_{18}N_6O_8S_3$
[6R-[6α,7β(R*)]]-7-[(Hydroxyphenylacetyl)amino]-8-oxo-3-[[1-(sulfomethyl)-1H-tetrazol-5-yl]thio]methyl]-5-thia-1-azabicyclo[4.2.0]oct-2-ene-2-carboxylic acid.
Antibacterial. *SmithKline Beecham Pharm.*

1989 Cefonicid Sodium
71420-79-6 1980
$C_{18}H_{16}N_6Na_2O_8S_3$
[6R-[6α,7β(R*)]]-7-[(Hydroxyphenylacetyl)amino]-8-oxo-3-[[1-(sulfomethyl)-1H-tetrazol-5-yl]thio]methyl]-5-thia-1-azabicyclo[4.2.0]oct-2-ene-2-carboxylic acid sodium salt.
SKF-75073; Cefodie; Mopnocid; Monocidur; Praticef. Antibacterial. *SmithKline Beecham Pharm.*

1990 Cefoperazone
62893-19-0 1981 263-749-4
$C_{25}H_{27}N_9O_8S_2$
[6R-[6α,7β(R*)]]-7-[[[[(4-Ethyl-2,3-dioxo-1-piperazinyl)carbonyl]amino](4-hydroxyphenyl)-acetyl]amino]-3-[[(1-methyl-1H-tetrazol-5-yl)thio]methyl]-8-oxo-5-thia-1-azabicyclo[4.2.0]oct-2-ene-2-carboxylic acid.
Antibacterial. mp = 169-171° (hydrated). *Roerig Div. Pfizer Pharm.; Toyama Chem. Co. Ltd.*

1991 Cefoperazone Sodium
62893-20-3 1981 263-751-5
$C_{25}H_{26}N_9NaO_8S_2$
[6R-[6α,7β(R*)]]-7-[[[[(4-Ethyl-2,3-dioxo-1-piperazinyl)carbonyl]amino](4-hydroxyphenyl)-acetyl]amino]-3-[[(1-methyl-1H-tetrazol-5-yl)thio]methyl]-8-oxo-5-thia-1-azabicyclo[4.2.0]oct-2-ene-2-carboxylic acid sodium salt.
CP-52640-2; T-1551; Bioperazone; Cefazone; Cefobid; Cefobine; Cefobis; Cefogram; Cefoneg; Cefosint; Dardum; Farecef; Kefazon; Novobiocyl; Pathozone; Peracef; Perocef; Tomabef. Antibacterial. *Roerig Div. Pfizer Pharm.; Toyama Chem. Co. Ltd.*

1992 Ceforanide
60925-61-3 1982
$C_{20}H_{21}N_7O_6S_2$
(6R-trans)-7-[[[2-(Aminomethyl)phenyl]acetyl]amino]-3-[[[1-(carboxymethyl)-1H-tetrazol-5-yl]thio]methyl]-8-oxo-5-thia-1-azabicyclo[4.2.0]oct-2-ene-2-carboxylic acid.
BL-S786; Precef [as sodium salt]. Antibacterial. mp >150° (dec). *Bristol-Myers Squibb Pharm. R&D.*

1993 Cefoselis
122841-10-5
$C_{19}H_{22}N_8O_6S_2$
(-)-5-Amino-2-[[(6R,7R)-7-[2-(2-amino-4-thiazolyl)-glyoxylamido]-2-carboxy-8-oxo-5-thia-1-azabicyclo-[4.2.0]oct-2-en-3-yl]methyl]-1-(2-hydroxyethyl)-pyrazolium hydroxide inner salt 7^2-(Z)-(O-methyloxime).
Cephalosporin antibiotic.

1994 Cefotaxime
63527-52-6 1983 264-299-1
$C_{16}H_{17}N_5O_7S_2$
[6R-[6α,7β(Z)]]-3-[(Acetyloxy)methyl]-7-[[(2-amino-4-thiazolyl)(methoxyimino)acetyl]amino]-8-oxo-5-thia-1-azabicyclo[4.2.0]oct-2-ene-2-carboxylic acid.

Cephalosporin antibacterial. Broad-spectrum. *Hoechst Roussel Pharm. Inc.; Roussel-UCLAF.*

1995 Cefotaxime Sodium
64485-93-4 1983 264-915-9
$C_{16}H_{16}N_5NaO_7S_2$
[6R-[6α,7β(Z)]]-3-[(Acetyloxy)methyl]-7-[[(2-amino-4-thiazolyl)(methoxyimino)acetyl]amino]-8-oxo-5-thia-1-azabicyclo[4.2.0]oct-2-ene-2-carboxylic acid sodium salt.
HR-756; RU-24756; Cefotax; Chemcef; Claforan; Pretor; Tolycar. Cephalosporin antibacterial. Broad-spectrum. $[\alpha]_D^{20}$ = +55 ± 2° (c = 0.8 in H_2O). *Hoechst Roussel Pharm. Inc.; Roussel-UCLAF.*

1996 Cefotetan
69712-56-7 1984 274-093-3
$C_{17}H_{17}N_7O_8S_4$
[6R-(6α,7α)]-7-[[[4-(2-Amino-1-carboxy-2-oxoethylidene)-1,3-dithietan-2-yl]carbonyl]amino]-7-methoxy-3-[[(1-methyl-1H-tetazol-5-yl)thio]methyl]-8-oxo-5-thia-1-azabicyclo[4.2.0]oct-2-ene-2-carboxylic acid.
Cefotan; ICI-156834. Antibacterial. Soluble in H_2O; LD_{50} (mmus iv) = 6350 mg/kg, (mmus ip) = 8120 mg/kg, (mmus sc,orl) > 10000 mg/kg, (mrat iv) = 8480, (mrat ip) = 8370 mg/kg, (mrat sc,orl) > 10000 mg/kg. *Zeneca Pharm.*

1997 Cefotetan Disodium
74356-00-6 1984 277-834-9
$C_{17}H_{15}N_7Na_2O_8S_4$
[6R-(6α,7α)]-7-[[[4-(2-Amino-1-carboxy-2-oxoethylidene)-1,3-dithietan-2-yl]carbonyl]amino]-7-methoxy-3-[[(1-methyl-1H-tetazol-5-yl)thio]methyl]-8-oxo-5-thia-1-azabicyclo[4.2.0]oct-2-ene-2-carboxylic acid disodium salt.
YM-09330; Apatef; Cefotan; Ceftenon; Cepan Darvilen. Antibacterial. LD_{50} (mmus iv) = 6350 mg/kg, (mmus ip) = 8120 mg/kg, (mmus sc,orl) > 10000 mg/kg, (mrat Iv) = 8480, (mrat ip) = 8370 mg/kg, (mrat sc,orl) > 10000 mg/kg. *Zeneca Pharm.*

1998 Cefotiam
61622-34-2 1985
$C_{18}H_{23}N_9O_4S_3$
(6R-trans)-7-[[(2-Amino-4-thiazolyl)-acetyl]amino]-3-[[[1-[2-(dimethylamino)ethyl]-1H-tetrazol-5-yl]-thio]methyl]-8-oxo-5-thia-1-azabicyclo[4.2.0]oct-2-ene-2-carboxylic acid.
SCE-963. Antibacterial. *Abbott Labs Inc.; Takeda Chem. Ind. Ltd.*

1999 Cefotiam Dihydrochloride
66309-69-1 1985 266-312-6
$C_{18}H_{25}Cl_2N_9O_4S_3$
(6R-trans)-7-[[(2-Amino-4-thiazolyl)-acetyl]amino]-3-[[[1-[2-(dimethylamino)ethyl]-1H-tetrazol-5-yl]-thio]methyl]-8-oxo-5-thia-1-azabicyclo[4.2.0]oct-2-ene-2-carboxylic acid dihydrochloride.
Abbott 48999; CGP-14221/E; Halospor; Pansporin; Pansporine; Spizef; Sporidyn. Cephalosporin antibacterial. Semisynthetic. Soluble in MeOH; slightly soluble in EtOH. *Abbott Labs Inc.; Takeda Chem. Ind. Ltd.*

2000 Cefoxazole
36920-48-6
$C_{21}H_{18}ClN_3O_7S$
(6R,7R)-7-[3-(o-Chlorophenyl)-5-methyl-4-isoxazole-carboxamido]-3-(hydroxymethyl)-8-oxo-5-thia-1-azabicyclo[4.2.0]oct-2-ene-2-carboxylic acid acetate (ester).
Cephalosporin antibiotic.

2001 Cefoxitin
35607-66-0 1986 252-641-2
$C_{16}H_{17}N_3O_7S_2$
(6R-cis)-3-[[(Aminocarbonyl)oxy]methyl]-7-methoxy-8-oxo-7-[(2-thienylacetyl)amino]-5-thia-1-azabicyclo[4.2.0]oct-2-ene-2-carboxylic acid.
Mefoxin. Antibacterial. mp = 149-150°; poorly soluble in H_2O, soluble in organic solvents; LD_{50} (rat orl) = 8980 mg/kg. *Merck & Co.Inc.*

2002 Cefoxitin Sodium
33564-30-6 1986 251-574-6
$C_{16}H_{16}N_3NaO_7S_2$
(6R-cis)-3-[[(Aminocarbonyl)oxy]methyl]-7-methoxy-8-oxo-7-[(2-thienylacetyl)amino]-5-thia-1-azabicyclo[4.2.0]oct-2-ene-2-carboxylic acid sodium salt.
Cefoxitin sodium salt; MK-306; Betacef; Farmoxin; Mefoxin; Mefoxitin; Merxin; Cenomycin. Antibacterial. $[\alpha]_{589nm}^{25}$ = +210° (c = 1 in MeOH); very soluble in H_2O; soluble in MeOH; sparingly soluble in EtOH, Me_2CO; insoluble in aromatic and aliphatic hydrocarbons; LD_{50} (mus iv) = 5.10 mg/kg, (rat iv) = 8.98 mg/kg, (dog iv) > 10 mg/kg. *Merck & Co.Inc.*

2003 Cefozopran
113359-04-9 1987
$C_{19}H_{17}N_9O_5S_2$
[6R-[6α,7β(Z)]]-1-[[7-[[-[[(5-Amino-1,2,4-thiadiazol-3-yl)(methoxyimino)acetyl]amino]-2-carboxy-8-oxo-5-thia-1-azabicyclo[4.2.0]oct-2-ene-3-yl]methyl]imidazo[1,2-b]pyridazinium inner salt.
Antibacterial. *Takeda Chem. Ind. Ltd.*

2004 Cefpimizole
84880-03-5 1988
$C_{28}H_{26}N_6O_{10}S_2$
[6R-[6α,7β(R*)]]-1-[[2-Carboxy-7-[[[[(5-carboxy-1H-imidazol-4-yl)carbonyl]amino]phenylacetyl]amino]-8-oxo-5-thia-1-azabicyclo[4.2.0]oct-2-en-3-yl]methyl]-4-(2-sulfoethyl)pyridinium hydroxide inner salt.
U-63196; AC-1370. Antibacterial. *Ajinomoto Co. Inc.*

2005 Cefpimizole Sodium
85287-61-2 1988
$C_{28}H_{25}N_6NaO_{10}S_2$
[6R-[6α,7β(R*)]]-1-[[2-Carboxy-7-[[[[(5-carboxy-1H-imidazol-4-yl)carbonyl]amino]phenylacetyl]amino]-8-oxo-5-thia-1-azabicyclo[4.2.0]oct-2-en-3-yl]methyl]-4-(2-sulfoethyl)pyridinium hydroxide inner salt monosodium salt.
U-63196E; Ajicef; Renilan. Antibacterial. $[\alpha]_D^{20}$ = -28.2° (c = 0.5 in H_2O); λ_m 257 nm (ε 22400 H_2O); soluble in H_2O; LD_{50} (mmus iv) = 2700 mg/kg, (mmus sc) = 8200 mg/kg, (mmus orl) > 15000 mg/kg, (fmus iv) = 2900 mg/kg, (fmus sc) = 6800 mg/kg, (fmus orl) > 15000 mg/kg, (mrat iv) = 4200 mg/kg, (mrat sc) = 12200 mg/kg, (mrat

orl) > 15000 mg /kg, (frat ip) = 3500 mg/kg, (frat sc) = 11500 mg/kg, (frat orl) > 15000 mg/kg. *Ajinomoto Co. Inc.*

2006 Cefpiramide
70797-11-4 1989
$C_{25}H_{24}N_8O_7S_2$
[6R-[6α,7β(R*)]]-7-[[[[(4-Hydroxy-6-methyl-3-pyridinyl)carbonyl]amino](4-hydroxyphenyl)-acetyl]amino]-3-[[(1-methyl-1H-tetrazol-5-yl)thio]methyl]-8-oxo-5-thia-1-azabicyclo[4.2.0]oct-2-ene-2-carboxylic acid.
Antibacterial. mp = 213-215°. *Sumitomo Pharm. Co. Ltd.*

2007 Cefpiramide Sodium
74849-93-7 1989
$C_{25}H_{23}N_8NaO_7S_2$
[6R-[6α,7β(R*)]]-7-[[[[(4-Hydroxy-6-methyl-3-pyridinyl)carbonyl]amino](4-hydroxyphenyl)-acetyl]amino]-3-[[(1-methyl-1H-tetrazol-5-yl)thio]methyl]-8-oxo-5-thia-1-azabicyclo[4.2.0]oct-2-ene-2-carboxylic acid sodium salt.
SM-1652; Wy-44635; Cefpiran; Suncefal; Sepatren. Antibacterial. *Sumitomo Pharm. Co. Ltd.*

2008 Cefpirome
84957-29-9 1990
$C_{22}H_{22}N_6O_5S_2$
[6R-[6α,7β(Z)]]-1-[[7-[[(2-Amino-4-thiazolyl)-methoxyimino)acetyl]amino]-2-carboxy-8-oxo-5-thia-1-azabicyclo[4.2.0]oct-2-ene-3-yl]methyl]-6,7-dihydro-5H-cyclopenta[b]pyrindinium inner salt.
HR-810. Antibacterial. LD_{50} (mus iv) = 1900-2400 mg/kg, (mus ip) = 2300-4200 mg/kg, (rat iv) = 1900-2150 mg/kg, (rat ip) = 5800-6550 mg/kg. *Hoechst AG.*

2009 Cefpirome Sulfate
98753-19-6 1990
$C_{22}H_{22}N_6O_5S_2.H_2SO_4$
[6R-[6α,7β(Z)]]-7-[[(2-Amino-4-thiazolyl)(methoxyimino)acetyl]amino]-2-carboxy-8-oxo-5-thia-1-azabicyclo[4.2.0]oct-2-ene-3-yl]methyl]-6,7-dihydro-5H-cyclopenta[b]pyridinium inner salt sulfate.
Broact; Cefrom; Keitin. Antibacterial. Dec 198-202°; $[\alpha]_D^{25}$ = -4.7° (c = 5 in H_2O); λ_m = 265 nm (ε 21100); soluble in aqueous buffer, pH 6.5 (>50g/100 ml). *Hoechst AG.*

2010 Cefpodoxime
80210-62-4
$C_{15}H_{17}N_5O_6S_2$
[6R-[6α,7β(Z)]]-7-[[(2-Amino-4-thiazolyl)methoxy-imino)acetyl]amino]-3-(methoxy-methyl)-8-oxo-5-thia-1-azabicyclo[4.2.0]oct-2-ene-2-carboxylic acid.
Antibacterial. *Sankyo Co. Ltd.*

2011 Cefpodoxime Proxetil
87239-81-4 1991
$C_{21}H_{27}N_5O_9S$
[6R-[6α,7β(Z)]]-7-[[(2-Amino-4-thiazolyl)methoxy-imino)acetyl]amino]-3-(methoxy-methyl)-8-oxo-5-thia-1-azabicyclo[4.2.0]oct-2-ene-2-carboxylic acid 1-[[(1-methylethoxy)carbonyl]oxy]ethyl ester.
CS-807; U-76252; Banan; Cefodox; Orelox; Otreon; Vantin. Antibacterial. LD_{50} (mmus sc) > 10000 mg/kg, (mmus ip) = 3502 mg/kg, (mmus orl) > 8000 mg/kg, (fmus sc) >10000 mg/kg, (fmus ip) = 2535 mg/kg, (fmus orl) > 8000 mg.kg, ((mrat sc) > 2000 mg/kg, (mrat ip) >4000 mg/kg, (mrat orl) > 4000 mg/kg, (frat sc) >2000 mg/kg, (frat ip) > 4000 mg/kg, (frat orl) > 4000 mg/kg. *Sankyo Co. Ltd.*

2012 Cefprozil [anhydrous]
92665-29-7 1992
$C_{18}H_{19}N_3O_5S$
[6R-[6α,7β(R*)]]-7-[[Amino(4-hydroxyphenyl)acetyl]-amino]-8-oxo-3-(1-propenyl)-5-thia-1-azabicyclo[4.2.0]oct-2-ene-2-carboxylic acid.
Antibacterial. *Bristol Laboratories.*

2013 Cefprozil Monohydrate
121123-17-9 1992
$C_{18}H_{19}N_3O_5S.H_2O$
[6R-[6α,7β(R*)]]-7-[[Amino(4-hydroxyphenyl)acetyl]amino]-8-oxo-3-(1-propenyl)-5-thia-1-azabicyclo[4.2.0]oct-2-ene-2-carboxylic acid monohydrate.
BMY-28100-03-800; Cefzil; Procef; BMY-28100 [Z form]; BMY-28167 [E form]; BBS-1067. Antibacterial. [Z Form]: mp = 218-220° (dec); λ_m 228, 279 nm (ε 12300, 9800 pH 7 phosphate buffer); [E Form]: mp = 230° (dec); λ_m 228, 292 nm (ε 13000, 16900 pH 7 phosphate buffer). *Bristol Laboratories.*

2014 Cefquinome
84957-30-2
$C_{23}H_{24}N_6O_5S_2$
[6R-[6α,7β(Z)]-7-[2-(2-Amino-4-thiazolyl)glyoxylamido]-2-carboxy-8-oxo-5-thia-1-azabicyclo[4.2.0]oct-2-en-3-yl]methyl]5,6,7,8-tetrahydroquinolinium hydroxide inner salt 7^2-(Z)-(O-methyloxime).
Antibacterial. *Hoechst Roussel Pharm. Inc.*

2015 Cefquinome Sulfate
118443-89-3
$C_{23}H_{24}N_6O_5S_2.H_2SO_4$
[6R-[6α,7β(Z)]-7-[2-(2-Amino-4-thiazolyl)glyoxylamido]-2-carboxy-8-oxo-5-thia-1-azabicyclo[4.2.0]oct-2-en-3-yl]methyl]5,6,7,8-tetrahydroquinolinium hydroxide inner salt 7^2-(Z)-(O-methyloxime) sulfate (1:1).
Antibacterial. *Hoechst Roussel Pharm. Inc.*

2016 Cefrotil
52231-20-6
$C_{20}H_{22}N_4O_4S$
(6R,7R)-3-Methyl-8-oxo-[2-[p-(1,4,5,6-tetrahydro-2-pyrimidinyl)phenyl]acetamido]-5-thia-1-aza-bicyclo[4.2.0]oct-2-ene-2-carboxylic acid.
Cephalosporin antibiotic.

2017 Cefroxadine
51762-05-1 1993 257-391-8
$C_{16}H_{19}N_3O_5S$
[6R-[6α,7β(R*)]]-7-[(Amino-1,4-cyclohexadien-1-ylacetyl)amino]-3-methoxy-8-oxo-5-thia-1-azabicyclo[4.2.0]oct-2-ene-2-carboxylic acid.
CGP-9000; Oraspor. Antibacterial. mp = 170° (dec); $[\alpha]_D^{20}$ = +87° (c = 1.093 in 0.1N HCl); λ_m 267 nm (ε 6100 0.1N HCl); LD_{50} (mus orl) > 6000 mg/kg, (mus ip) = 7090 mg/kg. *Ciba-Geigy Corp.*

2018 Cefsulodin
62587-73-9 1994
$C_{22}H_{20}N_4O_8S_2$
[6R-[6α,7β(R*)]]-4-(Aminocarbonyl)-1-[[2-carboxy-8-oxo-7-[(phenylsulfoacetyl)amino]-5-thia-1-azabicyclo-[4.2.0]oct-2-en-3-yl]methyl]pyridinium hydroxide inner salt monosodium salt.
Antibacterial. *Ciba-Geigy Corp.; Takeda Chem. Ind. Ltd.*

2019 Cefsulodin Sodium
52152-93-9 1994 257-692-4
$C_{22}H_{19}N_4NaO_8S_2$
[6R-[6α,7β(R*)]]-4-(Aminocarbonyl)-1-[[2-carboxy-8-oxo-7-[(phenylsulfoacetyl)amino]-5-thia-1-azabicyclo[4.2.0]-oct-2-en-3-yl]methyl]pyridinium inner salt sodium salt.
Cefsulodin sodium; 7-(α-sulphophenylacetamido)-3-(4'-carbamoylpyridinium)methyl-3-cephem-4-carboxylic acid sodium salt; sulcephalosporin; Abbot 46811; CGP-7174/E; SCE-129; Cefomonil; Monaspor; Pseudomonil; Pseudocef; Pyocefal; Takesulin; Tilmapor; Ulfar. Antibacterial. mp = 175° (dec); LD_{50} (mus orl) > 15000 mg/kg, (mus ip) > 4000 mg/kg. *Ciba-Geigy Corp.; Takeda Chem. Ind. Ltd.*

2020 Cefsumide
54818-11-0
$C_{17}H_{20}N_4O_6S_2$
(6R,7R)-7-[(2r)-2-Amino-2-(m-methanesulfonamido-phenyl)acetamido]-3-methyl-8-oxo-5-thia-1-azabicyclo[4.2.0]oct-2-ene-2-carboxylic acid.
Cephalosporin antibiotic.

2021 Ceftazidime
72558-82-8 1995
$C_{22}H_{22}N_6O_7S_2$
[6R-[6α,7β(Z)]]-4-(Aminocarbonyl)-1-[[7-[[(2-amino-4-thiazolyl)[(1-carboxy-1-methylethoxy)imino]acetyl]-amino]-2-carboxy-8-oxo-5-thia-1-azabicyclo[4.2.0]oct-2-en-3-yl]methyl]pyridinium inner salt sodium salt.
GR-20263. Cephalosporin antibiotic. *Eli Lilly & Co.; Glaxo Wellcome Inc.; SmithKline Beecham Pharm.*

2022 Ceftazidime Pentahydrate
78439-06-2 1995 276-715-9
$C_{22}H_{22}N_6O_7S_2.5H_2O$
[6R-[6α,7β(Z)]]-4-(Aminocarbonyl)-1-[[7-[[(2-amino-4-thiazolyl)[(1-carboxy-1-methylethoxy)imino]acetyl]-amino]-2-carboxy-8-oxo-5-thia-1-azabicyclo[4.2.0]oct-2-en-3-yl]methyl]pyridinium inner salt sodium salt pentahydrate.
Ceftim; Fortam; Fortaz; Fortum; Glazidim; Kefadim; Kefamin; Kefazim; Modacin; Panzid; Spectrum; Starcef; Tazicef; Tazidime. Cephalosporin antibiotic. λ_m = 257 nm ($E^{1\%}_{1\,cm}$ 348). *Eli Lilly & Co.; Glaxo Wellcome Inc.; SmithKline Beecham Pharm.*

2023 Cefteram
82547-58-8 1996
$C_{16}H_{17}N_9O_5S_2$
[6R-[6α,7β(Z)]]-7-[[(2-Amino-4-thiazolyl)(methyoxyimino)acetyl]amino]-3-[(5-methyl-2H-tetrazol-2-yl)methyl]-8-oxo-5-thia-1-azabicyclo[4.2.0]oct-2-en-2-carboxylic acid.
ceftetrame; Ro-19-5247; T-2525. Antibacterial. mp > 200°. *Toyama Chem. Co. Ltd.*

2024 Ceftezole
26973-24-0 1997
$C_{13}H_{12}N_8O_4S_3$
(6R-trans)-8-Oxo-7-[(1H-tetrazol-1-ylacetyl)amino]-5-thia-1-azabicyclo[4.2.0]oct-2-ene-2-carboxylic acid.
CTZ; CG-B3Q. Antibacterial. mp = 155° (dec); λ_m 273 nm ($E^{1\%}_{1\,cm}$ 274 pH 6.4). *Fujisawa Pharm. USA Inc.*

2025 Ceftibuten
97519-39-6 1998
$C_{15}H_{14}N_4O_6S_2$
[6R-[6α,7β(Z)]]-7-[[2-(2-Amino-4-thiazolyl)-4-carboxy-1-oxo-2-butenyl]amino]-8-oxo-5-thia-1-azabicyclo[4.2.0]oct-2-ene-2-carboxylic acid.
7432-S; Sch-39720. Antibacterial. *Shionogi & Co. Ltd.*

2026 Ceftiofur
80370-57-6 1999
$C_{19}H_{17}N_5O_7S_3$
[6R-[6α,7β(Z)]]-7-[[(2-Amino-4-thiazolyl)-(methoxyimino)acetyl]amino]-3-[[(2-furanylcarbonyl)thio]methyl]-8-oxo-5-thia-1-azabicyclo[4.2.0]oct-2-ene-2-carboxylic acid.
Antibacterial. *Sanofi Winthrop.*

2027 Ceftiofur Hydrochloride
103980-44-5 1999
$C_{19}H_{17}N_5O_7S_3.HCl$
[6R-[6α,7β(Z)]]-7-[[(2-Amino-4-thiazolyl)(methoxyimino)acetyl]amino]-3-[[(2-furanylcarbonyl)-thio]methyl]-8-oxo-5-thia-1-azabicyclo[4.2.0]oct-2-ene-2-carboxylic acid monohydrocholoride.
U-67279A. Antibacterial. *Sanofi Winthrop.*

2028 Ceftiofur Sodium
104010-37-9 1999
$C_{19}H_{16}N_5NaO_7S_3$
[6R-[6α,7β(Z)]]-7-[[(2-Amino-4-thiazolyl)-(methoxyimino)acetyl]amino]-3-[[(2-furanylcarbonyl)-thio]methyl]-8-oxo-5-thia-1-azabicyclo[4.2.0]oct-2-ene-2-carboxylic acid monosodium salt.
CM-31916; U-64279E; Excenel; Naxcel. Antibacterial. *Sanofi Winthrop.*

2029 Ceftiolene
77360-52-2
$C_{20}H_{18}N_8O_8S_3$
(6R,7R)-7-[2-(2-Amino-4-thiazolyl)glyoxylamido]-3-[(E)-2-[[4-(formylmethyl)-1,4,5,6-tetrahydro-5,6-dioxo-as-triazin-3-yl]thio]vinyl]-8-oxo-5-thia-1-azabicyclo[4.2.0]oct-2-ene-2-carboxylic acid 7^2-(Z)-(O-methyloxime).
Cephalosporin antibiotic.

2030 Ceftioxide
71048-88-9
$C_{16}H_{17}N_5O_8S_2$
(5S,6R,7R)-7-[2-(2-Amino-4-thiazolyl)glyoxylamido]-3-(hydroxymethyl)-8-oxo-5-thia-1-azabicyclo[4.2.0]oct-2-ene-2-carboxylic acid 7^2-(Z)-(O-methyloxime).
Cephalosporin antibiotic.

2031 Ceftizoxime
68401-81-0 2000
$C_{13}H_{13}N_5O_5S_2$
[6R-[6α,7β(Z)]]-7-[[(2-Amino-4-thiazolyl)-(methoxyimino)acetyl]amino]-8-oxo-5-thia-1-azabicyclo[4.2.0]oct-2-ene-2-carboxylic acid.
Antibacterial. *Fujisawa Pharm. USA Inc.*

2032 Ceftizoxime Sodium
68401-82-1 2000
$C_{13}H_{12}N_5NaO_5S_2$
[6R-[6α,7β(Z)]]-7-[[(2-Amino-4-thiazolyl)-(methoxyimino)acetyl]amino]-8-oxo-5-thia-1-azabicyclo[4.2.0]oct-2-ene-2-carboxylic acid sodium salt.
FK-749; FR-13479; SKF-88373; Cefizox; Ceftix; Ceftizon; Epocelin; Eposerin. Antibacterial. LD_{50} (rat, mice iv) = ≅ 6000 mg/kg. *Fujisawa Pharm. USA Inc.*

2033 Ceftriaxone
73384-59-5 2001 277-405-6
$C_{18}H_{18}N_8O_7S_3$
[6R-[6α,7β(Z)]]-7-[[(2-Amino-4-thiazolyl)(methoxyimino)acetyl]amino]-8-oxo-3-[[(1,2,5,6-tetrahydro-2-methyl-5,6-dioxo-1,2,4-triazin-3-yl)thio]methyl]-5-thia-1-azabicyclo[4.2.0]oct-2-en-2-carboxylic acid.
ceftriaxone. Antibacterial. *Hoffmann-LaRoche Inc.*

2034 Ceftriaxone Sodium
104376-79-6 2001
$C_{18}H_{16}N_8Na_2O_7S_3.3.5H_2O$
[6R-[6α,7β(Z)]]-7-[[(2-Amino-4-thiazolyl)-(methoxyimino)acetyl]amino]-8-oxo-3-[[(1,2,5,6-tetrahydro-2-methyl-5,6-dioxo-1,2,4-triazin-3-yl)thio]methyl]-5-thia-1-azabicyclo[4.2.0]oct-2-ene-2-carboxylic acid disodium salt hemiheptahydrate.
Ro-13-9904/001; Rocefin; Rocephin(e). Antibacterial. mp >155° (dec); $[\alpha]_D^{25}$ = -165° (c = 1 in H_2O); λ_m (H_2O) = 242, 272 nm (ε 32300, 29530); soluble in H_2O (40 g/100 ml); LD_{50} (mmus iv) = 3000 mg/kg, (mmus orl) > 10000 mg/kg, (fmus iv) = 2800 mg/kg, (fmus orl) > 10000 mg/kg, (mrat iv) = 2175 mg/kg, (mrat orl) > 10000 mg/kg, (frat iv) = 2175 mg/kg, (frat orl) > 10000 mg/kg. *Hoffmann-LaRoche Inc.*

2035 Cefuracetime
39685-31-9
$C_{17}H_{17}N_3O_8S$
(6R,7R)-7-[2-(2-Furyl)glyoxylamido]-3-(hydroxymethyl)-8-oxo-5-thia-1-azabicyclo[4.2.0]oct-2-ene-2-carboxylic acid 7^2-(Z)-(O-methyloxime).
Cephalosporin antibiotic.

2036 Cefuroxime
55268-75-2 2002 259-560-1
$C_{16}H_{16}N_4O_8S$
[6R-[6α,7β(Z)]]-3-[[(Aminocarbonyl)oxy]methyl]]-7-[[2-furanyl(methoxyimino)acetyl]amino]-8-oxo-5-thia-1-azabicyclo[4.2.0]oct-2-ene-2-carboxylic acid.
Antibacterial. $[\alpha]_D^{20}$ = +63.7° (c = 1.0 in 0.2 M phosphate buffer, pH 7); λ_m 274 nm (ε 17600 pH 6). *Glaxo Wellcome Inc.*

2037 Cefuroxime Axetil
64544-07-6
$C_{20}H_{22}N_4O_{10}S$
[6R-[6α,7β(Z)]]-3-[[(Aminocarbonyl)oxy]methyl]]-7-[[2-furanyl(methoxyimino)acetyl]amino]-8-oxo-5-thia-1-azabicyclo[4.2.0]oct-2-ene-2-carboxylic acid 1-acetoxyethyl ester.
CCI-15641; Ceftin; Cefurax; Cefazine; Elobact; Oraxim; Zinat; Zinnat. Antibacterial. *Glaxo Wellcome Inc.*

2038 Cefuroxime Pivoxetil
100680-33-9
$C_{23}H_{28}N_4O_{11}S$
[6R-[6α,7β(Z)]]-3-[[(Aminocarbonyl)oxy]methyl]]-7-[[2-furanyl(methoxyimino)acetyl]amino]-8-oxo-5-thia-1-azabicyclo[4.2.0]oct-2-ene-2-carboxylic acid (2,2-dimethyl-1-oxopropoxy)methyl ester.
Antibacterial. *Glaxo Wellcome Inc.*

2039 Cefuroxime Sodium
56238-63-2 2002 260-073-1
$C_{16}H_{15}N_4NaO_8S$
[6R-[6α,7β(Z)]]-3-[[(Aminocarbonyl)oxy]methyl]]-7-[[2-furanyl(methoxyimino)acetyl]amino]-8-oxo-5-thia-1-azabicyclo[4.2.0]oct-2-ene-2-carboxylic acid sodium salt.
Anaptivan; Biociclin; Biofurex; Bioxima; Cefamar; Cefoprim; Cefumax; Cefurex; Cefurin; Curocef; Curoxim; Duxima; Gibicef; Ipacef; Kefurox; Kesint; Lampsporin; Medoxim; Novocef; Spectrazole; Ultroxim; Zinacef. Antibacterial. $[\alpha]_D^{20}$ = +60° (c = 0.91 in H_2O); λ_m 274 nm (ε 17400 H_2O); freely soluble in H_2O; soluble in MeOH; very slightly soluble in EtOAc, Et_2O, C_6H_6, $CHCl_3$, octanol. *Glaxo Wellcome Inc.*

2040 Cefuzonam
82219-78-1 2003
$C_{16}H_{15}N_7O_5S_4$
[6R-[6α,7β(Z)]]-7-[[(2-Amino-4-thiazolyl)(methoxyimino)-acetyl]amino]-8-oxo-3-[(1,2,3-thiadiazol-5-ylthio)-methyl]-5-thia-1-azabicyclo[4.2.0]oct-2-ene-2-carboxylic acid.
CL-118523. Antibacterial. LD_{50} (fmus iv) = 4117 mg/kg, (fmus ip) = 6424 mg/kg) (fmus orl) > 10000 mg/kg, (mmus iv) > 4800 mg/kg, (mmus ip) = 6783 mg/kg, (mmus orl) > 10000 mg/kg, (frat iv) = 4281 mg/kg, (frat ip) > 8000 mg/kg, (frat orl) > 10000 mg/kg, (mrat iv) = 4222 mg/kg, (mrat ip) > 8000 mg/kg, (mrat orl) > 10000 mg/kg. *Am. Cyanamid; Takeda Chem. Ind. Ltd.*

2041 Cephacetrile
10206-21-0 233-508-8
$C_{13}H_{13}N_3O_6S$
(6R-trans)-3-[(Acetyloxy)-methyl]-7-[(cyanoacetyl)amino]-8-oxo-5-thia-1-azabicyclo[4.2.0]oct-2-ene-2-carboxylic acid.
Cefacetrile. Antibacterial. *CIBA plc.*

2042 Cephacetrile Sodium
23239-41-0 2019 245-513-2
$C_{13}H_{12}N_3NaO_6S$
(6R-trans)-3-[(Acetyloxy)-methyl]-7-[(cyanoacetyl)amino]-8-oxo-5-thia-1-azabicyclo[4.2.0]oct-2-ene-2-carboxylic acid monosodium salt.
Ciba 36278-Ba; Celospor; Vetimast. Antibacterial. A

semisynthetic cephalosporin. LD_{50} (mus iv) = 4500 ± 540 mg/kg, (mus sc) = 9100 ± 1500 mg/kg. *CIBA plc.*

2043 Cephalexin [anhydrous]
15686-71-2 2021
$C_{16}H_{17}N_3O_4S$
[6R-[6α,7β(R*)]]-7-[(Aminophenylacetyl)amino]-3-methyl-8-oxo-5-thia-1-azabicyclo[4.2.0]oct-2-ene-2-carboxylic acid.
Cefadros; Cefaloto; Cefanex; Cefaseptin; Ceporexine; Cex; Derantel; Efalexin; Farexin; Fergon 500; Garasin; Ibilex; Iwalexin; larixin; Lexibiotico; Llonexina; Madlexin; Mamalexin; Mecilex; Ohlexin; Oracocin; Rinesal; Sencephalin; Sintolexyn; Syncl; Taicelexin; Tokiolexin; Xahl; Alfaspoven [as sodium salt]. Antibacterial. λ_m = 260 nm (ε 7750). *Eli Lilly & Co.*

2044 Cephalexin Hydrochloride
105879-42-3 2021
$C_{16}H_{20}ClN_3O_5S$
[6R-[6α,7β(R*)]]-7-[(Aminophenylacetyl)amino]-3-methyl-8-oxo-5-thia-1-azabicyclo[4.2.0]oct-2-ene-2-carboxylic acid hydrochloride monohydrate.
LY-061188; Keftab. Antibacterial. LD_{50} (mus orl) = 1600-4500 mg/kg, (mus ip) = 400-1300 mg/kg, (rat orl) > 5000 mg/kg, (rat ip) > 3700 mg/kg. *Eli Lilly & Co.*

2045 Cephalexin Monohydrate
23325-78-2 2021 239-773-6
$C_{16}H_{17}N_3O_4S.H_2O$
[6R-[6α,7β(R*)]]-7-[(Aminophenylacetyl)amino]-3-methyl-8-oxo-5-thia-1-azabicyclo[4.2.0]oct-2-ene-2-carboxylic acid monohydrate.
Cefa-Iskia; Cefibacter; Ceporex; Ceporexin; Keforal; Keflet; Keflex; Oracef; Ortisporina; Sartosona; Servispor. Antibacterial. λ_m = 260 nm (ε = 7750); pKa = 5.2, 7.3; [monohydrate]: LD_{50} (rat orl) > 5000 mg/kg. *Bristol-Myers Oncology; Eli Lilly & Co.; Lemmon Co.*

2046 Cephaloglycin
22202-75-1 2023
$C_{18}H_{19}N_3O_6S.2H_2O$
[6R-[6α,7β(R*)]]-3-[(Acetyloxy)methyl]-7-[(aminophenylacetyl)amino]-8-oxo-5-thia-1-azabicyclo[4.2.0]oct-2-ene-carboxylic acid dihydrate.
Kafocin. Antibacterial. mp = 223-250° (dec); λ_m (2% DMF): 258 nm ($E^{1\%}_{1cm}$ 166). *Eli Lilly & Co.; Glaxo Wellcome Inc.; Merck & Co.Inc.*

2047 Cephaloglycin [anhydrous]
3577-01-3 2023 222-696-7
$C_{18}H_{19}N_3O_6S$
[6R-[6α,7β(R*)]]-3-[(Acetyloxy)methyl]-7-[(aminophenylacetyl)amino]-8-oxo-5-thia-1-azabicyclo[4.2.0]oct-2-ene-carboxylic acid.
Kefglycin. Antibacterial. *Eli Lilly & Co.; Glaxo Wellcome Inc.; Merck & Co.Inc.*

2048 Cephaloridine
50-59-9 2025 200-052-6
$C_{19}H_{17}N_3O_4S_2$
(6R-trans)-1-[[2-Carboxy-8-oxo-7[(2-thienylacetyl)amino]-5-thia-1-azabicyclo[4.2.0]oct-2-en-3-yl]-methyl]pyridinium inner salt.
1-[(7'-β-[2-(2-thienyl)acetamido]-8'-oxo-1'-aza-5'-thia-bicyclo[4.2.0]oct-2'-en-3'-yl)methyl]pyridinium 2'-carboxylate; N-[7-[(2-thienyl)acetamido]ceph-3-em-3-yl-methyl]pyridinium-4-carboxylate; Cefaloridin; Ceflorin; Cepaloridin; Cer; Faredina; Ceporan; Ceporin; Cilifor; Intrasporin; Keflodin; Lloncefal; Sefacin; Cepalorin; Deflorin; Kefspor; Loridine; Aliporina; Ampligram; Floridin. Antibacterial. $[\alpha]_D^{20}$ = 47.7° (H_2O, c = 1.25); LD_{50} (rat orl) = 2500-4000 mg/kg; λ_m = 239, 252 nm (ε 15160, 13950). *Eli Lilly & Co.; Glaxo Wellcome Inc.*

2049 Cephalosporin C
61-24-5 2026 200-501-6
$C_{16}H_{21}N_3O_8S$
[6R-[6α,7β(R*)]]-3-[(Acetyloxy)methyl]-7-[(5-amino-5-carboxy-1-oxopentyl)amino]-8-oxo-5-thia-1-azabicyclo[4.2.0]oct-2-ene-2-carboxylic acid.
Cephalosporin C; Averon-1; Cefalotin; Cemastin; Cephation; Ceporacin; Cepovenin; Chephalotin; Coaxin; Keflin; Lospoven; Microtin; Synclotin; Toricelocin. Antibacterial. *Merck & Co.Inc.; National Res. Dev. Corp.; Parke-Davis.*

2050 Cephalothin
153-61-7 2028 205-815-7
$C_{16}H_{16}N_2O_6S_2$
(6R-trans)-3-[(Acetyloxy)methyl]-8-oxo-7-[(2-thienyl-acetyl)amino]-5-thia-1-azabicyclo[4.2.0]oct-2-carboxylic acid.
7-(2-Thienylacetamido)cephalosporanic acid; 7-(thiophene-2-acetamido)cephalosporanic acid. Antibacterial. mp = 160-160.5°; $[\alpha]_D^{20}$ = 50° (CH_3CN, c = 1.03).

2051 Cephalothin Sodium
58-71-9 2028 200-394-6
$C_{16}H_{15}N_2NaO_6S_2$
(6R-trans)-3-[(Acetyloxy)methyl]-8-oxo-7-[(2-thienyl-acetyl)amino]-5-thia-1-azabicyclo[4.2.0]oct-2-carboxylic acid sodium salt.
Averon-1; Cefalotin; Cemastin; Cephation; Ceporacin; Cepovenin; Chephalotin; Coaxin; Keflin; Lospoven; Microtin; Synclotin; Toricelocin. Antibacterial. mp = 204-205°; $[\alpha]_D^{20}$ = +135° (c = 1.0 in H_2O); λ_m = 236, 260 nm (ε 12950, 9350); LD_{50} (mg/kg) (mus orl) > 20000 mg/kg, (mus ip) = 5670 mg/kg, (rat orl) > 10000 mg/kg, (rat ip) = 7716 mg/kg. *Eli Lilly & Co.; Glaxo Wellcome Inc.*

2052 Cephapirin
21593-23-7 244-466-5
$C_{17}H_{17}N_3O_6S_2$
(6R-trans)-3-[(Acetyloxy)methyl]-8-oxo-7-[[(4-pyridinylthio)acetyl]amino]-5-thia-1-azabicyclo[4.2.0]oct-2-ene
2-carboxylic acid.
Antibacterial. *Apothecon; Bristol-Myers Squibb Pharm. R&D.*

2053 Cephapirin Benzathine
97468-37-6
$C_{50}H_{54}N_8O_{12}S_4$
(6R-trans)-3-[(Acetyloxy)methyl]-8-oxo-7-[[(4-pyridinyl-thio)acetyl]amino]-5-thia-1-azabicyclo[4.2.0]oct-2-ene 2-carboxylic acid compound with N,N-bis(phenylmethyl))-1,2-ethanediamine (2:1).
Antibacterial. *Apothecon; Bristol-Myers Squibb Pharm. R&D.*

2054 Cephapirin Sodium
24356-60-3 2030 246-194-2
$C_{17}H_{16}N_3NaO_6S_2$
(6R-trans)-3-[(Acetyloxy)methyl]-8-oxo-7-[[(4-pyridinylthio)acetyl]amino]-5-thia-1-azabicyclo[4.2.0]oct-2-ene 2-carboxylic acid monosodium salt.
3-(hydroxymethyl)-8-oxo-7-[2-(4-pyridylthio)acetamido]-5-thia-1-azabicyclo[4.2.0]oct-2-ene 2-carboxylic acid acetate monosodium salt; 7-[α-(4-pyridylthio)-acetamido]cephalosporanic acid sodium salt; sodium 7-(pyrid-4-ylthioacetamido)cephalosporanate; Cephapirin sodium; BL-P-1322; Ambrocef; Brisfirina; Bristocef; Cefadyl; Cef-Lak; Cefatrexyl; Piricef; ToDay. Antibacterial. Soluble in H_2O. *Apothecon; Bristol-Myers Squibb Pharm. R&D.*

2055 Cephradine
38821-53-3 2032 254-137-8
$C_{16}H_{19}N_3O_4S$
[6R-[6α,7β(R*)]]-7-[(Amino-1,4-cyclohexadien-1-ylacetyl)amino]-3-methyl-8-oxo-5-thia-1-azabicyclo[4.2.0]oct-2-ene-2-carboxylic acid.
cefradine; Anspor; Velosef; SQ-11436; Cefradex; Cefrag; Cefro; Celex; Cesporan; Dimacef; Ecosporina; Easkacef; Lenzacef; Lisacef; Megacef; Samedrin; Sefril; Velocef. Antibacterial. mp = 140-142° (dec). *Apothecon; SmithKline Beecham Pharm.*

2056 Cephradine Dihydrate
31828-50-9 2032
$C_{16}H_{19}N_3O_4S.2H_2O$
[6R-[6α,7β(R*)]]-7-[(Amino-1,4-cyclohexadien-1-ylacetyl)amino]-3-methyl-8-oxo-5-thia-1-azabicyclo[4.2.0]oct-2-ene-2-carboxylic acid dihydrate.
cephadrine dihydrate; SQ-22022. Antibacterial. mp = 183-185°. *Apothecon; SmithKline Beecham Pharm.*

2057 Cephradine Monohydrate
75975-70-1
$C_{16}H_{19}N_3O_4S.H_2O$
[6R-[6α,7β(R*)]]-7-[(Amino-1,4-cyclohexadien-1-ylacetyl)amino]-3-methyl-8-oxo-5-thia-1-azabicyclo[4.2.0]oct-2-ene-2-carboxylic acid monohydrate.
Forticef. Antibacterial. *Apothecon; SmithKline Beecham Pharm.*

2058 Clavulanic Acid
58001-44-8 261-069-2
$C_8H_9NO_5$
3-(2-Hydroxyethylidene)-7-oxo-4-oxa-1-azabicyclo[3.2.0]heptane-2-carboxylic acid.
MM-14151. Antibacterial. A β-lactamase inhibitor. *SmithKline Beecham Pharm.*

2059 Clavulanic Acid Potassium Salt combined with Ticarcillin Disodium
116876-37-0
$C_{23}H_{22}KN_3Na_2O_{11}S_2$
Potassium 3-(2-hydroxyethylidene)-7-oxo-4-oxa-1-azabicyclo[3.2.0]heptane-2-carboxylate combined with ticarcillin sodium.
Betabactil; Timentin. Antibacterial. A β-lactamase inhibitor. *SmithKline Beecham Pharm.*

2060 Clemizole Penicillin
6011-39-8
$C_{35}H_{38}ClN_5O_4S$
Benzylpenicillin combined with 1-(p-chlorobenzyl)-2-(1-pyrrolidinylmethyl)benzimidazole.
Antibacterial.

2061 Clometicillin
1926-49-4 2445
$C_{17}H_{18}Cl_2N_2O_5S$
[2S-(2α,5α,6β)]-6-[[(3,4-Dichlorophenyl)methoxyacetyl]amino]-3,3-dimethyl-7-oxo-4-thia-1-azabicyclo[3.2.0]heptane-2-carboxylic acid.
clometacillin; clomethacillin; no. 356; penicillin 3566. Antibacterial. *Recherche et Ind. Therap.*

2062 Clometicillin Sodium Salt
2445
$C_{17}H_{17}Cl_2N_2NaO_5S$
[2S-(2α,5α,6β)]-6-[[(3,4-Dichlorophenyl)methoxyacetyl]amino]-3,3-dimethyl-7-oxo-4-thia-1-azabicyclo[3.2.0]heptane-2-carboxylic acid sodium salt.
Rixapen. Antibacterial. $[\alpha]_D$ = 210-220°. *Recherche et Ind. Therap.*

2063 Cloxacillin
642-78-4 2480 211-390-9
$C_{19}H_{18}ClN_3O_5S$
[2S-(2α,5α,6β)]-6-[[[3-(2-Chlorophenyl)-5-methyl-4-isoxazolyl]carbonyl]amino]-3,3-dimethyl-7-oxo-4-thia-1-azabicyclo[3.2.0]heptane-2-carboxylic acid.
Antibacterial. *Apothecon; SmithKline Beecham Pharm.*

2064 Cloxacillin Benzathine
23736-58-5 2480 245-855-2
$C_{54}H_{56}Cl_2N_8O_{10}S_2$
[2S-(2α,5α,6β)]-6-[[[3-(2-Chlorophenyl)-5-methyl-4-isoxazolyl]carbonyl]amino]-3,3-dimethyl-7-oxo-4-thia-1-azabicyclo[3.2.0]heptane-2-carboxylic acid benzathine salt.
Boviclox; Dry-Clox; Noroclox DC; Opticlox; Orbenin Dry Cow; Triclox. Antibacterial. *Apothecon; SmithKline Beecham Pharm.*

2065 Cloxacillin Sodium Monohydrate
7081-44-9 2480
$C_{19}H_{17}ClN_3NaO_5S.H_2O$
[2S-(2α,5α,6β)]-6-[[[3-(2-Chlorophenyl)-5-methyl-4-isoxazolyl]carbonyl]amino]-3,3-dimethyl-7-oxo-4-thia-1-azabicyclo[3.2.0]heptane-2-carboxylic acid sodium salt monohydrate.
sodium cloxacillin; BRL-1621; Bactopen; Cloxapen; Cloxypen; Ekvacillin; Gelstaph; Orbenin; Methoxillin-S; Prostaphlin-A; Staphobristol-250; Staphybiotic; Tegopen; Tepogen. Antibacterial. mp = 170° (dec); $[\alpha]_D^{20}$ = 163°; soluble in H_2O, polar organic solvents; LD_{50} (rat ip) = 1630 ± 112 mg/kg, (mus ip) = 1280 ± 50 mg/kg. *Apothecon; SmithKline Beecham Pharm.*

2066 Cyclacillin
3485-14-1 2769 222-470-8
$C_{15}H_{23}N_3O_4S$
[2S-(2α,5α,6β)]-6-[[(1-Aminocyclohexyl)carbonyl]amino]-3,3-dimethyl-7-oxo-4-thia-1-azabicyclo[3.2.0]heptane-2-carboxylic acid.

Wy-4508; Calthor; Citosarin; Cyclapen; Syngacillin; ciclacillin; Ultracillin; Vastcillin; Vatracin; Vipicil; Wyvital. Antibacterial. mp = 156-158° (dec), 182-183°; $[\alpha]_D^{25}$ = 268° (H_2O); soluble in H_2O (2.9 g/100 ml at 38°). *Wyeth-Ayerst Labs.*

2067 Dicloxacillin
3116-76-5 3134 221-488-3
$C_{19}H_{17}Cl_2N_3O_5S$
[2S-[2α,5α,6β(S*)]]-6-[[[3-(2,6-Dichlorophenyl)-5-methyl-4-isoxazolyl]carbonyl]amino]-3,3-dimethyl-7-oxo-4-thia-1-azabicyclo[3.2.0]heptane-2-carboxylic acid.
R-13423; BRL-1702; Maclicine. Antibacterial. *SmithKline Beecham Pharm.*

2068 Dicloxacillin Sodium
343-55-5 3134 206-444-3
$C_{19}H_{16}Cl_2N_3NaO_5S$
[2S-[2α,5α,6β]]-6- [[[3-(2,6-Dichlorophenyl)-5-methyl-4-isoxazolyl]carbonyl]amino]-3,3-dimethyl-7-oxo-4-thia-1-azabicyclo[3.2.0]heptane-2-carboxylic acid sodium salt.
sodium dicloxacillin; P-1011; Brispen; Constaphyl; Dichlor-Stapenor; Diclocil; Dycill; Dynapen; Noxaben; Pathocil; Pen-Sint; Stampen; Syntarpen; Veracillin. Antibacterial. Dec 222-225°; $[\alpha]_D^{20}$ = 127.2° (H_2O); soluble in H_2O, MeOH; less soluble in butanol; slightly soluble in Me_2CO and the usual organic solvents; LD_{50} (mus iv) = 900 mg/kg, (rat ip) = 630 mg/kg, (rat orl) > 5000 mg/kg. *SmithKline Beecham Pharm.*

2069 Dicloxacillin Sodium Monohydrate
13412-64-1 3134
$C_{19}H_{16}Cl_2N_3NaO_5.H_2O$
[2S-(2α,5α,6β)]-6-[[[3-(2,6-Dichlorophenyl)-5-methyl-4-isoxazolyl]carbonyl]amino]-3,3-dimethtyl-7-oxo-4-thia-1-azabicyclo[3.2.0]heptane-2-carboxylic acid sodium salt monohydrate.
sodium dicloxacillin monohydrate; P-1011; Brispen; Constphyl; Dichlor-Stapenor; Diclocil; Dycill; Dynapen; Noxaben; Pathocil; Pen-Sint; Stampen; Syntarpen; Veracillin. Antibacterial. Dec 222-225°; $[\alpha]_D^{20}$ = 127.2° (H_2O); soluble in H_2O, less soluble in BuOH, slightly soluble in Me_2CO and the usual organic solvents; LD_{50} (mus iv) = 900 mg/kg, (rat ip) = 630 mg/kg, (rat orl) > 5000 mg/kg. *SmithKline Beecham Pharm.*

2070 Epicillin
26774-90-3 3651 248-001-7
$C_{16}H_{21}N_3O_4S$
[2S-[2α,5α,6β(S*)]]-6-[(Amino-1,4-cyclohexadien-1-ylacetyl)amino]-3,3-dimethyl-7-oxo-4-thia-1-azabicyclo[3.2.0]heptane-2-carboxylic acid.
SQ-11302; Dexacilina; Dexacillin; Omnisan; Spectacillin. Antibacterial. Dec 202°. *Bristol-Myers Squibb Pharm. R&D.*

2071 Fenbenicillin
1926-48-3 4001
$C_{22}H_{22}N_2O_5S$
[2S-(2α,5α,6β)]-3,3-Dimethyl-7-oxo-6-[(phenoxyphenylacetyl)amino]-4-thia-1-azabicyclo[3.2.0]heptane-2-carboxylic acid.
Phenbenicillin. Antibacterial. *SmithKline Beecham Pharm.*

2072 Fenbenicillin Potassium Salt
1177-30-6 4001
$C_{22}H_{21}KN_2O_5S$
[2S-(2α,5α,6β)]-3,3-Dimethyl-7-oxo-6-[(phenoxyphenylacetyl)amino]-4-thia-1-azabicyclo[3.2.0]heptane-2-carboxylic acid potassium salt.
Penspek. Antibacterial. mp = 88=95°, dec 120-125°; readily soluble in H_2O; LD_{50} (mus iv) = 225 mg/kg, (mus ip) = 520 mg/kg, (mus orl) = 3000 mg/kg. *SmithKline Beecham Pharm.*

2073 Fibracillin
51154-48-4 257-021-5
$C_{26}H_{28}ClN_3O_6S$
D-6-[2-[2-(p-Chlorophenoxy)-2-methylpropionamido]-2-phenylacetamido]-3,3-dimethyl-7-oxo-4-thia-1-azabicyclo[3.2.0]heptane-2-carboxylic acid.
Antibacterial.

2074 Flomoxef
99665-00-6 4141
$C_{15}H_{18}F_2N_6O_7S_2$
(6R-cis)-7-[[[(Difluoromethyl)thio]acetyl]amino]-3-[[[1-(2-hydroxyethyl)-1H-tetrazol-5-yl]thio]methyl]-7-methoxy-8-oxo-5-oxa-1-azabicyclo[4.2.0]oct-2-ene-2-carboxylic acid.
Antibacterial. mp = 82.5-87.5°. *Shionogi & Co. Ltd.*

2075 Floxacillin
5250-39-5 4147 226-051-0
$C_{19}H_{17}ClFN_3O_5S$
[2S-[2α,5α,6β(S*)]-6-[[[3-(2-Chloro-6-fluorophenyl)-5-methyl-4-isoxazolyl]carbonyl]amino]-3,3-dimethyl-7-oxo-4-thia-1-azabicyclo[3.2.0]heptane-2-carboxylic acid.
Flucloxacillin; BRL-2039; Penplus; Staphlipen; Abboflox; Flupen; Floxapen; BRL-2039. Antibacterial. *SmithKline Beecham Pharm.*

2076 Floxacillin Sodium
34214-51-2
$C_{19}H_{16}ClFN_3NaO_5S.H_2O$
[2S-[2α,5α,6β(S*)]-6-[[[3-(2-Chloro-6-fluorophenyl)-5-methyl-4-isoxazolyl]carbonyl]amino]-3,3-dimethyl-7-oxo-4-thia-1-azabicyclo[3.2.0]heptane-2-carboxylic acid sodium salt.
Culpen; Floxapen; Ladropen; Stafoxil; Staphcil; Staphylex. Antibacterial. *SmithKline Beecham Pharm.*

2077 Fomidacillin
98048-07-8
$C_{24}H_{28}N_6O_{10}S$
(2S,5R,6R)-6-[(R)-2-(3,4-Dihydroxyphenyl)-2-(4-ethyl-2,3-dioxo-1-piperazinecarboxamido)acetamido]-6-formamido-3,3-dimethyl-7-oxo-4-thia-1-azabicyclo[3.2.0]heptane-2-carboxylic acid.
Antibacterial.

2078 Fumoxicillin
78186-33-1 278-865-0
$C_{21}H_{21}N_3O_6S$
(2S,5R,6R)-6-[(R)-2-(Furfurylideneamino)-2-(p-hydroxyphenyl)acetamido]-3,3-dimethyl-7-oxo-4-thia-1-azabicyclo[3.2.0]heptane-2-carboxylic acid.
FU-02. Antibacterial. *Farmatis S.R.L., Italy.*

2079 Fuzlocillin
66327-51-3
$C_{25}H_{26}N_6O_8S$
(2S,5R,6R)-6-[(2R)-2-[3-[(E)-Furfurylideneamino]-2-oxo-1-imidazolidinecarboxamido]-2-(p-hydroxyphenyl)acetamido]-3,3-dimethyl-7-oxo-4-thia-1-azabicyclo-[3.2.0]heptane-2-carboxylic acid.
Antibacterial.

2080 Gloximonam
90850-05-8
$C_{18}H_{25}N_5O_8S$
[[(2S,3S)-3-[(2-Amino-4-thiazolyl)glyoxylamido]-2-methyl-4-oxo-1-azetidinyl]oxy]acetic acid ester with tert-butyl glycolate 3^2-(Z)-(O-methyloxime).
SQ-82531. Antibacterial. *Bristol-Myers Squibb Pharm. R&D.*

2081 Hetacillin
3511-16-8 4706 222-512-5
$C_{19}H_{23}N_3O_4S$
[2S-[2α,5α,6β(S*)]]-6-(2,2-Dimethyl-5-oxo-4-phenyl-1-imidazolidinyl)-3,3-dimethyl-7-oxo-4-thia-1-azabicyclo[3.2.0]heptane-2-carboxylic acid.
6-(2,2-dimethyl-5-oxo-4-phenyl-1-imidazolidinyl)penicillanic acid; Hetacillin; phenazacillin; BRL-804; Penplenum; Versapen; Versatrex. Antibacterial. Dec 182.8-183.9°, 189.2-191.0°; $[\alpha]_D^{25}$ = +366° (C_5H_5N). *Bristol-Myers Squibb Pharm. R&D.*

2082 Hetacillin Potassium
5321-32-4 4706 226-182-3
$C_{19}H_{22}KN_3O_4S$
[2S-[2α,5α,6β(S*)]]-6-(2,2-Dimethyl-5-oxo-4-phenyl-1-imidazolidinyl)-3,3-dimethyl-7-oxo-4-thia-1-azabicyclo--[3.2.0]heptane-2-carboxylic acid potassium salt.
Uropen; Versapen K; HetacinK; Natacillin. Antibacterial. *Bristol-Myers Squibb Pharm. R&D.*

2083 Imipenem [anhydrous]
64221-86-9 4954 264-734-5
$C_{12}H_{17}N_3O_4S$
[5R-[5α,6α(R*)]]-6-(1-Hydroxyethyl)-3-[[2-[(iminomethyl)amino]ethyl]thio]-7-oxo-1-azabicyclo[3.2.0]hept-2-ene-2-carboxylic acid.
Antibacterial. Broad spectrum semi-synthetic antibiotic. *Merck & Co.Inc.*

2084 Imipenem Monohydrate
74431-23-5 4954
$C_{12}H_{17}N_3O_4S.H_2O$
[5R-[5α,6α(R*)]]-6-(1-Hydroxyethyl)-3-[[2-[(iminomethyl)amino]ethyl]thio]-7-oxo-1-azabicyclo[3.2.0]hept-2-ene-2-carboxylic acid mono-hydrate.
imipemide; MK-787; MK-0787; component of [in combination with cilastatin sodium]: Primaxin, Imipem, Tenacid, Tienam, Tracix, Zienam. Antibacterial. Broad spectrum semi-synthetic antibiotic. $[\alpha]_D^{25}$ = +86.8° (c = 0.05 in 0.1M phosphate, pH 7); λ_m 299 nm (ε 9670, 98% H_2O, NH_2OH ext); soluble in H_2O (1 g/100 ml), MeOH (0.5 g/100 ml), EtOH (0.02 g/100 ml), Me_2CO (< 0.01 g/100 ml), DMF (< 0.01 g/100 ml), DMSO (0.03 g/100 ml). *Merck & Co.Inc.*

2085 Lenampicillin
86273-18-9 5460
$C_{21}H_{23}N_3O_7S$
[2S-[2α,5α,6β(S*)]]-6-[(Aminophenylacetyl)amino]-3,3-dimethyl-7-oxo-4-thia-1-azabicyclo[3.2.0]heptane-2-carboxylic acid (5-methyl-2-oxo-1,3-dioxol-4-yl)methyl ester.
Antibacterial. Prodrug of ampicillin. Orally active. *Kanebo Pharm. Ltd.*

2086 Lenampicillin Hydrochloride
80734-02-6
$C_{21}H_{24}ClN_3O_7S$
[2S-[2α,5α,6β(S*)]]-6-[(Aminophenylacetyl)amino]-3,3-dimethyl-7-oxo-4-thia-1-azabicyclo[3.2.0]heptane-2-carboxylic acid (5-methyl-2-oxo-1,3-dioxol-4-yl)methyl ester hydrochloride (salt).
KB-1585; KBT-1585; Takacillin; Varacillin. Antibacterial. mp = 145°; LD50 (mrat orl) 10000 mg/kg, (mmus orl) = 8294 mg/kg, (mrat sc) = 4362 mg/kg, (mmus sc) = 3576 mg/kg, (mrat iv) = 976 mg/kg, (mmus iv) = 711 mg/kg, (dog orl) > 300 mg/kg. *Kanebo Pharm. Ltd.*

2087 Lenapenem
149951-16-6
$C_{18}H_{29}N_3O_5S$
(1R,5S,6S)-2-[(3S,5S)-5-[(R)-1-Hydroxy-3-(N-methylamino)propyl]pyrrolidin-3-ylthio]-6-[(R)-1-hydroxyethyl]-1-methyl-1-carbapen-2-em-3-carboxylic acid.
BO-2727. Penicillin antibacterial.

2088 Levopropylcillin
3736-12-7
$C_{18}H_{22}N_2O_5S$
(2S,5R,6R)-3,3-Dimethyl-7-oxo-6-[(2S)-phenoxybutyramido]-4-thia-1-azabicyclo[3.2.0]heptane-2-carboxylic acid.
Antibacterial. *Eli Lilly & Co.*

2089 Levopropylcillin Potassium
7245-75-2
$C_{18}H_{21}KN_2O_5S$
(2S,5R,6R)-3,3-Dimethyl-7-oxo-6-[(2S)-phenoxybutyramido]-4-thia-1-azabicyclo[3.2.0]heptane-2-carboxylic acid potassium salt.
BRL-284; P-248; 38389. Antibacterial. *Eli Lilly & Co.*

2090 Loracarbef [anhydrous]
76470-66-1
$C_{16}H_{16}ClN_3O_4$
[6R-[6α,7β(R*)]]-7-[(Aminophenyl)acetyl)amino]-3-chloro-8-oxo-1-azabicyclo[4.2.0]oct-2-ene-2-carboxylic acid.
Antibacterial. *Eli Lilly & Co.; Kyowa Hakko Kogyo Co. Ltd.*

2091 Loracarbef Monohydrate
121961-22-6 5606
$C_{16}H_{16}ClN_3O_4.H_2O$
[6R-[6α,7β(R*)]]-7-[(Aminophenyl)acetyl)amino]-3-chloro-8-oxo-1-azabicyclo[4.2.0]oct-2-ene-2-carboxylic acid monohydrate.
carbacefaclor; KT-3777; LY-163892 monohydrate; Lorabid. Antibacterial. mp = 232-238°; $[\alpha]_D^{20}$ = +27.5° (c = 1 in $CHCl_3$). *Eli Lilly & Co.; Kyowa Hakko Kogyo Co. Ltd.*

2092 Meropenem
96036-03-2 5960
$C_{17}H_{25}N_3O_5S$
[4R-[3(3S*,5S*),4α,5β,6β(R*)]]-3-[[5-[(Dimethylamino)-carbonyl]-3-pyrrolidinyl]thio]-6-(1-hydroxyethyl)-4-methyl-7-oxo-1-azabicyclo[3.2.0]hept-2-ene-2-carboxylic acid.
Antibacterial. *Zeneca Pharm.*

2093 Meropenem Trihydrate
119478-56-7 5960
$C_{17}H_{25}N_3O_5S.3H_2O$
[4R-[3(3S*,5S*),4α,5β,6β(R*)]]-3-[[5-[(Dimethylamino)-carbonyl]-3-pyrrolidinyl]thio]-6-(1-hydroxyethyl)-4-methyl-7-oxo-1-azabicyclo[3.2.0]hept-2-ene-2-carboxylic acid trihydrate.
ICI-194660; SM-7338; Merrem; Meronem. Antibacterial. *Zeneca Pharm.*

2094 Metampicillin
6489-97-0 5986 229-365-6
$C_{17}H_{19}N_3O_4S$
[2S-(2α,5α,6β)]-3,3-Dimethyl-6-[[(methyleneamino)-phenylacetyl]amino]-7-oxo-4-thia-1-azabicyclo[3.2.0]heptane-2-carboxylic acid.
methampicillin; Bonopen; Fedacilina; Micinovo; Pravacilin; Ruticina; Suvipen; Viderpen. Antibacterial. *E.R.A.S.M.E.*

2095 Methicillin
61-32-5 6047 200-505-8
$C_{17}H_{20}N_2O_6S$
(2S,5R,6R)-6-(2,6-Dimethoxybenzamido)-3,3-dimethyl-7-oxo-4-thia-1-azabicyclo[3.2.0]heptane-2-carboxylic acid.
Antibacterial. *Apothecon; SmithKline Beecham Pharm.*

2096 Methicillin Sodium
132-92-3 6047 205-083-9
$C_{17}H_{19}N_2NaO_6S$
[2S-(2α,5α,6β)]-6-[(2,6-Dimethoxybenzoyl)amino]-3,3-dimethyl-7-oxo-4-thia-1-azabicyclo[3.2.0]heptane-2-carboxylic acid monosodium salt.
dimethoxyphenecillin sodium; BRL-1241; X-1497; Azapen; Belfacillin; Celpillina; Celbenin; Cinopenil; Dimocillin; Flabelline; Penistaph; Staphcillin; SQ-16123. Antibacterial. mp = 196-197°; $[\alpha]_D^{20}$ = 230° (c = 5); λ_m = 281 nm ($E^{1\%}_{1cm}$ 55); soluble in H_2O (>300 mg/ml), EtOH, Me_2CO; insoluble in Et_2O, hydrocarbons; [dl-form]: mp = 230-232°; freely soluble in H_2O; [l-form]: mp = 238-239°; $[\alpha]_D^{24}$ = 218° (c = 0.01 H_2O). *Apothecon; SmithKline Beecham Pharm.*

2097 Methicillin Sodium Monohydrate
7246-14-2 6047
$C_{17}H_{19}N_2NaO_6S.H_2O$
(2S,5R,6R)-6-(2,6-Dimethoxybenzamido)-3,3-dimethyl-7-oxo-4-thia-1-azabicyclo[3.2.0]heptane-2-carboxylic acid sodium salt monohydrate.
Celbenin; Staphcilin; BRL-1241; SQ-16123; X-1497; Azapen; Belfacillin; Celpillina; Cinopenil; Dimocillin; Flabelline; Penistaph; Staphcillin. Antibacterial. mp = 196-197° (dec); $[\alpha]_D^{20}$ =230° (c = 5), 225° (c = 1); λ_m = 281 nm ($E^{1\%}_{1\ cm}$ 55); soluble in H_2O (> 30 g/100 ml at 20°), EtOH (4 g/100 ml), Et_2O (0.003 g/100 ml), Me_2CO (0.035 g/100 ml), $CHCl_3$ (0.006 g/100 ml), isooctane (< 0.003 g/100 ml). *Apothecon; SmithKline Beecham Pharm.*

2098 Mezlocillin
51481-65-3 6259 257-233-8
$C_{21}H_{25}N_5O_8S_2$
[2S-[2α,5α,6β(S*)]]-3,3-Dimethyl-6-[[[[[3-(methylsulfonyl)-2-oxo-1-imidazolidinyl]carbonyl]-amino]phenylacetyl]amino]-7-oxo-4-thia-1-azabicyclo[3.2.0]heptane-2-carboxylic acid.
Bay f 1353. Antibacterial. *Bayer Corp. Pharm. Div.*

2099 Mezlocillin Sodium
59798-30-0 6259
$C_{21}H_{24}N_5NaO_5S_2.H_2O$
[2S-[2α,5α,6β(S*)]]-3,3-Dimethyl-6-[[[[[3-(methylsulfonyl)-2-oxo-1-imidazolidinyl]carbonyl]-amino]phenylacetyl]amino]-7-oxo-4-thia-1-azabicyclo[3.2.0]heptane-2-carboxylic acid sodium salt monohydrate.
Baycipen; Baypen; Mezlin. Antibacterial. Soluble in H_2O, MeOH, DMF; insoluble in Me_2CO, EtOH. *Bayer Corp. Pharm. Div.*

2100 Moxalactam
64952-97-2 6369 265-287-9
$C_{20}H_{20}N_6O_9S$
7-[[Carboxy(4-hydroxyphenyl)acetyl]amino]-7-methoxy-3-[[(1-methyl-1H-tetrazol-5-yl)thio]methyl]-8-oxo-5-oxa-1-azabicyclo[4.2.0]oct-2-ene-2-carboxylic acid.
latamoxef; lamoxactam. Antibacterial. mp = 117-122°; $[\alpha]_D^{25}$= -15° (c = 0.216 MeOH); λ_m = 276 nm (ε 10200). *Eli Lilly & Co.; Shionogi & Co. Ltd.*

2101 Moxalactam Disodium
64953-12-4 6369 265-288-4
$C_{20}H_{18}N_6Na_2O_9S$
7-[[Carboxy(4-hydroxyphenyl)acetyl]amino]-7-methoxy-3-[[(1-methyl-1H-tetrazol-5-yl)thio]methyl]-8-oxo-5-oxa-1-azabicyclo[4.2.0]oct-2-ene-2-carboxylic acid disodium salt.
LY-12735; S-6059; Festamoxin; Moxalactam; Moxam; Shiomarin. Antibacterial. $[\alpha]_D^{22}$= -45° (H_2O); λ_m = 270 nm (ε 12000). *Eli Lilly & Co.; Shionogi & Co. Ltd.*

2102 Nafcillin
147-52-4 6438 205-690-9
$C_{21}H_{22}N_2O_5S$
[2S-(2α,5α,6β)]-6-[[(2-Ethoxy)-1-naphthalenynyl)carbonyl]amino]-3,3-dimethyl-7-oxo-4-thia-1-azabicyclo[3.2.0]heptane-2-carboxylic acid.
Antibacterial. *Apothecon; SmithKline Beecham Pharm.; Wyeth-Ayerst Labs.*

2103 Nafcillin Sodium
985-16-0 6438 213-574-4
$C_{21}H_{21}N_2NaO_5S$
[2S-(2α,5α,6β)]-6-[[(2-Ethoxy)-1-naphthalenyl)carbonyl]amino]-3,3-dimethyl-7-oxo-4-thia-1-azabicyclo[3.2.0]heptane-2-carboxylic acid monosodium salt.
Nafcillin sodium; Nafcil; Naftopen; Unipen. Antibacterial. *Apothecon; SmithKline Beecham Pharm.; Wyeth-Ayerst Labs.*

2104 Nafcillin Sodium Monohydrate
7177-50-6
$C_{21}H_{21}N_2NaO_5S.H_2O$
[2S-(2α,5α6β)]-6-[[(2-Ethoxy-1-naphthalenyl)-carbonyl]amino]-3,3-dimethyl-7-oxo-4-thia-1-azabicyclo[3.2.0]heptane-2-carboxylic acid monosodium salt.
Nafcil; Naftopen; Unipen. Antibacterial. *Apothecon; SmithKline Beecham Pharm.; Wyeth-Ayerst Labs.*

2105 Oxacillin
66-79-5 7036 200-635-5
$C_{19}H_{19}N_3O_5S$
[2S-(2α,5α,6β)]-3,3-Dimethyl-6-[[(5-methyl-3-phenyl-4-isoxazolyl)carbonyl]amino]-7-oxo-4-thia-1-azabicyclo[3.2.0]heptane-2-carboxylic acid.
oxazocilline. Antibacterial. *Apothecon; SmithKline Beecham Pharm.*

2106 Oxacillin Sodium
1173-88-2 7036 214-636-3
$C_{19}H_{18}N_3NaO_5S$
[2S-(2α,5α,6β)]-3.3-Dimethyl-6-[[(5-methyl-3-phenyl-4-isoxazolyl)carbonyl]amino]-7-oxo-4-thia-1-azabicyclo[3.2.0]heptane-2-carboxylic acid sodium salt.
penicillin P-12; sodium oxacillin; BRL-1400; Bactocill; Bristopen; Cryptocillin; Micropenin; Oxabel; Penstapho; Penstaphocid; Prostaphlin; Resistopen; Stapenor; Bactocil sodium. Antibacterial. mp = 188° (dec) $[\alpha]_D^{20}$ = +201° (c = 1 in H_2O); LD_{50} (rat orl) > 8000 mg/kg. *Apothecon; SmithKline Beecham Pharm.*

2107 Oxacillin Sodium Monohydrate
7240-38-2 7036
$C_{19}H_{18}N_3NaO_5S.H_2O$
[2S-(2α,5α,6β)]-3,3-Dimethyl-6-[[(5-methyl-3-phenyl-4-isoxazolyl)carbonyl]amino]-7-oxo-4-thia-1-azabicyclo[3.2.0]heptane-2-carboxylic acid sodium salt monohydrate.
oxazocilline; Stapenor; Stapenor Retard; penicillin P-12; sodium oxacillin; BRL-1400; Bactocill; Bristopen; Cryptocillin; Micropenin; Oxabel; Penstapho; Penstaphocid; Prostaphlin; Resistopen. Antibacterial. mp = 188° (dec); $[\alpha]_D^{20}$ = 201° (c = 1 H_2O); LD_{50} (rat orl) > 8000 mg/kg. *Apothecon; SmithKline Beecham Pharm.*

2108 Oxetacillin
53861-02-2
$C_{19}H_{23}N_3O_5S$
(2S,5R,6R)-6-[(R)-[4-(p-Hydroxyphenyl)-2,2-dimethyl-5-oxo-1-imidazolidinyl]]-3,3-dimethyl-7-oxo-4-thia-1-azabicyclo[3.2.0]heptane-2-carboxylic acid.
Antibacterial.

2109 Oximonam
90898-90-1
$C_{12}H_{15}N_5O_6S$
[[(2S,3S)-3-[(2-Amino-4-thiazolyl)glyoxylamido]-2-methyl-4-oxo-1-azetidinyl]oxy] acetate 3^2-(Z)-(O-methyloxime).
Antibacterial. *Bristol-Myers Squibb Pharm. R&D.*

2110 Oximonam Sodium
90849-08-4
$C_{12}H_{14}N_5NaO_6S$
[[(2S,3S)-3-[(2-Amino-4-thiazolyl)glyoxylamido]-2-methyl-4-oxo-1-azetidinyl]oxy] acetate 3^2-(Z)-(O-methyloxime) sodium salt.
SQ-82629. Antibacterial. *Bristol-Myers Squibb Pharm. R&D.*

2111 Panipenem
87726-17-8 7141
$C_{15}H_{21}N_3O_4S$
[5R-[3(S*),5α,6α(R*)]]-6-(1-Hydroxyethyl)-3-[[1-(1-iminoethyl)-3-pyrrolidinyl]thio]-7-oxo-1-azabicyclo[3.2.0]hept-2-ene-2-carboxylic acid.
CS-533; RS-533. Antibacterial. mp = 198-200° (dec); λ_m 298 nm (ε 10400 H_2O); LD_{50} (mmus iv) = 1700-2200 mg/kg, (fmus iv) = 1300-1700 mg/kg. *Sankyo Co. Ltd.*

2112 Penamecillin
983-85-7 7208 213-571-8
$C_{19}H_{22}N_2O_6S$
[2S-(2α,5α,6β)]-3,3-Dimethyl-7-oxo-6-[(phenylacetyl)amino]-4-thia-1-azabicyclo[3.2.0]heptane-2-carboxylic acid (acetyloxy) methyl ester.
Wy-20788; Havapen; penicillin G hydromethyl ester acetate; Wy-20788; acetoxymethyl benzylpenicillinate; benzylpenicillin acetoxymethyl ester. Antibacterial. mp = 106-108°; $[\alpha]_D^{20}$ = 154°. *Wyeth-Ayerst Labs.*

2113 Penethamate Hydriodide
808-71-9 7212 212-367-6
$C_{22}H_{32}IN_3O_4S$
[2S-(2α,5α,6β)]-3,3-Dimethyl-7-oxo-6-[(phenylacetyl)amino]-4-thia-1-azabicyclo[3.2.0]heptane-2-carboxylic acid 2-(diethylamino)ethyl ester monohydriodide.
ephicillin hydriodide; penethecillin; Alivin; Broncopen; Estopen; Neopenil; Pulmaxil N; Bronchocillin; Pulmo 500; Leocillin. Antibacterial. mp = 178-179°; slightly soluble in H_2O (0.96 g/100 ml at 20°). *Antibiotice S.A.*

2114 Penicillin G Aluminum
1406-06-0
$C_{48}H_{51}AlN_6O_{12}S_3$
[2S-(2α,5α,6β)]-3,3-Dimethyl-7-oxo-6-[(phenylacetyl)-amino]-4-thia-1-azabicyclo[3.2.0]heptane-2-carboxylic acid aluminum salt.
aluminum salt of penicillin G. Antibacterial.

2115 Penicillin G Benethamine
751-84-8 7220 212-029-8
$C_{31}H_{35}N_3O_4S$
[2S-(2α,5α,6β)]-3,3-Dimethyl-7-oxo-6-[(phenylacetyl)amino]-4-thia-1-azabicyclo[3.2.0]heptane-2-carboxylic acid compound with N-(phenylmethyl)benzeneethanamine (1:1).
benzylpenicillinic acid N-benzyl-β-phenylethylamine salt; benethamine penicillin G; Benapen; Betapen; Benetolin. Antibacterial. Semi-synthetic antibiotic prepared from penicillin G and N-benzylphenethylamine. mp = 146-147° (dec); slightly soluble in H_2O. *Glaxo Wellcome Inc.*

2116 Penicillin G Benzathine
1538-09-6 7221 216-260-5
$C_{48}H_{56}N_6O_8S_2$
[2S-(2α,5α,6β)]-3,3-Dimethyl-7-oxo-6-[(phenylacetyl)-amino]-4-thia-1-azabicyclo[3.2.0]heptane-2-carboxylic acid compound with N,N'-dibenzylethylenediamine (1:1).
Penicillin G Benzathine; Benethamine penicillin; Benepen; Betapen; Benetolin; Potassium penicillin G; Potassium benzylpenicillinate; Benzylpenicillinic acid potassium salt; Notaral; benzathine penicillin G; diamine penicillin; Beacillin; Megacillin suspension; Bicillin; Cillenta; Permapen; Duropenin; Dibencillin; DBED Penicillin; Penidural; Tardocillin; Dibencil; Lentopenil; Vicin Neolin; Pen-Di-Ben; Penidure; Moldamin; Extencilline; Longacilina; Longicil; Penadur; Penditan; Cepacilina. Antibiotic. mp = 123-124°; slightly soluble in H_2O; $[\alpha]_D^{25}$ = +206° (c = 0.105 in formamide). *Wyeth-Ayerst Labs.*

2117 Penicillin G Benzathine Tetrahydrate
41372-02-5
$C_{48}H_{56}N_6O_8S_2.4H_2O$
[2S-(2α,5α,6β)]-3,3-Dimethyl-7-oxo-6-[(phenylacetyl)-amino]-4-thia-1-azabicyclo[3.2.0]heptane-2-carboxylic acid compound with N,N'-dibenzylethylenediamine (1:1) tetrahydrate.
Bicillin; Bicillin L-A; Permapen; component of: Bicillin C-R, Flo-Cillin. Antibacterial. *Bristol-Myers Squibb Pharm. R&D; Roerig Div. Pfizer Pharm.; Wyeth-Ayerst Labs.*

2118 Penicillin G Benzhydralamine
1538-11-0 7222 216-261-0
$C_{48}H_{56}N_6O_8S_2$
[2S-(2α,5α,6β)]-3,3-Dimethyl-7-oxo-6-[(phenylacetyl)-amino]-4-thia-1-azabicyclo[3.2.0]heptane-2-carboxylic acid compound with N,N'-bis(phenylmethyl)-1,2-ethanediamine.
Beacillin; Megacillin Suspension; Bicillin; Cillenta; Permapen; Duropenin; Dibencillin; DBED Penicillin; Penidural; Tardocillin; Dibencil; Lentopenil; Vicin Neolin; Pen-Di-Ben; Penidure; Moldamin; Extencilline; Longfacilina; Longicil; Penadur; Penditan; Cepacilina. Antibacterial. mp = 123-124°; $[\alpha]_D^{25}$ = 206° (c = 0.105 formamide); soluble in H_2O (0.015 g/100 ml), C_6H_6 (0.038 g/100 ml), EtOH (0.52 g/100 ml), Me_2CO (0.15 g/100 ml), formamide (2.8 g/100 ml). *Wyeth-Ayerst Labs.*

2119 Penicillin G Calcium
1406-07-1 7223 215-795-1
$C_{32}H_{34}CaN_4O_8S_2$
[2S-(2α,5α,6β)]-3,3-Dimethyl-7-oxo-6-[(phenylacetyl)-amino]-4-thia-1-azabicyclo[3.2.0]heptane-2-carboxylic acid calcium salt.
calcium salt of penicillin G; calcium benzylpenicillinate; calcium penicillin G. Antibacterial. Freely soluble in H_2O, EtOH, glycerol, EtOAc, $CHCl_3$, Me_2CO.

2120 Penicillin G Chloroprocaine
575-52-0
$C_{29}H_{37}ClN_4O_6S$
[2S-(2α,5α,6β)]-3,3-Dimethyl-7-oxo-6-[(phenylacetyl)-amino]-4-thia-1-azabicyclo[3.2.0]heptane-2-carboxylic acid compound with 2-(diethylamino)ethyl 4-amino-2-chlorobenzoate (1:1).
Antibacterial.

2121 Penicillin G Hydrabamine
3344-16-9 7224 222-092-3
$C_{74}H_{100}N_6O_8S_2$
Benzylpencillinic acid N,N'-bis(dehydroabietyl)ethylenediamine double salt.
hydrabamine penicillin G; Compocillin. Antibacterial. Dec 171-174°; $[\alpha]_D^{25}$ = +115.3° (c = 10 in $CHCl_3$); solubilities (g/100 ml at 28°): H_2O 0.00075, MeOH 0.73, EtOH 0.52, iPrOH 0.17, isoamyl alcohol 0.31, cyclohexane 1.15, C_6H_6 6.00, C_7H_8 3.9, petroleum ether 0.0, isooctane 0.55. *Abbott Labs Inc.*

2122 Penicillin G Potassium
113-98-4 7225 204-038-0
$C_{16}H_{17}KN_2O_4S$
[2S-(2α,5α,6β)]-3,3-Dimethyl-7-oxo-6-[(phenylacetyl)-amino]-4-thia-1-azabicyclo[3.2.0]heptane-2-carboxylic acid monopotassium salt.
potassium penicillin G; potassium benzylpenicillinate; Notaral; Crytapen; Hipercilina; Pentid; Tabilin; Eskacillin; Forpen; Hylenta; Cosmopen; Falapen; Hyasorb; Cristapen; M-Cillin; Monopen; Megacillin Tablets. Antibacterial. mp = 214-217° (dec); $[\alpha]_D^{22}$ = 285-310° (c = 0.7); freely soluble in H_2O.

2123 Penicillin G Procaine
54-35-3 7226 200-205-7
$C_{29}H_{38}N_4O_6S$
[2S-(2α,5α,6β)]-3,3-Dimethyl-7-oxo-6-[(phenylacetyl)-amino]-4-thia-1-azabicyclo[3.2.0]heptane-2-carboxylic acid compound with 2-(diethylamino)ethyl 4-aminobenzoate (1:1).
benzylpenicillin procaine; Abbocillin-DC; Afsillin; Ampinpenicillin; Aquacillin; Aquasuspen; Avloprocil; Cilicaine; Crysticillin; Despacilina; Depocillin; Distaquaine; Dorsallin A.R.; Duracillin; Flo-cillin Aqueous; Hydracillin; Ilcocillin P; Kabipenin; Megapen; Mylipen; Ledercillin; Lenticillin; Mammacillin; Neoproc; Pentaquacaine G; Pen-50; Premocillin; Procanodia; Pro-Pen; Wycillin. Antibacterial. *Abbott Labs Inc.; Apothecon; Eli Lilly & Co.; Marion Merrell Dow Inc.; Roerig Div. Pfizer Pharm.; Schering-Plough Animal Health; Schering-Plough HealthCare Products; SmithKline Beecham Pharm.; Solvay Animal Health Inc.*

2124 Penicillin G Procaine Monohydrate
6130-64-9 7226
$C_{29}H_{38}N_4O_6S.H2O$
[2S-(2α,5α,6β)]-3,3-Dimethyl-7-oxo-6-[(phenylacetyl)-amino]-4-thia-1-azabicyclo[3.2.0]heptane-2-carboxylic acid compound with 2-(diethylamino)ethyl 4-aminobenzoate (1:1) monohydrate.
penicillin G compound with 2-(diethylamino)ethyl p-aminobenzoate (1:1) monohydrate; benzylpenicillin procaine; procaine benzylpenicillinate; procaine penicillin G; Abbocillin-DC; Afsillin; Ampinpenicillin; Aquacillin; Aquasuspen; Avloprocil; Cilicaine; Crysticillin; Despacilina; Depocillin; Distquaine; Dorsallin A.R.; Duracillin; Flo-Cillin Aqueous; Hydracillin; Ilcocillin P; Kabipenin; Ledercillin; Lenticilln; Megapen; Mylipen; Neoproc; Penaquacaine G; Pen-Fifty; Premocillin; Procanodia; Pro-Pen; Wycillin. Antibacterial. mp = 106-110° (dec); d = 1.255; soluble in H_2O (0.7 g/100 ml), more soluble in organic solvents; LD_{50} (mus sc) = 2300 mg/kg. *Abbott Labs Inc.; Apothecon; Eli Lilly & Co.; Marion Merrell Dow Inc.; Roerig Div. Pfizer Pharm.;*

Schering-Plough Animal Health; Schering-Plough HealthCare Products; SmithKline Beecham Pharm.; Solvay Animal Health Inc.

2125 Penicillin N
525-94-0 7227
$C_{14}H_{21}N_3O_6S$
[2S-[2α,5α,6β(S*)]]-6-[(5-Amino-5-carboxy-1-oxopentyl)amino]-3,3-dimethyl-7-oxo-4-thia-1-azabicyclo[3.2.0]heptane-2-carboxylic acid.
adicillin; cephalosporin N; adicillin; Synnematin B. Antibacterial. Soluble in H_2O; dextrorotary. *Eli Lilly & Co.*

2126 Penicillin O
87-09-2 7228
$C_{13}H_{18}N_2O_4S_2$
[2S-(2α,5α,6β)]-3,3-Dimethyl-7-oxo-6-[[(2-propenylthio)acetyl]amino]-4-thia-1-azabicyclo[3.2.0]heptane-2-carboxylic acid.
allomercaptomethylpenicillin; allylmercaptomethylpenicillinic acid; penicillin AT. Antibiotic produced by *Penicillium chrysogenum. Eli Lilly & Co.; Pharmacia & Upjohn.*

2127 Penicillin O Potassium
897-61-0 7228
$C_{13}H_{17}KN_2O_4S_2$
[2S-(2α,5α,6β)]-3,3-Dimethyl-7-oxo-6-[[(2-propenylthio)acetyl]amino]-4-thia-1-azabicyclo[3.2.0]heptane-2-carboxylic acid potassium salt.
potassium penicillin O; penicillin O potassium. Antibacterial. *Eli Lilly & Co.; Pharmacia & Upjohn.*

2128 Penicillin O Sodium
7177-54-0 7228 230-539-9
$C_{13}H_{17}N_2NaO_4S_2$
[2S-(2α,5α,6β)]-3,3-Dimethyl-7-oxo-6-[[(2-propenylthio)acetyl]amino]-4-thia-1-azabicyclo[3.2.0]heptane-2-carboxylic acid sodium salt.
Cer-O-Cillin Sodium. Antibacterial. *Eli Lilly & Co.; Pharmacia & Upjohn.*

2129 Penicillin V
87-08-1 7230 201-722-0
$C_{16}H_{18}N_2O_5S$
[2S-(2α,5α,6β)]-3,3-Dimethyl-7-oxo-6-[(phenoxyacetyl)amino]-4-thia-1-azabicyclo[3.2.0]heptane-2-carboxylic acid.
Acipen-V; Distaquaine V; Fenospen; Meropenin; Oracilline; Oratren; V-Cillin. Antibacterial. Dec at 120-128°; λ_m = 268, 274 nm (ε 1330, 1100); soluble in H_2O at pH 1.8 (25 mg/100 ml); soluble in polar organic solvents; practically insoluble in vegetable oils, liquid petrolatum. *Eli Lilly & Co.; Wyeth-Ayerst Labs.*

2130 Penicillin V Benzathine
5928-84-7 7231 227-667-2
$C_{48}H_{56}N_6O_{10}S_2$
[2S-(2α,5α,6β)]-3,3-Dimethyl-7-oxo-6-[(phenylacetyl)amino]-4-thia-1-azabicyclo-[3.2.0]heptane-2-carboxylic acid compound with N,N'-bis(phenylmethyl)-1,2-ethanediamine (2:1).
penicillin V DBED; benathine penicillin V; benzathine benzylpenicillin; Bicillin L-A; Falcopen-V; Ospen (syrup); Pen-Vee. Antibacterial. mp = 105-109°; poorly soluble in H_2O (0.0321 g/ 100 ml), more soluble in EtOH. *Wyeth-Ayerst Labs.*

2131 Penicillin V Benzathine Tetrahydrate
63690-57-3 7231
$C_{48}H_{56}N_6O_{10}S_2.4H_2O$
[2S-(2α,5α,6β)]-3,3-Dimethyl-7-oxo-6-[(phenoxyacetyl)-amino]-4-thia-1-azabicyclo[3.2.0]heptane-2-carboxylic acid compound with N,N'-bis(phenylmethyl)-1,2-ethanediamine (2:1) tetrahydrate.
penicillin V DBED; benzathine benzylpenicillin; Bicillin L-A; Falcopen-V; Ospen; Pen-Vee. Antibacterial. *Wyeth-Ayerst Labs.*

2132 Penicillin V Hydrabamine
6591-72-6 7232
$C_{74}H_{100}N_6O_{10}S_2$
N,N'-Bis(dehydroabietyl)ethylenediamine bis(phenoxymethylpenicillin).
hydrabamine penicillin V; Abbocillin V; Compocillin-V. Antibacterial. Solubilities (mg/ml): H_2O 0.05, MeOH 11.05, EtOH 5.8, isopropanol 1.75, isoamyl alcohol 6.85, cyclohexane 0.12, C_6H_6 1.4, C_7H_8 1.07, petroleum ether 0.06, isooctane 0.065, CCl_4 3.30, EtOAc 4.0, isoamyl acetate 4.0, isoamyl acetate 4.9, Me_2OH 10.2, methyl ethyl ketone 13.7, ether 0.095, ethylene chloride >20, dioxane 7.5. *Abbott Labs Inc.*

2133 Penicillin V Potassium
132-98-9 7230 205-086-5
$C_{16}H_{17}KN_2O_5S$
[2S-(2α,5α,6β)]-3,3-Dimethyl-7-oxo-6-[(phenoxyacetyl)-amino]-4-thia-1-azabicyclo[3.2.0]heptane-2-carboxylic acid monopotassium salt.
D-α-phenoxymethylpencillinate K salt; phenoxymethyl-penicillin Potassium; V-cillin-k; Compocillin-VK; Penvikal; Apsin VK; Arcacil; Beromycin 400; Cliacil; Distakaps V-k; Dowpen V-k; Fenoxypen; Oracil-VK; PVK; Suspen; Betapen V; phenoxymethylpenicillin potassium; V-Cillin-K; Compocillin-VK; Penvikal; Apsin VK; Arcacil; Beromycin 400; Cliacil; Distakaps V-K; Dowpen V-K; Fenoxypen; Oracil-VK; PVK; Suspen; Uticillin VK; Beromycin; Antibiocin; Arcasin; Betapen-VK; Calciopen K; Distaquaine V-K; DQV-K; Isocillin; Distaquaine V-K; Icipen; Ispenoral; Ledercillin VK; Megacillin Oral; Orapen; Ospeneff; Pedipen; Penagen; Pencompren; penicillin potassium phenoxymethyl; Pen-Vee-K; Pen-V-K; Qidpen VK; Robicillin VK; ocillin-VK; Roscopenin; Fenocin; Fenocin Forte; Penapar VK; Robicillin Vk; SK-Penicillin VK; Stabillin VK syrup 125; Stabillin VK syrup 62.5; SumapenVK; V-CIL-K; Veetids; Vepen; Pfizerpen VK; Uticillin VK; Beromycin; Antibiocin; Arcasin; Beromycin (Penicillin); Betapen-VK; Calciopen K; Distaquaine V-K; DqV-K; Isocillin; Icipen; Ispenoral; Ledercillin VK; Megacillin Oral; Orapen; Ospeneff; Pedipen; Penagen; Pencompren; Penicillin Potassium Phenoxymethyl; Pen-vee-k; Pen-V-K; Qidpen VK; Robicillin VK; Rocillin-VK; Roscopenin; Sk-penicillin VK; Stabillin VK Syrup 125; Stabillin VK Syrup 62.5; Sumapen VK; V-CIL-K; veetids; Vepen; Pfizerpen VK. Antibacterial. mp = 120-128° (dec); λ_m = 268, 274 nm (ε 1330 1100); soluble in H_2O (0.025 g/100 ml), polar organic solvents. *Apothecon; Eli Lilly & Co.; Lederle Labs.; Parke-Davis; Robins, A. H. Co.; SmithKline Beecham Pharm.; Wyeth-Ayerst Labs.*

2134 Penimepicycline
4599-60-4 7235 225-002-0
$C_{45}H_{56}N_6O_{14}S$
[2S-(2α,5α,6β)]-3,3-Dimethyl-7-oxo-6-[(phenylacetyl)-amino]-4-thia-1-azabicyclo[3.2.0]heptane-2-carboxylic acid compound with [4S-(4α,4aα,5aα,6β,12aα)]-4-(dimethylamino)-1,4,4a,5,5a,6,11,12a-octahydro-3,6,10,12,12a-pentahydroxy-N-[[4-(2-hydroxyethyl)-1-piperazinyl]-methyl]-6-methyl-1,11-dioxo-2-naphthacenecarboxamide (1:1).
penimepiciclina; Criseocil; Geotricyn; Mepenicycline; Olimpen; Penetracyne; Peniltetra; Prestociclina. Antibacterial. Dec above 143°; soluble in H_2O at 20° = 142 g/100 ml; $[\alpha]_D^{20}$ = -50.5° (c = 2 in MeOH). *E.R.A.S.M.E.*

2135 Phenethicillin Potassium
147-55-7 7369 205-691-4
$C_{17}H_{19}KN_2O_5S$
[2S-(2α,5α,6β)]-3,3-Dimethyl-7-oxo-6-[(1-oxo-2-phenoxypropyl)amino]-4-thia-1-azabicyclo[3.2.0]heptane-2-carbocylic acid monopotassium salt.
penicillin-152; penicillin MV; penicillin-152 potassium; potassium phenethicillin; Alfacilin; Alpen (obsolete); Brocsil; Broxil; Chemipen; Darcil; Dramcillin-S; Maxipen; Optipen; Oralopen; Peniplus; Penorale; Penova; Pensig; Syncillin; Synthecilline; Synthepen. Antibacterial. [dl-form]: Dec 230-232°; freely soluble in H_2O. *Bristol-Myers Squibb Pharm. R&D.*

2136 Piperacillin
61477-96-1 7616 262-811-8
$C_{23}H_{27}N_5O_7S$
[2S-[2α,5α,6β(S*)]]-6-[[[[(4-Ethyl-2,3-dioxo-1-piperazinyl)carbonyl]amino]phenylacetyl]amino]-3,3-dimethyl-7-oxo-4-thia-1-azabicyclo[3.2.0]heptane-2-carboxylic acid.
Antibacterial. *Lederle Labs.*

2137 Piperacillin Monohydrate
66258-76-2 7616
$C_{23}H_{27}N_5O_7S.H_2O$
[2S-[2α,5α,6β(S*)]]-6-[[[[(4-Ethyl-2,3-dioxo-1-piperazinyl)carbonyl]amino]phenylacetyl]amino]-3,3-dimethyl-7-oxo-4-thia-1-azabicyclo[3.2.0]heptane-2-carboxylic acid monohydrate.
Antibacterial. *Lederle Labs.*

2138 Piperacillin Sodium
59703-84-3 7616 261-868-6
$C_{23}H_{26}N_5NaO_7S$
[2S-[2α,5α,6β(S*)]]-6-[[[[(4-Ethyl-2,3-dioxo-1-piperazinyl)carbonyl]amino]phenylacetyl]amino]-3,3-dimethyl-7-oxo-4-thia-1-azabicyclo[3.2.0]heptane-2-carboxylic acid sodium salt.
Pipracil; Zosyn; T-1220; CL-227193. Antibacterial. mp = 183-185° (dec); LD_{50} (mus iv) = 5000 mg/kg, (rat iv) = 2700 mg/kg, (dog iv) > 6000 mg/kg, (mky iv) > 4000 mg/kg. *Lederle Labs.*

2139 Pirazmonam
108319-07-9
$C_{22}H_{24}N_{10}O_{12}S_2$
2-[[[(2-Amino-4-thiazolyl)[[1-[[3-(1,4-dihydro-5-hydroxy-4-oxopicolinamido)-2-oxo-1-imidazolidinyl]sulfonyl]carbamoly]-2-oxo-3-azetidinyl]carbamoyl]methylene]amino]oxy]-2-methylpropionic acid.
Antibacterial. *Bristol-Myers Squibb Pharm. R&D.*

2140 Pirazmonam Sodium
104393-00-2
$C_{22}H_{22}N_{10}Na_2O_{12}S_2$
2-[[[(2-Amino-4-thiazolyl)[[1-[[3-(1,4-dihydro-5-hydroxy-4-oxopicolinamido)-2-oxo-1-imidazolidinyl]sulfonyl]carbamoly]-2-oxo-3-azetidinyl]carbamoyl]methylene]amino]oxy]-2-methylpropionic acid disodium salt.
SQ-83360. Antibacterial. *Bristol-Myers Squibb Pharm. R&D.*

2141 Pirbenicillin
55975-92-3
$C_{24}H_{26}N_6O_5S$
(2S,5R,6R)-6-[(R)-2-[Isonicotinimidoylamino)acetamido]-2-phenylacetamido]-3,3-dimethyl-7-oxo-4-thia-1-azabicyclo[3.2.0]heptane-2-carboxylic acid.
Antibacterial. *Pfizer Inc.*

2142 Pirbenicillin Sodium
55162-26-0
$C_{24}H_{25}N_6NaO_5S$
(2S,5R,6R)-6-[(R)-2-[Isonicotinimidoylamino)acetamido]-2-phenylacetamido]-3,3-dimethyl-7-oxo-4-thia-1-azabicyclo[3.2.0]heptane-2-carboxylic acid sodium salt.
CP-33994-2. Antibacterial. *Pfizer Inc.*

2143 Piridicillin
69414-41-1
$C_{32}H_{35}N_5O_{11}S_2$
(2S,5R,6R)-6-[(R)-2-[6-[p-[Bis(2-hydroxyethyl)sulfamoyl]phenyl]-1,2-dihydro-2-oxonicotinamido]-2-(p-hydroxyphenyl)acetamido]-3,3-dimethyl-7-oxo-4-thia-1-azabicyclo[3.2.0]heptane-2-carboxylic acid.
Antibacterial. *Parke-Davis.*

2144 Piridicillin Sodium
69402-03-5
$C_{32}H_{34}N_5NaO_{11}S_2$
(2S,5R,6R)-6-[(R)-2-[6-[p-[Bis(2-hydroxyethyl)sulfamoyl]phenyl]-1,2-dihydro-2-oxonicotinamido]-2-(p-hydroxyphenyl)acetamido]-3,3-dimethyl-7-oxo-4-thia-1-azabicyclo[3.2.0]heptane-2-carboxylic acid sodium salt.
Antibacterial. *Parke-Davis.*

2145 Piroxicillin
82509-56-6
$C_{27}H_{28}N_8O_9S_2$
(2S,5R,6R)-6-[(R)-2-(p-Hydroxyphenyl)-2-[3-[4-hydroxy-2-(p-sulfamoylanilino)-5-pyrimidinyl]ureido]acetamido]-3,3-dimethyl-7-oxo-4-thia-1-azabicyclo[3.2.0]heptane-2-carboxylic acid.
Antibacterial.

2146 Pivampicillin
33817-20-8 7669 251-688-6
$C_{22}H_{29}N_3O_6S$
[2S-[2α,5α,6β(S*)]]-6-[(Aminophenylacetyl)amino]-3,3-dimethyl-7-oxa-4-thia-1-azabicyclo[3.2.1]heptane-2-carboxylic acid (2,2-dimethyl-1-oxopropoxy)-methyl ester.
ampicillin pivaloyloxymethyl ester; pivaloyloxymethyl ampicillinate; MK-191. Antibacterial. *LeoAB.*

2147 Pivampicillin Hydrochloride
26309-95-5 7669 247-604-2
$C_{22}H_{29}N_3O_6S.HCl$
[2S-[2α,5α,6β(S*)]]-6-[(Aminophenylacetyl)amino]-3,3-dimethyl-7-oxa-4-thia-1-azabicyclo[3.2.1]heptane-2-carboxylic acid (2,2-dimethyl-1-oxopropoxy)-methyl ester hydrochloride.
Pivampicillin hydrochloride; Alphacilina; Alphacillin; Berocillin; Centurina; Diancina; Inacilin; Maxifen; Pivatil; Pondocil; Pondocillin; Pondocillina; Sanguicillin. Antibacterial. mp = 155-156° (dec); $[\alpha]_D^{20}$ = 196° (c = 1 H_2O); λ_m = 268, 262, 256 nm ($E_{1\,cm}^{1\%}$ 3.9, 5.7, 6.3 H_2O); soluble in H_2O, less soluble in organic solvents; LD_{50} (mus orl) = 3340 mg/kg, (mus sc) = 3600 mg/kg, (rat orl) = 5000 mg/kg, (rat sc) = 4500 mg/kg. *LeoAB.*

2148 Pivampicillin Pamoate
39030-72-3
$C_{22}H_{29}N_3O_6S.HCl$
[2S-[2α,5α,6β(S*)]]-6-[(Aminophenylacetyl)amino]-3,3-dimethyl-7-oxa-4-thia-1-azabicyclo[3.2.1]heptane-2-carboxylic acid (2,2-dimethyl-1-oxopropoxy)-methyl ester hydrochloride.
Antibacterial. *LeoAB.*

2149 Pivampicillin Probenate
42190-91-0
$C_{35}H_{48}N_4O_{10}S_2$
[2S-[2α,5α,6β(S*)]]-6-[(Aminophenylacetyl)amino]-3,3-dimethyl-7-oxa-4-thia-1-azabicyclo[3.2.1]heptane-2-carboxylic acid (2,2-dimethyl-1-oxopropoxy)-methyl ester mono[4-[(dipropylamino)sulfonyl]benzoate] (1:1).
Antibacterial. *LeoAB.*

2150 Prazocillin
15949-72-1
$C_{19}H_{18}Cl_2N_4O_4S$
6-[1-(2,6-Dichlorophenyl)-4-methylpyrazole-5-carboxamido]-3,3-dimethyl-7-oxo-4-thia-1-azabicyclo-[3.2.0]heptane-2-carboxylic acid.
Antibacterial.

2151 Propicillin
551-27-9 8002 208-995-5
$C_{18}H_{22}N_2O_5S$
[2S-(2α,5α,6β)]-3,3-Dimethyl-7-oxo-6-[(1-oxo-2-phenoxybutyl)amino]-4-thia-1-azabicyclo[3.2.0]heptane-2-carboxylic acid.
levopropylcillin. Antibacterial. *SmithKline Beecham Pharm.*

2152 Propicillin Potassium
1245-44-9 8002 214-993-5
$C_{18}H_{21}KN_2O_5S$
[2S-(2α,5α,6β)]-3,3-Dimethyl-7-oxo-6-[(1-oxo-2-phenoxybutyl)amino]-4-thia-1-azabicyclo[3.2.0]heptane-2-carboxylic acid potassium salt.
BRL-284; PA-248; Baycillin; Brocillin; Cetacillin; Oricillin; Trecillin; Ultrapen. Antibacterial. mp ≅ 196° (dec); soluble in H_2O (8.3 g/100 ml), more soluble in EtOH. *Bayer Corp. Pharm. Div.; SmithKline Beecham Pharm.*

2153 Quinacillin
1596-63-0 8224 216-481-7
$C_{18}H_{14}N_4Na_2O_6S$
[2S-(2α,5α,6β)]-6-[[(3-Carboxy-2-quinoxalinyl)-carbonyl]amino]-3,3-dimethyl-7-oxo-4-thia-1-azabicyclo[3.2.0]heptane-2-carboxylic acid.
Antibacterial.

2154 Ritipenem
84845-57-8 8400
$C_{10}H_{12}N_2O_6S$
[5R-[5α,6α(R*)]]-3-[[(Aminocarbonyl)oxy]methyl]-6-(1-hydroxyethyl)-7-oxo-4-thia-1-azabicyclo-[3.2.0]hept-2-ene-2-carboxylic acid.
Antibacterial. *Farmitalia Carlo Erba Ltd.*

2155 Rotamicillin
55530-41-1
$C_{28}H_{31}N_5O_5S$
(2S,5R,6R)-3,3-Dimethyl-7-oxo-6-[(R)-2-phenyl-2-[2-[p-(1,4,5,6-tetrahydro-2-pyrimidinyl)phenyl]-acetamido]acetamido]-4-thia-1-azabicyclo[3.2.0]-heptane-2-carboxylic acid.
Antibacterial.

2156 Sanfetrinem
156769-21-0
$C_{14}H_{19}NO_5$
(1S,5S,8aS,8bR)-1,2,5,6,7,8,8a,8b-Octahydro-1-[(R)-1-hydroxyethyl]-5-methoxy-2-oxoazeto[2,1-a]isoindole-4-carboxylic acid.
Antibacterial. *Glaxo Wellcome Inc.*

2157 Sanfetrinem Sodium
141611-76-9
$C_{14}H_{18}NNaO_5$
(1S,5S,8aS,8bR)-1,2,5,6,7,8,8a,8b-Octahydro-1-[(R)-1-hydroxyethyl]-5-methoxy-2-oxoazeto[2,1-a]isoindole-4-carboxylic acid sodium salt.
GV-104326B. Antibacterial. *Glaxo Wellcome Inc.*

2158 Sanfetrinem Cilexetil
141646-08-4
$C_{23}H_{33}NO_8$
1-Hydroxyethyl-(1S,5S,8aS,8bR)-1,2,5,6,7,8,8a,8b-octahydro-1-[(R)-1-hydroxyethyl]-5-methoxy-2-oxo-azeto[2,1-a]-isoindole-4-carboxylic acid cyclohexyl carbonate (ester).
GV-118819X. Antibacterial. *Glaxo Wellcome Inc.*

2159 Sarmoxicillin
67337-44-4
$C_{21}H_{27}N_3O_6S$
Methoxymethyl (2S,5R,6R)-6-[4-(p-hydroxyphenyl)-2,2-dimethyl-5-oxo-1-imidazolidinyl]3,3-dimethyl-7-oxo-4-thia-1-azabicyclo[3.2.0]heptane-2-carboxylic acid.
Antibacterial. *Bristol-Myers Squibb Pharm. R&D.*

2160 Sulbactam
68373-14-8 9058 269-878-2
$C_8H_{11}NO_5S$
(2S-cis)-3,3-Dimethyl-7-oxo-4-thia-1-azabicyclo-[3.2.0]heptane-2-carboxylic acid.
CP-45899-2; Betamaze; penicillanic acid sulfone;

penicillanic acid 1,1-dioxide. A semisynthetic Antibiotic. A β-lactamase inhibitor. mp = 148-151°; $[\alpha]_D^{20}$ = 251° (c = 0.01 in pH 5 buffer). *Pfizer Inc.; Roerig Div. Pfizer Pharm.*

2161 Sulbactam Benzathine
83031-43-0 9058
$C_{32}H_{42}N_4O_{10}S_2$
(2S-cis)-3,3-Dimethyl-7-oxo-4-thia-1-azabicyclo-[3.2.0]heptane-2-carboxylic acid compound with N,N'-dibenzylethylenediamine.
CP-45899-99. Antibacterial. A β-lactamase inhibitor. *Pfizer Inc.*

2162 Sulbactam Pivoxil
69388-79-0 9058
$C_{14}H_{21}NO_7S$
(2S-cis)-3,3-Dimethyl-7-oxo-4-thia-1-azabicyclo[3.2.0]-heptane-2-carboxylic acid 4,4-dioxide (2,2-dimethyl-1-oxopropoxy)methyl ester.
penicillanic acid sulfone; penicillanic acid 1,1-dioxide; CP-45899; Unasyn® Oral. Antibacterial. A β-lactamase inhibitor. *Pfizer Inc.*

2163 Sulbactam Sodium
69388-84-7 9058 273-984-4
$C_8H_{10}NNaO_5S$
(2S-cis)-3,3-Dimethyl-7-oxo-4-thia-1-azabicyclo[3.2.0]-heptane-2-carboxylic acid sodium salt.
CP-45899-2; Sulperazone [with cefoperazone sodium]; Bethacil [with ampicillin sodium]; Loricin [with ampicillin sodium]; Unacid [with ampicillin sodium]; Unacim [with ampicillin sodium]; component of: Unasyn. Antibacterial. A β-lactamase inhibitor. *Roerig Div. Pfizer Pharm.*

2164 Sulbenicillin
41744-40-5 9059 255-528-6
$C_{16}H_{18}N_2O_7S_2$
3,3-Dimethyl-7-oxo-6-(2-phenyl-2-sulfoacetamido)-4-thia-1-azabicyclo[3.2.0]heptane-2-carboxylic acid.
sulfocillin. Antibacterial. *Takeda Chem. Ind. Ltd.*

2165 Sulbenicillin Disodium Salt
28002-18-8 9059 248-769-3
$C_{16}H_{16}N_2Na_2O_7S_2$
3,3-Dimethyl-7-oxo-6-(2-phenyl-2-sulfoacetamido)-4-thia-1-azabicyclo[3.2.0]heptane-2-carboxylic acid disodium salt.
Kedacillina; Sulpelin; Lilacillin. Antibacterial. *Takeda Chem. Ind. Ltd.*

2166 Sulopenem
120788-07-0
$C_{12}H_{15}NO_5S_3$
(5R,6S)-6-[(1R)-1-Hydroxyethyl]-7-oxo-3-[[(3S)-tetra-hydro-3-thienyl]thio]-4-thia-1-azabicyclo[3.2.0]hept-2-ene-2-carboxylic acid.
CP-70429. Antibacterial. *Pfizer Intl.*

2167 Sultamicillin
76497-13-7 9166
$C_{25}H_{30}N_4O_9S_2$
[2S-[2α(2R*,5S*),5α,6β(S*)]]-6-[(Aminophenylacetyl)-amino]-3,3-dimethyl-7-oxo-4-thia-1-azabicyclo[3.2.0]-heptane-2-carboxylic acid [[(3,3-dimethyl-7-oxo-4-thia-1-azabicyclo[3.2.0]hept-2-yl)carbonyl]oxy]methyl ester S,S-dioxide.
CP-49952; VD-1827; sultamicillin. Antibacterial. A β-lactamase inhibitor. Administered orally. A double ester of sulbactam and ampicillin. *Pfizer Inc.*

2168 Sultamicillin Tosylate
83105-70-8 9166
$C_{32}H_{38}N_4O_{12}S_3$
[2S-[2α(2R*,5S*),5α,6β(S*)]]-6-[(Aminophenylacetyl)-amino]-3,3-dimethyl-7-oxo-4-thia-1-azabicyclo[3.2.0]-heptane-2-carboxylic acid [[(3,3-dimethyl-7-oxo-4-thia-1-azabicyclo[3.2.0]hept-2-yl)carbonyl]oxy]methyl ester S,S-dioxide p-toluenesulfonate.
Bacimex; Bethacil orale; Unacid PD oral; Unacim orale; Unasyn. Antibacterial. *Pfizer Inc.*

2169 Suncillin
22164-94-9
$C_{16}H_{19}N_3O_7S_2$
3,3-Dimethyl-7-oxo-6-[2-phenyl-D-2-(sulfoamino)-acetamido]-4-thia-1-azabicyclo[3.2.0]heptane-2-carboxylic acid.
Antibacterial. *Bristol-Myers Squibb Pharm. R&D.*

2170 Suncillin Sodium
23444-86-2
$C_{16}H_{17}N_3Na_2O_7S_2$
3,3-Dimethyl-7-oxo-6-[2-phenyl-D-2-(sulfoamino)-acetamido]-4-thia-1-azabicyclo[3.2.0]heptane-2-carboxylic acid sodium salt.
BL-P1462. Antibacterial. *Bristol-Myers Squibb Pharm. R&D.*

2171 Talampicillin
47747-56-8 9204 256-332-3
$C_{24}H_{23}N_3O_6S$
[2S-[2α,5α,6β(S*)]]-6-[(Aminophenylacetyl)amino]-3,3-dimethyl-7-oxo-4-thia-1-azabicyclo[3.2.0]heptane-2-carboxylic acid 1,3-dihydro-3-oxo- 1-isobenzofuranyl ester.
Antibacterial. *SmithKline Beecham Pharm.; Yamanouchi U.S.A. Inc.*

2172 Talampicillin Hydrochloride
39878-70-1 9204
$C_{24}H_{23}N_3O_6S.HCl$
[2S-[2α,5α,6β(S*)]]-6-[(Aminophenylacetyl)amino]-3,3-dimethyl-7-oxo-4-thia-1-azabicyclo[3.2.0]heptane-2-carboxylic acid 1,3-dihydro-3-oxo-1-isobenzo furanyl ester hydrochloride.
Talat; Talpen; Yamacillin. Antibacterial. mp = 154-157° (dec). *SmithKline Beecham Pharm.; Yamanouchi U.S.A. Inc.*

2173 Tameticillin
56211-43-9
$C_{23}H_{33}N_3O_6S$
2-(Diethylamino)ethyl-(2S,5R,6R)-6-(2,6-dimethoxybenzamido)-3,3-dimethyl-7-oxo-4-thia-1-azabicyclo[3.2.0]heptane-2-carboxylate.
Antibacterial.

2174 Tazobactam
89786-04-9 9251
$C_{10}H_{12}N_4O_5S$
[2S-(2α,3β,5β)]-3-Methyl-7-oxo-3-(1H-1,2,3-triazol-1-ylmethyl)-4-thia-1-azabicyclo[3.2.0]heptane-2-carboxylic acid 4,4 dioxide.
YTR-830H; CL-298741. Antibacterial. A β-lactamase inhibitor. *Taiho.*

2175 Tazobactam Sodium
89785-84-2 9251
$C_{10}H_{11}N_4NaO_5S$
[2S-(2α,3β,5β)]-3-Methyl-7-oxo-3-(1H-1,2,3-triazol-1-ylmethyl)-4-thia-1-azabicyclo[3.2.0]heptane-2-carboxylic acid 4,4 dioxide sodium salt.
YTR-830; CL-307579; component of: Zosyn. Antibacterial. A β-lactamase inhibitor. mp > 170° (dec). *Lederle Labs.; Taiho.*

2176 Temocillin
66148-78-5 9288 266-184-1
$C_{16}H_{18}N_2O_7S_2$
[2S-(2α,5α,6β)]-6-[(Carboxy-3-thienylacetyl)amino]-6-methoxy-3,3-dimethyl-7-oxo-4-thia-1-azabicyclo[3.2.0]-heptane-2-carboxylic acid.
Antibacterial. *SmithKline Beecham Pharm.*

2177 Terdecamycin
113167-61-6
$C_{31}H_{43}N_3O_8$
4-Methyl-1-piperazinecarboxylic acid 7-ester with (-)-N-[1S,2R,3E,5E,7S,9E,11E,13S,15R,19R)-7,13-dihydroxy-1,4,10,19-tetramethyl-17,18-dioxo-16-oxabicyclo-[13.2.2]nonadeca-3,5,9,11-tetraen-2-yl]pyruvamide.
Antibacterial (veterinary).

2178 Thiphencillin
26552-51-2
$C_{16}H_{18}N_2O_4S_2$
3,3-Methyl-7-oxo-6-[2-(phenylthio)acetamido]-4-thia-1-azabicyclo[3.2.0]heptane-2-carboxylic acid.
Tifencillin. Antibacterial. *Eli Lilly & Co.*

2179 Thiphencillin Potassium
4803-45-6
$C_{16}H_{17}KN_2O_4S_2$
3,3-Methyl-7-oxo-6-[2-(phenylthio)acetamido]-4-thia-1-azabicyclo[3.2.0]heptane-2-carboxylic acid potassium salt.
Antibacterial. *Eli Lilly & Co.*

2180 Ticarcillin
34787-01-4 9568 252-213-5
$C_{15}H_{16}N_2O_6S_2$
[2S-[2α,5α,6β(S*)]]-6-[(Carboxy-3-thienylacetyl)amino]-3,3-dimethyl-7-oxo-4-thia-1-azabicyclo[3.2.0]heptane-2-carboxylic acid.
ticarcillin. Antibacterial. *SmithKline Beecham Pharm.*

2181 Ticarcillin Cresyl
59070-07-4
$C_{22}H_{22}N_2O_6S_2$
p-Tolyl-(R)-N-[2S,5R,6R)-2-carboxy-3,3-dimethyl-7-oxo-4-thia-1-azabicyclo[3.2.0]hept-6-yl)]-3-thiophenemalonamic acid.
Antibacterial. *SmithKline Beecham Pharm.*

2182 Ticarcillin Cresyl Sodium
59070-06-3
$C_{22}H_{21}N_2NaO_6S_2$
p-Tolyl-(R)-N-[2S,5R,6R)-2-carboxy-3,3-dimethyl-7-oxo-4-thia-1-azabicyclo[3.2.0]hept-6-yl)]-3-thiophene-malonamate sodium salt.
BRL-12594. Antibacterial. *SmithKline Beecham Pharm.*

2183 Ticarcillin Disodium
4697-14-7 9568 249-642-5
$C_{15}H_{14}N_2Na_2O_6S_2$
[2S-[2α,5α,6β(S*)]]-6-[(Carboxy-3-thienylacetyl)amino]-3,3-dimethyl-7-oxo-4-thia-1-azabicyclo[3.2.0]heptane-2-carboxylic acid disodium salt.
BRL-2288; Aerugipen; Monapen; Ticarpen; Ticillin. Antibacterial. Soluble in H_2O (>100 g/100 ml). *SmithKline Beecham Pharm.*

2184 Ticarcillin Monosodium
74682-62-5
$C_{15}H_{15}N_2NaO_6S_2$
[2S-[2α,5α,6β(S*)]]-6-[(Carboxy-3-thienylacetyl)amino]-3,3-dimethyl-7-oxo-4-thia-1-azabicyclo[3.2.0]heptane-2-carboxylic acid monosodium salt.
Antibacterial. *SmithKline Beecham Pharm.*

2185 Tigemonam
102507-71-1 9572
$C_{12}H_{15}N_5O_9S_2$
[S-(Z)]-[[[1-(2-Amino-4-thiazolyl)-2-[[2,2-dimethyl-4-oxo-1-(sulfooxy)-3-azetidinyl]amino]-2-oxoethylidene]amino]oxy]acetic acid.
Orally active antibacterial. Synthetic monosulfactam structurally similar to aztreonam, a monobactam. *Bristol-Myers Squibb Pharm. R&D.*

2186 Tigemonam Dicholine
102916-21-2 9572
$C_{22}H_{41}N_7O_{11}S_2$
[S-(Z)]-[[[1-(2-Amino-4-thiazolyl)-2-[[2,2-dimethyl-4-oxo-1-(sulfooxy)-3-azetidinyl]amino]-2-oxoethylidene]amino]oxy]acetic acid dicholine salt.
SQ-30836; Tigemen. Orally active antibacterial. Synthetic monosulfactam structurally similar to aztreonam, a monobactam. *Bristol-Myers Squibb Pharm. R&D.*

2187 Valspodar
121584-18-7
$C_{63}H_{111}N_{11}O_{12}$
Cyclo[[(2S,4R,6E)-4-methyl-2-(methylamino)-3-oxo-6-octenoyl]-L-valyl-N-methylglycyl-N-methyl-L-leucyl-L-valyl-N-methyl-L-leucyl-L-alanyl-D-alanyl-N-methyl-L-leucyl-N-methyl-L-leucyl-N-methyl-L-valyl].
PSC-833; Amdray. Chemosensitizer. Third-generation cyclosporine analog, an inhibitor of P-glycoprotein, the drug efflux pump. Multidrug resistance modulator coadministered with chemotherapeutic drugs in cancer treatment. *Novartis Pharm. Corp.; Sandoz Pharm. Corp.*

Leprostatic Antibiotics

2188 Acedapsone
77-46-3 20 201-028-8
$C_{16}H_{16}N_2O_4S$
N,N'-(Sulfonyldi-4,1-phenylene)bisacetamide.
Hansolar; CI-556; CN-1883; DADDS; PAM-MR-1165; Rodilone; diacetyldapsone; sulfadiamine; 1399F. Leprostatic; antimalarial. Sulfone antibiotic. mp = 289-292°; λ_m 256 284 nm (ε 25500 36200 MeOH); soluble in H_2O (0.0003 g/100 ml). *Parke-Davis.*

2189 Acetosulfone Sodium
128-12-1 73
$C_{14}H_{14}N_3NaO_5S_2$
N-(6-Sulfanilylmetanilyl)acetamide monosodium salt.
Promacetin; CI-100; IA-307; NSC-107528; acetosulphone; Internal Antiseptic No. 307; I.A. 307. Sulfone antibiotic. Leprostatic. Soluble in H_2O (3 g/100 ml); [free sulfonamide]: mp = 285°. *Parke-Davis.*

2190 Broxaldine
3684-46-6 222-971-1
$C_{17}H_{11}Br_2NO_2$
5,7-Dibromo-2-methyl-8-quinolinol benzoate ester.
A leprostatic antibacterial.

2191 Clofazimine
2030-63-9 2433 217-980-2
$C_{27}H_{22}Cl_2N_4$
3-(p-Chloroanilino)-10-(p-chlorophenyl)-2,10-dihydro-2-(isopropylimino)phenazine.
Lamprene; G-30320; NSC-141046. Antibacterial (tuberculostatic). Also leprostatic. mp = 210-212°; λ_m 284, 486 nm (abs. 1.30, 0.64, 0.01M HCl/MeOH); soluble in AcOH, DMF, $CHCl_3$ (6.6 g/100 ml), EtOH (0.14 g/100 ml), Et_2O (0.1 g/100 ml); insoluble in H_2O; LD_{50} (mus, rat, gpg orl) > 4000 mg/kg. *Ciba-Geigy Corp.*

2192 Dapsone
80-08-0 2885 201-248-4
$C_{12}H_{12}N_2O_2S$
4,4'-Sulfonyldianiline.
4,4'-diaminodiphenyl sulfone; diaphenylsulfone; NSC-6091; DDS; DADPS; 1358F; Avlosulfon; Croysulfone; Diphenasone; Disulone; Dumitone; Eporal; Novophone; Sulfona-Mae; Sulphadione; Udolac. Sulfone antibiotic. Leprostatic and dermatitis herpetiformis suppressant. Also used as a hardening agent with epoxy resins and, in veterinary medicine, as a coccidostat. mp = 175-176°, 180.5°; insoluble in H_2O; soluble in EtOH, MeOH, Me_2CO, dilute HCl. *I.G. Farben.*

2193 Diathymosulfone
5964-62-5 3037
$C_{32}H_{34}N_4O_4S$
Di-[4-(4-hydroxy-2-methyl-5-isopropylphenylazo)phenyl] sulfone.
thymol sulfone; Diatox. Antibacterial. Leprostatic. mp = 222-224°; λ_m 400 nm (EtOH); insoluble in H_2O; soluble in dioxane, Me_2CO, alkali solutions; less soluble in EtOH, Et_2O. *Lab. Laborec.*

2194 Ditophal
584-69-0
$C_{12}H_{14}O_2S_2$
S,S-Diethyl ester of 1,3-dithioisophthalic acid.
Etisul. A leprostatic antibiotic.

2195 Glucosulfone Sodium
554-18-7 4471 209-064-6
$C_{24}H_{34}N_2Na_2O_{18}S_3$
4,4'-Diaminophenylsulfone-N,N-di(dextrose sodium sulfate).
SN-166; 501-P; Protomin; Promin; Promanide; Angeli's Sulfone. Sulfone antibiotic; leprostatic. Soluble in H_2O; slightly soluble in EtOH; insoluble in Et_2O, C_6H_6, MeOH, EtOAc, C_5H_5N; LD_{50} (mus orl) = 3930 mg/kg. *Siegfried AG.*

2196 Hydnocarpic Acid
459-67-6 4797
$C_{16}H_{28}O_2$
(R)-2-Dyclopentene-1-undecanoic acid.
Isolated from Chaulmoogra oil. Antibacterial. Leprostatic. mp = 59-60°; $[\alpha]_D$ = 68.3° ($CHCl_3$); soluble in $CHCl_3$, sparingly soluble in other organic solvents; [dl-form]: mp = 59-59.5°; [sodium salt ($C_{16}H_{27}NaO_2$)]: soluble in H_2o, MLD (rat iv) = 100-125 mg/kg.

2197 Solasulfone
133-65-3 8859 205-116-7
$C_{30}H_{28}N_2Na_4O_{14}S_5$
1,1'-[Sulfonylbis(p-phenylimino)]bis-(3-phenyl-1,3-propanedisulfonic acid) tetrasodium salt.
phenopryldiasulfone sodium; tetrasodium 4,4-diphenylsulfone α,γ,α',γ'-tetra-sulfonate; solapsone; RP-3668; Cimedone; Sulphetrone. Sulfone antibiotic. Leprostatic. Crystallizes as a hydrate; very soluble in H_2O, insoluble in EtOH. *Glaxo Wellcome Inc.*

2198 Succisulfone
5934-14-5 9047 227-684-5
$C_{16}H_{16}N_2O_5$
4'-Sulfanilylsuccinanilic acid.
F 1500; 4-succinylamido-4'-aminodiphenyl sulfone; Fourneau 1500; Exosulfonyl. Antibacterial. Leprostatic. mp = 157°; soluble in ammonia. *Bayer Corp.; Eli Lilly & Co.*

2199 Sulfoxone Sodium
144-75-2 9141
$C_{14}H_{14}N_2Na_2O_6S_3$
Disodium [sulfonylbis(p-phenyleneimino)]-dimethanesulfinate.
aldesulfone sodium; Diasone; Diazon; Novotrone. Sulfone antibiotic. Leprostatic. Dec 263-265°; readily soluble in H_2O, slightly soluble in EtOH, insoluble in most organic solvents. *Abbott Labs Inc.*

Lincosamide Antibiotics

2200 Clindamycin
18323-44-9 2414 242-209-1
$C_{18}H_{33}ClN_2O_5S$
Methyl 7-chloro-6,7,8-trideoxy-6-(1-methyl-trans-4-propyl-L-2-pyrrolidinecarboxamido)-1-thio-L-threo-α-D-galactooctopyranoside.
Dalacine; Cleocin; U-21251; Antirobe; Dalacin C; Klimicin; Sobelin. Antibacterial. $[\alpha]_D$ = 214° ($CHCl_3$). *Pharmacia & Upjohn.*

2201 Clindamycin Hydrochloride Monohydrate
58207-19-5 2414
$C_{18}H_{33}ClN_2O_5S.H_2O$
Methyl 7-chloro-6,7,8-trideoxy-6-(1-methyl-trans-4-propyl-L-2-pyrrolidinecarboxamido)-1-thio-L-threo-α-D-galactooctopyranoside hydrochloride monohydrate.
Cleocin HCl; Dalacine. Antibacterial. mp = 141-143°; $[\alpha]_D$ = 144° (H_2O); pKa = 7.6; soluble in H_2O, C_5H_5N,

EtOH, DMF; LD_{50} (mus iv) = 245 mg/kg, (mus ip) = 361 mg/kg, (mus orl) = 2618 mg/kg. *Pharmacia & Upjohn.*

2202 Clindamycin Palmitate
36688-78-5 2414
$C_{34}H_{63}ClN_2O_6S$
Methyl 7-chloro-6,7,8-trideoxy-6-(1-methyl-trans-4-propyl-L-2-pyrrolidinecarboxamido)-1-thio-L-threo-α-D-galactooctopyranoside palmitate.
Antibacterial. *Pharmacia & Upjohn.*

2203 Clindamycin Palmitate Hydrochloride
25507-04-4 2414
$C_{34}H_{64}Cl_2N_2O_6S$
Methyl 7-chloro-6,7,8-trideoxy-6-(1-methyl-trans-4-propyl-L-2-pyrrolidinecarboxamido)-1-thio-L-threo-α-D-galactooctopyranoside palmitate hydrochloride.
Cleocin Pediatric. Antibacterial. *Pharmacia & Upjohn.*

2204 Clindamycin Phosphate
24729-96-2 2414
$C_{18}H_{34}ClN_2O_8PS$
Methyl 7-chloro-6,7,8-trideoxy-6-(1-methyl-trans-4-propyl-L-2-pyrrolidinecarboxamido)-1-thio-L-threo-α-D-galactooctopyranoside 2-(dihydrogen phosphate).
Cleocin Phosphate; Cleocin T; U-28,508; Dalacin T. Antibacterial. *Pharmacia & Upjohn.*

2205 Lincomycin
154-21-2 5525 205-824-6
$C_{18}H_{34}N_2O_6S$
Methyl 6,8-dideoxy-6-trans-(1-methyl-4-propyl-L-2-pyrrolidinecarboxamido)-1-thio-D-erythro-α-D-galactooctopyranoside.
U-10149; NSC-70731; lincolnensin; Lincolcina. Antibacterial. Produced by *Streptomyces lincolnensis var. lincolnensis*. Slightly soluble in H_2O; soluble in MeOH, EtOH, EtOAc, Me_2CO, $CHCl_3$. *Pharmacia & Upjohn.*

2206 Lincomycin Hydrochloride Monohydrate
7179-49-9 5525
$C_{18}H_{35}ClN_2O_6S.H_2O$
Methyl 6,8-dideoxy-6-trans-(1-methyl-4-propyl-L-2-pyrrolidinecarboxamido)-1-thio-D-erythro-α-D-galactooctopyranoside hydrochloride monohydrate.
Frademicina; Mycivin; Waynecomycin; Albiotic; Cillimycin; Lincomix; Lincocin [as hemihydrate]. Antibacterial. mp = 145-147°; $[\alpha]_D^{25}$ = 137° (H_2O); freely soluble in H_2O, MeOH, EtOH; sparingly soluble in most organic solvents; LD_{50} (mus, rat orl) = 4000 mg/kg, (mus,rat ip) = 1000 mg/kg. *Pharmacia & Upjohn.*

Macrolide Antibiotics

2207 Azithromycin [anhydrous]
83905-01-5 946
$C_{38}H_{72}N_2O_{12}$
[2R-(2R*,3S*,4R*,5R*,8R*,10R*,11R*,12S*,13S*,14R*)]-13-[(2,6-Dideoxy-3-C-methyl-3-O-methyl-α-L-ribo-hexopyranosyl)oxy]-2-ethyl-3,4,10-trihydroxy-3,5,6,8,10,12-heptamethyl-11-[[3,4,6-trideoxy-3-(dimethylamino)-β-D-xylo-hexopyranosyl]oxy]-1-oxa-6-azacyclopentadecan-15-one.
N-methyl-11-aza-10-deoxo-10-dihydroerythromycin A; 9-deoxo-9a-methyl-9a-aza-9a-homoetrythromycin A; CP-62993; XZ-450; Azitrocin; Sumamed; Trozocina; Zithromax; Zitromax. Macrolide antibiotic. mp = 113-115°; $[\alpha]_D^{20}$ = -37° (c = 1 in $CHCl_3$). *Pfizer Inc.*

2208 Berythromycin
527-75-3
$C_{37}H_{67}NO_{12}$
12-Deoxyerythromycin.
Abbott 24091. Macrolide antibiotic. Also antiamebic. *Abbott Labs Inc.*

2209 Carbomycin A
4564-87-8 1854
$C_{42}H_{67}NO_{16}$
(12S,13S)-9-Deoxy12,13-epoxy-12,13-dihydro-9-oxoleucomycin V 3-acetate 4^B-(3-methylbutanoate).
magnamycin A; deltamycin A_4; M-4209; NSC-51001. Macrolide antibiotic. Sixteen-member ring compound, similar to leucomycin and erythromycin. mp = 214°; $[\alpha]_D^{25}$= -58.6° ($CHCl_3$); λ_m 238, 327 nm ($E_{1cm}^{1\%}$ 185, 0.9, EtOH); pK_b = 7.2; soluble in H_2O (0.029 g/100 ml), MeOH (> 2 g/100 ml), EtOH (> 2 g/100 ml); LD_{50} (mus iv) = 550 mg/kg. *Pfizer Inc.*

2210 Carbomycin B
21238-30-2 1854
$C_{42}H_{67}NO_{15}$
9-Deoxy-9-oxo-leucomycinV 3-acetate 4^b-(3-methylbutanoate).
magnamycin B. Macrolide antibiotic. Sixteen-member ring compound, similar to leucomycin and erythromycin. mp = 141-144° (dec); $[\alpha]_D^{25}$= -35° (c = 1 $CHCl_3$); λ_m = 278 nm ($E_{1\,cm}^{1\%}$ 276 EtOH); soluble in EtOH (45 g/100 ml), H_2O (0.01g/100 ml). *Pfizer Inc.*

2211 Clarithromycin
81103-11-9 2400
$C_{38}H_{69}NO_{13}$
6-O-Methylerythromycin.
Abbott-56268; TE-031; Biaxin; Clathromycin; Klacid; Klaricid; Macladin; Naxy; Veclam; Zeclar. Macrolide antibiotic. mp = 217-220° (dec), 222-225°; λ_m = 288 nm ($CHCl_3$); $[\alpha]_D^{24}$= -90.4° (c = 1 $CHCl_3$); LD_{50} (mmus orl) = 2740 mg/kg, (mmus ip) = 1030, (mmus sc) > 5000 mg/kg, (fmus orl) = 2700 mg/kg, (fmus ip) = 850 mg/kg, (fmus sc) > 5000 mg/kg, (mrat orl) = 3470 mg/kg, (mrat ip) = 669 mg/kg, (mrat sc) > 5000 mg/kg, (frat orl) = 2700 mg/kg, (frat ip) = 753 mg/kg, (frat sc) > 5000 mg/kg. *Abbott Labs Inc.*

2212 Dirithromycin
62013-04-1 3418
$C_{42}H_{78}N_2O_{14}$
[9S(R)]-9-Deoxy-11-deoxy-9,11-[imino[2-(2-methoxyethoxy)ethylidene]oxy]-erythromycin.
KT-237216; AS-E 136; Dynabac; Noriclan; Nortron; Valodin. Macrolide antibiotic. mp = 186-189° (dec); pKa (66% aq CH_2F_2) = 9.0; LD_{50} (mus sc) = >1 mg/kg, (mus orl) > 1 mg/kg. *Boehringer Mannheim GmbH; Thomae GmbH Dr. Karl.*

2213 Erythromycin
114-07-8 3720 204-040-1
$C_{37}H_{67}NO_{13}$
(3S*,4S*,5S*,6R*,7R*,9R*,11R*,12R*,13S*,14R*)-4-[(2,6-Dideoxy-3-C-methyl-3-O-methyl-α-L-ribohexopyranosyl)oxy]-14-ethyl-7,12,13-trihydroxy-3,5,7,9,11,13-hexamethyl-6-[[3,4,6-trideoxy-3-(dimethylamino)-β-D-xylohexopyranosyl]-oxy]oxacyclotetradecane-2,10-dione.
A/T/S; Emgel; ERYC; Erycette; EryDerm; Erygel; Erymax; Ery-Tab; Erythro; Ilotycin; PCE; Erythromycin A; Abomacetin; Ak-Mycin; Aknin; E-Base; EMU; E-Mycin; Eritrocina; Erythromast 36; Erythromid; Erycin; Erycin; Erycinum; Ermycin; IndermRetcin; Staticin; Stiemycin; Torlamicina; component of: Benzamycin, Staticin, T-Stat. Macrolide antibiotic. Produced by *Streptomyces erytheus*. mp = 135-140°, 190-193°; $[\alpha]_D^{25}$ = -78° (c = 1.99, EtOH); λ_m 280 nm (pH 6.3); poorly soluble in H_2O (0.020 g/100 ml); freely soluble in MeOH, EtOH, Me_2CO; $CHCl_3$, EtOAc, CH_3CN; moderately soluble in Et_2O, ethylene dichloride. *Abbott Labs Inc.; Dermik Labs. Inc.; Eli Lilly & Co.; Glaxo Wellcome plc; Hoechst Roussel Pharm. Inc.; Ortho Pharm. Corp.; Parke-Davis; Westwood-Squibb Pharm. Inc.*

2214 Erythromycin Acistrate
96128-89-1 3721
$C_{57}H_{104}NO_{16}$
Erythromycin 2'-acetate stearate (salt).
Erasis; 2'-acetylerythromycin stearate. Macrolide antibiotic. *Kreussler Chemische-Fabrik.*

2215 Erythromycin Estolate
3521-62-8 3722 222-532-4
$C_{52}H_{97}NO_{18}S$
Erythromycin 2'-propionate dodecyl sulfate (salt).
Ilosone; Eritroger; Eromycin; Lauromicina; Neo-Erycinum; PELS; Roxomicina; Stellamicina; Eriscel; Estomicina; Eupragin; Marcoeritrex; Propiocine Enfant; Togiren. Macrolide antibiotic. mp = 135-140° (dec); soluble in H_2O (0.0024 g/100 ml); more soluble in EtOH, Me_2CO, $CHCl_3$; LD_{50} (rat orl) > 5000 mg/kg. *Eli Lilly & Co.*

2216 Erythromycin Ethylsuccinate
41342-53-4 3720
$C_{43}H_{75}NO_{16}$
Erythromycin 2'-(ethylsuccinate).
E.E.S.; EryPed; Pediamycin; Wyamycin E; Anamycin; Arpimycin; Durapaediat; E-Mycin e; Eryliquid; Erythro ES; Erythro-Holz; Erythroped; Esinol; Monomycin; Paediathrocin; Refkas; Sigapedil; component of: Pediazole. Macrolide antibiotic. mp = 109-110°; $[\alpha]_D$ = -42.5°. *Abbott Labs Inc.; Ross Products; Xoma Corp.*

2217 Erythromycin Glucoheptonate
23067-13-2 3723 245-407-6
$C_{44}H_{81}NO_{21}$
Erythromycin glucoheptonate (1:1) (salt).
Erythromycin gluceptate; Ilotycin Gluceptate. Macrolide antibiotic. mp = 95-140°; freely soluble in H_2O, EtOH, dioxane, Me_2CO, propylene glycol; insoluble in Et_2O, CCl_4, C_6H_6, C_7H_8. *Eli Lilly & Co.*

2218 Erythromycin Lactobionate
3847-29-8 3724 223-348-7
$C_{49}H_{89}NO_{25}$
Erythromycin mono(4-O-β-D-galactopyranosyl-D-gluconate) (salt).
Erythrocin Lactobionate. Macrolide antibiotic. mp = 145-150°; soluble in H_2O (20 g/100 ml); freely soluble in EtOH, slightly soluble in Et_2O. *Abbott Labs Inc.; Elkins-Sinn.*

2219 Erythromycin Propionate
134-36-1 3725 205-140-8
$C_{40}H_{71}NO_{14}$
Erythromycin 2'-propionate.
propionyl erythromicin; Propiocine. Macrolide antibiotic. mp = 122-126°; $[\alpha]_D^{25}$ = -81.6° (Me_2CO); slightly soluble in H_2O; readily soluble in MeOH, EtOH, Me_2CO, DMF, EtOAc; LD_{50} (mus orl) = 2870 mg/kg, (mus sc) > 5000 mg/kg, (rat orl) > 5000 mg/kg, (rat sc) > 5000 mg/kg. *Eli Lilly & Co.*

2220 Erythromycin Stearate
643-22-1 3726 211-396-1
$C_{55}H_{103}NO_{15}$
Erythromycin octadecanoate (salt).
Dowmycin E; Erypar; Erythrocin; Ethril; Wyamycin S; Abboticine; Bristamycin; Eratrex; Eryprim; Erythro S; Ethryn; Gallimycin; Meberyt; Pantomicina; Pfizer-E; SK-Erythromycin; Wemid. Macrolide antibiotic. Slightly soluble in EtOH, Et_2O, $CHCl_3$; insoluble in H_2O. *Abbott Labs Inc.; Bristol Myers Squibb Pharm. Ltd.; Marion Merrell Dow Inc.; Parke-Davis; Xoma Corp.*

2221 Flurithromycin
82664-20-8
$C_{37}H_{66}FNO_{13}$
(8S)-8-Fluoroerythromycin.
A macrolide antibacterial.

2222 Josamycin
16846-24-5 5280 240-871-6
$C_{42}H_{69}NO_{15}$
4-(Acetyloxy-6-[[3,6-dideoxy-4-O-[2,6-dideoxy-3-C-methyl-4-O-(3-methyl-1-oxobutyl)-α-L-ribohexopyranosyl]-3-(dimethylamino)-β-D-glucopyranosyl]oxy]-10-hydroxy-5-methoxy-9,16-dimethyl-2-oxooxacyclohexadeca-11,13-diene-7-acetaldehyde.
Iosalide; EN-141; Jomybel; Josamina; Leucomycin A_3. Macrolide antibiotic. mp = 130-133°; $[\alpha]_D^{25}$ = -70° (c = 1 EtOH); λ_m 232 nm ($E_{1\,cm}^{1\%}$ 325 0.001N HCl); insoluble in H_2O; very soluble in EtOH, MeOH, Me_2CO, $CHCl_3$, EtOAc, dioxane, dilute acids; soluble in BuOH, Et_2O, CCl_4, C_6H_6, C_7H_8; insoluble in petroleum ether, n-C_6H_{14}. *Yamanouchi U.S.A. Inc.*

2223 Josamycin Propionate
40922-77-8 5280 255-140-7
$C_{45}H_{73}NO_{16}$
Josamycin 10-propionate.
Josacine; Josamy; Josaxin; Wilprafen. Macrolide antibiotic. *Yamanouchi U.S.A. Inc.*

2224 Leucomycins
1392-21-8 5480
Kitasamycin; C-637; Ayermicina; Sineptina; Stereomycine; Syneptine. A macrolide antibiotic complex obtained from *Streptomyces kitasatoensis* and consisting of at least 6 components, leucomycin A_1, A_2, B_1, B_2, B_3 and B_4. Macrolide antibiotic. mp = 128-145°; $[\alpha]_D^{20}$ = -67.1 (EtOH); λ_m 232 285 nm ($e_{1\,cm}^{1\%}$ 228 8.6 EtOH); slightly soluble in H_2O, freely soluble in organic solvents; [Leucomycin A_1]: $[\alpha]_D^{25}$ = -66.0° ($CHCl_3$); λ_m 232 nm ($E_{1\,cm}^{1\%}$ 400 MeOH); [triacetylleucomycin A_1]: mp = 125-126°; $[\alpha]_D^{25}$ = -82.5° (c = 1.3 $CHCl_3$). *Toyo Jozo.*

2225 Lexithromycin
53066-26-5
$C_{38}H_{70}N_2O_{13}$
Erythromycin 9-(O-methyloxime).
Wy-48314. Macrolide antibiotic. *Xoma Corp.*

2226 Megalomicin
28022-11-9
$C_{44}H_{80}N_2O_{15}$
(3S,4S,5S,6R,7R,9R,11R,12R,13R,14R)-4-[(2,6-Dideoxy-3-C-methyl-α-L-ribohexopyranosyl)oxy]-14-ethyl-12,13-dihydroxy-3,5,7,9,11,13-hexamethyl-7-[[2,3,6-trideoxy-3-(dimethylamino)-α-L-ribohexopyranosyl]oxy]-6-[[3,4,6-trideoxy-3-(dimethylamino)-β-D-xylohexopyranosyl]-oxy]oxacyclotetradecane-2,10-dione.
Sch-13430. Macrolide antibiotic. *Schering-Plough HealthCare Products.*

2227 Megalomicin Potassium Phosphate
51481-68-6
$C_{44}H_{84}K_2N_2O_{23}P_2$
(3S,4S,5S,6R,7R,9R,11R,12R,13R,14R)-4-[(2,6-Dideoxy-3-C-methyl-α-L-ribohexopyranosyl)oxy]-14-ethyl-12,13-dihydroxy-3,5,7,9,11,13-hexamethyl-7-[[2,3,6-trideoxy-3-(dimethylamino)-α-L-ribohexopyranosyl]oxy]-6-[[3,4,6-trideoxy-3-(dimethylamino)-β-D-xylohexopyranosyl]oxy]oxacyclotetradecane-2,10-dione, compound with potassium dihydrogen phosphate (1:2).
Sch-13430.2KH_2PO_4. Macrolide antibiotic. *Schering-Plough HealthCare Products.*

2228 Midecamycin A_1
35457-80-8 6271 252-578-0
$C_{41}H_{67}NO_{15}$
Leucomycin V 3,4^B-dipropionate.
espinomycin A; mydecamycin; turimycin P3; SF-837; YL-704B1; Aboren; Medemycin; Midecin; Momicine; Myoxam; Normicina; Rubimycin. Macrolide antibiotic. mp = 155-156°; $[\alpha]_D^{23}$ = -67° (c = 1 EtOH); λ_m 232 nm ($E_{1\,cm}^{1\%}$ 325 EtOH); soluble in MeOH, EtOH, Me_2CO, $CHCl_3$, EtOAc, C_6H_6, Et_2O, dilute acids; insoluble in H_2O, C_6H_{14}, petroleum ether. *Meiji Seika Kaisha Ltd.*

2229 Midecamycin A_3
36025-69-1 6271
$C_{41}H_{65}NO_{15}$
9-Deoxy-9-oxoleucomycin V 3,4^B-dipropionate.
antibiotic SF-837A3. Macrolide antibiotic. mp = 122-125°; $[\alpha]_D^{22}$ = -44° (c = 1 EtOH); λ_m 280 nm ($E_D^{1\%}$295 EtOH); soluble in MeOH, EtOH, Me_2CO, $CHCl_3$, EtOAc, C_6H_6, Et_2O, dilute acid; insoluble in H_2O, C_6H_{14}, petroleum ether. *Meiji Seika Kaisha Ltd.*

2230 Miokamycin
55881-07-7 6291 259-879-6
$C_{45}H_{71}NO_{17}$
Leucomycin V 3^B,9-diacetate 3,4^B-dipropionate.
9,3-diacetylmidecamycin; MOM; ponsinomycin; Miocamycin. Macrolide antibiotic. Derivative of midecamycin mp = 220° (dec); $[\alpha]_D^{25}$ = -53° (c = 1.0 $CHCl_3$); $[\alpha]_D^{20}$ = -74° (c = 1.0 MeOH); λ_m 231 nm ($E_{1\,cm}^{1\%}$ 342 MeOH); soluble in MeOH, Me_2CO, $CHCl_3$; slightly soluble in H_2O. *Meiji Seika Kaisha Ltd.*

2231 Neutramycin
1404-08-6
$C_{34}H_{54}O_{14}$
Neutramycin.
AE-705W; LL-705W. Macrolide antibiotic. Antibiotic produced by *Streptomyces rimosus.*

2232 Oleandomycin
3922-90-5 6962 223-495-7
$C_{35}H_{61}NO_{12}$
Oleandomycin.
PA-105; Amimycin; Landomycin; Romicil. Macrolide antibiotic. Produced by *Streptomyces antibioticus.* Moderately soluble in H_2O; freely soluble in MeOH, EtOH, BuOH, Me_2CO; insoluble in C_6H_{14}, C_6H_6, CCl_4; λ_m 286-289 nm (MeOH). *Pfizer Inc.*

2233 Oleandomycin Hydrochloride
6696-47-5 6962
$C_{35}H_{62}ClNO_{12}$
Oleandomycin hydrochloride.
Macrolide antibiotic. mp = 134-135°; $[\alpha]_D^{25}$ = -54° (MeOH); freely soluble in H_2O; LD_{50} (mus orl) = 8200 mg/kg, (mus orl) = 600 mg/kg, (rat orl) > 10000 mg/kg, (rat iv) = 400 mg/kg. *Pfizer Inc.*

2234 Oleandomycin Phosphate
7060-74-4 6962 230-351-7
$C_{35}H_{64}NO_{16}P$
Oleandomycin phosphate (1:1).
Matromycin. Macrolide antibiotic. *Pfizer Inc.*

2235 Primycin A_1
47917-41-9 7931
$C_{55}H_{103}N_3O_{17}$
[5-[19-(α-D-Arabinofuranosyloxy)-35-butyl-10,12,14,16,18,22,26,30,34-nonahydroxy-3,5,21,33-tetramethyl-36-oxooxacyclohexatriaconta-4,20-dien-2-yl]-4-hydroxyhexyl]-guanidine.
Macrolide antibiotic. Major component from the mixture of > 20 macrolide antibiotics produced in the intestinal tract of the wax moth (*Galeria melonella).* Fairly soluble in MeOH; sparingly soluble in C_5H_5N, AcOH, H_2O; LD_{50} (mus ip) = 2.5 mg/kg, (gpg ip) = 5.0 mg/kg, (rat ip) = 10.0 mg/kg, (rbt ip) = 10.0 mg/kg. *Chinoin.*

2236 Repromicin
56689-42-0
$C_{31}H_{51}NO_8$
16-Ethyl-4-hydroxy-5,9,13,15-tetramethyl-2,10-dioxo-6-[[3,4,6-trideoxy-3-(dimethylamino)-β-D-xylo-hexopyranosyl]oxy]oxacyclohexadeca-11,3-diene-7-acetaldehyde.
Sch-16524. Macrolide antibiotic. *Schering-Plough HealthCare Products.*

2237 Rokitamycin
74014-51-0 8408
$C_{42}H_{69}NO_{15}$
[(4R,5S,6S,7R,9R,10R,11E,13E,16R)-7-(Formylmethyl)-4,10-dihydroxy-5-methoxy-9,16-dimethyl-2-oxooxacyclohexadeca-11,13-dien-6-yl]-3,6-dideoxy-4-O-(2,6-dideoxy-3-C-methyl-α-L-ribohexopyranosyl)-3-(dimethylamino)-β-D-glucopyranoside 4-butyrate 3-propionate.
rikamycin; M-19-Q; TMS-19Q; Ricamycin; Rokital. Macrolide antibiotic. mp = 116°; $[\alpha]_D^{2\cdot}$ = -71° (c = 1.0 $CHCl_3$); λ_m 232 nm (ε 28000, EtOH). *Meiji Seika Kaisha Ltd.*

2238 Rosaramicin
35834-26-5 8420 252-742-1
$C_{31}H_{51}NO_9$
3-Ethyl-7-hydroxy-2,8,12,16-tetramethyl-5,13-dioxo-9-[[3,4,6-trideoxy-3-(dimethylamino)-β-D-xylohexopyranosyl]oxy]-4,17-dioxabicyclo[14.1.0]heptadec-14-ene-10-acetaldehyde.
4'-deoxycirramycin A_1; antibiotic 67-694; juvenimicin A_3; rosamicin; M-4365A2; Sch-14947. Macrolide antibiotic. mp = 119-122°; $[\alpha]_D^{26}$ = -35° (EtOH); λ_m 240 nm (ε 14600 MeOH); slightly soluble in H_2O, Et_2O; freely soluble in MeOH, $CHCl_3$, $CHCl_3$, C_6H_6; LD_{50} (mus sc) = 625 mg/kg, (mus ip) = 350 mg/kg, (mus iv) = 155 mg/kg. *Schering-Plough HealthCare Products.*

2239 Roxithromycin
80214-83-1 8433
$C_{41}H_{76}N_2O_{15}$
Erythromycin 9-[O-[(2-methoxyethoxy)methyl]oxime].
RU-965; RU-28965; Assoral; Claramid; Forilin; Overal; Rossitrol; Rotramin; Rulid; Surlid. Macrolide antibiotic. $[\alpha]_D^{25}$ = -77.5° ± 2° (c = 0.45 $CHCl_3$). *Hoechst Roussel Pharm. Inc.*

2240 Spiramycin
8025-81-8 8904 232-429-6
Rovamycin.
IL-5902; RP-5337; NSC-55926; Selectomycin; Rovamicina; Leucomycin; Foromacidin; Provamycin; Rovamycin; Selectomycin; Sequamycin; Rovamicina; [Embonate]: Spira 200; [Hexanedioate]: spiramycin adipate; Stomamycin; Suanovil. Macrolide antibiotic. Mixture of macrolide antibiotics obtained from *Streptomyces ambofaciens* from the soil of Northern France. Its major components are the spiramycins I, II and III. $[\alpha]_D^{20}$ = -80° (MeOH); λ_m 231 nm (EtOH); soluble in most organic solvents; LD_{50} (rat orl) = 9400 mg/kg, (rat sc) = 1000 mg/kg, (rat iv) = 170 mg/kg. *Rhône-Poulenc Rorer Pharm. Inc.*

2241 Spiramycin I
24916-50-5 8904
$C_{43}H_{74}N_2O_{14}$
Foromacidin A. Macrolide antibiotic. mp = 134-137°; $[\alpha]_D^{20}$ = -96°; [triacetate]: mp = 140-142°; $[\alpha]_D^{20}$ = -92.5°. *Rhône-Poulenc Rorer Pharm. Inc.*

2242 Spiramycin II
24916-51-6 8904
$C_{45}H_{76}N_2O_{15}$
Foromacidin B. Macrolide antibiotic. mp = 130-133°; $[\alpha]_D^{20}$ = -86°; [diacetate]: mp = 156-160°; $[\alpha]_D^{20}$ = -98.4°. *Rhône-Poulenc Rorer Pharm. Inc.*

2243 Spiramycin III
24916-52-7 8904
$C_{46}H_{78}N_2O_{15}$
Foromacidin C. Macrolide antibiotic. mp = 128-131°; $[\alpha]_D^{20}$ = -83°; [diacetate]: mp = 140-142°; $[\alpha]_D^{20}$ = -90.4°. *Rhône-Poulenc Rorer Pharm. Inc.*

2244 Troleandomycin
2751-09-9 9899 220-392-9
$C_{41}H_{67}NO_{15}$
Triacetyloleandomycin.
oleandomycin triacetate ester; TAO; NSC-108166; Cyclamycin; Wytrion; Evramycin; Triocetin. Macrolide antibiotic. Dec 176°; $[\alpha]_D^{25}$ = -23° (MeOH); slightly soluble in H_2O (< 0.1 g/100 ml). *Roerig Div. Pfizer Pharm.*

Nitrofuran Antibiotics

2245 Furalazine
556-12-7
$C_9H_7N_5O_3$
3-Amino-6-[2-(5-nitro-2-furyl)vinyl]-as-triazine.
A nitrofuran antibiotic.

2246 Furaltadone
139-91-3 4315 205-384-5
$C_{13}H_{16}N_4O_6$
(±)-5-(Morpholinomethyl)-3-[(5-nitrofurfurylidene)amino]-2-oxazolidinone.
NF-260; Altafur; Altabactina; Furazolin; Ibifur; Medifuran; Nitraldone; Otifuril; Sepsinol; Ultrafur; Unifur; Valsyn. Antibiotic for urinary tract infections. mp = 206° (dec); soluble in H_2O (≅ 0.075 g/100 ml, 25°). *Norwich Eaton.*

2247 Furazolium Chloride
5118-17-2 4321
$C_9H_8ClN_3O_3S$
6,7-Dihydro-3-(5-nitro-2-furyl)-5H-imidazo[2,1-b]-thiazolium chloride.
NF-963; Novofur; Dermafur. Antibacterial. Dec > 250°; [free base]: mp = 171-172° (dec). *Norwich Eaton.*

2248 Furazolium Tartrate
17692-15-8 4321
$C_{13}H_{13}N_3O_9S$
6,7-Dihydro-3-(5-nitro-2-furyl)-5H-imidazo[2,1-b]-thiazolium hydrogen tartrate.
Antibacterial. *Norwich Eaton.*

2249 Nifuradine
555-84-0 6621
$C_8H_8N_4O_4$
1-[(5-Nitrofurfurylidene)amino]-2-imidazolidinone.
NF-246; NSC-6470; Renafur; oxafuradene. Antibacterial. mp = 261.5-263°; λ_m 387.5, 273 nm (ε 17550, 13200 H_2O); soluble in H_2O (0.008-0.010 g/100 ml); LD_{50} (rat orl) = 1681 mg/kg. *Norwich Eaton.*

2250 Nifuraldezone
3270-71-1 6622 221-890-9
$C_7H_6N_4O_5$
5-Nitro-2-furaldehyde semioxamazone.
NF-84; NSC-3184; Furamazone. Antibacterial. mp = 270° (dec). *Eaton Labs.*

2251 Nifuralide
54657-96-4
$C_{14}H_{13}N_5O_4S$
2-(Allylamino)-4-thiazole carboxylic acid [3-(5-nitro-2-furyl)allylidene]hydrazide.
Antibacterial.

2252 Nifuratel
4936-47-4 6623 225-576-2
$C_{10}H_{11}N_3O_5S$
5-[(Methylthio)methyl]-3-[(5-nitrofurfurylidene)amino]-2-oxazolidinone.
methylmercadone; Inimur; Macmiror; Magmilor; Omnes; Polmiror; Tydantil. Antibacterial; antifungal; antiprotozaol (used against trichomonas). mp = 182°. *Polichimica Sap.*

2253 Nifuratrone
19561-70-7
$C_7H_8N_2O_5$
N-(2-Hydroxyethyl)-α-(5-nitro-2-furyl)nitrone.
Antibacterial. *Dainippon Pharm.*

2254 Nifurdazil
5036-03-3
$C_{10}H_{12}N_4O_5$
1-(2-Hydroxyethyl)-3-[(5-nitrofurfurylidene)amino]-2-imidazolidinone.
NF-1010. Antibacterial.

2255 Nifurethazone
5580-25-6
$C_{10}H_{15}N_5O_4$
5-Nitro-2-furaldehyde 2-(2-dimethylaminoethyl)-semicarbazone.
Antibacterial.

2256 Nifurfoline
3363-58-4 6624 222-131-4
$C_{13}H_{15}N_5O_6$
3-(Morpholinomethyl)-1-[(5-nitrofurfurylidene)-amino]hydantoin.
Urbac. Antibacterial. mp = 194-196°, 206°. *Esteve Group.*

2257 Nifurimide
15179-96-1
$C_9H_{10}N_4O_4$
(±)-4-Methyl-1-[(5-nitrofurfurylidene)amino]-2-imidazolidinone.
NF-1120. Antibacterial.

2258 Nifurizone
26350-39-0
$C_{12}H_{13}N_5O_5$
1-(Methylcarbamoyl)-3-[[3-(5-nitro-2-furyl)allylidene]amino]-2-imidazolidinone.
CB-11380. Antibacterial.

2259 Nifurmazole
18857-59-5
$C_{11}H_{10}N_4O_6$
3-(Hydroxymethyl)-1-[[3-(5-nitro-2-furyl)allylidene]amino]-2-imidazolidinone.
CB-10615. Antibacterial; antiprotozoal, against trypanosomes.

2260 Nifurmerone
5579-95-3
$C_6H_4ClNO_4$
Chloromethyl 5-nitro-2-furyl ketone.
NF-71. Antibacterial. Antifungal.

2261 Nifuroquine
57474-29-0 6625 260-755-9
$C_{14}H_8N_2O_6$
4-(5-Nitro-2-furyl)quinaldic acid 1-oxide.
quinaldofur; Abimasten 100. Antibacterial. Used to treat mastitis in cows. mp = 190° (dec); insoluble in H_2O. *ABIC.*

2262 Nifuroxazide
965-52-6 6626 213-522-0
$C_{12}H_9N_3O_5$
p-Hydroxybenzoic acid 5-nitrofurfurylidene hydrazide.
RC-27109; Adral; Bacifurane; Diarlidan; Dicoferin; Ercefurol; Ercefuryl; Pentofuryl. Antibacterial. Used as an intestinal antiseptic. mp = 298°; insoluble in H_2O. *Robert et Carriere.*

2263 Nifuroxime
6236-05-1 6627 228-349-6
$C_5H_4N_2O_4$
5-Nitro-2-furaldehyde oxime.
Micofur; Mycofur; component of: Tricofuron. Antibacterial; topical anti-infective; antiprotozoal, against Trichomonas. Nitrofuran. mp = 156°, 163-164°; soluble in H_2O (0.1 g/100 ml at 25°), MeOH (8.9 g/100 ml), EtOH (3.9 g/100 ml). *Norwich Eaton.*

2264 Nifurpipone
24632-47-1
$C_{12}H_{17}N_5O_4$
4-Methyl-1-piperazineacetic acid (5-nitrofurfurylidene)-hydrazide.
NP; Rec-15-0122. Antibacterial.

2265 Nifurpirinol
13411-16-0 6628 236-503-9
$C_{12}H_{10}N_2O_4$
6-[2-(5-Nitro-2-furyl)vinyl]-2-pyridinemethanol.
Furanace; P-7138; furpyrinol; furpirinol. Antibacterial. Used to treat bacterial diseases in fish. mp = 170-171°; LD_{50} (eel orl) = 1780 mg/kg. *Dainippon Pharm.; Yamanouchi U.S.A. Inc.*

2266 Nifurprazine
1614-20-6 6629 216-563-2
$C_{10}H_8N_4O_3$
3-Amino-6-[2-(5-nitro-2-furyl)vinyl]pyridazine.
HB-115; furenapyridazin; Furenazin. Antibacterial, used topically. *Boehringer Mannheim GmbH.*

2267 Nifurprazine Hydrochloride
50832-74-1 6629
$C_{10}H_9ClN_4O_3$
3-Amino-6-[2-(5-nitro-2-furyl)vinyl]pyridazine hydrochloride.
Carofur. Antibacterial. mp = 290°. *Boehringer Mannheim GmbH.*

2268 Nifurquinazol
5055-20-9
$C_{16}H_{16}N_4O_5$
2,2'-[[2-(5-Nitro-2-furyl)-4-quinazolinyl]imino]diethanol.
NF-1088. Antibacterial.

2269 Nifurthiazole
3570-75-0
$C_8H_6N_4O_4S$
Formic acid 2-[4-(5-nitro-2-furyl)-2-thiazolyl]hydrazide.
AS-17665; NSC-525334. Antibacterial. *Abbott Labs Inc.*

2270 Nifurtimox
23256-30-6 6630 245-531-0
$C_{10}H_{13}N_3O_5S$
4-[(5-Nitrofurfurylidene)amino]-3-methylthiomorpholine 1,1-dioxide.
Bayer 2502; Lampit. Antibacterial, primarily antiprotozoal, against Trypanosoma. mp = 180-182°; LD_{50} (mus gvg) = 3720 mg/kg, (rat gvg) = 4050 mg/kg. *Bayer AG.*

2271 Nifurtoinol
1088-92-2 6631 214-126-0
$C_9H_8N_4O_6$
3-(Hydroxymethyl)-1-[(5-nitrofurfurylidene)amino]hydantoin.
Urfadyn. Antibacterial. Dec 270°; λ_m = 367.5, 265 nm (ε 17900, 12800 2% in DMF). *Norwich Eaton.*

2272 Nifurvidine
1900-13-6
$C_{11}H_9N_3O_4$
2-Methyl-6-[2-(5-nitro-2-furyl)vinyl]-4-pyrimidinol.
Antibacterial.

2273 Nifurzide
39978-42-2 6632 254-728-0
$C_{12}H_8N_4O_6S$
5-Nitro-2-thiophenecarboxylic acid [3-(5-nitro-2-furyl)allylidene] hydrazide.
Ricridene. Antibacterial. mp = 235-236°; LD_{50} (mus orl) = 3200 mg/kg. *Lipha Pharm. Inc.*

2274 Nitrofurantoin
67-20-9 6696 200-646-5
$C_8H_6N_4O_5$
1-[(5-Nitrofurfurylidene)amino]-hydantoin.
Macrobid; Macrodantin; Parfuran; Berkfurin; Chemiofuran; Cyantin; Cystit; FuaMed; Furachel; Furalan; Furadantin; Furadantine MC; Furadoin; Furantoina; Furobactina; Furophen T-Caps; Ituran; Trantoin; Urizept; Urodin; Urolong; Uro-Tablinen; Welfurin. Antibacterial. Used to treat mastitis in cows. Dec 270-272°; λ_m = 370 nm ($E^{1\%}_{1\,cm}$ 776); soluble in H_2O (0.019 g/100 ml at pH 7), EtOH 0.051 g/1oo ml), Me_2CO (0.51o g/100 ml), dmg (8 g/100 ml), glycerol (0.06 g/100 ml), polyethylene glycol (1.5 g/100 ml). *Eaton Labs.; Norwich Eaton; Parke-Davis; Procter & Gamble Pharm. Inc.*

2275 Nitrofurazone
59-87-0 6697 200-443-1
$C_6H_6N_4O_4$
5-Nitro-2-furaldehyde semicarbazone.
Aldomycin; Amifur; Furacin; Chemofuran; Furesol; Nifuzon; Nitrofural; Nitrozone; Furacinetten; Furacoccid; Furazol W; Mammex; Furaplast; Coxistat; Aldomycin; Nefco; Vabrocid; component of: Furacort, Furadex, Furea. Antibacterial. Used as a topical anti-infective. A nitrofuran. Dec 236-240°; λ_m = 260 375 nm; slightly soluble in H_2O (0.023 g/100 ml), EtOH (0.17 g/100 ml), proylene glycol (0.29 g/100 ml); insoluble in Et_2O; pH (saturated aqueous solution) 6.0-6.5; LD_{50} (rat orl) = 590 mg/kg, (rat sc) = 3000 mg/kg. *Norwich Eaton; Roberts Pharm. Corp.; SmithKline Beecham Animal Health.*

Miscellaneous Antibiotics

2276 Alafosfalin
60668-24-8 204 262-362-8
$C_5H_{13}N_2O_4P$
[(1R)-1-[(2S)-2-Aminopropionamido]ethyl]-phosphonic acid.
alaphosphin; Ro-3-7008. A phosphonic acid antibiotic mp = 295-296° (dec); $[\alpha]_D^{20}$ = -44.0° (c = 1 H_2O). *Hoffmann-LaRoche Inc.*

2277 Alexidine
22573-93-9 231 245-096-7
$C_{26}H_{56}N_{10}$
1,1-Hexamethylenebis[5-(2-ethylhexyl)-biguanide].
Win-21904; compound 904; Sterwin 904; Bisguanidine. Antibacterial; antiseptic. A guanidine derivative. [dihydrochloride]: mp = 220.6-223.4°. *Sterling Res. Labs.*

2278 Asperlin
30387-51-0
$C_{10}H_{12}O_5$
6-(1,2-Epoxypropyl)-5,6-dihydro-5-hydroxy-2H-pyran-2-one acetate.
Upjohn 224b; U-13933; NSC-93158. Antineoplastic antibiotic. Derived from *Aspergillus nidulans. Upjohn Ltd.*

2279 Avilamycin
11051-71-1
$C_{61}H_{88}Cl_2O_{32}$
O-(1R)-4-C-Acetyl-6-deoxy-2,3-O-methylene-D-galactopyranosylidene-(1→3-4)-2-O-(2-methyl-1-oxopropyl)-α-L-lyxopyranosyl-O-2,6-dideoxy-4-O-(3,5-dichloro-4-hydroxy-2-methoxy-6-methylbenzoyl)-β-D-arabinohexopyranosyl-(1→4)-O-2,6-dideoxy-D-arabinohexopyranosyl-(1→3)-O-6-deoxy-4-O-methylβ-D-galacto-pyranosyl-(1→4)-2,6-di-O-methyl-β-D-mannopyranoside.
Avilamycin A; Surmax; LY-048740. Antibacterial. *Eli Lilly & Co.*

2280 Batebulast
81907-78-0
$C_{19}H_{29}N_3O_2$
p-tert-Butylphenyl trans-4-(guanidinomethyl)-cyclohexanecarboxylate.
NCO-650. A trypsin inhibitor. Also inhibits methylation of phospholipids. Has antibacterial properties.

2281 Biphenamine
3572-52-9 1275 222-686-2
$C_{19}H_{23}NO_3$
2-(Diethylamino)ethyl-2-hydroxy-3-biphenylcarboxylate.
xenysalate. Local anesthetic; antibacterial; antifungal. Oily liquid; soluble in H_2O. *Wallace Labs.*

2282 Biphenamine Hydrochloride
5560-62-3 1275 226-930-9
$C_{19}H_{24}ClNO_3$
2-(Diethylamino)ethyl-2-hydroxy-3-biphenylcarboxylate hydrochloride.
Sebaclen; Sebaklen; component of: Alvinine Shampoo. Antibacterial; antifungal; anesthetic (topical). *Wallace Labs.*

2283 Bisdequalinium Diacetate
3785-44-2 223-252-5
$C_{44}H_{64}N_4O_4$
6,7,8,9,10,11,12,13,14,15,16,17,24,25,26,27,28,29,30,31,32,33-Docosahydro-35,37-dimethyl-5,34:18,23-diethenodibenzo[b,r][1,5,16,20]tetraazacyclotriacontine-23,24-diium diacetate.
Antibacterial.

2284 Carbadox
6804-07-5 229-879-0
$C_{11}H_{10}N_4O_4$
Methyl 3-(2-quinoxalinylmethylene)carbazate N^1,N^4-dioxide.
Mecadox; GS-6244. Antibacterial. *Pfizer Intl.*

2285 Chlorhexidine Phosphanilate
77146-42-0
$C_{34}H_{46}Cl_2N_{12}O_6P_2$
1,1'-Hexamethylenebis[5-(p-chlorophenyl)biguanide] (p-aminophenyl)phosphonate (1:2).
BMY-30120; CHP; WP-973. Antibacterial. *Bristol-Myers Squibb Pharm. R&D.*

2286 Chloroxylenol
88-04-0 2228 201-793-8
C_8H_9ClO
4-Chloro-3,5-xylenol.
parametaxylenol; Benzytol; Dettol. Antibacterial; antiseptic (topical, urinary); germicide. A phenol derivative. mp = 115°; bp = 246°; volatile with steam; slightly soluble in H_2O; more soluble in organic solvents.

2287 Clofoctol
37693-01-9 2439 253-632-6
$C_{21}H_{26}Cl_2O$
α-(2,4-Dichlorophenyl)-4-(1,1,3,3-tetramethylbutyl)-o-cresol.
Antibacterial.

2288 Coumermycin
4434-05-3
$C_{55}H_{59}N_5O_{20}$
5-Methylpyrrole-2-carboxylic acid diester with 3,3'-[(3-methylpyrrole-2,4-diyl)bis(carbonylimino)]bis[4-hydroxy-8-methyl-7-[(tetrahydro-3,4-dihydroxy-5-methoxy-6,6-dimethylpyran-2-yl)oxy]coumarin.
NSC-107412. Antibacterial.

2289 Coumermycin Sodium
87901-11-9
$C_{55}H_{57.6}N_5Na_{1.4}O_{20}$
5-Methylpyrrole-2-carboxylic acid 3,3' diester with 3,3'-[(3-methylpyrrole-2,4-diyl)bis(carbonylimino)]bis[7-(5,5-di-C-methyl-4-O-methyl-α-L-lyxopyranosyl)oxy]-4-hydroxy-8-methylcoumarin sodium salt (5:7).
Ro-5-4645/010. Antibacterial.

2290 Cycloserine
68-41-7 2820 200-688-4
$C_3H_6N_2O_2$
(+)-4-Amino-3-oxazolidinone.
orientomycin; PA-94; 106-7; Closina; Farmiserina; Micoserina; Oxamycin; Seromycin. Tuberculostatic. Antibiotic produced by *Streptomyces garyphalus sive orchidaceus*. Dec 155-156°; $[\alpha]_D^{23} = 116°$ (c = 1.17), $[\alpha]_{546}^{25} = 137°$ (c = 5 2N NaOH); λ_m 226 nm ($E_{1\,cm}^{1\%}$ 402); soluble in H_2O; slightly soluble in MeOH, propylene glycol. *Eli Lilly & Co.; Merck & Co.Inc.; Pfizer Intl.*

2291 Dequalinium Chloride
6707-58-0
$C_{30}H_{40}Cl_2N_4$
1,1'-Decamethylenebis(4-aminoquinaldinium) chloride.
Antibacterial agent.

2292 Dipyrithione
3696-28-4 223-024-5
$C_{10}H_8N_2O_2S_2$
2,2'-Dithiodipyridine 1,1'-dioxide.
OMDS. Antibacterial; antifungal. mp = 205°.

2293 Eperezolid
165800-04-4
$C_{18}H_{23}FN_4O_5$
N-[[(S)-3-[3-Fluoro-4-(4-glycoloyl-1-piperazinyl)phenyl]-2-oxo-5-oxazolidinyl]methyl]acetamide.
U-100592. Antibacterial.

2294 Fludalanine
35523-45-6 252-607-7
$C_3H_5DFNO_2$
3-Fluoro-D-alanine-2-d.
Antibacterial. *Merck & Co.Inc.*

2295 Fosfomycin
23155-02-4 245-463-1
$C_3H_7O_4P$
(-)-(1R,2S)-(1,2-Epoxypropyl)phosphonic acid.
Antibacterial. *Merck & Co.Inc.*

2296 Fosfomycin Tromethamine
78964-85-9 279-018-8
$C_7H_{18}NO_7P$
(1R,2S)-(1,2-Epoxypropyl)phosphonic acid compound with 2-amino-2-(hydroxymethyl)-1,3-propanediol (1:1).
Monurol; Z-1282. Antibacterial. *Zambon Group.*

2297 Fosmidomycin
66508-53-0
$C_4H_{10}NO_5P$
[3-(N-Hydroxyformamido)propyl]phosphonic acid.
Antibacterial.

2298 Fusidate Sodium
751-94-0 212-030-3
$C_{31}H_{47}NaO_6$
Sodium 3α,11α,16β-trihydroxy-29-nor-8α,9β,13α,14β-dammara-17(20),24-dien-21-oate 16-acetate.
fusidic acid sodium salt; Fucidine; SQ-16360. Antibacterial. *Bristol-Myers Squibb Pharm. R&D.*

2299 Fusidic Acid
6990-06-3 230-256-0
$C_{31}H_{48}O_6$
3α,11α,16β-Trihydroxy-29-nor-8α,9β,13α,14β-dammara-17(20),24-dien-21-oic acid 16-acetate.
SQ-16603. Antibacterial. *Bristol-Myers Squibb Pharm. R&D.*

2300 Haloprogin
777-11-7 212-286-6
$C_9H_4Cl_3IO$
3-Iodo-2-propynyl 2,4,5-trichlorophenyl ether.
M-1028; NSC-100071; component of: Halotex. Antibacterial. *Meiji Seika Kaisha Ltd.; Westwood-Squibb Pharm. Inc.*

2301 Hexedine
5980-31-4 4736
$C_{22}H_{45}N_3$
2,6-Bis(2-ethylhexyl)-hexahydro-7a-methyl-1H-imidazo[1,5-c]imidazole.
Sterisol; W-4701. Antibacterial. *Parke-Davis.*

2302 Letrazuril
103337-74-2
$C_{17}H_9Cl_2FN_4O_2$
(±)-[2,6-Dichloro-4-(4,5-dihydro-3,5-dioxo-as-triazin-2(3H)-yl)phenyl-(p-fluorophenyl)acetonitrile.
216. Used in the treatment of cryptosporidiosis. Antibacterial.

2303 Linezolid
165800-03-3
$C_{16}H_{20}FN_3O_4$
N-[[(S)-3-(3-Fluoro-4-morpholinophenyl)-2-oxo-5-oxazolidinyl]methyl]acetamide.
U-100766. Antibacterial. *Pharmacia & Upjohn.*

2304 Lombazole
60628-98-0 262-337-1
$C_{22}H_{17}ClN_2$
(±)-1-(α-4-Biphenylyl-o-chlorobenzyl)imidazole.
An imidazole. Antibacterial. *C.M. Ind.*

2305 Mequidox
16915-79-0
$C_{10}H_{10}N_2O_3$
3-Methyl-2-quinoxalinemethanol 1,4-dioxide.
GS-7443. Antibacterial. *Pfizer Intl.*

2306 Methenamine
100-97-0 6036 202-905-8
$C_6H_{12}N_4$
1,3,5,7-Tetraazatricyclo[3.3.1.1^{3,7}]decane.
Hexamethylenetetramine; Uritone; Urotropin; component of: Uro-Phosphate. Antibacterial. *ECR Pharm.; Parke-Davis.*

2307 Methenamine Anhydromethylenecitrate
6190-43-8 6038 228-236-1
$C_{13}H_{20}N_4O_7$
Hexamethylene anhydromethylenecitrate.
Helmitol; Fromanol; Citramin; Uropurgol; Urotropin New. Antibacterial.

2308 Methenamine Hippurate
5714-73-8 6039 227-206-5
$C_{15}H_{21}N_5O_3$
Hexamethylene monohippurate.
Hiprex; Urex. Antibacterial. *3M Pharm.; Merrell Pharm. Inc.*

2309 Methenamine Mandelate
587-23-5 6040 209-597-4
$C_{14}H_{20}N_4O_3$
Hexamethylenetetramine monomandelate.
Mandelamine; component of: Azo-Mandelamine. Antibacterial. *Parke-Davis.*

2310 Methenamine Sulfosalicylate
20480-93-7 6042
$C_{13}H_{18}N_4O_6S$
Hexamethylene sulfosalicylate.
Hexalet; Hexal; Sulfhexet. Antibacterial.

2311 Mupirocin
12650-69-0 6383
$C_{26}H_{44}O_9$
(E)-(2S,3R,4R,5S)-5-([(2S,3S,4S,5S)-2,3-Epoxy-5-hydroxy-4-methylhexyl]tetrahydro-3,4-dihydroxy-β-methyl-2H-pyran-2-crotonic acid ester with 9-hydroxynonanoic acid.
Bactroban; BRL-4910A; Bactoderm; Turixin; pseudomonic acid A. Antibacterial. mp = 77-78°; $[\alpha]_D^{20}$ = -19.3° (c = 1 MeOH); λ_m 222 nm (ε 14500 EtOH). *SmithKline Beecham Pharm.*

2312 Nitroxoline
4008-48-4 6734 223-662-4
$C_9H_6N_2O_3$
5-Nitro-8-quinolinol.
A-82. Antibacterial.

2313 Penoctonium Bromide
17088-72-1
$C_{26}H_{50}BrNO_2$
Diethyl(2-hydroxyethyl)octyl ammonium bromide dicyclopentylacetate.
Tested as an antibiotic. Potential chemotherapeutic agent.

2314 Pirlimycin
79548-73-5
$C_{17}H_{31}ClN_2O_5$
Methyl 7-chloro-6,7,8-trideoxy-6-cis-4-ethyl-L-pipecolamido)-1-thio-L-threo-α-D-galactooctopyranoside.
Antibacterial.

2315 Pirlimycin Hydrochloride
77495-92-2
$C_{17}H_{32}Cl_2N_2O_5.H_2O$
Methyl 7-chloro-6,7,8-trideoxy-6-cis-4-ethyl-L-pipecolamido)-1-thio-L-threo-α-D-galactooctopyranoside monohydrochloride monohydrate.
U-57930E. Antibacterial.

2316 Pirtenidine
103923-27-9
$C_{21}H_{38}N_2$
1,4-Dihydro-1-octyl-4-(octylimino)pyridine.
Antimicrobial. Used in the treatment of candidiasis. May have use in inhibiting dental plaque.

2317 Polynoxylin
9011-05-6 7735
$(C_4H_8N_2O_3)_n$
Poly[methyi[bis(hydroxymethyl)]ureylene]amer.
oxymetylurea; polynoxyline; Anaflex; Larex; Ponoxylan. Polymer condensation product of urea and formaldehyde with substantially linear chains, often delivered in aerosol spray or powder form. Antibacterial (topical). Amorphous powder; decomposes without melting; nearly insoluble in H_2O (0.28-0.31%).

2318 Quindecamine
19056-26-9 242-789-6
$C_{30}H_{38}N_4$
4,4'-(Decamethylenediimino)diquinaldine.
Antibacterial. *Marion Merrell Dow Inc.*

2319 Quindecamine Acetate
5714-05-6
$C_{34}H_{46}N_4O_4.H_2O$
4,4'-(Decamethylenediimino)diquinaldine diacetate dihydrate.
RMI-8090DJ. Antibacterial. *Marion Merrell Dow Inc.*

2320 Quindoxin
2423-66-7 219-352-3
$C_8H_6N_2O_2$
Quinoxaline 1,4-dioxide.
ICI-8173. Antibacterial. Used in animal husbandry. CAUTION: high degree of phototoxicity and mutagenicity.

2321 Roxarsone
121-19-7 204-453-7
$C_6H_6AsNO_6$
4-Hydroxy-3-nitrobenzenearsonic acid.
NSC-2101. Antibacterial.

2322 Satranidazole
56302-13-7
$C_8H_{11}N_5O_5S$
1-(1-Methyl-5-nitroimidazol-2-yl)-3-(methylsulfonyl)-2-imidazolidinone.
A nitroimidazole with high selective toxicity for anaerobic prokaryotes and eukaryotes. Antiamebic; antimicrobial; radiosensitizer.

2323 Spectinomycin
1695-77-8 8890 216-911-3
$C_{14}H_{24}N_2O_7$
Decahydro-4a,7,9-trihydroxy-2-methyl-6,8-bis(methylamino)-4H-pyrano[2,3-b][1,4]benzodioxin-4-one.
Aminoglycoside antibiotic.

2324 Spectinomycin Dihydrochloride Pentahydrate
22189-32-8 8890
$C_{14}H_{26}Cl_2N_2O_7.H_2O$
Decahydro-4a,7,9-trihydroxy-2-methyl-6,8-bis(methylamino)-4H-pyrano[2,3-b][1,4]benzodioxin-4-one dihydrochloride pentahydrate.
Aminoglycoside antibiotic.

2325 Spectinomycin Sulfate Tetrahydrate
64058-48-6 8890
$C_{14}H_{26}N_2O_{11}S.4H_2O$
Decahydro-4a,7,9-trihydroxy-2-methyl-6,8-bis(methylamino)-4H-pyrano[2,3-b][1,4]benzodioxin-4-one.
Aminoglycoside antibiotic.

2326 Stallimycin
636-47-5
$C_{22}H_{27}N_9O_4$
N-(2-Amidinoethyl)-4-formamido-1,1',1-trimethyl-N,4':N',4-ter[pyrrole-2-carboxamide].
Antibacterial. *Farmitalia Carlo Erba Ltd.*

2327 Stallimycin Hydrochloride
6576-51-8 229-505-6
$C_{22}H_{28}ClN_9O_4$
N-(2-Amidinoethyl)-4-formamido-1,1',1-trimethyl-N,4':N',4-ter[pyrrole-2-carboxamide] monohydrochloride.
Herperal; F.I. 6426. Antibacterial. *Farmitalia Carlo Erba Ltd.*

2328 Taurolidine
19388-87-5 9243 243-016-5
$C_7H_{16}N_4O_4S_2$
4,4'-Methylenebis(tetrahydro-1,2,4-thiadiazine 1,1-dioxide).
Antibacterial.

2329 Tetronasin 5930
75139-05-8 9384
$C_{35}H_{54}O_8$
4-Hydroxy-3-[(2S)-2-[(1S,2S,6R)-2-[(1E)-3-hydroxy-2-[(2R,3R,6S)-tetrahydro-3-methyl-6-[(1E,3S)-3-[(2R,3S,5R)-tetrahydro-5-[(1S)-1-methoxyethyl]-3-methyl-2-furyl]-1-butenyl]-2H-pyran-2-yl]propenyl]-6-methylcyclohexyl]propionyl]-2(5H)-furanone.
antibiotic M139603; ICI-139603; M-139603. Polyether-ionophore antibiotic (veterinary) produced by *Streptomyces longisporoflavus* NCIB 11426. Possesses a

biosynthetically rare acyl tetronic acid moiety. mp = 176-178°; $[\alpha]_D^{23}$= -82° (c = 0.2 in MeOH); λ_m = 234, 270 nm (ε 1300, 1100 in EtOH); pKa = 1.8 ±3 (in MeOH:H_2O 1:9); soluble in most organic solvents, insoluble in H_2O.

2330 Ticlatone
70-10-0
C_7H_4ClNOS
6-Chloro-1,2-benzisothiazolin-3-one.
FER-1443. Antibacterial; antifungal.

2331 Tilbroquinol
7175-09-9 230-533-6
$C_{10}H_8BrNO$
7-Bromo-5-methyl-8-quinolinol.
component of: Intetrix. An intestinal antiseptic. Antibiotic.

2332 Tiliquinol
5541-67-3 226-905-2
$C_{10}H_9NO$
5-Methyl-8-quinolinol.
component of: Intetrix. An intestinal antiseptic. Antibiotic.

2333 Tiodonium Chloride
38070-41-6
$C_{10}H_7Cl_2IS$
(p-Chlorophenyl)-2-thienyliodonium chloride.
DL-164. Antibacterial. *Marion Merrell Dow Inc.*

2334 Trospectomycin
88669-04-9 9917
$C_{17}H_{30}N_2O_7$
(2R,4aR,5aR,6S,7S,8R,9S,9aR,10aS)-2-Butyldecahydro-4a,7,9-trihydroxy-6,8-bis(methylamino)-4H-pyrano[2,3-b][1,4]benzodioxin-4-one.
Aminoglycoside antibiotic.

2335 Trospectomycin Sulfate Pentahydrate
88851-61-0 9917
$C_{17}H_{32}N_2O_{11}S.5H_2O$
(2R,4aR,5aR,6S,7S,8R,9S,9aR,10aS)-2Butyldecahydro-4a,7,9-trihydroxy-6,8-bis(methylamino)-4H-pyrano[2,3-b][1,4]benzodioxin-4-one sulfate (1:1) (salt) pentahydrate.
U-63366F. Aminoglycoside antibiotic.

2336 Tuberin
53643-53-1 9938
$C_{10}H_{11}NO_2$
N-[2-(4-Methoxyphenyl)ethenyl]formamide.
N-formyl-trans-p-methoxystyrylamine. Antibiotic (tuberculostatic). mp = 132-133°; λ_m 219, 285 nm ($E^{1\%}_{1\,cm}$ 870, 1710, MeOH); soluble in MeOH, EtOH, EtOAc, Me_2CO; moderately soluble in CCl_4, $CHCl_3$; sparingly soluble in H_2O, C_6H_6; insoluble in petroleum ether. *Inst. Phys. & Chem. Res.*

2337 Valnemulin
101312-92-9
$C_{31}H_{52}N_2O_5S$
[[2-[(R)-2-Amino-3-methylbutyramido]-1,1-dimethylethyl]thio]acetic acid 8-ester with (3aS,4R,5S,6S,8R,9R,9aR,10R)-octahydro-5,8-dihydroxy-4,6,9,10-tetramethyl-6-vinyl-3a,9-propano-3aH-cyclopentacycloocten-1(4H)-one.
Econor. Antibacterial (veterinary).

2338 Xibornol
13741-18-9 10210 237-312-3
$C_{18}H_{26}O$
6-Isobornyl-3,4-xylenol.
Antibacterial.

Polypeptide Antibiotics

2339 Amphomycin
1402-82-0 625 215-760-0
$C_{58}H_{91}N_{13}O_{20}$
R-Asp-MeAsp-Asp-Gly-Asp-Gly-Dab[a]-Val-Pro-Dab[b]-Pip[c] [R = (+)-3-anteisotridecenoic or (+)-3-isododecenoic acid; Dab[a] = D-erythro-α,β-diaminobutyric acid; Dab[b] = L-threo-α,β-diaminobutyric acid; Pip[c] = D-pipecolic acid, cyclized with the neighboring Dab]. Peptide antibiotic. Modified at the amino terminus with a fatty acid group. Active against gram positive bacteria. Produced by *Streptomyces canus. Bristol-Myers Squibb HIV Products.*

2340 Anthelmycin
1402-84-2
hikizimycin; 33876. A 4-aminohexylcytosine antibiotic which inhibits protein synthesis on both pro- and eukaryotic ribosomes by preventing the peptide bond-forming reaction. Antibiotic substance produced by *Streptomyces longissimus. Eli Lilly & Co.*

2341 Aspartocin
4117-65-1
$C_{42}H_{64}N_{12}O_{12}S_2$
Oxytocin 4-L-asparagine.
H-Cys-Tyr-Ile-Asn-Asn-Cys-Pro-Lue-Gly-NH_2-cyclic (1→6)-disulfide; A 8999. Antibiotic produced by *Streptomyces griseus.*

2342 Bacitracin
1405-87-4 965 215-786-2
$C_{66}H_{103}N_{17}O_{16}S$
Thiazole decapeptide.
Altracin; Ayfivin; Baciguent; Fortracin; Penitracin; Tropitracin; Zutracin; component of: Mycitracin. A peptide antibiotic complex produced by Bacillus *subtilis and licheniformis*. Soluble in H_2O, alcohol; practically insoluble in Et_2O, $CHCl_3$, Me_2CO; stable in acid solution; unstable in alkaline solution. *Merck & Co.Inc.; Pfizer Inc.; Upjohn Ltd.*

2343 Bacitracin Methylene Disalicylic Acid
1405-88-5 965
Bacitracin methylenebis[2-hydroxybenzoate].
bacitracin methylenedisalicylate. Peptide antibiotic. Soluble in H_2O.

2344 Bacitracin Zinc
1405-89-6 215-787-8
Bacitracins zinc complex.
component of: Cortisporin, Neo-Polycin, Neosporin, Polysporin. Peptide antibiotic. *Burroughs Wellcome Inc.; Marion Merrell Dow Inc.*

2345 Capreomycin
11003-38-6 1801
Capastat. Peptide antibiotic (tuberculostatic). Cyclic

peptide antibiotic similar to viomycin. Produced by *Streptomyces capreolus*. A mixture of capreomycins IA (25%), IB (67%), IIA (3%), IIB (6%). Soluble in H_2O, insoluble in most organic solvents. *Dista Products Ltd.; Eli Lilly & Co.*

2346 Capreomycin Disulfate
1405-36-3 1801
Caprocin; Ogostal. Peptide antibiotic (tuberculostatic). LD_{50} (mus iv) = 250 mg/kg, (mus sc) = 514 mg/kg, (rat iv) = 325 mg/kg, (rat sc) = 1191 mg/kg. *Eli Lilly & Co.*

2347 Capreomycin IA
37280-35-6 1801
$C_{25}H_{44}N_{14}O_8$
Peptide antibiotic (tuberculostatic). Similar to viomycin. Produced by *Streptomyces capreolus*. mp = 246-248° (dec); $[\alpha]_D^{22}$ = -21.9° (c = 0.5 H_2O); λ_m 269 nm (ε 24000 0.1N HCl), 268 nm (ε 23900 H_2O), 287 nm (ε 15900 0.1N NaOH). *Eli Lilly & Co.*

2348 Capreomycin IB
33490-33-4 1801
$C_{25}H_{44}N_{14}O_7$
Peptide antibiotic. Tuberculostatic. mp = 253-255° (dec); $[\alpha]_D^{22}$ = -44.6° (c = 0.55 H_2O); λ_m 268 nm (ε 22700 0.1N HCl), 268 nm (ε 22300 H_2O), 290 nm (ε 14400 0.1N NaOH). *Eli Lilly & Co.*

2349 Colistin
1066-17-7 2542 213-907-3
Cyclopolypeptide produced by *Bacillus colisitinus (Aerobacillus colistinus)*.
Polymyxin E.
Peptide antibiotic. Mixture of Colistin A, B and C. Colistin A is the same as Polymyxin E_1 ($C_{53}H_{100}N_{16}O_{13}$).

2350 Colistin Sodium Methanesulfonate
8068-28-8 2542 232-516-9
Peptide antibiotic.

2351 Colistin Sulfate
1264-72-8 2542 215-034-3
Peptide antibiotic.

2352 Dalfopristin
112362-50-2
$C_{34}H_{50}N_4O_9S$
(3R,4R,5E,10E,12E,14S,26R,26aS)-26-[[2-(Diethylamino)ethyl]sulfonyl]-8,9,14,15,24,25,26,26a-octahydro-14-hydroxy-3-isopropyl-4,12-dimethyl-3H-21,18-nitrilo-1H,22H-pyrrolo[2,1-c][1,8,4,19]-dioxodiazacyclotetracosine-1,7,16,22(4H,17H)-tetrone.
RP-54476. Peptide antibiotic. *Rhône-Poulenc Rorer Pharm. Inc.*

2353 Daptomycin
103060-53-3
$C_{72}H_{101}N_{17}O_{26}$
N-Decanoyl-L-tryptophyl-L-asparaginyl-L-aspartyl-L-threonylglycyl-L-ornithyl-L-aspartyl-D-alanyl-L-aspartylglycyl-D-seryl-threo-3-methyl-L-glutamyl-3-anthraniloyl-L-alanine ε_1-lactone.
LY-146032. Peptide antibiotic. *Eli Lilly & Co.*

2354 Enduracidin
11115-82-5 3619
Enradin [as monohydrochloride]. Peptide antibiotic. Cyclodepsipeptide antibiotic produced by *Streptomyces fungicidicus* comprising Enduracidins A and B, both 17-mers. [monohydrochloride]: Dec 234-238°; LD50 (mus orl) > 10000 mg/kg, (mus im) > 5000 mg/kg. *Takeda Chem. Ind. Ltd.*

2355 Enduracidin A Hydrochloride
33368-20-6 3619
$C_{107}H_{139}Cl_3N_{26}O_{31}$
component of: Enduracidin. Peptide antibiotic. mp = 240-245°; $[\alpha]_D^{23}$ = 92° (c = 0.5 DMF); λ_m 231, 272 nm (0.1N HCl). *Takeda Chem. Ind. Ltd.*

2356 Enduracidin B Hydrochloride
34765-98-5 3619
$C_{108}H_{141}Cl_3N_{26}O_{31}$
component of: Enduracidin. Peptide antibiotic. mp = 238-241°; $[\alpha]_D^{23}$ = 92° (c = 0.5 DMF); λ_m 231, 272 nm (0.1N HCl). *Takeda Chem. Ind. Ltd.*

2357 Enviomycin
33103-22-9 3636
$C_{25}H_{43}N_{13}O_{10}$
[[15-(3,6-Diamino-4-hydroxyhexanamido)-3-(hexahydro-2-imino-4-pyrimidinyl)-9,12-bis(hydroxymethyl-2,5,8,11,14-pentaoxo-1,4,7,10,13-pentaazacyclohexadec-6-ylidene]methyl]urea.
Tuberactinomycin N. Peptide antibiotic (tuberculostatic). [hydrochloride ($C_{25}H_{46}Cl_3N_{13}O_{10}$)]: mp > 245° (dec); $[\alpha]_D^{21}$ = -19.1°; λ_m 268 nm ($E_{1\,cm}^{1\%}$ 342 H_2O or 0.1N HCl), 288 nm ($E_{1\,cm}^{1\%}$ 215 0.1N NaOH); very soluble in H_2O, slightly soluble in organic solvents; LD_{50} (mus iv) = 485 mg/kg, (rat iv) = 680 mg/kg. *Toyo Brewing Company.*

2358 Fusafungine
1393-87-9 4335 215-737-5
S-314; Biofusal; Fusaloyos; Fusarine; Locabiotal. Peptide antibiotic. Antibiotic obtained from cultures of a *fusarium belomging to the Lateritium Wr.* section. mp = 125-129°; insoluble in H_2O, soluble in glycols and fats. *Biofarma A/S.*

2359 Gramicidin
1405-97-6 4553 215-790-4
CHO-Val-Gly-Ala-D-Leu-Ala-D-Val-Val-D-Val-Trp-D-Leu-Trp-D-Leu-Trp-D-Leu-Trp-$NHCH_2CH_2OH$.
Gramoderm; component of: Neo-Polycin, Neosporin Ophthalmic Solution, Spectrocin. Peptide antibiotic. *Bristol-Myers Squibb Pharm. R&D; Glaxo Wellcome Inc.; Marion Merrell Dow Inc.; Penick; Schering-Plough HealthCare Products.*

2360 Gramicidin S
113-73-5 4552
$C_{60}H_{92}N_{12}O_{10}$
Cyclo(L-valyl-L-ornithyl-L-leucyl-D-phenylalanyl-L-prolyl-L-valyl-L-ornithyl-L-leucyl-D-phenylalanyl-L-prolyl).
Peptide antibiotic.

2361 Mideplanin
122173-74-4
$C_{93}H_{109}Cl_2N_{11}O_{32}$
34-[(2-Acetamido-2-deoxy-β-D-glucopyranosyl)oxy]-15-amino-22,31-dichloro-56-[[2-deoxy-2-(8-methylnonanamido)-β-D-glucopyranosyl]oxy]-N-[3-(dimethylamino)-propyl]-2,3,16,17,18,19,35,36,37,38,48,49,50,50a-tetradecahydro-6,11,40,44-tetrahydroxy-42-(α-D-mannopyranosyloxy)-2,16,36,50,51,59-hexaoxo-1H,15H,34H-20,23:30,33-dietheno-3,18:35,48-bis(iminomethano)-4,8:10,14:25,28:43,47-tetrametheno-28H-[1,14,6,22]dioxadiazacyclooctacosio[4,5-m]-[10,2,16]benzoxadiazacyclotetracosine-38-carboxamide. MDL-62873. Peptide antibiotic.

2362 Mikamycin
11006-76-1 6275 234-244-6
Peptide antibiotic. Member of the streptogramin family of antibiotics, produced by *Streptomyces mitakaemsis*. Consists mainly of mikamycins A and B.

2363 Mikamycin A
21411-53-0 6275 244-376-6
see Virginiamycin M_1.

2364 Mikamycin B
3131-03-1 6275
$C_{45}H_{54}N_8O_{10}$
4-[4-(Dimethylamino)-N-methyl-L-phenylalanine]virginiamycin S_1.
mikamycin I_A; ostreigrycin B; streptogramin B; pristinamycin I_A; vernamycin B_α. Peptide antibiotic. Member of the streptogramin family of antibiotics. Dec 262-263°; λ_m = 209, 260, 305 nm ($E_{1\%}^{1\ cm}$ 605, 220, 101); $[\alpha]_{20}^{D}$ = -60.3° (c = 1.0 methanol); soluble in MeOH, EtOH, C_6H_6, Me_2CO, $CHCl_3$; nearly insoluble in H_2O, petroleum ether, C_6H_{14}; [monohydrate]: mp = 160°.

2365 Paldimycin
102426-96-0
U-70138. Peptide antibiotic. A *Streptomyces* peptide antibiotic, consisting of two components, Paldimycin A and Paldimycin B . *Pharmacia & Upjohn.*

2366 Paldimycin A
101411-70-5
$C_{44}H_{64}N_4O_{23}S_3$
2-Amino-5-[3-O-[2,6-dideoxy-4-C-[(1S)-1-hydroxyethyl]-3-O-methyl-α-L-lyxohexopyranosyl]-β-D-allopyranosyl]-5-hydroxy-3,6-dioxo-1-cyclohexene-1-carboxylic acid 4'-[3-[[(2R)-2-acetamido-2-carboxyethyl]thio]-2-[(dithiocarboxy)amino]buytrate] 6'-acetate.
Peptide antibiotic. *Pharmacia & Upjohn.*

2367 Paldimycin B
101411-71-6
$C_{43}H_{62}N_4O_{23}S_3$
2-Amino-5-[3-O-[2,6-dideoxy-4-C-[(1S)-1-hydroxyethyl]-3-O-methyl-α-L-lyxohexopyranosyl]-β-D-allopyranosyl]-5-hydroxy-3,6-dioxo-1-cyclohexene-1-carboxylic acid 4'-[3-[[(2R)-2-acetamido-2-carboxyethyl]thio]-2-[(dithiocarboxy)amino]buytrate] 6'-acetate.
Peptide antibiotic. *Pharmacia & Upjohn.*

2368 Polymyxin
1406-11-7 7734
Antibiotic complex produced by *Bacillus polymyxa*. Contains about 11 separate component polypeptide antibiotics, termed the polymyxins [hydrochloride salt]: dec 228-230°; $[\alpha]_D^{23}$ = -40° (c = 1.05); very soluble in H_2O, MeOH; solubility increases in higher alcohols; insoluble in other organic solvents, especially ethers, ketones, esters, hydrocarbons, chlorinated solvents. *Am. Cyanamid; Glaxo Wellcome Inc.*

2369 Polymyxin B
1404-26-8 7734 215-768-4
Peptide antibiotic. Mixture of Polymyxins B_1 and B_2. $[\alpha]_{5461}$ = -106.3°.

2370 Polymyxin B Sulfate
1405-20-5 7734 215-774-7
Aerosporin; Mastimyxin. Peptide antibiotic.

2371 Polymyxin B_1
4135-11-9 7734
$C_{56}H_{98}N_{16}O_{13}$
Peptide antibiotic. [pentahydrochloride salt ($C_{56}H_{103}Cl_5N_{16}O_{13}$)]: $[\alpha]_D^{25}$ = -85.11° (c = 2.33 75% EtOH).

2372 Polymyxin B_2
34503-87-2 7734
$C_{55}H_{96}N_{16}O_{13}$
Peptide antibiotic. $[\alpha]_{5461}^{22}$ = -112.4°.

2373 Polymyxin D_1
10072-50-1 7734
$C_{50}H_{93}N_{15}O_{15}$
Peptide antibiotic.

2374 Polymyxin D_2
34167-45-8 7734
$C_{49}H_{91}N_{15}O_{15}$
Peptide antibiotic.

2375 Pristinamycin
11006-76-1 7933 234-244-6
RP-7293; Pyostacine. Peptide antibiotic mixture productd by *Streptomyces prostinaespiralis*, consisting of pristinamycins I_A, I_B, I_C (identical to the vernamycins) and pristinamycins II_A, II_B (identical to the virginiamycins). *Rhône-Poulenc Rorer Pharm. Inc.*

2376 Quinupristin
120138-50-3
$C_{53}H_{67}N_9O_{10}S$
N-[(6R,9S,10R,13S,15aS,18R,22S,24aS)-22-[p-(Dimethylamino)benzyl]-6-ethyl-docosahydro-10,23-dimethyl-5,8,12,15,17,21,24-heptaoxo-13-phenyl-18-[[(3S)3-quinuclidinylthio]methyl]-12H-pyrido[2,1-f]pyrrolo-[2,1-l][1,4,7,10,13,16]oxapentaazacyclononadecin-9-yl]-3-hydroxy-picolinamide.
Derivative of virginiamycin S_1; RP-57669; Dalfopristin. Peptide antibiotic. *Rhône-Poulenc Rorer Pharm. Inc.*

2377 Ramoplanin
76168-82-6
Peptide antibiotic.

2378 Ramoplanin A_1
81988-87-6
$C_{118}H_{152}ClN_{21}O_{40}$
(S)-2-(3-Chloro-4-hydroxyphenyl)-N-[N^2-[(2Z,4Z)-2,4-octadienoyl]-(S)-asparaginyl-(2S,3S)-3-hydroxyasparaginyl-(R)-2-(p-hydroxyphenyl)-glycyl-(R)-ornithyl-(2R,3R)-allothreonyl-2-(p-hydroxyphenyl)glycyl-2-(p-hydroxyphenyl)-glycyl-(2S,3S)-allothreonyl-(S)-phenylalanyl-(R)-ornithyl-(S)-2-[p-[(2-O-α-D-manno-pyranosyl-α-D-mannopyranosyl)oxy]phenyl]-glycyl-(2R,3R)-allothreonyl-(S)-2-(p-hydroxyphenyl)glycylglycyl-(S)-leucyl-(R)-alanyl]-glycine-ψ_1-lactone.
Peptide antibiotic.

2379 Ramoplanin A'_1
124884-28-2
$C_{112}H_{14-}62ClN_{21}O_{35}$
(S)-2-(3-Chloro-4-hydroxyphenyl)-N-[N^2-[(2Z,4Z)-2,4-octadienoyl]-(S)-asparaginyl-(2S,3S)-3-hydroxyasparaginyl-(R)-2-(p-hydroxyphenyl)glycyl-(R)-ornithyl-(2R,3R)-allothreonyl-2-(p-hydroxyphenyl)glycyl-2-(p-hydroxyphenyl)-glycyl-(2S,3S)-allothreonyl-(S)-phenylalanyl-(R)-ornithyl-(S)-2-[p-(α-D-mannopyranosyloxy)phenyl]glycyl-(2R,3R)-allothreonyl-(S)-2-(p-hydroxyphenyl)glycylglycyl-(S)-leucyl-(R)-alanyl]glycine-ψ_2-lactone.
Peptide antibiotic.

2380 Ramoplanin A_2
81988-88-7
$C_{119}H_{154}ClN_{21}O_{40}$
(S)-2-(3-Chloro-4-hydroxyphenyl)-N-[N^2-[(2Z,4Z)-7-methyl-2,4-octadienoyl]-(S)-asparaginyl-(2S,3S)-3-hydroxyasparaginyl-(R)-2-(p-hydroxyphenyl)glycyl-(R)-ornithyl-(2R,3R)-allothreonyl-2-(p-hydroxyphenyl)glycyl-2-(p-hydroxyphenyl)-(S)-phenylalanylglycyl-(2S,3S)-allothreonyl-(S)-phenylalanyl-(R)-ornithyl-(S)-2-[p-[(2-O-α-D-mannopyranosyl-α-D-mannopyranosyl(oxy]phenyl]glycyl-(2R,3R)-allothreonyl-(S)-2-(p-(hydroxyphenyl)glycylglycyl-(S)-leucyl-(R)-alanyl]glycine ψ_1-lactone.
Peptide antibiotic.

2381 Ramoplanin A'_2
124884-29-3
$C_{113}H_{144}ClN_{21}O_{35}$
(S)-2-(3-Chloro-4-hydroxyphenyl)-N-[N^2-[(2Z,4Z)-7-methyl-2,4-octadienoyl]-(S)-asparaginyl-(2S,3S)-3-hydroxyasparaginyl-(R)-2-(p-hydroxyphenyl)glycyl-(R)-ornithyl(2R,3R)-allothreonyl-2-(p-hydroxyphenyl)glycyl-2-(p-hydroxyphenyl)glycyl(2S,3S)-allothreonyl-(S)-phenylalanyl-(R)-ornithyl-(S)-2-[p-(α-D-mannopyranosyloxy)phenyl]glycyl-(2R,3R)-allothreonyl-(S)-2-(p-hydroxyphenyl)glycylglycyl-(S)-leucyl-(R)-alanyl]glycine ψ_1-lactone.
Peptide antibiotic.

2382 Ramoplanin A_3
81988-89-8
$C_{120}H_{156}ClN_{21}O_{40}$
(S)-2-(3-Chloro-4-hydroxyphenyl)-N-[N^2-[(2Z,4Z)-8-methyl-2,4-nonadienoyl]-(S)-asparaginyl-(2S,3S)-3-hydroxyasparaginyl-(R)-2-(p-hydroxyphenyl)glycyl-(R)-ornithyl-(2R,3R)-allothreonyl-2-(p-hydroxyphenyl)glycyl-2-(p-hydroxyphenyl)glycyl-(2S,3S)-allothreonyl-(S)-phenylalanyl-(R)-ornithyl-(S)-2-[p-[(2-O-α-D-mannopyranosyl-α-D-mannopyranosyl)oxy]phenyl]glycyl-(2R,3R)-allothreonyl-(S)-2-(p-hydroxyphenyl)glycylglycyl-(S)-leucyl-(R)-alanyl]glycine ψ_1-lactone.
Peptide antibiotic.

2383 Ramoplanin A'_3
124884-30-6
$C_{114}H_{146}ClN_{21}O_{35}$
(S)-2-(3-Chloro-4-hydroxyphenyl)-N-[N^2-[(2Z,4Z)-8-methyl-2,4-nonadienoyl]-(S)-asparaginyl-(2S,3S)-3-hydroxyasparaginyl-(R)-2-(p-hydroxyphenyl)glycyl-(R)-ornithyl-(2R,3R)-allothreonyl-2-(p-hydroxyphenyl)glycyl-2-(p-hydroxyphenyl)glycyl-(2S,3S)-allothreonyl-(S)-phenylalanyl-(R)-ornithyl-(S)-2-[p-(α-D-mannopyranosyloxy)phenyl]glycyl-(2R,3R)-allothreonyl-(S)-2-(p-hydroxyphenyl)glycylglycyl-(S)-leucyl-(R)-alanyl]glycine ψ_1-lactone.
Peptide antibiotic.

2384 Ristocetin
1404-55-3 8398 215-770-5
[Ristocetin A]: $C_{95}H_{110}N_8O_{44}$
Ristomycin; Spontin; Riston. Peptide antibiotic obtained from *Nocardia lurida* .

2385 Teicoplanin
61036-64-4 9269
Targocid. Mixture of peptide antibiotics obtained from *Actinoplanes teichomyceticus*. Has six major components (Teicoplanins) . *Hoechst AG.*

2386 Teicoplanin $A_{2\text{-}1}$
91032-34-7
$C_{88}H_{95}Cl_2N_9O_{33}$
(3S,15R,18R,34R,35S,38S,48R,50aR)-34-([(2-Acetamido-2-deoxy-β-D-glucopyranosyl)oxy]-15-amino-22,31-dichloro-56-[[2-(Z)-4-decenamido-2-deoxy-β-D-glucopyranosyl]oxy]-2,3,16,17,18,19,35,36,37,38,48,49,50,50a-tetradecahydro-6,11,40,44-tetrahydroxy-42-(α-D-mannopyranosyloxy)-2,16,36,50,51,59-hexaoxo-1H,15H,34H,-20,23:30,33-dietheno-3,18:35,48-bis(iminomethano)-4,8:10,14:25,28:43,47-tetrametheno-28H-[1,14,6,22]-dioxadiazacyclooctacosino[4,5-m][10,2,16]benzoxadiazacyclotetracosine-38-carboxylic acid.
Peptide antibiotic. *Hoechst AG.*

2387 Teicoplanin $A_{2\text{-}2}$
91032-26-7
$C_{88}H_{97}Cl_2N_9O_{33}$
(3S,15R,18R,34R,35S,38S,48R,50aR0-34-[(2-Acetamido-2-deoxy-β-D-glucopyranosyl)oxy]-15-amino-22,31-dichloro-56-[[2-deoxy-2-(8-methylnonanonamido-β-D-glucopyranosl]oxy]-

2,3,16,17,18,19,35,36,37,38,48,49,50,50a-tetradecahydro-6,11,40,44-tetrahydroxy-42-(α-D-mannopyranosyloxy)-2,16,36,50,51,59-hexaoxo-1H,15H,34H,-20,23:30,33-dietheno-3,18:35,48-bis(iminomethano)-4,8:10,14:25,28:43,47-tetrametheno-28H-[1,14,6,22]dioxadiazacyclooctacosino[4,5-m][10,2,16]benzoxadiazacyclotetracosine-38-carboxylic acid.
Peptide antibiotic. *Hoechst AG.*

2388 Teicoplanin A_{2-3}
91032-36-9
$C_{88}H_{97}Cl_2N_9O_{33}$
(3S,15R,18R,34R,35S,38S,48R,50aR)-34-[(2-Acetamido-2-deoxy-β-D-glucopyranosyl)oxy]-15-amino-22,31-dichloro-56-[[2-decanamido-2-deoxy-β-D-glucopyranosyl)oxy]-2,3,16,17,18,19,35,36,37,38,48,49,50,50a-tetradecahydro-6,11,40,44-tetrahydroxy-42-(α-D-mannopyranosyloxy)-2,16,36,50,51,59-hexaoxo-1H,15H,34H-20,23:30,33-dietheno-3,18:35,48-bis(iminomethano)-4,8:10,14:25,28:43,47-tetrametheno-28H-[1,14,6,22]dioxadiazacyclooctacosino[4,5-m][10,2,16]benzoxadiazacyclotetracosine-38-carboxylic acid.
Peptide antibiotic. *Hoechst AG.*

2389 Teicoplanin A_{2-4}
91032-37-0
$C_{89}H_{99}Cl_2N_9O_{33}$
(3S,15R,18R,34R,35S,38S,48R,50aR)-34-[(2-Acetamido-2-deoxy-β-D-glucopyranosyl)oxy]-15-amino-22,31-dichloro-56-[[2-deoxy-2-(8-methyldecanamido)-β-D-glucopyranosyl]oxy]-2,3,16,17,18,19,35,36,37,38,48,49,50,50a-tetradecahydro-6,11,40,44-tetrahydroxy-42-(α-D-mannopyranosyloxy)-2,16,36,50,51,59-hexaoxo1H,15H,34H-20,23:30,33-dietheno-3,18:35,48-bis(iminomethano)-4,8:10,14:25,28:43,47-tetrametheno-28H-[1,14,6,22]dioxadiazacyclooctacosino[4,5-m][10,2,16]-benzoxadiazacyclotetracosine-38-carboxylic acid.
Peptide antibiotic. *Hoechst AG.*

2390 Teicoplanin A_{2-5}
91032-38-1
$C_{89}H_{99}Cl_2N_9O_{33}$
(3S,15R,18R,34R,35S,38S,48R,50aR)-34-[(2-Acetamido-2-deoxy-β-D-glucopyranosyl)oxy]-15-amino-22,31-dichloro-56-[[2-deoxy-2-(9-methyldecanamido)-β-D-glucopyranosyl]oxy]-2,3,16,17,18,19,35,36,37,38,48,49,50,50a-tetradecahydro-6,11,40,44-tetrahydroxy-42-(α-D-mannopyranosyloxy)-2,16,36,50,51,59-hexaoxo-1H,15H,34H-20,23:30,33-dietheno-3,18:35,48-bis(iminomethano)-4,8:10,14:25,28:43,47-tetrametheno-28H-[1,14,6,22]dioxadiazacyclooctacosino[4,5-m][10,2,16]benzoxadiazacyclotetracosine-38-carboxylic acid.
Peptide antibiotic. *Hoechst AG.*

2391 Teicoplanin A_{3-1}
93616-27-4
$C_{72}H_{68}Cl_2N_8O_{38}$
(3S,15R,18R,34R,35S,38S,48R,50aR)-34-[(2-Acetamido-2-deoxy-β-D-glucopyranosyl)oxy]-15-amino-22,31-dichloro-2,3,16,17,18,19,35,36,37,38,48,49,50,50a-tetradecahydro-6,11,40,44,56-pentahydroxy-42-(α-D-mannopyranosyloxy)-2,16,36,50,51,59-hexaoxo-1H,15H,34H-20,23:30,33-dietheno-3,18:35,48-bis(iminomethano)-4,8:10,14:25,28:43,47-tetrametheno-28H-[1,14,6,22]-dioxadiazacyclooctacosino[4,5-m][10,2,16]benzoxadiazacyclotetracosine-38-carboxylic acid.
Peptide antibiotic. *Hoechst AG.*

2392 Thiostrepton
1393-48-2 9502 215-734-9
$C_{72}H_{85}N_{19}O_{18}S_5$
Gargon; Thiactin; Bryamycin. Peptide antibiotic. Polypeptide antibiotic produced by *Streptomyces azureus*. Dec 246-256°; $[\alpha]^{23D}$ = -985.° (AcOH), -61° (dioxane), -20° (C_5H_5N); soluble in $CHCl_3$, dioxane, C_5H_5N, DMF, AcOH; insoluble in H_2O, MeOH, EtOH, nonpolar organic solvents; no uv maxima, shoulders at 225, 250, 280 nm ($E^{1\%}$: 520, 380, 255); [hemisuccinate]: mp = 200-220°. *Olin Mathieson.*

2393 Tuberactinomycin
11075-36-8 9935
Peptide antibiotic. Tuberculostatic. Polypeptide antibiotic obtained from *Streptomyces griseoverticillatus var. tuberacticus*. Composed of tuberactinomycins A, B, N and O. Tuberactinomycin B is identical to viomycin and tuberactinomycin N to enviomycin. [hydrochloride]: mp = 244-264° (dec); $[\alpha]_D^{25}$ = -31.5° (c = 1 H_2O); λ_m = 268 nm ($E^{1\%}_{1\,cm}$ 330 H_2O), 268.5 nm ($E^{1\%}_{1\,cm}$ 313 1N HCl), 285 nm ($E^{1\%}_{1\,cm}$ 206.5 0.1N NaOH); soluble in H_2O; poorly s.

2394 Tuberactinomycin A
33103-21-8 9935
$C_{25}H_{43}N_{13}O_{11}$
1-[(3R,4R)-4-Hydroxy-L-3,6-diaminohexanoic acid]viomycin.
Peptide antibiotic. Tuberculostatic.

2395 Tuberactinomycin O
33137-73-4 9935
$C_{25}H_{43}N_{13}O_9$
(R)-6-[L-2-(2-Amino-1,4,5,6-tetrahydro-4-pyrimidinyl)glycine]viomycin.
Peptide antibiotic. Tuberculostatic.

2396 Tyrocidine
8011-61-8 9967 232-375-3
Brevicidin [as hydrochloride]; Rapicidin [as hydrochloride]. A peptide antibiotic mixture from *Bacillus brevis*, it is a major constituent of tyrothricine and has three components, tyrocidines A, B and C [hydrochloride]: Dec 240°; $[\alpha]_D^{20}$ = -101° (c = 1.2 95% EtOH); soluble in EtOH, AcOH, C_5H_5N; slightly soluble in H_2O, Me_2CO; insoluble in Et_2O, $CHCl_3$, hydrocarbons. *Penick.*

2397 Tyrocidine A
1481-70-5 9967
$C_{66}H_{87}N_{13}O_{13}$
Peptide antibiotic. [hydrochloride]: mp = 240-242°; $[\alpha]_D^{25}$ = -111° (c = 1.37 50% EtOH); freely soluble in aqueous MeOH, EtOH; slightly soluble in MeOH, EtOH; insoluble in $CHCl_3$, Me_2CO, Et_2O.

2398 Tyrocidine B
865-28-1 9967
$C_{66}H_{88}N_{14}O_{13}$
Peptide antibiotic. [hydrochloride]: mp = 236-237°; $[\alpha]_D$ = -93.0° (c = 0.5 MeOH); has the same structure as tyrocidine A except that L-tryptophan replaces the L-phenylalanine.

2399 Tyrocidine C
3252-29-7 9967
$C_{70}H_{89}N_{15}O_{13}$
Peptide antibiotic. Has same structure as tyrocidine B except that D-tryptophan replaces the D-phenylalanine attached to L-asparagine.

2400 Tyrothricin
1404-88-2 9972 215-771-0
Coltirot; Martricin; Hydrotricine; Dermotricine; Tyroderm; Solutricine; Tyri 10. Polypeptide antibiotic mixture obtained from soil bacilli belonging to the *Tyrothrix* group of bacteria Dec 215-220°; poorly soluble in H_2O (0.21 g/100 ml); soluble in EtOH (2.8 g/100 ml), MeOH, propylene glycol, iPrOH (0.56 g/100 ml), C_6H_6 (0.30 g/100 ml), isooctane (0.0042 g/100 ml), CCl_4 (0.045 g/100 ml), EtOAc (0.265 g/100 ml), Me_2CO (0.68 g/100 ml), Et_2O (0.325 g/100 ml), dioxane (0.111 g/100 ml), $CHCl_3$ (0.16 g/100 ml); LD_{50} (mus sc) > 1500 mg/kg, (mus ip) = 100 mg/kg, (mus orl) > 3000 mg/kg.

2401 Vancomycin
1404-90-6 10066 215-772-6
$C_{66}H_{75}Cl_2N_9O_{24}$
Vancocin. A glycopeptide antibiotic. Amphoteric substance produced by *Streptomyces orientalis.* Inhibits bacterial monopeptide biosythesis.

2402 Vancomycin Hydrochloride
1404-93-9 10066
$C_{66}H_{76}Cl_3N_9O_{24}$
Lyphocin; Vancor. An amphoteric, glycoeptide antibiotic. Inhibits bacterial monopeptide biosythesis. White solid; λ_m 282 nm ($E^{1\%}_{1\ cm}$ 40 H_2O); soluble in H_2O (> 10 g/100 ml); moderately soluble in dilute MeOH; less soluble in higher alcohols, Me_2CO, Et_2O; LD_{50} (mus iv) = 489 mg/kg, (mus ip) = 1734 mg/kg, (mus sc, orl) = 5000 mg/kg.

2403 Viomycin
32988-50-4 10139 251-323-0
$C_{25}H_{43}N_{13}O_{10}$
Celiomycin; florimycin; tuberactinomycin B. Polypeptide antibiotic. Used as a tuberculostatic. Produced by *Streptomyces* spp. [hydrochloride]: mp = 270° (dec); $[\alpha]_D^{18}$ = -16.6° (c = 1 H_2O); λ_m 268 nm (log ε 4.5 H_2O), 268 nm (log ε 4.4 0.1N HCl), 285 nm (log ε 4.3 0.1N NaOH). *CIBA plc.*

2404 Viomycin Pantothenate Sulfate
1401-79-2 10139
$C_{23}H_{62}N_{14}O_{19}S$
Vionactan; Viothenat. Peptide antibiotic (tuberculostatic). mp = 242° (dec). *Ciba-Geigy Corp.; Grunenthal.*

2405 Viomycin Sulfate
37883-00-4 10139
$C_{25}H_{45}N_{13}O_{14}S$
Viocin. Peptide antibiotic (tuberculostatic). mp = 266° (dec); $[\alpha]_D^{18}$ = -29.5° (c = 1 H_2O); λ_m 268 nm (log ε 4.4 H_2O or 0.1N HCl or 0.1N NaOH); LD_{50} (mus iv) = 240 mg/kg, (mus sc) = 1750 mg/kg; soluble in H_2O, insoluble in most organic solvents. *CIBA plc.*

2406 Virginiamycin M_1
21411-53-0 10142 244-376-6
$C_{28}H_{35}N_3O_7$
mikamycin A; ostreogrycin A; pristinamycin II_A; staphylomycin M_1; streptogramin A; vernamycin A. Peptide antibiotic. mp = 165-167°; $[\alpha]_D$ = -190° ± 2° (c = 0.5 EtOH); λ_m 216 nm ($E^{1\%}_{1\ cm}$ 582 MeOH); soluble in Et_2O (0.1 g/100 ml), C_6H_6 (0.3 g/100 ml), EtOAc (0.5 g/100 ml); Me_2CO (2 g/100 ml), MeOH or EtOH (4 g/100 ml); dioxane and THF (5 g/100 ml); very soluble in $CHCl_3$, DMF; insoluble in H_2O, petroleum ether.

2407 Virginiamycin S_1
23152-29-6 10142 245-462-6
$C_{43}H_{49}N_7O_{10}$
staphylomycin S. Peptide antibiotic. mp = 240-242°; $[\alpha]_D^{20}$= -28° (c = 1 EtOH); λ_m 305 nm (log ε 3.85 EtOH); soluble in Et_2O (0.1 g/100 ml), MeOH (0.5 g/100 ml), EtOH (2.5 g/100 ml), C_6H_6 (2.5 g/100 ml), Me_2CO or EtOAc (3 g/100 ml), dioxane (4 g/100 ml); very soluble in $CHCl_3$, DMF; insoluble in H_2O, petroleum ether.

2408 Zinc Bacitracin
1405-89-6 10257 215-787-8
Bacitracin Zinc Complex.
Bacitracin zinc salt; Baciferm. Peptide antibiotic. Soluble in H_2O (0.23 - 0.45 g/100 ml at 28°), MeOH (0.65 g/100 ml), EtOH (0.20 g/100 ml), iPrOH (0.016 g/100 ml), EtOAc (0.13 g/100 ml), $CHCl_3$ (0.001 g/100 ml), petroleum ether (0.0025 g/100 ml).

Quinolone Antibiotics

2409 Alatrofloxacin
157182-32-6
$C_{26}H_{25}F_3N_6O_5$
7-[(1R,5S,6S)-6-[(S)-2-[(S)-2-Aminopropionamido]-propionamido]-3-azabicyclo[3.1.0]hex-3-yl]-1-(2,4-difluorophenyl)-6-fluoro-1,4-dihydro-4-oxo-1,8-naphthyridine-3-carboxylic acid.
Quinolone antibiotic. *Pfizer Intl.*

2410 Alatrofloxacin Mesylate
157605-25-9
$C_{27}H_{29}F_3N_6O_8S$
7-[(1R,5S,6S)-6-[(S)-2-[(S)-2-Aminopropionamido]-propionamido]-3-azabicyclo[3.1.0]hex-3-yl]-1-(2,4-difluorophenyl)-6-fluoro-1,4-dihydro-4-oxo-1,8-naphthyridine-3-carboxylic acid monomethanesulfonate.
CP-116517-27. Quinolone antibiotic. *Pfizer Intl.*

2411 Amifloxacin
86393-37-5 289-231-8
$C_{16}H_{19}FN_4O_3$
6-Fluoro-1,4-dihydro-1-(methylamino)-7-(4-methyl-1-piperazinyl)-4-oxo-3-quinolinecarboxylic acid.
Win-49375. Quinolone antibiotic. *Sterling Winthrop Inc.*

2412 Amifloxacin Mesylate
88036-80-0
$C_{17}H_{23}FN_4O_6S$
6-Fluoro-1,4-dihydro-1-(methylamino)-7-(4-methyl-1-piperazinyl)-4-oxo-3-quinolinecarboxylic acid monomethanesulfonate.
Win-49375-3. Quinolone antibiotic. *Sterling Winthrop Inc.*

2413 Balofloxacin
127294-70-6
$C_{20}H_{24}FN_3O_4$
(±)-1-Cyclopropyl-6-fluoro-1,4-dihydro-8-methoxy-7-[3-(methylamino)piperidino]-4-oxo-3-quinolinecarboxylic acid.
Quinolone antibacterial.

2414 Cinoxacin
28657-80-9 2369 249-133-8
$C_{12}H_{10}N_2O_5$
1-Ethyl-1,4-dihydro-4-oxo[1,3]dioxolo-[4,5-g]cinnoline-3-carboxylic acid.
Cinobac; 64716; Cinobac; Noxigram; Uronorm. Quinolone antibiotic. mp = 261-262° (dec); soluble in polar organic solvents; LD_{50} (rat orl) = 4160 mg/kg, (rat iv) = 900 mg/kg; [sodium salt ($C_{12}H_9N_2NaO_5$)]: soluble in aqueous solvents. *Eli Lilly & Co.*

2415 Ciprofloxacin
85721-33-1 2374
$C_{17}H_{18}FN_3O_3$
1-Cyclopropyl-6-fluoro-1,4-dihydro-4-oxo-7-(1-piperazinyl)-3-quinolinecarboxylic acid.
Cipro IV; Bay q 3939. Quinolone antibiotic. Dec 255-257°. *Bayer Corp. Pharm. Div.*

2416 Ciprofloxacin Monohydrochloride Monohydrate
86393-32-0 2374
$C_{17}H_{19}ClFN_3O_3.H_2O$
1-Cyclopropyl-6-fluoro-1,4-dihydro-4-oxo-7-(1-piperazinyl)-3-quinolinecarboxylic acid monohydrochloride monohydrate.
Ciloxan; Cipro; Bay o 9867 monohydrate; Baycip; Ciflox; Ciprinol; Ciprobay; Ciproxan; Ciproxin; Flociprin; Septicide; Velmonit. Quinolone antibiotic. mp = 318-320°. *Alcon Labs; Bayer Corp. Pharm. Div.*

2417 Clinafloxacin
105956-97-6 2413
$C_{17}H_{17}ClFN_3O_3$
(±)-7-(3-Amino-1-pyrrolidinyl)-8-chloro-1-cyclopropyl-6-fluoro-1,4-dihydro-4-oxo-3-quinolinecarboxylic acid.
CI-960. Quinolone antibiotic. mp = 253-258° (dec). *Parke-Davis.*

2418 Clinafloxacin Hydrochloride
105956-99-8 2413
$C_{17}H_{18}Cl_2FN_3O_3$
(±)-7-(3-Amino-1-pyrrolidinyl)-8-chloro-1-cyclopropyl-6-fluoro-1,4-dihydro-4-oxo-3-quinolinecarboxylic acid hydrochloride.
CI-960 HCl; AM-1091; CI-960; PD-127391. Quinolone antibiotic. mp = 263-265° (dec). *Parke-Davis.*

2419 Difloxacin
98106-17-3 3187
$C_{21}H_{19}F_2N_3O_3$
6-Fluoro-1-(p-fluorophenyl)-1,4-dihydro-7-(4-methyl-1-piperazinyl)-4-oxo-3-quinolinecarboxylic acid.
Quinolone antibiotic. *Abbott Labs Inc.*

2420 Difloxacin Hydrochloride
91296-86-5 3187
$C_{21}H_{20}ClF_2N_3O_3$
6-Fluoro-1-(p-fluorophenyl)-1,4-dihydro-7-(4-methyl-1-piperazinyl)-4-oxo-3-quinolinecarboxylic acid hydrochloride.
Abbott 56619; A-56619. Quinolone antibiotic. mp > 275°. *Abbott Labs Inc.*

2421 Droxacin
35067-47-1
$C_{14}H_{13}NO_4$
5-Ethyl-2,3,5,8-tetrahydro-8-oxofuro[2,3-g]quinoline-7-carboxylic acid.
Quinolone antibiotic. *Schering AG.*

2422 Droxacin Sodium
57363-13-0
$C_{14}H_{12}NNaO_4$
Sodium 5-ethyl-2,3,5,8-tetrahydro-8-oxofuro-[2,3-g]quinoline-7-carboxylate.
SH-263. Quinolone antibiotic. *Schering AG.*

2423 Enoxacin
74011-58-8 3625
$C_{15}H_{17}FN_4O_3$
1-Ethyl-6-fluoro-1,4-dihydro-4-oxo-7-(1-piperazinyl)-1,8-naphthyridine-3-carboxylic acid.
AT-2266; CI-919; PD-107779. Quinolone antibiotic. mp = 220-224°; LD_{50} (mmus iv) = 327 mg/kg, (mmus sc) = 1237 mg/kg, (mmus orl) > 5000 mg/kg, (fmus iv) = 391 mg/kg, (fmus sc) = 1320 mg/kg, (fmus orl) > 5000 mg/kg, (mrat iv) = 236 mg/kg, (mrat sc) > 2000 mg/kg, (mrat orl) > 5000 mg/kg, (frat iv) = 294 mg/kg, (frat orl) > 5000 mg/kg. *Rhône-Poulenc Rorer Pharm. Inc.*

2424 Enoxacin Sesquihydrate
84294-96-2 3625
$C_{15}H_{17}FN_4O_3.1.5H_2O$
1-Ethyl-6-fluoro-1,4-dihydro-4-oxo-7-(1-piperazinyl)-1,8-naphthyridine-3-carboxylic acid sesquihydrate.
Enoxen; Enoxor; Gyramid. Quinolone antibiotic. *Rhône-Poulenc Rorer Pharm. Inc.*

2425 Fleroxacin
79660-72-3 4137
$C_{17}H_{18}F_3N_3O_3$
6,8-Difluoro-1-(2-fluoroethyl)-1,4-dihydro-7-(4-methyl-1-piperazinyl)-4-oxo-3-quinolinecarboxylic acid.
Megalone; Ro-23-6240/000; AM-833; Megalocin;

Quinodis. Quinolone antibiotic. [hydrochloride]: mp = 269-271° (dec). *Hoffmann-LaRoche Inc.*

2426 Flumequine
42835-25-6 4172 255-962-6
$C_{14}H_{12}FNO_3$
9-Fluoro-6,7-dihydro-5-methyl-1-oxo-1H,5H-benzo[ij]quinolizine-2-carboxylic acid.
R-802; Apurone; Fantacin. Quinolone antibiotic. mp = 253-255°; insoluble in H_2O; soluble in EtOH, alkaline solutions. *3M Pharm.*

2427 Gatifloxacin
160738-57-8
$C_{19}H_{22}FN_3O_4$
(±)-1-Cyclopropyl-6-fluoro-1,4-dihydro-8-methoxy-7-(3-methyl-1-piperazinyl)-4-oxo-3-quinolinecarboxylic acid.
Quinolone antibacterial.

2428 Grepafloxacin
119914-60-2 4567
$C_{19}H_{22}FN_3O_3$
1-Cyclopropyl-6-fluoro-1,4-dihydro-5-methyl-7-(3-methyl-1-piperazinyl)-4-oxo-3-quinolinecarboxylic acid.
Quinolone antibiotic. [dihydrate]: mp = 190-192°. *Otsuka Am. Pharm. ; Parke-Davis.*

2429 dl-Grepafloxacin
146863-02-7 4567
$C_{19}H_{22}FN_3O_3$
(±)-1-Cyclopropyl-6-fluoro-1,4-dihydro-5-methyl-7-(3-methyl-1-piperazinyl)-4-oxo-3-quinolinecarboxylic acid.
Quinolone antibiotic. *Otsuka Am. Pharm. ; Parke-Davis.*

2430 dl-Grepafloxacin Hydrochloride
161967-81-3 4567
$C_{19}H_{23}ClFN_3O_3$
(±)-1-Cyclopropyl-6-fluoro-1,4-dihydro-5-methyl-7-(3-methyl-1-piperazinyl)-4-oxo-3-quinolinecarboxylic acid hydrochloride.
OPC-17116. Quinolone antibiotic. *Otsuka Am. Pharm. ; Parke-Davis.*

2431 Irloxacin
91524-15-1
$C_{16}H_{13}FN_2O_3$
1-Ethyl-6-fluoro-1,4-dihydro-4-oxo-7-pyrrol-1-yl-3-quinolinecarboxylic acid.
Quinolone antibacterial.

2432 Lomefloxacin
98079-51-7 5592
$C_{17}H_{19}F_2N_3O_3$
(±)-1-Ethyl-6,8-difluoro-1,4-dihydro-7-(3-methyl-1-piperazinyl)-4-oxo-3-quinolinecarboxylic acid.
SC-47111A. Quinolone antibiotic. mp = 239-240.5°; LD_{50} (mus iv) = 245.6 mg/kg, (mus orl) > 4000 mg/kg. *Searle G.D. & Co.*

2433 Lomefloxacin Hydrochloride
98079-52-8 5592
$C_{17}H_{20}ClF_2N_3O_3$
(±)-1-Ethyl-6,8-difluoro-1,4-dihydro-7-(3-methyl-1-piperazinyl)-4-oxo-3-quinolinecarboxylic acid hydrochloride.
SC-47111; Maxaquin; NY-198; Bareon; Chimono; Lomebact; Uniquin. Quinolone antibiotic. mp = 290-300° (dec). *Searle G.D. & Co.*

2434 Miloxacin
37065-29-5 6283
$C_{12}H_9NO_6$
5,8-Dihydro-5-methoxy-8-oxo-1,3-dioxolo[4,5-g]quinoline-7-carboxylic acid.
antibiotic AB 206; AB-206; Fuldazin. Quinolone antibiotic. mp = 264° (dec). *Sumitomo Pharm. Co. Ltd.*

2435 Nadifloxacin
124858-35-1 6430
$C_{19}H_{21}FN_2O_4$
(±)-9-Fluoro-6,7-dihydro-8-(4-hydroxypiperidino)-5-methyl-1-oxo-1H,5H-benzo[ij]quinolizine-2-carboxylic acid.
OPC-7251; jinofloxacin. Quinolone antibiotic. mp = 245-247° (dec); LD_{50} (mmus iv) = 376.5 mg/kg, (fmus iv) = 420.6 mg/kg, (mrat iv) = 225.7 mg/kg, (frat iv) = 240.5 mg/kg. *Otsuka Am. Pharm.*

2436 Nalidixic Acid
389-08-2 6446 206-864-7
$C_{12}H_{12}N_2O_3$
1-Ethyl-1,4-dihydro-7-methyl-4-oxo-1,8-naphthyridine-3-carboxylic acid.
Win-18320; NSC-82174; Cybis; NegGram; Wintomylon; Betaxina; Dixiben; Eucistin; Innoxalomn; Nalidicron; Nalitucsan; Narigix; Negram; Nevigramon; Nicelate; Nogram; Poleon; Specifin; Uriben; Uriclar; Uralgin; Urodixin; Uroman; Uroneg; Uropan. Quinolone antibiotic. mp = 229-230°; soluble in $CHCl_3$ (3.5 g/100 ml), C_7H_8 (0.16 g/100 ml), MeOH (0.13 g/100 ml), EtOH (0.09 g/100 ml), H_2O, Et_2O (0.01 g/100 ml); LD_{50} (mus orl) = 3300 mg/kg, (mus sc) = 500 mg/kg, (mus iv) = 176 mg/kg. *Sterling Winthrop Inc.*

2437 Nalidixic Acid Sodium Salt
15769-77-4
$C_{12}H_{11}N_2NaO_3$
Sodium 1-ethyl-1,4-dihydro-7-methyl-4-oxo-1,8-naphthyridine-3-carboxylate.
Win-18320-3. Quinolone antibiotic. *Sterling Winthrop Inc.*

2438 Norfloxacin
70458-96-7 6793 274-614-4
$C_{16}H_{18}FN_3O_3$
1-Ethyl-6-fluoro-1,4-dihydro-4-oxo-7-(1-piperazinyl)-3-quinolinecarboxylic acid.
Chibroxin; Noroxin; AM-715; MK-366; Baccidal; Barazan; Chibroxine; Chibroxol; Floxacin; Fulgram; Gonorcin; Lexinor; Noflo; Nolicin; Noracin; Noraxin; Norocin; Noroxin; Noroxine; Norxacin; Sebercim; Uroxacin; Utinor; Zoroxin. Quinolone antibiotic. mp = 220-221°; λ_m 274, 325, 336 nm ($A_{1\ cm}^{1\%}$ 1109, 437, 425, 0.1N NaOH); soluble in H_2O (0.028 g/100 ml), MeOH (0.098 gm/100 ml), EtOH (0.19 g/100 ml), Me_2CO (0.51 g/100 ml), $CHCl_3$ (0.55 g/100 ml), Et_2O (0.001 g/100 ml), C_6H_6 (0.015 g/100 ml), EtOAc (0.094 g/100 ml), octanol (0.51 g/100 ml), AcOH (34.0 g/100 ml); LD_{50} (mus orl) > 4000 mg/kg, (mus im) > 500 mg/kg, (mus iv) = 220 mg/kg, (rat orl) > 4000 mg/kg, (rat sc) = 1500 mg/kg, (rat im) > 500 mg/kg, (rat iv) = 270 mg/kg. *Merck & Co.Inc.*

2439 Ofloxacin
82419-36-1 6865
$C_{18}H_{20}FN_3O_4$
9-Fluoro-2,3-dihydro-3-methyl-10-(4-methyl-1-piperazinyl)-7-oxo-7H-pyrido[1,2,3-de]-1,4-benzoxazine-6-carboxylic acid.
Exocin; Floxin; Ocuflox; DL-8280; HOE-280; Flobacin; Floxil; Oflocet; Oflocin; Oxaldin; Tarivid; Visren. Quinolone antibiotic. mp = 250-257° (dec); LD_{50} (mmus orl) = 5450 mg/kg, (mmus iv) = 208 mg/kg, (mmus sc) > 10000 mg/kg, (fmus orl) = 5290 mg/kg, (fmus iv) = 233 mg/kg, (fmus sc) > 10000 mg/kg, (mrat orl) = 3590 mg/kg, (mrat iv) = 273 mg/kg, (mrat sc) = 7070 mg/kg, (frat orl) = 3750 mg/kg, (frat iv) = 276 mg/kg, (frat sc) = 9000 mg/kg. *Daiichi Seiyaku; Hoechst AG.*

2440 (S)-(-)-Ofloxacin
100986-85-4 6865
$C_{18}H_{20}FN_3O_4$
(S)-(-)-9-Fluoro-2,3-dihydro-3-methyl-10-(4-methyl-1-piperazinyl)-7-oxo-7H-pyrido[1,2,3-de]-1,4-benzoxazine-6-carboxylic acid.
Exocin; Floxin; Ocuflox; DL-8280; HOE-280; levofloxacin; DR-3355; Cravit. Quinolone antibiotic. mp = 225-227° (dec); $[\alpha]_D^{23}$ = -76.9° (c = 0.385 0.5N NaOH); LD_{50} (mmus orl) = 1881 mg/kg, (fmus orl) = 1803 mg/kg, (mrat orl) = 1478 mg/kg, (frat orl) = 1507 mg/kg. *Daiichi Seiyaku; Hoechst AG.*

2441 Oxolinic Acid
14698-29-4 7079 238-750-8
$C_{13}H_{11}NO_5$
5-Ethyl-5,8-dihydro-8-oxo-1,3-dioxolo[4,5-g]quinoline-7-carboxylic acid.
Utibid; W-4565; NSC-110364; Emyrenil; Inoxyl; Nidantin; Ossian; Oxoboi; Pietil; Prodoxol; Urinox; Uritrate; Uro-Alvar; Urotrate; Uroxin; Uroxol. Quinolone antibiotic. mp = 314-316° (dec); LD_{50} (mus orl) > 6000 mg/kg, (rat orl) > 2000 mg/kg. *Parke-Davis.*

2442 Pazufloxacin
127045-41-4 7188
$C_{16}H_{15}FN_2O_4$
(-)-(3S)-10-(1-Aminocyclopropyl)-9-fluoro-2,3-dihydro-3-methyl-7-oxo-7H-pyrido[1,2,3-de]-1,4-benzoxazine-6-carboxylic acid.
T-3761. Quinolone antibiotic. mp = 269-271.5°; $[\alpha]_D^{25}$ = -88.0° (c = 0.5 0.05N NaOH); LD_{50} (mmus iv) > 500 mg/kg. *Toyama Chem. Co. Ltd.*

2443 Pazufloxacin Methanesulfonate
163680-77-1 7188
$C_{17}H_{19}FN_2O_7S$
(-)-(3S)-10-(1-Aminocyclopropyl)-9-fluoro-2,3-dihydro-3-methyl-7-oxo-7H-pyrido[1,2,3-de]-1,4-benzoxazine-6-carboxylic acid monomethanesulfonate.
T-3762. Quinolone antibiotic. mp = 258-259° (dec); $[\alpha]_D^{20}$ = -64.2° (c = 1 1N NaOH); soluble in H_2O (> 20 g/100 ml at 25°). *Toyama Chem. Co. Ltd.*

2444 Pefloxacin
70458-92-3 7197 274-611-8
$C_{17}H_{20}FN_3O_3$
1-Ethyl-6-fluoro-1,4-dihydro-7-(4-methyl-1-piperazinyl)-4-oxo-3-quinolinecarboxylic acid.
EU-5306; 1589 RB; AM-725. Quinolone antibiotic. mp = 270-272° (dec); slightly soluble in H_2O; LD_{50} (mus iv) = 225 mg/kg, (mus orl) = 1000 mg/kg, (rat ip) = 1500 mg/kg, (rat orl) = 2500 mg/kg. *Dainippon Pharm.*

2445 Pefloxacin Methanesulfonate
70458-95-6 7197 274-613-9
$C_{18}H_{24}FN_3O_6S$
1-Ethyl-6-fluoro-1,4-dihydro-7-(4-methyl-1-piperazinyl)-4-oxo-3-quinolinecarboxylic acid monomethanesulfonate.
41 982RP. Quinolone antibiotic. *Dainippon Pharm.*

2446 Pefloxacin Methanesulfonate Dihydrate
149676-40-4 7197
$C_{18}H_{24}FN_3O_6S.2H_2O$
1-Ethyl-6-fluoro-1,4-dihydro-7-(4-methyl-1-piperazinyl)-4-oxo-3-quinolinecarboxylic acid monomethanesulfonate dihydrate.
41 982RP; 1589mRB; Peflacine; Peflox. A fluorinated quinolone antibiotic. Similar to norfloxacin. *Dainippon Pharm.*

2447 Pipemidic Acid
51940-44-4 7613 257-530-2
$C_{14}H_{17}N_5O_3$
8-Ethyl-5,8-dihydro-5-oxo-2-(1-piperazinyl)pyrido-[2,3-d]pyrimidine-6-carboxylic acid.
1489-RB; Filtrax; Memento 400; Pi-Coli; Pipeacid; Pipedac; Pipemid; Pipurin; Tractur; Uropimid; Urosten; Uroval. Quinolone antibiotic. mp = 253-255°; insoluble in Et_2O, C_6H_6; almost insoluble in H_2O, EtOH; slightly soluble in $CHCl_3$ (0.5 g/100 ml), MeOH (0.4 g/100 ml); LD_{50} (mus orl) = 4000 mg/kg, (mus ip) = 1000 mg/kg, (mus iv) = 50 mg/kg. *Dainippon Pharm.*

2448 Pipemidic Acid Trihydrate
72571-82-5 7613
$C_{14}H_{17}N_5O_3.3H_2O$
8-Ethyl-5,8-dihydro-5-oxo-2-(1-piperazinyl)pyrido-[2,3-d]pyrimidine-6-carboxylic acid trihydrate.
Deblaston; Dolcol; Pipram; Solupemid. Quinolone antibiotic. mp = 253-255°. *Dainippon Pharm.*

2449 Piromidic Acid
19562-30-2 7660 243-161-4
$C_{14}H_{16}N_4O_3$
8-Ethyl-5,8-dihydro-5-oxo-2-(1-pyrrolidinyl)pyrido-[2,3-d]pyrimidine-6-carboxylic acid.
PD-93; Bactramyl; Enterol; Gastrurol; Panacid; Pirudal; Purim; Reelon; Septural; Uropir. Quinolone antibiotic. mp = 314-316°; LD_{50} (mmus) = 287 mg/kg, (fmus iv) = 268 mg/kg, (mrat iv) = 177 mg/kg, (frat iv) = 158 mg/kg, (mus,rat orl,sc,ip) > 4000 mg/kg. *Dainippon Pharm.*

2450 Prulifloxacin
123447-62-1
$C_{21}H_{20}FN_3O_6S$
(±)-7-[4-[(Z)-2,3-Dihydroxy-2-butenyl]-1-piperazinyl]-6-fluoro-1-methyl-4-oxo-1H,4H-[1,3]thiazeto-[3,2-a]quyinoline-3-carboxylic acid cyclic carbonate.
NM-441. A prodrug of the quinolone carboxylic acid antibacterial agent NM394. Antibacterial.

2451 Rosoxacin
40034-42-2 8426 254-758-4
$C_{17}H_{14}N_2O_3$
1-Ethyl-1,4-dihydro-4-oxo-7-(4-pyridyl)-3-quinolinecarboxylioc acid.
Roxadyl; Win-35213; acrosoxacin; Eracin; Eradacil; Eradacin; Winuron. Quinolone antibiotic. mp = 290°. *Sterling Winthrop Inc.*

2452 Rufloxacin
101363-10-4 8448
$C_{17}H_{18}FN_3O_3S$
9-Fluoro-2,3-dihydro-10-(4-methyl-1-piperazinyl)-7-oxo-7H-pyrido[1,2,3-de]-1,4-benzothiazine-6-carboxylic acid.
MF-934. Quinolone antibiotic. Similar to ofloxacin. Used in treatment of urinary tract infections. *Mediolanum Farmaceutici S.p.A.*

2453 Rufloxacin Hydrochloride
106017-08-7 8448
$C_{17}H_{19}ClFN_3O_3S$
9-Fluoro-2,3-dihydro-10-(4-methyl-1-piperazinyl)-7-oxo-7H-pyrido[1,2,3-de]-1,4-benzothiazine-6-carboxylic acid hydrochloride.
ISF-09334; Qari; Monos; Tebraxin. Quinolone antibiotic. mp = 322-324°; LD_{50} (rat iv) = 285 mg/kg, (mus iv) = 224 mg/kg, (rbt orl) = 660 mg/kg, (mrat orl) = 631 mg/kg, (frat orl) = 501 mg/kg. *Mediolanum Farmaceutici S.p.A.*

2454 Sparfloxacin
110871-86-8 8884
$C_{19}H_{22}F_2N_4O_3$
5-Amino-1-cyclopropyl-7-(cis-3,5-dimethyl-1-piperazinyl)-6,8-difluoro-1,4-dihydro-4-oxo-3-quinolinecarboxylic acid.
Zagam; CI-978; AT-4140; Spara; PD-1315-1. Quinolone antibiotic. mp = 266-269° (dec). *Parke-Davis; Rhône-Poulenc Rorer Pharm. Inc.*

2455 Temafloxacin
108319-06-8 9284
$C_{21}H_{18}F_3N_3O_3$
(±)-1-(2,4-Difluorophenyl)-6-fluoro-1,4-dihydro-7-(3-methyl-1-piperazinyl)-4-oxo-3-quinolinecarboxylic acid.
T-1258; A-63004; Teflox; Temac; Omniflox. Quinolone antibiotic. *Abbott Labs Inc.*

2456 Temafloxacin Hydrochloride
105784-61-0 9284
$C_{21}H_{19}ClF_3N_3O_3$
(±)-1-(2,4-Difluorophenyl)-6-fluoro-1,4-dihydro-7-(3-methyl-1-piperazinyl)-4-oxo-3-quinolinecarboxylic acid hydrochloride.
Abbott 62254. Quinolone antibiotic. *Abbott Labs Inc.*

2457 Tosufloxacin
100490-36-6 9692
$C_{19}H_{15}F_3N_4O_3$
7-(3-Amino-1-pyrrolidinyl)-1-(2,4-difluorophenyl)-6-fluoro-1,4-dihydro-4-oxo-1,8-naphthyridine-3-carboxylic acid.
Abbott 61827; [as toluene sulfonate monohydrate]: tosufloxacin tosilate, A-3262, Ozex, Tosuxacin. Trifluorinated quinolone antibiotic. [toluene sulfonic acid monohydrate ($C_7H_8O_3S.H_2O$)]: mp = 258-260. *Abbott Labs Inc.*

2458 dl-Tosufloxacin
120382-07-2 9692
$C_{19}H_{15}F_3N_4O_3$
(±)-7-(3-Amino-1-pyrrolidinyl)-1-(2,4-difluorophenyl)-6-fluoro-1,4-dihydro-4-oxo-1,8-naphthyridine-3-carboxylic acid.
Abbott 61827. Trifluorinated quinolone antibiotic. *Abbott Labs Inc.*

2459 Tosufloxacin Hydrochloride
104051-69-6 9692
$C_{19}H_{16}ClF_3N_4O_3$
7-(3-Amino-1-pyrrolidinyl)-1-(2,4-difluorophenyl)-6-fluoro-1,4-dihydro-4-oxo-1,8-naphthyridine-3-carboxylic acid hydrochloride.
Abbott 60969. Quinolone antibiotic. mp = 247-250° (dec). *Abbott Labs Inc.*

2460 Trovafloxacin
147059-72-1 9919
$C_{20}H_{15}F_3N_4O_3$
7-[(1R,5S,6s)-6-Amino-3-azabicyclo[3.1.0]hex-3-yl]-1-(2,4-difluorophenyl)-6-fluoro-1,4-dihydro-4-oxo-1,8-naphthyridine-3-carboxylic acid.
CP-99219. Quinolone antibiotic. *Pfizer Intl.*

2461 Trovafloxacin Hydrochloride
146961-34-4 9919
$C_{20}H_{16}ClF_3N_4O_3$
7-[(1R,5S,6s)-6-Amino-3-azabicyclo[3.1.0]hex-3-yl]-1-(2,4-difluorophenyl)-6-fluoro-1,4-dihydro-4-oxo-1,8-naphthyridine-3-carboxylic acid hydrochloride.
Quinolone antibiotic. mp = 246° (dec). *Pfizer Intl.*

2462 Trovafloxacin Methanesulfonate
147059-75-4 9919
$C_{21}H_{19}F_3N_4O_6S$
7-[(1R,5S,6s)-6-Amino-3-azabicyclo[3.1.0]hex-3-yl]-1-(2,4-difluorophenyl)-6-fluoro-1,4-dihydro-4-oxo-1,8-naphthyridine-3-carboxylic acid monomethanesulfonate.
CP-99219-27. Quinolone antibiotic. *Pfizer Intl.*

Sulfonamide Antibiotics

2463 Acetyl Sulfamethoxypyrazine
3590-05-4 103 222-730-0
$C_{13}H_{14}N_4O_4S$
N^1-Acetyl-N^1-(3-methoxypyrazinyl)sulfanilamide.
Acetylazide. Sulfonamide antibiotic. mp = 199°. *Soc. Farmaceutici Italia.*

2464 Acetyl Sulfisoxazole
80-74-0 9125 201-305-3
$C_{13}H_{15}N_3O_4S$
N^1-Monoacetylsulfisoxazole.
Sulfonamide antibiotic. mp = 193-194°; soluble in H_2O (0.007 g/100 ml), MeOH (0.493 g/100 ml), 95% EtOH (0.570 g/100 ml), Et_2O (0.094 g/100 ml), $CHCl_3$ (2.9 g/100 ml). *Hoffmann-LaRoche Inc.*

2465 Benzylsulfamide
104-22-3 1182 203-186-3
$C_{13}H_{14}N_2O_2S$
N^4-Benzylsulfanilamide.
46 R.P.; RP-46; M&B-125; Septazine; Setazine; Chemodyn; Proseptazine. Sulfonamide antibiotic. Sulfanilamide prodrug. mp = 175°; slightly soluble in H_2O; soluble in EtOH; readily soluble in Me_2CO, dioxane. *May & Baker Ltd.*

2466 Chloramine B
127-52-6 2117 204-847-9
$C_6H_5ClNNaO_2S$
N-Chlorobenzenesulfonamide sodium salt.
sodium benzenesulfochloramine; Neomagnol. Sulfonamide antibiotic; topical antiseptic. Soluble in H_2O (5 g/100 ml), EtOH (4 g/100 ml); sparingly soluble in Et_2O, $CHCl_3$.

2467 Chloramine T
127-65-1 2118 204-854-7
$C_7H_7ClNNaO_2S$
N-Chloro-p-toluenesulfonamide trihydrate.
chloramine; Aktiven; Chloraseptine; Chlorazene; Chlorazone; Euclorina; Gansil; Halamid; Mianine; Tochlorine; Tolamine. Sulfonamide antibiotic; topical antiseptic. Fairly soluble in H_2O; insoluble in C_6H_6, $CHCl_3$, Et_2O.

2468 Dichloramine T
473-34-7 3098 207-462-4
$C_7H_6Cl_2NNaO_2S$
N,N-Dichloro-p-toluenesulfonamide.
Sulfonamide antibiotic. Used as a germicide. mp = 83°; almost insoluble in H_2O; soluble in C_6H_6 (100 g/100 ml), $CHCl_3$ (100 g/100 ml), CCl_4 (40 g/100 ml), AcOH; slightly soluble in petroleum ether. *Shell.*

2469 N^2-Formylsulfisomidine
795-13-1 4275
$C_{13}H_{14}N_4O_3S$
N-[4-[[(2,6-Dimethyl-4-pyrimidinyl)amino]-sulfonyl]phenyl]formamide.
formylsulfamethine; FAK III; Wometin. Sulfonamide antibiotic. mp = 248.5-250.5°. *VEB Farbenfabrik Wolfen.*

2470 N^4-β-D-Glucosylsulfanilamide
53274-53-6 4472
$C_{12}H_{18}N_2O_7S$
N^4-β-D-Glucosidosulfanilamide.
Prontoglucal; Prontoglukal. Sulfonamide antibiotic. mp = 204°; $[\alpha]_D^{22}$ = -117° (c = 0.9 C_5H_5N), $[\alpha]_D^{22}$ = -128° (c = 0.9 H_2O); [2,3,4,6-tetraacetate ($C_{20}H_{26}N_2O_{11}S$)]: mp = 204°; $[\alpha]_D^{22}$ = -81° (C_5H_5N); [N,N',2,3,4,6-hexaacetate ($C_{24}H_{30}N_2O_{13}S$)]: mp = 115°.

2471 Mafenide
138-39-6 5683 205-326-9
$C_7H_{10}N_2O_2S$
α-Amino-p-toluenesulfonamide.
Sulfamylon; NSC-34632; Marfanil; Mesudrin; Mesudin; Sulfamylon; Homosulfamine; Ambamide; Neofamid; Septicid; Emilene; Homonal; Paramenyl. Sulfonamide antibiotic. mp = 151-152°; soluble in dilute alkali, acid. *Sterling Winthrop Inc.*

2472 Mafenide Acetate
13009-99-9 5683 235-855-0
$C_9H_{14}N_2O_4S$
α-Amino-p-toluenesulfonamide acetate.
Mafatate; Mefamide. Sulfonamide antibiotic. LD_{50} (rat iv) = 2040 mg/kg, (mus iv) = 1580 mg/kg. *Sterling Winthrop Inc.*

2473 Mafenide Hydrochloride
138-37-4 5683 205-325-3
$C_7H_{11}ClN_2O_2S$
α-Amino-p-toluenesulfonamide hydrochloride.
Sulfonamide antibiotic. mp = 256°; LD_{50} (rat iv) = 1170 mg/kg, (mus iv) =900 mg/kg. *Sterling Winthrop Inc.*

2474 Mafenide Propionate
12001-72-8 5683
$C_{10}H_{16}N_2O_4S$
α-Amino-p-toluenesulfonamide propionate.
Sulfomyl. Sulfonamide antibiotic. mp = 158°; readily soluble in H_2O. *Sterling Winthrop Inc.*

2475 4'-(Methylsulfamoyl)sulfaniliamide
547-53-5 6202
$C_{13}H_{15}N_4O_3S_2$
N^1-Methyl-N^4-sulfanilylsulfanilamide.
DB87; Diseptal B; Neo Uliron. Sulfonamide antibiotic. mp = 141°; slightly soluble in H_2O; soluble in EtOH, Me_2CO; freely soluble in aqueous $NaHCO_3$. *I.G. Farben.*

2476 Noprylsulfamide
576-97-6 6777
$C_{15}H_{16}N_2Na_2O_8S_3$
N^4-(Disodium 1,3-disulfo-3-phenylpropyl)sulfanilamide.
RP-40; Solucin; Soluseptasine; Soluseptazine; Solusetazine; Sulphasolucin; Sulphasolutin. Sulfonamide antibiotic. Sulfanilamide prodrug. Soluble in H_2O (20 g/100 ml). *Rhône-Poulenc.*

2477 Phthalylsulfacetamide
131-69-1 7532 205-035-7
$C_{16}H_{14}N_2O_6S$
N^1-Acetyl-N^4-phthalylsulfanilamide.
Thalamyd; Talsigel; ftalicetimida; Enterocid; Enterosulfamid; Enterosulfon; Talecid; Rabalan; Sterathal. Sulfonamide antibiotic. mp = 196°; almost insoluble in H_2O, soluble in EtOH. *Bristol-Myers Squibb Pharm. R&D; Schering-Plough HealthCare Products.*

2478 Phthalylsulfathiazole
85-73-4 7533 201-627-4
$C_{17}H_{13}N_3O_5S_2$
4'-(2-Thiazolylsulfamoyl)phthalanilic acid.
AFI-Ftalyl; Entexidina; Ftalazol; Intestiazol; Sulfathalidine; Sulftalyl; Taleudron; Talidine; Thalazole; Ultratiazol. Sulfonamide antibiotic. mp = 272-277° (dec); insoluble in H_2O, $CHCl_3$; slightly soluble in EtOH, Et_2O; readily soluble in NaOH, KOH, NH_3 solutions; LD_{50} (mus ip) = 920 mg/kg. *E. Geistlich Sohne; Merck & Co.Inc.*

2479 Salazosulfadimidine
2315-08-4 8473 219-016-6
$C_{19}H_{17}N_5O_5S$
5-[p-[(4,6-Dimethyl-2-pyrimidinyl)sulfamoyl]phenylazo]salicylic acid.
salicylazosulfadimidine; salcylazosulfamethazine; Azudimidine. Sulfonamide antibiotic. mp = 207°.

2480 Succinylsulfathiazole
116-43-8 9046 204-141-0
$C_{13}H_{13}N_3O_5S_2.H_2O$
4'-(2-Thiazolylsulfamoyl)succinanilic acid.
Sulfasuxidine. Sulfonamide antibiotic. mp = 184-186°, 192-195°; slightly soluble in H_2O (0.021 g/100 ml); sparingly soluble in EtOH, Me_2CO; LD_{50} (mus ip) = 5700 mg/kg.

2481 Sulfabenzamide
127-71-9 9065 204-859-4
$C_{13}H_{12}N_2O_3S$
N^1-Benzoylsulfanilamide.
Sulfabenzide. Sulfonamide antibiotic. mp = 181.2-182.3°; poorly soluble in H_2O (0.0003 g/100 ml); soluble in EtOH (3 g/100 ml), Me_2CO (11.1 g/100 ml). *Monsanto Co.; Schering AG.*

2482 Sulfabromomethazine
116-45-0 9066 204-142-6
$C_{12}H_{13}BrN_4O_2S$
N^1-(5-Bromo-4,6-dimethyl-2-pyrimidinyl)sulfanilamide.
5-bromosulfamethazine; SN-3517. Sulfonamide antibiotic. Dec 250-252°; λ_m 238 272 nm (A 428 635 MeOH).

2483 Sulfacetamide
144-80-9 9067 205-640-6
$C_8H_{10}N_2O_3S$
N-Sulfanilylacetamide.
component of: Sultrin, Trysul. Sulfonamide antibiotic. mp = 182-184°; soluble in H_2O (0.67 g/100 ml at 20°), EtOH (6.6 g/100 ml), Me_2CO (14.3 g/100 ml); LD_{50} (dog orl) = 8000 mg/kg. *Ortho Pharm. Corp.; Savage Labs.*

2484 Sulfacetamide Sodium
6209-17-2 9067
$C_8H_9N_2NaO_3S.H_2O$
N-Sulfanilylacetamide sodium salt.
Bleph-10 Ophthalmic Ointment; Bleph-10 Ophthalmic Solution; Cetamide Ointment; Isopto Cetamide; Op-Sulfa 30; Sebizon; Sodium Sulamyd; Sulf 10; Sulfacet R; Ak-Sulf; Albucid; Antébor; Beocid-Puroptal; Bleph-10; Locula; Op-Sulfa; Prontamid; Sebizon; Sodium Sulamyd; Sulf-10; Sulten-10; component of: Blephamide, Blephamide S.O.P., Cetapred Ointment, FML-S, Isopto Cetapred, Metimyd, Optimyd, Vasocidin Ointment, Vasocidin Solution, Vasosulf. Sulfonamide antibiotic. mp = 257°; soluble in H_2O (66 g/100 ml); sparingly soluble in EtOH, Me_2CO. *Alcon Labs; Allergan Inc.; Broemmel Pharm.; CIBA Vision AG; Lemmon Co.; Schering-Plough HealthCare Products.*

2485 Sulfachlorpyridazine
80-32-0 9068 201-269-9
$C_{10}H_9ClN_4O_2S$
N^1-(6-Chloro-3-pyridazinyl)sulfanilamide.
Nefrosul; Sonilyn; Ciba 10370; Ba-10370; Cosulid; Cosumix. Sulfonamide antibiotic. *Am. Cyanamid.*

2486 Sulfachlorpyridazine Sodium Salt
23282-55-5 9068 245-553-0
$C_{10}H_9ClN_4NaO_2S$
N^1-(6-Chloro-3-pyridazinyl)sulfanilamide sodium salt.
Prinzone; Vetisulid. Sulfonamide antibiotic. *Am. Cyanamid.*

2487 Sulfachrysoidine
485-41-6 9069 207-617-6
$C_{13}H_{13}N_5O_4S$
3,5-Diamino-2-(p-sulfamoylphenylazo)benzoic acid.
Azo Compound No. 4; Rubiazol; Collubiazol. Sulfonamide antibiotic. mp > 300°.

2488 Sulfaclomide
4015-18-3
$C_{12}H_{13}ClN_4O_2S$
N^1-(5-Chloro-2,6-dimethyl-4-pyrimidinyl)-sulfanilamide.
A sulfonamide derivative. Antibacterial.

2489 Sulfacytine
17784-12-2 9070
$C_{12}H_{14}N_4O_3S$
N^1-(1-Ethyl-1,2-dihydro-2-oxo-4-pyrimidinyl)-sulfanilamide.
Renoquid; CI-636. Sulfonamide antibiotic. mp = 166.5-168°; λ_m 263 297 nm ($E^{1\%}_{1\,cm}$ 584 762 MeOH); soluble in H_2O (0.175 g/100 ml at pH 5 at 37°). *Glenwood Inc.; Parke-Davis.*

2490 Sulfadiazine
68-35-9 9071 200-685-8
$C_{10}H_{10}N_4O_2S$
N^1-2-Pyrimidinylsulfanilamide.
Coco-Diazine; Eskadiazine; Adiazine; Diazyl; Sulfolex; component of: Sulfonamide Duplex. Sulfonamide antibiotic. mp = 252-256°; soluble in H_2O (0.013 g/100 ml at pH 5.5, 37°, 0.2 g/100 ml at pH 7.5, 37°); sparingly soluble in EtOH, Me_2CO; freely soluble in dilute acid, alkali. *Eli Lilly & Co.; SmithKline Beecham Pharm.*

2491 Sulfadiazine Silver Salt
22199-08-2 9071 244-834-5
$C_{10}H_9AgN_4O_2S$
N^1-2-Pyrimidinylsulfanilamide monosilver(1+) salt.
silver sulfadiazine; Silvadene; Flamazine; Flammazine; component of: Boots SSD. Sulfonamide antibiotic. Anti-infective (topical). *Boots Pharmaceuticals Inc.; Hoechst Marion Roussel Inc.; Marion Merrell Dow Inc.*

2492 Sulfadiazine Sodium Salt
547-32-0 9071 208-919-0
$C_{10}H_9N_4NaO_2S$
N^1-2-Pyrimidinylsulfanilamide monoodium salt.
sulfadiazine sodium; soluble sulfadiazine. Sulfonamide antibiotic. Soluble in H_2O (50 g/100 ml). *Hoechst Marion Roussel Inc.*

2493 Sulfadicramide
115-68-4 9072 204-099-3
$C_{11}H_{14}N_2O_3S$
N'-(3,3-Dimethylacroyl)sulfanilamide.
Sulfonamide antibiotic. mp = 184-185°; slightly soluble in H_2O, Et_2O; freely soluble in EtOH, Me_2CO.

2494 Sulfadimethoxine
122-11-2 9073 204-523-7
$C_{12}H_{14}N_4O_4S$
N^1-(2,6-Dimethoxy-4-pyrimidinyl)sulfanilamide.
Agribon; Albon; Arnosulfan; Bactrovet; Diasulfa; Madribon; Maxulvet; Neostreptal; Sudine; Suldixine; Sulfabon; Sulxin; Sumbio; Ultrasulfon; component of: Rofenaid, Maxutrim, Prazil, Trivalbon. Sulfonamide antibiotic. mp = 201-203°; soluble in dilute HCl, $NaHCO_3$; soluble in H_2O (0.0046 g/100 ml at pH 4.10, 0.0295 g/100 ml at pH 6.7, 0.058 g/100 ml at pH 7.06, 5.17 g/100 ml at pH 8.71); LD_{50} (mus orl) > 10000 mg/kg. *Hoffmann-LaRoche Inc.; ICI; Oesterreiche Stickstoffwerke.*

2495 Sulfadimethoxine mixture with Trimethoprim
39469-82-4 9073
$C_{12}H_{14}N_4O_4S.C_{14}H_{18}N_4O_3$
Sulfadimethoxine/Trimethoprim.
Maxutrim; Prazil; Trivalbon. Sulfonamide antibiotic. *Hoffmann-LaRoche Inc.; Oesterreiche Stickstoffwerke.*

2496 Sulfadimethoxine Sodium Salt
1037-50-9 9073 213-859-3
$C_{12}H_{13}N_4NaO_4S$
N^1-(2,6-Dimethoxy-4-pyrimidinyl)sulfanilamide sodium salt.
Sulfonamide antibiotic. Freely soluble in H_2O. *Hoffmann-LaRoche Inc.; Oesterreiche Stickstoffwerke.*

2497 Sulfadimethoxine Sodium Salt mixture with Trimethoprim
131643-86-2 9073
$C_{12}H_{13}N_4NaO_4S.C_{14}H_{18}N_4O_3$
Sulfadimethoxine Sodium/Trimethoprim.
Sulfaprim; Vetiprim. Sulfonamide antibiotic. *Hoffmann-LaRoche Inc.; Oesterreiche Stickstoffwerke.*

2498 Sulfadoxine
2447-57-6 9074 219-504-9
$C_{12}H_{14}N_4O_4S$
N^1-(5,6-Dimethoxy-4-pyrimidinyl)sulfanilamide.
Fanasil; Fanzil; Ro-4-4393; component of: Fansidar. Sulfonamide antibiotic. mp = 190-194°; insoluble in Et_2O; slightly soluble in H_2O, EtOH, MeOH; soluble in dilute mineral acids, alkali hydroxides, carbonates; LD_{50} (mus orl) = 5000 mg/kg, (mus sc) = 2900 mg/kg, (mus ip) = 2900 mg/kg. *Hoffmann-LaRoche Inc.*

2499 Sulfadoxine mixture with Trimethoprim
39295-60-8 9074
$C_{12}H_{14}N_4O_4S.C_{14}H_{18}N_4O_3$
Sulfadoxine/Trimethoprim.
Animar; Borgal. Sulfonamide antibiotic. *Hoffmann-LaRoche Inc.*

2500 Sulfaethidole
94-19-9 9075 202-312-4
$C_{10}H_{12}N_4O_2S_2$
N^1-(5-Ethyl-1,3,4-thiadiazole-2-yl)sulfanilamide.
VK-55; Sul-Spantab; sulfaethylthiadiazole. Sulfonamide antibiotic. mp = 185.5-186°; soluble in H_2O (0.025 g/100 ml), MeOH (2.5 g/100 ml), EtOH (3.3 g/100 ml), Me_2CO (10 g/100 ml), Et_2O (0.74 g/100 ml), $CHCl_3$ (0.35 g/100 ml); insoluble in C_6H_6. *Schering AG.*

2501 Sulfaguanidine
57-67-0 9076 200-345-9
$C_7H_{10}N_4O_2S$
N^1-(Diaminomethylene)sulfanilamide.
RP-2275; Diacta; Ganidan; Guanicil; Resulfon; Shigatox. Sulfonamide antibiotic. Used in veterinary medicine to treat enteric infections. mp = 190-193°; soluble in H_2O (0.1 g/100 ml at 25°, 10 g/100 ml at 100°); sparingly soluble in EtOH, Me_2CO; freely soluble in dilute mineral acids; insoluble in NaOH solution; LD_{100} (mus ip) = 1000 mg/kg.

2502 Sulfaguanole
27031-08-9 9077 248-175-4
$C_{12}H_{15}N_5O_3S$
N^1-[(4,5-Dimethyl-2-oxazolyl)amidino]-sulfanilamide.
Enterocura. Sulfonamide antibiotic. Used, in veterinary medicine, to treat enteric infections. mp = 233-236°, 228-230°; insoluble in H_2O, soluble in NaOH solutions; LD_{50} (mus,rat orl) > 5000 mg/kg. *Nordmark.*

2503 Sulfalene
152-47-6 9078 205-804-7
$C_{11}H_{12}N_4O_3S$
N^1-(3-Methoxypyrazinyl)sulfanilamide.
Kelfizina; NSC-110433; Farmitalia 204/122; Dalysep; Kelfizine W; Longum; Policydal; Vetkelfizina. Sulfonamide antibiotic. mp = 176°; LD_{50} (mus orl) = 2164 mg/kg, (mus ip) = 1410 mg/kg. *Abbott Labs Inc.; Farmitalia Societa Farmaceutici.*

2504 Sulfalene mixture with Trimethoprim
50933-06-7 9078
$C_{11}H_{12}N_4O_3S.C_{14}H_{18}N_4O_3$
Sulfalene/trimethoprim.
Kelfiprim. Sulfonamide antibiotic. LD_{50} (mus orl) = 3500 mg/kg, (rat orl) = 3550 mg/kg. *Abbott Labs Inc.; Farmitalia Societa Farmaceutici.*

2505 Sulfaloxic Acid
14376-16-0 9080 238-348-2
$C_{16}H_{15}N_3O_7S$
4'-[[(Hydroxymethyl)carbamoyl]sulfamoyl]-phthalanilic acid.
sulphaloxic acid. Sulfonamide antibiotic. mp = 160-165°; soluble in dilute alkali. *Heyden Chem.*

2506 Sulfaloxic Acid Calcium Salt
59672-20-7 9080 261-850-8
$C_{32}H_{28}CaN_6O_{14}S_2$
4'-[[(Hydroxymethyl)carbamoyl]sulfamoyl]phthalanilic acid calcium salt (2:1).

Enteromide; Intestin-Euvernil. Sulfonamide antibiotic. *Heyden Chem.*

2507 Sulfamerazine
127-79-7 9081 204-866-2
$C_{11}H_{12}N_4O_2S$
N^1-(4-Methyl-2-pyrimidinyl)sulfanilamide.
RP-2632; Mesulfa; Percoccide; component of: Sulfonamide Duplex. Sulfonamide antibiotic. mp = 234-238°; λ_m 243 257 nm ($E^{1\%}_{1\ cm}$ 875 822 H_2O), 243 307 nm ($E^{1\%}_{1\ cm}$ 625 200 0.1M HCl), 271 nm ($E^{1\%}_{1\ cm}$ 835 EtOH); soluble in H_2O (0.035 g/100 ml at pH 5,5, 0.170 g/100 ml at pH 7.5); readily soluble in mineral acid and alkaline solutions; sparingly soluble in Me_2CO; slightly soluble in EtOH; insoluble in Et_2O, $CHCl_3$. *Eli Lilly & Co.; Merck & Co.Inc.*

2508 Sulfamerazine Sodium Salt
127-58-2 9081 204-851-0
$C_{11}H_{11}N_4NaO_2S$
N^1-(4-Methyl-2-pyrimidinyl)sulfanilamide sodium salt.
soluble sulfamerazine; Solumédine. Sulfonamide antibiotic. Soluble in H_2O (27.7 g/100 ml); slightly soluble in EtOH; insoluble in Et_2O, $CHCl_3$. *Eli Lilly & Co.; Merck & Co.Inc.*

2509 Sulfameter
651-06-9 9082 211-480-8
$C_{11}H_{12}N_4O_3S$
N^1-(5-Methoxy-2-pyrimidinyl)sulfanilamide.
AHR-857; sulfametorine; I-2586; Bayrena; Durenat; Kinecid; Kiron; Kirocid; Sulla, Supramid; Ultrax. Sulfonamide antibiotic. mp = 214-216°; λ_m 230 271 nm ($E^{1\%}$:1 cm 562 726); sparingly soluble in EtOH, H_2O, Et_2O; soluble in dilute acid, base; [sodium salt ($C_{11}H_{11}N_4NaO_3S$)]: LD_{50} (mus iv) = 1100 ± 200 mg/kg, (mus ip) = 1500 mg/kg, (mus orl) = 3000 mg/kg, (rat iv) = 1200 mg/kg, (rat ip) = 1100 mg/kg, (rat orl) = 1000 mg/kg. *Schering AG; SPOFA.*

2510 Sulfamethazine
57-68-1 9083 200-346-4
$C_{12}H_{14}N_4O_2S$
N^1-(4,6-Dimethyl-2-pyrimidinyl)sulfanilamide.
Calfspan Tablets; Sulka K Boluses; SulfaSURE SR Bolus; DiazilSulfadine; S-Dimidine; Dimidin-R; Neazina; Sulmet. Sulfonamide antibiotic. mp = 170-176°, 178-179°, 198-199°, 205-207°; λ_m 241 nm ($E^{1\%}_{1\ cm}$ 670 H_2O, pH 6.6), 243 257 nm ($E^{1\%}_{1\ cm}$ 765 776 0.01N NaOH), 241 297 nm ($E^{1\%}_{1\ cm}$ 561 266 0.01N HCl); soluble in H_2O (0.15 g/100 ml at 29°, 0.192 g/100 ml at 37°, pH 7.00); LD_{50} (mus ip) = 1060 mg/kg. *Fermenta Animal Health Co.; ICI; Inst. Chemioter.; Merck & Co.Inc.; Solvay Animal Health Inc.*

2511 Sulfamethazine Sodium Salt
1981-58-4 9083 217-840-0
$C_{12}H_{13}N_4NaO_2S$
N^1-(4,6-Dimethyl-2-pyrimidinyl)sulfanilamide sodium salt.
Intradine; Sulfoxine 33; Vesadin. Sulfonamide antibiotic. *Fermenta Animal Health Co.; ICI; Inst. Chemioter.; Merck & Co.Inc.; Solvay Animal Health Inc.*

2512 Sulfamethizole
144-82-1 9084 205-641-1
$C_9H_{10}N_4O_2S_2$
N^1-(5-Methyl-1,3,4-thiadiazol-2-yl)sulfanilamide.
Thiosulfil Forte; Famet; Lucosil; Methazol; Renasul; Rufol; Salimol; Sulfapyelon; Thidicur; Thiosulfil; Urolucosil; component of: Thiosulfil-A-Forte. Sulfonamide antibiotic. mp = 208°; soluble in H_2O (0.025 g/100 ml at pH 6.5, 20 g/100 ml at pH 7.5), MeOH (2.5 g/100 ml), EtOH (3,3 g/100 ml), Me_2CO (10 g/100 ml), Et_2O (0.07 g/100 ml), $CHCl_3$ (0.035 g/100 ml), insoluble in C_6H_6. *Lundbeck GmbH & Co.*

2513 Sulfamethomidine
3772-76-7 9085 223-219-5
$C_{12}H_{14}N_4O_3S$
N^1-(6-Methoxy-2-methyl-4-pyrimidinyl)sulfanilamide.
Duroprocin; Methofadin; Télémid. Sulfonamide antibiotic. mp = 146°. *Nordmark.*

2514 Sulfamethoxazole
723-46-6 9086 211-963-3
$C_{10}H_{11}N_3O_3S$
N^1-(5-Methyl-3-isoxazolyl)sulfanilamide.
Gantanol; Ro-4-2130; Sinomin; sulfisomezole; component of: Azo Gantanol, Bactrim, Cotrim, Septra, SMZ/TMP, Sulfatrim. Sulfonamide antibiotic used in the treatment of pneumocystis. mp = 167°; LD_{50} (mus orl) = 3662 mg/kg. *Apothecon; Glaxo Wellcome Inc.; Hoffmann-LaRoche Inc.; Lemmon Co.; Solvay UK Holding Co.Ltd.*

2515 Sulfamethoxazole mixture with Trimethoprim
8064-90-2 9086
$C_{10}H_{11}N_3O_3S.C_{14}H_{18}N_4O_3$
Sulfamethoxazole/trimethoprim.
co-trimopxazole; Abacin; Aop-Sulfatrim; Bactramin; Bactrim; Baktar; Chemotrim; Comox; Drylin; Eusaprim; Fectrim; Gantaprim; Gantrim; Imexim; Kepinol; Laratrim; Linaris; Microtrim; Nopil; Oraprim; Septra; Septrin; Sigaprim; Sulfotrim; Sulprim; Sumetrolim; Supracombin; Suprim; Teleprim; Thiocuran; Trigonyl; Trimesulf; Uroplus, Uro-Septra. Sulfonamide antibiotic. Used also to treat pneumocystis. LD_{50} (mus orl) = 5513 mg/kg. *Shionogi & Co. Ltd.*

2516 Sulfamethoxypyridazine
80-35-3 9087 201-272-5
$C_{11}H_{12}N_4O_3S$
N^1-(6-Methoxy-3-pyridazinyl)sulfanilamide.
Midicel; CL-13494; RP-7522; Kynex; Lederkyn; Midikel; Sulfalex; Sulfdurazin; Sultirene. Sulfonamide antibiotic. mp = 182-183°; soluble in H_2O (0.11 g/100 ml at pH 5, 37°, 0.12 g/100 ml at pH 6, 37°, 0.147 g/100 ml at pH 6.5, 37°); slightly soluble in EtOH, MeOH; more soluble in Me_2CO (2 g/100 ml) and DMF (100 g/100 ml); freely soluble in aqueous alkaline solutions; LD_{50} (mus orl) = 1750 mg/kg. *Am. Cyanamid; Parke-Davis.*

2517 Sulfamethoxypyridazine Acetyl
3568-43-2 222-664-2
$C_{13}H_{13}N_4NaO_4S$
N[1]-(6-Methoxy-3-pyridazinyl)sulfanilamide acetyl.
ND-1966; Midicel Acetyl. Sulfonamide antibiotic. *Am. Cyanamid; Parke-Davis.*

2518 Sulfamethoxypyridazine Sodium Salt
2577-32-4 9087 219-928-4
$C_{11}H_{11}N_4NaO_3S$
N[1]-(6-Methoxy-3-pyridazinyl)sulfanilamide sodium salt.
Davosin; Sulfoxine LA. Sulfonamide antibiotic. *Am. Cyanamid; Parke-Davis.*

2519 Sulfamethylthiazole
515-59-3 208-203-8
$C_{10}H_{11}N_3O_2S_2$
N[1]-(4-Methyl-2-thiazolyl)sulfanilamide.
Sulfonamide antibiotic. mp = 238-240°; soluble in H_2O (0.2 g/100 ml), insoluble in Et_2O, readily soluble in dilute mineral acids and solutions of alkali hydroxides and carbonates.

2520 Sulfamethylthiazole Sodium Salt
6101-28-6
$C_{10}H_{10}N_3NaO_2S_2$
N[1]-(4-Methyl-2-thiazolyl)sulfanilamide sodium salt.
Ultaseptyl. Sulfonamide antibiotic. Soluble in H_2O.

2521 Sulfametrole
32909-92-5 9089 251-288-1
$C_9H_{10}N_4O_3S_2$
N[1]-(4-Methoxy-1,2,5-thiadiazol-3-yl)sulfanilamide.
Sulfonamide antibiotic. mp = 149-150°. *OSSW.*

2522 Sulfametrole mixture with Trimethoprim
63749-94-0 9089
$C_9H_{10}N_4O_3S_2.C_{14}H_{18}N_4O_3$
Sulfametrole-trimethoprim.
Lidaprim; Maderan. Sulfonamide antibiotic. *OSSW.*

2523 Sulfamidochrysoidine
103-12-8 9092 203-081-2
$C_{12}H_{13}N_5O_2S$
4-[(2,4-Diaminophenyl)azo]benzenesulfonamide.
Sulfonamide antibiotic. *I.G. Farben; Winthrop.*

2524 Sulfamidochrysoidine Hydrochloride
33445-35-1 9092
$C_{12}H_{14}ClN_5O_2S$
4-[(2,4-Diaminophenyl)azo]benzenesulfonamide hydrochloride.
Pronotsil; Prontosil flavum; Prontosil rubrum; Rubiazol I; Septosan; Streptocide; Streptozon. Sulfonamide antibiotic. mp = 248-250°; soluble in H_2O (0.25 g/100 ml at 25°), EtOH, Me_2CO, fats, oils. *I.G. Farben; Winthrop.*

2525 Sulfamonomethoxine
1220-83-3
$C_{11}H_{12}N_4O_3S$
N[1]-(6-Methoxy-4-pyrimidinyl)sulfanilamide.
Sulfonamide antibiotic.

2526 Sulfamoxole
729-99-7 9093 211-982-7
$C_{11}H_{13}N_3O_3S$
N[1]-(4,5-Dimethyl-2-oxazolyl)sulfanilamide.
Justamil; Sulfmidil; Sulfuno; Tardamide. Sulfonamide antibiotic. mp = 193-194°; soluble in H_2O (0.085 g/100 ml), 0.01N HCl (0.163 g/100 ml), 0.01N NaOH (0.196 g/100 ml), MeOH (2.315 g/100 ml), $CHCl_3$ (0.240 g/100 ml); λ_m 210 250 270 nm (ε 740 546 857 5 mg/l MeOH); LD_{50} (mus orl) > 10000 mg/kg, (mus ip) = 1800 mg/kg, (rat orl) > 12500 mg/kg, (rat ip) = 2500 mg/kg. *Nordmark.*

2527 Sulfamoxole mixture with Trimethoprim
57197-43-0 9093
$C_{11}H_{13}N_3O_3S.C_{14}H_{18}N_4O_3$
Sulfamoxazole/trimethoprim.
co-trifamole; CN-3123; CoFram; Dibactil; Nevin; Supristol. Sulfonamide antibiotic. LD_{50} (mus orl) > 12000 mg/kg, (mus ip) = 1870 mg/kg, (rat orl) = 14000 mg/kg, (rat ip) = 2000 mg/kg. *Nordmark.*

2528 Sulfanilamide
63-74-1 9094 200-563-4
$C_6H_8N_2O_2S$
p-Aminobenzensulfonamide.
Sulfanilamide Vaginal Cream;1162-F; Albexan; Deseptyl; Prontalbin; Prontosil album; Prontylin; Septoplix; Streptocid album; Streptocide. Sulfonamide antibiotic. mp = 164.5-166.5°; λ_m 257 313 nm; soluble in H_2O (0.26 g/1oo ml at 10°, 0.75 g/100 ml at 25°; 1.70 g/100 ml at 40°, 4.0 g/100 ml at 60°, 47.7 g/100 ml at 100°), EtOH (2.7 g/100 ml), Me_2CO (20 g/100 ml), glycerol, propylene glycol, HCl, NaOH, KOH solutions; insoluble in C_6H_6, $CHCl_3$, Et_2O, petroleum ether; LD_{50} (dog orl) = 2000 mg/kg. *Lemmon Co.*

2529 4-Sulfanilamidosalicylic Acid
6202-21-7 9095 228-263-9
$C_{13}H_{12}N_2O_5S$
4-(p-Aminobenzenesulfonamido)-2-hydroxybenzoic acid.
Metrasil. Sulfonamide antibiotic. Dec 220°; soluble in H_2O at neutral pH. *Ward Blenkinsop.*

2530 Sulfanilate Zinc
31884-76-1
$C_{12}H_{12}N_2O_6S_2Zn.4H_2O$
Zinc 4-aminobenzenesulfonate zinc salt (2:1) tetrahydrate.
Nizin; component of: Op-Isophrin-Z. Sulfonamide antibiotic. *Broemmel Pharm.*

2531 N[4]-Sulfanilylanilamide
547-52-4 9100
$C_{12}H_{13}N_3O_4S_2$
4-(4'-Aminobenzenesulfonamido)benzene-sulfonamide.
DB-32; Diseptal C; Uliron C; Disulon; Neosanamid II; Albasil C; Disulfan. Sulfonamide antibiotic, used topically. mp = 133-134°; soluble in H_2O, MeOH, EtOH, Et_2O, NH_3, HCl; insoluble in petroleum ether, $CHCl_3$. *I.G. Farben.*

2532 Sulfanilylurea
547-44-4 9101 208-922-7
$C_7H_9N_3O_3S$
4-Amino-N-(aminocarbonyl)benzene-sulfonamide.
sulfaurea; Euvernil; Uractyl; Uramid; Urenil; Urosulfan. Sulfonamide antibiotic. mp = 146-148°; soluble in H_2O (0.81 g/100 ml at 37°); soluble in alkalies, forms a soluble sodium salt; [monohydrate]: mp = 1251-27°. *Ciba-Geigy Corp.*

2533 N-Sulfanilyl-3,4-Xylamide
120-34-3 9102 204-387-9
$C_{15}H_{16}N_2O_3S$
N^1-(3,4-Dimethylbenzoyl)sulfanilamide.
Geigy 867; Irgafen. Sulfonamide antibiotic. mp = 222-223°; sparingly soluble in H_2O. *Ciba-Geigy Corp.*

2534 Sulfanitran
122-16-7 9103
$C_{14}H_{13}N_3O_5S$
4'-[(p-Nitrophenyl)sulfamoyl]acetanilide.
NSC-77120; APNPS. Sulfonamide antibiotic. Also used as a coccidostat in poultry. mp = 239-240°, 264°; freely soluble in Me_2CO; soluble in hot EtOH, MeOH; sparingly soluble in H_2O, Et_2O.

2535 Sulfaperine
599-88-2 9104 209-976-4
$C_{11}H_{12}N_4O_2S$
N^1-(5-Methyl-2-pyrimidinyl)sulfanilamide.
Pallidin; Retardon; Rexulfa; Sintosulfa; Sulfatreis. Sulfonamide antibiotic. mp = 262-263°; sparingly soluble in EtOH, H_2O (0.04 g/100 ml at pH 5.5); soluble in aqueous solutions of acids and alkalies. *Merck & Co.Inc.*

2536 Sulfaphenazole
526-08-9 9105 208-384-3
$C_{15}H_{14}N_4O_2S$
N^1-(1-Phenylpyrazol-5-yl)sulfanilamide.
Sulfabid; Isarol V; Orisul; Orisulf. Sulfonamide antibiotic. mp = 179-183°; sparingly soluble in H_2O (0.15 g/100 ml at pH 7, 25°); more soluble in EtOH, MeOH, AcOH; LD_{50} (mus orl) = 5800 mg/kg; [sodium salt monohydrate ($C_{15}H_{13}N_4NaO_2S.H_2O$)]: soluble in H_2O. *CIBA plc; Purdue Pharma L.P.*

2537 Sulfaproxyline
116-42-7 9106 204-140-5
$C_{16}H_{18}N_2O_4S$
N^1-(4-Isopropoxybenzoyl)sulfanilamide.
sulphaproxyline; component of: Dosulfin. Sulfonamide antibiotic. mp = 172-173°. *Ciba-Geigy Corp.*

2538 Sulfapyrazine
116-44-9 9107
$C_{10}H_10N_4O_2S$
4-Amino-N-(pyrazinyl)benzene-sulfonamide.
Sulfonamide antibiotic. Dec 250-254°; slightly soluble in EtOH, Me_2CO; soluble in NaOH, KOH solutions, ammonia and mineral acids; almost insoluble in H_2O (0.0050 g/100 ml at 25°, 0.0052 g/100 ml at 37°); [sodium salt monohydrate ($C_{10}H_9N_4NaO_2S.H_2O$)]: freely soluble in H_2O (30 g/100 ml at 25°); very soluble in Me_2CO; soluble in EtOH; insoluble in Et_2O, $CHCl_3$.

2539 Sulfapyridine
144-83-2 9108 205-642-7
$C_{11}H_{11}N_3O_2S$
4-Amino-N,2-pyridinylbenzenesulfonamide.
N^1-2-pyridylsulfanilamide; M&B-693; Dagenan; Eubasin; Pyriamid; Coccoclase; Septipulmon. Sulfonamide antibiotic; dermatitis suppressant. Used in the treatment of dermatitis herpetiformis. mp = 190-191°; soluble in H_2O (0.029 g/100 ml), EtOH (0.227 g/100 ml), Me_2CO (1.54 g/100 ml); freely soluble in mineral acids, KOH, NaOH solutions; LD_{50} (mus orl) = 7500 mg/kg. *May & Baker Ltd.*

2540 Sulfapyridine Sodium Salt Monohydrate
127-57-1 9108 204-850-5
$C_{11}H_{10}N_3NaO_2S{\cdot}H_2O$
4-Amino-N-2-pyridinylbenzenesulfonamide sodium salt monohydrate.
soluble sulfapyridine; Izopiridina; Soludagenan, Sulfonamide antibiotic. Dermatitis suppressant. Used in the treatment of dermatitis herpetiformis. Soluble in H_2O (66.6 g/100 ml), EtOH (10 g/100 ml); LD_{50} (mus orl) = 2700 mg/kg. *May & Baker Ltd.*

2541 Sulfaquinoxaline
59-40-5 200-423-2
$C_{14}H_{12}N_4O_2S$
N^1-2-Quinoxalinylsulfanilamide.
component of: Sulquin 6-50 Concentrate. Sulfonamide antibiotic. *Solvay Animal Health Inc.*

2542 Sulfasalazine
599-79-1 9112 209-974-3
$C_{18}H_{14}N_4O_5S$
5-[[p-(2-Pyridylsulfamoyl)phenyl]azo]salicylic acid.
salazosulfapyridine; salicylazosulfapyridine; sulphasalazine; Azulfidine; Colo-Pleon; Salazopyrin. Anti-inflammatory (gastrointestinal). Sulfonamide; conjugate of 5-aminosalicylic acid and suphapyridine. Used in the treatment of ulcerative colitis and Crohn's disease. Dec 240-245°; λ_m = 237 ($E^{1\%}_{1cm}$ 658), 359 nm; slightly soluble in EtOH; nearly insoluble in H_2O, C_6H_6, $CHCl_3$, Et_2O. *Kabi Pharmacia Diagnostics; Pharmacia & Upjohn.*

2543 Sulfasomizole
632-00-8 9113 211-167-6
$C_{10}H_{11}N_3O_2S_2$
N^1-(3-Methyl-5-isothiazolyl)sulfanilamide.
sulphasomisole; Bidizole; Amidozol. Sulfonamide antibiotic. mp = 192-192.5°; soluble in H_2O (0.248 g/100 ml at pH 6.0, 2.26 g/100 ml at pH 7.0); [sodium salt monohydrate ($C_{10}H_{10}N_3NaO_2S2.H_2O$)]: dec 345°; freely soluble in H_2O. *May & Baker Ltd.*

2544 Sulfasymazine
1984-94-7 9114
$C_{13}H_{17}N_5O_2S$
N^1-(3-Methyl-1-phenylpyrazol-5-yl)sulfanilamide.
sulphsymazine; Symasul; Prosymasul. Sulfonamide

antibiotic. mp = 186.5-187.5°, 190-190.5°; soluble in H_2O (0.1 g/100 ml at pH 5.9). *Am. Cyanamid.*

2545 Sulfathiazole
72-14-0 9115 200-771-5
$C_9H_9N_3O_2S_2$
N^1-2-Thiazolylsulfanilamide.
RP-2090; M&B-760; Thiazamide; Cibazol; Enterobiocine; Duatok; Sulfamul; Sulfavitina; Sulzol; component of: Sultrin, Trysul. Sulfonamide antibiotic. mp = 202-202.5°; soluble in H_2O (0.06 g/100 ml at pH 6.03), EtOH (0.525 g/100 ml); soluble in Me_2CO, dliute mineral acids, KOH and NaOH solutions, ammonia, H_2O. *Ortho Pharm. Corp.; Savage Labs.*

2546 Sulfathiazole Sodium
144-74-1 9115 205-638-5
$C_9H_8N_3NaO_2S_2$
N^1-2-Thiazolylsulfanilamide sodium salt.
soluble sulfathiazole. Sulfonamide antibiotic. Soluble in H_2O (40 g/100 ml), EtOH (6.7 g/100 ml); LD_{50} (mus sc) = 1450 mg/kg, 1950 mg/kg. *Ortho Pharm. Corp.; Savage Labs.*

2547 Sulfathiourea
515-49-1 9116 208-201-7
$C_7H_9N_3O_2S_2$
1-Sulfanilyl-2-thiourea.
RP-2255; Badional; Fontamide. Sulfonamide antibiotic. Dec 171.5-172°; soluble in H_2O (1.1 g/100 ml).

2548 Sulfathiourea Sodium Salt
6101-34-4 9116
$C_7H_8N_3NaO_2S_2$
1-Sulfanilyl-2-thiourea sodium salt.
Sulfonamide antibiotic. Dec 245-245.5°; very soluble in H_2O.

2549 Sulfatolamide
1161-88-2 9117 214-600-7
$C_{14}H_{19}N_5O_4S_3$
1-Sulfanilyl-2-thiourea derivative of α-amino-p-toluenesulfonamide.
Marbadal. Sulfonamide antibiotic. mp = 179-181°; insoluble in Et_2O; slightly soluble in H_2O (< 0.78 g/100 ml); freely soluble in dilute HCl, NaOH, KOH solutions. *Schenley.*

2550 Sulfatroxazole
23256-23-7
$C_{11}H_{13}N_3O_3S$
N^1-(4,5-Dimethyl-3-isoxazolyl)sulfanilamide.
A sulfonamide derivative. Antibacterial.

2551 Sulfazamet
852-19-7 9118 212-707-3
$C_{16}H_{16}N_4O_2S$
N^1-(3-Methyl-1-phenylpyrazol-5-yl)-sulfanilamide.
sulfapyrazole; Vesulong. Sulfonamide antibiotic. mp = 195°, 181-182°. *CIBA plc.*

2552 Sulfisomidine
515-64-0 9124 208-204-3
$C_{12}H_{14}N_4O_2S$
N^1-(2,6-Dimethyl-4-pyrimidinyl)-sulfanilamide.
sulphasomidine; Elkosin; Elcosin; Elkosil; Domain; Aristamid. Sulfonamide antibiotic. mp = 243°; soluble in H_2O (0.12 g/100 ml at 15°, 0.30 g/100 ml at 30°, < 1.67 g/100 ml at 100°); slightly soluble in EtOH, Me_2CO; insoluble in C_6H_6, Et_2O, $CHCl_3$; freely soluble in dilute HCl and NaOH. *CIBA plc; Nordmark.*

2553 Sulfisoxazole
127-69-5 9125 204-858-9
$C_{11}H_{13}N_3O_3S$
N^1-(3,4-Dimethyl-5-isoxazolyl)sulfanilamide.
Gantrisin; Sulfalar; sulphafurazole; Sosol; Soxisol; Soxomide; Sulfazin; Sulfoxol; Sulsoxin; component of: Azo Gantrisin. Sulfonamide antibiotic. mp = 194°; soluble in H_2O (0.013 g/100 ml), EtOH; LD_{50} (mus orl) = 6800 mg/kg. *Hoffmann-LaRoche Inc.; Parke-Davis.*

2554 Sulfisoxazole Diethanolamine Salt
4299-60-9 9125 224-308-1
$C_{15}H_{24}N_4O_5S$
N^1-(3,4-Dimethyl-5-isoxazolyl)sulfanilamide diethanolamine salt.
sulfisoxazole diolamine; Suladrin; Sulfium. Sulfonamide antibiotic. Freely soluble in H_2O, soluble in EtOH. *Hoffmann-LaRoche Inc.; Parke-Davis.*

Sulfone Antibiotics

2555 Acedapsone
77-46-3 20 201-028-8
$C_{16}H_{16}N_2O_4S$
N,N'-(Sulfonyldi-4,1-phenylene)bisacetamide.
Hansolar; CI-556; CN-1883; DADDS; PAM-MR-1165; Rodilone; diacetyldapsone; sulfadiamine; 1399F. Leprostatic; antimalarial. Sulfone antibiotic. mp = 289-292°; λ_m 256 284 nm (ε 25500 36200 MeOH); soluble in H_2O (0.0003 g/100 ml). *Parke-Davis.*

2556 Acediasulfone
80-03-5 21 201-243-7
$C_{14}H_{13}N_2NaO_4S$
N-p-Sulfanilylphenylglycine sodium.
diaminosulfone-N-acetic acid; Sulfon-Cilag [as sodium salt]. Sulfone antibiotic. mp = 194°; soluble in MeOH, Me_2CO, dilute NaOH. *Cilag-Chemie Ltd.*

2557 Acetosulfone Sodium
128-12-1 73
$C_{14}H_{14}N_3NaO_5S_2$
N-(6-Sulfanilylmetanilyl)acetamide monosodium salt.
Promacetin; CI-100; IA-307; NSC-107528; acetosulphone; Internal Antiseptic No. 307; I.A. 307. Sulfone antibiotic. Leprostatic. Soluble in H_2O (3 g/100 ml); [free sulfonamide]: mp = 285°. *Parke-Davis.*

2558 Dapsone
80-08-0 2885 201-248-4
$C_{12}H_{12}N_2O_2S$
4,4'-Sulfonyldianiline.
4,4'-diaminodiphenyl sulfone; diaphenylsulfone; NSC-6091; DDS; DADPS; 1358F; Avlosulfon; Croysulfone; Diphenasone; Disulone; Dumitone; Eporal; Novophone; Sulfona-Mae; Sulphadione; Udolac. Sulfone antibiotic. Leprostatic and dermatitis herpetiformis suppressant. Also used as a hardening agent with epoxy resins and, in veterinary medicine, as a coccidostat. mp = 175-176°, 180.5°; insoluble in H_2O; soluble in EtOH, MeOH, Me_2CO, dilute HCl. *I.G. Farben.*

2559 Diathymosulfone
5964-62-5 3037
$C_{32}H_{34}N_4O_4S$
Di-[4-(4-hydroxy-2-methyl-5-isopropylphenylazo)-phenyl] sulfone.
thymol sulfone; Diatox; thymolated silver sulfone [as di-silver salt]. Antibacterial. Leprostatic. mp = 222-224°; λ_m 400 nm (EtOH); insoluble in H_2O; soluble in dioxane, Me_2CO, alkali solutions; less soluble in EtOH, Et_2O. *Lab. Laborec.*

2560 Glucosulfone Sodium
554-18-7 4471 209-064-6
$C_{24}H_{34}N_2Na_2O_{18}S_3$
4,4'-Diaminophenylsulfone-N,N-di(dextrose sodium sulfate).
SN-166; 501-P; Protomin; Promin; Promanide; Angeli's Sulfone. Sulfone antibiotic. Leprostatic. Soluble in H_2O; slightly soluble in EtOH; insoluble in Et_2O, C_6H_6, MeOH, EtOAc, C_5H_5N; LD_{50} (mus orl) = 3930 mg/kg. *Siegfried AG.*

2561 Solasulfone
133-65-3 8859 205-116-7
$C_{30}H_{28}N_2Na_4O_{14}S_5$
1,1'-[Sulfonylbis(p-phenylimino)]-bis-(3-phenyl-1,3-propanedisulfonic acid) tetrasodium salt.
phenopryldiasulfone sodium; tetrasodium 4,4-diphenyl-sulfone α,γ,α',γ'-tetra-sulfonate; solapsone; RP-3668; Cimedone; Sulphetrone. Sulfone antibiotic. Leprostatic. Crystallizes as a hydrate; very soluble in H_2O; insoluble in EtOH. *Glaxo Wellcome Inc.*

2562 Succisulfone
5934-14-5 9047 227-684-5
$C_{16}H_{16}N_2O_5$
4'-Sulfanilylsuccinanilic acid.
4-succinylamido-4'-aminodiphenyl sulfone; F 1500; Fourneau 1500; Exosulfonyl. Sulfone antibiotic. Leprostatic. mp = 157°; soluble in ammonia. *Bayer Corp. Pharm. Div.; Eli Lilly & Co.*

2563 Succisulfone 2,2'-Iminodiethanol Salt
547-36-4 9047 208-920-6
$C_{20}H_{27}N_3O_7S$
4'-Sulfanilylsuccinanilic acid 2,2'-iminodiethanol salt.
Sulfone antibiotic. *Theraplix.*

2564 Sulfanilate Zinc
31884-76-1 9096
$C_{12}H_{12}N_2O_6S_2Zn.4H_2O$
Zinc 4-aminobenzenesulfonate zinc salt (2:1) tetrahydrate.
Nizin; component of: Op-Isophrin-Z. Sulfonamide, sulfone antibiotic. *Broemmel Pharm.*

2565 Sulfanilic Acid
121-57-3 9096 204-482-5
$C_6H_7NO_3S.H_2O$
4-Aminobenzenesulfonic acid.
Sulfone antibiotic. Dec 288°, > 360°; soluble in H_2O (1.0 g/100 ml at 20°, 1.45 g/100 ml at 30°, 1.94 g/100 ml at 40°); insoluble in EtOH, C_6H_6, Et_2O; slightly soluble in hot MeOH.

2566 Sulfanilic Acid Monohydrate
6101-32-2 9096
$C_6H_7NO_3S.H_2O$
4-Aminobenzenesulfonic acid.
Sulfone antibiotic.

2567 Sulfanilic Acid Sodium Salt Dihydrate
6106-22-5 9096
$C_6H_7NO_3S.2H_2O$
Sodium 4-aminobenzenesulfonate dihydrate.
Sulfone antibiotic. Freely soluble in H_2O, soluble in hot MeOH.

2568 p-Sulfanilidobenzylamine
4393-19-5 9098
$C_{13}H_{14}N_2O_2S$
4-[(4-Aminophenyl)sulfonyl]benzene-methanamine.
Alphamide. Sulfone antibiotic. mp = 159°; [monohydro-chloride monohydrate ($C_{13}H_{15}ClN_2O_2S.H_2O$)]: mp = 195°; very soluble in H_2O; [dihydrochloride ($C_{13}H_{16}Cl_2N_2O_2S$)]: mp = 285°.

2569 Sulfoxone Sodium
144-75-2 9141
$C_{14}H_{14}N_2Na_2O_6S_3$
Disodium [sulfonylbis(p-phenyleneimino)]-dimethanesulfinate.
aldesulfone sodium; Diasone; Diazon; Novotrone. Sulfone antibiotic. Leprostatic. Used in treatment of dermatitis herpetiformis. Dec 263-265°; readily soluble in H_2O, slightly soluble in EtOH, insoluble in most organic solvents. *Abbott Labs Inc.*

2570 Thiazolsulfone
473-30-3 9444
$C_9H_9N_3O_2S_2$
2-Amino-5-sulfanilylthiazole.
4-aminophenyl-2'-aminothiazolyl-5'-sulfone; thiazosul-fone; thiazolesulfone; Promizole. Sulfone antibiotic. mp = 219-221° (dec); soluble in H_2O (0.03 - 0.04 g/100 ml at pH 6.5); freely soluble in Me_2CO, dioxane, 70% EtOH, dilute acids; moderatly soluble in EtOH, EtOAc, Et_2O. *Parke-Davis.*

Tetracycline Antibiotics

2571 Amicycline
5874-95-3
$C_{21}H_{23}N_3O_7$
9-Amino-4-(dimethylamino)-1,4,4a,5,5a,6,11,12a-octahydro-3,10,12,12a-tetrahydroxy-1,11-dioxo-2-naphthacenecarboxamide.
Tetracycline antibiotic.

2572 Apicycline
15599-51-6 772
$C_{30}H_{38}N_4O_{11}$
α-[4-(Dimethylamino)-1,4,4a,5,5a,6,11,12a-octahydro-3,6,10,12,12a-pentahydroxy-6-methyl-1,11-dioxo-2-naphthacenecarboxamido]-4-(2-hydroxyethyl)-1-piperazineacetic acid.
RIT-1140; Traserit. Tetracycline antibiotic. mp = 144.5° (dec); $[\alpha]_D$ = -123° (c = 0.5 MeOH), -133° (c = 0.33 H_2O). *Recherche et Ind. Therap.*

2573 Chlortetracycline
57-62-5 2245 200-341-7
$C_{22}H_{23}ClN_2O_8$
7-Chloro-4-(dimethylamino)-1,4,4a,5,5a,6,11,12a-octahydro-3,6,10,12,12a-pentahydroxy-6-methyl-1,11-dioxo-2-naphthacenecarboxamide.
7-chlorotetracycline; Acronize; Aureocina; Aureomycin; Biomitsin; Centraureo; Chrusomykine; Orospray. Tetracycline antibiotic; antiamebic; antiprotozoal. mp = 168-169°; $[\alpha]_D^{23}$= -275.0° (MeOH); λ_m = 230, 262.5 367.5 nm (0.1NHCl), 255 285 345 nm (0.1N NaOH); soluble in H_2O (0.05-0.06 g/100 ml); very soluble in aqueous solutions at pH > 8.5; freely soluble in cellosolves, dioxane, carbitol; soluble in MeOH, EtOH, BuOH, Me_2CO, EtOAc, C_6H_6; insoluble in Et_2O, petroleum ether. *Am. Cyanamid; Fermenta Animal Health Co.; Lederle Labs.*

2574 Chlortetracycline Hydrochloride
64-72-2 2245 200-591-7
$C_{22}H_{24}Cl_2N_2O_8$
7-Chloro-4-(dimethylamino)-1,4,4a,5,5a,6,11,12a-octahydro-3,6,10,12,12a-pentahydroxy-6-methyl-1,11-dioxo-2-naphthacenecarboxamide monohydrochloride.
Aureomycin; Fermycin Soluble; Aureociclina; Isphamycin. Tetracycline antibiotic; antiamebic; antiprotozoal. Dec > 210°; $[\alpha]_D^{23}$= -240°; soluble in H_2O (0.86 g/100 ml at 28°), MeOH (1.74 g/100 ml at 28°), EtOH (0.17 g/100 ml at 28°); insoluble in Me_2CO, Et_2O, $CHCl_3$, dioxane; LD_{50} (rat orl) = 10300 mg/kg. *Fermenta Animal Health Co.; Lederle Labs.*

2575 Clomocycline
1181-54-0 2448
$C_{23}H_{25}ClN_2O_9$
7-Chloro-4-(dimethylamino)-1,4,4a,5,5a,6,11,12a-octahydro-3,6,10,12,12a-pentahydroxy-N-(hydroxymethyl)-6-methyl-1,11-dioxo-2-naphthacenecarboxamide.
Megaclor. Tetracycline antibiotic. Dec 145-170°; soluble in H_2O (pH 6-8). *LeoAB.*

2576 Demeclocycline
127-33-3 2937 204-834-8
$C_{21}H_{21}ClN_2O_8$
7-Chloro-4-(dimethylamino)-1,4,4a,5,5a,6,11,12a-octahydro-3,6,10,12,12a-pentahydroxy-1,11-dioxo-2-naphthacene-carboxamide.
RP-10192; Biotercíclin; Declomycin; Deganol; Ledermycin; Periciclina. Tetracycline antibiotic. [sesquihydrate]: mp = 174-178° (dec); $[\alpha]_D^{25}$ =-258° (c = 0.5 0.1N H_2SO_4). *Am. Cyanamid; Merck & Co.Inc.; Olin Mathieson.*

2577 Demeclocycline Hydrochloride
64-73-3 2937 200-592-2
$C_{21}H_{22}Cl_2N_2O_8$
7-Chloro-4-(dimethylamino)-1,4,4a,5,5a,6,11,12a-octahydro-3,6,10,12,12a-pentahydroxy-1,11-dioxo-2-naphthacenecarboxamide hydrochloride.
Declomycin; Clortetrin; Demetraciclina; Detravis; Meciclin; Mexocine. Tetracycline antibiotic. LD_{50} (rat orl) = 2372 mg/kg. *Am. Cyanamid; Lederle Labs.; Merck & Co.Inc.; Olin Mathieson.*

2578 Doxycycline
564-25-0 3496 209-271-1
$C_{22}H_{24}N_2O_8$
4-(Dimethylamino)-1,4,4a,5,5a,6,11,12a-octahydro-3,5,10,12,12a-pentahydroxy-6-methyl-1,11-dioxo-2-naphthacene-carboxamide.
Monodox; Vibramycin. Tetracycline antibiotic. *Oclassen Pharm. Inc.; Pfizer Intl.*

2579 Doxycycline Hyclate
24390-14-5 3496
$C_{46}H_{56}Cl_2N_4O_{17}.H_2O$
4-(Dimethylamino)-1,4,4a,5,5a,6,11,12a-octahydro-3,5,10,12,12a-pentahydroxy-6-methyl-1,11-dioxo-2-naphthacenecarboxamide monohydrochloride compound with ethyl alcohol (2:1) monohydrate.
Doryx; Vibra-Tabs; Vivox; Azudoxat; Bassado; Clinofug; Diocimex; Doryx; Doxatet; Doxicrisol; Doxylar; Doxichel hyclate; Doxytem; Duradoxal; Granudoxy; Hydramycin; Mespafin; Nordox; Paldomycin; Retens; Ronaxan; Sigadoxin; Spanor; Tetradox; Unacil; Vibramycin Hyclate; Vibraveineuse; Vibravenös; Zadorin. Tetracycline antibiotic. dec 201°; $[\alpha]_D^{25}$= -110° (c = 1 in 0.01N HCl/MeOH); λ_m 267, 351 nm (log ε 4.24 4.12 0.01N HCl/MeOH); soluble in H_2O; LD_{50} (rat ip) = 262 mg/kg. *Apothecon; Bristol-Myers Squibb Pharm. R&D; Elkins-Sinn; Lemmon Co.; Parke-Davis; Pfizer Intl.*

2580 Doxycycline Hydrate
17086-28-1 3496
$C_{22}H_{24}N_2O_8.H_2O$
4-(Dimethylamino)-1,4,4a,5,5a,6,11,12a-octahydro-3,5,10,12,12a-pentahydroxy-6-methyl-1,11-dioxo-2-naphthacenecarboxamide monohydrate.
Monodox; Vibramycin; GS-3065; Jenacyclin; Supracyclin. Tetracycline antibiotic. *Oclassen Pharm. Inc.; Pfizer Intl.*

2581 Guamecycline
16545-11-2 4584 240-611-1
$C_{29}H_{38}N_8O_8$
N-[[4-(Amidinoamidino)-1-piperazinyl]methyl]-4-(dimethylamino)-1,4,4a,5,5a,6,11,12a-octahydro-3,6,10,12,12a-pentahydroxy-6-methyl-1,11-dioxo-2-naphthacenecarboxamide.
tetrabiguanide; xanthomycin; xantomicina. Tetracycline antibiotic. *Societa Prodiotti Antibiotici Italy.*

2582 Guamecycline Hydrochloride
13040-98-7 4584 235-913-5
$C_{29}H_{39}ClN_8O_8$
N-[[4-(Amidinoamidino)-1-piperazinyl]methyl]-4-(dimethylamino)-1,4,4a,5,5a,6,11,12a-octahydro-3,6,10,12,12a-pentahydroxy-6-methyl-1,11-dioxo-2-naphthacenecarboxamide hydrochloride.
Terratrex; Xantociclina. Tetracycline antibiotic. *Societa Prodiotti Antibiotici Italy.*

2583 Lymecycline
992-21-2 5658 213-592-2
$C_{29}H_{38}N_4O_{10}$
[4S-(4α,4aα,5aα,6β,12aα)]-N^6-[[[[4-(Dimethylamino)-1,4,4a,5,5a,6,11,12a-octahydro-3,6,10,12,12a-pentahydroxy-6-methyl-1,11-dioxo-2-naphthacenyl]carbonyl]amino]methyl]-L-lysine.
Armyl; Ciclolysal; Mucomycin; Tetralisal; Tertamyl; Tetralysal. Tetracycline antibiotic. [sodium salt ($C_{29}H_{37}N_4NaO_{10}$)]: λ_m 376 nm (MeOH). *Farmitalia Carlo Erba Ltd.*

2584 Meclocycline
2013-58-3 5818 217-938-3
$C_{29}H_{27}ClN_2O_{14}S$
[4S-(4α,4aα,5α,5aα,12aα)]-7-Chloro-4-(dimethylamino)-1,4,4a,5,5a,6,11,12a-octahydro-3,5,10,12,12a-pentahydroxy-6-methylene-1,11-dioxo-2-naphthacenecarboxamide.
GS-2989; NSC-78502; Meclan; Mecloderm; Meclosorb; Meclutin; Traumatociclina. Tetracycline antibiotic. Semisynthetic antibiotic related to tetracycline. λ_m = 245, 347 nm (log ε 4.34, 410 MeOH, 0.01 N HCl); LD_{50} (mus orl) > 5000 mg/kg, (mus ip) = 425 mg/kg. *Pfizer Intl.*

2585 Meclocycline 5-Sulfosalicylate
73816-42-9 5818 277-614-2
$C_{29}H_{25}ClN_2O_{13}S$
7-Chloro-4-(dimethylamino)-1,4,4a,5,5a,6,11,12a-octahydro-3,5,10,12,12a-pentahydroxy-6-methylene-1,11-dioxo-2-naphthacenecarboxamide mono (5-sulfosalicylate) (salt).
Meclan Cream; Meclan; Mecloderm; Meclosorb; Meclutin; Traumatociclina. Tetracycline antibiotic. λ_m 239 268 346 nm (log ε 4.46 4.07 4.11 0.01N HCl/MeOH). *Ortho Pharm. Corp.; Pfizer Inc.*

2586 Methacycline
914-00-1 6007 213-017-5
$C_{22}H_{22}N_2O_8$
4-(Dimethylamino)-1,4,4a,5,5a,6,11,12a-octahydro-3,5,10,12,12a-pentahydroxy-6-methylene-1,11-dioxo-2-naphthacenecarboxamide.
GS-2876; metacycline; Bialatan. Tetracycline antibiotic. *Pfizer Intl.*

2587 Methacycline Hydrochloride
3963-95-9 6007 223-568-3
$C_{22}H_{23}ClN_2O_8$
4-(Dimethylamino)-1,4,4a,5,5a,6,11,12a-octahydro-3,5,10,12,12a-pentahydroxy-6-methylene-1,11-dioxo-2-naphthacenecarboxamide hydrochloride.
Rondomycin; Adriamicina; Ciclobiotic; Germiciclin; Metadomus; Metilenbiotic; Londomycin; Optimycin; Physiomycine; Rindex. Tetracycline antibiotic. Dec 205°; soluble in H_2O; sparingly soluble in EtOH; insoluble in Et_2O, $CHCl_3$; λ_m 253 345 nm (log ε 4.37 4.19 0.01N HCl/MeOH); LD_{50} (rat ip) = 252 mg/kg, (mus ip) = 288 mg/kg. *Wallace Labs.*

2588 Minocycline
10118-90-8 6289
$C_{23}H_{27}N_3O_7$
4,7-Bis(dimethylamino)-1,4,4a,5,5a,6,11,12a-octahydro-3,10,12,12a-tetrahydroxy-1,1-dioxo-2-naphthacenecarboxamide.
Minocyn. Tetracycline antibiotic. Semi-synthetic. Effective against tetracycline-r esistant staphylococci. $[\alpha]_D^{25}$ = -166° (c = 0.524); λ_m 352 263 nm (4.16 4.23 0.1N HCl), 380 243 nm (log ε 4.30 4.38 0.1N NaOH). *Lederle Labs.; Parke-Davis.*

2589 Minocycline Hydrochloride
13614-98-7 6289 237-099-7
$C_{23}H_{28}ClN_3O_7$
4,7-Bis(dimethylamino)-1,4,4a,5,5a,6,11,12a-octahydro-3,10,12,12a-tetrahydroxy-1,1-dioxo-2-naphthacenecarboxamide hydrochloride.
Minocin; Vectrin; Klinomycin; Minomycin. Tetracycline antibiotic. *Lederle Labs.; Parke-Davis.*

2590 Nitrocycline
5585-59-1
$C_{21}H_{21}N_3O_9$
4-(Dimethylamino)-1,4,4a,5,5a,6,11,12a-octahydro-3,10,12,12a-tetrahydroxy-7-nitro-1,11-dioxo-2-naphthacenecarboxamide.
Tetracycline antibiotic.

2591 Oxytetracycline
6153-64-6 7111
$C_{22}H_{24}N_2O_9$
4-(Dimethylamino)-1,4,4a,5,5a,6,11,12a-octahydro-3,5,6,10,12,12a-hexahydroxy-6-methyl-1,11-dioxo-2-naphthacenecarboxamide.
glomycin; riomitsin; hydroxytetracycline. Tetracycline antibiotic. *C.M. Ind.; Pfizer Intl.; SmithKline Beecham Animal Health.*

2592 Oxytetracycline Dihydrate
79-57-2 7111 201-212-8
$C_{22}H_{24}N_2O_9.2H_2O$
4-(Dimethylamino)-1,4,4a,5,5a,6,11,12a-octahydro-3,5,6,10,12,12a-hexahydroxy-6-methyl-1,11-dioxo-2-naphthacenecarboxamide dihydrate.
OXTC; Terramycin; Abbocin; Berkmycen; Clinimycin; Imperacin; Oxatets; Oxydon; Oxymycin; Stecsolin; Stevasin; Terralon-LA; Unimycin. Tetracycline antibiotic. Dec 181-182°; $[\alpha]_D^{25}$ = -196.6°; (0.1N HCl), -2.1° (0.1N NaOH), 26.5° (MeOH); λ_m 249 276 353 nm ($E_{1\,cm}^{1\%}$ 240 322 301 pH 4.5 phosphate buffer); soluble in H_2O (3.14

g/100 ml at pH 1.2, 0.46 g/100 ml at pH 2.0, 0.14 g/100 ml at pH 3.0, 0.05 g/100 ml at pH 5.0, 0.07 g/100 ml at pH 6.0, 0.11 g/100 ml at pH 7.0, 3.86 g/100 ml at pH 9.0), absolute EtOH (1.2 g/100 ml), 95% EtOH (0.02 g/100 ml); [disodium salt ($C_{22}H_{22}N_2Na_2O_9.2H_2O$)]: soluble in EtOH (0.8 g/100 ml), MeOH (0.15 g/100 ml). *Abbott Labs Inc.; Pfizer Intl.; SmithKline Beecham Animal Health.*

2593 Oxytetracycline Hydrochloride
2058-46-0 7111 218-161-2
$C_{22}H_{25}ClN_2O_9$
4-(Dimethylamino)-1,4,4a,5,5a,6,11,12a-octahydro-3,5,6,10,12,12a-hexahydroxy-6-methyl-1,11-dioxo-2-napthacenecarboxamide monohydrochloride.
Alamycin; Aquacycline; Bio-Mycin; Duphacycline; Engemycin; Geomycin; Gynamousse; Macocyn; Medamycin; Occrycetin; Oxlopar; Oxybiocycline; Oxybiotic; Oxycyclin; Oxy-Dumocyclin; Oxyject; Oxylag; Oxypan; Oxytetracid; Oxytetrin; Oxy WS; Terrafungine; Terraject; Terramycin Hydrochloride; Tetramel; Tetran; Tetra-Tabilinen; Toxinal; Vendarcin; component of: Terra-Cortril. Tetracycline antibiotic; antirickettsial. Very soluble in H_2O; soluble in absolute ethanol; 95% EtOH. *Fermenta Animal Health Co.; Parke-Davis; Pfizer Inc.; TechAmerica.*

2594 Penimepicycline
4599-60-4 7235 225-002-0
$C_{45}H_{56}N_5O_{12}S$
4-(Dimethylamino)-1,4,4a,5,5a,6,11,12a-oxtahydro-3,6,10,12,12a-pentahydroxy-N[[4-(2-hydroxyethyl)-1-piperazinyl]methyl]-6-methyl-1,11-dioxo-2-naphthacenecarboxamide salt with phenoxymethylpenicillin.
penimepiciclina; Criseocil; Geotricyn; Olimpen; Penetracyne; Peniltetra; Prestociclina. Tetracycline antibiotic. Dec > 143°; soluble in H_2O (142.8 g/100 ml); $[\alpha]_D^{20}$ = -50.5° (c = 2 MeOH). *E.R.A.S.M.E.*

2595 Pipacycline
1110-80-1 7606 214-176-3
$C_{29}H_{38}N_4O_9$
4-Dimethylamino-1,4,4a,5,5a,6,11,12a-octahydro-3,6,10,12,12a-pentahydroxy-N-[[4-(2-hydroxyethyl)-1-piperazinyl]methyl]-6-methyl-1,11-dioxo-2-naphthacenecarboxamide.
mepicycline; mepiciclina; Ambra-Vena; Sieromicin; Valtomicina. Tetracycline antibiotic. Dec 162-163°; $[\alpha]_D^{20}$ = -195° (c = 0.5); λ_m = 286 355 nm (0.1N HCl); freely soluble in H_2O, MeOH, formamide; slightly soluble in EtOH, iPrOH; insoluble in Et_2O, C_6H_6, $CHCl_3$; LD_{50} (mus iv) = 188 mg/kg. *E.R.A.S.M.E.*

2596 Rolitetracycline
751-97-3 8411 212-031-9
$C_{27}H_{33}N_3O_8$
4-(Dimethylamino)-1,4,4a,5,5a,6,11,12a-octahydro-3,6,10,12,12a-pentahydroxy-6-methyl-1,11-dioxo-N-(1-pyrrolidinylmethyl)-2-naphthacenecarboxamide.
Syntetrin; SQ-15659; Reverin; Tetraverin; Transcycline. Tetracycline antibiotic. Dec 162-165°; soluble in H_2O (125 g/100 ml); freely soluble in EtOH, soluble in dilute acids, alkali. *Bristol-Myers Squibb Pharm. R&D.*

2597 Rolitetracycline Compound with Chloramphenicol Succinate
4154-10-3 8411
$C_{42}H_{49}Cl_2N_5O_{16}$
4-(Dimethylamino)-1,4,4a,5,5a,6,11,12a-octahydro-3,6,10,12,12a-pentahydroxy-6-methyl-1,11-dioxo-N-(1-pyrrolidinylmethyl)-2-naphthacenecarboxamide compound with chloramphenicol succinate.
cafrolicycline; gradocycline; levocycline; senociclin; Clorociclin; Crovicina; Metilcal; Proterciclina; Reicaf; Tecaf; Tetrafenicol. Tetracycline antibiotic. Dec 140-144°; very soluble in H_2O; insoluble in Et_2O, petroleum ether, C_6H_{14}. *Lab. ProTer.*

2598 Rolitetracycline Nitrate Sesquihydrate
26657-13-6 8411
$C_{27}H_{34}N_4O_{11}.1.5H_2O$
4-(Dimethylamino)-1,4,4a,5,5a,6,11,12a-octahydro-3,6,10,12,12a-pentahydroxy-6-methyl-1,11-dioxo-N-(1-pyrrolidinylmethyl)-2-naphthacenecarboxamide nitrate sesquihydrate.
Bristacin; Pyrrocycline-N; Syntetrex; Tetrim; Tetriv. Tetracycline antibiotic. LD_{50} (mus iv) = 91 mg/kg. *Bristol-Myers Squibb Pharm. R&D.*

2599 Sancycline
808-26-4 8500
$C_{21}H_{22}N_2O_7$
4-(Dimethylamino)-1,4,4a,5,5a,6,11,12a-octahydro-3,10,12,12a-tetrahydroxy-1,11-dioxo-2-naphthacenecarboxamide.
Bonomycin; GS-2147; NSC-51812; norcycline. Tetracycline antibiotic. [hydrochloride hemihydrate ($C_{21}H_{23}ClN_2O_7.0.5H_2O$)]: dec 215-220°; λ_m 267 347 nm (ε 19300 15500 0.01N HCl/MeOH), 217 268 343 nm (ε 13400 1900 14600 0.1N H_2SO_4). *Am. Cyanamid; Pfizer Intl.*

2600 Tetracycline
60-54-8 9337 200-481-9
$C_{22}H_{24}N_2O_8$
(4S,4aS,5aS,6S,12aS)-4-(Dimethylamino)-1,4,4a,5,5a,6,11,12a-octahydro-3,6,10,12,12a-pentahydroxy-6-methyl-1,11-dioxo-2-naphthacenecarboxamide.
tsiklomitsin; deschlorobiomycin; Abricycline; Ambramycin; Bio-Tetra; Cyclomycin; Dumocyclin; Liquamycin; Mysteclin-F; Talsutin; Tetradecin; component of: Mysteclin-F. Antiamebic; antibacterial; antirickettsial. Tetracycline antibiotic produced in *Streptomyces* species. [trihydrate]: Dec 170-175°; $[\alpha]_D^{25}$ = -257.9° (0.1 N HCl), -239° (MeOH); λ_m = 220, 268, 355 nm (ε 13000, 18040, 13320 0.1N HCl); soluble in H_2O (1.7 mg/ml), MeOH (> 20 mg/ml); LD_{50} (rat orl) = 707mg/kg, (mus orl) = 808 mg/kg. *Bristol-Myers Squibb Co; Pfizer Inc.*

2587 Methacycline Hydrochloride
3963-95-9 6007 223-568-3
$C_{22}H_{23}ClN_2O_8$
4-(Dimethylamino)-1,4,4a,5,5a,6,11,12a-octahydro-3,5,10,12,12a-pentahydroxy-6-methylene-1,11-dioxo-2-naphthacenecarboxamide hydrochloride.
Rondomycin; Adriamicina; Ciclobiotic; Germiciclin; Metadomus; Metilenbiotic; Londomycin; Optimycin; Physiomycine; Rindex. Tetracycline antibiotic. Dec 205°;

soluble in H_2O; sparingly soluble in EtOH; insoluble in Et_2O, $CHCl_3$; λ_m 253 345 nm (log ε 4.37 4.19 0.01N HCl/MeOH); LD_{50} (rat ip) = 252 mg/kg, (mus ip) = 288 mg/kg. *Wallace Labs.*

2588 Minocycline
10118-90-8 6289
$C_{23}H_{27}N_3O_7$
4,7-Bis(dimethylamino)-1,4,4a,5,5a,6,11,12a-octahydro-3,10,12,12a-tetrahydroxy-1,1-dioxo-2-naphthacenecarboxamide.
Minocyn. Tetracycline antibiotic. Semi-synthetic. Effective against tetracycline-resistant staphylococci. $[\alpha]_D^{25}$ = -166° (c = 0.524); λ_m 352 263 nm (4.16 4.23 0.1N HCl), 380 243 nm (log ε 4.30 4.38 0.1N NaOH). *Lederle Labs.; Parke-Davis.*

2589 Minocycline Hydrochloride
13614-98-7 6289 237-099-7
$C_{23}H_{28}ClN_3O_7$
4,7-Bis(dimethylamino)-1,4,4a,5,5a,6,11,12a-octahydro-3,10,12,12a-tetrahydroxy-1,1-dioxo-2-naphthacenecarboxamide hydrochloride.
Minocin; Vectrin; Klinomycin; Minomycin. Tetracycline antibiotic. *Lederle Labs.; Parke-Davis.*

2590 Nitrocycline
5585-59-1
$C_{21}H_{21}N_3O_9$
4-(Dimethylamino)-1,4,4a,5,5a,6,11,12a-octahydro-3,10,12,12a-tetrahydroxy-7-nitro-1,11-dioxo-2-naphthacenecarboxamide.
Tetracycline antibiotic.

2591 Oxytetracycline
6153-64-6 7111
$C_{22}H_{24}N_2O_9$
4-(Dimethylamino)-1,4,4a,5,5a,6,11,12a-octahydro-3,5,6,10,12,12a-hexahydroxy-6-methyl-1,11-dioxo-2-naphthacenecarboxamide.
glomycin; riomitsin; hydroxytetracycline. Tetracycline antibiotic. *C.M. Ind.; Pfizer Intl.; SmithKline Beecham Animal Health.*

2592 Oxytetracycline Dihydrate
79-57-2 7111 201-212-8
$C_{22}H_{24}N_2O_9.2H_2O$
4-(Dimethylamino)-1,4,4a,5,5a,6,11,12a-octahydro-3,5,6,10,12,12a-hexahydroxy-6-methyl-1,11-dioxo-2-naphthacenecarboxamide dihydrate.
OXTC; Terramycin; Abbocin; Berkmycen; Clinimycin; Imperacin; Oxatets; Oxydon; Oxymycin; Stecsolin; Stevasin; Terralon-LA; Unimycin. Tetracycline antibiotic. Dec 181-182°; $[\alpha]_D^{25}$ = -196.6°; (0.1N HCl), -2.1° (0.1N NaOH), 26.5° (MeOH); λ_m 249 276 353 nm ($E_{1\ cm}^{1\%}$ 240 322 301 pH 4.5 phosphate buffer); soluble in H_2O (3.14 g/100 ml at pH 1.2, 0.46 g/100 ml at pH 2.0, 0.14 g/100 ml at pH 3.0, 0.05 g/100 ml at pH 5.0, 0.07 g/100 ml at pH 6.0, 0.11 g/100 ml at pH 7.0, 3.86 g/100 ml at pH 9.0), absolute EtOH (1.2 g/100 ml), 95% EtOH (0.02 g/100 ml); [disodium salt ($C_{22}H_{22}N_2Na_2O_9.2H_2O$)]: soluble in EtOH (0.8 g/100 ml), MeOH (0.15 g/100 ml). *Abbott Labs Inc.; Pfizer Intl.; SmithKline Beecham Animal Health.*

2593 Oxytetracycline Hydrochloride
2058-46-0 7111 218-161-2
$C_{22}H_{25}ClN_2O_9$
4-(Dimethylamino)-1,4,4a,5,5a,6,11,12a-octahydro-3,5,6,10,12,12a-hexahydroxy-6-methyl-1,11-dioxo-2-napthacenecarboxamide monohydrochloride.
Alamycin; Aquacycline; Bio-Mycin; Duphacycline; Engemycin; Geomycin; Gynamousse; Macocyn; Medamycin; Occrycetin; Oxlopar; Oxybiocycline; Oxybiotic; Oxycyclin; Oxy-Dumocyclin; Oxyject; Oxylag; Oxypan; Oxytetracid; Oxytetrin; Oxy WS; Terrafungine; Terraject; Terramycin Hydrochloride; Tetramel; Tetran; Tetra-Tabilinen; Toxinal; Vendarcin; component of: Terra-Cortril. Tetracycline antibiotic; antirickettsial. Very soluble in H_2O; soluble in absolute ethanol; 95% EtOH. *Fermenta Animal Health Co.; Parke-Davis; Pfizer Inc.; TechAmerica.*

2594 Penimepicycline
4599-60-4 7235 225-002-0
$C_{45}H_{56}N_5O_{12}S$
4-(Dimethylamino)-1,4,4a,5,5a,6,11,12a-oxtahydro-3,6,10,12,12a-pentahydroxy-N[[4-(2-hydroxyethyl)-1-piperazinyl]methyl]-6-methyl-1,11-dioxo-2-naphthacenecarboxamide salt with phenoxymethylpenicillin.
penimepiciclina; Criseocil; Geotricyn; Olimpen; Penetracyne; Peniltetra; Prestociclina. Tetracycline antibiotic. Dec > 143°; soluble in H_2O (142.8 g/100 ml); $[\alpha]_D^{20}$= -50.5° (c = 2 MeOH). *E.R.A.S.M.E.*

2595 Pipacycline
1110-80-1 7606 214-176-3
$C_{29}H_{38}N_4O_9$
4-Dimethylamino-1,4,4a,5,5a,6,11,12a-octahydro-3,6,10,12,12a-pentahydroxy-N-[[4-(2-hydroxyethyl)-1-piperazinyl]methyl]-6-methyl-1,11-dioxo-2-naphthacenecarboxamide.
mepicycline; mepiciclina; Ambra-Vena; Sieromicin; Valtomicina. Tetracycline antibiotic. Dec 162-163°; $[\alpha]_D^{20}$= -195° (c = 0.5); λ_m = 286 355 nm (0.1N HCl); freely soluble in H_2O, MeOH, formamide; slightly soluble in EtOH, iPrOH; insoluble in Et_2O, C_6H_6, $CHCl_3$; LD_{50} (mus iv) = 188 mg/kg. *E.R.A.S.M.E.*

2596 Rolitetracycline
751-97-3 8411 212-031-9
$C_{27}H_{33}N_3O_8$
4-(Dimethylamino)-1,4,4a,5,5a,6,11,12a-octahydro-3,6,10,12,12a-pentahydroxy-6-methyl-1,11-dioxo-N-(1-pyrrolidinylmethyl)-2-naphthacene-carboxamide.
Syntetrin; SQ-15659; Reverin; Tetraverin; Transcycline. Tetracycline antibiotic. Dec 162-165°; soluble in H_2O (125 g/100 ml); freely soluble in EtOH, soluble in dilute acids, alkali. *Bristol-Myers Squibb Pharm. R&D.*

2597 Rolitetracycline Compound with Chloramphenicol Succinate
4154-10-3 8411
$C_{42}H_{49}Cl_2N_5O_{16}$
4-(Dimethylamino)-1,4,4a,5,5a,6,11,12a-octahydro-3,6,10,12,12a-pentahydroxy-6-methyl-1,11-dioxo-N-(1-pyrrolidinylmethyl)-2-naphthacenecarboxamide compound with chloramphenicol succinate.

cafrolicycline; gradocycline; levocycline; senociclin; Clorociclin; Crovicina; Metilcal; Proterciclina; Reicaf; Tecaf; Tetrafenicol. Tetracycline antibiotic. Dec 140-144°; very soluble in H_2O; insoluble in Et_2O, petroleum ether, C_6H_{14}. *Lab. ProTer.*

2598 Rolitetracycline Nitrate Sesquihydrate
26657-13-6 8411
$C_{27}H_{34}N_4O_{11}.1.5H_2O$
4-(Dimethylamino)-1,4,4a,5,5a,6,11,12a-octahydro-3,6,10,12,12a-pentahydroxy-6-methyl-1,11-dioxo-N-(1-pyrrolidinylmethyl)-2-naphthacenecarboxamide nitrate sesquihydrate.
Bristacin; Pyrrocycline-N; Syntetrex; Tetrim; Tetriv. Tetracycline antibiotic. LD_{50} (mus iv) = 91 mg/kg. *Bristol-Myers Squibb Pharm. R&D.*

2599 Sancycline
808-26-4 8500
$C_{21}H_{22}N_2O_7$
4-(Dimethylamino)-1,4,4a,5,5a,6,11,12a-octahydro-3,10,12,12a-tetrahydroxy-1,11-dioxo-2-naphthacenecarboxamide.
Bonomycin; GS-2147; NSC-51812; norcycline. Tetracycline antibiotic. [hydrochloride hemihydrate ($C_{21}H_{23}ClN_2O_7.0.5H_2O$)]: dec 215-220°; λ_m 267 347 nm (ε 19300 15500 0.01N HCl/MeOH), 217 268 343 nm (ε 13400 1900 14600 0.1N H_2SO_4). *Am. Cyanamid; Pfizer Intl.*

2600 Tetracycline
60-54-8 9337 200-481-9
$C_{22}H_{24}N_2O_8$
(4S,4aS,5aS,6S,12aS)-4-(Dimethylamino)-1,4,4a,5,5a,6,11,12a-octahydro-3,6,10,12,12a-pentahydroxy-6-methyl-1,11-dioxo-2-naphthacenecarboxamide.
tsiklomitsin; deschlorobiomycin; Abricycline; Ambramycin; Bio-Tetra; Cyclomycin; Dumocyclin; Liquamycin; Mysteclin-F; Talsutin; Tetradecin; component of: Mysteclin-F. Antiamebic; antibacterial; antirickettsial. Tetracycline antibiotic produced in *Streptomyces* species. [trihydrate]: Dec 170-175°; $[\alpha]_D^{25}$ = -257.9° (0.1 N HCl), -239° (MeOH); λ_m = 220, 268, 355 nm (ε 13000, 18040, 13320 0.1N HCl); soluble in H_2O (1.7 mg/ml), MeOH (> 20 mg/ml); LD_{50} (rat orl) = 707mg/kg, (mus orl) = 808 mg/kg. *Bristol-Myers Squibb Co.; Pfizer Inc.*

2601 Tetracycline Hydrochloride
64-75-5 9337 200-593-8
$C_{22}H_{25}ClN_2O_8$
(4S,4aS,5aS,6S,12aS)-4-(Dimethylamino)-1,4,4a,5,5a,6,11,12a-octahydro-3,6,10,12,12a-pentahydroxy-6-methyl-1,11-dioxo-2-naphthacenecarboxamide monohydrochloride.
Achro; Achromycin; Ala-Tet; Ambracyn; Ambramicina; Bristaciclina; Cefracycline; Criseociclina; Cyclopar; Diocyclin; Helvecyclin; Hostacyclin; Imex; Mediletten, Mephacyclin; Panmycin; Partrex; Polycycline; Purocyclina; Quadracyclin; Remicyclin; Riocyclin; Robitet; Ro-Cycline; Sanclomycine; Steclin; Sumycin; Supramycin; Sustamycin; Tefilin; Tetrabakat; Tetrabid; Tetrablet; tetrabon; Tetrachel; Tetracompren; Tetracyn; Tetrakap; Tetralution; Tetramavan; Tetramycin; Tetrosol; TetraSURE; Topicycline; Totomycin; Triphacyclin; Unicin; Vetquamycin-324. Tetracycline antibiotic; antiamebic; antiprotozoal. Dec 214°; $[\alpha]_D^{25}$ = -257.9° (c = 0.5 0.1N HCl); soluble in H_2O, MeOH, EtOH; insoluble in Et_2O, petroleum ether; LD_{50} (rat orl) = 6443 mg/kg. *Apothecon; Bristol-Myers Squibb Pharm. R&D; Fermenta Animal Health Co.; Lederle Labs.; Parke-Davis; Pharmacia & Upjohn; Robins, A. H. Co.*

2602 Tetracycline Phosphate Complex
1336-20-5 9337 215-646-0
(4S,4aS,5aS,6S,12aS)-4-(Dimethylamino)-1,4,4a,5,5a,6,11,12a-octahydro-3,6,10,12,12a-pentahydroxy-6-methyl-1,11-dioxo-2-naphthacenecarboxamide phosphate complex.
Tetrex; Tetrex BidCaps; Panmycin P; Telotrex; Tetradecin; Novum; Upcyclin; component of: Azotrex. Tetracycline antibiotic. Sparingly soluble in H_2O, slightly soluble in EtOH. *Bristol-Myers Squibb Pharm. R&D.*

Tuberculostatic Antibiotics

2603 p-Aminosalicylic Acid
65-49-6 498 200-613-5
$C_7H_7NO_3$
4-Amino-2-hydroxybenzoic acid.
Pamisyl; Rezipas. Antibacterial (tuberculostatic). mp = 150-151°; λ_m 265, 300 nm (0.1N HCl); soluble in H_2O (0.2 g/100 ml), EtOH (4.76 g/100 ml); slightly soluble in Et_2O; insoluble in C_6H_6; LD_{50} (mus orl) = 4000 mg/kg; [hydrochloride]: dec 224°. *Bristol-Myers Squibb Pharm. R&D; Parke-Davis.*

2604 p-Aminosalicylic Acid Hydrazide
6946-29-8 499 230-108-5
$C_7H_9N_3O_2$
4-Amino-2-hydroxybenzoic acid hydrazide.
Apacizin; Apacizina. Antibacterial (tuberculostatic). mp = 190-200°; slightly soluble in H_2O, more soluble in EtOH.

2605 p-Aminosalicylic Acid Potassium Salt
133-09-5 498 205-090-7
$C_7H_6KNO_3$
Potassium 4-amino-2-hydroxybenzoate.
Antibacterial (tuberculostatic). *Parke-Davis.*

2606 p-Aminosalicylic Acid Sodium Salt Dihydrate
6108-19-5 498
$C_7H_6NNaO_3.2H_2O$
Sodium 4-amino-2-hydroxybenzoate dihydrate.
Pamisyl Sodium. Antibacterial (tuberculostatic). Soluble in H_2O (50 g/100 ml); sparingly soluble in Me_2CO; insoluble in Et_2O, $CHCl_3$, C_6H_6. *Parke-Davis.*

2607 Benzoylpas
13898-58-3 1148
$C_{14}H_{11}NO_4$
4-(Benzoylamino)-2-hydroxybenzoic acid.
Antibacterial (tuberculostatic). mp = 260-261°. *Wander Pharma.*

2608 Benzoylpas Calcium Salt Pentahydrate
5631-00-5 1148
$C_{28}H_{20}CaN_2O_8.5H_2O$
Calcium 4-(benzoylamino)-2-hydroxybenzoate pentahydrate.
benzoylpas calcium; Benzapas; Benzacyl; Therapas. Antibacterial (tuberculostatic). *Wander Pharma.*

2609 Benzoylpas Sodium Salt
537-20-2 1148
$C_{14}H_{10}NNaO_4$
Sodium 4-(benzoylamino)-2-hydroxybenzoate.
BPAS. Antibacterial (tuberculostatic). *Wander Pharma.*

2610 5-Bromosalicylhydroxamic Acid
5798-94-7 1458
$C_7H_6BrNO_3$
5-Bromo-N,2-dihydroxybenzamide.
Brosalamid; Bromocyl. Antibacterial (tuberculostatic). Dec 232°; sparingly soluble in H_2O.

2611 Capreomycin
11003-38-6 1801
Capastat. Peptide antibiotic (tuberculostatic). Cyclic peptide antibiotic similar to viomycin. Produced by *Streptomyces capreolus*. A mixture of capreomycins IA (25%), IB (67%), IIA (3%), IIB (6%). Soluble in H_2O, insoluble in most organic solvents. *Dista Products Ltd.; Eli Lilly & Co.*

2612 Capreomycin Disulfate
1405-36-3 1801
Caprocin; Ogostal. Peptide antibiotic (tuberculostatic). LD_{50} (mus iv) = 250 mg/kg, (mus sc) = 514 mg/kg, (rat iv) = 325 mg/kg, (rat sc) = 1191 mg/kg. *Eli Lilly & Co.*

2613 Capreomycin IA
37280-35-6 1801
$C_{25}H_{44}N_{14}O_8$
Peptide antibiotic (tuberculostatic). Similar to viomycin. Produced by *Streptomyces capreolus*. mp = 246-248° (dec); $[\alpha]_D^{22}$ = -21.9° (c = 0.5 H_2O); λ_m 269 nm (ε 24000 0.1N HCl), 268 nm (ε 23900 H_2O), 287 nm (ε 15900 0.1N NaOH). *Eli Lilly & Co.*

2614 Capreomycin IB
33490-33-4 1801
$C_{25}H_{44}N_{14}O_7$
Peptide antibiotic. Tuberculostatic. mp = 253-255° (dec); $[\alpha]_D^{22}$ = -44.6° (c = 0.55 H_2O); λ_m 268 nm (ε 22700 0.1N HCl), 268 nm (ε 223000 H_2O), 290 nm (ε 14400 0.1N NaOH). *Eli Lilly & Co.*

2615 Clofazimine
2030-63-9 2433 217-980-2
$C_{27}H_{22}Cl_2N_4$
3-(p-Chloroanilino)-10-(p-chlorophenyl)-2,10-dihydro-2-(isopropylimino)-phenazine.
Lamprene; G-30320; NSC-141046. Antibacterial (tuberculostatic). Also leprostatic. mp = 210-212°; λ_m 284, 486 nm (abs. 1.30, 0.64, 0.01M HCl/MeOH); soluble in AcOH, DMF, $CHCl_3$ (6.6 g/100 ml), EtOH (0.14 g/100 ml), Et_2O (0.1 g/100 ml); insoluble in H_2O; LD_{50} (mus, rat, gpg orl) > 4000 mg/kg. *Ciba-Geigy Corp.*

2616 Cyacetacide
140-87-4 2751 205-437-2
$C_3H_5N_3O$
Cyanoacetic acid hydrazide.
Dictyzide; Mackreazid; Armazal; Reacid; Reazide; Hidacian; Leandin; Neohydrazid; Dictycide. Antibacterial (tuberculostatic). mp = 114.5-115°; freely soluble in H_2O. *Labs. O.M.*

2617 Cycloserine
68-41-7 2820 200-688-4
$C_3H_6N_2O_2$
(+)-4-Amino-3-oxazolidinone.
orientomycin; PA-94; 106-7; Closina; Farmiserina; Micoserina; Oxamycin; Seromycin. Tuberculostatic. Antibiotic produced by *Streptomyces garyphalus sive orchidaceus*. Dec 155-156°; $[\alpha]_D^{23}$ = 116° (c = 1.17), $[\alpha]_{546}^{25}$ = 137° (c = 5 2N NaOH); λ_m 226 nm ($E_{1\,cm}^{1\%}$ 402); soluble in H_2O; slightly soluble in MeOH, propylene glycol. *Eli Lilly & Co.; Merck & Co.Inc.; Pfizer Intl.*

2618 Dihydrostreptomycin
128-46-1 3222 204-888-2
$C_{21}H_{41}N_7O_{12}$
O-2-Deoxy-2-(methylamino)-α-L-glucopyranosyl-(1→2)-O-5-deoxy-3-C-(hydroxymethyl)-α-L-lyxofuranosyl-(1→4)-N,N'-bis(aminoiminomethyl)-D-streptamine.
DHSM; DST; Abiocine; Vibriomycin. Aminoglycoside antibiotic. Tuberculostatic. mp > 300°. *Am. Cyanamid; Bristol-Myers Squibb Co.; Heyden Chem.; Merck & Co.Inc.; Olin Res. Ctr.; Pfizer Inc.; Schenley; Takeda Chem. Ind. Ltd.*

2619 Dihydrostreptomycin Pantothenate
3563-84-6 3222 222-637-5
$C_{30}H_{58}N_8O_{17}$
O-2-Deoxy-2-(methylamino)-α-L-glucopyranosyl-(1→2)-O-5-deoxy-3-C-(hydroxymethyl)-α-L-lyxofuranosyl-(1→4)-N,N'-bis(aminoiminomethyl)-D-streptamine pantothenate.
Didrothenat; Pantostrep. Aminoglycoside antibiotic. Tuberculostatic. *Am. Cyanamid; Bristol-Myers Squibb Pharm. R&D; Heyden Chem.; Merck & Co.Inc.; Olin Res. Ctr.; Pfizer Inc.; Schenley; Takeda Chem. Ind. Ltd.*

2620 Dihydrostreptomycin Sesquisulfate
5490-27-7 3222 226-823-7
$C_{41}H_{88}N_{14}O_{36}$
O-2-Deoxy-2-(methylamino)-α-L-glucopyranosyl-(1→2)-O-5-deoxy-3-C-(hydroxymethyl)-α-L-lyxofuranosyl-(1→4)-N,N'-bis(aminoiminomethyl)-D-streptamine sulfate (2:3) (salt).
Didromycine; Double-mycin; Sol-Mycin; Streptomagna. Aminoglycoside antibiotic. Tuberculostatic. Dec 255-265°, 250°; $[\alpha]_D^{25}$ = -88.5° (c = 1); very soluble in H_2O; soluble in 50% MeOH/H_2O (crystals: 0.08 g/100 ml; powder: 10 g/100 ml); at 28° soluble in H_2O (> 2 g/100 ml), MeOH (0.035 g/100 ml), EtOH (0.010 g/100 ml). *Am. Cyanamid; Bristol-Myers Squibb Pharm. R&D; Heyden Chem.; Merck & Co.Inc.; Olin Res. Ctr.; Pfizer Inc.; Schenley; Takeda Chem. Ind. Ltd.*

2621 Enviomycin
33103-22-9 3636
$C_{25}H_{43}N_{13}O_{10}$
[[15-(3,6-Diamino-4-hydroxyhexanamido)-3-(hexahydro-2-imino-4-pyrimidinyl)-9,12-bis(hydroxymethyl-2,5,8,11,14-pentaoxo-1,4,7,10,13-pentaazacyclohexadec-6-ylidene]methyl]urea.
Tuberactinomycin N. Peptide antibiotic (tuberculostatic). [hydrochloride ($C_{25}H_{46}Cl_3N_{13}O_{10}$)]: mp > 245° (dec); $[\alpha]_D^{21}$ = -19.1°; λ_m 268 nm ($E_{1\,cm}^{1\%}$ 342 H_2O or 0.1N HCl), 288 nm ($E_{1\,cm}^{1\%}$ 215 0.1N NaOH); very soluble in H_2O, slightly soluble in organic solvents; LD_{50} (mus iv) = 485 mg/kg, (rat iv) = 680 mg/kg. *Toyo Brewing Company.*

2622 Ethambutol
74-55-5 3764 200-810-6
$C_{10}H_{24}N_2O_2$
(+)-2,2'-(Ethylenediimino)-di-1-butanol.
EMB. Antibacterial (tuberculostatic). mp = 87.5-88.8°; $[\alpha]_D^{25}$ =13.7° (c = 2 H_2O); soluble in $CHCl_3$, CH_2Cl_2; poorly soluble in C_6H_6, H_2O. *Lederle Labs.*

2623 Ethambutol Dihydrochloride
1070-11-7 3764 213-970-7
$C_{10}H_{26}Cl_2N_2O_2$
(+)-2,2'-(Ethylenediimino)-di-1-butanol dihydrochloride.
Myambutol; CL-40881; Dexambutol; Ebutol; Etibi; Etapiam; Myambutol; Mycobutol; Sural; Tibutol. Antibacterial (tuberculostatic). mp = 198.5-200.3°, 201.8-202.6°; $[\alpha]_D^{25}$= -7.6° (c = 2 H_2O); soluble in H_2O, DMSO; sparingly soluble in EtOH; poorly soluble in Me_2CO, $CHCl_3$. *Lederle Labs.*

2624 Ethionamide
536-33-4 3783 208-628-9
$C_8H_{10}N_2S$
2-Ethylthioisonicotinamide.
Trecator-SC; 1314-TH. Antibacterial (tuberculostatic). Dec 164-166°; sparingly soluble in H_2O, Et_2O; slightly soluble in MeOH, EtOH, propylene glycol; soluble in Me_2CO, dichloroethane, C_5H_5N. *Wyeth-Ayerst Labs.*

2625 Ftivazide
149-17-7
$C_{14}H_{13}N_3O_3$
Isonicotinic acid vanillylidenehydrazide.
A tuberculostatic antibiotic.

2626 Furonazide
3460-67-1 4330 222-411-6
$C_{12}H_{11}N_3O_2$
Isonicotinic acid α-methylfurfurylidene hydrazide.
INF; Furilazone; Clitizina; Menazone. Antibacterial (tuberculostatic). mp = 206°; λ_m 288, 276, 336 nm ($E_{1\,cm}^{1\%}$ 945, 588, 144 95% EtOH), 226, 273, 336 nm ($E_{1\,cm}^{1\%}$ 1147, 557, 133 0.1N NaOH); slightly soluble in H_2O, $CHCl_3$, Et_2O; soluble in Me_2CO, MeOH, DMF; less soluble in EtOH; LD_{50} (frat orl) = 2600 mg/kg, (mrat orl) = 2820 mg/kg.

2627 Glyconiazide
3691-74-5 4510 223-005-1
$C_{12}H_{13}N_3O_6$
D-Glucuronic acid σ-lactone 1-[(4-pyridinylcarboxyl)hydrazone].
Galatone; Gatalone; Glucazide; Gluconiazide; Gluronazide; Guidazide; Hydronsan; INH-G; Mycobactyl. Antibacterial (tuberculostatic). Dec 150-160°; freely soluble in H_2O, insoluble in cold EtOH, soluble in MeOH (1.2 g/100 ml at 66°). *University of California.*

2628 Isoniazid
54-85-3 5203 200-214-6
$C_6H_7N_3O$
Isonicotinic acid hydrazide.
Dinacrin; Ditubin; INH; Isolyn; Niconyl; Nydrazid; Rimifon; Tyvid; component of: Rifater. Antibacterial (tuberculostatic). mp = 171.4°; λ_m 266 nm ($E_{1\,cm}^{1\%}$ 378 H_2O), 265 nm ($E_{1\,cm}^{1\%}$ 420 0.1N HCl); soluble in H_2O (14 g/100 ml at 25°, 26 g/100 ml at 40°), EtOH (2 g/100 ml at 25°, 10 g/100 ml at 76°), $CHCl_3$ (0.1 g/100 ml); insoluble in Et_2O, C_6H_6; LD_{50} (mus ip) = 151 mg/kg, (mus iv) = 149 mg/kg. *Abbott Labs Inc.; Apothecon; Ciba-Geigy Corp.; Hoffmann-LaRoche Inc.; Marion Merrell Dow Inc.; Parke-Davis; Schering-Plough HealthCare Products; Sterling Winthrop Inc.*

2629 Isoniazid 4-Aminosalicylate
2066-89-9 5203 218-183-2
$C_{13}H_{14}N_4O_4$
Isonicotinic acid hydrazide 4-aminosalicylate.
pasiniazide; GEWO-399; Paraniazide; Dipasic. Antibacterial (tuberculostatic). mp = 142-144°; sparingly soluble in H_2O; λ_m = 272 303 nm ($E_{1\,cm}^{1\%}$ 550, 445). *Hoffmann-LaRoche Inc.*

2630 Isoniazid Methanesulfonate
13447-95-5 5203 236-605-3
$C_7H_9N_3O_4S$
Isonicotinic acid 2-(sulfomethyl)hydrazine.
Methaniazide. Antibacterial (tuberculostatic). Dec 187-189°. *Farmitalia Carlo Erba Ltd.*

2631 Isoniazid Methanesulfonate Calcium
6059-26-3 5203 227-987-2
$C_{14}H_{16}CaN_6O_8S_2$
Isonicotinic acid 2-(sulfomethyl)hydrazine calcium salt.
Neo-Tizide. Antibacterial (tuberculostatic). Dec 215-220°. *Farmitalia Carlo Erba Ltd.*

2632 Isoniazid Methanesulfonate Sodium
3804-89-5 5203 223-275-0
$C_7H_8N_3NaO_4S$
Isonicotinic acid 2-(sulfomethyl)hydrazine sodium salt.
Neoiscotin. Antibacterial (tuberculostatic). Dec 164-167°. *Farmitalia Carlo Erba Ltd.*

2633 Metazamide
14058-90-3
$C_{11}H_{12}N_2O_2$
1-(p-Methoxyphenyl)-5-methyl-4-imidazoline-2-one.
GPA-878. Tuberculostatic agent. Inhibits dehydrogenase activities.

2634 Metazide
1707-15-9
$C_{13}H_{14}N_6O_2$
Isonicotinic acid 2,2'-methylenedihydrazide.
Antibacterial (tuberculostatic).

2635 Morphazinamide
952-54-5 6355 213-460-4
$C_{10}H_{14}N_4O_2$
N-(Morpholinomethyl)pyrazinecarboxamide.
Morinamide; B-2310; Morfgazinamide. Antibacterial (tuberculostatic). mp = 118.5-119.5°; λ_m 269 317 nm (log ε 3.95, 2.77, EtOH); soluble in H_2O (33 g/100 ml), EtOH (3.3 g/100 ml), C_6H_6 (3.3 g/100 ml), $CHCl_3$ (40 g/100 ml). *Bracco Diagnostics Inc.*

2636 Morphazinamide Hydrochloride
1473-73-0 6355 216-013-1
$C_{10}H_{15}ClN_4O_2$
N-(Morpholinomethyl)pyrazinecarboxamide hydrochloride.
Morinamide hydrochloride; B-2311; Piazofolina; Piazolin. Antibacterial (tuberculostatic). mp = 196°; soluble in H_2O (50 g/100 ml), EtOH (0.29 g/100 ml), $CHCl_3$ (0.05 g/100 ml). *Bracco Diagnostics Inc.*

2637 Opiniazide
2779-55-7 6984
$C_{16}H_{15}N_3O_5$
5,6-Dimethoxyphthalaldehydic acid isonicotinoyl hydrazone.
saluside; saluzide; saliuzid; saluzid. Antibacterial (tuberculostatic). LD_{50} (gpg iv) = 1634 mg/kg.

2638 Phenyl Aminosalicylate
133-11-9 7426 205-092-8
$C_{13}H_{11}NO_3$
4-Amino-2-hydroxybenzoic acid phenyl ester.
fenamisal; Phenyl PAS; Pheny-PAS-Tebamin; Tebamin; Tebanyl. Antibacterial (tuberculostatic). mp = 153°, soluble in H_2O (0.012 g/100 ml), serum (0.0007 g/100 ml). *Rhône-Poulenc Rorer Pharm. Inc.*

2639 Protionamide
14222-60-7 8076 238-093-7
$C_9H_{12}N_2S$
2-Propylthioisonicotinamide.
TH-1321; RP-9778. Antibacterial (tuberculostatic). mp = 142°; soluble in EtOH, MeOH; slightly soluble in Et_2O, $CHCl_3$; insoluble in H_2O; LD_{50} (mus ip) = 1 mg/kg, (rat ip) = 1.32 mg/kg, (cat ip) > 1000 mg/kg. *Chimie et Atomistique.*

2640 Pyrazinamide
98-96-4 8140 202-717-6
$C_5H_5N_3O$
Pyrazinecarboxamide.
component of: Rifater. Antibacterial (tuberculostatic). mp = 189-191°; λ_m 269 nm ($E^{1\%}_{1\ cm}$ 660); soluble in H_2O (1.5 g/100 ml), MeOH (1.38 g/100 ml), absolute ethanol (0.57 g/100 ml), iPrOH (0.38 g/100 ml), Et_2O (0.1 g/100 ml), isooctane (0.001 g/100 ml), $CHCl_3$ (0.74 g/100 ml). *Marion Merrell Dow Inc.*

2641 Rifabutin
72559-06-9 8380
$C_{46}H_{62}N_4O_{11}$
(9S,12E,14S,15R,16S,17R,18R,19R,20S,21S,22E24Z06,-16,18,20-Tetrahydroxy-1'-isobutyl-14-methoxy-7,9,15,17,19,21,25-heptamethylspiro-[9,4-(epoxypentadeca[1,11,13]trienimino0-2H-furo[2',3':7,8]naphth[1,2-d]imidazole-2,4'-piperidine]-5,10,26(3H,9H)-trione 16-acetate.
Mycobutin; LM-427. Antibacterial (tuberculostatic). Very soluble in $CHCl_3$, soluble in MeOH, slightly soluble in EtOH, insoluble in H_2O; λ_m 493, 315, 274, 238 nm (MeOH). *Pharmacia & Upjohn.*

2642 Rifampin
13292-46-1 8382 236-312-0
$C_{43}H_{58}N_4O_{12}$
5,6,9,17,19,21-Hexahydroxy-23-methoxy-2,4,12,16,18,20,22-heptamethyl-8-[N-(4-methyl-1-piperazinyl)formimidoyl]-2,7-(epoxypentadeca[1,11,13]trienimino)naphtho[2,1-b]furan-1,11(2H)-dione 21-acetate.
refampicin; rifaldazine; rifamycin AMP; R/AMP; Abrifam; Eremfat; Rifa; Rifadin; Rifadine; Rifaprodin; Rifoldin; Rimactan; Rimactane; L-5103 Lepetit; Ba 41166/E; NSC-113926. Antibacterial (ansamycin). Tuberculostatic. Dec 183-188°; λ_m 237, 255,334, 475 nm (ε 33200, 32100, 27000, 15400 pH 7.38); freely soluble in CH_3Cl, DMSO; soluble in EtOAc, MeOH, THF; slightly soluble in H_2O, Me_2CO, CCl_4; LD_{50} (mus orl) = 885 mg/kg, (mus iv) = 260 mg/kg, (mus . *Ciba-Geigy Corp.; Marion Merrell Dow Inc.; Merrell Pharm. Inc.*

2643 Rifapentine
61379-65-5 8385 262-743-9
$C_{47}H_{64}N_4O_{12}$
3-[N-(4-Cyclopentyl-1-piperazinyl)formimidoyl]rifamycin.
MDL-473. Ansamycin antibacterial; tuberculostatic. mp = 179-180°; λ_m 475, 334 nm (ε 15200, 26700); LD_{50} (mus orl) > 2000 mg/kg, 3300 mg/kg, (mus ip) = 750 mg/kg, 710 mg/kg. *Marion Merrell Dow Inc.*

2644 Salinazid
495-84-1 8487
$C_{13}H_{11}N_3O_2$
1-Isonicotinoyl-2-salicylidinehydrazine.
Salizid. Antibacterial (tuberculostatic). mp = 232-233°, 251°; soluble in H_2O (0.005 g/100), EtOH (0.18 g/100 ml), propylene glycol (0.212 g/100 ml). *Parke-Davis.*

2645 Streptomycin
57-92-1 8983 200-355-3
$C_{21}H_{39}N_7O_{12}$
O-2-Deoxy-2-(methylamino)-α-L-glucopyranosyl-(1→2)-O-5-deoxy-3-C-formyl-α-L-lyxofuranosyl-(1→4)-N,N'-bis(aminoiminomethyl)-D-streptamine.
streptomycin A. Aminoglycoside antibiotic (tuberculostatic). [trihydrochloride ($C_{21}H_{42}Cl_3N_7O_{12}$)]: $[\alpha]_D^{26}$ = -84°; soluble in H_2O (> 2g/100 ml); MeOH (> 2 g/100 ml), EtOH (0.09 g/100 ml), iPrOH (0.012 g/100 ml), isoamyl alcohol (0.012 g/100 ml), petroleum ether (0.002 g/100 ml), CCl_4 (0.004 g/100 ml), Et_2O (0.001 g/100 ml). *Olin Res. Ctr.*

2646 Streptomycin Sesquisulfate
3810-74-0 8983 223-286-0
$C_{42}H_{84}N_{14}O_{36}S_3$
O-2-Deoxy-2-(methylamino)-α-L-glucopyranosyl-(1→2)-O-5-deoxy-3-C-formyl-α-L-lyxofuranosyl-(1→4)-N,N'-bis(aminoiminomethyl)-D-streptamine sulfate (2:3).
streptomycin sulfate; AgriStrep; Streptobrettin; Strycin; Vetstrep; component of: Intromycin. Aminoglycoside antibiotic (tuberculostatic). Soluble in H_2O (> 2 g/100 ml), MeOH (0.085 g/100 ml), EtOH (0.030 g/100 ml), iPrOH (0.001 g/100 ml), petroleum ether (0.0015 g/100 ml), CCl_4 (0.0035 g/100 ml), Et_2O (0.0035 g/100 ml). *Bristol-Myers Squibb Pharm. R&D; Enzon Inc.; Roerig Div. Pfizer Pharm.*

2647 Streptonicozid
5667-71-0 8985 227-128-1
$C_{54}H_{94}N_{20}O_{36}S_3$
4-Pyridinecarboxylic acid hydrazide hydrazone with O-2-deoxy-2-(methylamino)-α-L-glucopyranosyl-(1→2)-O-5-deoxy-3-C-formyl-α-L-lyxofuranosyl-(1→4)-N,N'-bis(aminoiminomethyl)-D-streptamine sulfate (2:3) (salt).
streptomyclidine isonicotinyl hydrazine sulfate; stretoniazide; Strazide; Streptohydrazid. Aminoglycoside antibiotic with tuberculostatic properties. Dec 230°; soluble in H_2O (> 2 g/100 ml), EtOH (0.0115 g/100 ml), CCl_4 (0.0025 g/100 ml), Et_2O (0.031 g/100 ml). *Pfizer Intl.*

2648 Subathizone
121-55-1 9030 204-480-4
$C_{10}H_{13}N_3O_2S_2$
p-Ethylsulfonylbenzaldehyde thiosemicarbazone.
Antibacterial (tuberculostatic). mp = 234-235° (dec). *Schenley.*

2649 Sulfoniazide
3691-81-4 9132
$C_{13}H_{11}N_3O_4S$
4-Pyridinecarboxylic acid [(3-sulfophenyl)methylene]hydrazide.
G-605. Antibacterial (tuberculostatic). Dec 250-253°; slightly soluble in H_2O; [sodium salt trihydrate ($C_{13}H_{10}N_3NaO_4S.3H_2O$; Sulfon-Niazone)]: soluble in H_2O (3 g/100 ml). *Tanake.*

2650 Terizidone
25683-71-0
$C_{14}H_{14}N_4O_4$
4,4'-[p-Phenylenebis(methyleneamino)]-di-(isoxazolidin-3-one).
terivalidin. A Schiff base of D-cycloserine. Antibacterial (tuberculostatic).

2651 Thiacetazone
104-06-3 9427 203-170-6
$C_{10}H_{12}N_4OS$
4'-Formylacetanilide thiosemicarbazone.
Amithiozone; thibone; Tb I-698; Berculon A; Conteben; Livazone; Mirizone Neustab; Panrone; Seroden; Tebethion; Thiocarbazil; Thioparamizone; Tibione; Tiobicina. Antibacterial (tuberculostatic). Dec 225-230°; λ_m 328 nm (EtOH); soluble in hot EtOH; sparingly soluble in cold EtOH; insoluble in H_2O, Me_2CO, C_6H_6, CCl_4, $CHCl_3$, CS_2, petroleum ether, other organic solvents; LD_{50} (mus sc) = 1000-2000 mg/kg.

2652 Tiocarlide
910-86-1 9593 213-006-5
$C_{23}H_{32}N_2O_2S$
4,4'-Bis(isopentyloxy)thiocarbanilide.
DATC; thiocarlide; DATC; Datanil; Disocarban; Isoxyl. Antibacterial (tuberculostatic). mp = 134-135°. *CIBA plc.*

2653 Tuberactinomycin
11075-36-8 9935
Peptide antibiotic (tuberculostatic). Polypeptide antibiotic obtained from *Streptomyces griseoverticillatus var. tuberacticus*. Composed of tuberactinomycins A, B, N and O. Tuberactinomycin B is identical to viomycin and tuberactinomycin N to enviomycin. [hydrochloride]: mp = 244-264° (dec); $[\alpha]_D^{25}$ = -31.5° (c = 1 H_2O); λ_m = 268 nm ($E^{1\%}_{1\,cm}$ 330 H_2O), 268.5 nm ($E^{1\%}_{1\,cm}$ 313 1N HCl), 285 nm ($E^{1\%}_{1\,cm}$ 206.5 0.1N NaOH); soluble in H_2O, slight.

2654 Tuberactinomycin A
33103-21-8 9935
$C_{25}H_{43}N_{13}O_{11}$
1-[(3R,4R)-4-Hydroxy-L-3,6-diaminohexanoic acid]-viomycin.
Peptide antibiotic. Tuberculostatic.

2655 Tuberactinomycin O
33137-73-4 9935
$C_{25}H_{43}N_{13}O_9$
(R)-6-[L-2-(2-Amino-1,4,5,6-tetrahydro-4-pyrimidinyl)glycine]viomycin.
Peptide antibiotic. Tuberculostatic.

2656 Tubercidin
69-33-0 9936 200-703-4
$C_{11}H_{14}N_4O_4$
7-β-D-Ribofuranosyl-7H-pyrrolo[2,3-d]pyrimidin-4-amine.
7-deazaadenosine; sparsamycin A; U-10071. Antibacterial (tuberculostatic); antifungal; antineoplastic. Antibiotic produced in *Streptomyces tubericidus*. Dec 247-248°; $[\alpha]_D^{17}$ = -67° (50% AcOH); λ_m 270 nm (ε 12,100 0.01N NaOH); soluble in H_2O (0.3 g/100 ml), MeOH (0.5 g/100 ml), EtOH (0.05 g/100 ml); insoluble in Me_2CO, EtOAc, $CHCl_3$, C_6H_6, petroleum ether; LD_{50} (mus.iv) = 45 mg/kg.

2657 Tuberin
53643-53-1 9938
$C_{10}H_{11}NO_2$
N-[2-(4-Methoxyphenyl)ethenyl]formamide.
N-formyl-trans-p-methoxystyrylamine. Antibacterial (tuberculostatic). mp = 132-133°; λ_m 219, 285 nm ($E^{1\%}_{1\,cm}$ 870, 1710, MeOH); soluble in MeOH, EtOH, EtOAc, Me_2CO; moderately soluble in CCl_4, $CHCl_3$; sparingly soluble in H_2O, C_6H_6; insoluble in petroleum ether. *Inst. Phys. & Chem. Res.*

2658 Verazide
93-47-0 10091
$C_{15}H_{15}N_3O_3$
1-Isonicotinoyl-2-veratrylidenehydrazine.

3,4-dimethoxybenzal isonicotinoyl-hydrazone. Antibacterial (tuberculostatic). Polypeptide antibiotic from various *Streptomyces* species. mp = 189-190°.

2659 Viomycin
32988-50-4 10139 251-323-0
$C_{25}H_{43}N_{13}O_{10}$
Celiomycin; florimycin; tuberactinomycin B. Peptide antibiotic. Tuberculostatic. [hydrochloride]: mp = 270° (dec); $[\alpha]_D^{18}$ = -16.6° (c = 1 H_2O); λ_m 268 nm (log ε 4.5 H_2O), 268 nm (log ε 4.4 0.1N HCl), 285 nm (log ε 4.3 0.1N NaOH). *CIBA plc.*

2660 Viomycin Pantothenate Sulfate
1401-79-2 10139
$C_{34}H_{62}N_{14}O_{19}$
Vionactan; Viothenat. Peptide antibiotic (tuberculostatic). mp = 242° (dec). *Ciba-Geigy Corp.; Grunenthal.*

2661 Viomycin Sulfate
37883-00-4 10139
$C_{25}H_{45}N_{13}O_{14}S$
Viocin. Peptide antibiotic (tuberculostatic). mp = 266° (dec); $[\alpha]_D^{18}$ = -29.5° (c = 1 H_2O); λ_m 268 nm (log ε 4.4 H_2O or 0.1N HCl or 0.1N NaOH); LD_{50} (mus iv) = 240 mg/kg, (mus sc) = 1750 mg/kg; soluble in H_2O, insoluble in most organic solvents. *CIBA plc.*

Antibacterial Adjuncts

2662 Betamipron
3440-28-6 1227
$C_{10}H_{11}NO_3$
N-Benzoyl-β-alanine.
CS-443. Antibacterial. A β-lactamase inhibitor. mp = 120°, 133°; readily soluble in hot H_2O, $CHCl_3$; soluble in EtOH, Et_2O, Me_2CO; LD_{50} (rat iv) > 3000 mg/kg. *Sankyo Co. Ltd.*

2663 Brobactam
26631-90-3 247-856-3
$C_8H_{10}BrNO_3S$
(2S,5R,6R)-6-Bromo-3,3-dimethyl-7-oxo-4-thia-1-azabicyclo[3.2.0]heptane-2-carboxylic acid.
A β-lactamase inhibitor, used as an adjunct to penicillin therapy.

2664 Cilastatin
82009-34-5 2331
$C_{16}H_{26}N_2O_5S$
[R-[(R*,S*(Z)]-7-[(2-Amino-2-carboxyethyl)thio]-2-[[(2,2-dimethylcyclopropyl)carbonyl]amino]-2-heptenoic acid.
MK-791. Used in combination with Imipenem, an antibacterial. A dipeptidase I inhibitor which prevents renal metabolism of penem and carbapenem antibiotics. *Merck & Co.Inc.*

2665 Cilastatin Sodium salt
81129-83-1 2331
$C_{16}H_{25}N_2NaO_5S$
[R-[(R*,S*(Z)]-7-[(2-Amino-2-carboxyethyl)thio]-2-[[(2,2-dimethylcyclopropyl)carbonyl]amino]-2-heptenoic acid acid sodium salt (1:1).
Cilastatin Sodium. Used in combination with Imipenem, an antibacterial. A dipeptidase I inhibitor which prevents renal metabolism of penem and carbapenem antibiotics. Very soluble in H_2O, MeOH. *Merck & Co.Inc.*

2666 Clavulanic acid
58001-44-8 261-069-2
$C_8H_9NO_5$
3-(2-Hydroxyethylidene)-7-oxo-4-oxa-1-azabicyclo[3.2.0]heptane-2-carboxylic acid.
MM-14151. Antibacterial. A β-lactamase inhibitor. *SmithKline Beecham Pharm.*

2667 Clavulanic Acid Potassium Salt combined with Ticarcillin Disodium
116876-37-0
$C_{23}H_{22}KN_3Na_2O_{11}S_2$
Potassium 3-(2-hydroxyethylidene)-7-oxo-4-oxa-1-azabicyclo[3.2.0]heptane-2-carboxylate combined with ticarcillin sodium.
Betabactil; Timentin. Antibacterial. A β-lactamase inhibitor. *SmithKline Beecham Pharm.*

2668 Sulbactam
68373-14-8 9058 269-878-2
$C_8H_{11}NO_5S$
(2S-cis)-3,3-Dimethyl-7-oxo-4-thia-1-azabicyclo[3.2.0]heptane-2-carboxylic acid.
CP-45899-2; Betamaze; penicillanic acid sulfone; penicillanic acid 1,1-dioxide. Antibacterial. A β-lactamase inhibitor. mp = 148-151°; $[\alpha]_D^{20}$ = 251° (c = 0.01 in pH 5 buffer). *Pfizer Inc.; Roerig Div. Pfizer Pharm.*

2669 Sulbactam Benzathine
83031-43-0 9058
$C_{32}H_{42}N_4O_{10}S_2$
(2S-cis)-3,3-Dimethyl-7-oxo-4-thia-1-azabicyclo-[3.2.0]heptane-2-carboxylic acid compound with N,N'-dibenzylethylenediamine.
CP-45899-99. Antibacterial. A β-lactamase inhibitor. *Pfizer Inc.*

2670 Sulbactam Pivoxil
69388-79-0 9058
$C_{32}H_{42}N_4O_{10}S_2$
(2S-cis)-3,3-Dimethyl-7-oxo-4-thia-1-azabicyclo-[3.2.0]heptane-2-carboxylic acid 4,4-dioxide (2,2-dimethyl-1-oxopropoxy)methyl ester.
penicillanic acid sulfone; penicillanic acid 1,1-dioxide; CP-45899; Unasyn® Oral. Antibacterial. A β-lactamase inhibitor. *Pfizer Intl.*

2671 Sulbactam Sodium
69388-84-7 9058 273-984-4
$C_8H_{10}NNaO_5S$
(2S-cis)-3,3-Dimethyl-7-oxo-4-thia-1-azabicyclo[3.2.0]heptane-2-carboxylic acid sodium salt.
CP-45899-2; Sulperazone [with cefoperazone sodium]; Bethacil [with ampicillin sodium]; Loricin [with ampicillin sodium]; Unacid [with ampicillin sodium]; Unacim [with ampicillin sodium]; component of: Unasyn. Antibacterial. A β-lactamase inhibitor. *Roerig Div. Pfizer Pharm.*

2672 Sultamicillin
76497-13-7 9166
$C_{25}H_{30}N_4O_9S_2$
[2S-[2α(2R*,5S*),5α,6β(S*)]]-6-[(Aminophenylacetyl)-amino]-3,3-dimethyl-7-oxo-4-thia-1-azabicyclo[3.2.0]heptane-2-carboxylic acid [[(3,3-dimethyl-7-oxo-4-thia-1-azabicyclo-[3.2.0]hept-2-yl)carbonyl]oxy]methyl ester S,S-dioxide.
CP-49952; VD-1827; sultamicillin. Antibacterial. A β-lactamase inhibitor. Administered orally. A double ester of sulbactam and ampicillin. *Pfizer Inc.*

2673 Sultamicillin Tosylate
83105-70-8 9166
$C_{32}H_{38}N_4O_{12}S_3$
[2S-[2α(2R*,5S*),5α,6β(S*)]]-6-[(Aminophenylacetyl)-amino]-3,3-dimethyl-7-oxo-4-thia-1-azabicyclo-[3.2.0]heptane-2-carboxylic acid [[(3,3-dimethyl-7-oxo-4-thia-1-azabicyclo[3.2.0]hept-2-yl)carbonyl]oxy]methyl ester S,S-dioxide tosylate.
Bacimex; Bethacil orale; Unacid PD oral; Unacim orale; Unasyn. Antibacterial. A β-lactamase inhibitor. *Pfizer Inc.*

2674 Tazobactam
89786-04-9 9251
$C_{10}H_{12}N_4O_5S$
[2S-(2α,3β,5β)]-3-Methyl-7-oxo-3-(1H-1,2,3-triazol-1-ylmethyl)-4-thia-1-azabicyclo[3.2.0]heptane-2-carboxylic acid 4,4 dioxide.
YTR-830H; CL-298741. Antibacterial. A β-lactamase inhibitor. *Taiho.*

2675 Tazobactam Sodium
89785-84-2 9251
$C_{10}H_{11}N_4NaO_5S$
[2S-(2α,3β,5β)]-3-Methyl-7-oxo-3-(1H-1,2,3-triazol-1-ylmethyl)-4-thia-1-azabicyclo[3.2.0]heptane-2-carboxylic acid 4,4 dioxide sodium salt.
YTR-830; CL-307579; component of: Zosyn. Antibacterial. A β-lactamase inhibitor. mp > 170° (dec). *Lederle Labs.; Taiho.*

Anticholinergics

2676 Adiphenine
64-95-9 160 200-599-0
$C_{20}H_{25}NO_2$
2-(Diethylamino)ethyl diphenylacetate.
Trasentine; Diphacil; Difacil. Smooth muscle relaxant. Anticholinergic. *CIBA plc.*

2677 Adiphenine Hydrochloride
50-42-0 160 200-036-9
$C_{20}H_{26}ClNO_2$
2-(Diethylamino)ethyl diphenylacetate hydrochloride.
NSC-129224; Trasentine hydrochloride; Difacil hydrochloride; Diphacil hydrochloride; Patrovina; Spasmolytin. Anticholinergic. Smooth muscle relaxant. mp = 113-114°; freely soluble in H_2O; poorly soluble in EtOH, Et_2O. *Ciba-Geigy Corp.*

2678 Adiphenine Methobromide
6113-04-8 160 228-077-8
$C_{21}H_{28}BrNO_2$
2-(Diethylamino)ethyl diphenylacetate methobromide.
adiphenine methyl bromide; Lunal. Smooth muscle relaxant, anticholinergic. LD_{50} (rbt iv) = 22.5-27.5 mg/kg. *CIBA plc.*

2679 Alverine
150-59-4 386 205-763-5
$C_{20}H_{27}N$
N-Ethyl-3,3'-diphenyldipropylamine.
Anticholinergic. $bp_{0.3}$ = 165-168°, bp_{13} = 210-215°.

2680 Alverine Citrate
5560-59-8 386 226-929-3
$C_{26}H_{35}NO_7$
N-Ethyl-3,3'-diphenyldipropylamine citrate.
NF XIII; Antispasmin; Calmabel; Gamatran; Profenine; Prophelan; Proverine; Spacolin; Spasmaverine; Spasmonal. Anticholinergic. Freely soluble in H_2O.

2681 Ambutonium
14007-49-9 406
$C_{20}H_{27}N_2O^+$
(3-Carbamoyl-3,3-diphenylpropyl)ethyldimethyl-ammonium.
Anticholinergic.

2682 Ambutonium Bromide
115-51-5 406 204-093-0
$C_{20}H_{27}BrN_2O$
(3-Carbamoyl-3,3-diphenylpropyl)ethyldimethyl-ammonium bromide.
R-100. Component of: Praxiten SP [with oxazepam]. Anticholinergic. mp = 228-229°.

2683 Aminopentamide
60-46-8 479 200-479-8
$C_{19}H_{24}N_2O$
α-[2-(Dimethylamino)propyl]-α-phenyl-benzeneacetamide.
BL-139; Centrine; Valeramide-OM. Anticholinergic. Also has antiemetic and anticonvulsant properties. [dl-form]: mp = 183-184°; insoluble in H_2O; [d-form]: mp = 136.5-137.5°; $[\alpha]_D^{23}$ = 98.9° (MeOH); [l-form]: mp = 136.5-137.5°; $[\alpha]_D^{23}$ = -101.9° (MeOH). *Bristol-Myers Squibb Co.*

2684 Aminopentamide Sulfate
60-46-8 479
$C_{19}H_{26}N_2O_5S$
α-[2-(Dimethylamino)propyl]-α-phenylbenzene-acetamide sulfate.
NND-1962. Anticholinergic. Also has antiemetic and anticonvulsant properties. [dl-form]: mp = 190-191° (dec); soluble in H_2O, EtOH; LD_{50} (mus iv) = 34.7 mg/kg, (mus orl) = 396 mg/kg; [dl-form acid sulfate]: mp = 185-187°; [medicinal grade]: mp = 178-181°; λ_m = 258.5 nm ($A_{1\ cm}^{1\%}$ 10.3 1% H_2SO_4); freely soluble in H_2O, EtOH; slightly soluble in $CHCl_3$; insoluble in Et_2O. *Bristol-Myers Squibb Co.*

2685 Amixetrine
24622-72-8 514
$C_{17}H_{27}NO$
1-[β-(Isopentyloxy)phenethyl]pyrrolidine.
Anti-inflammatory; anticholinergic. bp_2 = 121°; $n_D^{21.6}$ = 1.4978. *C.E.R.M.*

2686 Amixitrene Hydrochloride
24622-52-4 514 246-365-1
$C_{17}H_{28}ClNO$
1-[β-(Isopentyloxy)phenethyl]pyrrolidine monohydrochloride.
Somagest. Anti-inflammatory; anticholinergic. mp = 150°. *C.E.R.M.*

2687 Amprotropine
148-32-3 632
$C_{18}H_{29}NO_3$
3-(Diethylamino)-2,2-dimethylpropyl ester.
Anticholinergic.

2688 Amprotropine Phosphate
134-53-2 632
$C_{18}H_{32}NO_7P$
3-(Diethylamino)-2,2-dimethylpropyl ester phosphate (salt).
Syntropan. Anticholinergic. mp = 142-145°; soluble in H_2O; slightly soluble in EtOH; insoluble in $CHCl_3$, Et_2O.

2689 Anisotropine Methylbromide
80-50-2 709 201-285-6
$C_{17}H_{32}BrNO_2$
3α-Hydroxy-8-methyl-1αH,5αH-tropanium bromide 2-propylvalerate.
Valpin 50; Lytispasm; octatropine metylbromide. Anticholinergic. mp = 329°. *DuPont-Merck Pharm.; Endo Pharm. Inc.*

2690 Apoatropine
500-55-0 780 207-906-7
$C_{17}H_{21}NO_2$
1αH,5αH-Tropan-3α-ol atropate.
atropamine; atropyltropeine. Antispasmodic. Anticholinergic. Occurs in the root of *Atropa belladonna L. Solanaceae.* Can also be synthesized from atropine. CAUTION: can cause respiratory arrest. mp = 62°; insoluble in H_2O; soluble in EtOH, Et_2O, $CHCl_3$, C_6H_6, CS_2; slightly soluble in petroleum ether, isoamyl alcohol; LD_{50} (mus orl) = 160 mg/kg, (mus ip) = 14.1 mg/kg; [hydrochloride ($C_{17}H_{22}ClNO_2$)]: mp = 239°; soluble in hot H_2O; sparingly soluble in EtOH, Me_2CO; insoluble in Et_2O; [sulfate pentahydrate ($C_{34}H_{44}N_2O_8.5H_2O$)]: sparingly soluble in H_2O.

2691 Apoatropine Hydrochloride
5978-81-4 780 227-779-1
$C_{17}H_{22}ClNO_2$
1αH,5αH-Tropan-3α-ol atropate hydrochloride.
Anticholinergic, used as an antispasmodic. mp = 239°; soluble in hot H_2O, EtOH, Me_2CO; almost insoluble in Et_2O.

2692 Apoatropine Sulfate Pentahydrate
780
$C_{34}H_{44}N_2O_8.5H_2O$
1αH,5αH-Tropan-3α-ol atropate sulfate (2:1) (salt) pentahydrate.
Anticholinergic, used as an antispasmodic. Sparingly soluble in H_2O.

2693 Aprofene
3563-01-7
$C_{21}H_{27}NO_2$
2-Diethylaminoethyl 2,2-diphenylpropionate.
An irreversible antagonist for muscarinic receptors. Anticholinergic.

2694 Atromepine
428-07-9 5491
$C_{18}H_{25}NO_3$
(-)-3α-Tropanyl 2-methyl-2-phenylhydracrylate.
Levomepate; Dispan. Anticholinergic, used as an antispasmodic. [hydrochloride]: mp = 210-212°; $[\alpha]_D^{20}$ = -6.8°. *Gruppo Lepetit S.p.A.*

2695 Atropine
51-55-8 907 200-104-8
$C_{17}H_{23}NO_3$
1αH,5αH-Tropan-3α-ol (±)-tropate (ester).
tropine tropate; dl-tropyl tropate; atropine hyperduric [as mucate]. Anticholinergic. Parasympatholytic alkaloid from *Atropa belladona.* mp = 114-116°; soluble in H_2O (0.22 g/100 ml at 25°, 11.1 g/100 ml at 80°), EtOH (50 g/100 ml at 25°, 83.3 g/100 ml at 60°), glycerol (3.7 g/100 ml), Et_2O (4 g/100 ml), $CHCl_3$ (100 g/100 ml), C_6H_6; LD_{50} (rat orl) = 750 mg/kg.

2696 Atropine Oxide
4438-22-6 908
$C_{17}H_{23}NO_4$
1αH,5αH-Tropan-3α-ol (±)-tropate (ester) 8-oxide.
atropine-N-oxide; atropine aminoxide; genatropine; aminoxytropine tropate. Anticholinergic. mp = 127-128°, dec 135°; soluble in EtOH, $CHCl_3$; insoluble in Et_2O.

2697 Atropine Oxide Hydrochloride
4574-60-1 908 224-959-1
$C_{17}H_{24}ClNO_4$
1αH,5αH-Tropan-3α-ol (±)-tropate (ester) 8-oxide hydrochloride.
Tropinox, Xtro. Anticholinergic. mp = 192-193°.

2698 Atropine Sulfate [anhydrous]
55-48-1 908 200-235-0
$C_{34}H_{48}N_2O_{10}S$
1αH,5αH-Tropan-3α-ol (±)-tropate (ester) sulfate (2:1) (salt).
Atropisol; Atropisol® Ophthalmic Solution. Anticholinergic; mydriatic. Used in pre-anesthetic medication. Veterinary use is as an antispasmodic and as an antidote for organophosphorus insecticide poisoning. mp = 190-194°; soluble in H_2O (2.5 g/ml), less soluble in organic solvents; LD_{50} (rat orl) = 622 mg/kg.

2699 Atropine Sulfate Monohydrate
5908-99-6 908
$C_{34}H_{48}N_2O_{10}S.H_2O$
1αH,5αH-Tropan-3α-ol (±)-tropate (ester) sulfate (2:1) (salt) monohydrate.
Anticholinergic.

2700 Benactyzine
302-40-9 1055 206-123-8
$C_{20}H_{25}NO_3$
Diethyl(2-hydroxyethyl)amine benzilate.
Anticholinergic; antidepressant. mp = 51°. *Am. Cyanamid.*

2701 Benactyzine Hydrochloride
57-37-4 1055 200-324-4
$C_{20}H_{26}ClNO_3$
2-Diethylamino benzilate hydrochloride.
AY-5406-1; Actozine; Amizil; Arcadine; Cafron; Cedad; Cevanol; Fobex; Ibiotyzil; Lucidil; Nervacton; Neuroleptone; Nutinal; Parasan; Parpon; Suavitil; Tranquillin. Anticholinergic with antidepressant properties. mp = 177-178°; soluble in H_2O (14.9 g/100 ml at 25°), insoluble in Et_2O. *Am. Cyanamid.*

2702 Benapryzine
22487-42-9 1057
$C_{21}H_{27}NO_3$
2-(Ethylpropylamino)ethyl benzilate.
Anticholinergic. *Beecham Res. Labs. UK.*

2703 Benapryzine Hydrochloride
3202-55-9 1057
$C_{21}H_{28}ClNO_3$
2-(Ethylpropylamino)ethyl benzilate hydrochloride.
AP-1288; BRL-1288. Anticholinergic. mp = 164-166°; LD_{50} (mus sc) = 400 mg/kg, (mus orl) = 500 mg/kg. *Beecham Res. Labs. UK.*

2704 Benzetimide
119391-55-8 1104
$C_{23}H_{26}N_2O_2$
2-(1-Benzyl-4-piperidyl)-2-phenylglutarimide.
Anticholinergic. mp = 156-159°; [l-form]: mp = 180.5-182°; $[\alpha]_D^{20}$ = -124° ($CHCl_3$); LD_{50} (mus iv) = 38.5 mg/kg; [d-form]: dexetimide. *Janssen Pharm. Inc.*

2705 Benzetimide Hydrochloride
5633-14-7 1104 227-072-8
$C_{23}H_{27}ClN_2O_2$
2-(1-Benzyl-4-piperidyl)-2-phenylglutarimide hydrochloride.
Dioxatrine; R-4929; Spasmentral. Anticholinergic. Used in veterinary medicine as an antidiarrheal. mp = 299-301.5°; LD_{50} (rat orl) = 37.6 mg/kg, (rat iv) = 46.0 mg/kg, (mus orl) = 37.6 mg/kg, (mus iv) = 46.0 mg/kg. *Janssen Pharm. Inc.*

2706 Benzilonium Bromide
1050-48-2 1110 213-885-5
$C_{22}H_{28}BrNO_3$
1,1-Diethyl-3-[(hydroxydiphenylacetyl)oxy]pyrrolidinium bromide.
CI-379; PU-239; Minelsin; Portyn; Ulcoban; Minelcin; Minelco; Ortyn; Ortyn retard; Partyn; Pirbenina; Pyrbenine. Anticholinergic. mp = 203-204°; LD_{50} (rat orl) = 1.86 g/kg. *Parke-Davis.*

2707 Benztropine Mesylate
132-17-2 1156 205-048-8
$C_{22}H_{29}NO_4S$
3α-(Diphenylmethoxy)-1αH,5αH-tropane methanesulfonate.
Cogentin; Cogentinol; Cobrentin methanesulfonate. Anticholinergic. Antiparkinsonian. mp = 143°; λ_m = 259 nm (E_M 437); soluble in H_2O. *Apothecon; Merck & Co.Inc.*

2708 Benzydamine
642-72-8 1157 211-388-8
$C_{19}H_{23}N_3O$
1-Benzyl-3-[3-(dimethylamino)propoxy]-1H-indazole.
benzindamine. Analgesic; antipyretic; anti-inflammatory. Anticholinergic. $bp_{0.05}$ = 160°. *3M Pharm.; Angelini Pharm. Inc.*

2709 Benzydamine Hydrochloride
132-69-4 1157 205-076-0
$C_{19}H_{24}ClN_3O$
1-Benzyl-3-[3-(dimethylamino)propoxy]-1H-indazole monohydrochloride.
Difflam Oral Rinse; Difflam Cream; AP-1288; BRL-1288; AF-864; Afloben; Andolex; Benalgin; Benzyrin; Difflam; Dorinamin; Enzamin; Imotryl; Ririlim; Riripen; Salyzoron; Saniflor; Tamas; Tantum; Verax. Anticholinergic. Has analgesic, anti-inflammatory, and antipyretic properties. mp = 160°; λ_m = 306 nm ($E_{1\,cm}^{1\%}$ 160); soluble in H_2O, polar organic solvents; LD_{50} (rat orl) = 1050 mg/kg, (rat ip) = 100 mg/kg, (mus orl) = 515 mg/kg, (mus ip) = 110 mg/kg. *Angelini Pharm. Inc.*

2710 Bevonium
33371-53-8 1239
$[C_{22}H_{28}NO_3]^+$
2-(Hydroxymethyl)-1,1-dimethylpiperidinium benzilate.
Anticholinergic with antispasmodic and bronchodilating properties. Quaternary ammonium compound. *Grunenthal.*

2711 Bevonium Methyl Sulfate
5205-82-3 1239 226-001-8
$C_{23}H_{31}NO_7S$
2-(Hydroxymethyl)-1,1-dimethylpiperidinium methyl sulfate benzilate.
bevonium metilsulfate; CG-201; Acabel. Anticholinergic with antispasmodic and bronchodilating properties. mp = 134-135°. *Grunenthal.*

2712 Biperiden
514-65-8 1274 208-184-6
$C_{21}H_{29}NO$
α-5-Norbornen-2-yl-α-phenyl-1-piperidinepropanol.
KL-373; Biperiden; 3-piperidino-1-phenyl-1-bicycloheptenyl-1-propanol; Akineton. Anticholinergic. Antiparkinsonian. mp = 112-116°, 101°; slightly soluble in H_2O, EtOH; readily soluble in MeOH. *Knoll Pharm. Co.*

2713 Biperiden Hydrochloride
1235-82-1 1274 214-976-2
$C_{21}H_{30}ClNO$
α-5-Norbornen-2-yl-α-phenyl-1-piperidinepropanol hydrochloride.
Akineton hydrochloride; Akineton; Akinophyl. Anticholinergic. Antiparkinsonian. mp = 275° (dec), 238°; LD_{50} (mus orl) = 545 mg/kg, (mus iv) = 56 mg/kg. *Knoll Pharm. Co.*

2714 Biperiden Lactate
7085-45-2 1274 230-388-9
$C_{24}H_{35}NO_4$
α-5-Norbornen-2-yl-α-phenyl-1-piperidine-propanol lactate.
Akineton lactate. Anticholinergic. *Knoll Pharm. Co.*

2715 Bornaprine
20448-86-6
$C_{21}H_{31}NO_2$
3-(Diethylamino)propyl 2-phenyl-2-norbornane-carboxylate.
Anticholinergic.

2716 Botulin Toxin A
93384-43-1 1384 297-253-4
Botulin A; botulinium toxin type A; oculinum; AGN-191622; BoTox. A potent bacterial neurotoxin produced by *Clostridium botulinum*. Prevents release of acetylcholine. Death results from peripheral respiratory paralysis. Botulin A used in treatment of blepharospasm and achlasia. MLD (mus) = 0.0003 μg/kg.

2717 Butropium Bromide
29025-14-7 1569 249-375-4
$C_{28}H_{38}BrNO_4$
8-(p-Butoxybenzyl)-3α-hydroxy-1αH,5αH-tropanium bromide (-)-tropate.
Coliopan; BHB. Anticholinergic; antispasmodic. mp = 166-168°, 158-160°; $[\alpha]_D^{20}$ = -21.7° (c = 0.5 H_2O); soluble in AcOH, $CHCl_3$, DMF; less in EtOH, H_2O; insoluble in Me_2CO, Et_2O, C_6H_6; LD_{50} (mmus orl) = 1500 mg/kg, (mmus sc) = 660 mg/kg, (mmus iv) = 12.0 mg/kg. *Eisai Co. Ltd.*

2718 Butynamine
3735-65-7
$C_{10}H_{19}N$
N-tert-Butyl-N,1,1-trimethyl-2-propynylamine.
A muscarinic cholinergic receptor antagonist.

2719 Buzepide Metiodide
15351-05-0 1635 239-383-6
$C_{23}H_{31}IN_2O$
1-(3-Carbamoyl-3,3-diphenylpropyl)perhydro-1-methylazepinium iodide.
F.I. 6146; metazepium iodide; diphexamide methiodide. Anticholinergic. mp = 212-213° (dec); [buzepide]: mp = 141.5-143.5°. *N.V.Nederlandsche Comb. Chem. Ind.*

2720 Camylofine
54-30-8 1784 200-202-0
$C_{19}H_{32}N_2O_2$
N-[2-(Diethylamino)ethyl]-2-phenylglycine isopentyl ester.
acamylophenine; Adopon; Avadyl; Belosin; Navadyl; Novospasmin; Sintespasmil; Spasmocan. Anticholinergic. $bp_{1.5}$ = 174-178°, bp_4 = 165-180°. *Astra Sweden.*

2721 Camylofine Dihydrochloride
5892-41-1 1784 227-571-0
$C_{19}H_{34}Cl_2N_2O_2$
N-[2-(Diethylamino)ethyl]-2-phenylglycine isopentyl ester dihydrochloride.
Avacan. Anticholinergic. mp = 174-178°, 172°, 173°; soluble in H_2O; LD_{50} (mus orl) = 760 mg/kg, (mus sc) = 1350 mg/kg, (mus iv) = 49.2 mg/kg. *Astra Sweden.*

2722 Caramiphen
77-22-5 1820 201-013-6
$C_{18}H_{27}NO_2$
2-Diethylaminoethyl 1-phenylcyclopentane-1-carboxylate .
Anticholinergic. $bp_{0.07}$ = 112-115°. *Ciba-Geigy Corp.*

2723 Caramiphen Ethanedisulfonate
125-86-0 1820 204-759-0
$C_{38}H_{60}N_2O_{10}S_2$
2-Diethylaminoethyl 1-phenylcyclopentane-1-carboxylate ethanedisulfonate.
Alcopon; Taoryl; Toryn. Anticholinergic. mp = 115-116°; soluble in H_2O. *Ciba-Geigy Corp.*

2724 Caramiphen Hydrochloride
125-85-9 1820 204-758-5
$C_{18}H_{28}ClNO_2$
2-Diethylaminoethyl 1-phenylcyclopentane-1-carboxylate hydrochloride.
Panparnit; Parpanit. Anticholinergic; antitussive. mp = 145-146°; soluble in EtOH, slightly soluble in H_2O; LD_{50} (rat ip) = 209 mg/kg. *Ciba-Geigy Corp.*

2725 Chlorbenzoxamine
522-18-9 2125 208-323-0
$C_{27}H_{31}ClN_2O$
1-[2-(o-Chloro-α-phenylbenzyloxy)ethyl]-4-o-methylbenzylpiperazine.
Anticholinergic. $bp_{0.01}$= 234-236°, $bp_{0.005}$ = 235°.

2726 Chlorbenzoxamine Dihydrochloride
5576-62-5 2125 226-951-3
$C_{27}H_{33}Cl_3N_2O$
1-[2-(o-Chloro-α-phenylbenzyloxy)ethyl]-4-o-methylbenzylpiperazine dihydrochloride.
Libratar; Antiulcera Master; Gastomax. Anticholinergic. mp = 197-200°; soluble in MeOH, EtOH, $CHCl_3$; sparingly soluble in H_2O, Me_2CO; insoluble in CH_3CN, C_6H_6, Et_2O; LD_{50} (rat sc) = 4000 mg/kg, (rat iv) = 66 mg/kg, (mus orl) = 1400 mg/kg.

2727 Chlorphenoxamine
77-38-3 2234
$C_{18}H_{22}ClNO$
2-[1-(4-Chlorophenyl)-1-phenylethoxy]-N,N-dimethylethanamine.
Anticholinergic. $bp_{0.05}$ = 150-155°. *Marion Merrell Dow Inc.*

2728 Chlorphenoxamine Hydrochloride
562-09-4 2234 209-227-1
$C_{18}H_{24}Cl_3NO$
2-[1-(4-Chlorophenyl)-1-phenylethoxy]-N,N-dimethylethanamine hydrochloride.
Systral; 1766; Clorevan; Contristamine; Phenoxene; Phenoxine. Anticholinergic. mp = 128°; soluble in H_2O. *Marion Merrell Dow Inc.*

2729 Chlorphentermine
461-78-9 2235 207-314-9
$C_{10}H_{14}ClN$
4-Chloro-α,α-dimethylbenzeneethanamine.
clorfentermina; Lucofen; Teramine. Anorexic; CNS stimulant. Anticholinergic. Federally controlled substance (stimulant). bp_2 = 100-102°. *Parke-Davis.*

2730 Chlorphentermine Hydrochloride
151-06-4 2235 205-782-9
$C_{10}H_{15}Cl_2N$
4-Chloro-α,α-dimethylbenzeneethanamine hydrochloride.
Pre-Sate; S-62; W-2426; NSC-76098. Anorexic; anticholinergic. mp = 234°; soluble in H_2O (> 20 g/100 ml), LD_{50} (mus orl) = 270 mg/kg, (mus sc) = 267 mg/kg. *Parke-Davis.*

2731 Cimetropium Bromide
51598-60-8 2338
$C_{21}H_{28}BrNO_4$
8-(Cyclopropylmethyl)-6β,7β-epoxy-3α-hydroxy-1αH,5αH-tropanium bromide (-)-(S)-tropate.
DA-3177; N-cyclopropylmethylscopolamine bromide; Alginor. Anticholinergic. Spasmolytic with affinity for intestinal muscarinic receptors mp = 174°; $[\alpha]_D^{20}$ = -18.3° (c = 3). *DeAngeli.*

2732 Clidinium Bromide
3485-62-9 2412 222-471-3
$C_{22}H_{26}BrNO_3$
3-Hydroxy-1-methylquinuclidinium bromide benzilate.
Quarzan; Ro-2-3773; component of: Librax. Anticholinergic. mp = 240-241°. *Hoffmann-LaRoche Inc.*

2733 Cyclodrine
52109-93-0 2788
$C_{19}H_{29}NO_3$
2-Phenyl-2-(1-hydroxycyclopentyl)acetic acid β-diethylamino)ethyl ester.
Anticholinergic. *University of Michigan.*

2734 Cyclodrine Hydrochloride
78853-39-1 2788
$C_{19}H_{30}ClNO_3$
2-Phenyl-2-(1-hydroxycyclopentyl)acetic acid β-diethylamino)ethyl ester hydrochloride.
GT-92; Cyclopent. Anticholinergic. mp = 133-135°. *University of Michigan.*

2735 Cyclonium Iodide
6577-41-9 2806
$C_{22}H_{34}INO_2$
1-[(2-Cyclohexyl-2-phenyl-1,3-dioxolan-4-yl)methyl]-1-methylpiperidinium iodide.
oxapium iodide; ciclonium iodide; SH-100; Esperan. Antispasmodic. Anticholinergic. [α-form]: mp = 195-197°; [β-form]: mp = 150-152°; [both forms]: soluble in MeOH, EtOH, $CHCl_3$, poorly soluble in C_6H_6, C_7H_8, H_2O; LD_{50} (mus ip) = 106 mg/kg. *Toyama Chem. Co. Ltd.*

2736 Cyclopentolate Hydrochloride
5870-29-1 2815 227-521-8
$C_{17}H_{26}ClNO_3$
2-(Dimethylamino)ethyl 1-hydroxy-α-phenylcyclopentaneacetate hydrochloride.
Cyclogyl; Ak-Pentolate; Alnide; Mydplegic; Mydrilate; Zyklolat; component of: Cyclomydril. Anticholinergic (ophthalmic). Mydriatic. mp = 137-141°; freely soluble in H_2O, EtOH; insoluble in Et_2O. *Alcon Labs.*

2737 Cycrimine
77-39-4 2825 201-024-6
$C_{19}H_{29}NO$
α-Cyclopentyl-α-phenyl-1-piperidinepropanol.
Pagitane. Anticholinergic. *Eli Lilly & Co.*

2738 Cycrimine Hydrochloride
126-02-3 2825 204-764-8
$C_{19}H_{30}ClNO$
α-Cyclopentyl-α-phenyl-1-piperidinepropanol hydrochloride.
Pagitane hydrochloride. Anticholinergic. mp = 241-244° (dec); soluble in H_2O (0.6 g/100 ml), EtOH (2.0 g/100ml), $CHCl_3$ (3.0 g/100 ml). *Eli Lilly & Co.*

2739 Darifenacin
133099-04-4
$C_{28}H_{30}N$+27O2
(S)-1-[2-(2,3-Dihydro-5-benzofuranyl)ethyl]-α,α-diphenyl-3-pyrrolidineacetamide.
A muscarinic antagonist used as an anticholinergic agent.

2740 Deptropine
604-51-3 2958 210-069-0
$C_{23}H_{27}NO$
3-[10,11-Dihydro-5H-dibenzo[a,d]cyclohepten-5-yloxy]tropane.
dibenzheptropine. Anticholinergic; antihistaminic. [maleate]: mp = 133-136°.

2741 Deptropine Citrate
2169-75-7 2958 218-516-1
$C_{29}H_{35}NO_8$
3-[10,11-Dihydro-5H-dibenzo[a,d]cyclohepten-5-yloxy]tropane citrate.
BS-6987; Brontine. Anticholinergic; antihistaminic. LD_{50} (mus iv) = 32 mg/kg, (mus orl) = 300 mg/kg.

2742 Dexetimide
21888-98-2 2987
$C_{23}H_{26}N_2O_2$
(+)-2-(1-Benzyl-4-piperidyl)-2-phenylglutarimide.
dextrobenzetimide; dexbenzetimide. Anticholinergic; antiparkinsonian. mp = 181-183°; $[\alpha]_D^{20}$ = 125° ($CHCl_3$). *Abbott Laboratories Inc.*

2743 Dexetimide Hydrochloride
21888-96-0 2987 244-633-2
$C_{23}H_{27}ClN_2O_2$
(+)-2-(1-Benzyl-4-piperidyl)-2-phenylglutarimide hydrochloride.
R-16470; Tremblex. Anticholinergic; antiparkinsonian. mp = 270-275°; $[\alpha]_D^{20}$ = 125° (MeOH); LD_{50} (rat iv) = 45 mg/kg. *Abbott Laboratories Inc.*

2744 Dibutoline Sulfate
532-49-0 3082
$C_{30}H_{66}N_4O_8S$
2-[[(Dibutylamino)carbonyl]oxy]-N-ethyl-N,N-dimethylethanaminium sulfate (2:1).
dibuline sulfate. Anticholinergic. Unstable at 100° in H_2O.

2745 N-Butylscopolammonium Bromide
149-64-4 1624 205-744-1
$C_{21}H_{30}BrNO_4$
8-Butyl-6β,7β-epoxy-3α-hydroxy-1αH,5αH-tropanium bromide (-)-tropate.
scopolamine bromobutylate; hyoscine N-butyl bromide; Amisepan; Buscapina; Buscol; Buscolysin; Buscopan; Butylmin; Donopon; Monospan; Scobro; Scobron; Scobutil; Sparicon; Sporamin; Stibron; Tirantil. Anticholinergic; antispasmodic. mp = 142-144°; $[\alpha]_D^{20}$ = -20.8° (c = 3 H_2O); LD_{50} (mus iv) = 15.6 mg/kg, (mus ip) = 74 mg/kg, (mus sc) = 570 mg/kg, (mus orl) = 3000 mg/kg. *Boehringer Ingelheim GmbH.*

2746 Dicyclomine
77-19-0 3147 201-009-4
$C_{19}H_{35}NO_2$
2-(Diethylamino)ethyl [bicyclohexyl-1-carboxylate].
Dicycloverine. Anticholinergic. *Marion Merrell Dow Inc.*

2747 Dicyclomine Hydrochloride
67-92-5 3147 200-671-1
$C_{19}H_{36}ClNO_2$
2-(Diethylamino)ethyl [bicyclohexyl-1-carboxylate] hydrochloride.
Atumin; Benacol; Bentomine; Bentyl Hydrochloride; Bentylol Hydrochloride; Diocyl Hydrochloride; Di-Syntramine; Dyspas; Mamiesan; Merbentyl; Procyclomin; Wyovin Hydrochloride. Anticholinergic. mp = 164-166°; soluble in H_2O (25 g/100 ml). *Marion Merrell Dow Inc.*

2748 Diethazine
60-91-3 3157 200-491-3
$C_{18}H_{22}N_2S$
10-(2-Diethylaminoethyl)phenothiazine.
RP-2987; Deparkin; Dinezin; Dolisina; Eazaminum; Ethylemin; Parkazin. Anticholinergic; antiparkinsonian. Oily liquid; bp_{4-5} = 195-208°, $bp_{0.4-0.5}$ = 167-175°. *Rhône-Poulenc Rorer Pharm. Inc.*

2749 Diethazine Hydrochloride
341-70-8 3157 206-437-5
$C_{18}H_{23}ClN_2S$
10-(2-Diethylaminoethyl)phenothiazine hydrochloride.
Antipar; Aparkazin; Casantin; Diparcol; Latibon; Thiantan; Thiontan. Anticholinergic; antiparkinsonian. mp = 184-186°; soluble in H_2O (20 g/100 ml), EtOH (16.7 g/100 ml), $CHCl_3$ (20 g/100 ml); insoluble in Et_2O; LD_{50} (mus orl) = 450 mg/kg. *Rhône-Poulenc Rorer Pharm. Inc.*

2750 Difemerine
80387-96-8 3180
$C_{20}H_{25}NO_3$
2-(Dimethylamino)-1,1-dimethylethyl benzilate.
Anticholinergic; antispasmodic. mp = 78°. *Lab. UPSA.*

2751 Difemerine Hydrochloride
70280-88-5 3180 274-526-6
$C_{20}H_{26}ClNO_3$
2-(Dimethylamino)-1,1-dimethylethyl benzilate hydrochloride.
Luostyl. Anticholinergic; antispasmodic. mp = 182°. *Lab. UPSA.*

2752 Dihexyverine
561-77-3 3211
$C_{20}H_{35}NO_2$
2-Piperidinoethyl ester of bicyclohexyl-1-carboxylic acid.
dihexiverine; dicyclohexylacetic acid 2-piperidinoethyl ester. Anticholinergic.

2753 Dihexyverine Hydrochloride
5588-25-0 3211 226-996-9
$C_{20}H_{36}ClNO_2$
2-Piperidinoethyl ester of bicyclohexyl-1-carboxylic acid hydrochloride.
dihexiverine hydrochloride; JL-1078; Metaspas; Diverine; Olimplex; Seclin; Dispas; Neospasmina; Spasmodex; Spasmalex. Anticholinergic. mp = 175°; 200°; LD_{50} (mus ip) = 212 mg/kg.

2754 Diphemanil Methylsulfate
62-97-5 3361 200-552-4
$C_{21}H_{27}NO_4S$
4-(Diphenylmethylene)-1,1-dimethylpiperidinium methyl sulfate.
Prantal; Demotil; Diphenatil; Nivelona; Variton. Anticholinergic. mp = 194-195°; soluble in H_2O; LD_{50} (rat orl) = 1107 mg/kg, (mus orl) = 64 mg/kg, (gpg orl) = 404 mg/kg. *Schering-Plough Pharm.*

2755 N-(1,2-Diphenylethyl)nicotinamide
553-06-0 3382 209-031-6
$C_{20}H_{18}N_2O$
N-(1,2-Diphenylethyl)-3-pyridinecarboxamide.
nicofetamide; 1-nicotinylamino-1,2-diphenylethane; C-1065; Lyspamin. Anticholinergic. mp = 159°; soluble in 5N HCl; moderately soluble in MeOH, Me_2CO, Et_2O, C_6H_6; slightly soluble in EtOH; insoluble in H_2O. *Cilag-Chemie Ltd.*

2756 Dipiproverine
117-30-6 3399
$C_{20}H_{30}N_2O_2$
1-Piperidineethanol α-phenyl-1-piperidine acetate ester.
Anticholinergic. bp_1 = 188-190°.

2757 Dipiproverine Hydrochloride
2404-18-4 3399 219-293-3
$C_{20}H_{32}Cl_2N_2O_2$
1-Piperidineethanol α-phenyl-1-piperidine acetate ester dihydrochloride.

LD-935; Levospasme; Spasmonal. Anticholinergic. mp = 212-214°, 238°; soluble in H_2O (184 g/100 ml at 20°, 323 g/100 ml at 100°), iPrOH (5 g/100g at 20°, 12.5 g/100 g at 50°, 27 g/100 g at 80°).

2758 Diponium Bromide
2001-81-2 3402 217-891-9
$C_{20}H_{38}BrNO_2$
Triethyl(2-hydroxyethyl)ammonium bromide dicyclopentylacetate.
dipenine bromide; SA-267; Unospaston. Antispasmodic; anticholinergic. mp = 185-186°; freely soluble in H_2O; LD_{50} (mus orl) = 570 mg/kg, (mus ip) = 88 mg/kg, (mus iv) = 6.2 mg/kg, (rat orl) = 780 mg/kg, (rat iv) = 6.6 mg/kg. *Siegfried AG.*

2759 Domazoline
6043-01-2 227-931-7
$C_{14}H_{20}N_2O_2$
2-(3,6-Dimethoxy-2,4-dimethylbenzyl)-2-imidazole.
Anticholinergic. *Schering AG.*

2760 Domazoline Fumarate
35100-41-5
$C_{18}H_{24}N_2O_6$
2-(3,6-Dimethoxy-2,4-dimethylbenzyl)-2-imidazoline fumarate.
Anticholinergic. *Schering AG.*

2761 Elantrine
1232-85-5
$C_{20}H_{24}N_2$
11-[3-(Dimethylamino)propylidene]-5,6-dihydro-5-methylmorphanthridine.
EX-10-029-C; RMI-80029. Anticholinergic. *Marion Merrell Dow Inc.*

2762 Elucaine
25314-87-8
$C_{19}H_{23}NO_2$
α-[(Diethylamino)methyl]benzyl alcohol benzoate (ester).
Anticholinergic.

2763 Emepronium Bromide
3614-30-0 3598 222-786-6
$C_{20}H_{28}BrN$
Ethyl(3,3-diphenyl-1-methylpropyl)dimethylammonium bromide.
Cetiprin; Restenacht; Ripirin; Uro-Ripirin. Antispasmodic; anticholinergic. Muscarinic antagonist. mp = 204°. *Aktiebolaget. Recip.*

2764 Endobenzyline Bromide
52080-56-5 3612
$C_{20}H_{28}BrNO_3$
2-[(Bicyclo[2.2.1]hept-5-en-2-ylhydroxyphenylacetyl)oxy]-N,N,N-trimethylethanaminium bromide.
PC-1238; Ulcynendobenziline bromide. Anticholinergic. mp = 170°.

2765 Ethopropazine
522-00-9 3793 208-320-4
$C_{19}H_{24}N_2S$
10-[2-(Diethylamino)propyl] phenothiazine.
RP-3356; W-483; Isopthazine; Isothiazine; Parkin. Anticholinergic. mp = 53-55°; LD_{50} (mus orl) = 650 mg/kg. *Rhône-Poulenc Rorer Pharm. Inc.*

2766 Ethopropazine Hydrochloride
1094-08-2 3793 214-134-4
$C_{19}H_{25}ClN_2S$
10-[2-(Diethylamino)propyl] phenothiazine monohydrochloride.
Dibutil; Lysivane; Pardisol; Parphezein; Parphezin; Parsidol; Parsitan; Parsotil; Rodipal. Anticholinergic. Used as an antiparkinsonian and antiallergic. mp = 223-225°; soluble in H_2O (0.25 g/100 ml at 20°, 5 g/100 ml at 40°), EtOH (3.3 g/100 ml at 25°); sparingly soluble in Me_2CO; insoluble in Et_2O, C_6H_6. *Parke-Davis; Rhône-Poulenc Rorer Pharm. Inc.*

2767 Ethybenztropine
524-83-4 3802
$C_{22}H_{27}NO$
3α-(Diphenylmethoxy)-8-ethyl-1αh,5αh-nortropane .
UK-738; N-ethylbenztropine. Anticholinergic. Antibacterial. [hydrochloride (Ponalid)]: mp = 190-191°; [hydrobromide (Panolid)]: mp = 226-228°. *Boehringer Ingelheim GmbH; Sandoz Pharm. Corp.*

2768 Ethylbenzhydramine
642-58-0 3813
$C_{19}H_{25}NO$
2-(Diphenylmethoxy)-N,N-diethylethanamine.
etanautine. Anticholinergic; antiparkinsonian. $bp_{0.15}$ = 140-142°. *Ciba-Geigy Corp.; Parke-Davis.*

2769 Ethylbenzhydramine Hydrochloride
86-24-8 3813
$C_{19}H_{26}ClNO$
2-(Diphenylmethoxy)-N,N-diethylethanamine hydrochloride.
Antiparkin. Anticholinergic. Used as an antiparkinsonian. mp = 140°; soluble in H_2O, EtOH, Me_2CO, $CHCl_3$; slightly soluble in C_6H_6, Et_2O. *Ciba-Geigy Corp.*

2770 Etomidoline
21590-92-1 3928 244-463-9
$C_{23}H_{29}N_3O_2$
2-Ethyl-2,3-dihydro-3-[[4-[2-(1-piperidinyl)ethoxy]phenyl]amino]-1H-isoindol-1-one.
K-2680; Smedolin. Antispasmodic; anticholinergic. mp = 106-107°; LD_{50} (mus orl) = 176 mg/kg, (mus iv) = 44.8 mg/kg. *Farmitalia Carlo Erba Ltd.*

2771 Eucatropine
100-91-4 3943 202-900-0
$C_{17}H_{25}NO_3$
1,2,2,6-Tetramethyl-4-piperidyl mandelate.
Euphthalmine. Anticholinergic (ophthalmic). mp = 113°; insoluble in cold H_2O, soluble in organic solvents.

2772 Eucatropine Hydrochloride
536-93-6 3943 208-653-5
$C_{17}H_{26}ClNO_3$
1,2,2,6-Tetramethyl-4-piperidyl mandelate hydrochloride.
Anticholinergic (ophthalmic). mp = 183-184°; soluble in H_2O, EtOH (50 g/100 ml at 76°), $CHCl_3$; insoluble in Et_2O.

2773 Fenpiverinium Bromide
125-60-0 4032 204-744-9
$C_{22}H_{29}BrN_2O$
1-(4-Amino-4-oxo-3,3-diphenylbutyl)-1-methylpiperidinium bromide.
fenpipramide methobromide; 12494 Hoechst; Resantin. Antispasmodic; anticholinergic. mp = 177.5-178.5°, 216-216.5°; freely soluble in H_2O. *Hoechst AG.*

2774 Fentonium Bromide
5868-06-4 4048 227-520-2
$C_{31}H_{34}BrNO_4$
3α-Hydroxy-8-(p-phenylphenacyl)-1αH,5αH-tropanium bromide (-)-tropate.
phentonium bromide; FA-402; Z-326; Ketoscilium; Ulcesium. Anticholinergic and antispasmodic. mp = 203-205° (dec), 193-194°; $[\alpha]_D^{23}$ = -5.68° (c = 5 DMF), $[\alpha]_D^{25}$ = -4.7° (c = 5 DMF); LD_{50} (mus iv) = 12.1 mg/kg, (mus sc) > 400 mg/kg, (mus orl) > 400 mg/kg. *Whitefin Holding.*

2775 Flutropium Bromide
63516-07-4 4248
$C_{24}H_{29}BrFNO_3$
(8r)-8-(2-Fluoroethyl)-3α-hydroxy-1αH,5αH-tropanium bromide benzilate.
Ba-598Br; Flubron. Bronchodilator. Quaternary ammonium compound. Anticholinergic. mp = 192-193° (dec), 198-199°; LD_{50} (mmus iv) = 12.5 mg/kg, (mmus orl) = 760 mg/kg, (fmus iv) = 11.0 mg/kg, (fmus orl) = 810 mg/kg, (mrat iv) = 16.4 mg/kg, (mrat orl) = 830 mg/kg, (frat iv) = 18.4 mg/kg, (frat orl) = 740 mg/kg. *Boehringer Ingelheim GmbH.*

2776 Glycopyrrolate
596-51-0 4511 209-887-0
$C_{19}H_{28}BrNO_3$
3-Hydroxy-1,1-dimethylpyrrolidinium bromide α-cyclopentylmandelate.
copyrrolate; Robinul; AHR-504; Nodapton; Robanul; Tarodyl; Tarodyn. Anticholinergic. mp = 193.2-194.5°; soluble in H_2O; LD_{50} (fmus ip) = 107 mg/kg, (frat ip) = 196 mg/kg, (mrat orl) = 1150 mg/kg. *Whitehall-Robins.*

2777 Hemicholinium
16478-59-4 4676
$[C_{24}H_{34}N_2O_4]^{2+}$
2,2'-[1,1'-Biphenyl]-4,4'-diylbis-[2-hydroxy-4,4-dimethylmorpholinium].
Specific inhibitor of choline uptake by cholinergic presynaptic nerve terminals. Inhibits acetylcholine synthesis. A research tool used as acholinergic probe to deplete choline stores.

2778 Hemicholinium Dibromide
312-45-8 4676
$C_{24}H_{34}Br_2N_2O_4$
2,2'-[1,1'-Biphenyl]-4,4'-diylbis-[2-hydroxy-4,4-dimethylmorpholinium] dibromide.
HC-3; hemicholinium-3. Acetylcholine synthesis inhibitor. A research tool used as acholinergic probe to deplete choline stores. mp = 180°; moderately soluble in H_2O; soluble in EtOH; LD+si50(fmus ip) = 0.048-0.082 mg/kg. [monohydrate]: mp = 226-228° (dec).

2779 Hepzidine
1096-72-6
$C_{21}H_{25}NO$
4-(10,11-Dihydro-5H-dibenzo[a,d]cyclohepten-5-yloxy)-1-methylpiperidine.
Antidepressant; anticholinergic.

2780 Heteronium Bromide
7247-57-6 4709
$C_{18}H_{22}BrNO_3S$
(±)-3-Hydroxy-1,1-dimethylpyrrolidinium bromide α-phenyl-2-thiopheneglycolate.
Hetrum bromide; 31814. Anticholinergic. [α diastereoisomer]: mp = 210-211°; [β diastereoisomer]: mp = 182-184°. *Eli Lilly & Co.*

2781 Hexocyclium
6004-98-4 4744
$C_{19}H_{30}N_2O$
4-(β-Cyclohexyl-β-hydroxyphenethyl)-1,1-dimethylpiperazine.
Evipan Sodium; Hexanastab; Hexobarbitone-sodium; methenexyl sodium; Evipal sodium; 2,4,6(1H,3H,5H)-Pyrimidinetrione, 5-(1-cyclohexen-1-yl)-1,5-dimethyl-, sodium salt. Anticholinergic. *Sterling Winthrop Inc.*

2782 Hexocyclium Methosulfate
115-63-9 4744 204-097-2
$C_{21}H_{36}N_2O_5S$
4-(β-Cyclohexyl-β-hydroxyphenethyl)-1,1-dimethylpiperazinium methyl sulfate.
AMA-DE-1973; ND-1966; MI. Tral; Tral; Tralin. Anticholinergic. mp = 200-210°; λ_m = 252, 257, 263 nm ((0.1N H_2SO_4); very soluble in H_2O (50 g/100 ml), slightly soluble in $CHCl_3$, insoluble in Et_2O. *Abbott Labs Inc.*

2783 Homatropine
87-00-3 4766 201-716-8
$C_{16}H_{21}NO_3$
3α-Hydroxy-1αH,5αH-tropanium mandelate (ester).
mandelyltropeine; tropine mandelate. Anticholinergic (ophthalmic). [dl-form]: mp = 99-100°; soluble in organic solvents, slightly soluble in H_2O. *Alcon Labs; Ciba-Geigy Corp.*

2784 Homatropine Hydrobromide
51-56-9 4766 200-105-3
$C_{16}H_{22}BrNO_3$
3α-Hydroxy-1αH,5αH-tropanium mandelate (ester) hydrobromide.
Isopto Homatropine. Anticholinergic (ophthalmic). mp = 212°; soluble in H_2O (16.6 g/100 ml), EtOH (2.5 g/100 ml at 25°, 8.33 g/100 ml at 60°), $CHCl_3$ (0.24 g/100 ml); insoluble in ether. *Alcon Labs; Ciba-Geigy Corp.*

2785 Homatropine Methyl Bromide
80-49-9 4766 201-284-0
$C_{17}H_{24}BrNO_3$
3α-Hydroxy-8-methyl-1αH,5αH-tropanium bromide mandelate.
Arkitropin; Homapin; Malcotran; Mesopin; Novatrin; Novatropine; Sethyl; component of: Hycodan, Probilagol. Anticholinergic (ophthalmic). mp = 191-192° (dec); soluble in H_2O, slightly soluble in EtOH; LD_{50} (mus orl) =

1400 mg/kg, (mus ip) = 60 mg/kg. *DuPont-Merck Pharm.; Purdue Pharma L.P.*; 259l.

2786 Hydroxyamphetamine
1518-86-1 4855
$C_9H_{13}NO$
(±)-p-(2-Aminopropyl)phenol.
p-hydroxyphenylisopropylamine; α-methyltyramine; Paredrine; Paredrinex; Pulsoton. An α-adrenergic (ophthalmic) agonist. Used as a mydriatic agent. mp = 125-126°; soluble in H_2O, EtOH, $CHCl_3$, EtOAc; [iodide]: mp = 155°, soluble in H_2O, EtOH, Me_2CO; [hydrochloride]: mp = 171-172°; soluble in H_2O, EtOH; insoluble in Et_2O. *Allergan Inc.; SmithKline Beecham Pharm.*

2787 Hydroxyamphetamine Hydrobromide
306-21-8 4855 206-181-4
$C_9H_{14}BrNO$
(±)-p-(2-Aminopropyl)phenol hydrobromide.
Paredrine hydrobromide; component of: Paremyd. An α-adrenergic (ophthalmic) agonist. Used as a mydriatic agent. mp = 171-172°; soluble in H_2O, EtOH; insoluble in Et_2O. *Allergan Inc.; SmithKline Beecham Pharm.*

2788 Hyoscyamine
101-31-5 4907 202-933-0
$C_{17}H_{23}NO_3$
1αH,5αH-Tropan-3α-ol (-)-tropate (ester).
Cystospaz; l-tropine tropate; daturine; duboisine; l-hyoscyamine; Levsin. Anticholinergic. mp = 108.5°; $[\alpha]_D^{20}$ = -21° (EtOH); soluble in H_2O (0.36 g/100 ml), Et_2O (1.45 g/100 ml), C_6H_6 (0.67 g/100 ml), $CHCl_3$ (100 g/100 ml); freely soluble in EtOH. *Mallinckrodt Inc.*

2789 Hyoscyamine Hydrobromide
306-03-6 4907 206-174-6
$C_{17}H_{24}BrNO_3$
1αH,5αH-Tropan-3α-ol (-)-tropate (ester) hydrobromide.
Anticholinergic. mp = 152°; levorotatory; very soluble in H_2O, EtOH (33.3 g/100 ml), $CHCl_3$ (83.3 g/100 ml), Et_2O (0.04 g/100 ml). *Mallinckrodt Inc.*

2790 Hyoscyamine Sulfate
620-61-1 4907 210-644-6
$C_{18}H_{24}BrNO_4$
1αH,5αH-Tropan-3α-ol (-)-tropate (ester) sulfate (2:1) (salt).
Anticholinergic. Anticholinergic. *Ascher B.F. & Co.; Forest Pharm. Inc.; Robins, A. H. Co.; Schwarz Pharma Kremers Urban Co.; Wallace Labs; Zeneca Pharm.*

2791 Hyoscyamine Sulfate Dihydrate
6835-16-1 4907
$C_{34}H_{48}N_2O_{10}.2H_2O$
1αH,5αH-Tropan-3α-ol (-)-tropate (ester) sulfate (2:1) (salt) dihydrate.
Peptard; Egacene; Egazil Duretter. Anticholinergic. mp = 206°; $[\alpha]_D^{15}$ = -29° (c = 2); soluble in H_2O (200 g/100 ml), EtOH (20 g/100 ml); slightly soluble in $CHCl_3$, Et_2O. *Mallinckrodt Inc.*

2792 Idaverine
100927-13-7
$C_{24}H_{39}N_3O_3$
(+)-1-[4-[Ethyl-(p-methoxy-α-methylphenethyl)amino]butyryl]-N,N-dimethylisonipecotamide.
Anticholinergic. Muscarinic receptor antagonist.

2793 Ipratropium Bromide
66985-17-9 5089
$C_{20}H_{30}BrNO_3.H_2O$
(8r)-3α-Hydroxy-8-isopropyl-1αH,5αH-tropanium bromide monohydrate.
Sch-1000-Br-monohydrate; N-isopropylnoratropinium bromomethylate; 8-isopropylnoratropine methobromide; 3α-hydroxy-8-isopropyl-1αH,5αH-tropanium bromide tropate; Atem; Atrovent; Bitrop; Itrop; Narilet; Rinatec; component of: Combivent. An anticholinergic bronchodilator and antiarrhythmic. Quaternary ammonium compound. mp = 230-232°; soluble in H_2O, MeOH, EtOH; insoluble in Et_2O, $CHCl_3$, fluorohydrocarbons; LD_{50} (mmus orl) = 1001 mg/kg, (mmus iv) = 12.29 mg/kg, (mmus sc) = 300 mg/kg, (fmus orl) = 1083 mg/kg, (fmus iv) = 14.97 mg/kg, (fmus sc) = 340 mg/kg, (mrat orl) = 1663 mg/kg, (mrat iv) = 15.89 mg/kg, (frat orl) = 1779 mg/kg, (frat iv) = 15.70 mg/kg. *Boehringer Ingelheim Ltd.; Schering AG.*

2794 Ipratropium Bromide [anhydrous]
22254-24-6 5089 244-873-8
$C_{20}H_{30}BrNO_3$
(8r)-3α-Hydroxy-8-isopropyl-1αH,5αH-tropanium bromide.
N-isopropylnoratropinium bromomethylate; Sch-100. An anticholinergic bronchodilator and antiarrhythmic. Quaternary ammonium compound. *Boehringer Ingelheim Ltd.*

2795 Isopropamide
7492-32-2 5221 231-316-9
$[C_{22}H_{30}N_2O]^+$
(3-Carbamoyl-3,3-diphenylpropyl)diisopropylmethyl-ammonium.
Anticholinergic. mp = 84-86°. *SmithKline Beecham Pharm.*

2796 Isopropamide Iodide
71-81-8 5221 200-766-8
$C_{23}H_{33}IN_2O$
(3-Carbamoyl-3,3-diphenylpropyl)diisopropylmethyl-ammonium iodide.
Darbid; R-79; Priamide; Tyrimide; component of: Combid, Stelabid. Anticholinergic. Potent carbonic anhydrase inhibitor for control of glaucoma. mp = 198-201°, 189.0-191.5°; soluble in hot H_2O, MeOH, EtOH; insoluble in Et_2O. *SmithKline Beecham Pharm.*

2797 Lorglumide
97964-56-2
$C_{22}H_{32}Cl_2N_2O_4$
(±)-4-(3,4-Dichlorobenzamido)-N,N-dipentyl-glutaramic acid.
A selective cholecystokinin A (CCKA) receptor antagonist. Anticholinergic.

2798 Mazaticol
42024-98-6
$C_{21}H_{27}NO_3S_2$
6,6,9-Trimethyl-9-azabicyclo[3.3.1]non-3β-yl di-2-thienylglycolate.
PG-501. Potent inhibitor of muscarinic receptor binding. Anticholinergic; antiparkinsonian.

2799 Mecloxamine
5668-06-4 5823
$C_{19}H_{24}ClNO$
2-[1-(4-Chlorophenyl)-1-phenylethoxy]-N,N-dimethyl-1-propanamine.
Anticholinergic. Also has sedative and hypnotic properties. $bp_{0.6}$ = 154-160°. *Astra USA Inc.*

2800 Mecloxamine Citrate
5823
$C_{25}H_{32}ClNO_8$
2-[1-(4-Chlorophenyl)-1-phenylethoxy]-N,N-dimethyl-1-propanamine citrate.
component of: Melidorm. Sedative/hypnotic properties. Anticholinergic. mp = 120-124°. *Astra USA Inc.*

2801 Mepenzolate
25990-43-6 5893
$C_{20}H_{23}NO_3$
3-Hydroxy-1,1-dimethylpiperidinyl benzilate.
Anticholinergic. $bp_{0.03}$ = 175-176°. *Merrell Dow Pharm. Inc.*

2802 Mepenzolate Bromide
76-90-4 5893 200-992-7
$C_{21}H_{26}BrNO_3$
3-Hydroxy-1,1-dimethylpiperidinium bromide benzilate.
Cantil; Cantril; Gastropidil; Trancolon. Anticholinergic. Also reported to be hallucinogenic. mp = 228-229° (dec); LD_{50} (rat orl) = 742 ± 47 mg/kg. *Merrell Dow Pharm. Inc.*

2803 Metcaraphen
561-79-5 5996
$C_{20}H_{31}NO_2$
1-(3',4'-Dimethylphenyl)-1-cyclopentanecarboxylic acid 2-diethylaminoethyl ester.
Netrin. Anticholinergic. $bp_{0.05}$ = 126-128°. *Ciba-Geigy Corp.*

2804 Metcaraphen Hydrochloride
1950-31-8 5996
$C_{20}H_{32}ClNO_2$
1-(3',4'-Dimethylphenyl)-1-cyclopentanecarboxylic acid 2-diethylaminoethyl ester hydrochloride.
Anticholinergic. Soluble in H_2O. *Ciba-Geigy Corp.*

2805 Methantheline
5818-17-7 6025
$[C_{21}H_{23}NO_3]^+$
Diethyl(2-hydroxyethyl)methylammonium xanthene-9-carboxylate.
methanthelinium; Banthine. Anticholinergic. Used in veterinary medicine for *its antispasmodic and antisecretory effects. Roberts Pharm. Corp.; Searle G.D. & Co.*

2806 Methantheline Bromide
53-46-3 6025 200-176-0
$C_{21}H_{26}BrNO_3$
Diethyl(2-hydroxyethyl)methylammonium bromide xanthene-9-carboxylate.
methanthelinium bromide; Banthine Bromide; MTB-51; SC-2910; Avagal; Uldumont; Vagantin; Metaxan; Methanide; Xanteline; Gastron; Gastrosedan; Methanthine bromide; Vagamin; Metanyl; Doladene; Asabaine. Anticholinergic. mp = 175-176°; soluble in H_2O, EtOH; insoluble in Et_2O; λ_m = 246, 282 nm ($E_{1\,cm}^{1\%}$ 135 69 EtOH); LD_{50} (mus ip) = 76 mg/kg. *Roberts Pharm. Corp.; Searle G.D. & Co.*

2807 Methixene
4969-02-2 6059 225-610-6
$C_{20}H_{23}NS$
1-Methyl-3-(thioxanthen-9-ylmethyl)piperidine.
Smooth muscle relaxant. Anticholinergic. $bp_{0.07}$ = 171-175°. *Wander Pharma.*

2808 Methixene Hydrochloride
1553-34-0 6059 216-300-1
$C_{20}H_{24}ClNS$
1-Methyl-3-(thioxanthen-9-ylmethyl)piperidine hydrochloride.
Smooth muscle relaxant. Anticholinergic.

2809 Methixene Hydrochloride Monohydrate
7081-40-5 6059
$C_{20}H_{24}ClNS.H_2O$
1-Methyl-3-(thioxanthen-9-ylmethyl)piperidine hydrochloride monohydrate.
SJ-1977; NSC-78194; Cholinfall; Methixart; Methyloxan; Tremaril; Tremarit; Tremonil; Tremoquil; Trest. Smooth muscle relaxant. Anticholinergic. mp = 215-217°; λ_m = 268 nm (ε 10250 dilute HCl); soluble in H_2O, EtOH, $CHCl_3$. *Wander Pharma.*

2810 Methscopolamine Bromide
155-41-9 6084 205-844-5
$C_{18}H_{24}BrNO_4$
6β,7β-Epoxy-3α-hydroxy-8-methyl-1αH,5αH-tropanium bromide tropate (ester).
hyoscine methylbromide; Diopal; Holopon; Mescopil; Neo-Avagal; Pamine; Restropin. Anticholinergic. mp = 214-217° (dec); soluble in H_2O, EtOH. *Pharmacia & Upjohn.*

2811 Methylatropine Nitrate
52-88-0 907 200-156-1
$C_{18}H_{26}N_2O_6$
3α-Hydroxy-8-methyl-1αH,5αH-tropanium (±)-tropate nitrate (salt).
Eumydrin; Ekomine; Harvatrate; Metanite. Anticholinergic. mp = 163°; soluble in H_2O (1 g/ml), less soluble in EtOH, insoluble in organic solvents. *Hoechst Roussel Pharm. Inc.; Sterling Winthrop Inc.*

2812 Methylbenactyzium Bromide
3166-62-9 1055 221-628-3
$C_{21}H_{28}BrNO_3$
Diethyl(2-hydroxyethyl)methylammonium bromide benzilate.
benactyzine methobromide; Finalin; Spatomac.

Anticholinergic; antidepressant. mp = 169-170°. *Am. Cyanamid.*

2813 Metoquizine
7125-67-9
$C_{22}H_{27}N_5O$
3,5-Dimethyl-N-(4,6,6a,7,8,9,10,10a-octahydro-4,7-dimethylindolo[4,3-fg]quinolin-9-yl)pyrazole-1-carboxamide.
42406. Anticholinergic. *Eli Lilly & Co.*

2814 Nuvenzepine
96487-37-5
$C_{19}H_{20}N_4O_2$
6,11-Dihydro-11-(1-methylisonipecotoyl)-5H-pyrido[2,3-b][1,5]benzodiazepin-5-one.
DF-545. Selective M1 muscarinic receptor antagonist. Pirenzepine-related analog. Antimuscarinic.

2815 Octamylamine
502-59-0 6846 207-947-0
$C_{13}H_{29}N$
N-Isopentyl-1,5-dimethylhexylamine.
Octinum D; Octisamyl; Neo-Octon; Octometine. Anticholinergic; antispasmodic. bp_7 = 100-101°; [hydrochloride]: mp = 121°; soluble in H_2O, EtOH, Et_2O. *Knoll Pharm. Co.*

2816 Otenzepad
100158-38-1
$C_{24}H_{31}N_5O_2$
(±)-11-[[2-[(Diethylamino)methyl]piperidino]acetyl]-5,11-dihydro-6H-pyrido[2,3-b][1,4]benzodiazepin-6-one.
Muscarinic receptor antagonist. Anticholinergic.

2817 Otilonium Bromide
26095-59-0 247-457-4
$C_{29}H_{43}BrN_2O_4$
Diethyl(2-hydroxyethyl)methylammonium bromide p-[o-(oxtyloxy)benzamido]benzoate.
SP-63; Spasmomen. A muscarinic and tachykinin NK2 receptor antagonist and calcium channel blocker. Used for the symptomatic treatment of irritable bowel syndrome. Anticholinergic; spasmolytic agent.

2818 Oxybutynin
5633-20-5 7089
$C_{22}H_{31}NO_3$
4-(Diethylamino)-2-butynyl α-phenylcyclohexane-glycolate.
Anticholinergic. *Hoechst Marion Roussel Inc.*

2819 Oxybutynin Chloride
1508-65-2 7089 216-139-7
$C_{22}H_{32}ClNO_3$
4-(Diethylamino)-2-butynyl α-phenylcyclohexane-glycolate hydrochloride.
Ditropan; Dridase; MJ-4309-1; 5058; Cystrin; Pollakisu; Tropax. Anticholinergic. mp = 129-130°; LD_{50} (rat orl) = 1220 mg/kg. *Hoechst Marion Roussel Inc.*

2820 Oxyphencyclimine
125-53-1 7107 204-743-3
$C_{20}H_{28}N_2O_3$
(1,4,5,6-Tetrahydro-1-methyl-2-pyrimidinyl)methyl α-phenylcyclohexaneglycolate.
Antulcus; Caridan; Daricol; Setrol; Vio-Thene; Daricon; Naridan; Zamanil. Anticholinergic. *Pfizer Inc.*

2821 Oxyphencyclimine Hydrochloride
125-52-0 7107 204-742-8
$C_{20}H_{29}ClN_2O_3$
(1,4,5,6-Tetrahydro-1-methyl-2-pyrimidinyl)methyl α-phenylcyclohexaneglycolate monohydrochloride.
Daricon; component of: Vistrax. Anticholinergic. mp = 231-232° (dec); soluble in H_2O (1.2 g/100 ml). *Pfizer Inc.*

2822 Oxyphenonium Bromide
50-10-2 7109 200-010-7
$C_{21}H_{34}BrNO_3$
Diethyl(2-hydroxyethyl)methylammonium bromide α-phenylcyclohexane glycolate.
Antrenyl Duplex; Ba-5473; C-5473; Spasmophen. Anticholinergic. A proprietary preparation of oxyphenonium bromide. Used as a gastrointestinal sedative. mp = 189=194°; freely soluble in H_2O, sparingly soluble in EtOH. *CIBA plc.*

2823 Parapenzolate Bromide
5634-41-3 227-080-1
$C_{21}H_{26}BrNO_3$
4-Hydroxy-1,1-dimethylpiperidinium bromide benzilate.
Sch-3444. Anticholinergic. *Schering AG.*

2824 Pentapiperide
7009-54-3 7260 230-286-4
$C_{18}H_{27}NO_2$
1-Methyl-4-piperidyl 3-methyl-2-phenylvalerate ester.
C-4675 [as fumarate]. Anticholinergic; antispasmodic. [fumarate]: mp = 91-93°. *Cilag-Chemie Ltd.*

2825 Pentapiperium Methosulfate
7681-80-3 7260 231-678-8
$C_{20}H_{33}NO_6S$
1,1-Dimethyl-4-[(3-methyl-1-oxo-2-phenylpentyl)oxy]piperidinium methosulfate.
pentapiperium metilsulfate; pentapiperium mesylate; pentapiperium methylsulfonate; Hycholin; Quilene; Perium; Crylène. Anticholinergic; antispasmodic. mp = 110-112°. *Cilag-Chemie Ltd.*

2826 Penthienate
22064-27-3
$C_{17}H_{27}NO_3S$
2-[(Cyclopentylhydroxy-2-thienylacetyl)oxy]-N,N-diethylethanaminium bromide.
Anticholinergic. *Sterling Winthrop Inc.*

2827 Penthienate Bromide
60-44-6 7268 200-478-2
$C_{18}H_{30}BrNO_3S$
2-[(Cyclopentylhydroxy-2-thienylacetyl)oxy]-N,N-diethyl-N-methylethanaminium bromide.
Monodral bromide; Win-4369. Anticholinergic. mp = 124°; λ_m = 238 nm ($A^{1\%}_{1\,cm}$ 189); soluble in H_2O (0.2 g/ml), EtOH, $CHCl_3$; less soluble in non-polar organic solvents. *Sterling Winthrop Inc.*

2828 Phencarbamide
3735-90-8 7363 223-103-4
$C_{19}H_{24}N_2OS$
S-[2-(Diethylamino)ethyl] diphenylthiocarbamate.
Escorpal. Anticholinergic. mp = 48-49°; $bp_{0.01}$ = 120-126°; insoluble in H_2O; soluble in MeOH, EtOH, $CHCl_3$; less soluble in petroleum ether. *Bayer Corp.*

2829 Phencarbamide-1,5-naphthalene Disulfonate
72017-58-4 7363
$C_{29}H_{32}N_2O_7S_3$
S-[2-(Diethylamino)ethyl] diphenylthiocarbamate-1,5-naphthalene disulfonate.
Gelosedine; phencarbamide napadisilate. Anticholinergic. LD_{50} (rat orl) = 920 mg/kg. *Quinoderm Ltd.*

2830 PhencarbamideHydrochloride
58-13-9 7363 200-363-7
$C_{19}H_{25}ClN_2OS$
S-[2-(Diethylamino)ethyl] diphenylthiocarbamate hydrochloride.
Ba-1355; Escorpal. Anticholinergic. mp = 180-181°; soluble in H_2O; LD_{50} (rat orl) = 410 mg/kg. *Bayer Corp.*

2831 Phenglutarimide
1156-05-4 7377 214-587-8
$C_{17}H_{24}N_2O_2$
2,2-Diethylaminoethyl-2-phenylglutarimide.
Anticholinergic. mp = 125-127°. *Ciba-Geigy Corp.*

2832 Phenglutarimide Hydrochloride
1674-96-0 7377 216-819-3
$C_{17}H_{25}ClN_2O_2$
2,2-Diethylaminoethyl-2-phenylglutarimide hydrochloride.
Aturban; Aturbane; Aturbal. Anticholinergic. mp = 168-172° (dec); freely soluble in H_2O. *Ciba-Geigy Corp.*

2833 Pipenzolate
13473-38-6 7614 236-748-1
$C_{21}H_{25}NO_3$
1-Ethyl-3-hydroxypiperidinium benzilate.
Anticholinergic. *Marion Merrell Dow Inc.*

2834 Pipenzolate Bromide
125-51-9 7614 204-741-2
$C_{22}H_{28}BrNO_3$
1-Ethyl-3-hydroxy-1-methylpiperidinium bromide benzilate.
Piptal; JB-323; pibenzolone bromide. Anticholinergic. mp = 179-180°; soluble in H_2O. *Marion Merrell Dow Inc.*

2835 Piperidolate
82-98-4 7623 201-449-7
$C_{21}H_{25}NO_2$
1-Ethyl-3-piperidyl diphenylacetate.
Anticholinergic. $bp_{0.18}$ = 191-192°. *Marion Merrell Dow Inc.*

2836 Piperidolate Hydrochloride
129-77-1 7623 204-964-5
$C_{21}H_{26}ClNO_2$
1-Ethyl-3-piperidyl diphenylacetate hydrochloride.
Dactil; Crapinon. Anticholinergic. mp = 195-196°; soluble in H_2O. *Marion Merrell Dow Inc.*

2837 Piperilate
4546-39-8 7624
$C_{21}H_{25}NO_3$
1-Piperidineethanol benzilate.
Anticholinergic; antispasmodic.

2838 Piperilate Hydrochloride
4544-15-4 7624
$C_{21}H_{26}ClNO_3$
1-Piperidineethanol benzilate hydrochloride.
Daipisate; Norticon; Pensanate; Pipenate. Anticholinergic; antispasmodic. mp = 170-171°.

2839 Pirenzepine
28797-61-7 7646 249-228-4
$C_{19}H_{21}N_5O_2$
5,11-Dihydro-11-[(4-methyl-1-piperazinyl)acetyl]-6H-pyrido[2,3-b][1,4]benzodiazepin-6-one.
LS-519. Anticholinergic; antiulcerative. Gastric acid inhibitor. *Boehringer Ingelheim GmbH.*

2840 Pirenzepine Hydrochloride
29868-97-1 7646 249-907-5
$C_{19}H_{23}Cl_2N_5O_2$
5,11-Dihydro-11-[(4-methyl-1-piperazinyl)acetyl]-6H-pyrido[2,3-b][1,4]benzodiazepin-6-one dihydrochloride.
LS-59 Cl2; Duogastral;Durapirenz; Gasteril; Gastrozepin; Leblon; Maghen; Renzepin; Tabe; Ulcuforton; Ulcosan. Anticholinergic; antiulcerative. Gastric acid inhibitor. Soluble in H_2O, less soluble in MeOH, insoluble in Et_2O. *Boehringer Ingelheim GmbH.*

2841 Poldine
596-50-9
$C_{20}H_{23}NO_3$
2-(Hydroxymethyl)-1,1-dimethylpyrrolidine benzilate.
Anticholinergic. *McNeil Pharm.*

2842 Poldine Methylsulfate
545-80-2 7714 208-894-6
$C_{22}H_{29}NO_7S$
2-(Hydroxymethyl)-1,1-dimethylpyrrolidinium methyl sulfate benzilate.
Nacton; Nactate; IS-499; McN-R-726-47. Anticholinergic. Diuretic. mp = 154-155°; soluble in H_2O. *McNeil Pharm.*

2843 Pridinol
511-45-5 7922 208-128-0
$C_{20}H_{25}NO$
1,1-Diphenyl-3-piperidino-1-propanol.
ridinol; C-238. Anticholinergic; antiparkinsonian. mp = 120-121°; soluble in Me_2CO. *Glaxo Wellcome Inc.*

2844 Pridinol Hydrochloride
968-58-1 7922 213-529-9
$C_{20}H_{26}ClNO$
α,α-Diphenyl-1-piperidinepropanol hydrochloride.
Parks 12. Anticholinergic; antiparkinsonian. Dec 238°; soluble in EtOH; LD_{50} (mus iv) = 35 mg/kg, (mus ip) = 131°, (rat iv) = 33 mg/kg, (rat ip) = 91 mg/kg. *Glaxo Wellcome Inc.; Wellcome Foundation Ltd.*

2845 Pridinol Mesylate
6856-31-1 7922 229-953-2
$C_{21}H_{29}NO_4S$
α,α-Diphenyl-1-piperidinepropanol monomethanesulfonate.
pridinol methanesulfonate; Konlax; Loxeen; Lyseen; Mitanoline. Anticholinergic; antiparkinsonian. mp = 152.5-155°; sparingly soluble in H_2O. *Glaxo Wellcome Inc.; Wellcome Foundation Ltd.*

2846 Prifinium Bromide
4630-95-9 7923 225-051-8
$C_{22}H_{28}BrN$
3-(Diphenylmethylene)-1,1-diethyl-2-methylpyrrolidinium bromide.
pyrodifenium bromide; Padrin; Riabal. Antispasmodic. Peripheral cholinoceptor antagonist. mp = 216-218°; LD_{50} (mmus iv) = 11 mg/kg, (mmus ip) = 43 mg/kg, (mmus sc) =30 mg/kg, (mmus orl) = 330 mg/kg; [free base]: $bp_{0.15}$ = 183-185°.

2847 Procyclidine
77-37-2 7944 201-023-0
$C_{19}H_{29}NO$
α-Cyclohexyl-α-phenyl-1-pyrrolidinepropanol.
Skeletal muscle relaxant; anticholinergic; antiparkinsonian. mp = 85.5-86.5°;λ_m = 285.5 nm (ε 233 17% EtOH); [methosulfate ($C_{21}H_{35}NO_5S$)]: mp = 100°; soluble in H_2O, EtOH. *Eli Lilly & Co.; Glaxo Wellcome Inc.*

2848 Procyclidine Hydrochloride
1508-76-5 7944 216-141-8
$C_{19}H_{30}ClNO$
α-Cyclohexyl-α-phenyl-1-pyrrolidinepropanol hydrochloride.
Arpicolin; Kemadrin; Osnervan. Skeletal muscle relaxant; anticholinergic; antiparkinsonian. Dec 226-227°; soluble in H_2O (3 g/100 ml), organic solvents. *Eli Lilly & Co.; Glaxo Wellcome Inc.*

2849 Proglumide
6620-60-6 7958 229-567-4
$C_{18}H_{26}N_2O_4$
(±)-4-Benzamido-N,N-dipropylglutaramic acid.
xylamide; W-5219; CR-242; Milid; Milide; Nulsa; Promid. Anticholinergic. Cholecystokinin inhibitor. mp = 142-145°; LD_{50} (mus iv) = 2211-2649 mg/kg, (mus orl) = 7350-8861 mg/kg. *Wallace Labs.*

2850 Propantheline Bromide
50-34-0 7989 200-030-6
$C_{23}H_{30}BrNO_3$
2-(Hydroxyethyl)diisopropylmethylammonium bromide xanthene-9-carboxylate.
Pro-Banthine; Neo-Metantyl; Pantheline; Corrigast; Ercotina. Anticholinergic. mp = 159-161°; very soluble in H_2O, EtOH, $CHCl_3$; insoluble in Et_2O, C_6H_6. *Roberts Pharm. Corp.; Searle G.D. & Co.*

2851 Propenzolate
4354-45-4 7998
$C_{20}H_{29}NO_3$
(+)-1-Methyl-3-piperidyl (±)-α-phenylcyclohexaneglycolate.
Anticholinergic. *Lakeside BioTechnology; Marion Merrell Dow Inc.*

2852 Propenzolate Hydrochloride
1420-03-7 7998
$C_{20}H_{30}ClNO_3$
(+)-1-Methyl-3-piperidyl (±)-α-phenylcyclohexaneglycolate hydrochloride.
Delinal; NDR-263. Anticholinergic. mp = 222°. *Lakeside BioTechnology; Marion Merrell Dow Inc.*

2853 Propiverine
60569-19-9 8018
$C_{23}H_{29}NO_3$
1-Methyl-4-piperidyl diphenylpropoxyacetate.
Anticholinergic.

2854 Propiverine Hydrochloride
54556-98-8 8018
$C_{23}H_{30}ClNO_3$
1-Methyl-4-piperidyl diphenylpropoxyacetate hydrochloride.
P-4; Mictonorm; Mictonetten. Anticholinergic. Used in treatment of urinary incontinence. mp = 216-218°; sparingly soluble in H_2O, EtOH; LD_{50} (mmus iv) = 36 mg/kg, (mmus sc) = 223 mg/kg, (mmus orl) = 410 mg/kg, (fmus iv) = 36 mg/kg, (fmus sc) = 283 mg/kg, (fmus orl) = 323 mg/kg, (mrat iv) = 22 mg/kg, (mrat sc) = 1632 mg/kg, (mrat orl) = 1000 mg/kg, (frat iv) = 25 mg/kg, (frat sc) = 1411 mg/kg, (frat orl) = 1092 mg/kg.

2855 Propyromazine
145-54-0 8057 205-657-9
$C_{20}H_{23}BrN_2OS$
1-Methyl-1-(1-phenothiazin-10-ylcarbonylethyl)-pyrrolidinium bromide.
LD-335; SD-104-19; Diaspasmyl. Anticholinergic; antispasmodic. mp = 228-229°; [free base ($C_{19}H_{20}N_2OS$)]: mp = 94.5-95.5°. *Astra Chem. Ltd.*

2856 Scopolamine
51-34-3 8550 200-090-3
$C_{17}H_{21}NO_4$
6β,7β-Epoxy-1αH,5αH-tropan-3α-ol (-)-tropate (ester).
hyoscine; l-scopolamine; scopine tropate; S-(-)-tropate; tropic acid ester with scopine; Scop; Scopoderm-TTS; Transcop; Transderm Scop. Anticholinergic; antiemetic. Used in treatment of motion sickness. Tropane alkaloid found in *Scolinacea, particularly Datura metel l. Scopola carniolica. Constituent of impure duboisine, Duboisia myoporides.* Pure duboisine is l-hyoscyamine. Viscous liquid; decomposes; $[\alpha]_D^{20}$ = -28° (c = 2.7); soluble in H_2O (10.5 g/100 ml at 15°); freely soluble in hot H_2O, EtOH, Et_2O, $CHCl_3$, Me_2CO; sparingly soluble in C_6H_6, petroleum ether; [monohydrate]: mp = 59°. *Alcon Labs; Ciba-Geigy Corp.; Parke-Davis; Robins, A. H. Co.; Wallace Labs; Zeneca Pharm.*

2857 Scopolamine Hydrobromide
6533-68-2 8550 200-090-3
$C_{17}H_{22}BrNO_4.3H_2O$
6β,7β-Epoxy-1αH,5αH-tropan-3α-ol (-)-tropate (ester) hydrobromide trihydrate.
Scopolamine hydrobromide; Isopto Hyoscine; Transderm-Scop; component of: Barbidonna, Benacine, Donnagel, Donnatal, Donnazyme, Kinesed. Anticholinergic. Used in

treatment of motion sickness. mp = 195°; $[\alpha]_D^{25}$ = -24° to -26° (c = 5); λ_m = 246, 252, 258, 264 nm (A1%;ks$_{1\ cm}$ 3.5, 4.0, 4.5, 3.0 MeOH); soluble in H_2O (66 g/100 ml); EtOH (5 g/100 ml), $CHCl_3$; insoluble in Et_2O; LD_{50} (rat sc) = 3800 mg/kg. *Alcon Labs; Ciba-Geigy Corp.; Parke-Davis; Robins, A. H. Co.; Wallace Labs; Zeneca Pharm.*

2858 Scopolamine N-oxide
97-75-6 8551 202-606-2
$C_{17}H_{21}NO_5$
6β,7β-Epoxy-1αH,5αH-tropan-3α-ol (-)-tropate 8-oxide.
scopolamine aminoxide. Anticholinergic. mp = 80°; $[\alpha]_D^{20}$ = -14° (H_2O); [hydrobromide (Genoscopolamine)]: mp = 135-138°; $[\alpha]_D^{25}$ = -25° (c = 2); soluble in EtOH, Me_2CO. *Alcon Labs; Ciba-Geigy Corp.; Parke-Davis; Robins, A. H. Co.; Wallace Labs; Zeneca Pharm.*

2859 Secoverine
57558-44-8
$C_{22}H_{35}NO_2$
1-Cyclohexyl-4-[ethyl(p-methoxy-α-methylphenethyl)-amino]-1-butanone.
Selective muscarinic antagonist. Inhibits gastric secretion and gastric mobility. Spasmolytic.

2860 Stercuronium Iodide
30033-10-4
$C_{26}H_{43}IN_2$
(Cona-4,6-dienin-3β-yl)dimethylethyl-ammonium iodide.
Short-acting peripheral neuromuscular blocking agent. Muscarinic antagonist.

2861 Stramonium
8063-18-1 8975
Thorn apple; Jamestown weed; Jimpson weed; Jimson weed; stinkweed; devil's apple; apple of Peru. Anticholinergic. From dried leaves and flowering tops of *Datura stramonium L. Solanaceae.* Alkaloids including atropine, hyoscyamine, scopolamine.

2862 Sultroponium
15130-91-3 9170
$C_{20}H_{29}NO_6S$
endo-(±)-3-(3-Hydroxy-1-oxo-2-phenylpropoxy)-8-methyl-8-(3-sulfopropyl)-8-azoniabicyclo[3.2.1]octane inner salt.
8-(3-sulfopropyl)atropinium hydroxide inner salt; atropine sulfobetaine; sulfobetain of atropine; tropyl tropate sulfobetaine; A-118; Sultropan. Anticholinergic; antispasmodic. mp = 220° (dec); soluble in H_2O, warm EtOH; insoluble in Et_2O, Me_2CO, C_6H_6, $CHCl_3$; λ_m = 251.5, 257.5, 263.5 nm (H_2O).

2863 Telenzepine
80880-90-6 9270
$C_{19}H_{22}N_4O_2S$
4,9-Dihydro-3-methyl-4-[(4-methyl-1-piperazinyl)acetyl]-10H-thieno[3,4-b][1,5]benzodiazepin-10-one.
Antiulcerative; anticholinergic. Selective muscarinic M_1-receptor antagonist. Gastric acid inhibitor. mp = 263-264°. *Byk Gulden Lomberg GmbH.*

2864 Tematropium Methylsulfate
113932-41-5
$C_{21}H_{31}NO_8S$
3α-Hydroxy-8-methyl-1αH,5αH-tropanium methyl sulfate (salt) (±)-ethyl hydrogen phenylmalonate.
Anticholinergic. *Pharmos Corp.*

2865 Temiverine
173324-94-2
$C_{24}H_{35}NO_3$
(±)-4-Diethylamino-1,1-dimethylbut-2-yn-1-yl 2-cyclohexyl-2-hydroxy-2-phenylacetate.
p-INN; NS-21 [as hydrochloride monohydrate]. Used in treatment of urinary frequency and incontinence. Anticholinergic; calcium antagonist.

2866 Thiphenamil
82-99-5 9509
$C_{20}H_{25}NOS$
S-[2-(Diethylamino)ethyl] diphenylthioacetate.
Anticholinergic; smooth muscle relaxant. *Sterling Winthrop Inc.*

2867 Thiphenamil Hydrochloride
548-68-5 9509 208-955-7
$C_{20}H_{26}ClNOS$
S-[2-(Diethylamino)ethyl] diphenylthioacetate hydrochloride.
Thiphen; Trocinate. Anticholinergic; smooth muscle relaxant. mp = 129-130°; soluble in H_2O. *Sterling Winthrop Inc.*

2868 Tiemonium
6252-92-2 228-380-5
$C_{18}H_{25}NO_2S$
4-[3-Hydroxy-3-phenyl-3-(2-thienyl)propyl]-4-methylmorpholine.
Anticholinergic; antispasmodic. *C.E.R.M.*

2869 Tiemonium Iodide
144-12-7 9571 205-616-5
$C_{18}H_{24}INO_2S$
4-[3-Hydroxy-3-phenyl-3-(2-thienyl)propyl]-4-methylmorpholinium iodide.
TE-114; Visceralgina. Anticholinergic; antispasmodic. mp = 189-191°. *C.E.R.M.*

2870 Timepidium Bromide
35035-05-3 9583
$C_{17}H_{22}BrNOS_2$
3-(Di-2-thienylmethylene)-5-methoxy-1,1-dimethylpiperidinium bromide.
SA-504; Mepidium; Sesden. Anticholinergic. mp = 198-200°. *Tanabe Seiyaku Co. Ltd.*

2871 Tiotropium Bromide
139404-48-1 9598
$C_{19}H_{22}BrNO_4S_2$
(1α,2β,4β,5α,7β)-7-[(Hydroxydi-2-thienylacetyl)oxy]-9,9-dimethyl-3-oxa-9-azoniatricyclo[3.3.1.0^{2,4}]nonane bromide.
di-2-thienylglycolate; Ba-679-BR; Spiriva. A muscarinic antagonist used for treatment of obstructive airways disease. Bronchodilator; anticholinergic. mp = 218-220°. *Boehringer Ingelheim Ltd.*

2872 Tiquinamide
53400-67-2
$C_{11}H_{14}N_2S$
5,6,7,8-Tetrahydro-3-methylthio-8-quinolinecarboxamide.
WY-24081. Anticholinergic. *Wyeth-Ayerst Labs.*

2873 Tiquinamide Hydrochloride
53400-68-3
$C_{11}H_{15}ClN_2S$
5,6,7,8-Tetrahydro-3-methylthio-8-quinolinecarboxamide monohydrochloride.
WY-24081 HCl. Anticholinergic. *Wyeth-Ayerst Labs.*

2874 Tiquizium Bromide
71731-58-3 9602
$C_{19}H_{24}BrNS_2$
trans-3-(Di-2-thienylmethylene)octahydro-5-methyl-2H-quinolizinium bromide.
HSR-902; HS-902; Thiaton. Antispasmodic. Anticholinergic. mp = 278-281° (dec). *Hokoriku.*

2875 Tofenacin
15301-93-6 9641 239-338-0
$C_{17}H_{21}NO$
N-Methyl-2-[(o-methyl-α-phenylbenzyl)oxy]ethylamine.
Anticholinergic; antidepressant. $bp_{0.7}$ = 139-143°. *Brocades-Stheeman & Pharmacia.*

2876 Tofenacin Hydrochloride
10488-36-5 9641 234-011-9
$C_{17}H_{22}ClNO$
N-Methyl-2-[(o-methyl-α-phenylbenzyl)oxy]ethylamine hydrochloride.
BS-7331; Elamol; Tofacine. Anticholinergic; antidepressant. mp = 143-147°; freely soluble in H_2O, EtOH, MeOH, $CHCl_3$; slightly soluble in Me_2CO; insoluble in Et_2O; LD_{50} (mus orl) = 182 mg/kg, (mus sc) = 82 mg/kg, (mus ip) = 58 mg/kg, (mus iv) = 36.5 mg/kg, (rat ip) = 72 mg/kg, (gpg sc) = 92 mg/kg. *Brocades-Stheeman & Pharmacia.*

2877 Tolterodine
124937-51-5
$C_{22}H_{31}NO$
(+)-(R)-2-[α-[2-(Diisopropylamino)ethyl]benzyl]-p-cresol.
Detrol. Antimuscarinic. Used to treat urinary incontinence. *Pharmacia & Upjohn.*

2878 Toquizine
7125-71-5
$C_{23}H_{29}N_5O$
N,N,3,5,6-Pentamethyl-2-pyrazinamine.
Anticholinergic. *Eli Lilly & Co.*

2879 Triampyzine
6503-95-3
$C_9H_{15}N_3$
(Dimethylamino)trimethylpyrazine.
Anticholinergic.

2880 Triampyzine Sulfate
7082-30-6
$C_9H_{17}N_3O_4S$
N,N,3,5,6-Pentamethyl-2-pyrazinamine sulfate.
W-3976-B. Anticholinergic.

2881 Tricyclamol Chloride
3818-88-0 223-311-5
$C_{20}H_{32}ClNO$
(±)-1-(3-Cyclohexyl-3-hydroxy-3-phenylpropyl)-1-methylpyrrolidinium chloride.
Anticholinergic; antiparkinsonian. mp = 226-227° (dec); soluble in H_2O (3 g/100 ml); more soluble in EtOH, $CHCl_3$; insoluble in Et_2O. *Eli Lilly & Co.*

2882 Tridihexethyl
60-49-1 9794
$C_{19}H_{31}NO$
(3-Cyclohexyl-3-hydroxy-3-phenylpropyl)diethylamine.
Anticholinergic. Anti-inflammatory. *Am. Cyanamid; Lederle Labs.; Wallace Labs.*

2883 Tridihexethyl Chloride
4310-35-4 9794 224-323-3
$C_{21}H_{36}ClNO$
(3-Cyclohexyl-3-hydroxy-3-phenylpropyl)triethylammonium chloride.
Pathilon. Anticholinergic. *Am. Cyanamid; Lederle Labs.; Wallace Labs.*

2884 Tridihexethyl Iodide
125-99-5 9794 204-762-7
$C_{21}H_{36}INO$
(3-Cyclohexyl-3-hydroxy-3-phenylpropyl)triethylammonium iodide.
propethonium iodide; tridihexethide; 921 C; Claviton; component of: Milpath. Anticholinergic. mp = 179-184°; soluble in H_2O (1.1 g/100 ml); freely soluble in EtOH, $CHCl_3$; slightly soluble in Et_2O. *Am. Cyanamid; Lederle Labs.; Wallace Labs.*

2885 Trihexyphenidyl
144-11-6 9823 205-614-4
$C_{20}H_{31}NO$
α-Cyclohexyl-α-phenyl-1-piperidinepropanol.
Anticholinergic; antiparkinsonian. mp = 114.3-115°; [l-form]: mp = 112-113°; $[\alpha]_D^{20}$ = -25° (c = 0.4 EtOH). *Am. Cyanamid; Glaxo Wellcome plc.*

2886 Trihexyphenidyl Hydrochloride
52-49-3 9823 200-142-5
$C_{20}H_{32}ClNO$
α-Cyclohexyl-α-phenyl-1-piperidinepropanol hydrochloride.
Antitrem; Artane; benzhexol chloride; Aparkane; Broflex; Cyclodol; Pacitane; Paralest; Pargitan; Parkinane; Parkopan; Peragit; Pipanol; Sedrina; Tremin; Triphedinon; Triphenidyl; Tsiklodol. Anticholinergic; antiparkinsonian. mp = 258.5° (dec); soluble in H_2O (1.0 g/100 ml), EtOH (6 g/100 ml), $CHCl_3$ (5 g/100 ml); soluble in MeOH; slightly soluble in C_6H_6, Et_2O; [l-form]: mp = 264°; $[\alpha]_D^{20}$ = -30° (c = 0.4 $CHCl_3$). *Am. Cyanamid; Glaxo Wellcome plc.*

2887 Trimebutine
39133-31-8 9829 254-309-2
$C_{22}H_{29}NO_5$
3,4,5-Trimethoxybenzoic acid 2-(dimethylamino)-2-phenylbutyl ester.
Anticholinergic; antispasmodic. Opiate receptor agonist. Accelerates gastric emptying. mp = 78-80°; soluble in CH_2Cl_2. *Jouveinal.*

2888 Trimebutine Maleate
34140-59-5 9829 251-845-9
$C_{26}H_{33}NO_9$
3,4,5-Trimethoxybenzoic acid 2-(dimethylamino)-2-phenylbutyl ester maleate.
TM-906; Cerekinon; Debridat; Digerent; Foldox; Polibutin; Spabucol; Trimedat. Anticholinergic; antispasmodic. Opiate receptor agonist. Accelerates gastric emptying. mp = 105-106°. *Jouveinal.*

2889 Tropacine
6878-98-4 9903
$C_{22}H_{25}NO_2$
endo-α-Phenylbenzeneacetic acid-8-methyl-8-azabicyclo[3.2.1]oct-3-yl ester.
tropine diphenylacetate. Anticholinergic. [hydrochloride]: mp = 217-218°. *Ciba-Geigy Corp.*

2890 Tropatepine
27574-24-9
$C_{22}H_{23}NS$
3-Dibenzo[b,e]thiepin-11(6H)-ylidine-1αH,5αH-tropane.
SD-1248-17 [as hydrochloride]. Anticholinergic.

2891 Tropenzile
53834-53-0 9908
$C_{23}H_{27}NO_4$
(3-endo,6-exo)-α-Hydroxy-α-phenylbenzeneacetic acid 6-methoxy-8-methyl-8-azabicyclo[3.2.1]oct-3-yl ester.
tropenzilium bromide [as methyl bromide]; component of: Palerol [as hydrobromide]. Anticholinergic; antispasmodic. mp = 99-101°; [hydrochloride ($C_{23}H_{28}ClNO_4$)]: mp = 146-148°. *Sandoz Pharm. Corp.*

2892 Tropenzile Hydrochloride
9908
$C_{23}H_{28}ClNO_4$
(3-endo,6-exo)-α-Hydroxy-α-phenylbenzeneacetic acid 6-methoxy-8-methyl-8-azabicyclo[3.2.1]oct-3-yl ester hydrochloride.
Anticholinergic; antispasmodic. mp = 146-148°. *Sandoz.*

2893 Tropicamide
1508-75-4 9911 216-140-2
$C_{17}H_{20}N_2O_2$
N-Ethyl-2-phenyl-N-(4-pyridylmethyl)hydracrylamide.
Mydriacyl; Mydriaticum; component of: paremyd. Anticholinergic (ophthalmic). mp = 96-97°; λ_m = 254 nm (ε 5100 0.025 mg/ml in 0.1N HCl). *Alcon Labs; Allergan Inc.; Hoffmann-LaRoche Inc.*

2894 Trospium Chloride
10405-02-4 9918 233-875-4
$C_{25}H_{30}ClNO_3$
endo-3-[(Hydroxydiphenylacetyl)oxy]spiro[8-azonia-bicyclo[3.2.1]octane-8,1'-pyrrolidinium] chloride.
Relaspium; Spasmex. Anticholinergic; antispasmodic. mp = 255-257°; LD_{50} (mus iv) = 12.3 mg/kg. *Pfleger Dr.*

2895 Valethamate Bromide
90-22-2 10044 201-977-8
$C_{19}H_{32}BrNO_2$
N,N-Diethyl-N-methyl-2-(3-methyl-2-phenylvaleryloxy)ethylammonium bromide.
Resitan; Epidosin; Murel. Anticholinergic. mp = 100-101°; freely soluble in H_2O, EtOH; insoluble in Et_2O. *Kali-Chemie.*

2896 Vamicamide
132373-81-0 10052
$C_{18}H_{23}N_3O$
(±)-(R*)-α-[(R*)-2-(Dimethylamino)propyl]-α-phenyl-2-pyridineacetamide.
FK-176. Anticholinergic. mp = 156-157°, 132-134°; [(-)-(2R,4R) form]: mp = 104°; $[\alpha]_D$ = -103.1° (c = 0.5 MeOH); [(+)-(2S,4S) form]: mp = 103-104°; $[\alpha]_D$ = +103.3° (c = 0.5 MeOH). *Fujisawa Pharm. USA Inc.*

2897 Xenytropium Bromide
511-55-7 10208 208-129-6
$C_{30}H_{34}BrNO_3$
endo-(±)-8-([1,1'-Biphenyl]-4-ylmethyl)-3-(3-hydroxy-2-phenyl-1-oxopropoxy)-8-methyl-8-azoniabicyclo[3.2.1]octane bromide.
N-399; Gastropin; Gastripon. Anticholinergic; antispasmodic. Dec = 220-222°. *Licencia Budapest.*

2898 Zamifenacin
127308-82-1
$C_{27}H_{29}NO_3$
(R)-3-(Diphenylmethoxy)-1-[3,4-(methylenedioxy)phenethyl]piperidine.
UK-76654. Selective M3 muscarinic receptor antagonist. Anticholinergic.

Anticoagulants

2899 Acenocoumarol
152-72-7 29 205-807-3
$C_{19}H_{15}NO_6$
3-(α-Acetonyl-p-nitrophenyl)-4-hydroxycoumarin.
acenocoumarin; nicoumalone; G-23350; Minisintrom; Sinthrome; Sintrom. Anticoagulant. mp = 196-199°; insoluble in H_2O or organic solvents; LD_{50} (mus ip) = 114.7 mg/kg. *Ciba-Geigy Corp.*

2900 Ancrod
9046-56-4 671 232-933-6
Agkistrodon serine proteinase.
agkistrododon rhodostoma venom proteinase; A-38414; Abbott 38414; Arvin; Arwin; Venacil. Proteinase obtained from the venom of the Malan pit viper *Agkistrodon rhodostoma*. Acts specifically on fibrinogen. Used as an anticoagulant. Soluble in saline solution. *Abbott Labs Inc.*

2901 Anisindione
117-37-3 706 204-186-6
$C_{16}H_{12}O_3$
2-(p-Methoxyphenyl)indane-1,3-dione.

Miradon; SPE-2792; Unidone. Used as an anticoagulant. mp = 156-157°. *Schering AG.*

2902 Aprosulate Sodium
123072-45-7
$C_{27}H_{34}N_2Na_{16}O_{70}S_{16}$
N,N'-Trimethylenebis[actobionamide] hexadecakis(sodium sulfate) (ester).
Anticoagulant.

2903 Argatroban
74863-84-6 816
$C_{23}H_{36}N_6O_5S$
(2R,4R)-4-Methyl-1-[n^2-[(1,2,3,4-tetrahydro-3-methyl-8-quinolyl)sulfonyl]-L-arginyl]-pipecolic acid.
argipidin; MQPA. Synthetic thrombin inhibitor. Used as an anticoagulant. mp = 188-191°. *Mitsubishi Chem. Corp.*

2904 Argatroban Monohydrate
141396-28-3 816
$C_{23}H_{36}N_6O_5S.H_2O$
(2R,4R)-4-Methyl-1-[n^2-[(1,2,3,4-tetrahydro-3-methyl-8-quinolyl)sulfonyl]-L-arginyl]-pipecolic acid monohydrate.
Novastan; Slonnon; MCI-9038; MD-805; DK-7419; GN1600. Synthetic thrombin inhibitor. Used as an anticoagulant. mp = 176-180°; $[\alpha]_D^{27}$ = 76.1° (c = 1 0.2N HCl). *Mitsubishi Chem. Corp.*

2905 Bivalirudin
128270-60-0
$C_{98}H_{138}N_{24}O_{33}$
D-Phenylalanyl-L-prolyl-L-arginyl-L-prolylglycylglycylglycylglycyl-L-asparaginylglycyl-L-α-aspartyl-L-phenylalanyl-L-α-glutamyl-L-α-glutamyl-L-isoleucyl-L-prolyl-L-α-glutamyl-L-α-glutamyl-L-tyrosyl-L-leucine.
H-D-Phe-Pro-Arg-Pro-Gly-Gly-Gly-Gly-Asn-Gly-Asp-Phe-Glu-Glu-Ile-Pro-Glu-Glu-Tyr-Leu-OH; Angiomax; Hirulog; BG8967. Antithrombotic; anticoagulant. A 20-merpeptide. Used as an anticoagulant in conjunction with aspirin in patients with unstable angina undergoing percutaneous transluminal coronary angioplasty. *Biogen, Inc.*

2906 Bromindione
1146-98-1 1414
$C_{15}H_9BrO_2$
2-(p-Bromophenyl)-1,3-indanedione.
HL-255; MG-2555; Fluidane; Halinone. Used as an anticoagulant. mp = 137-139°. *U.S. Vitamin.*

2907 Clocoumarol
35838-63-2
$C_{21}H_{21}ClO_3$
3-[p-(2-Chloroethyl)-α-propylbenzyl]-4-hydroxycoumarin.
Anticoagulant.

2908 Clorindione
1146-99-2 2468 214-553-2
$C_{15}H_9ClO_2$
2-(p-Chlorophenyl)-1,3-indanedione.
G-25766; MG-2552; Indaliton. Used as an anticoagulant. mp = 145-146°; insoluble in H_2O, soluble in organic solvents; LD_{50} (mus orl) = 1220 mg/kg. *Ciba-Geigy Corp.*

2909 Coumetarol
4366-18-1 2633 224-455-1
$C_{21}H_{16}O_7$
3,3'-(2-Methoxyethylidene)bis(4-hydroxycoumarin).
cumetharol; cumethoxaethane; PH-137; Dicoumoxyl; Dicumoxane. Used as an anticoagulant. mp = 156-157°. *N.V. Amsterdamsche Chininefabriek.*

2910 Cyclocumarol
518-20-7 2786 208-248-3
$C_{20}H_{18}O_4$
3,4-Dihydro-2-methoxy-2-methyl-4-phenyl-2H,5H-pyrano[3,2-c][1]benzopyran-5-one.
anticoagulant No. 63; BL-5; Cumopyran; Cumopyrin. Used as an anticoagulant. mp = 166°; soluble in H_2O (1 g/100 ml); slightly soluble in EtOH, vegetable oil. *Abbott Labs Inc.; Wisconsin Alumni Res. Foundation.*

2911 Desirudin
120993-53-5 4754
$C_{287}H_{440}N_{80}O_{110}S_6$
63-Desulfohirudin.
CGP-39393. A recombinant variation on desulfatohirudin. Used as an anticoagulant. *Ciba-Geigy Corp.*

2912 Dextran Sulfate Sodium
9011-18-1 2993
Dextran sulfuric acid ester sodium salt.
Asuro; Colyonal; Dexulate; Dextrarine; MDS. A heparin-like polysaccharide. Used in veterinary medicine as an anticoagulant. Molecular weight range 4,000 - 500,000; up to 3 sulfate groups per glucose residue.

2913 Dicumarol
66-76-2 3140 200-632-9
$C_{19}H_{12}O_6$
3,3'-Methylenebis[4-hydroxycoumarin].
bishydroxycoumarin; Dicoumarol; Dicoumarin; Dicumol; Dufalone; Melitoxin. Used as an anticoagulant. mp = 297-293°; insoluble in H_2O, EtOH, Et_2O; slightly soluble in C_6H_6, C_5H_5N, $CHCl_3$; LD_{50} (rat orl) = 541.6 mg/kg. *Abbott Labs Inc.; Wisconsin Alumni Res. Foundation.*

2914 Diphenadione
82-66-6 3363 201-434-5
$C_{23}H_{16}O_3$
2-(Diphenylacetyl)-1,3-indandione.
diphacinone; U-1363; Dipaxin; Oragulant; Solvan; Didandin; Diphacin. Used as an anticoagulant and rodenticide. mp = 146-147°; insoluble in H_2O; slightly soluble in C_6H_6, EtOH; soluble in Me_2CO, AcOH; sodium salt is sparingly soluble in H_2O. *Pharmacia.*

2915 Ethyl Biscoumacetate
548-00-5 3818 208-940-5
$C_{22}H_{16}O_8$
Ethyl bis(4-hydroxy-2-oxo-2H-1-benzopyran-3-yl)-acetate.
B.O.E.A.; Dicumacyl; Pelentan; Stabilene; Tromexan; Tromexan Ethyl Acetate. Used as an anticoagulant. mp =

154-157°, 177-182°; insoluble in H_2O; soluble in Me_2CO (5 g/100 ml), C_6H_6; slightly soluble in EtOH, Et_2O; LD_{50} (mus orl) = 0.88 mg/kg, (mus ip) = 0.32 mg/kg, (rat orl) = 0.26 mg/kg, (rat ip) = 1.1 mg/kg. *Spojene.*

2916 Ethylidene Dicoumarol
1821-16-5 3858 217-342-3
$C_{20}H_{14}O_6$
3,3'-Ethylidenebis[4-hydroxy-2H-1-benzopyran-2-one].
Pertrombon. Used as an anticoagulant. mp = 178°; [dibenzoate ($C_{34}H_{22}O_8$)]: mp = 209-210°. *SPOFA.*

2917 Fluindarol
6723-40-6
$C_{16}H_9F_3O_2$
2-(α,α,α-Trifluoro-p-tolyl)indan-1,3-dione.
Anticoagulant.

2918 Fluindione
957-56-2 4168 213-484-5
$C_{15}H_9FO_2$
2-(p-Fluorophenyl)-1,3-indandione.
LM-123; Previscan. Used as an anticoagulant. mp = 120°; LD_{50} (mus orl) = 240 mg/kg. *Lipha Pharm. Inc.; U.S. Vitamin.*

2919 Heparin
9005-49-6 4685 232-681-7
Heparinic acid.
Arteven; Leparan. Anticoagulant. A glycosaminoglycan with anticoagulant properties. Binds to antithrombin III to accelerate the interaction of antithrombin III with coagulation factors. Catalyzes the inhibition of thrombin by heparin cofactor II. $[\alpha]_D^{20}$ = 55°.

2920 Heparin Calcium
37270-89-6 4685
Calcium heparinate.
Calciparine; Ecasolv. Used as an anticoagulant.

2921 Heparin Sodium
9041-08-1 4685
Sodium heparinate.
Ardeparin Sodium; Dalteparin Sodium; Fragmin; Heparin Fragment Kabi 2165; Heprinar; Hepsal; Lipo-Hepin; Liquémin; Lipo-Hepinette; Longheparin; Monoparin; Normiflo; Panheprin; Pularin; Liquaemin Sodium; Minihep; Thrombo-Hepin; Thromboliquine; Thrombophob; Tinzaparin Sodium; Unihep. Sodium salt of polymerized heparin obtained by heparinase (through nitrous acid degradation) from *Flavobacterium heparinum* (heparin lyase). Used as an anticoagulant. Molecular weight range 5500-6500 with 2.7 sulfate residues/disaccharide unit. $[\alpha]_D^{25}$ = 47° (c = 1.5 H_2O); soluble in H_2O (5 g/100 ml), saline solution; insoluble in EtOH, Me_2CO, C_6H_6, $CHCl_3$, Et_2O. *Pharmacia Hepar Inc.; Pharmacia.*

2922 Hexadimethrine Bromide
9011-04-5
$(C_{13}H_{30}Br_2N_2)_n$
N,N,N',N'-Tetramethyl-1,6-hexanediamine polymer with 1,3-dibromopropane.
Anticoagulant.

2923 Hirudin
8001-27-2 4754 232-279-1
Exhirud; Exhirudine; Hirudex. Anticoagulant extracted from the leach *Hirudo medicinalis.* Used as an anticoagulant. Single chain polypeptide with 65 amino acid residues.

2924 Lepirudin
138068-37-8
$C_{287}H_{440}N_{80}O_{111}S_6$
1-L-Leucine-2-L-threonine-63-desulfohirudin.
recombinant hirudin; r-hirudin; HBW-023; Leu-Thr-Tyr-Thr-Asp-Cys(6)-Thr-Glu-Ser-Gly-Asn-Leu-Cys(14)-Leu-Cys(16)-Glu-Gly-Ser-Asn-Val-Cys(22)-Gly-Gln-Gly-Asn-Lys-Cys(28)-Ile-Leu-Gly-Ser-Asp-Gly-Glu-Lys-Asn-Gln-Cys(39)-Val-Thr-Gly-Glu-Glu-Thr-Pro-Lys-Pro-Gln-Ser-His-Asn-Asp-Gly-Asp-Phe-Glu-Glu-Ile-Pro-Glu-Glu-Tyr-Leu-Gln [(6→14),(16→28),(22→39) cyclic disulfide. Recombinant hirudin (*hirudo medicinalis* isoform HV1). Produces parenteral anticoagulation for treatment of heparin-induced thrombocytopenia. Antithrombotic; anticoagulant.

2925 Lyapolate Sodium
25053-27-4 5645
PES; Peson
Sodium ethenosulfonate polymer.
sodium apolate; ethene sulfonic acid homopolymer. Used as an anticoagulant. *Hoechst Roussel Pharm. Inc.*

2926 Nafamostat
81525-10-2 6436
$C_{19}H_{17}N_5O_2$
6-Amidino-2-naphthyl-4-guanidinobenzoate.
nafamstat. Protease Inhibitor. Non-peptide enzyme inhibitor with inhibitory effects on trypsin, thrombin, kallikrein, plasmin, and complement-mediated hemolysis. *Torii Pharm. Co. Ltd.*

2927 Nafamostat Dimethanesulfonate
82956-11-4 6436
$C_{19}H_{17}N_5O_2.2CH_3SO_3H$
6-Amidino-2-naphthyl-4-guanidinobenzoate dimethanesulfonate.
nafamastat mesylate; FUT-175; Futhan. Used as an anticoagulant. Protease Inhibitor. Non-peptide enzyme inhibitor with inhibitory effects on trypsin, thrombin, kallikrein, plasmin, and complement-mediated hemolysis. mp = 217-220°, 260° (dec). *Torii Pharm. Co. Ltd.*

2928 Oxazidione
27591-42-0 7060 248-550-2
$C_{20}H_{19}NO_3$
2-(Morpholinomethyl)-2-phenyl-1,3-indanedione.
mofedione; LD-4610; Transidione; Amplidione. Used as an anticoagulant. mp = 88°; LD_{50} (mus orl) = 235 mg/kg; [hydrochloride]: mp = 150°. *Lab. Dausse.*

2929 Pentosan Polysulfate
37300-21-3 7276
$[C_5H_8O_{10}S_2]_{6-12}$
(1→4)-β-D-Xylan 2,3-bis(hydrogen sulfate).
xylan hydrogen sulfate; xylan polysulfate; CB-8061. Anti-inflammatory (interstitial cystitis); antithrombotic; anticoagulant. A semi-synthetic sulfated polyxylan polysulfate composed of β-D-xylopyranose residues with

properties similar to heparin. Molecular weight 1500-5000. *Wander Pharma.*

2930 Pentosan Polysulfate Sodium Salt
37319-17-8 7276
$[C_5H_6Na2O_{10}S_2]_{6-12}$
(1→4)-β-D-Xylan 2,3-bis(hydrogen sulfate) sodium salt.
SP54; PZ68; sodium pentosan polysulfate; sodium xylan polysulfate; Cartrophen; Elmiron; Fibrase; Fibrezym; Héoclar; Thrombocid. Anti-inflammatory (interstitial cystitis); antithrombotic; anticoagulant. Sodium salt of sulfated β-D-xylopyranose polymer. $[\alpha]_D^{20}$ = -57°; pH (10% aqueous solution) 6.0; N_D^{20} (10% aqueous solution) = 1.344; soluble in H_2O. *Wander Pharma.*

2931 Phenindione
83-12-5 7381 201-454-4
$C_{15}H_{10}O_2$
2-Phenyl-1,3-indanedione.
Hedulin; Indon; Pindione; Bindan; Dindevan; Dineval; Hemolidione; Indema; Fenilin; Fenhydren; Rectadione; Cronodione; Thrombasal. Used as an anticoagulant. mp = 149-151°; insoluble in cold H_2O; slightly soluble in warm H_2O; soluble in MeOH, EtOH, Et_2O, Me_2CO, C_6H_6, $CHCl_3$. *Marion Merrell Dow Inc.; Parke-Davis.*

2932 Phenprocoumon
435-97-2 7413 207-108-9
$C_{18}H_{16}O_3$
3-(α-Ethylbenzyl)-4-hydroxycoumarin.
Falithrom; Marcoumar; Marcumar; Liquamar. Used as an anticoagulant. mp = 179-180°. *Hoffmann-LaRoche Inc.*

2933 Phosvitin
9008-96-2 7522
A phosphoprotein of molecular weight ca. 40,000, isolated from egg yolk. Used as an anticoagulant.

2934 Picotamide
32828-81-2 7560 251-245-7
$C_{21}H_{20}N_4O_3$
4-Methoxy-N,N'-bis(3-pyridinylmethyl)-1,3-benzenedicarboxamide.
G-137; Plactidil [as monohydrate]. Antithrombotic; fibrinolytic; anticoagulant. Used as an anticoagulant. mp = 124°; LD_{50} (mus ip) = 1205 mg/kg. *Soc. Italo-Brit. L. Manetti.*

2935 Tioclomarol
22619-35-8 9594 245-132-1
$C_{22}H_{16}Cl_2O_4S$
3-[5-Chloro-α(p-chloro-β-hydroxyphenethyl)-2-thenyl]-4-hydroxycoumarin.
Apegmone. Used as an anticoagulant. mp = 104°. *Lipha Pharm. Inc.*

2936 Warfarin
81-81-2 10174 201-377-6
$C_{19}H_{16}O_4$
3-(α-Acetonylbenzyl)-4-hydroxycoumarin.
compound 42; WARF compound 42; Co-Rax; Rodex. Used as a rodenticide and an anticoagulant. Orally effective, fat soluble derivative of 4-hydroxycoumarin that induce hypocoagulability only *in vivo* by inducing the formation of structurally incomplete clotting factors. mp = 161°; λ_m = 308 nm (ε 13610 H_2O); soluble in Me_2CO, dioxane; slightly soluble in MeOH, EtOH, iPrOH, some oils; insoluble in H_2O, C_6H_6, C_6H_{12}.

2937 Warfarin Sodium
129-06-6 10174 204-929-4
$C_{19}H_{15}NaO_4$
3-(α-Acetonylbenzyl)-4-hydroxycoumarin sodium salt.
Coumadin; Panwarfin; Marevan; Prothromadin; Tintorane; Warfilone; Waran. Used as a rodenticide and an anticoagulant. Very soluble in H_2O, EtOH; slightly soluble in $CHCl_3$, Et_2O; LD_{50} (mrat orl) = 323 mg/kg, (frat orl) = 58 mg/kg, (mus orl) = 374 mg/kg, (rbt orl) ≅ 800 mg/kg, (mrat orl) = 100.3 mg/kg, (frat orl) = 8.7 mg/kg. *Abbott Labs Inc.; DuPont-Merck Pharm.*

Anticonvulsants

2938 Albutoin
830-89-7 218
$C_{10}H_{16}N_2OS$
3-Allyl-5-isobutyl-2-thiohydantoin.
BAX422Z; CO-ORD; Euprax. An anticonvulsant. mp = 210-211°.

2939 Aloxidone
526-35-2 315 208-389-0
$C_7H_9NO_3$
3-Ally-5-methyl-2,4-oxazolidinedione.
allomethadione; Malidone; Malazol. An anticonvulsant. $bp_{0.5}$ = 86-87°, $bp_{1.8}$ = 88-90°; slightly soluble in H_2O, soluble in EtOH. *Schering Plough Ltd.*

2940 Ameltolide
787-93-9
$C_{15}H_{16}N_2O$
4-Amino-2',6'-benzoxylidide.
LY-201116. An anticonvulsant. *Eli Lilly & Co.*

2941 Aminoglutethimide
125-84-8 460 204-756-4
$C_{13}H_{16}N_2O_2$
2-(p-Aminophenyl)-2-ethylglutarimide.
p-Aminoglutethimide; Ba-16038; NSC-330915; Cytadren; Elipten; Orimeten. Adrenocortical suppressant; anticonvulsant; antineoplastic agent. Aromatase inhibitor. Has been used in treatment of Cushing's syndrome and other adrenal hormone disorders and in palliative treatment of breast cancer. mp = 149-150°; insoluble in H_2O; poorly soluble in EtOAc, EtOH; readily soluble in other organic solvents; [hydrochloride]: mp = 223-225°, soluble in H_2O. *Ciba-Geigy Corp.*

2942 4-Amino-3-Hydroxybutyric Acid
352-21-6 464 206-518-5
$C_4H_9NO_3$
σ-Amino-β-hydroxybutyric acid.
buksamin; GABOB; Gabomade; Gamibetal. An anticonvulsant. [DL-form]: mp = 218° (dec); soluble in H_2O, sparingly soluble in organic solvents; [D(+)-form]: mp = 214° (dec); $[\alpha]_D^{20}$ = 18.3° (c = 2 H_2O); [L(-)-form]: mp = 212° (dec), 216-217° (dec); $[\alpha]_D^{2-}$ = -21.06° (c = 2 H_2O,

-20.7° (c = 1.83 H_2O). *Antonio Gallardo S.A.; Kaken Pharm. Co. Ltd.*

2943 Atolide
16231-75-7
$C_{18}H_{23}N_3O$
2-Amino-4'-(diethylamino)-o-benzotoluidide.
W-5733; Go-1213. An anticonvulsant. *Parke-Davis.*

2944 Atrolactamide
2019-68-3 904
$C_9H_{11}NO_2$
2-Hydroxy-2-phenylpropionamide.
M-144; Themisone. An anticonvulsant. [DL-form]: mp = 102°; freely soluble in H_2O; [L-form]: mp = 62.5-63.5°; $[\alpha]_D^{15}$ = 12.8° (c = 2.2 Me_2CO); freely soluble in H_2O; soluble in EtOH, Me_2CO.

2945 Avizafone
65617-86-9
$C_{22}H_{27}ClN_4O_3$
2'-Benzoyl-4'-chloro-2-[(S)-2,6-diaminohexanamido]-N-methylacetanilide.
Ro-03-7355/000; Pro-diazepam. Anticonvulsant; anxiolytic. *Hoffmann-LaRoche Inc.*

2946 Beclamide
501-68-8 1045 207-927-1
$C_{10}H_{12}ClNO$
N-Benzyl-3-chloropropionamide.
chloroethylphenamide; Chloracon; Hibicon; Neuracen; Nidrane; Posedrine; Seclar. An anticonvulsant. mp = 94°; slightly soluble in H_2O (0.005 - 0.01 g/100 ml); moderately soluble in EtOH, MeOH. *Am. Cyanamid.*

2947 Buramate
4663-83-6 1525
$C_{10}H_{13}NO_3$
2-Hydroxyethyl benzylcarbamate.
AC-601; NSC-30223; glycol benzylcarbamate; 2-hydroxyethyl benzylcarbamate; mono(benzylcarbamate); Hyamate. Antipsychotic; anticonvulsant. mp = 40°. *Ascher B.F. & Co.; Xttrium Labs. Inc.*

2948 Calcium Bromide
7789-41-5 1693 232-164-6
Br_2Ca
Calcium dibromide.
Anticonvulsant; sedative. Also used to treat hypocalcemia in veterinary medicine. mp = 730°; d_4^{25} = 3.353; soluble in H_2O, MeOH, EtOH, Me_2CO; insoluble in dioxane, $CHCl_3$, Et_2O.

2949 Carbamepazine
298-46-4 1826 206-062-7
$C_{15}H_{12}N_2O$
5H-Dibenz[b,f]azepine-5-carboxamine.
G-32883; Biston; Calepsin; Carbelan; Epitol; Finlepsin; Sirtal; Stazepine; Tegretal; Tegretol; Telesmin; Timonil. Analgesic; anticonvulsant. mp = 190-193°; Soluble in alcohol, Me_2CO, propylene glycol; nearly insoluble in H_2O; LD_{50} (mus orl) = 3750 mg/kg, (rat orl) = 4025 mg/kg. *Ciba-Geigy Corp.; Lemmon Co.*

2950 Cinromide
58473-74-8 2371
$C_{11}H_{12}BrNO$
(E)-m-Bromo-N-ethylcinnamamide.
BW-122U; Vumide. An anticonvulsant. mp = 89-90°; LD_{50} (mus ip) = 660 ± 28 mg/kg, (mus orl) = 2277 ± 250 mg/kg. *Glaxo Wellcome Inc.*

2951 Citenamide
10423-37-7
$C_{16}H_{13}NO$
5H-Dibenzo[a,d]cycloheptene-5-carboxamide.
AY-15613. An anticonvulsant. *Wyeth-Ayerst Labs.*

2952 Clomethiazole
533-45-9 2444 208-565-7
C_6H_8ClNS
5-(2-Chloroethyl)-4-methylthiazole.
S.C.T.Z.; chlorethiazol; chlormethiazole. A sedative/hypnotic; anticonvulsant. Oily liquid, d_4^{25} = 1.233; bp_7 = 92°; [hydrochloride]: mp = 130°; soluble in H_2O, EtOH; [methanedilsulfonate]: mp = 120°. *Hoffmann-LaRoche Inc.*

2953 Clomethiazole Ethanedisulfonate
1867-58-9 2444 217-483-0
$C_{14}H_{22}Cl_2N_2O_6S_4$
5-(2-Chloroethyl)-4-methylthiazole ethanedisulfonate.
SCTZ; Distraneurin; Hemineurin; Heminevrin. Anticonvulsant. Sedative/hypnotic. mp = 124°. *Hoffmann-LaRoche Inc.*

2954 Clonazepam
1622-61-3 2449 216-596-2
$C_{15}H_{10}ClN_3O_3$
5-(o-Chlorophenyl)-1,3-dihydro-7-nitro-2H-1,4-benzodiazepin-2-one.
Clonopin; Klonopin; Ro-5-4023; Iktorivil; Landsen; Rivotril. Antiepileptic agent with anxiolytic and antimanic properties. A benzodiazepine. Used as an an anticonvulsant. mp = 236.5-238.5°; λ_m = 248, 310 nm (ε 14500 11600 MeOH/iPrOH); soluble in Me_2CO (3.1 g/100 ml), $CHCl_3$ (1.5 g/100 ml), MeOH (0.86 g/100 ml), Et_2O (0.07 g/100 ml), C_6H_6 (0.05 g/100 ml), H_2O (< 0.01 g/100 ml); LD_{50} (mus orl) > 4000 mg/kg. *Hoffmann-LaRoche Inc.*

2955 Cyheptamide
7199-29-3 2828 230-570-8
$C_{16}H_{15}NO$
10,11-Dihydro-5H-dibenzo[a,d]cycloheptene-5-carboxamide.
AY-8682. An anticonvulsant. mp = 193-194°; soluble in $CHCl_3$; sparingly soluble in MeOH, Me_2CO; slightly soluble in EtOH, Et_2O; insoluble in H_2O; LD_{50} (mus orl) = 4200-5200 mg/kg, (mus ip) = 2400-2600 mg/kg. *Wyeth-Ayerst Labs.*

2956 Decimemide
14817-09-5 2909
$C_{19}H_{31}NO_4$
4-(Decyloxy)-3,5-dimethoxybenzamide.
EGYT-1050; Denegyt. An anticonvulsant. mp = 121-122°;

LD_{50} (mus orl) = 2950 mg/kg, (rat orl) = 1650 mg/kg. *EGYT*.

2957 Denzimol
73931-96-1
$C_{19}H_{20}N_2O$
(±)-α-(p-Phenethylphenyl)imidazole-1-ethanol.
Anticonvulsant.

2958 Dezinamide
91077-32-6
$C_{11}H_{11}F_3N_2O_2$
3-[(α,α,α)-m-Tolyloxy]-1-azetidinecarboxamide.
AHR-11748; AN-051. An anticonvulsant. *Athena Neurosciences Inc.*

2959 Diethadione
702-54-5 3155 211-867-1
$C_8H_{13}NO_3$
5,5-Diethyldihydro-2H-1,3-oxazine-2,4(3H)-dione.
Dioxone; Ledosten; Persisten; Tocèn. Anticonvulsant; analeptic. CNS stimulant. mp = 97-98°; LD_{50} (mus ip) = 52 mg/kg, (mus orl) = 130 mg/kg, (rat ip) = 31.6 mg/kg, (rat orl) = 70.5 mg/kg. *Gruppo Lepetit S.p.A.*

2960 Dimethadione
695-53-4 3260 211-781-4
$C_5H_7NO_3$
5,5-Dimethyl-2,4-oxazolidinedione.
BAX-1400Z; AC-1198; DMO; NSC-30152; Eupractone. Anticonvulsant. Metabolite of trimethadione. mp = 76-77°; LD_{50} (mus iv) = 450 mg/kg.

2961 Divalproex Sodium
76584-70-8 10049
$(C_{16}H_{31}NaO_4)_n$
Sodium hydrogen bis(2-propylvalerate) oligomer.
Depakote; Abbott 50711. Anticonvulsant; epileptic. *Abbott Labs Inc.*

2962 Doxenitoin
3254-93-1 3491 221-851-6
$C_{15}H_{14}N_2O$
5,5-Diphenyl-4-imidazolidinone.
SKF-2599; Glior. An anticonvulsant. mp = 183°, 185.5-186.5°; soluble in AcOH; moderately soluble in EtOH, EtOAc, C_6H_6, $CHCl_3$; insoluble in H_2O, petroleum ether; [hydrochloride]: mp = 205-206° (dec). *SmithKline Beecham Pharm.*

2963 Dulozafone
75616-02-3
$C_{20}H_{22}Cl_2N_2O_4$
2-[Bis(2-hydroxyethyl)amino]-4'-chloro-2'-(o-chlorobenzoyl)-N-methylacetanilide.
Anticonvulsant.

2964 Esuprone
91406-11-0
$C_{13}H_{14}O_5S$
7-Hydroxy-3,4-dimethylcoumarin ethanesulfonate.
Anticonvulsant.

2965 Etazepine
88124-27-0
$C_{17}H_{17}NO_2$
(±)-11-Ethoxy-5,11-dihydro-5-methyl-6H-dibenz[b,e]azepin-6-one.
Anticonvulsant.

2966 Eterobarb
27511-99-5 3758
$C_{16}H_{20}N_2O_5$
5-Ethyl-1,3-bis(methoxymethyl)-5-phenyl-barbituric acid.
Antilon; EX 12-095; RMI-16238; eterobarbital. An anticonvulsant. Structurally similar to phenobarbital. mp = 116-118°; LD_{50} (mus orl) = 470 mg/kg. *Marion Merrell Dow Inc.*

2967 Ethadione
520-77-4 3762 208-297-0
$C_7H_{11}NO_3$
3-Ethyl-5,5-dimethyl-2,4-oxazolidinedione.
3-Ethyl-5,5-dimethyl-2,4-diketooxazolidine; Dimedione; Petidiol; Didione; Petidion; Epinyl; Etydion; Petisan; Neo-Absentol. An anticonvulsant. mp = 76-77°. *Schering Plough Ltd.*

2968 Ethosuximide
77-67-8 3794 201-048-7
$C_7H_{11}NO_2$
2-Ethyl-2-methylsuccinimide.
Zarontin; CI-366; CN-10395; PM-671; NSC-64013; Atysmal; Capitus; Emeside; Epileo Petitmal; Ethymal; Mesentol; Pemal; Peptinimid; Petinimid; Petnidan; Pyknolepsinum; Simatin; Succimal; Suxilep; Suximal; Suxinutin. An anticonvulsant. mp = 64-65°; freely soluble in H_2O; LD_{50} (mus ip) = 1650 mg/kg, (mus orl) = 1750 mg/kg. *Parke-Davis.*

2969 Ethotoin
86-35-1 3795 201-665-1
$C_{11}H_{12}N_2O_2$
3-Ethyl-5-phenylimidazolidin-2,4-dione.
Peganone. An anticonvulsant. mp = 94°; sparingly soluble in cold H_2O; more soluble in hot H_2O; soluble in EtOH, Et_2O, C_6H_6. *Abbott Labs Inc.*

2970 Etiracetam
33996-58-6
$C_8H_{14}N_2O_2$
(±)-α-Ethyl-2-oxo-1-pyrrolidineacetamide.
Anticonvulsant.

2971 Felbamate
25451-15-4 3988 247-001-4
$C_{11}H_{14}N_2O_4$
2-Phenyl-1,3-propanediol dicarbamate.
Felbatol; W-554; Felbamyl; Taloxa; ADD-03055. An anticonvulsant. Structurally related to meprobamate. mp = 151-152°; sparingly soluble in H_2O, EtOH, MeOH, Me_2CO, $CHCl_3$; freely soluble in DMSO, DMF, 1-methyl-2-pyrrolidinone; LD_{50} (mus ip) = 4000 mg/kg. *Carter-Wallace.*

2972 Fluoresone
2924-67-6 4197 220-889-0
$C_8H_9FO_2S$
Ethyl p-fluorophenyl sulfone.
Bripadon; Caducid. Anticonvulsant; analgesic; anxiolytic. A tricyclic benzodiazepine derivative. mp = 41°; LD_{50} (mus orl) = 2500 mg/kg, 850 mg/kg, 542 mg/kg.

2973 Flurazepam
17617-23-1 4233 241-591-7
$C_{21}H_{23}ClFN_3O$
7-Chloro-1-[2-(diethylamino)ethyl]-5-(o-fluorophenyl)-1,3-dihydro-2H-1,4-benzodiazepin-2-one.
Felmane; Noctosom; Stauroderm. Anticonvulsant; sedative/hypnotic; muscle relaxant. A benzodiazepine. mp = 77-82°. *Hoffmann-LaRoche Inc.*

2974 Flurazepam Hydrochloride
1172-18-5 4233 214-630-0
$C_{21}H_{25}Cl_3FN_3O$
7-Chloro-1-[2-(diethylamino)ethyl]-5-(o-fluorophenyl)-1,3-dihydro-2H-1,4-benzodiazepin-2-one dihydrochloride.
Dalmane; Ro-5-6901; NSC-78559; Benozil; Dalmadorm; Dalmate; Dormodor; Felson; Insumin; Lunipax; Somlan. An anticonvulsant with sedative/hypnotic properties. Muscle relaxant. A benzodiazepine. mp = 190-220°; LD_{50} (mus ip) = 290 mg/kg, (mus orl) = 870 mg/kg, (mus iv) = 84 mg/kg. *Hoffmann-LaRoche Inc.*

2975 Fluzinamide
76263-13-3
$C_{12}H_{13}F_3N_2O_2$
N-Methyl-3-[(α,α,α-trifluoro-m-tolyl)oxy]-1-azetidinecarboxamide.
AHR-8559. An anticonvulsant. *Robins, A. H. Co.*

2976 Fosphenytoin Sodium
92134-98-0
$C_{16}H_{13}N_2Na_2O_6P$
3-(Hydroxymethyl)-5,5-diphenylhydantoin disodium phosphate (ester).
Cetebyx; ACC-9653-010 [as sodium salt]. An anticonvulsant. *Parke-Davis.*

2977 Gabapentin
60142-96-3 4343 262-076-3
$C_9H_{17}NO_2$
1-(Aminomethyl)cyclohexaneacetic acid.
Neurontin; CI-945; GOE-3450. An anticonvulsant. An amino acid related to gama-aminobutyric acid (GABA). mp = 162-166°, 165-167°; soluble in H_2O (> 10 g/100 ml at pH 7.4). *Parke-Davis.*

2978 Gacyclidine
68134-81-6
$C_{16}H_{25}NS$
1-[cis-2-Methyl-1-(2-thienyl)cyclohexyl]piperidine.
An NMDA antagonist. Antiepileptic and neuroprotective.

2979 Ganaxolone
38398-32-2
$C_{22}H_{36}O_2$
3α-Hydroxy-3-methyl-5α-pregnan-20-one.
Antiepileptic; anticonvulsant.

2980 Glutaurine
56488-60-9
$C_7H_{14}N_2O_6S$
N-(2-Sulfoethyl)-L-glutamine.
Antiepileptic; anticonvulsant.

2981 5-Hydroxytryptophan
56-69-9 4895 200-284-8
$C_{11}H_{12}N_2O_3$
5-Hydroxytryptophan.
5-HTP. Anticonvulsant; antidepressant; antimigraine.

2982 D-5-Hydroxytryptophan
4350-07-6 4895
$C_{11}H_{12}N_2O_3$
D-5-Hydroxytryptophan.
5-HTP. Anticonvulsant; antidepressant; antimigraine. $[\alpha]_D^{20}$ = 32.2° (H_2O).

2983 DL-5-Hydroxytryptophan
114-03-4 4895 204-039-6
$C_{11}H_{12}N_2O_3$
DL-5-Hydroxytryptophan.
5-HTP; Prétonine; Quietim. Anticonvulsant; antidepressant; antimigraine. mp = 298-300° (dec); λ_m = 278 nm (H_2O at pH 6); soluble in H_2O (1 g/100 ml at 5°, 5.5 g/100 ml at 100°), 50% EtOH (2.5 g/100 ml at 70°).

2984 L-5-Hydroxytryptophan
4350-09-8 4895 224-411-1
$C_{11}H_{12}N_2O_3$
L-5-Hydroxytryptophan.
5-HTP; oxitriptan; L-5HTP; Cincofarm; Levothym; Lévotinine; Oxyfan; Serotonyl; Telesol; Triptene; Tript-Oh. Anticonvulsant; antidepressant; antimigraine. $[\alpha]_D^{20}$ = -32.5° (H_2O), $[\alpha]_D^{20}$ = 16° (4N HCl).

2985 Ilepcimide
82857-82-7
$C_{15}H_{17}NO_3$
1-[(E)-3,4-(Methylenedioxy)cinnamoyl]piperidine.
Antiepilepsirine. An anticonvulsant. *Beijing Medical University/Ohio State University.*

2986 Lamotrigine
84057-84-1 5367 281-901-8
$C_9H_7Cl_2N_5$
3,5-Diamino-6-(2,3-dichlorophenyl)-as-triazine.
Lamictal; BW-430C; LTG. An anticonvulsant. mp = 216-218°; LD_{50} (mus orl) = 250 mg/kg. *Glaxo Wellcome Inc.*

2987 Levetiracetam
102767-28-2
$C_8H_{14}N_2O_2$
(S)-α-Ethyl-2-oxo-1-pyridolidineacetamide.
ucb-L059. Anticonvulsant. Anticonvulsant.

2988 Losigamone
112856-44-7
$C_{12}H_{11}ClO_4$
(5R*)-5-[(αS*)-o-Chloro-α-hydroxybenzyl]-4-methoxy-2(5H)furanone.
AO-33. Anticonvulsant; antiepileptic.

2989 Magnesium Bromide
7789-48-2 5695 232-170-9

Br_2Mg
Magnesium dibromide.
Anticonvulsant; sedative. Also used in organic synthesis. mp = 165° (dec); soluble in H_2O (333 g/100 ml), soluble in EtOH.

2990 Magnesium Sulfate [anhydrous]
7487-88-9 5731 231-298-2

$MgSO_4$
Sulfuric acid magnesium salt (1:1).
Kieserite [as monohydrate]. Magnesium replenisher. The heptahydrate is an anticonvulsant and cathartic. Also effective in termination of refractory ventricular tachyarrhythmias. Loss of deep tendon reflex is a sign of overdose. *Astra USA Inc.; Dow Chem. U.S.A.; McGaw Inc.*

2991 Magnesium Sulfate Heptahydrate
10034-99-8 5731

$MgSO_4.7H_2O$
Sulfuric acid magnesium salt (1:1) heptahydrate.
bitter salts; Epsom salts. Magnesium replenisher. Also effective in termination of refractory ventricular tachyarrhythmias. Loss of deep tendon reflex is a sign of overdose. d = 1.67; soluble in H_2O (71 g/100 ml at 20°, 91 g/100 ml at 40°), slightly soluble in EtOH. *Astra USA Inc.; Dow Chem. U.S.A.; McGaw Inc.*

2992 Mephenytoin
50-12-4 5898 200-012-8

$C_{12}H_{14}N_2O_2$
5-Ethyl-3-methyl-5-phenylhydantoin.
Mesantoin; Mesdontoin; 3-ethylnirvanol; Phenantoin; Sedantional; Gerot-Epilan; Sacerno; component of: Hydantal. An anticonvulsant. mp = 136-137°; insoluble in H_2O; LD_{100} (rat ip) = 270 mg/kg. *Sandoz Pharm. Corp.*

2993 Mephobarbital
115-38-8 5899 204-085-7

$C_{13}H_{14}N_2O_3$
5-Ethyl-1-methyl-5-phenylbarbituric acid.
Mebaral; Menta-Bal; Phemiton; Prominal; Isonal; component of: Mebroin. An anticonvulsant, sedative and hypnotic. The N-methyl analog of phenobarbital. mp = 176°; slightly soluble in cold H_2O; freely soluble in hot H_2O, EtOH. *Marion Merrell Dow Inc.; Sterling Winthrop Inc.*

2994 Metharbital
50-11-3 6029 200-011-2

$C_9H_{14}N_2O_3$
5,5-Diethyl-1-methylbarbituric acid.
Gemonil. An anticonvulsant. Analog of phenobarbital. mp = 155°; soluble in H_2O (0.12 g/100 ml), EtOH (4.3 g/100 ml), Et_2O (2.6 g/100 ml). *Abbott Labs Inc.*

2995 Methetoin
5696-06-0 6046

$C_{12}H_{14}N_2O_2$
5-Ethyl-1-methyl-5-phenylhydantoin.
Deltoin; N-3; NSC-524411. An anticonvulsant. mp = 210°. *Sandoz Pharm. Corp.*

2996 Methsuximide
77-41-8 6085 201-026-7

$C_{12}H_{13}NO_2$
N,2-Dimethyl-2-phenylsuccinimide.
Celontin; Petinutin; mesuximide. An anticonvulsant. Structurally related to ethosuximide. mp = 52-53°; $bp_{0.1}$ = 121-122°; freely soluble in MeOH, EtOH; LD_{50} (mus orl) = 1550 mg/kg. *Parke-Davis.*

2997 5-Methyl-5-(3-phenanthryl)hydantoin
3784-92-7 6184

$C_{18}H_{14}N_2O_2$
5-Methyl-5-(3-phenathrenyl)-2,4-imidazolidinedione.
An anticonvulsant. mp = 236-237°.

2998 5-Methyl-5-(3-phenanthryl)hydantoin Sodium Salt
74449-59-5 6184

$C_{18}H_{13}N_2NaO_2$
5-Methyl-5-(3-phenathrenyl)-2,4-imidazolidinedione sodium salt.
Bagrosin-Natrium. An anticonvulsant.

2999 3-Methyl-5-Phenylhydantoin
6846-11-3 6187

$C_{10}H_{10}N_2O_2$
3-Methyl-5-phenyl-2,4-imidazolidinedione.
Norantoin; Nuvarone. An anticonvulsant. mp = 162-163°.

3000 Milacemide
76990-56-2

$C_7H_{16}N_2O$
2-(Pentylamino)acetamide.
Antidepressant; anticonvulsant. *Searle G.D. & Co.*

3001 Milacemide Hydrochloride
76990-85-7

$C_7H_{17}ClN_2O$
2-(Pentylamino)acetamide monohydrochloride.
CP-1552S. Antidepressant; anticonvulsant. *Searle G.D. & Co.*

3002 Nabazenil
58019-65-1

$C_{35}H_{55}NO_3$
3-(1,2-Dimethylheptyl)-7,8,9,10-tetrahydro-6,69-trimethyl-6H-dibenzo[b,d]pyran-1-yl hexahydro-1H-azepine-1-butyrate.
SP-175. An anticonvulsant.

3003 Nafimidone Hydrochloride
70891-37-1

$C_{15}H_{13}ClN_2O$
2-Imidazol-1-yl-2'-acetonaphthone monohydrochloride.
An anticonvulsant. *C.M. Ind.; Syntex Labs. Inc.*

3004 Narcobarbital
125-55-3 6509

$C_{11}H_{15}BrN_2O_3$
5-(2-Bromo-2-propenyl)-1-methyl-5-(1-methylethyl)-2,4,6-(1H,3H,5H)-pyrimidinetrione.
enibomal. Sedative/hypnotic; anticonvulsant. Federally controlled substance (depressant). mp = 115°; sparingly soluble in H_2O; soluble in MeOH, EtOH, C_5H_5N. *Riedel de Haen (Chinosolfabrik).*

3005 Narcobarbital Sodium Salt
3329-16-6 6509 222-050-4
$C_{11}H_{14}BrN_2NaO_3$
5-(2-Bromo-2-propenyl)-1-methyl-5-(1-methylethyl)-2,4,6-(1H,3H,5H)-pyrimidinetrione sodium salt.
Eunarcon; Narcotal. Sedative/hypnotic. Anticonvulsant. Federally controlled substance (depressant). Soluble in H_2O. *Riedel de Haen (Chinosolfabrik).*

3006 Nimetazepam
2011-67-8 6641 217-931-5
$C_{16}H_{13}N_3O_3$
1,3-Dihydro-1-methyl-7-nitro-5-phenyl-2H-1,4-benzodiazepin-2-one.
1-methylnitrazepam; nimetazam; S-1530; Elimin; Hypnon. Skeletal muscle relaxant with anticonvulsant properties. The desmethyl derivative of nitrazepam. Federally controlled substance (depressant). mp = 156.5-157.5°; λ_m = 259, 308 nm (ε = 15800, 9600 MeOH); LD_{50} (mmus orl) = 910 mg/kg, (mmus ip) = 970 mg/kg, (mmus sc) = 1500 mg/kg, (fmus orl) = 750 mg/kg, (fmus ip) = 840 mg/kg, (fmus sc) = 1500 mg/kg, (mrat orl) = 1150 mg/kg, (mrat ip) = 970 mg/kg, (mrat sc) = 1000 mg/kg, (frat orl) = 970 mg/kg, (frat ip) = 980 mg/kg, (frat sc) = 1000 mg/kg. *Hoffmann-LaRoche Inc.; Sumitomo Pharm. Co. Ltd.*

3007 Nitrazepam
146-22-5 6667 205-665-2
$C_{15}H_{11}N_3O_3$
1,3-Dihydro-7-nitro-5-phenyl-2H-1,4benzodiazepin-2-one.
Mogadon; Ro-4-5360; Ro-5-3059; NSC-58775; Benzalin; Calsmin; Eatan; Eunoctin; Imeson; Insomin; Ipersed; Mogadan; Nelbon; Neuchlonic; Nitrados; Nitrenpax; Noctesed; Pelson; Radedorm; Remnos; Somnased; Somnibel; Somnite; Sonebon; Surem; Unisomnia. An anticonvulsant and sedative/hypnotic. A benzodiazepine. mp = 224-226°; λ_m = 277.5 nm ($e_{1\ cm}^{1\%}$ = 1500 0.1N H_2SO_4); soluble in EtOH, Me_2CO, EtOAc, $CHCl_3$; insoluble in H_2O, Et_2O, C_6H_6, C_6H_{14}; LD_{50} (rat orl) = 825 ± 80 mg/kg. *Hoffmann-LaRoche Inc.*

3008 Oxcarbazepine
28721-07-5 7063 249-188-8
$C_{15}H_{12}N_2O_2$
10,11-Dihydro-10-oxo-5H-dibenz[b,f]azepine-5-carboxamide.
oxacarbazepine; GP-47680; Trileptal. An anticonvulsant. mp = 215-216°. *Ciba-Geigy Corp.*

3009 Paramethadione
115-67-3 7161 204-098-8
$C_7H_{11}NO_3$
5-Ethyl-3,5-dimethyl-2,4-oxazolidinedione.
Paradione. An anticonvulsant. Liquid; d_4^{25} = 1.1180-1.1240; slightly soluble in H_2O; freely soluble in EtOH, $CHCl_3$, C_6H_6, Et_2O. *Abbott Labs Inc.*

3010 Phenacemide
63-98-9 7343 200-570-2
$C_9H_{10}N_2O_2$
(Phenylacetyl)urea.
Phenurone; phenacetylurea; Epiclase; Phacetur; Phetylureum. An anticonvulsant. mp = 212-216°; slighlty soluble in H_2O, EtOH, C_6H_6, $CHCl_3$, Et_2O; LD_{50} (mus orl) = 987 mg/kg, (rat orl) > 1781 mgkg. *Abbott Labs Inc.*

3011 Phenaglycodol
79-93-6 201-235-3
$C_{11}H_{15}ClO_2$
2-(p-Chlorophenyl)-3-methyl-2,3-butanediol.
Has anxiolytic and possibly antiepilpetic properties.

3012 Phenetharbital
357-67-5 7368 206-615-2
$C_{14}H_{16}N_2O_3$
5,5-Diethyl-1-phenyl-2,4,6(1H,3H,5H)-pyrimidinetrione.
phenidiemal; Fedibaretta; Pyrictal. An anticonvulsant. mp = 178°; freely soluble in hot H_2O.

3013 Pheneturide
90-49-3 7375 201-998-2
$C_{11}H_{14}N_2O_2$
2-Phenylbutyrylurea.
EPA; PBU; Benuride; ethylphenacemide. An anticonvulsant. [dl-form]: mp = 149-150°; [d-form]: mp = 168-169°; $[\alpha]_D^{17}$ = 54.0° (c = 1 EtOH), $[\alpha]_D^{22}$ = 53.8° (c = 1 Me_2CO), 48.2° (c = 1 dioxane); [l-form]: mp = 162-163°; $[\alpha]_D^{30}$ = -51.6° (c = 1 EtOH). *Labs. Sapos.*

3014 Phenobarbital
50-06-6 7386 200-007-0
$C_{12}H_{12}N_2O_3$
5-Ethyl-5-phenyl-2,4,6(1H,3H,5H)-pyrimidinetrione.
5-ethyl-5-phenylbarbituric acid; phenobarbitone; Agrypnal; Barbiphenyl; Barbipil; Eskabarb; Gardenal; Luminal; Phenobal; Solfoton; Talpheno; component of: Antrocol, Barbidonna, Bronkotabs, Chardonna-2, Donnatal, Donnazyme, Hydantal, Kinesed, Levsin PB Drops and Tablets, Quadrinal, Tedral. An anticonvulsant, sedative and hypnotic. A long-acting barbiturate. mp = 174-178°; λ_m = 240 nm ($A_{1\ cm}^{1\%}$ 431 pH10), 256 nm ($A_{1\ cm}^{1\%}$ 314 0.1N NaOH); soluble in H_2O (0.1 g/100 ml), EtOH (12.5 g/100 ml), $CHCl_3$ (2.5 g/100 ml), Et_2O (7.7 g/100 ml), C_6H_6 (0.14 g/100 ml); LD_{50} (mrat orl) = 1.1 mg/kg, (frat orl) = 2.3 mg/kg, (mrat derm) 2.5 mg/kg, (frat derm) 6.2 mg/kg. *ECR Pharm.; Knoll Pharm. Co.; Marion Merrell Dow Inc.; Parke-Davis; Robins, A. H. Co.; Sandoz Pharm. Corp.; Schwarz Pharma Kremers Urban Co.; SmithKline Beecham Pharm.; Sterling Winthrop Inc.; Wallace Labs; Zeneca Pharm.*

3015 Phenobarbital Sodium
57-30-7 7386 200-322-3
$C_{12}H_{11}N_2NaO_3$
5-Ethyl-5-phenyl-2,4,6(1H,3H,5H)-pyrimidinetrione sodium salt.
5-ethyl-5-phenylbarbituric acid sodium salt; Luminal Sodium; sol phenobarbital; sol phenobarbitone. Anticonvulsant sedative/hypnotic. A long-acting barbiturate. Soluble in H_2O (100 g/100 ml), EtOH (10 g/100 ml); insoluble in Et_2O, $CHCl_3$; LD_{50} (rat orl) = 660 mg/kg. *Sterling Winthrop Inc.; Wyeth-Ayerst Labs.*

3016 Phensuximide
86-34-0 7414 201-664-6
$C_{11}H_{11}NO_2$
(±)-N-Methyl-2-phenylsuccinimide.
Milontin; Lifene; Mirontin; Succitimal. An anticonvulsant. mp = 71-73°; slightly soluble in H_2O (≅ 0.42 g/100 ml at 25°); freely soluble in EtOH, MeOH; LD_{50} (mus orl) = 960 mg/kg. *Parke-Davis.*

3017 Phenylmethylbarbituric Acid
76-94-8 7457 200-994-8
$C_{11}H_{10}N_2O_3$
5-Methyl-5-phenyl-2,4,6(1H,3H,5H)-pyrimidinetrione.
Rutonal. An anticonvulsant; sedative/hypnotic. mp = 226°; insoluble in H_2O; soluble in EtOH, Et_2O; has water-soluble sodium salt. *Bayer Corp.*

3018 Phenytoin
57-41-0 7475 200-328-6
$C_{15}H_{12}N_2O_2$
5,5-Diphenylhydantoin.
diphenylhydantoin; Dilantin; Difhydan; Dihycon; Di-Hydan; Ekko; Hydantin; Hydantol; Lehydan; Lepitoin; Phenhydan; Zentropil; component of: Mebroin. Anticonvulsant; class IB antiarrhythmic. mp = 295-298°; insoluble in H_2O; soluble in EtOH (1.67 g/100 ml), Me_2CO (3.32 g/100 ml); LD_{50} (mus iv) = 92 mg/kg, (mus sc) = 110 mg/kg. *Parke-Davis; Sterling Winthrop Inc.*

3019 Phenytoin Sodium
630-93-3 7475 211-148-2
$C_{15}H_{11}N_2NaO_2$
5,5-Diphenylhydantoin sodium salt.
phenytoin soluble; Antisacer; Danten; Diphantoine; Diphenin; Diphenylan sodium; Epanutin; Minetoin; Tacosal; Solantyl; component of: Beuthanasia-D. Anticonvulsant; class IB antiarrhythmic. Soluble in EtOH (9.5 g/100 ml), H_2O (1.5 g/100 ml); insoluble in Et_2O, $CHCl_3$; LD_{50} (mus orl) = 490 mg/kg. *Elkins-Sinn; Schering-Plough Animal Health.*

3020 Phethenylate Sodium
510-34-9 7476
$C_{13}H_9N_2NaO_2S$
5-Phenyl-5-(2-thienyl)-2,4-imidazolidinedione monosodium salt.
Thiantoin Sodium; Thioantoine sodium. Anticonvulsant. Soluble in H_2O, EtOH; [free base]: mp = 256-257°.

3021 Pivagabine
69542-93-4
$C_9H_{17}NO_3$
4-[(2,2-Dimethyl-1-oxopropyl)amino]butanoic acid.
Tonerg. Modulates corticotropin releasing factor and has anxiolytic and antidepressant effects. Anticonvulsant.

3022 Potassium Bromide
7758-02-3 7780 231-830-3
BrK
Anticonvulsant; sedative. CAUTION: Large doses may cause CNS suppression. mp = 730°; d = 2.75; soluble in H_2O (66.6 g/100 ml at 25°, 100 g/100 ml at 100°), EtOH (0.4 g/100 ml), glycerol (21.7 g/100 ml).

3023 Primidone
125-33-7 7927 204-737-0
$C_{12}H_{14}N_2O_2$
5-Ethyldihydro-5-phenyl-4,6(1H,5H)-pyrimidinedione.
Mysoline; Neurosyn; Liskantin; Mylepsin; Resimatil; Sertan. An anticonvulsant. Analog of phenobarbital. mp = 281-282°; sparingly soluble in H_2O (0.06 g/100 ml); soluble in most organic solvents. *Fermenta Animal Health Co.; ICI; Wyeth-Ayerst Labs.*

3024 Progabide
62666-20-0 7955 263-679-4
$C_{17}H_{16}ClFN_2O_2$
4-[[α-(p-Chlorophenyl)-5-fluorosalicylidene]-amino]butyramide.
halogabide; Gabren; Gabrene; SL-76002. Smooth muscle relaxant; anticonvulsant. An antagonist of σ-aminobutyric acid with antiepileptic activity. mp = 133-135°, 142.5°; λ_m = 332, 250, 210 (ε 4200, 10800, 24000 MeOH); LD_{50} (mus ip) = 900 mg/kg. *Synthelabo Pharmacie.*

3025 Ralitoline
93738-40-0
$C_{13}H_{13}ClN_2O_2S$
(Z)-6'-Chloro-3-methyl-4-oxo-$\Delta^{2,\alpha}$thiazolidineaceto-o-toluidide.
CI-946. An anticonvulsant. *Parke-Davis.*

3026 Retigabine
150812-12-7
$C_{16}H_{18}FN_3O_2$
N-(2-Amino-4-(4-fluorobenzylamino)phenyl)carbamic acid ethyl ester.
D-23129; D-20443 [as dihydrochloride]. Anticonvulsant.

3027 Ropizine
3601-19-2
$C_{24}H_{26}N_4$
1-(Diphenylmethyl)-4-[[(6-methyl-2-pyridyl)-methylene]amino]piperazine.
SC-13504. An anticonvulsant. *Searle G.D. & Co.*

3028 Sabeluzole
104383-17-7 8462
$C_{22}H_{26}FN_3O_2S$
(±)-4-(2-Benzothiazolylmethylamino)-α-[(p-fluorophenoxy)methyl]-1-piperidineethanol.
R-58735; Reminyl. Anticonvulsant. Nootropic. mp = 101.7°. *Janssen Pharm. Inc.*

3029 Sodium Bromide
7647-15-6 8737 231-599-9
BrNa
Sedoneural. Sedative/hypnotic; anticonvulsant (veterinary). mp = 755°; d = 3.21; soluble in H_2O (91 g/100 ml), EtOH (6.25 g/100 ml), MeOH (16.6 g.100 ml); cyrstallizes as a dihydrate; LD_{50} (rat orl) = 3500 mg/kg.

3030 Solanum
8856
Bull nettle; Radical weed; sand-brier; horse-nettle
Air-dried ripe fuit of *Solanum carolinense L., Solanaceae.* Plant product, contains solanine and solanidine.

3031 Soretolide
130403-08-6
$C_{13}H_{14}N_2O_2$
2,6-Dimethyl-N-(5-methyl-3-isoxazolyl)benzamide.
D-2916. Antiepileptic.

3032 Stiripentol
49763-96-4 256-480-9
$C_{14}H_{18}O_3$
4,4-Dimethyl-1-[(3,4-methylenedioxy)phenyl]-1-penten-3-ol.
BCX-2600. An anticonvulsant. *Labs. Biocodex.*

3033 Strontium Bromide
10476-81-0 8997 233-969-5
Br_2Sr
Strontium dibromide.
An anticonvulsant. [hexahydrate]: mp = 88°, 643°; soluble in H_2O (285 g/100 ml), EtOH; insoluble in Et_2O, LD_{50} (rat ip) = 1000 mg/kg.

3034 Suclofenide
30279-49-3 9047 250-111-5
$C_{16}H_{13}ClN_2O_4S$
3-Chloro-4-(phenylsuccinimido)benzenesulfonamide.
CGP-8426; GS-385; Sulfalepsin. An anticonvulsant. mp = 205-207°; LD_{50} (mus orl) > 5000 mg/kg. *E. Geistlich Sohne.*

3035 Sulthiame
61-56-3 9167 200-511-0
$C_{10}H_{14}N_2O_4S_2$
p-(Tetrahydro-2H-1,2-thiazin-2-yl)benzenesulfonamide.
Conadil; Trolone; Riker 594; Elisal; Ospolot. An anticonvulsant. Carbonic anydrase inhibitor. mp = 180-182°; insoluble in cold H_2O; slightly soluble in boiling H_2O, EtOH. *3M Pharm.; Riker Labs.*

3036 Taltrimide
81428-04-8
$C_{13}H_{16}N_2O_4S$
N-Isopropyl-1,3-dioxo-2-isoindolineethanesulfonamide.
Anticonvulsant.

3037 Tetrantoin
52094-70-9 9372 257-658-9
$C_{12}H_{12}N_2O_2$
3',4'-Dihydrospiro[imidazolidine-4,2'-(1'H)-naphthalene]-2,5-dione.
Spirodon. An anticonvulsant. mp = 267-268°. *Cutter Labs.*

3038 Thiopental Sodium
71-73-8 9487 200-763-1
$C_{11}H_{17}N_2NaO_2S$
Sodium 5-ethyl-5-(1-methylbutyl)-2-thiobarbiturate.
Pentothal; thiomebumal sodium; penthiobarbital sodium; thiopentone sodium; thionembutal; Intraval Sodium; Nesdonal Sodium; Pentothal Sodium; Trapanal. Intravenous anesthetic; anticonvulsant. Ultra-short-acting barbiturate. Soluble in H_2O, EtOH; insoluble in C_6H_6, Et_2O, petroleum ether; LD_{50} (mus ip) = 149 mg/kg, (mus iv) = 78 mg/kg. *Abbott Labs Inc.*

3039 Tiagabine
115103-54-3 9557
$C_{20}H_{25}NO_2S_2$
(-)-(R)-1[4,4-Bis(3-methyl-2-thienyl)-3-butenyl] nipecotic acid.
Abbott 70569.1; Abbott 70569.HCl; ABT-569; NO-050328; NNC-05-0328; TGB. Anticonvulsant. *Abbott.*

3040 Tiagabine Hydrochloride
145821-59-6 9557
$C_{20}H_{26}ClNO_2S_2$
(-)-(R)-1[4,4-Bis(3-methyl-2-thienyl)-3-butenyl]nipecotic acid hydrochloride.
Gabitril; Tiabex. An anticonvulsant. mp = 192° (dec); $[\alpha]_D^{20}$ = -11°; soluble in H_2O (3 g/100 ml), insoluble in C_6H_{14}. *Abbott Labs Inc.*

3041 Tiletamine Hydrochloride
14176-50-2
$C_{12}H_{18}ClNOS$
2-(Ethylamino)-2-(2-thienyl)cyclohexanone hydrochloride.
CI-634; CL-399; CN-54521-2; component of: Telazol. An anticonvulsant. *Parke-Davis.*

3042 Topiramate
97240-79-4 9686
$C_{12}H_{21}NO_8S$
2,3:4,5-Di-O-isopropylidene-β-D-fructopyranose sulfamate.
Topamax; McN-4853; RWJ-17021. An anticonvulsant. mp = 125-126°; $[\alpha]_D^{23}$ = -34.0° (c = 0.4 MeOH). *McNeil Pharm.*

3043 Treptilamine
58313-74-9
$C_{20}H_{27}NO$
2-[(α-Tricyclo[2.2.1.0^{2,6}]hept-3-ylidenebenzyl)-oxy]triethylamine.
Spasmolytic.

3044 Trimethadione
127-48-0 9836 204-845-8
$C_6H_9NO_3$
3,5,5-Trimethyl-2,4-oxazolidinedione.
Tridione; Absentol; Epidione; Petidon; Ptimal. Anticonvulsant. An oxazolidinedione. mp = 46-46.5°; bp_5 ≅ 78°; soluble in H_2O (≅ 5g/100 ml); soluble in EtOH, C_6H_6, $CHCl_3$, Et_2O; insoluble in petroleum ether. *Abbott.*

3045 Valproic Acid
99-66-1 10049 202-777-3
$C_8H_{16}O_2$
Propylvaleric acid.
Depakene; 44089; Mylproin. An anticonvulsant. bp_{14} = 120-121°, bp_{20} = 128-130°; d_{25} = 0.904; slightly soluble in H_2O; LD_{50} (rat orl) = 670 mg/kg. *Abbott Labs Inc.; Solvay Pharm. Inc.*

3046 Valpromide
2430-27-5 10050 219-394-2
$C_8H_{17}NO$
2-Propylvaleramide.

dipropylacetamide; Dépamide. An anticonvulsant. mp = 125-126°; insoluble in H_2O. *Bofors.*

3047 Vigabatrin
60643-86-9 10114
$C_6H_{11}NO_2$
4-Amino-5-hexenoic acid.
Sabril; MDL-71754; GVG; RMI-71754. An anticonvulsant. mp = 209°; freely soluble in H_2O; LD_{50} (mus ip) > 2500 mg/kg. *Hoechst Marion Roussel Inc.*

3048 Zoniclezole Hydrochloride
121929-46-2
$C_{12}H_{11}Cl_2N_3O$
5-Chloro-3-(1-imidazol-1-ylethyl)-1,2-benzisoxazole monohydrochloride.
CGS-18416A. An anticonvulsant. *Ciba-Geigy Corp.*

3049 Zonisamide
68291-97-4 10323
$C_8H_8N_2O_3S$
1,2-Benzisoxazole-3-methane-sulfonamide.
CI-912; AD-810; PD-110843; Aleviatin; Exceglan; Excegram. An anticonvulsant. mp = 160-163°, 162-166°; sparingly soluble in H_2O, $CHCl_3$, C_6H_{14}; soluble in MeOH, EtOH, EtOAc, AcOH; LD_{50} (mus orl) = 1892 mg/kg, (mus sc) = 1273 mg/kg, (mus ip) = 699 mg/kg, (mus iv) = 604 mg/kg, (rat orl) = 2001 mg/kg, (rat sc) = 2569 mg/kg, (rat ip) = 733 mg/kg, (rat iv) = 748 mg/kg. *Parke-Davis.*

Antidepressants

3050 Adatanserin
127266-56-2
$C_{21}H_{31}N_5O$
N-[2-[4-(2-Pyrimidinyl)-1-piperazinyl]ethyl]-1-adamantanecarboxamide.
WY-50324. Antidepressant with anxiolytic properties. Adamantyl heteroarylpiperazine with dual serotonin 5-HT(1A) and 5-HT(2) activity. *Wyeth-Ayerst Labs.*

3051 Adatanserin Hydrochloride
144966-96-1
$C_{21}H_{32}ClN_5O$
N-[2-[4-(2-Pyrimidinyl)-1-piperazinyl]ethyl]-1-adamantanecarboxamide monohydrochloride.
WY-50324 HCl. Anxiolytic; antidepressant. An adamantyl heteroarylpiperazine with dual serotonin 5-HT(1A) and 5-HT(2) activity. *Wyeth-Ayerst Labs.*

3052 Adinazolam
37115-32-5 159
$C_{19}H_{18}ClN_5$
8-Chloro-1-[(dimethylamino)methyl]-6-phenyl-4H-s-triazolo[4,3-a][1,4]-benzodiazepine.
U-41123. Antidepressant with anxiolytic properties. Sedative. A triazolobenzodiazepine. Dimethylamino derivative of alprazolam. mp = 171-172.5°. *Ciba-Geigy Corp.; Pratt Pharm.; Upjohn Ltd.*

3053 Adinazolam Mesylate
57938-82-6 159
$C_{20}H_{22}ClN_5O_3S$
8-Chloro-1-[(dimethylamino)methyl]-6-phenyl-4H-s-triazolo[4,3-a][1,4]benzodiazepine monomethanesulfonate.
U-41123F; adinazolam monomethanesulfonate; Deracyn. Antidepressant with anxiolytic properties. Sedative/hypnotic. Tricyclic. mp = 230-244°. *Ciba-Geigy Corp.; Pharmacia & Upjohn; Upjohn Ltd.*

3054 Adrafinil
63547-13-7 168 264-303-1
$C_{15}H_{15}NO_3S$
2-[(Diphenylmethyl)sulfinyl]acetohydroxamic acid.
CRL-40028; Olmifon. Antidepressant. An α-adrenergic agonist. mp = 159-160°; soluble in H_2O (< 1g/l); LD_{50} (mus ip) < 2048 mg/kg, (mus orl) = 1950 mg/kg. *C.M. Ind.; Lab. Lafon France.*

3055 Alaproclate
60719-82-6
$C_{13}H_{18}ClNO_2$
2-(4-Chlorophenyl)-1,1-dimethylethyl ester DL-alanine.
GEA-654. Antidepressant. *Merck & Co.Inc.*

3056 Aletamine
4255-23-6
$C_{11}H_{15}N$
α-Allylphenethylamine.
Antidepressant. *Marion Merrell Dow Inc.*

3057 Aletamine Hydrochloride
4255-24-7
$C_{11}H_{16}ClN$
α-Allylphenethylamine hydrochloride.
NDR-5061A. Antidepressant. *Marion Merrell Dow Inc.*

3058 Alnespirone
138298-79-0
$C_{26}H_{38}N_2O_4$
(+)-(S)-N-[4-[(5-methoxy-3-chromanyl)propylamino]butyl]-1,1-cyclopentanediacetimide.
S 20499. An agonist of 5-HT1A receptors. Antidepressant.

3059 Amedalin
22136-26-1
$C_{19}H_{22}N_20$
3-Methyl-3-[3-(methylamino)propyl]-1-phenyl-2-indolinone.
Antidepressant. *Pfizer Inc.*

3060 Amedalin Hydrochloride
22232-73-1
$C_{19}H_{23}ClN_20$
3-Methyl-3-[3-(methylamino)propyl]-1-phenyl-2-indolinone monohydrochloride.
UK-3540-1. Antidepressant. *Pfizer Inc.*

3061 Amitriptyline
50-48-6 511 200-041-6
$C_{20}H_{23}N$
3-(10,11-Dihydro-5H-dibenzo-[a,d]cyclohepten-5-ylidene)-N,N-dimethyl-1-propanamine.

Antidepressant. Tricyclic. *Hoffmann-LaRoche Inc.; Merck & Co.Inc.*

3062 Amitriptyline Hydrochloride

549-18-8 511 208-964-6

$C_{20}H_{24}ClN$

3-(10,11-Dihydro-5H-dibenzo-[a,d]cyclohepten-5-ylidene)-N,N-dimethyl-1-propanamine hydrochloride.

Ro-4-1575; Adepril; Amineurin; Amitid; Amitril; Deprex; Domical; Elavil; Endep; Euplit; Laroxyl; Lentizol; Miketorin; Redomex; Saroten; Sarotex; Sylvemid; Triptizol; Tryptanol; Tryptizol. Antidepressant. Tricyclic. mp = 196-197°; freely soluble in H_2O, $CHCl_3$, alcohol; λ_m = 240 nm; pKa = 9.4; LD_{50} (rat orl) = 380 mg/kg. *Bristol-Myers Squibb Co.; Lemmon Co.; Merck & Co.Inc.; Parke-Davis; Roche Puerto Rico; Schering-Plough HealthCare Products.*

3063 Amitriptylinoxide

4317-14-0 512

$C_{20}H_{23}NO$

3-(10,11-Dihydro-5H-dibenzo-[a,d]cyclohepten-5-ylidene)-N,N-dimethyl-1-propanamine N-oxide.

amitriptyline N-oxide. Antidepressant. Tricyclic. *Dumex USA.*

3064 Amoxapine

14028-44-5 616 237-867-1

$C_{17}H_{16}ClN_3O$

2-Chloro-11-(1-piperazinyl)dibenz[b,f][1,4]oxazepine.

Asendin; Asendis; CL-67.772; Demolox; Moxadil. Antidepressant. Heterocyclic. mp = 175-176°; LD_{50} (mus ip) = 122 mg/kg, (mus orl) = 112 mg/kg. *Am. Cyanamid; Lederle Labs.*

3065 Aptazapine

71576-40-4

$C_{16}H_{19}N_3$

(±)-1,3,4,14b-Tetrahydro-2-methyl-2H,10H-pyrazino-[1,2-a]pyrrolo[2,1-c][1,4]benzodiazepine.

Antidepressant. Tricyclic.

3066 Aptazapine Maleate

71576-41-5

$C_{20}H_{23}N_3O_4$

(±)-1,3,4,14b-Tetrahydro-2-methyl-2H,10H-pyrazino-[1,2-a]pyrrolo[2,1-c][1,4]benzodiazepine maleate (1:1).

CSG-7525A. Antidepressant. Tricyclic.

3067 Azaloxan

72822-56-1

$C_{18}H_{25}N_3O_3$

(S)-1-[1-[2-(1,4-Benzodioxan-2-yl)ethyl]-4-piperidyl]-2-imidazolidinone.

Antidepressant.

3068 Azaloxan Fumarate

86116-60-1

$C_{22}H_{29}N_3O_7$

(S)-1-[1-[2-(1,4-Benzodioxan-2-yl)ethyl]-4-piperidyl]-2-imidazolidinone fumarate (1:1).

CGS-7135A. Antidepressant.

3069 Azepindole

26304-61-0

$C_{12}H_{14}N_2$

2,3,4,5-Tetrahydro-1H-[1,4]diazepino[1,2-a]indole.

McN-2453. Antidepressant. *McNeil Pharm.*

3070 Azipramine

58503-82-5

$C_{26}H72_6N_2$

1-[2-(Benzylmethylamino)ethyl]-6,7-dihydroindolo-[1,7-ab][1]benzazepine.

Antidepressant. *Pierrel S.p.A.*

3071 Azipramine Hydrochloride

57529-83-6

$C_{26}H_{27}ClN_2$

1-[2-(Benzylmethylamino)ethyl]-6,7-dihydroindolo-[1,7-ab][1]benzazepine monohydrochloride.

Pierrel-TQ 86. Antidepressant. *Pierrel S.p.A.*

3072 Befuraline

41717-30-0

$C_{20}H_{20}N_2O_2$

1-(2-Benzofuranylcarbonyl)-4-benzylpiperazine.

Antidepressant.

3073 Benactyzine

302-40-9 1055 206-123-8

$C_{20}H_{25}NO_3$

2-Diethylaminoethyl benzilate.

Antidepressant; anticholinergic. mp = 51°. *Am. Cyanamid.*

3074 Benactyzine Methobromide

3166-62-9 1055 221-628-3

$C_{21}H_{28}BrNO_3$

2-Diethylaminoethyl benzilate methobromide.

methylbenactyzium bromide; Finalin; Spatomac. Antidepressant; anticholinergic. mp = 169-170°. *Am. Cyanamid.*

3075 Benmoxine

7654-03-7 1072 231-619-6

$C_{15}H_{16}N_2O$

Benzoic acid 2-(1-phenylethyl)hydrazine.

benmoxin, Neuralex. Antidepressant. mp = 93-94°. *ICI.*

3076 Binedaline

60662-16-0 1266

$C_{19}H_{23}N_3$

1-[[2-(Dimethylamino)ethyl]-methylamino]-3-phenylindole.

Binodaline. Antidepressant. mp = 52-53°. *Siegfried AG.*

3077 Binedaline Hydrochloride

57647-35-5 1266 260-877-2

$C_{19}H_{24}ClN_3$

1-[[2-(Dimethylamino)ethyl]-methylamino]-3-phenylindole hydrochloride.

RU-39780; Sgd-Scha-1059; Ixprim. Antidepressant. mp = 195-196°; λ_m = 222, 263 mn (ε 28000, 14500); LD_{50} (mus orl) = 760 mg/kg, (rat orl) = 1380 mg/kg, (mus iv) = 54.0 mg/kg, (rat iv) = 27.2 mg/kg. *Siegfried AG.*

3078 Bipenamol
79467-22-4
$C_{14}H_{15}NOS$
o-[(α-Amino-o-tolyl)thio]benzyl alcohol.
Antidepressant.

3079 Bipenamol Hydrochloride
62220-58-0
$C_{14}H_{16}ClNOS$
o-[(α-Amino-o-tolyl)thio]benzyl alcohol hydrochloride.
BW-647U hydrochloride. Antidepressant.

3080 Brofaromine
63638-91-5
$C_{14}H_{16}BrNO_2$
4-(7-Bromo-5-methoxy-2-benzofuranyl)piperidine.
Antidepressant.

3081 Bupropion
34911-55-2 1523
$C_{13}H_{18}ClNO$
(±)-2-(tert-Butylamino)-3'-chloropropiophenone.
amfebutamon; amfebutamone. Anti-depressant with action similar to that of the tricyclic antidepressants. Pale yellow oil; bp = 52°; soluble in MeOH, EtOH, Me_2CO, Et_2O, C_6H_6. *Burroughs Wellcome Inc.*

3082 Bupropion Hydrochloride
31677-93-7 1523 250-759-9
$C_{13}H_{19}Cl_2NO$
(±)-2-(tert-Butylamino)-3'-chloropropiophenone hydrochloride.
Wellbatrin; Wellbutrin; Zyban. Antidepressant. Also used in smoking cessation therapy. mp = 233-234°; solubility (in H_2O) = 32 mg/ml, (in alcohol) = 193 mg/ml, (in 0.1N HCl) = 333 mg/ml; LD_{50}(mus ip) = 230 mg/kg, (rat ip) = 210 mg/kg, (mus orl) = 575 mg/kg, (rat orl) = 600 mg/kg. *Burroughs Wellcome Inc.*

3083 Butacetin
2109-73-1 1532
$C_{12}H_{17}NO_2$
4'-tert-Butoxyacetanilide.
BW-63-90; NSC-106564; Tromal. Antidepressant; analgesic. mp = 130°. *Burroughs Wellcome Inc.*

3084 Butriptyline
35941-65-2 1568
$C_{21}H_{27}N$
10,11-Dihydro-N,N,β-trimethyl-5H-dibenzo[a,d]cycloheptene-5-propanamine.
butriptylene. Antidepressant. Tricyclic. Oil; bp_1 = 180-185°. *Ayerst; Schering-Plough HealthCare Products.*

3085 Butriptyline Hydrochloride
5585-73-9 1568 226-983-8
$C_{21}H_{28}ClN$
10,11-Dihydro-N,N,β-trimethyl-5H-dibenzo[a,d]-cycloheptene-5-propanamine hydrochloride.
AY-62014; Evadene; Evadyne. Antidepressant. Tricyclic. mp = 188-190°; λ_m = 273, 270, 266 nm (ε 460, 441, 522, MeOH); soluble in H_2O; moderately soluble in aliphatic alcohols, $CHCl_3$; insoluble in Et_2O, paraffinic hydrocarbons; LD_{50} (mus orl) = 345 mg/kg, (mus ip) = 120 mg/kg. *Ayerst; Schering-Plough HealthCare Products.*

3086 Caroxazone
18464-39-6 1907 242-345-1
$C_{10}H_{10}N_2O_3$
2-Oxo-2H-1,3-benzoxazine-3(4H)-acetamide.
FI-6654; Timostenil. Antidepressant. mp = 203-205°. *Farmitalia Carlo Erba Ltd.*

3087 Cartazolate
34966-41-1
$C_{15}H_{22}N_4O_2$
Ethyl 4-(butylamino)-1-ethyl-1H-pyrazolo[3,4-b]-pyridine-5-carboxylate.
SQ-65396. Antidepressant. *Bristol-Myers Squibb Co.*

3088 Cericlamine
112922-55-1
$C_{12}H_{17}Cl_2NO$
(±)-3-(3,4-Dichlorophenyl)-2-(dimethylamino)-2-methyl-1-propanol.
Serotonin reuptake blocker. Possible use as an antidepressant.

3089 Ciclazindol
37751-39-6
$C_{17}H_{15}ClN_2O$
10-(m-Chlorophenyl)-2,3,4,10-tetrahydropyrimido [1,2-a]indol-10-ol.
WY-23409. Antidepressant. *Wyeth Labs.*

3090 Ciclopramine
33545-56-1
$C_{18}H_{20}N_2$
2,3,7,8-Tetrahydro-3-(methylamino)-1H-quino[1,8-ab][1]benzazepine.
Antidepressant.

3091 Cilobamine
69429-84-1
$C_{17}H_{23}Cl_2NO$
cis-2-(3,4-Dichlorophenyl)-3-(isopropylamino)bicyclo[2.2.2]octan-2-ol.
clobamine. Antidepressant. *Marion Merrell Dow Inc.*

3092 Cilobamine Mesylate
69429-85-2
$C_{18}H_{27}Cl_2NO_4S$
cis-2-(3,4-Dichlorophenyl)-3-(isopropylamino)bicyclo-[2.2.2]octan-2-ol methansulfonate (salt).
clobamine mesylate; RMI-81182EF. Antidepressant. *Marion Merrell Dow Inc.*

3093 Cimoxatone
73815-11-9
$C_{19}H_{18}N_2O_4$
α[p-[5-(Methoxymethyl)-2-oxo-3-oxazolidinyl]phenoxy]-m-tolunitrile.
Antidepressant.

3094 Cirazoline
59939-16-1
$C_{13}H_{16}N_2O$
2-[(o-Cyclopropylphenoxy)methyl]-2-imidazoline.
Possible MAO inhibitor. Antidepressant. An imidazoline.

3095 Citalopram
59729-33-8 2379 261-891-1
$C_{20}H_{21}FN_2O$
1-[3-(Dimethylamino)propyl]-1-(p-fluorophenyl)-5-phthalancarbonitrile.
LU-10-171; Bonitrile; Nitalapram. Antidepressant. A serotonin uptake inhibitor. $bp_{0.03}$ = 175-181°. *Kefalas A/S.*

3096 Citalopram Hydrobromide
59729-32-7 2379 261-890-6
$C_{20}H_{22}BrFN_2O$
1-[3-(Dimethylamino)propyl]-1-(p-fluorophenyl)-5-phthalancarbonitrile hydrobromide.
Cipramil. Antidepressant. A serotonin uptake inhibitor. mp = 182-183°. *Kefalas A/S.*

3097 Clodazon
4755-59-3
$C_{18}H_{20}ClN_3O$
5-Chloro-1-[3(dimethylamino)propyl]-3-phenyl-2-benzimidazolinone.
Antidepressant. *Sandoz Pharm. Corp.*

3098 Clodazon Hydrochloride
31959-88-3
$C_{18}H_{23}Cl_2N_3O_2$
5-Chloro-1-[3(dimethylamino)propyl]-3-phenyl-2-benzimidazolinone monohydrochloride monohydrate.
HUF-2446; AW-14'2446. Antidepressant. *Sandoz Pharm. Corp.*

3099 Clodazon Hydrochloride [anhydrous]
4913-61-5
$C_{18}H_{21}Cl_2N_3O$
5-Chloro-1-[3(dimethylamino)propyl]-3-phenyl-2-benzimidazolinone monohydrochloride [anhydrous].
Antidepressant. *Sandoz Pharm. Corp.*

3100 Clomipramine
303-49-1 2447 206-144-2
$C_{19}H_{23}ClN_2$
3-Chloro-5-[3-(dimethylamino)propyl]-10,11-dihydro-5H-dibenz[b,f]azepine.
chlorimipramine; G-34586. Antidepressant. A tricyclic used to treat obsessive-compulsive disorder. $bp_{0.03}$ = 160-170°. *Ciba-Geigy Corp.*

3101 Clomipramine Hydrochloride
17321-77-6 2447 241-344-3
$C_{19}H_{24}Cl_2N_2$
3-Chloro-5-[3-(dimethylamino)propyl]-10,11-dihydro-5H-dibenz[b,f]azepine monohydrochloride.
Anafranil; G-34586. Antidepressant. A tricyclic used to treat obsessive-compulsive disorder. mp = 189-190°. *Ciba-Geigy Corp.*

3102 Clorgiline
17780-72-2
$C_{13}H_{15}Cl_2NO$
N-[3-(2,4-Dichlorophenoxy)propyl]-N-methyl-2-propynylamine.
Clorgyline. MAO-A inhibitor. Antidepressant.

3103 Clovoxamine
54739-19-4
$C_{14}H_{21}ClN_2O_2$
4'-Chloro-5-methoxyvalerophenone (E)-O-(2-aminoethyl)oxime.
Antidepressant.

3104 Cotinine
486-56-6 2619 207-634-9
$C_{10}H_{12}N_2O$
1-Methyl-5-(3-pyridinyl)-2-pyrrolidinone.
Antidepressant. A viscous oil; bp_6 = 210-211°.

3105 Cotinine Fumarate
5695-98-7 2619
$(C_{10}H_{12}N_2O)_2.C_4H_4O_4$
(-)-1-Methyl-5-(3-pyridinyl)-2-pyrrolidinone fumarate (2:1).
Scotine. Antidepressant.

3106 Cyclindole
32211-97-5
$C_{14}H_{18}N_2$
3-(Dimethylamino)-1,2,3,4-tetrahydrocarbazole.
ciclindole; Win-27147-2. Antidepressant. *Sterling Winthrop Inc.*

3107 Cypenamine
15301-54-9
$C_{11}H_{15}N$
2-Phenyl-cyclopentylamine.
Antidepressant. *Marion Merrell Dow Inc.*

3108 Cypenamine Hydrochloride
5588-23-8
$C_{11}H_{16}ClN$
2-Phenyl-cyclopentylamine hydrochloride.
Antidepressant. Vitamin B_{12}. *Marion Merrell Dow Inc.*

3109 Cyprolidol
4904-00-1
$C_{21}H_{19}NO$
Diphenyl [2-(4-pyridinyl)cylclopropyl]methanol.
Antidepressant. Antihistaminic, antipruritic.

3110 Cyprolidol Hydrochloride
2364-72-9
$C_{21}H_{20}ClNO$
Diphenyl [2-(4-pyridinyl)cylclopropyl]methanol hydrochloride.
IN-1060; NSC-84973. Antidepressant.

3111 Cyproximide
15518-76-0
$C_{11}H_8ClNO_2$
1-(p-Chlorophenyl)-1,2-cyclopropanedicarboximide.
ciproximide; CL-53415. Antidepressant; antipsychotic.

3112 Daledalin
22136-27-2
$C_{19}H_{24}N_2$
3-Methyl-3-[3-(methylamino)propyl]-1-phenylindoline.
Antidepressant. Anesthetic. *Pfizer Inc.*

3113 Daledalin Tosylate
23226-37-1
$C_{26}H_{32}N_2O_3S$
3-Methyl-3-[3-(methylamino)propyl]-1-phenylindoline mono-p-toluenesulfonate.
UK-3557-15. Antidepressant. *Pfizer Inc.*

3114 Danitracen
31232-26-5
$C_{20}H_{21}NO$
9,10-Dihydro-10-(1-methyl-4-piperylidene)-9-anthrol.
Antidepressant.

3115 Dapoxetine
119356-77-3
$C_{21}H_{23}NO$
(+)-(S)-N,N-Dimethyl-α-[2-(1-naphthyloxy)ethyl]benzylamine.
LY-210448. Antidepressant. *Eli Lilly & Co.*

3116 Dapoxetine Hydrochloride
129938-20-1
$C_{21}H_{24}ClNO$
(+)-(S)-N,N-Dimethyl-α-[2-(1-naphthyloxy)ethyl]benzylamine hydrochloride.
LY-210448-HCl. Antidepressant. *Eli Lilly & Co.*

3117 Dazadrol
47029-84-5 256-292-7
$C_{15}H_{14}ClN_3O$
α-(p-Chlorophenyl)-α-2-imidazolin-2-yl-2-pyridinemethanol.
Antidepressant. *Schering-Plough HealthCare Products.*

3118 Dazadrol Maleate
25387-70-6
$C_{19}H_{18}ClN_3O_5$
α-(p-Chlorophenyl)-α-2-imidazolin-2-yl-2-pyridinemethanol maleate (1:1) (salt).
Sch-12650. Antidepressant. *Schering-Plough HealthCare Products.*

3119 Dazepinil
75991-50-3
$C_{17}H_{18}N_2$
(±)-4,5-Dihydro-2,3-dimethyl-4-phenyl-3H-1,3-benzodiazapine.
Antidepressant. *Hoechst Roussel Pharm. Inc.*

3120 Dazepinil Hydrochloride
75991-49-0
$C_{17}H_{18}ClN_2$
(±)-4,5-Dihydro-2,3-dimethyl-4-phenyl-3H-1,3-benzodiazapine monohydrochloride.
HRP-543; P-76-2543. Antidepressant. *Hoechst Roussel Pharm. Inc.*

3121 Demexiptiline
24701-51-7 2941
$C_{18}H_{18}N_2O$
5H-Dibenzo[a,d]cyclohepten-5-one O-[2-(methylamino)ethyl]oxime.
Antidepressant. Tricyclic. A colorless oil. *Bayer Corp.*

3122 Demexiptiline Hydrochloride
18059-99-9 2941 241-969-1
$C_{18}H_{19}ClN_2O$
5H-Dibenzo[a,d]cyclohepten-5-one O-[2-(methylamino)ethyl]oxime hydrochloride.
LM-2909; Deparon. Antidepressant. Tricyclic. mp = 232-233°; LD_{50} (rat orl) = 330 mg/kg, (rat iv) = 25 mg/kg. *Bayer Corp.*

3123 Depramine
303-54-8
$C_{19}H_{22}N_2$
5-[3-(Dimethylamino)propyl]-5H-dibenz[b,f]azepine.
balipramine; GP-31406. Antidepressant. Tricyclic.

3124 Desipramine
50-47-5 2966 200-040-0
$C_{18}H_{22}N_2$
10,11-Dihydro-N-(methylamino)propyl]-5H-dibenz[b,f]azepine.
desmethylimipramine; norimipramine. Antidepressant. Tricyclic. $bp_{0.02}$ = 172-174°; λ_m = 213, 252 nm (log ε 4.39, 3.93). *Lakeside BioTechnology.*

3125 Desipramine Hydrochloride
58-28-6 2966 200-373-1
$C_{18}H_{23}ClN_2$
10,11-Dihydro-N-(methylamino)propyl]-5H-dibenz[b,f]azepine monohydrochloride.
G-35020; JB-8181; EX-4355; RMI-9384A; NSC-114901; Norpramin; Nortrimil; Pertofran; Pertofrane. Antidepressant. Tricyclic. mp = 215-216°; soluble in H_2O; LD_{50} (mus orl) = 500 mg/kg, (rat orl) = 385 mg/kg, (mus ip) = 94 mg/kg, (rat ip) = 48 mg/kg. *Marion Merrell Dow Inc.; Rhône-Poulenc Rorer Pharm. Inc.*

3126 Dexamisole
14769-74-5 238-837-0
$C_{11}H_{12}N_2S$
(+)-2,3,5,6-Tetrahydro-6-phenylimidazo[2,1-b]thiazole.
R-12563 [as hydrochloride]. Antidepressant. *Abbott Laboratories Inc.*

3127 Deximafen
42116-77-8
$C_{11}H_{13}N_3$
(+)-2,3,5,6-Tetrahydro-5-(or 3)-phenyl-1H-imidazo[2,1-a]imidazole.
Antidepressant. *Abbott Laboratories Inc.*

3128 Dexnafenodone
92629-87-3
$C_{20}H_{23}NO$
(+)-(S)-2-[2-(Dimethylamino)ethyl]-3,4-dihydro-2-phenyl-1(2H)-naphthalenone.
Antidepressant.

3129 Dibenzepin
4498-32-2 3055
$C_{18}H_{21}N_3O$
10-[2-(Dimethylamino)ethyl]-5,10-dihydro-5-methyl-11-H-dibenzo[b,e][1,4]diazepin-11-one.
HF-1927. Antidepressant. Tricyclic. mp = 116-117°; $bp_{0.01}$ = 185°. *Sandoz Pharm. Corp.; Wander Pharma.*

3130 Dibenzepin Hydrochloride
315-80-0 3055 206-255-6
$C_{18}H_{22}ClN_3O$
10-[2-(Dimethylamino)ethyl]-5,10-dihydro-5-methyl-11-H-dibenzo[b,e][1,4]diazepin-11-one monohydrochloride.
Neodalit; Noveril. Antidepressant. Tricyclic. mp = 238°; λ_m (0.1N HCl) = 204, 220 nm (log ε 4.530, 4.458); pKa = 8.25; soluble in H_2O, alcohol, $CHCl_3$; LD_{50} (mus orl) = 215 mg/kg. *Sandoz Pharm. Corp.*

3131 Dimetacrine
4757-55-5 3258
$C_{20}H_{26}N_2$
10-[3-(Dimethylamino)propyl]-9,9-dimethylacridan.
Dimetacrino; Dimetacrinum; Dimethacrin. Antidepressant. Tricyclic. bp_1 = 200°. *Kefalas A/S; Siegfried AG.*

3132 Dimetacrine Hydrochloride
3258
$C_{20}H_{27}ClN_2$
10-[3-(Dimethylamino)propyl]-9,9-dimethylacridan hydrochloride.
Antidepressant. Tricyclic. mp = 151-154°. *Kefalas A/S; Siegfried AG.*

3133 Dimetacrine Tartrate
3759-07-7 3258 223-166-8
$C_{24}H_{32}N_2O_6$
10-[3-(Dimethylamino)propyl]-9,9-dimethylacridan tartrate.
SD-709; Istonil; Linostil. Antidepressant. Tricyclic. *Kefalas A/S; Siegfried AG.*

3134 Dimethazan
519-30-2 3261 208-266-1
$C_{11}H_{17}N_5O_2$
7-(2-Dimethylaminoethyl)theophylline.
Elidin. Antidepressant. mp = 95°.

3135 Dioxadrol
6495-46-1 3352
$C_{20}H_{23}NO_2$
2-(2,2-Diphenyl-1,3-dioxolan-4-yl)piperidine.
Rydar. CNS stimulant; antidepressant; smooth muscle relaxant. The dl-form of the free base is an antidepressant. The d-form, dexoxadrol is a CNS stimulant. Oily liquid. *Cutter Labs.*

3136 Dioxadrol Hydrochloride
3666-69-1 3352
$C_{20}H_{24}ClNO_2$
2-(2,2-Diphenyl-1,3-dioxolan-4-yl)piperidine hydrochloride.
CL-639C; dexoxadrol hydrochloride [d-form]; Relane [d-form]; levoxadrol hydrochloride [l-form]; Levoxan [l-form]. CNS stimulant; antidepressant; smooth muscle relaxant. The d-form is a CNS stimulant and analgesic; the l-form is a local anesthetic and a smooth muscle relaxant. mp = 256-260; LD_{50} (mus orl)= 240 mg/kg; [d-form hydrochloride]: mp = 254° (dec); $[\alpha]_D^{20}$ = 34° (c = 2 MeOH); LD_{50} (mus orl) = 340 mg/kg; [l-form hydrochloride]: mp = 248-254°; $[\alpha]_D^{20}$ = 34.5° (c = 2 MeOH); LD_{50} (mus orl) = 230 mg/kg. *Cutter Labs.*

3137 Dothiepin
204-031-2 3485 204-031-2
$C_{19}H_{21}NS$
3-Dibenzo[b,e]thiepin-11(6H)-ylidine-N,N-dimethyl-propanamine.
dosulepin; Dothep. Antidepressant. Tricyclic. $bp_{0.05}$ = 171-172°; mp = 55-57°. *SPOFA.*

3138 Dothiepin Hydrochloride
897-15-4 3485 212-978-8
$C_{19}H_{22}ClNS$
3-Dibenzo[b,e]thiepin-11(6H)-ylidine-N,N-dimethyl-propanamine hydrochloride.
dosulepin-hydrochloride; Altapin; Depresym; Prothiaden. Antidepressant. Tricyclic. mp = 218-221°; λ_m (MeOH) = 232, 260, 309 (log ε 4.41, 3.97, 3.53). *SPOFA.*

3139 Doxepin
1668-19-5 3492
$C_{19}H_{21}NO$
3-Dibenz[b,e]oxepin-11(6H)-ylidene-N,N-dimethyl-1-propanamine.
P-3693A; NSC-108160; Aponal; Curatin; Quitaxon. Antidepressant; antipruritic (veterinary). Tricyclic. Oily liquid; $bp_{0.03}$ = 154-157°; $bp_{0.2}$ = 260-270°; :D_{50} = (rat iv) = 16 mg/kg, (mus iv) = 26 mg.kg, (rat orl) = 147 mg/kg, (mus orl) = 135 mg/kg. *Fisons Pharm. Div.; Roerig Div. Pfizer Pharm.*

3140 Doxepin Hydrochloride
1229-29-4 3492 214-966-8
$C_{19}H_{22}ClNO$
3-Dibenz[b,e]oxepin-11(6H)-ylidene-N,N-dimethyl-1-propanamine hydrochloride.
Adapin; Aponal; Curatin; Quitaxon; Sinequin. Antidepressant. A cis-trans (approx. 1:5) mixture. Tricyclic. mp = 184-186°. *Fisons Pharm. Div.; Roerig Div. Pfizer Pharm.*

3141 Doxepin Hydrochloride E isomer
4698-39-9
$C_{19}H_{22}ClNO$
3-Dibenz[b,e]oxepin-11(6H)-ylidene-N,N-dimethyl-1-propanamine hydrochloride.
Antidepressant. Tricyclic. *Fisons Pharm. Div.; Roerig Div. Pfizer Pharm.*

3142 cis-Doxepin Hydrochloride
25127-31-5 3492
$C_{19}H_{22}ClNO$
cis-3-Dibenz[b,e]oxepin-11(6H)-ylidene-N,N-dimethyl-1-propanamine hydrochloride.
doxepin hydrochloride Z isomer; cidoxepin hydrochloride; P-4599. Antidepressant. Tricyclic. mp = 209-210.5°. *Fisons Pharm. Div.; Roerig Div. Pfizer Pharm.*

3143 trans-Doxepin Hydrochloride
3607-18-9 3492
$C_{19}H_{22}ClNO$
trans-3-Dibenz[b,e]oxepin-11(6H)-ylidene-N,N-dimethyl-1-propanamine hydrochloride.
Antidepressant. Tricyclic. mp = 192-193°. *Fisons Pharm. Div.; Roerig Div. Pfizer Pharm.*

3144 Duloxetine
116539-59-4 3518
$C_{18}H_{19}NOS$
(S)-N-Methyl-γ-(1-napthalenyloxy)-2-thiophenepropanamine.
LY-248686. Antidepressant. Dual serotonin and norepinephrine uptake inhibitor. *Eli Lilly & Co.*

3145 Duloxetine Hydrochloride
136434-34-9 3518
$C_{18}H_{20}ClNOS$
(S)-N-Methyl-γ-(1-napthalenyloxy)-2-thiophenepropanamine monohydrochloride.
Antidepressant. Dual serotonin and norepinephrine uptake inhibitor. White solid; pKa (DMF-H_2O, 66:34) = 9.6. *Eli Lilly & Co.*

3146 Eclanamine
71027-13-9
$C_{16}H_{22}Cl_2N_2O$
(±)-trans-2',4'-Dichloro-N-[2-(dimethylamino)-cyclopentyl]propionanilide.
Antidepressant. *Upjohn Ltd.*

3147 Eclanamine Maleate
71027-14-0
$C_{20}H_{26}Cl_2N_2O$+5
(±)-trans-2',4'-Dichloro-N-[2-(dimethylamino)-cyclopentyl]propionanilide maleate (1:1).
Antidepressant. *Upjohn Ltd.*

3148 Encyprate
2521-01-9
$C_{13}H_{17}NO_2$
Ethyl-N-benzcyclopropanecarbamate.
MO-1255; A-19757. Antidepressant. *Abbott Labs Inc.*

3149 Etoperidone
52942-31-1 3930
$C_{19}H_{28}ClN_5O$
1-[3-[4-(m-Chlorophenyl)-1-piperazinyl]propyl]-3,4-diethyl-Δ^2-1,2,4-triazolin-5-one.
Antidepressant. Liquid; $bp_{0.5}$ = 230°. *Angelini Pharm. Inc.*

3150 Etoperidone Hydrochloride
57775-22-1 3930 260-942-5
$C_{19}H_{29}Cl_2N_5O$
1-[3-[4-(m-Chlorophenyl)-1-piperazinyl]propyl]-3,4-diethyl-Δ^2-1,2,4-triazolin-5-one monohydrochloride.
AF-1191; McN-A-2673-11; ST-1191; Axiomin; Deprecer; Etonin; Etoran; Staff; Tropene. Antidepressant. mp = 197-198°; pH (2% aq soln) = 4.1 ± 0.2; soluble in H_2O, EtOH; slightly soluble in $CHCl_3$; practically insoluble in Me_2CO, C_6H_6, Et_2O; LD_{50} (mus orl) = 580 mg/kg, (rat orl) = 720 mg/kg, (mus ip) = 135 mg/kg, (rat ip) = 120 mg/kg. *Angelini Pharm. Inc.; McNeil Pharm.*

3151 Fantridone
17692-37-4
$C_{18}H_{20}N_2O$
5-[3-(Dimethylamino)propyl]-6(5H)-phenanthridinone.
Antidepressant. Tricyclic. *Searle G.D. & Co.*

3152 Fantridone Hydrochloride [anhydrous]
22461-13-8
$C_{18}H_{21}ClN_2O$
5-[3-(Dimethylamino)propyl]-6(5H)-phenanthridinone monohydrochloride.
Antidepressant. Tricyclic. *Searle G.D. & Co.*

3153 Fantridone Hydrochloride Monohydrate
24390-12-3
$C_{18}H_{23}ClN_2O_2$
5-[3-(Dimethylamino)propyl]-6(5H)-phenanthridinone monohydrochloride monohydrate.
AGN-616. Antidepressant. Tricyclic. *Searle G.D. & Co.*

3154 Febarbamate
13246-02-1 3983 236-226-3
$C_{20}H_{27}N_3O_6$
1-(3-Butoxy-2-hydroxypropyl)-5-ethyl-5-phenylbarbituric acid.
Go-560; G-Tril; Tymium. Antidepressant; thymoanaleptic. Syrup; cannot be crystallized; dec ≅196°; soluble in EtOH, dioxane; insoluble in H_2O; LD_{50} (mus orl) = 1065 mg/kg. *Labs. Sapos.*

3155 Femoxetine
59859-58-4 3993
$C_{20}H_{25}NO_2$
(+)-trans-3-[(p-Methoxyphenoxy)methyl]-1-methyl-4-phenylpiperidine.
Antidepressant. Serotonin uptake inhibitor. *Fabre Pierre; Ferrosan A/S.*

3156 Femoxetine Hydrochloride
56222-04-9 3993
$C_{20}H_{26}ClNO_2$
(+)-trans-3-[(p-Methoxyphenoxy)methyl]-1-methyl-4-phenylpiperidine hydrochloride.
FG-4963; Malexil. Antidepressant. Serotonin uptake inhibitor. LD_{50} (fmus iv) = 48 mg/kg, (fmus sc) = 941 mg/kg, (fmus orl) = 1408 mg/kg, (mmus iv) = 45 mg/kg, (mmus sc) = 723 mg/kg, (mmus orl) = 1687 mg/kg. *Fabre Pierre; Ferrosan A/S.*

3157 Fenmetozole
41473-09-0
$C_{10}H_{10}Cl_2N_2O$
2-[(3,4-Dichlorophenoxy)methyl]-2-imidazoline.
Narcotic antagonist. Used as an antidepressant. *Marion Merrell Dow Inc.*

3158 Fenmetozole Hydrochloride
23712-05-2
$C_{10}H_{11}Cl_3N_2O$
2-[(3,4-Dichlorophenoxy)methyl]-2-imidazoline monohydrochloride.
DH-524. Antidepressant; narcotic antagonist. *Marion Merrell Dow Inc.*

3159 Fenmetramide
5588-29-4
$C_{11}H_{13}NO_2$
5-Methyl-6-phenyl-3-morpholinone.
McN-1075; Feninetramide. Antidepressant. *McNeil Pharm.*

3160 Fenpentadiol
15687-18-0 4029 239-782-5
$C_{12}H_{17}ClO_2$
2-(4-Chlorophenyl)-4-methyl-2,4-pentanediol.
RD-292; Tredum. Antidepressant. mp = 76.5°; λ_m (0.1% in abs EtOH) = 228 mn; LD_{50} (mmus orl) = 940 mg/kg, (fmus orl) = 995 mg/kg, (mrat orl) = 1200 mg/kg, (frat orl) = 1250 mg/kg. *Lab. Albert Rolland.*

3161 Fezolamine
80410-36-2
$C_{20}H_{23}N_3$
1-[3-(Dimethylamino)propyl]-3,4-diphenylpyrazole.
Antidepressant. *Sterling Winthrop Inc.*

3162 Fezolamine Fumarate
80410-37-3
$C_{24}H_{27}N_3O_4$
1-[3-(Dimethylamino)propyl]-3,4-diphenylpyrazole fumarate.
Win-41528-2. Antidepressant. *Sterling Winthrop Inc.*

3163 Flerobuterol
82101-10-8
$C_{12}H_{18}FNO$
α-[(tert-Butylamino)methyl]-o-fluorobenzyl alcohol.
Antidepressant.

3164 Flibanserin
167933-07-5
$C_{20}H_{21}F_3N_4O$
1-[2-[4-(α,α,α-Trifluoro-m-tolyl)-1-piperazinyl]ethyl]-2-benzimidazolinone.
Antidepressant.

3165 Fluacizine
30223-48-4 4149
$C_{20}H_{21}F_3N_2OS$
10-[3-Diethylamino)propionyl)]-2-(trifluoromethyl)phenothiazine.
Fluoracizine; Fluoracizine; Ftoracizine; Phtorazisin. Tricyclic antidepressant. Phenothiazine derivative with psychotropic activity.

3166 Fluacizine Hydrochloride
4149
$C_{20}H_{22}CLF_3N_2OS$
10-[3-Diethylamino)propionyl)]-2-(trifluoromethyl)phenothiazine hydrochloride.
Fluoracyzine; Toracizin. Tricyclic antidepressant. Phenothiazine derivative with psychotropic activity. mp = 163-165°; soluble in H_2O, warm alcohols.

3167 Fluotracen
35764-73-9
$C_{21}H_{24}F_3N$
(±)-cis-9,10-Dihydro-N,N-trimethyl-2-(trifluoromethyl)-9-anthracenepropylamine.
Antidepressant; antipsychotic. *SmithKline Beecham Pharm.*

3168 Fluotracen Hydrochloride
57363-14-1
$C_{21}H_{25}ClF_3N$
(±)-cis-9,10-Dihydro-N,N-trimethyl-2-(trifluoromethyl)-9-anthracenepropylamine hydrochloride.
SKF-28175. Antidepressant; antipsychotic. *SmithKline Beecham Pharm.*

3169 Fluoxetine
54910-89-3 4222
$C_{17}H_{18}F_3NO$
(±)-N-Methyl-3-phenyl-3-[(α,α,α-trifluoro-p-tolyl)oxy]propylamine.
Antidepressant. Serotonin uptake inhibitor. *C.M. Ind.; Eli Lilly & Co.*

3170 Fluoxetine Hydrochloride
59333-67-4 4222 260-101-2
$C_{17}H_{19}ClF_3NO$
(±)-N-Methyl-3-phenyl-3-[(α,α,α-trifluoro-p-tolyl)oxy]propylamine hydrochloride.
LY-110140; Adofen; Fluctin; Fluoxeren; Fontex; Foxetin; Prozac; Reneuron. Antidepressant. Serotonin uptake inhibitor. mp = 158-159°; slightly soluble in H_2O (1-2 mg/ml), ethyl acetate, toluene, $CHCl_3$, C_6H_{14} (0.5-0.77 mg/ml); soluble in MeOH, ETOH (>100 mg/mg), acetonitrile (33-100 mg/ml), C_6H_6; λ_m = 227, 264, 268, 275 nm ($E^{1\%}_{1\,cm}$ 372, 29,29,22 MeOH); LD_{50} (rat orl) = 452 mg/kg. *C.M. Ind.; Eli Lilly & Co.*

3171 Fluparoxan
105182-45-4
$C_{10}H_{10}FNO_2$
(3aS,9aS)-5-Fluoro-2,3,3a,9a-tetrahydro-1H-[1,4]benzodioxino[2,3-c]pyrrole.
Antidepressant. *Glaxo Labs.*

3172 Fluparoxan Hydrochloride
111793-41-0
$C_{10}H_{11}ClFNO_2$
(3aS,9aS)-5-Fluoro-2,3,3a,9a-tetrahydro-1H-[1,4]benzodioxino[2,3-c]pyrrole hydrochloride.
GR-50360A. Antidepressant. *Glaxo Labs.*

3173 Fluvoxamine
54739-18-3 4251
$C_{15}H_{21}F_3N_2O_2$
5-Methoxy-4'-(trifluoromethyl)valerophenone (E)-*O*-(2-aminoethyl)oxime.
Antidepressant; anxiolytic; antiobsessional agent. Serotonin uptake inhibitor. *Philips-Duphar B.V.; Solvay Duphar Labs Ltd.*

3174 Fluvoxamine Maleate
61718-82-9 4251
$C_{19}H_{25}F_3N_2O_6$
5-Methoxy-4'-(trifluoromethyl)valerophenone (E)-*O*-(2-aminoethyl)oxime maleate.
DU-23000; MK-264; Dumirox; Faverin; Fevarin; Floxyfral; Luvox; Maveral. Antidepressant; anxiolytic; antiobsessional agent. Serotonin uptake inhibitor. mp = 120-121.5°. *Philips-Duphar B.V.; Solvay Pharm. Inc.*

3175 Gamfexine
7273-99-6
$C_{17}H_{27}N$
N,N-Dimethyl-γ-phenylcyclohexanepropylamine.
Win-1344. Antidepressant. *Sterling Winthrop Inc.*

3176 Guanoxyfen
13050-83-4
$C_{10}H_{15}N_3O$
(3-Phenoxypropyl)guanidine.
Antidepressant; antihypertensive. *Parke-Davis.*

3177 Guanoxyfen Sulfate
1021-11-0
$C_{20}H_{32}N_6O_6S$
(3-Phenoxypropyl)guanidine sulfate (2:1).
CI-515; CN-34799-5A; EA-166; HP-1598. Antidepressant; antihypertensive. *Parke-Davis.*

3178 Hepzidine
1096-72-6
$C_{21}H_{25}NO$
4-(10,11-Dihydro-5H-dibenzo[a,d]cyclohepten-5-yloxy)-1-methylpiperidine.
Antidepressant; anticholinergic.

3179 Imafen
53361-23-2
$C_{11}H_{13}N_3$
2,3,5,6-Tetrahydro-5-(or-3-)-phenyl-1H-imidazo[1,2a]imidazole.
Antidepressant. *Abbott Laboratories Inc.*

3180 Imafen Hydrochloride
53361-24-3
$C_{11}H_{14}ClN_3$
2,3,5,6-Tetrahydro-5-(or-3-)-phenyl-1H-imidazo[1,2a]imidazole monohydrochloride.
R-25540. Antidepressant. *Abbott Laboratories Inc.*

3181 Imiloxan
81167-16-0
$C_{14}H_{16}N_2O_2$
(±)-2-[(1-Ethyl-2-imidazolyl)methyl]-1,4-benzodioxane.
Antidepressant. *Syntex Labs. Inc.*

3182 Imiloxan Hydrochloride
86710-23-8
$C_{14}H_{17}ClN_2O_2$
(±)-2-[(1-Ethyl-2-imidazolyl)methyl]-1,4-benzodioxane hydrochloride.
RS-21361. Antidepressant. *Syntex Labs. Inc.*

3183 Imipramine
50-49-7 4955 200-042-1
$C_{19}H_{24}N_2$
10,11-Dihydro-5-(3-(dimethylamino)propyl)-5H-dibenz[b,f]azepine.
G-22355; Imizin. Antidepressant. Tricyclic. $bp_{0.1}$ = 160°.

3184 Imipramine Hydrochloride
113-52-0 4955 204-030-7
$C_{19}H_{25}ClN_2$
10,11-Dihydro-5-(3-(dimethylamino)propyl)-5H-dibenz[b,f]azepine hydrochloride.
Antideprin; Apo-Imipramine; Berkomine; Censtim; Censtin; Chrytemin; Deprinol; Dimipressin; DIPD; Dyna-Zina; Efuranol; Eupramin; Feinalmin; Imavate; Imidol; Imidobenzyle; Imilanyle; Imiprin; Imizin; Imizinum; Imiprin; Imavate; Impril; Intalpram; Iramil; Irmin; Janimine; Melipramine; Novo-pramine; Praminil; Presamine; Promiben; Pryleugan; SK-Pramine; Surplix; Timolet; Tipramine; Tofranil. Antidepressant. Tricyclic. mp = 174-175°; very soluble in H_2O, less soluble in organic solvents; LD_{50} (rat orl) = 490 mg/kg, (rat ip) = 90 mg/kg. *Abbott Labs Inc.; Ciba-Geigy Corp.*

3185 Imipramine N-Oxide
6829-98-7 4956 229-907-1
$C_{19}H_{24}N_2O$
10,11-Dihydro-5-(3-(dimethylamino)propyl)-5H-dibenz[b,f]azepine N-oxide.
imipraminoxide. Antidepressant. Tricyclic. mp = 120-123° (dec); soluble in MeOH, Et_2O, Me_2CO, C_6H_6; strongly hygroscopic. *Dumex USA.*

3186 Imipramine N-Oxide Hydrochloride
20438-98-6 4956
$C_{19}H_{25}ClN_2O$
10,11-Dihydro-5-(3-(dimethylamino)propyl)-5H-dibenz[b,f]azepine N-oxide hydrochloride.
Imiprex. Antidepressant. Tricyclic. mp = 167-174°; soluble in H_2O/EtOH/$CHCl_3$ mixture; nearly insoluble in Et_2O; LD_{50} (rat ip) = 90 mg/kg, (mus ip) = 150 mg/kg. *Dumex USA.*

3187 Imipramine Pamoate
10075-24-8 4955
$C_{42}H_{40}N_2O_6$
10,11-Dihydro-5-(3-(dimethylamino)propyl)-5H-dibenz[b,f]azepine pamoate.
Tofranil-PM. Antidepressant. Tricyclic. *Ciba-Geigy Corp.*

3188 Indalpine
63758-79-2 4965 264-445-4
$C_{15}H_{20}N_2$
3-[2-(4-Piperidyl)ethyl]indole.
LM-5008; Upstene. Antidepressant. A selective serotonin uptake inhibitor. [monohydrochloride ($C_{15}H_{21}ClN_2$)]: mp = 167°. *Mar-Pha Soc. Etud. Exploit. Marques.*

3189 Indeloxazine Hydrochloride
65043-22-3 4972
$C_{14}H_{18}ClNO_2$
(±)-2-[(Inden-7-yloxy)methyl]morpholine hydrochloride.
CI-874; YM-08054-1; Elen; Noin. Antidepressant; nootropic. Inhibits synaptosomal uptake of serotonin and noradrenaline. mp = 155-156°, 169-170°; LD_{50} (mus iv) = 47 mg/kg; [(+)-form]: mp = 112-113°; $[\alpha]_D^{21}$ = = 4.9° (c = 5 MeOH); [(-)-form]: mp = 142-142.5°; $[\alpha]_D^{20}$ = -4.9° (c = 5 MeOH). *Parke-Davis; Yamanouchi U.S.A. Inc.*

3190 Intriptyline
27466-27-9
$C_{21}H_{19}N$
4-(5H-Dibenzo[a,d]cyclohepten-5-ylidene)-N,N-dimethyl-s-butynylamine.
Antidepressant. Tricyclic. *Ayerst.*

3191 Intriptyline Hydrochloride
27466-29-1
$C_{21}H_{20}ClN$
4-(5H-Dibenzo[a,d]cyclohepten-5-ylidene)-N,N-dimethyl-s-butynylamine hydrochloride.
AY-22124. Antidepressant. *Ayerst.*

3192 Iprindole
5560-72-5 5091 226-933-5
$C_{19}H_{28}N_2$
5-[3-Dimethylamino)propyl]-6,7,8,9,10,11-hexahydro-5H-cyclooct[b]indole.
pramindole; Tertran. Antidepressant. *Am. Home Products.*

3193 Iprindole Hydrochloride
20432-64-8 5091 243-819-0
$C_{19}H_{29}ClN_2$
5-[3-Dimethylamino)propyl]-6,7,8,9,10,11-hexahydro-5H-cyclooct[b]indole hydrochloride.
WY-3263; Prondol; Galatur. Antidepressant. Tricyclic. mp = 146-147°. *Am. Home Products; Wyeth Labs.*

3194 Iproclozide
3544-35-2 5092 222-589-5
$C_{11}H_{15}ClN_2O_2$
(p-Chlorophenoxy)acetic acid 2-isopropylhydrazide.
PC-603; Sursum. Antidepressant. A monoamine oxidase inhibitor. mp = 93-94°.

3195 Iproniazid
54-92-2 5094 200-218-8
$C_9H_{13}N_3O$
Isonicotinic acid 2-isopropylhydrazide.
Antidepressant. Monoamine oxidase inhibitor. mp ≅ 112°; soluble in H_2O, EtOH; pH (aqueous solution) = 6.7; [Dihydrochloride]: Dec 227-228°; soluble in H_2O.

3196 Iproniazid Phosphate
305-33-9 5094 206-164-1
$C_9H_{16}N_3PO_5$
Isonicotinic acid 2-isopropylhydrazide phosphate.
Marsilid. Antidepressant. A monoamine oxidase inhibitor. mp = 175-184°; λ_m = 265 nm ($A^{1\%}_{1\,cm}$ 166 MeOH), 264 nm ($A^{1\%}_{1\,cm}$ 176 H_2O), 267 nm ($A^{1\%}_{1\,cm}$ 179 0.1N HCl), 244, 272, 308 nm ($A^{1\%}_{1\,cm}$ 113, 121, 135 0.1N KOH); soluble in H_2O (18.8 g/100 ml), MeOH (2.1 g/100 ml), EtOH (0.9 g/100 ml), $CHCl_3$ (0.06 g/100 ml); insoluble in Et_2O, C_6H_{14}.

3197 Isocarboxazid
59-63-2 5172 200-438-4
$C_{12}H_{13}N_3O_2$
5-Methyl-3-isoxazolecarboxylic acid 2-benzylhydrazide.
Ro-5-0831; Marplan. Antidepressant. A monoamine oxidase inhibitor. mp = 105-106°; barely soluble in H_2O (0.05%); slightly more soluble in 95% alcohol (1-2%), glycerol, propylene glycol. *Hoffmann-LaRoche Inc.*

3198 Ketipramine
796-29-2
$C_{19}H_{22}N_2O$
5-[3-(Dimethylamino)propyl]-5,11-dihydro-10H-dibenz[b,f]azepin-10-one.
Antidepressant. Tricyclic.

3199 Ketipramine Fumarate
17243-32-2
$C_{23}H_{26}N_2O_5$
5-[3-(Dimethylamino)propyl]-5,11-dihydro-10H-dibenz[b,f]azepin-10-one fumarate.
G-35259; Ketimipramine. Antidepressant. Tricyclic.

3200 Levoprotiline
76496-68-9
$C_{20}H_{23}NO$
(-)-(R)α-[(Methylamino)methyl]-9,10-ethanoanthracene-9(10H)-ethanol.
(-)-oxaprotiline; (-)-hydroxymaprotiline; CGP-12103A; 119. Inactive, R-(-)-enantiomer of racemic oxaprotiline, an antidepressant with noradrenaline uptake inhibiting activity. A tricyclic. *See* Oxaprotiline.

3201 Levosulpiride
23672-07-3 9163
$C_{15}H_{23}N_3O_4S$
(-)-N-[[(S)-1-Ethyl-2-pyrrolidinyl]methyl]-5-sulfamoyl-*o*-anisamide.
S(-)-sulpiride; Levobren; Levopraid. Antidepressant; antipsychotic; antiemetic. l-Form of Sulpiride, a dopamine receptor antagonist. mp = 185-187°.

3202 Lidanserin
73725-85-6
$C_{26}H_{31}FN_2O_4$
(±)-4-[3-[3-[4-(p-Fluorobenzoyl)piperidino]-propoxy]-4-methoxyphenyl]-2-pyrrolidinone.
Antidepressant.

3203 Litoxetine
86811-09-8
$C_{16}H_{19}NO$
4-(2-Naphthylmethoxy)piperidine.
SL-810385. Selective serotonin reuptake inhibitor. Antidepressant.

3204 Litracen
5118-30-9
$C_{20}H_{23}N$
9-(3-Methylaminopropylidene)-10,10-dimethyl-9,10-dihydroanthracene.
N-7049. A tricyclic. Antidepressant.

3205 Lofepramine
23047-25-8 5587 245-396-8
$C_{26}H_{27}ClN_2O$
4'-Chloro-2-[[3-(10,11-dihydro-5H-dibenz-[b,f]azepin-5-yl)propyl]methylamino]-acetophenone.
Lopramine. Tricyclic antidepressant. A psychotrophic drug related to Imipramine. mp = 104-106°; easily oxidized to desipramine and p-chlorobenzoic acid. *LeoAB.*

3206 Lofepramine Hydrochloride
26786-32-3 5587 248-002-2
$C_{26}H_{28}Cl_2N_2O$
4'-Chloro-2-[[3-(10,11-dihydro-5H-dibenz-[b,f]azepin-5-yl)propyl]methylamino]-acetophenone monohydrochloride.
Leo-640; WHR-2908A; Amplit; Gamanil; Gamonil; Timelit; Tymelyt. Tricyclic antidepressant. mp = 152-154°; soluble in alcohol, $CHCl_3$; practically insoluble in H_2O; LD_{50} (mus orl) > 2500 mg/kg, (rat orl) > 1000 mg/kg, (mus ip) > 920 mg/kg; (rat ip) > 1000 mg/kg, (mus/rat sc) > 1000 mg/kg.

3207 Lorapride
68677-06-5 272-057-1
$C_{14}H_{22}ClN_3O_3S$
5-Chloro-n'-[(1-ethyl-2-pyrrolidinyl)methyl]-2-methoxysulfanylzamide.
Antipsychotic; antidepressant; antiemetic. Sulpiride derivative.

3208 Lortalamine
70384-91-7
$C_{15}H_{17}ClN_2O_2$
(±)-(4aR*,10R*,10aS*)-8-Chloro-1,2,3,4,10,10a-hexahydro-2-methyl-4a,10-(iminoethano)-4aH-[1]benzopyrano[3,2-c]pyridin-12-one.
LM-1404. Antidepressant. Tetracyclic. *Lipha Pharm. Inc.*

3209 Mafoprazine
80428-29-1
$C_{22}H_{28}FN_3O_3$
4'-[3-[4-(o-Fluorophenyl)-1-piperazinyl]propoxy]-m-acetanisidide.
A phenothiazine with postsynaptic dopamine D2-receptor blocking and α_2-adrenoceptor-stimulating actions. Antipsychotic.

3210 Maprotiline
10262-69-8 5792 233-599-4
$C_{20}H_{23}N$
N-Methyl-9,10-ethanoanthracene-9(10H)-propylamine
Antidepressant. Tetracyclic. mp = 92-93°. *CIBA plc.*

3211 Maprotiline Hydrochloride
10347-81-6 5792 233-758-8
$C_{20}H_{24}ClN$
N-Methyl-9,10-ethanoanthracene-9(10H)-propylamine hydrochloride.
Ba-34276; Deprilept; Ludiomil; Psymion. Antidepressant. Tetracyclic. mp = 230-232°; soluble in MeOH, $CHCl_3$; slightly soluble in H_2O; insoluble in isooctane; LD_{50} (rat orl) = 900 mg/kg. *CIBA plc.*

3212 Mariptiline
60070-14-6
$C_{18}H_{18}N_2O$
1a,10b-Dihydrodibenzo[a,e]cyclopropa[c]cyclohepten-6(1H)one O-(2-aminoethyl)oxime.
Antidepressant. Tetracyclic.

3213 Maroxepin
65509-24-2
$C_{19}H_{19}NO$
2,3,4,5-Tetrahydro-3-methyl-1H-dibenz[2,3:6,7]oxepino[4,5-d]azepine.
Tetracyclic with high affinity for the invertebrate-specific octopamine receptor. A psychoactive agent with antidepressant and sedative properties.

3214 Medifoxamine
32359-34-5 5834 251-011-4
$C_{16}H_{19}NO_2$
(Dimethylamino)acetaldehyde diphenyl acetal.
Antidepressant. *Gerda.*

3215 Medifoxamine Fumarate
16604-45-8 5834 240-657-2
$C_{20}H_{23}NO_6$
(Dimethylamino)acetaldehyde diphenyl acetal fumarate.
LG-152; Cledial; Geraxyl. Antidepressant. mp = 128.5°. LD_{50} (rat orl) = 750 mg/kg. *Gerda.*

3216 Melitracen
5118-29-6 5866 225-858-5
$C_{21}H_{25}N$
N,N,10,10-Tetramethyl-δ9(10H),γ-anthracenepropylamine.
Antidepressant. Tricyclic. *Kefalas A/S.*

3217 Melitracen Hydrochloride
10563-70-9 5866 234-150-5
$C_{21}H_{26}ClN$
N,N,10,10-Tetramethyl-δ9(10H),γ-anthracenepropylamine hydrochloride.
U-24973A; Dixeran; Melixeran; Trausabun. Antidepressant. mp ≅ 246°; LD_{50} (mus iv) = 52 mg/kg. *Kefalas A/S.*

3218 Metapramine
21730-16-5 5991
$C_{16}H_{18}N_2$
10,11-Dihydro-5-methyl-10-(methylamino)-5H-dibenz[b,f]azepine.
RP-19560; Timaxel. Antidepressant. Hydrochloride used for injectable formulation; fumarate used for tablet formulations. [hydrochloride]: mp = 238-240°. *Rhône-Poulenc.*

3219 Methylbenactyzium Bromide
3166-62-9 1055 221-628-3
$C_{21}H_{28}BrNO_3$
Diethyl(2-hydroxyethyl)methylammonium bromide benzilate.
benactyzine methobromide; Finalin; Spatomac. Antidepressant; anticholinergic. mp = 169-170°. *Am. Cyanamid.*

3220 Metralindole
54188-38-4 6238
$C_{15}H_{17}N_3O$
2,4,5,6-Tetrahydro-9-methoxy-4-methyl-1H-3,4,6a-triazafluoranthene.
Tetracyclic antidepressant. A heterocyclic β-carboline derivative. mp = 164-165°.

3221 Metralindole Hydrochloride
53734-79-5 6238
$C_{15}H_{18}ClN_3O$
2,4,5,6-Tetrahydro-9-methoxy-4-methyl-1H-3,4,6a-triazafluoranthene hydrochloride.
Incazan. Tetracyclic antidepressant. A heterocyclic β-carboline derivative. mp = 305-308°; soluble in H_2O; LD_{50} (mus orl) = 445 mg/kg.

3222 Mianserin
24219-97-4 6260 246-088-6
$C_{18}H_{20}N_2$
1,2,3,4,10,14b-Hexahydro-2-methyldibenzo[c,f]pyrazino[1,2-a]azepine.
Tetracyclic antidepressant. Serotonin receptor antagonist. *Organon Inc.*

3223 Mianserin Hydrochloride
21535-47-7 6260 244-426-7
$C_{18}H_{21}ClN_2$
1,2,3,4,10,14b-Hexahydro-2-methyldibenzo[c,f]-pyrazino[1,2-a]azepine monohydrochloride.
GB-94; Org-GB-94; Athymil; Bolvidon; Lantanon; Norval; Tetramide; Tolvin; Tolvon. Tetracyclic antidepressant; antihistiminic. Serotonin receptor antagonist. mp = 282-284°; LD_{50} (mmus orl) = 356 mg/kg, (fmus orl) = 390 mg/kg, (mmus iv) = 32.5 mg/kg, (fmus iv) = 31.0 mg/kg. *Organon Inc.*

3224 Milacemide
76990-56-2
$C_7H_{16}N_2O$
2-(Pentylamino)acetamide.
Antidepressant; anticonvulsant. *Searle G.D. & Co.*

3225 Milacemide Hydrochloride
76990-85-7
$C_7H_{17}ClN_2O$
2-(Pentylamino)acetamide monohydrochloride.
CP-1552S. Antidepressant; anticonvulsant. *Searle G.D. & Co.*

3226 Milnacipran
92623-85-3 6281
$C_{15}H_{22}N_2O$
cis-(±)-2-(Aminomethyl)-N,N-diethyl-1-phenylcyclopropanecarboxamide.
midalcipran. Antidepressant. A serotonin and norepinephrine uptake inhibitor. *Fabre Pierre.*

3227 Milnacipran Hydrochloride
101152-94-7 6281
$C_{15}H_{23}ClN_2O$
cis-(±)-2-(Aminomethyl)-N,N-diethyl-1-phenylcyclopropanecarboxamide monohydrochloride.
F-2207. Antidepressant. A serotonin and norepinephrine uptake inhibitor. mp = 189-181°; LD_{50} (mus orl) = 237 mg/kg. *Fabre Pierre.*

3228 Minaprine
25905-77-5 6287 247-329-8
$C_{17}H_{22}N_4O$
4-[2-[(4-Methyl-6-phenyl-3-pyridazinyl)amino]-ethyl]morpholine.
Antidepressant. mp = 122°; insoluble in H_2O; slightly soluble in EtOH; soluble in $CHCl_3$. *Sanofi Winthrop.*

3229 Minaprine Hydrochloride
25953-17-7 6287
$C_{17}H_{24}Cl_2N_4O$
4-[2-[(4-Methyl-6-phenyl-3-pyridazinyl)amino]-ethyl]morpholine dihydrochloride.
Agr-1240; CB-30038; Brantur; Cantor. Antidepressant. mp = 182°. *Sanofi Winthrop.*

3230 Mirtazapine
61337-67-5 6295
$C_{17}H_{19}N_3$
1,2,3,4,10,14b-Hexahydro-2-methylpyrazino[2,1-a]-pyrido[2,3-c]benzazepine.
Org-3770; 6-Azamianserin; mepirzepine; Mirtazepine; Remeron. Tetracyclic antidepressant. An α_2-adrenergic blocker; analog of mianserin. mp = 114-116°. *Chinoin; Organon Inc.*

3231 Moclobemide
71320-77-9 6309
$C_{13}H_{17}ClN_2O_2$
p-Chloro-N-(2-morpholinoethyl)benzamide.
Ro-11-1163/000; Aurorix; Manerix; Moclaime. Antidepressant. A reversible monoamine oxidase A inhibitor. mp = 137°; LD_{50} (rat orl) = 707 mg/kg; [monohydrochloride]: mp = 208°. *C.M. Ind.; Hoffmann-LaRoche Inc.*

3232 Modaline
2856-74-8
$C_{10}H_{15}N_3$
2-Methyl-3-piperidinopyrazine.
Antidepressant. *Parke-Davis.*

3233 Modaline Sulfate
2856-75-9
$C_{10}H_{17}N_3O_4S$
2-Methyl-3-piperidinopyrazine sulfate (1:1).
W-3207B; NSC-89277. Antidepressant. *Parke-Davis.*

3234 Napactadine Hydrochloride
57166-13-9
$C_{14}H_{17}ClN_2$
N,N'-Dimethyl-2-napthaleneacetamidine monohydrochloride.
DL-588. Antidepressant. *Marion Merrell Dow Inc.*

3235 Napamezole
91524-14-0
$C_{14}H_{16}N_2$
2-[(3,4-Dihydro-2-napthyl)methyl]-2-imidazoline.
Antidepressant. *Sterling Winthrop Inc.*

3236 Napamezole Hydrochloride
87495-33-8
$C_{14}H_{17}ClN_2$
2-[(3,4-Dihydro-2-napthyl)methyl]-2-imidazoline monohydrochloride.
Win-51181-2. Antidepressant. *Sterling Winthrop Inc.*

3237 Napitane
148152-63-0
$C_{22}H_{25}NO_2$
(±)-(3R*)-3-Phenyl-1-[[(6R*)-6,7,8,9-tetrahydronaphtho[1,2-d]-1,3-dioxol-6-yl]methyl]pyrrolidine.
A-75200. Antidepressant. *Abbott Labs Inc.*

3238 Napitane Mesylate
149189-73-1
$C_{23}H_{29}NO_5S$
(±)-(3R*)-3-Phenyl-1-[[(6R*)-6,7,8,9-tetrahydronaphtho[1,2-d]-1,3-dioxol-6-yl]methyl]pyrrolidine methanesulfonate.
A-75200 Mesylate. Antidepressant. *Abbott Labs Inc.*

3239 Nefazodone
83366-66-9 6527
$C_{25}H_{32}ClN_5O_2$
2-[3-[4-(3-Chlorophenyl)-1-piperazinyl]propyl]-5-ethyl-2,4-dihydro-4-(2-phenoxyethyl)-3H-1,2,4-triazol-3-one.

Antidepressant. Selective serotonin 5-HT_2 receptor antagonist. mp = 83-83°. *Bristol-Myers Squibb Pharm. R&D; Mead Johnson Pharmaceuticals.*

3240 Nefazodone Hydrochloride
82752-99-6 6527

$C_{25}H_{33}Cl_2N_5O_2$

2-[3-[4-(3-Chlorophenyl)-1-piperazinyl]propyl]-5-ethyl-2,4-dihydro-4-(2-phenoxyethyl)-3H-1,2,4-triazol-3-one monohydrochloride.

BMY-13754; MJ-13754-1; Dutonin; Serazone. Antidepressant. Selective serotonin 5-HT_2 receptor antagonist. mp (crystals from iPrOH) = 186-87°, mp (crystals from EtOH) = 175-177°. *Bristol-Myers Squibb Pharm. R&D; Mead Johnson Pharmaceuticals.*

3241 Nefopam
13669-70-0 6529 237-148-2

$C_{17}H_{19}NO$

3,4,5,6-Tetrahydro-5-methyl-2-phenyl-1H-2,5-benzoxazocine.

Antidepressant; analgesic. A benzoxazocine; a centrally acting muscle relaxant. *Rexall SundownInc.; Riker Labs.*

3242 Nefopam Hydrochloride
23327-57-3 6529 245-585-5

$C_{17}H_{20}ClNO$

3,4,5,6-Tetrahydro-5-methyl-2-phenyl-1H-2,5-benzoxazocine hydrochloride.

R-738; Acupan; Ajan; Fenazoxine. Antidepressant; analgesic. A benzoxazocine; a centrally acting muscle relaxant. mp = 238-242°; LD_{50} (mus orl) = 119 mg/kg, (rat orl) = 178 mg/kg, (mus iv) = 44.5 mg/kg, (rat iv) = 28 mg/kg. *3M Pharm.; Rexall SundownInc.; Riker Labs.*

3243 Nialamide
51-12-7 6575 200-079-3

$C_{16}H_{18}N_4O_2$

Isonicotinic acid 2-[(2-benzylcarbamoyl)ethyl]hydrazide.

NF-XIII; Espril; Niamid; Niamidal; Niaquitil; Nuredal; Nyezin. Antidepressant. A monoamine oxidase inhibitor. mp = 152-153°; slightly soluble in H_2O, more soluble in acids; LD_{50} (mus orl) = 590 mg/kg, (mus ip) = 435 mg/kg, (mus iv) = 120 mg/kg. *Pfizer Inc.*

3244 Nisoxetine
53179-07-0

$C_{17}H_{21}NO_2$

(±)-3-(o-Methoxyphenoxy)-N-methyl-3-phenylpropylamine.

Compound 89218. Antidepressant. *Eli Lilly & Co.*

3245 Nitrafudam
64743-09-5

$C_{11}H_9N_3O_3$

5-(o-Nitrophenyl)-2-furamidine.

Antidepressant. *Norwich Eaton.*

3246 Nitrafudam Hydrochloride
57666-60-1

$C_{11}H_{10}ClN_3O_3$

5-(o-Nitrophenyl)-2-furamidine monohydrochloride.

F-853. Antidepressant. *Norwich Eaton.*

3247 Nomifensine
24526-64-5 6768

$C_{16}H_{18}N_2$

8-Amino-1,2,3,4-tetrahydro-2-methyl-4-phenylisoquinoline.

Antidepressant. A novel antidipressant distinguished by *its bicyclic structure. mp = 179-181°. Hoechst Roussel Pharm. Inc.*

3248 Nomifensine Maleate
32795-47-4 6768 251-223-7

$C_{20}H_{22}N_2O_4$

8-Amino-1,2,3,4-tetrahydro-2-methyl-4-phenylisoquinoline maleate (1:0).

HOE-984; Alival; Hostalival; Merital; Neurolene; Psicronizer. Antidepressant. A novel antidipressant distinguished by its bicyclic structure. mp = 199-201°; LD_{50} (rat orl) = 430 mg/kg, (rat iv) = 72 mg/kg, (mus sc) = 430 mg/kg. *Hoechst Roussel Pharm. Inc.*

3249 Nortriptyline
72-69-5 6812 200-788-8

$C_{19}H_{21}N$

3-(10,11-Dihydro-5H-dibenzo[a,d]cyclohepten-5-ylidene)-N-methyl-1-propanamine.

desitriptilina; desmethylamitriptyline; Avantyl; Aventyl; Noritren; Ateben; Psychostyl; Sesaval. Antidepressant. Tricyclic. *Merck & Co.Inc.*

3250 Nortriptyline Hydrochloride
894-71-3 6812 212-973-0

$C_{19}H_{22}ClN$

3-(10,11-Dihydro-5H-dibenzo[a,d]cyclohepten-5-ylidene)-N-methyl-1-propanamine hydrochloride.

Acetexa; Allegron; Altilev; Aventyl Hydrochloride; Nortrilen; Norzepine; Pamelor; Sensival; Vividyl. Antidepressant. Tricyclic. mp = 213-215°; λ_m = 240 nm (ε 13900); soluble in H_2O, EtOH, $CHCl_3$, insoluble in other organic solvents. *Eli Lilly & Co.; Merck & Co.Inc.; Sandoz Pharm. Corp.*

3251 Noxiptiline
3362-45-6 6821

$C_{19}H_{22}N_2O$

10,11-Dihydro-5H-dibenzo[a,d]cyclohepten-5-one *O*-[2-dimethyl-amino)ethyl]oxime.

noxiptilin; noxiptyline; Dibenzoxin. Antidepressant. Tricyclic. $bp_{0.05}$ = 160-164°. *Bayer Corp.; Pfizer Inc.*

3252 Noxiptiline Hydrochloride
4985-15-3 6821 225-638-9

$C_{19}H_{23}ClN_2O$

10,11-Dihydro-5H-dibenzo[a,d]cyclohepten-5-one *O*-[2-dimethyl-amino)ethyl]oxime monohydrochloride.

Bay-1521; Agedal; Nogedal. Antidepressant. Tricyclic. mp = 185-187°; LD_{50} (mus sc) = 240 mg/kg. *Bayer Corp.; Pfizer Inc.*

3253 Octamoxin
4684-87-1 6845

$C_8H_{20}N_2$

(1-Methylheptyl)hydrazine.

2-hydrazinooctane; octomoxine; Ximaol. Antidepressant. A monoamine oxidase inhibitor. *Octel Chem. Ltd.*

3254 Octamoxin Sulfate
3845-07-6 6845
$C_8H_{22}N_2O_4S$
(1-Methylheptyl)hydrazine sulfate.
Nimoal. Antidepressant. A monoamine oxidase inhibitor. mp = 78-80°. *Octel Chem. Ltd.*

3255 Octriptyline
47166-67-6
$C_{20}H_{21}N$
1a,10b-Dihydro-N-methyldibenzo[a,d]cyclopropa-[c]cycloheptene-$\delta^{6(1H),\gamma}$-propylamine.
Antidepressant. Tricyclic. *Searle G.D. & Co.*

3256 Octriptyline Phosphate
51481-67-5
$C_{20}H_{24}NPO_4$
1a,10b-Dihydro-N-methyldibenzo[a,d]cyclopropa-[c]cycloheptene-$\delta^{6(1H),\gamma}$-propylamine phosphate (1:1).
SC-27123. Antidepressant. Tricyclic. *Searle G.D. & Co.*

3257 Opipramol
315-72-0 6985 206-254-0
$C_{23}H_{29}N_3O$
4-[3-(5H-Dibenz[b,f]azepin-5-yl)propyl]-1-piperazineethanol.
Antidepressant; antipsychotic. Tricyclic. mp = 100-101°. *Ciba-Geigy Corp.; Rhône-Poulenc Rorer Pharm. Inc.*

3258 Opipramol Hydrochloride
909-39-7 6985 213-000-2
$C_{23}H_{31}Cl_2N_3O$
4-[3-(5H-Dibenz[b,f]azepin-5-yl)propyl]-1-piperazineethanol dihydrochloride.
G-33040; Dinsidon; Ensidon; Insidon; Nisidana. Antidepressant; antipsychotic. Tricyclic. mp = 229-230°; soluble in H_2O, EtOH; sparingly soluble in Me_2CO. *Ciba-Geigy Corp.; Rhône-Poulenc Rorer Pharm. Inc.*

3259 Oxaflozane
26629-87-8 7039 247-855-8
$C_{14}H_{18}F_3NO$
4-(1-Methylethyl)-2-[3-(trifluoromethyl)-phenyl]morpholine.
Antidepressant. A non-tryciclic antidepressant agent. Oil; $bp_{0.005}$ = 99°; n_D^{24}= 1.4751.

3260 Oxaflozane Hydrochloride
222629-86-7 7039
$C_{14}H_{19}ClF_3NO$
4-(1-Methylethyl)-2-[3-(trifluoromethyl)-phenyl]morpholine monohydrochloride.
CERM-1766; Conflictan. Antidepressant. A non-tryciclic antidepressant agent. mp = 164°; LD_{50} (mus orl) = 365 or 420 mg/kg, (mus iv) = 90 or 80 mg/kg.

3261 Oxaprotiline
56433-44-4
$C_{20}H_{23}NO$
(±)-α-[(Methylamino)methyl]-9,10-ethanoanthracene-9(10H)-ethanol.
hydroxymaprotiline; CGP-12104A [(+)-enantiomer]. Antidepressant. Noradrenaline uptake inhibitor. *Ciba-Geigy Corp.*

3262 Oxaprotiline Hydrochloride
39022-39-4
$C_{20}H_{24}ClNO$
(±)-α-[(Methylamino)methyl]-9,10-ethanoanthracene-9(10H)-ethanol hydrochloride.
C-49802B-Ba. Antidepressant. Noradrenaline uptake inhibitor. *Ciba-Geigy Corp.*

3263 Oxitriptyline
29541-85-3
$C_{19}H_{21}NO_2$
2-[(10,11-Dihydro-5H-dibenzo[a,d]cyclohepten-5-yl)oxy]-N,N-dimethylacetamide.
Tricyclic. Antidepressant.

3264 Oxypertine
153-87-7 7105 205-818-3
$C_{23}H_{29}N_3O_2$
5,6-Dimethoxy-2-methyl-3-[2-(4-phenyl-1-piperazinyl)ethyl]indole.
Win-18501-2; Forit; Integrin. Antidepressant. *Sterling Winthrop Inc.*

3265 Panuramine
80349-58-2
$C_{24}H_{25}N_3O_2$
1-Benzoyl-3-[1-(2-naphthylmethyl)-4-piperidyl]urea.
Wy-26002; 390. Selective inhibitor of 5-hydroxytryptamine uptake. Antidepressant.

3266 Paroxetine
61869-08-7 7175
$C_{19}H_{20}FNO_3$
(-)-(3S,4R)-4-[(p-Fluorophenyl)-3-[(3,4-methylenedioxy)phenoxy]methyl]piperidine.
BRL-29060; FG-7051; Aropax; Paxil; Seroxat. Antidepressant. Serotonin uptake inhibitor. [hydrochloride ($C_{19}H_{21}ClFNO_3$)]: mp = 118°; [hydrochloride hemihydrate ($C_{19}H_{21}ClFNO_3.0.5H_2O$)]: mp = 129-131°; [maleate]: mp = 136-138°; $[\alpha]_D$ = -87° (c = 5 EtOH); LD_{50} (mus sc) = 845 mg/kg, (mus orl) = 500 mg/k. *Beecham Res. Labs. UK; Ferrosan A/S; SmithKline Beecham Pharm.*

3267 Phenelzine
51-71-8 7366 200-117-9
$C_8H_{12}N_2$
β-Phenylethylhydrazine.
phenethylhydrazine; phenalzine. Antidepressant. A monoamine oxidase inhibitor. $bp_{0.1}$ = 74°; n_D^{20} = 1.5494; [hydrochloride]: mp = 174°. *Lakeside BioTechnology.*

3268 Phenelzine Dihydrogen Sulfate
156-51-4 7366 200-117-9
$C_8H_{14}N_2O_4S$
β-Phenylethylhydrazine sulfate (1:1).
phenelzine acid sulfate; W-1544a; Nardelzine; Nardil; Stinerval. Antidepressant. A monoamine oxidase inhibitor. Soluble in H_2O; LD_{50} (mus orl) = 156 mg/kg. *Lakeside BioTechnology; Parke-Davis.*

3269 Phenoxypropazine
3818-37-9 7411
$C_9H_{14}N_2O$
(1-Methyl-2-phenoxyethyl)hydrazine.
fenoxypropazine. Antidepressant. A monoamine oxidase inhibitor. $bp_{0.2}$ = 98-102°.

3270 Phenoxypropazine Maleate
3941-06-8 7411
$C_{13}H_{18}N_2O_5$
(1-Methyl-2-phenoxyethyl)hydrazine maleate (1:1).
Drazine. Antidepressant. A monoamine oxidase inhibitor. mp = 107-108°; very soluble in H_2O, less soluble in EtOH, iPrOH.

3271 Piberaline
39640-15-8 7547
$C_{17}H_{19}N_3O$
1-Benzyl-4-picolinoylpiperazine.
Trelibet. Antidepressant. *EGYT.*

3272 Pirandamine
42408-79-7
$C_{17}H_{23}NO$
1,3,4,9-Tetrahydro-N,N,1-trimethylindeno[2,1-c]pyran-1-ethylamine.
Antidepressant. *Ayerst.*

3273 Pirandamine Hydrochloride
42408-78-6
$C_{17}H_{24}ClNO$
1,3,4,9-Tetrahydro-N,N,1-trimethylindeno[2,1-c]pyran-1-ethylamine hydrochloride.
AY-23713. Antidepressant. *Ayerst.*

3274 Pirisudanol
33605-94-6 8175 251-591-9
$C_{16}H_{24}N_2O_6$
2-(Dimethylamino)ethyl [5-hydroxy-4-(hydroxymethyl)-6-methyl-3-pyridyl]methyl succinate.
Pyrisuccideanol. Antidepressant. *Soc. Etudes Sci. Ind. L'Ile de France.*

3275 Pirisudanol Dimaleate
53659-00-0 8175 258-690-6
$C_{24}H_{32}N_2O_{14}$
2-(Dimethylamino)ethyl [5-hydroxy-4-(hydroxymethyl)-6-methyl-3-pyridyl]methyl succinate maleate (1:2).
Mentium; Nadex; Pyrisuccideanol Dimaleate; Stivane. Antidepressant. mp = 134°. *Soc. Etudes Sci. Ind. L'Ile de France.*

3276 Pirlindole
60762-57-4
$C_{15}H_{18}N_2$
2,3,3a,4,5,6-Hexahydro-8-methyl-1H-pyrazino[3,2,1-jk]carbazole.
pyrazidole. A selective and reversible monoamine oxidase A inhibitor. Antidepressant.

3277 Pizotyline
15574-96-6 7671 239-632-9
$C_{19}H_{21}NS$
4-(9,10-Dihydro-4H-benzo[4,5]cyclohepta[1,2-b]thien-4-ylidene)-1-methylpiperidine.
BC-105; pizotifen; pizotifan; Litec. Antimigraine; appetite stimulant; antidepressant. Tricyclic. Serotonin antagonist (specific to migraine) structurally related to cyproheptadine. *Sandoz Pharm. Corp.*

3278 Pizotyline Hydrochloride
73391-87-4 7671
$C_{19}H_{22}ClNS$
4-(9,10-Dihydro-4H-benzo[4,5]cyclohepta[1,2-b]thien-4-ylidene)-1-methylpiperidine hydrochloride.
Anabolic; antimigraine; appetite stimulant; antidepressant. Tricyclic serotonin antagonist, specific to migraine. mp = 261-263° (dec). *Sandoz Pharm. Corp.*

3279 Pizotyline Malate
5189-11-7 7671 225-970-4
$C_{23}H_{27}NSO_5$
4-(9,10-Dihydro-4H-benzo[4,5]cyclohepta[1,2-b]thien-4-ylidene)-1-methylpiperidine compound with malic acid (1:1).
Sandomigran; Sanmigran; Sanomigran; Mosegor. Antimigraine; appetite stimulant; antidepressant. Tricyclic serotonin antagonist. mp = 185-186° (dec). *Sandoz Pharm. Corp.*

3280 Pridefine
5370-41-2
$C_{19}H_{21}N$
3-(Diphenylmethylene)-1-ethyl pyrrolidine.
Antidepressant.

3281 Pridefine Hydrochloride
23239-78-3
$C_{19}H_{22}ClN$
3-(Diphenylmethylene)-1-ethyl pyrrolidine hydrochloride.
AHR-1118. Antidepressant.

3282 Prolintane
493-92-5 7964 207-784-5
$C_{15}H_{23}N$
1-[1-(Phenylmethyl)butyl] pyrrolidine.
phenylpyrrolidinopentane; SP-732. CNS stimulant; antidepressant. $bp_{0.5}$= 105°, bp_{16} = 153°. *Boehringer Ingelheim GmbH; Thomae GmbH Dr. Karl.*

3283 Prolintane Hydrochloride
1211-28-5 7964 214-917-0
$C_{15}H_{24}ClN$
1-[1-(Phenylmethyl)butyl] pyrrolidine hydrochloride.
Katovit; Promotil. CNS stimulant; antidepressant. mp = 133-134°; LD_{50} (mus orl) = 257 mg/kg. *Boehringer Ingelheim GmbH; Thomae GmbH Dr. Karl.*

3284 Propizepine
10321-12-7 8019 233-705-9
$C_{17}H_{20}N_4O$
6,11-Dihydro-6-[2-(dimethylamino)-2-methylethyl]-5H-pyrido[2,3-b][1,5]benzodiazepin-5-one.
Antidepressant. Tricyclic. mp = 122°. *Lab. UPSA.*

3285 Propizepine Hydrochloride
14559-79-6 8019
$C_{17}H_{21}ClN_4O$
6,11-Dihydro-6-[2-(dimethylamino)-2-methylethyl]-5H-pyrido[2,3-b][1,5]benzodiazepin-5-one monohydrochloride.

UP-106; Vagran. Antidepressant. Tricyclic. mp = 235°. *Lab. UPSA.*

3286 Protriptyline
438-60-8 8088 207-119-9
$C_{19}H_{21}N$
N-Methyl-5H-dibenzo[a,d]cycloheptene-5-propylamine. amimetilina. Antidepressant. Tricyclic. *Merck & Co.Inc.*

3287 Protriptyline Hydrochloride
1225-55-4 8088 214-956-3
$C_{19}H_{22}ClN$
N-Methyl-5H-dibenzo[a,d]cycloheptene-5-propylamine hydrochloride.
MK-240; Concordin; Maximet; Triptil; Vivactil. Antidepressant. Tricyclic. mp = 169-171°; λ_m = 290 nm (ε = 13311); pKa = 8.2; soluble in H_2O; pKa = 8.2. *Merck & Co.Inc.*

3288 Quinupramine
31721-17-2 8267 250-780-3
$C_{21}H_{24}N_2$
10,11-Dihydro-5-(3-quinuclidinyl)-5H-dibenz[b,f]azepine.
LM-208; Kinupril; Kevopril. Tricyclic antidepressant. An analog of imipramine. mp = 150°. *Sogeras.*

3289 Quipazine
4774-24-7
$C_{13}H_{15}N_3$
2-(1-Piperazinyl)quinoline.
Antidepressant; oxytoxic. *Miles Inc.*

3290 Quipazine Maleate
5786-68-5 227-314-2
$C_{17}H_{19}N_3O_4$
2-(1-Piperazinyl)quinoline maleate (1:1).
MA-1291. Antidepressant; oxytoxic. *Bayer Corp. Pharm. Div.; Miles Inc.*

3291 Reboxetine
98769-81-4
$C_{19}H_{23}NO_3$
(±)-(2R*)-2-[(αR*)-α-(o-Ethoxyphenoxy)-benzyl]morpholine.
Selective noradrenaline reuptake inhibitor. Antidepressant. *Pharmacia & Upjohn.*

3292 Reboxetine Mesylate
98769-84-7
$C_{20}H_{27}NO_6S$
(±)-(2R*)-2-[(αR*)-α-(o-Ethoxyphenoxy)-benzyl]morpholine methanesulfonate.
PNU-155905E; FCE-20124; Edronax. Selective noradrenaline reuptake inhibitor. Antidepressant. *Pharmacia & Upjohn.*

3293 Ritanserin
87051-43-2 8399
$C_{27}H_{25}F_2N_3OS$
6-[2-[4-[Bis(p-fluorophenyl)methylene]piperidino]ethyl]-7-methyl-5H-thiazolo[3,2-a]pyrimidin-5-one.
R-55667; Tiserton. Anxiolytic; antidepressant. A selective serotonin receptor antagonist. mp = 145.5°; LD_{50} (mmus iv) = 28.2 mg/kg, (mmus orl) = 626 mg/kg, (fmus iv) = 28.2 mg/kg, (fmus orl) = 993 mg/kg, (mrat iv) = 20.0 mg/kg, (mrat orl) = 856 mg/kg, (frat iv) = 22.2 mg/kg, (frat orl) = 515 mg/kg, (mdog iv) = 24.1 mg/kg, (mdog orl) = 1280 mg/kg, (fdog iv) = 33.2 mg/kg, (fdog orl) = 640-1280 mg/kg. *Janssen Pharm. Inc.*

3294 Rolicyprine
2829-19-8 8409
$C_{14}H_{16}N_2O_2$
L-trans-(+)-5-Oxo-N-(2-phenylcyclopropyl)-2-pyrrolidinecarboxamide.
EX-4883; RMI-83027; Cypromin; Rolicypram. Antidepressant. mp = 144-147°; $[\alpha]_D^{25}$ = +104.28° (DMF); LD_{50} (rat orl) = 88-104 mg/kg. *Colgate-Palmolive; Lakeside BioTechnology; Marion Merrell Dow Inc.; Schering AG.*

3295 Rolipram
61413-54-5 8410 262-771-1
$C_{16}H_{21}NO_3$
4-[3-(Cyclopentyloxy)-4-methoxyphenyl]-2-pyrrolidinone.
ZK-62711. Antidepressant. A selective type-IV inhibitor of cyclic AMP phosphodiesterase; a pyrrolidone antidepressant. mp = 132°. *Schering AG.*

3296 Roxindole
112192-04-8 8432
$C_{23}H_{26}N_2O$
3-[4-(3,6-Dihydro-4-phenyl-1(2H)-pyridylbutyl]-indol-5-ol.
Dopamine D_2 receptor agonist. Used as an antidepressant.

3297 Roxindole Hydrochloride
108050-82-4 8432
$C_{23}H_{27}ClN_2O$
3-[4-(3,6-Dihydro-4-phenyl-1(2H)-pyridinyl)butyl]-1H-indol-5-ol hydrochloride.
EMD-38362. A D_2 receptor agonist. mp = 274°.

3298 Roxindole Mesylate
119742-13-1 8432
$C_{24}H_{30}N_2O_4S$
3-[4-(3,6-Dihydro-4-phenyl-1(2H)-pyridylbutyl]indol-5-ol monomethanesulfonate.
EMD-49980. Dopamine D_2 receptor agonist. Used as an antidepressant.

3299 Seproxetine
126924-38-7
$C_{16}H_{16}F_3NO$
(S)-3-Phenyl-3-[(α,α,α-trifluoro-p-tolyl)oxy]propylamine.
Antidepressant. *Eli Lilly & Co.*

3300 Seproxetine Hydrochloride
127685-30-7
$C_{16}H_{17}ClF_3NO$
(S)-3-Phenyl-3-[(α,α,α-trifluoro-p-tolyl)oxy]propylamine hydrochloride.
LY-215229. Antidepressant. *Eli Lilly & Co.*

3301 Sertraline
79617-96-2 8612
$C_{17}H_{17}Cl_2N$
(1S,4S)-4-(3,4-Dichlorophenyl)-1,2,3,4-tetrahydro-N-methyl-1-naphthalenamine.

Antidepressant. Selective serotonin uptake inhibitor. *Pfizer Inc.; Roerig Div. Pfizer Pharm.*

3302 Sertraline Hydrochloride
79559-97-0 8612
$C_{17}H_{18}Cl_3N$
(1S,4S)-4-(3,4-Dichlorophenyl)-1,2,3,4-tetrahydro-N-methyl-1-naphthalenamine hydrochloride.
Cp-51974-1; Lustral; Zoloft. Antidepressant. Serotonin uptake inhibitor. mp = 243-245°; $[\alpha]_D^{23}$ = 38° (c = 2 MeOH). *Pfizer Inc.; Roerig Div. Pfizer Pharm.*

3303 Setiptiline
57262-94-9 260-653-4
$C_{19}H_{19}N$
1,2,3,4-Tetrahydro-2-methyl-1H-dibenzo[3,4:6,7]-cyclohepta[1,2-c]pyridine.
MO-8282; Org-8282; Tecipul [as maleate]; 272. A tetracyclic. Antidepressant.

3304 Sibutramine
106650-56-0 8629
$C_{17}H_{26}ClN$
(±)-1-(p-Chlorophenyl)-α-isobutyl-N,N-dimethylcyclobutanemethylamine.
Anorexic; antidepressant. A monoamine reuptake inhibitor. *Boots Pharmaceuticals Inc.*

3305 Sibutramine Hydrochloride [anhydrous]
84485-00-7 8629
$C_{17}H_{27}Cl_2N$
(±)-1-(p-Chlorophenyl)-α-isobutyl-N,N-dimethylcyclobutanemethylamine hydrochloride.
Anorexic; antidepressant. Monoamine reuptake inhibitor. *Boots Co.; Boots Pharmaceuticals Inc.*

3306 Sibutramine Hydrochloride Monohydrate
125494-59-9 8629
$C_{17}H_{27}Cl_2N$
(±)-1-(p-Chlorophenyl)-α-isobutyl-N,N-dimethylcyclobutanemethylamine hydrochloride monohydrate.
BTS-54524; Reductil; Meridia. Anorexic; antidepressant. Monoamine reuptake inhibitor. mp = 193-195.5°. *Boots Co.; Boots Pharmaceuticals Inc.*

3307 Sulpiride
15676-16-1 9163 239-753-7
$C_{15}H_{23}N_3O_4S$
N-[(1-Ethyl-2-pyrrolidinyl)methyl]-2-methoxy-5-sulfamoylbenzamide.
Abilit; Aiglonyl; Coolspan; Dobren; Dogmatil; Dogmatyl; Dolmatil; Guastil; Meresa; Miradol; Mirbanil; Misulvan; Neogama; Omperan; Pyrikappl; Sernevin; Splotin; Sulpitil; Sursumid; Synedil; Trilan; Levosulpiride [as l-form]; S(-)-Sulpiride [as l-form]; Levobren [as l-form]; Levopraid [as l-form]. Antidepressant; antipsychotic; antiemetic. Dopamine D_2- and D_3-receptor antagonist. White, odorless crystalline powder; mp = 178-180°; poorly soluble in MeOH; nearly insoluble in H_2O, Et_2O, $CHCl_3$, C_6H_6; pKa_1 = 10.19; $[\alpha]_D^{25}$ = -66.8° (c = 0.5 DMF); LD_{50} (mus ip) = 170 mg/kg, (mus orl) = 2250 mg/kg. *Delagrange; Ravizza; Soc. Etudes Sci. Ind. L'Ile de France.*

3308 Suritozole
110623-33-1
$C_{10}H_{10}FN_3S$
3-(m-Fluorophenyl-1,4-dimethyl-Δ^2-1,2,4-triazoline-5-thione.
MDL-26479. Antidepressant. *Marion Merrell Dow Inc.*

3309 Talsupram
21489-20-3
$C_{20}H_{25}NS$
1,3-Dihydro-N,3,3-trimethyl-1-phenylbenzo(c)thiophen-1-propylamine.
Selective noradrenaline uptake inhibitor. Antidepressant.

3310 Tametraline
52795-02-5
$C_{17}H_{19}N$
(1R,4S)-1,2,3,4-Tetrahydro-N-methyl-4-phenyl-1-napthylamine.
Antidepressant. *Pfizer Inc.*

3311 Tametraline Hydrochloride
52760-47-1
$C_{17}H_{20}ClN$
(1R,4S)-1,2,3,4-Tetrahydro-N-methyl-4-phenyl-1-napthylamine hydrochloride.
CP-24441-1. Antidepressant. *Pfizer Inc.*

3312 Tampramine
83166-17-0
$C_{23}H_{24}N_4$
11-[3-(Dimethylamino)propyl]-6-phenyl-11H-pyrido[2,3-b][1,4]benzodiazepine.
Antidepressant. Tricyclic.

3313 Tampramine Fumarate
83166-18-1
$C_{27}H_{28}N_4O_4$
11-[3-(Dimethylamino)propyl]-6-phenyl-11H-pyrido[2,3-b][1,4]benzodiazepine fumarate (1:1).
AHR-9377. Antidepressant. Tricyclic.

3314 Tandamine
42408-80-0
$C_{18}H_{26}N_2S$
1-[2-Dimethyamino)ethyl]-9-ethyl-1,2,3,9-tetrahydro-1-methylthiopyrano[3,4-b]-indole.
Antidepressant. Tricyclic.

3315 Tandamine Hydrochloride
58167-78-5
$C_{18}H_{27}ClN_2S$
1-[2-Dimethyamino)ethyl]-9-ethyl-1,2,3,9-tetrahydro-1-methylthiopyrano[3,4-b]-indole monohydrochloride.
AY-23946. Antidepressant. Tricyclic.

3316 Tandospirone
87760-53-0 9219
$C_{21}H_{29}N_5O_2$
(3aα,4β,7β,7aα)-Hexahydro-2-[4-[4-(2-pyrimidinyl)-1-piperazinyl]butyl]-4,7-methano-1H-isoindole-1,3(2H)-dione.
Anxiolytic; antidepressant. A serotonin receptor agonist. mp = 112-113.5°. *Pfizer Inc.; Sumitomo Pharm. Co. Ltd.*

3317 Tandospirone Citrate
112457-95-1 9219
$C_{27}H_{37}N_5O_9$
(1R*,2S*,3R*,4S*)-N-[4-[4-(2-Pyrimidinyl)-1-piperazinyl]-butyl]-2,3-norbornanedicarboximide citrate (1:1).
SM-3997. Anxiolytic; antidepressant. A serotonin receptor agonist. mp = 169.5-170°. *Pfizer Inc.; Sumitomo Pharm. Co. Ltd.*

3318 Teniloxazine
62473-79-4
$C_{16}H_{19}NO_2S$
(±)-2-[[(α-2-Thienyl-o-tolyl)oxy]methyl]morpholine.
Y-8894 [as maleate]. Antidepressant; nootropic; antianoxic.

3319 Thiazesim
5845-26-1 9440
$C_{19}H_{22}N_2OS$
5-[2-(Dimethylamino)ethyl]-2,3-dihydro-2-phenyl-1,5-benzothiazepin-4(5H)-one.
thiazenone; tiazesim. Antidepressant. *Bristol-Myers Squibb Pharm. R&D; Olin Mathieson.*

3320 Thiazesim Hydrochloride
3122-01-8 9440
$C_{19}H_{23}ClN_2OS$
5-[2-(Dimethylamino)ethyl]-2,3-dihydro-2-phenyl-1,5-benzothiazepin-4(5H)-one monohydrochloride.
tiazesim hydrochloride; SQ-10469; Altinil. Antidepressant. mp = 222-224°. *Bristol-Myers Squibb Pharm. R&D; Olin Mathieson.*

3321 Thozalinone
655-05-0 9521
$C_{11}H_{12}N_2O_2$
2-(Dimethylamino)-5-phenyl-2-oxazolin-4-one.
CL-39808; Stimsen; thozalinone; tozalinone. Antidepressant. mp = 133-136°. *Am. Cyanamid.*

3322 Tianeptine
66981-73-5 9560
$C_{21}H_{25}ClN_2O_4S$
7-[(3-Chloro-6,11-dihydro-6-methyl-dibenzo[c,f][1,2]-thiazepin-11-yl)amino]-heptanoic acid S,S-dioxide.
Antidepressant. A tricyclic compound with psychostimulant, antiulcerative, and anitemetic properties. *Sci. Union et Cie France.*

3323 Tianeptine Sodium Salt
30123-17-2 9560 250-059-3
$C_{21}H_{24}ClN_2NaO_4S$
Sodium 7-[(3-Chloro-6,11-dihydro-6-methyldibenzo[c,f][1,2]thiazepin-11-yl)amino] heptanoate S,S-dioxide.
Stablon. Antidepressant. A tricyclic compound with psychostimulant, antiulcerative, and anitemetic properties. mp = 180°. *Sci. Union et Cie France.*

3324 Tifemoxone
39754-64-8 254-620-3
$C_{11}H_{13}NO_2S$
Tetrahydro-6-(phenoxymethyl)-2H-1,3-oxazine-2-thione.
Putative antidepressant.

3325 Tiflucarbine
89875-86-5
$C_{16}H_{17}FN_2S$
9-Ethyl-4-fluoro-7,8,9,10-tetrahydro-1-methyl-6H-pyrido[4,3-b]thieno[3,2-e]indole.
BAY-P-4495. Protein kinase C/calmodulin inhibitor. Serotonin (5-HT) receptor antagonist. Antidepressant.

3326 Tisocromide
35423-51-9
$C_{19}H_{30}N_2O_6S$
N-[3-(Dimethylamino)-1,3-dimethylbutyl]-6,7-dimethoxy-2,1-benzoxathian-3-carboxamide 1,1-dioxide.
Antidepressant; antioxidant; antihypoxic.

3327 Toloxatone
29218-27-7 9659 249-522-2
$C_{11}H_{13}NO_3$
5-(Hydroxymethyl)-3-m-tolyl-2-oxazolidinone.
MD-69276; Humoryl; Perenum. Antidepressant. A reversible monoamine oxidase A inhibitor. mp = 76°; LD_{50} (mus orl) = 1850 mg/kg, 1500 mg/kg. *Delalande; Delandale Labs. Ltd.*

3328 Tomoxetine
83015-26-3
$C_{17}H_{21}NO$
(-)-N-Methyl-3-(o-tolyloxy)propylamine.
Antidepressant.

3329 Tomoxetine Hydrochloride
82248-59-7
$C_{17}H_{22}ClNO$
(-)-N-Methyl-3-(o-tolyloxy)propylamine hydrochloride.
Ly-139603. Antidepressant. *Eli Lilly & Co.*

3330 Tranylcypromine
155-09-9 9708 205-841-9
$C_9H_{11}N$
(±)-trans-2-Phenylcyclopropanamine.
SKF-trans-385. Antidepressant. A monoamine oxidase inhibitor. Liquid; $bp_{1.5-1.6}$ = 79-80°; [hydrochloride]: mp = 164-166°. *SmithKline Beecham Pharm.*

3331 Tranylcypromine Sulfate
13492-01-8 9708 236-807-1
$C_{18}H_{24}N_2O_4S$
(±)-trans-2-Phenylcyclopropamine sulfate (2:1).
Parnate; Tylciprine. Antidepressant. A monoamine oxidase inhibitor. Soluble in H_2O; less soluble in EtOH, Et_2O; insoluble in $CHCl_3$. *SmithKline Beecham Pharm.*

3332 Trazium Esilate
97110-59-3
$C_{19}H_{18}ClN_3O_4S$
1-(p-Chlorophenyl)-1,2-dihydro-1-hydroxy-as-triazino[6,1-a]isoquinolin-5-ium ethanesulfonate.
EGYT-3615. As-triazino isoquinolinium salt. Antidepressant. Tetrabenazine, yohimbine potentiation antagonist.

3333 Trazodone
19794-93-5 9712 243-317-1
$C_{19}H_{24}ClN_5O$
2-[3-[4-m-Chlorophenyl)-1-piperazinyl]propyl]-s-triazolo-[4,3-a]pyridin-3(2H)-one.

Antidepressant. mp = 86-87°; pKa (50% EtOH) = 6.14. *Angelini Francesco.*

3334 Trazodone Hydrochloride

25332-39-2 9712 246-855-5

$C_{19}H_{25}Cl+2N_5O$

2-[3-[4-m-Chlorophenyl)-1-piperazinyl]propyl]-s-triazolo-[4,3-a]pyridin-3(2H)-one monohydrochloride.

AF-1161; Bimaran; Desyrel; Molipaxin; Pragmazone; Thombran; Tombran; Trazolan; Trittico. Antidepressant. mp = 233°; soluble in $CHCl_3$, less soluble in polar solvents, practically insoluble in common organic solvents; λ_m = 211, 246, 274, 312 nm (ε 50100, 11730, 3840, 3820); LD_{50} (mus iv) = 96 mg/kg. *Angelini Francesco; Lemmon Co.; Mead Johnson Pharmaceuticals.*

3335 Trebenzomine

23915-73-3

$C_{12}H_{17}NO$

(±)-(cis or trans)-N,N,2-Trimethyl-3-chromanamine.

Antidepressant. *Parke-Davis.*

3336 Trebenzomine Hydrochloride

23915-74-4

$C_{12}H_{18}ClNO$

(±)-(cis or trans)-N,N,2-Trimethyl-3-chromanamine hydrochloride.

CI-686-HCl. Antidepressant. *Parke-Davis.*

3337 Trimipramine

739-71-9 9852 212-008-3

$C_{20}H_{26}N_2$

5-[3-(Dimethylamino)-2-methylpropyl]-10,11-dihydro-5H-dibenz[b,f]azepine.

RP-7162; trimeprimine; trimeproprimine; Rhotrimine; Surmontil. Antidepressant. Tricyclic. mp = 45°. *Wyeth Labs.*

3338 Trimipramine Maleate

521-78-8 9852 208-318-3

$C_{24}H_{30}N_2$

5-[3-(Dimethylamino)-2-methylpropyl]-10,11-dihydro-5H-dibenz[b,f]azepine maleate (1:1).

Stangyl (tablets); Surmontil (tablets). Antidepressant. Tricyclic. mp = 142°; soluble in $CHCl_3$; slightly soluble in H_2O, EtOH; practically insoluble in Et_2O. *Wyeth Labs.*

3339 Trimipramine Methanesulfonate

9852

$C_{13}H_{19}NO_5S$

5-[3-(Dimethylamino)-2-methylpropyl]-10,11-dihydro-5H-dibenz[b,f]azepine methanesulfonate.

Stangyl (ampules); Surmontil (ampules). Antidepressant. Tricyclic. *Wyeth Labs.*

3340 Venlafaxine

93413-69-5 10079

$C_{17}H_{27}NO_2$

(±)-1-[α-[(Dimethylamino)methyl]-p-methoxybenzyl]cyclohexanol.

Venlafexine. Antidepressant. A serotonin noradrenaline reuptake inhibitor. mp = 102-104°; [(-)-form)]: $[\alpha]_D^{25}$ = -27.1° (c = 1.04 in 95% EtOH); [(+)-form]: $[\alpha]_D^{25}$ = +27.6° (c = 1.07 in 95% EtOH). *Am. Home Products; Wyeth-Ayerst Labs.*

3341 Venlafaxine Hydrochloride

99300-78-4 10079

$C_{17}H_{28}ClNO_2$

(±)-1-[α-[(Dimethylamino)methyl]-p-methoxybenzyl]cyclohexanol hydrochloride.

WY-4530; WY-45651 [(-)-form]; WY-45655 [(+)-form]; Effexor. Antidepressant. A serotonin noradrenaline reuptake inhibitor. mp = 215-217°; soluble in H_2O (57.2 g/100 ml); [(+) or (-) form]: mp = 240-240.5°; [(-)-form)]: $[\alpha]_D^{25}$ = +4.6° (c = 1.0 in EtOH); [(+)-form]: $[\alpha]_D^{25}$ = -4.7° (c = 0.945 in EtOH). *Am. Home Products; Wyeth-Ayerst Labs.*

3342 Viloxazine

46817-91-8 10116 256-281-7

$C_{13}H_{19}NO_3$

2-[(o-Ethoxyphenoxy)methyl]morpholine.

ICI-58834. Antidepressant. *ICI.*

3343 Viloxazine Hydrochloride

35604-67-2 10116 252-638-6

$C_{13}H_{20}ClNO_3$

2-[(o-Ethoxyphenoxy)methyl]morpholine hydrochloride.

Catatrol; Vivilan; Vicilan; Vivarint. Antidepressant. mp = 185-186°; LD_{50} (mus orl) = 1000 mg/kg, (mus iv) = 60 mg/kg. *ICI .*

3344 Viqualine

72714-74-0

$C_{20}H_{26}N_2O$

6-Methoxy-4-[3-[(3R,4R)-3-vinyl-4-piperidyl]propyl]quinoline.

PK-5078. Used in treatment of alcoholism. Serotonin releaser and uptake inhibitor; antidepressant.

3345 Zimeldine

56775-88-3 10254

$C_{16}H_{17}BrN_2$

(Z)-3-[1-(p-Bromophenyl)-3-(dimethylamino)propenyl]pyridine.

H-102/09; Zimelidine. Antidepressant. An antidepressant that inhibits membranal 5-hydroxytryptamine uptake. *Astra Hassle AB.*

3346 Zimeldine Hydrochloride

61129-30-4 10254 262-279-7

$C_{16}H_{21}BrCl_2N_2O$

(Z)-3-[1-(p-Bromophenyl)-3-(dimethylamino)propenyl]pyridine dihydrochloride monohydrate.

Normud; Zelmid; Zimelidine Hydrochloride.

Antidepressant. Inhibits membranal 5-hydroxytryptamine (serotonin) uptake. mp = 193°. *Astra Hassle AB; Astra Sweden; Merck & Co.Inc.*

3347 Zimeldine Hydrochloride [anhydrous]

60525-15-7 10254

$C_{16}H_{19}BrCl_2N_2$

(Z)-3-[1-(p-Bromophenyl)-3-(dimethylamino)propenyl]pyridine dihydrochloride.

Antidepressant. Inhibits membranal 5-hydroxytryptamine (serotonin) uptake. *Astra Hassle AB.*

3348 Zoloperone
52867-74-0
$C_{22}H_{24}FN_3O_3$
4-(p-Fluorophenyl)-5-[2-[4-(o-methoxyphenyl)-1-piperazinyl]ethyl]-4-oxazolin-2-one.
LR-511. Free base, sertraline [79617-96-2]. An antidepressant. Neuroleptic; antipsychotic.

3349 Zometapine
51022-73-2
$C_{14}H_{15}ClN_4$
4-(m-Chlorophenyl)-1,6,7,8-tetrahydro-1,3-dimethylpyrazolo[3,4-e][1,4]diazepine.
CI-781. Antidepressant. *Parke-Davis.*

Antidiabetics

3350 Acarbose
56180-94-0 15 260-030-7
$C_{25}H_{43}NO_{18}$
O-4,6-Dideoxy-4-[[(1S,4R,5R,6S)-4,5,6-trihydroxy-3-(hydroxymethyl)-2-cyclohexen-1-yl]amino]-α-D-glucopyranosyl-(1→4)-O-α-D-glucopyrranosyl-(1→4)-D-glucose.
Precose; Bay g 5421; Glucobay. An α-glucosidase inhibitor with antidiabetic activity. $[\alpha]_D^{18}$ = 165° (c = 0.4 H_2O). *Bayer AG.*

3351 Acetohexamide
968-81-0 59 213-530-4
$C_{15}H_{20}N_2O_4S$
4-acetyl-N-[(cyclohexylamino)-carbonyl]-benzenesulfonamide.
Dimelor; Dymelor; Dimelin; Ordimel; Tsiklamid; Cyclamide. Antidiabetic agent. An oral agent for treatment of hypoglycemia. mp = 188-190°; soluble in C_5H_5N; slightly soluble in EtOH, $CHCl_3$; insoluble in H_2O, Et_2O. *Eli Lilly & Co.*

3352 Amlintide
122384-88-7
$C_{165}H_{261}N_{51}O_{55}S_2$
H-Lys-Cys-Asn-Thr-Ala-Thr-Cys-Ala-Thr-Gln-Arg-Leu-Ala-Asn-Phe-Leu-Val-His-Ser-Ser-Asn-Asn-Phe-Gly-Ala-Ile-Leu-Ser-Ser-Thr-Asn-Val-Gly-Ser-Asn-Thr-Tyr-NH_2 cyclic 2-7 disulfide.
AC-001. Antidiabetic agent. Synthetic peptide. *Amylin Pharm. Inc.*

3353 Bemetizide
1824-52-8 217-357-5
$C_{15}H_{16}ClN_3O_4S_2$
6-Chloro-3,4-dihydro-3-(α-methylbenzyl)-2H-1,2,4-benzothiadiazine-7-sulfonamide 1,1-dioxide.
Antidiabetic.

3354 Benfosformin
52658-53-4
$C_9H_{12}N_5Na_2O_3P$
Disodium [(benzylamidino)amidino]phosphoramidate monohydrate.
Antidiabetic agent.

3355 Benfosformin [anhydrous]
35282-33-8
$C_9H_{12}N_5Na_2O_3P.H_2O$
Disodium [(benzylamidino)amidino]phosphoramidate.
Antidiabetic agent.

3356 Buformin
692-13-7 1500 211-726-4
$C_6H_{15}N_5$
1-Butylbiguanide.
DBV; W-37. Antidiabetic agent. Very soluble in H_2O; [nitrate]: mp = 125-126° (dec). *U.S. Vitamin.*

3357 Buformin Hydrochloride
1190-53-0 1500 214-723-6
$C_6H_{15}N_5$
1-Butylbiguanide hydrochloride.
Andere; Biforon; Bigunal; Bufonamin; Bulbonin; Diabrin; Dibetos; Gliporal; Insulamin; Krebon; Panformin; Silubin; Sindiatil; Tidemol; Ziavetine. Antidiabetic agent. mp = 174-177°; soluble in H_2O, EtOH; LD_{50} (mus ip) = 380 mg/kg. *U.S. Vitamin.*

3358 Butoxamine
2922-20-5
$C_{15}H_{23}NO_3$
α-[1-(tert-Butylamino)ethyl]-2,5-dimethoxybenzyl alcohol.
Butaxamine. Antidiabetic agent. *Glaxo Wellcome Inc.*

3359 Butoxamine Hydrochloride
5696-15-1 227-169-5
$C_{15}H_{24}ClNO_3$
α-[1-(tert-Butylamino)ethyl]-2,5-dimethoxybenzyl alcohol hydrochloride.
Butaxamine hydrochloride; BW-64-9; NSC-106565. Antidiabetic agent. *Glaxo Wellcome Inc.*

3360 Camiglibose
132438-21-2
$C_{13}H_{25}NO_9$
Methyl-6-deoxy-6-[(2R,3R,4R,5S)-3,4,5-trihydroxy-2-(hydroxymethyl)piperidino]-α-D-glucopyranoside sesquihydrate.
MDL-73945. Antidiabetic agent. *Marion Merrell Dow Inc.*

3361 Carbutamide
339-43-5 1881 206-424-4
$C_{11}H_{17}N_3O_3S$
4-Amino-N-[(butylamino)carbonyl]benzenesulfonamide.
1-Butyl-3-sulfanilylurea; Invenol; aminophenurobutane; BZ-55; U-6987; Bucarban; Glucidoral; Glucofren; Invenol; Nadisan. Antidiabetic agent. mp = 144-145°; soluble in H_2O; LD_{50} (mus, sc) = 3 g/kg. *Pharmacia & Upjohn.*

3362 Chlorpropamide
94-20-2 2239 202-314-5
$C_{10}H_{13}ClN_2O_3S$
4-Chloro-N-[(propylamino)-carbonyl]benzenesulfonamide.
N-propyl-N'-(p-chlorobenzenesulfonyl)urea; P-607; Adiaben; Asucrol; Catanil; Chloronase; Diabechlor; Diabenal; Diabetoral; Diabinese; Melitase; Millinese; Oradian; Stabinol. Antidiabetic agent. mp = 127-129°;

λ_m = 232.5 nm (ε 16500 0.01N HCl); soluble in H_2O (2.2 mg/ml pH 6) insoluble at higher pH; soluble in EtOH, $CHCl_3$; less soluble in C_6H_6, Et_2O; LD_{50} (rat ip) = 580 mg/kg. *Pfizer Intl.*

3363 Ciglitazone
74772-77-3
$C_{18}H_{23}NO_3S$
(±)-5-[p-[(1-Methylcyclohexyl)methoxy]benzyl]-2,4-thiazolidinedione.
Add-3878; U-63287. Antidiabetic agent. *Takeda Chem. Ind. Ltd.*

3364 Clomoxir
88431-47-4
$C_{14}H_{17}ClO_3$
(±)-2-[5-(p-Chlorophenyl)pentyl]glycidic acid.
Megaclor. Carnitine palmitoyl transferase inhibitor. Orally active antidiabetic agent.

3365 Darglitazone
141200-24-0
$C_{23}H_{20}N_2O_4S$
(±)-5-[p-[3-(5-Methyl-2-phenyl-4-oxazolyl)propionyl]benzyl]-2,4-thiazolidinedione.
Antidiabetic agent. Oral hypoglycemic agent. *Pfizer Intl.*

3366 Darglitazone Sodium
141683-98-9
$C_{23}H_{19}NaN_2O_4S$
(±)-5-[p-[3-(5-Methyl-2-phenyl-4-oxazolyl)propionyl]-benzyl]-2,4-thiazolidinedione sodium salt.
CP-86325-2. Antidiabetic agent. Oral hypoglycemic agent. *Pfizer Intl.*

3367 Deriglidole
122830-14-2
$C_{16}H_{21}N_3$
(+)-1,2,4,5-Tetrahydro2-(2-imidazolin-2-yl)-2-propylpyrrolo[3,2,1-hi]imdazole.
Antidiabetic agent.

3368 Dibotin
114-86-3 7376 204-057-4
$C_{10}H_{15}N_5$
N-(2-Phenylethyl)imidodicarbonimidic diamide.
1-phenethylbiguanide; phenformin; fenformin; fenormin; PEDG; β-PEBG; Dipar. Antidiabetic agent. Known to cause lactic acidosis. *Ciba-Geigy Corp.*

3369 Englitazone
109229-58-5
$C_{20}H_{19}NO_3S$
(-)-5-[[(2R)-2-Benzyl-6-chromanyl]methyl]-2,4-thiazolidinedione.
Antidiabetic agent. *Pfizer Intl.*

3370 Englitazone Sodium
109229-57-4
$C_{20}H_{18}NNaO_3S$
(-)-5-[[(2R)-2-Benzyl-6-chromanyl]methyl]-2,4-thiazolidinedione sodium salt.
CP-72467-2. Antidiabetic agent. *Pfizer Intl.*

3371 Etoformin
45086-03-1
$C_8H_{19}N_5$
1-Butyl-2-ethylbiguanide.
Antidiabetic agent. *Schering AG.*

3372 Etoformin Hydrochloride
53597-26-5
$C_8H_{20}ClN_5$
1-Butyl-2-ethylbiguanide monohydrochloride.
SH E 199. Antidiabetic agent. *Schering AG.*

3373 Gliamilide
51876-98-3
$C_{23}H_{33}N_5O_5S$
endo-1-[[4-[2-(2-Methoxynicotinamido)ethyl]-piperidino]sulfonyl]-3-(5-norbornen-2-ylmethyl)urea.
CP-27634. Antidiabetic agent. *Pfizer Intl.*

3374 Glibenese
29094-61-9 4451 249-427-6
$C_{21}H_{27}N_5O_4S$
1-Cyclohexyl-3-[[p-[2-(5-methylpyrazinecarboxamido)ethyl]phenyl]sulfonyl]urea.
Glucotrol; Glucotrol XL; CP-28720; K-4024. Antidiabetic agent. mp = 208-209°; LD_{50} (rat ip) = 1.2 g/kg, (mus ip) > 3 g/kg. *Apothecon; Pratt Pharm.*

3375 Glibornuride
26944-48-9 4447 248-124-6
$C_{18}H_{26}N_2O_4S$
endo,endo-1-[(1R)-(2-Hydroxy-3-bornyl)]-3-(p-tolylsulfonyl)urea.
Glutril; Ro-6-4563; Gluboride. Antidiabetic agent. mp = 192-195°; $[\alpha]_D$= 63.8° (EtOH). *Hoffmann-LaRoche Inc.*

3376 Glicetanile
24455-58-1
$C_{23}H_{25}ClN_4O_4S$
5'-Chloro-2-[p-[(5-isobutyl-2-pyrimidinyl)sulfamoyl]phenyl]-o-acetanisidide.
glydanile. Antidiabetic agent. *Schering AG.*

3377 Glicetanile Sodium
24428-71-5
$C_{23}H_{24}ClN_4NaO_4S$
5'-Chloro-2-[p-[(5-isobutyl-2-pyrimidinyl)sulfamoyl]phenyl]-o-acetanisidide monosodium salt.
Glydanile sodium. Antidiabetic agent. *Schering AG.*

3378 Gliflumide
35273-88-2
$C_{25}H_{29}FN_4O_4S$
(-)-(S)-N-(5-Fluoro-2-methoxy-α-methylbenzyl)-2-[p-[(5-isobutyl-2-pyrimidinyl)sulfamoyl]phenyl]acetamide.
SH3.1168. Antidiabetic agent.

3379 Glipalamide
37598-94-0
$C_{12}H_{15}N_3O_3S$
(±)-5-Methyl-N-(p-tolylsulfonyl)-2-pyrazoline-1-carboxamide.
Antidiabetic.

3380 Glipizide
29094-61-9 4451 249-427-6
$C_{21}H_{27}N_5O_4S$
1-Cyclohexyl-3-[[p-[2-(5-methylpyrazinecarboxamido)-ethyl]phenyl]sulfonyl]urea.
Glucotrol; Glucotrol XL; CP-28720; K-4024. Antidiabetic agent. mp = 208-209°; LD_{50} (rat ip) = 1.2 g/kg, (mus ip) > 3 g/kg. *Apothecon; Pratt Pharm.*

3381 Gliquidone
33342-05-1 4452 251-463-2
$C_{27}H_{33}N_3O_6S$
1-Cyclohexyl-3-[[p-[2-(3,4-dihydro-7-methoxy-4,4-dimethyl-1,3-dioxo-2(1H)-isoquinolyl)ethyl]-phenyl]sulfonyl]urea.
ARDF 26; Glurenorm. Antidiabetic agent. mp = 180-182°. *Boehringer Ingelheim GmbH.*

3382 Glisamuride
52430-65-6
$C_{23}H_{31}N_5O_4S$
1-Methyl-3-[p-[[3-(4-methylcyclohexyl)ureido]-sulfonyl]phenethyl]-1-(2-pyridyl)urea.
Antidiabetic.

3383 Glisolamide
24477-37-0
$C_{20}H_{26}N_4O_5S$
1-Cyclohexyl-3-[[p-[2-(5-methyl-3-isoxazolecarboxamido)ethyl]phenyl]sulfonyl]urea.
Antidiabetic.

3384 Glisoxepid
25046-79-1 4453 246-579-5
$C_{20}H_{27}N_5O_5S$
N-[2-[4-[[[[(Hexahydro-1H-azepin-1-yl)amino]carbonyl]amino]sulfonyl]phenyl]ethyl]-5-methyl-3-isoxazolecarboxamide.
BS-4231; RP-22410; Pro-Diaban. Antidiabetic agent. mp = 189°; LD_{50} (rat orl) >10 g/kg, (rat iv) = 196 mg/kg, (mus orl) > 10 g/kg, (mus iv) = 283 mg/kg, (cat orl) > 4 g/kg, (dog orl) > 2 g/kg. *Bayer Corp.*

3385 Glucagon
16941-32-5
His-Ser-Gln-Gly-Thr-Phe-Thr-Ser-Asp-Tyr-Ser-Lys-Tyr-Leu-Asp-Ser-Arg-Arg-Ala-Gln-Asp-Phe-Val-Gln-Trp-Leu-Met-Asn-Thr.
Glukagon; hyperglycemic-glycogenolytic factor; HG-factor; HGF. Antidiabetic agent. Polypetpide hormone produced in the pancreas. Insoluble in H_2O (pH 3-9.5).

3386 Glurenorm
33342-05-1 4452 251-463-2
$C_{27}H_{33}N_3O_6S$
1-Cyclohexyl-3-[[p-[2-(3,4-dihydro-7-methoxy-4,4-dimethyl-1,3-dioxo-2(1H)-isoquinolyl)ethyl]phenyl]sulfonyl]urea.
ARDF 26; Glurenorm; . Antidiabetic agent. mp = 180-182°. *Boehringer Ingelheim GmbH.*

3387 Glyburide
10238-21-8 4486 233-570-6
$C_{23}H_{28}ClN_3O_5S$
5-Chloro-N-[2-[4-[[[(cyclohexylamino)carbonyl]amino]-sulfonyl]phenyl]ethyl]-2-methoxybenzamide.
glybenzcyclamide; glibenclamide; HB-419; U-26452; Abbenclamide; Adiab; Azuglucon; Bastiverit; Dia-basan; Diabeta; Diabiphage; Daonil; Duraglucon; Wuglucon; Gilemal; Gliben-Puren N; Glidiabet; Glimidstada; Glubate; Glucoremed; Gluco-Tabinen; Glycolande; Gl. Antidiabetic agent. Second generation sulfonylurea with hypoglycemic activity. mp = 169-170°, 172-174°; poorly soluble in H_2O, soluble in organic solvents; LD_{50} (mus orl) > 20 g/kg, (mus ip) > 12.5 g/kg, (mus sc) > 20 g/kg. (rat orl) > 20 g/kg, rat ip) > 12.5 g/kg, (rat sc) > 20 g/kg. *Abbott Labs Inc.*

3388 Glyhexamide
451-71-8 4516
$C_{16}H_{22}N_2O_3S$
N-[(Cyclohexylamino)carbonyl]-2,3-dihydro-1H-indene-5-sulfonamide.
SQ-15860; Subose; NSC-106960. Antidiabetic agent. mp = 153-155°. *Bristol-Myers Squibb Co.*

3389 Glymidine
339-44-6 4517 206-426-5
$C_{13}H_{15}N_{34}S$
N-[5-(2-Methoxyethoxy)-2-pyrimidinyl]-benzenesulfonamide.
glycodiazine. Antidiabetic agent. mp = 152-154°; soluble in EtOH (9.1 g/l), C_7H_8 (6.7 g/l). *Schering AG.*

3390 Glyoctamide
1038-59-1
$C_{16}H_{24}N_2O_3S$
1-[(p-Chlorophenyl)sulfonyl]-3-[p-(dimethylamino)-phenyl]urea.
Antidiabetic agent. *Hoechst Roussel Pharm. Inc.*

3391 Glyparamide
5581-42-0
$C_{15}H_{16}ClN_3O_3S$
1-[(p-Chlorophenyl)sulfonyl]-3-[p-(dimethylamino)-phenyl]urea.
P-1306. Antidiabetic agent. *Pfizer Intl.*

3392 Hydroxyhexamide
3168-01-2
$C_{13}H_{18}N_2O_4S$
1-[(p-Hydroxyphenyl)sulfonyl]-3-cyclohexylurea.
Metabolite of acetohexamide, a hypoglycemic agent.

3393 Insulin
9004-10-8 5011 232-672-8
Hypurin [bovine]; Iletin II [procine]; Velosulin [porcine]; insulin (prb) [human, recombinant]; Huminsulin [human, recombinant]; Humulin [human, recombinant]; Humulina [human, recombinant]; insulin emp [human, semi-synthetic]; Biohulin [human, semi-synthetic]; Novolin [human, semi-synthetic]; Orgasuline [human, semi-synthetic]; Insulatard; Lente; Novolin; NPH; NPH Iletin; Protamine; Actrapid; Initard; Isotard; Isophane; Mixtard; Monotard; Protamine Zinc; Protaphane; Ultralente; Ultratard. Antidiabetic agent. Growth regulator. Soluble in dilute acids and alkalies. *Eli Lilly & Co.*

3394 Insulin$_{131}$I
37294-43-2 5011
Radio-iodinated insulin.
Imusay-131. Radioactive diagnostic agent. Used to study

insulin binding factors in insulin-resistant sera. *Abbott Labs Inc.*

3395 Insulin Argine
68859-20-1 5011
$C_{269}H_{407}N_{73}O_{79}S_6$
30^Ba-L-Arginine-30_Bb-L-arginineinsulin (human).
Antidiabetic agent.

3396 Insulin Aspart
116094-23-6 5011
$C_{256}H_{381}N_{65}O_{79}S_6$
28_B-L-Aspartic acid-insulin (human).
Insulin X14; INA-X14; B28-Asp-Insulin. Antidiabetic agent. *Novo Nordisk Pharm. Inc.*

3397 Insulin Defalan, Bovine
51798-72-2 5011 257-427-2
$C_{247}H_{372}N_{64}O_{75}S_6$
1_B-de(L-Phenylalanine)insulin (bovine).
Antidiabetic agent.

3398 Insulin Defalan, Porcine
11091-62-6 5011 234-312-5
$C_{247}H_{372}N_{64}O_{75}S_6$
1_B-de(L-Phenylalanine)insulin (porcine).
Antidiabetic agent.

3399 Insulin Detemir
169148-63-4
$C_{267}H_{402}N_{64}O_{76}S_6$
29_B-(N_6-Myristoyl-L-lysine)-30_B-de-L-threonineinsulin (human).
NN-304. Antidiabetic agent. *Novo Nordisk Pharm. Inc.*

3400 Insulin Glargine
160337-95-1 5011
$C_{267}H_{404}N_{72}O_{78}S_6$
21_A-Glycine-30Ba-L-arginine-30_Bb-L-arginineinsulin (human).
Antidiabetic agent.

3401 Insulin, Biphasic Isophane
8063-29-4 5011
Neutral insulin.
insulin, biphasic isophane. Antidiabetic agent. Sterile suspension of bovine insulin complexed with protamine in a solution of porcine insulin, or a sterile suspension of human insulin complexed with protamine in a solubion of human insulin. A hormone isolated from the pancreas; used in the treatment of diabetes. *Eli Lilly & Co.*

3402 Insulin, Dalanated
9004-12-0 5011
SN-44. Insulin derivative: C-terminal alanine removed from B chain of insulin. Antidiabetic agent. *Eli Lilly & Co.*

3403 Insulin, Human
11061-68-0 5011 234-279-7
$C_{257}H_{383}N_{65}O_{77}S_6$
Human insulin.
Humulin; Novolin; Velosulin. Antidiabetic agent. The natural antidiabetic principle produced by the human pancreas. *Christiaens S.A.; Novo Nordisk Pharm. Inc.*

3404 Insulin, Isophane
9004-17-5 5011
Isophane insulin.
NPH insulin; neutral protein Hagedorn insulin; NPH; NPH Iletin. Antidiabetic agent. Crystallized form consisting of protamine, insulin, and zinc. Activity begins 3-4 hours after sc injection; lasts 18-28 hours. pH = 7.1-7.4. *Christiaens S.A.; Novo Nordisk Pharm. Inc.*

3405 Insulin, Lispro
133107-64-9 5011
28_B-L-Lysine-29_B-L-prolineinsulin (human).
LY-275585. Antidiabetic agent. *Eli Lilly & Co.*

3406 Insulin, Neutral
9004-14-2 5011
Antidiabetic agent. A neutral, buffered solution of porcine insulin. *Eli Lilly & Co.*

3407 Insulin, Protamine Zinc
9004-17-5 5011
Protamine zinc insulin injection.
insulin zinc protamine injection; PZI insulin. Antidiabetic agent. An aqueous suspension. Activity begins 4-6 hours after sc injection; lasts 38 hours. pH = 7.1-7.4.

3408 Insulin, Zinc
8049-62-5 5011 232-469-4
Lente Iletin; Monotard; Semilente Insulin; Semilente Iletin. Antidiabetic agent. Injectable form. *Eli Lilly & Co.*

3409 Linogliride
75358-37-1
$C_{16}H_72N_4O$
N-(1-Methyl-2-pyrrolidinylidene)-N'-phenyl-4-morpholinecarboxamidine.
Antidiabetic agent. *McNeil Pharm.*

3410 Linogliride Fumarate
78782-47-5
$C_{20}H_{26}N_4O_5$
N-(1-Methyl-2-pyrrolidinylidene)-N'-phenyl-4-morpholinecarboxamidine fumarate.
McN-3935. Antidiabetic agent. *McNeil Pharm.*

3411 Meglitinide
54870-28-9
$C_{17}H_{16}ClNO_4$
p-[2-(5-Chloro-o-anisamido)ethyl]benzoid acid.
A benzoic acid analog, hypoglycaemic agent. Antidiabetic.

3412 Metformin
657-24-9 6001 211-517-8
$C_4H_{11}N_5$
N,N-Dimethylbiguanide.
N,N-dimethylbiguanide; N'-dimethylguanylguanidine; LA-6023; DMGG; Glucinan [as p-chlorophenoxyacetate]; metformin pamoate [as emboate]; Stagid [as emboate]. Antidiabetic agent. *Bristol-Myers Squibb Co.*

3413 Metformin Hydrochloride
1115-70-4 6001 214-230-6
$C_4H_{12}ClN_5$
N,N-Dimethylimidocarbonimidic diamide monohydrochloride.
Diabex; Diaformin; Diabetosan; Glucophage;

Metiguanide. Antidiabetic agent. mp = 218-220°, 232°; soluble in H_2O, EtOH; insoluble in Et_2O, $CHCl_3$; LD_{50} (rat orl) = 1000 mg/kg, (rat sc) = 300 mg/kg. *Bristol-Myers Squibb Co.*

3414 Methyl Palmoxirate
69207-52-9
Antidiabetic agent.

3415 Miglitol
72432-03-2 276-661-6
$C_8H_{17}NO_5$
(2R,3R,4R,5S)-1-(2-Hydroxyethyl)-2-(hydroxymethyl)-3,4,5-piperidinetriol.
Glyset; Bay m 1099. α-glucosidase inhibitor. Antidiabetic. *Bayer AG; Pharmacia & Upjohn.*

3416 Monotard, Human
11070-73-8 5011 234-291-2
Insulin zinc, human.
Antidiabetic agent.

3417 Naglivan
122575-28-4
$C_{22}H_{46}N_4O_3S+2V$
Bis[2-amino-3-mercapto-N-octylpropionamidato(1-)-S]oxovanadium.
An organic vanadyl compound. Antidiabetic.

3418 Palmoxirate Sodium
79069-97-9
$C_{17}H_{31}NaO_3.2H_2O$
Sodium (±)-2-tetradecylglycidate dihydrate.
McN-3802-21-98. Antidiabetic agent. *McNeil Pharm.*

3419 Palmoxiric Acid
68170-97-8
$C_{17}H_{32}O_3$
(±)-2-Tetradecylglycidic acid.
McN-3802. Antidiabetic agent. *McNeil Pharm.*

3420 Phenformin Hydrochloride
834-28-6 7376 212-637-3
$C_{10}H_{16}ClN_5$
N-(2-Phenylethyl)imidodicarbonimidic diamide monohydrochloride.
Azucaps; DBI; DBI-TD; Debeone-DT; Debinyl; Dibein; Feguanide; Glucopostin; Insoral; Lentobetic; Meltrol; Normoglucina. Antidiabetic agent. Known to cause lactic acidosis. mp = 175-178°; soluble in H_2O; LD_{50} (mus orl) = 450 mg/kg, (mus iv) = 19 mg/kg. *Ciba-Geigy Corp.*

3421 Pioglitazone
111025-46-8 7605
$C_{19}H_{20}N_2O_3S$
(±)-5-[p-[2-[(5-Ethyl-2-pyridyl)ethoxy]benzyl]-2,4-thiazolidinedione.
AD-4833. Antidiabetic agent. Insulin sensitizer. mp = 183-184°. *Takeda Chem. Ind. Ltd.*

3422 Pioglitazone Hydrochloride
112529-15-4 7605
$C_{19}H_{21}ClN_2O_3S$
(±)-5-[p-[2-[(5-Ethyl-2-pyridyl)ethoxy]benzyl]-2,4-thiazolidinedione monohydrochloride.
U-72107A. Antidiabetic agent. Insulin sensitizer. mp = 193-194°. *Takeda Chem. Ind. Ltd.*

3423 Pirogliride
62625-18-7
$C_{16}H_{22}N_4$
N-(1-Methyl-2-pyrrolidinylidene)-N'-phenyl-1-pyrrolidinecarboxamidine.
Antidiabetic agent. *McNeil Pharm.*

3424 Pirogliride Tartrate
62625-19-8
$C_{20}H_{28}N_4O_6$
N-(1-Methyl-2-pyrrolidinylidene)-N'-phenyl-1-pyrrolidinecarboxamidine L-(+)-tartrate(1:1).
McN-3495. Antidiabetic agent. *McNeil Pharm.*

3425 Pramlintide
151126-32-8
$C_{171}H_{267}N_{51}O_{53}S_2$
H-Lys-Cys(-Cys7)-Asn-Thr-Ala-Thr-Cys-Ala-Thr-Gln-Arg-Leu-Ala-Asn-Phe-Leu-Val-His-Ser-Ser-Asn-Asn-Phe-Gly-Pro-Ile-Leu-Pro-Pro-Thr-Asn-Val-Gly-Ser-Asn-Thr-Tyr-NH_2.
AC0137. Antidiabetic agent. *Amylin Pharm. Inc.*

3426 Proinsulin Human
67422-14-4
Antidiabetic agent. *Schering-Plough Pharm.*

3427 Redul
3459-20-9 4517 222-399-2
$C_{13}H_{14}N_3NaO_4S$
N-[5-(2-Methoxyethoxy)-2-pyrimidinyl]benzenesulfonamide sodium salt.
Glymidine sodium salt; SH-717; Glyconormal; Gondafon; Lycanol. Antidiabetic agent. mp = 221-226°; soluble in H_2O (75%), less soluble in EtOH; LD_{50} (rat orl) = 2.85 g/kg, (rat iv) = 2.00 g/kg, (mus orl) = 5.30 g/kg, (mus iv) = 1.48 g/kg. *Schering AG.*

3428 Rosiglitazone
122320-73-4
$C_{18}H_{19}N_3O_3S$
(±)-5-[p-[2-(Methyl-2-pyridylamino)ethoxy]benzyl]-2,4-thiazolidinedione .
Avandia; Brl 49653; 5-((4-(2-Methyl-2-(pyridinylamino)-ethoxy)phenyl)methyl)-2,4-thiazolidinedione-2-butenedioate; Rosiglitazone. Antidiabetic; safe if used in combination with metformin hydrochloride. *Bristol-Myers Squibb Pharm. R&D; SmithKline Beecham Pharm.*

3429 Rosiglitazone Maleate
155141-29-0
$C_{22}H_{23}N_3O_7S$
(±)-5-[p-[2-(Methyl-2-pyridylamino)ethoxy]benzyl]-2,4-thiazolidinedione maleate (1:1).
Avandia; BRL-49653C. Antidiabetic; safe if used in combination with metformin hydrochloride. *Bristol-Myers Squibb Pharm. R&D; SmithKline Beecham Pharm.*

3430 Secretin
1393-25-5 8564 215-733-3
$C_{130}H_{220}N_{44}O_{41}$ [porcine secretin]
Duodenal mucosal hormone.
secretine; sekretolin; Secretin-Kabi. A strongly basic

polypeptide gastrointestinal hormone produced in the duodenum and jejunum of many animal species. Stimulates exocrine pancreatic secretion and lowers blood sugar level. Pancreatic hormone. *KabiVitrum AB.*

3431 Seglitide
81377-02-8
$C_{44}H_{56}N_8O_7$
Cyclo(N-methyl-L-alanyl-L-tyrosyl-D-tryptophyl-L-lysyl-L-valyl-L-phenylalanyl) .
Antidiabetic agent. *Merck & Co.Inc.*

3432 Seglitide Acetate
99248-33-6
$C_{46}H_{60}N_8O_9$
Cyclo(N-methyl-L-alanyl-L-tyrosyl-D-tryptophyl-L-lysyl-L-valyl-L-phenylalanyl) monoacetate (salt).
MK-678. Antidiabetic agent. *Merck & Co.Inc.*

3433 Tolanase
1156-19-0 9644 214-588-3
$C_{14}H_{21}N_3O_3S$
N-[[(Hexahydro-1H-azepin-1-yl)amino]carbonyl]-4-methylbenzenesulfonamide.
Tolazamide; Tolazolamide; U-17835; Diabewas; Norglycin; Tolanase; Tolinase; NSC-70762. Antidiabetic agent. mp = 170-173°. *Pharmacia & Upjohn.*

3434 Tolbutamide
64-77-7 9646 200-594-3
$C_{12}H_{18}N_2O_3S$
1-Butyl-3-(p-tolylsulfonyl)urea.
D-860; U-2043; Artosin; Diaben; Diasulfon; Dolipol; Glyconon; Ipoglicone; Mobenol; Orabet; Orinase; Oterben; Pramidex; Rastinon; Tolbusal. Antidiabetic agent. Used in veterinary medicine as a hypoglycemic agent. mp = 128.5-129.5°. *Pharmacia & Upjohn.*

3435 Tolbutamide Sodium
473-41-6 9646
$C_{12}H_{17}N_2NaO_3S$
1-Butyl-3-(p-tolylsulfonyl)urea monosodium salt.
Orinase Diagnostic. Antidiabetic agent. Used in veterinary medicine as a hypoglycemic agent. mp = 130-133°, [tetrahydrate]: mp = 41-43°. *Pharmacia & Upjohn.*

3436 Tolpyrramide
5588-38-5
$C_{12}H_{16}N_2O_3S$
N-p-Tolylsulfonyl-1-pyrrolidinecarboxamide.
NSC-106572. Antidiabetic agent.

3437 Troglitazone
97322-87-7 9898
$C_{24}H_{27}NO_5S$
(±)-all rac-5-[p-[(6-Hydroxy-2,5,7,8-tetramethyl-2-chromanyl)methoxy]benzyl]-2,4-thiazolidinedione.
CS-045; CI-991; GR-92132X; romglizone. Antidiabetic agent. Insulin sensitizer. mp = 184-186°. *Pfizer Inc.; Sankyo Co. Ltd.*

3438 Zopolrestat
110703-94-1 10325
$C_{19}H_{12}F_3N_3O_3S$
3,4-Dihydro-4-oxo-3-[[5-(trifluoromethyl)-2-benzothiazolyl]methyl]-1-phthalazineacetic acid.
CP-73850. Antidiabetic agent. Aldose reductase inhibitor used in treatment of diabetic neuropathy. mp = 197-198°. *Pfizer Inc.*

Antidiarrheal Agents

3439 Acetorphan
81110-73-8 72
$C_{21}H_{23}NO_4S$
(±)-N-[2-[(Acetylthio)methyl]-1-oxo-3-phenylpropyl]glycine phenylmethyl ester.
Tiorfan; ecadotril [as (S)-form]; sinorphan [as (S)-form]. The racemate is an antidiarrheal, the (S)-form is an antihypertensive. Antisecretory enkephalinase inhibitor. mp = 89°; [(S)-form]: mp = 71°; $[\alpha]_D^{25}$ = -24.1° (c = 1.2 MeOH); LD_{50} (mus iv) > 100 mg/kg. *Bioproject.*

3440 Acetyltannic Acid
1397-74-6 107 215-741-7
Diacetyltannic acid.
tannyl acetate; Acetannin; Tannigen. Interstinal astringent, used as an antidiarrheal. Slightly soluble in H_2O, EtOH; soluble in EtOAc; incompatible with hydroxides, carbonates, alkalies, iron salts.

3441 Alkofanone
7527-94-8 254
$C_{21}H_{19}NO_3S$
3-[(4-Aminophenyl)sulfonyl]-1,3-diphenyl-1-propanone.
Nu-404; Ro-2-0404; Alfone. Antidiarrheal. mp = 221-223° (dec); slightly soluble in Me_2CO, dioxane; insoluble in most other organic solvents. *Hoffmann-LaRoche Inc.*

3442 Almasilate
71205-22-6
$Al_2MgO_8Si_2.xH_2O$
Magnesium aluminosilicate.
Antidiarrheal.

3443 Aluminum Salicylates, Basic
375
$(C_7H_5O_3)_n.Al(OH)_{3-n}.xH_2O$
Alunozal [as monosalicylate ($C_7H_7AlO_5$)]; Baluvet [as disalicylate ($C_{14}H_{11}AlO_7.H_2O$)]. Sparingly soluble in H_2O. *Soc. Chim. des Usines du Rhône.*

3444 Bismuth Subsalicylate
14882-18-9 1327 238-953-1
$C_7H_5BiO_4$
2-Hydroxybenzoic acid bismuth (3+) salt.
basic bismuth salicylate; oxo(salicylato)bismuth; Bismogenol Tosse Inj.; Stabisol. Antidiarrheal; antacid; antiulcerative. Also used as a lupus erythematosus suppressant. Insoluble in H_2O, EtOH. *Mobay.*

3445 Catechin
154-23-4 1950 205-825-1
$C_{15}H_{14}O_6$
(2R-trans)-2-(3,4-Dihydroxyphenyl)-3,4-dihydro-2H-1-benzopyran-3,5,7-triol.
catechol; 3,3',4',5,7-flavanpentol; catechinic acid; catechuic acid; dexcyanidanol; cyanidol; Catergen. Antidiarrheal. Hepatoprotectant. mp = 93-96°, 175-177°; $[\alpha]_D^{18}$ = 16 to 18.4°.

3446 dl-Catechin
7295-85-4 1950 230-731-2
$C_{15}H_{14}O_6$
(±)-(2R-trans)-2-(3,4-Dihydroxyphenyl)-3,4-dihydro-2H-1-benzopyran-3,5,7-triol.
3,3',4',5,7-flavanpentol; dl-catechol; catechinic acid; catechuic acid; cyanidol; Catergen. Antidiarrheal. Hepatoprotectant. mp = 212-216°; slightly soluble in cold H_2O, Et_2O; soluble in hot H_2O, EtOH, Me_2CO, AcOH; insoluble in C_6H_6, $CHCl_3$, petroleum ether.

3447 l-Catechin
18829-70-4 1950 242-611-7
$C_{15}H_{14}O_6$
l-(2R-trans)-2-(3,4-Dihydroxyphenyl)-3,4-dihydro-2H-1-benzopyran-3,5,7-triol.
catechol; l-catechol; 3,3',4',5,7-flavanpentol; catechinic acid; catechuic acid; (-)-cyanidanol-3; cyanidol; Catergen. Antidiarrheal. Hepatoprotectant. mp = 93-96°, 175-177°; $[\alpha]_D$ = -16.8°.

3448 Difenoxin
28782-42-5 3183
$C_{28}H_{28}N_2O_2$
1-(3-Cyano-3,3-diphenylpropyl)-4-phenylisonipecotic acid.
McN-JR-15403-11; difenoxilic acid; difenoxylic acid; Lyspafen; R-15403 [as hydrochloride]. Active metabolite of diphenoxylate. Antiperistaltic, antidiarrheal. [hydrochloride]: mp = 290°, slightly soluble in H_2O (0.023 g/100 ml), $CHCl_3$, THF, dimethylacetamide, DMSO; LD_{50} (rat orl) = 149 mg/kg. *Abbott Laboratories Inc.*

3449 Diphenoxylate
915-30-0 3371 213-020-1
$C_{30}H_{32}N_2O_2$
Ethyl-1-(3-cyano-3,3-diphenylpropyl)-4-phenylisonipecotate.
R-1132. *Abbott Laboratories Inc.*

3450 Diphenoxylate Hydrochloride
3810-80-8 3371 223-287-6
$C_{30}H_{33}ClN_2O_2$
Ethyl-1-(3-cyano-3,3-diphenylpropyl)-4-phenylisonipecotate monohydrochloride.
component of: Lomotil, Diarsed, Reasec. Antiperistaltic; antdiarrheal. mp = 220.5-222°; λ_m = 252, 258, 264 nm; soluble in AcOH (50 g/100 ml), DMF (50 g/100 ml), $CHCl_3$ (36 g/100 ml), MeOH (> 5 g/100 ml), EtOH (0.3 g/100 ml), H_2O (0.08 g/100 ml), C_6H_{14} (0.05 g/100 ml). *Abbott Laboratories Inc.; Searle G.D. & Co.*

3451 Igmesine
140850-73-3
$C_{23}H_{29}N$
(+)-α-[(E)-Cinnamyl]-N-(cyclopropylmethyl)-α-ethyl-N-methylbenzylamine.
Antidiarrheal.

3452 Lidamidine
66871-56-5 5504
$C_{11}H_{16}N_4O$
1-(Methylamidino)-3-(2,6-xylyl)urea.
Has antisecretory; antimotility properties. Antiperistaltic, antdiarrheal. *Rorer.*

3453 Lidamidine Hydrochloride
65009-35-0 5504 265-307-6
$C_{11}H_{17}ClN_4O$
1-(Methylamidino)-3-(2,6-xylyl)urea monohydrochloride.
WHJR-1142A; Lidarral; Smodin. Has antisecretory; antimotility properties. Antiperistaltic, antdiarrheal. mp = 194-197°; λ_m = 262, 271 nm (ε 626, 524 H_2O); soluble in H_2O (15.35 g/100 ml at 25°), MeOH (29.79 g/100 ml), EtOH (8.85 g/100 ml), $CHCl_3$ (0.46 g/100 ml), C_6H_{14} (0.001 g/100 ml); LD_{50} (mmus orl) = 260 mg/kg, (mrat orl) = 267 mg/kg, (frat orl) = 160 mg/kg, (mus iv) = 56 mg/kg. *Rorer.*

3454 Loperamide
53179-11-6 5601 258-416-5
$C_{29}H_{33}ClN_2O_2$
4-(p-Chlorophenyl)-4-hydroxy-N,N-dimethyl-α,α-diphenyl-1-piperidinebutyramide.
Antiperistaltic; antidiarrheal. *Janssen Pharm. Inc.; Lemmon Co.; McNeil Pharm.; Rhône-Poulenc Rorer Pharm. Inc.*

3455 Loperamide Hydrochloride
34552-83-5 5601 252-082-4
$C_{29}H_{34}Cl_2N_2O_2$
4-(p-Chlorophenyl)-4-hydroxy-N,N-dimethyl-α,α-diphenyl-1-piperidinebutyramide monohydrochloride.
Imodium; Imodium A-D; Maalox Antidiarrheal; R-18553; PJ-185; R-18553; Arret; Blox; Brek; Dissenten; Fortasec; Imosec; Imossel; Lopemid; Lopemin; Loperyl; Suprasec; Tebloc. Antiperistaltic; antidiarrheal. mp = 222-223°; λ_m = 253, 259, 265, 273 nm (ε 532, 648, 581, 233, 0.1N HcL/iPrOH 10:90 v/v); soluble in H_2O (0.14 g/100 ml pH 1.7, 0.008 g/100 ml pH 6.1, < 0.001 g/100 ml pH 7.9), MeOH (28.6 g/100 ml), EtOH (5.37 g/100 ml), iPrOH (1.1 g/100 ml), CH_2Cl_2 (35.1 g/100 ml), Me_2CO (0.20 g/100 ml), EtOAc (0.035 g/100 ml), Et_2O (0.001 g/100 ml), C_6H_{14} (0.001 g/100 ml), C_7H_8 (0.001 g/100 ml), DMF (10.3 g/100 ml), THF (0.32 g/100 ml), DMSO (20.5 g/100 ml), propylene glycol (5.64 g/100 ml); LD_{50} (mus sc) = 75 mg/kg, (mus ip) = 28 mg/kg, (mus orl) = 105 mg/kg, (rat orl) = 185 mg/kg. *Janssen Pharm. Inc.; Lemmon Co.; McNeil Pharm.; Rhône-Poulenc Rorer Pharm. Inc.*

3456 Mebiquine
23910-07-8 5810
$C_{10}H_{10}BiNO_3$
Dihydroxy(6-methyl-8-quinolinolato)bismuth.
DV-1; Arrétyl. Antidiarrheal. Insoluble in H_2O, soluble in strong acids; LD_{50} (mus orl) > 10000 mg/kg. *Ugine Kuhlmann.*

3457 Rolgamidine
66608-04-6
$C_9H_{16}N_4O$
trans-N-(Diaminomethylene)-2,5-dimethyl-3-pyrroline-1-acetamide.
WY-25021. Antidiarrheal. *Wyeth-Ayerst Labs.*

3458 Trillium

9825

Beth root; Indian Balm; ground lily; birthroot. Dried rhizomet of *Trillium erectum* L. Astringent used to treat diarrhea.

3459 Urarigenin

466-09-1 10035 207-373-0

$C_{23}H_{34}O_4$

(3β,5α)-3,14-Dihydroxycard-20(22)-enolide.

odorigeni. Antidiarrheal. mp = 240-256°; $[\alpha]_D^{20}$ = 10.5° (c = 1.056 EtOH); [3-O-acetate]: mp = 262-266°; $[\alpha]_D^{22}$ = 4.6° (c = 1.09 $CHCl_3$).

3460 Uzarin

20231-81-6 10035 243-616-7

$C_{35}H_{54}O_{14}$

(3β,5α)-3-[(6-O-β-D-Glucopyranosyl-β-D-glucopyranosyl)oxy]-14-hydroxycard-20(22)-enolide.

Antidiarrheal. mp = 206-208°, 266-270°; $[\alpha]_D^{20}$ = -27° c = 1.075 C_5H_5N), $[\alpha]_D^{19}$ = -1.4° (c = 0.85 MeOH); λ_m = 217 nm (log ε 4.23); soluble in C_5H_5N, methyl cellosolve; sparingly soluble in H_2O; insoluble in Et_2O, $CHCl_3$.

3461 Zaldaride

109826-26-8 10243

$C_{26}H_{28}N_4O_2$

(±)-1-[1-[(4-Methyl-4H,6Hpyrrolo[1,2-α][4,1]benzoxazepin-4-ylmethyl]-4-piperidyl]-2-benzimidazolinone.

Ion channel; calmodulin blocker. Antidiarrheal. mp = 173-175°. *Ciba-Geigy Corp.*

3462 Zaldaride Maleate

109826-27-9 10243

$C_{30}H_{32}N_4O_6$

(±)-1-[1-[(4-Methyl-4H,6Hpyrrolo[1,2-α][4,1]benzoxazepin-4-ylmethyl]-4-piperidyl]-2-benzimidazolinone maleate.

CGS-9343B; ZY-17617B. Ion channel; calmodulin blocker. Antidiarrheal. mp = 189-190°. *Ciba-Geigy Corp.*

Antidiuretics

3463 Argipressin Tannate

113-79-1 204-035-4

8-L-Argininevasopressin tannate.

vasopressin tannate; Pitressin Tannate; CI-107. Antidiuretic. *Parke-Davis.*

3464 Desmopressin

16679-58-6 2969 240-726-7

$C_{46}H_{66}N_{14}O_{13}S_2$

1-(3-Mercaptopropionic acid)-8-D-arginine-vasopressin.

Adiuretin SD; DAV Ritter; DDAVP; Desmospray; Minirin. Antidiuretic. $[\alpha]_D^{25}$ = 85.5° ± 2°. *Rhône-Poulenc Rorer Pharm. Inc.*

3465 Desmopressin Acetate

62357-86-2 2969

$C_{48}H_{68}N_{14}O_{14}S_2.3H_2O$

1-(3-Mercaptopropionic acid)-8-D-arginine-vasopressin monoacetate (salt) trihydrate.

DDAVP; Stimate; Octostim. Antidiuretic. *Centeon L.L.C.; Rhône-Poulenc Rorer Pharm. Inc.*

3466 Felypressin

56-59-7 3992 200-282-7

$C_{46}H_{65}N_{13}O_{11}S_2$

2-(L-Phenylalanine)-8-L-lysine-vasopressin.

PLV-2; Octapressin. Antidiuretic and vasoconstrictor. *Sandoz Pharm. Corp.*

3467 Lypressin

50-57-7 5661 200-050-5

$C_{46}H_{65}N_{13}O_{12}S_2$

L-Cysteinyl-L-tyrosyl-L-phenylalanyl-L-glutaminyl-L-asapraginyl-L-cysteinyl-L-prolyl-L-lysylglycinamide cyclic (1→6)-disulfide.

vasopressin, lysine form; 8-L-lysine-vasopressin; L-8; Diapid; Postacton; Syntopressin. Antidiuretic and vasopressor.

3468 Ornipressin

3397-23-7 7001 222-253-8

$C_{45}H_{63}N_{13}O_{12}S_2$

8-Ornithine-vasopressin.

POR-8. Antidiuretic. *Sandoz Pharm. Corp.*

3469 Oxycinchophen

485-89-2 7091 207-624-4

$C_{16}H_{11}NO_3$

3-Hydroxy-2-phenyl-4-quinolinecarboxylic acid.

3-hydroxy-2-phenylcinchoninic acid; 3-hydroxychinchophen; HCP; Fenidrone; Magnofenyl; Magnophenyl; Oxinofen; Reumalon. Antidiuretic; uricosuric. Dec 206-207°; sparingly soluble in H_2O, Et_2O; soluble in AcOH, alkalies, hot EtOH, C_6H_6. *Chemo Puro.*

3470 Pituitary, Posterior

7666

Pituitary extract (posterior).

Pituamin; Di-Sipidin; Pituitrin. Oxytocic; antidiuretic. Desiccated hypophysis. From the posterior lobe of pituitary body of domesticated animals. Contains oxytocin, vasopressin. Partially soluble in H_2O. *Parke-Davis.*

3471 Terlipressin

14636-12-5 9310 238-680-8

$C_{52}H_{74}N_{16}O_{15}S_2$

N-[N-(N-Glycylglycyl)glycyl]-8-l-lysine-vasopressin.

Glypressin. Analog of lypressin. Antidiuretic and vasopressor. [diacetate pentahydrate (Glycylpressin)]: $[\alpha]_D^{25}$ = -82° (c = 0.2 in 1M AcOH).

3472 Vasopressin

9034-50-8 10073

$C_{46}H_{65}N_{15}O_{12}S_2$

β-Hypophamine.

Pitressin; Antidiuretic hormone; beta-hypophamine; Leiormone; Tonephin; Vasophysin. Antidiuretic; vasopressor hormone. Hemostatic. Obtained from posterior lobe of the pituitary of healthy domestic animals or by synthesis. Two vasopressins, which differ in the amino acid at position 8, have been isolated:arginine vasopressin and lysine vasopressin. *Parke-Davis.*

3473 Vasopressin Injection
11000-17-2 234-236-2

$C_{46}H_{65}N_{13}O_{12}S_2$

Antidiuretic; vasopressor hormone. Hemostatic. *Parke-Davis.*

3474 Vasopressin, Arginine form
113-79-1 10073 204-035-4

$C_{46}H_{65}N_{15}O_{12}S_2$

8-L-Arginine-vasopressin.

arginine vasopressin; argipressin; rindervasopressin; antidiuretic hormone; β-hypophamine; Leiormone; Tonephin; Vasophysin. Antidiuretic; vasopressor hormone. Hemostatic. *Parke-Davis.*

3475 Vasopressin, Lysine form
50-57-7 5661 200-050-5

$C_{46}H_{65}N_{13}O_{12}S_2$

8-L-Lysine-vasopressin.

Diapid; Lypressin; Phe^3-Lys^8-oxytocin; Pitressin; Postaction; Schweine-Vasopressin; Syntopressin. Antidiuretic; vasopressor hormone. Hemostatic. See *Lypressin. Parke-Davis.*

Antidotes

3476 Acetylcysteine
616-91-1 89 210-498-3

$C_5H_9NO_3S$

L-α-Acetamido-β-mercaptopropionic acid.

N-acetyl-L-cysteine; 5052; NSC-111180; Airbron; Broncholysin; Brunac; Fabrol; Fluatox; Fluimucil; Fluimucetin; Fluprowit; Mucocedyl; Mucolator; Mucolyticum; Mucomyst; Muco Sanigen; Mucosolvin; Mucret; NAC; Neo-Fluimucil; Parvolex; Respaire; Tixair. Mucolytic; corneal vulnerary; antidote to acetaminophen poisoning. mp = 109-110°; LD_{50} (rat orl) = 5050 mg/kg. *Bristol Laboratories; DuPont-Merck Pharm.; Mead Johnson Labs.; Mead Johnson Pharmaceuticals.*

3477 Albumen
216

Egg white; dried egg white; Albumin. Antidote to mercury poisoning. Yellow powder, decomposes in moist air; solution coagulates at 61°.

3478 p-Aminopropiophenone
70-69-9 490 200-742-7

PAPP.

Antidote to cyanide poisoning. mp = 140°; bp = 482°; soluble in H_2O, EtOH; [hydrochloride ($C_9H_{12}ClNO$)]: mp = 198-199°; soluble in H_2O; LD_{50} (dog iv) = 7 mg/kg. *Pfalz & Bauer.*

3479 Asoxime Chloride
34433-31-3 870

$C_{14}H_{16}Cl_2N_4O_3$

1-[[[4-(Aminocarbonyl)pyridino]methoxy]methyl]-2-[(hydroxyimino)methyl]pyridinium dichloride.

HI-6. Cholinesterase reactivator. Used as an antidote for organophosphorus poisoning. [monohydrate]: mp = 145-147°; λ_m = 350, 300, 270, 218 nm (log ε 3.093, 4.018, 3.966, 4.181 H_2O). *E. Merck.*

3480 Cupric Sulfate
7758-99-8 2722

$CuSO_4.5H_2O$

Copper (II) sulfate (1:1) pentahydrate.

bluestone; blue vitriol; Roman vitriol; Salzburg vitriol. Antiseptic; antifungal(topical); antidote to phosphorus. Used in veterinary medicine as a mineral source, anthelmintic, emetic, fungicide. Occurs as the mineralchalcanthite. Loses $2H_2O$ at 30°, 2 more at 110°, and becomes anhydrous by 250°; $d_4^{15.6}$ = 2.286; soluble in H_2O, MeOH, glycerol; slightly soluble in EtOH; LD_{50} (rat orl) = 960 mg/kg.

3481 Cupric Sulfate [anhydrous]
7798-98-7 2722

CuO_4S

Copper (II) sulfate (1:1).

Antiseptic; antifungal (topical); antidote to phosphorus. Occurs as the mineral hydrocyanite. Dec above 560°; d = 3.6; hygroscopic; soluble in H_2O; practically insoluble in alcohol.

3482 Cysteamine
60-23-1 2848 200-463-0

C_2H_7NS

2-Aminoethanethiol.

MEA; L-1573; mercamine; Becaptan; Lambratene. Used as a radioprotective agent and to produce ulcers in rats. Antiurolithic. Antidote to acetaminophen poisoning. mp = 97-98.5°; freely soluble in H_2O; LD_{50} (mus orl) = 625 mg/kg, (mus ip) = 250 mg/kg. *Parke-Davis.*

3483 Cysteamine Hydrochloride
156-57-0 2848 205-858-1

C_2H_8ClNS

2-Aminoethanethiol hydrochloride.

Cl-9148. Used as a radioprotective agent and to produce ulcers in rats. Antiurolithic. Antidote to acetaminophen poisoning. mp = 70.2-70.7°; soluble in H_2O, EtOH; LD_{50} (rat ip) = 231.9 mg/kg, (rbt iv) = 149.5 mg/kg. *Parke-Davis.*

3484 Deferoxamine
70-51-9 2914

$C_{25}H_{48}N_6O_8$

N-[5-[3-[(5-Aminopentyl)hydroxycarbamoyl]-propion-amido]pentyl]-3-[[5-(N-hydroxy-acetamido)pentyl]carbamoyl]-propionohydroxamic acid.

NSC-527604; Desferioxamine B. Chelating agent for iron. mp = 138-140°; soluble in H_2O (1.2 g/100 ml at 20°). *Ciba-Geigy Corp.*

3485 Deferoxamine Hydrochloride
1950-39-6 2914 217-767-4

$C_{25}H_{49}ClN_6O_8$

N-[5-[3-[(5-Aminopentyl)hydroxycarbamoyl]-propion-amido]pentyl]-3-[[5-(N-hydroxy-acetamido)pentyl]-carbamoyl]-propionohydroxamic acid monohydrochloride.

Ba-29837. Chelating agent for iron. mp = 172-175°. *Ciba-Geigy Corp.*

3486 Deferoxamine Methanesulfonate
138-14-7 2914 205-314-3
$C_{26}H_{52}N_6O_{11}S$
N-[5-[3-[(5-Aminopentyl)hydroxycarbamoyl]propionamido]pentyl]-3-[[5-(N-hydroxyacetamido)pentyl]-carbamoyl]-propionohydroxamic acid monomethanesulfonate.
Desferal mesylate; Ba-33112; DFOM; Desferal. Chelating agent for iron. mp = 148-149°; soluble in H_2O (> 20 g/100 ml). *Ciba-Geigy Corp.*

3487 Dicobalt Edetate
36499-65-7
$C_{10}H_{12}Co_2N_2O_8$
Cobalt(2+)-[(ethylenedinitrilo)tetraacetato]cobaltate(2-).
Kelocyanor Antidote for cyanide poisoning.

3488 Dimercaprol
59-52-9 3255
$C_3H_8OS_2$
2,3-Dimercapto-1-propanol.
BAL in Oil; British Anti-Lewisite; Dicaptol; Sulfactin. Antidote to gold, mercury and arsenic poisoning. d_4^{25} = 1.2385; $bp_{0.2}$ = 60°, $bp_{5.6}$ = 100°, bp_{15} = 120°, bp_{25} = 130°, bp_{40} = 140°; soluble in H_2O (8.7 g/100 ml); LD_{50} (rat im) = 86.7 mg/kg. *Becton Dickinson Microbiology Systems; E. I. DuPont de Nemours Inc.*

3489 2,3-Dimercapto-1-Propanesulfonic Acid
74-61-3 3256
$C_3H_8O_3S_3$
2,3-Dithiolpropanesulfonic acid.
Antidote for heavy metal poisoning.

3490 2,3-Dimercapto-1-Propanesulfonic Acid Sodium salt
4076-02-2 3256 223-796-3
$C_3H_7NaO_3S_3$
Sodium 2,3-dithiolpropanesulfonate.
DMPS; unitiol; Dimaval. Antidote for heavy metal poisoning. mp = 235°; LD_{50} (mus ip) = 982.8 mg/kg.

3491 Ditiocarb Sodium
148-18-5 3443 205-710-6
$C_5H_{10}NNaS_2.3H_2O$
Sodium diethyldithiocarbamate.
dithiocarb; DTC; DDC; DEDC; DDTC; DeDTC; Imuthiol. Immunomodulator, chelating agent and antidote, especially for nickel and cadmium poisoning. Used to counter platinum toxicity is pateints tretaed with cis-Platin or to treat Wilson's disease. [trihydrate]: mp = 90-92°, 94-102°; freely soluble in H_2O, EtOH, MeOH, Me_2CO, insoluble in Et_2O, C_6H_6; λ_m = 257, 290 nm (ε 1200, 13000 H_2O); LD_{50} (rat orl) = 2830 mg/kg, (mus orl) = 1870 mg/kg, (mus iv) > 1000 mg/kg.

3492 Edrophonium Chloride
116-38-1 3562 204-138-4
$C_{10}H_{16}ClNO$
Ethyl(m-hydroxyphenyl)dimethylammonium chloride.
Antirex; Enlon; Reversol; Tensilon; edrophone bromide [as bromide]; Ro-2-3198 [as bromide]; component of: Enlon Plus. Antidote to curare alkaloids. Also used as a diagnostic aid in myasthenia gravis. Cholinesterase inhibitor. mp = 162-163° (dec); soluble in H_2O, EtOH, insoluble in Et_2O, $CHCl_3$; [bromide ($C_{10}H_{16}BrNO$)]: mp = 151-152°; soluble in H_2O (> 10 g/100 ml), EtOH, insoluble in Et_2O. *Anaquest; ICN Pharm. Inc.; Organon Inc.*

3493 Folinic Acid
58-05-9 4254 200-361-6
$C_{20}H_{23}N_7O_7$
N-[p-[[(2-Amino-5-formyl-5,6,7,8-tetrahydro-4-hydroxy-6-pteridinyl)methyl]amino]benzoyl]-glutamic acid.
5-formyltetrahydrofolate; CF; citrovorum factor; leucovorin. Antianemic (folate deficiency); antidote to folic acid antagonists. Antineoplastic agent. mp = 240-250° (dec); λ_m = 282 nm (%T = 27.0 for 10 mg/l in 0.1N NaOH). *Glaxo Wellcome Inc.; Res. Corp.*

3494 Folinic Acid Calcium Salt Pentahydrate
6035-45-6 4254
$C_{20}H_{21}CaN_7O_7.5H_2O$
Calcium N-p-[[[(6RS)-2-amino-5-formyl-5,6,7,8-tetrahydro-4-hydroxy-6-pteridinyl]methyl]amino]-benzoyl]-L-glutamate (1:1) pentahydrate.
leucovorin calcium pentahydrate; Wellcovorin; NSC-3590; calcium folinate; Folaren; Foliben; Lederfolat; Lederfolin; Leucovorin; Leucosar; Rescufolin; Rescuvolin; Tonofolin. Antianemic (folate deficiency); antidote to folic acid antagonists. Freely soluble in H_2O, insoluble in EtOH; $[\alpha]_D^{21}$ = 14.9° (c = 1 H_2O). *Elkins-Sinn; Glaxo Wellcome plc; Immunex Corp.*

3495 l-Folinic Acid Calcium Salt
80433-71-2 4254
$C_{20}H_{21}CaN_7O_7$
Calcium N-p-[[[(6RS)-2-amino-5-formyl-5,6,7,8-tetrahydro-4-hydroxy-6-pteridinyl]methyl]amino]-benzoyl]-L-glutamate (1:1).
calcium (6S) folinate; calcium levofolinate; levoleucovorin calcium; CL-307782; Elvorine. Antianemic (folate deficiency); antidote to folic acid antagonists. $[\alpha]_D^{21}$ = -15.1° (c = 1.82). *Elkins-Sinn; Glaxo Wellcome plc; Immunex Corp.*

3496 Fomepizole
7554-65-6 4257 231-445-0
$C_4H_6N_2$
4-Methylpyrazole.
4-MP. Alcohol dehydrogenase inhibitor. Antidote to MeOH and ethylene glycol poisoning. mp = 15.5-18.5°; bp_{18} = 98.5-99.5°; bp_{730} = 204-205°; λ_m = 220 nm (log ε 3.47 EtOH), 226 nm (log ε 3.65 6N HCl); soluble in H_2O, EtOH; LD_{50} (7 day) (mus iv) = 312 mg/kg, (rat iv) = 312 mg/kg, (mus orl) = 640 mg/kg, (rat orl) = 534 mg/kg.

3497 Leucovorin Calcium
1492-18-8 4254 216-082-8
$C_{20}H_{21}CaN_7O_7$
Calcium N-p-[[[(6RS)-2-amino-5-formyl-5,6,7,8-tetrahydro-4-hydroxy-6-teridinyl]methyl]amino]benzoyl]-L-glutamate (1:1).
leucovorin calcium; Wellcovorin; NSC-3590. Antianemic (folate deficiency); antidote to folic acid antagonists. Soluble in H_2O, insoluble in EtOH; $[\alpha]_D^{21}$ = 14.9° (c = 1 H_2O). *Elkins-Sinn; Glaxo Wellcome plc; Immunex Corp.*

3498 Methionine
63-68-3 6053 200-562-9
$C_5H_{11}NO_2S$
L-2-Amino-4-(methylthio)butyric acid.
L-methionine; Met; M; Acimethin. Hepatoprotectant and antidote for acetaminophen poisoning. Urinary acidifier. An essential amino acid. mp = 280-282° (dec); $[\alpha]_D^{25}$ = -8.11° (c = 0.8), $[\alpha]_D^{20}$ = 23.40° (c = 5.0 3N HCl).

3499 D-Methionine
348-67-4 6053 206-483-6
$C_5H_{11}NO_2S$
D-2-Amino-4-(methylthio)butyric acid.
D-methionine. Hepatoprotectant and antidote for acetaminophen poisoning. Urinary acidifier. $[\alpha]_d^{25}$ = 8.12° (c = 0.8), $[\alpha]_D^{25}$ = -21.18° (c = 0.8 0.2N HCl).

3500 DL-Methionine
59-51-8 6053 200-432-1
$C_5H_{11}NO_2S$
DL-2-Amino-4-(methylthio)butyric acid.
racemethionine; DL-methionine; Amurex; Banthionine; Dyprin; Lobamine; Metione; Pedameth; Urimeth. Hepatoprotectant and antidote for acetaminophen poisoning. Urinary acidifier. mp = 281° (dec); d = 1.340; soluble in H_2O (1.82 g/100 ml at 0°, 3.38 g/100 ml at 25°, 6.07 g/100 ml at 60°, 10.52 g/100 ml at 75°, 17.60 g/100 ml at 100°); slightly soluble in EtOH; insoluble in Et_2O.

3501 Methylene Blue
61-73-4 6137 200-515-2
$C_{16}H_{18}ClN_3S.3H_2O$
3,7-Bis(dimethylamino)phenothiazin-5-ium chloride trihydrate.
methylthioninium chloride; tetramethylthionine chloride; solvent blue 8; C.I. Basic Blue 9 trihydrate; Swiss Blue; Urolene Blue; CI-52015. Antimethemoglobinemic; antidote to cyanide poisoning. Used in veterinary medicine as an antiseptic, disinfectant and anitdote to cyanide and nitrate. [trihydrate]: Green crystals; λ_m = 668, 609 nm; soluble in H_2O (4 g/100 ml), EtOH (1.5 g/100 ml); insoluble in Et_2O.

3502 Neostigmine
59-99-4 6553
$[C_{12}H_{19}N_2O_2]^+$
(m-Hydroxyphenyl)trimethylammonium dimethylcarbamate.
synstigmin; proserine. Cholinergic; miotic; antidote for curare poisoning. *Apothecon; Elkins-Sinn; Hoffmann-LaRoche Inc.; ICN Pharm. Inc.*

3503 Neostigmine Bromide
114-80-7 6553 204-054-8
$C_{12}H_{19}BrN_2O_2$
(m-Hydroxyphenyl)trimethylammonium bromide dimethylcarbamate.
Juvastigmin (tabl.); Neoesserin; Neostigmin (tabl.); Normastigmin (tabl.); Prostigmin. Cholinergic; miotic; antidote for curare poisoning. mp = 167° (dec); soluble in H_2O (100 g/100 ml), EtOH. *Apothecon; Elkins-Sinn; Hoffmann-LaRoche Inc.; ICN Pharm. Inc.*

3504 Neostigmine Methylsulfate
51-60-5 6553 200-109-5
$C_{13}H_{22}N_2O_6S$
(m-Hydroxyphenyl)trimethylammonium methyl sulfate dimethylcarbamate.
Intrastigmina; Juvastigmin (amp.); Metastigmin; Neostigmin (inj.); Normastigmin (amp.); Prostigmin (amp.); Stiglyn. Cholinergic; miotic; antidote for curare poisoning. Cholinesterase inhibitor. mp = 142-145°; soluble in H_2O (10 g/100 ml), less soluble in EtOH; LD_{50} (mus iv) = 0.16 mg/kg, (mus sc) = 0.42 mg/kg, (mus orl) = 7.5 mg/kg. *Apothecon; Elkins-Sinn; Hoffmann-LaRoche Inc.; ICN Pharm. Inc.*

3505 Potassium Nitrite
7758-09-0 7816 231-832-4
KNO_2
Nitrous acid potassium salt.
Vasodilator; antidote for cyanide poisoning. d = 1.915; mp = 441° (dec); soluble in H_2O (285 g/100 ml), EtOH; LD_{50} (rbt orl) = 199 mg/kg.

3506 Protamine Sulfate
9009-65-8
Protamine sulphate. Antidote to heparin. *Elkins-Sinn.*

3507 Sodium Nitrite
7632-00-0 8793 231-555-9
$NaNO_2$
Nitrous acid sodium salt.
erinitrit. Antidote to cyanide poisoning; vasodilator. Also used as a reagent in manufacture of inorganic and organic compounds, as well as in other industrial processes. mp = 217°; dec above 320°; d = 2.17; soluble in H_2O; slightly soluble in EtOH; aqueous solution is alkaline; oxidizes to nitrate in air; LD_{50} (rat orl) = 180 mg/kg.

3508 Sodium Thiosulfate
7772-98-7 8844 231-867-5
$Na_2S_2O_3$
Disodium thiosulfate.
Sulfactol; sodium hyposulfite; hypo; antichlor; Sodothiol; Sulfothiorine; Ametox; component of: Tinver. Antidote to cyanide poisoning. d = 1.69; mp = 48°; soluble in H_2O, insoluble in EtOH; LD_{50} (rat iv) > 2500 mg/kg. *Pilkington Barnes Hind; Sterling Winthrop Inc.*

3509 Succimer
304-55-2 9034 206-155-2
$C_4H_6O_4S_2$
meso-2,3-Dimercaptosuccinic acid.
DMS; DMSA; DIM-SA; Ro-1-7977; Chemet. Chelating agent used as an antidote in cases of heavy metal poisoning. Diagnostic aid (the m99 isotope as a radioactive imaging agent). mp = 192-194°; LD_{50} (mus ip) > 3000 mg/kg. *Hoffmann-LaRoche Inc.*

3510 Tacrine
321-64-2 9199 206-291-2
$C_{13}H_{14}N_2$
9-Amino-1,2,3,4-tetrahydroacridine.
Nootropic. Respiratory stimulant. Centrally active anticholinesterase. Cognition adjuvant and antidote for curare poisoning. mp = 183-184°. *Parke-Davis.*

3511 Tacrine Hydrochloride
1684-40-8 9199 216-867-5
$C_{13}H_{15}ClN_2$
9-Amino-1,2,3,4-tetrahydroacridine monohydrochloride.
Cognex; CI-970; THA. Nootropic. Respiratory stimulant. Centrally active anticholinesterase. Cognition adjuvant and antidote for curare poisoning. mp = 283-284°; soluble in H_2O. *Parke-Davis.*

3512 Tiopronin
1953-02-2 9597 217-778-4
N-(2-mercapto-1-oxopropyl)glycine.
Acadione; BRN 1859822; Capen; Captimer; CCRIS 1935; Epatiol; Meprin (detoxicant); Mercaptopropionylglycine; (2-Mercaptopropionyl)glycine; Mucolysin; N-(2-Mercapto-1-oxopropyl)glycine; N-(2-Mercaptopropionyl)-glycine; Sutilan; Thiola; Thiolpropionamidoacetic acid; Thiopronin; Thiopronine; Thiosol; Tioglis; Tiopronin; Tiopronine; Tiopronino; Tioproninum; Vincol; . Mucolytic; expectorant. Used as an antidote in heavy metal poisoning. mp = 95-97°; LD_{50} (mus iv) = 2100 mg/kg. *Boehringer Ingelheim Ltd.*

Antidyskinetics

3513 Amantadine
768-94-5 389 212-201-2
$C_{10}H_{17}N$
Tricyclo[3.3.1.1^{3,7}]decan-1-amine.
Antiviral; antidyskinetic; antiparkinsonian. mp = 160-190°, 180-192°; sparingly soluble in H_2O. *Apothecon; DuPont-Merck Pharm.; Solvay Pharm. Inc.*

3514 Amantadine Hydrochloride
665-66-7 389 211-560-2
$C_{10}H_{18}ClN$
Tricyclo[3.3.1.1^{3,7}]decan-1-amine hydrochloride.
Symadine; Symmetrel; EXP-105-1; NSC-83653; Amazolon; Mantadix; Mantadan; Mantadine; Midantan; Mydantane; Virofral. Antiviral; antidyskinetic; antiparkinsonian. Dec 360°; soluble in H+2O (> 5 g/100 ml), insoluble in Et_2O; LD_{50} (mus orl) = 700 mg/kg, (rat orl) = 1275 mg/kg. *Apothecon; DuPont-Merck Pharm.; Solvay Pharm. Inc.*

3515 Cabergoline
81409-90-7 1637
$C_{26}H_{37}N_5O_2$
1-[(6-Allylergolin-8β-yl)carbonyl]-1-[3-(dimethylamino)propyl]-3-ethylurea.
FCE 21336; Dostinex. Antidyskinetic; antihyperprolactinemic. Dopamine receptor agonist; prolactin inhibitor. mp = 102-104°; LD_{50} (mus orl) > 400 mg/kg; [diphosphate ($C_{26}H_{43}N_5O_{10}P_2$)]: mp = 153-155°. *Farmitalia Carlo Erba Ltd.*

3516 Clonidine
4205-90-7 2450 224-119-4
$C_9H_9Cl_2N_3$
2-[(2,6-Dichlorophenyl)imino]imidazolidine.
Catapres-TTS; ST-155-BS. An α_2-adrenergic agonist used as an antihypertensive and antidyskinetic. mp = 130°. *Boehringer Ingelheim GmbH.*

3517 Clonidine Hydrochloride
4205-91-8 2450 224-121-5
$C_9H_{10}Cl_3N_3$
2-[(2,6-Dichlorophenyl)imino]imidazolidine hydrochloride.
Catapres; ST-155; component of: Combipres. An α_2-adrenergic agonist used as an antihypertensive and antidyskinetic. mp = 305°; soluble in H_2O (7.7 g/100 ml at 20°, 16.6 g/100 ml at 60°), MeOH (17.25 g/100 ml), EtOH (4 g/100 ml), $CHCl_3$ (0.02 g/100 ml); λ_m = 213, 271, 302 nm (ε 8290, 713, 340 H_2O); LD_{50} (mus orl) = 328 mg/kg, (mus iv) = 18 mg/kg, (rat orl) = 270 mg/kg, (rat iv) = 29 mg/kg. *Boehringer Ingelheim GmbH; Parke-Davis.*

3518 Haloperidol
52-86-8 4629 200-155-6
$C_{21}H_{23}ClFNO_2$
4-[4-(p-Chlorophenyl)-4-hydroxypiperidino]-4'-fluorobutyrophenone.
R-1625; McN-JR-1625; Aloperidin; Bioperidolo; Brotopon; Dozic; Einalon S; Eukystol; Haldol; Halosten; Keselan; Linton; Peluces; Serenace; Serenase; Sigaperidol. Antidyskinetic and antipsychotic. Used to treat Gilles de la Tourette's disease. A Butyrophenone. mp = 148-149.4°; λ_m = 247, 221 nm (ε 13300, 15000 HCl/MeOH); soluble in H_2O (0.0014 g/100 ml); freely soluble in MeOH, $CHCl_3$, Me_2CO, C_6H_6; LD_{50} (rat orl) = 165 mg/kg, (mus ip) = 60 mg/kg. *Abbott Laboratories Inc.; Lemmon Co.; McNeil Pharm.*

3519 Haloperidol Decanoate
74050-97-8 4629 277-679-7
$C_{31}H_{41}ClFNO_3$
4-[4-(p-Chlorophenyl)-4-hydroxypiperidino]-4'-fluorobutyrophenone decanoate.
KD-16; R-13762; Haldol Decanoate; Halomonth; Neoperidole. Antidyskinetic (in Gilles de la Tourette's disease); antipsychotic. A Butyrophenone. *Abbott Laboratories Inc.*

3520 Haloperidol Hydrochloride
1511-16-6 4629
$C_{21}H_{24}Cl_2FNO_2$
4-[4-(p-Chlorophenyl)-4-hydroxypiperidino]-4'-fluorobutyrophenone hydrochloride.
Antidyskinetic; antipsychotic. Used to treat Gilles de la Tourette's disease. mp = 226-227.5°; soluble in H_2O (0.3 g/100 ml). *Lemmon Co.; McNeil Pharm.*

3521 Pimozide
2062-78-4 7589 218-171-7
$C_{28}H_{29}F_2N_3O$
1-[1-[4,4-Bis(p-fluorophenyl)butyl]-4-piperidyl]-2-benzimidazolinone.
KD-136; Haldol decanoate; Halomonth; Neoperidole. Antipsychotic; antidyskinetic. *Abbott Laboratories Inc.; Lemmon Co.*

3522 Selegiline
14611-51-9 8569
$C_{13}H_{17}N$
(-)-(R)-N-α-Dimethyl-N-2-propynylphenethylamine.
(-)-deprenil; L-deprenyl; deprenyl [as (±)-form]; phenylisopropyl-N-methylpropinylamine [as (±)-form].

Antidyskinetic; antiparkinsonian. Monoamine oxidase B inhibitor. Related to pargyline. $bp_{0.8}$ = 92-93°; $[\alpha]_D^{20}$ = -11.2°; [(dl-form)]: bp_5 = 103-110°; n_D^{20} = 1.5224. *Chinoin; Somerset Pharm. Inc.*

3523 Selegiline Hydrochloride
14611-52-0 8569
$C_{13}H_{18}ClN$
(-)-(R)-N-α-Dimethyl-N-2-propynylphenethylamine hydrochloride.
deprenyl; Eldepryl; Déprényl; Eldéprine; Eldepryl; Jumex; Movergan; Plurimen; E-250 [as (±)-form hydrochloride]. Antidyskinetic; antiparkinsonian. Monoamine oxidase B inhibitor. Related to pargyline. mp = 141-142°; $[\alpha]_D^{25}$ = -10.8° (c = 6.48 H_2O); LD_{50} (rat iv) = 81 mg/kg, (rat sc) = 280 mg/kg; [(±)-form hydrochloride]: LD_{50} (rat iv) = 63 mg/kg, (rat sc) = 126 mg/kg, (rat orl) = 385 mg/kg. *Chinoin; Somerset Pharm. Inc.*

3524 Tetrabenazine
58-46-8 9325 200-383-6
$C_{19}H_{27}NO_3$
1,3,4,6,7,11b-Hexahydro-3-isobutyl-9,10-dimethoxy-2H-benzo[a]quinolizin-2-one.
Ro-1-9569. Antidyskinetic; antipsychotic. A dopamine depleting agent; a tricyclic. mp = 125-126°; [oxime]: mp = 158°. *Hoffmann-LaRoche Inc.*

3525 Tetrabenazine Hydrochloride
2105-47-7 9325
$C_{19}H_{28}ClNO_3$
1,3,4,6,7,11b-Hexahydro-3-isobutyl-9,10-dimethoxy-2H-benzo[a]quinolizin-2-one hydrochloride.
Antipsychotic; antidyskinetic. A dopamine depleting agent; a tricyclic. mp = 208-210°; soluble in hot H_2O, insoluble in Me_2CO; λ_m = 230, 284 nm (ε 7780, 3320 EtOH). *Hoffmann-LaRoche Inc.*

3526 Tetrabenazine Methanesulfonate
804-53-5 9325
$C_{19}H_{27}NO_3$
1,3,4,6,7,11b-Hexahydro-3-isobutyl-9,10-dimethoxy-2H-benzo[a]quinolizin-2-one methanesulfonate.
Nitoman. Antidyskinetic; antipsychotic. A dopamine depleting agent; a tricyclic. mp = 126-130°; soluble in EtOH, sparingly soluble in H_2O, nearly insoluble in Et_2O. *Hoffmann-LaRoche Inc.*

3527 Tiapride
51012-32-9 9561 256-907-9
$C_{15}H_{24}N_2O_4S$
N-[2-(Diethylamino)ethyl]-5-(methylsulfonyl)-o-anisamide.
FLC 1374. Antidyskinetic. mp = 123-125°. *Chemo Puro.*

3528 Tiapride Hydrochloride
51012-33-0 9561 256-908-4
$C_{15}H_{25}ClN_2O_4S$
N-[2-(Diethylamino)ethyl]-5-(methylsulfonyl)-o-anisamide hydrochloride.
Gramalil; Italprid; Luxoben; Sereprile; Tiapridal; Tiapridex. Antidyskinetic. *Chemo Puro.*

Antieczema Agents

3529 Ascomycin
NSC-106410. Used to treat eczema.

3530 Coal Tar
2487
Clinitar; Psorigel; T/Gel; pixalbol
Used as a topical antieczematic. Black liquid/semisolid; insoluble in H_2O, soluble in most organic solvents.

3531 Evening Primrose Oil
3953
Efamol; Efamst; Epogam
Seed oil of the evening primrose, *Oenothera Biennis L. Onagraceae*. Major constituents are linolenic acid (92%) and σ-linolenic acid (9%). Used as a dietary supplement and in treatment of atopic eczema and mastaglia. d_{15} = 0.9283.

3532 Fluprednidene
2193-87-5
$C_{22}H_{27}FO_5$
9-Fluoro-11β,17,21-trihydroxy-16-methylenepregan-1,4-diene-3,20-dione.
Antieczematic.

3533 σ-Linolenic Acid
506-26-3 5531
Gamolenic acid; GLA; Viacutan
(Z,Z,Z)-6,9,12-Octadecatrienoic acid.
A prostaglandin precursor, used as a topical antieczematic. [hexabromide ($C_{18}H_{30}Br_6O_2$)]: mp = 201-202°.

3534 Tacrolimus
104987-11-3 9200
$C_{44}H_{69}NO_{12}.H_2O$
(-)-(3S,4R,5S,8R,9E12S,14S,15R,16R,18R,19R26aS)-8-allyl-5,6,8,11,12,13,14,15,16,17,18,19,24,25,26,26a-Hexadecahydro-5,19-dihydroxy-3-[(E)-2-[(1R,3R,4R)-4-hydroxy-3-methoxycyclohexyl]-1-methylvinyl]-14,16-dimethoxy-4,10,12,18-tetramethyl-15,19-epoxy-3H-pyrido[2,1-c][1,4]oxaazacyclotricosine-1,7,20,21(4H,23H)-tetrone)monohydrate.
Prograf; Protopic; FK-506; FR-900506. Immunosuppressant. Used to treat eczema. [monohydrate]: mp = 127-129°; $[\alpha]_D^{23}$ = -84.4° (c = 1.02 $CHCl_3$); soluble in MeOH, EtOH, Me_2CO, EtOAc, $CHCl_3$; Et_2O; sparingly soluble in C_6H_{14}, petroleum ether; insoluble in H_2O; LD_{50} (mus ip) > 200 mg/kg, (mrat iv) = 57 mg/kg, (mrat orl) = 134 mg/kg, (frat iv) = 23.6 mg/kg, (frat orl) = 194 mg/kg. *Fujisawa Pharm. USA Inc.*

Antiemetics

3535 Acetylleucine Monoethanolamine
149-90-6 96
$C_{10}H_{22}N_2O_4$
N-Acetyl-DL-leucine compound with 2-aminoethanol.
RP-7452; Tanganil. Antiemetic; antivertigo agent. mp = 150°; soluble in H_2O (> 20g/100 ml). *Rhône-Poulenc Rorer Pharm. Inc.*

3536 Alizapride
59338-93-1 246 261-710-6
$C_{16}H_{21}N_5O_2$
N-[(1-Allyl-2-pyrrolidinyl)methyl]-6-methoxy-1H-benzotriazole-5-carboxamide.
Antipsychotic; antiemetic. A benzamide. mp = 139°; LD_{50} (mus iv) = 92.7 mg/kg.

3537 Alizapride Hydrochloride
59338-87-3 246
$C_{16}H_{22}ClN_5O_2$
N-[(1-Allyl-2-pyrrolidinyl)methyl]-6-methoxy-1H-benzotriazole-5-carboxamide hydrochloride.
Nausilen; Plitican; Vergentan. Antipsychotic; antiemetic. A benzamide. mp = 206-208°.

3538 Alosetron
122852-42-0
$C_{17}H_{18}N_4O$
2,3,4,5-Tetrahydro-5-methyl-2-[(5-methylimidazol-4-yl)methyl]-1H-pyrido[4,3-b]indol-1-one.
2,3,4,5-Tetrahydro-5-methyl-2-((5-methyl-1H-imidazol-4-yl)methyl)-1H-pyrido(4,3-b)indol-1-one; GR 68755; Lotronex. Antiemetic. Used to treat irritable bowel syndrome in women. *Glaxo Wellcome Inc.*

3539 Alosetron Hydrochloride
122852-69-1
$C_{17}H_{19}ClN_4O$
2,3,4,5-Tetrahydro-5-methyl-2-[(5-methylimidazol-4-yl)methyl]-1H-pyrido[4,3-b]indol-1-one monohydrochloride.
GR-68755C. Antiemetic. Used to treat irritable bowel syndrome in women. *Glaxo Wellcome Inc.*

3540 Azasetron
123040-69-7 933
$C_{17}H_{20}ClN_3O_3$
(±)-6-Chloro-3,4-dihydro-4-methyl-3-oxo-N-3-quinuclidinyl-2H-1,4-benzoxazine-8-carboxamide.
nazasetron. Antiemetic. Specific serotonin $5HT_3$ receptor antagonist. *Yoshitomi.*

3541 Azasetron Hydrochloride
141922-90-9 933
$C_{17}H_{21}Cl_2N_3O_3$
(±)-6-Chloro-3,4-dihydro-4-methyl-3-oxo-N-3-quinuclidinyl-2H-1,4-benzoxazine-8-carboxamide monohydrochloride.
Y-25130; Serotone. Antiemetic. Specific serotonin $5HT_3$ receptor antagonist. mp= 281° (dec), 305° (dec); LD_{50} (mrat iv) = 135 mg/kg, (frat iv) = 132 mg/kg. *Yoshitomi.*

3542 Batanopride
102670-46-2
$C_{17}H_{26}ClN_3O_3$
4-Amino-5-chloro-N-[2-(diethylamino)ethyl]-2-[(1-methylacetonyl)oxy]benzamide.
Antiemetic. *Bristol-Myers Squibb Pharm. R&D.*

3543 Batanopride Hydrochloride
102670-59-7
$C_{17}H_{27}Cl_2N_3O_3$
4-Amino-5-chloro-N-[2-(diethylamino)ethyl]-2-[(1-methylacetonyl)oxy]benzamide hydrochloride.
BMY-25801-01. Antiemetic. *Bristol-Myers Squibb Pharm. R&D.*

3544 Bemesetron
40796-97-2
$C_{15}H_{17}Cl_2NO_2$
1αH,5αH-Tropan-3α-yl 3,5-dichlorobenzoate.
MDL-72222. Antiemetic. *Marion Merrell Dow Inc.*

3545 Benzquinamide
63-12-7 1154
$C_{22}H_{32}N_2O_5$
2-(Acetoxy)-N,N-diethyl-1,3,4,6,7,11b-hexahydro-9,10-dimethoxy-2H-benzo[a]quinolizine-3-carboxamide.
BZQ; NSC-64375; P-2647; Emete-Con; Emeticon; Promecon; Quantril. Antipsychotic; antiemetic. A tricyclic. mp = 130-131.5°; LD_{50} (rat orl) = 990 mg/kg, (mus ip) = 376 mg/kg. *Pfizer Intl.; Roerig Div. Pfizer Pharm.*

3546 Bromopride
4093-35-0 1454 223-842-2
$C_{14}H_{22}BrN_3O_2$
4-Amino-5-bromo-N-[2-(diethylamino)ethyl]-o-anisamide.
Emepride; Emoril; Viadil. Antiemetic. *Teikoku Hormone Mfg. Co. Ltd.*

3547 Bromopride Hydrochloride
52423-56-0 1454 257-906-6
$C_{14}H_{23}ClBrN_3O_2$
4-Amino-5-bromo-N-[2-(diethylamino)ethyl]-o-anisamide hydrochloride.
Cascapride; Plesium; Praiden; Valopride; Viaben. Antiemetic. *Teikoku Hormone Mfg. Co. Ltd.*

3548 Buclizine
82-95-1 1484 201-448-1
$C_{28}H_{33}ClN_2$
1-(p-tert-Butylbenzyl)-4-(p-chloro-α-phenylbenzyl)piperazine.
hitabutyzyne; histabutizine; UCB-4445; Buclifen; Longifene; Posdel; Postafen; Vibazine. Antiemetic. $bp_{0.001}$ = 217-220°. *Pfizer Intl.; Zeneca Pharm.*

3549 Buclizine Hydrochloride
129-74-8 1484 204-962-4
$C_{28}H_{35}Cl_3N_2$
1-(p-tert-Butylbenzyl)-4-(p-chloro-α-phenylbenzyl)piperazine dihydrochloride.
Vibazine; Bucladin S; Buclina; Softran. Antiemetic. mp = 230-240°. *Pfizer Intl.; Zeneca Pharm.*

3550 Cerium Oxalate
139-42-4 2045 205-362-5
$C_6Ce_2O_{12}$
Cerium(III) oxalate.
Sedemesis. Antiemetic. Insoluble in H_2O, soluble in mineral acids.

3551 Chlorpromazine
50-53-3 2238 200-045-8
$C_{17}H_{19}ClN_2S$
2-Chloro-10-[3-(dimethylamino)propyl]phenothiazine.
2601-A; HL-5746; RP-4560; SKF-2601-A; Aminazine;

Ampliactil; Amplictil; Chlorderazin; Chlropromados; Elmarin; Esmind; Fenactil; Novomazina; Promactil; Promazil; Proma; Prozil; Plegomazin; Sanopron; Thorazine; Wintermin. Antiemetic; antipsychotic. Phenothiazine. $bp_{0.8}$ = 200-205°. *Elkins-Sinn; KV Pharm.; Labaz S.A.; Parke-Davis; Rhône-Poulenc Rorer Pharm. Inc.*

3552 Chlorpromazine Hydrochloride
69-09-0 2238 200-701-3
$C_{17}H_{20}Cl_2N_2S$
2-Chloro-10-[3-(dimethylamino)propyl]phenothiazine hydrochloride.
clorazepate potassium; dipotassium clorazepate; Chloractil; Chlorazin; Gen-xene; Hebanil; Hibanil; Hibernal; Klorpromex; Largactil; Largaktyl; Marazine; Megaphen; Novo-Clopate; Nu-Clopate; Promacid; Promapar; Propaphenin; Sonazine; Taroctyl; Thorazine; Torazina; Tranzene. Antiemetic; antipsychotic. Tranquilizer (veterinary). A phenothiazine. Dec 179-180°; pH (5% aq) = 4.0-5.5; soluble in H_2O (40 g/100ml), MeOH, EtOH, $CHCl_3$; insoluble in C_6H_6, Et_2O; LD_{50} (rat orl) = 2225 mg/kg. *DDSA Pharm. Ltd.; Elkins-Sinn; KV Pharm.; Labaz S.A.; Parke-Davis; Pharmacia & Upjohn; Rhône-Poulenc Rorer Pharm. Inc.; SmithKline Beecham Pharm.*

3553 Cilansetron
120635-74-7
$C_{20}H_{21}N_3O$
(-)-(R)-56,9,10-Tetrahydro-10-[(2-methylimidazol-1-yl)methyl]-4H-pyrido[3,2,1-jk]carbazol-11-(8H)-one.
Antiemetic. 5-HT_3 receptor antagonist.

3554 Cipropride
68475-40-1
$C_{17}H_{25}N_3O_4S$
N-[[-(Cyclopropylmethyl)-2-pyrrolidinyl]methyl]-5-sulfamoyl-o-anisamide.
Antiemetic.

3555 Clebopride
55905-53-8 2404 259-885-9
$C_{20}H_{24}ClN_3O_2$
4-Amino-N-(1-benzyl-4-piperidyl)-5-chloro-o-anisamide.
Cleboril. Antispasmodic; antiemetic. Selective D2 receptor antagonist. mp = 194-195°. *Anphar; Grupo Farmaceutico Almirall S.A.*

3556 Clebopride Hydrochloride
2404
$C_{20}H_{25}Cl_2N_3O_2.H_2O$
4-Amino-N-(1-benzyl-4-piperidyl)-5-chloro-o-anisamide hydrochloride monohydrate.
Antiemetic with antispasmodic properties. Selective D2 receptor antagonist. mp = 217-219°; LD_{50} (mus orl) > 1000 mg/kg. *Grupo Farmaceutico Almirall S.A.*

3557 Clebopride Malate
57645-91-7 2404 260-874-6
$C_{24}H_{30}ClN_3O_7$
4-Amino-N-(1-benzyl-4-piperidyl)-5-chloro-o-anisamide malate.
Amicos; Clanzol; Clast; Cleboril; Cleprid; Motilex. Antiemetic with antispasmodic properties. Selective D2 receptor antagonist. *Anphar; Grupo Farmaceutico Almirall S.A.*

3558 Cyclizine
82-92-8 2779 201-445-5
$C_{18}H_{22}N_2$
1-(Diphenylmethyl)-4-methylpiperazine.
Compound 47-83; Marzine; Marezine; Nautazine; Ne-Devomit. Antihistaminic. mp = 105.5-107.5°; λ_m = 269, 263, 258, 225 nm (ε 540, 742, 694, 11300).1N HCl); soluble in $CHCl_3$ (1.1 g/ml), Et_2O, EtOH 0.17 g/ml, H_2O < 0.001 g/ml; LD_{50} (mus orl) = 147 mg/kg. *Glaxo Wellcome Inc.*

3559 Cyclizine Hydrochloride
303-25-3 2779 206-136-9
$C_{18}H_{23}ClN_2$
1-(Diphenylmethyl)-4-methylpiperazine hydrochloride.
Valoid. Antiemetic. Soluble in H_2O. *Glaxo Wellcome Inc.*

3560 Cyclizine Lactate
5897-19-8 2779
$C_{21}H_{28}N_2O_3$
1-(Diphenylmethyl)-4-methylpiperazine monolactate.
Injection. *Glaxo Wellcome Inc.*

3561 Dimenhydrinate
523-87-5 3252 208-350-8
$C_{24}H_{28}ClN_5O_3$
8-Chlorotheophylline compound with 2-(diphenylmethoxy)-N,N-dimethylethylamine (1:1).
Dommanate; chloranautine; Amosyt; Anautine; Andramine; Antemin; Diamarin; Dimate; Dramamine; Dramarin; Dramocen; Dramyl; Emedyl; Emes; Epha; Gravol; Menhydrinate; Reidamine; Removine; Travel-Gum; Travelin; Travelmin; Vomex A; Xamamina; Faston. Antiemetic. mp = 102-107°; soluble in H_2O (3 mg/ml), more soluble in organic solvents. *Forest Pharm. Inc.; Wyeth-Ayerst Labs.*

3562 Diphenidol
972-02-1 3369 213-540-9
$C_{21}H_{27}NO$
α,α-Diphenyl-1-piperidinebutanol.
SK&F-478. mp = 104-105°; LD_{50} (rat sc) = 50 mg/kg. *SmithKline Beecham Pharm.*

3563 Diphenidol Hydrochloride
3254-89-5 3369 221-850-0
$C_{21}H_{28}ClNO$
α,α-Diphenyl-1-piperidinebutanol hydrochloride.
SK&F-478-A; Vontrol; Ansmin; Cefadol; Celmidol; Difenidolin; Maniol; Mecalmin; Pineroro; Satanolon; Tenesdol; Wansar. mp = 212-214°; soluble in H_2O, MeOH, $CHCl_3$; insoluble in Et_2O, C_6H_6, petroleum ether; LD_{50} (mus orl) = 430 mg/kg, (mus ip) = 105 mg/kg, (mus iv) = 37 mg/kg, (rat orl) = 515 mg/kg, (rat ip) = 82 mg/kg, (rat iv) = 29 mg/kg. *SmithKline Beecham Pharm.*

3564 Diphenidol Pamoate
26363-46-2 3369
$C_{65}H_{70}N_2O_8$
α,α-Diphenyl-1-piperidinebutanol compound with 4,4'-methylenebis[3-hydroxy-2-naphthoic acid] (2:1).
SK&F-478-J. *SmithKline Beecham Pharm.*

3565 Dolasetron
115956-12-2 3471
$C_{19}H_{20}N_2O_3$
Indole-3-carboxylic acid ester with (8r)-hexahydro-8-hydroxy-2,6-methano-2H-quinolizin-3(4H)-one.
MDL-73147. Antiemetic; antimigraine. Serotonin receptor antagonist. *Hoechst Marion Roussel Inc.*

3566 Dolasetron Mesylate
115956-13-3 3471
$C_{20}H_{24}N_2O_6S$
Indole-3-carboxylic acid ester with (8r)-hexahydro-8-hydroxy-2,6-methano-2H-quinolizin-3(4H)-one monomethanesulfonate.
Anxemet; MDL-73147EF. Antiemetic; antimigraine. Serotonin receptor antagonist. mp = 278°. *Hoechst Marion Roussel Inc.*

3567 Domperidone
57808-66-9 3476 260-968-7
$C_{22}H_{24}ClN_5O_2$
5-Chloro-1-[1-[3-(2-oxo-1-benzimidazolinyl)propyl]-4-piperidyl]-2-benzimidazolinone.
Motilium; R-33812; Euciton; Evoxin; Gastronorm; Mod; Nauzelin; Peridon; Peridys. Antiemetic. mp = 242.5°. *Abbott Laboratories Inc.*

3568 Dronabinol
1972-08-3 9349
$C_{21}H_{30}O_2$
(6aR,10aR)-6a,7,8,10a-Tetrahydro-6,6,9-trimethyl-3-pentyl-6H-dibenzo[b,d]pyran-1-ol.
Marinol; NSC-134454; tetrahydrocannabinol; (-)-Δ^1-3,4-trans tetrahydrocannabinol; Δ^9-THC; Δ^1-THC; QCD-84924. Antiemetic. The (-)-Δ^1-3,4-trans form is used as an antiemetic for cancer patients and as a topical medication for hypertensive glaucomas. $bp_{0.02}$ = 200°; $[\alpha]_D^{20}$ = -150° (c = 0.53 $CHCl_4$); λ_m = 283, 276 nm (log ε 3.21, 3.20 EtOH); LD_{50} (mrat orl) = 800 mg/kg, (frat orl) = 730 mg/kg. *Unimed Pharm. Inc.*

3569 Exepanol
77416-65-0
$C_{11}H_{15}NO_2$
(±)-cis-2,3,4,5-Tetrahydro-3-(methylamino)-1-benzoxepin-5-ol.
Antiemetic.

3570 Fludorex
15221-81-5
$C_{11}H_{14}F_3NO$
N-Methyl-β-methoxy-m-(trifluoromethyl)-phenethylamine.
Win-11464. Antiemetic; anorexic. *Sterling Winthrop Inc.*

3571 Flumeridone
75444-64-3 278-211-4
$C_{22}H_{23}ClFN_5O_2$
5-Chloro-1-[1-[3-(5-chloro-2,3-dihydro-2-oxo-1-benzimidazolinyl)propyl]-4-piperidyl]-2-benzimidazolinone.
R-45486. Antiemetic.

3572 Galdansetron
116684-92-5
$C_{18}H_{19}N_3O$
(3R)-2,3-Dihydro-9-methyl-3-[(5-methylimidazol-4-yl)methyl]carbazol-4(1H)-one.
GR-81225X. Antiemetic. *Glaxo Wellcome Inc.*

3573 Galdansetron Hydrochloride
156712-35-5
$C_{18}H_{20}ClN_3O$
(3R)-2,3-Dihydro-9-methyl-3-[(5-methylimidazol-4-yl)methyl]carbazol-4(1H)-one monohydrochloride.
GR-81225C. Antiemetic. *Glaxo Wellcome Inc.*

3574 Granisetron
109889-09-0 4557
$C_{18}H_{24}N_4O$
1-Methyl-N-(9-methyl-endo-9-azabicyclo[3.3.1]non-3-yl)-1H-imidazole-3-carboxamide.
BRL-43694. Antiemetic. Specific serotonin $5HT_3$ receptor antagonist. Used in treatment of nausea in cancer patients. *SmithKline Beecham Pharm.*

3575 Granisetron Hydrochloride
107007-99-8 4557
$C_{18}H_{25}ClN_4O$
1-Methyl-N-(9-methyl-endo-9-azabicyclo[3.3.1]non-3-yl)-1H-imidazole-3-carboxamide hydrochloride.
BRL-43694A; Kytril. Antiemetic. Specific serotonin $5HT_3$ receptor antagonist. Used in treatment of nausea in cancer patients. mp = 290-292°. *SmithKline Beecham Pharm.*

3576 Lerisetron
143257-98-1
$C_{18}H_{20}N_4$
1-Benzyl-2-(1-piperazinyl)benzimidazole.
Antiemetic. 2-Piperazinylbenzimidazole derivative. Specific 5HT3-receptor antagonist.

3577 Lorapride
68677-06-5 272-057-1
$C_{14}H_{22}ClN_3O_3S$
5-Chloro-n^1-[(1-ethyl-2-pyrrolidinyl)methyl]-2-methoxysulfanylzamide.
Antipsychotic; antidepressant; antiemetic. Sulpiride derivative.

3578 Lurosetron
128486-54-4
$C_{17}H_{17}FN_4O$
6-Fluoro-2,3,4,5-tetrahydro-5-methyl-2-[(5-methylimidazol-4-yl)methyl]-1H-pyrido-[4,3-b]indol-1-one.
Antiemetic. *Glaxo Wellcome plc.*

3579 Lurosetron Mesylate
143486-90-2
$C_{18}H_{21}FN_4O_4S$
6-Fluoro-2,3,4,5-tetrahydro-5-methyl-2-[(5-methylimidazol-4-yl)methyl]-1H-pyrido-[4,3-b]indol-1-one monomethanesulfonate.
GR-87442N. Antiemetic. *Glaxo Wellcome plc.*

3580 Meclizine
569-65-3 5817 209-323-3
$C_{25}H_{27}ClN_2$
1-(p-Chloro-α-phenylbenzyl)-4-(m-methylbenzyl)piperazine.
meclozine; parachloramine. Antiemetic. bp_2= 230°. *KV Pharm.; Pfizer Intl.; Roerig Div. Pfizer Pharm.*

3581 Meclizine Hydrochloride
31884-77-2 5817 214-164-8
$C_{25}H_{29}Cl_3N_2.2H_2O$
1-(p-Chloro-α-phenylbenzyl)-4-(m-methylbenzyl)piperazine dihydrochloride monohydrate.
UCB-5062; Ancolan; Antivert; Bonamine; Bonine; Calmonal; Diadril; Histametizine; Navicalm; Neo-Istafene; Peremesin; Postafene; Sabari; Sea-Legs; Veritab. Antiemetic. Insoluble in H_2O (0.1 g/100 ml); freely soluble in $CHCl_3$, C_5H_5N. *KV Pharm.; Pfizer Intl.; Roerig Div. Pfizer Pharm.*

3582 Methallatal
115-56-0 6011 204-095-1
$C_{10}H_{14}N_2O_2S$
5-Ethyldihydro-5-(2-methyl-2-propenyl)-2-thioxo-4,6(1H,5H)-pyrimidinedione.
V-12; Mosidal. Antiemetic. mp = 160-161°; insoluble in H_2O; sodium salt is soluble in H_2O, EtOH. *Abbott Labs Inc.*

3583 Metoclopramide
364-62-5 6226 206-662-9
$C_{14}H_{22}ClN_3O_2$
4-Amino-5-chloro-N-[2-(diethylamino)ethyl]-2-methoxybenzamide.
Metoclopramide; 4-Amino-5-chloro-N-[2-(diethylamino)-ethyl]-o-anisamide.HCl.H2O; AHR-3070-C; Cerucal; Clopromate; Draclamid; Emperal; Eucil; Gastrese; Gastrobid; Gastromax; Gastrosil; Gastro-tablinen; Gastrotem; Gastro-Timelets; Maxeran; Maxolon; MCP-ratiopharm; Meclopran; Metamide; Metocobil; Metramid; Parmid; Paspertin; Plasil; Regla. Antiemetic. Has neuroleptic activity. A benzamide. mp = 146-148°; soluble in H_2O (0.2 g/100 ml), more soluble in organic solvents. *Abbott Labs Inc.; Apothecon; Lemmon Co.; SmithKline Beecham Pharm.; Whitehall-Robins.*

3584 Metoclopramide Hydrochloride
54143-57-6 6226 206-662-9
$C_{14}H_{23}Cl_2N_3O_2.H_2O$
4-Amino-5-chloro-N-[2-(diethylamino)ethyl]-2-methoxybenzamide monohydrochloride monohydrate.
Maxolon; Reglan; AHR-3070-C; Emetid; Gastronerton; Primperan; AHR-3070-C; Cerucal; Clopromate; Draclamid; Emperal; Eucil; Gastrese; Gastrobid; Gastromax; Gastrosil; Gastro-tablinen; Gastrotem; Gastro-Timelets; Maxeran; Maxolon; MCP-ratiopharm; Meclopran; Metamide; Metoclol; Metocobil; Metramid; Moriperan; Mygdalon; Parmid; Paspertin; Peraprin; Plasil; Pramiel; Reglan. Antiemetic. Has neuroleptic activity. A benzamide. mp = 182.5-184°; slightly soluble in H_2O, more soluble in organic solvents. *Abbott Labs Inc.; Apothecon; Lemmon Co.; SmithKline Beecham Pharm.; Whitehall-Robins.*

3585 Metopimazine
14008-44-7 6233 237-818-4
$C_{22}H_{27}N_3O_3S_2$
1-[3-[2-(Methylsulfonyl-10H-phenothiazin-10-yl]propyl]isonipecotamide.
EXP 999; RP-9965; Vogalene. Antiemetic. mp = 170-171°; LD_{50} (rat orl) = 976 mg/kg, (rat sc) = 1080 mg/kg. *DuPont-Merck Pharm.; Rhône-Poulenc.*

3586 Mobenzoxamine
65329-79-5
$C_{30}H_{35}FN_2O_3$
4'-Fluoro-4-[4-[2-[(p-methoxy-α-phenylbenzyl)oxy]ethyl]-1-piperazinyl]butyrophenone.
Antiemetic.

3587 Mociprazine
56693-13-1 260-340-2
$C_{22}H_{32}N_2O_3$
α-[[(1-Ethynylcyclohexyl)oxy]methyl]-4-(o-methoxyphenyl)-1-piperazineethanol.
CERM-3517. Antiemetic.

3588 Nabilone
51022-71-0 6427
$C_{24}H_{36}O_3$
(±)-3-(1,1-Dimethylheptyl-6,6aβ,7,8,10,10aα-hexahydro-1-hydroxy-6,6-dimethyl-9H-dibenzo[b,d]pyran-9-one.
Cesamet; Compd. 109514; LY-109514. Antiemetic. A synthetic cannabinoid with antiemetic, antiglaucoma and CNS activity. mp = 159-160°; λ_m=207, 280 nm (ε 47000, 250 EtOH). *Eli Lilly & Co.*

3589 Naboctate
74912-19-9
$C_{33}H_{53}NO_3$
(±)-7,8,9,10-Tetrahydro-6,6,9-trimethyl-3-(1-methyloctyl)6H-dibenzo[b,d]pyran-1-yl 4-(diethylamino)-butyrate.
Antiemetic; antiglaucoma agent.

3590 Naboctate Hydrochloride
73747-21-4
$C_{33}H_{54}ClNO_3$
(±)-7,8,9,10-Tetrahydro-6,6,9-trimethyl-3-(1-methyloctyl)-6H-dibenzo[b,d]pyran-1-yl 4-(diethylamino)butyrate hydrochloride.
SP-325. Antiglaucoma agent; antiemetic.

3591 Nonabine
16985-03-8
$C_{25}H_{33}NO_2$
7-(1,2-Dimethylheptyl)-2,2-dimethyl-4-(4-pyridiyl)-2H-1-benzopyran-5-ol.
BRL-4664. Antiemetic.

3592 Ondansetron
116002-70-1 6979
$C_{18}H_{19}N_3O$
(±)-2,3-Dihydro-9-methyl-3-[(2-methylimidazol-1-yl)methyl]carbazol-4(1H)-one.
Anxiolytic; antiemetic; antischizophrenic. Specific serotonin receptor ($5HT_3$) antagonist. mp = 231-232°;

[3S-form]: $[\alpha]_D^{25}$ = -14° (c = 0.19, MeOH); [3R-form]: $[\alpha]_D^{24}$ = +16° (c = 0.34, MeOH). *Glaxo Wellcome Inc.*

3593 Ondansetron Hydrochloride
103639-04-9 6979
$C_{18}H_{20}ClN_3O.H_2O$
(±)-2,3-Dihydro-9-methyl-3-[(2-methylimidazol-1-yl)methyl]carbazol-4(1H)-one monohydrochloride monohydrate.
Zofran; GR-38032F. Anxiolytic; antiemetic; anti-schizophrenic. Serotonin receptor antagonist. mp = 178.5-179.5°. *Glaxo Wellcome Inc.*

3594 Oxypendyl
5585-93-3 7104
$C_{20}H_{26}N_4OS$
4-[3-(10H-Pyrido[3,2-b][1,4]benzothiazin-10-yl)-propyl]piperazin-1-ylethanol.
oxipendyl; D-706. Antiemetic. bp_1 ≅ 263°; bp_6 = 280-300°.

3595 Oxypendyl Hydrochloride
17297-82-4 7104
$C_{20}H_{28}Cl_2N_4OS$
4-[3-(10H-Pyrido[3,2-b][1,4]benzothiazin-10-yl)-propyl]piperazin-1-ylethanol dihydrochloride.
Antiemetic. mp = 231-232°; soluble in H_2O (1.2 g/100 ml).

3596 Palonosetron
135729-56-5
$C_{19}H_{22}N_2O$
2,4,5,6-Tetrahydro-2-[(3S)-3-quinuclidinyl]-1H-benz[de]isoquinolin-1-one.
Antiemetic. *Syntex Labs. Inc.*

3597 Palonosetron Hydrochloride
135729-55-4
$C_{19}H_{23}ClN_2O$
2,4,5,6-Tetrahydro-2-[(3S)-3-quinuclidinyl]-1H-benz[de]isoquinolin-1-one hydrochloride.
RS-25259-197. Antiemetic. *Syntex Labs. Inc.*

3598 Pancopride
121243-20-7
$C_{18}H_{24}ClN_3O_2$
4-Amino-5-chloro-α-cyclopropyl-N-3-quinuclidinyl-o-anisamide.
Anxiolytic; antiemetic; CNS stimulant. *Grupo Farmaceutico Almirall S.A.*

3599 (±)-Pancopride
121650-80-4
$C_{18}H_{24}ClN_3O_2$
(±)-4-Amino-5-chloro-α-cyclopropyl-N-3-quinuclidinyl-o-anisamide.
Anxiolytic; antiemetic; peristaltic stimulant. *Grupo Farmaceutico Almirall S.A.*

3600 Pipamazine
84-04-8 7607 201-512-9
$C_{21}H_{24}ClN_3OS$
10-[3-(4-Carbamoylpiperidino)propyl]-2-chlorophenothiazine.
Mornidine; Nausidol; SC-9387. Antiemetic. mp = 139°; [hydrochloride]: mp = 196-197°. *Searle G.D. & Co.*

3601 Prochlorperazine
58-38-8 7942 200-379-4
$C_{20}H_{24}ClN_3S$
2-Chloro-10-[3-(4-methyl-1-piperazinyl)propyl]-phenothiazine.
chlormeprazine; prochlorpemazine; prochlorperazine; Bayer A173; RP-6140; SKF-4657; Compazine; Eskatrol. Antiemetic; antipsychotic. For vertigo. *Bayer AG; Rhône-Poulenc Rorer Pharm. Inc.; SmithKline Beecham Pharm.*

3602 Prochlorperazine Edisylate
1257-78-9 7942 215-019-1
$C_{22}H_{30}ClN_3O_6S_3$
2-Chloro-10-[3-(4-methyl-1-piperazinyl)propyl]-phenothiazine 1,2-ethanedisulfonate (1:1).
prochlorperazine ethanedisulfonate; Novamin; Tementil; Compazine Injection; Compazine Syrup. Antiemetic; antipsychotic. A phenothiazine, used in treatment of vertigo. *Rhône-Poulenc Rorer Phàrm. Inc.; SmithKline Beecham Pharm.; Wyeth-Ayerst Labs.*

3603 Prochlorperazine Maleate
84-02-6 7942 201-511-3
$C_{28}H_{32}ClN_3O_8S$
2-Chloro-10-[3-(4-methyl-1-piperazinyl)propyl]-phenothiazine maleate (1:2).
Buccastem; Compazine; Meterazine; Stemetil; Vertigon; component of: Combid. Antiemetic; antipsychotic. A phenothiazine, used in treatment of vertigo. mp = 228°; slightly soluble in H_2O (0.1 g/100 ml), MeOH, EtOH; insoluble in C_6H_6, Et_2O, $CHCl_3$; LD_{50} (mus sc) = 400 mg/kg, (mus ip) = 120 mg/kg, (mus iv) = 90 mg/kg, (mus orl) = 400 mg/kg. *Apothecon; Rhône-Poulenc Rorer Pharm. Inc.; SmithKline Beecham Pharm.*

3604 Promethazine
60-87-7 7970 200-489-2
$C_{17}H_{20}N_2S$
10-[2-(Dimethylamino)propyl]phenothiazine.
proazamine; RP-3277. Antihistaminic. Also has antiemetic and CNS depressant properties. mp = 60°; bp_3 = 190-192°. *Rhône-Poulenc; Wyeth Labs.*

3605 Promethazine Hydrochloride
58-33-3 7970 200-375-2
$C_{17}H_{21}ClN_2S$
10-[2-(Dimethylamino)propyl]phenothiazine hydrochloride.
Anergan 25; Anergan 50; Phenergan; Mepergan; Phanergan D; Phenergan VC; RP-3389; Atosil; Dorme; Duplamin; Fellozine; Fenazil; Genphen; Hiberna; Lergigan; Phencen; Phenergan; Prorex; Prothazine; Provigan; Remsed. Antihistaminic. Also has antiemetic and CNS depressant properties. mp = 230-232°; λ_m = 249, 297 nm (ε 28770, 3400 H_2O); soluble in H_2O, EtOH, $CHCl_3$; insoluble in Me_2CO, EtOAc, Et_2O; LD_{50} (mus iv) = 55 mg/kg. *Rhône-Poulenc; Wyeth Labs.*

3606 Promethazine Teoclate
17693-51-5 7970 241-691-0
$C_{17}H_{20}N_2S \cdot C_7H_7ClN_4O_2$
10-[2-(Dimethylamino)propyl]phenothiazine compound with 8-chlorotheophylline.
Avomine. Antihistaminic. Also has antiemetic and CNS depressant properties. *Rhône-Poulenc; Wyeth Labs.*

3607 Pyridoxine Hydrochloride

58-56-0 8166 200-386-2

$C_8H_{12}ClNO_3$

5-Hydroxy-6-methyl-3,4-pyridinedimethanol hydrochloride.

Beesix; Hexa-Betalin; Hexavibex; Bonasanit; Hexabione hydrochloride; Pyridipca; Pyridox; Bécilan; Benadon; Hexermin; component of: Bendectin, Spondylonal. Vitamin, vitamin source. Enzyme cofactor. May improve hematopoiesis in patients with sideroblastic anemias. mp = 205-212° (dec); λ_m = 290 nm (ε 8400 0.1N HCl), 253, 325 nm (ε 3700, 7100, pH 7); soluble in H_2O (22.2 g/100 ml), EtOH (1.1 g/100 ml), propylene glycol; sparingly soluble in Me_2CO; insoluble in Et_2O, $CHCl_3$. *BASF Corp.; Eli Lilly & Co.; Forest Pharm. Inc.; General Aniline; Lederle Labs.; Marion Merrell Dow Inc.; Merck & Co.Inc.; Parke-Davis.*

3608 Scopolamine

51-34-3 8550 200-090-3

$C_{17}H_{21}NO_4$

[7(S)-(1α,2β,4β,5α,7β)]-(Hydroxymethyl)benzene-acetic acid 9-methyl-3-oxa-9-azatricyclo-[3.3.1.0^{2,4}]non-7-yl ester.

hyoscine; l-scopolamine; scopine tropate; S-(-)-tropate; tropic acid ester with scopine; Scop; Scopoderm-TTS; Transcop; Transderm Scop. Anticholinergic; antiemetic. For motion sickness. Tropane alkaloid found in *Scolinacea*. Constituent of impure duboisine. Pure duboisine is l-hyoscyamine. Viscous liquid; decomposes; $[\alpha]_D^{20}$ = -28° (c = 2.7); soluble in H_2O (10.5 g/100 ml at 15°); EtOH, Et_2O, $CHCl_3$, Me_2CO; sparingly soluble in C_6H_6, petroleum ether; [monohydrate]: mp = 59°. *Alcon Labs; Ciba-Geigy Corp.; Parke-Davis; Robins, A. H. Co.; Wallace Labs; Zeneca Pharm.*

3609 Scopolamine Hydrobromide

114-49-8 8550 204-050-6

$C_{17}H_{22}BrNO_4.3H_2O$

[7(S)-(1α,2β,4β,5α,7β)]-(Hydroxymethyl)benzene-acetic acid 9-methyl-3-oxa-9-azatricyclo-[3.3.1.0^{2,4}]non-7-yl ester hydrobromide trihydrate.

Scopolamine hydrobromide trihydrate; scopolammonium bromide; Scopos; Sea Legs; Wellcome Brand Scopolamine Hydrobromide Injection; Scopolaminium bromide; Scopolammonium bromide; Scopos; Sereen; Triptone; 1αH,5αH-Tropan-3α-ol, 6β,7β-epoxy-, (-)-tropate (ester), hydrobromide; Beldavrin; Euscopol; Hydroscine hydrobromide; Hyocine F hydrobromide; Hyoscine bromide; Hyoscine hydrobromide; L-Hyoscine hydrobromide; component of: Benacine. Antiemetic. mp = 195°; $[\alpha]_D^{25}$ = -25° (c = 5); λ_m = 246, 252, 258, 264 nm ($A_{1\ cm}^{1\%}$ 3.5, 4.0, 4.5, 3.0 MeOH); soluble in H_2O (0.66 g/ml), less soluble in organic solvents; LD_{50} (rat sc) = 3800 mg/kg. *Alcon Labs; Ciba-Geigy Corp.; Parke-Davis; Robins, A. H. Co.; Wallace Labs; Wellcome Foundation Ltd.; Zeneca Pharm.*

3610 Sulpiride

15676-16-1 9163 239-753-7

$C_{15}H_{23}N_3O_4S$

N-[(1-Ethyl-2-pyrrolidinyl)methyl]-2-methoxy-5-sulfamoylbenzamide.

Abilit; Aiglonyl; Coolspan; Dobren; Dogmatil; Dogmatyl; Dolmatil; Guastil; Meresa; Miradol; Mirbanil; Misulvan; Neogama; Omperan; Pyrikappl; Sernevin; Splotin; Sulpitil; Sursumid; Synedil; Trilan; Levosulpiride [as l-form]; S(-)-Sulpiride [as l-form]; Levobren [as l-form]; Levopraid [as l-form]. Antidepressant; antipsychotic; antiemetic. Dopamine D_2- and D_3-receptor antagonist. mp = 178-180°; poorly soluble in MeOH; nearly insoluble in H_2O, Et_2O, $CHCl_3$, C_6H_6; pKa_1 = 10.19; $[\alpha]_D^{25}$ = -66.8° (c = 0.5 DMF); LD_{50} (mus ip) = 170 mg/kg, (mus orl) = 2250 mg/kg. *Delagrange; Ravizza; Soc. Etudes Sci. Ind. L'Ile de France.*

3611 Thiethylperazine

1420-55-9 9449 215-819-0

$C_{22}H_{29}N_3S_2$

2-(Ethylthio)-10-[3-(4-methyl-1-piperazinyl)propyl]-phenthiazine.

Antiemetic. mp = 62-64°; $bp_{0.01}$ = 227°. *Sandoz Pharm. Corp.*

3612 Thiethylperazine Dihydrochloride

9449

$C_{22}H_{31}Cl_2N_3S_2$

2-(Ethylthio)-10-[3-(4-methyl-1-piperazinyl)propyl]-phenthiazine dihydrochloride.

Antiemetic. mp = 214-216°. *Sandoz Pharm. Corp.*

3613 Thiethylperazine Dimalate

52239-63-1 9449 257-780-2

$C_{30}H_{41}N_3O_{10}S_2$

2-(Ethylthio)-10-[3-(4-methyl-1-piperazinyl)propyl]-phenthiazine malate.

Norzine. Antiemetic. mp = 139°. *Sandoz Pharm. Corp.*

3614 Thiethylperazine Dimaleate

1179-69-7 9449 214-648-9

$C_{30}H_{37}N_3O_8S_2$

2-(Ethylthio)-10-[3-(4-methyl-1-piperazinyl)propyl]-phenthiazine dimaleate.

GS-95; NSC-130044; Torecan maleate; Toresten; Tresten. Antiemetic. mp = 188-190° (dec). *Sandoz Pharm. Corp.*

3615 Thioproperazine

316-81-4 9494 206-262-4

$C_{22}H_{30}N_4O_2S_2$

N,N-Dimethyl-10-[3-(4-methyl-1-piperazinyl)propyl]-phenothiazine-2-sulfonamide.

RP-7843; SKF-5883. Antipsychotic; antiemetic. A phenothiazine. mp = 140°; [fumarate]: mp = 182°. *Rhône-Poulenc; SmithKline Beecham Pharm.*

3616 Thioproperazine Mesylate

2347-80-0 9494 219-074-2

$C_{24}H_{38}N_4O_8S_4$

N,N-Dimethyl-10-[3-(4-methyl-1-piperazinyl)propyl]-phenothiazine-2-sulfonamide dimethanesulfonate.

thioperazine dimethanesulfonate; Majeptil; Vontil. Antipsychotic; antiemetic. A phenothiazine. *Rhône-Poulenc; SmithKline Beecham Pharm.*

3617 Trimethobenzamide

138-56-7 9839 205-332-1

$C_{21}H_{28}N_2O_5$

N-[p-[2-(Dimethylamino)ethoxy]benzyl]-3,4,5-trimethoxybenzamide.

Antiemetic. *Hoffmann-LaRoche Inc.*

3618 Trimethobenzamide Hydrochloride
554-92-7 9839 209-075-6
$C_{21}H_{29}ClN_2O_5$
N-[p-[2-(Dimethylamino)ethoxy]benzyl]-3,4,5-trimethoxybenzamide monohydrochloride.
Ro-2-9578; Anaus; Tigan; Xametina. Antiemetic. mp = 187.5-190°; freely soluble in H_2O (> 50 g/100 ml). *Hoffmann-LaRoche Inc.*

3619 Tropisetron
89565-68-4 9914
$C_{17}H_{20}N_2O_2$
1αH,5αH-Tropan-3α-yl indole-3-carboxylate.
ICS-205-930. Antiemetic. Serotonin receptor antagonist. mp = 201-202°. *Sandoz Pharm. Corp.*

3620 Tropisetron Hydrochloride
105826-92-4 9914
$C_{17}H_{21}ClN_2O_2$
1αH,5αH-Tropan-3α-yl indole-3-carboxylate hydrochloride.
Navoban; Novaban. Antiemetic. Serotonin receptor antagonist. mp = 283-285°. *Sandoz Pharm. Corp.*

3621 Zacopride
90182-92-6
$C_{15}H_{20}ClN_3O_2$
4-Amino-5-chloro-N-3-quinuclidinyl-o-anisamide.
Antiemetic; peristaltic stimulant. *Robins, A. H. Co.,.*

3622 Zacopride Hydrochloride
99617-34-2
$C_{15}H_{21}Cl_2N_3O_2$
4-Amino-5-chloro-N-3-quinuclidinyl-o-anisamide monohydrochloride monohydrate.
AHR-11190-B. Antiemetic; peristaltic stimulant. *Robins, A. H. Co.,.*

Antiestrogens

3623 Centchroman
78994-24-8 2018
$C_{30}H_{35}NO_3$
(trans)-1-[2-[p-(7-Methoxy-2,2-dimethyl-3-phenyl-4-chromanyl)phenoxy]ethyl]pyrrolidine.
Antiestrogen. Estrogen antagonist. mp = 99-101°. *Rexall SundownInc.*

3624 Centchroman Hydrochloride
51023-56-4 2018
$C_{30}H_{36}ClNO_3$
(trans)-1-[2-[p-(7-Methoxy-2,2-dimethyl-3-phenyl-4-chromanyl)phenoxy]ethyl]pyrrolidine hydrochloride.
67/20 CDRI. Antiestrogen. Estrogen antagonist. mp = 165-166°; λ_m = 232, 278 nm (ε 3701 at 278 nm MeOH); soluble in $CHCl_3$ (10 g/100 ml), Me_2CO (5 g/100 ml), EtOH (1.67 g/100 ml), MeOH (5 g/100 ml); insoluble in H_2O; LD_{50} (mus ip) = 400 mg/kg. *Rexall SundownInc.*

3625 Clometherone
5591-27-5
$C_{22}H_{31}ClO_2$
6α-Chloro-16α-methylpregn-4-ene-3,2-dione.
38000. Antiestrogen. *Eli Lilly & Co.*

3626 Clomiphene
911-45-5 2446 213-008-6
$C_{26}H_{28}ClNO$
2-[p-(2-Chloro-1,2-diphenylvinyl)phenoxy] triethylamine.
clomifene; chloramiphene; MRL-41. Antiestrogen. Gonad stimulating principal. Synthetic estrogen agonist-antagonist. *Lemmon Co.; Merrell Pharm. Inc.; Serono Labs Inc.*

3627 Clomiphene Citrate
50-41-9 2446 200-035-3
$C_{32}H_{36}ClNO_8$
2-[p-(2-Chloro-1,2-diphenylvinyl)phenoxy]-triethylamine citrate (1:1).
clomifene citrate; MRL-41; MER-41; NSC-35770; Clomid; Clomphid; Clomivid; Clostilbegyt; Dyneric; Ikaclomine; Pergotime; Serophene; zuclomiphene [cis-form]; enclomiphene [trans-form]. Gonad stimulating principal. Antiestrogen. mp = 116.5-118°; slightly soluble in H_2O, $CHCl_3$, EtOH; freely soluble in MeOH; insoluble in Et_2O. *Lemmon Co.; Merrell Pharm. Inc.; Serono Labs Inc.*

3628 Delmadinone
15262-77-8 2930 239-306-6
$C_{21}H_{25}ClO_3$
6-Chloro-17-hydroxypregna-1,4,6-triene-3,20-dione.
Δ^1-chlormadinone. Progestogen with antiestrogenic and antiandrogenic activity. *Pharmacia & Upjohn.*

3629 Delmadinone Acetate
13698-49-2 2930 237-219-8
$C_{23}H_{27}ClO_4$
6-Chloro-17-hydroxypregna-1,4,6-triene-3,20-dione acetate.
RS-1301; Δ^1-chlormadinone acetate; Delminal; Estrex; Tardastrex; Tarden; Zenadrex. Progestogen with antiestrogenic and antiandrogenic activity. mp = 168-170°; $[\alpha]_D$ = -83° ($CHCl_3$); λ_m = 229, 258, 297 nm (log ε 4.00, 4.00, 4.03 EtOH). *Pharmacia & Upjohn.*

3630 Enclomiphene
15690-57-0
$C_{26}H_{28}ClNO$
(E)-2-[p-(2-Chloro-1,2-diphenylvinyl)phenoxy]-triethylamine.
Antiestrogen. *Marion Merrell Dow Inc.*

3631 Levormeloxifene
78994-23-7
$C_{30}H_{35}NO_3$
(-)-1-[2-[4-[(3R,4R)-7-Methoxy-2,2-dimethyl-3-phenyl-4-chromanyl)phenoxy]ethyl]-pyrrolidine.
L-enantiomer of the racemic compound ormeloxifene, a selective estrogen receptor modulator. Estrogen receptor agonist.

3632 Nafoxidine
1845-11-0
$C_{29}H_{31}NO_2$
1-[2-[p-(3,4-Dihydro-6-methoxy-2-phenyl-1-naphthyl)phenoxy]ethyl]pyrrolidine.
Antiestrogen. *Pharmacia & Upjohn.*

3633 Nafoxidine Hydrochloride
1847-63-8
$C_{29}H_{32}ClNO_2$
1-[2-[p-(3,4-Dihydro-6-methoxy-2-phenyl-1-naphthyl)phenoxy]ethyl]pyrrolidine hydrochloride.
U-11100A; NSC-70735. Antiestrogen. *Pharmacia & Upjohn.*

3634 Nitromifene
10448-84-7
$C_{27}H_{28}N_2O_4$
1-[2-[p-[α-(p-Methoxyphenyl)-β-nitrostyryl]phenoxy]ethyl]pyrrolidine.
Antiestrogen. *Parke-Davis.*

3635 Nitromifene Citrate
5863-35-4
$C_{33}H_{36}N_2O_{11}$
1-[2-[p-[α-(p-Methoxyphenyl)-β-nitrostyryl]phenoxy]-ethyl]pyrrolidine citrate (1:1).
Antiestrogen. *Parke-Davis.*

3636 Panomifene
77599-17-8
$C_{25}H_{24}F_3NO_2$
(E)-2-[[2-[p-(3,3,3-Trifluoro-1,2-diphenylpropenyl)-phenoxy]ethyl]amino]ethanol.
EGIS-5660. An analog of tamoxifen. Antiestrogen.

3637 Raloxifene
84449-90-1 8281
$C_{28}H_{27}NO_4S$
6-Hydroxy-2-(p-hydroxyphenyl)benzo-[b]thien-3-yl-p-(2-piperidinoethoxy)phenyl ketone.
keoxifene; LY-139481. Antiosteoporotic; antiestrogenic. Used to treat osteoporosis. mp = 143-147°; λ_m = 290 nm (ε 34000 EtOH). *Eli Lilly & Co.*

3638 Raloxifene Hydrochloride
82640-04-8 8281
$C_{28}H_{28}ClNO_4S$
6-Hydroxy-2-(p-hydroxyphenyl)benzo-[b]thien-3-yl-p-(2-piperidinoethoxy)phenyl ketone hydrochloride.
keoxifene hydrochloride; LY-156758. Antiestrogen; used to treat osteoporosis. Nonsteroidal estrogen receptor mixed agonist- antagonist. mp = 258°; λ_m = 286 nm (ε 32800 EtOH). *Eli Lilly & Co.*

3639 Tamoxifen
10540-29-1 9216 234-118-0
$C_{26}H_{29}NO$
(Z)-2-[p-(1,2-Diphenyl-1-butenyl)phenoxy]-N,N-dimethylethylamine.
Antiestrogen. Antineoplastic. Nonsteroidal estrogen antagonist used in prevention and palliative treatment of breast cancer. mp = 96-98°; [cis-form]: mp = 72-74°. *Zeneca Pharm.*

3640 Tamoxifen Citrate
54965-24-1 9216 259-415-2
$C_{32}H_{37}NO_8$
(Z)-2-[p-(1,2-Diphenyl-1-butenyl)phenoxy]-N,N-dimethylethylamine citrate.
ICI-46474 citrate; Kessar; Noltam; Nolvadex; Nourytam; Tamofen; Tomaxasta; Zemide; TMX; ICI-47699 [as cis-form citrate]. Antiestrogen. Antineoplastic. Nonsteroidal estrogen antagonist used in prevention and palliative treatment of breast cancer. mp = 140-142°; slightly soluble in H_2O; more soluble in EtOH, MeOH, Me_2CO; LD_{50} (mus ip) = 200 mg/kg, (mus iv) = 62.5 mg/kg, (mus orl) = 3000-6000 mg/kg, (rat ip) = 600 mg/kg), (rat iv) = 62.5 mg/kg, (rat orl) = 1200-2500 mg/kg; [cis-form citrate]: mp = 126-128°. *ICI ; Zeneca Pharm.*

3641 Toremifene
89778-26-7 9688
$C_{26}H_{28}ClNO$
2-[p-[(Z)-4-Chloro-1,2-diphenyl-1-butenyl]phenoxy]-N,N-dimethylethylamine.
Antiestrogen and antineoplastic. mp = 108-110°. *Farmos Group Ltd.*

3642 Toremifene Citrate
89778-27-8 9688
$C_{32}H_{36}ClNO_8$
2-[p-[(Z)-4-Chloro-1,2-diphenyl-1-butenyl]phenoxy]-N,N-dimethylethylamine citrate.
FC-1157a; Fareston. Antiestrogen and antineoplastic. mp = 160-162°. *Farmos Group Ltd.*

3643 Trioxifene
63619-84-1
$C_{30}H_{31}NO_3$
3,4-Dihydro-2-(p-methoxyphenyl)-1-naphthyl-p-[2-(1-pyrrolidinyl)ethoxy] phenyl ketone.
Antiestrogen. *Eli Lilly & Co.*

3644 Trioxifene Mesylate
68307-81-3
$C_{31}H_{35}NO_6S$
3,4-Dihydro-2-(p-methoxyphenyl)-1-naphthyl-p-[2-(1-pyrrolidinyl)ethoxy] phenyl ketone methanesulfonate.
Compound 133314. Antiestrogen. *Eli Lilly & Co.*

3645 Zindoxifene
86111-26-4
$C_{21}H_{21}NO_4$
1-Ethyl-2-(p-hydroxyphenyl)-3-methylindol-5-ol diacetate (ester).
D-16726. Antiestrogen; antineoplastic.

Antifibrotics

3646 Aminobenzoate Potassium
138-84-1 7766 205-338-4
$C_7H_6KNO_2$
p-Aminobenzoic acid potassium salt.
potassium p-aminobenzoate; KPABA; Potoba. Antifibrotic; analgesic. Used in idiopathic pulmonary fibrosis. Soluble in H_2O; nearly insoluble in Et_2O. *Glenwood Inc.*

3647 Safironil
134377-69-8
$C_{15}H_{23}N_3O_4$
N,N'-Bis(3-methoxypropyl)-2,4-pyridinecarboxamide.
Antifibrotic.

Antifungals

3648 Acrosorcin
7527-91-5 128 231-389-7
$C_{25}H_{28}N_{2O}2$
4-Hexylresorcinol compound with 9-aminoacridine.
Akrinol. Antifungal. *Schering-Plough HealthCare Products.*

3649 Aliconazole
63824-12-4 264-498-3
$C_{18}H_{13}Cl_3N_2$
(Z)-1-[2,4-Dichloro-β-(p-chlorophenyl)cinnamyl]-imidazole.
Antifungal.

3650 Alteconazole
93479-96-0
$C_{17}H_{12}Cl_3N_3O$
cis-1-[2-(p-Chlorophenyl)-3-(2,4-dichlorophenyl)-2,3-epoxypropyl]-1H-1,2,4-triazole.
Antifungal.

3651 Ambruticin
58857-02-6
$C_{28}H_{42}O_6$
6-[2-[2-[5-(6-Ethyl-3,6-dihydro-5-methyl-2H-pyran-2-yl)-3-methyl-1,4-hexadienyl]-3-methylcyclopropyl]vinyl]-tetrahydro-4,5-dihydroxy-2H-pyran-2-acetic acid.
W7783; SMP-78 Acid S. Antifungal. *Parke-Davis.*

3652 Amorolfine
78613-35-1 612
$C_{21}H_{35}NO$
(±)-cis-2,6-Dimethyl-4-[2-methyl-3-(p-tert-pentylphenyl)propyl]morpholine.
Ro-14-4767/000; Loceryl. Antimycotic, used as a topical antifungal. $bp_{0.036}$ = 134°. *Hoffmann-LaRoche Inc.*

3653 Amorolfine Hydrochloride
78613-38-4 612
$C_{21}H_{36}ClNO$
(±)-cis-2,6-Dimethyl-4-[2-methyl-3-(p-tert-pentylphenyl)propyl]morpholine hydrochloride.
Ro-14-4767/002. Antimycotic, used as a topical antifungal. *Hoffmann-LaRoche Inc.*

3654 Amphotericin B
1397-89-3 627 215-742-2
$C_{47}H_{73}NO_{17}$
[1R-1R*,3S*,5R*,6R*,9R*,11R*,15S*,16R*,17R*,18S*,19E,21E,23E,25E,27E,29E,31E,33R*,35S*,36R*,37S*)]33-[(3-Amino-3,6-dideoxy-β-D-mannopyranosyl)oxy]-1,3,5,6,9,11,17,37-octahydroxy-15,16,18-trimethyl-13-oxo-14,39-dioxabicyclo[33.3.1]nonatriaconta-19,21,23,25,27,29,31-heptaene-36-carboxylic acid.
Amphocin; Fungizone; Ambisome; Amphozone; Fungilin; Ampho-Moronal; component of: Mysteclin-F. Antifungal. mp > 170° (dec); λ_m = 406, 382, 363, 345 nm (MeOH); $[\alpha]_D^{24}$ = 333° (acidic DMF), -33.6° (0.1N MeOH/HCl); insoluble in H_2O at pH 6-7, at pH 2 or pH 11, 0.001 g/100 ml); soluble in DMF (0.2 - 0.4 g/100 ml), DMSO (3-4 g/100 ml); LD_{50} (mus ip) = 88 mg/kg, (mus iv) = 4 mg/kg. *Apothecon; Bristol-Myers Squibb Co.; Pharmacia & Upjohn.*

3655 Azaconazole
60207-31-0 262-102-3
$C_{12}H_{11}Cl_2N_3O_2$
1-[[2-(2,4-Dichlorophenyl)-1,3-dioxolan-2-yl]methyl]-1H-1,2,4-triazole.
Azoconazole; R-28644. Antifungal. *Abbott Labs. Inc.*

3656 Azaserine
115-02-6 932 204-061-6
$C_5H_7N_3O_4$
L-Serine diazoacetate (ester).
CI-337; CN-15,757; P-165; NSC-742. Antifungal with antineoplastic activity. mp = 146-162° (dec); $[\alpha]_D^{27.5}$ = -0.5° (H_2O, pH 5.18, c = 8.46); λ_m = 250.5 nm ($E^{1\%}_{1\,cm}$ 1140, pH 7), 252 nm ($E^{1\%}_{1\,cm}$ 1230 0.1N NaOH); soluble in H_2O, less soluble in organic solvents; LD_{50} (mus or. *Parke-Davis.*

3657 Basifungin
127785-64-2
$C_{60}H_{92}N_8O_{11}$
N-[(2R,3R)-2-Hydroxy-3-methylvaleryl]-N-methyl-L-valyl-L-phenylalanyl-N-methyl-L-phenylalanyl-L-prolyl-L-alloisoleucyl-N-methyl-L-valyl-L-leucyl-3-hydroxy-N-methyl-L-valine α_1-lactone.
LY-295337; NK-204; R106-1. Antifungal. *Eli Lilly & Co.*

3658 Becliconazole
112893-26-2
$C_{18}H_{12}Cl_2N_2O$
(±)-1-[o-Chloro-α-(5-chloro-2-benzofuranyl)benzyl]imidazole.
Antifungal agent.

3659 Bifonazole
60628-96-8 1260 262-336-6
$C_{22}H_{18}N_2$
(±)-1-(p,α-Diphenylbenzyl)imidazole.
Mycospor; Bay h 4502; Amycor; Azolmen; Bedriol; Mycosporan. Antifungal. mp = 142°; soluble in EtOPH, MeOH, DMF, DMSO, H_2O (0.1 mg/100 ml at pH 6); LD_{50} (mus orl) = 2629 mg/kgm (rat orl) = 2854 mg/kg. *Bayer AG.*

3660 Biphenamine
3572-52-9 1275 222-686-2
$C_{19}H_{23}NO_3$
2-(Diethylamino)ethyl-2-hydroxy-3-biphenyl-carboxylate .
xenysalate. Local anesthetic; antibacterial; antifungal. Oily liquid; soluble in H_2O. *Wallace Labs.*

3661 Biphenamine Hydrochloride
5560-62-3 1275 226-930-9
$C_{19}H_{24}ClNO_3$
2-(Diethylamino)ethyl-2-hydroxy-3-biphenyl-carboxylate hydrochloride.
Sebaclen; Sebaklen; component of: Alvinine Shampoo. Antibacterial; antifungal; anesthetic (topical). *Wallace Labs.*

3662 Bispyrithione Magsulfex
67182-81-4
$C_{10}H_8MgN_2O_6S_3.3H_2O$
(2,2'-Dithiopyridine 1,1'-dioxide)sulfatomagnesium trihydrate.
Omadine MDS. Antifungal. *Olin Res. Ctr.*

3663 Blastomycin
1362-89-6
Antifungal agent.

3664 Bromosalicylchloranilide
3679-64-9 1457 222-957-5
$C_{13}H_9BrClNO_2$
5-Bromo-N-(4-chlorophenyl)-2-hydroxybenzamide.
Multifungin. Antifungal. mp = 238-243°. *Knoll Pharm. Co.*

3665 Buclosamide
575-74-6 1485 209-390-9
$C_{11}H_{14}ClNO_2$
N-Butyl-4-chlorosalicylamide.
Antifungal. mp = 90-92°. *Hoechst AG.*

3666 Buclosamide combination with Salicylic Acid
75199-98-3 1485
$C_{18}H_{20}ClNO_3$
N-Butyl-4-chlorosalicylamide combination with o-hydroxybenzoic acid.
Jadit. Antifungal. *Hoechst AG.*

3667 Butenafine
101828-21-1 1547
$C_{23}H_{27}N$
N-(p-tert-Butylbenzyl)-N-methyl-1-naphthalenemethylamine.
Squalene epoxidase inhibitor. Antifungal. *Mitsui Toatsu; Penederm Inc.*

3668 Butenafine Hydrochloride
101827-46-7 1547
$C_{23}H_{28}ClN$
N-(p-tert-Butylbenzyl)-N-methyl-1-naphthalenemethylamine hydrochloride.
Mentax; KP-363. Squalene epoxidase inhibitor. Antifungal. mp = 200-202°; slightly soluble in H_2O; very soluble in MeOH, EtOH, CH_2Cl_2, $CHCl_3$. *Mitsui Toatsu; Penederm Inc.*

3669 Butoconazole
64872-76-0 1561
$C_{19}H_{17}Cl_3N_2S$
(±)-1-[4-(p-Chlorophenyl)-2-[(2,6-dichlorophenyl)thio]butyl]imidazole.
Topical antifungal. mp = 68-70.5°. *Syntex Labs. Inc.*

3670 Butoconazole Nitrate
64872-77-1 1561
$C_{19}H_{18}Cl_3N_3O_3S$
(±)-1-[4-(p-Chlorophenyl)-2-[(2,6-dichlorophenyl)thio]butyl]imidazole mononitrate.
Femstat; Gynomyk; RS-35887; RS-35887-00-10-3. Topical antifungal. mp = 162-163°; LD_{50} (musorl) > 3200 mg/kg, (mrat orl) > 3200 mg/kg, (frat orl) = 1720 mg/kg, (mus ip) > 1600 mg/kg, (mrat ip) = 940, (frat ip) = 940 mg/kg. *Syntex Labs. Inc.*

3671 Calcium Propionate
4075-81-4 1745 223-795-8
$C_6H_{10}CaO_4$
Propionic acid calcium salt.
Antifungal. Soluble in H_2O; slightly soluble in MeOH, EtOH; insoluble in organic solvents.

3672 Calcium Undecylenate
$C_{22}H_{38}CaO_4$
Undecylenic acid calcium salt.
calcium 10-undecenoate. Antifungal.

3673 Candicidin
1403-17-4 1789 215-763-7
Vanobid; Levorin; Candimon; NSC-94219. The major of four components of Candicin is Candicidin D (levorin A_2). Topical antifungal. λ_m = 403, 380 ($E^{1\%}_{1\,cm}$ 1150), 360 nm; insoluble in H_2O, most organic solvents; soluble in DMF, DMSO, AcOH; LD_{50} (mus ip) = 14 mg/kg. *Key Pharm.*

3674 Carbol-Fuchsin
8052-17-3
Antifungal, applied as a topical solution.

3675 Chlordantoin
5588-20-5 2130 226-995-3
$C_{11}H_{17}Cl_3N_2O_2S$
5-(1-Ethylpentyl)-3-[(trichloromethyl)thio]hydantoin.
Clodantoin; component of: Sporostacin. Antifungal. *Ortho Pharm. Corp.*

3676 Chlormidazole
3689-76-7 2156 222-998-9
$C_{15}H_{13}ClN_2$
1-p-Chlorobenzyl-2-methylbenzimidazole.
Clomidazole. Antifungal. [monohydrate]: mp = 67-68°.

3677 Chlormidazole Hydrochloride
74298-63-8 2156 277-804-5
$C_{15}H_{14}Cl_2N_2$
1-p-Chlorobenzyl-2-methylbenzimidazole hydrochloride.
H-115; Diamyceline; Futrican. Antifungal. mp = 227-228°.

3678 Chlorphenesin
104-29-0 2230 203-192-6
$C_9H_{11}ClO_3$
3-(4-Chlorophenoxy)-1,2-propanediol.
Adermykon; Mycil. Antifungal. mp = 77-79°; poorly soluble in H_2O (< 1 g/100 ml). *BDH Laboratory Supplies.*

3679 Ciclopirox
29342-05-0 2325 249-577-2
$C_{12}H_{17}NO_2$
6-Cyclohexyl-1-hydroxy-4-methyl-2-(1H)-pyridone.
Loprox; HOE-296b. Antifungal. mp = 144°. *Hoechst Roussel Pharm. Inc.*

3680 Ciclopirox Olamine
41621-49-2 2325 255-464-9
$C_{14}H_{24}N_2O_3$
6-Cyclohexyl-1-hydroxy-4-methyl-2-(1H)-pyridone compound with 2-aminoethanol (2:1).
Loprox; HOE-296; Batrafen; Brumixol; Ciclochem; Micoxolamina; Mycoster; Terit. Antifungal. LD_{50} (mus orl) = 2858 mg/kg, (rat orl) = 3290 mg/kg. *Hoechst Roussel Pharm. Inc.*

3681 Cilofungin
79404-91-4
$C_{49}H_{71}N_7O_{17}$
(4R,5R)-4,5-Dihydroxy-N^2-[p-(octoyloxy)benzoyl]-L-ornithyl-L-threonyl-trans-4-hydroxy-L-prolyl-(S)-4-hydroxy-4-(p-hydroxyphenyl)-L-threonyl-L-threonyl-(3S,4S)-3-hydroxy-4-methyl-L-proline cyclic (6→1)-peptide.
LY-121019. Antifungal. *Eli Lilly & Co.*

3682 Cisconazole
104456-79-3
$C_{19}H_{15}F_3N_2OS$
2-(±)-cis-1-[[3-[(2,6-Difluorobenzyl)oxy]-5-fluoro-2,3-dihydrobenzo[b]thien-2-yl]methyl]imidazole.
Sch-35852. Antifungal. *Schering-Plough HealthCare.*

3683 Climbazole
38083-17-9
$C_{15}H_{17}ClN_2O_2$
1-(p-Chlorophenoxy)-1-imidazol-1-yl-3,3-dimethyl-2-butanone.
Antifungal.

3684 Clotrimazole
23593-75-1 2478 245-764-8
$C_{22}H_{17}ClN_2$
1-(o-Chloro-α,α-diphenylbenzyl)imidazole.
Femcare; Gyne-Lotrimin; Lotrimin; Lotrimin AF Cream; Lotrimin AF Solution; Lotrimin Jock-Itch Cream; Lotrimin Jock-Itch Lotion; Mycelex; Mycelex 7; Mycelex G; Mycelex OTC; Mycelex Troche; Veltrim; BAY 5907; Canesten; Canifug; Empecid; Monobaycuten; Mycofug; Mycosporin; Pedisafe; Rimazole; Tibatin; Trimysten; component of: Lotrimax, Lotrisone, Otomax. Antifungal. mp = 147-149°; slighlty soluble in H_2O, C_6H_6, C_7H_8; soluble in Me_2CO, $CHCl_3$, EtOAc, DMF; LD_{50} (mmus orl) = 923 mg/kg, (rat orl) = 708 mg/kg; [hydrochloride]: mp = 159°. *Bayer AG; Bayer Corp., Ag. Div., Animal Health; C.M. Ind.; Key Pharm.; Lemmon Co.; Schering-Plough HealthCare Products.*

3685 Cloxyquin
130-16-5 2483 204-978-1
C_9H_6ClNO
5-Chloroquinolin-8-ol.
cloxiquine; Chlorisept. Antiseptic; antifungal; antibacterial. A quinoline. mp = 130°; sparingly soluble in cold dilute HCl; [hydrochloride]: mp = 256-258°.

3686 Coparaffinate
8001-60-3 2582
isopar. A mixture of H_2O-insoluble isoparafffinic acids partially neutralized with hydroxybenzyl dialiphatic amines. Marketed as a 17% ointment with 4% TiO_2 in an ointment base. Antifungal. d = 0.970 - 0.980; imiscible with H_2O, soluble in EtOH.

3687 Croconazole
77175-51-0 2429
$C_{18}H_{15}ClN_2O$
[1-[1-[o-[(m-Chlorobenzyl)oxy]phenyl]vinyl]imidazole.
croconazole. Antifungal. mp = 72-73°; soluble in EtOAC. *Shionogi & Co. Ltd.*

3688 Croconazole Monohydrochloride
77174-66-4 2429
$C_{18}H_{16}Cl_2N_2O$
[1-[1-[o-[(m-Chlorobenzyl)oxy]phenyl]vinyl]imidazole hydrochloride.
croconazole hydrochloride; 710674-S; Pilzcin. Antifungal. mp = 148.5-150°; LD_{50} (rat sc) = 7000 mg/kg, (rat orl) = 2500 mg/kg. *Shionogi & Co. Ltd.*

3689 Cuprimyxin
28069-65-0
$C_{26}H_{18}CuN_4O_8$
Bis(6-methoxy-1-phenazinol 5,10-dioxidato)copper.
Unitop; Ro-7-4488/1. Antifungal. *Hoffmann-LaRoche Inc.*

3690 Democonazole
70161-09-0
$C_{19}H_{15}Cl_3N_2O_2$
(E)-1-[2,4-Dichloro-β-[2-(p-chlorophenoxy)ethoxy]styryl]imidazole.
Antifungal.

3691 Denofungin
11056-13-6
U-28009. Antibiotic produced by *Streptomyces hygroscopicus* vairant. Antibacterial and antifungal. *Pharmacia & Upjohn.*

3692 Dermostatin
11120-15-3 2960
$C_{41}H_{66}O_{11}$
Dermostatin.
Virdofulvin, Dermastatin. Antifungal. Mixture of dermostatin A (43%) and dermostatin B (57%). %). A polyene antibiotic produced by *streptomyces viridogriseus*. Sinters at 180°, darkens at 200°; $[\alpha]_D$ = -82° (c = 0.2 MeOH); λ_m = 383, 282, 223 nm ($E^{1\%}_{1\ cm}$ 1000, 212, 130); soluble in aq. MeOH; [acetate]: mp = 146-147°; $[\alpha]^{23}_D$= -59.8° (c = 1.37 $CHCl_3$).

3693 Diamthazole Dihydrochloride
136-96-9 3033 205-270-5
$C_{15}H_{24}ClN_3OS$
6-(2-Diethylaminoethoxy)-2-dimethylaminobenzothiazole dihydrochloride.
Dimazole dihydrochloride; Asterol dihydrochloride; Atelor; Ro-2-2453. Antifungal. mp = 240-243°; freely soluble in H_2O, MeOH, EtOH. *Hoffmann-LaRoche Inc.*

3694 Dipyrithione
3696-28-4 223-024-5
$C_{10}H_8N_2O_2S_2$
2,2'-Dithiodipyridine 1,1'-dioxide.
OMDS. Antibacterial; antifungal. mp = 205°.

3695 Doconazole
59831-63-9
$C_{26}H_{22}Cl_2N_2O_3$
cis-1-[[4-[(4-Biphenyloxy)methyl]-2-(2,4-dichlorophenyl)-1,3-dioxolan-2-yl]methyl]imidazole.
R-34000. Antifungal. *Abbott Laboratories Inc.*

3696 Ebselen
60940-34-3
$C_{13}H_9NOSe$
2-Phenyl-1,2-benzisoselenazolin-3-one.
Antifungal; antioxidant.

3697 Econazole
27220-47-9 3550 248-341-6
$C_{18}H_{15}Cl_3N_2O$
1-[2,4-Dichloro-β-[(p-chlorobenzyl)oxy]phenethyl]imidazole.
Antifungal. mp = 86.8°. *Abbott Laboratories Inc.; IC.M. Ind.*

3698 Econazole Nitrate
24169-02-6 3550 246-053-5
$C_{18}H_{16}Cl_3N_3O_4$
1-[2,4-Dichloro-β-[(p-chlorobenzyl)oxy]phenethyl]-imidazole nitrate.
R-14827; Epi-Pevaryl; Gyno-Pevaril; Ifenec; Micofugal; Micogin; Palavale; Pargin; Pevaryl; Spectazole. Antifungal. mp = 162°; slightly soluble in H_2O, common organic solvents; LD_{50} (mus orl) = 462.7 mg/kg, (rat orl) = 667.7 mg/kg. *Abbott Laboratories Inc.*

3699 Econazole Nitrate (±)
68797-31-9 3550 272-295-6
$C_{18}H_{16}Cl_3N_3O_4$
(±)-1-[2,4-Dichloro-β-[(p-chlorobenzyl)oxy]phenethyl]-imidazole nitrate.
Ecostatin; Spectazole; SQ-13050. Antifungal. *Bristol-Myers Squibb Co.; C.M. Ind.; Ortho Pharm. Corp.*

3700 Enilconazole
35554-44-0 3622 252-615-0
$C_{14}H_{14}Cl_2N_2O$
(±)-1-[β-(Allyloxy)-2,4-dichlorophenethyl]imidazole.
R-23979; imazalil; Clinafarm; Imaverol. Antifungal. Insoluble in H_2O, slightly soluble in organic solvents. *Abbott Laboratories Inc.*

3701 Ethonam Nitrate
15037-55-5
$C_{16}H_{19}N_3O_5$
Ethyl 1-(1,2,3,4-tetrahydro-1-naphthyl)imidazole-5-carboxylate mononitrate.
R-10.100. Antifungal. *Abbott Laboratories Inc.*

3702 Exalamide
53370-90-4 3955 258-504-3
$C_{13}H_{19}NO_2$
o-(Hexyloxy)benzamide.
HBA; Hyperan. Antifungal (topical). Derivative of salicylamide. mp = 71°; soluble in MeOH, Me_2CO, $CHCl_3$, C_6H_6; slightly soluble in Et_2O; nearly insoluble in H_2O.

3703 Fenticonazole
72479-26-6 4047
$C_{24}H_{20}Cl_2N_2OS$
(±)-1-[2,4-Dichloro-β-[[p-(phenylthio)benzyl]oxy]phenethyl]imidazole.
Antifungal. *Recordati Corp.*

3704 Fenticonazole Nitrate
73151-29-8 4047 277-302-6
$C_{24}H_{21}Cl_2N_3O_4S$
(±)-1-[2,4-Dichloro-β-[[p-(phenylthio)benzyl]oxy]phenethyl]imidazole mononitrate.
Rec-15/1476; Falvin; Fentiderm; Lomexin. Antifungal. mp = 136°; λ_m = 252 nm (ε 13894); soluble in H_2O (< 0.01 g/100 ml), Et_2O (< 0.01 g/100 ml), EtOH (3 g/100 ml), MeOH (10 g/100 ml), $CHCl_3$ (30 g/100 ml), DMF (60 g/100 ml); LD_{50} (mus ip) = 1191 mg/kg, (mrat ip) = 440 mg/kg, (frat ip) = 309 mg/kg, (mus, rat orl) > 3000 mg/kg. *Recordati Corp.*

3705 Filipin
11078-21-0 4123
$C_{35}H_{58}O_{11}$
4,6,8,10,12,14,16,27-Octahydroxy-3-(1-hydroxyethyl)-17,28-dimethyloxacyclooctacosa-17,19,21,23,25-pentaen-2-one.
U-5956; NSC-3364. Polyene antibiotic complex containing mainly Filipin II, Filipin III and Filipin IV. Antifungal. mp = 195-205°; $[\alpha]_D^{22}$ = -148.3° (c = 0.89 MeOH); λ_m = 322, 338, 355 nm ($E_{1\,cm}^{1\%}$ 910, 1360, 1330 MeOH); freely soluble in DMG, C_5H_5N; soluble in MeOH, EtOH, BuOH, iPrOH, Et_2O, AcOH; insoluble in H_2O, $CHCl_3$.

3706 Filipin III
480-49-9 4123
$C_{35}H_{58}O_{11}$
4,6,8,10,12,14,16,27-Octahydroxy-3-(1-hydroxyethyl)-17,28-dimethyloxacyclooctacosa-17,19,21,23,25-pentaen-2-one.
15-deoxylagosin. Major constituent of Filipin; isomeric with Filipin IV. Antifungal. mp = 163-180°; $[\alpha]_D^{25}$ = -245° (c = 0.8 DMF); λ_m = 243, 308, 321, 337, 354 nm ($E_{1\,cm}^{1\%}$ 62, 413, 851, 1368, 1343, MeOH).

3707 Fluconazole
86386-73-4 4158
$C_{13}H_{12}F_2N_6O$
2,4-Difluoro-α,α-bis(1H-1,2,4-triazol-1-ylmethyl)benzyl alcohol.
Diflucan; UK-49858; Biozolene; Elazor; Triflucan. Antifungal. mp = 138-140°. *Roerig Div. Pfizer Pharm.*

3708 Flucytosine
2022-85-7 4161 217-968-7
$C_4H_4FN_3O$
4-Amino-5-fluoro-2(1H)-pyrimidinone.
5-fluorocytosine; Ancobon; Ro-2-9915; Ancotil; Alcobon. Antifungal. mp = 295-297°; λ_m = 285 nm (ε 8900 0.1N HCl); soluble in H_2O (1.5 g/100 mlat 25°); LD_{50} (mus orl, sc) > 2000 mg/kg, (mus ip) = 1190 mg/kg, (mus iv) = 500 mg/kg. *Hoffmann-LaRoche Inc.*

3709 Flutrimazole
119006-77-8 4247
$C_{22}H_{16}F_2N_2$
1-[o-Fluoro-α-(p-fluorophenyl)-α-phenylbenzyl]imidazole.
UR-4056. Antifungal. mp = 164-167°, 161-163°; LD_{50} (mmus orl) > 1000 mg/kg, (mmus ip) > 2000 mg/kg, (fmus orl) > 1000 mg/kg, (fmus ip) > 2000 mg/kg, mrat orl) = 808 mg/kg, (mrat ip) = 1079 mg/kg, (frat orl) = 1214 mg/kg, (frat ip) = 1446 mg/kg. *Uriach.*

3710 Fungichromin
6834-98-6 4312 229-913-4
$C_{35}H_{58}O_{12}$
15-Hydroxyfilipin III.
Antibiotic A 246; cogomycin; lagosin; pentamycin; Cantricin. Polyene macrolide antibiotic, related to Filipin. Antifungal. mp = 157-162°; $[\alpha]_D^{20}$ = -227.7° (c = 0.53 DMF); λ_m = 357, 338, 322 nm ($E_{1\,cm}^{1\%}$ 1231, 1250, 786, MeOH); LD_{50} (mus orl) = 1624 mg/kg, (mus ip) = 33.3 mg/kg.

3711 Fungimycin
1404-87-1
NC-1968; WX-2412. Antibiotic produced by *Streptomyces coelicolo var. aminopholus.* Antifungal. *Parke-Davis.*

3712 Griseofulvin
126-07-8 4571 204-767-4
$C_{17}H_{17}ClO_6$
7-Chloro-2',4,6-trimethoxy-6'β-methylspiro[benzofuran-2(3H),1'-[2-cyclohexene]-3,4'-dione.
Fulvicin Bolus; Fulvicin-P/G; Fulvicin-U/F; Grifulvin V; Grisactin; Gris-PEG; Fulvidex; Grysio; amudane; Curling factor; Fulcin; Fulvicin; Grifulvin; Griséfulin; Grisovin; Lamoryl; Likuden; Neo-Fulcin; Polygris; Poncyl-FP; Spirofulvin; Sporostatin. Antifungal. mp = 220°; $[\alpha]_D^{17}$ = 370° ($CHCl_3$, saturated); λ_m = 286, 325 nm; soluble in DMF (12-14 g/100 ml); slightly soluble in MeOH, EtOH, Me_2CO, C_6H_6, $CHCl_3$, EtOAc, AcOH; insoluble in H_2O, petroleum ether. *Allergan Herbert; Ortho Pharm. Corp.; Schering-Plough Animal Health; Schering-Plough HealthCare Products; Wyeth-Ayerst Labs.*

3713 Hachimycin
1394-02-1 4616
Trichomycin; Cabimicina; Trichonat. Heptaene macrolide antibiotic produced by *Streptomyces hachijoensis.* Antifungal and antiprotozoal against Trichomonas. Yellow crystals; forms a water-soluble sodium salt; LD_{50} (mus ip) = 5 mg/kg.

3714 Halethazole
15599-36-7 4622
$C_{19}H_{21}ClN_2OS$
5-Chloro-2-[p-(2-diethylaminoethoxy)phenyl]-benzothiazole.
haletazole; Episol. Antiseptic and antifungal. mp = 93-94°; [citrate]: mp = 167°. *Crookes Healthcare Ltd.*

3715 Hamycin
1403-71-0 4641
Primamycin. Antibiotic produced by *Streptomyces pimprina.* Antifungal. Dec 160°; $[\alpha]_D^{25}$ = 216°; λ_m = 383 nm ($E_{1\,cm}^{1\%}$ 916 MeOH); soluble in C_5H_5N, collidine; insoluble in H_2O and other organic solvents; LD_{50} (mus iv) = 6.16 mg/kg, 1.20 mg/kg; [Hamycin A]: mp > 300°; $[\alpha]_D^{24}$ = 181.1° (c = 0.6 DMF); λ_m = 380 nm ($E_{1\,cm}^{1\%}$ 989). *Hindustan Antibiotics Ltd.*

3716 Hexetidine
141-94-6 4741 205-513-5
$C_{21}H_{45}N_3$
5-Amino-1,3-bis(2-ethylhexyl)hexahydro-5-methylpyrimidine.
Sterisil; Glypesin; Hexigel; Hexocil; Hexoral; Hextril; Oraldene; Steri/Sol. Antifungal. Liquid; d_{20}^{20} = 0.8889; $bp_{0.4}$ = 160°; soluble in petroleum Et_2O, MeOH, C_6H_6, Me_2CO, EtOH, C_6H_{14}, $CHCl_3$. *Parke-Davis.*

3717 Isoconazole
27523-40-6 5176 248-508-3
$C_{18}H_{14}Cl_4N_2O$
1-[2,4-Dichloro-β-[(2,6-dichlorobenzyl)oxy]-phenethyl]imidazole.
Antibacterial and antifungal. *Abbott Laboratories Inc.; C.M. Ind.*

3718 Isoconazole Nitrate
24168-96-5 5176 246-051-4
$C_{18}H_{15}Cl_4N_3O_4$
1-[2,4-Dichloro-β-[(2,6-dichlorobenzyl)oxy]-phenethyl]imidazole nitrate.
R-15454; Fazol; Gyno-Travogen; Travogen; Travogyn. Antibacterial and antifungal. mp = 182-183°. *Abbott Laboratories Inc.; C.M. Ind.*

3719 Itraconazole
84625-61-6 5262
$C_{35}H_{38}Cl_2N_8O_4$
(±)-1-sec-Butyl-4-p-[4-[p-[[(2R*,4S*)-2-(2,4-dichlorophenyl)-2-(1H-1,2,4-triazol-1-ylmethyl)-1,3-dioxolan-4-yl]methoxy]phenyl]-1-piperazinyl]phenyl]-Δ^2-1,2,4-triazolin-5-one.
Sporanox; oriconazole; R-51211; Itrizole; Triasporin. Antifungal. Orally active. Structurally related to ketoconazole. mp = 166.2°; insoluble in H_2O; LD_{50} (14 day) (mus orl) > 320 mg/kg, (rat orl) > 320 mg/kg, (dog orl) > 200 mg/kg. *Abbott Laboratories Inc.*

3720 Kalafungin
11048-15-0
$C_{16}H_{12}O_6$
3,3a,5,11b-Tetrahydro-7-hydroxy-5-methyl-2H-furo[3,2-b]naphtho[2,3-d]pyran-2,6,11-trione.
U-19718; NSC-137443. Antibiotic produced by *Streptomyces tanashiensis strain kala.* Antifungal.

3721 Ketoconazole
65277-42-1 5313 265-667-4
$C_{26}H_{28}Cl_2N_4O_4$
(±)-cis-1-Acetyl-4-[p-[[2-(2,4-dichlorophenyl)-2-(imidazol-1-ylmethyl)-1,3-dioxolan-4-yl]methoxy]phenyl]piperazine.
Nizoral, R-41400; Ketoisdin; Fungarest; Fungoral; Ketoderm; Orifungal M; Panfungol. Antifungal. Orally active, broad-spectrum antimycotic. mp = 146°; LD_{50} (mus iv) = 44 mg/kg, (mus orl) = 702 mg/kg, (rat iv) = 86 mg/kg, (rat orl) = 227 mg/kg, (gpg iv) = 28 mg/kg, (gpg

orl) = 202 mg/kg, (dog iv) = 49 mg/kg, (dog orl) = 780 mg/kg. *Abbott Laboratories Inc.*

3722 Lanoconazole
101530-10-3 5370
$C_{14}H_{10}ClN_3S_2$
(±)-α-[(E)-4-(o-Chlorophenyl)-1,3-dithiolan-2-ylidene]imidazole-1-acetonitrile.
latoconazole; TJN-318; NND-318; Astat. Antifungal. mp = 141.5°; LD_{50} (mmus orl) = 3224 mg/kg, (mmus ip) = 2158 mg/kg, (fmus orl) = 2715 mg/kg, (fmus ip) = 1743 mg/kg, (mrat orl) = 993 mg/kg, (mrat ip) = 1655 mg/kg, (frat orl) = 652 mg/kg, (frat ip) = 2596 mg/kg, mus, rat sc) > 5000 mg/kg. *Nihon Nohyaku Co. Ltd.*

3723 Loflucarban
790-69-2 5589 212-336-7
$C_{13}H_9Cl_2FN_2S$
3,5-Dichloro-4'-fluorothiocarbanilide.
Fluonilid. Antifungal. mp = 148°; soluble in ethyl oleate, isopropyl myristate. *Madan.*

3724 Lomofungin
26786-84-5
$C_{15}H_{10}N_2O_6$
Methyl 6-formyl-4,7,9-trihydroxy-1-phenazine-carboxylate.
Antibiotic produced by *Streptomyces lomonodensis var. lomondensis.* Antifungal.

3725 Lucensomycin
13058-67-8 5621 235-950-7
$C_{36}H_{53}NO_{13}$
lucimycin; Antibiotic FI 1163; FI-1163; Etruscomicina; Etruscomycin. Antibiotic produced by *Streptomyces lucensis.* Antifungal. $[\alpha]_D^{20}$ = 296° (C_5H_5N), 50° (MeOH - 0.1N HCl); λ_m = 218, 278, 290, 303, 318 nm ($E^{1\%}_{1\,cm}$ 300, 370, 780, 1170, 1098); insoluble in H_2O, EtOH, non-polar solvents; soluble in C_5H_5N, DMF; LD_{50} (mus orl) = 1163 mg/kg. *Farmitalia Carlo Erba Ltd.*

3726 Lydimycin
10118-85-1
$C_{10}H_{14}N_2O_3S$
5-(Hexahydro)-2-oxo-1H-thieno[3,4-d]imidazol-4-yl-2-pentenoic acid.
U-15965. Antibiotic produced by *Streptomyces lydicus.* Antifungal.

3727 Mepartricin
11121-32-7 5891
mixture (≅ 1:1) of mepartricin A and mepartricin B; SPA-S-160; SN 654; Partricin methyl ester; methylpartricin; Ipertrofan; Orofungin; Tricandil; Tricangine. Antifungal and antiprotozoal (trichomonas); also used to treat benign prostatic hypertrophy. Methyl ester of the heptaene macrolide antibiotic complex, partricin. λ_m = 401, 378, 359, 340 nm; slightly soluble in H_2O, Et_2O, C_6H_6, petroleum ether; soluble in ROH, C_5H_5N, DMF, DMSO, Me_2CO; LD_{50} (mus orl) > 2000 mg/kg, (mus ip) = 200 mg/kg.

3728 Mepartricin A
62534-68-3 5891
$C_{60}H_{88}N_2O_{10}$
40-Demethyl-3,7-dideoxo-3,7-dihydroxy-N^{47}-methyl-5-oxocandicidin D methyl ester cyclic 15,19-hemiacetal gedamycin methyl ester.
Antifungal and antiprotozoal against Trichomonas; also used to treat benign prostatic hypertrophy. mp = 145-149° (dec); λ_m = 400, 377, 357, 339, 287, 240, 234, 204 nm (ε 79326, 92454, 68094, 51685, 14199, 24612, 26505, 16092 MeOH).

3729 Mepartricin B
62534-69-4 5891
$C_{59}H_{86}N_2O_{19}$
SPA-S-222 [as sodium lauryl sulfate complex]; Montricin [as sodium lauryl sulfate complex]. Antifungal and antiprotozoal against Trichomonas; also used to treat benign prostatic hypertrophy. mp = 154-158°; λ_m = 402, 379, 359, 340, 285. 233. 204 nm (ε 81101, 94729, 64171, 41558, 16196, 23835, 21696, MeOH).

3730 Mercurobutol
498-73-7 207-869-7
$C_{10}H_{13}ClHgO$
4-tert-Butyl-2-chloromercuriphenol.
Antifungal. Antifungal agent.

3731 Metacresol
108-39-4 2645 203-577-9
C_7H_8O
3-Hydroxytoluene.
3-methylphenol. Topical antiseptic and antifungal. d_4^{20}= 1.034; mp = 11-12°; bp = 202°; soluble in H_2O (2.5 g/100 ml), EtOH, $CHCl_3$, Et_2O; LD_{50} (rat orl) = 2020 mg/kg.

3732 Metipirox
29342-02-7
$C_7H_9NO_2$
1-Hydroxy-4,6-dimethyl-2(1H)-pyridone.
Antimycotic pyridone derivative.

3733 Miconazole
22916-47-8 6266 245-324-5
$C_{18}H_{14}Cl_4N_2O$
1-[2,4-Dichloro-β-[(2,4-dichlorobenzyl]oxy]phenethyl]imidazole.
Monistat IV. Antifungal. *Abbott Laboratories Inc.; C.M. Ind.*

3734 Miconazole Nitrate
22832-87-7 6266 245-256-6
$C_{18}H_{15}Cl_4N_3O_4$
1-[2,4-Dichloro-β-[(2,4-dichlorobenzyl]oxy]phenethyl]imidazole nitrate.
R-14889; Antifungal Cream; Lotrimin AF Powder; Lotrimin AF Powder Aerosol; Lotrimin AF Spray Liquid; Loptrimin AF Jock-Itch Powder Aerosol; Micatin; Monistat Cream and Suppositories; Monistat-Derm; Zeasorb-AF; Aflorix; Albistat; Andergin; Brentan; Conoderm; Conofite; Daktar; Daktarin; Deralbine; Dermonistat; Epi-Monistat; Florid; Fungiderm; Fungisdin; Gyno-Daktarin; Gyno-Monistat; Miconal; Ecobi; Micotef; Monistat; Prilagin; Vodol. Antifungal. mp = 170.5, 184-185°; [(+)-form nitate]: mp = 135.3°; $[\alpha]_D^{20}$ = 59° (MeOH); [(-)-form nitate]: mp = 135°; $[\alpha]_D^{20}$ = - 58° (MeOH). *Abbott Laboratories Inc.; C.M. Ind.; Carrington Labs Inc.;*

Lemmon Co.; Ortho Pharm. Corp.; Schering-Plough Pharm.; Stiefel Labs Inc.

3735 Monensin
17090-79-8 6329 241-154-0
$C_{36}H_{62}O_{11}$
2-[2-Ethyloctahydro-3'-methyl-5'-[tetrahydro-6-hydroxy-6-(hydroxymethyl)-3,5-dimethyl-2H-pyran-2-yl][2,2'-bifuran-5-yl]]-9-hydroxy-β-methoxy-α,σ,2,8-tetramethyl-1,6-dioxapsiro[4.5]decan-7-butanoic acid.
A-3823A; 63714; monensic acid. Antibiotic produced by *Streptomyces cinnamonensis*. Antifungal; antiprotozoal. mp = 103-105°; $[\alpha]_D$ = 47.7°; slightly soluble in H_2O, more soluble in hydrocarbons, very soluble in organic solvents; LD_{50} (mus orl) = 43.8 ± 5.2 mg/kg, (chicks orl) = 284 ± 47 mg/kg. *Eli Lilly & Co.*

3736 Monensin Sodium
22373-78-0 6329 244-941-7
$C_{36}H_{61}NaO_{11}$
Sodium 2-[2-ethyloctahydro-3'-methyl-5'-[tetrahydro-6-hydroxy-6-(hydroxymethyl)-3,5-dimethyl-2H-pyran-2-yl][2,2'-bifuran-5-yl]]-9-hydroxy-β-methoxy-α,σ,2,8-tetramethyl-1,6-dioxapsiro[4.5]decan-7-butanoate.
Rumensin; Romensin; Coban. Antibiotic produced by *Streptomyces cinnamonensis*. Antifungal; antiprotozoal. mp = 267-269°; $[\alpha]_D$ = 57.3°; slightly soluble in H_2O, soluble in hydrocarbons, organic solvents. *Eli Lilly & Co.*

3737 Naftifine
65472-88-0 6442
$C_{21}H_{21}N$
(E)-N-Cinnamyl-N-methyl-1-naphthalenemethylamine.
Exoderil; Naftin; Naftifungin. Antifungal. $bp_{0.015}$ = 162-167°. *Allergan Herbert; C.M. Ind.; Sandoz Pharm. Corp.*

3738 Naftifine Hydrochloride
65473-14-5 6442
$C_{21}H_{22}ClN$
(E)-N-Cinnamyl-N-methyl-1-naphthalenemethylamine hydrochloride.
AW-105-843; SN-105-843; Exoderil; Naftin. Antifungal. mp = 177°. *Allergan Herbert; C.M. Ind.; Sandoz Pharm. Corp.*

3739 Natamycin
7681-93-8 6513 231-683-5
$C_{33}H_{47}NO_{13}$
22-[(3-Amino-3,6-dideoxy-β-D-mannopyranosyl)oxy]-1,3,26-trihydroxy-12-methyl-10-oxo-6,11,28-trioxatricyclo[22.3.1.$0^{5,7}$]-octacosa-8,14,16,18,20-pentaene-25-carboxylic acid.
Natacyn; CL-12625; Antibiotic A-5283; Pimaricin; tennecetin; Mycophyt; Myuprozine; Pimafucin; Synogil. Antifungal. Dec 280-300°; $[\alpha]_D^{20}$ = 278° (c = 1 AcOH); λ_m = 220, 280, 290, 303, 318 nm (ε 21300, 26630, 52930, 83220, 76230, MeOH/AcOH); insoluble in most organic solvents; LD_{50} (mrat orl) = 2730 mg/kg, (frat orl) = 4670 mg/kg. *Alcon Labs; Am. Cyanamid.*

3740 Neomycin Undecylenate
1406-04-8 6543 215-793-0
Neomycin undecenate.
Neodecyllin; component of: Otalgine. Aminoglycoside antibiotic; antifungal. *Penick; Purdue Pharma L.P.*

3741 Neticonazole
130726-68-0
$C_{17}H_{22}N_2OS$
(E)-1-[(2-(Methylthio)-1-[o-(pentyloxy)phenyl]vinyl]-imidazole.
SS-717. Antifungal.

3742 Nifuratel
4936-47-4 6623 225-576-2
$C_{10}H_{11}N_3O_5S$
5-[(Methylthio)methyl]-3-[(5-nitrofurfurylidene)amino]-2-oxazolidinone.
Macmiror; Magmilor; Pilmiror; Tydantil; methylmercadone; Inimur; Omnes. Antibacterial; antiprotozoal (against Trichomonas); antifungal. mp = 182°. *Polichimica Sap.*

3743 Nifurmerone
5579-95-3
$C_6H_4ClNO_4$
Chloromethyl 5-nitro-2-furyl ketone.
NF-71. Antibacterial. Antifungal.

3744 Nitralamine
71872-90-7
$C_{10}H_{13}ClN_2O_2S$
2-[[o-Chloro-α-(nitromethyl)benzyl]thio]-ethylamine.
Antifungal. *Searle G.D. & Co.*

3745 Nitralamine Hydrochloride
1432-75-3
$C_{10}H_{14}Cl_2N_2O_2S$
2-[[o-Chloro-α-(nitromethyl)benzyl]thio]-ethylamine monohydrochloride.
SC-12350. Antifungal. *Searle G.D. & Co.*

3746 Nystatin
1400-61-9 6834 215-749-0
Mycostatin; Mycostatin Pastilles; Nystex; O-V Statin; Fungicidin; Biofanal; Diastatin; Candex; Candio-Hermal; Moronal; Nystavescent; component of: Mycolog II, Myco-Triacet II, Mytrex, Nystaform, Nystaform HC, Panolog Cream, Terrastatin. Mixture of Nystatin A_1, A_3 and A_3, biologically active polyene antibiotics. Produced by *Streptomyces* spp. Antifungal. Dec 250°; $[\alpha]_D^{25}$ = -10° (AcOH), 21° (C_5H_5N), 12° (DMF), -7° (0.1N HCl/EtOH); λ_m = 290, 307, 322 nm; soluble in H_2O (0.4 g/100 ml), MeOH (1.12 g/100 ml), EtOH (0.12 g/100 ml), CCl_4 (0.123 g/100 ml), $CHCl_3$ (0.048 g/100 ml), C_6H_6 (0.028 g/100 ml), ethylene glycol (0.875 g/100 ml); LD_{50} (mus ip) ≅ 200 mg/kg. *Apothecon; Bayer AG; Bristol-Myers Oncology; Bristol-Myers Squibb Co.; Lederle Labs.; Lemmon Co.; Pfizer Inc.; Savage Labs; Solvay Animal Health Inc.*

3747 Nystatin A_1
34786-70-4 6834
$C_{46}H_{75}NO_{17}$
Major component of Nystatin. Antifungal. *Apothecon; Bayer AG; Bristol-Myers Oncology; Bristol-Myers Squibb Co.; Lederle Labs.; Lemmon Co.; Pfizer Inc.; Savage Labs; Solvay Animal Health Inc.*

3748 Octanoic Acid
124-07-2 1808 204-677-5
$C_8H_{16}O_2$
n-Octanoic acid.
caprylic acid. Antifungal. mp = 6.7°; bp = 239.7°; d_4^{20} = 0.910; slightly soluble in H_2O (0.068 g/100 ml); freely soluble in EtOH, $CHCl_3$, Et_2O, CS_2, AcOH, petroleum ether; LD_{50} (rat orl) = 10080 mg/kg. *Crookes Healthcare Ltd.; Standard Oil Co. Indiana.*

3749 Oligomycin A
579-13-5 6970 209-437-3
$C_{45}H_{74}O_{11}$
One of four major components of the macrolide antibiotic complex produced by *Streptomyces diastatochromogenes*. Possibly antifungal. mp = 140-141° and 150-151°; λ_m = 225 nm (ε 20,000 EtOH); soluble in H_2O (0.002 g/100 ml), Et_2O (28 g/100 ml), C_6H_6 (6 g/100 ml), petroleum ether (0.02 g/100 ml), EtOH (25 g/100 ml), AcOH (37.5 g/100 ml), Me_2CO (85 g/100 ml). *Wisconsin Alumni Res. Foundation.*

3750 Oligomycin B
11050-94-5 6970 234-275-5
$C_{45}H_{72}O_{12}$
28-Oxooligomycin A.
One of four major components of the macrolide antibiotic complex produced by *Streptomyces diastatochromogenes*. Possibly antifungal. *Wisconsin Alumni Res. Foundation.*

3751 Oligomycin C
11052-72-5 6970 234-276-0
$C_{45}H_{74}O_{10}$
12-Deoxyoligomycin A.
One of four major components of the macrolide antibiotic complex produced by *Streptomyces diastatochromogenes*. Possibly antifungal. *Wisconsin Alumni Res. Foundation.*

3752 Oligomycin D
1404-59-7 6970
$C_{44}H_{72}O_{11}$
26-Demethyloligomycin A.
rutamycin; A-272; RR-32705. A component of the macrolide antibiotic complex produced by *Streptomyces* spp. Antifungal. mp = 116-119°; $[\alpha]_D^{20}$ = -62° (c = 1.36 $CHCl_3$). *Wisconsin Alumni Res. Foundation.*

3753 Omoconazole
74512-12-2 6978
$C_{20}H_{17}Cl_3N_2O_2$
(Z)-1-[2,4-Dichloro-β-[2-(p-chlorophenoxy)ethoxy]-α-methylstyryl]imidazole .
CM-8282. Antifungal. mp = 89-90°. *Siegfried AG.*

3754 Omoconazole Nitrate
83621-06-1 6978
$C_{20}H_{18}Cl_3N_3O_5$
(Z)-1-[2,4-Dichloro-β-[2-(p-chlorophenoxy)ethoxy]-α-methylstyryl]imidazole mononitrate.
10 80 07; Sgd-12878; Fangorex; Fongarex. Antifungal. mp = 118-120°, 122.5°. *Siegfried AG.*

3755 Orconazole
66778-37-8
$C_{18}H_{15}Cl_3N_2O$
(±)-1-[p-Chloro-β-[(2,6-dichlorobenzyl)oxy]phenethyl]-imidazole.
Antifungal. *Abbott Laboratories Inc.*

3756 Orconazole Nitrate
66778-38-9
$C_{18}H_{16}Cl_3N_3O_4$
(±)-1-[p-Chloro-β-[(2,6-dichlorobenzyl)oxy]phenethyl]-imidazole mononitrate.
R-15556. Antifungal. *Abbott Laboratories Inc.*

3757 Oxiconazole Nitrate
64211-46-7 7071 264-730-3
$C_{18}H_{14}Cl_4N_4O_4$
2',4'-Dichloro-2-imidazol-1-ylacetophenone (Z)-[O-(2,4-dichlorobenzyl)oxime] mononitrate.
Oxistat; Sgd-301-76; Ro-13-8996; ST-813; Gyno-Myfungar; Myfungar; Oceral; Oxistat. Antifungal. mp = 137-138°. *Glaxo Wellcome plc; Siegfried AG.*

3758 Oxifungin
64057-48-3
$C_{13}H_{12}N_4O$
1,2-Dihydro-3-(phenoxymethyl)pyrido[3,4-e]as-triazine.
Antifungal.

3759 Oxifungin Hydrochloride
55242-74-5
$C_{13}H_{13}ClN_4O$
1,2-Dihydro-3-(phenoxymethyl)pyrido[3,4-e]as-triazine monohydrochloride.
EU-3421. Antifungal.

3760 Parconazole
61400-59-7
$C_{17}H_{16}Cl_2N_2O_3$
cis-1-[[2-(2,4-Dichlorophenyl)-4-[(2-propynyloxy)-methyl]-1,3-dioxolan-2-yl]methyl]imidazole.
Antifungal. *Abbott Laboratories Inc.*

3761 Parconazole Hydrochloride
62973-77-7 263-777-7
$C_{17}H_{17}Cl_3N_2O_3$
cis-1-[[2-(2,4-Dichlorophenyl)-4-[(2-propynyloxy)-methyl]-1,3-dioxolan-2-yl]methyl]imidazole monohydrochloride.
R-39500. Antifungal. *Abbott Laboratories Inc.*

3762 Pecilocin
19504-77-9 7193 243-116-9
$C_{17}H_{25}NO_3$
[R-(E,E,E)]-1-(8-Hydroxy-6-methyl-1-oxo-2,4,6-dodecatrienyl)-2-pyrrolidone.
Supral; Variotin. Antibiotic obtained from *Paecyllomyces varioti Banier var. antibioticus*. Antifungal. $[\alpha]_D^{28}$ = -5.68° (MeOH); freely soluble in MeOH, EtOH, Me_2CO, EtOAc, C_6H_6, Et_2O, $CHCl_3$, C_5H_5N, dioxane, AcOH, slightly soluble in H_2O, petroleum ether; λ_m = 318, 324 nm ($E_{1\,cm}^{1\%}$ 1198 MeOH); [monohydrate]: mp = 41.5-42.5°; λ_m = 320 nm (ε 46000). *Nippon Kayaku Co. Ltd.*

3763 Perimycin
11016-07-2 7310 234-247-2
Aminomycin; fungimycin; WX-2412; NC-1968. Polyene antifungal antibiotic complex isoated from *Streptomyces coelicolo var. aminophilus* and containing, as a major component, Perimycin A ($C_{59}H_{88}N_2O_{17}$). Antifungal. λ_m = 383 nm ($E^{1\%}_{1\ cm}$ 1000 MeOH); soluble in aqueous Me_2CO, C_5H_5N; THF and dioxane, MeOH, DMF, DMSO; insoluble in H_2O, petroleum ether, EtOAc, C_6H_6. *Warner-Lambert.*

3764 Pirtenidine
103923-27-9
$C_{21}H_{38}N_2$
1,4-Dihydro-1-octyl-4-(octylimino)pyridine.
Used in the treatment of candidiasis. May have use in inhibiting dental plaque. Antimicrobial.

3765 Potash Sulfurated
39365-88-3 7762
(K_2S_x)
Mixture of potassium and sulfides.
liver of sulfur; sulfurated potassa; hepar sulfuris. Sulfide source (skin diseases). Decomposes on exposure to air; soluble in H_2O; slightly soluble in alcohol.

3766 Potassium Iodide
7681-11-0 7809 231-659-4
IK
Jodid; Thyro-Block; Thyrojod; component of: Mudrane Tablets, Mudrane-2 Tablets, quadrinal. Antifungal; expectorant; iodine supplement. Used in veterinary medicine (orally) to treat goiter, actinobacillosis, actinomycosis, idodine deficiency, lead or mercury poisoning. d = 3.12; mp = 680°; soluble in H_2O (142.8 g/100 ml at 20°, 200 g/100 ml at 100°), EtOH (4.5 g/100 ml at 20°, 12.5 g/100 ml at 76°), absolute EtOH (1.96 g/100 ml), MeOH (12.5 g/100 ml), Me_2CO (1.3 g/100 ml), glycerol (50 g/100 ml), ethylene glycol (40 g/100 ml); LD_{50} (rat iv) = 285 mg/kg. *ECR Pharm.; Knoll Pharm. Co.; Wallace Labs.*

3767 Proclonol
14088-71-2 237-934-5
$C_{16}H_{14}Cl_2O$
Bis(p-chlorophenyl)cyclopropylmethanol.
R-8284. Antifungal; anthelmintic. *Abbott Laboratories Inc.*

3768 Propionic Acid
79-09-4 8010 201-176-3
$C_3H_6O_2$
Propionic acid.
methylacetic acid; ethylformic acid. Antifungal. mp = -21.5°; bp = 141.1°; miscible with H_2O, soluble in organic solvents; LD_{50} (rat orl) = 4290 mg/kg. *E. I. DuPont de Nemours Inc.*

3769 Pyrithione
1121-30-8 8178 214-328-9
C_5H_5NOS
1-Hydroxy-2(1H)-pyridinethione.
PTO; Omadine. Antifungal. *Olin Mathieson; Procter & Gamble Pharm. Inc.*

3770 Pyrithione Zinc
13463-41-7 8178 236-671-3
$C_{10}H_8N_2O_2S_2Zn$
Bis[1-hydroxy-2(1H)-pyridinethionato]zinc.
zinc pyrithione; zinc pyridinethione; bis(2-pyridylthio)zinc 1,1'-dioxide; zinc omadine; Danex; Sebulon Shampoo; ZNP Bar; Desquaman; component of: Head and Shoulders. Antibacterial; antifungal; antiseborrheic. *Allergan Herbert; Herbert; Lederle Labs.; Olin Res. Ctr.; Stiefel Labs Inc.; Westwood-Squibb Pharm. Inc.*

3771 Pyrrolnitrin
1018-71-9 8202 213-812-7
$C_{10}H_6Cl_2N_2O_2$
3-Chloro-4-(3-chloro-2-nitrophenyl)pyrrole.
52230; NSC-107654; Pyroace; PN. Antifungal. mp = 124.5°; λ_m = 252 nm (ε 7500); slightly soluble in H_2O, petroleum ether; soluble in MeOH, EtOH, BuOH, Me_2CO, EtOAC, C_6H_6, $CHCl_3$, CCl_4, C_5H_5N, AcOH; LD_{50} (rat ip) = 68 mg/kg, (rbt ip) = 105 mg/kg. *Eli Lilly & Co.*

3772 Rilopirox
104153-37-9
$C_{19}H_{16}ClNO_4$
6-[[p-(p-Chlorophenoxy)phenoxy]methyl]-1-hydroxy-4-methyl-2(1H)-pyridone.
Antimycotic hydroxypyridone derivative. Antifungal.

3773 Salicylanilide
87-17-2 8482 201-727-8
$C_{13}H_{11}NO_2$
2-Hydroxy-N-phenylbenzamide.
Salinidol. Topical antifungal. mp = 135.8-136.2°; slightly soluble in H_2O; freely soluble in EtOH, Et_2O, $CHCl_3$, C_6H_6. *Dow Chem. U.S.A.*

3774 Sanguinarium Chloride
5578-73-4
$C_{20}H_{14}ClNO_4$
13-Methyl-[1,3]benzodioxolo[5,6-c]-1,3-dioxolo[4,5-l]-phenthridinium chloride.
Sanguinarine chloride. Antifungal; anti-inflammatory; antimicrobial. *Atrix Labs.*

3775 Saperconazole
110588-57-3 8510
$C_{35}H_{38}F_2N_8O_4$
(±)-1-sec-Butyl-4-[p-[4-[p-[[(2R*,4S*)-2-(2,4-difluorophenyl)-2-(1H-1,2,4-triazol-1-ylmethyl)-1,3-dioxolan-4-yl]methoxy]phenyl]-1-piperazinyl]phenyl]-Δ^2,1,2,4-triazolin-5-one.
R-66905. Antifungal. mp = 189.5°, poorly soluble in H_2O. *Abbott Laboratories Inc.*

3776 Scopafungin
11056-18-1
U-29479; NSC-107041. Antibiotic produced by *Streptomyces hygroscopicus* variant. Antifungal.

3777 Selenium Sulfide
7488-56-4 8580 231-303-8
SeS_2
Selenium disulfide.
Exsel; Seleen; Selsun; Selsun Blue. Antiseborrheic (topi-

cal); antifungal. Insoluble in H_2O; LD_{50} (rat orl) = 138 mg/kg. *Abbott Labs Inc.; Allergan Herbert; Ross Products.*

3778 Sertaconazole
99592-32-2 8610
$C_{20}H_{15}Cl_3N_2OS$
(±)-1-[2,4-Dichloro-β-[(7-chlorobenzo[b]thien-3-yl]-piperidino]ethyl]-2-imidazolidinone.
FI-7045. Antifungal. mp = 146-147°. *Ferrer.*

3779 Sertaconazole Nitrate
99592-39-9 8610
$C_{20}H_{16}Cl_3N_4O_4S$
(±)-1-[2,4-Dichloro-β-[(7-chlorobenzo[b]thien-3-yl]-piperidino]ethyl]-2-imidazolidinone nitrate.
FI-7056; Dermofix; Zalain. Antifungal. mp = 158-160°; soluble in EtOH (1.7 g/100 ml), $CHCl_3$ (1.5 g/100 ml); Me_2CO (0.95 g/100 ml), n-octanol (0.069 g/100 ml); insoluble in H_2O; λ_m = 302.3 ($A^{1\%}_{1\,cm}$ 79.8), 292.9, 260.3 nm. *Ferrer.*

3780 Siccanin
22733-60-4 8630
$C_{22}H_{30}O_3$
(13aS)-1,2,3,4,4aβ,5,6,6a,11bβ,13bβ-Decahydro-4,4,6aβ,9-tetramethyl13H-benzo[a]furo[2,3,4-mn]-xanthen-11-ol.
Tackle. From *Helminthosporium siccans*. Antifungal. mp = 139-140°; $[\alpha]^{20}_D$ = -136° (c = 2 $CHCl_3$); λ_m = 210, 285 nm (ε 45690, 171 EtOH); very soluble in $CHCl_3$, DMF, C_6H_6; soluble in Me_2CO, Et_2O, EtOAc, EtOH; insoluble in H_2O; LD_{50} (mus orl) > 6000 mg/kg, (mus sc,ip) > 3000 mg/kg, (rat orl) > 1000 mg/kg, (rat sc,ip) >600 mg/kg. *Sankyo Co. Ltd.*

3781 Sinefungin
58944-73-3
$C_{15}H_{23}N_7O_5$
6,9-Diamino-1-(6-amino-9H-purin-9-yl)-1,5,6,7,8,9-hexadeoxy-β-D-ribodecafuranuroic acid.
compound 57926. Antibiotic produced by *Streptomyces grisolus*. Antifungal. *Eli Lilly & Co.*

3782 Sodium Propionate
137-40-6 8816 205-290-4
$C_3H_5NaO_2$
Propionic acid sodium salt, anhydrous.
component of: Prophyllin. Antifungal. Soluble in H_2O (100 g/100 ml at 25°, 153.8 g/100 ml at 100°), EtOH (4.16 g/100 ml at 25°). *Rystan Co. Inc.*

3783 Sulbentine
350-12-9 9061 206-497-2
$C_{17}H_{18}N_2S_2$
3,5-Dibenzyltetrahydro-2H-1,3,5-thiadiazine-2-thione.
D 47; Fungiplex; dibenzthione; Refungine. Antifungal. mp = 101-102°.

3784 Sulconazole
61318-90-9 9062
$C_{18}H_{15}Cl_3N_2S$
(±)-1-[2,4-Dichloro-β-[(p-chlorobenzyl)thio]-phenethyl]imidazole.
Antifungal. *C.M. Ind.; Syntex Labs. Inc.*

3785 Sulconazole Nitrate
61318-91-0 9062
$C_{18}H_{16}Cl_3N_3O_3S$
(±)-1-[2,4-Dichloro-β-[(p-chlorobenzyl)yhio]phenethyl]-imidazole mononitrate.
Exelderm; RS-44872; RS-44872-00-10-3; Myk; Sulcosyn. Antifungal. mp = 130.5-132°. *C.M. Ind.; Syntex Labs. Inc.*

3786 Tenonitrozole
3810-35-3 9292 223-282-9
$C_8H_5N_3O_3S_2$
N-(5-Nitro-2-thiazolyl)-2-thiophenecarboxamide.
TC-109; thenitrazole; Atrican; Moniflagon. Antifungal; antiprotozoal against Trichomonas. Synthesized from 2-thenoyl chloride and 2-amino-5-nitrothiazole. mp = 255-256°. *Chantereau.*

3787 Terbinafine
91161-71-6 9299
$C_{21}H_{25}N$
(E)-N-(6,6-Dimethyl-2-hepten-4-ynyl)-N-methyl-1-naphthalenemethylamine.
Lamasil; SF-86-327. Antifungal. [hydrochloride]: mp = 195-198°; LD_{50} (mus orl) = 4000 mg/kg, (mus iv) = 393 mg/kg, (rat orl) = 4000 mg/kg, (rat iv) = 213 mg/kg. *Sandoz Pharm. Corp.*

3788 Terconazole
67915-31-5 9303 267-751-6
$C_{26}H_{31}Cl_2N_5O_3$
cis-1-[p-[[2-(2,4-Dichlorophenyl)-2-(1H-1,2,4-triazol-4-ylmethyl)-1,3-dioxolan-4-yl]methoxy]phenyl]-4-isopropylpiperazine.
Terazol Cream & Suppositories; Triaconazole; R-42470; Fungistat; Gyno-Terazol; Terazol; Tercospor. Antifungal. mp = 126.3°. *Abbott Laboratories Inc.; Ortho Pharm. Corp.*

3789 Thiram
137-26-8 9510 205-286-2
$C_6H_{12}N_2S_4$
Bis(dimethylthiocarbamoyl) disulfide.
Rezifilm; SQ-1489; NSC-1771; TMTD; ENT-987; Thiurad; Thylate; Fernasan; Nomersan; Pomarsol; Tersan; Tuads, Arasan. Antifungal. mp = 155-156°; d = 1.29; insoluble in H_2O; soluble in EtOH, Et_2O (0.2 g/100 ml), Me_2CO (1.2 g/100 ml), C_6H_6 (2.5 g/100 ml); LD_{50} (rat orl) = 640 mg/kg. *Bristol-Myers Squibb Co.*

3790 Ticlatone
70-10-0
C_7H_4ClNOS
6-Chloro-1,2-benzisothiazolin-3-one.
FER-1443. Antibacterial; antifungal.

3791 Tioconazole
65899-73-2 9595 265-973-8
$C_{16}H_{13}Cl_3N_2OS$
1-[2,4-Dichloro-β[(2-chloro-3-thenyl)oxy]phenethyl]imidazole.
Vagistat; Vagistat-1; UK-20349; Fungibacid; Gyno-Trosyd; Trosyd; Trosyl; Zoniden. Antifungal. [hydrochloride ($C_{16}H_{14}Cl_4N_2OS$)]: mp = 168-170°. *Mead Johnson Nutritionals; Pfizer Inc.*

3792 Tolciclate
50838-36-3 9647 256-792-5
$C_{20}H_{21}NOS$
O-(1,2,3,4-Tetrahydro-1,4-methanonaphthalen-6-yl) m,N-dimethylthiocarbanilate.
K 9147; Fungifos; Kilmicen; Tolmicen. Antifungal. mp = 92-94°; soluble in C6H14 (1.49 g/100 ml), n-octanol (2.39 g/100 ml); insoluble in H_2O; LD_{50} (mus orl) = 4000 mg/kg, (rat orl) = 6000 mg/kg, (dog orl) = 5000 mg/kg. *Farmitalia Carlo Erba Ltd.*

3793 Tolindate
27877-51-6 9652
$C_{18}H_{19}NOS$
O-5-Indanyl m,N-dimethylthiocarbanilate.
Dalnate. Antifungal. mp = 94-95°. *U.S. Vitamin.*

3794 Tolnaftate
2398-96-1 9656 219-266-6
$C_{19}H_{17}NOS$
O-2-Naphthyl m,N-dimethylthiocarbanilate.
Aftate; Dr. Scholl's Athlete's Foot Spray; Tinactin; Tritin; Sch-10144; component of: Tinavet; naphthiomate T; Sch-10144; Chinofungin; Dungistop; Hi-Alarzin; Sporiline; Timoped; Tonoftal; Tniaderm. Antifungal. mp = 110.5-111.5°; insoluble in H_2O; sparingly soluble in MeOH, EtOH; soluble in $CHCl_3$ (66 g/100 ml), Me_2CO (12.5 g/100 ml), CCl_4 (10 g/100 ml); LD_{50} (mus orl) > 10000 mg/kg, (mus sc) > 6000 mg/kg, (rat oprl) > 6000 mg/kg, (rat sc) > 4000 mg/kg. *Schering-Plough Pharm.*

3795 Triacetin
102-76-1 9721 203-051-9
$C_9H_{14}O_6$
Glyceryl triacetate.
Enzactin; Fungacetin. Antifungal. d_4^{25} = 1.1562; mp = -78°; bp = 258-260°; bp_{40} = 172°; soluble in H_2O (7.1 g/100 ml); miscible with EtOH, Et_2O, $CHCl_3$; slightly soluble in CS_2; LD_{50} (mus iv) = 1600 ± 81 mg/kg. *Eastman Chem. Co.; Whitehall-Robins.*

3796 Triafungin
55242-77-8
$C_{13}H_{10}N_4$
3-Benzylpyriso[3,4-c]-as-triazine.
EU-3325. Antifungal.

3797 2,4,6-Tribromo-m-cresol
4619-74-3 9742 204-278-6
$C_7H_5Br_3O$
2,4,6-Tribromo-3-methylphenol.
Micatex. Antiseptic; antifungal (topical). Phenol. mp = 84°.

3798 Tubercidin
69-33-0 9936 200-703-4
$C_{11}H_{14}N_4O_4$
7-β-D-Ribofuranosyl-7H-pyrrolo[2,3-d]pyrimidin-4-amine.
7-deazaadenosine; sparsamycin A; U-10071. Antibacterial (tuberculostatic); antifungal; antineoplastic. Antibiotic produced in *Streptomyces tubericidus.* Dec 247-248°; $[\alpha]_D^{17}$ = -67° (50% AcOH); λ_m 270 nm (ε 12,100 0.01N NaOH); soluble in H_2O (0.3 g/100 ml), MeOH (0.5 g/100 ml), EtOH (0.05 g/100 ml); insoluble in Me_2CO, EtOAc, $CHCl_3$, C_6H_6, petroleum ether; LD_{50} (mus iv) 45 mg/kg.

3799 Ujothion
1219-77-8 9978
$C_{12}H_{14}N_2O_2S_2$
5-Benzyldihydro-6-thioxo-1,3,5-thiadiazine-3(4H)-acetic acid.
Antifungal. mp = 152°.

3800 Undecylenic Acid
112-38-9 9983 203-965-8
$C_{11}H_{20}O_2$
10-Undecenoic acid.
Declid; Renselin; Sevinon; component of: Fulvidex. Antifungal. d_4^{24}= 0.9072, d_{25}^{25} = 0.9102, d_{45}^{45} = 0.8993, $d_4^{79.9}$= 0.8653; mp = 24.5°; bp = 275°, bp_{182} = 232-235°, bp_{130} = 230-235°, bp_{100} = 213.5°, bp_{90} = 198-200°, bp_{15} = 168.3°; $bp_{1.0}$ = 131°; insoluble in H_2O; soluble in EtOH, $CHCl_3$, Et_2O; LD_{50} (mus orl) = 8150 mg/kg, (mus ip) = 960 mg/kg. *Schering-Plough Animal Health.*

3801 Valconazole
56097-80-4
$C_{16}H_{18}Cl_2N_2O_2$
(±)-2-(2,4-Dichlorophenoxy)-1-imidazol-1-yl-4,4-dimethyl-3-pentanone.
Broad-spectrum triazole antifungal.

3802 Viridin
3306-52-3 10144 221-987-6
$C_{20}H_{16}O_6$
1β-Hydroxy-2+[b-methoxy-18-norandrosta-5,8,11,13-tetraeno[6,5,4-bc]furan-3,7,17,trione.
Antifungal. mp = 245° (dec), 222-224°, 200-205° (dec); $[\alpha]_D^{19}$= -224°; λ_m = 242, 300 nm (log ε 4.49 4.22); soluble in H_2O, $CHCl_3$; sparingly soluble in CS_2, CCl_4; insoluble in Et_2O; [β-isomer]: mp = 240-245° (dec); $[\alpha]_D^{16}$= -224°; λ_m = 243, 300 nm (log ε 4.45, 4.25).

3803 Viridofulvin
1405-00-1
Antibiotic, antifungal from *Streptomyces viridogriseus.*

3804 Zinc Propionate
557-28-8 10286 209-167-6
$C_6H_{10}O_4Zn$
Propionic acid zinc salt.
Antifungal. Used as fungicide on adhesive tape. Soluble in H_2O (32 g/100 ml at 15°), EtOH (2.8 g/100 ml at 15°, 17.2 g/100 ml at 76°.

3805 Zinc Undecylenate
557-08-4 9983 209-155-0
$C_{22}H_{38}O_4Zn$
Zinc 10-undecenoate.
Antifungal. mp = 115-116°.

3806 Zinoconazole
84697-21-2
$C_{15}H_{11}Cl_3N_4S$
5-Chloro-2-thienylimidazolylmethyl ketone.
Antifungal. *Searle G.D. & Co.*

3807 Zinoconazole Hydrochloride
80168-44-1
$C_{15}H_{12}Cl_4N_4S$
5-Chloro-2-thienylimidazolylmethyl ketone hydrochloride.
SC-38390. Antifungal. *Searle G.D. & Co.*

Antiglaucoma Agents

3808 Acetazolamide
59-66-5 50 200-440-5
$C_4H_6N_4O_3S_2$
N-(5-Sulfamoyl-1,3,4-thiadiazol-2-yl)acetamide.
acetazoleamide; carbonic anhydrase inhibitor 6063; acetamox; atenezol; cidamex; defiltran; diacarb; diamox; didoc; diluran; diureticum-Holzinger; diuriwas; diutazol; donmox; edemox; fonurit; glaupax; glupax; natrionex; nephramid; vetamox (sodium salt). Diuretic; antiglaucoma agent. Carbonic anhydrase inhibitor. mp = 258-259°; sparingly soluble in H_2O; pKa = 7.2. *Lederle Labs.*

3809 Acetazolamide Sodium
1424-27-7 50
$C_4H_6N_4NaO_3S_2$
N-(5-Sulfamoyl-1,3,4-thiadiazol-2-yl)acetamide monosodium salt.
Diamox Parenteral; Vetamox. Carbonic anhydrase inhibitor, used as diuretic, in treatment of glaucoma. *Lederle Labs.*

3810 Alprenoxime Hydrochloride
121009-30-1
$C_{15}H_{23}ClN_2O_2$
1-(o-Allylphenoxy)-3-(isopropylamino)-2-propanone oxime monohydrochloride.
HGP-5; CDDD 1815. Antiglaucoma agent. *Pharmos Corp.*

3811 Befunolol
39552-01-7 1050
$C_{16}H_{21}NO_4$
7-[2-Hydroxy-3-(isopropylamino)propoxy]-2-benzofuranyl methyl ketone.
A β-adrenergic blocker. Used as an antiglaucoma agent. mp = 115°; LD_{50} (mus iv) = 100-105 mg/kg. *Kakenyaku Kako.*

3812 Befunolol Hydrochloride
39543-79-8 1050
$C_{16}H_{22}ClNO_4$
7-[2-Hydroxy-3-(isopropylamino)propoxy]-2-benzofuranyl methyl ketone hydrochloride.
BFE-60; Benfuran; Bentos; Bentox; Glauconex. A β-adrenergic blocker. Used as an antiglaucoma agent. mp = 163°. *Kakenyaku Kako.*

3813 R-(+)-Befunolol Hydrochloride
66685-79-8 1050
$C_{16}H_{22}ClNO_4$
(R)-(+)-7-[2-Hydroxy-3-(isopropylamino)propoxy]-2-benzofuranyl methyl ketone hydrochloride.
A β-adrenergic blocker. Used as an antiglaucoma agent. mp = 151°; $[\alpha]_D$ = 15.3° (c = 1 MeOH). *Kakenyaku Kako.*

3814 (S)-(-)-Befunolol Hydrochloride
66717-59-7 1050
$C_{16}H_{22}ClNO_4$
(S)-(-)-7-[2-Hydroxy-3-(isopropylamino)propoxy]-2-benzofuranyl methyl ketone hydrochloride.
A β-adrenergic blocker. Antiglaucoma agent. mp = 151-152°; $[\alpha]_D$ = -125.5° (c = 1 MeOH). *Kakenyaku Kako.*

3815 Betaxolol
63659-18-7 1229
$C_{18}H_{29}NO_3$
(±)-1-[p-[2-(Cyclopropylmethoxy)ethyl]phenoxy]-3-(isopropylamino)-2-propanol.
Antianginal agent with antihypertensive and antiglaucoma properties. Cardioselective β_1 adrenergic blocker. mp = 70-72°. *Alcon Labs; Synthelabo Pharmacie.*

3816 Betaxolol Hydrochloride
63659-19-8 1229 264-384-3
$C_{18}H_{30}ClNO_3$
(±)-1-[p-[2-(Cyclopropylmethoxy)ethyl]phenoxy]-3-(isopropylamino)-2-propanol hydrochloride.
SLD-212; SL-75.212; Betoptic; Betoptima; Kerlone. Antianginal agent with antihypertensive and antiglaucoma properties. Cardioselective β_1 adrenergic blocker. mp = 116°; LD_{50} (mus orl) = 94 mg/kg, (mus iv) = 37 mg/kg. *Alcon Labs; Synthelabo Pharmacie.*

3817 Brimonidine
59803-98-4 1399
$C_{11}H_{10}BrN_5$
5-Bromo-6-(2-imidazolin-2-ylamino)quinoxaline.
UK-14304; AGN-190342. Adrenergic (ophthalmic); used to treat glaucoma. An α_2-adrenoceptor agonist. mp = 252°. *Allergan Inc.*

3818 Brimonidine Tartrate
70359-46-5 1399
$C_{15}H_{16}BrN_5O_6$
5-Bromo-6-(2-imidazolin-2-ylamino)quinoxaline D-tartrate.
Alphagan; UK-14304-18; AGN 190342-LF. Adrenergic (ophthalmic); used to treat glaucoma. An α_2-adrenoceptor agonist. mp = 207.5°. *Allergan Inc.*

3819 Brinzolamide
138890-62-7
$C_{12}H_{21}N_3O_5S_3$
(R)-4-(Ethylamino)-3,4-dihydro-2-(3-methoxypropyl)-2H-thieno[3,2-e]-1,2-thiazine-6-sulfonamide1,1-dioxide.
AL-4862. antiglaucoma agent. *Alcon Labs.*

3820 Bupranolol
14556-46-8 1521
$C_{14}H_{22}ClNO_2$
1-(tert-Butylamino)-3-[(6-chloro-m-tolyl)-oxy]-2-propanol.
bupranol; Ophtorenin. Antianginal agent with antiarrhythmic, antihypertensive and antiglaucoma properties. A β-adrenergic blocker.

3821 Bupranolol Hydrochloride
15148-80-8 1521 239-208-3
$C_{14}H_{23}Cl_2NO_2$
1-(tert-Butylamino)-3-[(6-chloro-m-tolyl)-oxy]-2-propanol hydrochloride.

B-1312; KL-255; Betadran; Betadrenol; Looser; Panimit. Antianginal agent with antiarrhythmic, antihypertensive and antiglaucoma properties. A β-adrenergic blocker. mp = 220-222°.

3822 Carteolol
51781-06-7 1917
$C_{16}H_{24}N_2O_3$
5-[3-[(1,1-Dimethylethyl)amino]-2-hydroxypropyl]-3,4-dihydro-2(1H)-quinolinone.
Antianginal agent with antiarrhythmic, antihypertensive and antiglaucoma properties. A β-adrenergic blocker. *Abbott Labs Inc.; Otsuka Am. Pharm.*

3823 Carteolol Hydrochloride
51781-21-6 1917 257-415-7
$C_{16}H_{25}ClN_2O_3$
5-[3-[(1,1-Dimethylethyl)amino]-2-hydroxypropyl]-3,4-dihydro-2(1H)-quinolinone hydrochloride.
Carteolol hydrochloride, Abbott 43326; OPC-1085; Arteoptic; Caltidren; Carteol; Cartrol; Endak; Mikelan; Optipress; Tenalet; Tenalin; Teoptic. Antianginal agent with antiarrhythmic, antihypertensive and antiglaucoma properties. A β-adrenergic blocker. mp = 278°; LD_{50} (mrat orl) = 1380 mg/kg, (mrat iv) = 158 mg/kg, (mrat ip) = 380 mg/kg, (mmus orl) = 810 mg/kg, (mmus iv) = 54.5 mg/kg, (mmus ip) = 380 mg/kg. *Abbott Labs Inc.; Otsuka Am. Pharm.*

3824 Colforsin
66575-29-9 266-410-9
$C_{22}H_{34}O_7$
(3R,4aR,5S,6S,6aS,10S,10aR,10bS)-Dodecahydro-5,6,10,10b-tetrahydroxy-3,4a,7,7,10a-pentamethyl-3-vinyl-1H-naphtho[2,1-b]pyran-1-one 5-acetate.
HL-362; L-75-1362B. Antiglaucoma agent. *Hoechst Roussel Pharm. Inc.*

3825 Dapiprazole Hydrochloride
72822-13-0 2884
$C_{19}H_{28}ClN_5$
5,6,7,8-Tetrahydro-3-[2-(4-o-tolyl-1-piperazinyl)ethyl]-s-triazolo[4,3-a]pyridine hydrochloride.
AF-2139; Glamidolo; Reversil; Rev-Eyes. An α_1-adrenergic blocking agent with antiglaucoma and miotic activity. mp = 206-207°; LD_{50} (mus ip) = 260 mg/kg. *Angelini Pharm. Inc.*

3826 Dichlorphenamide
120-97-8 3127 204-440-6
$C_6H_6Cl_2N_2O_4S_2$
4,5-Dichloro-m-benzenedisulfonamide.
Daranide; Antidrasi; Oratrol. Carbonic anhydrase inhibitor used to treat glaucoma. mp = 239-241°, 228.5-229°; insoluble in H_2O, soluble in alklaine solutions. *Merck & Co.Inc.*

3827 Dipivefrin
52365-63-6 3400
$C_{19}H_{29}NO_5$
(±)-3,4-Dihydroxy-α-[(methylamino)methyl]benzyl alcohol 3,4-dipivalate.
dipivalyl epinephrine; DPE. Adrenergic (ophthalmic); used to treat glaucoma. mp = 146-147°. *Alcon Labs; Allergan Inc.*

3828 Dipivefrin Hydrochloride
64019-93-8 3400 264-609-5
$C_{19}H_{30}ClNO_5$
(±)-3,4-Dihydroxy-α-[(methylamino)methyl]benzyl alcohol 3,4-dipivalate hydrochloride.
Diopine; d Epifrin; Diphemin; Pivalephrine; Propine. An α-adrenergic (ophthalmic) agonist. Used as an antiglaucoma agent. mp = 158-159°; soluble in H_2O, EtOH. *Alcon Labs; Allergan Inc.*

3829 Dorzolamide
120279-96-1 3484
$C_{10}H_{16}N_2O_4S_3$
(4S,6S)-4-(Ethylamino)-5,6-dihydro-6-methyl-4H-thieno[2,3-b]thiopyran-2-sulfonamide 7,7-dioxide.
Adrenergic (ophthalmic); a carbonic anhydrase inhibitor, used to treat glaucoma. *Merck & Co.Inc.*

3830 Dorzolamide
120279-96-1 3484
$C_{10}H_{16}N_2O_4S_3$
(4S,6S)-4-(Ethylamino)-5,6-dihydro-6-methyl-4H-thieno[2,3-b]thiopyran-2-sulfonamide 7,7-dioxide.
Adrenergic (ophthalmic); used to treat glaucoma. *Merck & Co.Inc.*

3831 Dorzolamide Hydrochloride
130693-82-2 3484
$C_{10}H_{17}ClN_2O_4S_3$
(4S,6S)-4-(Ethylamino)-5,6-dihydro-6-methyl-4H-thieno[2,3-b]thiopyran-2-sulfonamide 7,7-dioxide hydrochloride.
Trusopt; MK-507; component of: Cosopt. Adrenergic (ophthalmic); a carbonic anhydrase inhibitor, used to treat glaucoma. mp = 283-285°; $[\alpha]_D^{24}$ = -8.34° (c = 1 MeOH); soluble in H_2O. *Merck & Co.Inc.*

3832 Epinephrine
51-43-4 3656 200-098-7
$C_9H_{13}NO_3$
(-)-3,4-Dihydroxy-α-[(methylamino)methyl]benzyl alcohol.
l-methylaminoethanolcatechol; adrenalin; levorenen; Bronkaid Mist; Epifrin; Epiglaufrin; Eppy; Glaucon; Glauposine; Primatene Mist; Simplene; Sus-phrine; Suprarenaline; component of: Citanest Forte. Bronchodilator; cardiostimulant; mydriatic; antiglaucoma. Endogenous catecholamine with combined α- and β-agonist activity. Principal sympathomimetic hormone produced by the adrenal medulla. mp = 211-212°; dec 215°; $[\alpha]_D^{25}$ = -50.0° to -53.5° (in 0.6N HCl); slightly soluble in H_2O, EtOH; soluble in aqueous solutions of mineral acids; insoluble in aqueous solutions of ammonia and alkali carbonates; insoluble in Et_2O, Me_2CO, oils; LD_{50} (mus ip) = 4 mg/kg. *Alcon Labs; Allergan Inc.; Astra Sweden; Bristol-Myers Squibb Co.; CIBA Vision AG; Elkins-Sinn; Evans Medical Ltd.; Parke-Davis; Sterling Health U.S.A.; Whitehall-Robins; Wyeth-Ayerst Labs.*

3833 Epinephrine d-Bitartrate
51-42-3 3656 200-097-1
$C_{13}H_{19}NO_9$
(-)-3,4-Dihydroxy-α-[(methylamino)methyl]benzyl alcohol (+)-tartrate (1:1) salt.
Asmatane Mist; Asthmahaler Epitrate; Bronitin Mist;

Bronkaid Mist Suspension; Epitrate; Medihaler-Epi; Primatene Mist Suspension; Suprarenin; component of: Asthmahaler, E-Pilo. Bronchodilator; cardiostimulant; mydriatic; antiglaucoma. Adrenergic (opthalmic). mp = 147-154° (some decomposition); soluble in H_2O (30 g/100 ml); slightly soluble in EtOH. *3M Pharm.; CIBA Vision Corp.; Menley & James Labs Inc.; Sterling Health U.S.A.; Whitehall-Robins; Wyeth-Ayerst Labs.*

3834 Epinephrine Hydrochloride
55-31-2 3656
$C_9H_{14}ClNO_3$
(-)-3,4-Dihydroxy-α-[(methylamino)methyl]benzyl alcohol hydrochloride.
Adrenalin; Epifrin; Glaucon; Suprarenin. Bronchodilator; cardiostimulant; mydriatic; antiglaucoma. An α-adrenergic agonist. *Alcon Labs; Allergan Inc.; Astra Chem. Ltd.; Bristol-Myers Squibb Co.; CIBA Vision AG; Elkins-Sinn; Forest Pharm. Inc.; Parke-Davis; Sterling Health U.S.A.; Whitehall Labs. Inc.; Wyeth-Ayerst Labs.*

3835 dl-Epinephrine
329-65-7 3656 206-347-6
$C_{20}H_{32}N_2O_6$
(±)-α-[1-(Methylamino)ethyl]benzenemethanol sulfate salt (2:1).
racepinefrine; racepinephrine. An α-adrenergic agonist. Used as a bronchodilator, cardiostimulant, mydriatic and antiglaucoma agent. Slightly soluble in H_2O, EtOH. *Alcon Labs; Allergan Inc.; Astra Chem. Ltd.; Bristol-Myers Squibb Co.; CIBA Vision AG; Elkins-Sinn; Forest Pharm. Inc.; Parke-Davis; Sterling Health U.S.A.; Whitehall Labs. Inc.; Wyeth-Ayerst Labs.*

3836 d-Epinephrine Hydrochloride
329-63-5 3656 206-346-0
$C_{20}H_{33}ClN_2O_6$
(±)-α-[1-(Methylamino)ethyl]benzenemethanol hydrochloride salt.
Asthmanefrin; Vaponefrin. An α-adrenergic agonist. Used as a bronchodilator, cardiostimulant, mydriatic and antiglaucoma agent. mp = 157°, soluble in H_2O; sparingly soluble in EtOH. *Alcon Labs; Allergan Inc.; Astra Chem. Ltd.; Bristol-Myers Squibb Co.; CIBA Vision AG; Elkins-Sinn; Forest Pharm. Inc.; Parke-Davis; Sterling Health U.S.A.; Whitehall Labs. Inc.; Wyeth-Ayerst Labs.*

3837 Falintolol
90581-63-8
$C_{12}H_{24}N_2O_2$
Cyclopropyl methyl ketone (±)-(EZ)-O-[3-(tert-butyl-amino)-2-hydroxypropyl]oxime.
A beta-adrenergic antagonist used for treatment of glaucoma.

3838 Latanoprost
130209-82-4 5387
$C_{26}H_{40}O_5$
Isopropyl (Z)-7-[(1R,2R,3R,5S)-3,5-dihydroxy-2-[(3R)-3-hydroxy-5-phenylpentyl]-cyclopentyl]-6-heptenoate.
Xalatan; PHXA41; XA41. Prostaglandin. Antiglaucoma agent. $[\alpha]_D^{20}$= 31.57° (c = 0.91 CH_3CN). *Kabi Pharmacia Diagnostics; Pharmacia & Upjohn.*

3839 Levobunolol
47141-42-4 5488
$C_{17}H_{25}NO_3$
(-)-5-[3-(tert-Butylamino)-2-hydroxypropoxy]-3,4-dihydro-1(2H)-naphthalenone.
l-bunolol; W-6421A. Adrenergic β-receptor; antiglaucoma agent. LD_{50} (mrat orl) = 700 mg/kg, (mrat iv) = 25 mg/kg, (frat orl) = 800 mg/kg, (frat iv) = 28 mg/kg, (mmus orl) = 1530 mg/kg, (mmus iv) = 78 mg/kg, (fmus orl) = 1220 mg/kg, (fmus iv) = 84 mg/kg, (mhmtr orl) = 435 mg/kg, (fhmtr orl) = 500 mg/kg; (dog orl) = 100 mg/kg. *Allergan Inc.*

3840 Levobunolol Hydrochloride
27912-14-7 5488 248-725-3
$C_{17}H_{26}ClNO_3$
(-)-5-[3-(tert-Butylamino)-2-hydroxypropoxy]-3,4-dihydro-1(2H)-naphthalenone hydrochloride.
Vitagan; Gotensin; Betagan; W-7000A. Adrenergic β-receptor; antiglaucoma agent. mp = 209-211°; $[\alpha]_{598}^{24}$ = -19.6° ± 0.7° (c = 2.90 MeOH); λ_m = 221. 253. 310 nm (ε 24700, 9000. 2400. NaOH). *Allergan Inc.*

3841 Methazolamide
554-57-4 6031 209-066-7
$C_5H_8N_4O_3S_2$
N-(4-Methyl-2-sulfamoyl-Δ^2-1,3,4-thiadiazolin-5-ylidene)acetamide.
Neptazane. Carbonic anhydrase inhibitor, used as diuretic, in treatment of glaucoma. mp = 213-214°; λ_m = 254 nm (log ε 3.66 95% EtOH), 247 nm (log ε3.61 0.1N NaOH). *Lederle Labs.*

3842 Metipranolol
22664-55-7 6221 245-151-5
$C_{17}H_{27}NO_4$
(±)-1-(4-Hydroxy-2,3,5-trimethylphenoxy)-3-(isopropylamino)-2-propanol 4-acetate.
OptiPranolol; BM01.004; VUFB6453; Betamet; methypranol; trimepranol; Betanol; Disorat; Glauline; Glausyn; Turoptin. A β-adrenergic blocker. Antihypertensive agent also used to treat glaucoma. mp = 105-107°, 108.5-110.5°; λ_m = 278, 274 nm ($A_{1\ cm}^{1\%}$ 51.3, 50.5 MeOH); freely soluble in EtOH, $CHCl_3$, C_6H_6; slightly soluble in Et_2O; insoluble in H_2O; LD_{50} (mus iv) = 31 mg/kg. *Bausch & Lomb Vision Care Div.; SPOFA.*

3843 Metipranolol Hydrochloride
36592-77-5 6221
$C_{17}H_{28}ClNO_4$
(±)-1-(4-Hydroxy-2,3,5-trimethylphenoxy)-3-(isopropylamino)-2-propanol 4-acetate hydrochloride.
Betamann; Optipranolol. A β-adrenergic blocker. Antihypertensive; antiglaucoma agent. Soluble in H_2O. *Bausch & Lomb Vision Care Div.; SPOFA.*

3844 Naboctate Hydrochloride
73747-21-4
$C_{33}H_{54}ClNO_3$
(±)-7,8,9,10-Tetrahydro-6,6,9-trimethyl-3-(1-methyloctyl)-6H-dibenzo[b,d]pyran-1-yl 4-(diethylamino)butyrate hydrochloride.
SP-325. Antiglaucoma agent; antiemetic.

3845 Pilocarpine
92-13-7 7578 202-128-4
$C_{11}H_{16}N_2O_2$
3-Ethyldihydro-4-[(1-methyl-1H-imidazol-5-yl)methyl]-2(3H)-furanone.
Ocusert Pilo. Antiglaucoma agent; cholinergic (ophthalmic); miotic. mp = 34°; bp_5 = 260°; $[\alpha]_D^{18}$ = 106° (c = 2); soluble in H_2O, EtOH, $CHCl_3$; insoluble in Et_2O, C_6H_6, petr ether. *Alcon Labs; CIBA Vision AG; Wyeth Labs.*

3846 Pilocarpine Hydrochloride
54-71-7 7578 200-212-5
$C_{11}H_{17}ClN_2O_2$
3-Ethyldihydro-4-[(1-methyl-1H-imidazol-5-yl)methyl]-2(3H)-furanone hydrochloride.
Adsorbocarpine; Almocarpine; Isopto Carpine; component of: E-Pilo, Pilocar, Pilopine HS Gel. Antiglaucoma agent; cholinergic (ophthalmic). Also a miotic. mp = 204-205°; $[\alpha]_D^{18}$ = 91° (c = 2); freely soluble in H_2O, EtOH; insoluble in Et_2O, $CHCl_3$. *Alcon Labs; CIBA Vision AG; Wyeth Labs.*

3847 Pindolol
13523-86-9 7597 236-867-9
$C_{14}H_{20}N_2O_2$
1-(Indol-4-yloxy)-3-(isopropylamino)-2-propanol.
LB-46; Visken; prinodolol; Betapindol; Blocklin L; Calvisken; Decreten; Durapindol; Glauco-Visken; Pectobloc; Pinbetol; Pynastin. Antianginal agent with antiarrhythmic, antihypertensive and antiglaucoma properties. A β-adrenergic blocker. mp = 171-163°. *Salix Pharm. Inc.*

3848 Pirenoxine
1043-21-6 7645 213-872-4
$C_{16}H_8N_2O_5$
1-Hydroxy-5-oxo-5H-pyrido[3,2-a]phenoxazine-3-carboxylic acid.
Clarvisan [as hydrochloride]. Used in the treatment of cataracts. Antioxidant. mp = 247-248° (dec).

3849 Pirnabine
19825-63-9
$C_{19}H_{24}O_3$
7,8,9,10-Tetrahydro-3,6,6,9-tetramethyl-6H-dibenzo[b,d]pyran-1-ol acetate.
SP-304. Antiglaucoma agent.

3850 (±)-Pirnabine
68298-00-0
$C_{19}H_{24}O_3$
(±)-7,8,9,10-Tetrahydro-3,6,6,9-tetramethyl-6H-dibenzo[b,d]pyran-1-ol acetate.
Antiglaucoma agent.

3851 Timolol
91524-16-2 9585
$C_{13}H_{24}N_4O_3S.1/2H_2O$
(S)-1-(tert-Butylamino)-3-[(4-morpholino-1,2,5-thiadiazol-3-yl)oxy]-2-propanol hemihydrate.
Antianginal agent with antiarrhythmic (class II), antihypertensive and antiglaucoma properties. A β-adrenergic blocker. [(±) form]: mp = 71.5-72.5°. *Merck & Co.Inc.*

3852 Timolol Maleate
26921-17-5 9585 248-111-5
$C_{17}H_{28}N_4O_7S$
(S)-1-(tert-Butylamino)-3-[(4-morpholino-1,2,5-thiadiazol-3-yl)oxy]-2-propanol maleate (1:1).
Blocadren; Timoptic; Timoptol; MK-950; Aquanil; Betim; Proflax; Temserin; Tenopt; Timacar; Timacor; component of: Cosopt, Timolide. Antianginal agent with antiarrhythmic (class II), antihypertensive and antiglaucoma properties. A β-adrenergic blocker. mp = 201.5-202.5°; $[\alpha]_{405}^{24}$ = -12.0° (c = 5 1N HCl); $[\alpha]_D^{25}$ = -4.2°; λ_m = 294 nm ($A_{1\ cm}^{1\%}$ 200 0.1N HCl); soluble in EtOH, MeOH; poorly soluble in $CHCl_3$, cyclohexane; insoluble in isooctane, Et_2O. *Merck & Co.Inc.*

3853 Unoprostone
120373-36-6 9984
$C_{22}H_{38}O_5$
(+)-(Z)-7-[(1R,2R,3R,5S)-3,5-Dihydroxy-2-(3-oxodecyl)cyclopentyl]-5-heptenoic acid.
Prostaglandin. Antiglaucoma agent; used in treatent of intraocular hypertension. *Ueno Fine Chem. Industry Ltd.*

3854 Unoprostone Isopropyl Ester
120373-24-2 9984
$C_{25}H_{44}O_5$
Isopropyl (+)-(Z)-7-[(1R,2R,3R,5S)-3,5-dihydroxy-2-(3-oxodecyl)cyclopentyl]-5-heptenoate.
UF-021; Rescula. Prostaglandin. Antiglaucoma agent; used in treatent of intraocular hypertension. *Ueno Fine Chem. Industry Ltd.*

Antigonadotropins

3855 Danazol
17230-88-5 2875 241-270-1
$C_{22}H_{27}NO_2$
17α-Pregna-2,4-dien-20-yno[2,3-d]isoxazol-17-ol.
Danocrine; Chronogyn; Bonzol; Cyclomen; Danol; Danovaol; Ladogal; Winobanin; Win-17757. Anterior pituitary suppressant; anabolic steroid derivative of ethisterone with mild androgenic side effects. An antogonadotropin. mp = 224.4-226.8°; $[\alpha]_D^{25}$ = 7.5° (EtOH), $[\alpha]_D^{25}$ = 21.9° ($CHCl_3$); λ_m = 286 nm (ε 11300 EtOH). *Sterling Winthrop Inc.*

3856 Gestrinone
16320-04-0 4423
$C_{21}H_{24}O_2$
13-Ethyl-17-hydroxy-18,19-dinor-17α-pregna-4,9,11-trien-20-yn-3-one.
R-2323; RU-2323; A-46745; Dimetriose; Dimetrose; Nemestran; Tridomose. Progestogen. An antogonadotropin. mp = 154°; $[\alpha]_D^{20}$ = 84.6° (c = 0.41 MeOH). *Hoechst Roussel Pharm. Inc.*

3857 Paroxypropione
70-70-2 7176 200-743-2
$C_9H_{10}O_2$
4'-Hydroxypropiophenone.
B-360; H 365; NSC-2834; p-propionylphenol; P.O.P.; Profenone; Frenantol; Frenohypon; Paroxon; Possipion; Hypostat. A pituitary antogonadotropin. mp = 149°;

soluble in H_2O (0.0345 g/100 ml at 15°, 3.3 g/100 ml at 100°); freely soluble in EtOH, Et_2O.

3858 Ramorelix
127932-90-5
$C_{74}H_{95}ClN_{16}O_{18}$
1-[N-Acetyl-3-(2-naphthyl)-D-alanyl-p-chloro-D-phenylalanyl-D-tryptophyl-L-seryl-L-tyrosyl-O-(6-deoxy-α-L-mannopyranosyl)-D-seryl-L-leucyl-L-arginyl-L-prolyl]semicarbazide.
Hoe-013. Luteinizing-hormone-releasing hormone antagonist. Antigonadotropin. *Hoechst Roussel Pharm. Inc.*

Antigout Agents

3859 Allopurinol
315-30-0 287 206-250-9
$C_5H_4N_4O$
1H-Pyrazolo[3,4-d]pyrimidin-4-ol.
HPP; BW-15658; NSC-1390; Adenock; Aloral; Alositol; Sllo-puren; Allozym; Allurtal; Anoprolin; Anzief; Apulonga; Apurol; Apurin; Bleminol; Bloxanth; Caplenal; Cellidrin; Cosuric; Dabroson; Embarin; Epidropal; Foligan; Geapur; Gichtex; Hamarin; Hexanurat; Ketanrift; Ketobun-A; Ledopur; Lopurin; Lysuron; Miniplanor; Monarch; Nektrohan; Remid; Riball; Sigapurol; Suspendol; Takanarumin; Urbol; Uricemil; Uripurinol; Urobenyl; Urosin; Urtias; Xanturat; Zyloprim; Zyloric. Xanthine oxidase inhibitor. Used to treat gout. mp > 350°; λ_m = 257 nm (ε 7200 0.1N NaOH), 250 nm (ε 7600 0.1N HCl), 252 nm (ε 7600 MeOH); soluble in H_2O (0.048 g/100 ml), $CHCl_3$ (0.060 g/100 ml), EtOH (0.030 g/100 ml), DMSO (0.46 g/100 ml), n-octanol (< 0.001 g/100 ml). *Glaxo Wellcome Inc.; Knoll Pharm. Co.*

3860 Amflutizol
82114-19-0
$C_{11}H_7F_3N_2O_2S$
4-Amino-3-(α,α,α-trifluoro-m-tolyl)-5-isothiazolecarboxylic acid.
LY-141894. Gout suppressant. *Eli Lilly & Co.*

3861 Carprofen
53716-49-7 1912 258-712-4
$C_{15}H_{12}ClNO_2$
(±)-6-Chloro-α-methylcarbazole-2-acetic acid.
Ro-20-5720/000; C-5720; Imadyl; Rimadyl. Anti-inflammatory, used to treat gout. mp = 197-198°; LD_{50} (mus orl) = 400 mg/kg. *Hoffmann-LaRoche Inc.*

3862 Colchicine
64-86-8 2536 200-598-5
$C_{22}H_{25}NO_6$
[N-(5,6,7,9-Tetrahydro-1,2,3,10-tetramethoxy-9-oxobenzo[a]heptalen-7-yl]acetamide.
component of: Colbenemid. Used to treat gout. mp = 142-150°, 157°; $[\alpha]_D^{17}$ = -429° (c = 1.72), $[\alpha]_D^{17}$ = -121° (c = 0.9 $CHCl_3$); λ_m = 350.5, 243 nm (log ε 4.22 4.47 EtOH); soluble in H_2O (4.5 g/100 ml), Et_2O (0.45 g/100 ml), C_6H_6 (1 g/100 ml); freely soluble in EtOH, $CHCl_3$; insoluble in petrolerum ether; LD_{50} (rat iv) = 1.6 mg/kg, (mus iv) = 4.13 mg/kg. *Merck & Co.Inc.*

3863 Oxypurinol
2465-59-0 219-570-9
$C_5H_4N_4O_2$
1H-Pyazolo[3,4-d]pyrimidine-4,6-diol.
NSC-76239. Xanthine oxidase inhibitor. Used to treat gout.

3864 Probenecid
57-66-9 7934 200-344-3
$C_{13}H_{19}NO_4S$
p-(Dipropylsulfamoyl)benzoic acid.
Benemid; Probecid; Proben; component of: Colbenemid, Polycillin-PRB. Uricosuric. Used to treat gout. mp = 194-196°; λ_m = 242.5 nm (0.1N NaOH); soluble in $CHCl_3$, insoluble in H_2O; LD_{50} (rat orl) = 1600 mg/kg. *Apothecon; Merck & Co.Inc.; Wyeth-Ayerst Labs.*

3865 Sulfinpyrazone
57-96-5 9121 200-357-4
$C_{23}H_{20}N_2O_3S$
1,2-Diphenyl-4-[2-(phenylsulfinyl)ethyl]-3,5-pyrazolidinedione.
G-28315; Anturan; Anturane; Anturano; Enturen. Uricosuric and antithrombotic. Used to treat gout. A nonsteroidal anti-inflammatory agent. Inhibits cyclooxygenase. Prolongs circulating platelet survival. mp = 136-137°; λ_m = 255 nm (1N NaOH); soluble in EtOAc, $CHCl_3$; slightly soluble in H_2O, EtOH, Et_2O, mineral oils; [d-form]: mp = 130-133°; $[\alpha]^2_D$ = 67.1° (c = 2.04 EtOH), $[\alpha]_D^{25}$ = 109.3 (c = 0.5 $CHCl_3$); [l-form]: mp = 130-133°; $[\alpha]_D^{23}$ = -64.2° (c = 2.14 EtOH), $[\alpha]_D^{26}$ = -104.5° (c = 0.5 $CHCl_3$). *Ciba-Geigy Corp.*

Antihistaminics

3866 Acrivastine
87848-99-5 129
$C_{22}H_{24}N_2O_2$
(E)-6-[(E)-3-(1-Pyrrolidinyl)-1-p-tolypropenyl]-2-pyridineacrylic acid.
BW-825C; BW-0270C; BW-A825C; Semprex; component of: Semprex D, Duact. Antihistaminic. Non-sedating histamine H_1-receptor antagonist. Analog of triprolidine. mp = 222° (dec). *Glaxo Wellcome Inc.*

3867 Alloclamide
5486-77-1 263
$C_{16}H_{23}ClN_2O_2$
2-(Allyloxy)-4-chloro-N-[2-(diethylamino)ethyl]-benzamide.
component of: Dégryp [as hydrochloride]. Antitussive; antihistaminic. [hydrochloride] mp = 125-127°; soluble in EtOH; LD_{50} (mus orl)= 740 mg/kg. *C.E.R.M.*

3868 Antazoline
91-75-8 718
$C_{17}H_{19}N_3$
2-[(N-Benzylanilino)methyl]-2-imidazol.
phenazoline; imidamine; 5512-M; Antistine; Antistin; Histostab; Antastan; Antasten; Antihistal; Azalone; Ben-a-hist. Antihistaminic. mp = 120-122°. *Ciba-Geigy Corp.*

3869 Antazoline Hydrochloride
2508-72-7 718 219-719-8
$C_{17}H_{20}ClN_3$
4,5-Dihydro-N-phenyl-N-(phenylmethyl)-1H-imidazole-2-methanamine hydrochloride.
Antistin; Dibistin; 2-(N-benzylanilinomethyl)-2-imidazoline; phenazoline; imidamine; 5512-M; antistine; Histostab; Antastan; Antasten; Antihistal; Azalone; Ben-a-hist; Fenazolina; Histazine. Antihistaminic. mp = 237-241°; λ_m = 242 nm ($E^{1\%}_{1\ cm}$ 495-515); soluble in H_2O (2.5 g/100 ml), more soluble in EtOH, less soluble in organic solvents. *Ciba-Geigy Corp.*

3870 Antazoline Phosphate
154-68-7 718 205-831-4
$C_{17}H_{19}N_3.H_3PO_4$
4,5-Bihydro-N-phenyl-N-(phenylmethyl))-1H-imidazole-2-methanamine phosphate.
Vasocon-A. Antihistaminic. mp = 194-198°; soluble in H_2O, less soluble in organic solvents. *Ciba-Geigy Corp.*

3871 Astemizole
68844-77-9 891 272-441-9
$C_{28}H_{31}FN_4O$
1-(p-Fluorobenzyl)-2-[[1-(p-methoxyphenethyl)-4-piperidyl]amino]benzimidazole.
Hismanal; R-43512; Astemisan; Histamen; Histaminos; Histazol; Kelp; Laridal; Metodik; Novo-Nastizol A; Paralergin; Retolen; Waruzol. Antiallergic; antihistaminic. Non-sedating type histamine H_1-receptor antagonist. mp = 149.1°; soluble in organic solvents, insoluble in H_2O; λ_m = 219, 249, 286 nm (ε 27250, 6480, 8634 EtOH), 209, 277 nm (ε 57889, 18073, 0.1N HCl). *Janssen Pharm. Ltd.*

3872 Azatadine
3964-81-6 934
$C_{20}H_{22}N_2$
6,11-Dihydro-11-(1-methyl-4-piperidinylidene)-5H-benzo[5,6]cyclohepta[1,2-b]pyridine.
Antihistaminic. mp = 124-126°.

3873 Azatadine Dimaleate
3978-86-7 934
$C_{28}H_{30}N_2O_8$
6,11-Dihydro-11-(1-methyl-4-piperidinylidene)-5H-benzo[5,6]cyclohepta[1,2-b]pyridine maleate (1:2).
Optimine; Trinalin; Sch-10649; Atoramin; Bonamid; Idulian; Zadine. Antihistaminic. mp = 152-154°.

3874 Azelastine
58581-89-8 939
$C_{22}H_{24}ClN_3O$
4-[(4-Chlorophenyl)methyl]-2-(hexahydro-1-methyl-1H-azepin-4-yl)-1(2H)-phthalazinone.
Antiallergic; antiasthmatic; antihistaminic. Orally active histamine H_1 receptor antagonist. Oil, soluble in CH_2Cl_2; gives a crystalline monohydrate. *Asta-Werke AG; Carter-Wallace.*

3875 Azelastine Hydrochloride
79307-93-0 939
$C_{22}H_{25}Cl_2N_3O$
4-[(4-Chlorophenyl)methyl]-2-(hexahydro-1-methyl-1H-azepin-4-yl)-1(2H)-phthalazinone hydrochloride.
A-5610; W-2979M; E-0659; Allergodil; Astelin; Azeptin; Rhinolast. Antiallergic; antiasthmatic; antihistaminic. Orally active H_2-histamine receptor antagonist. mp = 225-229°; LD_{50} (mmus iv) = 36.5 mg/kg, (mmus ip) = 56.4 mg/kg, (mmus sc) = 63.0 mg/kg, (mmus orl) = 124 mg/kg, (fmus iv) = 35.5 mg/kg, (fmus ip) = 42.8 mg/kg, (fmus sc) = 54.2 mg/kg, (fmus orl) = 139 mg/kg, (mrat iv) = 26.9 mg/kg, (mrat ip) = 4. *Asta-Werke AG; Carter-Wallace.*

3876 Bamipine
4945-47-5 983
$C_{19}H_{24}N_2$
4-(N-Benzylanilino)-1-methylpiperidine.
Soventol. Antihistaminic. mp = 115°. *KV Pharm.*

3877 Bamipine Dihydrochloride
61732-85-2 983
$C_{19}H_{26}Cl_2N_2$
4-(N-Benzylanilino)-1-methylpiperidine dihydrochloride.
Taumidrine. Antihistaminic. mp = 189° (dec). *KV Pharm.*

3878 Barmastine
99156-66-8
$C_{27}H_{29}N_7O_2$
3-[2-[4-[(3-(2-Furfuryl)-3H-imidazo[4,5-b]pyridin-2-yl)amino]piperidino]ethyl]-2-methyl-4H-pyrido[1,2-a]pyrimidin-4-one.
R-57959. Antihistaminic. *Janssen Pharm. Inc.*

3879 Bietanautine
6888-11-5 1253
$C_{35}H_{41}N_9O_9$
1,2,3,6-Tetrahydro-1,3-dimethyl-2,6-dioxo-7H-purine-7-acetic acid compound with 2-(diphenylmethoxy)-N,N-dimethylethanamine (2:1).
Etanautine; Nautamine. Antihistaminic; antiemetic; antiparkinsonian. mp = 168-170°; soluble in EtOH, sparingly soluble in H_2O. *Delagrange.*

3880 Bromodiphenhydramine
118-23-0 1439
$C_{17}H_{20}BrNO$
2-[(p-Bromo-α-phenylbenzyl)oxy]-N,N-dimethylethylamine.
Ambodryl; Bromo-benadryl; Deserol; Histabromamine. Antihistaminic. *Parke-Davis.*

3881 Bromodiphenhydramine Hydrochloride
1808-12-4 1439
$C_{17}H_{21}BrClNO$
2-[(p-Bromo-α-phenylbenzyl)oxy]-N,N-dimethylethylamine hydrochloride.
Ambodryl hydrochloride. Antihistaminic. mp = 144-145°. *Parke-Davis.*

3882 Brompheniramine
86-22-6 1467
$C_{16}H_{19}BrN_2$
2-[p-Bromo-α-[2-(dimethylamino)ethyl]-benzyl]pyridine.
parabromdylamine. Antihistaminic. $bp_{0.5}$ = 147-152°; soluble in dilute acids. *ECR Pharm.; Schering-Plough Pharm.; Whitehall-Robins.*

3883 Brompheniramine Maleate
980-71-2 1467
$C_{20}H_{23}BrN_2O_4$
2-[p-Bromo-α-[2-(dimethylamino)ethyl]benzyl]pyridine maleate.
Allent; Dimetapp; Dimetane; Dristan; DrixoralLodrane; Ilvin; Nagemid; Symptom 3; Veltane. Antihistaminic. mp = 132-134°; soluble in H_2O. *ECR Pharm.; Schering-Plough Pharm.; Whitehall-Robins.*

3884 Bufenadrine
604-74-0
$C_{21}H_{29}NO$
2-[(o-tert-Butyl-α-phenylbenzyl)oxy]-N,N-dimethylethylamine.
Antihistaminic.

3885 Bufrolin
54867-56-0
$C_{18}H_{16}N_2O_6$
6-Butyl-1,4,7,10-tetrahydro-4,10-dioxo-1,7-phenanthraline-2,8-dicarboxylic acid.
An antiallergic and antihistaminic drug.

3886 Carbinoxamine
486-16-8 1845 222-498-0
$C_{16}H_{19}ClN_2O$
2-[p-Chloro-α-[2-(dimethylamino)ethoxy]benzyl]pyridine.
Paracarinoxamine. Antihistaminic. Antihistaminic. $bp_{0.1}$ = 158-162°. *McNeil Consumer Products Co.; Ross Products.*

3887 Carbinoxamine Maleate
3505-38-2 1845
$C_{20}H_{23}ClN_2O_5$
2-[p-Chloro-α-[2-(dimethylamino)ethoxy]benzyl]pyridine maleate.
Clistin; Rondec; Rondec DM; Allergefon; Ciberon; Hislosine; Lergefin; Polistin T-Caps. Antihistaminic. mp = 117-119°; soluble in H_2O, EtOH, $CHCl_3$; LD_{50} (mus ip) = 166 mg/kg. *McNeil Consumer Products Co.; Ross Products.*

3888 Carebastine
90729-42-3
$C_{32}H_{37}NO_4$
p-[4-[4-(Diphenylmethoxy)piperidino]butyryl]-α-methylhydratropic acid.
Antiallergic; antihistaminic.

3889 Cetirizine
83881-51-0 2063
$C_{21}H_{25}ClN_2O_3$
(±)-[2-[4-(p-Chloro-α-phenylbenzyl)-1-piperazinyl]ethoxy]acetic acidetirizine.
Antihistaminic. mp = 110-115°. *Pfizer Inc.*

3890 Cetirizine Hydrochloride
83881-52-1 2063
$C_{21}H_{27}Cl_3N_2O_3$
(±)-[2-[4-(p-Chloro-α-phenylbenzyl)-1-piperazinyl]ethoxy]acetic acidetirizine hydrochloride.
Zyrtec; P 071; Alerlisin; Formistin; Reactine; Virlix; Zirtek; Zyrlex. Antihistaminic. mp = 225°. *Pfizer Inc.*

3891 Cetoxime
25394-78-9 2065
$C_{15}H_{17}N_3O$
2-N-Benzylanilinoacetamidoxime.
Antihistaminic. mp = 107-108°. *Boots Pharmaceuticals Inc.*

3892 Cetoxime Hydrochloride
22204-29-1 2065
$C_{15}H_{18}ClN_3O$
2-N-Benzylanilinoacetamidoxime hydrochloride.
Febramine. Antihistaminic. mp = 164-165°. *Boots Pharmaceuticals Inc.*

3893 Chlorcyclizine
82-93-9 2128
$C_{18}H_{21}ClN_2$
1-(p-Chloroα-phenylbenzyl)-4-methylpiperazine.
Antihistaminic. bp0.1-0.15 = 137-145°. *Abbott Labs Inc.; Glaxo Wellcome plc.*

3894 Chlorcyclizine Hydrochloride
1620-21-9 2128
$C_{18}H_{22}Cl_2N_2$
1-(p-Chloroα-phenylbenzyl)-4-methylpiperazine monohydrochloride.
Di-paralene; Mantadil; Perazil; Histantin. Antihistaminic. mp = 226-227°; soluble in H_2O (500 g/l), EtOH (90 g/l), $CHCl_3$ (250 g/l); insoluble in Et_2O, C_6H_6; [dihydrochloride]: LD_{50} (mus ip) = 137 mg/kg. *Abbott Labs Inc.; Glaxo Wellcome plc.*

3895 Chloropyramine
59-32-5 2214
$C_{16}H_{20}ClN_3$
2-[(p-Chlorobenzyl)(2-dimethylaminoethyl)amino]pyridine.
Halopyramine; Chloropyribenzamine. Antihistaminic. $bp_{0.2}$ = 154-155°. *Ciba-Geigy Corp.*

3896 Chlorothen
148-65-2 2220
$C_{14}H_{18}ClN_3S$
2-[(5-Chloro-2-thenyl)[2-(dimethylamino)ethyl]-amino]pyridine.
chloropyrilene; chloromethapyrilene; chlorothenylpyramine. Antihistaminic. $bp_{1.0}$ = 155-156°. *Monsanto Co.*

3897 Chlorothen Citrate
148-64-1 2220
$C_{20}H_{26}ClN_3O_7S$
2-[(5-Chloro-2-thenyl)[2-(dimethylamino)ethyl]-amino]pyridine citrate.
chlorothenium citrate; Tagafen. Antihistaminic. mp = 112-116°; λ_m = 240 nm ($E^{1\%}_{1\,cm}$ 390-410); soluble in H_2O; soluble in H_2O (30 g/l), EtOH (15 g/l); insoluble in non-polar solvents; LD_{50} (mus ip) = 105 mg/kg. *Monsanto Co.*

3898 Chlorpheniramine
132-22-9 2232 205-054-0
$C_{16}H_{19}ClN_2$
2-[p-Chloro-α-[2-(dimethylamino)ethyl]benzyl]pyridine.
chlorprophenpyridamine; chlorphenamine; Haynon.

Antihistaminic. $bp_{1.0}$ = 142°. *Schering-Plough HealthCare Products.*

3899 Chlorpheniramine Maleate

113-92-8 2232 204-037-5

$C_{20}H_{23}ClN_2O_4$

2-[p-Chloro-α-[2-(dimethylamino)ethyl]benzyl]pyridine maleate (1:1).

chlorphenamine hydrogen maleate; Allergisan; Antagonate; Chlor-Trimeton; Chlor-Tripolon; Cloropiril; C-Meton; Histadur; Histaspan; Lorphen; Piriton; Pyridamal-100; Teldrin; component of: Allarest, A.R.M; Azimycin, Cerose-DM, Children's Tylenol Cold, Comhist, Contac, Coricidin, Corilin, Deconamine, Demazin, Diathal, Dristan Cold, Drize, Fedahist, Histabid Duracap, Histalet Forte, Histaspan Plus, Hycomine, Intensin, Isoclor, Metrevet, Naldecon, Novahistine, Ornade, PediaCare Cold-Cough, PV Tussin Syrup, Rhinex D-Lay, Ru-Tuss, Robitussin, Sinarest, Sine-Off, Sudafed Plus, Theraflu, Triaminic, Trind, Trind-DM, Tussar-2, Tussar DM, Tussar SF, TussarViro-Med, Tylenol Cold and Flu, Tylenol Cold Medication. Antihistaminic. mp = 130-135°; λ_m = 261 nm (ε 5760 H_2O); soluble in EtOH (330 g/l), $CHCl_3$ (240 g/l), H_2O (160 g/l), MeOH (130 g/l); less soluble in non-polar solvents; LD_{50} (mus orl) = 162 mg/kg. *Apothecon; Ciba-Geigy Corp.; Eli Lilly & Co.; Forest Pharm. Inc.; Glaxo Wellcome plc; KV Pharm.; McNeil Pharm.; Mead Johnson Nutritionals; Menley & James Labs Inc.; Nippon Shinyaku Japan; Rhône-Poulenc Rorer Pharm. Inc.; Sandoz Pharm. Corp.; Schering-Plough HealthCare Products; Schering-Plough Pharm.; SmithKline Beecham Pharm.; Solvay Animal Health Inc.; Sterling Health U.S.A.; Whitehall Labs. Inc.; Wyeth-Ayerst Labs.*

3900 d-Chlorpheniramine Maleate

2438-32-6 2232 219-450-6

$C_{20}H_{23}ClN_2O_4$

γ-(4-Chlorophenyl)-N,N-dimethyl-2-pyridinepropanamine maleate.

Fortamine; Isomerine; Phenamin; Phendextro; Polamin; Polaramine; Polaronil; Sensidyn. Antihistaminic. mp = 113-115°; $[\alpha]_D^{25}$ = 44.3° (c = 1 DMF). *Ciba-Geigy Corp.; DuPont-Merck Pharm.; ECR Pharm.; Eli Lilly & Co.; Forest Pharm. Inc.; Glaxo Wellcome plc; Lemmon Co.; Marion Merrell Dow Inc.; McGaw Inc.; Nippon Shinyaku Japan; Rhône-Poulenc Rorer Pharm. Inc.; Savage Labs; Schering-Plough HealthCare Products; SmithKline Beecham Pharm.; Solvay Animal Health Inc.; Sterling Health U.S.A.; Whitehall Labs. Inc.; Wyeth-Ayerst Labs.*

3901 d-Chlorpheniramine

25523-97-1 2232 247-073-7

$C_{16}H_{19}ClN_2$

γ-(4-Chlorophenyl)-N,N-dimethyl-2-pyridine-propanamine.

dexchlorpheniramine; d-chlorpheniramine. Antihistaminic. Oily liquid; $[\alpha]_D^{25}$ = 49.8° (c = 1 DMF). *Ciba-Geigy Corp.; DuPont-Merck Pharm.; ECR Pharm.; Eli Lilly & Co.; Forest Pharm. Inc.; Glaxo Wellcome plc; Lemmon Co.; Marion Merrell Dow Inc.; McGaw Inc.; Nippon Shinyaku Japan; Rhône-Poulenc Rorer Pharm. Inc.; Savage Labs; Schering-Plough HealthCare Products; SmithKline Beecham Pharm.; Solvay Animal Health Inc.; Sterling Health U.S.A.; Whitehall Labs. Inc.; Wyeth-Ayerst Labs.*

3902 Cinnarizine

298-57-7 2365 206-064-8

$C_{26}H_{28}N_2$

1-Cinnamyl-4-(diphenylmethyl)piperazine.

cinnipirine; R-516; R-1575; 516-MD; Aplactan; Aplexal; Apotomin; Artate; Carecin; Cerebolan; Cerepar; Cinaperazine; Cinazyn; Cinnacet; Cinnageron; Corathiem; Denapol; Dimitron; Eglen; Folcodal; Giganten; Glanil; Hilactan; Ixterol; Katoseran; Labyrin; Midronal; Mitronal; Olamin; Processine; Sedatromin; Sepan; Siptazin; Spaderizine; Stugeron; Stutgeron; Stutgin; Toliman; component of: Emesazine. Antihistaminic; vasodilator (peripheral, cerebral). Calcium channel blocker with antiallergic and antivasoconstricting activity. [hydrochloride]: mp = 192° (dec); soluble in H_2O (20 g/l). *Abbott Laboratories Inc.*

3903 Clemastine

15686-51-8 2405

$C_{21}H_{26}ClNO$

(+)-(2R)-2-[2-[[(R)-p-Chloro-α-methyl-α-phenylbenzyl]oxy]ethyl]-1-methylpyrrolidine.

meclastine. Antihistaminic. $bp_{0.02}$ = 154°; $[\alpha]_D^{20}$ = 33.6° (EtOH). *Lemmon Co.; Sandoz Pharm. Corp.; SmithKline Beecham Pharm.*

3904 Clemastine Fumarate

14976-57-9 2405

$C_{25}H_{30}ClNO_5$

(+)-(2R)-2-[2-[[(R)-p-Chloro-α-methyl-α-phenylbenzyl]oxy]ethyl]-1-methylpyrrolidine fumarate.

Tavist; HS-592; Aloginan; Alphamin; Anhistan; Fuluminol; Inbestan; Kinotomin; Lacretin; Lecasol; Maikohis; Mallermin-F; Marsthine; Masletine; Piloral; Reconin; Tavegil; Tavegyl; Telgin-G; Trabest; Xolamin. Antihistaminic. mp = 177-178°; $[\alpha]_D^{21}$ = 16.9° (MeOH); LD_{50} (mus orl) = 730 mg/kg, (mus iv) = 43 mg/kg, (rat orl) =3550 mg/kg, (rat iv) = 82 mg/kg. *Lemmon Co.; Sandoz Pharm. Corp.; SmithKline Beecham Pharm.*

3905 Clemizole

442-52-4 2406 207-133-5

$C_{19}H_{20}ClN_3$

1-[(4-Chlorophenyl)methyl]-2-(1-pyrrolidinylmethyl)-1H-benzimidazole.

Neopenyl; Lergopenin; Depocural. Antihistaminic. mp = 167°; forms water-soluble salts. *Schering AG.*

3906 Clemizole Hydrochloride

1163-36-6 2406

$C_{19}H_{21}Cl_2N_3$

1-[(4-Chlorophenyl)methyl]-2-(1-pyrrolidinylmethyl)-1H-benzimidazole hydrochloride.

AL-20; Allercur; Histacuran; Klemidox; Reactrol. Antihistaminic. mp = 239-241°. *Schering AG.*

3907 Clobenzepam

1159-93-9 2420

$C_{17}H_{18}ClN_3O$

7-Chloro-10-[2-(dimethylamino)ethyl]-5,10-dihydro-11H-dibenzo[b,e][1,4]-diazepin-11-one.

Antihistaminic. mp = 165-166°; λ_m = 230 nm (ε 32,740). *Parke-Davis.*

3908 Clobenzepam Hydrochloride
2726-03-6 2420
$C_{17}H_{19}Cl_2N_3O$
7-Chloro-10-[2-(dimethylamino)ethyl]-5,10-dihydro-11H-dibenzo[b,e][1,4]-diazepin-11-one monohydrochloride.
Tarpan. Antihistaminic. mp = 225-233°; LD_{50} (mus orl) = 330 mg/kg. *Parke-Davis.*

3909 Clobenztropine
5627-46-3 2422
$C_{21}H_{24}ClNO$
3-[(p-Chloro-α-phenylbenzyl)oxy]tropane.
Teprin. Antihistaminic. [hydrochloride]: mp = 215-217°; soluble in H_2O, EtOH; [hydrobromide]: mp = 197-200°; soluble in H_2O, EtOH; [methobromide]: mp = 245-248°; soluble in H_2O, EtOH; [methochloride]: mp = 261-263°; soluble in H_2O, EtOH. *Schenley.*

3910 Clocinizine
298-55-5 2428
$C_{26}H_{27}ClN_2$
1-(p-Chloro-α-phenylbenzyl)-4-cinnamylpiperazine.
Denoral. Antihistaminic. [hydrochloride]: mp = 200-201°. *Abbott Laboratories Inc.*

3911 Closiramine
47135-88-6
$C_{18}H_{21}ClN_2$
8-Chloro-11-[2-(dimethylamino)ethyl]-6,11-dihydro-5H-benzo[5,6]cyclohepta[1,2-b]pyridine.
Antihistaminic. *Schering AG.*

3912 Closiramine Aceturate
23256-09-9
$C_{22}H_{28}ClN_3O_3$
8-Chloro-11-[2-(dimethylamino)ethyl]-6,11-dihydro-5H-benzo[5,6]cyclohepta[1,2-b]pyridine compound with N-acetylglycine.
Sch-12169. Antihistaminic. *Schering AG.*

3913 Cromolyn
16110-51-3 2658 240-279-8
$C_{23}H_{12}O_{11}$
5,5'-[(2-Hydroxytrimethylene)dioxy]bis[4-oxo-4H-1-benzopyran-2-carboxylic acid].
cromoglycic acid; Duracroman. Antiallergic; prophylactic antiasthmatic. mp = 241-242° (dec); [monohydrate]: mp = 216-217°. *Bausch & Lomb Pharm. Inc.; Fisons plc.*

3914 Cromolyn Disodium Salt
15826-37-6 2658 239-926-7
$C_{23}H_{14}Na_2O_{11}$
Sodium 5,5'-[(2-hydroxytrimethylene)dioxy]bis[4-oxo-4H-1-benzopyran-2-carboxylate].
cromolyn sodium; disodium cromoglycate; DSCG; FPL-670; Aarane; Alercrom; Alerion; Allergocrom; Colimune; Crolom; Cromovet; Fivent; Gastrocom; Gastrofrenal; Inostral; Intal; Introl; Irtan; Lomudal; Lomupren; Lomusol; Lomuspray; Nalcrom; Nalcron; Nasalcrom; Nasmil; Opticrom; Rynacrom; Sofro; Vividrin. Antiallergic; prophylactic antiasthmatic. Soluble in H_2O, insoluble in organic solvents; LD_{50} (rat orl) > 8000 mg/kg. *Bausch & Lomb Pharm. Inc.; Fisons plc.*

3915 Cycliramine
47128-12-1
$C_{18}H_{19}ClN_2$
4-(p-Chloro-α-2-pyridylbenzylidene)-1-methylpiperidine.
Antihistaminic. *Schering AG.*

3916 Cycliramine Maleate
5781-37-3
$C_{22}H_{23}ClN_2O_4$
4-(p-Chloro-α-2-pyridylbenzylidene)-1-methylpiperidine maleate.
Prolergic; Sch-2544; NSC-70933. Antihistaminic. *Schering AG.*

3917 Cyproheptadine
129-03-3 2842 204-928-9
$C_{21}H_{21}N$
4-(5H-Dibenzo[a,d]cyclohepten-5-ylidene)-1-methyl piperidine.
Antihistaminic; antipruritic. mp = 112.3-113.3°. *Merck & Co.Inc.*

3918 Cyproheptadine Hydrochloride Sesquihydrate
41354-29-4 2842
$C_{21}H_{22}ClN.1.5H_2O$
4-(5H-Dibenzo[a,d]cyclohepten-5-ylidene-1-methylpiperidine hydrochloride sesquihydrate.
Anarexol; Antegan; Ifrasarl; Nuran; Periactin; Vimicon; Cipractin; Peritol; Periactin. Antihistaminic; antipruritic; used for the treatment of coughs; colds. Dec 252.6-253.6°; λ_m = 224, 285 nm ($E^{1\%}_{1\,cm}$ 1656 355 0.1N H_2SO_4); soluble in MeOH (66.6 g/100 ml), $CHCl_3$ (6.25 g/100 ml), EtOH (2.88 g/100 ml), H_2O (0.36 g/100 ml); insoluble in Et_2O; LD_{50} (mus orl) = 74.2 mg/kg. *Merck & Co.Inc.*

3919 Dacemazine
518-61-6 189
$C_{16}H_{16}N_2OS$
10-(N,N-Dimethylglycyl)phenothiazine.
Ahistan. Antihistaminic. Antineoplastic agent. mp = 144-145°; [hydrochloride]: mp = 230-231°. *Searle G.D. & Co.*

3920 Deptropine
604-51-3 2958 210-069-0
$C_{23}H_{27}NO$
endo-3-[(10,11-Dihydro-5H-dibenzo[a,d]cyclohepten-5-yl)oxy]-8-methyl-8-azabicyclo[3.2.1]octane.
dibenzheptropine. Anticholinergic; antihistaminic. [maleate]: mp = 133-136°.

3921 Deptropine Citrate
2169-75-7 2958 218-516-1
$C_{29}H_{35}NO_8$
endo-3-[(10,11-Dihydro-5H-dibenzo[a,d]cyclohepten-5-yl)oxy]-8-methyl-8-azabicyclo[3.2.1]octane citrate.
BS-6987; Brontine. Antihistaminic; anticholinergic. LD_{50} (mus iv) = 32 mg/kg, (mus orl) = 300 mg/kg.

3922 Dexbrompheniramine
132-21-8 1467
$C_{16}H_{19}BrN_2$
(+)-2-[p-Bromo-α-[2-(dimethylamino)ethyl]benzyl pyridine.

parabromdylamine. Antihistaminic. $bp_{0.5}$ = 147-152°. *Schering-Plough HealthCare Products; Schering-Plough Pharm.*

3923 Dexbrompheniramine Maleate
2391-03-9 1467
$C_{20}H_{23}BrNBr_2O_4$
(+)-2-[p-Bromo-α-[2-(dimethylamino)ethyl]benzyl pyridine maleate.
Disomer; Disophrol; Drixoral; Dimegan; Dimetane; Dimotane; Ilvin; Nagemid; Symptom 3; Veltane. Antihistaminic. mp = 132-134°; soluble in H_2O, less soluble in EtOH. *Schering-Plough HealthCare Products; Schering-Plough Pharm.*

3924 Dimethindene
5636-83-9 3265
$C_{20}H_{24}N_2$
2-[1-[2-[2-(Dimethylamino)ethyl]inden-3-yl]-ethyl]pyridine.
dimethpyrindene. Antihistaminic. *Ciba-Geigy Corp.; Marion Merrell Dow Inc.*

3925 Dimethindene Maleate
3614-69-5 3265 222-789-2
$C_{24}H_{28}N_2O_4$
2-[1-[2-[2-(Dimethylamino)ethyl]inden-3-yl]ethyl]pyridine maleate.
Forhistal maleate; Su-6518; Fenistil; Fenostil; Forhistal; NSC-107677. Antihistaminic. mp = 159-161°; LD_{50} (rat iv) = 26.8 mg/kg, (rat orl) = 618.2 mg/kg. *Ciba-Geigy Corp.; Marion Merrell Dow Inc.*

3926 Diphenhydramine
58-73-1 3367
$C_{17}H_{21}NO$
2-(Diphenylmethoxy)-N,N-dimethylethylamine.
benzhydramine. Antihistaminic. Also used in preventing motion sickness. $bp_{2.0}$ = 150-165°. *Bristol-Myers Squibb HIV Products; Elkins-Sinn; Glaxo Wellcome Inc.; McNeil Consumer Products Co.; Parke-Davis; Schering-Plough Pharm.; SmithKline Beecham Pharm.; Sterling Health U.S.A.; Whitehall-Robins; Wyeth-Ayerst Labs.*

3927 Diphenhydramine Citrate
88637-37-0 3367
$C_{23}H_{29}NO_8$
2-(Diphenylmethoxy)-N,N-dimethylethylamine citrate.
Excedrin. Antihistaminic. Also used in preventing motion sickness. *Bristol-Myers Squibb HIV Products.*

3928 Diphenhydramine Hydrochloride
147-24-0 3367
$C_{17}H_{22}ClNO$
2-(Diphenylmethoxy)-N,N-dimethylethylamine hydrochloride.
Actifed; Alledryl; Allergina; Amidryl; Bagodryl; Bax; Benadryl; Bena-Fedrin; Benocten; Benodine; Benylin; Benzantin; Dibondrin; Dihydral; Diphantine; Dolestan; Fenylhist; Halbmond; Histacyl; Noctomin; S 8; Sedopretten; Sekundal-D; Syntedril; Wehydryl; component of : Benacine, Benylin, Caladryl, Contac, Coricidin, Midol, Sleep-eze, Tylanol PM, Ziradryl. Antihistaminic. Also used in preventing motion sickness. mp = 166-170°; soluble in H_2O (1 g/ml), EtOH (0.5 g/ml), $CHCl_3$ (0.5 g/ml), Me_2CO (0.02 g/ml); insoluble in non-polar solvents; LD_{50} (rat orl) = 500 mg/kg. *Elkins-Sinn; Glaxo Wellcome Inc.; McNeil Consumer Products Co.; Parke-Davis; Schering-Plough Pharm.; SmithKline Beecham Pharm.; Sterling Health U.S.A.; Whitehall-Robins; Wyeth-Ayerst Labs.*

3929 Diphenylpyraline
147-20-6 3390 205-049-3
$C_{19}H_{23}NO$
4-(Diphenylmethoxy)-1-methylpiperidine.
Antihistaminic. *3M Pharm.; SmithKline Beecham Pharm.*

3930 Diphenylpyraline Hydrochloride
132-18-3 3390 205-049-3
$C_{19}H_{24}ClNO$
4-(Diphenylmethoxy)-1-methylpiperidine hydrochloride.
Diafen; Hispril; Histryl; histryl Spansule Capsules; Anginosan; Belfene; Dayfen; Histyn; Kolton; Kolton Jelly; Lergobine. Antihistaminic. mp = 206°; soluble in H_2O, EtOH; insoluble in organic solvents. *3M Pharm.*

3931 Diphenylpyraline Theoclate
606-90-6 3390
$C_{26}H_{30}ClN_5O_3$
4-Diphenylmethoxy-1-methylpiperidine compound with 8-chlorotheophylline.
Kolton; Mepedyl; Piprinhydrinate. Antihistaminic. mp = 174-176°; poorly soluble in H_2O, soluble in EtOH. *Boots Co.*

3932 Dorastine
21228-13-7
$C_{20}H_{22}ClN_3$
8-Chloro-2,3,4,5-tetrahydro-2-methyl-5-[2-(6-methyl-3-pyridyl)ethyl]-1H-pyrido[4,3-b]indole.
Antihistaminic. *Hoffmann-LaRoche Inc.*

3933 Dorastine Hydrochloride
21228-28-4
$C_{20}H_{24}Cl_3N_3$
8-Chloro-2,3,4,5-tetrahydro-2-methyl-5-[2-(6-methyl-3-pyridyl)ethyl]-1H-pyrido[4,3-b]indole dihydrochloride.
Ro-5-9110/1. Antihistaminic. *Hoffmann-LaRoche Inc.*

3934 Doxylamine
469-21-6 3497 207-414-2
$C_{17}H_{22}N_2O$
2-[α-[2-(Dimethylamino)ethoxy]-α-methylbenzyl]pyridine.
Antihistaminic. Sedative/hypnotic. $bp_{0.5}$ = 137-141°; soluble in acids. *Marion Merrell Dow Inc.; Pfizer Intl.; Whitehall Labs. Inc.*

3935 Doxylamine Succinate
562-10-7 3497 209-228-7
$C_{21}H_{28}N_2O_5$
2-[α-[2-(Dimethylamino)ethoxy]-α-methylbenzyl]pyridine succinate.
Decapryn Succinate; Unisom; Mereprine; Alsodorm; Gittalun; Hoggar N; Sedaplus; component of: Robitussin Night Time Cold Formula. Sedative/hypnotic; antihistaminic. mp = 100-104°; soluble in H_2O (100 g/100 ml), EtOH (50 g/100 ml), $CHCl_3$ (50 g/100 ml); slightly soluble in C_6H_6, Et_2O; LD_{50} (mus orl) = 470 mg/kg, (mus iv) = 62 mg/kg, (mus sc) = 460 mg/kg, (rbt

orl) = 250 mg/kg, (rbt iv) = 49 mg/kg, (mrat sc) = 440 mg/kg, (frat sc) = 445 mg/kg. *Marion Merrell Dow Inc.; Pfizer Intl.; Whitehall Labs. Inc.*

3936 Ebastine
90729-43-4 3534
$C_{32}H_{39}NO_2$
4'-tert-Butyl-4-[4-(diphenylmethoxy)piperidino]-butyrophenone.
LAS W-090; RP-64305; Bastel; Ebastel. Antihistaminic. Non-sedating histamine H_1-receptor antagonist. [fumarate]: mp = 197-198°. *Grupo Farmaceutico Almirall S.A.*

3937 Efletirizine
150756-35-7
$C_{21}H_{24}F_2N_2O_3$
[2-[4-[Bis(p-fluorophenyl)methyl]-1-piperazinyl]-ethoxy]acetic acid.
Antihistaminic.

3938 Embramine
3565-72-8 3596
$C_{18}H_{22}BrNO$
2-[(p-Bromo-α-methyl-α-phenylbenzyl)oxy]-N,N-dimethylethylamine.
Antihistaminic. $bp_{0.3}$ = 135-140°. *SmithKline Beecham Pharm.*

3939 Embramine Hydrochloride
13977-28-1 3596 237-757-3
$C_{18}H_{23}BrClNO$
2-[(p-Bromo-α-methyl-α-phenylbenzyl)oxy]-N,N-dimethylethylamine hydrochloride.
Embelin; Mebryl; Mebrophenhydramine; Bromadryl. Antihistaminic. mp = 150-152°; LD_{50} (mus orl) = 330 mg/kg, (musiv) = 80 mg/kg. *SmithKline Beecham Pharm.*

3940 Emedastine
87233-61-2 3597
$C_{17}H_{26}N_4O$
1-(2-Ethoxyethyl)-2-(hexahydro-4-methyl-1H-1,4-diazepin-1-yl)benzimidazole.
Antiallergic; antihistaminic; prophylactic antiasthmatic. Histamine H_1-receptor antagonist. *Kanebo Pharm. Ltd.*

3941 Emedastine
87233-61-2 3597
$C_{17}H_{26}N_4O$
1-(2-Ethoxyethyl)-2-(hexahydro-4-methyl-1H-1,4-diazepin-1-yl)benzimidazole.
Antiallergic; antihistaminic; prophylactic antiasthmatic. Histamine H_1-receptor antagonist. *Kanebo Pharm. Ltd.*

3942 Emedastine Difumarate
87233-62-3 3597
$C_{25}H_{34}N_4O_9$
1-(2-Ethoxyethyl)-2-(hexahydro-4-methyl-1H-1,4-diazepin-1-yl)benzimidazole fumarate (1:2).
AL-3432A; KB-2413; LY-188695. Antiallergic; antihistaminic; prophylactic antiasthmatic. Histamine H_1-receptor antagonist. mp = 148-151°; LD_{50} (gpg orl) = 744 mg/kg. *Kanebo Pharm. Ltd.*

3943 Epinastine
80012-43-7 3655
$C_{16}H_{15}N_3$
3-Amino-9,13b-dihydro-1H-dibenz[c,f]imidazo[1,5-a]-azepine.
WAL-801. Antihistaminic. Adrenergic. mp = 205-208°; [hydrobromide]: mp = 284-286°. *Boehringer Ingelheim GmbH.*

3944 Epinastine Hydrochloride
80012-44-8 3655
$C_{16}H_{16}ClN_3$
3-Amino-9,13b-dihydro-1H-dibenz[c,f]imidazo[1,5-a]-azepine hydrochloride.
WAL-801CL; Alesion. Antihistaminic. Adrenergic. mp = 273-275°; soluble in H_2O; LD_{50} (mrat orl) = 314 mg/kg, (mrat iv) = 17 mg/kg, (frat orl) = 192 mg/kg, (frat iv) = 22 mg/kg. *Boehringer Ingelheim GmbH.*

3945 Etymemazine
523-54-6 3938
$C_{20}H_{26}N_2S$
10[3-(Dimethylamino)-2-methylpropyl]-2-ethylphenothiazine.
RP-6484; ethylisobutrazine; Diquel. Antihistaminic. *Rhône-Poulenc Rorer Pharm. Inc.*

3946 Etymemazine Hydrochloride
3737-33-5 3938
$C_{20}H_{27}ClN_2S$
10[3-(Dimethylamino)-2-methylpropyl]-2-ethylphenothiazine hydrochloride.
RP-6484; Nuital; Sergetyl. Antihistaminic. mp = 160-163°; soluble in H_2O, EtOH, MeOH, Me_2CO, $CHCl_3$; insoluble in Et_2O. *Rhône-Poulenc Rorer Pharm. Inc.*

3947 Fenethazine
522-24-7 4013
$C_{16}H_{18}N_2S$
10(2-Dimethylaminoethyl)phenothiazine.
RP-3015; SC-1627; Anergan; Anergen; Ethysine; Etisine; Lisergan; Lysergan; Phenethazinum; Rutergan. Antihistaminic. bp_1 = 183-187°; [hydrochloride]: mp = 201-201.5°; LD_{50} (mus ip) = 115-120 mg/kg. *Rhône-Poulenc.*

3948 Fexofenadine
83799-24-0 4113
$C_{32}H_{39}NO_4$
(±)-p-[1-Hydroxy-4-[4-(hydroxydiphenyl-methyl)piperidino]butyl]-α-methylhydratropic acid.
carboxyterfenadine; terfenadine carboxylate; bufexamac. Antihistaminic. A proprietary preparation of bufexamac in a cream base; used in the treatment of eczema. mp = 142-143°. *Hoechst Marion Roussel Inc.*

3949 Fexofenadine Hydrochloride
138452-21-8 4113
$C_{32}H_{40}ClNO_4$
(±)-p-[1-Hydroxy-4-[4-(hydroxydiphenyl-methyl)piperidino]butyl]-α-methylhydratropic acid hydrochloride.
Telfast; MDL-16455A. Antihistaminic. *Hoechst Marion Roussel Inc.*

3950 Hexopyrronium Bromide
3734-12-1
$C_{20}H_{30}BrNO_3$
1,1-Dimethyl-3-hydroxypyrrolidinium bromide.
Antihistaminic.

3951 Histapyrrodine
493-80-1 4757
$C_{19}H_{24}N_2$
1-(2-N-Benzylanilinoethyl)pyrrolidine.
Antihistaminic. bp_1 = 198-205°.

3952 Histapyrrodine Hydrochloride
6113-17-3 4757
$C_{19}H_{25}ClN_2$
1-(2-N-Benzylanilinoethyl)pyrrolidine hydrochloride.
Calcistin; Domistan; Luvistin. Antihistaminic. mp = 196-197°; soluble in H_2O (20 g/l).

3953 N-Hydroxyethylpromethazine Hydrochloride
2090-54-2 4870
$C_{19}H_{25}ClN_2OS$
N-(2-Hydroxyethyl)-N,N,α-trimethyl-10H-phenothiazine-10-ethanaminium chloride.
Aprobit. Antihistaminic. mp = 233°.

3954 Hydroxyzyne
68-88-2 4897
$C_{21}H_2$77ClN2O2
2-[2-[4-(p-Chloroα-phenylbenzyl)-1-piperazinyl]ethoxy]ethanol.
UCB-4492; Tran-Q; Tranquizine. Anxiolytic; antihistaminic; minor tranquillizer. *Pfizer Inc.*

3955 Hydroxyzyne Hydrochloride
2192-20-3 4897 218-586-3
$C_{21}H_{29}Cl_3N_2O_2$
2-[2-[4-(p-Chloroα-phenylbenzyl)-1-piperazinyl]ethoxy]ethanol dihydrochloride.
Alamon; Atarax; Aterax; Durrax; Orgatrax; Quiess; QYS; Vistaril Parenteral; Marax. Anxiolytic; antihistaminic. Has been used as a minor tranquillizer. H_1 receptor antagonist. mp = 193°; soluble in H_2O (<70 g/100 ml), $CHCl_3$ (6 g/100 ml), Me_2CO (0.2 g/100 ml), Et_2O (< 0.01 g/100 ml); LD_{50} (rat ip) = 126 mg/kg, (rat orl) = 950 mg/kg. *Elkins-Sinn; Forest Pharm. Inc.; KV Pharm.; Pfizer Inc.; Roerig Div. Pfizer Pharm.*

3956 Hydroxyzyne Pamoate
10246-75-0 4897 233-582-1
$C_{44}H_{43}ClN_2O_8$
2-[2-[4-(p-Chloroα-phenylbenzyl)-1-piperazinyl]ethoxy]-ethanol 4,4'-methylenebis[3-hydroxy-2-naphthoate].
Equipose; Masmoran; Paxisitil; Vistaril pamoate. Anxiolytic; antihistaminic; minor tranquillizer. H_1-receptor antagonist. Insoluble in H_2O. *Pfizer Inc.*

3957 Iproheptine
13946-02-6 6243
$C_{11}H_{25}N$
N-Isopropyl-1,5-dimethylhexylamine.
Metron; Metron S. Antihistaminic. bp_{23} = 84-85°.

3958 Isopromethazine
303-14-0 5219
$C_{17}H_{20}N_2S$
N,N,β-Trimethyl-10H-phenothiazine-10-ethanamine.
Lilly 01526; RP-4460; Isophenergan; Fen-Bridal. Antihistaminic. [hydrochloride]: mp = 193-194°. *Rhône-Poulenc.*

3959 Isothipendyl
482-15-5 5248 214-957-9
$C_{16}H_{19}N_3S$
10-(2-Dimethylaminopropyl)-10H-pyrido[3,2-b]-[1,4]benzothiazine.
Antihistaminic. $bp_{0.4}$ = 171-174°; LD_{50} (mus ip) = 65 mg/kg, (mus orl) = 18 mg/kg.

3960 Isothipendyl Hydrochloride
1225-60-1 5248
$C_{16}H_{20}ClN_3S$
10-(2-Dimethylaminopropyl)-10H-pyrido[3,2-b]-[1,4]benzothiazine hydrochloride.
Andantol; Andanton; Nilergex. Antihistaminic. Soluble in H_2O.

3961 Levocabastine
79547-78-7 5489
$C_{26}H_{29}FN_2O_2$
(-)-trans-1-[cis-4-Cyano-4-(p-fluorophenyl)cyclohexyl]-3-methyl-4-phenylisonipecotic acid.
(-)-cabastine. Antihistaminic. *Abbott Laboratories Inc.*

3962 Levocabastine Hydrochloride
79516-68-0 5489
$C_{26}H_{30}ClFN_2O_2$
(-)-trans-1-[cis-4-Cyano-4-(p-fluorophenyl)cyclohexyl]-3-methyl-4-phenylisonipecotic acid monohydrochloride.
Livostin; R-50547; Levophta. Antihistaminic. *Abbott Laboratories Inc.*

3963 Linetastine
159776-68-8
$C_{35}H_{40}N_2O_6$
1-[5'-(3-Methoxy-4-ethoxycarbonyloxyphenyl)-2',4'-pentadienoyl aminoethyl]-4-diphenyl-methoxypiperidine.
TMK-688. Antihistaminic; leukotriene antagonist. Antihistaminic; leukotriene antagonist.

3964 Loratadine
79794-75-5 5608
$C_{22}H_{23}ClN_2O_2$
Ethyl 4-(8-chloro-5,6-dihydro-11H-benzo-[5,6]cyclo-hepta[1,2-b]pyridin-11-ylidine)-1-piperidinecarboxylate.
Sch-29851; Claritin; Clarityn; Lisino. Antihistaminic. mp = 134-136°. *Schering AG.*

3965 Loxanast
69915-62-4
$C_{14}H_{26}O_2$
cis-4-Isohexyl-1-methylcyclohexanecarboxylic acid.
Antihistaminic.

3966 Mebhydrolin
524-81-2 5809
$C_{19}H_{20}N_2$
5-Benzyl-1,3,4,5-tetrahydro-2-methyl-2H-pyrido-[4,3-b]indole.
Incidal. Antihistaminic. mp = 95°; bp_1 = 207-215°; insoluble in H_2O, soluble in polar organic solvents. *Bayer AG; Schenley.*

3967 Mebhydrolin Napadisylate
6153-33-9 5809 228-170-3
$C_{48}H_{48}N_4O_6S_2$
5-Benzyl-1,3,4,5-tetrahydro-2-methyl-2H-pyrido-[4,3-b]indole 1,5-naphthalenedisulfonate (salt).
Diazoline; Fabahistin; Omeril. Antihistaminic. mp = 280° (dec); insoluble in H_2O. *Bayer AG; Schenley.*

3968 Medrylamine
524-99-2 5839
$C_{18}H_{23}NO_2$
2-(p-Methoxy-α-phenylbenzyloxy)-N,N-dimethylethylamine.
Histaphen; Histaphene; Postafen; Salve. Antihistaminic. [hydrochloride]: mp = 141°. *UCB Pharma.*

3969 Mequitamium Iodide
101396-42-3
$C_{21}H_{25}IN_2S$
(±)-1-Methyl-3-(phenothiazin-10-ylmethyl)quinuclidinium iodide.
LG-30435. The (+)-(S)-enantiomer 1β is 10x more potent than (-)-(R)-enantiomer 1α as a histamine antagonist; the two enantiomers show the same antimuscarinic activity in vitro. Antihistaminic; antiallergic; antimuscarinic.

3970 Mequitazine
29216-28-2 5911 249-521-7
$C_{20}H_{22}N_2S$
10-(3-Quinuclidinylmethyl)phenothiazine.
LM-209; Butix; Metaplexan; Mircol; Primalan; Zesulan. Antihistaminic. mp = 130-131°.

3971 Methafurylene
531-06-8 6010
$C_{14}H_{19}N_3O$
N-(2-Furanylmethyl)-N',N'-dimethyl-N-2-pyridinyl-1,2-ethanediamine.
Antihistaminic. $bp_{0.2}$ = 117.5-118°; [hydrochloride]: mp = 117-119°; [dihydrogen citrate]: mp = 95-97°. *Rhône-Poulenc.*

3972 Methafurylene Fumarate
5429-41-4 6010
$C_{18}H_{23}N_3O_5$
N-(2-Furanylmethyl)-N',N'-dimethyl-N-2-pyridinyl-1,2-ethanediamine fumarate.
F-151; Foralamin. Antihistaminic. *Rhône-Poulenc.*

3973 Methaphenilene
493-78-7 6026
$C_{15}H_{20}N_2S$
N,N-Dimethyl-N'-(α-thenyl)-N'-phenethylenediamine.
00836; RP-2740; W-50 base; Diatrin base. Antihistaminic. bp_7 = 183-185°.

3974 Methaphenilene Hydrochloride
7084-07-3 6026
$C_{15}H_{21}ClN_2S$
N,N-Dimethyl-N'-(α-thenyl)-N'-phenethylenediamine hydrochloride.
Diatrin; Enstamine; Nilhistin. Antihistaminic. mp = 186-187°; soluble in H_2O, EtOH; LD_{50} (mus ip) = 117 mg/kg.

3975 Methapyrilene Fumarate
33032-12-1 6027
$C_{40}H_{50}N_6O_{12}S_2$
2-[[2-(Dimethylamino)ethyl]-2-thenylaminopyridine hydrochloride fumarate (2:3).
Antihistaminic. mp = 135-136°. *Abbott Labs Inc.; Eli Lilly & Co.*

3976 Methapyrilene Hydrochloride
135-23-9 6027
$C_{14}H_{20}ClN_3S$
2-[[2-(Dimethylamino)ethyl]-2-thenylaminopyridine hydrochloride.
Antihistaminic. mp = 162°; λ_m = 238 nm ($E^{1\%}_{1\ cm}$ 623); soluble in H_2O (2 g/ml), EtOH (0.2 g/ml), $CHCl_3$ (0.3 g/ml); insoluble in non-polar solvents. *Abbott Labs Inc.; Eli Lilly & Co.*

3977 Methapyriline
91-80-5 6027 202-099-8
$C_{14}H_{19}N_3S$
2-[[2-(Dimethylamino)ethyl]-2-thenylaminopyridine.
AH-42; Thenylene; Pyrathyn; Semikon; Thionylan; Histadyl; Restryl; Rest-on; Sleepwell; Paradormalene; Pyrinistab; Pyrinistol; Lullamin. Antihistaminic. bp_3 = 173-175°; LD_{50} (rat orl) = 375 mg/kg, (mus orl) = 182 mg/kg, (mus iv) = 19.8 mg/kg, (gpg orl) = 375 mg/kg. *Abbott Labs Inc.; Eli Lilly & Co.*

3978 p-Methyldiphenhydramine
19804-27-4 6130
$C_{18}H_{23}NO$
N,N-Dimethyl-2-[(4-methylphenyl)phenylmethoxy]-ethanamine.
Neo-Benodine; Toladryl. Antihistaminic. $bp_{0.1}$ = 143°; [hydrochloride]: mp = 150-152°. *Parke-Davis.*

3979 Mifentidine
83184-43-4
$C_{13}H_{16}N_4$
N-(p-Imidazol-4-ylphenyl)-N'-isopropylformamidine.
DA-4577. A novel histamine H2-receptor antagonist. Antihistaminic.

3980 Mizolastine
108612-45-9
$C_{24}H_{25}FN_6O$
2-[[1-[1-(p-Fluorobenzyl-2-benzimidazolyl]-4-piperidyl]methylamino]-4(3H)-pyrimidinone.
SL-850324. Selective histamine H1 receptor antagonist. Antihistiminic.

3981 Moxastine
3572-74-5 6370
$C_{18}H_{23}NO$
2-(1,1-Diphenylethoxy)-N,N-dimethylethylamine.
Spofa 325; Alfadryl; α-methyldiphenhydramine.

Antihistaminic. $bp_{0.15}$ = 129-136°; [hydrochloride]: mp = 168°.

3982 Noberastine
110588-56-2
$C_{17}H_{21}N_5O$
3-(5-Methylfurfuryl)-2-(4-piperidiylamino)-3H-imidazo[4,5-b]pyridine.
R-64947. Antihistaminic. *Janssen Pharm. Inc.*

3983 Orphenadrine
83-98-7 7007 201-509-2
$C_{18}H_{23}NO$
N,N-Dimethyl-2-[(o-methyl-α-phenylbenzyl)-oxy]ethylamine.
BS-5930; Biorphen; Brocasipal. Antihistaminic; skeletal muscle relaxant. bp_{12} = 195°. *3M Pharm.; Forest Pharm. Inc.*

3984 Orphenadrine Citrate
4682-36-4 7007 225-137-5
$C_{24}H_{31}NO_8$
N,N-Dimethyl-2-[(o-methyl-α-phenylbenzyl)-oxy]ethylamine citrate (1:1).
Banflex; Disipal; Norflex; X-Otag; component of: Norgesic. Antihistaminic. Skeletal muscle relaxant with anticonvulsant properties. *3M Pharm.; Forest Pharm. Inc.*

3985 Orphenadrine Hydrochloride
341-69-5 7007
$C_{18}H_{24}ClNO$
N,N-Dimethyl-2-[(o-methyl-α-phenylbenzyl)oxy]ethylamine hydrochloride.
Disipal; Mephenamin. Antihistaminic. mp = 156-157°; soluble in H_2O, EtOH, $CHCl_3$; insoluble in Me_2CO, C_6H_6, Et_2O. *3M Pharm.*

3986 Oxatomide
60607-34-3 7058 262-320-9
$C_{27}H_{30}N_4O$
1-[3-[4-(Diphenylmethyl)-1-piperazinyl]propyl]-2-benzimidazolinone.
R-35443; Celtect; Cobiona; Dasten; Tinset. Antihistaminic; antiallergic; antiasthmatic. mp = 153.6°; LD_{50} (rat orl) > 2560 mg/kg, (rat iv) = 30 mg/kg, (gpg orl) = 320 mg/kg, (gpg iv) = 27 mg/kg, (mus orl) >2560 mg/kg, (mus iv) = 27 mg/kg. *Abbott Laboratories Inc.*

3987 Oxomemazine
3689-50-7 7080 222-996-8
$C_{18}H_{22}N_2O_2S$
10-[3-(Dimethylamino)-2-methylpropyl]phenothiazine 5,5-dioxide.
RP-6847; [hydrochloride]: Doxergan; Imakol. Selective muscarinic M_1 receptor antagonist. Useful as an antihistaminic. mp = 115°; [hydrochloride]: mp = 250°.

3988 Phenindamine
82-88-2 7380 209-320-7
$C_{19}H_{19}N$
2,3,4,9-Tetrahydro-2-methyl-9-phenyl-1H-indeno[2,1-c]pyridine.
Nu-1504. Antihistaminic. mp = 91°; d = 1.17. *Hoffmann-LaRoche Inc.; Solvay Pharm. Inc.*

3989 Phenindamine Tartrate
569-59-5 7380
$C_{23}H_{25}NO_6$
2,3,4,9-Tetrahydro-2-methyl-9-phenyl-1H-indeno[2,1-c]pyridine tartrate.
Thephorin; PV Tussin Syrup; PV Tussin Tablet; Pernovin. Antihistaminic. mp = 165-167°; soluble in H_2O (25 g/l), insoluble in EtOH; LD_{50} (rat orl) = 280 mg/kg. *Hoffmann-LaRoche Inc.; Solvay Pharm. Inc.*

3990 Pheniramine
86-21-5 7383
$C_{16}H_{20}N_2$
2-[α]-2-Dimethylaminoethyl]benzyl]pyridine .
prophenpyridamine; propheniramine; Avil; Tripoton. Antihistaminic. bp_{13} = 181°; d = 1.0081; insoluble in H_2O; soluble in dilute acids, organic solvents. *Alcon Labs; Whitehall-Robins.*

3991 Pheniramine Maleate
132-20-7 7383 205-051-4
$C_{20}H_{24}N_2O_4$
2-[α]-[2-Dimethylaminoethyl]benzyl]pyridine bimaleate.
Dristan Nasal Spray; Naphcon A; Daneral, Daneral SA; Inhiston; Trimeton. Antihistaminic. mp = 107°; soluble in H_2O, EtOH; less soluble in organic solvents. *Alcon Labs; Whitehall-Robins.*

3992 Phenyltoloxamine
92-12-6 7469
$C_{17}H_{21}NO$
N,N-Dimethyl-2-(α-phenyl-o-toloxy)ethylamine.
Antihistaminic. $bp_{0.1}$ = 141-144°. *Apothecon; Ascher B.F. & Co.; Parke-Davis; Schwarz Pharma Kremers Urban Co.*

3993 Phenyltoloxamine Citrate
1176-08-5 7469
$C_{23}H_{29}NO_8$
N,N-Dimethyl-2-(α-phenyl-o-toloxy)ethylamine citrate.
Kutrase; Mobigesic; Naldecon; Sinubid. Antihistaminic. mp = 138-140°; soluble in H_2O. *Apothecon; Ascher B.F. & Co.; Parke-Davis; Schwarz Pharma Kremers Urban Co.*

3994 Pirdonium Bromide
35620-67-8
$C_{22}H_{30}BrNO$
1,1-Dimethyl-2-[[(p-methyl-α-phenylbenzyl)oxy]methyl]piperidinium bromide.
A hydrophilic histamine H1-receptor antagonist.

3995 Promethazine
60-87-7 7970 200-489-2
$C_{17}H_{20}N_2S$
10-[2-(Dimethylamino)propyl]phenothiazine.
proazamine; RP-3277. Antihistaminic. Also has antiemetic and CNS depressant properties. mp = 60°; bp_3 = 190-192°. *Rhône-Poulenc; Wyeth Labs.*

3996 Promethazine Hydrochloride
58-33-3 7970 200-375-2
$C_{17}H_{21}ClN_2S$
10-[2-(Dimethylamino)propyl]phenothiazine hydrochloride.

Anergan 25; Anergan 50; Phenergan; Mepergan; Phanergan D; Phenergan VC; RP-3389; Atosil; Dorme; Duplamin; Fellozine; Fenazil; Genphen; Hiberna; Lergigan; Phencen; Phenergan; Prorex; Prothazine; Provigan; Remsed. Antihistaminic. Also has antiemetic and CNS depressant properties. mp = 230-232°; λ_m = 249, 297 nm (ε 28770, 3400 H_2O); soluble in H_2O, EtOH, $CHCl_3$; insoluble in Me_2CO, EtOAc, Et_2O; LD_{50} (mus iv) = 55 mg/kg. *Rhône-Poulenc; Wyeth Labs.*

3997 Promethazine Teoclate
17693-51-5 7970 241-691-0
$C_{17}H_{20}N_2S{\cdot}C_7H_7ClN_4O_2$
10-[2-(Dimethylamino)propyl]phenothiazine compound with 8-chlorotheophylline.
Avomine. Antihistaminic. Also has antiemetic and CNS depressant properties. *Rhône-Poulenc; Wyeth Labs.*

3998 Pyribenzamine
91-81-6 9868
$C_{16}H_{21}N_3$
2-[Benzyl[2-(dimethylamino)]ethyl]amino]pyridine.
Tripelennamine. Antihistaminic. $bp_{0.1}$ = 138-142°; soluble in H_2O. *Ciba-Geigy Corp.; Solvay Animal Health Inc.*

3999 Pyribenzamine Citrate
6138-56-3 9868
$C_{22}H_{29}N_3O_7$
2-[Benzyl[2-(dimethylamino)]ethyl]amino]pyridine citrate.
Antihistaminic; antiallergic. mp = 106-110°; soluble in H_2O, EtOH; less soluble in non-polar solvents. *Ciba-Geigy Corp.; Solvay Animal Health Inc.*

4000 Pyribenzamine Hydrochloride
154-69-8 9868
$C_{16}H_{22}ClN_3$
2-[Benzyl[2-(dimethylamino)]ethyl]amino]pyridine hydrochloride.
Re Covr; Dehistin; Azaron; Pyribenzamine; PBZ; Vetobenzamina; Vetibenzamine. Antihistaminic. mp = 192-193°; λ_m = 244, 305 nm (ε 14470, 4780 H_2O); soluble in H_2O (1.3 g/ml), less soluble in organic solvents; LD_{50} (mus ip) = 47 mg/kg. *Ciba-Geigy Corp.; Solvay Animal Health Inc.*

4001 Pyrilamine
91-84-9 8168
$C_{17}H_{23}N_3O$
2-[[2-(Dimethylamino)ethyl](p-methoxybenzyl)amino]pyridine.
mepyramine; pyranisamine; RP-2786. Antihistaminic. bp_5 = 201°; LD_{50} (mus orl) = 312 mg/kg. *Bristol-Myers Squibb HIV Products; Hoechst Roussel Pharm. Inc.; Schering-Plough Pharm.; Sonus Pharm. Inc.; Sterling Health U.S.A.; Whitehall-Robins.*

4002 Pyrilamine Maleate
59-33-6 8168 200-422-7
$C_{21}H_{27}N_3O_5$
2-[[2-(Dimethylamino)ethyl](p-methoxybenzyl)amino]pyridine maleate.
Histavet-P; Pymafed; Histalet Forte; Midol; PV Tussin Syrup; Robitussin; Antamine; Antisan; Dorantamin; Enrumay; Histalon; Histan; Histapyran; Histatex; Neo-Antergan; Paraminyl; Parmal; Pyramal; Stamine; Stangen; Thylogen. Antihistaminic. mp = 100-101°; λ_m = 244 nm ($E^{1\%}_{1\,cm}$ 420); soluble in H_2O (2.5 g/ml), less soluble in organic solvents; LD_{50} (mus orl) = 338 mg/kg. *Bristol-Myers Squibb HIV Products; Hoechst Roussel Pharm. Inc.; Schering-Plough Pharm.; Sonus Pharm. Inc.; Sterling Health U.S.A.; Whitehall-Robins.*

4003 Pyroxamine
7009-68-9
$C_{18}H_{20}ClNO$
3-[(p-Chlorochloro-α-phenylbenzyl)oxy]-1-methylpyrrolidine .
Antihistaminic. *Whitehall-Robins.*

4004 Pyroxamine Maleate
5560-75-8
$C_{22}H_{24}ClNO_5$
3-[(p-Chlorochloro-α-phenylbenzyl)oxy]-1-methylpyrrolidine maleate.
AHR-224; NSC-64540. Antihistaminic. *Whitehall-Robins.*

4005 Pyrrobutamine
91-82-7 8196
$C_{20}H_{22}ClN$
1-[σ-(p-Chlorobenzyl]cinnamyl]pyrrolidine.
Antihistaminic. mp = 48-49°; $bp_{0.3}$ = 190-195°; d^{25} = 1.1052; λ_m = 360, 243 nm (α 112, 9500 95% EtOH). *Eli Lilly & Co.*

4006 Pyrrobutamine Phosphate
135-31-9 8196
$C_{20}H_{28}ClNO_8P_2$
1-[σ-(p-Chlorobenzyl]cinnamyl]pyrrolidine phosphate.
Pyronil. Antihistaminic. mp = 129-130°; soluble in H_2O (10 g/100ml), EtOH; less soluble in organic solvents; LD_{50} (mus orl) = 1116 mg/kg, (mus sc) = 1270 mg/kg, (mus im) = 837 mg/kg, (mus iv) = 53.5 mg/kg, (gpg sc) = 1241 mg/kg, (gpg im) = 625 mg/kg, (gpg orl) = 992 mg/kg. *Eli Lilly & Co.*

4007 Quifenadine
10447-39-9
$C_{20}H_{23}NO$
α,α-Diphenyl-3-quinuclidinemethanol.
phencarol [as hydrochloride]; fencarol [as hydrochloride]. Derivative of quinuclidylarylcarbinol. Antihistaminic.

4008 Rocastine
91833-77-1
$C_{13}H_{19}N_3OS$
(±)-2-[2-(Dimethylamino)ethyl]-3,4-dihydro-4-methylpyrido[3,2-f]-1,4-oxazepine-5(2H)-thione.
Antihistaminic. *Whitehall-Robins.*

4009 Rocastine Hydrochloride
99617-35-3
$C_{13}H_{20}ClN_3OS.H_2O$
(±)-2-[2-(Dimethylamino)ethyl]-3,4-dihydro-4-methylpyrido[3,2-f]-1,4-oxazepine-5(2H)-thione monohydrochloride monohydrate.
AHR-11325-D; . Antihistaminic. *Whitehall-Robins.*

4010 Rotoxamine
5560-77-0 1845
$C_{16}H_{19}ClN_2O$
(-)-2-[p-Chloro-α[2-(dimethylamino)ethoxy]-benzyl]pyridine.
l-carbinoxamine; McN-R-73-Z. Antihistaminic. $bp_{0.5}$ = 143-144°; $[\alpha]_D^{25}$ = -6.8° (c = 2 MeOH). *McNeil Consumer Products Co.*

4011 Rotoxamine Tartrate
49746-00-1 1845
$C_{20}H_{25}ClN_2O_7$
(-)-2-[p-Chloro-α[2-(dimethylamino)ethoxy]benzyl]-pyridine tartrate.
Twiston. Antihistaminic. mp = 143-145°; $[\alpha]_D^{25}$ = 37.2° (c = 20 MeOH). *McNeil Consumer Products Co.*

4012 Rupatadine
158876-82-5
$C_{26}H_{26}ClN_3$
8-Chloro-6,11-dihydro-11-[1-[(5-methyl-3-pyridinyl)-methyl]-4-piperidinylidene]-5H-benzo[5,6]-cyclohepta[1,2b]pyridine .
UR-12592. Orally active dual antagonist of histamine and platelet-activating factor.

4013 Setastine
64294-95-7 8619
$C_{22}H_{28}ClNO$
1-[2-[(p-Chloro-α-methyl-α-phenylbenzyl)oxy]-ethyl]hexahydro-1H-azepine.
Antihistaminic. *EGYT.*

4014 Setastine Hydrochloride
59767-13-4 8619
$C_{22}H_{29}Cl_2NO$
1-[2-[(p-Chloro-α-methyl-α-phenylbenzyl)oxy]-ethyl]hexahydro-1H-azepine hydrochloride.
EGIS-2062; EGYT-2062; Loderix. Antihistaminic. *EGYT.*

4015 Talastine
16188-61-7 9205
$C_{19}H_{21}N_3O$
2-[2-(Dimethylamino)ethyl]-4-benzyl-1(2H)phthalazinone.
Antihistaminic. $bp_{0.3}$ = 215-222°.

4016 Talastine Hydrochloride
16188-76-4 9205
$C_{19}H_{22}ClN_3O$
2-[2-(Dimethylamino)ethyl]-4-benzyl-1(2H)phthalazinone hydrochloride.
HL-2186; Ahanon. Antihistaminic. mp = 178°; LD_{50} (mus ip) = 116 mg/kg.

4017 Tazifylline
79712-55-3
$C_{23}H_{32}N_6O_3S$
(±)-7-[2-Hydroxy-3-[4-[3-(phenylthio)propyl]-1-piperazinyl]propyl]theophylline.
Antihistaminic.

4018 Tazifylline Hydrochloride
79712-53-1
$C_{23}H_{34}ClN_6O_3S$
(±)-7-[2-Hydroxy-3-[4-[3-(phenylthio)propyl]-1-piperazinyl]propyl]theophylline dihydrochloride.
RS-49014. Antihistaminic.

4019 Temelastine
86181-42-2
$C_{21}H_{24}BrN_5O$
2-[[4-(5-Bromo-3-methyl-2-pyridyl)butyl]amino]-5-[(6-methyl-3-pyridyl)methyl]-4(1H)-pyrimidinone.
SKF-93944. Antihistaminic. *SmithKline Beecham Pharm.*

4020 Terfenadine
50679-08-8 9307
$C_{32}H_{41}NO_2$
α[4-(1,1-Dimethylethyl)phenyl]-4-(hydroxydiphenylmethyl)-1-piperidinebutanol.
Seldane; MDL-9918; Allerplus; Cyater; Nebralin; Teldane; Teldanex; Terdin; Terfex; Ternadin; Triludan. Antihistaminic. mp = 146.5 - 148.5°; λ_m = 260 nm (A 660.4, MeOH), 260 nm (A 671.4 EtOH), 260 nm (A 762.2 CH_2Cl_2); soluble in H_2O (1 mg/100 ml), EtOH (3.78 g/100 ml), MeOH (3.75 g/100 ml), C_6H_{14} (34 mg/100 ml), 0.1M HCl (12 mg/100 ml), 0.1M citric acid (110 mg/100 ml), 0.1M tartaric acid (110 mg/100 ml); LD_{50} (rat mus gpg orl) > 2000 mg/kg. *Merrell Pharm. Inc.*

4021 Thenaldine
86-12-4 9413 201-651-5
$C_{17}H_{22}N_2S$
1-Methyl-4-N-2-thenylanilinopiperidine.
thenalidine; Sandostene. Antihistaminic; antipruritic. The tartrate is used as an antihistaminic. mp = 95-97°; $bp_{0.02}$ = 158-160°; [tartrate]: mp = 170-172°. *Sandoz Pharm. Corp.*

4022 Thenyldiamine
91-79-2 9416
$C_{14}H_{19}N_3S$
2-[[2-(Dimethylamino)ethyl]-3-thenylamino]-pyridine.
Win-2848; Thenfadil; dethylandiamine. Antihistaminic. bp_1 = 169-172°.

4023 Thenyldiamine Hydrochloride
958-93-0 9416
$C_{14}H_{20}ClN_3S$
2-[[2-(Dimethylamino)ethyl]-3-thenylamino]pyridine hydrochloride.
Antihistaminic. mp = 169.5-170; soluble in H_2O (< 20 g/100 ml); LD_{50} (rat orl) = 525 mg/kg.

4024 Thiazinamium
2338-21-8 9441 219-051-7
$C_{18}H_{23}N_2S^+$
Trimethyl(1-methyl-2-phenothiazin-10-ylethyl)ammonium.
Antihistaminic.

4025 Thiazinamium Chloride
4320-13-2
$C_{18}H_{23}ClN_2S$
Trimethyl(1-methyl-2-phenothiazin-10-ylethyl)-ammonium chloride.
WY-460E. Antihistaminic; antiallergic.

4026 Thiazinamium Metilsulfate
58-34-4 9441
$C_{18}H_{23}ClN_2S$
Trimethyl(1-methyl-2-phenothiazin-10-ylethyl)-ammonium chloride.
WY-460E; RP-3554; Multergan; Padisal. Antihistaminic; antiallergic. mp = 206-210° (dec); soluble in H_2O (10 g/100 ml), EtOH; less soluble in Me_2CO; insoluble in Et_2O, C_6H_6. *Wyeth-Ayerst Labs.*

4027 Thonzylamine
91-85-0 9513
$C_{16}H_{22}N_4O$
2-[[2-(Dimethylamino)ethyl](p-methoxybenzyl)-amino]pyrimidine.
Neohetramine. Antihistaminic. $bp_{2.2}$ = 185-187°; LD_{50} (gpg orl) = 493 mg/kg. *Parke-Davis.*

4028 Thonzylamine Hydrochloride
63-56-9 9513
$C_{16}H_{23}ClN_4O$
2-[[2-(Dimethylamino)ethyl](p-methoxybenzyl)amino]pyrimidine hydrochloride.
Super Anahist; Neohetramine hydrochloride; Tonamil. Antihistaminic. mp = 173-176°; soluble in H_2O, EtOH, $CHCl_3$; insoluble in Et_2O. *Parke-Davis.*

4029 Tolpropamine
5632-44-0 9662 227-071-2
$C_{18}H_{23}N$
N,N-Dimethyl-3-phenyl-3-p-tolylpropamine.
Pragman Gelee [as hydrochloride]. Topical antihistaminic; antipruritic. [hydrochloride]: mp = 182-184°. *Hoechst AG.*

4030 Triprolidine
486-12-4 9877
$C_{19}H_{22}N_2$
(E)-2-[3-(1-Pyrrolidinyl)-1-p-toluenepropenyl]pyridine .
Antihistaminic. mp = 59-61°; λ_m = 236, 285 nm (ε 15300, 6800 EtOH). *Glaxo Wellcome Inc.; Sterling Health U.S.A.*

4031 Triprolidine Hydrochloride
6138-79-0 9877
$C_{19}H_{23}ClN_2$
(E)-2-[3-(1-Pyrrolidinyl)-1-p-toluenepropenyl]pyridine monohydrochloride .
Actifed. Antihistaminic. *Glaxo Wellcome Inc.; Sterling Health U.S.A.*

4032 Triprolidine Hydrochloride [anhydrous]
550-70-9 9877
$C_{19}H_{23}ClN_2.H_2O$
(E)-2-[3-(1-Pyrrolidinyl)-1-p-toluenepropenyl]pyridine monohydrochloride monohydrate.
295C51; Actidil; Actidilon; Pro-Actidil; Pro-Entra; Venen. Antihistaminic. mp = 116-118°; λ_m = 235, 283 nm (ε 15000, 7400 EtOH); moderately soluble in H_2O, EtOH, MeOH. *Glaxo Wellcome Inc.; Sterling Health U.S.A.*

4033 Tritoqualine
14504-73-5 9894
$C_{26}H_{32}N_2O_8$
7-Amino-4,5,6-triethoxy-3-(5,6,7,8-tetrahydro-4-methoxy-6-methyl-1,3-dioxolo[4,5-g]isoquinoli-5-yl)phthalide.
L-554; Hypostamine; tritocaline; Inhibostamin; Livalfa. Antihistaminic. mp = 183°.

4034 Zolamine
553-13-9 10319
$C_{15}H_{21}N_3OS$
2-[[2-(Dimethylamino)ethyl]-(p-methoxybenzyl)amino]thiazole.
Local anesthetic. Antihistaminic. Isostere of pyrilamine. bp_7 = 217-219°. *Schering-Plough HealthCare Products.*

4035 Zolamine Hydrochloride
1155-03-9 10319
$C_{15}H_{22}ClN_3OS$
N-[(4-Methoxyphenyl)methyl]-N',N'-dimethyl-N-2-thiazolyl-1,2-ethanediamine hydrochloride.
194-B; WI 291; component of: Otodyne. Antihistaminic. Isostere of pyrilamine with local anesthetic properties. mp = 167.5-167.8°; soluble in H_2O. *Schering-Plough HealthCare Products.*

Antihyperlipoproteinemics

4036 Acetiromate
2260-08-4
$C_{15}H_9I_3O_5$
4-(4-Hydroxy-3-iodophenoxy)-3,5-diiodobenzoic acid acetate.
Hypolipidemic.

4037 Acifran
72420-38-3 112
$C_{12}H_{10}O_4$
(±)-4,5-Dihydro-5-methyl-4-oxo-5-phenyl-2-furoic acid.
AY-25712; Reductol. Hypolipidemic. mp = 176°; λ_m = 281 nm (ε 7960 MeOH); LD_{50} (rat orl) = 3 g/kg; [(-) isomer]: mp = 87-89°; $[\alpha]_D^{25}$ = -144.7° (MeOH, c = 2.0). *Ayerst.*

4038 Acipimox
51037-30-0 113
$C_6H_6N_2O_3$
5-Methylpyrazinecarboxylic acid 4-oxide.
K-9321; Olbemox; Olbetam. Hypolipidemic. mp = 177-180°; LD_{50} (mus orl) =3500 mg/kg. *Farmitalia Carlo Erba Ltd.*

4039 Aluminum Clofibrate
24818-79-9 2437 246-477-0
$C_{20}H_{21}AlCl_2O_7$
Di-[2-(4-chlorophenoxy)-2-methylpropionato] hydroxyaluminum.
Alufibrate; Atherolip; Atherolipin. Hypolipidemic agent. A pharmaceutical used in the treatment of arteriosclerosis.

4040 Atorvastatin
134523-00-5 897
$C_{33}H_{35}FN_2O_5$
(βR,δR)-2-(p-Fluorophenyl)-β,δ-dihydroxy-5-isopropyl-3-phenyl-4-(phenylcarbamoyl)pyrrole-1-heptanoic acid.
CI-981. Hypolipidemic agent. Hydroxymethylglutarate co-enzyme A reductase inhibitor. *Parke-Davis.*

4041 Atorvastatin Calcium
134523-03-8 897
$C_{66}H_{68}CaF_2N_4O_{10}$
Calcium (βR,δR)-2-(p-fluorophenyl)-β,δ-dihydroxy-5-isopropyl-3-phenyl-4-(phenylcarbamoyl)pyrrole-1-heptanoate.
probenecid-colchicine. Hypolipidemic. Hydroxymethylglutarate co-enzyme A reductase inhibitor. $[\alpha]_D$ = -7.4° (DMSO c = 1). *Parke-Davis.*

4042 Beclobrate
55937-99-0 1046
$C_{22}H_{23}ClO_3$
Ethyl (±)-2-[[α-(p-chlorophenyl)-p-tolyl]oxy]-2-methylbutyrate.
Sgd-24774; Beclipur; Turec. Hypolipidemic agent. bp0.01-0.1 = 200-204°; LD_{50} (mus orl) = 8 g/kg.

4043 Beloxamide
15256-58-3
$C_{18}H_{21}NO_2$
N-(Benzyloxy)-N-(3-phenylpropyl)acetamide.
W-1372. Hypolipidemic agent. *Wallace Labs.*

4044 Benfluorex
23602-78-0 1066
$C_{19}H_{20}F_3NO_2$
2-[[α-Methyl-m-(trifluoromethyl)phenethyl]amino]ethanol benzoate (ester).
S-780; SE-780; Minolip. Hypolipidemic. Colorless oil.

4045 Benfluorex Hydrochloride
23642-66-2 1066
$C_{19}H_{21}ClF_3NO_2$
2-[[α-Methyl-m-(trifluoromethyl)phenethyl]amino]ethanol benzoate (ester) monohydrochloride (salt).
S-992; JP-992; Mediator; Mediaxal. Hypolipidemic agent. mp = 161-162°. *C.M. Ind.*

4046 Bezafibrate
41859-67-0 1240 255-567-9
$C_{19}H_{20}ClNO_4$
2-[4-[2-(4-Chlorobenzoyl)amino]-ethyl]phenoxy]-2-methylpropanoic acid.
BM-15075; Befizal; Bezalip; Bezatol; Cedur; Difaterol. Hypolipidemic agent. A fibric acid; reduces plasma levels of very-low-density lipoprotein. mp = 186°. *Boehringer Mannheim GmbH; C.M. Ind.*

4047 Binifibrate
69047-39-8 1267
$C_{25}H_{23}ClN_2O_7$
2-(p-Chlorophenoxy) 2-methylpropionic acid ester with 1,3-dinicotinoyloxy-2-propanol.
WAC-104; Biniwas. Hypolipidemic agent. mp = 100°; LD_{50} (mus, rat orl) > 4000 mg/kg.

4048 Boxidine
10355-14-3
$C_{19}H_{20}F_3NO$
1-[2-[[4'-(Trifluoromethyl)[1,1'-biphenyl]-4-yl]oxy]ethyl]-pyrrolidine.
CL-65205. Hypolipidemic agent.

4049 Carnitine
461-06-3 1898
$C_7H_{15}NO_3$
L-(3-Carboxy-2-hydroxypropyl)trimethylammonium hydroxide inner salt.
3-Hydroxy-4-trimethylammoniobutanoate; γ-Trimethyl-β-hydroxybutyrobetaine; levocarnitine; vitamin B_7; Cardiogen; Carnitene; Carnicor; Carnum; Carrier; Miocor; Miotonal; Vitacarn. Hypolipidemic agent. mp = 197-198° (dec); $[\alpha]_D^{30}$ = -23.9° (H_2O c = 0.86); soluble in H_2O, hot EtOH; insoluble in Me_2CO, Et_2O, C_6H_6.

4050 Cerivastatin
145599-86-6
$C_{26}H_{34}FNO_5$
(+)-(3R,5S,6E)-7-[4-(p-Fluorophenyl)-2,6-diisopropyl-5-(methoxymethyl)-3-pyridyl]-3,5-dihydroxy-6-heptenoic acid.
Hypolipidemic agent. Hydroxymethylglutarate co-enzyme A reductase inhibitor. *Bayer AG.*

4051 Cerivastatin Sodium
143201-11-0
$C_{26}H_{33}FNNaO_5$
(+)-Sodium (3R,5S,6E)-7-[4-(p-fluorophenyl)-2,6-diisopropyl-5-(methoxymethyl)-3-pyridyl]-3,5-dihydroxy-6-heptenoate.
Bay w 6228; Baycol. Hypolipidemic agent. Hydroxymethylglutarate co-enzyme A reductase inhibitor. *Bayer AG.*

4052 Cetaben
55986-43-1
$C_{23}H_{39}NO_2$
p-(Hexadecylamino)benzoic acid.
Hypolipidemic agent.

4053 Cetaben Sodium
64059-66-1
$C_{23}H_{38}NNaO_2$
Sodium (p-hexadecylamino)benzoate.
CL-203821. Hypolipidemic agent.

4054 Cholestyramine Resin
11041-12-6 2257
Cholestyramin.
colestyramine; Cholybar; Duolite AP-143 Resin; Questran; Questran Light; Dowex 1-X2-Cl; MK-135; Cuemid; Quantalan. Ion exchange resin, binds bile acids. Used as a hypolipidemic agent. Bile acid sequestrant. Ion exchange resin, insoluble in H_2O, organic solvents. *Bristol Laboratories; Parke-Davis; Rohm & Haas Co.*

4055 Choloxin
137-53-1 9555 205-301-2
$C_{15}H_{10}I_4NNaO_4$
Dextrothyroxine sodium.
D-thyroxine sodium; Biotirmone; Choloxin; Detyroxin; Dethyrona; Dextroid; Dynothel; Eulipos; Sodium D-

thyroxine. Hypolipidemic agent. Antihyperlipoproteinemic. *Boots Pharmaceuticals Inc.; Johnson R. W. Pharm. Res. Institute; Knoll Pharm. Co.*

4056 Chondroitin 4-Sulfate
24967-93-9 2270
Chondroitin 4-(sodium sulfate).
ORG-10172. Hypolipidemic agent. Mucopolysaccharide; major constituent of the cartilagenous tissue in the body; Antihyperlipo-proteinemic agent. $[\alpha]^D$ = -28° to -32°. *Diosynth BV.*

4057 Chondroitin Sulfate
9007-28-7 2270 232-696-9
Chondroitin sulfuric acid.
Chonsurid; Structum. Hypolipidemic agent. Mucopolysaccharide; major constituent of the cartilagenous tissue in the body; Antihyperlipoproteinemic agent. *Diosynth BV.*

4058 Ciprofibrate
52214-84-3 2373
$C_{13}H_{14}Cl_2O_3$
2-[4-(2,2-Dichlorocyclopropyl)phenoxy]-2-methylpropionic acid.
Win-35833; Ciprol; Lipanor; Modalim. Hypolipidemic agent. A fibric acid; reduces plasma triglycerides by lowering plasma levels of very-low-density lipoprotein. Chemicaly related to clofibrate. mp = 114-116°. *Sterling Winthrop Inc.*

4059 Clinofibrate
30299-08-2 2415
$C_{28}H_{36}O_6$
2,2'-(4,4'-Cyclohexylidinediphenoxy)-2,2'-dimethyldibutyric acid.
S-8527; Lipoclin. Hypolipidemic agent. mp = 143-146°; insoluble in H_2O, soluble in organic solvents; LD_{50} (mmus orl) = 1800 mg/kg, (mmusip) = 255 mg/kg, (mmus sc) = 410 mg/kg, (mrat orl) > 4000 mg/kg, (mrat ip) = 205 mg/kg, (mrat sc) = 2200 mg/kg.

4060 Clofibrate
637-07-0 2436 211-277-4
$C_{12}H_{15}ClO_3$
Ethyl 2-(4-chlorophenoxy)-2-methylpropionate.
ethyl-α-p-chlorophenoxyisobutyrate; clofibric acid; Abitrate; Atromid S; Amotril; Anparton; Apolan; Artevil; Ateculon; Arteriosan; Atheropront; Atromidin; Bioscleran; Claripex; Clobren-SF; Clofinit; CPIB; Hyclorate; Liprinal; Neo-Atromid; Normet; Normolipol; Recolip; Regelan; Serotinex; Sklerolip; Sklerepmexe; Sklero-Tablinene; Ticlobran; Xyduril. Hypolipidemic agent. A fibric acid; reduces plasma triglycerides by lowering plasma levels of very-low-density lipoprotein. bp_{20} = 148-150°; insoluble in H_2O, soluble in organic solvents; LD_{50} (rat orl) = 1.65 g/kg. *ICI Chem. & Polymers Ltd.; Wyeth-Ayerst Labs.*

4061 Clofibric Acid
882-09-7 2437 212-925-9
$C_{10}H_{11}ClO_3$
2-(4-Chlorophenoxy)-2-methylpropanoic acid.
chlorophibrinic acid; Arteriohom; Regulipid. Hypolipidemic agent. Antihyperlipoproteinemic. mp = 118-119°; LD_{50} (rat orl) = 897 mg/kg. *Wyeth-Ayerst Labs.*

4062 Clofibride
26717-47-5
$C_{16}H_{22}ClNO_4$
2-(p-Chlorophenoxy)-2-methylpropionic acid ester with 4-hydroxy-N,N-dimethylbutyramide.
Antihyperlipoproteinemic.

4063 Colestipol
26658-42-4 2538
N-(2-Aminoethyl)-N'-[(2-aminoethyl]amino]ethyl]-1,2-ethanediamine polymer with (chloromethyl)oxirane.
Hypolipidemic agent. Bile acid sequestrant. *Pharmacia & Upjohn.*

4064 Colestipol Hydrochloride
37296-80-3 2538
N-(2-Aminoethyl)-N'-[(2-aminoethyl]amino]ethyl]-1,2-ethanediamine polymer with (chloromethyl)oxirane hydrochloride salt.
U-26597A; Cholestabyl; Lestid; Colestid; Colestid. Hypolipidemic. Bile acid sequestrant. Basic anion-exchange resin, highly cross-linked, insoluble. LD_{50} (rat orl) > 1000 mg/kg, (rat ip) > 4000 mg/kg. *Pharmacia & Upjohn.*

4065 Colextran
9015-73-0 2980
Dextran 2-(diethylamino)ethyl ether.
diethylaminoethyl dextran; Detaxtran; DEAE-dextran; basic Dextran. Hypolipidemic agent. Soluble in H_2O, saline. *Pharmacia & Upjohn.*

4066 Doxercalciferol
1α-Hydroxyvitamin D(2).
Hectorol. Antihyperphosphatemic; calcium regulator. A vitamin D prohormone. Suppresses intact parathyroid hormone (iPTH) with minimal increase in serum calcium and phosphorus. Used to treat hypercalcemia and hyperphosphatemia in dialysis patients; treatment of secondary hyperparathyroidism. *Bone-Care Intl.*

4067 Dulofibrate
61887-16-9
$C_{16}H_{14}Cl_2O_3$
p-Chlorophenyl 2-(p-chlorophenoxy)-2-methylpropionate.
Antihyperlipoproteinemic.

4068 Eniclobrate
60662-18-2
$C_{24}H_{24}ClNO_3$
3-Pyridylmethyl (±)-2-[[α-(p-chlorophenyl)-p-tolyl]oxy]-2-methylbutyrate.
Antihyperlipoproteinemic.

4069 Ethyl Icosapentate
73310-10-8
$C_{22}H_{34}O_2$
Ethyl all-cis-5,8,11,14,17-icosapentaenoic acid.
Antihyperlioproteinemic.

4070 Etofibrate
31637-97-5 3923
$C_{18}H_{18}ClNO_5$
2-Hydroxyethyl nicotinate 2-(p-chlorophenoxy)-2-methylpropionate (ester).
ethofibrate; Lipo-Merz. Hypolipidemic agent. mp = 100°.

4071 Fenofibrate
49562-28-9 4019
$C_{20}H_{21}ClO_4$
Isopropyl 2-[p-(p-chlorobenzoyl)phenoxy]-2-methylpropionate.
Procetofen; Procetofene; LF-178; Ankebin; Elasterin; Fenobrate; Fenotard; Lipanthyl; Lipantil; Lipidil; Lipoclar; Lipofene; Liposit; Lipsin; Nolipax; Procetoken; Protolipan; Secalip. Hypolipidemic agent. A fibric acid; reduces plasma triglycerides by lowering plasma levels of very-low-density lipoprotein. mp = 80-81°; insoluble in H_2O; slightly soluble in EtOH, MeOH; more soluble in non-polar solvents; LD_{50} (mus orl) = 1600 mg/kg.

4072 Fluvastatin
93957-54-1 4250
$C_{24}H_{26}FNO_4$
(±)-(3R*,5S*,6E)-7-(3-p-Fluorophenyl)-1-isopropylindol-2-yl]-3,5-dihydroxy-6-heptenoic acid.
Hypolipidemic agent. Hydroxymethylglutarate co-enzyme A reductase inhibitor. *Sandoz Pharm. Corp.*

4073 Fluvastatin Sodium
93957-55-2 4250
$C_{24}H_{25}FNNaO_4$
Sodium (±)-(3R*,5S*,6E)-7-(3-p-fluorophenyl)-1-isopropylindol-2-yl]-3,5-dihydroxy-6-heptenoate.
XU-62320; Lescol; fluindostatin. Hypolipidemic agent. Hydroxymethylglutarate co-enzyme A reductase inhibitor. mp = 194-197°. *Sandoz Pharm. Corp.*

4074 Gamma Oryzanol
11042-64-1 7016
$C_{40}H_{58}O_4$
9,19-Cyclo-9β-lanost-24-en-3β-ol 4-hydroxy-3-methoxy cinnamate.
OZ; γ-OZ; γ-orizanol; Caclate; Gammajust 50; Gamma-OZ; Gammariza; Gammatsul; Guntrin; Hi-Z; Maspiron; Oliver; Oryvita; Oryzaal; Thiaminogen. Hypolipidemic agent with antiulcerative properties. mp = 135-137°; λ_m = 216, 231, 291, 315 nm (heptane). *Toyo Koatsu Co. Ltd.*

4075 Gemcadiol
35449-36-6
$C_{14}H_{30}O_2$
2,2,9,9-Tetramethyl-1,10-decanediol.
CI-720. Hypolipidemic agent. *Parke-Davis.*

4076 Gemfibrozil
25812-30-0 4394 247-280-2
$C_{15}H_{22}O_3$
2,2-Dimethyl-5-(2,5-xylyloxy)valeric acid.
CI-719; Decrelip; Genlip; Gevilon; Lipur; Lopid; Lopizid. Hypolipidemic agent. Serum lipid regulator. A fibric acid; reduces plasma levels of very-low-density lipoprotein. mp = 61-63°; $bp_{0.02}$ = 158-159°; LD_{50} (mus orl) = 3162 mg/kg, (rat orl) = 4786 mg/kg. *Parke-Davis.*

4077 Glunicate
80763-86-6
$C_{36}H_{28}N_6O_{10}$
2-Deoxy-2-nicotinamido-β-D-glucopyranose 1,3,4,6-tetranicotinate.
Has hypolipidemic properties.

4078 Halofenate
26718-25-2
$C_{19}H_{17}ClF_3NO_4$
(p-Chlorophenyl)[(α,α,α-trifluoro-m-tolyl)oxy]acetic acid ester with N-(2-hydroxyethyl)acetamide.
Lipivas. Uricosuric; antihyperlipoproteinemic.

4079 Icosapent
10417-94-4 3572
$C_{20}H_{30}O_2$
(all Z)-5,8,11,14,17-Eicosapentaenoic acid.
Hypolipidemic agent. Prostaglandin and thromboxane precursor. Colorless oil.

4080 Itanoxone
58182-63-1
$C_{17}H_{13}ClO_3$
2-[p-(o-Chlorophenyl)phenacyl]acrylic acid.
Antihyperlipoproteinemic agent.

4081 Levocarnitine
541-15-1 1898 208-768-0
$C_7H_{15}NO_3$
L-(3-Carboxy-2-hydroxypropyl)trimethylammonium hydroxide inner salt.
3-Hydroxy-4-trimethylammoniobutanoate; γ-Trimethyl-β-hydroxybutyrobetaine; levocarnitine; vitamin B_7; Cardiogen; Carnitene; Carnicor; Carnitor; Carnum; Carrier; Miocor; Miotonal; Vitacarn. Hypolipidemic agent. Carnitine replenisher. Dec 197-198°; $[\alpha]_D^{30}$ = -23.9° (c = 0.86 H_2O); soluble in H_2O, hot EtOH; insoluble in Me_2CO, Et_2O, C_6H_6. *Sigma-Tau Pharm. Inc.*

4082 Lifibrate
22204-91-7
$C_{20}H_{21}Cl_2NO_4$
1-Methyl-4-piperidyl glyoxylate 2-[bis(p-chlorophenyl)-acetal].
42-348. Hypolipidemic agent. *Sandoz Pharm. Corp.*

4083 Lovastatin
75330-75-5 5616
$C_{24}H_{36}O_5$
[1S-[1α(R*),3α,7β,8β(2S*,4S*),8aβ]]-2-Methylbutanoic acid 1,2,3,7,8,8a-hexahydro-3,7-dimethyl-8-[2-(tetra-hydro-4-hydroxy-6-oxo-2H-pyran-2-yl]ethyl]-1-naphthal-enyl ester.
6α-methylcompactin; mevinolin; monacolin K; MK-803; Lovalip; Mevacor; Mevinacor; Mevlor; Sivlor. Antihyperlipoproteinemic. HMG-CoA reductase inhibitor; used as an antihypercholesterolemic agent. A fungal metabolite. mp = 174°; $[\alpha]_D^{25}$ = 323° (c = 0.5 CH_3CN); λ_m = 231, 238, 247 nm ($A^{1\%}$ 532, 621, 418); almost insoluble in H_2O, soluble in organic solvents; LD_{50} (mus orl) > 1000 mg/kg. *Merck & Co.Inc.*

4084 Magnesium Clofibrate
14613-30-0 2437
$C_{20}H_{20}Cl_2MgO_6$
Bis[2-(p-chlorophenoxy)-2-methylpropionato]magnesium.
UR-112; Clomag. Hypolipidemic. Decreases rate of cholesterol synthesis. mp = 326-328°; soluble in H_2O (0.45 g/100 ml), EtOH (7 g/100 ml), $CHCl_3$ (0.02 g/100 ml).

4085 Meglutol
503-49-1 5852
$C_6H_{10}O_5$
3-Hydroxy-3-methylglutaric acid.
CB-337; dicrotalic acid; medroglutaric acid; HMG; HMGA; Lipoglutaren; Mevalon. Hypolipidemic agent. mp = 108-109°; soluble in H_2O; LD_{50} (mus orl) = 7.33 g/kg, (mus ip) - 3.23 g/kg. *Hoechst Roussel Pharm. Inc.*

4086 Melinamide
14417-88-0 5864
$C_{26}H_{41}NO$
N-(α-Methylbenzyl)linoleamide.
MBLA; AC-223; Artes. Hypolipidemic agent. mp < 4°; $bp_{0.03}$ = 200-215°; $bp_{0.07}$ = 200-204°. *Sumitomo Pharm. Co. Ltd.*

4087 Mevastatin
73573-88-3 6251
$C_{23}H_{34}O_5$
(1S,7S,8S,8aR)-1,2,3,7,8,8a-Hexahydro-7-methyl-8-[2-[(2R,4R)-tetrahydro-4-hydroxy-6-oxo-2H-pyran-2-yl]ethyl]-1-naphthyl (S)-2-methylbutyrate.
Antihyperlipoproteinemic. HMG-CoA reductase inhibitor; used as an antihypercholesterolemic agent. mp = 152°; $[\alpha]_D^{22}$ = 283° (Me_2CO c = 0.48); λ_m = 230, 237, 246 nm (log ε 4.28, 4.30, 4.11). *Sankyo Co. Ltd.*

4088 Nafenopin
3771-19-5
$C_{20}H_{22}O_3$
2-Methyl-2-[4-(1,2,3,4-tetrahydro-1-naphthyl)phenoxy]propionic acid.
Su-13437; TPIA; ch 13-437; ciba 13437 su; c 13437 su. Hypolipidemic agent. mp = 117-118°. *CIBA plc.*

4089 Neomycin
1404-04-2 6542 215-766-3
Mycifradin; Fradiomycin; Neomin; Neolate; Neomas; Pimavecort; Vonamycin Powder V. Aminoglycoside antibiotic. Antibiotic produced by *Streptomyces fradiae*. Consists of Neomycins A, B and C. Has a hypolipidemic effect when administered orally, probably through the formation of insoluble complexes with bile acids in the intestine. Insoluble in organic solvents; slightly soluble in H_2O (< 25 g/100 ml). *Merck & Co.Inc.; Pharmacia & Upjohn.*

4090 Niacin
59-67-6 6612 200-441-0
$C_6H_5NO_2$
3-Pyridinecarboxylic acid.
Nicotinic acid; Niac; Nicamin; Nicobid; Nicolar; Wampocap; P.P. Factor; Akotin; Daskil; Niacor; Nicacid; Nicangin; Niconacid; NicoSpan. Vitamin (enzyme cofactor). At high doses, decreases hepatic secretion of very-high-density lipoproteins as a result of reduced triglyceride synthesis. Used in treatment of hypercholesterolemias and hypertriglyceridemias. mp = 236.6°; λ_m = 263 nm; soluble in H_2O (1.67 g/100 ml); freely soluble in H_2O at 100°, EtOH at 76°; insoluble in Et_2O; LD_{50} (raty sc) = 5000 mg/kg. *Abbott Labs Inc.; Apothecon; Forest Pharm. Inc.; Marion Merrell Dow Inc.; Rhône-Poulenc Rorer Pharm. Inc.; Wallace Labs.*

4091 Nicanartine
150443-71-3
$C_{23}H_{33}NO_2$
(5-(3,5-Di-tert-butyl)-4-hydroxyphenyl-1-(3-pyridyl)-2-oxapentane.
MRZ-3/124. Antioxidant with atherosclerotic plaque-reducing actiivity.

4092 Niceritrol
5868-05-3 6581
$C_{29}H_{24}N_4O_8$
3-Pyridinecarboxylic acid 2,2-bis[[(3-pyridinylcarbonyl)-oxy]methyl]-1,3-propanediyl ester.
pentaerythritol tetranicotinate; 8-AL; Perycit; Bufor. Hypolipidemic agent. mp = 160-162°, 163-164°; LD_{50} (mus orl) > 20 g/kg, (mus sc) > 5 g/kg; (mus ip) > 5 g/kg, (rat orl) >20 g/kg, (rat sc) >5 g/kg, (rat ip) > 5 g/kg, (rbt orl) > 10 g/kg, (rbt ip) > 5 g/kg.

4093 Nicofibrate
31980-29-7 6604
$C_{16}H_{16}ClNO_3$
3-Pyridiylmethyl 2-(p-chlorophenoxy))-2-methylpropionate.
clofenpyride. Hypolipidemic agent. mp = 48-49°; $bp_{0.4}$ = 180°. [hydrochloride salt]: mp = 115.5-118.5°. *Merck & Co.Inc.*

4094 Oxiniacic Acid
2398-81-4 7075
$C_6H_5NO_3$
Nicotinic acid 1-oxide.
3-carboxypyridine N-oxide. Hypolipidemic agent. mp = 244-245°; slightly soluble in H_2O, AcOH; less soluble in EtOH; insoluble in non-polar organic solvents; λ_m = 220, 260 nm (ε 22400, 10200 0.1N H_2SO_4).

4095 Pantethine
16816-67-4 7144
$C_{22}H_{42}N_4O_8S_2$
D-Bis(N-pantothenyl)-2-aminoethyl)-disulfide.
Lipodel; Pantetina; Panthecin; Pantomin; Pantosin. Hypolipidemic agent. $[\alpha]_D^{27}$ = 13.5° (H_2O c = 3.75); soluble in H_2O, EtOH; insoluble in other organic solvents.

4096 Pimetine
3565-03-5
$C_{16}H_{26}N_2$
4-Benzyl-1-[2-(dimethylamino)ethyl]piperidine.
Hypolipidemic agent.

4097 Pimetine Hydrochloride
4991-68-8
$C_{16}H_{28}Cl_2N_2$
4-Benzyl-1-[2-(dimethylamino)ethyl]piperidine dihydrochloride.
Hypolipidemic agent.

4098 Pirifibrate
55285-45-5 7650
$C_{17}H_{18}ClNO_4$
[6-(Hydroxymethyl)-2-pyridyl]methyl 2-(p-chlorophen-oxy)-2-methylpropionate.

EL-466; Bratenol. Hypolipidemic agent. mp = 46°; LD_{50} (mus ip) = 915-1098 mg/kg. *Hoechst Roussel Pharm. Inc.*

4099 Pirinixic acid
50892-23-4
$C_{14}H_{14}ClN_3O_2S$
[[4-Chloro-6-(2,3-xylidino)-2-pyrimidinyl]thio] acetic acid.
WY-14643. Peroxisome proliferator; produces a significant hepatomegaly and induces the peroxisomal fatty acid beta-oxidation enzyme system together with profound proliferation of peroxisomes in hepatic parenchymal cells. Antihyperlipoproteinemic.

4100 Pirinixil
65089-17-0
$C_{16}H_{19}ClN_4O_2S$
2-[[4-Chloro-6-(2,3-xylidino)-2-pyrimidinyl]thio]-N-(2-hydroxyethyl)acetamide.
BR-931. Peroxisome proliferator. Antihyperlipoproteinemic.

4101 Pirozadil
54110-25-7 7662
$C_{27}H_{29}NO_{10}$
2,6-Pyridinediyldimethylenebis(3,4,5-trimethoxybenzoate).
722-D; Pemix. Hypolipidemic agent. mp = 119-126°; soluble in $CHCl_3$, dioxane, CH_3CN; insoluble in Et_2O, H_2O.

4102 Polidexide
56227-39-5
Poly-[2-(diethylamino)ethyl] polyglycerylenedextran.
Secholex. An ion-exchange resin that acts as a bile acid-sequestering agent. Antihyperlipoproteinemic.

4103 Polidexide Sulfate
63494-82-6
Dextran 2-(diethylamino)ethyl 2-[[2-(diethylamino)-ethyl]diethylamino]ethyl ester sulfate epichlorohydrin crosslinked.
An ion-exchange resin that acts as a bile acid-sequestering agent. Antihyperlipoproteinemic.

4104 Pravastatin
81093-37-0
$C_{23}H_{36}O_7$
(+)-(3R,5R)-3,5-Dihydroxy-7-[(1S,2S,6R,8S,8aR)-6-hydroxy-2-methyl-8-[(S)-2-methylbutyryloxy]-1,2,6,7,8,8a-hexahydro-1-naphthyl]heptanoic acid.
3β-hydroxycompactin; eptastatin. Antihyperlipoproteinemic. HMG-CoA reductase inhibitor; used as an antihypercholesterolemic agent. Active metabolite of mevastatin. *Sankyo Co. Ltd.*

4105 Pravastatin Sodium
81131-70-6 7984
$C_{23}H_{35}NaO_7$
Sodium (+)-(3R,5R)-3,5-dihydroxy-7-[(1S,2S,6R,8S,8aR)-6-hydroxy-2-methyl-8-[(S)-2-methylbutyryloxy]-1,2,6,7,8,8a-hexahydro-1-naphthyl]heptanoate.
3β-hydroxycompactin sodium salt; eptastatin sodium; CS-514; SQ-31000; Elisor; Lipostat; Mevalotin; Oliprevin; Pravachol; Praváselect; Selectin; Selipran; Vasten. Antihyperlipoproteinemic. HMG-CoA reductase inhibitor; used as an antihypercholesterolemic agent. Active metabolite of mevastatin. λ_m = 230, 237, 245 nm. *Sankyo Co. Ltd.*

4106 Probucol
23288-49-5 7935 245-560-9
$C_{31}H_{48}O_2S_2$
Acetone bis(3,5-di-t-butyl-4-hydroxyphenyl) mercaptole.
DH-581; Lorelco; Lurselle; Sinlestal. Synthetic lipophilic antioxidant. For treatment of hypercholesterolemia. mp = 124.5-126°, 125-126.5°. *Merrell Pharm. Inc.*

4107 Ronifibrate
42597-57-9 8414
$C_{19}H_{20}ClNO_5$
3-Hydroxypropyl nicotinate 2-(p-chlorophenoxy)-2-methylpropionate (ester).
I-612; Cloprane. Hypolipidemic agent. LD_{50} (mus orl) = 4.08 g/kg. *Yamanouchi U.S.A. Inc.*

4108 Simvastatin
79902-63-9 8686
$C_{25}H_{38}O_5$
2,2-Dimethyl butyric acid 8-ester with (4R,6R)-6-[2-[(1S,2S,6R,8S,8aR)-1,2,6,7,8,8a-hexahydro-8-hydroxy-2,6-dimethyl-1-naphthyl]ethyl]tetrahydro-4-hydroxy-2H-pryan-2-one.
MK-733; synvinolin (formerly); Denan; Liponorm; Lodalès; Simovil; Sivastin; Zocor; Zocord. Antihyperlipoproteinemic. HMG-CoA reductase inhibitor; used as an antihypercholesterolemic agent. Synthetic analog of lovastatin. mp = 135-138°. *Merck & Co.Inc.*

4109 β-Sitosterol
83-46-5 8697 201-480-6
$C_{29}H_{50}O$
(3β)-Stigmast-5-en-3-ol.
α-dihydrofucosterol; α-phytosterol; cinchol; cupreol; rhammol; quebrachol; sitosterin; Cytellin; Harzol; Prostasal; Sito-Lande. Anticholesterolemic. For treatment of prostatic adenoma. Plant sterol similar to cholesterol; lowers plasma concentrations of low-density-lipoprotein. mp = 140°; $[\alpha]_D^{25}$ = -41° (c = 2 $CHCl_3$). *Eli Lilly & Co.*

4110 Sorbinicate
6184-06-1 228-230-9
$C_{42}H_{32}N_6O_{12}$
D-Glucitol hexanicotinate.
Nicotinic acid derivative. Antilipolytic; antihyperlipoproteinemic; antiarteriosclerotic.

4111 Sultosilic Acid
57775-26-5 9169
$C_{13}H_{12}O_7S_2$
2,5-Dihydroxybenzenesulfonic acid 5-p-toluenesulfonate.
Hypolipidemic agent.

4112 Sultosilic Acid Piperazine Salt
57775-27-6 9169
$C_{17}H_{22}N_2O_7S_2$
2,5-Dihydroxybenzenesulfonic acid 5-p-toluenesulfonate piperazine salt.
diethylenediamine sultosylate; piperazine sultosylate; A-585; Mimedran. mp = 74°; LD_{50} (dog ip) = 605 mg/kg, (mrat ip) = 833.6 mg/kg, (frat ip) = 1272 mg/kg, (rat orl) > 11 g/kg.

4113 Tazasubrate
79071-15-1
$C_{18}H_{17}NO_3S_2$
(±)-α-[(6-Ethoxy-2-benzothiazolyl)thio]hydratropic acid.
EMD-34853. Lipid lowering agent.

4114 Telmesteine
122946-43-4
$C_9H_{11}NO_4S$
(-)-3-Ethyl hydrogen (R)-3,4-thiazolidinedicarboxylate.
Antihyperlipoproteinemic.

4115 Terbufibrol
56488-59-6
$C_{20}H_{24}O_5$
p-[3-(p-tert-Butylphenoxy)-2-hydroxypropoxy]benzoic acid.
Antihyperlipoproteinemic.

4116 Theofibrate
54504-70-0 9420
$C_{19}H_{21}ClN_4O_5$
2-(p-Chlorophenoxy) 2-methylpropionic acid ester with 7-(2-hydroxyethyl)theophylline.
Duolip; ML-1024; etofylline clofibrate. Hypolipidemic agent with antilipemic, antithrombotic and platelet aggregation inhibitory acitvity. mp = 133-135°; insoluble in H_2O, EtOH; soluble in Me_2CO, $CHCl_3$ and hot alcohols; LD_{50} (mus orl) = 11.7 mg/kg, (dog orl) > 10.0 g/kg, (rat orl) = 17.0 g/kg. *MerckleL. GmbH.*

4117 D-Thyroxine
51-49-0 9555 200-102-7
$C_{15}H_{11}I_4NO_4$
D-O-(4-Hydroxy-3,5-diiodophenyl)-3,5-diiodotyrosine.
dextrothyroxine; Debetrol. L-form is a thyroid hormone; D-form is antihyperlipoproteinemic. Optical isomer of the endogenous hormone L-thyroxine. Can produce modest lowering of plasma low-density lipoprotein. CAUTION: may cause serious cardiac toxicity. Dec 237°; $[\alpha]^{21}_{546}$ = 2.97° (c = 3.7 NaOH/EtOH). *Astra USA Inc.; Baxter Healthcare Systems; Forest Pharm. Inc.; Knoll Pharm. Co.*

4118 Tiadenol
6964-20-1 9556
$C_{14}H_{30}O_2S_2$
2,2'-(Decamethylenedithio)diethanol.
LL-1558; Delipid; Eulip; Finlipol; Tiaden; Tiaterol. Hypolipidemic agent. mp = 69.5; λ_m = 212 nm; insoluble in H_2O; soluble in EtOH, $CHCl_3$. *Eastman Kodak.*

4119 Tibric Acid
37087-94-8
$C_{14}H_{18}ClNO_4S$
2-Chloro-5-[(cis-3,5-dimethylpiperidino)sulfonyl]benzoic acid.
CP-18524; CAS RN 24358-29-0. Hypolipidemic agent. *Pfizer Inc.*

4120 Tizoprolic Acid
30709-69-4
$C_7H_9NO_2S$
2-Propyl-5-thiazolecarboxylic acid.
Antihyperlipoproteinemic.

4121 Tocofibrate
50465-39-9
$C_{39}H_{59}ClO+4$
2,5,7,8-Tetramethyl-2-(4,8,12-trimethyltridecyl)-6-chromanyl 2-(p-chlorophenoxy)-2-methylpropionate.
Antyhyperlipoproteinemic.

4122 Treloxinate
30910-27-1
$C_{16}H_{12}Cl_2O_4$
Methyl-2,10-dichloro-12H-dibenzo[d,g][1,3]dioxocin-6-carboxylate.
Hypolipidemic agent. *Marion Merrell Dow Inc.*

4123 Triparanol
78-41-1 9867
$C_{27}H_{32}ClNO_2$
2-p-Chlorophenyl-1-[p-(2-diethylaminoethoxy)phenyl]-1-p-tolylethanol.
MER-2p; Trianel; Hipocolestina; Triparin; Acosterina; Metasclene; Diticyl; Drenaren; Clotrox; Tropalin; Trikosterol; Valip; Verdiana; Metasqualene; Sclane. Hypolipidemic agent. Antilipemic. mp = 102-104°; insoluble in H_2O, soluble in EtOH.

4124 Xenbucin
959-10-4 10205
$C_{16}H_{16}O_2$
(±)-α-Ethyl-4-biphenylacetic acid.
Liosol; Liposana; MG-1559; Maggioni 1559. Hypolipidemic. mp = 123-125°; insoluble in H_2O, soluble in most organic solvents. *Maggioni Farmaceutici S.p.A.*

Antihyperphosphatemics

4125 Aluminum Hydroxide
21645-51-2 355 244-492-7
H_3AlO_3
Aluminum hydroxide.
Amphojel; Dialume; Simeco; component of: Arthritis Pain Formula Maximum Strength, Calcitrel, Camalox, Gelusil, Kestomatin, Kudrox, Maalox, Maalox HRF, Maalox Plus, Simeco Suspension, Tricreamalate, Trsiogel, Wingel. Antacid with antihyperphosphatemic properties. White bulky amorphous powder; insoluble in H_2O, soluble in alkaline or acid solutions. *Wyeth Labs.*

4126 Aluminum Hydroxychloride
1327-41-9 356 215-477-2
$Al_2Cl(OH)_5{\cdot}2H_20$
Aluminum chlorohydrate.
basic aluminum chloride; aluminum chlorohydroxide; Astringen; Chlorhydrol; Hyperdrol; Locron; Phosphonorm. Astringent; antihyperphosphatemic. Used in antiperspirants. Anhidrotic. *Elizabeth Arden.*

4127 Sevelamer Hydrochloride
182683-00-7
$(C_3H_7N)_m.(C_3H_5ClO)_n.xHCl$
Allylamine polymer with 1-chloro-2,3-epoxypropane hydrochloride.
RenaGel; GT16-026A. Antihyperphosphatemic. *Dow Chem. U.S.A.*

Antihypertensives

4128 Acebutolol
37517-30-9 16 253-539-0
$C_{18}H_{28}N_2O_4$
(±)-3'-Acetyl-4'-[2-hydroxy-3-(1-methylethylamino)-propoxy]butyranilide.
Monitan; Sectral; Prent. Antianginal agent. Class II antiarrhythmic agent. Cardioselective β-adrenergic blocking agent. Has antiarrhythmic and antihypertensive properties. mp = 119-123°. *Wyeth Labs.*

4129 Acebutolol Hydrochloride
34381-68-5 16 251-980-3
$C_{18}H_{29}ClN_2O_4$
(±)-3'-Acetyl-4'-[2-hydroxy-3-(1-methylethylamino)-propoxy]butyranilide hydrochloride.
IL-17803A; M&B-17803A; Acetanol; Neptall; Sectral. Antianginal agent. Class II antiarrhythmic agent. Cardioselective β-adrenergic blocking agent. Has antiarrhythmic and antihypertensive properties. mp = 141-143°; soluble in H_2O (200 mg/ml), EtOH (70 mg/ml). *Wyeth Labs.*

4130 Ajmaline
4360-12-7 194 224-439-4
$C_{20}H_{26}N_2O_2$
Ajmalan-17,20-diol.
rauwolfine; Gilurytmal; Cardiorythmine; Ritmos; Tachmalin. Antiarrhythmic agent with antihypertensive properties. [MeOH solvate]: mp = 158-160°; $[\alpha]_D^{18}$ = 131° (c = 0.4 $CHCl_3$); [anhydrous form]: mp = 205-207°; $[\alpha]_D^{20}$ = 144° (c = 0.8 $CHCl_3$); λ_m = 247, 295 nm (log ε 3.94, 3.49 EtOH); soluble in EtOH, MeOH, Et_2O, $CHCl_3$; poorly soluble in H_2O.

4131 Alacepril
74258-86-9 202
$C_{20}H_{26}N_2O_5S$
N-[1-[(S)-3-Mercapto-2-methylpropionyl]-L-prolyl]-3-phenyl-L-alanine acetate (ester).
DU-1219; Cetapril. Angiotensin-converting enzyme inhibitor. Used to treat hypertension. mp= 155-156°; $[\alpha]_D^{25}$ = -81.3° (c = 1.02 EtOH); LD_{50} (rat orl) > 5000 mg/kg, (rat sc) > 3000 mg/kg, (rat ip) ≅ 2000 mg/kg, (mus orl) > 5000 mg/kg, (mus sc) > 3000 mg/kg, (mus ip) ≅ 3000 mg/kg. *Dainippon Pharm.*

4132 Alfuzosin
81403-80-7 237
$C_{19}H_{27}N_5O_4$
(±)-N-[3-[(4-Amino-6,7-dimethoxy-2-quinazolinyl)methylamino]propyl]tetrahydro-2-furamide.
SL-77499. Antihypertensive agent. An α_1 andrenoreceptor antagonist used in the treatment of benign prostatic hyperplasia. *Synthelabo Pharmacie.*

4133 Alfuzosin Hydrochloride
81403-68-1 237
$C_{19}H_{28}ClN_5O_4$
(±)-N-[3-[(4-Amino-6,7-dimethoxy-2-quinazolinyl)methylamino]propyl]tetrahydro-2-furamide monohydrochloride.
SL-77499-10; Alfoten; Urion; Xatral. Antihypertensive agent. An α_1 andrenoreceptor antagonist used in the treatment of benign prostatic hyperplasia. mp = 225°, 235° (dec). *Synthelabo Pharmacie.*

4134 Alprenolol
13655-52-2 321 237-140-9
$C_{15}H_{23}NO_2$
1-(o-Allylphenoxy)-3-(isopropylamino)-2-propanol.
H 56/28. Antianginal agent. antiarrhythmic agent. Cardioselective β-adrenergic blocking agent. Has antiarrhythmic and antihypertensive properties. *C.M. Ind.; ICI.*

4135 Alprenolol Hydrochloride
13707-88-5 321 237-244-4
$C_{15}H_{24}ClNO_2$
1-(o-Allylphenoxy)-3-(isopropylamino)-2-propanol hydrochloride.
Applobal; Aprobal; Aptine; Aptol Duriles; Gubernal; Regletin; Yobir. Antianginal agent. Class II antiarrhythmic agent. Cardioselective β-adrenergic blocking agent. Has antiarrhythmic and antihypertensive properties. mp = 107-109°; LD_{50} (mus orl) = 278.0 mg/kg, (rat orl) = 597.0 mg/kg, (rbt orl) = 337.3 mg/kg. *C.M. Ind.; ICI.*

4136 Althiazide
5588-16-9 326 226-994-8
$C_{11}H_{14}ClN_3O_4S_3$
3-[(Allylthio)methyl-6-chloro-3,4-dihydro-2H-1,2,4-benzothiadiazine-7-sulfonamide 1,1-dioxide.
P-1779; Altizide; Aldactazine. Antihypertensive; diuretic. mp = 206-207°. *Pfizer Inc.*

4137 Ambuside
3754-19-6 405 223-158-4
$C_{13}H_{16}ClN_3O_5S_2$
N'-Allyl-4-chloro-6-[(3-hydroxy-2-butenylidene)-amino]-m-benzenedisulfonamide.
2-allylsulfamoyl-2-chloro-4-(3-hydroxy-2-butenylidine)-aniline; EX-4810; RMI-83047; Hydrion; Novohydrin. Anti-hypertensive; diuretic. mp = 205-207°; λ_m = 343 nm (ε 32900). *Marion Merrell Dow Inc.*

4138 γ-Amino butyric Acid
56-12-2 450 200-258-6
$C_4H_9NO_2$
4-Aminobutanoic acid.
piperidic acid; GABA; gammalon. Antihypertensive. mp = 202° (dec to form pyrrolidone and H_2O); freely soluble in H_2O, poorly soluble in organic solvents; [hydrochloride ($C_4H_{10}ClNO_2$)]: mp - 135-136°; [ethyl ester ($C_6H_{13}NO_2$)]: bp_{12} = 76°.

4139 Amlodipine
88150-42-9 516
$C_{20}H_{25}ClN_2O_5$
3-Ethyl-5-methyl (±)-2-[(2-aminoethoxy)methyl]-4-(o-chlorophenyl)-1,4-dihydro-6-methyl-3,5-pyridine-dicarboxylate.
Antianginal agent. Dihydropyridine calcium channel blocker. Has antihypertensive properties. *Ciba-Geigy Corp.; Pfizer Inc.*

4140 Amlodipine Besylate
111470-99-6 516
$C_{26}H_{31}ClN_2O_8S$
3-Ethyl-5-methyl (±)-2-[(2-aminoethoxy)methyl]-4-(o-chlorophenyl)-1,4-dihydro-6-methyl-3,5-pyridinedi-carboxylate monobenzenesulfonate.
Norvasc; UK-48340-26; Antacal; Istin; Monopina; component of: Lotrel. Antianginal agent. Dihydropyridine calcium channel blocker. Has antihypertensive properties. mp = 178-179°. *Ciba-Geigy Corp.; Pfizer Inc.*

4141 Amlodipine Maleate
88150-47-4 516
$C_{24}H_{29}ClN_2O_9$
3-Ethyl-5-methyl (±)-2-[(2-aminoethoxy)methyl]-4-(o-chlorophenyl)-1,4-dihydro-6-methyl-3,5-pyridine-dicarboxylate maleate.
UK-48340-11. Antianginal agent. Dihydropyridine calcium channel blocker. Has antihypertensive properties. *Pfizer Inc.*

4142 Amosulalol
85320-68-9 614
$C_{18}H_{24}N_2O_5S$
(±)-5-[1-Hydroxy-2-[[2-(o-methoxyphenoxy)-ethyl]amino]ethyl]-o-toluenesulfonamide.
An α_1-adrenergic blocking agent with antihypertensive activity. *Yamanouchi U.S.A. Inc.*

4143 Amosulalol Hydrochloride
93633-92-2 614
$C_{18}H_{25}ClN_2O_5S$
(±)-5-[1-Hydroxy-2-[[2-(o-methoxyphenoxy)-ethyl]amino]ethyl]-o-toluene-sulfonamide hydrochloride.
YM-09538; Lowgan. An α_1-adrenergic blocking agent with antihypertensive activity. mp = 158-160°; [R(-)-form]: mp = 158°; $[\alpha]_D^{20}$ = -30.4° (c = 1 MeOH); [S(+)-form]: mp = 158°; $[\alpha]_D^{20}$ = 30.7° (c = 1 MeOH). *Yamanouchi U.S.A. Inc.*

4144 Amoxydramine Camsilate
15350-99-9 239-382-0
$C_{27}H_{37}NO_6S$
2-(Diphenylmethoxy)-N,N-dimethylethylamine-N-oxide 2-oxo-10-bornanesulfonate.
Antihypertensive agent.

4145 Aranidipine
86780-90-7 806
$C_{19}H_{20}N_2O_7$
(±)-Acetyl methyl 1,4-dihydro-2,6-dimethyl-4-(o-nitro-phenyl)-3,5-pyridinedicarboxylate.
MPC-1304. Antihypertensive agent. Calcium channel blocker. *Maruko Seiyaku.*

4146 Arfalasin
60173-73-1
$C_{48}H_{67}N_{13}O_{11}$
1-Succinamic acid-5-L-valine-8-(L-2-phenylglycine)-angiotensin II.
HOE-409. An antihypertensive agent related to saralasin. Angiotensin II blocker.

4147 Arotinolol
68377-92-4 827
$C_{15}H_{21}N_3O_2S_3$
(±)-5-[2-[[3-(tert-Butylamino)-2-hydroxypropyl]thio]-4-thiazolyl]-2-thiophenecarboxamide.
Antianginal agent. Also antihypertensive and antiarrhythmic. Possesses both α- and β-adrenergic blocking activity. A propanolamine derivative. mp = 148-149°. *Sumitomo Pharm. Co. Ltd.*

4148 Arotinolol Hydrochloride
68377-91-3 827
$C_{15}H_{22}ClN_3O_2S_3$
(±)-5-[2-[[3-(tert-Butylamino)-2-hydroxypropyl]thio]-4-thiazolyl]-2-thiophenecarboxamide hydrochloride.
S-596; ARL; Almarl. Antianginal; antihypertensive; antiarrhythmic. Possesses both α- and β-adrenergic blocking activity. A propanolamine derivative. mp = 234-235.5°; LD_{50} (mus iv) = 86 mg/kg, (mus ip) = 360 mg/kg, (mus orl) > 5000 mg/kg. *Sumitomo Pharm. Co. Ltd.*

4149 Atenolol
29122-68-7 892 249-451-7
$C_{14}H_{22}N_2O_3$
2-[p-[2-Hydroxy-3-(isopropylamino)propoxy]-phenyl]acetamide.
ICI-66082; AteHexal; Atenol; Cuxanorm; Ibinolo; Myocord; Prenormine; Seles Beta; Selobloc; Teno-basan; Tenoblock; Tenormin; Uniloc. Antianginal; class II antiarrhythmic. Cardioselective β-adrenergic blockinger. Has antiarrhythmic and antihypertensive properties. mp = 146-148°, 150-152°; λ_m = 225, 275, 283 nm (MeOH); very soluble in MeOH; soluble in AcOH, DMSO; less soluble in Me_2CO, dioxane; insoluble in CH_3CN, EtOAc, $CHCl_3$; LD_{50} (mus orl) = 2000 mg/kg, (mus iv) = 98.7 mg/kg, (rat orl) = 3000 mg/kg, (rat iv) = 59.24 mg/kg. *Apothecon; C.M. Ind.; ICI ; Lemmon Co.; Zeneca Pharm.*

4150 Azamethonium Bromide
306-53-6 929 206-186-1
$C_{13}H_{33}Br_2N_3$
[(Methylimino)diethylene]bis(ethyldimethylammonium bromide).
pentamethazine dibromide; Präparat 9295; Ciba 9295; Pendoimid; Azameton; Azamethone; Ganlion; Pentaméthazène. Antihypertensive agent. Ganglion blocking agent. mp = 212-215°; freely soluble in H_2O; LD_{50} (mus orl) = 2500 mg/kg, (mus iv) = 60 mg/kg, (rbt orl) = 3000 mg/kg, (rbt sc) = 160 mg/kg, (rbt iv) = 75 mg/kg. *Ciba-Geigy Corp.*

4151 Azelnidipine
123524-52-7
$C_{33}H_{34}N_4O_6$
3-[1-(Diphenylmethyl)-3-azetidinyl] 5-isopropyl (±)-2-amino-1,4-dihydro-6-methyl-4-(m-nitrophenyl)-3,5-pyridinedicarboxylate.
Calcium channel blocker. Antihypertensive.

4152 Azepexole
36067-73-9
$C_9H_{15}N_3O$
2-Amino-6-ethyl-5,6,7,8-tetrahydro-4H-oxazolo[4,5-d]-azepine.
Antihypertensive agent.

4153 Barnidipine
104713-75-9 1031
$C_{27}H_{29}N_3O_6$
(+)-(3'S,4S)-1-Benzyl-3-pyrrolidinyl methyl 1,4-dihydro-2,6-dimethyl-4-(m-nitrophenyl)-3,5-pyridinedicarboxylate.
Mepirodipine. Antianginal agent. Dihydropyridine calcium channel blocker. Has antihypertensive properties. mp = 137-139°; $[\alpha]_D^{20}$ = 64.8 (c = 1 MeOH). *Yamanouchi U.S.A. Inc.*

4154 Barnidipine Hydrochloride
1031
$C_{27}H_{30}ClN_3O_6$
(+)-(3'S,4S)-1-Benzyl-3-pyrrolidinyl methyl 1,4-dihydro-2,6-dimethyl-4-(m-nitrophenyl)-3,5-pyridinedicarboxylate hydrochloride.
YM-09730-5; Hypoca. Antianginal agent. Dihydropyridine calcium channel blocker. Has antihypertensive properties. mp = 226-228°; $[\alpha]_D^{20}$ = 116.4° (c = 1 MeOH); insoluble in H_2O; LD_{50} (mrat orl) = 105 mg/kg, (frat orl) = 113 mg/kg. *Yamanouchi U.S.A. Inc.*

4155 Benazepril
86541-75-5 1058
$C_{24}H_{28}N_2O_5$
(3S)-3-[[(1S)-1-Carboxy-3-phenylpropyl]amino]-2,3,4,5-tetrahydro-2-oxo-1H-1-benzazepine-1-acetic acid 3-ethyl ester.
CGS-14824A; Briem; Cibacen; Cibacène; Lotensin. Angiotensin-converting enzyme inhibitor. Used to treat hypertension. mp= 148-149°; $[\alpha]_D$ = -159° (c = 1.2 EtOH). *Ciba-Geigy Corp.*

4156 Benazepril Hydrochloride
86541-74-4 1058
$C_{24}H_{29}ClN_2O_5$
(3S)-3-[[(1S)-1-Carboxy-3-phenylpropyl]amino]-2,3,4,5-tetrahydro-2-oxo-1H-1-benzazepine-1-acetic acid 3-ethyl ester hydrochloride.
Lotensin; CGS-14824A HCl; component of Lotensin-HCT, Lotrel, Lotrel capsules. Angiotensin-converting enzyme inhibitor. Used to treat hypertension. mp = 188-190°; $[\alpha]_D$ = -141° (c = 0.9 EtOH). *Ciba-Geigy Corp.*

4157 Benazeprilate
86541-78-8 1058
$C_{22}H_{24}N_2O_5$
(3S)-3-[[(1S)-1-Carboxy-3-phenylpropyl]amino]-2,3,4,5-tetrahydro-2-oxo-1H-1-benzazepine-1-acetic acid.
CGS-14831. Angiotensin-converting enzyme inhibitor. Used to treat hypertension. mp = 270-272°; $[\alpha]_D$ = -200.5° (c = 1 in 3% aq. NaOH). *Ciba-Geigy Corp.*

4158 Bendroflumethazide
73-48-3 1064 200-800-1
$C_{15}H_{14}F_3N_3O_4S_2$
3-Benzyl-3,4-dihydro-6-(trifluoromethyl)-2H-1,2,4-benzothiadiazine-7-sulfonamide 1,1-dioxide.
benzylhydroflumethiazide; benzydroflumethiazide; bendrofluazide; Naturetin; Corzide; Rautrax N; Rauzide; Aprinox; Benzy-Rodiuran; Berkozide; Bristuric; Bristuron; Centyl; Flumersil; Naturetin; Naturine; Neo-Naclex; Naigaril; Nikion; Orsile; Pluryle; Plusuril; Poliuron; Relan Beta; Salures; Sinesalin; Sodiuretic; Urlea. Diuretic; antihypertensive. A thiazide. mp = 224.5=225.5°, 221-223°; λ_m = 208, 273, 326 nm ($E_{1\,cm}^{1\%}$ 745, 565, 96 MeOH); soluble in Me_2CO, EtOH; insoluble in H_2O, $CHCl_3$, C_6H_6, Et_2O. *Apothecon; Bristol-Myers Squibb Co.*

4159 Benidipine
105979-17-7 1071
$C_{28}H_{31}N_3O_6$
(±)-(R*)-3-[(R*)-1-Benzyl-3-piperidyl] methyl 1,4-dihydro-2,6-dimethyl-4-(m-nitrophenyl)-3,5-pyridinedicarboxylate.
Antihypertensive. Calcium channel blocker. *Kyowa Hakko Kogyo Co. Ltd.*

4160 Benidipine Hydrochloride
91599-74-5 1071
$C_{28}H_{32}ClN_3O_6$
(±)-(R*)-3-[(R*)-1-Benzyl-3-piperidyl] methyl 1,4-dihydro-2,6-dimethyl-4-(m-nitrophenyl)-3,5-pyridinedicarboxylate hydrochloride(α form).
KW-3049; Coniel. Antihypertensive. Calcium channel blocker. mp = 199.4-200.4°; λ_m = 238, 359 nm (ε = 28000, 6680 EtOH); soluble in H_2O (0.19 g/100 ml); MeOH (6.9 g/100 ml), EtOH (2.2 g/100 g); $CHCl_3$ (0.16 g/100 g), Me_2CO (0.13 g/100 g), EtOAc (0.0056 g/100 g); C_7H_8 (0.0019 g/100 g), C_7H_{14} (0.00009 g/100 g); LD_{50} (mus orl) = 218 mg/kg. *Kyowa Hakko Kogyo Co. Ltd.*

4161 Benzoclidine
16852-81-6
$C_{14}H_{17}NO_2$
3-Quinuclidinol benzoate.
Used as an antihypertensive agent.

4162 Benzthiazide
91-33-8 1155 202-061-0
$C_{15}H_{14}ClN_3O_4S_3$
3-[(Benzylthio)methyl]-6-chloro-2H-1,2,4-benzothiadiazine-7-sulfonamide 1,1-dioxide.
Fovane; Exna; Aquatag; Dihydrex; Diucen; Edemex; ExNa; Exosalt; Freeuril; HyDrine; Lemazide; Proaqua; Urese; component of: Dytide. Diuretic. A thiazide. mp = 231-232°, 238-239°; insoluble in H_2O; soluble in alkaline solutions; LD_{50} (rat orl) >10000 mg/kg, (rat iv) = 422 mg/kg, (mus orl) > 5000 mg/kg, (mus iv) = 410 mg/kg. *Robins, A. H. Co.*

4163 Benzylhydrochlorothiazide
1824-50-6 1171
$C_{14}H_{14}ClN_3O_4S_2$
3-Benzyl-3,4-dihydro-6-chloro-2H-1,2,4-thiadiazine-7-sulfonamide 1,1-dioxide.
Behyd. Diuretic; antihypertensive. mp = 260-262°, 269°.

4164 Betaxolol
63659-18-7 1229
$C_{18}H_{29}NO_3$
(±)-1-[p-[2-(Cyclopropylmethoxy)ethyl]phenoxy]-3-(isopropylamino)-2-propanol.
Antianginal agent with antihypertensive and antiglaucoma properties. Cardioselective β_1 adrenergic blocker. mp = 70-72°. *Alcon Labs; Synthelabo Pharmacie.*

4165 Betaxolol Hydrochloride
63659-19-8 1229 264-384-3
$C_{18}H_{30}ClNO_3$
(±)-1-[p-[2-(Cyclopropylmethoxy)ethyl]phenoxy]-3-(isopropylamino)-2-propanol hydrochloride.
SLD-212; SL-75.212; Betoptic; Betoptima; Kerlone. Antianginal agent with antihypertensive and antiglaucoma properties. Cardioselective β_1 adrenergic blocker. mp = 116°; LD_{50} (mus orl) = 94 mg/kg, (mus iv) = 37 mg/kg. *Alcon Labs; Synthelabo Pharmacie.*

4166 Bethanidine
55-73-2 1233
$C_{10}H_{15}N_3$
1-Benzyl-2,3-dimethylguanidine.
Antihypertensive. Adrenergic neuron blocking agent. mp = 195-197°. *Glaxo Wellcome Inc.*

4167 Bethanidine Sulfate
114-85-2 1233 204-056-9
$C_{20}H_{32}N_6SO_4$
1-Benzyl-2,3-dimethylguanidine sulfate.
BW-467-C-60; NSC-106563; Benzaidin; Bendogen; Benzoxine, Betaling; Betanidol; Esbatal; Eusmanid; Hypersin; Tenathan. Antihypertensive. Adrenergic neuron blocking agent. LD_{50} (mus iv) = 12 mg/kg, (mus ip) = 150 mg/kg, (mus sc) = 260 mg/kg. *Glaxo Wellcome Inc.*

4168 Bevantolol
59170-23-9 1238
$C_{20}H_{27}NO_4$
(±)-1-[(3,4-Dimethoxyphenethyl)amino]-3-(m-toloxy)-2-propanol.
Antianginal agent. Cardioselective β-adrenergic blocking agent. Has antiarrhythmic and antihypertensive properties. *Parke-Davis.*

4169 Bevantolol Hydrochloride
42864-78-8 1238
$C_{20}H_{28}ClNO_4$
(±)-1-[(3,4-Dimethoxyphenethyl)amino]-3-(m-toloxy)-2-propanol hydrochloride.
Vantol; CI-775; Ranestol; Sentiloc. Antianginal agent. Cardioselective β-adrenergic blocking agent. Has antiarrhythmic and antihypertensive properties. mp = 137-138°. *Parke-Davis.*

4170 Bietaserpine
53-18-9 1254 200-165-0
$C_{39}H_{53}N_3O_9$
Methyl 1-[2-(diethylamino)ethyl]-18β-hydroxy-11,17α-dimethoxy-3β,20α-yohimban-16β-carboxylate.
DL-152; S-1210; 1-[2-(diethylamino)ethyl]reserpine; diethylaminoreserpine. Antihypertensive. Cyclic AMP phosphodiesterase inhibitor. $[\alpha]_D^{17}$ = -121° (c = 2 $CHCl_3$); soluble in polar organic solvents. *Dautreville & Lebas.*

4171 Bietaserpine Bitartrate
1111-44-0 1254 214-180-5
$C_{43}H_{59}N_3O_{15}$
Methyl 1-[2-(diethylamino)ethyl]-18β-hydroxy-11,17α-dimethoxy-3β,20α-yohimban-16β-carboxylate bitartrate (salt).
DL-152; Tensibar. Antihypertensive. Cyclic AMP phosphodiesterase inhibitor. mp = 145-150° (dec); LD_{50} (mus orl) = 620 mg/kg, (mus ip) = 430 mg/kg, (mus iv) = 215 mg/kg. *Dautreville & Lebas.*

4172 Bisoprolol
66722-44-9 1336
$C_{18}H_{31}NO_4$
(±)-1-[[α(2-Isopropoxyethoxy)-p-tolyl]oxy]-3-(isopropylamino)-2-propanol.
EMD-33512. Antihypertensive. Cardioselective β_1-adrenergic blocker. *E. Merck.*

4173 Bisoprolol Hemifumarate
104344-23-2 1336
$C_{20}H_{33}NO_6$
(±)-1-[[α(2-Isopropoxyethoxy)-p-tolyl]oxy]-3-(isopropylamino)-2-propanol hemifumarate.
Concor; Detensiel; Emvoncor; Emcor; Eurtadal; Isoten; Monocor; Soprol; Zebeta; component of: Ziac. Antihypertensive. Cardioselective β_1-adrenergic blocker. mp = 100°; soluble in EtOH. *E. Merck.*

4174 Bopindolol
62658-63-3 1362
$C_{23}H_{28}N_2O_3$
(±)-1-(tert-Butylamino)-3-[(2-methylindol-4-yl)oxy]-2-propanol benzoate (ester).
Antihypertensive. Non-selective β-adrenergic blocker. Soluble in Et_2O, CH_2Cl_2; LD_{50} (mus iv) = 17 mg/kg. *Sandoz Pharm. Corp.*

4175 Bopindolol Maleate
62658-64-4 1362
$C_{27}H_{32}N_2O_7$
(±)-1-(tert-Butylamino)-3-[(2-methylindol-4-yl)oxy]-2-propanol benzoate (ester) maleate.
Sandonorm. Antihypertensive. Non-selective β-adrenergic blocker. *Sandoz Pharm. Corp.*

4176 Bopindolol Malonate
82857-38-3 1362
$C_{26}H_{32}N_2O_7$
(±)-1-(tert-Butylamino)-3-[(2-methylindol-4-yl)oxy]-2-propanol benzoate (ester) malonate.
LT-31-200; Wandonorm. Antihypertensive. Non-selective β-adrenergic blocker. *Sandoz Pharm. Corp.*

4177 Budralazine
36798-79-5 1492
$C_{14}H_{16}N_4$
4-Methyl-3-penten-2-one (1-phthalazinyl)hydrazone.
DJ-1461; Buterazine. An α-adrenergic agonist. Used as an antihypertensive agent. Direct-acting vasodilator with central sympathoinhibitory activity. A derivative of hydralazine. mp = 132-133°; λ_m = 208, 240, 289, 357 nm (ε = 27000, 89000, 20000, 15000 MeOH); LD_{50} (mus orl) = 1820 mg/kg, (mus ip) = 4020 mg/kg, (rat orl) = 620 mg/kg, (rat ip) = 3570 mg/kg. *Daiichi Seiyaku.*

4178 Bufeniode
22103-14-6 1495 244-781-8
$C_{19}H_{23}I_2NO_2$
4-Hydroxy-3,5-diiodo-α-[1[(1-methyl-3-phenylpropyl)-amino]ethyl] benzyl alcohol.
HF-241; diiodobuphenine; Diastal; Proclival. Peripheral vasodilator, used as an antihypertensive agent. mp (slow

heating) = 185° (some decomposition); mp (fast heating) = 212°; LD_{50} (mus ip) > 600 mg/kg, (mus orl) > 2000 mg/kg. *Lab. Houde.*

4179 Bufuralol
54340-62-4 1504 259-112-5
$C_{16}H_{23}NO_2$
α-[(tert-Butylamino)methyl]-7-ethyl-2-benzofuranmethanol.
A β-adrenergic blocker with peripheral vasodilating activity. Antianginal with antihypertensive properties. *Hoffmann-LaRoche Inc.*

4180 Bufuralol Hydrochloride
59652-29-8 1504
$C_{16}H_{24}ClNO_2$
α-[(tert-Butylamino)methyl]-7-ethyl-2-benzofuranmethanol hydrochloride.
Ro-3-4787; Angium. A β-adrenergic blocker with peripheral vasodilating activity. Antianginal with antihypertensive properties. mp = 146°; LD_{50} (mus iv) = 29.7 mg/kg, (mus ip) = 88.0 mg/kg, (mus orl) = 177 mg/kg, (rat sc) = 1400 mg/kg, (rat orl) = 750 mg/kg; [(+)-hydrochloride]: mp = 122-123°; $[\alpha]^{20}_{365}$ = 135.0° (c = 1 EtOH); [(-)-hydrochloride]: mp = 122-123°; $[\alpha]^{20}_{365}$ = -136.0° (c = 1 EtOH). *Hoffmann-LaRoche Inc.*

4181 Bunazosin
80755-51-7 1512
$C_{19}H_{27}N_5O_3$
1-(4-Amino-6,7-dimethoxy-2-quinazolinyl)-4-butylhexahydro-1H-1,4-diazepine.
Antihypertensive. Adrenergic blocker. *Eisai Co. Ltd.*

4182 Bunazosin Hydrochloride
52712-76-2 1512
$C_{19}H_{28}ClN_5O_3$
1-(4-Amino-6,7-dimethoxy-2-quinazolinyl)-4-butyl-hexahydro-1H-1,4-diazepine hydrochloride.
E-643; Detantol. Antihypertensive. Adrenergic blocker. mp = 280-282°. *Eisai Co. Ltd.*

4183 Bunitrolol
34915-68-9 1514
$C_{14}H_{20}N_2O_2$
o-[3-(tert-Butylamino)-2-hydroxypropoxy]benzonitrile.
Ko-1366. Antianginal agent with antiarrhythmic and antihypertensive properties. A β-adrenergic blocker. *Boehringer Ingelheim GmbH.*

4184 Bunitrolol Hydrochloride
23093-74-5 1514 245-427-5
$C_{14}H_{21}ClN_2O_2$
o-[3-(tert-Butylamino)-2-hydroxypropoxy]benzonitrile hydrochloride.
Betriol; Stresson. Antianginal agent with antiarrhythmic and antihypertensive properties. A β-adrenergic blocker. mp = 163-165°; LD_{50} (mus orl) = 1344-1440 mg/kg, (mus ip) = 264-265 mg/kg, (rat orl) = 639-649 mg/kg, (rat ip) = 222-225 mg/kg. *Boehringer Ingelheim GmbH.*

4185 Bupranolol
14556-46-8 1521
$C_{14}H_{22}ClNO_2$
1-(tert-Butylamino)-3-[(6-chloro-m-tolyl)-oxy]-2-propanol.
bupranol; Ophtorenin. Antianginal agent with antiarrhythmic, antihypertensive and antiglaucoma properties. A β-adrenergic blocker.

4186 Bupranolol Hydrochloride
15148-80-8 1521 239-208-3
$C_{14}H_{23}Cl_2NO_2$
1-(tert-Butylamino)-3-[(6-chloro-m-tolyl)-oxy]-2-propanol hydrochloride.
B-1312; KL-255; Betadran; Betadrenol; Looser; Panimit. Antianginal agent with antiarrhythmic, antihypertensive and antiglaucoma properties. β-Adrenergic blocker. mp = 220-222°.

4187 Butanserin
87051-46-5
$C_{24}H_{26}FN_3O_3$
3-[4-[4-(p-Fluorobenzoyl)piperidino]butyl]-2,4(1H,3H)-quinazolinedione.
R-53393. Antihypertensive. 5-Hydroxytryptamine S2-antagonist. Selective α_1-adrenoceptor antagonist.

4188 Buthiazide
2043-38-1 1554 218-048-8
$C_{11}H_{16}ClN_3O_4S_2$
6-Chloro-3,4-dihydro-3-isobutyl-2H-1,2,4-benzothiadiazine-7-sulfonamide 1,1-dioxide.
Su-6187; S-3500; Eunephran; Saltucin; Modenol. Diuretic; antihypertensive. A thiazide. mp= 228°, 241-245°. *Searle G.D. & Co.*

4189 Butofilolol
64552-17-6 1563
$C_{17}H_{26}FNO_3$
(±)-2'-[3-(tert-Butylamino)-2-hydroxy-propoxyl-5'-fluoro-butyrophenone.
CM-6805. Antihypertensive. A β-adrenergic blocker. mp = 88-89°. *C.M. Ind.*

4190 Butofilolol Maleate
88606-96-6 1563 289-431-5
$C_{21}H_{30}FNO_7$
(±)-2'-[3-(tert-Butylamino)-2-hydroxy propoxyl-5'-fluorobutyrophenone maleate.
Cafide. Antihypertensive. A β-adrenergic blocker. *C.M. Ind.*

4191 Cadralazine
64241-34-5 1669
$C_{12}H_{21}N_5O_3$
Ethyl-6-[ethyl(2-hydroxypropyl)amino]-3-pyridazine-carbazate.
DC-826; ISF-2469; Cadral; Cadraten; Cadrilan; Presmode. Peripheral vasodilator related to hydralazine. Used as an antihypertensive agent. mp = 160-162°; λ_m = 248, 340 nm (ε = 22100, 2250); soluble in H_2O (130 mg/100 ml), HCl (23500 mg/100 ml), DMSO (932300 mg/ml), MeOH (2100 mg/100 ml), dioxane (1860 mg/100 ml), $CHCl_3$ (850 mg/100 ml); insoluble in Et_2O, C_6H_6, C_6H_{12}; LD_{50} (rat iv) = 259 mg/kg, (rat orl) = 2060 mg/kg, (dog iv) ≅ 4000 mg/kg, (dog orl) > 2000 mg/kg, (mus ip) = 700 mg/kg.

4192 Candesartan
139481-59-7 1788
$C_{24}H_{20}N_6O_3$
2-Ethoxy-1-[p-(o-1H-tetrazol-5-ylphenyl)-benzyl]-7-benzimidazole-carboxylic acid.
CV-11974. Non-peptidic angiotensin II type-1 receptor antagonist. Used as an antihypertensive. mp = 183-185°. *Takeda Chem. Ind. Ltd.*

4193 Candesartan Cilexetil
145040-37-5 1788
$C_{33}H_{34}N_6O_6$
(±)-1-Hydroxyethyl 2-ethoxy-1-[p-(o-1H-tetrazol-5-yl-phenyl)benzyl]-7-benzimidazolecarboxylate cyclohexyl-carbonate (ester).
TCV-116. Nonpeptidic angiotensin II type-1 receptor antagonist. Used as an antihypertensive. Ester prodrug of Candesartan. mp = 163° (dec). *Takeda Chem. Ind. Ltd.*

4194 Captopril
62571-86-2 1817 263-607-1
$C_9H_{15}NO_3S$
1-[(2S)-3-Mercapto-2-methylpropionyl]-L-proline.
Capoten; SQ-14225; Acediur; Acepril; Aceplus; Alopresin; Acepress; Capoten; Captolane; Captoril; Cesplon; Dilabar; Garranil; Hipertil; Lopirin; Lopril; Tensobon; Tensoprel; component of: Capozide, Acezide, Captea, Ecazide. Angiotensin-converting enzyme inhibitor. Used to treat hypertension. Orally active peptidyldipeptide hydrolase inhibitor. mp = 103-104°, 86°, 87-88°, 104-105°; $[\alpha]_D^{22}$ = -131.0° (c = 1.7 EtOH); freely soluble in H_2O, EtOH, $CHCl_3$, CH_2Cl_2; LD_{50} (mus iv) = 1040 mg/kg, (mus orl) = 6000 mg/kg. *Apothecon; Bristol-Myers Squibb Co.; Squibb E.R. & Sons.*

4195 Carazolol
57775-29-8 1822 260-945-1
$C_{18}H_{22}N_2O_2$
1-(Carbazol-4-yloxy)-3-(isopropylamino)-2-propanol.
BM-51052; Conducton; Suacron. A β-adrenergic blocker with antihypertensive, antianginal and antiarrhythmic activity. Used for treatment of stress in pigs (veterinary). [hydrochloride]: mp = 234-235°. *Boehringer Mannheim GmbH.*

4196 Carmoxirole
98323-83-2 1893
$C_{24}H_{26}N_2O_2$
3-[4-(3,6-Dihydro-4-phenyl-1(2H)-pyridyl)butyl]-indole-5-carboxylic acid.
Antihypertensive. Selective dopamine D_2-receptor antagonist. mp = 284-285°. *E. Merck.*

4197 Carmoxirole Hydrochloride
115092-85-8 1893
$C_{24}H_{27}ClN_2O_2$
3-[4-(3,6-Dihydro-4-phenyl-1(2H)-pyridyl)butyl]-indole-5-carboxylic acid hydrochloride.
EMD-45609. Selective dopamine D_2-receptor antagonist. Used as an antihypertensive agent. mp = 298-299°; λ_m = 242, 281 nm (MeOH). *E. Merck.*

4198 Carteolol
51781-06-7 1917
$C_{16}H_{24}N_2O_3$
5-[3-[(1,1-Dimethylethyl)amino]-2-hydroxypropyl]-3,4-dihydro-2(1H)-quinolinone.
Antianginal agent with antiarrhythmic, antihypertensive and antiglaucoma properties. A β-adrenergic blocker. *Abbott Labs Inc.; Otsuka Am. Pharm.*

4199 Carteolol Hydrochloride
51781-21-6 1917 257-415-7
$C_{16}H_{25}ClN_2O_3$
5-[3-[(1,1-Dimethylethyl)amino]-2-hydroxypropyl]-3,4-dihydro-2(1H)-quinolinone hydrochloride.
Carteolol hydrochloride, Abbott 43326; OPC 1085; Arteoptic; Caltidren; Carteol; Cartrol; Endak; Mikelan; Optipress; Tenalet; Tenalin; Teoptic. Antianginal agent with antiarrhythmic, antihypertensive and antiglaucoma properties. A β-adrenergic blocker. mp = 278°; LD_{50} (mrat orl) = 1380 mg/kg, (mrat iv) = 158 mg/kg, (mrat ip) = 380 mg/kg, (mmus orl) = 810 mg/kg, (mmus iv) = 54.5 mg/kg, (mmus ip) = 380 mg/kg. *Abbott Labs Inc.; Otsuka Am. Pharm.*

4200 Carvediol
72956-09-3 1924
$C_{24}H_{26}N_2O_4$
(±)-1-(Carbazol-4-yloxy)-3-[[2-(o-methoxyphen-oxy)ethyl]amino]-2-propanol.
Coreg; BM-14190; DQ-2466; Dilatrend; Dimitone; Eucardic; Kredex; Querto. Antianginal; antihypertensive. Nonselective β-adrenergic blocker with vasodilating properties. mp = 114-115°. *Boehringer Mannheim GmbH; SmithKline Beecham Pharm.*

4201 Celiprolol
56980-93-9 2007 260-497-7
$C_{20}H_{33}N_3O_4$
3-[3-Acetyl-4-[3-(tert-butylamino)-2-hydroxypropoxy]-phenyl]-1,1-diethylurea.
N-[3-acetyl-4-(3'-tert-butylamino-2'-hydroxy)propoxy]-phenyl-N'-diethylurea; ST-1396. Antianginal; antihypertensive agent. Cardioselective β_1 adrenergic blocker. mp = 110-112°. *Hoechst Marion Roussel Inc.*

4202 Celiprolol Hydrochloride
57470-78-7 2007 260-752-2
$C_{20}H_{34}ClN_3O_4$
3-[3-Acetyl-4-[3-(tert-butylamino)-2-hydroxypropoxy]phenyl]-1,1-diethylurea monohydrochloride.
Celectol; Corliprol; Selecor; Selectol. Antianginal; antihypertensive agent. Cardioselective β_1 adrenergic blocker. mp = 197-200° (dec); soluble in H_2O (15.1 g/100 ml), MeOH (18.2 g/100 ml), EtOH (1.61 g/100 ml), $CHCl_3$ (0.42 g/100 ml); λ_m = 221, 324 nm ($E_{1\%}$ 652, 57 H_2O), 231, 324 nm [$E_{1\%}$ 660, 60 0.01N HCl), 231, 324 nm ($E_{1\%}$ 640, 60 0.01N NaOH), 232, 329 nm ($E_{1\%}$ 775, 58 MeOH); LD_{50} (mmus iv) = 56.2 mg/kg, (mmus orl) = 1834 mg/kg, (mrat iv) = 68.3 mg/kg, (mrat orl) = 3826 mg/kg. *Hoechst Marion Roussel Inc.*

4203 Ceronapril
111223-26-8 2038
$C_{21}H_{33}N_2O_6P$
1-[(2S)-6-Amino-2-hydroxyhexanoyl]-L-proline hydrogen (4-phenylbutyl)phosphonate (ester).
SQ-29852; Ceranapril. Angiotensin-converting enzyme inhibitor. Used to treat hypertension. mp = 190-195°; $[\alpha]_D$ = -47.5° (c = 1 MeOH). *Bristol-Myers Squibb Co.*

4204 Cetamolol
34919-98-7 2060
$C_{16}H_{26}N_2O_4$
(±)-2-[o-[3-(tert-Butylamino)-2-hydroxypropoxy]phenoxy]-N-methylacetamide.
Cardioselective β_2-adrenergic blocker. Used as an antihypertensive. mp = 96-97°. *ICI.*

4205 Cetamolol Hydrochloride
77590-95-5 2060 278-729-0
$C_{16}H_{27}ClN_2O_4$
(±)-2-[o-[3-(tert-Butylamino)-2-hydroxypropoxy]-phenoxy]-N-methylacetamide hydrochloride.
AI-27303; ICI-72222; Betacor. Cardioselective β_2-adrenergic blocker. Used as an antihypertensive. *ICI.*

4206 Chlorisondamine Chloride
69-27-2 2151
$C_{14}H_{20}Cl_6N_2$
4,5,6,7-Tetrachloro-2-(2-dimethylaminoethyl)-2-methylisoindolinium chloride methochloride.
chlorisondamine dimethochloride; Su-3088; Ecolid; Ecolid chloride. Antihypertensive. Ganglion blocker. mp = 258-265°; soluble in H_2O, EtOH, MeOH. *Ciba-Geigy Corp.*

4207 Chlorothiazide
58-94-6 2221 200-404-9
$C_7H_6ClN_3O_4S_2$
6-Chloro-2H-1,2,4-benzothiadiazine-7-sulfonamide 1,1-dioxide.
Chlotride; Diuril; Diuril Boluses; Aldoclor; Diupres; Diuril Lyovac [as sodium salt]; Lyovac Diuril [as sodium salt]. Diuretic; antihypertensive. A thiazide. mp = 342.5-343°; soluble in DMSO, DMF; less soluble in MeOH, C_5H_5N; insoluble in Et_2O, $CHCl_3$, C_6H_6; poorly soluble in H_O. *Merck & Co.Inc.*

4208 Chlorthalidone
77-36-1 2246 201-022-5
$C_{14}H_{11}ClN_2O_4S$
2-Chloro-5-(1-hydroxy-3-oxo-1H-isoindolinyl)-benzenesulfonamide.
Hygroton; Thalitone; Combipres; Demi-Regroton; Regroton; Tenoretic; G-33182; NSC-69200. Diuretic; antihypertensive. Nonthiazide compound with a similar mechanism of action to the thiazide diuretics. mp= 224-226°; λ_m = 220 nm (MeOH); soluble in H_2O (12 mg/ml at 20°, 27 mg/ml at 37°), slightly soluble in EtOH, Et_2O. *Boehringer Ingelheim GmbH; KV Pharm.; Parke-Davis; Rhône-Poulenc Rorer Pharm. Inc.; Zeneca Pharm.*

4209 Cianergoline
74627-35-3
$C_{19}H_{22}N_4O$
(α-RS)-α-Cyano-6-methylergoline-8β-propionamide.
Has antihypertensive properties. Acts as a dopaminergic agonist.

4210 Cicletanine
89943-82-8 2323
$C_{14}H_{12}ClNO_2$
(±)-3-(p-Chlorophenyl)-1,3-dihydro-6-methylfuro[3,4-c]pyridin-7-ol.
(±)-BN-1270; Win-90,000; cicletanide; cycletanide. Antihypertensive. *Sterling Winthrop Inc.*

4211 Cicletanine Hydrochloride
82747-56-6 2323
$C_{14}H_{13}Cl_2NO_2$
(±)-3-(p-Chlorophenyl)-1,3-dihydro-6-methylfuro[3,4-c]pyridin-7-ol hydrochloride.
BN-1270; Coverine; Justar; Secletan; Tenstaten. Antihypertensive. mp = 219-228°. *Sterling Winthrop Inc.*

4212 Ciclosidomine
66564-16-7 2326
$C_{13}H_{20}N_4O_3$
N-(Cyclohexylcarbonyl)-3-morpholinosydnone imine.
Peripheral vasodilator similar to molsidomine. Used as an antihypertensive. *Boehringer Ingelheim GmbH.*

4213 Ciclosidomine Hydrochloride
26209-07-4 2326
$C_{13}H_{21}ClN_4O_3$
N-(Cyclohexylcarbonyl)-3-morpholinosydnone imine hydrochloride.
PR-G-138-CL; Neopres. Peripheral vasodilator similar to molsidomine. Used as an antihypertensive. mp = 187°. *Boehringer Ingelheim GmbH.*

4214 Cilazapril
92077-78-6 2332
$C_{22}H_{31}N_3O_5.H_2O$
(1S,9S)-9-[[(S)-1-Carboxy-3-phenylpropyl]amino]-octahydro-10-oxo-6H-pyridazino[1,2-a]-[1,2]diazepine-1-carboxylic acid 9-ethyl ester monohydrate.
Inhibace; Ro-31/2848/006; Dynorm; Initiss; Justor; Vascase. Antihypertensive. Angiotensin-converting enzyme inhibitor. mp = 95-97°; $[\alpha]_D^{20}$ = -62.51° (c = 1 EtOH). *Hoffmann-LaRoche Inc.*

4215 Cilazaprilat
90139-06-3 2332
$C_{20}H_{27}N_{35}$
N-(1S,9S)-1-Carboxy-10-oxoperhydropyridazino-[1,2-α][1,2]diazepine-9-yl-4-phenyl-L-homoalanine.
Ro-31-3113. Antihypertensive. Angiotensin-converting enzyme inhibitor. mp = 242°; $[\alpha]_D^{20}$= -74.7° (c = 0.5 0.1M NaOH). *Hoffmann-LaRoche Inc.*

4216 Cilnidipine
102106-21-8 2334
$C_{27}H_{28}N_2O_7$
(E)-Cinnamyl 2-methoxyethyl 1,4-dihydro-2,6-dimethyl-4-(m-nitrophenyl)-3,5-pyridine dicarboxylate.
Antihypertensive. Dihydropyridine calcium channel blocker. *Fujirebio Inc.*

4217 (±)-Cilnidipine
132203-70-4 2334
$C_{27}H_{28}N_2O_7$
(±)-(E)-Cinnamyl 2-methoxyethyl 1,4-dihydro-2,6-dimethyl-4-(m-nitrophenyl)-3,5-pyridinedicarboxylate.
(±)-FRC-8653. Antihypertensive. Dihydropyridine calcium channel blocker. mp = 115.5-116.6°; LD_{50} (mmus orl) > 5000 mg/kg, (mmus sc) > 5000 mg/kg, (mmus ip) = 1845 mg/kg, (fmus orl) > 5000 mg/kg, (fmus sc) > 5000 mg/kg, (fmus ip) = 2353 mg/kg, (mrat orl) > 5000 mg/kg, (mrat sc) > 5000 mg/kg, (mrat ip) = 441 mg/kg, (frat o. *Fujirebio Inc.*

4218 Cilutazoline
104902-08-1
$C_{14}H_{18}N_2O$
2-[[(6-Cyclopropyl-m-tolyl)oxy]methyl]-2-imidazoline.
Thought to be antihypertensive.

4219 Clentiazem
96125-53-0 2408
$C_{22}H_{25}ClN_2O_4S$
(+)-(2S,3S)-8-Chloro-5-[2-(dimethylamino)ethyl]-2,3-dihydro-3-hydroxy-2-(p-methoxyphenyl)-1,5-benzothiazepin-4(5H)-one acetate (ester).
The 8-chloro derivative of diltiazem. A calcium channel blocker used for *its antihypertensive properties. Tanabe Seiyaku Co. Ltd.*

4220 Clentiazem Maleate
96128-92-6 2408
$C_{26}H_{29}ClN_2O_8S$
(+)-(2S,3S)-8-Chloro-5-[2-(dimethylamino)ethyl]-2,3-dihydro-3-hydroxy-2-(p-methoxyphenyl)-1,5-benzothiazepin-4(5H)-one acetate (ester) maleate (1:1).
TA-3090; Logna. The 8-chloro derivative of diltiazem. A calcium channel blocker used for its antihypertensive properties. mp = 160.5-161.5°; $[\alpha]_D^{20}$ = 76.5° (c = 1 MeOH). *Tanabe Seiyaku Co. Ltd.*

4221 Clevidipine
166432-28-6
$C_{21}H_{23}Cl_2NO_6$
(±)-Hydroxymethyl methyl 4-(2,3-dichlorophenyl))-1,4-dihydro-2,6-dimethyl-3,5-pyridinedicarboxylate butyrate (ester).
Calcium channel blocker used as an antihypertensive agent.

4222 Clonidine
4205-90-7 2450 224-119-4
$C_9H_9Cl_2N_3$
2-[(2,6-Dichlorophenyl)imino]imidazolidine.
Catapres-TTS; ST-155-BS. An α_2-adrenergic agonist used as an antihypertensive and antidyskinetic. mp = 130°. *Boehringer Ingelheim GmbH.*

4223 Clonidine Hydrochloride
4205-91-8 2450 224-121-5
$C_9H_{10}Cl_3N_3$
2-[(2,6-Dichlorophenyl)imino]imidazolidine hydrochloride.
Catapres; ST-155; component of: Combipres. An α_2-adrenergic agonist used as an antihypertensive and antidyskinetic. mp = 305°; soluble in H_2O (7.7 g/100 ml at 20°, 16.6 g/100 ml at 60°), MeOH (17.25 g/100 ml), EtOH (4 g/100 ml), $CHCl_3$ (0.02 g/100 ml); λ_m = 213, 271, 302 nm (ε 8290, 713, 340 H_2O); LD_{50} (mus orl) = 328 mg/kg, (mus iv) = 18 mg/kg, (rat orl) = 270 mg/kg, (rat iv) = 29 mg/kg. *Boehringer Ingelheim GmbH; Parke-Davis.*

4224 Clopamide
636-54-4 2454 211-261-7
$C_{14}H_{20}ClN_3O_3S$
4-Chloro-N-(2,6-dimethylpiperidino)-3-sulfamoylbenzamide.
chlosudimeprimyl; DT-327; Adurix; Aquex; Brinaldix. Antihypertensive; diuretic. A thiazide. [hydrazine derivative]: mp = 244-246°. *Sandoz Pharm. Corp.*

4225 Cryptenamine Tannates
2674
Unitensen tannate.
Ester alkaloids from Vreatrum species. Contains proveratrine A and B, germitrine, neogermitrine and germidine. Antihypertensive. Soluble in EtOH, slightly soluble in H_2O. *Irwin Neissler.*

4226 Cyclopenthiazide
742-20-1 2813 212-012-5
$C_{13}H_{18}ClN_3O_4S_2$
6-Chloro-3-(cyclopentylmethyl)-3,4-dihydro-2H-1,2,4-benzothiadiazine-7-sulfonamide 1,1-dioxide.
Su-8341; NSC-107679; cyclomethiazide; tsiklometiazid; Su-8341; Navidrex; Navidrix; Salimid. Diuretic. mp= 230°; LD_{50} (rat iv) = 142 mg/kg, (mus iv) = 232 mg/kg. *Ciba-Geigy Corp.*

4227 Cyclothiazide
2259-96-3 2822 218-859-7
$C_{14}H_{16}ClN_3O_4S_2$
6-Chloro-3,4-dihydro-3-(5-norbornen-2-yl)-2H-1,2,4-benzothiadiazine-7-sulfonamide 1,1-dioxide.
Lilly 35483; Aquirel; Anhydron; Doburil; Fluidil. Diuretic; antihypertensive. mp = 234°. *Eli Lilly & Co.*

4228 Debrisoquin
1131-64-2 2901 214-470-1
$C_{10}H_{13}N_3$
3,4-Dihydro-2(1H)-isoquinolinecarboxamidine.
isocaramidine. Antihypertensive. Monoamine oxidase inhibitor. *Hoffmann-LaRoche Inc.*

4229 Debrisoquin Sulfate
581-88-4 2901 209-472-4
$C_{20}H_{28}N_6O_4S$
3,4-Dihydro-2(1H)-isoquinolinecarboxamidine sulfate (2:1).
Declinax; Ro-5-3307/1. Antihypertensive. Monoamine oxidase inhibitor. mp = 278-280°, 284-285°, 266-268°; soluble in H_2O; LD_{50} (neonate rat orl) = 88 ± 18 mg/kg, (rat orl) = 1580 ± 163 mg/kg. *Hoffmann-LaRoche Inc.*

4230 Delapril
83435-66-9 2928
$C_{26}H_{32}N_2O_5$
Ethyl (S)-2-[[(S)-1-[(carboxymethyl)-2-indanylcarbamoyl]-ethyl]amino]-4-phenylbutyrate.
Alindapril; Indalapril. Angiotensin-converting enzyme inhibitor. Antihypertensive. *Takeda Chem. Ind. Ltd.*

4231 Delapril Hydrochloride
83435-67-0 2928
$C_{26}H_{33}ClN_2O_5$
Ethyl (S)-2-[[(S)-1-[(carboxymethyl)-2-indanylcarbamoyl]ethyl]amino]-4-phenylbutyrate monohydrochloride.
REV-6000A; CV-3317; Adecut; Cupressin. Angiotensin-converting enzyme inhibitor. Used to treat hypertension. mp = 166-170°; $[\alpha]_D^{22}$ = 18.5° (c = 1 MeOH). *Takeda Chem. Ind. Ltd.*

4232 Deserpidine
131-01-1 2964 205-004-8
$C_{32}H_{38}N_2O_8$
Methyl 18β-hydroxy-17α-methoxy-3β,20α-yohimban-16β-carboxylate 3,4,5-trimethoxybenzoate (ester).
Harmonyl; Raunormine; canescine; recanescine; 11-desmethoxyreserpine; component of: Enduronyl. Antihypertensive. [α form]: mp = 228-232°; [β form]: mp = 230-232°; [σ form]: mp= 138° and 226-232°; $[\alpha]_D^{20}$= -163° (c = 0.5 C_5H_5N); λ_m = 218, 272, 290 nm (log ε = 4.79, 4.26, 4.07 EtOH); [nitrate]: mp = 254-260°; [oxalate]: mp = 239-243°. *Abbott Labs Inc.; Ciba-Geigy Corp.; Penick; Roussel-UCLAF.*

4233 Deserpidine Hydrochloride
6033-69-8 2964
$C_{32}H_{39}ClN_2O_8$
Methyl 18β-hydroxy-17α-methoxy-3β,20α-yohimban-16β-carboxylate 3,4,5-trimethoxybenzoate (ester) hydrochloride (1:1).
Antihypertensive. mp = 253-256°. *Abbott Labs Inc.; Ciba-Geigy Corp.; Penick; Roussel-UCLAF.*

4234 Dexlofexidine
81447-79-2
$C_{11}H_{12}Cl_2N_2O$
(+)-(S)-2[1-(2,6-Dichlorophenoxy)ethyl]-2-imidazoline.
An α_2-adrenoceptor agonist. Used as an antihypertensive.

4235 Diazoxide
364-98-7 3051 206-668-1
$C_8H_7ClN_2O_2S$
7-Chloro-3-methyl-2H-1,2,4-benzothiadiazine 1,1-dioxide.
SRG-95213; Sch-6783; NSC-64198; Eudemine injection; Proglicem; Hyperstat; Hypertonalum; Mutabase; Proglycem. Antihypertensive. A potent vasodilator. An ATP-dependent potassium-channel opener. A benzothiadiazine derivative similar in structure to the thiazides. mp = 330-331°; λ_m = 268 nm (ε = 11300 MeOH); soluble in EtOH, insoluble in H_2O. *Schering-Plough HealthCare Products.*

4236 Dihydralazine
484-23-1 3212 207-605-0
$C_8H_{10}N_6$
1,4-Dihydrazinophthalazine.
Vasodilator with antihypertensive properties. mp = 180° (dec); LD_{50} (rat ip) = 206 mg/kg. *Ciba-Geigy Corp.*

4237 Dihydralazine Sulfate
7327-87-9 3212 230-808-0
$C_8H_{12}N_6O_4S.2.5H_2O$
1,4-Dihydrazinophthalazine hydrogen sulfate.
dihydralazine sulfate hemipentahydrate; Depressan; Dihyzin; Nepresol; Nepréssol; dihydralazine mesylate [as methanesulfonate]; Nepresol Inject [as methanesulfonate]. Vasodilator with antihypertensive properties. [sulfate]: mp = 233° (dec). *Ciba-Geigy Corp.*

4238 Dilevalol
75659-07-3 3245
$C_{19}H_{24}N_2O_3$
(-)-5-[(1R)-1-Hydroxy-2-[[(1R)-1-methyl-3-phenylpropyl]amino]ethyl]salicylamide.
R,R-labetalol; Sch-19927; Dilevalon; Levadil; Unicard. Non-selective β-adrenergic blocker with vasodilating and antihypertensive properties. Active isomer of labetalol. $[\alpha]_D$ = -21.7°. *Schering-Plough HealthCare Products.*

4239 Dilevalol Hydrochloride
75659-08-4 3245
$C_{19}H_{25}ClN_2O_3$
(-)-5-[(1R)-1-Hydroxy-2-[[(1R)-1-methyl-3-phenylpropyl]amino]ethyl]salicylamide monohydrochloride.
Sch-19927-HCl. Non-selective β-adrenergic blocker with vasodilating and antihypertensive properties. mp = 133-134° (dec), 192-193.5° (dec); $[\alpha]_D^{26}$ = -30.6° (c = 1.0 EtOH). *Schering-Plough HealthCare Products.*

4240 Dioxadilol
80743-08-4
$C_{16}H_{25}NO_4$
(±)-1-(1,4-Benzodioxan-2-ylmethoxy)-3-(tert-butylamino)-2-propanol.
Has beta adrenolytic activity. Antihypertensive; antianginal; antiarrhythmic; less potent than propranolol.

4241 Doxazosin
74191-85-8 3489
$C_{23}H_{25}N_5O_5$
1-(4-Amino-6,7-dimethoxy-2-quinazolinyl)-4-(1,4-benzodioxan-2-ylcarbonyl)piperazine.
UK-33274. Antihypertensive. Also used in the treatment of benign prostatic hyperplasia. Selective α-adrenergic blocker related to Prazosin. [monohydrochloride]: mp = 289-290°. *Pfizer Intl.; Roerig Div. Pfizer Pharm.*

4242 Doxazosin Monomethanesulfonate
77883-43-3 3489
$C_{24}H_{29}N_5O_8S$
1-(4-Amino-6,7-dimethoxy-2-quinazolinyl)-4-(1,4-benzodioxan-2-ylcarbonyl)piperazine monomethanesulfonate.
Cardura; UK-33274-27; doxazosin mesylate; Alfadil; Cardenalin; Cardular; Cardura; Cardran; Diblocin; Normothen; Supressin. Antihypertensive. Also used in the treatment of benign prostatic hyperplasia. Selective α-adrenergic blocker related to Prazosin. *Pfizer Intl.; Roerig Div. Pfizer Pharm.*

4243 Efonidipine
111011-63-3 3566
$C_{34}H_{38}N_3O_7P$
2-(N-Benzylanilino)ethyl (±)-1,4-dihydro-2,6-dimethyl-4-(m-nitrophenyl)-5-phosphononicotinate cyclic 2,2-dimethyltrimethylene ester.
Dihydropyridine calcium channel blocker. Antihypertensive. mp = 169-170°, 155-156°. *Nissan Kenzai Co. Ltd.*

4244 Efonidipine Hydrochloride
111011-53-1 3566
$C_{34}H_{39}ClN_3O_7P$
2-(N-Benzylanilino)ethyl (±)-1,4-dihydro-2,6-dimethyl-4-(m-nitrophenyl)-5-phosphononicotinate cyclic 2,2-dimethyltrimethylene ester hydrochloride.
Dihydropyridine calcium channel blocker. Used as an antihypertensive agent. Forms an ethanol solvate (NZ-105; Landel), mp = 151° (dec). LD_{50} (mus orl) > 600 mg/kg; [(S) form]: mp = 190-192°; $[\alpha]_D^{25}$ = 7.0° (c = 0.50 $CHCl_3$); [(R) form]: mp = 190-192°; $[\alpha]_D^{25}$ = -7.0° (c = 0.5 $CHCl_3$). *Nissan Kenzai Co. Ltd.*

4245 Enalapril
75847-73-3 3605
$C_{20}H_{28}N_2O_5$
1-[N-[(S)-1-Carboxy-3-phenylpropyl]-L-alanyl]-L-proline 1'-ethyl ester.
Angiotensin-converting enzyme inhibitor. Used to treat hypertension. Orally active peptidyldipeptide hydrolase inhibitor. *Merck & Co.Inc.*

4246 Enalapril Maleate
76095-16-4 3605 278-375-7
$C_{24}H_{32}N_2O_9$
1-[N-[(S)-1-Carboxy-3-phenylpropyl]-L-alanyl]-L-proline 1'-ethyl ester maleate (1:1).
Enacard; Renitec; Vasotec; MK-421; Amprace; Bitensil; Cardiovet; Enaloc; Enapren; Glioten; Hipoartel; Innovace; Lotrial; Olivin; Pres; Reniten; Renivace; Xanef; component of: Vaseretic, Acesistem, Co-Renitec, Innozide, Renacor, Xynertec. Angiotensin-converting enzyme inhibitor. Used to treat hypertension. mp = 143-144.5°; $[\alpha]_D^{25}$ = -42.2° (c = 1 MeOH); soluble in H_2O (2.5 g/100 ml), EtOH (8 g/100 ml), MeOH (20 g/100 ml). *Merck & Co.Inc.*

4247 Enalaprilat
84680-54-6 3606 278-459-3
$C_{18}H_{24}N_2O_5.2H_2O$
1-[N-[(S)-1-Carboxy-3-phenylpropyl]-L-alanyl]-L-proline dihydrate.
enalaprilic acid; MK-422; Vasotec Injection; Vasotec IV. Angiotensin-converting enzyme inhibitor. Used to treat hypertension. mp = 148-151°; $[\alpha]_D$ = -67.0° (0.1M HCl). *Merck & Co.Inc.*

4248 Enalaprilat [anhydrous]
76420-72-9 3606
$C_{18}H_{24}N_2O_5$
1-[N-[(S)-1-Carboxy-3-phenylpropyl]-L-alanyl]-L-proline.
enalaprilic acid. Angiotensin-converting enzyme inhibitor. Active metabolite of enalipril. Used to treat hypertension. *Merck & Co.Inc.*

4249 Endralazine Monomethanesulfonate
65322-72-7 3617
$C_{15}H_{19}N_5O_4S$
6-Benzoyl-5,6,7,8-tetrahydropyrido[4,3-c]-pyridazin-3(2H)-one hydrazone monomethanesulfonate.
Migranal; endralazine mesylate; BQ-22-708; Miretilan. Antihypertensive agent with peripheral vasodilating properties. mp = 185-188° (dec). *Sandoz Pharm. Corp.*

4250 Epanolol
86880-51-5 3641
$C_{20}H_{23}N_3O_4$
(±)-N-[2-[[3-(o-Cyanophenoxy)-2-hydroxypropyl]amino]ethyl]-2-(p-hydroxyphenyl)acetamide.
ICI-141292; Visacor. Antianginal; antihypertensive agent. Cardioselective β_1-adrenergic blocker with sympathomimetic activity. mp = 118-120°. *ICI.*

4251 Epithiazide
1764-85-8 3661 217-181-9
$C_{10}H_{11}ClF_3N_3O_4S_2$
6-Chloro-3,4-dihydro-3[[(2,2,2-trifluoroethyl)thio]-methyl]-2H-1,2,4-benzothiadiazine-7-sulfonamide 1,1-dioxide.
P-2105; NSC-108164; Thiaver. Diuretic; antihypertensive. mp = 206-207°. *Pfizer Inc.*

4252 Eprosartan
133040-01-4 3669
$C_{23}H_{24}N_2O_4S$
(E)-2-Butyl-1-(p-carboxybenzyl)-α-2-thenylimidazole-5-acrylic acid.
SK&F-108566. Non-peptidic angiotensin II receptor antagonist. Used as an antihypertensive. mp = 260-261°. *SmithKline Beecham Pharm.*

4253 Eprosartan Mesylate
144143-96-4 3669
$C_{24}H_{28}N_2O_7S_2$
(E)-2-Butyl-1-(p-carboxybenzyl)-α-2-thenylimidazole-5-acrylic acid monomethanesulfonate.
SK&F-108566J; Eprosartan methanesulfonate. Non-peptidic angiotensin II receptor antagonist. Used as an antihypertensive. *SmithKline Beecham Pharm.*

4254 Ethiazide
1824-58-4 3778 217-358-0
$C_9H_{12}ClN_3O_4S_2$
6-Chloro-3-ethyl-3,4-dihydro-2H-1,2,4-benzothiadiazine-7-sulfonamide 1,1-dioxide.
acthiazidum; Hypertane. Diuretic. mp = 269-270°. *Abbott Labs Inc.; Ciba-Geigy Corp.; Lederle Labs.; Merck & Co.Inc.; Parke-Davis; Searle G.D. & Co.; Solvay Duphar Labs Ltd.; Solvay Pharm. Inc.; Squibb E.R. & Sons; Wallace Labs.*

4255 Ethomoxane
16509-23-2
$C_{15}H_{23}NO_3$
(±)-2-(Butylaminomethyl)-8-ethoxy-1,4-benzodioxan.
An α-adrenoceptor antagonist used as an antihypertensive.

4256 Ethomoxane Hydrochloride
6038-78-4
$C_{15}H_{24}ClNO_3$
(±)-2-(Butylaminomethyl)-8-ethoxy-1,4-benzodioxan hydrochloride.
An α-adrenoceptor antagonist used as an antihypertensive.

4257 Eticlopride
84226-12-0
$C_{17}H_{25}ClN_2O_3$
(-)-(S)-5-Chloro-3-ethyl-N-[(1-ethyl-2-pyrrolidinyl)methyl]-6-methoxysalicylamide.
Dopamine receptor antagonist; antihypertensive.

4258 Fantofarone
114432-13-2 3975
$C_{13}H_{38}N_2O_5S$
1-[[p-[3-[(3,4-Dimethoxyphenethyl)methylamino]propoxy]phenyl]-sulfonyl]-2-isopropylindolizine.
SR-33557. A calcium channel blocker used as an antihypertensive agent. mp = 82-83°; d = 1.21 g/ml; soluble in H_2O (0.06 g/100 ml), organic solvents. *Sanofi Winthrop.*

4259 Felodipine
86189-69-7 3991
$C_{18}H_{19}Cl_2NO_4$
Ethyl methyl 4-(2,3-dichlorophenyl)-1,4-dihydro-2,6-dimethyl-3,5-pyridinedicarboxylate.
Plendil; H 154/82; Agon; Feloday; Flodil; Hydac; Munobal; Prevex; Splendil. Antianginal; antihypertensive agent. Dihydropyridine calcium channel blocker sold as the racemate. mp= 145°. *Astra Chem. Ltd.; Merck & Co.Inc.*

4260 dl-Felodipine
72509-76-3 3991
$C_{18}H_{19}Cl_2NO_4$
(±) Ethyl methyl 4-(2,3-dichlorophenyl)-1,4-dihydro-2,6-dimethyl-3,5-pyridinedicarboxylate.
Plendil; H 154/82; Agon; Feloday; Flodil; Hydac; Munobal; Prevex; Splendil. Antianginal; antihypertensive agent. Dihydropyridine calcium channel blocker sold as the racemate. mp= 145°. *Astra Chem. Ltd.; Merck & Co.Inc.*

4261 Fenoldopam
67227-56-9 4020
$C_{16}H_{16}ClNO_3$
6-Chloro-2,3,4,5-tetrahydro-1-(p-hydroxyphenyl)-1H-3-benzazepine-7,8-diol.
SKF-82526. Antihypertensive. Dopamine D_1 receptor agonist. [hydrobromide]: mp = 277° (dec). *SmithKline Beecham Pharm.*

4262 Fenoldopam Mesylate
67227-57-0 4020
$C_{17}H_{20}ClNO_6S$
6-Chloro-2,3,4,5-tetrahydro-1-(p-hydroxyphenyl)-1H-3-benzazepine-7,8-diol monomethanesulfonate (salt).
fenoldopam monomethanesulfonate; SKF-82526J; Corlopam. Dopamine agonist used as an antihypertensive agent. mp = 274° (dec). *SmithKline Beecham Pharm.*

4263 Fenquizone
20287-37-0 4039 243-689-5
$C_{14}H_{12}ClN_3O_3S$
(±)-7-Chloro-1,2,3,4-tetrahydro-4-oxo-2-phenyl-6-quinazolinesulfonamide.
M.G. 13054. Diuretic. mp > 310°; insoluble in H_2O. *Maggioni Farmaceutici S.p.A.*

4264 Fenquizone Monopotassium
52246-40-9 4039
$C_{14}H_{11}ClKN_3O_3S$
(±)-7-Chloro-1,2,3,4-tetrahydro-4-oxo-2-phenyl-6-quinazolinesulfonamide monopotassium salt.
Idrolone. Diuretic. Soluble in H_2O. *Maggioni Farmaceutici S.p.A.*

4265 Flosequinan
76568-02-0 4146
$C_{11}H_{10}FNO_2S$
7-Fluoro-1-methyl-3-(methylsulfinyl)-4(1H)-quinolone.
BTS 49 465; flosequinon; Manoplax. Acts as an arterial and venous vasodilator. Used as an antihypertensive agent; a cardiotonic. mp = 226-228°. *Boots Co.*

4266 Flufylline
82190-91-8
$C_{21}H_{24}FN_5O_3$
7-[2-[4-(p-Fluorobenzoyl)piperidino]ethyl]theophylline.
Antihypertensive.

4267 Flusoxolol
84057-96-5
$C_{22}H_{30}FNO_4$
(S)-1-[p-[2-[(p-Fluorophenethyl)oxy]ethoxy]phenoxy]-3-(isopropylamino)-2-propanol.
Antihypertensive; antianginal; antiarrhythmic. A selective β_1-adrenoceptor agonist.

4268 Flutonidine
28125-87-3
$C_{10}H_{12}FN_3$
2-(5-Fluoro-o-toluidino)-2-imidazoline.
ST-600. An α_2-adrenergic agonist used as an antihypertensive.

4269 Fosinopril
98048-97-6 4282
$C_{30}H_{46}NO_7P$
(4S)-4-Cyclohexyl-1-[[(R)-[(S)-1-hydroxy-2-methylpropoxy](4-phenylbutyl)phosphinyl]acetyl-L-proline propionate (ester).
fosenopril. Antihypertensive. Angiotensin-converting enzyme inhibitor. [diacid (SQ-27519)]: mp = 149-153°; $[\alpha]_D$= -24° (c = 1 MeOH). *Squibb E.R. & Sons.*

4270 Fosinopril Sodium
88889-14-9 4282
$C_{30}H_{45}NNaO_7P$
(4S)-4-Cyclohexyl-1-[[(R)-[(S)-1-hydroxy-2-methylpropoxy](4-phenylbutyl)phosphinyl]acetyl-L-proline propionate (ester) sodium salt.
SQ-28555; Monopril; Acecor; Secorvas; Staril. Antihypertensive. Angiotensin-converting enzyme inhibitor. *Mead Johnson Labs.; Mead Johnson Pharmaceuticals.*

4271 Furnidipine
138661-03-7
$C_{21}H_{24}N_2O_7$
(±)-Methyl tetrahydrofurfuryl-1,4-dihydro-2,6-dimethyl-4-(o-nitrophenyl)-3,5-pyridinedicarboxylate.
A calcium channel blocker. Antihypertensive.

4272 Furosemide
54-31-9 4331 204-822-2
$C_{12}H_{11}ClN_2O_5S$
4-Chloro-N-furfuryl-5-sulfamoylanthranlic acid.
Diuretic salt; Disal; Lasix; Frumil; LB-502; Aisemide; Beronald; Desdemin; Discoid; Diural; Dryptal; Durafurid; Errolon; Eutensin; Frusetic; Frusid; Fulsix; Fuluvamide; Furesis; Furo-Puren; Furosedon; Hydro-rapid; Impugan; Katlex; Lasilix; Lowpston; Macasirool; Mirfat; Nicorol; Odemase; Oedemex; Profemin; Rosemide; Rusyde; Trofurit; Urex. Diuretic. Furosemide used with controlled release potassium chloride to control edema. mp = 206°; λ_m = 288, 276, 336 nm ($E^{1\%}_{1\ cm}$ 945, 588, 144 95% EtOH); soluble in Me_2CO, MeOH, DMF; less soluble in EtOH, H_2O, $CHCl_3$, Et_2O; LD_{50} (frat orl) = 2600 mg/kg, (mrat orl) = 2820 mg/kg. *Astra USA Inc.; Elkins-Sinn; Fermenta Animal Health Co.; Hoechst Roussel Pharm. Inc.; Parke-Davis; Rhône-Poulenc Rorer Pharm. Inc.*

4273 Guabenxan
19889-45-3
$C_{10}H_{13}N_3O_2$
(1,4-Benzodioxan-6-ylmethyl)guanidine.
Antihypertensive.

4274 Guanabenz
5051-62-7 4585 225-750-8
$C_8H_8Cl_2N_4$
[(2,6-Dichlorobenzylidene)amino]guanidine.
Wy-8678; NSC-68982. A centrally acting α_2-adrenergic agonist used as an antihypertensive agent. mp = 227-229° (dec). *Sandoz Pharm. Corp.; Wyeth Labs.*

4275 Guanabenz Monoacetate
23256-50-0 4585 245-534-7
$C_{10}H_{12}Cl_2N_4O_2$
[(2,6-Dichlorobenzylidene)amino]guanidine monoacetate.
Wytensin; Wy-8678 acetate; Rexitene; Tenelid. A centrally acting α_2-adrenergic agonist used as an antihypertensive agent. mp = 192.5° (dec); soluble in H_2O (1100 mg/100 ml), EtOH (5000 mg/ml), propylene glycol (10000 mg/100 ml), $CHCl_3$ (60 mg/100 ml), EtOAc (100 mg/100 ml). *Sandoz Pharm. Corp.; Wyeth Labs.*

4276 Guanacline
1463-28-1 4586
$C_9H_{18}N_4$
[2-(3,6-Dihydro-4-methyl-1(2H)-pyridyl)ethyl]-guanidine.
cyclazenin; FBA-1464. Antihypertensive. Administration can cause a marked loss of neurons in the ganglia of the peripheral sympathetic nervous system. *Bayer AG.*

4277 Guanacline Monosulfate
1562-71-6 4586 216-344-1
$C_9H_{20}N_4O_4S$
[2-(3,6-Dihydro-4-methyl-1(2H)-pyridyl)ethyl]-guanidine sulfate.
B-1464; Leron. Antihypertensive. Administration can cause a marked loss of neurons in the ganglia of the peripheral sympathetic nervous system. mp = 185-186° (dec). *Bayer AG.*

4278 Guanadrel
40580-59-4 4587
$C_{10}H_{19}N_3O_2$
(1,4-Dioxaspiro[4.5]dec-2-ylmethyl)guanidine.
Orally active postganglionic sympathetic inhibitor. Used as an antihypertensive agent. *Cutter Labs; Fisons plc.*

4279 Guanadrel Sulfate
22195-34-2 4587
$C_{20}H_{40}N_6O_8S$
(1,4-Dioxaspiro[4.5]dec-2-ylmethyl)guanidine sulfate.
Hylorel; CL-1388R; U-28288D; Anarel. Orally active postganglionic sympathetic inhibitor. Used as an antihypertensive agent. mp = 213.5-215°. *Cutter Labs; Fisons plc.*

4280 Guanazodine
32059-15-7 4588
$C_9H_{20}N_4$
[(Octahydro-2-azocinyl)methyl]guanidine.
Antihypertensive agent. A hypotensive agent. Structurally related to guanethidine. *EGYT.*

4281 Guanazodine Sulfate Monohydrate
4588
$C_9H_{24}N_4O_5S$
[(Octahydro-2-azocinyl)methyl]guanidine sulfate monohydrate.
EGYT-739; Calnegyt; Sanegyt. Antihypertensive agent. A hypotensive agent. Structurally related to guanethidine. mp = 239-241°; LD_{50} (mus orl) = 2450 mg/kg, (mus iv) = 165 mg/kg, (mus sc) = 700 mg/kg. *EGYT.*

4282 Guanclofine
55926-23-3
$C_9H_{12}Cl_2N_4$
[2-(2,6-Dichloroanilino)ethyl]guanidine.
Adrenergic neurone-blocking agent; antihypertensive.

4283 Guanethidine
55-65-2 4589 200-241-3
$C_{10}H_{22}N_4$
[2-(Hexahydro-1(2H)-azocinyl)ethyl]guanidine.
Su-5864; Eutensol; Dopom; Octatensine; oktadin; oktatenzin; Oktatensin; Sanotensin; Abapresin. Antihypertensive agent. Adrenergic neurone-blocking agent with antihypertensive properties. *Ciba-Geigy Corp.*

4284 Guanethidine Monosulfate
645-43-2 4589 211-442-0
$C_{10}H_{24}N_4O_4$
[2-(Hexahydro-1(2H)-azocinyl)ethyl]guanidine sulfate (1:1).
Ismelin sulfate; component of: Esimil. Antihypertensive. Adrenergic neurone-blocking agent with antihypertensive properties. mp = 276-281°. *Ciba-Geigy Corp.*

4285 Guanethidine Sulfate
60-02-6 4589 200-452-0
$C_{20}H_{46}N_8O_4$
[2-(Hexahydro-1(2H)-azocinyl)ethyl]guanidine sulfate (2:1).
Su-5864; NSC-29863; Guethine; Iporal; Isobarin; Ismelin. Antihypertensive agent. Adrenergic neurone-blocking agent with antihypertensive properties. *Ciba-Geigy Corp.*

4286 Guanfacine
29110-47-2 4590 249-442-8
$C_9H_9Cl_2N_3O$
N-Amidino-2-(2,6-dichlorophenyl)acetamide.
Centrally active α_2-adrenergic agonist used as an antihypertensive. mp = 225-227°. *Robins, A. H. Co.; Sandoz Pharm. Corp.*

4287 Guanfacine Hydrochloride
29110-48-3 4590 249-443-3
$C_9H_{10}Cl_3N_3O$
N-Amidino-2-(2,6-dichlorophenyl)acetamide hydrochloride.
BS-100-141; LON-798; Estulic; Tenex. Centrally active α_2-adrenergic agonist; used as an antihypertensive. mp = 213-216°; LD_{50} (mus orl) = 165 mg/kg. *Robins, A. H. Co.; Sandoz Pharm. Corp.*

4288 Guanochlor
5001-32-1 4595 225-667-7
$C_9H_{12}Cl_2N_4O$
[[2-(2,6-Dichlorophenoxy)ethyl]amino]guanidine.
Antihypertensive agent. *Pfizer Inc.*

4289 Guanochlor Sulfate
551-48-4 4595 208-996-0
$C_{18}H_{26}Cl_4N_8O_6S$
[[2-(2,6-Dichlorophenoxy)ethyl]amino]guanidine sulfate (2:1).
Vatensol; NSC-108163. Antihypertensive agent. mp = 214°. *Pfizer Inc.*

4290 Guanoxabenz
24047-25-4 4597
$C_8H_8Cl_2N_4O$
1-[(2,6-Dichlorobenzylidene)amino]-3-hydroxyguanidine.
43-663. Antihypertensive agent. *Sandoz Pharm. Corp.*

4291 Guanoxabenz Hydrochloride
23256-40-8 4597 245-532-6
$C_8H_9Cl_3N_4O$
1-[(2,6-Dichlorobenzylidene)amino]-3-hydroxyguanidine hydrochloride.
Benzerial. Antihypertensive agent. mp = 173-175°. *Sandoz Pharm. Corp.*

4292 Guanoxan
2165-19-7 4598
$C_{10}H_{13}N_3O_2$
(1,4-Benzodioxan-2-ylmethyl)guanidine.
2-guanidinomethyl-1,4-benzodioxan. Antihypertensive agent. mp = 164-165°. *Pfizer Inc.*

4293 Guanoxan Sulfate
5714-04-5 4598
$C_{20}H_{28}N_6O_8$
(1,4-Benzodioxan-2-ylmethyl)guanidine sulfate (2:1).
Envacar; 3-01003. Antihypertensive agent. *Pfizer Inc.*

4294 Hexamethonium
60-26-4 4724
$[C_{12}H_{30}N_2]^+$
N,N,N,N',N',N'-Hexamethyl-1,6-hexanediaminium.
hexathonide; hexamethone; Hexathide [as iodide]; Vegolysen-T [as tartrate]. Ganglion blocker used as an antihypertensive agent. *May & Baker Ltd.*

4295 Hexamethonium Bromide
55-97-0 4724 200-249-7
$C_{12}H_{30}Br_2N_2$
N,N,N,N',N',N'-Hexamethyl-1,6-hexanediaminium bromide.
Bistrium bromide; Esametina; Gangliostat; Hexameton bromide; Hexanium bromide; Simpatoblock; Vegolysen; Vegolysin. Ganglion blocker used as an antihypertensive agent. mp = 274-276°; soluble in H_2O, EtOH; insoluble in Me_2CO, $CHCl_3$, Et_2O; hygroscopic; pH (10% aq solution) = 5.5-6.5. *May & Baker Ltd.*

4296 Hexamethonium Chloride
60-25-3 4724 200-465-1
$C_{12}H_{30}Cl_2N_2$
N,N,N,N',N',N'-Hexamethyl-1,6-hexanediaminium chloride.
Bistrium chloride; Chloor-Hexaviet; Esomid chloride; Hestrium chloride; Hexameton chloride; Hexone chloride; Hiohex chloride; Methium chloride; Meton. Ganglion blocker used as an antihypertensive agent. mp = 289-292° (dec); soluble in H_2O, EtOH; insoluble in $CHCl_3$, Et_2O; hygroscopic; pH (10% aq solution) = 3.8. *May & Baker Ltd.*

4297 Hydracarbazine
3614-47-9 4798 222-788-7
$C_5H_7N_5O$
6-Hydrazino-6-pyridazinecarboxamide.
Normatensyl. Diuretic; antihypertensive. mp = 249-250° (dec). *Chimie et Atomistique.*

4298 Hydralazine
86-54-4 4800 201-680-3
$C_8H_8N_4$
1-Hydrazinophthalazine.
Apresoline; Hypophthalin; Hipoftalin; C-5968; Präparat 5968; Ciba-5968; 1-hydrazinophthalazine. Antihypertensive agent. mp= 172-173°; soluble in 2N AcOH (33.3 g/100 ml), warm MeOH (8.4 g/100 ml); LD_{50} (mus orl) = 122 mg/kg, (mus ip) = 101 mg/kg, (rat orl) = 90 mg/kg, (rat ip) = 40 mg/kg. *Ciba-Geigy Corp.; Medco Res. Inc.; Solvay Pharm. Inc.*

4299 Hydralazine Hydrochloride
304-20-1 4800 206-151-0
$C_8H_9ClN_4$
1-Hydrazinophthalazine hydrochloride.
Apresoline hydrochloride; Lopres; component of: Apresazide, BiDil, H.H. 25/25, H.H. 50/50, Ser-Ap-Es, Unipres. Antihypertensive agent. mp = 273° (dec); soluble in H_2O (3.01 g/100 ml at 15°, 4.42 g/100 ml at 25°), EtOH (0.2 g/100 ml); slightly soluble in Et_2O; λ_m = 211, 240, 260, 304, 315 nm. *Ciba-Geigy Corp.; Medco Res. Inc.; Solvay Pharm. Inc.*

4300 Hydralazine Polystirex
Sulfonated diethenylbenzene-ethenylbenzene copolymer complex with 1-hydrazinophthalazine.
Antihypertensive agent. A polymer complex. *Fisons plc.*

4301 Hydrochlorothiazide
58-93-5 4822 200-403-3
$C_7H_{8C}lN_3O_4S_2$
6-Chloro-3,4-dihydro-2H-1,2,4-benzothiadiazine-7-sulfonamide 1,1-dioxide.
chlorsulthiadil; Esidrex; Dichlotride; HydroDIURIL; Hydrozide; Oretic; Thiuretic; Acuretic; Aldactazide; Aldoril; Apresazide; Caplaril; Capozide; Dyazide; Esimil; H.H. 25/25; H.H. 50/50; Hydropres; Hyzaar; Inderide; Lopressor HCT; Lotensin HCT; Maxzide; Moduretic; Prinzide; Ser-Ap-Es; Timolide; Unipres; Vaseretic; Ziac; component of: Micardis (with telmisartan). A diuretic used in the treatment of hypertension. A thiazide. mp= 273-275°; λ_m = 317, 271, 226 nm ($A^{1\%}_{1\,cm}$ 130, 654, 1280 MeOH/HCl); soluble in MeOH, EtOH, Me_2CO; insoluble in H_2O; LD_{50} (mus iv) = 590 mg/kg, (mus orl) > 8000 mg/kg. *Abbott Labs Inc.; Ciba-Geigy Corp.; Lederle Labs.; Lemmon Co.; Merck & Co.Inc.; Parke-Davis; Searle G.D. & Co.; Solvay Pharm. Inc.; Squibb E.R. & Sons; Wallace Labs; Wyeth Labs.*

4302 Hydroflumethiazide
135-09-1 4830 200-203-6
$C_8H_8F_3N_3O_4S_2$
3,4-Dihydro-6-(trifluoromethyl)-2H-1,2,4-benzothiadiazine-7-sulfonamide 1,1-dioxide.
Diumide-K; Diucardin; Saluron; Salutensin; dihydroflumethiazide; methforylthiazidine; metflorylthiazidine; Bristab; Bristurin; Di-Ademil; Diucardin; Elodrine; Finuret; Hydol; Hydrenox; Leodrine; NaClex; Olmagran; Rodiuran; Rontyl; Sisuril; Vergonil. A diuretic used in the treatment of hypertension. A thiazide. mp = 272-273°; λ_m = 272.5 nm (log ε 4.286 MeOH); soluble in Me_2CO (> 100 mg/ml), MeOH (58 mg/ml), CH_3CN (43 mg/ml), H_2O (0.3 mg/ml), Et_2O (0.2 mg/ml), C_6H_6 (< 0.1 mg/ml); pK_1 = 8.9, pK_2 = 10.7; LD_{50} (mus orl) > 8000 mg/kg, (mus iv) = 750 mg/kg, (mus ip) = 6280 mg/kg. *Apothecon; Roberts Pharm. Corp.; Wyeth-Ayerst Labs.*

4303 Idazoxan
79944-58-4
$C_{11}H_{12}N_2O_2$
(±)-2-(1,4-Benzodioxan-2-yl)-2-imidazoline.
An alpha(2)-adrenoceptor antagonist with antihypertensive properties.

4304 Idrapril
127420-24-0
$C_{11}H_{18}N_2O_5$
(1S,2R)-2-[[(Hydroxycarbamoyl)methyl]methylcarbamoyl]-cyclohexanecarboxylic acid.
An ACE inhibitor. Used as an antihypertensive agent.

4305 Iganidipine
119687-33-1
$C_{28}H_{38}N_4O_6$
(±)-3-(4-Allyl-1-piperazinyl)-2,2-dimethylpropyl methyl 1,4-dihydro-2,6-dimethyl-4-(m-nitrophenyl)-3,5-pyridinedicarboxylate.
A calcium antagonist with antihypertensive properties.

4306 Imidapril
89371-37-9 4947
$C_{20}H_{27}N_3O_6$
(S)-3-(N-[(S)-1-Ethoxycarbonyl-3-phenylpropyl]-L-alanyl)-1-methyl-2-oxoimidazoline-4-carboxylic acid.
Antihypertensive. Angiotensin-converting enzyme inhibitor. mp = 139-140°; $[\alpha]_D^{20}$ = -71.7° (c = 0.5 EtOH). *Tanabe Seiyaku Co. Ltd.*

4307 Imidapril Hydrochloride
89396-94-1 4947
$C_{20}H_{28}ClN_3O_6$
(S)-3-(N-[(S)-1-Ethoxycarbonyl-3-phenylpropyl]-L-alanyl)-1-methyl-2-oxoimidazoline-4-carboxylic acid mono-hydrochloride.
TA-6366; Novaloc; Tanapril. Antihypertensive. Angiotensin-converting enzyme inhibitor. mp = 214-216° (dec); $[\alpha]_D^{20}$ = -64.1° (c = 0.5 EtOH). *Tanabe Seiyaku Co. Ltd.*

4308 Imidaprilat
89371-44-8 4947
$C_{18}H_{23}N_3O_6$
(S)-3-(N-[(S)-1-Carboxyl-3-phenylpropyl]-L-alanyl)-1-methyl-2-oxoimidazoline-4-carboxylic acid.
imidaprilate. Antihypertensive. Angiotensin-converting enzyme inhibitor. mp = 239-241°; $[\alpha]_D^{19}$ = -88.4° (c = 1 5% $NaHCO_3$). *Tanabe Seiyaku Co. Ltd.*

4309 Indapamide
26807-65-8 4969 248-012-7
$C_{16}H_{16}ClN_3O_3S$
4-Chloro-N-(2-methyl-1-indolinyl)-3-sulfamoylbenzamide.
Lozol; S-1520; SE-1520; Bajaten; Damide; Fludex; Indaflex; Indamol; Ipamix; Natrilix; Noranat; Tandix; Veroxil; Pressural [as hemihydrate]. Diuretic; antihypertensive. Nonthiazide compound with a similar mechanism of action to the thiazide diuretics. mp = 160-162°; LD_{50} (rat ip) = 393-421 mg/kg; (rat iv) = 394-440 mg/kg, (rat orl) > 3000 mg/kg, (mus ip) = 410-564 mg/kg, (mus iv) = 577-635 mg/kg, (mus orl) > 3000 mg/kg, (gpg ip) = 347-416 mg/kg, (gpg iv) = 272-358 mg/kg, (gpg orl) > 3000 mg/kg. *Apothecon; Rhône-Poulenc Rorer Pharm. Inc.*

4310 Indenolol
60607-68-3 4974 262-323-5
$C_{15}H_{21}NO_2$
1-[1H-Inden-4 (or -7)-yloxy]-3-(isopropylamino)-2-propanol.
Sch-28316Z; YB-2. Antiarrhythmic; antihypertensive; antianginal agent. Non-selective β-adrenergic blocker. A 1:2 tautomeric mixture of the 4- and 7- indenyloxy isomers. mp = 88-89°. *Schering-Plough HealthCare Products; Yamanouchi U.S.A. Inc.*

4311 Indenolol Hydrochloride
81789-85-7 4974
$C_{15}H_{22}ClNO_2$
1-[1H-Inden-4 (or -7)-yloxy]-3-[(1-methylethyl)amino]-2-propanol hydrochloride.
Pulsan; Securpres. Antihypertensive; antiarrhythmic; antianginal. Non-selective β-adrenergic blocker. mp = 147-148°; LD_{50} (mus iv) = 26 mg/kg. *Schering AG; Yamanouchi U.S.A. Inc.*

4312 Indoramin
26844-12-2 5000 248-041-5

$C_{22}H_{25}N_3O$

N-[1-(2-Indol-3-ylethyl)-4-piperidyl]benzamide.

Wy-21901. An α_1-adrenergic blocking agent with antihypertensive and bronchodilating activity. mp = 208-210°. *Wyeth-Ayerst Labs.*

4313 Indoramin Hydrochloride
38821-52-2 5000 254-136-2

$C_{22}H_{26}ClN_3O$

N-[1-(2-Indol-3-ylethyl)-4-piperidyl]benzamide hydrochloride.

Wy-21901 HCl; Baratol; Doralese; Vidora; Wydora; Wypres; Wypresin. An α_1-adrenergic blocking agent with antihypertensive and bronchodilating activity. mp = 230-232°, 258-260°. *Wyeth-Ayerst Labs.*

4314 Irbesartan
138402-11-6 5097

$C_{25}H_{28}N_6O$

2-Butyl-3-[p-(o-1H-tetrazol-5-ylphenyl)benzyl]-1,3-diazaspiro[4.4]non-1-en-4-one.

BMS-186295; SR-47436. Non-peptidic angiotensin II type-1 receptor antagonist. Used as an antihypertensive. mp = 180-181°. *Bristol-Myers Squibb Pharm. R&D; Sanofi Winthrop.*

4315 Irindalone
96478-43-2

$C_{24}H_{29}FN_4O$

(+)-(1R,3S)-1-[2-[4-[3-(p-Fluorophenyl)-1-indanyl]-1-piperazinyl]ethyl]-2-imidazolidinone.

A new S2-serotonergic antagonist with antihypertensive activity.

4316 Isradipine
75695-93-1 5260

$C_{19}H_{21}N_3O_5$

Isopropyl methyl (±)-4-(4-benzofurazanyl)-1,4-dihydro-2,6-dimethyl-3,5-pyridinedicarboxylate.

isrodipine; DynaCirc; PN-200-110; Clivoten; Dynacrine; Esradin; Lomir; Prescal; Rebriden. Antihypertensive; antianginal. A dihydropyridine calcium channel blocker. mp = 168-170°; [(S)-(+)-form (PN-205-033)]: mp = 142°; $[\alpha]_D^{20}$ = 6.7° (c = 1.5 EtOH); [(R)-(-)-form (PN-205-034)]: mp = 140°; $[\alpha]_D^{20}$ = 6.7° (c = 1.5 EtOH). *Sandoz Pharm. Corp.*

4317 Ketanserin
74050-98-9 5307 277-680-2

$C_{22}H_{22}FN_3O_3$

3-[2-[4-(p-Fluorobenzoyl)piperidino]ethyl]-2,4-(1H,3H)-quinazolinedione.

R-41468. Specific serotonin $5HT_2$-receptor antagonist used as an antihypertensive agent. mp = 227-235°; soluble in H_2O (0.001 g/100 ml), EtOH (0.038 g/100 ml), DMF (2.34 g/100 ml). *Janssen Pharm. Inc.*

4318 Ketanserin Tartrate
83846-83-7 5307 281-062-8

$C_{26}H_{28}FN_3O_9$

3-[2-[4-(p-Fluorobenzoyl)piperidino]ethyl]-2,4-(1H,3H)-quinazolinedione tartrate.

R-49945; Ket; Perketan; Serepress; Sufrexal. Specific serotonin $5HT_2$-receptor antagonist used as an antihypertensive agent. *Janssen Pharm. Inc.*

4319 Labetalol
36894-69-6 5341 253-258-3

$C_{19}H_{24}N_2O_3$

5-[1-Hydroxy-2-[(1-methyl-3-phenylpropyl)amino]-ethyl]salicylamide.

ibidomide; Dilevalol [as (R,R)-isomer]. Competitive α- and β-adrenergic receptor antagonist. Used as an antihypertensive. *Glaxo Wellcome Inc.; Key Pharm.*

4320 Labetalol Hydrochloride
32780-64-6 5341 251-211-1

$C_{19}H_{25}ClN_2O_3$

5-[1-Hydroxy-2-[(1-methyl-3-phenylpropyl)amino]-ethyl]salicylamide hydrochloride.

Normodyne; Trandate; Sch-15719W; AH-5158A; Amipress; Ipolab; Labelol; Labracol; Presdate; Pressalolo; Vescal. Competitive α- and β-adrenergic receptor antagonist. Used as an antihypertensive. mp = 187-189°; soluble in H_2O, EtOH; insoluble in Et_2O, $CHCl_3$; LD_{50} (mmus ip) = 114 mg/kg, (mmus iv) = 47 mg/kg, (mmus orl) = 1450 mg/kg, (fmus ip) = 120 mg/kg, (fmus iv) = 54 mg/kg, (fmus orl) = 1800 mg/kg, (mrat ip) = 113 mg/kg, (mrat iv) = 60 mg/kg, (mrat orl) = 4550 mg/kg, (frat ip) = 107 mg/kg, (frat iv) = 53 mg/kg, (frat ol) = 4000 mg/kg. *C.M. Ind.; Glaxo Wellcome Inc.; Key Pharm.*

4321 Lacidipine
103890-78-4 5344

$C_{26}H_{33}NO_6$

4-[o-[(E)-2-Carboxyvinyl]phenyl]-1,4-dihydro-2,6-dimethyl-3,5-pyridinedicarboxylic acid 4-tert-butyldiethyl ester.

GR-43659X; GX-1048; Caldine; Lacipil; Lacirex; Motens. Dihydropyridine calcium channel blocker used as an antihypertensive agent. mp = 174-175°. *Glaxo Wellcome plc.*

4322 Lercanidipine
100427-26-7 5469

$C_{36}H_{41}N_3O_6$

(±)-2-[(3,3-Diphenylpropyl)methylamino]-1,1-dimethylethyl methyl 1,4-dihydro-2,6-dimethyl-4-(m-nitrophenyl)-3,5-pyridinedicarboxylate.

masnidipine. Dihydropyridine calcium channel blocker used as an antihypertensive agent. *Recordati Corp.*

4323 Lercanidipine Hydrochloride
132866-11-6 5469

$C_{36}H_{42}ClN_3O_6$

(±)-2-[(3,3-Diphenylpropyl)methylamino]-1,1-dimethylethyl methyl 1,4-dihydro-2,6-dimethyl-4-(m-nitrophenyl)-3,5-pyridinedicarboxylate hydrochloride.

Rec-15-2375; R-75. Dihydropyridine calcium channel blocker used as an antihypertensive agent. [hemihydrate]: mp = 119-123°; LD_{50} (mus ip) = 83 mg/kg, (mus orl) = 657 mg/kg. *Recordati Corp.*

4324 Libenzapril
109214-55-3

$C_{18}H_{25}N_3O_5$

N-[(3S)-1-(Carboxymethyl)-2,3,4,5-tetrahydro-2-oxo-1H-1-benzazepin-3-yl]-L-lysine.

CGS-16617. Antihypertensive. Angiotensin-converting enzyme inhibitor. *Ciba-Geigy Corp.*

4325 Lisinopril
83915-83-7 5540
$C_{21}H_{31}N_3O_5.2H_2O$
1-[N^2[(S)-1-Carboxy-3-phenylpropyl]-L-lysyl]-L-proline dihydrate.
MK-521; RS-10029; Acerbon; Alapril; Carace; Coric; Novatec; Prinil; Prinivil; Tensopril; Vivatec; Zestril; component of: Prinzide. Antihypertensive. Angiotensin-converting enzyme inhibitor. λ_m = 246, 254, 258, 261, 267 nm ($A^{1\%}_{1\ cm}$ 4.0, 4.5, 5.1, 5.1, 3.7 0.1N NaOH), 246 253, 258, 264, 267 nm ($A^{1\%}_{1\ cm}$ 3.2, 3.9, 4.5, 3.0, 2.8 0.1NHCl); $[\alpha]^{25}_{405}$ = -120° (c = 1 0.25M Zn(OAc)$_2$ pH 6.4), [α]25. *Merck & Co.Inc.*

4326 Lofexidine
31036-80-3 5588
$C_{11}H_{12}Cl_2N_2O$
2-[1-(2,6-Dichlorophenoxy)ethyl]-2-imidazoline.
Vasoactive agent used as an antihypertensive drug. An alpha 2-adrenoreceptor agonist. mp = 126-128°. *Marion Merrell Dow Inc.*

4327 Lofexidine Hydrochloride
21498-08-8 5588
$C_{11}H_{13}Cl_3N_2O$
2-[1-(2,6-Dichlorophenoxy)ethyl]-2-imidazoline hydrochloride.
MDL-14042; Baq-168; Britlofex; Lofetensin; Loxacor. Vasoactive agent used as an antihypertensive drug. An alpha 2-adrenoreceptor agonist. mp = 221-223°, 230-232°; very soluble in H_2O, EtOH; slightly soluble in iPrOH; insoluble in Et_2O; LD_{50} (mus orl) = 74-147 mg/kg, (mus iv) = 8-18 mg/kg, (rat orl) = 74-147 mg/kg, (rat iv) = 8-18 mg/kg, (dog orl) = 74-147 mg/kg, (dog iv) = 8-818. *Marion Merrell Dow Inc.*

4328 Losartan
114798-26-4 5613
$C_{22}H_{23}ClN_6O$
2-Butyl-4-chloro-1-[p-(o-1H-tetrazol-5-ylphenyl)benzyl]-imidazole-5-methanol.
Non-peptide angiotensin II receptor antagonist used as an antihypertensive agent. mp = 183.5-184.5°. *DuPont-Merck Pharm.; Merck & Co.Inc.*

4329 Losartan Monopotassium Salt
124750-99-8 5613
$C_{22}H_{22}ClKN_6O$
2-Butyl-4-chloro-1-[p-(o-1H-tetrazol-5-ylphenyl)benzyl]-imidazole-5-methanol monopotassium salt.
Cozaar; DuP-753; Du Pont 753; DUP-753; MK-954; component of: Hyzaar. Non-peptide angiotensin II receptor antagonist used as an antihypertensive agent. *DuPont-Merck Pharm.; Merck & Co.Inc.*

4330 Manidipine
89226-50-6 5786
$C_{35}H_{38}N_4O_6$
2-[4-(Diphenylmethyl)-1-piperazinyl]ethyl methyl (±)-1,4-dihydro-2,6-dimethyl-4-(m-nitrophenyl)-3,5-pyridinecarboxylate.
franipidine; manidipine 6300. Vasodilator; antihypertensive. Dihydropyridine calcium channel blocker used as an antihypertensive agent. mp = 125-128°. *Takeda Chem. Ind. Ltd.*

4331 Manidipine Dihydrochloride
89226-75-5 5786
$C_{35}H_{40}Cl_2N_4O_6$
2-[4-(Diphenylmethyl)-1-piperazinyl]ethyl methyl (±)-1,4-dihydro-2,6-dimethyl-4-(m-nitrophenyl)-3,5-pyridinecarboxylate dihydrochloride.
CV-4093; Calslot. Vasodilator; antihypertensive. Dihydropyridine calcium channel blocker used as an antihypertensive agent. mp [α form]: = 157-163°; mp [β form]: = 174-180°; mp [monohydrate]: = 167-170°; LD_{50} (mmus sc) = 387 mg/kg, (mmus ip) = 62.2 mg/kg, (mmus orl) = 190 mg/kg, (fmus sc) = 340 mg/kg), (fmus ip) = 68.0 mg/kg, (fmus orl) = 171 mg/kg, (mrat sc) = 2.

4332 Mebutamate
64-55-1 5813 200-587-5
$C_{10}H_{20}N_2O_4$
2-sec-Butyl-2-methyl-1,3-propanediol dicarbamate.
Capla; Dormate; dicamoylmethane; W-583; Butatensin; Carbuten; Mebutina; Prean; Sigmafon; Vallene; Mega; No-Press; Axiten; Ipotensivo; component of: Caplaril. Antihypertensive agent. mp = 77-79°; soluble in organic solvents, slightly soluble in H_2O (0.1 g/100 ml). *Wallace Labs.*

4333 Mecamylamine
60-40-2 5814 200-476-1
$C_{11}H_{21}N$
N,2,3,3-Tetramethyl-2-norbornamine.
mecamine. Antihypertensive agent. Ganglionic blocker. Oily liquid; $bp_{4.0}$ = 72°; slightly soluble in H_2O. *Merck & Co. Inc.*

4334 Mecamylamine Hydrochloride
826-39-1 5814 212-555-8
$C_{11}H_{22}ClN$
N,2,3,3-Tetramethyl-2-norbornamine hydrochloride.
Inversine; Mevasine. Antihypertensive agent. Ganglionic blocker. mp - 245.5-246.5°; soluble in H_2O (21.2 g/100 ml), EtOH (8.2 g/100 ml), glycerol (10.4 g/100 ml), iPrOH (2.1 g/100 ml). *Merck & Co.Inc.*

4335 Mepindolol
23694-81-7 5901 245-831-1
$C_{15}H_{22}N_2O_2$
1-[Isopropylamino]-3-[(2-methyl-indol-4-yl)oxy]-2-propanol.
SH-E-222 [as sulfate salt]; Betagon [as sulfate salt]; Corindolan [as sulfate salt]; Mepicor [as sulfate salt]. Antianginal; antihypertensive agent. A β-adrenergic blocker. mp = 100-102°, 95-97°. *Sandoz Pharm. Corp.*

4336 Methyclothiazide
135-07-9 6086 205-172-2
$C_9H_{11}Cl_2N_3O_4S_2$
6-Chloro-3-(chloromethyl)-3,4-dihydro-2-methyl-2H-1,2,4-benzothiadiazine-7-sulfonamide 1,1-dioxide.
Enduron; Aquatensen; Enduronyl; Eutron; NSC-110431; Duretic; Naturon. Diuretic; antihypertensive. A thiazide. mp= 225°; λ_m = 226, 267, 311 nm (ε 39300, 21250, 3300 MeOH); very soluble in Me_2CO, C_5H_5N; less

soluble in MeOH, EOH; insoluble in H_2O, $CHCl_3$, C_6H_6. *Abbott Labs Inc.; Carter-Wallace; Wallace Labs.*

4337 Methyl 4-Pyridyl Ketone Thiosemicarbazone
3115-21-7 6196
$C_8H_{10}N_4S$
2-[1-(4-Pyridinyl)ethylidene]hydrazine-carbothioamide.
Antihypertensive agent. mp = 229-231° (dec). *Schenley.*

4338 Methyl 4-Pyridyl Ketone Thiosemicarbazone Hydrochloride
2260-13-1 6196
$C_8H_{11}ClN_4S$
2-[1-(4-Pyridinyl)ethylidene]hydrazine-carbothioamide hydrochloride.
Depreton. Antihypertensive agent. *Schenley.*

4339 Methyldopa
555-30-6 6132 209-089-2
$C_{10}H_{13}NO_4$
3-Hydroxy-α-methyl-L-tyrosine.
AMD; α-methyldopa; MK-351; Aldomet; Aldometil; Aldomine; Dopamet; Dopegyt; Elanpres; Equibar; Lederdopa; Medomet; Medopa; Medopren; Methoplain; Sembrina; Presinol. Antihypertensive agent. Activates central α_2-adrenergic receptors. [l-form sesquihydrate]: mp = 300° (dec); $[\alpha]_D^{23}$ = -4.0° ± 0.5° (c = 1 0.1N HCl); λ_m = 281 nm (ε 2780); soluble in H_2O (1 g/100 ml at 25°), insoluble in common organic solvents; [D-form]: soluble in H_2O (1.8 g/100 ml). *Merck & Co.Inc.*

4340 Methyldopa Ethyl Ester Hydrochloride
2508-79-4 6132 219-720-3
$C_{12}H_{18}ClNO_4$
3-Hydroxy-α-methyl-L-tyrosine ethyl ester hydrochloride.
methyldopate hydrochloride. Antihypertensive agent. Activates central α_2-adrenergic receptors. Soluble in H_2O (1-30 g/100 ml at 25°). *Merck & Co.Inc.*

4341 Meticrane
1084-65-7 6220 214-112-4
$C_{10}H_{13}NO_4S_2$
7-Methylthiochroman-7-sulfonamide 1,1-dioxide.
SD-17102; Arresten; Fontilix. Diuretic; antihypertensive. mp= 236-237°. *Soc. Ind. Fabric. Antiboit.*

4342 Metipranolol
22664-55-7 6221 245-151-5
$C_{17}H_{27}NO_4$
(±)-1-(4-Hydroxy-2,3,5-trimethylphenoxy)-3-(isopropyl-amino)-2-propanol 4-acetate.
OptiPranolol; BM01.004; VUFB6453; Betamet; methypranol; trimepranol; Betanol; Disorat; Glauline; Glausyn; Turoptin. A β-adrenergic blocker. Antihypertensive agent also used to treat glaucoma. mp = 105-107°, 108.5-110.5°; λ_m = 278, 274 nm ($A_{1\,cm}^{1\%}$ 51.3, 50.5 MeOH); freely soluble in EtOH, $CHCl_3$, C_6H_6; slightly soluble in Et_2O; insoluble in H_2O; LD_{50} (mus iv) = 31 mg/kg. *Bausch & Lomb Pharm. Inc.; SPOFA.*

4343 Metipranolol Hydrochloride
36592-77-5 6221
$C_{17}H_{28}ClNO_4$
(±)-1-(4-Hydroxy-2,3,5-trimethylphenoxy)-3-(isopropylamino)-2-propanol 4-acetate hydrochloride.
Betamann; Optipranolol. A β-adrenergic blocker. Antihypertensive agent also used to treat glaucoma. Soluble in H_2O. *Bausch & Lomb Pharm. Inc.; SPOFA.*

4344 Metolazone
17560-51-9 6231 241-539-3
$C_{16}H_{16}ClN_3O_3S$
7-Chloro-1,2,3,4-tetrahydro-2-methyl-4-oxo-3-o-tolyl-6-quinazolinesulfonamide.
Mykrox; Zaroxolyn; SR-720-22; Diulo; Metenix;Oldren; Xuret. Diuretic; antihypertensive. Nonthiazide compound with a similar mechanism of action to the thiazide diuretics. mp= 252-254°; LD_{50} (mus orl) > 5000 mg/kg, (mus ip) > 1500 mg/kg. *Fisons plc.*

4345 Metoprolol
37350-58-6 6235 253-483-7
$C_{15}H_{25}NO_3$
1-(Isopropylamino)-3-[p-(2-methoxyethyl)phenoxy]-2-propanol.
CGP-2175; H-93/26. Antianginal; class II antiarrhythmic; antihypertensive. A β-adrenergic blocker. Lacks inherent sympathomimetic activity. *Apothecon; Astra Chem. Ltd.; Ciba-Geigy Corp.; Lemmon Co.*

4346 Metoprolol Fumarate
119637-66-0 6235
$C_{34}H_{54}N_2O_{10}$
1-(Isopropylamino)-3-[p-(2-methoxyethyl)phenoxy]-2-propanol fumarate (2:1) (salt).
Lopressor OROS; CGP-2175C. Antianginal; class II antiarrhythmic; antihypertensive. A β-adrenergic blocker. Lacks inherent sympathomimetic activity. Insoluble in EtOAc, Me_2CO, Et_2O, C_7H_{16}. *Ciba-Geigy Corp.*

4347 Metoprolol Succinate
98418-47-4 6235
$C_{34}H_{56}N_2O_{10}$
1-(Isopropylamino)-3-[p-(2-methoxyethyl)phenoxy]-2-propanol succinate (2:1) (salt).
Toprol XL; H-93/26 succinate. Antianginal; class II antiarrhythmic; antihypertensive. A β-adrenergic blocker. Lacks inherent sympathomimetic activity. *Apothecon; Astra Chem. Ltd.; Astra Sweden; Ciba-Geigy Corp.; Lemmon Co.*

4348 Metoprolol Tartrate
56392-17-7 6235 260-148-9
$C_{34}H_{56}N_2O_{12}$
1-(Isopropylamino)-3-[p-(2-methoxyethyl)phenoxy]-2-propanol (2:1) dextro-tartrate salt.
CGP-2175E; HCTCGP 2175E; Beloc; Betaloc; Lopressor; Lopresor; Prelis; Seloken; Selopral; Selo-Zok; component of: Lopressor HCT. Antianginal; class II antiarrhythmic; antihypertensive. A β-adrenergic blocker. Lacks inherent sympathomimetic activity. Soluble in MeOH (50 g/100 ml), H_2O (> 100 g/100 ml), $CHCl_3$ (49.6 g/100 ml), Me_2CO (0.11 g/100 ml), CH_3CN (0.089 g/100 ml); insoluble in C_6H_{14}; LD_{50} (fmus iv) = 118 mg/kg, (fmus orl) = 3090 mg/kg, (mrat iv) ≅ 90 mg/kg, (mrat orl) = 3090

mg/kg, (mrat iv) ≅ 90 mg/kg, (mrat orl) = 3090 mg/kg. *Apothecon; Astra Chem. Ltd.; Ciba-Geigy Corp.; Lemmon Co.; Sandoz Nutrition.*

4349 dl-Metyrosine
620-30-4 6248 210-635-7
$C_{10}H_{13}NO_3$
(±)-α-Methyl-L-tyrosine.
Demser. Antipheochromocytoma. A tyrosine hydroxylase inhibitor used as an antihypertensive in pheochromocytoma. Dec 320°, 330-332°; soluble in H_2O (0.057 g/100 ml at 25°). *Merck & Co.Inc.*

4350 l-Metyrosine
672-87-7 6248 211-599-5
$C_{10}H_{13}NO_3$
(-)-α-Methyl-L-tyrosine.
Demser; L-α-MT; MK-781. Antipheochromocytoma. A tyrosine hydroxylase inhibitor used as an antihypertensive in pheochromocytoma. mp = 310-315°. *Merck & Co.Inc.*

4351 Mibefradil
116644-53-2 6261
$C_{29}H_{38}FN_3O_3$
(1S,2S)-[2-[[3-(2-Benzimidazolyl)propyl]methylamino]-ethyl-6-fluoro-1,2,3,4-tetrahydro-1-isopropyl-2-naphthyl methoxyacetate.
Calcium channel blocker; antihypertensive. Blocks both L- and T-type calcium channels with a more selective blockade of T-type channels. *Hoffmann-LaRoche Inc.*

4352 Mibefradil Dihydrochloride
116666-63-8 6261
$C_{29}H_{40}Cl_2FN_3O_3$
(1S,2S)-[2-[[3-(2-Benzimidazolyl)propyl]methylamino]ethyl]-6-fluoro-1,2,3,4-tetrahydro-1-isopropyl-2-naphthyl methoxyacetate dihydrochloride.
Posicor; Ro-40-5967/001. Calcium channel blocker with antihypertensive activity. Blocks both L- and T-type calcium channels with a more selective blockade of T-type channels. mp = 128°; soluble in H_2O. *Hoffmann-LaRoche Inc.*

4353 Minoxidil
38304-91-5 6290 253-874-2
$C_9H_{15}N_5O$
2,4-Diamino-6-piperidinopyrimidine 3-oxide.
PDP; U-10858; Alopexil; Alostil; Loniten; Lonolox; Minoximen; Normoxidil; Pierminox; Prexidil; Regaine; Rogaine; Tricoxidil. Antialopecia agent with antihypertensive properties. Orally active. A piperidinopyrimidine derivative that activates potassium channels, producing vascular smooth muscle hyperpolarization and relaxation. mp = 248°, 259-261° (dec); λ_m = 230, 261, 285 nm (ε 35210, 11210, 11790 EtOH); 232, 280 nm (ε 26350, 28350 0.1N H_2SO_4), 231, 261.5, 285 nm (ε 36100, 11400, 12040 0.1N KOH); soluble in propylene glycol (7.5 g/100 ml), MeOH (4.4 g/100 ml), EtOH (2.9 g/100 ml), iPrOH (0.67 g/100 ml), DMSO (0.65 g/100 ml), H_2O (0.22 g/100 ml), $CHCl_3$ (0.05 g/100 ml); slightly soluble in Me_2CO, EtOAc, Et_2O, C_6H_6, CH_3CN; LD_{50} (rat iv) = 49 mg/kg, (mus iv) = 51 mg/kg. *Pharmacia & Upjohn.*

4354 Moexipril
103775-10-6
$C_{25}H_{30}N_2O_7$
(3S)-2-[(2s)-N-[(1S)-1-Carboxy-3-phenylpropyl]alanyl]-1,2,3,4-tetrahydro-6,7-dimethoxy-3-isoquinolinecarboxylic acid 2 ethyl ester.
RS-10029. Antihypertensive. Angiotensin-converting enzyme inhibitor. *Schwarz Pharma Kremers Urban Co.*

4355 Moexipril Hydrochloride
82586-52-5
$C_{27}H_{35}ClN_2O_7$
(3S)-2-[(2S)-N-[(1S)-1-Carboxy-3-phenylpropyl]-L-alanyl]-1,2,3,4-tetrahydro-6,7-dimethoxy-3-isoquinolinecarboxylic acid 2-ethyl ester hydrochloride.
Univasc; SPM-925; CI-925; RS-10085-197. Antihypertensive. Angiotensin-converting enzyme inhibitor. *Schwarz Pharma Kremers Urban Co.*

4356 Moexiprilat
103775-14-0
$C_{25}H_{30}N_2O_7$
(3S)-2-[(2S)-N-[(1S)-1-Carboxy-3-phenylpropyl]alanyl]-1,2,3,4-tetrahydro-6,7-dimethoxy-3-isoquinolinecarboxylic acid.
Antihypertensive. Angiotensin-converting enzyme inhibitor. *Schwarz Pharma Kremers Urban Co.*

4357 Mopidralazine
75841-82-6
$C_{14}H_{19}N_5O$
4-[6-(2,5-Dimethylpyrrol-1-yl)amino]-3-pyrazinyl]morpholine.
MDL-899 [as hydrochloride]. A pyrrolylpyridazinamine antihypertensive.

4358 Moprolol
5741-22-0 6345 227-254-7
$C_{13}H_{21}NO_3$
1-(Isopropylamino)-3-(o-methoxyphenoxy)-2-propanol.
Antihypertensive agent with antianginal properties. mp = 82-83°; [l-form (levomoprolol)]: mp = 78-80°; [α] = -5.5° ± 0.2°. *ICI.*

4359 Moprolol Hydrochloride
27058-84-0 6345 248-195-3
$C_{13}H_{22}ClNO_3$
1-(Isopropylamino)-3-(o-methoxyphenoxy)-2-propanol hydrochloride.
SD-1601; Omeral. Antihypertensive agent with antianginal properties. mp = 110-112°; [l-form hydrochloride (Levotensin)]: mp = 121-123°; $[\alpha]_D^{25}$ = -16.3° (c = 5.0 EtOH). *ICI.*

4360 Moveltipril
85856-54-8 6368
$C_{19}H_{30}N_2O_5S$
(-)-1-[(2S)-3-Mercapto-2-methylpropionyl]-L-proline ester with N-(cyclohexylcarbonyl)-thio-D-alanine.
altiopril. Antihypertensive. Angiotensin-converting enzyme inhibitor. mp = 113-116°; $[\alpha]_D$ = 14.2° (c = 1.05 MeOH). *Chugai Pharm. Co. Ltd.*

4361 Moveltipril Calcium
85921-53-5 6368
$C_{38}H_{58}CaN_4O_{10}S_2$
(-)-1-[(2S)-3-Mercapto-2-methylpropionyl]-L-proline ester with N-(cyclohexylcarbonyl)thio-D-alanine calcium salt (2:1).
MC-838; Lowpres. Antihypertensive. Angiotensin-converting enzyme inhibitor. mp ≅ 190°; $[\alpha]_D^{20}$ = -48° to -52° (c = 1 MeOH); very soluble in H_2O, MeOH; soluble in EtOH, $CHCl_3$; insoluble in Me_2CO, EtOAc; LD_{50} (mmus orl) > 10.0 g/kg, (mmus ip) = 2.1 g/kg, (mmus sc) = 3.0 g/kg, (fmus orl) > 10 g/kg, (fmus ip) = 2.3 g/kg, (fmus sc) = 3.8 g/kg, (mrat orl) > 10.0 g/kg, (mrat ip) = 1.3 g/kg, (mrat sc) = 3.4 g/kg, (frat orl) > 10.0 g/kg, (frat ip) = 1.3 g/kg, (frat sc) = 3.9 g/kg, (mdog orl) > 6.0 g/kg, (fdog orl) > 6.0 g/kg. *Chugai Pharm. Co. Ltd.*

4362 Moxonidine
75438-57-2 6375
$C_9H_{12}ClN_5O$
4-Chloro-5-(2-imidazolin-2-ylamino)-6-methoxy-2-methylpyrimidine.
BDF-5895; Cynt; Physiotens. Vasoconstrictor; antihypertensive. An α-adrenergic agonist. mp = 217-219° (dec). *Beiersdorf AG.*

4363 Moxonidine Hydrochloride
75438-58-3 6375
$C_9H_{13}Cl_2N_5O$
4-Chloro-5-(2-imidazolin-2-ylamino)-6-methoxy-2-methylpyrimidine hydrochloride.
Vasoconstrictor; antihypertensive. An α-adrenergic agonist. mp = 189°. *Beiersdorf AG.*

4364 Muzolimine
55294-15-0 6397 259-573-2
$C_{11}H_{11}Cl_2N_3O$
3-Amino-1-(3,4-dichloro-α-methyl)benzyl-2-pyrazolin-3-one.
Edrul®; BAY g 2821. Diuretic. mp = 127-129°; LD_{50} (mus orl) = 1794 mg/kg, (dog orl) = 2000 mg/kg, (rbt orl) = 1250 mg/kg, (rat orl) = 1559 mg/kg. *Bayer AG.*

4365 Nadolol
42200-33-9 6431 255-706-3
$C_{17}H_{27}NO_4$
1-(tert-Butylamino)-3-[(5,6,7,8-tetrahydro-cis-6,7-dihydroxy-1-naphthyl)oxy]-2-propanol.
Corgard; SQ-11725; Anabet; Solgol. Antihypertensive and antianginal agent. Class II antiarrhythmic. A β-adrenergic blocker. mp = 124-136°; λ_m = 270, 278 nm ($E^{1\%}_{1\,cm}$ 37.5, 39.1, MeOH); pKa = 9.67; poorly soluble in organic solvents; insoluble in Me_2CO, C_6H_6, Et_2O, C_6H_{14}; LD_{50} (rat orl) = 5300 mg/kg, (mus orl) = 4500 mg/kg. *Apothecon; Bristol-Myers Squibb Co.*

4366 Naftopidil
57149-07-2 6443
$C_{24}H_{28}N_2O_3$
4-(o-Methoxyphenyl)-α-[(1-naphthyloxy)methyl]-1-piperazineethanol.
KT-611; Avishot; Flivas. An α_1 adrenergic blocker and serotonin $5HT_{1A}$ receptor agonist. Used as an antihypertensive agent. Also used in treatment of BPH. mp = 125-126°, 125-129°; insoluble in H_2O; LD_{50} (mus orl) = 1300 mg/kg, (rat orl) = 6400 mg/kg. *Boehringer Ingelheim Ltd.; Boehringer Mannheim GmbH; C.M. Ind.*

4367 (±)-Naftopidil
132295-16-0 6443
$C_{24}H_{28}N_2O_3$
(±)-4-(o-Methoxyphenyl)-α-[(1-naphthyloxy)methyl]-1-piperazineethanol.
KT-611; Avishot; Flivas. An α_1 adrenergic blocker and serotonin $5HT_{1A}$ receptor agonist. Used as an antihypertensive agent. Also used in treatment of BPH. mp = 125-126°, 125-129°; insoluble in H_2O; LD_{50} (mus orl) = 1300 mg/kg, (rat orl) = 6400 mg/kg. *Boehringer Mannheim GmbH.*

4368 Naftopidil Dihydrochloride
57149-08-3 6443
$C_{24}H_{30}Cl_2N_2O_3$
4-(o-Methoxyphenyl)-α-[(1-naphthyloxy)methyl]-1-piperazineethanol dihydrochloride.
An α_1 adrenergic blocker and serotonin $5HT_{1A}$ receptor agonist. Used as an antihypertensive agent. Also used in treatment of BPH. mp = 212-213°. *Boehringer Mannheim GmbH.*

4369 Nebivalol
99200-09-6 6519
$C_{22}H_{25}F_2NO_4$
α,α'-(Iminodimethylene)bis[6-fluoro-2-chromanmethanol].
R-65824; dl-nebivolol; narbivolol; Nebilet. Antihypertensive agent. A β-adrenergic blocker. *Janssen Pharm. Inc.*

4370 Nicardipine
55985-32-5 6579 259-932-3
$C_{26}H_{29}N_3O_6$
2-(Benzylmethylamino) ethyl methyl 1,4-dihydro-2,6-dimethyl-4-(m-nitrophenyl)-3,5-pyridinedicarboxylate.
Antianginal; antihypertensive; vasodilator. Dihydropyridine calcium channel blocker. Has antihypertensive properties. *Syntex Labs. Inc.; Yamanouchi U.S.A. Inc.*

4371 Nicardipine Hydrochloride
54527-84-3 6579 259-198-4
$C_{26}H_{30}Cl9N_3O_6$
2-(Benzylmethylamino) ethyl methyl 1,4-dihydro-2,6-dimethyl-4-(m-nitrophenyl)-3,5-pyridinedicarboxylate monohydrochloride.
YC-93; RS-69216; RS-69216-XX-07-0; Barizin; Bionicard; Cardene; Dacarel; Lecibral; Lescodil; Loxen; Nerdipina; Nicant; Nicardal; Nicarpin; Nicapress; Nicodel; Nimicor; Perdipina; Perdipine; Ranvil; Ridene; Rycarden; Rydene; Vasodin; Vasonase. Antianginal; antihypertensive agent. Dihydropyridine calcium channel blocker. Has antihypertensive properties. [α form]: mp = 179-181°; [β form]: mp = 168-170°; LD_{50} (mrat orl) = 634 mg/kg), (mrat iv) = 18.1 mg/kg, (frat orl) = 557 mg/kg, (frat iv) = 25.0 mg/kg, (mmus orl) = 634 mg/kg, (mmus iv) = 20.7 mg/kg, (fmus orl) = 650 mg/kg, (fmus iv) = 19.9 mg. *Syntex Labs. Inc.; Yamanouchi U.S.A. Inc.*

4372 Nifedipine
21829-25-4 6617 244-598-3
$C_{17}H_{18}N_2O_6$
Dimethyl 1,4-dihydro-2,6-dimethyl-4-(o-nitrophenyl)-3,5-pyridinedicarboxylate.
Bay a 1040; Adalat; Adalate; Adapress; Aldipin; Alfadat; Anifed; Aprical; Bonacid; Camont; Chronadalate; Citilat; Coracten; Cordicant; Cordilan; Corotrend; Duranifin; Ecodipi; Hexadilat; Introcar; Kordafen; Nifedicor; Nifedin; Nifelan; Nifelat; Nifensar XL; Orix; Oxcord; Pidilat; Procardia; Sepamit; Tibricol; Zenusin. Coronary vasodilator used as an antianginal agent. Dihydropyridine calcium channel blocker. mp = 172-174°; λ_m = 340, 235 nm (ε 5010, 21590 MeOH), 338, 238 nm (ε 5740, 20600 0.1N HCl), 340, 238 nm (5740, 20510 0.1N NaOH); soluble in Me_2CO (25.0 g/100 ml), CH_2Cl_2 (16 g/100ml), $CHCl_3$ (14 g/100 ml), EtOAc (5 g/100 ml), MeOH (2.6 g/100 ml), EtOH (1.7 g/100 ml); LD_{50} (mus orl) = 494 mg/kg, (mus iv) = 4.2 mg/kg, (rat orl) = 1022 mg/kg, (rat iv) = 15.5 mg/kg. *Bayer AG; Miles Inc.; Pfizer Inc.; Pratt Pharm.*

4373 Nilvadipine
75530-68-6 6637
$C_{19}H_{19}N_3O_6$
5-Isopropyl 3-methyl 2-cyano-1,4-dihydro-6-methyl-4-(m-nitrophenyl)-3,5-pyridinedicarboxylate.
nivadipine; nivaldipine; CL-287389; FK-235; FR-34235; SK&F-102362; Escor; Nivadil. Antianginal; antiarrhythmic agent. Dihydropyridine calcium channel blocker. Has antihypertensive properties. mp = 148-150°; [(+) form]: $[\alpha]_D^{20}$ = 222.42° (c = 1 MeOH); [(-) form]: $[\alpha]_D^{20}$ = -219.62° (c= 1 MeOH). *Fujisawa Pharm. USA Inc.; SmithKline Beecham Pharm.*

4374 Nipradilol
81486-22-8 6655
$C_{15}H_{22}N_2O_6$
8-[2-Hydroxy-3-(isopropylamino)propoxy]-3-chromanol.
Nipradolol; K-351; Hypadil. Antianginal; antihypertensive. β-adrenergic blocker; vasodilator. mp = 107-116°, 110-122°; LD_{50} (mus iv) = 74.0 mg/kg, (rat iv) = 73.0 mg/kg, (mus orl) = 540 mg/kg. *Kowa Chem. Ind. Co. Ltd.*

4375 Nisoldipine
63675-72-9 6658 264-407-7
$C_{20}H_{24}N_2O_6$
1,4-Dihydro-2,6-dimethyl-4-(2-nitrophenyl)-3,5-pyridinedicarboxylic acid methyl 2-methylpropyl ester.
Bay k 5552; Baymycard; Nisocor; Norvasc; Sular; Syscor; Zadipina. Antianginal; antihypertensive. Calcium channel blocker. Dihydropyridine calcium channel blocker. mp = 151-152°. *Bayer AG; Miles Inc.; Zeneca Pharm.*

4376 Nitrendipine
39562-70-4 6669 254-513-1
$C_{18}H_{20}N_2O_6$
(±)-Ethyl methyl-1,4-dihydro-2,6-dimethyl-4-(m-nitrophenyl)-3,5-pyridinedicarboxylate.
Baypress; Bay e 5009; Bayotensin; Bylotensin; Deiten; Nidrel. Dihydropyridine calcium channel blocker used as an antihypertensive agent. mp = 158°; insoluble in H_2O; LD_{50} (mus iv) = 39 mg/kg, (mus orl) = 2540 mg/kg, (rat iv) = 12.6 mg/kg, (rat orl) > 10000 mg/kg. *Bayer AG.*

4377 Olmidine
22693-65-8
$C_9H_{10}N_2O_3$
3,4-(Methylenedioxy)mandelamidine.
dl-mandelamidine. Antihypertensive. Inhibits adrenergic transmission.

4378 Oxprenolol
6452-71-7 7086 229-257-9
$C_{15}H_{23}NO_3$
1-(o-Allyloxyphenoxy)-3-isopropylamino-2-propanol .
Antianginal; antihypertensive; class IV antiarrhythmic. A β-adrenergic blocker. mp = 78-80°. *Bayer AG; Ciba-Geigy Corp.*

4379 Oxprenolol Hydrochloride
6452-73-9 7086 229-260-5
$C_{15}H_{24}ClNO_3$
1-(o-Allyloxyphenoxy)-3-isopropylamino-2-propanol hydrochloride.
Ba-39089; Coretal; Laracor; Paritane; Slow-Pren; Trasicor; Trasacor. Antianginal; antihypertensive; class IV antiarrhythmic. A β-adrenergic blocker. mp = 107-109°. *Bayer AG; Ciba-Geigy Corp.*

4380 Paraflutizide
1580-83-2 7157 216-426-7
$C_{14}H_{13}ClFN_3O_4S_2$
6-Chloro-3,4-dihydro-3-(p-fluorobenzyl)-2H-1,2,4-benzothiadiazine-7-sulfonamide 1,1-dioxide.
LD-3612. Diuretic. mp= 238-240°.

4381 Pargyline
555-57-7 7172 209-101-6
$C_{11}H_{13}N$
N-Methyl-N-2-propynylbenzylamine.
MO-911; A-19120; Eudatin; Supirdyl. Monoamine oxidase inhibitor with antihypertensive properties. bp_{11} = 96-97°. *Abbott Labs Inc.*

4382 Pargyline Hydrochloride
306-07-0 7172 206-175-1
$C_{11}H_{14}ClN$
N-Methyl-N-2-propynylbenzylamine hydrochloride.
Eutonyl; A-19120; MO-911; NSC-43798; component of: Eutron. Monoamine oxidase inhibitor with antihypertensive properties. mp = 154-155°; readily soluble in H_2O. *Abbott Labs Inc.*

4383 Pempidine
79-55-0 7207 201-211-2
$C_{10}H_{21}N$
1,2,2,6,6-Pentamethylpiperidine.
Ganglion blocking agent with antihypertensive properties. bp = 147°; [p-toluenesulfonate]: mp = 162-163°. *May & Baker Ltd.*

4384 Pempidine Tartrate
546-48-5 7207 208-902-8
$C_{14}H_{27}NO_6$
1,2,2,6,6-Pentamethylpiperidine tartrate.
M&B-4486; Pempidil; Pempiten; Perolysen; Tenormal; Tensinol; Tensoral. Ganglion blocking agent with antihypertensive properties. mp = 160°; soluble in EtOH, moderately soluble in H_2O. *May & Baker Ltd.*

4385 Penbutolol
38363-40-5 7209
$C_{18}H_{29}NO_2$
(S)-1-(tert-Butylamino)-3-(o-cyclopentylphenoxy)-2-propanol.
Antianginal; antihypertensive; antiarrhythmic. A β-adrenergic calcium channel blocker with antiarrhythmic activity. mp = 68-72°; $[\alpha]_D^{20}$ = -11.5° (c = 1 MeOH); soluble in MeOH, EtOH, $CHCl_3$. *Hoechst AG.*

4386 Penbutolol Sulfate
38363-32-5 7209 253-906-5
$C_{36}H_{40}N_2O_8S$
(S)-1-(tert-Butylamino)-3-(o-cyclopentylphenoxy)-2-propanol sulfate (salt) (2:1).
HOE-893d; HOE-39-893d; Betapressin; Levatol; Paginol. Antianginal; antihypertensive; antiarrhythmic. β-Adrenergic calcium channel blocker. mp = 216-218° (dec); $[\alpha]_D^{20}$ = -24.6° (c = 1 MeOH). *Hoechst AG.*

4387 Pentacynium Bis(methyl sulfate)
3810-83-1 7243
$C_{29}H_{45}N_3O_9S_2$
4-[2-[(5-Cyano-5,5-diphenylpentyl)methylamino]ethyl]-4-methylmorpholinium bis(methylsulfate).
pentacyone mesylate; Presidal. Ganglion blocking agent with antihypertensive properties. mp = 173-175°; soluble in H_2O. *Glaxo Wellcome Inc.*

4388 Pentamethonium Bromide
541-20-8 7253 208-771-7
$C_{11}H_{28}Br_2N_2$
N,N,N,N',N',N'-Hexamethyl-1,5-pentanediaminium bromide.
C-5; Penthonium; Lytensium. Ganglion blocking agent with antihypertensive properties.

4389 Pentolinium Tartrate
52-62-0 7274 200-146-7
$C_{23}H_{42}N_2O_{12}$
1,1'-(1,5-Pentanediyl)bis[1-methyl-pyrrolidinium] salt with [R-(R*, R*)]-2,3-dihydroxybutanedioic acid.
pentapyrrolidinium bitartrate; M&B-2050A; Ansolysen tartrate; Ansolysen bitartrate; Pentilium. Ganglion blocking agent with antihypertensive properties. mp = 203° (dec); freely soluble in H_2O (250 g/100 ml); soluble in EtOH (0.12 g/100 ml); insoluble in Et_2O, $CHCl_3$. *May & Baker Ltd.*

4390 Pentopril
82924-03-6
$C_{18}H_{23}NO_5$
Ethyl (αR,σR,2S)-2-carboxy-α,σ-dimethyl-δ-oxo-1-indoline valerate.
CGS-13945. Antihypertensive. Angiotensin-converting enzyme inhibitor. *Ciba-Geigy Corp.*

4391 Perindopril
82834-16-0
$C_{19}H_{32}N_2O_5$
(2S,3aS,7aS)-1-[(S)-N-[(S)-1-Carboxybutyl]alanyl]hexahydro-2-indoline carboxylic acid, 1-ethyl ester.
S-9490; McN-A-2833; Coversyl. Ace inhibitor. *Ortho-McNeill; Solvay Pharm. Inc.*

4392 Perindopril tert-Butylamine
107133-36-8 7311
$C_{23}H_{43}N_3O_5$
(2S,3aS,7aS)-1-[(S)-N-[(S)-1-Carboxybutyl]alanyl]hexahydro-2-indolinecarboxylic acid tert-butylamine salt.
perindopril ebumine; perinodpril erbimune; S-9490-3; McN-A-2833-109; Aceon; Coversum; Coversyl; Procaptan. Angiotensin-converting enzyme inhibitor. Used to treat hypertension. *McNeil Pharm.*

4393 Phenactropinium Bromide
3784-89-2 7346
$C_{24}H_{28}ClNO_4$
3-[(Hydroxyphenylacetyl)oxy]-8-methyl-8-(2-oxo-2-phenylethyl)-8-azoniabicyclo[3.2.1]octane chloride.
N-phenacylhomatropinium chloride; Trophenium. Ganglion blocking agent with antihypertensive properties. mp = 195-197°. *Smith T&H.*

4394 Pheniprazine
55-52-7 7382 200-236-6
$C_9H_{14}N_2$
(1-Methyl-2-phenylethyl)hydrazine.
PIH. Antihypertensive agent. [D-form]: bp_{10} = 135-138°; $[\alpha]_D^{25}$ = 4.5° (c = 5 MeOH); [L-form]: bp_{10} = 135-138°; $[\alpha]_D^{25}$ = -4.5° (c = 5 MeOH). *Lakeside BioTechnology.*

4395 (±)-Pheniprazine
52031-11-5 7382
$C_9H_{14}N_2$
(±)-(1-Methyl-2-phenylethyl)hydrazine.
Antihypertensive agent. $bp_{0.5}$ = 82-86°. *Lakeside BioTechnology.*

4396 (±)-Pheniprazine Hydrochloride
54779-57-6 7382
$C_9H_{15}ClN_2$
(±)-(1-Methyl-2-phenylethyl)hydrazine hydrochloride.
JB-516; Catral; Catron; Catroniazid; Cavodil. Antihypertensive agent. mp = 124-125°; [D-form]: mp = 152-154°, 148-149°; $[\alpha]_D^{25}$ = 12.8° (c = 5 H_2O), 13.8° (c = 1 H_2O); [l-form]: mp = 152-154°, 148-149°; $[\alpha]_D^{25}$ = -12.5° (c = 5 H_2O), -14.0° (c = 1 H_2O). *Lakeside BioTechnology.*

4397 Phentolamine
50-60-2 7417 200-053-1
$C_{17}H_{19}N_3O$
m-[N-(2-Imidazolin-2-ylmethyl)-p-toluidino]phenol.
Regitine; C-7337. An α-adrenergic blocker used as an antihypertensive agent. Used in the diagnosis and treatment of pheochromocytoma. mp = 174-175°. *Ciba-Geigy Corp.*

4398 Phentolamine Hydrochloride
73-05-2 7417 200-793-5
$C_{17}H_{20}ClN_3O$
m-[N-(2-Imidazolin-2-ylmethyl)-p-toluidino]phenol hydrochloride.
Regitine hydrochloride. An α-adrenergic blocker with antihypertensive properties. Used in the diagnosis and treatment of pheochromocytoma. mp = 239-240°; soluble in H_2O (2 g/100 ml), EtOH (1.43 g/100 ml); slightly soluble in $CHCl_3$; insoluble in Me_2CO, EtOAc; LD_{50} (rat

iv) = 75 mg/kg, (rat sc) = 275 mg/kg, (rat orl) = 1250 mg/kg. *Ciba-Geigy Corp.*

4399 Phentolamine Methanesulfonate
65-28-1 7417 200-604-6
$C_{18}H_{23}N_3O_4S$
m-[N-(2-Imidazolin-2-ylmethyl)-p-toluidino]phenol monomethanesulfonate.
phentolamine mesylate; Regitine; Rogitine. An α-adrenergic blocker used as an antihypertensive agent. Also used in the diagnosis and treatment of pheochromocytoma. mp = 177-181°; soluble in H_2O (2 g/100 ml), EtOH (4.35 g/100 ml), $CHCl_3$ (0.15 g/100 ml). *Ciba-Geigy Corp.*

4400 Piclonidine
72467-44-8 276-672-6
$C_{14}H_{17}Cl_2N_3O$
N-(2,6-Dichlorophenyl)-4,5-dihydro-N-(tetrahydro-2H-pyran-2-yl)-1H-Imidazol-2-amine.
LR-99853. Clonidine analog. Antihypertensive.

4401 Pildralazine
64000-73-3 7577
$C_8H_{15}N_5O$
(±)-1-[(6-Hydrazino-3-pyridazinyl)methylamino]-2-propanol.
propyldazine; propildazine. Peripheral vasodilator with antihypertensive activity. *I.S.F.*

4402 Pildralazine Dihydrochloride
56393-22-7 7577
$C_8H_{17}Cl_2N_5O$
(±)-1-[(6-Hydrazino-3-pyridazinyl)methylamino]-2-propanol dihydrochloride.
ISF-2123; Atensil. Peripheral vasodilator with antihypertensive activity. mp = 206-209° (dec); LD_{50} (mus ip) = 357 mg/kg, (mus orl) = 1170 mg/kg, (rat ip) = 355 mg/kg, (rat orl) = 1230 mg/kg. *I.S.F.*

4403 Pinacidil
85371-64-8 7592
$C_{13}H_{19}N_5.H_2O$
(±)-2-Cyano-1-(4-pyridyl)-3-(1,2,2-trimethylpropyl)guanidine monohydrate.
Pindac; P-1134; [anhydrous form] CAS 60560-33-0. Potassium channel opening vasodilator used as an antihypertensive agent. mp= 164-165°; LD_{50} (mus orl) = 600 mg/kg, (rat orl) = 570 mg/kg. *Eli Lilly & Co.*

4404 Pindolol
13523-86-9 7597 236-867-9
$C_{14}H_{20}N_2O_2$
1-(Indol-4-yloxy)-3-(isopropylamino)-2-propanol.
LB-46; Visken; prinodolol; Betapindol; Blocklin L; Calvisken; Decreten; Durapindol; Glauco-Visken; Pectobloc; Pinbetol; Pynastin. Antianginal agent with antiarrhythmic, antihypertensive and antiglaucoma properties. A β-adrenergic blocker. mp = 171-163°. *Sandoz Pharm. Corp.*

4405 Piperoxan
59-39-2 7631
$C_{14}H_{19}NO_2$
2-Piperidinomethyl-1,4-benzodioxan.
benzodioxane; Benodaine. An α-adrenergic blocker used as an antihypertensive agent. Also used as a diagnostic aid and a treatment for pheochromocytoma. bp_{17} = 193°. *Rhône-Poulenc Rorer Pharm. Inc.*

4406 Piperoxan Hydrochloride
135-87-5 7631 205-222-3
$C_{14}H_{20}ClNO_2$
2-Piperidinomethyl-1,4-benzodioxan hydrochloride.
compd. 933F; Fourneau 933. An α-adrenergic blocker used as an antihypertensive agent. Also used as a diagnostic aid and a treatment for pheochromocytoma. [dl-form]: mp = 232-234°; λ_m = 275 nm ($E^{1\%}_{1\,cm}$ 82); freely soluble in H_2O, iPrOH (10.8 mg/g). *Rhône-Poulenc Rorer Pharm. Inc.*

4407 Polythiazide
346-18-9 7744 206-468-4
$C_{11}H_{13}ClF_3N_3O_4S_3$
6-Chloro-3,4-dihydro-2-methyl-3-[[(2,2,2-trifluoromethyl)thio]methyl]-2H-1,2,4-benzothiadiazine-7-sulfonamide 1,1-dioxide.
Renese; P-2525; NSC-108161; Drenusil; Nephril. Diuretic; antihypertensive. A thiazide. mp= 202.5°; soluble in MeOH, Me_2CO; insoluble in H_2O, $CHCl_3$. *Pfizer Inc.*

4408 Prazosin
19216-56-9 7897 242-885-8
$C_{19}H_{21}N_5O_4$
1-(4-Amino-6,7-dimethoxy-2-quinazolinyl)-4-(2-furoyl)-piperazine.
furazosin. An α_1-adrenergic blocking agent used as an antihypertensive agent and also in treatment of BPH. mp = 278-280°. *Brocades-Stheeman & Pharmacia; Pfizer Intl.*

4409 Prazosin Hydrochloride
19237-84-4 7897 242-903-4
$C_{19}H_{22}ClN_5O_4$
1-(4-Amino-6,7-dimethoxy-2-quinazolinyl)-4-(2-furoyl)piperazine hydrochloride.
Minipress; CP-12299-1; Alpress LP; Duramipress; Eurex; Hypovase; Peripress; Sinetens. An α_1-adrenergic blocking agent used as an antihypertensive agent and also in treatment of BPH. Soluble in Me_2CO (0.72 mg/100 ml), MeOH (640 mg/100 ml), EtOH (84 mg/100 ml), DMF (130 mg/100 ml), dimethylacetamide (120 mg/100 ml), H_2O (140 mg/100 ml at pH 3.5), $CHCl_3$ (4.1 mg/100 ml); λ_m = 246, 329 nm (a_M = 137 ± 3, 27.6 ± 0.3 MeOH/1% HCl). *Brocades-Stheeman & Pharmacia; Pfizer Inc.*

4410 Pronetalol
54-80-8 7974
$C_{15}H_{19}NO$
2-Isopropyl-1-(naphth-2-yl)ethanol.
pronethalol; nethalide. Antianginal; antihypertensive; antiarrhythmic. A β-adrenergic blocker. mp = 108°. *ICI ; Wyeth-Ayerst Labs.*

4411 Pronetalol Hydrochloride
51-02-5 7974
$C_{15}H_{20}ClNO$
2-Isopropylamino-1-(naphth-2-yl)ethanol hydrochloride.
pronethalol hydrochloride; ICI-38174; AY-6204; Alderlin. Antianginal; antihypertensive; antiarrhythmic. A β-

adrenergic blocker. mp = 184°; LD_{50} (mus orl) = 512 mg/kg, (mus iv) = 46 mg/kg. *ICI ; Wyeth-Ayerst Labs.*

4412 Propranolol
525-66-6 8025 208-378-0
$C_{16}H_{21}NO_2$
1-(Isopropylamino)-3-(1-naphthyloxy)-2-propanol.
Avlocardyl; Euprovasin. Antianginal; antihypertensive; antiarrhythmic (class II) agent. A β-adrenergic blocker. mp = 96°. *ICI ; Parke-Davis; Quimicobiol; Wyeth-Ayerst Labs.*

4413 Propranolol Hydrochloride
318-98-9 8025 206-268-7
$C_{16}H_{22}ClNO_2$
1-(Isopropylamino)-3-(1-naphthyloxy)-2-propanol hydrochloride.
AY-64043; ICI-45520; NSC-91523; Angilol; Apsolol; Bedranol; Beprane; Berkolol; Beta-Neg; Beta-Tablinen; Beta-Timelets; Cardinol; Caridolol; Deralin; Dociton; Duranol; Efektolol; Elbol; Frekven; Inderal; Indobloc; Intermigran; Kemi S; Oposim; Prano-Puren; Prophylux; Propranur; Pylapron; Rapynogen; Sagittol; Servanolol; Sloprolol; Sumial; Tesnol. Antianginal; antihypertensive; antiarrhythmic (class II) agent. A β-adrenergic blocker. mp = 163-164°; soluble in H_2O, alcohol; insoluble in Et_2O, C_6H_6, EtOAc; LD_{50} (mus orl) = 565 mg/kg, (mus iv) = 22 mg/kg, (mus ip) = 107 mg/kg. *ICI ; Parke-Davis; Quimicobiol; Wyeth Labs.*

4414 Protoveratrines
8086
Provell [mixture of protoveratines A and B]; Tensatrin [mixture of protoveratines A and B]; Veralba [mixture of protoveratines A and B]; Protalba [protoveratine A]; veratetrine [protoveratine B]; neoprotoveratrine [protoveratine B]. Antihypertensive agent. Extract from the rhizome of *Veratrum album L., Liliaceae,* consists primarily of protoveratrine A and protoveratrine B. A group II sodium channel toxin. May enhance release of acetylcholine from nerve terminals. mp = 266-267° (dec); $[\alpha]_D^{25}$= -8.5° (c= 1.99 $CHCl_3$); insoluble in H_2O, soluble in polar organic solvents; LD_{50} (mus iv) = 0.048 mg/kg.

4415 Quinapril
85441-61-8 8233
$C_{25}H_{30}N_2O_5$
(S)-2-[(S)-N-[(S)-1-Carboxy-3-phenylpropyl]alanyl]-1,2,3,4-tetrahydro-3-isoquinolinecarboxylic acid 1-ethyl ester.
component of: Accuretic; Acequide; Koretic. Antihypertensive. Angiotensin-converting enzyme inhibitor. Orally active peptidyldipeptide hydrolase inhibitor. *Parke-Davis.*

4416 Quinapril Hydrochloride
82586-55-8 8233
$C_{25}H_{31}ClN_2O_5$
(S)-2-[(S)-N-[(S)-1-Carboxy-3-phenylpropyl]alanyl]-1,2,3,4-tetrahydro-3-isoquinolinecarboxylic acid 1-ethyl ester monohydrochloride.
CI-906; PD-109452-2; Accupril; Accuprin; Accupro; Acequin; Acuitel; Korec; Quinazil; component of: Accuretic; Acequide; Korectic. Antihypertensive. Angiotensin-converting enzyme inhibitor. mp = 120-130°, 119-121.5°; $[\alpha]_D^{23}$ = 14.5° (c = 1.2 EtOH), $[\alpha]_D^{25}$ = 15.4° (c = 2.0 MeOH); LD_{50} (mmus orl) = 1739 mg/kg, (mmus iv) = 504 mg/kg, (fmus orl) = 1840 mg/kg, (fmus iv) = 523 mg/kg, (mrat orl) = 4280 mg/kg, (mrat iv) = 158 mg/kg, (frat orl) = 3541 mg/kg, (frat iv) = 107 mg/kg. *Parke-Davis.*

4417 Quinaprilat
82768-85-2 8233
$C_{23}H_{26}N_2O_5$
(S)-2-[(S)-N-[(S)-1-Carboxy-3-phenylpropyl]alanyl]-1,2,3,4-tetrahydro-3-isoquinolinecarboxylic acid.
CI-928. Antihypertensive. Angiotensin-converting enzyme inhibitor. mp = 166-168°; $[\alpha]_D^{23}$ = 20.9° (c = 1 MeOH). *Parke-Davis.*

4418 Quinazosin
15793-38-1
$C_{17}H_{23}N_5O_2$
2-(Allyl-1-piperazinyl)-4-amino-6,7-dimethyoxyquinazoline.
Antihypertensive. Quinazoline derivative. *Pfizer Inc.*

4419 Quinazosin Hydrochloride
7262-00-2
$C_{17}H_{25}Cl_2N_5O_2$
2-(Allyl-1-piperazinyl)-4-amino-6,7-dimethyoxyquinazoline dihydrochloride.
CP-11332-1. Antihypertensive. Quinazoline derivative. *Pfizer Inc.*

4420 Quinethazone
73-49-4 8240 200-801-7
$C_{10}H_{12}ClN_3O_3S$
7-Chloro-2-ethyl-1,2,3,4-tetrahydro-4-oxo-6-quinazolinesulfonamide.
CL-36010; Hydromox; Aquamox. Diuretic; antihypertensive. Nonthiazide compound with a similar mechanism of action to the thiazide diuretics. mp = 250-252°; soluble in Me_2CO, EtOH. *Am. Cyanamid.*

4421 Ramipril
87333-19-5 8283
$C_{23}H_{32}N_2O_5$
(2S,3aS,6aS)-1-[(S)-N-[(S)-1-Carboxy-3-phenylpropyl]alanyl]octahydrocyclopenta[b]-pyrrole-2-carboxylic acid.
Altace; HOE-498; Cardace; Delix; Pramace; Quark; Ramace; Triatec; Tritace; Unipril; Vesdil. Antihypertensive. Angiotensin-converting enzyme inhibitor. Orally active peptidyldipeptide hydrolase inhibitor. mp = 109°; $[\alpha]_D^{24}$ = 33.2° (c= 1 0.1N ethanolic HCl); LD_{50} (mmus iv) = 1194 mg/kg, (mmus orl) = 10933 mg/kg, (fmus iv) = 1158 mg/kg, (fmus orl) = 10048 mg/kg, (mrat iv) = 687 mg/kg, (mrat orl) > 10000 mg/kg, (frat iv) = 608 mg/kg, (fr. *Hoechst Roussel Pharm. Inc.*

4422 Raubasine
483-04-5 8291 207-589-5
$C_{21}H_{24}N_2O_3$
16,17-Didehydro-19α-methyloxayohimban-16-carboxylic acid methyl ester.
tetrahydroserpentine; ajmalicine; Circolene; Hydrosarpan; Isoarteril; Lamuran. An α_1-adrenergic blocker isolated from the bark of Corynanthe johimbe K. Sachum., Rubiaceae. Used as an antihypertensive agent. Used as an antihypertensive agent and also as a cerebral

and peripheral vasodilator with anti-ischemic properties. mp = 257° (dec); $[\alpha]_D^{20}$ = -60° (c = 0.5 $CHCl_3$); $[\alpha]_D^{20}$ = -39° (c = 0.25 MeOH)λ_m = 227, 292 nm (log ε = 4.62 3.79 EtOH); [hydrochloride]: mp = 290° (dec); sparingly soluble in H_2O; $[\alpha]_D^{20}$ = -17° (c = 0.5 MeOH); [hydrobromide]: mp = 295-296°.

4423 Remikiren
126222-34-2
$C_{33}H_{50}N_4O_6S$
(αS)-α-[(αS)-α-[(tert-Butylsulfonyl)methyl]cinnamamido]-N-[(1S,2R,3S)-1-(cyclohexylnethyl)-3-cyclopropyl-2,3-dihydroxypropyl]imidazole-4-propionamide.
Ro-42-5892. Renin inhibitor. Antihypertensive.

4424 Rescimetol
73573-42-9 8310
$C_{33}H_{38}N_2O_8$
(3β,16β,17α,18β(E),20α)-18-[[3-(4-Hydroxy-3-methoxyphenyl)-1-oxo-2-propenyl]oxy]-11,17-dimethoxyyohimban-1-6carboxylic acid methyl ester.
CD-3400; WHO-4939; Toscarna. Antihypertensive agent. Analog of rescinnamine. mp = 259-260°. *Nippon Chemiphar.*

4425 Rescinnamine
24815-24-5 8311 246-471-8
$C_{35}H_{42}N_2O_9$
(3β,16β,17α,18β,20α)-11,17-Dimethoxy-18-[[1-oxo-3-(3,4,5-trimethoxyphenyl)-2-propenyl]oxy-3,20-yohimban-16-carboxylic acid methyl ester.
reserpinine; Anaprel; Apoterin S; Cartric; Cinnaloid; Moderil. Antihypertensive agent. Found in *Rauwolfia serpentina Beth., Apocynaceae.* Related to reserpine. mp = 238-239° (in vacuo); $[\alpha]_D^{24}$ = -97° (c = 1 $CHCl_3$); λ_m = 228, 302 nm (log ε 4.79, 4.48 MeOH); insoluble in H_2O; moderately soluble in C_6H_6, MeOH, $CHCl_3$, other organic solvents. *Pfizer Inc.*

4426 Reserpine
50-55-5 8314 200-047-9
$C_{33}H_{40}N_2O_9$
Methyl 18β-hydroxy-11,17α-dimethoxy-3β,20α-yohimban-16β-carboxylate 3,4,5-trimethoxybenzoate (ester).
Crystoserpine; Eskaserp; Rau-Sed; Reserpoid; Rivasin; Sandril; Sedaraupin; Serpasil Serpasol; Serpine; Serpiloid; component of: Demi-Regroton, Diupres, Diutensen-R, Hydropres, Metatensin, Naquival, Regroton, Renese R, Salutensin, Ser-Ap-Es, Unipres. Antihypertensive agent. Found in *Rauwolfia serpentina* spp (snakeroot). Interferes with norepinephrine storage, thereby reducing vascular smooth muscle tone as well as venous tone. mp = 264-265°, 277-277.5° (dec); $[\alpha]_D^{23}$ = -118° ($CHCl_3$); $[\alpha]_D^{26}$ = -164° (c = 0.96 C_5H_5N); $[\alpha]_D^{26}$ = -168° (c = 0.624 DMF); λ_m = 216, 267, 295 nm (ε 61700, 17000, 10200, $CHCl_3$); sparingly soluble in H_2O; freely soluble in $CHCl_3$, CH_2Cl_2, AcOH; soluble in C_6H_6, EtOAc, EtOH (0.55 g/100 ml), Et_2O; [hydrochloride hydrate]: mp = 224° (dec). *3M Pharm.; Bristol-Myers Squibb Co.; Ciba-Geigy Corp.; Eli Lilly & Co.; Marion Merrell Dow Inc.; Merck & Co.Inc.; Pfizer Inc.; Rhône-Poulenc Rorer Pharm. Inc.; Roberts Pharm. Corp.; Schering-Plough HealthCare Products; Solvay Pharm. Inc.; Wallace Labs.*

4427 Rilmenidene
54187-04-1 8388 259-021-0
$C_{10}H_{16}N_2O$
2-[(Dicyclopropylmethyl)amino]-2-oxazoline.
oxaminozoline; S-3341; [phosphate]: Hyperium; S-3341-3. An α_2-adrenoceptor agonist used as an antihypertensive agent. mp= 106-107°; [phosphate]: soluble in H_2O (19 g/100 ml), MeOH (7 g/100 ml), $CHCl_3$, EtOH (0.7 g/100 ml); LD_{50} (mus orl) = 375 mg/kg, (rat orl) = 295 mg/kg; [fumarate]: mp = 170°. *Sci. Union et Cie France.*

4428 Sampatrilat
129981-36-8
$C_{26}H_{40}N_4O_9S$
N-[[1-[(S)-3-[(S)-6-Amino-2-methanesulfonamidohexanamido]-2-carboxypropyl]-cyclopentyl]carbonyl]-L-tyrosine.
A novel dual inhibitor of both angiotensin-converting enzyme (ACE) and neutral endopeptidase (NEP). Antihypertensive.

4429 Saralasin
34273-10-4 8518
$C_{42}H_{65}N_{13}O_{10}$
N-[1-[N-[N-[N-[N-[N^2-(N-Methylglycyl)-L-arginyl]-L-valyl]-L-tyrosyl]-L-valyl]-L-histidyl]-L-prolyl]-L-alanine.
Sar-Arg-Val-Tyr-Val-His-Pro-Ala; 1-sar-8-ala-angiotensin II. Antihypertensive. Peptide angiotensin receptor blocking agent; specific antagonist of angiotensin II. Proposed as a diagnostic aid in identifying angiotensin-dependent hypertension. *Norwich Eaton.*

4430 Saralasin Hydrated Acetate
39698-78-7 8518
$C_{42}H_{65}N_{13}O_{10}.xC_2H_4O_2.xH_2O$
N-[1-[N-[N-[N-[N-[N^2(N-Methylglycyl-L-arginyl]-L-valyl]-L-tyrosyl]-L-valyl]-L-histidyl]-L-prolyl]-L-alanine acetate (salt) hydrate.
P-113; Sarenin; Sar-Arg-Val-Tyr-Val-His-Pro-Ala. A specific antagonist of angiotensin II. Used as an antihypertensive agent. Also used for diagnosis of renin-dependent hypertension. mp = 256°; soluble in H_2O; LD_{50} (mmus iv) = 1171 mg/kg. *Norwich.*

4431 Semotiadil
116476-13-2 8590
$C_{29}H_{32}N_2O_6S$
(R)-2-[2-[3-[[2-(1,3-Benzodioxol-5-yloxy)ethyl]methylamino]propoxy]-5-methoxyphenyl]-4-methyl-2H-1,4-benzothiazin-3(4H)-one.
sesamodil; DS-4823. Antianginal; antihypertensive; antiarrhythmic. Benzothiazine calcium antagonist. 8590.

4432 Semotiadil Fumarate
116476-14-3 8590
$C_{33}H_{36}N_2O_{10}S$
(R)-2-[2-[3-[[2-(1,3-Benzodioxol-5-yloxy)ethyl]methylmino]propoxy]-5-methoxyphenyl]-4-methyl-2H-1,4-benzothiazin-3(4H)-one fumarate.
SD-3211. Antianginal; antihypertensive; antiarrhythmic. Benzothiazine calcium antagonist. mp = 134-135°; $[\alpha]_D^{25}$ = 195° (DMSO). 8590.

4433 Sodium Nitroprusside
14402-89-2 8794 238-373-9

$C_5FeN_6Na_2O$
Disodium pentacyanonitrosylferrate(2-).
sodium nitroferricyanide; sodium nitroprussiate; Nipruss; Nipride [dihydrate]; Nitropress [dihydrate]. Antihypertensive. A potent, directly acting vasodilator used intravenously for treatment of hypertensive emergencies. Also used as a reagent in the detection of many organic compounds and alkali sulfides. *Abbott Labs Inc.; Elkins-Sinn; Hoffmann-LaRoche Inc.*

4434 Sodium Nitroprusside Dihydrate
13755-38-9 8794

$C_5FeN_6Na_2O.2H_2O$
Disodium pentacyanonitrosylferrate(2-) dihydrate.
Nipride; Nitropress. Antihypertensive. Soluble in H_2O (43 g/100 ml), slightly soluble in EtOH. *Abbott Labs Inc.; Elkins-Sinn; Hoffmann-LaRoche Inc.*

4435 Sotalol
3930-20-9 8876

$C_{12}H_{20}N_2O_3S$
4'-[1-Hydroxy-2-(isopropylamino)ethyl]methanesulfonanilide.
Antianginal; class II and III antiarrhythmic; antihypertensive. A β-adrenergic blocker used as an antihypertensive agent. λ_m = 242.2, 275.2 nm ($CHCl_3$). *Berlex Labs Inc.; Bristol-Myers Squibb Co.*

4436 Sotalol Hydrochloride
959-24-0 8876 213-496-0

$C_{12}H_{21}ClN_2O_3S$
4'-[1-Hydroxy-2-(isopropylamino)ethyl]-methanesulfonanilide monohydrochloride.
MJ-1999; Beta-Cardone; Betapace; Darob; Sotacor; Sotalex. Antianginal; class II and III antiarrhythmic; antihypertensive. A β-adrenergic blocker. mp = 206.5-207°, 218-219°; soluble in H_2O, less soluble in $CHCl_3$; LD_{50} (mmus orl) = 2600 mg/kg, (mmus ip) = 670 mg/kg, (mrat orl) = 3450 mg/kg, (mrat ip) = 680 mg/kg, (rbt orl) = 1000 mg/kg, (dog ip) = 330 mg/kg. *Berlex Labs Inc.; Bristol-Myers Squibb Co.*

4437 Spirapril
83647-97-6 8905

$C_{22}H_{30}N_2O_5S_2.1/2H_2O$
(8S)-7-[(S)-N-[(S)-1-Carboxy-3-phenylpropyl]alanyl]-1,4-dithia-7-azaspiro[4.4]nonane-8-carboxylic acid 1-ethyl ester hemihydrate.
Antihypertensive. Angiotensin-converting enzyme inhibitor. $[\alpha]_D^{26}$ = -29.5° (c = 0.2 EtOH). *Schering-Plough HealthCare Products.*

4438 Spirapril Hydrochloride
94841-17-5 8905

$C_{22}H_{31}ClN_2O_5S_2$
(8S)-7-[(S)-N-[(S)-1-Carboxy-3-phenylpropyl]alanyl]-1,4-dithia-7-azaspiro[4.4]nonane-8-carboxylic acid 1-ethyl ester monohydrochloride.
Sch-33844; TI-211-950; Renormax; Renpress; Sandopril. Antihypertensive. Angiotensin-converting enzyme inhibitor. mp = 192-194° (dec); $[\alpha]_D^{26}$ = -11.2° (c = 0.4 EtOH). *Schering-Plough HealthCare Products.*

4439 Spiraprilat
83602-05-5 8905

$C_{20}H_{26}N_2O_5S_2$
(8S)-7-[(S)-N-[(S)-1-Carboxy-3-phenylpropyl]alanyl]-1,4-dithia-7-azaspiro[4.4]nonane-8-carboxylic acid.
Sch-33861; spiraprilic acid. Antihypertensive. Angiotensin-converting enzyme inhibitor. mp = 163-165° (dec); $[\alpha]_D^{26}$ = 4.1° (c = 0.4 EtOH). *Schering-Plough HealthCare Products.*

4440 Sulfinalol
66264-77-5 9120

$C_{20}H_{27}NO_4S$
4-Hydroxy-α-[[[3-(p-methoxyphenyl)-1-methylpropyl]-amino]methyl]-3-(methylsulfinyl)benzyl alcohol.
A β-adrenergic blocker used as an antihypertensive agent. *Sterling Winthrop Inc.*

4441 Sulfinalol Hydrochloride
63251-39-8 9120 264-046-5

$C_{20}H_{28}ClNO_4S$
4-Hydroxy-α-[[[3-(p-methoxyphenyl)-1-methylpropyl]amino]methyl]-3-(methylsulfinyl)benzyl alcohol hydrochloride.
Win-408087; Perifadil. A β-adrenergic blocker used as an antihypertensive agent. mp = 172-175°. *Sterling Winthrop Inc.*

4442 Syrosingopine
84-36-6 9193 201-527-0

$C_{35}H_{42}N_2O_{11}$
18-[[4-[(Ethoxycarbonyl)oxy]-3,5-dimethoxybenzoyl]-oxy]-11,17-dimethoxyyohimban-16-carboxylic acid methyl ester.
syringopine; Su-3118; Isotense; Londomin; Raunova; Seniramin; Singoserp; Siringina. Antihypertensive agent. mp = 175-179°. *Ciba-Geigy Corp.*

4443 Talinolol
57460-41-0 9208

$C_{20}H_{33}N_3O_3$
(±)-1-[p-[3-(tert-Butylamino)-2-hydroxypropoxy]phenyl]-3-cyclohexylurea.
02-115; Cordanum. Antiarrhythmic agent with antihypertensive properties. A β-adrenergic blocker related to practolol. mp = 142-144°; LD_{50} (rat orl) = 1180 mg/kg, (rat ip) = 54.3 mg/kg, (rat iv) = 29.7 mg/kg, (mus orl) = 593 mg/kg, (mus ip) = 74.7 mg/kg, (mus iv) = 25.0 mg/kg. *Ciba-Geigy Corp.*

4444 Teclothiazide
4267-05-4 9261 224-253-3

$C_8H_7Cl_4N_3O_4S_2$
6-Chloro-3,4-dihydro-3-(trichloromethyl)-2H-1,2,4-benzothiadiazine-7-sulfonamide 1,1-dioxide.
Diuretic. A thiazide. mp = 300-303°, 287°.

4445 Teclothiazide Potassium
5306-80-9 9261 226-157-7

$C_8H+67Cl_4KN_3O_4S_2$
6-Chloro-3,4-dihydro-3-(trichloromethyl)-2H-1,2,4-benzothiadiazine-7-sulfonamide 1,1-dioxide potassium salt.
PS-207; K-33; Depleil. Diuretic. A thiazide. LD_{50} (mus ip) = 4.75 g/kg.

4446 Temocapril
111902-57-9 9287
$C_{23}H_{28}N_2O_5S_2$
(+)-(2S,6R)-6-[[(1S)-1-Carboxy-3-phenylpropyl]-amino]tetrahydro-5-oxo-2-(2-thienyl)-1,4-thiazepine-4(5H)-acetic acid 6-ethyl ester.
Angiotensin-converting enzyme inhibitor. Used to treat hypertension. mp =168°; $[\alpha]^{23}$ = 40° (c = 1.1 DMF). *Sankyo Co. Ltd.*

4447 Temocapril Hydrochloride
110221-44-8 9287
$C_{23}H_{29}ClN_2O_5S_2$
(+)-(2S,6R)-6-[[(1S)-1-Carboxy-3-phenylpropyl]-amino]tetrahydro-5-oxo-2-(2-thienyl)-1,4-thiazepine-4(5H)-acetic acid 6-ethyl ester mono-hydrochloride.
CS-622. Angiotensin-converting enzyme inhibitor. Used to treat hypertension. mp = 187° (dec); $[\alpha]_D^{25}$ = 47.7° (c = 1 DMF); LD_{50} (mus orl) > 5000 mg/kg, (rat orl) > 5000 mg/kg, (dog orl) > 800 mg/kg. *Sankyo Co. Ltd.*

4448 Temocaprilate
110221-53-9 9287
$C_{21}H_{24}N_2O_5S_2$
(+)-(2S,6R)-6-[[(1S)-1-Carboxy-3-phenylpropyl]amino]-tetrahydro-5-oxo-2-(2-thienyl)-1,4-thiazepine-4(5H)-acetic acid.
temocaprilat; RS-5139. Angiotensin-converting enzyme inhibitor. Used to treat hypertension. mp = 246° (dec); $[\alpha]_D^{25}$ = 63.4° (c = 1 DMF). *Sankyo Co. Ltd.*

4449 Teprotide
35115-60-7
$C_{53}H_{76}N_{14}O_{12}$
5-Oxo-L-prolyl-L-tryptophyl-L-prolyl-L-arginyl-L-prolyl-L-glutaminyl-L-isoleucyl-L-prolyl-L-proline.
SQ-20881. Angiotensin-converting enzyme inhibitor. Used to treat hypertension. *Bristol-Myers Squibb Co.*

4450 Terazosin [anhydrous]
63074-08-8 9297
$C_{19}H_{25}N_5O_4$
1-(4-Amino-6,7-dimethoxy-2-quinazolinyl)-4-(tetrahydro-2-furoyl)piperazine.
An α_1-adrenergic blocker related to prazosin. Used as an antihypertensive agent and in treatment of BPH. mp = 272.6-274°; λ_m = 212, 245, 330 nm (a 65.7, 127.5, 24.0 H_2O); soluble in MeOH (3.37 g/100 ml), H_2O (2.97 g/100 ml), EtOH (0.41 g/100 ml), $CHCl_3$ (0.12 g/100 ml), Me_2CO (.1 mg/100 ml); insoluble in C_6H_{14}. *Abbott Labs Inc.*

4451 Terazosin Hydrochloride Dihydrate
70024-40-7 9297
$C_{19}H_{26}ClN_5O_4$
1-(4-Amino-6,7-dimethoxy-2-quinazolinyl)-4-(tetrahydro-2-furoyl)piperazine hydrochloride dihydrate.
Abbott-45975; Heitrin; Hytracin; Hytrin; Hytrinex; Itrin; Urodie; Vasocard; Vasomet; Vicard. Used to treat hypertension and benign antiprostatic hypertrophy. An α_1-adrenergic blocker related to prazosin. mp = 271-274°; soluble in H_2O (2.42 g/100 ml); LD_{50} (mrat iv) = 277 mg/kg, (frat iv) = 293 mg/kg; [anhydrous hydrochloride]: mp = 278-279°; soluble in H_2O (76.12 g/100 ml); LD_{50} (mus iv) = 259.3 mg/kg. *Abbott Labs Inc.*

4452 Tertatolol
34784-64-0 9318
$C_{16}H_{25}NO_2S$
(±)-1-(tert-Butylamino)-3-(thiochroman-8-yloxy)-2-propanol.
A non-selective β-adrenergic blocker. Used as an antihypertensive agent. mp = 70-72°. *Sci. Union et Cie France.*

4453 Tertatolol Hydrochloride
33580-30-2 9318 251-578-8
$C_{16}H_{26}ClNO_2S$
(±)-1-(tert-Butylamino)-3-(thiochroman-8-yloxy)-2-propanol hydrochloride.
S-2395; SE-2395; Artex; Artexal; Prenalex. A non-selective β-adrenergic blocker. Used as an antihypertensive agent. mp = 180-183°; LD_{50} (rat iv) = 40 mg/kg, (rat ip) = 90 mg/kg, (mus iv) = 37 mg/kg, (mus ip) = 120 mg/kg. *Sci. Union et Cie France.*

4454 Tiamenidine
31428-61-2 9558
$C_8H_{10}ClN_3S$
2-[(2-Chloro-4-methyl-3-thienyl)amino]-2-imidazoline.
HOE-440; Symcor Base TTS; Thiamendidine. An α-adrenoceptor agonist related to clonidine and used as an antihypertensive agent. mp = 152°. *Hoechst Roussel Pharm. Inc.*

4455 Tiamenidine Hydrochloride
51274-83-0 9558 257-100-4
$C_8H_{11}Cl_2N_3S$
2-[(2-Chloro-4-methyl-3-thienyl)amino]-2-imidazoline hydrochloride.
HOE-42-440; Sundralen; Symcor; component of: Symcorad. An α-adrenoceptor agonist related to clonidine and used as an antihypertensive agent. mp = 228-229°; LD_{50} (rat iv) = 40 mg/kg, (mus iv) = 45 mg/kg, (mus sc) = 170 mg/kg, (mus orl) = 400 mg/kg. *Hoechst Roussel Pharm. Inc.*

4456 Tibalosin
63996-84-9
$C_{21}H_{27}NOS$
(±)-erythro-2,3-Dihydro-α-[1-[(4-phenyl-butyl)amino]ethyl]benzo[b]thiophene-5-methanol.
CP-804-S. Heterocyclic aminoalcohol related to ifenprodil. Antagonist acting at the polyamine site of the N-methyl-D-aspartate (NMDA) subtype of glutamate receptor. Antihypertensive; alpha 1-adrenoceptor antagonist.

4457 Tilisolol
85136-71-6 9579
$C_{17}H_{24}N_2O_3$
(±)-4-[3-(tert-Butylamino)-2-hydroxypropoxy]-2-methyl-isocarbostyril.
Antiarrhythmic; antihypertensive. A β-adrenergic blocker. *Nisshin Denka K.K.*

4458 Tilisolol Hydrochloride
62774-96-3 9579
$C_{17}H_{25}ClN_2O_3$
(±)-4-[3-(tert-Butylamino)-2-hydroxypropoxy]-2-methylisocarbostyril hydrochloride.
N-696; Selecal. Antiarrhythmic; antihypertensive. A β-adrenergic blocker. mp = 203-205°; LD_{50} (mmus orl) = 1393 mg/kg, (mmus sc) = 1219 mg/kg, (mmus ip) = 578 mg/kg, (mmus iv) = 74.3 mg/kg, (fmus orl) = 1290 mg/kg, (fmus sc) = 1245 mg/kg, (fmus ip) = 557 mg/kg, (fmus iv) = 104.7 mg/kg, (mrat orl) = 145 mg/kg, (mrat sc) = . *Nisshin Denka K.K.*

4459 Timolol
91524-16-2 9585
$C_{13}H_{24}N_4O_3S.1/2H_2O$
(S)-1-(tert-Butylamino)-3-[(4-morpholino-1,2,5-thiadiazol-3-yl)oxy]-2-propanol hemihydrate.
Antianginal agent with antiarrhythmic (class II), antihypertensive and antiglaucoma properties. A β-adrenergic blocker. [(±) form]: mp = 71.5-72.5°. *Merck & Co.Inc.*

4460 Timolol Maleate
26921-17-5 9585 248-111-5
$C_{17}H_{28}N_4O_7S$
(S)-1-(tert-Butylamino)-3-[(4-morpholino-1,2,5-thiadiazol-3-yl)oxy]-2-propanol maleate (1:1).
Blocadren; Timoptic; Timoptol; MK-950; Aquanil; Betim; Proflax; Temserin; Tenopt; Timacar; Timacor; component of: Cosopt, Timolide. Antianginal agent with antiarrhythmic (class II), antihypertensive and antiglaucoma properties. A β-adrenergic blocker. mp = 201.5-202.5°; $[\alpha]^{24}_{405}$ = -12.0° (c = 5 1N HCl), $[\alpha]^{25}_{D}$ = -4.2°; λ_m = 294 nm ($A^{1\%}_{1\ cm}$ 200 0.1N HCl); soluble in EtOH, MeOH; poorly soluble in $CHCl_3$, cyclohexane; insoluble in isooctane, Et_2O. *Merck & Co.Inc.*

4461 Todralazine
14679-73-3 9640
$C_{11}H_{12}N_4O_2$
Ethyl 3-(1-phthalazinyl)carbazate.
ecarazine; carboethoxyphthalazinohydrazine.
Antihypertensive agent. *Polfa.*

4462 Todralazine Hydrochloride
3778-76-5 9640
$C_{11}H_{13}ClN_4O_2$
Ethyl 3-(1-phthalazinyl)carbazate hydrochloride.
CEPH; BT-621; Apiracohl; Apredor; Apride; Atapren; Binazin; Illcut; Propat. Antihypertensive agent. LD_{50} (mus ip) = 500 mg/kg. *Polfa.*

4463 Toliprolol
2933-94-0 9653 220-905-6
$C_{13}H_{21}NO_2$
1-(Isopropylamino)-3-(m-tolyloxy)-2-propanol.
MHIP. Antianginal; antihypertensive agent. A β-adrenergic blocker. mp = 75-76°, 79°. *Boehringer Ingelheim GmbH; ICI.*

4464 Toliprolol Hydrochloride
306-11-6 9653 206-177-2
$C_{13}H_{22}ClNO_2$
1-(Isopropylamino)-3-(m-tolyloxy)-2-propanol hydrochloride.
ICI-45763; Ko-592; Doberol; Sinorytmal. Antianginal; antihypertensive agent. A β-adrenergic blocker. mp = 120-121°; λ_m = 270 nm ($E^{1\%}_{1\ cm}$ 498 H_2O). *Boehringer Ingelheim GmbH; ICI.*

4465 Tolonidine
4201-22-3 9657
$C_{10}H_{12}ClN_3$
2-(2-Chloro-p-toluidino)-2-imidazoline.
ST-375. Antihypertensive agent with direct α-sympatheticomimetic properties. Structurally related to clonidine. mp = 148-150°. *Boehringer Ingelheim GmbH.*

4466 Tolonidine Nitrate
57524-15-9 9657 260-785-2
$C_{10}H_{13}ClN_4O_3$
2-(2-Chloro-p-toluidino)-2-imidazoline nitrate (salt).
CERM-10137; Euctan. Antihypertensive agent with direct α-sympatheticomimetic properties. Structurally related to clonidine. mp = 162-164°; LD_{50} (mmus orl) = 160 mg/kg, (mmus iv) = 21.25 mg/kg, (mrat orl) = 420 mg/kg, (mrat iv) = 42 mg/kg. *Boehringer Ingelheim GmbH.*

4467 Trandolapril
87679-37-6 9703
$C_{24}H_{34}N_2O_5$
(2S,3aR,7aS)-1-[(S)-N-[(S)-1-Carboxy-3-phenylpropyl]-alanyl]hexahydro-2-indolinecarboxylic acid 1-ethyl ester.
RU-44570; Odrik; Gopten. Angiotensin-converting enzyme inhibitor. Antihypertensive. *Roussel-UCLAF.*

4468 Trandolaprilate
87679-71-8 9703
$C_{22}H_{30}N_2O_5$
(2S,3aR,7aS)-1-[(S)-N-[(S)-1-Carboxy-3-phenylpropyl]-alanyl]hexahydro-2-indolinecarboxylic acid.
RU-44403; trandolaprilat. Angiotensin-converting enzyme inhibitor. Antihypertensive. *Roussel-UCLAF.*

4469 Trequinsin
79855-88-2
$C_{24}H_{27}N_3O_3$
2,3,6,7-Tetrahydro-2-(mesitylimino)-9,10-dimethoxy-3-methyl-4H-pyrimido[6,1-a]isoquinoline-4-one.
HL-725 [as hydrochloride]. Selective phosphodiesterase 3 inhibitor. Antihypertensive vasodilator; antithromotic.

4470 Trichlormethiazide
133-67-5 9754 205-118-8
$C_8H_8Cl_3N_3O_4S_2$
6-Chloro-3-(dichloromethyl)-3,4-dihydro-2H-1,2,4-thiadiazine-7-sulfonamide 1,1-dioxide.
3-dichloromethylhydrochlorothiazide; hydrochlorothiazide; trichloromethiazide; Achletin; Anatran; Anistadin; Aponorin; Carvacron; Diurese; Esmarin; Fluitran; Fluitran; Flutra; Intromene; Kubacron; Metahydrin; Metatensin; Naqua; Naquasone; Naquival; Salirom; Tachionin; Tolcasone; Triflumen. Diuretic; antihypertensive. A thiazide. mp= 248-250°, 266-273°; soluble in H_2O (0.8 mg/ml), EtOH (21 mg/ml), MeOH (60 mg/ml); LD_{50} (rat orl) > 20000 mg/kg. *Marion Merrell Dow Inc.; Merrell Pharm. Inc.; Schering-Plough HealthCare Products; Schering-Plough Pharm.*

4471 Trimazosin
35795-16-5 9828 252-732-7
$C_{20}H_{29}N_5O_6$
2-Hydroxy-2-methylpropyl 4-(4-amino-6,7,8-trimethoxy-2-quinazolinyl)-1-piperazine carboxylate.
An α_1-adrenergic receptor antagonist with anti-hypertensive activity. mp = 158-159°. *Pfizer Inc.*

4472 Trimazosin Hydrochloride Monohydrate
53746-46-6 9828
$C_{20}H_{30}ClN_5O_6.H_2O$
2-Hydroxy-2-methylpropyl 4-(4-amino-6,7,8-trimethoxy-2-quinazolinyl)-1-piperazine carboxylate monohydrochloride monohydrate.
CP-19106-1; Cardovar; Supres. An α_1-adrenergic receptor antagonist with antihypertensive activity. mp = 166-169° (dec). *Pfizer Inc.*

4473 Trimethaphan Camsylate
68-91-7 9837 200-696-8
$C_{32}H_{40}N_2O_5S_2$
(+)-1,3-Dibenzyldecahydro-2-oxoimidazo[4,5-c]-thieno[1,2-a]thiolium 2-oxo-10-bornane-sulfonate (1:1).
Arfonad; Nu-2222. Antihypertensive agent. Ganglionic blocker. mp = 245° (dec); $[\alpha]_D^{20}$ = 22.0° (c = 4 H_2O); soluble in H_2O (20 g/100 ml), EtOH (50 g/100ml); slightly soluble in Me_2CO, Et_2O. *Hoffmann-LaRoche Inc.*

4474 Trimethidinium Methosulfate
14149-43-0 9838 237-994-2
$C_{19}H_{42}N_2O_8S_2$
1,3,8,8-Tetramethyl-3-[3-(trimethylammonio)-propyl]-3-azoniabicyclo[3.2.1]octane bis(methyl sulfate).
Ganglion blocking agent with antihypertensive properties. mp = 192-193°. *Boehringer Ingelheim GmbH.*

4475 Tripamide
73803-48-2 9866
$C_{16}H_{20}ClN_3O_3S$
4-Chloro-N-(endo-hexahydro-4,7-methanoisoindolin-2-yl)-3-sulfamoylbenzamide.
toripamide; ADR-033; E-614; Normonal. Diuretic; antihypertensive.

4476 Tyrosinase
9002-10-2 9969 232-653-4
A copper-containing enzyme with antihypertensive properties.

4477 Urapidil
34661-75-1 10003 252-130-4
$C_{20}H_{29}N_5O_3$
6-[[3-[4-(o-Methoxyphenyl)-1-piperazinyl]propyl]-amino]-1,3-dimethyluracil.
B-66256; Ebrantil; Eupressyl; Mediatensyl; Uraprene. An α_1-adrenergic antagonist used as an antihypertensive agent. Derivative of uracil. mp = 156-158°; λ_m = 237, 268 nm (ε = 11000, 26700 MeOH); LD_{50} (mmus orl) = 750 mg/kg, (mmus iv) = 260 mg/kg, (mrat orl) = 550 mg/kg, (mrat iv) = 145 mg/kg. *Byk Gulden Lomberg GmbH.*

4478 Valsartan
137862-53-4 10051
$C_{24}H_{29}N_5O_3$
N-[p-(o-1H-Tetrazol-5-ylphenyl)benzyl]-N-valeryl-L-valine.
CGP-48933. A non-peptide angiotensin II AT_1-receptor antagonist. Used as an antihypertensive agent. mp = 116-117°. *Ciba-Geigy Corp.*

4479 Xipamide
14293-44-8 10212 238-216-4
$C_{15}H_{15}ClN_2O_4S$
4-Chloro-5-sulfamoyl-2',6'-salicyloxilidide.
MJF-10938; Be-1293; Bei-1293; Aquaphor; Chronexan; Diurexan; Lumitens. Diuretic; antihypertensive. mp = 256°. *Beiersdorf AG.*

4480 Zofenopril Calcium
81938-43-4
$C_{44}H_{44}CaN_2O_8S_4$
(4S)-N-[(s)-3-Mercapto-2-methylpropionyl]-4-(phenylthio)-L-proline benzoate (ester) calcium salt.
SQ-26991; Zoprace. Angiotensin-converting enzyme inhibitor. For hypertension. *Bristol-Myers Squibb Co.*

4481 Zolasartan
145781-32-4
$C_{24}H_{20}BrClN_6O_3$
1-[[3-Bromo-2-(o-1H-tetrazol-5-ylphenyl)-5-benzo-furanyl]methyl]-2-butyl-4-chloroimidazole-5-carboxylic acid.
GR-117289. Selective, potent, orally active, long-acting nonpeptide angiotensin II type 1 receptor antagonist. Analog of losartan. Antihypertensive.

Antihyperthyroids

4482 2-Amino-4-methylthiazole
1603-91-4 470 216-505-6
$C_4H_6N_2S$
4-Methyl-2-thiazolamine.
Nomortiroide. Antihyperthyroid. mp = 45-46°; bp_{20} = 124-126°, $bp_{0.4}$ = 70°; very soluble in H_2O, EtOH, Et_2O.

4483 2-Aminothiazole
96-50-4 501 202-511-6
$C_3H_4N_2S$
2-Thiazolamine.
Abadol; Basedol. Thyroid inhibitor. mp = 93°; soluble in hot H_2O; slightly soluble in cold H_2O, EtOH, Et_2O; LD_{50} (rat orl) = 480 mg/kg. *Mallinckrodt Inc.; Monsanto Co.; Squibb E.R. & Sons.*

4484 Carbimazole
22232-54-8 1844 244-854-4
$C_7H_{10}N_2O_2S$
Ethyl -3-methyl-2-thioimidazoline-1-carboxylate.
Neo-mercazole; Neo-Thyreostat. Thyroid inhibitor. mp = 122-125°; λ_m = 291 nm (H_2O), 227, 291 nm (($E^{1\%}_{1\,cm}$ 557, 0.1N H_2SO_4); soluble in H_2O 0.2 g/100 ml at 20°), EtOH (2.0 g/100 ml at 20°), Et_2O (0.3 g/100 ml at 20°), $CHCl_3$ (33.3 g/100 ml at 20°), Me_2CO (5.88 g/100 ml at 20°).

4485 3,5-Dibromo-L-tyrosine Monohydrate
300-38-9 3079 206-091-5
$C_9H_9Br_2NO_3.H_2O$
β-(3,5-Dibromo-4-hydroxyphenyl)alanine.
Biotiren; Bromotiren. Thyroid inhibitor. Dec 245°; $[\alpha]_D^{20}$ = 1.3° (c = 5 4% HCl); soluble in H_2O (0.4 g/100 ml at 25°, 3.3 g/100 ml at 100°), slightly soluble in EtOH, insoluble in Et_2O; [DL-form]: dec 245°; soluble in H_2O (0.18 g/100 ml at 20°).

4486 3,5-Diiodotyrosine
66-02-4 3236 200-620-3
$C_9H_9I_2NO_3$
3,5-Diodo-4-hydroxy-β-phenylalanine.
iodogorgoic acid; Agontan. Thyroid inhibitor. [L-form]: Dec 213°; $[\alpha]_D^{20}$ = 2.89° (c = 4.92 4% HCl), $[\alpha]_D^{20}$ = 2.27° (c = 4.54 in 25% NH_3); soluble in H_2O (0.02 g/100 ml at 0°, 0.06 g/100 ml at 25°, 0.19 g/100 ml at 50°, 0.56 g/100 at 75°, 1.7 g/100 ml at 100°); [DL-form]: dec 200°; soluble in H_2O (0.015 g/1oo ml at 0°, 0.034 g/100 ml at 25°, 0.077 g/100 ml at 50°). *Basic Inc.*

4487 Iodine
7553-56-2 5034 231-442-4
I_2
Iodine.
Antihyperthyroid and topical anti-infective. Also used, particularly in veterinary medicine, to treat goiter. Halogen. CAUTION: Overexposure could cause irritation of eyes and nose, lacrimation, headache, tight chest, skin burns or rash, cutaneous hypersensitivity. Highly corrosive on the GI tract. mp = 113.6°, bp = 185.2°; d^{25} = 4.93; soluble in H_2O (0.017 g/100 ml), C_6H_6 (14.09 g/100 g), CS_2 (16.5 g/100 g), EtOH (21.43 g/100 g), Et_2O (25.2 g/100 g), C_6H_{12} (2.7 g/100 g), CCl_4 (2.6 g/100 g at 35°); incompatible with tannins, alkaloids, starch.

4488 Methimazole
60-56-0 6049 200-482-4
$C_4H_6N_2S$
1-Methylimidazole-2-thiol.
Tapazole; mercazolyl; thiamazole; Basolan; Danantizol; Favistan; Frentirox; Mercazole; Metazolo; Thacapzol; Thycapsol; Strumazol. Antihyperthyroid. mp = 146-148°; bp = 280° (dec); λ_m = 211, 251.5 nm ($E_{1cm}^{1\%}$ 593, 1528, 0.1N H_2SO_4); freely soluble in H_2O; soluble in EtOH, $CHCl_3$; sparingly soluble in Et_2O, C_6H_6, petroleum ether. *Eli Lilly & Co.*

4489 Methylthiouracil
56-04-2 6210 200-252-3
$C_5H_6N_2OS$
2,3-Dihydro-6-methyl-2-thioxo-4(1H)-pyrimidinone.
4-methyluracil; 6-methyl-2-thiouracil; NSC-9378; NSC-193526; Alkiron; Antibason; Basecil; Basethyrin; Methiacil; Methicil; Methiocil; Methylthiouracil; Muracil; Muracin; MTU; Orcanon; Prostrumyl; Strumacil; Thimecil; Thiomecil; Thiomidil; Thioryl; Thiothymin; Thiothyron; Thiuryl; Thyreonorm; Thyreostat; Thyreostat I; Thyril; Tiomeracil; Tiorale M; Tiotiron; USAF EK-6454. Antineoplastic agent. Thyroid inhibitor. mp = 326-331° (dec); slightly soluble in H_2O at 25° (0.67 g/100 ml at 100°), Et_2O, EtOH, Me_2CO; insoluble in C_6H_6, $CHCl_3$; MLD (rbt orl) = 2500 mg/kg.

4490 Propylthiouracil
51-52-5 8054 200-103-2
$C_7H_{10}N_2OS$
6-Propyl-2-thiouracil.
Propacil; Propycil; Propyl-Thyracil; Thyreostat II. Antihyperthyroid. mp = 219-221°; λ_m = 275, 214 nm (ε 15800, 15600, MeOH), 315.5, 260, 207.5 (ε 10900, 10700, 15400 MeOH/KOH); soluble in H_2O (0.11 g/100 ml at 20°, 1.1 g/100 ml at 100°), EtOH 1.67 g/100 ml), Me_2CO (1.67 g/100 ml); insoluble in Et_2O, C_6H_6, $CHCl_3$. *Lederle Labs.*

4491 Sodium Perchlorate
7601-89-0 8798 231-511-9
$ClNaO_4$
Perchloric acid sodium salt.
Irenat. Thyroid inhibitor. d = 2.02; dec → 130°; very soluble in H_2O.

4492 Thibenzazoline
6028-35-9 9446
$C_9H_{10}N_2O_2S$
1,3-Dihydro-1,3-bis(hydroxymethyl)-2H-benzimidazole-2-thione.
Thyreocordon. Antihyperthyroid. mp = 160-162°; soluble in dilute alkali.

4493 Thiobarbital
77-32-7 9455 201-020-4
$C_8H_{12}N_2O_2S$
5,5-Diethyldihydro-2-thioxo-4,6(1H,5H)-pyrimidine-dione.
Ibition. Antihyperthyroid. mp = 180°; soluble in H_2O (1.14 g/100 ml at 100°), EtOH, $CHCl_3$, Et_2O, Me_2CO, NH_3, alkaline solutions; sparingly soluble in C_7H_8; insoluble in C_6H_6.

4494 2-Thiouracil
141-90-2 9504 205-508-8
$C_4H_4N_2OS$
2,3-Dihydro-2-thioxo-4(1H)-pyrimidinone.
Deracil. Thyroid depressant. Used in treatment of hyperthyroidism, angina and congestive heart failure. Slightly soluble in H_2O (0.050 g/100 ml); insoluble in EtOH, Et_2O, acids; LD_{50} (rat ip) = 1500 mg/kg.

Antihypotensives

4495 Amezinium Methyl Sulfate
30578-37-1 412 250-248-0
$C_{12}H_{15}N_3O_5S$
4-Amino-6-methoxy-1-phenylpyridazinium methyl sulfate.
LU-1631; Regulton; Risumic; Supratonin. Antihypotensive. Sympathomimetic with vascular and cardiac activity. mp = 1676° (dec); LD_{50} (mus orl) = 1630 mg/kg, (mus iv) = 40.4 mg/kg, (rat orl) = 1410 mg/kg, (rat iv) = 45.5 mg/kg. *BASF Corp.*

4496 Angiotensin Amide
53-73-6 689 200-182-3
$C_{49}H_{70}N_{14}O_{11}$
N-[1-[N-[N-[N-[N-(N^2-L-Asparaginyl-L-arginyl)-L-valyl]-L-tyrosyl]-L-valyl]-L-histidyl]-L-prolyl]-3-phenyl-L-alanine.
NSC-107678; Asn-Arg-Val-Tyr-Val-His-Pro-Phe; 5-valine angiotensin II amide; Hypertensin; Ipertensina. Vasoconstrictor; antihypotensive. *Parke-Davis.*

4497 Ciclafrine
55694-98-9
$C_{15}H_{21}NO_2$
m-1-Oxo-4-azaspiro[4.6]undec-2-ylphenol.
Antihypotensive. *Parke-Davis.*

4498 Ciclafrine Hydrochloride
51222-36-7
$C_{15}H_{22}ClNO_2$
m-1-Oxo-4-azaspiro[4.6]undec-2-ylphenol hydrochloride.
W-43026A; Go-3026A. Antihypotensive. *Parke-Davis.*

4499 Dimetofrine
22950-29-4 3314 245-348-6
$C_{11}H_{17}NO_4$
4-Hydroxy-3,5-dimethoxy-α-[(methylamino)methyl]-benzyl alcohol.
dimethophrine; dimetrophine. An α-adrenergic agonist. Used as an antihypotensive. mp = 178° (dec). *Zambeletti.*

4500 Dimetofrine Hydrochloride
22775-12-8 3314 245-212-6
$C_{11}H_{18}ClNO_4$
4-Hydroxy-3,5-dimethoxy-α-[(methylamino)methyl]-benzyl alcohol hydrochloride.
Pressamina. An α-adrenergic agonist. Used as an antihypotensive agent. mp = 171-173°. *Zambeletti.*

4501 Dopamine
51-61-6 3479 200-110-0
$C_8H_{11}NO_2$
4-(2-Aminoethyl)pyrocatechol.
3-hydroxytyramine. Adrenergic; used as an antihypotensive agent and cardiotonic. Endogenous catecholamine with α- and β-adrenergic activity. *Astra USA Inc.; Elkins-Sinn; Parke-Davis.*

4502 Dopamine Hydrochloride
62-31-7 3479 200-527-8
$C_8H_{12}ClNO_2$
4-(2-Aminoethyl)pyrocatechol hydrochloride.
Dopastat; ASL-279; Cardiosteril; Dynatra; Inovan; Inotropin. Adrenergic; used as an antihypotensive and cardiotonic. Dec 241°; freely soluble in H_2O; soluble in EtOH, MeOH; insoluble in Et_2O, $CHCl_3$, C_6H_6, C_7H_8, petroleum ether; [hydrobromide]: mp = 210-214° (dec). *Astra USA Inc.; Elkins-Sinn; Parke-Davis.*

4503 Etifelmin
341-00-4 3909
$C_{17}H_{19}N$
2-(Diphenylmethylene)butylamine.
2-(diphenylmethylene)butylamine; etifelmine; EDPA; Na III. CNS stimulant; antihypotensive. *Giulini.*

4504 Etifelmin Hydrochloride
1146-95-8 3909
$C_{17}H_{20}ClN$
2-(Diphenylmethylene)butylamine hydrochloride.
etifelmine hydrochloride; Tensinase D; component of: Gilutensin [with nicotine]. CNS stimulant; antihypotensive. mp = 232°; soluble in H_2O. *Giulini.*

4505 Etilefrin
709-55-7 3911 211-910-4
$C_{10}H_{15}NO_2$
α-[(Ethylamino)methyl]-m-hydroxybenzyl alcohol.
ethylphenylephrine; etiladrianol. Antihypotensive. Adrenomimetic. *Labs. Fher S.A.*

4506 dl-Etilefrin
10128-36-6 3911 233-359-9
$C_{10}H_{15}NO_2$
(±)-α-[(Ethylamino)methyl]-m-hydroxybenzyl alcohol.
Antihypotensive. Adrenomimetic. mp = 147-148°. *Labs. Fher S.A.*

4507 dl-Etilefrin Hydrochloride
534-87-2 3911
$C_{10}H_{16}ClNO_2$
(±)-α-[(Ethylamino)methyl]-m-hydroxybenzyl alcohol hydrochloride.
Apocretin; Circupon; Effontil; Effortil; Effortilvet; Efortil; Ethyl Adrianol; Eti-Puren; Kertasin; Pulsamin; Tonus-Forte. Antihypotensive. Adrenomimetic. mp = 121°; freely soluble in H_2O, soluble in EtOH, insoluble in $CHCl_3$. *Labs. Fher S.A.*

4508 Gepefrine
18840-47-6 4409
$C_9H_{13}NO$
(+)-(S)-m-(2-Aminopropyl)phenol.
Sympathomimetic isomer of hydroxyamphetamine. Used as an antihypotensive agent. mp = 155-158°; $[\alpha]_D^{25}$ = 31.8° (c = 2 MeOH). *Helopharm.*

4509 Gepefrine Tartrate
60763-48-6 4409 262-417-6
$C_{13}H_{19}NO_7$
(+)-(S)-m-(2-Aminopropyl)phenol tartrate.
Pressionorm; Wintonin. Sympathomimetic isomer of hydroxyamphetamine. Used as an antihypotensive agent. *Helopharm.*

4510 Metaraminol
54-49-9 5993
$C_9H_{13}NO_2$
(-)-α-(1-Aminoethyl)-m-hydroxybenzyl alcohol.
metaradrine; Pressonex. An α-adrenergic agonist; used as an antihypotensive agent. [oxalate dihydrate ($C_{11}H_{15}NO_6$)]: mp = 190°; $[\alpha]_D^{20}$ = -21.66°; soluble in H_2O. *Merck & Co.Inc.; Sterling Winthrop Inc.*

4511 Metaraminol Bitartrate
33402-03-8 5993 251-502-3
$C_{13}H_{19}NO_8$
(-)-α-(1-Aminoethyl)-m-hydroxybenzyl alcohol tartrate (1:1) salt.
Aramine; Icoral B; Pressorol. An α-adrenergic agonist;

used as an antihypotensive agent. mp = 176-177°; freely soluble in H_2O. *Merck & Co.Inc.; Sterling Winthrop Inc.*

4512 Metaraminol Hydrochloride
5967-52-2 5993
$C_9H_{14}ClNO_2$
(-)-α-(1-Aminoethyl)-m-hydroxybenzyl alcohol hydrochloride.
Aramine; Pressonex Bitartrate. An α-adrenergic agonist; used as an antihypotensive agent. $[\alpha]_D^{20}$ = -19.75°; soluble in H_2O; LD_{50} (mus ip) = 440 mg/kg. *Merck & Co.Inc.; Sterling Winthrop Inc.*

4513 Methoxamine
390-28-3 6067 206-867-3
$C_{11}H_{17}NO_3$
(±)-α-(1-Aminoethyl)-2,5-dimethoxybenzyl alcohol.
2,5-dimethoxynorephedrine. α-Adrenergic agonist; vasoconstrictor. Used as an antihypotensive. *Burroughs Wellcome Inc.; Glaxo Wellcome Inc.*

4514 Methoxamine Hydrochloride
61-16-5 6067 200-499-7
$C_{11}H_{18}ClNO_3$
(±)-α-(1-Aminoethyl)-2,5-dimethoxybenzyl alcohol hydrochloride.
Vasoxine; Vasoxyl; Vasylox. α-Adrenergic agonist (vasoconstrictor). Used as an antihypotensive. mp = 212-216°; soluble in H_2O (40 g/100 ml), EtOH (8.4 g/100 ml); insoluble in Et_2O, C_6H_6; $CHCl_3$. *Burroughs Wellcome Inc.; Glaxo Wellcome Inc.*

4515 Midodrine
42794-76-3 6272 255-945-3
$C_{12}H_{18}N_2O_4$
(±)-2-Amino-N-(β-hydroxy-2,5-dimethoxyphenethyl)acetamide.
St-1085. α-Adrenergic; vasoconstrictor. Used as an antihypotensive. *OSSW; Roberts Pharm. Corp.*

4516 Midodrine Hydrochloride
3092-17-9 6272
$C_{12}H_{19}ClN_2O_4$
(±)-2-Amino-N-(β-hydroxy-2,5-dimethoxyphenethyl)acetamide monohydrochloride.
Pro-Amatine; A-4020 Linz; St. Peter 224; ST-1085; Alphamine; Amatine; Gutron; Hipertan; Metligene; Midamine. Vasopressor used as an antihypotensive agent. An α-adrenergic agonist. mp = 192-193°. *OSSW; Roberts Pharm. Corp.*

4517 Norepinephrine
51-41-2 6788 200-096-6
$C_8H_{11}NO_3$
(-)-α-(Aminomethyl)-3,4-dihydroxybenzyl alcohol.
Noradrenaline; levarterenol; Adrenor; Levophed. Vasopressor used as an antihypotensive agent. An α-adrenergic agonist. mp = 216.5-218° (dec); $[\alpha]_D^{25}$ = -37.3° (c = 5 H_2O). *Sterling Winthrop Inc.*

4518 Norepinephrine d-Bitartrate
69815-49-2 6788
$C_{12}H_{17}NO_9$
(-)-α-(Aminomethyl)-3,4-dihydroxybenzyl alcohol tartrate (1:1) salt monohydrate.
Levophed; Levarterenol bitartrate; Aktamin; Binodrenal. Vasopressor used as an antihypotensive agent. An α-adrenergic agonist. mp = 102-104°; $[\alpha]_D^{25}$ = -10.7° (c = 1.6 H_2O); freely soluble in H_2O. *Sterling Winthrop Inc.*

4519 Norepinephrine Hydrochloride
329-56-6 6788 206-345-5
$C_8H_{12}ClNO_3$
(-)-α-(Aminomethyl)-3,4-dihydroxybenzyl alcohol hydrochloride.
Aterenol. Vasopressor used as an antihypotensive agent. An α-adrenergic agonist. mp = 145.2=146.4°; $[\alpha]_D^{25}$ = -40° (c = 6); soluble in H_2O. *Sterling Winthrop Inc.*

4520 Pholedrine
370-14-9 7485 206-725-0
$C_{10}H_{15}NO$
4-[2-(Methylamino)propyl]phenol.
Knoll H_{75}. Vasopressor used as an antihypotensive agent. An α-adrenergic agonist. mp = 162-163°; slightly soluble in H_2O; soluble in EtOH, Et_2O. *Knoll Pharm. Co.*

4521 Pholedrine Sulfate
6114-26-7 7485 228-083-0
$C_{20}H_{32}N_2O_6S$
4-[2-(Methylamino)propyl]phenol sulfate.
Paredrinol; Pulsotyl; Veritol. Vasopressor used as an antihypotensive agent. An α-adrenergic agonist. Dec 320-323°; soluble in H_2O; LD_{50} (rat sc) = 500 mg/kg. *Knoll Pharm. Co.*

4522 Synephrine
94-07-5 9189 202-300-9
$C_9H_{13}NO_2$
(RS)-1-(4-Hydroxyphenyl)-2-(methylamino)ethanol.
Oxedrine; Analeptin; Ethaphene; Parasympatol; Simpalon; Synephrin; Synthenate. Vasopressor; anti-hypotensive agent. An α-adrenergic agonist. mp = 184-185°. *Boehringer Ingelheim GmbH.*

4523 Synephrine Hydrochloride
5985-28-4 9189 227-804-6
$C_9H_{14}ClNO_2$
(RS)-1-(4-Hydroxyphenyl)-2-(methylamino)ethanol hydrochloride.
Oxedrine hydrochloride. Vasopressor; antihypotensive agent. An α-adrenergic agonist. mp = 151-152°; soluble in H_2O. *Boehringer Ingelheim GmbH.*

4524 Synephrine Tartaric Acid Monoester
6414-49-9 9189
$C_{13}H_{17}NO_7$
(RS)-1-(4-Hydroxyphenyl)-2-(methylamino)ethanol tartrate (1:1) ester.
Neupentedrin; Pentedrin. Used as a vasopressor and antihypotensive agent. An α-adrenergic agonist. *Boehringer Ingelheim GmbH.*

4525 Synephrine Tartrate
16589-24-5 9189 240-647-8
$C_{22}H_{32}N_2O_{10}$
(RS)-1-(4-Hydroxyphenyl)-2-(methylamino)ethanol tartrate (2:1) salt.
Corvasymton; Simpadren; Sympathol. An α-adrenergic agonist. Used as a vasopressor and antihypotensive agent. mp = 188-190° (dec); soluble in H_2O, EtOH. *Boehringer Ingelheim GmbH.*

4526 Theodrenaline
13460-98-5
$C_{17}H_{21}N_5O_5$
7-[2-[2-(3,4-Dihydroxyphenyl)-2-hydroxyethylamino]ethyl]theophylline.
Akrinor. Theophylline derivative. Betamimetic catecholamine. Vasoconstrictor. Used to treat hypotension.

Antihypothyroids

4527 Levothyroxine Sodium
55-03-8 5497
$C_{15}H_{10}I_4NNaO_4$
O-(4-Hydroxy-3,5-diiodophenyl)-3,5-diiodo-L-tyrosine monosodium salt.
L-thyroxine sodium salt; Levothroid; Synthroid Sodium; Eltroxin; Euthyrox; Laevoxin; Letter; Levaxin; Levothyrox; Oroxine; Thyroxevan. Thyroid hormone. Used in treatment of hyperthyroidism. Sodium salt of the amino acid L-thyroxine. Obtained from thyroid gland of domesticated animals or synthesized. *Astra USA Inc.; Forest Pharm. Inc.; Knoll Pharm. Co.*

4528 Levothyroxine Sodium Pentahydrate
25416-65-3 5497
$C_{15}H_{10}I_4NNaO_4 \cdot 5H_2O$
O-(4-Hydroxy-3,5-diiodophenyl)-3,5-diiodo-L-tyrosine sodium salt pentahydrate.
L-thyroxine sodium salt pentahydrate. Thyroid hormone. Used in treatment of hyperthyroidism. Sodium salt of the amino acid L-thyroxine. From thyroid gland of domesticated animals or synthesized. d = 2.381; $[\alpha]_D^{20}$ = -4.4° (c= 3 70% EtOH); slightly soluble in H_2O (0.015 g/100 ml); more soluble in EtOH; very slightly soluble in $CHCl_3$, Et_2O. *Astra USA Inc.; Forest Pharm. Inc.; Knoll Pharm. Co.*

4529 Liothyronine
6893-02-3 5535 228-120-0
$C_{15}H_{12}I_3NO_4$
L-3-[4-(4-Hydroxy-3-iodophenoxy)-3,5-diiodophenyl]alanine.
T-3. Thyroid hormone. Dec 236-237°; $[\alpha]_D^{29.5}$ = 21.5° (c = 4.75 in a mixture of 1 part 1N HCl and 2 parts EtOH); insoluble in H_2O, EtOH, propylene glycol; soluble in dilute alkalies. *SmithKline Beecham Pharm.*

4530 Liothyronine Hydrochloride
6138-47-2 5535 228-120-0
$C_{15}H_{13}ClI_3NO+4$
L-3-[4-(4-Hydroxy-3-iodophenoxy)-3,5-diiodophenyl]alanine hydrochloride.
Thybon. Thyroid hormone. Dec 202-203°; $[\alpha]_D^{29.5}$ = 21.5° (c = 4.75 HCl/EtOH). *SmithKline Beecham Pharm.*

4531 Liothyronine Sodium
55-06-1 5535 200-223-5
$C_{15}H_{11}I_3NNaO_4$
Monosodium L-3-[4-(4-hydroxy-3-iodophenoxy)-3,5-diiodophenyl]alanine.
liothyroniinde sodium salt; sodium L-triiodothyronine; Cytomel; Cytobin; Cytomine; Cyomel; Cynomel; Cytomel; Tertroxin; Triostat; Triothyrone. Thyroid hormone. *SmithKline Beecham Pharm.*

4532 Thyroid
9551
Dried thyroid.
NSC-26492; Tiroidina; Thyradin; Thyrocrine; Tyroidina. Thyroid hormone. Dried and powdered thyroid gland of domesticated animals.

4533 Thyroidin
9552
Iodothyrin.
Thyroid hormone. An extract of the thyroid gland, diluted with (e.g.) milk sugar.

4534 D-Thyroxine
51-49-0 9555 200-102-7
$C_{15}H_{11}I_4NO_4$
D-O-(4-Hydroxy-3,5-diiodophenyl)-3,5-diiodotyrosine.
dextrothyroxine; Debetrol. The L-form is a thyroid hormone; the D-form is antihyperlipoproteinemic. Optical isomer of the endogenous hormone L-thyroxine. Can produce modest lowering of plasma low-density lipoprotein. CAUTION: may cause serious cardiac toxicity. Dec 237°; $[\alpha]_{546}^{21}$ = 2.97° (c = 3.7 NaOH/EtOH). *Astra USA Inc.; Baxter Healthcare Systems; Forest Pharm. Inc.; Knoll Pharm. Co.*

4535 DL-Thyroxine
300-30-1 9555 206-088-9
$C_{15}H_{11}I_4NO_4$
(±)-O-(4-Hydroxy-3,5-diiodophenyl)-3,5-diiodotyrosine.
DL-thyroxine. The L-form is a thyroid hormone, the D-form is antihyperlipoproteinemic. Thyroid hormone. Dec 230-231°; insoluble in H_2O and most organic solvents. *Astra USA Inc.; Baxter Healthcare Systems; Forest Pharm. Inc.; Knoll Pharm. Co.*

4536 L-Thyroxine
51-48-9 9555 200-101-1
$C_{15}H_{11}I_4NO_4$
L-O-(4-Hydroxy-3,5-diiodophenyl)-3,5-diodo-L-tyrosine.
The L-form is a thyroid hormone, the D-form is antihyperlipoproteinemic. Dec 235-236°; $[\alpha]_{546}^{25}$ = -3.2° (90.66 g in 6.07 g of 0.5N NaOH and 13.03 g EtOH), $[\alpha]_D^{20}$ = -4.4° (c = 3 0.13N NaOH/70% EtOH). *Astra USA Inc.; Baxter Healthcare Systems; Forest Pharm. Inc.; Knoll Pharm. Co.*

4537 D-Thyroxine Sodium Salt
137-53-1 9555 205-301-2
$C_{15}H_{10}I_4NNaO_4$
D-O-(4-Hydroxy-3,5-diiodophenyl)-3,5-diiodotyrosine sodium salt.
dextrothyroxine sodium; Biotirmone; Choloxin; Detyroxin; Dethyrona; Dextroid; Dynothel; Eulipos. The L-form is a thyroid hormone, the D-form is

antihyperlipoproteinemic. Thyroid hormone. *Astra USA Inc.; Baxter Healthcare Systems; Forest Pharm. Inc.; Knoll Pharm. Co.*

4538 Thyroxine
7488-70-2 9555
$C_{15}H_{11}I_4NO_4$
O-(4-Hydroxy-3,5-diiodophenyl)-3,5-diiodotyrosine.
3,5,3',5'-tetraiodothyronine. The L-form is a thyroid hormone; the D-form is antihyperlipoproteinemic. *Astra USA Inc.; Baxter Healthcare Systems; Forest Pharm. Inc.; Knoll Pharm. Co.*

4539 Tiratricol
51-24-1 9603 200-086-1
$C_{14}H_9I_3O_4$
[4-(4-Hydroxy-3-iodophenoxy)-3,5-diiodophenyl]-acetic acid.
Triac; Triacana. Antihypothyroid. mp = 65°, 180-183°. *Glaxo Labs.*

4540 TSH
9002-71-5 9931 232-664-4
Thyroid-stimulating hormone.
thyrotropin; thyrotropic hormone; thyreotrophic hormone; TTH; Dermathycin; Thytropar. Thyrotropic hormone. Diagnostic aid (thyroid function). A glycoprotein produced by anterior lobe of pituitary gland. Stimulates production of thyroxine, raising metabolic rate. Inactivated by heating, proteolysis, oxidizing agents.

Anti-inflammatories, Nonsteroidal

4541 Aceclofenac
89796-99-6 19
$C_{16}H_{13}Cl_2NO_4$
2-[(2,6-Dichlorophenyl)amino]benzeneacetic acid carboxymethyl ester.
PR-82/3; Airtal; Biofenac; Tresquim. Anti-inflammatory; analgesic. Arylacetic acid derivative. mp = 149-150°; λ_m = 275 nm (log ε 4.14 EtOH). *Ajinomoto Co. Inc.*

4542 Acemetacin
53164-05-9 27 258-403-4
$C_{21}H_{18}ClNO_6$
1-(4-Chlorobenzoyl)-5-methoxy-2-methyl-1H-indole-3-acetic acid carboxymethyl ester.
TV-1322; Acemix; Emflex; Rantudil; Rheumibis; Solart. Anti-inflammatory. Derivative of indomethacin. mp = 150-153°; LD_{50} (mmus orl) = 55.5 mg/kg, (mmus iv) = 34.1 mg/kg. *Troponwerke Dinklage.*

4543 Acetaminosalol
118-57-0 46 204-261-3
$C_{15}H_{13}NO_4$
4'-Hydroxyacetanilide salicylate.
p-acetamidophenyl salicylate; p-acetylaminophenol salicylic acid ester; phenestal; Phenosal; Salophen. Analgesic; antipyretic; anti-inflammatory. mp = 187°; nearly insoluble in petroleum ether, cold H_2O; more soluble in warm H_2O; soluble in alcohol, Et_2O, C_6H_6; incompatible with alkalies and alkaline solutions.

4544 Acexamic Acid
57-08-9 43 200-310-8
$C_8H_{15}NO_3$
6-(Acetylamino)hexanoic acid.
acetaminocaproic acid; ε-acetamidocaproic acid; N-acetyl-6-aminohexanoic acid; CY-153. Anti-inflammatory. mp = 104-105.5° (also reported as 112°). *Meiji Seika Kaisha Ltd.*

4545 Acexamic Acid Sodium Salt
7234-48-2 43 230-635-0
$C_8H_{14}NNaO_3$
6-(Acetylamino)hexanoic acid monosodium salt.
Plastenan. Anti-inflammatory.

4546 Acexamic Acid Zinc Salt
70020-71-2 43
$C_{16}H_{28}N_2O_6Zn$
6-(Acetylamino)hexanoic acid zinc salt.
zinc acexamate; ε-acetamidocaproic acid zinc salt; Copinal. Antiulcerative. Free base and sodium salt have anti-inflammatory properties. [free acid]: mp = 104-105.5°, 112°. *Rowa-Wagner.*

4547 S-Adenosylmethionine
29908-03-0 155 249-946-8
$C_{15}H_{22}N_6O_5S$
(3S)-5'-[(3-Amino-3-carboxypropyl)methylsulfonio]-5'-deoxyadenosine inner salt.
active methionine; ademetionine; AdoMet; SAMe; Donamet; S. Amet; Gumbaral [as disulfate ditosylate]; Samyr [as disulfate ditosylate]. Anti-inflammatory; hepatoprotectant. Endogenous methyl donor involved in ezymatic transmethylation reactions. Used in treatment of chronic liver disease. [chloride ($C_{15}H_{23}ClN_6O_5S$)]: λ_m 260 nm (H_2O); $[\alpha]_D^{25}$ = 32°, (c = 3.3 H_2O); [disulfate ditosylate (Gumbaral, Samyr, $C_{29}H_{42}N_6O_{19}S_5$)]: LD_{50} (mus iv) = 560 mg/kg, (mus ip) = 2500 mg/kg, (mus orl) > 6000 mg/kg. *Ajinomoto Co. Inc.; Merck & Co.Inc.; Yamaza Shoyu.*

4548 Alclofenac
22131-79-9 220 244-795-4
$C_{11}H_{11}ClO_3$
3-Chloro-4-(2-propenyloxy)benzeneacetic acid.
W-7320; Allopydin; Epinal; Medifenac; Mervan; Neoston; Prinalgin; Reufenac; Zumaril. Analgesic; antipyretic; anti-inflammatory. Arylacetic acid derivative. mp = 92-93°; LD_{50} (rat orl) = 1050 mg/kg, (rat sc) = 600 mg/kg, (rat ip) = 555 mg/kg. *Continental Pharma Inc.*

4549 Alminoprofen
39718-89-3 308 254-604-6
$C_{13}H_{17}NO_2$
p-[(2-Methylalyly)amino]hydratropic acid.
EB-382; Minalfene. Anti-inflammatory; analgesic. mp = 107°; LD_{50} (mus orl) = 2400 mg/kg.

4550 Amfenac
51579-82-9 413
$C_{15}H_{13}NO_3$
2-Amino-3-benzoylphenylacetic acid.
Anti-inflammatory. Arylacetic acid derivative. mp = 121-123 (dec); LD_{50} (mus orl) = 615 mg/kg, (rat orl) = 311 mg/kg. *Robins, A. H. Co.*

4551 Amfenac Sodium
61618-27-7 413
$C_{15}H_{14}NNaO_4$
Sodium (2-amino-3-benzoylphenyl)acetate monohydrate.
AHR-5850D; Fenamate; Fenazox. Anti-inflammatory. Arylacetic acid derivative. mp = 254-255.5°.

4552 3-Amino-4-hydroxybutyric Acid
589-44-6 463
$C_4H_9NO_3$
γ-Hydroxy-β-aminobutyric acid.
GOBAB. Anti-inflammatory; antifungal; antiseptic. mp = 216°, 228°, 233°. *Kaken Pharm. Co. Ltd.*

4553 Amixetrine
24622-72-8 514
$C_{17}H_{27}NO$
1-[β-(Isopentyloxy)phenethyl]pyrrolidine.
Anti-inflammatory; anticholinergic. bp_2 = 121°; $n_D^{21.6}$ = 1.4978. *C.E.R.M.*

4554 Amixitrene Hydrochloride
24622-52-4 514 246-365-1
$C_{17}H_{28}ClNO$
1-[β-(Isopentyloxy)phenethyl]pyrrolidine monohydrochloride.
Somagest. Anti-inflammatory; anticholinergic. mp = 150°. *C.E.R.M.*

4555 Ammonium Salicylate
528-94-9 584 208-444-9
$C_7H_9NO_3$
2-Hydroxybenzoic acid monoammonium salt.
salicylic acid monoammonium salt; Salicyl-Vasogen. Analgesic. Used topically to loosen psoriatic scales. Soluble in H_2O (100 g/100 ml), EtOH (33.3 g/100 ml).

4556 Ampiroxicam
99464-64-9 629
$C_{20}H_{21}N_3O_7S$
(±)-4-(1-Hydroxyethoxy)-2-methyl-N-2-pyridyl-2H-1,2-benzothiazine-3-carboxamide ethyl carbonate (ester).
CP-65703; Flucam; Nasil. Anti-inflammatory. A thiazinecarboxamide; prodrug of piroxicam. mp = 159-161°; LD_{50} (mrat orl) = 1798 mg/kg, (frat orl) = 747 mg/kg. *Pfizer Inc.*

4557 Amtolmetin Guacil
87344-06-7 637
$C_{24}H_{24}N_2O_5$
N-[(1-Methyl-5-p-toluoylpyrrol-2-yl)acetyl]glycine o-methoxyphenyl ester.
ST-679; MED-15; amtolmethin guacil. Analgesic; anti-inflammatory. Ester prodrug of tolmetin. Arylacetic acid derivative. mp = 117-120°; soluble in common organic solvents; LD_{50} (mmus ip) = 1370 mg/kg, (mrat ip) = 1100 mg/kg, (mmus orl) > 1500 mg/kg, (mrat orl) = 1450 mg/kg. *Sigma-Tau Pharm. Inc.*

4558 Anakinra
143090-92-0 5022
$C_{759}H_{1186}N_{208}O_{232}S_{10}$
N^2-L-Methionyl-interleukin 1 receptor antagonist (human isoform x reduced).
Antril. Anti-inflammatory. Nonglycosylated, recombinant human IL-1ra, molecular weight 17 kDa. Inhibits interleukin-1. Used in treatment of sepsis syndrome. *Synergen Inc.*

4559 Anirolac
66635-85-6
$C_{16}H_{15}NO_4$
(±)-5-p-Anisoyl-2,3-dihydro-1H-pyrrolizine-1-carboxylic acid.
RS-37326. Anti-inflammatory; analgesic. *Syntex Labs. Inc.*

4560 Anitrazafen
63119-27-7
$C_{18}H_{17}N_3O_2$
5,6-Bis(p-Methoxyphenyl)-3-methyl-as-triazine.
LY-122512. Anti-inflammatory (topical). *Eli Lilly & Co.*

4561 Apazone
13539-59-8 768 236-913-8
$C_{16}H_{20}N_4O_2$
5-(Dimethylamino)-9-methyl-2-propyl-1H-pyrazolo[1,2-a][1,2,4]benzotriazine-1,3(2H)-dione.
azapropazone; AHR-3018; Mi-85; NSC-102824; Cinnamin; Sinnamin. Anti-inflammatory; analgesic. A pyrazolone. mp = 228°. *Siegfried AG.*

4562 Apazone Dihydrate
22304-30-9 768
$C_{16}H_{24}N_4O_4$
5-(Dimethylamino)-9-methyl-2-propyl-1H-pyrazolo[1,2-a][1,2,4]benzotriazine-1,3(2H)-dione dihydrate.
MI-85Di; Azapren; Prolixan; Rheumox; Tolyprin. Anti-inflammatory; analgesic. A pyrazolone. mp = 247-248°. *Siegfried AG.*

4563 Aspirin
50-78-2 886 200-064-1
$C_9H_8O_4$
Acetylsalicylic acid.
o-carboxyphenyl acetate; 2-(acetyloxy)benzoic acid; acetate salicylic acid; salicylic acid acetate; Acenterline; Aceticyl; Acetosal; Acetosalic Acid; Acetosalin; Acetylin; Acetyl-SAL; Acimetten; Acylpyrin; Arthrisin, A.S.A; Asatard; Aspro; Asteric; Caprin; Claradin; Colfarit; Contrheuma retard; Duramax; ECM; Ecotrin; Empirin; Encaprin; Endydol; Entrophen; Enterosarine; Helicon; Levius; Longasa; Measurin; Neuronika; Platet; Rhodine; Salacetin; Salcetogen; Saletin; Solrin; Solpyron; Xaxa; Alka-seltzer; Anacin; Ascriptin; Bufferin; Coricidin D; Darvon compound; Excedrin; Gelprin; Robaxisal; Vanquish; Ascoden-30; Coricidin; Norgesic; Persistin; Supac; Triaminicin; acetophen; acidum acetylsalicylicum; acetilum acidulatum; Acetonyl; Adiro; acenterine; acetosal; acetosalic acid; acetosalin; acetylin; acetylsal; acylpyrin;asteric; caprin; colfarit; entrophen; enterosarine; rhodine; salacetin; salcetogen; saletin; acesal; acetilsalicilico; acetisal; acetonyl; asagran; asatard; aspalon; aspergum; aspirdrops; AC 5230; benaspir; entericin; extren; bialpirinia; contrheuma retard; Crystar; Delgesic; Dolean ph 8; enterophen; globoid; idragin. Analgesic; antipyretic; anti-inflammatory. Also has platelet aggregation inhibiting, antithrombotic, and antirheumatic properties. *See* Acetylsalicylic Acid. mp = 135; d = 1.40; λ_m = 229, 277 nm ($E^{1\%}_{1cm}$ 484, 68 in 0.1N H_2SO_4); pH (25°) = 3.49; somewhat soluble in H_2O (1 g/300 ml at 25°, 1 g/100 ml at 37°); soluble in EtOH (1 g/5 ml), $CHCl_3$ (1 g/17 ml), Et_2O (1 g/10 ml). *Bayer AG;*

Boots Pharmaceuticals Inc.; Bristol-Myers Squibb Co.; Eli Lilly & Co.; Parke-Davis; Schering-Plough Pharm.; SmithKline Beecham Pharm.; Sterling Health U.S.A.; Sterling Winthrop Inc.; Upjohn Ltd.

4564 Atiprimod
123018-47-3
$C_{22}H_{44}N_2$
2-[3-(Diethylamino)propyl]-8,8-dipropyl-2-azaspiro[4,5]decane.
Antirheumatic agent with anti-inflammatory properties. *SmithKline Beecham Pharm.*

4565 Atiprimod Dihydrochloride
130065-61-1
$C_{22}H_{46}Cl_2N_2$
2-[3-(Diethylamino)propyl]-8,8-dipropyl-2-azaspiro[4,5]-decane dihydrochloride.
SK&F-106615-A2. Antirheumatic agent with anti-inflammatory properties. *SmithKline Beecham Pharm.*

4566 Atiprimod Dimaleate
183063-72-1
$C_{30}H_{52}N_2O_8$
2-[3-(Diethylamino)propyl]-8,8-dipropyl-2-azaspiro[4,5]-decane maleate (1:2).
SKF-106615-A12. Antirheumatic agent with anti-inflammatory properties. *SmithKline Beecham Pharm.*

4567 Bendazac
20187-55-7 1061
$C_{16}H_{14}N_2O_3$
[[1-(Phenylmethyl)-1H-indazol-3-yl]oxy]acetic acid.
bendazolic acid; bindazac; AF-983; Dogalina; Versus; Zildasac. Anti-inflammatory. Used in veterinary medicine for treatment of cataracts. mp = 160°; λ_m = 306 nm ($E^{1\%}_{1cm}$ 191); LD_{50} (mus iv) = 380 mg/kg, (mus orl) = 1105 mg/kg; soluble in $CHCl_3$, Me_2CO; nearly insoluble in H_2O. *Angelini Pharm. Inc.*

4568 Bendazac [lysine]
81919-14-4
[(1-Benzyl-1H-indazol-3-yl)oxy]acetic acid.
AF-1934. Anti-inflammatory. *Angelini Pharm. Inc.*

4569 Benorylate
5003-48-5 1074 225-674-5
$C_{17}H_{15}NO_5$
4-Acetamidophenyl salicylate acetate.
benorilate; fenasprate; WIN-11450; Benoral; Benortan; Quinexin; Salipran. Analgesic; antipyretic; anti-inflammatory. Derivative of salicylic acid. mp = 175-176°; LD_{50} (mus orl) = 2000 mg/kg, (rat orl) = 10000 mg/kg, (mus ip) = 1255 mg/kg, (rat ip) = 1830 mg/kg. *Sterling Winthrop Inc.*

4570 Benoxaprofen
51234-28-7 1075 257-069-7
$C_{16}H_{12}ClNO_3$
2-(4-Chlorophenyl)-α-methyl-5-benzoxazoleacetic acid.
Compound 90459; Coxigon; Opren; Oraflex; Uniprofen. Anti-inflammatory; analgesic. Arylpropionic acid derivative. mp = 189-190°; LD_{50} (mus orl) = 800 mg/kg. *Eli Lilly & Co.*

4571 Benzpiperylon
53-89-4 1152 200-187-0
$C_{22}H_{25}N_3O$
4-Benzyl-2-(1-methyl-4-piperidyl)-5-phenyl-4-pyrazolin-3-one.
benzpiperilone; benzpiperylone; KB-95; Benzometan; Humedil; Reublonil; Telon. Analgesic; anti-inflammatory. A pyrazolone. Dec 181-183°; pk_1 = 6.73, pK_2 = 9.13; LD_{50} (mus iv) = 160 mg/kg, (mus orl) = 1880 mg/kg. *Sandoz Pharm. Corp.*

4572 Benzydamine
642-72-8 1157 211-388-8
$C_{19}H_{23}N_3O$
1-Benzyl-3-[3-(dimethylamino)propoxy]-1H-indazole.
benzindamine. Analgesic; antipyretic; anti-inflammatory. Anticholinergic. $bp_{0.05}$ = 160°. *3M Pharm.; Angelini Pharm. Inc.*

4573 Benzydamine Hydrochloride
132-69-4 1157 205-076-0
$C_{19}H_{24}ClN_3O$
1-Benzyl-3-[3-(dimethylamino)propoxy]-1H-indazole monohydrochloride.
Difflam Oral Rinse; Difflam Cream; AP-1288; BRL-1288; AF-864; Afloben; Andolex; Benalgin; Benzyrin; Difflam; Dorinamin; Enzamin; Imotryl; Ririlim; Riripen; Salyzoron; Saniflor; Tamas; Tantum; Verax. Analgesic; antipyretic; anti-inflammatory. Anticholinergic. mp = 160°; λ_m = 306 nm ($E^{1\%}_{1cm}$ 160); soluble in H_2O, EtOH, $CHCl_3$, n-BuOH; LD_{50} (mus ip) = 110 mg/kg, (rat ip) = 100 mg/kg, (mus orl) = 515 mg/kg, (rat orl) = 1050 mg/kg. *Angelini Pharm. Inc.*

4574 Bermoprofen
72619-34-2 1199
$C_{18}H_{16}O_4$
(±)-10,11-Dihydro-α,8-dimethyl-11-oxodibenz[b,f]-oxepin-2-acetic acid.
AD-1590; AJ-1590; Dibenon. Anti-inflammatory; analgesic; antipyretic. Prostaglandin synthetase inhibitor; a derivative of arylpropionic acid derivative. mp = 128-129°; LD_{50} (mus orl) = 500 mg/kg, (rat orl) = 147 mg/kg. *Dainippon Pharm.*

4575 α-Bisabolol
515-69-5 1281 208-205-9
$C_{15}H_{26}O$
(R*,R*)-α,4-Dimethyl-α-(4-methyl-3-pentenyl)cyclohex-3-ene-1-methanol.
Camilol; Dragosantol; Hydagen B. Anti-inflammatory. Also used as an ingredient in cosmetics. A sesqueterpine isolated from the essential oils of a variety of plants. bp_{12} = 155-157°; d^{23}_D = 0.9223; n^{23}_D = 1.4917; miscible with alcohols, oils, lipophilic substances.

4576 (-)-Epi-α-Bisabolol
78148-59-1 1281
$C_{15}H_{26}O$
(R*,R*)-(-)-α,4-Dimethyl-α-(4-methyl-3-pentenyl)cyclohex-3-ene-1-methanol.
anymol. Anti-inflammatory. $[\alpha]^{20}_D$ = -67.6°.

4577 (-)-α-Bisabolol
23089-26-1 1281 245-423-3
$C_{15}H_{26}O$
[S-(R*,R*)]-α,4-Dimethyl-α-(4-methyl-3-pentenyl)cyclohex-3-ene-1-methanol.
[S-(R*,R*)]-α-bisabolol; levomenol; Kamillosan. Anti-inflammatory. Active principle in natural extracts. A sesquiterpene. bp_{12} = 153°; d^{20} = 0.9211; n_D^{20} = 1.4936; $[\alpha]_D$ = -55.7°; LD_{50} (mus orl) = 11350 mg/kg, (rat orl) = 14850 mg/kg.

4578 (+)-Epi-α-Bisabolol
76738-75-5 1281
$C_{15}H_{26}O$
(R*,R*)-(+)-α,4-Dimethyl-α-(4-methyl-3-pentenyl)cyclohex-3-ene-1-methanol.
Anti-inflammatory. $[\alpha]_D^{20}$= + 67.4°.

4579 (+)-α-Bisabolol
23178-88-3 1281
$C_{15}H_{26}O$
[R-(R*,R*)]-α,4-Dimethyl-α-(4-methyl-3-pentenyl)cyclohex-3-ene-1-methanol.
(+)-α-bisabolol. Anti-inflammatory. $bp_{1.0}$ = 120-122°; d^{20} = 0.9213; n_D^{20}= 1.4919; $[\alpha]_D^{20}$= + 57.04°.

4580 (±)-α-Bisabolol
25428-43-7 1281 246-973-7
$C_{15}H_{26}O$
(R*,R*)-(±)-α,4-Dimethyl-α-(4-methyl-3-pentenyl)cyclohex-3-ene-1-methanol.
Anti-inflammatory.

4581 Bromelain
9001-00-7 1409
bromelin; Ananase; Extranase; Inflamen; Traumanase. Anti-inflammatory. A concentrate of proteolytic enzymes derived from the pineapple plant *Ananas sativus.* Two protein-digesting, milk-clotting enzymes, one from pineapple fruit juice and one from the plant stem. Also used for tenderizing meat, chill-proofing beer, in production of protein hydrolyzates. λ_m (stem) = 280 nm ($A_{1cm}^{1\%}$ 20.1). *Pineapple Res. Inst.*

4582 Bromfenac
91714-94-2 1411
$C_{15}H_{12}BrNO_3$
2-Amino-3-(p-bromobenzoyl)phenyl]-acetic acid.
AHR-10282. Analgesic; anti-inflammatory. Prostaglandin synthetase inhibitor; analog of amfenac, an arylacetic acid derivative. *Robins, A. H. Co.*

4583 Bromfenac Sodium
120638-55-3 1411
$C_{15}H_{11}BrNNaO_3.1/2H_2O$
Sodium [2-amino-3-(p-bromobenzoyl)phenyl]-acetate sesquihydrate.
AHR-10282B. Analgesic; anti-inflammatory. Arylacetic acid derivative. mp = 284-286° (dec); pKa = 4.29; soluble in H_2O, MeOH, dilute base; insoluble in $CHCl_3$, dilute acid. *Robins, A. H. Co.*

4584 Bromosaligenin
2316-64-5 1460 219-026-0
$C_7H_7BrO_2$
5-Bromo-2-hydroxybenzyl alcohol.
5-bromosalignenin; Bromsalizol. Anti-inflammatory. Derivative of salicylic acid. mp = 109°; soluble in H_2O, EtOH, Et_2O, EtOAc, olive oil; moderately soluble in $CHCl_3$, C_6H_6.

4585 Broperamole
33144-79-5
$C_{15}H_{18}BrN_5O$
1-[3-[5-(m-Bromophenyl)-2H-tetrazol-2-yl]propionyl]-piperidine.
TR-2378. Anti-inflammatory. *Miles Inc.*

4586 Bucloxic Acid
32808-51-8 1486 251-231-0
$C_{16}H_{19}ClO_3$
3-(3-Chloro-4-cyclohexylbenzoyl)propionic acid.
bucloxonic acid. Anti-inflammatory. Arylpropionic acid derivative. mp = 163°; λ_m = 255 mn (ε 15500 in EtOH); LD_{50} (mus orl) = 900 mg/kg, (rat orl) =120 mg/kg, (mus ip) = 1100 mg/kg, (rat ip) = 210 mg/kg. *Clin-Byla France.*

4587 Bucloxic Acid Calcium Salt
32808-53-0 1486 251-232-6
$C_{32}H_{36}CaCl_2O_6$
Calcium 3-chloro-4-cyclohexyl-γ-oxobenzenebutyrate.
CB-804; Esfar. Anti-inflammatory. Arylpropionic acid derivative. LD_{50} (mus orl) = 1700 mg/kg, (rat orl) =175 mg/kg, (mus ip) = 1700 mg/kg, (rat ip) = 200 mg/kg. *Clin-Byla France.*

4588 Bucolome
841-73-6 1487 212-666-1
$C_{14}H_{22}N_2O_3$
5-Butyl-1-cyclohexylbarbituric acid.
BCP; Paramidin. Anti-inflammatory. mp = 84°; $bp_{0.8}$ = 185-197°. *Takeda Chem. Ind. Ltd.*

4589 Bufexamac
2438-72-4 1497 219-451-1
$C_{12}H_{17}NO_3$
2-(p-Butoxyphenyl)acetohydroxamic acid.
CP-1044-J3; Droxarol; Droxaryl; Feximac; Malipuran; Mofenar; Norfemac; Parfenac; Parfenal. Anti-inflammatory; analgesic; antipyretic. Arylacetic acid derivative. mp = 153-155°; nearly insoluble in H_2O; LD_{50} (mus orl) > 8 g/mg. *C.M. Ind.; Pfizer Inc.*

4590 Bufezolac
50270-32-1
$C_{21}H_{22}N_2O_2$
1-Isobutyl-3,4-diphenylpyrazole-5-acetic acid.
Anti-inflammatory agent.

4591 Bumadizone Calcium Salt Hemihydrate
69365-73-7 1507
$C_{38}H_{42}CaN_4O_6.1/2H_2O$
Butylmalonic acid mono(1,2-diphenylhydrazide) calcium salt (2:1) hemihydrate.
Bumaflex; Eumotol; Rheumatol. Analgesic; antipyretic; anti-inflammatory. Arylbutyric acid derivative. Dec 154°;

soluble in $CHCl_3$; EtOH, Et_2O; slightly soluble in H_2O; LD_{50} (mus orl) = 2500 mg/kg, (rat orl) = 1250 mg/kg, (mus iv) = 258 mg/kg, (rat iv) = 263 mg/kg. *Ciba-Geigy Corp.*

4592 Butibufen
55837-18-8 1557 259-849-2
$C_{14}H_{20}O_2$
2-(p-Isobutylphenyl)butyric acid.
Butilopan. Anti-inflammatory Arylbutyric acid derivative. mp = 51-53°; LD_{50} (mus orl) = 810 mg/kg.

4593 Butixirate
19992-80-4 10205 243-454-7
$C_{28}H_{33}NO_2$
(±)-α-Ethyl-4-biphenylacetic acid compound with trans-4-phenylcyclohexylamine (1:1).
xenbucin phenylcyclohexylamine; MG-5771; Flectar. Analgesic; antirheumatic; anti-inflammatory. mp ≅ 225°; LD_{50} (mus ip) = 183 μM/kg. *Maggioni Farmaceutici S.p.A.*

4594 Calcium Acetylsalicylate
69-46-5 1684 200-707-6
$C_{18}H_{14}CaO_8$
2-(Acetyloxy)benzoic acid calcium salt.
salicylic acid acetate calcium salt; acetylsalicylic acid calcium salt; calcium aspirin; soluble aspirin; Ascal; Cal-Aspirin; Dispril; Disprin; Kalmopyrin; Kalsetal; Solaspin; Tylcalsin. Analgesic; antipyretic; anti-inflammatory. Derivative of salicylic acid. Soluble in H_2O (16.6 g/100 ml), EtOH (1.25 g/100 ml). *Lee Labs.*

4595 Calcium Acetylsalicylate Carbamide
5749-67-7 1684
$C_{19}H_{18}CaN_2O_9$
2-(Acetyloxy)benzoic acid calcium salt complex with urea.
carbaspirin calcium; carbasalate calcium; acetylsalicylic acid calcium salt complex with urea; urea calcium acetylsalicylate; carbaspirin calcium; Alcacyl; Calurin; Solupsan. Analgesic; antipyretic; anti-inflammatory. Derivative of salicylic acid. Dec 243-245°; soluble in H_2O (23.1 g/100 ml at 37°), pH = 4.8. *Lee Labs.*

4596 Carprofen
53716-49-7 1912 258-712-4
$C_{15}H_{12}ClNO_2$
(±)-6-Chloro-α-methylcarbazole-2-acetic acid.
Ro-20-5720/000; C-5720; Imadyl; Rimadyl. Anti-inflammatory, used to treat gout. mp = 197-198°; LD_{50} (mus orl) = 400 mg/kg. *Hoffmann-LaRoche Inc.*

4597 Celecoxib
169590-42-5
$C_{17}H_{14}F_3N_3O_2S$
4-[5-(4-Methylphenyl)-3-(trifluoromethyl)-1H-pyrazol-1-yl] benzenesulfonamide.
Anti-inflammatory, used as an NSAID, especially in pain control in arthritis. A COX-2 inhibitor. *Searle G.D. & Co.*

4598 Cicloprofen
36950-96-6 253-287-1
$C_{16}H_{14}O_2$
α-Methylfluorene-2-acetic acid.
SQ-20824. Anti-inflammatory. *Bristol-Myers Squibb Co.*

4599 Cinmetacin
20168-99-4 2353 243-555-6
$C_{21}H_{19}NO_4$
1-Cinnamoyl-5-methoxy-2-methylindole-3-acetic acid.
Cindomet; Indolacin. Anti-inflammatory. Arylacetic acid derivative. mp = 170-172°, 164-165°; LD_{50} (mus ip) = 360 mg/kg, (rat ip) = 590 mg/kg, (mus orl) = 750 mg/kg, (rat orl) = 1020 mg/kg. *Sumitomo Pharm. Co. Ltd.*

4600 Cintazone
2056-56-6
$C_{22}H_{22}N_2O_2$
2-Pentyl-6-phenyl-1H-pyrazolo[1,2-a]cinnoline-1,3(2H)-dione.
cinnopentazone; AHR-3015; Scha-306; NSC-102825. Anti-inflammatory. *Robins, A. H. Co.*

4601 Clidanac
34148-01-1 2411
$C_{16}H_{19}ClO_2$
6-Chloro-5-cyclohexyl-1-indancarboxylic acid.
TAI-284; Britai; Indanal. Anti-inflammatory; antipyretic. Arylcarboxylic acid derivative. mp = 150.5-152.5°; LD_{50} (rat orl) = 41 mg/kg. *Bristol-Myers Squibb Co.*

4602 Cliprofen
51022-75-4
$C_{14}H_{11}ClO_3S$
3-Chloro-4-(-2-thenoyl)hydratropic acid.
R-12160. Anti-inflammatory. *Cilag-Chemie Ltd.*

4603 Clofexamide
1223-36-5
$C_{14}H_{21}ClN_2O_2$
2-(p-Chlorophenoxy)-N-[2-(diethylamino)ethyl]-acetamide.
ANP 246. Anti-inflammatory.

4604 Clofezone
60104-29-2
$C_{14}H_{21}ClN_2O_2.C_{19}H_{20}N_2O_2.2H_2O$
Equimolar mixture of clofexamide and phenylbutazone.

4605 Clopirac
42779-82-8 2459 255-938-5
$C_{14}H_{14}ClNO_2$
1-(4-Chlorophenyl)-2,5-dimethyl-1H-pyrrole-3-acetic acid.
BRL-13856; CP-172AP; Clopiran. Anti-inflammatory. Arylacetic acid derivative. *Beecham Res. Labs. UK; Continental Pharma Inc.; Pfizer Inc.*

4606 Cloximate
58832-68-1
$C_{14}H_{19}ClN_2O_3$
2-(Dimethylamino)ethyl(E)-[[(p-chloro-α-methylbenzylidene)amino]oxy]acetate.
anti-inflammatory.

4607 Delmetacin
16401-80-2
$C_{18}H_{15}NO_3$
1-Benzoyl-2-methylindole-3-acetic acid.
Anti-inflammatory; antipyretic; analgesic.

4608 Dexibuprofen
51146-56-6 4925
$C_{13}H_{18}O_2$
(+)-(S)-p-Isobutylhydratropic acid.
Analgesic; anti-inflammatory. Cyclooxygenase inhibitor. *See* Ibuprophen. *Merck & Co.Inc.*

4609 Dexibuprofen Lysine
141505-32-0 4925
$C_{19}H_{34}N_2O_5$
(+)-(S)-p-Isobutylhydratropic acid L-lysine salt monohydrate.
Ibuprofen S-form L-lysine salt; ML-223; MK-223; Doctrin. Analgesic; anti-inflammatory. S-Form of ibuprofen L-lysine salt. Cyclooxygenase inhibitor. *Merck & Co.Inc.*

4610 Dexibuprofen Lysine [anhydrous]
113403-10-4 4925
$C_{19}H_{3N2}O_4$
(+)-(S)-p-Isobutylhydratropic acid L-lysine salt.
L-669445. Analgesic; anti-inflammatory. S-Form of ibuprofen L-lysine salt. Cyclooxygenase inhibitor. *Merck & Co.Inc.*

4611 Dexindoprofen
53086-13-8 4999 258-351-2
$C_{17}H_{15}NO_3$
(+)-(S)-p-(1-Oxo-2-isoindolinyl)hydratropic acid.
(+)-form indoprofen; Nedius. Analgesic; anti-inflammatory. *See* indoprofen. mp = 205-207°; $[\alpha]_D^{20}$ = +48° (c = 0.05 DMSO); LD_{50} (rat iv) = 31.98 mg/kg, (rat orl) = 33.75 mg/kg. *Chugai Pharm. Co. Ltd.*

4612 Diclofenac
15307-86-5 3132 239-348-5
$C_{14}H_{11}Cl_2NO_2$
2-[(2,6-Dichlorophenyl)amino]benzeneacetic acid.
Anti-inflammatory. Arylacetic acid derivative. mp = 156-158°. *Ciba-Geigy Corp.*

4613 Diclofenac Sodium
15307-79-6 3132 239-346-4
$C_{14}H_{10}Cl_2NNaO_2$
2-[(2,6-Dichlorophenyl)amino]benzeneacetic acid monosodium salt.
GP-45840; Allvoran; Assaren; Benfofen; Cataflam; Delphimix; Dichronic; Diclobenin; Diclo-Phlogont; Diclo-Puren; Diclord; Dicloreum; Dolobasan; Duravolten; Evofenac; Effekton; Kriplex; Neriodin; Novapirina; Primofenac; Prophenatin; Rhumalgan; Tsudohmin; Valetan; Voldal; Voltaren; Voltarol; Xenid. Anti-inflammatory. Arylacetic acid derivative. mp = 283-284°; λ_m = 283 nm (ε 10500 MeOH); soluble in H_2O, MeOH; less soluble in Me_2CO, CH_3CN, cyclohexane; LD_{50} (mus orl) 390 mg/kg, (rat orl) = 150 mg/kg. *Ciba-Geigy Corp.*

4614 Difenamizole
20170-20-1 3181
$C_{20}H_{22}N_4O$
2-(Dimethylamino)-N-(1,3-diphenylpyrazol-5-yl)propionamide.
AP-14; Pasalin. Analgesic; anti-inflammatory. A pyrazole. mp = 123-128°, 120-122°; soluble in Me_2CO, $CHCl_3$, C_6H_6; nearly insoluble in H_2O; LD_{50} (mus iv) = 103 mg/kg, (mus ip) = 186 mg/kg, (mus sc) = 525 mg/kg, (mus orl) = 560 mg/kg. *Takeda Chem. Ind. Ltd.*

4615 Difenpiramide
51484-40-3 3184 257-235-9
$C_{16}H_{19}N_2O$
N-Pyridin-2-yl[1,1'-biphenyl]-4-acetamide.
diphenpyramide; Z-876; Difenax. Anti-inflammatory. mp = 122-124°; LD_{50} (mmus orl) = 2590 mg/kg, (mrat orl) = 2075 mg/kg, (mmus ip) = 1421 mg/kg, (mrat ip) = 1396 mg/kg. *Zambeletti.*

4616 Diflumidone
22736-85-2
$C_{14}H_{11}F_2NO_3S$
3'-Benzoyl-1,1-difluoromethanesulfonanilide.
Anti-inflammatory. *3M Company; Lab. Bago S.A.*

4617 Diflumidone Sodium
22737-01-5
$C_{14}H_{10}F_2NNaO_3S$
3'-Benzoyl-1,1-difluoromethanesulfonanilide sodium salt.
BA-4164-8; MBR-4164-8. Anti-inflammatory. *3M Company; Lab. Bago S.A.*

4618 Diflunisal
22494-42-4 3190 245-034-9
$C_{13}H_8F_2O_3$
2',4'-Difluoro-4-hydroxy-3-biphenylcarboxylic acid.
MK-647; Adomal; Difludol; Dolisal; Dolobid; Dolobis; Flovacil; Fluniget; Fluodonil; Flustar. Analgesic; anti-inflammatory. mp = 210-211°; sparingly soluble in H_2O; LD_{50} (fmus orl) = 439 mg/kg. *E. Merck; N.V. Philips.*

4619 Diftalone
21626-89-1
$C_{16}H_{12}N_2O_2$
Phthalazino[2,3-b]phthalazine-5,12(7H,14H)-dione.
L-5418. Anti-inflammatory. *Marion Merrell Dow Inc.*

4620 Ditazole
18471-20-0 3432 242-353-5
$C_{19}H_{20}N_2O_3$
2,2'-[(4,5-Diphenyl-2-oxazolyl)imino]bisethanol.
ditazol; diethylphenazol; S-222; Ageroplas. Anti-inflammatory. mp = 96-98°; LD_{50} (rat orl) = 11380 mg/kg, (rat ip) = 7770 mg/kg. *Serono Labs Inc.*

4621 Droxicam
90101-16-9 3512
$C_{16}H_{11}N_3O_5S$
5-Methyl-3-(2-pyridyl)-2H,5H-1,3-oxazino[5,6-c]-[1,2]benzothiazine-2,4(3H)-dione 6,6-dioxide.
E-3128; Dobenam; Droxar; Ombolan. Anti-inflammatory. A thiazinecarboxamide, prodrug of piroxicam. mp = 259-261°; LD_{50} (mmus orl) = 6192 mg/kg, (fmus orl) = 8841 mg/kg, (mrat orl) = 1434 mg/kg, (frat orl) = 1994 mg/kg. *Provesan S.A.*

4622 Eltenac
72895-88-6
$C_{12}H_9Cl_2NO_2S$
4-(2,6-Dichloroanilino)-3-thiopheneacetic acid.
An NSAID.

4623 Emorfazone
38957-41-4 3603 254-220-9
$C_{11}H_{17}N_3O_3$
4-Ethoxy-2-methyl-5-(4-morpholinyl)-3(2H)-pyridazinone.
M-73101; Nandron; Pentoyl. Anti-inflammatory; analgesic. mp = 89-91°; LD_{50} (mus ip) = 700 mg/kg. *Robert et Carriere.*

4624 Enfenamic Acid
23049-93-6 3620
$C_{15}H_{15}NO_2$
N-Phenethylanthranilic acid.
RH-8; Tromaril. Anti-inflammatory; analgesic. Aminoarylcarboxylic acid derivative. mp = 116-117°; insoluble in H_2O; LD_{50} (mus ip) = 575 mg/kg, (mus orl) > 2000 mg/kg. *CSIR New Delhi.*

4625 Enolicam
59755-82-7
$C_{17}H_{12}Cl_3NO_4S$
3',4',7'-Trichloro-2,3-dihydro-5-hydroxy-1-benzothiepin-4-carboxanilide 1,1-dioxide.
Anti-inflammatory; antirheumatic.

4626 Enolicam Sodium [anhydrous]
59756-39-7
$C_{17}H_{11}Cl_3NNaO_4S$
3',4',7'-Trichloro-2,3-dihydro-5-hydroxy-1-benzothiepin-4-carboxanilide 1,1-dioxide monosodium salt.
CGS-5391B. Anti-inflammatory; antirheumatic. *Ciba-Geigy Corp.*

4627 Enolicam Sodium Monohydrate
73574-69-3
$C_{17}H_{13}Cl_3NNaO_5S$
3',4',7'-Trichloro-2,3-dihydro-5-hydroxy-1-benzothiepin-4-carboxanilide 1,1-dioxide monosodium salt monohydrate.
Anti-inflammatory; antirheumatic. *Ciba-Geigy Corp.*

4628 Epirizole
18694-40-1 3659 242-507-1
$C_{11}H_{14}N_4O_2$
4-Methoxy-2-(5-methoxy-3-methylpyrazol-1-yl)-6-methylpyrimidine.
mepirizole; DA-398; Mebron. Analgesic; anti-inflammatory; antipyretic. mp = 90-92°; slightly soluble in H_2O; soluble in dilute acids, EtOH, C_6H_6, dichloroethane, Et_2O, Me_2CO; LD_{50} (mus orl) = 820 mg/kg. *Daiichi Pharm. Corp.*

4629 Esculamine
2908-75-0
$C_{15}H_{19}NO_6$
8-[[Bis(2-hydroxyethyl)amino]methyl]-6,7-dihydroxy-4-methylcoumarin.
Anti-inflammatory.

4630 Etersalate
62992-61-4 3759 263-780-3
$C_{19}H_{19}NO_6$
2-(Acetyloxy)benzoic acid 2-[4-(acetylamino)phenoxy]-ethyl ester.
eterilate; etherylate; eterylate; Daital. Analgesic; anti-inflammatory; antipyretic. Derivative of salicylic acid. mp = 139-141°; LD_{10} (rat orl) = 7000 mg/kg.

4631 Etodolac
41340-25-4 3920
$C_{17}H_{21}NO_3$
1,8-Diethyl-1,3,4,9-tetrahydropyranol[3,4-b]indole-1-acetic acid.
etodolic acid; AY-24236; Edolan; Lodine; Ramodar; Tedolan; Ultradol; Zedolac. Anti-inflammatory; analgesic. Arylacetic acid derivative. mp = 145-148°. *Am. Home Products; Wyeth-Ayerst Labs.*

4632 Etofenamate
30544-47-9 3922 250-231-8
$C_{18}H_{18}F_3NO_4$
2-(2-Hydroxyethoxy)ethyl-N-(α,α,α-trifluoro-m-tolyl)-anthranilate.
B-577; Bay-d-1107; TV-485; WHR-5020; Bayrogel; Glasel; Rheumon gel; Traumon Gel. Anti-inflammatory; analgesic. Percutaneously active antiphlogistic agent. Aminoarylcarboxylic acid derivative. Viscous oil; thermolabile at 180°; $bp_{0.001}$ = 130-135°; λ_m = 286 nm ($E^{1\%}_{1cm}$ 423 in MeOH); soluble in lower alcohols, organic solvents; barely soluble in H_2O (0.16 mg/100 ml at 22°); LD_{50} (mrat orl) = 292 mg/kg, (mrat iv) = 140 mg/kg, (mrat ip) = 373 mg/kg, (mrat sc) = 643 mg/kg, (frat orl) = 470 mg/kg, (frat iv) = 226 mg/kg, (frat ip) = 397 mg/kg, (frat sc) = 568 mg/kg. *Farbenfabriken Bayer AG; Troponwerke Dinklage.*

4633 Felbinac
5728-52-9 3989 227-233-2
$C_{14}H_{12}O_2$
4-Biphenylacetic acid.
CL-83544; L-141; LJC-10141; BPAA; Napageln; Traxam. Anti-inflammatory; analgesic. A derivative of arylacetic acid. A metabolite of fenbufen. mp = 164-165°; LD_{50} = 164 mg/kg. *Lederle Labs.*

4634 Felbinac Ethyl Ester
14062-23-8 3989
$C_{16}H_{16}O_2$
4-Biphenylacetic acid ethyl ester.
LM-001; Diatec. Anti-inflammatory; analgesic. Arylacetic acid derivative. *Lederle Labs.*

4635 Fenamole
5467-78-7 226-780-4
$C_7H_7N_5$
5-Amino-1-phenyl-1H-tetrazol.
AL-0559; P-463; PAT; NSC-25413. Anti-inflammatory.

4636 Fenbufen
36330-85-5 4003 252-979-0
$C_{16}H_{14}O_3$
3-(4-Biphenylylcarbonyl)propionic acid.
CL-82204; Bufemid; Cinopal; Cinopol; Lederfen. Anti-inflammatory. mp = 185-187°; LD_{50} (rat orl) = 200-720 mg/kg. *Am. Cyanamid; Lederle Labs.*

4637 Fenclofenac
34645-84-6 252-126-2
$C_{14}H_{10}Cl_2O_3$
2-(2,4-Dichlorophenoxy)benzeneacetic acid.
Rx-67408; R-67408; Flenac. Anti-inflammatory. Arylacetic acid derivative. *Norwich Eaton.*

4638 Fenclorac
36616-52-1
$C_{14}H_{16}Cl_2O_2$
α,3-Dichloro-4-cyclohexyl)benzeneacetic acid.
WHR-539. Anti-inflammatory. Arylacetic acid derivative.

4639 Fenclozic Acid
17969-20-9 4010
$C_{11}H_8ClNO_2S$
2-(p-Chlorophenyl)-4-thiazoleacetic acid.
acidum fenclozicum; ICI-54450; Myalex. Anti-inflammatory. Arylacetic acid derivative. mp = 155-156°; LD_{50} (mus iv) = 14.5 mg/kg, (mus orl) = 950 mg/kg; soluble in common organic solvents; nearly insoluble in H_2O. *ICI.*

4640 Fendosal
53597-27-6 4012
$C_{25}H_{19}NO_3$
5-(4,5-Dihydro-2-phenyl-3H-benz[e]indol-3-yl)salicylic acid.
P71-0129; Alnovin. Anti-inflammatory. Derivative of salicylic acid. mp = 223-225° (dec); LD_{50} (mus orl) = 740 mg/kg, (rat orl) = 450 mg/kg. *Hoechst Roussel Pharm. Inc.*

4641 Fenflumizole
73445-46-2
$C_{23}H_{18}F_2N_2O_2$
2-(2,4-Difluorophenyl)-4,5-bis(p-methoxyphenyl)imidazole.
Anti-inflammatory.

4642 Fenoprofen
31879-05-7 4021 250-850-3
$C_{15}H_{14}NO_3$
(±)-m-Phenoxyhydratropic acid.
Lilly-53858. Anti-inflammatory; analgesic. Arylpropionic acid derivative. Oil; $bp_{0.11}$ = 168-171°; n_D^{25} = 1.5742; pKa = 7.3. *Eli Lilly & Co.*

4643 Fenoprofen Calcium
53746-45-5 4021
$C_{30}H_{30}CaO_8$
Calcium (±)-m-phenoxyhydratropate dihydrate.
Lilly-69323; Fenopron; Fepron; Feprona; Nalfon; Nalgesic; Progesic. Anti-inflammatory; analgesic. Arylpropionic acid derivative. Soluble in H_2O (2.5 mg/ml), more soluble in EtOH, less soluble in organic solvents; pKa = 4.5; aqueous solutions sensitive to uv light; LD_{50} (mus orl) = 800 mg/kg. *Eli Lilly & Co.*

4644 Fenpipalone
21820-82-6
$C_{17}H_{22}N_2O_2$
5-[2-(3,6-Dihydro-4-phenyl-1(2H)-pyridinyl)ethyl]-3-methyl-2-oxazolidinone.
AHR-1680. Anti-inflammatory. *Robins, A. H. Co.*

4645 Fentiazac
18046-21-4 4045 241-958-1
$C_{17}H_{12}ClNO_2S$
4-(4-Chlorophenyl)-2-phenyl-5-thiazoleacetic acid.
BR-700; CH-800; WY-21894; Donorest; Flogene; Norvedan. Anti-inflammatory. Arylacetic acid derivative. mp = 161-162°; LD_{50} (rat orl) = 661 mg/kg. *Wyeth Labs.*

4646 Fepradinol
63075-47-8 4052
$C_{12}H_{19}NO_2$
(±)-α-[[(2-Hydroxy-1,1-dimethylethyl)amino]-methyl]benzyl alcohol.
Dalgen [as hydrochloride]; Flexidol [as hydrochloride]. Anti-inflammatory. mp = 142-143°; (also reported as 140-142°, 139°); λ_m = 250.5, 256.4, 262.0 nm (log ε 2.17, 2.28, 2.25 in 0.1N HCl).

4647 Feprazone
30748-29-9 4053 250-324-3
$C_{20}H_{20}N_2O_2$
4-(3-Methyl-2-butenyl)-1,2-diphenyl-3,5-pyrazolidinedione.
phenylprenazone; prenazone; DA-2370; Analud; Methrazone; Zepelin. Anti-inflammatory. A pyrazolone. mp = 156.5; λ_m = 264 nm (log ε 4.19 in EtOH); pKa 5.09; LD_{50} (mmus ip) = 408.8 mg/kg, (mmus orl) = 1067 mg/kg; soluble in Me_2CO, $CHCl_3$, DMF; sparingly soluble in Et_2O, MeOH, EtOH, cyclohexane; nearly insoluble in 10% HCl, 10% AcOH, H_2O. *DeAngeli.*

4648 Flazalone
21221-18-1
$C_{19}H_{19}F_2NO_2$
p-Fluorophenyl 4-(p-fluorophenyl)-4-hydroxy-1-methyl-3-piperidyl ketone.
NSC-102629. Anti-inflammatory. *3M Pharm.*

4649 Florifenine
83863-79-0
$C_{23}H_{22}F_3N_3O_2$
2-(1-Pyrrolidinyl)ethyl-N-[7-(trifluoromethyl)-4-quinolyl]anthranilate.
Anti-inflammatory.

4650 Flosulide
80937-31-1
$C_{16}H_{13}F_2NO_4S$
N-[6-(2,4-Difluorophenoxy)-1-oxo-5-indanyl]methanesulfonamide.
Anti-inflammatory. COX-2 inhibitor.

4651 Flufenamic Acid
530-78-9 4167 208-494-1
$C_{14}H_{10}F_3NO_2$
N-(α,α,α-Trifluoro-m-tolyl)anthranilate.
CI-440; CN-27544; INF-1837; NSC-82699; Achless; Ansatin; Arlef; Fullsafe; Meralen; Paraflu; Parlef; Ristogen; Sastridex; Surika; Tecramine. Anti-inflammatory; analgesic. Aminoarylcarboxylic acid derivative. mp = 125°; LD_{50} (mus orl) = 715 mg/kg. *Parke-Davis.*

4652 Flufenamic Acid Aluminum Salt
61891-34-7 4167
$C_{42}H_{27}AlF_9N_3O_6$
N-(α,α,α-Trifluoro-m-tolyl)anthranilate aluminum salt (3:1).
aluminum flufenamate; Alfenamin; Opyrin. Anti-inflammatory; analgesic. Aminoarylcarboxylic acid derivative. *Parke-Davis.*

4653 Flufenamic Acid Butyl Ester
67330-25-0 4167
$C_{18}H_{18}F_3NO_2$
N-(α,α,α-Trifluoro-m-tolyl)anthranilate butyl ester.
ufenamate; Fenazol; Combec. Anti-inflammatory; analgesic. Aminoarylcarboxylic acid derivative. *Parke-Davis.*

4654 Flumizole
36740-73-5
$C_{18}H_{15}F_3N_2O_2$
4,5-Bis(p-methoxyphenyl)-2-(trifluoromethyl)imidazole.
CP-22665. Anti-inflammatory. *Pfizer Inc.*

4655 Flunixin
38677-85-9 4182
$C_{14}H_{11}F_3N_2O_2$
2-(2-Methyl-3-trifluoromethylanilino)nicotinic acid.
Sch-14714. Analgesic; anti-inflammatory; antipyretic. Cyclooxygenase inhibitor. mp = 226-228°; pKa' = 5.28. *Schering-Plough HealthCare Products.*

4656 Flunixin Meglumine
42461-84-7 4182 255-836-0
$C_{21}H_{28}F_3N_3O_7$
2-(2-Methyl-3-trifluoromethylanilino)nicotinic acid compound with 1-deoxy-1-(methylamino)-D-glucitol (1:1).
Banamine; Finadyne. Analgesic; anti-inflammatory; antipyretic. Cyclooxygenase inhibitor. mp = 135-139°; soluble in H_2O. *Schering-Plough HealthCare Products.*

4657 Flunoxaprofen
66934-18-7 4183
$C_{16}H_{12}FNO_3$
(+)-2-(p-Fluorophenyl-α-methyl-5-benzoxazoleacetic acid.
RV-12424; Priaxim. Anti-inflammatory. Arylpropionic acid derivative. mp = 162-164°; $[\alpha]_D^{20}$ = +50° (c = 2% DMF); LD_{50} (mus orl) 1200 mg/kg. *Eli Lilly & Co.; Ravizza.*

4658 Fluquazone
37554-40-8
$C_{16}H_{10}ClF_3N_2O$
6-Chloro-4-phenyl-1-(2,2,2-trifluoroethyl-2(1H)-quinazolinone.
EN-970. Anti-inflammatory.

4659 Flurbiprofen
5104-49-4 4234 225-827-6
$C_{15}H_{13}FO_2$
2-Fluoro-α-methyl[1,1'-biphenyl]-4-acetic acid.
BTS-18322; U-27182; Adfeed; Ansaid; Antadys; Cebutid; Froben; Flurofen; Ocufen; Stayban; Zepolas. Anti-inflammatory; analgesic. Arylpropionic acid derivative. mp = 110-111°. *Boots Pharmaceuticals Inc.; C.M. Ind.; Upjohn Ltd.*

4660 Fluretofen
56917-29-4
$C_{14}H_9F$
4'-Ethynyl-2-fluorobiphenyl.
Compound 93819. Anti-inflammatory; antithrombotic. *Eli Lilly & Co.*

4661 Flutiazin
7220-56-6
$C_{14}H_8F_3NO_2S$
8-(Trifluoromethyl)phenothiazine-1-carboxylic acid.
Anti-inflammatory. *SmithKline Beecham Pharm.*

4662 Furobufen
38873-55-1
$C_{16}H_{12}O_4$
γ-Oxo-2-dibenzofuranbutyric acid.
Anti-inflammatory. *Wyeth-Ayerst Labs.*

4663 Furofenac
56983-13-2
$C_{12}H_{14}O_3$
2-Ethyl-2,3-dihydro-5-benzofuranacetic acid.
Anti-inflammatory.

4664 Gentisic Acid
490-79-9 4404 207-718-5
$C_7H_6O_4$
2,5-Dihydroxybenzoic acid.
5-hydroxysalicylic acid. Analgesic; anti-inflammatory. Occurs in gentian root. Derivative of salicylic acid. mp = 199-200°; pK (25°) = 2.93; soluble in H_2O (1 part in 200 at 5°), EtOH, Et_2O; nearly insoluble in CS_2, $CHCl_3$, C_6H_6. *Monsanto Co.; Schering AG.*

4665 Gentisic Acid Sodium Salt
4955-90-2 4404 207-718-5
$C_7H_5NaO_4$
2,5-Dihydroxybenzoic acid sodium salt.
sodium gentisate; Gentinatre; Gentisod; Legential; Gentisine UCB. Analgesic; anti-inflammatory. Soluble in H_2O. *Monsanto Co.; Schering AG.*

4666 Glucametacin
52443-21-7 4456 257-923-9
$C_{25}H_{27}ClN_2O_8$
2-[[[1-(4-Chlorobenzoyl)-5-methoxy-2-methyl-1H-indol-3-yl]acetyl]amino]-2-deoxy-D-glucose.
glucametacine; glucamethacin; indomethacin glucosamide; Euminex [as monohydrate]; Teorema [as monohydrate]; Teoremac [as monohydrate]. Anti-inflammatory. Arylacetic acid derivative.

4667 Glycol Salicylate
87-28-5 4509 201-737-2
$C_9H_{10}O_4$
2-Hydroxybenzoic acid 2-hydroxyethyl ester.
2-hydroxyethyl salicylate; monoglycol salicylate; ethylene glycol monosalicylate; GL-7; Glysal; Norgesic; Phlogont (salve); Spirosal. Counterirritant; anti-inflammatory (topical). Derivative of salicylic acid. bp_{12} = 169-172°; soluble in H_2O (10.1 g/100 ml); more soluble in organic solvents.

4668 Guacetisal
55482-89-8 886 259-663-1
$C_{16}H_{14}O_5$
2-(Acetyloxy)benzoic acid 2-methoxyphenyl ester.
Broncaspin; Guaiaspir. Analgesic; antipyretic; anti-inflammatory. See aspirin.

4669 Guaiazulene
489-84-9 4581 207-701-2
$C_{15}H_{18}$
7-Isopropyl-1,4-dimethylazulene.
S-guaiazulene; AZ-8; AZ-8 Beris; Eucazulen; Kessazulen; Vaumigan. Anti-inflammatory; antiulcerative. *See* sodium gualenate. bp_{10} = 165-170°.

4670 Ibufenac
1553-60-2 4924 216-302-2
$C_{12}H_{16}O_2$
4-(2-Methylpropyl)benzeneacetic acid.
RD-11654; Dytransin; Ibunac. Analgesic; anti-inflammatory. Arylacetic acid derivative. mp = 85-87°; slightly soluble in H_2O; soluble in organic solvents; LD_{50} (mus orl) = 1800 mg/kg. *Boots Pharmaceuticals Inc.*

4671 Ibuprofen
15687-27-1
$C_{13}H_{18}O_2$
α-2-(p-Isobutylphenyl)propionic acid.
Anti-inflammatory. Arylpropionic acid derivative. *Boots Pharmaceuticals Inc.*

4672 (±)-Ibuprofen
58560-75-1 4925 239-784-6
$C_{13}H_{18}O_2$
(±)-2-(p-Isobutylphenyl)propionic acid.
RD-13621; U-18753; Adran; Advil; Anco; Amibufen; Anflagen; Apsifen; Artril 300; Bluton; Brufen; Brufort; Buburone; Butylenin; Dansida; Dentigoa; Dolgin; Dolgirid; Dolgit; Dolocyl; Dolo-Dolgit; Ebufac; Emodin; Epobron; Femadon; Fenbid; Gynofug; Haltran; Ibu-Attritin; Ibumetin; Ibuprocin; Ibutad; Opturem; Pediaprofen; Proflex; Prontalgin; Recidol; Roidenin; Seclodin; Suspren; Tabalon; Trendar; Urem. Anti-inflammatory. mp = 75-77°; fairly insoluble in H_2O; soluble in most organic solvents; LD_{50} (mus ip) = 495 mg/kg, (mus orl) = 1255 mg/kg. *Boots Pharmaceuticals Inc.; Bristol-Myers Squibb Co.; Ciba-Geigy Corp.; McNeil Pharm.; Sterling Health U.S.A.; Upjohn Ltd.; Whitehall Labs. Inc.*

4673 Ibuprofen Aluminum
61054-06-6
$C_{26}H_{35}AlO_5$
(±)-Hydroxybis(p-isobutylhydratropato)-aluminum.
U-18573G; Motrin-A. Anti-inflammatory. *Upjohn Ltd.*

4674 Ibuprofen Lysine Salt
4925
$C_{19}H_{32}N_2O_4$
(±)-2-(p-Isobutylphenyl)propionate lysine salt.
MK-223 [S-form]; Solprofen. Anti-inflammatory. *Boots Pharmaceuticals Inc.*

4675 Ibuprofen Methylglucamine Salt
135861-34-6 4925
$C_{13}H_{18}O_2$
(±)-2-(p-Isobutylphenyl)propionate methylglucamine salt.
Artrene. Anti-inflammatory. *Boots Pharmaceuticals Inc.*

4676 Ibuprofen Piconol
112017-99-9
$C_{19}H_{23}NO_2$
2-Pyridylmethyl (±)-p-isobutylhydratropate.
Be-100; U-18573G. Anti-inflammatory (topical). *Hisamitsu Pharm. Co. Ltd.; Upjohn Ltd.*

4677 Ibuproxam
53648-05-8 4926 258-683-8
$C_{13}H_{19}NO_2$
p-Isobutylhydratropohydroxamic acid.
G-277; Ibudros. Anti-inflammatory. Hydroxylamine derivative of ibuprofen. mp = 119-121°; LD_{50} (mus orl) > 2000 mg/kg; soluble in MeOH, EtOH, Et_2O; nearly insoluble in H_2O, petroleum ether.

4678 Imidazole Salicylate
36364-49-5 4949
$C_{10}H_{10}N_2O_3$
Salicylic acid compound with imidazole (1:1).
salizolo; IFT-182; Flogozen; Selezen. Anti-inflammatory; analgesic; antipyretic. Derivative of salicylic acid. Also used in heat-sensitive copying materials. mp = 123-124°; λ_m = 300 mn ($E^{1\%}_{1cm}$ 182.5); soluble in H_2O (>100 mg/ml); LD_{50} (mmus sc) = 763 mg/kg, (mmus iv) = 422 mg/kg, (mmus orl) = 1121 mg/kg. *Italfarmaco S.p.A.*

4679 Indomethacin
53-86-1 4998 200-186-5
$C_{19}H_{16}ClNO_4$
1-(p-Chlorobenzoyl)-5-methoxy-2-methylindol-3-acetic acid.
Amuno; Argun; Artracin; Artrinovo; Bonidin; Catlep; Chibro-Amuno; Chrono-Indicid; Confortid; Dolcidium; Durametacin; Elmetacin; Idomethine; Imbrilon; Inacid; Indacin; Indocid; Indocin; Indomed; Indomee; Indomethine; Indomod; Indo-Phlogont; Indoptic; Indoptol; Indorektal; Indo-Tablinen; Indoxen; Inflazon; Infrocin; Inteban SP; Lausit; Mezolin; Mikametan; Mobilan; Rheumacin LA; Tannex; Vonum; Liometacen [meglumine salt]. Anti-inflammatory; analgesic; antipyretic. Arylacetic acid derivative which blocks prostaglandin biosynthesis. mp = 155-162°; λ_m = 230, 260, 319 mn (ε 10800, 16200, 6290 EtOH); pKa = 4.5; soluble in EtOH, Et_2O, Me_2CO, castor oil; nearly insoluble in H_2O; decomposes in strong alkalai; LD_{50} (rat ip) = 13 mg/kg. *Gruppo Lepetit S.p.A.; Lemmon Co.; Merck & Co.Inc.; Sumitomo Pharm. Co. Ltd.*

4680 Indomethacin Sodium Trihydrate
74252-25-8 4998
$C_{19}H_{21}ClNNaO_7$
Sodium 1-(p-chlorobenzoyl)-5-methoxy-2-methylindol-3-acetic acid trihydrate.
Indocin IV; Osmosin. Anti-inflammatory; analgesic; antipyretic. Arylacetic acid derivative. pH (1% aqueous solution) = 8.4; soluble in MeOH, H_2O; slightly soluble in $CHCl_3$, Me_2CO. *Merck & Co.Inc.*

4681 Indoprofen
31842-01-0 4999 250-833-0
$C_1H_7H_{15}NO_3$
p-(1-Oxo-2-isoindolinyl)hydratropic acid.
IPP; K-4277; Bor-ind; Flosin; Flosint; Isindone; Praxis;

Reumofene. Analgesic; anti-inflammatory. Arylacetic acid derivative. mp = 213-214°; LD_{50} (rat orl) = 61 mg/kg. *Farmitalia Carlo Erba Ltd.*

4682 Intrazole
15992-13-9
$C_{17}H_{12}ClN_5O$
1-(p-Chlorobenzoyl)-3-(1H-tetrazol-5-ylmethyl)indole.
BL-R743. Anti-inflammatory.

4683 Isofezolac
50270-33-2 5188 256-512-1
$C_{23}H_{18}N_2O_2$
1,3,4-Triphenylpyrazole-5-acetic acid.
LM-22102; Sofenac. Anti-inflammatory; analgesic; antipyretic. Prostaglandin sythetase inhibitor. Arylacetic acid derivative. mp = 200°; LD_{50} (mus orl) = 215 mg/kg, (rat orl) = 13 mg/kg.

4684 Isonixin
57021-61-1 5209 260-521-6
$C_{14}H_{14}N_2O_2$
2-Hydroxy-2',6'-nicotinoxylidide.
Nixyn. Analgesic; anti-inflammatory. Aminoarylcarboxylic acid derivative. mp = 266-267°; soluble in $CHCl_3$, strong alkali; nearly insoluble in acids, H_2O; LD_{50} (mmus orl) >7000 mg/kg, (mus, rat ip) > 2000 mg/kg. *Hermes (GB) Ltd.*

4685 Isoxepac
55453-87-7 5257
$C_{16}H_{12}O_4$
6,11-Dihydro-11-oxodibenz[b,e]oxepin-2-acetic acid.
oxepinac; HP-549; P-720549; Artil. Anti-inflammatory. Arylacetic acid derivative. mp = 131-132.5°; LD_{50} (rat orl) = 199 mg/kg. *Hoechst Roussel Pharm. Inc.*

4686 Isoxicam
34552-84-6 5528 252-084-5
$C_{14}H_{13}N_3O_5S$
4-Hydroxy-2-methyl-N-(5-methyl-3-isoxazolyl)-2H-1,2-benzothiazine-3-carboxamide.
W-8495; Floxicam; Maxicam; Pacyl; Vectren. Anti-inflammatory. A thiazinecarboxamide. mp = 265-271° (dec). *CK Witco Chem. Corp.; Parke-Davis.*

4687 Ketoprofen Lysine Salt
57469-78-0 5316
$C_{22}H_{28}O_5$
m-Benzoylhydratropic acid lysine salt.
Artrosilene. Anti-inflammatory; analgesic. Arylpropionic acid derivative. *Rhône-Poulenc Rorer Pharm. Inc.; Wyeth-Ayerst Labs.*

4688 Ketorolac
74103-06-3 5318
$C_{15}H_{13}NO_3$
(±)-5-Benzoyl-2,3-dihydro-1H-pyrrolizine-1-carboxylic acid.
RS-37619. Analgesic; anti-inflammatory. Prostaglandin biosynthesis inhibitor. Arylcarboxylic acid derivative. mp = 160-161°; λ_m = 245, 312 nm (ε 7080, 17400 MeOH); pKa = 3.5; LD_{50} (mus orl) 200 mg/kg. *Merck & Co.Inc.; Syntex Labs. Inc.*

4689 Ketorolac Tromethamine
74103-07-4 5318
$C_{19}H_{24}N_2O_6$
(±)-5-Benzoyl-2,3-dihydro-1H-pyrrolizine-1-carboxylic acid compound with 2-amino-2-(hydroxymethyl)-1,3-propanediol (1:1).
Acular; Dolac; Lixidol; Tarazyn; Toradol; Toratex. Analgesic; anti-inflammatory. Arylcarboxylic acid derivative. *Syntex Labs. Inc.*

4690 Lofemizole
65571-68-8 265-818-4
$C_{10}H_9ClN_2$
4-(p-Chlorophenyl)-5-methylimidazole.
Anti-inflammatory; analgesic; antipyretic. *Farmatis S.R.L., Italy.*

4691 Lofemizole Hydrochloride
70169-80-1
$C_{10}H_{10}Cl_2N_2$
4-(p-Chlorophenyl)-5-methylimidazole monohydrochloride.
Anti-inflammatory; analgesic; antipyretic. *Farmatis S.R.L., Italy.*

4692 Lonaprofen
41791-49-5 255-555-3
$C_{14}H_{13}ClO_3$
Methyl-2-[(1-chloro-2-naphthyloxy]propionate.
Analgesic; anti-inflammatory.

4693 Lonazolac
53808-88-1 5596 258-791-5
$C_{17}H_{13}ClN_2O_2$
3-(4-Chlorophenyl)-1-phenyl-1H-pyrazole-4-acetic acid.
Anti-inflammatory. Arylacetic acid derivative. mp = 150-151°; λ_m = 281 nm (ε 24800 MeOH); pKa 4.3; LD_{50} (mmus iv) = 195 mg/kg. *Byk Gulden Lomberg GmbH; Wyeth Labs.*

4694 Lonazolac
75821-71-5 5596 278-322-8
$C_{34}H_{24}CaCl_2N_4O_4$
3-(4-Chlorophenyl)-1-phenyl-1H-pyrazole-4-acetic acid calcium salt.
Argun L; Irritren. Anti-inflammatory. Arylacetic acid derivative. mp = 270-290° (dec); λ_m = 280 nm (ε 46400 0.1N HCl); LD_{50} (mmus orl) = 670 mg/kg, (fmus orl) = 845 mg/kg, (mrat orl) = 730 mg/kg, (frat orl) = 1000 mg/kg. *Byk Gulden Lomberg GmbH; Wyeth Labs.*

4695 Lornoxicam
70374-39-9 5612
$C_{13}H_{10}ClN_3O_4S_2$
6-Chloro-4-hydroxy-2-methyl-N-2-pyridyl-2H-thieno-[2,3-e]-1,2-thiazine-3-carboxamide 1,1-dioxide.
chlortenoxicam; Ro-13-9297; TS-110. Anti-inflammatory; analgesic. Cyclooxygenase inhibitor; structurally similar to tenoxicam. A thiazinecarboxamide. mp = 225-230° (dec); λ_m = 371 nm; pKa_2 = 4.7; LD_{50} (mus, rat, dog, rbt, mky orl) > 10 mg/kg. *Hoffmann-LaRoche Inc.*

4696 Losmiprofen
74168-08-4
$C_{17}H_{15}ClO_4$
(±)-2-[[3-(p-Chlorobenzoyl)-o-tolyl]oxy]propionic acid.
Analgesic; anti-inflammatory.

4697 Lotifazole
71119-10-3
$C_{12}H_9Cl_3N_2O_2S$
2',2',2-Trichloroethyl 4-phenyl-2-thiazolecarbamate.
F-1686. Anti-inflammatory.

4698 Loxoprofen
68767-14-6 5619
$C_{15}H_{18}O_3$
(±)-p-[(2-Oxocyclopentyl)methyl]hydratropic acid.
Nonsteroidal anti-inflammatory prodrug; analgesic. Arylpropionic acid derivative. Active metabolite is the trans-cycloohydroxypentane. Oil; $bp_{0.3}$ = 190-195°; mp = 108.5-111°. *Sankyo Co. Ltd.*

4699 Loxoprofen Sodium
80382-23-6 5619
$C_{15}H_{17}NaO_3$
(±)-p-[(2-Oxocyclopentyl)methyl]hydratropate sodium salt.
CS-600; Loxonin. Nonsteroidal anti-inflammatory prodrug; analgesic. Arylpropionic acid derivative. Active metabolite is the trans-cycloohydroxypentane. *Sankyo Co. Ltd.*

4700 Lysine Acetylsalicylate
62952-06-1 5668 263-769-3
$C_{15}H_{22}N_2O_6$
DL-Lysine mono[2-(acetyloxy)benzoate].
aspirin DL-lysine; aspirin lysine salt; lysine monosalicylate acetate; LAS; Aspidol; Delgesic; Flectadol; Lysal; Quinvet; Venopirin; Vetalgina. Analgesic; antipyretic; anti-inflammatory. Water soluble, injectable aspirin derivative. mp = 154-156°; soluble in H_2O; slightly soluble in EtOH; insoluble in organic solvents.

4701 Mabuprofen
82821-47-4 280-048-9
$C_{15}H_{23}NO_2$
(±)-N-(2-Hydroxyethyl)-2-[4-(2-methylpropyl)phenyl]-propionamide.
Analgesic; anti-inflammatory.

4702 Meclofenamate Sodium
6385-02-0
$C_{29}H_{27}ClN_2O_{14}S$
[4S-(4α,4aα,5α,5aα,12aα)]-7-Chloro-4-(dimethylamino)-1,4,4a,5,5a,6,11,12a-octahydro-3,5,10,12,12a-pentahydroxy-6-methylene-1,11-dioxo-2-naphthacenecarboxamide.
Meclan; Mecloderm; Meclosorb; Meclutin; Traumatociclina. Analgesic; anti-inflammatory; antipyretic. Aminoarylcarboxylic acid derivative. *Parke-Davis.*

4703 Meclofenamate Sodium
6385-02-0 5819 228-983-3
$C_{14}H_{12}Cl_2NNaO_3$
2-[(2,6-Dichloro-3-methylphenyl)amino]benzoate sodium salt monohydrate.
Lenidolor; Meclodol; Meclomen; Movens. Anti-inflammatory; antipyretic. Aminoarylcarboxylic acid derivative. mp = 289-291°; soluble in H_2O (15 mg/ml), pH = 8.7. *Parke-Davis.*

4704 Meclofenamic Acid
644-62-2 5819 211-419-5
$C_{14}H_{11}Cl_2NO_2$
2-[(2,6-Dichloro-3-methylphenyl)amino]benzoic acid.
meclophenamic acid; CI-583; INF-4668; Arquel. Anti-inflammatory; antipyretic. Aminoarylcarboxylic acid derivative. mp = 257-259°, 248-250°; slightly soluble in H_2O (0.03 mg/ml); more soluble in 0.1N NaOH (28 mg/ml); pH (saturated aqueous solution) 6.9. *Parke-Davis.*

4705 Meloxicam
71125-38-7 5869
$C_{14}H_{13}N_3O_4S$
4-Hydroxy-2-methyl-N-(5-methyl-2-thiazolyl)-2H-1,2-benzothiazine-3-carboxamide 1,1-dioxide.
Metacam. Anti-inflammatory. mp = 254° (dec); pKa = 4.08 in H_2O; LD_{50} (mus orl) = 470 mg/kg. *Boehringer Ingelheim Ltd.; Thomae GmbH Dr. Karl.*

4706 Mesalamine
89-57-6 5964 201-919-1
$C_7H_7NO_3$
5-Aminosalicylic acid.
5-amino-2-hydroxybenzoic acid; m-aminosalicylic acid; fisalamine; mesalazine; 5-ASA; Asacol; Asacolitin; Claversal; Lixacol; Mesasal; Pentasa; Rowasa; Salofalk. Anti-inflammatory (gastrointestinal). Derivative of salicylic acid. Used in the treatment of ulcerative colitis. Also used in the manufacture of light sensitive paper, azo and sulfur dyes. Dec ≅ 280°; slightly soluble in cold H_2O, alcohol; more soluble in hot H_2O; soluble in HCl. *C.M. Ind.; Ciba-Geigy Corp.; Solvay Pharm. Inc.*

4707 Meseclazone
29053-27-8
$C_{11}H_{10}ClNO_3$
7-Chloro-3,3a-dihydro-2-methyl-2H,9H-isoxazolo[3,2-b][1,3]benzoxazin-9-one.
W-2395. Anti-inflammatory. *Wallace Labs.*

4708 Metanixin
4394-04-1
$C_{14}H_{14}N_2O_2$
2-(2,6-Xylidino)nicotinic acid.
Anilinonicotinic acid derivative. Anti-inflammatory.

4709 Metbufen
63472-04-8
$C_{17}H_{16}O_3$
3-(4-Biphenylylcarbonyl)-2-methylpropionic acid.
Anti-inflammatory.

4710 Metiazinic Acid
13993-65-2 6219 237-795-0
$C_{15}H_{13}NO_2S$
10-Methylphenothiazine-2-acetic acid.
methiazic acid; methiazinic acid; metiazic acid; RP-16091; Soridermal; Soripal. Anti-inflammatory. Arylacetic acid derivative. mp = 146°; λ_m = 253, 305 nm (0.1N

NaOH); soluble in Me_2CO, Et_2O, $CHCl_3$; forms a H_2O-soluble sodium salt; LD_{50} (mus orl) = 800 mg/kg. *Rhône-Poulenc Rorer Pharm. Inc.*

4711 Mipragoside
131129-98-1
$C_{76}H_{137}N_3O_{31}$
N-(II^3-N-Acetylneuraminosylgangliotetraosyl)ceramide isopropyl ester.
AGF-44. Ester derivative of ganglioside GM1. Used in the treatment of allergic conjunctivitis. Antinociceptive; anti-inflammatory.

4712 Mofebutazone
2210-63-1 6312 218-641-1
$C_{13}H_{16}N_2O_2$
4-Butyl-1-phenyl-3,5-pyrazolidinedione.
monophenylbutazone; Arcomonol Tablets; Mobutazon; Mobuzon; Mofesal; Monazon; Monobutyl; Monorheumetten; Reumatox. Anti-inflammatory. A pyrazolone. mp = 102-103°; λ_m = 240, 275 nm ($E^{1\%}_{1cm}$ 443,245 EtOH); LD_{50} (mus iv) = 600 mg/kg.

4713 Mofezolac
78967-07-4 6314
$C_{19}H_{17}NO_5$
3,4-Bis(p-methoxyphenyl)-5-isoxazoleacetic acid.
N-22. Anti-inflammatory; analgesic. Prostaglandin biosynthase inhibitor. A derivative of arylacetic acid. mp = 147.5°; λ_m = 236 mn (ε 18300 MeOH); soluble in organic solvents; slightly soluble in H_2O; LD_{50} (mmus orl) = 1528 mg/kg, (mmus ip) = 275 mg/kg, (mmus sc) = 612 mg/kg.

4714 Morazone
6536-18-1 6349 229-447-1
$C_{23}H_{27}N_3O_2$
4-[(3-Methyl-2-phenyl-4-morpholino)methyl]antipyrine.
R-445; Tarugan. Analgesic; anti-inflammatory; antipyretic. A pyrazolone. mp = 149-150°; soluble in $CHCl_3$, MeOH, Me_2CO; slightly soluble in Et_2O. *Ravensberg.*

4715 Morazone Hydrochloride
50321-35-2 6349 256-540-4
$C_{23}H_{28}ClN_3O_2$
4-[(3-Methyl-2-phenyl-4-morpholino)methyl]antipyrine monohydrochloride.
Analgesic; anti-inflammatory; antipyretic. A pyrazolone. mp = 171-172° (dec); soluble in H_2O. *Ravensberg.*

4716 Morniflumate
65847-85-0 6620
$C_{19}H_{20}F_3N_3O_3$
2-Morpholinoethyl 2-(α,α,α-trifluoro-m-toluidino)-nicotinate.
niflumic acid β-morpholinoethyl ester; UP-164; Flomax; Nifluril (suppositories). Anti-inflammatory. *See* niflumic acid. *Hexachemie.*

4717 Morpholine Salicylate
147-90-0 6362 205-703-8
$C_{11}H_{15}NO_4$
Tetrahydro-2H-1,4-oxazine salicylate.
Retarcyl; Deposal. Analgesic; antipyretic; anti-inflammatory. mp = 110-111°; soluble in H_2O, alcohols, EtOAc, Me_2CO, C_6H_6, $CHCl_3$; nearly insoluble in C_7H_8, xylene, petroleum ether, Et_2O, CCl_4.

4718 Nabumetone
42924-53-8 6428
$C_{15}H_{16}O_2$
4-(6-Methoxy-2-naphthyl)-2-butanone.
BRL-14777; Arthaxan; Balmox; Consolan; Nabuser; Relafen; Relifen; Relifex. Anti-inflammatory; analgesic. Nonacidic, lipophillic prodrug, metabolized in vivo to 6-methoxy-2-napthylacetic acid, a prostaglandin sythesis inhibitor. mp = 80°. *Beecham Res. Labs. UK; C.M. Ind.*

4719 1-Naphthyl Salicylate
550-97-0 6501
$C_{17}H_{12}O_3$
2-Hydroxybenzoic acid 1-naphthalenyl ester.
α-naphthyl salicylate; α naphthol salicylate; Alphol. Anti-infective; anti-inflammatory. Derivative of salicylic acid. A phenol. mp = 83°; insoluble in H_2O; soluble in EtOH, Et_2O, oils.

4720 Naproxen
22204-53-1 6504 244-838-7
$C_{14}H_{14}O_3$
(+)-6-Methoxy-α-methyl-2-napthaleneacetic acid.
MNPA; RS-3540; Bonyl; Diocodal; Dysmenalgit; Equiproxen; Floginax; Laraflex; Laser; Naixan; Napren; Naprium; Naprius; Naprosyn; Naprosyne; Naprux; Naxen; Nycopren; Panoxen; Prexan; Proxen; Proxine; Reuxen; Veradol; Xenar. Anti-inflammatory; analgesic; antipyretic. Arylpropionic acid derivative. mp = 152-154°; $[\alpha]_D$ = +66° (c = 1 in $CHCl_3$); nearly insoluble in H_2O; soluble in organic solvents; LD_{50} (mus orl) = 1234 mg/kg, (rat orl) = 534 mg/kg, (rat ip) = 575 mg/kg. *Syntex Labs. Inc.*

4721 Naproxen Piperazine
70981-66-7 6504
$C_{32}H_{38}N_2O_6$
(+)-6-Methoxy-α-methyl-2-napthaleneacetate piperazine salt (2:1).
piproxen; Numidan. Anti-inflammatory; analgesic; antipyretic. Arylpropionic acid derivative. *Syntex Labs. Inc.*

4722 Naproxen Sodium
26159-34-2 6504 247-486-2
$C_{14}H_{13}NaO_3$
(-)-Sodium 6-methoxy-α-methyl-2-napthalene-acetate.
RS-3560; Aleve; Anaprox; Apranax; Axer; Alfa; Flanax; Gynestrel; Miranax; Primeral; Synflex. Anti-inflammatory; analgesic; antipyretic. Arylpropionic acid derivative. mp = 244-246°; $[\alpha]_D$ = -11° (in MeOH). *Syntex Labs. Inc.*

4723 Naproxol
26159-36-4
$C_{14}H_{16}O_2$
(-)-6-Methoxy-β-methyl-2-napthaleneethanol.
RS-4034. Anti-inflammatory; analgesic; antipyretic. *Syntex Labs. Inc.*

4724 Nicaraven
79455-30-4
$C_{15}H_{16}N_4O_2$
(±)-N,N'-Propylenebis[nicotinamide].
Hydroxyl free radical scavenger.

4725 Niflumic Acid
4394-00-7 6620 224-516-2
$C_{13}H_9F_3N_2O_2$
2-[3-(Trifluoromethyl)anilino]nicotinic acid.
UP-83; Actol; Forenol; Landruma; Nifluril. Anti-inflammatory. Aminoarylcarboxylic acid derivative. mp = 204°; LD_{50} (rat orl) = 370 mg/kg, (rat ip) = 155 mg/kg. *Lab. UPSA.*

4726 Nimesulide
51803-78-2 6640 257-431-4
$C_{13}H_{12}N_2O_5S$
N-(4-Nitro-2-phenoxyphenyl)methonesulfonamide.
R-805; Aulin; Flogovital; Mesulid; Nisulid. Anti-inflammatory. Prostaglandin synthetase and platelet aggregation inhibitor. mp = 143-144.5°; LD_{50} (rat orl) = 324 mg/kg. *C.M. Ind.*

4727 Niometacin
16426-83-8 240-479-5
$C_{18}H_{16}N_2O_4$
5-Methoxy-2-methyl-1-nicotinoylindole-3-acetic acid.
Anti-inflammatory.

4728 Nitraquazone
56739-21-0
$C_{16}H_{13}N_3O_4$
3-Ethyl-1-(m-nitrophenyl)-2,4[1H,3H]-quinazolindione.
TVX-2706. Selective phosphodiesterase inhibitor. Anti-inflammatory.

4729 Olsalazine
15722-48-2 6976
$C_{14}H_{10}N_2O_6$
3,3'-Azobis(6-hydroxybenzoic acid).
C.I. Mordant Yellow 5; 5,5'-azobis(salicylic acid); azodisal. Anti-inflammatory (gastrointestinal); antiulcerative. Derivative of salicylic acid. Dimer of mesalamine. Used in the treatment of ulcerative colitis. Mordant dye for wool. *Ciba-Geigy Corp.; Kabi Pharmacia Diagnostics.*

4730 Olsalazine Sodium
6054-98-4 6976 227-975-7
$C_{14}H_8N_2Na_2O_6$
Disodium 3,3'-azobis(6-hydroxybenzoic acid).
sodium azodisalicylate; azodisal sodium; disodium azodisalicylate; C.I. 14130; Ph-CJ-91B; Dipentum; C.I. Mordant Yellow 5 disodium salt. Anti-inflammatory (gastrointestinal); antiulcerative. Derivative of salicylic acid. Used in the treatment of ulcerative colitis. Mordant dye for wool. Yellow powder; soluble in H_2O; moderately soluble in EtOH. *Ciba-Geigy Corp.; Kabi Pharmacia Diagnostics; Pharmacia & Upjohn.*

4731 Orgotein
9016-01-7 9177 232-771-6
Artolasi; Ormetein; Ontosein; Oximorm; Palosein; Peroxinorm. Anti-inflammatory; antirheumatic. Water soluble protein congeners derived from red blood cells, liver, and other tissues. Molecular weight 33000. Copper-Zinc chelate having superoxide dismutase activity. Member of the superoxide dismutase family, a group of naturally occurring enzymes that act as free oxygen radical scavengers to protect against the effects of biologically generated superoxide oxygen radicals by dismutation to hydrogen peroxide. LD_{50} (mus sc) > 5800 mg/kg, (rat sc) > 400 mg/kg, (mus ip) > 60 mg/kg, (rat ip) > 284 mg/kg, (mus iv) > 4000 mg/kg.

4732 Orpanoxin
60653-25-0
$C_{13}H_{11}ClO_4$
5-(4-Chlorophenyl)-β-hydroxy-
furanpropanoic acid.
F-776. Anti-inflammatory. *Norwich Eaton.*

4733 Oxaceprol
33996-33-7 7035 251-780-6
$C_7H_{11}NO_4$
trans-1-Acetyl-4-hydroxy-L-proline.
CO-61; AHP-200; Jonctum. Anti-inflammatory; vulnerary. Derivative of hydroproline. mp = 133-134°, 126-128°; $[\alpha]_D^{20}$ = -116.5° (c = 3.2); $[\alpha]_D^{18}$ = -119.5° (c = 3.75); soluble in alcohol, H_2O, MeOH; insoluble in $CHCl_3$. *Richardson-Merrell.*

4734 Oxametacin
27035-30-9 7048 248-179-6
$C_{19}H_{17}ClN_2O_4$
1-(p-Chlorobenzoyl)-5-methoxy-2-methylindole-3-
acetohydroxamic acid.
indoxamic acid; Dinulcid; Flogar. Anti-inflammatory. Derivative of indomethacin, a derivative of arylacetic acid. mp = 181-182°; soluble in common organic solvents; LD_{50} (rat orl) = 96 mg/kg.

4735 Oxaprozin
21256-18-8 7057 244-296-1
$C_{18}H_{15}NO_3$
4,5-Diphenyl- 2-oxazolepropanoic acid.
WY-21743; Alvo; Daypro; Durapro; Duraprost; Oxapro. Anti-inflammatory. Arylpropionic acid derivative. mp = 160.5-161.5°. *Wyeth Labs.*

4736 Oxepinac
55689-65-1
$C_{16}H_{12}O_4$
6,11-Dihydro-11-oxodibenz[b,e]oxepin-3-acetic acid.
Anti-inflammatory.

4737 Oxindanac
68548-99-2
$C_{17}H_{14}O_4$
(±)-5-Benzoyl-6-hydroxy-1-indancarboxylic acid.
Anti-inflammatory.

4738 Oxolamine
959-14-8 7078 213-493-4
$C_{14}H_{19}N_3O$
N,N-Diethyl-3-phenyl-1,2,4-oxadiazole-
5-ethanamine.
Anti-inflammatory (respiratory). Liquid; $bp_{0.04}$ = 127°. *Angelini Francesco; SmithKline Beecham Pharm.*

4739 Oxolamine Citrate
1949-20-8 7078 217-760-6
$C_{20}H_{27}N_3O_8$
N,N-Diethyl-3-phenyl-2-hydroxy-1,2,3-propanetricarboxylate-1,2,4-oxadiazole-5-ethanamine.
AF-438; SKF-9976; Bredon; Broncatar; Flogobron; Oxarmin; Perebron; Prilon. Anti-inflammatory (respiratory). Used in inflammatory conditions of the respiratory tract. Crystals; λ_m (aqueous solution) = 239 mn (ε 260) , 273 nm; 283 nm; slightly soluble in H_2O, EtOH. *Angelini Francesco; SmithKline Beecham Pharm.*

4740 Oxyphenbutazone
129-20-4 7106 204-936-2
$C_{19}H_{20}N_2O_3$
4-Butyl-1-(p-hydroxyphenyl)-2-phenyl-3,5-pyrazolidinedione.
p-hydroxyphenylbutazone; G-27202; Californit; Crovaril; Flogitolo; Flogoril; Frabel; Neo-Farmadol; Oxalid; Rapostan; Tandacote; Tandearil; Visubutina. Anti-inflammatory; antirheumatic. A pyrazolone. mp = 124-125°. *Ciba-Geigy Corp.*

4741 Oxyphenbutazone Monohydrate
7081-38-1 7106
$C_{19}H_{22}N_2O_4$
4-Butyl-1-(p-hydroxyphenyl)-2-phenyl-3,5-pyrazolidinedione monohydrate.
Imbun; Phlogistol; Phlogase; Phlogont. Anti-inflammatory; antirheumatic. A pyrazolone. mp = 96°; soluble in EtOH, MeOH, $CHCl_3$, C_6H_6, Et_2O. *Ciba-Geigy Corp.*

4742 Paranyline
1729-61-9 7163
$C_{21}H_{16}N_2$
4-(9H-Fluoren-9-ylidenemethyl)benzenecarboximidamide.
renytoline; 9-(p-guanylbenzal)fluorene; MER-27. Anti-inflammatory. *Marion Merrell Dow Inc.*

4743 Paranyline Hydrochloride
5585-60-4 7163
$C_{21}H_{17}ClN_2$
4-(9H-Fluoren-9-ylidenemethyl)benzenecarboximidamide monohydrochloride.
Anti-inflammatory. mp = 308°. *Marion Merrell Dow Inc.*

4744 Parsalmide
30653-83-9 7177 250-274-2
$C_{14}H_{18}N_2O_2$
5-Amino-N-butyl-2-(2-propynyloxy)benzamide.
MY-41-6; Sinovial; Parsal. Anti-inflammatory; analgesic. Derivative of salicylic acid. mp = 83-85°; λ_m = 220, 327 nm (in absolute alcohol), 284 nm (log ε 3.39 in H_2O, pH 2), 313 nm (log ε 3.40 in H_2O, pH 8.5); soluble in many organic solvents; slightly soluble in H_2O; nearly insoluble in petroleum ether, cyclohexane.

4745 Pegorgotein
155773-57-2 9177
PEG-SOD; Win-22118; Dismutase. Anti-inflammatory; antirheumatic. Orgotein polyethylene glycol conjugate; molecular weight 71-105 kDa. Member of the superoxide dismutase family of enzymes. *Sterling Winthrop Inc.*

4746 Pentosan Polysulfate
37300-21-3 7276
$[C_5H_8O_{10}S_2]_{6-12}$
(1→4)-β-D-Xylan 2,3-bis(hydrogen sulfate).
xylan hydrogen sulfate; xylan polysulfate; CB-8061. Anti-inflammatory (interstitial cystitis); antithrombotic; anticoagulant. A semi-synthetic sulfated polyxylan polysulfate composed of β-D-xylopyranose residues with properties similar to heparin. Molecular weight 1500-5000. *Wander Pharma.*

4747 Pentosan Polysulfate Sodium
116001-96-8 7276
$[C_5H_6Na_2O_{10}S_2]_n$
Xylan hydrogen sulfate sodium salt.
sodium pentosan polysulfate; sodium xylan polysulfate; PZ-68; SP-54; Cartrophen; Elmiron; Fibrase; Fibrezym; Hémoclar; Thrombocid. Anti-inflammatory (interstitial cystitis); antithrombotic. n_D^{20} (10% aqueous solution) = 1.344; $[\alpha]_D^{20}$ = -57°; pH (10% aqueous solution) 6.0; soluble in H_2O.

4748 Pentosan Polysulfate Sodium Salt
37319-17-8 7276
$[C_5H_6Na2O_{10}S_2]_{6-12}$
(1→4)-β-D-Xylan 2,3-bis(hydrogen sulfate) - sodium salt.
SP54; PZ68; sodium pentosan polysulfate; sodium xylan polysulfate; Cartrophen; Elmiron; Fibrase; Fibrezym; Héoclar; Thrombocid. Anti-inflammatory (interstitial cystitis); antithrombotic; anticoagulant. Sodium salt of sulfated β-D-xylopyranose polymer. $[\alpha]_D^{20}$ = -57°; pH (10% aqueous solution) 6.0; N_D^{20} (10% aqueous solution) = 1.344; soluble in H_2O. *Wander Pharma.*

4749 Perisoxal
2055-44-9 7318
$C_{16}H_{20}N_2O_2$
3-(2-Piperidino-1-hydroxyethyl)-5-phenylisoxazole.
Anti-inflammatory; analgesic. mp = 107-108°.

4750 Perisoxal Citrate
2139-25-5 7318
$C_{38}H_{48}N_4O_9$
3-(2-Piperidino-1-hydroxyethyl)-5-phenylisoxazole citrate (2:1).
Isoxal. Anti-inflammatory; analgesic. mp = 135-145°; LD_{50} (mus sc) = 416 mg/kg.

4751 Phenbutazone Sodium Glycerate
34214-49-8
$C_{21}H_{27}N_2NaO_5$
4-Butyl-3-hydroxy-1,2-diphenyl-3-pyrazolin-5-one sodium salt compound with glycerol (1:1).
G-23872. Anti-inflammatory. A pyrazolone.

4752 Phenyl Acetylsalicylate
134-55-4 7424 205-147-6
$C_{15}H_{12}O_4$
2-(Acetyloxy)benzoic acid phenyl ester.
acetylphenylsalicylate; Acetylsalol; Spiroform; Vesipyrin. Analgesic; anti-inflammatory; antipyretic. mp = 97°; bp_{11} = 198°; insoluble in H_2O; soluble in alcohol, Et_2O.

4753 Phenyl Salicylate
118-55-8 7464 204-259-2
$C_{13}H_{10}O_3$
2-Hydroxybenzoic acid phenyl ester.
Salol. Analgesic; anti-inflammatory; antipyretic. Also used in the manufacture of plastics, laquers, adhesives, waxes, polishes; used in suntan oils and creams. Has some light absorbing and plasticizer properties. mp = 41-43°; bp_{12} = 173°; d = 1.25; slightly soluble in H_2O (1 g/6670 ml); more soluble in alcohol (17 g/100 ml), C_6H_6 (67 g/100 ml), amyl alcohol (20 g/100 ml), liquid paraffin (10 g/100 ml), almond oil (25 g/100 ml); soluble in Me_2CO, .

4754 Piketoprofen
60576-13-8 7576
$C_{22}H_{20}N_2O_2$
m-Benzoyl-N-(4-methyl-2-pyridyl)hydratropamide.
Calmatel (aerosol). Anti-inflammatory (topical). Derivative of arylpropionic acid; derivative of ketoprofen. Oil; soluble in CH_2Cl_2, EtOH; insoluble in H_2O.

4755 Piketoprofen Hydrochloride
59512-37-7 7576
$C_{22}H_{21}ClN_2O_2$
m-Benzoyl-N-(4-methyl-2-pyridyl)hydratropamide monohydrochloride.
Calmatel (cream). Analgesic; anti-inflammatory (topical). Derivative of arylpropionic acid; derivative of ketoprofen. mp = 180-182°.

4756 Pipebuzone
27315-91-9 7610 248-398-7
$C_{25}H_{32}N_4O_2$
4-Butyl-4-[(4-methyl-1-piperazinyl)methyl]-1,2-diphenyl-3,5-pyrazolidinedione.
LD-4644; Elarzone. Anti-inflammatory; analgesic; antipyretic. Derivative of phenylbutazone, an anti-inflammatory agent. A pyrazolone. mp = 129°.

4757 Pirazolac
71002-09-0 7643 275-102-3
$C_{17}H_{12}ClFN_2O_2$
4-(4-Chlorophenyl)-1-(4-fluorophenyl)-1H-pyrazole-3-acetic acid.
ZK-76604. Anti-inflammatory. Arylacetic acid derivative. mp = 148-149°, 135-136°. *Schering AG.*

4758 Pirfenidone
53179-13-8
$C_{12}H_{11}NO$
5-Methyl-1-phenyl-2(1H)pyridone.
AMR-69. Analgesic; anti-inflammatory; antipyretic.

4759 Piroxicam
36322-90-4 7661 252-974-3
$C_{15}H_{13}N_3O_4S$
4-Hydroxy-2-methyl-N-2-pyridinyl-2H-1,2-benzothiazine-3-carboxamide 1,1-dioxide.
CP-16171; Artroxicam; Baxo; Bruxicam; Caliment; Erazon; Feldene; Flogobene; Geldene; Impromtal; Larapam; Pirkam; Piroflex; Reudene; Riacen; Roxicam; Roxiden; Sasulen; Solocalm; Zunden. Anti-inflammatory. A thiazinecarboxamide. mp = 198-200°; pKa = 6.3 (2:1 dioxane:H_2O); LD_{50} (mus orl) = 360 mg/kg. *Pfizer Inc.*

4760 Piroxicam Cinnamate
87234-24-0 7661
$C_{24}H_{19}N_3O_5S$
4-Hydroxy-2-methyl-N-2-pyridinyl-2H-1,2-benzothiazine-3-carboxamide 1,1-dioxide cinnamic acid ester.
cinnoxicam; SPA-S-510; Sinartol; Zelis; Zen. Anti-inflammatory. A thiazinecarboxamide. *Pfizer Inc.*

4761 Piroxicam Compound with β-Cyclodextrin
121696-62-6 7661
$C_{57}H_{83}N_3O_{39}S$
4-Hydroxy-2-methyl-N-2-pyridinyl-2H-1,2-benzothiazine-3-carboxamide 1,1-dioxide compound with β-cyclodextrin.
Brexin; Cicladol; Cycladol. Anti-inflammatory; analgesic; antirheumatic. A thiazinecarboxamide. *Pfizer Inc.*

4762 Piroxicam Olamine
85056-47-9 7661
$C_{17}H_{20}N_4O_5S$
4-Hydroxy-2-methyl-N-2-pyridinyl-2H-1,2-benzothiazine-3-carboxamide 1,1-dioxide compound with 2-aminoethanol.
CP-16171-85. Anti-inflammatory; analgesic. A thiazinecarboxamide. *Pfizer Inc.*

4763 Pirprofen
31793-07-4 7663 250-805-8
$C_{13}H_{14}ClNO_2$
3-Chloro-4-(-3-pyrrolin-1-yl)hydratropic acid.
Su-21524; Rangasil 400; Rengasil; Seflenyl. Anti-inflammatory. Arylpropionic acid derivative. mp = 98-100°. *CIBA plc.*

4764 Pranoprofen
52549-17-4 7890
$C_{15}H_{13}NO_3$
α-Methyl-5H-[1]benzopyrano[2,3-b]pyridine-7-acetic acid.
Y-8004; Niflan. Anti-inflammatory. Arylpropionic acid derivative. mp = 182-183°; LD_{50} (mmus orl) = 447.3 mg/kg, (mrat orl) = 87.3 mg/kg. *Yoshitomi.*

4765 Prifelone
69425-13-4
$C_{19}H_{24}O_2S$
3,5-Di-tert-butyl-4-hydroxyphenyl 2-thienyl ketone.
R-830; R-830T; S-16820. Anti-inflammatory (dermatologic). *3M Pharm.*

4766 Proglumetacin
57132-53-3 7957
$C_{46}H_{58}ClN_5O_8$
3-[4-(2-Hydroxyethyl)-1-piperazinyl]propyl DL-4-benzamido-N,N-dipropylglutaramate 1-(p-chlorobenzoyl)-5-methoxy-2-methylindole-3-acetate (ester).
Anti-inflammatory. Derivative of indomethacin, a derivative of arylacetic acid.

4767 Proglumetacin Dimaleate
59209-20-4 7957
$C_{54}H_{66}ClN_5O_{12}$
3-[4-(2-Hydroxyethyl)-1-piperazinyl]propyl DL-4-benzamido-N,N-dipropylglutaramate 1-(p-chlorobenzoyl)-5-methoxy-2-methylindole-3-acetate (ester) dimaleate.

protacine; CR-604; Afloxan; Miridacin; Protaxon; Proxil. Anti-inflammatory. Derivative of indomethacin, a derivative of arylacetic acid. mp = 146-148°; LD_{50} (mmus orl) = 262 mg/kg, (mrat orl) = 450 mg/kg.

4768 Propyphenazone
479-92-5 8056 207-539-2
$C_{14}H_{18}N_2O$
4-Isopropyl-2,3-dimethyl-1-phenyl-3-pyrazolin-5-one.
4-isopropylphenazone; isopropylphenazone; Budirol; Causyth; Cibalgina; Eufibron; Isopropchin. Analgesic; antipyretic; anti-inflammatory. A pyrazolone. mp = 103°; soluble in H_2O (0.24 g/100 g at 16.5°), organic solvents. *Hoffmann-LaRoche Inc.*

4769 Proquazone
22760-18-5 8059 245-203-7
$C_{18}H_{18}N_2O$
1-Isopropyl-7-methyl-4-phenyl-2(1H)-quinazolinone.
RU-43-715; Sandoz 43-715; Anthrex; Biarison. Anti-inflammatory. mp = 137-138°; soluble in $CHCl_3$; insoluble in H_2O. *Sandoz Pharm. Corp.*

4770 Prospidium Chloride
23476-83-7
$C_{18}H_{36}Cl_4N_4O_2$
3,12-Bis(3-chloro-2-hydroxypropyl)-3-,12-diazonia-dispiro[5.2.5.2]hexadecane dichloride.
Anti-inflammatory; antirheumatic; antineoplastic; cytostatic.

4771 Protizinic Acid
13799-03-6 8077 237-453-0
$C_{17}H_{17}NO_3S$
7-Methoxy-α,10-dimethylphenothiazine-2-acetic acid.
17190-RP; Pirocrid. Anti-inflammatory. Leukokinin receptor antagonist. Arylpropionic acid derivative. mp = 124-125°. *Rhône-Poulenc Rorer Pharm. Inc.*

4772 Ramifenazone
3615-24-5 8282 222-791-3
$C_{14}H_{19}N_3O$
4-Isopropylamino-2,3-dimethyl-1-phenyl-3-pyrazolin-5-one.
4-isopropylaminoantipyrine; isopropylaminophenazone; isopyrin; component of: Tomanol (with phenylbutazine). Analgesic; anti-inflammatory; antipyretic. Has been evaluated as an antimigraine in combination with phenylbutazone. A pyrazolone. mp = 80°; LD_{50} (mus ip) = 843 mg/kg, (mus orl) = 1070 mg/kg.

4773 Rofecoxib
162011-90-7
$C_{17}H_{14}O_4S$
4-[4-(Methylsulfonyl)phenyl]-3-phenyl-2(5H)-furanone.
MK-966; Vioxx. Analgesic; anti-inflammatory. A COX-2 specific inhibitor used in the treatment of osteoarthritis. Sparingly soluble in Me_2CO; slightly soluble in MeOH, isopropyl acetate; very slightly soluble in EtOH; practically insoluble in octanol; insoluble in H_2O. *Merck & Co.Inc.*

4774 Romazarit
109543-76-2
$C_{15}H_{16}ClNO_4$
2-[[2-(4-Chlorophenyl)-4-methyl-5-oxazoly]methoxy]-2-methylpropionic acid.
Ro-31-3948/000. Anti-inflammatory; antirheumatic. *Hoffmann-LaRoche Inc.*

4775 Salacetamide
487-48-9 8471 207-656-9
$C_9H_9NO_3$
N-Acetyl-2-hydroxybenzamide.
N-acetylsalicylamide; acetsalicylamide; Actylamide. Analgesic; anti-inflammatory; antipyretic. Derivative of salicylic acid. mp = 148°.

4776 Salcolex
28038-04-2
$C_{24}H_{46}MgN_2O_{12}S \cdot 4H_2O$
Choline salicylate (salt) compound with magnesium sulfate (2:1) tetrahydrate.
Analgesic; anti-inflammatory; antipyretic. *Wallace Labs.*

4777 Salcolex [anhydrous]
54194-00-2
$C_{24}H_{46}MgN_2O_{16}S$
Choline salicylate (salt) compound with magnesium sulfate (2:1).
Analgesic; anti-inflammatory; antipyretic. *Wallace Labs.*

4778 Salicylamide
65-45-2 8480 200-609-3
$C_7H_7NO_2$
2-Hydroxybenzamide.
Cidal; Salamid; Samid; Salizell; Salymid; Urtosal. Analgesic. Derivative of salicylic acid. mp = 140°; soluble in H_2O (20 mg/ml), soluble in organic solvents; pH (saturated aqueous solution) ≅ 5; LD_{50} (mus orl) = 1.4 g/kg. *Lemmon Co.*

4779 Salicylamide O-Acetic Acid
25359-22-6 8481
$C_9H_9NO_4$
[2-(Aminocarbonyl)phenoxy]acetic acid.
o-(carbamylphenoxy)acetic acid. Analgesic; anti-inflammatory; antipyretic. Derivative of salicylic acid. mp = 221°; soluble in aqueous alkali. *Yoshitomi.*

4780 Salicylamide O-Acetic Acid Sodium Salt
3785-82-8 8481
$C_9H_8NNaO_4$
[2-(Aminocarbonyl)phenoxy]acetate sodium salt (1:1).
Salizell ampules. Analgesic; anti-inflammatory; antipyretic. Derivative of salicylic acid. mp = 212-215°. *Yoshitomi.*

4781 Salicylsulfuric Acid
89-45-2 8486
$C_7H_6O_6S$
2-(Sulfoxy)benzoic acid.
salicylic acid, acid sulfate; salicylic acid sulfuric acid ester. Analgesic; anti-inflammatory. Derivative of salicylic acid.

4782 Salicylsulfuric Acid Sodium Salt
6155-64-2 8486
$C_7H_5NaO_6S$
2-(Sulfoxy)benzoate monosodium salt.
sodium salicylsulfate; Salcyl; Salcylix. Analgesic; anti-inflammatory. Derivative of salicylic acid. Soluble in H_2O; insoluble in organic solvents.

4783 Salnacedin
87573-01-1
$C_{12}H_{13}NO_5S$
N-Acetyl-L-cysteine salicylate.
G-201,SCY. Analgesic; anti-inflammatory. *Genta Inc.*

4784 Salsalate
552-94-3 8491 209-027-4
$C_{14}H_{10}O_5$
2-Hydroxybenzoic acid 2-carboxyphenyl ester.
disalicylic acid; salicyloxysalicylic acid; salicylsalicylic acid; NSC-49171; Disalcid; Disalgesic; Mono-Gesic; Salflex. Analgesic; anti-inflammatory. Nonacetylated aspirin analog. mp = 148-149°; nearly insoluble in H_2O; sparingly soluble in organic solvents. *3M Pharm.; Boehringer Mannheim GmbH; Fisons Pharm. Div.*

4785 Seclazone
29050-11-1
$C_{10}H_8ClNO_3$
7-Chloro-3,3a-dihydro-2-methyl-2H,9H-isoxazolo[3,2-b][1,3]benzoxazin-9-one.
W-2352. Anti-inflammatory; uricosuric. *Wallace Labs.*

4786 Sermetacin
57645-05-3 260-873-0
$C_{22}H_{21}ClN_2O_6$
N-[[1-(4-Chlorobenzoyl)-5-methoxy-2-methyl-1H-indol-3-yl]acetyl]-L-serine.
SH-G-318-AB. Anti-inflammatory. *Schering AG.*

4787 Sodium Gualenate
6223-35-4 4581 228-309-8
$C_{21}H_{21}N_3O_6$
5-Isopropyl-3,8-dimethyl-1-azulenesulfonic acid sodium salt.
guaiazulene 3-sulfonate sodium salt; guaiazulene soluble; Azulon. Anti-inflammatory; antiulcerative. *See* guaiazulene.

4788 Sudoxicam
34042-85-8
$C_{13}H_{11}N_3O_4S_2$
4-Hydroxy-2-methyl-N-2-thiazolyl-2H-1,2-benzothiazine-3-carboxamide 1,1-dioxide.
CP-15973. Anti-inflammatory. *Pfizer Inc.*

4789 Sulfasalazine
599-79-1 9112 209-974-3
$C_{18}H_{14}N_4O_5S$
5-[[p-(2-Pyridylsulfamoyl)phenyl]azo]salicylic acid.
salazosulfapyridine; salicylazosulfapyridine; sulphasalazine; Azulfidine; Colo-Pleon; Salazopyrin. Anti-inflammatory (gastrointestinal). Sulfonamide; conjugate of 5-aminosalicylic acid and suphapyridine. Used in the treatment of ulcerative colitis and Crohn's disease. Dec 240-245°; λ_m = 237 ($E^{1\%}_{1cm}$ 658), 359 nm; slightly soluble in EtOH; nearly insoluble in H_2O, C_6H_6, $CHCl_3$, Et_2O. *Kabi Pharmacia Diagnostics; Pharmacia & Upjohn.*

4790 Sulfinpyrazone
57-96-5 9121 200-357-4
$C_{23}H_{20}N_2O_3S$
1,2-Diphenyl-4-[2-(phenylsulfinyl)ethyl]-3,5-pyrazolidinedione.
G-28315; Anturan; Anturane; Anturano; Enturen. Uricosuric and antithrombotic. Used to treat gout. A nonsteroidal anti-inflammatory agent. Inhibits cyclooxygenase. Prolongs circulating platelet survival. mp = 136-137°; λ_m = 255 nm (1N NaOH); soluble in EtOAc, $CHCl_3$; slightly soluble in H_2O, EtOH, Et_2O, mineral oils; [d-form]: mp = 130-133°; $[\alpha]^{2}_{D}$ = 67.1° (c = 2.04 EtOH), $[\alpha]^{25}_{D}$ = 109.3 (c = 0.5 $CHCl_3$); [l-form]: mp = 130-133°; $[\alpha]^{23}_{D}$ = -64.2° (c = 2.14 EtOH), $[\alpha]^{26}_{D}$ = -104.5° (c = 0.5 $CHCl_3$). *Ciba-Geigy Corp.*

4791 Sulindac
38194-50-2 9155 253-819-2
$C_{20}H_{17}FO_3S$
cis-5-Fluoro-2-methyl-1-[(p-methylsulfinyl)-phenyl]benzylidene]indene-3-acetic acid.
MK-231; Aflodac; Algocetil; Arthrocine; Artribid; Citireuma; Clinoril; Clisundac; Imbaral; Reumofil; Reumyl; Sudac; Sulinol; Sulreuma. Anti-inflammatory. Arylacetic acid derivative. mp = 182-185°; λ_m = 327, 285, 256, 226 nm ($E^{1\%}_{1cm}$ 375, 420, 410, 540 0.1N HCl); slightly soluble in H_2O (0.3 g/100 ml, pH 7), similarly soluble in polar organic solvents; pKa (25°) = 4.7. *Merck & Co.Inc.*

4792 Suprofen
40828-46-4 9180 255-096-9
$C_{14}H_{12}O_3S$
2-Thenoylhydratropic acid.
sutoprofen; R-25061; Masterfen; Profenol; Srendam; Supranol; Suprol; Sulproltin; Suprocil; Topalgic. Anti-inflammatory; analgesic. Prostaglindin biosynthesis inhibitor. A derivative of arylpropionic acid. mp = 124.3°; pKa = 3.91; λ_m = 266, 292 nm (ε 15700, 15600 0.01N HCL-90% iPrOH); soluble in alcohols, Et_2O, $CHCl_3$, Me_2CO, polyethylene glycol, alkalies; slightly soluble in H_2O; nearly insoluble in n-hexane; LD_{50} (mus orl) = 590 mg/kg, (dog orl) = 160 mg/kg, (rat orl) = 353 mg/kg, (gpg orl) = 280 mg/kg. *Abbott Laboratories Inc.; Alcon Labs; McNeil Pharm.*

4793 Suxibuzone
27470-51-5 9185 248-477-6
$C_{24}H_{26}N_2O_6$
4-Butyl-4-(hydroxymethyl)-1,2-diphenyl-3,5-pyrazolidinedione hydrogen succinate (ester).
AE-17; Calibéne; Danilon; Flogos; Solurol. Anti-inflammatory. A pyrazolone; prudrug of phenylbutazone. mp = 126-127°; soluble in most organic solvents; insoluble in H_2O; LD_{50} (mus orl) = 3.06 mg/kg. *Esteve Group.*

4794 Talmetacin
67489-39-8
$C_{27}H_{20}ClNO_6$
(±)-1-(p-Chlorobenzoyl)-5-methoxy-2-methylindole-3-acetic acid ester with 3-hydroxyphthalide.
BA-7605-06. Analgesic; anti-inflammatory; antipyretic. *Lab. Bago S.A.; Resfar S.R.L.*

4795 Talniflumate
66898-62-2 9214
$C_{21}H_{13}F_3N_2O_4$
Phthalidyl 2-(α,α,α-trifluoro-m-toluidino)nicotinate.
BA-7602-06; Somalgen. Analgesic; anti-inflammatory. Derivative of niflumic acid, an anti-inflammatory agent. An aminoarylcarboxylic acid derivative. mp = 165-166°; λ_m = 287, 357 nm (ε 25600, 7800 $CHCl_3$); LD_{50} (rat orl) = 1200 mg/kg. *Lab. Bago S.A.*

4796 Talosalate
66898-60-0
$C_{17}H_{12}O_6$
1,3-Dihydro-3-oxo-1-isobenzofuranyl ester 2-(acetoxy)-benzoic acid.
BA-7604-02. Analgesic; anti-inflammatory. *Lab. Bago S.A.*

4797 Tebufelone
112018-00-5
$C_{20}H_{28}O_2$
3',5'-Di-tert-butyl-4'-hydroxy-5-hexynophenone.
NE-11740. Analgesic; anti-inflammatory. *Norwich Eaton.*

4798 Tenidap
120210-48-2 9290
$C_{14}H_9ClN_2O_3S$
(Z)-5-Chloro-3-(α-hydroxy-2-thenylidene)-2-oxo-1-indolinecarboxamide.
CP-66248. Anti-inflammatory. Used in the treatment of osteoarthritis and rheumatoid arthritis. mp = 230° (dec). *Pfizer Inc.*

4799 Tenidap Sodium
119784-94-0 9290
(Z)-5-Chloro-3-(α-hydroxy-2-thenylidene)-2-oxo-1-indolinecarboxamide monosodium salt.
CP-66248-2. Anti-inflammatory. Used in the treatment of osteoarthritis and rheumatoid arthritis. mp = 237-238°. *Pfizer Inc.*

4800 Tenoxicam
59804-37-4 9293
$C_{13}H_{11}N_3O_4S_2$
4-Hydroxy-2-methyl-N-2-pyridyl-2H-thieno[2,3-e]-1,2-thiazene-3-carboxamide 1,1-dioxide.
Ro-12-0068; Alganex; Dolmen; Liman; Mobiflex; Rexalgan; Tilatil; Tilcotil. Anti-inflammatory; analgesic. A thiazinecarboxamide. mp = 209-213° (dec). *Hoffmann-LaRoche Inc.*

4801 Terofenamate
29098-15-5 9312 249-434-4
$C_{17}H_{17}Cl_2NO_3$
Ethoxymethyl N-(2,6-dichloro-m-tolyl)anthranilate.
etoclofene; Etofen. Anti-inflammatory; analgesic. Aminoarylcarboxylic acid derivative. mp = 73-73°; LD_{50} (mus orl) = 918 mg/kg, (rat orl) = 307 mg/kg, (mus ip) = 300 mg/kg, (rat ip) = 274 mg/kg. *Lusofarmico.*

4802 Tesicam
21925-88-2
$C_{16}H_{12}ClN_2O_3$
4'-Chloro-1,2,3,4-tetrahydro-1,3-dioxo-4-isoquinolinecarboxanilide.
CP-13608. Anti-inflammatory. *Pfizer Inc.*

4803 Tesimide
35423-09-7
$C_{16}H_{15}NO_2$
4-Benzylidine-5,6,7,8-tetrahydro-1,3(2H,4H)-isoquinolinedione.
Anti-inflammatory. *Parke-Davis.*

4804 Tetriprofen
28168-10-7
$C_{15}H_{18}O_2$
p-1-Cyclohexen-1-ylhydratropic acid.
47-210 [as sodium]. Analgesic; anti-inflammatory.

4805 Thiazolinobutazone
54749-86-9 9443 259-319-0
$C_{22}H_{26}N_4O_2S$
4-Butyl-1,2-diphenylpyrazolidine-3,5-dione compound with 4,5-dihydrothiazol-2-amine (1:1).
phenylbutazone 2-amino-2-thiazoline salt; TZB; LAS-11871; Fordonal. Anti-inflammatory. A pyrazolone. mp = 164-166°; LD_{50} (mus orl) = 1425 mg/kg, (rat orl) = 1650 mg/kg.

4806 Tiaprofenic Acid
33005-95-7 9562 251-329-3
$C_{14}H_{12}O_3S$
5-Benzoyl-α-methyl-2-thiopheneacetic acid.
FC-3001; RU-15060; Suralgan; Surgam. Anti-inflammatory. Arylpropionic acid derivative. mp = 96°. *C.M. Ind.; Roussel-UCLAF.*

4807 Tiaramide
32527-55-2 9564 251-083-7
$C_{15}H_{18}ClN_3O_3S$
4-[(5-Chloro-2-oxo-3-benzothiazolyl)acetyl]-1-piperazineethanol.
tialamide. Antiasthmatic; antiallergic; anti-inflammatory Arylacetic acid derivative. pKa = 6.2. *Fujisawa Pharm. USA Inc.; Rhône-Poulenc Rorer Pharm. Inc.*

4808 Tiaramide Hydrochloride
35941-71-0 9564 252-802-7
$C_{15}H_{18}Cl_2N_3O_3S$
4-[(5-Chloro-2-oxo-3-benzothiazolyl)acetyl]-1-piperazineethanol monohydrochloride.
NTA-194; FK-1160; Solantal. Antiasthmatic; antiallergic; anti-inflammatory Arylacetic acid derivative. mp = 159-161°; pH (10% aqueous solution) = 3.4-3.7; soluble in H_2O; slightly soluble in organic solvents; LD_{50} (mmus iv) = 178 mg/kg, (mmus ip) = 298 mg/kg, (mmus orl) = 564 mg/kg, (mrat iv) = 203 mg/kg, (mrat ip) = 540 mg/kg, (mrat orl) = 3600 . *Fujisawa Pharm. USA Inc.; Rhône-Poulenc Rorer Pharm. Inc.*

4809 Tiflamizole
62894-89-7
$C_{17}H_{10}F_6N_2O_2S$
4,5-Bis(p-fluorophenyl)-2-[91,1,2,2-tetrafluoroethyl)sulfonyl]imidazole.
Thioimidazole derivative. Cyclooxygenase inhibitor in prostaglandin synthesis. Anti-inflammatory; antiarthritic.

4810 Timegadine
71079-19-1 275-184-0
$C_{20}H_{23}N_5S$
1-Cyclohexyl-2-(2-methyl-4-quinolyl)-3-(2-thiazolyl)guanidine.
SR-1368. A tri-substituted guanidine derivative that inhibits both arachidonate cyclo-oxygenase and lipoxygenase activity. Used in treatment of rheumatoid arthritis. Anti-inflammatory.

4811 Tinoridine
24237-54-5 9590 246-102-0
$C_{17}H_{20}N_2O_2S$
Ethyl 2-amino-6-benzyl-4,5,6,7-tetrahydrothieno-[2,3-c]pyridine-3-carboxylate.
Y-3642. Analgesic; anti-inflammatory; antipyretic. Arylcarboxylic acid derivative. mp = 112-113°; slightly soluble in H_2O; LD_{50} (mus orl) = 5400 mg/kg, (rat orl) > 10200 mg/kg, (mus ip) = 1600 mg/kg, (rat ip) = 1250 mg/kg. *Yoshitomi.*

4812 Tinoridine Hydrochloride
25913-34-2 9590
$C_{17}H_{21}ClN_2O_2S$
Ethyl 2-amino-6-benzyl-4,5,6,7-tetrahydrothieno-[2,3-c]pyridine-3-carboxylate hydrochloride.
Dimaten, Noflamin. Analgesic; anti-inflammatory; antipyretic. Arylcarboxylic acid derivative. mp = 234-235° (dec); slightly soluble in MeOH; less soluble in H_2O, Et_2O, Me_2CO, C_6H_6; LD_{50} (mus orl) = 1601 mg/kg. *Yoshitomi.*

4813 Tiopinac
61220-69-7
$C_{16}H_{12}O_3S$
6,11-Dihydro-11-oxodibenzo[b,e]thiepin-3-acetic acid.
RS-40974. Analgesic; anti-inflammatory; antipyretic. *Syntex Labs. Inc.*

4814 Tioxaprofen
40198-53-6 254-834-7
$C_{18}H_{13}Cl_2NO_3S$
2-[[4,5-Bis-(p-chlorophenyl)-2-oxazolyl]thio]-propionic acid.
EMD-26644. Anti-inflammatory; antithrombotic; antimycotic.

4815 Tolfenamic Acid
13710-19-5 9650 237-264-3
$C_{14}H_{12}ClNO_2$
N-(3-Chloro-o-tolyl)anthranilic acid.
GEA-6414; Clotam; Tolfedine; Tolfine. Anti-inflammatory; analgesic Prostaglandin biosynthesis inhibitor. Derivative of aminoarylcarboxylic acid. Related structurally to mefanamic and flufenamic acids, anti-inflammatory agents. mp = 207-207.5°. *Gea A/S; Parke-Davis.*

4816 Tolmetin
26171-23-3 9655 247-497-2
$C_{15}H_{15}NO_3$
1-Methyl-5-p-toluoylpyrrole-2-acetic acid.
McN-2559. Anti-inflammatory. Arylacetic acid derivative. mp = 155-157°. *McNeil Pharm.*

4817 Tolmetin Sodium [anhydrous]
35711-34-3 252-687-3
$C_{15}H_{14}NNaO_3$
Sodium 1-methyl-5-p-toluoylpyrrole-2-acetic acid.
Anti-inflammatory. *McNeil Pharm.*

4818 Tolmetin Sodium Dihydrate
64490-92-2 9655
$C_{15}H_{18}NNaO_5$
Sodium 1-methyl-5-p-toluoylpyrrole-2-acetic acid dihydrate.
McN-2559-21-98; Reutol; Tolectin; Tolmene. Anti-inflammatory. Arylacetic acid derivative. *McNeil Pharm.*

4819 Tomoxiprole
76145-76-1
$C_{21}H_{20}N_2O$
3-Isopropyl-2-(p-methoxyphenyl)-3H-naphth[1,2-d]imidazole.
MDL-035. Selectively inhibits cyclooxygenase-2. Anti-inflammatory.

4820 Tribuzone
13221-27-7 236-191-4
$C_{22}H_{24}N_2O_3$
4-(4,4-Dimethyl-3-oxopentyl)-1,2-diphenyl-3,5-pyrazolidinedione.
Anti-inflammatory.

4821 Triflumidate
24243-89-8
$C_{17}H_{14}F_3NO_5S$
Ethyl-m-benzoyl-N-[(trifluoromethyl)sulfonyl]carbanilate.
BA-4223; MBR-4223. Anti-inflammatory.

4822 Tropesin
65189-78-8 9909 265-607-7
$C_{28}H_{24}ClNO_6$
(±)-1-(4-Chlorobenzoyl)-5-methoxy-2-methyl-1H-indole-3-acetic acid 2-carboxy-2-phenyl ester.
Repanidal. Anti-Inflammatory; analgesic. Tropic acid ester of indomethacin, an arylacetic acid derivative. mp ≅ 130; LD_{50} (frat orl) = 140 mg/kg. *SPOFA.*

4823 Vedaprofen
71109-09-6 275-196-6
$C_{19}H_{22}O_2$
(±)-4-Cyclohexyl-α-methyl-1-naphthaleneacetic acid.
CERM-10202; PM-150. Anti-Inflammatory; analgesic. Used in veterinary medicine. *PCAS.*

4824 Ximoprofen
56187-89-4 10211 260-041-7
$C_{15}H_{19}NO_3$
4-[3-(Hydroxyimino)cyclohexyl]-α-methylbenzene-acetic acid.
XIFAM; 13832-JL. Anti-inflammatory. Arylpropionic acid derivative. mp = 178°.

4825 Zaltoprofen
89482-00-8 10244
$C_{17}H_{14}O_3S$
(±)-10,11-Dihydro-α-methyl-10-oxodibenzo[b,f]thiepin-2-acetic acid.
CN-100. Anti-inflammatory. Arylpropionic acid derivative. (S)-Form possesses anti-inflammatory activity. mp = 130-133°; [(S) form]: mp = 129.5-131°; $[\alpha]_D^{25}$ = +32.4° (c = 1 in $CHCl_3$); soluble in Me_2CO, $CHCl_3$, MeOH; slightly soluble in EtOH, C_6H_6; nearly insoluble in H_2O, cyclohexane. *Nippon Chemiphar.*

4826 Zidometacin
62851-43-8 263-740-5
$C_{19}H_{16}N_4O_4$
1-(4-Azidobenzoyl)-5-methoxy-2-methyl-1H-indole-3-acetic acid.
Anti-inflammatory. *Pierrel S.p.A.*

4827 Zileuton
111406-87-2 10253
$C_{11}H_{12}N_2O_2S$
(±)-1-(1-Benzo[b]thien-2-ylethyl)-1-hydroxyurea.
A-64077; Abbott-64077; Leutrol; Zyflo. Antiasthmatic; anti-inflammatory. Inhibitor of 5-lipoxygenase, the initial enzyme in the biosynthesis of leukotrienes from arachidonic acid. mp = 157-158°. *Abbott Labs Inc.*

4828 Zomepirac
33369-31-2 10322 251-474-2
$C_{15}H_{14}ClNO_3$
5-(4-Chlorobenzoyl)-1,4-dimethyl-1H-pyrrole-2-acetic acid.
Analgesic; anti-inflammatory. Arylacetic acid derivative. mp = 178-179°. *McNeil Pharm.*

4829 Zomepirac Sodium [anhydrous]
64092-48-4 10322 264-669-2
$C_{15}H_{13}ClNNaO_3$
5-(4-Chlorobenzoyl)-1,4-dimethyl-1H-pyrrole-2-acetic acid sodium salt.
Analgesic; anti-inflammatory. Arylacetic acid derivative. *McNeil Pharm.*

4830 Zomepirac Sodium Dihydrate
64092-49-5 10322
$C_{15}H_{17}ClNNaO_5$
5-(4-Chlorobenzoyl)-1,4-dimethyl-1H-pyrrole-2-acetic acid sodium salt dihydrate.
McN-2783-21-98; Zomax; Zomaxin; Zopirac. Analgesic; anti-inflammatory. Arylacetic acid derivative. mp = 295-296°. *McNeil Pharm.*

Anti-inflammatories, Steroidal

4831 21-Acetoxypregnenolone
566-78-9 77 209-298-9
$C_{23}H_{34}O_4$
(3β)-21-(Acetyloxy)-3-hydroxypregn-5-en-20-one.
prebediolone acetate; A.O.P.; Acetoxanon; Artixone acetate. Anti-inflammatory. mp = 184-185°; very slightly soluble in Et_2O, C_5H_{12}; soluble in $CHCl_3$, C_7H_8. *Schering-Plough HealthCare Products.*

4832 Alclometasone
67452-97-5 221
$C_{22}H_{29}ClO_5$
7α-Chloro-11β,17,21-trihydroxy-16α-methylpregna-1,4-diene-3,20-dione.
7α-chloro-16α-methylprednisolone. Anti-inflammatory (topical). Non-fluorinated corticosteroid with low systemic effects. mp = 176-179°; $[\alpha]_D^{26}$ = +47.5° (c = 3 DMF); λ_m = 242 nm (ε 15500 MeOH). *Glaxo Labs.; Schering-Plough HealthCare Products.*

4833 Alclometasone Dipropionate
66734-13-2 221 266-464-3
$C_{28}H_{37}ClO_7$
7α-Chloro-11β,17,21-trihydroxy-16α-methylpregna-1,4-diene-3,20-dione 17,21-dipropionate.
Sch-22219; Aclovate; Vaderm. Anti-inflammatory (topical). mp = 212-216°; $[\alpha]_D^{26}$ = +42.6° (c = 3 DMF); λ_m = 242 nm (ε 15600 MeOH). *Glaxo Labs.; Schering-Plough HealthCare Products.*

4834 Algestone
595-77-7 238 209-869-2
$C_{21}H_{30}O_4$
16α,17-Dihydroxypregn-4-ene-3,20-dione.
alphasone; 16α,17-dihydroxyprogesterone. Anti-inflammatory. mp = 225°; $[\alpha]_D^{22}$ = +95° (c = 0.81 $CHCl_3$); λ_m = 240 nm (ε 16600). *Olin Res. Ctr.; Searle G.D. & Co.*

4835 Algestone Acetonide
4968-09-6 238 225-608-5
$C_{24}H_{34}O_4$
16α,17-Dihydroxypregn-4-ene-3,20-dione cyclic acetal with acetone.
alphasone acetonide. Anti-inflammatory (topical). A progestational steroid. mp = 210°; $[\alpha]_D^{20}$ = 137° (c = 0.7 $CHCl_3$). *Olin Res. Ctr.; Parke-Davis; Searle G.D. & Co.*

4836 Amcinafal
3924-70-7 223-497-8
$C_{28}H_{35}FO_6$
(R)-9-Fluoro-11β,16α,17,21-tetrahydroxypregna-1,4-diene-3,20-dione cyclic 16,17-acetal with 3-pentanone.
SQ-15102. Anti-inflammatory. *Bristol-Myers Squibb Co.*

4837 Amcinafide
7332-27-6
$C_{29}H_{33}FO_6$
(R)-9-Fluoro-11β,16α,17,21-tetrahydroxypregna-1,4-diene-3,20-dione cyclic 16,17-acetal with acetophenone.
SQ-15112. Anti-inflammatory. *Bristol-Myers Squibb Co.*

4838 Ampiroxicam
99464-64-9 629
$C_{20}H_{21}N_3O_7S$
(±)-4-(1-Hydroxyethoxy)-2-methyl-N-2-pyridyl-2H-1,2-benzothiazine-3-carboxamide ethyl carbonate (ester).
CP-65703; Flucam; Nasil. Anti-inflammatory. A thiazinecarboxamide; prodrug of piroxicam. mp = 159-161°; LD_{50} (mrat orl) = 1798 mg/kg, (frat orl) = 747 mg/kg. *Pfizer Inc.*

4839 Beclomethasone
4419-39-0 1047 224-585-9
$C_{22}H_{29}ClO_5$
9-Chloro-11β,17,21-trihydroxy-16β-methylpregna-1,4-diene-3,20-dione.
9α-chloro-16β-methylprednisolone; beclometasone. Inhalant form used as an antiasthmatic. Also has antiallergic and topical anti-inflammatory properties. Glucocorticoid. *Glaxo Wellcome Inc.; Key Pharm.; Merck & Co.Inc.; Schering AG; Schevico.*

4840 Beclomethasone Dipropionate
5534-09-8 1047 226-886-0
$C_{28}H_{37}ClO_7$
9-Chloro-11β,17,21-trihydroxy-16β-methyl-pregna-1,4-diene-3,20-dione 17,21-dipropionate.
Sch-8020W; Aerobec; Aldecin; Anceron; Andion; Beclovent Inhaler; Beclacin; Belcoforte; Belcomet; Belchlorhinol; Becloval; Becodisks; Beconase; Becotide; Clenil-A; Entyderma; Inalone O; Inalone R; Korbutone; Propaderm; Rino-Clenil; Sanasthmax; Sanasthymyl; Vancenase; Vanceril; Viarex; Viarox. Antiallergic; antiasthmatic (inhalant); anti-inflammatory. Glucocorticoid. mp = 117-120°; $[\alpha]_D$ = 98° (c = 1 dioxane); λ_m = 238 nm (ε 15990 EtOH). *Glaxo Labs.; Key Pharm.; Merck & Co.Inc.; Schering-Plough HealthCare Products.*

4841 Budesonide
51333-22-3 1490 257-139-7
$C_{25}H_{34}O_6$
(R,S)-11β,16α,17,21-Tetrahydroxypregna-1,4-diene-3,20-dione cyclic 16,17-acetal with butyraldehyde.
S-1320; Bidien; Budeson; Cortivent; Micronyl; Preferid; Pulmicort; Rhinocort; Spirocort. Anti-inflammatory; antiasthmatic. Non-halogenated glucocorticoid related to triamcinolone hexacetonide with a high ratio of topical to systemic activity. A mixture of two isomers in which the S-isomer varies between 40-51%. mp = 221-232° (dec); $[\alpha]_D^{25}$= 98.9 (c = 0.28 CH_2Cl_2). *Astra Chem. Ltd.; Bofors.*

4842 R-Budesonide
51372-29-3 1490
$C_{25}H_{34}O_6$
11β,16α(R),17,21-Tetrahydroxypregna-1,4-diene-3,20-dione cyclic 16,17-acetal with butyraldehyde.
Anti-inflammatory.

4843 S-Budesonide
51372-28-2 1490
$C_{25}H_{34}O_6$
11β,16α(S),17,21-Tetrahydroxypregna-1,4-diene-3,20-dione cyclic 16,17-acetal with butyraldehyde.
Anti-inflammatory.

4844 Clobetasol
25122-41-2 2423 246-633-8
$C_{22}H_{28}ClFO_4$
(11β,16β)-21-Chloro-9-fluoro-11,17-dihydroxy-16-methylpregna-1,4-diene-3,20-dione.
Anti-inflammatory. Glucocorticoid. *Glaxo Labs.*

4845 Clobetasol Propionate
25122-46-7 2423 246-634-3
$C_{25}H_{32}ClFO_5$
(11β,16β)-21-Chloro-9-fluoro-11,17-dihydroxy-16-methylpregna-1,4-diene-3,20-dione 17-propionate.
CCl-4725; GR-2/925; Clobesol; Dermoval; Dermovate; Dermoxin; Dermoxinale; Temovate. Anti-inflammatory. Glucocorticoid. mp = 195.5-197°; $[\alpha]_D$ = +103.8° (c = 1.04 dioxane); λ_m = 237 nm (ε 15000 EtOH). *Glaxo Labs.*

4846 Clobetasone
54063-32-0 2424 258-953-5
$C_{22}H_{26}ClFO_4$
(16β)-21-Chloro-9-fluoro-17-hydroxy-16-methylpregna-1,4-diene-3,11,20-trione.
21-chloro-11-dehydrobetamethasone. Anti-inflammatory. Glucocorticoid. *Glaxo Labs.*

4847 Clobetasone Butyrate
25122-57-0 2424 246-635-9
$C_{26}H_{32}ClFO_5$
(16β)-21-Chloro-9-fluoro-17-hydroxy-16-methylpregna-1,4-diene-3,11,20-trione 17-butyrate.
CCl-5537; GR-2/1214; Emovate; Eumovate; Molivate. Anti-inflammatory. Glucocorticoid. mp = 90-100°. *Glaxo Labs.*

4848 Cloticasone
87556-66-9
$C_{22}H_{27}ClF_2O_4S$
S-(Chloromethyl) 6α,9-difluro-11β,17-dihydroxy-16α-methyl-3-oxoandrosta-1,4-diene-17β-carbothioate.
Anti-inflammatory. *Glaxo Labs.*

4849 Cloticasone Propionate
80486-69-7
$C_{25}H_{31}ClF_2O_5S$
S-(Chloromethyl) 6α,9-difluro-11β,17-dihydroxy-16α-methyl-3-oxoandrosta-1,4-diene-17β-carbothioate 17-propionate.
CCl-18773. Anti-inflammatory. *Glaxo Labs.*

4850 Cormethasone
35135-68-3
$C_{22}H_{25}F_3O_4$
6,6,9-Trifluoro-11β,17,21-trihydroxy-16α-methylpregna-1,4-diene-3,20-dione.
Anti-inflammatory (topical). *Syntex Labs. Inc.*

4851 Cormethasone Acetate
35135-67-2
$C_{24}H_{29}F_3O_6$
6,6,9-Trifluoro-11β,17,21-trihydroxy-16α-methylpregna-1,4-diene-3,20-dione 21-acetone.
RS-3694R. Anti-inflammatory (topical). *Syntex Labs. Inc.*

4852 Cortodoxone
152-58-9 205-805-2
$C_{21}H_{30}O_4$
17,21-Dihydroxypregn-4-ene-3,20-dione.
SKF-3050; NSC-18317. Anti-inflammatory. *SmithKline Beecham Pharm.*

4853 Deflazacort
14484-47-0 2916 238-483-7
$C_{25}H_{31}NO_6$
(11β,16β)-21-(Acetyloxy)-11-hydroxy-2'-methyl-5'H-pregna-1,4-dieno[16,17-doxazole-3,20-dione.
oxazacort; azacort; DL-458-IT; L-5458; MDL-458; Calcort; Deflan; Dezacor; Flantadin; Lantadin. Anti-

inflammatory; glucocorticoid. Systemic corticosteroid; oxazoline derivative of prednisolone. mp = 255-256.5°; $[\alpha]_D$ = +62.3° (c = 0.5 $CHCl_3$); λ_m = 241-242 nm ($E^{1\%}_{1cm}$ 352.5 MeOH); LD_{50} (mus orl) = 5200 mg/kg. *Gruppo Lepetit S.p.A.; Marion Merrell Dow Inc.*

4854 Desonide
638-94-8 2973 211-351-6
$C_{24}H_{32}NO_6$
(11β,16α)-11,21-Dihydroxy-16,17-[(1-methylethylidene)-bis(oxy)]pregna-1,4-diene-3,20-dione.
prednacinolone; D-2083; Locapred; Sterax; Steroderm; Topifug; Tridesilon. Anti-inflammatory. mp = 274-275°, 263-266°; $[\alpha]^{25}_D$ = +123° (c = 0.5 DMF); λ_m = 242 nm ($E^{1\%}_{1cm}$ 356); LD_{50} (mus sc) = 93 mg/kg. *Am. Cyanamid; Galderma Labs Inc.; Miles Inc.; Squibb E.R. & Sons.*

4855 Dexamethasone
50-02-2 2986 200-003-9
$C_{22}H_{29}FO_5$
9-Fluoro-11β,17,21-trihydroxy-16α-methylpregna-1,4-diene-3,20-dione.
hexadecadrol; Aeroseb-Dex; Corson; Cortisumman; Decacort; Decaderm; Decadron; Decalix; Decasone; Dekacort; Deltafluorene; Deronil; Deseronil; Dexacortal; Dexacortin; Dexafarma; Dexa-Mamallet; Dexameth; Dexamonozon; Dexapos; Dexa-sine; Dexasone; Dexinoral; Dinormon; Fluormone; Isopto-Dex; Lokalison F; Loverine; Luxazone; Maxidex; Millicorten; Pet-Derm III; component of: Azimycin, Azium, Deronil, Dexacidin, Fulvidex, Maxitrol, Naquasone, Tobradex, Tresaderm. Anti-inflammatory; glucocorticoid; diagnostic aid (Cushing's Syndrome, depression). mp = 262-264°, 268-271°; $[\alpha]^{25}_D = {}_7.5°$ (in dioxane); somewhat soluble in H_2O (0.1 mg/ml 25°); soluble in Me_2CO, EtOH, $CHCl_3$. *Alcon Labs; Herbert; Iolab; Merck & Co.Inc.; Organon Inc.; Schering-Plough HealthCare Products; Solvay Pharm. Inc.; Upjohn Ltd.*

4856 Dexamethasone 21-Acetate
1177-87-3 2986 214-646-8
$C_{24}H_{31}FO_6$
9-Fluoro-11β,17,21-trihydroxy-16α-methyl-pregna-1,4-diene-3,20-dione
21-acetate.
Dalalone DP; Dalalone LA; Decadronal; Decadron-LA; Dectancyl; Dexacortisyl. Anti-inflammatory; glucocorticoid; diagnostic aid (Cushing's Syndrome, depression). mp = 215-221°, 229-231°, 238-240°; $[\alpha]^{25}_D$ = +73° ($CHCl_3$); λ_m = 239 nm (ε 14900). *Forest Pharm. Inc.; Merck & Co.Inc.*

4857 Dexamethasone 21-Phosphate Disodium Salt
2392-39-4 2986 219-243-0
$C_{22}H_{28}FNa_2O_8P$
9-Fluoro-11β,17,21-trihydroxy-16α-methyl-pregna-1,4-diene-3,20-dione 21-phosphate disodium salt.
dexamethasone sodium; dexamethasone 21-phosphate disodium salt; dexamethasone 21-(dihydrogen phosphate) disodium salt; Ak-Dex; Baldex; Colvasone; Dalalone; Decadron; Dexabene; Dezone; Fortecortin; Hexadrol; Maxidex Ointment; Oradexon; Orgadrone; Solu-Deca; component of: NeoDecadron. Anti-inflammatory; glucocorticoid (antiasthmatic); diagnostic aid (Cushing's Syndrome, depression). Used in diagnosis of Cushing's syndrome and depression. mp = 233-235°; $[\alpha]_D$ = 257° (H_2O); $[\alpha]^{25}_D$ = 74° ± 4° (c = 1); λ_m = 238-239 nm (ε 14000); soluble in H_2O. *Alcon Labs; Elkins-Sinn; Forest Pharm. Inc.; Merck & Co.Inc.; Organon Inc.*

4858 Dexamethasone Acefurate
83880-70-0 2986
$C_{29}H_{33}FO_8$
9-Fluoro-11β,17,21-trihydroxy-16α-methyl-pregna-1,4-diene-3,20-dione 17-(2-furoate).
Sch-31353. Anti-inflammatory; glucocorticoid; diagnostic aid (Cushing's Syndrome, depression). *Schering-Plough HealthCare Products.*

4859 Dexamethasone Acetate Monohydrate
55821-90-3 2986
$C_{24}H_{33}FO_7$
9-Fluoro-11β,17,21-trihydroxy-16α-methyl-pregna-1,4-diene-3,20-dione 21-acetate monohydrate.
Dalalone DP; Dalalone LA; Decadron-LA. Anti-inflammatory; glucocorticoid; diagnostic aid (Cushing's Syndrome, depression). *Forest Pharm. Inc.; Merck & Co.Inc.*

4860 Dexamethasone Diethylaminoacetate
2986
$C_{28}H_{41}FNO_6$
9-Fluoro-11β,17,21-trihydroxy-16α-methyl-pregna-1,4-diene-3,20-dione
21-diethylaminoacetate.
Solu-Forte-Cortin. Anti-inflammatory; glucocorticoid; diagnostic aid (Cushing's Syndrome, depression). *Merck & Co.Inc.*

4861 Dexamethasone Dimethylbutyrate
2986
$C_{28}H_{39}FO_6$
9-Fluoro-11β,17,21-trihydroxy-16α-methyl-pregna-1,4-diene-3,20-dione
21-(3,3-dimethylbutyrate).
dexamethasone tert-butylacetate; Decadron TBA. Anti-inflammatory; glucocorticoid; diagnostic aid (Cushing's Syndrome, depression). *Merck & Co.Inc.*

4862 Dexamethasone Dipropionate
55541-30-5 2986 259-699-8
$C_{28}H_{37}FO_7$
9-Fluoro-11β,17,21-trihydroxy-16α-methyl-pregna-1,4-diene-3,20-dione
17,21-dipropionate.
ST12; THS-101; Methaderm. Anti-inflammatory; glucocorticoid; diagnostic aid (Cushing's Syndrome, depression). *Innothera.*

4863 Dexamethasone Isonicotinate
2265-64-7 2986 218-866-5
$C_{28}H_{32}FNO_6$
9-Fluoro-11β,17,21-trihydroxy-16α-methyl-pregna-1,4-diene-3,20-dione
21-(4-pyridinecarboxylate).
dexamethasone 21-isonicotinate; Auxiloson; Ausixone; Voren. Anti-inflammatory; glucocorticoid; diagnostic aid (Cushing's Syndrome, depression). mp = 250-252°; $[\alpha]^{27}_D$ = +183.5° (in dioxane). *Merck & Co.Inc.*

4864 Dexamethasone Palmitate
14899-36-6 2986
$C_{38}H_{37}FO_7$
9-Fluoro-11β,17,21-trihydroxy-16α-methylpregna-1,4-diene-3,20-dione 21-palmitate.
Limethasone. Anti-inflammatory; glucocorticoid; diagnostic aid (Cushing's Syndrome, depression). *Merck & Co.Inc.*

4865 Diflorasone
2557-49-5 3186 219-875-7
$C_{22}H_{28}F_2O_5$
(6α,11β,16β)-6,9-Difluoro-11,17,21-trihydroxy-16-methylpregna-1,4-diene-3,20-dione.
6α,9α-difluoro-16β-methylprednisolone. Anti-inflammatory (topical); glucocorticoid. The 16β-analog of flumethasone, a glucocorticoid. *Merck & Co.Inc.; Pfizer Inc.; Upjohn Ltd.*

4866 Diflorasone Diacetate
33564-31-7 3186 251-575-1
$C_{26}H_{32}F_2O_7$
(6α,11β,16β)-17,21-Bis(acetyloxy)-6,9-difluoro-11-hydroxy-16-methyl-pregna-1,4-diene-3,20-dione.
U-24865; Dermaflor; Diacort; Difulal; Florone; Maxiflor; Psorcon; Soriflor. Anti-inflammatory (topical); glucocorticoid. mp = 221-223 (dec); $[\alpha]_D$ = +61° (in $CHCl_3$); λ_m = 238 nm (ε 17250). *ABIC; Dermik Labs. Inc.; Merck & Co.Inc.; Pfizer Inc.; Upjohn Ltd.*

4867 Diflucortolone
2607-06-9 3189
$C_{22}H_{28}F_2O_4$
(6α,11β,16α)-6,9-Difluoro-11,21-dihydroxy-16-methylpregna-1,4-diene-3,20-dione.
Anti-inflammatory. Glucocorticoid. The 9α-fluoro derivative of fluocortolone. mp = 240-244°, 248-249°; $[\alpha]_D^{22}$ = +111° (in MeOH); λ_m = 237 nm (ε 16600). *Schering-Plough HealthCare Products.*

4868 Diflucortolone Pivalate
15845-96-2 3189
$C_{27}H_{36}F_2O_5$
(6α,11β,16α)-6,9-Difluoro-11,21-dihydroxy-16-methylpregna-1,4-diene-3,20-dione 21-pivalate.
SH-968; Neribas; Neriforte; Nerisona; Nerisone; Temetex; Texmeten. Anti-inflammatory. Glucocorticoid. mp = 195-195.5°; $[\alpha]_D^{22}$ = +100.8° (in dioxane); LD_{50} (mus orl) > 4000 mg/kg, (rat orl) = 3100 mg/kg, (mus sc) = 180 mg/kg, (rat sc) = 13 mg/kg, (mus ip) = 450 mg/kg, (rat ip) = 98 mg/kg. *Schering-Plough HealthCare Products.*

4869 Difluprednate
23674-86-4 3194 245-815-4
$C_{27}H_{34}F_2O_7$
(6α,11β)-21-(Acetyloxy)-6,9-difluoro-11-hydroxy-17-(1-oxobutoxy)pregna-1,4-diene-3,20-dione.
CM-9155; W-6309; Epitopic; Myser. Anti-inflammatory. mp = 191-194°; $[\alpha]_D^{22}$ = +31.7° (c = 0.5 dioxane); λ_m = 237-238 nm ($E_{1cm}^{1\%}$ = 320). *CK Witco Chem. Corp.*

4870 Drocinonide
36637-22-6 219-093-6
$C_{24}H_{35}FO_6$
9-Fluoro-11β,16α,17,21-tetrahydroxy-5α-pregnane-3,20-dione cyclic 16,17-acetal with acetone.
Anti-inflammatory. *Bristol-Myers Squibb Co.*

4871 Endrysone
35100-44-8 252-362-6
$C_{22}H_{30}O_3$
(6α,11β)-11-Hydroxy-6-methylpregna-1,4-diene-3,20-dione.
endrisone. Anti-inflammatory (topical, ophthalmic). *Lark S.p.A.*

4872 Enoxolone
471-53-4 3628 207-444-6
$C_{30}H_{46}O_4$
(3β,20β)-3-Hydroxy-11-oxoolean-12-en-29-oic acid.
glycyrrhetic acid; 18β-glycyrrhetinic acid; uralenic acid; Arthrodont; Biosone; P.O. 12. Anti-inflammatory (topical). mp = 296°; $[\alpha]_D^{21}$ = +86° (in EtOH); soluble in $CHCl_3$, dioxane, C_5H_5N, AcOH; nearly insoluble in petroleum ether. *Farmitalia Carlo Erba Ltd.*

4873 Fluazacort
19888-56-3 4151 243-400-2
$C_{25}H_{30}FNO_6$
(11β,16β)-21-(Acetyloxy)-9-fluoro-11-hydroxy-2'-methyl-5'H-pregna-1,4-dieno[17,16-d]oxazole-3,20-dione.
L-6400; Azacortid. Anti-inflammatory. Neuroleptic and antipsychotic. mp = 252-255°; $[\alpha]_D$ = +54.8 (c = 0.5 $CHCl_3$); λ_m = 238-240 nm ($E_{1cm}^{1\%}$ 330 MeOH). *Gruppo Lepetit S.p.A.; Marion Merrell Dow Inc.*

4874 Flumethasone
2135-17-3 4173 218-370-9
$C_{22}H_{28}F_2O_5$
(6α,11β,16α)-6,9-Difluoro-11,17,21-trihydroxy-16-methylpregna-1,4-diene-3,20-dione.
flumetasone; 6α-fluorodexamethazone; U-10974; NSC-5402; Aniprome; Cortexilar; Flucort; Methagon. Glucocorticoid; anti-inflammatory. *Upjohn Ltd.*

4875 Flumethasone Acetate
4173
$C_{24}H_{30}F_2O_6$
(6α,11β,16α)-6,9-Difluoro-11,17,21-trihydroxy-16-methylpregna-1,4-diene-3,20-dione 21-acetate.
flumetasone 1-acetate. Glucocorticoid; anti-inflammatory. mp = 260-264°; $[\alpha]_D$ = 91° (in EtOH); λ_m = 237 nm (log ε 4.16). *Upjohn Ltd.*

4876 Flumethasone Pivalate
2002-29-1 4173 218-370-9
$C_{27}H_{36}F_2O_6$
(6α,11β,16α)-6,9-Difluoro-11,17,21-trihydroxy-16-methylpregna-1,4-diene-3,20-dione 21-pivalate.
flumetasone 21-pivalate; NSC-107680; Locacorten (obsolete); Locorten; Lorinden; Losalen. Glucocorticoid; anti-inflammatory. *Upjohn Ltd.*

4877 Fluocinolone Acetonide
67-73-2 4185 200-668-5
$C_{24}H_{30}72O_6$
(6α,11β,16α)-6,9-Difluoro-11,21-dihydroxy-16,17-[(1-methylethylidene)bis(oxy)]pregna-1,4-diene-3,20-dione.
NSC-92339; Coriphate; Cortiplastol; Dermalar; Fluonid; Fluovitef; Fluvean; Fluzon; Jellin; Localyn; Synalar; Synamol; Synandone; Synemol; Synotic; Synsac; component of: N-Synalar. Glucocoriticoid; anti-inflammatory. mp = 265-266°; $[\alpha]_D$ = +95° (in $CHCl_3$); λ_m = 238 nm (log ε 4.21). *Lemmon Co.; Syntex Labs. Inc.*

4878 Fluocinonide
356-12-7 4186 206-597-6
$C_{26}H_{32}F_2O_7$
(6α,11β,16α)-21-(Acetyloxy)-6,9-difluoro-11-hydroxy-16,17-[(1-methylethylidine)bis(oxy)]pregna-1,4-diene-3,20-dione cyclic 16,17-acetal with acetone.
fluocinolide (obsolete); fluocinolide acetate (obsolete); fluocinolone acetonide acetate; NSC-101791; Biscosal; Dermaplus; Lidex; Metosyn; Synalate (obsolete); Straderm; Topsym; Topsymin; Topsyne; Topsyn. Anti-inflammatory; glucocorticoid. mp = 308-311°; $[\alpha]_D$= 83° (in $CHCl_3$); λ_m = 237 nm (log ε 4.18). *Olin Res. Ctr.; Pharm. Res. Products; Syntex Labs. Inc.*

4879 Fluocortin
33124-50-4 4187 251-383-8
$C_{22}H_{27}FO_5$
(6α,11β,16α)-6-Fluoro-11-hydroxy-16-methyl-3,20-dioxopregna-1,4-dien-21-oic acid.
Anti-inflammatory. *Schering AG.*

4880 Fluocortin Butyl
41767-29-7 4187 255-543-8
$C_{26}H_{35}FO_5$
(6α,11β,16α)-6-Fluoro-11-hydroxy-16-methyl-3,20-dioxopregna-1,4-dien-21-oic acid butyl ester.
SH K 203; Varlane; Vaspit. Anti-inflammatory. The butyl ester derivative of fluocortolone-21-acid, a metabolite of fluocortolone, a glucocorticoid. mp = 191.5°; $[\alpha]_D^{25}$ = +136° (c = 0.5 $CHCl_3$); λ_m = 242 nm (ε 16800 in MeOH); soluble in $CHCl_3$, EtOH; nearly insoluble in ethyl ether; LD_{50} (mus,rat orl,sc) > 4000 mg/kg. *Schering AG.*

4881 Fluorometholone
426-13-1 4213 207-041-5
$C_{22}H_{29}FO_4$
(6α,11β)-9-Fluoro-11,17-dihydroxy-6-methylpregna-1,4-diene-3,20-dione.
fluormetholon; Cortilet; Delmeson; Efflumidex; Fluaton; Flumetholon; Fluor-Op; FML; FML Forte; FML Liquifilm; FML S.O.P.; Loticort; Oxylone; Ursnon; component of: FML-S Liquifilm, Neo-Oxylone. Glucocorticoid; anti-inflammatory. mp = 292-303°. *Allergan Inc.; Iolab; Upjohn Ltd.*

4882 Fluorometholone Acetate
3801-06-7 4213 223-270-3
$C_{24}H_{31}FO_5$
(6α,11β)-9-Fluoro-11,17-dihydroxy-6-methylpregna-1,4-diene-3,20-dione 17-acetate.
U-17323; Flarex; component of: Tobrasone. Glucocorticoid; anti-inflammatory. mp = 230-232°; $[\alpha]_D$ = +28° (in $CHCl_3$). *Am. Cyanamid; Upjohn Ltd.*

4883 Flurandrenolide
1524-88-5 4232 216-196-8
$C_{24}H_{33}F_2O_6$
(6α,11β,16α)-6-Fluoro-11,21-dihydroxy-16,17-[(1-methylethylidene)bis(oxy)]pregn-4-ene-3,20-dione.
fluorandrenolone; flurandrenolone; flurandrenolone acetate (obsolete); fludroxycortide; Cordran; Drenison; Drocort; Haelan; Sermaka. Glucocorticoid; anti-inflammatory. mp = 247-255°; $[\alpha]_D^{25}$ = +140-150° (in $CHCl_3$); λ_m = 236 nm (log ε 4.17). *Eli Lilly & Co.; Syntex Labs. Inc.*

4884 Fluticasone
90566-53-3 4244
$C_{25}H_{31}F_3O_5S$
S-Fluoromethyl 6α,9-difluoro-11β,17-dihydroxy-16α-methyl-3-oxoandrosta-1,4-diene-17β-carbothioic acid.
Antiallergic; anti-inflammatory. Derivative of flumethasone. *Glaxo Labs.*

4885 Fluticasone Propionate
80474-14-2 4244
$C_{22}H_{27}F_3O_4S$
S-Fluoromethyl 6α,9-difluoro-11β,17-dihydroxy-16α-methyl-3-oxoandrosta-1,4-diene-17β-carbothioate 17-propionate.
CCI-18781; Cutivate; Flixonase; Flixotide; Flovent; Flunase. Antiallergic; anti-inflammatory. mp = 272-273° (dec); $[\alpha]_D$ = +30° (c = 0.35). *Glaxo Labs.*

4886 Halcinonide
3093-35-4 4621 221-439-6
$C_{24}H_{32}ClFO_5$
(11β,16α)-21-Chloro-9-fluoro-11-hydroxy-16,17-[(1-methylethylidene)bis(oxy)]pregn-4-ene-3,20-dione.
SQ-18566; Halciderm; Halcimat; Halog. Anti-inflammatory (topical). mp = 264-265° (dec); $[\alpha]_D^{25}$ = +155° (in $CHCl_3$); λ_m = 238 nm (ε 16400 MeOH); soluble in Me_2CO; $CHCl_3$, DMSO; slightly soluble in C_6H_6, EtOH, Et_2O, MeOH; insoluble in H_2O, 0.1M HCl, 0.1M NaOH; LD_{50} (mus ip) = 150 mg/kg. *Bristol-Myers Squibb Co.*

4887 Halobetasol
98651-66-2 4625
$C_{22}H_{27}ClF_2O_4$
21-Chloro-6α,9-difluoro-11β,17-dihydroxy-16β-methylpregna-1,4-diene-3,20-dione.
ulobetasol. Anti-inflammatory. *Ciba-Geigy Corp.*

4888 Halobetasol Propionate
66852-54-8 4625
$C_{25}H_{31}ClF_2O_5$
(6α,11β,16β)-21-Chloro-6,9-difluoro-11-hydroxy-16-methy-17-(1-oxopropoxy)pregna-1,4-diene-3,20-dione.
ulobetasol propionate; BMY-30056; CGP-14458; Ultravate. Anti-inflammatory. Trihalogenated corticosteroid structurally related to clobetasol. mp = 220-221°. *Ciba-Geigy Corp.; Westwood-Squibb Pharm. Inc.*

4889 Halometasone
50629-82-8 4628 256-664-9
$C_{22}H_{27}ClF_2O_5$
(6α,11β,16α)-2-Chloro-6,9-difluoro-11,17,21-trihydroxy-16-methypregna-1,4-diene-3,20-dione.
2-chloroflumethasone; C-48401-Ba; Sicorten [monohydrate]. Anti-inflammatory (topical); antipruritic. Synthetic corticosteroid. mp = 220-222° (dec). *Ciba-Geigy Corp.*

4890 Halopredone
57781-15-4 4630 260-953-5
$C_{21}H_{25}BrF_2O_5$
(6β,11β)-2-Bromo-6,9-difluoro-11,17,21-trihydroxy-pregna-1,4-diene-3,20-dione.
Anti-inflammatory (topical). *Pierrel S.p.A.*

4891 Halopredone Acetate
57781-14-3 4630 260-951-4
$C_{25}H_{29}BrF_2O_7$
(6β,11β)-2-Bromo-6,9-difluoro-11,17,21-trihydroxy-pregna-1,4-diene-3,20-dione 17,21-diacetate.
Haloart; Topicon. Anti-inflammatory (topical). mp = 290-292°; $[\alpha]_D^{24}$ = -36° (c = 1 $CHCl_3$); λ_m = 246 nm (ε 12500 MeOH). *Pierrel S.p.A.*

4892 Isoflupredone
338-95-4 5190 206-422-3
$C_{21}H_{27}FO_5$
(11β)-9-Fluoro-11,17,21-trihydroxypregna-1,4-diene-3,20-dione.
9-fluoroprednisolone. Anti-inflammatory (veterinary). mp = 263-266° (dec), 274-275° (dec); $[\alpha]_D^{23}$ = +108° (c = 0.611 EtOH), $[\alpha]_D^{23}$ = + 94° (in alcohol); λ_m = 240 nm (ε 15800 EtOH). *CIBA plc; Schering-Plough HealthCare Products.*

4893 Isoflupredone Acetate
338-98-7 5190 206-423-9
$C_{23}H_{29}FO_6$
(11β)-9-Fluoro-11,17,21-trihydroxypregna-1,4-diene-3,20-dione 21-acetate.
U-6013; Predef. Anti-inflammatory (veterinary). mp = 244-246° (dec); $[\alpha]_D^{23}$ = +108° (c = 0.735 dioxane); λ_m = 240 nm (ε 16250 EtOH). *CIBA plc; Schering-Plough HealthCare Products; Upjohn Ltd.*

4894 Loteprednol
129260-79-3 5614
$C_{21}H_{27}ClO_4$
(11β,17α)-11-Hydroxy-3-oxoandrosta-1,4-diene-17-carboxyl.
Anti-inflammatory (topical). *Otsuka Am. Pharm.*

4895 Loteprednol Etabonate
82034-46-6 5614
$C_{24}H_{31}ClO_7$
(11β,17α)-17-[(Ethyoxycarbonyl)oxy]-11-hydroxy-3-oxoandrosta-1,4-diene-17-carboxylic acid chloromethyl ester.
CDDD-5604; HGP-1; P-5604; Lenoxin. Anti-inflammatory (topical). Ophthalmic corticosteroid. mp = 220.5-223.5°; very slightly soluble in H_2O (0.00005 g/100 ml at 25°); more soluble in 50% propylene glycol and H_2O (0.0037 g/100 ml); lipophilicity (log K) = 3.04. *Otsuka Am. Pharm. ; Xenon Vision.*

4896 Mazipredone
13085-08-0 5802
$C_{26}H_{38}N_2O_4$
(11β)-11,17-Dihydroxy-21-(4-methyl-1-piperazinyl)pregna-1,4-diene-3,20-dione.
Anti-inflammatory. Dec 199°. *Pfizer Inc.*

4897 Mazipredone Hydrochloride
60-39-3 5802 200-475-6
$C_{26}H_{39}ClN_2O_4$
(11β)-11,17-Dihydroxy-21-(4-methyl-1-piperazinyl)pregna-1,4-diene-3,20-dione monohydrochloride.
Depresolone. Anti-inflammatory. Dec 246°; soluble in H_2O. *Pfizer Inc.*

4898 Meclorisone
4732-48-3
$C_{22}H_{28}Cl_2O_2$
9,11β-Dichloro-17,21-dihydroxy-16α-methylpregna-1,4-diene-3,20-dione.
Anti-inflammatory (topical). *Schering-Plough HealthCare Products.*

4899 Meclorisone Dibutyrate
10549-91-4 234-132-7
$C_{30}H_{40}Cl_2O_6$
9,11β-Dichloro-17,21-dihydroxy-16α-methylpregna-1,4-diene-3,20-dione dibutyrate.
Sch-11572. Anti-inflammatory (topical). *Schering-Plough HealthCare Products.*

4900 Methylprednisolone Suleptanate
90350-40-6
$C_{33}H_{48}NNaO_{10}S$
(6α,11β)-11,17-Dihydroxy-6-methyl-21-[[8-[methyl(2-sulfoethyl)amino]-1,8-dioxooctyl]oxy]pregna-1,4-diene-3,20-dione monosodium salt.
Medrosol. Anti-inflammatory; glucocorticoid. *Upjohn Ltd.*

4901 Mometasone Furoate
83919-23-7 6324
$C_{27}H_{30}Cl_2O_6$
(11β,16α)-9,21-Dichloro-17-[(2-furanylcarbonyl)oxy]-11-hydroxy-16-methylpregna-1,4-diene-3,20-dione.
Sch-32088; Elocon. Anti-inflammatory (topical). Topical corticosteroid. mp = 218-220°; $[\alpha]_D^{26}$ = + 58.3° (in dioxane); λ_m = 247 nm (ε 26300 in MeOH). *Schering-Plough HealthCare Products.*

4902 Nicocortonide
65415-41-0 265-754-7
$C_{31}H_{37}NO_7$
(11β)-14,17-[2-Butenylidenebis(oxy)]-11-hydroxy-21-[(4-pyridinylcarbonyl)oxy]-pregn-4-ene-3,20-dione.
Anti-inflammatory (steroidal).

4903 Prednazate
5714-75-0
$C_{46}H_{58}ClN_3O_9S$
11β,17,21-Trihydroxypregna-1,4-diene-3,20-dione 21-(hydrogen succinate) compound with 4-[3-(2-chlorophenothiazin-10-yl)propyl]-1-piperazineethanol (1:1).
prednisolone compound with perphenazine; Sch-6620.

Anti-inflammatory. Corticosteroid combined with an antipsychotic. *Schering-Plough HealthCare Products.*

4904 Prednazoline
6693-90-9
$C_{35}H_{47}N_2O_9P$
11β,17,21-Trihydroxypregna-1,4-diene-3,20-dione 21-(di-hydrogen phosphate) compound with 2-[(2-isopropylphenoxy)methyl]-2-imidazoline.
prednisolone phosphate compound with fenoxazoline. Anti-inflammatory. Corticosteroid combined with a sympathomimetic.

4905 Resocortol
76675-97-3
$C_{22}H_{36}O_4$
11β,17α-Dihydroxy-17-propionylandrost-4-en-3-one.
Anti-inflammatory.

4906 Resocortol Butyrate
76738-96-0
$C_{26}H_{38}O_5$
11β,17α-Dihydroxy-17-propionylandrost-4-en-3-one 17-butyrate.
Anti-inflammatory.

4907 Rimexolone
49697-38-3 8392
$C_{24}H_{34}O_3$
(11β,16α,17β)-11-Hydroxy-16,17dimethyl-17-(1-oxopropylandrosta)-1,4-diene-3,20-dione.
trimexolone; Org-6216; Rimexel; Vexol. Anti-inflammatory (local). mp = 258-268°; $[\alpha]_D$ = + 100° (c = 0.92 C_5H_5N); λ_m = 244 nm (ε 14600). *Akzo Chemie ; Organon Inc.*

4908 Tixocortol
61951-99-3 9623
$C_{24}H_{30}O_4S$
(11β)-11,17-Dihydroxy-21-mercaptopregn-4-ene-3,20-dione.
Anti-inflammatory. Dec 220-221°; λ_m = 241 nm (ε16500 95% EtOH). *Jouveinal.*

4909 Tixocortol Pivalate
55560-96-8 9623 259-706-4
$C_{26}H_{38}O_5S$
(11β)-21-[(2,2-Dimethyl-1-oxopropyl)thio]-11,17-dihydroxypregn-4-ene-3,20-dione.
JO-1016; Pivalone; Rectovalone; Tiovalon. Anti-inflammatory. mp = 195-200°; $[\alpha]_D^{20}$ = +145° (c = 1 dioxane); λ_m = 229 nm (log ε 4.259 MeOH). *Jouveinal.*

4910 Triclonide
26849-57-0
$C_{24}H_{28}Cl_3FO_4$
(6α,11β,16α)-9,11,21-Trichloro-6-fluoro-16,17-[(1-methylethylidene)bis(oxy)]pregna-1,4-diene-3,20-dione.
RS-4464. Anti-inflammatory. *Syntex Labs. Inc.*

Antimalarials

4911 Acedapsone
77-46-3 20 201-028-8
$C_{16}H_{16}N_2O_4S$
N,N'-(Sulfonyldi-4,1-phenylene)bisacetamide.
Hansolar; CI-556; CN-1883; DADDS; PAM-MR-1165; Rodilone; diacetyldapsone; sulfadiamine; 1399F. Leprostatic; antimalarial. Sulfone antibiotic. mp = 289-292°; λ_m 256 284 nm (ε 25500 36200 MeOH); soluble in H_2O (0.0003 g/100 ml). *Parke-Davis.*

4912 Amodiaquine
86-42-0 609 201-669-3
$C_{20}H_{22}ClN_3O$
4-[(7-Chloro-4-quinolyl)amino]-α-(diethylamino)-o-cresol.
SN-10751. Antimalarial; antiprotozoal against Toxoplasma. mp = 208° (dec). *Parke-Davis.*

4913 Amodiaquine Dihydrochloride Dihydrate
6398-98-7 609
$C_{20}H_{24}Cl_3N_3O.2H_2O$
4-[(7-Chloro-4-quinolyl)amino]-α-(diethylamino)-o-cresol dihydrochloride dihydrate.
CAM-AQ1; Camoquin; Flavoquine; Miaquin. Antimalarial; antiprotozoal. Dec 150-160°; λ_m = 342 nm ($E_{1\ cm}^{1\%}$ 349, MeOH), 341.5 nm ($E_{1\ cm}^{1\%}$ = 389 0.1N HCl), 342 nm ($E_{1\ cm}^{1\%}$ 396 0.1N NaOH); soluble in H_2O; sparingly soluble in EtOH; insoluble in Et_2O, $CHCl_3$, C_6H_6. *Parke-Davis.*

4914 Amopyroquine
550-81-2
$C_{20}H_{20}ClN_5$
4-[(7-Chloro-4-quinolyl)amino]-α-1-pyrrolidinyl-o-cresol.
Antimalarial agent.

4915 Amquinate
17230-85-2
$C_{18}H_{24}N_2O_3$
Methyl 7-(diethylamino)-4-hydroxy-6-propyl-3-quinolinecarboxylate.
Antimalarial.

4916 Arteether
75887-54-6 853
$C_{17}H_{28}O_5$
[3R-(3α,5aβ,6β,8aβ,9α,10α,12β,12aR*)]-10-Ethoxydecahydro-3,6,9-trimethyl-3,12-epoxy-12H-pyrano[4,3-j]-1,2-benzodioxepin.
dihydroartemisinin ethyl ether; SM-227; dihydro-qinghaosu ethyl ether. Antimalarial. mp = 80-82°; $[\alpha]_D^{21}$ = 154.5° (c = 1.0 $CHCl_3$).

4917 Arteflene
123407-36-3
$C_{19}H_{18}F_6O_3$
(1S,4R,5R,8S)-4-[(Z)-2,4-Bis(trifluoromethyl)styryl]-4,8-dimethyl-2,3-dioxabicyclo[3.3.1]-nonan-7-one.
Ro-42-1611. Antimalarial. *Hoffmann-LaRoche Inc.*

4918 Artemether
71963-77-4 854
$C_{16}H_{26}O_5$
[3R-(3α,5aβ,6β,8aβ,9α,10α,12β,12aR*)]-Decahydro-10-methoxy-3,6,9-trimethyl-3,12-epoxy-12H-pyrano[4,3-j]-1,2-dibenzidioxepin.
dihydroartemisinin methyl ether; SM-224. Antimalarial. mp = 86-88°; $[\alpha]_D^{19.5}$ = 171° (c = 2.59 $CHCl_3$); LD_{50} (mus im) = 263 mg/kg.

4919 Artemisinin
63968-64-9 856
$C_{15}H_{22}O_5$
(3R,5aS,6R,8aS,9R,12S,12aR)-Octahydro-3,6,9-trimethyl-3,12-epoxy-12H-pyrano[4,3-j]-1,2-benzodiozepin-10(3H)-one.
artemisine; arteannuin; huanghuahaosu; QHS; qinghaosu; qing hau sau. Antimalarial. mp = 156-157°; $[\alpha]_D^{17}$ = 66.3° (c = 1.64 $CHCl_3$); soluble in aprotic solvents; LD_{50} (mus orl) = 5105 mg/kg, 4228 mg/kg, (mus im) = 2800 mg/kg, 3840 mg/kg, (mus ip) = 1558 mg/kg, (rat orl) = 5576 mg/kg, (rat im) = 2571 mg/kg.

4920 Artesunate
88495-63-0 857
$C_{19}H_{28}O_8$
(3R,5aS,6R,8aS,9R,10S,12R,12aR)-Decahydro-3,6,9-trimethyl-3,12-epoxy-12H-pyrano[4,3-j]-1,2-benzodiozepin-10-ol hydrogen succinate.
artesunic acid; dihydroqinghasu hemsuccinate. Antimalarial. [sodium salt (SM-804)]: LD_{50} (mus iv) = 520 mg/kg, (mus im) = 475 mg/kg.

4921 Atovaquone
95233-18-4 898
$C_{22}H_{19}ClO_3$
2-[trans-4-(p-Chlorophenyl)cyclohexyl]-3-hydroxy-1,4-naphthoquinone.
Mepron; 566C80; 566C; BW-566C; BW-556C-80; Acuvel; Wellvone. Antimalarial; antipneumocystic. Antiprotozoal against Toxoplasma. mp = 216-219°. *Glaxo Wellcome Inc.*

4922 Bebeerine
477-60-1 1042
$C_{36}H_{38}N_2O_6$
1'-α-6,6'-Dimethoxy-2,2'-dimethyltubocuraran-7',12'-diol.
d-bebeerine; chondodendrine; pelosine. Antimalarial. mp = 215°; $[\alpha]_D^{20}$ = 345.7° (c = 0.4 1N HCl); soluble in C_6H_6, $CHCl_3$, C_5H_5N; [hydrochloride]: mp = 260°; $[\alpha]_D^{20}$ = 294° (c = 0.7); soluble in H_2O, EtOH.

4923 Berberine
2086-83-1 1192 218-229-1
$[C_{20}H_{18}NO_4]^+$
5,6-Dihydro-9,10-dimethoxybenzo[g]-1,3-benzodioxolo[5,6-a]quinolizinium.
umbellatine. Antipyretic; antimalarial; antibacterial. Bitter stomachic. Alkaloid isolated from *Hydrastis candensis L., Berberidaceae*; also found in other plants. mp = 145°; λ_m = 265, 343 nm; [sulfate trihydrate]: LD_{50} (mus ip) = 24.3 mg/kg.

4924 Chirata
2104
Chiretta; chirayita; bitter stick; East Indian Balmony. Extract from the dired plant *Swertia (Ophelia) chirata.* Antimalarial.

4925 Chlorguanide
500-92-5 2138 207-915-6
$C_{11}H_{16}ClN_5$
1-(p-Chlorophenyl)-5-isopropylbiguanide.
chloroguanide; proguanil. Antimalarial. mp = 129°. *Rhône-Poulenc Rorer Pharm. Inc.*

4926 Chlorguanide Hydrochloride
637-32-1 2138 211-283-7
$C_{11}H_{17}Cl_2N_5$
1-(p-cClorophenyl)-5-isopropylbiguanide hydrochloride.
Paludrine; M-4888; RP-3359; SN-12837; Diguanyl; Drinupal; Guanatol; Palusil; Tirian. Antimalarial. mp = 243-244°; λ_m = 259 nm (EtOH); soluble in EtOH; slightly soluble in H_2O; insoluble in $CHCl_3$, Et_2O; LD_{50} (rat orl) = 200 mg/kg. *Eli Lilly & Co.; Zeneca Pharm.*

4927 Chloroquine
54-05-7 2215 200-191-2
$C_{18}H_{26}ClN_3$
7-Chloro-4-[[4-(diethylamino)-1-methylbutyl]amino]-quinoline.
Aralen; SN-7618; RP-3377; Artrichin; Bemaphate; Capquin; Nivaquine B; Resoquine; Reumachlor; Sanoquin; nivaquine [as sulfate]. Antiamebic; antimalarial; antirheumatic. Also used as a lupus erythematosus suppressant. mp = 87°. *Sterling Winthrop Inc.*

4928 Chloroquine Dihydrochloride
3545-67-3 2215 222-592-1
$C_{18}H_{28}Cl_3N_3$
7-Chloro-4-[[4-(diethylamino)-1-methylbutyl]amino]-quinoline dihydrochloride.
Aralen hydrochloride. Antiamebic; antimalarial. Also an antirheumatic agent. *Sterling Winthrop Inc.*

4929 Chloroquine Diphosphate
50-63-5 2215 200-055-2
$C_{18}H_{32}ClN_3P_2O_8$
7-Chloro-4-[[4-(diethylamino)-1-methylbutyl]amino]-quinoline diphosphate.
Aralen phosphate; Arechin; Avloclor; Imagon; Malaquin; Resochin; Tresochin. Antiamebic; antimalarial; antirheumatic; lupus erythematosus suppressant. mp = 193-195°, 215-218°; soluble in H_2O, insoluble in organic solvents. *Sterling Winthrop Inc.*

4930 Chlorproguanil
537-21-3 2237 208-660-3
$C_{11}H_{15}Cl_2N_5$
1-(3,4-Dichlorophenyl)-5-isopropylbiguanide.
M-5943. Antimalarial. *ICI.*

4931 Chlorproguanil Hydrochloride
15537-76-5 2237
$C_{11}H_{16}Cl_3N_5$
1-(3,4-Dichlorophenyl)-5-isopropylbiguanide hydrochloride.

Lapudrine. Antimalarial. mp = 246-247°; soluble in H_2O (1 g/100 ml). *ICI.*

4932 Cinchona
2343
Calisaya bark; Peruvian bark; Cinchona bark; Jesuit's bark. Antimalarial. Dried bark of cinchocha species. Consists of about 35% alkaloids.

4933 Cinchonidine
485-71-2 2345 207-622-3
$C_{19}H_{22}N_2O$
(8α,9R)-Cinchonan-9-ol.
cinchovatine; α-quinidine. Antimalarial. mp = 210°; $[\alpha]_D^{20}$ = -109.2° (EtOH); soluble in EtOH, $CHCl_3$; moderately soluble in Et_2O; LD_{50} (rat ip) = 206 mg/kg.

4934 Cinchonine
118-10-5 2346 204-234-6
$C_{19}H_{22}N_2O$
(9S)-Cinchonan-9-ol.
Antimalarial. mp = 265°; $[\alpha]_D$ = 229° (EtOH); insoluble in H_2O; soluble in EtOH (1.67 g/100 ml at 25°, 4 g/100 ml at 76°), $CHCl_3$ (0.91 g/100 ml), Et_2O (0.2 g/100 ml); LD_{50} (rat ip) = 152 mg/kg.

4935 Clociguanil
3378-93-6
$C_{12}H_{15}Cl_2N_5O$
4,6-Diamino-1-[(3,4-dichlorobenzyl)oxy]-1,2-dihydro-2,2-dimethyl-s-triazine.
Antimalarial.

4936 Cycloguanil
516-21-2 2790
$C_{11}H_{14}ClN_5$
4,6-Diamino-1-(p-chlorophenyl)-1,2-dihydro-2,2-dimethyl-s-triazine.
chlorazine; chlorguanide triazine; TCl; M-10580; D-20. Antimalarial. mp = 146°; λ_m = 241 nm (log ε 4.11 H_2O). *Parke-Davis.*

4937 Cycloguanil Pamoate
609-78-9 2790
$C_{45}H_{44}Cl_2N_{10}O_6$
4,6-Diamino-1-(p-chlorophenyl)-1,2-dihydro-2,2-dimethyl-s-triazine compound (2:1) with 4,4'-methylenebis[3-hydroxy-2-naphthoic acid].
cycloguanil embonate; CI-501; CN-14,329-23A; PAM-MR-807-23a; NSC-77830; Camolar. Antimalarial. mp = 231-234°; soluble in H_2O (0.003 g/100 ml). *Parke-Davis.*

4938 Enpiroline Phosphate
66364-74-7
$C_{19}H_{21}F_6N_2O_5P$
(±)-(R*,R*)-α-[2-(trifluoromethyl)-6-(α,α,α-trifluoro-p-tolyl)-4-pyridyl]-2-piperidinemethanol phosphate (1:1) (salt).
WR-180409. Antimalarial. *Walter Reed Army Inst. of Res.*

4939 Gentiopicrin
20831-76-9 4403 244-070-2
$C_{16}H_{20}O_9$
(5R-trans)-5-Ethenyl-6-(β-D-glucopyranosyloxy)-5,6-dihydro-1H,3H-pyrano[3,4-c]pyran-1-one.
gentiopicroside. Antimalarial. Isolated from *Gentiana lutea L., Gentianaceae.* mp = 191°; $[\alpha]_D^{20}$ = -199° (RtOH); λ_m = 270 nm (log ε 3.96, EtOH).

4940 Halofantrine
69756-53-2 4626 274-104-1
$C_{26}H_{30}Cl_2F_3NO$
1,3-Dichloro-α-[2-(dibutylamino)ethyl]-6-(trifluoromethyl)-9-phenanthrenemethanol.
Halfan. Antimalarial. [β-glycerophosphate]: mp = 60-65°. *SmithKline Beecham Pharm.; Walter Reed Army Inst. of Res.*

4941 Halofantrine Hydrochloride
36167-63-2 4626 252-895-4
$C_{26}H_{31}Cl_3F_3NO$
1,3-Dichloro-α-[2-(dibutylamino)ethyl]-6-(trifluoromethyl)-9-phenanthrenemethanol hydrochloride.
WR-171669; SKF-102886; Halfan. Antimalarial. mp = 93-96°, 203-204°. *SmithKline Beecham Pharm.; Walter Reed Army Inst. of Res.*

4942 Hydroxychloroquine
118-42-3 4863 204-249-8
$C_{18}H_{26}ClN_3O$
2-[[4-[(7-Chloro-4-quinolyl)amino]-pentyl]ethylamino]ethanol.
oxychloroquine; oxichloroquine. Antimalarial; lupus erythematosus suppressant. Antirheumatic agent. Also an antiamebic, antimalarial. Used in combination with cyclophosphamide and azathioprine in treatment of rheumatoid arthritis. mp = 89-91°. *Apothecon; Sterling Winthrop Inc.*

4943 Hydroxychloroquine Sulfate
747-36-4 4863 212-019-3
$C_{18}H_{28}ClN_3O_5S$
2-[[4-[(7-Chloro-4-quinolyl)amino]-pentyl]ethylamino]ethanol sulfate (1:1) (salt).
Plaquenil Sulfate; Ercoquin; Quensyl. Antirheumatic agent. Also antiamebic, antimalarial. Also has activity as a lupus erythematosus suppressant. Two forms, mp = 198° and 240°; freely soluble in H_2O, insoluble in most organic solvents. *Apothecon; Sterling Winthrop Inc.*

4944 Mefloquine
53230-10-7 5845
$C_{17}H_{16}F_6N_2O$
(DL-erythro-α-2-Piperidyl-2,8-bis(trifluoromethyl)-4-quinolinemethanol.
WR-142490; Ro-21-5998. Antimalarial. *Hoffmann-LaRoche Inc.; Walter Reed Army Inst. of Res.*

4945 Mefloquine Hydrochloride
51773-92-3 5845
$C_{17}H_{17}ClF_6N_2O$
(DL-erythro-α-2-Piperidyl-2,8-bis(trifluoromethyl)-4-quinolinemethanol hydrochloride.
Lariam; Ro-21-5998/001; Fansimef [as combination with sulfadoxine and pyrimethamine]. Antimalarial. mp = 259-260° (dec); slightly soluble in H_2O; soluble in EtOH, EtOAc. *Hoffmann-LaRoche Inc.; Walter Reed Army Inst. of Res.*

4946 Menoctone
14561-42-3

$C_{24}H_{32}O_3$
2-(8-Cyclohexyloctyl)-3-hydroxy-1,4-naphthoquinone.
Win-11530; NSC-103336. Antimalarial. *Sterling Winthrop Inc.*

4947 3-Methylarsacetin
25384-21-8 6102

$C_9H_{12}AsNO_4$
[4-(Acetylamino)-3-methylphenyl]arsonic acid.
Orsudan. Antimalarial. Dec 306°; soluble in hot H_2O, MeOH; poorly soluble in EtOH, organic solvents.

4948 Mirincamycin
31101-25-4

$C_{19}H_{35}ClN_2O_5S$
Methyl 7-chloro-6,7,8-trideoxy-6-(cis-4-pentyl-L-2-pyrrolidinecarboxamido)-1-thio-L-threo-α-D-galacto-octapyranoside with methyl 7-chloro-6,7,8-trideoxy-6-(trans-4-pentyl-L-2-pyrrolidinecarboxamido)-1-thio-L-threo-α-D-galacto-octapyranoside.
Antibacterial; antimalarial.

4949 Mirincamycin Hydrochloride
8063-91-0

$C_{19}H_{36}Cl_2N_2O_5S$
Methyl 7-chloro-6,7,8-trideoxy-6-(cis-4-pentyl-L-2-pyrrolidinecarboxamido)-1-thio-L-threo-α-D-galacto-octapyranoside with methyl 7-chloro-6,7,8-trideoxy-6-(trans-4-pentyl-L-2-pyrrolidinecarboxamido)-1-thio-L-threo-α-D-galacto-octapyranoside monhydrochloride.
U-24729A. Antibacterial; antimalarial.

4950 Pamaquine
491-92-9 7134 211-224-5

$C_{19}H_{29}N_3O$
N^1,N^1-Diethyl-N^4-(6-methoxy-8-quinolinyl)-1,4-pentanediamine.
Plasmochin; Aminoquin; Praequine; Béprochine; Gamefar; Quipenyl; Plasmoquine. Antimalarial. $bp_{0.03}$ = 175-180°; bp_1 = 182-194°.

4951 Pamaquine Naphthoate
635-05-2 7134

$C_{42}H_{45}N_3O_7$
N^1,N^1-Diethyl-N^4-(6-methoxy-8-quinolinyl)-1,4-pentanediamine naphthoate.
pamaquine pamoate; pamoquine ebonate. Antimalarial. Nearly insoluble in H_2O; soluble in alcohol, Me_2CO.

4952 Pentaquine
86-78-2

$C_{18}H_{27}N_3O$
8-(5-Isopropylaminoamylamino)-6-methoxyquinoline.
Antimalarial.

4953 Pentaquine Phosphate
5428-64-8

$C_{18}H_{30}N_3O_5P$
8-(5-Isopropylaminoamylamino-6-methoxy quinoline phosphate.
An 8-aminoquinoline. Antimalarial.

4954 Plasmocid
551-01-9 7680

$C_{17}H_{25}N_3O$
N,N-Diethyl-N'-(6-methoxy-8-quinolinyl)-1,3-propanediamine.
710-F; SN-3115; Fourneau 710; Antimalarine; Rhodoquine. Antimalarial. $bp_{1.0}$ = 182°; d_4^{24} = 1.0569; [dihydrochloride]: mp = 218-220°; [diphosphate]: mp = 169-171°.

4955 Primaquine
90-34-6 7925 201-987-2

$C_{15}H_{21}N_3O$
8-[(4-Amino-1-methylbutyl)amino]-6-methoxyquinoline.
SN-13272. Antimalarial. $bp_{0.2}$ = 175-179°; soluble in Et_2O; [oxalate]: mp = 182.5-185°.

4956 Primaquine Diphosphate
63-45-6 7925 200-560-8

$C_{15}H_{27}N_3O_9P$
8-[(4-Amino-1-methylbutyl)amino]-6-methoxyquinoline phosphate.
Antimalarial. mp = 197-198°; moderately soluble in H_2O.

4957 Pyrimethamine
58-14-0 8169 200-364-2

$C_{12}H_{13}ClN_4$
2,4-Diamino-5-(p-chlorophenyl)-6-ethylpyrimidine.
RP-4753; Chloridin; Daraprim; Malocide; Tinduring; component of: Fansidar. Antimalarial; antiprotozoal, targeting Toxoplasma. mp = 233-234°, 240-242°; insoluble in H_2O; soluble in EtOH (0.9 g/100 ml at 25°, 2.5 g/100 ml at 75°), dilute HCl (0.5 g/100 ml); sparingly soluble in propylene glycol, dimethylacetamide at 70°. *Glaxo Wellcome Inc.; Hoffmann-LaRoche Inc.; Rhône-Poulenc Rorer Pharm. Inc.*

4958 Quinacrine
83-89-6 8225 201-508-7

$C_{23}H_{30}ClN_3O$
6-Chloro-9-[[4-(diethylamino)-1-methylbutyl]amino]-2-methoxyacridine.
Mepacrine; Atabrine. Anthelmintic; antimalarial. Targets cestodes. *Sterling Winthrop Inc.*

4959 Quinacrine Dihydrochloride Dihydrate
6151-30-0 8225

$C_{23}H_{31}Cl_2N_3O.2H_2O$
6-Chloro-9-[[4-(diethylamino)-1-methylbutyl]amino]-2-methoxyacridine dihydrochloride dihydrate.
Atabrine hydrochloride; RP-866; SN-390. Anthelmintic; antimalarial. Targets cestodes. Dec 248-250°; soluble in H_2O (2.8 g/100 ml); slightly soluble in EtOH, MeOH; insoluble in C_6H_6, Et_2O, Me_2CO. *Sterling Winthrop Inc.*

4960 Quinacrine Methanesulfonate Monohydrate
6598-46-5 8225

$C_{25}H_{38}ClN_3O_7S_2.H_2O$
6-Chloro-9-[[4-(diethylamino)-1-methylbutyl]amino]-2-methoxyacridine monomethanesulfonate monohydrate.
Anthelmintic; antimalarial. Targets cestodes. Soluble in H_2O (33 g/100 ml at 15°), EtOH (2.8 g/100 ml). *Sterling Winthrop Inc.*

4961 Quinidine
56-54-2 8244 200-279-0
$C_{20}H_{24}N_2O_2$
(8R,9S)-6'-Methoxycinchonan-9-ol.
Pitayin; conquinine, β-quinine; Coccinine; Conchinine; Cin-quin; Quinicardine; Quinidex; Cardioquin; Quindine; Quinaglute; conquinine; pitayine; (+)-quinidine; β-quinidine; Kinidin; (+)-Quindine. Class 1A antiarrhythmic agent. Antimalarial. A dextrorotatory stereoisomer of quinine. Found in cinchona bark. mp = 174-175°; $[\alpha]_D^{15}$ = 230° (c = 1.8 $CHCl_3$), $[\alpha]_D^{17}$ = 258° (EtOH), $[\alpha]_D^{17}$ = 322° (c = 1.6 2M HCl); soluble in H_2O (0.05 g/100 ml at 25°, 0.125 g/100 ml at 100°), EtOH (2.78 g/100 ml), Et_2O (1.78. *Eli Lilly & Co.; Merrell Pharm. Inc.; Parke-Davis.*

4962 Quinidine Gluconate
7054-25-3 8244 230-333-9
$C_{26}H_{36}N_2O_9$
(8R,9S)-6'-Methoxycinchonan-9-ol gluconate.
gluconic acid quinidine salt; Duraquin; Quinaglute. Class IA antiarhythmic; antimalarial. mp = 175-176.5°; soluble in H_2O (11.1 g/100 ml), EtOH (1.67 g/100 ml). *Berlex Labs Inc.; Parke-Davis.*

4963 Quinidine Hydrogen Sulfate Tetrahydrate
6151-39-9 8244
$C_{20}H_{24}N_2O.H_2SO_4.4H_2O$
(8R,9S)-6'-Methoxycinchonan-9-ol hydrogen sulfate tetrahydrate.
quinidine bisulfate; Chinidin-Duriles; Kiditard; Kinichron; Kinidin Durules; Quiniduran. Antimalarial. Soluble in H_2O. *Eli Lilly & Co.; Merrell Pharm. Inc.; Parke-Davis.*

4964 Quinidine Polygalacturonate
7681-28-9 8244
$(C_{20}H_{24}N_2O.C_6H_{10}O_7.H_2O)_x$
(8R,9S)-6'-Methoxycinchonan-9-ol polygalacturonate.
Cardioquin; Galactoquin; Naticardina. Antimalarial. mp = 180° (dec); soluble in hot alcohols, H_2O; LD_{50} (rat orl) = 3200 ±350, (mus orl) = 2680 ±210.

4965 Quinine
130-95-0 8245 205-003-2
$C_{20}H_{24}N_2O_2$
(8S,9R)-6'-Methoxycinchonan-9-ol.
Skeletal muscle relaxant; antimalarial. The primary alkaloid of Cinchona species. An optical isomer of quinidine. mp = 177° (dec); $[\alpha]_D^{15}$ = -169° (c = 2 97% EtOH), $[\alpha]_D^{17}$ = -117° (c = 1.5 $CHCl_3$), $[\alpha]_D^{15}$ = -285° (c = 0.4M 0.1N H_2SO_4); soluble in H_2O (0.05 g/100 ml at 25°, 0.13 g/100 ml at 100°), EtOH (125 g/100 ml), C_6H_6 (1.25 g/100 ml at 25°, 5.5 g/100 ml at 50°), $CHCl_3$ (83.3 g/100 ml), Et_2O (0.4 g/100 ml), glycerol (5 g/100 ml); insoluble in petroleum ether.

4966 Quinine Sulfate Dihydrate
6119-70-6 8245 200-046-3
$C_{40}H_{50}N_4O_8S.2H_2O$
(8α,9R)-6'-Methoxycinchonan-9-ol sulfate (2:1) (salt) dyhydrate.
Coco-Quinine; Quinamm; Quine; Quinate; Quinsan; component of: Quinamm. Antimalarial. $[\alpha]_D^{15}$ = -220° (c = 5 0.5N HCl); soluble in H_2O (0.12 g/100 ml at 20°, 3.12 g/100 ml at 100°), EtOH (0.83 g/100 ml at 20°, 10 g/100 ml at 75°); slightly soluble in $CHCl_3$, Et_2O. *Eli Lilly & Co.; Merrell Pharm. Inc.; Parke-Davis.*

4967 Quinine Tannate
1407-83-6 215-805-4
Quinine tannate.
Less bitter than quinine sulfate. Antimalarial.

4968 Quinocide
525-61-1 8252
$C_{15}H_{21}N_3O$
8-[(4-Aminopentyl)amino]-6-methoxyquinoline.
chinocide; khinocyde. Antimalarial. $bp_{1.0}$ = 183-186°; mp = 46°; [hydrochloride]: mp = 224-224,.5°; [dihydrochloride]: mp = 227-227.5°; [diphosphate]: mp ≅ 175°.

4969 Quinoline
91-22-5 8253 202-051-6
C_9H_7N
Benzo[b]pyridine.
Leucoline; chinoleine; 1-benzazine. Antimalarial. mp = -15°; bp = 237.7°, bp_{100} = 163.2°, bp_{40} = 136.7°, bp_{20} = 119.8°, bp_{10} = 103.8°, bp_5 = 89.6°, bp_1 = 59.7°; soluble in H_2O, EtOH, Et_2O, CS_2; LD_{50} (rat orl) = 460 mg/kg.

4970 Sodium Arsenate Dibasic
7778-43-0 8720 231-902-4
$AsHNa_2O_4$
Disodium arsenate.
Formerly used as an antimalarial. [heptahydrate]: mp = 57°; d = 1.87; soluble in H_2O 76.9 g/100 ml), glycerol, slightly soluble in EtOH; LD_{75} (rat ip) = 14-18 mg/kg.

4971 Tebuquine
74129-03-6
$C_{26}H_{25}Cl_2N_3O$
3-[(tert-Butylamino)methyl]-4'-chloro-5-[(7-chloro-4-quinolyl)amino]-2-biphenylol.
CI-897; WR-228258. Antimalarial. *Parke-Davis.*

Antimanics

4972 Lithium Acetate
546-89-4 5544 208-914-3
$C_2H_3LiO_2$
Acetic acid lithium salt.
Quilonorm; Quilonum. Antimanic. mp = 286° (dec); freely soluble in H_2O, EtOH.

4973 Lithium Carbonate
554-13-2 5552 209-062-5
CLi_2O_3
Carbonic acid dilithium salt.
dilithium carbonate; CP-15467-61; NSC-16895; Camcolit; Carbolith; Carbolithium; Ceglution; Eskalith; Hypnorex; Limas; Liskonum; Lithane; Lithicarb; Lithobid; Lithonate; Lithotabs; Phasal; Plenur; Priadel; Quilonorm-retard; Quilonum-retard; Téralithe. Antimanic. mp = 720° ± 1°; d = 2.11; soluble in H_2O (1.28 g/100 ml at 20°, 0.71 g/100 ml at 100°), dilute acids; insoluble in EtOH; LD_{50} (rat orl) = 710 mg/kg. *Bayer Corp. Pharm. Div.; Ciba-*

Geigy Corp.; Miles Inc.; SmithKline Beecham Pharm.; Solvay Pharm. Inc.

4974 Lithium Chloride
7447-41-8 5553 231-212-3
ClLi
Hydrochloric acid lithium salt.
Antimanic. mp = 605°, bp = 1382°.

4975 Lithium Citrate
919-16-4 5555 213-045-8
$C_6H_5Li_3O_7$
Propanetricarboxylic acid 2-hydroxy-trilithium salt tetrahydrate.
trilithium citrate tetrahydrate; Cibalith-S; Litarex; Lithonate S. Antimanic; antidepressant. Deliquescent; soluble in H_2O (66.6 g/100 ml). *Ciba-Geigy Corp.*

4976 Lithium Hydroxide
1310-65-2 5560 215-183-4
H_3LiO_2
Lithium hydroxide monohydrate.
lithium hydrate. Antimanic. [anhydrous form]: mp = 471°; d = 2.54; soluble in H_2O, slightly soluble in EtOH; [monohydrate]: d_4^{20} = 1.51; soluble in H_2O (10.7 g/100 ml at 0°, 109 g/100 at 100°), slightly soluble in EtOH.

4977 Lithium Sulfate
10377-48-7 5569 233-820-4
Li_2O_4S
Sulfuric acid lithium salt.
dilthium sulfate; Lithiophor; Lithium-Duriles. Antimanic. [monohydrate]: d = 2.06; soluble in H_2O (0.38 g/100 ml), insoluble in EtOH.

Antimethemoglobinemics

4978 Ergot
3702
Dried sclerotia of the fungus *Claviceps purpurea (Fries) Tul., Hypocreaceae*. Vasoconstrictor, specific to migraine. Antimigraine. *Miles Inc.*

4979 Methylene Blue
61-73-4 6137 200-515-2
$C_{16}H_{18}ClN_3S.3H_2O$
3,7-Bis(dimethylamino)phenothiazin-5-ium chloride trihydrate.
methylthioninium chloride; tetramethylthionine chloride; solvent blue 8; C.I. Basic Blue 9 trihydrate; Swiss Blue; Urolene Blue; CI-52015. Antimethemoglobinemic; antidote to cyanide poisoning. Used in veterinary medicine as an antiseptic, disinfectant and anitdote to cyanide and nitrate. [trihydrate]: Green crystals; λ_m = 668, 609 nm; soluble in H_2O (4 g/100 ml), EtOH (1.5 g/100 ml); insoluble in Et_2O.

4980 Naratriptan
121679-13-8
$C_{17}H_{25}N_3O_2$
N-Methyl-3-(1-methyl-4-piperidinyl)indole-5-ethanesulfonamide.
Antimigraine. *Glaxo Labs.*

Antimigraines

4981 Almotriptan
154323-57-6
$C_{17}H_{25}N_3O_2S$
1-[[[3-[2-(Dimethylamino)ethyl]indol-5-yl]methyl]-sulfonyl]pyrrolidine.
LAS-31416. Antimigraine. *Grupo Farmaceutico Almirall S.A.*

4982 Alniditan
152317-89-0
$C_{17}H_{26}N_4O$
(-)-2-[[3-[[(R)-2-Chromanylmethyl]amino]propyl]amino]-1,4,5,6-tetrahydropyrimidine.
Antimigraine. *Abbott Laboratories Inc.*

4983 Alniditan Dihydrochloride
155428-00-5
$C_{17}H_{28}Cl_2N_4O$
(-)-2-[[3-[[(R)-2-Chromanylmethyl]amino]propyl]amino]-1,4,5,6-tetrahydropyrimidine dihydrochloride.
R-91274. Antimigraine. *Abbott Laboratories Inc.*

4984 Alpiropride
81982-32-3 319 279-867-4
$C_{21}H_{23}Cl_2N_3O$
(±)-N-[(1-Allyl-2-pyrrolidinyl)methyl]-4-amino-5-(methylsulfamoyl)-o-anisamide.
RIV-2093; Revistel. Dopamine D_2 receptor antagonist. Antimigraine. mp = 168.5-169°; LD_{50} (mmus iv) = 44 mg/kg, (mmus ip) = 184 mg/kg, (mmus sc) = 204 mg/kg, (mmus orl) = 3600 mg/kg. *Synthelabo Pharmacie.*

4985 Avitriptan
151140-96-4
$C_{22}H_{30}N_6O_3S$
3-[3-[4-(5-Methoxy-4-pyrimidinyl)-1-piperazinyl]propyl]-N-methylindoile-5-methanesulfonamide.
Antimigraine. *Bristol-Myers Squibb Co.*

4986 Avitriptan Fumarate
171171-42-9
$C_{26}H_{34}N_6O_7S$
3-[3-[4-(5-Methoxy-4-pyrimidinyl)-1-piperazinyl]propyl]-N-methylindoile-5-methanesulfonamide fumarate.
BMS-180048; BMS-180048-02. Antimigraine. *Bristol-Myers Squibb Co.*

4987 Dihydroergotamine
511-12-6 3217 208-123-3
$C_{33}H_{37}N_5O_5$
9,10-Dihydro-12'-hydroxy-2'-methyl-5'-(phenylmethyl)ergotoman-3',6',18-trione.
Divegal [as tartrate]. Antiadrenergic; antimigraine. α-Adrenergic blocker with venoconstrictor activity. Also binds to serotonin $5HT_1$-receptors. mp = 239°; $[\alpha]_D^{20}$ = -64°, $[\alpha]_{546}^{20}$ = -79° (c = 0.5 C_5H_5N); insoluble in H_2O; sparingly soluble in EtOH, MeOH, $CHCl_3$, C_6H_6; Dec 210-215°. *Sandoz Pharm. Corp.*

4988 Dihydroergotamine Methanesulfonate
6190-39-2 3217 228-235-6
$C_{34}H_{41}N_5O_8S$
9,10-Dihydro-12'-hydroxy-2'-methyl-5'-(phenylmethyl)-ergotoman-3',6',18-trione monomethanesulfonate.

DHE-45; Agit; Angionorm; Dergotamine; DET MS; D.H.E. 45; Diergo; Dihydergot; Dirgotarl; Endophleban; Ergomimet; Ergont; Ergotonin; Ikaran; Migranal; Morena; Orstanorm; Séglor; Tonopres; Verladyn. Antiadrenergic; antimigraine. α-Adrenergic blocker with venoconstrictor activity. Also binds to serotonin $5HT_1$-receptors. mp = 230-235°; moderatley soluble in H_2O. *Sandoz Pharm. Corp.*

4989 Dolasetron
115956-12-2 3471
$C_{19}H_{20}N_2O_3$
Indole-3-carboxylic acid ester with (8r)-hexahydro-8-hydroxy-2,6-methano-2H-quinolizin-3(4H)-one.
MDL-73147. Antiemetic; antimigraine. Serotonin receptor antagonist. *Hoechst Marion Roussel Inc.*

4990 Dolasetron Mesylate
115956-13-3 3471
$C_{20}H_{24}N_2O_6S$
Indole-3-carboxylic acid ester with (8r)-hexahydro-8-hydroxy-2,6-methano-2H-quinolizin-3(4H)-one monomethanesulfonate.
Anxemet; MDL-73147EF. Antiemetic; antimigraine. Serotonin receptor antagonist. mp = 278°. *Hoechst Marion Roussel Inc.*

4991 Eletriptan
143322-58-1
$C_{22}H_{26}N_2O_2S$
3-[[(R)-1-Methyl-2-pyrrolidinyl]methyl]-5-[2-(phenylsulfonyl)ethyl]indole.
UK-116044. Antimigraine.

4992 Ergocornine
564-36-3 3685 209-272-7
$C_{31}H_{39}N_5O_5$
(5'α)-12'-Hydroxy-2',5'-bis(1-methylethyl)ergotaman-3',6',18-trione.
Antimigraine. Dec 181°; $[\alpha]_D^{20}$ = -110° (C_5H_5N), -175° ($CHCl_3$); λ_m = 311 nm (log ε 3.91 MeOH); soluble in Me_2CO, $CHCl_3$, EtOAc; slightly soluble in EtOH, MeOH; insoluble in H_2O; [phosphate]: mp = 190-195° (dec); [ethanesulfonate]: mp = 209° (dec). *Sandoz Pharm. Corp.*

4993 Ergocryptine
511-10-4 3689 208-122-8
$C_{32}H_{41}N_5O_5$
ergokryptine. Antimigraine. Mixture of two isomers, α-ergocryptine and β-ergocryotine. *Sandoz Pharm. Corp.*

4994 α-Ergocryptine
511-09-1 3689 208-121-2
$C_{32}H_{41}N_5O_5$
(5'α)-12'-Hydroxy-2'-(1-methylethyl)-5'-(2-methylpropyl)-ergotaman-3',6',18-trione.
Antimigraine. mp = 212° (dec); $[\alpha]_D^{20}$= -120° (C_5H_5N), -198° ($CHCl_3$); λ_m = 241, 312.5 nm (log ε 4.31 3.95 MeOH); freely soluble in EtOH, $CHCl_3$; insoluble in H_2O. *Sandoz Pharm. Corp.*

4995 β-Ergocryptine
20315-46-2 3689 243-728-6
$C_{32}H_{41}N_5O_5$
[5'α(S)]-12'-Hydroxy-2'-(1-methylethyl)-5'-(1-methylpropyl)ergotaman-3',6',18-trione.
Antimigraine. mp = 173° (dec); $[\alpha]_D^{20}$ = -98° (c = 0.5 C_5H_5N), -179° (c = 0.5 $CHCl_3$); λ_m = 312 nm (log ε 3.93 MeOH). *Sandoz Pharm. Corp.*

4996 Ergotamine
113-15-5 3703 204-023-9
$C_{33}H_{35}N_5O_5$
(5'α)-12'-Hydroxy-2'-methyl-5'-(phenylmethyl)-ergotoman-3',6',18-trione.
Ergoton-A [as succinate]. Analgesic, specific to migraine. Antimigraine. An alkaloid from ergot. Used in medicine as a vasoconstrictor. Dec 212-214°; $[\alpha]_D^{20}$= -160° ($CHCl_3$); insoluble in H_2O, petroleum ether; soluble in MeOH (1.4 g/100 ml), EtOH (0.33 g/100 ml), Me_2CO (0.67 g/100 ml); freely soluble in $CHCl_3$, C_5H_5N, AcOH; soluble in EtOAc; slightly soluble in. *3M Pharm.; Fisons plc; Organon Inc.; Parke-Davis; Sandoz Pharm. Corp.*

4997 Ergotamine Tartrate
379-79-3 3703 206-835-9
$C_{70}H_{76}N_{10}O_{16}$
12'-Hydroxy-2'-methyl-5'α-(phenylmethyl)ergotaman-3',6',18-trione [R-(R*,R*)]-2,3-dihydroxybutanedioate (2:1) salt.
Ergate; Ergomar; Ergostat; Ergotartrat; Exmigra; Femergin; Gynergin; Lingraine; Lingran; Medihaler Ergotamine; component of: Cafergot, Wigraine. Analgesic, specific to migraine. Antimigraine. Used as an oxytocic (veterinary). Serotonin receptor agonist. mp = 203° (dec); $[\alpha]_D^{25}$= -125 to -155° (c = 0.4 $CHCl_3$); soluble in H_2O (2 mg/ml), EtOH; LD_{50} (rat iv) = 80 mg/kg. *3M Pharm.; Fisons Pharm. Div.; Organon Inc.; Parke-Davis; Sandoz Pharm. Corp.*

4998 Fampridine
504-24-5 3974 207-987-9
$C_5H_6N_2$
4-Aminopyridine.
4-AP; EL-970; 4-pyridylamine; 4-pyridinamine; γ-aminopyridine. Analgesic; antimigraine. Potassium channel blocker. Prolongs action potential in demyelinated nerve fibers. Used in treatment of multiple sclerosis. mp = 158-159°; soluble in H_2O, EtOH; less soluble in non-polar organic solvents.

4999 Flumedroxone
15687-21-5
$C_{22}H_{29}F_3O_3$
17-Hydroxy-6α-(trifluoromethyl)pregn-4-en-3,20-dione.
Antimigraine.

5000 Flumedroxone Acetate
987-18-8 4171 213-577-0
$C_{24}H_{31}F_3O_4$
17-Hydroxy-6α-(trifluoromethyl)pregn-4-en-3,20-dione acetate.
WG-537; Demigran. Antimigraine. mp ≅ 206°; $[\alpha]_D^{20}$ = 30°; λ_m = 234 nm (ε 15600 EtOH). *Lovens Komiske Fabrik AS.*

5001 Fonazine
7456-24-8 4260 231-229-6
$C_{19}H_{25}N_3O_2S_2$
10-[2-(Dimethylamino)propyl]-N,N-dimethylphenothiazine-2-sulfonamide.
dimethothiazine; dimethiotazine; dimetiotazine; RP-8599. Antimigraine. Serotonin inhibitor. [hydrochloride ($C_{19}H_{26}ClN_3O_2S_2$)]: mp = 214° (dec). *Rhône-Poulenc Rorer Pharm. Inc.*

5002 Fonazine Methanesulfonate
7455-39-2 4260
$C_{20}H_{29}N_3O_5S_3$
10-[2-(Dimethylamino)propyl]-N,N-dimethyl-phenothiazine-2-sulfonamide monomethanesulfonate.
IL-6302 mesylate; 8599 R.P. mesylate; Banistyl; Bonpac; Calsekin; Migristène; Neomestine; Promaquid; Yoristen. Antimigraine. Serotonin inhibitor. *Rhône-Poulenc Rorer Pharm. Inc.*

5003 Iprazochrome
7248-21-7
$C_{12}H_{16}N_4O_3$
3-Hydroxy-1-isopropyl-5,6-indolinedione 5-semicarbazone.
Serotonin antagonist used to treat migraine.

5004 Lisuride
18016-80-3 5541 241-925-1
$C_{20}H_{26}N_4O$
3-(9,10-Didehydro-6-methylergolin-8α-yl)-1,1-diethylurea.
methylergol carbamide; lysuride. Antimigraine; prolactin inhibitor; antiparkinsonian. Dopamine D_2 receptor agonist. mp = 186°; $[\alpha]_D^{20}$ = 313° (c = 0.60 C_5H_5N).

5005 Lisuride Maleate
19875-60-6 5541 243-387-3
$C_{24}H_{30}N_4O_5$
3-(9,10-Didehydro-6-methylergolin-8α-yl)-1,1-diethylurea maleate.
Apodel; Cuvalit; Dopergin; Eunal; Lysenyl; Revanil. Antimigraine; prolactin inhibitor; antiparkinsonian. Dopamine D_2 receptor agonist mp = 200° (dec); $[\alpha]_D^{20}$ = 288° (c = 0.5 MeOH); λ_m = 313 nm (MeOH); LD_{50} (mus iv) = 14.4 mg/kg.

5006 Lomerizine
101477-55-8 5593
$C_{27}H_{30}F_2N_2O_3$
1-[Bis(p-Fluorophenyl)methyl]-4-(2,3,4-trimethoxy-benzyl)piperazine.
Antimigraine. Diphenylpiperazine calcium channel blocker. Selective cerebral vasodilator. *Kanebo Pharm. Ltd.*

5007 Lomerizine Hydrochloride
101477-54-7 5593
$C_{27}H_{32}Cl_2F_2N_2O_3$
1-[Bis(p-Fluorophenyl)methyl]-4-(2,3,4-trimethoxy-benzyl)piperazine dihydrochloride.
KB-2796. Antimigraine. Diphenylpiperazine calcium channel blocker. Selective cerebral vasodilator. mp = 214-218° (dec), 204-207° (dec); LD_{50} (mus orl) = 300 mg/kg. *Kanebo Pharm. Ltd.*

5008 Metergotamine
22336-84-1
$C_{34}H_{37}N_5O_5$
1-Methylergotamine.
MY-25 [as bitartrate]. Analgesic; antimigraine.

5009 Methysergide
361-37-5 6217 206-644-0
$C_{21}H_{27}N_3O_2$
(+)-9,10-Didehydro-N-[1-(hydroxymethyl)propyl]-1-methyl-D-lysergamide.
UML-491. Vasoconstrictor, specific to migraine. Antimigraine. Serotonin receptor antagonist. mp = 194-196°; $[\alpha]_D^{20}$ = -45° (c = 0.5 C_5H_5N); [dimaleate]: mp = 165°; soluble in MeOH, less soluble in H_2O, insoluble in EtOH. *Sandoz Pharm. Corp.*

5010 Methysergide Hydrogen Maleate
129-49-7 6217 204-950-9
$C_{25}H_{31}N_3O_6$
(+)-9,10-Didehydro-N-[1-(hydroxymethyl)propyl]-1-methyl-D-lysergamide hydrogen maleate.
Sansert; Deseril; Désernil. Serotonin receptor agonist. Vasoconstrictor, specific to migraine. Antimigraine. *Sandoz Pharm. Corp.*

5011 Naratriptan Hydrochloride
143388-64-1
$C_{17}H_{27}Cl_2N_3O_2$
N-Methyl-3-(1-methyl-4-piperidinyl)indole-5-ethane-sulfonamide monohydrochloride.
GR-85548A. Antimigraine. *Glaxo Labs.*

5012 Oxetorone
26020-55-3 7068 247-411-3
$C_{21}H_{21}NO_2$
3-Benzofuro[3,2-c][1]benzoxepin-6(12H)-ylidene-N,N-dimethyl-1-propanamine.
Antimigraine. Novel serotonin and histamine antagonist. *Labaz S.A.*

5013 Oxetorone Fumarate
34522-46-8 7068
$C_{25}H_{25}NO_6$
3-Benzofuro[3,2-c][1]benzoxepin-6(12H)-ylidene-N,N-dimethyl-1-propanamine fumarate.
L-6257; Nocertone; Oxedix. Antimigraine. Novel serotonin and histamine antagonist. mp = 160°. *Labaz S.A.*

5014 Pizotyline
15574-96-6 7671 239-632-9
$C_{19}H_{21}NS$
4-(9,10-Dihydro-4H-benzo[4,5]cyclohepta[1,2-b]thien-4-ylidene)-1-methylpiperidine.
BC-105; pizotifen; pizotifan; Litec. Antimigraine; appetite stimulant; antidepressant. Tricyclic. Serotonin antagonist (specific to migraine) structurally related to cyproheptadine. *Sandoz Pharm. Corp.*

5015 Pizotyline Hydrochloride
73391-87-4 7671
$C_{19}H_{22}ClNS$
4-(9,10-Dihydro-4H-benzo[4,5]cyclohepta[1,2-b]thien-4-ylidene)-1-methylpiperidine hydrochloride.
Anabolic; antimigraine; appetite stimulant;

antidepressant. Tricyclic serotonin antagonist, specific to migraine. mp = 261-263° (dec). *Sandoz Pharm. Corp.*

5016 Pizotyline Malate
5189-11-7 7671 225-970-4
$C_{23}H_{27}NSO_5$
4-(9,10-Dihydro-4H-benzo[4,5]cyclohepta[1,2-b]thien-4-ylidene)-1-methylpiperidine compound with malic acid (1:1).
Sandomigran; Sanmigran; Sanomigran; Mosegor. Antimigraine; appetite stimulant; antidepressant. Tricyclic serotonin antagonist. mp = 185-186° (dec). *Sandoz Pharm. Corp.*

5017 Rizatriptan
144034-80-0
$C_{15}H_{19}N_5$
3-[2-(Dimethylamino)ethyl]-5-(1H-1,2,4-triazol-1-ylmethyl)indole.
Antimigraine. *Merck & Co.Inc.*

5018 Rizatriptan Benzoate
145202-66-0
$C_{22}H_{25}N_5O_2$
3-[2-(Dimethylamino)ethyl]-5-(1H-1,2,4-triazol-1-ylmethyl)indole monobenzoate.
MK-0462. Antimigraine. *Merck & Co.Inc.*

5019 Rizatriptan Sulfate
159776-67-7
$C_{30}H_{40}N_{10}O_4S.H_2O$
3-[2-(Dimethylamino)ethyl]-5-(1H-1,2,4-triazol-1-ylmethyl)indole sulfate (2:1) monohydrate.
MK-A462. Antimigraine. *Merck & Co.Inc.*

5020 Sergolexole
108674-86-8
$C_{26}H_{36}N_2O_3$
trans-4-Methoxycyclohexyl 1-isopropyl-6-methylergoline-8β-carboxylate.
Antimigraine. *Eli Lilly & Co.*

5021 Sergolexole Maleate
108674-87-9
$C_{30}H_{40}N_2O_7$
trans-4-Methoxycyclohexyl 1-isopropyl-6-methylergoline-8β-carboxylate maleate (1:1).
LY-281067. Antimigraine. *Eli Lilly & Co.*

5022 Sumatriptan
103628-46-2 9172
$C_{14}H_{21}N_3O_2S$
3-[2-(Dimethylamino)ethyl]-N-methylindole-5-methanesulfonamide.
GR-43175. Serotonin $5HT_1$ receptor agonist. Antimigraine. mp = 169-171°. *Glaxo Wellcome Inc.*

5023 Sumatriptan Succinate
103628-48-4 9172
$C_{18}H_{27}N_3O_6S$
3-[2-(Dimethylamino)ethyl]-N-methylindole-5-methanesulfonamide succinate (1:1).
Imigran; Imitrex; GR-43175C. Serotonin $5HT_1$ receptor agonist. Antimigraine. mp = 165-166°. *Glaxo Wellcome Inc.*

5024 Zatosetron
123482-22-5
$C_{19}H_{25}ClN_2O_2$
5-Chloro-2,3-dihydro-2,2-dimethyl-N-1αH,5αH-tropan-3α-yl-7-benzofurancarboxamide.
LY-277359. Antimigraine. *Eli Lilly & Co.*

5025 Zatosetron Maleate
123482-23-5
$C_{23}H_{29}ClN_2O_6$
5-Chloro-2,3-dihydro-2,2-dimethyl-N-1αH,5αH-tropan-3α-yl-7-benzofurancarboxamide maleate (1:1).
LY-277359 maleate. Antimigraine. *Eli Lilly & Co.*

5026 Zolmitriptan
139264-17-8
$C_{16}H_{21}N_3O_2$
(S)-4-[[3-[2-(Dimethylamino)ethyl]indol-5-yl]methyl]-2-oxazolidinone.
Zomig; 311C90. Antimigraine. *Glaxo Wellcome Inc.; Zeneca Pharm.*

Antineoplastics

5027 Acitretin
55079-83-9 114 259-474-4
$C_{21}H_{26}O_3$
(all E)-9-(4-Methoxy-2,3,6-trimethylphenyl)-3,7-dimethyl-2,4,6,8-nonatrienoic acid.
Soriatane; etretin; Neotigason; Ro-10-1670/000. Antipsoriatic. An aromatic retinoid. The primary active metabolite of etretinate. Used in the treatment of psoriasis and other dermatologic disorders. Shown to inhibit angiogenesis and to be effective in prevention of tumor development. mp = 228-230°; LD_{50} (mus ip) > 400 mg/kg (1 day), = 700 (10days), = 700 (20 days). *Hoffmann-LaRoche Inc.*

5028 Acivicin
42228-92-2
$C_5H_7ClN_2O_3$
(αS,5S)-α-Amino-3-chloro-2-isoxazoleacetic-5-acetic acid.
AT-125; U-42126. Antineoplastic agent. *Upjohn Ltd.*

5029 Aclarubicin
57576-44-0 115
$C_{42}H_{53}NO_{15}$
2-Ethyl-1,2,3,4,6,11-hexahydro-2,5,7-trihydroxy-6,11-dioxo-4-[[2,3,6-trideoxy-4-O-[2,6-dideoxy-4-O[(2R-trans)-tetrahydro-6-methyl-5-oxo-2H-pyran-2-yl]-α-L-lyxohexopyranosyl]-3-(dimethylamino)-α-L-lyxohexopyranosyl]oxy]-1-naphthacene carboxylic acid methyl ester.
Aclacinomycin A; NSC-208734; antibiotic MA 144A1; Jaclacin. Antineoplastic agent. mp = 151-153° (dec); $[\alpha]_D^{24}$ = -11.5° (CH_2Cl_2 c = 1); λ_m = 229.5, 259, 289.5, 431 nm ($E_{1\,cm}^{1\%}$ 550, 326, 135, 161 MeOH), 229.5, 258.5, 290, 431 nm ($E_{1\,cm}^{1\%}$ 571, 338, 130, 161 0.1N HCl), 239, 287, 523 nm ($E_{1\,cm}^{1\%}$ 450, 113, 127 0.1N NaOH); soluble in $CHCl_3$, EtOAc; insoluble in non-polar organic solvents;

LD_{50} mus ip) = 22.6 mg/kg, (mus iv) = 33.7 mg/kg. *Bristol-Myers Squibb Pharm. R&D.*

5030 Acodazole
79152-85-5
$C_{20}H_{19}N_5O$
N-Methyl-N-[4-[(7-methyl-1H-imidazo[4,5-f]quinolin-9-yl)amino]phenylacetamide.
Antineoplastic agent.

5031 Acodazole Hydrochloride
55435-65-9
$C_{20}H_{20}ClN_5O$
N-Methyl-N-[4-[(7-methyl-1H-imidazo[4,5-f]quinolin-9-yl)amino]phenylacetamide monohydrochloride.
NSC-305884. Antineoplastic agent.

5032 Aconiazide
13410-86-1
$C_{15}H_{13}N_3O_4$
Isonicotinic acid [0-(carboxymethoxy)benzylidene]-hydrazide.
Antineoplastic agent.

5033 Acronine
7008-42-6
$C_{20}H_{19}NO_3$
3,12-Dihydro-6-methoxy-3,3,12-trimethyl-7H-pyrano[2,3-c]acridin-7-one.
Acromycine; Acronine; Acronycine; Compound 42339; NCI-C01536; NSC-403169. Antineoplastic agent. *Eli Lilly.*

5034 Adozelesin
110314-48-2
$C_{30}H_{22}N_4O_4$
N-[2-[(4,5,8,8a-Tetrahydro-7-methyl-4-oxocyclopropa[c]pyrrolo[3,2-e]indol-2(1h)-yl)carbonyl]-1H-indol-5-yl]benzofurancarboxamide.
U-73975. Antineoplastic agent. *Upjohn Ltd.*

5035 Alanosine
5854-93-3 207
$C_3H_7N_3O_4$
(-)-(S)-2-Amino-3-(hydroxynitrosamino)propionic acid.
Antibiotic isolated from *Streptomyces alanosinicus.* Antineoplastic agent. Dec 190°; $[\alpha]_D$ = 8° (1N HCl), -46° (0.1N NaOPH), -37.8° (H_2O); λ_m 228 nm ($E^{1\%}_{1\,cm}$ 505 0.1N HCl), 250 nm ($E^{1\%}_{1\,cm}$ 630 0.1N NaOH); slightly soluble in H_2O, insoluble in the common organic solvents; LD_{50} (mus ip) = 600 mg/kg, (mus iv) = 300 mg/kg. *Gruppo Lepetit S.p.A.*

5036 Aldesleukin
110942-02-4 5020
Interleukin-2.
T-cell growth factor; TCGF; Thymocyte Stimulating Factor; IL-2. Antineoplastic; antiviral agent.

5037 Alestramustine
139402-18-9
$C_{26}H_{36}Cl_2N_2O_4$
Estradiol 3-[bis(2-chloroethyl)carbamate] 17-ester with L-alanine.
Antineoplastic agent.

5038 Alitretinoin
5300-03-8
$C_{20}H_{28}O_2$
(2E,4E,6Z,8E)-3,7-Dimethyl-9-(2,6,6-trimethyl-1-cyclohexen-1-yl)-2,4,6,8-nonatrienoic acid.
Panretin; Panrexin; 9-cis-retinoic acid; NSC-659772; LG-100057; LGD-1057; ALRT-1057; AGN-192013; . Antineoplastic used in the treatment of Kaposi's sacroma and acute promyelocytic leukemia. *Ligand.*

5039 Altretamine
645-05-6 328
$C_9H_{18}N_6$
N,N,N',N',N,N-Hexamethyl-1,3,5-triazine-2,4,6-triamine.
Hexalen; HMM; ENT050852; Hexastat; NSC-13875. Antineoplastic agent. mp = 172-174°; λ_m = 226 nm (ε 49400 EtOH); LD_{50} (rat orl) = 350 mg/kg, (gpg orl)= 255 mg/kg. *U.S. Bioscience Corp.*

5040 Ambamustine
85754-59-2
$C_{29}H_{39}Cl_2FN_4O_4S$
N-[3-[m-[Bis-(2-chloroethyl)amino]phenyl]-N-[3-(p-fluorophenyl)-L-alanyl]-L-alanyl]-L-methionine ethyl ester.
Antineoplastic agent.

5041 Ambazone
539-21-9 208-713-0
$C_8H_{11}N_7S.H_2O$
p-Benzoquinone amidinohydrazone thiosemicarbazone hydrate.
A membrane-active antitumor agent.

5042 Ambomycin
1402-81-9
NSC-53397. Antineoplastic antibiotic produced by *Streptomyces ambofaciens. Pfizer Intl.*

5043 Ametantrone Acetate
70711-40-9
$C_{26}H_{36}N_4O_8$
1,4-Bis[[2-[(2-hydroxyethyl)amino]ethyl]amino]-9,10-anthracenedione diacetate.
CI-881; NSC-287513. Antineoplastic agent. *Parke-Davis.*

5044 Aminoglutethimide
125-84-8 460 204-756-4
$C_{13}H_{16}N_2O_2$
2-(p-Aminophenyl)-2-ethylglutarimide.
Cytadren; Elipten; Orimeten; p-Aminoglutethimide; Ba-16038; NSC-330915. Adrenocortical suppressant; anticonvulsant; antineoplastic agent. Aromatase inhibitor. Has been used in treatment of Cushing's syndrome and other adrenal hormone disorders and in palliative treatment of breast cancer. mp = 149-150°; insoluble in H_2O; poorly soluble in EtOAc, EtOH; readily soluble in other organic solvents; [hydrochloride]: mp = 223-225°, soluble in H_2O. *Ciba-Geigy Corp.*

5045 Aminopterin
54-62-6 493 200-209-9
$C_{19}H_{20}N_8O_5$
N-[p-[[(2,4-Diamino-6-pteridinyl)methyl]amino]benzoyl]glutamic acid.

Experimental antitumor agent. Also used as a rodenticide. λ_m 261, 282, 373 nm (log ε 4.41, 4.39, 3.91 0.1N NaOH). *Am. Cyanamid.*

5046 Aminopterin Sodium
58602-66-7 493
$C_{19}H_{18}N_8Na_2O_5$
Sodium N-[p-[[(2,4-diamino-6-pteridinyl)methyl]amino]-benzoyl]glutamate.
Experimental antitumor agent. Also used as a rodenticide. *Am. Cyanamid.*

5047 Amonafide
69408-81-7
$C_{16}H_{17}N_3O_2$
3-Amino-N-[2-(dimethylamino)ethyl]naphthalimide.
Antitumor agent.

5048 Amrubicin
110267-81-7
$C_{25}H_{25}NO_9$
(±)-(7S,9S)-9-Acetyl-9-amino-7-[(2-deoxy-β-D-erythropentopyranosyl)oxy]-7,8,9,10-tetrahydro-6,11-dihydroxy-5,12-naphthacenedione.
Experimental antitumor agent.

5049 Amsacrine
51264-14-3 635 257-094-3
$C_{21}H_{19}N_3O_3S$
N-[4-(9-Acridinylamino)-3-methoxyphenyl]methanesulfonamide.
m-AMSA; CI-880; SN-11841; Amerkin; Amsidine; Amsidyl; Lamasine; NSC-249992. Antineoplastic agent. LD_{50} (mus orl) = 810 mg/m^2.

5050 Amsidyl
51264-14-3 635 257-094-3
$C_{21}H_{19}N_3O_3S$
N-[4-(9-Acridinylamino)-3-methoxyphenyl]methanesulfonamide.
amsacrine; m-AMSA; CI-880; SN-11841; Amerkin; Amsidine; Amsidyl; Lamasine; NSC-249992. Antineoplastic agent. LD_{50} (mus orl) = 810 mg/m^2. *Parke-Davis.*

5051 Anastrozole
120511-73-1 667
$C_{17}H_{19}N_5$
α,α,α',α'-Tetramethyl-5-(1H-1,2,4-triazol-1-ylmethyl)-1,3-benzenediacetonitrile.
Arimidex; ICI-D-1033; ZD-1033. Antineoplastic agent. Aromatase inhibitor. Used in breast cancer treatment. mp = 81-82°. *ICI ; Zeneca Pharm.*

5052 Anaxirone
77658-97-0 278-745-8
$C_{11}H_{15}N_3O_5$
Tris(2,3-epoxypropyl)bicarbamimide.
Experimental antitumor agent.

5053 Ancitabine Hydrochloride
10212-25-6 670
$C_9H_{11}N_3O_4$
(2R,3R,3aS,9aR)-2,3,3a,9a-Tetrahydro-3-hydroxy-6-imino-6H-furo[2',3';4,5]oxazol[3,2-a]pyrimidine-2-methanol.
cyclocytidine; cyclocytidine hydrochloride; cycloCMP hydrochloride; NSC-145668; OCTD hydrochloride; 2,2'-o-Cyclocytidine; 2, 2'-Anhydro-1β-D-arabinofuranosylcytosine hydrochloride. Antineoplastic agent. mp = 248-250°; $[\alpha]_D^{23}$ = -21.8° (H_2O, c = 2.0); λ_m = 262, 231 nm (ε 10600, 9400, pH 1-7). *Merck & Co.Inc.*

5054 Ancyte
2608-24-4 7635
$C_{12}H_{22}N_2O_8S_2$
N,N'-bis(3-methanesulfonyloxypropionyl)-piperazine.
1,4-bis[3-[(methylsulfonyl)oxy]-1-oxopropyl]piperazine; NSC-47774; Ancyte; A-20968; Piposulfan. Antineoplastic agent. mp = 175-177°. *Abbott Labs Inc.*

5055 Anthramycin
4803-27-4 724
$C_{16}H_{17}N_3O_4$
3-(5,10,11,11a)-Tetrahydro-9,11-dihydroxy-8-methyl-5-oxo-1H-pyrrolo[2,1-c][1.4]benzodiazepin-2-yl-2-propenamide.
NRRL-3143. Antineoplastic agent. Isolated from *Streptomyces refuineus* and *Streptomyces spadicogriseus.* mp = 188-194°; λ_m = 235, 333 nm (ε 18200, 31800, CH_3CN); $[\alpha]_D^{25}$ = + 930°. *Hoffmann-LaRoche Inc.*

5056 Apigenin
520-36-5 773 208-292-3
$C_{15}H_{10}O_5$
5,7-Dihydroxy-2-(4-hydroxyphenyl)-4H-1-benzopyran-4-one.
apigenine; pelargidenon 1449; Versulin; 4',5,7-Trihydroxyflavone; 5,7,4'-Trihydroxyflavone; Naringenin chalcone. A yellow dyestuff obtained by decomposing apiine, a glucoside found in parsley. A MAP kinase inhibitor. Also inhibits the proliferation of malignant tumor cells. mp = 345-350°; λ_m = 269, 340 nm (ε 18800, 20900 EtOH); insoluble in H_2O, soluble in EtOH.

5057 Asparaginase
9015-68-3 871
L-Asparaginase aminohydrolase.
colaspase; L-asnase; E.C. 3.5.1.1.; MK-965; Crasnitin; Elspar; Kidrolase; Leunase; NSC-109229. Antineoplastic agent. $[\alpha]_D^{20}$ = -31°; λ_m = 278 nm ($A^{12\%}_{1\,cm}$ 7.1 0.03M sodium phosphate at pH 7.3); soluble in H_2O, insoluble in organic solvents.

5058 Asperlin
30387-51-0
$C_{10}H_{12}O_5$
6-(1,2-Epoxypropyl)-5,6-dihydro-5-hydroxy-2H-pyran-2-one acetate.
Upjohn 224b; U-13933; NSC-93158. Antineoplastic antibiotic. Derived from *Aspergillus nidulans. Upjohn Ltd.*

5059 Asulacrine
80841-47-0
$C_{24}H_{24}N_4O_4S$
9-[2-Methoxy-4-(methylsulfonylamino)anilino]-N,5-dimethylacridine-4-carboxamide.
CI-921; NSC-343499. Potential antitumor agent.

5060 Atrimustine
75219-46-4
$C_{41}H_{47}Cl_2NO_6$
Estradiol 3-benzoate 17-glycolate 4-[p-[bis(2-chloroethyl)amino]phenyl]-butyrate.
Used as an antitumor agent.

5061 Azacitidine
320-67-2 923 206-280-2
$C_8H_{12}N_4O_5$
4-Amino-1-β-D-ribofuranosyl-1,3,5-triazin-2(1H)-one.
Azacytidine; 5 AZC; 5-AC; 5-AZCR; Antibiotic U 18496; ladakamycin; U-18496; WR-183027; mylosar; Azacitidine; 5-Azacytidine; 5-AzaC; NSC-102816. Antineoplastic agent. An RNA/DNA antimetabolite. Used as an antineoplastic agent. mp = 228-230° (dec); $[\alpha]_D^{20}$ = 40° (c = 1 H_2O); λ_m = 241 nm (ε 8767 H_2O); LD_{50} (mus orl) = 572 mg/kg.

5062 Azaserine
115-02-6 932 204-061-6
$C_5H_7N_3O_4$
L-Serine diazoacetate (ester).
CI-337; CN-15757; P-165; NSC-742. Antifungal with antineoplastic activity. mp = 146-162° (dec); $[\alpha]_D^{27.5}$ = -0.5° (H_2O, pH 5.18, c = 8.46); λ_m = 250.5 nm ($E^{1\%}_{1\,cm}$ 1140, pH 7), 252 nm ($E^{1\%}_{1\,cm}$ 1230 0.1N NaOH); soluble in H_2O, less soluble in organic solvents; LD_{50} (mus or. *Parke-Davis.*

5063 Azetepa
125-45-1
$C_8H_{14}N_5OPS$
P,P-Bis(1-aziridinyl)-N-ethyl-N-1,3,4-thiadiazol-2-yl phoshpinic amide.
CL-25477; NSC-64826. Antineoplastic agent.

5064 Azimexon
64118-86-1 264-679-7
$C_{19}H_{14}N_4O$
1-[(1-(2-Cyano-1-aziridinyl)-1-methylethyl]-2-aziridine-carboxamide.
Immunomodulator. Used in cancer treatment.

5065 Azotomycin
7644-67-9
$C_{17}H_{23}N_7O_8$
6-Diazo-N-(6-diazo-N-L-α-glutamyl-5-oxo-L-norleucyl)-5-oxo-L-norleucine.
Diazomycin B; Duazomycin B; NSC-56654. Antineoplastic antibiotic produced by *Streptomyces ambofaciens. Pfizer Intl.*

5066 Batimastat
130370-60-4 1036
$C_{23}H_{31}N_4O_3S_2$
[2R-[1(S*),2R*,3S*]]-N^4-Hydroxy-N^1-[2-(methyl-amino)-2-oxo-1-(phenylmethyl)ethyl]-2-(2-methylpropyl)-3-[(2-thienylthio)-methyl]butanediamide.
BB-94. Antineoplastic agent adjunct. Antimetastatic agent. mp = 236-238°. *British Biotechnology Ltd.*

5067 Benaxibine
27661-27-4
$C_{12}H_{15}NO_6$
p-(D-Xylosylamino)benzoic acid.
Potential antitumor agent.

5068 Bendamustine
16506-27-7
$C_{16}H_{21}Cl_2N_3O_22$
5-[Bis(2-chloroethyl)amino]-1-methyl-2-benzimidazo-lebutyric acid.
Antineoplastic agent.

5069 Benzodepa
1980-45-6 1118
$C_{12}H_{16}N_{3O}3P$
[Bis(1-aziridinyl)phosphinyl]carbamic acid phenylmethyl ester.
AB-103; Dualar; NSC-37096. Antineoplastic agent. mp = 134-135°; insoluble in H_2O, soluble in organic solvents.

5070 Besigomsin
58546-54-6
$C_{23}H_{28}O_7$
(+)-(6S,7S, Biar-R)-5,6,7,8-tetrahydro-1,2,3,13-tetra-methoxy-6,7-dimethylbenzo[3,4]cycloocta-[1,2-f][1,3]benzodioxol-6-ol.
A lignan component of Schizandra fruits which inhibits development of preneoplastic lesions.

5071 Bisantrene
78186-34-2 1284
$C_{22}H_{22}N_8$
9,10-Anthracenedicarboxaldehyde bis[(4,5-dihydro-1H-imidazol-2-yl)hydrazone].
Antineoplastic agent.

5072 Bisantrene Hydrochloride
71439-68-4 1284
$C_{22}H_{24}Cl_2N_8$
9,10-Anthracenedicarboxaldehyde bis[(4,5-dihydro-1H-imidazol-2-yl)hydrazone] dihydrochloride.
CL-216942; ADAH; ADCA; ; NSC-337776; Orange Crush; Zantrëne. Antineoplastic agent. mp = 288-289° (hemihydrate); λ_m = 260, 415 nm (ε 72700, 16300 H_2O).

5073 Bisnafide
144849-63-8
$C_{32}H_{28}N_6O_8$
[R-(R*,R*)]-2,2'-[1,2-Ethanediylbis[imino(1-methyl-2,1-ethanediyl)]]bis[5-nitro1H-benz[d,e]isoquinoline-1,3(2H)-dione.
Antineoplastic agent. *DuPont-Merck Pharm.*

5074 Bisnafide Dimesylate
145124-30-7
$C_{34}H_{36}N_6O_{14}S$
2,2'-[1,2-Ethanediylbis[imino(1-methyl-2,1-ethanediyl)]]bis[5-nitro1H-benz[d,e]-isoquinoline-1,3(2H)-dione [R-(R*,R*)] dimethanesulfonate.
VersaLuma; DMP-840. Antineoplastic agent. *DuPont-Merck Pharm.*

5075 Bizelesin
129655-21-6
$C_{43}H_{36}Cl_2N_8O_5$
[S-(R*,R*)]-6, 6'-[Carbonylbis(imino-1H-indole-5,2-diyl-carbonyl)]bis[8-(chloromethyl)-3,6,7,8-tetrahydro-1-methyl-benzo[1,2-b:4,3-b']dipyrrol-4-ol.
U-77779; NSC-615291. Antineoplastic. *Upjohn Ltd.*

5076 Bleomycin
11056-06-7 1351 232-925-2
$C_{55}H_{84}N_{17}O_{21}S_3$
Bleo; NSC-125066. An antineoplastic antibiotic produced by *Streptomyces verticillus*. Very soluble in H_2O, MeOH; slightly soluble in EtOH; Practically insoluble in Me_2OH, EtOAc, ether; λ_m = 244-248, 289-294 nm ($E^{1\%}_{1\ cm}$ 121-148, 102-121). *Bristol-Myers Oncology.*

5077 Bleomycin Sulfate
9041-93-4 1351 232-925-2
$C_{55}H_{84}N_{17}O_{21}S_3$
N^1-[3-(Dimethylsulfonio)-propyl]bleomycinamide.
Blenoxane. Antineoplastic agent. A mixture of glycopeptide antibiotics isolated from a strain of *Streptomyces verticillus* and converted into sulfates. *Bristol-Myers Oncology.*

5078 Brequinar
96187-53-0 1394
$C_{23}H_{15}F_2NO_2$
6-Fluoro-2-(2'-fluoro-[1,1'-biphenyl]-4-yl)-3-methyl-4-quinolinecarboxylic acid.
Biphenquinate; BPQ. Antineoplastic agent; immunosuppressant. Dihydroorotate dehydrogenase inhibitor. mp = 315-317°; soluble in H_2O, DMF. *E. I. DuPont de Nemours Inc.*

5079 Brequinar Sodium
96201-88-6 1394
$C_{23}H_{14}F_2NNaO_2$
6-Fluoro-2-(2'-fluoro-[1,1'-biphenyl]-4-yl)-3-methyl-4-quinolinecarboxylic acid sodium salt.
Dup-785; NSC-368390. Antineoplastic agent; immunosuppressant. Dihydroorotate dehydrogenase inhibitor. mp > 360°; soluble in H_2O. *E. I. DuPont de Nemours Inc.*

5080 Bromebric Acid
5711-40-0
$C_{11}H_9BrO_4$
(E)-3-p-Anisoyl-3-bromoacrylic acid.
Antitumor agent.

5081 Bromebric Acid Sodium Salt
21739-91-3
$C_{11}H_8BrNaO_4$
(E)-3-p-Anisoyl-3-bromoacrylic acid sodium salt.
Cytembena. Antitumor agent.

5082 Bropirimine
56741-95-8
$C_{10}H_8BrN_3O$
2-Amino-5-bromo-6-phenyl-4(3H)-pyrimidinone.
U-54461. Antiviral agent with antitumor activity. *Upjohn Ltd.*

5083 Brosuximide
22855-57-8
$C_{10}H_8BrNO_2$
2-(m-Bromophenyl)succimide.
An antineoplastic agent.

5084 Broxuridine
59-14-3
$C_9H_{11}BrN_2O_5$
5-Bromo-2'-deoxyuridine.
bromodeoxyuridine; 5-bromodeoxyuridine; bromouracil deoxyriboside; 5-bromouracil; NSC-38297; 5-bromouracil-2-deoxyriboside; broxuridine; BDU; 5-BDU; BRUDR; BUDR. Antineoplastic agent.

5085 Budotitane
85969-07-9
$C_{24}H_{28}O_6Ti$
Diethoxybis(1-phenyl-1,3-butanedionato)titanium.
A titanium-containing antineoplastic agent.

5086 Cactinomycin
8052-16-2 1642
Actinomycin C.
Sanamycin; NSC-18268. Antineoplastic agent.

5087 Capecitabine
154361-50-9
$C_{15}H_{22}FN_3O_6$
[1-(5-Deoxy-β-D-ribofuranosyl)-5-fluoro-1,2-dihydro-2-oxo-4-pyrimidinyl]-carbamic acid pentyl ester.
Ro-09-1978/000. Antineoplastic agent. *Hoffmann-LaRoche Inc.*

5088 Caracemide
81424-67-1
$C_6H_{11}N_3O_4$
N-[(Methylamino)carbonyl]-N-[(methylamino)carbonyl]oxy] acetamide.
NSC-253272. Antineoplastic agent.

5089 Carbetimer
82230-03-3
N-137. Antineoplastic agent. 2,5-furandione polymer with ethylene, reaction product with ammonia.

5090 Carboplatin
41575-94-4 1870
$C_6H_{12}N_2O_4Pt$
cis-Diammine(1,1-cyclobutanedicarboxylato)platinum.
Paraplatin; JM-8; NSC-241240. Antineoplastic agent. Soluble in H_2O; LD_{50} (mus ip) = 150 mg/kg, (mus iv) = 140 mg/kg; (rat iv) = 85 mg/kg. *Bristol-Myers Oncology.*

5091 Carboquone
24279-91-2 1872
$C_{15}H_{19}N_3O_5$
2,5-Bis(1-aziridinyl)-3-(2-hydroxy-1-methoxyethyl)-6-methyl-p-benzoquinone carbamate ester.
carbazilquinone, Esquinone; NSC-134679. Antineoplastic agent. mp = 202° (dec); insoluble in H_2O, slightly soluble in organic solvents; LD_{50} (mus orl) = 30.8 mg/kg, (mus iv) = 6.09 mg/kg, (mus ip) = 3.84 mg/kg; (rat orl) = 28.0 mg/kg, (rat iv) = 3.88 mg/kg, (rat ip) = 3.16 mg/kg.

5092 Carmofur
61422-45-5 1892
$C_{11}H_{16}FN_3O_3$
5-Fluoro-N-hexyl-3,4-dihydro-2,4-dioxo-1(2H)-pyrimidinecaroxamide.
HCFU; Mifurol; Yamaful. Antineoplastic agent. Orally active, cytostatic. mp = 110-111°; λ_m = 258 nm (ε 11,600 $CHCl_3$).

5093 Carmustine
154-93-8 1894 205-838-2
$C_5H_9Cl_2N_3O_2$
N,N'-Bis(2-chloroethyl)-N-nitrosourea.
Bischloroethyl nitrosourea; BiCNU; BCNU; N,N'-bis(2-chloroethyl)-N-nitrosourea; N,N-Bis (2-chloroethyl)-N-nitrosourea; Bis(2-chloroethyl)nitrosourea; Becenun; Carmubris; Carmustin; FDA 0345; Gliadel; Nitrumon; NSC-409962; NSC-409962; SK-27702; SRI-1720. Antineoplastic agent. mp = 30-32°; soluble in H_2O (4 mg/ml), more soluble in organic solvents; LD_{50} (mus orl) = 19-25 mg/kg, (mus ip) = 26 mg/kg, (mus sc)= 24 mg/kg, (rat orl) = 30-34 mg/kg.

5094 Carubicin
50935-04-1 1921
$C_{26}H_{27}NO_{10}$
(8S-cis)-8-Acetyl-10-[(3-amino-2,3,6-trideoxy-α-L-lyxo-hexopyranosyl)oxy]-7,8,9,10-tetrahydro-1,6,8,11-tetra-hydroxy-5,12-naphthacenedione .
Carminomicin I; Carminomycin; Carminomycin I; Carubicin; CARMINOMYCIN; Karminomitsin; NSC-180024. Antineoplastic agent. Anthracycline antibiotic isolated from Actinomadura carminata.

5095 Carubicin Hydrochloride
52794-97-5 1921
$C_{26}H_{28}ClNO_{10}$
(8S-cis)-8-Acetyl-10-[(3-amino-2,3,6-trideoxy-α-L-lyxo-hexopyranosyl)oxy]-7,8,9,10-tetrahydro-1,6,8,11-tetra-hydroxy-5,12-naphthacenedione hydrochloride.
Carminomycin hydrochloride; Carubicin hydrochloride; Karminomycin hydrochloride. Antineoplastic agent. Anthracycline antibiotic isolated from Actinomadura carminata. $[\alpha]_D^{20}$ = 289°; λ_m = 236, 255, 462, 478, 492 nm; soluble in H_2O, MeOH; insoluble in organic solvents; LD_{50} (mus orl) = 7.3 mg/kg, (mus iv) = 1.3 mg/kg, (mus sc) = 3.7 mg/kg.

5096 Carzelesin
119813-10-4
$C_{41}H_{37}ClN_6O_5$
(S)-N-[2-[[1-(Chloromethyl)-1,6-dihydro-8-methyl-5-[[(phenylamino)carbonyl]oxy]benzo[1,2-b;4,3-b']-dipyrrol-3(2H)-yl]carbonyl]-1H-indol-5-yl]-6-diethyl-amino)-2-benzofurancarboxamide.
U-80244. Antineoplastic agent.

5097 Cemadotin
159776-69-9
$C_{35}H_{56}N_6O_5$
N,N-Dimethyl-L-valyl-L-valyl-N-methyl-L-valyl-L-prolyl-N-benzyl-L-prolinamide.
Potential antitumor agent.

5098 Chlorambucil
305-03-3 2116
$C_{14}H_{19}Cl_2NO_2$
4-[Bis(2-chloroethyl)amino]benzenebutanoic acid.
Leukeran Tablets; 4-[p-[bis(2-chloroethyl)amino]phenyl-]butyric acid; Ambochlorin; Leukeran; chloraminophene; CB-1348; NSC-3088. Antineoplastic agent. A proprietary formulation of chlorambucil; for treatment of chronic lymphocytic leukemia, Hodgkins disease, certain forms of non-Hodgkins lymphoma, Walderstroms macroglobuliremia and advanced ovarian adenocarcinoma. mp = 64-66°; soluble in Et_2O, alcohol, $CHCl_3$, Me_2OH; insoluble in H_2O; LD_{50} (rat ip) = 17.7 mg/kg. *Glaxo Wellcome Inc.*

5099 Chlornaphazine
494-03-1 2157
$C_{14}H_{15}Cl_2N$
N,N-Bis(2-chloroethyl)-2-naphthylamine.
CB-1048; R-48; Cloronaftina; Erysan; NSC-62209. Antineoplastic agent. mp = 54°; bp = 210°; poorly soluble in H_2O (< 0.1 mg/l), more soluble in organic solvents.

5100 Chromomycin A3
7059-24-7 2295
$C_{57}H_{82}O_{26}$
3β-O-(4-O-acetyl-2,6-dideoxy-3-C-methyl-α-L-arabino-hexopyranosyl)-7-methylolivomycin D.
Aburamycin B; Chromomycin; Toyomycin; NSC-58514. Antineoplastic agent. mp = 185° (dec); $[\alpha]_D^{23}$ = -55° (EtOH); λ_m = 230, 281, 304, 318, 330, 412 nm (log ε 4.39, 4.72, 3.85, 3.92, 3.84, 4.07); LD_{50} (mus iv) = 1.85 mg/kg, (mus ip) = 1.7 mg/kg.

5101 Ciaftalan Zinc
14320-04-8
$C_{32}H_{16}N_8Zn$
(SP-4-1)-[Phthalocyaninato(2-)-N^{29},N^{30},N^{31},N^{32}]zinc.
Used in photodynamic therapeutic treatment of cancer.

5102 Cirolemycin
11056-12-5
U-12241. Antineoplastic antibiotic produced by *Streptomyces bellus.*

5103 Cisplatin
15663-27-1 2378 239-733-8
$Cl_2H_6N_2Pt$
cis-Diamminedichloroplatinum.
cis-diamminedichloroplatinum; cis-platinum II; (SP-4-2)-Diamminedichloroplatinum; cis-DDP; CACP; CPDC; DDP; Briplatin; Cismaplat; Cisplatyl; Citoplatino; Lederplatin; Neoplatin; Platamine; Platinex; Platiblastin; Platinol; Platinoxan; Platistin; Platosin; Rand; NSC-119875. Antineoplastic agent. mp = 270° (dec); soluble in H_2O (253 mg/100g), insoluble in organic solvents; LD_{50} (gpg ip) = 9.7 mg/kg. *Bristol-Myers Oncology; Lederle Labs.*

5104 Cladribine
4291-63-8 2397
$C_{10}H_{12}ClN_5O_3$
2-Chloro-2'-deoxyadenosine.
leustatin; RWJ-26251; NSC-105014; cladribine; 2-CdA; CldAdo. Antineoplastic agent. mp = 220° (softens); $[\alpha]_D^{25}$ =

-18.8° (DMF, c = 1.0); λ_m = 265 nm (0.1n NaOH), 265 nm (0.1N HCl). *Ortho Biotech Inc.*

5105 Clanfenur
51213-99-1
$C_{16}H_{15}ClFN_3O_2$
1-(p-Chlorophenyl)-3-(6-fluoro-N,N-dimethyl-anthraniloyl)urea.
Antineoplastic.

5106 Condyline
518-28-5 7704 208-250-4
$C_{22}H_{22}O_8$
[5R-(5α,5aβ,8aα,9α)]-5,8,8a,9-Tetrahydro-9-hydroxy-5-(3,4,5-trimethoxyphenyl)-furo[3',4':6,7]naphtho[2,3-d]-1,3-dioxol-6(5aH)-one.
Bisoprolol; podofilox; podophyllotoxin; NSC-24818. mp = 114-118°, 183-184°; $[\alpha]_D^{20}$ = -132.7° ($CHCl_3$); soluble in H_2O (120 mg/l), more soluble in organic solvents; LD_{50} (rat iv) = 8.7 mg/kg, (rat ip) = 15 mg/kg. *E. Merck.*

5107 Crisnatol
96389-68-3
$C_{23}H_{23}NO_2$
2-[(6-Chrysenylmethyl)amino]-2-methyl-1,3-propanediol.
BW-A770U. Antineoplastic agent. *Glaxo Wellcome Inc.*

5108 Crisnatol Mesylate
96389-69-4
$C_{24}H_{27}NO_5S$
2-[(6-Chrysenylmethyl)amino]-2-methyl 1,3-propanediol methanesulfonate (salt).
BW-A770U mesylate. Antineoplastic agent. *Glaxo Wellcome Inc.*

5109 Cyclophosphamide
50-18-0 2816 200-015-4
$C_7H_{15}Cl_2N_2O_2P$
(Bis(chloro-2-ethyl)amino)-2-tetrahydro-3,4,5,6-oxaza-phosphorine-1,3,2-oxide-2 hydrate.
Asta-B-518; Clafen; Claphene; Cyclophosphamid; Cyclophosphamide; Cyclophosphamidum; Cyclophosphan; Cyclophosphane; Cyclostin; Cytophosphan; Cytoxan; NSC-26271; CB-4564; CP; CPA; CTX; CY; Endoxan; Endoxan R; Endoxan-Asta; Endoxana; Endoxanal; Endoxane; Enduxan; Genoxal; Hexadrin; Mitoxan; Neosar; NCI-C04900; Procytox; Semdoxan; Sendoxan; Senduxan; SK-20501; Zyklophosphamid. Immunosuppressant; antineoplastic agent. Listed as a known carcinogen. mp = 41-45°; soluble in H_2O (40 g/l), less soluble in oganic solvents; LD_{50} (rat orl) = 94 mg/kg. *Bristol-Myers Oncology; Pharmacia & Upjohn.*

5110 Cyclophosphamide Monohydrate
6055-19-2 2816
$C_7H_{15}Cl_2N_2O_2P.H_2O$
N,N-Bis(2-chloroethyl)tetrahydro-2H-1,3,2-oxazaphosphorin-2-amine 2-oxide monohydrate.
Cycloblastin; Cyclostin; Endoxan; Procytox; Sendoxan; Cytoxan; NSC-26271. Immunosuppressant; antineoplastic agent. mp = 41-45°; soluble in H_2O (40 g/l), less soluble in oganic solvents; LD_{50} (rat orl) = 94 mg/kg. *Degussa-Hüls Corp.*

5111 Cytarabine
147-94-4 2853 205-705-9
$C_9H_{13}N_3O_5$
4-Amino-1-β-D-arabinofuranosyl-2(1H)-pyrimidinone.
1-β-D-arabinofuranosylcytosine; β-cytosine arabino-side; CHX-3311; U-19920; Alexan; Arabitin; Aracytidine; Aracytine; Ara-C; Cytosar; Cytosar U; Erpalfa; Iretin; Udicil; NSC-287459. Antineoplastic agent; antiviral agent. mp = 212-213°; $[\alpha]_D^{23}$ = 158° (c = 0.5, H_2O); λ_m = 281, 212.5 nm (ε 13171, 10230, pH 2). *Ciba-Geigy Corp.; Pharmacia & Upjohn.*

5112 Dacarbazine
4342-03-4 2866 224-396-1
$C_6H_{10}N_6O$
5-(3,3-Dimethyl-1-triazenyl)-1H-imidazole-4-carboxamide.
(Dimethyltriazeno)imidazolecarboxamide; Dacarbazine; Deticene; NSC-45388; Dimethyltriazenoimidazole-carboxamide; DIC; DTIC; DTIC-Dome; DTIE; Imidazole carboxamide; ICDMT; ICDT; NCI-C04717. Antineoplastic agent. May be a carcinogen. Dec (explosive) 250-255°; λ_m = 237 nm (ε 11200 pH 7); insoluble in H_2O (10 mg/100 ml), organic solvents; LD_{50} (rat orl) = 2147 mg/kg. *Bayer Corp. Pharm. Div.*

5113 Dactinomycin
50-76-0 2867
$C_{62}H_{86}N_{12}O_{16}$
Di-.xi.-lactone N,N'-[(2-amino-4,6-dimethyl-3-oxo-3H-phenoxazine-1,9-diyl)bis[carbonylimino[2-(1-hydroxy-ethyl)-1-oxo-2,1-ethanediyl]imino[2-(1-methylethyl)-1-oxo-2,1-ethanediyl]-1,2-pyrrolidinediylcarbonyl-(methylimino)(1-oxo-2,1-ethanediyl]bis[N-methyl]-L-valine.
Actinomycin 7; Actinomycindioic D acid, dilactone; NSC-3053; Actactinomycin A IV; Actinomycin AIV; Actinomycin D; Actinomycin IV; Actinomycin X 1; actinomycin[thr-val-pro-sar-meval]; Cosmegen; C1; Dactinomycin; Dactinomycin D; Dilactone actinomycin D acid; Dilactone actinomycindioic D acid; HBF-386; Lyovac cosmegen; meractinomycin; NCI-C04682; Oncostatin K. Antineoplastic agent. Antibiotic from *Streptomyces parvullus.* mp = 241-243° (dec); $[\alpha]_D^{28}$ = -315° (MeOH, c = 0.25); λ_m = 244, 441 nm ($A_{1\ cm}^{1\%}$ 281, 206); soluble in organic solvents; light sensitive; LD_{50} (mus orl) = 13.0 mg/kg, (rat orl) = 7.2 mg/kg. *Merck & Co.Inc.*

5114 Datelliptium Chloride
105118-14-7
$C_{23}H_{28}ClN_3O$
2-[2-(Diethylamino)ethyl]-9-hydroxy-5,11-dimethyl-6H-pyrido[4,3-b]carbazolium chloride.
Antineoplastic .

5115 Daunorubicin
20830-81-3 2890
$C_{27}H_{29}NO_{10}$
8-Acetyl-10-[(3-amino-2,3,6-trideoxy-alpha-L-lyxo-hexopyranosyl)oxy]-7,8,9,10-tetrahydro-6,8,11-tri-hydroxy-1-methoxy-,(8S-cis)-5,12-Naphthacenedione monohydrochloride.
Antibiotics from Streptomyces coeruleorubidus; Cerubidine; Daunoblastin; Daunomycin; NSC-83142; Daunomycin, hydrochloride; Daunorubicin hydro-

chloride; NDC-0082-4155; Ondena; Rubidomycin hydrochloride; RP-13057; RUBIDOMYCIN. Antineoplastic agent. mp = 208-209°; LD_{50} (mus iv) = 20 mg/kg, (mis ip) = 5 mg/kg, (rat iv) = 13 mg/kg, (rat ip) = 8 mg/kg. *Rhône-Poulenc Rorer Pharm. Inc.; Wyeth-Ayerst Labs.*

5116 Daunorubicin Hydrochloride

23541-50-6 2890

$C_{27}H_{30}ClNO_{10}$

8-Acetyl-10-[(3-amino-2,3,6-trideoxy-alpha-L-lyxo-hexopyranosyl)oxy]-7,8,9,10-tetrahydro-6,8,11-trihydroxy-1-methoxy-,(8S-cis)-5,12-Naphthacenedione.

Cerubidine; Daunoblastin; Daunomycin hydrochloride; DAUNORUBICIN HCL; NSC-82151; NDC-0082-4155; Ondena; Rubidomycin hydrochloride; RP-13057 hydrochloride; RUBIDOMYCIN; RUBOMYCIN C. Antineoplastic agent. mp = 188-190° (dec); $[\alpha]_D^{20}$ = 248° (MeOH, c = 0.05-0.10); soluble in H_2O, polar organic solvents; insoluble in non-polar organic solvents; λ_m = 234, 252, 290, 480, 495, 532 nm ($E_{1\,cm}^{1\%}$ = 665 462, 153, 214, 218, 112, MeOH); LD_{50} (mus iv)= 26 mg/kg. *Rhône-Poulenc Rorer Pharm. Inc.; Wyeth-Ayerst Labs.*

5117 Decitabine

2353-33-5

$C_8H_{12}N_4O_4$

4-Amino-1-(2-deoxy-β-D-erythropentofuranosyl)-3,5-triazin-2(1H)-one.

5-Aza-2'-deoxycytidine; NSC-127716; Decitabine. Antineoplastic agent. *Pharmachemie.*

5118 Defosfamide

3733-81-1 2917

$C_9H_{20}Cl_3N_2O_3P$

N,N-Bis(2-chloroethyl)-N'-(3-hydroxypropyl)phosphorodiamidic acid 2-chloroethyl ester.

B-612; B 612-Asta; Defosfamid; Defosfamide; Desmofosfamide; Desmophosphamidum; Mitarson; NSC-40627. Antineoplastic agent. Liquid; sg = 1.3675; insoluble in H_2O, soluble in organic solvents.

5119 Demecolcine

477-30-5 2938

$C_{21}H_{25}NO_5$

Deacetyl-N-methylcolchicine.

C-12669; Alkaloid H 3, from Colchicum autumnale; C-12669; Ciba 12669 A; Colcemide; Colchamine; NSC-3096; Colchicine, N-deacetyl-N-methyl-; Kolchamin; Kolkamin; Omaine; Reichstein's F; Santavy's substance F; Substance F. Antineoplastic. mp = 186°; $[\alpha]_D^{20}$ = -129.0° ($CHCl_3$ c = 1); λ_m = 245 355 nm (log ε 4.55 4.24 EtOH); soluble in acid solutions, organic solvents.

5120 Denileukin Diftitox

173146-27-5

$C_{2560}H_{4038}N_{678}O_{799}S_{17}$

N-L-Methionyl-387-L-histidine-388-L-alanine-1-388-toxin (*Cornebacterium diphtheriae* strain C7) (388 → 2') protein with 2-133-interleukin 2 (human clone pTIL2-21a).

Ontak; LY335348; DAB_{389}IL2. Antineoplastic agent. *Ligand; Seragen.*

5121 Desrazoxane

24584-09-6 8295

$C_{11}H_{16}N_4O_4$

(+)-(S)-4,4'-Propylenedi-2,6-piperazinedione.

(+)-razoxane; dexrazoxane; (+)-4,4'-propylenedi-2,6-piperazinedione; (+)-(3,5,3',5'-tetraoxo)-1,2-dipiperazinopropane; ICI-59118; ICRF-159; NSC-129943; ADR-529; ICRF-187; NSC-169780; Zinecard; Cardioxane [hydrochloride]; Eucardion [hydrochloride]. Antianginal agent. Cardioprotectant. The racemate is used as an antineoplastic. mp = 193°; $[\alpha]_D$ = 11.35° (c = 5, DMF); soluble in H_2O (10 mg/ml), 0.1N HCl (35-43 mg/ml), NaOH (6.7-10 mg/ml), EtOH (1 mg/ml), MeOH (7.1-10 mg/ml). *Pharmacia & Upjohn.*

5122 Detorubicin

66211-92-5

$C_{33}H_{39}NO_{14}$

2-(Diethyl acetal) glyoxylic acid 3^2 ester doxorubicin b. Antineoplastic.

5123 Dexormaplatin

96392-96-0

$C_6H_{14}Cl_4N_2Pt$

(+)-trans-Tetrachloro(1,2-cyclohexanediamine)platinum. U-78938. Antineoplastic agent. *Pharmacia & Upjohn.*

5124 Dezaguanine

41729-52-6

$C_6H_6N_4O$

6-Amino-1,5-dihydro-4H-imidazo[4,5-c]pyridin-4-one.

CI-908; ICN-4221; NSC-261726. Antineoplastic agent. *Parke-Davis.*

5125 Dezaguanine Mesylate

87434-82-0

$C_7H_{10}N_4O_4S$

6-Amino-1,5-dihydro-4H-imidazo[4,5-c]pyridin-4-one methanesulfonate.

CI-908 mesylate; PD 90695-73. Antineoplastic agent. *Parke-Davis.*

5126 Diaziquone

57998-68-2 3044

$C_{16}H_{20}N_4O_6$

[2,5-Bis(1-aziridinyl)-3,6-dioxo-1,4-cyclohexadiene-1,4-diyl]bis-carbamic acid diethyl ester.

AZQ; CI-904; NSC-182986. Antineoplastic agent. mp = 230° (dec); λ_m = 340 nm (log ε 4.17); soluble in H_2O (0.5 mg/ml); LD_{50} (mus iv) = 30.9 mg/m^2. *Parke-Davis.*

5127 Dinaline

58338-59-3

$C_{13}H_{13}N_3O$

2',4-Diaminobenzanilide.

Antineoplastic.

5128 Ditercalinium Chloride

74517-42-3

$C_{46}H_{50}Cl_2N_6O_2$

2,2'-([4,4'-Bipiperidine]-1,1'-diyldiethylene)bis[10-methoxy-7H-pyrido[4,3-c]cabrazolium] dichloride.

Antineoplastic.

5129 Docetaxel
148408-66-6
$C_{43}H_{53}NO_{14}.3H_2O$
[2aR-[2aα,4β,4aβ,6β,9α(αR*,βS*),11α,12α,12aα,12bα]]-β-12b-(Acetyloxy)-12-(benzoyloxy)-2a,3,4,4a,5,6,9,10,11,12,12a,12b-dodecahydro-4,6,11-trihydroxy-4a,8,13,13-tetramethyl-5-oxo-7,11-methano-1H-cyclodeca[3,4]-benz[1,2-b]oxet-9-yles [[(1,1-dimethylethoxy)carbonyl]-amino]-α-hydroxy-benzenepropanoic acid trihydrate.RP-56976; taxotere. Antineoplastic agent. Binds to tubulin and inhibits depolymerization of microtubules. *Rhône-Poulenc Rorer Pharm. Inc.*

5130 Docetaxel [anhydrous]
114977-28-5 3458
$C_{43}H_{53}NO_{14}$
[2aR-[2aα,4β,4aβ,6β,9α(αR*,βS*),11α,12α,12aα,12bα]]-β-12b-(Acetyloxy)-12-(benzoyloxy)-2a,3,4,4a,5,6,9,-10,11,12,12a,12b-dodecahydro-4,6,11-trihydroxy-4a,8,13,13-tetramethyl-5-oxo-7,11-methano-1H-cyclodeca[3,4]benz[1,2-b]oxet-9-yles [[(1,1-dimethyl-ethoxy)-carbonyl]amino]-α-hydroxy-benzenepropanoic acid.
taxotere; docetaxel; N-debenzoyl-N-tert-butoxycarbonyl-10-deacetyl taxol. Antineoplastic agent. Binds to tubulin and inhibits deplymerization of microtubules. mp = 232°; $[\alpha]_D$ = -36° (EtOH c= 0.74); λ_m = 230, 275, 283 nm (ε 14800, 1730, 1670).

5131 Doxifluridine
3094-09-5 3493
$C_9H_{11}FN_2O_5$
5'-Deoxy-4-fluorouridine.
5'-DFUR; 5'-dFUrd; Ro-21-9738; Flutron; Furtulon. Antineoplastic agent. mp = 189-190°, 186-188°, 192-193°; $[\alpha]_D^{25}$= 18.4° (H_2O c = 0.419); λ_m = 268-269 nm(ε 8550); LD_{50} (mus iv 14 day) > 1000 mg/kg, (rat iv 14 day) > 2000 mg/kg, (mrat orl) = 3471 mg/kg, (frat orl) = 3390 mg/kg, (mmus orl) > 5000 mg/kg, (fmus orl) > 5000 mg/kg. *Hoffmann-LaRoche Inc.*

5132 Doxorubicin
23214-92-8 3495
$C_{27}H_{29}NO_{11}$
α-3b-Glycoloyl-1,2,3,4,6,11-hexahydro-3,5,12-tri-hydroxy-10-methoxy-6,11-dioxo-1a-naphthacenyl 3-amino-2,3,6-trideoxy-L-lyxo-hexopyranoside.
adriblastina; FI-106. Antineoplastic agent. mp = 229-231°. *Farmitalia Societa Farmaceutici.*

5133 Doxorubicin Hydrochloride
25316-40-9 3495 246-818-3
$C_{27}H_{30}ClNO_{11}$
α-3b-Glycoloyl-1,2,3,4,6,11-hexahydro-3,5,12-tri-hydroxy-10-methoxy-6,11-dioxo-1a-naphthacenyl 3-amino-2,3,6-trideoxy-L-lyxo-hexopyranoside hydrochloride.
Adriacin; Adriamycin hydrochloride; Adriamycin, hydrochloride (8CI); Adriblastin; Adriblastina; NSC-123127; Adriblastina; ADM hydrochloride; ADR; DOX HCl; FI-106; FI-6804; Hydroxydaunorubicin hydrochloride. Antineoplastic agent. mp = 204-205°; $[\alpha]_D^{20}$= 248° (MeOH c= 0.1); λ_m = 233, 252, 288, 479, 496, 529 nm; soluble in H_2O, polar organic solvents; LD_{50} (mus iv) = 21.1 mg/kg. *Farmitalia Societa Farmaceutici.*

5134 Duazomycin
1403-47-0
$C_8H_{11}N_3O_4$
N-Acetyl-6-diazo-5-oxo L-norleucine.
A-10270A; Acetyl-DON; Diazomycin A; Duazomycin (USAN); Duazomycin A; NSC-51097. Antineoplastic agent. An antibiotic from *Streptomyces ambofaciens. Pfizer Inc.*

5135 Ecomustine
98383-18-7
$C_{10}H_{18}ClN_3O_6$
Methyl 3-[3-(2-chloroethyl)-3-nitrosoureido]-2,3-dideoxy-α-D-arabinohexapyranoside.
Antineoplastic.

5136 Edaravone
89-25-8 6809
$C_{10}H_{10}N_2O$
2 4-Dihydro-5-methyl-2-phenyl-3H-pyrazol-3-one.
NSC-12; NSC-2629; NSC-26139; C.I. Developer 1; Developer Z; Methylphenylpyrazolone; Norphenazone; NCI-C03952. Antineoplastic agent. Also used in treatment of stroke. mp = 129-130°; bp_{265} = 287°. *Pfalz & Bauer.*

5137 Edatrexate
80576-83-6 3553
$C_{22}H_{25}N_7O_5$
N-[4-[1-[(2,4-Diamino-6-pteridinyl)-methyl]propyl]-benzoyl]-L-glutamic acid.
10-EdAM; 10-Ethyl-10-dezaz-aminopterin; Psyllium; NSC-626715; CGP-30694. Antineoplastic agent. λ_m = 255, 370 nm (ε 30731, 7582 pH 12). *Ciba-Geigy Corp.*

5138 Edelfosine
70641-51-9
$C_{27}H_{58}NO_6P$
(±)-2-Methoxy-3-(octadecyloxy)propyl hydrogen phosphate choline hydroxide inner salt.
Antineoplastic.

5139 Eflornithine
67037-37-0 3564
$C_6H_{12}F_2N_2O_2$
2-(Difluoromethyl)-DL-ornithine.
α-difluoromethylornithine; DFMO; RFI-71782; MDL-71782-A; Ornidyl. Antineoplastic; antipneumocystic; antiprotozoal (trypanosoma). Irreversibly inhibits ornithine decarboxylase. Has an antiproliferative effect on tumor cells. Used to remove facial hair; also effective against sleeping sickness. *Marion Merrell Dow Inc.*

5140 Eflornithine Hydrochloride
96020-91-6 3564
$C_6H_{13}ClF_2N_2O_2.H_2O$
2-(Difluoromethyl)-DL-ornithine hydrochloride monohydrate.
Ornidyl; MDL-71782A. Antineoplastic; anti-pneumocystic; antiprotozoal (trypanosoma). Irreversibly inhibits ornithine decarboxylase. Has an antiproliferative effect on tumor cells. mp = 183°. *Marion Merrell Dow Inc.*

5141 Elacridar
143664-11-3
$C_{34}H_{33}N_3O_5$
N-[4-(2-(3,4-Dihydro-6,7-dimethoxy-2(1h)-isoquinolyl)ethyl]phenyl]-9,10-dihydro-5-methoxy-9-oxo-4-acridinecarboxamide.
Antineoplastic agent (adjunct). *Glaxo Wellcome Inc.*

5142 Elacridar Hydrochloride
143851-98-3
$C_{34}H_{34}ClN_3O_5$
N-[4-(2-(3,4-Dihydro-6,7-dimethoxy-2(1h)-isoquinolyl)ethyl]phenyl]-9,10-dihydro-5-methoxy-9-oxo-4-acridinecarboxamide hydrochloride.
GF-120918A. Antineoplastic agent (adjunct). *Glaxo Wellcome Inc.*

5143 Elliptinium Acetate
58337-35-2 3590
$C_{20}H_{20}N_2O_3$
9-Hydroxy-2,5,11-trimethyl-6H-pyrido[4,3-b]-carbazolium acetate.
Celiptium; 9-Hydroxy-2-methylellipticinium acetate; Ellipticine analog; H9M2E; NMHE; NSC-264137. Antineoplastic agent.

5144 Elmustine
60784-46-5
$C_5H_{10}ClN_3O_3$
1-(2-Chloroethyl)-3-(2-hydroxyethyl)-1-nitrosourea.
Antineoplastic.

5145 Elsamitrucin
97068-30-9
$C_{33}H_{35}NO_{13}$
10-[[2-O-(2-Amino-2,6-dideoxy-3-O-methyl-α-D-galacto-pyranosyl)-6-deoxy-3-C-methyl-β-D-galactopyranosyl]-oxy]-6-hydroxy-1-methyl-benzo[h][1]benzopyrano-[5,4,3-cde][1]benzopyran-5,12-dione.
Chartreusin analog; Elsamicin; NSC-369327. Antineoplastic agent. *Bristol-Myers Squibb Pharm. R&D.*

5146 Emitefur
110690-43-2 3601
$C_{28}H_{19}FN_4O_8$
m-[[3-(Ethoxymethyl)-5-fluoro-3,6-dihydro-2,6-dioxo-1(2H)-pyrimidinyl]carbonyl]benzoic acid 2-ester with 2,6-dihydroxynicotinonitrile benzoate.
Antineoplastic. mp = 162-164°; LD_{50} (mus orl) > 5000 mg/kg, (frat orl) = 1850 mg/kg, (mrat orl) = 1934 mg/kg.

5147 Eniluracil
59989-18-3
$C_6H_4N_2O_2$
5-Ethynyl-2,4(1H,3H)-pyrimidinedione.
776C85. Antineoplastic agent (adjunct). *Glaxo Wellcome Inc.*

5148 Enloplatin
111523-41-2
$C_{13}H_{22}N_2O_5Pt$
[1,1-Cyclobutanedicarboxylato(2-)](tetrahydro-4H-pyran-4,4-dimethanamine-N,N'-platinum, (SP-4-2).
CL-287110. Antineoplastic agent.

5149 Enocitabine
55726-47-1 3624
$C_{31}H_{55}N_3O_6$
N-(1-β-D-Arabinofuranosyl-1,2-dihydro-2-oxo-4-pyrimidinyl)docosanamide.
NSC-239336. Antineoplastic agent. mp = 141-142°; $[\alpha]_D$ = 70° (c = 1, THF, 22°); λ_m = 216, 248, 303 nm (ε 16400, 15200, 8200 iPrOH).

5150 Enpromate
10087-89-5
$C_{22}H_{23}NO_2$
1,1-Diphenyl-2-propynyl cyclohexancarbamate.
NSC-112682. Antineoplastic agent. *Eli Lilly & Co.*

5151 Epipropidine
5696-17-3
$C_{16}H_{28}N_2O_2$
1,1'-Bis(2,3-epoxypropyl)-4,4'-bipiperidine.
Epipropidine; Eponate; Epoxypropidine; Lilly Res. No. 28002; Lilly 28002; NSC-56308. Antineoplastic agent. *Eli Lilly & Co.*

5152 Epirubicin
56420-45-2 3660
$C_{27}H_{29}NO_{11}$
(1S,3S)-Glycoloyl-1,2,3,4,6,11-hexahydro-3,5,12-trihydroxy-10-methoxy-6,11-dioxo-1-naphthacenyl 3-amino-2,3,6-trideoxy-α-L-arabino hexopyranoside.
pidorubicin; 4'-epidoxorubicin; 4'-epi-DX; IMI-28. Antineoplastic agent. Analog of the anthracycline antibiotic doxorubicin differing only in the position of the C_4 hydroxyl of the sugar moiety. *Farmitalia Carlo Erba Ltd.*

5153 Epirubicin Hydrochloride
56390-09-1 3660
$C_{27}H_{30}ClNO_{11}$
(1S,3S)-Glycoloyl-1,2,3,4,6,11-hexahydro-3,5,12-trihydroxy-10-methoxy-6,11-dioxo-1-naphthacenyl 3-amino-2,3,6-trideoxy-α-L-arabino hexopyranoside hydrochloride.
Farmorubicin; Pharmorubicin. Antineoplastic agent. Analog of the anthracycline antibiotic doxorubicin differing only in the position of the C_4 hydroxyl of the sugar moiety. mp = 185° (dec); $[\alpha]_D^{20}$ = 274° (MeOH c = 0.01). *Farmitalia Carlo Erba Ltd.*

5154 Erbulozole
124784-31-2
$C_{24}H_{27}N_3O_5S$
Ethyl (±)-cis-p-[[[2-(imidazol-1-ylmethyl)-2-(p-methoxyphenyl)-1,3-dioxolan-4-yl]methyl]thio]carbanilate.
Antineoplastic agent (adjunct). *Janssen Pharm. Inc.*

5155 Esorubicin
63521-85-7
$C_{27}H_{29}NO_{10}$
(2S-(2α(8R*,10R*),4β,6β))-10-[(4-Aminotetrahydro-6-methyl-2H-pyran-2-yl)oxy]-7,8,9,10-tetrahydro-6,8,11-trihydroxy-8-(hydroxyacetyl)-1-methoxy-5,12-naphthacenedione.
Deoxydoxorubicin. Antineoplastic agent. *Farmitalia Carlo Erba Ltd.*

5156 Esorubicin Hydrochloride
63950-06-1
$C_{27}H_{30}ClNO_{10}$
(2S-(2α(8R*,10R*),4β,6β))-10-[(4-Aminotetrahydro-6-methyl-2H-pyran-2-yl)oxy]-7,8,9,10-tetrahydro-6,8,11-trihydroxy-8-(hydroxyacetyl)-1-methoxy-5,12-naphthacenedione hydrochloride.
4'-Deoxydoxorubicin hydrochloride; Escorubicin hydrochloride; IMI 58 ; NSC-267469. Antineoplastic agent. *Farmitalia Carlo Erba Ltd.*

5157 Estramustine
2998-57-4 3749
$C_{23}H_{31}Cl_2NO_3$
Estradiol 3-[bis(2-chloroethyl)carbamate].
Ro-22-2296/000; Leo 275; NSC-89201. Antineoplastic agent (bound to nitrogen mustard). An estradiol. mp = 104-105°; $[\alpha]_D^{20}$ = 50° (dioxane); λ_m = 271, 277 nm (EtOH). *Hoffmann-LaRoche Inc.*

5158 Estramustine Phosphate
4891-15-0 3749 225-512-3
$C_{23}H_{31}Cl_2NO_3$
Estra-1,3,5(10)-triene-3,17β-diol 3-[bis(2-chloroethyl)-carbamate].
Ro-21-8837; Estracyt. Antineoplastic agent (bound to nitrogen mustard). An estradiol. mp = 155°; $[\alpha]_D^{20}$ = 30° (dioxane); soluble in H_2O. *Hoffmann-LaRoche Inc.*

5159 Estramustine Phosphate Sodium
52205-73-9 3749 257-735-7
$C_{23}H_{30}Cl_2NNa_2O_6P$
Estradiol 3-[bis(2-chloroethyl)carbamate] 17-(dihydrogen phosphate) disodium salt.
Ro-21-8837/001; Emcyt; NSC-89199. Antineoplastic agent (bound to nitrogen mustard). An estradiol. *Hoffmann-LaRoche Inc.*

5160 Etoglucid
1954-28-5 3926 217-784-7
$C_{12}H_{22}O_6$
1,2:15,16-Diepoxy-4,7,10,13-tetraoxa-hexadecane.
ICI-32865; ethoglucid; Ayerst 62013; Diglycidyl-triethylene glycol; Epodyl; Ethoglucid; Etoglucid; Etoglucide; NSC-80439; ICI-32865; Oxirane, 2,2'-(2,5,8,11-tetraoxadodecane-1,12-diyl)bis-; Triethylene glycol diglycidyl ether; Triethylene glycol, bis(2,3-epoxypropyl) ether; TDE. Antineoplastic agent. mp = -15°; $bp_{0.005}$ = 140°; d^{20} = 1.1312. *ICI.*

5161 Etoposide
33419-42-0 3931
$C_{29}H_{32}O_{13}$
[5R-[5α,5aβ,8aα,9β(R*)]]-9-[(4,6-O-Ethylidene-β-D-glucopyranosyl)oxy]-5,8,8a,9-tetrahydro-5-(4-hydroxy-3,5-dimethoxyphenyl)-furo[3',4':6,7]naphtho[2,3-d]-1,3-dioxol-6(5aH)-one.
Toposar; VePesid; VP-16-213; NSC-141540; Epipodophyllotoxin VP-16213; EPEG; Lastet; Vepesid J. Antineoplastic agent. Semisynthetic. A derivative of podophyllotoxin. Related to teniposide. mp = 236-251°; $[\alpha]_D^{20}$ = -110.5° ($CHCl_3$ c = 0.6); λ_m = 283 nm (ε 4245 MeOH); pKa = 9.8. *Bristol-Myers Oncology.*

5162 Etoposide Phosphate
117091-64-2
$C_{29}H_{33}O_{16}P$
[5R-[5α,5aβ,8aα,9β(R*)]]-5-[3,5-Dimethoxy-4-(phosphonooxy)phenyl]-9-[(4,6-O-ethylidene-β-D-glucopyranosyl)oxy]-5,8,8a,9-tetrahydro-furo[3',4':6,7]-naphtho[2,3-d]-1,3-dioxol-6(5aH)-one.
BMY-40481. Antineoplastic agent. *Bristol-Myers Squibb Pharm. R&D.*

5163 Etoprine
18588-57-3
$C_{12}H_{12}Cl_2N_4$
2,4-Diamino-5-(3,4-dichlorophenyl)-6-ethylpyrimidine.
Ethodichlorophen; NSC-3062. Antineoplastic agent.

5164 Fadrozole
102676-47-1 3969
$C_{14}H_{13}N_3$
(±)-p-(5,6,7,8-Tetrahydroimidazo[1,5-a]pyridin-5-yl)-benzonitrile.
Antineoplastic agent. Aromatase inhibitor. mp = 117-118°. *Ciba-Geigy Corp.*

5165 Fadrozole Hydrochloride
102676-96-0 3969
$C_{14}H_{14}ClN_3$
(±)-p-(5,6,7,8-Tetrahydroimidazo[1,5-a]pyridin-5-yl)-benzonitrile monohydrochloride.
CGS-16949A. Antineoplastic agent. Aromatase inhibitor. mp = 231-233°; soluble in H_2O. *Ciba-Geigy Corp.*

5166 Fazarabine
65886-71-7
$C_8H_{12}N_4O_5$
4-Amino-1-β-D-arabinofuranosyl-s-triazin-2(1H)-one.
ara-AC; 5-azacytosine arabinoside; Kymarabine; NSC-281272. Antineoplastic agent. Canine contagious hepatitis vaccine.

5167 Fenretinide
65646-68-6 4040
$C_{26}H_{33}NO_2$
N-(4-Hydroxyphenyl)retinamide.
all-trans-4'-hydroxyretinanilide. Antineoplastic. mp = 173-175°, 162-163°, 178-181°; λ_m = 370 nm (ε 44500 $CHCl_3$), λ_m = 362 nm (ε 47900 MeOH). *McNeil Pharm.*

5168 Floxuridine
50-91-9 4148
$C_9H_{11}FN_2O_5$
2'-Deoxy-5-fluorouridine.
FUDR; NSC-26740. Antiviral; antineoplastic agent. mp = 150-151°; λ_m = 268 nm (ε 7570, pH 7.2), 270 nm (ε 6480, pH 14); $[\alpha]_D$ = 37° (H_2O), 48.6° (DMF). *Hoffmann-LaRoche Inc.*

5169 Fludarabine
21679-14-1 4162
9-β-D-Arabinofuranosyl-2-fluoroadenine.
2-Fluoro Ara-A; NSC-118218; 2-Fluoroadenine arabinoside. Antineoplastic agent. mp = 260°; $[\alpha]_D^{25}$ = 17° (EtOH c = 0.1); λ_m = 262 nm (ε 13,200 pH 1), 261 nm (ε

14,800 pH 7), 262 nm (ε 15,000 pH 13); poorly soluble in H_2O, organic solvents. *Berlex Labs Inc.*

5170 Fludarabine Phosphate
75607-67-9 4162
$C_{10}H_{13}FN_5O_7P$
9-β-D-Arabinofuranosyl-2-fluoroadenine 5'-(dihydrogen-phosphate).
2-F-ara-AMP; NSC-328002; Fludara; NSC-312887. Antineoplastic agent. Soluble in H_2O. *Berlex Labs Inc.*

5171 Fluorouracil
51-21-8 4219 200-085-6
$C_4H_3FN_2O_2$
5-Fluoro-2,4(1H,3H)-Pyrimidinedione.
Fluorouracil; FU; 5-FU; Adrucil; Efudex; Fluoroplex; Ro-2-9757; Arumel; Carzonal; Effluderm (free base); Efudix; Fluoroblastin; Fluracil; Fluri; Fluril; Kecimeton; Timazin; U-8953; Ulup; 5-Flouracyl; efurix; fluracilum; ftoruracil; queroplex; NSC-19893. Antineoplastic agent. mp = 282-283° (dec); λ_m = 265-266 nm (ε 7070 0.1N HCl); insoluble in H_2O; LD_{50} (rat orl) = 230 mg/kg. *Allergan Herbert; Hoffmann-LaRoche Inc.; Pharmacia & Upjohn.*

5172 Flurocitabine
40505-45-1
$C_9H_{10}FN_3O_4$
(2R,3R,3aS,9aR)-7-Fluoro-2,3,3a,9a-tetrahydro-3-hydroxy-6-imino-6H-furo[2',3':4,5]oxazolo[3,2-a]pyrimidine-2-methanol.
Ro-21-0702; AAFC; RN-37717-21-8; NSC-166641; Anhydro-arabinosyl-5-fluoro-cytosine; Cyclo cytidine, 5-fluoro-; Cyclo FC; 5-fluoro-O,2,2'-cyclocytidine; 5-F-Anhydro-ara-C hydrochloride. Antineoplastic agent. *Hoffmann-LaRoche Inc.*

5173 Folinic Acid
58-05-9 4254 200-361-6
$C_{20}H_{23}N_7O_7$
N-[p-[[(2-Amino-5-formyl-5,6,7,8-tetrahydro-4-hydroxy-6-pteridinyl)methyl]amino]benzoyl]-glutamic acid.
5-formyltetrahydrofolate; CF; citrovorum factor; leucovorin. Antianemic (folate deficiency); antidote to folic acid antagonists. Antineoplastic agent. mp = 240-250° (dec); λ_m = 282 nm (%T = 27.0 for 10 mg/l in 0.1N NaOH). *Glaxo Wellcome Inc.; Res. Corp.*

5174 Fosfestrol
522-40-7 4279 208-328-8
$C_{18}H_{22}O_8P_2$
α,α'-Diethyl-(E)-4,4'-stilbenediol bis(dihydrogen phosphate).
DESDP; diethylstilbesterol bisphosphate; diethylstilbestryl bisphosphate; stilbestrol diphosphate; Stilphostrol. Estrogen; antineoplastic (hormonal). mp = 204-206° (dec); poorly soluble in H_2O. *Asta-Werke AG; Bayer AG; Bayer Corp. Pharm. Div.; Miles Inc.*

5175 Fosfestrol Disodium Salt
5965-09-3 4279 227-746-1
$C_{18}H_{20}Na_2O_8P_2$
α,α'-Diethyl-(E)-4,4'-stilbenediol bis(dihydrogen phosphate) disodium salt.
Estrogen; antineoplastic (hormonal). mp = 230°; soluble in H_20. *Asta-Werke AG; Bayer AG; Bayer Corp. Pharm. Div.; Miles Inc.*

5176 Fosfestrol Tetrasodium Salt
4719-75-9 4279 225-209-6
$C_{18}H_{18}Na_{28}P_2$
α,α'-Diethyl-(E)-4,4'-stilbenediol bis(dihydrogen phosphate) tetrasodium salt.
st52-asta; ST-51; Cytonal; Honvan; Honvol; Stilbostatin. Estrogen; antineoplastic (hormonal). *Asta-Werke AG; Bayer AG; Bayer Corp. Pharm. Div.; Miles Inc.*

5177 Fosquidone
114517-02-1
$C_{28}H_{22}NO_6P$
Benzyl (±)-5,8,13,14-tetrahydro-14-methyl-8,13-dioxobenz[5,6]isoindolo[2,1-b]isoquinolin-9-yl hydrogen phosphate.
GR-63178K. Antineoplastic agent. *Glaxo Wellcome plc.*

5178 Fostriecin
87810-56-8
$C_{19}H_{27}O_9P$
5,6-Dihydro-6-[3,4,6,13-tetrahydroxy-3-methyl-1,7,9,11-tridecatetraenyl]2H-pyran-2-one 4-(hydrogen phosphate. Antineoplastic agent. *Parke-Davis.*

5179 Fostriecin Sodium
87860-39-7
$C_{19}H_{26}NaO_9P$
5,6-Dihydro-6-[3,4,6,13-tetrahydroxy-3-methyl-1,7,9,11-tridecatetraenyl]2H-pyran-2-one 4-(sodium hydrogen phosphate.
CI-920. Antineoplastic agent. *Parke-Davis.*

5180 Fotemustine
92118-27-9 4285
$C_9H_{19}ClN_3O_5P$
(±) Diethyl [1-[3-(2-chloroethyl)-3-nitrosoureido]-ethyl]phosphonate.
S-10036; Muphoran. Antineoplastic agent. mp = 85°.

5181 Gallium Nitrate
13494-90-1 4364
GaN_3O_9
Gallium(III) nitrate (1:3), anhydrous.
gallium trinitrate; nitric acid, gallium salt; Ganite; NSC-15200. Antineoplastic; antihypercalcemic. Calcium regulator. Bone resorption inhibitor. Soluble in H_2O, EtOH, Et_2O; LD_{50} (mus iv) = 55 mg/kg, (rat iv) = 46 mg/kg, (rbt iv) = 43 mg/kg.

5182 Galocitabine
124012-42-6
$C_{19}H_{22}FN_3O_8$
N-[1-(5-Deoxy-β-D-ribofuranosyl)-5-fluoro-1,2-dihydro-2-oxo-4-pyrimidinyl]-3,4,5-trimethoxybenzamide.
Antineoplastic.

5183 Gemcitabine
95058-81-4 4392
$C_9H_{11}F_2N_3O_4$
2'-Deoxy-2',2'-difluorocytidine.
LY-188011;dFdC; dFdCyd; Gemzar; NSC-613327. Antineoplastic agent. $[\alpha]_{365}$ = 425° (MeOH c = 0.96);

λ_m = 234, 268 nm (ε 7810, 8560 EtOH); LD_{10} (rat iv) = 200 mg/m^2. *Eli Lilly & Co.*

5184 Gemcitabine Hydrochloride
122111-03-9 4392
$C_9H_{12}ClF_2N_3O_4$
2'-Deoxy-2',2'-difluorocytidine hydrochloride.
mp = 287-292°; $[\alpha]_D$ = 48°; $[\alpha]_{365}$ = 257.9° (D_2O c = 1); λ_m = 232, 268 nm (ε 7960, 9360, H_2O).

5185 Gold Au-198
10043-49-9 4533
^{198}Au
Gold-198.
colloidal gold (^{198}Au); Aurcoloid; Aureotrope; Auroscan-198. Antineoplastic agent. Radioactive agent (β- and γ-emitter) use as a diagnostic reagent in liver imaging. *Abbott Labs Inc.; Bristol-Myers Squibb Pharm. R&D; Mallinckrodt Inc.*

5186 Hexaprofen
24645-20-3
$C_{15}H_{20}O_2$
p-Cyclohexylhydratropic acid.
Reported to be antimetastatic in Lewis lung tumors.

5187 Hydroxyurea
127-07-1 4896 204-821-7
$CH_4N_2O_2$
N-(Aminocarbonyl)hydroxylamine.
Biosupressin; Hidrix; Hydrea; Hydreia; N-carbamoyl-hydroxylamine; Hydroxylurea; NSC-32065; Hydroxyurea; Hydura; Hydurea; HU; Litaler; Litalir; N-Hydroxyurea; NCI-C04831; Onco-carbide; Oxyurea; SK 22591; SQ-1089; Urea,. mp = 133-136°; soluble in H_2O, EtOH. *Bristol-Myers Oncology.*

5188 Idarubicin
58957-92-9 4931
$C_{26}H_{27}NO_9$
(7S-cis)-9-Acetyl-7-[(3-amino-2,3,6-trideoxy-α-L-lyxo-hexopyranosyl)oxy]-7,8,9,10-tetrahydro- 6,9,11-trihydroxy-5,12-naphthacenedione.
Daunomycin, 4-demethoxy-; Idarubicin; 4-DMD; 4-Demethoxydaunorubicin; NSC-256439. Antineoplastic agent. *Farmitalia Carlo Erba Ltd.*

5189 Idarubicin Hydrochloride
57852-57-0 4931
$C_{26}H_{28}ClNO_9$
(7S-cis)-9-Acetyl-7-[(3-amino-2,3,6-trideoxy-α-L-lyxo-hexopyranosyl)oxy]-7,8,9,10-tetrahydro- 6,9,11-tri-hydroxy-5,12-naphthacenedione hydrochloride.
Daunomycin, 4-demethoxy-, hydrochloride; Idarubicin hydrochloride; 4-DMD HCl; 4-Demethoxydaunorubicin hydrochloride; NSC-256439; Idamycin; Zavedos. Antineoplastic agent. mp = 183-185°, 172-174°; $[\alpha]_D^{20}$ = 205°, 188° (MeOH c = 0.1). *Farmitalia Carlo Erba Ltd.*

5190 Ifosfamide
3778-73-2 4937 223-237-3
$C_7H_{15}Cl_2N_2O_2P$
3-(2-Chloroethyl)-2-[(2-chloroethyl)amino]tetrahydro-2H-1,3,2-oxazaphosphorine 2-oxide.
Isophosphamide; Ifex; Iphosphamid; Isoendoxan; Cyfos; Holoxan; Mitoxana; Naxamide; A-4942; Asta-Z-4942; NSC-109724; Holoxan 1000; Ifosfamid; Iphosphamid; Isofosfamide; MJF-9325; Z-4942. Antineoplastic agent. Cytostatic related to cyclophosphamide. mp = 39-41°; LD_{50} (rat ip) = 150 mg/kg, 160 mg/kg. *Bristol-Myers Oncology.*

5191 Ilmofosine
89315-55-9
$C_{26}H_{56}NO_5PS$
(±)-3,5-Dioxa-9-thia-4-phosphapentacosan-1-aminium 4-hydroxy-7-(methoxymethyl)-N,N,N-trimethyl hydroxide inner salt 4-oxide.
BM-41440. *Boehringer Mannheim GmbH.*

5192 Imexon
59643-91-3
$C_4H_5N_3O$
4-Imino-1,3-diazabicyclo[3.1.0]hexan-2-one.
Antineoplastic.

5193 Improsulfan
13425-98-4 4962
$C_8H_{19}NO_6S_2$
3,3'-Iminobis(1-propanol) dimethanesulfonate.
Improsulfan tosylate; Improsulfan; NSC-102627; Yoshi 864; Bis(3-mesyloxypropyl)amine hydrochloride; Compound 864; IPD; IPD hydrochloride. mp = 94-95°; LD_{50} (rat iv) = 75 mg/kg.

5194 Interferon α
74899-72-2 5016 232-710-3
Alpha interferon.
alfa-interferon; alferon; OFN-α; LeIF; leukocyte interferon; lymphoblastoid interferon; Berfor [α-2C]; Alpha 2 [α-2C]; NSC-339140 [α-n1]; Suniferon [α-n1]; Wellferon [α-n1]; Alferon [α-n3]. Antineoplastic; antiviral; immunomodulator. In combination with nevirapine, has been used to treat hepatitis C successfully. Family of homologous proteins that inhibit cell proliferation and viral replication; modulates immune response. Produced by human lymphoblastoid cells induced with *Sendai* virus. MW = 18-20 kD.

5195 Interferon α-2a
76543-88-9 5016
$C_{860}H_{1353}N_{227}O_{255}S_9$
Interferon alfa-2a.
interferon αA; IFN-αA; Ro-22-8181; Canferon; Roferon-A; Ro-22-8181; roferon A. Antineoplastic; antiviral; immunomodulator. Alpha interferon, natural (injectable form); used for the treatment of genital warts. *Hoffmann-LaRoche Inc.; Interferon Sciences Inc.*

5196 Interferon α-2b
99210-65-8 5016
$C_{860}H_{1353}N_{229}O_{255}S_9$
Interferon alfa-2b.
INF-α_2; Sch-30500; YM-14090; Cibian; Introna; Intron A; Viraferon. Antineoplastic; antiviral; immunomodulator. *Schering-Plough HealthCare Products.*

5197 Interferon β
74899-71-1 5017 232-710-3
Beta interferon.

fibroblast interferon; FIF; IFN-β; Feron; Fiblaferon; Frone; Nafron. Antineoplastic; antiviral; immunomodulator.

5198 Interferon β-1a
145258-61-3 5017
$C_{908}H_{1406}N_{246}O_{252}S_7$
Interferon beta-1 (human fibroblast protein moiety).
Neoferon. Antineoplastic; antiviral; immunomodulator. A glycosylate polypeptide of 166 amino acid residues. From cultured Chinese Hamster ovary containing the engineered gene for human interferon beta. *Biogen, Inc.*

5199 Interferon β-1b
145155-23-3 5017
$C_{903}H_{1399}N_{245}O_{252}S_5$
17-L-Serine-2-166 interferon β1 (human fibroblast reduced).
Betaseron. Antineoplastic; antiviral; immunomodulator. A non-glycosylated polypeptide of 165 amino acid residues. Produced by *E. Coli. Berlex Labs Inc.*

5200 Interferon γ
9008-11-1 5018 232-710-3
Gamma interferon.
IFN-γ; immune IFN; ImIFN; type II interferon; S-6810 [as cys-tyr-cyc-interferon γ]; Actimmune [as cys-tyr-cyc-interferon γ]; Gammaferon [as cys-tyr-cyc-interferon γ]; Immuneron [as cys-tyr-cyc-interferon γ]; Polyferon [as cys-tyr-cyc-interferon γ]. Antiviral; antineoplastic; immunomodulator. Lymphokine produced by T-cells, structurally unrelated to interferon-α or -β. Molecular weight 16,000-25,000 daltons.

5201 Interferon γ-1a
98059-18-8 5018
$C_{761}H_{1206}N_{214}O_{225}S_6$
Interferon gamma-a.
Antineoplastic; antiviral; immunomodulator. Polypeptide of 146 amino acid residues. Produced in *E coli* K-12 (C600) by expression interferon gamma cDNA derived from human splenic lymphosyte mRNA. *Berlex Labs Inc.*

5202 Interferon γ-1b
98059-61-1 5018
$C_{734}H_{1166}N_{204}O_{204}S_5$
N^2-L-Methionyl-1-139-interferon-γ (human lymphocyte protein moiety reduced).
Actimmune; Immukin. Antineoplastic; antiviral; immunomodulator. *Genentech Inc.*

5203 Iodothiouracil
5984-97-4
$C_4H_3IN_2OS$
5-Iodo-2-thiouracil.
Antineoplastic, used with iodine isotopes for treatment and diagnosis.

5204 Iproplatin
62928-11-4
$C_6H_{20}Cl+62N_2O_2Pt$
ab-Dichloro-cd-dihydroxy-df-bis(isopropylamine)-platinum.
CHIP; Iproplatin (USAN); NSC-256927; JM-9. Antineoplastic agent. *Bristol-Myers Squibb Pharm. R&D.*

5205 Irinotecan
97682-44-5 5104
$C_{33}H_{38}N_4O_6$
(+)-7-Ethyl-10-hydroxycamptothecine 10-[1,4'-bipiperidine]-1'-carboxylate.
Antineoplastic agent. DNA topoisomerase I inhibitor. mp = 222-223°. *Pharmacia & Upjohn.*

5206 Irinotecan Hydrochloride Trihydrate
136572-09-3 5104
$C_{33}H_{39}ClN_4O_6.3H_2O$
(+)-7-Ethyl-10-hydroxycamptothecine 10-[1,4'-bipiperidine]-1'-carboxylate monohydrochloride trihydrate.
Camptosar; U-101440E. Antineoplastic agent. mp = 256.5°; $[\alpha]_D^{20}$ = 67.7° (H_2O, c = 1); λ_m = 221, 254, 359, 372 nm (ε 53800, 36600, 26200, 25300, EtOH); LD_{50} (mus ip) = 177.5 mg/kg, (mus orl) = 765.3 mg/kg. *Pharmacia & Upjohn.*

5207 Letrozole
112809-51-5 5474
$C_{17}H_{11}N_5$
4,4'-(1H-1,2,4-Triazol-1-ylmethylene)dibenzonitrile.
CGS-20267. Antineoplastic agent. Aromatase inhibitor. Related to fadrozole. mp = 181-183°. *Ciba-Geigy Corp.*

5208 Liarozole Hydrochloride
145858-50-0
$C_{17}H_{14}Cl_2N_4$
(±)-5-(m-Chloro-α-imidazol-1-ylbenzyl)benzimidazole monhydrochloride.
R-75251. Antineoplastic agent. *Janssen Pharm. Inc.*

5209 Lobaplatin
135558-11-1
$C_9H_{18}N_2O_3Pt$
cis-[trans-1,2-Cyclobutanebis(methylamine)][(S)-lactato-O^1,O^1]platinum.
D-19466. Platinum-based cytostatic drug. Antineoplastic.

5210 Lometrexol
106400-81-1
$C_{21}H_{25}N_5O_6$
N-[p-[2-[(R)-2-Amino-3,4,5,6,7,8-hexahydro-4-oxopyrido[2,3-d]pyrimidin-6-yl]ethyl]benzoyl]-L-glutamic acid.
LY-264618. Antineoplastic agent. *Eli Lilly & Co.*

5211 Lometrexol Sodium
120408-07-3
$C_{21}H_{23}Na_2N_5O_6$
N-[p-[2-[(R)-2-Amino-3,4,5,6,7,8-hexahydro-4-oxopyrido[2,3-d]pyrimidin-6-yl]ethyl]benzoyl]-L-glutamic acid disodium salt.
LY-264618 disodium. Antineoplastic agent. *Eli Lilly & Co.*

5212 Lomustine
13010-47-4 5594 235-859-2
$C_9H_{16}ClN_3O_2$
N-(2-Chloroethyl)-N'-cyclohexyl-N-nitroso-urea.
RB-1509; Belustine; Cecenu; CeeNU; Chloroethylcyclohexylnitrosourea; CiNu; CCNU; ICIG-1109; NSC-79037; NCI-C04740; SRI-2200. Antineoplastic agent. CAUTION: This substance may be a

carcinogen. mp = 90°; soluble in H_2O, organic solvents; LD_{50} (mus orl) = 51 mg/kg, (mus ip) = 56 mg/kg, (mus sc) = 61 mg/kg. *Bristol-Myers Oncology.*

5213 Lonidamine
50264-69-2 5598
$C_{15}H_{10}Cl_2N_2O_2$
1-(2,4-Dichlorobenzyl)-1H-imidazole-3-carboxylic acid.
Antineoplastic agent (adjunct). mp = 207°; soluble in MeOH, AcOH; LD_{50} (mus orl)= 900 mg/kg, (mus ip) = 435 mg/kg, (rat orl) = 1700 mg/kg, (rat ip) = 525 mg/kg.

5214 Losoxantrone
88303-60-0
$C_{22}H_{27}N_5O_4$
7-Hydroxy-2-[2-[(2-hydroxyethyl)amino]ethyl]-5-[[2-[(2-hydroxyethyl)amino]ethyl]amino]anthra-[1,9-cd]-pyrazol-6(2H)-one.
biantrazole. Antineoplastic agent. *DuPont-Merck Pharm.*

5215 Losoxantrone Hydrochloride
88303-61-1
$C_{22}H_{29}Cl_2N_5O_4.0.5H_2O$
7-Hydroxy-2-[2-[(2-hydroxyethyl)amino]ethyl]-5-[[2-[(2-hydroxyethyl)amino]ethyl]amino]anthra-[1,9-cd]-pyrazol-6(2H)-one dihydrochloride hemihydrate.
DUP-941; NSC-357885. Antineoplastic agent. *DuPont-Merck Pharm.*

5216 Lurtotecan
149882-10-0
$C_{28}H_{30}N_4O_6$
(8S)-8-Ethyl-2,3-dihydro-8-hydroxy-15-[(4-methyl-1-piperazinyl)methyl]-11H-p-dioxino[2,3-g]pyrano-[3',4':6,7]indolizino[1,2-b]quinoline-9,12(8H,14H)-dione.
Antineoplastic agent. DNA topoisomerase I inhibitor. *Glaxo Wellcome Inc.*

5217 Lurtotecan Dihydrochloride
155773-58-3
$C_{28}H_{32}Cl_2N_4O_6$
(8S)-8-Ethyl-2,3-dihydro-8-hydroxy-15-[(4-methyl-1-piperazinyl)methyl]-11H-p-dioxino[2,3-g]pyrano-[3',4':6,7]indolizino[1,2-b]quinoline-9,12(8H,14H)-dione dihydrochloride.
GI147211C.

5218 Mafosfamide
88859-04-5
$C_9H_{19}Cl_2N_2O_5PS_2$
(±)-2-[[2-[Bis(2-chloroethyl)amino]tetrahydro-2H-1,3,2-oxazaphosphorin-4-yl]thio]ethanesulfonic acid P-cis-oxide.
Antineoplastic; immunosuppressant. A cyclophosphamide.

5219 Mannomustine Hydrochloride
551-74-6
$C_{10}H_{24}Cl_4N_2O_4$
1,6-bis[(2-Chloroethyl)amino]-1,6-dideoxy-D-mannitol hydrochloride.
BCM; Degranol; Degranol chinoin; Dimesylmannitol; Mannitol mustard dihydrochloride; NSC-9698; Mannitol nitrogen mustard; Mannogranol; Mannomustine dihydrochloride. Antineoplastic agent.

5220 Mannosulfan
7518-35-6
$C_{10}H_{22}O_{14}S_4$
D-Mannitol 1,2,5,6-tetramethanesulfonate.
R-52. Antineoplastic.

5221 Masoprocol
27686-84-6 6786
meso-4,4'-(2,3-Dimethyltetramethylene)dipyrocatachol.
CHX-100; meso-NDGA; Actinex. Antineoplastic agent. Used as an antioxidant for fats and oils in foods. mp = 185-186°; λ_m = 283, 218 nm (ε 6660, 13400 MeOH); poorly soluble in H_2O, more soluble in EtOH, insoluble in non-polar organic solvents. *Schwarz Pharma Kremers Urban Co.*

5222 Maytansine
35846-53-8 5800
$C_{34}H_{46}ClN_3O_{10}$
N-Acetyl-N-methyl alanine 6-ester with 11-chloro-6,21-dihydroxy-12,20-dimethoxy-2,5,9,16-tetramethyl-4,24-dioxa-9,22-diazatetracyclo[19.3.1.1(10,24).0(3,5)]-hexacosa-10,12,14[26] 16,18-pentaene-8,23-dione.
Maitansine; Maysanine; Maytansin; Maytansine; MTS; NSC-153858. Antineoplastic agent. mp= 171-172°; $[\alpha]_D^{26}$= -145° ($CHCl_3$ c = 0.055); λ_m = 233, 254, 282, 290 nm (ε 29800, 27200, 5690, 5520, EtOH); LD_{50} (rat sc) = 0.48 mg/kg. *Bristol-Myers Squibb Pharm. R&D.*

5223 Mechlorethamine
51-75-2 5815 200-120-5
$C_5H_{11}Cl_2N$
N,N-Bis(2-chloroethyl)methylamine.
MBA; Nitrogen Mustard; Mechloroethamine; HN-2; Mustine Note; Dichloren; Caryolysine. Antineoplastic agent. mp = -60°; bp_{18} = 87°; d_4^{25}= 1.118; slightly soluble in H_2O, soluble in organic solvents; LD_{50} (rat iv) = 1.1 mg/kg (hydrochloride). *Merck & Co.Inc.*

5224 Mechlorethamine Hydrochloride
55-86-7 5815
$C_5H_{12}Cl_3N$
N,N-Bis(2-chloroethyl)methylamine monohydrochloride.
Azotoyperite; C-6866; Caryolysine hydrochloride; Chloramin hydrochloride; Chlorethamine; Chlorethazine; Chlormethine hydrochloride; Chlormethinum; Dema; Dichloren hydrochloride; Dichloromethyldiethylamine hydrochloride; Dimitan; Embechine; Embichin hydrochloride; NSC-762; Embikhine; Erasol hydrochloride; Erasol-Ido; HN2 hydrochloride; Mechlorethamine hydrochloride; Mitoxine; Mustargen hydrochloride; Mustine hydrochloride; MBA hydrochloride; N-Lost; Nitol; Nitol takeda; Nitrogen mustard hydrochloride; Nitrogranulogen hydrochloride; NCI-C56382; NM; SK-101. Antineoplastic agent. *Merck & Co.Inc.*

5225 Medorubicin
64314-52-9
$C_{26}H_{27}NO_{10}$
(7S-cis)-7-[(3-Amino-2,3,6-trideoxy-α-L-lyxo-hexopyranosyl)oxy]-7,8,9,10-tetrahydro-6,9,11-trihydroxy-9-(hydroxyacetyl)-5,12-naphthacenedione.
Medorubicin; 4-DMA; 4-Demethoxyadriamycin. Antineoplastic agent.

5226 Medorubicin Hydrochloride
64363-63-9
$C_{26}H_{28}ClNO_{10}$
(7S-cis)-7-[(3-Amino-2,3,6-trideoxy-α-L-lyxo-hexopyranosyl)oxy]-7,8,9,10-tetrahydro-6,9,11-trihydroxy-9-(hydroxyacetyl)-5,12-naphthacenedione hydrochloride.
Medorubicin hydrochloride; 4-DMA HCl; 4-Demethoxyadriamycin HCl; NSC-256438. Antineoplastic agent.

5227 Melphalan
148-82-3 5871 205-726-3
$C_{13}H_{18}Cl_2N_2O_2$
4-[Bis(2-chloroethyl)amino]-L-phenylalanine.
CB-3025; alanine nitrogen mustard; L-phenylalanine mustard hydrochloride; L-PAM; melfalan; L-sarcolysine; Alkeran; Sarcoclorin; NSC-8806 [as hydrochloride]. Antineoplastic agent. Carcinogen. The intravenous formulation is used for the treatment of localized malignant melanoma of the extremities and localized soft tissue sarcoma of the extremities by regional arterial perfusion. In tablet formulation, used for the palliative treatment of multiple myeloma and advanced ovarian adrenocarcinoma. mp = 182-183°; $[\alpha]_D^{25}$= 7.5° (1.0N HCl); $[\alpha]_D^{22}$= -31.5° (MeOH, c = 0.67); insoluble in H_2O, soluble in EtOH; LD_{50} (rat ip) = 4.5 mg/kg. *Wellcome Foundation Ltd.*

5228 DL-Melphalan
5871
$C_{13}H_{18}Cl_2N_2O_2$
4-[Bis(2-chloroethyl)amino]-phenylalanine.
merphalan; sarcolysine. Antineoplastic agent. Carcinogen. mp = 180-181°.

5229 D-Melphalan
5871
$C_{13}H_{18}Cl_2N_2O_2$
4-[Bis(2-chloroethyl)amino]-D-phenylalanine.
D-sarcolysine; medphalan; CB-3026. Antineoplastic agent. Carcinogen. mp = 181.5-182° (dec); $[\alpha]_D^{21}$= -7.5° (c = 1.26 in 1.0 N HCl).

5230 Menogaril
71628-96-1 5881
$C_{28}H_{31}NO_{10}$
4-(Dimethylamino)-3,4,5,6,11,12,13,14-octahydro-3,5,8,10,13-pentahydroxy-11-methoxy-6,13-dimethyl 2,6-epoxy-2H-naphthaceno[1,2-b]oxocin-9,16-dione.
U-52047; 7(R)-O-Methylnogarol; 7-O-Methylnogarol; 7-OMEN; NSC-269148. Antineoplastic agent. mp = 247-249° (dec), 250-254° (dec); $[\alpha]_D^{20}$ = 857° ($CHCl_3$ c = 0.112), 867° ($CHCl_3$ c = 0.045); 958° ($CHCl_3$ c = 0.163); λ_m = 235, 251, 257, 290, 479 (ε 41200, 25500, 24150, 10500, 15530 EtOH). *Upjohn Ltd.*

5231 Mequinol
150-76-5 205-769-8
$C_7H_8O_2$
4-Methoxyphenol.
Antioxidant. Studied for its utility in treatment of malignant melanomas.

5232 6-Mercaptopurine
6112-76-1 5919
$C_5H_4N_4S$
1,7-Dihydro-6H-purine-6-thione monohydrate.
6-purinethiol hydrate; purine-6-thiol monohydrate; NSC-755. Antineoplastic; immunosuppressant. Dec 313-314°; λ_m = 230, 312 nm (ε 14000, 19600 0.1N NaOH); insoluble in H_2O, organic solvents; slightly soluble in EtOH; LD_{50} (mus ip) = 157 mg/kg.

5233 6-Mercaptopurine [anhydrous]
50-44-2 5919 200-037-4
$C_5H_4N_4S$
1,7-Dihydro-6H-purine-6-thione.
purine-6-thiol; thiohypoxanthine; 3H-purine-6-thiol; 6MP; 6-purinethiol; 6-thioxopurine; 7-mercapto-1,3,4,6-tetrazaindene; U-4748; Ismipur; Leukeran; Leukerin; Leupurin; Mercaleukim; Mercaleukin; Mercaptopurine; Mercapurin; Mern; Puri-Nethol; Purimethol; Purinethiol. Antineoplastic; immunosuppressant.

5234 Metamelfalan
1088-80-8
$C_{13}H_{18}Cl_2N_2O_2$
3-[m-[Bis(2-chloroethyl)amino]phenyl]-L-alanine.
L-m-Sarcolysin; NSC-67781. Antineoplastic agent.

5235 Metesind
138384-68-6
$C_{23}H_{24}N_4O_3S$
4-[[α-[(2-Aminobenz[c,d]indol-6-yl)methylamino]-p-tolyl]sulfonyl]morpholine.
Antineoplastic agent. Thymidylate synthase inhibitor. *Agouron Pharm. Inc.*

5236 Metesind Glucuronate
157182-23-5
$C_{29}H_{34}N_4O_{10}S$
4-[[α-[(2-Aminobenz[c,d]indol-6-yl)methylamino]-p-tolyl]sulfonyl]morpholine mono-D-glucuronate.
AG-331. Antineoplastic agent. Thymidylate synthase inhibitor. *Agouron Pharm. Inc.*

5237 Methotrexate
59-05-2 6065 200-413-8
$C_{20}H_{22}N_8O_5$
L-(+)-N-[[(2,4-Diamino-6-pteridinyl)methyl]methylamino]benzoyl]glutamic acid.
4-amino-10-methylfolic acid; amethopterin; methylaminopterin; CL-14377; EMT-25299; NSC-740; MTX; Emtexate; Ledertrexate; Metatrexan; A-Methopterin; Methopterin; Mexate; Rheumatrex. Antineoplastic; antirheumatic. Folic acid antagonist. [monohydrate]: mp = 185-204°; λ_m = 244, 307 nm (0.1N HCl); soluble in aqueous solutions with some decomposition; LD_{50} (rat orl) = 135 mg/kg. *Bristol-Myers Oncology; Lederle Labs.*

5238 Methylthiouracil
56-04-2 6210 200-252-3
$C_5H_6N_2OS$
2,3-Dihydro-6-methyl-2-thioxo-4(1H)-pyrimidinone.
4-methyluracil; 6-methyl-2-thiouracil; NSC-9378; NSC-193526; Alkiron; Antibason; Basecil; Basethyrin; Methiacil; Methicil; Methiocil; Methylthiouracil; Muracil; Muracin; MTU; Orcanon; Prostrumyl; Strumacil;

Thimecil; Thiomecil; Thiomidil; Thioryl; Thiothymin; Thiothyron; Thiuryl; Thyreonorm; Thyreostat; Thyreostat I; Thyril; Tiomeracil; Tiorale M; Tiotiron; USAF EK-6454. Antineoplastic agent. Thyroid inhibitor. mp = 326-331° (dec); slightly soluble in H_2O at 25° (0.67 g/100 ml at 100°), Et_2O, EtOH, Me_2CO; insoluble in C_6H_6, $CHCl_3$; MLD (rbt orl) = 2500 mg/kg.

5239 Metoprine
7761-45-7
$C_{11}H_{10}Cl_2N_4$
5-(3,4-Dichlorophenyl)-6-methyl-2,4-pyrimidine-diamine.
NSC-19494; BW-197U; BW-50197; DDMP; Methodichlorophen; SK-5265; U-197; NSC-7364. Antineoplastic agent.

5240 Meturedepa
1661-29-6 6245
$C_{11}H_{22}N_3O_3P$
Ethyl [bis(2,2-dimethyl-1-aziridinyl)phosphinyl]-carbamate.
AB-132; Turloc; NSC-51325. Antineoplastic agent. mp = 119-121°.

5241 Miltefosine
58066-85-6 6285
$C_{21}H_{46}NO_4P$
Choline hydroxide hexadecyl hydrogen phosphate inner salt.
hexadecylphosphorylcholine; D-18506; HPC; Miltex. Antineoplastic. mp = 232-234° (dec); LD_{50} (rat orl) = 246 mg/kg.

5242 Miproxifene
129612-87-9
$C_{29}H_{35}NO_2$
(Z)-α-[p-[2-(Dimethylamino)ethoxy]phenyl]-α'-ethyl-4'-isopropyl-4-stilbenol.
TAT-59 [as phosphate]. Antineoplastic.

5243 Mithramycin
18378-89-7 7696 232-455-8
$C_{22}H_{38}O_5$
[2S-[2α,3β(1R*,3R*,4S*)]]-6-[[2,6-Dideoxy-3-O-(2,6-dideoxy-β-D-arabino-hexopyranosyl)-β-D-arabino-hexopyranosyl]oxy]-2-[(O-2,6-dideoxy-3-C-methyl-β-D-ribo-hexopyranosyl-(1→4)-O-2,6-dideoxy-α-D-lyxo-hexopyranosyl-(1→3)-2,6-dideoxy-β-D-arabino-hexopyranosyl)oxy]-3-(3,4-dihydroxy-1-methoxy-2-oxopentyl)-3,4-dihydro-8,9-dihydroxy-7-methyl-1(2H)-anthracenone.
A-2371; Antibiotic LA-7017; Aurelic acid; Aureolic acid; Mithracin; Mithramycin A; Mitramycin; PA-144; NSC-24559; plicamycin. Antineoplastic agent. mp = 180-183°; $[\alpha]_D^{20}$ = 51° (c = 0.4 EtOH); soluble in H_2O, polar organic solvents; less soluble in Et_2O, C_6H_6; LD_{50} (rat iv) = 1.74 mg/kg. *Bayer Corp. Pharm. Div.*

5244 Mitindomide
10403-51-7
$C_{14}H_{12}N_2O_4$
(3aα,3bβ,4α,4aβ,7aβ,8α,8aβ,8bα)-3a,3b,4,4a,7a,8,8a,8b-Octahydro-4,8-ethenopyrrolo[3',4':3,4]cyclobut[1,2-f]isoindole-1,3,5,7(2H,6H)tetrone.
Benzenebismaleimide adduct; Mitindomide; NSC-284356. Antineoplastic agent. *National Cancer Inst.*

5245 Mitobronitol
488-41-5 6298
$C_6H_{12}Br_2O_4$
1,6-Dibromo-1,6-dideoxy-D-mannitol.
D-Dibromomannitol; Dibromannit; Dibromannitol; Dibromomannitol; Mielobromol; NSC-94100; Mitobronitol; Myelobromol; NCI-C04762; R-54; 1,6-Dibromo-D-mannitol; 1,6-Dibromo-1,6-dideoxy-D-mannitol; 1,6-Dibromomannitol. Antineoplastic agent. mp = 176-178°.

5246 Mitocarcin
11056-14-7
Antineoplastic agent. Antibiotic derived from *Streptomyces* species.

5247 Mitocromin
11043-98-4
B-35251; NSC-77471. Antineoplastic agent. Antibiotic derived from *Streptomyces viridochromogenes* species.

5248 Mitogillin
1403-99-2
Antineoplastic agent. Antibiotic produced by *Aspergillus restrictus*.

5249 Mitoguazone
459-86-9 6299
$C_5H_{12}N_8$
1,1'-[(Methylethanediylidene)dinitrilo]-diguanidine.
Antineoplastic agent. mp = 225° (dec); λ_m = 283 nm (ε 38400 pH 1), 325 nm (ε 33500 pH 11).

5250 Mitolactol
10318-26-0 6300
$C_6H_{12}Br_2O_4$
1,6-Dibromo-1,6-dideoxy-D-galactitol.
Dibromdulcit; Dibromdulcitol; Dibromodulcitol; Dibromogalactitol; DBD; Elobromol; NSC-104800; NCI-C04795. Antineoplastic agent. mp = 187-188°; LD_{50} (rat orl) = 1400 mg/kg, (rat ip) = 470 mg/kg.

5251 Mitomalcin
11043-99-5
NSC-113233. Antineoplastic agent. Antibiotic produced by *Streptomyces malayensis. Pfizer Intl.*

5252 Mitomycin
50-07-7 6301 200-008-6
$C_{15}H_{18}N_4O_5$
(1aS,8S,8aR,8bS)-6-Amino-1,1a,2,8,8a,8b-hexahydro-8-(hydroxymethyl)-8a-methoxy-5-methylazirino-[2',3':3,4]-pyrrolo[1,2-a]indole-4,7-dione carbamate (ester).
mitomycin C; MMC; Ametycine; Mitocin-C; Mutamycin; NSC-26980. Antineoplastic agent. One of the mitomycins, a group of antitumor antibiotics from *Streptomyces caespitosus*. mp > 360°; soluble in H_2O, organic solvents; λ_m = 216, 360, 560 nm ($E^{1\%}_{1\,cm}$ 742, 742, 0.06 MeOH); LD_{50} (mus iv) = 5 mg/kg. *Bristol-Myers Oncology.*

5253 Mitonafide
54824-17-8
$C_{16}H_{15}N_3O_4$
N-[2-(Dimethylamino)ethyl]-3-nitronaphthalimide.
Non-cationic tricyclic aromatic carboxamide cytotoxic agent. Antineoplastic.

5254 Mitopodozide
1508-45-8 7703
$C_{24}H_{30}N_2O_8$
Podophyllic acid 2-ethylhydrazide.
SPI-77; NSC-72274. Antineoplastic agent. $[\alpha]_D$= -154° ($CHCl_3$ c = 0.5).

5255 Mitoquidone
91753-07-0
$C_{20}H_{13}NO_2$
5,14-Dihydrobenz[5,6]isoindolo[2,1-b]isoquinoline-8,13-dione.
GR-30921. A pentacyclic pyrroloquinone. Antineoplastic.

5256 Mitosper
11056-15-8
NSC-117032. Antineoplastic agent. *State of Michigan Department of Public Health.*

5257 Mitotane
53-19-0 6302 200-166-6
$C_{14}H_{10}Cl_4$
1,1-Dichloro-2-(o-chlorophenyl)-2-(p-chlorophenyl)ethane.
Lysodren; o,p'-DDE; NSC-38721. Antineoplastic agent. mp = 76-78°; soluble in EtOH, isooctane, CCl_4. *Bristol-Myers Oncology.*

5258 Mitoxantrone
65271-80-9 6303
$C_{22}H_{28}N_4O_6$
1,4-Dihydroxy-5,8-bis[[2-[(2-hydroxyethyl)amino]-ethyl]amino]-9,10-anthracenedione.
Dihydroxyanthraquinone; DHAQ; Mitoxantrone [as free base]; NSC-279836. Antineoplastic agent. mp = 160-162°; λ_m = 244, 279, 525, 620, 660 nm (log ε 4.64, 4.31, 3.70, 4.37, 4.38 EtOH); sparingly soluble in H_2O, EtOH; insoluble in organic solvents. *Immunex Corp.*

5259 Mitoxantrone Dihydrochloride
70476-82-3 6303
$C_{22}H_{30}Cl_2N_4O_6$
1,4-Dihydroxy-5,8-bis[[2-[(2-hydroxyethyl)amino]-ethyl]amino]-9,10-anthracenedione dihydrochloride.
DHAQ; CL-232315; NSC-279836; Novantrone; Immunex. Antineoplastic agent. mp = 203-205°; λ_m = 241, 273, 608, 658 nm (ε 41000, 12000, 19200, 20900 H_2O); sparingly soluble in H_2O, MeOH; insoluble in organic solvents. *Immunex Corp.*

5260 Mitozolomide
85622-95-3 287-943-3
$C_7H_7ClN_6O_2$
3-(2-Chloroethyl)-3,4-dihydro-4-oxoimidazo[5,1-d]-astetrazine-8-carboxamide.
M&B-39565; NSC-353451. Antitumor bicyclic imidazotetrazine; chloroethylating agent. Antineoplastic.

5261 Mivobulin Isethionate
126268-81-3
$C_{17}H_{19}N_5O_2.C_2H_6O_4S$
Ethyl (S)-5-amino-1,2-dihydro-2-methyl-3-phenylpyrido-[3,4-b]pyrazine-7-carbamate mono(2-hydroxyethane-sulfonate).
BW-B1090U; Mivacron; CI-980. Antineoplastic agent. Microtubule inhibitor. *Parke-Davis.*

5262 Mofarotene
125533-88-2
$C_{29}H_{39}NO_2$
4-[2-[p-[(E)-2-(5,6,7,8-Tetrahydro-5,5,8,8-tetramethyl-2-naphthyl)propenyl]phenoxy]ethyl]-morpholine.
Ro-40-8757. Retinoid analog. Antineoplastic.

5263 Momordicine
$C_9H_{14}N_4O_4$
N-(Ethoxycarbonyl)-3-(4-morpholinyl)sydnone imine.
morsydomine; SIN-10; Corvaton; Corvasal; Molsidolat; Morial; Motazomin; Molsidain; Molsidaine. Antianginal.

5264 Mopidamol
13665-88-8 6344
$C_{19}H_{31}N_7O_4$
2,2',2,2'''-[(4-Piperidinopyrimido[5,4-d]pyrimidine-2,6-diyl)dintrilo]tetraethanol.
RA-233; Rapenton. Antineoplastic agent. Platelet aggregation inhibitor with antimetastatic properties. mp = 157-158°.

5265 Moss Starch
1402-10-4 5503 215-755-3
$C_{18}H_{32}O_{15}$
Lichenin.
Antineoplastic agent. Soluble in H_2O; $[\alpha]_D$ = 18.4°.

5266 Mycophenolic Acid
24280-93-1 6408
$C_{17}H_{20}O_6$
(E)-6-(1,3-Dihydro-4-hydroxy-6-methoxy-7-methyl-3-oxo-5-isobenzofuranyl)-4-methyl-4-hexenoic acid.
Melbex; NSC-129185; Lilly 68618; Melbex; MPA. Antineoplastic agent. mp = 141°; insoluble in H_2O, soluble in organic solvents; LD_{50} (mus orl) = 2500 mg/kg, (mus iv) = 550 mg/kg, (rat orl) = 700 mg/kg, (rat iv) = 450 mg/kg. *Eli Lilly & Co.*

5267 Myleran
55-98-1 1529 200-250-2
$C_6H_{14}O_6S_2$
1,4-Butanediol dimethanesulfonate.
AN-33501; Busulfan; Busulphan; Buzulfan; Citosulfan; CB-2041; GT-2041; GT-41; Leucosulfan; Mablin; Mielevcin; Mielosan; Mielucin; Milecitan; Mileran; Misulban; NSC-750; Mitostan; Myeloleukon; Myelosan; Mylecytan; Myleran; NCI-C01592; Sulphabutin; X-149; 1,4-Bis(methanesulfonyloxy)butane; 2041 C.B. Antineoplastic agent. mp = 114-118°; insoluble in H_2O; soluble in organic solvents; LD_{50} (mus iv) = 1.8 mg/kg. *Glaxo Wellcome Inc.*

5268 Mylosar
320-67-2 923 206-280-2
$C_8H_{12}N_4O_5$
4-Amino-1-β-D-ribofuranosyl-1,3,5-triazin-2(1H)-one.
5-azacytidine; ladakamycin; U-18496; NSC-102816. Antineoplastic agent. mp = 228-230°; $[\alpha]_D^{25}$ = 39° (H_2O, c = 1); λ_m = 241 nm (ε 8767 H_2O), 249 nm (ε 3077 0.01N HCl), 223 nm (ε 24200 0.01N KOH); LD_{50} (mus ip) = 115.9 mg/kg, (mus orl) = 572.3 mg/kg. *Pharmacia & Upjohn.*

5269 Nedaplatin
95734-82-0
$C_2H_8N_2O_3Pt$
cis-Diammine(glycolato-O^1,O^2)platinum.
Antineoplastic agent.

5270 Nemorubicin
108852-90-0
$C_{32}H_{37}NO_{13}$
(1S,3S)-3-Glycoloyl-1,2,3,4,6,11-hexahydro-3,5,12-trihydroxy-10-methoxy-6,11-dioxo-1-naphthacenyl 2,3,6-trideoxy-3-[(S)-2-methoxymorpholino]-α-L-lyxo-hexopyranoside.
Antineoplastic agent.

5271 Nigrin
3930-19-6 8986 223-501-8
$C_{25}H_{22}N_4O_8$
5-Amino-6-(7-amino-5,8-dihydro-6-methoxy-5,8-dioxo-2-quinolyl)-4-(2-hydroxy-3,4-dimethoxyphenyl)-3-methyl-picolinic acid.
streptonigrin; Abbott Crystalline antibiotic; Antibiotic from *Streptomyces flocculus*; AO50165L302; NSC-45383; Bruneomycin; Nigrin; Rufochromomycin; Rufocromomycin; Streptonigran; Streptonigrin; STP; STREPTONIGRIN; 5278 R. P. Antineoplastic agent. mp= 262-263° (dec); λ_m = 248, 375-380 nm (ε 38400, 17400, MeOH); slightly soluble in H_2O, alcohols; more soluble in dioxane, C_5H_5N, DMF. *Pfizer Inc.*

5272 Nimustine
42471-28-3 6645
$C_9H_{13}ClN_6O_2$
3-[(-4-Amino-2-methyl-5-pyrimidinyl)methyl]-1-(2-chloroethyl)-1-nitrosourea.
9-methylfolic acid; Bremfol. Antineoplastic agent. mp = 125° (dec).

5273 Nimustine Hydrochloride
55661-38-6 6645
$C_9H_{14}Cl_2N_6O_2$
3-[(-4-Amino-2-methyl-5-pyrimidinyl)methyl]-1-(2-chloroethyl)-1-nitrosourea hydrochloride.
ACNU; CS-439 HCl; Nidran hydrochloride; NSC-245382. Antineoplastic. λ_m = 245 nm ($E_{1\ cm}^{1\%}$ = 480-510 0.04N HCl); soluble in EtOH, less soluble in other organic solvents; LD_{50} (mus iv) = 62 mg/kg, (rat iv) = 46 mg/kg.

5274 Nitracrine Hydrochloride
6514-85-8 6661
$C_{18}H_{22}Cl_2N_4O_2$
9-[[3-(Dimethylamino)propyl]amino]-1-nitroacridine dihydrochloride.
C-283; Ledacrine; Nitracrine dihydrochloride; NSC-247561. Antineoplastic agent. mp = 134-135°; insoluble in H_2O, soluble in organic solvents.

5275 Nocodazole
31430-18-9
$C_{14}H_{11}N_3O_3S$
[[5-(2-Thienylcarbonyl)-1H-benzimidazol-2-yl]]carbamic acid methyl ester.
Oncodazole; R-17934; NSC-238159. Antineoplastic agent. *Abbott Laboratories Inc.*

5276 Nogalamycin
1404-15-5 6767
$C_{39}H_{49}NO_{16}$
(2α,3β,4α,5β,6α,11β,13α,14α)-(+)-11-[(6-Deoxy-3-C-methyl-2,3,4-tri-O-methyl-α-L-mannopyranosyl)-oxy]-4-(dimethylamino)-3,4,5,6,9,11,12,13,14,16-decahydro-3,5,8,10,13-pentahydroxy-6,13-dimethyl-9,16-dioxo-2,6-epoxy-2H-naphthaceno[1,2-b]oxocin-14-carboxylic acid methyl ester.
Antibiotic from *Streptomyces nogalater* var. *nogalater*; Antibiotic 205t3; Nogalamycin (8CI); U-15167; NSC-70845. mp = 195-196° (dec); $[\alpha]_D^{25}$ = 425° ($CHCl_3$ c = 0.11); λ_m = 236, 258, 292 nm (ε 52360, 24755, 9890 EtOH); insoluble in H_2O, alcohols, soluble in organic solvents; LD_{50} (mus iv) = 11.75 mg/kg, (mus ip) = 4.79 mg/kg. *Rybar Labs. Ltd.*

5277 Oltipraz
64224-21-1 264-736-6
$C_8H_6N_2S_3$
4-Methyl-5-(pyrazinyl)-3H-1,2-dithiole-3-thione.
RP-35972. Anticarcinogen. An an antischistosomal drug with chemoprotective properties.

5278 Ormaplatin
62816-98-2
$C_6H_{14}Cl_4N_2Pt$
(±)-trans-Tetrachloro(1,2-cyclohexanediamine)platinum.
U-77233; tetraplatin; NSC-363812. Antineoplastic agent. *Pharmacia & Upjohn.*

5279 Oxaliplatin
61825-94-3 7044
$C_8H_{12}N_2O_4Pt$
[(1R,2R)-1,2-Cyclohexanediamine-N,N']-[oxolato(2-)-O,O']platinum.
1,2-Cyclohexanediamine, platinum complex, (1R-trans)-(9CI); Trans-l-diaminocyclohexane oxalatoplatinum; oxalatoplatin; oxalatoplatinum; Pt-(oxalato)(trans-l-dach); l-OHP; RP-54780; OXALIPLATIN; DACPLAT; ELOXATIN; NSC-266046. Antineoplastic agent. Third generation platinum complex. Soluble in H_2O (7.9 mg/ml).

5280 Oxisuran
27302-90-5
$C_8H_9NO_2S$
(Methylsulfinyl)methyl-2-pyridiyl ketone.
Ismisupren; Ketone, (methylsulfinyl)methyl 2-pyridyl; NSC-356716; 2-[(methylsulfinyl)acetyl]pyridine; W 6495. Antineoplastic agent. *Parke-Davis.*

5281 Paclitaxel
33069-62-4 7117
$C_{47}H_{51}NO_{14}$
[2aR-[2aα,4β,4aβ,6β,9α-(R*,βS*),11α,12α,12aα,12bα]]-β-(Benzoylamino)-α-hydroxybenzenepropanoic acid 6,12b-bis(acetyloxy)-12-(benzoyloxy)- 2a,3,4,4a,5,6,9,10,11,12,12a,12b-dodecahydro-4,11-dihydroxy-4a,8,13,13-tetramethyl-5-oxo-7,11-methano-1H-cyclodeca[3,4]benz-[1,2-b]oxet-9-yl ester.
Taxol®; NSC-125973. Antineoplastic agent. Microtubule inhibitor; used in treatment of ovarian cancer. mp = 213-216° (dec); $[\alpha]_D^{20}$ = -49° (MeOH); λ_m = 227, 273 nm (ε 29800 1700 MeOH). *Bristol-Myers Oncology.*

5282 Pazelliptine
65222-35-7
$C_{22}H_{27}N_5$
10-[[3-(Diethylamino)propyl]amino]-6-methyl-5H-pyrido[3',4':4,5]pyrrolo[2,3-g]isoquinoline.
Antineoplastic agent.

5283 Pegaspargase
130167-69-0 871
(Monomethoxypolyethylene glycol succinimidyl)$_{7a}$-L-asparaginase.
Oncaspar. Antineoplastic agent. Reaction product of asparaginase wth succinic anhydride. Used in treatment of acute leukemia. *Enzon Inc.*

5284 Peldesine
133432-71-0
$C_{12}H_{11}N_5O$
2-Amino-3,5-dihydro-7-(3-pyridylmethyl)-4H-pyrrolo-[3,2-d]pyrimidin-4-one.
BCX-34. Antineoplastic; antipsoriatic agent. *BioCryst Pharm. Inc.*

5285 Peliomycin
1404-20-2
$C_{46}H_{76}O_{14}$
NSC-76455. Antineoplastic agent.

5286 Penoctonium Bromide
17088-72-1
$C_{26}H_{50}BrNO_2$
Diethyl(2-hydroxyethyl)octyl ammonium bromide dicyclopentylacetate.
Tested as an antibacterial agent. Potential chemotherapeutic agent.

5287 Pentamustine
73105-03-0
$C_8H_{16}ClN_3O_2$
1-(2-Chloroethyl)-3-neopentyl-1-nitrosourea.
NCNU. Antineoplastic agent. *National Foundation for Cancer Res.*

5288 Pentostatin
53910-25-1 7277
$C_{11}H_{16}N_4O+4$
(R)-3-(2-Deoxy-β-D-erythro-pentofuranosyl)-3,6,7,8-tetrahydroimidazo[4,5-d][1,3]diazepin-8-ol.
Co-V; Co-Vidarabine; CI-825; CL-67310465; Covidarabine; Deaminase inhibitor; NSC-218321; Pentostatin; PD-ADI; Vira A deaminase inhibitor; 2'-Dexoycoformycin; 2'-DCF; Nipent. Antineoplastic agent. Potentiator. mp = 220-225°, 204-209.5°; λ_m = 282 nm (ε 8000 pH 7), 283 nm (ε 7970 pH 11), 283 nm (ε 7570 → 3143 over 6.5 hours pH 2); $[\alpha]_D^{25}$ = 76.4° (H_2O c = 1), $[\alpha]_D^{23}$ = 73.0° (pH 7 buffer c = 1). *Parke-Davis.*

5289 Peplomycin
68247-85-8 7286
$C_{61}H_{88}N_{18}O_{21}S_2$
N^1-[3-[[(S)-(α-Methylbenzyl)]amino]propyl]bleomycinamide.
Antineoplastic agent. Bleomycin derivative with cytostatic acitvity and less pulmonary toxicity than bleomycin. *Bristol-Myers Oncology.*

5290 Peplomycin Sulfate
70384-29-1 7286
$C_{61}H_{88}N_{18}O_{21}S_2.H_2SO_4$
N^1-[3-[[(S)-(α-Methylbenzyl)]amino]propyl]bleomycinamide sulfate (1:1) (salt).
NK-631. Antineoplastic agent. Bleomycin derivative with cytostatic acitvity and less pulmonary toxicity than bleomycin. mp = 196-198°; $[\alpha]_{436}^{25}$ = -2.0° (H_2O c = 1); soluble in H_2O, MeOH, AcOH, DMSO, DMF; insoluble in less polar solvents; LD_{50} (rat sc) = 234 mg/kg, (rat ip)= 208 mg/kg, (rat iv) = 245 mg/kg, (mus sc) = 88 mg/kg, (mus ip) 85 mg/kg,. *Bristol-Myers Oncology.*

5291 Perfosfamide
62435-42-1 7303
$C_7H_{15}Cl_2N_2O_4P$
(±)-cis-2-[Bis(2-chloroethyl)amino]tetrahydro-2H-1,3,2-oxazaphosphorin-4-yl hydroperoxide.
P-oxide; Pergamid;4-HC; NSC-181815. Antineoplastic agent. mp = 107=108°; LD_{50} (rat iv) = 115 mg/kg, (rat ip) = 131 mg/kg, (mus iv) = 235 mg/kg, (mus ip) = 181 mg/kg. *Scios Nova Inc.*

5292 Pipobroman
54-91-1
$C_{10}H_{16}Br_2N_2O_2$
1,4-Bis(3-bromopropionyl)-piperazine.
Amedel; N,N'-bis(3-bromopropionyl)piperazine; Vercyte; NSC-25154; A-8103. Antineoplastic agent. *Abbott Labs Inc.*

5293 Piposulfan
2608-24-4 7635
$C_{12}H_{22}N_2O_8S_2$
1,4-Dihydracryloylpiperazine dimethanesulfonate.
A-20968; Ancyte; 1,4-Dihydracryloylpiperazine, dimethanesulfonate; NSC-47774. Antineoplastic agent. mp = 175-177°. *Abbott Labs Inc.*

5294 Pirarubicin Hydrochloride
72496-41-4 7642
$C_{32}H_{38}ClNO_{12}$
(8S,10S)-10-[[3-Amino-2,3,6-trideoxy-4-O-(2R-tetra-hydro-2H-pyran-2-yl-α-L-lyxopyranosyl]oxy]-8-glycoloyl-7,8,9,10-tetrahydro-6,8,11-trihydroxy-1-methoxy-5,12-naphthacenedione.
THP-adriamycin HCl; Pirarubicin HCl; NSC-654509. Antineoplastic agent.

5295 Piritrexim
72732-56-0 7654
$C_{17}H_{19}N_5O_2$
2,4-Diamino-6-(2,5-dimethoxybenzyl)-5-methylpyrido[2,3-d]pyrimidine.
BW-301U. Antiproliferative agent. mp = 252-254°. *Glaxo Wellcome Inc.*

5296 Piritrexim Isethionate
79483-69-5 7654
$C_{17}H_{19}N_5O_2.C_2H_6O_4S$
2,4-Diamino-6-(2,5-dimethoxybenzyl)-5-methylpyrido-[2, 3-d]pyrimidine mono(2-hydroxyethanesulfonate).
BW-301U isethionate. Antiproliferative agent. A proprietary preparation of chlorpheniramine maleate; antihistaminic. LD_{50} (rat orl) = 764 mg/kg; LD_{90} (rat orl) = 1572 mg/kg. *Glaxo Wellcome Inc.*

5297 Piroxantrone
91441-23-5
$C_{21}H_{25}N_5O_4$
5-[(3-Aminopropyl)amino]-7,10-dihydroxy-2-[2-[(2-hydroxyethyl)amino]ethyl]anthra[1,9-cd]pyrazol-6(2H)-one.
Antineoplastic agent. *Parke-Davis.*

5298 Piroxantrone Hydrochloride
105118-12-5
$C_{21}H_{27}Cl_2N_5O_4$
5-[(3-Aminopropyl)amino]-7,10-dihydroxy-2-[2-[(2-hydroxyethyl)amino]ethyl]anthra[1,9-cd]pyrazol-6(2H)-one dihydrochloride.
CI-942. Antineoplastic agent. *Parke-Davis.*

5299 Podophyllotoxin
518-28-5 7704 208-250-4
$C_{22}H_{22}O_8$
[5R-(5α,5aβ,8aα,9α)]-5,8,8a,9-Tetrahydro-9-hydroxy-5-(3,4,5-trimethoxyphenyl)-furo[3',4':6,7]naphtho[2,3-d]-1,3-dioxol-6(5aH)-one.
podofilox; podophyllotoxin; NSC-24818; Bisoprolol; Condolyne. Antiviral agent. Precursor of antineoplastic agents etoposide and teniposide, found in rhizomes of *Podophyllum peltatum L. Podophyllaceae* of North America. mp = 114-118°, 183-184° (after drying); $[\alpha]_D^{20}$ = -132.7° ($CHCl_3$); soluble in H_2O (120 mg/l), more soluble in organic solvents; LD_{50} (rat iv) = 8.7 mg/kg, (rat ip) = 15 mg/kg. *E. Merck.*

5300 Porfimer Sodium
87806-31-3 7755
Photofrin II.
CL-184116. Antineoplastic agent; radioprotector; radiosensitizer.

5301 Porfiromycin
801-52-5 7756
$C_{16}H_{20}N_4O_5$
6-Amino-1,1a,2,8,8a,8b-hexahydro-8-(hydroxymethyl)-8a-methoxy-1,5-dimethyl-azirino[2',3':3,4]pyrrolo[1,2-a]indole-4,7-dione carbamate (ester).
methyl mitomycin C; methylmitomycin; N-methylmitomycin C; porphyromycin; Regamycin; NSC-56410; U-14743. Antineoplastic agent. Dec 201-202°; $[\alpha]_D^{25}$ = 275° (c = 0.1% MeOH); λ_m = 217, 360, 555 nm (ε 24600, 23000, 209 MeOH); slightly soluble in H_2O, moderately soluble in polar organic solvents; insoluble in hydrocarbon solvents. *Pharmacia & Upjohn.*

5302 Prednimustine
29069-24-7 7900
$C_{35}H_{45}Cl_2NO_6$
11β,17-Dihydroxy-21-[4-[4-[bis(2-chloroethyl)amino]-phenyl]-1-oxobutoxy].
Leo 1031; Sterecyt; NSC-171345; NSC-134087. Antineoplastic agent. mp = 163-164°; $[\alpha]_D^{24}$ = 92.9° ($CHCl_3$ c = 1.06). *SmithKline Beecham Pharm.*

5303 Procarbazine
671-16-9 7938
$C_{12}H_{19}N_3O$
N-(1-Methylethyl)-4-[(2-methylhydrazino)methyl]-benzamide.
MIH; Ro-4-6467; Ibenzmethyzine; NSC-77213. Antineoplastic agent. [hydrobromide]: dec 216-217°. *Hoffmann-LaRoche Inc.*

5304 Procarbazine Hydrochloride
366-70-1 7938
$C_{12}H_{20}ClN_3O$
N-(1-Methylethyl)-4-[(2-methylhydrazino)methyl]-benzamide monohydrochloride.
ibenzmethyzin; MIH; Ro-4-6467; ibenzmethyzine hydrochloride; Matulane; MBH; MIH hydrochloride; PCB hydrochloride; Natulan Hydrochloride; Nathulane; Natulan; Natulanar; IBZ; Ro-4-6467; NCI-C01810; PCB hydrochloride; Ro 4 6467/1. Antineoplastic agent. May be a carcinogen. mp = 223-226°; LD_{50} (rat orl) = 785 mg/kg. *Hoffmann-LaRoche Inc.*

5305 Proresid
1508-45-8 7703
$C_{24}H_{30}N_2O_8$
Podophyllic acid 2-ethylhydrazide.
SPI-77; NSC-72274. Antineoplastic agent. $[\alpha]_D$ = -154° (c = 0.5 $CHCl_3$).

5306 Puromycin
53-79-2 8130
$C_{22}H_{29}N_7O_5$
3'-(α-Amino-p-methoxyhydrocinnamamido)-3'-deoxy-N,N-dimethyladenosine.
CL-13900; P-638; 3123-L; Stylomycin. Antineoplastic. Antiprotozoal against Trypanosoma. mp = 175.5-177°; $[\alpha]_D^{25}$ = -11° (EtOH); λ_m = 275 nm (ε 20300 0.13N NaOH), 267.5 nm (ε 19500 0.1N HCl); LD_{50} (mus iv) = 350 mg/kg, (mus ip) = 525 mg/kg; (mus orl) = 675 mg/kg. *Am. Cyanamid; ICN Pharm. Inc.*

5307 Puromycin Hydrochloride
58-58-2 8130
$C_{22}H_{31}Cl_2N_7O_5$
3'-(α-Amino-p-methoxyhydrocinnamamido)-3'-deoxy-N,N-dimethyladenosine dihydrochloride.
CL-16536; NSC-3055; CL-13900 dihydrochloride; P-638 dihydrochloride; Stylomycin dihydrochloride; 3123L, dihydrochloride. Antineoplastic. Antiprotozoal against Trypanosoma. *ICN Pharm. Inc.*

5308 Pyrazofurin
30868-30-5
$C_9H_{13}N_3O_6$
4-Hydroxy-3-β-D-ribofuranosylpyrazole-5-carboxamide. 1H-Pyrazole-5-carboxamide, 4-hydroxy-3-β-D-ribofuranosyl-; NSC-143095; Pirazofurin; β-Pyrazomycin; Pyrazomycin; Pyrozofurin; Przf; Pyrazofurin; Pyrazomycin; 47599. Antineoplastic agent. *Eli Lilly & Co.*

5309 Raltitrexed
112887-68-0 9684
$C_{21}H_{22}N_4O_6S$
N-[5-[[3,4-Dihydro-2-methyl-4-oxo-6-quinazolinyl)methyl]methylamino]-2-thenoyl]-L-glutamic acid.
Tomudex; ZD-1694. Antineoplastic agent. Thymidylate synthase inhibitor, used in treatment of advanced colorectal cancer. Monohydrate is soluble in H_2O, mp = 180-184°. *Zeneca Pharm.*

5310 (±)-Razoxane
21416-87-5 8295
$C_{11}H_{16}N_4O_4$
4,4'-(1-Methyl-1,2-ethanediyl)bis-2,6-piperazinedione. Razoxin; (±)-4,4'-propylenedi-2,6-piperazinedione; (±)-(3,5,3',5'-tetraoxo)-1,2-dipiperazinopropane; ICI-59118; ICRF-159; NSC-129943. Razoxin is an anticancer preparation containing razoxane. mp = 237-239°. *ICI Chem. & Polymers Ltd.*

5311 Retelliptine
72238-02-9
$C_{25}H_{42}N_4O$
1-[[3-(Diethylamino)propyl]amino]-9-methoxy-5,11-dimethyl-6H-pyrido[4,3-b]carbazole.
Antineoplastic agent.

5312 Retinoic Acid
302-79-4 8333 206-129-0
$C_{20}H_{28}O_2$
[all-E]-3,7-Dimethyl-9-(2,6,6-trimethyl-1-cyclohexen-1-yl)-2,4,6,8-nonatrienoic acid.
all-trans-Retinoic acid; vitamin A acid; tretinoin; NSC-122578; Aberel; Airol; Aknoten; Cordes Vas; Dermairol; Epi-Aberel; Eudyna; Retin-A; Vesanoid; Vesnaroid. Antineoplastic agent. Also a keratolytic, used to treat acne. mp = 180-182°; λ_m = 351 nm (ε 45000 MeOH); LD_{50} (10 day) (mus ip) = 790 mg/kg, (mus orl) = 2200 mg/kg, (rat ip) = 790 mg/kg, (rat orl) = 2000 mg/kg. *Degussa Ltd.; Hoffmann-LaRoche Inc.*

5313 Riboprine
7724-76-7
$C_{15}H_{21}N_5O_4$
N-(3-Methyl-2-butenyl)adenosine.
IPA; NSC-105546. Antineoplastic agent. *Roswell Park Memorial Inst.*

5314 Rituximab
174722-31-7
Immunoglobulin G1 disulfide with human-mouse monoclonal IDEC-C2B8 kappa-chain, dimer.
IDEC-C2B8; IDEC-102. Antineoplastic agent. Microtubule inhibitor. *IDEC Pharm. Corp.*

5315 Rodorubicin
96497-67-5
$C_{48}H_{64}N_2O_{17}$
(1S,3R,4R)-3-Ethyl-1,2,3,4,6,11-hexahydro-3,5,10,12-tetrahydroxy-6,11-dioxo-4-[[2,3,6-trideoxy-3-(dimethylamino)-α-L-lyxohexopyranosyl]oxy]-1-naphthacenyl O-3,6-dideoxy-α-L-erythrohexopyranos-4-ulosyl(1→4)-O-2,6-dideoxy-α-L-lyxohexopyranosyl-(1→4)-2,3,6-trideoxy-3-(dimethylamino)-α-L-lyxohexopyranoside 2,3'-anhydride.
Antineoplastic agent.

5316 Rogletimide
121840-95-7
$C_{12}H_{14}N_2O_2$
(±)-2-Ethyl-2-(4-pyridyl)glutarimide.
Antineoplastic agent. *U.S. Bioscience Corp.*

5317 Safingol
15639-50-6
$C_{18}H_{39}NO_2$
(2S,3S)-2-Amino-1,3-octadecanediol.
Antineoplastic (adjunct); antipsoriatic. *Sphinx Pharm. Corp.*

5318 Safingol Hydrochloride
139755-79-6
$C_{18}H_{40}ClNO_2$
(2S,3S)-2-Amino-1,3-octadecanediol hydrochloride.
Antineoplastic (adjunct); antipsoriatic. *Sphinx Pharm. Corp.*

5319 Samarium Sm-153 Lexidronam Pentasodium
154427-83-5
$C_6H_{17}N_2O_{12}P_4{}^{153}Sm$
[N,N'O^P,OP;ks',O^{P}',o^{P}''']-[[[1,2-ethanediyl-bis[nitrilobis(methylene)]]tetrakis[phosphonato]](8-)-Samarate-(5-)-^{153}Sm.
Antineoplastic agent. *Cytogen Corp.*

5320 Sedoxantrone Trihydrochloride
119221-49-7
$C_{21}H_{30}Cl_3N_5OS$
5-[(2-Aminoethyl)amino]-2-[2-(diethylamino)ethyl]-2H-[1]benzothiopyrano[4,3,2-cd]indazol-8-ol trihydrochloride.
CI-958. Antineoplastic agent. DNA topoisomerase II inhibitor. *Parke-Davis.*

5321 Semustine
13909-09-6
$C_{10}H_{18}ClN_3O_2$
1-(2-Chloroethyl)-3-(4-methylcyclohexyl)-1-nitrosourea.
methyl CCNU; trans-Methyl-CCNU; Lomustine, methyl; MeCCNU; NSC-95441. Antineoplastic agent.

5322 Simtrazene
5579-27-1
$C_{14}H_{16}N_4$
1,4-Dimethyl-1,4-diphenyl-2-tetrazene.
CL-26193; NSC-83799. Antineoplastic agent.

5323 Sobuzoxane
98631-95-9 8708
$C_{22}H_{34}N_4O_{10}$
4,4'-Ethylenebis[1-(hydroxymethyl)-2,6-piperazinedione] bis(isobutyl carbonate).
Antineoplastic agent. mp = 128-130°, 132-133°; insoluble in H_2O; LD_{50} (mmus ip) = 807 mg/kg, (mmus sc) = 400 mg/kg, (fmus ip) = 960 mg/kg, (fmus sc) = 673 mg/kg, (mrat ip)= 877 mg/kg, (mrat sc) = 3025 mg/kg, (frat ip) = 567 mg/kg, (frat sc) = 2821 mg/kg; reported as >5000 mg/kg orl in all species.

5324 Sodium Iodide, Radioactive
7790-26-3 8778
^{131}INa
Sodium iodide-^{131}I.
sodium iodide, radioactive; sodium radio-iodide; Iodotope; Oriodide; Radiocaps-131; Theriodide-131; Tracervial-131. Antineoplastic; diagnostic aid (radioactive agent). Used to test thyroid function. Radioactive iodine (^{131}I) is a β- and σ-emitter with a halflife of 8 days. *Abbott Labs Inc.*

5325 Sodium Phenylbutyrate
1716-12-7
$C_{10}H_{11}NaO_2$
Sodium 4-phenylbutyrate.
PBA; Buphenyl; TriButyrate. A short-chain aromatic fatty acid that inhibits cell proliferation and induces apoptosis. Also used in treatment of ornithine transcarbamylase deficiency as a vehicle for waste nitrogen excretion in patients with inborn errors of urea synthesis. Antineoplastic, hematopoietic and antihyperammonemic. Used to reduce levels of ammonia in the blood.

5326 Sparfosate Sodium
66569-27-5
$C_6H_8NNa_2O_8P$
N-(Phosphonoacetyl)-L-aspartic acid disodium salt.
CI-882. Antineoplastic agent. *Parke-Davis.*

5327 Sparfosic Acid
51321-79-0
$C_6H_{10}NO_8P$
N-(Phosphonoacetyl)-L-aspartic acid.
Antineoplastic agent. *Parke-Davis.*

5328 Spirogermanium
41992-22-7 8916
$C_{17}H_{38}Cl_2GeN_2$
2-[3-(Dimethylamino)propyl]-8,8-diethyl-2-aza-8-germaspiro[4.5]decane dihydrochloride.
Spirogermanium 32; NSC-192965. Antineoplastic agent. mp = 287-288°; LD_{50} (mus orl) = 324 mg/kg. *Unimed Pharm. Inc.*

5329 Spirogermanium Dihydrochloride
41992-23-8 8916
$C_{17}H_{36}GeN_2$
2-[3-(Dimethylamino)propyl]-8,8-diethyl-2-aza-8-germa-spiro[4.5]decane.
spirogermanium ; NSC-192965. Antineoplastic agent. mp = 287-288°; LD_{50} (mus orl) = 324 mg/kg. *Unimed Pharm. Inc.*

5330 Spiromustine
56605-16-4
$C_{14}H_{23}Cl_2N_3O_2$
3-[2-[Bis(2-chloroethyl)amino]ethyl]-1,3-diazaspiro-[4.5]decane-2,4-dione.
Spirohydantoin mustard; SHM; NSC-172112. Antineoplastic agent.

5331 Spiroplatin
74790-08-2
$C_8H_{18}N_2O_4PtS$
cis[1,1-Cyclohexanebis(methylamine)](sulfato)platinum.
NSC-311056. Antineoplastic agent. *Bristol-Myers Oncology.*

5332 Streptonigrin
3930-19-6 8986 223-501-8
$C_{54}H_{94}N_{20}O_{36}S_3$
4-Pyridine carboxylic acid hydrazide hydrazone with O-2-deoxy-2-(methylamino)-α-L-glucopyranosyl-(1→2)-O-5-deoxy-3-C-formyl-α-L-lyxofuranosyl-(1→4)-N,N'-bis(aminoiminomethyl)-D-streptamine sulfate (2:3) (salt).
streptoniazide; Strazide; Streptohydrazid; NSC-56748; NSC-83950; NSC-45383; A 050165L302; Abbott Crystalline antibiotic; Antibiotic from *Streptomyces flocculus*; Bruneomycin; Nigrin; Rufochromomycin; Rufocromomycin; Streptonigran; STP; 5278 R. P. Antineoplastic agent. mp = 262-263°; dec 275°; λ_m = 248, 375-380 nm (ε 38400, 17400 MeOH); slightly soluble in H_2O, more soluble in polar organic solvents. *Pfizer Inc.*

5333 Streptozocin
18883-66-4 8991 242-646-8
$C_8H_{15}N_3O_7$
2-Deoxy-2-[[(methylnitrosoamino)carbonyl]amino]-D-glucose.
NSC-37917; NSC-85998; Streptozoticin; Streptozotocin, Pure; STZ; U-9889; Zanosar. Antineoplastic agent. mp = 115° (dec); soluble in H_2O, lower alcohols, ketones; λ_m = 228 nm (ε 6360); LD_{50} (mus ip) = 360 mg/kg. *Pharmacia & Upjohn.*

5334 Sufosfamide
37753-10-9
$C_8H_{18}ClN_2O_5PS$
2-[[3-(2-Chloroethyl)tetrahydro-2H-1,3,2-oxazaphos-phorin-2-yl]amino]ethanol methanesulfonate (ester).
An oxazaphosphorine derivative. Antineoplastic; immunosuppressant.

5335 Sulofenur
110311-27-8
$C_{16}H_{15}ClN_2O_3S$
1-(p-Chlorophenyl)-3-(5-indanylsulfonyl)urea.
LY-186641. Antineoplastic agent. *Eli Lilly & Co.*

5336 Tallimustine
115308-98-0
$C_{32}H_{38}Cl_2N_{10}O_4$
N-(2-Amidinoethyl)-4-[p-[bis(2-chloroethyl)amino]benzamido]-1,1',1-trimethyl-N,4':N',4-ter[pyrrole-2-carboxamide.
PNU-152241; FCE-24517. Distamycin-A derivative. Antineoplastic; antitumor.

5337 Tamoxifen
10540-29-1 9216 234-118-0
$C_{26}H_{29}NO$
(Z)-2-[p-(1,2-Diphenyl-1-butenyl)phenoxy]-N,N-dimethylethylamine.
Antiestrogen. Antineoplastic. Nonsteroidal estrogen antagonist for prevention, palliative treatment of breast cancer. mp = 96-98°; [cis-form]: mp = 72-74°. *Zeneca Pharm.*

5338 Tamoxifen Citrate
54965-24-1 9216 259-415-2
$C_{32}H_{37}NO_8$
(Z)-2-[p-(1,2-Diphenyl-1-butenyl)phenoxy]-N,N-dimethylethylamine citrate.
ICI-46474 citrate; Kessar; Noltam; Nolvadex; Nourytam; Tamofen; Tomaxasta; Zemide; TMX; ICI-47699 [as cis-form citrate]. Antiestrogen. Antineoplastic. Nonsteroidal estrogen antagonist used in prevention and palliative treatment of breast cancer. mp = 140-142°; slightly soluble in H_2O; more soluble in EtOH, MeOH, Me_2CO; LD_{50} (mus ip) = 200 mg/kg, (mus iv) = 62.5 mg/kg, (mus orl) = 3000-6000 mg/kg, (rat ip) = 600 mg/kg), (rat iv) = 62.5 mg/kg, (rat orl) = 1200-2500 mg/kg; [cis-form citrate]: mp = 126-128°. *ICI ; Zeneca Pharm.*

5339 Tauromustine
85977-49-7
$C_7H_{15}ClN_4O_4S$
1-(2-Chloroethyl)-3-[2-(dimethylsulfamoyl)ethyl]-1-nitrosourea.
TCNU. Chemotherapeutic; antitumor. A nitrosourea.

5340 Tecogalan Sodium
134633-29-7
DS-4152. Antineoplastic agent (adjunct). Fermentation Product of *Arthrobacter* sp. AT-25. *Daiichi Pharm. Corp.*

5341 Tegafur
17902-23-7 9267
$C_8H_9FN_2O_3$
5-Fluoro-1-(tetrahydro-2-furanyl)-2,4(1H,3H)-pyrimidinedione.
Sunfural; FT-207; MJF-12264; Citofur; Coparogin; Exonal; Fental; NSC-148958; Franrose; Ftorafur; Fulaid; Fulfeel; Furafluor; Furofutran; Futraful; Lamar; Lifril; Neberk; Nitobanil; Riol; Sinoflurol; Tefsiel C. Antineoplastic agent. mp = 164-165°; λ_m = 270 nm (ε 8460 pH 2, 8050 pH 7, 6700 pH 12); soluble in H_2O, EtOH, DMF; insoluble in Et_2O; LD_{50} (mus orl 3 day) = 900 mg/kg, (mus orl) = 750 mg/kg, (mus ip) = 1150 mg/kg. *Asahi Chem. Industry.*

5342 Teloxantrone
91441-48-4
$C_{23}H_{29}N_5O_6$
7,10-Dihydroxy-2-[2-[(2-hydroxyethyl)amino]ethyl]-5-[[2-(methylamino)ethyl]amino]-anthra[1,9-cd]pyrazol-6(2H)-one acetate (salt) hydrobromide (10:5:21).
Moxantrazole; DUP-937; NSC-355644. Antineoplastic agent. *DuPont-Merck Pharm.*

5343 Teloxantrone Hydrochloride
132937-88-3
$C_{23}H_{30}ClN_5O_6$
7,10-Dihydroxy-2-[2-[(2-hydroxyethyl)amino]ethyl]-5-[[2-(methylamino)ethyl]amino]-anthra[1,9-cd]pyrazol-6(2H)-one acetate (salt) hydrobromide (10:5:21) hydroxide.
NSC-355644. Antineoplastic agent. *DuPont-Merck Pharm.*

5344 Temoporfin
122341-38-2
$C_{44}H_{32}N_4O_4$
3,3',3,3'''-(7,8-Dihydroporphyrin-5,10,15,20-tetrayl)tetraphenol.
EF9. Antineoplastic agent. *Scotia Pharm. Ltd.*

5345 Temozolomide
85622-93-1 9289
$C_6H_6N_6O_2$
3,4-Dihydro-3-methyl-4-oxoimidazo[5,1-d]-as-tetrazine-8-carboxamide.
M&B-39831; Methazolastone; NSC-362856. Antineoplastic agent. mp = 212° (dec); λ_m = 327 nm (EtOH). *May & Baker Ltd.*

5346 Teniposide
29767-20-2 9291
$C_{32}H_{32}O_{13}S$
[5R-[5α,5aβ,8aα,9β(R*)]]-5,8,8a,9-Tetrahydro-5-(4-hydroxy-3,5-dimethoxyphenyl)-9-[[4,6-O-(2-thienylmethylene)-β-D-glucopyranosyl]oxy]furo[3',4':6,7]-naphtho[2,3-d]-1,3-dioxol-6(5aH)-one.
ETP; VM-26; Vehem-Sandoz; Vumon; NSC-362856. Antineoplastic agent. Semi-synthetic derivative of podophyllotoxin. mp = 242-246°; $[\alpha]_D^{20}$ = -107° ($CHCl_3$/MeOH, 9:1); λ_m = 283 nm ($E_{1\,cm}^{1\%}$ 64.1, MeOH). *Sandoz Pharm. Corp.*

5347 Teroxirone
59653-73-5
$C_{12}H_{15}N_3O_6$
(RS,RS,SR)-1,3,5-Tris(2,3-epoxypropyl)-s-triazine-2,4,6-(1H,3H,5H)-trione.
α-triglycidyl isocyanurate; αTGI; NSC-296934. Antineoplastic agent.

5348 Teslac
968-93-4 9321 213-534-6
$C_{19}H_{24}O_3$
13-Hydroxy-3-oxo-13,17-secoandrosta-1,4-dien-17-oic acid δ-lactone.
testolactone; SQ-9538; Fludestrin; NSC-12173; NSC-23759. Antineoplastic agent. mp = 218-219°; $[\alpha]_D^{23}$ = -46° (c = 1.24 $CHCl_3$); λ_m = 242 nm (ε 15800, EtOH). *Bristol-Myers Oncology.*

5349 Thalidomide
50-35-1 9390 200-031-1
$C_{13}H_{10}N_2O_4$
N-(2,6-Dioxo-3-piperidyl)phthalimide.
Kevadon; K-17; NSC-66847; Distaval; Softenon; Sedalis; Talimol; Pantosediv; Neurosedyn; Contergan. Formerly used as a sedative/hypnotic but discontinued because of its marked teratogenicity. Has recently been shown to be effective against bone cancer. mp = 269-271°; λ_m 220, 300 nm (pH 7); poorly soluble in H_2O (0.005 g/100 ml); sparingly soluble in H_2O, MeOH, EtOH, EtOAc, AcOH; very soluble in C_5H_5N, DMF, dioxane; insoluble in Et_2O, $CHCl_3$, C_6H_6. *Grunenthal; Marion Merrell Dow Inc.*

5350 Thiamiprine
5581-52-2 9434
$C_9H_8N_8O_2S$
2-Amino-6-[(1-methyl-4-nitro-1H-imidazol-5-yl)thio]-purine.
BW-57-323; Guaneran; NSC-38887. Antineoplastic agent. mp > 200° (dec); λ_m = 320 nm (pH 1), 315 nm (pH 11). *Glaxo Wellcome Inc.*

5351 Thioguanine
154-42-7 9473 205-827-2
$C_5H_5N_5S.xH_2O$
2-Amino-1,7-dihydro-6H-purine-6-thione.
2-aminopurine-6(1H)-thione; tioguanine; Tabloid; NSC-752. Antineoplastic agent. mp > 360°. *Glaxo Wellcome Inc.*

5352 Thioguanine Hemihydrate
5580-03-0
$C_5H_5N_5S.0.5H_2O$
2-Amino-1,7-dihydro-6H-purine-6-thione.
Lanvis; BW-5071; Tabloid; Thioguanine; Tioguanin; Tioguanine; TG; Wellcome U3B; X 27; NSC-752. mp > 360°.

5353 Thiotepa
52-24-4 9805 200-135-7
$C_6H_{12}N_3PS$
1,1',1''-Phosphinothioylidynetris-aziridine.
CBC-806495; Girostan; NCI-C01649; Oncotepa; Oncothio-tepa; SK-6882; STEPA; Tespa; NSC-6396; Tespamin; Tespamine; Thio-tepa; Thio-tepa S; Thio-Tep; Thio-Tepa; Thiofozil; Thiotef; Thiotepa; Thioplex; Tifosyl; Tio-tef; Tiofosfamid; Tiofosyl; Tiofozil; TESPA; TIO TEF; TSPA. Antineoplastic agent. mp = 51°; soluble in H_2O (190 g/l), organic solvents; LD_{50} (rat iv) = 15 mg/kg. *Immunex Corp.*

5354 Thymalfasin
62304-98-7
$C_{129}H_{215}N_{33}O_{55}$
N-Acetyl-L-seryl-L-α-aspartyl-L-alanyl-L-alanyl-L-valyl-L-α-aspartyl-L-threonyl-L-seryl-L-seryl-L-α-glutamyl-L-isoleucyl-L-threonyl-L-threonyl-L-lysyl-L-α-aspartyl-L-leucyl-L-lysyl-L-α-glutamyl-L-lysyl-L-lysyl-L-α-glutamyl-L-valyl-L-valyl-L-α-glutamyl-L-α-glutamyl-L-alanyl-L-α-glutamyl-L-asparagine.
thymosin-α1; Zadaxin. Antineoplastic agent. Also used in hepatitis treatment, vaccine enhancement and treatment of infectious diseases. *SciClone Pharm. Inc.*

5355 Tiazofurin
60084-10-8
$C_9H_{12}N_2O_5S$
2-β-D-Ribofuranosyl-4-thiazolecarboxamide.
CI-909; NSC-286193. Antineoplastic agent. *Parke-Davis.*

5356 Tirapazamine
27314-97-2
$C_7H_6N_4O_2$
3-Amino-1,2,4-benzotriazine 1,4-dioxide.
WIN-59075; SR-4233; Triazone; SR-259075. Antineoplastic agent. A hypoxic cytotoxin, used in combination with cisplatin.

5357 Topotecan
123948-87-8 9687
$C_{23}H_{23}N_3O_5$
(S)-10-[(Dimethylamino)methyl]-4-ethyl-4,9-dihydroxy-1H-pyrano[3',4':6,7]indolizino-[1,2-b]quinoline-3,14(4H,12H)-dione.
Antineoplastic agent. DNA topoisomerase I inhibitor. soluble in H_2O (< 1 mg/ml). *SmithKline Beecham Pharm.*

5358 Topotecan Hydrochloride
119413-54-6 9687
$C_{23}H_{24}ClN_3O_5$
(S)-10-[(Dimethylamino)methyl]-4-ethyl-4,9-dihydroxy-1H-pyrano[3',4':6,7]indolizino-[1,2-b]quinoline-3,14(4H,12H)-dione monohydrochloride.
SK&FS-104864-A. Antineoplastic agent. DNA topoisomerase I inhibitor. *SmithKline Beecham Pharm.*

5359 Triaziquone
68-76-8 9733 200-692-6
$C_{12}H_{13}N_3O_2$
2,3,5-Tris(1-aziridinyl)-2,5-cyclohexadiene-1,4-dione.
A 163; Bayer 3231; Oncoredox; Oncovedex; Prenimon; Riker 601; Trenimon; Treninon; NSC-29215; Triaziquinone; Triaziquon; Triethyleniminobenzoquinone; tris(aziridinyl)-p-benzoquinone; TEIB; 10257 R.P. Antineoplastic agent. mp = 162-163°; poorly soluble in H_2O, soluble in oganic solvents. *Bayer Corp. Pharm. Div.*

5360 Trichlormethine
555-77-1 9777
$C_6H_{12}Cl_3N$
2,2',2-Trichlorotriethylamine.
trimustine; Hn3; Lekamin; R 47; Sinalost; SK-100; Trichlormethine; Trillekamin; Trimitan; Trimustine; NSC-260424; Tris-N-lost; TS-160. Antineoplastic agent. mp= -4°; bp_{15} = 144°; d_4^{25} = 1.2347; slightly soluble in H_2O, soluble in organic solvents.

5361 Triciribine Phosphate
61966-08-3
$C_3H_4Cl_3NO_2$
2,2,2-Trichloroethanol carbamate.
Compralgyl; trichloroethyl urethan; 2,2,2-trichloroethyl carbamate; carbamic acid trichloroethyl ester; Voluntal. Antineoplastic agent. *Bayer Corp. Pharm. Div.*

5362 Triethylenemelamine
51-18-3 9803
$C_9H_{12}N_6$
2,4,6-Tris(1-aziridinyl)-1,3,5-triazine.
DRP-859025; ENT-25296; M-9500; Persistol; Persistol HOE-1/193; R-246; SK-1133; NSC-9706; Tem-Simes; Tretamin; Tretamine; Triamelin; Triaziridinyl triazine; Triethanomelamine; Triethylenemelamine; Tris-(ethyleneimino)triazine; Trisaziridinyltriazine; TAT; TEM; TET. Antineoplastic agent. Can be synthesized from ethylenimine and cynuric chloride. mp = 39° (dec); soluble in H_2O (40%), less soluble in organic solvents; LD_{50} (mus ip) = 2.8 mg/kg, (mus orl) = 15 mg/kg, (rat ip) - 1.0 mg/kg, (rat orl) = 13 mg/kg.

5363 Trimetrexate
52128-35-5 9851
$C_{19}H_{23}N_5O_3$
5-Methyl-6-[[(3,4,5-trimethoxyphenyl)amino]methyl]-2,4-quinazolinediamine.
TMQ; CI-898; NSC-249008. Antineoplastic agent. *Parke-Davis.*

5364 Trimetrexate Glucuronate
82952-64-5 9851
$C_{25}H_{33}N_5O_{10}$
5-Methyl-6-[[(3,4,5-trimethoxyphenyl)amino]methyl]-2,4-quinazolinediamine mono-D-glucuronate.
NSC-249008. Antineoplastic agent. Soluble in H_2O (> 50 mg/ml). *Parke-Davis.*

5365 Trofosfamide
22089-22-1 9897
$C_9H_{18}Cl_3N_2O_2P$
N,N,3-Tris(2-chloroethyl)tetrahydro-2H-1,3,2-oxazaphosphorin-2-amine 2-oxide.
Z-4828; Ixoten; trilophosphamide; NSC-109723. Antineoplastic agent. mp = 50-51°; $[\alpha]_D^{25}$ = -28.6° (MeOH, c = 2.0); LD_{50} (mus ip) = 212 mg/kg. *Asta-Werke AG.*

5366 Tubulozole
84697-22-3
$C_{23}H_{23}Cl_2N_3O_4S$
Ethyl (±)-cis-p-[[[2-(2,4-dichlorophenyl)-2-(imidazol-1-ylmethyl)-1,3-dioxolan-4-yl]methyl]thio]carbanilate.
R-46846; NSC-376450. Antineoplastic agent. Microtubule inhibitor. *Abbott Laboratories Inc.*

5367 Tubulozole Hydrochloride
83529-08-2
$C_{23}H_{24}Cl_3N_3O_4S$
Ethyl (±)-cis-p-[[[2-(2,4-dichlorophenyl)-2-(imidazol-1-ylmethyl)-1,3-dioxolan-4-yl]methyl]thio]carbanilate monohydrochloride.
NSC-376450.

5368 Turosteride
137099-09-3
$C_{27}H_{45}N_3O_3$
1-(4-Methyl-3-oxo-4-aza-5 alpha-androstane-17 beta-carbonyl)-1,3-diisopropylurea.
FCE-26073. A potent and selective inhibitor of 5 alpha-reductase, the enzyme responsible for the conversion of testosterone to 5 alpha-dihydrotestosterone. Antineoplastic; antitumor.

5369 Ubenimex
58970-76-6 9973 261-529-2
$C_{16}H_{24}N_2O_4$
(-)-N-[(2S,3R)-3-Amino-2-hydroxy-4-phenylbutyryl]-L-leucine.
Bestatin; NK-421; NSC-265489. Immunomodulator. Antineoplastic agent. A dipeptide antitumor antibiotic obtained from *Streptomyces olivoreticuli.* mp = 233-236°; $[\alpha]_D^{20}$ = -15.5° (c = 1.0 1N HcL); λ_m 241.5, 248,253,258,264.5, 268 nm ($E_{1\,cm}^{1\%}$ 3.8, 4.0, 5.0, 6.0, 4.6, 2.7); soluble in AcOH, DMSO, MeOH; less soluble in H_2O; insoluble in EtOAc, C_6H_6, C_6H_1. *Microbiochem. Res. Found.*

5370 Uracil Mustard
66-75-1 9986 200-631-3
$C_8H_{11}Cl_2N_3O_2$
5-[Bis(2-chloroethyl)amino]uracil.
U-8344; Aminouracil mustard; Chlorethaminacil; CB-4835; Demethyldopan; Desmethyldopan; NSC-34462; ENT 50439; Nordopan; NCI-C04820; NSC-34462; SK-19849; U 8344; U-8344; Uracil nitrogen mustard; Uracillost; Uracilmostaza; Uramustin; Uramustine. Antineoplastic agent. mp = 206° (dec); slightly soluble in H_2O; λ_m = 257 nm (ε 5675, 0.01N H_2SO_4 in 95% EtOH); LD_{50} (rat ip)= 1.25-2.5 mg/kg.

5371 Uredepa
302-49-8 10011
$C_7H_{14}N_3O_3P$
Ethyl [bis(1-aziridinyl)phosphinyl]carbamate.
Avinar; AB 100; AB-100; bis(ethylenimido)phosphorylurethan; NSC-37095. Antineoplastic agent. mp = 88-90°; soluble in H_2O. *Armour Pharm. Co. Ltd.*

5372 Urethane
51-79-6 10013
$C_3H_7NO_2$
Ethyl carbamate.
ethyl urethan; NSC-746. Antineoplastic agent. mp = 48-50°; bp = 182-184°; soluble in H_2O (2 g/ml), soluble in organic solvents; MLD (mus ip) = 2.1-2.2 g/kg.

5373 Valrubicin
56124-62-0
$C_{34}H_{36}F_3NO_{13}$
2-((2S,4S)-1,2,3,4,6,11-hexahydro-2,5,12-trihydroxy-7-methoxy-6,11-dioxo-4-((2,3,6-trideoxy-3-((trifluoroacetylamino)-α-L-lysohexopyranoxyl)oxy)-2-naphthacenyl)-2-oxoethyl pentanoate.
Trifluoroacetyladriamycin-14-valerate; Valstar; AD 32; Antibiotic AD 32; NSC-246131; N-Trifluoroacetyldoxorubicin 14-valerate; N-Trifluoroacetyladriamycin 14-valerate; Trifluoroacetyladriamycin-14-valerate; Valrubicin; Valstar. Antineoplastic. *Medeva.*

5374 Valspodar
121584-18-7
$C_{63}H_{111}N_{11}O_{12}$
Cyclo[[(2S,4R,6E)-4-methyl-2-(methylamino)-3-oxo-6-octenoyl]-L-valyl-N-methylglycyl-N-methyl-L-leucyl-L-valyl-N-methyl-L-leucyl-L-alanyl-D-alanyl-N-methyl-L-leucyl-N-methyl-L-leucyl-N-methyl-L-valyl].
PSC-833; Amdray. Chemosensitizer. Third-generation cyclosporine analog, an inhibitor of P-glycoprotein, the drug efflux pump. Multidrug resistance modulator coadministered with chemotherapeutic drugs in cancer treatment. *Novartis Pharm. Corp.; Sandoz Pharm. Corp.*

5375 Vapreotide
103222-11-3
$C_{57}H_{70}N_{12}O_9S_2$
D-Phenylalanyl-L-cysteinyl-L-tyrosyl-D-tryptophyl-L-lysyl-L-valyl-L-cysteinyl-L-tryptophanamide cyclic (2-7) disulfide.
RC-160; BMY-41606. Antineoplastic agent. *Bristol-Myers Squibb Pharm. R&D.*

5376 Vercyte
54-91-1 7634
$C_{10}H_{16}Br_2N_2O_2$
1,4-Bis(3-bromopropionyl)-piperazine.
Amedel; pipobroman; NSC-25154; A-1803; A-8103. Antineoplastic agent. mp = 106-107°. *Abbott Labs Inc.*

5377 Verteporfin
129497-78-5
$C_{14}H_{22}N_2O.HCl.H_2O$
2-(Diethylamino)-N-(2,6-dimethylphenyl)acetamide hydrochloride monohydrate.
Lignocaine hydrochloride; Versicane; Lidesthesin; Lignavet; Odontalg; Sedagul; Xylocard; Xyloneural. Local anesthetic. *May & Baker Ltd.; Rhône-Poulenc Rorer Pharm. Inc.*

5378 Vinblastine
865-21-4 10119
$C_{46}H_{58}N_4O_9$
Vincaleukoblastine.
vinblastine; NSC-49842. Antineoplastic agent. mp = 211-216°; $[\alpha]_D^{26}$ = 42° ($CHCl_3$); insoluble in H_2O, soluble in organic solvents. *Eli Lilly & Co.*

5379 Vinblastine Sulfate
143-67-9 10119 205-606-0
$C_{46}H_{60}N_4O_{13}S$
Vincaleukoblastine sulfate.
vinblastine sulfate; 29060-LE; Velsar; Belvan, VLB; Exal; Vincaleukoblastine sulfate (1:1) (salt); VLB monosulfate; Velbe; NSC-49842. Antineoplastic agent. mp = 284-285°; $[\alpha]_D^{26}$= -28° (c = 1.01 in MeOH). *Eli Lilly & Co.*

5380 Vincamine
1617-90-9 10120 216-576-3
$C_{21}H_{26}N_2O_3$
(3α,14β,16α)-14,15-Dihydro-14-hydroxy-eburnamenine-14-carboxylic acid methyl ester.
Angiopac; Arteriovinca; Devincan; Equipur; Novicet; Ocu-vinc; Oxygeron; Perval; Pervincamine; Pervone; Sostenil; Tripervan; Vincadar; Vincafarm; Vincafolina; Vincafor; Vincagil; Vincalen; Vincamidol; Vincapront; Vinvasaunier; Vincimax; Vinodrel Retard; Vraap; Cetal [as hydrochloride]; Esberidin [as hydrochloride]; NSC-91998. Antineoplastic agent. Vasodilator (cerebral). Alkaloid isolated from *vinca minor*. Other alkaloids from the vincamine fraction are vincine, vincaminine, and vincinine. Occurs naturally in d-form. mp = 232-233°; $[\alpha]_D^{23}$= +41° (C_5H_5N); λ_m = 225, 278 nm (log ε 4.14, 3.61); LD_{50} (mus iv) = 75 mg/kg, (mus sc) > 1000 mg/kg, (mus orl) = 1000 mg/kg; [(±)-form]: mp = 228-229°.

5381 Vincristine
57-22-7 10124 200-318-1
$C_{46}H_{56}N_4O_{10}$
22-Oxo-vincaleukoblastine.
leucristine; VCR; LCR. Antitumor alkaloid from *Vinca rosea*. Related to vinblastine. mp = 218-220°; $[\alpha]_D^{25}$= +17°; $[\alpha]_D^{25}$= +26.2° (ethylene chloride); pKa = 5.0; λ_m = 220, 225, 296 nm (log a_m 4.65, 4.21, 4.18); LD_{50} (mus ip) = 5.2 mg/kg.

5382 Vincristine Sulfate
2068-78-2 10124 218-190-0
$C_{46}H_{56}N_4O_{10}.H_2SO_4$
22-Oxo-vincaleukoblastinesulfate (1:1) (salt) (9CI).
Kyocristine; Leurocristine sulfate; Lilly 37231; LCR; Oncovin; Onkovin; Vincristine sulfate; Vincristine, sulfate; Vincrisul; VCR sulfate; 37231; NSC-67574. Antineoplastic agent. Antitumor alkaloid *from Vinca rosea. Eli Lilly & Co.*

5383 Vindesine
53643-48-4 10125
$C_{43}H_{55}N_5O_7$
3-Carbamoyl-4-deacetyl-3-de(methoxy-carbonyl)vincaleukoblastine.
Compound 112531; NSC-245467. Antineoplastic agent. Synthetic derivative of vinblastine, a *Catharanthus* alkaloid. mp= 230-232°; $[\alpha]_D^{25}$= 39.4° (MeOH c = 1); λ_m = 214, 266, 288, 296 nm (ε 53400, 17450, 13950, 12500 MeOH). *Eli Lilly & Co.*

5384 Vindesine Sulfate
59917-39-4 10125 261-984-7
$C_{43}H_{57}N_5O_{11}S$
3-Carbamoyl-4-deacetyl-3-de(methoxy-carbonyl)vincaleukoblastine sulfate.
Eldisine; LY-099094; NSC-245467. Antineoplastic agent. mp > 250°; LD_{50} (mus iv) = 6.3 mg/kg, (rat iv) = 2.0 mg/kg, (mus ip) = 8.8 mg/kg. *Eli Lilly & Co.*

5385 Vinepidine
68170-69-4
$C_{46}H_{56}N_4O_9$
(4'S)-4'-Deoxyleurocristine.
LY-119863. Antineoplastic agent. *Eli Lilly & Co.*

5386 Vinepidine Sulfate
83200-11-7
$C_{46}H_{58}N_4O_{13}S$
(4'S)-4'-Deoxyleurocristine sulfate.
Antineoplastic agent. *Eli Lilly & Co.*

5387 Vinflunine
162652-95-1
$C_{45}H_{54}F_2N_4O_8$
20',20'-Difluoro-3',4'-dihydrovinorelbine.
A fluorinated Vinca alkaloid. Antineoplastic.

5388 Vinfosiltine
123286-00-0
$C_{51}H_{72}N_5O_{10}P$
[23(S)]-4-Deacetyl-3-de(methoxycarbonyl)-3-[(2-methyl-1-phosphonopropyl)carbamoyl]vincaleukoblastine diethyl ether.
S-12363. Vinca alkaloid derivative. Antineoplastic.

5389 Vinglycinate
865-24-7
$C_{48}H_{63}N_5O_9$
4-Deacetylvincaleukoblastine 4-(N,N-dimethylglycinate) (ester).
Antineoplastic agent. *Eli Lilly & Co.*

5390 Vinglycinate Sulfate
7281-31-4
$C_{48}H_{63}N_5O_9.1.5H_2SO_4$
4-Deacetylvincaleukoblastine 4-(N,N-dimethylglycinate) (ester) sulfate (2:3) salt.
Antineoplastic agent. *Eli Lilly & Co.*

5391 Vinleucinol
81571-28-0
$C_{51}H_{69}N_5O_9$
[23(1S,2S)]-4-Deacetyl-3-[(1-carboxy-2-methylbutyl)carbamoyl]-3-de(methoxy-carbonyl)-vincaleukoblastine ethyl ether.
VileE; vinblastine-isoleucinate. Vinca alkaloid derivative. Antineoplastic.

5392 Vinleurosine
23360-92-1
$C_{46}H_{56}N_4O_9$
Leurosine.
Leurosine; Lilly 32645; NSC-528004. Antineoplastic agent. Alkaloid isolated from Madagascar periwinkle. *Eli Lilly & Co.*

5393 Vinleurosine Sulfate
1404-95-1
$C_{46}H_{58}N_4O_{13}S$
Leurosine sulfate.
Leurosine sulfate; Lilly 32645; NSC-528004. Antineoplastic agent. Sulfate salt of alkaloid isolated from Madagascar periwinkle. *Eli Lilly & Co.*

5394 Vinorelbine
71486-22-1 10127
$C_{45}H_{54}N_4O_8$
3',4'-Didehydro-4'-deoxy-8'-norvinca-leukoblastine.
Antineoplastic agent. $[\alpha]_D^{20} = 52.4°$ ($CHCl_3$ c = 0.3); λ_m = 215, 268, 282,293, 310 nm (ε 3700, 11000, 9500, 7600, 4400 EtOH). *Glaxo Wellcome Inc.*

5395 Vinorelbine Tartrate
125317-39-7 10127
$C_{53}H_{66}N_4O_{20}$
3',4'-Didehydro-4'-deoxy-8'-norvincaleukoblastine L-(+)-tartrate.
Navelbine. Antineoplastic agent. Soluble in H_2O, EtOH. *Glaxo Wellcome Inc.*

5396 Vinrosidine
15228-71-4
$C_{46}H_{58}N_4O_9$
4'-Deoxy-3'-hydroxy vincaleucoblastine.
Antineoplastic agent. Alkaloid isolated from Vinca rosea Linne. *Eli Lilly & Co.*

5397 Vinrosidine Sulfate
18556-44-0
$C_{46}H_{58}N_4O_9.xH_2SO_4$
4'-Deoxy-3'-hydroxy vincaleucoblastine sulfate.
36781. Antineoplastic agent. Sulfate salt of alkaloid isolated from *Vinca rosea* Linne. *Eli Lilly & Co.*

5398 Vintriptol
81600-06-8
$C_{56}H_{68}N_6O_9$
[23(S)]-4-Deacetyl-3-[(1-carboxy-2-indol-3-ylethyl)carbamoyl]-3-de(methoxycarbonyl)vincaleukoblastine diethyl ether.
Vinblastine tryptophan ester. Vinca alkaloid derivative; a tryptophan ester of vinblastine. Antineoplastic.

5399 Vinzolidine Sulfate
67699-41-6
$C_{48}H_{60}ClN_5O_{13}S$
Methyl (3R,5S,7R,9S)-9-[3'-(2-chloroethyl)-6,7-didehydro-4β-hydroxy-16-methoxy-1-methyl-2',4'-dioxo-2β,3β,5α,12β,19α-spiro[aspidospermidine-3,5'-oxazolidin]-15-yl]-5-ethyl-1,4,5,6,7,8,9,10-octahydro-5-hydroxy-2H-3,7-methanoazacycloundecino[5,4-b]indole-9-carboxylate -4'-acetate (ester) sulfate (1:1) (salt).
LY-104208. Antineoplastic agent. *Eli Lilly & Co.*

5400 Vorozole
129731-10-8
$C_{16}H_{13}ClN_6$
(+)-(S)-6-(p-Chloro-α-1H-1,2,4-triazol-1-ylbenzyl)-1-methyl-1H-benzotriazole.
R-83842. Antineoplastic agent. *Janssen Pharm. Inc.*

5401 Zeniplatin
111490-36-9
$C_{11}H_{20}N_2O_6Pt$
cis-[2,2-Bis(aminomethyl)-1,3-propanediol]-(1,1-cyclo-butanedicarboxylato)platinum.
CL-286558. Antineoplastic agent.

5402 Zinostatin
9014-02-2
Neocarzinostatin.
Antineoplastic agent. Polypeptide derived from *Streptomyces carzinostaticus* var. *Bristol-Myers Squibb Pharm. R&D.*

5403 Zinostatin Stimalamer
123760-07-6
Neocarzinostatin stimalamer.
Antineoplastic agent. Derivative of polypeptide derived from *Streptomyces carzinostaticus* var. *Bristol-Myers Squibb Pharm. R&D.*

5404 Zorubicin
54083-22-6 10326
$C_{34}H_{35}N_3O_{10}$
Benzoic acid hydrazide 3-hydrazone with daunorubicin.
Antineoplastic agent. *Rhône-Poulenc.*

5405 Zorubicin Hydrochloride
36508-71-1 10326
$C_{34}H_{36}ClN_3O_{10}$
Benzoic acid hydrazide 3-hydrazone with daunorubicin monohydrochloride.
RP-22050 hydrochloride; NSC-164011. Antineoplastic agent. $[\alpha]_D^{20} = -50°$; λ_m = 232.5, 253, 480, 495 nm (ε 40225, 35300, 10480, 10300 MeOH); LD_{50} (mus sc) = 13.66 mg/kg, (mus ip) = 4.42 mg/kg, (mus iv) = 8.50 mg/kg. *Rhône-Poulenc.*

Antineutropenics

5406 Daniplestim
161753-30-6

$C_{564}H_{909}N_{161}O_{166}S_5$

14-L-Alanine-18-L-isoleucine-25-L-histidine-29-L-arginine-32-L-asparagine-37-L-proline-42-L-serine-45-L-methionine-51-L-arginine-55-L-threonine-59-L-leucine-62-L-valine-67-L-histidine-69-L-glutamic acid-73-glycine-76-L-alanine-79-L-arginine-82-L-glutamine.

Hematopoietic stimulant; antineutropenic. *Searle G.D. & Co.*

5407 Filgrastim
121181-53-1 4558

$C_{845}H_{1339}N_{223}O_{243}S_9$

N-L-Methionyl-colony-stimulating factor (human clone 1034).

Neupogen; r-metHuG-CSF; recombinant methionyl human G-CSF. Hematopoietic stimulant; antineutropenic. A glycoprotein that stimulates neutrophil development. *Amgen Inc.*

5408 Granulocyte Colony Stimulating Factor
143011-72-7 4558

Hematopoietic growth factor.

CSF-β; G-CSF; GM-DF; MGI-2; pluripoietin. Hematopoietic stimulant; antineutropenic. Stimulates development of neutrophils. Enhances the functional activities of the mature end-cell. A glycoprotein.

5409 Granulocyte-Macrophage Colony Stimulating Factor
83869-56-1 4559

Colony-stimulating factor 2.

CSF-2; CSFα; GM-CSF; NIF-T. Hematopoietic stimulant; antineutropenic. Promotes the proliferation and development of early erythroid, megakayocytic and eosinophilic progenitor cells. Inhibits neutrophil migration. Enhances activities of mature end-cells.

5410 Lenograstim
135968-09-1 4558

Granocyte.

rG-CSF; Neutrogrin. Hematopoietic stimulant; antineutropenic. Component 1 [135968-09-1] and component 2 [130120-54-6]. Produced in Chinese hamster ovary cells by recombinant DNA technology. Resembles human G-CSF. *Chugai Pharm. Co. Ltd.*

5411 Milodistim
137463-76-4

$C_{1336}H_{2116}N_{362}O_{410}S_{13}$

Colony stimulating factor 2 (human clone pHG25 protein moiety reduced).

Pixykine; PIXY321. Hematopoietic stimulant; antineutropenic. *Immunex Corp.*

5412 Molgramostim
99283-10-0 4559

$C_{639}H_{1007}N_{171}O_{196}S_8$

Colony stimulating factor 2 (human clone pHG25 protein moiety reduced).

Sch-39300. Hematopoietic stimulant; antineutropenic. Non-glycosylated protein. Produced in *E. coli* by recombinant DNA technology. *Schering-Plough Pharm.*

5413 Muplestim
148641-02-5

$C_{670}H_{1074}N_{186}O_{199}S_5$

Human interleukin 3.

SDZ ILE 964. Hematopoietic stimulant; antineutropenic. *Sandoz Pharm. Corp.*

5414 Nartograstim
134088-74-7 4558

1-(N-L-Methionyl-L-alanine)-3-L-threonine-4-L-tyrosine-5-L-arginine-17-L-serine colony stimulating factor (human clone 1034).

marograstim; KW-2228; Noyap. Hematopoietic stimulant; antineutropenic. A glycoprotein that stimulates neutrophil development. GSF-mutein produced in *E. coli* by recombinat DNA technology.

5415 Regramostim
127757-91-9 4559

$C_{637}H_{1003}N_{171}O_{187}S_8$

Colony stimulating factor 2 (human clone pCSF-1 protein moiety reduced), glycoform GMC 89-107.

GMC-89-107; GM-CSF; rhGm-CSF. Hematopoietic stimulant; antineutropenic. A glycoprotein produced in Chinese hamster ovary cells by recombinant DNA technology. Molecular weight 21-34 kDa. *Sandoz Pharm. Corp.*

5416 Sargramostim
123774-72-1 4559

$C_{639}H_{1002}N_{168}O_{196}S_8$

Colony stimulating factor 2 (human clone pHG$_{25}$ protein moiety), 23-L-leucine.

Leukine; B161.012; rhu GM-CSF. Hematopoietic stimulant; antineutropenic. A variably glycoprotein produced in yeast by recombinant DNA technology. Molecular weight 15.5-19.5 kDa. *Immunex Corp.*

Antiosteoporetics

5417 Alendronic Acid
66376-36-1 228

$C_4H_{13}NO_7P_2$

(4-Amino-1-hydroxybutylidene)bisphosphonic acid.

ABDP. Calcium regulator. mp = 233-235° (dec). *Inst. Gentili S.p.A.; Merck & Co.Inc.*

5418 Alendronic Acid Trisodium Salt
121268-17-5 228

$C_4H_{12}NNaO_7P_2.3H_2O$

Sodium trihydrogen (4-amino-1hydroxybutylidene)-diphosphonate trihydrate.

Fosamax; MK-217; G-704650; Adronat; Alendros; Onclast. Calcium regulator. *Inst. Gentili S.p.A.; Merck & Co.Inc.*

5419 Clodronic Acid
10596-23-3 2432 234-212-1

$CH_4Cl_2O_6P_2$

(Dichloromethylene)diphosphonic acid.

Cl_2MDP; DMDP. Calcium regulator. mp = 249-251°. *Procter & Gamble Pharm. Inc.*

5420 Clodronic Acid Disodium Salt
22560-50-5 2432 245-078-9
$CH_2ClNa_2O_6P_2$
Sodium (dichloromethylene)diphosphonate.
DClMDP; Bonefos; Clasteon; Difoafonal; Loron; Mebonat; Ossiten; Ostac. Calcium regulator. *Procter & Gamble Pharm. Inc.*

5421 Etidronic Acid
2809-21-4 3908 220-552-8
$C_2H_8O_7P_2$
(1-Hydroxyethylidene)diphosphonic acid.
EHDP. Calcium regulator. Very soluble in H_2O, insoluble in AcOH. *Procter & Gamble Pharm. Inc.*

5422 Etidronic Acid Disodium Salt
7414-83-7 3908 231-025-7
$C_2H_6Na_2O_7P_2$
Disodium dihydrogen (1-hydroxyethylidene)-diphosphonate.
Didronel; Calcimux; Diphos; Etidron. Calcium regulator. Bone resorption inhibitor. Soluble in H_2O. *Procter & Gamble Pharm. Inc.*

5423 Fosfomycin Calcium Salt Monohydrate
26016-98-8 247-408-7
$C_3H_5CaPO_4$
Calcium (2R-cis)-(3-methyloxiranyl)-phosphonate salt (1:1).
Calcium fosfomycin; Calcium phosphomycin; Calcium phosphonomycin; Fosfomycin calcium; Fosfomycin calcium salt; Fosmicin. Polypeptide antibiotic. A broad-spectrum antibiotic potentially useful for the treatment of vancomycin-resistant infections.

5424 Idoxifene
116057-75-1
$C_{28}H_{30}INO$
1-[2-[p-[(E)-β-Ethyl-α-(p-iodophenyl)styryl]phenoxy]ethyl]pyrrolidine.
CB-7432; SB-223030. Antineoplastic; hormone (replacement therapy, estrogen receptor antagonist). Used to treat and prevcent osteoporosis. *SmithKline Beecham Pharm.*

5425 Olpadronic Acid
63132-39-8
$C_5H_{15}NO_7P_2$
[3-(Dimethylamino)-1-hydroxypropylidene] diphosphonic acid.
An N-containing bisphosphonate that suppresses bone resorption by inhibiting isopentenyl pyrophosphate isomerase/farnesyl pyrophosphate synthase activity. Antiosteoporotic.

5426 Pamidronic Acid
40391-99-9 7135 254-905-2
$C_3H_{11}NO_7P_2$
(3-Amino-1-hydroxypropylidene)diphosphonic acid.
ADP; AHPrBP. Bone resorption inhibitor. *Ciba-Geigy.*

5427 Pamidronic Acid Disodium Salt
57248-88-1 7135 260-647-1
$C_3H_9NNa_2O_7P_2.5H_2O$
Disodium dihydrogen (3-amino-1-hydroxypropylidene)-diphosphonate pentahydrate.
Aredia; CGP-23339AE; Aminomux. Bone resorption inhibitor. *Ciba-Geigy Corp.*

5428 Raloxifene
84449-90-1 8281
$C_{28}H_{27}NO_4S$
6-Hydroxy-2-(p-hydroxyphenyl)benzo[b]thien-3-yl-p-(2-piperidinoethoxy)phenyl ketone.
keoxifene; LY-139481. Antiosteoporotic; antiestrogenic. Used to treat osteoporosis. mp = 143-147°; λ_m = 290 nm (ε 34000 EtOH). *Eli Lilly & Co.*

5429 Raloxifene Hydrochloride
82640-04-8 8281
$C_{28}H_{28}ClNO_4S$
6-Hydroxy-2-(p-hydroxyphenyl)benzo[b]thien-3-yl-p-(2-piperidinoethoxy)phenyl ketone hydrochloride.
keoxifene hydrochloride; LY-156758. Antiestrogen; used to treat osteoporosis. Nonsteroidal estrogen receptor mixed agonist-antagonist. mp = 258°; λ_m = 286 nm (ε 32800 EtOH). *Eli Lilly & Co.*

5430 Risedronic Acid
105462-24-6 8396
$C_7H_{11}NO_7P_2$
[1-Hydroxy-2-(3-pyridyl)ethylidene]diphosphonic acid.
Calcium regulator, used as a bone resorption inhibitor. *Merck & Co.Inc.; Procter & Gamble Pharm. Inc.*

5431 Risedronic Acid Monosodium Salt
115436-72-1 8396
$C_7H_{10}NNaO_7P_2$
Sodium trihydrogen [1-hydroxy-2-(3-pyridyl)ethylidene]-diphosphonate.
Actonel; risedronate sodium; NE-58095. Calcium regulator; bone resorption inhibitor in treatment of osteoporosis. *Merck & Co.Inc.; Procter & Gamble Pharm. Inc.*

5432 Sodium Fluoride
7681-49-4 8762 231-667-8
FNa
Hydrofluoric acid sodium salt.
Florinse; Minute-Gel; Neutra-Care; Pediaflor; Chemifluor; Dentalfluoro; Duraphat; Fuoros; Luride-SF; Villiaumite; Florocid; Flura-Drops; Karidium; Lemoflur; Ossalin; Ossin; Osteo-F; Osteoflur; Slow-Fluoride; Zymafluor. Dental caries prophylactic. d = 2.78; mp = 993°; bp = 1704°; soluble in H_2O (4.0 g/100 ml at 15°, 4.3 g/100 ml at 25°, 5.0 g/100 ml at 100°), insoluble in EtOH; LD_{50} (rat orl) = 180 mg/kg. *Ross Products.*

5433 Teriparatide
52232-67-4
$C_{181}H_{291}N_{55}O_{51}S_2$
L-Seryl-L-valyl-L-seryl-L-α-glutamyl-L-isoleucyl-L-methionyl-L-histidyl-L-asparaginyl-L-leucylglcyl-L-lysyl-L-histidyl-L-leucyl-L-asparaginyl-L-seryl-L-methionyl-L-α-glutaminyl-L-arginyl-L-valyl-L-α-glutamyl-L-tryptophyl-L-leucyl-L-arginyl-L-lysyl-L-lysyl-L-leucyl-L-glutaminyl-L-α-aspartyl-L-valyl-L-histidyl-L-asparaginyl-L-phenylalanine.
LY-333334; hPTH 1-34; H-Ser-Val-Ser-Glu-Ile-Gln-Leu-Met-His-Asn-Leu-Gly-Lys-His-Leu-Asn-Ser-Met-Glu-Arg-Val-Glu-Trp-Leu-Arg-Lys-Lys-Leu-Gln-Asp-Val-His-Asn-Phe-OH. Bone resorption inhibitor. Osteoporosis therapy adjunct. *Rhône-Poulenc Rorer Pharm. Inc.*

5434 Teriparatide Acetate
99294-94-7 9309
$C_{181}H_{291}N_{55}O_{51}S_2.xH_2O.yC_2H_4O_2$
L-Seryl-L-valyl-L-seryl-L-α-glutamyl-L-isoleucyl-L-methionyl-L-histidyl-L-asparaginyl-L-leucylglcyl-L-lysyl-L-histidyl-L-leucyl-L-asparaginyl-L-seryl-L-methionyl-L-α-glutaminyl-L-arginyl-L-valyl-L-α-glutamyl-L-tryptophyl-L-leucyl-L-arginyl-L-lysyl-L-lysyl-L-leucyl-L-glutaminyl-L-α-aspartyl-L-valyl-L-histidyl-L-asparaginyl-L-phenylalanine acetate (salt) hydrate.
MN-10T; Parathar; hPTH 1-34 acetate; H-Ser-Val-Ser-Glu-Ile-Gln-Leu-Met-His-Asn-Leu-Gly-Lys-His-Leu-Asn-Ser-Met-Glu-Arg-Val-Glu-Trp-Leu-Arg-Lys-Lys-Leu-Gln-Asp-Val-His-Asn-Phe-OH.$xH_2O.yCH_3COOH$. Bone resorption inhibitor. Osteoporosis therapy adjunct. Synthetic polypeptide; the 1-34 fragment of human parathyroid hormone. Also used as a diagnostic aid. *Rhône-Poulenc Rorer Pharm. Inc.*

5435 Tiludronate Disodium
149845-07-8
$C_7H_7ClNa_2O_6P_2S$
Disodium dihydrogen [[(p-chlorophenyl)thio]methylene]-diphosphonate.
SR-41319B. Used for treatment of and intervention in osteoporosis and treatment of Paget's disease. *Sanofi Winthrop.*

5436 Tiludronic Acid
89987-06-4 9582
$C_7H_9ClO_6P_2S$
[[(p-Chlorophenyl)thio]methylene]-diphosphonic acid.
SR-41319; ACPMD; Cl-TMBP; ME-3737; Skelid. Used for treatment of and intervention in osteoporosis and treatment of Paget's disease. Has anti-inflammatory activity. $PK_1 = 10.85$, $pK_2 = 6.90$, $pK_3 = 2.95$, $pK_4 = 1.30$; [di-tert-butylamine salt]: mp = 253° (dec). *Sanofi Winthrop.*

5437 Zoledronate Disodium
165800-07-7
$C_5H_8N_2Na_2O_7P_2.4H_2O$
Disodium dihydrogen (1-hydroxy-2-imidazol-1-ylethylidene)diphosphonate tetrahydrate.
CGP-42446A. Osteoporosis therapy adjunct. *Ciba-Geigy Corp.*

5438 Zoledronate Trisodium
165800-08-8
$C_{25}H_{35}N_{10}Na_{15}O_{35}P_{10}.2H_2O$
Trisodium hydrogen (1-hydroxy-2-imidazol-1-ylethylidene)diphosphonate hydrate (5:2).
CGP-42446B. Osteoporosis therapy adjunct. *Ciba-Geigy Corp.*

5439 Zoledronic Acid
165800-06-6
$C_5H_{10}N_2O_7P_2.H_2O$
(1-Hydroxy-2-imidazol-1-ylethylidene)diphosphonic acid monohydrate.
CGP-42446. Osteoporosis therapy adjunct. *Ciba-Geigy Corp.*

Antipagetics

5440 Alendronic Acid
66376-36-1 228
$C_4H_{13}NO_7P_2$
(4-Amino-1-hydroxybutylidene)bisphosphonic acid.
ABDP. Calcium regulator. mp = 233-235° (dec). *Inst. Gentili S.p.A.; Merck & Co.Inc.*

5441 Alendronic Acid Trisodium Salt
121268-17-5 228
$C_4H_{12}NNaO_7P_2.3H_2O$
Sodium trihydrogen (4-amino-1-hydroxybutylidene)-diphosphonate trihydrate.
Fosamax; MK-217; G-704650; Adronat; Alendros; Onclast. Calcium regulator. *Inst. Gentili S.p.A.; Merck & Co.Inc.*

5442 Elcatonin
60731-46-6 3578 262-393-7
$C_{148}H_{244}N_{42}O_{47}$
1-Butyric acid-7-(L-2-aminobutyric acid)-26-L-aspartic acid-27-L-valine-29-L-alaninecalcitonin (salmon).
carbocalcitonin; HC-58; Calcinil; Carbicalcin; Elcitonin; Turbocalcin. Calcium regulator. Used to treat Paget's disease. *Toyo Jozo.*

5443 Etidronic Acid
2809-21-4 3908 220-552-8
$C_2H_8O_7P_2$
(1-Hydroxyethylidene)diphosphonic acid.
EHDP. Calcium regulator. Very soluble in H_2O, insoluble in AcOH. *Procter & Gamble Pharm. Inc.*

5444 Etidronic Acid Disodium Salt
7414-83-7 3908 231-025-7
$C_2H_6Na_2O_7P_2$
Disodium dihydrogen (1-hydroxyethylidene)-diphosphonate.
Didronel; Calcimux; Diphos; Etidron. Calcium regulator. Bone resorption inhibitor. Soluble in H_2O. *Procter & Gamble Pharm. Inc.*

5445 Pamidronic Acid
40391-99-9 7135 254-905-2
$C_3H_{11}NO_7P_2$
(3-Amino-1-hydroxypropylidene)diphosphonic acid.
ADP; AHPrBP. Bone resorption inhibitor. *Ciba-Geigy Corp.*

5446 Pamidronic Acid Disodium Salt
57248-88-1 7135 260-647-1
$C_3H_9NNa_2O_7P_2.5H_2O$
Disodium dihydrogen (3-amino-1-hydroxypropylidene)-diphosphonate pentahydrate.
Aredia; CGP-23339AE; Aminomux. Bone resorption inhibitor. *Ciba-Geigy Corp.*

5447 Teriparatide
52232-67-4
$C_{181}H_{291}N_{55}O_{51}S_2$
L-Seryl-L-valyl-L-seryl-L-α-glutamyl-L-isoleucyl-L-methionyl-L-histidyl-L-asparaginyl-L-leucylglcyl-L-lysyl-L-histidyl-L-leucyl-L-asparaginyl-L-seryl-L-methionyl-L-α-glutaminyl-L-arginyl-L-valyl-L-α-glutamyl-L-tryptophyl-L-leucyl-L-arginyl-L-lysyl-L-lysyl-L-leucyl-L-glutaminyl-L-α-

aspartyl-L-valyl-L-histidyl-L-asparaginyl-L-phenylalanine. LY-333334; hPTH 1-34; H-Ser-Val-Ser-Glu-Ile-Gln-Leu-Met-His-Asn-Leu-Gly-Lys-His-Leu-Asn-Ser-Met-Glu-Arg-Val-Glu-Trp-Leu-Arg-Lys-Lys-Leu-Gln-Asp-Val-His-Asn-Phe-OH. Bone resorption inhibitor. Osteoporosis therapy adjunct. *Rhône-Poulenc Rorer Pharm. Inc.*

5448 Teriparatide Acetate
99294-94-7 9309
$C_{181}H_{291}N_{55}O_{51}S_2.xH_2O.yC_2H_4O_2$
L-Seryl-L-valyl-L-seryl-L-α-glutamyl-L-isoleucyl-L-methionyl-L-histidyl-L-asparaginyl-L-leucylglcyl-L-lysyl-L-histidyl-L-leucyl-L-asparaginyl-L-seryl-L-methionyl-L-α-glutaminyl-L-arginyl-L-valyl-L-α-glutamyl-L-tryptophyl-L-leucyl-L-arginyl-L-lysyl-L-lysyl-L-leucyl-L-glutaminyl-L-α-aspartyl-L-valyl-L-histidyl-L-asparaginyl-L-phenylalanine acetate (salt) hydrate.
MN-10T; Parathar; hPTH 1-34 acetate; H-Ser-Val-Ser-Glu-Ile-Gln-Leu-Met-His-Asn-Leu-Gly-Lys-His-Leu-Asn-Ser-Met-Glu-Arg-Val-Glu-Trp-Leu-Arg-Lys-Lys-Leu-Gln-Asp-Val-His-Asn-Phe-OH.$xH_2O.yCH_3COOH$. Bone resorption inhibitor. Osteoporosis therapy adjunct. Synthetic polypeptide; the 1-34 fragment of human parathyroid hormone. Also used as a diagnostic aid. *Rhône-Poulenc Rorer Pharm. Inc.*

5449 Tiludronate Disodium
149845-07-8
$C_7H_7ClNa_2O_6P_2S$
Disodium dihydrogen [[(p-chlorophenyl)thio]methylene]-diphosphonate.
SR-41319B. Used for treatment of and intervention in osteoporosis and treatment of Paget's disease. *Sanofi Winthrop.*

5450 Tiludronic Acid
89987-06-4 9582
$C_7H_9ClO_6P_2S$
[[(p-Chlorophenyl)thio]methylene]diphosphonic acid.
SR-41319; ACPMD; Cl-TMBP; ME-3737; Skelid. For treatment of osteoporosis and Paget's disease. [di-tert-butylamine salt]: mp = 253° (dec). *Sanofi Winthrop.*

5451 Zoledronate Disodium
165800-07-7
$C_5H_8N_2Na_2O_7P_2.4H_2O$
Disodium dihydrogen (1-hydroxy-2-imidazol-1-ylethylidene)diphosphonate tetrahydrate.
CGP-42446A. Osteoporosis therapy adjunct. *Ciba-Geigy Corp.*

5452 Zoledronate Trisodium
165800-08-8
$C_{25}H_{35}N_{10}Na_{15}O_{35}P_{10}.2H_2O$
Trisodium hydrogen (1-hydroxy-2-imidazol-1-ylethylidene)diphosphonate hydrate (5:2).
CGP-42446B. Osteoporosis therapy adjunct. *Ciba-Geigy Corp.*

5453 Zoledronic Acid
165800-06-6
$C_5H_{10}N_2O_7P_2.H_2O$
(1-Hydroxy-2-imidazol-1-ylethylidene)diphosphonic acid monohydrate.
CGP-42446. Osteoporosis therapy adjunct. *Ciba-Geigy Corp.*

Antiparkinsonians

5454 Amantadine
768-94-5 389 212-201-2
$C_{10}H_{17}N$
Tricyclo[3.3.1.1^{3,7}]decan-1-amine.
Antiviral; antidyskinetic; antiparkinsonian. mp = 160-190°, 180-192°; sparingly soluble in H_2O. *Apothecon; DuPont-Merck Pharm.; Solvay Pharm. Inc.*

5455 Amantadine Hydrochloride
665-66-7 389 211-560-2
$C_{10}H_{18}ClN$
Tricyclo[3.3.1.1^{3,7}]decan-1-amine hydrochloride.
Symadine; Symmetrel; EXP-105-1; NSC-83653; Amazolon; Mantadix; Mantadan; Mantadine; Midantan; Mydantane; Virofral. Antiviral; antidyskinetic; antiparkinsonian. Dec 360°; soluble in H+2O (> 5 g/100 ml), insoluble in Et_2O; LD_{50} (mus orl) = 700 mg/kg, (rat orl) = 1275 mg/kg. *Apothecon; DuPont-Merck Pharm.; Solvay Pharm. Inc.*

5456 Benserazide
322-35-0 1079
$C_{10}H_{15}N_3O_5$
DL-Serine 2-[(2,3,4-trihydroxyphenyl)hydrazide.
component of: co-beneldopa, Madopa. Peripheral decarboxylase inhibitor. Antiparkinsonian. Used in combination with levodopa to treat Parkinson's disease. *Hoffmann-LaRoche Inc.*

5457 Benserazide Hydrochloride
14919-77-8 1079 238-991-9
$C_{10}H_{16}ClN_3O_5$
DL-Serine 2-[(2,3,4-trihydroxyphenyl)hydrazide hydrochloride.
Ro-4-4602. Peripheral decarboxylase inhibitor. Antiparkinsonian. mp = 146-148°; soluble in H_2O. *Hoffmann-LaRoche Inc.*

5458 Benztropine
86-13-5
$C_{21}H_{25}NO$
3α-(Diphenylmethoxy)-1αH,5αH-tropane.
Anticholinergic. Antiparkinsonian. *Apothecon; Merck & Co.Inc.*

5459 Benztropine Mesylate
132-17-2 1156 205-048-8
$C_{22}H_{29}NO_4S$
3α-(Diphenylmethoxy)-1αH,5αH-tropane methanesulfonate.
Cogentin; Cogentinol; Cobrentin methanesulfonate. Anticholinergic. Antiparkinsonian. mp = 143°; λ_m = 259 nm (E_M 437); soluble in H_2O. *Apothecon; Merck & Co.Inc.*

5460 Bietanautine
6888-11-5 1253
$C_{35}H_{41}N_9O_9$
1,2,3,6-Tetrahydro-1,3-dimethyl-2,6-dioxo-7H-purine-7-acetic acid compound with 2-(diphenylmethoxy)-N,N-dimethylethanamine (2:1).
Etanautine; Nautamine. Antihistaminic; antiemetic;

antiparkinsonian. mp = 168-170°; soluble in EtOH, sparingly soluble in H_2O. *Delagrange.*

5461 Biperiden
514-65-8 1274 208-184-6
$C_{21}H_{29}NO$
α-5-Norbornen-2-yl-α-phenyl-1-piperidinepropanol.
KL-373; Biperiden; 3-piperidino-1-phenyl-1-bicycloheptenyl-1-propanol; Akineton. Anticholinergic. Antiparkinsonian. mp = 112-116°, 101°; slightly soluble in H_2O, EtOH; readily soluble in MeOH. *Knoll Pharm. Co.*

5462 Biperiden Hydrochloride
1235-82-1 1274 214-976-2
$C_{21}H_{30}ClNO$
α-5-Norbornen-2-yl-α-phenyl-1-piperidinepropanol hydrochloride.
Akineton hydrochloride; Akineton; Akinophyl. Anticholinergic. Antiparkinsonian. mp = 275° (dec), 238°; LD_{50} (mus orl) = 545 mg/kg, (mus iv) = 56 mg/kg. *Knoll Pharm. Co.*

5463 Biperiden Lactate
7085-45-2 1274 230-388-9
$C_{24}H_{35}NO_4$
α-5-Norbornen-2-yl-α-phenyl-1-piperidinepropanol lactate.
Akineton lactate. Anticholinergic. *Knoll Pharm. Co.*

5464 Bromocriptine
25614-03-3 1437 247-128-5
$C_{32}H_{40}BrN_5O_5$
2-Bromo-12'-hydroxy-2'-(1-methylethyl)-5'-(2-methylpropyl)-5'α-ergotaman-3',6',18-trione.
CB-154. Prolactin inhibitor. Antiparkinsonian. Dopamine receptor agonist. mp = 215-218° (dec); $[\alpha]_D^{20}$ = -195° (c = 1 CH_2Cl_2); LD_{50} (rbt orl) > 1000 mg/kg, (rbt iv) = 12 mg/kg. *Sandoz Pharm. Corp.*

5465 Bromocriptine Mesylate
22260-51-1 1437 244-881-1
$C_{33}H_{44}BrN_5O_8S$
2-Bromo-12'-hydroxy-2'-(1-methylethyl)-5'-(2-methylpropyl)-5'α-ergotaman-3',6',18-trione monomethanesulfonate (salt).
bromocriptine methanesulfonate; CB-154 mesylate; Parlodel; Bagren; Pravidel. Prolactin inhibitor used to treat Parkinson's disease. Dopamine receptor agonist. mp = 192-196° (dec); $[\alpha]_D^{20}$ = 95° (c = 1 MeOH/CH_2Cl_2); soluble in MeOH (91 g/100 ml), EtOH (2.3 g/100 ml), H_2O (0.08 g/100 ml), $CHCl_3$ (0.045 g/100ml)C_6H_6 (< 0.01 g/100 ml). *Sandoz Pharm. Corp.*

5466 Budipine
57982-78-2 1491 261-062-4
$C_{21}H_{27}N$
1-tert-Butyl-4,4-diphenylpiperidine.
Antiparkinsonian. Indirect dopaminergic activity with weak antimuscarinic action. *Byk Gulden Lomberg GmbH.*

5467 Budipine Hydrochloride
63661-61-0 1491 264-388-5
$C_{21}H_{28}ClN$
1-tert-Butyl-4,4-diphenylpiperidine hydrochloride.
Parkinsan. Antiparkinsonian. Indirect dopaminergic activity with weak antimuscarinic action. LD_{50} (mmus orl) = 120 mg/kg, (mmus iv) = 33 mg/kg, (rat orl) = 165 mg/kg, (rat iv) = 28 mg/kg. *Byk Gulden Lomberg GmbH.*

5468 Carbidopa
28860-95-9 1843 249-271-9
$C_{10}H_{14}N_2O_4.H_2O$
(-)-L-α-Hydrazino-3,4-dihydroxy-α-methylhydrocinnamic acid monohydrate.
Lodosyn; Lodosin; HMD; MK-486; component of: Sinemet. Decarboxylase inhibitor. Antiparkinsonian. mp = 203-205° (dec), 208°; $[\alpha]_D$ = -17.3 (MeOH); [DL-form]: mp = 206-208° (dec); λ_m = 282.5 nm (ε 2940 MeOH). *DuPont-Merck Pharm.; Lemmon Co.*

5469 Carmantadine
38081-67-3 253-774-9
$C_{14}H_{21}NO_2$
1-(1-Amantadinyl)-2-azetidinecarboxylic acid.
Sch-15427. Antiparkinsonian. *Schering-Plough HealthCare Products.*

5470 Ciladopa
80109-27-9
$C_{21}H_{26}N_2O_4$
(-)-(S)-2-[4-(β-Hydroxy-3,4-dimethoxyphenethyl)-1-piperazinyl]-2,4,6-cycloheptatrien-1-one.
Dopaminergic agent. Antiparkinsonian. *Wyeth-Ayerst Labs.*

5471 Ciladopa Hydrochloride
83529-09-3
$C_{21}H_{27}ClN_2O_4$
(-)-(S)-2-[4-(β-Hydroxy-3,4-dimethoxyphenethyl)-1-piperazinyl]-2,4,6-cycloheptatrien-1-one hydrochloride.
Tremerase; AY-27110. Dopaminergic agent. Antiparkinsonian. *Wyeth-Ayerst Labs.*

5472 Dexetimide
21888-98-2 2987
$C_{23}H_{26}N_2O_2$
(+)-2-(1-Benzyl-4-piperidyl)-2-phenylglutarimide.
dextrobenzetimide; dexbenzetimide. Anticholinergic; antiparkinsonian. mp = 181-183°; $[\alpha]_D^{20}$ = 125° ($CHCl_3$). *Abbott Laboratories Inc.*

5473 Dexetimide Hydrochloride
21888-96-0 2987 244-633-2
$C_{23}H_{27}ClN_2O_2$
(+)-2-(1-Benzyl-4-piperidyl)-2-phenylglutarimide hydrochloride.
R-16470; Tremblex. Anticholinergic; antiparkinsonian. mp = 270-275°; $[\alpha]_D^{20}$ = 125° (MeOH); LD_{50} (rat iv) = 45 mg/kg. *Abbott Laboratories Inc.*

5474 Diethazine
60-91-3 3157 200-491-3
$C_{18}H_{22}N_2S$
10-(2-Diethylaminoethyl)phenothiazine.
RP-2987; Deparkin; Dinezin; Dolisina; Eazaminum; Ethylemin; Parkazin. Anticholinergic; antiparkinsonian. Oily liquid; bp_{4-5} = 195-208°, $bp_{0.4-0.5}$ = 167-175°. *Rhône-Poulenc Rorer Pharm. Inc.*

5475 Diethazine Hydrochloride
341-70-8 3157 206-437-5
$C_{18}H_{23}ClN_2S$
10-(2-Diethylaminoethyl)phenothiazine hydrochloride.
Antipar; Aparkazin; Casantin; Diparcol; Latibon; Thiantan; Thiontan. Anticholinergic; antiparkinsonian. mp = 184-186°; soluble in H_2O (20 g/100 ml), EtOH (16.7 g/100 ml), $CHCl_3$ (20 g/100 ml); insoluble in Et_2O; LD_{50} (mus orl) = 450 mg/kg. *Rhône-Poulenc Rorer Pharm. Inc.*

5476 Diprobutine
61822-36-4
$C_{10}H_{23}N$
1,1-Dipropylamine.
Antiparkinsonian. A noncompetitive blocker of nicotinic acetylcholine receptors.

5477 Dopamantine
39907-68-1 254-697-3
$C_{19}H_{25}NO_3$
N-(3,4-Dihydroxyphenyl)ethyl)-1-adamantane-carboxamide.
Sch-15507. Antiparkinsonian. *Schering-Plough HealthCare Products.*

5478 Doreptide
90104-48-6
$C_{17}H_{24}N_4O_3$
(2S)-N-[(αR)-α-[(Carbamoylmethyl)carbamoyl]-α-ethylbenzyl]-2-pyrrolidinecarboxamide.
1-propyl-2-phenyl-1-2-aminobutanoylglycinamide.
Antiparkinsonian. L-dopa potentiating agent.

5479 Droxidopa
23651-95-8 3513
$C_9H_{11}NO_5$
(-)-Threo-3-(3,4-dihydroxyphenyl))-L-serine.
L-threo-DOPS; L-DOPS; SM-5688; Dops. Antiparkinsonian. mp = 232-235°, 229-232°; $[\alpha]_D^{20}$ = -39° (c = 1 in 1N HCl), -42.0° (c = 1 in 1N HCl). *Parke-Davis.*

5480 Endomide
4582-18-7
$C_{17}H_{28}N_2O_2$
(1R,2S,3S,4S)-N,N,N',N'-Tetraethyl-5-norbornene-2,3-dicarboxamide.
Antiparkinsonian.

5481 Entacapone
130929-57-6
$C_{14}H_{15}N_3O_5$
(E)-α-Cyano-N,N-diethyl-3,4-dihydroxy-5-nitrocinnam-amide.
Antiparkinsonian. Catechol-O-methyltransfersase inhibitor.

5482 Ethopropazine
522-00-9 3793 208-320-4
$C_{19}H_{24}N_2S$
10-[2-(Diethylamino)propyl]phenothiazine.
RP-3356; W-483; Isothazine; Isothiazine; Parkin. Anticholinergic; antiparkinsonian. mp = 53-55°; LD_{50} (mus orl) = 650 mg/kg. *Parke-Davis.*

5483 Ethopropazine Hydrochloride
1094-08-2 3793 214-134-4
$C_{19}H_{25}ClN_2S$
10-[2-(Diethylamino)propyl]phenothiazine hydrochloride.
Dibutil; Lysivane; Pardisol; Parphezein; Parphezin; Parsidol; Parsitan; Parsotil; Rodipal. Anticholinergic; antiparkinsonian. mp = 223-225°; soluble in H_2O (0.25 g/100 ml at 20°, 5 g/100 ml at 40°), EtOH (3.3 g/100 ml at 25°); sparingly soluble in Me_2CO; insoluble in Et_2O, C_6H_6. *Parke-Davis; Rhône-Poulenc Rorer Pharm. Inc.*

5484 Ethylbenzhydramine
642-58-0 3813
$C_{19}H_{25}NO$
2-(Diphenylmethoxy)-N,N-diethylethanamine.
etanautine. Anticholinergic; antiparkinsonian. $bp_{0.15}$ = 140-142°. *Ciba-Geigy Corp.; Parke-Davis.*

5485 Ethylbenzhydramine Hydrochloride
86-24-8 3813
$C_{19}H_{26}ClNO$
2-(Diphenylmethoxy)-N,N-diethylethanamine hydrochloride.
Antiparkin. Anticholinergic; antiparkinsonian. mp = 140°; freely soluble in H_2O; soluble in EtOH, Me_2CO, $CHCl_3$; slightly soluble in C_6H_6, Et_2O. *Parke-Davis.*

5486 Ethylbenzhydramine Methyl Iodide
5982-52-5 3813 227-792-2
$C_{20}H_{28}INO$
2-(Diphenylmethoxy)-N,N-diethylethanamine methyl iodide.
Metropin. Anticholinergic; antiparkinsonian. *Parke-Davis.*

5487 Ipidacrine
62732-44-9
$C_{12}H_{16}N_2$
9-Amino-2,3,5,6,7,8-hexahydro-1H-cyclopent[b]-quinoline.
NIK-247. Nootropic. Antiparkinsonian. Improves scopolamine-induced amnesia. Under study for use in treatment of Alzheimer's disease.

5488 Lazabemide
103878-84-8 5407
$C_8H_{10}ClN_3O$
N-(2-Aminoethyl)-5-chloropicolinamide.
Ro-19-6327. Antiparkinsonian. Selective and reversible monoamine oxidase B inhibitor. *Hoffmann-LaRoche Inc.*

5489 Lazabemide Hydrochloride
103878-83-7 5407
$C_8H_{11}Cl_2N_3O$
N-(2-Aminoethyl)-5-chloropicolinamide hydrochloride.
Ro-19-6327/000. Antiparkinsonian. Selective and reversible monoamine oxidase B inhibitor. mp = 193-195°; LD_{50} (mus orl) = 1000-2000 mg/kg. *Hoffmann-LaRoche Inc.*

5490 Levodopa
59-92-7 5490 200-445-2
(-)-3-(3,4-Dihydroxyphenyl)-L-alanine.
L-dopa; Bendopa; Dopar; Larodopa; Levopa; Deadopa; Dopaflex; Dopal; Dopaidan; Dopalina; Doparkine;

Doparl; Dopasol; Dopaston; Dopastral; Cidandopa; Doprin; Eldopal; Eldopar; Eldopatec; Eurodopa; Maipedopa; Laradopa; Maipedopa; Ledopa; Parda; Levopa; Veldopa; Weldopa; component of: Madopa, Sinemet. Antiparkinsonian. Naturally occurring isomer of dopa, the biological precursor of catecholamine. mp = 276-278° (dec), 284-286°; $[\alpha]_D^{13}$ = -13.1° (c = 5.12 in 1N HCl); λ_m = 220.5, 280 nm (log ε 3.79, 3.42 0.001N HCl); soluble in H_2O (0.165 g/100 ml); insoluble in EtOH, C_6H_6, EtOAc, $CHCl_3$; LD_{50} (mus orl) = 3650 mg/kg, (rat orl) = 4000 mg/kg, (rbt orl) = 609 mg/kg. *DuPont-Merck Pharm.; Hoffmann-LaRoche Inc.; ICN Pharm. Inc.; Lemmon Co.; Roberts Pharm. Corp.; SmithKline Beecham Pharm.*

5491 Lometraline
39951-65-0
$C_{13}H_{18}ClNO$
8-Chloro-1,2,3,4-tetrahydro-5-methoxy-N,N-dimethyl-1-naphthylamine.
Antipsychotic. Antiparkinsonian. *Pfizer Inc.*

5492 Lometraline Hydrochloride
34552-78-8
$C_{13}H_{19}Cl_2NO$
8-Chloro-1,2,3,4-tetrahydro-5-methoxy-N,N-dimethyl-1-naphthylamine hydrochloride.
CP-14368-1. Antipsychotic. Antiparkinsonian. *Pfizer Inc.*

5493 Mazaticol
42024-98-6
$C_{21}H_{27}NO_3S_2$
6,6,9-Trimethyl-9-azabicyclo[3.3.1]non-3β-yl di-2-thienylglycolate.
PG-501. Potent inhibitor of muscarinic receptor binding. Anticholinergic; antiparkinsonian.

5494 Mofegiline
119386-96-8 6313
$C_{11}H_{13}F_2N$
(E)-2-(Fluoromethylene)-4-(p-fluorophenyl)-butylamine.
Antiparkinsonian. Selective and irreversible monoamine oxidase B inhibitor. *Marion Merrell Dow Inc.*

5495 Mofegiline Hydrochloride
120635-25-8 6313
$C_{11}H_{14}ClF_2N$
(E)-2-(Fluoromethylene)-4-(p-fluorophenyl)-butylamine hydrochloride.
MDL-72974A. Antiparkinsonian. Selective and irreversible monoamine oxidase B inhibitor. mp = 131°. *Marion Merrell Dow Inc.*

5496 Naxagolide
88058-88-2
$C_{15}H_{21}NO_2$
(+)-(4aR,10bR)-3,4,4a,5,6,10b-Hexahydro-4-propyl-2H-naphth[1,2-b]-1,4-oxazin-9-ol.
Antiparkinsonian. Dopamine agonist. *Merck & Co.Inc.*

5497 Naxagolide Hydrochloride
99705-65-4
$C_{15}H_{22}ClNO_2$
(+)-(4aR,10bR)-3,4,4a,5,6,10b-Hexahydro-4-propyl-2H-naphth[1,2-b]-1,4-oxazin-9-ol hydrochloride.
MK-458; L-647339. Antiparkinsonian. Dopamine agonist. *Merck & Co.Inc.*

5498 Pareptide
61484-38-6
$C_{14}H_{26}N_4O_3$
N-[D-1-[(Carbamoylmethyl)carbamoyl]-3-methylbutyl]-N-methyl-L-2-pyrrolidinecarboxamide.
Antiparkinsonian. Melanotropin-inhibiting factor analog. *Wyeth-Ayerst Labs.*

5499 Pareptide Sulfate
61484-39-7
$C_{28}H_{54}N_8O_{10}S$
N-[D-1-[(Carbamoylmethyl)carbamoyl]-3-methylbutyl]-N-methyl-L-2-pyrrolidinecarboxamide sulfate (2:1).
AY-24856. Antiparkinsonian. Melanotropin-inhibiting factor analog. *Wyeth-Ayerst Labs.*

5500 Pergolide
66104-22-1 7304
$C_{19}H_{26}N_2S$
8β-[(Methylthio)methyl]-6-propylergoline.
Dopamine agonist. Antiparkinsonian. mp = 206-209°. *Eli Lilly & Co.*

5501 Pergolide Methanesulfonate
66104-23-2 7304
$C_{20}H_{30}N_2O_3S_2$
8β-[(Methylthio)methyl]-6-propylergoline monomethanesulfonate.
pergolide mesylate; Permax; Celance; LY-127809. Dopamine agonist. Antiparkinsonian. mp = 225°; λ_m = 279 nm (ε 6980 H_2O), 281 nm (ε 6993 EtOlH); $[\alpha]_D^{20}$ = -18.0° to -23.0° (c = 1 DMF); sparingly soluble in DMF, MeOH; slightly soluble in H_2O, 0.01N HCl, $CHCl_3$, CH_3CN, CH_2Cl_2, EtOH; very slightly soluble in Me_2CO; insoluble in Et_2O. *Eli Lilly & Co.*

5502 Piroheptine
16378-21-5 7658
$C_{22}H_{25}N$
3-(10,11-Dihydro-5H-dibenzo[a,d]cyclohepten-5-ylidene)-1-ethyl-2-methylpyridine.
Antiparkinsonian. bp_4 = 167°; λ_m = 240 nm (ε 12100 EtOH). *Fujisawa Pharm. USA Inc.*

5503 Piroheptine Hydrochloride
16378-22-6 7658
$C_{22}H_{26}ClN$
3-(10,11-Dihydro-5H-dibenzo[a,d]cyclohepten-5-ylidene)-1-ethyl-2-methylpyridine hydrochloride.
Trimol. Antiparkinsonian. mp = 250-253°; LD_{50} (mmus orl) = 253 mg/kg, (mmus iv) = 19 mg/kg, (mmus ip) = 95 mg/kg, (mmus sc) = 109 mg/kg, (mrat orl) = 600 mg/kg, (mrat iv) = 17 mg/kg, (mrat ip) = 110 mg/kg, (mrat sc) = 330 mg/kg. *Fujisawa Pharm. USA Inc.*

5504 Pramipexole
104632-26-0 7885
$C_{10}H_{17}N_3S$
(S)-2-Amino-4,5,6,7-tetrahydro-6-(propylamino)benzothiazole.
U-98528E; SUD919CL2Y. Dopamine D_2-receptor

agonist; antischizophrenic; antidepressant. Antiparkinsonian. *Boehringer Ingelheim GmbH.*

5505 Pramipexole Hydrochloride
104632-25-9 7885
$C_{10}H_{18}ClN_3S$
(S)-2-Amino-4,5,6,7-tetrahydro-6-(propylamino)benzothiazole hydrochloride.
SND-19. Dopamine D_2-receptor agonist; antischizophrenic; antidepressant. Antiparkinsonian. mp = 296-298°; $[\alpha]_D^{20}$ = -67.2° (c = 1 MeOH). *Boehringer Ingelheim GmbH.*

5506 Pridinol
511-45-5 7922 208-128-0
$C_{20}H_{25}NO$
α,α-Diphenyl-1-piperidinepropanol.
ridinol; C-238. Anticholinergic; antiparkinsonian. mp = 120-121°; soluble in Me_2CO. *Glaxo Wellcome Inc.; Wellcome Foundation Ltd.*

5507 Pridinol Hydrochloride
968-58-1 7922 213-529-9
$C_{20}H_{26}ClNO$
α,α-Diphenyl-1-piperidinepropanol hydrochloride.
Parks 12. Anticholinergic; antiparkinsonian. Dec 238°; soluble in EtOH; LD_{50} (mus iv) = 35 mg/kg, (mus ip) = 131°, (rat iv) = 33 mg/kg, (rat ip) = 91 mg/kg. *Glaxo Wellcome Inc.; Wellcome Foundation Ltd.*

5508 Pridinol Mesylate
6856-31-1 7922 229-953-2
$C_{21}H_{29}NO_4S$
α,α-Diphenyl-1-piperidinepropanol monomethanesulfonate.
pridinol methanesulfonate; Konlax; Loxeen; Lyseen; Mitanoline. Anticholinergic; antiparkinsonian. mp = 152.5-155°; sparingly soluble in H_2O. *Glaxo Wellcome Inc.; Wellcome Foundation Ltd.*

5509 Procyclidine
77-37-2 7944 201-023-0
$C_{19}H_{29}NO$
α-Cyclohexyl-α-phenyl-1-pyrrolidinepropanol.
Skeletal muscle relaxant; anticholinergic; antiparkinsonian. mp = 85.5-86.5°;λ_m = 285.5 nm (ε 233 17% EtOH); [methosulfate ($C_{21}H_{35}NO_5S$)]: mp = 100°; soluble in H_2O, EtOH. *Eli Lilly & Co.; Glaxo Wellcome Inc.*

5510 Procyclidine Hydrochloride
1508-76-5 7944 216-141-8
$C_{19}H_{30}ClNO$
α-Cyclohexyl-α-phenyl-1-pyrrolidinepropanol hydrochloride.
Arpicolin; Kemadrin; Osnervan. Anticholinergic. Skeletal muscle relaxant. Antiparkinsonian. Dec 226-227°; soluble in H_2O (3 g/100 ml), organic solvents. *Eli Lilly & Co.; Glaxo Wellcome Inc.*

5511 Prodipine
31314-38-2 7949
$C_{20}H_{25}N$
1-Isopropyl-4,4-diphenylpiperidine hydrochloride.
CNS stimulating agent with antidepressant properties. Used as an antiparkinsonian. Monoamine oxidase inhibitor. $bp_{0.01}$ = 117-125°. *Byk Gulden Lomberg GmbH.*

5512 Prodipine Hydrochloride
31314-39-3 7949
$C_{20}H_{26}ClN$
1-Isopropyl-4,4-diphenylpiperidine hydrochloride.
Anthen. CNS stimulating agent with antidepressant properties. Used as an antiparkinsonian. Monoamine oxidase inhibitor. mp = 267°. *Byk Gulden Lomberg GmbH.*

5513 Quinelorane
97466-90-5
$C_{14}H_{22}N_4$
(-)-(5aR,9aR)-2-Amino-5,5a,6,78,9,9a,10-octahydro-6-propylpyrido-2,3-g]quinazoline.
Antihypertensive. Antiparkinsonian. *Eli Lilly & Co.*

5514 Quinelorane Hydrochloride
97548-97-5
$C_{14}H_{23}ClN_4$
(-)-(5aR,9aR)-2-Amino-5,5a,6,78,9,9a,10-octahydro-6-propylpyrido-2,3-g]quinazoline dihydrochloride.
LY-163502. Antihypertensive. Antiparkinsonian. *Eli Lilly & Co.*

5515 Rasagiline
136236-51-6
$C_{12}H_{13}N$
(R)-N-2-Propynyl-1-indanamine.
Antiparkinsonian. Monoamine oxidase B inhibitor. *Lemmon Co.*

5516 Rasagiline Methanesulfonate
161735-79-1
$C_{13}H_{17}NO_3S$
(R)-N-2-Propynyl-1-indanamine monomethanesulfonate.
TVP-1012. Antiparkinsonian. Monoamine oxidase B inhibitor. *Lemmon Co.*

5517 Ropinirole
91374-21-9 8416
$C_{16}H_{24}N_2O$
4-[2-(Dipropylamino)ethyl]-2-indolinone.
SKF-101468. Antiparkinsonian. Dopamine D_2 receptor agonist. *SmithKline Beecham Pharm.*

5518 Ropinirole Hydrochloride
91374-20-8 8416
$C_{16}H_{25}ClN_2O$
4-[2-(Dipropylamino)ethyl]-2-indolinone monohydrochloride.
SKF-101468-A. Antiparkinsonian. Dopamine D_2 receptor agonist. mp = 241-243°. *SmithKline Beecham Pharm.*

5519 Selegiline
14611-51-9 8569
$C_{13}H_{17}N$
(-)-(R)-N-α-Dimethyl-N-2-propynylphenethylamine.
(-)-deprenil; L-deprenyl; deprenyl [as (±)-form]; phenylisopropyl-N-methylpropinylamine [as (±)-form]. Antidyskinetic; antiparkinsonian. Monoamine oxidase B

inhibitor. Related to pargyline. $bp_{0.8}$ = 92-93°; $[\alpha]_D^{20}$ = -11.2°; [(dl-form)]: bp_5 = 103-110°; n_D^{20} = 1.5224. *Chinoin; Somerset Pharm. Inc.*

5520 Selegiline Hydrochloride
14611-52-0 8569
$C_{13}H_{18}ClN$
(-)-(R)-N-α-Dimethyl-N-2-propynylphenethylamine hydrochloride.
deprenyl; Eldepryl; Déprényl; Eldéprine; Eldepryl; Jumex; Movergan; Plurimen; E-250 [as (±)-form hydrochloride]. Antidyskinetic; antiparkinsonian. Monoamine oxidase B inhibitor. Related to pargyline. mp = 141-142°; $[\alpha]_D^{25}$ = -10.8° (c = 6.48 H_2O); LD_{50} (rat iv) = 81 mg/kg, (rat sc) = 280 mg/kg; [(±)-form hydrochloride]: LD_{50} (rat iv) = 63 mg/kg, (rat sc) = 126 mg/kg, (rat orl) = 385 mg/kg. *Chinoin; Somerset Pharm. Inc.*

5521 Talipexole
101626-70-4 9209
$C_{10}H_{15}N_3S$
6-Allyl-2-amino-5,6,7,8-tetrahydro-4H-thiazolo[4,5-d]azepine.
An α_2-adrenergic and dopamine D_2 receptor agonist. Used as an antiparkinsonian. *Boehringer Ingelheim GmbH.*

5522 Talipexole Hydrochloride
36085-73-1 9209
$C_{10}H_{17}Cl_2N_3S$
6-Allyl-2-amino-5,6,7,8-tetrahydro-4H-thiazolo[4.5-d]azepine hydrochloride.
B-HT-920; Domin. Antiparkinsonian. A α_2-adrenoceptor and dopamine D_2-receptor agonist. mp = 245° (dec). *Boehringer Ingelheim GmbH.*

5523 Terguride
37686-84-3 9308 253-624-2
$C_{20}H_{28}N_4O$
1,1-Diethyl-3-(6-methylergolin-8α-yl)urea.
TDHL. Prolactin inhibitor. Antiparkinsonian. Ergot derivative with dopamine agonist and antagonist activity. mp = 203-204° (dec), 205-207° (dec); $[\alpha]_D^{20}$ = 30° (c = 1 C_5H_5N); 29° (c = 0.2 C_5H_5N); λ_m = 292, 281, 224 nm (log ε 3.72, 3.81, 4.42 C_5H_5N); insoluble in H_2O. *SPOFA.*

5524 Terguride Hydrogen Maleate
37686-85-4 9308 253-625-8
$C_{24}H_{32}N_4O_5$
1,1-Diethyl-3-(6-methylergolin-8α-yl)urea maleate.
SH-406; VUFB-6638; ZK-31224; Dironyl; Mysalfon. Prolactin inhibitor. Antiparkinsonian. Ergot derivative with dopamine agonist and antagonist activity. mp = 190-191°; [monohydrate]: mp = 150-153°; $[\alpha]_D^{20}$ = -15.0° (c = 0.1 H_2O); soluble in H_2O (1.26 g/100 ml). *SPOFA.*

5525 Tolcapone
134308-13-7
$C_{14}H_{11}NO_5$
3,4-Dihydroxy-4'-methyl-5-nitrobenzophenone.
Ro-40-7592. Antihyperprolactinemic; antiparkinsonian. Catechol-O-methyltransfersase inhibitor.

5526 Tricyclamol Chloride
3818-88-0 223-311-5
$C_{20}H_{32}ClNO$
(±)-1-(3-Cyclohexyl-3-hydroxy-3-phenylpropyl)-1-methylpyrrolidinium chloride.
Anticholinergic; antiparkinsonian. mp = 226-227° (dec); soluble in H_2O (3 g/100 ml); more soluble in EtOH, $CHCl_3$; insoluble in Et_2O. *Eli Lilly & Co.*

5527 Trihexyphenidyl
144-11-6 9823 205-614-4
$C_{20}H_{31}NO$
α-Cyclohexyl-α-phenyl-1-piperidine-propanol.
Anticholinergic; antiparkinsonian. mp = 114.3-115°; [l-form]: mp = 112-113°; $[\alpha]_D^{20}$ = -25° (c = 0.4 EtOH). *Am. Cyanamid; Glaxo Wellcome plc.*

5528 Trihexyphenidyl Hydrochloride
52-49-3 9823 200-142-5
$C_{20}H_{32}ClNO$
α-Cyclohexyl-α-phenyl-1-piperidine-propanol hydrochloride.
Antitrem; Artane; benzhexol chloride; Aparkane; Broflex; Cyclodol; Pacitane; Paralest; Pargitan; Parkinane; Parkopan; Peragit; Pipanol; Sedrina; Tremin; Triphedinon; Triphenidyl; Tsiklodol. Anticholinergic; antiparkinsonian. mp = 258.5° (dec); soluble in H_2O (1.0 g/100 ml), EtOH (6 g/100 ml), $CHCl_3$ (5 g/100 ml), soluble in MeOH; slightly soluble in C_6H_6, Et_2O; [l-form]; mp = 264°; $[\alpha]_D^{20}$ = -30° (c = 0.4 $CHCl_3$). *Am. Cyanamid; Glaxo Wellcome plc.*

Antipheochromocytoma Agents

5529 Isaxonine
4214-72-6 5118
$C_7H_{11}N_3$
2-(Isopropylamino)pyrimidine.
Accelerates the rate of peripheral nerve regeneration. Explored as a nerve-regenerating drug in treatment of peripheral neuropathy, but dropped because it is associated with onset of hepatitis. mp = 27-28°; bp_{12} = 92-93°.

5530 Metyrosine
672-87-7 6248 211-599-5
$C_{10}H_{13}NO_3$
(-)-α-Methyl-L-tyrosine.
Demser; L-α-MT; MK-781. A tyrosine hydroxylase inhibitor used as an antihypertensive in pheochromocytoma. mp = 310-315°. *Merck & Co.Inc.*

5531 dl-Metyrosine
620-30-4 6248 210-635-7
$C_{10}H_{13}NO_3$
(±)-α-Methyl-L-tyrosine.
Demser. A tyrosine hydroxylase inhibitor used as an antihypertensive in pheochromocytoma. Antipheochromocytoma. Dec 320°, 330-332°; soluble in H_2O (0.057 g/100 ml at 25°). *Merck & Co.Inc.*

5532 Phenoxybenzamine
59-96-1 7409 200-446-8
$C_{18}H_{22}ClNO$
N-(2-Chloroethyl)-N-(1-methyl-2-phenoxyethyl)-benzylamine.
bensylyt; 688-A. Antihypertensive. Antipheochromocytoma. α-Adrenergic blocker. CAUTION: The hydrochloride may be a carcinogen. mp = 38-40°; soluble in C_6H_6. *SmithKline Beecham Pharm.*

5533 Phenoxybenzamine Hydrochloride
63-92-3 7409 200-569-7
$C_{18}H_{23}Cl_2NO$
N-(2-Chloroethyl)-N-(1-methyl-2-phenoxyethyl)-benzylamine hydrochloride.
Dibenzyline; Dibenzylin; Dibenyline; Dibenzyran. Antihypertensive. Antipheochromocytoma. α-Adrenergic blocker. CAUTION: The hydrochloride may be a carcinogen. mp = 137.5-140°; soluble in EtOH, propylene glycol; sparingly soluble in H_2O. *SmithKline Beecham Pharm.*

5534 Phentolamine
50-60-2 7417 200-053-1
$C_{17}H_{19}N_3O$
m-[N-(2-Imidazolin-2-ylmethyl)-p-toluidino]phenol.
Regitine; C-7337. An α-adrenergic blocker used as an antihypertensive agent. Used in the diagnosis and treatment of pheochromocytoma. mp = 174-175°. *Ciba-Geigy Corp.*

5535 Phentolamine Hydrochloride
73-05-2 7417 200-793-5
$C_{17}H_{20}ClN_3O$
m-[N-(2-Imidazolin-2-ylmethyl)-p-toluidino]phenol hydrochloride.
Regitine hydrochloride. An α-adrenergic blocker with antihypertensive properties. Used in the diagnosis and treatment of pheochromocytoma. mp = 239-240°; soluble in H_2O (2 g/100 ml), EtOH (1.43 g/100 ml); slightly soluble in $CHCl_3$; insoluble in Me_2CO, EtOAc; LD_{50} (rat iv) = 75 mg/kg, (rat sc) = 275 mg/kg, (rat orl) = 1250 mg/kg. *Ciba-Geigy Corp.*

5536 Phentolamine Methanesulfonate
65-28-1 7417 200-604-6
$C_{18}H_{23}N_3O_4S$
m-[N-(2-Imidazolin-2-ylmethyl)-p-toluidino]phenol monomethanesulfonate.
phentolamine mesylate; Regitine; Rogitine. An α-adrenergic blocker used as an antihypertensive agent. Also used in the diagnosis and treatment of pheochromocytoma. mp = 177-181°; soluble in H_2O (2 g/100 ml), EtOH (4.35 g/100 ml), $CHCl_3$ (0.15 g/100 ml). *Ciba-Geigy Corp.*

Antipneumocystics

5537 Atovaquone
95233-18-4 898
$C_{22}H_{19}ClO_3$
2-[trans-4-(p-Chlorophenyl)cyclohexyl]-3-hydroxy-1,4-naphthoquinone.
Mepron; 566C80; 566C; BW-566C; BW-556C-80; Acuvel; Wellvone. Antimalarial; antipneumocystic. Antiprotozoal against Toxoplasma. mp = 216-219°. *Glaxo Wellcome Inc.*

5538 Eflornithine
67037-37-0 3564
$C_6H_{12}F_2N_2O_2$
2-(Difluoromethyl)-DL-ornithine.
α-difluoromethylornithine; DFMO; RFI-71782; MDL-71782-A; Ornidyl. Antineoplastic; antipneumocystic; antiprotozoal (trypanosoma). Irreversibly inhibits ornithine decarboxylase. Has an antiproliferative effect on tumor cells. Used to remove facial hair; also effective against sleeping sickness. *Marion Merrell Dow Inc.*

5539 Eflornithine Hydrochloride
96020-91-6 3564
$C_6H_{13}ClF_2N_2O_2.H_2O$
2-(Difluoromethyl)-DL-ornithine hydrochloride monohydrate.
Ornidyl; MDL-71782A. Antineoplastic; antipneumocystic; antiprotozoal (trypanosoma). Irreversibly inhibits ornithine decarboxylase. Has an antiproliferative effect on tumor cells. mp = 183°. *Marion Merrell Dow Inc.*

5540 Pentamidine
100-33-4 7254 202-841-0
$C_{19}H_{24}N_4O_2$
4,4'-(Pentamethylenedioxy)dibenzamidine.
Antipneumocystic. Also used as antiprotozoal (trypanosoma, leishmania). Dec 186°. *Fujisawa Pharm. USA Inc.; Rhône-Poulenc Rorer Pharm. Inc.*

5541 Pentamidine Dimethanesulfonate
6823-79-6 7254 229-898-4
$C_{21}H_{32}N_4O_8S_2$
4,4'-(Pentamethylenedioxy)dibenzamidine dimethanesulfonate.
pentamidine mesylate; Lomodine. Antifungal and antiprotozoal: targets Leishmania, Trypanosoma. Antipneumocystic. *Fujisawa Pharm. USA Inc.; May & Baker Ltd.; Rhône-Poulenc Rorer Pharm. Inc.*

5542 Pentamidine Isethionate
140-64-7 7254 205-424-1
$C_{23}H_{36}N_4O_{10}S_2$
4,4'-(Pentamethylenedioxy)dibenzamidine isethionate.
pentamidine isetionate; MB-800, RP-2512; Aeropent; Banambax; NebuPent; Pentacarinat; Pentam; Pentam 300; Pneumopent. Antifungal and antiprotozoal: targets Leishmania, Trypanosoma. Antipneumocystic. mp = 180°; soluble in H_2O (10 g/100 ml at 25°, 25 g/100 ml at 100°); soluble in glycerol; slightly soluble in EtOH; insoluble in Et_2O, Me_2CO, $CHCl_3$. *Fujisawa Pharm. USA Inc.; May & Baker Ltd.; Rhône-Poulenc Rorer Pharm. Inc.*

5543 Sulfamethoxazole
723-46-6 9086 211-963-3
$C_{10}H_{11}N_3O_3S$
N^1-(5-Methyl-3-isoxazolyl)sulfanilamide.
Gantanol; Ro-4-2130; Sinomin; sulfisomezole; component of: Azo Gantanol, Bactrim, Cotrim, Septra, SMZ/TMP, Sulfatrim. Sulfonamide antibiotic used in the treatment of pneumocystis. mp = 167°; LD_{50} (mus orl) =

3662 mg/kg. *Apothecon; Glaxo Wellcome Inc.; Hoffmann-LaRoche Inc.; Lemmon Co.; Solvay UK Holding Co.Ltd.*

Antiprostatic Hypertrophy Agents

5544 Alfuzosin
81403-80-7 237
$C_{19}H_{27}N_5O_4$
(±)-N-[3-[(4-Amino-6,7-dimethoxy-2-quinazolinyl)-methylamino]propyl]tetrahydro-2-furamide.
SL-77499. Antihypertensive agent. An α_1 andrenoreceptor antagonist used in the treatment of benign prostatic hyperplasia. *Synthelabo Pharmacie.*

5545 Alfuzosin Hydrochloride
81403-68-1 237
$C_{19}H_{28}ClN_5O_4$
(±)-N-[3-[(4-Amino-6,7-dimethoxy-2-quinazolinyl)-methylamino]propyl]tetrahydro-2-furamide monohydrochloride.
SL-77499-10; Alfoten; Urion; Xatral. Antihypertensive agent. An α_1 andrenoreceptor antagonist used in the treatment of benign prostatic hyperplasia. mp = 225°, 235° (dec). *Synthelabo Pharmacie.*

5546 Doxazosin
74191-85-8 3489
$C_{23}H_{25}N_5O_5$
1-(4-Amino-6,7-dimethoxy-2-quinazolinyl)-4-(1,4-benzodioxan-2-ylcarbonyl)piperazine.
UK-33274. Antihypertensive. Also used in the treatment of benign prostatic hyperplasia. Selective α-adrenergic blocker related to Prazosin. [monohydrochloride]: mp = 289-290°. *Pfizer Intl.; Roerig Div. Pfizer Pharm.*

5547 Doxazosin Monomethanesulfonate
77883-43-3 3489
$C_{24}H_{29}N_5O_8S$
1-(4-Amino-6,7-dimethoxy-2-quinazolinyl)-4-(1,4-benzodioxan-2-ylcarbonyl)piperazine monomethanesulfonate.
Cardura; UK-33274-27; doxazosin mesylate; Alfadil; Cardenalin; Cardular; Cardura; Cardran; Diblocin; Normothen; Supressin. Antihypertensive. Also used in the treatment of benign prostatic hyperplasia. Selective α-adrenergic blocker related to Prazosin. *Pfizer Intl.; Roerig Div. Pfizer Pharm.*

5548 Epristeride
119169-78-7 3668
$C_{25}H_{37}NO_3$
17β-(tert-Butylcarbamoyl)androsta-3,5-diene-3-carboxylic acid.
SK&F-105657. A 5α-reductase inhibitor. Used in treatment of benign prostatic hypertrophy. mp = 242-249°. *SmithKline Beecham Pharm.*

5549 Finasteride
98319-26-7 4125
$C_{23}H_{36}N_2O_2$
N-tert-Butyl-3-oxo-4-aza-5α-androst-1-ene 17β-carboxamide.
Propecia; Proscar; MK-906; Chibro-Proscar; Finastid; Prostide. A 5α-reductase inhibitor (testosterone → dihydrotestosterone converting enzyme). Formerly used as a treatment for benign prostatic hypertrophy. Also reported to have antialopecia properties and is now used to treat alopecia. mp = 257°, 252-254°; $[\alpha]_D$ = -59° (c = 1 MeOH); freely soluble in $CHCl_3$, DMSO, MeOH, EtOH, n-PrOH; sparingly soluble in propylene glycol, polyethylene glycol 400; very slightly soluble in H_2O, acids, bases. *Merck & Co.Inc.*

5550 Gestonorone Caproate
1253-28-7 4422 215-010-2
$C_{26}H_{38}O_4$
17-Hydroxy-19-norpregn-4-ene-3,20-dione hexanoate.
SH-582; NSC-84054; gestronol caproate; Depostat. Progestogen. Used in treatment of benign prostatic hypertrophy. mp = 123-124°; $[\alpha]_D$ = 13° ($CHCl_3$); λ_m = 239 nm (ε 17540). *Schering-Plough HealthCare Products.*

5551 Mepartricin
11121-32-7 5891
mixture (≅ 1:1) of mepartricin A and mepartricin B; SPA-S-160; SN 654; Partricin methyl ester; methylpartricin; Ipertrofan; Orofungin; Tricandil; Tricangine. Antifungal; antiprotozoal (trichomonas); also used to treat benign prostatic hypertrophy. Methyl ester of the heptaene macrolide antibiotic complex, partricin. λ_m = 401, 378, 359, 340 nm; slightly soluble in H_2O, Et_2O, C_6H_6, petroleum ether; soluble in ROH, C_5H_5N, DMF, DMSO, Me_2CO; LD_{50} (mus orl) > 2000 mg/kg, (mus ip) = 200 mg/kg.

5552 Mepartricin A
62534-68-3 5891
$C_{60}H_{88}N_2O_{10}$
40-Demethyl-3,7-dideoxo-3,7-dihydroxy-N^{47}-methyl-5-oxocandicidin D methyl ester cyclic 15,19-hemiacetal gedamycin methyl ester.
Antifungal; antiprotozoal against Trichomonas; also used to treat benign prostatic hypertrophy. mp = 145-149° (dec); λ_m = 400, 377, 357, 339, 287, 240, 234, 204 nm (ε 79326, 92454, 68094, 51685, 14199, 24612, 26505, 16092 MeOH).

5553 Mepartricin B
62534-69-4 5891
$C_{59}H_{86}N_2O_{19}$
SPA-S-222 [as sodium lauryl sulfate complex]; Montricin [as sodium lauryl sulfate complex]. Antifungal; antiprotozoal against Trichomonas; also used to treat benign prostatic hypertrophy. mp = 154-158°; λ_m = 402, 379, 359, 340, 285. 233. 204 nm (ε 81101, 94729, 64171, 41558, 16196, 23835, 21696, MeOH).

5554 Naftopidil
57149-07-2 6443
$C_{24}H_{28}N_2O_3$
4-(o-Methoxyphenyl)-α-[(1-naphthyloxy)methyl]-1-piperazineethanol.
KT-611; Avishot; Flivas. An α_1 adrenergic blocker and serotonin $5HT_{1A}$ receptor agonist. Used as an antihypertensive agent. Also used in treatment of BPH. mp = 125-126°, 125-129°; insoluble in H_2O; LD_{50} (mus orl) = 1300 mg/kg, (rat orl) = 6400 mg/kg. *Boehringer Ingelheim Ltd.; Boehringer Mannheim GmbH; C.M. Ind.*

5555 (±)-Naftopidil
132295-16-0 6443
$C_{24}H_{28}N_2O_3$
(±)-4-(o-Methoxyphenyl)-α-[(1-naphthyloxy)methyl]-1-piperazineethanol.
KT-611; Avishot; Flivas. An α_1 adrenergic blocker and serotonin $5HT_{1A}$ receptor agonist. Used as an antihypertensive agent. Also used in treatment of BPH. mp = 125-126°, 125-129°; insoluble in H_2O; LD_{50} (mus orl) = 1300 mg/kg, (rat orl) = 6400 mg/kg. *Boehringer Mannheim GmbH.*

5556 Naftopidil Dihydrochloride
57149-08-3 6443
$C_{24}H_{30}Cl_2N_2O_3$
4-(o-Methoxyphenyl)-α-[(1-naphthyloxy)methyl]-1-piperazineethanol dihydrochloride.
An α_1 adrenergic blocker and serotonin $5HT_{1A}$ receptor agonist. Used as an antihypertensive agent. Also used in treatment of BPH. mp = 212-213°. *Boehringer Mannheim GmbH.*

5557 Osaterone
105149-04-0 7019
$C_{20}H_{25}ClO_4$
(+)-6-Chloro-17-hydroxy-2-oxapregna-4,6-diene-3,2-dione.
antiandrogen (androgen receptor antagonist). Used in treatment of benign prostatic hypertrophy. mp = 218-221°. *Teikoku Hormone Mfg. Co. Ltd.*

5558 Osaterone Acetate Ester
105149-00-6 7019
$C_{22}H_{27}ClO_5$
(+)-6-Chloro-17-hydroxy-2-oxapregna-4,6-diene-3,2-dione acetate (ester).
2-oxochlormadinone acetate; TZP-4238. antiandrogen (androgen receptor antagonist). Used in treatment of benign prostatic hypertrophy. mp = 253-255°. *Teikoku Hormone Mfg. Co. Ltd.*

5559 Oxendolone
33765-68-3 7065
$C_{20}H_{30}O_2$
16β-Ethyl-17β-hydroxyestr-4-en-3-one.
TSAA-291. antiandrogen (androgen receptor antagonist). Used in treatment of benign prostatic hypertrophy. mp = 152-153°; $[\alpha]_D$ = 41° (c = 1.0 EtOH); λ_m = 240 nm (ε 15800 EtOH); LD_{50} (rat mus orl) > 10000 mg/kg, (rat mus im) = 5000-10000 mg/kg, (rat mus ip) = 5000 - 10000 mg/kg. *Takeda Chem. Ind. Ltd.*

5560 Sitogluside
474-58-8
$C_{35}H_{60}O_6$
3β-(α-D-Glucopyranosyloxy)stigmast-5-ene.
BSSG; EU-4906; AW-10; WA-184. Antiprostatic hypertrophy.

5561 Tamsulosin
106133-20-4 9217
$C_{20}H_{28}N_2O_5S$
(-)-(R)-5-[2-[[2-(o-Ethoxyphenoxy)ethyl]amino]propyl]-2-methoxybenzenesulfonamide.
amsulosin. Specific α_1-adrenoceptor antagonist. Used in treatment of benign prostatic hypertrophy. *Boehringer Ingelheim Pharm. Inc.; Yamanouchi U.S.A. Inc.*

5562 dl-Tamsulosin Hydrochloride
80223-99-0 9217
$C_{20}H_{29}ClN_2O_5S$
(±)-(R)-5-[2-[[2-(o-Ethoxyphenoxy)ethyl]amino]propyl]-2-methoxybenzenesulfonamide hydrochloride.
LY-253351; YM-617; Amsulosin. Used in treatment of benign prostatic hypertrophy. Specific α_1-adrenoceptor antagonist. mp = 254-256°. *Boehringer Ingelheim Pharm. Inc.; Yamanouchi U.S.A. Inc.*

5563 (R)-Tamsulosin Hydrochloride
106463-17-6 9217
$C_{20}H_{29}ClN_2O_5S$
(-)-(R)-5-[2-[[2-(o-Ethoxyphenoxy)ethyl]amino]propyl]-2-methoxybenzenesulfonamide hydrochloride.
LY-253351; R-(-)-YM-12617; YM-12617-1; YM-617; Harnal. Specific α_1-adrenoceptor antagonist. Used in treatment of benign prostatic hypertrophy. mp = 228-230°; $[\alpha]_D^{24}$ = -4.0° (c = 0.35 MeOH). *Boehringer Ingelheim Pharm. Inc.; Yamanouchi U.S.A. Inc.*

5564 (S)-Tamsulosin Hydrochloride
106463-19-8 9217
$C_{20}H_{29}ClN_2O_5S$
(+)-(S)-5-[2-[[2-(o-Ethoxyphenoxy)ethyl]amino]propyl]-2-methoxybenzenesulfonamide hydrochloride.
YM-12617-2. Specific α_1-adrenoceptor antagonist. Used in treatment of benign prostatic hypertrophy. mp = 228-230°; $[\alpha]_D^{24}$ = +4.2° (c = 0.36 MeOH). *Boehringer Ingelheim Pharm. Inc.; Yamanouchi U.S.A. Inc.*

5565 Terazosin [anhydrous]
63074-08-8 9297
$C_{19}H_{25}N_5O_4$
1-(4-Amino-6,7-dimethoxy-2-quinazolinyl)-4-(tetrahydro-2-furoyl)piperazine.
An α_1-adrenergic blocker related to prazosin. Used as an antihypertensive agent and in treatment of benign prostatic hypertrophy. mp = 272.6-274°; λ_m = 212, 245, 330 nm (a 65.7, 127.5, 24.0 H_2O); soluble in MeOH (3.37 g/100 ml), H_2O (2.97 g/100 ml), EtOH (0.41 g/100 ml), $CHCl_3$ (0.12 g/100 ml), Me_2CO (.1 mg/100 ml); insoluble in C_6H_{14}. *Abbott Labs Inc.*

5566 Terazosin Hydrochloride Dihydrate
70024-40-7 9297
$C_{19}H_{26}ClN_5O_4$
1-(4-Amino-6,7-dimethoxy-2-quinazolinyl)-4-(tetrahydro-2-furoyl)piperazine hydrochloride dihydrate.
Abbott-45975; Heitrin; Hytracin; Hytrin; Hytrinex; Itrin; Urodie; Vasocard; Vasomet; Vicard. Used in treatment of hypertension and benign antiprostatic hypertrophy. An α_1-adrenergic blocker related to prazosin. mp = 271-274°; soluble in H_2O (2.42 g/100 ml); LD_{50} (mrat iv) = 277 mg/kg, (frat iv) = 293 mg/kg; [anhydrous hydrochloride]: mp = 278-279°; soluble in H_2O (76.12 g/100 ml); LD_{50} (mus iv) = 259.3 mg/kg. *Abbott Labs Inc.*

Antiprotozoals

5567 Acetarsone
97-44-9 49 202-582-3

$C_8H_{10}AsNO_5$

N-Acetyl-4-hydroxy-m-arsanilic acid.

Stovarsol; acetarsol; acetphenarsine; Ehrlich 594; Fourneau 190; F-190; Amarsan; Arsaphen; Dynarsan; Goyl; Kharophen; Limarsol; Malagride; Gynoplix; Oralcid; Devegan; Orarsan; Osarsal; Osvarsan; Paroxyl; Sanogyl; Spirocid; S.V.C.; Monargan; Ginarsol; Stovarsolan; Realphene [as calcium salt]; Bistovol [as bismuth salt]. Antisyphilitic; antiprotozoal against Trichomonas. Dec 240-250°; slightly soluble in H_2O; MLD (rbt orl) = 125-150 mg/kg, (cat orl) = 150-175 mg/kg. *Abbott Labs Inc.; Rhône-Poulenc Rorer Pharm. Inc.*

5568 Acetarsone Diethylamine Salt
534-33-8 49 208-597-1

$C_{12}H_{21}AsN_2O_5$

N-Acetyl-4-hydroxy-m-arsanilic acid diethylamine salt.

Acetarsin; Acetilarsano; Acetylarsan; Arsaphenan; Golarsyl; Syntharsol. Antiprotozoal (Trichomonas). Antisyphilitic. *Abbott Labs Inc.; Rhône-Poulenc Rorer Pharm. Inc.*

5569 Acranil
1684-42-0 123 216-868-0

$C_{21}H_{28}Cl_3N_3O_2$

1-[(6-Chloro-2-methoxy-9-acridinyl)amino]-3-(diethylamino)-2-propanol dihydrochloride.

SKF-16214-A2; SN-186. Antiprotozoal (Giardia). mp = 237-239° (dec); soluble in H_2O; [free base ($C_{21}H_{26}ClN_3O_2$)]: mp = 105-107°; sparingly soluble in Et_2O. *SmithKline Beecham Pharm.*

5570 Aminitrozole
140-40-9 430 205-414-7

$C_5H_5N_3O_3S$

N-(5-Nitro-2-thiazolyl)acetamide.

CL-5279; acinitrazole; Tritheon; Trichorad; Trichoral; Gynofon; Enheptin-A; Pleocide. Antiprotozoal (Trichomonas). mp = 264-265°; soluble in aqueous NH_3, NaOH. *Am. Cyanamid.*

5571 Amodiaquine
86-42-0 609 201-669-3

$C_{20}H_{22}ClN_3O$

4-[(7-Chloro-4-quinolyl)amino]-α-(diethylamino)-o-cresol.

SN-10751. Antimalarial; antiprotozoal against Toxoplasma. mp = 208° (dec). *Parke-Davis.*

5572 Amodiaquine Hydrochloride
69-44-3 609 200-706-0

$C_{20}H_{23}Cl_2N_3O$

4-[(7-Chloro-4-quinolinyl)amino]-α-(diethylamino)-o-cresol hydrochloride.

Camoquin hydrochloride; CAM-AQ1; Flavoquine, Miaquin. Antimalarial; antiprotozoal against Toxoplasma. Dec 150-160°; λ_m = 342 nm ($E^{1\%}_{1cm}$ 349 MeOH), 341.5 nm ($E^{1\%}_{1cm}$ = 389 H_2O), 342 nm ($E^{1\%}_1$ 396 0.1N HCl); soluble in H_2O; sparingly soluble in EtOH; very slightly soluble in C_6H_6, $CHCl_3$, Et_2O; [diihydrochloride hemihydrate]: mp = 243°, slightly soluble in H_2O, EtOH. *Parke-Davis.*

5573 Anisomycin
22862-76-6 708 245-269-7

$C_{14}H_{19}NO_4$

1,4,5-Trideoxy-1,4-imino-5-(4-methoxyphenyl)-D-xylopentitol 3-acetate.

Flagecidin. Antiprotozoal against Trichomonas. mp = 140-141°; $[\alpha]_D^{23}$ = -30° (MeOH); λ_m = 224, 277, 283 nm (ε 10800. 1800, 1600); soluble in H_2O, EtOH, MeOH, EtOAc, Me_2CO, $CHCl_3$; insoluble in C_6H_6; [hydrochloride ($C_{14}H_{20}ClO_4$)]: mp = 187-188°; very soluble in H_2O; [deacetylanisomycin ($C_{12}H_{17}NO_3$)]: mp = 176-179°; $[\alpha]_D^{25}$ = -20.0° (MeOH). *Pfizer Inc.*

5574 Atovaquone
95233-18-4 898

$C_{22}H_{19}ClO_3$

2-[trans-4-(p-Chlorophenyl)cyclohexyl]-3-hydroxy-1,4-naphthoquinone.

Mepron; 566C80; 566C; BW-566C; BW-556C-80; Acuvel; Wellvone. Antimalarial; antipneumocystic. Antiprotozoal against Toxoplasma. mp = 216-219°. *Glaxo Wellcome Inc.*

5575 Azanidazole
62973-76-6 930

$C_{10}H_{10}N_6O_2$

(E)-2-Amino-4-[2-(1-methyl-5-nitroimidazol-2-yl)vinyl]pyrimidine.

nitromidine; F-4; Triclose. Antiprotozoal against Trichomonas. mp = 232-235°; soluble in DMF, DMSO; slightly soluble in dioxane, Me_2CO; LD_{50} (mus orl) = 5100 mg/kg, (mus ip) = 590 mg/kg, (rat orl) = 7600 mg/kg, (rat ip) = 860 mg/kg. *Chemoterapico.*

5576 Bamnidazole
31478-45-2 250-650-6

$C_7H_{10}N_4O_4$

2-Methyl-5-nitroimidazole-1-ethanol carbamate.

RP-20578. Antiprotozoal against Trichomonas. *Rhône-Poulenc Rorer Pharm. Inc.*

5577 Benznidazole
22994-85-0 1114

$C_{12}H_{12}N_4O_3$

N-Benzyl-2-nitroimidazole-1-acetamide.

Ro-7-1051; Radanil. Antiprotozoal against Trypanosoma. mp = 188.5-190°; λ_m = 313 nm (ε 7600 EtOH); soluble in H_2O (0.04 g/100ml at 37°). *Hoffmann-LaRoche Inc.*

5578 Carnidazole
42116-76-7 1897 255-663-0

$C_8H_{12}N_4O_3S$

O-Methyl [2-(2-methyl-5-nitroimidazol-1-yl)ethyl]thiocarbamate.

R-25831; Spartrix. Antiprotozoal against Trichomonas. mp = 142.4°. *Abbott Laboratories Inc.*

5579 Chlortetracycline
57-62-5 2245 200-341-7

$C_{22}H_{23}ClN_2O_8$

7-Chloro-4-(dimethylamino)-1,4,4a,5,5a,6,11,12a-octahydro-3,6,10,12,12a-pentahydroxy-6-methyl-1,11-dioxo-2-naphthacenecarboxamide.

7-chlorotetracycline; Acronize; Aureocina; Aureomycin; Biomitsin; Centraureo; Chrusomykine; Orospray.

Tetracycline antibiotic; antiamebic; antiprotozoal. mp = 168-169°; $[\alpha]_D^{23}$ = -275.0° (MeOH); λ_m = 230, 262.5 367.5 nm (0.1NHCl), 255 285 345 nm (0.1N NaOH); soluble in H_2O (0.05-0.06 g/100 ml); very soluble in aqueous solutions at pH > 8.5; freely soluble in cellosolves, dioxane, carbitol; soluble in MeOH, EtOH, BuOH, Me_2CO, EtOAc, C_6H_6; insoluble in Et_2O, petroleum ether. *Am. Cyanamid; Fermenta Animal Health Co.; Lederle Labs.*

5580 Chlortetracycline Bisulfate
2245
7-Chloro-4-(dimethylamino)-1,4,4a,5,5a,6,11,12a-octahydro-3,6,10,12,12a-pentahydroxy-6-methyl-1,11-dioxo-2-naphthacenecaroxamide bisulfate.
Tetracycline antibiotic; antiamebic.

5581 Chlortetracycline Hydrochloride
64-72-2 2245 200-591-7
$C_{22}H_{24}Cl_2N_2O_8$
7-Chloro-4-(dimethylamino)-1,4,4a,5,5a,6,11,12a-octahydro-3,6,10,12,12a-pentahydroxy-6-methyl-1,11-dioxo-2-naphthacenecarboxamide monohydrochloride.
Aureomycin; Fermycin Soluble; Aureociclina; Isphamycin. Tetracycline antibiotic; antiamebic; antiprotozoal. Dec > 210°; $[\alpha]_D^{23}$ = -240°; soluble in H_2O (0.86 g/100 ml at 28°), MeOH (1.74 g/100 ml at 28°), EtOH (0.17 g/100 ml at 28°); insoluble in Me_2CO, Et_2O, $CHCl_3$, dioxane; LD_{50} (rat orl) = 10300 mg/kg. *Fermenta Animal Health Co.; Lederle Labs.*

5582 Doxycycline Calcium
94088-85-4 3496 302-088-9
$C_{44}H_{46}CaN_2O_8$
[4S-(4α,4aα,5α,5aα,6α,12aα)]-4-(Dimethylamino)-1,4,4a,5,5a,6,11,12a-octahydro3,5,10,12,12a-pentahydroxy-6-methyl-1,11-dioxo-2-naphthacenecarboxamide calcium (2:1) (salt).
Antibacterial; antiamebic.

5583 Eflornithine
67037-37-0 3564
$C_6H_{12}F_2N_2O_2$
2-(Difluoromethyl)-DL-ornithine.
α-difluoromethylornithine; DFMO; RFI-71782; MDL-71782-A; Ornidyl. Antineoplastic; antipneumocystic; antiprotozoal (trypanosoma). Irreversibly inhibits ornithine decarboxylase. Has an antiproliferative effect on tumor cells. Used to remove facial hair; also effective against sleeping sickness. *Marion Merrell Dow Inc.*

5584 Eflornithine Hydrochloride
96020-91-6 3564
$C_6H_{13}ClF_2N_2O_2.H_2O$
2-(Difluoromethyl)-DL-ornithine hydrochloride monohydrate.
Ornidyl; MDL-71782A. Antineoplastic; antipneumocystic; antiprotozoal (trypanosoma). Irreversibly inhibits ornithine decarboxylase. Has an antiproliferative effect on tumor cells. mp = 183°. *Marion Merrell Dow Inc.*

5585 Ethylstibamine
1338-98-3 3896
$C_{10}H_{19}N_2O_3Sb$
Dihydroxyphenylstibine oxide compound with diethylamine.
Bayer 693; Neostibosan; Stibosamine; Astaril. Antiprotozoal against Trypanosoma. Freely soluble in H_2O.

5586 Flubendazole
31430-15-6 4154 250-624-4
$C_{16}H_{12}FN_3O_3$
Methyl 5-(p-fluorobenzoyl)-2-benzimidazolecarbamate.
R-17889; Flubenol; Flumoxal; Flumoxane; Fluvermal. Anthelmintic; antiprotozoal. mp = 260°; LD_{50} (mus, rat, gpg orl) > 2560 mg/kg. *Abbott Laboratories Inc.*

5587 Flunidazole
4548-15-6
$C_{11}H_{10}FN_3O_3$
2-(p-Fluorophenyl)-5-nitroimidazole-1-ethanol.
Anthelmintic; antiprotozoal. *Merck & Co.Inc.*

5588 Furazolidone
67-45-8 4320 200-653-3
$C_8H_7N_3O_5$
3-[(5-Nitrofurfurylidene)amino]-2-oxazolidinone.
NF-180; Furovag; Furox; Furoxane; Furoxone; Giarlam; Giardil; Medaron; Neftin; Nicolen; Nifulidone; Ortazol; Roptazol; Tikofuran; Topazone; component of: Tricofuron. Antiseptic (topical); anti-infective (topical); nitrofuran antiprotozoal (topical) used against trichomonas. mp = 256-257°; soluble in H_2O (0.004 g/100 ml); decomposed by alkali. *Norwich Eaton; Roberts Pharm. Corp.; SmithKline Beecham Animal Health; Solvay Animal Health Inc.*

5589 Hachimycin
1394-02-1 4616
Trichomycin; Cabimicina; Trichonat. Heptaene macrolide antibiotic produced by *Streptomyces hachijoensis*. Antifungal and antiprotozoal against Trichomonas. Yellow crystals; forms a water-soluble sodium salt; LD_{50} (mus ip) = 5 mg/kg.

5590 Halofuginone Hydrobromide
64924-67-0 4627
$C_{16}H_{18}Br_2ClN_3O_3$
(±)-trans-7-Bromo-6-chloro-3-[3-(3-hydroxy-2-piperidyl)-acetonyl]-4(3H)-quinazolinone monohydrobromide.
Stenorol; RU-19110. Antiprotozoal - coccidiostat mp = 247° (dec). *Roussel-UCLAF.*

5591 Hydroxystilbamidine
495-99-8 4892 207-811-0
$C_{16}H_{16}N_4O$
2-Hydroxy-4,4'-stilbenedicarboxamidine.
Antiprotozoal against Leishmania. mp = 235°; LD_{50} (mus iv) = 27 mg/kg, (mus sc) = 140 mg/kg. *May & Baker Ltd.*

5592 Hydroxystilbamidine Isethionate
533-22-2 4892 208-557-3
$C_{20}H_{28}N_4O_9S$
2-Hydroxy-4,4'-stilbenedicarboxamidine bis(2-hydroxyethanesulfonate) (salt).
Antiprotozoal against Leishmania. mp = 286° (dec); soluble in H_2O, EtOH (1.0 g/100 ml); insoluble in Et_2O. *May & Baker Ltd.*

5593 Imidocarb
27885-92-3 4951 248-711-7
$C_{19}H_{20}N_6O$
3,3'-di-2-Imidazolin-2-ylcarbanilide.
Antiprotozoal against Babesia. *Wander Pharma.*

5594 Imidocarb Hydrochloride
5318-76-3 4951 226-179-7
$C_{19}H_{21}ClN_6O$
3,3'-di-2-Imidazolin-2-ylcarbanilide hydrochloride.
4A65; imizol [dipropionate]; Imizad Equine Injection [dipropionate]. Antiprotozoal against Babesia. mp = 350° (dec); LD_{50} (mus sc) = 107 mg/kg, (rat sc) = 150 mg/kg. *Wander Pharma.*

5595 Ipronidazole
14885-29-1 5095 238-957-3
$C_7H_{11}N_3O_2$
2-Isopropyl-1-methyl-5-nitroimidazole.
Ipropran; Ro-7-1554; NSC-109212. Antiprotozoal against Histomonas. mp = 60°; LD_{50} (poult orl) = 640 ± 25 mg/kg; [hydrochloride]: mp = 177-182°; soluble in H_2O. *Hoffmann-LaRoche Inc.*

5596 Lauroguadine
135-43-3 5398
$C_{20}H_{36}N_6O$
1,1'-[4-(Dodecyloxy)-m-phenylene]diguanidine.
P7. Antiprotozoal (Trichomonas). [dihydrochloride monohydrate; P-7; Farmidril]]: dec 250°; soluble in H_2O (0.2 g/100 ml at 25°, 10 g/100 ml at 100°), MeOH (84.2 g/100 at 25°). *Farmitalia Carlo Erba Ltd.*

5597 Levofuraltadone
3795-88-8
$C_{13}H_{16}N_4O_6$
(-)-5-(Morpholinomethyl)-3-[(5-nitrofurfurylidene)amino]-2-oxazolidinone.
NF-602; NSC-527986. Antiprotozoal; antibacterial.

5598 Melarsomine
128470-15-5
$C_{13}H_{21}AsN_8S_2$
bis(2-Aminoethyl) p-[(4,6-diamino-s-triazin-2-yl)amino]dithiobenzenearsonite.
RM 340 [as dihydrochloride]; Cymelarsan [as dihydrochloride]; CyMel [as dihydrochloride]. A melaminophenyl arsenical trypanocide (veterinary). Antiprotozoal.

5599 Melarsonyl
37526-80-0
$C_{13}H_{13}AsN_6O_4S_2$
2-[p-[(4,6-Diamino-s-triazin-2-yl)amino]phenyl]-1,3,2-dithiarsolane-4,5-dicarboxylic acid.
A melaminophenyl arsenical trypanocide (veterinary). Antiprotozoal.

5600 Melarsonyl Potassium
13355-00-5 236-405-6
$C_{13}H_{11}AsK_2N_6O_4S_2$
Potassium 2-[p-[(4,6-diamino-s-triazin-2-yl)-amino]phenyl]-1,3,2-dithiarsolane-4,5-dicarboxylate.
RP-9955; mel W; trimelarsan. A melaminophenyl arsenical trypanocide (veterinary). Antiprotozoal.

5601 Melarsoprol
494-79-1 5856 207-793-4
$C_{12}H_{15}AsN_6OS_2$
2-[p-(4,6-Diamino-s-triazin-2-ylamino)phenyl]-1,3,2-dithiarsolane-4-methanol.
RP-3854; Mel B; Arsobal. Antiprotozoal against Trypanosoma. Soluble in propylene glycol; insoluble in H_2O, MeOH, EtOH.

5602 Mepartricin
11121-32-7 5891
mixture (≅ 1:1) of mepartricin A and mepartricin B; SPA-S-160; SN 654; Partricin methyl ester; methylpartricin; Ipertrofan; Orofungin; Tricandil; Tricangine. Antifungal; antiprotozoal (trichomonas); also used to treat benign prostatic hypertrophy. Methyl ester of the heptaene macrolide antibiotic complex, partricin. λ_m = 401, 378, 359, 340 nm; slightly soluble in H_2O, Et_2O, C_6H_6, petroleum ether; soluble in ROH, C_5H_5N, DMF, DMSO, Me_2CO; LD_{50} (mus orl) > 2000 mg/kg, (mus ip) = 200 mg/kg.

5603 Mepartricin A
62534-68-3 5891
$C_{60}H_{88}N_2O_{10}$
40-Demethyl-3,7-dideoxo-3,7-dihydroxy-N^{47}-methyl-5-oxocandicidin D methyl ester cyclic 15,19-hemiacetal gedamycin methyl ester.
Antifungal; antiprotozoal against Trichomonas; also used to treat benign prostatic hypertrophy. mp = 145-149° (dec); λ_m = 400, 377, 357, 339, 287, 240, 234, 204 nm (ε 79326, 92454, 68094, 51685, 14199, 24612, 26505, 16092 MeOH).

5604 Mepartricin B
62534-69-4 5891
$C_{59}H_{86}N_2O_{19}$
SPA-S-222 [as sodium lauryl sulfate complex]; Montricin [as sodium lauryl sulfate complex]. Antifungal; antiprotozoal against Trichomonas; also used to treat benign prostatic hypertrophy. mp = 154-158°; λ_m = 402, 379, 359, 340, 285. 233. 204 nm (ε 81101, 94729, 64171, 41558, 16196, 23835, 21696, MeOH).

5605 Metronidazole
443-48-1 6242 207-136-1
$C_6H_9N_3O_3$
2-Methyl-5-nitroimidazole-1-ethanol.
Metro Cream & Gel; Protostat; Satric; Bayer 5360; RP-8823; NSC-50364; Arilin; Clont; Deflamon; Elyzol; Flagyl; Fossyol; Gineflavir; Klion; MetroGel; Metrolag; Metrolyl; Metrotop; Orvagil; Rathimed; Sanatrichom; Trichazol; Tricocet; Trichocide; Tricho Cordes; Tricho-Gynaedron; Trivazol; Vagilen; Vagimid; Zadstat; component of: Flagyl I.V. RTU, Metro I.V. Antiprotozoal against Trichomonas. antiamebic and antibacterial. mp =158-160°; soluble in H_2O (1.0 g/100 ml), EtOH (0.5 g/100 ml), Et_2O (< 0.05 g/100 ml), $CHCl_3$ (< 0.05 g/100 ml); sparingly soluble in DMF. *Bayer Corp. Pharm. Div.; Elkins-Sinn; Galderma Labs Inc.; Lemmon Co.; McGaw Inc.; Ortho Pharm. Corp.; Savage Labs; SCS Pharm.*

5606 Metronidazole Hydrochloride
69198-10-3 6242
$C_6H_{10}ClN_3O_3$
2-Methyl-5-nitroimidazole-1-ethanol.
Flagyl I.V.; SC-326421. Antiprotozoal against Trichomonas. antiamebic and antibacterial. *SCS Pharm.*

5607 Metronidazole Phosphate
73334-05-1
$C_6H_{10}N_3O_6P$
2-Methyl-5-nitroimidazole-1-ethanol dihydrogen phosphate (ester).
U-54555. Antiprotozoal against Trichomonas. antiamebic and antibacterial.

5608 Misonidazole
13551-87-6 236-931-6
$C_7H_{11}N_3O_4$
α-(Methoxymethyl)-2-nitroimidazole-1-ethanol.
Ro-7-0582. Antiprotozoal against Trichomonas. *Hoffmann-LaRoche Inc.*

5609 Monensin
17090-79-8 6329 241-154-0
$C_{36}H_{62}O_{11}$
2-[2-Ethyloctahydro-3'-methyl-5'-[tetrahydro-6-hydroxy-6-(hydroxymethyl)-3,5-dimethyl-2H-pyran-2-yl][2,2'-bifuran-5-yl]]-9-hydroxy-β-methoxy-α,σ,2,8-tetramethyl-1,6-dioxapsiro[4.5]decan-7-butanoic acid.
A-3823A; 63714; monensic acid. Antibiotic produced by *Streptomyces cinnamonensis*. Antifungal, antibiotic and antiprotozoal. mp = 103-105°; $[\alpha]_D$ = 47.7°; slightly soluble in H_2O, more soluble in hydrocarbons, very soluble in organic solvents; LD_{50} (mus orl) = 43.8 ± 5.2 mg/kg, (chicks orl) = 284 ± 47 mg/kg. *Eli Lilly & Co.*

5610 Monensin Sodium
22373-78-0 6329 244-941-7
$C_{36}H_{61}NaO_{11}$
Sodium 2-[2-ethyloctahydro-3'-methyl-5'-[tetrahydro-6-hydroxy-6-(hydroxymethyl)-3,5-dimethyl-2H-pyran-2-yl][2,2'-bifuran-5-yl]]-9-hydroxy-β-methoxy-α,σ,2,8-tetramethyl-1,6-dioxapsiro[4.5]decan-7-butanoate.
Rumensin; Romensin; Coban. Antibiotic produced by *Streptomyces cinnamonensis*. Antifungal, antibiotic and antiprotozoal. mp = 267-269°; $[\alpha]_D$ = 57.3°; slightly soluble in H_2O, more soluble in hydrocarbons, very soluble in organic solvents. *Eli Lilly & Co.*

5611 Moxipraquine
23790-08-1
$C_{24}H_{38}N_4O_2$
4-[6-[(6-Methoxy-8-quinolyl)amino]hexyl]-α-methyl-1-piperazinepropanol.
349-C59. A novel 8-aminoquinolone compound with antitrypanosomal activity and significant fetal toxicity in rats and rabbits. Antiprotozoal.

5612 Moxnidazole
52279-59-1
$C_{13}H_{18}N_6O_5$
3-[[(1-Methyl-5-nitroimidazol-2-yl)methylene]amino]-5-(morpholinomethyl)-2-oxazolidinone.
SH-240. Antiprotozoal (Trichomonas). *Schering AG.*

5613 Nifuratel
4936-47-4 6623 225-576-2
$C_{10}H_{11}N_3O_5S$
5-[(Methylthio)methyl]-3-[(5-nitrofurfurylidene)amino]-2-oxazolidinone.
Macmiror; Magmilor; Polmiror; Tydantil; Omnes; methylmercadone; Inimur. Antibacterial; antifungal; antiprotozoal, against Trichomonas. mp = 182°. *Polichimica Sap.*

5614 Nifuroxime
6236-05-1 6627 228-349-6
$C_5H_4N_2O_4$
5-Nitro-2-furaldehyde oxime.
Micofur; Mycofur; component of: Tricofuron. Antibacterial; topical anti-infective; antiprotozoal, against Trichomonas. Nitrofuran. mp = 156°, 163-164°; soluble in H_2O (0.1 g/100 ml at 25°), MeOH (8.9 g/100 ml), EtOH (3.9 g/100 ml). *Norwich Eaton.*

5615 Nifursemizone
5579-89-5
$C_8H_{10}N_4O_4$
5-Nitro-2-furaldehyde 2-ethylsemicarbazone.
NF-161. Antihistomonad. Used for poultry.

5616 Nifursol
16915-70-1 240-963-6
$C_{12}H_7N_5O_9$
3,5-Dinitrosalicylic acid (5-nitrofurfurylidene-(hydrazide).
Antihistomonad. Used for poultry.

5617 Nifurtimox
23256-30-6 6630 245-531-0
$C_{10}H_{13}N_3O_5S$
4-[(5-Nitrofurfurylidene)amino]-3-methylthiomorpholine 1,1-dioxide.
Bayer 2502; Lampit. Antibacterial, primarily antiprotozoal, against Trypanosoma. mp = 180-182°; LD_{50} (mus gvg) = 3720 mg/kg, (rat gvg) = 4050 mg/kg. *Bayer AG.*

5618 Nimorazole
6506-37-2 6644 229-394-4
$C_9H_{14}N_4O_3$
4-[2-(5-Nitroimidazol-1-yl)ethyl]morpholine.
K-1900; Acterol; Esclama; Naxofem; Naxogin; Nulogyl. Antiprotozoal against Trichomonas. mp = 110-111°; soluble in H_2O, EtOH, Me_2CO, $CHCl_3$; LD_{50} (mus orl) = 1530 mg/kg. *Merck & Co.Inc.*

5619 Nitarsone
98-72-6 6659 202-695-8
$C_6H_6AsNO_5$
p-Nitrobenzenearsonic acid.
NSC-5085. Antihistomonad. Dec 298-300°; slightly soluble in H_2O, EtOH at 25°; more soluble in warm H_2O, EtOH. *U.S. Government.*

5620 N-Methylglucamine Antimonate
133-51-7 6154 228-506-9
$C_7H_{18}NO_8Sb$
1-Deoxy-1-(methylamino)-D-glucitol antimonate.
RP-2168; Glucantim; Glucantime; Protostib.

Antiprotozoal against Leishmania. Soluble in H_2O (35 g/100 ml); insoluble in EtOH, Et_2O, $CHCl_3$. *Rhône-Poulenc Rorer Pharm. Inc.*

5621 Oxophenarsine Hydrochloride
538-03-4 7082 208-682-3
$C_6H_7AsClNO_2$
2-Amino-4-arsenosophenol hydrochloride.
Mapharsen; Ehrlich 5; Arseno 39; Arsenoxide; Maspharside; Mapharsal; Fontarsan; Arsenosan; Oxiarsolan. Antiprotozoal against Trypanosoma. Soluble in H_2O, EtOH, MeOH. *Parke-Davis.*

5622 Partricin
11096-49-4 7181
Ayfactin; SPA-S-132. Heptaene macrolide antibiotic complex produced by *Streptomyces Aureofaciens* NRRL 3878. Antiprotozoal. *SPA.*

5623 Pentamidine
100-33-4 7254 202-841-0
$C_{19}H_{24}N_4O_2$
4,4'-(Pentamethylenedioxy)dibenzamidine.
Antifungal and antiprotozoal: targets Leishmania, Trypanosoma. Antipneumocystic. Dec 186°; [dihydrochloride ($C_{19}H_{26}Cl_2N_4O_2$]: mp = 232-234°; LD_{50} (mus iv) = 28 mg/lg, (mus sc) = 64 mg/kg. *Fujisawa Pharm. USA Inc.; May & Baker Ltd.; Rhône-Poulenc Rorer Pharm. Inc.*

5624 Pentamidine Dimethanesulfonate
6823-79-6 7254 229-898-4
$C_{21}H_{32}N_4O_8S_2$
4,4'-(Pentamethylenedioxy)dibenzamidine dimethanesulfonate.
pentamidine mesylate; Lomodine. Antifungal and antiprotozoal: targets Leishmania, Trypanosoma. Antipneumocystic. *Fujisawa Pharm. USA Inc.; May & Baker Ltd.; Rhône-Poulenc Rorer Pharm. Inc.*

5625 Pentamidine Isethionate
140-64-7 7254 205-424-1
$C_{23}H_{36}N_4O_{10}S_2$
4,4'-(Pentamethylenedioxy)dibenzamidine isethionate.
pentamidine isetionate; MB-800, RP-2512; Aeropent; Banambax; NebuPent; Pentacarinat; Pentam; Pentam 300; Pneumopent. Antifungal and antiprotozoal: targets Leishmania, Trypanosoma. Antipneumocystic. mp = 180°; soluble in H_2O (10 g/100 ml at 25°, 25 g/100 ml at 100°); soluble in glycerol; slightly soluble in EtOH; insoluble in Et_2O, Me_2CO, $CHCl_3$. *Fujisawa Pharm. USA Inc.; May & Baker Ltd.; Rhône-Poulenc Rorer Pharm. Inc.*

5626 Propamidine
104-32-5 7981 203-195-2
$C_{17}H_{20}N_4O_2$
4,4'-(Trimethylenedioxy)dibenzamidine.
4,4'-diamidino-α,ω-diphenoxypropane; 4,4'-(trimethylenedioxy)dibenzamidine. Antiprotozoal (Trypanosoma); antiamebic; anti-infective (topical, veterinary). Used as an antiprotozoal against Trypanosoma and Babesia and an antiamebic. *May & Baker Ltd.*

5627 Propamidine Isethionate
140-63-6 7981 205-423-6
$C_{21}H_{32}N_4O_{10}S_2$
4,4'-(Trimethylenedioxy)dibenzamidine ethanesulfonic acid.
M&B-782; Brolene Drops. Topical anti-infective; antiamebic. Used as an antiprotozoal against Trypanosoma and Babesia. mp ≅ 235°; soluble in H_2O (20 g/100 ml), EtOH (3 g/100 ml), glycerol; insoluble in Et_2O, $CHCl_3$. *May & Baker Ltd.*

5628 Puromycin
53-79-2 8130
$C_{22}H_{29}N_7O_5$
(S)-3'-[[2-Amino-3-(4-methoxyphenyl)-1-oxopropyl]amino]-3'-deoxy-N,N-dimethyladenosine.
CL-13900; P-638; 3123-L; Stylomycin. Antineoplastic. Antiprotozoal against Trypanosoma. mp = 175.5-177°; $[\alpha]_D^{25}$ = -11° (EtOH); λ_m = 275 nm (ε 20300 0.13N NaOH), 267.5 nm (ε 19500 0.1N HCl); LD_{50} (mus iv) = 350 mg/kg, (mus ip) = 525 mg/kg, (mus orl) = 675 mg/kg. *Am. Cyanamid; ICN Pharm. Inc.*

5629 Puromycin Hydrochloride
58-58-2 8130
$C_{22}H_{31}Cl_2N_7O_5$
(S)-3'-[[2-Amino-3-(4-methoxyphenyl)-1-oxopropyl]amino]-3'-deoxy-N,N-dimethyladenosine dihydrochloride.
CL-16536; NSC-3055; CL-13900 dihydrochloride; P-638 dihydrochloride; Stylomycin dihydrochloride; 3123L, dihydrochloride. Antineoplastic. Antiprotozoal against Trypanosoma. *ICN Pharm. Inc.*

5630 Pyrimethamine
58-14-0 8169 200-364-2
$C_{12}H_{13}ClN_4$
2,4-Diamino-5-(p-chlorophenyl)-6-ethylpyrimidine.
RP-4753; Chloridin; Daraprim; Malocide; Tinduring; component of: Fansidar. Antimalarial; antiprotozoal, targeting Toxoplasma. mp = 233-234°, 240-242°; insoluble in H_2O; soluble in EtOH (0.9 g/100 ml at 25°, 2.5 g/100 ml at 75°), dilute HCl (0.5 g/100 ml); sparingly soluble in propylene glycol, dimethylacetamide at 70°. *Glaxo Wellcome Inc.; Hoffmann-LaRoche Inc.; Rhône-Poulenc Rorer Pharm. Inc.*

5631 Quinapyramine
20493-41-8 8234
$C_{17}H_{22}N_6$
4-Amino-6-[(2-amino-1,6-dimethyl-4(1H)-pyrimidinylidene)amino]-1,2-dimethylquinolinium conjugate monoacid.
M-7555; Antrycide. Antiprotozoal against Trypanosoma. *ICI.*

5632 Quinapyramine Chloride
23609-65-6 8234 245-784-7
$C_{17}H_{22}Cl_2N_6$
4-Amino-6-[(2-amino-1,6-dimethyl-4(1H)-pyrimidinylidene)amino]-1,2-dimethylquinolinium dichloride.
Antiprotozoal against Trypanosoma. Trypanosoma. mp = 316-317° (dec); LD_{50} (mus iv) = 10-15 mg/kg. *ICI.*

5633 Quinapyramine Sulfate
23609-66-7 8234
$C_{19}H_{28}N_6O_8S_2$
4-Amino-6-[(2-amino-1,6-dimethyl-4(1H)-pyrimidinylidene)amino]-1,2-dimethylquinolinium dimethosulfate.
Antiprotozoal against Trypanosoma. Trypanosoma. mp = 255-256°; freely soluble in H_2O; LD_{50} (mus iv) = 10-15 mg/kg. *ICI* .

5634 Ronidazole
7681-76-7 8413 231-675-1
$C_6H_8N_4O_4$
1-Methyl-7-nitroimidazole-2-methanol carbamate ester.
MCMN; Dugro; Ridzol. Antimicrobial; antiprotozoal. mp = 167-169°; soluble in H_2O (0.29 g/100 ml at 25°); freely soluble in Me_2CO; soluble in MeOH, EtOH, $CHCl_3$, EtOAc. *Merck & Co.Inc.*

5635 Secnidazole
3366-95-8 8562 222-134-0
$C_7H_{11}N_3O_3$
α,2-Dimethyl-5-nitroimidazole-1-ethanol.
PM-185184; RP-14539; Flagentyl. Antiamebic; antiprotozoal against Trichomonas. mp = 76°. *Rhône-Poulenc Rorer Pharm. Inc.*

5636 Silver Picrate
146-84-9 8670 205-682-5
$C_6H_2AgN_3O_7$
2,4,67-Trinitrophenol silver salt.
silver trinitrophenolate; Picragol; Picrotol. Antiprotozoal against Trichomonas. Soluble in H_2O (2 g/100 ml), sparingly soluble in EtOH, Me_2CO, glycerol; insoluble in $CHCl_3$, Et_2O.

5637 Sodium Antimonylgluconate
16307-91-5
$C_6H_8NaO_7Sb$
Triostam. Sodium salt of a trivalent antimony derivative of gluconic acid. Antileishmanial (antiprotozoal); antischistosomal (anthelmintic).

5638 Sodium Stibogluconate
16037-91-5 742
$C_{12}H_{17}Na_3O_{17}Sb_2.9H_2O$
2,4:2',4'-O-(Oxydistibylidyne)bis[D-gluconic acid] Sb,Sb'-dioxide trisodium salt nonahydrate.
Pentostam; Solustibosan. Antiprotozoal against Leishmania. Freely soluble in H_2O. *Glaxo Wellcome Inc.*

5639 Stilbamidine
122-06-5 8970 204-519-5
$C_{16}H_{16}N_4$
4,4'-(1,2-Ethanediyl)bisbenzenecarboximidamide.
4,4'-stilbenedocarboxamidine. Antiprotozoal against Leishmania and Trypanosoma. [dihydrochloride ($C_{16}H_{18}Cl_2N_4$)]: LD_{50} (mus iv) = 31 mg/kg, (mus sc) = 180 mg/kg. *May & Baker Ltd.*

5640 Stilbamidine Isethionate
140-59-0 8970
$C_{20}H_{28}N_4O_8S_2$
4,4'-(1,2-Ethanediyl)bisbenzenecarboximidamide isethionate.
M&B-744. Antiprotozoal against Leishmania and Trypanosoma. Dec 290°; λ_m = 330 nm ($E^{1\%}_{1\ cm}$ = 750); soluble in H_2O (33-40 g/100 ml), MeOH (1.5 g/100 ml). *May & Baker Ltd.*

5641 Sulnidazole
51022-76-5
$C_9H_{14}N_4O_3S$
O-Methyl [2-(2-ethyl-5-nitroimidazol-1-yl)ethyl]thiocarbamate.
R-26412. Antiprotozoal against Trichomonas. *Cilag-Chemie Ltd.*

5642 Suramin Sodium
129-46-4 9181 204-949-3
$C_{51}H_{34}N_6Na_6O_{23}S_6$
Hexasodium 8,8'-[ureylenebis[m-phenylenecarbonylimino(4-methyl-m-phenylene)carbonylimino]]di-1,3,5-naphthalenetrisulfonate.
suramin hexasodium; Bayer 205; Fourneau 309; Antrypol; Germanin; Moranyl; Naganol; Naphuride. Antineoplastic; antiviral. Anthelmintic against nematodes; antiprotozoal against Trypanosoma. Freely soluble in H_2O; poorly soluble in EtOH; insoluble in Et_2O, $CHCl_3$, petroleum ether; LD_{50} (mus iv) ≅ 620 mg/kg. *Bayer AG; Parke-Davis.*

5643 Tenonitrozole
3810-35-3 9292 223-282-9
$C_8H_5N_3O_3S_2$
N-(5-Nitro-2-thiazolyl)-2-thiophenecarboxamide.
TC-109; thenitrazole; Atrican; Moniflagon. Antifungal; antiuprotozoal against Trichomonas. mp = 255-256°. *Chantereau.*

5644 Tinidazole
19387-91-8 9588 243-014-4
$C_8H_{13}N_3O_4S$
1-[2-(Ethylsulfonyl)ethyl]-2-methyl-5-nitroimidazole.
CP-12574; Fasigin; Fasigyn; Pletil; Simplotan; Sorquetan; Tricolam; Trimonase. Antiamebic; antifungal; antiprotozoal against Giardia, Trichomonas. mp = 127-128°; LD_{50} (mus orl) > 3600 mg/kg, (mus ip) > 2000 mg/kg. *Pfizer Inc.*

5645 Trypan Red
574-64-1 9924 209-372-0
$C_{32}H_{19}N_6Na_5O_{15}S_5$
4,4'-[(3-Sulfo[1,1'-biphenyl]-4,4'-diyl)bis(azo)]bis[3-amino-2,7-naphthalenedisulfonic acid] pentasodium salt.
CI-22850. Antiprotozoal against Trypanosoma. Has been used as a trypanocide. Soluble in H_2O, insoluble in EtOH.

5646 Tryparsamide
554-72-3 9925 209-070-9
$C_8H_{10}AsN_2NaO_4$
[4-[(2-Amino-2-oxoethyl)amino]phenyl]arsonic acid monosodium salt.
Glyphenarsine; Tryparsone; Tryponarsyl; Trypothane. Antiprotozoal against Trypanosoma. Soluble in H_2O (50 g/100 ml); slightly soluble in EtOH; insoluble in Et_2O, $CHCl_3$.

5647 Urea Stibamine
1340-35-8 10010
MF Unknown
Carbostibamide.
Anthelmintic, targeting nematodes and Schistosoma; antiprotozoal against Leishmania. Chemical composition uncertain. Active principle thought to be a substituted urea: sym-diphenylcarbamido-4,4-distibinic acid. Soluble in H_2O, partly soluble in EtOH, Et_2O. *Bristol-Myers Squibb Co.*

Antipruritics

5648 Ammonium Lactate
515-98-0 561 208-214-8
$C_3H_9NO_3$
2-Hydroxypropanoic acid mono-ammonium salt.
BMS-186091. Antipruritic. mp = 91-94°; soluble in H_2O, glycerol, MeOH; insoluble in higher alcohols, Et_2O, Me_2CO, EtOAc. *Westwood-Squibb Pharm. Inc.*

5649 Camphor
76-22-2 1779 200-945-0
$C_{10}H_{16}O$
1,7,7-Trimethylbicyclo[2.2.1]heptan-2-one.
2-bornanone; 2-camphanone; norcamphane; gum camphor; Japan camphor; Formosa camphor; laurel camphor; spirit of camphor; component of: Campho-Phenique Cold Sore Gel, Campho-Phenique Liquid, Heet, Minut-Rub, Pazo Ointment, Sarna. Topical antipruritic; antiseptic. Plasticizer for cellulose nitrate, other explosives and lacquers, insecticides, moth and mildew proofing, tooth powders, flavoring, embalming, pyrotechnics, intermediate. CAUTION: Ingestion or injection may cause nausea, vomiting, vertigo, mental confusion, delirium, clonic convulsions, coma, respiratory failure, death. mp = 179.75°; bp = 204°; $[\alpha]_D^{25}$ = 41° - 43° (c= 10 EtOH); d_4^{25} = 0.992; soluble in H_2O (0.125 g/100 ml), EtOH (100 g/100 ml), Et_2O (100 g/100 ml), $CHCl_3$ (200 g/100 ml), C_6H_6 (250 g/100 ml), Me_2CO (250 g/100 ml); LD_{50} (mus ip) = 3000 mg/kg. *Bristol-Myers Squibb Co.; Sterling Health U.S.A.; Stiefel Labs Inc.; Whitehall Labs. Inc.*

5650 Cyproheptadine
129-03-3 2842 204-928-9
$C_{21}H_{21}N$
4-(5H-Dibenzo[a,d]cyclohepten-5-ylidene-1-methylpiperidine .
Antihistaminic; antipruritic. mp = 112.3-113.3°. *Merck & Co.Inc.*

5651 Cyproheptadine Hydrochloride Monohydrate
6032-06-0 2842
$C_{21}H_{22}ClN \cdot H_2O$
4-(5H-Dibenzo[a,d]cyclohepten-5-ylidene-1-methylpiperidine hydrochloride monohydrate.
Antihistaminic; antipruritic. mp = 214-216°; soluble in H_2O (0.5 g/100 ml). *Merck & Co.Inc.*

5652 Cyproheptadine Hydrochloride Sesquihydrate
41354-29-4 2842
$C_{21}H_{22}ClN.1.5H_2O$
4-(5H-Dibenzo[a,d]cyclohepten-5-ylidene-1-methylpiperidine hydrochloride sesquihydrate.
Anarexol; Antegan; Ifrasarl; Nuran; Periactin; Vimicon; Cipractin; Peritol; Periactin. Antihistaminic; antipruritic. Used for the treatment of coughs, colds. Dec 252.6-253.6°; λ_m = 224, 285 nm ($E_{1\,cm}^{1\%}$ 1656 355 0.1N H_2SO_4); soluble in MeOH (66.6 g/100 ml), $CHCl_3$ (6.25 g/100 ml), EtOH (2.88 g/100 ml), H_2O (0.36 g/100 ml); insoluble in Et_2O; LD_{50} (mus orl) = 74.2 mg/kg. *Merck & Co.Inc.*

5653 Dichlorisone
7008-26-6 3099 230-283-8
$C_{21}H_{26}Cl_2O_4$
9α,11β-Dichloro-17α,21-dihydroxypregna-1,4-diene-3,20-dione.
Diloderm; Disoderm. Topical antipruritic. Dec 238-241°; $[\alpha]_D^{20}$ = 134° (C_5H_5N); λ_m = 237 nm (ε 15400 MeOH). *Schering-Plough HealthCare Products.*

5654 Dichlorisone 21-Acetate
79-61-8 3099 201-213-3
$C_{23}H_{28}Cl_2O_5$
9α,11β-Dichloro-17α,21-dihydroxypregna-1,4-diene-3,2o-dione 21-acetate.
Astroderm. Topical antipruritic. Dec 246-253°; $[\alpha]_D^{25}$ = 162° (dioxane); λ_m = 237 nm (ε 15000 MeOH). *Schering-Plough HealthCare Products.*

5655 Glycine
56-40-6 4500 200-272-2
$C_2H_5NO_2$
Aminoacetic acid.
Gly; G; Glycolixir; glycocoll; Gyn-Hydralin; Glycosthène; component of: Corilin. Antipruritic. mp = 290°; d = 1.1607; soluble in H_2O (25.0 g/100 ml at 25°, 39.1 g/100 ml at 50°, 54.4 g/100 ml at 75°, 67.2 g/100 ml at 100°), EtOH (0.06 g/100 ml, C_5H_5N (0.60 g/100 ml); insoluble in Et_2O. *Bristol-Myers Squibb Co.; McGaw Inc.; Schering-Plough HealthCare Products.*

5656 Halometasone
50629-82-8 4628 256-664-9
$C_{22}H_{27}ClF_2O_5$
2-Chloro-6α,9-difluoro-11β,17,21-trihydroxy-16α-methylpregna-1,4-diene-3,20-dione.
2-chloroflumethasone; C-48401-Ba; Sicorten [monohydrate]. Anti-inflammatory (topical); antipruritic. Synthetic corticosteroid. mp = 220-222° (dec). *Ciba-Geigy Corp.*

5657 3-Hydroxycamphor
10373-81-6 4862
3-Hydroxy-1,7,7-trimethylbicyclo[2.2.1]heptan-2-one.
oxycamphor; Oxaphor. Antipruritic. mp = 205-206°; soluble in H_2O (2 g/100 ml); very soluble in EtOH, $CHCl_3$, Et_2O.

5658 Levomenthol
2216-51-5 218-690-9
$C_{10}H_{20}O$
(-)-(1R,3R,4S)-Menthol.
Antipruritic.

5659 Menthol
89-78-1 5882 201-939-0
$C_{10}H_{20}O$
5-Methyl-2-(1-methylethyl)cyclohexanol.
Fisherman's Friend Lozenges; Therapeutic Mineral Ice; hexahydrothymol; peppermint camphor; component of: Dermoplast, Minut-Rub, Robitussin Cough Drops, Sarna, Theragesic. Antipruritic. mp = 41-43°; bp = 212°; $[\alpha]_D^{18}$ = -50° (c = 10 EtOH); slightly soluble in H_2O; very soluble in EtOH, $CHCl_3$, Et_2O, petroleum ether; LD_{50} (rat orl) = 3180 mg/kg. *Bristol-Myers Squibb Co.; Mission Pharmacal Co.; Robins, A. H. Co.; Stiefel Labs Inc.*

5660 Mesulfen
135-58-0 5977 205-202-4
$C_{14}H_{12}S_2$
2,7-Dimethylthianthrene.
mesulphen; Mitigal; Odylen; Sudermo; Peligal; Neosulfine. Ectoparasiticide; antipruritic; scabicide. mp = 123°; bp_3 = 184°; bp_{14} = 228-231°; freely soluble in Me_2CO, $CHCl_3$, Et_2O, petroleum ether; moderatley soluble in EtOH, EtOAc; insoluble in H_2O.

5661 Methdilazine
1982-37-2 6034 217-841-6
$C_{18}H_{20}N_2S$
10[(1-Methyl-3-pyrrolidinyl)methyl]phenothiazine.
Tacaryl. Antipruritic. mp = 87-88°. *Westwood-Squibb Pharm. Inc.*

5662 Methdilazine Hydrochloride
1229-35-2 6034 214-967-3
$C_{18}H_{21}ClN_2S$
10[(1-Methyl-3-pyrrolidinyl)methyl]phenothiazine hydrochloride.
Tacaryl hydrochloride; Dilosyn; Disyncran. Antipruritic. mp = 187.5-189°; LD_{50} (rat orl) = 320 mg/kg. *Westwood-Squibb Pharm. Inc.*

5663 Phenol
108-95-2 7390 203-632-7
C_6H6_O
Hydroxybenzene.
Carbolic acid; phenic acid; phenylic acid; phenyl hydroxide; oxybenzene; component of: Anbesol, Campho-Phenique Cold Sore Gel, Campho-Phenique Gel, Campho-Phenique Liquid. Local anesthetic. Antipruritic. CAUTION: Toxicity can arise from cutaneous absorption, ingestion, or inhalation. mp = 40.85°; bp = 812°; d = 1.071; soluble in H_2O (6.6 g/100 ml), C_6H_6 (8.3 g/100 ml); very soluble in EtOH, $CHCl_3$, Et_2O, CS_2; insoluble in petroleum ether; LD_{50} (rat orl) = 530 mg/kg. *Robins, A. H. Co.; Sterling Health U.S.A.; Whitehall Labs. Inc.*

5664 Polidocanol
3055-99-0 7717 221-284-4
$C_{12}H_{35}(OCH_2CH_2)_nOH$ (average polymer, n = 9)
α-Dodecyl-ω-hydroxypoly(oxy-1,2-ethanediyl).
polyethylene glycol (9) monodecyl ether; dodecyl alcohol polyoxyethylene ether; hydroxypolyethoxydodecane; laureth 9; polyoxyethylene lauryl ether; polyethylene glycol monododecyl ether; Aethoxysklerol; Aetoxisclerol; Atlas G-4829; Hetoxol L-9; Lipal 9LA; Thesit. Sclerosing agent; antipruritic; anesthetic (topical). Also used as a solvent, non-ionic emulsifier, pharmaceutic aid (surfactant), spermatacide. Soluble in H_2O, EtOH, C_7H_8; miscible with hot mineral, natural and synthetic oils; miscible with fats and fatty alcohols; LD_{50} (mus orl) = 1170 mg/kg, (mus iv) 125 mg/kg. *Kreussler Chemische-Fabrik.*

5665 Thenaldine
86-12-4 9413 201-651-5
$C_{17}H_{22}N_2S$
1-Methyl-4-N-2-thenylanilinopiperidine.
thenalidine; Sandostene. Antihistaminic; antipruritic. The tartrate is used as an antihistaminic. mp = 95-97°; $bp_{0.02}$ = 158-160°; [tartrate]: mp = 170-172°. *Sandoz Pharm. Corp.*

5666 Tolpropamine
5632-44-0 9662 227-071-2
$C_{18}H_{23}N$
N,N-Dimethyl-3-phenyl-3-p-tolylpropamine.
Pragman Gelee [as hydrochloride]. Topical antihistaminic; antipruritic. [hydrochloride]: mp = 182-184°. *Hoechst AG.*

5667 Trimeprazine
84-96-8 9834 201-577-3
$C_{18}H_{22}N_2S$
10-[3-(Dimethylamino)-2-methylpropyl]-phenothiazine.
alimemazine; methylpromazine; Bayer 1219. Antipruritic. mp = 68°; $bp_{0.3}$ = 150-175°. *Rhône-Poulenc Rorer Pharm. Inc.*

5668 Trimeprazine Tartrate
4330-99-8 9834 224-368-9
$C_{40}H_{50}N_4O_6S_2$
10-[3-(Dimethylamino)-2-methylpropyl]phenothiazinetartrate.
Temaril16; Panectyl; Repeltin; Temaril; Theralene; Vallergan. Antipruritic. Soluble in H_2O, slightly soluble in EtOH. *Rhône-Poulenc Rorer Pharm. Inc.*

Antipsoriatics

5669 Acitretin
55079-83-9 114 259-474-4
$C_{21}H_{26}O_3$
(all E)-9-(4-Methoxy-2,3,6-trimethylphenyl)-3,7-dimethyl-2,4,6,8-nonatrienoic acid.
Soriatane; etretin; Neotigason; Ro-10-1670/000. Antipsoriatic. An aromatic retinoid. The primary active metabolite of etretinate. Used in the treatment of psoriasis and other dermatologic disorders. Shown to inhibit angiogenesis and to be effective in prevention of tumor development. mp = 228-230°; LD_{50} (mus ip) > 400 mg/kg (1 day), = 700 (10days), = 700 (20 days). *Hoffmann-LaRoche Inc.*

5670 Ammonium Salicylate
528-94-9 584 208-444-9
$C_7H_9NO_3$
2-Hydroxybenzoic acid monoammonium salt.
salicylic acid monoammonium salt; Salicyl-Vasogen. Analgesic. Used topically to loosen psoriatic scales. Soluble in H_2O (100 g/100 ml), EtOH (33.3 g/100 ml).

5671 Anthralin
1143-38-0 723 214-538-0
$C_{14}H_{10}O_3$
1,8-Dihydroxy-9-anthrone.
Anthra-Derm; DrithoCreme; Drithoscalp; Lasan; dithranol; Batidrol; Cignolin; Cigthranol; Psodadrate; Psoriacide. Antipsoriatic. mp = 176-181°; insoluble in H_2O; freely soluble in $CHCl_3$; soluble in Me_2CO, C_6H_6, C_5H_5N, oils; slightly soluble in EtOH, Et_2O, AcOH. *Dermik Labs. Inc.; Stiefel Labs Inc.*

5672 Anthralin Triacetate
16203-97-7 723 240-333-0
$C_{20}H_{16}O_6$
1,8,9-Triacetoxyanthracene.
Exolan; 1,8,9-triacetoxyanthracene. Antipsoriatic. *Dermik Labs. Inc.; Stiefel Labs Inc.*

5673 Azaribine
2169-64-4 937
$C_{14}H_{17}N_3O_9$
2-β-D-Ribofuranosyl-as-triazine-3,5(2H,4H)-dione 2',3',5'-triacetate.
Triazure; CB-304; NSC-67239. Antipsoriatic. LD_{50} (mus orl) = 7800 mg/kg, (rat orl) = 12000 mg/kg; [free alcohol]: mp = 160-161°; $[\alpha]_D^{24}$ = -132° (C_5H_5N); λ_m = 262 nm (ε 6100 H_2O). *Grunenthal.*

5674 6-Azauridine
54-25-1 937 200-199-6
$C_8H_{11}N_3O_6$
2-β-D-Ribofuranosyl-1,2,4-triazine-3,5(2H,4H)-dione.
AzUR; Rib-Azauracil. Antipsoriatic. The better antipsoriatic is the triacetate (azaribine). mp = 160-161°; $[\alpha]_D^{24}$ = -132° (C_5H_5N); λ_m 262 nm (ε 6100 H_2O).

5675 Bergaptene
484-20-8 1196 207-604-5
$C_{12}H_8O_4$
4-Methoxy-7H-furo[3,2-g][1]benzopyran-7-one.
bergapten; bergaptan; heraclin; majudin; 5-MOP; Psoraderm. Antipsoriatic. mp = 188°; insoluble in H_2O; slightly soluble in AcOH, $CHCl_3$, C_6H_6; soluble in absolute EtOH (1.66 g/100 ml).

5676 Butantrone
75464-11-8
$C_{18}H_{16}O_4$
10-Butyryl-1,8-dihydroxyanthrone.
An antipsoriatic drug.

5677 Calcipotriene
112965-21-6 1679
$C_{27}H_{40}O_3$
(1α,3β,5Z,7E,22E,24S)-24-Cyclopropyl-9,10-secochola-5,7,10(19),22-tetraene-1,3,24-triol.
MC-903; calciptriol; Daivonex; Dovonex; component of: Dovex. Vitamin, vitamin source. Antipsoriatic. mp = 166-168°; λ_m 264 nm (ε 17200 96% EtOH). *LeoAB; Westwood-Squibb Pharm. Inc.*

5678 Cedefingol
35301-24-7
$C_{20}H_{41}NO_3$
N-[(1S,2S)-2-Hydroxy-1-(hydroxymethyl)heptadecyl]-acetamide.
SPC-101210. Antineoplastic (adjunct); antipsoriatic. *Sphinx Pharm. Corp.*

5679 Chrysarobin
491-59-8 2312
$C_{15}H_{12}O_3$
1,8-Hydroxy-3-methyl-9-anthrone.
purified Goa powder; purified araroba. Formerly used as an antipsoriatic. mp = 203.4-204°; slightly soluble in H_2O; soluble in EtOH (0.26 g/100 ml), C_6H_6 (3.3 g/100 ml), $CHCl_3$ (66.6 g/100 ml), Et_2O (0.62 g/100 ml), CS_2 (0.55 g/100 ml).

5680 Cycloheximide
66-81-9 2797
$C_{15}H_{23}NO_4$
3-[(R)-2-[(1S,3S,5S)-3,5-Dimethyl-2-oxocyclohexyl]-2-hydroxyethyl]glutarimide.
U-4527. Antipsoriatic. mp = 119.5-121°, 115-116°; $[\alpha]_D^{29}$ = -3.38° (c = 9.47 MeOH), $[\alpha]_D^{25}$ = 6.8° (c = 2 H_2O); soluble in H_2O (2.1 g/100 ml), amyl alcohol (7 g/100 ml), $CHCl_3$, Et_2O, Me_2CO; MeOH, EtOH; LD_{50} (mus iv) = 150 mg/kg. *Pharmacia & Upjohn.*

5681 Domoprednate
66877-67-6
$C_{26}H_{36}O_5$
11α,17α-Dihydroxy-D-homopregna-1,4-diene-3,20-dione.
Antipsoriatic.

5682 Enazadrem Phosphate
132956-22-0
$C_{18}H_{28}N_3O_5P$
4,6-Dimethyl-2-[(6-phenylhexyl)amino]-5-pyrimidinol phosphate.
CP-70490-09. Antipsoriatic. *Pfizer Inc.*

5683 Etretinate
54350-48-0 3935 259-119-3
$C_{23}H_{30}O_3$
Ethyl (all E)-9-(4-methoxy-2,3,6-trimethylphenyl)-3,7-dimethyl-2,4,6,8-nonateraenoate.
Ro-10-9359; Tegison; Tigason. Antipsoriatic. mp = 104-105°; LD_{50} (mus ip 1 day) > 4000 mg/kg, (mus ip 20 day) = 1176 mg/kg, (rat ip 20 day) > 2000 mg/kg, (mus orl 20 day) > 2000 mg/kg, (rat orl 20 day) > 4000 mg/kg. *Hoffmann-LaRoche Inc.*

5684 Lexacalcitol
131875-08-6
$C_{29}H_{48}O_4$
(5Z,7E,20R)-20-[(4-Ethyl-4-hydroxyhexyl)oxy]-9,10-secopregna-5,7,10(19)-tiene-1α,3β-diol.

KH-1060. Vitamin D analog with antiproliferative properties. Antipsoriatic.

5685 Liarozole Fumarate
145858-52-2
$C_{46}H_{38}Cl_2N_8O_{12}$
(±)-5-(m-Chloro-α-imidazol-1-ylbenzyl)benzimidazole fumarate (2:3).
Liazal; R-85246. Antipsoriatic. *Janssen Pharm. Inc.*

5686 Lonapalene
91431-42-4 5595
$C_{16}H_{15}ClO_6$
6-Chloro-2,3-dimethoxy-1,4-naphthalenediol diacetate.
RS-43179. Antipsoriatic. mp = 93-94°, 87.5-89.5°. *Syntex Intl. Ltd.*

5687 Maxacalcitol
103909-75-7
$C_{26}H_{42}O_4$
1α 25-Dihydroxy-22-oxacalcitriol.
Vitamin D analog with antiproliferative properties. Antipsoriatic.

5688 Peldesine
133432-71-0
$C_{12}H_{11}N_5O$
2-Amino-3,5-dihydro-7-(3-pyridylmethyl)-4H-pyrrolo[3,2-d]pyrimidin-4-one.
BCX-34. Antineoplastic; antipsoriatic agent. *BioCryst Pharm. Inc.*

5689 Pyrogallol
87-66-1 8184 201-762-9
$C_6H_6O_3$
1,2,3-Benzenetriol.
pyrogallic acid. Antipsoriatic. Catechol-O-methyltransferase inhibitor. mp = 131-133°; bp = 309°; d = 1.45; soluble in H_2O (58.8 g/100 ml), EtOH (76.9 g/100 ml), Et_2O (62.5 g/100 ml); slightly soluble in C_6H_6, $CHCl_3$, CS_2; LD_{50} (rbt orl) = 1600 mg/kg.

5690 Pyrogallol Monoacetate
1330-51-4 8184
$C_8H_8O_4$
1-Acetoxy-2,3-dihydroxybenzene.
Eugallol. Antipsoriatic. Catechol-O-methyltransferase inhibitor. Soluble in H_2O, EtOH, $CHCl_3Et_2O$, Me_2CO, castor oil.

5691 Pyrogallol Triacetate
525-52-0 8184 208-374-9
$C_{12}H_{12}O_6$
1,2,3-Triacetoxybenzene.
acetpyrogall; Lenigallol. Antipsoriatic. Catechol-O-methyltransferase inhibitor. mp = 165°; soluble in EtOH, slightly soluble in H_2O.

5692 Safingol
15639-50-6
$C_{18}H_{39}NO_2$
(2S,3S)-2-Amino-1,3-octadecanediol.
Antineoplastic; antipsoriatic agent. *Sphinx Pharm. Corp.*

5693 Safingol Hydrochloride
139755-79-6
$C_{18}H_{40}ClNO_2$
(2S,3S)-2-Amino-1,3-octadecanediol hydrochloride.
Antineoplastic (adjunct); antipsoriatic. *Sphinx Pharm. Corp.*

5694 Tacalcitol
57333-96-7 9197
$C_{27}H_{44}O_3$
(+)-(5Z,7E,24R)-9,10-Secocholesta-5,7,10(19)-triene-1α,3β,24-triol.
Bonalfa; TV-02. Antipsoriatic. λ_m = 265 nm (EtOH).

5695 Tazarotene
118292-40-3 9249
$C_{21}H_{21}NO_2S$
Ethyl 6-[(4,4dimethylthiochroman-6-yl)ethynyl]nicotinate.
Zorac; AGN-190168. Keratolytic. Used to treat acne and psoriasis. White solid. *Allergan Inc.*

5696 Tepoxalin
103475-41-8
$C_{20}H_{20}ClN_3O_3$
5-(p-Chlorophenyl)-1-(p-methoxyphenyl)-N-methylpyrazole-3-propionohydroxamic acid.
ORF-20485; RWJ-20485. Antipsoriatic. *Johnson R. W. Pharm. Res. Institute; Ortho Pharm. Corp.*

5697 Ticolubant
154413-61-3
$C_{23}H_{19}Cl_2NO_3S$
(E)-6-[[(2,6-Dichlorophenyl)thio]methyl]-3-(phenethyloxy)-2-pyridineacrylic acid.
SB-209247. Antipsoriatic. *SmithKline Beecham Pharm.*

Antipsychotics

5698 Acetophenazine
2752-68-0 70 2751-68-0
$C_{23}H_{29}N_3O_2S$
10-[3-[4-(2-Hydroxyethyl)-1-piperazinyl]propyl]phenothiazin-2-yl methyl ketone.
Antipsychotic. A phenothiazine. *Schering-Plough HealthCare Products.*

5699 Acetophenazine Maleate
5714-00-1 70 227-202-3
$C_{31}H_{37}N_3O_{10}S$
10-[3-[4-(2-Hydroxyethyl)-1-piperazinyl]propyl]-phenothiazin-2-yl methyl ketone maleate (1:2).
acetophenazine dimaleate; Sch-6673; NSC-70600; Tindal. Antipsychotic. A phenothiazine. mp = 167-168.5°; LD_{50} (rat orl) = 433 mg/kg. *Schering-Plough HealthCare Products.*

5700 Alentemol
112891-97-1
$C_{19}H_{25}NO$
(+)-2-(Dipropylamino)-2,3-dihydrophenalen-5-ol.
Antipsychotic. A dopamine agonist. *Pharmacia & Upjohn; Upjohn Ltd.*

5701 Alentemol Hydrobromide
112892-81-6
$C_{19}H_{25}BrNO$
(+)-2-(Dipropylamino)-2,3-dihydrophenalen-5-ol hydrobromide.
U-68553B. Antipsychotic. A dopamine agonist. *Pharmacia & Upjohn; Upjohn Ltd.*

5702 Alizapride
59338-93-1 246 261-710-6
$C_{16}H_{21}N_5O_2$
6-Methoxy-N-[[1-(2-propenyl)-2-pyrrolidinyl]-methyl]-1H-benzotriazole-5-carboxamide.
Antipsychotic; antiemetic. A benzamide. mp = 139°; LD_{50} (mus iv) = 92.7 mg/kg.

5703 Alizapride Hydrochloride
59338-87-3 246
$C_{16}H_{22}ClN_5O_2$
6-Methoxy-N-[[1-(2-propenyl)-2-pyrrolidinyl]methyl]-1H-benzotriazole-5-carboxamide monohydrochloride.
Nausilen; Plitican; Vergentan. Antipsychotic; antiemetic. A benzamide. mp = 206-208°.

5704 Alpertine
27076-46-6
$C_{25}H_{31}N_3O_4$
Ethyl-5,6-dimethoxy-3-[2-(4-phenyl-1-piperazinyl)ethyl]indole-2-carboxylate.
Win-31665. Antipsychotic. *Sterling Health U.S.A.*

5705 Amisulpride
71675-85-9 508 275-831-7
$C_{17}H_{27}N_3O_4S$
4-Amino-N-[(1-ethyl-2-pyrrolidinyl)methyl]-5-(ethylsulfonyl)-o-anisamide.
aminosultopride; DAN-2163; Socian; Solian. Antipsychotic. A neuroleptic agent, analog of sulpiride, sultopiride, a benzamide. mp = 126-127°; LD_{50} (mus iv) 56 mg/kg, (mus ip) 175 mg/kg, (mus sc) 1030 mg/ks.

5706 Azaperone
1649-18-9 931 216-715-8
$C_{19}H_{22}FN_3O$
4'-Fluoro-4-[4-(2-pyridinyl)-1-piperazinyl]butyrophenone.
R-1929; Stresnil; Suicalm. Sedative (veterinary); tranquilizer (veterinary). mp = 73-75°. *Janssen Pharm. Ltd.*

5707 Batelapine
95634-82-5
$C_{16}H_{20}N_6$
2-Methyl-5-(4-methyl-1-piperazinyl)-11H-s-triazolo[1,5-c]benzodiazepine.
Antipsychotic. *Ciba-Geigy Corp.*

5708 Batelapine Maleate
120360-10-3
$C_{20}H_{24}N_6O_4$
2-Methyl-5-(4-methyl-1-piperazinyl)-11H-s-triazolo[1,5-c]benzodiazepine maleate (1:1).
CGS-13429A. Antipsychotic. *Ciba-Geigy Corp.*

5709 Benperidol
2062-84-2 1077 218-172-2
$C_{22}H_{24}FN_3O_2$
1-[1-[4-(4-Fluorophenyl)-4-oxobutyl]-4-piperidinyl]-1,3-dihydro-2H-benzimidazol-2-one.
McN-JR-4584; R-4584; Anquil; Frenactyl; Frénactil; Glianimon. Antipsychotic. A butyrophenone. mp = 170-172°. *Janssen Pharm. Ltd.*

5710 Benperidol Hydrochloride
74298-73-0 1077 277-807-1
$C_{22}H_{25}ClFN_3O_2$
1-[1-[4-(4-Fluorophenyl)-4-oxobutyl]-4-piperidinyl]-1,3-dihydro-2H-benzimidazol-2-one monohydrochloride.
Antipsychotic. A butyrophenone. [hydrochloride hydrate]: mp = 134-142°. *Janssen Pharm. Ltd.*

5711 Benzindopyrine Hydrochloride
5585-71-7
$C_{22}H_{21}ClN_2$
1-Benzyl-3-[2-(4-pyridyl)ethyl]indole monohydrochloride.
IN-461; NSC-17789. Antipsychotic.

5712 Benzquinamide
63-12-7 1154
$C_{22}H_{32}N_2O_5$
2-(Acetoxy)-N,N-diethyl-1,3,4,6,7,11b-hexahydro-9,10-dimethoxy-2H-benzo[a]quinolizine-3-carboxamide.
BZQ; NSC-64375; P-2647; Emete-Con; Emeticon; Promecon; Quantril. Antipsychotic; antiemetic. A tricyclic. mp = 130-131.5°; LD_{50} (rat orl) = 990 mg/kg, (mus ip) = 376 mg/kg. *Pfizer Intl.; Roerig Div. Pfizer Pharm.*

5713 Biriperone
41510-23-0
$C_{24}H_{26}FN_3O$
(±)-4'-Fluoro-4-(3,4,6,7,12,12a-hexahydropyrazino[1',2':1,6]pyrido[3,4-b]indol-2(1H)-yl)butyrophenone.
Centbutindole. Antipsychotic; comparable to trifluoperazine.

5714 Blonanserin
132810-10-7
$C_{23}H_{30}FN_3$
2-(4-Ethyl-1-piperazinyl)-4-(p-fluorophenyl)-5,6,7,8,9,10-hexahydrocycloocta[b]pyridine.
Antipsychotic agent.

5715 Brofoxine
21440-97-1 244-389-7
$C_{10}H_{10}BrN_2$
6-Bromo-1,4-dihydro-4,4-dimethyl-2H-3,1-benzoxazin-2-one.
FI-6820. Antipsychotic. *Farmitalia Societa Farmaceutici.*

5716 Bromerguride
83455-48-5
$C_{20}H_{25}BrN_4O$
3-(2-Bromo-9,10-didehydro-6-methylergolin-8α-yl)-1,1-diethylurea.
A dopamine-antagonistic ergot derivative with antipsychotic properties.

5717 Bromperidol
10457-90-6 1466 233-943-3
$C_{21}H_{23}BrFNO_2$
4-[4-(p-Bromophenyl)-4-hydroxypiperidino]-4'-fluorobutyrophenone.
R-11333; Azurene; Impromen; Tesoprel. Antipsychotic. A bromine analog of haloperidol. mp = 155-158°; λ_m = 245 mn; pKa = 8.6-8.7; soluble in H_2O, 0.1M acids. *Cilag-Chemie Ltd.*

5718 Bromperidol Decanoate
75067-66-2 1466 278-065-1
$C_{31}H_{41}BrFNO_3$
4-[4-(p-Bromophenyl)-4-hydroxypiperidino]-4'-fluorobutyrophenone decanoate.
R-46541. Antipsychotic. A bromine analog of haloperidol. *Janssen Pharm. Ltd.*

5719 Buramate
4663-83-6 1525
$C_{10}H_{13}NO_3$
2-Hydroxyethyl benzylcarbamate.
AC-601; NSC-30223; glycol benzylcarbamate; 2-hydroxyethyl benzylcarbamate; mono(benzylcarbamate); Hyamate. Antipsychotic; anticonvulsant. mp = 40°. *Ascher B.F. & Co.; Xttrium Labs. Inc.*

5720 Butaclamol
51152-91-1
$C_{25}H_{31}NO$
(±)-3α-tert-Butyl-2,3,4,4aβ,8,9,13bα,14-octahydro-1H-benzo[6,7]-cyclohepta[1,2,3-de]pyrido[2,1-]isoquinolin-3-ol.
Antipsychotic.

5721 Butaclamol Hydrochloride
36504-94-6
$C_{25}H_{32}ClNO$
(±)-3α-tert-Butyl-2,3,4,4aβ,8,9,13bα,14-octahydro-1H-benzo[6,7]-cyclohepta[1,2,3-de]pyrido[2,1-]isoquinolin-3-ol hydrochloride.
AY-23038. Antipsychotic.

5722 Butaperazine
653-03-2 1543 211-493-9
$C_{24}H_{31}N_3OS$
1-[10-[3-(4-Methyl-1-piperazinyl)propyl]phenothiazin-2-yl]-1-butanone.
AHR-3000; Bayer-1362; Riker-595; butyrylperazine; Repoise; Tyrylen. Antipsychotic. A phenothiazine. $bp_{0.05}$ = 270-280°. *3M Pharm.; Bayer AG.*

5723 Butaperazine Dimaleate
1063-55-4 1543 213-900-5
$C_{32}H_{39}N_3O_8S$
1-[10-[3-(4-Methyl-1-piperazinyl)propyl]phenothiazin-2-yl]-1-butanone dimaleate (1:2).
Randolectil. Antipsychotic. A phenothiazine. [maleate]: mp = 180-182°. *3M Pharm.; Bayer AG.*

5724 Carphenazine
2622-30-2 1910 220-072-9
$C_{24}H_{31}N_3O_2S$
1-[10-[3-[4-(2-Hydroxyethyl)-1-piperazinyl]propyl]phenothiazin-2-yl]-1-propanone.
Proketazine. Antipsychotic. A phenothiazine. *Am. Home Products; Schering-Plough HealthCare Products; Wyeth Labs.*

5725 Carphenazine Maleate
2975-34-0 1910 221-019-2
$C_{32}H_{39}N_3O_{10}S$
1-[10-[3-[4-(2-Hydroxyethyl)-1-piperazinyl]propyl]phenothiazin-2-yl]-1-propanone dimaleate (1:2).
Wy-2445; NSC-71755. Antipsychotic. A phenothiazine. mp = 175-177°. *Am. Home Products; Schering-Plough HealthCare Products; Wyeth Labs.*

5726 Carpipramine
5942-95-0 1911 227-700-0
$C_{28}H_{38}N_4O$
1'-[3-(10,11-Dihydro-5H-dibenz[b,f]azepin-5-yl)propyl]-(1,4'-bipiperidine)-4'-carboxamide.
Bay-b-4343-b. Antipsychotic. A tricyclic. *Yoshitomi.*

5727 Carpipramine Dihydrochloride
7075-03-8 1911 230-372-1
$C_{28}H_{41}Cl_2N_4O_2$
1'-[3-(10,11-Dihydro-5H-dibenz[b,f]azepin-5-yl)propyl]-(1,4'-bipiperidine)-4'-carboxamide dihydrochloride monohydrate.
PZ-1511; Defekton; Prazinil. Antipsychotic. A tricyclic. mp = 260°; sparingly soluble in H_2O; LD_{50} (rat iv)= 37 mg/kg, (rat ip) 76 mg/kg; (rat orl) 1025 mg/kg. *Yoshitomi.*

5728 Chlorproethazine
84-01-5 2236 201-510-8
$C_{19}H_{23}ClN_2S$
2-Chloro-10-(3-diethylaminopropyl)phenothiazine.
RP-4909. Muscle relaxant; antipsychotic. A phenothiazine. bp_1 = 225-240°. *Rhône-Poulenc Rorer Pharm. Inc.*

5729 Chlorproethazine Hydrochloride
4611-02-3 2236 225-018-8
$C_{19}H_{24}ClN_2S$
2-Chloro-10-(3-diethylaminopropyl)phenothiazine hydrochloride.
Neuriplege. Muscle relaxant; antipsychotic. A phenothiazine. mp = 178°; soluble in H_2O (1.67 g/100 ml), EtOH (0.33 g/100 ml), $CHCl_3$ (20 g/100 ml); insoluble in Me_2CO, C_6H_6, Et_2O. *Rhône-Poulenc Rorer Pharm. Inc.*

5730 Chlorpromazine
50-53-3 2238 200-045-8
$C_{17}H_{19}ClN_2S$
2-Chloro-10-[3-(dimethylamino)propyl]phenothiazine.
2601-A; HL-5746; RP-4560; SKF-2601-A; Aminazine; Ampliactil; Amplictil; Chlorderazin; Chlropromados; Elmarin; Esmind; Fenactil; Novomazina; Promactil; Promazil; Proma; Prozil; Plegomazin; Sanopron; Thorazine; Wintermin. Antiemetic; antipsychotic. A phenothiazine. Oily liquid; $bp_{0.8}$ = 200-205°. *Elkins-Sinn; KV Pharm.; Labaz S.A.; Parke-Davis; Rhône-Poulenc Rorer Pharm. Inc.*

5731 Chlorpromazine Hydrochloride
69-09-0 2238 200-701-3
$C_{17}H_{20}Cl_2N_2S$
2-Chloro-10-[3-(dimethylamino)propyl]phenothiazine hydrochloride.

clorazepate potassium; dipotassium clorazepate; Chloractil; Chlorazin; Gen-xene; Hebanil; Hibanil; Hibernal; Klorpromex; Largactil; Largaktyl; Marazine; Megaphen; Novo-Clopate; Nu-Clopate; Promacid; Promapar; Propaphenin; Sonazine; Taroctyl; Thorazine; Torazina; Tranzene. Antiemetic; antipsychotic. Used in veterinary medicine as a tranquilizer. A phenothiazine. Dec 179-180°; pH (5% aq) = 4.0-5.5; soluble in H_2O (40 g/100ml), MeOH, EtOH, $CHCl_3$; insoluble in C_6H_6, Et_2O; LD_{50} (rat orl) = 2225 mg/kg. *DDSA Pharm. Ltd.; Elkins-Sinn; KV Pharm.; Labaz S.A.; Parke-Davis; Pharmacia & Upjohn; Rhône-Poulenc Rorer Pharm. Inc.; SmithKline Beecham Pharm.*

5732 Chlorprothixene
113-59-7 2241 204-032-8
$C_{18}H_{18}ClNS$
(Z)-2-Chloro-N,N-dimethylthioxanthene-$\delta^{9,\gamma}$-propylamine.
N-714; Ro-4-04033; Taractan; Truxal; Truxaletten; Tarasan. Antipsychotic. A thioxanthene. mp = 97-98°; insoluble in H_2O; soluble in organic solvents; LD_{50} (rat orl) = 380 mg/kg. *Am. Cyanamid; Hoffmann-LaRoche Inc.; Kissei.*

5733 Cinperene
14796-24-8
$C_{25}H_{28}N_2O_2$
2-(1-Cinnamyl-4-piperidyl)-2-phenylglutarimide.
R-5046. Antipsychotic. *Janssen Pharm. Ltd.*

5734 Cintriamide
5588-21-6
$C_{12}H_{15}NO_4$
3,4,5-Trimethoxycinnamamide.
Cintramide. Antipsychotic. *Lederle Labs.*

5735 Cinuperone
82117-51-9
$C_{23}H_{24}FN_3O$
4'-Fluoro-4-[4-(3-isoquinolyl)-1-piperazinyl]butyrophenone.
Sigma antagonist, used as an antipsychotic.

5736 Clocapramine
47739-98-0 2427
$C_{28}H_{37}ClN_4O$
1'-[3-(3-Chloro-10,11-dihydro-5H-dibenz[b,f]azepin-5-yl)propyl][1,4'-bipiperidine]-4'-carboxaminde.
3-chlorocarpipramine; clocarpramine; CCP. Antipsychotic. A tricyclic. *Yoshitomi.*

5737 Clocapramine Dihydrochloride Monohydrate
60789-62-0 2427
$C_{28}H_{41}Cl_3N_4O_2$
1'-[3-(3-Chloro-10,11-dihydro-5H-dibenz[b,f]azepin-5-yl)propyl][1,4'-bipiperidine]-4'-carboxaminde dihydrochloride monohydrate.
Y-4153; Clofekton. Antipsychotic. A tricyclic. mp = 267°; soluble in H_2O, AcOH; practically insoluble in Et_2O, Me_2CO; LD_{50} (mus ip) = 160 mg/kg, (rat ip) = 125 mg/kg, (mus orl) = 2550 mg/kg, (rat orl) = 6800 mg/kg. *Yoshitomi.*

5738 Clomacran
5310-55-4 2441
$C_{18}H_{21}ClN_2$
2-Chloro-9-[3-(dimethylamino)propyl]acridan.
SK&F-14336D. Antipsychotic. A tricyclic. *SmithKline Beecham Pharm.*

5739 Clomacran Phosphate
22199-46-8 2441
$C_{18}H_{24}ClN_2O_4P$
2-Chloro-9-[3-(dimethylamino)propyl]acridan phosphate (1:1).
SK&F-14336; Devryl; Olaxin. Antipsychotic. A tricyclic. LD_{50} (mus orl) = 222 mg/kg, (rat orl) = 350 mg/kg. *SmithKline Beecham Pharm.*

5740 Clopenthixol
982-24-1 2455 213-566-0
$C_{22}H_{25}ClN_2OS$
4-[3-(2-Chlorothioxanthen-9-ylidene)propyl]-1-piperazineethanol.
NSC-64087; α-clopenthixol [cis(Z)-form]; zuclopenthixol [cis(Z)-form]. Antipsychotic. A thioxanthene neuroleptic. Colorless syrup; sparingly soluble in Et_2O; readily soluble in MeOH. *Kefalas A/S.*

5741 Clopenthixol Dihydrochloride
633-59-0 2455 211-194-3
$C_{22}H_{27}Cl_3N_2OS$
4-[3-(2-Chlorothioxanthen-9-ylidene)propyl]-1-piperazineethanol dihydrochloride.
AY-62021; N-746; Ciatyl; Cisordinol [cis(Z)-form]; Clopixol [cis(Z)-form]; Sordenac; Sordinol. Antipsychotic. A thioxanthene neuroleptic. mp = 250-260°; mp (cis(Z)-form) = 250-260°; soluble in H_2O, sparingly soluble in EtOH; LD_{50} (mus iv) = 111 mg/kg. *Kefalas A/S.*

5742 Clopimozide
53179-12-7
$C_{28}H_{28}ClF_2N_3O$
1-[1-[4,4-Bis(p-fluorophenyl)butyl]-4-piperidyl]-5-chloro-2-benzimidazolinone.
R-29764. Antipsychotic. *Janssen Pharm. Ltd.*

5743 Clopipazan
60085-78-1
$C_{19}H_{18}ClNO$
4-(2-Chlorozanthen-9-ylidene)-1-methylpiperidine.
Antipsychotic. *SmithKline Beecham Pharm.*

5744 Clopipazan Mesylate
60086-22-8
$C_{20}H_{22}ClNO_4S$
4-(2-Chlorozanthen-9-ylidene)-1-methylpiperidine methanesulfonate.
SK&F-69634. Antipsychotic. *SmithKline Beecham Pharm.*

5745 Cloroperone
61764-61-2
$C_{22}H_{23}ClFNO_2$
4-[4-(p-Chlorobenzoyl)piperidino]-4'-fluorobutyrophenone.
Antipsychotic.

5746 Cloroperone Hydrochloride
55695-56-2
$C_{22}H_{24}Cl_2FNO_2$
4-[4-(p-Chlorobenzoyl)piperidino]-4'-fluorobutyrophenone hydrochloride.
AHR-6134. Antipsychotic.

5747 Clorotepine
13448-22-1
$C_{19}H_{21}ClN_2S$
1-(8-Chloro-10,11-dihydrodibenzo[b,f]thiepin-10-yl)-4-methylpiperazine.
Antipsychotic agent.

5748 Clospirazine
24527-27-3 2474 246-298-8
$C_{22}H_{24}ClN_3OS$
8-[3-(2-Chloro-10H-phenothiazin-10-yl)propyl]-1-thia-4,8-diazaspiro[4.5]decan-3-one.
spiclomazine. Antipsychotic. A phenothiazine. mp = 154-156°. *Yoshitomi.*

5749 Clospirazine Hydrochloride
27007-85-8 2474
$C_{22}H_{25}Cl_2N_3OS$
8-[3-(2-Chloro-10H-phenothiazin-10-yl)propyl]-1-thia-4,8-diazaspiro[4.5]decan-3-one hydrochloride.
APY-606; Diceplon; Disperon. Antipsychotic. A phenothiazine. mp = 262-264°; sparingly soluble in MeOH, H_2O $CHCl_3$; LD_{50} (rat ip) = 2950 mg/kg, (rat sc) > 5490 mg/kg, (rat orl) = 4000 mg/kg. *Yoshitomi.*

5750 Clothiapine
2058-52-8 2476 218-162-8
$C_{18}H_{18}ClN_3S$
2-Chloro-11-(4-methyl-1-piperazinyl)dibenzo[b,f][1,4]thiazepine.
HF-2159; Entumine; Etumine. Antipsychotic. A tricyclic. mp = 118-120°. *Sandoz Pharm. Corp.*

5751 Clothixamide
4177-58-6
$C_{24}H_{28}ClN_3OS$
4-[3-(2-Chlorothioxanthen-9-ylidene)propyl]-N-methyl-1-piperazinepropionamide.
Antipsychotic. *Pfizer Inc.*

5752 Clothixamide Maleate
4434-20-2
$C_{32}H_{36}ClN_3O_9S$
4-[3-(2-Chlorothioxanthen-9-ylidene)propyl]-N-methyl-1-piperazinepropionamide maleate.
P-4385B; NSC-78714. Antipsychotic. *Pfizer Inc.*

5753 Cloxypendyl
15311-77-0
$C_{20}H_{25}ClN_4OS$
4-[3-(3-Chloro-10H-pyrido[3,2-b][1,4]benzothiazin-10-yl)propyl]-1-piperazineethanol.
Antipsychotic agent.

5754 Clozapine
5786-21-0 2484 227-313-7
$C_{18}H_{19}ClN_4$
8-Chloro-11-(4-methyl-1-piperazinyl)-5H-dibenzo[b,e][1,4]diazepine.
HF-1854; Clozaril; Leponex. An atypical antipsychotic. A tricyclic. mp = 183-184°; λ_m 215,230, 261,297 nm (ε 27400, 25800, 16800, 10500 MeOH); LD_{50} (rat iv) = 58 mg/kg, (rat orl) = 260 mg/kg. *Sandoz Pharm. Corp.*

5755 Cyamemazine
3546-03-0 2753 222-594-2
$C_{19}H_{22}N_3S$
10-[3-Dimethylamino)-2-methylpropyl]phenothiazine-1-carbonitrile.
RP-7204; TH-2602; cyamepromezine; Kyamepromazine; Ciamatil; Tercian. Antipsychotic. A phenothiazine. Yellow oil: $bp_{0.2-0.5}$ = 205-220°; yellow powder: mp = 89-96°; practically insoluble in H_2O; soluble in EtOH and organic solvents. *Rhône-Poulenc Rorer Pharm. Inc.*

5756 Cyclophenazine
17692-26-1
$C_{23}H_{26}F_3N_3S$
10-[3-(4-Cyclopropyl-1-piperazinyl)propyl]-2-(trifluoromethyl)phenothiazine.
Antipsychotic. *Eli Lilly & Co.*

5757 Cyclophenazine Hydrochloride
15686-74-5
$C_{23}H_{28}Cl_2F_3N_3S$
10-[3-(4-Cyclopropyl-1-piperazinyl)propyl]-2-(trifluoromethyl)phenothiazine dihydrochloride.
Antipsychotic. *Eli Lilly & Co.*

5758 Dextromoramide Tartrate
2922-44-3 2997 220-870-7
$C_{29}H_{38}N_2O_8$
(+)-4-[2-Methyl-4-oxo-3,3-diphenyl-4-(1-pyrrolidinyl)butyl]morpholine bitartrate.
dextromoramide bitartrate; Dimorlin Tartrate. Analgesic, narcotic. Neurotropic. Used as an antipsychotic. Federally controlled substance (opiate). Dec 189-192°; solubility (w/v) at 25°: in H_2O 20%, $CHCl_3$ 30%, MeOH 40%, EtOH 100%, Me_2CO 100%. *Janssen Pharm. Inc.; SmithKline Beecham Pharm.*

5759 Didrovaltrate
18296-45-2
$C_{22}H_{32}O_8$
1,4a,5,7a-Tetrahydro-16,-dihydroxyspiro[cyclopenta[c]pyran-7(6H)-,2'-oxirane]-4-methanol 6-acetate 1,4-diisovalerate.
Sedative, antispasmodic.

5760 Dixyrazine
2470-73-7 3450 219-591-3
$C_{24}H_{33}N_2O_2S$
2-[2-[4-[2-Methyl-3-(10H-phenothiazin-10-yl)propyl]-1-piperazinyl]ethoxy]ethanol.
UCB-3412; Esocalm; Esucos. Antipsychotic. A phenothiazine.

5761 Droperidol
548-73-2 3505 208-957-8
$C_{22}H_{22}FN_3O_2$
1-[1-[3-(p-Fluorobenzoyl)propyl]-1,2,3,6-tetrahydro-4-pyridyl]-2-benzimidazolinone.
McN-JR-4749; R-4749; dehydrobenzperidol; Dridol; Droleptan; Inapsine; component of: Innovar, Thalamonal.

Antipsychotic; Tranquilizer (veterinary). A Butyrophenone. mp = 145-146.5°; λ_m = 245, 280 nm (ε 15600, 7500); soluble in $CHCl_3$ and DMF; barely soluble in C_6H_6, EtOH, H_2O; pKa = 7.64; heat and light sensitive; LD_{50} (mus sc) 125 mg/kg, (mus ip) = 43 mg/kg. *Astra USA Inc.; DuPont-Merck Pharm.; Janssen Pharm. Ltd.*

5762 Etazolate
51022-77-6
$C_{14}H_{19}N_5O_2$
Ethyl 1-ethyl-4-(isopropylidenehydrazino)-1H-pyrazolo[3,4-b]pyridine-5-carboxylate.
Antipsychotic. *Bristol-Myers Squibb Co.*

5763 Etazolate Hydrochloride
35838-58-5
$C_{14}H_{20}ClN_5O_2$
Ethyl 1-ethyl-4-(isopropylidenehydrazino)-1H-pyrazolo[3,4-b]pyridine-5-carboxylate monohydrochloride.
SQ-20009. Antipsychotic. *Bristol-Myers Squibb Co.*

5764 Fananserin
127625-29-0
$C_{23}H_{24}FN_3O_2S$
2-[3-[4-(p-Fluorophenyl)-1-piperazinyl]propyl]-2H-naphtho-[1,8-cd]-isothiazole 1,1-dioxide.
RP-62203. *Rhône-Poulenc Rorer Pharm. Inc.*

5765 Fenimide
60-45-7
$C_{13}H_{15}NO_2$
3-Ethyl-2-methyl-2-phenylsuccinimide.
CI-419; PM-1807. Antipsychotic. *Parke-Davis.*

5766 Fluanisone
1480-19-9 4150 216-038-8
$C_{21}H_{21}FN_2O_2$
4'-Fluoro-4-[4-(o-methoxyphenyl)-1-piperazinyl]buterophenone.
MD-2028; R-2028; R2167; haloanisone; Sedalande. Antipsychotic. A Butyrophenone. mp = 67-68°; insoluble in H_2O, soluble in organic solvents; LD_{50} (mus ip) = 200 mg/kg.

5767 Flucindole
40594-09-0
$C_{14}H_{16}F_2N_2$
3-(Dimethylamino)-6,8-difluoro-1,2,3,4-tetrahydrocarbazole.
Win-35150. Antipsychotic. *Sterling Winthrop Inc.*

5768 Flumezapine
61325-80-2
$C_{17}H_{19}FN_4S$
7-Fluoro-2-methyl-4-(4-methyl-1-piperazinyl)-10H-thieno[2,3-b][1,5]benzodiazepine.
LY-120363. Antipsychotic. *Eli Lilly & Co.*

5769 Flupentixol
2709-56-0 4224 220-304-9
$C_{23}H_{25}F_3N_2OS$
2-Trifluoromethyl-9-[3-[4-(2-hydroxyethyl)piperazin-1-yl]propylidene]thioxanthene.
N-7009; LC-44; flupenthixol. Antipsychotic. A neuroleptic agent related structurally to thiothixene, a thioxanthene. *Kefalas A/S; SmithKline Beecham Pharm.*

5770 Flupentixol Decanoate
30909-51-4 4224 250-385-6
$C_{33}H_{47}F_3N_2O_4S_2$
2-Trifluoromethyl-9-[3-[4-(2-hydroxyethyl)piperazin-1-yl]propylidene]thioxanthene decanoate.
LU-5-110; Depixol; Fluanxol, Dépot; Viscoleo. Antipsychotic. A neuroleptic agent related structurally to thiothixene, a thioxanthene. *Kefalas A/S; SmithKline Beecham Pharm.*

5771 Flupentixol Dihydrochloride
2413-38-9 4224 219-321-4
$C_{23}H_{27}CL_2F_3N_2OS$
2-Trifluoromethyl-9-[3-[4-(2-hydroxyethyl)piperazin-1-yl]propylidene]thioxanthene dihydrochloride.
Emergil; Fluanxol; Siplarol; Metamin. Antipsychotic. A neuroleptic agent related structurally to thiothixene, a thioxanthene. *Kefalas A/S; SmithKline Beecham Pharm.*

5772 Fluperlapine
67121-76-0
$C_{19}H_{20}FN_3$
3-Fluoro-6-(4-methyl-1-piperazinyl)morphanthridine.
Antipsychotic.

5773 Fluphenazine
69-23-8 4226 200-702-9
$C_{22}H_{26}F_3N_3OS$
4-[3-[2-(Trifluoromethyl)-10H-phenothiazin-10-yl]propyl]-1-piperazineethanol.
S-94; SQ-4918. Antipsychotic. A phenothiazine. Viscous oil; $b_{0.5}$ = 268274°; $bp_{0.3}$ = 250-252°. *Bristol-Myers Squibb Co.; Olin Res. Ctr.; SmithKline Beecham Pharm.*

5774 Fluphenazine Decanoate
5002-47-1 4226 225-672-4
$C_{32}H_{44}F_3N_3O_2S$
4-[3-[2-(Trifluoromethyl)-10H-phenothiazin-10-yl]propyl]-1-piperazineethanol decanoate.
SQ-10733; QD-10733; Dapotum D; Lyogen; Depot; Modecate; Prolixin Decanoate; Siqualine. Antipsychotic. A phenothiazine. Viscous liquid; mp = 30-32°; insoluble in H_2O; very soluble in organic solvents. *SmithKline Beecham Pharm.*

5775 Fluphenazine Enanthate
2746-81-8 4226 220-385-0
$C_{29}H_{38}F_3N_3O_2S$
2-[4-[3-[2-(Trifluoromethyl)phenothiazin-10-yl]propyl]-1-piperazinyl]ethyl heptanoate.
SQ-16144; Moditen Enanthate; Prolixin Enanthate. Antipsychotic. A phenothiazine. Viscous liquid or oily solid. *Bristol-Myers Squibb Co.; Pfizer Inc.; SmithKline Beecham Pharm.*

5776 Fluphenazine Hydrochloride
146-56-5 4226 205-674-1
$C_{22}H_{28}Cl_2F_3N_3OS$
4-[3-[2-(Trifluoromethyl)-10H-phenothiazin-10-yl]propyl]-1-piperazineethanol dihydrochloride.
Anatensol; Dapotum; Lyogen; Moditen; Omca; Pacinol; Permitil; Prolixin; Siqualone; Tensofin; Valamina.

Antipsychotic. A phenothiazine. mp = 235-237°. *Bristol-Myers Squibb Co.; Schering-Plough HealthCare Products; SmithKline Beecham Pharm.*

5777 Fluspiperone
54965-22-9
$C_{23}H_{25}F_2N_3O_2$
8-[3-(p-Fluorobenzoyl)propyl-1-(p-fluorophenyl)-1,3,8-triazaspiro[4.5]decan-4-one.
R-28930. Antipsychotic. *Janssen Pharm. Ltd.*

5778 Fluspirilene
1841-19-6 4241 217-418-6
$C_{29}H_{31}F_2N_3O$
8-[4,4-Bis(p-fluorophenyl)butyl]-1-phenyl-1,3,8-triazaspiro-[4.5]decan-4-one.
McN-JR-6218; R-6218; Imap; Redeptin. Antipsychotic. mp = 187.5-190°; soluble in H_2O (0.015 mg/ml); LD_{50} (rat im) 146 mg/kg. *Janssen Pharm. Ltd.*

5779 Flutroline
70801-02-4
$C_{27}H_{25}F_3N_2O$
(±)-8-Fluoro-α,5-bis(p-fluorophenyl)-1,3,4,5-tetrahydro-2H-pyrido[4,3-b]indole-2-butanol.
CP-36584. Antipsychotic. *Pfizer Inc.*

5780 Gevotroline
107266-06-8
$C_{19}H_{20}FN_3$
8-Fluoro-2,3,4,5-tetrahydro-2-[3-(3-pyridyl)propyl-1H-pyrido[4,3-b]indole.
Antipsychotic. *Wyeth-Ayerst Labs.*

5781 Gevotroline Hydrochloride
112243-58-0
$C_{19}H_{21}ClFN_3$
8-Fluoro-2,3,4,5-tetrahydro-2-[3-(3-pyridyl)propyl-1H-pyrido[4,3-b]indole mono-hydrochloride.
WY-47384. Antipsychotic. *Wyeth-Ayerst Labs.*

5782 Halopemide
59831-65-1
$C_{21}H_{22}ClFN_4O_2$
N-[2-[4-(Chloro-2-oxo-1-benzimidazolinyl)piperidino]-ethyl]-p-fluorobenzenamide.
R-34301. Antipsychotic. *Janssen Pharm. Ltd.*

5783 Haloperidol
52-86-8 4629 200-155-6
$C_{21}H_{23}ClFNO_2$
4-[4-(p-Chlorophenyl)-4-hydroxypiperidino]-4'-fluorobutyrophenone.
R-1625; McN-JR-1625; Aloperidin; Bioperidolo; Brotopon; Dozic; Einalon S; Eukystol; Haldol; Halosten; Keselan; Linton; Peluces; Serenace; Serenase; Sigaperidol. Antidyskinetic; antipsychotic. Used to treat Gilles de la Tourette's disease. A Butyrophenone. mp = 148-149.4°; λ_m = 247, 221 nm (ε 13300, 15000 HCl/MeOH); soluble in H_2O (0.0014 g/100 ml); freely soluble in MeOH, $CHCl_3$, Me_2CO, C_6H_6; LD_{50} (rat orl) = 165 mg/kg, (mus ip) = 60 mg/kg. *Abbott Laboratories Inc.; Lemmon Co.; McNeil Pharm.*

5784 Haloperidol Decanoate
74050-97-8 4629 277-679-7
$C_{31}H_{41}ClFNO_3$
4-[4-(p-Chlorophenyl)-4-hydroxypiperidino]-4'-fluorobutyrophenone decanoate.
KD-16; R-13762; Haldol Decanoate; Halomonth; Neoperidole. Antidyskinetic (in Gilles de la Tourette's disease); antipsychotic. A Butyrophenone. *Abbott Laboratories Inc.*

5785 Iloperidone
133454-47-4
$C_{24}H_{27}FN_2O_4$
4'-[3-[4-(6-Fluoro-1,2-benzisoxazol-3-yl)piperidino]propoxy]-3'-methoxyacetophenone.
HP-873. Antipsychotic. *Hoechst Roussel Pharm. Inc.; Ilon Labs.*

5786 Imiclopazine
7224-08-0 4945
$C_{25}H_{32}ClN_5OS$
1-[2-[4-[3-(2-Chlorophenothiazin-10-yl)propyl]-1-piperazinyl]ethyl]-3-methyl-2-imidazolidinone.
chlorimpiphenine. Antipsychotic. A phenothiazine. $bp_{0.01}$ = 260°. *Astra Sweden.*

5787 Imiclopazine Dihydrochloride
7414-95-1 4945
$C_{25}H_{34}Cl_3N_5OS$
1-[2-[4-[3-(2-Chlorophenothiazin-10-yl)propyl]-1-piperazinyl]ethyl]-3-methyl-2-imidazolidinone dihydrochloride.
P-4241; Ponsital. Antipsychotic. A phenothiazine. *Astra Sweden.*

5788 Imidoline
5588-31-8
$C_{13}H_{18}ClN_3O$
1-(m-Chlorophenyl)-3-[2-(dimethylamino)ethyl]-2-imidazolidinone.
Antipsychotic. *Lederle Labs.*

5789 Imidoline Hydrochloride
7303-78-8
$C_{13}H_{19}Cl_2N_3O$
1-(m-Chlorophenyl)-3-[2-(dimethylamino)ethyl]-2-imidazolidinone monohydrochloride.
CL-48156. Antipsychotic. *Lederle Labs.*

5790 Lenperone
24678-13-5 246-399-7
$C_{22}H_{23}F_2NO_2$
4'-Fluoro-4-(p-fluorobenzoyl)piperidino]butyrophenone.
AHR-2277 [as hydrochloride]. Antipsychotic. *Robins, A. H. Co.*

5791 Lithium Carbonate
554-13-2 5552 209-062-5
CLi_2O_3
Carbonic acid dilithium salt.
dilithium carbonate; CP-15467-61; NSC-16895; Camcolit; Carbolith; Carbolithium; Ceglution; Eskalith; Hypnorex; Limas; Liskonum; Lithane; Lithicarb; Lithobid; Lithonate; Lithotabs; Phasal; Plenur; Priadel; Quilonorm-retard; Quilonum-retard; Téralithe. Antimanic. mp = 720° ± 1°; d = 2.11; soluble in H_2O (1.28 g/100 ml at 20°, 0.71

g/100 ml at 100°), dilute acids; insoluble in EtOH; LD_{50} (rat orl) = 710 mg/kg. *Bayer Corp. Pharm. Div.; Ciba-Geigy Corp.; Miles Inc.; SmithKline Beecham Pharm.; Solvay Pharm. Inc.*

5792 Lithium Citrate
6080-58-6 5555
$C_6H_{13}Li_3O_{11}$
Trilithium citrate tetrahydrate.
Cibalith-S; Litarex; Lithonate S. Antimanic; antidepressant. Deliquesces on exposure to moist air; loses H_2O at 105°; soluble in H_2O; pH 8. *Ciba-Geigy Corp.*

5793 Lithium Hydroxide
1310-66-3 5560 215-183-4
H_3LiO_2
Lithium hydroxide monohydrate.
lithium hydrate. Antimanic. [anhydrous form]: mp = 471°; d = 2.54; soluble in H_2O, slightly soluble in EtOH; [monohydrate]: d_4^{20}= 1.51; soluble in H_2O (10.7 g/100 ml at 0°, 109 g/100 at 100°), slightly soluble in EtOH.

5794 Lometraline
39951-65-0
$C_{13}H_{18}ClNO$
8-Chloro-1,2,3,4-tetrahydro-5-methoxy-N,N-dimethyl-1-naphthylamine.
Antipsychotic; antiparkinsonian. *Pfizer Inc.*

5795 Lometraline Hydrochloride
34552-78-8
$C_{13}H_{19}Cl_2NO$
8-Chloro-1,2,3,4-tetrahydro-5-methoxy-N,N-dimethyl-1-naphthylamine hydrochloride.
CP-14368-1. Antipsychotic; antiparkinsonian. *Pfizer Inc.*

5796 Lorapride
68677-06-5 272-057-1
$C_{14}H_{22}ClN_3O_3S$
5-Chloro-n^1-[(1-ethyl-2-pyrrolidinyl)methyl]-2-methoxysulfanylzamide.
Antipsychotic; antidepressant; antiemetic. Sulpiride derivative.

5797 Lysergide
50-37-3 5665 200-033-2
$C_{20}H_{25}N_3O$
N,N-Diethyllysergamide.
delysid; D-lysergic acid dethylamide; Ergine; LSA; LSD; LSD-25; Lysergsaure diethylamid; Lysergamide; Lysergide; Lysergic acid amide; Lysergic acid diethylamide. Used in biochemical research as an antagonist to serotonin. Has been used experimentally as an adjunct in the study and treatment of mental disorders. A controlled substance (hallucinogen). mp = 80-85°; $[\alpha]_D^{20}$= +17° (c = 0.5 in C_5H_5N); λ_m (in EtOH) = 311 nm; LD_{50} (rat iv) = 16.5 mg/kg. *Eli Lilly & Co.; Farmitalia Societa Farmaceutici.*

5798 Mazindol
22232-71-9 5801 244-857-0
$C_{16}H_{13}ClN_2O$
5-(p-Chlorophenyl)-2,5-dihydro-3H-imidazo[2,1-a]-isoindol-5-ol.
SaH-42548; 42548; Magrilon; Mazanor; Mazildene; Sanorex; Terenac; Teronac. Anorexic; CNS stimulant. Federally controlled substance (stimulant). mp = 215-217°, 198-199°; λ_m = 223, 268.5, 272 nm (ε 19000, 4400, 4400 95% EtOH); insoluble in H_2O, soluble in EtOH. *Am. Home Products; Sandoz Pharm. Corp.; Wyeth-Ayerst Labs.*

5799 Melperone
3757-80-2 5870
$C_{16}H_{22}FNO$
4'-Fluoro-4-(-methylpiperidino)butyrophenone.
FG-5111; methylperone; flubuperone. Antipsychotic. A neuroleptic agent related structurally to haloperidol, a butyrophenone antipsychotic. $bp_{0.1}$ = 120-125°. *Ferrosan A/S.*

5800 Melperone hydrochloride
1622-79-3 5870 216-599-9
$C_{16}H_{23}ClFNO$
4'-Fluoro-4-(-methylpiperidino)butyrophenone monohydrochloride.
FG-5111; Buronil; Eunerpan. Antipsychotic. A neuroleptic agent related structurally to haloperidol, a butyrophenone antipsychotic. mp = 209-211°; LD_{50} (rat orl) 330 mg/kg, (rat iv) = 35 mg/kg. *Ferrosan A/S.*

5801 Mepazine
60-89-9 5892 200-490-8
$C_{19}H_{22}N_2S$
10-[1-(Methyl-3-piperidyl)methyl]-phenothiazine.
mesapin; pecazine; MPMP; P-391; III-2318; Paxital; Lacumin; Pacatal; Nothiazine; Pacatol. Antipsychotic. A phenothiazine. bp_4 = 230-235°. *Promonta.*

5802 Mepazine Acetate
24360-97-2 5892 246-207-1
$C_{21}H_{26}N_2O_2S$
10-[1-(Methyl-3-piperidyl)methyl]phenothiazine acetate (1:1).
Pecazine. Antipsychotic. Tranquilizer (veterinary). A phenothiazine. *Promonta.*

5803 Mesoridazine
5588-33-0 5970
$C_{21}H_{26}N_2OS_2$
10-[2-(1-Methyl-2-piperidyl)ethyl]-2-methylsulfinyl-phenothiazine.
TPS-23; thioridazine-2-sulfoxide. Antipsychotic. A dopamine receptor blocking agent, a phenothiazine antipsychotic; analog of thioridazine. Oily product. *Sandoz Pharm. Corp.*

5804 Mesoridazine Besylate
32672-69-8 5970
$C_{27}H_{32}N_2O_4S_3$
10-[2-(1-Methyl-2-piperidyl)ethyl]-2-methylsulfinyl-phenothiazine monobenzenesulfonate.
mesoridazine bezenesulfonate; NC-123; Lidanar; Lidanil; Serentil. Antipsychotic. A dopamine receptor blocking agent, a phenothiazine antipsychotic; analog of thioridazine. LD_{50} (mus iv) = 33 mg/kg, (mus orl) = 360 mg/kg. *Sandoz Pharm. Corp.*

5805 Methoxypromazine
61-01-8 6080 200-497-6
$C_{18}H_{22}N_2OS$
2-Methoxy-N,N-dimethyl-10H-phenothiazine-10-propanamine.
methopromazine; RP-4632. Antipsychotic. A phenothiazine. mp = 44-48°.

5806 Methoxypromazine Maleate
3403-42-7 6080 222-277-9
$C_{22}H_{26}N_2O_5S$
2-Methoxy-N,N-dimethyl-10H-phenothiazine-10-propanamine maleate (1:1).
Mopazine; Tentone; Vetomazine. Antipsychotic. A phenothiazine. mp = 141-145°; somewhat soluble in H_2O; more soluble in MeOH, $CHCl_3$, DMF; less soluble in EtOH, C_6H_6, Et_2O.

5807 Metiapine
5800-19-1
$C_{19}H_{21}N_3S$
2-Methyl-11-(4-methyl-1-piperazinyl)dibenzo[b,f][1,4]-thiazepine.
Antipsychotic. *Marion Merrell Dow Inc.*

5808 Metofenazate
388-51-2 6228
$C_{31}H_{36}ClN_3O_5S$
2-[4-[3-(2-Chlorophenothiazin-10-yl)propyl]-1-piperazinyl]ethyl 3,4,5-trimethoxybenzoate.
methophenazine; perphenazine 3,4,5-trimethoxybenzoate. Antipsychotic. A phenothiazine.

5809 Metofenazate Fumarate
522-23-6 6228
$C_{39}H_{44}ClN_3O_{13}S$
2-[4-[3-(2-Chlorophenothiazin-10-yl)propyl]-1-piperazinyl]ethyl 3,4,5-trimethoxybenzoate difumarate.
Frenolon. Antipsychotic. A phenothiazine. mp = 107-107°.

5810 Milenperone
59831-64-0 261-947-5
$C_{22}H_{23}ClFN_3O_2$
5-Chloro-1-[3-[4-(p-fluorobenzoyl)piperidino]propyl]-2-benzimidazolinone.
R-34009. Antipsychotic. *Janssen Pharm. Inc.*

5811 Milipertine
24360-55-2
$C_{24}H_{31}N_3O_3$
5,6-Dimethoxy-3-[2-[4-(o-methoxyphenyl)-1-piperazinyl]ethyl]-2-methylindole.
Win-18935. Antipsychotic. *Sterling Res. Labs.*

5812 Molindone
7416-34-4 6315
$C_{16}H_{24}N_2O_2$
3-Ethyl-6,7-dihydro-2-methyl-5-(morpholinomethyl)-indol-4(5H)-one.
Antipsychotic. An indole derivative. mp = 180-181°. *Endo Pharm. Inc.*

5813 Molindone Hydrochloride
15622-65-8 6315
$C_{16}H_{25}ClN_2O_2$
3-Ethyl-6,7-dihydro-2-methyl-5-(morpholinomethyl)-indol-4(5H)-one monohydrochloride.
EN-1733A; Lindone; Moban. Antipsychotic. An indole derivative. LD_{50} (rat orl) = 261 mg/kg. *Abbott Labs Inc.; DuPont-Merck Pharm.; Endo Pharm. Inc.*

5814 Moperone
1050-79-9 6343 213-887-6
$C_{22}H_{26}FNO_2$
4'-Fluoro-4-(4-hydroxy-4-p-tolylpiperidino)-butyrophenone.
R-1658. Antipsychotic. A butyrophenone. mp = 118-119.5°; λ_m =246.5 (ε 12200). *Janssen Pharm. Inc.*

5815 Moperone Hydrochloride
3871-82-7 6343 223-392-7
$C_{22}H_{27}ClFNO_2$
4'-Fluoro-4-(4-hydroxy-4-p-tolylpiperidino)-butyrophenone monohydrochloride.
Luvatren. Antipsychotic. A butyrophenone. mp = 216-218°. *Janssen Pharm. Inc.*

5816 Mosapramine
89419-57-7 6365
$C_{28}H_{35}ClN_4O$
(±)-1'-[3-(3-Chloro-10,11-dihydro-5H-dibenz-[b,f]azepine-5-yl)propyl]hexahydrospiro-[imidazo[1,2-a]pyridine-3(2H),4'-piperidin]-2-one.
clospipramine. Antipsychotic. A dopamine receptor antagonist with adrenoceptor blocking activity, metabolite of clocapramine, a tricyclic antipsychotic. *Yoshitomi.*

5817 Mosapramine Dihydrochloride
98043-60-8 6365
$C_{28}H_{37}Cl_3N_4O$
(±)-1'-[3-(3-Chloro-10,11-dihydro-5H-dibenz-[b,f]azepine-5-yl)propyl]hexahydrospiro-[imidazo[1,2-a]pyridine-3(2H),4'-piperidin]-2-one dihydrochloride.
Y-516; Cremin. Antipsychotic. A dopamine receptor antagonist with adrenoceptor blocking activity, metabolite of clocapramine, a tricyclic antipsychotic. mp = 271°; LD_{50} (mus orl) = 1008 mg/kg, (mus ip) = 74 mg/kg, (mus sc) = 1147 mg/kg. *Yoshitomi.*

5818 Naranol
22292-91-7
$C_{18}H_{21}NO_2$
8,9,10,11,11a,12-Hexahydro-8,10-dimethy-7aH-napthol[1',2':5:6]pyranol[3,2-c]pyridin-7a-ol.
Antipsychotic. *Parke-Davis.*

5819 Naranol Hydrochloride
34256-91-2
$C_{18}H_{22}ClNO_2$
8,9,10,11,11a,12-Hexahydro-8,10-dimethy-7aH-napthol[1',2':5:6]pyranol[3,2-c]pyridin-7a-ol hydrochloride.
W-5494A. Antipsychotic. *Parke-Davis.*

5820 Neflumozide
86636-93-3
$C_{22}H_{23}FN_4O_2$
1-[1-[3-(6-Fluoro-1,2-benzisoxazole-3-yl)propyl]-4-piperidyl]-2-benzimidazolinone.
Antipsychotic. *Hoechst Roussel Pharm. Inc.*

5821 Neflumozide Hydrochloride
86015-38-5
$C_{22}H_{24}ClFN_4O_2$
1-[1-[3-(6-Fluoro-1,2-benzisoxazole-3-yl)propyl]-4-piperidyl]-2-benzimidazolinone monohydrochloride.
HRP-913; P79-3913. Antipsychotic. *Hoechst Roussel Pharm. Inc.*

5822 Nemonapride
93664-94-9 6533
$C_{21}H_{26}ClN_3O_2$
(±)-cis-N-(1-Benzyl-2-methyl-3-pyrrolidinyl)-5-chloro-4-(methylamino)-o-anisamide.
emonapride; YM-09151-2; Emilace. Selective dopaminde D_2 receptor antagonist; bezamide antipsychotic. mp = 152-153°. *Yamanouchi U.S.A. Inc.*

5823 Nerisopam
102771-12-0
$C_{18}H_{19}N_3O_2$
1-(4-Aminophenyl)-4-methyl-7,8-dimethoxy-5H-2,3-benzodiazepine.
GYKI-52322. A psychoactive 5H-2,3-benzodiazepine with neuroleptic properties. Anxiolytic; antipsychotic.

5824 Ocaperidone
129029-23-8
$C_{24}H_{25}FN_4O_2$
3-[2-[4-(6-Fluoro-1,2-benzisoxazol-3-yl)piperidino]ethyl]-2,9-dimethyl-4H-pyrido[1,2-a]pyridimin-4-one.
Antipsychotic. A benzisoxazole.

5825 Olanzapine
132539-06-1 6959
$C_{17}H_{20}N_4S$
2-Methyl-4-(4-methyl-1-piperazinyl)-10H-thieno[2,3-b][1,5]benzodiazepine.
LY-170053; Lanzac. Antipsychotic. A serotonin and dopamine receptor antagonist with anticholinergic activity; a tricyclic antipsychotic. mp = 195°. *Eli Lilly & Co.*

5826 Opipramol
315-72-0 6985 206-254-0
$C_{23}H_{29}N_3O$
4-[3-(5H-Dibenz[b,f]azepin-5-yl)propyl]-1-piperazineethanol.
Antidepressant; antipsychotic. Tricyclic. mp = 100-101°. *Ciba-Geigy Corp.; Rhône-Poulenc Rorer Pharm. Inc.*

5827 Opipramol Hydrochloride
909-39-7 6985 213-000-2
$C_{23}H_{31}Cl_2N_3O$
4-[3-(5H-Dibenz[b,f]azepin-5-yl)propyl]-1-piperazineethanol dihydrochloride.
G-33040; Dinsidon; Ensidon; Insidon; Nisidana. Antidepressant; antipsychotic. Tricyclic. mp = 229-230°; soluble in H_2O, EtOH; sparingly soluble in Me_2CO. *Ciba-Geigy Corp.; Rhône-Poulenc Rorer Pharm. Inc.*

5828 Oxaflumazine
16498-21-8 7040
$C_{26}H_{32}F_3N_3O_2S$
10-[3-[4-(2-m-Dioxanylethyl)-1-piperazinyl]propyl]-2-(trifluoromethyl)phenothiazine.
Antipsychotic. A phenothiazine. *S.I.F.A.*

5829 Oxaflumazine Disuccinate
7450-97-7 7040 231-219-1
$C_{34}H_{44}F_3N_3O_{10}S$
10-[3-[4-(2-m-Dioxanylethyl)-1-piperazinyl]propyl]-2-(trifluoromethyl)phenothiazine disuccinate.
SD-270-31; Oxaflumine. Antipsychotic. A phenothiazine. mp = 136-138°; λ_m (MeOH) = 259, 310 nm (log ε 4.53, 3.58); LD_{50} (mus iv) = 94 mg/kg, (mus ip) = 175 mg/kg, (mus orl) = 919 mg/kg. *S.I.F.A.*

5830 Oxiperomide
5322-53-2
$C_{20}H_{23}N_3O_3$
1-[1-(2-Phenoxyethyl)-4-piperidyl]-2-benzimidazolinone.
R-4714. Antipsychotic. *Janssen Pharm. Ltd.*

5831 Panamesine
139225-22-2
$C_{23}H_{26}N_2O_6$
(5S)-(-)-[4-Hydroxy-4-(3,4-benzodioxol-5-yl)piperidin-1-ylmethyl]-3-(4-methoxyphenyl)-oxazolidin-2-one.
EMD-57445. High affinity sigma ligand; atypical antipsychotic.

5832 Penfluridol
26864-56-2 7213 248-074-5
$C_{28}H_{27}ClF_5NO$
1-[4,4-Bis(p-flurophenyl)butyl]-4-(4-chloro-α,α,α-trifluoro-m-tolyl)-4-piperidinol.
McN-JR-16341; R-16341; Semap. Antipsychotic. mp = 105-107°; slightly soluble in H_2O; LD_{50} (rat orl) = 87 mg/kg. *Abbott Laboratories Inc.; McNeil Pharm.*

5833 Pentiapine Maleate
81382-52-7
$C_{19}H_{21}N_5O_4S$
5-(4-Methyl-1-piperazinyl)imidazo[2,1-b][1,3,5]-benzothiadiazepine maleate (1:1).
CGS-10746B. Antipsychotic. *Ciba-Geigy Corp.*

5834 Perazine
84-97-9 7294 201-578-9
$C_{20}H_{25}N_3S$
10-[3-(4-Methyl-1-piperazinyl)propyl]-10H-phenothiazine.
P-725. Antipsychotic. A phenothiazine. mp = 51-53°; $bp_{0.001}$ = 160-170°. *Rhône-Poulenc Rorer Pharm. Inc.*

5835 Pericyazine
2622-26-6 7306 220-071-3
$C_{21}H_{23}N_3OS$
10-[3-(4-Hydroxypiperidino)propyl]-phenothiazine-2-carbonitrile.
periciazine; propericiazine; RP-8909; SKF-20716; Aolept; Nelactil; Neuleptil. Antipsychotic. A psychotherapeutic phenothiazine. mp = 116-117°; λ_m = 232.5, 271.5 (ε 4319, 4503); LD_{50} (rat orl) = 395 mg/kg. *Rhône-Poulenc Rorer Pharm. Inc.*

5836 Perimethazine
13093-88-4 7309 236-009-3
$C_{22}H_{28}N_2O_2S$
1-[3-(2-Methoxyphenothiazin-10-yl)-2-methylpropyl]-4-piperidinol.
perimetazine; AN-1317; RP-9159; Leptryl. Antipsychotic. A phenothiazine. mp = 137-138°. *Rhône-Poulenc Rorer Pharm. Inc.*

5837 Perospirone
150915-41-6
$C_{23}H_{30}N_4O_2S$
N-[4-[4-(1,2-Benzisothiazol-3-yl)-1-piperazinyl]butyl]-1,2-cis-cyclohexanedicarboximide.
SM-9018. A novel seratonin 5-HT2, dopamine D2 antagonist. Antipsychotic.

5838 Perphenazine
58-39-3 7323 200-381-5
$C_{21}H_{26}ClN_3OS$
4-[3-(2-Chlorophenothiazin-10H-phenothiazine-10-yl)propyl]-1-piperazineethanol.
chloriprozine; PZC; Sch-3940; Decentan; Fentazin; Perphenan; Trilafon; Trilifan; component of: Etrafon, Triavil. Antipsychotic. mp = 94-100°; $bp_{0.15}$ = 214-218°; insoluble in H_2O, soluble in organic solvents. *Schering AG.*

5839 Pimozide
2062-78-4 7589 218-171-7
$C_{28}H_{29}F_2N_3O$
1-[1-[4,4-Bis(p-Flurophenyl)butyl]-4-piperidyl]-2-benzimidazolinone.
McN-JR-6238; R-6238; Orap; Opiran. Antipsychotic. Has been used in treatment of schizophrenia and in management of Tourette's syndrome. mp = 214-218°; insoluble in H_2O (<0.0001 mg/100 ml); soluble in dilute acids (<0.05 mg/100 ml); weak base with pKa = 7.32. *Janssen Pharm. Ltd.; Lemmon Co.*

5840 Pinoxepin
14008-66-3
$C_{23}H_{27}ClN_2O_2$
(Z)-4-[3-(2-Chlorodibenz[b,e]oxepin-11-(6H)-ylidene)propyl]-1-piperazineethanol.
Antipsychotic. *Pfizer Inc.*

5841 Pinoxepin Hydrochloride
14008-46-9
$C_{23}H_{29}Cl_3N_2O_2$
(Z)-4-[3-(2-Chlorodibenz[b,e]oxepin-11-(6H)-ylidene)propyl]-1-piperazineethanol dihydrochloride.
P-5227. Antipsychotic. *Pfizer Inc.*

5842 Pipamperone
1893-33-0 7608
$C_{21}H_{30}FN_3O_2$
1'-[3-(p-Fluorobenzoyl)propyl][1,4'-bipiperidine]-4'-carboxamide.
R-3345; floropipamide. Antipsychotic. A buterophenone. *Abbott Laboratories Inc.*

5843 Pipamperone Hydrochloride
244-68-2 7608
$C_{21}H_{32}Cl_2FN_3O_2$
1'-[3-(p-Fluorobenzoyl)propyl][1,4'-bipiperidine]-4'-carboxamide dihydrochloride.
Dipiperon; Piperonil; Propitan. Antipsychotic. A buterophenone. mp = 124.5-126.0°. *Abbott Laboratories Inc.*

5844 Piperacetazine
3819-00-9 7615 223-312-0
$C_{24}H_{30}N_2O_2S$
1-[10-[3-[4-(2-Hydroxyethyl)-1-piperidinyl]propyl]-10H-phenothiazin-2-yl]ethanone.
PC-1421; Psymod; Quide. Antipsychotic. A phenothiazine. [hydrochloride]: mp = 100-110°. *Marion Merrell Dow Inc.; Searle G.D. & Co.*

5845 Pipotiazine
39860-99-6 7636 254-659-6
$C_{24}H_{33}N_3O_3S_2$
10-[3-[4-(2-Hydroxyethyl)piperidino]propyl]-N,N-dimethylphenothiazine-2-sulfonamide.
RP-19366; pipothiazine; Piportil. Antipsychotic. A phenothiazine. LD_{50} (mus ip) = 108 mg/kg, (mus sc) = 360 mg/kg, (mus orl) = 440 mg/kg. *Rhône-Poulenc Rorer Pharm. Inc.*

5846 Pipotiazine Palmitate
37517-26-3 7636 253-536-4
$C_{40}H_{63}N_3O_4S_2$
10-[3-[4-(2-Hydroxyethyl)piperidino]propyl]-N,N-dimethylphenothiazine-2-sulfonamide palmitate (ester).
RP-19552; IL-19552; pipotiazine palmitic ester; Piportil L4. Antipsychotic. A phenothiazine. *Rhône-Poulenc Rorer Pharm. Inc.*

5847 Pipotiazine Undecylenate
22178-11-6 7636 244-819-3
$C_{35}H_{51}N_3O_4S_2$
10-[3-[4-(2-Hydroxyethyl)piperidino]propyl]-N,N-dimethylphenothiazine-2-sulfonamide undecylenate (ester).
RP-19551; pipotiazine undecylenic ester; Piportil M2. Antipsychotic. A phenothiazine. *Rhône-Poulenc Rorer Pharm. Inc.*

5848 Piquindone
78541-97-6
$C_{15}H_{22}N_2O$
(±)-trans-3-Ethyl-1,4s,5,6,7,8,8a,9-octahydro-2,6-dimethyl-4H-pyrrolo[2,3-g]isoquinolin-4-one.
Antipsychotic. *Hoffmann-LaRoche Inc.*

5849 Piquindone Hydrochloride
83784-19-4
$C_{15}H_{27}ClN_2O_3$
(±)-trans-3-Ethyl-1,4s,5,6,7,8,8a,9-octahydro-2,6-dimethyl-4H-pyrrolo[2,3-g]isoquinolin-4-one monohydrochloride dihydrate.
Ro-22-1319/003. Antipsychotic. *Hoffmann-LaRoche Inc.*

5850 Preclamol
85966-89-8
$C_{14}H_{21}NO$
(-)-(S)-m-(1-Propyl-3-piperidyl)phenol.
3-PPP. Partial dopamine autoreceptor agonist. Antipsychotic.

5851 Prochlorperazine
58-38-8 7942 200-379-4
$C_{20}H_{24}ClN_3S$
2-Chloro-10-[3-(4-methyl-1-piperazinyl)-propyl]phenothiazine.
chlormeprazine; prochlorpemazine; prochlorperazine; Bayer A173; RP-6140; SKF-4657; Compazine; Eskatrol. Antiemetic; antipsychotic. A phenothiazine, used in treatment of vertigo. *Bayer AG; Rhône-Poulenc Rorer Pharm. Inc.; SmithKline Beecham Pharm.*

5852 Prochlorperazine Edisylate
1257-78-9 7942 215-019-1
$C_{22}H_{32}ClN_3O_6S_3$
2-Chloro-10-[3-(4-methyl-1-piperazinyl)-propyl]phenothiazine 1,2-ethanedisulfonate (1:1).
prochlorperazine ethanedisulfonate; Novamin; Tementil; Compazine Injection; Compazine Syrup. Antiemetic; antipsychotic. A phenothiazine, used in treatment of vertigo. *Rhône-Poulenc Rorer Pharm. Inc.; SmithKline Beecham Pharm.; Wyeth-Ayerst Labs.*

5853 Prochlorperazine Maleate
84-02-6 7942 201-511-3
$C_{28}H_{32}ClN_3O_8S$
2-Chloro-10-[3-(4-methyl-1-piperazinyl)propyl]-phenothiazine maleate (1:2).
Buccastem; Compazine; Meterazine; Stemetil; Vertigon; component of: Combid. Antiemetic; antipsychotic. A phenothiazine, used in treatment of vertigo. mp = 228°; slightly soluble in H_2O (0.1 g/100 ml), MeOH, EtOH; insoluble in C_6H_6, Et_2O, $CHCl_3$; LD_{50} (mus sc) = 400 mg/kg, (mus ip) = 120 mg/kg, (mus iv) = 90 mg/kg, (mus orl) = 400 mg/kg. *Apothecon; Rhône-Poulenc Rorer Pharm. Inc.; SmithKline Beecham Pharm.*

5854 Prochlorperazine Mesylate
51888-09-6 7942 257-495-3
$C_{22}H_{30}ClN_3)_6S_3$
2-Chloro-10-[3-(4-methyl-1-piperazinyl)propyl]-phenothiazine methanesulfonate (2:1).
prochlorperazine dimethanesulfonate. Antiemetic; antipsychotic. A phenothiazine, used in treatment of vertigo. *Rhône-Poulenc Rorer Pharm. Inc.; SmithKline Beecham Pharm.*

5855 Promazine
58-40-2 7966 200-382-0
$C_{17}H_{20}N_2S$
10-[3-(Dimethylamino)propyl]phenothiazine.
RP-3276; Wy-1094. Antipsychotic; tranquilizer (veterinary). A phenothiazine. Oily liquid; $bp_{0.3}$ = 203-210°. *Rhône-Poulenc Rorer Pharm. Inc.*

5856 Promazine Hydrochloride
53-60-1 7966 200-179-7
$C_{17}H_{21}ClN_2S$
10-[3-(Dimethylamino)propyl]phenothiazine monohydrochloride.
Liranol; Promwill; Prazine; Prolactyl; Sparine; Talofen; component of: Ketaset Plus (veterinary). Antipsychotic; tranquilizer (veterinary). A phenothiazine. Dec 181°; soluble in H_2O, EtOH, $CHCl_3$, practically insoluble in Et_2O, C_6H_6. *DuPont-Merck Pharm.; Forest Pharm. Inc.; Rhône-Poulenc Rorer Pharm. Inc.; Wyeth-Ayerst Labs.*

5857 Prothipendyl
303-69-5 8073
$C_{16}H_{19}N_3S$
N,N-Dimethyl-10H-pyrido[3,2-b][1,4]benzo-thiazine-10-propanamine.
Antipsychotic. A tricyclic. $bp_{0.7}$ = 217-219°; $bp_{0.5}$ = 195-198°. *Olin Res. Ctr.*

5858 Prothipendyl Hydrochloride
1225-65-6 8073 214-958-4
$C_{16}H_{20}ClN_3S$
N,N-Dimethyl-10H-pyrido[3,2-b][1,4]benzo-thiazine-10-propanamine mono-hydrochloride.
D-206; Dominal; Tolnate. Antipsychotic. A tricyclic. mp = 108-112°; freely soluble in H_2O, MeOH; practically insoluble in Et_2O. *Olin Res. Ctr.*

5859 Prothixene
2622-24-4
$C_{18}H_{19}NS$
N,N-Dimethylthioxanthene-$\Delta^{9,\gamma}$-propylamine.
prooxin. Ataraxic; tranquilizer; antipsychotic. LD_{50} (mus iv) = 280 mg/kg.

5860 Psilocybin
520-52-5 8111 208-294-4
$C_{12}H_{17}N_2O_4P$
3-(2-Dimethylaminoethyl)indol-4-yl dihydrogen phosphate.
psylocybin; psilocybine; indocybin; CY-39. Psychomimetic. The major component of Tenanacatl, the mushroom *Psilocybe mexicana Heim, Agaricacea*. A federally controlled substance (hallucinogen). mp = 220-228°, 185-195°; λ_m = 220, 267, 290 nm (Log ε 4.6, 3.8, 3.6 MeOH); pH (50% aqueous EtOH) = 5.2; somewhat soluble in hot H_2O, MeOH; barely soluble in EtOH; nearly insoluble in C_6H_6, $CHCl_3$. *Sandoz Pharm. Corp.*

5861 Remoxipride
80125-14-0 8301
$C_{16}H_{23}BrN_2O_3$
(-)-(S)-3-Bromo-N-[(1-ethyl-2-pyrrolidinyl)methyl]-2,6-dimethoxybenzamide.
Antipsychotic. A specific dopamine D_2-receptor antagonist; a benzamide antipsychotic. $[\alpha]_D^{20}$ = -64° (c = 2 EtOH). *Astra USA Inc.*

5862 Remoxipride Hydrochloride Monohydrate
82935-42-0 8301
$C_{16}H_{26}BrClN_2O_4$
(-)-(S)-3-Bromo-N-[(1-ethyl-2-pyrrolidinyl)methyl]-2,6-dimethoxybenzamide monohydrochloride monohydrate.
A-33547; FLA-731; Roxiam. Antipsychotic. A specific dopamine D_2-receptor antagonist; a benzamide antipsychotic. Loss of H_2O at 105°; mp = 107°; $[\alpha]_D^{20}$ = -11° (c = 2 in H_2O); λ_m (H_2O) = 286 nm (ε 2280); soluble in H_2O, EtOH, CH_2Cl_2, Me_2CO; LD_{50} (rat ip) = 338 mg/kg. *Astra USA Inc.*

5863 Rilapine
79781-95-6
$C_{22}H_{20}ClN_3$
(Z)-2-Chloro-10-(4-methyl-1-piperazinyl)-5H-dibenzo[a,d]cycloheptene-$\Delta^{5,\alpha}$-acetonitrile.
Has high affinity for the serotonin 5-HT1C receptor. Neuroleptic; atypical antipsychotic.

5864 Rimcazole
75859-04-0
$C_{21}H_{27}N_3$
9-[3-cis-3,5-Dimethyl-1-piperazinyl)propyl]-carbazole.
Antipsychotic.

5865 Rimcazole Hydrochloride
75859-03-9
$C_{21}H_{29}Cl_2N_3$
9-[3-cis-3,5-Dimethyl-1-piperazinyl)propyl]-carbazole dihydrochloride.
BW-234U. Antipsychotic.

5866 Risperidone
106266-06-2 8397
$C_{23}H_{27}FN_4O_2$
3-[2-[4-(6-Fluoro-1,2-benzisoxazol-3-yl)piperidino]ethyl]-6,7,8,9-tetrahydro-2-methyl-4H-pyrido[1,2-a]pyridimin-4-one.
R-64766; Risperdal. Antipsychotic. A combined serotonin and dopamine receptor antagonist, a benzisoxazole antipsychotic. mp = 170°; LD_{50} (rat iv) = 29.7 mg/kg, (dog iv) = 14.1 mg/kg, (rat orl), 82.1= mg/kg, (dog orl) = 18.3 mg/kg. *Janssen Pharm. Ltd.*

5867 Seperidol
10457-91-7
$C_{22}H_{22}ClF_4NO_2$
4-[4-(4-Chloro-α,α,α-trifluoro-m-tolyl)-4-hydroxypiperidinol-4'-fluorobutyrophenone.
Antipsychotic. *Abbott Laboratories Inc.*

5868 Seperidol Hydrochloride
17230-87-4
$C_{22}H_{23}Cl_2F_4NO_2$
4-[4-(4-Chloro-α,α,α-trifluoro-m-tolyl)-4-hydroxypiperidinol-4'-fluorobutyrophenone hydrochloride.
R-9298. Antipsychotic. *Abbott Laboratories Inc.*

5869 Setoperone
86487-64-1
$C_{21}H_{24}FN_3O_2S$
6-[2-[4-p-Fluorobenzoyl)piperidinyl]ethyl]-2,3-dihydro-7-methyl-5H-thiazolo[3,2-α]pyrimidin-5-one.
R-52245. Antipsychotic. *Abbott Laboratories Inc.*

5870 Spiperone
749-02-0 8903 212-024-0
$C_{23}H_{26}FN_3O_2$
8-[3-(p-Fluorobenzoyl)propyl-1-phenyl-1,3,8-triazaspiro[4.5]decan-4-one.
R-5147; Spiropitan. Antipsychotic. A butyrophenone. mp = 190-193.6°. *Abbott Laboratories Inc.*

5871 Sulforidazine
14759-06-9 9137 238-818-7
$C_{21}H_{26}N_2O_2S_2$
10-[2-(1-Methyl-2-piperidinyl)ethyl]-2-methylsulfonylphenothiazine.
TPN-12; thioridazine-2-sulfone; Imagotan; Inofal. Antipsychotic. A dopamine receptor blocker; metabolite of thioridazine and mesoridazine; a phenothiazine antipsychotic. mp = 121-123°. *Sandoz Pharm. Corp.*

5872 Sultopride
53583-79-2 9168 258-641-9
$C_{17}H_{26}N_2O_4S$
N-[(1-Ethyl-2-pyrrolidiny)methyl]-5-(ethylsulfonyl)-o-anisamide.
Antipsychotic. A dopamine D_2-receptor antagonist; a benzamide antipsychotic.

5873 Sultopride Hydrochloride
23694-17-9 9168 245-829-0
$C_{17}H_{27}ClN_2O_4S$
N-[(1-Ethyl-2-pyrrolidiny)methyl]-5-(ethylsulfonyl)-o-anisamide monohydrochloride.
LIN-1418; Barnetil; Barnotil. Antipsychotic. A dopamine D_2-receptor antagonist; a benzamide antipsychotic. mp = 181-182°.

5874 Teflutixol
55837-23-5
$C_{23}H_{26}F_4N_2OS$
4-[3-[6-Fluoro-2-(trifluoromethyl)-9H-thioxanthen-9-yl]propyl]-1-piperazineethanol.
Lu-10-022. Thioxanthene derivative. Antipsychotic.

5875 Tenilapine
82650-83-7
$C_{17}H_{16}N_4S_2$
(E)-5-(4-Methyl-1-piperazinyl)-9H-dithieno[3,4-b:3',4'-3]-azepine-$\Delta^{9,\alpha}$-acetonitrile.
An atypical neuroleptic. Antipsychotic.

5876 Tetrabenazine
58-46-8 9325 200-383-6
$C_{19}H_{27}NO_3$
1,3,4,6,7,11b-Hexahydro-3-isobutyl-9,10-dimethoxy-2H-benzo[a]quinolizin-2-one.
Ro-1-9569. Antidyskinetic and antipsychotic. A dopamine depleting agent; a tricyclic. Studied for treatment of hyperdyskinetic disorders. mp = 125-126°; [hydrochloride]: mp = 208-210°; λm = 230, 284 nm (ε 7780, 3920); soluble in hot H_2O; nearly insoluble in Et_2O; [oxime]: mp = 158°. *Hoffmann-LaRoche Inc.*

5877 Tetrabenazine Methanesulfonate
804-53-5 9325
$C_{19}H_{27}NO_3$
1,3,4,6,7,11b-Hexahydro-3-isobutyl-9,10-dimethoxy-2H-benzo[a]quinolizin-2-one methanesulfonate.
Nitoman. Antidyskinetic; antipsychotic. A dopamine depleting agent; a tricyclic. mp = 126-130°; soluble in EtOH, sparingly soluble in H_2O, nearly insoluble in Et_2O. *Hoffmann-LaRoche Inc.*

5878 Thiopropazate
84-06-0 9493 201-513-4
$C_{23}H_{28}ClN_3O_2S$
4-[3-(2-Chlorophenothiazin-10-yl)propyl]-1-piperazineethanol acetate.
Antipsychotic. A phenothiazine. $bp_{0.1}$ = 214-218°; soluble in Et_2O.

5879 Thiopropazate Hydrochloride
146-28-1 9493 205-666-8
$C_{23}H_{30}Cl_3N_3O_2S$
4-[3-(2-Chlorophenothiazin-10-yl)propyl]-1-piperazineethanol acetate dihydrochloride.
Dartal; Dartalan. Antipsychotic. A phenothiazine. Dec 223-229°; freely soluble in H_2O; less soluble in organics.

5880 Thioproperazine
316-81-4 9494 206-262-4
$C_{22}H_{30}N_4O_2S_2$
N,N-Dimethyl-10-[3-(4-methyl-1-piperazinyl)propyl]phenothiazine-2-sulfonamide.
RP-7843; SKF-5883. Antipsychotic; antiemetic. A phenothiazine. mp = 140°; [fumarate]: mp = 182°. *Rhône-Poulenc; SmithKline Beecham Pharm.*

5881 Thioproperazine Mesylate
2347-80-0 9494 219-074-2
$C_{24}H_{38}N_4O_8S_4$
N,N-Dimethyl-10-[3-(4-methyl-1-piperazinyl)propyl]phenothiazine-2-sulfonamide dimethanesulfonate.
thioperazine dimethanesulfonate; Majeptil; Vontil. Antipsychotic; antiemetic. A phenothiazine. *Rhône-Poulenc; SmithKline Beecham Pharm.*

5882 Thioridazine
50-52-2 9497 200-044-2
$C_{21}H_{26}N_2S_2$
10-[2-(1-Methyl-2-piperidyl)ethyl]-2-(methylthio)phenothiazine.
Mellaril-S. Antipsychotic; sedative. A dopamine receptor blocker; parent compound of sulforidazine and mesoridazine. A phenothiazine derivative. mp = 72-74°; $bp_{0.02}$ - 230°; λ_m 263, 314 nm (ε 38172, 4595 EtOH), 230, 263 nm (ε 20939, 45954 0.1N HCl), 313 nm (ε 5226 0.1N NaOH); insoluble in H_2O, soluble in EtOH 16.6 g/100 ml), $CHCl_3$ 123 g/100 ml), Et_2O (33.3 g/100 ml); L. *Sandoz Pharm. Corp.*

5883 Thioridazine Hydrochloride
130-61-0 9497 204-992-8
$C_{21}H_{27}ClN_2S_2$
10-[2-(1-Methyl-2-piperidyl)ethyl]-2-(methylthio)phenothiazine monohydrochloride.
TP-21; Aldazine; Mallorol; Mellaril; Melleretten; Melleril; Novoridazine; Orsanil; Ridazin; Stalleril. Antipsychotic, used also as a sedative/hypnotic. A phenothiazine dopamine receptor blocker. mp = 158-160°; λ_m = 262, 310 nm (ε 41842, 3215 H_2O), 264, 310 nm (ε 41598, 3256 EtOH), 264 305 nm (ε 42371 5495 0.1N HCl), 263 nm (ε 18392 0.1N NaOH); soluble in H_2O (11.1 g/100 ml), EtOH (10 g/100 ml), MeOH, $CHCl_3$ (20 g/100 ml). *Sandoz Pharm. Corp.*

5884 Thiothixene
5591-45-7 9503 227-001-0
$C_{23}H_{29}N_3O_2S_2$
N,N-Dimethyl-9-[3-(4-methyl-1-piperazinyl)-propylidene]thioxanthene-2-sulfonamide.
P-4657B; NSC-108165; tiotixene; Navane; Orbinamon. Antipsychotic. Available as the *cis isomer, the trans* isomer or the mixture of isomers; a thioxanthene. [cis isomer]: mp = 147-149°; λ_m = 228, 260, 310 nm (log ε 4.6, 4.3, 3.9 MeOH); more potent than the rans isomer; LD_{50} (rat ip) = 55 mg/kg; [trans isomer]: mp = 123-125°; λ_m = 229, 252, 301 nm (log ε 4.5, 4.2, 3.9 MeOH); LD_5. *Lemmon Co.; Pfizer Inc.; Roerig Div. Pfizer Pharm.*

5885 Thiothixene [z]
3313-26-6 9503
$C_{23}H_{29}N_3O_2S_2$
(Z)-N,N-Dimethyl-9-[3-(4-methyl-1-piperazinyl)-propylidene]thioxanthene-2-sulfonamide.
Antipsychotic. A thioxanthene. *Roerig Div. Pfizer Pharm.*

5886 Thiothixene Hydrochloride
22189-31-7 9503
$C_{23}H_{35}Cl_2N_3O_4S_2$
N,N-Dimethyl-9-[3-(4-methyl-1-piperazinyl)-propylidene]thioxanthene-2-sulfonamide dihydrochloride dihydrate.
CP-12252-1; Navane Hydrochloride. Antipsychotic. A thioxanthene. *Pfizer Inc.; Roerig Div. Pfizer Pharm.*

5887 Thiothixene Hydrochloride [anhydrous (z)]
49746-04-5 9503
$C_{23}H_{31}Cl_2N_3O_2S_2$
(Z)-N,N-Dimethyl-9-[3-(4-methyl-1-piperazinyl)-propylidene]thioxanthene-2-sulfonamide dihydrochloride.
Antipsychotic. A thioxanthene. *Pfizer Inc.*

5888 Thiothixene Hydrochloride [anhydrous]
58513-59-0 9503
$C_{23}H_{31}Cl_2N_3O_2S_2$
N,N-Dimethyl-9-[3-(4-methyl-1-piperazinyl)-propylidene]thioxanthene-2-sulfonamide dihydrochloride.
Antipsychotic. A thioxanthene. *Pfizer Inc.*

5889 Thiothixene Hydrochloride [z]
49746-09-0 9503
$C_{23}H_{35}Cl_2N_3O_4S_2$
(Z)-N,N-Dimethyl-9-[3-(4-methyl-1-piperazinyl)-propylidene]thioxanthene-2-sulfonamide dihydrochloride dihydrate.
Antipsychotic. A thioxanthene. *Pfizer Inc.*

5890 Timelotem
96306-34-2
$C_{17}H_{18}FN_3S$
(±)-10-Fluoro-1,2,3,4,4a,5-hexahydro-3-methyl-7-(2-thienyl)pyrazino[1,2-a][1,4]benzodiazepine.
An atypical neuroleptic. Antipsychotic.

5891 Timeperone
57648-21-2 9584 260-880-9
$C_{22}H_{24}FN_3OS$
4-Fluoro-4-[4-(2-thioxo-1-benzimidazolinyl)piperidino]-butyrophenone.

DD-3480; Tolopelon. Antipsychotic. A butyrophenone derivative with neuroleptic activity. mp = 201-203°; λ_m (EtOH) = 226.5, 246, 309 nm; slightly soluble in H_2O; LD_{50} (rat orl) = 232 mg/kg, (mus orl) = 478 mg/kg. *Daiichi Pharm. Corp.*

5892 Tioperidone
52618-67-4
$C_{25}H_{32}N_4O_2S$
3-[4-[4-[o-(Propylthio)phenyl]-1-piperazinyl]butyl]-2,4(1H,3H)-quinazolinedione.
Antipsychotic. *Parke-Davis.*

5893 Tioperidone Hydrochloride
52618-68-5
$C_{25}H_{33}ClN_4O_2S$
3-[4-[4-[o-(Propylthio)phenyl]-1-piperazinyl]butyl]-2,4(1H,3H)-quinazolinedione monohydrochloride.
CI-787. Antipsychotic. *Parke-Davis.*

5894 Tiospirone
87691-91-6
$C_{24}H_{32}N_4O_2S$
N-[4-[4-(1,2-Benzisothiazol-3-yl)-1-piperazinyl]butyl]-1,1-cyclopentanediacetimide.
Antipsychotic.

5895 Tiospirone Hydrochloride
87691-92-7
$C_{24}H_{33}ClN_4O_2S$
N-[4-[4-(1,2-Benzisothiazol-3-yl)-1-piperazinyl]butyl]-1,1-cyclopentanediacetimide monohydrochloride.
BMY-13859-1. Antipsychotic.

5896 Trifluoperazine
117-89-5 9811 204-219-4
$C_{21}H_{24}F_3N_3S$
10-[3-(4-Methyl-1-piperazinyl)propyl]-2-(trifluoromethyl)phenothiazine.
Antipsychotic, used also as a sedative/hypnotic. A phenothiazine. $bp_{0.6}$ = 202-210°; λ_m 258, 307.5 (log ε 4.50 3.50 EtOH); LD_{50} (rat orl) = 542.7 mg/kg, (mus orl) = 424.0 mg/kg. *Apothecon; SmithKline Beecham Pharm.*

5897 Trifluoperazine Dihydrochloride
440-17-5 9811 207-123-0
$C_{21}H_{26}Cl_2F_3N_3S$
10-[3-(4-Methyl-1-piperazinyl)propyl]-2-(trifluoromethyl)phenothiazine dihydrochloride.
triftazin; triphthasine; Eskazinyl; Eskazine; Jatroneural; Modalina; Stelazine; Terfluzine. Antipsychotic, used also as a sedative/hypnotic. A phenothiazine. mp = 242-243°; freely soluble in H_2O, insoluble in Et_2O, C_6H_6; hygroscopic. *Labaz S.A.; SmithKline Beecham Pharm.*

5898 Trifluperidol
749-13-3 9813 749-13-3
$C_{22}H_{23}F_4NO_2$
4'-Fluoro-4-[4-hydroxy-4-(α,α,α-trifluoro-m-tolyl)piperidino]butyrophenone.
McN-JR-2498; flumoperone. Antipsychotic. A butyrophenone. *Abbott Laboratories Inc.; McNeil Pharm.*

5899 Trifluperidol Hydrochloride
2062-77-3 9813 218-170-1
$C_{22}H_{24}ClF_4NO_2$
4'-Fluoro-4-[4-hydroxy-4-(α,α,α-trifluoro-m-tolyl)piperidino]butyrophenone monohydrochloride.
R-2489; Psicoperidol; Psychoperidol; Triperidol. Antipsychotic. A butyrophenone. mp = 200-201°; soluble in H_2O; LD_{50} (rat sc) = 70 mg/kg. *Abbott Laboratories Inc.; McNeil Pharm.*

5900 Triflupromazine
146-54-3 9814 205-673-6
$C_{18}H_{19}F_3N_2S$
10-[3-(Dimethylamino)propyl]-2-(trifluoromethyl)-phenothiazine.
fluopromazine. Antipsychotic. A phenothiazine. $bp_{0.7}$ = 162-164°; n_D^{23} = 1.580. *SmithKline Beecham Pharm.*

5901 Triflupromazine Hydrochloride
445-54-3 9814 214-149-6
$C_{18}H_{20}ClF_3N_2S$
10-[3-(Dimethylamino)propyl]-2-(trifluoromethyl)-phenothiazine monohydrochloride.
fluopromazine; Adazine; Fluorofen; Psyquil; Syquil; Vespral; Vesprin; Vetame. Antipsychotic. A phenothiazine. Dec 173-174°; λ_m = 255, 305 nm (E 700, 90); soluble in H_2O, EtOH, Me_2CO. *SmithKline Beecham Pharm.*

5902 Zetidoline
51940-78-4
$C_{16}H_{22}ClN_3O$
1-(m-Chlorophenyl)-3-[2-(3,3-dimethyl-1-azetidinyl)ethyl]-2-imidazolidinone.
DL-308-IT. Specific dopamine D2 receptor antagonist. Antipsychotic.

5903 Ziprasidone
146939-27-7 10304
$C_{21}H_{21}ClN_4OS$
5-[2-[4-(1,2-Benzisothiazol-3-yl)-1-piperazinyl]ethyl]-6-chloro-1,3-dihydro-2H-indol-2-one.
CP-88059. Antipsychotic. A dopamine D_2 and serotonin 5-HT_2 antagonist. *Pfizer Inc.*

5904 Ziprasidone Hydrochloride Monohydrate
138982-67-9 10304
$C_{21}H_{24}Cl_2N_4O_2S$
5-[2-[4-(1,2-Benzisothiazol-3-yl)-1-piperazinyl]ethyl]-6-chloro-1,3-dihydro-2H-indol-2-one monohydrochloride monohydrate.
CP-88059-1. Antipsychotic. A dopamine D_2 and serotonin 5-HT_2 antagonist. [hemihydrate]: mp > 300°. *Pfizer Inc.*

5905 Zotepine
26615-21-4 10327
$C_{18}H_{18}ClNOS$
2-[(8-Chlorodibenzo[b,f]thiepin-10-yl)oxy]-N,N-dimethylethylamine.
Lodopin; Nipolept. Antipsychotic. A tricyclic enol-ether compound with psychotropic and neuroleptic activity. mp = 90-91°; λ_m (95% EtOH) = 266 nm; LD_{50} (mus orl) = 108 mg/kg, (rat orl) = 458 mg/kg, (mus iv) = 43.3 mg/kg,

(mus ip)= 40.0 mg/kg, (mus sc) = 84.9 mg/kg. *Fujisawa Pharm. USA Inc.*

Antipyretics

5906 Acetaminophen

103-90-2 45 203-157-5

$C_8H_9NO_2$

4'-Hydroxyacetanilide.

p-hydroxyacetanilide; p-acetamidophenol; p-aceaminophenol; N-acetyl-p-aminophenol; paracetamol; Abensanil; Acamol; Acetalgin; Alpiny; Amadil; Anaflon; Anhiba; Apamide; APAP; Banesin; Ben-u-ron; Bickie-mol; Calpol; Captin; Claratal; Cetadol; Dafalgan; Datril; Dirox; Disprol; Doliprane; Dolprone; Dymadon; Enelfa; Eneril; Eu-Med; Exdol; Febrilex; Finimal; Gelocatil; Hedex; Homoolan; Korum; Momentum; Naprinol; Nebs; Nobedon; Ortensan; Pacemol; Paldesic; Panadol; Panaleve; Panasorb; Panets; Panodil; Parelan; Paraspen; Parmol; Pasolind N; Phenaphen; Salzone; Tabalgin; Tapar; Tempra; Tralgon; Tylenol; Valadol; component of: Actifed Plus, Allerest Sinus Pain Formula, Anexsia, Aspirin-Free Anacin, Children's Tylenol Cold Tablets, Contac Cough & Sore Throat Formula, Contac Jr Non-drowsy Formula, Contac Nighttime Cold Medicine, Contac Severe Cold Formula, Coricidin, Darvoset-N, Dristan Cold (Multisymptom), Dristan Cold (No Drowsiness), Empracet, Endecon, Gemnisyn, Headache Strength Allerest, Hycomine Compound, Hy-Phen, Intensin, Liquiprin, Maximum Strength Sine-Aid, Midol, Midol Maximum Strength, Midol PMS, Naldegesic, Naldetuss, Ornex, Percocet, Percogesic with Codeine, Propacet, Quiet World, Rhinex D-Lay Tablets, Sinarest, Sine-Off Maxiumum Strength Allergy/Sinus, Sine-Off Maximum Strength No Drowsiness Formula Caplets, Sinubid, St Joseph's Cold Tablets for Children, Sudafed Sinus, Supac, Teen Midol, TheraFlu, Tylenol Allergy Sinus, Tylenol Cold and Flu Multi-Symptom, Tylenol Cold Medication Caplets, Liquid and Tablets, Tylenol Cold Night TIme Liquid, Tylenol Cold No Drowsiness, Tylenol PM Tablets and Caplets, Tylenol with Codeine, Tylox, Vanquish, Vicodin, Wygesic, Zydone . Analgesic; antipyretic; anti-inflammatory. mp = 169-170.5°; d_4^{21} = 1.293; λ_m = 250 nm (ε 13800 in EtOH); slightly soluble in cold H_2O, Et_2O; more soluble in hot H_2O; soluble in alcohol, dimethylformamide, ethylene dichloride, Me_2CO, EtOAc; nearly insoluble in petroleum ether, pentane, C_6H_6; LD_{50} (mus orl) = 338 mg/kg, (mus ip) = 500 mg/kg. *Bristol-Myers Squibb Co.; Forest Pharm. Inc.; McNeil Pharm.; Mead Johnson Nutritionals; Novo Nordisk Pharm. Inc.; Parke-Davis; Robins, A. H. Co.; Sterling Health U.S.A.; Warner-Lambert.*

5907 Acetaminosalol

118-57-0 46 204-261-3

$C_{15}H_{13}NO_4$

4'-Hydroxyacetanilide salicylate.

p-acetamidophenyl salicylate; p-acetylaminophenol salicylic acid ester; phenestal; Phenosal; Salophen. Analgesic; antipyretic; anti-inflammatory. mp = 187°; nearly insoluble in petroleum ether, cold H_2O; more soluble in warm H_2O; soluble in alcohol, Et_2O, C_6H_6; incompatible with alkalies and alkaline solutions.

5908 Acetanilide

103-84-4 47 203-150-7

C_8H_9NO

N-Phenylacetamide.

antifebrin; acetylaniline; acetylaminobenzene. Antipyretic; analgesic. mp = 113-115°; bp = 304-305°; d_4^{15}= 1.219; pK (28°) = 13.0; soluble in H_2O, alcohol, $CHCl_3$, Me_2CO, glycerol, dioxane, Et_2O, C_6H_6; nearly insoluble in petroleum ether; LD_{50} (rat ig) = 800 mg/kg.

5909 Aconite

8063-12-5 118

Monkshood; Wolf's Bane; Friar's Cowl; Mouse Bane. Antipyretic. Dried tuberous root of *Aconitum napellus L. Ranunculaceae.* Contains aconine, napelline, picraconitine; aconitic acid, itaconic acid, succinic and malonic acids, levulose and fats. Quite toxic. Used as antipyretic.

5910 Alclofenac

22131-79-9 220 244-795-4

$C_{11}H_{11}ClO_3$

3-Chloro-4-(2-propenyloxy)benzeneacetic acid.

W-7320; Allopydin; Epinal; Medifenac; Mervan; Neoston; Prinalgin; Reufenac; Zumaril. Analgesic; antipyretic; anti-inflammatory. Arylacetic acid derivative. mp = 92-93°; LD_{50} (rat orl) = 1050 mg/kg, (rat sc) = 600 mg/kg, (rat ip) = 555 mg/kg. *Continental Pharma Inc.*

5911 Aluminum Bis(acetylsalicylate)

23414-80-1 338

$C_{18}H_{15}AlO_9$

Bis[2-(acetyloxy)benzoato-O']hydroxyaluminum.

aluminum diacetylsalicylate; acetylsalicylate aluminum; aluminum diaspirin; aluminum aspirin; monohydroxyaluminum diacetylsalicylate; monohydroxyaluminum bis(acetylsalicylate); Rumasal. Analgesic; antipyretic. *See* dihydroxyaluminum acetylsalicylate. Nearly insoluble in H_2O, alcohol, Et_2O; decomposes in dilute acids or alkalies and alkali carbonates.

5912 Aminochlorthenoxazin

3567-76-8 455

$C_{12}H_{11}ClN_2O_2$

6-Amino-2-(2-chloroethyl)-2,3-dihydro-4H-1,3-benzoxazin-4-one.

component of: Dereuma. Antipyretic; analgesic. mp = 164°; nearly insoluble in H_2O; LD_{50} (mus ip) 1950 mg/kg, (mus orl) 10000 mg/kg.

5913 Aminochlorthenoxazin Hydrochloride

3443-15-0 455

$C_{12}H_{12}Cl_2N_2O_2$

6-Amino-2-(2-chloroethyl)-2,3-dihydro-4H-1,3-benzoxazin-4-one monohydrochloride.

A-350; ICI-350. Antipyretic; analgesic. mp = 209-210°; soluble in H_2O; LD_{50} (mus ip) 920 mg/kg, (mus orl) 2250 mg/kg, (mus iv) 290.

5914 Aminophenazone Cyclamate

747-30-8 495

$C_{20}H_{30}N_4O_4S$

4-Dimethylamino-2,3-dimethyl-1-phenyl-3-pyrazolin-5-one cyclohexylsulfamate.

aminopyrine cyclohexylsulfamate. Analgesic; antipyretic. *See* aminopyrine.

5915 Aminopyrine
58-15-1 495 200-365-8
$C_{13}H_{17}N_3O$
4-Dimethylamino-2,3-dimethyl-1-phenyl-3-pyrazolin-5-one.
amidopyrine; aminophenazone; dimethylamino-phenyldimethylpyrazone; Amidofebrin; Amidopyrazoline; Anafebrina; Brufaneuxol; Dimapyrin; Dipirin; Febrinina; Itamidone; Mamallet-A; Netsusarin; Novamidon; Piridol; Polinalin; Pyradone; Pyramidon. Analgesic; antipyretic. CAUTION: may cause agranulocytosis. mp = 107-109°; soluble in alcohol (1 g/1.5 ml), C_6H_6 (1 g/12 ml), $CHCl_3$ (1 g/1 ml), Et_2O (1g/13 ml), H_2O (1g/18 ml); LD_{50} (rat orl) = 1700 mg/kg.

5916 Aminopyrine Bicamphorate
94442-12-3 495
$C_{33}H_{49}N_3O_9$
4-Dimethylamino-2,3-dimethyl-1-phenyl-3-pyrazolin-5-one bicamphorate.
Pyramidon bicamphorate. Analgesic; antipyretic. mp = 94°; soluble in H_2O with gradual decomposition; soluble in alcohol.

5917 Aminopyrine Salicylate
603-57-6 495 210-049-1
$C_{20}H_{23}N_3O_4$
4-Dimethylamino-2,3-dimethyl-1-phenyl-3-pyrazolin-5-one salicylate.
Pyramidon salicylate. Analgesic; antipyretic. mp = 70°; soluble in H_2O (1 g/16 ml), alcohol (1 g/6 ml).

5918 Antipyrine Acetylsalicylate
569-84-6 757
$C_{20}H_{20}N_2O_5$
2,3-Dimethyl-1-phenyl-3-pyrazolin-5-one acetylsalicylate.
Acetopyrine; Acopyrine; Acetasol. Analgesic. Has been used as an antipyretic and analgesic in veterinary medicine. mp = 63-65°; soluble in hot H_2O, $CHCl_3$, alcohol; sparingly soluble in Et_2O, cold H_2O (1 g/400 ml).

5919 Aspirin
50-78-2 886 200-064-1
$C_9H_8O_4$
Acetylsalicylic acid.
o-carboxyphenyl acetate; 2-(acetyloxy)benzoic acid; acetate salicylic acid; salicylic acid acetate; Acenterline; Aceticyl; Acetosal; Acetosalic Acid; Acetosalin; Acetylin; Acetyl-SAL; Acimetten; Acylpyrin; Arthrisin, A.S.A; Asatard; Aspro; Asteric; Caprin; Claradin; Colfarit; Contrheuma retard; Duramax; ECM; Ecotrin; Empirin; Encaprin; Endydol; Entrophen; Enterosarine; Helicon; Levius; Longasa; Measurin; Neuronika; Platet; Rhodine; Salacetin; Salcetogen; Saletin; Solrin; Solpyron; Xaxa; Alka-seltzer; Anacin; Ascriptin; Bufferin; Coricidin D; Darvon compound; Excedrin; Gelprin; Robaxisal; Vanquish; Ascoden-30; Coricidin; Norgesic; Persistin; Supac; Triaminicin; acetophen; acidum acetylsalicylicum; acetilum acidulatum; Acetonyl; Adiro; acenterine; acetosal; acetosalic acid; acetosalin; acetylin; acetylsal; acylpyrin;asteric; caprin; colfarit; entrophen; enterosarine; rhodine; salacetin; salcetogen; saletin; acesal; acetilsalicilico; acetisal; acetonyl; asagran; asatard; aspalon; aspergum; aspirdrops; AC 5230; benaspir; entericin; extren; bialpirinia; contrheuma retard; Crystar; Delgesic; Dolean ph 8; enterophen; globoid; idragin. Analgesic; antipyretic; anti-inflammatory. Also has platelet aggregation inhibiting, antithrombotic, and antirheumatic properties. *See* acetylsalicylic acid. mp = 135; d = 1.40; λ_m = 229, 277 nm ($E^{1\%}_{1cm}$ 484, 68 in 0.1N H_2SO_4); pH (25°) = 3.49; somewhat soluble in H_2O (1 g/300 ml at 25°, 1 g/100 ml at 37°); soluble in alcohol (1 g/5 ml), $CHCl_3$ (1 g/17 ml), Et_2O (1 g/10 ml). *Boots Pharmaceuticals Inc.; Bristol-Myers Squibb Co.; Eli Lilly & Co.; Parke-Davis; Schering-Plough Pharm.; SmithKline Beecham Pharm.; Sterling Health U.S.A.; Sterling Winthrop Inc.; Upjohn Ltd.*

5920 Benorylate
5003-48-5 1074 225-674-5
$C_{17}H_{15}NO_5$
4-Acetamidophenyl salicylate acetate.
benorilate; fenasprate; WIN-11450; Benoral; Benortan; Quinexin; Salipran. Analgesic; antipyretic; anti-inflammatory. Derivative of salicylic acid. mp = 175-176°; LD_{50} (mus orl) = 2000 mg/kg, (rat orl) = 10000 mg/kg, (mus ip) = 1255 mg/kg, (rat ip) = 1830 mg/kg. *Sterling Winthrop Inc.*

5921 Benzydamine
642-72-8 1157 211-388-8
$C_{19}H_{23}N_3O$
1-Benzyl-3-[3-(dimethylamino)propoxy]-1H-indazole.
benzindamine. Analgesic; antipyretic; anti-inflammatory. Anticholinergic. $bp_{0.05}$ = 160°. *3M Pharm.; Angelini Pharm. Inc.*

5922 Benzydamine Hydrochloride
132-69-4 1157 205-076-0
$C_{19}H_{24}ClN_3O$
1-Benzyl-3-[3-(dimethylamino)propoxy]-1H-indazole monohydrochloride.
Difflam Oral Rinse; Difflam Cream; AP-1288; BRL-1288; AF-864; Afloben; Andolex; Benalgin; Benzyrin; Difflam; Dorinamin; Enzamin; Imotryl; Ririlim; Riripen; Salyzoron; Saniflor; Tamas; Tantum; Verax. Analgesic; antipyretic; anti-inflammatory. Anticholinergic. mp = 160°; λ_m = 306 nm ($E^{1\%}_{1cm}$ 160); soluble in H_2O, EtOH, $CHCl_3$, n-butanol; LD_{50} (mus ip) = 110 mg/kg, (rat ip) = 100 mg/kg, (mus orl) = 515 mg/kg, (rat orl) = 1050 mg/kg. *Angelini Pharm. Inc.*

5923 Berberine
2086-83-1 1192 218-229-1
$[C_{20}H_{18}NO_4]^+$
5,6-Dihydro-9,10-dimethoxybenzo[g]-1,3-benzodioxolo[5,6-a]quinolizinium.
umbellatine. Antipyretic; antimalarial; antibacterial. Bitter stomachic. Alkaloid isolated from *Hydrastis candensis L., Berberidaceae*; also found in other plants. mp = 145°; λ_m = 265, 343 nm; [sulfate trihydrate]: LD_{50} (mus ip) = 24.3 mg/kg.

5924 Bermoprofen
72619-34-2 1199
$C_{18}H_{16}O_4$
(±)-10,11-Dihydro-α,8-dimethyl-11-oxodibenz[b,f]-oxepin-2-acetic acid.

AD-1590; AJ-1590; Dibenon. Anti-inflammatory; analgesic; antipyretic. Prostaglandin synthetase inhibitor; a derivative of arylpropionic acid derivative. mp = 128-129°; LD_{50} (mus orl) = 500 mg/kg, (rat orl) = 147 mg/kg. *Dainippon Pharm.*

5925 Bromoacetanilide
103-88-8 1420 203-154-9
C_8H_8BrNO
N-(4-Bromophenyl)acetamide.
p-bromoacetanalide; monobromoacetanalide; bromoanalide; Antisepsin; Asepsin; Bromoantifebrin. Analgesic; antipyretic. mp = 168°; d = 172; nearly insoluble in cold H_2O; slightly soluble in hot H_2O; soluble in C_6H_6; $CHCl_3$, EtOAc; moderately soluble in alcohol. *Monsanto Co.*

5926 Bufexamac
2438-72-4 1497 219-451-1
$C_{12}H_{17}NO_3$
2-(p-Butoxyphenyl)acetohydroxamic acid.
CP-1044-J3; Droxarol; Droxaryl; Feximac; Malipuran; Mofenar; Norfemac; Parfenac; Parfenal. Anti-inflammatory; analgesic; antipyretic. Arylacetic acid derivative. mp = 153-155°; nearly insoluble in H_2O; LD_{50} (mus orl) > 8 g/mg. *C.M. Ind.; Pfizer Inc.*

5927 Bumadizone
3583-64-0 1507 222-710-1
$C_{19}H_{22}N_2O_3$
Butylmalonic acid mono(1,2-diphenylhydrazide).
bumadizon. Analgesic; antipyretic; anti-inflammatory. Main product of hydrolysis of phenylbutazone. mp = 116-117° (also as 77-79°); λ_m = 234,264 nm (ε 16200, 3700 in 0.1 N NaOH). *Ciba-Geigy Corp.*

5928 Bumadizone Calcium Salt Hemihydrate
69365-73-7 1507
$C_{38}H_{42}CaN_4O_6.1/2H_2O$
Butylmalonic acid mono(1,2-diphenylhydrazide) calcium salt (2:1) hemihydrate.
Bumaflex; Eumotol; Rheumatol. Analgesic; antipyretic; anti-inflammatory. Arylbutyric acid derivative. Dec 154°; soluble in $CHCl_3$; EtOH, Et_2O; slightly soluble in H_2O; LD_{50} (mus orl) = 2500 mg/kg, (rat orl) = 1250 mg/kg, (mus iv) = 258 mg/kg, (rat iv) = 263 mg/kg. *Ciba-Geigy Corp.*

5929 Calcium Acetylsalicylate
69-46-5 1684 200-707-6
$C_{18}H_{14}CaO_8$
2-(Acetyloxy)benzoic acid calcium salt.
salicylic acid acetate calcium salt; acetylsalicylic acid calcium salt; calcium aspirin; soluble aspirin; Ascal; Cal-Aspirin; Dispril; Disprin; Kalmopyrin; Kalsetal; Solaspin; Tylcalsin. Analgesic; antipyretic; anti-inflammatory. Derivative of salicylic acid. Soluble in H_2O (16.6 g/100 ml), EtOH (1.25 g/100 ml). *Lee Labs.*

5930 Calcium Acetylsalicylate Carbamide
5749-67-7 1684
$C_{19}H_{18}CaN_2O_9$
2-(Acetyloxy)benzoic acid calcium salt complex with urea.
carbaspirin calcium; carbasalate calcium; acetylsalicylic acid calcium salt complex with urea; urea calcium acetylsalicylate; carbaspirin calcium; Alcacyl; Calurin; Solupsan. Analgesic; antipyretic; anti-inflammatory. Derivative of salicylic acid. Dec 243-245°; soluble in H_2O (23.1 g/100 ml at 37°), pH = 4.8. *Lee Labs.*

5931 Chlorthenoxazine
132-89-8 2247 205-082-3
$C_{10}H_{10}ClNO_2$
2-(2-Chloroethyl)-2,3-dihydro-4H-1,3-benzoxazin-4-one.
chlorthenoxazin; AP-67; Apirazin; Ossapirina; Ossazone; Piroxina; Reumagrip; Valmorin; Valtorin; component of: Trigatan, Fiobrol. Antipyretic; analgesic. mp = 146-147° (dec); λ_m = 297.5 nm; soluble in $CHCl_3$.

5932 Choline Salicylate
2016-36-6 2265 217-948-8
$C_{12}H_{19}NO_4$
2-(Hydroxyethyl)trimethylammonium salicylate.
choline salicylic acid salt; salicylic acid choline salt; Actasal; Arthropan; Artrobione; Audax; Mundisal. Analgesic; antipyretic. mp = 49-50°; soluble in H_2O, polar organic solvents, insoluble in non-polar organic solvents. *Omnium Chim.*

5933 Clidanac
34148-01-1 2411
$C_{16}H_{19}ClO_2$
6-Chloro-5-cyclohexyl-1-indancarboxylic acid.
TAI-284; Britai; Indanal. Anti-inflammatory; antipyretic. Arylcarboxylic acid derivative. mp = 150.5-152.5°; LD_{50} (rat orl) = 41 mg/kg. *Bristol-Myers Squibb Co.*

5934 Dihydroxyaluminum Acetylsalicylate
53230-06-1 3226
$C_9H_9AlO_6$
Dihydroxy(acetylsalicylato)aluminum.
dihydroxyaluminum aspirin. Analgesic; antipyretic. May contain one or more molecules of $Al(OH)_3$. Stable in aqueous suspension at neutral pH; decomposes below pH 4. *Keystone Chemurgic.*

5935 Dipyrocetyl
486-79-3 3413 207-641-7
$C_{11}H_{10}O_6$
2,3-Bis(acetyloxy)benzoic acid.
2,3-diacetoxybenzoic acid; Artromialgina; Movirene. Analgesic; antipyretic. mp = 148-150°, 146-170°; soluble in organic solvents; nearly insoluble in H_2O. *Hoechst Roussel Pharm. Inc.*

5936 Dipyrone
5907-38-0 3414
$C_{13}H_{18}N_3NaO_5S$
Sodium (antipyrinylmethylamino)-methanesulfonate monohydrate.
sodium methylaminoantipyrine methanesulfonate; methylmelubrin; methampyrone; metamizol; analgin; sulpyrin; Alginodia; Algocalmin; Bonpyrin; Conmel; Divarine; Dolazon; D-Pron; Dya-Tron; Espyre; Farmolisina; Feverall; Fevonil; Keypyrone; Metilon; Minalgin; Narone; Nartate; Nevralgina; Nolotil; Novacid; Novaldin; Novalgin; Novemina; Novil; Paralgin; Pyralgin; Pyril; Pyrilgin; Pyrojec; Tega-Pyrone; Unagen. Analgesic; antipyretic. Soluble in H_2O; less soluble in EtOH; nearly

insoluble in Et_2O, Me_2CO, C_6H_6, $CHCl_3$. *Farmitalia Societa Farmaceutici; Hoechst Roussel Pharm. Inc.; Sterling Winthrop Inc.*

5937 Dipyrone [anhydrous]
68-89-3 3414
$C_{13}H_{16}N_3NaO_4S$
Sodium (antipyrinylmethylamino)-methanesulfonate.
Analgesic; antipyretic. *Farmitalia Societa Farmaceutici; Hoechst Roussel Pharm. Inc.; Sterling Winthrop Inc.*

5938 Dipyrone Magnesium Salt
6150-97-6 3414
$C_{26}H_{32}MgN_6NaO_8S_2$
Sodium (antipyrinylmethylamino)-methanesulfonate magnesium salt (2:1).
Magnopyrol. Analgesic; antipyretic. *Farmitalia Societa Farmaceutici; Hoechst Roussel Pharm. Inc.; Sterling Winthrop Inc.*

5939 Epirizole
18694-40-1 3659 242-507-1
$C_{11}H_{14}N_4O_2$
4-Methoxy-2-(5-methoxy-3-methylpyrazol-1-yl)-6-methylpyrimidine.
mepirizole; DA-398; Mebron. Analgesic; anti-inflammatory; antipyretic. mp = 90-92°; slightly soluble in H_2O; soluble in dilute acids, EtOH, C_6H_6, dichloroethane, Et_2O, Me_2CO; LD_{50} (mus orl) = 820 mg/kg. *Daiichi Pharm. Corp.*

5940 Etersalate
62992-61-4 3759 263-780-3
$C_{19}H_{19}NO_6$
2-(Acetyloxy)benzoic acid 2-[4-(acetylamino)phenoxy]ethyl ester.
eterilate; etherylate; eterylate; Daital. Analgesic; anti-inflammatory; antipyretic. Derivative of salicylic acid. mp = 139-141°; LD_{10} (rat orl) = 7000 mg/kg.

5941 Flunixin
38677-85-9 4182
$C_{14}H_{11}F_3N_2O_2$
2-(2-Methyl-3-trifluoromethylanilino)nicotinic acid.
Sch-14714. Analgesic; anti-inflammatory; antipyretic. Cyclooxygenase inhibitor. mp = 226-228°; pKa' = 5.28. *Schering-Plough HealthCare Products.*

5942 Flunixin Meglumine
42461-84-7 4182 255-836-0
$C_{21}H_{28}F_3N_3O_7$
2-(2-Methyl-3-trifluoromethylanilino)nicotinic acid compound with 1-deoxy-1-(methylamino)-D-glucitol (1:1).
Banamine; Finadyne. Analgesic; anti-inflammatory; antipyretic. Cyclooxygenase inhibitor. mp = 135-139°; soluble in H_2O. *Schering-Plough HealthCare Products.*

5943 Guacetisal
55482-89-8 886 259-663-1
$C_{16}H_{14}O_5$
2-(Acetyloxy)benzoic acid 2-methoxyphenyl ester.
Broncaspin; Guaiaspir. Analgesic; antipyretic; anti-inflammatory. See aspirin.

5944 Imidazole Salicylate
36364-49-5 4949
$C_{10}H_{10}N_2O_3$
Salicylic acid compound with imidazole (1:1).
salizolo; IFT-182; Flogozen; Selezen. Anti-inflammatory; analgesic; antipyretic. Derivative of salicylic acid. Also used in heat-sensitive copying materials. mp = 123-124°; λ_m = 300 mn ($E^{1\%}_{1cm}$ 182.5); soluble in H_2O (>100 mg/ml); LD_{50} (mmus sc) = 763 mg/kg, (mmus iv) = 422 mg/kg, (mmus orl) = 1121 mg/kg. *Italfarmaco S.p.A.*

5945 Indomethacin
53-86-1 4998 200-186-5
$C_{19}H_{16}ClNO_4$
1-(p-Chlorobenzoyl)-5-methoxy-2-methylindol-3-acetic acid.
Amuno; Argun; Artracin; Artrinovo; Bonidin; Catlep; Chibro-Amuno; Chrono-Indicid; Confortid; Dolcidium; Durametacin; Elmetacin; Idomethine; Imbrilon; Inacid; Indacin; Indocid; Indocin; Indomed; Indomee; Indomethine; Indomod; Indo-Phlogont; Indoptic; Indoptol; Indorektal; Indo-Tablinen; Indoxen; Inflazon; Infrocin; Inteban SP; Lausit; Mezolin; Mikametan; Mobilan; Rheumacin LA; Tannex; Vonum; Liometacen [meglumine salt]. Anti-inflammatory; analgesic; antipyretic. Arylacetic acid derivative which blocks prostaglandin biosynthesis. mp = 155-162°; λ_m = 230, 260, 319 mn (ε 10800, 16200, 6290 EtOH); pKa = 4.5; soluble in EtOH, Et_2O, Me_2CO, castor oil; nearly insoluble in H_2O; decomposes in strong alkalai; LD_{50} (rat ip) = 13 mg/kg. *Gruppo Lepetit S.p.A.; Lemmon Co.; Merck & Co.Inc.; Sumitomo Pharm. Co. Ltd.*

5946 Indomethacin Sodium Trihydrate
74252-25-8 4998
$C_{19}H_{21}ClNNaO_7$
Sodium 1-(p-chlorobenzoyl)-5-methoxy-2-methylindol-3-acetic acid trihydrate.
Indocin IV; Osmosin. Anti-inflammatory; analgesic; antipyretic. Arylacetic acid derivative. pH (1% aqueous solution) = 8.4; soluble in MeOH, H_2O; slightly soluble in $CHCl_3$, Me_2CO. *Merck & Co.Inc.*

5947 Isofezolac
50270-33-2 5188 256-512-1
$C_{23}H_{18}N_2O_2$
1,3,4-Triphenylpyrazole-5-acetic acid.
LM-22102; Sofenac. Anti-inflammatory; analgesic; antipyretic. Prostaglandin sythetase inhibitor. Arylacetic acid derivative. mp = 200°; LD_{50} (mus orl) = 215 mg/kg, (rat orl) = 13 mg/kg.

5948 p-Lactophenetide
539-08-2 5355 208-708-3
$C_{11}H_{15}NO_3$
N-(4-Ethoxyphenyl)-2-hydroxy-propanamide.
4'-ethoxylactanilide; Fenolactine; Lactophenin; Phenolactine. Analgesic; antipyretic mp = 117-118°; somewhat soluble in H_2O (1 g/330 ml cold, 1 g/55 ml hot); more soluble in alcohol (1 g/8.5 ml); slightly soluble in Et_2O, petroleum ether.

5949 Lofemizole
65571-68-8 265-818-4
$C_{10}H_9ClN_2$
4-(p-Chlorophenyl)-5-methylimidazole.
Anti-inflammatory; analgesic; antipyretic. *Farmatis S.R.L., Italy.*

5950 Lofemizole Hydrochloride
70169-80-1
$C_{10}H_{10}Cl_2N_2$
4-(p-Chlorophenyl)-5-methylimidazole monohydrochloride.
Anti-inflammatory; analgesic; antipyretic. *Farmatis S.R.L., Italy.*

5951 Lysine Acetylsalicylate
62952-06-1 5668 263-769-3
$C_{15}H_{22}N_2O_6$
DL-Lysine mono[2-(acetyloxy)benzoate].
aspirin DL-lysine; aspirin lysine salt; lysine monosalicylate acetate; LAS; Aspidol; Delgesic; Flectadol; Lysal; Quinvet; Venopirin; Vetalgina. Analgesic; antipyretic; anti-inflammatory. Water soluble, injectable aspirin derivative. mp = 154-156°; soluble in H_2O; slightly soluble in EtOH; insoluble in organic solvents.

5952 Magnesium Acetylsalicylate
132-49-0 5689 205-062-4
$C_{18}H_{14}MgO_8$
2-(Acetyloxy)benzoic acid magnesium salt.
magnesium aspirin; Apyron; Fyracyl; Magisal; Magnespirin; Novacetyl. Analgesic; anti-inflammatory. Soluble in H_2O; less soluble in organic solvents.

5953 Meclofenamate Sodium
6385-02-0
$C_{29}H_{27}ClN_2O_{14}S$
[4S-(4α,4aα,5α,5aα,12aα)]-7-chloro-4-(dimethylamino)-1,4,4a,5,5a,6,11,12a-octahydro-3,5,10,12,12a-pentahydroxy-6-methylene-1,11-dioxo-2-naphthacenecarboxamide.
Meclan; Mecloderm; Meclosorb; Meclutin; Traumatociclina. Analgesic; anti-inflammatory; antipyretic. Aminoarylcarboxylic acid derivative. *Parke-Davis.*

5954 Meclofenamate Sodium
6385-02-0 5819 228-983-3
$C_{14}H_{12}Cl_2NNaO_3$
2-[(2,6-Dichloro-3-methylphenyl)amino]benzoate sodium salt monohydrate.
Lenidolor; Meclodol; Meclomen; Movens. Anti-inflammatory; antipyretic. Aminoarylcarboxylic acid derivative. mp = 289-291°; soluble in H_2O (15 mg/ml), pH = 8.7. *Parke-Davis.*

5955 Meclofenamic Acid
644-62-2 5819 211-419-5
$C_{14}H_{11}Cl_2NO_2$
2-[(2,6-Dichloro-3-methylphenyl)amino]benzoic acid.
meclophenamic acid; CI-583; INF-4668; Arquel. Anti-inflammatory; antipyretic. Aminoarylcarboxylic acid derivative. mp = 257-259° (also as 248-250°); slightly soluble in H_2O (0.03 mg/ml); more soluble in 0.1N NaOH (28 mg/ml); pH (saturated aqueous solution) 6.9. *Parke-Davis.*

5956 Morazone
6536-18-1 6349 229-447-1
$C_{23}H_{27}N_3O_2$
4-[(3-Methyl-2-phenyl-4-morpholino)methyl]antipyrine.
R-445; Tarugan. Analgesic; anti-inflammatory; antipyretic. A pyrazolone. mp = 149-150°; soluble in $CHCl_3$, MeOH, Me_2CO; slightly soluble in Et_2O. *Ravensberg.*

5957 Morazone Hydrochloride
50321-35-2 6349 256-540-4
$C_{23}H_{28}ClN_3O_2$
4-[(3-Methyl-2-phenyl-4-morpholino)methyl]antipyrine monohydrochloride.
Analgesic; anti-inflammatory; antipyretic. A pyrazolone. mp = 171-172° (dec); soluble in H_2O. *Ravensberg.*

5958 Morpholine
110-91-8 6362 203-815-1
C_4H_9NO
Tetrahydro-2H-1,4-oxazine.
diethylene oximide; diethylene imidoxide. Analgesic; antipyretic; anti-inflammatory. Also used as solvent for resins, waxes, casein, dyes. CAUTION: Overexposure may cause visual disturbance, nose irritation, coughing, respiratory irritation, eye and skin irritation, liver and kidney damage. Liquid; mp = -4.9°; bp_{760} = 128.9°; bp_6 = 20.0°; d_D^{20} = 1.007; n_D^{20} = 1.4540; miscible with H_2O, many organic solvents, oils; LD_{50} (frat orl) = 1050 mg/kg.

5959 Morpholine Salicylate
147-90-0 6362 205-703-8
$C_{11}H_{15}NO_4$
Tetrahydro-2H-1,4-oxazine salicylate.
Retarcyl; Deposal. Analgesic; antipyretic; anti-inflammatory. mp = 110-111°; soluble in H_2O, alcohols, EtOAc, Me_2CO, C_6H_6, $CHCl_3$; nearly insoluble in C_7H_8, xylene, petroleum ether, Et_2O, CCl_4.

5960 Naproxen
22204-53-1 6504 244-838-7
$C_{14}H_{14}O_3$
(+)-6-Methoxy-α-methyl-2-napthaleneacetic acid.
MNPA; RS-3540; Bonyl; Diocodal; Dysmenalgit; Equiproxen; Floginax; Laraflex; Laser; Naixan; Napren; Naprium; Naprius; Naprosyn; Naprosyne; Naprux; Naxen; Nycopren; Panoxen; Prexan; Proxen; Proxine; Reuxen; Veradol; Xenar. Anti-inflammatory; analgesic; antipyretic. Arylpropionic acid derivative. mp = 152-154°; $[\alpha]_D$ = +66° (c = 1 in $CHCl_3$); nearly insoluble in H_2O; soluble in organic solvents; LD_{50} (mus orl) = 1234 mg/kg, (rat orl) = 534 mg/kg, (rat ip) = 575 mg/kg. *Syntex Labs. Inc.*

5961 Naproxen Piperazine
70981-66-7 6504
$C_{32}H_{38}N_2O_6$
(+)-6-Methoxy-α-methyl-2-napthaleneacetate piperazine salt (2:1).
piproxen; Numidan. Anti-inflammatory; analgesic; antipyretic. Arylpropionic acid derivative. *Syntex Labs. Inc.*

5962 Naproxen Sodium
26159-34-2 6504 247-486-2
$C_{14}H_{13}NaO_3$
(-)-Sodium 6-methoxy-α-methyl-2-napthaleneacetate.
RS-3560; Aleve; Anaprox; Apranax; Axer; Alfa; Flanax; Gynestrel; Miranax; Primeral; Synflex. Anti-inflammatory; analgesic; antipyretic. Arylpropionic acid derivative. mp = 244-246°; $[\alpha]_D$ = -11° (in MeOH). *Syntex Labs. Inc.*

5963 Naproxol
26159-36-4
$C_{14}H_{16}O_2$
(-)-6-Methoxy-β-methyl-2-napthaleneethanol.
RS-4034. Anti-inflammatory; analgesic; antipyretic. *Syntex Labs. Inc.*

5964 Nifenazone
2139-47-1 6619 218-387-1
$C_{17}H_{16}N_4O_2$
N-Antipyrinylnicotinamide.
N-nicotinoylaminoantipyrine; Dolongan; Nicopyron; Nikofezon; Niprazine; Phenicazone; Reupiron; Thylin. Analgesic; antipyretic. mp = 252-253° (also as 256-258°); slightly soluble in H_2O, Me_2CO, ethylacetate, Et_2O; soluble in hot H_2O, alcohol, $CHCl_3$, dilute acids.

5965 5'-Nitro-2'-propoxyacetanilide
553-20-8 6726
$C_{11}H_{14}N_2O_4$
N-(5-Nitro-2-propoxyphenyl)acetamide.
Falimint. Antipyretic; analgesic. mp = 102.5-103.5°. *Schwartz's Essencefabriken.*

5966 Phenacetin
62-44-2 7344 200-533-0
$C_{10}H_{13}NO_2$
N-(4-Ethoxyphenyl)acetamide.
p-acetophenetitide; p-ethoxyacetanilide; para-acetphenetidin; acetophenetidin; acetphenetidin; Phenin; component of: P-A-C compound, APC Tablets. Analgesic; antipyretic. CAUTION: may be a carcinogen. mp = 134-135°; soluble in H_2O (1 g in 1310 ml cold, 82 ml boiling), EtOH (1 g in 15 ml cold, 2.8 ml boiling), Et_2O (1 g/90 ml), $CHCl_3$ (1 g/14 ml); LD_{50} (rat orl) = 1650 mg/kg. *Marion Merrell Dow Inc.; Monsanto Co.; Upjohn Ltd.*

5967 Phenicarbazide
103-03-7 7378 203-072-3
1-Phenylsemicarbazide.
2-phenylhydrazinecarboxamide; Cryogenine; Kryogenin. Antipyretic. mp = 172°; soluble in hot H_2O, alcohol, MeOH, Me_2CO; very slightly soluble in cold H_2O, Et_2O, C_6H_6, ligroin.

5968 Phenocoll
103-97-9 7388 203-163-8
$C_{10}H_{14}N_2O_2$
2-Amino-4'-ethoxyacetanilide.
phenokoll; Phenamine. Antipyretic; analgesic. mp = 100.5°. *Schering-Plough HealthCare Products.*

5969 Phenocoll Salicylate
140-47-6 7388
$C_{17}H_{20}N_2O_5$
2-Amino-4'-ethoxyacetanilide salicylate.
Salocoll. Antipyretic; analgesic. Soluble in H_2O (0.5% cold, 5% hot). *Schering-Plough HealthCare Products.*

5970 Phenopyrazone
3426-01-5 7401 222-324-3
$C_{15}H_{12}N_2O_2$
1,4-Diphenylpyrazolidine-3,5-dione.
Analgesic; antipyretic. mp = 233-234°. *Knoll Pharm. Co.*

5971 Phenyl Acetylsalicylate
134-55-4 7424 205-147-6
$C_{15}H_{12}O_4$
2-(Acetyloxy)benzoic acid phenyl ester.
acetylphenylsalicylate; Acetylsalol; Spiroform; Vesipyrin. Analgesic; anti-inflammatory; antipyretic. mp = 97°; bp_{11} = 198°; insoluble in H_2O; soluble in alcohol, Et_2O.

5972 Phenyl Salicylate
118-55-8 7464 204-259-2
$C_{13}H_{10}O_3$
2-Hydroxybenzoic acid phenyl ester.
Salol. Analgesic; anti-inflammatory; antipyretic. Used in suntan oils and creams. Has some light absorbing properties. mp = 41-43°; bp_{12} = 173°; d = 1.25; slightly soluble in H_2O (1 g/6670 ml); more soluble in alcohol (17 g/100 ml), C_6H_6 (67 g/100 ml), amyl alcohol (20 g/100 ml), liquid paraffin (10 g/100 ml), almond oil (25 g/100 ml); soluble in Me_2CO, $CHCl_3$, ET_2O, oils, glycerol.

5973 Pipebuzone
27315-91-9 7610 248-398-7
$C_{25}H_{32}N_4O_2$
4-Butyl-4-[(4-methyl-1-piperazinyl)methyl]-1,2-diphenyl-3,5-pyrazolidinedione.
LD-4644; Elarzone. Anti-inflammatory; analgesic; antipyretic. Derivative of phenylbutazone, an anti-inflammatory agent. A pyrazolone. mp = 129°.

5974 Pirfenidone
53179-13-8
$C_{12}H_{11}NO$
5-Methyl-1-phenyl-2(1H)pyridone.
AMR-69. Analgesic; anti-inflammatory; antipyretic.

5975 Propacetamol
66532-85-2 7976 266-390-1
$C_{14}H_{20}N_2O_3$
N,N-Diethylglycine 4-(acetylamino)phenyl ester.
4-acetamidophenyl (diethylamino)acetate; N,N-diethylglycine ester with 4'-hydroxyacetanilide. Analgesic; antipyretic. Injectable prodrug (acetaminophen). *Hexachemie.*

5976 Propacetamol Hydrochloride
66532-86-3 7976
$C_{14}H_{22}ClN_2O_3$
N,N-Diethylglycine 4-(acetylamino)phenyl ester hydrochloride.
UP-34101; Pro-Dafalgan. Analgesic; antipyretic. Injectable prodrug of acetaminophen. mp = 228°; soluble in H_2O. *Hexachemie.*

5977 Propyphenazone
479-92-5 8056 207-539-2
$C_{14}H_{18}N_2O$
4-Isopropyl-2,3-dimethyl-1-phenyl-3-pyrazolin-5-one.
4-isopropylphenazone; isopropylphenazone; Budirol; Causyth; Cibalgina; Eufibron; Isopropchin. Analgesic; antipyretic; anti-inflammatory. A pyrazolone. mp = 103°; soluble in H_2O (0.24 g/100 g at 16.5°), organic solvents. *Hoffmann-LaRoche Inc.*

5978 Ramifenazone
3615-24-5 8282 222-791-3
$C_{14}H_{19}N_3O$
4-Isopropylamino-2,3-dimethyl-1-phenyl-3-pyrazolin-5-one.
4-isopropylaminoantipyrine; isopropylaminophenazone; isopyrin; component of: Tomanol (with phenylbutazine). Analgesic; anti-inflammatory; antipyretic. Has been evaluated as an antimigraine in combination with phenylbutazone. A pyrazolone. mp = 80°; LD_{50} (mus ip) = 843 mg/kg, (mus orl) = 1070 mg/kg.

5979 Salacetamide
487-48-9 8471 207-656-9
$C_9H_9NO_3$
N-Acetyl-2-hydroxybenzamide.
N-acetylsalicylamide; acetsalicylamide; Actylamide. Analgesic; anti-inflammatory; antipyretic. Derivative of salicylic acid. mp = 148°.

5980 Salcolex
28038-04-2
$C_{24}H_{46}MgN_2O_{12}S{\cdot}4H_2O$
Choline salicylate (salt) compound with magnesium sulfate (2:1) tetrahydrate.
Analgesic; anti-inflammatory; antipyretic. *Wallace Labs.*

5981 Salcolex [anhydrous]
54194-00-2
$C_{24}H_{46}MgN_2O_{16}S$
Choline salicylate (salt) compound with magnesium sulfate (2:1).
Analgesic; anti-inflammatory; antipyretic. *Wallace Labs.*

5982 Salicylamide O-Acetic Acid
25359-22-6 8481
$C_9H_9NO_4$
[2-(Aminocarbonyl)phenoxy]acetic acid.
o-(carbamylphenoxy)acetic acid. Analgesic; anti-inflammatory; antipyretic. Derivative of salicylic acid. mp = 221°; soluble in aqueous alkali. *Yoshitomi.*

5983 Salicylamide O-Acetic Acid Sodium Salt
3785-82-8 8481
$C_9H_8NNaO_4$
[2-(Aminocarbonyl)phenoxy]acetate sodium salt (1:1).
Salizell ampules. Analgesic; anti-inflammatory; antipyretic. Derivative of salicylic acid. mp = 212-215°. *Yoshitomi.*

5984 Sodium Salicylate
54-21-7 8819 200-198-0
$C_7H_5NaO_3$
2-Hydroxybenzoic acid monosodium salt.
Alysine; Idocyl; Enterosalicyl; Enterosalil. Analgesic; antipyretic. Used as a preservative. Soluble in H_2O (1.1 g/ml), less soluble in organic solvents; incompatible withh ferric salts, lime water, spirit nitrous Et_2O, mineral acids, iodine, lead acetate, silver nitrate, sodium phosphate; LD_{50} (rat ip)= 780 mg/kg. *Marion Merrell Dow Inc.*

5985 Talmetacin
67489-39-8
$C_{27}H_{20}ClNO_6$
(±)-1-(p-Chlorobenzoyl)-5-methoxy-2-methylindole-3-acetic acid ester with 3-hydroxyphthalide.
BA-7605-06. Analgesic; anti-inflammatory; antipyretic. *Lab. Bago S.A.; Resfar S.R.L.*

5986 Tetrandrine
518-34-3 9369
$C_{38}H_{42}N_2O_6$
(1β)-6,6',7,12-Tetramethoxy-2,2'-dimethylberbaman.
Phaenthine [l-form]. Analgesic; antipyretic. Found in root of *Staphania tetandra s. Moore, Menispermaceae.* Present in Chinese drug han-fang-chi. mp = 217-218°; [l-form]: mp = 210°; $[\alpha]_D^{26}$ = +252.4° ($CHCl_3$); $[\alpha]_D^{26}$ = -278° ($CHCl_3$); nearly insoluble in H_2O, petroleum ether; soluble in Et_2O and some other organic solvents.

5987 Tetrydamine
17289-49-5
$C_9H_{15}N_3$
4,5,6,7-Tetrahydro-2-methyl-3-(methylamino)-2H-indazol.
POLI-67. Analgesic; anti-inflammatory. *Polichimica Sap.*

5988 Tinoridine
24237-54-5 9590 246-102-0
$C_{17}H_{20}N_2O_2S$
Ethyl 2-amino-6-benzyl-4,5,6,7-tetrahydro-thieno[2,3-c]pyridine-3-carboxylate.
Y-3642. Analgesic; anti-inflammatory; antipyretic. Arylcarboxylic acid derivative. mp = 112-113°; slightly soluble in H_2O; LD_{50} (mus orl) = 5400 mg/kg, (rat orl) > 10200 mg/kg, (mus ip) = 1600 mg/kg, (rat ip) = 1250 mg/kg. *Yoshitomi.*

5989 Tinoridine Hydrochloride
25913-34-2 9590
$C_{17}H_{21}ClN_2O_2S$
Ethyl 2-amino-6-benzyl-4,5,6,7-tetrahydro-thieno[2,3-c]pyridine-3-carboxylate hydrochloride.
Dimaten; Noflamin. Analgesic; anti-inflammatory; antipyretic. Arylcarboxylic acid derivative. mp = 234-235° (dec); slightly soluble in MeOH; less soluble in H_2O, Et_2O, Me_2CO, C_6H_6; LD_{50} (mus orl) = 1601 mg/kg. *Yoshitomi.*

5990 Tiopinac
61220-69-7
$C_{16}H_{12}O_3S$
6,11-Dihydro-11-oxodibenzo[b,e]thiepin-3-acetic acid.
RS-40974. Analgesic; anti-inflammatory; antipyretic. *Syntex Labs. Inc.*

Antirickettsials

5991 Chloramphenicol

56-75-7 2120

$C_{11}H_{12}Cl_2N_2O_5$

[R-(R*,R*)]-2,2-Dichloro-N-[2-hydroxy-1-(hydroxymethyl)-2-(4-nitrophenyl)ethyl]acetamide.

Ak-Chlor; Amphicol; Anacetin; Aquamycetin; Chemicetina; Chloramex; Chlorasol; Chloricol; Chlorocid; Chloromycetin; Chloroptic; Cloramfen; Clorocyn; Enicol; Farmicetina; Fenicol; Globenicol; Intramycetin; Kemicetine; Leukomycin; Micoclorina; Mychel; Mycinol; Novomycetin; Ophthochlor; Pantovernil; Paraxin; Quemicetina; Ronphenil; Sintomicetina; Sno Phenicol; Synthomycetin; Tevcocin; Tifomycine; Veticol; Viceton. Antibacterial; antirickettsial. A broad spectrum antibiotic obtained from cultures of the soil bacterium *streptomyces venezuelae*. mp = 150.5-151.5°; $[\alpha]_D^{27}$= 18.6° (c = 4.86 EtOH), $[\alpha]_D^{25}$= -25.5° (EtOAc); λ_m 278 nm ($E_D^{1\%}$298); soluble in H_2O (0.25 g/100 ml at 25°), propylene glycol (15.1 g/100 ml at 25°); very soluble in MeOH, EtOH, BuOH, EtOAc; Me_2CO; fairly soluble in Et_2O; insoluble in C_6H_6, petroleum ether. *Allergan Inc.; Chinoin; Fermenta Animal Health Co.; Fujisawa Pharm. USA Inc.; Parke-Davis; Schering-Plough Animal Health; TechAmerica.*.

5992 Chloramphenicol Arginine Succinate

34327-18-9 2120

$C_{21}H_{30}Cl_2N_6O_{10}$

[R-(R*,R*)]-2,2-Dichloro-N-[2-hydroxy-1-(hydroxymethyl)-2-(4-nitrophenyl)ethyl]acetamide monosuccinate arginine.

chloramphenicol monosuccinate arginine salt; Paraxin Succinate A. Antibacterial; antirickettsial. A broad spectrum antibiotic obtained from cultures of the soil bacterium *streptomyces venezuelae*. mp = 135-145° (dec). *Chinoin.*

5993 Chloramphenicol Monosuccinate Sodium Salt

982-57-0 2120 213-568-1

$C_{15}H_{15}Cl_2N_2NaO_8$

[R-(R*,R*)]-2,2-Dichloro-N-[2-hydroxy-1-(hydroxymethyl)-2-(4-nitrophenyl)ethyl]acetamide monosuccinate sodium.

chloramphenicol sodium succinate; protophenicol. Antibacterial; antirickettsial. A broad spectrum antibiotic obtained from cultures of the soil bacterium *streptomyces venezuelae*. Freely soluble in H_2O (50 g/100 ml). *Chinoin.*

5994 Chloramphenicol Palmitate

530-43-8 2120 208-477-9

$C_{27}H_{42}Cl_2N_2O_6$

[R-(R*,R*)]-2,2-Dichloro-N-[2-hydroxy-1-(hydroxymethyl)-2-(4-nitrophenyl) propyl ester hexadecanoic acid.

Chlorambon; Chloropal; Clorolifarina. Antibacterial; antirickettsial. A broad spectrum antibiotic obtained from cultures of the soil bacterium *streptomyces venezuelae*. mp = 90°; $[\alpha]_D^{26}$= 24.6° (c = 5 EtOH); λ_m 271 nm ($E_{1\ cm}^{1\%}$ 179 EtOH); slightly soluble in H_2O (0.105 g/100 ml at 28°), petroleum ether (0.0225 g/100 ml); freely soluble in MeOH, EtOH, $CHCl_3$, Et_2O, C_6H_6. *Chinoin; Parke-Davis.*

5995 Chloramphenicol Pantothenate Complex

31342-36-6 2120

$C_{62}H_{80}CaCl_8N_{10}O_{30}$

N-(2,4-Dihydroxy-3,3-dimethyl-1-oxobutyl)-β-alanine calcium salt (2:1) with [R-(R*,R*)]-2,2-Dichloro-N-[2-hydroxy-1-(hydroxymethyl)-2-(4-nitrophenyl)ethyl]-acetamide (1:4).

chloramphenicol pantothenate calcium complex (4:1); Pantothenic acid calcium salt (2:1) compound with D-threo-(-)-2,2-dichloro-N[β-hydroxy-α-(hydroxymethyl)-p-nitrophenyl]acetamide (1:4); Pantofenicol. Antibacterial; antirickettsial. A broad spectrum antibiotic obtained from cultures of the soil bacterium *streptomyces venezuelae*. *Chinoin; Eprova AG; Pluriquimica, Portugal.*

5996 Oxytetracycline Hydrochloride

2058-46-0 7111 218-161-2

$C_{22}H_{25}ClN_2O_9$

4-(Dimethylamino)-1,4,4a,5,5a,6,11,12a-octahydro-3,5,6,10,12,12a-hexahydroxy-6-methyl-1,11-dioxo-2-napthacenecarboxamide monohydrochloride.

Alamycin; Aquacycline; Bio-Mycin; Duphacycline; Engemycin; Geomycin; Gynamousse; Macocyn; Medamycin; Occrycetin; Oxlopar; Oxybiocycline; Oxybiotic; Oxycyclin; Oxy-Dumocyclin; Oxyject; Oxylag; Oxypan; Oxytetracid; Oxytetrin; Oxy WS; Terrafungine; Terraject; Terramycin Hydrochloride; Tetramel; Tetran; Tetra-Tabilinen; Toxinal; Vendarcin; component of: Terra-Cortril. Tetracycline antibiotic; antirickettsial. Very soluble in H_2O; soluble in absolute ethanol; 95% EtOH. *Fermenta Animal Health Co.; Parke-Davis; Pfizer Inc.; TechAmerica.*

5997 p-Aminobenzoic Acid

150-13-0 443 205-753-0

$C_7H_7NO_2$

4-Aminobenzoic acid.

para-aminobenzoic acid; vitamin B_x; bacterial vitamin H^1; chromotrichia factor; antichromotrichia factor; trichochromogenic factor; anticantic vitamin; PABA; Amben; Paraminol; Sunbrella; component of: Pabanol, PreSun. Ultra-violet screen; antirickettsial (formerly). A naturally occurring B complex factor. Found in baker's yeast (5-6 ppm) and brewer's yeast (10-100 ppm). mp = 187-187.5°; λ_m 266 nm ($E_{1\ cm}^{1\%}$ 1070 H_2O), 288 nm ($E_{1\ cm}^{1\%}$ 137 iPrOH); soluble in H_2O (0.59 g/100 mlat 25°, 1.1 g/100 ml at 100°), EtOH (12.5 g/100 ml), Et_2O (2 g/100 ml), EtOAc, AcOH; slightly soluble in C_6H_6; insoluble in petroleum ether; pKa = 4.65, 4.80; pH (0.5%) = 3.5; LD_{50} (mus orl) = 2850 mg/kg, (rat orl) > 6000 mg/kg, (rbt iv) = 2000 mg/kg, (rbt orl) = 1830 mg/kg. *Aktieselskabet Pharmacia; Dey Labs; DuPont-Merck Pharm.; Heyden Chem.; ICN Pharm. Inc.*

5998 Tetracycline

60-54-8 9337 200-481-9

$C_{22}H_{24}N_2O_8$

(4S,4aS,5aS,6S,12aS)-4-(Dimethylamino)-1,4,4a,5,5a,6,11,12a-octahydro-3,6,10,12,12a-pentahydroxy-6-methyl-1,11-dioxo-2-naphthacenecarboxamide.

tsiklomitsin; deschlorobiomycin; Abricycline; Ambramycin; Bio-Tetra; Cyclomycin; Dumocyclin; Liquamycin; Mysteclin-F; Talsutin; Tetradecin; component of: Mysteclin-F. Antiamebic; antibacterial; antirickettsial. Tetracycline antibiotic produced in *Streptomyces* species. [trihydrate]: Dec 170-175°; $[\alpha]_D^{25}$=

-257.9° (0.1 N HCl), -239° (MeOH); λ_m = 220, 268, 355 nm (ε 13000, 18040, 13320 0.1N HCl); soluble in H_2O (1.7 mg/ml), MeOH (> 20 mg/ml); LD_{50} (rat orl) = 707mg/kg, (mus orl) = 808 mg/kg. *Bristol-Myers Squibb Co.; Pfizer Inc.*

Antiseborrheics

5999 Chloroxine
773-76-2 2227 212-258-3
$C_9H_5Cl_2NO$
5,7-Dichloro-8-quinolinol.
Capitrol. Antiseborrheic. mp = 179-180°; soluble in C_6H_6, Me_2CO; slightly soluble in cold alcohol, AcOH; readily soluble in sodium and potassium hydroxides and in acids, forming yellow solutions. *Westwood-Squibb Pharm. Inc.*

6000 3-O-Lauroylpyroxidoxol Diacetate
1562-13-6 5401
$C_{24}H_{37}NO_6$
5-Lauroyloxy-6-methyl-3,4-pyridinedimethanol diacetate.
Epixine; Rosamit. Antiseborrheic. mp = 44°; practically insoluble in H_2O; soluble in Et_2O, $CHCl_3$, EtOH, ethylene dichloride. *Soc. Belge des Labs. Labaz.*

6001 Piroctone
50650-76-5 7657
$C_{14}H_{23}NO_2$
1-Hydroxy-4-methyl-6-(2,4,4-trimethylpentyl)-2(1H)-pyridone.
Antiseborrheic. mp = 108°. *Hoechst AG.*

6002 Piroctone Olamine
68890-66-4 7657 272-574-2
$C_{16}H_{30}N_2O_3$
1-Hydroxy-4-methyl-6-(2,4,4-trimethylpentyl)-2(1H)-pyridone compound with 2-aminoethanol (1:1).
piroctone ethanolamine salt; Octopirox. Antiseborrheic. *Hoechst AG.*

6003 Pyrithione Zinc
13463-41-7 8178 236-671-3
$C_{10}H_8N_2O_2S_2Zn$
Bis[1-hydroxy-2(1H)-pyridinethionato]zinc.
zinc pyrithione; zinc pyridinethione; bis(2-pyridylthio)zinc 1,1'-dioxide; zinc omadine; Danex; Sebulon Shampoo; ZNP Bar; Desquaman; component of: Head and Shoulders. Antibacterial; antifungal; antiseborrheic. *Allergan Herbert; Herbert; Lederle Labs.; Olin Res. Ctr.; Stiefel Labs Inc.; Westwood-Squibb Pharm. Inc.*

6004 Resorcinol
1008-46-3 8323 203-585-2
$C_6H_6O_2$
1,3-Benzenediol.
m-dihydroxybenzene; resorcin; component of: Acnomel, Rezamid, Sulforcin. Keratolytic; antiseborrheic. CAUTION: Irritating to skin, mucous membranes; absorption can cause methemoglobinemia, cyanosis, convulsions, death. mp = 109-111°; soluble in H_2O, alcohol; freely soluble in Et_2O, glycerol; slightly soluble in $CHCl_3$; protect from light; incompatible with acetanilide, albumin, alkalies, antipyrine, camphor, ferric salts, menthol, spirit nitrous ether, urethan. *Dermik Labs. Inc.; Menley & James Labs Inc.*

6005 Resorcinol Monoacetate
102-29-4 8323 203-022-0
$C_8H_8O_3$
1,3-Benzenediol monoacetate.
acetylresorcinol; Euresol. Keratolytic; antiseborrheic. CAUTION: Irritating to skin, mucous membranes; absorption can cause methemoglobinemia, cyanosis, convulsions, death. Oily liquid; bp 283° with decomposition; miscible with alcohol, C_6H_6, $CHCl_3$, Me_2CO; soluble in solutions of alkali hydroxides. *Knoll Pharm. Co.*

6006 Selenium Sulfide
7488-56-4 8580 231-303-8
SeS_2
Selenium disulfide.
Exsel; Seleen; Selsun; Selsun Blue. Antiseborrheic (topical); antifungal. Insoluble in H_2O; LD_{50} (rat orl) = 138 mg/kg. *Abbott Labs Inc.; Allergan Herbert; Herbert; Ross Products.*

6007 Tioxolone
4991-65-5 9600 225-653-0
$C_7H_4O_3S$
6-Hydroxy-1,3-benzoxathiol-2-one.
thioxolone; Camyna; Stepin; component of: Psoil [with hydrocortisone]. Antiseborrheic. Used to treat acne. mp = 160°; nearly insoluble in H_2O; soluble in EtOH, iPrOH, propylene glycol, Et_2O, C_6H_6, C_7H_8; hydrolyzed by alkali. *Boehringer Ingelheim GmbH; Sterling Winthrop Inc.; Tillots Pharma; Winthrop.*

Antiseptics

6008 Acetomeroctol
584-18-9 63 209-534-0
$C_{16}H_{24}HgO_3$
(Acetato-O)[2-hydroxy-5-(1,1,3,3-tetramethylbutyl)-phehyl]mercury.
Merbak. Antiseptic; topical anti-infective. Phenol. mp = 158°; nearly insoluble in H_2O; soluble in alcohol, $CHCl_3$; sparingly soluble in C_6H_6.

6009 Alexidine
22573-93-9 231 245-096-7
$C_{26}H_{56}N_{10}$
1,1-Hexamethylenebis[5-(2-ethylhexyl)biguanide].
Win-21904; compound 904; Sterwin 904; Bisguanidine. Antibacterial; antiseptic. A guanidine derivative. [dihydrochloride]: mp = 220.6-223.4°. *Sterling Res. Labs.*

6010 Aluminum Acetate Solution
139-12-8 332 205-354-1
$C_6H_9AlO_6$
Acetic acid aluminum salt.
Burow's solution; Buro-Sol Concentrate; Domeboro; component of: Otic Domeboro. Antiseptic (topical); astringent. Colorless liquid; d 1.002; pH (1:20 aqueous solution) = 4.2. *Doak Pharmacal Co. Inc.; Miles Inc.*

6011 Aluminum Subacetate Solution
142-03-0 380 205-518-2
$C_4H_7AlO_5$
Bis(acetato-O)hydroxyaluminum.
Essigsäure Tonerde; basic aluminum acetate. Antiseptic (topical); astringent. Also used as mordant in fabric dyeing, fireproofing, and printing; used in antiperspirantes and embalming fluids. CAUTION: Ingestion may cause severe nausea, vomiting, diarrhea, melena, hematemesis. Colorless liquid; d = 1.045; gradually becomes turbid and colloidal; basic salt precipitates out.

6012 Aluminum Sulfate
10043-01-3 381 233-135-0
$Al_2O_{12}S_3$
Aluminum sulfate (2:3).
alunogenite; cake alum [as octadecahydrate]; patent alum [as octadecahydrate]; component of: Bluboro. Antiseptic. Occurs as the mineral alunogenite. Soluble in H_2O; nearly insoluble in alcohol. *Herbert.*

6013 Amantanium Bromide
58158-77-3 390
$C_{25}H_{46}BrNO_2$
Decyl(2-hydroxyethyl)dimethylammonium bromide 1-adamantanecarboxylate.
Amantol. Antiseptic. Quaternary ammonium compound. mp = 182-184°; LD_{50} (mus orl) = 910 mg/kg. *Colgate-Palmolive; Rotta Pharm.*

6014 Ambazone
6011-12-7 395 208-713-0
$C_8H_{11}N_7S$
p-Benzoquinone amidinohydrazone thiosemicarbazone hydrate.
Inversal; Primal; Promassol. Antibacterial; antiseptic. Guanidine. Dec 195°; sparingly soluble in H_2O; moderately soluble in alcohol, Me_2CO; freely soluble in DMF, dilute acids. *Bayer AG.*

6015 3-Amino-4-hydroxybutyric Acid
589-44-6 463
$C_4H_9NO_3$
γ-Hydroxy-β-aminobutyric acid.
GOBAB. Antiseptic; anti-inflammatory; antifungal. mp = 216°, 228°, 233°. *Kaken Pharm. Co. Ltd.*

6016 Aminoquinuride
3811-56-1 496
$C_{21}H_{20}N_6O$
1,3-Bis(4-amino-2-methyl-6-quinolyl)urea.
aminoquincarbamide; aminochinuride; aminokinuride; Surfen. Antiseptic. Quinoline. Dec 255° (effervescence). *I.G. Farben.*

6017 Ammonium Benzoate
1863-63-4 521 217-468-9
$C_7H_9NO_2$
Benzoic acid ammonium salt.
Used in medicine as a urinary anti-infective. Also used as a preservative for latex and glue. mp = 198°; d = 1.26; soluble in H_2O (0.21 g/ml), less soluble in organic solvents; incompatible with ferric salts, alkali hydroxides, carbamates. *Hooker Chem.*

6018 Balsam of Peru
8007-00-9 232-352-8
Peruvian balsam; Indian balsam; China oil; Black balsam; Honduras balsam; Surinam balsam. An oleoresin obtained from the bark of *Myroxylon pereinaeg.* It contains esters of cinnamic and benzoic acids. Antiseptic, used in medicine and perfumery.

6019 Benzalkonium Chloride
8001-54-5 1086 264-151-6
Alkylbenzyldimethylammonium chloride.
Benirol; BTC; Capitol; Cequartyl; Drapolene; Drapolex; Enuclen; Germinol; Germitol; ; Osvan; Paralkan; Roccal; Rodalon; Zephiran Chloride; Zephirol. Antiseptic. A mixture of alkyldimethylbenzylammonium chlorides used as a topical antiseptic and udder wash (veterinary). Very soluble in H_2O, alcohol, Me_2CO; slightly soluble in C_6H_6; almost insoluble in Et_2O; LD_{50} (rat orl) = 400 mg/kg; incompatible with anionic detergents and nitrates. *Alcon Labs; Ortho Pharm. Corp.; Sterling Winthrop Inc.*

6020 Benzethonium Chloride
121-54-0 1103 204-479-9
$C_{27}H_{42}ClNO_2$
Benzyldimethyl[2-[2-[p-(1,1,3,3-tetramethylbutyl)-phenoxy]ethoxy]ethyl]ammonium chloride.
Phemerol Chloride. Topical antiseptic (veterinary); topical anti-infective. Quaternary ammonium compound. CAUTION: Ingestion may cause vomiting, collapse, convulsions, coma. mp = 164-166°; soluble in H_2O, alcohol, Me_2CO, $CHCl_3$; incompatible with soap, anionic detergents; LD_{50} (rat) = 420 mg/kg. *Parke-Davis.*

6021 Benzoxiquine
86-75-9 1143 201-697-6
$C_{16}H_{11}NO_2$
8-Quinolol benzoate.
benzoxyline; NSC-3951; Dioxyline. Antiseptic. Quinoline. mp = 118-120°; soluble in alcohol, Et_2O. *Labs. Franca Inc.*

6022 Benzoxonium Chloride
19379-90-9 1144 243-008-1
$C_{23}H_{42}ClNO_2$
Benzyldodecylbis(2-hydroxyethyl)ammonium chloride.
D-301; ZY-15021; Absonal V; Bialcol; Bradophen; Orofar. Antiseptic. Quaternary ammonium compound. mp = 107-109°; soluble in H_2O, alcohol, C_6H_6, C_7H_8, C_6H_6. LD_{50} (rat orl) = 750 mg/kg. *CIBA plc.*

6023 Bisdiqualinium Chloride
52951-36-7 1288
$C_{40}H_{58}Cl_2N_4$
1,1'-Decamethylene-4,4'-(1,10-decamethylenediimino)bis[quinaldinium chloride].
Salvizol. Antiseptic; disinfectant. Quaternary ammonium compound.

6024 Bismuth Iodide Oxide
7787-63-5 1309 232-126-9
BiIO
Bismuth oxyiodide.
basic bismuth iodide; bismuthyl iodide; bismuth subiodide; iodooxobismuthine. Antiseptic. Halogen-containing compound. Reddish powder or crystals;

d = 7.92; nearly insoluble in H_2O, alcohol, $CHCl_3$; soluble in HCl.

6025 Bismuth Iodosubgallate
138-58-9 1310 205-333-7
$C_7H_6BiIO_6$
Hydroxyiodo[(3,4,5-trihydroxybenzoyl)oxy]-bismuthine.
(gallato)hydroxyiodobismuth; bismuth oxyiodogallate; Airoform; Airogen. Antiseptic; anti-infective. Halogen containing compound. Decomposed by H_2O, acids; nearly insoluble in alcohol, Et_2O, $CHCl_3$; soluble in solutions of alkali hydroxides; light sensitive.

6026 Bismuth Tribromophenate
5175-83-7 1333 225-958-9
$C_{18}H_6BiBr_9O_3$
Bismuth tris(2,4,6-tribromophenoxide).
bismuth tribromophenol; tribromophenobismuth; Sigmaform; Xeroform. Antiseptic; anti-infective. Halogen containing compound. Stable below 120°; slightly soluble in H_2O, alcohol, $CHCl_3$, vegetable oils; decomposed by strong alkalies and strong acid.

6027 Bithionol
97-18-7 1343 202-565-0
$C_{12}H_6Cl_4O_2S$
2,2'-thiobis(4,6-dichlorophenol).
XL-7; Actamer; Bithin; Lorothidol. Topical anti-infective, used in veterinary medicine as an anthelmintic and antiseptic. mp = 188°; d_4^{25} = 1.73; insoluble in H_2O, soluble in organic solvents. *I.G. Farben; Sterling Winthrop Inc.*

6028 Bithionol Sodium
6385-58-6 1343
$C_{12}H_6Cl_4Na_2O_2S$
2,2'-thiobis(4,6-dichlorophenol) sodium salt.
bithionate sodium; Vancide BN. Topical anti-infective. *I.G. Farben.*

6029 Bithionol Sulfoxide
844-26-8 1343
$C_{12}H_6Cl_42O_3S$
2,2'-thiobis(4,6-dichlorophenol) sulfoxide.
BTS; Bitin-s; Disto-5. Topical anti-infective. *I.G. Farben.*

6030 Boric Acid
10043-35-3 1364 233-139-2
H_3BO_3
Orthoboric acid.
boracic acid; Borofax. Antiseptic; astringent. mp 171°; somewhat soluble in H_2O; more soluble in alcohol; LD_{50} (rat orl) = 5.14 g/kg.

6031 Bornyl Chloride
464-41-5 1369 207-350-5
$C_{10}H_{17}Cl$
endo-2-Chloro-1,7,7-trimethylbicyclo[2.2.1]heptane.
pinene hydrochloride; terpene hydrochloride; turpentine camphor. Antiseptic. Halogen containing compound. mp = 132°; bp = 207-208°; nearly insoluble in H_2O; soluble in alcohol, Et_2O.

6032 Broxyquinoline
521-74-4 1474 208-317-8
$C_9H_5Br_2NO$
5,7-Dibromoquinolin-8-ol.
Brodiar; Broxykinolin; Colepur; Fenilor; Intensopan. Antiseptic. Quinoline. mp = 196°; d = 2.189; soluble in $CHCl_3$, alcohol, C_6H_6, AcOH; slightly soluble in Et_2O; nearly insoluble in H_2O.

6033 Cadexomer Iodine
94820-09-4 1646
Iodosorb. Antiseptic; antiulcerative; vulnerary. Product of reaction of dextrin with epichlorohydrin coupled with ion-exchange groups and iodine. *Perstorp AB.*

6034 Cadmium Salicylate
19010-79-8 1661 242-749-8
$C_{14}H_{10}CdO_6$
Cadmium disalicylate.
Antiseptic. Phenol. mp = 242° (dec); slightly soluble in cold H_2O, MeOH, EtOH; soluble in boiling H_2O.

6035 Calcium Iodate
7789-80-2 1719 232-191-3
CaI_2O_6
Iodic acid calcium salt.
Laurarite. Antiseptic. Nutritional source of iodine in foods and feedstuffs. Stable below 540°; d_4^{14}= 4.519; sensitive to reducing agents; slightly soluble in H_2O; more soluble in aqueous solutions of iodides and amino acids; soluble in nitric acid; insoluble in alcohol.

6036 Calcium Peroxide
1305-79-9 1736 215-139-4
CaO_2
Calcium dioxide.
Antiseptic. Peroxide. Also used as a stabilizer for rubber. Commercial product usually contains 60% CaO_2 with some $Ca(OH)_2$ and $CaCO_3$. Dec in moist air; slightly soluble in H_2O; soluble in acids with formation of H_2O. *DuPont Pharm. Co.*

6037 Camphor
76-22-2 1779 200-945-0
$C_{10}H_{16}O$
1,7,7-Trimethylbicyclo[2.2.1]heptan-2-one.
2-bornanone; 2-camphanone; norcamphane; gum camphor; Japan camphor; Formosa camphor; laurel camphor; spirit of camphor; component of: Campho-Phenique Cold Sore Gel, Campho-Phenique Liquid, Heet, Minut-Rub, Pazo Ointment, Sarna. Topical antipruritic; antiseptic. Plasticizer for cellulose nitrate, other explosives and lacquers, insecticides, moth and mildew proofing, tooth powders, flavoring, embalming, pyrotechnics, intermediate. CAUTION: Ingestion or injection may cause nausea, vomiting, mental confusion, delirium, clonic convulsions, coma, respiratory failure, death. mp = 179.75°; bp = 204°; $[\alpha]_D^{25}$ = 41° - 43° (c= 10 EtOH); d_4^{25} = 0.992; soluble in H_2O (0.125 g/100 ml), EtOH (100 g/100 ml), Et_2O (100 g/100 ml), $CHCl_3$ (200 g/100 ml), C_6H_6 (250 g/100 ml), Me_2CO (250 g/100 ml); LD_{50} (mus ip) = 3000 mg/kg. *Bristol-Myers Squibb Co.; Sterling Health U.S.A.; Stiefel Labs Inc.; Whitehall Labs. Inc.*

6038 Carvacrol
499-75-2 1923 207-889-6
$C_{10}H_{14}O$
2-Methyl-5-(1-methylethyl)phenol.
2-hydroxy-p-cymene; isopropyl o-cresol; isothymol. Anthelmintic. Targets nematodes. Used as a general disinfectant. Phenol found in the oil of origanum, thyme, marjoram, and summer savory. An essential oil. Liquid; mp 0°; bp_{760} = 237-238°; bp_{18} = 118-122°; bp_3 = 93°; d_4^{20}= 0.976; d_{25}^{25} = 0.9751; n_D^{20}= 1.52295; λ_m = 277.5 (log ε 3.262 in EtOH); volatile with steam; nearly insoluble in H_2O; soluble in alcohol, Et_2O; LD_{50} (rbt orl) = 100 mg/kg.

6039 Cetalkonium Chloride
122-18-9 2059 204-526-3
$C_{25}H_{46}ClN$
Benzylhexadecyldimethylammonium chloride.
cetyldimethylbenzylammonium chloride; Acetoquate CDAC; Acquat CDAC; Ammonyx G; Ammonyx T; Banicol; Cetol; Zettyn. Anti-infective; topical. Quaternary ammonuim compound. mp = 59°; soluble in H_2O, alcohol, Me_2CO, EtOAc, propylene glycol, sorbitol solutions, glycerol, Et_2O, CCl_4; pH of aqueous solutions 7.2. *I.G. Farben; ICI ; Sterling Winthrop Inc.; Zeeland Chem. Inc.*

6040 Cethexonium Bromide
1794-74-7 2061
$C_{24}H_{50}BrNO$
Hexadecyl(2-hydroxycyclohexyl)dimethylammonium bromide.
Biocidan. Antiseptic. Quaternary ammonuim compound. mp = 75°; soluble in H_2O, alcohol, $CHCl_3$; practically insoluble in petroleum ether.

6041 Cethexonium Chloride
58703-78-9
$C_{24}H_{50}ClNO$
Hexadecyl(2-hydroxycyclohexyl)dimethylammonium chloride.
Antiseptic. Quaternary ammonuim compound.

6042 Cetylpyridinium Chloride
6004-24-6 2074
$C_{21}H_{40}ClNO$
1-Hexadecylpyridinium chloride monohydrate.
Halset. Antiseptic (topical); disinfectant. Quaternary ammonuim compound. mp = 77-78°; soluble in H_2O, alcohol, $CHCl_3$; slightly soluble in C_6H_6, Et_2O; LD_{50} (rat sc) = 250 mg/kg, (rat ip) = 6 mg/kg, (rat iv) = 30 mg/kg, (rat orl) = 200 mg/kg.

6043 Cetylpyridinium Chloride [anhydrous]
123-03-5 2074 204-593-9
$C_{21}H_{38}ClN$
1-Hexadecylpyridinium chloride.
Ceepryn; Cepacol; Dobendan; Medilave; Merocet; Pristacin; Pyrisept. Antiseptic; disinfectant. Quaternary ammonuim compound. *Hexcel; Marion Merrell Dow Inc.*

6044 Chlorhexidine
55-56-1 2140 200-238-7
$C_{22}H_{30}Cl_2N_{10}$
1,1'-Hexamethylenebis[5-(p-chlorophenyl)biguanide].
Antiseptic; disinfectant. A bisbiguanide with bacteriostatic activity. mp = 134°; strong alkaline reaction; soluble in H_2O at 20°. *ICI.*

6045 Chlorhexidine Acetate
56-95-1 2140 200-302-4
$C_{26}H_{38}Cl_2N_{10}O_4$
1,1'-Hexamethylenebis[5-(p-chlorophenyl)biguanide] diacetate.
Chlorasept 2000; Nolvasan. Antiseptic; disinfectant. A bisbiguanide with bacteriostatic activity. mp = 154-155°; neutral reaction; soluble in H_2O at 20° (1.9 g/100 ml); aqueous solutions decompose above 70°; soluble in alcohol, glycerol; propylene glycol; polyethylene glycols; LD_{50} (mus orl) = 2 g/kg. *ICI.*

6046 Chlorhexidine Gluconate
18472-51-0 2140 242-354-0
$C_{34}H_{54}Cl_2N_{10}O_{14}$
1,1'-Hexamethylenebis[5-(p-chlorophenyl)biguanide] di-D-gluconate.
Chlorhexamed; Bacticlens; Corsodyl; Gingisan; Hibiclens; Hibidil; Hibiscrub; Hibital; Hibitane; Peridex; pHiso-Med; Plac Out; Rotersept; Secalan; Sterilon; Unisept; component of: Hibistat. Antiseptic; disinfectant. A bisbiguanide with bacteriostatic activity. soluble in H_2O at 20°; LD_{50} (mus iv) 22 mg/kg, (mus orl) 1800 mg/kg. *Boots Pharmaceuticals Inc.; ICI ; Poythress.*

6047 Chlorhexidine Hydrochloride
3697-42-5 2140 223-026-6
$C_{22}H_{32}Cl_4N_{10}$
1,1'-Hexamethylenebis[5-(p-chlorophenyl)biguanide] dihydrochloride.
AY-5312; Lisium. Anti-infective (topical). Dec 260-262°; soluble in H_2O at 20° (0.06 g/ 100 ml). *ICI.*

6048 Chloroazodin
502-98-7 2171 207-955-4
$C_2H_4Cl_2N_6$
N,N''-Dichlorodiazenedicarboximidamide.
chlorazodin; dichloroazodicarbonamidine; Azochloramine. Antiseptic. Local anesthetic (veterinary). Dec explosively at 155°; sparingly soluble in H_2O, alcohol; slightly soluble in other organic solvents; all solutions decompose on exposure to light.

6049 Chlorocresol
59-50-7 2184 200-431-6
C_7H_7ClO
4-Chloro-m-cresol.
3-methyl-4-chlorophenol; parachlorometacresol; 6-chloro-m-cresol. Antiseptic; disinfectant. mp = 55.5°; bp = 235°; somewhate soluble in H_2O; more soluble in hot H_2O; Freely soluble in organic solvents. *Kalle BV.*

6050 Chloroxylenol
88-04-0 2228 201-793-8
C_8H_9ClO
4-Chloro-3,5-xylenol.
parametaxylenol; Benzytol; Dettol. Antibacterial; antiseptic (topical, urinary); germicide. A phenol derivative. mp = 115°; bp = 246°; volatile with steam; slightly soluble in H_2O; more soluble in organic solvents.

6051 Chlorquinaldol
72-80-0 2243 200-789-3
$C_{10}H_7Cl_2NO$
5,7-Dichloro-2-methylquinolin-8-ol.
hydroxydichloroquinaldinol; chloroquinaldol; Afungil; Quesil; Siogène; Gyno-Sterosan; Gynotherax; Saprosan; Sterosan; Steroxin; Siosteran. Antiseptic. Quinoline. mp = 114-115° (with slight decomposition); λ_m = 326 nm ($A^{1\%}_{1\,cm}$ 170 EtOH); nearly insoluble in H_2O; more soluble in organic solvents. *Ciba-Geigy Corp.*

6052 Clofucarban
369-77-7 2438 206-724-5
$C_{14}H_9Cl_2F_3N_2O$
N-(4-Chlorophenyl)-N'-[4-chloro-3-(trifluoromethyl)phenyl]urea.
halocarban; Irgasan CF3; Irgosan CF3. Antiseptic; disinfectant. Halogen containing compound. mp = 214-215°; insoluble in H_2O; soluble in organic solvents. *Am. Cyanamid; Ciba-Geigy Corp.*

6053 Clorophene
120-32-1 2469 204-385-8
$C_{13}H_{11}ClO$
2-Benzyl-4-chlorophenol.
clorofene; NSC-59989; Septiphene; Santophen 1. Antiseptic. Phenol. Used in disinfectant preparations. mp = 48.5°; $bp_{3.5}$ = 160-162°; $d^{55}_{15.5}$ = 1.186-1.190. *Astra Draco AB.*

6054 Cloxyquin
130-16-5 2483 204-978-1
C_9H_6ClNO
5-Chloroquinolin-8-ol.
cloxiquine; Chlorisept. Antiseptic; antifungal; antibacterial. A quinoline. mp = 130°; sparingly soluble in cold dilute HCl; [hydrochloride]: mp = 256-258°.

6055 Creosote, Coal Tar
8001-58-9 2641 232-287-5
Coal tar creosote.
coal tar; wash oil. Disinfectant. Also used as a wood preservative and insecticide. Distillate of coal tar produced by the high temperature carbonization of bituminous coal. It consists primarily of aromatic hydrocarbons, tar acids and tar bases. CAUTION: Readily absorbed through GI tract and skin. Overexposure can cause GI irritation and congestion. Direct contact may cause irritation, burning, itching, erythema, papular and vesicular eruptions, keratoconjunctivitis. Systemic poisoning can cause salivation, vomiting, respiratory difficulties, vertigo, headache, loss of pupillary reflexes, hypothermia, cyanosis, mild convulsions. Carcinogenic. Translucent brown to black oily liquid; flammable; flash point = 165°F (75°C); ignition temperature = 637°F (335°C); practically insoluble in H_2O.

6056 Creosote, Wood
8021-39-4 2642 232-419-1
Wood creosote.
wood creosote; beechwood creosote; Creosote. Antiseptic; expectorant. Liquid obtained from wood tars by distillation; composed mainly of guaiacol and creosol. Colorless or yellow oily liquid; D^{25}_{25} > 1.076; bp 203-220°; somewhat soluble in H_2O; soluble in glycerol, glacial acetic acid, fixed alkali hydroxied solutions; incompatible with acacia, albumin, oxidizers, and cupric, ferric, gold and silver salts.

6057 Cresol
1319-77-3 2645 215-293-2
C_7H_8O
Cresylic acid.
cresylol; tricresol. Antiseptic; disinfectant. A mixture of three isomeric cresols in which the m-isomer predominates. Obtained from coal tar. CAUTION: Poisonous. Has corrosive action on tissues. Potential symptoms of overexposure are CNS effects; skin and eye burns; dermatitis; liver, kidney and lung damage. Acute exposure can lead to muscular weakness, gastroenteric disturbances, severe depression, collapse, and death. Liquid; d^{25}_{25} = 1.030-1.038; distills between 195-205°; slightly soluble in H_2O; miscible with alcohol, C_6H_6, Et_2O, glycerol, petroleum ether; soluble in fixed alkali hydroxide solutions; [m-cresol]: LD_{50} (rat orl) = 2.02 g/kg.

6058 m-Cresyl Acetate
122-46-3 2651
$C_9H_{10}O_2$
Acetic acid 3-methylphenyl ester.
m-tolyl acetate; acetic acid ester; acetic acid m-cresol ester; acetylmetacresol; metacresol acetate; Cresatin; Cresatin Metacresylacetate; Cresatin-Sulzberger; Metacresylacetate-Sulzberger; Kresatin. Antiseptic (topical); antifungal. Oily liquid; bp = 212°; bp_{13} = 99°; d^{26}_4= 1.048; volatile with steam; λ_m + 262.5, 269.5 nm (in MeOH); practically insoluble in H_2O, glycerol; miscible with alcohol, H_2O, $CHCl_3$, petroleum ether, C_6H_6; soluble in petrolatum (5%), cottonseed oil.

6059 Cupric Sulfate
7758-99-8 2722
$CuSO_4.5H_2O$
Copper (II) sulfate (1:1) pentahydrate.
bluestone; blue vitriol; Roman vitriol; Salzburg vitriol. Antiseptic; antifungal(topical); antidote to phosphorus. Used in veterinary medicine as a mineral source, anthelmintic, emetic, fungicide. Occurs as the mineralchalcanthite. Loses $2H_2O$ at 30°, 2 more at 110°, and becomes anhydrous by 250°; $d^{15.6}_4$= 2.286; soluble in H_2O, MeOH, glycerol; slightly soluble in EtOH; LD_{50} (rat orl) = 960 mg/kg.

6060 Cupric Sulfate [anhydrous]
7798-98-7 2722
CuO_4S
Copper (II) sulfate (1:1).
Antiseptic; antifungal (topical); antidote to phosphorus. Occurs as the mineral hydrocyanite. Dec above 560°; d = 3.6; hygroscopic; soluble in H_2O; practically insoluble in alcohol.

6061 Dequalinium Chloride
522-51-0 2959 208-330-9
$C_{30}H_{40}Cl_2N_4$
1,1'-(1,10-Decanediyl)bis[4-amino-2-methylquinolinium chloride].
BADQ-10; decamine; dekamiln; Decatylen; Dekadin; Dequadin Chloride; Dequafungen; Dequavet; Dequavagyn; Eriosept; Evazol; Grocreme; Labosept;

Optipect; Phylletten; Polycidine; Sorot; component of: Efisol, Gargilon, Gramipan, Hexalyse, Micrin. Antiseptic; disinfectant. Quaternary ammonuim compound. mp = 326° (dec); soluble in H_2O (1 g/200 ml at 25°). *Allen & Hanbury.*

6062 Dibromopropamidine
496-00-4 3073
$C_{17}H_{18}Br_2N_4O_2$
4,4'-(Trimethylenedioxy)bis(3-bromobenzamidine).
dibrompropamidine. Antiseptic; antiamebic. Used as a preservative in cosmetics. *May & Baker Ltd.*

6063 Dibromopropamidine Isethionate
614-87-9 3073 210-399-5
$C_{21}H_{30}Br_2N_4O_{10}S_2$
4,4'-(Trimethylenedioxy)bis(3-bromobenzamidine) di(2-hydroxyethanesulfonate) (ester).
dibrompropamidine isethionate; Brolene Ointment; Brulidine. Antiseptic; antiamebic. Also used as a preservative in cosmetics. mp = 226°; soluble in H_2O (0.5 g/ml), EtOH (1.6 g/100 ml), glycerol; insoluble in Et_2O, $CHCl_3$, petroleum ether; incompatible with chlorides, sulfates, many organic anions (forms sparingly soluble salts). *May & Baker Ltd.*

6064 Disiquonium Chloride
68959-20-6 273-403-4
$C_{27}H_{60}ClNO_3$
Didecylmethyl[3-(trimethoxysilyl)propyl]ammonium chloride.
Antiseptic. Quaternary ammonuim compound.

6065 Dodecarbonium Chloride
100-95-8 3466 202-904-2
$C_{23}H_{41}ClN_2O$
N-[2-(Dodecylamino)-2-oxoethyl]-N,N-dimethylbenzenemethanaminium chloride.
Straminol; Urolocide. Antiseptic; disinfectant. Quaternary ammonuim compound. mp = 147-148°; soluble in H_2O, alcohol; insoluble in Et_2O, Me_2CO, C_6H_6; LD_{50} (rat orl) = 100 mg/kg. *Patchem AG.*

6066 Domiphen Bromide
538-71-6 3474 208-702-0
$C_{22}H_{40}BrNO$
Dodecyldimethyl(2-phenoxyethyl)ammonium bromide.
NSC-39415; PDDB; phenododecinium bromide; Bradosol Bromide; Oradol; Modicare; Neo-Bradoral. Antiseptic; anti-infective (topical). Quaternary ammonuim compound. mp = 112-113°; Freely soluble in warm H_2O; Soluble in EtOH, Me_2CO, EtOAc, $CHCl_3$; very slightly soluble in C_6H_6; incompatible with soap. *CIBA plc.*

6067 Ethylhydrocupreine
522-60-1 3855 208-333-5
$C_{21}H_{28}N_2O_2$
(8α,9R)-6'-Ethoxy-10,11-dihydrocinchonan-9-ol.
hydrocupreine ethyl ether; Numoquin, Optoquine. Antiseptic. Quinoline. mp = 123-128°; $[\alpha]_D^{25}$ = -136° (EtOH); insoluble in H_2O; soluble in organic solvents.

6068 Ethylhydrocupreine Hydrochloride
3413-58-9 3855 222-302-3
$C_{21}H_{29}ClN_2O_2$
(8α,9R)-6'-Ethoxy-10,11-dihydrocinchonan-9-ol monohydrochloride.
Neumolisina. Antiseptic. Quinoline. mp = 252-1254sg; $[\alpha]_D^{21}$ = -123.6° (H_2O); soluble in H_2O, alcohol, $CHCl_3$; sparingly soluble in dry Me_2CO; practically insoluble in Et_2O; light sensitive.

6069 Ethylparaben
120-47-8 3883 204-399-4
$C_9H_{10}O_3$
Ethyl p-hydroxybenzoate.
Nipagin A; Ethyl Parasept; Solbrol A. Antifungal; antiseptic. Used as a preservative for pharmaceuticals. mp = 116°; bp = 297-298°; soluble in alcohol, Et_2O, H_2O.

6070 Euprocin
1301-42-4 3950
$C_{24}H_{34}N_2O_2$
(8α,9R)-10,11-Dihydro-6'-(3-methylbutoxy)cinchonan-9-ol.
hydrocupreine isopentyl ether; isoamylhydrocupreine; isopentylhydrocupreine; $O^{6'}$-isopentylhydrocupreine; eucupreine; Eucupin. Antiseptic; local anesthetic. A quinoline. mp = 152°; soluble in EtOH, Et_2O, $CHCl_3$; insoluble in H_2O; LD_{50} (mus sc) = 300 mg/kg, (rbt iv) = 13 mg/kg. *Schering-Plough HealthCare Products.*

6071 Euprocin Hydrochloride
18984-80-0 3950
$C_{24}H_{38}Cl_2N_2O_3$
(8α,9R)-10,11-Dihydro-6'-(3-methylbutoxy)cinchonan-9-ol dihydrochloride monohydrate.
isoamylhydrocupreine hydrochloride; isopentylhydrocupreine dihydrochloride; WI-287; component of: Otodyne. Antiseptic; local anesthetic. Quinoline. Freely soluble in EtOH, soluble in H_2O (6.6 g/100 ml). *Schering AG.*

6072 Fenticlor
97-24-5 4046 202-568-7
$C_{12}H_8Cl_2O_2S$
2,2'-Thiobis[4-chlorophenol].
S-7; NSC-4112; Novex. Anti-infective (topical). Phenol used as a fungicide, especially against *Monosporidium apiospermum.* mp = 175°; soluble in aqueous NaOH solutions, alcohol, hot C_6H_6. *I.G. Farben.*

6073 Fludazonium Chloride
53597-28-7
$C_{26}H_{20}Cl_5FN_2O_2$
1-[2,4-Dichloro-β-[(2,4-dichlorobenzyl)oxy]phenethyl]-3-(p-fluorophenacyl)imidazolium chloride.
R-23633. Antiseptic. *Janssen Pharm. Inc.*

6074 Fluorosalan
4776-06-1 4216 225-322-0
$C_{14}H_8Br_2F_3NO_2$
3,5-Dibromo-2-hydroxy-N-[3-(trifluoromethyl)phenyl]-benzamide.
flusalan; Fluorophene. Antiseptic; disinfectant.

6075 Furazolidone
67-45-8 4320 200-653-3

$C_8H_7N_3O_5$

3-[(5-Nitrofurfurylidene)amino]-2-oxazolidinone.

NF-180; Furovag; Furox; Furoxane; Furoxone; Giarlam; Giardil; Medaron; Neftin; Nicolen; Nifulidone; Ortazol; Roptazol; Tikofuran; Topazone; component of: Tricofuron. Antiseptic (topical); anti-infective (topical); nitrofuran antiprotozoal (topical) used against trichomonas. mp = 256-257°; soluble in H_2O (0.004 g/100 ml); decomposed by alkali. *Norwich Eaton; Roberts Pharm. Corp.; SmithKline Beecham Animal Health; Solvay Animal Health Inc.*

6076 Fursalan
15686-77-8

$C_{12}H_{13}Br_2NO_3$

3,5-Dibromo-N-(tetrahydrofurfuryl)salycylamide.

Disinfectant.

6077 Gallacetophenone
528-21-2 4360 208-430-2

$C_8H_8O_4$

1-(2,3,4-Trihydroxyphenyl)ethanone.

Alizarine yellow C. Antiseptic. mp = 173°; λ_m (MeOH) = 237,296 (ε 8560, 12500); soluble in H_2O, alcohol, Et_2O, solution of sodium acetate. *Eastman Kodak.*

6078 Halimide®
19014-05-2 4624 242-754-5

$C_{22}H_{40}ClN$

Dodecylbenzyltrimethylammonium chloride.

Antiseptic; disinfectant; cationic surface active agent. Quaternary ammonuim compound. *Aktieselskabet Pharmacia.*

6079 Halquinol
8067-69-4 4637

5,7-Dichloro-8-quinolinol mixture with 5-chloro-8-quinolinol and 7-chloro-8-quinolinol.

chlorquinol; SQ-16401; CHQ; Halquivet; Quinolor; Quixalin; Quixalud; Tarquinor. Anti-infective (topical). Quinoline. *Olin Mathieson.*

6080 Hexachlorophene
70-30-4 4716 200-733-8

$C_{13}H_6Cl_6O_2$

2,2'-Methylenebis[3,4,6-trichlorophenol].

AT-7; G-11; Bilevon; Dermadex; Exofene; Hexosan; pHisohex; Soy-Dome; Surgi-Cen; Surofene. Antiseptic; disinfectant; anthelmintic (flukicide, veterinary). Phenol used mainly in the manufacture of soaps. mp = 164-165°; practically insoluble in H_2O; soluble in common organic solvents; LD_{50} (mrat orl) = 66 mg/kg, (frat orl) = 57 mg/kg. *Miles Inc.; Sterling Winthrop Inc.*

6081 Hydrargaphen
14235-86-0 4805 238-107-1

$C_{33}H_{24}Hg_2O_6S_2$

Phenylmercuric 3,3'-methylenebis(2-naphthalene-sulfonate).

phenylmercuric Fixtan; Conotrane; Fibrotan; Hydraphen; Penotrane; P.M.F.; Versotrane; Septotan. Antiseptic; anti-infective (topical). Mercurial compound. Practically insoluble in H_2O; forms colloidal solutions in alkali methal dinaphthylmethane sulfonates; colloid tends to absorb at interfaces and form charged hydrated aggregates; LD_{50} (mus orl) = 80 mg/kg. *Ward Blenkinsop.*

6082 Hydrastine
118-08-1 4806 204-233-0

$C_{21}H_{21}NO_6$

[S-(R*,S*)]-6,7-Dimethoxy-3-(5,6,7,8-tetrahydro-6-methyl-1,3-dioxolo[4,5-g]isoquinolin-5-yl)-1(3H)-isobenzo-furanone.

l-β-hydrastine. Antiseptic. A quinoline naturally occurring in l-β-form in *Hydrastitis canadensis L. Ranunculaceae* together with berberine and canadine. Hydrochloride formerly used as a uterine hemostatic. mp = 132°; mp [hydrochloride] = 116°; $[\alpha]_D^{20}$ = -50° (c = 0.3 abs alcohol); λ_m = 202, 218, 238, 298, 316 mn (log ε 4.79, 4.53, 4.15, 3.86, 3.63 EtOH); pK = 7.8; soluble in Me_2CO, C_6H_6; insoluble in H_2O; hydrochloride is very soluble in H_2O.

6083 Hydrogen Peroxide
7722-84-1 4839 231-765-0

H_2O_2

Hydrogen dioxide.

hydroperoxide; Albone; Hioxyl; Lensan A; Mirasept; Oxysept; Pegasyl. Antiseptic (topical). Peroxide. Marketed as a solution in H_2O in concentrations of 3-90% by weight. Also used in rocket propulsion, as a dough conditioner, and as a maturing and bleaching agent in food. CAUTION: Symptoms of overexposure include irritation of eyes, nose and throat; corneal ulceration; erythema, vesicles on skin; bleaching of hair.

6084 8-Hydroxyquinoline
148-24-3 4890 205-711-1

C_9H_7NO

8-Quinolol.

oxoquinoline; hydroxybenzopyridine; oxybenzopyridine; phenopyridine; oxychinolin; oxine. Antiseptic. Quinoline. Also used as a fungistat, a chelating agent in the determination of trace metals. mp = 76°; bp 267°; nearly insoluble in H_2O, Et_2O; soluble in Me_2CO, alcohol, $CHCl_3$, C_6H_6, aqueous mineral acids; LD_{50} (mus ip) = 48 mg/kg.

6085 8-Hydroxyquinoline Aluminum Sulfate
153-77-5 4890

$C_{27}H_{24}AlN_3O_{15}S_3$

8-Quinolol aluminum sulfate.

Nyxolan; Aloxyn. Anthelmintic. Used as antiperspirant, deodorant.

6086 8-Hydroxyquinoline Sulfate
134-31-6 4890 205-137-1

$C_{18}H_{16}N_2O_6S$

8-Quinolol sulfate (2:1) salt.

Chinosol. Antiseptic. Quinoline. Also used as a fungistat, a chelating agent in the determination of trace metals. mp = 175-178°; soluble in H_2O; slightly soluble in glycerol, alcohol; insoluble in Et_2O.

6087 Ichthammol
8029-68-3 4929 232-439-0

Ammonium bituminosulfonate.

ammonium bithiolcium; ammonium ichthosulfonate; ammonium sulfobituminate; ammonium sulfoichthyolate; bitumol; bituminol; ichthammonium; ichthosulfol;

Amsubit; Bitulan; Ichthosauran; Ichthymall; Hirathiol; Ichden; Ichtammon; Ichthadone; Ichthymall; Ichthyol; Ichthysalle; Ichthalum; Ichthium; Ichthosan; Ichthosauran; Ichthynat; Ichthyopon; Ichtopur; Leukochthol; Lithol; Petrosulpho; Perichthol; Piscarol; Pisciol; Adnexol [injectable form]. Antiseptic (topical). A product of a distillate from mineral deposits (bituminous schists); contains saturated and unsaturated hydrocarbons, nitrogenous acids, thiophene derivatives. Miscible with H_2O, glycerol, propylene glycol, fats, oils, carbowaxes, lanolin; partially soluble in alcohol, Et_2O. *Mallinckrodt Inc.*

6088 Ictasol
12542-33-5
$C_{28}H_{36}Na_2O_6S_3$ (tentative)
Ictasol.
Disinfectant. A product of a distillate from mineral deposits (bituminous schists).

6089 Iodic Acid
7782-68-5 5032 231-962-1
HIO_3
Antiseptic; astringent; disinfectant. Halogen containing compound. mp = 110° (dec); d^0_4 = 4.629; soluble in H_2O (269 g/ 100 ml at 20°, 295 g/100 ml at 40°), HNO_3, dilute EtOH; insoluble in absolute EtOH, Et_2O, $CHCl_3$; darkens on exposure to light.

6090 Iodine
7553-56-2 5034 231-442-4
I_2
Iodine.
Antihyperthyroid; topical anti-infective. Also used, particularly in veterinary medicine, to treat goiter. Halogen. CAUTION: Overexposure could cause irritation of eyes and nose, lacrimation, headache, tight chest, skin burns or rash, cutaneous hypersensitivity. Highly corrosive on the GI tract. mp = 113.6°, bp = 185.2°; d^{25} = 4.93; soluble in H_2O (0.017 g/100 ml), C_6H_6 (14.09 g/100 g), CS_2 (16.5 g/100 g), EtOH (21.43 g/100 g), Et_2O (25.2 g/100 g), C_6H_{12} (2.7 g/100 g), CCl_4 (2.6 g/100 g at 35°); incompatible with tannins, alkaloids, starch.

6091 Iodine Monochloride
7790-99-0 5038 232-236-7
CII
Iodine chloride.
Wijs' chloride. Anti-infective (topical). Halogen containing compound. CAUTION: Attacks the skin, forming dark, painful patches. mp [α form]: = 27.2°; mp [β form]: = 13.9°; bp [β form]: = 97° (dec); d^{29}_4 [β form]: = 3.10; soluble in H_2O, alcohol, Et_2O, CS_2, AcOH.

6092 Iodine Trichloride
865-44-1 5041 212-739-8
Cl_3I
Iodine chloride.
Anti-infective (topical). Halogen containing compound. Also used as a chlorinating and oxidizing agent. CAUTION: Corrosive to human skin. Concentrated solutions are strongly irritating. mp 33°; d^{-4} = 3.203; volatile at room temperature.

6093 Iodochlorhydroxyquin
130-26-7 5052 204-984-4
C_9H_5ClINO
5-Chloro-7-iodo-8-quinolinol.
clioquinol; chloroiodoquin; iodochlorohydroxyquinoline; iodochloroxyquinoline; Amebil; Alchloquin; Amoenol; Bactol; Barquinol; Budoform; Chinoform; Clioquinol; Cliquinol; Cort-Quin; Eczecidin; Enteroquinol; Entero-Septol; Entero-Vioform; Enterozol; Entrokin;Hi-Eneterol; Iodoenterol; Nioform; Nystaform; Quinambicide; Quin-O-Crème; Rheaform Boluses; Rometin; Vioform; Vioformio; component of: Domeform-HC, Formtone-HC, Lidaform-HC, Nystaform, Nystaform-HC, Racet, Vioform-Hydrocortisone. Antiamebic; topical anti-infective. Also used as an intestinal anti-infective (veterinary). Quinoline. CAUTION: Has been linked with occurrence of subacute myelo-optic neuropathe. Dec 178-179°; λ_m = 266 nm ($A^{1\%}_{1\,cm}$ 1120 in 0.1 N MeOHic NaOH), 269 nm ($A^{1\%}_{1\,cm}$ 1120 (MeOH/KOH), 255 nm ($A^{1\%}_{1\,cm}$ 1570 EtOH); slightly soluble in $CHCl_3$, AcOH; nearly insoluble in H_2O, cold alcohol, Et_2O; LD_{50} (cat orl) = 400 mg/kg. *Bayer AG; Ciba-Geigy Corp.; Dermik Labs. Inc.; Lemmon Co.; Marion Merrell Dow Inc.*

6094 Iodoform
75-47-8 5054 200-874-5
CHI_3
Triiodomethane.
Antiseptic (topical). Halogen containing compound. mp 120°; dec at high temperature with the evolution of iodine; volatile with steam; slightly soluble in H_2O; more soluble in common organic solvents; LD_{50} (mus sc) = 1.6 mmoles/kg.

6095 Isomerol [a]
7256-12-6
C_7H_6HgO
2-Methyl-7-oxa-8-mercurabicyclo[4.2.0]octa-1,3,5-triene.
Parahydrecin; component: of Unguentine. Antiseptic. *Norwich Eaton.*

6096 Isomerol [b]
7256-12-7
C_7H_6HgO
4-Methyl-7-oxa-8-mercurabicyclo[4.2.0]octa-1,3,5-triene.
Parahydrecin; component: of Unguentine. Antiseptic. *Norwich Eaton.*

6097 Jothion
Calisaya bark; Peruvian bark; Cinchona bark. Dried stem or bark of various species of cinchona. Contains up to 35 alkaloids such as quinine, cinchotannic, quinic and quinovic acids. Used as an antimalarial.

6098 Lauralkonium Chloride
19486-61-4 205-351-5
$C_{29}H_{44}ClNO_2$
Benzyl[2-[p-(lauroyl)phenoxy]ethyl]dimethylammonium chloride.
Antiseptic.

6099 Laurolinium Acetate
146-37-2 5399 205-668-9
$C_{24}H_{38}N_2O_2$
4-Amino-1-dodecylquialdinium acetate.

Laurodin. Antiseptic. Quaternary ammonuim compound. mp = 170-171°; soluble in H_2O; LD_{50} (mus orl) = 131± 36.2 mg/kg, (mus sc) = 30.2 ± 5.6 mg/kg, (mus ip) = 2.3 ± 0.2 mg/kg. *Allen & Hanbury*.

6100 Magnesium Peroxide
14452-57-4 5717 238-438-1
MgO_2
Magnesium dioxide.
magnesium perhydrol; magnesium superoxol. Antiseptic; antacid. Peroxide. The commercial product contains 15-25% MgO_2, the balance being $Mg(OH)_2$. Insoluble in H_2O; gradually dec in H_2O with liberation of O_2; soluble in dilute acids, forming H_2O_2.

6101 Mecetronium Ethylsulfate
3006-10-8 221-106-5
$C_{22}H_{49}NO_4S$
Ethylhexadecyldimethylammonium ethyl sulfate.
mecetronium etilsulfate. Antiseptic. *Asahi Chem. Industry*.

6102 Meralein Sodium
4386-35-0 5912 224-498-6
$C_{19}H_9HgI_2NaO_5$
(3',6'-Dihydroxy-2',7'-diiodospiro[3H-2,1-benzoxathiole-3,9'-[9H]xanthen]-4'-yl)hydroxymercury S,S-dioxide monosodium salt.
sodium meralein; Merodicein. Antiseptic (topical). Mercurial compound. Soluble in H_2O. *Hynson Westcott & Dunning*.

6103 Merbromin
129-16-8 5914 204-933-6
$C_{20}H_8Br_2HgNa_2O_6$
(2',7'-Dibromo-3',6'-dihydroxy-3-oxospiro[isobenzofuran-1(3H),9'-[9H]xanthen]-4'-yl)hydroxymercury disodium salt.
mercurochrome; dibromohydroxymercurifluorescein disodium salt; no. 220 sol; Mercurochrome-220 Soluble; Chromargyre; Planochrome; Flavurol; D.M.O.F.; Mercurophage; Mercurocol; Gallochrome; Gynochrome; Mercurome; Asceptichrome; Mercuranine. Antibacterial; antiseptic (veterinary). Mercurial compound. Soluble in H_2O; incompatible with acids, alkaloidal salts, most local anesthetics. *Hynson Westcott & Dunning*.

6104 Mercufenol Chloride
90-03-9 5920 201-962-6
C_6H_5ClHgO
Chloro(o-hydroxyphenyl)mercury.
2-chloromercuriophenol; U-7743; Myringacaine Drops; Salicresin Fluid; component of: Mercresin. Anti-infective (topical); disinfectant. CAUTION: poisonous. Prepared from phenol and mercuric acetate. mp = 150152°; slightly soluble in cold H_2O; moderately soluble in boiling H_2O; freely soluble in alcohol, hot C_6H_6; sparingly soluble in $CHCl_3$. *Upjohn Ltd*.

6105 Mercuric Chloride
7487-94-7 5926 231-299-8
Cl_2Hg
Mercury bichloride.
mercury chloride; mercury dichloride; corrosive sublimate; mercury perchloride; corrosive mercury chloride. Antiseptic (topical); disinfectant. Mercurial compound. CAUTION: *Violent poison*. May be fatal if swallowed. mp = 277°; d = 5.4; volatilizes at 300°; soluble in H_2O and many other common solvents; incompatible with formates, sulfites, albumin, gelatin, alkalies, alkaloid salts, ammonia, lime H_2O, antimony, arsenic, bromides, borax, carbonates, reduced iron, copper, iron, lead, silver salts, tannic acid, vegetable astringents.

6106 Mercuric Sulfide, Red
1344-48-5 5945 215-696-3
HgS
Mercury (II) sulfide.
vermillion; Chinese red; C.I. Pigment Red; C.I. 77766. Antiseptic. Mercurial compound that occurs in nature as the mineral cinnabar. Prepared from mercuric acetate, ammonium thiocyanate, glacial acetic acid and hydrogen sulfide. Blackens on exposure to light; practically insoluble in H_2O.

6107 Mercurophen
52486-78-9 5947
$C_6H_4HgNNaO_4$
Sodium 4-(hydroxymercuri)-2-nitrophenolate.
Antiseptic; disinfectant. Mercurial compound. CAUTION: *Poisonous*. Soluble in hot H_2O.

6108 Mercurous Acetate
631-60-7 5948 211-161-3
$C_6H_6Hg_2O_4$
Dimercury di(acetate).
mercury acetate. Antiseptic. Mercurial compound. Somewhat soluble in H_2O, dilute AcOH; insoluble in alcohol, Et_2O; aqueous solutions decompose quickly in light and heat.

6109 Mercurous Chloride
10112-91-1 5951 233-307-5
Hg_2Cl_2
Dimercury dichloride.
calomel; mild mercury chloride; mercury monochloride; mercury subchloride; precipité blanc; Calogreen. Antiseptic; cathartic; diuretic; antisyphilitic. Mercurial compound. CAUTION: Excessive doses can cause mercury poisoning. Sublimes at 400-500°; d = 7.15; practically insoluble in H_2O; insoluble in alcohol, Et_2O; incompatible with bromides, iodides, alkali chlorides, sulfates, sulfites, carbonates, hydroxides, lime H_2O, acacia, ammonia, golden antimony sulfide, cocaine, suanides, copper salts, hydrogen peroxide, iodine, iodoform, lead salts, silver salts, soap, sulfides.

6110 Mercurous Iodide
15385-57-6 5953 239-409-6
Hg_2I_2
Dimercury diiodide.
yellow mercury iodide; mercury protoiodide. Antiseptic; antibacterial. Mercurial compound. CAUTION: Mercury yellow or green when combined with a soluble iodide forms a highly *poisonous* mercuric iodide. mp = 290°; d = 7.70; insoluble in H_2O, alcohol, Et_2O; soluble in solutions of mercurous or mercuric nitrates; incompatible with soluble iodides; sensitive to light.

6111 Metabromsalan
2577-72-2 219-933-1
$C_{13}H_9Br_2NO_2$
3,5-Dibromosalicylanilide.
NSC-526280. Disinfectant.

6112 Methenamine Tetraiodine
12001-65-9 6043
$C_6H_{12}I_4N_4$
Hexamethylenetetramine tetraiodide.
Iodoformine; Mirion; Siomine. Antiseptic; Iodine source. Prepared from potassium mercuric iodide and methenamine. Reddish powder; deflagrates at 138°; nearly insoluble in H_2O; slightly soluble in alcohol, $CHCl_3$, Et_2O, carbon disulfide; soluble in aqueous solutions of sodium or potassium iodides, sodium thiosulfate, dilute HCl; decomposition likely in aqueous solution.

6113 2-(Methoxymethyl)-5-nitrofuran
586-84-5 6075 209-586-4
$C_6H_7NO_4$
2-Nitro-2-furfurylmethyl ether.
Furbenal; Furaspor. Antiseptic; antifungal. A nitrofuran. Prepared through nitration of furfuryl mether ether in acetic anhydride. Oily liquid; bp_3 = 104-105°; bp_4 = 114-117°; d_{20}^{20} = 1.283; n^{20}_D = 1.5325-1.5343; miscible with EtOH; soluble in H_2O.

6114 Methylbenzethonium Chloride
1320-44-1 6103 246-675-7
$C_{28}H_{45}ClNO_4$
Benzyldimethyl[2-[2-[[4-(1,1,3,3-tetramethylbutyl)tolyl]ethoxy]ethyl]ammonium chloride monohydrate.
Delvan; Diaparene chloride; Hyamine 10X. Anti-infective (topical). Quaternary ammonuim compound. mp = 161-163°; freely soluble in H_2O, alcohol, Cellosolve; $CHCl_3$; hot C_6H_6. *Miles Inc.; Sterling Winthrop Inc.*

6115 Myrtol
8002-55-9 6423
Gelomyrtol.
Fraction of the volatile oil from *Myrtus communis L, Myrtaceae distilling between 160-180°, consisting chiefly of eualyptol and dextro*-pinene with some camphor. Liquid; d 0.895; $[\alpha]_D^{20}$= +10°; n_D^{20} 1.465; freely soluble in alcohol, Et_2O.

6116 1-Naphthyl Salicylate
550-97-0 6501
$C_{17}H_{12}O_3$
2-Hydroxybenzoic acid 1-naphthalenyl ester.
α-naphthyl salicylate; α naphthol salicylate; Alphol. Anti-infective; anti-inflammatory. Derivative of salicylic acid. A phenol. mp = 83°; insoluble in H_2O; soluble in EtOH, Et_2O, oils.

6117 2-Naphthyl Salicylate
613-78-5 6502 210-355-5
$C_{17}H_{12}O_3$
2-Hydroxybenzoic acid 2-naphthalenyl ester.
α-naphthol salicylate; Betol; Naphthalol; Naphthosalol; Salinaphthol. Antiseptic; anti-inflammatory. Phenol. mp = 95°; insoluble in H_2O, glycerol; sparingly soluble in cold alcohol; soluble in C_6H_6, Et_2O, boiling alcohol.

6118 Negatol®
9011-02-3 6531
Hydroxymethylbenzenesulfonic acid polymer with formaldehyde.
Albocresil; Albothyl; Negatan. Antiseptic. High molecular weight colloidal product prepared by reacting sulfonated m-cresol and formaldehyde. Soluble in H_2O forming colloidal solutions; pH (5% w/v) 1.0. *Eli Lilly & Co.*

6119 Nidroxyzone
405-22-1 6615 206-970-3
$C_8H_{10}N_4O_5$
5-Nitro-furaldehyde-2-(2-hydroxyethyl)semicarbazone.
NF-67; Furadroxyl. Antiseptic. Nitrofuran. mp = 214-216° (dec); slightly soluble in H_2O (1:2000). *Norwich Eaton.*

6120 Nifuroxime
6236-05-1 6627 228-349-6
$C_5H_4N_2O_4$
5-Nitro-2-furaldehyde oxime.
Micofur; Mycofur; component of: Tricofuron. Antibacterial; topical anti-infective; antiprotozoal, against Trichomonas. Nitrofuran. mp = 156°, 163-164°; soluble in H_2O (0.1 g/100 ml at 25°), MeOH (8.9 g/100 ml), EtOH (3.9 g/100 ml). *Norwich Eaton.*

6121 Nitrofurazone
59-87-0 6697 200-443-1
$C_6H_6N_4O_4$
5-Nitro-2-furaldehyde semicarbazone.
Aldomycin; Amifur; Furacin; Chemofuran; Furesol; Nifuzon; Nitrofural; Nitrozone; Furacinetten; Furacoccid; Furazol W; Mammex; Furaplast; Coxistat; Aldomycin; Nefco; Vabrocid; component of: Furacort, Furadex, Furea. Antibacterial. Used as a topical anti-infective. A nitrofuran. Dec 236-240°; λ_m = 260 375 nm; slightly soluble in H_2O (0.023 g/100 ml), EtOH (0.17 g/100 ml), proylene glycol (0.29 g/100 ml); insoluble in Et_2O; pH (saturated aqueous solution) 6.0-6.5; LD_{50} (rat orl) = 590 mg/kg, (rat sc) = 3000 mg/kg. *Norwich Eaton; Roberts Pharm. Corp.; SmithKline Beecham Animal Health.*

6122 Nitromersol
133-58-4 6707 205-112-5
$C_7H_5HgNO_3$
5-Methyl-2-nitro-7-oxa-8-mercurabicyclo[4.2.0]octa-1,3,5-triene.
Metaphen. Antiseptic; anti-infective (topical). Mercurial compound. Anhydride of 4-nitro-3-hydroxymercuri-o-cresol. Insoluble in H_2O; nearly insoluble in Me_2CO, alcohol, Et_2O, aqueous sodium carbonate solutions; soluble in alkali, ammonia by opening anhydride ring and forming a salt; soluble in boiling glacial acetic acid.

6123 Noxythiolin
15599-39-0 6822 239-679-5
C_3H_8OS
1-Hydroxymethyl-3-methyl-2-thiourea.
noxytiolin; Noxyflex-S. Antiseptic. mp = 88-90°; soluble in H_2O; LD_{50} (mus orl) > 3 g/kg. *E. Geistlich Sohne.*

6124 Octenidine
71251-02-0 6852
$C_{36}H_{62}N_4$
N,N'-(1,10-Decanediyldi-1(4H)-pyridinyl-4-ylidene)bis [1-octanamine].
Win-41464; Win-41464-6 [as disaccharin]. Antiseptic. *Sterling Winthrop Inc.*

6125 Octenidine Hydrochloride
70775-75-6 6852 274-861-8
$C_{36}H_{64}Cl_2N_4$
N,N'-(1,10-Decanediyldi-1(4H)-pyridinyl-4-ylidene)bis [1-octanamine] dihydrochloride.
Win-41464-2; Neo Kodan; Octeniderm; Octenisept. Antiseptic. mp = 215-217°. *Sterling Winthrop Inc.*

6126 Ornidazole
16773-42-5 7000 240-826-0
$C_7H_{10}ClN_3O_3$
α-(Chloromethyl)-2-methyl-5-nitro-1H-imidazole-1-ethanol.
Ro-7-0207; Madelen; Ornidal; Tiberal. Anti-infective. mp = 77-78°; λ_m = 288, 312 nm (ε 3720, 9150 in 2-propanol); pKa = 2.4 ± 0.1; LD_{50} (mus orl) > 2000 mg/kg, (mus ip) > 2000 mg/kg. *Hoffmann-LaRoche Inc.*

6127 Oxychlorosene
8031-14-9 7090
$C_{20}H_{35}ClO_4S$
Monoxychlorosene.
Clorpactin; Clorpactin XCB. Antiseptic. A buffered organic hypochlorous acid derivative with slightly acid pH. Halogen containing compound. Aqueous solutions are unstable and should be freshly prepared. *Guardian Labs.*

6128 Oxychlorosene Sodium
52906-84-0 7090
$C_{20}H_{35}ClO_4NaS$
Monoxychlorosene sodium.
Clorpactin WCS. Antiseptic. Halogen containing compound. *Guardian Labs.*

6129 Parachlorophenol
106-48-9 2206 203-402-6
C_6H_5ClO
4-Chlorophenol.
p-chlorophenol. Antiseptic. CAUTION: Irritating to skin. May cause tremors, convulsions, dyspnea, coma. mp = 43.2-43.7°; bp = 220°; d_4^{78} = 1.2238; n_D^{50} = 1.5419; n_D^{40} = 1.5579; sparingly soluble in H_2O, liquid petrolatum; soluble in alcohol, glycerin, Et_2O, $CHCl_3$, fixed and volatile oils; LD_{50} (rat orl) = 670 mg/kg. *Dow Chem. U.S.A.; Merrell Pharm. Inc.*

6130 Parachlorophenol, Camphorated
8003-18-7
Anti-infective, topical (dental).

6131 Phenoctide
78-05-7 7389 201-078-0
$C_{27}H_{42}ClNO$
N,N-Diethyl-N-[2-[4-(1,1,3,3-tetramethylbutyl)phenoxy]ethyl]benzenemethanaminium chloride.
Octaphen. Anti-infective (topical). Quaternary ammonuim compound. mp = 112-114°, 95°. *Ward Blenkinsop.*

6132 Phenoline
Carbolic camphor. Phenol with camphor (*phenol cum camphorae B.P.*).

6133 Phenosalyl
$C_{23}H_{29}NO_3.HCl$
1-(3-Hydroxy-3-phenylpropyl)-4-phenyl-4-piperidine carboxylic acid ethyl ester hydrochloride.
Lealgin; Operidine. Analgesic and narcotic.

6134 2-Phenoxyethanol
122-99-6 7410 204-589-7
$C_8H_{10}O_2$
1-Hydroxy-2-phenoxyethane.
ethylene glycol monophenyl ether; β-hydroxyethyl phenyl ether; Phenoxethol; Phenoxetol; Phenyl Cellosolve. Antiseptic. Oily liquid; d_{20}^{20} = 1.1094; d_4^{22} = 1.102; mp = 14°; bp_{80} = 165°; bp_{25} = 137°; n_D^{20} = 1.534; flash point = 250°; soluble in H_2O; freely soluble in alcohol, Et_2O, NaOH solutions; LD_{50} (rat orl) = 1.26 mg/kg.

6135 Phenylmercuric Borate
102-98-7 7456 203-068-1
$C_5H_7BHgO_3$
(Dihydrogen borato)phenylmercury.
phenylmercuric borate; phenylmercury borate; Famosept; Merfen. Antiseptic, topical. mp = 112-113°; soluble in H_2O, alcohol, glycerol.

6136 Phenylmercuric Nitrate, Basic
55-68-5 7455
$C_{12}H_{11}Hg_2NO_4$
Nitratophenylmercury.
merphenyl nitrate; (nitrato-O)-phenylmercury; Phe-Mer-Nite; Phenmerzyl Nitrate. Antiseptic. Pharmaceutic aid; bactericide; germicide. mp = 187-190° (dec); moderately soluble in H_2O, alcohol, glycerol; practically insoluble in organic solvents; LD_{50} = (mus sc) 0.045 mg/kg, (mus iv) 0.027 mg/kg. *Marion Merrell Dow Inc.; Schering AG.*

6137 Picloxydine
5636-92-0 7553 227-084-3
$C_{24}H_2H_4Cl_2N_{10}$
1,1'-[1,4-Piperazinediyl-bis(imidocarbonyl)]bis[3-(p-chlorophenyl)guanidine].
Antibacterial (topical); antiseptic. A heterocyclic biguanidine with antibacterial activity. *Nicholas Labs. Ltd.*

6138 Plastosol
The specific substance derived from plasma. It is an inactive form of plasmin which, when activated, by natural activators such as streptokinase or urokinase, has the property of lysing fibrinogen, fibrin, and other proteins.

6139 Potassium Permanganate
7722-64-7 7824 231-760-3
$KMnO_4$
Permanganic potassium salt.
chameleon mineral. Anti-infective (topical).

Permanganate. CAUTION: Explosive when brought into contact with organic or other readily oxidizable substances, either in solution or dry state. Dilute solutions are mildly irritating; concentrated solutions are caustic. Dec 240°; d = 2.7; soluble in H_2O; decomposed by alcohol and many other organic solvents, concentrated acids, reducing agents; incompatible with alcohol, arsenites, bromides, iodides, hydrochlorid acid, charcoal, organic substances, ferrous or mercuic salts, hypophosphites, hyposulfites, sulfites, peroxides, oxalates; LD_{50} (rat orl) = 1090 mg/kg.

6140 Potassium Tetraiododimercurate(II)
7783-33-7 7862 231-990-4
HgI_4K_2
Dipotassium tetraiodomercurate.
mercuric potassium iodide. Anti-infective (topical); disinfectant. Mercurial compound. CAUTION: *Poisonous.* Soluble in H_2O, alcohol, Et_2O, Me_2CO.

6141 Potassium Triiododimercurate(II) Solution
22330-18-3 7870 244-913-4
Potassium triiodomercurate.
Mercuric potassium iodide solution; potassium mercuriiodide solution; solution potassium iodohydrargyrate; Channing's solution; Thoulet's solution. Antiseptic. Mercurial compound. Also used reagent for alkaloids. CAUTION: *Poisonous.*

6142 Povidone-Iodine
25655-41-8 7880
$(C_6H_9NO)_n.xI$
1-Vinyl-2-pyrrolidinone polymer compound with iodine.
Betadine; Betaisodona; Braunol; Braunosan H Disadine DP; Disphex; Efodine; Inadine; Isodine; Proviodine; Traumasept; Videne; PVP-I; PVP-Iodine; PVP-Iodine, 30-06. Anti-infective, topical. An iodophor. Soluble in alcohol, H_2O; practically insoluble in $CHCl_3$, CCl_4, Et_2O, solvent hexane, Me_2CO. *Intl. Specialty Products; Purdue Pharma L.P.*

6143 Propamidine
104-32-5 7981 203-195-2
$C_{17}H_{20}N_4O_2$
4,4'-(Trimethylenedioxy)dibenzamidine.
4,4'-diamidino-α,ω-diphenoxypropane. Antiprotozoal (Trypanosoma); antiamebic; anti-infective (topical, veterinary). *May & Baker Ltd.*

6144 Propamidine Isethionate
140-63-6 7981 205-423-6
$C_{21}H_{32}N_4O_{10}S_2$
4,4'-(Trimethylenedioxy)dibenzamidine ethanesulfonic acid.
M&B-782; Brolene Drops. Topical anti-infective; antiamebic. Used as an antiprotozoal against Trypanosoma and Babesia. mp 235°; soluble in H_2O, glycerol, 95% alcohol; practically insoluble in Et_2O, $CHCl_3$, oils. *May & Baker Ltd.*

6145 β-Propiolactone
57-57-8 8005 200-340-1
$C_3H_4O_2$
2-Oxetanone.
NSC-21626; hydracrylic acid β-lactone; β-propionolactone; propanolide; Betaprone. Disinfectant. A versatile reagent in organic synthesis. CAUTION: Overexposure can cause skin irritation, blistering and burns; corneal opacity; dysuria; hematuria. May be a carcinogen. Liquid; mp = -33.4°; bp_{760} = 150° (dec); bp_{20} = 60°; bp_{10} = 51°; d_4^{20}= 1.1420; n_D^{20}= 1.4131; flash point = 70° (158°F); soluble in H_2O; miscible with alcohol, Me_2CO, Et_2O, $CHCl_3$. *Forest Pharm. Inc.*

6146 Silver Bromide
7785-23-1 8649 232-076-8
AgBr
Silver bromide.
Topical anti-infective; astringent. Silver compound. mp = 432°; d = 6.47; slightly soluble in H_2O (0.0000135 g/100 ml at 25°); insoluble in EtOH, most acids; moderately soluble in 10% NH_4OH (0.33 g/100 ml at 12°); soluble in solutions of alkali cyanides; sparingly soluble in solutions of thiocyanides or thiosulfates; soluble in 220 parts saturated NaCl, in 35 parts of staurated KBr; slightly soluble in ammonium carbonate solutions.

6147 Silver Fluoride
7775-41-9 8657 231-895-8
AgF
Silver monofluride.
argentous fluoride. Antiseptic. Silver compound. d = 5.852; mp = 435°; bp 1150°; soluble in H_2O when freshly prepared; becomes insoluble in moist air because of basic fluoride formation; forms several hydrates; soluble in HF, NH_3, CH_3CN.

6148 Silver Lactate
128-00-7 8660
$C_3H_5AgO_3$
2-Hydroxypropanoic acid sliver salt (1:1).
Topical anti-infective; astringent. Silver compound. Soluble in H_2O (6.6 g/100 ml); slightly soluble in alcohol; light sensitive.

6149 Silver Nitrate
7761-88-8 8661 231-853-9
$AgNO_3$
Nitric acid silver salt.
Antiseptic. Silver compound. CAUTION: Poisonous. mp = 212°; d = 4.35; dec 440°; soluble in boiling H_2O, boiling alcohol, ammonia H_2O; slightly soluble in alcohol, Me_2CO, Et_2O; pH (aqueous solution) = 6.

6150 Silver Protein
9015-51-4 8671
argentoproteinum; protargin; silver proteinate; silver nucleate; silver nucleinate; argentum vitellinum [mild]; Argyrol [mild]; Silvol [mild]; albumose silver [strong]; Protargol [strong]. Antiseptic. Group of compounds characterized as colloidal combinations of silver and protein. Compound(s) prepared from a silver salt with gelatin, serum albumin, casein or peptone. Soluble in H_2O; nearly insoluble in alcohol, Et_2O, $CHCl_3$. *Iolab.*

6151 Sodium Borate
1303-96-4 8733 232-160-4
$B_4Na_2O_7$
Sodium diborate.
sodium pyroborate; sodium tetraborate; borax;

component of: Collyrium Eye Wash. Antiseptic (vet); detergent (vet); astringent (vet); pharmaceutic aid (alkalizer). Bright orange crystals. Used as an oxidizing agent in synthetic chemistry. Also used as a topical anti-infective. Slowly soluble in H_2O. *Wyeth-Ayerst Labs.*

6152 Sodium Hypochlorite
8007-59-8 8773 231-668-3
ClNaO
Sodium hypochlorite.
aqueous solution: Eau de Labarraque, Clorox, Dazzle; diluted soda solution: modified Dakin's solution. Anti-infective; antiseptic, topical (vet). A diluted soda solution used as an antiseptic for wound irrigation (vet). Halogen containing compound. mp = 18°; decomposed by CO_2 in air; anhydrous form is explosive; soluble in H_2O; remarkably stable in aqueous solution.

6153 Sodium Iodate
7681-55-2 8776 231-672-5
$INaO_3$
Sodium iodate.
Antiseptic (mucous membranes). Halogen containing compound. d = 4.28; soluble in H_2O; insoluble in alcohol.

6154 Strontium Peroxide
1314-18-7 9010 215-224-6
O_2Sr
Strontium peroxide.
Antiseptic. Peroxide. Nearly insoluble in H_2O; forms hydrogen peroxide with dilute acids; gradually dec on exposure to air.

6155 Sulfadiazine Silver Salt
22199-08-2 9071 244-834-5
$C_{10}H_9AgN_4O_2S$
N^1-2-Pyrimidinylsulfanilamide monosilver(1+) salt.
silver sulfadiazine; Silvadene; Flamazine; Flammazine; component of: Boots SSD. Sulfonamide antibiotic. Anti-infective (topical). *Boots Pharmaceuticals Inc.; Hoechst Marion Roussel Inc.; Marion Merrell Dow Inc.*

6156 Sulfadimethoxine
122-11-2 9073 204-523-7
$C_{12}H_{14}N_4O_4S$
N^1-(2,6-Dimethoxy-4-pyrimidinyl)sulfanilamide.
Agribon; Albon; Arnosulfan; Bactrovet; Diasulfa; Madribon; Maxulvet; Neostreptal; Sudine; Suldixine; Sulfabon; Sulxin; Sumbio; Ultrasulfon; component of: Rofenaid, Maxutrim, Prazil, Trivalbon. Sulfonamide antibiotic. mp = 201-203°; soluble in dilute HCl, $NaHCO_3$; soluble in H_2O (0.0046 g/100 ml at pH 4.10, 0.0295 g/100 ml at pH 6.7, 0.058 g/100 ml at pH 7.06, 5.17 g/100 ml at pH 8.71); LD_{50} (mus orl) > 10000 mg/kg. *Hoffmann-LaRoche Inc.; ICI; Oesterreiche Stickstoffwerke.*

6157 Symclosene
87-90-1 9188 201-782-8
$C_3Cl_3N_3O_3$
1,3,5-Trichloro-s-triazine-2,4,6(1H,3H,5H)-trione.
trichloroiminocyanuric acid; trichloroisocyanuric acid; NSC-405124; ACL-85; Chloreal. Anti-infective (topical). Halogen containing compound. A chlorinating agent, disinfectant, and industrial deodorant. component of: household cleaners. CAUTION: Irritating to eyes, skin, mucous membranes. mp = 246-247° (dec); pH 4.4; slightly soluble in H_2O; soluble in chlorinated and highly polar solvents. *Grace W.R. & Co.; Monsanto Co.*

6158 α-Terpineol
98-55-5 9316 202-680-6
$C_{10}H_{18}O$
α,α,4-Trimethyl-3-Cyclohexene-1-methanol.
p-menth-1-en-8-ol. Antiseptic. One of three isomers of terpineol. Liquid; [dl-form]: bp_3 = 85°; d^{15} = 0.9386; n_4^{20} = 1.4831.

6159 Thimerfonate Sodium
5964-24-9 9450 227-741-4
$C_8H_9HgNaO_3S_2$
Ethyl(hydrogen p-mercaptobenzenesulfonato)mercury sodium salt.
Sulfo-Merthiolate. Anti-infective (topical). Mercurial compound. Soluble in H_2O. *Eli Lilly & Co.*

6160 Thimerosol
54-64-8 9451 200-210-4
$C_9H_9HgNaO_2S$
Ethyl(sodium o-mercaptobenzoato)mercury.
sodium ethylmercurithiosalicylate; thiomersalate; mercurothiolate; Merthiolate; Merzonin; Vitaseptol; component of: Collyrium Eye Wash. Anti-infective (topical); pharmaceutic aid (preservative). Mercurial compound. Soluble in H_2O; less soluble in alcohol; practically insoluble in Et_2O, C_6H_6. *Eli Lilly & Co.; Wyeth-Ayerst Labs.*

6161 Thiosalan
15686-78-9
$C_{13}H_8Br_3NOS$
3,4,5'-Tribromo-2-mercaptobenzanilide.
Disinfectant.

6162 Tibezonium Iodide
54663-47-7 9565 259-284-1
$C_{28}H_{32}IN_3S_2$
Diethylmethyl[(2-[[4-[p-(phenylthio)phenyl]-3H-1,5-benzodiazepin-2-yl]thio]ethyl]-ammonium iodide.
thiabenzazonium iodide; Rec-15-0691; Antoral. Antibacterial. Quaternary ammonuim compound. mp = 162°; LD_{50} (mus olr) > 10000 mg/kg, (rat orl) = 9000 mg/kg; (mus ip) = 42 mg/kg, (rat ip) = 35 mg/kg. *Recordati Corp.*

6163 Tibrofan
15686-72-3
$C_{11}H_6Br_3NOS$
4,4'-Tribromo-2-thiophenecarboxyanilide.
Disinfectant.

6164 2,4,6-Tribromo-m-cresol
4619-74-3 9742 204-278-6
$C_7H_5Br_3O$
2,4,6-Tribromo-3-methylphenol.
Micatex. Antiseptic; antifungal (topical). Phenol. mp = 84°.

6165 Tribromsalan
87-10-5 9747 201-723-6
$C_{13}H_8Br_3NO_2$
3,4',5-Tribromosalicylanilide.
TBS; Temasept IV; Tuasol. Disinfectant; bacteriostat. mp = 227-228°; practically insoluble in H_2O; soluble in hot Me_2CO, DMF. *Colgate-Palmolive; Dow Chem. U.S.A.; Sogeras.*

6166 3',4',5'-Trichlorosalicylanilide
642-84-2 9775 211-391-4
$C_{13}H_8Cl_3NO_2$
5-Chloro-N-(3,4-dichlorophenyl)-2-hydroxybenzamide.
Anobial. Phenol used as an antiseptic and deodorant in soaps and cosmetics. mp = 246-248°; forms a H_2O soluble sodium salt. *Ciba-Geigy Corp.*

6167 Triclobisonium Chloride
79-90-3 9785 201-232-7
$C_{36}H_{74}Cl_2N_2$
N,N,N',N'-Tetramethyl-N,N'-bis[1-methyl-3-(2,2,6-trimethylcyclohexyl)propyl]-1,6-hexanediaminium dichloride.
Ro-5-0810/1; Triburon. Antiseptic. Quaternary ammonuim compound. mp =243-253° (dec); soluble in H_2O, $CHCl_3$, alcohol. *Hoffmann-LaRoche Inc.*

6168 Triclosan
3380-34-5 9790 222-182-2
$C_{12}H_7Cl_3O_2$
5-Chloro-2-(2,4-dichlorophenoxy)phenol.
CH-3635; Aquasept; Gamophen; Irgasan DP 300; Sapoderm; SterZac. Antiseptic; disinfectant. Halogen containing compound. mp = 54-57.3°; pKa = 7.9; insoluble in H_2O; soluble in alkaline solutions and many organic solvents. *Ciba-Geigy Corp.*

6169 Troclosene Potassium
2244-21-5 9896 218-828-8
$C_3Cl2KN_3O_3$
1,3-Dichloro-s-triazine-2,4,6(1H,3H,5H)trione potassium salt.
potassium troclosene; potassium dichloroisocyanurate; potassium salt of dichloroisocyanuric acid; ACL-59. Anti-infective (topical). Halogen containing compound. Source of Cl in solid bleach and detergent formulations. *Monsanto Co.*

6170 Urea Hydrogen Peroxide
124-43-6 10007 204-701-4
$CH_6N_2O_3$
Urea compound with hydrogen peroxide.
carbamide peroxide; hydrogen peroxide carbamide; Debrox; Exterol; Gly-Oxide; Hyperol; Ortizon [obsolete]; Perhydrit; Perhydrol-Urea. Antiseptic. Usually contains 35% H_2O_2. Dec in air to urea, oxygen and H_2O; soluble in H_2O; partly decomposed by alcohol or Et_2O to H_2O_2 and urea.

6171 Yellow Precipitate
21908-53-2 5936 244-654-7
HgO
Mercuric oxide.
mercuric oxide, yellow. Antiseptic, topical (opthalmic). CAUTION: poisonous. Incompatible with reducing agents; light sensitive.

6172 Zinc Permanganate
23414-72-4 10281 245-646-6
Mn_2O_8Zn
Permanganic acid zinc salt (2:1).
Antiseptic; astringent. Permanganate. Deteriorates on exposure to light, air; soluble in H_2O (33 g/100 ml); decomposed by alcohol.

6173 Zinc Peroxide
1314-22-3 10282 215-226-7
O_2Zn
Zinc superoxide.
ZPO. Antiseptic (topical); astringent. Dec above 150°; soluble in dilute acids, liberating H_2O_2; insoluble in but decomposed by H_2O.

6174 Zinc Salicylate
16283-36-6 10288 240-380-7
$C_{14}H_{10}O_6Zn$
2-Hydroxybenzoic acid zinc salt (2:1).
zinc disalicylate. Astringent; antiseptic; topical protectant. [trihydrate]: Soluble in H_2O, EtOH.

6175 Zinc Tannate
8011-65-2 10295 232-377-4
Zinc oxide and tannins.
sal barnit. Astringent; antiseptic. Practically insoluble in H_2O; soluble in dilute acids.

Antispasmodics

6176 Alibendol
26750-81-2 243 247-960-9
$C_{13}H_{17}NO_4$
5-Allyl-N-(2-hydroxyethyl)-3-methoxysalicylamide.
EB-1856; FC-54; H-3774; Cebera. Choleretic and antispasmodic. mp = 95°; λ_m = 316, 218 nm (EtOH); LD_{50} (mmus orl) > 3000 mg/kg, (mmus sc) > 2000 mg/kg, (mmus iv) = 217 mg/kg. *Roussel-UCLAF.*

6177 Ambucetamide
519-88-0 403 208-278-7
$C_{17}H_{28}N_2O_2$
2-(Dibutylamino)-2-(p-methoxyphenyl)acetamide.
A-16; Dibutamide; Bersen; Meritin. Antispasmodic. mp = 125-127°, 134°; insoluble in H_2O; soluble in EtOH, iPrOH, AcOH.

6178 Aminopromazine
58-37-7 487 200-378-9
$C_{19}H_{25}N_3S$
10-[2,3-Bis(dimethylamino)propyl]phenothiazine.
proquamezine; RP-3828; Tetrameprozine. Antispasmodic. A Phenothiazine. *Rhône-Poulenc Rorer Pharm. Inc.*

6179 Aminopromazine Fumarate
3688-62-8 487 222-987-9
$C_{42}H_{54}N_6O_4S_2$
10-[2,3-Bis(dimethylamino)propyl]phenothiazine fumarate (2:1).

Lispamol; Lorusil; Spamol. Antispasmodic. Phenothiazine derivative dec 166-170°; soluble in H_2O (9.0 g/100 ml), MeOH (5 g/100 ml), EtOH (0.5 g/100 ml); slightly soluble in iPrOH, Me_2CO; insoluble in C_6H_6, Et_2O. *Rhône-Poulenc Rorer Pharm. Inc.*

6180 Apoatropine
500-55-0 780 207-906-7
$C_{17}H_{21}NO_2$
endo-α-Methylenebenzeneacetic acid 8-methyl-8-azabicyclo[3.2.1]oct-3-yl ester.
atropamine; atropyltropeine. Antispasmodic. Anticholinergic. Occurs in the root of *Atropa belladonna L. Solanaceae.* mp = 62°; insoluble in H_2O; soluble in EtOH, Et_2O, $CHCl_3$, C_6H_6, CS_2; slightly soluble in petroleum ether, isoamyl alcohol; LD_{50} (mus orl) = 160 mg/kg, (mus ip) = 14.1 mg/kg; [hydrochloride ($C_{17}H_{22}ClNO_2$)]: mp = 239°; soluble in hot H_2O; sparingly soluble in EtOH, Me_2CO; insoluble in Et_2O; [sulfate pentahydrate ($C_{34}H_{44}N_2O_8.5H_2O$)]: sparingly soluble in H_2O.

6181 Bencianol
85443-48-7 287-240-1
$C_{28}H_{22}O_6$
(2R,3S)-3',4'-(Diphenylmethylene)dioxy]-3,5,7-flavan triol.
ZY-15051. antispasmogenic. Adrenoceptor antagonist.

6182 Bevonium Methyl Sulfate
5205-82-3 1239 226-001-8
$C_{23}H_{31}NO_7S$
2-(Hydroxymethyl)-1,1-dimethylpiperidinium methyl sulfate benzilate.
bevonium metilsulfate; CG-201; Acabel. Anticholinergic with antispasmodic and bronchodilating properties. mp = 134-135°. *Grunenthal.*

6183 Bietamiverine
479-81-2 1252 207-538-7
$C_{19}H_{30}N_2O_2$
2-Diethylaminoethyl α-phenyl-1-piperidineacetate hydrochloride.
Antispasmodic. bp_1 - 65°; d_4^{25} = 1.0184. *Nordmark.*

6184 Bietamiverine Dihydrochloride
2691-4-65 1252
$C_{19}H_{32}Cl_2N_2O_2$
2-Diethylaminoethyl α-phenyl-1-piperidineacetate dihydrochloride.
Antispasmodic. mp = 194-195°. *Nordmark.*

6185 Bietamiverine Hydrochloride
1477-10-7 1252
$C_{19}H_{31}ClN_2O_2$
2-Diethylaminoethyl α-phenyl-1-piperidineacetate monohydrochloride.
Novosparol. Antispasmodic. mp = 187-189°. *Nordmark.*

6186 Butaverine
55837-14-4 1544
$C_{18}H_{27}NO_2$
Butyl 3-phenyl-3-(1-piperidyl)propionate.
butamiverine. Antispasmodic.

6187 Butaverine Hydrochloride
17824-89-4 1544
$C_{18}H_{28}ClNO_2$
Butyl 3-phenyl-3-(1-piperidyl)propionate Hydrochloride.
Espasmo-Gemora; Gemora. Antispasmodic. mp = 170°.

6188 Butropium Bromide
29025-14-7 1569 249-375-4
$C_{28}H_{38}BrNO_4$
8-(p-Butoxybenzyl)-3α-hydroxy-1αH-tropanium bromide (-)-tropate.
BHB; Coliopan. Anticholinergic; antispasmodic. mp = 166-168°, 158-160°; $[\alpha]_D^{20}$ =.-21.7° (c = 0.5 H_2O); freely soluble in AcOH; soluble in $CHCl_3$, DMF; slightly soluble in H_2O, 0.1N HCl, 0.1N NaOH; insoluble in Me_2CO, Et_2O, C_6H_6; LD_{50} (mmus orl) = 1500 mg/kg, (mmus sc) = 66- mg/kg, (mmus iv) = 12.0 mg/kg. *Eisai Co. Ltd.*

6189 N-Butylscopolammonium Bromide
149-64-4 1624 205-744-1
$C_{21}H_{30}BrNO_4$
8-Butyl-6β,7β-epoxy-3α-hydroxy-1αH,5αH-tropanium bromide (-)-tropate.
scopolamine bromobutylate; Amisepan; Buscapina; Buscol; Buscolysin; Buscopan; Butylmin; Donopon; Monospan; Scobro; Scobron; Scobutil; Sparicon; Sporamin; Stibron; Tirantil; hyoscine N-butyl bromide. Antispasmodic. mp = 142-144°; $[\alpha]_D^{20}$ = -20.8° (c = 3 H_2O); LD_{50} (mus iv) = 15.6 mg/kg, (mus ip) = 74 mg/kg, (mus sc) = 570 mg/kg, (mus orl) = 3000 mg/kg. *Boehringer Ingelheim GmbH.*

6190 Caroverine
23465-76-1 1906
$C_{23}H_{27}N_3O_2$
1-[2-(Dimethylamino)ethyl]-3-(p-methoxybenzyl)-2(1H)-quinoxalinone.
Spasmium. Antispasmodic. mp = 69°; $bp_{0.01}$ = 202°; [hydrochloride ($C_{22}H_{28}ClN_3O_3$)]: dec 188°. *Donau Pharm.*

6191 Cimetropium Bromide
51598-60-8 2338
$C_{21}H_{28}BrNO_4$
[7(S)-(1α,2β,4β,5α,7β)]-9-(Cyclopropylmethyl)-7-(3-hydroxy-1-oxo-2-phenylpropoxy)-9-methyl-3-oxa-9-azoniatricyclo[3.3.1.o^{2,4}]nonane bromide.
DA-3177; N-cyclopropylmethylscopolamine bromide; Alginor. Antispasmodic. Spasmolytic with affinity for intestinal muscarinic receptors. mp = 174°; $[\alpha]_D^{20}$ = -18.3° (c = 3). *DeAngeli.*

6192 Cinnamedrine
90-86-8 2357 202-021-2
$C_{19}H_{23}NO$
α-[1-(Cinnamylmethylamino)ethyl]benzyl alcohol.
N-cinnamylephedrine. Smooth muscle relaxant; antispasmodic. Reduces epinephrine-induced autonomic activity. mp = 72-78°; [dl-hydrochloride]: mp = 180-185°; soluble in H_2O (1.25 g/100 ml at 20°), EtOH (10 g/100 ml at 20°), $CHCl_3$ (2 g/100 ml); sparingly soluble in Et_2O, C_6H_6. *Sterling Winthrop Inc.; Winthrop.*

6193 Clebopride
55905-53-8 2404 259-885-9
$C_{20}H_{24}ClN_3O_2$
4-Amino-N-(1-benzyl-4-piperidyl)-5-chloro-o-anisamide.
Cleboril. Antispasmodic; antiemetic. Selective D2 receptor antagonist. mp = 194-195°; [hydrochloride monohydrate ($C_{20}H_{25}Cl_2N_3O_2.H2O$)]: mp = 217-219°; LD_{50} (mus orl) > 1000 mg/kg. *Anphar; Grupo Farmaceutico Almirall S.A.*

6194 Clebopride Malate
57645-91-7 2404 260-874-6
$C_{24}H_{30}ClN_3O_7$
4-Amino-N-(1-benzyl-4-piperidyl)-5-chloro-o-anisamide malate.
Amicos; Clanzol; Clast; Cleboril; Clepríd; Motilex. Antiemetic with antispasmodic properties. Selective D2 receptor antagonist. *Anphar; Grupo Farmaceutico Almirall S.A.*

6195 Cyclonium Iodide
6577-41-9 2806
$C_{22}H_{34}INO_2$
1-[(2-Cyclohexyl-2-phenyl-1,3-dioxolan-4-yl)methyl]-1-methylpiperidinium iodide.
oxapium iodide; ciclonium iodide; SH-100; Esperan. Antispasmodic. Anticholinergic. [α-form]: mp = 195-197°; [β-form]: mp = 150-152°; [both forms]: soluble in MeOH, EtOH, $CHCl_3$, poorly soluble in C_6H_6, C_7H_8, H_2O; LD_{50} (mus ip) = 106 mg/kg. *Toyama Chem. Co. Ltd.*

6196 Difemerine
80387-96-8 3180
$C_{20}H_{25}NO_3$
2-(Dimethylamino)-1,1-dimethylethyl benzilate.
Anticholinergic and antispasmodic. mp = 78-78.3°. *Lab. UPSA.*

6197 Difemerine Hydrochloride
70280-88-5 3180 274-526-6
$C_{20}H_{26}ClNO_3$
2-(Dimethylamino)-1,1-dimethylethyl benzilate hydrochloride.
Luostyl. Anticholinergic and antispasmodic. mp = 182°. *Lab. UPSA.*

6198 Diisopromine
5966-41-6 3239 227-752-4
$C_{21}H_{29}N$
N,N-Diisopropyl-3,3-diphenylpropylamine.
dispromine. Antispasmodic. *Abbott Laboratories Inc.; N.V.Nederlandsche Comb. Chem. Ind.*

6199 Diisopromine Hydrochloride
24358-65-4 3239 246-201-9
$C_{21}H_{30}ClN$
N,N-Diisopropyl-3,3-diphenylpropylamine hydrochloride.
Agofell; Bilagol. Antispasmodic. mp = 175-176°; λ_m = 254, 259.5, 268.8 nm; soluble in H_2O, MeOH, EtOH, $CHCl_3$; insoluble in Et_2O. *Abbott Laboratories Inc.; N.V.Nederlandsche Comb. Chem. Ind.*

6200 Dioxaphetyl Butyrate
467-86-7 3354 207-402-7
$C_{22}H_{27}NO_3$
Ethyl 4-morpholino-2,2-diphenylbutyrate.
Amidalgon; Spasmoxal; Spasmoxale. Antispasmodic; narcotic analgesic. Federally controlled substance (opiate). [hydrochloride ($C_{22}H_{28}ClNO_3$)]: mp = 168-169°. *Sterling Winthrop Inc.; Winthrop.*

6201 Diponium Bromide
2001-81-2 3402 217-891-9
$C_{20}H_{38}BrNO_2$
Triethyl(2-hydroxyethyl)ammonium bromide dicyclopentylacetate.
dipenine bromide; SA-267; Unospaston. Antispasmodic; anticholinergic. mp = 185-186°; freely soluble in H_2O; LD_{50} (mus orl) = 570 mg/kg, (mus ip) = 88 mg/kg, (mus iv) = 6.2 mg/kg, (rat orl) = 780 mg/kg, (rat iv) = 6.6 mg/kg. *Siegfried AG.*

6202 Drofenine
1679-76-1 3501
$C_{20}H_{31}NO_2$
2-(Diethylamino)ethyl α-phenylcyclohexaneacetate.
hexahydroadiphenine. Antispasmodic. Selective muscarinic M_1-receptor antagonist. $bp_{0.15}$ = 158°. *CIBA plc.*

6203 Drofenine Hydrochloride
548-66-3 3501 208-954-1
$C_{20}H_{32}ClNO_2$
2-(Diethylamino)ethyl α-phenylcyclohexaneacetate hydrochloride.
Trasentine-A; Trasentine 6-H. Antispasmodic. Selective muscarinic M_1-receptor antagonist. mp = 145-147°; freely soluble in H_2O; sparingly soluble in EtOH, Et_2O; LD_{50} (mus iv) = 65.6 mg/kg. *CIBA plc.*

6204 Emepronium Bromide
3614-30-0 3598 222-786-6
$C_{20}H_{28}BrN$
Ethyl(3,3-diphenyl-1-methylpropyl)dimethylammonium bromide.
Cetiprin; Restenacht; Ripirin; Uro-Ripirin. Antispasmodic; anticholinergic. Muscarinic antagonist. mp = 204°. *Aktiebolaget. Recip.*

6205 Ethaverine
486-47-5 3773 207-633-3
$C_{24}H_{29}NO_4$
1-(3,4-Diethoxybenzyl)-6,7-diethoxyisoquinoline.
ethylpapaverine; Dyscural. Antispasmodic. L-type calcium channel inhibitor. mp = 99-101°; insoluble in H_2O; slightly soluble in EtOH, $CHCl_3$, Et_2O.

6206 Ethaverine Hydrochloride
985-13-7 3773 213-573-9
$C_{24}H_{30}ClNO_4$
1-(3,4-Diethoxybenzyl)-6,7-diethoxyisoquinoline hydrochloride.
Barbonin, Circubid; Diquinol; Ethabid; Isovex; Laverin; Perparin; Perperine. Antispasmodic. L-type calcium channel inhibitor. mp = 186-188°; soluble in H_2O (2.5 g/100 ml).

6207 Etomidoline
21590-92-1 3928 244-463-9
$C_{23}H_{29}N_3O_2$
2-Ethyl-3-(β-piperidino-p-phenetidino)phthalimidine.
K-2680; Smedolin. Anticholinergic and antispasmodic. mp = 106-107°; LD_{50} (mus orl) = 176 mg/kg, (mus iv) = 44.8 mg/kg. *Farmitalia Carlo Erba Ltd.*

6208 Feclemine
3590-16-7 3986
$C_{24}H_{42}N_2$
2-(α-Cyclohexylbenzyl)-N,N,N',N'-tetraethyl-1,3-propanediamine.
UCB-1545; Licaran; Spasmexan. Antispasmodic. *UCB Pharma.*

6209 Feclemine Hydrochloride
115097-93-3 3986
$C_{24}H_{43}ClN_2$
2-(α-Cyclohexylbenzyl)-N,N,N',N'-tetraethyl-1,3-propanediamine hydrochloride.
Licabile. Antispasmodic. mp < 70°; $bp_{0.05}$ = 143°. *UCB Pharma.*

6210 Fenalamide
4551-59-1 3995 224-917-2
$C_{19}H_{30}N_2O_3$
Ethyl N-[2-(diethylamino)ethyl]-2-ethyl-2-phenylmalonamate.
Spasmamide. Skeletal muscle relaxant. Used as an antispasmodic. bp_3 = 182-188°; insoluble in H_2O, soluble in EtOH, MeOH, EtOAc, C_6H_6, $CHCl_3$, Et_2O, mineral acids; [hydrobromide]: mp = 74-80° (dec); [hydrochloride]: mp = 71-74°. *Marion Merrell Dow Inc.; Schering-Plough HealthCare Products.*

6211 Fenoverine
37561-27-6 4023 253-552-1
$C_{26}H_{25}N_3O_3S$
10-[(4-Piperonyl-1-piperazinyl)acetyl]phenothiazine.
Spasmopriv. Antispasmodic. mp = 141-142°; LD_{50} (mus orl) = 1500 mg/kg, (mus ip) = 2500 mg/kg.

6212 Fenpiprane
3540-95-2 4031
$C_{20}H_{25}N$
1-(3,3-Diphenylpropyl)piperidine.
component of: Aspasan. Antiallergic; antispasmodic. mp = 41-42.5°; bp_8 = 210-220°. *Hoechst Roussel Pharm. Inc.; Sterling Winthrop Inc.; Winthrop-Stearns.*

6213 Fenpiprane Hydrochloride
3329-14-4 4031 222-049-9
$C_{20}H_{26}ClN$
1-(3,3-Diphenylpropyl)piperidine hydrochloride.
component of: Efosin. Antiallergic; antispasmodic. mp = 216-217°. *Hoechst Roussel Pharm. Inc.; Sterling Winthrop Inc.; Winthrop-Stearns.*

6214 Fenpiverinium Bromide
125-60-0 4032 204-744-9
$C_{22}H_{29}BrN_2O$
1-(3-Carbamoyl-3,3-diphenylpropyl)-1-methylpiperidinium bromide.
fenpipramide methobromide; 12494 Hoechst; Resantin. Antispasmodic; anticholinergic. mp = 177.5-178.5°, 216-216.5°; freely soluble in H_2O. *Hoechst AG.*

6215 Fentonium Bromide
5868-06-4 4048 227-520-2
$C_{31}H_{34}BrNO_4$
3α-Hydroxy-8-(p-phenylphenacyl)-1αH,5αH-tropanium bromide (-)-tropate.
phentonium bromide; FA-402; Z-326; Ketoscilium; Ulcesium. Anticholinergic and antispasmodic. mp = 203-205° (dec), 193-194°; $[\alpha]_D^{23}$ = -5.68° (c = 5 DMF), $[\alpha]_D^{25}$ = -4.7° (c = 5 DMF); LD_{50} (mus iv) = 12.1 mg/kg, (mus sc) > 400 mg/kg, (mus orl) > 400 mg/kg. *Whitefin Holding.*

6216 Flavoxate
15301-69-6 4135 239-337-5
$C_{24}H_{25}NO_4$
2-Piperidinoethyl 3-methyl-4-oxo-2-phenyl-4H-1-benzopyran-8-carboxylate.
2-piperidinoethyl 3-methylflavone-8-carboxylate. Antispasmodic; smooth muscle relaxant. A flavone derivative. Slightly soluble in H_2O (0.001 g/100 ml at 37°); soluble in EtOH, $CHCl_3$; LD_{50} (rat orl) = 1110 mg/kg, (rat iv) = 20.8 mg/kg. *Recordati Corp.; Seceph; SmithKline Beecham Pharm.*

6217 Flavoxate Hydrochloride
3717-88-2 4135 223-066-4
$C_{24}H_{26}ClNO_4$
2-Piperidinoethyl 3-methyl-4-oxo-2-phenyl-4H-1-benzopyran-8-carboxylate hydrochloride.
DW-61; NSC-114649; Rec-7-0040; Bladderon; Genurin; Patricin; Spasuret; Urispas. Antispasmodic; smooth muscle relaxant. A flavone derivative. mp = 232-234°; LD_{50} (rat iv) = 27.4 mg/kg. *Recordati Corp.; Seceph.*

6218 Flavoxate Succinate
28782-19-6 4135 249-217-4
$C_{28}H_{31}NO_8$
2-Piperidinoethyl 3-methyl-4-oxo-2-phenyl-4H-1-benzopyran-8-carboxylate succinate.
Antispasmodic; smooth muscle relaxant. A flavone derivative. Soluble in H_2O (33.7 g/100 ml at 37°). *Recordati Corp.; Seceph.*

6219 Flopropione
2295-58-1 4142 218-942-8
$C_9H_{10}O_4$
1-(2,4,6-Trihydroxyphenyl)-1-propanone.
phloropropiophenone; RP-13907; Argobyl; Cospanon; Flopion; Gallepronin; Gasstenon; Labroda; Labrodax; Pasmus; Profenon; Spamorin; Spasmoril; Supanate; Supazlun. Antispasmodic. mp = 175-176°; soluble in EtOH, Et_2O, EtOAc, hot H_2O; poorly soluble in cold H_2O.

6220 Gluconic Acid
526-95-4 4464 208-401-4
$C_6H_{12}O_7$
D-Gluconic acid.
dextronic acid; maltonic acid; glyconic acid; glycogenic acid; pentahydroxycaproic acid; Almora [as magnesium salt dihydrate ($C_{12}H_{22}MgO_{14}.2H_2O$)]; Ultra-Mg [as magnesium salt dihydrate]. Antispasmodic. mp = 131°; d_4^{25} = 1.24; $[\alpha]_D^{20}$ = -6.7° (c = 1); freely soluble in H_2O,

slightly soluble in EtOH, insoluble in Et_2O and most other organic solvents. *Pfizer Intl.*

6221 Hexasonium Iodide
3569-59-3
$C_{18}H_{27}IO_2S$
(2-Hydroxyethyl)dimethyl sulfonium iodide α-phenyl cyclohexaneacetate.
Spasmolytic.

6222 Hydramitrazine
13957-36-3 4803 237-736-9
$C_{11}H_{23}N_7$
4,6-Bis(diethylamino)-1,3,5-triazin-2(1H)-one hydrazone.
Meladrazine. Antispasmodic. *CIBA plc.*

6223 Hydramitrazine Tartrate
20423-87-4 4803 243-808-0
$C_{15}H_{29}N_7O_6$
4,6-Bis(diethylamino)-1,3,5-triazin-2(1H)-one hydrazone tatrtate.
Lisidonil. Antispasmodic. *CIBA plc.*

6224 Hymecromone
90-33-5 4903 201-986-7
$C_{10}H_8O_3$
7-Hydroxy-4-methyl-2H-1-benzopyran-2-one.
4-methylumbelliferone; imecromone; 4-MU; Bilcolic; Biliton H; Cantabilin; Cantabiline; Cholonerton; Cholspasmin; Cumarote C; Himecol; Medilla; Mendiaxon. Antispasmodic; choleretic. mp = 185-186°, 194-195°; λ_m = 221, 251, 322.5 nm (MeOH); soluble in MeOH, AcOH; slightly soluble in Et_2O, $CHCl_3$, insoluble in cold H_2O. *Bayer Corp.*

6225 Hymecromone O,O-Diethyl-phosphorothioate
299-45-6 4903
$C_{14}H_{17}O_5PS$
7-Hydroxy-4-methyl-2H-1-benzopyran-2-one O,O-diethyl phosphorothioate.
E-838; Potasan. Antispasmodic; choleretic. Acts as a cholinesterase inhibitor (cf. Parathion) and is used as an insecticide, especially effective against the Colorado beetle. mp = 38°; $bp_{1.0}$ = 210° (dec); d_4^{38} = 1.260; sparingly soluble in H_2O; slightly soluble in petroleum ether; soluble in most other organic solvents; LD_{50} (rat ip) = 15 mg/kg. *Bayer Corp.*

6226 Leiopyrrole
5633-16-9 5457
$C_{23}H_{28}N_2O$
N,N-Diethyl-2-[2-(2-methyl-5-phenyl-1H-pyrrol-1-yl)phenoxy]ethanamine.
DV-714. Antispasmodic. bp_{13} = 232°.

6227 Leiopyrrole Hydrochloride
14435-78-0 5457
$C_{23}H_{29}ClN_2O$
N,N-Diethyl-2-[2-(2-methyl-5-phenyl-1H-pyrrol-1-yl)phenoxy]ethanamine hydrochloride.
Leioplegil. Antispasmodic. mp = 138°; readily soluble in H_2O.

6228 Mebeverine
3625-06-7 5808 222-830-4
$C_{25}H_{35}NO_5$
4-[Ethyl(p-methoxy-α-methylphenethyl)amino]butyl veratrate.
Smooth muscle relaxant; local anesthetic. Used as an antispasmodic. Binds to α- and β-adrenoceptors and inhibits phosphodiesterase activity. *N.V. Philips.*

6229 Mebeverine Hydrochloride
2753-45-9 5808 220-400-0
$C_{25}H_{36}ClNO_5$
4-[Ethyl(p-methoxy-α-methylphenethyl)amino]butyl veratrate hydrochloride.
CSAG 144; Colofac; Duspatalin; Duspatal. Antispasmodic. Smooth muscle relaxant. A pyrazolone. mp = 125-127°, 129-131°. *N.V. Philips.*

6230 Moxaverine
10539-19-2 6371 234-117-5
$C_{20}H_{21}NO_2$
3-Ethyl-6,7-dimethoxy-1-(phenylmethyl)-isoquinoline.
Antispasmodic. Cyclic adenosine monophosphate phosphodiesterase inhibitor. mp = 78-79°. *Orgamol S.A.*

6231 Moxaverine Hydrochloride
1163-37-7 6371 214-607-5
$C_{20}H_{22}ClNO_2$
3-Ethyl-6,7-dimethoxy-1-(phenylmethyl)isoquinoline hydrochloride.
Eupaverin; Eupaverina; Kollateral; Sorbosan. Antispasmodic. Cyclic adenosine monophosphate phosphodiesterase inhibitor. mp = 214°, 208-210° (dec); sparingly soluble in cold H_2O; soluble in hot H_2O, EtOH, other organic solvents. *Orgamol S.A.*

6232 Nafiverine
5061-22-3 6439 225-766-5
$C_{34}H_{38}N_2O_4$
α-Methyl-1-naphthaleneacetic acid 1,4-piperazinediyldi-2,1-ethanediyl ester.
Naftidan. Antispasmodic. [dihydrochloride ($C_{34}H_{40}Cl_2N_2O_4$)]: mp = 220-221°. *DeAngeli.*

6233 Octamylamine
502-59-0 6846 207-947-0
$C_{13}H_{29}N$
6-Methyl-N-(3-methylbutyl)-2-heptanamine.
Octinum D; Octisamyl; Neo-Octon; Octometine. Anticholinergic; antispasmodic. bp_7 = 100-101°; [hydrochloride ($C_{13}H_{30}ClN$)]: mp = 121°; soluble in H_2O, EtOH, Et_2O. *Knoll Pharm. Co.*

6234 Octaverine
549-68-8 6851 208-976-1
$C_{23}H_{27}NO_5$
6,7-Dimethoxy-1-(3,4,5-triethoxyphenyl)-isoquinoline.
oktaverine. Antispasmodic. Insoluble in H_2O; [hydrochloride ($C_{23}H_{28}ClNO_5$)]: mp = 199-200°; sparingly soluble in H_2O (0.2 g/100 ml). *Asta Chem. Fabrik.*

6235 Pentapiperide
7009-54-3 7260 230-286-4
$C_{18}H_{27}NO_2$
α-(1-Methylpropyl)benzeneacetic acid 1-methyl-4-piperidinyl ester.
C-4675 [as fumarate]. Anticholinergic; antispasmodic. [fumarate]: mp = 91-93°. *Cilag-Chemie Ltd.*

6236 Pentapiperium Methosulfate
7681-80-3 7260 231-678-8
$C_{20}H_{33}NO_6S$
α-(1-Methylpropyl)benzeneacetic acid 1-methyl-4-piperidinium methylsulfate.
pentapiperium metilsulfate; pentapiperium mesylate; pentapiperium methylsulfonate; Hycholin; Quilene; Perium; Crylène. Anticholinergic; antispasmodic. mp = 110-112°. *Cilag-Chemie Ltd.*

6237 Phenamacide Hydrochloride
31031-74-0 7350
$C_{13}H_{20}ClNO_2$
(±)-α-Aminobenzeneacetic acid 3-methylbutyl ester hydrochloride.
Aklonin. Antispasmodic. mp = 154°.

6238 Phloroglucinol
108-73-6 7482 203-611-2
$C_6H_6O_3$
1,3,5-Benzenetriol.
phloroglucin; Dilospan S; Spasfon-Lyoc. Antispasmodic. mp = 218°; soluble in H_2O (1 g/100 ml), EtOH (10 g/100 ml), C_5H_5N (200 g/100 ml); LD_{50} (mus orl) = 4550 mg/kg, (mus ip) = 4050 mg/kg, (mus sc) = 5520 mg/kg. *Hooker Chem.*

6239 Pinaverium Bromide
53251-94-8 7595 258-450-0
$C_{26}H_{41}Br_2NO_4$
4-[(2-Bromo-4,5-dimethoxyphenyl)methyl]-4-[2-[2-(6,6-dimethylbicyclo[3.1.1]hept-2-yl)ethoxy]ethyl]morpholinium bromide.
LAT-1717; Dicetel. Antispasmodic; only weakly anticholinergic. mp = 181°, 170°; LD_{50} (mus orl) = 1400 mg/kg, (mus iv) = 66 mg/kg, 37 ± 2.4 mg/kg. *Societe Berri-Balzac.*

6240 Piperilate
4546-39-8 7624
$C_{21}H_{25}NO_3$
α-Hydroxy-α-phenylbenzeneacetic acid 2-(1-piperidinyl)ethyl ester.
1-piperidineethanol benzilate; Panpurol [as ethyl bromide ($C_{23}H_{30}BrNO_3$)]. Anticholinergic; antispasmodic.

6241 Piperilate Hydrochloride
4544-15-4 7624
$C_{21}H_{26}ClNO_3$
α-Hydroxy-α-phenylbenzeneacetic acid 2-(1-piperidinyl)ethyl ester hydrochloride.
Daipisate; Norticon; Pensanate; Pipenale. Anticholinergic; antispasmodic. mp = 170-171°.

6242 Pipoxolan Hydrochloride
18174-58-8 7637
$C_{22}H_{26}ClNO_3$
5,5-Diphenyl-2-(2-piperidinoethyl)-1,3-dioxolan-4-one hydrochloride.
BR-18; Rowapraxin. Skeletal muscle relaxant used as an antispasmodic. mp = 207-209°; soluble in H_2O; LD_{50} (rat orl) = 1500 mg/kg, (rat iv) = 60 mg/kg, (rat ip) = 130 mg/kg, (rat sc) > 300 mg/kg, (mus orl) = 700 mg/kg, (mus iv) = 35 mg/kg, (mus ip) = 130 mg/kg. *Rowa Ltd.*

6243 Pitofenone
54063-52-4
$C_{22}H_{25}NO_4$
Methyl o-[p-(2-piperidinoethoxy)benzoyl]benzoate.
Antispasmodic.

6244 Pramiverin
14334-40-8 7887
$C_{21}H_{27}N$
N-(1-Methylethyl)-4,4-diphenylcyclohexanamine.
pramiverine; primaverine; propaminodiphen. Anticholinergic; antispasmodic. mp = 70°; $bp_{0.05}$ ≅ 165°. *E. Merck.*

6245 Pramiverin Hydrochloride
14334-41-9 7887 238-284-5
$C_{21}H_{28}ClN$
N-(1-Methylethyl)-4,4-diphenylcyclohexanamine hydrochloride.
EMD-9806; HSP-2986; Monoverin; Sistalgin. Anticholinergic; antispasmodic. mp = 230°, 234-237°; soluble in H_2O (0.3 g/100 ml), EtOH (4 g/100 ml), $CHCl_3$ (5 g/100 ml); insoluble in Et_2O; LD_{50} (14 days) (mus orl) = 346 mg/kg, (mus iv) = 25 mg/kg, (rat orl) = 623 mg/kg, (rat iv) = 26 mg/kg. *E. Merck.*

6246 Prifinium Bromide
4630-95-9 7923 225-051-8
$C_{22}H_{28}BrN$
3-(Diphenylmethylene)-1,1-diethyl-2-methylpyrrolidinium bromide.
pyrodifenium bromide; Padrin; Riabal. Antispasmodic. Peripheral cholinoceptor antagonist. mp = 216-218°; LD_{50} (mmus iv) = 11 mg/kg, (mmus ip) = 43 mg/kg, (mmus sc) =30 mg/kg, (mmus orl) = 330 mg/kg; [free base]: $bp_{0.15}$ = 183-185°.

6247 Propyromazine
145-54-0 8057 205-657-9
$C_{20}H_{23}BrN_2OS$
1-Methyl-1-(1-phenothiazin-10-ylcarbonylethyl)-pyrrolidinium bromide.
LD-335; SD-104-19; Diaspasmyl. Anticholinergic; antispasmodic. mp = 228-229°; [free base ($C_{19}H_{20}N_2OS$)]: mp = 94.5-95.5°. *Astra Chem. Ltd.*

6248 Proxazole
5696-09-3 8090
$C_{17}H_{25}N_3O$
N,N-Diethyl-3-(1-phenylpropyl)-1,2,4-oxadiazole-5-ethanamine.
Aerbron; propoxaline. Antispasmodic. Smooth muscle relaxant used as an anticonvulsant. $bp_{0.2}$ = 132°; [nitrate]: mp = 127-128°. *Angelini Francesco.*

6249 Proxazole Citrate
132-35-4 8090 205-059-8
$C_{23}H_{33}N_3O_8$
N,N-Diethyl-3-(1-phenylpropyl)-1,2,4-oxadiazole-5-ethanamine citrate.
AF-634; Flou; Pirecin; Toness. Antispasmodic. Smooth muscle relaxant. LD_{50} (rat ip) = 39 mg/kg, (rat orl) = 60 mg/kg. *Angelini Pharm. Inc.*

6250 Prozapine
3426-08-2 8093 222-325-9
$C_{21}H_{27}N$
1-(3,3-Diphenylpropyl)hexahydro-1H-azepine.
hexadiphane. Antispasmodic. bp_1 = 170-174°. *N.V.Nederlandsche Comb. Chem. Ind.*

6251 Racefemine
22232-57-1 8273 244-856-5
$C_{18}H_{23}NO$
(±)-α-Methyl-N-(1-methyl-2-phenoxyethyl)benzeneethanamine.
CB-3697; [fumarate]: Dysmalgine. Antispasmodic. Smooth muscle relaxant. [isomer 1]: $bp_{0.05}$ = 132-135°; fumarate mp = 162°; hydrochloride mp = 156-157°; [isomer 2]: hydrochloride mp = 167-168°; [isomer 3]: $[\alpha]_D^{21}$ = 41° C = 0.01 EtOH); fumarate mp = 164-165°; hydrochloride mp = 186-187°, $[\alpha]_D^{21}$ = 22° (c = 0.01 EtOH); [isomer 4]: $[\alpha]_D^{22}$ = 22° (c = 0.005 EtOH), hydrochloride mp = 160-163°; $[\alpha]_D^{20}$ = 22° (c = 0.01 EtOH); [isomer 5]: $[\alpha]_D^{24}$ = -24° (c = 0.01 EtOH), hydrochloride mp = 189-190°; $[\alpha]_D^{21}$ = -22° (c = 0.01 EtOH); [isomer 6]: $[\alpha]_D^{24}$ = -41° (c = 0.01 EtOH); fumarate mp = 164-164.5° $[\alpha]_D^{22}$ = -15° (c = 0.01 EtOH); hydrochloride mp = 186-187°; $[\alpha]_D^{21}$ = -21.5° (c = 0.01 EtOH). *Clin-Byla France.*

6252 Rociverine
53716-44-2 8406 258-711-9
$C_{20}H_{37}NO_3$
1-Hydroxy-[1,1'-bicyclohexyl]-2-carboxylic acid 2-(diethylamino)-1-methylethyl ester.
LG-30158; Rilaten. Antispasmodic. Has neurotropic and myotropic activities. $bp_{0.1}$ = 148-150°; insoluble in H_2O; soluble in EtOH, Et_2O, C_6H_6, $CHCl_3$. *Guidotti.*

6253 Sintropium Bromide
79467-19-9 8693
$C_{19}H_{36}BrNO_2$
(endo,syn)-8-Methyl-8-(1-methylethyl)-3-[(1-oxo-2-propylpentyl)oxy]-8-azoniabicyclo[3.2.1]octane bromide.
VAL-480; VAL-4000. Anticholinergic; antispasmodic. mp > 280° (dec); LD_{50} (mus ip) = 77 mg/kg. *Valeas.*

6254 Spasmolytol
25333-96-4 8888
$C_{14}H_{21}Br_2NO_2$
2-[(3,5-Dibromo-2-methoxyphenyl)methoxy]-N,N-diethylethanamine.
A-124. Antispasmodic. bp_3 = 185-195°; [hydrochloride ($C_{14}H_{22}Br_2ClNO_2$)]: mp = 124-127°. *Hynson Westcott & Dunning.*

6255 Stilonium Iodide
77257-42-2
$C_{22}H_{30}INO$
Triethyl[2-[(E)-(p-styrylphenoxy)ethyl]ammonium iodide.
Elvetil. Antispasmodic. *Maggioni Farmaceutici S.p.A.*

6256 Sultroponium
15130-91-3 9170
$C_{20}H_{29}NO_6S$
endo-(±)-3-(3-Hydroxy-1-oxo-2-phenylpropoxy)-8-methyl-8-(3-sulfopropyl)-8-azoniabicyclo[3.2.1]octane inner salt.
8-(3-sulfopropyl)atropinium hydroxide inner salt; atropine sulfobetaine; sulfobetain of atropine; tropyl tropate sulfobetaine; A-118; Sultropan. Anticholinergic; antispasmodic. mp = 220° (dec); soluble in H_2O, warm EtOH; insoluble in Et_2O, Me_2CO, C_6H_6, $CHCl_3$; λ_m = 251.5, 257.5, 263.5 nm (H_2O).

6257 Terflavoxate
86433-40-1
$C_{26}H_{29}NO_4$
1,1-Dimethyl-2-piperidinoethyl 3-methyl-4-oxo-2-phenyl-4H-1-benzopyran-8-carboxylate.
REC-15/2053. Smooth muscle relaxant. Bladder spasmolytic.

6258 Tiemonium Iodide
144-12-7 9571 205-616-5
$C_{18}H_{24}INO_2S$
4-[3-Hydroxy-3-phenyl-3-(2-thienyl)propyl]-4-methylmorpholinium iodide.
TE-114; Visceralgina. Anticholinergic; antispasmodic. mp = 189-191°. *C.E.R.M.*

6259 Tigloidine
495-83-0 9574 207-810-5
$C_{13}H_{21}NO_2$
2-Methyl-2-butenoic acid [1α,3α(E),5α]-8-methyl-8-azabicyclo[3.2.1]oct-3-yl ester.
3β-tigloyloxytropane. Antispasmodic. [hydrobromide ($C_{13}H_{22}BrNO_2$)]: mp = 234-235°; soluble in $CHCl_3$.

6260 Tiquizium Bromide
71731-58-3 9602
$C_{19}H_{24}BrNS_2$
trans-3-(Di-2-thienylmethylene)octahydro-5-methyl-2H-quinolizinium bromide.
HSR-902; HS-902; Thiaton. Antispasmodic. Anticholinergic. mp = 278-281° (dec). *Hokoriku.*

6261 Tiropramide
55837-29-1 9607
$C_{28}H_{41}N_3O_3$
α-(Benzoylamino)-4-[2-(diethylamino)ethoxy]-N,N-dipropylbenzenepropanamide.
CR-605. Antispasmodic. Smooth muscle relaxant. Derivative of tyrosine. May inhibit cAMP catabolism. mp = 65-67°; LD_{50} (rat iv) = 33.9 mg/kg. *Rotta Pharm.*

6262 Tiropramide Hydrochloride
53567-47-8 9607
$C_{28}H_{42}ClN_3O_3$
α-(Benzoylamino)-4-[2-(diethylamino)ethoxy]-N,N-dipropylbenzenepropanamide hydrochloride.

Alfospas; Maiorad. Antispasmodic. Smooth muscle relaxant. *Rotta Pharm.*

6263 Tizanidine
51322-75-9 9624

$C_9H_8ClN_5S$

5-Chloro-4-(2-imidazolin-2-ylamino)-2,1,3-benzothiadiazole.

DS-103-282; Sirdalud; Ternelin. Antispasmodic; centrally acting skeletal muscle relaxant. α_2 adrenoceptor agonist; centrally active myotonolytic. Has been investigated for antitremor activity. mp = 221-223°; LD_{50} (mus orl) = 235 mg/kg. *Athena Neurosciences Inc.; Wander Pharma.*

6264 Tizanidine Hydrochloride
64461-82-1 9624

$C_9H_9Cl_2N_5S$

5-Chloro-4-(2-imidazolin-2-ylamino)-2,1,3-benzothiadiazole monohydrochloride.

AN021; DS-103-282; Zanaflex. Antispasmodic; centrally acting skeletal muscle relaxant. α_2 adrenoceptor agonist; centrally active myotonolytic. Has been investigated for antitremor activity. *Athena Neurosciences Inc.; Wander Pharma.*

6265 Trepibutone
41826-92-0 9718

$C_{16}H_{22}O_6$

2,4,5-Triethoxy-σ-oxybenzenebutanoic acid.

AA-149; Supacal. Choleretic; antispasmodic. mp = 150-151°; LD_{50} (mmus orl) = 1340 mg/kg, (mmus ip) = 530 mg/kg, (mrat orl) = 2450 mg/kg, (mrat ip) = 410 mg/kg. *Takeda Chem. Ind. Ltd.*

6266 Tricromyl
85-90-5 9791 201-641-0

$C_{10}H_8O_2$

3-Methyl-4H-1-benzoxopyran-4-one.

methylchromone; Cromonalgina. Antispasmodic; coronary vasodilator. Originated from an ancient Egyptian drug now termed *bezr el khelda*. mp = 68°; λ_m = 304 nm (in EtOH). *Lab. Franc. Chimiother.*

6267 Trifolium
9818

Meadow clover; red clover; purple clover; cow clover. Dried inflorescence of *Trifolium pratense L. Leguminosae.* Antispasmodic and expectorant.

6268 Trimebutine
39133-31-8 9829 254-309-2

$C_{22}H_{29}NO_5$

3,4,5-Trimethoxybenzoic acid 2-(dimethylamino)-2-phenylbutyl ester.

Anticholinergic; antispasmodic. Opiate receptor agonist. Accelerates gastric emptying. mp = 78-80°; soluble in CH_2Cl_2. *Jouveinal.*

6269 Trimebutine Maleate
34140-59-5 9829 251-845-9

$C_{26}H_{33}NO_9$

3,4,5-Trimethoxybenzoic acid 2-(dimethylamino)-2-phenylbutyl ester maleate.

TM-906; Cerekinon; Debridat; Digerent; Foldox; Polibutin; Spabucol; Trimedat. Anticholinergic; antispasmodic. Opiate receptor agonist. Accelerates gastric emptying. mp = 105-106°. *Jouveinal.*

6270 N,N,1-Trimethyl-3,3-diphenylpropylamine
13957-55-6 9843

$C_{18}H_{23}N$

N,N,α-Trimethyl-σ-phenylbenzenepropanamine.

Recipavrin. Antispasmodic. [l-form hydrochloride]: mp = 180-182°; $[\alpha]_D^{20}$ = -43.3° (c = 1.04 H_2O); [d-form hydrochloride]: mp = 179-181°; $[\alpha]_D^{20}$ = 43.1° (c = 0.53 H_2O); [nitrate]: mp 118-120°.

6271 Tropenzile
53834-53-0 9908

$C_{23}H_{27}NO_4$

(3-endo,6-exo)-α-Hydroxy-α-phenylbenzeneacetic acid 6-methoxy-8-methyl-8-azabicyclo[3.2.1]oct-3-yl ester.

tropenzilium bromide [as methyl bromide]; component of: Palerol [as hydrobromide]. Anticholinergic; antispasmodic. mp = 99-101°; [hydrochloride ($C_{23}H_{28}ClNO_4$)]: mp = 146-148°. *Sandoz Pharm. Corp.*

6272 Xenytropium Bromide
511-55-7 10208 208-129-6

$C_{30}H_{34}BrNO_3$

endo-(±)-8-([1,1'-Biphenyl]-4-ylmethyl)-3-(3-hydroxy-2-phenyl-1-oxopropoxy)-8-methyl-8-azoniabicyclo[3.2.1]octane bromide.

N-399; Gastropin; Gastripon. Antispasmodic; anticholinergic. dec 220-222°. *Licencia Budapest.*

Antisyphilitics

6273 Acetarsone
97-44-9 49 202-582-3

$C_8H_{10}AsNO_5$

N-Acetyl-4-hydroxy-m-arsanilic acid.

Stovarsol; acetarsol; acetphenarsine; Ehrlich 594; Fourneau 190; F-190; Amarsan; Arsaphen; Dynarsan; Goyl; Kharophen; Limarsol; Malagride; Gynoplix; Oralcid; Devegan; Orarsan; Osarsal; Osvarsan; Paroxyl; Sanogyl; Spirocid; S.V.C.; Monargan; Ginarsol; Stovarsolan; Realphene [as calcium salt]; Bistovol [as bismuth salt]. Antisyphilitic; antiprotozoal against Trichomonas. Dec 240-250°; slightly soluble in H_2O; MLD (rbt orl) = 125-150 mg/kg, (cat orl) = 150-175 mg/kg. *Abbott Labs Inc.; Rhône-Poulenc Rorer Pharm. Inc.*

6274 Acetarsone Diethylamine Salt
534-33-8 49 208-597-1

$C_{12}H_{21}AsN_2O_5$

N-Acetyl-4-hydroxy-m-arsanilic acid diethylamine salt.

Acetarsin; Acetilarsano; Acetylarsan; Arsaphenan; Golarsyl; Syntharsol. Antiprotozoal against Trichomonas. Antisyphilitic. *Abbott Labs Inc.; Rhône-Poulenc Rorer Pharm. Inc.*

6275 Arsacetin
618-22-4 829 210-541-6

$C_8H_{10}AsNO_4$

[4-(Acetylamino)phenyl]arsonic acid.

N-acetylarsanilic acid. Antisyphilitic. [sodium salt tetrahydrate ($C_8H_9AsNNaO_4.4H_2O$)]: soluble in H_2O (10 g/100 ml at 25°, 33.3 g/100 ml at 100°); LD_{50} (rbt iv) = 550 mg/kg.

6276 Arsphenamine
139-93-5 851 205-386-6
$C_{12}H_{14}As_2Cl_2N_2O_2$
4,4'-(1,2-Diarsenediyl)bis[2-aminophenol] dihydrochloride.
Ehrlich 606; Arsaminol; Kharsivan; Salvarsan; Sanluol. Used formerly as an antisyphilitic. Soluble in H_2O, EtOH, glycerol; slightly soluble in $CHCl_3$, Et_2O; LD_{100} (rat iv) = 140 mg/kg. *Hoechst AG.*

6277 Bismuth Butylthiolaurate
53897-25-9 1301
$C_{16}H_{33}BiO_4S$
2-(Butylthio)dodecanoic acid bismuth basic salt.
Bisspecia; Neocardyl. Used formerly as an antisyphilitic. Insoluble in H_2O, soluble in oils.

6278 Bismuth Chloride Oxide
7787-59-9 1303 232-122-7
BiClO
Bismuth oxychloride.
basic bismuth chloride; bismuthyl chloride; bismuth subchloride; pearl white; blanc d'Espagne; blanc de perle; Chlorbismol. Antisyphilitic. d = 7.72; melts at low red heat; insoluble in H_2O; soluble in HCl, HNO_3.

6279 Bismuth Ethyl Camphorate
52951-37-8 1304
$C_{36}H_{57}BiO_{12}$
d-Camphoric acid ethyl ester bismuth salt.
bismuth(III) salt of d-camphoric acid ethyl ester. Used formerly as an antisyphilitic. mp = 61-67°; insoluble in H_2O; soluble in $CHCl_3$, Et_2O, ethylene dichloride; MLD (rat mi) = 250 mg/kg.

6280 Bismuth Potassium Tartrate
5798-41-4 1318 227-345-1
Potassium bismuth tartrate.
potassium bismuthotartrate; potassium bismuthyl tartrate; tartaric acid bismuth complex potassium salt. Used formerly as an antisyphilitic. Soluble in H_2O (50 g/100 ml), insoluble in organic solvents. *Searle G.D. & Co.*

6281 Bismuth Sodium Tartrate
31586-77-3 1321 250-719-0
Sodium bismuth tartrate.
sodium bismuthyl tartrate; tartaric acid bismuth complex, sodium salt; Natrol; Tartrol. Used formerly as an antisyphilitic. Soluble in H_2O (33 g/100 ml), insoluble in organic solvents. *Searle G.D. & Co.*

6282 d-Camphocarboxylic Acid Basic Bismuth salt
4154-53-4 1778 242-404-1
$C_{33}H_{46}Bi_2O_{11}$
Basic bismuth d-camphocarboxylate.
Angimuth. Used formerly as an antisyphilitic. Insoluble in H_2O; soluble in MeOH, Et_2O, C_6H_6. *Abbott Labs Inc.*

6283 Dichlorophenarsine
455-83-4 3121
$C_6H_6AsCl_2NO$
3-Amino-4-hydroxyphenyldichloroarsine.
Used formerly as an antisyphilitic. *Parke-Davis.*

6284 Dichlorophenarsine Hydrochloride
536-29-8 3121
$C_6H_7AsCl_3NO$
3-Amino-4-hydroxyphenyldichloroarsine hydrochloride.
Dichlor-Mapharsen; dichlorophenarsinammonium chloride. Used formerly as an antisyphilitic. mp = 200°; readily soluble in H_2OLD_{50} (mus ip) = 41 mg/kg. *Parke-Davis.*

6285 Ethanearsonic Acid
507-32-4 3768
$C_2H_7AsO_3$
Ethylarsonic acid.
ethylarsinic acid. Used formerly as an antisyphilitic. mp = 99.5°; soluble in H_2O (70 g/100 ml at 27°, 112 g/100 ml at 40°), 95% EtOH (39.4 g/100 ml at 25°).

6286 Ethanearsonic Acid Disodium Salt
5982-55-8 3768
$C_2H_5AsNa_2O_3$
Sodium ethylarsonate.
Mon-Arsone. Used formerly as an antisyphilitic. Very soluble in H_2O.

6287 Mercuric Benzoate
583-15-3 5924 209-499-1
$C_{14}H_{10}HgO_4$
Mercury(II) benzoate.
Used formerly as an antisyphilitic. Soluble in H_2O (1.1 g/100 ml at 25°, 2.5 g/100 ml at 100°), slightly soluble in EtOH.

6288 Mercurous Chloride
10112-91-1 5951 233-307-5
Hg_2Cl_2
Dimercury dichloride.
calomel; mild mercury chloride; mercury monochloride; mercury subchloride; precipité blanc; Calogreen. Antiseptic; cathartic; diuretic; antisyphilitic. Used as a cathartic, local antiseptic and desiccant in veterinary medicine. Mercurial compound. CAUTION: Excessive doses can cause mercury poisoning. Sublimes at 400-500°; d = 7.15; practically insoluble in H_2O; insoluble in alcohol, Et_2O; incompatible with bromides, iodides, alkali chlorides, sulfates, sulfites, carbonates, hydroxides, lime H_2O, acacia, ammonia, golden antimony sulfide, cocain.

6289 Sodium Arsanilate
127-85-5 8719 204-869-9
$C_6H_7AsNNaO_3$
(p-Aminophenyl)arsonic acid sodium salt.
arsanilic acid sodium salt; sodium aminarsonate; sodium anilarsonate; Arsamin; Atoxyl; Nuarsol; Protoxyl; Soamin; Sonate; Piglet Pro-Gen V; Trypoxyl. Used formerly as an antisyphilitic. Soluble in H_2O (16.6 g/100 ml), EtOH (1 g/100 ml).

6290 Sodium Arsphenamine
1936-28-3 8722
$C_{12}H_{10}As_2N_2Na_2O_2$
4,4'-Arsenobis[2-aminophenol] disodium salt.
arsphenamine sodium. Used formerly as an antisyphilitic. Freely soluble in H_2O.

6291 Sozoiodole-Mercury
515-43-5 8881
$C_6H_2HgI_2O_4S$
4-Hydroxy-3,5-diiodobenzenesulfonic acid mercury salt.
Used formerly as an antisyphilitic. Orange-yellow powder, insoluble in H_2OEtOH, Et_2O, glycerol; soluble in NaCl or KI solutions.

6292 Sulfarsphenamine
618-82-6 9111 210-564-1
$C_{14}H_{14}As_2N_2Na_2O_8S_2$
[1,2-Diarsenediylbis[(6-hydroxy-3,1-phenylene)imino]]bismethanesulfonic acid disodium salt.
sulfarsenobenzene; Karsulphan; Myosalvarsan; Myarsenol; Metarsenobillon; Thiosarmine. Used formerly as an antisyphilitic. Very soluble in H_2O, slightly soluble in EtOH.

Antithrombocythemics

6293 Anagrelide
68475-42-3 665
$C_{10}H_7Cl_2N_3O$
6,7-Dichloro-1,5-dihydroimidazo[2,1-b]quinazolin-2(3H)-one.
Antithrombotic; antithrombocythemic. *Bristol-Myers Squibb Co.; Roberts Pharm. Corp.*

6294 Anagrelide Hydrochloride
58579-51-4 665
$C_{10}H_8Cl_3N_3O$
6,7-Dichloro-1,5-dihydroimidazo[2,1-b]quinazolin-2(3H)-one hydrochloride.
BL-4162A; BMY-26538-01; Agrelin; Agrylin. Antithrombotic; antithrombocythemic. mp >280°. *Bristol-Myers Squibb Co.; Roberts Pharm. Corp.*

Antithrombotics

6295 Anagrelide
68475-42-3 665
$C_{10}H_7Cl_2N_3O$
6,7-Dichloro-1,5-dihydroimidazo[2,1-b]quinazolin-2(3H)-one.
Antithrombotic; antithrombocythemic. *Bristol-Myers Squibb Co.; Roberts Pharm. Corp.*

6296 Anagrelide Hydrochloride
58579-51-4 665
$C_{10}H_8Cl_3N_3O$
6,7-Dichloro-1,5-dihydroimidazo[2,1-b]quinazolin-2(3H)-one hydrochloride.
BL-4162A; BMY-26538-01; Agrelin; Agrylin. Antithrombotic; antithrombocythemic. mp >280°. *Bristol-Myers Squibb Co.; Roberts Pharm. Corp.*

6297 Antithrombin III
52014-67-2
The glycoprotein antithrombin obtained from human plasma.
Kybernin; Thrombate III. Antithrombotic. *Bayer Corp. Pharm. Div.; Hoechst Roussel Pharm. Inc.*

6298 Argatroban
74863-84-6 816
$C_{23}H_{36}N_6O_5S$
(2R,4R)-4-Methyl-1-[n^2-[(1,2,3,4-tetrahydro-3-methyl-8-quinolyl)sulfonyl]-L-arginyl]pipecolic acid.
argipidin; MQPA. Synthetic thrombin inhibitor. Used as an anticoagulant. mp = 188-191°. *Mitsubishi Chem. Corp.*

6299 Argatroban Monohydrate
141396-28-3 816
$C_{23}H_{36}N_6O_5S.H_2O$
(2R,4R)-4-Methyl-1-[n^2-[(1,2,3,4-tetrahydro-3-methyl-8-quinolyl)sulfonyl]-L-arginyl]pipecolic acid monohydrate.
Novastan; Slonnon; MCI-9038; MD-805; DK-7419; GN1600. Synthetic thrombin inhibitor. Used as an anticoagulant. mp = 176-180°; $[\alpha]_D^{27}$ = 76.1° (c = 1 0.2N HCl). *Mitsubishi Chem. Corp.*

6300 Aspirin
50-78-2 886 200-064-1
$C_9H_8O_4$
Acetylsalicylic acid.
o-carboxyphenyl acetate; 2-(acetyloxy)benzoic acid; acetate salicylic acid; salicylic acid acetate; Acenterline; Aceticyl; Acetosal; Acetosalic Acid; Acetosalin; Acetylin; Acetyl-SAL; Acimetten; Acylpyrin; Arthrisin, A.S.A; Asatard; Aspro; Asteric; Caprin; Claradin; Colfarit; Contrheuma retard; Duramax; ECM; Ecotrin; Empirin; Encaprin; Endydol; Entrophen; Enterosarine; Helicon; Levius; Longasa; Measurin; Neuronika; Platet; Rhodine; Salacetin; Salcetogen; Saletin; Solrin; Solpyron; Xaxa; Alka-seltzer; Anacin; Ascriptin; Bufferin; Coricidin D; Darvon compound; Excedrin; Gelprin; Robaxisal; Vanquish; Ascoden-30; Coricidin; Norgesic; Persistin; Supac; Triaminicin; acetophen; acidum acetylsalicylicum; acetilum acidulatum; Acetonyl; Adiro; acenterine; acetosal; acetosalic acid; acetosalin; acetylin; acetylsal; acylpyrin;asteric; caprin; colfarit; entrophen; enterosarine; rhodine; salacetin; salcetogen; saletin; acesal; acetilsalicilico; acetisal; acetonyl; asagran; asatard; aspalon; aspergum; aspirdrops; AC 5230; benaspir; entericin; extren; bialpirinia; contrheuma retard; Crystar; Delgesic; Dolean ph 8; enterophen; globoid; idragin. Analgesic; antipyretic; anti-inflammatory. Also has platelet aggregation inhibiting, antithrombotic, and antirheumatic properties. Can inhibit the synthesis of platelet thromboxane A_2, thereby inhibiting platelet aggregation. mp = 135; d = 1.40; λ_m = 229, 277 nm ($E^{1\%}_{1cm}$ 484, 68 0.1N H_2SO_4); pH (25°) = 3.49; slightly soluble in H_2O (1 g/300 ml at 25°, 1 g/100 ml at 37°); soluble in alc (1 g/5 ml), $CHCl_3$ (1 g/17 ml), Et_2O (≅1 g/10 ml); LD_{50} (mus orl) = 1100 mg/kg. *Boots Pharmaceuticals Inc.; Bristol-Myers Squibb Co.; Eli Lilly & Co.; Parke-Davis; Schering-Plough Pharm.; SmithKline Beecham Pharm.; Sterling Health U.S.A.; Sterling Winthrop Inc.; Upjohn Ltd.*

6301 Beciparcil
130782-54-6
$C_{12}H_{13}NO_3S_2$
p-[(5-Thio-β-D-xylopyranosyl)thio]benzonitrile.
Antithrombotic.

6302 Bivalirudin
128270-60-0
$C_{98}H_{138}N_{24}O_{33}$
D-Phenylalanyl-L-prolyl-L-arginyl-L-prolylglycylglycylglycylglycyl-L-asoaraginylglycyl-L-α-aspartyl-L-phenylalanyl-L-α-glutamyl-L-α-glutamyl-L-tyrosyl-L-leucine.
H-D-Phe-Pro-Arg-Pro-Gly-Gly-Gly-Gly-Asn-Gly-Asp-Phe-Glu-Glu-Ile-Pro-Glu-Glu-Tyr-Leu-OH; Hirulog; BG8967. Antithrombotic; anticoagulant. A 20-merpeptide. *Biogen, Inc.*

6303 Cilostazol
73963-72-1 2335
$C_{20}H_{27}N_5O_2$
6-[4-(1-Cyclohexyl-1H-tetrazol-5-yl)butoxy]-3,4-dihydroxycarbostyril.
OPC-13013; OPC-21; Pletaal. Antithrombotic; vasodilator; platelet inhibitor. Antithrombotic. For treatment of intermittent claudication. mp = 169.4-170.3°; λ_m = 257 nm (ε 15200 MeOH); freely soluble in AcOH, $CHCl_3$, DMSO; insoluble in Et_2O, H_2O, 0.1N HCl, 0.1N NaOH. *Otsuka Am. Pharm.*

6304 Clopidogrel
113665-84-2 2457
$C_{16}H_{16}ClN_2OS$
Methyl (+)-(S)-α-(o-chlorophenyl)-6,7-dihydrothieno[3,2-c]pyridine-5(4H)-acetate.
SR-25990. Platelet inihibtor; antithrombotic. $[\alpha]_d^{20}$ = 51.52° (c = 1.61 MeOH). *Sanofi Winthrop.*

6305 Clopidogrel Hydrogen Sulfate
135046-48-9 2457
$C_{16}H_{18}ClN_2O_5S_2$
Methyl (+)-(S)-α-(o-chlorophenyl)-6,7-dihydrothieno[3,2-c]pyridine-5(4H)-acetate sulfate (1:1).
SR-25990C; Plavix. Platelet inihibtor; antithrombotic. mp = 184°; $[\alpha]_D^{20}$ = 55.10° (c = 1.891 MeOH). *Sanofi Winthrop.*

6306 Cloricromen
68206-94-0 2467
$C_{20}H_{26}ClNO_5$
Ethyl [[8-chloro-3-[2-(diethylamino)ethyl]-4-methyl-2-oxo-2H-1-benzopyran-7-yl]oxy]acetate.
AD_6; 8-chlorocarbochromen. Antithrombotic; vasodilator (coronary). Related to coumarin. mp = 147-148°. *Fidia Pharm.*

6307 Cloricromen Hydrochloride
74697-28-2 2467
$C_{20}H_{27}Cl_2NO_5$
Ethyl [[8-chloro-3-[2-(diethylamino)ethyl]-4-methyl-2-oxo-2H-1-benzopyran-7-yl]oxy]acetate hydrochloride.
Cromocap; Proendotel. Vasodilator; antithrombotic. mp = 219-220°. *Fidia Pharm.*

6308 Dalteparin
9005-49-6 2870
Fragment of heparin.
Tedelparin. Antithrombotic. Low-molecular-weight fragment of heparin prepared by nitrous acid depolymerization of porcine mucosal heparin. Average molecular weight 4000-6000 Da. Hemorrhagic effects lower than those of heparin. *Kabi Pharmacia Diagnostics.*

6309 Dalteparin Sodium
9041-08-1 2870
Sodium salt of heparin fragment.
Kabi 2165; FR-860; Boxol; Fragmin. Antithrombotic. Sodium salt of a low-molecular-weight fragment of heparin prepared by nitrous acid depolymerization of porcine mucosal heparin. Average molecular weight 4000-6000 Da. *Kabi Pharmacia Diagnostics.*

6310 Daltroban
79094-20-5 2871
$C_{16}H_{16}ClNO_4S$
[p-[2-(p-Chlorobenzenesulfonamido)ethyl]phenyl]-acetic acid.
SKF-96148; BM-13505. Antithrombotic; immunosuppressant. Thromboxane synthetase inhibitor. *Boehringer Mannheim GmbH; SmithKline Beecham Pharm.*

6311 Danaparoid Sodium
57459-72-0
ORG-10172. Antithrombotic. Mixture of sodium salts of heparin sulfate and chondroitin sulfate. *Diosynth BV.*

6312 Dazoxiben
78218-09-4
$C_{12}H_{12}N_2O_3$
p-(2-Imidazol-1-ylethoxy)benzoic acid.
Antithrombotic. *Pfizer Intl.*

6313 Dazoxiben Hydrochloride
74226-22-5
$C_{12}H_{13}ClN_2O_3$
p-(2-Imidazol-1-ylethoxy)benzoic acid hydrochloride.
UK-37248-01. Antithrombotic. *Pfizer Intl.*

6314 Defibrotide
83712-60-1 2915
Fraction P; defibrinotide; Dasovas; Noravid; Prociclide. Antithrombotic. Polydeoxyribonucleotides from bovine lung; molecular weight between 15000 and 30000. *Crinos.*

6315 Dipyridamole
58-32-2 3410 200-374-7
$C_{24}H_{40}N_8O_4$
2,2',2'',2'''-[(4,8-Dipiperidinylpyrimido[5,4-d]pyrimidine-2,6-diyl)dinitrilo]tetraethanol.
NSC-515776; RA-8; Anginal; Cardoxil; Cleridium; Coridil; Coronarine; Curantyl; Dipyridan; Gulliostin; Natyl; Peridamol; Persantine; Piroan; Prandiol; Protangix. Vasodilator (coronary). Phosphodiesterase inhibitor. Decreases platelet aggregation to damaged endothelium. mp = 163°; slightly soluble in H_2O; soluble in MeOH, EtOH, $CHCl_3$; slightly soluble in Me_2CO, C_6H_6, EtOAc; LD_{50} (rat orl) = 8400 mg/kg, (rat iv) = 208 mg/kg. *Boehringer Ingelheim Ltd.; Thomae GmbH Dr. Karl.*

6316 Efegatran
105806-65-3
$C_{21}H_{32}N_6O_3$
N-Methyl-D-phenylalanyl-N-[(1S)-1-formyl-4-guanidino-butyl]-L-prolinamide.
LY-294468. Antithrombotic. *Eli Lilly & Co.*

6317 Efegatran Sulfate
126721-07-1
$C_{21}H_{34}N_6O_7$
N-Methyl-D-phenylalanyl-N-[(1S)-1-formyl-4-guanidino-butyl]-L-prolinamide sulfate (1:1).
LY-294468 sulfate. Antithrombotic. *Eli Lilly & Co.*

6318 Enoxaparin Sodium
9041-08-1 3626
Sodium salt of depolymerized heparin.
Clexane; Ultraparin; Lovenox Injection; RP-54563; PK-10169. Antithrombotic. A low-molecular-weight fragment of heparin prepared from the benzylic ester of porcine mucosal heparine. Has a 4-eno pyranosuronate sodium group on the non-reducing end of the polymer. Molecular weight between 3500 and 5500 Da. *Rhône-Poulenc Rorer Pharm. Inc.*

6319 Fluretofen
56917-29-4
$C_{14}H_9F$
4'-Ethynyl-2-fluorobiphenyl.
compound 93819. Antithrombotic. *Eli Lilly & Co.*

6320 Ifetroban
143443-90-7
$C_{25}H_{32}N_2O_5$
o-[[(1S,2R,3S,4R)-3-[4-(pentylcarbamoyl)-2-oxazolyl]-7-oxabicyclo[2.2.1]hept-2-yl]methyl]-hydrocinnamic acid.
BMS-180291. Antithrombotic. *Bristol-Myers Squibb Pharm. R&D.*

6321 Ifetroban Sodium
156715-37-6
$C_{25}H_{31}N_2NaO_5$
Sodium o-[[(1S,2R,3S,4R)-3-[4-(pentylcarbamoyl)-2-oxazolyl]-7-oxabicyclo[2.2.1]hept-2-yl]methyl]-hydrocinnamate.
BMS-180291-02. Antithrombotic. *Bristol-Myers Squibb Pharm. R&D.*

6322 Iliparcil
137214-72-3
$C_{16}H_{18}O_6S$
4-Ethyl-7-[(5-thio-β-D-xylopyranosyl)oxy]coumarin.
Antithrombotic.

6323 Iloprost
78919-13-8 4940
$C_{22}H_{32}O_4$
(E)-(3aS,4R,5R,6aS)-Hexahydro-5-hydroxy-4-[(E)-(3S,4RS)-3-hydroxy-4-methyl-1-octen-6-ynyl]-$\Delta^{2(1H),\delta}$-pentalenevaleric acid.
ciloprost; ZK-36374; [tromethamine salt ($C_{26}H_{43}NO_7$)]: Endoprost; Ilomedin. Antithrombotic; peripheral vasodilator. Colorless oil. *Schering AG.*

6324 Indobufen
63610-08-2 4991 264-364-4
$C_{18}H_{17}NO_3$
(±)-2-[p-(1-Oxo-2-isoindolinyl)phenyl]butyric acid.
K-2930; Ibustrin. Antithrombotic. mp = 182-184°. *CIBA plc.*

6325 Inicarone
39178-37-5
$C_{17}H_{15}NO_2$
2-Isopropyl-3-benzofuranyl 4-pyridyl ketone.
Antithrombotic.

6326 Inogatran
155415-08-0
$C_{21}H_{38}N_6O_4$
N-[(1R)-2-Cyclohexyl-1-[[(2S)-2-[(3-guanidinopropyl)-carbamoyl]piperidino]carbonyl]ethyl]glycine.
Thrombin inhibitor. Antithrombotic.

6327 Integrelin
157630-07-4 5013
A fibrinogen receptor antagonist used as an antithrombotic. Isolated from the venom of the southeastern pigmy rattlesnake. A synthetic, disulfide linked heptapeptide based on barbourin.

6328 Isbogrel
89667-40-3 5120
$C_{18}H_{19}NO_2$
(E)-7-Phenyl-7-(3-pyridyl)-6-heptenoic acid.
CV-4151. Antithrombotic. Thromboxane synthetase inhibitor. mp = 114-115°. *Takeda Chem. Ind. Ltd.*

6329 Israpafant
117279-73-9
$C_{28}H_{29}ClN_4S$
(±)-4-(o-Chlorophenyl)-2-(p-isobutylphenethyl)-6,9-dimethyl-66H-thieno[3,2-f]-s-triazolo[4,3-a]-[1,4]diazepine.
Antithrombotic. Platelet-activating factor (PAF) antagonist.

6330 Lamifiban
144412-49-7 5362
$C_{24}H_{28}N_4O_6$
[[1-[N-(p-Amidinobenzoyl)-L-tyrosyl]-4-piperidyl]oxy]-acetic acid.
Ro-44-9883/000. Antithrombotic. Specific nonpeptide platelet fibrinogen receptor (GPIIb/IIIa) antagonist. mp > 200° (dec); $[\alpha]_D^{20}$ = 29.8° (c = 0.86 1N HCl). *Hoffmann-LaRoche Inc.*

6331 Lamifiban Trifluoracetate Salt
144412-50-0 5362
$C_{26}H_{29}F_3N_4O_8$
[[1-[N-(p-Amidinobenzoyl)-L-tyrosyl]-4-piperidyl]oxy]-acetic acid trifluoroacetate salt.
Antithrombotic. Specific nonpeptide platelet fibrinogen receptor (GPIIb/IIIa) antagonist. mp = 125-130° (dec); LD_{50} (mus iv) = 250 mg/kg. *Hoffmann-LaRoche Inc.*

6332 Lamoparan
5366
Low molecular weight heparinoid.
Org-10172. Antithrombotic. Derived from porcine intestinal mucosa. A mixture of sulfated

glycosaminoglycans. Mean molecular weight is 6500. Activity similar to that of heparin, but with a lower hemorrhagic effect. $[\alpha]_D^{20}$ = 30° to 70°.

6333 Lefradafiban
149503-79-7
$C_{23}H_{25}N_3O_6$
(3S,5S)-5-[[[4'-(Carboxyamidino)-4-biphenylyl]oxy]-methyl]-2-oxo-3-pyrrolidineacetic acid di-methyl ester.
Prodrug of fradafiban, a nonpeptide platelet glycoprotein IIb/IIIa antagonist.

6334 Lepirudin
138068-37-8
$C_{287}H_{440}N_{80}O_{111}S_6$
1-L-Leucine-2-L-threonine-63-desulfohirudin.
recombinant hirudin; r-hirudin; HBW-023; Leu-Thr-Tyr-Thr-Asp-Cys(6)-Thr-Glu-Ser-Gly-Asn-Leu-Cys(14)-Leu-Cys(16)-Glu-Gly-Ser-Asn-Val-Cys(22)-Gly-Gln-Gly-Asn-Lys-Cys(28)-Ile-Leu-Gly-Ser-Asp-Gly-Glu-Lys-Asn-Gln-Cys(39)-Val-Thr-Gly-Glu-Glu-Thr-Pro-Lys-Pro-Gln-Ser-His-Asn-Asp-Gly-Asp-Phe-Glu-Glu-Ile-Pro-Glu-Glu-Tyr-Leu-Gln [(6→14),(16→28),(22→39) cyclic disulfide. Recombinant hirudin (*hirudo medicinalis* isoform HV1). Produces parenteral anticoagulation for treatment of heparin-induced thrombocytopenia. Antithrombotic; anticoagulant.

6335 Linotroban
120824-08-0
$C_{14}H_{15}NO_5S_2$
5-[2-(Phenylsulfonylamino)ethyl]-thienyloxy-acetic acid.
HN-11500. Thromboxane A2 receptor antagonist. Antithrombotic.

6336 Melagatran
159776-70-2
$C_{22}H_{31}N_5O_4$
N-[(R)-[[(2S)-2-[(p-Amidinobenzyl)carbamoyl]-1-azetidinyl]carbonyl]cyclohexylmethyl]glycine.
A synthetic, low molecular-weight thrombin inhibitor. Antithrombotic.

6337 Modipafant
122957-06-6
$C_{34}H_{29}ClN_6O_3$
(+)-(R)-4-(o-Chlorophenyl)-1,4-dihydro-6-methyl-2-[p-(2-methyl-1H-imidazo[4,5-c]pyridin-1-yl)phenyl]-5-(2-pyridylcarbamoyl)nicotinate.
UK-80067. Platelet-activating factor (PAF) antagonist. Antithrombotic.

6338 Nadroparin Calcium
6434
Calcium salt of heparin depolymerized by nitrous acid degradation.
CY-16. Antithrombotic. Molecular weight between 4000 and 5000.

6339 Nafazatrom
59040-30-1 261-571-1
$C_{16}H_{16}N_2O_2$
3-Methyl-1-[2-(2-naphthyloxy)ethyl]-2-pyrazolin-5-one.
Bay g 6575. Synthetic pyrazolinone derivative. Leukotriene synthesis inhibitor; lipoxygenase inhibitor. Antimetastatic; antithrombotic.

6340 Napsagatran
159668-20-9
$C_{26}H_{34}N_6O_6S.H_2O$
N-[N^4-[[(3S-1-Amidino-3-piperidyl]methyl]-n^2-(2-naphthylsulfonyl)-L-asparaginyl]-N-cyclopropylglycine monohydrate.
Ro-46-6240/010. Antithrombotic. *Hoffmann-LaRoche Inc.*

6341 Naroparcil
120819-70-7
$C_{19}H_{17}NO_4S_2$
p-[p-[(5-Thio-β-D-xylopyranosyl)thio]benzoyl]benzonitrile.
A β-D-Xyloside analog. Antithrombotic.

6342 Nictindole
36504-64-0 253-070-1
$C_{17}H_{16}N_2O_3$
[2-(1-Methylethyl)-1H-indol-3-yl]-3-pyridinyl-methanone.
L-8027. Thromboxane inhibitor. Antithrombotic.

6343 Orbofiban Acetate
165800-05-5
$C_{17}H_{23}N_5O4.\ 0.25H_2O$
N-[[(3S)-1-(p-Amidinophenyl)-2-oxo-3-pyrrolidinyl]-carbamoyl]-β-alanine ethyl ester monoacetate quadrantihydrate.
SC-57099B. Antithrombotic; platelt aggregation inhibitor. *Searle G.D. & Co.*

6344 Ozagrel
82571-53-7 7115
$C_{13}H_{12}N_2O_2$
(E)-p-(Imidazol-1-ylmethyl)cinnamic acid.
OKY-046. Antianginal; antithrombotic. Thromboxane synthetase inhibitor. mp = 223-224°. *Kissei; Ono Pharm.*

6345 Ozagrel Hydrochloride
78712-43-3 7115
$C_{13}H_{13}ClN_2O_2$
(E)-p-(Imidazol-1-ylmethyl)cinnamic acid hydrochloride.
Antianginal; antithrombotic. Thromboxane synthetase inhibitor. mp = 214-217°. *Kissei; Ono Pharm.*

6346 Ozagrel Sodium
7115
$C_{13}H_{11}N_2NaO_2$
Sodium (E)-p-(Imidazol-1-ylmethyl)cinnamic acid.
Cataclot; Xanbon. Antianginal; antithrombotic. Thromboxane synthetase inhibitor. LD_{50} (mmus iv) = 1940 mg/kg, (mmus orl) = 3800 mg/kg, (mmus sc) = 2450 mg/kg, (fmus iv) = 1580 mg/kg, (fmus orl) = 3600 mg/kg, (fmus sc) = 2100 mg/kg, (mrat iv) = 1150 mg/kg, (mrat orl) = 5900 mg/kg, (mrat sc) = 2300 mg/kg, (frat iv) = 1300 mg/kg, (frat orl) = 5700 mg/kg, (frat sc) = 2250 mg/kg. *Kissei; Ono Pharm.*

6347 Pentosan Polysulfate
37300-21-3 7276
$[C_5H_8O_{10}S_2]_{6-12}$
(1→4)-β-D-Xylan 2,3-bis(hydrogen sulfate).
xylan hydrogen sulfate; xylan polysulfate; CB-8061. Anti-inflammatory (interstitial cystitis); antithrombotic; anticoagulant. A semi-synthetic sulfated polyxylan polysulfate composed of β-D-xylopyranose residues with

properties similar to heparin. Molecular weight 1500-5000. *Wander Pharma.*

6348 Pentosan Polysulfate Sodium Salt
37319-17-8 7276
$[C_5H_6Na2O_{10}S_2]_{6-12}$
(1→4)-β-D-Xylan 2,3-bis(hydrogen sulfate) sodium salt.
SP54; PZ68; sodium pentosan polysulfate; sodium xylan polysulfate; Cartrophen; Elmiron; Fibrase; Fibrezym; Héoclar; Thrombocid. Anti-inflammatory (interstitial cystitis); antithrombotic; anticoagulant. Sodium salt of sulfated β-D-xylopyranose polymer. $[\alpha]_D^{20}$ = -57°; N_D^{20} (10% aqueous solution) = 1.344; soluble in H_2O. *Wander Pharma.*

6349 Picotamide
32828-81-2 7560 251-245-7
$C_{21}H_{20}N_4O_3$
4-Methoxy-N,N'-bis(3-pyridinylmethyl)-1,3-benzenedicarboxamide.
G-137; Plactidil [as monohydrate]. Antithrombotic; fibrinolytic; anticoagulant. mp = 124°; LD_{50} (mus ip) = 1205 mg/kg. *Soc. Italo-Brit. L. Manetti.*

6350 Picotamide Hydrate
80530-63-8 7560
$C_{21}H_{20}N_4O_3.H_2O$
4-Methoxy-N,N'-bis(3-pyridinylmethyl)-1,3-benzenedicarboxamide monohydrate.
Plactidil. Antithrombotic. *Soc. Italo-Brit. L. Manetti-H. Roberts.*

6351 Plafibride
63394-05-8 7673 264-121-2
$C_{16}H_{22}ClN_3O_4$
1-[2-(p-Chlorophenoxy)-2-methylpropionyl]-3-(morpholinomethyl)urea.
ITA-104; Idonor; Perifunal. Antithrombotic. mp = 100-102°; soluble in Me_2CO; slightly soluble in EtOH; insoluble in H_2O, petroleum ether; LD_{50} (mus orl) = 3569 mg/kg, (rat orl) > 4000 mg/kg, (gpg orl) = 2168 mg/kg. *Investigacion Tecnica y Aplicada.*

6352 Ramatroban
116649-85-5
$C_{21}H_{21}FN_2O_4S$
(3R)-3-(4-Fluorophenylsulfonamido)-1,2,3,4-tetrahydro-9-carbazo lepropanoic acid.
BAY- u3405. A thromboxane A2 receptor antagonist. Antithrombotic antiasthmatic. *Bayer AG.*

6353 Reviparin Sodium
9041-08-1 8336
Sodium salt of heparin.
Antithrombotic. Heparin from porcine intestinal mucosa depolymerized by nitrous acid depolymerization. Molecular weight between 3500 and 4500.

6354 Ridogrel
110140-89-1 8379
$C_{18}H_{17}F_3N_2O_3$
(E)-5-[[[α-3-Pyridyl-m-(trifluoromethyl)benzylidene]-amino]oxy]valeric acid.
R-68070. Thromboxane synthetase inhibitor. Used as an antithrombotic. mp = 70.3°. *Janssen Pharm. Inc.*

6355 Rolafagrel
89781-55-5
$C_{14}H_{12}N_2O_2$
5,6-Dihydro-7-imidazol-1-yl)-naphthoic acid.
FCE-22178. Thromboxane inhibitor. Used in amelioration of progressive kidney disease. Antithrombotic.

6356 Roxifiban Acetate
176022-59-6
$C_{23}H_{33}N_5O_8$
(2S)-3-[2-[(5R)-3-(p-Amidinophenyl)-2-isoxazolin-5-yl]acetamido]-2-(carboxyamino)propionic acid 2-butyl-methyl ester monoacetate.
DMP-754. Fibrinogen receptor antagonist; used as an antithrombotic. *DuPont-Merck Pharm.*

6357 Rupatadine
158876-82-5
$C_{26}H_{26}ClN_3$
8-Chloro-6,11-dihydro-11-[1-[(5-methyl-3-pyridinyl)methyl]-4-piperidinylidene]-5H-benzo[5,6]-cyclohepta[1,2b]pyridine .
UR-12592. Orally active dual antagonist of histamine and platelet-activating factor.

6358 Sarpogrelate
125926-17-2
$C_{24}H_{31}NO_6$
(±)-2-(Dimethylamino)-1-[[o-(m-methoxyphenethyl)-phenoxy]ethyl hydrogen succinate.
MCI-9042 [as hydrochloride]; Anplag [as hydrochloride]. A serotonin 2A (5-HT2A) receptor antagonist. Antiplatelet agent. Antithromobotic.

6359 Satigrel
111753-73-2
$C_{20}H_{19}NO_4$
4-Cyano-5,5-bis(p-methoxyphenyl)-4-pentenoic acid.
E-5510. Antiplatelet aggregation agent. Antithrombotic.

6360 Sibrafiban
172927-65-0
$C_{20}H_{28}N_4O_6$
Ethyl (Z)-[[1-N-[(p-hydroxyamidino)benzoyl]-L-alanyl]-4-piperidyl]oxy]acetate.
Ro-48-3657/001. Fibrinogen receptor antagonist and platelet aggregation inhibitor; used as an antithrombotic. *Hoffmann-LaRoche Inc.*

6361 Sulfinpyrazone
57-96-5 9121 200-357-4
$C_{23}H_{20}N_2O_3S$
1,2-Diphenyl-4-[2-(phenylsulfinyl)ethyl]-3,5-pyrazolidinedione.
G-28315; Anturan; Anturane; Anturano; Enturen. Uricosuric and antithrombotic. Used to treat gout. A nonsteroidal anti-inflammatory agent. Inhibits cyclooxygenase. Prolongs circulating platelet survival. mp = 136-137°; λ_m = 255 nm (1N NaOH); soluble in EtOAc, $CHCl_3$; slightly soluble in H_2O, EtOH, Et_2O, mineral oils; [d-form]: mp = 130-133°; $[\alpha]^2{}_D$ = 67.1° (c = 2.04 EtOH), $[\alpha]_D^{25}$ = 109.3 (c = 0.5 $CHCl_3$); [l-form]: mp = 130-133°; $[\alpha]_D^{23}$ = -64.2° (c = 2.14 EtOH), $[\alpha]_D^{26}$ = -104.5° (c = 0.5 $CHCl_3$). *Ciba-Geigy Corp.*

6362 Sulodexide
57821-29-1
Glucorono-2-amino-2-deoxyglucoglucan sulfate.
3GS. A glycosaminoglycan. Highly purified preparation of a fast-moving heparin fraction as well as dermatan-sulfate. Used in treatment of peripheral arterial disease, cardiovascular events, postphlebitic syndrome and albuminuria in nephropathy. Antithrombotic; profibrinolytic.

6363 Taprostene
108945-35-3 9227
$C_{24}H_{30}O_5$
α-[(2Z,3aR,4R,5R,6aS)-4-[(1E,3S)-3-Cyclohexyl-3-hydroxy-propenyl]hexahydro-5-hydroxy-2H-cyclopenta[b]furan-2-ylidene]-m-toluic acid.
Antithrombotic; platelet aggregation inhibitor. A prostacyclin analog. *Grunenthal.*

6364 Taprostene Sodium Salt
87440-45-7 9227
$C_{24}H_{29}NaO_5$
α-[(2Z,3aR,4R,5R,6aS)-4-[(1E,3S)-3-Cyclohexyl-3-hydroxy-propenyl]hexahydro-5-hydroxy-2H-cyclopenta[b]furan-2-ylidene]-m-toluic acid sodium salt.
CG-4203; Rheocyclan. Antithrombotic; platelet aggregation inhibitor. A prostacyclin analog. $[\alpha]_D^{22} = 249°$ (c = 0.68 MeOH); LD_{50} (mus iv) = 164 mg/kg, (rat iv) = 20 mg/kg. *Grunenthal.*

6365 Teopranitol
81792-35-0
$C_{16}H_{22}N_6O_7$
1,4:3,6-Dianhydro-2-deoxy-2-[[3-(1,2,3,6-tetrahydro-1,3-dimethyl-2,6-dioxopurin-7-yl)propyl]amino]-L-iditol 5-nitrate.
An organic nitrate that stimulates the release of a prostacyclin (PGI2)-like antiplatelet activity. Shows vasodilating and antiplatelet activity.

6366 Terbogrel
149979-74-8
$C_{23}H_{27}N_5O_2$
(5E)-6-[m-(3-tert-Butyl-2-cyanoguanidino)phenyl]-6-(3-pyridyl)-5-hexenoic acid.
Omega-disubstituted alkenoic acid derivative derived from samixogrel. Combined thromboxane A2 receptor antagonist-thromboxane A2 synthase inhibitor. Antithrombotic.

6367 Ticlopidine
55142-85-3 9569 259-498-5
$C_{14}H_{14}ClNS$
5-(o-Chlorobenzyl)-4,5,6,7-tetrahydrothieno-[3,2-c]-pyridine.
Platelet aggregation inhibitor; used as antithrombotic. *Sanofi Winthrop.*

6368 Ticlopidine Hydrochloride
53885-35-1 9569 258-837-4
$C_{14}H_{15}Cl_2NS$
5-(o-Chlorobenzyl)-4,5,6,7-tetrahydrothieno-[3,2-c]-pyridine hydrochloride.
Anagregal; Caudaline; Panaldine; Ticlid; Ticlodox; Ticlodone; Ticlosin; Tiklid; 53-32C; 4-C-32. Platelet aggregation inhibitor; used as antithrombotic. mp = 190°; λ_m = 214, 268, 295 nm ($A_{1\ cm}^{1\%}$ = 303.8, 1.314 2 H_2O); soluble in H_2O, EtOH, MeOH, $CHCl_3$; insoluble in Et_2O; LD_{50} (mus iv) = 55 mg/kg, (mus orl) > 300 mg/kg. *C.M. Ind.; Sanofi Winthrop.*

6369 Tinzaparin Sodium
9041-08-1 9592
Sodium salt of heparin depolymerized by heparinase degradation; molecular weight between 1500 and 10000. Antithrombotic. *LeoAB; Novo Nordisk Pharm. Inc.*

6370 Tioxaprofen
40198-53-6 254-834-7
$C_{18}H_{13}Cl_2NO_3S$
2-[[4,5-Bis-(p-chlorophenyl)-2-oxazolyl]thio]-propionic acid.
EMD-26644. Anti-inflammatory; antithrombotic; antimycotic.

6371 Tirofiban
144494-65-5 9605
$C_{22}H_{36}N_2O_5S$
N-(Butylsulfonyl)-4-[4-(4-piperidyl)butoxy]-L-phenylalanine.
Fibrinogen receptor antagonist. Used as an antithrombotic and in the treatment of unstable angina. mp = 223-225°. *Merck & Co.Inc.*

6372 Tirofiban Hydrochloride
53567-47-8 9605
$C_{22}H_{37}ClN_2O_5S.H_2O$
N-(Butylsulfonyl)-4-[4-(4-piperidyl)butoxy]-L-phenylalanine hydrochloride.
Aggrastat; MK-383; L-700462. Fibrinogen receptor antagonist. Used as an antithrombotic and in the treatment of unstable angina. *Merck & Co.Inc.*

6373 Tirofiban Hydrochloride Monohydrate
150915-40-5 9605
$C_{22}H_{37}ClN_2O_5S.H_2O$
N-(Butylsulfonyl)-4-[4-(4-piperidyl)butoxy]-L-phenylalanine hydrochloride monohydrate.
L-700462; MK-383; Aggrastat. Fibrinogen receptor antagonist. Used as an antithrombotic and in the treatment of unstable angina. mp = 131-132°; $[\alpha]_D^{25}$ = -14.4° (c = 0.92 MeOH). *Merck & Co.Inc.*

6374 Trequinsin
79855-88-2
$C_{24}H_{27}N_3O_3$
2,3,6,7-Tetrahydro-2-(mesitylimino)-9,10-dimethoxy-3-methyl-4H-pyrimido[6,1-a]isoquinoline-4-one.
HL-725 [as hydrochloride]. Selective phosphodiesterase 3 inhibitor. Antihypertensive vasodilator; antithromotic.

6375 Tretoquinol
21650-42-0 9719
$C_{19}H_{23}NO_5$
(±)-1,2,3,4-Tetrahydro-1-(3,4,5-trimethoxybenzyl)-6,7-isoquinolinediol.
trimethoquinol; trimetoquinol; AQ-110 [as l-form hydrochloride]; Inolin [as l-form hydrochloride]; Vems [as l-form hydrochloride]. A catecholamine. A highly potent β_2-adrenoceptor and site-selective thromboxane A2/prostaglandin H2 receptor ligand. *l*-Form hydrochloride acts as a bronchodilator. Dec 224.5-226°.

6376 Trifenagrel
84203-09-8
$C_{25}H_{25}N_3O$
2-[o-[2-(Dimethylamino)ethoxy]phenyl]-4,5-diphenylimidazole.
BW-325U. Antithrombotic. *Glaxo Wellcome Inc.*

6377 Triflusal
322-79-2 9817 206-297-5
$C_{10}H_7F_3O_4$
α,α,α-Trifluoro-2,4-cresotic acid acetate.
UR-1501; Disgren. Inhibitor of platelet aggregation, used as an antithrombotic. mp = 110-112°, 120-122°; soluble in EtOH, insoluble in H_2O; LD_{50} (mus orl) = 437 mg/kg, (mus ip) = 380 mg/kg, (rat orl) = 402 mg/kg, (ra ip) = 217 mg/kg. *Uriach.*

Antitussives

6378 Alloclamide
5486-77-1 263
$C_{16}H_{23}ClN_2O_2$
2-(Allyloxy)-4-chloro-N-[2-(diethylamino)ethyl]benzamide.
component of: Dégryp [as hydrochloride]. Antitussive; antihistaminic. [hydrochloride] mp = 125-127°; soluble in EtOH; LD_{50} (mus orl)= 740 mg/kg. *C.E.R.M.*

6379 Amicibone
23271-63-8 416
$C_{22}H_{31}NO_3$
Benzyl-1-[2-(hexahydro-1H-azepin-1-yl)ethyl]-2-oxocyclohexanecarboxylate.
Biotussial; Pectipront. Antitussive.

6380 Benproperine
2156-27-6 1078
$C_{21}H_{27}NO$
1-[2-(2-Benzylphenoxy)-1-methylethyl]piperidine.
Antitussive. $bp_{0.2}$ = 159-161°. *Aktieselskabet Pharmacia.*

6381 Benproperine Pamoate
64238-92-2 1078 264-745-5
$C_{65}H_{70}N_2O_8$
1-[2-(2-Benzylphenoxy)-1-methylethyl]piperidinium 3,3'-dihydroxy-4,4'-methylenedi-2-naphthoate.
benproperine emboate; benzproperine amboate; Tussafug. Antitussive. *Aktieselskabet Pharmacia.*

6382 Benproperine Phosphate
19428-14-9 1078 243-050-0
$C_{21}H_{30}NO_5P$
1-[2-(2-Benzylphenoxy)-1-methylethyl]piperidine trihydrogen phosphate.
ASA-158/5; Blascorid; Pirexyl. Antitussive. mp = 150-152°; LD_{50} (mus ip) = 192 mg/kg; (mus orl) = 1365 mg/kg. *Aktieselskabet Pharmacia.*

6383 Benzonatate
104-31-4 1127 203-194-7
$C_{30}H_{53}NO_{11}$
2,5,8,11,14,17,20,23,26-Nonaoxactacosan-28-yl p-(butylamino)benzoate.
benzononatine; Exangit; Tessalon; Ventussin. Antitussive. Oil; soluble in most organic solvents; insoluble in aliphatic hydrocarbons. *CIBA plc.*

6384 Bibenzonium
59866-76-1
$C_{19}H_{26}NO^+$
[2-(1,2-Diphenylethoxy)ethyltrimethyl]ammonium.
Antitussive.

6385 Bibenzonium Bromide
15585-70-3 1244 239-643-9
$C_{19}H_{26}BrNO$
[2-(1,2-Diphenylethoxy)ethyltrimethyl]ammonium bromide.
ES-132; Lysobex; Lysibex; Lysbex; Medipectol; Sedobex; Thoragol. Antitussive. mp = 144-147°; soluble in H_2O, EtOH; insoluble in other organic solvents. *Eprova AG.*

6386 Bromoform
75-25-2 1441 200-854-6
$CHBr_3$
Tribromomethane.
Antitussive; sedative. Federally controlled substance. CAUTION: Dangerous substance; irratates skin, eyes, respiratory system; depresses central nercous system; causes liver damage. May be habit forming. Liquid; bp = 149-150°; mp = 7.5°; d_D^{15} = 2.9035; n_D^{15} = 1.6005; soluble in H_2O (0.125 g/100 ml); miscible with EtOH, C_6H_6, $CHCl_3$, Et_2O, Me_2CO, oils; incompatible with caustic alkalies; protect from light; LD_{50} (mus sc) = 1814 mg/kg.

6387 Butamirate
18109-80-3 1539 242-005-2
$C_{18}H_{29}NO_3$
2-[2-(Diethylamino)ethoxy]ethyl 2-phenylbutyrate.
butamyrate. Antitussive. Liquid; bp_1 = 140-155°; practically insoluble in H_2O; soluble in EtOH, Me_2CO, Et_2O. *Abbott Labs Inc.; Hommel GmbH.*

6388 Butamirate Citrate
18109-81-4 1539 242-006-8
$C_{24}H_{37}NO_{10}$
2-[2-(Diethylamino)ethoxy]ethyl 2-phenylbutyrate citrate (1:1).
Abbott-36581; HH-197; Acodeen; Panatus; Sincodix; Sinecond. Antitussive. mp = 75°. *Abbott Labs Inc.; Hommel GmbH.*

6389 Butetamate
14007-64-8 1551 237-817-9
$C_{16}H_{25}NO_2$
2-(Diethylamino)ethyl 2-phenylbutyrate.
butethamate; HH-105. Antitussive. bp_{11} = 167-169°; n_D^{20} = 1.4909. *Hommel GmbH.*

6390 Butetamate Citrate
3639-12-1 1551
$C_{22}H_{33}NO_9$
2-(Diethylamino)ethyl 2-phenylbutyrate citrate (1:1).
Abuphenine; Convenil; Hicoseen; Phenesis; Phenetin. Antitussive. mp = 109-110°; soluble in EtOH. *Hommel GmbH.*

6391 Camenthol
A mixture of camphor and menthol for inhalation. *See* camphor, menthol.

6392 Caramiphen
77-22-5 1820 201-013-6
$C_{18}H_{27}NO_2$
2-Diethylaminoethyl 1-phenylcyclopentane-1-carboxylate.
Anticholinergic. $bp_{0.07}$ = 112-115°. *Ciba-Geigy Corp.*

6393 Caramiphen Ethanedisulfonate
125-86-0 1820 204-759-0
$C_{38}H_{60}NO_{10}S_2$
2-Diethylaminoethyl 1-phenylcyclopentane-1-carboxylate ethanedisulfonate (2:1).
caramiphen edisilate; Alcopon; Taoryl; Toryn; component of: Tuss-Ornade (with phenylpropanolamine hydrochloride). Antitussive. Anticholinergic mp = 115-116°; soluble in H_2O, EtOH, pharmaceutical syrups. *Ciba-Geigy Corp.*

6394 Caramiphen Hydrochloride
125-85-9 1820 204-758-5
$C_{18}H_{28}ClNO_2$
2-Diethylaminoethyl 1-phenylcyclopentane-1-carboxy-late hydrochloride.
Panparnit; Parpanit. Anticholinergic; antitussive. mp = 145-146°; soluble in EtOH; less soluble in H_2O; LD_{50} (rat ip) = 209 mg/kg. *Ciba-Geigy Corp.*

6395 Carbetapentane
77-23-6 1840
$C_{20}H_{31}NO_3$
2-[2-(Dimethylamino)ethoxy]ethyl 1-phenylcyclopentane-carboxylate.
pentoxyverine; pentoxiverine; Atussil. Antitussive. $bp_{0.01}$ = 165-170°. *Pfizer Inc.*

6396 Carbetapentane Citrate
23142-01-0 1840
$C_{26}H_{39}NO_{10}$
2-[2-(Dimethylamino)ethoxy]ethyl 1-phenylcyclopentane-carboxylate citrate (1:1).
UCB-2543; Antees; Calnathal; Carbetane; Cossym; Fustpentane; Germapect; Pencal; Sedotussin; Toclase; Tosnore; Tuclase. Antitussive. mp = 93°; soluble in H_2O, $CHCl_3$, EtOH, Me_2CO, EtOAc; practically insoluble in Et_2O, petroleum ether; C_6H_6. *Pfizer Inc.*

6397 Chlophedianol
791-35-5 2106 212-340-9
$C_{17}H_{20}ClNO$
2-Chloro-α-[2-(dimethylamino)-ethyl]benzhydrol.
clofedanol; Tussistop. Antitussive. mp = 120°; LD_{50} (mus iv) = 70 mg/kg. *Bayer AG.*

6398 Chlophedianol Hydrochloride
511-13-7 2106 208-124-9
$C_{17}H_{21}Cl_2NO$
2-Chloro-α-[2-(dimethylamino)-ethyl]benzhydrol hydrochloride.
clofedanol hydrochloride; SL-501; Coldrin; Pectolitan; Refugal; Ulone; Ulo. Antitussive. mp = 190-191°; soluble in H_2O; EtOH; sparingly soluble in Et_2O, C_6H_6, EtOAc; LD_{50} (rat orl) = 350 mg/kg, (mus sc) = 95 mg/kg. *3M Pharm.; Bayer AG.*

6399 Clobutinol
14860-49-2 2425 238-926-4
$C_{14}H_{22}ClNO$
p-Chloro-α-[2-(dimethylamino)-1-methylethyl]-α-methylphenyl alcohol.
Antitussive. bp_{12} = 179-180°. *Thomae GmbH Dr. Karl.*

6400 Clobutinol Hydrochloride
1215-83-4 2425 214-931-7
$C_{14}H_{23}Cl_2NO$
p-Chloro-α-[2-(dimethylamino)-1-methylethyl]-α-methylphenyl alcohol hydrochloride.
KAT-256; Biotertussin; Silomat. Antitussive. mp = 169-170°; soluble in H_2O; LD_{50} (mus orl) = 600 mg/kg, (mus ip) = 130 mg/kg. *Thomae GmbH Dr. Karl.*

6401 Cloperastine
3703-76-2 2456 223-042-3
$C_{20}H_{24}ClNO$
1-[2-[(p-Chloro-α-phenylbenzyl)oxy]ethyl]piperidine.
Antitussive. $bp_{0.06}$ = 172-174°; $bp_{0.15}$ = 178-179°. *Yoshitomi.*

6402 Cloperastine Hydrochloride
14984-68-0 2456 239-067-8
$C_{20}H_{25}Cl_2NO$
1-[2-[(p-Chloro-α-phenylbenzyl)oxy]ethyl]piperidine hydrochloride.
Hustazol; Nitossil; Novotusil; Seki. Antitussive. mp = 147.9°. *Yoshitomi.*

6403 Codeine N-Oxide
3688-65-1 2527 222-988-4
$C_{18}H_{21}NO_4$
7,8-Didehydro-4,5α-epoxy-3-methoxy-17-methyl morphinan-6α-ol 17-oxide.
Gencodeine; genkodein; Codeigene. Antitussive. Federally controlled substance (opium derivative). mp = 231-232°.

6404 Codeine Polystirex
Sulfonated styrene-divinylbenzene copolymer complex with 7,8-didehydro-4,5α-epoxy-3-methoxy-17-methyl morphinan-6α-ol.
component of: Penntuss. Antitussive. *Fisons plc.*

6405 Codoxime
7125-76-0 230-425-9
$C_{20}H_{24}N_2O_5$
[[(4,5α-Epoxy-3-methoxy-17-methylmorphinan-6-ylidene)amino]oxy]acetic acid.
Antitussive. *Marion Merrell Dow Inc.*

6406 Cyclexanone
15301-52-7 2775
$C_{16}H_{25}NO_2$
2(Cyclopent-1-enyl)-2-(2-morphinolinoethyl)cyclopentanone.
pentethylcyclanone. Antitussive. $bp_{0.035}$ = 112°. *CIBA plc.*

6407 Cyclexanone Hydrochloride
2775
$C_{16}H_{25}ClNO_2$
2(Cyclopent-1-enyl)-2-(2-morphinolinoethyl)-cyclopentanone hydrochloride.
Exopan; Exopon. Antitussive. mp = 209°. *CIBA plc.*

6408 Dextromethorphan
125-71-3 8274 204-752-2
$C_{18}H_{25}NO$
3-Methoxy-17-methyl-9α,13α-morphinan.
racemethorphan [as d-form]. Antitussive. Federally controlled substance. *See* racemethorphan. *Hoffmann-LaRoche Inc.*

6409 Dextromethorphan Hydrobromide
6700-34-1 8274
$C_{18}H_{28}BrNO_2$
3-Methoxy-17-methyl-9α,13α-morphinan hydrobromide monohydrate.
Ro-1-5470/5, demorphan hydrobromide; Benylin DM; Canfodion; Cosylan; Hihustan M; Methorate; PediaCare 1; Romilar; St Joseph Syrup; component of: Cerose-DM, Chloraseptic DM, Contac Cough Formula, Contac Cough & Sore Throat Formula, Contac Jr Non-drowsy Formula, Contac Nighttime Cold Medicine, Contac Severe Cold Formula, Coricidin, Dimacol, Endotussin-NN, Endotussin-NN Pediatric, Nalde. Antitussive. Federally controlled substance. *See* racemethorphan. mp = 122-124°; $[\alpha]_D^{20}$ = 27.6° (c = 1.5 H_2O); soluble in H_2O (1.5 g/100 ml), EtOH; less soluble in organic solvents. *Ciba-Geigy Corp.; Hoffmann-LaRoche Inc.; McNeil Pharm.; Parke-Davis; Schering-Plough Pharm.; Upjohn Ltd.*

6410 Dextromethorphan Hydrobromide [anhydrous]
125-69-9 8274 204-750-1
$C_{18}H_{26}BrNO$
3-Methoxy-17-methyl-9α,13α-morphinan hydrobromide.
Ro-1-5470. Antitussive. Federally controlled substance. *See* racemethorphan. mp = 124-126°. *Hoffmann-LaRoche Inc.*

6411 Dextromethorphan Polistirex
Sulfonated styrene-divinylbenzene copolymer complex with 3-methoxy-17-methyl-9α,13α,14α-morphinan.
Antitussive.

6412 Dimemorfan
36309-01-0 3251 252-963-3
$C_{18}H_{25}N$
(+)-3,17-Dimethylmorphinan.
d-3-methyl-N-morphinan. Antitussive. Oil; $bp_{0.3}$ = 130-136°; or crystals mp = 90-93°. *Yamanouchi U.S.A. Inc.*

6413 Dimemorfan Phosphate
36304-84-4 3251 252-958-6
$C_{18}H_{28}NO_4P$
(+)-3,17-Dimethylmorphinan phosphate.
Astomin. Antitussive. mp = 267-269°; $[\alpha]_D^{23}$ = +25.7° (c = 0.5 in MeOH); soluble in glacial acetic acid; sparingly soluble in H_2O; MeOH; practically insoluble in EtOH, Me_2CO, $CHCl_3$, C_6H_6, Et_2O; LD_{50} (mus sc) = 223 mg/kg, (mus orl) = 475 mg/kg. *Yamanouchi U.S.A. Inc.*

6414 Dimethoxanate
477-93-0 3272 207-520-9
$C_{19}H_{22}N_2O_3S$
2-(2-Dimethylaminoethoxy)ethyl phenothiazine-10-carboxylate.
Tussidin; Cotrane; component of: Cothera Syrup. Antitussive. *Am. Home Products.*

6415 Dimethoxanate Hydrochloride
518-63-8 3272 208-255-1
$C_{19}H_{23}ClN_2O_3S$
2-(2-Dimethylaminoethoxy)ethyl phenothiazine-10-carboxylate.
Antitussive. mp = 161-163 (dec). *Am. Home Products.*

6416 Dropropizine
17692-31-8 3507 241-683-7
$C_{13}H_{20}N_2O_2$
3-(4-Phenyl-1-piperazinyl)-1,2-propanediol.
UCB-1967; Ribex; levodropropizine [S-form]; Danka [S-form]; Levotuss [S-form]; Rapitux [S-form]. Antitussive. Cough suppressive phenylpiperazine derivative. mp = 105-108°; LD_{50} (rat iv) = 200 mg/kg; (rat orl) = 750 mg/kg; [S-form]: mp = 98-100°; $[\alpha]_D^{25}$ = -10° (c = 1.0 in EtOH); LD_{50} (rat orl) = 886.6 mg/kg, (rat ip) = 401.3 mg/kg. *Labs. Franca Inc.*

6417 Drotebanol
3176-03-2 3511
$C_{19}H_{27}NO_4$
3,4-Dimethoxy-17-methylmorphinan-6β,14-diol.
oxymethebanol; Metebanyl. Antitussive. Federally controlled substance (opium derivative). mp = 165.5-166.5°; slightly soluble in H_2O; more soluble in organic solvents; LD_{50} (mus orl) = 1300 mg/kg, (mus sc) = 1500 mg/kg, (mus iv) = 91 mg/kg. *Sankyo Co. Ltd.*

6418 Eprazinone
10402-90-1 3666 233-873-3
$C_{24}H_{32}N_{2O2}$
3-[4-(β-Ethoxyphenylethyl)-1-piperazinyl]-2-methylpropiophenone.
Antitussive.

6419 Eprazinone Dihydrochloride
10402-53-6 3666 233-872-8
$C_{24}H_{34}Cl_2N_{2O2}$
3-[4-(β-Ethoxyphenylethyl)-1-piperazinyl]-2-methylpropiophenone dihydrochloride.
746-CE; Eftapan; Mucitux. Antitussive. mp = 201° or 160°; LD_{50} (mus orl) = 729 mg/kg, (mus iv) = 38 mg/kg.

6420 Ethyl Dibunate
5560-69-0 3834 226-931-4
$C_{20}H_{28}O_3S$
Ethyl 3,6-di-tert-butylnaphthalenesulfonate.
dibunate ethyl; NDR-304; Neodyne. Antitussive. mp = 138-139°. *Marion Merrell Dow Inc.*

6421 Eucalyptol
470-82-6 3940 207-431-5
$C_{10}H_{18}O$
1,3,3-Trimethyl-2-oxabicyclo[2.2.2]octane.
1,8-epoxy-p-menthane; cineole; cajeputol. Pharmaceutic aid (flavor). Chief constituent of oil of eucaluptus. bp =

176-177°; mp = 1.5°; n_D^{20} = 1.455-1.460; insoluble in H_2O; miscible with EtOH, $CHCl_3$, Et_2O, AcOH, oils.

6422 Fominoben
18053-31-1 4258 241-964-4
$C_{21}H_{24}ClN_3O_3$
3'-Chloro-α-[methyl[(morpholinocarbonyl)methyl]amino]-o-benzotoluidide.
Antitussive; respiratory stimulant. mp = 122.5-123°. *Boehringer Ingelheim Ltd.; Thomae GmbH Dr. Karl.*

6423 Fominoben Hydrochloride
24600-36-0 4258 246-344-7
$C_{21}H_{25}Cl_2N_3O_3$
3'-Chloro-α-[methyl[(morpholinocarbonyl)methyl]amino]-o-benzotoluidide hydrochloride.
PB-89; Finaten; Noleptan; Oleptan; Terion; Tussirama. Antitussive; respiratory stimulant. mp = 206-208° (dec); soluble in H_2O (0.1 mg/100 ml); LD_{50} (mus ip) = 630 mg/kg, (rat ip) = 1201 mg/kg, (mus orl) = 2200 mg/kg, (rat orl) = 1250 mg/kg. *Boehringer Ingelheim Ltd.; Thomae GmbH Dr. Karl.*

6424 Guaiapate
852-42-6 4580 212-713-6
$C_{18}H_{29}NO_4$
1-[2-[2-[2-(o-methoxyphenoxy)ethoxy]ethoxy]ethyl]piperidine.
MG-5454; Klamar. Antitussive. Basic ether derivative of guaiacol. Liquid; $bp_{0.5}$ = 190-193°; LD_{50} (mus ip) = 82 mg/kg. *Maggioni Farmaceutici S.p.A.*

6425 Heptabarbital
509-86-4 4689 208-107-6
$C_{13}H_{18}N_2O_3$
5-(Cyclohepten-1-yl)-5-ethylbarbituric acid.
heptabarb; heptabarbitone; Heptadorm; Medomin. Sedative/hypnotic. Federally controlled substance (depressant). CAUTION: may be habit forming. mp = 174°; λ_m = 218.5, 254 nm (0.2N NaOH); sparingly soluble in H_2O; soluble in EtOH (4.0 g/100 ml), Me_2CO (5.7 g/100 ml), $CHCl_3$ (1.4 g/100 ml). *Ciba-Geigy Corp.*

6426 Hydrocodone Polystirex
Sulfonated stryene-divinylbenzene copolymer complex with 4,5α-epoxy-3-methoxy-17-methyl morphinan-6-one. component of: Tussionex. Antitussive. *Fisons plc.*

6427 Isoaminile
77-51-0 5124 201-033-5
$C_{16}H_{24}N_2$
4-Dimethylamino-2-isopropyl-2-phenylvaleronitrile.
Aprecon; Dimyril; Nullatuss. Antitussive. Liquid; bp_3 = 138-146°. *Kali-Chemie.*

6428 Isoaminile Citrate
28416-66-2 5124 249-011-4
$C_{22}H_{32}N_2O_7$
4-Dimethylamino-2-isopropyl-2-phenylvaleronitrile citrate.
Perocan. Antitussive. mp = 63-64°. *Kali-Chemie.*

6429 Isoaminile Cyclamate
10075-36-2 5124 233-207-1
$C_{22}H_{37}N_3O_3S$
4-Dimethylamino-2-isopropyl-2-phenylvaleronitrile cyclamate.
Mucalan; Peracon. Antitussive. *Kali-Chemie.*

6430 Levodropropizine
99291-25-5
$C_{13}H_{20}N_2O_2$
(-)-(S)-3-(4-Phenyl-1-piperazinyl)-1,2-propanediol.
LVDP. Antitussive. S-(-)-enantiomer of dropropizine.

6431 Levomethorphan
125-70-2 8274 204-751-7
$C_{18}H_{25}NO$
(-)-3-Methoxy-N-methylmorphinan.
Antitussive. Federally controlled substance (opiate). *See* racemethorphan. *Hoffmann-LaRoche Inc.*

6432 Levomethorphan Hydrobromide
125-68-8 8274
$C_{18}H_{30}BrNO_3$
(-)-3-Methoxy-N-methylmorphinan hydrobromide dihydrate.
l-form of racemethorphan; Ro-1-5470/6; Ro-1-7788. Antitussive. Federally controlled substance (opiate). *See* racemethorphan. mp = 124-126°; $[\alpha]_D^{20}$ = -26.3°. *Hoffmann-LaRoche Inc.*

6433 Levopropoxyphene
2338-37-6 5495 2338-37-6
$C_{22}H_{29}NO_2$
(1R,2S)-1-Benzyl-3-dimethylamino-2-methyl-1-phenylpropyl propionate.
l-propoxyphene. Antitussive. mp = 75-76°; $[\alpha]_D^{25}$ = -68.2° (c = 0.6 in $CHCl_3$). *Eli Lilly & Co.*

6434 Levopropoxyphene Napsylate [anhydrous]
5714-90-9 5495 227-209-1
$C_{32}H_{39}NO_6S$
2-Naphthalenesulfonic acid compound with (-)-α-[2-(dimethylamino)-1-methylethyl]-α-phenylphenethyl propanoate (1:1).
Antitussive. *Eli Lilly & Co.*

6435 Levopropoxyphene Napsylate Monohydrate
55557-30-7 5495
$C_{32}H_{37}NO_5S$
2-Naphthalenesulfonic acid compound with (-)-α-[2-(dimethylamino)-1-methylethyl]-α-phenylphenethyl propanoate (1:1) monohydrate.
levopropoxyphene 2-napthalenesulfonate; Contratuss; Letusin; Novrad. Antitussive. LD_{50} (frat orl) 1455 mg/kg. *Eli Lilly & Co.*

6436 Medazomide
300-22-1
$C_6H_9N_3O_2$
1,4,5,6-Tetrahydro-1-methyl-6-oxo-3-pyridazinecarboxamide.
Possible antitussive agent.

6437 Meprotixol
4295-63-0
$C_{19}H_{23}NO_2S$
9-[3-(Dimethylamino)propyl]-2-methoxy-thioxanthene-9-ol.
N-7020. Antitussive. May also have antirheumatic activity.

6438 Moguisteine
119637-67-1
$C_{16}H_{21}NO_5S$
Ethyl (±)-2-[(o-methoxyphenoxy)methyl]-β-oxo-3-thiazolidinepropionate.
BBR-2173. A peripherally acting non-narcotic antitussive.

6439 Morclofone
31848-01-8 6350 250-838-8
$C_{21}H_{24}ClNO_5$
4'-Chloro-3,5-dimethoxy-4-(2-morpholinoethoxy)-benzophenone.
dimecolphenone; K-3712; Medicil; Nitux; Plausitin. Antitussive. mp = 91-92°; LD_{50} (mus orl) = 522 mg/kg. *Farmitalia Carlo Erba Ltd.*

6440 Moxazocine
58239-89-7
$C_{18}H_{25}NO_2$
(-)-(2R,6S,11R)-3-(Cyclopropylmethyl)-1,2,3,4,5,6-hexahydro-11-methoxy-6-methyl-2,6-methano-3-benzazocin-8-ol.
levo-BL-4566. Analgesic; antitussive. *Bristol Myers Squibb Pharm. Ltd.*

6441 Noscapine
128-62-1 6815 204-899-2
$C_{22}H_{23}NO_7$
[S-(R*,S*)]-6,7-Dimethoxy-3-(5,6,7,8-tetrahydro-4-methoxy-6-methyl-1,3-dioxolo[4,5-g]isoquinolin-5-yl)-1(3H)-Isobenzofuranone.
narcotine; l-α-narcotine; narcosine; methoxyhydrastine; opian; opianine; NSC-5366; Coscopin; Coscotabs; Capval; Longatin; Lyobex; Narcompren; Narcotussin; Nectadon; Nicolane; Nipaxon; Noscapalin; Terbenol; Tusscapine; Vadebex; dl-narcotine [dl-form]; gnoscopine [dl-form]. Antitussive. Opium alkaloid. Occurs in *Papaver somniferum*. mp = 176°; d = 1.395; sublimes at 150-160° (11 mm pressure); pK = 7.8; λ_m= 209, 291, 309-310 nm (log ε = 4.86, 3.60, 3.69); insoluble in oils; slightly soluble in NH_4OH and hot alkali solutions; salts formed with acids are dextrarotory and unstable in H_2O. *Fisons plc.*

6442 Noscapine Camphorsulfonate
6815
$C_{32}H_{43}N_{13}O_{12}S_6$
[S-(R*,S*)]-6,7-Dimethoxy-3-(5,6,7,8-tetrahydro-4-methoxy-6-methyl-1,3-dioxolo[4,5-g]-isoquinolin-5-yl)-1(3H)-Isobenzofuranone camphorsulfonate.
374-JL; Tulisan. Antitussive. Contains 36% camphosulfonic acid. mp = 188-191°; $[\alpha]_D^{33}$= +32.7 (c = 4.56 in H_2O); soluble in H_2O, EtOH; less soluble in organic solvents.

6443 Noscapine Hydrochloride
912-60-7 6815 213-014-9
$C_{22}H_{24}ClNO_7$
[S-(R*,S*)]-6,7-Dimethoxy-3-(5,6,7,8-tetrahydro-4-methoxy-6-methyl-1,3-dioxolo[4,5-g]isoquinolin-5-yl)-1(3H)-Isobenzofuranone hydrochloride.
Antitussive. Soluble in H_2O.

6444 Oxeladin
468-61-1 7064 207-412-1
$C_{20}H_{33}NO_3$
2-(2-Diethylaminoethoxy)ethyl 2-ethyl-2-phenylbutyrate.
Antitussive. Oil; $bp_{0.5}$ = 150°; soluble in dilute HCl, EtOH, Me_2CO, Et_2O, C_7H_8; practically insoluble in H_2O; stable in acids; unstable in alkalies. *British Drug Houses.*

6445 Oxeladin Citrate
52432-72-1 7064 257-910-8
$C_{26}H_{41}NO_{10}$
2-(2-Diethylaminoethoxy)ethyl 2-ethyl-2-phenylbutyrate citrate (1:1).
Pectamol; Petcamon; Paxeladine; Silopentol. Antitussive. mp = 90-91°; soluble in H_2O. *British Drug Houses.*

6446 Oxolamine
959-14-8 7078 213-493-4
$C_{14}H_{19}N_3O$
N,N-Diethyl-3-phenyl-1,2,4-oxadiazole-5-ethanamine.
Anti-inflammatory (respiratory). Liquid; $bp_{0.04}$ = 127°. *Angelini Francesco; SmithKline Beecham Pharm.*

6447 Oxolamine Citrate
1949-20-8 7078 217-760-6
$C_{20}H_{27}N_3O_8$
N,N-Diethyl-3-phenyl-2-hydroxy-1,2,3-propane-tricarboxylate-1,2,4-oxadiazole-5-ethanamine.
AF-438; SKF-9976; Bredon; Broncatar; Flogobron; Oxarmin; Perebron; Prilon. Anti-inflammatory (respiratory). Used in inflammatory conditions of the respiratory tract. Crystals; λ_m (aqueous solution) = 239 mn (ε 260) , 273 nm; 283 nm; slightly soluble in H_2O, EtOH. *Angelini Francesco; SmithKline Beecham Pharm.*

6448 Pemerid
50432-78-5
$C_{15}H_{32}N_2O$
4-[3-Dimethylamino)propoxyl]-1,2,2,6,6-pentamethyl-piperidine.
Antitussive. *Parke-Davis.*

6449 Pemerid Nitrate
34114-01-7
$C_{15}H_{34}N_4O_7$
4-[3-Dimethylamino)propoxyl]-1,2,2,6,6-pentamethyl-piperidine dinitrate.
W-2394A. Antitussive. *Parke-Davis.*

6450 Pholcodine
509-67-1 7484 208-102-9
$C_{23}H_{30}N_2O_4$
(5α,6α)-7,8-Didehydro-4,5-epoxy-17-methyl-3-[2-(4-morpholinyl)ethoxy]morphinan-6-ol.
morpholinylethylmorphine; β-morpholinylethylmorphine; homocodeine; Codylin; Ethnine; Galenphol; Galphol;

Memine; Pectolin; Weifacodine. Antitussive. Federally controlled substance (opiate). [monohydrate]: mp = 91°; $[\alpha]_D^{20}$ = -95° (c = 2 EtOH); slightly soluble in H_2O (2.0 g/100 ml); more soluble in organic solvents; LD_{50} (mus sc) = 540 mg/kg. *Lab. Dausse; Purdue Pharma L.P.*

6451 Picoperine
21755-66-8 7558
$C_{19}H_{25}N_3$
1-[2-[N-(2-Pyridylmethyl)anilino]ethyl]piperidine.
picoperidamine; TAT-3. Antitussive. Liquid; bp_4 = 195-196°. *Takeda Chem. Ind. Ltd.*

6452 Picoperine Hydrochloride
24699-40-9 7558
$C_{19}H_{26}ClN_3$
1-[2-[N-(2-Pyridylmethyl)anilino]ethyl]piperidine hydrochloride.
Coben. Antitussive. mp = 183-185°; soluble in H_2O, EtOH; less soluble in organic solvents; LD_{50} (mus sc) = 210 mg/kg, (mus ip) = 85 mg/kg, (mus iv) = 17 mg/kg, (mus orl) = 240 mg/kg. *Takeda Chem. Ind. Ltd.*

6453 Picoperine Tripalmitate
24656-22-2 7558
$C_{67}H_{121}N_3O_6$
1-[2-[N-(2-Pyridylmethyl)anilino]ethyl]piperidine tripalmitate.
Coben P. Antitussive. mp = 57-60°; soluble in EtOH, $CHCl_3$; less soluble in dioxane, Me_2CO, C_6H_6; nearly insoluble in C_6H_{14}, H_2O, LD_{50} (mus orl) = 1900 mg/kg. *Takeda Chem. Ind. Ltd.*

6454 Pipazethate
2167-85-3 7609 218-508-8
$C_{21}H_{25}N_3O_3S$
2-(2-Piperidinoethoxy)ethyl 10H-pyrido[3,2-b]-[1,4]benzothiadiazine-10-carboxylate.
1-azaphenothiazine-10-carboxylic acid 2-(2-piperidinoethoxy)ethyl ester; pipazetate; D-254; SKF-70230-A; SQ-15874. Antitussive. *Bristol-Myers Squibb Pharm. R&D; Degussa-Hüls Corp.; SmithKline Beecham Pharm.*

6455 Pipazethate Hydrochloride
6056-11-7 7609 227-980-4
$C_{21}H_{26}ClN_3O_3S$
2-(2-Piperidinoethoxy)ethyl 10H-pyrido[3,2-b]-[1,4]benzothiadiazine-10-carboxylate hydrochloride.
pipazetate hydrochloride; Lenopect; Selvigon; Selvjgon; Theratuss. Antitussive. mp = 160-161°; soluble in H_2O, MeOH; insoluble in non-polar organic solvents; LD_{50} (rat orl) = 560 mg/kg. *Degussa-Hüls Corp.*

6456 Piperidione
77-03-2 7622 200-999-5
$C_9H_{15}NO_2$
3,3-Diethyl-2,4-piperidinedione.
dihyprylone; Sedulon; Tusseval. Antitussive; sedative/hypnotic. mp = 102-107°; soluble in H_2O, EtOH, $CHCl_3$.

6457 Prenoxdiazine
47543-65-7
$C_{23}H_{27}N_3O$
1-[2-[3-(2,2-Diphenylethyl)-1,2,4-oxadiazol-5-yl]-ethyl]piperidene.
Libexin. Antitussive. *Chinoin.*

6458 Prenoxdiazine Hydrochloride
982-43-4 7918 213-567-6
$C_{23}H_{28}ClN_3O$
1-[2-[3-(2,2-Diphenylethyl)-1,2,4-oxadiazol-5-yl]ethyl]piperidine monohydrochloride.
HK-256; Libexin; Lomapect; Tibexin. Antitussive. mp = 192-193°; LD_{50} (mus orl) = 920 mg/kg, (mus iv) = 34 mg/kg. *Chinoin.*

6459 Promedol
64-39-1 7968 200-583-3
$C_{17}H_{25}NO_2$
1,2,5-Trimethyl-4-phenyl-4-propionyloxypiperidine.
dimethylmeperidine; α-promedol [α isomer]; isopromedol [β isomer hydrochloride]; trimeperidine [γ isomer hydrochloride]; γ-promedol [γ isomer]. Analgesic, narcotic; antitussive. Federally controlled substance (opiate). [hydrochloride, α isomer]: mp = 153-154°; [hydrochloride, β isomer]: mp = 183-184°; [hydrochloride, γ isomer]: mp = 222-223°.

6460 Promolate
3615-74-5 222-797-6
$C_{16}H_{23}NO_4$
2-Morpholinoethyl 2-methyl-2-phenoxypropionate.
morphethylbutyne. Antitussive.

6461 Proxorphan
69815-38-9
$C_{19}H_{25}NO_2$
(-)-(4aR,5R,10bS)-13-(Cyclopropylmethyl)-4,4a,5,6-tetrahydro-3H-5,10b-(iminoethano)-1H-naptho[1,2-c]-pyran-9-ol.
Analgesic; antitussive. *Bristol-Myers Squibb Pharm. R&D.*

6462 Proxorphan Tartrate
69815-39-0
$C_{42}H_{56}NO_{10}$
(-)-(4aR,5R,10bS)-13-(Cyclopropylmethyl)-4,4a,5,6-tetrahydro-3H-5,10b-(iminoethano)-1H-naptho[1,2-c]-pyran-9-ol D-(-)-tartrate (2:1) (salt).
BL-5572M. Analgesic; antitussive. *Bristol-Myers Squibb Pharm. R&D.*

6463 Racemethorphan
510-53-2 8274 208-114-4
$C_{18}H_{25}NO$
(±)-3-Methoxy-17-methylmorphinan.
deoxydihydrothebacodine; methorphan; Ro-1-5470 [as hydrobromide]. Antitussive. Federally controlled substance (opiate). *See* dextromethorphan, levomethorphan. [hydrobromide]: mp = 124-126°. *Hoffmann-LaRoche Inc.*

6464 Sodium Dibunate
14992-59-7 8753 239-079-3
$C_{18}H_{23}NaO_3S$
Sodium 2,6-di-tert-butyl-1(or -3)-napthalenesulfonate.
L-1633l; 1633 Labaz; Becantal; Becantex; Dibunafon;

Keuten; Linctussal. Antitussive. Mixture of at least two isomers. Dec > 300°; slightly soluble in cold, more soluble in hot H_2O; soluble in MeOH; less soluble in EtOH.

6465 Suxemerid
47662-15-7
$C_{24}H_{44}N_2O_4$
Bis(1,2,2,6,6-pentamethyl-4-piperidyl) - succinate.
Antitussive. *Parke-Davis.*

6466 Suxemerid Sulfate
34144-82-6
$C_{24}H_{48}N_2O_{12}S_2$
Bis(1,2,2,6,6-pentamethyl-4-piperidyl) succinate sulfate (1:2).
W-2180. Antitussive. *Parke-Davis.*

6467 Tipepidine
5169-78-8 9601
$C_{15}H_{17}NS_2$
3-(Di-2-thienylmethylene)-1-methylpiperidine.
tipedine; AT-327; CR-662. Antitussive. mp = 64-65°; bp_{4-5} = 178-184°; LD_{50} (mus ip) = 294 mg/kg; (mus orl) = 867 mg/kg. *Tanabe Seiyaku Co. Ltd.*

6468 Tipepidine Citrate Monohydrate
9601
$C_{21}H_{27}NO_8S_2$
3-(Di-2-thienylmethylene)-1-methylpiperidine citrate monohydrate.
bithiodine. Antitussive. mp = 139-139°; soluble in H_2O, EtOH, propylene glycol. *Tanabe Seiyaku Co. Ltd.*

6469 Tipepidine Hibenzate
31139-87-4 9601
$C_{29}H_{27}NO_4S_2$
3-(Di-2-thienylmethylene)-1-methylpiperidine hibenzate.
Asverin; Sotal. Antitussive. mp = 187-190°. *Tanabe Seiyaku Co. Ltd.*

6470 Vadocaine
72005-58-4
$C_{18}H_{28}N_2O_2$
(±)-6'-Methoxy-2-methyl-1-piperidinepropiono-2',4'-xylidide.
OR-K-242. Local anesthetic; antitussive. Anilide derivative.

6471 Xyloxemine
1600-19-7
$C_{23}H_{33}NO_2$
2-[2-(Di-2,6-xylylmethoxy)ethoxy]-N,N-dimethylethylamine.
BS-6748 [as hydrochloride]. Antitussive.

6472 Zipeprol
34758-83-3 10303 252-191-7
$C_{23}H_{32}N_2O_3$
4-(2-Methoxy-2-phenylethyl)-α-(methoxyphenylmethyl)-1-piperazineethanol.
Antitussive. mp = 83°. *Hexcel.*

6473 Zipeprol Dihydrochloride
34758-84-4 10303 252-192-2
$C_{23}H_{34}Cl_2N_2O_3$
4-(2-Methoxy-2-phenylethyl)-α-(methoxyphenylmethyl)-1-piperazineethanol dihydrochloride.
CERM-3024; Antituxil-Z; Citizeta; Mirsol; Respilene; Respirase; Zitoxil. Antitussive. mp = 231°; LD_{50} (mus orl) = 301 mg/kg. *Hexcel.*

Antiulceratives

6474 Aceglutamide Aluminum
12607-92-0 25
$C_{35}H_{59}Al_3N_{10}O_{24}$
Pentakis (N^2-acetl-L-glutaminato)terahydroxytrialuminum.
Glumal; KW-110. Antiulcerative. mp = 221° (dec); soluble in H_2O; insoluble in MeOH, EtOH, Me_2CO; LD_{50} (mmus orl) = 14.3 g/kg, (mmus ip) = 5.0 g/kg, (mmus iv) = 0.46 g/kg, (mrat orl) > 14.5 g/kg, (mrat ip) = 4.2 g/kg, (mrat iv) = 0.40 g/kg. *Kyowa Hakko Kogyo Co. Ltd.*

6475 Acetoxolone
6277-14-1 76 228-475-1
$C_{32}H_{48}O_5$
(20β)-3β-(Acetyloxy)-11-oxoolean-12-en-29-oic acid.
acetylglycyrrhetinic acid; glycyrrhetic acid acetate. Antiulcerative. mp= 322-325°; $[\alpha]_D^{20}$= 141°. *Dott. Inverni & Della Beffa.*

6476 Acetoxolone Aluminum Salt
29728-34-5 76 249-815-5
$C_{96}H_{141}AlO_{15}$
(20β)-3β-(Acetyloxy)-11-oxoolean-12-en-29-oic acid aluminum salt.
Oriens. Antiulcerative. mp = 286-290°; $[\alpha]_D^{20}$= 126° ±2° (c = 1 $CHCl_3$); insoluble in H_2O; LD_{50} (mrat orl) > 3300 mg/kg. *Dott. Inverni & Della Beffa.*

6477 Acexamic Acid Zinc Salt
70020-71-2 43
$C_{16}H_{28}N_2O_6Zn$
6-(Acetylamino)hexanoic acid zinc salt.
zinc acexamate; ε-acetamidocaproic acid zinc salt; Copinal. Antiulcerative. Free base and sodium salt have anti-inflammatory properties. [free acid]: mp = 104-105.5°, 112°. *Rowa-Wagner.*

6478 Aldioxa
5579-81-7 224 226-964-4
$C_4H_7AlN_4O_5$
Dihydroxy[(2-hydroxy-5-oxo-2-imidazolin-4-yl)-ureato]aluminum.
aluminum dihydroxy allantoinate; ALDA; RC-172; Alanetorin; Alusa; Arlanto; Ascomp; Chlokale; Isalon; Nische; Peptilate. Astringent; keratolytic. Antiulcerative. mp = 230°, insoluble in polar and non-polar solvents. *ICI Americas Inc.*

6479 Arbaprostil
55028-70-1 808
$C_{21}H_{34}O_5$
(E,Z)-(1R,2R,3R)-7-[3-Hydroxy-2-[(3R)-(3-hydroxy-3-methyl-1-octenyl)]-5-oxocyclopentyl]-5-heptenoic acid.

15-methylprostaglandin E_2; U-42842; Arbacet. Gastric antisecretory agent. Antiulcerative. Prostaglandin. [(15S)-methyl ester ($C_{22}H_{36}O_5$)]: $[\alpha]_D$ = -79° (c = 1.3 $CHCl_3$); λ_m = 278 nm (ε 25250 EtOH); [(15R)-methyl ester ($C_{22}H_{36}O_5$)]: $[\alpha]_D$ = -74° (c = 1.0 $CHCl_3$); λ_m = 278 nm (ε 25200 EtOH). *Pharmacia & Upjohn.*

6480 Balsalazide

80573-04-2

$C_{17}H_{15}N_3O_6$

(E)- 5-((4-(((2-carboxyethyl)amino)carbonyl)phenyl)azo)-2-hydroxybenzoic acid.

4-Aminobenzoyl-b-alanine; Balsalazide; Balsalazida; Balsalazide; Balsalazido; Balsalazidum; (E)-5-((4-(((2-Carboxyethyl)amino)carbonyl)phenyl)azo)-2-hydroxybenzoic acid. Antiulcerative. A prodrug that releases 5-aminosalicylic acid into the colonic lumen. Used in the treatment of ulcerative colitis.

6481 Balsalazide Disodium

150399-21-6

?$C_{17}H_{13}N_3Na_2O_6$

(E)- 5-((4-(((2-carboxyethyl)amino)carbonyl)phenyl)azo)-2-hydroxybenzoic acid disodium salt.

Colazal; Colazide. Antiulcerative. A prodrug that releases 5-aminosalicylic acid into the colonic lumen. Used in the treatment of ulcerative colitis. Pharm; Salix.

6482 Benexate

78718-52-2

$C_{65}H_{98}ClN_3O_{39}$

Benzyl 2-[trans-4-(guanidinomethyl)cyclohexylcarbonyloxy]benzoate hydrochloride β-cyclodextrin clathrate.

A formulated-deliverable form of benexate, an antiulcerative.

6483 Benexate Hydrochloride

78718-25-9 1065

$C_{23}H_{28}ClN_3O_4$

trans-2-[[[4-[[(Aminoiminomethyl)amino]methyl]-cyclohexyl]carbonyl]oxy]benzoic acid phenmethyl ester monohydrochloride.

Antiulcerative. Synthetic protease inhibitor. mp = 83°. *Nippon Chemiphar.*

6484 Bergenin

477-90-7 1197

$C_{14}H_{16}O_9$

3,4,4a,10b-Tetrahydro-3,4,8,10-tetrahydroxy-2-(hydroxymethyl)-9-methoxypyrano[3,2-c][2]-benzopyran-6H-one.

Antiulcerative. mp = 238°; λ_m 275 220 nm (log ε 3.92 4.42); $[\alpha]_D^{18}$ = -37.7° (c = 1.96 EtOH), $[\alpha]_D^{24}$ = -45.3° (c = 0.51 H_2O); freely soluble in H_2O, soluble in EtOH; [monohydrate]: mp = 140°, slightly soluble in H_2O, freely soluble in EtOH.

6485 Cadexomer Iodine

94820-09-4 1646

Iodosorb. Antiseptic; antiulcerative; vulnerary. Product of reaction of dextrin with epichlorohydrin coupled with ion-exchange groups and iodine. *Perstorp AB.*

6486 Carbenoxolone

5697-56-3 1839 227-174-2

$C_{34}H_{50}O_7$

3β-Hydroxy-11-oxoolean-12-en-30-oic acid hydrogen succinate.

carbenoxalone. Antiulcerative. Anti-inflammatory gluccocorticoid related to enoxolone; a gastro-intestinal sedative. mp = 291-294°; $[\alpha]_D^{20}$= 128° ($CHCl_3$). *Biorex Labs Ltd.*

6487 Carbenoxolone Sodium

7421-40-1 1839 231-044-0

$C_{34}H_{48}Na_2O_7$

3β-Hydroxy-11-oxoolean-12-en-30-oic acid hydrogen succinate disodium salt.

carbenoxalone sodium; Biogastrone; Bioplex; Bioral; Duogastrone; Neogel; Pyrogastrone; Sanodin; Ulcus-Tablinen. Antiulcerative. Anti-inflammatory gluccocorticoid related to enoxolone; a gastro-intestinal sedative. Soluble in H_2O; LD_{50} (mmus ip) = 120 mg/kg, (mmus iv) = 198 mg/kg, (mrat orl) = 3200 mg/kg. *Biorex Labs Ltd.*

6488 Cetraxate

34675-84-8 2067

$C_{17}H_{23}NO_4$

p-Hydroxyhydrocinnamic acid trans-4-(aminomethyl)-cyclohexanecarboxylate.

Antiulcerative. Derivative of tranexamic acid. mp = 200-280°. *Daiichi Seiyaku.*

6489 Cetraxate Hydrochloride

27724-96-5 2067

$C_{17}H_{24}ClNO_4$

p-Hydroxyhydrocinnamic acid trans-4-(aminomethyl)-cyclohexanecarboxylate hydrochloride.

DV-1006; Neuer. Antiulcerative. Derivative of tranexamic acid. mp = 238-240°. *Daiichi Seiyaku.*

6490 Cimetidine

51481-61-9 2337 257-232-2

$C_{10}H_{16}N_6S$

2-Cyano-1-methyl-3-[2-[[(5-methylimidazol-4-yl)methyl]thio]ethyl]guanidine.

SKF-92334; Acibilin; Acinil; Cimal; Cimetag; Cimetum; Edalene; Dyspamet; Eureceptor; Gastromet; Peptol; Tagamet; Tametin; Tratul; Ulcedin; Ulcedine; Ulcerfen; Ulcimet; Ulcofalk; Ulcomedina; Ulcomet; Ulhys. Antiulcerative. Histamine H_2-receptor antagonist. mp = 141-143°; soluble in H_2O (11.4 g/l 37°); LD_{50} (mus orl) = 2600 mg/kg, (mus iv) = 150 mg/kg, (mus ip) = 470 mg/kg, (rat orl) = 5000 mg/kg, (rat iv) = 106 mg/kg, (rat ip) = 650 mg/kg. *Apothecon; Lemmon Co.; SmithKline Beecham Pharm.*

6491 Cimetidine Hydrochloride

70059-30-2 2337 274-297-2

$C_{10}H_{17}ClN_6S$

2-Cyano-1-methyl-3-[2-[[(5-methylimidazol-4-yl)methyl]thio]ethyl]guanidine monohydrochloride.

Tagamet Injection; Tagamet Liquid; Aciloc; Biomag; Brumetidina; Notul. Antiulcerative. Histamine H_2-receptor antagonist. *Apothecon; Lemmon Co.; SmithKline Beecham Pharm.*

6492 Colloidal Bismuth Subcitrate
57644-54-9 2549 260-872-5
$C_9H_{11}IN_2O_5$
Tripotassium dicitrato bismuthate.
De-Nol; CBS; De-Noltab; Duosol; Ulcerone. Antiulcerative. A cytoprotective, polymeric bismuth citrato complex. soluble in H_2O, dilute alkali. *Gist-Brocades Intl.*

6493 Deboxamet
34024-41-4
$C_{12}H_{14}N_2O_3$
5-Methoxy-2-methylindole-3-acetohydroxamic acid.
Antiulcerative.

6494 Ebrotidine
100981-43-9 3536
$C_{14}H_{17}BrN_6O_2S_3$
p-Bromo-N-[(E)-[[2-[[[2-[(diaminomethylene)amino]4-thiazolyl]methyl]thio]ethyl]amino]methylene]-benzenesulfonamide.
FI-3542; Ebrodin; Ulsanic. Antiulcerative. Histamine H_2 receptor antagonist. mp = 142.5-146°.

6495 Ecabet
33159-27-2 3538
$C_{20}H_{28}O_5S$
13-Isopropyl-12-sulfopodocarpa-8,11,13-trien-15-oic acid.
Antiulcerative. [hemihydrate] $[\alpha]_D^{25}$= 72.4° (c= 2.5 EtOH).

6496 Ecabet Sodium
86408-72-2 3538
$C_{20}H_{27}NaO_5S$
13-Isopropyl-12-sulfopodocarpa-8,11,13-trien-15-oic acid sodium salt.
TA-2711; Gastron. Antiulcerative. [pentahydrate] mp > 300°; $[\alpha]_D^{20}$= 59.4° (c = 0.5).

6497 Egualen
99287-30-6
$C_{15}H_{18}O_3S$
3-Ethyl-7-isopropyl-1-azulenesulfonic acid.
Antiulcerative.

6498 Enisoprost
81026-63-3
$C_{22}H_{36}O_5$
(±)-Methyl(Z)-7-[(1R,2R,3R)-3-hydroxy-2-[(E)-(4RS)-4-hydroxy-4-methyl-1-octenyl]-5-oxocyclopentyl]-4-heptenoate.
SC-34301. Antiulcerative. *Searle G.D. & Co.*

6499 Enprostil
73121-56-9 3629
$C_{23}H_{28}O_6$
Methyl 7-[(1R*,2R*,3R*)-3-hydroxy-2-[(E)-(3R*)-3-hydroxy-4-phenoxy-1-butenyl]-5-oxocyclopentyl]-4,5-heptadienoate.
RS-84135; Camleed; Fundyl; Gardrin; Gardrine; Syngard. Antiulcerative. Gastric antisecretory. Prostaglandin. mp < 46°; slightly soluble in H_2O; soluble in EtOH, propylene glycol, propylene carbonate; λ_m = 220, 265, 271, 277 nm (log ε 4.01, 3.14, 3.24, 3.16 MeOH). *Syntex Labs. Inc.*

6500 Esaprazole
64204-55-3 3736
$C_{12}H_{23}N_3O$
N-Cyclohexyl-1-piperazineacetamide.
hexaprazol; hexaprazole; exaprazole; C-63; Prazol; [hydrochloride] CO-1063. Antiulcerative. mp = 111-112°; $bp_{0.5}$ = 190°; LD_{50} (mus orl) = 1974 mg/kg, (mus iv) = 271 mg/kg, rat orl) = 3900 mg/kg. *Camillo-Corvi.*

6501 Famotidine
76824-35-6 3972
$C_8H_{15}N_7O_2S_3$
[1-Amino-3-[[[2-[(diaminomethylene)amino]-4-thiazolyl]-metyhl]thio]propylidene]sulfamide.
MK-208; YM-11170; Amfamox; Dispromil; Famodil; Famodine; Famosan; Famoxal; Fanosin; Fibonel; Ganor; Gaster; Gastridin; Gastropen; Ifada; Lecedil; Motiax; Muclox; Nulcerin; Pepcid®; Pepcid AC; Pepcidina; Pepcidine; Pepdine; Pepdul; Peptan; Ulcetrax; Ulfamid; Ulfinol. Antiulcerative. Histamine H_2 receptor antagonist. Used for short-term treatment of active duodenal ulcers. mp = 163-164°; soluble in DMF (80 g/100 ml), AcOH (50 g/100 ml), MeOH (0.3 g/100 ml), H_2O (0.1 g/100 ml); insoluble in EtOH, EtOAc, $CHCl_3$; LD_{50} (mus iv) = 244.4 mg/kg. *Johnson & Johnson-Merck Consumer Pharm.; Merck & Co.Inc.*

6502 Gefarnate
51-77-4 4385 200-121-0
$C_{27}H_{44}O_2$
trans-3,7-Dimethyl-2,6-octadienyl-5,9,13-trimethyl-4,8,12-tetradectrienoate.
DA 688; geranyl farnesyl acetate; Alsanate; Arsanyl; Dixnalate; Gefanil; Gefarnyl; Gefulcer; Osteol; Salanil; Zackal. Antiulcerative. $bp_{0.05}$ = 165-168°; λ_m = 204 nm ($E^{1\%}_{1\ cm}$ 486); soluble in EtOH, Et_2O, Et_2O, DMF, Me_2CO; oils; insoluble in H_2O, formamide, ethylene glycol, propylene glycol, glycerol. *Ist. De Angeli.*

6503 Guaiazulene
489-84-9 4581 207-701-2
$C_{15}H_{18}$
1,4-Dimethyl-7-(1-methylethyl)azulene.
S-guaiazulene; AZ 8; AZ 8 Beris; Eucazulen; Kessazulen; Vaumigan. Anti-inflammatory; antiulcerative. Blue oil; bp_{10} = 165-170°.

6504 Irsogladine
57381-26-7 5113
$C_9H_7Cl_2N_5$
2,4-Diamino-6-(2,5-dichlorophenyl)-s-triazine.
dicloguamine. Antiulcerative. mp = 268-269°. *Nippon Shinyaku Japan.*

6505 Irsogladine Maleate
84504-69-8 5113
$C_{13}H_{11}Cl_2N_5O_4$
2,4-Diamino-6-(2,5-dichlorophenyl)-s-triazine maleate.
MN-1695; Gaslon. Antiulcerative. mp = 205° (dec); LD_{50} (mmus orl) = 6035 mg/kg, (mmus sc) = 2841 mg/kg, (mmus ip) = 775 mg/kg, (fmus orl) = 5697 mg/kg, (fmus sc) = 3216 mg/kg, (fmus ip) = 1006 mg/kg, (mrat orl) = 3898 mg/kg, (mrat sc) = 1600 mg/kg, (mrat ip) = 558

mg/kg, (frat orl) = 2917 mg/kg, (frat sc) = 1524 mg/kg, (frat ip) = 545 mg/kg. *Nippon Shinyaku Japan.*

6506 Isotiquimide
56717-18-1
$C_{11}H_{14}N_2S$
(±)-5,6,7,8-Tetrahydro-4-methylthio-8-quinolinecarboxamide.
WY-24377. Antiulcerative. *Wyeth-Ayerst Labs.*

6507 Lansoprazole
103577-45-3 5373
$C_{16}H_{14}F_3N_3O_2S$
2-[[[3-Methyl-4-(2,2,2-trifluoroethoxy)-2-pyridyl]methyl]-sulfinyl]benzimidazole.
Prevacid; Prevacid(TAP); AG-1749; A-65006; Agopton; Lanzor; Ogast; Takepron; Zoton. Antiulcerative. Gastric proton pump inhibitor. mp = 178-182° (dec). *Takeda Chem. Ind. Ltd.*

6508 Lavoltidine
76956-02-0
$C_{19}H_{29}N_5O_2$
1-Methyl-5-[[3-[(α-piperidino-m-tolyl)oxy]propy]amino]-1H-1,2,4-triazole-3-methanol.
Loxotidine; AH23844. Antiulcerative. Histamine H_2 receptor antagonist. *Glaxo Labs.*

6509 Lavoltidine Succinate
86160-82-9
$C_{42}H_{64}N_{10}O_8$
1-Methyl-5-[[3-[(α-piperidino-m-tolyl)oxy]propy]amino]-1H-1,2,4-triazole-3-methanol succinate (salt) (2:1).
Loxotidine succinate; AH23844A. Antiulcerative. Histamine H_2 receptor antagonist. *Glaxo Labs.*

6510 Leminoprazole
104340-86-5
$C_{19}H_{23}N_3OS$
(±)-2-[[o-(Isobutylmethylamino)benzyl]sulfinyl]benzimidazole.
NC-1300-O-3. Antiulcerative. H^+,K(+)-ATPase (acid pump) inhibitor. Stimulates synthesis and secretion of mucus.

6511 Lozilurea
71475-35-9
$C_{10}H_{13}ClN_2O$
1-(m-Chlorobenzyl)-3-ethylurea.
ITA-312. Antiulcerative.

6512 Meciadanol
65350-86-9 265-710-7
$C_{16}H_{16}O_6$
(2R,3S)-3-Methoxy-3',4',5,7-flavanterol.
Zyma S.A., Nyon, Switzerland. Antiulcerative. A catechin bioflavonoid with gastric cycoprotective effects.

6513 Misoprostol
59122-46-2 6297
$C_{22}H_{38}O_5$
(±)-Methyl (1R,2R,3R)-3-hydroxy-2-[(E)-(4RS)-4-hydroxy-4-methyl-1-octenyl]-5-oxocyclopentaneheptanoate.
Cytotec; SC-29333. Antiulcerative. Cytoprotective prostaglandin PGE_1 analog. Soluble in H_2O; LD_{50} (rat ip) = 40-62 mg/kg, (rat orl) = 81-100 mg/kg, (mus ip) = 70-160 mg/kg, (mus orl) = 27-138 mg/kg. *Searle G.D. & Co.*

6514 Niperotidine
84845-75-0 284-304-0
$C_{20}H_{26}N_4O_5S$
N-[2-[[5-[(Dimethylamino)methyl]furfuryl]thio]ethyl]-2-nitro-N'-piperonyl-1,1-ethenediamine.
Antiulcerative. Histamine H2-receptor antagonist. Associated liver toxicity.

6515 Nizatidine
76963-41-2 6758
$C_{12}H_{21}N_5O_2S_2$
N-[2-[[[2-[(Dimethylamino)methyl]-4-thiazolyl]methyl]thio]ethyl]-N'-methyl-2-nitro-1,1-ethenediamine.
Axid; LY-139037; ZE-101; ZL-101; Calmaxid; Cronizat; Distaxid; Gastrax; Naxidine; Nizax; Nizaxid; Zanizal. Antiulcerative. Histidne H_2 receptor antagonist. mp = 130-132°; λ_m = 240, 325 nm (ε 8400, 19600 MeOH); 260, 314 (ε 11820, 15790 H_2O); soluble in $CHCl_3$ (> 100 mg/ml), MeOH (50-100 mg/ml); less soluble in H_2O (10-33 mg/ml); poorly soluble in EtOAc (1-2 mg/ml), iPrOH (3.3-5 mg/ml); insoluble in C_6H_6 (<0.5 mg/ml), Et_2O (< 0.5 mg/ml), $C_8H_{17}OH$ (< 0.5 mg/ml). *Eli Lilly & Co.*

6516 Nolinium Bromide
40759-33-9
$C_{15}H_{11}BrCl_2N_2$
2-(3,4-Dichloroanilino)quinolizinium bromide.
EU-2972. Antiulcerative.

6517 Olsalazine
15722-48-2 6976
$C_{14}H_{10}N_2O_6$
3,3'-Azobis(6-hydroxybenzoic acid).
C.I. Mordant Yellow 5; 5,5'-azobis(salicylic acid); azodisal. Anti-inflammatory (gastrointestinal); antiulcerative. Derivative of salicylic acid. Dimer of mesalamine. Used in the treatment of ulcerative colitis. Mordant dye for wool. *Ciba-Geigy Corp.; Kabi Pharmacia Diagnostics.*

6518 Olsalazine Sodium
6054-98-4 6976 227-975-7
$C_{14}H_8N_2Na_2O_6$
Disodium 3,3'-azobis(6-hydroxybenzoic acid).
sodium azodisalicylate; azodisal sodium; disodium azodisalicylate; C.I. 14130; Ph-CJ-91B; Dipentum; C.I. Mordant Yellow 5 disodium salt. Anti-inflammatory (gastrointestinal); antiulcerative. Derivative of salicylic acid. Used in the treatment of ulcerative colitis. Mordant dye for wool. Yellow powder; soluble in H_2O; moderately soluble in EtOH. *Ciba-Geigy Corp.; Kabi Pharmacia Diagnostics; Pharmacia & Upjohn.*

6519 Omeprazole
73590-58-6 6977
$C_{17}H_{19}N_3O_3S$
5-Methoxy-2-[[(4-methoxy-3,5-dimethyl-2-pyridyl)methyl]sulfinyl]benzimidazole.
Prilosec; Antra; Gastroloc; Losec; Mepral; Mopral; Omepral; Omeprazen; Parizac; Pepticum. Antiulcerative. Gastric antisecretory agent (proton pump inhibitor). Used

in treatment of Zollinger-Ellison syndrome. mp = 156°; LD_{50} (mus iv) = 80 mg/kg, (mus orl) > 4000 mg/kg, (rat iv) > 50 mg/kg, (rat orl) > 4000 mg/kg. *Astra Hassle AB; Astra Sweden; Merck & Co.Inc.*

6520 Omeprazole Sodium
95510-70-6 6977
$C_{17}H_{18}N_3NaO_3S$
5-Methoxy-2-[[(4-methoxy-3,5-dimethyl-2-pyridyl)methyl]sulfinyl]benzimidazole sodium salt.
Losec Sodium; H 168/68. Antiulcerative. Gastric antisecretory agent; proton pump inhibitor. *Astra Hassle AB; Astra Sweden; Merck & Co.Inc.*

6521 Ornoprostil
70667-26-4 7003
$C_{23}H_{38}O_6$
Methyl (-)-(1R,2R,3R)-3-hydroxy-2-[(E)-(3S,5S)-3-hydroxy-5-methyl-1-nonenyl]-ε,5-dioxocyclopentaneheptanoate.
ronoprost; ONO-1308; OU-1308; Alloca; Ronok. Antiulcerative. Gastric antisecretory. *Ono Pharm.*

6522 Pantoprazole
102625-70-7 7146
$C_{16}H_{15}F_2N_3O_4S$
5-(Difluoromethoxy)-2-[[(3,4-dimethoxy-2-pyridyl)methyl]sulfinyl]benzimidazole.
SK&F-96022; BY-1023. Antiulcerative. Gastric proton pump inhibitor. mp = 139-140° (dec); [sodium salt]: mp >130° (dec); λ_m = 289 (ε 16400). *Byk Gulden Lomberg GmbH; SmithKline Beecham Pharm.*

6523 Picoprazole
78090-11-6
$C_{17}H_{17}N_3O_3S$
Methyl 6-methyl-2-[[(3-methyl-2-pyridyl)methyl]sulfinyl]-5-benzimidazolecarboxylate.
H-149/94. A benzimidazole derivative that suppresses acid secretion through inhibition of (H+/K+)-ATPase. Antiulcerative; cyctoprotectant.

6524 Pifarnine
56208-01-6 7575
$C_{27}H_{40}N_2O_2$
1-Piperonyl-4-(3,7,11-trimethyl-2,6,10-dodecatrienyl)piperazine.
U-27; Pifazin. Antiulcerative. Non-cholinergic gastric secretory inhibitor. λ_m = 287 nm ($E^{1\%}_{1\,cm}$ 94.6 EtOH); d^{20} = 1.013-1.015; soluble in most organic solvents, insoluble in H_2O; LD_{50} (mus orl) = 2175 mg/kg, (mus iv) = 41.6 mg/kg, (mus ip) = 500 mg/kg, (rat orl)= 2610 mg/kg, (rat iv) = 33.3 mg/kg. *Pierrel S.p.A.*

6525 Pipratecol
15534-05-1 239-578-6
$C_{19}H_{24}N_2O_4$
α-(3,4-Dihydroxyphenyl)-4-(2-methoxyphenyl)-1-piperazineethanol.
711-SE. Antiulcerative. Substituted thienoimidazole H+/K(+)-ATPase inhibitor.

6526 Pirenzepine
28797-61-7 7646 249-228-4
$C_{19}H_{21}N_5O_2$
5,11-Dihydro-11-[(4-methyl-1-piperazinyl)acetyl]-6H-pyrido[2,3-b][1,4]benzodiazepin-6-one.
LS-519. Anticholinergic; antiulcerative. Gastric acid inhibitor. *Boehringer Ingelheim GmbH.*

6527 Pirenzepine Hydrochloride
29868-97-1 7646 249-907-5
$C_{19}H_{23}Cl_2N_5O_2$
5,11-Dihydro-11-[(4-methyl-1-piperazinyl)acetyl]-6H-pyrido[2,3-b][1,4]benzodiazepin-6-one dihydrochloride.
LS-59 Cl2; Duogastral;Durapirenz; Gasteril; Gastrozepin; Leblon; Maghen; Renzepin; Tabe; Ulcuforton; Ulcosan. Anticholinergic; antiulcerative. Gastric acid inhibitor. Soluble in H_2O, less soluble in MeOH, insoluble in Et_2O. *Boehringer Ingelheim GmbH.*

6528 Plaunotol
64218-02-6 7692
$C_{20}H_{34}O_2$
(2Z,6E)-2-[(3E)-4,8-Dimethyl-3,7-nonadienyl]-6-methyl-2,6-octadiene-1,8-diol.
CS-684; Kelnac. Antiulcerative. Soluble in organic solvents, insoluble in H_2O; LD_{50} (mmus orl) = 8800 µl/kg, (fmus orl) = 8100 µl/kg, (mrat orl) = 10900 µl/kg, (frat orl) = 11200 µl/kg. *Sankyo Co. Ltd.*

6529 Polaprezinc
107667-60-7 7712
$(C_9H_{12}N_4O_3Zn)_n$
[N-β-Alanyl-L-histidinato(2-)-N,N^N,O^α]zinc.
zinc L-carnosine; Z-103; Promac. Antiulcerative. Has antioxidant and gastroprotective properties. Insoluble in H_2O; LD_{50} (mmus ip) = 220 mg/kg, (mmus sc) = 758 mg/kg, (mmus orl) = 1269 mg/kg, (fmus ip) = 165 mg/kg, (fmus sc) = 874 mg/kg, (fmus orl) = 1331 mg/kg, (mrat ip) = 405 mg/kg, (mrat sc) > 5000 mg/kg, (mrat orl) = 8441 mg/kg, (frat ip) = . *Hamari Chem. Ltd.*

6530 Rabeprazole Sodium
117976-90-6 8272
$C_{18}H_{20}N_3NaO_3S$
2-[[[4-(3-Methoxypropoxy)-3-methyl-2-pyridinyl]methyl]sulfinyl]benzimidazole sodium salt.
LY-307640 sodium; E-3810. Antiulcerative. Gastric proton pump inhibitor. mp = 140-141° (dec). *Eisai Co. Ltd.; Eli Lilly & Co.*

6531 Rabeprazole Sodium
117976-90-6 8272
$C_{18}H_{20}N_3NaO_3S$
2-[[[4-(3-Methoxypropoxy)-3-methyl-2-pyridinyl]methyl]sulfinyl]benzimidazole sodium salt.
LY-307640 sodium; E-3810. Antiulcerative. Gastric proton pump inhibitor. mp = 140-141° (dec). *Eisai Co. Ltd.; Eli Lilly & Co.*

6532 Ramixotidine
84071-15-8
$C_{16}H_{21}N_3O_3S$
N-[2-[[5-[(Dimethylamino)methyl]furfuryl]thio]ethyl]nicotinamide 1-oxide.
CM-57755 [as dihydrochloride]. Antiulcerative. Histamine H2-receptor antagonist with gastric antisecretory effects.

6533 Ranitidine
66357-35-5 8286 266-332-5
$C_{13}H_{22}N_4O_3S$
N-[2-[[5-[(Dimethylamino)methyl]furfuryl]thio]ethyl]-N'-methyl-2-nitro-1,1-ethenediamine.
Antiulcerative. Histamine H_2 receptor antagonist; inhibits gastric secretion. mp = 69-70°. *Glaxo Labs.*

6534 Ranitidine Hydrochloride
66357-59-3 8286 266-333-0
$C_{13}H_{23}ClN_4O_3S$
N-[2-[[5-[(Dimethylamino)methyl]furfuryl]thio]ethyl]-N'-methyl-2-nitro-1,1-ethenediamine hydrochloride.
AH-19065; Azantac; Melfax; Noctone; Raniben; Ranidil; Raniplex; Sostril; Taural; Terposen; Trigger; Ulcex; Ultidine; Zantac; Zantic. Antiulcerative. Histamine H_2 receptor antagonist; inhibits gastric secretion. mp = 133-134°; soluble in H_2O, AcOH; less soluble in EtOH, MeOH; insoluble in organic solvents. *Glaxo Labs.*

6535 Rebamipide
90098-04-7 8296
$C_{19}H_{15}ClN_2O_4$
(±)-α-(p-Chlorobenzamido)-1,2-dihydro-2-oxo-4-quinolinepropionic acid.
OPC-12759; Mucosta; proamipide. Antiulcerative. Gastric cytoprotectant. [hemihydrate]: mp = 299-290° (dec); [(-)-form]: mp = 304-306°; [(-)-form]: $[\alpha]_D^{20}$ = -116.7 (c = 1.0 in DMF); [(+)-form]: $[\alpha]_D^{20}$ = +116.9 (c = 1.0 DMF). *Otsuka Am. Pharm.*

6536 Remiprostol
110845-89-1
$C_{25}H_{36}O_5$
(±)-Methyl-(Z)-7-[(1R,2R,3R)-2-[(1E,5E)-(4RS)-6-(1-cyclopenten-1-yl)-4-hydroxy-4-methyl-1,5-hexadienyl]-3-hydroxy-5-oxocyclopentyl]-5-heptenoate.
SC-48834. Antiulcerative. *Searle G.D. & Co.*

6537 Rioprostil
77287-05-9 8395
$C_{21}H_{38}O_4$
(2R,3R,4R)-4-Hydroxy-2-(7-hydroxyheptyl)-3-[(E)-(4RS)-(4-hydroxy-4-methyl-1-octenyl)]cyclopentanone.
TR-4698; ORF-15927; RWJ-15927; Bay o 6893; Rostil. Antiulcerative. Prostaglandin. Gastric antisecretory agent. $[\alpha]_D$ = -58.6° (c = 1 $CHCl_3$). *Bayer Corp. Pharm. Div.; Miles Inc.*

6538 Rosaprostol
56695-65-9 8419 260-341-8
$C_{18}H_{34}O_3$
(1RS,2SR,5RS)-2-Hexyl-5-hydrocyclopentane-heptanoic acid.
2-hexyl-5-hydroxycyclopentaneheptanoic acid; C-83; IBI-C83; Rosal. Gastric antisecretory; cytoprotectant. Antiulcerative. Prostaglandin. Oil; [sodium salt ($C_{18}H_{33}N_O3$)]: LD_{50} (mus orl) = 3000 mg/kg, (rat orl) > 5000 mg/kg. *Ist. Biochim.*

6539 Rotraxate
92071-51-7 8428
$C_{17}H_{23}NO_3$
p-[[trans-4-(Aminomethyl)cyclohexyl]carbonyl]hydrocinnamic acid.
traxaprone. Antiulcerative. Gastric antisecretory and cytoprotectant. *Teijin Ltd.*

6540 Rotraxate Hydrochloride
82085-94-7 8428
$C_{17}H_{24}ClNO_3$
p-[[trans-4-(Aminomethyl)cyclohexyl]carbonyl]-hydrocinnamic acid monohydrochloride.
TEI-5103; TG-51; Cumelon. Antiulcerative. Gastric antisecretory and cytoprotectant. Similar to cetraxate. mp = 245° (dec), 221-227°; LD_{50} (mrat orl) = 9800 mg/kg, (mrat ip) = 862 mg/kg, (mrat sc) = 5000 mg/kg, (frat orl) = 9800 mg/kg, (frat ip) = 835 mg/kg, (frat sc) = 5000 mg/kg. *Teijin Ltd.*

6541 Roxatidine
78273-80-0 8431
$C_{19}H_{28}N_2O_4$
N-[3-[(α-Piperidino-m-tolyl)oxy]propyl]-glycolamide.
Rosoxacin; HOE-062. Antiulcerative. Histamine H_2 receptor antagonist. mp = 59-60°; LD_{50} (mus orl)= 1000 mg/kg. *Hoechst Roussel Pharm. Inc.*

6542 Roxatidine Acetate
78628-28-1 8431
$C_{19}H_{28}N_2O_4$
N-[3-[(α-Piperidino-m-tolyl)oxy]propyl]glycolamide acetate (ester).
HOE-062; aceroxatidine. Antiulcerative. Histidine H_2 receptor antagonist. mp = 59-60°, LD_{50} (mmus orl) = 1000 mg/kg. *Hoechst Roussel Pharm. Inc.*

6543 Roxatidine Acetate Hydrochloride
93793-83-0 8431
$C_{19}H_{29}ClN_2O_4$
N-[3-[(α-Piperidino-m-tolyl)oxy]propyl]glycolamide acetate (ester) hydrochloride.
HOE-760; pifatidine; TZU-0460; Altat; Gastralgin; Neo H2; Roxit. Antiulcerative. Histidine H_2 receptor antagonist. mp = 145-146°. *Hoechst Roussel Pharm. Inc.*

6544 Sofalcone
64506-49-6 8850
$C_{27}H_{30}O_6$
[5-[(3-Methyl-2-butenyl)oxy]-2-[p-[(3-methyl-2-butenyl)oxy]cinnamoyl]phenoxy]-acetic acid.
Su-88; Solon. Antiulcerative. mp = 143-144°; LD_{50} (mus, rat orl) > 10 g/kg. *Taisho.*

6545 Somatostatin
38916-34-6 8863 254-186-5
$C_{76}H_{104}N_{18}O_{19}S_2$
Ala-Gly-Cys-Lys-Asn-Phe-Phe-Trp-Lys-Thr-Phe-Thr-Ser-Cys (Cys-Cys disulfide).
GH-RIF; growth hormone-release inhibiting factor; somatotropin release inhibiting factor; SRIF; SRIF-14. Gastric antisecretory agent. Used for treatment of hemorrhaging of gastro-duodenal ulcers. Growth hormone inhibitor. Studied for use as an antidiabetic. A cyclic tetradecapeptide which inhibits the release of growth hormone, insulin and glucagon. *Genentech Inc.*

6546 Somatostatin Acetate
8863

$C_{76}H_{104}N_{18}O_{19}S_2 \cdot C_2H_4O_2$
Ala-Gly-Cys-Lys-Asn-Phe-Phe-Trp-Lys-Thr-Phe-Thr-Ser-Cys (Cys-Cys disulfide) acetate.
SRIF-A; Aminopan; Modustatina; Somatofalk; Stilamin. Gastric antisecretory agent. Used for treatment of hemorrhaging of gastro-duodenal ulcers. Growth hormone inhibitor. Studied for use as an antidiabetic. A cyclic tetradecapeptide which inhibits the release of growth hormone, insulin and glucagon. *Genentech Inc.*

6547 Spizofurone
72492-12-7 8918

$C_{12}H_{10}O_3$
5-Acetylspiro[benzofuran-2(3H),1'-cyclopropan]-3-one.
AG-629; Maon. Antiulcerative. mp = 102-104°, 106-107°. *Takeda Chem. Ind. Ltd.*

6548 Sucralfate
54182-58-0 9049 259-018-4

$C_{12}H_mAl_{16}O_nS_8$
Sucrose octakis(hydrogen sulfate) aluminum complex.
Carafate; Antepsin; Citogel; Hexagastron; Keal; Succosa; Sucralfin; Sucrate; Sugast; Sulcrate; Ulcar; Ulcerlmin; Ulcogant. Gastrointestinal antiulcerative. Insoluble in H_2O, EtOH; soluble in dilute HCl and NaOH solutions. *Chugai Pharm. Co. Ltd.; Hoechst Marion Roussel Inc.*

6549 Sucrosofate
57680-56-5

$C_{12}H_{22}O_{35}S_8$
Sucrose octakis(sulfuric acid).
Antiulcerative. *Marion Merrell Dow Inc.*

6550 Sucrosofate Potassium
76578-81-9

$C_{12}H_{14}K_8O_{35}S_8.7H_2O$
Sucrose octakis(potassium sulfate) heptahydrate.
Agent M-01. Antiulcerative. *Marion Merrell Dow Inc.*

6551 Sulfasalazine
599-79-1 9112 209-974-3

$C_{18}H_{14}N_4O_5S$
5-[[p-(2-Pyridylsulfamoyl)phenyl]azo]salicylic acid.
salazosulfapyridine; salicylazosulfapyridine; sulphasalazine; Azulfidine; Colo-Pleon; Salazopyrin. Anti-inflammatory (gastrointestinal). Sulfonamide; conjugate of 5-aminosalicylic acid and suphapyridine. Used in the treatment of ulcerative colitis and Crohn's disease. Dec 240-245°; λ_m = 237 ($E^{1\%}_{1cm}$ 658), 359 nm; slightly soluble in EtOH; nearly insoluble in H_2O, C_6H_6, $CHCl_3$, Et_2O. *Kabi Pharmacia Diagnostics; Pharmacia & Upjohn.*

6552 Sulglicotide
54182-59-1
Sulfuric polyester of a glycopeptide isolated from pig duodenum.
sulglycotide. A sulfoglycopeptide. Antiulcerative; cytoprotectant.

6553 Telenzepine
80880-90-6 9270

$C_{19}H_{22}N_4O_2S$
4,9-Dihydro-3-methyl-4-[(4-methyl-1-piperazinyl)acetyl]-10H-thieno[3,4-b][1,5]benzodiazepin-10-one.
Antiulcerative; anticholinergic. Selective muscarinic M_1-receptor antagonist. Gastric acid inhibitor. mp = 263-264°. *Byk Gulden Lomberg GmbH.*

6554 Teprenone
6809-52-5 9296

$C_{23}H_{38}O$
6,10,14,18-Tetramethyl-5,9,13,17-nonadecatetraene-2-one, mixture of (5E,9E,13E) and (5Z,9E,13E) isomers.
geranylgeranylacetone; GGA; E-0671; E36U31; Selbex. Antiulcerative. $bp_{0.01}$ = 155-160°; $d_4^{20.5}$ = 0.9081.

6555 Tolimidone
41964-07-2

$C_{11}H_{10}N_2O_2$
5-(m-Toloxy)-2(1H)-pyrimidinone.
CP-26154. Antiulcerative. *Pfizer Inc.*

6556 Trapencaine
104485-01-0

$C_{22}H_{34}N_2O_3$
(±)-trans-2-(1-Pyrrolidinyl)cyclohexyl m-(pentyloxy)-carbanilate.
pentacaine. Local anesthetic; antiulcerative; gastroprotective.

6557 Trimoprostil
69900-72-7 9853

$C_{23}H_{38}O_4$
(Z)-7-[(1R,2R,3R)-2-[(E)-(3R)-3-Hydroxy-4,4-dimethyl-1-octenyl]-3-methyl-5-oxocyclopentyl]-5-heptenoic acid.
Ro-21-6937/000; TM-PGE_2; Ulstar. Antiulcerative. Synthetic prostaglandin E_2 analog with gastric antisecretory activity. $[\alpha]_D$ = -51.54° (c = 1 $CHCl_3$); LD_{50} mus orl) = 41 mg/kg, (mis ip) = 70 mg/kg, (mus sc) = 68 mg/kg, (rat orl) = 23 mg/kg, (rat ip) = 21 mg/kg, (rat sc) = 29 mg/kg. *Hoffmann-LaRoche Inc.*

6558 Tritiozine
35619-65-9 9889 252-645-4

$C_{14}H_{19}NO_4S$
4-(3,4,5-Trimethoxythiobenzoyl)morpholine.
Trithiozine; sulmetozine; ISF-2001; Tresanil. Antiulcerative. Non-cholinergic gastric secretion inhibitor. Gastric cytoprotectant. mp = 141-143°; LD_{50} (mus ip)= 2000 mg/kg. *I.S.F.*

6559 Troxipide
99777-81-8 9921

$C_{15}H_{22}N_2O_4$
(±)-(3,4,5-Trimethoxy-N-3-piperidylbenzamide.
KU-54; Aplace. Antiulcerative. mp= 179-185°; soluble in EtOH; LD_{50} (mrat orl) = 500 mg/kg, (mrat sc) > 4150 mg/kg, (mrat ip) = 340 mg/kg, (frat orl) = 2100 mg/kg, (frat sc) > 4150 mg/kg, (frat ip) = 340 mg/kg, (mmus orl) = 2200 mg/kg, (mmus sc) = 1600 mg/kg, (mmus ip) = 300 mg/kg, (fmus orl) = 2000 mg/kg, (fmus sc) = 1550 mg/kg, (fmus ip) = 305 mg/kg. *Kyorin Pharm. Co. Ltd.*

6560 Zolimidine
1222-57-7 10320 214-947-4

$C_{14}H_{12}N_2O_2S$
2-[4-(Methylsulfonyl)phenyl]imidazo[1.2-a]pyridine.
zoliridine; Solimidin. Antiulcerative. Non-cholinergic gastroprotective agent. mp = 242-244°; LD_{50} (rat orl) =

3710 mg/kg; [hydrochloride]: LD_{50} (mus ip) = 800 mg/kg. *Selvi.*

Antivirals

6561 Abacavir
136470-78-5
$C_{14}H_{18}N_6O$
4-[2-Amino-6-(cyclopropylamino)-9H-purin-9-yl]-2-cyclopentene-1-methanol, (1S, cis).
Antiviral agent. *Glaxo Wellcome Inc.*

6562 Abacavir Succinate
168146-84-7
$C_{18}H_{24}N_6O_5$
4-[2-Amino-6-(cyclopropylamino)-9H-purin-9-yl]-2-cyclopentene-1-methanol, (1S, cis), butanedioate (1:1) (salt).
Antiviral agent. *Glaxo Wellcome Inc.*

6563 Abacavir Sulfate
188062-50-2
$C_{14}H_{20}N_6O_5S$
(1S,4R)-4-(2-Amino-6-(cyclopropylamino)-9H-purin-9-yl)-2-cyclopentene-1-methanol.
Abacavir sulfate; 1592U89 sulfate; ABC sulfate; Abacavir sulfate; DRG-0257; Ziagen; (component of) Trizivir; component of: trizivir. Antiviral. Used alone or in combination with other antiretroviral agents for the treatment of HIV-1 infection. *Glaxo Wellcome Inc.*

6564 Acedoben
556-08-1 209-114-7
$C_9H_9NO_3$
p-Carboxyacetanilide.
p-(acetamino)benzoic acid; 4-(acetylamino)benzoic acid; N-acetyl-p-aminobenzoic acid. Examined as an antiviral agent.

6565 Acemannan
110042-95-0 26
Polymanoacetate.
Carraklenz Wound & Skin Cleanser; Carrisyn; Snow & Sun Sports Gel; component of: Carraklenz Incontinence Skin Care Kit, Moisture Barrier Cream with Zinc, Moisture Guard, Skin Balm, Snow & Sun Sunburn Spray. Antiviral agent. Immunomodulator. From mucilage of *Aloe barbadensis* (aloe vera). A polydispersed, acetylated, linear mannan. MW: 1-2 x 10^6 daltons. *Carrington Labs Inc.*

6566 Acyclovir
59277-89-3 148 261-685-1
$C_8H_{11}N_5O_3$
2-Amino-1,9-dihydro-9-[(2-hydroxyethoxy)methyl]-6H-purin-6-one.
Azone; Acycloguanosine; BW-248U; Wellcome 248U; Acicloftal; Cargosil; Laurocapram; Poviral; Virorax; Zovirax; Vipral; Aciclovir; Acyclo-V; Zyclir. Antiviral agent. Antiviral used in treatment of herpes virus. A nucleoside analog that is preferentially taken up by infected cells and then inhibits viral DNA synthesis by interfering with transcription. mp = 256.5-257°; LD_{50} (mus orl) >10,000 mg/kg. *Glaxo Wellcome Inc.*

6567 Acyclovir Sodium
69657-51-8 148
$C_8H_{10}N_5NaO_3$
2-Amino-1,9-dihydro-9-[(2-hydroxyethoxy)methyl]-6H-purin-6-one monosodium salt.
Antiviral agent. *Glaxo Wellcome Inc.*

6568 Adefovir Dipivoxil
142340-99-6
$C_{20}H_{32}N_5O_8P$
[[[2-(6-Amino-9H-purin-9-yl)ethoxy]methyl]phosphinylidene]bis(oxymethylene) 2,2-dimethylpropanoate.
Antiviral agent. *Gilead Sciences Inc.*

6569 Afovirsen
151356-08-0
$C_{192}H_{250}N_{57}O_{107}P_{19}S_{19}$
2'-Deoxy-P-thiocytidylyl-(5'→3')-P-thiothymidylyl-(5'→3')-2'-deoxy-P-thioguanylyl-(5'→3')-2'-deoxy-P-thiocytidylyl-(5'→3')-P-thiothymidylyl-(5'→3')-2'-deoxy-P-thiocytidylyl-(5'→3')-2'-deoxy-P-thiocytidylyl-(5'→3')-P-thiothymidylyl-(5'→3')-P-thiothymidylyl-(5'→3')-2'-deoxy-P-thiocytidylyl-(5'→3')-P-thiothymidylyl-(5'→3')-2'-deoxy-P-thioadenylyl-(5'→3')-2'-deoxy-P-thiocytidylyl-(5'→3')-(5'→3')-2'-deoxy-P-thiocytidylyl-(5'→3')-P-thiothymidylyl-(5'→3')-P-thiothymidylyl-(5'→3')-2'-deoxy-P-thiocytidylyl-(5'→3')-2'-deoxy-P-thioguanylyl-(5'→3')-P-thiothymidylyl-(5'→3')-thymidine.
An antisense phosphorothioate oligodeoxynucleotide with antiviral properties.

6570 Almurtide
61136-12-7
$C_{18}H_{30}N_4O_{11}$
2-Acetamido-3-O-[[[(1S)-1-[[(1R)-1-carbamoyl-3-carboxypropyl]carbamoyl]ethyl]carbamoyl]methyl]-2-deoxy-D-glucopyranose.
Compound with some antiviral properties. Reported to prevent viral oncogenesis.

6571 Alovudine
25526-93-6
$C_{10}H_{13}FN_2O_4$
3'-Deoxy-3'-fluorothymidine.
Antiviral agent.

6572 Alvircept Sudotox
137487-62-8
$C_{2600}H_{4130}N_{748}O_{812}S_{10}$
Antiviral agent. Synthetic protein of 59,187 daltons.

6573 Amantadine
768-94-5 389 212-201-2
$C_{10}H_{17}N$
Tricyclo[3.3.1.1^{3,7}]decan-1-amine.
Antiviral; antidyskinetic; antiparkinsonian. mp = 160-190°, 180-192°; sparingly soluble in H_2O. *Apothecon; DuPont-Merck Pharm.; Solvay Pharm. Inc.*

6574 Amantadine Hydrochloride
665-66-7 389 211-560-2
$C_{10}H_{18}ClN$
Tricyclo[3.3.1.1^{3,7}]decan-1-amine hydrochloride.
Symadine; Symmetrel; EXP-105-1; NSC-83653; Amazolon; Mantadix; Mantadan; Mantadine; Midantan;

Mydantane; Virofral. Antiviral; antidyskinetic; antiparkinsonian. Dec 360°; soluble in H+2O (> 5 g/100 ml), insoluble in Et_2O; LD_{50} (mus orl) = 700 mg/kg, (rat orl) = 1275 mg/kg. *Apothecon; DuPont-Merck Pharm.; Solvay Pharm. Inc.*

6575 Amidapsone
3569-77-5
$C_{13}H_{13}N_3O_3S$
(p-Sulfanilylphenyl)urea.
Antiviral for poultry.

6576 Amprenavir
161841-49-9
$C_{25}H_{35}N_3O_6S$
(3S)-Tetrahydro-3-furyl [(αS)-α-[(1R-1-hydroxy-2-(N[1]-isobutylsulfanilamido)ethyl]phenethyl]carbamate.
Agenerase; VX-478; 141W94; KVX-478. HIV-protease inhibitor. *Glaxo Wellcome Inc.; Vertex.*

6577 Aranotin
19885-51-9
$C_{20}H_{18}N_2O_7S_2$
5-(Acetyloxy)-5,5a,13,13a-tetrahydro-13-hydroxy-8H,16H-7a,15a-epidithio-7H,15H-bisoxepino[3',4':4,5]-pyrrolo[1,2-a:1',2'-d]pyrazine-7,15-dione.
Antiviral agent. *Eli Lilly & Co.*

6578 Arildone
56219-57-9 260-066-3
$C_{20}H_{29}ClO_4$
4-[6-(2-Chloro-4-methoxyphenoxy)hexyl]-3,5-heptanedione.
Win-38020. Antiviral agent. *Sterling Winthrop Inc.*

6579 Atevirdine
136816-75-6
$C_{21}H_{25}N_5O_2$
1-[3-(Ethylamino)-2-pyridinyl]-4-[(5-methoxy-1H-indol-2-yl)carbonyl piperazine.
Antiviral agent.

6580 Atevirdine Mesylate
138540-32-6
$C_{22}H_{29}N_5O_5S$
1-[3-(Ethylamino)-2-pyridinyl]-4-[(5-methoxy-1H-indol-2-yl)carbonyl piperazine monomethanesulfonate.
U-87201E. Antiviral agent.

6581 Avridine
35607-20-6
$C_{43}H_{90}N_2O_2$
2,2'-[[3-(Dioctadecylamino)propyl]imino]bisethanol.
CP-20961. Antiviral agent. *Pfizer Inc.*

6582 Betasizofiran
39464-87-4 254-464-6
$(C_{24}H_{40}O_{20})_n$
Scleroglucan.
poly[→3(O-β-D-glucopyranosyl-(1→3)-O-[β-D-glucopyranosyl-(1→6)-O-β-D-glucopyranosyl-(1→3)-O-β-D-glucopyranosyl-(1→]. Reported to have antiviral properties. Produced by *Sclerotium rolfsii.* Molecular weight about 5 x 10^6.

6583 Brivudine
69304-47-8
$C_{11}H_{13}BrN_2O_5$
(E)-5-(2-Bromovinyl)-2'-deoxyuridine.
Antiviral agent.

6584 Bropirimine
56741-95-8
$C_{10}H_8BrN_3O$
2-Amino-5-bromo-6-phenyl-4(3H)-pyrimidinone.
U-54461. Antiviral with antitumor activity. *Upjohn Ltd.*

6585 Buciclovir
86304-28-1
$C_9H_{13}N_5O_3$
(R)-9-(3,4-Dihydroxybutyl)guanine.
An antiviral agent.

6586 Celgosivir Hydrochloride
141117-12-6
$C_{12}H_{22}ClNO_5$
Octahydro-1,7,8-trihydroxy-6-indolizinyl butanoate hydrochloride.
Antiviral agent. *Hoechst Marion Roussel Inc.*

6587 Cicloxolone
52247-86-6
$C_{38}H_{56}O_7$
3β-Hydroxy-11-oxoolean-12-en-30-oic acid hydrogen cis-1,2-cyclohexanedicarboxylate.
Investigated for its antiviral activity.

6588 Cidofovir
149394-66-1
$C_8H_{14}N_3O_6P.2H_2O$
[[2-(4-Amino-2-oxo-1(2H)-pyrimidinyl)-1-(hydroxymethyl)ethoxy]methyl] phosphonic acid dihydrate, (S), dihydrate.
Antiviral agent. *Gilead Sciences Inc.*

6589 Cidofovir [anhydrous]
113852-37-2 2329
$C_8H_{14}N_3O_6P$
[[2-(4-Amino-2-oxo-1(2H)-pyrimidinyl)-1-(hydroxymethyl)ethoxy]methyl] phosphonic acid, (S).
(S)-HPMPC; GS-504; Vistide. Antiviral agent. DNA synthesis inhibitor. Active against cytomegalovirus and herpe simplex virus. mp = 260° (dec); $[\alpha]_D^{2\cdot} = -97.3°$ (H_2O, c = 0.8); [Monohydrate (pH 2)]: λ_m = 279 nm (ε 13000). *Gilead Sciences Inc.*

6590 Cipamfylline
132210-43-6
$C_{13}H_{17}N_5O_2$
8-Amino-1,3-bis(cyclopropylmethyl)-3,7-dihydro-1H-purine-2,6-dione.
Antiviral agent. *SmithKline Beecham Pharm.*

6591 Citenazone
21512-15-2
$C_7H_6N_4S_2$
5-Formyl-2-thiophenecarbonitrile thiosemicarbazone.
HOE-105. Antiviral developed as chemoprophylactic against smallpox, chicken pox. *Hoechst AG.*

6592 Cytarabine
147-94-4 2853 205-705-9
$C_9H_{13}N_3O_5$
4-Amino-1-β-D-arabinofuranosyl-2(1H)-pyrimidinone.
1-β-D-arabinofuranosylcytosine; β-cytosine arabino-side; CHX-3311; U-19920; Alexan; Arabitin; Aracytidine; Aracytine; Ara-C; Cytosar; Cytosar U; Erpalfa; Iretin; Udicil; NSC-287459. Antineoplastic agent; antiviral agent. A cytotoxic drug. mp = 212-213°; $[\alpha]_D^{23}$ = 158° (c = 0.5, H_2O); λ_m = 281, 212.5 nm (ε 13171, 10230, pH 2). *Ciba-Geigy Corp.; Pharmacia & Upjohn.*

6593 Cytarabine Hydrochloride
69-74-9 200-713-9
$C_9H_{13}N_3O_5$
4-Amino-1-β-D-arabinofuranosyl-2(1H)-pyrimidinone monohydrochloride.
NSC-63878. Antiviral agent. A cytotoxic drug. Also has antiviral activity. *Pharmacia & Upjohn.*

6594 Delavirdine
136817-59-9 2929
$C_{22}H_{28}N_6O_3S$
1-[3-[(1-Methylethyl)amino]-2-pyridinyl]-4-[5-[(methylsulfonyl)amino]-1H-indol-2-yl]carbonylpiperazine.
U-90152. Antiviral agent. mp = 226-228°.

6595 Delavirdine Mesylate
147221-93-0 2929
$C_{23}H_{32}N_6O_6S_2$
1-[3-[(1-Methylethyl)amino]-2-pyridinyl]-4-[5-[(methylsulfonyl)amino]-1H-indol-2-yl]carbonyl-piperazine monomethanesulfonate.
U-90152S; Rescriptor. Antiviral agent. *Bristol-Myers Squibb HIV Products.*

6596 Denotivir
51287-57-1
$C_{18}H_{14}ClN_3O_2S$
5-Benzamido-4'-chloro-3-methyl-4-isothiazole-carboxanilide.
Antiviral agent.

6597 Desciclovir
84408-37-7
$C_8H_{11}N_5O_2$
2-[(2-Amino-9H-purin-9-yl)methoxyethanol.
BW-A515U. Antiviral agent.

6598 Didanosine
69655-05-6 3148
$C_{19}H_{35}NO_2$
2',3'-Dideoxyinosine.
Videx. Antiviral agent. mp = 160-163°; λ_m = 248 nm (pH 2), 254 nm (pH 12). *Bristol-Myers Squibb HIV Products.*

6599 Disoxaril
87495-31-6
$C_{20}H_{26}N_2O_3$
5-[7-[4-(4,5-Dihydro-2-oxazolyl)phenoxy]heptyl]-3-methylisoxazole.
Win-51711. Antiviral agent. *Sterling Winthrop Inc.*

6600 Droxinavir
159910-86-8
$C_{29}H_{51}N_5O_4$
N-Methylglycyl-N-[3-[[[(1,1-dimethylethyl)amino]-carbonyl](3-methylbutyl)amino]-2-hydroxy-1-(phenylmethyl)propyl]-3-methyl-L-valinamide.
Antiviral agent.

6601 Droxinavir Hydrochloride
155662-50-3
$C_{29}H_{52}ClN_5O_4$
N-Methylglycyl-N-[3-[[[(1,1-dimethylethyl)amino]-carbonyl](3-methylbutyl)amino]-2-hydroxy-1-(phenylmethyl)propyl]-3-methyl-L-valinamide monohydrochloride.
Antiviral agent.

6602 Edoxudine
15176-29-1 3561 239-226-1
$C_{11}H_{16}N_2O_5$
2'-Deoxy-5-ethyluridine.
EDU; EUDR; RWJ-15817; ORF-15817; Aedurid; Edurid. Antiviral agent. Active against Herpes Simplex. mp = 152-153°; λ_m = 267 nm (ε 9610, pH 2), 267 nm (ε 7280 pH 1). *Ortho Pharm. Corp.*

6603 Enviradene
80883-55-2
$C_{19}H_{21}N_3O_2S$
1-[(1-Methylethyl)sulfonyl]-6-(1-phenyl-1-propenyl)-1H-benzimidazol-2-amine.
Antiviral agent. *Eli Lilly & Co.*

6604 Enviroxime
72301-79-2 3637
$C_{17}H_{18}N_4O_3S$
6-[(Hydroxyimino)phenylmethyl]-1-[(1-methylethyl)-sulfonyl]-1H-benzimidazol-2-amine.
LY-122772. Antiviral agent. Inhibitor of rhinovirus propagation. mp = 198-199°; λ_m = 254, 285 nm (ε 20800, 13200, MeOH). *Eli Lilly & Co.*

6605 Epervudine
60136-25-6
$C_{12}H_{18}N_2O_5$
2'-Deoxy-5-isopropylurisine.
Antiviral.

6606 Famciclovir
104227-87-4 3971
$C_{14}H_{19}N_5O_4$
2-[2-(2-Amino-9H-purin-9-yl)ethyl]-1,3-propanediol diacetate (ester).
Famvir; FCV; BRL-42810. Antiviral agent. mp = 102-104°; λ_m 222, 244, 309 nm (ε 27500, 4890, 7160 MeOH); soluble in H_2O (>25 g/100 ml), Me_2CO, MeOH; less soluble in EtOH, iPrOH. *SmithKline Beecham Pharm.*

6607 Famotine
18429-78-2
$C_8H_{15}N_7O_2S_3$
1-[(4-Chlorophenoxy)methyl]-3,4-dihydroisoquinoline.
Pepcid; Pepcid PM; Amfamox; Pepcidine; YM-11170; MK-208; Amfamox; Dispromil; Famodil; Famodine; Famosan; Famoxal; Fanosin; Fibonel; Ganor; Gaster;

Gastridin; Gastropen; Ifada; Lecedil; Motiax; Muclox; Nulcerin; Pepcidina; Pepdine; Pepdul; Peptan; Ulcetrax; Ulfamid; Ulfinol. Antiviral agent.

6608 Famotine Hydrochloride
10500-82-0
$C_{16}H_{15}Cl_2NO$
1-[(4-Chlorophenoxy)methyl]-3,4-dihydroisoquinoline monohydrochloride.
UK-2054. Antiviral agent.

6609 Felvizumab
167747-20-8
immunoglobulin G1 (human-mouse monoclonal), γ chain antirespiratory syncitial virus. Antiviral agent. Monoclonal antibody.

6610 Fiacitabine
69123-90-6
$C_9H_{11}FIN_3O_4$
4-Amino-1-(2-deoxy-2-fluoro-β-D-arabinofuranosyl)-5-iodo-2(1H)-pyrimidinone.
Antiviral agent. *Oclassen Pharm. Inc.*

6611 Fialuridine
69123-98-4 4114
$C_9H_{10}FIN_2O_5$
1-(2-Deoxy-2-fluoro-β-D-arabinofuranosyl)-5-iodo-2,4(1H,3H)-pyrimidinedione.
FIAU. Antiviral agent. Active against hepatitis B. mp = 216-217°.

6612 Floxuridine
50-91-9 4148
$C_9H_{11}FN_2O_5$
2'-Deoxy-5-fluorouridine.
FUDR; NSC-26740. Antiviral; antineoplastic agent. mp = 150-151°; λ_m = 268 nm (ε 7570, pH 7.2), 270 nm (ε 6480, pH 14); $[\alpha]_D$ = 37° (H_2O), 48.6° (DMF). *Hoffmann-LaRoche Inc.*

6613 Fomivirsen
144245-52-3
Antiviral agent. Antisense antiviral agent used in treatment of cytomegalovirus. *Isis Pharm. Inc.*

6614 Fomivirsen Sodium
160369-77-7
Antiviral agent. Antisense antiviral agent used in treatment of cytomegalovirus. *Isis Pharm. Inc.*

6615 Fosarilate
73514-87-1 277-523-8
$C_{17}H_{28}ClO_5P$
6-(2-Chloro-4-methoxyphenoxy)hexylphosphonic acid diethyl ester.
Antiviral agent. *Sterling Res. Labs.*

6616 Foscarnet Sodium
63585-09-1 4277
$C_{17}H_{28}ClO_5P$
Dihydroxyphosphinecarboxylic acid oxide trisodium salt.
trisodium phosphonoformate; Foscavir; Triapten; EHB-776. Antiviral agent. mp > 250°; LD_{50} (mus ip) = 384-768 mg/kg. *Astra USA Inc.; Lederle Labs.*

6617 Fosfonet Disodium [anhydrous]
36983-81-0 253-297-6
$C_2H_3Na_2O_5P$
Phosphonoacetic acid disodium salt.
Abbott 38642. Antiviral agent.

6618 Fosfonet Disodium Monohydrate
54870-27-8
$C_2H_3Na_2O_5P.H_2O$
Phosphonoacetic acid disodium salt monohydrate.
Abbott 38642. Antiviral agent.

6619 Fosfonoacetic Acid
4408-78-0 224-558-1
$C_2H_3Na_2O_5P$
Phosphonoacetic acid.
Antiviral agent. *Abbott Labs Inc.*

6620 Ganciclovir
82410-32-0 4374
$C_9H_{13}N_5O_4$
2-Amino-1,9-dihydro-9-[[2-hydroxy-1-(hydroxymethyl)ethoxy]methyl]-6H-purin-6-one.
Cytovene; RS-21592; BW-759U. Antiviral agent. mp = 248-249° (dec); λ_m = 254 nm (ε 12880 MeOH); soluble in H_2O (0.43 g/100 ml at pH 7); LD_{50} (mus ip) = 1000-2000 Mg/kg. *Syntex Intl. Ltd.*

6621 Ganciclovir Sodium
107910-75-8 4374
$C_9H_{12}N_5NaO_4$
2-Amino-1,9-dihydro-9-[[2-hydroxy-1-(hydroxymethyl)ethoxy]methyl]-6H-purin-6-one monosodium salt.
RS-21592 sodium. Antiviral agent. *Syntex Intl. Ltd.*

6622 Herpid
67-68-5 3308 200-664-3
C_2H_6OS
Dimethyl sulfoxide.
DMSO; SQ-9453; DMS-70; DMS-90; Demavet; Demeso; Dolicur; Domoso; Dromisol; Gamasol 90; Hyadur; Kemsol; Rimso-50; Sclerosol; Somipront; Syntexan. Antiviral agent. mp = 18.5°; bp = 189°; soluble in H_2O, organic solvents; LD_{50} (rat orl) = 19690 mg/kg.

6623 Ibacitabine
611-53-0 210-269-8
$C_9H_{12}IN_3O_4$
2'-Deoxy-5-iodocytidine.
Antiviral; antiherpes agent.

6624 Idoxuridine
54-42-2 4934 200-207-8
$C_9H_{11}IN_2O_5$
2'-Deoxy-5-iodouridine.
SK&F-14287; NSC-39661; 5IUDR; IDU; Allergan 211; Dendrid; Emanil; Herpe-Gel; Herplex; Idexur; Idoxene; Idulea; Iduridin; Kerecid; Ophthalmadine; Stoxil; Virudox. Antiviral agent. mp = 160° (dec), 190-195°, 240°; $[\alpha]_D^{25}$ = 7.4° (H_2O, c = 0.108); λ_m = 288 nm (log ε 3.87, H_2O); LD_{50} (mus ip) = 1800 mg/kg. *Allergan Inc.; SmithKline Beecham Pharm.*

6625 Indinavir
180683-37-8
$C_{36}H_{47}N_5O_4.H_2O$
2,3,5-Trideoxy-N-(2,3-dihydro-2-hydroxy-1H-inden-1-yl)-5-[2-[[(1,1-dimethylethyl)amino]carbonyl]-4-(3-pyridinylmethyl)-1-piperazinyl]-2-(phenylmethyl)-D-erythro-pentonamide monohydrate.
Antiviral agent. *Merck & Co.Inc.*

6626 Indinavir Sulfate
157810-81-6 4979
$C_{36}H_{47}N_5O_4.H_2O$
2,3,5-Trideoxy-N-(2,3-dihydro-2-hydroxy-1H-inden-1-yl)-5-[2-[[(1,1-dimethylethyl)amino]carbonyl]-4-(3-pyridinylmethyl)-1-piperazinyl]-2-(phenylmethyl)-D-erythropentonamide monohydrate sulfate (1:1) salt.
Crixivan; MK-639; L-735,524. Antiviral agent. mp = 153-154°, 167.5-168°; $[\alpha]_D^{22}$ = 24.1° ($CHCl_3$, c = 0.0133); soluble in H_2O (> 0.15 g/100 ml pH 4). *Merck & Co.Inc.*

6627 Inosine Pranobex
36703-88-5 5006 253-162-1
$C_{52}H_{78}N_{10}O_{17}$
Inosine-2-hydroxypropyldimethylammonium 4-acetamidobenzoate (1:3).
methisoprinol; NP-113; NPT-10381; Aviral; Delimmun; Imunoviral; Inosiplex; Isoprinosin; Isoprinosina; Isoprinosine; Isoviral; Methisoprinol; Modimmunal; Pranosina; Pranosine; Viruxan. Antiviral agent; immunomodulator. An immunostimulant. Soluble in H_2O; LD_{50} (mus,rat orl,ip) > 4000 mg/kg. *Newport.*

6628 Interferon α-2a
76543-88-9 5016
$C_{860}H_{1353}N_{227}O_{255}S_9$
Interferon alfa-2a.
interferon αA; IFN-αA; Ro-22-8181; Canferon; Roferon-A; Ro-22-8181; roferon A. Antineoplastic; antiviral; immunomodulator. Inhibits viral replication. Alpha interferon, natural (injectable form); used for the treatment of genital warts. *Hoffmann-LaRoche Inc.; Interferon Sciences Inc.*

6629 Interferon γ
9008-11-1 5018 232-710-3
Gamma interferon.
IFN-γ; immune IFN; ImIFN; type II interferon; S-6810 [as cys-tyr-cyc-interferon γ]; Actimmune [as cys-tyr-cyc-interferon γ]; Gammaferon [as cys-tyr-cyc-interferon γ]; Immuneron [as cys-tyr-cyc-interferon γ]; Polyferon [as cys-tyr-cyc-interferon γ]. Antiviral; antineoplastic; immunomodulator. Lymphokine produced by T-cells, structurally unrelated to interferon-α or -β (binds to a different cell surface receptor). Molecular weight 16,000-25,000 daltons.

6630 Kethoxal
27762-78-3 5310
$C_6H_{12}O_4$
1,1-Dihydroxy-3-ethoxy-2-butanone.
U-2032. Antiviral agent. bp = 145°; soluble in H_2O, C_6H_6, EtOH. *Upjohn Ltd.*

6631 Lamivudine
134678-17-4 5365
$C_8H_{11}N_3O_3S$
4-Amino-1-[2-(hydroxymethyl)-1,3-oxathiolan-5-yl]-2(1H)pyrimidinone, (2R, cis).
Epivir; GR-109714X. Antiviral agent. mp = 160-162°; $[\alpha]_D^{21}$ = -132° (MeOH, c = 1.08). *Glaxo Wellcome Inc.*

6632 Lobucavir
127759-89-1
$C_{11}H_{15}N_5O_3$
[1R-(1α,2β,3α)]-2-Amino-9-[2,3-bis(hydroxymethyl)-cyclobutyl]-1,9-dihydro-6H-purin-6-one.
BMS-180194; SQ-34514. Antiviral agent. *Bristol-Myers Squibb HIV Products.*

6633 Loviride
147362-57-0
$C_{17}H_{16}Cl_2N_2O_2$
(±)-2-(6-Acetyl-m-toluidino)-2-(2,6-dichlorophenyl)acetamide.
R-89439. Antiviral agent. *Janssen Pharm. Inc.*

6634 Lysozyme Chloride
9001-63-2 5671 232-620-4
N-Acetylmuramide glycanohydrolase hydrochloride.
muramidase hydrochloride; Acdeam; Antalzyme; Immunozima; Lanzyme; Leftose; Likinozym; Lisozima; Murazyme; Neutase; Neuzyme; Toyolysom-DS. Mucolytic enzyme, acts as an antiviral agent.

6635 Memotine
18429-69-1
$C_{17}H_{17}NO_2$
3,4-Dihydro-1-(4-methoxyphenoxy)methyl isoquinoline.
Antiviral agent. *Pfizer Inc.*

6636 Memotine Hydrochloride
10540-97-3
$C_{17}H_{18}ClNO_2$
3,4-Dihydro-1-(4-methoxyphenoxy)methyl isoquinoline dihydrochloride.
UK-2371. Antiviral agent. *Pfizer Inc.*

6637 Methisazone
1910-68-5 6057 217-616-2
$C_{10}H_{10}N_4OS$
2-(1,2-Dihydro-1-methyl-2-oxo-3H-indol-3-ylidene hydrazinecarbothioamide.
BW-33-T-57; NSC-69811; N-methylisatin 3-thiosemicarbazone; Marboran; Viruzona. Antiviral agent. mp = 245°.

6638 Moroxydine
3731-59-7 6354 223-093-1
$C_6H_{13}N_5O$
4-Morpholinecarboximidoylguanidine.
SKF-8898-A; ABOB; Virusmin. Antiviral agent.

6639 Nelfinavir
159989-64-7
$C_{32}H_{45}N_3O_4S$
N-(1,1-Dimethylethyl)decahydro-2-[2-hydroxy-3-[(3-hydroxy-2-methylbenzoyl)amino]-4-(phenylthio)butylisoquinoline carboxamide.
Antiviral agent.

6640 Nelfinavir Mesylate
159989-65-8
$C_{33}H_{49}N_3O_7S_2$
N-(1,1-Dimethylethyl)decahydro-2-[2-hydroxy-3-[(3-hydroxy-2-methylbenzoyl)amino]-4-(phenylthio)butyl-isoquinoline carboxamide monomethanesulfonate.
Antiviral agent.

6641 Netivudine
84558-93-0
$C_{12}H_{16}N_2O_6$
1-β-D-Arabinofuranosyl-5-(1-propynyl)uracil.
882C87. Antiviral agent.

6642 Nevirapine
129618-40-2 6573
$C_{15}H_{14}N_4O$
11-Cyclopropyl-5,11-dihydro-4-methyl- 6H-dipyrido-[3,2-b:2',3'-e][1,4]diazepin-6-one.
Viramune. Antiviral agent. Immunomodulator. Non-nucleoside reverse transcriptase inhibitor, specific to HIV-1. In combination with α-interferon, has been used to treat hepatitis C successfully. mp = 247-249°; soluble in H_2O at pH < 3, almost insoluble at pH 7. *Boehringer Ingelheim Pharm. Inc.*

6643 Oseltamivir
204255-11-8
$C_{16}H_{31}N_2PO_8$
Ethyl (3R,4R,5S)-4-acetamido-5-amino-3-(1-ethylprop-oxy)-1-cyclohexene-1-carboxylate phosphate (1:1).
Tamiflu; Ro 64-0796/002. Antiviral. For the prophylaxis of influenza virus. Neuraminidase inhibitor. Used to treat influenza. *Hoffmann-LaRoche Inc.*

6644 Palinavir
154612-39-2
$C_{41}H_{52}N_6O_5$
[2S-[1[1R*(R*),2S*]-2α,4α-N-[1-[[[3-[2-[[(1,1-Dimethyl-ethyl)amino]carbonyl]-4-(4-pyridinylmethoxy)-1-piperidinyl]-2-hydroxy-1-(phenylmethyl)propyl]-amino]carbonyl]-2-methylpropyl]-2-quinolinecarboxamide.
BILA 2011 BS. Antiviral agent. *Boehringer Ingelheim Pharm. Inc.*

6645 Penciclovir
39809-25-1 7210
$C_{10}H_{15}N_5O_3$
2-Amino-1,9-dihydro-9-[4-hydroxy-3-(hydroxymethyl)-butyl]-6H-purin-6-one.
BRL-39123. Antiviral agent. Carba analog of ganciclovir. Active against several herpes viruses. mp = 275-277°; λ_m = 253 nm (ε 11500, H_2O); soluble in H_2O (0.17 g/100 ml at 20°, pH 7). *SmithKline Beecham Pharm.*

6646 Pirodavir
124436-59-5
$C_{21}H_{27}N_3O_3$
4-[2-[1-(6-Methyl-3-pyridazinyl)-4-piperidinyl]ethoxy]-benzoic acid ethyl ester.
R-77975. Antiviral agent. *Janssen Pharm. Inc.*

6647 Podophyllotoxin
518-28-5 7704 208-250-4
$C_{22}H_{22}O_8$
[5R-(5α,5aβ,8aα,9α)]-5,8,8a,9-Tetrahydro-9-hydroxy-5-(3,4,5-trimethoxyphenyl)-furo[3',4':6,7]naphtho[2,3-d]-1,3-dioxol-6(5aH)-one.
podofilox; podophyllotoxin; NSC-24818; Bisoprolol; Condolyne. Antiviral agent. Precursor of antineoplastic agents etoposide and teniposide, found in rhizomes of *Podophyllum peltatum L. Podophyllaceae* of North America. mp = 114-118°, 183-184° (after drying); $[\alpha]_D^{20}$ = -132.7° ($CHCl_3$); soluble in H_2O (120 mg/l), more soluble in organic solvents; LD_{50} (rat iv) = 8.7 mg/kg, (rat ip) = 15 mg/kg. *E. Merck.*

6648 Raluridine
119644-22-3
$C_9H_{10}ClFN_2O_4$
5-Chloro-2',3'-dideoxy-3'-fluorouridine.
935U83. Antiviral agent. *Glaxo Wellcome Inc.*

6649 Ribavirin
36791-04-5 8365
$C_8H_{12}N_4O_5$
1-β-D-Ribofuranosyl-1H-1,2,4-triazole-3-carboxamide.
Virazole. Antiviral agent. A broad spectrum antiviral nucleoside. mp = 166-168°, 174-176°; $[\alpha]_D^{25}$ = -36.5° LD_{50} (mus ip) = 1300 mg/kg, (rat orl) = 5300 mg/kg. *ICN Pharm. Inc.*

6650 Rimantadine
13392-28-4 8390
$C_{12}H_{21}N$
α-Methyl-tricyclo[3.3.1.1^{3,7}]-decane-1-methanamine.
remantidin; remantidine. Antiviral agent. Derivative of adamantane.

6651 Rimantadine Hydrochloride
1501-84-4 8390
$C_{12}H_{22}ClN$
α-Methyl-tricyclo[3.3.1.1^{3,7}]-decane-1-methanamine hydrochloride.
remantadin(e) hydrochloride; EXP-126; Flumadine; Meradan; Meradane; Roflual. Antiviral agent. mp = 373-375°.

6652 Ritonavir
155213-67-5 8402
$C_{37}H_{48}N_6O_5S_2$
10-Hydroxy-2-methyl-5-(1-methylethyl)-1-[2-(1-methylethyl)-4-thiazolylmethyl ester of 2,4,7,12-tetraazatridecan-13-oic acid.
Norvir; A-84538; Abbott 84538; ABT-538. Antiviral agent. *Abbott Labs Inc.*

6653 Rociclovir
108436-80-2
$C_{15}H_{25}N_5O_3$
2-Amino-9-[[2-isopropoxy-1-(isopropoxymethyl)-ethoxy]methyl]purine.
Antiviral agent.

6654 Saquinavir
127779-20-8 8516
$C_{38}H_{50}N_6O_5$
N^1-[3-[3-[[(1,1-Dimethylethyl)amino]carbonyl]octahydro-2(1H)-isoquinolinyl]-2-hydroxy-1-(phenylmethyl)propyl]-2-[(2-quinolinylcarbonyl)amino]butanediamide.
Invirase. Antiviral agent. $[\alpha]_D^{20}$ = -55.9° (MeOH, c = 0.5); soluble in H_2O (0.22 g/100 ml at 21°). *Hoffmann-LaRoche Inc.*

6655 Saquinavir Mesylate
149845-06-7 8516
$C_{39}H_{54}N_6O_8S$
N^1-[3-[3-[[(1,1-Dimethylethyl)amino]carbonyl]octahydro-2(1H)-isoquinolinyl]-2-hydroxy-1-(phenylmethyl)propyl]-2-[(2-quinolinylcarbonyl)amino]butanediamide monomethanesulfonate.
Ro-31-8959/003; Invirase. Antiviral agent. *Hoffmann-LaRoche Inc.*

6656 Somantadine
79594-24-4
$C_{14}H_{25}N$
α,α-Dimethyl tricyclo[3.3.1.1^{3,7}]decan-1-amine.
Antiviral agent. *Fisons Pharm. Div.*

6657 Somantadine Hydrochloride
68693-30-1
$C_{14}H_{26}ClN$
α,α-Dimethyl tricyclo[3.3.1.1^{3,7}]decan-1-amine monohydrochloride.
Antiviral agent. *Fisons Pharm. Div.*

6658 Sorivudine
77181-69-2 8875
$C_{11}H_{13}BrN_2O_6$
1-β-D-Arabinofuranosyl-5-(2-bromoethenyl)-2,4(1H,3H)-pyrimidinedione.
5-bromovinyl-araU; Brovavir; Bravavir; BV-araU; BVAU; YN-72; SQ-32756; Usevir. Antiviral agent. mp = 182°, 195-200° (dec); $[\alpha]_D^{25}$ = 0.5° (1N NaOH); LD_{50} (mus ip) = 3300 mg/kg, (mus sc) > 5000 mg/kg, (mus orl) > 10,000 mg/kg. *Bristol-Myers Squibb HIV Products.*

6659 Statolon
11006-77-2 8957
Vistatolon; NSC-71901. Antiviral agent. Polysaccharide. Effective prophylactically against a wide range of viruses in animals. Produced by *Penicillium stoloniferum. Eli Lilly & Co.*

6660 Stavudine
3056-17-5 8958
$C_{10}H_{12}N_2O_4$
2',3'-Dihydro-3'-deoxythymidine.
d4T; BMY-27857; Zerit. Antiviral agent. mp = 165-166°; $[\alpha]_D^{25}$ = -39.4° (H_2O c = 0.701); λ_m = 266 nm (ε 10149 H_2O). *Bristol-Myers Squibb HIV Products.*

6661 Steffimycin
11033-34-4
U-20661. Antibiotic with antiviral activity, produced by *Streptomyces steffisburgensis. Upjohn Ltd.*

6662 Stoxil
54-42-2 4934 200-207-8
$C_9H_{11}IN_2O_5$
2'-Deoxy-5-iodouridine.
Herplex; Dendrid. Antiviral agent. mp = 160° (dec), 190-195°, 240°; $[\alpha]_D^{25}$ = 7.4° (H_2O, c = 0.108); λ_m = 288 nm (log ε 3.87, H_2O); LD_{50} (mus ip) = 1800 mg/kg.

6663 Telinavir
143224-34-4
$C_{33}H_{44}N_6O_5$
[1S-[1R*(R*),2S*]]-N^1-[3-[[[(1,1-Dimethylethyl)amino]-carbonyl](2-methylpropyl)amino]-2-hydroxy-1-(phenylmethyl)propyl]-2-[(2-quinolinylcarbonyl)-amino]butanediamide.
SC-52151. Antiviral agent. *Searle G.D. & Co.*

6664 Tilorone
27591-97-5 9581
$C_{25}H_{34}N_2O_3$
2,7-Bis[2-(diethylamino)ethoxy]-9H-fluoren-9-one.
bis-DEAE-fluorenone. Antiviral. *Marion Merrell Dow Inc.*

6665 Tilorone Hydrochloride
27591-69-1 9581
$C_{25}H_{36}Cl_2N_2O_3$
2,7-Bis[2-(diethylamino)ethoxy]-9H-fluoren-9-one dihydrochloride.
NSC-143969. Antiviral agent. mp = 235-237°; λ_m = 269 nm ($E_{1\,cm}^{1\%}$ = 1600, H_2O); LD_{50} (mus orl) = 959 mg/kg, (mus ip) = 145 mg/kg, (rat orl) = 852 mg/kg, (rat ip) = 244 mg/kg.

6666 Tivirapine
137332-54-8
$C_{16}H_{20}ClN_3S$
(S)-8-Chloro-4,5,6,7-tetrahydro-5-methyl-6-(3-methyl-2-butenyl)imidazo[4,5,1-jk][1,4]benzodiazepine-2(1H)-thione.
TIBO-R-86183. Non-nucleoside reverse transcriptase inhibitor.

6667 Trecovirsen Sodium
170274-79-0
$C_{16}H_{15}ClN_2OS$
Gem91. Antiviral agent. *Hybridon Inc.*

6668 Tromantadine
53783-83-8 9901 258-770-0
$C_{16}H_{28}N_2O_2$
N-1-Adamantyl-2-[2-(dimethylamino)ehoxy]acetamide.
Antiviral agent.

6669 Trovirdine
149488-17-5
$C_{13}H_{13}BrN_4S$
1-(5-Bromo-2-pyridyl)-3-[2-(2-pyridyl)ethyl]-2-thiourea.
LY-300046-HCl. HIV reverse transcriptase non-nucleoside inhibitor.

6670 Valacyclovir
124832-26-4 10039
$C_{17}H_{21}NO_3$
2-[(2-Amino-1,6-dihydro-6-oxo-9H-purin-9-yl)methoxy ethyl ester of L-valine.

256U87; valACV; valaciclovir. Antiviral agent. Prodrug of acyclovir. *Glaxo Wellcome Inc.*

6671 Valacyclovir Hydrochloride
124832-27-5 10039
$C_{17}H_{22}ClNO_3$
2-[(2-Amino-1,6-dihydro-6-oxo-9H-purin-9-yl)methoxy ethyl ester of L-valine monohydrochloride.
256U87 hydrochloride; 256U; BW-256U87; BW-256; Valtrex. Antiviral agent. λ_m = 252.8 nm (ε 8250, H_2O); soluble in H_2O (17.4 g/100 ml). *Glaxo Wellcome Inc.*

6672 Vidarabine
24356-66-9
$C_{10}H_{13}N_5O_4.H_2O$
9-β-D-Arabinofuranosyl-9H-purin-6-amine monohydrate.
Vira-A. Antiviral agent. *Parke-Davis.*

6673 Vidarabine Anhydrous
5536-17-4 10113 226-893-9
$C_{10}H_{14}N_5O_7P$
9-β-D-Arabinofuranosyl-9H-purin-6-amine.
arabinosyladenine; adenine arabinoside; spongoadenosine; ara-A; CI-673; Arasena-A; Vira-A. Antiviral agent. mp = 257-257.5°; $[\alpha]_D^{27}$ = -0.5° (c = 0.25); λ_m = 259 nm (ε 13400 pH 7); LD_{50} (mus ip) = 4677 mg/kg, (mus orl) > 7950 mg/kg. *Parke-Davis.*

6674 Vidarabine Phosphate
29984-33-6 249-990-8
$C_{10}H_{13}N_5O_4$
9-β-D-Arabinofuranosyl-9H-purin-6-amine phosphate.
CI-808. Antiviral agent. *Parke-Davis.*

6675 Vidarabine Sodium Phosphate
71002-10-3
$C_{10}H_{13}N_5O_4$
9-β-D-Arabinofuranosyl-9H-purin-6-amine phosphate disodium salt.
CI-808 sodium. Antiviral agent. *Parke-Davis.*

6676 Xenazoic Acid
1174-11-4 10204
$C_{23}H_{21}NO_4$
p-[(α-Ethoxy-p-phenylphenacyl)amino]benzoic acid.
CV-58903; SKF-8318; xanalamine; CV-58903; SKF-8318; Xenovis. Antiviral agent. mp = 192° (dec).

6677 Zalcitabine
7481-89-2 10242
$C_{19}H_{19}NOS$
2',3'-Dideoxycytidine.
Hivid; NSC-606170. Antiviral. mp = 215-217°; $[\alpha]_D^{25}$ = 81° (H_2O, c = 0.635); λ_m = 280 nm (ε 17720 0.1N HCl), 270 nm (ε 8410 0.1NNaOH). *Hoffmann-LaRoche Inc.*

6678 Zanamivir
139110-80-8
$C_{12}H_{20}N_4O_7$
5-(Acetylamino)-4-[(aminoiminomethyl)amino]-2,6-anhydro-3,4,5-trideoxy-D-glycero-D-galacto-non-2-enonic acid.
GR-121167X; Relenza. A neuraminidase inhibitor. Used to treat influenza. Antiviral agent. *Glaxo Wellcome plc.*

6679 Zidovudine
30516-87-1 10252
$C_{10}H_{13}N_5O_4$
3'-Azido-3'-deoxythymidine.
AZT; BW-A509U; Compound S; 3'-azidothymidine; Zidovudine; Retrovir; Azidothymidine; ZVD; ZDV; 3-azido-3-deoxythymidine. Antiviral agent. Pyrimidine nucleoside analog. Reverse transcriptase inhibitor. mp = 106-112°; soluble in H_2O (1-5 g/100 ml at 17°); more soluble in organic solvents; λ_m = 266.5 nm (ε 11650); LD_{50} = (mmus orl) = 3568 mg/kg; (fmus orl) = 3084 mg/kg; (mrat orl) = 3084 mg/kg; (frat orl) 3683 mg/kg; (all species iv) > 750 mg/kg. *Glaxo Wellcome Inc.*

6680 Zinviroxime
72301-78-1
$C_{17}H_{18}N_4O_3S$
6-[(Hydroxyimino)phenylmethyl]-1-[(1-methylethyl)-sulfonyl]-1H-benzimidazol-2-amine.
Viroxime. Antiviral agent. *Eli Lilly & Co.*

Anxiolytics

6681 Abecarnil
111841-85-1 2
$C_{24}H_{24}N_2O_4$
Isopropyl-6-(benzyloxy)-4-(methoxymethyl)-9H-pyrido[3,4-b]indole-3-carboxylate.
ZK-112119. Anxiolytic. mp = 150-151°. *Schering AG.*

6682 Acepromazine
61-00-7 32 200-496-0
$C_{19}H_{22}N_2OS$
10-[3-(Dimethylamino)propyl]-10H-phenothiazin-2-yl methyl ketone.
acetazine; acetopromazine; acetylpromazine; 1522-CB; Vetranquil. Sedative, used as a veterinary tranquillizer. $bp_{0.5}$ = 220-240°.

6683 Acepromazine Maleate
3598-37-6 32 222-748-9
$C_{23}H_{26}N_2O_5S$
10-[3-(Dimethylamino)propyl]-10H-phenothiazin-2-yl methyl ketone maleate.
Atravet; Calmivet; Notensil; Plegicil; Sedalin; Soprontin. Sedative, used as a veterinary tranquillizer. mp = 135-136°; soluble in H_2O; LD_{50} (rat orl) = 130 mg/kg), (rat iv) = 70 mg/kg (calculated as free base).

6684 Aceprometazine
13461-01-3 236-661-9
$C_{19}H_{22}N_2OS$
10-[2-(Dimethylamino)propyl]phenothiazin-2-yl methyl ketone.
Anxiolytic.

6685 Adatanserin
127266-56-2
$C_{21}H_{31}N_5O$
N-[2-[4-(2-Pyrimidinyl)-1-piperazinyl]ethyl]-1-adamantanecarboxamide.

WY-50324. Antidepressant with anxiolytic properties. Adamantyl heteroarylpiperazine with dual serotonin 5-HT(1A) and 5-HT(2) activity. *Wyeth-Ayerst Labs.*

6686 Adatanserin Hydrochloride
144966-96-1
$C_{21}H_{32}ClN_5O$
N-[2-[4-(2-Pyrimidinyl)-1-piperazinyl]ethyl]-1-adamant-anecarboxamide monohydrochloride.
WY-50324 HCl. Anxiolytic; antidepressant. Adamantyl heteroarylpiperazine with dual serotonin 5-HT(1A) and 5-HT(2) activity. *Wyeth-Ayerst Labs.*

6687 Adinazolam
37115-32-5 159
$C_{19}H_{18}ClN_5$
8-Chloro-1-[(dimethylamino)methyl]-6-phenyl-4H-s-tri-azolo[4,3-a][1,4]benzo-diazepine.
U-41123. Antidepressant with anxiolytic properties. Sedative. A triazolobenzodiazepine. Dimethylamino derivative of alprazolam. mp = 171-172.5°. *Ciba-Geigy Corp.; Pratt Pharm.; Upjohn Ltd.*

6688 Adinazolam Mesylate
57938-82-6 159
$C_{20}H_{22}ClN_5O_3S$
8-Chloro-1-[(dimethylamino)methyl]-6-phenyl-4H-s-triazolo[4,3-a][1,4]benzodiazepine monomethanesulfonate.
U-41123F; adinazolam monomethanesulfonate; Deracyn. Antidepressant with anxiolytic properties. Sedative/hypnotic. Tricyclic. mp = 230-244°. *Ciba-Geigy Corp.; Pharmacia & Upjohn; Upjohn Ltd.*

6689 Alpidem
82626-01-5 318
$C_{21}H_{23}Cl_2N_3O$
6-Chloro-2-(p-chlorophenyl)-N,N-dipropylimidazo-[1.2-a]pyridine-3-acetamide.
SL 80.0342-00. Anxiolytic. mp = 140-141°. *Synthelabo Pharmacie.*

6690 Alprazolam
28981-97-7 320 249-349-2
$C_{17}H_{13}ClN_4$
8-Chloro-1-methyl-6-phenyl-4H-s-triazolo[4,3-a][1,4]-benzodiazepine.
D-65MT; U-31889; Alplax; Tafil; Tranquinal; Tranquinal; Trankimazin; Xanax; Xanor. Anxiolytic. Sedative/hypnotic. A benzodiazepine. mp = 228-228.5°; λ_m = 222 nm (ε 40250 EtOH); soluble in EtOH, insoluble in H_2O; LD_{50} (mus orl) = 1020 mg/kg, (mus ip) = 540 mg/kg, (rat orl) > 2000 mg/kg, (rat ip) = 610 mg/kg. *Pharmacia & Upjohn.*

6691 Amperozide
75558-90-6 619
$C_{23}H_{29}F_2N_3O$
4-[4,4-Bis(p-fluorophenyl)butyl]-N-ethyl-1-piperazinecarboxamide.
FG-5606; Hogpax. A piperazine derivative with effects on stress-induced disorders. Used in veterinary medicine as an antiaggressive.

6692 Amperozide Hydrochloride
75529-73-6 619
$C_{23}H_{30}ClF_2N_3O$
4-[4,4-Bis(p-fluorophenyl)butyl]-N-ethyl-1-piperaiznecarboxamide hydrochloride.
Used in veterinary medicine as an antiaggressive. mp = 177-178°.

6693 Avizafone
65617-86-9
$C_{22}H_{27}ClN_4O_3$
2'-Benzoyl-4'-chloro-2-[(S)-2,6-diaminohexanamido]-N-methylacetanilide.
Ro-03-7355/000; Pro-diazepam. Anticonvulsant; anxiolytic. *Hoffmann-LaRoche Inc.*

6694 Azacyclonol
115-46-8 925 204-092-5
$C_{18}H_{21}NO$
α,α-Diphenyl-4-piperidinemethanol.
Diphenyl-4-piperidylmethanol; MER-17; Psychosan; Frenoton; Ataractan; Calmeran. Anxiolytic. mp = 160-161°; LD_{50} (mus orl) = 650 mg/kg. *Marion Merrell Dow Inc.*

6695 Azacyclonol Hydrochloride
1798-50-1 925 217-284-9
$C_{18}H_{22}ClNO$
α,α-Diphenyl-4-piperidinemethanol hydrochloride.
Frenquel. Anxiolytic. mp = 283-285°; soluble in H_2O. *Marion Merrell Dow Inc.*

6696 Azapetine
146-36-1 205-667-3
1-Allyl-2,7-dihydro-3,4:5,6-dibenzazepine.
Ilidar [as phosphate]. Anxiolytic. *Sonus Pharm. Inc.*

6697 Benzoctamine
17243-39-9 1117
$C_{18}H_{19}N$
N-Methyl-9,10-ethanoanthracene-9(10H)-methylamine.
Skeletal muscle relaxant; sedative/hypnotic; anxiolytic. *Ciba-Geigy Corp.*

6698 Benzoctamine Hydrochloride
10085-81-1 1117 233-216-0
$C_{18}H_{20}ClN$
N-Methyl-9,10-ethanoantrhacene-9(10H)-methylamine hydrochloride.
Ba-30803; Tacitin. Skeletal muscle relaxant; sedative/hypnotic; anxiolytic. mp = 320-322°; LD_{50} (rat orl) = 700 ± 170 mg/kg. *Ciba-Geigy Corp.*

6699 Binospirone
102908-59-8
$C_{20}H_{26}N_2O_4$
(±)-N-[2-[(1,4-Benzodioxan-2-ylmethyl)amino]ethyl]-1,1-cyclopentanediacetimide.
Anxiolytic; skeletal muscle relaxant. *Marion Merrell Dow Inc.*

6700 Binospirone Mesylate
124756-23-6
$C_{21}H_{30}N_2O_7S$
(±)-N-[2-[(1,4-Benzodioxan-2-ylmethyl)amino]ethyl]-1,1-cyclopentanediacetimide monomethanesulfonate.

MDL-73005EF. Anxiolytic; skeletal muscle relaxant. *Marion Merrell Dow Inc.*

6701 Bretazenil
84379-13-5
$C_{19}H_{20}BrN_3O_3$
tert-Butyl (S)-8-bromo-11,12,13,13a-tetrahydro-9-oxo-9H-imidazo[1.5-a]pyrrolo[2,1-c][1,4]-benzodiazepine-1-carboxylate.
Ro-16-6028/000. Anxiolytic. *Hoffmann-LaRoche Inc.*

6702 Bromazepam
1812-30-2 1406 217-322-4
$C_{14}H_{10}BrN_3O$
7-Bromo-1,3-dihydro-5-(2-pyridyl)-2H-1,4-benzodiazepin-2-one.
7-Bromo-5-(2-pyridyl)-3H-1,4-benzodiazepin-2(1H)-one; Ro-5-3350; Compendium; Creosedin; Durazanil; Lexomil; Lexotan; Lexotanil; Normoc. Anxiolytic, used as a minor tranquillizer. A benzodiazepine with anxiolytic properties. A Federally controlled substance (depressant). mp = 237-238°; LD_{50} (rat orl) = 3050±405 mg/kg. *Roche Products Ltd.; Roche Puerto Rico.*

6703 Buspirone
36505-84-7 1528 253-072-2
$C_{21}H_{31}N_5O_2$
N-[4-[4-(2-Pyrimidinyl)-1-piperazinyl]butyl]-1,1-cyclopentanediacetamide hydrochloride.
Serotonin receptor agonist. Non benzodiazepine anxiolytic, used as a minor tranquillizer. An azapirone. *Mead Johnson Pharmaceuticals.*

6704 Buspirone Hydrochloride
33386-08-2 1528 251-489-4
$C_{21}H_{32}ClN_5O_2$
N-[4-[4-(2-Pyrimidinyl)-1-piperazinyl]butyl]-1,1-cyclopentanediacetamide hydrochloride.
Ansial; Ansiced; Axoren; Bespar; Buspar; Buspimem; Buspinol; Buspisal; Censpar; Lucelan; Narol; Travin. Non-benzodiazepine anxiolytic, used as a minor tranquillizer. An azapirone. Serotonin receptor agonist. mp = 201.5-202.5°; LD_{50} (rat ipr) = 136 mg/kg. *Mead Johnson Pharmaceuticals.*

6705 Camazepam
36104-80-0 1773 252-866-6
$C_{19}H_{18}ClN_3O_3$
7-Chloro-1,3-dihydro-3-hydroxy-1-methyl-5-phenyl-2H-1,4-benzodiazepin-2-one dimethylcarbamate (ester).
SB-5833; Albego. Anxiolytic. mp = 173-174°; soluble in EtOH, less soluble in H_2O; LD_{50} (mus orl) = 970 mg/kg, (rat orl) > 4000 mg/kg.

6706 Canbisol
56689-43-1
$C_{24}H_{38}O_3$
(±)-3-(1,1-Dimethylheptyl)-6aβ,7,8,9,10,10aα-hexahydro-6,6-dimethyl-6H-dibenzo[b,d]pyran-1,9-diol.
A cannabinoid-related anxiolytic.

6707 Captodiamine
486-17-9 1816 207-629-1
$C_{21}H_{29}NS_2$
2-[(p-Butylthio)-α-phenylbenzylthio]-N,N-dimethylethylamine.
p-butylthiodiphenlmethyl 2-dimethlaminoethyl sulfide; captodiame; captodiam; captodramin. Anxiolytic.

6708 Captodiamine Hydrochloride
904-04-1 1816 212-992-4
$C_{21}H_{30}ClNS_2$
2-[(p-Butylthio)-α-phenylbenzylthio]-N,N-dimethylethylamine hydrochloride.
captodiame hydrochloride; Covatine; Covatix; Suvren. Anxiolytic. mp = 131-132°; LD_{50} (mus ip 96 hr) = 180 mg/kg, (mus orl 96 hr) = 1630 mg/kg, (rat ip 96 hr) = 343 mg/kg, (rat orl 96 hr) = 3800 mg/kg.

6709 Carburazepam
59009-93-7 261-555-4
$C_{17}H_{16}ClN_3O_2$
7-Chloro-1,2,3,5-tetrahydro-1-methyl-2-oxo-5-phenyl-4H-1,4-benzodiazepine-4-carboxamide.
(S)-uxepam. An anxiolytic structurally related to chlordiazepoxide.

6710 Chlordiazepoxide
58-25-3 2132 200-371-0
$C_{16}H_{14}ClN_3O$
7-Chloro-2-(methylamino)-5-phenyl-3H-1,4-benzodiazepin 4-oxide.
Abboxide; Librelease; Libritabs; Limbitrol; Menrium; methaminodiazepoxide; Clopoxide; Helogaphen; Multum; Risolid; Silibrin; Tropium. Anxiolytic. Sedative/hypnotic. A benzodiazepine. mp = 236-236.5°. *Abbott Labs Inc.; Hoffmann-LaRoche Inc.; Roche Puerto Rico.*

6711 Chlordiazepoxide Hydrochloride
438-41-5 2132 207-117-8
$C_{16}H_{15}Cl_2N_3O$
7-Chloro-2-(methylamino)-5-phenyl-3H-1,4-benzodiazepin 4-oxide hydrochloride.
A-Poxide; Librium; SK-Lygen; Librax; Ro-5-0690; NSC-115748; Ansiacal; Balance; Benzodiapin; Cebrum; Corax; Disarim; Elenium; Equibral; Labican; Lentotran; Librium; O.C.M.; Psichial; Psicoterina; Reliberan; Seren Vita; Viansin. Anxiolytic. Sedative/hypnotic. A benzodiazepine. mp = 213°; soluble in H_2O (0.15 g/100 ml). *Abbott Labs Inc.; Hoffmann-LaRoche Inc.; Parke-Davis; Roche Puerto Rico; SmithKline Beecham Pharm.*

6712 Chlormezanone
80-77-3 2155 201-307-4
$C_{11}H_{12}ClNO_3S$
2-(p-Chlorophenyl)tetrahydro-3-methyl-4H-1,3-thiazin-4-one 1,1-dioxide.
dichloromezanone; chlormethazanone; Alinam; Banabin-Sintyal; Fenarol; Lobak; Mio-Sed; Rexan; Rilansyl; Rilaquil; Rilasol; Supotran; Suprotran; Tanafol; Trancote; Trancopal; Transanate. Anxiolytic; skeletal muscle relaxant. mp = 116.2-118.2°; soluble in H_2O (< 0.25 g/100 ml), EtOH (< 1 g/100 ml). *Sterling Winthrop Inc.*

6713 Chlorpromazine Hydrochloride
69-09-0 2238 200-701-3
$C_{17}H_{20}Cl_2N_2S$
2-Chloro-10-[3-(dimethylamino)propyl]phenothiazine hydrochloride.
clorazepate potassium; dipotassium clorazepate; Chloractil; Chlorazin; Gen-xene; Hebanil; Hibanil;

Hibernal; Klorpromex; Largactil; Largaktyl; Marazine; Megaphen; Novo-Clopate; Nu-Clopate; Promacid; Promapar; Propaphenin; Sonazine; Taroctyl; Thorazine; Torazina; Tranzene. Antiemetic; antipsychotic. Used in veterinary medicine as a tranquilizer. A phenothiazine. Dec 179-180°; pH (5% aq) = 4.0-5.5; soluble in H_2O (40 g/100ml), MeOH, EtOH, $CHCl_3$; insoluble in C_6H_6, Et_2O; LD_{50} (rat orl) = 2225 mg/kg. *DDSA Pharm. Ltd.; Elkins-Sinn; KV Pharm.; Labaz S.A.; Parke-Davis; Pharmacia & Upjohn; Rhône-Poulenc Rorer Pharm. Inc.; SmithKline Beecham Pharm.*

6714 Ciclotizolam
58765-21-2
$C_{20}H_{18}BrClN_4S$
2-Bromo-4-(o-chlorophenyl)-9-cyclohexyl-6H-thieno[3,2-f]-s-triazolo[4,3-a][1,4]diazepine.
We-973-BS. Anxiolytic.

6715 Clazolam
7492-29-7
$C_{18}H_{17}ClN_2O$
2-Chloro-5,9,10,14b-tetrahydro-5-methylisoquino[2,1-d][1,4]-benzodiazepin-6(7H)-one.
41-123. Anxiolytic; anticonvulsant. Used as a minor tranquillizer. *Sandoz Pharm. Corp.*

6716 Clobazam
22316-47-8 2417 244-908-7
$C_{16}H_{13}ClN_2O_2$
7-Chloro-1-methyl-5-phenyl-1H-1,5-benzodiazepine-2,4(3H,5H)-dione.
Urbanyl; HR-376; H-4723; LM-2717; Frisium; Urbadan. Anxiolytic; anticonvulsant. Used as a minor tranquillizer. mp = 166 -168°. *Hoechst Roussel Pharm. Inc.*

6717 Clorazepate Dipotassium
57109-90-7 2465 260-565-6
$C_{16}H_{11}ClK_2N_2O_4$
Potassium 7-chloro2,3-dihydro-2-oxo-5-phenyl-1H-1,4-benzodiazepine-3-carboxylate compound with KOH (1:1).
Tranxene; 4306 CB; Abbott 35616; Belseren; Mendon; Tranxilene; Tranxilium; Transene. Anxiolytic; anticonvulsant. Used as a minor tranquillizer. Poorly soluble in EtOH; insoluble in Et_2O, $CHCl_3$; λ_m = 231, 311 nm (ε 33500, 2450 H_2O); LD_{50} (mus orl) = 700 mg/kg, (mus ip) = 290 mg/kg, (rat orl) > 1000 mg/kg. *Abbott Labs Inc.; Clin-Byla France.*

6718 Clorazepate Monopotassium
5991-71-9 2465 227-817-7
$C_{16}H_{10}ClKN_2O_3$
Potassium 7-chloro2,3-dihydro-2-oxo-5-phenyl-1H-1,4-benzodiazepine-3-carboxylate.
4311 CB; Abbott 39083; Azene. Anxiolytic. Used as a minor tranquillizer. *Abbott Labs Inc.; Clin-Midy.*

6719 Clorazepic Acid
20432-69-3 2465
$C_{16}H_{11}ClN_2O_3$
7-Chloro-2,3-dihydro-2-oxo-5-phenyl-1H-1,4-benzodiazepine-3-carboxylic acid.
Anxiolytic. Used as a minor tranquillizer. *Abbott Labs Inc.; Clin-Midy.*

6720 Clotiazepam
33671-46-4 2477 251-627-3
$C_{16}H_{15}ClN_2OS$
5-(o-Chlorophenyl)-7-ethyl-1,3-dihydro-1-methyl-2H-thieno[2,3-e]-1,4-diazepin-2-one.
Y-6047; Clozan; Rise; Rize; Rizen; Tienor; Trecalmo; Veratran. A thienodiazepine anxiolytic. mp = 105-106°; LD_{50} (mus orl) = 636 mg/kg, (mus ip) = 440 mg/kg.

6721 Cloxazolam
24166-13-0 2481
$C_{17}H_{14}Cl_2N_2O_2$
10-Chloro-11b-(o-chlorophenyl)-2,3,7,11b-tetrahydro-oxazolo[3,2-d][1,4]benzodiazepin-6(5H)-one.
CS-370; Betavel; Enadel; Lubalix; Olcadil; Sepazon; Tolestan. Anxiolytic. Federally controlled substance (depressant). mp = 202-204°; soluble in glacial AcOH; sparingly soluble in $CHCl_3$; slightly soluble in Me_2CO, dehydrated EtOH, EtOAc, C_6H_6; insoluble in H_2O; LD_{50} (mus, orl) = 3300 mg/kg, (mus ip) > 2000 mg/kg.

6722 Cyclarbamate
5779-54-4 2772 227-302-7
$C_{21}H_{24}N_2O_4$
1,1-Cyclopentanedimethanol dicarbanilate.
cyclopentaphene; C-1428; BSM-906M; Calmalone; Casmalon. Anxiolytic; skeletal muscle relaxant. mp = 151-152°; insoluble in H_2O; slightly soluble in EtOH, glycerol, propylene glycol. *Lab Cassenne Marion.*

6723 Cyclobenzaprine
303-53-7 2782 206-145-8
$C_{20}H_{21}N$
N,N-Dimethyl-5H-dibenzo[a,d]cycloheptene-$\Delta^{5,\sigma}$-propylamine.
MK-130; Ro-4-1577; RP-9715. Anxiolytic; skeletal muscle relaxant. bp_1 = 175-180°; λ_m = 224, 289 nm (log ε = 4.57, 4.02). *Apothecon; Hoffmann-LaRoche Inc.; Merck & Co.Inc.*

6724 Cyclobenzaprine Hydrochloride
6202-23-9 2782 228-264-4
$C_{20}H_{22}ClN$
N,N-Dimethyl-5H-dibenzo[a,d]cycloheptene-$\Delta^{5,\sigma}$-propylamine hydrochloride.
Flexeril; Flexiban. Anxiolytic; skeletal muscle relaxant. mp = 216-218°; soluble in H_2O (> 20 g/100ml); soluble in MeOH, EtOH; less soluble in iPrOH, $CHCl_3$, CH_2Cl_2; insoluble in hydrocarbons; λ_m = 226, 295 nm (ε 52300, 12000); LD_{50} (mus iv) = 35 mg/kg, (mus orl) = 250 mg/kg. *Apothecon; Hoffmann-LaRoche Inc.; Merck & Co.Inc.*

6725 Delorazepam
2894-67-9
$C_{15}H_{10}Cl_2N_2O$
7-Chloro-5-(o-chlorophenyl)-1,3-dihydro-2H-1,4-benzodiazepin-2-one.
Anxiolytic.

6726 Demoxepam
963-39-3 213-515-2
$C_{15}H_{11}ClN_2O_2$
7-Chloro-1,3-dihydro-5-phenyl-2H-1,4-benzodiazepin-2-one 4-oxide.

Ro-5-2092; NSC-46007. Anxiolytic, used as a minor tranquillizer. *Hoffmann-LaRoche Inc.*

6727 Deramciclane
120444-71-5
$C_{20}H_{31}NO$
N,N-Dimethyl-2-[[(1R,2S,4R)-2-phenyl-2-bornyl]oxy]ethylamine.
Anxiolytic.

6728 Dexmedetomidine
113775-47-6 5830
$C_{13}H_{16}N_2$
(+)-4-[(S)-α,2,3-Trimethylbenzyl]imidazole.
MPV-1440; dexmedetomidine; (S)-medetomidine. Anxiolytic with sedative; analgesic properties. *Farmos Group Ltd.*

6729 Diazepam
439-14-5 3042 207-122-5
$C_{16}H_{13}ClN_2O$
7-Chloro-1,3-dihydro-1-methyl-5-phenyl-2H-1,4-benzodiazepin-2-one.
Alupram; Valium; Valrelease; LA-111; Ro- 5-2807; Wy-3467; NSC-77518; Apaurin; Atensine; Atilen; Bialzepam; Calmpose; Ceregulart; Dialar; Diazemuls; Dipam; Eridan; Eurosan; Evacalm; Faustan; Gewacalm; Horizon; Lamra; Lembrol; Levium; Mandrozep; Neurolytril; Noan; Novazam; Paceum; Pacitran; Paxate; Paxel; Pro-Pam; Q-Pam; Relanium; Sedapam; Seduxen; Servizepam; Setonil; Solis; Stesolid; Tranquase; Tranquo-Puren; Tranquo-Tablinen; Unisedil; Valaxona; Valiquid; Valium; Valium Injectable; Valrelease; Vival; Vivol. Anxiolytic, skeletal muscle relaxant and sedative/hypnotic. Also used as an intravenous anesthetic. Pharmaceutical preparation for the treatment of depression. A benzodiazepine. mp = 125-126°; soluble in DMF, $CHCl_3$, C_6H_6; Me_2CO, EtOH; slightly soluble in H_2O; LD_{50} (rat orl) = 710 mg/kg. *Berk Pharm. Ltd.; Hoffmann-LaRoche Inc.*

6730 Divaplon
90808-12-1
$C_{17}H_{17}N_3O_2$
6-Ethyl-7-methoxy-5-methylimidazo[1,2-a]pyrimidin-2-yl phenyl ketone.
GABA receptor agonist. Anxiolytic.

6731 Emylcamate
78-28-4 3604 201-101-4
$C_7H_{15}NO_2$
3-Methyl-3-pentanol carbamate.
methyl diethyl carbinol urethan; tert-hexanol; KABI 925; MK-250; JD-91; Nuncital; Restetal; Statran; Striatran. Anxiolytic. Antibacterial. mp = 56 - 58.5°; $bp_{1.0}$ = 35°; soluble in H_2O (0.4 g/100 ml); freely soluble in EtOH, Et_2O, C_6H_6, glycol ethers. *E. Merck; Kabi Pharmacia Diagnostics.*

6732 Enciprazine
68576-86-3 3610
$C_{23}H_{32}N_2O_6$
(±)-4-(o-Methoxyphenyl)-α-[(3,4,5-trimethoxyphenoxy)methyl]-1-piperazineethanol.
Anxiolytic, used as a minor tranquillizer. *C.M. Ind.; Wyeth-Ayerst Labs.*

6733 Enciprazine Hydrochloride
68576-88-5 3610 271-509-5
$C_{23}H_{34}Cl_2N_2O_6$
(±)-4-(o-Methoxyphenyl)-α-[(3,4,5-trimethoxy-phenoxy)methyl]-1-piperazineethanol dihydrochloride.
WY-48624; D-13112. Anxiolytic, used as a minor tranquillizer. mp = 196-197°. *C.M. Ind.; Wyeth-Ayerst Labs.*

6734 Enpiprazole
31729-24-5
$C_{16}H_{21}ClN_4$
1-(o-Chlorophenyl)-4-[2-(1-methylpyrazol-4-yl)-ethyl]piperazine.
Anxiolytic.

6735 Ethyl Loflazepate
29177-84-2 3867 249-489-4
$C_{18}H_{14}ClFN_2O_3$
Ethyl 7-chloro-5-(o-fluorophenyl)-2,3-dihydro-2-oxo-1H-1,4-benzodiazepine-3-carboxylate.
CM-6912; Meilax; Victan. Anxiolytic. mp = 193-194°. *Hoffmann-LaRoche Inc.*

6736 Etifoxine
21715-46-8 3910
$C_{17}H_{17}ClN_2O$
6-Chloro-2-(ethylamino)-4-methyl-4-phenyl-4H-3,1-benzoxazine.
36-801; HOE-36801. Psychotropic agent with anxiolytic andanticonvulsant activity. mp = 90-92°; λ_m = 273 nm (ε 21200 EtOH); LD_{50} (mus orl) = 12000 mg/kg. *Hoechst AG.*

6737 Etifoxine Hydrochloride
56776-32-0 3910 260-380-0
$C_{17}H_{18}Cl_2N_2O$
6-Chloro-2-(ethylamino)-4-methyl-4-phenyl-4H-3,1-benzoxazine hydrochloride.
Stresam. Psychotropic agent with anxiolytic andanticonvulsant activity. mp = 150-151°. *Hoechst AG.*

6738 Etizolam
40054-69-1 3919
$C_{17}H_{15}ClN_4S$
4-(o-Chlorophenyl)-2-ethyl-9-methyl-6H-thieno[3,2-f]-s-triazolo[4,3-a][1,4]diazepine.
Y-7131; Depas. Anxiolytic. May affect monoamine metabolism in the brain. mp = 147-148°; LD_{50} (mrat orl) = 3619 mg/kg, (mrat ip) = 865, (frat orl) = 3509 mg/kg, (frat ip) = 825 mg/kg, (mmus orl) = 4358 mg/kg, (mmus ip) = 830 mg/kg, (fmus orl) = 4258 mg/kg, (fmus ip) = 783 mg/kg, (rat mus sc) > 5000 mg/kg. *Yoshitomi.*

6739 Flesinoxan
98206-10-1 4138
$C_{22}H_{26}FN_3O_4$
(+)-(S)-p-Fluoro-N-[2-[4-[2-(hydroxymethyl)-1,4-benzodioxan-5-yl]-1-piperazinyl]ethyl]-benzamide.
Anxiolytic. Serotonin $5-HT_{1A}$ receptor agonist. A phenylpiperazine derivative. *Solvay Duphar Labs Ltd.*

6740 Flesinoxan Hydrochloride
98205-89-1 4138
$C_{22}H_{27}ClFN_3O_4$
(+)-(S)-p-Fluoro-N-[2-[4-[2-(hydroxymethyl)-1,4-benzodioxan-5-yl]-1-piperazinyl]ethyl]benzamide hydrochloride.
DU-29373. Anxiolytic. Serotonin 5-HT_{1A} receptor agonist. A phenylpiperazine derivative. mp = 184.5 - 185.5°, 183 - 184°; $[\alpha]_D$ = 25° (c = 1 MeOH); freely soluble in H_2O, sparingly soluble in EtOH. *Solvay Duphar Labs Ltd.*

6741 Fludiazepam
3900-31-0 4164
$C_{16}H_{12}ClFN_2O$
7-Chloro-5-(o-fluorophenyl)-1,3-dihydro-1-methyl-2H-1,4-benzodiazepin-2-one.
ID-540; Ro-5-3438; Erispan. Anxiolytic. mp = 88-92°, 69-72°; LD_{50} (mus orl) = 910 mg/kg, (mus ip) = 360 mg/kg, (mus sc) = 1150 mg/kg. *Hoffmann-LaRoche Inc.*

6742 Fluoresone
2924-67-6 4197 220-889-0
$C_8H_9FO_2S$
Ethyl p-fluorophenyl sulfone.
Bripadon; Caducid. Anticonvulsant; analgesic; anxiolytic. A tricyclic benzodiazepine derivative. mp = 41°; LD_{50} (mus orl) = 2500 mg/kg, 850 mg/kg, 542 mg/kg.

6743 Flutazolam
27060-91-9 4243
$C_{19}H_{18}ClFN_2O_3$
10-Chloro-11b-(o-fluorophenyl)-2,3,7,11b-tetrahydro-oxazolo[3,2-d][1,4]benzodiazepin-6(5H)-one.
MS-4101; Ro-7-6102; Coreminal. Anxiolytic. mp = 142-147°; soluble in $CHCl_3$, EtOH; less soluble in Me_2CO, C_6H_6, MeOH; insoluble in H_2O. *Hoffmann-LaRoche Inc.*

6744 Flutoprazepam
25967-29-7 4245
$C_{19}H_{16}ClFN_2O$
7-Chloro-1-(cyclopropylmethyl)-5-(o-fluorophenyl)-1,3-dihydro-2H-1,4-benzodiazepin-2-one.
KB-509; ID-1937; Restar; Restas. Anxiolytic. A benzodiazepine derivative. mp = 118-122°; LD_{50} (mmus orl) = 2640 mg/kg, (mmus ip) = 2400 mg/kg, (fmus orl) = 2430 mg.kg, (fmus ip) = 2110 mg/kg, (mrat orl) = 13760 mg/kg, (mrat ip) = 2460 mg/kg, (frat orl) = 10060 mg/kg, (frat ip) = 2230 mg/kg, (mus, rat sc) > 5000 mg/kg. *Sumitomo Pharm. Co. Ltd.*

6745 Fluvoxamine
54739-18-3 4251
$C_{15}H_{21}F_3N_2O_2$
5-Methoxy-4'-(trifluoromethyl)valerophenone (E)-O-(2-aminoethyl)oxime.
Antidepressant; anxiolytic; antiobsessional agent. Serotonin uptake inhibitor. *Philips-Duphar B.V.; Solvay Duphar Labs Ltd.*

6746 Fluvoxamine Maleate
61718-82-9 4251
$C_{19}H_{25}F_3N_2O_6$
5-Methoxy-4'-(trifluoromethyl)valerophenone (E)-O-(2-aminoethyl)oxime maleate.
DU-23000; MK-264; Dumirox; Faverin; Fevarin; Floxyfral; Luvox; Maveral. Antidepressant; anxiolytic; antiobsessional agent. Serotonin uptake inhibitor. mp = 120-121.5°. *Philips-Duphar B.V.; Solvay Pharm. Inc.*

6747 Gepirone
83928-76-1
$C_{19}H_{29}N_5O_2$
3,3-Dimethyl-1-[4-[4-(2-pyrimidinyl)-1-piperazinyl]butyl]glutarimide.
Anxiolytic. *Bristol Myers Squibb Pharm. Ltd.*

6748 Gepirone Hydrochloride
83928-66-9
$C_{19}H_{26}ClN_5O_2$
3,3-Dimethyl-1-[4-[4-(2-pyrimidinyl)-1-piperazinyl]butyl]glutarimide monohydrochloride.
BMY-13805-1. Anxiolytic. *Bristol Myers Squibb Pharm. Ltd.*

6749 Girisopam
82230-53-3
$C_{18}H_{17}ClN_2O_2$
1-(m-Chlorophenyl)-7,8-dimethoxy-4-methyl-5H-2,3-benzodiazepine.
Anxiolytic.

6750 Glemanserin
107703-78-6
$C_{20}H_{25}NO$
1-Phenethyl-α-phenyl-4-piperidinemethanol.
MDL-11939. Anxiolytic. *Marion Merrell Dow Inc.*

6751 (±)-Glemanserin
132553-86-7
$C_{20}H_{25}NO$
(±)-1-Phenethyl-α-phenyl-4-piperidinemethanol.
MDL-11939. Anxiolytic. *Marion Merrell Dow Inc.*

6752 Glutamic Acid
56-86-0 4477 200-293-7
$C_5H_9NO_4$
(S)-2-Aminopentanedioic acid.
L-glutamic acid; α-aminoglutaric acid; 1-aminopropane-1,3-dicarboxylic acid; Glutacid; Glutaminol; Glutaton. The magnesium salt hydrobroimde is used as an anxiolytic. mp = 160° (dec); $[\alpha]_D^{22.4}$ = 31.4° (6N HCl); soluble in H_2O (8.6 g/l at 25°), insoluble in organic solvents.

6753 Halazepam
23092-17-3 4619 245-425-4
$C_{17}H_{12}ClF_3N_2O$
7-Chloro-1,3-dihydro-5-phenyl-1-(2,2,2-trifluoroethyl)-2H-1,4-benzodiazepin-2-one.
Paxipam; Sch-12041. Anxiolytic with sedative and hypnotic properties. A benzodiazepine. mp = 164-166°; LD_{50} (mus orl) > 4000 mg/kg. *Schering-Plough HealthCare Products.*

6754 Hydroxyphenamate
50-19-1 4884 200-017-5
$C_{11}H_{15}NO_3$
2-Phenyl-1,2-butanediol 1-carbamate.
AI-0361; P-301; NSC-108034; Oxyfenamate; Listica.

Anxiolytic used as a minor tranquillizer. mp= 55-56.5°; soluble in H_2O (2.5 g/100 ml); LD_{50} (mus orl) = 830 mg/kg. *Armour Pharm. Co. Ltd.*

6755 Hydroxyzyne
68-88-2 4897 200-693-1
$C_{21}H_{27}ClN_2O_2$
2-[2-[4-(p-Chloroα-phenylbenzyl)-1-piperazinyl]ethoxy]ethanol.
UCB-4492; Tran-Q; Tranquizine. Anxiolytic; antihistaminic. Has been used as a minor tranquillizer. H_1 receptor antagonist. *Pfizer Inc.*

6756 Hydroxyzyne Hydrochloride
2192-20-3 4897 218-586-3
$C_{21}H_{29}Cl_3N_2O_2$
2-[2-[4-(p-Chloroα-phenylbenzyl)-1-piperazinyl]ethoxy]ethanol dihydrochloride.
Alamon; Atarax; Aterax; Durrax; Orgatrax; Quiess; QYS; Vistaril Parenteral; Marax. Anxiolytic; antihistaminic. Has been used as a minor tranquillizer. H_1 receptor antagonist. mp = 193°; soluble in H_2O (<70 g/100 ml), $CHCl_3$ (6 g/100 ml), Me_2CO (0.2 g/100 ml), Et_2O (< 0.01 g/100 ml); LD_{50} (rat ip) = 126 mg/kg, (rat orl) = 950 mg/kg. *Elkins-Sinn; Forest Pharm. Inc.; KV Pharm.; Pfizer Inc.; Roerig Div. Pfizer Pharm.*

6757 Hydroxyzyne Pamoate
10246-75-0 4897 233-582-1
$C_{44}H_{43}ClN_2O_8$
2-[2-[4-(p-Chloroα-phenylbenzyl)-1-piperazinyl]ethoxy]ethanol 4,4'-methylenebis[3-hydroxy-2-naphthoate].
Equipose; Masmoran; Paxisitil; Vistaril pamoate. Anxiolytic; antihistaminic. Has been used as a minor tranquillizer. H_1 receptor antagonist. Insoluble in H_2O. *Pfizer Inc.*

6758 Iclazepam
57916-70-8
$C_{21}H_{21}ClN_2O_2$
7-Chloro-1-[2-(cyclopropylmethoxy)ethyl]-1,3-dihydro-5-phenyl-2H-1,4-benzodiazepin-2-one.
Anxiolytic.

6759 Ipsapirone
95847-70-4 5096
$C_{19}H_{23}N_5O_3S$
2-[4-[4-(2-Pyrimidinyl)-1-piperazinyl]butyl]-1,2-benzoisothiazolin-3-one 1,1-dioxide.
Isapirone. A serotonin (5-hydroxytryptamine, 5-HT_1) receptor antagonist. Non-benzodiazepine anxiolytic. mp = 137-138°. *Bayer AG.*

6760 Ipsapirone Hydrochloride
92589-98-5 5096
$C_{19}H_{24}ClN_5O_3S$
2-[4-[4-(2-Pyrimidinyl)-1-piperazinyl]butyl]-1,2-benzoisothiazolin-3-one 1,1-dioxide monohydrochloride.
Bay q 7821; TVX Q 7821. Non-benzodiazepine anxiolytic. A serotonin (5-hydroxytryptamine, 5-HT_1) receptor agonist. mp = 221-222°. *Bayer AG.*

6761 Itasetron
123258-84-4
$C_{16}H_{20}N_4O_2$
2-Oxo-N-1αH,5αH-tropan-3α-yl-1-benzimidazoline-1-carboxamide.
U-98079-A; DAU6215CL. Anxiolytic; antidepressant; antiemetic. *Boehringer Ingelheim GmbH.*

6762 Ketazolam
27223-35-4 5308 248-346-3
$C_{20}H_{17}ClN_2O_3$
11-Chloro-8,12b-dihydro-2,8-dimethyl-12b-phenyl-4H-[1,3]-oxazino[3,2-d][1,4]benzodiazepine-4,7(6H)dione.
U-28774; Anseren; Ansieten; Anxon; Contamex; Loftran; Unakalm. Anxiolytic used as a minor tranquillizer. Anesthetic. Federally controlled substance (depressant). mp = 182-183.5°; λ_m = 202, 241 nm (ε 40600, 18400 EtOH). *Pharmacia & Upjohn.*

6763 Lesopitron
132449-46-8 5470
$C_{15}H_{21}ClN_6$
2-[4-[4-(4-Chloropyrazol-1-yl)butyl]-1-piperazinyl]pyrimidine.
Anxiolytic. Selective serotonin 5HT_{1A} receptor agonist. *Esteve Group.*

6764 Lesopitron Dihydrochloride
132449-89-9 5470
$C_{15}H_{23}Cl_3N_6$
2-[4-[4-(4-Chloropyrazol-1-yl)butyl]-1-piperazinyl]-pyrimidine dihydrochloride.
E-4424. Anxiolytic. Selective serotonin 5HT_{1A} receptor agonist. mp = 194-197.5°. *Esteve Group.*

6765 Lopirazepam
42863-81-0 255-974-1
$C_{14}H_9Cl_2N_3O_2$
7-Chloro-5-(o-chlorophenyl)-1,3-dihydro-3-hydroxy-2H-pyrido[3,2-e]-1,4-diazepin-2-one.
D-12524. A benzodiazepine. Sedative; anxiolytic.

6766 Loprodiol
2209-86-1 7247 218-635-9
$C_5H_{10}Cl_2O_2$
2,2-Bis(chloromethyl)-1,3-propanediol.
pentaerythritol dichlorohydrin; Dispranol. Tranquillizer; muscle relaxant. mp = 79-80°, 65°. 83°, 95°; bp_{12} = 158.5-160°. *Heyden Chem.*

6767 Lorazepam
846-49-1 5609 212-687-6
$C_{15}H_{10}Cl_2N_2O_2$
7-Chloro-5-(o-chlorophenyl)-1,3-dihydro-3-hydroxy-2H-1,4-benzodiazepin-2-one.
Almazine; Ativan; WY-4036; Emotival; Lorax; Lorsilan; Pro Dorm; Psicopax; Punktyl; Quait; Securit; Sedatival; Sedazin; Somagerol; Tavor; Temesta; Wypax. Anxiolytic. Pharmaceutical preparation for the treatment of anxiety. mp = 166-168°; λ_m = 229 nm (1N NaOH), 233 nm (1N HCl); soluble in H_2O (0.008 g/100 ml), $CHCl_3$ (0.33 g/100 ml), EtOH (1.4 g/100 ml), propylene glycol (1.6 g/100 ml), EtOAc (3.0 g/100 ml); LD_{50} (mus orl) = 3178 mg/kg, (rat orl) > 5000 mg/kg. *Wyeth-Ayerst Labs.*

6768 Lorzafone
59179-95-2
$C_{18}H_{17}Cl_2N_3O_3$
2-(2-Aminoacetamido)-4'-chloro-2'-(o-chlorobenzoyl)-N-methylacetanilide .
Minor tranquillizer. *Eli Lilly & Co.*

6769 Lorzafone Monohydrate
81603-65-8
$C_{18}H_{17}Cl_2N_3O_3.H_2O$
2-(2-Aminoacetamido)-4'-chloro-2'-(o-chlorobenzoyl)-N-methylacetanilide monohydrate.
LY-123508. Minor tranquillizer. *Eli Lilly & Co.*

6770 Loxapine
1977-10-2 5617 217-835-3
$C_{18}H_{18}ClN_3O$
2-Chloro-11-(4-methyl-1-piperazinyl)dibenz[b,f][1,4]oxazepine.
Loxitane IM; SUM-3170; CL-62362; oxilapine; S-805. Minor tranquillizer. mp = 109-110°; LD_{50} (mus orl) = 65 mg/kg. *Lederle Labs.*

6771 Loxapine Succinate
27833-64-3 5617 248-682-0
$C_{22}H_{24}ClN_3O_5$
2-Chloro-11-(4-methyl-1-piperazinyl)dibenz[b,f][1,4]oxazepine succinate.
Daxolin; Loxitane Capsules; CL-71563; Loxapac; Loxitane. Minor tranquillizer. *Bayer AG; Lederle Labs.*

6772 Magnesium Glutamate Hydrobromide
53459-38-4 4477 258-566-1
$C_{10}H_{17}BrMgN_2O_8.H_2O$
Magnesium bromoglutamate.
Psicosoma; Psycho-Soma; Psychoverlan. Anxiolytic. mp = 225-227°; d_{20}^{20} = 1.4601; soluble in H_2O (2.05 g/100 ml at 25°); [d-form]: $[\alpha]_D^{20}$ = -30.5° (c = 1 in 6N HCl).

6773 Meclonazepam
58662-84-3
$C_{16}H_{12}ClN_3O_3$
(+)-(S)-5-(o-Chlorophenyl)-1,3-dihydro-3-methyl-7-nitro-2H-1,4-benzodiazepin-2-one.
Minor tranquillizer.

6774 Mecloralurea
1954-79-6 5822 217-786-8
$C_4H_7Cl_3N_2O_2$
1-Methyl-3-(2,2,2-trichloro-1-hydroxyethyl)urea.
trichloroethylolmethylurea; Heraldium. Anxiolytic. mp = 135-140°; soluble in H_2O, AcOH, Et_2O, EtOH, C_5H+5N; insoluble in $CHCl_3$, hydrocarbons. *Centre d'Etudes l'Ind. Pharm.*

6775 Medazepam
2898-12-6 5829 220-783-4
$C_{16}H_{15}ClN_2$
7-Chloro-2,3-dihydro-1-methyl-5-phenyl-1H-1,4-benzodiazepine.
Ansilan; Diepin; Medazepol; Megasedan; Narsis; Nobrium; Psiquium; Resmit; Rudotel; Tranquilax. Minor tranquillizer. mp = 95-97°; LD_{50} (mus orl) = 1070 mg/kg, (mus ip) = 360 mg/kg. *Hoffmann-LaRoche Inc.*

6776 Medazepam Hydrochloride
2898-11-5 5829 220-782-9
$C_{16}H_{15}ClN_2$
7-Chloro-2,3-dihydro-1-methyl-5-phenyl-1H-1,4-benzodiazepine monohydrochloride.
Ro 5-4556. Minor tranquillizer. *Hoffmann-LaRoche Inc.*

6777 Mephenoxalone
70-07-5 5896 200-723-3
$C_{11}H_{13}NO_4$
5-[(o-Methylphenoxy)methyl]-2-oxazolidinone.
metoxadone; methoxydon; methoxydone; methoxadone; AHR-233; OM-518; Control-Om; Dorsiflex; Dorsilon; Ekilan; Lenetran; Placidex; Riself; Tranpoise; Trepidone; Xerene. Tranquillizer. Skeletal muscle relaxant; anxiolytic. mp = 143-145°; insoluble in H_2O; LD_{50} (rat orl) = 3820 ± 17 mg/kg. *Marion Merrell Dow Inc.; Robins, A. H. Co.*

6778 Mepiprazole
20326-12-9 5902
$C_{16}H_{21}ClN_4$
1-(m-Chlorophenyl)-4-[2-(5-methylpyrazol-3-yl)ethyl]piperazine.
EMD-16923. Psychotropic agent with CNS-depressant properies. Tranquillizer. mp = 106°. *E. Merck.*

6779 Mepiprazole Dihydrochloride
20344-15-4 5902
$C_{16}H_{23}Cl_3N_4$
1-(m-Chlorophenyl)-4-[2-(5-methylpyrazol-3-yl)ethyl]piperazine dihydrochloride.
H-4007; Psigodal. Psychotropic agent with CNS-depressant properies. Tranquillizer. mp = 234°. *E. Merck.*

6780 Meprobamate
57-53-4 5908 200-337-5
$C_9H_{18}N_2O_4$
2-Methyl-2-propyl-1,3-propanediol dicarbamate.
Equanil; Meprospan; Meprotabs; Miltown; Tamate; Appetrol; Deprol; Equagesic; Micrainin; Milpath; Milprem; Miltrate; PMB-200; PMB-400; procalmadiol; procalmidol; Amosene; Andaxin; Aneural; Artolon; Atraxin; Ayeramate; Bamo 400; Biobamat; Calmiren; Cirpon; Cyrpon; Ecuanil; Fas-Cile 200; Gadexyl; Holbamate; Kesso-Bamate; Klort; Mar-Bate; Mepavlon; Meposed; Meprin; Meprindon; Meprobam; Mepr; Meprotabs; Meproten; Meprotil; Meptran; Mesmar; Miltaun; Morbam; My-trans; Nervonus; Oasil; Panediol; Perequil; Pertranquil; Placidon; Probamyl; Promate; Quaname; Quanil; Reostral; Restenil; Restran; Sowell; Trankvilan; Tranlisant; Tranquilan; Tranquiline; Urbilat. Anxiolytic with sedative and hypnotic properties. A propanediol carbamate. mp = 104-106°; soluble in H_2O (0.34 g/100 ml), more soluble in organic solvents; stable in dilute acid or alkali; LD_{50} (mus ip) = 800 mg/kg. *Lederle Labs.; Marion Merrell Dow Inc.; Wallace Labs; Wyeth-Ayerst Labs.*

6781 Metaclazepam
65517-27-3 5980
$C_{18}H_{18}BrClN_2O$
7-Bromo-5-(o-chlorophenyl)-2,3-dihydro-2-(methoxymethyl)-1-methyl-1H-1,4-benzodiazepine.
brometazepam; metuclazepam; Ka-2547; KC-2547. Anxiolytic. Benzodiazepam derivative. *Kali-Chemie.*

6782 Metaclazepam Hydrochloride

61802-93-5 5980 263-234-4

$C_{18}H_{19}BrCl_2N_2O$

7-Bromo-5-(o-chlorophenyl)-2,3-dihydro-2-(methoxymethyl)-1-methyl-1H-1,4-benzodiazepine hydrochloride.

Talis. Anxiolytic. Benzodiazepam derivative. Synthesized from 2-amino-2',5-dichlorobenzophenone. mp= 193-196°; LD_{50} (mus orl) = 1578 mg/kg. *Kali-Chemie.*

6783 Mexazolam

31868-18-5 6254

$C_{18}H_{16}Cl_2N_2O_2$

10-Chloro-11b-(o-chlorophenyl)-2,3,7,11b-tetrahydro-3-methyloxazolo[3,2-d][1,4]benzodiazepin-6(5H)-one.

CS-386; Melex. Anxiolytic. mp = 172-175°; LD_{50} (mmus orl) = 4687 mg/kg, (mmus ip or sc) > 6000 mg/kg, (fmus orl) = 4571 mg/kg, (fmus ip or sc) > 6000 mg/kg, (mrat orl) = 810 mg/kg, (mrat ip or sc) > 4000 mg/kg, (frat orl) = 4500 mg/kg, (frat ip or sc) > 4000 mg/kg.

6784 Mirisetron

135905-89-4

$C_{24}H_{31}N_3O_2$

1-Cyclohexyl-1,4-dihydro-4-oxo-N-1αH,5αH-tropan-3α-yl-3-quinolinecarboxamide.

Anxiolytic. *Wyeth-Ayerst Labs.*

6785 Mirisetron Maleate

148611-75-0

$C_{28}H_{35}N_3O_6$

1-Cyclohexyl-1,4-dihydro-4-oxo-N-1αH,5αH-tropan-3α-yl-3-quinolinecarboxamide maleate (1:1).

WAY-SEC-579. Anxiolytic. *Wyeth-Ayerst Labs.*

6786 Motrazepam

29442-58-8

$C_{17}H_{15}N_3O_4$

1,3-Dihydro-1-(methoxymethyl)-7-nitro-5-phenyl-2H-1,4-benzodiazepin-2-one.

Tolhart; Tolosate; Toloxyn; Tolserol; Tolseron; Tolulexin; Tolulox; Tolyspaz; Walconesin. Anxiolytic. *Wallace Labs.*

6787 Nerisopam

102771-12-0

$C_{18}H_{19}N_3O_2$

1-(4-Aminophenyl)-4-methyl-7,8-dimethoxy-5H-2,3-benzodiazepine.

GYKI-52322. A psychoactive 5H-2,3-benzodiazepine with unique neuroleptic properties. Anxiolytic; antipsychotic.

6788 Nisobamate

25269-04-9

$C_{13}H_{26}N_2O_4$

2-(Hydroxymethyl)-2,3-dimethylpentyl isopropylcarbamate carbamate (ester).

W-1015. Anxiolytic with sedative/hypnotic properties. *Wallace Labs.*

6789 Nordazepam

1088-11-5 6784 214-123-4

$C_{15}H_{11}ClN_2O$

7-Chloro-1,3-dihydro-5-phenyl-2H-1,4-benzodiazepin-2-one.

desmethyldiazepam; nordiazepam; DMDZ; A-101; Ro-5-2180; Calmday; Madar; Nordaz; Praxadium; Stilny. Anxiolytic. mp = 216-217°; λ_m 313 nm ($E^{1\%}_{1\,cm}$ 82, $CHCl_3$); slightly soluble in EtOH, $CHCl_3$; insoluble in H_2O; LD_{50} (mus orl) = 1300 mg/kg, 2750 = mg/kg, (mus ip) > 400 mg/kg, (rat orl) > 5200 mg/kg. *Hoffmann-LaRoche Inc.*

6790 Nortetrazepam

10379-11-0

$C_{15}H_{15}ClN_2O$

7-Chloro-5-(1-cyclohexen-1-yl)-1,3-dihydro-2H-1,4-benzodiazepin-2-one.

CB-4260. Anxiolytic.

6791 Ocinaplon

96604-21-6

$C_{17}H_{11}N_5O$

2-Pyridyl 7-(4-pyridyl)pyrazolo[1,5-a]pyrimidin-3-yl ketone.

CL-273547. Anxiolytic.

6792 Ondansetron

116002-70-1 6979

$C_{18}H_{19}N_3O$

(±)-2,3-Dihydro-9-methyl-3-[(2-methylimidazol-1-yl)methyl]carbazol-4(1H)-one.

Anxiolytic; antiemetic; antischizophrenic. Specific serotonin receptor ($5HT_3$) antagonist. Studied as an antiemetic. mp = 231-232°; [3S-form]: $[\alpha]_D^{25}$ = -14° (c = 0.19, MeOH); [3R-form]: $[\alpha]_D^{24}$ = +16° (c = 0.34, MeOH). *Glaxo Wellcome Inc.*

6793 Ondansetron Hydrochloride

103639-04-9 6979

$C_{18}H_{20}ClN_3O.H_2O$

(±)-2,3-Dihydro-9-methyl-3-[(2-methylimidazol-1-yl)methyl]carbazol-4(1H)-one monohydrochloride monohydrate.

Zofran; GR-38032F. Anxiolytic; antiemetic; antischizophrenic. Serotonin receptor antagonist. mp = 178.5-179.5°. *Glaxo Wellcome plc.*

6794 Oxanamide

126-93-2 7053

$C_8H_{15}NO_2$

2,3-Epoxy-2-ethylhexanamide.

Quiactin. Anxiolytic. mp = 90-91°; soluble in H_2O (1 g/100 ml); LD_{50} (mus ip) = 720 mg/kg, (mus orl) = 1220 mg/kg, (rat orl) = 1360 mg/kg.

6795 Oxazepam

604-75-1 7059 210-076-9

$C_{15}H_{11}ClN_2O_2$

7-Chloro-1,3-dihydro-3-hydroxy-5-phenyl-2H-1,4-benzodiazepin-2-one.

Wy-3498; Adumbran; Aplakil; Azutranquil; Bonare; Durazepam; Enidrel; Hilong; Isodin; Lederpam; Limbial; Serax; Nesontil; Noctazepam; Oxanid; Oxa-puren; Praxiten; Propax; Quilibrex; Rondar; Serax; Serenal; Serenid; Serepax; Seresta; Sigacalm; Sobril; Tazepam; Uskan; Zaxopam. Anxiolytic. A benzodiazepine. mp = 205-206°; insoluble in H_2O; soluble in EtOH, $CHCl_3$, dioxane; LD_{50} (mus, rat orl) > 5010 mg/kg. *Abbott Labs Inc.; Wyeth-Ayerst Labs.*

6796 Oxazolam
24143-17-7 7061 246-032-0
$C_{18}H_{17}ClN_2O_2$
10-Chloro-2,3,7,11b-tetrahydro-2-methyl-11b-phenyl-Oxazolo[3,2-d][1,4]benzodiazepin-6(5H)-one.
oxazolazepam; Hializan; Serenal; Tranquit. Anxiolytic. A benzodiazepine. mp = 186-188°; soluble in $CHCl_3$; slightly soluble in EtOH; nearly insoluble in H_2O.

6797 Pagoclone
133737-32-3
$C_{23}H_{22}ClN_3O_2$
(+)-2-(7-Chloro-1,8-naphthyridin-2-yl)-3-(5-methyl-2-oxohexyl)phthalimidine.
IP-456; RP-62955. Anxiolytic. *Janssen Pharm. Inc.*

6798 Panadiplon
124423-84-3
$C_{18}H_{17}N_5O_2$
3-(5-Cyclopropyl-1,2,4-oxadiazol-3-yl)-5-isopropylimidazo[1,5-a]quinoxalin-4(5H)-one.
U-78875; FG-10571. Anxiolytic.

6799 Pancopride
121243-20-7
$C_{18}H_{24}ClN_3O_2$
4-Amino-5-chloro-α-cyclopropyl-N-3-quinuclidinyl-o-anisamide.
Anxiolytic; antiemetic; CNS stimulant. *Grupo Farmaceutico Almirall S.A.*

6800 (±)-Pancopride
121650-80-4
$C_{18}H_{24}ClN_3O_2$
(±)-4-Amino-5-chloro-α-cyclopropyl-N-3-quinuclidinyl-o-anisamide.
Anxiolytic; antiemetic; CNS stimulant. *Grupo Farmaceutico Almirall S.A.*

6801 Pazinaclone
103255-66-9 7187
$C_{25}H_{23}ClN_4O_4$
(±)-8-[[2-(7-Chloro-1,8-naphthyridin-2-yl)-3-oxo-1-isoindolinyl]acetyl]-1,4-dioxa-8-azaspiro[4,5]decane.
A-77000; DN-2327. Anxiolytic. Pharmaceutical preparation for the treatment of constipation. mp = 238-239°. *Takeda Chem. Ind. Ltd.; TAP Pharm. Inc.*

6802 Pentabamate
5667-70-9
$C_8H_{16}N_2O_4$
3-Methyl-2,4-pentanedioldicarbamate.
Anxiolytic. *Farbenfabriken Bayer AG.*

6803 Phenaglycodol
79-93-6 201-235-3
$C_{11}H_{15}ClO_2$
2-(p-Chlorophenyl)-3-methyl-2,3-butanediol.
Has anxiolytic and possibly antiepilpetic properties.

6804 Phenprobamate
673-31-4 7412 211-606-1
$C_{10}H_{13}NO_2$
3-Phenyl-1-propanol carbamate.
proformiphen; MH-532; Extacol; Palmita; Spantol; Anseprон; Gamaquil; Quamaquil. Anxiolytic; skeletal muscle relaxant. mp = 101-104°; soluble in EtOH, $CHCl_3$, propylene glycol, ethylenediamine, DMF; less soluble in Et_2O; insoluble in H_2O; LD_{50} (mus orl) = 840 mg/kg. *Siegfried AG.*

6805 Pinazepam
52463-83-9 7596 257-934-9
$C_{18}H_{13}ClN_2O$
7-Chloro-1,3-dihydro-5-phenyl-1-(2-propynyl)-2H-1,4-benzodiazepin-2-one.
Z-905; Domar; Duna. Anxiolytic. mp = 140-142°; LD_{50} (mus orl) = 1355 mg/kg, 670 mg/kg (mus ip) = 266 mg/kg, (rat orl) = 5819 mg/kg, (rat ip) = 622 mg/kg. *Zambeletti.*

6806 Pipequaline
77472-98-1
$C_{22}H_{24}N_2$
2-Phenyl-4-[2-(4-piperidyl)ethyl]quinoline.
PK-8165; 45319-RP. A putative mixed agonist/antagonist at benzodiazepine receptors. Anxiolytic.

6807 Pirenperone
75444-65-4 278-213-5
$C_{23}H_{24}FN_3O_2$
3-[2-[4-(p-Fluorobenzoyl)piperidino]ethyl]-2-methyl-4H-pyrido[1,2-a]pyrimidin-4-one.
R-47465. Anxiolytic. *Janssen Pharm. Inc.*

6808 Pivoxazepam
55299-10-0
$C_{20}H_{19}ClN_2O_3$
7-Chloro-1,3-dihydro-3-hydroxy-5-phenyl-2H-1,4-benzodiazepin-2-one pivalate(ester).
Anxiolytic.

6809 Poskine
585-14-8
$C_{20}H_{25}NO_5$
3-[2-Phenyl-2-(propionyloxymethyl)acetyloxy]-6,7-epoxytropane.
Anxiolytic.

6810 Prazepam
2955-38-6 7895 220-975-8
$C_{19}H_{17}ClN_2O$
7-Chloro-1-(cyclopropylmethyl)-1,3-dihydro-5-phenyl-2H-1,4-benzodiazepin-2-one.
W-4020; Centrax; Demetrin; Lysanxia; Prazene; Sedapran; Settima; Trepidan; Verstran. Anxiolytic; sedative/hypnotic. mp = 145-146°. *Parke-Davis.*

6811 Premazepam
57435-86-6
$C_{15}H_{15}N_3O$
3,7-Dihydro-6,7-dimethyl-5-phenylpyrrolo[3,4-e]-1,4-diazepin-2(1H)-one.
Anxiolytic.

6812 Procymate
13931-64-1 7945
$C_{10}H_{19}NO_2$
1-Cyclohexylpropyl carbamate.
T-3033; Equipax. Anxiolytic. mp = 128-129°. *Matieres Colorantes.*

6813 Proflazeparn
52829-30-8
$C_{18}H_{16}ClFN_2O_3$
7-Chloro-1-(2,3-dihydroxypropyl)-5-(o-fluorophenyl)-1,3-dihydro-2H-1,4-benzodiazepin-2-one.
Anxiolytic. Used as a topical antiseptic.

6814 Ripazepam
26308-28-1
$C_{15}H_{16}N_4O$
1-Ethyl-4,6-dihydro-3-methyl-8-phenylpyrazolo[4,3-e]-[1,4]diazepin-5(1H)-one.
Cl-683. Anxiolytic.

6815 Ritanserin
87051-43-2 8399
$C_{27}H_{25}F_2N_3OS$
6-[2-[4-[Bis(p-fluorophenyl)methylene]piperidino]ethyl]-7-methyl-5H-thiazolo[3,2-a]pyrimidin-5-one.
R-55667; Tiserton. Anxiolytic; antidepressant. A selective serotonin receptor antagonist. mp = 145.5°; LD_{50} (mmus iv) = 28.2 mg/kg, (mmus orl) = 626 mg/kg, (fmus iv) = 28.2 mg/kg, (fmus orl) = 993 mg/kg, (mrat iv) = 20.0 mg/kg, (mrat orl) = 856 mg/kg, (frat iv) = 22.2 mg/kg, (frat orl) = 515 mg/kg, (mdog iv) = 24.1 mg/kg, (mdog orl) = 1280 mg/kg, (fdog iv) = 33.2 mg/kg, (fdog orl) = 640-1280 mg/kg. *Janssen Pharm. Inc.*

6816 Roxoperone
2804-00-4
$C_{19}H_{23}FN_2O_5$
8-[3-(p-Fluorobenzoyl)propyl]-2-methyl-2,8-diazaspiro[4.5]decane-1,3-dione.
Anxiolytic.

6817 Serazapine
115313-22-9
$C_{22}H_{23}N_3O_2$
(±) Methyl 1,3,4,16b-tetrahydro-2-methyl-2H,10H-indolo-[2,1-c]pyrazino[1.2-a][1,4]benzodiazepin-16-carboxylate.
Anxiolytic. *Ciba-Geigy Corp.*

6818 Serazapine Hydrochloride
117581-05-2
$C_{22}H_{24}ClN_3O_2$
(±) Methyl 1,3,4,16b-tetrahydro-2-methyl-2H,10H-indolo[2,1-c]pyrazino[1.2-a][1,4]benzodiazepin-16-carboxylate monohydrochloride.
CGS-15040A. Anxiolytic. *Ciba-Geigy Corp.*

6819 Sulazepam
2898-13-7
$C_{16}H_{13}ClN_2S$
7-Chloro-1,3-dihydro-1-methyl-5-phenyl-2H-1,4-benzodiazepine-2-thione.
W-3676. Anxiolytic used as a minor tranquillizer. Antibacterial. *Parke-Davis.*

6820 Suriclone
53813-83-5 9182 258-794-1
$C_{20}H_{20}ClN_5O_3S_2$
4-Methylpiperazine-1-carboxylic acid ester with (±)-6-(7-chloro-1,8-naphthyridin-2-yl)-2,3,6,7-tetrahydro-7-hydroxy-5H-p-dithiino[2,3-c]pyrrol-5-one.
RP-31264; Celexane; Clexane; Suril. Anxiolytic. mp = 280°. *Rhône-Poulenc Rorer Pharm. Inc.*

6821 Taclamine
34061-33-1
$C_{21}H_{23}N$
2,3,4,4a,8,9,3b,14-Octahydro-1H-benzo[6,7]cyclohepta[1,2,3-de]pyrido[2,1-a]isoquinoline.
Anxiolytic used as a minor tranquillizer. *Wyeth-Ayerst Labs.*

6822 Taclamine Hydrochloride
34061-34-2
$C_{21}H_{24}ClN$
2,3,4,4a,8,9,3b,14-Octahydro-1H-benzo[6,7]cyclohepta-[1,2,3-de]pyrido[2,1-a]isoquinoline hydrochloride.
AY-22214. Anxiolytic used as a minor tranquillizer. *Wyeth-Ayerst Labs.*

6823 Tandospirone
87760-53-0 9219
$C_{21}H_{29}N_5O_2$
(3aα,4β,7β,7aα)-Hexahydro-2-[4-[4-(2-pyrimidinyl)-1-piperazinyl]butyl]-4,7-methano-1H-isoindole-1,3(2H)-dione.
Anxiolytic; antidepressant. A serotonin receptor agonist. mp = 112-113.5°. *Pfizer Inc.; Sumitomo Pharm. Co. Ltd.*

6824 Tandospirone Citrate
112457-95-1 9219
$C_{27}H_{37}N_5O_9$
(3aα,4β,7β,7aα)-Hexahydro-2-[4-[4-(2-pyrimidinyl)-1-piperazinyl]butyl]-4,7-methano-1H-isoindole-1,3(2H)-dione 2-hydroxy-1,2,3-propanetricarboxylate (1:1).
SM-3997. Serotonin receptor agonist. Anxiolytic; antidepressant. mp = 169.5-170°; [hydrochloride]: mp = 227-229°. *Pfizer Inc.; Sumitomo Pharm. Co. Ltd.*

6825 Temazepam
846-50-4 9285 212-688-1
$C_{16}H_{13}ClN_2O_2$
7-Chloro-1,3-dihydro-3-hydroxy-1-methyl-5-phenyl-2H-1,4-benzodiazepin-2-one.
oxydiazepam; Wy-3917; ER-115; K-3917; Ro-5-5354; Euhypnos; Euipnos; Gelthix; Levanxene; Levanxol; Normison; Perdorm; Planum; Remestan; Restoril. Sedative/hypnotic; anxiolytic. A benzodiazepine. mp = 119-121°. *Hoffmann-LaRoche Inc.; Sandoz Pharm. Corp.*

6826 Tofisopam
22345-47-7 9642 244-922-3
$C_{22}H_{26}N_2O_4$
1-(3,4-Dimethoxyphenyl)-5-ethyl-7,8-dimethoxy-4-methyl-5H-2,3-benzodiazepine.
EGYT-341; Grandaxin; Seriel. Anxiolytic. mp = 156-157°; λ_m = 310, 272,239 nm (ε 16100, 11200, 26300 MeOH). *EGYT.*

6827 Triflubazam
22365-40-8
$C_{17}H_{13}F_3N_2O_2$
1-Methyl-5-phenyl-7-(trifluoromethyl)-1H-1,5-benzodiazepine-2,4(3H,5H)-dione.
WE352; ORF-8063. Anxiolytic used as a minor tranquillizer. *C. H. Boehringer Sohn.*

6828 Trimetozine
635-41-6 9850 211-236-0
$C_{14}H_{19}NO_5$
4-(3,4,5-Trimethoxybenzoyl)morpholine.
Abbott-22370; PS-2383; NSC-62939; V-7; Opalène; Trioxazine. Anxiolytic with sedative/hypnotic properties. mp = 120-122°; slightly soluble in H_2O, EtOH. *Abbott Labs Inc.*

6829 Trimetozine
635-41-6 9850 211-236-0
$C_{14}H_{19}NO_5$
4-(3,4,5-Trimethoxybenzoyl)morpholine.
Abbott-22370; PS-2383; NSC-62939; Opalene; Trioxazine. Anxiolytic with sedative-hypnotic properties. mp = 120-122°; slightly soluble in H_2O, EtOH. *Abbott Labs Inc.*

6830 Tuclazeparn
51037-88-8
$C_{17}H_{16}Cl_2N_2O$
7-Chloro-5-(o-chlorophenyl)-2,3-dihydro-1-methyl-1H-1,4-benzodiazepine-2-methanol.
Anxiolytic.

6831 Tybamate
4268-36-4 9960 224-254-9
$C_{13}H_{26}N_2O_4$
2-(Hydroxymethyl)-2-methylpentylbutylcarbamate carbamate.
Solacen; W713; Nospan; Tybatran. Anxiolytic used as a minor tranquillizer. mp = 49-51°; $bp_{0.006}$ = 150-152°. *Wallace Labs.*

6832 Uldazepam
28546-58-9
$C_{18}H_{15}Cl_2N_3O$
2-[(Allyloxy)amino]-7-chloro-5-(o-chlorophenyl)-3H-1,4-benzodiazepine.
U-31920. Anxiolytic with sedative/hypnotic properties.

6833 Valnoctamide
4171-13-5 10048 224-033-7
$C_8H_{17}NO$
2-Ethyl-3-methylvaleramide.
Axiquel; McN-X-181; NSC-32363; Nirvanil. Anxiolytic. mp = 113.5-114°. *McNeil Pharm.*

6834 Zalospirone
114298-18-9
$C_{24}H_{29}N_5O_2$
3a,4,4a,6a,7,7a-Hexahydro-2-[4-[4-(2-pyrimidinyl)-1-piperazinyl]butyl]-4,7-etheno-1H-cyclobut[f]-isoindole-1,3(2H)-dione.
WY-47846. Anxiolytic. *Wyeth-Ayerst Labs.*

6835 Zalospirone Hydrochloride
114374-97-9
$C_{24}H_{30}ClN_5O_2$
3a,4,4a,6a,7,7a-Hexahydro-2-[4-[4-(2-pyrimidinyl)-1-piperazinyl]butyl]-4,7-etheno-1H-cyclobut[f]-isoindole-1,3(2H)-dione hydrochloride.
WY-47846HCl. Anxiolytic. *Wyeth-Ayerst Labs.*

Aromatase Inhibitors

6836 Aminoglutethimide
125-84-8 460 204-756-4
$C_{13}H_{16}N_2O_2$
2-(p-Aminophenyl)-2-ethylglutarimide.
p-Aminoglutethimide; Ba-16038; NSC-330915; Cytadren; Elipten; Orimeten. Adrenocortical suppressant; anticonvulsant; antineoplastic agent. Aromatase inhibitor. Has been used in treatment of Cushing's syndrome and other adrenal hormone disorders and in palliative treatment of breast cancer. mp = 149-150°; insoluble in H_2O; poorly soluble in EtOAc, EtOH; readily soluble in other organic solvents; [hydrochloride]: mp = 223-225°, soluble in H_2O. *Ciba-Geigy Corp.*

6837 Anastrozole
120511-73-1 667
$C_{17}H_{19}N_5$
α,α,α',α'-Tetramethyl-5-(1H-1,2,4-triazol-1-ylmethyl)-m-benzenediacetonitrile.
Arimidex; ICI-D-1033; ZD-1033. Aromatase inhibitor. Used as an antineoplastic agent. mp = 81-82°. *ICI ; Zeneca Pharm.*

6838 Atamestane
96301-34-7
$C_{20}H_{26}O_2$
1-Methylandrosta-1,4-diene-3,17-dione.
Aromatase inhibitor.

6839 Exemestane
107868-30-4
$C_{20}H_{24}O_2$
6-Methyleneandrosta-1,4-diene-3,17-dione.
Aromatase inhibitor.

6840 Fadrozole
102676-47-1 3969
$C_{14}H_{13}N_3$
(±)-p-(5,6,7,8-Tetrahydroimidazo[1,5-a]pyridin-5-yl)benzonitrile.
Antineoplastic agent. Aromatase inhibitor. mp = 117-118°. *Ciba-Geigy Corp.*

6841 Fadrozole Hydrochloride
102676-31-3 3969
$C_{14}H_{14}ClN_3$
p-(5,6,7,8-Tetrahydroimidazo[1,5-a]pyridin-5-yl)benzonitrile hydrochloride.
CGS-16949A. Aromatase inhibitor. mp = 231-233°; soluble in H_2O. *Ciba-Geigy Corp.*

6842 (±)-Fadrozole Hydrochloride
102676-96-0 3969
$C_{14}H_{14}ClN_3$
(±)-p-(5,6,7,8-Tetrahydroimidazo[1,5-a]pyridin-5-yl)-benzonitrile hydrochloride.
CGS-16949A. Aromatase inhibitor. Antineoplastic agent. mp = 117-118°. *Ciba-Geigy Corp.*

6843 Formestane
566-48-3 4267
$C_{19}H_{26}O_3$
4-Hydroxyandrost-4-ene-3,17-dione.

CGP-32349; 4OHA; Lentaron. Aromatase inhibitor. mp = 199-202°, 203.5-206°; λ_m = 278 nm (ε 11030 EtOH); $[\alpha]_D^{20}$ = 181° (c = 7.7 $CHCl_3$). *Ciba-Geigy Corp.*

6844 Letrozole
112809-51-5 5474
$C_{17}H_{11}N_5$
4,4'-(1H-1,2,4-Triazol-1-ylmethylene)dibenzonitrile.
CGS-20267. Antineoplastic agent. Aromatase inhibitor. Related to fadrozole. mp = 181-183°. *Ciba-Geigy Corp.*

Astringents

6845 Alcloxa
1317-25-5 215-262-3
$C_4H_9Al_2ClN_4O_7$
Chlorotetrahydroxy[(2-hydroxy-5-oxo-2-imidazolin-4-yl)-ureato]dialuminum.
aluminum chlorhydroxy allantoinate; ALCA; RC-173. Astringent; keratolytic. *ICI Americas Inc.*

6846 Aldioxa
5579-81-7 224 226-964-4
$C_4H_7AlN_4O_5$
Dihydroxy[(2-hydroxy-5-oxo-2-imidazolin-4-yl)ureato]aluminum.
aluminum dihydroxy allantoinate; ALDA; RC-172; Alanetorin; Alusa; Arlanto; Ascomp; Chlokale; Isalon; Nische; Peptilate. Astringent; keratolytic. Antiulcerative. mp = 230°, insoluble in polar and non-polar solvents. *ICI Americas Inc.*

6847 Alkannin
517-88-4 253 208-245-7
$C_{16}H_{16}O_5$
(S)-5,8-Dihydroxy-2-(1-hydroxy-4-methyl-3-pentenyl)-1,4-naphthalenedione.
anchusa acid; anchusin; alkanna red; alkanet extract; C.I. Natural Red 20; C.I. 75530; [(+)-form]: shikonin; [(±)-form]: shikalkin. Astringent. mp = 149°; $[\alpha]_{Cd}^{20}$ = -165° (C_6H_6), -226° ($CHCl_3$), -254° ± 7° ($CHCl_3$); sparingly soluble in H_2O, soluble in most organic solvents; LD_{50} (mmus orl) = 3000 ± 1000 mg/kg, (fmus orl) 3100 ± 100 mg/kg, (rat orl) > 1000 mg/kg.

6848 Aluminum Acetate Solution
8006-13-1 332
$C_6H_9AlO_6$
A solution of about 5% neutral aluminum acetate.
Burow's Solution; Domeboro. Astringent and antiseptic. d ≅ 1.002.

6849 Aluminum Acetotartrate
333
Consists of about 70% basic aluminum acxetate and 30% tartaric acid.
Essitol. Astringent and antiseptic. Slowly soluble in cold H_2O, insoluble in EtOH.

6850 Aluminum Ammonium Sulfate
7784-25-0 335 232-055-3
$AlH_4NO_8S_2$
Burnt ammonium alum; exsiccated ammonium alum. Astringent. d = 1.65; mp = 94.5°; soluble in H_2O (14.3 g/100 ml at 20°, 200 g/100 ml at 100°, freely soluble in glycerol, insoluble in EtOH.

6851 Aluminum Chlorate
15477-33-5 347 239-499-7
$AlCl_3O_9$
Chloric acid aluminum salt.
Astringent; antiseptic. Occurs a hexahydrate and a nonahydrate. Freely soluble in H_2O, soluble in EtOH.

6852 Aluminum Chlorate Nonahydrate
7784-15-8 347
$AlCl_3O_9.9H_2O$
Chloric acid aluminum salt.
Mallebrin. Astringent and antiseptic. Freely soluble in H_2O, soluble in EtOH.

6853 Aluminum Chloride
7446-70-0 348 231-208-1
$AlCl_3$
Aluminum chloride.
Strong irritant.

6854 Aluminum Chloride Hexahydrate
7784-13-6 348
$AlCl_3.6H_2O$
Aluminum chloride hexahydrate.
Aluwets; Anhydrol; Driclor; component of: Drysol, Xerac AC. Astringent. Soluble in H_2O (111 g/100 ml), EtOH (25 g/100 ml), Et_2O, glycerol, propylene glycol. *Person & Covey Inc.*

6855 Aluminum Hydroxychloride
1327-41-9 356 215-477-2
$Al_2Cl(OH)_5 \cdot 2H_20$
Aluminum chlorohydrate.
basic aluminum chloride; aluminum chlorohydroxide; Astringen; Chlorhydrol; Hyperdrol; Locron; Phosphonorm. Astringent; antihyperphosphatemic. Used in antiperspirants. Anhidrotic. Soluble in H_2O (55 g/100 ml). *Elizabeth Arden.*

6856 Aluminum β-Naphtholdisulfonate
1300-81-8 363
$C_{30}H_{18}Al_2O_{21}S_6$
2-Hydroxynaphthalenedisulfonic acid aluminum salt.
Alumnol. Astringent and antiseptic. Soluble in H_2O (66 g/100 ml), soluble in glycerol, slightly soluble in EtOH, insoluble in Et_2O.

6857 Aluminum Potassium Sulfate
10043-67-1 373 233-141-3
$AlKO_8S_2$
burnt alum; exsiccated alum. Astringent. Soluble in H_2O (5 g/100 ml at 20°, 100 g/100 ml at 100°), insoluble in EtOH; [dodecahydrate (alum; potassium alum; Kalinite; alum flour; alum meal; cube alum)]: mp = 92.5°; d = 1.725; soluble in H_2O (13.9 g/100 ml at 20°, 330 g/100 ml at 100+s.

6858 Aluminum Sodium Sulfate
10102-71-3 378 233-277-3
$AlNaO_8S_2$
sodium alum [as dodecahydrate]; soda alum [as

dodecahydrate]. Astringent. [dodecahydrate]: mp ≅ 60°; d = 1.61; soluble in H_2O (100 g/100 ml), insoluble in EtOH.

6859 Aluminum Subacetate Solution
8000-61-1 380
Contains about 8% aluminum diacetate.
Essigsäure Tonerde. Astringent. Clear colorless liquid, d = 1.045.

6860 Ammonium Ferric Sulfate
10138-04-2 549 233-382-4
$FeH_4NO_8S_2$
Ferric ammonium sulfate.
ferric alum [as dodecahydrate]; iron alum [as dodecahydrate]. Astringent. [dodecahydrate]: mp ≅ 37°; d = 1.71; very soluble in H_2O, insoluble in EtOH.

6861 Baicalein
491-67-8 971
$C_{15}H_{10}O_5$
5,6,7-Trihydroxy-2-phenyl-4H-1-benzopyran-4-one.
noroxylin; Isolated from roots of *Scutellaria baicalensis*. Astringent. dec 264-265°; λ_m 324 276 nm (log ε 4.18 4.42 EtOH); soluble in EtOH, MeOH, Et_2O, Me_2CO, EtOAc, hot AcOH; sparingly soluble in $CHCl_3$, nitrobenzene; insoluble in H_2O.

6862 Bismuth Oxide
1304-76-3 1314 215-134-7
Bi_2O_3
Bismuth trioxide.
bismuthous oxide; bismuth yellow. Astringent. Insoluble in H_2O, soluble in HCl or HNO_3.

6863 Bismuth Subgallate
22650-86-8 1325
$C_7H_5BiO_6$
Basic bismuth gallate.
gallic acid bismuth basic salt; B.S.G.; Dermatol. Topical protectant; astringent; antacid. Insoluble in H_2O, EtOH, $CHCl_3$, Et_2O; soluble in dilute alkaline solutions, hot mineral acids.

6864 Bismuth Tannate
1330
Tanbismuth.
tannic acid bismuth derivative. Topical protectant; astringent. Contains about 36% bismuth. Insoluble in H_2O, EtOH, Et_2O.

6865 Boric Acid
10043-35-3 1364 233-139-2
H_3BO_3
Boric acid.
boracic acid; orthoboric acid; Borofax; component of: Bluboro, Borofax, Collyrium Eye Wash, Collyrium Fresh-Eye Drops. Astringent and antiseptic. mp ≅ 171°; soluble in H_2O (5.5 g/100 ml at 20°, 25 g/100 ml at 100°), EtOH (5.5 g/100 ml at 20°, 16.6 g/100 ml at 76°), glycerol (25 g/100 ml); LD_{50} (rat orl) = 5140 mg/kg. *Allergan Herbert; Glaxo Wellcome Inc.; Wyeth-Ayerst Labs.*

6866 Calcium Hydroxide
1305-62-0 1716 215-137-3
CaH_2O_2
Calcium hydroxide.
calcium hydrate; slaked lime. Astringent. d = 2.08 - 2.34; slightly soluble in H_2O, soluble in glycerol; LD_{50} (rat orl) = 7340 mg/kg.

6867 Cupric Citrate
866-82-0 2702 212-752-9
$C_6H_4Cu_2O_7$
2-Hydroxy-1,2,3-propanetricarboxylic acid copper salt (1:2).
Cuprocitrol. Astringent and antiseptic. Slightly soluble in H_2O; soluble in NH_4OH, dilute mineral acids.

6868 Dichloroacetic Acid
79-43-6 3100 201-207-0
$C_2H_2Cl_2O_2$
Dichloroethanoic acid.
DCA. Keratolytic; astringent. bp = 193-194°; d_4^{20}= 1.563; two crystalline forms, mp = 9.7° and -4°; soluble in H_2O, EtOH, Et_2O; LD_{50} (rat orl) = 2820 mg/kg; [ethyl ester ($C_4H_6Cl_2O_2$)]: bp = 158.3-158.7°; d_4^{20}= 1.282; slightly soluble in.

6869 Ferric Chloride
7705-08-0 4061 231-729-4
Cl_3Fe
Flores martis. The hexahydrate is an astringent. mp ≅ 300°; bp ≅ 316°; d^{25} = 2.90; readily soluble in H_2O, EtOH, Et_2O, Me_2CO, slightly soluble in CS_2, insoluble in EtOAc; [hexahydrate]: mp ≅ 37°; readily soluble in H_2O, EtOH, Et_2O, Me_2CO; LD_{50} (mus iv) = 0.24 mg/kg.

6870 Formic Acid
64-18-6 4268 200-579-1
CH_2O_2
Ameisensäure. Counter-irritant and astringent. bp = 100.8°; mp = 8.4°; d^{20} = 1.220; miscible with H_2O, Et_2O, Me_2CO, EtOAc, MeOH, EtOH; partially soluble in C_6H_6, C_7H_8; LD_{50} (mus iv) = 145 mg/kg, (mus orl) = 1100 mg/kg.

6871 Gallic Acid
149-91-7 4363 205-749-9
$C_7H_6O_5$
3,4,5-Trihydroxybenzoic acid.
Formerly used as an astringent. mp = 235-240° (dec), 258-265° (dec), 225-230° (dec); soluble in H_2O (1.15 g/100 ml at 20°, 33 g/100 ml at 100°), EtOH (16.6 g/100 ml), Et_2O (1 g/100 ml), glycerol (10 g/100 ml), Me_2CO (20 g/100 ml); insoluble in C_6H_6, $CHCl_3$, petroleum ether; LD_{50} (rbt orl) = 5000 mg/kg. *Mallinckrodt Inc.*

6872 Iodic Acid
7782-68-5 5032 231-962-1
HIO_3
Antiseptic; astringent; disinfectant. Halogen containing compound. mp = 110° (dec); d^0_4 = 4.629; soluble in H_2O (269 g/ 100 ml at 20°, 295 g/100 ml at 40°), HNO_3, dilute

EtOH; insoluble in absolute EtOH, Et_2O, $CHCl_3$; darkens on exposure to light.

6873 Lead Acetate
301-04-2 5411 206-104-4
$C_4H_6O_4Pb$
Acetic acid lead salt (2:1).
neutral lead acetate; normal lead acetate; sugar of lead; salt of Saturn. Astringent. [trihydrate]: d = 2.55; mp = 75°; soluble in H_2O (62.5 g/100 ml at 20°, 200 g/100 ml at 100°), EtOH 3.3 g/100 ml; freely soluble in glycerol; LD_{50} (rat ip) = 235.5 mg/kg.

6874 Methionic Acid
503-40-2 6052 207-966-4
$CH_4O_6S_2$
Methanedisulfonic acid.
Antiperspirant; the aluminum salt is used as an astringent. mp = 96-100°; [aluminum salt ($C_3H_6Al_2O_{18}S_6$)]: soluble in H_2O.

6875 Silver Bromide
7785-23-1 8649 232-076-8
AgBr
Silver bromide.
Topical anti-infective; astringent. Silver compound. mp = 432°; d = 6.47; slightly soluble in H_2O (0.0000135 g/100 ml at 25°); insoluble in EtOH, most acids; moderately soluble in 10% NH_4OH (0.33 g/100 ml at 12°); soluble in solutions of alkali cyanides; sparingly soluble in solutions of thiocyanides or thiosulfates; soluble in 220 parts saturated NaCl, in 35 parts of staurated KBr; slightly soluble in ammonium carbonate solutions.

6876 Silver Lactate
128-00-7 8660
$C_3H_5AgO_3$
2-Hydroxypropanoic acid sliver salt (1:1).
Astringent; topical anti-infective. Silver compound. Soluble in H_2O (6.6 g/100 ml); slightly soluble in alcohol; light sensitive.

6877 Sodium Formate
141-53-7 8765 205-488-0
$CHNaO_22$
Formic acid sodium salt.
Astringent. d = 1.92; mp = 253°; soluble in H_2O (77 g/100 ml), glycerol; slightly soluble in EtOH.

6878 Tannic Acid
1401-55-4 9221 215-753-2
Extracted from the bark and fruit of many plants.
Tannin; gallotannin; gallotannic acid. Astringent. Soluble in H_2O (285 g/100 ml), glycerol (100 g/100 ml at 60°), EtOH, Me_2CO; insoluble in C_6H_6, $CHCl_3$, Et_2O, petr ether, CS_2, CCl_64; LD_{100} (mus orl) = 6000 mg/kg.

6879 Tannoform
9010-29-1 9222
Methyleneditannin.
tannin-formaldehyde; Helgotan. Astringent. mp ≅ 230° (dec); insoluble in H_2O; soluble in EtOH, alkaline solutions.

6880 White Lotion
Astringent and topical protectant.

6881 Witch Hazel
4638
Distillate from dormant twigs of *Hamamelis virginiana*. Dickenson's Witch Hazel Formula. Astringent. *Dickinson E. E. Co.*

6882 Zinc Acetate
557-34-6 10256 209-170-2
$C_4H_6O_4Zn.2H_2O$
Acetic acid zinc salt dihydrate.
Emetic and astringent. mp = 237°; d = 1.735; soluble in H_2O (43.4 g/100 at 25°, 62.5 g/100 ml at 100°), EtOH (3.3 g/100 ml at 25°, 100 g/100 ml at 76°); LD_{50} (rat orl) = 2460 mg/kg.

6883 Zinc Carbonate
3486-35-9 10260 222-477-6
CO_3Zn
Carbonic acid zinc salt (1:1).
Astringent, antiseptic and topical protectant. Soluble in H_2O ((0.001 g/100 ml at 15°); soluble in dilutes minerals acids, alkalies.

6884 Zinc Chloride
7646-85-7 10261 231-592-0
Cl_2Zn
Hydrochloric acid zinc salt (2:1).
butter of zinc. Astringent. d^{25} = 2.907; mp ≅ 290°; bp = 732°; soluble in H_2O (432 g/100 ml at 25°, 614 g/100 ml at 100°), 2% HCl (400 g/100 ml), EtOH (77 g/100 ml), glycerol (50 g/100 ml); freely soluble in Me_2CO; LD_{50} (rat iv) = 60-90 mg/kg.

6885 Zinc Iodide
10139-47-6 10269 233-396-0
I_2Zn
Hydriodic acid zinc salt (2:1).
Astringent and topical antiseptic. d^{25} = 4.74; mp ≅ 446°; bp ≅ 625° (dec); soluble in H_2O (330 g/100 ml at 20°, 500 g/100 ml at 100°), glycerol (50 g/100 ml); freely soluble in EtOH, Et_2O.

6886 Zinc Oxide
1314-13-2 10279 215-222-5
OZn
flowers of zinc; philosopher's wool; zinc white; C.I. Pigment White 4; C.I. 77947. Topical protectant. Astringent. d = 5.67; d_4^{20} = 5.607; nearly insoluble in H_2O; soluble in dilute acetic acid, mineral acids, ammonia, ammonium carbonate or dilute AcOH. CAUTION: overexposure to fumes can lead to illness.

6887 Zinc Permanganate
23414-72-4 10281 245-646-6
Mn_2O_8Zn
Permanganic acid zinc salt (2:1).
Antiseptic; astringent. Permanganate. Usually sold in 95% pure form. Deteriorates on exposure to light, air; soluble in H_2O (33 g/100 ml); decomposed by alcohol.

6888 Zinc Peroxide
1314-22-3 10282 215-226-7
O_2Zn
Zinc superoxide.
ZPO. Antiseptic (topical); astringent. Dec above 150°; soluble in dilute acids, liberating H_2O_2; insoluble in but decomposed by H_2O.

6889 Zinc p-Phenolsulfonate
127-82-2 10283 204-867-8
$C_{12}H_10O_8S_2Zn$
p-Hydroxybenzensulfonic acid zinc salt (2:1).
zinc sulfocarbolate; zinc sulfophenate; Phenozin. Astringent. Soluble in H_2O (62.5 g/100 ml at 20°, 250 g/100 ml at 100°), EtOH (55.6 g/100 ml).

6890 Zinc Salicylate
16283-36-6 10288 240-380-7
$C_{14}H_{10}O_6Zn$
2-Hydroxybenzoic acid zinc salt (2:1).
zinc disalicylate. Astringent, antiseptic; topical protectant. [trihydrate]: Soluble in H_2O, EtOH.

6891 Zinc Sulfate
7733-02-0 10293 231-793-3
$ZnSO_4$
Sulfuric acid zinc salt (1:1).
Verazinc; white vitriol; zinc vitriol; Kreatol; Optraex; Solvezink; Solvazinc; Zincaps; Zincate; Zincomed; Z-Span; component of: VasoClear A, Zincfrin. Electrolyte replenisher. Zinc supplement. Astringent. Soluble in H_2O, insoluble in EtOH. *Alcon Labs; CIBA Vision AG; Forest Pharm. Inc.*

6892 Zinc Sulfate Heptahydrate
7446-20-0 10293
$ZnSO_4.7H_2O$
Sulfuric acid zinc salt (1:1) heptahydrate.
Op-Thal-Zin; Redeema. Electrolyte replenisher. Zinc supplement. Astringent. d = 1.97; mp = 100°; soluble in H_2O (166.6 g/100 ml), glycerol (40 g/100 ml); insoluble in EtOH. *Alcon Labs; CIBA Vision AG; Forest Pharm. Inc.*

6893 Zinc Tannate
8011-65-2 10295 232-377-4
Zinc oxide and tannins.
sal barnit. Astringent; antiseptic. Practically insoluble in H_2O; soluble in dilute acids.

Benzodiazepine Antagonists

6894 Flumazenil
78755-81-4 4169
$C_{15}H_{14}FN_3O_3$
Ethyl 8-fluoro-5,6-dihydro-5-methyl-6-oxo-4H-imidazo[1,5-a][1,4]benzodiazepin-3-carboxylate.
Ro-15-1788/000; flumazepil; Anexate; Lanexat; Mazicon; Romazicon. Benzodiazepine antagonist. Used for reversal of sedation and reversal of neurological deficits in patients with hepatic encephalopathy. mp = 201-203°; LD_{50} (mus ip) = 4000 mg/kg, (mus orl) = 4300 mg/kg, (rat ip) = 4300 mg/kg, (rat orl) = 6000 mg/kg. *Hoffmann-LaRoche Inc.*

6895 Sarmazenil
78771-13-8
$C_{15}H_{14}ClN_3O_3$
Ethyl 7-chloro-5,6-dihydro-5-methyl-6-oxo-4H-imidazo-[1,5-a][1,4]benzodiazepine-3-carboxylate.
Ro-15-3505. Benzodiazepine-receptor partial inverse agonist.

β-Adrenergic Agonists

6896 Albuterol
18559-94-9 217 242-424-0
$C_{13}H_{21}NO_3$
2-(tert-Butylamino)-1-(4-hydroxy-3-hydroxymethylphenyl)ethanol.
salbutamol; Proventil Inhaler; Ventalin Inhaler. Bronchodilator; tocolytic. Ephedrine derivative. mp = 151°, 157-158°; soluble in most organic solvents. *Allen & Hanbury; Apothecon; Glaxo Wellcome Inc.; Key Pharm.*

6897 Albuterol Sulfate
51022-70-9 217 256-916-8
$C_{26}H_{44}N_2O_{10}S$
2-(tert-Butylamino)-1-(4-hydroxy-3-hydroxymethylphenyl)ethanol sulfate (2:1).
Sch-13949W Sulfate; Aerolin; Asmaven; Broncovaleas; Cetsim; Cobutolin; Ecovent; Loftan; Proventil; Salbumol; Salbutard; Salbutine; Salbuvent; Sultanol; Ventelin; Ventodiscks; Ventolin; Volma. Bronchodilator; tocolytic. Ephedrine derivative. *Allen & Hanbury; Apothecon; Glaxo Labs.; Key Pharm.; Lemmon Co.; Schering AG.*

6898 Alifedrine
78756-61-3
$C_{18}H_{27}NO_2$
1-Cyclohexyl-3-[[(αS,βR)-β-hydroxy-α-methylphenethyl]amino]-1-propanone.
A novel beta-adrenergic partial agonist.

6899 Bambuterol
81732-65-2 980
$C_{18}H_{29}N_3O_5$
(±)-5-[2-(tert-Butylamino)-1-hydroxyethyl]-m-phenylene bis(dimethylcarbamate).
terbutaline bisdimethylcarbamate. Bronchodilator. Ester prodrug of the β_2-adrenergic agonist terbutaline. Ephedrine derivative. *Astra Draco AB.*

6900 Bambuterol Hydrochloride
81732-46-9 980
$C_{18}H_{30}ClN_3O_5$
(±)-5-[2-(tert-Butylamino)-1-hydroxyethyl]-m-phenylene bis(dimethylcarbamate) monohydrochloride.
KWD-2183; Bambec. Bronchodilator. Ephedrine derivative. *Astra Draco AB.*

6901 Bitolterol
30392-40-6 1344
$C_{28}H_{31}NO_5$
4-[2-(tert-Butylamino)-1-hydroxyethyl]-o-phenylene di-p-toluate.
Bronchodilator. α_2-adrenergic agonist; a diester of N-tert-butylnorepinephrine. Ephedrine derivative. *Sterling Winthrop Inc.*

6902 Bitolterol Mesylate
30392-41-7 1344 250-177-5
$C_{29}H_{35}NO_8S$
4-[2-(tert-Butylamino)-1-hydroxyethyl]-o-phenylene di-p-toluate methanesulfonate (salt).
bitoterol methanesulfonate; Win-32784; Biterol; Effectin; Tornalate. Bronchodilator. α_2-adrenergic agonist; a diester of N-tert-butylnorepinephrine. Ephedrine derivative. mp = 170-172°; soluble in DMSO. *Sterling Winthrop Inc.*

6903 Bornaprolol
66451-06-7
$C_{19}H_{29}NO_2$
1-(Isopropylamino)-3-(o-2-exo-norbornylphenoxy)-2-propanol.
A beta-adrenergic antagonist.

6904 Carbuterol
34866-47-2 1882 252-257-5
$C_{13}H_{21}N_3O_3$
[5-[2-[(1,1-Dimethylethyl)amino]-1-hydroxyethyl]-2-hydroxyphenyl]urea.
Bronchodilator. β-Adrenergic agonist related to isoproterenol; selective for airway smooth muscle receptors. Ephedrine derivative. mp = 174-176°; LD_{50} (mus iv) = 32.8 mg/kg, (rat iv) = 77.2 mg/kg, (mus orl) = 3134.6 mg/kg.

6905 Carbuterol Hydrochloride
34866-46-1 1882 252-255-4
$C_{13}H_{22}ClN_3O_3$
[5-[2-[(1,1-Dimethylethyl)amino]-1-hydroxyethyl]-2-hydroxyphenyl]urea monohydrochloride.
SKF-40383; Bronsecur; Pirem. Bronchodilator. Ephedrine derivative. mp = 204-207° (dec). *SmithKline Beecham Pharm.*

6906 Clenbuterol
37148-27-9 2407 253-366-0
$C_{12}H_{18}N_2O$
4-Amino-α-[(tert-butylamino)methyl]-3,5-dichlorobenzyl alcohol.
NAB-365; Monores. Bronchodilator. Substituted phenylethanolamine with β_2 sympathomimetic activity. Ephedrine derivative. *Thomae GmbH Dr. Karl.*

6907 Clenbuterol Hydrochloride
21898-19-1 2407 244-643-7
$C_{12}H_{19}ClN_2O$
4-Amino-α-[(tert-butylamino)methyl]-3,5-dichlorobenzyl alcohol monohydrochloride.
NAB-365Cl; Spiropent; Ventipulmin. Bronchodilator. β-Adrenergic agonist. Ephedrine derivative. mp = 174-175°; soluble in H_2O, MeOH, EtOH; slightly soluble in $CHCl_3$; insoluble in C_6H_6; LD_{50} (mus orl) = 176 mg/kg, (rat orl) = 315 mg/kg, (gpg orl) = 67.1, (mus iv) = 27.6 mg/kg. *Thomae GmbH Dr. Karl.*

6908 Clorprenaline
3811-25-4 2470 223-291-8
$C_{11}H_{16}ClNO$
o-Chloro-α-[(isopropylamino)methyl]benzyl alcohol.
N-(β-o-chlorophenyl-β-hydroxyethyl)isopropylamine; isoprophenamine; isoprofenamine. Bronchodilator. β-Adrenergic agonist. Ephedrine derivative. *Eli Lilly & Co.*

6909 Clorprenaline Hydrochloride
6933-90-0 2470 230-058-4
$C_{11}H_{17}Cl_2NO$
o-Chloro-α-[(isopropylamino)methyl]benzyl alcohol hydrochloride.
Bronchodilator. β-Adrenergic agonist. Ephedrine derivative. *Eli Lilly & Co.*

6910 Clorprenaline Hydrochloride Monohydrate
5588-22-7 2470 209-612-4
$C_{11}H_{19}Cl_2NO_2$
o-Chloro-α-[(isopropylamino)methyl]benzyl alcohol hydrochloride monohydrate.
Broncon; Clopinerin; Conselt; Fusca; Kalutein; Pentadoll; Restanolon. Bronchodilator. β-Adrenergic agonist. Ephedrine derivative. mp = 163-164°. *Eli Lilly & Co.*

6911 Denopamine
71771-90-9 2943
$C_{18}H_{23}NO_4$
(-)-(R)-α-[[(3,4-Dimethoxyphenethyl)amino]methyl]-p-hydroxybenzyl alcohol.
TA-064; Carguto; Kalgut. Cardiotonic. Selective β_1-adrenoceptor agonist with positive inotropic activity. [l-form hydrochloride]: mp = 138-139.5°; $[\alpha]_D^{25}$ = -38.0° (c = 1 MeOH); [dl-form hydrochloride]: mp = 164-167°. *Tanabe Seiyaku Co. Ltd.*

6912 Dioxethedrin
497-75-6 3356 207-849-8
$C_{11}H_{17}NO_3$
α-(1-Ethylaminoethyl)protocatechuyl alchohol.
dioxethedrine; component of: Bexol. Bronchodilator. β-Adrenergic agonist. Ephedrine derivative.

6913 Dioxethedrin Hydrochloride
497-75-6 3356
$C_{11}H_{18}ClNO_3$
α-(1-Ethylaminoethyl)protocatechuyl alchohol hydrochloride.
dioxethedrine hydrochloride. Bronchodilator. β-Adrenergic agonist. mp = 212-214°.

6914 Dobutamine
34368-04-2 3456
$C_{18}H_{23}NO_3$
(±)-4-[2-[[3-(p-Hydroxyphenyl)-1-methylpropyl]-amino]ethyl]pyrocatechol.
Compound 81929. Cardiotonic. A β_1-adrenoceptor agonist derived from dopamine. *Eli Lilly & Co.*

6915 Dobutamine Hydrochloride
49745-95-1 3456 256-464-1
$C_{18}H_{24}ClNO_3$
(±)-4-[2-[[3-(p-Hydroxyphenyl)-1-methylpropyl]-amino]ethyl]pyrocatechol hydrochloride.
Dobutrex; Inotrex; 46236. Cardiotonic. A β_1-adrenoceptor agonist derived from dopamine. mp = 184-186°; λ_m = 281, 223 nm (ε 4768, 14400, MeOH); LD_{50} (mus iv) = 73 mg/kg. *Astra USA Inc.; Eli Lilly & Co.*

6916 Dobutamine Lactobionate
104564-71-8
$C_{30}H_{45}NO_{15}$
(±)-4-[2-[[3-(p-Hydroxyphenyl)-1-methylpropyl]amino]-ethyl]pyrocatechol lactobionate (salt).
LY-207506. Cardiotonic. A β_1-adrenoceptor agonist derived from dopamine. *Eli Lilly & Co.*

6917 Dopexamine
86197-47-9 3482
$C_{22}H_{32}N_2O_2$
4-[2-[[6-(Phenethylamino)hexyl]amino]-ethyl]pyrocatechol.
FPL-60278; Dopacard. Cardiotonic. Dopamine receptor agonist. Has little or no α- or β-adrenoceptor activity. *Fisons Pharm. Div.*

6918 Dopexamine Hydrochloride
86484-91-5 3482
$C_{22}H_{34}Cl_2N_2O_2$
4-[2-[[6-(Phenethylamino)hexyl]amino]ethyl]-pyrocatechol dihydrochloride.
FPL-60278AR. Cardiotonic. Dopamine receptor agonist. Has little or no α- or β-adrenoceptor activity. [hydrobromide]: mp = 227-228°. *Fisons Pharm. Div.*

6919 Doxaminol
55286-56-1
$C_{26}H_{29}NO_3$
6,11-Dihydro-N-(2-hydroxy-3-phenoxypropyl)-N-methyldibenz[b,e]oxepin-11-ethylamine.
A β-adrenergic agonist. Has cardiotonic properties.

6920 Ephedrine Hydrochloride
134-71-4 3645 205-153-9
$C_{10}H_{16}ClNO$
α-[1-(Methylamino)ethyl]benzenemethanol hydrochloride.
racephedrine hydrochloride [dl-form]; Ephetonin [dl-form]. Adrenergic (bronchodilator). α- and β-adrenergic agonist. Found in Ma Huang, other Ephedra species. [(dl)-form]: mp = 187-188°; soluble in H_2O (25 g/100 ml); slightly soluble in EtOH; nearly insoluble in Et_2O. *Bristol Laboratories; Eli Lilly & Co.; Glaxo Labs.; Parke-Davis; Poythress; Sterling Winthrop Inc.; Whitehall Labs. Inc.*

6921 Ephedrine Sulfate
134-27-5 3645
$C_{10}H_{17}NO_4S$
α-[1-(Methylamino)ethyl]benzenemethanol sulfate.
racephedrine sulfate [dl-form]. Adrenergic (bronchodilator). α- and β-adrenergic agonist. Occurs in Ma Huang and other Ephedra species. mp (dl-form) = 247°; soluble in H_2O, EtOH; pH 6.

6922 dl-Ephedrine
90-81-3 3645 202-017-0
$C_{10}H_{15}NO$
(±)-α-[1-(Methylamino)ethyl]benzenemethanol.
1-phenyl-2-methylaminopropanol; racemic ephedrine; racephedrine. Adrenergic (bronchodilator). α- and β-adrenergic agonist. Occurs in Ma Huang and other Ephedra species. The d-isomer is used as a decongestant. mp = 79°; soluble in H_2O, EtOH, Et_2O, $CHCl_3$, oils.

6923 l-Ephedrine
299-42-3 3645 206-080-5
$C_{10}H_{15}NO$
L-erythro-2-(Methylamino)-1-phenylpropan-1-ol.
l-ephedrine; (-)-ephedrine; l-α-[1-(Methylamino)-ethyl]benzenemethanol; component of: Bena-Fedrin, Bronkotabs, Mudrane GG Elixir, Primatene, Quadrinal, Tedral. Bronchodilator. An α-adrenergic agonist. Waxy solid; may contain up to 1/2 mole H_2O; mp = 34°; bp = 255°; somewhat soluble in H_2O (5 g/100 ml); more soluble in EtOH (500 g/100 ml), $CHCl_3$, Et_2O, oils. *Eli Lilly & Co.; Knoll Pharm. Co.; Parke-Davis; Whitehall-Robins; Hybridon Inc.; Minerals Tech.*

6924 l-Ephedrine Hydrochloride
50-98-6 3645 200-074-6
$C_{10}H_{16}ClNO$
L-erythro-2-(Methylamino)-1-phenylpropan-1-ol hydrochloride.
Ephedral; Sanedrine; component of: Amesec, Bena-Fedrin, Mudrane GG Elixir, Primatene Tablets, Quadrinal, Quibron Plus, Tedral. An α-adrenergic agonist. Bronchodilator. mp = 216-220°; $[\alpha]_D^{25}$ = -33° to -35.5° (c = 5); soluble in H_2O (33.3 g/100 ml), EtOH (7.1 g/100 ml); nearly insoluble in Et_2O, $CHCl_3$. *Chemoterapico; Eli Lilly & Co.; Hybridon Inc.; Knoll Pharm. Co.; Parke-Davis; Whitehall-Robins.*

6925 l-Ephedrine Sulfate
134-72-5 3645 205-154-4
$C_{20}H_{32}N_2O_6$
L-erythro-2-(Methylamino)-1-phenylpropan-1-ol sulfate (2:1) salt.
Isofedrol; component of: Bronkaid, Pazo Ointment, Pazo Suppository. Bronchodilator. Adrenergic agonist. mp = 245° (dec); $[\alpha]_D^{25}$ = -29.5° to -32.0° (c = 5); soluble in H_2O (83 g/100 ml), EtOH (1.05 g/100 ml); freely soluble in hot EtOH. *Boehringer Mannheim GmbH; Bristol-Myers Squibb HIV Products; Eli Lilly & Co.; Sterling Health U.S.A.*

6926 Epinephrine
51-43-4 3656 200-098-7
$C_9H_{13}NO_3$
(R)-4-[1-Hydroxy-2-(methylamino)ethyl]-1,2-benzenediol.
l-methylaminoethanolcatechol; adrenalin; levorenen; Bronkaid Mist; Epifrin; Epiglaufrin; Eppy; Glaucon; Glauposine; Primatene Mist; Simplene; Sus-phrine; Suprarenaline; component of: Citanest Forte. Bronchodilator; cardiostimulant; mydriatic; antiglaucoma. Endogenous catecholamine with combined α- and β-agonist activity. Principal sympathomimetic hormone produced by the adrenal medulla. mp = 211-212°; dec 215°; $[\alpha]_D^{25}$ = -50.0° to -53.5° (in 0.6N HCl); slightly soluble in H_2O, EtOH; soluble in aqueous solutions of mineral acids; insoluble in aqueous solutions of ammonia and alkali carbonates; insoluble in Et_2O, Me_2CO, oils; LD_{50} (mus ip) = 4 mg/kg. *Alcon Labs; Allergan Inc.; Astra Sweden; Bristol-Myers Squibb Co.; CIBA Vision AG; Elkins-Sinn; Evans Medical Ltd.; Parke-Davis; Sterling Health U.S.A.; Whitehall-Robins; Wyeth-Ayerst Labs.*

6927 Epinephrine d-Bitartrate
51-42-3 3656 200-097-1
$C_{13}H_{19}NO_9$
(R)-4-[1-Hydroxy-2-(methylamino)ethyl]-1,2-benzenediol [R-(R*,R*)]-2,3-dihydroxybutanedioate (1:1) (salt).
Asmatane Mist; Asthmahaler Epitrate; Bronitin Mist; Bronkaid Mist Suspension; Epitrate; Medihaler-Epi; Primatene Mist Suspension; Suprarenin; component of: Asthmahaler, E-Pilo. Bronchodilator; cardiostimulant; mydriatic; antiglaucoma. Adrenergic (opthalmic). mp = 147-154° (some decomposition); soluble in H_2O (30 g/100 ml); slightly soluble in EtOH. *3M Pharm.; CIBA Vision Corp.; Menley & James Labs Inc.; Sterling Health U.S.A.; Whitehall-Robins; Wyeth-Ayerst Labs.*

6928 Epinephrine Hydrochloride
55-31-2 3656
$C_9H_{14}ClNO_3$
(-)-3,4-Dihydroxy-α-[(methylamino)methyl]benzyl alcohol hydrochloride.
Adrenalin; Epifrin; Glaucon; Suprarenin. Bronchodilator; cardiostimulant; mydriatic; antiglaucoma. An α-adrenergic agonist. *Alcon Labs; Allergan Inc.; Astra Chem. Ltd.; Bristol-Myers Squibb Co.; CIBA Vision AG; Elkins-Sinn; Forest Pharm. Inc.; Parke-Davis; Sterling Health U.S.A.; Whitehall Labs. Inc.; Wyeth-Ayerst Labs.*

6929 dl-Epinephrine
329-65-7 3656 206-347-6
$C_{20}H_{32}N_2O_6$
(±)-α-[1-(Methylamino)ethyl]benzenemethanol.
racepinefrine; racepinephrine. An α-adrenergic agonist. Bronchodilator, cardiostimulant, mydriatic, antiglaucoma. Slightly soluble in H_2O, EtOH. *Alcon Labs; Allergan Inc.; Astra Chem. Ltd.; Bristol-Myers Squibb Co.; CIBA Vision AG; Elkins-Sinn; Forest Pharm. Inc.; Parke-Davis; Sterling Health U.S.A.; Whitehall Labs. Inc.; Wyeth-Ayerst Labs.*

6930 dl-Epinephrine Hydrochloride
329-63-5 3656 206-346-0
$C_{20}H_{33}ClN_2O_6$
(±)-α-[1-(Methylamino)ethyl]benzenemethanol hydrochloride salt.
Asthmanefrin; Vaponefrin. An α-adrenergic agonist. Used as a bronchodilator, cardiostimulant, mydriatic and antiglaucoma agent. mp = 157°, soluble in H_2O; sparingly soluble in EtOH. *Alcon Labs; Allergan Inc.; Astra Chem. Ltd.; Bristol-Myers Squibb Co.; CIBA Vision AG; Elkins-Sinn; Forest Pharm. Inc.; Parke-Davis; Sterling Health U.S.A.; Whitehall Labs. Inc.; Wyeth-Ayerst Labs.*

6931 Epinephryl Borate
5579-16-8 226-961-8
$C_9H_{12}BNO_4$
(R)-2-Hydroxy-α-[(methylamino)methyl]-1,3,2-benzodioxaborole-5-methanol.
Epinal; Eppy/N. Adrenergic (opthalmic). *Alcon Labs; Pilkington Barnes Hind.*

6932 Etafedrine
7681-79-0 3752 231-677-2
$C_{12}H_{19}NO$
α-[1-(Ethylmethylamino)ethyl]benzenemethanol.
Menetryl; Novedrine; 451. Bronchodilator. β-Adrenergic agonist. Ephedrine derivative. *Marion Merrell Dow Inc.*

6933 Etafedrine Hydrochloride
5591-29-7 3752 227-000-5
$C_{12}H_{20}ClNO$
α-[1-(Ethylmethylamino)ethyl]benzenemethanol hydrochloride.
Nethamine; component of: Nethaphyl. Bronchodilator. β-Adrenergic agonist. Ephedrine derivative. mp = 183-184°; soluble in H_2O (67 g/100 ml), EtOH (12.5 g/100 ml). *Marion Merrell Dow Inc.*

6934 Ethylnorepinephrine
536-24-3 3880
$C_{10}H_{15}NO_3$
4-(2-Amino-1-hydroxybutyl)-1,2-benzenediol.
ethylnoradrenaline; ethylnorsprarenin; E.N.E; E.N.S. Bronchodilator. Ephedrine derivative. *Sterling Winthrop Inc.*

6935 Ethylnorepinephrine Hydrochloride
3198-07-0 3880
$C_{10}H_{16}ClNO_3$
4-(2-Amino-1-hydroxybutyl)-1,2-benzenediol hydrochloride.
Bronkephrine. Bronchodilator. Ephedrine derivative. Dec 199-200°; soluble in H_2O. *Sterling Winthrop Inc.*

6936 Fenoterol
13392-18-2 4022
$C_{17}H_{21}NO_4$
5-[1-Hydroxy-2-[[2-(4-hydroxyphenyl)-1-methylethyl]amino]ethyl]-1,3-benzenediol.
TH-1165. Bronchodilator; tocolytic. β-Adrenergic agonist. Ephedrine derivative. *Boehringer Ingelheim Pharm. Inc.*

6937 Fenoterol Hydrobromide
1944-12-3 4022
$C_{17}H_{22}BrNO_4$
5-[1-Hydroxy-2-[[2-(4-hydroxyphenyl)-1-methylethyl]amino]ethyl]-1,3-benzenediol hydrobromide.
TH-1165a; Airum; Berotec; Dosberotec; Partusisten. Bronchodilator; tocolytic. β-Adrenergic agonist. Ephedrine derivative. mp = 222-223°; LD_{50} (mus sc) = 1100 mg/kg, (mus orl) = 1990 mg/kg. *Boehringer Ingelheim Pharm. Inc.*

6938 Fenspiride
5053-06-5 4041 225-751-3
$C_{15}H_{20}N_2O_2$
8-(2-Phenylethyl)-1-oxa-3,8-diazaspiro[4.5]decan-2-one.
decaspiride; DESP. Bronchodilator. α-Adrenergic blocker. bp_2 = 126-127°. *Marion Merrell Dow Inc.*

6939 Fenspiride Hydrochloride
5053-08-7 4041 225-752-9
$C_{15}H_{20}N_2O_2$
8-(2-Phenylethyl)-1-oxa-3,8-diazaspiro[4.5]decan-2-one monohydrochloride.
NAT-333; NDR-5998A; Decaspir; Fluiden; Pneumorel; Respiride; Tefencia; Viarespan. Bronchodilator. α-Adrenergic blocker. Dec 232-233° (dec); soluble in H_2O; LD_{50} (mus iv) = 106 mg/kg, (rat orl) = 437 mg/kg. *Marion Merrell Dow Inc.*

6940 Formoterol
73573-87-2 4272
$C_{19}H_{24}N_2O_4$
(±)-2'-Hydroxy-5'-[(RS)-1-hydroxy-2-[[(RS)-p-methoxy-α-methylphenyl]amino]ethyl]formanilide.
CGP-25827A. Antiasthmatic. α-Adrenoceptor stimulating catecholamine analog with selective bronchodilator activity. Ephedrine derivative. *Yamanouchi U.S.A. Inc.*

6941 Formoterol Fumarate Dihydrate
43229-80-7 4272
$C_{42}H_{56}N_4O_{14}$
(±)-2'-Hydroxy-5'-[(RS)-1-hydroxy-2-[[(RS)-p-methoxy-α-methylphenyl]amino]ethyl]formanilide fumarate dihydrate.
BD-40A; Atock; Foradil. Antiasthmatic. α-Adrenoceptor stimulating catecholamine analog with selective bronchodilator activity. Ephedrine derivative. mp = 138-140°; LD_{50} (mrat orl) = 3130 mg/kg, (mrat iv) = 98 mg/kg, (mrat sc) = 1000 mg/kg, (mrat ip) = 170 mg/kg. *Yamanouchi U.S.A. Inc.*

6942 Hexoprenaline
3215-70-1 4745
$C_{22}H_{32}N_2O_6$
α,α'-[Hexamethylene-bis(iminomethylene)]bis(3,4-dihydroxybenzyl alcohol).
BYK-1512. Bronchodilator; tocolytic. β-Adrenergic agonist. Ephedrine derivative. [hemihydrate]: mp = 162-165°. *Lentia; Oesterreiche Stickstoffwerke; OSSW; Savage Labs.*

6943 Hexoprenaline Dihydrochloride
4323-43-7 4745 224-354-2
$C_{22}H_{34}Cl_2N_2O_6$
α,α'-[Hexamethylene-bis(iminomethylene)]bis(3,4-dihydroxybenzyl alcohol) dihydrochloride.
ST-1512; Ipradol. Bronchodilator; tocolytic. β-Adrenergic agonist. Ephedrine derivative. mp = 197.5-198°. *Lentia; OSSW; Savage Labs.*

6944 Hexoprenaline Sulfate
32266-10-7 4745 250-974-8
$C_{22}H_{34}N_2O_{10}S$
α,α'-[Hexamethylene-bis(iminomethylene)]bis(3,4-dihydroxybenzyl alcohol) sulfate.
Bronalin; Delaprem; Etoscol; Gynipral; Ipradol; Leanol. Bronchodilator; tocolytic. β-Adrenergic agonist. Ephedrine derivative. mp = 222-228°. *Lentia; Oesterreiche Stickstoffwerke; OSSW; Savage Labs.*

6945 Ibopamine
66195-31-1 4921 266-229-5
$C_{17}H_{25}NO_4$
4-[2-(Methylamino)ethyl]-o-phenylene diisobutyrate.
SB-7505; 3,4-di-o-isobutyryl epinine. Cardiotonic. Inotropic agent with dopaminergic and adrenergic agonist activities. *Simes S.p.A.; SmithKline Beecham Pharm.*

6946 Ibopamine Hydrochloride
75011-65-3 4921 278-056-2
$C_{17}H_{26}Cl5NO_4$
4-[2-(Methylamino)ethyl]-o-phenylene diisobutyrate hydrochloride.
SB-7505; Inopamil; Scandine. Cardiotonic. Inotropic agent with dopaminergic and adrenergic agonist activities. mp = 132°. *Simes S.p.A.; SmithKline Beecham Pharm.*

6947 Isoetharine
530-08-5 5185 208-472-1
$C_{13}H_{21}NO_3$
3,4-Dihydroxy-α-[1-(isopropylamino)propyl]benzyl alcohol.
etyprenaline; isoetarine; Win-3406; Dilabron; Neoisuprel. Bronchodilator. Ephedrine derivative. *I.G. Farben; Sterling Winthrop Inc.*

6948 Isoetharine Hydrochloride
2576-92-3 5185 219-927-9
$C_{13}H_{22}ClNO_3$
3,4-Dihydroxy-α-[1-(isopropylamino)propyl]benzyl alcohol hydrochloride.
Asthmalitan; Bronkosol; Numotac. Bronchodilator. Ephedrine derivative. mp = 212-213° (dec).

6949 Isoetharine Mesylate
7279-75-6 5185 230-695-8
$C_{14}H_{25}NO_6S$
3,4-Dihydroxy-α-[1-(isopropylamino)propyl]benzyl alcohol methanesulfonate (salt).
isoetarine mesilate; Bronkometer. Bronchodilator. Ephedrine derivative. *I.G. Farben; Sterling Winthrop Inc.*

6950 Isoproterenol
7683-59-2 5236 231-687-7
$C_{11}H_{17}NO_3$
3,4-Dihydroxy-α-[(isopropylamino)methyl]benzyl alcohol.
epinephrine isopropyl homolog; isoprenaline; dihydroxyphenylethanolisopropylamine; N-isopropylnoradrenaline; A-21; Aludrine; Aleudrin; Asiprenol; Asmalar; Assiprenol; Bellasthman; Isonorin; Isorenin; Isopropydrin; Isuprel; Isupren; Neodrenal; Neo-Epinine; Norisodrine; Novodrin; Proternol; Respifral; Saventrine; Vapo-N-Iso. Bronchodilator; anticholingergic. β-Adrenergic agonist; catecholamine derivative. Ephedrine derivative. [(dl)-form]: mp = 155.5°; [(l)-form]: mp = 164-165°; $[\alpha]_D^{19}$ = -45.0° (c = 2 in 2N HCl); pKa = 8.64; LD_{50} (mrat orl) = 3675 mg/kg. *3M Pharm.; Abbott Labs Inc.; Eli Lilly & Co.; Fisons plc; Parke-Davis; Sterling Winthrop Inc.*

6951 Isoproterenol Hydrochloride
51-30-9 5236 200-089-8
$C_{11}H_{18}ClNO_3$
3,4-Dihydroxy-α-[(isopropylamino)methyl]benzyl alcohol hydrochloride.
Aerolone; Aerotrol; Euspiran; Isomenyl; Isovon; Mistarel; Suscardia; component of: Aerolone Solution, Duo-Medihaler. Bronchodilator; anticholingergic. Ephedrine derivative. [(dl)-form]: mp = 170-171°; [(l)-form]: deco 162-164°; $[\alpha]_D^{19}$ = -50°; soluble in H_2O; less soluble in EtOH; nearly insoluble in many organic solvents; LD_{50} (rat orl) 2220 mg/kg. *3M Pharm.; Abbott Labs Inc.; Eli Lilly & Co.; Fisons plc; Parke-Davis; Sterling Winthrop Inc.*

6952 Isoproterenol Sulfate [anhydrous]
299-95-6 5236 206-085-2
$C_{22}H_{36}N_2O_{10}S$
3,4-Dihydroxy-α-[(isopropylamino)methyl]benzyl alcohol sulfate (2:1) salt.

isoprenaline sulfate. Bronchodilator; anticholingergic. Ephedrine derivative.

6953 L-Isoproterenol D-Bitartrate
54750-10-6 5236 259-322-7
$C_{17}H_{23}NO_9$
L-3,4-Dihydroxy-α-[(isopropylamino)methyl]benzyl alcohol D-Bitartrate.
Isolevin [as dihydrate]. Bronchodilator; anticholinergic. Ephedrine derivative. [dihydrate]: mp = 80-83° (sinters at 78°); $[\alpha]_D^{19}$ = -14.9° (c = 2 .31).

6954 L-Tretoquinol
30418-38-3 9719
$C_{19}H_{23}NO_5$
L-1,2,3,4-Tetrahydro-1-(3,4,5-trimethoxybenzyl)-6,7-isoquinolinediol.
trimethoquinol. Bronchodilator.

6955 L-Tretoquinol Hydrochloride
18559-63-2 9719
$C_{19}H_{24}ClNO_5$
L-1,2,3,4-Tetrahydro-1-(3,4,5-trimethoxybenzyl)-6,7-isoquinolinediol monohydrochloride.
AQ-110; Inolin; Vems. Bronchodilator. Soluble in H_2O, EtOH; [dl-form]: dec 224.5-226°.

6956 Mabuterol
56341-08-3 5674
$C_{13}H_{18}ClF_3N_2O$
4-Amino-α-[(tert-butylamino)methyl]-3-chloro-5-(trifluoromethyl)benzyl alcohol.
ambuterol. Bronchodilator. Orally active β-adrenergic agonist related to clenbuterol. Ephedrine derivative. *Boehringer Ingelheim Ltd.; Thomae GmbH Dr. Karl.*

6957 dl-Mabuterol Hydrochloride
95656-48-7 5674
$C_{13}H_{19}Cl_2F_3N_2O$
4-Amino-α-[(tert-butylamino)methyl]-3-chloro-5-(trifluoromethyl)benzyl alcohol hydrochloride.
KF-868; PB-868Cl; Broncholin. Bronchodilator. Ephedrine derivative. mp = 205-206°; somewhat soluble in H_2O; LD_{50} (mmus ip) = 60.3 mg/kg, (mmus orl) = 220.8 mg/kg.

6958 Metaproterenol
586-06-1 5992 209-569-1
$C_{11}H_{17}NO_3$
1-(3,5-Dihydroxyphenyl)-2-isopropylaminoethanol.
orciprenaline. Bronchodilator. Ephedrine derivative. mp = 100°. *Boehringer Ingelheim Ltd.*

6959 Metaproterenol Polistirex
586-06-1
Sulfonated styrene-divinylbenzene copolymer complex with 1-(3,5-dihydroxyphenyl)-2-isopropylaminoethanol.
Bronchodilator. *Fisons plc.*

6960 Metaproterenol Sulfate
5874-97-5 5992 227-539-6
$C_{22}H_{36}N_2O_{10}S$
1-(3,5-Dihydroxyphenyl)-2-isopropylaminoethanol sulfate (2:1).
orciprenaline sulfate; TH-152; Alotec; Alupent; Metaprel; Novasmasol. Bronchodilator. Ephedrine derivative. mp = 202-203°; LD_{50} (rat orl) = 42 mg/kg. *Boehringer Ingelheim Ltd.; Sandoz Pharm. Corp.*

6961 Methoxyphenamine
93-30-1 6077 202-237-7
$C_{11}H_{17}NO$
2-Methoxy-N,α-dimethylbenzene-ethanamine.
Bronchodilator. β-Adrenergic agonist. Oil; bp_2 = 97-99°. *Nippon Shinyaku Japan.*

6962 Methoxyphenamine Hydrochloride
5588-10-3 6077 226-993-2
$C_{11}H_{18}ClNO$
2-Methoxy-N,α-dimethylbenzene-ethanamine hydrochloride.
Proasma; Orthoxine; component of: Orthoxicol. Bronchodilator. β-Adrenergic agonist. mp = 129-131°; soluble in H_2O, EtOH, $CHCl_3$; slightly soluble in Et_2O, C_6H_6; pH (5% aqueous solution) = 5.3-5.7. *Hexcel; Nippon Shinyaku Japan; Upjohn Ltd.*

6963 Oxyfedrine
15687-41-9 7096
$C_{19}H_{23}NO_3$
3-[[(αS,βR)-β-Hydroxy-α-methylphenethyl]amino]-3'-methoxypropiophenone.
oxyphedrine. Cardiotonic; antianginal. Used to treat cardiac insufficiency. Partial β-adrenergic agonist with coronary vasodilating and positive inotropic effects. *Degussa-Hüls Corp.*

6964 DL-Oxyfedrine Hydrochloride
16648-69-4 7096 240-696-5
$C_{19}H_{24}ClNO_3$
(DL)-3-[[β-Hydroxy-α-methylphenethyl]amino]-3'-methoxypropiophenone hydrochloride.
Cardiotonic; antianginal. Used to treat cardiac insufficiency. Partial β-adrenergic agonist with coronary vasodilating and positive inotropic effects. mp = 173-175°; LD_{50} (mus iv) = 34 mg/kg. *Degussa-Hüls Corp.*

6965 L-Oxyfedrine Hydrochloride
16777-42-7 7096 240-828-1
$C_{19}H_{24}ClNO_3$
(L)-3-[[β-Hydroxy-α-methylphenethyl]amino]-3'-methoxypropiophenone hydrochloride.
D-563; Ildamen; Modacor. Cardiotonic; antianginal. Used to treat cardiac insufficiency. Partial β-adrenergic agonist with coronary vasodilating and positive inotropic effects. mp = 192-194°; LD_{50} (mus iv) = 29 mg/kg. *Degussa-Hüls Corp.*

6966 Pirbuterol
38677-81-5 7644
$C_{12}H_{20}N_2O_3$
α^6-[(tert-Butylamino)methyl]-3-hydroxy-2,6-pyridine-dimethanol.
Bronchodilator. Analog of albuterol, a β_2-adrenergic agonist. Ephedrine derivative. *Pfizer Inc.*

6967 Pirbuterol Acetate
65652-44-0 7644 265-862-4
$C_{14}H_{24}N_2O_5$
α^6-[(tert-Butylamino)methyl]-3-hydroxy-2,6-pyridinedimethanol monoacetate.
CP-24314-14; Maxair; Spirolair. Bronchodilator. Ephedrine derivative. *3M Pharm.; Pfizer Inc.*

6968 Pirbuterol Hydrochloride
38029-10-6 7644 253-751-3
$C_{12}H_{22}Cl_2N_2O_3$
α^6-[(tert-Butylamino)methyl]-3-hydroxy-2,6-pyridinedimethanol dihydrochloride.
CP-24314-1; Broncocor; Exirel. Bronchodilator. Ephedrine derivative. mp = 182° (dec). *Pfizer Inc.*

6969 Prenalterol
57526-81-5 7917 260-791-5
$C_{12}H_{19}NO_3$
(-)-(S)-1-(p-Hydroxyphenoxy)-3-(isopropylamino)-2-propanol.
A β_1-adrenergic agonist, used as a cardiotonic. mp = 127-128°; $[\alpha]_D^{20}$ = -1° ± 1°, $[\alpha]_{Hg}^{20}$ = 2° ± 1° (c = 0.940 MeOH). *Ciba-Geigy Corp.*

6970 Prenalterol Hydrochloride
61260-05-7 7917 262-676-5
$C_{12}H_{10}ClNO_3$
(-)-(S)-1-(p-Hydroxyphenoxy)-3-(isopropylamino)-2-propanol hydrochloride.
H133/22; CGP-7760B; Hyprenan; Varebian. A β_1-adrenergic agonist, used as a cardiotonic. *Ciba-Geigy Corp.*

6971 Procaterol
72332-33-3 7939 276-590-0
$C_{16}H_{22}N_2O_3$
(±)-erythro-8-Hydroxy-5-[1-hydroxy-2-(isopropylamino)butyl]carbostyril.
Bronchodilator. Symapthomimetic amine with selective β_2-adrenergic agonist activity. Ephedrine derivative. *Otsuka Am. Pharm.*

6972 Procaterol Hydrochloride
59828-07-8 7939
$C_{16}H_{23}ClN_2O_3$
(±)-erythro-8-Hydroxy-5-[1-hydroxy-2-(isopropylamino)-butyl]carbostyril.
Cl-888; OPC-2009; Lontermin; Masacin; Meptin; Onsukil; Pro-Air; Procadil; Promaxol; Propulum. Bronchodilator. Ephedrine derivative. [hemihydrate]: mp = 193-197° (dec); soluble in MeOH; slightly soluble in EtOH; nearly insoluble in Me_2CO, Et_2O, EtOAc, $CHCl_3$, C_6H_6; LD_{50} (mrat orl) = 2600 mg/kg; (mrat iv) = 80 mg/kg. *Otsuka Am. Pharm. ; Parke-Davis.*

6973 Protokylol
136-69-6 8082 205-255-3
$C_{18}H_{21}NO_5$
4-[2-[[2-(1,3-Benzodioxol-5-yl)-1-methylethyl]amino]-1-hydroxyethyl]-1,2-benzenediol.
Bronchodilator. β-Adrenergic agonist. Ephedrine derivative. *Lakeside BioTechnology.*

6974 Protokylol Hydrochloride
136-70-9 8082 205-254-8
$C_{18}H_{22}ClNO_5$
4-[2-[[2-(1,3-Benzodioxol-5-yl)-1-methylethyl]amino]-1-hydroxyethyl]-1,2-benzenediol hydrochloride.
JB-251; Caytine; Ventaire. Bronchodilator. β-Adrenergic agonist. Ephedrine derivative. mp = 126-127°; soluble in H_2O; LD_{50} (rat orl) 940 mg/kg. *Lakeside BioTechnology.*

6975 Reproterol
54063-54-6 8307 258-956-1
$C_{18}H_{23}N_5O_5$
7-[3-[[2-(3,5-Dihydroxyphenyl)-2-hydroxyethyl]amino]-propyl]-3,7-dihydro-1,3-dimethyl-1H-purine-2,6-dione.
D-1959. Bronchodilator. Theophylline derivative with selective β_2 receptor activity. Ephedrine derivative. *Degussa-Hüls Corp.*

6976 Reproterol Hydrochloride
13055-82-8 8307 235-942-3
$C_{18}H_{24}ClN_5O_5$
7-[3-[[2-(3,5-Dihydroxyphenyl)-2-hydroxyethyl]amino]-propyl]-3,7-dihydro-1,3-dimethyl-1H-purine-2,6-dione monohydrochloride.
D-1959-HCl; W-2946M; Asmaterolo; Bronchodil; Bronchospasmin. Bronchodilator. Ephedrine derivative. mp = 249-250°; LD_{50} (mus iv) = 148 mg/kg, (mus orl) > 10000 mg/kg. *Degussa-Hüls Corp.; Wallace Labs.*

6977 Rimiterol
32953-89-2 8393 251-305-2
$C_{12}H_{17}NO_3$
(R*,S*)-4-(Hydroxy-2-piperidinylmethyl)-1,2-benzenediol.
Bronchodilator. Ephedrine derivative. mp = 203-204°. *SmithKline Beecham Pharm.*

6978 Rimiterol Hydrobromide
31842-61-2 8393 250-834-6
$C_{12}H_{18}BrNO_3$
(R*,S*)-4-(Hydroxy-2-piperidinylmethyl)-1,2-benzenediol hydrobromide.
R-798; WG-253; Asmaten; Pulmadil. Bronchodilator. Ephedrine derivative. mp = 220° (dec). *3M Pharm.; SmithKline Beecham Pharm.*

6979 Ritodrine
26652-09-5 8401 247-879-9
$C_{17}H_{21}NO_3$
erythro-p-Hydroxy-α-[1-[(p-hydroxyphenethyl)amino]-ethyl]benzyl alcohol.
DU-21220. Smooth muscle relaxant; tocolytic agent. A β-adrenergic agonist. mp = 88-90°. *Astra Chem. Ltd.; Astra Sweden; Philips-Duphar B.V.; Teva Pharm. (USA).*

6980 Ritodrine Hydrochloride
23239-51-2 8401 245-514-8
$C_{17}H_{22}ClNO_3$
erythro-p-Hydroxy-α-[1-[(p-hydroxyphenethyl)amino]-ethyl]benzyl alcohol hydrochloride.
Pre-Par; Yutopar; DU-21220; Miolene; Prempar; Utemerin; Utopar. Smooth muscle relaxant. A β-adrenergic agonist and tocolytic. mp = 193-195° (dec); λ_m = 267.5 nm (ε 3310). *Astra Chem. Ltd.; Astra Sweden; Philips-Duphar B.V.; Teva Pharm. (USA).*

6981 Salmefamol
18910-65-1 242-662-5
$C_{19}H_{25}NO_4$
α-[(p-Methoxy-α-methylphenylethylamino)methyl]-4-hydroxy-m-xylene-α,α'-diol.
AH-3923. beta Adrenergic agonist; bronchodilator.

6982 Salmeterol
89365-50-4 8489
$C_{25}H_{37}NO_4$
(±)-4-Hydroxy-α'-[[[6-(4-phenylbutoxy)hexyl]amino]methyl]-m-xylene-α,α'-diol.
GR-33343X. Bronchodilator. Structural analog of albuterol. $β_2$-Adrenergic agonist. Ephedrine derivative. mp = 75.5-76.5°. *Glaxo Labs.*

6983 Salmeterol Xinafoate
49749-08-3 8489
$C_{36}H_{45}NO_7$
(±)-4-Hydroxy-α'-[[[6-(4-phenylbutoxy)hexyl]amino]methyl]-m-xylene-α,α'-diol 1-hydroxy-2-napthoate (salt).
GR-3343G; Arial; Salmetedur; Serevent. Bronchodilator. Ephedrine derivative. mp = 137-138°; soluble in MeOH; slightly soluble in EtOH, $CHCl_3$, iPrOH; sparingly soluble in H_2O. *Glaxo Labs.*

6984 Soterenol
13642-52-9 8877
$C_{12}H_{20}N_2O_4S$
2'-Hydroxy-5'-[1-hydroxy-2-(isopropylamino)ethyl]methanesulfonamide.
MJ-1992. Bronchodilator. β-Adrenergic agonist. Ephedrine derivative.

6985 Soterenol Hydrochloride
14816-67-2 8877 238-889-4
$C_{12}H_{21}ClN_2O_4S$
2'-Hydroxy-5'-[1-hydroxy-2-(isopropylamino)ethyl]methanesulfonamide monohydrochloride.
Bronchodilator. β-Adrenergic agonist. Ephedrine derivative. mp = 195.5-196.5° (dec); LD_{50} (mus iv) = 41 mg/kg, (mus ip) = 315 mg/kg, (mus orl) = 660 mg/kg.

6986 Terbutaline
23031-25-6 9302 245-385-8
$C_{12}H_{19}NO_3$
5-[2-[(1,1-Dimethylethyl)amino]-1-hydroxyethyl]-1,3-benzenediol.
Bronchodilator; tocolytic. β-Adrenergic agonist. Ephedrine derivative. mp = 119-120°. *Astra Draco AB; Ciba-Geigy Corp.; Merrell Pharm. Inc.*

6987 Terbutaline Sulfate
23031-32-5 9302 245-386-3
$C_{24}H_{40}N_2O_{10}S$
5-[2-[(1,1-Dimethylethyl)amino]-1-hydroxyethyl]-1,3-benzenediol sulfate (2:1) (salt).
KWD-2019; Brethaire; Brethine; Bricanyl; Butaliret; Monovent; Terbasmin; Terbul. Bronchodilator; tocolytic. β-Adrenergic agonist. Ephedrine derivative. mp = 246-248°; $λ_m$ = 276 nm ($A^{1\%}_{1cm}$ 67.6 0.1 N HCl); pKa_1 = 8.8, pKa_2 = 10.1, pKa_3 = 11.2; soluble in H_2O (>2.0 g/100 ml), 0.1N HCl (>2.0 g/100 ml), 0.1N NaOH (>2.0 g/100 ml), EtOH (0.012 g/100 ml), 10% EtOH (>2.0 . *Astra Draco AB; Ciba-Geigy Corp.; Marion Merrell Dow Inc.; Merrell Pharm. Inc.*

6988 Theodrenaline
13460-98-5
$C_{17}H_{21}N_5O_5$
7-[2-[2-(3,4-Dihydroxyphenyl)-2-hydroxyethylamino]-ethyl]theophylline.
Akrinor. Theophylline derivative. Betamimetic catecholamine. Vasoconstrictor. Used to treat hypotension.

6989 Trecadrine
90845-56-0
$C_{27}H_{29}NO$
(1R,2S)-α-[1-[[2-(10,11-Dihydro-5H-dibenzo[a,d-]cyclohexen-5-ylidine)ethyl]methylamino]ethyl]-benzyl alcohol.
beta 3 Adrenergic agonist with potential antidiabetic properties.

6990 Tretoquinol
21650-42-0 9719
$C_{19}H_{23}NO_5$
(±)-1,2,3,4-Tetrahydro-1-(3,4,5-trimethoxybenzyl)-6,7-isoquinolinediol.
trimethoquinol; trimetoquinol; AQ-110 [as l-form hydrochloride]; Inolin [as l-form hydrochloride]; Vems [as l-form hydrochloride]. A catecholamine. A highly potent $β_2$-adrenoceptor and site-selective thromboxane A2/prostaglandin H2 receptor ligand. *l*-Form hydrochloride acts as a bronchodilator. Dec 224.5-226°.

6991 Tulobuterol
41570-61-0 9942
$C_{12}H_{18}ClNO$
2-Chloro-α-[[(1,1-dimethylethyl)-amino]ethyl]benzenemethanol.
α-[(tert-Butylamino)methyl]-o-chlorobenzyl alcohol; HN-078; HOKU-81. Bronchodilator. Used in the treatment of asthma. A β-adrenergic receptor agonist, related structurally to terbutaline, an ephedrine derivative. mp = 89-91°; LD_{50} (mmus orl) = 305 mg/kg, (mrat orl) = 850 mg/kg, (mrbt orl) = 563 mg/kg, (mmus sc) = 170 mg/kg, (mrat sc) = 417 mg/kg, (mrbt sc) = 164 mg/kg.

6992 Tulobuterol Hydrochloride
56776-01-3 9942
$C_{12}H_{19}Cl_2NO$
2-Chloro-α-[[(1,1-dimethylethyl)-amino]methyl]benzenemethanol hydrochloride.
Atenos; Berachin; Brelomax; Bremax; Hokunalin; Respacal. Bronchodilator. Used in the treatment of asthma. A β-adrenergic receptor agonist. Ephedrine derivative. mp = 161-163°.

6993 Xamoterol
81801-12-9 10189
$C_{16}H_{25}N_3O_5$
(±)-N-[2-[[2-Hydroxy-3-(p-hydroxyphenoxy)propyl-]amino]ethyl]-4-morpholinecarboxamide.
Cardiotonic. Partial β-adrenergic agonist with positive inotropic activity. *ICI.*

6994 Xamoterol Hemifumarate
73210-73-8 10189 277-319-9
$C_{36}H_{54}N_6O_{14}$
(±)-N-[2-[[2-Hydroxy-3-(p-hydroxyphenoxy)-propyl]amino]ethyl]-4-morpholine-carboxamide fumarate (2:1) (salt).
Carwin; Corwin; Xamtol. Cardiotonic. Partial β-adrenergic agonist with positive inotropic activity. mp = 168-169° (dec). *ICI.*

β-Adrenergic Antagonists

6995 Acebutolol
37517-30-9 16 253-539-0
$C_{18}H_{28}N_2O_4$
(±)-3'-Acetyl-4'-[2-hydroxy-3-(1-methylethylamino)propoxy]butyranilide.
Monitan; Sectral; Prent. Antianginal agent. Class II antiarrhythmic agent. Cardioselective β-adrenergic blocking agent. Has antiarrhythmic and antihypertensive properties. mp = 119-123°. *Wyeth Labs.*

6996 Acebutolol Hydrochloride
34381-68-5 16 251-980-3
$C_{18}H_{29}ClN_2O_4$
(±)-3'-Acetyl-4'-[2-hydroxy-3-(1-methylethylamino)propoxy]butyranilide hydrochloride.
IL-17803A; M&B-17803A; Acetanol; Neptall; Sectral. Antianginal agent. Class II antiarrhythmic agent. Cardioselective β-adrenergic blocking agent. Has antiarrhythmic and antihypertensive properties. mp = 141-143°; soluble in H_2O (20 g/100 ml), EtOH (7 g/100ml). *Wyeth Labs.*

6997 Alprenolol
13655-52-2 321 237-140-9
$C_{15}H_{23}NO_2$
1-(o-Allylphenoxy)-3-(isopropylamino)-2-propanol.
H 56/28. Antianginal agent. antiarrhythmic agent. Cardioselective β-adrenergic blocking agent. Has antiarrhythmic and antihypertensive properties. *C.M. Ind.; ICI.*

6998 Alprenolol Hydrochloride
13707-88-5 321 237-244-4
$C_{15}H_{24}ClNO_2$
1-(o-Allylphenoxy)-3-(isopropylamino)-2-propanol hydrochloride.
Applobal; Aprobal; Aptine; Aptol Duriles; Gubernal; Regletin; Yobir. Antianginal agent. Class II antiarrhythmic agent. Cardioselective β-adrenergic blocking agent. Has antiarrhythmic and antihypertensive properties. mp = 107-109°; LD_{50} (mus orl) = 278.0 mg/kg, (rat orl) = 597.0 mg/kg, (rbt orl) = 337.3 mg/kg. *C.M. Ind.; ICI.*

6999 Amosulalol
85320-68-9 614
$C_{18}H_{24}N_2O_5S$
(±)-5-[1-Hydroxy-2-[[2-(o-methoxyphenoxy)ethyl]amino]ethyl]-o-toluenesulfonamide.
An α_1-adrenergic blocking agent with antihypertensive activity. *Yamanouchi U.S.A. Inc.*

7000 Amosulalol Hydrochloride
93633-92-2 614
$C_{18}H_{25}ClN_2O_5S$
(±)-5-[1-Hydroxy-2-[[2-(o-methoxyphenoxy)ethyl]amino]ethyl]-o-toluene-sulfonamide hydrochloride.
YM-09538; Lowgan. An α_1-adrenergic blocking agent with antihypertensive activity. mp = 158-160°; [R(-)-form]: mp = 158°; $[\alpha]_D^{20}$ = -30.4° (c = 1 MeOH); [S(+)-form]: mp = 158°; $[\alpha]_D^{20}$ = 30.7° (c = 1 MeOH). *Yamanouchi U.S.A. Inc.*

7001 Arnolol
87129-71-3
$C_{14}H_{23}NO_3$
(±)-3-Amino-1-[p-(2-methoxyethyl)phenoxy]-3-methyl-2-butanol.
A beta blocker.

7002 Arotinolol
68377-92-4 827
$C_{15}H_{21}N_3O_2S_3$
(±)-5-[2-[[3-(tert-Butylamino)-2-hydroxypropyl]thio]-4-thiazolyl]-2-thiophenecarboxamide.
Antianginal agent. Also antihypertensive and antiarrhythmic. Possesses both α- and β-adrenergic blocking activity. A propanolamine derivative. mp = 148-149°. *Sumitomo Pharm. Co. Ltd.*

7003 Arotinolol Hydrochloride
68377-91-3 827
$C_{15}H_{22}ClN_3O_2S_3$
(±)-5-[2-[[3-(tert-Butylamino)-2-hydroxypropyl]thio]-4-thiazolyl]-2-thiophenecarboxamide hydrochloride.
S-596; ARL; Almarl. Antianginal agent. Also antihypertensive and antiarrhythmic. Possesses both α- and β-adrenergic blocking activity. A propanolamine derivative. mp = 234-235.5°; LD_{50} (mus iv) = 86 mg/kg, (mus ip) = 360 mg/kg, (mus orl) > 5000 mg/kg. *Sumitomo Pharm. Co. Ltd.*

7004 Atenolol
29122-68-7 892 249-451-7
$C_{14}H_{22}N_2O_3$
2-[p-[2-Hydroxy-3-(isopropylamino)propoxy]phenyl]acetamide.
ICI-66082; AteHexal; Atenol; Cuxanorm; Ibinolo; Myocord; Prenormine; Seles Beta; Selobloc; Teno-basan; Tenoblock; Tenormin; Uniloc. Antianginal; class II antiarrhythmic agent. Cardioselective β-adrenergic blocking agent. Has antiarrhythmic and antihypertensive properties. mp = 146-148°, 150-152°; λ_m = 225, 275, 283 nm (MeOH); very soluble in MeOH; soluble in AcOH, DMSO; less soluble in Me_2CO, dioxane; insoluble in CH_3CN, EtOAc, $CHCl_3$; LD_{50} (mus orl) = 2000 mg/kg, (mus iv) = 98.7 mg/kg, (rat orl) = 3000 mg/kg, (rat iv) = 59.24 mg/kg. *Apothecon; C.M. Ind.; ICI ; Lemmon Co.; Zeneca Pharm.*

7005 Befunolol
39552-01-7 1050
$C_{16}H_{21}NO_4$
7-[2-Hydroxy-3-(isopropylamino)propoxy]-2-benzofuranyl methyl ketone.
A β-adrenergic blocker. Used as an antiglaucoma agent.

mp = 115°; LD_{50} (mus iv) = 100-105 mg/kg. *Kakenyaku Kako.*

7006 Befunolol Hydrochloride
39543-79-8 1050
$C_{16}H_{22}ClNO_4$
7-[2-Hydroxy-3-(isopropylamino)propoxy]-2-benzofuranyl methyl ketone hydrochloride.
BFE-60; Benfuran; Bentos; Bentox; Glauconex. A β-adrenergic blocker. Used as an antiglaucoma agent. mp = 163°. *Kakenyaku Kako.*

7007 R-(+)-Befunolol Hydrochloride
66685-79-8 1050
$C_{16}H_{22}ClNO_4$
(R)-(+)-7-[2-Hydroxy-3-(isopropylamino)propoxy]-2-benzofuranyl methyl ketone hydrochloride.
A β-adrenergic blocker. Used as an antiglaucoma agent. mp = 151°; $[\alpha]_D$ = 15.3° (c = 1 MeOH). *Kakenyaku Kako.*

7008 S-(-)-Befunolol Hydrochloride
66717-59-7 1050
$C_{16}H_{22}ClNO_4$
(S)-(-)-7-[2-Hydroxy-3-(isopropylamino)propoxy]-2-benzofuranyl methyl ketone hydrochloride.
A β-adrenergic blocker. Used as an antiglaucoma agent. mp = 151-152°; $[\alpha]_D$ = -125.5° (c = 1 MeOH). *Kakenyaku Kako.*

7009 Betaxolol
63659-18-7 1229
$C_{18}H_{29}NO_3$
(±)-1-[p-[2-(Cyclopropylmethoxy)ethyl]phenoxy]-3-(isopropylamino)-2-propanol.
Antianginal agent with antihypertensive and antiglaucoma properties. Cardioselective β_1 adrenergic blocker. mp = 70-72°. *Alcon Labs; Synthelabo Pharmacie.*

7010 Betaxolol Hydrochloride
63659-19-8 1229 264-384-3
$C_{18}H_{30}ClNO_3$
(±)-1-[p-[2-(Cyclopropylmethoxy)ethyl]phenoxy]-3-(isopropylamino)-2-propanol hydrochloride.
SLD-212; SL-75.212; Betoptic; Betoptima; Kerlone. Antianginal agent with antihypertensive and antiglaucoma properties. Cardioselective β_1 adrenergic blocker. mp = 116°; LD_{50} (mus orl) = 94 mg/kg, (mus iv) = 37 mg/kg. *Alcon Labs; Synthelabo Pharmacie.*

7011 Bevantolol
59170-23-9 1238
$C_{20}H_{27}NO_4$
(±)-1-[(3,4-Dimethoxyphenethyl)amino]-3-(m-toloxy)-2-propanol.
Antianginal agent. Cardioselective β-adrenergic blocking agent. Has antiarrhythmic and antihypertensive properties. *Parke-Davis.*

7012 Bevantolol Hydrochloride
42864-78-8 1238
$C_{20}H_{28}ClNO_4$
(±)-1-[(3,4-Dimethoxyphenethyl)amino]-3-(m-toloxy)-2-propanol hydrochloride.
Vantol; CI-775; Ranestol; Sentiloc. Antianginal agent. Cardioselective β-adrenergic blocking agent. Has antiarrhythmic and antihypertensive properties. mp = 137-138°. *Parke-Davis.*

7013 Bisoprolol
66722-44-9 1336
$C_{18}H_{31}NO_4$
(±)-1-[[α(2-Isopropoxyethoxy)-p-tolyl]oxy]-3-(isopropylamino)-2-propanol.
EMD-33512. Antihypertensive. Cardioselective β_1-adrenergic blocker. *E. Merck.*

7014 Bisoprolol Hemifumarate
104344-23-2 1336
$C_{20}H_{33}NO_6$
(±)-1-[[α(2-Isopropoxyethoxy)-p-tolyl]oxy]-3-(isopropylamino)-2-propanol hemifumarate.
Concor; Detensiel; Emvoncor; Emcor; Eurtadal; Isoten; Monocor; Soprol; Zebeta; component of: Ziac. Antihypertensive. Cardioselective β_1-adrenergic blocker. mp = 100°; soluble in EtOH. *E. Merck.*

7015 Bopindolol
62658-63-3 1362
$C_{23}H_{28}N_2O_3$
(±)-1-(tert-Butylamino)-3-[(2-methylindol-4-yl)oxy]-2-propanol benzoate (ester).
Antihypertensive. Non-selective β-adrenergic blocker. Soluble in Et_2O, CH_2Cl_2; LD_{50} (mus iv) = 17 mg/kg. *Sandoz Pharm. Corp.*

7016 Bopindolol Maleate
62658-64-4 1362
$C_{27}H_{32}N_2O_7$
(±)-1-(tert-Butylamino)-3-[(2-methylindol-4-yl)oxy]-2-propanol benzoate (ester) maleate.
Sandonorm. Antihypertensive. Non-selective β-adrenergic blocker. *Sandoz Pharm. Corp.*

7017 Bopindolol Malonate
82857-38-3 1362
$C_{26}H_{32}N_2O_7$
(±)-1-(tert-Butylamino)-3-[(2-methylindol-4-yl)oxy]-2-propanol benzoate (ester) malonate.
LT-31-200; Wandonorm. Antihypertensive. Non-selective β-adrenergic blocker. *Sandoz Pharm. Corp.*

7018 Brefonalol
104051-20-9
$C_{22}H_{28}N_2O_2$
(±)-6-[2-[(1,1-Dimethyl-3-phenylpropyl)amino]-1-hydroxy-ethyl]-3,4-dihydrocarbostyril.
A beta-adrenergic blocker.

7019 Broxaterol
76596-57-1 278-494-4
$C_9H_{15}BrN_2O_2$
(±)-3-Bromo-α-[(tert-butylamino)methyl]-5-isoxazolemethanol.
A beta-adrenergic receptor antagonist.

7020 Bucumolol
58409-59-9 1489
$C_{17}H_{23}NO_4$
8-[(3-tert-Butylamino)-2-hydroxypropoxy]-5-methylcoumarin.

Antianginal agent. A β-adrenergic blocker, antianginal and antiarrhythmic properties. *Sankyo Co. Ltd.*

7021 Bucumolol Hydrochloride
30073-40-6 1489
$C_{17}H_{24}ClNO_4$
8-[(3-tert-Butylamino)-2-hydroxypropoxy]-5-methylcoumarin hydrochloride.
CS-359; Bucumarol. Antianginal agent. A β-adrenergic blocker with antianginal and antiarrhythmic properties. mp = 226-228°; LD_{50} (mmus orl) = 676 mg/kg, (mmus iv) = 33.1 mg/kg, (fmus orl) = 692 mg/kg, (fmus iv) = 31.6 mg/kg. *Sankyo Co. Ltd.*

7022 Bufetolol
53684-49-4 1496
$C_{18}H_{29}NO_4$
1-(tert-Butylamino)-3-[o-[(tetrahydrofurfuryl)oxy]phenoxy]-2-propanol.
Antianginal agent. A β-adrenergic blocker with antianginal and antiarrhythmic properties. $bp_{0.07}$ = 180-186°. *Yoshitomi.*

7023 Bufetolol Hydrochloride
35108-88-4 1496 252-369-4
$C_{18}H_{30}ClNO_4$
1-(tert-Butylamino)-3-[o-[(tetrahydrofurfuryl)-oxy]phenoxy]-2-propanol hydrochloride.
Y-6124; Adobiol. Antianginal agent. A β-adrenergic blocker with antianginal and antiarrhythmic properties. mp = 153.5-157°, 151-154°; soluble in H_2O, MeOH, AcOH; slightly soluble in C_6H_6; insoluble in Et_2O; LD_{50} (mus orl) = 409 mg/kg, (mus sc) = 507 mg/kg, (rat orl) = 1142 mg/kg, (rat sc) = 1904 mg/kg. *Yoshitomi.*

7024 Bufuralol
54340-62-4 1504 259-112-5
$C_{16}H_{23}NO_2$
α-[(tert-Butylamino)methyl]-7-ethyl-2-benzofuranmethanol.
A β-adrenergic blocker with peripheral vasodilating activity. Antianginal with antihypertensive properties. *Hoffmann-LaRoche Inc.*

7025 Bufuralol Hydrochloride
59652-29-8 1504
$C_{16}H_{24}ClNO_2$
α-[(tert-Butylamino)methyl]-7-ethyl-2-benzofuranmethanol hydrochloride.
Ro-3-4787; Angium. A β-adrenergic blocker with peripheral vasodilating activity. Antianginal with antihypertensive properties. mp = 146°; LD_{50} (mus iv) = 29.7 mg/kg, (mus ip) = 88.0 mg/kg, (mus orl) = 177 mg/kg, (rat sc) = 1400 mg/kg, (rat orl) = 750 mg/kg; [(+)-hydrochloride]: mp = 122-123°; $[\alpha]_{365}^{20}$ = 135.0° (c = 1 EtOH); [(-)-hydrochloride]: mp = 122-123°; $[\alpha]_{365}^{20}$ = -136.0° (c = 1 EtOH). *Hoffmann-LaRoche Inc.*

7026 Bunitrolol
34915-68-9 1514
$C_{14}H_{20}N_2O_2$
o-[3-(tert-Butylamino)-2-hydroxypropoxy]benzonitrile.
Ko-1366. Antianginal; antiarrhythmic; antihypertensive. A β-adrenergic blocker. *Boehringer Ingelheim GmbH.*

7027 Bunitrolol Hydrochloride
23093-74-5 1514 245-427-5
$C_{14}H_{21}ClN_2O_2$
o-[3-(tert-Butylamino)-2-hydroxypropoxy]benzonitrile hydrochloride.
Betriol; Stresson. Antianginal agent with antiarrhythmic and antihypertensive properties. A β-adrenergic blocker. mp = 163-165°; LD_{50} (mus orl) = 1344-1440 mg/kg, (mus ip) = 264-265 mg/kg, (rat orl) = 639-649 mg/kg, (rat ip) = 222-225 mg/kg. *Boehringer Ingelheim GmbH.*

7028 Bupranolol
14556-46-8 1521
$C_{14}H_{22}ClNO_2$
1-(tert-Butylamino)-3-[(6-chloro-m-tolyl)-oxy]-2-propanol.
1-(6-chloro-3-methylphenoxy-3-tert-butylaminopropan-2-ol; bupranol; Ophtorenin. Antianginal agent with antiarrhythmic, antihypertensive and antiglaucoma properties. A β-adrenergic blocker.

7029 Bupranolol Hydrochloride
15148-80-8 1521 239-208-3
$C_{14}H_{23}Cl_2NO_2$
1-(tert-Butylamino)-3-[(6-chloro-m-tolyl)-oxy]-2-propanol hydrochloride.
B-1312; KL-255; Betadran; Betadrenol; Looser; Panimit. Antianginal agent with antiarrhythmic, antihypertensive and antiglaucoma properties. A β-adrenergic blocker. mp = 220-222°.

7030 Butidrine
55837-18-8 1558 259-849-2
$C_{16}H_{25}NO$
5,6,7,8-Tetrahydro-α-[[(1-isopropyl)amino]methyl]-2-naphthalenemethanol.
Antiarrhythmic agent. A β-adrenergic blocker. *Holding Ceresia.*

7031 Butidrine Hydrochloride
1506-12-3 1558
$C_{16}H_{26}ClNO$
5,6,7,8-Tetrahydro-α-[[(1-isopropyl)amino]methyl]-2-naphthalene-methanol hydrochloride.
butydrine hydrochloride; Betabloc; Recetan. Antiarrhythmic agent. A β-adrenergic blocker. mp = 129-130°; LD_{50} (mus iv) = 20.2 mg/kg, (mus orl) = 235 mg/kg. *Holding Ceresia.*

7032 Butofilolol
64552-17-6 1563
$C_{17}H_{26}FNO_3$
(±)-2'-[3-(tert-Butylamino)-2-hydroxypropoxyl-5'-fluorobutyrophenone.
CM-6805. Antihypertensive. A β-adrenergic blocker. mp = 88-89°. *C.M. Ind.*

7033 Butofilolol Maleate
88606-96-6 1563 289-431-5
$C_{21}H_{30}FNO_7$
(±)-2'-[3-(tert-Butylamino)-2-hydroxypropoxyl-5'-fluorobutyrophenone maleate.
Cafide. Antihypertensive. A β-adrenergic blocker. *C.M. Ind.*

7034 Carazolol
57775-29-8 1822 260-945-1
$C_{18}H_{22}N_2O_2$
1-(Carbazol-4-yloxy)-3-(isopropylamino)-2-propanol.
BM-51052; Conducton; Suacron. A β-adrenergic blocker; antihypertensive; antianginal; antiarrhythmic. Used for treatment of stress in pigs (veterinary). [hydrochloride]: mp = 234-235°. *Boehringer Mannheim GmbH.*

7035 Carteolol
51781-06-7 1917
$C_{16}H_{24}N_2O_3$
5-[3-[(1,1-Dimethylethyl)amino]-2-hydroxypropyl]-3,4-dihydro-2(1H)-quinolinone.
Antianginal agent with antiarrhythmic, antihypertensive and antiglaucoma properties. A β-adrenergic blocker. *Abbott Labs Inc.; Otsuka Am. Pharm.*

7036 Carteolol Hydrochloride
51781-21-6 1917 257-415-7
$C_{16}H_{25}ClN_2O_3$
5-[3-[(1,1-Dimethylethyl)amino]-2-hydroxypropyl]-3,4-dihydro-2(1H)-quinolinone hydrochloride.
Carteolol hydrochloride, Abbott 43326; OPC-1085; Arteoptic; Caltidren; Carteol; Cartrol; Endak; Mikelan; Optipress; Tenalet; Tenalin; Teoptic. Antianginal agent with antiarrhythmic, antihypertensive and antiglaucoma properties. A β-adrenergic blocker. mp = 278°; LD_{50} (mrat orl) = 1380 mg/kg, (mrat iv) = 158 mg/kg, (mrat ip) = 380 mg/kg, (mmus orl) = 810 mg/kg, (mmus iv) = 54.5 mg/kg, (mmus ip) = 380 mg/kg. *Abbott Labs Inc.; Otsuka Am. Pharm.*

7037 Carvediol
72956-09-3 1924
$C_{24}H_{26}N_2O_4$
(±)-1-(Carbazol-4-yloxy)-3-[[2-(o-methoxyphenoxy)-ethyl]amino]-2-propanol.
Coreg; BM-14190; DQ-2466; Dilatrend; Dimitone; Eucardic; Kredex; Querto. Antianginal; antihypertensive. Non-selective β-adrenergic blocker with vasodilating activity. mp = 114-115°. *Boehringer Mannheim GmbH; SmithKline Beecham Pharm.*

7038 Celiprolol
56980-93-9 2007 260-497-7
$C_{20}H_{33}N_3O_4$
3-[3-Acetyl-4-[3-(tert-butylamino)-2-hydroxypropoxy]-phenyl]-1,1-diethylurea.
ST-1396. Antianginal; antihypertensive agent. Cardioselective $β_1$ adrenergic blocker. mp = 110-112°. *Hoechst Marion Roussel Inc.*

7039 Celiprolol Hydrochloride
57470-78-7 2007 260-752-2
$C_{20}H_{34}ClN_3O_4$
3-[3-Acetyl-4-[3-(tert-butylamino)-2-hydroxypropoxy]-phenyl]-1,1-diethylurea monohydrochloride.
Celectol; Corliprol; Selecor; Selectol. Antianginal; antihypertensive agent. Cardioselective $β_1$ adrenergic blocker. mp = 197-200° (dec); soluble in H_2O (15.1 g/100 ml), MeOH (18.2 g/100 ml), EtOH (1.61 g/100 ml), $CHCl_3$ (0.42 g/100 ml); $λ_m$ = 221, 324 nm ($E_{1\%}$ 652, 57 H_2O), 231, 324 nm ($E_{1\%}$ 660, 60 0.01N HCl), 231, 324 nm ($E_{1\%}$ 640, 60 0.01N NaOH), 232, 329 nm ($E_{1\%}$ 775, 58 MeOH); LD_{50} (mmus iv) = 56.2 mg/kg, (mmus orl) = 1834 mg/kg, (mrat iv) = 68.3 mg/kg, (mrat orl) = 3826 mg/kg. *Hoechst Marion Roussel Inc.*

7040 Cetamolol
34919-98-7 2060
$C_{16}H_{26}N_2O_4$
(±)-2-[o-[3-(tert-Butylamino)-2-hydroxypropoxy]phenoxy]-N-methylacetamide.
Cardioselective $β_2$-adrenergic blocker. Used as an antihypertensive. mp = 96-97°. *ICI.*

7041 Cetamolol Hydrochloride
77590-95-5 2060 278-729-0
$C_{16}H_{27}ClN_2O_4$
(±)-2-[o-[3-(tert-Butylamino)-2-hydroxypropoxy]-phenoxy]-N-methylacetamide hydrochloride.
AI-27303; ICI-72222; Betacor. Cardioselective $β_2$-adrenergic blocker. Used as an antihypertensive. *ICI.*

7042 Cloranolol
39563-28-5 2464
$C_{13}H_{19}Cl_2NO_2$
1-(tert-Butylamino)-3-(2,5-dichlorophenoxy)-2-propanol.
A β-adrenergic blocker used as an antiarrhythmic agent. mp = 82-83°. *Res. Inst. Pharm. Chem.*

7043 Cloranolol Hydrochloride
54247-25-5 2464 259-044-6
$C_{13}H_{20}Cl_3NO_2$
1-(tert-Butylamino)-3-(2,5-dichlorophenoxy)-2-propanol hydrochloride.
GYKI-40199; Tobanum. A β-adrenergic blocker used as an antiarrhythmic agent. mp = 210-212°. *Res. Inst. Pharm. Chem.*

7044 Dilevalol
75659-07-3 3245
$C_{19}H_{24}N_2O_3$
(-)-5-[(1R)-1-Hydroxy-2-[[(1R)-1-methyl-3-phenylpropyl]-amino]ethyl]salicylamide.
R,R-labetalol; Sch-19927; Dilevalon; Levadil; Unicard. Non-selective β-adrenergic blocker with vasodilating and antihypertensive properties. Active isomer of labetalol. $[α]_D$ = -21.7°. *Schering-Plough HealthCare Products.*

7045 Dilevalol Hydrochloride
75659-08-4 3245
$C_{19}H_{25}ClN_2O_3$
(-)-5-[(1R)-1-Hydroxy-2-[[(1R)-1-methyl-3-phenylpropyl]-amino]ethyl]salicylamide monohydrochloride.
Sch-19927-HCl. Non-selective β-adrenergic blocker with vasodilating and antihypertensive properties. mp = 133-134° (dec), 192-193.5° (dec); $[α]_D^{26}$ = -30.6° (c = 1.0 EtOH). *Schering-Plough HealthCare Products.*

7046 Epanolol
86880-51-5 3641
$C_{20}H_{23}N_3O_4$
(±)-N-[2-[[3-(o-Cyanophenoxy)-2-hydroxypropyl]-amino]ethyl]-2-(p-hydroxyphenyl)acetamide.
ICI-141292; Visacor. Antianginal; antihypertensive agent. Cardioselective $β_1$-adrenergic blocker with sympathomimetic activity. mp = 118-120°. *ICI.*

7047 Esmolol
103598-03-4 3741
$C_{16}H_{25}NO_4$
(±)-Methyl p-[2-hydroxy-3-(isopropylamino)propoxy]-hydrocinnamate.
Class II antiarrhythmic. Ultra-short-acting β-adrenergic blocker. mp = 48-50°. *Am. Hospital Supply.*

7048 Esmolol Hydrochloride
81161-17-3 3741
$C_{16}H_{26}ClNO_4$
(±)-Methyl p-[2-hydroxy-3-(isopropylamino)propoxy]-hydrocinnamate hydrochloride.
ASL-8052; Brevibloc. Class II antiarrhythmic. Ultra-short-acting β-adrenergic blocker. mp = 85-86°. *Am. Hospital Supply.*

7049 Falintolol
90581-63-8
$C_{12}H_{24}N_2O_2$
Cyclopropyl methyl ketone (±)-(EZ)-O-[3-(tert-butylamino)-2-hydroxypropyl]oxime.
A beta-adrenergic antagonist; for treatment of glaucoma.

7050 Idropranolol
27581-02-8
$C_{16}H_{23}NO_2$
1-[(5,6-Dihydro-1-naphthyl)oxy]-3-(isopropylamino)-2-propanol.
A β-adrenergic blocker.

7051 Indenolol
60607-68-3 4974 262-323-5
$C_{15}H_{21}NO_2$
1-[1H-Inden-4 (or -7)-yloxy]-3-(isopropylamino)-2-propanol.
Sch-28316Z; YB-2. Antiarrhythmic; antihypertensive; antianginal agent. Non-selective β-adrenergic blocker. A 1:2 tautomeric mixture of the 4- and 7- indenyloxy isomers. mp = 88-89°. *Schering-Plough HealthCare Products; Yamanouchi U.S.A. Inc.*

7052 Indenolol Hydrochloride
81789-85-7 4974
$C_{15}H_{22}ClNO_2$
1-[1H-Inden-4 (or -7)-yloxy]-3-[(1-methylethyl)amino]-2-propanol hydrochloride.
Pulsan; Securpres. Antihypertensive; antiarrhythmic; antianginal. Non-selective β-adrenergic blocker. mp = 147-148°; LD_{50} (mus iv) = 26 mg/kg. *Schering AG; Yamanouchi U.S.A. Inc.*

7053 Labetalol
36894-69-6 5341 253-258-3
$C_{19}H_{24}N_2O_3$
5-[1-Hydroxy-2-[(1-methyl-3-phenylpropyl)amino]-ethyl]salicylamide.
ibidomide. Competitive α- and β-adrenergic receptor antagonist. Used as an antihypertensive. The (R,R) isomer is dilevalol. *Glaxo Wellcome Inc.; Key Pharm.*

7054 Labetalol Hydrochloride
32780-64-6 5341 251-211-1
$C_{19}H_{25}ClN_2O_3$
5-[1-Hydroxy-2-[(1-methyl-3-phenylpropyl)amino]ethyl]s-alicylamide hydrochloride.
Normodyne; Trandate; Sch-15719W; AH-5158A; Amipress; Ipolab; Labelol; Labracol; Presdate; Pressalolo; Vescal. Competitive α- and β-adrenergic receptor antagonist. Used as an antihypertensive. mp = 187-189°; soluble in H_2O, EtOH; insoluble in Et_2O, $CHCl_3$; LD_{50} (mmus ip) = 114 mg/kg, (mmus iv) = 47 mg/kg, (mmus orl) = 1450 mg/kg, (fmus ip) = 120 mg/kg, (fmus iv) = 54 mg/kg, (fmus orl) = 1800 mg/kg, (mrat ip) = 113 mg/kg, (mrat iv) = 60 mg/kg, (mrat orl) = 4550 mg/kg, (frat ip) = 107 mg/kg, (frat iv) = 53 mg/kg, (frat ol) = 4000 mg/kg. *C.M. Ind.; Glaxo Wellcome Inc.; Key Pharm.*

7055 Landiolol
133242-30-5
$C_{25}H_{39}N_3O_8$
(-)-[(S)-2,2-Dimethyl-1,3-dioxolan-4-yl]methyl p-[(S)-2-hydroxy-3-[[2-(4-morpholinecarboxamido)-ethyl]amino]propoxy]hydrocinnamate.
An ultra-short acting, selective beta blocker.

7056 Levobunolol
47141-42-4 5488
$C_{17}H_{25}NO_3$
(-)-5-[3-(tert-Butylamino)-2-hydroxypropoxy]-3,4-dihydro-1(2H)-naphthalenone.
l-bunolol; W-6421A. Adrenergic β-receptor; antiglaucoma agent. LD_{50} (mrat orl) = 700 mg/kg, (mrat iv) = 25 mg/kg, (frat orl) = 800 mg/kg, (frat iv) = 28 mg/kg, (mmus orl) = 1530 mg/kg, (mmus iv) = 78 mg/kg, (fmus orl) = 1220 mg/kg, (fmus iv) = 84 mg/kg, (mhmtr orl) = 435 mg/kg, (fhmtr orl) = 500 mg/kg; (dog orl) = 100 mg/kg. *Allergan Inc.*

7057 Mepindolol
23694-81-7 5901 245-831-1
$C_{15}H_{22}N_2O_2$
1-[Isopropylamino]-3-[(2-methyl-indol-4-yl)oxy]-2-propanol.
SH-E-222 [as sulfate salt]; Betagon [as sulfate salt]; Corindolan [as sulfate salt]; Mepicor [as sulfate salt]. Antianginal; antihypertensive agent. A β-adrenergic blocker. mp = 100-102°, 95-97°. *Sandoz Pharm. Corp.*

7058 Metipranolol
22664-55-7 6221 245-151-5
$C_{17}H_{27}NO_4$
(±)-1-(4-Hydroxy-2,3,5-trimethylphenoxy)-3-(isopropyl-amino)-2-propanol 4-acetate.
OptiPranolol; BM01.004; VUFB6453; Betamet; methypranol; trimepranol; Betanol; Disorat; Glauline; Glausyn; Turoptin. A β-adrenergic blocker. Antihypertensive agent also used to treat glaucoma. mp = 105-107°, 108.5-110.5°; λ_m = 278, 274 nm ($A^{1\%}_{1\,cm}$ = 51.3, 50.5 MeOH); freely soluble in EtOH, $CHCl_3$, C_6H_6, slightly soluble in Et_2O, insoluble in H_2O; LD_{50} (mus iv) = 31 mg/kg. *Bausch & Lomb Pharm. Inc.; SPOFA.*

7059 Metipranolol Hydrochloride
36592-77-5 6221
$C_{17}H_{28}ClNO_4$
(±)-1-(4-Hydroxy-2,3,5-trimethylphenoxy)-3-(isopropy-lamino)-2-propanol 4-acetate hydrochloride.
Betamann; Optipranolol. A β-adrenergic blocker. Antihypertensive agent also used to treat glaucoma. Soluble in H_2O. *Bausch & Lomb Pharm. Inc.; SPOFA.*

7060 Metoprolol
37350-58-6 6235 253-483-7
$C_{15}H_{25}NO_3$
1-(Isopropylamino)-3-[p-(2-methoxyethyl)phenoxy]-2-propanol.
CGP-2175; H-93/26. Antianginal; class II antiarrhythmic; antihypertensive. A β-adrenergic blocker. Lacks inherent sympathomimetic activity. *Apothecon; Astra Chem. Ltd.; Ciba-Geigy Corp.; Lemmon Co.*

7061 Metoprolol Fumarate
119637-66-0 6235
$C_{34}H_{54}N_2O_{10}$
1-(Isopropylamino)-3-[p-(2-methoxyethyl)phenoxy]-2-propanol fumarate (2:1) (salt).
Lopressor OROS; CGP-2175C. Antianginal; class II antiarrhythmic; antihypertensive. A β-adrenergic blocker. Lacks inherent sympathomimetic activity. Insoluble in EtOAc, Me_2CO, Et_2O, C_7H_{16}. *Ciba-Geigy Corp.*

7062 Metoprolol Succinate
98418-47-4 6235
$C_{34}H_{56}N_2O_{10}$
1-(Isopropylamino)-3-[p-(2-methoxyethyl)phenoxy]-2-propanol succinate (2:1) (salt).
Toprol XL; H-93/26 succinate. Antianginal; class II antiarrhythmic; antihypertensive. β-adrenergic blocker. Lacks inherent sympathomimetic activity. *Apothecon; Astra Chem. Ltd.; Astra Sweden; Ciba-Geigy Corp.; Lemmon Co.*

7063 Metoprolol Tartrate
56392-17-7 6235 260-148-9
$C_{34}H_{56}N_2O_{12}$
1-(Isopropylamino)-3-[p-(2-methoxyethyl)phenoxy]-2-propanol (2:1) dextro-tartrate salt.
CGP-2175E; HCTCGP 2175E; Beloc; Betaloc; Lopressor; Lopresor; Prelis; Seloken; Selopral; Selo-Zok; component of: Lopressor HCT. Antianginal; class II antiarrhythmic; antihypertensive. A β-adrenergic blocker. Lacks inherent sympathomimetic activity. Soluble in MeOH (50 g/100 ml), H_2O (> 100 g/100 ml), $CHCl_3$ (49.6 g/100 ml), Me_2CO (0.11 g/100 ml), CH_3CN (0.089 g/100 ml); insoluble in C_6H_{14}; LD_{50} (fmus iv) = 118 mg/kg, (fmus orl) = 3090 mg/kg, (mrat iv) ≅ 90 mg/kg, (mrat orl) = 3090 mg/kg, (mrat iv) ≅ 90 mg/kg, (mrat orl) = 3090 mg/kg. *Apothecon; Astra Chem. Ltd.; Ciba-Geigy Corp.; Lemmon Co.; Sandoz Nutrition.*

7064 Moprolol
5741-22-0 6345 227-254-7
$C_{13}H_{21}NO_3$
1-(Isopropylamino)-3-(o-methoxyphenoxy)-2-propanol.
Antihypertensive agent with antianginal properties. mp = 82-83°; [l-form (levomoprolol)]: mp = 78-80°; [α] = -5.5° ± 0.2°. *ICI.*

7065 Moprolol Hydrochloride
27058-84-0 6345 248-195-3
$C_{13}H_{22}ClNO_3$
1-(Isopropylamino)-3-(o-methoxyphenoxy)-2-propanol hydrochloride.
SD-1601; Omeral. Antihypertensive agent with antianginal properties. mp = 110-112°; [l-form hydrochloride (Levotensin)]: mp = 121-123°; $[\alpha]_D^{25}$ = -16.3° (c = 5.0 EtOH). *ICI.*

7066 Nadolol
42200-33-9 6431 255-706-3
$C_{17}H_{27}NO_4$
1-(tert-Butylamino)-3-[(5,6,7,8-tetrahydro-cis-6,7-dihydroxy-1-naphthyl)oxy]-2-propanol.
Corgard; SQ-11725; Anabet; Solgol. Antihypertensive and antianginal agent. Class II antiarrhythmic. A β-adrenergic blocker. mp = 124-136°; λ_m = 270, 278 nm ($E_{1\ cm}^{1\%}$ 37.5, 39.1, MeOH); pKa = 9.67; poorly soluble in organic solvents; insoluble in Me_2CO, C_6H_6, Et_2O, C_6H_{14}, LD_{50} (rat orl) = 5300 mg/kg, (mus orl) = 4500 mg/kg. *Apothecon; Bristol-Myers Squibb Co.*

7067 Nadoxolol
54063-51-3 6432
$C_{14}H_{16}N_2O_3$
3-Hydroxy-4-(1-naphthyloxy)butyramidoxime.
antiarrhythmic agent. Antihypertensive, antianginal. Used for the treatment of cardiovascular disorders. A β-adrenergic blocker. *Orsymonde.*

7068 Nadoxolol Hydrochloride
35991-93-6 6432 252-825-2
$C_{14}H_{17}ClN_2O_3$
3-Hydroxy-4-(1-naphthyloxy)butyramidoxime hydrochloride.
LL-1530; Bradyl. Antihypertensive; antianginal. Used for the treatment of cardiovascular disorders. Antiarrhythmic agent. A β-adrenergic blocker. mp = 188°; soluble in H_2O, EtOH, MeOH; insoluble in Et_2O; LD_{50} (mus iv) = 180 mg/kg, (mus orl) = 1000 mg/kg. *Orsymonde.*

7069 Nebivalol
99200-09-6 6519
$C_{22}H_{25}F_2NO_4$
α,α'-(Iminodimethylene)bis[6-fluoro-2-chromanmethanol].
R-65824; dl-nebivolol; narbivolol; Nebilet. Antihypertensive agent. A β-adrenergic blocker. *Janssen Pharm. Inc.*

7070 Nifenalol
7413-36-7 6618 231-023-6
$C_{11}H_{16}N_2O_3$
α-[(Isopropylamino)methyl]-p-nitrobenzyl alcohol.
isophenethanol; INPEA. Antiarrhythmic agent; antianginal. A β-adrenergic blocker. mp = 98°. *Selvi.*

7071 Nifenalol Hydrochloride
5704-60-9 6618 227-194-1
$C_{11}H_{17}ClN_2O_3$
α-[(Isopropylamino)methyl]-p-nitrobenzyl alcohol hydrochloride.
Inpea. antiarrhythmic agent with antianginal properties. A β-adrenergic blocker. mp = 181°. *Selvi.*

7072 Nipradilol
81486-22-8 6655
$C_{15}H_{22}N_2O_6$
8-[2-Hydroxy-3-(isopropylamino)propoxy]-3-chromanol.
Nipradolol; K-351; Hypadil. Antianginal; antihypertensive agent. A β-adrenergic blocker with vasodilating activity.

mp = 107-116°, 110-122°; LD_{50} (mus iv) = 74.0 mg/kg, (rat iv) = 73.0 mg/kg, (mus orl) = 540 mg/kg. *Kowa Chem. Ind. Co. Ltd.*

7073 Oxprenolol
6452-71-7 7086 229-257-9
$C_{15}H_{23}NO_3$
1-(o-Allyloxyphenoxy)-3-isopropylamino-2-propanol.
Antianginal; antihypertensive; class IV antiarrhythmic. β-adrenergic blocker. mp = 78-80°. *Bayer AG; Ciba-Geigy Corp.*

7074 Oxprenolol Hydrochloride
6452-73-9 7086 229-260-5
$C_{15}H_{24}ClNO_3$
1-(o-Allyloxyphenoxy)-3-isopropylamino-2-propanol hydrochloride.
Ba-39089; Coretal; Laracor; Paritane; Slow-Pren; Trasicor; Trasacor. Antianginal; antihypertensive; class IV antiarrhythmic. A β-adrenergic blocker. mp = 107-109°. *Bayer AG; Ciba-Geigy Corp.*

7075 Pacrinolol
65655-59-6
$C_{23}H_{28}N_2O_4$
(-)-p-[3-[(3,4-Dimethoxyphenethyl)amino]-2-hydroxy-propoxy]-β-methylcinnamonitrile.
Hoe-224A. A highly cardioselective beta-sympathicolytic with significant and long acting blood pressure lowering properties. A β-adrenergic blocker.

7076 Pafenolol
75949-61-0
$C_{18}H_{31}N_3O_3$
(±)-1-[p-[2-Hydroxy-3-(isopropylamino)propoxy]-phenethyl]-3-isopropylurea.
Selective adrenergic beta 1-blocking agent.

7077 Pargolol
47082-97-3
$C_{16}H_{23}NO_3$
1-(tert-Butylamino)-3-[o-(2-propynyloxy)phenoxy]-2-propanol.
beta Adrenergic blocker.

7078 Penbutolol
38363-40-5 7209
$C_{18}H_{29}NO_2$
(S)-1-(tert-Butylamino)-3-(o-cyclopentylphenoxy)-2-propanol.
Antianginal; antihypertensive; antiarrhythmic; β-adrener-gic calcium channel blocker. mp = 68-72°; $[\alpha]_D^{20}$ = -11.5° (c = 1 MeOH); soluble in MeOH, EtOH, $CHCl_3$. *Hoechst AG.*

7079 Penbutolol Sulfate
38363-32-5 7209 253-906-5
$C_{36}H_{40}N_2O_8S$
(S)-1-(tert-Butylamino)-3-(o-cyclopentylphenoxy)-2-propanol sulfate (salt) (2:1).
HOE-893d; HOE-39-893d; Betapressin; Levatol; Paginol. Antianginal; antihypertensive; antiarrhythmic. A β-adrenergic calcium channel blocker with antiarrhythmic activity. mp = 216-218° (dec); $[\alpha]_D^{20}$ = -24.6° (c = 1 MeOH). *Hoechst AG.*

7080 Penirolol
58503-83-6
$C_{15}H_{22}N_2O_2$
o-[2-Hydroxy-3-(tert-pentylamino)propoxy]benzonitrile.
A β-adrenergic blocker.

7081 Pindolol
13523-86-9 7597 236-867-9
$C_{14}H_{20}N_2O_2$
1-(Indol-4-yloxy)-3-(isopropylamino)-2-propanol.
LB-46; Visken; prinodolol; Betapindol; Blocklin L; Calvisken; Decreten; Durapindol; Glauco-Visken; Pectobloc; Pinbetol; Pynastin. Antianginal agent with antiarrhythmic, antihypertensive and antiglaucoma properties. A β-adrenergic blocker. mp = 171-163°. *Sandoz Pharm. Corp.*

7082 Pirepolol
69479-26-1
$C_{21}H_{32}N_4O_5$
(±)-6-[[2-[[3-(p-Butoxyphenoxy)-2-hydroxypropyl]amino]-ethyl]amino]-1,3-dimethyluracil.
A β-adrenergic blocker.

7083 Practolol
6673-35-4 7882 229-712-1
$C_{14}H_{22}N_2O_3$
4'-[2-Hydroxy-3-(isopropylamino)propoxy]acetanilide.
AY-21011; ICI-50172; Dalzic; Eraldin. Antiarrhythmic. A β-adrenergic blocker. mp = 134-136°; soluble in i-PrOH; [hydrochloride monhydrate]: mp = 140-142°; [R(+) form]: mp = 130-131.5°; $[\alpha]_{365}^{25}$ = 4.3° (EtOH), $[\alpha]_{578}^{25}$ = 3.5° (EtOH); [R(+) hydrochloride]: $[\alpha]_{436}^{25}$ = 26.0° (EtOH),. *ICI ; Wyeth-Ayerst Labs.*

7084 Procinolol
27325-36-6
$C_{15}H_{23}NO_2$
1-(o-Cyclopropylphenoxy)-3-(isopropylamino)-2-propanol.
SD-2124-01. A β-adrenergic blocker.

7085 Pronetalol
54-80-8 7974
$C_{15}H_{19}NO$
2-Isopropyl-1-(naphth-2-yl)ethanol.
pronethalol; nethalide. Antianginal; antihypertensive; antiarrhythmic. A β-adrenergic blocker. mp = 108°. *ICI ; Wyeth-Ayerst Labs.*

7086 Pronetalol Hydrochloride
51-02-5 7974
$C_{15}H_{20}ClNO$
2-Isopropylamino-1-(naphth-2-yl)ethanol hydrochloride.
pronethalol hydrochloride; ICI-38174; AY-6204; Alderlin. Antianginal; antihypertensive; antiarrhythmic. A β-adrenergic blocker. mp = 184°; LD_{50} (mus orl) = 512 mg/kg, (mus iv) = 46 mg/kg. *ICI ; Wyeth-Ayerst Labs.*

7087 Propranolol
525-66-6 8025 208-378-0
$C_{16}H_{21}NO_2$
1-(Isopropylamino)-3-(1-naphthyloxy)-2-propanol.
Avlocardyl; Euprovasin. Antianginal; antihypertensive;

antiarrhythmic (class II) agent. A β-adrenergic blocker. mp = 96°. *ICI ; Parke-Davis; Quimicobiol; Wyeth-Ayerst Labs.*

7088 Propranolol Hydrochloride
318-98-9 8025 206-268-7
$C_{16}H_{22}ClNO_2$
1-(Isopropylamino)-3-(1-naphthyloxy)-2-propanol hydrochloride.
1-[(1-methylethyl)amino]-3-(1-naphthalenyloxy)-2-propanol; AY-64043; ICI-45520; NSC-91523; Angilol; Apsolol; Bedranol; Beprane; Berkolol; Beta-Neg; Beta-Tablinen; Beta-Timelets; Cardinol; Caridolol; Deralin; Dociton; Duranol; Efektolol; Elbol; Frekven; Inderal; Indobloc; Intermigran; Kemi S; Oposim; Prano-Puren; Prophylux; Propranur; Pylapron; Rapynogen; Sagittol; Servanolol; Sloprolol; Sumial; Tesnol. Antianginal; antihypertensive; antiarrhythmic (class II) agent. A β-adrenergic blocker. mp = 163-164°; soluble in H_2O, alcohol; insoluble in Et_2O, C_6H_6, EtOAc; LD_{50} (mus orl) = 565 mg/kg, (mus iv) = 22 mg/kg, (mus ip) = 107 mg/kg. *ICI ; Parke-Davis; Quimicobiol; Wyeth Labs.*

7089 Ridazolol
83395-21-5
$C_{15}H_{18}Cl_2N_4O_3$
(±)-4-Chloro-5-[[2-[[3-(o-chlorophenoxy)-2-hydroxypropyl]amino]ethyl]amino]-3(2H)-pyridazinone.
A β-adrenergic blocker.

7090 Soquinolol
61563-18-6
$C_{17}H_{26}N_2O_3$
5-[3-(tert-Butylamino)-2-hydroxypropoxy]-3,4-dihydro-2(1H)-isoquinoline-carboxaldehyde.
We-704; Sertum. Highly potent non-subtype-selective β-adrenergic receptor blocker.

7091 Sotalol
3930-20-9 8876
$C_{12}H_{20}N_2O_3S$
4'-[1-Hydroxy-2-(isopropylamino)-ethyl]methanesulfonanilide.
Antianginal; class II and III antiarrhythmic; antihypertensive. A β-adrenergic blocker. λ_m = 242.2, 275.2 nm ($CHCl_3$). *Berlex Labs Inc.; Bristol-Myers Squibb Co.*

7092 Sotalol Hydrochloride
959-24-0 8876 213-496-0
$C_{12}H_{21}ClN_2O_3S$
4'-[1-Hydroxy-2-(isopropylamino)-ethyl]methanesulfonanilide monohydrochloride.
MJ-1999; Beta-Cardone; Betapace; Darob; Sotacor; Sotalex. Antianginal with class II and III antiarrhythmic properties. Antihypertensive. A β-adrenergic blocker. mp = 206.5-207° (has also been reported as 218-219°); soluble in H_2O, less soluble in $CHCl_3$; LD_{50} (mmus orl) = 2600 mg/kg, (mmus ip) = 670 mg/kg, (mrat orl) = 3450 mg/kg, (mrat ip) = 680 mg/kg, (rbt orl) = 1000 mg/kg, (dog ip) = 330 mg/kg. *Berlex Labs Inc.; Bristol-Myers Squibb Co.*

7093 Spirendolol
65429-87-0
$C_{21}H_{31}NO_3$
(±)-4'-[3-(tert-Butylamino)-2-hydroxypropoxy]spiro[cyclohexane-1,2'-indan]-1'-one.
A specific β-2 adrenergic receptor antagonist.

7094 Sulfinalol
66264-77-5 9120
$C_{20}H_{27}NO_4S$
4-Hydroxy-α-[[[3-(p-methoxyphenyl)-1-methylpropyl]amino]methyl]-3-(methylsulfinyl)benzyl alcohol.
A β-adrenergic blocker used as an antihypertensive agent. *Sterling Winthrop Inc.*

7095 Sulfinalol Hydrochloride
63251-39-8 9120 264-046-5
$C_{20}H_{28}ClNO_4S$
4-Hydroxy-α-[[[3-(p-methoxyphenyl)-1-methylpropyl]amino]methyl]-3-(methylsulfinyl)benzyl alcohol hydrochloride.
Win-408087; Perifadil. A β-adrenergic blocker used as an antihypertensive agent. mp = 172-175°. *Sterling Winthrop Inc.*

7096 Talinolol
57460-41-0 9208
$C_{20}H_{33}N_3O_3$
(±)-1-[p-[3-(tert-Butylamino)-2-hydroxypropoxy]phenyl]-3-cyclohexylurea.
02-115; Cordanum. Antiarrhythmic agent with antihypertensive properties. A β-adrenergic blocker related to practolol. mp = 142-144°; LD_{50} (rat orl) = 1180 mg/kg, (rat ip) = 54.3 mg/kg, (rat iv) = 29.7 mg/kg, (mus orl) = 593 mg/kg, (mus ip) = 74.7 mg/kg, (mus iv) = 25.0 mg/kg. *Ciba-Geigy Corp.*

7097 Teoprolol
65184-10-3 265-600-9
$C_{23}H_{30}N_6O_4$
3,7-Dihydro-7-[3-[[2-hydroxy-3-[(2-methyl-1H-indol-4-yl)oxy]propyl]amino]butyl]-1,3-dimethyl-1H-purine-2,6-dione.
A β-adrenergic blocker.

7098 Tertatolol
34784-64-0 9318
$C_{16}H_{25}NO_2S$
(±)-1-(tert-Butylamino)-3-(thiochroman-8-yloxy)-2-propanol.
A non-selective β-adrenergic blocker. Used as an antihypertensive agent. mp = 70-72°. *Sci. Union et Cie France.*

7099 Tertatolol Hydrochloride
33580-30-2 9318 251-578-8
$C_{16}H_{26}ClNO_2S$
(±)-1-(tert-Butylamino)-3-(thiochroman-8-yloxy)-2-propanol hydrochloride.
S-2395; SE-2395; Artex; Artexal; Prenalex. A non-selective β-adrenergic blocker. Used as an antihypertensive agent. mp = 180-183°; LD_{50} (rat iv) = 40 mg/kg, (rat ip) = 90 mg/kg, (mus iv) = 37 mg/kg, (mus ip) = 120 mg/kg. *Sci. Union et Cie France.*

7100 Tienoxolol
90055-97-3
$C_{21}H_{28}N_2O_5S$
(±)-Ethyl 2-[3-(tert-butylamino)-2-hydroxypropoxy]-5-(2-thienocarboxamido)benzoate.
A diuretic beta-blocking agent.

7101 Tilisolol
85136-71-6 9579
$C_{17}H_{24}N_2O_3$
(±)-4-[3-(tert-Butylamino)-2-hydroxypropoxy]-2-methylisocarbostyril.
Antiarrhythmic; antihypertensive. A β-adrenergic blocker. *Nisshin Denka K.K.*

7102 Tilisolol Hydrochloride
62774-96-3 9579
$C_{17}H_{25}ClN_2O_3$
(±)-4-[3-(tert-Butylamino)-2-hydroxypropoxy]-2-methylisocarbostyril hydrochloride.
N-696; Selecal. Antiarrhythmic; antihypertensive. A β-adrenergic blocker. mp = 203-205°; LD_{50} (mmus orl) = 1393 mg/kg, (mmus sc) = 1219 mg/kg, (mmus ip) = 578 mg/kg, (mmus iv) = 74.3 mg/kg, (fmus orl) = 1290 mg/kg, (fmus sc) = 1245 mg/kg, (fmus ip) = 557 mg/kg, (fmus iv) = 104.7 mg/kg, (mrat orl) = 145 mg/kg, (mrat sc) = . *Nisshin Denka K.K.*

7103 Timolol
91524-16-2 9585
$C_{13}H_{24}N_4O_3S.1/2H_2O$
(S)-1-(tert-Butylamino)-3-[(4-morpholino-1,2,5-thiadiazol-3-yl)oxy]-2-propanol hemihydrate.
Antianginal; antiarrhythmic (class II); antihypertensive; antiglaucoma agent. A β-adrenergic blocker. [(±) form]: mp = 71.5-72.5°. *Merck & Co.Inc.*

7104 Timolol Maleate
26921-17-5 9585 248-111-5
$C_{17}H_{28}N_4O_7S$
(S)-1-(tert-Butylamino)-3-[(4-morpholino-1,2,5-thiadiazol-3-yl)oxy]-2-propanol maleate (1:1).
Blocadren; Timoptic; Timoptol; MK-950; Aquanil; Betim; Proflax; Temserin; Tenopt; Timacar; Timacor; component of: Cosopt, Timolide. Antianginal agent with antiarrhythmic (class II), antihypertensive and antiglaucoma properties. A β-adrenergic blocker. mp = 201.5-202.5°; $[\alpha]_{405}^{24}$ = -12.0° (c = 5 1N HCl), $[\alpha]_D^{25}$ = -4.2°; λ_m = 294 nm ($A_{1\ cm}^{1\%}$ 200 0.1N HCl); soluble in EtOH, MeOH; poorly soluble in $CHCl_3$, cyclohexane, insoluble in isooctane, Et_2O. *Merck & Co.Inc.*

7105 Toliprolol
2933-94-0 9653 220-905-6
$C_{13}H_{21}NO_2$
1-(Isopropylamino)-3-(m-tolyloxy)-2-propanol.
MHIP. Antianginal; antihypertensive agent. A β-adrenergic blocker. mp = 75-76°, 79°. *Boehringer Ingelheim GmbH; ICI.*

7106 Toliprolol Hydrochloride
306-11-6 9653 206-177-2
$C_{13}H_{22}ClNO_2$
1-(Isopropylamino)-3-(m-tolyloxy)-2-propanol hydrochloride.
ICI-45763; Ko-592; Doberol; Sinorytmal. Antianginal; antihypertensive agent. A β-adrenergic blocker. mp = 120-121°; λ_m = 270 nm ($E_{1\ cm}^{1\%}$ 498 H_2O). *Boehringer Ingelheim GmbH; ICI.*

7107 Xibenolol
81584-06-7 10209
$C_{15}H_{25}NO_2$
(±)-1-(tert-Butylamino)-3-(2,3-xylyloxy)-2-propanol.
Antiarrhythmic agent. Non-selective β-adrenergic blocker. mp = 57°; $bp_{0.7}$ = 134-136°; λ_m = 271.2, 274, 279.3 nm (ε 10800, 10700, 11100 EtOH). *Teikoku Hormone Mfg. Co. Ltd.*

7108 Xibenolol Hydrochloride
59708-57-5 10209
$C_{15}H_{26}ClNO_2$
(±)-1-(tert-Butylamino)-3-(2,3-xylyloxy)-2-propanol hydrochloride.
D-32; Selapin; Rhythminal; Rythminal. Antiarrhythmic agent. Non-selective β-adrenergic blocker. mp = 135-137°. *Teikoku Hormone Mfg. Co. Ltd.*

Bone Resorption Inhibitors

7109 Alendronic Acid
66376-36-1 228
$C_4H_{13}NO_7P_2$
(4-Amino-1-hydroxybutylidene)bisphosphonic acid.
ABDP. Calcium regulator. mp = 233-235° (dec). *Inst. Gentili S.p.A.; Merck & Co.Inc.*

7110 Alendronic Acid Trisodium salt
121268-17-5 228
$C_4H_{12}NNaO_7P_2.3H_2O$
Sodium trihydrogen (4-amino-1hydroxybutylidene)diphosphonate trihydrate.
Fosamax; MK-217; G-704650; Adronat; Alendros; Onclast. Calcium regulator. *Inst. Gentili S.p.A.; Merck & Co.Inc.*

7111 Clodronic Acid
10596-23-3 2432 234-212-1
$CH_4Cl_2O_6P_2$
(Dichloromethylene)diphosphonic acid.
Cl_2MDP; DMDP. Calcium regulator. mp = 249-251°. *Procter & Gamble Pharm. Inc.*

7112 Clodronic Acid Disodium Salt
22560-50-5 2432 245-078-9
$CH_2Cl_2Na_2O_6P_2$
Sodium (dichloromethylene)diphosphonate.
DCIMDP; Bonefos; Clasteon; Difoafonal; Loron; Mebonat; Ossiten; Ostac. Calcium regulator. *Procter & Gamble Pharm. Inc.*

7113 Etidronic Acid
2809-21-4 3908 220-552-8
$C_2H_8O_7P_2$
(1-Hydroxyethylidene)diphosphonic acid.
EHDP. Calcium regulator. Very soluble in H_2O, insoluble in AcOH. *Procter & Gamble Pharm. Inc.*

7114 Etidronic Acid Disodium Salt
7414-83-7 3908 231-025-7
$C_2H_6Na_2O_7P_2$
Disodium dihydrogen (1-hydroxyethylidene)diphosphonate.
Didronel; Calcimux; Diphos; Etidron. Calcium regulator. Bone resorption inhibitor. Soluble in H_2O. *Procter & Gamble Pharm. Inc.*

7115 Pamidronic Acid
40391-99-9 7135 254-905-2
$C_3H_{11}NO_7P_2$
(3-Amino-1-hydroxypropylidene)-diphosphonic acid.
ADP; AHPrBP. Bone resorption inhibitor. *Ciba-Geigy Corp.*

7116 Pamidronic Acid Disodium salt
57248-88-1 7135 260-647-1
$C_3H_9NNa_2O_7P_2.5H_2O$
Disodium dihydrogen (3-amino-1-hydroxy-propylidene)diphosphonate pentahydrate.
Aredia; CGP-23339AE; Aminomux. Bone resorption inhibitor. *Ciba-Geigy Corp.*

7117 Risedronic Acid
105462-24-6 8396
$C_7H_{11}NO_7P_2$
[1-Hydroxy-2-(3-pyridyl)ethylidene] diphosphonic acid.
Calcium regulator, used as a bone resorption inhibitor. *Merck & Co.Inc.; Procter & Gamble Pharm. Inc.*

7118 Risedronic Acid Monosodium Salt
115436-72-1 8396
$C_7H_{10}NNaO_7P_2$
Sodium trihydrogen [1-hydroxy-2-(3-pyridyl)-ethylidene]diphosphonate.
Actonel; risedronate sodium; NE-58095. Calcium regulator, used as a bone resorption inhibitor in treatment of osteoporosis. *Merck & Co.Inc.; Procter & Gamble Pharm. Inc.*

7119 Tiludronate Disodium
149845-07-8
$C_7H_7ClNa_2O_6P_2S$
Disodium dihydrogen [[(p-chlorophenyl)thio]-methylene]diphosphonate.
SR-41319B. Used for treatment of and intervention in osteoporosis and treatment of Paget's disease. *Sanofi Winthrop.*

7120 Tiludronic Acid
89987-06-4 9582
$C_7H_9ClO_6P_2S$
[[(p-Chlorophenyl)thio]methylene] diphosphonic acid.
SR-41319; ACPMD; Cl-TMBP; ME-3737; Skelid. Used for treatment of and intervention in osteoporosis and treatment of Paget's disease. [di-tert-butylamine salt]: mp = 253° (dec). *Sanofi Winthrop.*

Bradycardic Agents

7121 Ivabradine
155974-00-8
$C_{27}H_{36}N_2O_5$
3-[3-[[[(7S)-3,4-Dimethoxybicyclo[4.2.0]octa-1,3,5-trien-7-yl]methyl]methylamino]propyl]-1,3,4,5-terahydro-7,8-dimethoxy-2H-3-benzazepin-2-one.
S-16257. Bradycardiac agent. Direct sinus node inhibitor.

7122 Zatebradine
85175-67-3 10245
$C_{26}H_{36}N_2O_5$
3-[3-[(3,4-Dimethoxyphenethyl)methylamino]propyl]-1,3,4,5-tetrahydro-7,8-dimethoxy-2H-3-benzazepin-2-one.
Antianginal agent. Specific bradycardic agent; sinus node inhibitor. *Boehringer Ingelheim GmbH.*

7123 Zatebradine Hydrochloride
91940-87-3 10245
$C_{26}H_{37}ClN_2O_5$
3-[3-[(3,4-Dimethoxyphenethyl)methylamino]propyl]-1,3,4,5-tetrahydro-7,8-dimethoxy-2H-3-benzazepin-2-one hydrochloride.
UL-FS-49. Antianginal agent. Specific bradycardic agent; sinus node inhibitor. mp = 168°, 185° (2 crystalline modifications); soluble in H_20. *Boehringer Ingelheim GmbH.*

Bradykinin Antagonists

7124 Bradycor
140661-97-8 1385
$C_{128}H_{194}N_{40}O_{28}S_2$
Dimer of D-Arg^0-[Hyp^3, Cys^6, D-Phe^7, Leu^8]-bradykinin joined by a bisuccinimdohexane linker via the sulfhydryl moieties of the cysteine residues.
Deltibant; CP-0127. A Bradykinin antagonist. Used in the treatment of systemic inflammatory response syndrome. *Cortech Inc.*

7125 Icatibant
130308-48-4
$C_{59}H_{89}N_{19}O_{13}S$
(R)-Arginyl-(S)-arginyl-(S)-prolyl-(2S,4R)-(4-hydroxyprolyl)glycyl-(S)-[3-(2-thienyl)alanyl]-(S)-seryl-(R)-[(1,2,3,4-tetrahydro-acetate (salt).
A Bradykinin antagonist. Used in the treatment of systemic inflammatory response syndrome. *Hoechst Roussel Pharm. Inc.*

7126 Icatibant Acetate
138614-30-9
$C_{59}H_{89}N_{19}O_{13}S.xC_2H_4O_2$
(R)-Arginyl-(S)-arginyl-(S)-prolyl-(2S,4R)-(4-hydroxyprolyl)glycyl-(S)-[3-(2-thienyl)alanyl]-(S)-seryl-(R)-[(1,2,3,4-tetrahydro-3-isoquinolyl)carbonyl]-(2S,3aS,7aS)-[(hexahydro-2-indolinyl)carbonyl]-(S)-arginine acetate (salt).
HOE-140. A Bradykinin antagonist. Used in the treatment of systemic inflammatory response syndrome. *Hoechst Roussel Pharm. Inc.*

Bronchodilators

7127 Acefylline
652-37-9 22 211-490-2
$C_9H_{10}N_4O_4$
1,2,3,6-Tetrahydro-1,3-dimethyl-2,6-dioxopurine-7-acetic acid.
carboxymethyltheophylline; 7-theophylline acetic acid; Aminodal [as sodium salt]. Diuretic; cardiotonic; bronchodilator. Xanthine derivative. mp = 271°; [sodium salt]: mp > 300°. *E. Merck.*

7128 Acefylline Clofibrol
70788-27-1
$C_{19}H_{21}ClN_4O_5$
2-(p-Chlorophenoxy)-2-methylpropyl 1,2,3,6-tetrahydro-1,3-dimethyl-2,6-dioxopurine-7-acetate.
Bronchodilator.

7129 Acefylline Piperazine
18428-63-2 23 242-614-3
$C_{13}H_{20}N_6O_4$
1,2,3,6-Tetrahydro-1,3-dimethyl-2,6-dioxopurine-7-acetic acid compound with piperazine.
piperazine theophylline ethanoate; piperazine theophylline-7-acetate; acepifylline; Dynaphylline; Etaphydel; Etaphylline; Etafillina. Bronchodilator. Undefined mixture of 2:1 and 1:1 salts containing 75% theophylline acetic acid and 25% anhydrous piperazine. Active metabolite is acefylline, a xanthine derivative.

7130 Acefylline Piperazine
18833-13-1 23 242-614-3
$C_{22}H_{30}N_{10}O_8$
Piperazine 7-theophyllineacetate.
Bronchodilator.

7131 Acefylline Sodium Salt
837-27-4 22 212-652-5
$C_9H_9N_4NaO_4$
Sodium 1,2,3,6-tetrahydro-1,3-dimethyl-2,6-dioxopurine-7-acetate.
Aminodal. Diuretic; cardiotonic; bronchodilator. Xanthine derivative. mp > 300°. *E. Merck.*

7132 Albuterol
18559-94-9 217 242-424-0
$C_{13}H_{21}NO_3$
2-(tert-Butylamino)-1-(4-hydroxy-3-hydroxymethyl-phenyl)ethanol.
salbutamol; Proventil Inhaler; Ventalin Inhaler. Bronchodilator; tocolytic. Ephedrine derivative. mp = 151°, 157-158°; soluble in most organic solvents. *Allen & Hanbury; Apothecon; Glaxo Wellcome Inc.; Key Pharm.*

7133 Albuterol Sulfate
51022-70-9 217 256-916-8
$C_{26}H_{44}N_2O_{10}S$
2-(tert-Butylamino)-1-(4-hydroxy-3-hydroxymethyl-phenyl)ethanol sulfate (2:1).
Sch-13949W Sulfate; Aerolin; Asmaven; Broncovaleas; Cetsim; Cobutolin; Ecovent; Loftan; Proventil; Salbumol; Salbutard; Salbutine; Salbuvent; Sultanol; Ventelin; Ventodiscks; Ventolin; Volma. Bronchodilator; tocolytic. Ephedrine derivative. *Allen & Hanbury; Apothecon; Glaxo Labs.; Key Pharm.; Lemmon Co.; Schering AG.*

7134 Ambuphylline
5634-34-4 404 227-077-5
$C_{11}H_{19}N_5O_3$
Theophylline compound with 2-amino-2-methyl-1-propanol (1:1).
2-amino-2-methyl-1-propanol compound with theophylline; bufylline (formerly); theophylline aminoisobutanol; Butaphyllamine; Buthoid; component of: Nethaphyl. Bronchodilator; diuretic; smooth muscle relaxant. Xanthine derivative. mp = 254-256°; soluble in H_2O (55 g/100 ml); LD_{50} (mus orl) = 600 mg/kg. *Marion Merrell Dow Inc.*

7135 Aminophylline
317-34-0 485 206-264-5
$C_{16}H_{24}N_{10}O_4$
3,7-Dihydro-1,3-dimethyl-1H-purine-2,6-dione compound with ethylenediamine (2:1).
theophylline compound with ethylenediamine (2:1); theophylline ethylenediamine; theophyllamine; Aminocardol; Aminodur; Aminophyllin; Cardiofilina; Cardiomin; Cardophylin; Carena; Diophyllin; Euphyllin CR; Genophyllin; Grifomin; Inophylline; Metaphyllin; Minaphol; Pecram; Peterphyllin; Phyllindon; Phyllocontin; Pylcardin; Rectalad Aminophylline; Somophyllin; Stenovasan; Tefamin; Theodrox; Theolamine; Theomin; Theophyldine; component of: Amesec, Mudrane GG Tablets, Mudrane GG-2 Tablets, Mudrane Tablets, Mudrane-2 Tablets. Smooth muscle relaxant, used as a bronchodilator. Xanthine derivative. CAUTION: Causes acute poisoning: restlessness, anorexia, nausea, fever, vomiting, dehydration, tremors, delirium, coma; may cause cardiovascular and respiratory collapse, shock, cyanosis, death. [Dihydrate]: soluble in H_2O (20 g/100 ml); insoluble in EtOH, Et_2O; LD_{50} (mus orl) = 540 mg/kg. *ECR Pharm.; Elkins-Sinn; Fisons plc; Glaxo Labs.; Purdue Pharma L.P.; Searle G.D. & Co.; Wallace Labs.*

7136 Azanator
37885-92-8 253-694-4
$C_{18}H_{18}N_2O$
5-(1-Methyl-4-piperidylidene)-5H-[1]benzopyrano[2,3-b]-pyridine.
Bronchodilator. *Schering AG.*

7137 Azanator Maleate
39624-65-2
$C_{22}H_{22}N_2O_5$
5-(1-Methyl-4-piperidylidene)-5H-[1]benzopyrano[2,3-b]-pyridine maleate (1:1).
Sch-15280. Bronchodilator. *Schering AG.*

7138 Bambuterol
81732-65-2 980
$C_{18}H_{29}N_3O_5$
(±)-5-[2-(tert-Butylamino)-1-hydroxyethyl]-m-phenylene bis(dimethylcarbamate).
terbutaline bisdimethylcarbamate. Bronchodilator. Ester prodrug of the β_2-adrenergic agonist terbutaline. Ephedrine derivative. *Astra Draco AB.*

7139 Bambuterol Hydrochloride
81732-46-9 980
$C_{18}H_{30}ClN_3O_5$
(±)-5-[2-(tert-Butylamino)-1-hydroxyethyl]-m-phenylene bis(dimethylcarbamate) monohydrochloride.

KWD-2183; Bambec. Bronchodilator. Ephedrine derivative. *Astra Draco AB.*

7140 Bamifylline
2016-63-9 982
$C_{20}H_{27}N_5O_3$
8-Benzyl-7-[2-[ethyl-(2-hydroxyethyl)amino]ethyl]-theophylline.
benzetamophylline; 8012 CB; Bamiphylline. Bronchodilator. Xanthine derivative. mp = 80-80.5°. *Christiaens S.A.*

7141 Bamifylline Hydrochloride
20684-06-4 982 243-967-6
$C_{20}H_{28}ClN_5O_3$
8-Benzyl-7-[2-[ethyl-(2-hydroxyethyl)amino]ethyl]-theophylline monohydrochloride.
AC-3810; BAX-2739Z; Briofil; Trentadil. Bronchodilator. Xanthine derivative. mp = 185-186°; LD_{50} (mus orl) = 246 mg/kg, (rat orl) = 1139 mg/kg, (mus ip) = 89 mg/kg, (rat ip) = 131 mg/kg, (mus iv) = 67 mg/kg, (rat iv) = 65 mg/kg. *Christiaens S.A.*

7142 Bevonium
33371-53-8 1239
$[C_{22}H_{28}NO_3]^+$
2-(Hydroxymethyl)-1,1-dimethylpiperidinium benzilate.
Anticholinergic with antispasmodic and bronchodilating properties. A quaternary ammonium. *Grunenthal.*

7143 Bevonium Methyl Sulfate
5205-82-3 1239 226-001-8
$C_{23}H_{31}NO_7S$
2-(Hydroxymethyl)-1,1-dimethylpiperidinium methyl sulfate benzilate.
bevonium metilsulfate; CG-201; Acabel. Anticholingergic; antipsasmodic; bronchodilator. Quaternary ammonium compound. mp = 134-135°. *Grunenthal.*

7144 Bitolterol
30392-40-6 1344
$C_{28}H_{31}NO_5$
4-[2-(tert-Butylamino)-1-hydroxyethyl]-o-phenylene di-p-toluate.
Bronchodilator. α_2-adrenergic agonist; a diester of N-tert-butylnorepinephrine. Ephedrine derivative. *Sterling Winthrop Inc.*

7145 Bitolterol Mesylate
30392-41-7 1344 250-177-5
$C_{29}H_{35}NO_8S$
4-[2-(tert-Butylamino)-1-hydroxyethyl]-o-phenylene di-p-toluate methanesulfonate (salt).
bitoterol methanesulfonate; Win-32784; Biterol; Effectin; Tornalate. Bronchodilator. α_2-adrenergic agonist; a diester of N-tert-butylnorepinephrine. Ephedrine derivative. mp = 170-172°; soluble in DMSO. *Sterling Winthrop Inc.*

7146 Butaprost
69648-38-0
$C_{24}H_{40}O_5$
Methyl (1R,2R,3R)-3-hydroxy-2-[(1E,4R)-4-hydroxy-4-(1-propylcyclobutyl)-1-butenyl]-5-oxocyclopentane-heptanoate.
TR-4979; Bay q 4218. Bronchodilator. *Miles Inc, Bayer Corp. Pharm. Div.*

7147 Carbuterol
34866-47-2 1882 252-257-5
$C_{13}H_{21}N_3O_3$
[5-[2-[(1,1-Dimethylethyl)amino]-1-hydroxyethyl]-2-hydroxyphenyl]urea.
Bronchodilator. β-Adrenergic agonist related to isoproterenol, with selectivity for airway smooth muscle receptors. Ephedrine derivative. mp = 174-176°; LD_{50} (mus iv) = 32.8 mg/kg, (rat iv) = 77.2 mg/kg, (mus orl) = 3134.6 mg/kg. *SmithKline Beecham Pharm.*

7148 Carbuterol Hydrochloride
34866-46-1 1882 252-255-4
$C_{13}H_{22}ClN_3O_3$
[5-[2-[(1,1-Dimethylethyl)amino]-1-hydroxyethyl]-2-hydroxyphenyl]urea monohydrochloride.
SKF-40383; Bronsecur; Pirem. Bronchodilator. Ephedrine derivative. mp = 204-207° (dec). *SmithKline Beecham Pharm.*

7149 Choline Theophyllinate
4499-40-5 2266 224-798-7
$C_{12}H_{21}N_5O_3$
2-Hydroxy-N,N,N-trimethylethanammonium salt with 3,7-dihydro-1,3-dimethyl-1H-purine-2,6-dione (1:1).
theophylline cholinate; theophylline salt of choline; oxtriphylline; oxytrimethylline; Cholinophylline; Choledyl; Filoral; Sabidal; Soliphylline; Teokolin; Teofilcolina; Theoxylline; component of: Brondecon, Cholinophylline Magnesium Glycinate (43%). Bronchodilator. Xanthine derivative. *See* monotheamin, amylobarbital. Soluble in H_2O. *Parke-Davis.*

7150 Clenbuterol
37148-27-9 2407 253-366-0
$C_{12}H_{18}N_2O$
4-Amino-α-[(tert-butylamino)methyl]-3,5-dichlorobenzyl alcohol.
NAB-365; Monores. Bronchodilator. Substituted phenylethanolamine with β_2 sympathomimetic activity. Ephedrine derivative. *Thomae GmbH Dr. Karl.*

7151 Clenbuterol Hydrochloride
21898-19-1 2407 244-643-7
$C_{12}H_{19}ClN_2O$
4-Amino-α-[(tert-butylamino)methyl]-3,5-dichlorobenzyl alcohol monohydrochloride.
NAB-365Cl; Spiropent; Ventipulmin. Bronchodilator. β-Adrenergic agonist. Ephedrine derivative. mp = 174-175°; soluble in H_2O, MeOH, EtOH; slightly soluble in $CHCl_3$; insoluble in C_6H_6; LD_{50} (mus orl) = 176 mg/kg, (rat orl) = 315 mg/kg, (gpg orl) = 67.1, (mus iv) = 27.6 mg/kg. *Thomae GmbH Dr. Karl.*

7152 Clorprenaline
3811-25-4 2470 223-291-8
$C_{11}H_{16}ClNO$
o-Chloro-α-[(isopropylamino)methyl]benzyl alcohol.
isoprophenamine; isoprofenamine. Bronchodilator. β-Adrenergic agonist. Ephedrine derivative. *Eli Lilly & Co.*

7153 Clorprenaline Hydrochloride
6933-90-0 2470 230-058-4
$C_{11}H_{17}Cl_2NO$
o-Chloro-α-[(isopropylamino)methyl]benzyl alcohol hydrochloride.
Bronchodilator. β-Adrenergic agonist. Ephedrine derivative. *Eli Lilly & Co.*

7154 Clorprenaline Hydrochloride Monohydrate
5588-22-7 2470 209-612-4
$C_{11}H_{19}Cl_2NO_2$
o-Chloro-α-[(isopropylamino)methyl]benzyl alcohol hydrochloride monohydrate.
Broncon; Clopinerin; Conselt; Fusca; Kalutein; Pentadoll; Restanolon. Bronchodilator. β-Adrenergic agonist. Ephedrine derivative. mp = 163-164°. *Eli Lilly & Co.*

7155 Colterol
18866-78-9
$C_{12}H_{19}NO_3$
(±)-α-[(tert-Butylamino)methyl]-3,4-dihydroxybenzyl alcohol.
Bronchodilator. *Sterling Winthrop Inc.*

7156 Colterol Mesylate
17605-73-1
$C_{13}H_{23}NO_6S$
(±)-α-[(tert-Butylamino)methyl]-3,4-dihydroxybenzyl alcohol methanesulfonate (salt).
Win-5563-3. Bronchodilator. *Sterling Winthrop Inc.*

7157 Darodipine
72803-02-2
$C_{19}H_{21}N_3O_5$
Diethyl 4-(4-benzofurazanyl)-1,4-dihydro-2,6-dimethyl-3,5-pyridinedicarboxylate.
PY-108-068. Antihypertensive; bronchodilator; vasodilator. Dihydropyridine calcium antagonist. *Sandoz Pharm. Corp.*

7158 Demelverine
13977-33-8
$C_{17}H_{21}N$
N-Methyldiphenethylamine.
Bronchodilator.

7159 Dioxethedrin
497-75-6 3356 207-849-8
$C_{11}H_{17}NO_3$
α-(1-Ethylaminoethyl)protocatechuyl alchohol.
dioxethedrine; component of: Bexol. Bronchodilator. β-Adrenergic agonist. Ephedrine derivative.

7160 Dioxethedrin Hydrochloride
497-75-6 3356
$C_{11}H_{18}ClNO_3$
α-(1-Ethylaminoethyl)protocatechuyl alchohol hydrochloride.
dioxethedrine hydrochloride. Bronchodilator. β-Adrenergic agonist. mp = 212-214°.

7161 Doxaprost
51953-95-8
$C_{21}H_{36}O_4$
(1R*,2R*)-2-[(E)-3-Hydroxy-3-methyl-1-octyl]-5-oxocyclopentaneheptanoic acid.
AY-24559. Bronchodilator.

7162 Doxofylline
69975-86-6 3494 274-239-6
$C_{11}H_{14}N_4O_4$
7-(1,3-Dioxolan-2-ylmethyl)-3,7-dihydro-1,3-dimethyl-1H-purine-2,6-dione.
doxophylline; doxofilline; ABC-12/3; Ansimar; Maxivent; Ventax. Bronchodilator. Xanthine derivative. mp = 144-145.5°; soluble in H_2O, Me_2CO, EtOAc, C_6H_6, $CHCl_3$, dioxane, hot EtOHs; nearly insoluble in Et_2O, petroleum ether; LD_{50} (mus orl) = 841 mg/kg, (rat orl) = 1022.4 mg/kg, (mus iv) = 215.6 mg/kg, (rat ip) = 445 mg/kg. *Roberts Pharm. Corp.*

7163 Dyphylline
479-18-5 3529 207-526-1
$C_{10}H_{14}N_4O_4$
7-(2,3-Dihydroxypropyl)-3,7-dihydro-1,3-dimethyl-1H-purine-2,6-dione.
diprophylline; glyphylline; glyfyllin; AFI-phyllin; Astmamasit; Athmolysin; Astrophyllin; Circair; Coronarin; Cor-Theophyllin; Dilor; Hiphyllin; Hyphylline; Lufyllin; Neophyl; Neostenovasan; Neothylline; Neotilina; Neo-Vasophylline; Neutrafil; Neutraphylline; Prophyllen; Silbephylline; Solufilin; Solufyllin; Theal ampules; Thefulan; Neutraphylline; Prophyllen; Silbephylline; Solufilin; Solufyllin; Theal ampules; Thefylan; component of: Dilor G, Neothylline-GG. Bronchodilator. Xanthine derivative. mp = 158°; λ_m = 273 nm ($A^{1\%}_{1\,cm}$ 361 in 0.001% H_2O); soluble in H_2O (30 g/100 ml), less soluble in organic solvents; LD_{50} (mus orl) = 3400 mg/kg, (mus sc) = 1430 mg/kg. *Carter-Wallace; ICN Pharm. Inc.; Lemmon Co.; Savage Labs.*

7164 Enprofylline
41078-02-8 255-201-8
$C_8H_{10}N_4O_2$
3,7-Dihydro-3-propyl-1H-purine-2,6-dione.
enprofylline; D-4028. Bronchodilator. Xanthine derivative. *Astra Sweden.*

7165 Ephedrine
90-81-3 3645 202-017-0
$C_{10}H_{15}NO$
α-[1-(Methylamino)ethyl]benzenemethanol.
1-phenyl-2-methylaminopropanol; racemic ephedrine [dl-form]; racephedrine [dl-form]. Adrenergic (bronchodilator). α- and β-adrenergic agonist. Occurs in Ma Huang and other Ephedra species. The d-isomer is used as a decongestant. [dl-form]: mp = 79°; soluble in H_2O, EtOH, Et_2O, $CHCl_3$, oils.

7166 Ephedrine Hydrochloride
134-71-4 3645 205-153-9
$C_{10}H_{16}ClNO$
α-[1-(Methylamino)ethyl]benzenemethanol hydrochloride.
racephedrine hydrochloride [dl-form]; Ephetonin [dl-form]. Adrenergic (bronchodilator). α- and β-adrenergic agonist. Found in Ma Huang and other Ephedra species. [dl-form]: mp = 187-188°; soluble in H_2O (25 g/100 ml); somewhat soluble in EtOH; nearly insoluble in Et_2O; pH 6. *Bristol Laboratories; Eli Lilly & Co.; Glaxo Labs.; Parke-Davis; Poythress; Sterling Winthrop Inc.; Whitehall Labs. Inc.*

7167 Ephedrine Sulfate
134-72-5 3645 205-154-4
$C_{10}H_{17}NO_4S$
α-[1-(Methylamino)ethyl]benzenemethanol sulfate.
racephedrine sulfate [dl-form]. Adrenergic (bronchodilator). α- and β-adrenergic agonist. Occurs in Ma Huang and other Ephedra species. [dl-form]: mp = 247°; soluble in H_2O, EtOH; pH 6. *Boehringer Mannheim GmbH.*

7168 l-Ephedrine
299-42-3 3645 206-080-5
$C_{10}H_{15}NO$
L-erythro-2-(Methylamino)-1-phenylpropan-1-ol.
l-ephedrine; (-)-ephedrine; l-α-[1-(Methylamino)ethyl]-benzenemethanol; component of: Bena-Fedrin, Bronkotabs, Mudrane GG Elixir, Primatene, Quadrinal, Tedral. Bronchodilator. An α-adrenergic agonist. Waxy solid; may contain up to 1/2 mole H_2O; mp = 34°; bp = 255°; somewhat soluble in H_2O (5 g/100 ml); more soluble in EtOH (500 g/100 ml), $CHCl_3$, Et_2O, oils. *Chemoterapico; Eli Lilly & Co.; Hybridon Inc.; Knoll Pharm. Co.; Parke-Davis; Whitehall-Robins;*

7169 l-Ephedrine Hydrochloride
50-98-6 3645 200-074-6
$C_{10}H_{16}ClNO$
L-erythro-2-(Methylamino)-1-phenylpropan-1-ol hydrochloride.
Ephedral; Sanedrine; component of: Amesec, Bena-Fedrin, Mudrane GG Elixir, Primatene Tablets, Quadrinal, Quibron Plus, Tedral. An α-adrenergic agonist. Bronchodilator. mp = 216-220°; $[\alpha]_D^{25}$ = -33° to -35.5° (c = 5); soluble in H_2O (33.3 g/100 ml), EtOH (7.1 g/100 ml); nearly insoluble in Et_2O, $CHCl_3$. *Chemoterapico; Eli Lilly & Co.; Hybridon Inc.; Knoll Pharm. Co.; Parke-Davis; Whitehall-Robins.*

7170 l-Ephedrine Sulfate
134-72-5 3645 205-154-4
$C_{20}H_{32}N_2O_6$
L-erythro-2-(Methylamino)-1-phenylpropan-1-ol sulfate (2:1) salt.
Isofedrol; component of: Bronkaid, Pazo Ointment, Pazo Suppository. Bronchodilator. Adrenergic agonist. mp = 245° (dec); $[\alpha]_D^{25}$ = -29.5° to -32.0° (c = 5); soluble in H_2O (83 g/100 ml), EtOH (1.05 g/100 ml); freely soluble in hot EtOH. *Boehringer Mannheim GmbH; Bristol-Myers Squibb HIV Products; Eli Lilly & Co.; Sterling Health U.S.A.*

7171 Epinephrine
51-43-4 3656 200-098-7
$C_9H_{13}NO_3$
(R)-4-[1-Hydroxy-2-(methylamino)ethyl]-1,2-benzenediol.
l-methylaminoethanolcatechol; adrenalin; levorenen; Bronkaid Mist; Epifrin; Epiglaufrin; Eppy; Glaucon; Glauposine; Primatene Mist; Simplene; Sus-phrine; Suprarenaline; component of: Citanest Forte. Bronchodilator; cardiostimulant; mydriatic; antiglaucoma. Endogenous catecholamine with combined α- and β-agonist activity. Principal sympathomimetic hormone produced by the adrenal medulla. mp = 211-212°; dec 215°; $[\alpha]_D^{25}$ = -50.0° to -53.5° (in 0.6N HCl); slightly soluble in H_2O, EtOH; soluble in aqueous solutions of mineral acids; insoluble in aqueous solutions of ammonia and alkali carbonates; insoluble in Et_2O, Me_2CO, oils; LD_{50} (mus ip) = 4 mg/kg. *Alcon Labs; Allergan Inc.; Astra Sweden; Bristol-Myers Squibb Co.; CIBA Vision AG; Elkins-Sinn; Evans Medical Ltd.; Parke-Davis; Sterling Health U.S.A.; Whitehall-Robins; Wyeth-Ayerst Labs.*

7172 Epinephrine d-Bitartrate
51-42-3 3656 200-097-1
$C_{13}H_{19}NO_9$
(R)-4-[1-Hydroxy-2-(methylamino)ethyl]-1,2-benzenediol [R-(R*,R*)]-2,3-dihydroxybutanedioate (1:1) (salt).
Asmatane Mist; Asthmahaler Epitrate; Bronitin Mist; Bronkaid Mist Suspension; Epitrate; Medihaler-Epi; Primatene Mist Suspension; Suprarenin; component of: Asthmahaler, E-Pilo. Bronchodilator; cardiostimulant; mydriatic; antiglaucoma. Adrenergic (opthalmic). mp = 147-154° (some decomposition); soluble in H_2O (30 g/100 ml); slightly soluble in EtOH. *3M Pharm.; CIBA Vision Corp.; Menley & James Labs Inc.; Sterling Health U.S.A.; Whitehall-Robins; Wyeth-Ayerst Labs.*

7173 Epinephrine Hydrochloride
55-31-2 3656
$C_9H_{14}ClNO_3$
(-)-3,4-Dihydroxy-α-[(methylamino)methyl]benzyl alcohol hydrochloride.
Adrenalin; Epifrin; Glaucon; Suprarenin. Bronchodilator; cardiostimulant; mydriatic; antiglaucoma. An α-adrenergic agonist. *Alcon Labs; Allergan Inc.; Astra Chem. Ltd.; Bristol-Myers Squibb Co.; CIBA Vision AG; Elkins-Sinn; Forest Pharm. Inc.; Parke-Davis; Sterling Health U.S.A.; Whitehall Labs. Inc.; Wyeth-Ayerst Labs.*

7174 dl-Epinephrine
329-65-7 3656 206-347-6
$C_{20}H_{32}N_2O_6$
(±)-α-[1-(Methylamino)ethyl]benzenemethanol sulfate salt (2:1).
racepinefrine; racepinephrine. An α-adrenergic agonist. Used as a bronchodilator, cardiostimulant, mydriatic and antiglaucoma agent. Slightly soluble in H_2O, EtOH. *Alcon Labs; Allergan Inc.; Astra Chem. Ltd.; Bristol-Myers Squibb Co.; CIBA Vision AG; Elkins-Sinn; Forest Pharm. Inc.; Parke-Davis; Sterling Health U.S.A.; Whitehall Labs. Inc.; Wyeth-Ayerst Labs.*

7175 dl-Epinephrine Hydrochloride
329-63-5 3656 206-346-0
$C_{20}H_{33}ClN_2O_6$
(±)-α-[1-(Methylamino)ethyl]benzenemethanol hydrochloride salt.
Asthmanefrin; Vaponefrin. An α-adrenergic agonist. Used as a bronchodilator, cardiostimulant, mydriatic and antiglaucoma agent. mp = 157°, soluble in H_2O; sparingly soluble in EtOH. *Alcon Labs; Allergan Inc.; Astra Chem. Ltd.; Bristol-Myers Squibb Co.; CIBA Vision AG; Elkins-Sinn; Forest Pharm. Inc.; Parke-Davis; Sterling Health U.S.A.; Whitehall Labs. Inc.; Wyeth-Ayerst Labs.*

7176 Epinephryl Borate
5579-16-8 226-961-8
$C_9H_{12}BNO_4$
(R)-2-Hydroxy-α-[(methylamino)methyl]-1,3,2-benzodioxaborole-5-methanol.

Epinal; Eppy/N. Adrenergic (opthalmic). *Alcon Labs; Pilkington Barnes Hind.*

7177 Eprozinol
32665-36-4 3670 251-146-9
$C_{22}H_{30}N_2O_2$
4-(β-Methoxyphenethyl)-α-epin-4(1H)-one.
Bronchodilator. Ephedrine derivative.

7178 Eprozinol Dihydrochloride
27588-43-8 3670
$C_{22}H_{32}Cl_2N_2O_2$
4-(β-Methoxyphenethyl)-α-epin-4(1H)-one dihydrochloride.
Alecor; Brovel; Eupnéron. Bronchodilator. mp = 164°; soluble in H_2O, EtOH; LD_{50} (mus orl) = 500 mg/kg.

7179 Etafedrine
7681-79-0 3752 231-677-2
$C_{12}H_{19}NO$
α-[1-(Ethylmethylamino)ethyl]benzenemethanol.
Menetryl; Novedrine. Bronchodilator. β-Adrenergic agonist. Ephedrine derivative. *Marion Merrell Dow Inc.*

7180 Etafedrine Hydrochloride
5591-29-7 3752 227-000-5
$C_{12}H_{20}ClNO$
α-[1-(Ethylmethylamino)ethyl]benzenemethanol hydrochloride.
Nethamine; component of: Nethaphyl. Bronchodilator. β-Adrenergic agonist. Ephedrine derivative. mp = 183-184°; soluble in H_2O (67 g/100 ml), EtOH (12.5 g/100 ml). *Marion Merrell Dow Inc.*

7181 Etamiphyllin
314-35-2 3754 206-244-6
$C_{13}H_{21}N_5O_2$
7-[2-(Diethylamino)ethyl]-3,7-dihydro-1,3-dimethyl-1H-purine-2,6-dione.
etamiphylline; dietamiphylline; millophyline. Bronchodilator. Xanthine derivative. Waxy solid; mp = 75°; soluble in H_2O, Me_2CO; slightly soluble in EtOH, Et_2O.

7182 Etamiphyllin Camphorsulfonate
19326-29-5 3754
$C_{23}H_{37}N_5O_6S$
7-[2-(Diethylamino)ethyl]-3,7-dihydro-1,3-dimethyl-1H-purine-2,6-dione camphorsulfonate.
Camphophyline. Bronchodilator. Xanthine derivative. mp = 174°.

7183 Etamiphyllin Heparinate
59547-58-9 3754
7-[2-(Diethylamino)ethyl]-3,7-dihydro-1,3-dimethyl-1H-purine-2,6-dione heparinate.
Milhérine. Bronchodilator. Xanthine derivative.

7184 Etamiphyllin Hydrochloride
17140-68-0 3754 206-244-6
$C_{13}H_{22}ClN_5O_2$
7-[2-(Diethylamino)ethyl]-3,7-dihydro-1,3-dimethyl-1H-purine-2,6-dione monohydrochloride.
Bronchodilator. Xanthine derivative. mp = 239-241°; soluble in H_2O.

7185 Etamiphyllin Iodomethylate
3754
7-[2-(Diethylamino)ethyl]-3,7-dihydro-1,3-dimethyl-1H-purine-2,6-dione iodomethylate.
Iodaphyline; Iodafilina; Jodo-Metil-Fillina. Bronchodilator. Xanthine derivative.

7186 Ethylnorepinephrine
536-24-3 3880
$C_{10}H_{15}NO_3$
4-(2-Amino-1-hydroxybutyl)-1,2-benzenediol.
ethylnoradrenaline; ethylnorsprarenin; E.N.E; E.N.S. Bronchodilator. Ephedrine derivative. *Sterling Winthrop Inc.*

7187 Ethylnorepinephrine Hydrochloride
3198-07-0 3880
$C_{10}H_{16}ClNO_3$
4-(2-Amino-1-hydroxybutyl)-1,2-benzenediol hydrochloride.
Bronkephrine. Bronchodilator. Ephedrine derivative. Dec 199-200°; soluble in H_2O. *Sterling Winthrop Inc.*

7188 Etofylline
519-37-9 3924 208-269-8
$C_9H_{12}N_4O_3$
3,7-Dihydro-7-(2-hydroxyethyl)-1,3-dimethyl-1H-purine-2,6-dione.
oxyethyltheophylline; Oxytheonyl; Oxphylline; Phyllocormin N. Bronchodilator. Pharmacological action as of theophylline, a xanthine derivative. *See* theophylline. mp = 158°; soluble in H_2O; moderately soluble in EtOH; pH (5% aqueous solution) = 6.5-7.0.

7189 Fenoterol
13392-18-2 4022
$C_{17}H_{21}NO_4$
5-[1-Hydroxy-2-[[2-(4-hydroxyphenyl)-1-methylethyl]amino]ethyl]-1,3-benzenediol.
TH-1165. Bronchodilator; tocolytic. β-Adrenergic agonist. Ephedrine derivative. *Boehringer Ingelheim Pharm. Inc.*

7190 Fenoterol Hydrobromide
1944-12-3 4022
$C_{17}H_{22}BrNO_4$
5-[1-Hydroxy-2-[[2-(4-hydroxyphenyl)-1-methylethyl]amino]ethyl]-1,3-benzenediol hydrobromide.
TH-1165a; Airum; Berotec; Dosberotec; Partusisten. Bronchodilator; tocolytic. β-Adrenergic agonist. Ephedrine derivative. mp = 222-223°; LD_{50} (mus sc) = 1100 mg/kg, (mus orl) = 1990 mg/kg. *Boehringer Ingelheim Pharm. Inc.*

7191 Fenprinast
75184-94-0
$C_{16}H_{16}ClN_5O$
4-(p-Chlorobenzyl)-1,4,6,7-tetrahydro-6,6-dimethyl-9H-imidazo[1,2-a]purin-9-one.
Bronchodilator. *Bristol-Myers Squibb Pharm. R&D.*

7192 Fenprinast Hydrochloride
77482-47-4
$C_{16}H_{19}Cl_2N_5O_2$
4-(p-Chlorobenzyl)-1,4,6,7-tetrahydro-6,6-dimethyl-9H-imidazo[1,2-a]purin-9-one monohydrochloride monohydrate.

MJ-13401-1-3. Bronchodilator (antiallergic). *Bristol-Myers Squibb Pharm. R&D.*

7193 Fenspiride
5053-06-5 4041 225-751-3
$C_{15}H_{20}N_2O_2$
8-(2-Phenylethyl)-1-oxa-3,8-diazaspiro[4.5]decan-2-one.
decaspiride; DESP. Bronchodilator. α-Adrenergic blocker. *Marion Merrell Dow Inc.*

7194 Fenspiride Hydrochloride
5053-08-7 4041 225-752-9
$C_{15}H_{21}ClN_2O_2$
8-(2-Phenylethyl)-1-oxa-3,8-diazaspiro[4.5]decan-2-one monohydrochloride.
NAT-333; NDR-5998A; Decaspir; Fluiden; Pneumorel; Respiride; Tefencia; Viarespan. Bronchodilator. α-Adrenergic blocker. Dec 232-233° (dec); soluble in H_2O; LD_{50} (mus iv) = 106 mg/kg, (rat orl) = 437 mg/kg. *C.M. Ind.; Marion Merrell Dow Inc.*

7195 Flutropium Bromide
63516-07-4 4248
$C_{24}H_{29}BrFNO_3$
(8r)-8-(2-Fluoroethyl)-3α-hydroxy-1αH,5αH-tropanium bromide benzilate.
Ba-598Br; Flubron. Bronchodilator. Quaternary ammonium compound. Anticholinergic. mp = 192-193° (dec), 198-199°; LD_{50} (mmus iv) = 12.5 mg/kg, (mmus orl) = 760 mg/kg, (fmus iv) = 11.0 mg/kg, (fmus orl) = 810 mg/kg, (mrat iv) = 16.4 mg/kg, (mrat orl) = 830 mg/kg, (frat iv) = 18.4 mg/kg, (frat orl) = 740 mg/kg. *Boehringer Ingelheim GmbH.*

7196 Formoterol
73573-87-2 4272
$C_{19}H_{24}N_2O_4$
(±)-2'-Hydroxy-5'-[(RS)-1-hydroxy-2-[[(RS)-p-methoxy-α-methylphenyl]amino]ethyl]formanilide.
CGP-25827A. Antiasthmatic. α-Adrenoceptor stimulating catecholamine analog with selective bronchodilator activity. Ephedrine derivative. *Yamanouchi U.S.A. Inc.*

7197 Formoterol Fumarate Dihydrate
43229-80-7 4272
$C_{42}H_{56}N_4O_{14}$
(±)-2'-Hydroxy-5'-[(RS)-1-hydroxy-2-[[(RS)-p-methoxy-α-methylphenyl]amino]ethyl]formanilide fumarate dihydrate.
BD-40A; Atock; Foradil. Antiasthmatic. α-Adrenoceptor stimulating catecholamine analog with selective bronchodilator activity. Ephedrine derivative. mp = 138-140°; LD_{50} (mrat orl) = 3130 mg/kg, (mrat iv) = 98 mg/kg, (mrat sc) = 1000 mg/kg, (mrat ip) = 170 mg/kg. *Yamanouchi U.S.A. Inc.*

7198 Guaithylline
5634-38-8 4582 227-078-0
$C_{17}H_{22}N_4O_6$
Theophylline compound with 3-(o-methoxyphenoxy)-1,2-propanediol.
guaifylline; Eclabron. Bronchodilator; expectorant. Xanthine derivative. *See* guaifenesin. *Degussa-Hüls Corp.; Forest Pharm. Inc.; Key Pharm.; Robins, A. H. Co.; Xttrium Labs. Inc.*

7199 Hexoprenaline
3215-70-1 4745
$C_{22}H_{32}N_2O_6$
α,α'-[Hexamethylene-bis(iminomethylene)]bis(3,4-dihydroxybenzyl alcohol).
BYK-1512. Bronchodilator; tocolytic. β-Adrenergic agonist. Ephedrine derivative. [hemihydrate]: mp = 162-165°. *Lentia; Oesterreiche Stickstoffwerke; OSSW; Savage Labs.*

7200 Hexoprenaline Dihydrochloride
4323-43-7 4745 224-354-2
$C_{22}H_{34}Cl_2N_2O_6$
α,α'-[Hexamethylene-bis(iminomethylene)]bis(3,4-dihydroxybenzyl alcohol) dihydrochloride.
ST-1512; Ipradol. Bronchodilator; tocolytic. β-Adrenergic agonist. Ephedrine derivative. mp = 197.5-198°. *Lentia; OSSW; Savage Labs.*

7201 Hexoprenaline Sulfate
32266-10-7 4745 250-974-8
$C_{22}H_{34}N_2O_{10}S$
α,α'-[Hexamethylene-bis(iminomethylene)]bis(3,4-dihydroxybenzyl alcohol) sulfate.
Bronalin; Delaprem; Etoscol; Gynipral; Ipradol; Leanol. Bronchodilator; tocolytic. β-Adrenergic agonist. Ephedrine derivative. mp = 222-228°. *Lentia; Oesterreiche Stickstoffwerke; OSSW; Savage Labs.*

7202 Hoquizil
21560-59-8
$C_{19}H_{26}N_4O_5$
2-Hydroxy-2-methylpropyl 4-(6,7-dimethoxy-4-quinazolinyl)-1-piperazinecarboxylate.
Bronchodilator. A quinazoline derivative. *Pfizer Inc.*

7203 Hoquizil Hydrochloride
23256-28-2
$C_{19}H_{27}ClN_4O_5$
2-Hydroxy-2-methylpropyl 4-(6,7-dimethoxy-4-quinazolinyl)-1-piperazinecarboxylate monohydrochloride.
CP-14185-1. Bronchodilator. A quinazoline derivative. *Pfizer Inc.*

7204 Ipratropium Bromide
66985-17-9 5089
$C_{20}H_{30}BrNO_3.H_2O$
(8r)-3α-Hydroxy-8-isopropyl-1αH,5αH-tropanium bromide monohydrate.
Sch-1000-Br-monohydrate; Atem; Atrovent; Bitrop; Itrop; Narilet; Rinatec; component of: Combivent. An anticholinergic bronchodilator and antiarrhythmic. Quaternary ammonium compound. mp = 230-232°; soluble in H_2O, MeOH, EtOH; insoluble in Et_2O, $CHCl_3$, fluorohydrocarbons; LD_{50} (mmus orl) = 1001 mg/kg, (mmus iv) = 12.29 mg/kg, (mmus sc) = 300 mg/kg, (fmus orl) = 1083 mg/kg, (fmus iv) = 14.97 mg/kg, (fmus sc) = 340 mg/kg, (mrat orl) = 1663 mg/kg, (mrat iv) = 15.89 mg/kg, (frat orl) = 1779 mg/kg, (frat iv) = 15.70 mg/kg. *Boehringer Ingelheim Ltd.; Schering AG.*

7205 Ipratropium Bromide [anhydrous]
22254-24-6 5089 244-873-8
$C_{20}H_{30}BrNO_3$
(8r)-3α-Hydroxy-8-isopropyl-1αH,5αH-tropanium bromide monohydrate.

N-isopropylnoratropinium bromomethylate; Sch-100. Anticholinergic bronchodilator; antiarrhythmic. Quaternary ammonium compound. *Boehringer Ingelheim Ltd.*

7206 Isbufylline
90162-60-0
$C_{11}H_{16}N_4O_2$
7-Isobutyltheophylline.
A new antibronchospastic xanthine; used as a bronchodilator.

7207 Isoetharine
530-08-5 5185 208-472-1
$C_{13}H_{21}NO_3$
3,4-Dihydroxy-α-[1-(isopropylamino)propyl]benzyl alcohol.
etyprenaline; isoetarine; Win-3406; Dilabron; Neoisuprel. Bronchodilator. Ephedrine derivative. *I.G. Farben; Sterling Winthrop Inc.*

7208 Isoetharine Hydrochloride
2576-92-3 5185 219-927-9
$C_{13}H_{22}ClNO_3$
3,4-Dihydroxy-α-[1-(isopropylamino)propyl]benzyl alcohol hydrochloride.
Asthmalitan; Bronkosol; Numotac. Bronchodilator. Ephedrine derivative. mp = 212-213° (dec). *I.G. Farben; Parke-Davis; Sterling Winthrop Inc.*

7209 Isoetharine Mesylate
7279-75-6 5185 230-695-8
$C_{14}H_{25}NO_6S$
3,4-Dihydroxy-α-[1-(isopropylamino)propyl]benzyl alcohol methanesulfonate (salt).
isoetarine mesilate; Bronkometer. Bronchodilator. Ephedrine derivative. *I.G. Farben; Sterling Winthrop Inc.*

7210 Isoproterenol
7683-59-2 5236 231-687-7
$C_{11}H_{17}NO_3$
3,4-Dihydroxy-α-[(isopropylamino)methyl]benzyl alcohol.
epinephrine isopropyl homolog; isoprenaline; dihydroxyphenylethanolisopropylamine; N-isopropylnoradrenaline; A-21; Aludrine; Aleudrin; Asiprenol; Asmalar; Assiprenol; Bellasthman; Isonorin; Isorenin; Isopropydrin; Isuprel; Isupren; Neodrenal; Neo-Epinine; Norisodrine; Novodrin; Proternol; Respifral; Saventrine; Vapo-N-Iso. Bronchodilator; anticholingergic. β-Adrenergic agonist; catecholamine derivative. Ephedrine derivative. [(dl)-form]: mp = 155.5°; [(l)-form]: mp = 164-165°; $[\alpha]_D^{19}$ = -45.0° (c = 2 in 2N HCl); pKa = 8.64; LD_{50} (mrat orl) = 3675 mg/kg. *3M Pharm.; Abbott Labs Inc.; Eli Lilly & Co.; Fisons plc; Parke-Davis; Sterling Winthrop Inc.*

7211 Isoproterenol Hydrochloride
51-30-9 5236 200-089-8
$C_{11}H_{18}ClNO_3$
3,4-Dihydroxy-α-[(isopropylamino)methyl]benzyl alcohol hydrochloride.
Aerolone; Aerotrol; Euspiran; Isomenyl; Isovon; Mistarel; Suscardia; component of: Aerolone Solution, Duo-Medihaler. Bronchodilator; anticholingergic. Ephedrine derivative. [(dl)-form]: mp = 170-171°; [(l)-form]: dec 162-164°; $[\alpha]_D^{19}$ = -50°; soluble in H_2O; less soluble in EtOHs; nearly insoluble in some organic solvents; LD_{50} (rat orl) 2220 mg/kg. *3M Pharm.; Abbott Labs Inc.; Eli Lilly & Co.; Fisons plc; Parke-Davis; Sterling Winthrop Inc.*

7212 Isoproterenol Sulfate
6700-39-6 5236
$C_{22}H_{40}N_2O_{12}S$
3,4-Dihydroxy-α-[(isopropylamino)methyl]benzyl alcohol sulfate (2:1) salt dihydrate.
Aludrin; Isomist; Propal; Luf-Iso; Medihaler Iso. Bronchodilator; anticholingergic. Ephedrine derivative. [(dl)-form]: mp = 128° (with some decomposition); soluble in H_2O (25 g/100 ml); slightly soluble in EtOH; nearly insoluble in many organic solvents. *3M Pharm.; Abbott Labs Inc.; Carter-Wallace.*

7213 Isoproterenol Sulfate [anhydrous]
299-95-6 5236 206-085-2
$C_{22}H_{36}N_2O_{10}S$
3,4-Dihydroxy-α-[(isopropylamino)methyl]benzyl alcohol sulfate (2:1) salt.
isoprenaline sulfate. Bronchodilator; anticholingergic. Ephedrine derivative.

7214 L-Isoproterenol D-Bitartrate
54750-10-6 5236 259-322-7
$C_{17}H_{23}NO_9$
L-3,4-Dihydroxy-α-[(isopropylamino)methyl]benzyl alcohol D-Bitartrate.
Isolevin [as dihydrate]. Bronchodilator; anticholinergic. Ephedrine derivative. [dihydrate]: mp = 80-83° (sinters at 78°); $[\alpha]_D^{19}$ = -14.9° (c = 2 .31).

7215 Lomifylline
10226-54-7 233-547-0
$C_{13}H_{18}N_4O_3$
7-(5-Oxohexyl)theophylline.
Xanthine derivative. Bronchodilator.

7216 Mabuterol
56341-08-3 5674
$C_{13}H_{18}ClF_3N_2O$
4-Amino-α-[(tert-butylamino)methyl]-3-chloro-5-(trifluoromethyl)benzyl alcohol.
ambuterol. Bronchodilator. Orally active β-adrenergic agonist related to clenbuterol. Ephedrine derivative. *Boehringer Ingelheim Ltd.; Thomae GmbH Dr. Karl.*

7217 dl-Mabuterol Hydrochloride
95656-48-7 5674
$C_{13}H_{19}Cl_2F_3N_2O$
4-Amino-α-[(tert-butylamino)methyl]-3-chloro-5-(trifluoromethyl)benzyl alcohol hydrochloride.
KF-868; PB-868Cl; Broncholin. Bronchodilator. Ephedrine derivative. mp = 205-206°; somewhat soluble in H_2O; LD_{50} (mmus ip) = 60.3 mg/kg, (mmus orl) = 220.8 mg/kg. *Boehringer Ingelheim Ltd.; Thomae GmbH Dr. Karl.*

7218 Medibazine
53-31-6 5831 200-168-7
$C_{25}H_{26}N_2O_2$
1-(Diphenylmethyl)-4-piperonylpiperazine.
S-4105. Vasodilator (coronary); bronchodilator. N-substituted piperazine derivative.

7219 Medibazine Dihydrochloride
96588-03-3 5831
$C_{25}H_{28}Cl_2N_2O_2$
1-(Diphenylmethyl)-4-piperonylpiperazine dihydrochloride.
Vialibran. Vasodilator (coronary); bronchodilator. N-substituted piperazine derivative. mp = 288°. *Sci. Union et Cie France.*

7220 Metaproterenol
586-06-1 5992 209-569-1
$C_{11}H_{17}NO_3$
1-(3,5-Dihydroxyphenyl)-2-isopropylaminoethanol.
orciprenaline. Bronchodilator. Ephedrine derivative. mp = 100°. *Boehringer Ingelheim Ltd.*

7221 Metaproterenol Polistirex
586-06-1
Sulfonated styrene-divinylbenzene copolymer complex with 1-(3,5-dihydroxyphenyl)-2-isopropylaminoethanol.
Bronchodilator. *Fisons plc.*

7222 Metaproterenol Sulfate
5874-97-5 5992 227-539-6
$C_{22}H_{36}N_2O_{10}S$
1-(3,5-Dihydroxyphenyl)-2-isopropylaminoethanol sulfate (2:1).
orciprenaline sulfate; TH-152; Alotec; Alupent; Metaprel; Novasmasol. Bronchodilator. Ephedrine derivative. mp = 202-203°; LD_{50} (rat orl) = 42 mg/kg. *Boehringer Ingelheim Ltd.; Sandoz Pharm. Corp.*

7223 Methoxyphenamine
93-30-1 6077 202-237-7
$C_{11}H_{17}NO$
2-Methoxy-N,α-dimethylbenzeneethanamine.
Bronchodilator. β-Adrenergic agonist. Oil; bp_2 = 97-99°. *Nippon Shinyaku Japan.*

7224 Methoxyphenamine Hydrochloride
5588-10-3 6077 226-993-2
$C_{11}H_{18}ClNO$
2-Methoxy-N,α-dimethylbenzeneethanamine hydrochloride.
Proasma; Orthoxine; component of: Orthoxicol. Bronchodilator. β-Adrenergic agonist. mp = 129-131°; soluble in H_2O, EtOH, $CHCl_3$; slightly soluble in Et_2O, C_6H_6; pH (5% aqueous solution) = 5.3-5.7. *Hexcel; Nippon Shinyaku Japan; Upjohn Ltd.*

7225 Methylephedrine
552-79-4 6145 209-022-7
$C_{11}H_{17}NO$
erythro-α-[1-(Dimethylamino)ethyl]-benzyl alcohol.
N-methylephedrine; N,N-dimethylnorephedrine; 2-dimethylamino-1-phenylpropanol. Analeptic. CNS stimulant. Ephedrine derivative. [dl-form]: mp = 63.5-64.5°; soluble in most solvents; [dl-form hydrochloride]: mp = 207-208°; [d-form]: mp = 87-87.5°; $[\alpha]_D^{20}$ = 29.2° (c = 4 MeOH); [d-form hydrochloride]: mp = 192°; $[\alpha]_D^{20}$ = 30.1°; [l-form]: mp = 87-88°; $[\alpha]_D$ = -29.5° (c = 4.5 MeOH); [l-form hydrochloride]: mp = 192°; $[\alpha]_D^{20}$ = -29.8° (c = 4.6); soluble in H_2O, less soluble in EtOH, poorly soluble in Me_2CO.

7226 Metiprenaline
1212-03-9
$C_{12}H_{19}NO_3$
α-[(Isopropylamino)methyl]vanillyl alcohol.
See isoproterenol. Bronchodilator.

7227 Mexafylline
80294-25-3
$C_{14}H_{18}N_4O_2$
3-(3-Cyclohexen-1-ylmethyl)-1,8-dimethylxanthine.
Theophylline derivative. Bronchodilator.

7228 Motapizone
90697-57-7
$C_{12}H_{12}N_4OS$
(±)-4,5-Dihydro-6-(4-imidazole-1-yl-2-thienyl)-5-methyl-3(2H)-pyridazinone.
Specific phosphodiesterase inhibitor with some bronchodilating and vasodilating activities.

7229 Nestifylline
116763-36-1
$C_{11}H_{14}N_4O_2S_2$
7-(1,3-Dithiolan-2-ylmethyl)theophylline.
Theophylline derivative. Bronchodilator.

7230 Nisbuterol
60734-87-4
$C_{22}H_{27}NO_6$
(±)-α-[(tert-Butylamino)methyl]-3,4-dihydroxybenzyl alcohol 3-acetate 4-p-anisate.
Bronchodilator. *Sterling Winthrop Inc.*

7231 Nisbuterol Mesylate
60734-88-5
$C_{23}H_{31}NO_9S$
(±)-α-[(tert-Butylamino)methyl]-3,4-dihydroxybenzyl alcohol 3-acetate 4-p-anisate methanesulfonate (salt).
Win-34886. Bronchodilator. *Sterling Winthrop Inc.*

7232 Nolpitantium Besilate
155418-06-7
$C_{43}H_{50}Cl_2N_2O_5S$
1-[2-[(S)-3-(3,4-Dichlorophenyl)-1-[(m-isopropoxyphenyl)acetyl]-3-piperidyl]ethyl]-4-phenylquinuclidinium benzenesulfonate.
SR-140333. Tachykinin NK1 receptor antagonist. Bronchodilator.

7233 Oxitropium Bromide
30286-75-0 7077 250-113-6
$C_{19}H_{26}NO_4$
(8r)-6β,7β-Epoxy-8-ethyl-3α-hydroxy-1αH,5αH-tropanium bromide (-)-tropate.
Ba-253; Ba-253-BR-L; Oxivent; Tersigat; Ventilat. Bronchodilator. Anticholinergic. Quaternary ammonium compound. mp = 203-204° (dec); $[\alpha]_D^{21}$ = -25° (c = 2 in H_2O). *Boehringer Ingelheim Ltd.*

7234 Picumeterol
130641-36-0
$C_{21}H_{29}Cl_2N_3O_2$
(-)-(R)-4-Amino-3,5-dichloro-α-[[[6-[2-(2-pyridyl)ethoxy]hexyl]amino]methyl]-benzyl alcohol.
GR-114297X. Bronchodilator. *Glaxo Labs.*

7235 Picumeterol Fumarate
130641-37-1
$C_{46}H_{62}Cl_4N_6O_8$
(-)-(R)-4-Amino-3,5-dichloro-α-[[[6-[2-(2-pyridyl)-ethoxy]hexyl]amino]methyl]benzyl alcohol fumarate (2:1) (salt).
GR-114297A. Bronchodilator. *Glaxo Labs.*

7236 Piquizil
21560-58-7
$C_{19}H_{26}N_4O_4$
Isobutyl 4-(6,7-dimethoxy-4-quinazolinyl)-1-piperazinecarboxylate.
Bronchodilator. A quinazoline derivative. *Pfizer Inc.*

7237 Piquizil Hydrochloride
23256-26-0
$C_{19}H_{27}ClN_4O_4$
Isobutyl 4-(6,7-dimethoxy-4-quinazolinyl)-1-piperazinecarboxylate monohydrochloride.
CP-12521-1. Bronchodilator. A quinazoline derivative. *Pfizer Inc.*

7238 Pirbuterol
38677-81-5 7644
$C_{12}H_{20}N_2O_3$
α^6-[(tert-Butylamino)methyl]-3-hydroxy-2,6-pyridinedimethanol.
Bronchodilator. Analog of albuterol, a β_2-adrenergic agonist. Ephedrine derivative. *Pfizer Inc.*

7239 Pirbuterol Acetate
65652-44-0 7644 265-862-4
$C_{14}H_{24}N_2O_5$
α^6-[(tert-Butylamino)methyl]-3-hydroxy-2,6-pyridinedimethanol monoacetate.
CP-24314-14; Maxair; Spirolair. Bronchodilator. Ephedrine derivative. *3M Pharm.; Pfizer Inc.*

7240 Pirbuterol Hydrochloride
38029-10-6 7644 253-751-3
$C_{12}H_{22}Cl_2N_2O_3$
α^6-[(tert-Butylamino)methyl]-3-hydroxy-2,6-pyridinedimethanol dihydrochloride.
CP-24314-1; Broncocor; Exirel. Bronchodilator. Ephedrine derivative. mp = 182° (dec). *Pfizer Inc.*

7241 Procaterol
72332-33-3 7939 276-590-0
$C_{16}H_{22}N_2O_3$
(±)-erythro-8-Hydroxy-5-[1-hydroxy-2-(isopropylamino)butyl]carbostyril.
Bronchodilator. Symapthomimetic amine with selective β_2-adrenergic agonist activity. Ephedrine derivative. *Otsuka Am. Pharm.*

7242 Procaterol Hydrochloride
59828-07-8 7939
$C_{16}H_{23}ClN_2O_3$
(±)-erythro-8-Hydroxy-5-[1-hydroxy-2-(isopropylamino)butyl]carbostyril.
CI-888; OPC-2009; Lontermin; Masacin; Meptin; Onsukil; Pro-Air; Procadil; Promaxol; Propulum. Bronchodilator. Ephedrine derivative. [hemihydrate]: mp = 193-197° (dec); soluble in MeOH; slightly soluble in EtOH; nearly insoluble in Me_2CO, Et_2O, EtOAc, $CHCl_3$, C_6H_6; LD_{50} (mrat orl) = 2600 mg/kg; (mrat iv) = 80 mg/kg. *Otsuka Am. Pharm. ; Parke-Davis.*

7243 Protokylol
136-69-6 8082 205-255-3
$C_{18}H_{21}NO_5$
4-[2-[[2-(1,3-Benzodioxol-5-yl)-1-methylethyl]amino]-1-hydroxyethyl]-1,2-benzenediol.
Bronchodilator. β-Adrenergic agonist. Ephedrine derivative. *Lakeside BioTechnology.*

7244 Protokylol Hydrochloride
136-70-9 8082 205-254-8
$C_{18}H_{22}ClNO_5$
4-[2-[[2-(1,3-Benzodioxol-5-yl)-1-methylethyl]amino]-1-hydroxyethyl]-1,2-benzenediol hydrochloride.
JB-251; Caytine; Ventaire. Bronchodilator. β-Adrenergic agonist. Ephedrine derivative. mp = 126-127°; soluble in H_2O; LD_{50} (rat orl) 940 mg/kg. *Lakeside BioTechnology.*

7245 Proxyphylline
603-00-9 8092 210-028-7
$C_{10}H_{14}N_4O_3$
3,7-Dihydro-7-(2-hydroxypropyl)-1,3-dimethyl-1H-purine-2,6-dione.
Brontyl; Proxy-Retardoral; Purophyllin; Spantin; Spasmolysin; Thean; Theon. Bronchodilator; vasodilator. Smooth muscle relaxant. Xanthine derivative. mp = 135-136°; soluble in H_2O (100 g/100 ml), absolute EtOH (7 g/100 ml); more soluble in boiling EtOH; pH (5% aqueous solution) = 5.0-7.0.

7246 Pseudoephedrine
3645
$C_{10}H_{15}NO$
DL-Threo-2-(methylamino)-1-phenylpropan-1-ol.
dl-pseudoephedrine. Adrenergic agonist. mp = 118°.

7247 d-Pseudoephedrine
90-82-4 3645 202-018-6
$C_{10}H_{15}NO$
[S-(R*,R*)]-α-[1-(methylamino)ethyl]benzenemethanol.
d-isoephedrine; d-+ly-ephedrine. Decongestant. α-adrenergic agonist. mp = 119°; $[\alpha]_D^{20}$ = 51° (c = 0.6 in EtOH); pH (0.5% aqueous solution) = 10.8; sparingly soluble in H_2O; soluble in EtOH, Et_2O.

7248 d-Pseudoephedrine Hydrochloride
345-78-8 3645 206-462-1
$C_{10}H_{16}ClNO$
[S-(R*,R*)]-α-[1-(methylamino)ethyl]benzenemethanol hydrochloride.
Galpseud; Novafed; Rhinalair; Otrinol; Sinufed; Sudafed; Symptom 2; PediaCare Decongestant Drops, Actifed, Actifed Plus, Actifed with Codeine Cough Syrup, Advil

Cold & Sinus, Allent, Alleract, Allerest; component of: Brexin EX, Brexin LA, Children's Tylenol Cold Tablets, Congestac, Contac Jr Non-drowsy Formula, Contac Nighttime Cold Medicine, Deconamine; componetnt of Dimacol, Dorcol, Dristan Allergy, Dristan Sinus, Fedahist, Histalet Syrup, Histalet X T. Decongestant. Adrenergic agonist. mp = 181-182°; $[\alpha]_D^{20}$ = +62° (c = 0.8); λ_m = 208, 215, 257, 264 mn (ε 8300, 161, 201, 161, EtOH); pKa = 9.22; pH (0.5% aqueous solution) = 5.9; soluble in H_2O; LD_{50} (mus ip) = 165 mg/kg. *Boots Pharmaceuticals Inc.; Burroughs Wellcome Inc.; Fisons plc; Marion Merrell Dow Inc.; McNeil Pharm.; Menley & James Labs Inc.; Rhône-Poulenc Rorer Pharm. Inc.; Robins, A. H. Co.; Ross Products; Sandoz Pharm. Corp.; Savage Labs; Schwarz Pharma Kremers Urban Co.; SmithKline Beecham Pharm.; Whitehall Labs. Inc.; Wyeth-Ayerst Labs.*

7249 d-Pseudoephedrine Sulfate
7460-12-0 3645 231-243-2
$C_{20}H_{32}N_2O_6S$
[S-(R*,R*)]-α-[1-(Methylamino)ethyl]benzenemethanol sulfate (2:1).
Sch-4855; Afrin Tablets; Afrinol; Chlor-Trimeton ND; Drixoral ND; Duration Tablets; component of: Chlor-Trimeton Decongestant, Claritin D, Disophrol; Drixoral Allergy/Sinus, Drixoral Cold & Allergy, Drixoral Cold & Flu, Polaramine Expectorant, Trinalin. The d-form is a nasal decongestant. An α-adrenergic agonist. Sympathomimetic. Used in veterinary medicine as an antihypotensive, mydriatic, antiallergic, and CNS stimulant. *ICI ; Schering-Plough HealthCare Products; Schering-Plough Pharm.*

7250 Quazodine
4015-32-1
$C_{12}H_{14}N_2O_2$
4-Ethyl-6,7-dimethoxyquinazoline.
MJ-1988. Cardiotonic; bronchodilator. A quinazoline derivative.

7251 Quinterenol
13757-97-6
$C_{14}H_{18}N_2O_2$
8-Hydroxy-α-[(isopropylamino)methyl]-5-quinolinemethanol.
quinprenaline. Bronchodilator. *Pfizer Inc.*

7252 Quinterenol Sulfate
13758-23-1
$C_{28}H_{38}N_4O_8S$
8-Hydroxy-α-[(isopropylamino)methyl]-5-quinolinemethanol sulfate (2:1).
quinprenaline sulfate; CP-10303-8. Bronchodilator. *Pfizer Inc.*

7253 Reproterol
54063-54-6 8307 258-956-1
$C_{18}H_{23}N_5O_5$
7-[3-[[2-(3,5-Dihydroxyphenyl)-2-hydroxyethyl]amino]propyl]-3,7-dihydro-1,3-dimethyl-1H-purine-2,6-dione.
D-1959. Bronchodilator. Theophylline derivative with selective β_2 receptor activity. Ephedrine derivative. *Degussa-Hüls Corp.*

7254 Reproterol Hydrochloride
13055-82-8 8307 235-942-3
$C_{18}H_{24}ClN_5O_5$
7-[3-[[2-(3,5-Dihydroxyphenyl)-2-hydroxyethyl]amino]propyl]-3,7-dihydro-1,3-dimethyl-1H-purine-2,6-dione monohydrochloride.
D-1959-HCl; W-2946M; Asmaterolo; Bronchodil; Bronchospasmin. Bronchodilator. Ephedrine derivative. mp = 249-250°; LD_{50} (mus iv) = 148 mg/kg, (mus orl) > 10000 mg/kg. *Degussa-Hüls Corp.; Wallace Labs.*

7255 Revatropate
149926-91-0
$C_{19}H_{27}NO_4S$
(R)-3-quinuclidinyl (S)-β-hydroxy-α-[2-(R)-methylsulfinyl]ethyl]hydratropate.
A novel antimuscarinic agent with selectivity for M1 and M3 receptors. Bronchodilator.

7256 Rimiterol
32953-89-2 8393 251-305-2
$C_{12}H_{17}NO_3$
(R*,S*)-4-(Hydroxy-2-piperidinylmethyl)-1,2-benzenediol.
Bronchodilator. Ephedrine derivative. mp = 203-204°. *SmithKline Beecham Pharm.*

7257 Rimiterol Hydrobromide
31842-61-2 8393 250-834-6
$C_{12}H_{18}BrNO_3$
(R*,S*)-4-(Hydroxy-2-piperidinylmethyl)-1,2-benzenediol hydrobromide.
R-798; WG-253; Asmaten; Pulmadil. Bronchodilator. Ephedrine derivative. mp = 220° (dec). *3M Pharm.; SmithKline Beecham Pharm.*

7258 Rispenzepine
96449-05-7
$C_{19}H_{20}N_4O_2$
(±)-6,11-Dihydro-11-(1-methylnipecotoyl)-5H-pyrido[2,3-b][1,5]benzodiazepin-5-one.
Muscarinic M1/M3 receptor antagonist. Bronchodilator.

7259 Salmefamol
18910-65-1 242-662-5
$C_{19}H_{25}NO_4$
α-[(p-Methoxy-α-methylphenylethylamino)methyl]-4-hydroxy-m-xylene-α,α'-diol.
AH-3923. beta Adrenergic agonist; bronchodilator.

7260 Salmeterol
89365-50-4 8489
$C_{25}H_{37}NO_4$
(±)-4-Hydroxy-α'-[[[6-(4-phenylbutoxy)hexyl]amino]methyl]-m-xylene-α,α'-diol.
GR-33343X. Bronchodilator. Structural analog of albuterol. β_2-Adrenergic agonist. Ephedrine derivative. mp = 75.5-76.5°. *Glaxo Labs.*

7261 Salmeterol Xinafoate
49749-08-3 8489
$C_{36}H_{45}NO_7$
(±)-4-Hydroxy-α'-[[[6-(4-phenylbutoxy)hexyl]amino]methyl]-m-xylene-α,α'-diol 1-hydroxy-2-napthoate (salt).
GR-3343G; Arial; Salmetedur; Serevent. Bronchodilator.

Ephedrine derivative. mp = 137-138°; soluble in MeOH; slightly soluble in EtOH, $CHCl_3$, iPrOH; sparingly soluble in H_2O. *Glaxo Labs.*

7262 Saredutant
142001-63-6
$C_{31}H_{35}Cl_2N_3O_2$
N-[(S)-β-[2-(4-acetamido-4-phenylpiperidino)ethyl]-3,4-dichlorophenethyl]-N-methylbenzamide.
SR-48968. A potent and specific nonpeptide tachykinin NK2 receptor antagonist. Reduces histamine-induced airway hyperresponsiveness. Bronchodilator.

7263 Soterenol
13642-52-9 8877
$C_{12}H_{20}N_2O_4S$
2'-Hydroxy-5'-[1-hydroxy-2-(isopropylamino)ethyl]methanesulfonamide.
MJ-1992. Bronchodilator. β-Adrenergic agonist. Ephedrine derivative.

7264 Soterenol Hydrochloride
14816-67-2 8877 238-889-4
$C_{12}H_{21}ClN_2O_4S$
2'-Hydroxy-5'-[1-hydroxy-2-(isopropylamino)ethyl]methanesulfonamide monohydrochloride.
Bronchodilator. β-Adrenergic agonist. Ephedrine derivative. mp = 195.5-196.5° (dec); LD_{50} (mus iv) = 41 mg/kg, (mus ip) = 315 mg/kg, (mus orl) = 660 mg/kg.

7265 Sulfonterol
42461-79-0
$C_{14}H_{23}NO_4S$
α-[(tert-Butylamino)methyl]-4-hydroxy-3-[(methylsylfonyl)methyl]benzyl alcohol.
Bronchodilator. *SmithKline Beecham Pharm.*

7266 Sulfonterol Hydrochloride
42461-78-9
$C_{14}H_{24}ClNO_4S$
α-[(tert-Butylamino)methyl]-4-hydroxy-3-[(methylsylfonyl)methyl]benzyl alcohol hydrochloride.
SKF-53705-A. Bronchodilator. *SmithKline Beecham Pharm.*

7267 Suloxifen
25827-12-7
$C_{18}H_{24}N_2OS$
N-[2-(Diethylamino)ethyl]-s,s-diphenylsulfoximine.
Bronchodilator. *Goedecke, Germany; Parke-Davis.*

7268 Suloxifen Oxalate
25827-13-8
$C_{20}H_{26}N_2O_4S$
N-[2-(Diethylamino)ethyl]-s,s-diphenylsulfoximine oxalate (1:1).
Go-1733; W-6439A. Bronchodilator. *Goedecke, Germany; Parke-Davis.*

7269 Terbutaline
23031-25-6 9302 245-385-8
$C_{12}H_{19}NO_3$
5-[2-[(1,1-Dimethylethyl)amino]-1-hydroxyethyl]-1,3-benzenediol.
Bronchodilator; tocolytic. β-Adrenergic agonist. Ephedrine derivative. mp = 119-120°. *Astra Draco AB; Ciba-Geigy Corp.; Merrell Pharm. Inc.*

7270 Terbutaline Sulfate
23031-32-5 9302 245-386-3
$C_{24}H_{40}N_2O_{10}S$
5-[2-[(1,1-Dimethylethyl)amino]-1-hydroxyethyl]-1,3-benzenediol sulfate (2:1) (salt).
KWD-2019; Brethaire; Brethine; Bricanyl; Butaliret; Monovent; Terbasmin; Terbul. Bronchodilator; tocolytic. β-Adrenergic agonist. Ephedrine derivative. mp = 246-248°; λ_m = 276 nm ($A^{1\%}_{1cm}$ 67.6 0.1 N HCl); pKa_1 = 8.8, pKa_2 = 10.1, pKa_3 = 11.2; soluble in H_2O (>2.0 g/100 ml), 0.1N HCl (>2.0 g/100 ml), 0.1N NaOH (>2.0 g/100 ml), EtOH (0.012 g/100 ml), 10% EtOH (>2.0 . *Astra Draco AB; Ciba-Geigy Corp.; Marion Merrell Dow Inc.; Merrell Pharm. Inc.*

7271 Theobromine
83-67-0 9418 201-494-2
$C_7H_8N_4O_2$
3,7-Dihydro-3,7-dimethyl-1H-purine-2,6-dione.
3,7-dimethylxanthine. Diuretic; bronchodilator; cardiotonic. Principle alkaloid of the cacao bean. Also found in coca nuts and tea. A methylxanthine CNS stimulant. mp = 357°; sublimes at 290-295°; very slightly soluble in H_2O, EtOH; soluble in fixed alkali hydroxides, concentrated acids; moderately soluble in ammonia; insoluble in C_6H_6, Et_2O, $CHCl_3$, CCl_4.

7272 1-Theobromineacetic Acid
5614-56-2 9419 227-034-0
$C_9H_{10}N_4O_4$
2,3,6,7-Tetrahydro-3,7-dimethyl-2,6-dioxo-1H-purine-1-acetic acid.
Bronchodilator. Xanthine derivative. mp = 260°.

7273 1-Theobromineacetic Acid Sodium Salt
32245-40-2 9419
$C_9H_9N_4NaO_4$
2,3,6,7-Tetrahydro-3,7-dimethyl-2,6-dioxo-1H-purine-1-acetic acid sodium (salt).
sodium theobromine acetate; Técarine. Bronchodilator. Xanthine derivative. Soluble in H_2O.

7274 Theobromine Calcium Salicylate
8065-51-8 9418
$C_{21}H_{18}CaN_4O_8$
3,7-Dihydro-3,7-dimethyl-1H-purine-2,6-dione
calcium salt mixture (1:1) with calcium salicylate.
Theocalcin; Calcium Diuretin. Diuretic; bronchodilator; cardiotonic. Xanthine derivative. Slightly soluble in H_2O.

7275 Theobromine Sodium Acetate
8002-88-8 9418
$C_9H_{11}N_4NaO_4.H_2O$
3,7-Dihydro-3,7-dimethyl-1H-purine-2,6-dione sodium salt mixture with sodium acetate (1:1), containing 1 H_2O.
Thesodate. Diuretic; bronchodilator; cardiotonic. Xanthine derivative. Soluble in H_2O.

7276 Theobromine Sodium Salicylate
8048-31-5 9418
$C_{12}H_{13}N_4NaO_5.H_2O$
3,7-Dihydro-3,7-dimethyl-1H-purine-2,6-dione sodium salt mixture with sodium salicylate (1:1), containing 1 H_2O.
Diuretin. Diuretic; bronchodilator; cardiotonic. Xanthine derivative. Soluble in H_2O; sparingly soluble in cold EtOH; pH 10.

7277 Theophylline
58-55-9 9421 200-385-7
$C_7H_8N_4O_2$
3,7-Dihydro-1,3-dimethyl-1H-purine-2,6-dione.
1,3-Dimethylxanthine; Accurbron; Aerobin; Aerolate; Afonilm; Armophylline; Austyn; Bilordyl; Brochoretard; Bronkodyl; Cétraphylline; Constant-T; Duraphyl; Duraphyllin; Diffumal; Elixophyllin; Etheophyl; Euphyllin; Euphylong; LaBID; Lasma; Physpan; Pro-Vent; PulmiDur; Pulmo-Timelets; Respbid; Slo-Bid; Slo-Phyllin; Solosin; Somophyllin-CRT; Somophyllin-T; Sustaire; Talotren; Tesona; Theobid; Theo-24; Theobid Duracap; Theoclear; Theochron; Theo-Dur; Theograd; Theolair; Theon; Theophyl; Theoplus; Theo-Sav; Theostat; Theovent; Unifyl; Uniphyl; Uniphyllin; Xanthium; component of: Bronkotabs, Dicurin Procaine, Mudrane GG Elixer, Primatene Tablets, Quibron, Quibron Plus, Quibron-T/SR, Slo-Phyllin GG, Tedral, Theolair Plus, Theo-Organdin. Bronchodilator. Xanthine derivative with diuretic, cardiac stimulant and smooth muscle relaxant activities; isomeric with theobromine; small amounts found in tea. *3M Pharm.; Boehringer Ingelheim Ltd.; Bristol Laboratories; Ciba-Geigy Corp.; Eli Lilly & Co.; Fisons plc; Forest Pharm. Inc.; Glaxo Wellcome plc; Key Pharm.; Norwich Eaton; Poythress; Purdue Pharma L.P.; Rhône-Poulenc Rorer Pharm. Inc.; Savage Labs; Schering AG; Searle G.D. & Co.; Sterling Winthrop Inc.; Wallace Labs; Whitehall Labs. Inc.*

7278 Theophylline Ethanolamine
573-41-1 9421 209-355-8
$C_9H_{15}N_5O_3$
3,7-Dihydro-1,3-dimethyl-1H-purine-2,6-dione compound with 2-aminoethanol (1:1).
theophylline 2-aminoethanol; theophylline olamine; Monotheamin. Bronchodilator. Xanthine derivative. Soluble in H_2O.

7279 Theophylline Isopropanolamine
5600-19-1 9421 227-017-8
$C_{10}H_{17}N_5O_3$
3,7-Dihydro-1,3-dimethyl-1H-purine-2,6-dione compound with 1-amino-2-propanol (1:1).
Oxyphyllin. Bronchodilator. Xanthine derivative. Soluble in H_2O; pH (aqueous solution) = 9.20.

7280 Theophylline Lysine Salt
9421
3,7-Dihydro-1,3-dimethyl-1H-purine-2,6-dione compound with lysine (1:1) (salt).
Paidomal. Bronchodilator. Xanthine derivative.

7281 Theophylline Monohydrate
5967-84-0 9421
$C_7H_{10}N_4O_3$
3,7-Dihydro-1,3-dimethyl-1H-purine-2,6-dione monohydrate.
Bronchodilator. Pharmaceutic necessity for Aminophylline Injection. Xanthine derivative. mp = 270-274°; pKa = 8.77; λ_m = 274 nm (in 0.1N NaOH); somewhat soluble in H_2O (8.3 g/100 ml), EtOH (1.25 g/100 ml), $CHCl_3$ (0.91 g/100 ml); soluble in hot H_2O, alkali hydroxides, ammonia, dilute HCl, dilute HNO_3; sparingly soluble in Et_2O.

7282 Theophylline Sodium Acetate
8002-89-9 9421
3,7-Dihydro-1,3-dimethyl-1H-purine-2,6-dione compound with sodium acetate (1:1), containing 1 H_2O.
Theocin. Bronchodilator. Xanthine derivative.

7283 Theophylline Sodium Glycinate
8000-10-0 9421
$C_9H_{12}N_5NaO_4$
3,7-Dihydro-1,3-dimethyl-1H-purine-2,6-dione monosodium salt with glycine.
sodium theophyllin-7-ylglycinate; Biophylline; Englate; Glytheonate; Panophylline; Pemophyllin. Smooth muscle relaxant, used as a bronchodilator. Xanthine derivative. Dec 180°; d^{20} = 1.05; soluble in H_2O (18 g/100 ml); pH (saturated solution) = 8.7-9.1. *Marion Merrell Dow Inc.*

7284 Tibenelast Sodium
105102-18-9
$C_{13}H_{13}NaO_4S$
Sodium 5,6-diethoxybenzo[b]thiophene-2-carboxylate.
LY-186655. A bronchodilator and antiasthmatic. Phosphodiesterase inhibitor. *Eli Lilly & Co.*

7285 Tiotropium
136310-93-5 9598
$[C_{19}H_{22}NO_4S_2]^+$
(1α,2β,4β,5α,7β)-7-[(Hydroxydi-2-thienylacetyl)oxy]-9,9-dimethyl-3-oxa-9-azoniatricycol[3.3.1.0^{2,4}]nonane.
Ba-679. Bronchodilator. Muscarinic receptor antagonist. Quaternary ammonium compound. *Boehringer Ingelheim Ltd.*

7286 Tiotropium Bromide
139404-48-1 9598
$C_{19}H_{22}BrNO_4S_2$
(1α,2β,4β,5α,7β)-7-[(Hydroxydi-2-thienylacetyl)oxy]-9,9-dimethyl-3-oxa-9-azoniatricyclo[3.3.1.0^{2,4}]nonane bromide.
di-2-thienylglycolate; Ba-679-BR; Spiriva. A muscarinic antagonist used for treatment of obstructive airways disease. Bronchodilator; anticholinergic. mp = 218-220°. *Boehringer Ingelheim Ltd.*

7287 Tretoquinol
21650-42-0 9719
$C_{19}H_{23}NO_5$
(±)-1,2,3,4-Tetrahydro-1-(3,4,5-trimethoxybenzyl)-6,7-isoquinolinediol.
trimethoquinol; trimetoquinol; AQ-110 [as l-form hydrochloride]; Inolin [as l-form hydrochloride]; Vems [as l-form hydrochloride]. A catecholamine. A highly potent β_2-adrenoceptor and site-selective thromboxane A2/prostaglandin H2 receptor ligand. *l*-Form hydrochloride acts as a bronchodilator. Dec 224.5-226°.

7288 L-Tretoquinol
30418-38-3 9719
$C_{19}H_{23}NO_5$
L-1,2,3,4-Tetrahydro-1-(3,4,5-trimethoxybenzyl)-6,7-isoquinolinediol.
trimethoquinol. Bronchodilator.

7289 L-Tretoquinol Hydrochloride
18559-63-2 9719
$C_{19}H_{24}ClNO_5$
L-1,2,3,4-Tetrahydro-1-(3,4,5-trimethoxybenzyl)-6,7-isoquinolinediol monohydrochloride.
AQ-110; Inolin; Vems. Bronchodilator. Soluble in H_2O, EtOH; [dl-form]: dec 224.5-226°.

7290 Tulobuterol
41570-61-0 9942
$C_{12}H_{18}ClNO$
2-Chloro-α-[[(1,1-dimethylethyl)amino]-ethyl]benzenemethanol.
α-[(tert-Butylamino)methyl]-o-chlorobenzyl alcohol; HN-078; HOKU-81. Bronchodilator. Used in the treatment of asthma. A β-adrenergic receptor agonist, related structurally to terbutaline, an ephedrine derivative. mp = 89-91°; LD_{50} (mmus orl) = 305 mg/kg, (mrat orl) = 850 mg/kg, (mrbt orl) = 563 mg/kg, (mmus sc) = 170 mg/kg, (mrat sc) = 417 mg/kg, (mrbt sc) = 164 mg/kg.

7291 Tulobuterol Hydrochloride
56776-01-3 9942
$C_{12}H_{19}Cl_2NO$
2-Chloro-α-[[(1,1-dimethylethyl)amino]-methyl]benzenemethanol hydrochloride.
Atenos; Berachin; Brelomax; Bremax; Hokunalin; Respacal. Bronchodilator. Used in the treatment of asthma. A β-adrenergic receptor agonist. Ephedrine derivative. mp = 161-163°.

7292 Verofylline
66172-75-6
$C_{12}H_{18}N_4O_2$
(±)-1,8-Dimethyl-3-(2-methylbutyl)xanthine.
CK-0383. Bronchodilator; antiasthmatic. *Berlex Labs Inc.*

7293 Xanoxate Sodium
41147-04-0
$C_{17}H_{13}NaO_5$
Sodium 7-isopropoxy-9-oxoxanthene-2-carboxylate.
RS-6818. Bronchodilator. *Syntex Labs. Inc.*

7294 Xanoxic Acid
33459-27-7
$C_{17}H_{14}O_5$
7-Isopropoxy-9-oxoxanthene-2-carboxylic acid.
Bronchodilator. *Syntex Labs. Inc.*

7295 Zardaverine
101975-10-4
$C_{12}H_{10}F_2N_2O_3$
6-[4-(Difluoromethoxy-3-methoxyphenyl]-3(2H)-pyridazinone.
Dual phosphodiesterase 3/4 inhibitor. Inhibits the proliferation of human peripheral blood mononuclear cells. Bronchodilator; antiasthmatic.

7296 Zindotrine
56383-05-2
$C_{11}H_{15}N_5$
8-Methyl-6-piperidino-s-triazolo[4,3-b]pyridazine.
MDL-257. Bronchodilator. *Marion Merrell Dow Inc.*

7297 Zinterol
37000-20-7
$C_{19}H_{26}N_2O_4S$
5'-[2-[(α,α-Dimethylphenethyl)amino]-1-hydroxyethyl]-2'-hydroxymethanesulfonanilide.
Bronchodilator. *Bristol Myers Squibb Pharm. Ltd.*

7298 Zinterol Hydrochloride
38241-28-0
$C_{19}H_{27}ClN_2O_4S$
5'-[2-[(α,α-Dimethylphenethyl)amino]-1-hydroxyethyl]-2'-hydroxymethanesulfonanilide hydrochloride.
MJ-9184-1. Bronchodilator. *Bristol Myers Squibb Pharm. Ltd.*

Calcium Channel Blockers

7299 Amlodipine
88150-42-9 516
$C_{20}H_{25}ClN_2O_5$
3-Ethyl-5-methyl (±)-2-[(2-aminoethoxy)methyl]-4-(o-chlorophenyl)-1,4-dihydro-6-methyl-3,5-pyridinedicarboxylate.
Antianginal agent. Dihydropyridine calcium channel blocker. Has antihypertensive properties. *Ciba-Geigy Corp.; Pfizer Inc.*

7300 Amlodipine Besylate
111470-99-6 516
$C_{26}H_{31}ClN_2O_8S$
3-Ethyl-5-methyl (±)-2-[(2-aminoethoxy)methyl]-4-(o-chlorophenyl)-1,4-dihydro-6-methyl-3,5-pyridinedicarboxylate monobenzenesulfonate.
Norvasc; UK-48340-26; Antacal; Istin; Monopina; component of: Lotrel. Antianginal agent. Dihydropyridine calcium channel blocker. Has antihypertensive properties. mp = 178-179°. *Ciba-Geigy Corp.; Pfizer Inc.*

7301 Amlodipine Maleate
88150-47-4 516
$C_{24}H_{29}ClN_2O_9$
3-Ethyl-5-methyl (±)-2-[(2-aminoethoxy)methyl]-4-(o-chlorophenyl)-1,4-dihydro-6-methyl-3,5-pyridinedicarboxylate maleate.
UK-48340-11. Antianginal agent. Dihydropyridine calcium channel blocker. Has antihypertensive properties. *Pfizer Inc.*

7302 Aranidipine
86780-90-7 806
$C_{19}H_{20}N_2O_7$
(±)-Acetyl methyl 1,4-dihydro-2,6-dimethyl-4-(o-nitrophenyl)-3,5-pyridine-dicarboxylate.
MPC-1304. Antihypertensive agent. Calcium channel blocker. *Maruko Seiyaku.*

7303 Azelnidipine
123524-52-7
$C_{33}H_{34}N_4O_6$
3-[1-(Diphenylmethyl)-3-azetidinyl] 5-isopropyl (±)-2-amino-1,4-dihydro-6-methyl-4-(m-nitrophenyl)-3,5-pyridinedicarboxylate.
Antihypertensive. Calcium channel blocker.

7304 Barnidipine
104713-75-9 1031
$C_{27}H_{29}N_3O_6$
(+)-(3'S,4S)-1-Benzyl-3-pyrrolidinyl methyl 1,4-dihydro-2,6-dimethyl-4-(m-nitrophenyl)-3,5-pyridinedicarboxylate.
Mepirodipine. Antianginal agent. Dihydropyridine calcium channel blocker. Has antihypertensive properties. mp = 137-139°; $[\alpha]_D^{20}$ = 64.8 (c = 1 MeOH). *Yamanouchi U.S.A. Inc.*

7305 Barnidipine Hydrochloride
1031
$C_{27}H_{30}ClN_3O_6$
(+)-(3'S,4S)-1-Benzyl-3-pyrrolidinyl methyl 1,4-dihydro-2,6-dimethyl-4-(m-nitrophenyl)-3,5-pyridinedicarboxylate hydrochloride.
YM-09730-5; Hypoca. Antianginal agent. Dihydropyridine calcium channel blocker. Has antihypertensive properties. mp = 226-228°; $[\alpha]_D^{20}$ = 116.4° (c = 1 MeOH); insoluble in H_2O; LD_{50} (mrat orl) = 105 mg/kg, (frat orl) = 113 mg/kg. *Yamanouchi U.S.A. Inc.*

7306 Bencyclane
2179-37-5 1060 218-547-0
$C_{19}H_{31}NO$
3-[(1-Benzylcycloheptyl)oxy]-N,N-dimethylpropylamine.
benzcyclan. Vasodilator. bp_3 = 146-156°. *EGYT.*

7307 Bencyclane Fumarate
14286-84-1 1060 238-204-9
$C_{23}H_{35}NO_5$
3-[(1-Benzylcycloheptyl)oxy]-N,N-dimethylpropylamine fumarate.
EGYT-201; Angiociclan; Dantrium; Dilangio; Fludilat; Fluxema; Halidor; Vasorelax. Vasodilator. mp = 131-133°; soluble in H_2O (1 g/100 ml at 25°); readily soluble in EtOH; slightly soluble in Me_2CO; λ_m (pH 3.4-6.6) = 207 nm; LD_{50} (mus orl) = 445.6 mg/kg, (mus iv) = 49.9 mg/kg, (mus ip) = 132 mg/kg, (mus sc) = 203 mg/kg. *EGYT.*

7308 Benidipine
105979-17-7 1071
$C_{28}H_{31}N_3O_6$
(±)-(R*)-3-[(R*)-1-Benzyl-3-piperidyl] methyl 1,4-dihydro-2,6-dimethyl-4-(m-nitrophenyl)-3,5-pyridinedicarboxylate.
Antihypertensive. Calcium channel blocker. *Kyowa Hakko Kogyo Co. Ltd.*

7309 Benidipine Hydrochloride
91599-74-5 1071
$C_{28}H_{32}ClN_3O_6$
(±)-(R*)-3-[(R*)-1-Benzyl-3-piperidyl] methyl 1,4-dihydro-2,6-dimethyl-4-(m-nitrophenyl)-3,5-pyridinedicarboxylate hydrochloride(α form).
KW-3049; Coniel. Antihypertensive. Calcium channel blocker. mp = 199.4-200.4°; λ_m = 238, 359 nm (ε = 28000, 6680 EtOH); soluble in H_2O (0.19 g/100 ml); MeOH (6.9 g/100 ml), EtOH (2.2 g/100 g); $CHCl_3$ (0.16 g/100 g), Me_2CO (0.13 g/100 g), EtOAc (0.0056 g/100 g); C_7H_8 (0.0019 g/100 g), C_7H_{14} (0.00009 g/100 g); LD_{50} (mus orl) = 218 mg/kg. *Kyowa Hakko Kogyo Co. Ltd.*

7310 Bepridil
64706-54-3 1188
$C_{24}H_{34}N_2O$
1-[2-(N-Benzylanilino)-1-(isobutoxymethyl)ethyl]-pyrrolidine.
Antianginal. Calcium channel blocker with antianginal and antiarrhythmic (class IV) properties. $bp_{0.1}$ = 184°; $bp_{0.5}$ = 192°; n_D^{20} = 1.5538. *McNeil Pharm.; Wallace Labs.*

7311 Bepridil Hydrochloride
74764-40-2 1188
$C_{24}H_{37}ClN_2O_3$
1-[2-(N-Benzylanilino)-1-(isobutoxymethyl)ethyl]-pyrrolidine monohydrochloride monohydrate.
CERM-1978; Angopril; Bepadin; Cordium; Vascor. Antianginal. Calcium channel blocker with antianginal and antiarrhythmic (class IV) properties. mp 91°; LD_{50} (mus orl) = 1955 mg/kg, (mus iv) = 23.5 mg/kg. *C.M. Ind.; McNeil Pharm.; Wallace Labs.*

7312 Bometolol
65008-93-7
$C_{25}H_{32}N_2O_7$
(±)-8-(Acetonyloxy)-5-[3-[(3,4-dimethoxy-phenethyl)amino]-2-hydroxypropoxy]-3,4-dihydrocarbostyril.
A calcium channel blocker with antiarrhythmic properties.

7313 Brinazarone
89622-90-2
$C_{25}H_{32}N_2O_2$
p-[3-(tert-Butylamino)propoxy]phenyl 2-isopropyl-3-indolizinyl ketone.
A calcium channel blocker and potential antiarrhythmic agent.

7314 Cilnidipine
102106-21-8 2334
$C_{27}H_{28}N_2O_7$
(E)-Cinnamyl 2-methoxythyl 1,4-dihydro-2,6-dimethyl-4-(m-nitrophenyl)-3,5-pyridinedicarboxylate.
Antihypertensive. Dihydropyridine calcium channel blocker. *Fujirebio Inc.*

7315 (±)-Cilnidipine
132203-70-4 2334
$C_{27}H_{28}N_2O_7$
(±)-(E)-Cinnamyl 2-methoxeythyl 1,4-dihydro-2,6-dimethyl-4-(m-nitrophenyl)-3,5-pyridinedicarboxylate.
(±)-FRC-8653. Antihypertensive. Dihydropyridine calcium channel blocker. mp = 115.5-116.6°; LD_{50} (mmus orl) > 5000 mg/kg, (mmus sc) > 5000 mg/kg, (mmus ip) = 1845 mg/kg, (fmus orl) > 5000 mg/kg, (fmus sc) > 5000 mg/kg, (fmus ip) = 2353 mg/kg, (mrat orl) > 5000 mg/kg, (mrat sc) > 5000 mg/kg, (mrat ip) = 441 mg/kg, (frat o. *Fujirebio Inc.*

7316 Cinnarizine
298-57-7 2365 206-064-8
$C_{26}H_{28}N_2$
1-Cinnamyl-4-(diphenylmethyl)piperazine.
cinnipirine; R-516; R-1575; 516-MD; Aplactan; Aplexal; Apotomin; Artate; Carecin; Cerebolan; Cerepar; Cinaperazine; Cinazyn; Cinnacet; Cinnageron; Corathiem; Denapol; Dimitron; Eglen; Folcodal; Giganten; Glanil; Hilactan; Ixterol; Katoseran; Labyrin; Midronal; Mitronal; Olamin; Processine; Sedatromin; Sepan; Siptazin; Spaderizine; Stugeron; Stutgeron; Stutgin; Toliman; component of: Emesazine. Antihistaminic; vasodilator (peripheral, cerebral). Calcium channel blocker with antiallergic and antivasoconstricting activity. [hydrochloride]: mp = 192° (dec); soluble in H_2O (20 g/l). *Abbott Laboratories Inc.*

7317 Clentiazem
96125-53-0 2408
$C_{22}H_{25}ClN_2O_4S$
(+)-(2S,3S)-8-Chloro-5-[2-(dimethylamino)ethyl]-2,3-dihydro-3-hydroxy-2-(p-methoxyphenyl)-1,5-benzothiazepin-4(5H)-one acetate (ester).
The 8-chloro derivative of diltiazem. A calcium channel blocker used for *its antihypertensive properties. Tanabe Seiyaku Co. Ltd.*

7318 Clentiazem Maleate
96128-92-6 2408
$C_{26}H_{29}ClN_2O_8S$
(+)-(2S,3S)-8-Chloro-5-[2-(dimethylamino)ethyl]-2,3-dihydro-3-hydroxy-2-(p-methoxyphenyl)-1,5-benzothiazepin-4(5H)-one acetate (ester) maleate (1:1).
TA-3090; Logna. The 8-chloro derivative of diltiazem. A calcium channel blocker used for its antihypertensive properties. mp = 160.5-161.5°; $[\alpha]_D^{20}$ = 76.5° (c = 1 MeOH). *Tanabe Seiyaku Co. Ltd.*

7319 Diltiazem
42399-41-7 3247 255-796-4
$C_{22}H_{26}N_2O_4S$
(+)-5-[2-(Dimethylamino)ethyl]-cis-2,3-dihydro-3-hydroxy-2-(p-methoxyphenyl)-1,5-benzothiazepin-4(5H)-one acetate (ester).
Antianginal; antihypertensive; antiarrhythmic (class IV). Calcium channel blocker with coronary vasodilating properties. *Bristol Myers Squibb Pharm. Ltd.; Forest Pharm. Inc.; Hoechst Marion Roussel Inc.; Lemmon Co.; Rhône-Poulenc Rorer Pharm. Inc.; Shionogi & Co. Ltd.; Tanabe Seiyaku Co. Ltd.*

7320 d-cis-Diltiazem Hydrochloride
33286-22-5 3247 251-443-3
$C_{22}H_{27}ClN_2O_4S$
(+)-5-[2-(Dimethylamino)ethyl]-cis-2,3-dihydro-3-hydroxy-2-(p-methoxyphenyl)-1,5-benzothiazepin-4(5H)-one acetate (ester) monohydrochloride.
CRD-401; RG-83606; Adizem; Altiazem; Anginyl; Angizem; Britiazim; Bruzem; Calcicard; Cardizem; Citizem; Cormax; Deltazen; Diladel; Dilpral; Dilrene; Dilzem; Dilzene; Herbesser; Masdil; Tildiem. Antianginal; antihypertensive; class IV antiarrhythmic. Calcium channel blocker with coronary vasodilating properties. mp = 207.5-212°; $[\alpha]_D^{24}$ = +98.3 ± 1.4° (c = 1.002 in MeOH); soluble in H_2O, MeOH, $CHCl_3$; slightly soluble in absolute EtOH; practically insoluble in C_6H_6; LD_{50} (mmus iv) = 61 mg/kg, (mmus sc) = 260 mg/kg, (mmus orl) = 740 mg/kg, (fmus iv) = 58 mg/kg, (fmus sc) = 280 mg/kg, (fmus orl) = 640 mg/kg, (mrat iv) = 38 mg/kg, (mrat sc) = 520 mg/kg, (mrat orl) = 560 mg/kg, (frat iv) = 39 mg/kg, (frat sc) = 550 mg/kg, (frat orl) = 610 mg/kg. *Bristol Myers Squibb Pharm. Ltd.; Forest Pharm. Inc.; Hoechst Marion Roussel Inc.; Lemmon Co.; Rhône-Poulenc Rorer Pharm. Inc.; Tanabe Seiyaku Co. Ltd.*

7321 Diltiazem Malate
144604-00-2 3247
$C_{26}H_{32}N_2O_9S$
(+)-5-[2-(Dimethylamino)ethyl]-cis-2,3-dihydro-3-hydroxy-2-(p-methoxyphenyl)-1,5-benzothiazepin-4(5H)-one acetate (ester) (S)-malate (1:1) monohydrochloride.
MK-793. Antianginal; antihypertensive; class IV antiarrhythmic. Calcium channel blocker with coronary vasodilating activity. Class IV antiarrhythmic. *Bristol Myers Squibb Pharm. Ltd.; Forest Pharm. Inc.; Hoechst Marion Roussel Inc.; Lemmon Co.; Rhône-Poulenc Rorer Pharm. Inc.*

7322 Efonidipine
111011-63-3 3566
$C_{34}H_{38}N_3O_7P$
2-(N-Benzylanilino)ethyl (±)-1,4-dihydro-2,6-dimethyl-4-(m-nitrophenyl)-5-phosphononicotinate cyclic 2,2-dimethyltrimethylene ester.
Dihydropyridine calcium channel blocker. Used as an antihypertensive agent. mp = 169-170°, 155-156°. *Nissan Kenzai Co. Ltd.*

7323 Efonidipine Hydrochloride
111011-53-1 3566
$C_{34}H_{39}ClN_3O_7P$
2-(N-Benzylanilino)ethyl (±)-1,4-dihydro-2,6-dimethyl-4-(m-nitrophenyl)-5-phosphononicotinate cyclic 2,2-dimethyltrimethylene ester hydrochloride.
Dihydropyridine calcium channel blocker. Used as an antihypertensive agent. Forms an ethanol solvate (NZ-105; Landel), mp = 151° (dec). LD_{50} (mus orl) > 600 mg/kg; [(S) form]: mp = 190-192°; $[\alpha]_D^{25}$ = 7.0° (c = 0.50 $CHCl_3$); [(R) form]: mp = 190-192°; $[\alpha]_D^{25}$ = -7.0° (c = 0.5 $CHCl_3$). *Nissan Kenzai Co. Ltd.*

7324 Elgodipine
119413-55-7 3587
$C_{29}H_{33}FN_2O_6$
2-[(p-Fluorobenzyl)methylamino]ethyl isopropyl (±)-1,4-dihydro-2,6-dimethyl-4-[2,3-(methylenedioxy)phenyl]-3,5-pyridinedicarboxylate.
Antianginal agent. A dihydropyridine calcium channel blocker. *Inst. Invest. Desarr.; Quimicobiol.*

7325 Elgodipine Hydrochloride
121489-04-1 3587
$C_{29}H_{34}ClFN_2O_6$
2-[(p-Fluorobenzyl)methylamino]ethyl isopropyl (±)-1,4-dihydro-2,6-dimethyl-4-[2,3-(methylenedioxy)phenyl]-3,5-pyridinedicarboxylate hydrochloride.
IQB-875. Antianginal agent. A dihydropyridine calcium channel blocker. mp = 194-195°; LD_{50} (mus ip) = 30-40 mg/kg. *Inst. Invest. Desarr.; Quimicobiol.*

7326 Etafenone
90-54-0 3753 202-002-9
$C_{21}H_{27}NO_2$
2'-[2-(Diethylamino)ethoxy]-3-phenyl]propylphenone.
LG-11457. Vasodilator (coronary). bp_{30} = 264-268°. *Guidotti.*

7327 Etafenone Hydrochloride
2192-21-4 3753
$C_{21}H_{28}ClNO_2$
2'-[2-(Diethylamino)ethoxy]-3-phenyl]propylphenone monohydrochloride.
heptaphenone; Asmedol; Baxacor; Corodilan; Dialicor; Pagano-Cor; Relicor. Vasodilator (coronary). mp = 129-130°; LD_{50} (rat orl) = 716 mg/kg, (rat iv) = 20.8 mg/kg. *Guidotti.*

7328 Fantofarone
114432-13-2 3975
$C_{13}H_{38}N_2O_5S$
1-[[p-[3-[(3,4-Dimethoxyphenethyl)methylamino]propoxy]phenyl]sulfonyl]-2-isopropylindolizine.
SR-33557. A calcium channel blocker used as an antihypertensive agent. mp = 82-83°; d = 1.21 g/ml; soluble in H_2O (0.06 g/100 ml), organic solvents. *Sanofi Winthrop.*

7329 Felodipine
86189-69-7 3991
$C_{18}H_{19}Cl_2NO_4$
Ethyl methyl 4-(2,3-dichlorophenyl)-1,4-dihydro-2,6-dimethyl-3,5-pyridinedicarboxylate.
Plendil; H 154/82; Agon; Feloday; Flodil; Hydac; Munobal; Prevex; Splendil. Antianginal; antihypertensive agent. Dihydropyridine calcium channel blocker sold as the racemate. mp= 145°. *Astra USA Inc.; Merck & Co.Inc.*

7330 dl-Felodipine
72509-76-3 3991
$C_{18}H_{19}Cl_2NO_4$
(±) Ethyl methyl 4-(2,3-dichlorophenyl)-1,4-dihydro-2,6-dimethyl-3,5-pyridinedicarboxylate.
Plendil; H 154/82; Agon; Feloday; Flodil; Hydac; Munobal; Prevex; Splendil. Antianginal; antihypertensive agent. Dihydropyridine calcium channel blocker sold as the racemate. mp= 145°. *Astra USA Inc.; Merck & Co.Inc.*

7331 Fendiline
13042-18-7 4011 235-915-6
$C_{23}H_{25}N$
N-(3,3-Diphenylpropyl)-α-methylbenzylamine.
Vasodilator (coronary). Calcium blocking agent. bp_1 = 183-187°. *Chinoin.*

7332 Fendiline Hydrochloride
13636-18-5 4011 237-121-5
$C_{23}H_{25}N$
N-(3,3-Diphenylpropyl)-α-methylbenzylamine hydrochloride.
HK-137; Cordan; Fendilar; Sensit. Vasodilator (coronary). mp = 204-205°; slightly soluble in H_2O; soluble in MeOH, EtOH, $CHCl_3$; LD_{50} (mus iv) = 14.5 mg/kg, (mus orl) = 950 mg/kg. *Chinoin.*

7333 Flunarizine
52468-60-7 4179 257-937-5
$C_{26}H_{26}F_2N_2$
(E)-1-[Bis-(p-fluorophenyl)methyl]-4-cinnamylpiperazine.
Vasodilator (peripheral, cerebral). Calcium channel blocker. Fluoronated derivative of cinnarizine. Also binds to α-adrenoceptors. *Janssen Pharm. Inc.*

7334 Flunarizine Hydrochloride
30484-77-6 4179 250-216-6
$C_{26}H_{28}Cl_2F_2N_2$
(E)-1-[Bis-(p-fluorophenyl)methyl]-4-cinnamylpiperazine dihydrochloride.
R-14950; Dinaplex; Flugeral; Flunagen; Flunarl; Fluxarten; Gradient; Issium; Mondus; Sibelium. Vasodilator (peripheral, cerebral). Calcium channel blocker. mp = 251.5°. *Janssen Pharm. Inc.*

7335 Furnidipine
138661-03-7
$C_{21}H_{24}N_2O_7$
(±)-Methyl tetrahydrofurfuryl-1,4-dihydro-2,6-dimethyl-4-(o-nitrophenyl)-3,5-pyridinedicarboxylate.
A calcium channel blocker. Antihypertensive.

7336 Gallopamil
16662-47-8 4369
$C_{28}H_{40}N_2O_5$
5-[(3,4-Dimethoxyphenethyl)methylamino]-2-isopropyl-2-(3,4,5-trimethoxyphenyl)-valeronitrile.
methoxyverapamil; D-600. Antianginal agent. Calcium channel blocking agent, related to verapamil. Pale yellow viscous oil. *Knoll Pharm. Co.*

7337 Gallopamil Hydrochloride
16662-46-7 4369 240-704-7
$C_{28}H_{40}N_2O_5$
5-[(3,4-Dimethoxyphenethyl)methylamino]-2-isopropyl-2-(3,4,5-trimethoxyphenyl)-valeronitrile hydrochloride.
Algoclor; Procorum. Antianginal agent. Calcium channel blocking agent, related to verapamil. mp = 145-148°. *Knoll Pharm. Co.*

7338 d-Gallopamil Hydrochloride
38176-09-9 4369
$C_{28}H_{40}N_2O_5$
d-5-[(3,4-Dimethoxyphenethyl)methylamino]-2-isopropyl-2-(3,4,5-trimethoxyphenyl)-valeronitrile hydrochloride.
Algoclor; Procorum. Antianginal agent. Calcium channel blocking agent, related to verapamil. mp = 160.5=161.5°; $[\alpha]_D^{25}$ = 11.7° (c = 5.02 EtOH). *Knoll Pharm. Co.*

7339 l-Gallopamil Hydrochloride
36222-39-6 4369
$C_{28}H_{40}N_2O_5$
l-5-[(3,4-Dimethoxyphenethyl)methylamino]-2-isopropyl-2-(3,4,5-trimethoxyphenyl)-valeronitrile hydrochloride.
Algoclor; Procorum. Antianginal agent. Calcium channel blocking agent, related to verapamil. mp = 160.5-161.5°; $[\alpha]_D^{25}$ = -11.7° (c = 5.04 EtOH). *Knoll Pharm. Co.*

7340 Isradipine
75695-93-1 5260
$C_{19}H_{21}N_3O_5$
Isopropyl methyl (±)-4-(4-benzofurazanyl)-1,4-dihydro-2,6-dimethyl-3,5-pyridinedicarboxylate.
isrodipine; DynaCirc; PN 200-110; Clivoten; Dynacrine; Esradin; Lomir; Prescal; Rebriden. Antihypertensive; antianginal; dihydropyridine calcium channel blocker. mp = 168-170°; [(S)-(+)-form (PN-205-033)]: mp = 142°; $[\alpha]_D^{20}$ = 6.7° (c = 1.5 EtOH); [(R)-(-)-form (PN-205-034)]: mp = 140°; $[\alpha]_D^{20}$ = 6.7° (c = 1.5 EtOH). *Sandoz Pharm. Corp.*

7341 Lacidipine
103890-78-4 5344
$C_{26}H_{33}NO_6$
4-[o-[(E)-2-Carboxyvinyl]phenyl]-1,4-dihydro-2,6-dimethyl-3,5-pyridinedicarboxylic acid 4-tert-butyldiethyl ester.
GR-43659X; GX-1048; Caldine; Lacipil; Lacirex; Motens. Dihydropyridine calcium channel blocker used as an antihypertensive agent. mp = 174-175°. *Glaxo Wellcome plc.*

7342 Lemildipine
125729-29-5
$C_{20}H_{22}Cl_2N_2O_6$
3-Isopropyl 5-methyl (±)-4-(2,3-dichlorophenyl)-1,4-dihydro-2-(hydroxymethyl)-6-methyl-3,5-pyridinedicarboxylate carbamate (ester).
Dihydropyridine calcium channel blocker.

7343 Lercanidipine
100427-26-7 5469
$C_{36}H_{41}N_3O_6$
(±)-2-[(3,3-Diphenylpropyl)methylamino]-1,1-dimethylethyl methyl 1,4-dihydro-2,6-dimethyl-4-(m-nitrophenyl)-3,5-pyridinedicarboxylate.
masnidipine. Dihydropyridine calcium channel blocker used as an antihypertensive agent. *Recordati Corp.*

7344 Lercanidipine Hydrochloride
132866-11-6 5469
$C_{36}H_{42}ClN_3O_6$
(±)-2-[(3,3-Diphenylpropyl)methylamino]-1,1-dimethylethyl methyl 1,4-dihydro-2,6-dimethyl-4-(m-nitrophenyl)-3,5-pyridinedicarboxylate hydrochloride.
Rec-15-2375; R-75. Dihydropyridine calcium channel blocker used as an antihypertensive agent. [hemihydrate]: mp = 119-123°; LD_{50} (mus ip) = 83 mg/kg, (mus orl) = 657 mg/kg. *Recordati Corp.*

7345 Levemopamil
101238-51-1
$C_{23}H_{30}N_2$
(-)-S-2-Isopropyl-5-(methylphenylamino)-2-phenylvaleronitrile.
(S)-emopamil. A phenylalkylamine calcium channel blocker with antagonistic action on serotonin 5-HT2 receptors.

7346 Levosemotiadil
116476-16-5
$C_{29}H_{32}N_2O_6S$
(-)-(S)-2-[5-Methoxy-2-[3-[methyl[2-[3,4-(methylenedioxy)phenoxy]ethyl]amino]propoxy]phenyl]-4-methyl-2H-1,4-benzothiazin-3(4H)-one.
SD-3212 [as fumarate]. A benzothiazine calcium channel antagonist.

7347 Lidoflazine
3416-26-0 5507 222-312-8
$C_{30}H_{35}F_2N_3O$
4-[4,4-Bis(p-fluorophenyl)butyl]-1-piperazineaceto-2',6'-xylidide.
McN-JR-7094; R-7904; Angex; Clinium; Klinium; Ordiflazine; Corflazine. Coronary vasodilator; antianginal agent. Used for treatment of angina pectoris. mp = 159-161°; soluble in $CHCl_3$, less soluble in other organic solvents, insoluble in H_2O. *Abbott Laboratories Inc.; Janssen Pharm. Inc.; McNeil Pharm.*

7348 Lomerizine
101477-55-8 5593
$C_{27}H_{30}F_2N_2O_3$
1-[Bis(4-Fluorophenyl)methyl]-4-[(2,3,4-trimethyoxyphenyl)methyl]methyl]piperazine.
Antimigraine; vasodilator (cerebral). Diphenylpiperazine calcium channel blocker; selective cerebral vasodilator.

7349 Lomerizine Dihydrochloride
101477-54-7 5593
$C_{27}H_{32}Cl_2F_2N_2O_3$
1-[Bis(p-fluorophenyl)methyl]-4-(2,3,4-trimethoxybenzyl)piperazine dihydrochloride.
KB-2796. Antimigraine. Diphenylpiperazine calcium channel blocker. Selective cerebral vasodilator. mp = 214-218° (dec), 204-207° (dec); LD_{50} (mus orl) = 300 mg/kg. *Kanebo Pharm. Ltd.*

7350 Manidipine
89226-50-6 5786
$C_{35}H_{38}N_4O_6$
2-[4-(Diphenylmethyl)-1-piperazinyl]ethyl methyl (±)-1,4-dihydro-2,6-dimethyl-4-(m-nitrophenyl)-3,5-pyridinecarboxylate.
franidipine. Vasodilator; antihypertensive. Dihydropyridine calcium channel blocker used as an antihypertensive agent. mp = 125-128°. *Takeda Chem. Ind. Ltd.*

7351 Manidipine 6300
120092-68-4 5786
$C_{35}H_{38}N_4O_6$
2-[4-(Diphenylmethyl)-1-piperazinyl]ethyl methyl (±)-1,4-dihydro-2,6-dimethyl-4-(m-nitrophenyl)-3,5-pyridinecarboxylate.
Vasodilator; antihypertensive. Dihydropyridine calcium channel blocker used as an antihypertensive agent. mp = 125-128°. *Takeda Chem. Ind. Ltd.*

7352 Manidipine Dihydrochloride
89226-75-5 5786
$C_{35}H_{39}ClN_4O_6$
2-[4-(Diphenylmethyl)-1-piperazinyl]ethyl methyl (±)-1,4-dihydro-2,6-dimethyl-4-(m-nitrophenyl)-3,5-pyridinecarboxylate hydrochloride.
CV-4093; Calslot. Vasodilator; antihypertensive. Dihydropyridine calcium channel blocker used as an antihypertensive agent. mp [α form]: = 157-163°; mp [β form]: = 174-180°; mp [monohydrate]: = 167-170°; LD_{50} (mmus sc) = 387 mg/kg, (mmus ip) = 62.2 mg/kg, (mmus orl) = 190 mg/kg, (fmus sc) = 340 mg/kg), (fmus ip) = 68.0

mg/kg, (fmus orl) = 171 mg/kg, (mrat sc) = 222 mg/kg, (mrat ip) = 66.5 mg/kg, (mrat orl) = 247 mg/kg, (frat sc) = 199 mg/kg, (frat ip) = 48.8 mg/kg, (frat orl) = 156 mg/kg); [α-form]: mp = 157-163°; [β-form]: mp = 174-180°; [β-form monohydrate]: mp = 167-170°. *Takeda Chem. Ind. Ltd.*

7353 Mesudipine
62658-88-2

$C_{19}H_{24}N_2O_4S$

Diethyl 1',4'-dihydro-2',6'-dimethyl-2-(methylthio)-[3,4'-bipyridine]-3',5'-dicarboxylate.

Calcium channel blocker.

7354 Mibefradil
116644-53-2 6261

$C_{29}H_{38}FN_3O_3$

(1S,2S)-[2-[[3-(2-Benzimidazolyl)propyl]methylamino]ethyl-6-fluoro-1,2,3,4-tetrahydro-1-isopropyl-2-naphthyl methoxyacetate.

Calcium channel blocker with antihypertensive activity. Blocks both L- and T-type calcium channels with a more selective blockade of T-type channels. *Hoffmann-LaRoche Inc.*

7355 Mibefradil Dihydrochloride
116666-63-8 6261

$C_{29}H_{40}Cl_2FN_3O_3$

(1S,2S)-[2-[[3-(2-Benzimidazolyl)propyl]methylamino]ethyl]-6-fluoro-1,2,3,4-tetrahydro-1-isopropyl-2-naphthyl methoxyacetate dihydrochloride.

Posicor; Ro-40-5967/001. Calcium channel blocker with antihypertensive activity. Blocks both L- and T-type calcium channels with a more selective blockade of T-type channels. mp = 128°; soluble in H_2O. *Hoffmann-LaRoche Inc.*

7356 Nexopamil
136033-49-3

$C_{24}H_{40}N_2O_3$

(2S)-5-(Hexylmethylamino)-2-isopropyl-2-(3,4,5-trimethoxyphenyl)valeronitrile.

LU-49938. Serotonin 5-HT2 receptor and calcium channel antagonist. A verapamil derivative.

7357 Nicardipine
55985-32-5 6579 259-932-3

$C_{26}H_{29}N_3O_6$

2-(Benzylmethylamino) ethyl methyl 1,4-dihydro-2,6-dimethyl-4-(m-nitrophenyl)-3,5-pyridinedicarboxylate.

Antianginal; antihypertensive; vasodilator. Dihydropyridine calcium channel blocker. Has antihypertensive properties. *Syntex Labs. Inc.; Yamanouchi U.S.A. Inc.*

7358 Nicardipine Hydrochloride
54527-84-3 6579 259-198-4

$C_{26}H_{30}Cl9N_3O_6$

2-(Benzylmethylamino) ethyl methyl 1,4-dihydro-2,6-dimethyl-4-(m-nitrophenyl)-3,5-pyridinedicarboxylate monohydrochloride.

YC-93; RS-69216; RS-69216-XX-07-0; Barizin; Bionicard; Cardene; Dacarel; Lecibral; Lescodil; Loxen; Nerdipina; Nicant; Nicardal; Nicarpin; Nicapress; Nicodel; Nimicor; Perdipina; Perdipine; Ranvil; Ridene; Rycarden; Rydene; Vasodin; Vasonase. Antianginal; antihypertensive agent. Dihydropyridine calcium channel blocker. Has antihypertensive properties. [α form]: mp = 179-181°; [β form]: mp = 168-170°; LD_{50} (mrat orl) = 634 mg/kg), (mrat iv) = 18.1 mg/kg, (frat orl) = 557 mg/kg, (frat iv) = 25.0 mg/kg, (mmus orl) = 634 mg/kg, (mmus iv) = 20.7 mg/kg, (fmus orl) = 650 mg/kg, (fmus iv) = 19.9 mg. *Syntex Labs. Inc.; Yamanouchi U.S.A. Inc.*

7359 Nifedipine
21829-25-4 6617 244-598-3

$C_{17}H_{18}N_2O_6$

Dimethyl 1,4-dihydro-2,6-dimethyl-4-(o-nitrophenyl)-3,5-pyridinedicarboxylate.

Bay a 1040; Adalat; Adalate; Adapress; Aldipin; Alfadat; Anifed; Aprical; Bonacid; Camont; Chronadalate; Citilat; Coracten; Cordicant; Cordilan; Corotrend; Duranifin; Ecodipi; Hexadilat; Introcar; Kordafen; Nifedicor; Nifedin; Nifelan; Nifelat; Nifensar XL; Orix; Oxcord; Pidilat; Procardia; Sepamit; Tibricol; Zenusin. Coronary vasodilator used as an antianginal agent. Dihydropyridine calcium channel blocker. mp = 172-174°; λ_m = 340, 235 nm (ε 5010, 21590 MeOH), 338, 238 nm (ε 5740, 20600 0.1N HCl), 340, 238 nm (5740, 20510 0.1N NaOH); soluble in Me_2CO (25.0 g/100 ml), CH_2Cl_2 (16 g/100ml), $CHCl_3$ (14 g/100 ml), EtOAc (5 g/100 ml), MeOH (2.6 g/100 ml), EtOH (1.7 g/100 ml); LD_{50} (mus orl) = 494 mg/kg, (mus iv) = 4.2 mg/kg, (rat orl) = 1022 mg/kg, (rat iv) = 15.5 mg/kg. *Bayer AG; Miles Inc.; Pfizer Inc.; Pratt Pharm.*

7360 Niludipine
22609-73-0 245-120-6

$C_{25}H_{34}N_2O_8$

Bis(2-propoxyethyl) 1,4-dihydro-2,6-dimethyl-4-(3-nitrophenyl)-,3,5-Pyridinedicarboxylate.

Bay a 7168. Voltage-sensitive calcium channel blocker with antianginal activity.

7361 Nilvadipine
75530-68-6 6637

$C_{19}H_{19}N_3O_6$

5-Isopropyl 3-methyl 2-cyano-1,4-dihydro-6-methyl-4-(m-nitrophenyl)-3,5-pyridinedicarboxylate.

nivadipine; nivaldipine; CL-287389; FK-235; FR-34235; SK&F-102362; Escor; Nivadil. Antianginal; antiarrhythmic agent. Dihydropyridine calcium channel blocker. Has antihypertensive properties. mp = 148-150°; [(+) form]: $[\alpha]_D^{20}$ = 222.42° (c = 1 MeOH); [(-) form]: $[\alpha]_D^{20}$ = -219.62° (c= 1 MeOH). *Fujisawa Pharm. USA Inc.; SmithKline Beecham Pharm.*

7362 Nimodipine
66085-59-4 6643 266-127-0

$C_{21}H_{26}N_2O_7$

Isopropyl 2-methoxyethyl 1,4-dihydro-2,6-dimethyl-4-(m-nitrophenyl)-3,5-pyridinedicarboxylate.

BAY e 9736; Admon; Nimotop; Periplum. Vasodilator (cerebral). Dihydropyridine calcium channel blocker. mp = 125°; LD_{50} (mus orl) = 3562, (rat orl) = 6599, (mus iv) = 33 mg/kg, (rat iv) = 16 mg/kg; [(+)-form]: $[\alpha]_D^{20}$ = +7.9° (c = 0.439 in dioxane); [(-)-form]: $[\alpha]_D^{20}$ = -7.93° (c = 0.374 in dioxane). *Bayer AG; Miles Inc.*

7363 Nisoldipine
63675-72-9 6658 264-407-7
$C_{20}H_{24}N_2O_6$
1,4-Dihydro-2,6-dimethyl-4-(2-nitrophenyl)-3,5-pyridinedicarboxylic acid methyl 2-methylpropyl ester.
Bay k 5552; Baymycard; Nisocor; Norvasc; Sular; Syscor; Zadipina. Antianginal; antihypertensive agent. Calcium channel blocker. Dihydropyridine calcium channel blocker. mp = 151-152°. *Bayer AG; Miles Inc.; Zeneca Pharm.*

7364 Nitrendipine
39562-70-4 6669 254-513-1
$C_{18}H_{20}N_2O_6$
(±)-Ethyl methyl-1,4-dihydro-2,6-dimethyl-4-(m-nitrophenyl)-3,5-pyridine-dicarboxylate.
Baypress; Bay e 5009; Bayotensin; Bylotensin; Deiten; Nidrel. Dihydropyridine calcium channel blocker used as an antihypertensive agent. mp = 158°; insoluble in H_2O; LD_{50} (mus iv) = 39 mg/kg, (mus orl) = 2540 mg/kg, (rat iv) = 12.6 mg/kg, (rat orl) > 10000 mg/kg. *Bayer AG.*

7365 Otilonium Bromide
26095-59-0 247-457-4
$C_{29}H_{43}BrN_2O_4$
Diethyl(2-hydroxyethyl)methylammonium bromide p-[o-(oxtyloxy)benzamido]benzoate.
SP-63; Spasmomen. A muscarinic and tachykinin NK2 receptor antagonist and calcium channel blocker. Used for the symptomatic treatment of irritable bowel syndrome. Anticholinergic; spasmolytic agent.

7366 Oxodipine
90729-41-2
$C_{19}H_{21}NO_6$
Ethyl methyl 1,4-dihydro-2,6-dimethyl-4-[2,3-(methylenedioxy)phenyl]-3,5-pyridinedicarboxylate.
Calcium channel blocker.

7367 Palonidipine
96515-73-0
$C_{29}H_{34}FN_3O_6$
(±)-3-(Benzylmethylamino)-2,2-dimethylpropyl methyl 4-(2-fluoro-5-nitrophenyl)-1,4-dihydro-2,6-dimethyl-3,5-pyridinedicarboxylate.
TC-81 [as hydrochloride]. Calcium channel blocker.

7368 Perhexiline
6621-47-2 7305 229-569-5
$C_{19}H_{35}N$
2-(2,2-Dicyclohexylethyl)piperidine.
perhexilene. Vasodilator (coronary); diuretic. Calcium blocking agent. [hydrochloride]: mp= 243-245.5°. *Marion Merrell Dow Inc.*

7369 Perhexiline Maleate
6724-53-4 7305 229-775-5
$C_{23}H_{39}NO_4$
2-(2,2-Dicyclohexylethyl)piperidine maleate (1:1).
perhexilene maleate; Pexid. Vasodilator (coronary); diuretic. Calcium blocking agent. mp = 188.5-191°; LD_{50} (rat orl) > 7000 mg/kg, (mus orl) = 4370 mg/kg. *Marion Merrell Dow Inc.*

7370 Piprofurol
40680-87-3 255-035-6
$C_{26}H_{33}NO_6$
α-[2-(4-Hydroxyphenyl)ethyl]-4,7-dimethoxy-6-[2-(1-piperidinyl)ethoxy]-5-benzofuranmethanol.
A benzofuran chalcon derivative. Calcium channel blocker.

7371 Pranidipine
99522-79-9
$C_{25}H_{24}N_2O_6$
(E)-Cinnamyl methyl (±)-1,4-dihydro-2,6-dimethyl-4-(m-nitrophenyl)-3,5-pyridinedicarboxylate.
OPC-13340. A dihydropyridine. Calcium channel blocker.

7372 Prenylamine
390-64-7 7919 206-869-4
$C_{24}H_{27}N$
N-(3,3-Diphenylpropyl)-α-methylphenethylamine.
B-436; Elecor. Vasodilator (coronary). mp = 36.5-37.5°. *Hoechst AG.*

7373 Prenylamine Lactate
69-43-2 7919 200-705-5
$C_{27}H_{33}NO_3$
N-(3,3-Diphenylpropyl)-α-methylphenethylamine lactate.
Angormin; Bismetin; Carditin-Same; Coredamin; Corontin; Crepasin; Daxauten; Hostaginan; Incoran; Irrorin; Lactamin; Plactamin; Reocorin; Roinin; Seccidin; Sedolaton; Segontin; Synadrin. Vasodilator (coronary). mp = 140-142°; λ_m = 260 nm ($E^{1\%}_{1cm}$ 170 in $CHCl_3$); sparingly soluble in H_2O (0.5%); soluble in organic solvents. *Hoechst AG.*

7374 Riodipine
71653-63-9
$C_{18}H_{19}F_2NO_5$
Dimethyl 4-[o-(difluoromethoxy)phenyl]-1,4-dihydro-2,6-dimethyl-3,5-pyridinedicarboxylate.
ryodipine. Derivative of phenylpyridine. Calcium channel blocker.

7375 Ronipamil
85247-77-4
$C_{32}H_{48}N_2$
(±)-2-[3-(Methylphenethylamino)propyl]-2-phenyltetradecanenitrile.
A long acting phenylalkylamine derivative. Vasodilator (coronary); calcium antagonist.

7376 Semotiadil
116476-13-2 8590
$C_{29}H_{32}N_2O_6S$
(R)-2-[2-[3-[[2-(1,3-Benzodioxol-5-yloxy)ethyl]-methylamino]propoxy]-5-methoxyphenyl]-4-methyl-2H-1,4-benzothiazin-3(4H)-one.
sesamodil; DS-4823. Antianginal; antihypertensive; antiarrhythmic. Benzothiazine calcium antagonist. 8590.

7377 Semotiadil Fumarate
116476-14-3 8590
$C_{33}H_{36}N_2O_{10}S$
(R)-2-[2-[3-[[2-(1,3-Benzodioxol-5-yloxy)ethyl]-methylamino]propoxy]-5-methoxyphenyl]-4-methyl-2H-1,4-benzothiazin-3(4H)-one fumarate.

SD-3211. Antianginal; antihypertensive; antiarrhythmic. Benzothiazine calcium antagonist. mp = 134-135°; $[\alpha]_D^{25}$ = 195° (DMSO). 8590.

7378 Teludipine
108687-08-7
$C_{28}H_{38}N_2O_6$
4-[o-[(E)-2-Carboxyvinyl]phenyl]-2-[(dimethyl-amino)methyl]-1,4-dihydro-6-methyl-3,5-pyridinecarboxylic acid 4-tert-butyl diethyl ester.
Calcium channel blocker. Antihypertensive. *Glaxo Wellcome Inc.*

7379 Teludipine Hydrochloride
108700-03-4
$C_{28}H_{39}ClN_2O_6$
4-[o-[(E)-2-Carboxyvinyl]phenyl]-2-[(dimethyl-amino)methyl]-1,4-dihydro-6-methyl-3,5-pyridinecarboxylic acid 4-tert-butyl diethyl ester monohydrochloride.
GR-53992B(GX-1296b). Calcium channel blocker. Antihypertensive. *Glaxo Wellcome Inc.*

7380 Terodiline
15793-40-5 9311
$C_{20}H_{27}N$
N-tert-Butyl-1-methyl-3,3-diphenylpropylamine.
Coronary vasodilator used as an antianginal agent. Used also in treatment of urinary incontinence. Calcium antagonist with anticholinergic and vasodilatory activity. Liquid; $bp_{1.0}$ = 130-132°. *Marion Merrell Dow Inc.*

7381 Terodiline Hydrochloride
7082-21-5 9311
$C_{20}H_{28}ClN$
N-tert-Butyl-1-methyl-3,3-diphenylpropylamine hydrochloride.
Bicor; Mictrol; Micturin; Micturol. Coronary vasodilator used as an antianginal agent. Used also in treatment of urinary incontinence. Calcium antagonist with anticholinergic and vasodilatory activity. mp = 178-180°; soluble in EtOH, slightly soluble in Et_2O. *Marion Merrell Dow Inc.*

7382 Valperinol
64860-67-9
$C_{16}H_{27}NO_4$
(2R*,4R*,4aS*,5R*,7S*,7aR*,8R*)-Hexahydro-4-methoxy-8-methyl-7a(piperidinomethyl)-2,5-methanocyclopenta-m-dioxin-7-ol.
Calcium antagonist.

7383 Verapamil
52-53-9 10083 200-145-1
$C_{27}H_{38}N_2O_4$
5-[(3,4-Dimethoxyphenethyl)methylamino]-2-(3,4-dimethoxyphenyl)-2-isopropylvaleronitrile.
D-365; CP-16533-1; iproveratril. Antianginal; class IV antiarrhythmic agent. Coronary vasodilator with calcium channel blocking activity. $bp_{0.001}$ = 243-246°; insoluble in H_2O; slightly soluble in C_6H_6, C_6H_{16}, Et_2O; soluble in EtOH, MeOH, Me_2CO, EtOAc, $CHCl_3$. *Bristol-Myers Squibb Co.*

7384 Verapamil Hydrochloride
152-11-4 10083 205-800-5
$C_{27}H_{39}ClN_2O_4$
5-[(3,4-Dimethoxyphenethyl)methylamino]-2-(3,4-dimethoxyphenyl)-2-isopropylvaleronitrile hydrochloride.
Arapamyl; Berkatens; Calan; Cardiagutt; Cardibeltin; Cordilox; Dignover; Drosteakard; Geangin; Isoptin; Quasar; Securon; Univer; Vasolan; Veracim; Veramex; Veraptin; Verelan; Verexamil. Antianginal; class IV antiarrhythmic agent. Coronary vasodilator with calcium channel blocking activity. mp = 138.5-140.5°; soluble in H_2O (7 g/100 g, 83 mg/ml), EtOH (26 mg/ml), propylene glycol (93 mg/ml), MeOH (> 100 mg/ml), iPrOH (4.6 mg/ml), EtOAc (1.0 mg/ml), DMF (> 100 mg/ml), CH_2Cl_2 (> 100 mg/ml), C_6H_{14} (0.001 mg/ml); LD_{50} (rat iv) = 16 mg/kg, (mus iv) = 8 mg/kg. *Knoll Pharm. Co.; Lederle Labs.; Parke-Davis; Searle G.D. & Co.*

Calcium Regulators

7385 Calcifediol
19356-17-3 1677
$C_{27}H_{44}O_2.H_2O$
9,10-Secocholesta-5,7,10(19)-triene-3,25-diol monohydrate.
Calderol; U-32070E; 25-hydroxy Vitamin D_3; 25-HCC; Dedrogyl; Didrogyl; Hidroferol. he principle circulating form of Vitamin D_3. Calcium regulator. λ_m 265 nm (ε 18000 EtOH). *Organon Inc.*

7386 Calcitonin
9007-12-9 1680 232-693-2
$C_{151}H_{226}N_{40}O_{45}S_3$
H-Cys-Ser-Asn-Leu-Ser-Thr-Cys-Val-Leu-Gly-Lys-Leu-Ser-Gln-Glu-Leu-His-LysLeu-Gln-Thr-Tyr-Pro-Arg-Thr-Asn-Thr-Gly-Ser-Gly-Thr-Pro-NH_2 (1→7 cyclic disulfide).
Calcimar; Cibacalcin; Thyrocalcitonin; TCA; TCT. A polypeptide hormone, secreted in the thyroid gland, that lowers the calcium concentration in the plasma of mammals. Acts to oppose the bone and renal effects of parathyroid hormone, thereby inhibiting bone resorption of calcium. Other effects are hypocalcemia, hypophosphatemia, decreased urinary calicum. Calcium regulator. *Astra Chem. Ltd.; Ciba-Geigy Corp.*

7387 Calcitonin, Human Synthetic
21215-62-3 1680 244-276-2
$C_{151}H_{226}N_{40}O_{45}S_3$
H-Cys-Ser-Asn-Leu-Ser-Thr-Cys-Val-Leu-Gly-Lys-Leu-Ser-Gln-Glu-Leu-His-LysLeu-Gln-Thr-Tyr-Pro-Arg-Thr-Asn-Thr-Gly-Ser-Gly-Thr-Pro-NH_2 (1→7 cyclic disulfide).
Cibacalcin. A polypeptide hormone that lowers the calcium concentration in the plasma of mammals. Calcium regulator. *Astra Chem. Ltd.; Ciba-Geigy Corp.*

7388 Calcitonin, Porcine
12321-44-7 1680 235-585-3
Calcitar; Calcitare; Staporos. A polypeptide hormone that lowers the calcium concentration in the plasma of mammals. Calcium regulator. *Astra Chem. Ltd.; Ciba-Geigy Corp.*

7389 Calcitonin, Salmon Synthetic
47931-85-1 1680 256-342-8
$C_{145}H_{240}N_{44}O_{48}S_2$
H-Cys-Gly-Asn-Leu-Ser-Thr-Cys-Met-Leu-Gly-Thr-Tyr-Thr-Gln-Asp-Phe-Asn-Lys-Phe-His-Thr-Phe-Pro-Gln-Thr-Ala-Ile-Gly-Val-Gly-Ala-Pro-NH_2 (1→7) cyclic disulfide.
Salcatonin; Calciben; Calcimar; Calsyn; Calsynar; Catonin; Karil; Miacalcic; Miacalcin; Miadenil; Osteocalcin; Prontocalcin; Rulicalcin; Salmotonin; Stalcin; Tonocalcin. A polypeptide hormone that lowers the calcium concentration in the plasma of mammals. Calcium regulator. *Astra Chem. Ltd.; Ciba-Geigy Corp.*

7390 Calcitriol
32222-06-3 1681 250-963-8
$C_{27}H_{44}O_3$
(5Z,7E)-9,10-Secocholesta-5,7,10(19)-triene-1α,3β,25-triol.
1α,25-dihydroxycholecalciferol; 1α,25-dihydroxyvitamin D_3; Calcijex; Rocaltrol; Toptriol; Ro-21-5535. Calcium regulator; vitamin (antirachitic). Biologically active form of vitamin D_3, involved in intestinal calcium transport and bone calcium resorption, formed by hydroxylation of vitamin D_3 in liver and kidney. mp = 111-115°; λ_m = 264 nm (ε 19000 EtOH); $[\alpha]_D^{25}$ = 48° (MeOH); slightly soluble in MeOH, EtOH, EtOAc, THF; air, light sensitive. *Abbott Labs Inc.; Hoffmann-LaRoche Inc.*

7391 Clodronic Acid
10596-23-3 2432 234-212-1
$CH_4ClO_6P_2$
(Dichloromethylene)diphosphonic acid.
Cl_2MDP; DMDP. Calcium regulator. mp = 249-251°. *Procter & Gamble Pharm. Inc.*

7392 Clodronic Acid Disodium Salt
22560-50-5 2432 245-078-9
$CH_2ClNa_2O_6P_2$
Sodium (dichloromethylene)diphosphonate.
DClMDP; Bonefos; Clasteon; Difoafonal; Loron; Mebonat; Ossiten; Ostac. Calcium regulator. *Procter & Gamble Pharm. Inc.*

7393 Dihydrotachysterol
67-96-9 3223
$C_{28}H_{46}O$
9,10-Secoergosta-5,7,22-trien-3β-ol.
Hytakerol; AT-10; Antitanil; Calcamine; Dygratyl; Dihydral; Parterol; Tachyrol. Calcium regulator. mp = 125-127°; $[\alpha]_D^{22}$ = 07.5° ($CHCl_3$); λ_m = 242, 251, 261 nm ($E^{1\%}_{1\,cm}$ 870, 1010, 650); insoluble in H_2O, readily soluble in organic solvents. *Sterling Winthrop Inc.*

7394 Doxercalciferol
1α-Hydroxyvitamin D(2).
Hectorol. Antihyperphosphatemic; calcium regulator. A vitamin D prohormone. Suppresses intact parathyroid hormone (iPTH) with minimal increase in serum calcium and phosphorus. Used to treat hypercalcemia and hyperphosphatemia in dialysis patients. Used for treatment of secondary hyperparathyroidism. *Bone-Care Intl.*

7395 Elcatonin
60731-46-6 3578 262-393-7
$C_{148}H_{244}N_{42}O_{47}$
1-Butyric acid-7-(L-2-aminobutyric acid)-26-L-aspartic acid-27-L-valine-29-L-alanine-calcitonin (salmon).
carbocalcitonin; HC-58; Calcinil; Carbicalcin; Elcitonin; Turbocalcin. Calcium regulator. Used to treat Paget's disease. *Toyo Jozo.*

7396 Etidronic Acid
2809-21-4 3908 220-552-8
$C_2H_8O_7P_2$
(1-Hydroxyethylidene)diphosphonic acid.
EHDP. Calcium regulator. Very soluble in H_2O, insoluble in AcOH. *Procter & Gamble Pharm. Inc.*

7397 Etidronic Acid Disodium Salt
7414-83-7 3908 231-025-7
$C_2H_6Na_2O_7P_2$
Disodium dihydrogen (1-hydroxyethylidene)-diphosphonate.
Didronel; Calcimux; Diphos; Etidron. Calcium regulator. Bone resorption inhibitor. Soluble in H_2O. *Procter & Gamble Pharm. Inc.*

7398 Falecalcitriol
83805-11-2
$C_{27}H_{38}F_6O_3$
(+)-(5Z,7E)-26,26,26,27,27,27-Hexafluoro-9,10-seco-cholesta-5,7,10(19)-triene-1α,3β,25-triol.
Flocalcitrol. Calcium regulator.

7399 Gallium Nitrate
13494-90-1
GaN_3O_9
Gallium(III) nitrate (1:3), anhydrous.
gallium trinitrate; nitric acid, gallium salt; Ganite; NSC-15200. Antineoplastic; antihypercalcemic. Calcium regulator. Bone resorption inhibitor.

7400 Gallium Nitrate Nonahydrate
135886-70-3
$GaN_3O_9.9H_2O$
Nitric acid gallium salt nonahydrate.
NSC-15200. Calcium regulator. Bone resorption inhibitor.

7401 Ipriflavone
35212-22-7 5090
$C_{18}H_{16}O_3$
7-Isopropoxyflavone.
FL-113; TC-80; Iprosten; Osten; Osteofix; Yambolap. Calcium regulator. Bone resorption inhibitor. mp = 115-117°. *Chinoin.*

7402 Levosimendan
141505-33-1
$C_{14}H_{12}N_6O$
Mesoxalonitrile (-)-[p[(R)-1,4,5,6-tetrahydro-4-methyl-6-oxo-3-pyridazinyl]phenyl]-hydrazone.
OR-1259. Calcium sensitizer. Acts as a Positive inotropic agent.

7403 Oxidronic Acid
15468-10-7 7074
$CH_6O_7P_2$
(Hydroxymethylene)diphosphonic acid.
HMDP. Calcium regulator. The ^{99m}Tc complex is used as a diagnostic aid. *Procter & Gamble Pharm. Inc.*

7404 Oxidronic Acid Sodium Salt
14255-61-9 7074
$CH_5NaO_7P_2$
(Hydroxymethylene)diphosphonic acid sodium salt.
Calcium regulator. The ^{99m}Tc complex is a diagnostic aid. mp = 297-300°. *Procter & Gamble Pharm. Inc.*

7405 Parathyroid Hormone
9002-64-6 7168
Paroidin; Parathormone; PTH. Blood calcium regulator. *Parke-Davis.*

7406 Piridronic Acid
75755-07-6
$C_7H_{11}NO_6P_2$
[2-(2-Pyridinyl)ethylidene]bisphosphonic acid.
Calcium regulator.

7407 Piridronic Acid Sodium Salt
100188-33-8
$C_7H_{10}NNaO_6P_2$
[2-(2-Pyridinyl)ethylidene]bisphosphonic acid monosodium salt.
NE-97221. Calcium regulator.

7408 Potassium Phosphate, Dibasic
7758-11-4 7828 231-834-5
HK_2O_4P
Dipotassium hydrogen phosphate.
dipotassium phosphate; dikalium phosphate; DKP; Isolyte. Cathartic; calcium regulator. Also used as a buffering agent. Soluble in H_2O (150 g/100 ml at 25°); slightly soluble in EtOH. *McGaw Inc.*

7409 Risedronic Acid
105462-24-6 8396
$C_7H_{11}NO_7P_2$
[1-Hydroxy-2-(3-pyridyl)ethylidene]diphosphonic acid.
Calcium regulator, used as a bone resorption inhibitor. *Merck & Co.Inc.; Procter & Gamble Pharm. Inc.*

7410 Risedronic Acid Monosodium Salt
115436-72-1 8396
$C_7H_{10}NNaO_7P_2$
Sodium trihydrogen [1-hydroxy-2-(3-pyridyl)-ethylidene]diphosphonate.
Actonel; risedronate sodium; NE-58095. Calcium regulator, used as a bone resorption inhibitor in treatment of osteoporosis. *Merck & Co.Inc.; Procter & Gamble Pharm. Inc.*

7411 Secalciferol
55721-11-4
$C_{27}H_{44}O_3$
(5Z,7E,24R)-9,10-Secocholesta-5,7,10(19)-triene-β,24,25-triol.
Osteo D. Calcium regulator. *Teva Pharm. (USA).*

7412 Simendan
131741-08-7
$C_{14}H_{12}N_6O$
(±)-(R,S)-[[4-(1,4,5,6-Tetrahydro-4-methyl-6-oxo-3-pyridazinyl)phenyl]hydrazone.
Calcium sensitizer. Positive inotropic agent. *See* levosimendan.

7413 Sodium Sulfate [anhydrous]
7757-82-6 8829 231-820-9
Na_2SO_4
Sulfuric acid disodium salt.
mirabilite; thenardite; salt cake; component of: Colyte. Calcium regulator. Also used as a laxative-cathartic. mp = 800°; d = 2.7; soluble in H_2O (50 g/100 ml at 33°; solubility decreases with temperature to 41.6 g/100 ml at 100°); insoluble in EtOH. *Schwarz Pharma Kremers Urban Co.*

7414 Sodium Sulfate Decahydrate
7727-73-3 8829
$Na_2SO_4.10H_2O$
Sulfuric acid disodium salt decahydrate.
Glauber's salt; component of: Colyte. Calcium regulator. Laxative-cathartic. mp = 32.4°; d = 1.46; loses all H_2O at 100°; soluble in H_2O (66 g/100 ml at 25°, 30 g/100 ml at 15°); soluble in glycerol; insoluble in EtOH. *Schwarz Pharma Kremers Urban Co.*

7415 Teriparatide
52232-67-4
$C_{181}H_{291}N_{55}O_{51}S_2$
L-Seryl-L-valyl-L-seryl-L-α-glutamyl-L-isoleucyl-L-methionyl-L-histidyl-L-asparaginyl-L-leucylglcyl-L-lysyl-L-histidyl-L-leucyl-L-asparaginyl-L-seryl-L-methionyl-L-α-glutaminyl-L-arginyl-L-valyl-L-α-glutamyl-L-tryptophyl-L-leucyl-L-arginyl-L-lysyl-L-lysyl-L-leucyl-L-glutaminyl-L-α-aspartyl-L-valyl-L-histidyl-L-asparaginyl-L-phenylalanine.
LY-333334; hPTH 1-34; H-Ser-Val-Ser-Glu-Ile-Gln-Leu-Met-His-Asn-Leu-Gly-Lys-His-Leu-Asn-Ser-Met-Glu-Arg-Val-Glu-Trp-Leu-Arg-Lys-Lys-Leu-Gln-Asp-Val-His-Asn-Phe-OH. Bone resorption inhibitor. Osteoporosis therapy adjunct. *Rhône-Poulenc Rorer Pharm. Inc.*

7416 Teriparatide Acetate
99294-94-7 9309
$C_{181}H_{291}N_{55}O_{51}S_2.xH_2O.yC_2H_4O_2$
L-Seryl-L-valyl-L-seryl-L-α-glutamyl-L-isoleucyl-L-methionyl-L-histidyl-L-asparaginyl-L-leucylglcyl-L-lysyl-L-histidyl-L-leucyl-L-asparaginyl-L-seryl-L-methionyl-L-α-glutaminyl-L-arginyl-L-valyl-L-α-glutamyl-L-tryptophyl-L-leucyl-L-arginyl-L-lysyl-L-lysyl-L-leucyl-L-glutaminyl-L-α-aspartyl-L-valyl-L-histidyl-L-asparaginyl-L-phenylalanine acetate (salt) hydrate.
MN-10T; Parathar; hPTH 1-34 acetate; H-Ser-Val-Ser-Glu-Ile-Gln-Leu-Met-His-Asn-Leu-Gly-Lys-His-Leu-Asn-Ser-Met-Glu-Arg-Val-Glu-Trp-Leu-Arg-Lys-Lys-Leu-Gln-Asp-Val-His-Asn-Phe-OH.$xH_2O.yCH_3COOH$. Bone resorption inhibitor. Osteoporosis therapy adjunct. Synthetic polypeptide; the 1-34 fragment of human parathyroid hormone. Also used as a diagnostic aid. *Rhône-Poulenc Rorer Pharm. Inc.*

Carbonic Anhydrase Inhibitors

7417 Acetazolamide
59-66-5 50 200-440-5
$C_4H_6N_4O_3S_2$
N-(5-Sulfamoyl-1,3,4-thiadiazol-2-yl)acetamide.
acetazoleamide; carbonic anhydrase inhibitor 6063; acetamox; atenezol; cidamex; defiltran; diacarb; diamox; didoc; diluran; diureticum-Holzinger; diuriwas; diutazol; donmox; edemox; fonurit; glaupax; glupax; natrionex; nephramid; vetamox (sodium salt). Diuretic; antiglaucoma agent. Carbonic anhydrase inhibitor. mp = 258-259°; sparingly soluble in H_2O; pKa = 7.2. *Lederle Labs.*

7418 Acetazolamide Sodium
1424-27-7 50
$C_4H_6N_4NaO_3S_2$
N-(5-Sulfamoyl-1,3,4-thiadiazol-2-yl)acetamide monosodium salt.
Diamox Parenteral; Vetamox. Carbonic anhydrase inhibitor, used as diuretic, in treatment of glaucoma. *Lederle Labs.*

7419 Butazolamide
16790-49-1 1545
$C_6H_{10}N_4O_3S_2$
N-[5-(Aminosulfonyl)-1,3,4-thiadiazol-2-yl]butanamide.
SKF-4965; Butamide. Carbonic anhydrase inhibitor, used as diuretic, in treatment of glaucoma mp = 260-262° (dec). *Am. Cyanamid.*

7420 Dichlorphenamide
120-97-8 3127 204-440-6
$C_6H_6Cl_2N_2O_4S_2$
4,5-Dichloro-m-benzenedisulfonamide.
Daranide; Antidrasi; Oratrol. Carbonic anhydrase inhibitor used to treat glaucoma. mp = 239-241°, 228.5-229°; insoluble in H_2O, soluble in alklaine solutions. *Merck & Co.Inc.*

7421 Dorzolamide
120279-96-1 3484
$C_{10}H_{16}N_2O_4S_3$
(4S,6S)-4-(Ethylamino)-5,6-dihydro-6-methyl-4H-thieno[2,3-b]thiopyran-2-sulfonamide 7,7-dioxide.
Adrenergic (ophthalmic); a carbonic anhydrase inhibitor, used to treat glaucoma. *Merck & Co.Inc.*

7422 Dorzolamide Hydrochloride
130693-82-2 3484
$C_{10}H_{17}ClN_2O_4S_3$
(4S,6S)-4-(Ethylamino)-5,6-dihydro-6-methyl-4H-thieno[2,3-b]thiopyran-2-sulfonamide 7,7-dioxide hydrochloride.
Trusopt; MK-507; component of: Cosopt. Adrenergic (ophthalmic); a carbonic anhydrase inhibitor, used to treat glaucoma. mp = 283-285°; $[\alpha]_D^{24}$ = -8.34° (c = 1 MeOH); soluble in H_2O. *Merck & Co.Inc.*

7423 Ethoxzolamide
452-35-7 3801 207-199-5
$C_9H_{10}N_2O_3S_2$
6-Ethoxy-2-benzothiazolesulfonamide.
ethoxyzolamide; Cardrase; Ethamide; Glaucotensil; Redupresin. Carbonic anhydrase inhibitor, used as diuretic, in treatment of glaucoma mp = 188-190.5°. *Upjohn Ltd.*

7424 Flumethiazide
148-56-1 4174 205-717-4
$C_8H_6F_3N_3O_4S_2$
6-(Trifluoromethyl)-2H-1,24-benzothiadiazine-7-sulfonamide 1,1-dioxide.
trimfluoromethylthiazide; Ademol; Fludemil. Carbonic anhydrase inhibitor. mp = 305.4-307.8° (dec); λ_m = 278 nm ($E^{1\%}_{1\ cm}$ 335); sparingly soluble in H_2O (5 g/100 ml at 100° with dec); soluble in MeOH, EtOH, DMF; insoluble in EtOAc, MEK, C_6H_6, C_7H_8; some dec in boiling H_2O. *Olin Res. Ctr.*

7425 Methazolamide
554-57-4 6031 209-066-7
$C_5H_8N_4O_3S_2$
N-(4-Methyl-2-sulfamoyl-Δ^2-1,3,4-thiadiazolin-5-ylidene)acetamide.
Neptazane. Carbonic anhydrase inhibitor, used as diuretic, in treatment of glaucoma. Federally controlled substance (stimulant). mp = 213-214°; λ_m = 254 nm (log ε 3.66 95% EtOH), 247 nm (log ε3.61 0.1N NaOH). *Lederle Labs.*

Cardiotonics

7426 Acefylline
652-37-9 22 211-490-2
$C_9H_{10}N_4O_4$
1,2,3,6-Tetrahydro-1,3-dimethyl-2,6-dioxopurine-7-acetic acid.
carboxymethyltheophylline; 7-theophylline acetic acid; Aminodal [as sodium salt]. Diuretic; cardiotonic; bronchodilator. Xanthine derivative. mp = 271°; [sodium salt]: mp > 300°. *E. Merck.*

7427 Acefylline Sodium Salt
837-27-4 22 212-652-5
$C_9H_9N_4NaO_4$
Sodium 1,2,3,6-tetrahydro-1,3-dimethyl-2,6-dioxopurine-7-acetate.
Aminodal. Diuretic; cardiotonic; bronchodilator. Xanthine derivative. mp > 300°. *E. Merck.*

7428 Acetyldigitoxin
1111-39-3 90 214-178-4
$C_{43}H_{66}O_{14}$
(3β,5β)-3-[(O-3-O-Acetyl-2,6-dideoxy-β-D-ribohexopyranosyl-(1→4)-2,6-dideoxy-β-D-ribohexopyranosyl-(1→4)-2,6-dideoxy-β-D-ribohexopyranosyl)oxy]-14-hydroxycard-20(22)-enolide.
Acylanid. Cardiotonic. The α and β forms differ in the position of the acetyl group. [α-form]: mp = 217-221°; $[\alpha]_D^{20}$= 5° (c = 0.7 C_5H_5N); slightly soluble in $CHCl_3$; [β-form]: dec 225°; $[\alpha]_D^{20}$ = 16.7 (C_5H_5N); soluble in $CHCl_3$ (11-14 g/100 ml), MeOH (0.67 g/100 ml), amyl alcohol (0.45 g/100 ml); insoluble in H_2O (0.0005 g/100 ml), Et_2O. *Sandoz Pharm. Corp.*

7429 Acrihellin
67696-82-6 266-909-1
$C_{29}H_{38}O_7$
3β,5,14-Trihydroxy-19-oxo-5β-bufa-20,22-dienolide 3-(3-methylcrotonate).
Cardiotonic. *Carter-Wallace.*

7430 Actodigin
36983-69-4
$C_{29}H_{49}O_9$
3β-(β-D-Glucopyranosyloxy)-14,23-dihydroxy-24-nor-5β,14β-chol-20(22)-en-21-oic acid σ-lactone.
AY-22241. Cardiotonic. *Wyeth-Ayerst Labs.*

7431 Adibendan
100510-33-6
$C_{16}H_{14}N_4O$
5,7-Dihydro-7,7-dimethyl-2-(4-pyridyl)pyrrolo[2,3-f]-benzimidazol-6(3H)-one.
A cardiotonic agent with highly selective phosphodiesterase III inhibitory properties.

7432 2-Amino-4-picoline
695-34-1 486 211-780-9
$C_6H_8N_2$
4-Methyl-2-pyridinamine.
aminopicoline; W-45; component of: Ascensil, Askensil. Analgesic; cardiac stimulant. mp = 100-100.5°; bp_{11} = 115-117°; soluble in H_2O, alcohols, dimethylformamide, coal tar bases; slightly soluble in petroleum ether, aliphatic hydrocarbons. *Raschig GmbH.*

7433 2-Amino-4-picoline Camphorsulfonate
12261-97-1 486
$C_{16}H_{24}N_2O_4S$
4-Methyl-2-pyridamine camphorsulfonate.
aminopicoline camphorsulfonate; Piricardio; Varunax. Analgesic; cardiotonic. *Raschig GmbH.*

7434 2-Amino-4-picoline Hydrochloride
2403-84-1 486
$C_6H_9ClN_2$
4-Methyl-2-pyridamine monohydrochloride.
Analgesic; cardiotonic. mp = 176-177°; freely soluble in H_2O, EtOH. *Raschig GmbH.*

7435 Amrinone
60719-84-8 634 262-390-0
$C_{10}H_9N_3O$
5-Amino[3,4'-bipyridin]-6(1H)-one.
Inocor; Win-40680; Vesistol; Cartonic; Wincoram. Cardiotonic. Phosphodiesterase inhibitor. mp = 294-297° (dec). *Sterling Winthrop Inc.*

7436 Apovincamine
4880-92-6
$C_{21}H_{24}N_2O_2$
Methyl(3α,16α)-eburnamenine-14-carboxylate.
Reported to enhance cardiovascular activity. Cardiotonic.

7437 Bemarinone
92210-43-0
$C_{11}H_{12}N_2O_3$
5,6-Dimethyl-4-methyl-2(1H)-quinazolinone.
Cardiotonic (positive inotropic, vasodilator). A quinazoline. *Ortho Pharm. Corp.*

7438 Bemarinone Hydrochloride
101626-69-1
$C_{11}H_{13}ClN_2O_3$
5,6-Dimethyl-4-methyl-2(1H)-quinazolinone monohydrochloride.
ORF-16600; RWJ-16600. Cardiotonic (positive inotropic, vasodilator). A quinazoline. *Ortho Pharm. Corp.*

7439 Bemoradan
112018-01-6
$C_{13}H_{13}N_3O_3$
7-(1,4,5,6-Tetrahydro-4-methyl-6-oxo-3-pyridazinyl)-2H-1,4-benzoxazin-3-(4H)-one.
ORF-22867. Cardiotonic. *Ortho Pharm. Corp.*

7440 Benafentrine
35135-01-4
$C_{23}H_{27}N_3O_3$
cis-4'-(1,2,3,4,4a,10b-Hexahydro-8,9-dimethoxy-2-methylbenzo[c][1,6]naphthyridin-6-yl)-acetanilide.
Phosphoroesterase inhibitor. Cardiotonic.

7441 Benfurodil Hemisuccinate
3447-95-8 1070 222-367-8
$C_{19}H_{18}O_7$
2-(1-Hydroxyethyl)-β-(hydroxymethyl)-3-methyl-5-benzofuranacrylic acid γ-lactone hydrogen succinate.
4091-CB; benzofurodil; Eucilat; Eudilat. Cardiotonic; vasodilator. mp = 144°; soluble in alkaline solutions; LD_{50} (mus orl) = 550 mg/kg. *Clin-Byla France.*

7442 Bucladesine
362-74-3 1483
$C_{18}H2_4N_5O_8P$
N-(9-β-D-Ribofuranosyl-9H-(purin-6-yl)butyramide cyclic 3',5'-(hydrogen phosphate) 2'-butyrate.
DBcAMP. Cardiotonic. Vasodilating cyclic nucleotide that mimics the action of cyclic AMP. *Daiichi Pharm. Corp.*

7443 Bucladesine Barium Salt
18837-96-2 1483
$C_{36}H_{46}BaN_{10}O_{16}P_2$
N-(9-β-D-Ribofuranosyl-9H-(purin-6-yl)butyramide cyclic 3',5'-(hydrogen phosphate) 2'-butyrate barium (2:1) (salt).
Cardiotonic. Vasodilating cyclic nucleotide that mimics the action of cyclic AMP. *Daiichi Pharm. Corp.*

7444 Bucladesine Sodium Salt
16980-89-5 1483 241-059-4
$C_{18}H2_3N_5NaO_8P$
N-(9-β-D-Ribofuranosyl-9H-(purin-6-yl)butyramide cyclic 3',5'-(hydrogen phosphate) 2'-butyrate sodium salt.
DC-2797; Actosin. Cardiotonic. Vasodilating cyclic nucleotide that mimics the action of cyclic AMP. λ_m = 270 nm (EtOH/ammonium acetate). *Daiichi Pharm. Corp.*

7445 Buquineran
59184-78-0
$C_{20}H_{29}N_5O_3$
1-Butyl-3-[1-(6,7-dimethoxy-4-quinazolinyl)-4-piperidyl]urea.
A phosphodiesterase inhibitor, used as a cardiotonic.

7446 Butopamine
66734-12-1
$C_{18}H_{23}NO_3$
(R)-p-Hydroxy-α-[[[(R)-3-(p-hydroxyphenyl)-1-methylpropyl]amino]methyl]benzyl alcohol.
Compound LY-131126. A β-adrenergic agonist with strong positive chronotropic properties. *Eli Lilly & Co.*

7447 Camphotamide
4876-45-3 1782 225-484-2
$C_{21}H_{32}N_2O_5S$
3-Diethylcarbamoyl-1-methylpyridinium camphorsulfonate.
camphetamide; camphramine; Tonicorine. CNS stimulant; cardiac stimulant. mp = 174-175°; soluble in H_2O, EtOH, Et_2O; insoluble in C_6H_6, petroleum ether. *Soc. Franc. Recherches Biochim.*

7448 Carbazeran
70724-25-3
$C_{18}H_{24}N_4O_4$
1-(6,7-Dimethoxy-1-phthalazinyl)-4-piperidyl ethylcarbamate.
UK-31557. Cardiotonic. Phosphodiesterase inhibitor. *Pfizer Intl.*

7449 Cariporide
159138-80-4
$C_{12}H_{17}N_3O_3S$
N-(Diaminomethylene)-4-isopropyl-3-(methylsulfonyl)benzamide.
Inhibits Na^+/H^+ exchange. Cardioprotectant during myocardial ischemia. cardioprotective .

7450 Carsatrin
125363-87-3
$C_{25}H_{26}F_2N_6O_S$
(±)-4-[Bis(p-fluorophenyl)methyl]-α-[(9H-purin-6-ylthio)methyl]-1-piperazineethanol.
Cardiotonic. A positive inotropic agent. Probable ion channel modulator. *Johnson R. W. Pharm. Res. Institute.*

7451 Carsatrin Succinate
132199-13-4
$C_{29}H_{32}F_2N_6O_5S$
(±)-4-[Bis(p-fluorophenyl)methyl]-α-[(9H-purin-6-ylthio)methyl]-1-piperazineethanol succinate (1:1) (salt).
RWJ-24517. Cardiotonic. A positive inotropic agent. Probable ion channel modulator. *Johnson R. W. Pharm. Res. Institute.*

7452 Cilostamide
68550-75-4
$C_{20}H_{26}N_2O_3$
N-Cyclohexyl-4-[(1,2-dihydroxy-2-oxo-6-quinolyl)oxy]-N-methylbutyramide.
Phosphodiesterase III inhibitor.

7453 Convallatoxin
508-75-8 2575 208-086-3
$C_{29}H_{42}O_{10}$
(3β,5β)-3-[(6-Deoxy-α-L-mannopyranosyl)-oxy]-5,14-dihydroxy-19-oxocard-20(22)-enolide.
strophanthidin α-L-rhamnoside; Convallaton; Corglykon; Korglykon. Cardiotonic. Found in blossoms of lilly of the valley (*Convallaria majalis L., Liliaceae*). mp = 235-242°; $[\alpha]_D^{22}$ = -1.7° ± 3° (c = 0.65 MeOH), $[\alpha]_D^{25}$ = -9.4° ± 3° (c = 0.72 dioxane); soluble in EtOH, Me_2CO; slightly soluble in H_2O (0.05 g/100 ml), $CHCl_3$, EtOAc; LD_{50} (mus ip) = 10.0 mg/gk, (rat ip) = 16.0 mg/kg; [tri-O-acetate (C35H48O13)]: mp = 215-238°; $[\alpha]_D^{25}$ = -5.5° ± 2° (c = 0.962 $CHCl_3$). *Hoechst AG.*

7454 Cymarin
508-77-0 2830 208-087-9
$C_{30}H_{44}O_9$
3β-[(2,6-Dideoxy-3-O-methyl-β-D-ribopyranosyl)-oxy]-5β,14-dihydroxy-19-oxocard-20(22)-enolide.
K-strophanthin-α; Alvonal MR. Cardiotonic. mp = 148°; $[\alpha]_D^{20}$ = 39.2° (MeOH), $[\alpha]_D^{22}$ = 39.0° (c = 1.7 $CHCl_3$); soluble in MeOH, $CHCl_3$; insoluble in H_2O; LD_{50} (rat iv) = 24.8 ± 1.8 mg/kg; [sesquihydrate]: mp = 184-185°; [monoacetylcymarin ($C_{32}H_{46}O_{10}$)]: mp = 175-176°; $[\alpha]_D^{22}$ = 45.1° (EtOH), soluble in $CHCl_3$, insoluble in H_2O.

7455 Denbufylline
57076-71-8
$C_{16}H_{24}N_4O_3$
7-acetonyl-1,3-dibutylxanthine.
Phosphodiesterase inhibitor.

7456 Denopamine
71771-90-9 2943
$C_{18}H_{23}NO_4$
(-)-(R)-α-[[(3,4-Dimethoxyphenethyl)amino]methyl]-p-hydroxybenzyl alcohol.
TA-064; Carguto; Kalgut. Cardiotonic. Selective β_1-adrenoceptor agonist with positive inotropic activity. [l-form hydrochloride]: mp = 138-139.5°; $[\alpha]_D^{25}$ = -38.0° (c = 1 MeOH); [dl-form hydrochloride]: mp = 164-167°. *Tanabe Seiyaku Co. Ltd.*

7457 Deslanoside
17598-65-1 2967 241-568-1
$C_{47}H_{74}O_{19}$
3-[(O-β-D-Glucopyranosyl-(1→4)-O-3,6-dideoxy-β-D-ribohexopyranosyl-(1→4)-O-2,6-dideoxy-β-D-ribohexopyranosyl-(1→4)-12,14-dihydroxycard-20(22)-enolide.
deacetylanatoside C; Cedilanid D; Desace; Desaci; Lanimerck (ampuls); Purpurea glycoside C. Cardiotonic. From the leaves of *Digitalis lanata*. dec 265-268°; $[\alpha]_D^{20}$ = 12° (c = 1.084 75% EtOH); soluble in H_2O (0.02 g/100 ml), MeOH (0.5 g/100 ml), EtOH (0.04 g/100 ml); slightly soluble in $CHCl_3$; insoluble in Et_2O. *Sandoz Pharm. Corp.*

7458 Digitalin

752-61-4 3200 212-036-6

$C_{36}H_{56}O_{14}$

(3β,5β,16β)-3-[(6-Deoxy-4-O-β-D-glucopyranosyl-3-O-methyl-β-D-galactopyranosyl)oxy]-14,16-dihydroxycard-20(22)-enolide.

digitalinum verum; digitalinum true; Schmiedeberg's digitalin; Diginorgin. Cardiotonic. From the seeds of *Digitalis purpurea L. Scrophulariaceae and the roots of Adenium hoghel.* mp = 240-243°; $[\alpha]_D^{20}$ = -1.1° (c = 0.894 MeOH); slightly soluble in H_2O, $CHCl_3$, Et_2O; soluble in EtOH.

7459 Digitalis

3201

Digitfortis; Digiglusin; Foxglove; Fairy Gloves; purple foxglove; Digitora; Neodigitalis; Pil-Digis. Cardiotonic. From the leaves of *Digitalis purpurea L. Scrophulariaceae*. CAUTION: The therapeutic dose is close to the toxic dose. Can cause anorexia, nausea, salivation, vomiting, diarrhea, headache, drowsiness, disorientation, delirium, hallucinations, death. *Eli Lilly & Co.; Parke-Davis.*

7460 Digitoxin

71-63-6 3206 200-760-5

$C_{41}H_{64}O_{13}$

3-[(O-2,6-Dideoxy-β-D-ribohexopyranosyl-(1→4)-O-2,6-dideoxy-β-D-ribohexopyranosyl-(1→4)-2,6-dideoxy-β-D-ribohexopyranosyl)oxy]-14-hydroxycard-20(22)-enolide.

Cardidigin; Crystodigin; Digisidin; Unidigin; digitophyllin; Cardigin; Carditoxin; Coramedan; Cristapurat; Digicor; Digilong; Digimerck; Digimed; Digipural; Ditaven; Digisidin; Digitaline Nativelle; Lanatoxin; Myodigin; Purodigin; Purpurid; Tradigal. Cardiotonic. Secondary glycoside from the dried leaves of *Digitalis purpurea L. Scrophulariaceae*. mp = 256-257°; $[\alpha]_D^{20}$ = 4.8° (c = 1.2 dioxane); soluble in $CHCl_3$ (2.5 g/100 ml), EtOH (1.67 g/100 ml), EtOAc (0.25 g/100 ml), Me_2CO, amyl alcohol, C_5H_5N; sparingly soluble in Et_2O, petroleum ether, H_2O 0.001 g/100 ml at 20°; LD_{50} (gpg orl) = 60 mg/kg, (cat orl) = 0.18 mg/kg. *Eli Lilly & Co.; Marion Merrell Dow Inc.; Sterling Winthrop Inc.*

7461 Digoxin

20830-75-5 3210 244-068-1

$C_{41}H_{64}O_{14}$

3-[(O-2,6-Dideoxy-β-D-ribohexopyranosyl-(1→4)-O-2,6-dideoxy-β-D-ribohexopyranosyl-(1→4)-2,6-dideoxy-β-D-ribohexopyranosyl)oxy]-14-hydroxycard-20(22)-enolide.

Lanoxicaps; Lanoxin; Cordioxil; Davoxin; Digacin; Dilanacin; Dixina; Dokim; Dynamos; Eudigox; Lanacordin; Lanicor; Lenoxicaps; Lenoxin; Longdigox; NeoDioxanin; Rougoxin; Stillacor; Vanoxin. Cardiotonic. Secondary glycoside from *Digitalis purpurea L. Scrophulariaceae*. Dec = 230-265°; $[\alpha]_{Hg}^{25}$ = 13.4 - 13.8° (c = 10 C_5H_5N); λ_m = 220 nm (ε 12800 EtOH); soluble in EtOH, C_5H_5N; insoluble in $CHCl_3$, Me_2CO, EtOAc, H_2O, Et_2O. *Glaxo Wellcome Inc.; Wyeth-Ayerst Labs.*

7462 Dioxyline Phosphate

5667-46-9

Paveril phosphate.

Cardiotonic. *Eli Lilly & Co.*

7463 Dobutamine

34368-04-2 3456

$C_{18}H_{23}NO_3$

(±)-4-[2-[[3-(p-Hydroxyphenyl)-1-methylpropyl]amino]ethyl]pyrocatechol.

Compound 81929. Cardiotonic. A β_1-adrenoceptor agonist derived from dopamine. *Eli Lilly & Co.*

7464 Dobutamine Hydrochloride

49745-95-1 3456 256-464-1

$C_{18}H_{24}ClNO_3$

(±)-4-[2-[[3-(p-Hydroxyphenyl)-1-methylpropyl]amino]ethyl]pyrocatechol hydrochloride.

Dobutrex; Inotrex; 46236. Cardiotonic. A β_1-adrenoceptor agonist derived from dopamine. mp = 184-186°; λ_m = 281, 223 nm (ε 4768, 14400, MeOH); LD_{50} (mus iv) = 73 mg/kg. *Astra USA Inc.; Eli Lilly & Co.*

7465 Dobutamine Lactobionate

104564-71-8

$C_{30}H_{45}NO_{15}$

(±)-4-[2-[[3-(p-Hydroxyphenyl)-1-methylpropyl]amino]ethyl]pyrocatechol lactobionate (salt).

LY-207506. Cardiotonic. A β_1-adrenoceptor agonist derived from dopamine. *Eli Lilly & Co.*

7466 Docarpamine

74639-40-0 3457

$C_{21}H_{30}N_2O_8S$

(-)-(S)-2-Acetamido-N-(3,4-dihydroxyphenethyl)-4-(methylthio)butyramide bis(ethyl carbonate) (ester).

TA-870; Tanadopa. Cardiotonic. Dopamine prodrug. mp = 85-90°, 105-108°; $[\alpha]_D^{20}$ = -15.6° (c = 2 MeOH); slightly soluble in H_2O, soluble in EtOH; LD_{50} (mrat sc) = 1000-1400 mg/kg, (frat sc) = 1000 mg/kg, (rat,dog orl) > 2000 mg/kg. *Tanabe Seiyaku Co. Ltd.*

7467 Dopamine

51-61-6 3479 200-110-0

$C_8H_{11}NO_2$

4-(2-Aminoethyl)pyrocatechol.

3-hydroxytyramine. Adrenergic; used as an antihypotensive agent and cardiotonic. Endogenous catecholamine with α- and β-adrenergic activity. *Astra USA Inc.; Elkins-Sinn; Parke-Davis.*

7468 Dopamine Hydrochloride

62-31-7 3479 200-527-8

$C_8H_{12}ClNO_2$

4-(2-Aminoethyl)pyrocatechol hydrochloride.

Dopastat; ASL-279; Cardiosteril; Dynatra; Inovan; Inotropin. Adrenergic; used as an antihypotensive and cardiotonic. Dec 241°; freely soluble in H_2O; soluble in EtOH, MeOH; insoluble in Et_2O, $CHCl_3$, C_6H_6, C_7H_8, petroleum ether; [hydrobromide]: mp = 210-214° (dec). *Astra USA Inc.; Elkins-Sinn; Parke-Davis.*

7469 Dopexamine

86197-47-9 3482

$C_{22}H_{32}N_2O_2$

4-[2-[[6-(Phenethylamino)hexyl]amino]ethyl]pyrocatechol.

FPL-60278; Dopacard. Cardiotonic. Dopamine receptor

agonist. Has little or no α- or β-adrenoceptor activity. *Fisons Pharm. Div.*

7470 Dopexamine Hydrochloride
86484-91-5 3482

$C_{22}H_{34}Cl_2N_2O_2$

4-[2-[[6-(Phenethylamino)hexyl]amino]-ethyl]pyrocatechol dihydrochloride.

FPL-60278AR. Cardiotonic. Dopamine receptor agonist. Has little or no α- or β-adrenoceptor activity. [hydrobromide]: mp = 227-228°. *Fisons Pharm. Div.*

7471 Doxaminol
55286-56-1

$C_{26}H_{29}NO_3$

6,11-Dihydro-N-(2-hydroxy-3-phenoxypropyl)-N-methyldibenz[b,e]oxepin-11-ethylamine.

A β-adrenergic agonist. Has cardiotonic properties.

7472 Enoximone
77671-31-9 3627

$C_{12}H_{12}N_2O_2S$

4-Methyl-5-[p-(methylthio)benzoyl]-4-imidazolin-2-one.

Perfan; MDL-17043; fenoximone; RMI-17043; Perfane. Cardiotonic. Selective phosphodiesterase inhibitor with vasodilating and positive inotropic activity. mp = 255-258° (dec). *Marion Merrell Dow Inc.*

7473 Erythrophleine
36150-73-9 3728

$C_{24}H_{39}NO_5$

[1S-(1α,4aα,4bβ,7E,8β,8aα,9α,10aβ)]-Tetradecahydro-9-hydroxy-1,4a,8-trimethyl-7-[2-[2-(methylamino)ethoxy]-2-oxoethylidene]-1-phenanthrenecarboxylic acid methyl ester.

norcassamidine. Cardiotonic. From the bark of *Erythrophleum guineense G. Don., Leguminossae*. mp = 115°; $[\alpha]_D^{20}$ = -22.5° (c = 0.65 EtOH); soluble in H_2O, EtOH.

7474 Etomoxir
124083-20-1

$C_{17}H_{23}ClO_4$

Ethyl-(+)-(R)-2-[6-(p-chlorophenoxy)hexyl]glycidate.

Cardiotonic.

7475 Fenalcomine
34616-39-2 3996

$C_{20}H_{27}NO_2$

α-Ethyl-p-[2-[(α-methylphenethyl)amino]ethoxy]-benzyl alcohol.

Cardiotonic. Local anesthetic. *LaRoche-Navarron.*

7476 Fenalcomine Hydrochloride
34535-83-6 3996 252-075-6

$C_{20}H_{28}ClNO_2$

α-Ethyl-p-[2-[(α-methylphenethyl)amino]ethoxy]-benzyl alcohol hydrochloride.

Cordoxene. Cardiotonic. Local anesthetic. *LaRoche-Navarron.*

7477 Flosequinan
76568-02-0 4146

$C_{11}H_{10}FNO_2S$

7-Fluoro-1-methyl-3-(methylsulfinyl)-4(1H)-quinolone.

BTS 49 465; flosequinon; Manoplax. Acts as an arterial and venous vasodilator. Used as an antihypertensive. Also used as a cardiotonic. mp = 226-228°. *Boots Co.*

7478 Gitalin
1405-76-1 4437 215-784-1

Cardiotonic. An extract of *Digitalis purpurea L. Sacrophulariaceae*. It is a mixture of digitoxin; gitoxin and gitaloxin (16-formylgitoxin, Cristaloxine). Readily soluble in EtOH, slightly soluble in H_2O.

7479 Gitaloxin
3261-53-8

$C_{42}H_{64}O_{15}$

Gitoxin 16-formate.

Cardiotonic.

7480 Gitoformate
7685-23-6

$C_{46}H_{64}O_{19}$

Gitoxin-3',3,3',4',16-pentaformate.

Cardiotonic.

7481 Gitoxin
4562-36-1 4441 224-934-5

$C_{41}H_{64}O_{14}$

(3β,5β,16β)-3-[(O-2,6-Dideoxy-β-D-ribohexopyranosyl-(1→4)-O-2,6-dideoxy-β-D-ribohexopyranosyl-(1→4)-2,6-dideoxy-β-D-ribohexopyranosyl)oxy]-14,16-dihydroxycard-20(22)-enolide.

anhydrogitalin; bigitalin; pseudodigitoixn. Cardiotonic. dec 285°; $[\alpha]_{546}^{20}$ = 3.5° (c = 1.02 C_5H_5N); λ_m = 315, 415, 495, 530 nm ($E_{1\,cm}^{1\%}$ 275, 285, 430, 505 98% H_2SO_4); insoluble in $CHCl_3$, EtOAc, Me_2CO; soluble in C_5H_5N, EtOH. *VEB Arzneimittelwerk.*

7482 Glycocyamine
352-97-6 4505 206-529-5

$C_3H_7N_3O_2$

N-(Aminoiminomethyl)glycine.

guanidineacetic acid; guanidoacetic acid. Used in combination with betaine as a cardiotonic. dec 280-284°; soluble in H_2O.

7483 Heptaminol
372-66-7 4691 206-758-0

$C_8H_{19}NO$

6-Amino-2-methyl-2-heptanol.

Cardiotonic. bp_7 = 92-93°. *Bilhuber.*

7484 Heptaminol Hydrochloride
543-15-7 4691 208-837-5

$C_8H_{20}ClNO$

6-Amino-2-methyl-2-heptanol hydrochloride.

RP-2831; Cortensor; Eoden; Hept-a-myl; Heptylon. Cardiotonic. mp = 150°; freely soluble in H_2O; soluble in EtOH; insoluble in Me_2CO, C_6H_6, Et_2O. *Bilhuber.*

7485 Hydrastinine
6592-85-4 4807 229-533-9

$C_{11}H_{13}NO_3$

5,6,7,8-Tetrahydro-6-methyl-1,3-dioxolo[4,5-g]isoquinolin-5-ol.

The hydrochloride salt is used as a cardiotonic and uterine hemostatic. Has dopamine receptor blocking activity. mp = 117°; soluble in EtOH, $CHCl_3$, hot H_2O, insoluble in cold H_2O.

7486 Ibopamine
66195-31-1 4921 266-229-5

$C_{17}H_{25}NO_4$

4-[2-(Methylamino)ethyl]-o-phenylene diisobutyrate.

SB-7505; 3,4-di-o-isobutyryl epinine. Cardiotonic. Inotropic agent with dopaminergic and adrenergic agonist activities. *Simes S.p.A.; SmithKline Beecham Pharm.*

7487 Ibopamine Hydrochloride
75011-65-3 4921 278-056-2

$C_{17}H_{26}Cl5NO_4$

4-[2-(Methylamino)ethyl]-o-phenylene diisobutyrate hydrochloride.

SB-7505; Inopamil; Scandine. Cardiotonic. Inotropic agent with dopaminergic and adrenergic agonist activities. mp = 132°. *Simes S.p.A.; SmithKline Beecham Pharm.*

7488 Imazodan
84243-58-3

$C_{13}H_{12}N_4O$

4,5-Dihydro-6-(p-imidazol-1-ylphenyl)-3(2H)-pyridazinone.

Cardiotonic. Selective type-III phosphodiesterase inhibitor.

7489 Imazodan Hydrochloride
89198-09-4

$C_{13}H_{13}ClN_4O$

4,5-Dihydro-6-(p-imidazol-1-ylphenyl)-3(2H)-pyridazinone monohydrochloride.

Cardiotonic. Selective type-III phosphodiesterase inhibitor.

7490 Indolidan
100643-96-7

$C_{14}H_{15}N_3O_2$

3,3-Dimethyl-5-(1,4,5,6-tetrahydro-6-oxo-3-pyridazinyl)-2-indolinone.

LY-195115. Cardiotonic. A potent, selective inhibitor of cyclic nucleotide phosphodiesterase. Positive inotropic agent. *Eli Lilly & Co.*

7491 Isomazole
86315-52-8

$C_{14}H_{13}N_3O_2S$

2-[2-Methoxy-4-(methylsulfinyl)phenyl]-1H-imidazo[4,5-c]pyridine.

Cardiotonic. A partial phosphodiesterase inhibitor and calcium sensitizer. Positive inotropic agent. *Eli Lilly & Co.*

7492 Isomazole Hydrochloride
87359-33-9

$C_{14}H_{14}ClN_3O_2S$

2-[2-Methoxy-4-(methylsulfinyl)phenyl]-1H-imidazo[4,5-c]pyridine monohydrochloride.

LY-175326. Cardiotonic. A partial phosphodiesterase inhibitor and calcium sensitizer. Positive inotropic agent. *Eli Lilly & Co.*

7493 Lanatoside A
17575-20-1 5368 241-544-0

$C_{49}H_{76}O_{19}$

(3β,5β)-3-[(O-β-D-Glucopyranosyl-(1→4)-O-3-O-acetyl-2,6-dideoxy-β-D-ribohexopyranosyl-(1→4)-O-2,6-dideoxy-β-D-ribohexopyranosyl-(1→4)-2,6-dideoxy-β-D-ribohexopyranosyl)oxy]-14-hydroxycard-20(22)-enolide.

digilanide A; Adigal. Cardiotonic. One of a family of glycosides from various species of *Digitalis*. dec 245-248°; $[\alpha]_D^{20}$ = 31.6° (c = 1.92 95% EtOH), $[\alpha]_D^{20}$ = 23.2° (c = 3.8 dioxane); soluble in MeOH (5 g/100 ml), EtOH (2.5 g/100 ml), $CHCl_3$ (0.44 g/100 ml), H_2O (0.006 g/100 ml).

7494 Lanatoside B
17575-21-2 5368 241-545-6

$C_{49}H_{76}O_{20}$

(3β,5β)-3-[(O-β-D-Glucopyranosyl-(1→4)-O-3-O-acetyl-2,6-dideoxy-β-D-ribohexopyranosyl-(1→4)-O-2,6-dideoxy-β-D-ribohexopyranosyl-(1→4)-2,6-dideoxy-β-D-ribohexopyranosyl)oxy]-14, 16-dihydroxycard-20(22)-enolide.

digilanide B. Cardiotonic. One of a family of glycosides from various species of *Digitalis*. dec 245-248°; $[\alpha]_D^{20}$ = 36.7° (c = 1.88 95% EtOH), $[\alpha]_D^{20}$ = 31.8° (c = 1.8 dioxane); soluble in MeOH (5 g/100 ml), EtOH (2.5 g/100 ml), $CHCl_3$ (0.44 g/100 ml); nearly insoluble in H_2O.

7495 Lanatoside C
17575-22-3 5368 241-546-1

$C_{49}H_{76}O_{20}$

(3β,5β)-3-[(O-β-D-Glucopyranosyl-(1→4)-O-3-O-acetyl-2,6-dideoxy-β-D-ribohexopyranosyl-(1→4)-O-2,6-dideoxy-β-D-ribohexopyranosyl-(1→4)-2,6-dideoxy-β-D-ribohexopyranosyl)oxy]-12,14-dihydroxycard-20(22)-enolide.

digilanide C; Allocor; Cedilanid; Ceglunat; Celadigal; Cetosanol; Lanimerck (suppositories). Cardiotonic. One of a family of glycosides from various species of *Digitalis*. dec 248-250°; $[\alpha]_D^{20}$ = 33.4 - 33.7° (c = 2 EtOH); soluble in MeOH (0.005 g/100 ml), $CHCl_3$ (0.05 g/100 ml); freely soluble in C_5H_5N, dioxane; insoluble in Et_2O, petroleum ether.

7496 Lanatoside D
17575-31-2 5368

$C_{49}H_{76}O_{21}$

(3β,5β)-3-[(O-β-D-Glucopyranosyl-(1→4)-O-3-O-acetyl-2,6-dideoxy-β-D-ribohexopyranosyl-(1→4)-O-2,6-dideoxy-β-D-ribohexopyranosyl-(1→4)-2,6-dideoxy-β-D-ribohexopyranosyl)oxy]-12,14,16-trihydroxycard-20(22)-enolide.

digilanide D. Cardiotonic. One of a family of glycosides from various species of *Digitalis*. dec 242-250°; $[\alpha]_D^{20}$ = 40.5° c = 5.95 MeOH); λ_m = 220 nm (log ε 4.16).

7497 Levdobutamine
61661-06-1

$C_{18}H_{23}NO_3$

4-[2-[[(S)-3-(p-Hydroxyphenyl)-1-methylpropyl]amino]ethyl]pyrocatechol.

LY-206243. Cardiotonic. A β_1-adrenoceptor agonist derived from dopamine. *Eli Lilly & Co.*

7498 Levdobutamine Lactobionate
129388-07-4
$C_{30}H_{45}NO_{15}$
4-[2-[[(S)-3-(p-Hydroxyphenyl)-1-methylpropyl]amino]ethyl]pyrocatechol lactiobionate (1:1) (salt).
LY-206243 lactobionate. Cardiotonic. A β_1-adrenoceptor agonist derived from dopamine. *Eli Lilly & Co.*

7499 Loprinone
106730-54-5 5605
$C_{14}H_{10}N_4O$
1,2-Dihydro-5-imidazo[1,2-a]pyridin-6-yl-6-methyl-2-oxo-3-pyridinecarbonitrile.
olprinone. Cardiotonic. Phosphodiesterase inhibitor. Positive inotropic agent. mp > 300°; [hydrochloride monohydrate ($C_{14}H_{11}ClN_4O$)]: mp > 300°; LD_{50} (mrat orl) = 7804 mg/kg, (mrat iv) = 176 mg/kg, (mrat sc) = 2133 mg/kg, (frat orl) >10000 mg/kg, (frat iv) = 240 mg/kg, (frat sc) = 2890 mg/kg, (mmus orl) > 10000 mg/kg, (mmus iv) = 242 mg/kg, (mmus sc) = 3898 mg/kg, (fmus orl) > 10000 mg/kg, (fmus iv) = 269 mg/kg, (fmus sc) = 4479 mg/kg. *Eisai Co. Ltd.*

7500 Medorinone
88296-61-1
$C_9H_8N_2O$
5-Methyl-1,6-naphthyridin-2(1H)-one.
Win-49016. Cardiotonic. Phosphodiesterase inhibitor. Positive inotropic agent. *Sterling Winthrop Inc.*

7501 Meribendan
119322-27-9
$C_{15}H_{14}N_6O$
4,5-Dihydro-5-methyl-6-(2-pyrazol-3-yl-5-benzimidazolyl)-3(2H)-pyridazinone.
PDE-III inhibitor. Cardiotonic.

7502 Milrinone
78415-72-2 6284 278-903-6
$C_{12}H_9N_3O$
1,6-Dihydro-2-methyl-6-oxo[3,4'-bipyridine]-5-carbonitrile.
Win-47203; Corotrope. Cardiotonic. Phosphodiesterase inhibitor. mp > 300°. *Sterling Winthrop Inc.*

7503 Milrinone Lactate
100286-97-3 6284
$C_{12}H_9N_3O.xC_3H_6O_3$
1,6-Dihydro-2-methyl-6-oxo[3,4'-bipyridine]-5-carbonitrile lactate.
Primacor. Cardiotonic. Phosphodiesterase inhibitor. *Sterling Winthrop Inc.*

7504 Neriifolin
466-07-9 6559 207-372-5
$C_{30}H_{46}O_8$
(3β,5β)-3-[(6-Deoxy-3-O-methyl-α-L-glucopyranosly)oxy]-14-hydroxycard-20(22)-enolide.
Cardiotonic. Cardiac glycoside isolated from *Thevitia neriifolia*. mp = 218-225°, 208°; $[\alpha]_D^{23}$ = -50.2° (MeOH); λ_m = 217 nm (log ε 4,1 MeOH).

7505 Neriifolin 2'-Acetate
25633-33-4 6559
$C_{32}H_{48}O_9$
(3β,5β)-3-[(6-Deoxy-3-O-methyl-α-L-glucopyranosly)oxy]-14-hydroxycard-20(22)-enolide 2'-acetate.
cerberin; veneniferin; monoacetylneriifolin. Cardiotonic. Cardiac glycoside. mp = 212-215°; $[\alpha]_D^{19}$ = -82° ($CHCl_3$); soluble in EtOH, AcOH, $CHCl_3$, Et_2O; insoluble in H_2O.

7506 Oleandrin
465-16-7 6963 207-361-5
$C_{32}H_{48}O_9$
(3β,5β,16β)-16-(Acetyloxy)-3-[(2,6-dideoxy-3-O-methyl-α-L-arabinohexopyranosyl)oxy]-14-hydroxycard-20(22)-enolide.
neriolin; Corrigen; Folinerin. Diuretic; cardiotonic. Glycoside. From the leaves of *Nerium oleander L. Apocynaceae* (Laurier rose). mp = 250°; $[\alpha]_D^{25}$ = -48.0° (c = 1.3 MeOH); λ_m= 220 nm (log ε 4.20); insoluble in H_2O, soluble in EtOH, $CHCl_3$; [desacetyloleandrin]: mp = 238-240°; $[\alpha]_D^{18}$ = -24.9°.

7507 Ouabain
630-60-4 7031 211-139-3
$C_{29}H_{44}O_12$
3-[(6-Deoxy-α-L-mannopyranosyl)oxy]-1,5,11,14,19-pentahydroxycard-20(22)-enolide.
G-strophanthin; Gratus strophanthin; acocantherin. Cardiotonic. Cardiac glycoside. From the seeds of *Strophanthus gratus*. [octahydrate (Purostrophan; Strodival; Strophoperm)]: dec 190°; $[\alpha]_D^{25}$ = -31° to =33.5° (c = 1); soluble in H_2O (1.3 g/100 ml at 25°, 20 g/100 ml at 100°), EtOH (1.0 g/100 ml at 25°, 20 g/100 ml at 75°), amyl alcohol, dioxane; slightly soluble in Et_2O, $CHCl_3$, EtOAc; LD_{50} (rat iv) = 14 mg/kg.

7508 Oxyfedrine
15687-41-9 7096
$C_{19}H_{23}NO_3$
3-[[(αS,βR)-β-Hydroxy-α-methylphenethyl]amino]-3'-methoxypropiophenone.
oxyphedrine. Cardiotonic; antianginal. Used to treat cardiac insufficiency. Partial β-adrenergic agonist with coronary vasodilating and positive inotropic effects. *Degussa-Hüls Corp.*

7509 DL-Oxyfedrine Hydrochloride
16648-69-4 7096 240-696-5
$C_{19}H_{24}ClNO_3$
(DL)-3-[[β-Hydroxy-α-methylphenethyl]amino]-3'-methoxypropiophenone hydrochloride.
Cardiotonic; antianginal. Used to treat cardiac insufficiency. Partial β-adrenergic agonist with coronary vasodilating and positive inotropic effects. mp = 173-175°; LD_{50} (mus iv) = 34 mg/kg. *Degussa-Hüls Corp.*

7510 L-Oxyfedrine Hydrochloride
16777-42-7 7096 240-828-1
$C_{19}H_{24}ClNO_3$
(L)-3-[[β-Hydroxy-α-methylphenethyl]amino]-3'-methoxypropiophenone hydrochloride.
D-563; Ildamen; Modacor. Cardiotonic; antianginal.

Used to treat cardiac insufficiency. Partial β-adrenergic agonist with coronary vasodilating and positive inotropic effects. mp = 192-194°; LD_{50} (mus iv) = 29 mg/kg. *Degussa-Hüls Corp.*

7511 Pelrinone
94386-65-9

$C_{12}H_{11}N_5O$
1,4-Dihydro-2-methyl-4-oxo-6[(3-pyridylmethyl)amino]-5-pyrimidinecarbonitrile.
Cardiotonic. Phosphodiesterase III inhibitor. *Wyeth-Ayerst Labs.*

7512 Pelrinone Hydrochloride
89232-84-8

$C_{12}H_{12}ClN_5O$
1,4-Dihydro-2-methyl-4-oxo-6[(3-pyridylmethyl)amino]-5-pyrimidinecarbonitrile monohydrochloride.
Myotrope; AY-28768. Cardiotonic. Phosphodiesterase III inhibitor. *Wyeth-Ayerst Labs.*

7513 Pimobendan
74150-27-9 7588

$C_{19}N_{18}N_4O_2$
(±)-4,5-Dihydro-6-[2-(p-methoxyphenyl)-5-benzimidazolyl]-5-methyl-3(2H)-pyridazinone.
UDCG-115. Cardiotonic. A putative calcium sensitizer and phosphodiesterase inhibitor. *Boehringer Ingelheim Pharm. Inc.; Thomae GmbH Dr. Karl.*

7514 Pimobendan Hydrochloride
77469-98-8 7588

$C_{19}N_{19}ClN_4O_2$
(±)-4,5-Dihydro-6-[2-(p-methoxyphenyl)-5-benzimidazolyl]-5-methyl-3(2H)-pyridazinone hydrochloride.
Cardiotonic. A putative calcium sensitizer and phosphodiesterase inhibitor. mp = 311° (dec); LD_{50} (mus orl) = 600 mg/kg. *Boehringer Ingelheim Pharm. Inc.; Thomae GmbH Dr. Karl.*

7515 Piroximone
84490-12-0

$C_{11}H_{11}N_3O_2$
4-Ethyl-5-isonicotinoyl-4-imidazolin-2-one.
MDL-19205. Cardiotonic. *Marion Merrell Dow Inc.*

7516 Prenalterol
57526-81-5 7917 260-791-5

$C_{12}H_{19}NO_3$
(-)-(S)-1-(p-Hydroxyphenoxy)-3-(isopropylamino)-2-propanol.
A $β_1$-adrenergic agonist, used as a cardiotonic. mp = 127-128°; $[α]_D^{20}$ = -1° ± 1°, $[α]_{Hg}^{20}$ = 2° ± 1° (c = 0.940 MeOH). *Ciba-Geigy Corp.*

7517 Prenalterol Hydrochloride
61260-05-7 7917 262-676-5

$C_{12}H_{10}ClNO_3$
(-)-(S)-1-(p-Hydroxyphenoxy)-3-(isopropylamino)-2-propanol hydrochloride.
H133/22; CGP-7760B; Hyprenan; Varebian. A $β_1$-adrenergic agonist; cardiotonic. *Ciba-Geigy Corp.*

7518 Prinoxodan
111786-07-3

$C_{13}H_{14}N_4O_2$
3,4-Dihydro-3-methyl-6-(1,4,5,6-tetrahydro-6-oxo-3-pyridazinyl)-2(1H)-quinazolinone.
RGW-2938. Cardiotonic. An orally effective positive inotropic and cardiac vasodilator agent. *Rhône-Poulenc Rorer Pharm. Inc.*

7519 Proscillaridin
466-06-8 8060 207-370-4

$C_{30}H_{42}O_8$
[(6-Deoxy-α-L-mannopyranosyl)oxy]-14-hydroxybufa-4,20,22-trienolide.
Tradenal; A-32686; 2936; proscillaridin A; desglucotrnasvaaline; Caradrin; Cardion; Carmazon; Proscillan; Prostosin; Proszine; Protasin; Purosin-TC; Sandoscill; Scillacrist; Simeon; Solestril; Stellarid; Talucard; Talusin; Urgilan; Wirnesin. Cardiotonic. A cardiac glycoside. mp = 219-222°; $[α]_D^{20}$ = -91.5° (MeOH); LD_{50} (mrat orl) = 56 mg/kg, (frat orl) = 76 mg/kg. *Knoll Pharm. Co.*

7520 Proscillaridin 4-Methyl Ether
33396-37-1 8060 251-493-6

$C_{31}H_{44}O_8$
[(6-Deoxy-α-L-mannopyranosyl)oxy]-14-hydroxybufa-4,20,22-trienolide 4-methyl ether.
meproscillarin; Clift. Cardiotonic. A cardiac glycoside. mp = 213-217°; $[α]_D^{20}$ = -94° (MeOH); $λ_m$ = 297 nm (log ε 3.79 (MeOH), 355 nm (log ε 4.65 1N KOH/MeOH); soluble in MeOH, EtOH, THF, dioxane; slightly soluble in $CHCl_3$, CH_2Cl_2, Me_2CO; insoluble in H_2O, non-polar organic solvents. *Knoll Pharm. Co.*

7521 Quazinone
70018-51-8

$C_{11}H_{10}ClN_3O$
(R)-6-Chloro-1,5-dihydro-3-methylimidazo[2,1-b]-quinazolin-2(3H)-one.
Ro-13-6438/006. Cardiotonic. A positive inotropic agent with vasodilating activity. *Hoffmann-LaRoche Inc.*

7522 Quazodine
4015-32-1

$C_{12}H_{14}N_2O_2$
4-Ethyl-6,7-dimethoxyquinazoline.
MJ-1988. Cardiotonic; bronchodilator. A quinazoline derivative.

7523 Resibufogenin
465-39-4 8315

$C_{24}H_{32}O_4$
14,15β-Epoxy-3β-hydroxy-5β-bufa-20,22-dienolide.
Respigon. Cardiotonic. Cytotoxic component of toad venom. mp = 113-140°, 155-168°; $[α]_D^{22}$ = -7.1° (c = 1.259 $CHCl_3$); [hydrochloride]: dec 230-232°; $[α]_D^{15}$ = 15.1° (c = 0.530 $CHCl_3$); $λ_m$ = 298 nm (log ε 3.74 EtOH).

7524 Scillaren
11003-70-6 8543

A mixture of the glycosides, Scillaren A and Scillaren B. Cardiotonic. $[α]_D^{20}$ = -25° to -35° (c = 2 75% EtOH);

soluble in H_2O (0.033 g/100 ml), EtOH, MeOH (20 g/100 ml); insoluble in Et_2O, $CHCl_3$; LD_{50} (cat iv) = 0.18 - 0.62 mg/kg; LD (rbt orl) = 0.95 mg/gk, LD (rat sc) = 10 mg/kg.

7525 Scillarenin

465-22-5 8544

$C_{24}H_{32}O_4$

(3β)-3,14-Dihydroxybufa-4,20,22-trienolide.

mp = 232-238°; $[\alpha]_D^{20}$ = -16.8° (c = 0.357 MeOH); $[\alpha]_D^{20}$ = 17.9° (c = 0.39 $CHCl_3$); λ_m = 300 nm (log ε 3.72); LD (cat iv) = 0.1567 mg/kg; [3-acetate ($C_{26}H_{34}O_5$)]: mp = 240-243°; $[\alpha]_D^{20}$ = -23.4° (c = 1.365 $CHCl_3$).

7526 Strophanthin

11005-63-3 9016 234-239-9

K-strophanthin; K-strophanthoside. A mixture of α-glucose (19%), β-glucose (19%), cymarose (15%) and strophanthidin (47%). Ouabain is G-strophanthin. Soluble in H_2O, dilute EtOH; insoluble in $CHCl_3$, Et_2O, C_6H_6; MLD (cat iv) = 0.11 mg/kg, (rat iv) = 9.4 mg/kg.

7527 Sulmazole

73384-60-8 9159 277-406-1

$C_{14}H_{13}N_3O_2S$

2-[2-Methoxy-4-(methylsulfinyl)phenyl]-3H-imidazo-[4,5-b]pyridine.

AR-L; 115BS; Vardax. Cardiotonic. Non-glycoside, non-adrenergic inotropic. mp = 203-205°; LD_{50} (mus orl) = 560 mg/kg, (mus iv) = 163 mg/kg. *Thomae GmbH Dr. Karl.*

7528 Tazolol

39832-48-9

$C_9H_{16}N_2O_2S$

2-(±)-1-(Isopropylamino)-3-(2-thiazolyloxy)-2-propanol.

Cardiotonic. *Syntex Intl. Ltd.*

7529 Tazolol Hydrochloride

38241-39-3

$C_9H_{17}ClN_2O_2S$

2-(±)-1-(Isopropylamino)-3-(2-thiazolyloxy)-2-propanol monohydrochloride.

RS-6245. Cardiotonic. *Syntex Intl. Ltd.*

7530 Theobromine

83-67-0 9418 201-494-2

$C_7H_8N_4O_2$

3,7-Dihydro-3,7-dimethyl-1H-purine-2,6-dione.

3,7-dimethylxanthine. Diuretic; bronchodilator; cardiotonic. Principle alkaloid of the cacao bean. Also found in coca nuts and tea. A methylxanthine CNS stimulant. mp = 357°; sublimes at 290-295°; very slightly soluble in H_2O, EtOH; soluble in fixed alkali hydroxides, concentrated acids; moderately soluble in ammonia; insoluble in C_6H_6, Et_2O, $CHCl_3$, CCl_4.

7531 Toborinone

143343-83-3

$C_{21}H_{24}N_2O_5$

(±)-6-[3-(3,4-Dimethoxybenzylamino)-2-hydroxypropoxy]-2(1H)-quinolinone.

OPC-18790. A novel cardiotonic, positive inotropic agent with an inhibitory action on phosphodiesterase.

7532 Tolafentrine

139308-65-9

$C_{28}H_{31}N_3O_4S$

(-)-4'-(cis-1,2,3,4,4a,10b-Hexahydro-8,9-dimethoxy-2-methylbenzo[c][1,6]naphthyridin-6-yl)-p-toluenesulfonanilide.

Type III/IV-selective phosphodiesterase inhibitor.

7533 Vesnarinone

81840-15-5 10105

$C_{22}H_{25}N_3O_4$

1-(1,2,3,4-Tetrahydro-2-oxo-6-quinolyl)-4-veratroylpiperazine.

OPC-8212; Arkin; pieranometazine. Cardiotonic. Operates through the sodium and potassium rectifying channels and has limited phosphodiesterase inhibiting activity. mp = 238.1-239.5°, 238.1-239.8°; λ_m = 271 nm (ε 25100 MeOH); soluble in AcOH (18.7 g/100 ml), $CHCl_3$ (14.2 g/100 ml), benzyl alcohol (5.9 g/100 ml), DMSO (2.51 g/100 ml); DMF (1.18 g/100 ml), dioxane (0.17 g/100 ml); MeOH (0.12 g/100 ml), Me_2CO (0.064 g/100 ml), EtOH (0.04 g/100 ml), H_2O (0.002 g/100 ml). *Otsuka Am. Pharm.*

7534 Xamoterol

81801-12-9 10189

$C_{16}H_{25}N_3O_5$

(±)-N-[2-[[2-Hydroxy-3-(p-hydroxyphenoxy)-propyl]amino]ethyl]-4-morpholine-carboxamide.

1-(4-hydroxyphenoxy)-3-[2-(4-morpholinocarboxamido)-ethylamino-2-propanol. Partial β-adrenergic agonist with positive inotropic activity. Has been used as a cardiotonic. *ICI.*

7535 Xamoterol Hemifumarate

73210-73-8 10189 277-319-9

$C_{36}H_{54}N_6O_{14}$

(±)-N-[2-[[2-Hydroxy-3-(p-hydroxyphenoxy)-propyl]amino]ethyl]-4-morpholinecarboxamide fumarate (2:1) (salt).

Carwin; Corwin; Xamtol. Cardiotonic. Partial β-adrenergic agonist with positive inotropic activity. mp = 168-169° (dec). *ICI.*

CCK Antagonists

7536 Devazepide

103420-77-5

$C_{25}H_{20}N_4O_2$

(S)-N-(2,3-Dihydro-1-methyl-2-oxo-5-phenyl-1H-1,4-benzodiazepin-3-yl)indole-2-carboxamide.

MK-329; L-364718. Cholecystokinin inhibitor. *Merck & Co.Inc.*

7537 Dexloxiglumide

119817-90-2

$C_{21}H_{30}Cl_2N_2O_5$

(R)-4-(3,4-Dichlorobenzamido)-N-(3-methoxypropyl)-N-pentylglutaramic acid.

Cholecystokinin inhibitor.

7538 Lintitript
136381-85-6
$C_{20}H_{14}ClN_3O_3S$
1-[[2-(4-(2-Chlorophenyl)thiazol-2-yl)aminocarbonyl] indolyl] acetic acid.
SR-27897. Cholecystokinin receptor antagonist.

7539 Loxiglumide
107097-80-3 5618
$C_{21}H_{30}Cl_2N_2O_5$
(±)-4-(3,4-Dichlorobenzamido)-N-(3-methoxypropyl)-N-pentylglutaramic acid.
CR-1505. Cholecystokinin A inhibitor, used as a gastroprokinetic agent. mp = 113-115°; slightly soluble in H_2O (0.01 g/100 ml). *Rotta Pharm.*

7540 Proglumide
6620-60-6 7958 229-567-4
$C_{18}H_{26}N_2O_4$
(±)-4-Benzamido-N,N-dipropyl-glutaramic acid.
xylamide; W-5219; CR-242; Milid; Milide; Nulsa; Promid. Anticholinergic. Cholecystokinin inhibitor. mp = 142-145°; LD_{50} (mus iv) = 2211-2649 mg/kg, (mus orl) = 7350-8861 mg/kg. *Wallace Labs.*

Chelating Agents

7541 Deferiprone
30652-11-0
$C_7H_9NO_2$
3-Hydroxy-1,2-dimethyl-4(1H)-pyridone.
L1. Chelating agent. Used to treat thalassemia.

7542 Deferoxamine
70-51-9 2914
$C_{25}H_{48}N_6O_8$
N-[5-[3-[(5-Aminopentyl)hydroxycarbamoyl]propionamido]pentyl]-3-[[5-(N-hydroxyacetamido)pentyl]-carbamoyl]propionohydroxamic acid.
NSC-527604; Desferioxamine B. Chelating agent for iron. mp = 138-140°; soluble in H_2O (1.2 g/100 ml at 20°). *Ciba-Geigy Corp.*

7543 Deferoxamine Hydrochloride
1950-39-6 2914 217-767-4
$C_{25}H_{49}ClN_6O_8$
N-[5-[3-[(5-Aminopentyl)hydroxycarbamoyl]propionamido]pentyl]-3-[[5-(N-hydroxyacetamido)pentyl]-carbamoyl]propionohydroxamic acid monohydrochloride.
Ba-29837. Chelating agent for iron. mp = 172-175°. *Ciba-Geigy Corp.*

7544 Deferoxamine Methanesulfonate
138-14-7 2914 205-314-3
$C_{26}H_{52}N_6O_{11}S$
N-[5-[3-[(5-Aminopentyl)hydroxycarbamoyl]propionamido]pentyl]-3-[[5-(N-hydroxyacetamido)pentyl]-carbamoyl]propionohydroxamic acid monomethanesulfonate.
Desferal mesylate; Ba-33112; DFOM; Desferal. Chelating agent for iron. mp = 148-149°; soluble in H_2O (> 20 g/100 ml). *Ciba-Geigy Corp.*

7545 Ditiocarb Sodium
148-18-5 3443 205-710-6
$C_5H_{10}NNaS_2.3H_2O$
Sodium diethyldithiocarbamate.
dithiocarb; DTC; DDC; DEDC; DDTC; DeDTC; Imuthiol. Immunomodulator, chelating agent and antidote, especially for nickel and cadmium poisoning. Used to counter platinum toxicity is pateints tretaed with cis-Platin or to treat Wilson's disease. [trihydrate]: mp = 90-92°, 94-102°; freely soluble in H_2O, EtOH, MeOH, Me_2CO, insoluble in Et_2O, C_6H_6; λ_m = 257, 290 nm (ε 1200, 13000 H_2O); LD_{50} (rat orl) = 2830 mg/kg, (mus orl) = 1870 mg/kg, (mus iv) > 1000 mg/kg.

7546 Edetate Calcium Disodium
62-33-9 3555 200-529-9
$C_{10}H_{12}CaN_2Na_2O_8.xH_2O$
Disodium[(ethylenedinitrilo)tetraacetato]calciate(2-) hydrate.
EDTA calcium; edathamiol calcium disodium; sodium calciumedetate; Calcitetracemate Disodium; Calcium Disodium Versenate; Ledclair; Mosatil; Antallin; Sormetal; Versene CA. Complexes lead ions. Soluble in H_2O.

7547 Edetate Disodium
139-33-3 3556 205-358-3
$C_{10}H_{14}N_2Na_2O_8.2H_2O$
Disodium(ethylenedinitrilo)tetraacetatedihydrate.
disodium edathamil; EDTA disodium; tetracemate disodium; disodium edetate; Chelaplex III; Endrate disodium; Sequestrene NA2; sodium versenate; Titriplex III; Versene Disodium Salt. mp = 252° (dec); soluble in H_2O; LD_{50} (rat orl) = 2000 mg/kg. *Dennis Martin.*

7548 Edetate Sodium
64-02-8 3557 200-573-9
$C_{10}H_{12}N_2Na_4O_8$
Tetrasodium(ethylenedinitrilo)tetraacetate.
EDTA tetrasodium; tetrasodium edetate; Endrate Tetrasodium; Questex; Versene; Sequestrene; Tetrine; Kalex; Trilon B; Komplexon; Nullapon; Aquamollin; Complexone; Distol 8; Irgalon; Calsol; Syntes 12a; Tyclarosol; Nervanaid B. mp > 300°; very soluble in H_2O, partially soluble in EtOH. *Dennis Martin.*

7549 Edetate Trisodium
150-38-9 3558 205-758-8
$C_{10}H_{13}N_2Na_3O_8$
Trisodium hydrogen(ethylenedinitrilo)tetraacetate.
trisodium edetate; Limclair; Versene-9; Sequestrene NA3. mp > 300°; very soluble in H_2O.

7550 Gluconolactone
90-80-2 4465 202-016-5
$C_6H_{10}O_6$
D-Gluconic acid δ-lactone.
Fujiglucon. dec 153°; $[\alpha]_D^{20}$ = 61.7° (c = 1); soluble in H_2O (59 g/100 ml), EtOH (1 g/100 ml); insoluble in Et_2O.

7551 Penicillamine
52-67-5 7214 200-148-8
$C_5H_{11}NO_2S$
D-3-Mercaptovaline.
Depen; Cuprimine; β-Thiovaline; Cuprenil; Depamine; Mercaptyl; Pendramine; Perdolat; Sufirtan; Trolovol. Antirheumatic agent. Also used as a copper chelator in treatment of Wilson's disease. mp = 202-206°; $[\alpha]_D^{25}$ = -63° (c = 1 C_5H_5N); LD_{50} (rat orl) > 10000 mg/kg, (rat ip) > 660 mg/kg. *Merck & Co.Inc.; Wallace Labs.*

7552 Penicillamine Hydrochloride
2219-30-9 7214 218-727-9
$C_5H_{12}ClNO_2S$
3-Mercaptovaline hydrochloride.
Distamine; Metalcaptase. Antirheumatic agent. Also used as a copper chelator in treatment of Wilson's disease. mp = 177.5° (dec); $[\alpha]_D^{25}$ = -63° (1N NaOH); soluble in H_2O, EtOH; LD_{50} (mus iv) = 2289 mg/kg.

7553 DL-Penicillamine
52-66-4 7214 200-147-2
$C_5H_{11}NO_2S$
DL-3-Mercaptovaline.
Antirheumatic agent. Also used as a copper chelator in treatment of Wilson's disease. mp = 201° (dec); LD_{50} (rat orl) = 365 mg/kg.

7554 DL-Penicillamine Hydrochloride
22572-05-0 7214 245-093-0
$C_5H_{12}ClNO_2S$
DL-3-Mercaptovaline hydrochloride.
Antirheumatic agent. Also used as a copper chelator in treatment of Wilson's disease. mp = 145-148°.

7555 L-Penicillamine
1113-41-3 7214 214-203-9
$C_5H_{11}NO_2S$
L-3-Mercaptovaline.
Antirheumatic agent. Also used as a copper chelator in treatment of Wilson's disease. mp = 190-194°; $[\alpha]_D^{25}$ = 63° (1N NaOH); LD_{50} (rat ip) = 350 mg/kg.

7556 Pentetate Calcium Trisodium
12111-24-9 7265 235-169-1
$C_{14}H_{18}CaN_3Na_3O_{10}$
Trisodium [N,N-bis[2-[bis(carboxymethyl)amino]ethyl]-glycinato(5-)]calciate(3-).
pentacin; Calcium Chel 330; Ditripentat; Penthamil. Soluble in H_2O, insoluble in EtOH; LD_{50} (rat ip) = 3800 mg/kg.

7557 Pentetic Acid
67-43-6 7266 200-652-8
$C_{14}H_{23}N_3O_{10}$
diethylenetriaminepentaacetic acid.
diethylenetriamine pentaacetic acid; DTPA. *Carbide & Carbon Chem.*

7558 Succimer
304-55-2 9034 206-155-2
$C_4H_6O_4S_2$
meso-2,3-Dimercaptosuccinic acid.
DMS; DMSA; DIM-SA; Ro-1-7977; Chemet. Chelating agent used as an antidote in cases of heavy metal poisoning. Diagnostic aid (the m99 isotope as a radioactive imaging agent). mp = 192-194°; LD_{50} (mus ip) > 3000 mg/kg. *Hoffmann-LaRoche Inc.*

7559 Trientine
112-24-3 9796 203-950-6
$C_6H_{18}N_4$
Triethylenetetramine.
trien; TETA; TECZA. Chelating agent used in treatment of Wilson's disease. mp = 12°, bp = 266-267°; d^{15} = 0.9817; soluble in H_2O, EtOH; LD_{50} (rat orl) = 2500 mg/kg. *Merck & Co.Inc.*

7560 Trientine Dihydrochloride
38260-01-4 9796 253-854-3
$C_6H_{20}Cl_2N_4$
Triethylenetetramine dihydrochloride.
Cuprid; Syprine. Chelating agent used in treatment of Wilson's disease. mp = 115-118°. *Merck & Co.Inc.*

7561 Zinc Acetate, Basic
82279-57-0
$C_{12}H_{18}O_{13}Zn_4$
Hexakis(μ-acetato)-μ4-oxotetrazinc.
Used for the treatment and prophylaxis of hepatic copper toxicosis; for maintenance therapy in patients initially treated with a chelating agent. Formerly used as an emetic.

Cholelitholytics

7562 Chenodiol
474-25-9 2096 207-481-8
$C_{26}4H_{40}O_4$
3α,7α-Dihydroxy-5β-cholan-24-oic acid.
Chenix; CDC; Chendol; Chenocedon; Chenocol; Chenodex; Chenofalk; Chenossil; Chenosäure; Cholanorm; Fluibil; Hekbilin; Kebilis; Ulmenide. Anticholelithogenic. mp = 119°; $[\alpha]_D^{20}$ = 11.5° (dioxane); soluble in MeOH, EtOH; Me_2CO, AcOH, Et_2O, EtOAc; insoluble in H_2O, C_6H_6, petroleum ether; [diformate ($C_{25}H_{40}O_6$)]: mp = 172°; [methyl ester ($C_{25}H_{42}O_4$)]: mp = 90-91°; $[\alpha]_D^{25}$ = 20°. *C.M. Ind.; Solvay Pharm. Inc.*

7563 Cicloxilic Acid
57808-63-6
$C_{13}H_{16}O_3$
cis-2-Hydroxy-2-phenylcyclopentaneacetate.
Affects bile flow and lipid composition.

7564 Methyl tert-Butyl Ether
1634-04-4 6111 216-653-1
$C_5H_{12}O$
2-Methoxy-2-methylpropane.
MTBE. Anticholelithogenic. Used therapeutically to dissolve cholesterol calculi. mp = -109°; bp = 55.2°; d_4^{20} = 0.7404; soluble in H_2O (4.8 g/100 ml); LC_{50} (mus 15 min) = 140.8 mg/l of atmosphere. *Res. Corp.*

7565 Monoctanoin
502-54-5 6335
$C_{11}H_{22}O_4$
Glycerol 1-octanoate.
octanoic acid 2,3-dihydroxypropyl ester; caprylic acid α-monoglyceride; α-monocaprylin. Anticholelithogenic. mp = 39.5-40.5°. *Stokely-Van Camp.*

7566 Ursodiol
128-13-2 10026 204-879-3
$C_{24}H_{40}O_4$
3α,7β-Dihydroxy-5β-cholan-24-oic acid.
ursodeoxycholic acid; Actigall; Arsacol; Cholit-Ursan; Delursan; Desol; Destolit; Deursil; Litursol; Lyeton; Paptarom; Solutrat; Urdes; Ursacol; Urso; Ursobilin; Ursochol; Ursodamor; Ursofalk; Ursolvan. Anticholelithogenic. mp = 203°; $[\alpha]_D^{20}$ = 57° (c = 2 EtOH); freely soluble in EtOH, AcOH; soluble in $CHCl_3$, Et_2O; insoluble in H_2O; LD_{50} (mus iv) = 100 mg/kg, 260 mg/kg, (mus sc = 6000 mg/kg, (mus ip) = 1200 mg/kg, (rat sc) = 2000 mg/kg, (rat ip) = 1000 mg/kg, (rat iv) = 310 mg/kg; [Diformate ($C_{26}H_{40}O_6$)]: mp = 170°; [diacetate ($C_{28}H_{44}O_6$)]: mp = 98-102°. *C.M. Ind.; Ciba-Geigy Corp.*

7567 Ursulcholic Acid
88426-32-8
$C_{24}H_{40}O_{10}S_2$
3α,7β-Dihydroxy-5β-cholan-24-oic acid bis(hydrogen sulfate).
Anticholelithogenic.

Choleretics

7568 Alibendol
26750-81-2 243 247-960-9
$C_{13}H_{17}NO_4$
5-Allyl-N-(2-hydroxyethyl)-3-methoxysalicylamide.
EB-1856; FC-54; H-3774; Cebera. Choleretic; antispasmodic. mp = 95°; λ_m = 316, 218 nm (EtOH); LD_{50} (mmus orl) > 3000 mg/kg, (mmus sc) > 2000 mg/kg, (mmus iv) = 217 mg/kg. *Roussel-UCLAF.*

7569 Anethole Trithione
532-11-6 683 208-528-5
$C_{10}H_7OS_2$
5-(p-Methoxyphenyl)-3H-1,2-dithiole-3-thione.
trithioanethole; Heporal; Mucinol; Trithio; Sulfralem; Tiotrifar; Felviten; Sulfogal; Sulfarlem. Choleretic. mp = 111°; insoluble in H_2O; soluble in C_5H_5N; $CHCl_3$, C_6H_6, dioxane; CS_2; slightly soluble in Et_2O, Me_2CO, EtOAc, AcOH, EtOH, petroleum ether; [oxime ($C_{10}H_9NO_2S_2$)]: mp = 170°; soluble in dioxane; [methiodide]: mp = 189°.

7570 Azintamide
1830-32-6 945 217-384-2
$C_{10}H_{14}ClN_3OS$
2-[(6-Chloro-3-pyridazinyl)thio]-N,N-diethylacetamide.
ST-9067; Oragallin. Choleretic. mp = 97-98°; freely soluble in EtOAc, $CHCl_3$, C_6H_6, Me_2CO; soluble in H_2O (0.5 g/100 ml); λ_m = 316, 306, 258 nm (ε 14040, 14920, 14210 EtOH); LD_{50} (mus orl) = 2340 mg/kg, (rat orl) = 1550 mg/kg. *Lentia.*

7571 Cholic Acid
81-25-4 2258 201-337-8
$C_{24}H_{40}O_5$
3α,7α,12α-Trihydroxy-5β-cholan-24-oic acid.
cholalic acid; Colalin. Choleretic. mp = 198°; $[\alpha]_D^{20}$ = 37° (c = 0.6 EtOH); soluble in H_2O (0.028 g/100 ml), EtOH (3.1 g/1oo ml), Et_2O (0.12 g/100 ml), $CHCl_3$ (0.51 g/100 ml), C_6H_6 (0.03 g/100 ml), Me_2CO (2.8 g/100 ml), AcOH (15.2 g/100 ml); [methyl ester ($C_{25}H_{42}O_5$)]: mp = 155-156°; [ethyl ester ($C_{26}H_{44}O_5$)]: mp = 162-163°; [sodium salt ($C_{24}H_{39}NaO_5$)]: soluble in H_2O (> 56/9 g/100 ml).

7572 Cicrotoic Acid
25229-42-9 2327 246-739-4
$C_{10}H_{16}O_2$
β-Methylcyclohexaneacrylic acid.
AD-106; Accroibile. Choleretic. mp = 85-86°; λ_m 219 nm (ε 12000); LD_{50} (mus orl) = 1925 mg/kg, (rat orl) = 2900 mg/kg. *Soc. Chim. Org. Biol.*

7573 Clanobutin
30544-61-7 2398 250-232-3
$C_{18}H_{18}ClNO_4$
4-[p-Chloro-N-(p-methoxyphenyl)benzamido]-butyric acid.
Bykahepar. Choleretic. mp = 115-116°; soluble in H_2O (1.39 g/100 ml at pH 7); LD_{50} (rat orl) > 2000 mg/kg, (rat iv) = 570 mg/kg. *Byk Gulden Lomberg GmbH.*

7574 Cyclobutyrol
512-16-3 2784 208-138-5
$C_{10}H_{18}O_3$
α-Ethyl-1-hydroxycyclohexaneacetic acid.
Choleretic. mp = 81-82°; bp_{24} = 164°, bp_{16} = 167-170°; slightly soluble in H_2O, petroleum ether; very soluble in EtOH, Me_2CO, dioxane, $CHCl_3$, Et_2O. *Lab. Jacques Logeais.*

7575 Cyclobutyrol Sodium Salt
1130-23-0 2784 214-458-6
$C_{10}H_{17}NaO_3$
Sodium α-ethyl-1-hydroxycyclohexaneacetate.
Bilimix; Bis-Bil; Colepan; Dimene; Epa-Bon; Hebucol; Maricolene; Yachicol; Tri-Bil; Tribilina. Choleretic. mp = 299-300°. *Lab. Jacques Logeais.*

7576 Cyclovalone
579-23-7 2823 209-438-9
$C_{22}H_{22}O_5$
2,6-Divanillylidenecyclohexanone.
Beveno; Divanil; DVC. Choleretic. mp = 178-179°; soluble in H_2O, EtOH. *Chem.-Pharm. Fabrik.*

7577 Cynarine
30964-13-7 2835
$C_{25}H_{24}O_{12}$
1-Carboxy-4,5-dihydroxy-1,3-cyclohexylene-3,4-dihydroxycinnamate.

Cynarin; Cinarine; Lisrocol; Plemocil. Choleretic. Active principle of artichoke. mp = 225-227°; $[\alpha]_D^{25}$ = -59° (c = 2 MeOH); λ_m 326 nm ($E_{1\ cm}^{1\%}$ 616 MeOH); soluble in H_2O, AcOH, EtOH, MeOH. *Farmitalia Carlo Erba Ltd.*

7578 Dehydrocholic Acid

81-23-2 2922 201-335-7

$C_{24}H_{34}O_5$

3,7,12-Trioxo-24-cholanic acid.

Acolen; Bilidren; Bilostat; Cholan-DH; Cholagon; Cholepatin; Chologon; Decholin; Dehychol; Deidrocolico Vita; Didrocolo; Erebile; Felacrinos; Procholon. Choleretic. mp = 237°; $[\alpha]_D^{20}$ = 26° (c = 1.4 EtOH); soluble in H_2O (0.02 g/100 ml), EtOH (0.33 g/100 ml), Et_2O (0.05 g/100 ml); $CHCl_3$ (0.94 g/100 ml), C_6H_6 (0.1 g/100 ml), Me_2CO (0.78 g/100 ml), EtOAc (0.74 g/100 ml), AcOH (0.74 g/100 ml). *Merck & Co.Inc.*

7579 Dehydrocholic Acid Sodium Salt

145-41-5 2922 205-652-1

$C_{24}H_{33}NaO_5$

Sodium 3,7,12-trihydroxy-5β-cholan-24-oate.

Carachol; Dycholium; Suprachol. Choleretic. Very soluble in H_2O. *Merck & Co.Inc.*

7580 Deoxycholic Acid

83-44-3 2946 201-478-5

$C_{24}H_{40}O_4$

3α,12α-Dihydroxy-5β-cholan-24-oic acid.

desoxycholic acid. Choleretic. mp = 176-178°; $[\alpha]_D^{20}$ = 55° (EtOH); soluble in H_2O (0.0-24 g/100 ml at 15°), EtOH (22.1 g/100 ml), Et_2O (0.12 g/100 ml), $CHCl_3$ (0.29 g/100 ml), C_6H_6 (0.012 g/100 ml), Me_2CO (1.05 g/100 ml), AcOH (0.91 g/100 ml); [sodium salt ($C_{24}H_{39}NaO_4$, sodium deoxycholate)]: soluble in H_2O (> 33.3 g/100 ml at 15°). *Armour Pharm. Co. Ltd.*

7581 Dimecrotic Acid

7706-67-4 3248

$C_{12}H_{14}O_4$

2,4-Dimethoxy-β-methylcinnamic acid.

Choleretic. mp = 149°. *Unicler.*

7582 Dimecrotic Acid Magnesium Salt

54283-65-7 3248

$C_{24}H_{26}MgO_8$

Magnesium 2,4-dimethoxy-β-methylcinnamate.

Visobil; Hepadial. Choleretic. mp = 135°; LD_{50} (mus ip) = 1.3 mg/kg, (rat ip) = 1.0 mg/kg. *Unicler.*

7583 α-Ethylbenzyl Alcohol

93-54-9 3816 202-256-0

$C_9H_{12}O$

α-Ethylbenzenemethanol.

SH-261; Ejibil; Livonal; Phenycholon; Phenicol; Phenychol; Felicur; Felitrope. Choleretic. bp = 219°, bp15:Lks = 107°, bp_3 = 78°; d_4^{25} = 0.9915; λ_m = 250, 260 nm (ε 173, 114); soluble in MeOH, EtOH, Et_2O, C_6H_6, C_7H_8; LD_{50} (rat orl) = 1580 mg/kg.

7584 Exiproben

26281-69-6 3959

$C_{16}H_{24}O_5$

o-[3-(Hexyloxy)-2-hydroxypropoxy]-benzoic acid.

Choleretic. *Boehringer Ingelheim GmbH.*

7585 Exiproben Sodium salt

3478-44-2 3959 222-454-0

$C_{16}H_{23}NaO_5$

Sodium o-[3-(hexyloxy)-2-hydroxypropoxy]-benzoate.

DCH-21; Etopalin. Choleretic. mp = 147°. *Boehringer Ingelheim GmbH.*

7586 Febuprol

3102-00-9 3985 221-454-8

$C_{13}H_{20}O_3$

1-Butoxy-3-phenoxy-2-propanol.

H-33; K-10033; Valbil; Valbilan. Choleretic. bp_{11} = 165°, $bp_{1.0}$ = 125-132°; d_4^{20} = 1.027; LD_{50} (mus orl) = 3050 mg/kg, (mus ip) = 436 mg/kg, (rat orl) = 2370 mg/kg, (rat ip) = 400 mg/kg. *Klinge Pharma GmbH.*

7587 Fencibutirol

5977-10-6 4009 227-773-9

$C_{16}H_{22}O_3$

(±)-α-Ethyl-1-hydroxy-4-phenylcyclohexane-acetic acid.

α-(1-hydroxy-4-phenylcyclohexyl)butyric acid; Verecolene; MG-4833; Biligen; Hepasil. Choleretic. mp = 157°. *Maggioni Farmaceutici S.p.A.*

7588 Fenipentol

583-03-9 4016 209-493-9

$C_{11}H_{16}O$

α-Butylbenzyl alcohol.

α-butylbenzenemethanol; phenylbutylcarbinol; phenylpentanol; PC 1; Ph BC; Pancoral. Choleretic. bp_{12} = 123-124°; insoluble in H_2O, soluble in organic solvents; LD_{50} (mus ip) = 1030 mg/kg, (mus orl) = 3100 mg/kg. *Thomae GmbH Dr. Karl.*

7589 Florantyrone

519-95-9 4143 208-279-2

$C_{20}H_{14}O_3$

σ-Oxo-8-fluoranthenebutyric acid.

Kanchol. Choleretic. mp = 208°; soluble in MeOH, EtOH, $NaHCO_3$ solution. *Miles Inc.; Searle G.D. & Co.*

7590 Hymecromone

90-33-5 4903 201-986-7

$C_{10}H_8O_3$

7-Hydroxy-4-methyl-2H-1-benzo-pyran-2-one.

4-methylumbelliferone; imecromone; 4-MU; Bilcolic; Biliton H; Cantabilin; Cantabiline; Cholonerton; Cholspasmin; Cumarote C; Himecol; Medilla; Mendiaxon. Antispasmodic; choleretic. mp = 185-186°, 194-195°; λ_m = 221, 251, 322.5 nm (MeOH); soluble in MeOH, AcOH; slightly soluble in Et_2O, $CHCl_3$, insoluble in cold H_2O. *Bayer Corp.*

7591 Hymecromone O,O-Diethylphosphorothioate
299-45-6 4903
$C_{14}H_{17}O_5PS$
7-Hydroxy-4-methyl-2H-1-benzopyran-2-one O,O-diethyl phosphorothioate.
E-838; Potasan. Antispasmodic; choleretic. Acts as a cholinesterase inhibitor (cf. Parathion) and is used as an insecticide, especially effective against the Colorado beetle. mp = 38°; $bp_{1.0}$ = 210° (dec);d_4^{38}= 1.260; sparingly soluble in H_2O, slightly soluble in petroleum ether, soluble in most other organic solvents; LD_{50} (rat ip) = 15 mg/kg. *Bayer Corp.*

7592 Menbutone
3562-99-0 5878 222-631-2
$C_{15}H_{14}O_4$
3-(4-Methoxy-1-naphthoyl)propionic acid.
Ictéryl. Choleretic. mp = 172-173°. *Searle G.D. & Co.*

7593 Menbutone Magnesium Salt
16643-66-6 5878 240-690-2
$C_{30}H_{26}MgO_8$
Magnesium 3-(4-methoxy-1-naphthoyl)propionate.
Hepalande. Choleretic. *Searle G.D. & Co.*

7594 3-(o-Methoxyphenyl)-2-phenylacrylic Acid
25333-25-9 6078 246-856-0
$C_{16}H_{14}O_3$
α-[(2-Methoxyphenyl)methylene]benzeneacetic acid.
Choleretic. [trans-form]: mp = 184°; [cis-form]: mp = 131.5°.

7595 3-(o-Methoxyphenyl)-2-phenylacrylic Acid Magnesium Salt
14222-61-8 6078
$C_{32}H_{26}MgO_6$
Magnesium α-[(2-methoxyphenyl)methylene]-benzeneacetate.
AN-1022; Bilcrine. Choleretic.

7596 Metochalcone
18493-30-6 6225 242-377-6
$C_{18}H_{18}O_4$
1-(2,4-Dimethoxyphenyl)-3-(4-methoxyphenyl)-2-propen-1-one.
2',4,4'-trimethoxychalcone; CB-1314; Lesidrin; Vesidril; Vesidryl. Choleretic; diuretic. mp = 97°.

7597 Moquizone
19395-58-5 6347 243-021-2
$C_{20}H_{21}N_3O_3$
2,3-Dihydro-1-(morpholinoacetyl)-3-phenyl-4(1H)-quinazolinone.
Rec-14-0127. Choleretic. mp = 135-137°, 128-130°. *Seceph.*

7598 Moquizone Hydrochloride
19395-78-9 6347 243-022-8
$C_{20}H_{22}ClN_3O_3$
2,3-Dihydro-1-(morpholinoacetyl)-3-phenyl-4(1H)-quinazolinone hydrochloride.
Peristil. Choleretic. mp = 210-214°; soluble in H_2O (12.2 g/100 ml at 25°); LD_{50} mus orl) = 1155 mg/kg, (mus iv) = 237 mg/kg, (rat orl) = 2135 mg/kg, (rat iv) = 146 mg/kg. *Seceph.*

7599 Osalmid
526-18-1 7018 208-385-9
$C_{13}H_{11}NO_3$
4'-Hydroxysalicylanilide.
oksafenamide; oxaphenamide; Driol; Jestmin; Kanochol; Saryuurin; Yoshicol. Choleretic. mp = 179°; insoluble in cold H_2O, AcOH; slightly soluble in warm H_2O, C_6H_6, C_7H_8; freely soluble in MeOH, EtOH, Me_2CO, Et_2O; [diacetate ($C_{17}H_{15}NO_5$)]: mp = 151°.

7600 Ox Bile Extract
7062
Sodium choleate.
Purified oxgall; oxgall; Bi-Ketolan; Bilein; Bilicholan; Cholatol; Crescefel; Desicol; Doxychol; Glycotauro; Panoxolin; Plebilin. Choleretic. Very soluble in H_2O, EtOH.

7601 4,4'-Oxydi-2-butanol
821-33-0 7094 212-475-3
$C_8H_{18}O_3$
3,3'-Dihydroxydibutyl ether.
DHBE; Diskin; Dyskinébyl; Dis-Cinil; Colenormol. Choleretic; antispasmodic. Soluble in H_2O.

7602 Piprozolin
17243-64-0 7639 241-280-6
$C_{14}H_{22}N_2O_3S$
Ethyl 3-ethyl-4-oxo-5-piperidino-$\Delta^{2,\alpha}$-thiazolidineacetate.
W-3699; Gö-919; Coleflux; Epsyl; Probilin; Secrebil. Choleretic. mp = 86-87°; insoluble in H_2O, soluble in dilute acids and most organic solvents; λ_m = 245, 285 nm (ε 8200 20000 MeOH); LD_{50} (mus orl) = 1070 mg/kg, (rat orl) = 3256 mg/kg. *Warner-Lambert.*

7603 4-Salicoylmorpholine
3202-84-4 8485
$C_{11}H_{13}NO_3$
4-(2-Hydroxybenzoyl)morpholine.
L-1102; Tardisal. Choleretic. mp = 175°; soluble in H_2O (0.41 g/100 ml), EtOH (3.3 g/100 ml), Et_2O (0.22 g/100 ml). *Dow Chem. U.S.A.*

7604 Sincalide
25126-32-3 8689 246-639-0
$C_{49}H_{62}N_{10}O_{16}S_3$
L-Aspartyl-L-tyrosyl-L-methionylglycyl-L-tryptophyl-L-methionyl-L-aspartylphenyl-L-alaninamide hydrogen sulfate (ester).
Asp-Tyr(SO_3H)-Met-Gly-Trp-Met-Asp-Phe-NH_2; Kinevac; SQ-19844. Choleretic. $[\alpha]_D^{23}$ = -18.4° (c = 0.7 in 1N NH_4OH); λ_m = 280 288 nm (ε 4850 4230 0.1N NaOH). *Bristol-Myers Squibb Co; Squibb E.R. & Sons.*

7605 Taurocholic Acid
81-24-3 9242 201-336-2
$C_{26}H_{45}NO_7S$
2-[[(3α,7α,12α-Trihydroxy-24-oxo-5β-cholan-24-yl]-amino]ethanesulfonic acid.

cholaic acid; cholyltaurine; N-choloyltaurine. Choleretic. dec 125°; $[\alpha]_D^{18}$ = 38.8° (c = 2 EtOH); freely soluble in H_2O; soluble in EtOH; insoluble in Et_2O, EtOAc; LD_{50} (rat orl) = 380 mg/kg; [sodium salt ($C_{26}H_{44}NNaO_7S$)]: dec 230°; $[\alpha]_D^{20}$ = 24° (c = 3); freely soluble in H_2O, EtOH; [barium salt ($C_{52}H_{88}BaN_2O_{14}S_2$)]: dec 225-227°; $[\alpha]_D^{20}$ = 25.6°.

7606 Tocamphyl
5634-42-4 9630 227-081-7
$C_{23}H_{37}NO_6$
1-(p,α-Dimethylbenzyl)camphorate compound with 2,2'-iminodiethanol (1:1).
Biliphorine; Hepatoxane; Syncuma. Choleretic. Soluble in H_2O. *Chemiewerk Homburg.*

7607 Trepibutone
41826-92-0 9718
$C_{16}H_{22}O_6$
3-(2,4,5-Triethoxybenzoyl)propionic acid.
AA-149; Supacal. Choleretic; antispasmodic. mp = 150-151°; LD_{50} (mmus orl) = 1340 mg/kg, (mmus ip) = 530 mg/kg, (mrat orl) = 2450 mg/kg, (mrat ip) = 410 mg/kg. *Takeda Chem. Ind. Ltd.*

7608 Vanitiolide
17692-71-6 10071 241-690-5
$C_{12}H_{15}NO_3S$
4-(Thiovanilloyl)morpholine.
Bildux. Choleretic.

Cholinergics

7609 Aceclidine
6109-70-2 8266 228-071-5
$C_9H_{15}NO_2$
dl-1-Azabicyclo[2.2.2]octan-3-ol acetate (ester).
dl-3-quinuclidinol acetate; 3-acetoxyquinuclidine; 3-quinuclidinyl acetate. Cholinergic agent. $bp_{0.4}$ = 73-74°, bp_{11} = 113-115°. *Hoffmann-LaRoche Inc.*

7610 Acetylcholine Bromide
66-23-9 87 200-622-4
$C_7H_{16}BrNO_2$
2-Acetyloxy-N,N,N-trimethylethanaminium bromide.
Pragmoline; Tonocholin B. Cholinergic agent. Also has miotic properties. Very soluble in cold H_2O, decomposed by hot H_2O, soluble in EtOH, insoluble in Et_2O.

7611 Acetylcholine Chloride
60-31-1 88 200-468-8
$C_7H_{16}ClNO_2$
2-Acetyloxy-N,N,N-trimethylethanaminium chloride.
Acecoline; Arterocoline; Miochol; Ovisot. Cholinergic; cardiac depressant; miotic; vasodilator (peripheral). mp = 149-152°; very soluble in cold H_2O, decomposed by hot H_2O, soluble in EtOH, insoluble in Et_2O; LD_{50} (rat sc) = 250 mg/kg. *Iolab.*

7612 Aclatonium Napadisilate
55077-30-0 116
$C_{30}H_{46}N_2O_{14}S_2$
2-[2-(Acetyloxy)-1-oxopropoxyl]-N,N,N-trimethylethanaminium 1,5-naphthalenedisulfonate (2:1).
TM-723; Abovis. Cholinergic agent. mp = 189-191°; LD_{50} (mus orl) = 15000 mg/kg; (mus sc) = 820 mg/lg, (dog orl) > 10000 mg/kg. *Toyama Chem. Co. Ltd.*

7613 Benzpyrinium Bromide
587-46-2 1153
$C_{15}H_{17}BrN_2O_2$
1-Benzyl-3-hydroxypyridinium bromide dimethylcarbamate.
benzstigminum bromidum; Stigmenene bromide; Stigmonene bromide. Cholinergic agent. mp = 114-115°; freely soluble in H_2O, EtOH; insoluble in $Et_2O\lambda_m$ = 269 nm ($E_{1\,cm}^{1\%}$ = 136). *Warner-Lambert.*

7614 Bethanechol Chloride
590-63-6 1232 209-686-8
$C_7H_{17}ClN_2O_2$
(2-Hydroxypropyl)trimethylammonium chloride carbamate.
Myotonachol; Urecholine; Mechothane; Myocholine; Mictone; Myotonine chloride; Uro-Carb. Cholinergic agent. Used to stimulate contraction of urinary bladder and gastrointestinal tract. mp = 218-219° (dec); soluble in H_2O (166.6 g/100 ml), EtOH (8.0 g/100 ml). *Glenwood Inc.; Merck & Co.Inc.*

7615 Carbachol
51-83-2 1823 200-127-3
$C_6H_{15}ClN_2O_2$
2-[(Aminocarbonyl)oxy]-N,N,N-trimethylethanaminium chloride.
choline chloride carbamate; Carbastat Intraocular; Isopto Carbachol; Miostat; Carcholin; Moryl; Doryl; Coletyl; Lentin. Cholinergic agent. Also has miotic properties. mp = 200-203°, 204-205°, 210-212°; soluble in H_2O (100 g/100 ml), MeOH (10 g/100 ml); insoluble in $CHCl_3$, Et_2O; LD_{50} (mus orl) = 15 mg/kg, (mus iv) = 0.3 mg/kg. *Alcon Labs; CIBA Vision AG.*

7616 Carpronium Chloride
13254-33-6 1913 236-243-6
$C_8H_{18}ClNO_2$
(3-Carboxypropyl)trimethylammonium chloride methyl ester.
Actinomin; Furozin. Cholinergic agent. mp = 126°, very soluble in H_2O, MeOH, EtOH; soluble in Me_2CO; insoluble in Et_2O.

7617 Demecarium Bromide
56-94-0 2936 200-301-9
$C_{32}H_{52}Br_2N_4O_4$
(m-Hydroxyphenyl)trimethylammonium bromide decamethylenebis[methylcarbamate] (2:1).
Humorsol; BC-48; Tosmilen. Cholinergic agent (ophthalmic). Cholinesterase inhibitor. mp = 162-167° (dec); freely soluble in H_2O, EtOH; less soluble in Me_2CO; insoluble in Et_2O. *Merck & Co.Inc.*

7618 Dexpanthenol
81-13-0 2988 201-327-3
$C_9H_{19}NO_4$
D-(+)-2,4-Dihydroxy-N-(3-hydroxypropyl)-3,3-dimethylbutyramide.
D-Panthenol 50; Ilopan; Motilyn; pantothenol; pantothenyl alcohol; Alcopan-250; Intrapan; Pantenyl; Panthoderm; Bepanthen; Cozyme; Urupan; component of: Ilopan-Choline. Cholinergic agent. The dl-form is used as a vitamin. $bp_{0.02}$ = 118-120°; d_{20}^{20} = 1.2; $[\alpha]_D^{20}$ = 29.5° (c = 5); freely soluble in H_2O, EtOH, MeOH; slightly soluble in Et_2O. *Abbott Labs Inc.; BASF Corp.; Hoffmann-LaRoche Inc.; Savage Labs.*

7619 Diisopropyl Paraoxon
3254-66-8 3242
$C_{12}H_{18}NO_6P$
Bis(1-methylethyl) phosphoric acid 4-nitrophenyl ester.
Miopticol; Propicol. Cholinergic agent (ophthalmic). *Albright & Wilson Americas Inc.*

7620 Echothiophate Iodide
513-10-0 3549 208-152-1
$C_9H_{23}INO_3PS$
(2-Mercaptoethyl)trimethylammonium iodide S-ester with O,O-diethyl phosphorothioate.
Phospholine iodide; 217-MI; ecothiopate iodide; Diethoxyphosphinylthiocholine iodide. Cholinergic agent (ophthalmic). Cholinesterase inhibitor. Used in treatment of glaucoma. mp = 138°, 124-124.5°; soluble in H_2O, EtOH, $CHCl_3$. *Wyeth-Ayerst Labs.*

7621 Edrophonium Chloride
116-38-1 3562 204-138-4
$C_{10}H_{16}ClNO$
Ethyl(m-hydroxyphenyl)dimethylammonium chloride.
Antirex; Enlon; Reversol; Tensilon; edrophone bromide [as bromide]; Ro-2-3198 [as bromide]; component of: Enlon Plus. Antidote to curare alkaloids. Also used as a diagnostic aid in myasthenia gravis. Cholinesterase inhibitor. mp = 162-163° (dec); soluble in H_2O, EtOH, insoluble in Et_2O, $CHCl_3$; [bromide (edrophone bromide; $C_{10}H_{16}BrNO$)]: mp = 151-152°; soluble in H_2O (> 10 g/100 ml), EtOH, insoluble in Et_2O. *Anaquest; ICN Pharm. Inc.; Organon Inc.*

7622 Eptastigmine
101246-68-8 3672
$C_{21}H_{33}N_3O_2$
(3aS-cis)-1,2,3,3a,8,8a-Hexahydro-1,3a,8-trimethyl-pyrrolo[2,3-b]indol-5-yl heptylcarbamic acid ester.
N-demethyl-N-heptylphysostigmine; heptylstigmine. Cholinergic agent. mp = 60-64°, 58-62°; λ_m = 303, 253 nm (ε 3300 14200 MeOH). *Consiglio Nazionale delle Ricerche.*

7623 Eptastigmine Tartrate
121652-76-4 3672
$C_{25}H_{39}N_3O_8$
(3aS-cis)-1,2,3,3a,8,8a-Hexahydro-1,3a,8-trimethylpyrrolo[2,3-b]indol-5-yl heptylcarbamic acid ester tartrate (salt).
MF-201; N-Demethyl-N-heptylphysostigmine tartate. Cholinergic agent. Cholinesterase inhibitor. mp = 122-123°; soluble in H_2O (70 g/100 ml); LD_{50} (mus ip) = 35 mg/kg. *Consiglio Nazionale delle Ricerche.*

7624 Eseridine
25573-43-7 3740 247-111-2
$C_{15}H_{21}N_3O_3$
(4aS,9aS)-2,3,4,4a,9,9a-Hexahydro-2,4a,9-trimethyl-1,2-oxazino[6,5-b]indol-6-ylmethylcarbamate.
physostigmine aminoxide; eserine aminoxide; eserine oxide. Cholinergic agent. mp = 129°; $[\alpha]_D^{15}$ = -175° (EtOH); soluble in EtOH, $CHCl_3$, C_6H_6, Et_2O, petroleum ether; Me_2CO, dilute acids; insoluble in H_2O.

7625 Eseridine Salicylate
5995-96-0 3740 227-835-5
$C_{22}H_{27}N_3O_6$
(4aS,9aS)-2,3,4,4a,9,9a-Hexahydro-2,4a,9-trimethyl-1,2-oxazino[6,5-b]indol-6-ylmethylcarbamate salicylate.
Geneserine 3. Cholinergic agent.

7626 Furtrethonium
7618-86-2 4334
$[C_8H_{14}NO]^+$
N,N,N-Trimethyl-2-furanmethanaminium.
furfuryltrimethylammonium; furtrimethonium; Furmethide. Cholinergic agent. The iodide is used as a sympathomimetic. [benzenesulfonate]: mp = 134-135°; soluble in H_2O, EtOH. *SmithKline Beecham Pharm.*

7627 Furtrethonium Iodide
541-64-0 4334 208-789-5
$C_8H_{14}INO$
N,N,N-Trimethyl-2-furanmethanaminium iodide.
Furamon; Furanol. Cholinergic agent, also used as a sympathomimetic. mp = 116-117°, 118-120°; soluble in H_2O, EtOH; insoluble in C_6H_6. *SmithKline Beecham Pharm.*

7628 Isoflurophate
55-91-4 5192 200-247-6
$C_6H_{14}FO_3P$
Diisopropyl fluorophosphonate.
Floropryl; fluostigmine; isofluorphate; DFP; Diflupyl; Dyflos; Fluropryl. Cholinergic agent (ophthalmic). d = 1.055; mp = -82°; bp_5 = 46°, bp_9 = 62°, bp_{760} = 183° (calc.); soluble in H_2O (154 g/100 ml); LD_{50} (mus sc) = 3.71 mg/kg, (mus orl) = 36.8 mg/kg. *Merck & Co.Inc.*

7629 Methacholine Chloride
62-51-1 6003 200-537-2
$C_8H_{18}ClNO_2$
(2-Hydroxypropyl)trimethylammonium chloride acetate.
Provocholine; Amechol. Cholinergic agent. Also a parasympathomimetic bronchoconstrictor used as a diagnostic aid in cases of suspected bronchial asthma. mp = 172-173°; freely soluble in H_2O, EtOH, $CHCl_3$; insoluble in Et_2O. *Hoffmann-LaRoche Inc.*

7630 Muscarine
300-54-9 6389 206-094-1
$[C_9H_{20}NO_2]^+$
[2S-(2α,4β,5α)]-Tetrahydro-4-hydroxy-N,N,N,5-tetramethyl-2-furanmethanaminium.

Cholinergic agent. [chloride]: mp = 180-181°; $[\alpha]_D^{25}$ = 8.1° (c = 3.5 EtOH); soluble in H_2O, EtOH; less soluble in Et_2O, $CHCl_3$, Me_2CO; LD_{50} (mus iv) = 0.23 mg/kg.

7631 Neostigmine

59-99-4 6553

$[C_{12}H_{19}N_2O_2]^+$

3-[[(Dimethylamino)carbonyl]oxy]-N,N,N-trimethylbenzenaminium.

synstigmin; proserine. Cholinergic; miotic; antidote for curare poisoning. *Apothecon; Elkins-Sinn; Hoffmann-LaRoche Inc.; ICN Pharm. Inc.*

7632 Neostigmine Bromide

114-80-7 6553 204-054-8

$C_{12}H_{19}BrN_2O_2$

3-[[(Dimethylamino)carbonyl]oxy]-N,N,N-trimethylbenzenaminium bromide.

Juvastigmin (tabl.); Neoesserin; Neostigmin (tabl.); Normastigmin (tabl.); Prostigmin. Cholinergic agent. Also a miotic and used as an antidote for curare poisoning. mp = 167° (dec); soluble in H_2O (100 g/100 ml), EtOH. *Apothecon; Elkins-Sinn; Hoffmann-LaRoche Inc.; ICN Pharm. Inc.*

7633 Neostigmine Methylsulfate

51-60-5 6553 200-109-5

$C_{13}H_{22}N_2O_6S$

3-[[(Dimethylamino)carbonyl]oxy]-N,N,N-trimethylbenzenaminium methyl sulfate.

Intrastigmina; Juvastigmin (amp.); Metastigmin; Neostigmin (inj.); Normastigmin (amp.); Prostigmin (amp.); Stiglyn. Cholinergic; miotic; antidote for curare poisoning. Cholinesterase inhibitor. mp = 142-145°; soluble in H_2O (10 g/100 ml), less soluble in EtOH; LD_{50} (mus iv) = 0.16 mg/kg, (mus sc) = 0.42 mg/kg, (mus orl) = 7.5 mg/kg. *Apothecon; Elkins-Sinn; Hoffmann-LaRoche Inc.; ICN Pharm. Inc.*

7634 Nicotine

54-11-5 6611 200-193-3

$C_{10}H_{14}N_2$

(S)-3-(1-Methyl-2-pyrrolidinyl)pyridine.

Habitrol; Nicabate; Nicoderm; Nicolan; Nicopatch; Nicotell TTS; Nicotinell; Tabazur. Smoking cessation adjunct. Ganglionic stimulant which also stimulates the CNS. Also used as an insecticide and fumigant. Severe nicotine poisoning can cause nausea, salivation, abdominal pain, vomiting, diarrhea, cold sweat, dizziness, mental confusion, weakness, drop in blood pressure and pulse, convulsions, paralysis of respiratory muscles. From dried leaves of *Nicotiana tabacum and N. Rustica*. Oily liquid; bp_{745} = 247° (partial dec), bp_{17} = 123-125°; n_D^{20} = 1.5282; d_D^{20} = 1.0097; $[\alpha]_D^{20}$ = -169°; miscible with H_2O; soluble in alcohol, $CHCl_3$, Et_2O, petroleum ether, kerosene, oils.

7635 Oxapropanium Iodide

541-66-2 7056

$C_7H_{16}INO_2$

(1,3-Dioxolan-4-ylmethyl)trimethylammonium iodide.

vasodilatateur 2249-F; 2249-F; Dilvasene. Cholinergic agent. mp = 158-160°; soluble in H_2O, EtOH (40 g/100 ml) at 76°; sparingly soluble in cold EtOH, Et_2O, $CHCl_3$, C_6H_6; [free base]; bp_{21} = 68°. *Rhône-Poulenc Rorer Pharm. Inc.*

7636 Oxotremorine

70-22-4 7085 200-728-0

$C_{15}H_{23}NO_3$

1-[4-(1-Pyrrolidinyl)-2-butynyl]-2-pyrrolidinone.

Potent muscarinic agent. Produces symptoms similar to those of parkinsonism. Used as a research tool in the study of antiparkinsonian drugs. Cholinergic. Liquid; $bp_{0.6}$ = 124°; n_D^{25} = 1.5156.

7637 Physostigmine

57-47-6 7540 200-332-8

$C_{15}H_{21}N_3O_2$

(3aS-cis)-1,2,3,3a,8,8a-Hexahydro-1,3a,8-trimethylpyrrolo[2,3-b]indol-5-ol methylcarbamate (ester).

Cogmine; Eserine. Cholinergic agent (cholinesterase inhibitor). Also a miotic. Obtained from Calabar beans (*Physostigma venenosum*). mp = 105-106°; $[\alpha]_D^{17}$ = -76° (c = 1.3 $CHCl_3$), $[\alpha]_D^{25}$ = -120° (C_6H_6); soluble in EtOH, C_6H_6, $CHCl_3$; slightly soluble in H_2O; pKa_1 = 6.12, pka_2 = 12.24; may oxidize to form eserine blue ($C_{26}H_{31}N_5O_2$); LD_{50} (mus orl) = 4.5 mg/kg.

7638 Physostigmine Salicylate

57-64-7 7540 200-343-8

$C_{22}H_{27}N_3O_5$

(3aS-cis)-1,2,3,3a,8,8a-Hexahydro-1,3a,8-trimethylpyrrolo[2,3-b]indol-5-ol methylcarbamate (ester) salicylate.

Antilirium; Isopto Eserine. Cholinergic agent (cholinesterase inhibitor). Also used as a miotic. mp = 185-187°; λ_m = 239, 252, 303 nm (log ε 4.09, 4.04, 3.78 MeOH); soluble in H_2O (1.33 g/100 ml at 25°, 6.25 g/100 ml at 80°), EtOH (6.25 g/100 ml at 25°, 20 g/100 ml at 76°), $CHCl_3$ (16.6 g/100 ml), Et_2O (0.4 g/100 ml); LD_{50} (mus ip) = 0.64 mg/kg. *Alcon Labs; Forest Pharm. Inc.*

7639 Physostigmine Sulfate

64-47-1 7540 200-585-4

$C_{30}H_{44}N_6O_8S$

(3aS-cis)-1,2,3,3a,8,8a-Hexahydro-1,3a,8-trimethylpyrrolo[2,3-b]indol-5-ol methylcarbamate (ester) sulfate.

Eserine sulfate. Cholinergic agent (cholinesterase inhibitor). Also a miotic. mp = 140°; soluble in EtOH (250 g/100 ml), H_2O (25 g/100 ml), Et_2O (0.83 g/100 ml). *CIBA Vision AG.*

7640 Quilostigmine

139314-01-5

$C_{23}H_{27}N_3O_2$

(3aS,8aR)-1,2,3,3a,8,8a-Hexahydro-1,3a,8-trimethylpyrrolo[2,3-b]indol-5-yl 3,4-dihydro-2(1H)-isoquinoline-carboxylate.

HP-290; NXX-066. Cholinergic agent. *Hoechst Marion Roussel Inc.*

7641 Troxypyrrolium Tosilate
3612-98-4
$C_{25}H_{35}NO_8S$
1-Ethyl-1-(2-hydroxyethyl)pyrrolidinium p-toluenesulfonate 3,4,5-trimethoxybenzoate.
Choline uptake inhibitor.

7642 Xanomeline
131986-45-3 10190
$C_{14}H_{23}N_3OS$
3-[4-(Hexyloxy)-1,2,5-thiadiazol-3-yl]-1,2,5,6-tetrahydro-1-methylpyridine.
Cholinergic; nootropic; selective muscarinic M_2-receptor agonist. *Eli Lilly & Co.; Novo Nordisk Pharm. Inc.*

7643 Xanomeline Oxalate
141064-23-5 10190
$C_{16}H_{25}N_3O_5S$
3-[4-(Hexyloxy)-1,2,5-thiadiazol-3-yl]-1,2,5,6-tetrahydro-1-methylpyridine oxalate.
Cholinergic; nootropic; selective muscarinic M_2-receptor agonist. mp = 148°. *Eli Lilly & Co.; Novo Nordisk Pharm. Inc.*

7644 Xanomeline Tartrate
152854-19-8 10190
$C_{18}H_{29}N_3O_7S$
3-[4-(Hexyloxy)-1,2,5-thiadiazol-3-yl]-1,2,5,6-tetrahydro-1-methylpyridine tartrate.
LY-246708 tartrate; NNC-11-0232; Lomeron; Memcor. Cholinergic; nootropic. Selective muscarinic M_2-receptor agonist. mp = 95.5°. *Eli Lilly & Co.; Novo Nordisk Pharm. Inc.*

Cholinesterase Inhibitors

7645 Ambenonium Chloride
115-79-7 396 204-107-5
$C_{28}H_{42}Cl_4N_4O_2$
[Oxalylbis(iminoethylene)]bis[(o-chlorobenzene)-diethylammonium] dichloride.
Win-8077; Mysuran; Mytelase chloride. Cholinesterase inhibitor. mp = 196-199°; freely soluble in H_2O. *Sterling Winthrop Inc.*

7646 Demecarium Bromide
56-94-0 2936 200-301-9
$C_{32}H_{52}Br_2N_4O_4$
(m-Hydroxyphenyl)trimethylammonium bromide decamethylenebis[methylcarbamate] (2:1).
Humorsol; BC-48; Tosmilen. Cholinergic agent (ophthalmic). Cholinesterase inhibitor. mp = 162-167° (dec); freely soluble in H_2O, EtOH; less soluble in Me_2CO; insoluble in Et_2O. *Merck & Co.Inc.*

7647 Distigmine Bromide
15876-67-2 3426 240-013-0
$C_{22}H_{32}Br_2N_4O_4$
3-Hydroxy-1-methylpyridinium bromide hexamethylenebis-(N-methylcarbamate).
BC-51; Ubretid. Cholinesterase inhibitor. dec 149°. *Oesterreiche Stickstoffwerke.*

7648 Echothiophate Iodide
513-10-0 3549 208-152-1
$C_9H_{23}INO_3PS$
(2-Mercaptoethyl)trimethylammonium iodide S-ester with O,O-diethyl phosphorothioate.
Phospholine iodide; 217-MI; ecothiopate iodide; Diethoxyphosphinylthiocholine iodide. Cholinergic agent (ophthalmic). Cholinesterase inhibitor. Used in treatment of glaucoma. mp = 138°, 124-124.5°; soluble in H_2O, EtOH, $CHCl_3$. *Wyeth-Ayerst Labs.*

7649 Edrophonium Chloride
116-38-1 3562 204-138-4
$C_{10}H_{16}ClNO$
Ethyl(m-hydroxyphenyl)dimethylammonium chloride.
Antirex; Enlon; Reversol; Tensilon; edrophone bromide [as bromide]; Ro-2-3198 [as bromide]; component of: Enlon Plus. Antidote to curare alkaloids. Also used as a diagnostic aid in myasthenia gravis. Cholinesterase inhibitor. mp = 162-163° (dec); soluble in H_2O, EtOH, insoluble in Et_2O, $CHCl_3$; [bromide (edrophone bromide; $C_{10}H_{16}BrNO$)]: mp = 151-152°; soluble in H_2O (> 10 g/100 ml), EtOH, insoluble in Et_2O. *Anaquest; ICN Pharm. Inc.; Organon Inc.*

7650 Eptastigmine
101246-68-8 3672
$C_{21}H_{33}N_3O_2$
N-Demethyl-N-heptylphysostigmine.
heptylphysostigmine; heptylstigmine. Cholinesterase inhibitor. mp = 60-64°, 58-62°; λ_m = 303 253 nm (ε 3300 14200 MeOH). *Consiglio Nazionale delle Ricerche.*

7651 Eptastigmine Tartrate
121652-76-4 3672
$C_{25}H_{39}N_3O_8$
(3aS-cis)-1,2,3,3a,8,8a-Hexahydro-1,3a,8-trimethyl-pyrrolo[2,3-b]indol-5-yl heptylcarbamic acid ester tartrate (salt).
MF-201; N-Demethyl-N-heptylphysostigmine tartate. Cholinergic agent. Cholinesterase inhibitor. mp = 122-123°; soluble in H_2O (70 g/100 ml); LD_{50} (mus ip) = 35 mg/kg. *Consiglio Nazionale delle Ricerche.*

7652 Galanthamine
357-70-0 4357
$C_{17}H_{21}NO_3$
(4aS,6R,8aS)-4a,5,9,10,11,12-Hexahydro-3-methoxy-11-methyl-6H-benzofuro[3a,3,2-ef][2]benzazepin-6-ol.
galantamine; lycoremine; Jilkon. Cholinesterase inhibitor. mp = 126-127°; $[\alpha]_D^{20}$ = -118.8° (c = 1.378 EtOH); fairly soluble in hot H_2O; soluble in EtOH, Me_2CO, $CHCl_3$; insoluble in Et_2O, C_6H_6; [hydrochloride ($C_{17}H_{22}ClNO_3$)]: mp = 256-257° (dec); sparingly soluble in H_2O, EtOH, Me_2CO.

7653 Galanthamine Hydrobromide
1953-04-4 4357 217-780-5
$C_{17}H_{22}BrNO_3$
(4aS,6R,8aS)-4a,5,9,10,11,12-Hexahydro-3-methoxy-11-methyl-6H-benzofuro[3a,3,2-ef][2]benzazepin-6-ol hydrobromide.

Nivalin. Cholinesterase inhibitor. dec 246-247°; $[\alpha]_D^{20}$= -93.1° (c = 0.1015 H_2O); LD_{50} (mus iv) = 5.2 ± 0.2 mg/kg.

7654 Neostigmine
59-99-4 6553
$[C_{12}H_{19}N_2O_2]^+$
3-[[(Dimethylamino)carbonyl]oxy]-N,N,N-trimethylbenzenaminium.
synstigmin; proserine. Cholinergic; miotic; antidote for curare poisoning. *Apothecon; Elkins-Sinn; Hoffmann-LaRoche Inc.; ICN Pharm. Inc.*

7655 Neostigmine Bromide
114-80-7 6553 204-054-8
$C_{12}H_{19}BrN_2O_2$
3-[[(Dimethylamino)carbonyl]oxy]-N,N,N-trimethylbenzenaminium bromide.
Juvastigmin (tabl.); Neoesserin; Neostigmin (tabl.); Normastigmin (tabl.); Prostigmin. Cholinergic agent. Cholinesterase inhibitor. Also a miotic and used as an antidote for curare poisoning. mp = 167° (dec); soluble in H_2O (100 g/100 ml), EtOH. *Apothecon; Elkins-Sinn; Hoffmann-LaRoche Inc.; ICN Pharm. Inc.*

7656 Neostigmine Methylsulfate
51-60-5 6553 200-109-5
$C_{13}H_{22}N_2O_6S$
3-[[(Dimethylamino)carbonyl]oxy]-N,N,N-trimethylbenzenaminium methyl sulfate.
Intrastigmina; Juvastigmin (amp.); Metastigmin; Neostigmin (inj.); Normastigmin (amp.); Prostigmin (amp.); Stiglyn. Cholinergic; miotic; antidote for curare poisoning. Cholinesterase inhibitor. mp = 142-145°; soluble in H_2O (10 g/100 ml), less soluble in EtOH; LD_{50} (mus iv) = 0.16 mg/kg, (mus sc) = 0.42 mg/kg, (mus orl) = 7.5 mg/kg. *Apothecon; Elkins-Sinn; Hoffmann-LaRoche Inc.; ICN Pharm. Inc.*

7657 Physostigmine
57-47-6 7540 200-332-8
$C_{15}H_{21}N_3O_2$
(3aS-cis)-1,2,3,3a,8,8a-Hexahydro-1,3a,8-trimethylpyrrolo[2,3-b]indol-5-ol methylcarbamate (ester).
Cogmine; Eserine. Cholinergic agent (cholinesterase inhibitor). Also a miotic. Obtained from Calabar beans (*Physostigma venenosum*). mp = 105-106°; $[\alpha]_D^{17}$= -76° (c = 1.3 $CHCl_3$), $[\alpha]_D^{25}$= -120° (C_6H_6); soluble in EtOH, C_6H_6, $CHCl_3$; slightly soluble in H_2O; pKa_1 = 6.12, pka_2 = 12.24; may oxidize to form eserine blue ($C_{26}H_{31}N_5O_2$); LD_{50} (mus orl) = 4.5 mg/kg.

7658 Physostigmine Salicylate
57-64-7 7540 200-343-8
$C_{22}H_{27}N_3O_5$
(3aS-cis)-1,2,3,3a,8,8a-Hexahydro-1,3a,8-trimethylpyrrolo[2,3-b]indol-5-ol methylcarbamate (ester) salicylate.
Antilirium; Isopto Eserine. Cholinergic agent (cholinesterase inhibitor). Also a miotic. mp = 185-187°; λ_m = 239, 252, 303 nm (log ε 4.09, 4.04, 3.78 MeOH); soluble in H_2O (1.33 g/100 ml at 25°, 6.25 g/100 ml at 80°), EtOH (6.25 g/100 ml at 25°, 20 g/100 ml at 76°), $CHCl_3$ (16.6 g/100 ml), Et_2O (0.4 g/100 ml); LD_{50} (mus ip) = 0.64 mg/kg. *Alcon Labs; Forest Pharm. Inc.*

7659 Physostigmine Sulfate
64-47-1 7540 200-585-4
$C_{30}H_{44}N_6O_8S$
(3aS-cis)-1,2,3,3a,8,8a-Hexahydro-1,3a,8-trimethylpyrrolo[2,3-b]indol-5-ol methylcarbamate (ester) sulfate.
Eserine sulfate. Cholinergic agent (cholinesterase inhibitor). Also a miotic. mp = 140°; soluble in EtOH (250 g/100 ml), H_2O (25 g/100 ml), Et_2O (0.83 g/100 ml). *CIBA Vision AG.*

7660 Pyridostigmine
155-97-5 8161
$C_9H_{13}N_2O_2^+$
3-Hydroxy-1-methylpyridinium dimethylcarbamate.
Cholinesterase Inhibitor. *Hoffmann-LaRoche Inc.; ICN Pharm. Inc.; Organon Inc.*

7661 Pyridostigmine Bromide
101-26-8 8161 202-929-9
$C_9H_{13}BrN_2O_2$
3-Hydroxy-1-methylpyridinium bromide dimethylcarbamate.
Mestinon; Regonol; Kalymin; Ro-1-5130. Cholinesterase Inhibitor. Used in treatment of myasthenia gravis and as a pre-exposure antidote to chemical warfare agents. mp = 152-154°; soluble in H_2O, EtOH; insoluble in Et_2O, Me_2CO, C_6H_6. *Hoffmann-LaRoche Inc.; ICN Pharm. Inc.; Organon Inc.*

7662 Tacrine
321-64-2 9199 206-291-2
$C_{13}H_{14}N_2$
9-Amino-1,2,3,4-tetrahydroacridine.
Centrally active anticholinesterase. Nootropic. Respiratory stimulant. Cognition adjuvant and antidote for curare poisoning. mp = 183-184°. *Parke-Davis.*

7663 Tacrine Hydrochloride
1684-40-8 9199 216-867-5
$C_{13}H_{15}ClN_2$
9-Amino-1,2,3,4-tetrahydroacridine monohydrochloride.
Cognex; CI-970; THA. Nootropic. Respiratory stimulant. Centrally active anticholinesterase. Cognition adjuvant and antidote for curare poisoning. mp = 283-284°; soluble in H_2O. *Parke-Davis.*

Cholinesterase Reactivators

7664 Asoxime Chloride
34433-31-3 870
$C_{14}H_{16}Cl_2N_4O_3$
1-[[[4-(Aminocarbonyl)pyridino]methoxy]-methyl]-2-[(hydroxyimino)methyl]pyridinium dichloride.
HI-6. Cholinesterase reactivator. Used as an antidote for organophosphorus poisoning. [monohydrate]: mp = 145-147°; λ_m = 350, 300, 270, 218 nm (log ε 3.093, 4.018, 3.966, 4.181 H_2O). *E. Merck.*

7665 Diacetyl Monoxime
57-71-6
$C_4H_7NO_2$
2,3-Butanedione monoxime.
Cholinesterase reactivator. mp = 75-58°; bp = 185-186°.

7666 Obidoxime Chloride
114-90-9 6835 204-059-5
$C_{14}H_{16}Cl_2N_4O_3$
1,1'-[Oxybis(methylene)]-bis[4-(hydroxyimino)methyl]-pyridinium dichloride.
BH-6; LüH-6; Toksobidin; Toxogonin. Cholinesterase reactivator. [syn-isomer]: mp = 235-236° (dec); [anti-form]: mp = 218-220° (dec); [both forms]: freely soluble in H_2O; LD_{50} (mus orl) > 2240 mg/kg; [dibromide ($C_{14}H_{16}Br_2N_4O_3$)]: mp = 202-203° (dec). *E. Merck.*

7667 Pralidoxime Chloride
51-15-0 7884 200-080-9
$C_7H_9ClN_2O$
2-[(Hydroxyimino)methyl]-1-methylpyridinium chloride.
2-PAM chloride; Protopam chloride. Cholinesterase reactivator. Used as an antidote in cases of poisoning by nerve gas and cholinesterase inhibitory insecticides. mp = 235-238° (dec); insoluble in Me_2CO; soluble in iPrOH (0.09 g/100 ml), EtOH (0.89 g/100 ml), MeOH (8.5 g/100 ml), H_2O (65.5 g/100 ml); LD_{50} (rbt iv) = 95 mg/kg, (mus iv) = 115°; (mus ip) = 205 mg/kg, (mus orl) = 4100 mg/kg. *U.S. Government.*

7668 Pralidoxime Iodide
94-63-3 7884 202-349-6
$C_7H_9IN_2O$
2-[(Hydroxyimino)methyl]-1-methylpyridinium iodide.
2-PAM. Cholinesterase reactivator. Used as an antidote in cases of poisoning by nerve gas and cholinesterase inhibitory insecticides. mp = 225-226°; very soluble in H_2O; soluble in EtOH (4.8 g/100 ml at 25°); insoluble in Et_2O, Me_2CO; LD_{50} (mus iv) = 140-178 mg/kg, (mus ip) = 136-260 mg/kg, (mus sc) = 290-340 mg/kg, (mus orl) = 1500-4000 mg/kg. *U.S. Government.*

7669 Pralidoxime Mesylate
154-97-2 7884 205-839-8
$C_8H_{12}N_2O_4S$
2-[(Hydroxyimino)methyl]-1-methylpyridinium mesylate.
Contrathion. Cholinesterase reactivator. Used as an antidote in cases of poisoning by nerve gas and cholinesterase inhibitory insecticides. mp = 155°; soluble in H_2O (50 g/100 ml); LD_{50} (mus iv) = 118-122°; (mus ip) = 216 mg/kg, (mus orl) = 3700 mg/kg, (rat iv) = 109 mg/kg, (rat ip) = 262 mg/kg. *U.S. Government.*

7670 Trimedoxime Bromide
56-97-3 200-304-5
$C_{18}H_{18}Br_2N_4O_3$
1,1'-(1,3-Propanediyl)bis[4-[(hydroxyimino)methyl]pyridinium dibromide.
TMB-4; C-434. Cholinesterase reactivator. Analog of obidoxime.

CNS Stimulants

7671 Amfonelic Acid
15180-02-6
$C_{18}H_{16}N_2O_3$
7-Benzyl-1-ethyl-4-oxo-1,8-naphthyridine-3-carboxylic acid.
Win-25978; NSC-100638. CNS stimulant. *Sterling Winthrop Inc.*

7672 Amineptene
57574-09-1 429 260-818-0
$C_{22}H_{27}NO_2$
7-[10,11-Dihydro-5H-dibenzo[a,d]cyclohepten-5-yl)-amino]heptanoic acid.
CNS stimulant. *Sci. Union et Cie France.*

7673 Amineptine Hydrochloride
30272-08-3 429 250-107-3
$C_{22}H_{28}ClNO_2$
7-[10,11-Dihydro-5H-dibenzo[a,d]cyclohepten-5-yl)-amino]heptanoic acid hydrochloride.
C S-1694; Maneon; Survector. CNS stimulant. mp = 226-230°.. *Union et Cie France.*

7674 Amiphenazole
490-55-1 505 207-713-8
$C_9H_9N_3S$
2,4-Diamino-5-phenylthiazole.
Daptazole; DAPT; Phenamizole; Dizol; Daptazile; Fenamizol; 5-Phenyl-2,4-thiazolediamine. CNS stimulant. A proprietary preparation of amiphenazole; a central nervous stimulant used as a narcotic antagonist and, in veterinary medicine as a barbiturate and morphine antagonist. mp = 163-164° (dec); [hydrobromide] mp> 250°; soluble in H_2O; [benzene sulfonate] mp = 261-262° (dec); insoluble in H_2O, most organic solvents. *Nicholas Labs. Ltd.*

7675 Amphetamine
300-62-9 623 206-096-2
$C_9H_{13}N$
(±)-α-Methylphenethylamine.
Actedron; Allodene; Adipan; Sympatedrine; Psychedrine; Isomyn; Isoamyne; Mecodrin; Norephedrane; Novydrine; Elastonon; Ortedrine; Phenedrine; Profamina; Propisamine; Sympamine; Simpatedrin. Anorexic; CNS stimulant. Psychomotor stimulant. d_4^{25} = 0.913; bp = 200-203°, bp_{13} = 82-85°; soluble in H_2O, EtOH, Et_2O; LD_{50} (rat sc) = 180 mg/kg. *Interco Fribourg.*

7676 Amphetaminil
17590-01-1 624 241-560-8
$C_{17}H_{18}N_2$
N-(α-Methylphenethyl)-2-phenylglycinonitrile.
AN 1; Aponeuron. CNS stimulant (psychotropic). mp = 85-87°; [hydrochloride] mp - 134-16°.

7677 Ampyzine
5214-29-9
$C_6H_9N_3$
(Dimethylamino)pyrazine.
CNS stimulant. *Parke-Davis.*

7678 Ampyzine Sulfate
7082-29-3
$C_6H_{11}N_3O_4S$
(Dimethylamino)pyrazine sulfate (1:1).
W-3580B. CNS stimulant. *Parke-Davis.*

7679 Azabon
1150-20-5
$C_{14}H_{20}N_2O_2S$
4-(3-Azabicyclo[3.2.2]non-3-ylsulfonylbenzeneamine.
CNS stimulant.

7680 Bemegride
64-65-3 1054 200-588-0
$C_8H_{13}NO_2$
3-Ethyl-3-methylglutarimide.
NP-13; Megimide; Mikedimide; Eukraton; Malysol. Respiratory stimulant. CNS stimulant, used as an antidote for barbiturate poisoning. mp = 127°; soluble in H_2O, Me_2CO; LD_{50} (mus iv) = 18.8 mg/kg, (rat iv) = 17.0 mg/kg. *Abbott Labs Inc.*

7681 Benzphetamine
156-08-1 1151
$C_{17}H_{21}N$
N-Benzyl-N,α-dimethylphenethylamine.
Anorexic; CNS stimulant. $bp_{0.02}$ = 127°; insoluble in H_2O; soluble in MeOH, EtOH, Et_2O, C_6H_6, $CHCl_3$, Me_2CO. *Pharmacia & Upjohn.*

7682 Benzphetamine Hydrochloride
5411-22-3 1151 226-489-2
$C_{17}H_{22}ClN$
N-Benzyl-N,α-dimethylphenethylamine hydrochloride.
Didrex; Inapetyl. Anorexic; CNS stimulant. mp = 129-130°; soluble in H_2O, EtOH; dextrorotatory. *Pharmacia & Upjohn.*

7683 Brucine
357-57-3 1476 206-614-7
$C_{23}H_{26}N_2O_4$
2,3-Dimethoxystrychnidine-10-one.
10,11-dimethoxystrychnine. CNS stimulant. Toxic alkaloid, related to strychnine. Found in Strychnos seeds. mp = 178°; $[\alpha]_D$ = -127° ($CHCl_3$), -85° (abs. EtOH); λ_m = 263, 301 nm (log ε 4.09, 3.93 EtOH); soluble in MeOH (1.25 g/ml), EtOH (0.77 g/ml), $CHCl_3$ (0.2 g/ml), EtOAc (0.04 g/ml), glycerol (2.78 g/100 ml), C_6H_6 (1 g/100 ml), Et_2O (0.53 g/100 ml), H_2O (0.75 g/l at 25°, 1.3 g/l at 100°); LD_{50} (rat orl) = 1 mg/kg.

7684 Brucine Sulfate
4845-99-2 1476 225-432-9
$C_{46}H_{54}N_4O_{12}S$.7H2O
2,3-Dimethoxystrychnidine-10-one sulfate (2:1) heptahydrate.
CNS stimulant. Toxic alkaloid, related to strychnine. Soluble in H_2O (1.3 g/100 ml at 25°, 10 g/100 ml at 100°), EtOH (0.95 g/100 ml), $CHCl_3$ (0.59 g/100 ml).

7685 Bucuculline
485-49-4 1250 207-619-7
$C_{20}H_{17}NO_6$
[(R-(R*,S*)]-6-(5,6,7,8-Tetrahydro-6-methyl-1,3-dioxolo[4,5-g]isoquinolin-5-yl)furo[3,4-e]-1,3-benzodioxol-8(6H)-one.
Analeptic stimulant. Found in (d-form) *Dicentra cucullaria.* mp = 215°, 177°; λ_m 225, 296, 324 nm (ε 36700, 6390, 5870 acidified EtOH); soluble in C_6H_6, $CHCl_3$, EtOAc; nearly insoluble in alcohol, Et_2O.

7686 Caffeine
58-08-2 1674 200-362-1
$C_8H_{10}N_4O_2$
3,7-Dihydro-1,3,7-trimethyl-1H-purine-2,6-dione.
methyltheobromine; theine; 1,3,7-trimethylxanthine; 1,3,7-trimethyl-2,6-dioxopurine; coffeine; 1,3,7-trimethylxanthine; thein; guaranine; methyltheobromine; NoDoz; Anacin; Cafergot; Dasin; DHCplus; Excedrin; Hycomine; Miudol; Neuranidal; Norgesic; Phensal; Propoxyphene Compound 65; SK-65 Compound; Synalgos; Vanquish; Wigraine. CNS stimulant. Used in beverages and medicines (CNS stimulant). A methylxanthine. Found in coffee, tea, mate+a leaves, cola nuts. mp = 238°; sublimes 178°; d_4^{18} = 1.23; soluble in H_2O, EtOH, Et_2O, Me_2CO, C_6H_6; LD_{50} (mmus orl) = 127 mg/kg, (fmus orl) = 230 mg/kg, (mhtr orl) = 230 mg/kg, (fhtr orl) = 249 mg/kg, (mrat orl) = 355 mg/kg, (frat orl) = 247 mg/kg, (mrbt orl) = 246 mg/kg, (frbt orl) = 224 mg/kg. *3M Pharm.; Bristol-Myers Squibb HIV Products; E. I. DuPont de Nemours Inc.; Eli Lilly & Co.; Lemmon Co.; Marion Merrell Dow Inc.; Organon Inc.; Purdue Pharma L.P.; SmithKline Beecham Pharm.; Sterling Health U.S.A.; Whitehall-Robins; Wyeth-Ayerst Labs.*

7687 Caffeine Monohydrate
5743-12-4 1674
$C_8H_{10}N_4O_2.H_2O$
3,7-Dihydro-1,3,7-trimethyl-1H-purine-2,6-dione monohydrate.
CNS stimulant. Used in beverages and medicines (CNS stimulant). A methylxanthine. Dehydration is complete at 80°. *3M Pharm.; Bristol-Myers Squibb HIV Products; E. I. DuPont de Nemours Inc.; Eli Lilly & Co.; Lemmon Co.; Marion Merrell Dow Inc.; Organon Inc.; Purdue Pharma L.P.; SmithKline Beecham Pharm.; Sterling Health U.S.A.; Whitehall-Robins; Wyeth-Ayerst Labs.*

7688 Caffeine, Citrated
69-22-7 1674
$C_8H_{10}N_4O_2.C_6H_8O_7$
Caffeine citrate.
citrated caffeine. CNS stimulant. Used in beverages and medicines (CNS stimulant). A methylxanthine. Soluble in warm H_2O. *3M Pharm.; Bristol-Myers Squibb HIV Products; E. I. DuPont de Nemours Inc.; Eli Lilly & Co.; Lemmon Co.; Marion Merrell Dow Inc.; Organon Inc.; Purdue Pharma L.P.; SmithKline Beecham Pharm.; Sterling Health U.S.A.; Whitehall-Robins; Wyeth-Ayerst Labs.*

7689 Camphotamide
4876-45-3 1782 225-484-2
$C_{21}H_{32}N_2O_5S$
3-Diethylcarbamoyl-1-methylpyridinium camphorsulfonate.

camphetamide; camphramine; Tonicorine. CNS stimulant; cardiac stimulant. mp = 174-175°; soluble in H_2O, EtOH, Et_2O; insoluble in C_6H_6, petroleum ether. *Soc. Franc. Recherches Biochim.*

7690 Cathinone
71031-15-7 1954

$C_9H_{11}NO$

(-)-α-Aminopropiophenone.

CNS stimulant. Alkaloid extracted from the leaves of Khat. $[\alpha]_D$= -46.8° (MeOH, c = 0.24); [hydrochloride] mp = 189-190°.

7691 Chlorphentermine
461-78-9 2235 207-314-9

$C_{10}H_{14}ClN$

4-Chloro-α,α-dimethylbenzeneethanamine.

clorfentermina; Lucofen; Teramine. Anorexic; CNS stimulant. Anticholinergic. Federally controlled substance (stimulant). bp_2 = 100-102°. *Parke-Davis.*

7692 Chlorphentermine Hydrochloride
151-06-4 2235 205-782-9

$C_{10}H_{15}Cl_2N$

4-Chloro-α,α-dimethylbenzeneethanamine hydrochloride.

Pre-Sate; S-62; W-2426; NSC-76098. Anorexic; CNS stimulant. Anticholinergic. Federally controlled substance (stimulant). mp = 234°; soluble in H_2O (> 20 g/100 ml), LD_{50} (mus orl) = 270 mg/kg, (mus sc) = 267 mg/kg. *Parke-Davis.*

7693 Clortermine
10389-73-8 2472

$C_{10}H_{14}ClN$

o-Chloro-α,α-dimethylphenethylamine.

Anorexic; CNS stimulant. Federally controlled substance (stimulant). bp_{16}= 116-118°. *Ciba-Geigy Corp.*

7694 Clortermine Hydrochloride
10389-72-7 2472

$C_{10}H_{15}Cl_2N$

o-Chloro-α,α-dimethylphenethylamine hydrochloride.

Su-10568; Voranil. Anorexic; CNS stimulant. Federally controlled substance (stimulant). mp = 245-246°; LD_{50} (rat orl) =332 ± 23 mg/kg. *Ciba-Geigy Corp.*

7695 Cyprodenate
15585-86-1

$C_{13}H_{25}NO_2$

2-(Dimethylamino)ethylcyclohexane-propionate.

A psychotonic brain stimulant.

7696 Deanol
108-01-0 2900 203-542-8

$C_4H_{11}NO$

2-(Dimethylamino)ethanol.

N,N-dimethyl-2-hydroxyethylamine. CNS stimulant. Studied for treatment of hyperkinesia, involuntary movement disorders. bp_{758} = 135°; d_4^{20}= 0.8866; miscible with H_2O, EtOH, Et_2O.

7697 Deanol Aceglumate
3342-61-8 2900 222-085-5

$C_{11}H_{22}N_2O_6$

2-(Dimethylamino)ethanol aceglumate.

Cleregil; Risatarun. CNS stimulant. Soluble in H_2O. *Interco Fribourg.*

7698 Deanol Acetamidobenzoate
3635-74-3 2900 222-858-7

$C_{13}H_{20}N_2O_4$

2-(Dimethylamino)ethanol acetamidobenzoate.

Deaner; Pabenol. CNS stimulant. mp = 159-161.5°; soluble in H_2O. *Riker Labs.*

7699 Deanol Bitartrate
5988-51-2 2900 227-809-3

$C_8H_{17}NO_7$

2-(Dimethylamino)ethanol bitartrate.

Liparon. CNS stimulant. Soluble in H_2O. *Interco Fribourg.*

7700 Deanol Hemisuccinate
2900

$C_8H_{17}NO_5$

2-(Dimethylamino)ethanol hemisuccinate.

Tonibral; Rischiaril. CNS stimulant. Soluble in H_2O, EtOH; less soluble in C_6H_6, $CHCl_3$. *Interco Fribourg.*

7701 Demanyl Phosphate
6909-62-2 2935

$C_4H_{12}NO_4P$

Mono[2-(dimethylamino)ethyl]phosphoric acid ester.

phosphoryldimethylcolamine; P-DMEA; Panclar. CNS stimulant; psychotonic. mp = 175-176°; [monohydrate] mp = 78-81°. *Ciba-Geigy Corp.*

7702 Dexoxadrol
4741-41-7

$C_{20}H_{23}NO_2$

(+)-2-(2,2-Diphenyl-1,3-dioxolan-4-yl)piperidine.

d-dioxadrol. CNS stimulant; analgesic. *Upjohn Ltd.*

7703 Dexoxadrol Hydrochloride
631-06-1

$C_{20}H_{24}ClNO_2$

(+)-2-(2,2-Diphenyl-1,3-dioxolan-4-yl)piperidine hydrochloride.

CL-911C; U-22559A; NSC-526062; d-dioxadrol hydrochloride. CNS stimulant; analgesic. *Upjohn Ltd.*

7704 Dextroamphetamine
51-64-9 623 200-112-1

$C_9H_{13}N$

(+)-α-Methylphenethylamine.

NSC-73713. CNS stimulant; anorexic. d_4^{25}= 0.913; bp 200-203°, bp_{13} = 82-85°; soluble in H_2O, EtOH, Et_2O; LD_{50} (rat sc) = 180 mg/kg. *Fisons plc.*

7705 Dextroamphetamine Phosphate
7528-00-9 623

$C_9H_{16}NO_4P$

(+)-α-Methylphenethylamine phosphate.

Actemin; Aktedron; Monophos; Profetamine phosphate;

Racephen; Raphetamine phosphate. CNS stimulant; anorexic. mp = 300° (dec); soluble in H_2O, EtOH; insoluble in C_6H_6, $CHCl_3$, Et_2O. *SmithKline Beecham Pharm.*

7706 Dextroamphetamine Sulfate
51-63-8 2996 200-111-6
$C_{18}H_{28}N_2O_4S$
(+)-α-Methylphenethylamine sulfate.
Dexedrine; Dexamyl; Eskatrol; Dexampex; Dexedrine sulfate; Afatin; Dexamphetamine; d-Amfetasul; Domafate; Obesedrin; Dexten; Maxiton; Sympamin; Simpamina-D; Albemap; Dadex; Ardex; Dexalone; Amsustain; Betafedrina; d-Betafedrine; Diocurb; component of: Carboxyphen, Bontril. CNS stimulant; anorexic. Used in veterinary medicine as a sympathomimetic. Federally controlled substance (stimulant). mp > 300°; $[\alpha]_D^{20}$ = 21.8° (c = 2); soluble in H_2O (10 g/100 ml), EtOH (0.2 g/100 ml); LD_{50} (mus orl) = 10 mg/kg. *SmithKline Beecham Pharm.*

7707 Diethadione
702-54-5 3155 211-867-1
$C_8H_{13}NO_3$
5,5-Diethyldihydro-2H-1,3-oxazine-2,4(3H)-dione.
Dioxone; Ledosten; Persisten; Tocèn. Anticonvulsant; analeptic. CNS stimulant. mp = 97-98°; LD_{50} (mus ip) = 52 mg/kg, (mus orl) = 130 mg/kg, (rat ip) = 31.6 mg/kg, (rat orl) = 70.5 mg/kg. *Gruppo Lepetit S.p.A.*

7708 Diethylpropion
90-84-6 3175 202-019-1
$C_{13}H_{19}NO$
2-(Diethylamino)propiophenone.
α-benzoyltriethylamine; amfepramone. Anorexic; CNS stimulant. *3M Pharm.; Marion Merrell Dow Inc.*

7709 Diethylpropion Hydrochloride
134-80-5 3175 205-156-5
$C_{13}H_{20}ClNO$
2-(Diethylamino)propiophenone hydrochloride.
Tenuate; Tepanil; Anfamon; Anorex; Danylen; Dobesin; Frekentine; Keramik; Keramin; Magrene; Modulor; Moderatan; Parabolin; Prefamone; Regenon; Tenuate Dospan; Tylinal. Anorexic; CNS stimulant. mp= 168° (dec). *3M Pharm.; Marion Merrell Dow Inc.*

7710 Difluanine
5522-39-4
$C_{28}H_{33}F_2N_3$
1-(2-Anilinoethyl)-4-[4,4-bis(p-fluorophenyl)butyl]-piperazine.
CNS stimulant. *McNeil Pharm.*

7711 Difluanine Hydrochloride
5522-33-8
$C_{22}H_{36}Cl_3F_2N_3$
1-(2-Anilinoethyl)-4-[4,4-bis(p-fluorophenyl)butyl]-piperazine trihydrochloride.
McN-JR-7242-11; R-7242. CNS stimulant. *McNeil Pharm.*

7712 Dimorpholamine
119-48-2 3317 204-328-7
$C_{20}H_{38}N_4O_4$
N,N'-1,2-Ethanediylbis[n-butyl-4-morpholine-carboxamide].
TH-1064; Amipan T; Théraleptique; Theraptique. CNS stimulant (analeptic). Respiratory stimulant. mp = 41-42°; $bp_{0.4}$ = 229°; soluble in H_2O (0.5 g/ml); LD_{50} (mus iv) = 54 mg/kg, (mus ip) = 80 mg/kg, (mus sc) = 104 mg/kg, (mus orl) = 380 mg/kg. *ICI.*

7713 Dioxadrol
6495-46-1 3352
$C_{20}H_{23}NO_2$
2-(2,2-Diphenyl-1,3-dioxolan-4-yl)piperidine.
Rydar. CNS stimulant; antidepressant; smooth muscle relaxant. The dl-form of the free base is an antidepressant. The d-form, dexoxadrol is a CNS stimulant. *Cutter Labs.*

7714 Dioxadrol Hydrochloride
3666-69-1 3352
$C_{20}H_{24}ClNO_2$
2-(2,2-Diphenyl-1,3-dioxolan-4-yl)piperidine hydrochloride.
CL-639C; dexoxadrol hydrochloride [d-form]; Relane [d-form]; levoxadrol hydrochloride [l-form]; Levoxan [l-form]. CNS stimulant; antidepressant; smooth muscle relaxant. The d-form is a CNS stimulant and analgesic; the l-form is a local anesthetic and a smooth muscle relaxant. mp = 256-260; LD_{50} (mus orl)= 240 mg/kg; [d-form hydrochloride]: mp = 254° (dec); $[\alpha]_D^{20}$ = 34° (c = 2 MeOH); LD_{50} (mus orl) = 340 mg/kg; [l-form hydrochloride]: mp = 248-254°; $[\alpha]_D^{20}$ = 34.5° (c = 2 MeOH); LD_{50} (mus orl) = 230 mg/kg. *Cutter Labs.*

7715 Doxapram
309-29-5 3488 206-216-3
$C_{24}H_{30}N_2O_2$
1-Ethyl-4-(2-morpholinoethyl)-3,3-diphenyl-2-pyrrolidinone.
Respiratory stimulant. CNS stimulant. Studied for use in treating bronchopulmonary disorders. *Robins, A. H. Co.*

7716 Doxapram Hydrochloride Monohydrate
7081-53-0 3488
$C_{24}H_{31}ClN_2O_2.H_2O$
1-Ethyl-4-(2-morpholinoethyl)-3,3-diphenyl-2-pyrrolidinone monohydrochloride monohydrate.
Dopram; AHR-619; Doxapril; Stimulexin. Respiratory stimulant. CNS stimulant. mp = 217-219°; soluble in H_2O; slightly soluble in EtOH, $CHCl_3$; LD_{50} (rat orl) = 261 mg/kg. *Robins, A. H. Co.*

7717 Ethamivan
304-84-7 3765 206-157-3
$C_{12}H_{17}NO_3$
N,N-Diethylvanillamide.
vanillic acid diethylamide; vanillic diethylamide; NSC-406087; Cardiovanil; Emivan; Vandid. CNS; respiratory stimulant. mp = 95-95.5°; LD_{50} (rat ip) = 28 mg/kg. *3M Pharm.; Oesterreiche Stickstoffwerke.*

7718 Etifelmin
341-00-4 3909
$C_{17}H_{19}N$
2-(Diphenylmethylene)butylamine.
2-(diphenylmethylene)butylamine; etifelmine; EDPA; Na III. CNS stimulant; antihypotensive. *Giulini.*

7719 Etifelmin Hydrochloride
1146-95-8 3909
$C_{17}H_{20}ClN$
2-Ethyl-3,3-diphenyl-2-propenylamine hydrochloride.
etifelmine hydrochloride; Tensinase D; component of: Gilutensin [with nicotine]. CNS stimulant; antihypotensive. mp = 232°; soluble in H_2O. *Giulini.*

7720 Etilamfetamine
457-87-4 3809
$C_{11}H_{17}N$
N-Ethyl-α-methylphenethylamine.
N-Ethylamphetamine; Adiparthrol; Apetinil. Anorexic; CNS stimulant. Federally controlled substance (stimulant). bp_{14} = 104.5-106°; [hydrochloride, d-form]: mp = 154-156°; $[\alpha]_D^{15}$ = 17.2° (c = 2 H_2O); [hydrochloride, l-form]: mp = 155-156°; $[\alpha]_D^{25}$ = -17.3° (c = 2 H_2O). *Sterling Winthrop Inc.*

7721 Etryptamine
2235-90-7 3937
$C_{12}H_{16}N_2$
3-(2-Aminobutyl)indole.
mp = 97-99°; λ_m = 220.5, 281, 289.5 (ε 20500, 6000, 6500). *Pharmacia & Upjohn.*

7722 Etryptamine Acetate
118-68-3 3937 204-268-1
$C_{14}H_{20}N_2O_2$
3-(2-Aminobutyl)indole monoacetate.
U-17312E; NSC-63963; Monase. CNS stimulant. mp = 165-166°. *Pharmacia & Upjohn.*

7723 Fencamfamine
1209-98-9 4006
$C_{15}H_{21}N$
3-Phenyl-N-ethyl-2-norbornanamine.
Euvitol. CNS stimulant. mp = 150-152°. *E. Merck.*

7724 Fencamfamine Hydrochloride
2240-14-4 4006 218-805-2
$C_{15}H_{22}ClN$
3-Phenyl-N-ethyl-2-norbornanamine hydrochloride.
Altimina; Sicoclor. CNS stimulant. mp = 278-279°; freely soluble in H_2O, less soluble in organic solvents; λ_m = 296 nm ($E^{1\%}_{1\,cm}$ 398 H_2O); LD_{50} (mus orl) = 418 mg/kg, (mus ip) = 82 mg/kg, (rat orl) = 508 mg/kg, (rat ip) = 93 mg/kg. *E. Merck.*

7725 Fenethylline
3736-08-1 4014
$C_{18}H_{23}N_5O_2$
7-[2-[(α-Methylphenethyl)amino]methyl]theophylline.
theophyllineethylamphetamine. CNS stimulant. *Degussa-Hüls Corp.*

7726 Fenethylline Hydrochloride
1892-80-4 4014 217-580-8
$C_{18}H_{24}ClN_5O_2$
7-[2-[(α-Methylphenethyl)amino]methyl]theophylline monohydrochloride.
H-814; Captagon. CNS stimulant. mp = 227-229°, 237-239°; [d-form or l-form]: mp = 246-247°. *Degussa-Hüls Corp.*

7727 Fenozolone
15302-16-6 4028 239-339-6
$C_{11}H_{12}N_2O_2$
2-Ethylamino-4-oxo-5-phenyl-2-oxazoline.
LD-3394; Ordinator. CNS stimulant. mp = 148°; λ_m = 221 nm (log ε 4.42 EtOH); LD_{50} (mus orl) = 425 mg/kg, (mus ip) = 175 mg/kg. *Lab. Dausse.*

7728 Flubanilate
847-20-1
$C_{14}H_{19}F_3N_2O_2$
Ethyl N-[2-(dimethylamino)ethyl]-m-(trifluoromethyl)-carbanilate.
CNS stimulant.

7729 Flubanilate Hydrochloride
967-48-6
$C_{14}H_{20}ClF_3N_2O_2$
Ethyl N-[2-(dimethylamino)ethyl]-m-(trifluoromethyl)-carbanilate monohydrochloride.
CNS stimulant.

7730 Flurothyl
333-36-8 4236
$C_4H_4F_6O$
Bis(2,2,2)trifluoroethyl ether.
Flurotyl; SK&F-6539; Indiklon; Idoklon. CNS stimulant. bp = 63.9°; d_4^{20} = 1.41; insoluble in H_2O, soluble in alcohol. *Anaquest.*

7731 Hexacyclonate Sodium
7009-49-6 4717
$C_9H_{15}NaO_3$
Sodium 1-(hydroxymethyl)cyclohexaneacetate.
Neuryl. CNS stimulant. mp = 106-108°; soluble in H_2O, MeOH, EtOH; less soluble in Et_2O, Me_2CO. *Warner-Lambert.*

7732 Hexacyclonic Acid
7491-42-1 4717
$C_9H_{16}O_3$
1-(Hydroxymethyl)cyclohexaneacetic acid.
CNS stimulant. *Warner-Lambert.*

7733 Homocamfin
535-86-4 4768
$C_{10}H_{16}O$
3-Methyl-5-(1-methylethyl)-2-cyclohexen-1-one.
Cyclosal; Hexetone. CNS stimulant. Generally supplied as a mixture with 2.5 parts of sodium salicylate. bp_{18} = 127-128°, bp_{16} = 124-126°, bp_{15} = 120-122°, $bp_{11.5}$ = 113-116°, bp_9 = 108-110°; slightly soluble in H_2O; soluble in EtOH, C_6H_6, Et_2O.

7734 Hopantenic Acid
18679-90-8 4781

$C_{10}H_{19}NO_5$
D-(+)-4-(2,4-Dihydroxy-3,3-dimethylbutyramido)-butyric acid.
D-homopantothenic acid. CNS stimulant. $[\alpha]_D^{20}$ = 23.8°; LD_{50} (mmus ip) = 850 mg/kg, (mmus sc) = 2063 mg/kg, (mmus orl) = 6297 mg/lg, (fmus ip) = 954 mg/kg, (fmus sc) = 2495 mg/kg, (fmus orl) = 7935 mg/kg, (mrat ip) = 1575 mg/kg, (mrat sc) = 5940 mg/kg, (mrat orl) = 16810 mg/kg, (frat ip) = 1458 mg/kg, (frat sc) = 7348 mg/kg, (frat orl) = 13350 mg/kg. *Tanabe Seiyaku Co. Ltd.*

7735 Indriline
7395-90-6

$C_{19}H_{21}N$
N,N-Dimethyl-1-phenylindene-1-ethylamine .
CNS stimulant.

7736 Indriline Hydrochloride
2988-32-1

$C_{19}H_{22}ClN$
N,N-Dimethyl-1-phenylindene-1-ethylamine hydrochloride.
MJ-1986. CNS stimulant.

7737 Lubeluzole
144665-07-6

$C_{22}H_{25}F_2N_3O_2S$
(+)-(S)-4-(2-Benzothiazolylmethylamino)-α-[(3,4-difluorophenoxy)methyl]-1-piperidineethanol.
R-86926. CNS stimulant. Used in treatment of stroke. *Abbott Laboratories Inc.*

7738 Mazindol
22232-71-9 5801 244-857-0

$C_{16}H_{13}ClN_2O$
5-(p-Chlorophenyl)-2,5-dihydro-3H-imidazo[2,1-a]isoindol-5-ol.
SaH-42548; 42548; Magrilon; Mazanor; Mazildene; Sanorex; Terenac; Teronac. Anorexic; CNS stimulant. Federally controlled substance (stimulant). mp = 215-217°, 198-199°; λ_m = 223, 268.5, 272 nm (ε 19000, 4400, 4400 95% EtOH); insoluble in H_2O, soluble in EtOH. *Am. Home Products; Sandoz Pharm. Corp.; Wyeth-Ayerst Labs.*

7739 Meclofenoxate
51-68-3 5820 200-116-3

$C_{12}H_{16}ClNO_3$
2-(Dimethylamino)ethyl(p-chlorophenoxy)acetate .
centrophenoxine; meclofenoxane; acephen; ANP-235; Analux; Cetrexin; Proseryl. CNS stimulant. Also used as a plant growth regulator. *CNRS.*

7740 Meclofenoxate Hydrochloride
3685-84-5 5820 222-975-3

$C_{12}H_{17}Cl_2NO_3$
2-(Dimethylamino)ethyl(p-chlorophenoxy)acetate hydrochloride.
Cellative; Clocete; Lucidril; Methocynal; Proserout; Brenal; Marucotol; Helfergin. CNS stimulant. Also used as a plant growth regulator. mp = 135-139°; soluble in H_2O, iPrOH, Me_2CO; insoluble in C_6H_6, Et_2O, $CHCl_3$; LD_{50} (mus iv) = 330 mg/kg, (mus ip) = 845 mg/kg, (mus orl) = 1750 mg/kg. *CNRS.*

7741 Mefexamide
1227-61-8 5844 214-963-1

$C_{15}H_{24}N_2O_3$
N-[2-(Diethylamino)ethyl]-2-(p-methoxyphenoxy)-acetamide.
mexephenamide; NP-297; Mefexadyne; Timodyne. CNS stimulant. [hydrochloride]: mp = 112°.

7742 Mesocarb
34262-84-5

$C_{18}H_{18}N_4O_2$
3-(α-Methylphenethyl)-N-(phenylcarbamoyl)-sydnone imine.
sidnocarb; sydnocarb; sydnocarbum. A specific inhibitor of in vitro synaptosomal catecholamine uptake. CNS stimulant.

7743 Methylephedrine
552-79-4 6145 209-022-7

$C_{11}H_{17}NO$
erythro-α-[1-(Dimethylamino)ethyl]-benzyl alcohol.
N-methylephedrine; N,N-dimethylnorephedrine; 2-dimethylamino-1-phenylpropanol. Analeptic. CNS stimulant. Ephedrine derivative. [dl-form]: mp = 63.5-64.5°; soluble in most solvents; [dl-form hydrochloride]: mp = 207-208°; [d-form]: mp = 87-87.5°; $[\alpha]_D^{20}$= 29.2° (c = 4 MeOH); [d-form hydrochloride]: mp = 192°; $[\alpha]_D^{20}$ = 30.1°; [l-form]: mp = 87-88°; $[\alpha]_D$ = -29.5° (c = 4.5 MeOH); [l-form hydrochloride]: mp = 192°; $[\alpha]_D^{20}$= -29.8° (c = 4.6); soluble in H_2O, less soluble in EtOH, poorly soluble in Me_2CO.

7744 Methylphenidate
113-45-1 6186 204-028-6

$C_{14}H_{19}NO_2$
Methyl α-phenyl-2-piperidine acetate.
Ritalin. CNS stimulant. Psychomotor stimulant. $bp_{0.6}$ = 135-137°; insoluble in H_2O, petroleum ether; soluble in EtOH, EtOAc, Et_2O. *Ciba-Geigy Corp.*

7745 Methylphenidate Hydrochloride
298-59-9 6186 206-065-3

$C_{14}H_{20}ClNO_2$
Methyl α-phenyl-2-piperidine acetate hydrochloride.
Ritalin hydrochloride. CNS stimulant. Psychomotor stimulant. mp = 224-226°; soluble in H_2O, EtOH, $CHCl_3$; LD_{50} (mus orl) = 190 mg/kg. *Ciba-Geigy Corp.*

7746 Modafinil
68693-11-8 6311

$C_{15}H_{15}NO_2S$
2-[(Diphenylmethyl)sulfinyl]acetamide.
CRC-40476; CEP-1538; 2-(benzhydrylsulfinyl)acetamide; Provigil; DEP 1538; CRC 40476. CNS stimulant and α_1-adrenergic agonist. Antinarcoleptic. Used for treatment of narcoplesy and hypersomnia. mp = 164-166°. 3038; *Lab. Lafon France.*

7747 Nicotine

54-11-5 6611 200-193-3

$C_{10}H_{14}N_2$

(S)-3-(1-Methyl-2-pyrrolidinyl)pyridine.

Habitrol; Nicabate; Nicoderm; Nicolan; Nicopatch; Nicotell TTS; Nicotinell; Tabazur. Smoking cessation adjunct. Ganglionic stimulant which also stimulates the CNS. Also used as an insecticide and fumigant. Severe nicotine poisoning can cause nausea, salivation, abdominal pain, vomiting, diarrhea, cold sweat, dizziness, mental confusion, weakness, drop in blood pressure and pulse, convulsions, paralysis of respiratory muscles. From dried leaves of *Nicotiana tabacum and N. Rustica*. Oily liquid; bp_{745} = 247° (partial dec), bp_{17} = 123-125°; n_D^{20} = 1.5282; d_D^{20} = 1.0097; $[\alpha]_D^{20}$ = -169°; miscible with H_2O; soluble in alcohol, $CHCl_3$, Et_2O, petroleum ether, kerosene, oils.

7748 Nikethamide

59-26-7 6635 200-418-5

$C_{10}H_{14}N_2O$

N,N-Diethyl-3-pyridinecarboxamide.

N,N-diethylnicotinamide; Anacardone; Astrocar; Carbamidal; Cardamine; Cardiamid; Cardimon; Coracon; Coractiv N; Coramine; Cordiamin; Corediol; Cormed; Cormid; Corvitol; Corvotone; Dynacoryl; Ecoran; Inicardio; Niamine; Nicamide; Nicor; Nicorine; Nikardin; Pyricardyl; Salvacard; Stimulin; Ventramine. CNS; respiratory stimulant. mp = 24-26°; bp = 296-300°, bp_{10} = 158-159°, bp_3 = 128-129°, $bp_{0.4}$ = 115°; d_4^{25} = 1.058-1.066; miscible with H_2O, Et_2O, $CHCl_3$, Me_2CO, EtOH; LD_{50} (rat ip) = 272 mg/kg. *Abbott Labs Inc.; Ciba-Geigy Corp.*

7749 Pemoline

2152-34-3 7206 218-438-8

$C_9H_8N_2O_2$

2-Imino-5-phenyl-4-oxazolidinone.

Cylert; NSC-25159; phenylisohydantoin; azoxodone; PIO; LA-956; YH-1; Azoksodon; Deltamine; Hyton Asa; Kethamed; Nitan; Pioxol; Pondex; Senior; Sigmadyn; Stimul; Tradon; Volital. CNS stimulant. Psychomotor stimulant. mp = 256-257° (dec), insoluble in H_2O, Et_2O, Me_2CO, dilute HCl; soluble in EtOH, propylene glycol; LD_{50} (rat orl)= 500 mg/kg. *Abbott Labs Inc.*

7750 Pentylenetetrazole

54-95-5 7283 200-219-3

$C_6H_{10}N_4$

6,7,8,9-Tetrahydro-5H-tetrazolo[1,5-a]azepine.

Cardiazol; Metrazol; pentetrazol; Cenalene-M; Cenazol; Coranormol; Corazole; Corvasol; Deumacard; Gewazol; Korazol; Phrenazol; Ventrazol. CNS stimulant. mp = 57-60°; soluble in H_2O, organic solvents; LD_{50} (rat ip) = 62 mg/kg, (rat sc) = 85 ± 2 mg/kg. *Knoll Pharm. Co.*

7751 Phenmetrazine

134-49-6 7385 205-143-4

$C_{11}H_{15}NO$

3-Methyl-2-phenylmorpholine.

A-66. Anorexic; CNS stimulant. Federally controlled substance (stimulant). bp_{12} = 138-140°, bp_1 = 104°. *Boehringer Ingelheim GmbH.*

7752 Phenmetrazine Hydrochloride

1707-14-8 7385 216-950-6

$C_{11}H_{16}ClNO$

3-Methyl-2-phenylmorpholine hydrochloride.

Preludin; Marsin; Neo-Zine. Anorexic; CNS stimulant. Federally controlled substance (stimulant). mp = 182°; soluble in H_2O (250 g/100 ml), 95% EtOH (50 g/100 ml), $CHCl_3$ (50 g/100 ml); poorly soluble in Et_2O. *Boehringer Ingelheim GmbH.*

7753 Phentermine

122-09-8 7415 204-522-1

$C_{10}H_{15}N$

α,α-Dimethylphenethylamine.

Inoamin. Anorexic; CNS stimulant. Systemic appetite suppressant. Psychomotor stimulant. bp_{750} = 205°; bp_{21} = 100°. *Fisons plc; Lemmon Co.; Marion Merrell Dow Inc.; SmithKline Beecham Pharm.; Wyeth-Ayerst Labs.*

7754 Phentermine Hydrochloride

1197-21-3 7415 214-821-9

$C_{10}H_{16}ClN$

α,α-Dimethylphenethylamine hydrochloride.

Adipex-P; Fastin; Wilpo; Obermine Black & Yellow. Anorexic; CNS stimulant. Systemic appetite suppressant. Psychomotor stimulant. mp = 198°. *Lemmon Co.; SmithKline Beecham Pharm.*

7755 Picrotoxin

124-87-8 7570 204-716-6

$C_{30}H_{34}O_{13}$

Cocculin.

Fish berry; Cocculine. CNS; respiratory stimulant. Isolated from seed of *Animirta cocculus*. Used as an antidote for barbiturate poisoning and as a fish poison. CAUTION: Extremely poisonous. mp = 203°; $[\alpha]_D^{16}$ = -29.3° (c = 4 EtOH); soluble in H_2O (0.285 g/100 ml at 25°, 20 g/100 ml at 100°), EtOH (7.4 g/100 ml at 25°, 33 g/100 ml at 778°); sparingly soluble in Et_2O, $CHCl_3$; LD_{50} (mus ip) = 7.2 mg/kg; highly toxic to fish. *Indofine Chem. Co.; Pfalz & Bauer; TCI America.*

7756 Pipradol

467-60-7 7638 207-394-5

$C_{18}H_{21}NO$

α,α-Diphenyl-2-piperidinemethanol.

MRD-108. CNS stimulant. *Marion Merrell Dow Inc.*

7757 Pipradol Hydrochloride

71-78-3 7638 200-764-7

$C_{18}H_{22}ClNO$

α,α-Diphenyl-2-piperidinemethanol hydrochloride.

Detaril; Meratrans; Meratonic; Stimolag Fortis. CNS stimulant. mp = 308-309° (dec); soluble in hot H_2O (1.6 g/100 ml). *Marion Merrell Dow Inc.*

7758 Prolintane

493-92-5 7964 207-784-5

$C_{15}H_{23}N$

1-[α-(Phenylmethyl)butyl] pyrrolidine.

phenylpyrrolidinopentane; SP-732. CNS stimulant;

antidepressant. $bp_{0.5}$= 105°, bp_{16} = 153°. *Boehringer Ingelheim GmbH; Thomae GmbH Dr. Karl.*

7759 Prolintane Hydrochloride

1211-28-5 7964 214-917-0

$C_{15}H_{24}ClN$

1-(α-Propylphenethyl)pyrrolidine hydrochloride.

Katovit; Promotil. CNS stimulant; antidepressant. mp = 133-134°; LD_{50} (mus orl) = 257 mg/kg. *Boehringer Ingelheim GmbH; Thomae GmbH Dr. Karl.*

7760 Pyrovalerone

3563-49-3 8194

$C_{16}H_{23}NO$

4'-Methyl-2-(1-pyrrolidinyl)valerophenone.

CNS stimulant. $bp_{0.08}$ = 104°. *Sandoz Pharm. Corp.*

7761 Pyrovalerone Hydrochloride

1147-62-2 8194 214-556-9

$C_{16}H_{24}ClNO$

4'-Methyl-2-(1-pyrrolidinyl)valerophenone hydrochloride.

F-1983; Centroton; Thymergix. CNS stimulant. mp = 178°; LD_{50} (mus orl) = 350 mg/kg. *Sandoz Pharm. Corp.*

7762 Securinine

5610-40-2 8565

$C_{13}H_{15}NO_2$

Securinan-11-one.

From leaves and roots of *Securinega suffruticosa, Euphorbiaceae. Component of the Phyllanthus discoideus* leaf. GABAA receptor blocker. Used in traditional Chinese medicine to treat aplastic anemia. Has antimalarial and antibacterial properties CNS stimulant. mp = 142-143°; $[\alpha]_D^{20}$= - 1042° (c = 1 alcohol); λ_m = 256, 330 mn (log ε 4.27, 3.30 alcohol); [hydrochloride]: mp = 230°; [nitrate]: mp = 205°; $[\alpha]_D^{20}$= -312° (c = 1 alcohol); LD_{50} (mus iv) = 3.5 ±0.9 mg/kg.

7763 Siagoside

100345-64-0

$C_{73}H_{129}N_3O_{30}$

N-(II^3-N-Acetylneuraminosylgangliotetraosyl)ceramid intramolecular ester.

Ganglioside. May attenuate some morphological and functional deficits related to striatal damage after acute cerebral ischemia.

7764 Strychnine

57-24-9 9020 200-319-7

$C_{21}H_{22}N_2O_2$

Strychnidin-10-one.

Veterinary: Has been used as a tonic and CNS stimulant. Found commonly in seeds of *Strychnos nux-vomica L., Loganiaceae and beans of S. Ignatti.* CAUTION: Extremely poisonous. mp = 275-285°; d^{18} = 1.359; $[\alpha]_D^{18}$= -104.3°; soluble in EtOH (0.55 g/100 ml), $CHCl_3$ (15 g/100 ml), C_6H_6 (0.67 g/100 ml), MeOH (0.4 g/100 ml), C_5H_5N (1.2 g/100 ml); LD_{50} (rat, slow infusion) = 0.96 mg/kg; [hydrochloride dihydrate]: soluble in H_2O (25 mg/ml), alcohol (1.25 mg/kg); insoluble in Et_2O; pH (0.01 M) = 5.4.

7765 Tenamfetamine

51497-09-7

$C_{10}H_{13}NO_2$

(±)-α-Methyl-3,4-(methylenedioxy)phenethylamine.

CNS stimulant.

7766 Theobromine

83-67-0 9418 201-494-2

$C_7H_8N_4O_2$

3,7-Dihydro-3,7-dimethyl-1H-purine-2,6-dione.

3,7-dimethylxanthine. Diuretic; bronchodilator; cardiotonic. Principle alkaloid of the cacao bean. Also found in coca nuts and tea. A methylxanthine CNS stimulant. mp = 357°; sublimes at 290-295°; very slightly soluble in H_2O, EtOH; soluble in fixed alkali hydroxides, concentrated acids; moderately soluble in ammonia; insoluble in C_6H_6, Et_2O, $CHCl_3$, CCl_4.

7767 Theophylline

58-55-9 9421 200-385-7

$C_7H_8N_4O_2$

3,7-Dihydro-1,3-dimethyl-1H-purine-2,6-dione.

1,3-Dimethylxanthine; Accurbron; Aerobin; Aerolate; Afonilm; Armophylline; Austyn; Bilordyl; Brochoretard; Bronkodyl; Cétraphylline; Constant-T; Duraphyl; Duraphyllin; Diffumal; Elixophyllin; Etheophyl; Euphyllin; Euphylong; LaBID; Lasma; Physpan; Pro-Vent; PulmiDur; Pulmo-Timelets; Respbid; Slo-Bid; Slo-Phyllin; Solosin; Somophyllin-CRT; Somophyllin-T; Sustaire; Talotren; Tesona; Theobid; Theo-24; Theobid Duracap; Theoclear; Theochron; Theo-Dur; Theograd; Theolair; Theon; Theophyl; Theoplus; Theo-Sav; Theostat; Theovent; Unifyl; Uniphyl; Uniphyllin; Xanthium; component of: Bronkotabs, Dicurin Procaine, Mudrane GG Elixer, Primatene Tablets, Quibron, Quibron Plus, Quibron-T/SR, Slo-Phyllin GG, Tedral, Theolair Plus, Theo-Organdin. Bronchodilator. Xanthine derivative with diuretic, cardiac stimulant and smooth muscle relaxant activities; isomeric with theobromine; small amounts found in tea. *3M Pharm.; Boehringer Ingelheim Ltd.; Bristol Laboratories; Ciba-Geigy Corp.; Eli Lilly & Co.; Fisons plc; Forest Pharm. Inc.; Glaxo Wellcome plc; Key Pharm.; Norwich Eaton; Poythress; Purdue Pharma L.P.; Rhône-Poulenc Rorer Pharm. Inc.; Savage Labs; Schering AG; Searle G.D. & Co.; Sterling Winthrop Inc.; Wallace Labs; Whitehall Labs. Inc.*

Contraceptives

7768 Desogestrel

54024-22-5 2971

$C_{22}H_{30}O$

13-Ethyl-11-methylene-18,19-dinor-17α-pregn-4-en-20-yn-17-ol.

Org-2969; Cyclosa; Dicromil; Marvelon 150/320; Mercilon; Ortho-Cept; Oviol; Varnoline; component of: Desogen. Progestogen with low androgenic potency. In combination with an estrogen (ethinylestradiol), used as an oral contraceptive [71138-35-7]. mp = 109-110°; $[\alpha]_D^{20}$= 55° ($CHCl_3$). *Organon Inc.*

7769 Ethinyl Estradiol

57-63-6 3780 200-342-2

$C_{20}H_{24}O_2$

19-Nor-17α-pregna-1,3,5(10)-trien-20-yn-3,17-diol.

Diogyn E; Estinyl; NSC-10973; component of: Brevicon, Demulen, Desogen, Estopherol, Estostep, Levlen, Loestrin, Lo/Ovral, ModiCon, Nordette, Norethrin 1/35E, Norlestrin, Ortho-Cyclen, Ortho-Novum, Ovcon, Ovral, Tri-Levlen, Triphasil. Estrogen. In combination with a progestogen (Desogestrel, Gestodene; Lynestrenol or Norgestimate), used as an oral contraceptive [71138-35-7], [109852-02-0], [8064-76-4], [79871-54-8], respectively. mp = 141-146°; $[\alpha]_D^{25}$ = 0 ± 1° (dioxane); λ_m 281 nm (ε 2040 ± 60 EtOH); insoluble in H_2O; soluble in EtOH (16.6 g/100 ml), Et_2O (25 g/100 ml), Me_2CO (20 g/100 ml), dioxane (25 g/100 ml), $CHCl_3$ (5 g/100 ml); LD_{50} (rat orl) = 2952 mg/kg, (mus orl) = 1737 mg/kg. *Astra Chem. Ltd.; Berlex Labs Inc.; Eprova AG; Marion Merrell Dow Inc.; Mead Johnson Labs.; Organon Inc.; Parke-Davis; Pfizer Intl.; Roberts Pharm. Corp.; Schering-Plough HealthCare Products; Searle G.D. & Co.; Syntex Intl. Ltd.; Wyeth-Ayerst Labs.*

7770 Ethynodiol

1231-93-2 3905 214-971-5

$C_{20}H_{28}O_2$

19-Nor-17α-pregn-4-en-20-yne-3β,17-diol.

ED. Progestogen. *Searle G.D. & Co.*

7771 Ethynodiol Diacetate

297-76-7 3905 206-044-9

$C_{24}H_{32}O_4$

19-Nor-17α-pregn-4-en-20-yne-3β,17-diol diacetate.

SC-11800; Femulen; Luteonorm; Luto-Metrodiol; Metrodiol; component of: Demulen, Metrulen, Ovulen, Luteolas, Ovaras, Conova, Miniluteolas. Progestogen. Used in combination with an estrogen (mestranol, ethinyl estradiol) as an oral contraceptive. mp - 126-127°; $[\alpha]_D$ = -72.5° ($CHCl_3$). *Searle G.D. & Co.*

7772 Gestodene

60282-87-3 4421 262-145-8

$C_{21}H_{26}O_2$

13-Ethyl-17-hydroxy-18,19-dinor-17α-pregna-4,15-dien-20-yn-3-one.

SH B 331; component of: Femodene, Femovan, Ginoden, Gynera, Milvane, Minulet, Monodie, Phaeva, Triminulet. Progestogen. Used in combination with ethinyl estradiol [109852-02-0] as an oral contraceptive. mp = 197.9°. *Schering AG.*

7773 Lynestrenol

52-76-6 5659 200-151-4

$C_{20}H_{28}O$

19-Nor-17α-pregn-4-en-20-yn17β-ol.

NSC-37725; Exluton; Exlutona; Exlutena; Orgametril; Orgametil; component of: Anacyclin; Fysionorm; Minilyn; Noracyclin; Ovanon; Ovoresta; Yermonil; Lyndiol. Progestogen. Used in combination with estrogens (ethinyl estradiol, mestranol), used as an oral contraceptive [109852-02-0], [8015-14-3] respectively. mp = 158-160°; $[\alpha]_D$ = -13° ($CHCl_3$).

7774 Medroxyprogesterone

520-85-4 5838 208-298-6

$C_{22}H_{32}O_3$

17-Hydroxy-6α-methylpregn-4-en-3,20-dione.

Progestogen. Orally active progestogen used with estrogens (eg. ethinyl estradiol) in oral contraceptives. Used in veterinary medicine for estrus regulation. mp = 220-223.5°; $[\alpha]_D^{25}$ = 75° ($CHCl_3$); λ_m = 241 nm (ε 16000 EtOH). *Farmitalia Carlo Erba Ltd.; Pharmacia & Upjohn.*

7775 Medroxyprogesterone Acetate

71-58-9 5838

$C_{24}H_{34}O_4$

17α-Hydroxy-6α-methylprogesterone 17-acetate.

MAP; Amen; Clinovir; Curretab; Cycrin; Depo-Clinovir; Depo-Provera; Farlutal; Gestapuran; G-Farlutal; Hysron; Lutoral; Nadigest; Nidaxin; Oragest; Perlutex; Prodasone; Provera; Sodelut G; Veramix; component of: Provest. Orally active progestogen once used with estrogens in oral contraceptives. Used in veterinary medicine for estrus regulation. Progestogen. mp = 207-209°; $[\alpha]_D$ = 61° ($CHCl_3$); λ_m = 240 nm (ε 15900 EtOH). *Farmitalia Carlo Erba Ltd.; Pharmacia & Upjohn; Solvay Pharm. Inc.; Wyeth-Ayerst Labs.*

7776 Mestranol

72-33-3 5976 200-777-8

$C_{21}H_{26}O_2$

3-Methoxy-19-nor-17α-pregna-1,3,5(10)-trien-20-yn-17-ol.

17α-ethynylestradiol 3-methyl ether; EE_3ME; 33355; Menophase; Norquen; Ovastol; component of: Enovid, Norethrin 1/50M, Norinyl, Norquen, Ortho-Novum, Ovulen. Estrogen. Used in combination with a progestogen as an oral contraceptive. May be a carcinogen. mp = 150-151°; λ_m = 279, 287.5 nm ($E_{1\,cm}^{1\%}$ 82, 14.4 MeOH); insoluble in H_2O; poorly soluble in MeOH; soluble in EtOH, $CHCl_3$, Et_2O, Me_2CO, dioxane. *Ortho Pharm. Corp.; Roberts Pharm. Corp.; Searle G.D. & Co.; Syntex Intl. Ltd.*

7777 Nomegestrol

58691-88-6

$C_{21}H_{28}O_3$

17-Hydroxy-6-methyl-19-norpregna-4,6-diene-3,20-dione.

TX-066 [as acetate]; Lutenyl [as acetate]; Uniplant [as acetate]; Thermex; Monaco. A synthetic progestin with a high affinity for the progesterone receptor used, in the acetate form, as a subdermal contraceptive. Nonandrogenic progestogen.

7778 Norethindrone

68-22-4 6790 200-681-6

$C_{20}H_{26}O_2$

17-Hydroxy-19-nor-17α-pregn-4-en-20-yn-3-one.

anhydrohydroxynorprogesterone; 19-norethisterone; nor-pregneninolone; NSC-9564; Conludag; Menzol; Micronor; Micronovum; Mini-Pe; mini-pill; Norcolut; Noriday; Norluten; Norlutin; Nor-QD; Primolut N; Utovlan; component of: Binovum, Brevicon, Brevinor, Conceplan, Modicon, Neocon 1/35, Norimin, Norinyl 1/35, Norquentiel, Ortho-Novum 1/35, Ortho-Novum 7/7/7, Ovcon, Ovysmen, Synphase, Tri-Norinyl, Trinovum, Norinyl-1, Ortho-Novin 1/50, Ortho-Novum 1/50. Used in combination with estrogens as an oral contraceptive. Progestogen. mp = 203-204°; $[\alpha]_D^{20}$ = -31.7° ($CHCl_3$); λ_m = 240 nm (log ε 4.24). *Syntex Labs. Inc.*

7779 Norethindrone Acetate
51-98-9 6790 200-132-0
$C_{22}H_{28}O_3$
17-Hydroxy-19-nor-17α-pregn-4-en-20-yn-3-one acetate.
Aygestin; Milligynon; Norlutate; Primolut-Nor; component of: Anovlar, Estrostep, Etalontin, Gynovlar, Loestrin, Minovlar, Norlestrin, Primosiston. Used in combination with estrogens as an oral contraceptive. Progestogen. mp = 161-162°; λ_m = 240 nm (ε 18690). *Bristol-Myers Squibb Pharm. R&D; Parke-Davis; Schering AG; Syntex Labs. Inc.; Wyeth-Ayerst Labs.*

7780 Norethynodrel
68-23-5 6791 200-682-1
$C_{20}H_{26}O_2$
17-Hydroxy-19-nor-17α-pregn-5(10)-en-20-yn-3-one.
SC-4642; NSC-15432; component of: Conovid E, Enavid, Enovid. Progestogen. Used in combination with estrogens as an oral contraceptive. mp = 169-170°; $[\alpha]_D$ = 108° (c = 1 $CHCl_3$). *Searle G.D. & Co.*

7781 Norgestimate
35189-28-7 6796
$C_{23}H_{31}NO_3$
(+)-13-Ethyl-17-hydroxy-18,19-dinor-17α-pregn-4-en-20-yn-3-one oxime acetate.
ORF-10131; RWJ-10131; D-138; component of: Ortho-Cyclen, Cilest, Ortho Tri-Cyclen, Ortrel, TriCilest. Progestogen. Used in combination with an estrogen (ethinyl estradiol) as an oral contraceptive. mp = 214-218°; $[\alpha]_D^{25}$ = 110°. *Ortho Pharm. Corp.*

7782 Norgestrel
6533-00-2 6797 229-433-5
$C_{21}H_{28}O_2$
(±)-13-Ethyl-17α-hydroxy-18,19-dinorpregn-4-en-20-yn-3-one .
WY-3707; Neogest; Ovrette; component of: Lo/Ovral, Ovral, Stediril. Progestogen. Used in combination with an estrogen (ethinyl estradiol) as an oral contraceptive. The levorotatory isomer is biologically active. mp = 205-207°; λ_m = 241 nm (ε 16700 EtOH). *Wyeth-Ayerst Labs.*

7783 (-)-Norgestrel
797-63-7 6797
$C_{21}H_{28}O_2$
(-)-13-Ethyl-17α-hydroxy-18,19-dinorpregn-4-en-20-yn-3-one.
levonorgestrel; D-norgestrel; dexnorgestrel; Wy-5104; Microlut; Microval; Norgeston; Norplant; component of: Levlen, Logynon, Microgynon, Nordette, Ovran, Ovranette, Tetragynon, Tri-Levlen, Trinordiol, Triphasil. Progestogen. Used in implants as an oral contraceptive. The levorotatory isomer is biologically active. mp = 235-237°; $[\alpha]_D^{20}$ = -32.4° (c = 0.496 $CHCl_3$); λ_m = 241 nm (ε 16770 MeOH). *Berlex Labs Inc.; Wyeth-Ayerst Labs.*

7784 (+)-Norgestrel
797-64-8 6797
$C_{21}H_{28}O_2$
(8α,9β,10α,13α,14β)-13-Ethyl-17-hydroxy-18,19-norpregn-4-en-20-yn-3-one.
dextronorgestrel. Progestogen. The levorotatory isomer is biologically active. mp = 238-242°; $[\alpha]_D^{25}$ = 40.7° ($CHCl_3$). *Wyeth-Ayerst Labs.*

Cytoprotectants

7785 Aceglutamide Aluminum
12607-92-0 25
$C_{35}H_{59}Al_3N_{10}O_{24}$
Pentakis (N^2-acetl-L-glutaminato)terahydroxytrialuminum.
Glumal; KW-110. Antiulcerative. mp = 221° (dec); soluble in H_2O; insoluble in MeOH, EtOH, Me_2CO; LD_{50} (mmus orl) = 14.3 g/kg, (mmus ip) = 5.0 g/kg, (mmus iv) = 0.46 g/kg, (mrat orl) > 14.5 g/kg, (mrat ip) = 4.2 g/kg, (mrat iv) = 0.40 g/kg. *Kyowa Hakko Kogyo Co. Ltd.*

7786 Acetoxolone
6277-14-1 76 228-475-1
$C_{32}H_{48}O_5$
(20β)-3β-(Acetyloxy)-11-oxoolean-12-en-29-oic acid.
acetylglycyrrhetinic acid; glycyrrhetic acid acetate. Antiulcerative. mp= 322-325°; $[\alpha]_D^{20}$= 141°. *Dott. Inverni & Della Beffa.*

7787 Acetoxolone Aluminum Salt
29728-34-5 76 249-815-5
$C_{96}H_{141}AlO_{15}$
(20β)-3β-(Acetyloxy)-11-oxoolean-12-en-29-oic acid aluminum salt.
Oriens. Antiulcerative. mp = 286-290°; $[\alpha]_D^{20}$= 126° ±2° (c = 1 $CHCl_3$); insoluble in H_2O; LD_{50} (mrat orl) > 3300 mg/kg. *Dott. Inverni & Della Beffa.*

7788 Benexate Hydrochloride
78718-25-9 1065
$C_{23}H_{28}ClN_3O_4$
trans-2-[[[4-[[(Aminoiminomethyl)amino]methyl]-cyclohexyl]carbonyl]oxy]benzoic acid phenmethyl ester monohydrochloride.
Antiulcerative. Synthetic protease inhibitor. mp = 83°. *Nippon Chemiphar.*

7789 Carbenoxolone
5697-56-3 1839 227-174-2
$C_{34}H_{50}O_7$
3β-Hydroxy-11-oxoolean-12-en-30-oic acid hydrogen succinate.
carbenoxalone. Antiulcerative. Anti-inflammatory gluccocorticoid related to enoxolone; a gastro-intestinal sedative. mp = 291-294°; $[\alpha]_D^{20}$= 128° ($CHCl_3$). *Biorex Labs Ltd.*

7790 Carbenoxolone Sodium
7421-40-1 1839 231-044-0
$C_{34}H_{48}Na_2O_7$
3β-Hydroxy-11-oxoolean-12-en-30-oic acid hydrogen succinate disodium salt.
carbenoxalone sodium; Biogastrone; Bioplex; Bioral; Duogastrone; Neogel; Pyrogastrone; Sanodin; Ulcus-Tablinen. Antiulcerative. Anti-inflammatory gluccocorticoid related to enoxolone; a gastro-intestinal sedative. Soluble in H_2O; LD_{50} (mmus ip) = 120 mg/kg, (mmus iv) = 198 mg/kg, (mrat orl) = 3200 mg/kg. *Biorex Labs Ltd.*

7791 Cetraxate
34675-84-8 2067
$C_{17}H_{23}NO_4$
p-Hydroxyhydrocinnamic acid trans-4-(aminomethyl)-cyclohexanecarboxylate.
Antiulcerative. Derivative of tranexamic acid. mp = 200-280°. *Daiichi Seiyaku.*

7792 Cetraxate Hydrochloride
27724-96-5 2067
$C_{17}H_{24}ClNO_4$
p-Hydroxyhydrocinnamic acid trans-4-(aminomethyl)-cyclohexanecarboxylate hydrochloride.
DV-1006; Neuer. Antiulcerative. Derivative of tranexamic acid. mp = 238-240°. *Daiichi Seiyaku.*

7793 Guaiazulene
489-84-9 4581 207-701-2
$C_{15}H_{18}$
1,4-Dimethyl-7-(1-methylethyl)azulene.
S-guaiazulene; AZ 8; AZ 8 Beris; Eucazulen; Kessazulen; Vaumigan. Anti-inflammatory; antiulcerative. Blue oil; bp_{10} = 165-170°.

7794 Irsogladine
57381-26-7 5113
$C_9H_7Cl_2N_5$
2,4-Diamino-6-(2,5-dichlorophenyl)-s-triazine.
dicloguamine. Antiulcerative. mp = 268-269°. *Nippon Shinyaku Japan.*

7795 Irsogladine Maleate
84504-69-8 5113
$C_{13}H_{11}Cl_2N_5O_4$
2,4-Diamino-6-(2,5-dichlorophenyl)-s-triazine maleate.
MN-1695; Gaslon. Antiulcerative. mp = 205° (dec); LD_{50} (mmus orl) = 6035 mg/kg, (mmus sc) = 2841 mg/kg, (mmus ip) = 775 mg/kg, (fmus orl) = 5697 mg/kg, (fmus sc) = 3216 mg/kg, (fmus ip) = 1006 mg/kg, (mrat orl) = 3898 mg/kg, (mrat sc) = 1600 mg/kg, (mrat ip) = 558 mg/kg, (frat orl) = 2917 mg/kg, (frat sc) = 1524 mg/kg, (frat ip) = 545 mg/kg. *Nippon Shinyaku Japan.*

7796 Nitecapone
116313-94-1
$C_{12}H_{11}NO_6$
3-(3,4-Dihydroxy-5-nitrobenzylidene)-2,4-pentanedione.
OR-462. Nitrocatechol-type catechol-O-methyl-transferase inhibitor. Antioxidant; cytoprotectant.

7797 Picoprazole
78090-11-6
$C_{17}H_{17}N_3O_3S$
Methyl 6-methyl-2-[[(3-methyl-2-pyridyl)methyl]sulfinyl]-5-benzimidazolecarboxylate.
H 149/94. A benzimidazole derivative that suppresses acid secretion through inhibition of (H+/K+)-ATPase. Antiulcerative; cyctoprotectant.

7798 Plaunotol
64218-02-6 7692
$C_{20}H_{34}O_2$
(2Z,6E)-2-[(3E)-4,8-Dimethyl-3,7-nonadienyl]-6-methyl-2,6-octadiene-1,8-diol.
CS-684; Kelnac. Antiulcerative. Soluble in organic solvents, insoluble in H_2O; LD_{50} (mmus orl) = 8800 µl/kg, (fmus orl) = 8100 µl/kg, (mrat orl) = 10900 µl/kg, (frat orl) = 11200 µl/kg. *Sankyo Co. Ltd.*

7799 Polaprezinc
107667-60-7 7712
$(C_9H_{12}N_4O_3Zn)_n$
[N-β-Alanyl-L-histidinato(2-)-N,N^N,O^α]zinc.
zinc L-carnosine; Z-103; Promac. Antiulcerative. Has antioxidant and gastroprotective properties. Insoluble in H_2O; LD_{50} (mmus ip) = 220 mg/kg, (mmus sc) = 758 mg/kg, (mmus orl) = 1269 mg/kg, (fmus ip) = 165 mg/kg, (fmus sc) = 874 mg/kg, (fmus orl) = 1331 mg/kg, (mrat ip) = 405 mg/kg, (mrat sc) > 5000 mg/kg, (mrat orl) = 8441 mg/kg, (frat ip) = . *Hamari Chem. Ltd.*

7800 Rebamipide
90098-04-7 8296
$C_{19}H_{15}ClN_2O_4$
(±)-α-(p-Chlorobenzamido)-1,2-dihydro-2-oxo-4-quinolinepropionic acid.
OPC-12759; Mucosta; proamipide. Antiulcerative. Gastric cytoprotectant. [hemihydrate]: mp = 299-290° (dec); [(-)-form]: mp = 304-306°; [(-)-form]: $[\alpha]_D^{20}$ = -116.7 (c = 1.0 in DMF); [(+)-form]: $[\alpha]_D^{20}$ = +116.9 (c = 1.0 DMF). *Otsuka Am. Pharm.*

7801 Sofalcone
64506-49-6 8850
$C_{27}H_{30}O_6$
[5-[(3-Methyl-2-butenyl)oxy]-2-[p-[(3-methyl-2-butenyl)oxy]cinnamoyl]phenoxy]acetic acid.
Su-88; Solon. Antiulcerative. mp = 143-144°; LD_{50} (mus, rat orl) > 10 g/kg. *Taisho.*

7802 Spizofurone
72492-12-7 8918
$C_{12}H_{10}O_3$
5-Acetylspiro[benzofuran-2(3H),1'-cyclopropan]-3-one.
AG-629; Maon. Antiulcerative. mp = 102-104°, 106-107°. *Takeda Chem. Ind. Ltd.*

7803 Sucralfate
54182-58-0 9049 259-018-4
$C_{12}H_mAl_{16}O_nS_8$
Sucrose octakis(hydrogen sulfate) aluminum complex.
Carafate; Antepsin; Citogel; Hexagastron; Keal; Succosa; Sucralfin; Sucrate; Sugast; Sulcrate; Ulcar; Ulcerlmin; Ulcogant. Gastrointestinal antiulcerative. Insoluble in H_2O, EtOH; soluble in dilute HCl and NaOH solutions. *Chugai Pharm. Co. Ltd.; Hoechst Marion Roussel Inc.*

7804 Sulglicotide
54182-59-1
Sulfuric polyester of a glycopeptide isolated from pig duodenum.
sulglycotide. A sulfoglycopeptide. Antiulcerative; cytoprotectant.

7805 Teprenone
6809-52-5 9296
$C_{23}H_{38}O$
6,10,14,18-Tetramethyl-5,9,13,17-nonadecatetraene-2-one, mixture of (5E,9E,13E) and (5Z,9E,13E) isomers.

geranylgeranylacetone; GGA; E-0671; E36U31; Selbex. Antiulcerative. $bp_{0.01}$= 155-160°; $d_4^{20.5}$= 0.9081.

7806 Troxipide

99777-81-8 9921

$C_{15}H_{22}N_2O_4$

(±)-(3,4,5-Trimethoxy-N-3-piperidylbenzamide.

KU-54; Aplace. Antiulcerative. mp= 179-185°; soluble in EtOH; LD_{50} (mrat orl) = 500 mg/kg, (mrat sc) > 4150 mg/kg, (mrat ip) = 340 mg/kg, (frat orl) = 2100 mg/kg, (frat sc) > 4150 mg/kg, (frat ip) = 340 mg/kg, (mmus orl) = 2200 mg/kg, (mmus sc) = 1600 mg/kg, (mmus ip) = 300 mg/kg, (fmus orl) = 2000 mg/kg, (fmus sc) = 1550 mg/kg, (fmus ip) = 305 mg/kg. *Kyorin Pharm. Co. Ltd.*

7807 Zolimidine

1222-57-7 10320 214-947-4

$C_{14}H_{12}N_2O_2S$

2-[4-(Methylsulfonyl)phenyl]imidazo[1.2-a]pyridine.

Solimidin. Antiulcerative. Non-cholinergic gastroprotective agent. mp = 242-244°; LD_{50} (rat orl) = 3710 mg/kg; [hydrochloride]: LD_{50} (mus ip) = 800 mg/kg. *Selvi.*

Decongestants

7808 Amidephrine

3354-67-4 418

$C_{10}H_{16}N_2O_3S$

3'-[1-Hydroxy-2-(methylamino)ethyl] methanesulfonanilide.

amidefrine; MJ-1996. Vasoconstrictor; decongestant (nasal); adrenergic. α-adrenergic receptor agonist. mp = 159-161°. *Mead Johnson Labs.*

7809 Amidephrine Mesylate

1421-68-7 418

$C_{11}H_{20}N_2O_6S_2$

3'-[1-Hydroxy-2-(methylamino)ethyl] methanesulfonanilide monomethane-sulfonate (salt).

amidephrine monomethanesulfonate; amidefrine mesilate; MJ-5190; Dircol; Fentrinol; Nalde. Vasoconstrictor; decongestant (nasal); adrenergic. α-adrenergic receptor agonist. mp = 207-209°; LD_{50} (frat orl) = 13-36 mg/kg. *Mead Johnson Labs.*

7810 Cafaminol

30924-31-3 1671 250-390-3

$C_{11}H_{17}N_5O_3$

3,7-Dihydro-8-[(2-hydroxyethyl)methylamino]-1,3,7-trimethyl-1H-purine-2,6-dione.

methylcoffanolamine; Rhinoptil. Decongestant (nasal). Alkanolamine derivative of caffeine. mp = 162-164°; soluble in H_2O (6 g/100 ml); pH (aqueous solution) = 6.9; LD_{50} (mmus sc) = 700 mg/kg.

7811 Cyclopentamine

102-45-4 2808

$C_9H_{19}N$

N,α-Dimethylcyclopentaneethylamine .

cyclopentadrine; Clopane; Sinos; Cyklosal; Cyclonarol. An α-adrenergic agonist. Has vasoconstrictor properties and is used as a nasal decongestant. bp_{30} = 83-86°; n_D^{25}= 1.4500. *Eli Lilly & Co.*

7812 Cyclopentamine Hydrochloride

3459-06-1 2808 208-681-8

$C_9H_{20}ClN$

N,α-Dimethylcyclopentaneethylamine hydrochloride.

Clopane Hydrochloride; component of: Aerolone Solution. Adrenergic (vasoconstrictor); nasal decongestant. mp = 113-115°; soluble in H_2O. *Eli Lilly & Co.*

7813 d-Ephedrine

321-98-2 3645 206-293-3

$C_{10}H_{15}NO$

D-erythro-2-(Methylamino)-1-phenylpropan-1-ol.

(+)-ephedrine; d-pseudoephedrine; d-+ly-ephedrine; d-isoephedrine. Decongestant. Adrenergic agonist. A stereoisomer of ephedrine. mp = 119°; $[\alpha]_D^{20}$= +51° (c = 0.6 in EtOH); pH (0.5% aqueous solution) = 10.8; sparingly soluble in H_2O; soluble in EtOH, Et_2O. *Eli Lilly & Co.*

7814 d-Ephedrine Hydrochloride

24221-86-1 3645 246-090-7

$C_{10}H_{16}ClNO$

D-erythro-2-(Methylamino)-1-phenylpropan-1-ol hydrochloride.

(+)-ephedrine hydrochloride; d-pseudoephedrine hydrochloride; Galpseud; Novafed; Otrinol; Rhinalair; Sinufed; Sudafed; Symptom 2. Decongestant. Adrenergic agonist. A stereoisomer of ephedrine. mp = 181-182°; $[\alpha]_D^{20}$= +62° (c = 0.8); λ_m = 208, 251, 257, 264 nm (ε 8300, 161, 201, 161 EtOH); pKa = 9.22; pH (0.5% aqueous solution) = 5.9; soluble in H_2O, EtOH, $CHCl_3$; LD_{50} (mus ip) = 165 mg/kg. *Burroughs Wellcome Inc.; Marion Merrell Dow Inc.*

7815 d-Ephedrine Sulfate

3645

$C_{20}H_{32}N_2O_6$

D-erythro-2-(Methylamino)-1-phenylpropan-1-ol sulfate (2:1) salt.

Afrinol. Decongestant. Adrenergic agonist. A stereoisomer of ephedrine. *Schering AG.*

7816 Fenoxazoline

4846-91-7 4025 225-437-6

$C_{13}H_{18}N_2O$

4,5-Dihydro-2-[[2-(1-methylethyl)phenoxy]methyl]-1H-imidazole.

phenoxazoline. CNS Stimulant. An α-adrenergic agonist with sympathomimetic properties. *Lab. Dausse.*

7817 Fenoxazoline Hydrochloride

21370-21-8 4025

$C_{13}H_{19}ClN_2O$

4,5-Dihydro-2-[[2-(1-methylethyl)phenoxy]methyl]-1H-imidazole monohydrochloride.

Aturgyl; Snup. An α-adrenergic agonist with sympathomimetic properties. CNS Stimulant. mp = 174°; soluble in H_2O, EtOH. *Lab. Dausse.*

7818 Indanazoline
40507-78-6 4967
$C_{12}H_{15}N_3$
2-(4-Indanylamino)-2-imidazoline.
Decongestant (nasal); vasoconstrictor. mp = 109-113°. *Nordmark.*

7819 Indanazoline Hydrochloride
40507-80-0 4967 254-945-0
$C_{12}H_{16}ClN_3$
2-(4-Indanylamino)-2-imidazoline hydrochloride.
EV-A-16; Farial. An α-adrenergic agonist. Used as a decongestant and vasoconstrictor. mp = 182-184°; LD_{50} (mmus orl) = 179 mg/kg, (mmus iv) = 22.3 mg/kg, (fmus orl) = 233 mg/kg, (fmus iv) = 26.9 mg/kg, (mrat orl) = 481 mg/kg, (mrat iv) = 16.3 mg/kg, (frat orl) = 542 mg/kg, (frat iv) = 17.6 mg/kg. *Nordmark.*

7820 Levmetamfetamine
33817-09-3
$C_{10}H_{15}N$
(-)-(R)-N,α-Dimethylphenethylamine.
Nasal decongestant.

7821 Levopropylhexedrine
6192-97-8 8045 228-245-0
$C_{10}H_{21}N$
(-)-N,α-Dimethylcyclohexaneethylamine.
l-propylhexedrine. Adrenergic (vasoconstrictor); decongestant. *See* propylhexedrine. Oily liquid; bp_9 = 80-81°; n_D^{20} = 1.4590. *SmithKline Beecham Pharm.*

7822 Metizoline
17692-22-7 6223
$C_{13}H_{14}N_2S$
2-[(2-Methylbenzo[b]thien-3-yl)methyl]-2-imidazoline.
benazoline. An α-adrenergic agonist. Used as a vasoconstrictor and nasal decongestant. mp = 156-157°. *E. Merck; Marion Merrell Dow Inc.*

7823 Metizoline Hydrochloride
5090-37-9 6223 225-811-9
$C_{13}H_{15}ClN_2S$
2-[(2-Methylbenzo[b]thien-3-yl)methyl]-2-imidazoline monohydrochloride.
EX-10-781; RMI-10482A; Ellsyl; Elsyl; Eunasin. An α-adrenergic agonist. Used as a vasoconstrictor and nasal decongestant. mp = 244-246°; slightly soluble in H_2O; soluble in alcohols; nearly insoluble in $CHCl_3$; LD_{50} (mus ip) = 49 mg/kg; (mus ip) = 9.1 mg/kg; (mus orl) = 155 mg/kg; (rat orl) = 74 mg/kg. *E. Merck; Marion Merrell Dow Inc.; Merck & Co.Inc.*

7824 Naphazoline
835-31-4 6455 212-641-5
$C_{14}H_{14}N_2$
2-(1-Naphthylmethyl)-2-imidazoline.
Privine. Used as a vasoconstrictor and nasal decongestant. α-Adrenergic agonist (vasoconstrictor). *Alcon Labs; Bristol-Myers Squibb HIV Products; CIBA Vision AG; Ciba-Geigy Corp.; Pilkington Barnes Hind; Ross Products.*

7825 Naphazoline Hydrochloride
550-99-2 6455 208-989-2
$C_{14}H_{15}ClN_2$
2-(1-Naphthylmethyl)-2-imidazoline monohydrochloride.
Ak-Con; Albalon; Allerest Eye Drops; Clera; Coldan; Comfort Eye Drops; Degest-2; Iridina Due; Naphcon; Niazol; Opcon; Privine Hydrochloride; Rhinatin; Rhinoperd; Sanorin; Sanorin-Spofa; Strictylon; Vasoclear; Vasocon; component of: Abalon-A Liquifilm, Clear Eyes, Naphcon-A, VasoClear A, Vasocon-A. Used as a vasoconstrictor and nasal decongestant. α-Adrenergic agonist (vasoconstrictor). mp = 255-260°; λ_m = 223, 270, 280, 287, 291 nm ($E_{1cm}^{1\%}$ 3622, 239, 286, 196, 198 EtOH); pKa (25°C) = 10.35; soluble in H_2O (40 g/100 ml), EtOH; slightly soluble in $CHCl_3$; insoluble in C_6H_6, Et_2O; LD_{50} (rat sc) = 385 mg/kg. *Alcon Labs; Allergan Inc.; Bristol-Myers Squibb HIV Products; CIBA Vision AG; Ciba-Geigy Corp.; Fisons plc; Iolab; Lovens Komiske Fabrik AS; Patchem AG; Ross Products.*

7826 Nemazoline
130759-56-7
$C_{10}H_{11}N_3$
2-(4-Amino-3,5-dichlorobenzyl)-2-imidazonline.
Nasal decongestant. *Schering-PloughHealth Care.*

7827 Octodrine
543-82-8 6854 208-851-1
$C_8H_{19}N$
1,5-Dimethylhexylamine.
SK&F-51; Vaporpac. Used as a nasal decongestant; local anesthetic. An α-adrenergic (vasoconstrictor) agonist. Viscous liquid; [dl-form]: bp = 154-156°; [hydrochloride]: soluble in H_2O; LD_{50} (mus ip) = 59 mg/kg, (rat ip) = 41.5 mg/kg. *SmithKline Beecham Pharm.*

7828 Oxymetazoline
1491-59-4 7100 216-079-1
$C_{16}H_{24}N_2O$
3-[(4,5-Dihydro-1H-imidazol-2-yl)methyl]-6-(1,1-dimethylethyl)-2,4-dimethylphenol.
H-990; Hazol; Navasin; Nezeril; Rhinofrenol; Rhinolitan; Sinerol; component of: Drixin. An α-adrenergic agonist (vasoconstrictor). Used as a nasal decongestant. mp = 181-183°. *Boehringer Ingelheim GmbH; Bristol-Myers Squibb HIV Products; E. Merck; Menley & James Labs Inc.; Schering-Plough Pharm.; Sterling Health U.S.A.; Whitehall-Robins.*

7829 Oxymetazoline Hydrochloride
2315-02-8 7100 219-015-0
$C_{16}H_{25}ClN_2O$
3-[(4,5-Dihydro-1H-imidazol-2-yl)methyl]-6-(1,1-dimethylethyl)-2,4-dimethylphenol monohydrochloride.
Sch-9384; Afrazine; Afrin; Iliadin; Nafrine; Nasivin; Oxilin; Sinex; Allerest 12 Hour Nasal Spray; 4-Way Nasal 12 Hour Spray; Dristan Long-Lasting Nasal Mist; Duration; Neo-Synephrine 12 Hour; Neo-Synephrine 12 Hour NTZ; Nostrilla; Ocuclear; Sch-9384; Afrazine; Iliadin; Nafrine; Nasivin; Oxilin; Sinex; component of: Benzedrex Nasal Spray 12 Hour, Drixin. An α-adrenergic agonist. Used as a nasal decongestant. mp = 300-303° (dec); soluble in H_2O, EtOH; insoluble in Et_2O, $CHCl_3$, C_6H_6; LD_{50} (mus orl) = 10 mg/kg. *Boehringer Ingelheim*

Ltd.; E. Merck; Fisons plc; Schering AG; Schering-Plough Pharm.; Sterling Winthrop Inc.; Whitehall Labs. Inc.

7830 Phenylephrine

59-42-7 7440 200-424-8

$C_9H_{13}NO_2$

(R)-3-Hydroxy-α-[(methylamino)methyl]-benzenemethanol.

Mydriatic; decongestant. α-Adrenergic agonist. mp = 169-172°.

7831 Phenylephrine Hydrochloride

61-76-7 7440 200-517-3

$C_9H_{14}ClNO_2$

(R)-m-Hydroxy-α[(methylamino)methyl]benzyl alcohol hydrochloride.

metaoxedrin; Adrianol; Ak-Dilate; Ak-Nefrin; Alcon Efrin; Biomydrin; Isophrin; m-Sympatol; Mezaton; Mydfrin; Neophryn; Neo-Synephrine; Nostril; Pyracort D; Prefrin; component of: Afrin 4-Way Nasal Spray Regular and Menthol, Anaplex HD, Benzedrex Nasal Spray Regular, Cerose-DM, Cyclomydril, Dristan Cold Multi-Symptom, Dristan Nasal Spray, Entex Capsules and Liquid, Histalet Forte, Hycomine Compound, Isopto Frin, Naldecon, Novahistine, Op-Isophrin-Z, Phenergan VC, Prefrin Liquifilm, Preparation H Cream, Preparation H Ointment, PV Tussin Syrup, Relief, RU-Tuss, Tympagesic, Vasosulf, Zincfrin. Mydriatic; decongestant. α-Adrenergic agonist. mp = 140-145°; $[\alpha]_D^{25}$ = -46.2° to -47.2°; soluble in H_2O, EtOH; LD_{50} (rat ip) = 17 ± 1.1 mg/kg, (rat sc) = 33.0 ± 2 mg/kg. *Alcon Labs; Boehringer Ingelheim Ltd.; Boots Pharmaceuticals Inc.; Bristol-Myers Squibb Co; CIBA Vision AG; DuPont-Merck Pharm.; Fisons plc; Marion Merrell Dow Inc.; Norwich Eaton; Parke-Davis; Robins, A. H. Co.; Schering AG; Schering-Plough Pharm.; Solvay Animal Health Inc.; Sterling Winthrop Inc.; Whitehall Labs. Inc.; Wyeth-Ayerst Labs.*

7832 Phenylpropanolamine

14838-15-4 7461 238-900-2

$C_9H_{13}NO$

(1RS,2SR)-2-Amino-1-phenylpropan-1-ol.

(±)-norephedrine; dl-norephedrine. Decongestant; anorexic. Sympathomimetic amine. [(dl)-form]: mp = 101-101.5°. *Knoll Pharm. Co.*

7833 Phenylpropanolamine Hydrochloride

154-41-6 7461 205-826-7

$C_9H_{14}ClNO$

(±)-α-(1-Aminoethyl)benzenemethanol hydrochloride.

1-phenyl-2-amino-1-propanol hydrochloride; mydriatin; Kontexin; Monydrin; Obestat; Propadrine; component of: A.R.M., Chlor-Trimeton Allergy/Sinus/Headache, Comtrex Liquigels, Comtrex Non-Drowsy Liquigels, Contac, Contac 12 Hour Caplets, Contac 12 Hour Capsules, Contac Severe Cold Maximum Strength, Coricidin D, Coricidin Sinus, Corsym Capsules, Demazin, Demilets, Diet Gard, Dimetapp Allergy Sinus, Dimetapp Cold and Allergy, Dimetapp Cold and Cough; Dimetapp DM, Dimetapp Elixir, Dimetapp Extentabs, Dimetapp Liquigels, Dimetapp 4 Hour Tablets, Entex Capsules and Liquid, Histalet Forte, Hycomine, Hycomine Pediatric, Naldecon, Naldecon CX, Ornade, Rhinex D-Lay Tablets, Robitussin CF, Sine-Off Sinus Medicine Tablets Aspirin Formula, Sinubid, Teldrin Timed Release Allergy Capsules, Triaminic, Trind, Trind-DM, Tuss-Ornade. Anorexic and decongestant. Psychomotor stimulant. Used in veterinary medicine as a bronchodilator and nasal decongestant. Sympathomimetic amine. Related to norpseudoephedrine. mp = 190-194°; soluble in H_2O, EtOH; insoluble in Et_2O, $CHCl_3$, C_6H_6; pKa = 9.44 ± 0.04; LD_{50} (rat orl) = 1490 mg/kg; [(+)-form hydrochloride]: mp = 171-172°; $[\alpha]_D^{25}$ = 32° (H_2O); [(±)-form base]: mp = 101-101.5. *Apothecon; Boots Pharmaceuticals Inc.; Bristol-Myers Squibb HIV Products; DuPont-Merck Pharm.; Fisons plc; Lemmon Co.; Mead Johnson Nutritionals; Menley & James Labs Inc.; Parke-Davis; Procter & Gamble Pharm. Inc.; Robins, A. H. Co.; Sandoz Pharm. Corp.; Schering-Plough Pharm.; SmithKline Beecham Pharm.; Solvay Pharm. Inc.; Taiho; Whitehall-Robins.*

7834 Propylhexedrine

101-40-6 8045 222-741-0

$C_{10}H_{21}N$

(±)-N,α-Dimethylcyclohexaneethylamine.

hexahydrodesoxyephedrine; Benzedrex; Dristan Inhaler. Adrenergic (vasoconstrictor); decongestant. Oily liquid; bp_{760} = 205°; bp_{20} = 92-93°; [(d)-form]: bp_{10} = 82-83°; [(l)-form]: bp_9 = 80-81°; d_4^{25} = 0.8501; n_D^{20} = 1.4600; [(d)-form]: n_D^{20} = 1.4588; [(l)-form]: n_D^{20} = 1.4590; slightly soluble inH_2O; miscible with EtOH, $CHCl_3$, Et_2O. *Menley & James Labs Inc.; SmithKline Beecham Pharm.; Whitehall Labs. Inc.*

7835 Propylhexedrine Hydrochloride

6192-95-9 8045 228-246-6

$C_{10}H_{22}ClN$

(±)-N,α-Dimethylcyclohexaneethylamine monohydrochloride.

Adrenergic (vasoconstrictor); decongestant. Dec 127-128°; soluble in H_2O. *SmithKline Beecham Pharm.*

7836 l-Propylhexedrine Ethylphenylbarbiturate

4388-82-3 8045 224-504-7

$C_{22}H_{33}N_3O_3$

l-N,α-Dimethylcyclohexaneethylamine ethylphenylbarbiturate.

barbexaclone; Maliasin. An α-adrenergic agonist (vasoconstrictor). Used as a mydriatic and nasal decongestant. *See* propylhexedrine. *Menley & James Labs Inc.; SmithKline Beecham Pharm.; Whitehall-Robins.*

7837 Tetrahydrozoline

84-22-0 9358 201-522-3

$C_{13}H_{16}N_2$

2-(1,2,3,4-Tetrahydro-1-naphthyl)-2-imidazoline.

tetryzoline. Adrenergic (vasoconstrictor); decongestant. An α-adrenergic agonist. *Alcon Labs; Marion Merrell Dow Inc.; Pfizer Inc.; Ross Products; Wyeth-Ayerst Labs.*

7838 Tetrahydrozoline Hydrochloride

522-48-5 9358 208-329-3

$C_{13}H_{17}ClN_2$

2-(1,2,3,4-Tetrahydro-1-naphthyl)-2-imidazoline hydrochloride.

Murine Plus; Rhinopront; Soothe; Tinarhinin; Tyzanol; Tyzine; Visine; Yxin; component of: Collyrium Fresh-Eye Drops. An α-adrenergic agonist (vasoconstrictor). Used as a nasal decongestant. mp = 256-257° (dec); λ_m = 264.5, 271.5 nm ($A_{1\,cm}^{1\%}$ 17.5, 15.5); freely soluble in H_2O, EtOH;

slightly soluble in $CHCl_3$; insoluble in Et_2O. *Alcon Labs; Key Pharm.; Marion Merrell Dow Inc.; Pfizer Inc.; Ross Products; Wyeth-Ayerst Labs.*

7839 Tramazoline
1082-57-1 9702 214-105-6
$C_{13}H_{17}N_3$
4,5-Dihydro-N-(5,6,7,8-tetrahydro-1-naphthalenyl)-1H-imidazol-2-amine.
An α-adrenergic agonist. Used as a nasal decongestant. mp = 142-143°. *Boehringer Ingelheim GmbH; Thomae GmbH Dr. Karl.*

7840 Tramazoline Hydrochloride Monohydrate
3715-90-0 9702 223-064-3
$C_{13}H_{20}ClN_3O$
4,5-Dihydro-N-(5,6,7,8-tetrahydro-1-naphthalenyl)-1H-imidazol-2-amine monohydrochloride monohydrate.
KB-227; Biciron; Ellatun; Rhinaspray; Rhinogutt; Rhinospray; Rinogutt; Towk. An α-adrenergic agonist (vasoconstrictor). Used as a nasal decongestant. mp = 172-174°; soluble in H_2O; LD_{50} (mus orl) = 195 mg/kg. *Boehringer Ingelheim GmbH; Thomae GmbH Dr. Karl.*

7841 Tuaminoheptane
123-82-0 9934 204-655-5
$C_7H_{17}N$
1-Methylhexylamine.
2-heptanamine; 2-aminoheptane; Tuamine. Decongestant (nasal); adrenergic (vasoconstrictor). α-Adrenergic agonist. Topical vasoconstrictor. Volatile liquid; bp_{760} = 142-144°; d_D^{25} = 0.7600-0.7660; n_D^{25} = 1.4150-1.4200; pH (1% aqueous solution) = 11.45; slightly soluble in H_2O; soluble in EtOH, Et_2O, petroleum ether, $CHCl_3$, C_6H_6. *Eli Lilly & Co.*

7842 Tuaminoheptane Sulfate
6411-75-2 9934 229-113-5
$C_{14}H_{36}N_2O_4S$
Bis[(1-methylhexyl)ammonium] sulfate.
2-heptanamine sulfate (2:1); Heptedrine. Decongestant (nasal); adrenergic (vasoconstrictor). α-Adrenergic agonist. Topical vasoconstrictor. Soluble in H_2O; pH (1% aqueous solution) = 5.4. *Eli Lilly & Co.*

7843 Tymazoline
24243-97-8 9965
$C_{14}H_{20}N_2O$
2-(2-Isopropyl-5-methylphenoxymethyl)-2-imidazoline.
Pernazene. An α-adrenergic agonist. Used as a nasal decongestant. [hydrochloride]: mp = 215-217°, 223.5-225°. *CIBA plc.*

7844 Xylometazoline
526-36-3 10219 208-390-6
$C_{16}H_{24}N_2$
2-[[4-(1,1-Dimethylethyl)-2,6-dimethylphenyl]-methyl]-4,5-dihydro-1H-imidazole.
Otrivin. Decongestant; adrenergic (vasoconstrictor). α-Adrenergic agonist. Used as a nasal decongestant. mp = 131-133°. *Ciba-Geigy Corp.; Sterling Winthrop Inc.*

7845 Xylometazoline Hydrochloride
1218-35-5 10219 214-936-4
$C_{16}H_{25}ClN_2$
2-[[4-(1,1-Dimethylethyl)-2,6-dimethylphenyl]-methyl]-4,5-dihydro-1H-imidazole monohydrochloride.
Neo-Rinoleina; Neo-Synephrine II; Novorin; Olynth; Otriven; Otrivin hydrochloride; Otrix; Therapin; Xymelin. Nasal decongestant and adrenergic (vasoconstrictor). An α-Adrenergic agonist. Soluble in H_2O (< 3 g/100 ml), MeOH, EtOH; nearly insoluble in C_6H_6, Et_2O. *CIBA plc; Ciba-Geigy Corp.; Sterling Winthrop Inc.*

Depigmentors

7846 Captamine Hydrochloride
13242-44-9 236-221-6
$C_4H_{12}ClNS$
2-(Dimethylamino)ethanethiol hydrochloride.
NSC-45463. Depigmentor. mp = 158-160°. *Schering-Plough HealthCare Products.*

7847 Hydroquinine
522-66-7 4852
$C_{20}H_{26}N_2O_2$
(8α,9R)-10,11-Dihydro-6'-methoxycinchonan-9-ol).
dihydroquinine. Depigmentor. An alkaloid of cinchona. Stereoisomer of hydroquinidine. mp = 172°; $[\alpha]_D^{18}$ = -142° (EtOH), $[\alpha]_D^{20}$ = -236° (c = 0.82 0.1N H_2SO_4); almost insoluble in H_2O (0.029 g/100 ml); freely soluble in EtOH, Me_2CO, $CHCl_3$, Et_2O, petroleum ether; [hydrochloride hemihydrate ($C_{20}H_{27}ClN_2O_2.0.5H_2O$)]: mp = 208°; $[\alpha]_D^{21}$ = -124° (c = 1.1); nearly insoluble in Et_2O; freely soluble in H_2O, EtOH, MeOH, Me_2CO; pH (0.005 M) = 5.85.

7848 Hydroquinone
123-31-9 4853 204-617-8
$C_6H_6O_2$
1,4-Benzenediol.
Black & White Bleaching Cream; Eldopaque Forte; Eldoquin Forte; Solaquin Forte; Eldopacque; Tecquinol; quinol; Aida; component of: Artra. Depigmentor. mp = 172-175°; bp = 285°; d = 1.3280; soluble in H_2) (7.1 g/100 ml), EtOH, Et_2O; slightly soluble in C_6H_6; LD_{50} (rat orl) = 320 mg/kg. *Eastman Chem. Co.; ICN Pharm. Inc.; Schering-Plough HealthCare Products; Schering-Plough Pharm.*

7849 Monobenzone
103-16-2 6331 203-083-3
$C_{13}H_{12}O_2$
4-Benzyloxyphenol.
hydroquinone monobenzyl ether; Benoquin; Depigman; Pigmex; Benzoquin; Agerite. Depigmentor. mp = 121-123°; insoluble in cold H_2O; soluble in hot H_2O (→ 1 g/100 ml); soluble in EtOH, Et_2O, C_6H_6; LD_{50} (rat ip) = 4500 mg/kg.

Dermatitis Suppressants

7850 Dapsone
80-08-0 2885 201-248-4
$C_{12}H_{12}N_2O_2S$
4,4'-Sulfonyldianiline.
4,4'-diaminodiphenyl sulfone; diaphenylsulfone; NSC-6091; DDS; DADPS; 1358F; Avlosulfon; Croysulfone; Diphenasone; Disulone; Dumitone; Eporal; Novophone; Sulfona-Mae; Sulphadione; Udolac. Sulfone antibiotic. Leprostatic and dermatitis herpetiformis suppressant. Also used as a hardening agent with epoxy resins and, in veterinary medicine, as a coccidostat. mp = 175-176°, 180.5°; insoluble in H_2O; soluble in EtOH, MeOH, Me_2CO, dilute HCl. *I.G. Farben.*

7851 Sulfapyridine
144-83-2 9108 205-642-7
$C_{11}H_{11}N_3O_2S$
4-Amino-N,2-pyridinylbenzenesulfonamide.
N^1-2-pyridylsulfanilamide; M&B-693; Dagenan; Eubasin; Pyriamid; Coccoclase; Septipulmon. Sulfonamide antibiotic; dermatitis suppressant. Used in the treatment of dermatitis herpetiformis. mp = 190-191°; soluble in H_2O (0.029 g/100 ml), EtOH (0.227 g/100 ml), Me_2CO (1.54 g/100 ml); freely soluble in mineral acids, KOH, NaOH solutions; LD_{50} (mus orl) = 7500 mg/kg. *May & Baker Ltd.*

7852 Sulfapyridine Sodium Salt Monohydrate
127-57-1 9108 204-850-5
$C_{11}H_{10}N_3NaO_2S \cdot H_2O$
4-Amino-N-2-pyridinylbenzenesulfonamide sodium salt monohydrate.
soluble sulfapyridine; Izopiridina; Soludagenan. Sulfonamide antibiotic. Dermatitis suppressant. Used in the treatment of dermatitis herpetiformis. Soluble in H_2O (66.6 g/100 ml), EtOH (10 g/100 ml); LD_{50} (mus orl) = 2.7 mg/kg. *May & Baker Ltd.*

Diagnostic Aids

7853 Glycerol
56-81-5 4493 200-289-5
$C_3H_8O_3$
1,2,3-Propanetriol.
glycerine; glycerin; tryhydroxypropane; incorporation factor; IFP; Bulbold; Cristal; Glyceol Opthalgan; Osmoglyn. Osmotic diuretic. Diagnostic aid (opthalmic). Used to reduced intraocular pressure and vitreous volume for ocular surgery. Syrupy liquid; mp = 178°; bp1.0:sk = 125.5; n_D^{25} = 1.4730; d_{25}^{25} = 1.24910; miscible with H_2O, alcohol; insoluble in C_6H_6, $CHCl_3$, CCl_4, petroleum ether, oils; LS_{50} (rat orl) > 20 ml/kg, (rat iv) = 4.4 ml/kg.

7854 Metyrapone
54-36-4 6246 200-206-2
$C_{14}H_{14}N_2O$
2-Methyl-1,2-di-3-pyridyl-1-propanone.
methopyrapone; mepyrapone; metopyrone; methbipyranone; Metopirone; Metroprione. Diagnostic aid. Used in the determination of pituitary function, adrenocortical insufficiency, and Cushings disease (hypercortisolism). mp = 50-51°. *Ciba-Geigy Corp.*

7855 Metyrapone Tartrate
908-35-0 6246
$C_{14}H_{14}N_2O.2C_4H_6O_6$
2-Methyl-1,2-di-3-pyridyl-1-propanone tartrate (1:2).
Su-4885; Metopirone Ditartrate. Diagnostic aid. Used in the determination of pituitary function, adrenocortical insufficiency, and Cushings disease (hypercortisolism). *Ciba-Geigy Corp.*

Digestive Aids

7856 Amylase
9000-92-4 640 232-567-7
Digestive aid. Usually refers to α-amylase, an enzyme derived from bacteria.

7857 α-Amylase, Bacterial
9000-85-5 640 232-560-9
Digestive aid. Generally derived from bacteria such as *Bacillus subtilis.* The amylases are enzymes which catalyze the hydrolysis of α-1→4 glucosidic linkages in polysaccharides.

7858 α-Amylase, Porcine
9000-90-2 640 232-565-6
Buclamase; Maxilase. Digestive aid. Enzyme derived from swine pancrease. Molecular weight ≅ 45,000.

7859 α-Amylase, Sweet Potato
9000-91-3 640 232-566-1
Digestive aid. Enzyme from sweet potato, molecular weight ≅ 152,000.

7860 Lipase
9001-62-1 5536 232-619-9
Triacylglycerol lipase.
Digestive aid. An esterase which hydrolyzes fats, producing fatty acids and glycerol.

7861 Pancreatin
8049-47-6 7137 232-468-9
Diastase vera; Creon; Pancrease; Pancrex-Vet; Pankrotanon; Panzytrat; Zypanar. Digestive aid. Used in veterinary medicine in treatment of pancreatic enzyme deficiency. Derived from pancreas of hog or ox. Contains the enzymes amylopsin, steapsin, and trypsin. Converts strach into soluble carbohydrates. Insoluble in EtOH; partly soluble in H_2O; highest activity in neutral or slightly alkaline media.

7862 Pancrelipase
53608-75-6 7138 258-659-7
Accelerase; Cotazym; Ilozyme; Ku-Zyme HP; Pancrease; Viokase. Digestive aid. Enzyme concentrate containing primarily lipase, and also amylase and proteases.

7863 Papain
9001-73-4 7148 232-627-2
vegetable pepsin; Arbuz; Caroid; Nematolyt; Papayotin; Summetrin; Tromasin; Velardon; Vermizym; component of: Panafil. Proteolytic enzyme. Digestive aid. A proteolytic enzyme isolated from the fruit and leaves of

Carica papaya. Used as a digestive aid, also as a debridant, an anthelmintic targeting Nematodes and as an agent which can prevent adhesions. λ_m = 278 nm ($A^{1\%}_{1\,cm}$ 25.0); insoluble in most organic solvents. *Rystan Co. Inc.; Sterling Winthrop Inc.*

7864 Pepsin
9001-75-6 7289 232-629-3
Puerzym. Digestive aid. Has been used to treat deficiency of pepsin secretion. The principal digestive enzyme in gastric juice; controls the degradation of proteins to proteoses and peptones hydrolyzing only peptide bonds. $[\alpha]_D^{26}$= -64.5° (pH 4.6); freely soluble in H_2O; insoluble in EtOH, $CHCl_3$, Et_2O.

7865 Rennin
9001-98-3 8303 232-645-0
Chymosin; rennase; lab; abomasal enzyme. Digestive aid. The milk-clotting enzyme from the stomach of the calf. Secreted as prorennin which is converted to rennin in acid. Rennet, a dried extract containing rennin is used in themanufacture of cheese and rennet casein.

Diuretics

7866 Acefylline
652-37-9 22 211-490-2
$C_9H_{10}N_4O_4$
1,2,3,6-Tetrahydro-1,3-dimethyl-2,6-dioxopurine-7-acetic acid.
carboxymethyltheophylline; 7-theophylline acetic acid; Aminodal [as sodium salt]. Diuretic; cardiotonic; bronchodilator. Xanthine derivative. mp = 271°; [sodium salt]: mp > 300°. *E. Merck.*

7867 Acetazolamide
59-66-5 50 200-440-5
$C_4H_6N_4O_3S_2$
N-(5-Sulfamoyl-1,3,4-thiadiazol-2-yl)acetamide.
acetazoleamide; carbonic anhydrase inhibitor 6063; acetamox; atenezol; cidamex; defiltran; diacarb; diamox; didoc; diluran; diureticum-Holzinger; diuriwas; diutazol; donmox; edemox; fonurit; glaupax; glupax; natrionex; nephramid; vetamox (sodium salt). Diuretic; antiglaucoma agent. Carbonic anhydrase inhibitor. mp = 258-259°; sparingly soluble in H_2O; pKa = 7.2. *Lederle Labs.*

7868 Acetazolamide Sodium
1424-27-7 50
$C_4H_6N_4NaO_3S_2$
N-(5-Sulfamoyl-1,3,4-thiadiazol-2-yl)acetamide monosodium salt.
Diamox Parenteral; Vetamox. Carbonic anhydrase inhibitor, used as diuretic, in treatment of glaucoma. *Lederle Labs.*

7869 Alipamide
3184-59-6
$C_9H_{12}ClN_3O_3S$
4-Chloro-3-sulfamoylbenzoic acid 2,2-dimethylhydrazide.
CI-546; CN-38474; D-1721. Diuretic; antihypertensive. *Parke-Davis.*

7870 Althiazide
5588-16-9 326 226-994-8
$C_{11}H_{14}ClN_3O_4S_3$
3-[(Allylthio)methyl-6-chloro-3,4-dihydro-2H-1,2,4-benzothiadiazine-7-sulfonamide 1,1-dioxide.
P-1779; Altizide; Aldactazine. Diuretic; antihypertensive. mp = 206-207°. *Pfizer Inc.*

7871 Amanozine
537-17-7 388
$C_9H_9N_5$
N-Phenyl-1,3,5-triazine-2,4-diamine.
N-phenylformoguanamine; W-1191-2; Urofort. Diuretic. mp= 235-236°; [hydrochloride]: mp = 258-260°. *Richter.*

7872 Ambuphylline
5634-34-4 404 227-077-5
$C_{11}H_{19}N_5O_3$
Theophylline compound with 2-amino-2-methyl-1-propanol (1:1).
2-amino-2-methyl-1-propanol compound with theophylline; bufylline (formerly); theophylline aminoisobutanol; Butaphyllamine; Buthoid; component of: Nethaphyl. Bronchodilator; diuretic; smooth muscle relaxant. Xanthine derivative. mp = 254-256°; soluble in H_2O (55 g/100 ml); LD_{50} (mus orl) = 600 mg/kg. *Marion Merrell Dow Inc.*

7873 Ambuside
3754-19-6 405 223-158-4
$C_{13}H_{16}ClN_3O_5S_2$
N^1-Allyl-4-chloro-6-[(3-hydroxy-2-butenylidene)amino]-m-benzenedisulfonamide.
EX-4810; RMI-83047; Hydrion; Novohydrin. Diuretic; antihypertensive. mp = 205-207°; λ_m = 343 nm (ε 32900). *Marion Merrell Dow Inc.*

7874 Amiloride
2609-46-3 426 220-024-7
$C_6H_8ClN_7O$
N-Amidino-3,5-diamino-6-chloropyrazine-carboxamide.
guanamprazine; amipramidin; amipramizide. Diuretic. Aldosterone antagonist. Potassium conserving agent. mp = 240.5-241.5°. *Merck & Co.Inc.*

7875 Amiloride Hydrochloride
17440-83-4 426
$C_6H_9Cl_2N_7O.2H_2O$
N-Amidino-3,5-diamino-6-chloropyrazine-carboxamide monohydrochloride dihydrate.
Amilorin; Frumil; Midamide; Midamor; Moduretic; N-Amidino-3,5-diamino-6-chloropyrazinecarboxamide monohydrochloride dihydrate; Amikal; Colectril; Modamide. Diuretic. Aldosterone antagonist. Potassium conserving agent. mp = 285-288°; λ_m = 212, 285, 362 nm ($E^{1\%}_{1\,cm}$ 642, 555, 617 H_2O); slightly soluble in H_2O, EtOH; insoluble in organic solvents; freely soluble in DMSO. *Merck & Co.Inc.*

7876 Aminometradine
642-44-4 471 211-384-6
$C_9H_{13}N_3O_2$
1-Allyl-6-amino-3-ethyluracil.

Mictine; Katapyrin; Mincard; Catapyrin. Diuretic. mp = 75-115°, [anhydrous]: mp = 143-144°. *Searle G.D. & Co.*

7877 Amisometradine
550-28-7 507 208-980-3
$C_9H_{13}N_3O_2$
6-Amino-3-1-(2-methylallyl)-2,4(1H,3H)-pyrimidinedione.
aminoisometradine; Rolicton. Diuretic. mp = 175°; soluble in H_2O (2 g/100 ml), EtOH, Me_2CO; insoluble in Et_2O; LD_{50} (mus orl) = 610 mg/kg, (mus ip) = 415 mg/kg. *Searle G.D. & Co.*

7878 Ammonium Acetate
631-61-8 520 211-162-9
$C_2H_7NO_2$
Acetic acid ammonium salt.
Mindererus's spirit. Diuretic. mp = 114°; d = 1.07; soluble in H_2O, EtOH; less soluble in Me_2CO.

7879 Anaritide
95896-08-5
$C_{112}H_{175}N_{39}O_{35}S_3$
H-Arg-Ser-Ser-Cys-Phe-Gly-Glt-Arg-Met-Asp-Arg-Ile-Gly-Ala-Gln-Ser-Gly-Leu-Gly-Cys-Asn-Ser-Phe-Arg-Tyr-OH.
atriopeptid-21 (rat), N-L-arginyl-8-L-methionine-21a-L-phenylalanine-21b-L-arginine-21c-L-tyrosine; Wyeth 47,663. Diuretic; antihypertensive. *Wyeth Labs.*

7880 Anaritide Acetate
104595-79-1
$C_{112}H_{175}N_{39}O_{35}S_3.xC_2H_4O_2$
H-Arg-Ser-Ser-Cys-Phe-Gly-Glt-Arg-Met-Asp-Arg-Ile-Gly-Ala-Gln-Ser-Gly-Leu-Gly-Cys-Asn-Ser-Phe-Arg-Tyr-OH.xCH_3COOH.
atriopeptid-21 (rat), N-L-arginyl-8-L-methionine-21a-L-phenylalanine-21b-L-arginine-21c-L-tyrosine, acetate (salt). Diuretic; antihypertensive. *Wyeth Labs.*

7881 Arbutin
497-76-7 812 207-850-3
$C_{12}H_{16}O_7$
4-Hydroxyphenyl-β-D-glucopyranoside.
hydroquinone glucose; arbutoside; ursin; Uvasol. Diuretic; urinary anti-infective. Found in leaves from species of *Saxifragacea, Rosacea, Ericaceae*. mp = 165°, 199.5-200°; $[\alpha]_D^{25}$= -64° (c = 3); soluble in H_2O, EtOH; highly hygroscopic.

7882 Azolimine
40828-45-3
$C_{10}H_{11}N_3O$
2-Imino-3-methyl-1-phenyl-4-imidazolidinone.
CL-90748. Diuretic.

7883 Azosemide
27589-33-9 953 248-549-7
$C_{12}H_{11}ClN_6O_2S_2$
2-Chloro-5-(1H-tetrazol-5-yl)-N^4-2-thenyl-sulfanilamide.
Ple-1053; Diart; Diurapid; Luret. Diuretic. mp = 218-221°. *Boehringer Mannheim GmbH.*

7884 Bemitradine
88133-11-3
$C_{15}H_{17}N_5O$
5-Amino-8-(2-ethoxyethyl)-7-phenyl-s-triazolo-[1,5-c]pyrimidine.
SC-33643. Diuretic; antihypertensive . *Searle G.D. & Co.*

7885 Bendroflumethazide
73-48-3 1064 200-800-1
$C_{15}H_{14}F_3N_3O_4S_2$
3-Benzyl-3,4-dihydro-6-(trifluoromethyl)-2H-1,2,4-benzothiadiazine-7-sulfonamide 1,1-dioxide.
benzylhydroflumethiazide; benzydroflumethiazide; bendrofluazide; Naturetin; Corzide; Rautrax N; Rauzide; Aprinox; Benzy-Rodiuran; Berkozide; Bristuric; Bristuron; Centyl; Flumersil; Naturetin; Naturine; Neo-Naclex; Naigaril; Nikion; Orsile; Pluryle; Plusuril; Poliuron; Relan Beta; Salures; Sinesalin; Sodiuretic; Urlea. Diuretic; antihypertensive. A thiazide. mp = 224.5=225.5°, 221-223°; λ_m = 208, 273, 326 nm ($E^{1\%}_{1\,cm}$ 745, 565, 96 MeOH); soluble in Me_2CO, EtOH; insoluble in H_2O, $CHCl_3$, C_6H_6, Et_2O. *Apothecon; Bristol-Myers Squibb Co.*

7886 Benzthiazide
91-33-8 1155 202-061-0
$C_{15}H_{14}ClN_3O_4S_3$
3-[(Benzylthio)methyl]-6-chloro-2H-1,2,4-benzo-thiadiazine-7-sulfonamide 1,1-dioxide.
Fovane; Exna; Aquatag; Dihydrex; Diucen; Edemex; ExNa; Exosalt; Freeuril; HyDrine; Lemazide; Proaqua; Urese; component of: Dytide. Diuretic. A thiazide. mp = 231-232°, 238-239°; insoluble in H_2O; soluble in alkaline solutions; LD_{50} (rat orl) >10000 mg/kg, (rat iv) = 422 mg/kg, (mus orl) > 5000 mg/kg, (mus iv) = 410 mg/kg. *Robins, A. H. Co.*

7887 Benzylhydrochlorothiazide
1824-50-6 1171
$C_{14}H_{14}ClN_3O_4S_2$
3-Benzyl-3,4-dihydro-6-chloro-2H-1,2,4-thiadiazine-7-sulfonamide 1,1-dioxide.
Behyd. Diuretic; antihypertensive. mp = 260-262°, 269°.

7888 Besulpamide
90992-25-9
$C_{15}H_{16}ClN_3O_3S$
1-(4-Chloro-3-sulfamoylbenzamido)-2,4,6-trimethylpyridinium hydroxide inner salt.
Has diuretic properties.

7889 Brocrinat
72481-99-3
$C_{15}H_9BrFNO_4$
[[7-Bromo-3-(o-fluorophenyl)-1,2-benzisoxazol-6-yl]-oxy]acetic acid.
HP-522; HP-3522; P-78-3522. Diuretic. *Hoechst Roussel Pharm. Inc.*

7890 Bumetanide
28395-03-1 1508 249-004-6
$C_{17}H_{20}N_2O_5S$
3-(Butylamino)-4-phenoxy-5-sulfamoylbenzoic acid.
Bumex; PF-1593; Ro-10-6338; Burinex; Fontego; Fordiuran; Lixil; Lunetoron. A high-ceiling, or loop

diuretic. mp = 230-231°; LD_{50} (mus iv) = 330 mg/kg. *Hoffmann-LaRoche Inc.*

7891 Butazolamide
16790-49-1 1545
$C_6H_{10}N_4O_3S_2$
N-[5-(Aminosulfonyl)-1,3,4-thiadiazol-2-yl]butanamide.
SKF-4965; Butamide. Carbonic anhydrase inhibitor, used as diuretic, in treatment of glaucoma mp = 260-262° (dec). *Am. Cyanamid.*

7892 Buthiazide
2043-38-1 1554 218-048-8
$C_{11}H_{16}ClN_3O_4S_2$
6-Chloro-3,4-dihydro-3-isobutyl-2H-1,2,4-benzothiadiazine-7-sulfonamide 1,1-dioxide.
Su-6187; S-3500; Eunephran; Saltucin; Modenol. Diuretic; antihypertensive. A thiazide. mp= 228°, 241-245°. *Searle G.D. & Co.*

7893 Canrenoate Potassium
2181-04-6 1795 218-554-9
$C_{22}H_{29}KO_4$
Potassium 17-hydroxy-3-oxo-17α-pregna-4,6-diene-21-carboxylate.
SC-14266; Kanrenol; Soldactone; Venactone. Diuretic. Aldosterone antagonist. *Searle G.D. & Co.*

7894 Canrenoic Acid
4138-96-9 223-963-0
$C_{22}H_{30}O_4$
17-Hydroxy-3-oxo-17α-pregna-4,6-diene-21-carboxylic acid.
Diuretic. Aldosterone antagonist. *Searle G.D. & Co.*

7895 Canrenone
976-71-6 1795 213-554-5
$C_{22}H_{28}O_3$
17-Hydroxy-3-oxo-17α-pregna-4,6-diene-21-carboxylic acid σ lactone.
SC-9376; Phanurane. Diuretic. Aldosterone antagonist. mp = 149-151°; $[\alpha]_D$ = 24.5° ($CHCl_3$); λ_m = 283 nm (ε 26700). *Searle G.D. & Co.*

7896 Chloraminophenamide
121-30-2 2119 204-463-1
$C_6H_8ClN_3O_4S_2$
4-Amino-6-chloro-1,3-benzenedisulfonamide.
Idorese. Diuretic. mp = 251-252°; λ_m = 223.5-224.5, 265-266, 312-314 nm (ε 41776, 18633, 3874); slightly soluble in H_2O. *Merck & Co.Inc.*

7897 Chlorazanil
500-42-5 2123 207-904-6
$C_9H_8ClN_5$
2-Amino-4-p-chloroanilino-s-triazine.
chlorazinil; ASA-226; Diurazine; Triazurol; Orpizin; Daquin; Neo-Urofort; Neurofort. Diuretic. mp = 233-234°, 256-258°. *3M Pharm.*

7898 Chlorazanil Hydrochloride
2019-25-2 2123 217-962-4
$C_9H_9Cl_2N_5$
2-Amino-4-p-chloroanilino-s-triazine hydrochloride.
Daquin; Doclizid T; Orpidan. Diuretic. mp = 227-228°. *3M Pharm.*

7899 Chlormerodrin
62-37-3 2154 200-530-4
$C_5H_{11}ClHgN_2O_2$
[3-(Chloromercuri)-2-methoxypropyl)]-urea.
chlormeroprin; Mercloran; Neohydrin; Katonil; Mercoral; Diurone; Percapyl; Merilid; Oricur. Diuretic. mp = 152-153°; soluble in H_2O (1.1 g/100 ml), EtOH (1.1 g/100 ml); poorly soluble in $CHCl_3$; LD_{50} (rat orl) = 82 mg/kg. *Marion Merrell Dow Inc.; Parke-Davis.*

7900 Chlorothiazide
58-94-6 2221 200-404-9
$C_7H_6ClN_3O_4S_2$
6-Chloro-2H-1,2,4-benzothiadiazine-7-sulfonamide 1,1-dioxide.
Chlotride; Diuril; Diuril Boluses; Aldoclor; Diupres; Diuril Lyovac [as sodium salt]; Lyovac Diuril [as sodium salt]. Diuretic; antihypertensive. A thiazide. mp = 342.5-343°; soluble in DMSO, DMF; less soluble in MeOH, C_5H_5N; insoluble in Et_2O, $CHCl_3$, C_6H_6; poorly soluble in H_O. *Merck & Co.Inc.*

7901 Chlorthalidone
77-36-1 2246 201-022-5
$C_{14}H_{11}ClN_2O_4S$
2-Chloro-5-(1-hydroxy-3-oxo-1H-isoindolinyl)-benzenesulfonamide.
Hygroton; Thalitone; Combipres; Demi-Regroton; Regroton; Tenoretic; G-33182; NSC-69200. Diuretic; antihypertensive. Nonthiazide compound with a similar mechanism of action to the thiazide diuretics. mp= 224-226°; λ_m = 220 nm (MeOH); soluble in H_2O (12 mg/ml at 20°, 27 mg/ml at 37°), slightly soluble in EtOH, Et_2O. *Boehringer Ingelheim GmbH; KV Pharm.; Parke-Davis; Rhône-Poulenc Rorer Pharm. Inc.; Zeneca Pharm.*

7902 Clazolimine
40828-44-2
$C_{10}H_{10}ClN_3O$
1-(p-Chlorophenyl)-2-imino-3-methyl-4-imidazolidinone.
CL-88893. Diuretic.

7903 Clofenamide
671-95-4 2434 211-588-5
$C_6H_7ClN_2O_4S_2$
4-Chloro-m-benzenedisulfonamide.
chlorphenamide; Salco; Saltron; Soluran; Aquedux; Haflutan. Diuretic. mp = 206-207°; soluble in hot H_2O, EtOH; less soluble in cold solvents.

7904 Clopamide
636-54-4 2454 211-261-7
$C_{14}H_{20}ClN_3O_3S$
4-Chloro-N-(2,6-dimethylpiperidino)-3-sulfamoylbenzamide.
chlosudimeprimyl; DT-327; Adurix; Aquex; Brinaldix. Antihypertensive; diuretic. A thiazide. [hydrazine derivative]: mp = 244-246°. *Sandoz Pharm. Corp.*

7905 Clorexolone
2127-01-7 2466 218-342-6
$C_{14}H_{17}ClN_2O_3S$
6-Chloro-2-cyclohexyl-3-oxo-5-isoindolinesulfonamide.
M&B-8430; RP-12833; Flonatril; Nefrolan. Diuretic. mp = 266-268°; poorly soluble in H_2O (1.6 mg/100 ml). *Marion Merrell Dow Inc.*

7906 Cyclopenthiazide
742-20-1 2813 212-012-5
$C_{13}H_{18}ClN_3O_4S_2$
6-Chloro-3-(cyclopentylmethyl)-3,4-dihydro-2H-1,2,4-benzothiadiazine-7-sulfonamide 1,1-dioxide.
Su-8341; NSC-107679; cyclomethiazide; tsiklometiazid; Su-8341; Navidrex; Navidrix; Salimid. Diuretic. mp= 230°; LD_{50} (rat iv) = 142 mg/kg, (mus iv) = 232 mg/kg. *Ciba-Geigy Corp.*

7907 Diapamide
3688-85-5
$C_9H_{11}ClN_2O_3S$
4-Chloro-N-methyl-3-(methylsulfamoyl)benzamide.
CI-456; CN-36,337; D-1593. Diuretic; antihypertensive. *Parke-Davis.*

7908 Disulfamide
671-88-5 3427 211-585-9
$C_7H_9ClN_2O_4S_2$
5-Chlorotoluene-2,4-disulfonamide.
disulphamide; Disamide; Natirene 25. Diuretic. mp = 260°; λ_m = 285 nm (ε 805 EtOH); insoluble in H_2O; soluble in EtOH (1.89 - 2.23 g/100 ml), iPrOH (0.35 g/100 ml), $CHCl_3$ (0.001 g/100 ml). *BDH Laboratory Supplies.*

7909 Epithiazide
1764-85-8 3661 217-181-9
$C_{10}H_{11}ClF_3N_3O_4S_2$
6-Chloro-3,4-dihydro-3[[(2,2,2-trifluoroethyl)thio]-methyl]-2H-1,2,4-benzothiadiazine-7-sulfonamide 1,1-dioxide.
P-2105; NSC-108164; Thiaver. Diuretic; antihypertensive. mp = 206-207°. *Pfizer Inc.*

7910 Ethacrynate Sodium
6500-81-8 3761
$C_{13}H_{11}Cl_2NaO_4$
Sodium [2,3-dichloro-4-(2-methylenebutyryl)-phenoxy]acetate.
Edecrin sodium; Lyovac Sodium Edecrin. Diuretic. λ_m = 225 nm (ε 15287 H_2O); soluble in H_2O (< 9 g/100 ml). *Merck & Co.Inc.*

7911 Ethacrynic Acid
58-54-8 3761 200-384-1
$C_{13}H_{12}Cl_2O_4$
[2,3-Dichloro-4-(2-methylenebutyryl)phenoxy]-acetic acid.
Edecril; Edecrin; MK-595; Crinuril; Endecril; Hydromedin; Reomax; Taladren; Uregit. A high-ceiling, or loop diuretic. mp = 121-122°; sparingly soluble in H_2O, soluble in $CHCl_3$; LD_{50} (mus iv) = 176 mg/kg, (mus orl) = 627 mg/kg. *Merck & Co.Inc.*

7912 Ethiazide
1824-58-4 3778 217-358-0
$C_9H_{12}ClN_3O_4S_2$
6-Chloro-3-ethyl-3,4-dihydro-2H-1,2,4-benzothiadiazine-7-sulfonamide 1,1-dioxide.
acthiazidum; Hypertane. Diuretic. mp = 269-270°. *Abbott Labs Inc.; Ciba-Geigy Corp.; Lederle Labs.; Lemmon Co.; Merck & Co.Inc.; Parke-Davis; Searle G.D. & Co.; Solvay Pharm. Inc.; Squibb E.R. & Sons; Wallace Labs.*

7913 Ethoxzolamide
452-35-7 3801 207-199-5
$C_9H_{10}N_2O_3S_2$
6-Ethoxy-2-benzothiazolesulfonamide.
ethoxyzolamide; Cardrase; Ethamide; Glaucotensil; Redupresin. Carbonic anhydrase inhibitor, used as diuretic, in treatment of glaucoma mp = 188-190.5°. *Upjohn Ltd.*

7914 Etozolin
73-09-6 3934 200-794-0
$C_{13}H_{20}N_2O_3S$
[3-Methyl-4-oxo-5-(1-piperidinyl)-2-thiazolidinylidene]acetic acid ethyl ester.
Go-787; W-2900A; Elkapin. Diuretic. mp = 140°; λ_m = 283, 243 nm (log ε 4.32, 4.0 MeOH); LD_{50} (mus ip) = 1.210 g/kg, (rat ip) = 1.575 g/kg; [hydrochloride]: mp = 158-159°. *Warner-Lambert.*

7915 Famotidine
76824-35-6 3972
$C_8H_{15}N_7O_2S_3$
[1-Amino-3-[[[2-[(diaminomethylene)amino]-4-thiazolyl]metyhl]thio]propylidene]sulfamide.
MK-208; YM-11170; Amfamox; Dispromil; Famodil; Famodine; Famosan; Famoxal; Fanosin; Fibonel; Ganor; Gaster; Gastridin; Gastropen; Ifada; Lecedil; Motiax; Muclox; Nulcerin; Pepcid®; Pepcid AC; Pepcidina; Pepcidine; Pepdine; Pepdul; Peptan; Ulcetrax; Ulfamid; Ulfinol. Antiulcerative. Histamine H_2 receptor antagonist. Used for short term treatment of active duodenal ulcers. mp = 163-164°; soluble in DMF (80 g/100 ml), AcOH (50 g/100 ml), MeOH (0.3 g/100 ml), H_2O (0.1 g/100 ml); insoluble in EtOH, EtOAc, $CHCl_3$; LD_{50} (mus iv) = 244.4 mg/kg. *Johnson & Johnson-Merck Consumer Pharm.; Merck & Co.Inc.*

7916 Fenquizone
20287-37-0 4039 243-689-5
$C_{14}H_{12}ClN_3O_3S$
(±)-7-Chloro-1,2,3,4-tetrahydro-4-oxo-2-phenyl-6-quinazolinesulfonamide.
M.G. 13054. Diuretic. mp > 310°; insoluble in H_2O. *Maggioni Farmaceutici S.p.A.*

7917 Fenquizone Monopotassium
52246-40-9 4039
$C_{14}H_{11}ClKN_3O_3S$
(±)-7-Chloro-1,2,3,4-tetrahydro-4-oxo-2-phenyl-6-quinazolinesulfonamide monopotassium salt.
Idrolone. Diuretic. Soluble in H_2O. *Maggioni Farmaceutici S.p.A.*

7918 Furosemide

54-31-9 4331 204-822-2

$C_{12}H_{11}ClN_2O_5S$

4-Chloro-N-furfuryl-5-sulfamoyl-anthranlic acid.

Diuretic salt; Disal; Lasix; Frumil; LB-502; Aisemide; Beronald; Desdemin; Discoid; Diural; Dryptal; Durafurid; Errolon; Eutensin; Frusetic; Frusid; Fulsix; Fuluvamide; Furesis; Furo-Puren; Furosedon; Hydro-rapid; Impugan; Katlex; Lasilix; Lowpston; Macasirool; Mirfat; Nicorol; Odemase; Oedemex; Profemin; Rosemide; Rusyde; Trofurit; Urex. Diuretic. Furosemide used with controlled release potassium chloride to control edema. A high-ceiling, or loop diuretic. mp = 206°; λ_m = 288, 276, 336 nm ($E^{1\%}_{1\,cm}$ 945, 588, 144 95% EtOH); soluble in Me_2CO, MeOH, DMF; less soluble in EtOH, H_2O, $CHCl_3$, Et_2O; LD_{50} (frat orl) = 2600 mg/kg, (mrat orl) = 2820 mg/kg. *Astra Sweden; Elkins-Sinn; Fermenta Animal Health Co.; Hoechst Roussel Pharm. Inc.; Parke-Davis; Rhône-Poulenc Rorer Pharm. Inc.*

7919 Furterene

7761-75-3

$C_{10}H_9N_7O$

2,4,7-Triamino-6-(2-furyl)-pteridine.

Diuretic.

7920 Glycerol

56-81-5 4493 200-289-5

$C_3H_8O_3$

1,2,3-Propanetriol.

glycerine; glycerin; tryhydroxypropane; incorporation factor; IFP; Bulbold; Cristal; Glyceol Opthalgan; Osmoglyn. Osmotic diuretic. Diagnostic aid (opthalmic). Used to reduced intraocular pressure and vitreous volume for ocular surgery. Syrupy liquid; mp = 178°; bp1.0:sk = 125.5; n_D^{25} = 1.4730; d_{25}^{25} = 1.24910; miscible with H_2O, alcohol; insoluble in C_6H_6, $CHCl_3$, CCl_4, petroleum ether, oils; LS_{50} (rat orl) > 20 ml/kg, (rat iv) = 4.4 ml/kg.

7921 Hydracarbazine

3614-47-9 4798 222-788-7

$C_5H_7N_5O$

6-Hydrazino-6-pyridazinecarboxamide.

Normatensyl. Diuretic; antihypertensive. mp = 249-250° (dec). *Chimie et Atomistique.*

7922 Hydrochlorothiazide

58-93-5 4822 200-403-3

$C_7H_{8C}lN_3O_4S_2$

6-Chloro-3,4-dihydro-2H-1,2,4-benzothiadiazine-7-sulfonamide 1,1-dioxide.

chlorsulthiadil; Esidrex; Dichlotride; HydroDIURIL; Hydrozide; Oretic; Thiuretic; Acuretic; Aldactazide; Aldoril; Apresazide; Caplaril; Capozide; Dyazide; Esimil; H.H. 25/25; H.H. 50/50; Hydropres; Hyzaar; Inderide; Lopressor HCT; Lotensin HCT; Maxzide; Moduretic; Prinzide; Ser-Ap-Es; Timolide; Unipres; Vaseretic; Ziac. Diuretic. A thiazide. mp= 273-275°; λ_m = 317, 271, 226 nm ($A^{1\%}_{1cm}$ 130, 654, 1280 MeOH/HCl); soluble in MeOH, EtOH, Me_2CO; insoluble in H_2O; LD_{50} (mus iv) = 590 mg/kg, (mus orl) > 8000 mg/kg. *Abbott Labs Inc.; Ciba-Geigy Corp.; Lederle Labs.; Lemmon Co.; Merck & Co.Inc.; Parke-Davis; Searle G.D. & Co.; Solvay Pharm. Inc.; Squibb E.R. & Sons; Wallace Labs; Wyeth Labs.*

7923 Hydroflumethiazide

135-09-1 4830 200-203-6

$C_8H_8F_3N_3O_4S_2$

3,4-Dihydro-6-(trifluoromethyl)-2H-1,2,4-benzothiadiazine-7-sulfonamide 1,1-dioxide.

Diumide-K; Diucardin; Saluron; Salutensin; dihydroflumethiazide; methforylthiazidine; metflorylthiazidine; Bristab; Bristurin; Di-Ademil; Diucardin; Elodrine; Finuret; Hydol; Hydrenox; Leodrine; NaClex; Olmagran; Rodiuran; Rontyl; Sisuril; Vergonil. Diuretic with antihypertensive properties. A thiazide. mp = 272-273°; λ_m = 272.5 nm (log ε 4.286 MeOH); soluble in Me_2CO (> 100 mg/ml), MeOH (58 mg/ml), CH_3CN (43 mg/ml), H_2O (0.3 mg/ml), Et_2O (0.2 mg/ml), C_6H_6 (< 0.1 mg/ml); LD_{50} (mus orl) > 8000 mg/kg, (mus iv) = 750 mg/kg, (mus ip) = 6280 mg/kg. *Apothecon; Roberts Pharm. Corp.; Wyeth-Ayerst Labs.*

7924 Indacrinone

57296-63-6

$C_{18}H_{14}Cl_2O_4$

(±)-[(6,7-Dichloro-2-methyl-1-oxo-2-phenyl-5-indanyl)oxy]acetic acid.

MK-196. Diuretic; antihypertensive. *Merck & Co.Inc.*

7925 Indapamide

26807-65-8 4969 248-012-7

$C_{16}H_{16}ClN_3O_3S$

4-Chloro-N-(2-methyl-1-indolinyl)-3-sulfamoylbenzamide.

Lozol; S-1520; SE-1520; Bajaten; Damide; Fludex; Indaflex; Indamol; Ipamix; Natrilix; Noranat; Tandix; Veroxil; Pressural [as hemihydrate]. Diuretic; antihypertensive. Nonthiazide compound with a similar mechanism of action to the thiazide diuretics. mp = 160-162°; LD_{50} (rat ip) = 393-421 mg/kg; (rat iv) = 394-440 mg/kg, (rat orl) > 3000 mg/kg, (mus ip) = 410-564 mg/kg, (mus iv) = 577-635 mg/kg, (mus orl) > 3000 mg/kg, (gpg ip) = 347-416 mg/kg, (gpg iv) = 272-358 mg/kg, (gpg orl) > 3000 mg/kg. *Apothecon; Rhône-Poulenc Rorer Pharm. Inc.*

7926 Isosorbide

652-67-5 5244 211-492-3

$C_6H_{10}O_4$

1,4:3,6-Dianhydro-D-glucitol.

1,4:3,6-dianhydrosorbitol; AT-101; NSC-40725; Hydronol; Ismotic; Isobide. Osmotic diuretic. mp = 61-64°; $[\alpha]_D$ = +44°. *Alcon Labs.*

7927 Mannitol

69-65-8 5788 200-711-8

$C_6H_{14}O_6$

D-Mannitol.

mannite; manna sugar; cordycepic acid; SDM-25; Diosmol; Manicol; Mannidex; Osmitrol; Osmosal; Resectisol. Osmotic diuretic; also used as a renal function diagnostic aid. Used as a pharmaceutical excipient and flavoring agent. mp= 166-168°; $bp_{3.5}$ = 290-295°; d^{20} = 1.52; $[\alpha]_D^{20}$= 23° (borax solution); soluble in H_2O (182 g/l), EtOH (12 g/l). *Astra Sweden; Baxter Healthcare Systems; ICI Americas Inc.; McGaw Inc.; Zeneca Pharm.*

7928 Mefruside
7195-27-9 5847 230-562-4
$C_{13}H_{19}ClN_2O_5S_2$
4-Chloro-N'-methyl-N'-(tetrahydro-2-methylfurfuryl)-m-benzenedisulfonamide.
BAY-1500; Baycaron. Diuretic. The l-form is the more active diuretic. [dl-form]: mp= 149-150°; [d-form]: mp = 146°; $[\alpha]_{578}^{20}$ = +5.4° (c = 2.026 MeOH); [l-form]: mp = 146°; $[\alpha]_{578}^{20}$= -5.5° (c = 2.100 MeOH). *Bayer Corp.*

7929 Meralluride
8069-64-5 5913
$C_{16}H_{24}HgN_6O_8$
N-[[3-(Hydroxymercuri)-2-methoxypropyl]-carbamoyl]succinamic acid compound with theophylline.
Diuretic. *Marion Merrell Dow Inc.*

7930 Meralluride Sodium
129-99-7 5913
$C_{16}H_{23}HgN_6NaO_8$
N-[[3-(Hydroxymercuri)-2-methoxypropyl]carbamoyl]succinamate sodium compound with theophylline.
Mercuhydrin; Dilurgen; Mercardan; Mercuretin. Diuretic. Soluble in hot H_2O, AcOH; insoluble in EtOH, $CHCl_3$, Et_2O; LD_{50} (rat sc) = 28 ± 7 mg/kg. *Marion Merrell Dow Inc.*

7931 Mercamphamide
127-50-4 5915
$C_{14}H_{25}HgNO_5$
3-[[3-(Hydroxymercuri)-2-methoxypropyl]carbamoyl]-1,2,2-trimethylcyclopentanecarboxylic acid.
Diuretic. Soluble in EtOH, slightly soluble in H_2O.

7932 Mercaptomerin
20223-84-1 5918
$C_{16}H_{27}HgNO_6S$
[3-(3-Carboxy-2,2,3-trimethylcyclopentanecarboxamido)-2-methoxypropyl](hydrogen mercaptoacetato)mercury.
Diuretic. *Am. Home Products.*

7933 Mercaptomerin Sodium
21259-76-7 5918 244-298-2
$C_{16}H_{25}HgNNa_2O_6S$
[3-(3-Carboxy-2,2,3-trimethylcyclopentanecarboxamido)-2-methoxypropyl](hydrogen mercaptoacetato)mercury disodium salt.
Diucardyn sodium; Thiomerin sodium. Diuretic. mp = 150-155° (dec); soluble in H_2O, EtOH; insoluble in Et_2O, C_6H_6, $CHCl_3$. *Am. Home Products.*

7934 Mercumallylic Acid-Theophylline Sodium
8018-15-3 5921
$C_{21}H_{21}HgN_4NaO_8$
[3-(3-Carboxy-2-oxo-2H-1-benzopyran-8-yl)-2-methoxy-propyl]hydroxymercurate(1-) sodium compound with theophylline.
mercumatilin sodium; Cumertilin sodium. Diuretic. Very soluble in H_2O; LD_{50} (rat iv) = 9.8 mg Hg/kg, (rat orl) = 238 mg Hg/kg. *Endo Pharm. Inc.*

7935 Mercumatilin
86-36-2 5921
$C_{14}H_{14}HgO_6$
[3-(3-Carboxy-2-oxo-2H-1-benzopyran-8-yl)-2-methoxy-propyl]hydroxymercurate(1-) hydrogen.
mercumallylic acid. Diuretic. mp = 155-160°, 197°; slightly soluble in H_2O, EtOH, $CHCl_3$; insoluble in Et_2O. *Endo Pharm. Inc.*

7936 Mercurophylline
8012-34-8 5915
$C_{21}H_{32}HgN_5NaO_7$
Sodium 3-[3-(hydroxymercuri)-2-methoxypropyl]-camphoramate compound with theophylline.
Mercamphamide-theophylline; Novurit. Diuretic.

7937 Mersalyl
492-18-2 5962 207-748-9
$C_{13}H_{16}HgNNaO_6$
Sodium o-[(3-hydroxymercuri-2-methoxypropyl)-carbamoyl]phenoxyacetate.
Salyrgan; mercuramide; Mercusal; Mersalin. Diuretic. Soluble in H_2O (1 g/ml), less soluble in organic solvents; LD_{50} (rat iv) = 17.7 mg/kg, (mus iv) = 72.6 mg/kg. *Ciba-Geigy Corp.; Hoechst AG.*

7938 Methalthiazide
5611-64-3
$C_{12}H_{16}ClN_3O_4S_3$
3-[(Allylthio)methyl]-6-chloro-3,4-dihydro-2-methyl-2H-1,2,4-benzothiadiazine-7-sulfonamide 1,1-dioxide.
P-2530. Diuretic; antihypertensive. A thiazide. *Pfizer Inc.*

7939 Methazolamide
554-57-4 6031 209-066-7
$C_5H_8N_4O_3S_2$
N-(4-Methyl-2-sulfamoyl-Δ^2-1,3,4-thiadiazolin-5-ylidene)acetamide.
Neptazane. Carbonic anhydrase inhibitor, used as diuretic, in treatment of glaucoma. mp = 213-214°; λ_m = 254 nm (log ε 3.66 95% EtOH), 247 nm (log ε3.61 0.1N NaOH). *Lederle Labs.*

7940 Methyclothiazide
135-07-9 6086 205-172-2
$C_9H_{11}Cl_2N_3O_4S_2$
6-Chloro-3-(chloromethyl)-3,4-dihydro-2-methyl-2H-1,2,4-benzothiadiazine-7-sulfonamide 1,1-dioxide.
Enduron; Aquatensen; Enduronyl; Eutron; NSC-110431; Duretic; Naturon. Diuretic; antihypertensive. A thiazide. mp= 225°; λ_m = 226, 267, 311 nm (ε 39300, 21250, 3300 MeOH); very soluble in Me_2CO, C_5H_5N; less soluble in MeOH, EOH; insoluble in H_2O, $CHCl_3$, C_6H_6. *Abbott Labs Inc.; Carter-Wallace; Wallace Labs.*

7941 Metiamide
34839-70-8
$C_9H_{16}N_4S_2$
1-Methyl-3-[2-[[(5-methylimidazol-4-yl)methyl]-thio]ethyl]-2-thiourea.
SK&F-92058. Histamine H_2-receptor Antagonist. Used as an antiulcerative. *SmithKline Beecham Pharm.*

7942 Meticrane

1084-65-7 6220 214-112-4

$C_{10}H_{13}NO_4S_2$

7-Methylthiochroman-7-sulfonamide 1,1-dioxide.

SD-17102; Arresten; Fontilix. Diuretic; antihypertensive. mp= 236-237°. *Soc. Ind. Fabric. Antiboit.*

7943 Metochalcone

18493-30-6 6225 242-377-6

$C_{18}H_{18}O_4$

1-(2,4-Dimethoxyphenyl)-3-(4-methoxyphenyl)-2-propen-1-one.

2',4,4'-trimethoxychalcone; CB-1314; Lesidrin; Vesidril; Vesidryl. Choleretic; diuretic. mp = 97°.

7944 Metolazone

17560-51-9 6231 241-539-3

$C_{16}H_{16}ClN_3O_3S$

7-Chloro-1,2,3,4-tetrahydro-2-methyl-4-oxo-3-o-tolyl-6-quinazolinesulfonamide.

Mykrox; Zaroxolyn; SR-720-22; Diulo; Metenix;Oldren; Xuret. Diuretic; antihypertensive. Nonthiazide compound with a similar mechanism of action to the thiazide diuretics. mp= 252-254°; LD_{50} (mus orl) > 5000 mg/kg, (mus ip) > 1500 mg/kg. *Fisons plc.*

7945 7-Morpholinomethyltheophylline

5089-89-4 6363 225-808-2

$C_{12}H_{17}N_5O_3$

3,7-Dihydro-1,3-dimethyl-7-(4-morpholinylmethyl)-1H-purine-2,6-dione.

Xanturil. Diuretic. mp = 177°.

7946 Muzolimine

55294-15-0 6397 259-573-2

$C_{11}H_{11}Cl_2N_3O$

3-Amino-1-(3,4-dichloro-α-methyl)benzyl-2-pyrazolin-3-one.

Edrul®; BAY g 2821. Diuretic. mp = 127-129°; LD_{50} (mus orl) = 1794 mg/kg, (dog orl) = 2000 mg/kg, (rbt orl) = 1250 mg/kg, (rat orl) = 1559 mg/kg. *Bayer Corp.*

7947 Niravoline

130610-93-4

$C_{22}H_{25}N_3O_3$

N-Methyl-2-(m-nitrophenyl)-N-[(1S,2S)-2-(1-pyrrolidinyl)-1-indanyl]acetamide.

RU-51599. A +lk-opioid receptor agonist. Diuretic; aquaretic. *Roussel-UCLAF.*

7948 Oleandrin

465-16-7 6963 207-361-5

$C_{32}H_{48}O_9$

16β-(Acetyloxy)-3β-[(2,6-dideoxy-3-O-methyl-α-L-arabino-hexopyranosyl)oxy]-14-hydroxy-5β-card-20(22)-enolide.

neriolin; Corrigen; Folinerin. Diuretic; cardiotonic. Glycoside. From the leaves of *Nerium oleander L. Apocynaceae* (Laurier rose). mp = 250°; $[\alpha]_D^{25}$= -48.0° (c = 1.3 MeOH); λ_m= 220 nm (log ε 4.20); insoluble in H_2O, soluble in EtOH, $CHCl_3$; [desacetyloleandrin]: mp = 238-240°; $[\alpha]_D^{18}$= -24.9°.

7949 Oxmetidine

72830-39-8

$C_{19}H_{21}N_5O_3S$

2-[[2-[[(5-Methylimidazol-4-yl)methyl]thio]ethyl]amino]-5-piperonyl-4-(1H)-pyrimidinone.

Histamine H_2-receptor Antagonist. Used as an antiulcerative. *SmithKline Beecham Pharm.*

7950 Oxmetidine Hydrochloride

63204-23-9

$C_{19}H_{23}Cl_2N_5O_3S$

2-[[2-[[(5-Methylimidazol-4-yl)methyl]thio]ethyl]amino]-5-piperonyl-4-(1H)-pyrimidinone dihydrochloride.

SK&F-92994-A_2. Histamine H_2-receptor Antagonist. Used as an antiulcerative. *SmithKline Beecham Pharm.*

7951 Oxmetidine Mesylate

84455-52-7

$C_{21}H_{29}N_5O_9S_3$

2-[[2-[[(5-Methylimidazol-4-yl)methyl]thio]ethyl]amino]-5-piperonyl-4-(1H)-pyrimidinone dimethanesulfonate.

SK&F-92994-J_2. Antiulcerative. Histamine H_2-receptor antagonist. *SmithKline Beecham Pharm.*

7952 Oxycinchophen

485-89-2 7091 207-624-4

$C_{16}H_{11}NO_3$

3-Hydroxy-2-phenyl-4-quinolinecarboxylic acid.

3-hydroxy-2-phenylcinchoninic acid; 3-hydroxy-chinchophen; HCP; Fenidrone; Magnofenyl; Magnophenyl; Oxinofen; Reumalon. Antidiuretic; uricosuric. Dec 206-207°; sparingly soluble in H_2O, Et_2O; soluble in AcOH, alkalies, hot EtOH, C_6H_6. *Chemo Puro.*

7953 Ozolinone

56784-39-5 260-383-7

$C_{11}H_{16}N_2O_3S$

(Z)-3-Methyl-4-oxo-5-piperidino-$\Delta^{2,\alpha}$-thiazolidineacetic acid.

Goedecke 382. Diuretic. *Goedecke; Parke-Davis.*

7954 Pamabrom

606-04-2 7133 210-103-4

$C_{11}H_{18}BrN_5O_3$

8-Bromotheophylline compound with 2-amino-2-methyl-1-propanol.

Midol; Premsyn PMS; Sunril. Diuretic. Formulated with acetaminophen and pyrilamine maleate. mp= 300° (dec); soluble in H_2O (> 30 g/100 ml). *Sterling Health U.S.A.*

7955 Paraflutizide

1580-83-2 7157 216-426-7

$C_{14}H_{13}ClFN_3O_4S_2$

6-Chloro-3,4-dihydro-3-(p-fluorobenzyl)-2H-1,2,4-benzothiadiazine-7-sulfonamide 1,1-dioxide.

LD-3612. Diuretic. mp= 238-240°.

7956 Penflutizide

1766-91-2 217-186-6

$C_{13}H_{18}F_3N_3O_4S_2$

3,4-Dihydro-3-pentyl-6-trifluoromethyl)-2H-1,2,4-benzothiadiazine-7-sulfonamide 1,1-dioxide.

A photosensitive compound. A thiazide diuretic.

7957 Perhexiline
6621-47-2 7305 229-569-5
$C_{19}H_{35}N$
2-(2,2-Dicyclohexylethyl)piperidine.
perhexilene. Vasodilator (coronary); diuretic. Calcium blocking agent. [hydrochloride]: mp= 243-245.5°. *Marion Merrell Dow Inc.*

7958 Perhexiline Maleate
6724-53-4 7305 229-775-5
$C_{23}H_{39}NO_4$
2-(2,2-Dicyclohexylethyl)piperidine maleate (1:1).
perhexilene maleate; Pexid. Vasodilator (coronary); diuretic. Calcium blocking agent. mp = 188.5-191°; LD_{50} (rat orl) > 7000 mg/kg, (mus orl) = 4370 mg/kg. *Marion Merrell Dow Inc.*

7959 Piretanide
55837-27-9 7647 259-852-9
$C_{17}H_{18}N_2O_5S$
4-Phenoxy-3-(1-pyrrolidinyl))-5-sulfamoylbenzoic acid.
Arlix; HOE-118; S-73-4118; Arelix; Diumax; Eurelix; Tauliz. Diuretic. mp = 225-227°; fluoresces at 366 nm; LD_{50} (rat orl) = 5601 mg/kg, (mus orl) = 3672 mg/kg. *Hoechst Roussel Pharm. Inc.*

7960 Polythiazide
346-18-9 7744 206-468-4
$C_{11}H_{13}ClF_3N_3O_4S_3$
6-Chloro-3,4-dihydro-2-methyl-3-[[(2,2,2-trifluoromethyl)thio]methyl]-2H-1,2,4-benzothiadiazine-7-sulfonamide 1,1-dioxide.
Renese; P-2525; NSC-108161; Drenusil; Nephril. Thiazide diuretic; antihypertensive. mp= 202.5°; soluble in MeOH, Me_2CO; insoluble in H_2O, $CHCl_3$. *Pfizer Inc.*

7961 Potassium Bitartrate
868-14-4 7776 212-769-1
$C_4H_5KO_6$
[R-(R*,R*)]-2,3-Dihydroxybutanedioic acid monopotassium salt.
argol; potassium hydrogen tartrate; cream of tartar; cremor tartari; faecula; faecla. Diuretic; laxative; cathartic. A crystalline crust deposited on the sides of the vat in which grape juice has been fermented; it contains 40-70% tartaric acid, principally as potassium hydrogen tartrate. Used as a laxative, cathartic and diuretic. Soluble in H_2O (0.62 g/100 ml at 25°, 6.25 g/100 ml at 100°), EtOH (0.011 g/100 ml), dilute mineral acids, alkaline solutions.

7962 Potassium Carbonate
584-08-7 7781 209-529-3
CK_2O_3
Carbonic acid potassium salt.
salt of tartar; pearl ash. Diuretic. d = 2.29; mp= 891°; soluble in H_2O (1 g/ml); insoluble in EtOH, organic solvents.

7963 Potassium nitrate
7757-79-1 7815 231-818-8
KNO_3
Nitric acid potassium salt.
saltpeter, niter. Diuretic. mp = 333°; d = 2.11; soluble in H_2O (350 mg/ml), insoluble in organic solvents; LD_{50} (rbt orl) = 1.17 g/kg.

7964 Protheobromine
50-39-5 8072 200-034-8
$C_{10}H_{14}N_4O_3$
1-(2-Hydroxypropyl)-3,7-dihydro-3,7-dimethyl-1H-purine-2,6-dione.
Tebe; Bonicor. Diuretic. mp= 140-142°; soluble in H_2O, $CHCl_3$, EtOH; insoluble in Et_2O; LD_{50} (mus sc) = 580 mg/kg.

7965 Quinethazone
73-49-4 8240 200-801-7
$C_{10}H_{12}ClN_3O_3S$
7-Chloro-2-ethyl-1,2,3,4-tetrahydro-4-oxo-6-quinazolinesulfonamide.
CL-36010; Hydromox; Aquamox. Diuretic; antihypertensive. Nonthiazide compound with a similar mechanism of action to the thiazide diuretics. mp = 250-252°; soluble in Me_2CO, EtOH. *Am. Cyanamid.*

7966 Ranitidine
66357-35-5 8286 266-332-5
$C_{13}H_{22}N_4O_3S$
N-[2-[[5-[(Dimethylamino)methyl]furfuryl]thio]ethyl]-N'-methyl-2-nitro-1,1-ethenediamine.
Antiulcerative. Histamine H_2 receptor antagonist; inhibits gastric secretion. mp = 69-70°. *Glaxo Labs.*

7967 Ranitidine Bismuth Citrate
128345-62-0 8286
$C_{19}H_{27}BiN_4O_{10}S$
N-[2-[[5-[(Dimethylamino)methyl]furfuryl]thio]ethyl]-N'-methyl-2-nitro-1,1-ethenediamine compound with bismuth (3+) citrate (1:1).
Ranitidine bismutrex; GR-122311X; Pylorid. Histamine H_2-receptor Antagonist. Used as an antiulcerative. *Glaxo Wellcome plc.*

7968 Ranitidine Hydrochloride
66357-59-3 8286 266-333-0
$C_{13}H_{23}ClN_4O_3S$
N-[2-[[5-[(Dimethylamino)methyl]furfuryl]thio]ethyl]-N'-methyl-2-nitro-1,1-ethenediamine hydrochloride.
AH-19065; Azantac; Melfax; Noctone; Raniben; Ranidil; Raniplex; Sostril; Taural; Terposen; Trigger; Ulcex; Ultidine; Zantac; Zantic. Antiulcerative. Histamine H_2 receptor antagonist; inhibits gastric secretion. mp = 133-134°; soluble in H_2O, AcOH; less soluble in EtOH, MeOH; insoluble in organic solvents. *Glaxo Labs.*

7969 Sodium Citrate
68-04-2 8746 200-675-3
$C_6H_5Na_3O_7$
Trisodium citrate.
Citrosodine; Citnatin; Cystemme; Urisal. Systemic alkalizer; diuretic; expectorant; sudorific. Used in veterinary medicine as an anticoagulant in blood collection. [dihydrate]: soluble in H_2O; insoluble in EtOH; pH 8; LD_{50} (rat ip) = 1548 mg/kg; [pentahydrate]: not as stable as the dihydrate.

7970 Spironolactone
52-01-7 8917 200-133-6
$C_{24}H_{32}O_4S$
17-Hydroxy-7α-mercapto-3-oxo-17α-pregn-4-ene-21-carboxylic acid σ-lactone acetate.
Abbolactone; Aldactone; Aldactazide; SC-9420; Aldace; Aldopur; Almatol; Altex; Aquareduct; Deverol; Diatensec; Dira; Duraspiron; Euteberol; Lacalmin; Lacdene; Laractone; Nefurofan; Osiren; Osyrol; Sagisal; Sincomen; Spiretic; Spiroctan; Spiroderm; Spirolone; Spiro-Tablinen; Supra-Puren; Suracton; Urusonin; Verospiron; Xenalon. Diuretic. Aldosterone antagonist. Potassium sparing. Used for edema in cirrhosis of the liver, nephrotic syndrome, congestive heart failure, potentiation of thiazide and loop diuretics, hypertension and Conn's syndrome. mp = 134-135°, 201-202°; $[\alpha]_D^{20}$ = -33.5° ($CHCl_3$); λ_m = 238 nm (ε 20200); insoluble in H_2O, soluble in most organic solvents. *Abbott Labs Inc.; Parke-Davis; Searle G.D. & Co.*

7971 Spiroxasone
6673-97-8
$C_{24}H_{34}O_3S$
4',5'-Dihydro-7α-mercaptospiro[androst-4-ene-17,2'-(3'H)-furan]-3-one acetate.
Diuretic. *Merck & Co.Inc.*

7972 Sufotidine
80343-63-1
$C_{20}H_{31}N_5O_3S$
1-[m-[3-[[1-Methyl-3-[(methylsulfonyl)methyl]-1H-1,2,4-triazol-5-yl]amino]propoxy]benzyl]piperidine.
AH-25352X. Histamine H_2-receptor Antagonist. Used as an antiulcerative. *Glaxo Wellcome plc.*

7973 Teclothiazide
4267-05-4 9261 224-253-3
$C_8H_7Cl_4N_3O_4S_2$
6-Chloro-3,4-dihydro-3-(trichloromethyl)-2H-1,2,4-benzothiadiazine-7-sulfonamide 1,1-dioxide.
Diuretic. A thiazide. mp = 300-303°, 287°.

7974 Teclothiazide Potassium
5306-80-9 9261 226-157-7
$C_8H+67Cl_4KN_3O_4S_2$
6-Chloro-3,4-dihydro-3-(trichloromethyl)-2H-1,2,4-benzothiadiazine-7-sulfonamide 1,1-dioxide potassium salt.
PS-207; K-33; Depleil. Diuretic. A thiazide. LD_{50} (mus ip) = 4.75 g/kg.

7975 Theobromine
83-67-0 9418 201-494-2
$C_7H_8N_4O_2$
3,7-Dihydro-3,7-dimethyl-1H-purine-2,6-dione.
3,7-dimethylxanthine. Diuretic; bronchodilator; cardiotonic. Principle alkaloid of the cacao bean. Also found in coca nuts and tea. A methylxanthine CNS stimulant. mp = 357°; sublimes at 290-295°; very slightly soluble in H_2O, EtOH; soluble in fixed alkali hydroxides, concentrated acids; moderately soluble in ammonia; insoluble in C_6H_6, Et_2O, $CHCl_3$, CCl_4.

7976 Theobromine Calcium Salicylate
8065-51-8 9418
$C_{21}H_{18}CaN_4O_8$
3,7-Dihydro-3,7-dimethyl-1H-purine-2,6-dione calcium salt mixture (1:1) with calcium salicylate.
Theocalcin; Calcium Diuretin. Diuretic; bronchodilator; cardiotonic. Xanthine derivative. Amorphous powder, partially soluble in H_2O.

7977 Ticrynafen
40180-04-9 9570 254-826-3
$C_{13}H_8Cl_2O_4S$
[2,3-Dichloro-4-(2-thienylcarbonyl)phenoxy]acetic acid.
tienylic acid; tienilic acid; thienylic acid; ANP-3624; CE-3624; SKF-62698; Difluorex; Selacryn. Diuretic; uricosuric; antihypertensive. mp = 148-149°, 157°; LD_{50} (mus iv) = 225 mg/kg, (mus orl) = 1275 mg/kg. *SmithKline Beecham Pharm.*

7978 Tienoxolol
90055-97-3
$C_{21}H_{28}N_2O_5S$
(±)-Ethyl 2-[3-(tert-butylamino)-2-hydroxypropoxy]-5-(2-thienocarboxamido)benzoate.
A diuretic beta-blocking agent.

7979 Tifluadom
81656-30-6
$C_{22}H_{20}FN_3OS$
(±)-N-[[5-(o-Fluorophenyl)-2,3-dihydro-1-methyl-1H-1,4-benzodiazepin-2-yl]methyl]-3-thiophenecarboxamide.
KC-5103. A benzodiazepine. kappa opioid receptor agonist. Analgesic; diuretic.

7980 Tiotidine
69014-14-8
$C_{10}H_{16}N_8S_2$
2-Cyano-1-[2-[[[2-[(diaminomethylene)amino]-4-thiazolyl]methyl]thio]ethyl]-3-methylguanidine.
ICI-125211. Histamine H_2-receptor Antagonist. Used as an antiulcerative. *ICI.*

7981 Tizolemide
56488-58-5
$C_{11}H_{14}ClN_3O_3S_2$
2-Chloro-5-[4-hydroxy-3-methyl-2-(methylimino)-4-thiazolidinyl]benzenesulfonamide.
HOE-740. A sulphonamide. Diuretic. *Hoechst Roussel Pharm. Inc.*

7982 Torsemide
56211-40-6 9690
$C_{16}H_{20}N_4O_3S$
1-Isopropyl-3-[(4-m-toluidino-3-pyridyl)sulfonyl]urea.
BM02.015; AC-4464; JDL-464; Demadex; Toradiur; Torem; Unat. Diuretic. mp = 163-164°. *Boehringer Mannheim GmbH; Christiaens S.A.*

7983 Triamterene
396-01-0 9731 206-904-3
$C_{12}H_{11}N_7$
2,4,7-Triamino-6-phenylpteridine.
Dyrenium; Dyazide; SK&F-8542; NSC-77625; Ademin; Ademine; pterophene; pterofen; Jatropur; Teriam;

Triteren; Urocaudal. Diuretic. Aldosterone antagonist. Potassium sparing. mp= 316°, 327°; λ_m = 356 nm (ε 21000 4.5% HCOOH). *SmithKline Beecham Pharm.*

7984 Trichlormethiazide

133-67-5 9754 205-118-8

$C_8H_8Cl_3N_3O_4S_2$

6-Chloro-3-(dichloromethyl)-3,4-dihydro-2H-1,2,4-thiadiazine-7-sulfonamide 1,1-dioxide.

3-dichloromethylhydrochlorothiazide; hydrochlorothiazide; trichloromethiazide; Achletin; Anatran; Anistadin; Aponorin; Carvacron; Diurese; Esmarin; Fluitran; Fluitran; Flutra; Intromene; Kubacron; Metahydrin; Metatensin; Naqua; Naquasone; Naquival; Salirom; Tachionin; Tolcasone; Triflumen. Diuretic; antihypertensive. A thiazide. mp= 248-250°, 266-273°; soluble in H_2O (0.8 mg/ml), EtOH (21 mg/ml), MeOH (60 mg/ml); LD_{50} (rat orl) > 20000 mg/kg. *Marion Merrell Dow Inc.; Merrell Pharm. Inc.; Schering-Plough HealthCare Products; Schering-Plough Pharm.*

7985 Triflocin

13422-16-7

$C_{13}H_9F_3N_2O_2$

4-(α,α,α-Trifluoro-m-toluidino)nicotinic acid.

C-65562. Diuretic.

7986 Tripamide

73803-48-2 9866

$C_{16}H_{20}ClN_3O_3S$

4-Chloro-N-(endo-hexahydro-4,7-methanoisoindolin-2-yl)-3-sulfamoylbenzamide.

toripamide; ADR-033; E-614; Normonal. Diuretic; antihypertensive.

7987 Ularitide

118812-69-4

$C_{145}H_{234}N_{52}O_{44}S_3$

L-Threonyl-L-alanyl-L-prolyl-L-arginyl-L-seryl-L-leucyl-L-arginyl-L-arginyl-L-seryl-L-seryl-L-cysteinyl-L-phenylalanylglycylglycyl-L-arginyl-L-methionyl-L-aspartyl-L-arginyl-L-isoleucylglycyl-L-alanyl-L-glutaminyl-L-serylglycyl-L-leucylglycyl-L-cysteinyl-L-asparaginyl-L-seryl-L-phenylalanyl-L-arginyl-L-tyrosine cyclic-(11→27)-disulfide.

urodilatin; CDD-95-126; ANP-95-126; H-Thr-Ala-Pro-Arg-Ser-Leu-Arg-Arg-Ser-Ser-Cys(11)-Phe-Gly-Gly-Arg-Met-Asp-Arg-Ile-Gly-Ala-Gln-Ser-Gly-Leu-Gly-Cys(27)-Asn-Ser-Phe-Arg-Tyr-OH cyclic-(11→27)-disulfide. A natriuretic peptide that exerts strong diuretic and natriuretic effects when infused intravenously. Used in the treatment of acute renal failure.

7988 Urea

57-13-6 10005 200-315-5

CH_4N_2O

Carbamide.

Elaqua XX; Nutraplus; Ureaphil; Aquacare; Panafil; Aquadrate; Basodexan; Hyanit; Keratinamin; Onychomal; Pastaron; Ureophil; Urepearl. Osmotic diuretic. mp = 132.7°; soluble in H_2O (1 g/ml), EtOH (5 g/100 ml), MeOH (16 g/100 ml); insoluble in $CHCl_3$, Et_2O. *Abbott Labs Inc.; Galderma Labs Inc.; Menley & James Labs Inc.; Pfanstiehl Labs Inc.; Rystan Co. Inc.*

7989 Xipamide

14293-44-8 10212 238-216-4

$C_{15}H_{15}ClN_2O_4S$

4-Chloro-5-sulfamoyl-2',6'-salicyloxilidide.

MJF-10938; Be-1293; Bei-1293; Aquaphor; Chronexan; Diurexan; Lumitens. Diuretic; antihypertensive. mp = 256°. *Beiersdorf AG.*

7990 Zaltidine Hydrochloride

90274-23-0

$C_8H_{12}Cl_2N_6S$

[4-(2-Methylimidazol-5-yl)-2-thiazolyl]guanidine dihydrochloride.

CP-57361-01. Antiulcerative. Histamine H_2-receptor antagonist. *Pfizer Inc.*

Dopamine Receptor Agonists

7991 Alentemol

112891-97-1

$C_{19}H_{25}NO$

(+)-2-(Dipropylamino)-2,3-dihydrophenalen-5-ol.

Antipsychotic. A dopamine agonist. *Pharmacia & Upjohn; Upjohn Ltd.*

7992 Alentemol Hydrobromide

112892-81-6

$C_{19}H_{25}BrNO$

(+)-2-(Dipropylamino)-2,3-dihydrophenalen-5-ol hydrobromide.

U-68553B. Antipsychotic. A dopamine agonist. *Pharmacia & Upjohn; Upjohn Ltd.*

7993 Bromocriptine

25614-03-3 1437 247-128-5

$C_{32}H_{40}BrN_5O_5$

2-Bromo-12'-hydroxy-2'-(1-methylethyl)-5'-(2-methylpropyl)ergotaman-3',6',18-trione.

CB-154. Prolactin inhibitor. Used to treat Parkinson's disease. mp = 215-218°; $[\alpha]_D^{20}$ = -195° (c = 1 CH_2Cl_2); LD_{50} (rbt orl) > 1000 mg/kg, (rbt iv) = 12.0 mg/kg. *Sandoz Pharm. Corp.*

7994 Bromocriptine Mesylate

22260-51-1 1437 244-881-1

$C_{33}H_{44}BrN_5O_8S$

2-Bromo-12'-hydroxy-2'-(1-methylethyl)-5'-(2-methylpropyl)-5'α-ergotaman-3',6',18-trione monomethanesulfonate (salt).

bromocriptine methanesulfonate; CB-154 mesylate; Parlodel; Bagren; Pravidel. Prolactin inhibitor used to treat Parkinson's disease. Dopamine receptor agonist. mp = 192-196° (dec); $[\alpha]_D^{20}$ = 95° (c = 1 MeOH/CH_2Cl_2); soluble in MeOH (91 g/100 ml), EtOH (2.3 g/100 ml), H_2O (0.08 g/100 ml), $CHCl_3$ (0.045 g/100ml)C_6H_6 (< 0.01 g/100 ml). *Sandoz Pharm. Corp.*

7995 Cabergoline

81409-90-7 1637

$C_{26}H_{37}N_5O_2$

1-[(6-Allylergolin-8β-yl)carbonyl]-1-[3-(dimethylamino)propyl]-3-ethylurea.

FCE 21336; Dostinex. Antidyskinetic; antihyperprolactinemic. Dopamine receptor agonist; prolactin inhibitor. mp = 102-104°; LD_{50} (mus orl) > 400 mg/kg; [diphosphate ($C_{26}H_{43}N_5O_{10}P_2$)]: mp = 153-155°. *Farmitalia Carlo Erba Ltd.*

7996 Carmoxirole
98323-83-2 1893

$C_{24}H_{26}N_2O_2$
3-[4-(3,6-Dihydro-4-phenyl-1(2H)-pyridyl)butyl]indole-5-carboxylic acid.
Antihypertensive. Selective dopamine D_2-receptor antagonist. mp = 284-285°. *E. Merck.*

7997 Carmoxirole Hydrochloride
115092-85-8 1893

$C_{24}H_{27}ClN_2O_2$
3-[4-(3,6-Dihydro-4-phenyl-1(2H0-pyridinyl)butyl]-1H-indole-5-carboxylic acid hydrochloride.
EMD-45609. Selective dopamine D_2-receptor antagonist. Used as an antihypertensive agent. mp = 298-299°; λ_m = 242, 281 nm (MeOH). *E. Merck.*

7998 Dopexamine
86197-47-9 3482

$C_{22}H_{32}N_2O_2$
4-[2-[[6-(Phenethylamino)hexyl]amino]ethyl]-pyrocatechol.
FPL-60278; Dopacard. Cardiotonic. Dopamine receptor agonist. Has little or no α- or β-adrenoceptor activity. *Fisons Pharm. Div.*

7999 Dopexamine Hydrochloride
86484-91-5 3482

$C_{22}H_{34}Cl_2N_2O_2$
4-[2-[[6-(Phenethylamino)hexyl]amino]ethyl]-pyrocatechol dihydrochloride.
FPL-60278AR. Cardiotonic. Dopamine receptor agonist. Has little or no α- or β-adrenoceptor activity. [hydrobromide]: mp = 227-228°. *Fisons Pharm. Div.*

8000 Etisulergine
64795-23-9

$C_{19}H_{28}N_4O_2S$
N,N-Diethyl-N'-(6-methylergolin-8α-yl)sulfamide.
Dopamine agonist.

8001 Fenoldopam
67227-56-9 4020

$C_{16}H_{16}ClNO_3$
6-Chloro-2,3,4,5-tetrahydro-1-(p-hydroxyphenyl)-1H-3-benzazepine-7,8-diol.
SKF-82526. Antihypertensive. Dopamine D_1 receptor agonist. [hydrobromide]: mp = 277° (dec). *SmithKline Beecham Pharm.*

8002 Fenoldopam Mesylate
67227-57-0 4020

$C_{17}H_{20}ClNO_6S$
6-Chloro-2,3,4,5-tetrahydro-1-(p-hydroxyphenyl)-1H-3-benzazepine-7,8-diol monomethanesulfonate (salt).
fenoldopam monomethanesulfonate; SKF-82526J; Corlopam. Dopamine agonist used as an antihypertensive agent. mp = 274° (dec). *SmithKline Beecham Pharm.*

8003 Ibopamine
66195-31-1 4921 266-229-5

$C_{17}H_{25}NO_4$
4-[2-(Methylamino)ethyl]-o-phenylene diisobutyrate.
SB-7505; 3,4-di-o-isobutyryl epinine. Cardiotonic. Inotropic agent with dopaminergic and adrenergic agonist activities. *Simes S.p.A.; SmithKline Beecham Pharm.*

8004 Ibopamine Hydrochloride
75011-65-3 4921 278-056-2

$C_{17}H_{26}Cl5NO_4$
4-[2-(Methylamino)ethyl]-o-phenylene diisobutyrate hydrochloride.
SB-7505; Inopamil; Scandine. Cardiotonic. Inotropic agent with dopaminergic and adrenergic agonist activities. mp = 132°. *Simes S.p.A.; SmithKline Beecham Pharm.*

8005 Lisuride
18016-80-3 5541 241-925-1

$C_{20}H_{26}N_4O$
3-(9,10-Didehydro-6-methylergolin-8α-yl)-1,1-diethylurea.
methylergol carbamide; lysuride. Antimigraine; prolactin inhibitor; antiparkinsonian. Dopamine D_2 receptor agonist. mp = 186°; $[\alpha]_D^{20}$ = 313° (c = 0.60 C_5H_5N).

8006 Lisuride Maleate
19875-60-6 5541 243-387-3

$C_{24}H_{30}N_4O_5$
3-(9,10-Didehydro-6-methylergolin-8α-yl)-1,1-diethylurea maleate.
Apodel; Cuvalit; Dopergin; Eunal; Lysenyl; Revanil. Antimigraine; prolactin inhibitor; antiparkinsonian. Dopamine D_2 receptor agonist mp = 200° (dec); $[\alpha]_D^{20}$ = 288° (c = 0.5 MeOH); λ_m = 313 nm (MeOH); LD_{50} (mus iv) = 14.4 mg/kg.

8007 Mergocriptine
81968-16-3

$C_{33}H_{43}N_5O_5$2-methyl-alpha-ergocryptine
2-Methyl-α-ergocryptine.
CBM-36-733. A long-acting ergot derivative with an agonistic action on dopamine D1 and D2 receptors.

8008 Naxagolide
88058-88-2

$C_{15}H_{21}NO_2$
(+)-(4aR,10bR)-3,4,4a,5,6,10b-Hexahydro-4-propyl-2H-naphth[1,2-b]-1,4-oxazin-9-ol.
Antiparkinsonian. Dopamine agonist. *Merck & Co.Inc.*

8009 Naxagolide Hydrochloride
99705-65-4

$C_{15}H_{22}ClNO_2$
(+)-(4aR,10bR)-3,4,4a,5,6,10b-Hexahydro-4-propyl-2H-naphth[1,2-b]-1,4-oxazin-9-ol hydrochloride.
DL-588. Antiparkinsonian. Dopamine agonist. *Merck & Co.Inc.*

8010 Pergolide
66104-22-1 7304

$C_{19}H_{26}N_2S$
8β-[(Methylthio)methyl]-6-propylergoline.

Dopamine agonist. Used in treatment of Parkinson's disease. mp = 206-209°. *Eli Lilly & Co.*

8011 Pergolide Methanesulfonate
66104-23-2 7304
$C_{20}H_{30}N_2O_3S_2$
8β-[(Methylthio)methyl]-6-propylergoline monomethanesulfonate.
pergolide mesylate; Permax; Celance; LY-127809. Dopamine agonist. Antiparkinsonian. mp = 225°; λ_m = 279 nm (ε 6980 H_2O), 281 nm (ε 6993 EtOIH); $[\alpha]_D^{20}$ = -18.0° to -23.0° (c = 1 DMF); sparingly soluble in DMF, MeOH; slightly soluble in H_2O, 0.01N HCl, $CHCl_3$, CH_3CN, CH_2Cl_2, EtOH; very slightly soluble in Me_2CO; insoluble in Et_2O. *Eli Lilly & Co.*

8012 Pramipexole
104632-26-0 7885
$C_{10}H_{17}N_3S$
(S)-2-Amnio-4,5,6,7-tetrahydro-6-(propylamino)-benzothiazole.
U-98528E; SUD919CL2Y. Dopamine D_2-receptor agonist; antiparkinsonian; antischizophrenic; anti-depressant. [dihydrochloride]: mp = 296-298°; $[\alpha]_D^{20}$ = -67.2° (MeOH c = 1). *Boehringer Ingelheim Pharm. Inc.*

8013 Preclamol
85966-89-8
$C_{14}H_{21}NO$
(-)-(S)-m-(1-Propyl-3-piperidyl)phenol.
3-PPP. Partial dopamine autoreceptor agonist. Antipsychotic.

8014 Proterguride
77650-95-4
$C_{22}H_{32}N_4O$
1,1-Diethyl-3-(6-propylergolin-8α-yl)urea.
Analog of ergoline; derivative of terguride. Partial dopamine receptor agonist.

8015 Quinagolide
87056-78-8 8226
$C_{20}H_{33}N_3O_3S$
(±)-N,N-Diethyl-N'-[(3R*,4aR*,10aS*)-1,2,3,4,4a,5,-10,10a-octahydro-6-hydroxy-1-propylbenzo[g]quinolin-3-yl]sulfamide.
Prolactin inhibitor. Dopamine D_2 receptor agonist. mp = 122.5-124°. *Sandoz Pharm. Corp.*

8016 Quinagolide Hydrochloride
94424-50-7 8226
$C_{20}H_{34}ClN_3O_3S$
(±)-N,N-Diethyl-N'-[(3R*,4aR*,10aS*)-1,2,3,4,4a,5,-10,10a-octahydro-6-hydroxy-1-propylbenzo[g]quinolin-3-yl]sulfamide hydrochloride (salt).
CV-205-502; SDZ-205-502; Norprolac. Prolactin inhibitor. Dopamine D_2 receptor agonist. mp = 234-236°. *Sandoz Pharm. Corp.*

8017 Ropinirole
91374-21-9 8416
$C_{16}H_{24}N_2O$
4-[2-(Dipropylamino)ethyl]-2-indolinone.
SKF-101468. Antiparkinsonian. Dopamine D_2 receptor agonist. *SmithKline Beecham Pharm.*

8018 Ropinirole Hydrochloride
91374-20-8 8416
$C_{16}H_{25}ClN_2O$
4-[2-(Dipropylamino)ethyl]-2-indolinone monohydrochloride.
SKF-101468-A. Antiparkinsonian. Dopamine D_2 receptor agonist. mp = 241-243°. *SmithKline Beecham Pharm.*

8019 Roxindole
112192-04-8 8432
$C_{23}H_{26}N_2O$
3-[4-(3,6-Dihydro-4-phenyl-1(2H)-pyridylbutyl]-indol-5-ol.
Dopamine D_2 receptor agonist. Used as an antidepressant.

8020 Roxindole Hydrochloride
108050-82-4 8432
$C_{23}H_{27}ClN_2O$
3-[4-(3,6-Dihydro-4-phenyl-1(2H)-pyridinyl)butyl]-1H-indol-5-ol hydrochloride.
EMD-38362. A D_2 receptor agonist. mp = 274°.

8021 Roxindole Mesylate
119742-13-1 8432
$C_{24}H_{30}N_2O_4S$
3-[4-(3,6-Dihydro-4-phenyl-1(2H)-pyridylbutyl]indol-5-ol monomethanesulfonate.
EMD-49980. Dopamine D_2 receptor agonist; antidepressant.

8022 Talipexole
101626-70-4 9209
$C_{10}H_{15}N_3S$
6-Allyl-2-amino-5,6,7,8-tetrahydro-4H-thiazolo[4,5-d]-azepine.
An α_2-adrenergic and dopamine D_2 receptor agonist; antiparkinsonian. *Boehringer Ingelheim GmbH.*

8023 Talipexole Dihydrochloride
36085-73-1 9209
$C_{10}H_{16}ClN_3S$
6-Allyl-2-amino-5,6,7,8-tetrahydro-4H-thiazolo[4,5-d]-azepine dihydrochloride.
B-HT-920; Domin. An α_2-adrenergic and dopamine D_2 receptor agonist. Used as an antiparkinsonian. mp = 245° (dec). *Boehringer Ingelheim GmbH.*

8024 Tienocarbine
75458-65-0
$C_{15}H_{16}N_2S$
7,8,9,10-Tetrahydro-1,9-dimethyl-6H-pyrido[4,3-b]-thieno[3,2-e]indole.
Mixed dopamine agonist-antagonist.

8025 Vanoxerine
67469-69-6
$C_{28}H_{32}F_2N_2O$
1-[2-[Bis(p-fluorophenyl)methoxy]ethyl]-4-(3-phenylpropyl)piperazine.
GBR-12909. Selective dopamine uptake inhibitor.

Dopamine Receptor Antagonists

8026 Alentemol
112891-97-1
$C_{19}H_{25}NO$
(+)-2-(Dipropylamino)-2,3-dihydrophenalen-5-ol.
Antipsychotic. A dopamine agonist. *Pharmacia & Upjohn; Upjohn Ltd.*

8027 Alentemol Hydrobromide
112892-81-6
$C_{19}H_{25}BrNO$
(+)-2-(Dipropylamino)-2,3-dihydrophenalen-5-ol hydrobromide.
U-68553B. Antipsychotic. A dopamine agonist. *Pharmacia & Upjohn; Upjohn Ltd.*

8028 Berupipam
150490-85-0
$C_{19}H_{19}BrClNO_2$
(+)-(5S)-(5-Bromo-2,3-dihydro-7-benzofuranyl)-8-chloro-2,3,4,5-tetrahydro-3-methyl-1H-3-benzazepin-7-ol.
A dopamine D1 receptor antagonist.

8029 Brocresine
555-65-7
$C_7H_8BrNO_2$
5-[(Aminooxy)methyl]-2-bromophenol.
NSD-1055. A DOPA decarboxylase inhibitor.

8030 Bromocriptine
25614-03-3 1437 247-128-5
$C_{32}H_{40}BrN_5O_5$
2-Bromo-12'-hydroxy-2'-(1-methylethyl)-5'-(2-methylpropyl)-5'α-ergotaman-3',6',18-trione.
CB-154. Dopamine receptor agonist. Prolactin inhibitor and antiparkinsonian. mp = 215-218°; $[\alpha]_D^{20}$ = -195° (c = 1 CH_2Cl_2); LD_{50} (rbt orl) > 1000 mg/kg, (rbt iv) = 12.0 mg/kg. *Sandoz Pharm. Corp.*

8031 Bromocriptine Mesylate
22260-51-1 1437 244-881-1
$C_{33}H_{44}BrN_5O_8S$
2-Bromo-12'-hydroxy-2'-(1-methylethyl)-5'-(2-methylpropyl)-5'α-ergotaman-3',6',18-trione monomethanesulfonate (salt).
bromocriptine methanesulfonate; CB-154 mesylate; Parlodel; Bagren; Pravidel. Prolactin inhibitor used to treat Parkinson's disease. Dopamine receptor agonist. mp = 192-196° (dec); $[\alpha]_D^{20}$ = 95° (c = 1 $MeOH/CH_2Cl_2$); soluble in MeOH (91 g/100 ml), EtOH (2.3 g/100 ml), H_2O (0.08 g/100 ml), $CHCl_3$ (0.045 g/100ml)C_6H_6 (< 0.01 g/100 ml). *Sandoz Pharm. Corp.*

8032 Cabergoline
81409-90-7 1637
$C_{26}H_{37}N_5O_2$
N-[3-(Dimethylamino)propyl]-N-[(ethylamino)carbonyl]-6-(2-propenyl)-ergoline-8β-carboxamide.
FCE 21336; Dostinex. Antidyskinetic; antihyperprolactinemic. Dopamine receptor agonist; prolactin inhibitor. mp = 102-104°; LD_{50} (mus orl) > 400 mg/kg; [diphosphate]: mp = 153-155°. *Farmitalia Carlo Erba Ltd.*

8033 Cabergoline Diphosphate
85329-89-1 1637
$C_{26}H_{43}N_5P_2O_{10}$
N-[3-(Dimethylamino)propyl]-N-[(ethylamino)carbonyl]-6-(2-propenyl)-ergoline-8β-carboxamide diphosphate.
Dopamine receptor agonist; prolactin inhibitor. mp = 153-155°. *Farmitalia Carlo Erba Ltd.*

8034 Carmoxirole
98323-83-2 1893
$C_{24}H_{26}N_2O_2$
3-[4-(3,6-Dihydro-4-phenyl-1(2H)-pyridyl)butyl]indole-5-carboxylic acid.
Antihypertensive. Selective dopamine D_2-receptor antagonist. mp = 284-285°. *E. Merck.*

8035 Carmoxirole Hydrochloride
115092-85-8 1893
$C_{24}H_{27}ClN_2O_2$
3-[4-(3,6-Dihydro-4-phenyl-1(2H0-pyridinyl)butyl]-1H-indole-5-carboxylic acid hydrochloride.
EMD-45609. Selective dopamine D_2-receptor antagonist. Antihypertensive. mp = 298-299°; λ_m = 242, 281 nm (MeOH). *E. Merck.*

8036 Diclofensine
67165-56-4
$C_{17}H_{17}Cl_2NO$
(±)-4-(3,4-Dichlorophenyl)-1,2,3,4-tetrahydro-7-methoxy-2-methylisoquinoline.
Dopamine uptake inhibitor. GABA-receptor agonist.

8037 Dopexamine
86197-47-9 3482
$C_{22}H_{32}N_2O_2$
4-[2-[[6-(Phenethylamino)hexyl]amino]ethyl]-pyrocatechol.
FPL-60278; Dopacard. Cardiotonic. Dopamine receptor agonist. Little or no α-,β-adrenoceptor activity. *Fisons Pharm. Div.*

8038 Dopexamine Hydrochloride
86484-91-5 3482
$C_{22}H_{34}Cl_2N_2O_2$
4-[2-[[6-(Phenethylamino)hexyl]amino]ethyl]-pyrocatechol dihydrochloride.
FPL-60278AR. Cardiotonic. Dopamine receptor agonist. Has little or no α- or β-adrenoceptor activity. [hydrobromide]: mp = 227-228°. *Fisons Pharm. Div.*

8039 Fenoldopam
67227-56-9 4020
$C_{16}H_{16}ClNO_3$
6-Chloro-2,3,4,5-tetrahydro-1-(p-hydroxyphenyl)-1H-3-benzazepine-7,8-diol.
SKF-82526. Antihypertensive. Dopamine D_1 receptor agonist. [hydrobromide]: mp = 277° (dec). *SmithKline Beecham Pharm.*

8040 Fenoldopam Mesylate
67227-57-0 4020
$C_{17}H_{20}ClNO_6S$
6-Chloro-2,3,4,5-tetrahydro-1-(p-hydroxyphenyl)-1H-3-benzazepine-7,8-diol monomethanesulfonate (salt).

fenoldopam monomethanesulfonate; SKF-82526J; Corlopam. Dopamine agonist used as an antihypertensive agent. mp = 274° (dec). *SmithKline Beecham Pharm.*

8041 Ibopamine
66195-31-1 4921 266-229-5
$C_{17}H_{25}NO_4$
4-[2-(Methylamino)ethyl]-o-phenylene diisobutyrate.
SB-7505; 3,4-di-o-isobutyryl epinine. Cardiotonic. Inotropic agent with dopaminergic and adrenergic agonist activities. *Simes S.p.A.; SmithKline Beecham Pharm.*

8042 Ibopamine Hydrochloride
75011-65-3 4921 278-056-2
$C_{17}H_{26}Cl5NO_4$
4-[2-(Methylamino)ethyl]-o-phenylene diisobutyrate hydrochloride.
SB-7505; Inopamil; Scandine. Cardiotonic. Inotropic agent with dopaminergic and adrenergic agonist activities. mp = 132°. *Simes S.p.A.; SmithKline Beecham Pharm.*

8043 Indatraline
86939-10-8
$C_{16}H_{15}Cl_2N$
(±)-trans-3-(3,4-Dichlorophenyl)-N-methyl-1-indanamine.
Lu-19-005. A dopamine reuptake blocker.

8044 Lisuride
18016-80-3 5541 241-925-1
$C_{20}H_{26}N_4O$
3-(9,10-Didehydro-6-methylergolin-8α-yl)-1,1-diethylurea.
methylergol carbamide; lysuride. Antimigraine; prolactin inhibitor; antiparkinsonian. Dopamine D_2 receptor agonist. mp = 186°; $[\alpha]_D^{20}$= 313° (c = 0.60 C_5H_5N).

8045 Lisuride Maleate
19875-60-6 5541 243-387-3
$C_{24}H_{30}N_4O_5$
3-(9,10-Didehydro-6-methylergolin-8α-yl)-1,1-diethylurea maleate.
Apodel; Cuvalit; Dopergin; Eunal; Lysenyl; Revanil. Antimigraine; prolactin inhibitor; antiparkinsonian. Dopamine D_2 receptor agonist mp = 200° (dec); $[\alpha]_D^{20}$= 288° (c = 0.5 MeOH); λ_m = 313 nm (MeOH); LD_{50} (mus iv) = 14.4 mg/kg.

8046 Mezilamine
50335-55-2
$C_{11}H_{18}ClN_5S_2$
4-Chloro-2-(methylamino)-6-(4-methyl-1-piperazinyl)-5-(methylthio)pyridimidine.
Dopamine receptor antagonist.

8047 Naxagolide
88058-88-2
$C_{15}H_{21}NO_2$
(+)-(4aR,10bR)-3,4,4a,5,6,10b-Hexahydro-4-propyl-2H-naphth[1,2-b]-1,4-oxazin-9-ol.
Antiparkinsonian. Dopamine agonist. *Merck & Co.Inc.*

8048 Naxagolide Hydrochloride
99705-65-4
$C_{15}H_{22}ClNO_2$
(+)-(4aR,10bR)-3,4,4a,5,6,10b-Hexahydro-4-propyl-2H-naphth[1,2-b]-1,4-oxazin-9-ol hydrochloride.
MK-458; L-647339. Antiparkinsonian. Dopamine agonist. *Merck & Co.Inc.*

8049 Nemonapride
93664-94-9 6533
$C_{21}H_{26}ClN_3O_2$
(±)-cis-N-(1-Benzyl-2-methyl-3-pyrrolidinyl)-5-chloro-4-(methylamino)-o-anisamide.
emonapride; YM-09151-2; Emilace. Selective dopaminde D_2 receptor antagonist; bezamide antipsychotic. mp = 152-153°. *Yamanouchi U.S.A. Inc.*

8050 Odapipam
131796-63-9
$C_{19}H_{20}ClNO_2$
(+)-(S)-8-Chloro-5-(2,3-dihydrobenzofuran-7-yl)-7-hydroxy-3-methyl-2,3,4,5-tetrahydro-1H-3-benzazepine.
NNC-756. Dopamine receptor antagonist.

8051 Pergolide
66104-22-1 7304
$C_{19}H_{26}N_2S$
8β-[(Methylthio)methyl]-6-propylergoline.
Dopamine agonist. Antiparkinsonian. mp = 206-209°. *Eli Lilly & Co.*

8052 Pergolide Methanesulfonate
66104-23-2 7304
$C_{20}H_{30}N_2O_3S_2$
8β-[(Methylthio)methyl]-6-propylergoline monomethanesulfonate.
pergolide mesylate; Permax; Celance; LY-127809. Dopamine agonist. Antiparkinsonian. mp = 225°; λ_m = 279 nm (ε 6980 H_2O), 281 nm (ε 6993 EtOIH); $[\alpha]_D^{20}$= -18.0° to -23.0° (c = 1 DMF); sparingly soluble in DMF, MeOH; slightly soluble in H_2O, 0.01N HCl, $CHCl_3$, CH_3CN, CH_2Cl_2, EtOH; very slightly soluble in Me_2CO; insoluble in Et_2O. *Eli Lilly & Co.*

8053 Picobenzide
51832-87-2
$C_{15}H_{16}N_2O$
3,4-Dimethyl-N-(4-pyridylmethyl)benzamide.
M-14012-4; Dosetil. Dopamine receptor antagonist.

8054 Piflutixol
54341-02-5
$C_{24}H_{25}F_4NOS$
1-[3-[6-Fluoro-2-(trifluoromethyl)thioxanthen-9-ylidene]propyl]-4-piperidineethanol.
Dopamine receptor antagonist.

8055 Pramipexole
104632-26-0 7885
$C_{10}H_{17}N_3S$
(S)-2-Amino-4,5,6,7-tetrahydro-6-(propylamino)-benzothiazole.

U-98528E; SUD919CL2Y. Dopamine D_2-receptor agonist; antischizophrenic; antidepressant. Antiparkinsonian. *Boehringer Ingelheim GmbH.*

8056 Pramipexole Hydrochloride
104632-25-9 7885

$C_{10}H_{18}ClN_3S$

(S)-2-Amino-4,5,6,7-tetrahydro-6-(propylamino)-benzothiazole hydrochloride.

SND-19. Dopamine D_2-receptor agonist; antischizophrenic; antidepressant. Antiparkinsonian. mp = 296-298°; $[\alpha]_D^{20}$ = -67.2° (c = 1 MeOH). *Boehringer Ingelheim GmbH.*

8057 Prosulpride
68556-59-2 271-484-0

$C_{16}H_{25}N_3O_4S$

N-[(1-Propyl-2-pyrrolidinyl)methyl]-5-sulfamoyl-o-anisamide.

Dopamine receptor antagonist.

8058 Quinagolide
87056-78-8 8226

$C_{20}H_{33}N_3O_3S$

(±)-N,N-Diethyl-N'-[(3R*,4aR*,10aS*)-1,2,3,4,4a,5,10,10a-octahydro-6-hydroxy-1-propylbenzo[g]quinolin-3-yl]sulfamide.

Prolactin inhibitor. Dopamine D_2 receptor agonist. mp = 122.5-124°. *Sandoz Pharm. Corp.*

8059 Quinagolide Hydrochloride
94424-50-7 8226

$C_{20}H_{34}ClN_3O_3S$

(±)-N,N-Diethyl-N'-[(3R*,4aR*,10aS*)-1,2,3,4,4a,5,10,10a-octahydro-6-hydroxy-1-propylbenzo[g]quinolin-3-yl]sulfamide hydrochloride.

CV-205-502; SDZ-205-502; Norprolac. Prolactin inhibitor. Dopamine D_2 receptor agonist. mp = 234-236°. *Sandoz Pharm. Corp.*

8060 Ropinirole
91374-21-9 8416

$C_{16}H_{24}N_2O$

4-[2-(Dipropylamino)ethyl]-2-indolinone.

SKF-101468. Antiparkinsonian. Dopamine D_2 receptor agonist. *SmithKline Beecham Pharm.*

8061 Ropinirole Hydrochloride
91374-20-8 8416

$C_{16}H_{25}ClN_2O$

4-[2-(Dipropylamino)ethyl]-2-indolinone monohydrochloride.

SKF-101468-A. Antiparkinsonian. Dopamine D_2 receptor agonist. mp = 241-243°. *SmithKline Beecham Pharm.*

8062 Roxindole
112192-04-8 8432

$C_{23}H_{26}N_2O$

3-[4-(3,6-Dihydro-4-phenyl-1(2H)-pyridylbutyl]-indol-5-ol.

Dopamine D_2 receptor agonist. Used as an antidepressant.

8063 Roxindole Hydrochloride
108050-82-4 8432

$C_{23}H_{27}ClN_2O$

3-[4-(3,6-Dihydro-4-phenyl-1(2H)-pyridinyl)butyl]-1H-indol-5-ol hydrochloride.

EMD-38362. A dopamine D_2 receptor agonist. mp = 274°.

8064 Talipexole
101626-70-4 9209

$C_{10}H_{15}N_3S$

6-Allyl-2-amino-5,6,7,8-tetrahydro-4H-thiazolo[4,5-d]-azepine.

An α_2-adrenergic and dopamine D_2 receptor agonist. Used as an antiparkinsonian. *Boehringer Ingelheim GmbH.*

8065 Talipexole Hydrochloride
36085-73-1 9209

$C_{10}H_{16}ClN_3S$

6-Allyl-2-amino-5,6,7,8-tetrahydro-4H-thiazolo[4.5-d]-azepine hydrochloride.

B-HT-920; Domin. An α_2-adrenergic and dopamine D_2 receptor agonist. Used as an antiparkinsonian. mp = 245° (dec). *Boehringer Ingelheim GmbH.*

8066 Tienocarbine
75458-65-0

$C_{15}H_{16}N_2S$

7,8,9,10-Tetrahydro-1,9-dimethyl-6H-pyrido[4,3-b]-thieno[3,2-e]indole.

Mixed dopamine agonist-antagonist.

8067 Tilozepine
42239-60-1

$C_{17}H_{18}ClN_3S$

7-Chloro-4-(4-methyl-1-piperazinyl)-10H-thieno[3,2-c]-[1]benzazepine.

Dopamine receptor antagonist.

8068 Tropapride
76352-13-1

$C_{23}H_{28}N_2O_3$

N-(8-Benzyl-1αH,5αH-nortropan-3β-yl)-o-veratramide.

MD-790501. A dopamine D2 receptor antagonist with sodium salt-dependent binding. A benzamide derivative.

8069 Vanoxerine
67469-69-6

$C_{28}H_{32}F_2N_2O$

1-[2-[Bis(p-fluorophenyl)methoxy]ethyl]-4-(3-phenylpropyl)piperazine.

GBR-12909. Selective dopamine uptake inhibitor.

8070 Zetidoline
51940-78-4

$C_{16}H_{22}ClN_3O$

1-(m-Chlorophenyl)-3-[2-(3,3-dimethyl-1-azetidinyl)ethyl]-2-imidazolidinone.

DL-308-IT. Specific dopamine D2 receptor antagonist. Antipsychotic.

Ectoparasiticides

8071 Amitraz

33089-61-1 510 251-375-4

$C_{19}H_{23}N_3$

N-Methyl-N'-2,4-xylyl-N-(N-2,4-xylylformimidoyl)-formamidine.

U-36059; BTS-27419; ENT-27967; BAAM; Mitaban; Mitac; Taktic. Ectoparasiticide; scabicide. mp = 86-87°; insoluble in H_2O, soluble in most organic solvents; LD_{50} (mus orl) = 1600 mg/kg; (rat orl) = 400 mg/kg. *Boots Pharmaceuticals Inc.*

8072 Benzyl Benzoate

120-51-4 1162 204-402-9

$C_{14}H_{12}O_2$

Benzoic acid phenylmethyl ester.

Benylate; Ascabin; Venzonate; Ascabiol. Ectoparasiticide. mp = 21°; bp = 323-324°, bp_{16} = 189=191°, $bp_{4.5}$ = 156°; d_4^{25}= 1.118; insoluble in H_2O, glycerol; soluble in organic solvents; LD_{50} (rat orl) = 1700 mg/kg, (mus orl) = 1400 mg/kg, (rbt orl) = 1800 mg/kg, (gpg orl) = 1000 mg/kg. *Sterling Winthrop Inc.*

8073 Carbaryl

63-25-2 1831 200-555-0

$C_{12}H_{11}NO_2$

1-Naphthyl methylcarbamate.

ENT-23969; OMS-29; UC-7744; Arylam; Carylderm; Clinicide; Derbac; Dicarbam; Ravyon; Seffein; Sevin. Ectoparasiticide. mp = 142°; d_{20}^{20} = 1.232; soluble in H_2O (0.012 g/100 ml), DMF, Me_2CO, cyclohexanone, isophorone; LD_{50} (rat orl) = 250 mg/kg. *Union Carbide Corp.*

8074 Crotamiton

483-63-6 2661 207-596-3

$C_{13}H_{17}NO$

N-Ethyl-o-crotonotoluidide.

Crotamitex; Eurax; Euraxil; Veteusan; component of: Eurax. Ectoparasiticide; scabicide. bp_{13} = 163-165°; soluble in MeOH, EtOH. *Westwood-Squibb Pharm. Inc.*

8075 DDT

50-29-3 2898 200-024-3

$C_{14}H_9Cl_5$

1,1,1-Trichloro-2,2-bis(p-chlorophenyl)ethane.

clofenotane; dicophaner; pentachlorin; Agritan; Gesapon; Gesarex; Gesarol; Guesapon; Neocid. Pediculicide; ectoparasiticide. mp = 108.5-109°; λ_m 236 nm (95% EtOH); insoluble in H_2O; soluble in Me_2CO (58 g/100 ml), C_6H_6 (78 g/100 ml), benzyl benzoate (42 g/100 ml), CCl_4 (45 g/100 ml), C_6H_5Cl (74 g/100 ml), cyclohexanone (116 g/100 ml), EtOH (2 g/100 ml), Et_2O (28 g/100 ml), iPrOH (3 g/100 ml), C_5H_5N, dioxane; LD_{50} (mrat orl) =113 mg/kg, (frat orl) = 118 mg/kg. *Ciba-Geigy Corp.*

8076 Dixanthogen

502-55-6 3449 207-944-4

$C_6H_{10}O_2S_4$

O,O-Diethyl dithiobis(thioformate).

bisethylxanthogen; diethyl xanthogenate; ethylxanthic disulfide; preparation K; EXD; Auligen; Aulinogen; Bexide; Herbisan; Lenisarin; Sulfasan. Pediculicide; ectoparasiticide. mp = 28-32°; soluble in EtOH (2 g/100 ml); freely soluble in C_6H_6, Et_2O, petroleum ether, oils; insoluble in H_2O; LD_{50} (rat orl) = 480 mg/kg.

8077 Doramectin

117704-25-3 3483

$C_{50}H_{74}O_{14}$

(2aE,4E,8E)-(5'S,6S,6'R,7S,11R, 13S, 15S,17aR,20R,20aR,20bS)-6'-Cyclohexyl-5',6,6',7,10,11,14,15,17a,20,20a,20b-dodecahydro-20,20b-dihydroxy-5',6,8,19-tetramethyl-17-oxospiro[11,15-methano-2H,13H,17H-furo[4,3,2-pq]-[2,6]-benzodioxacyclooctadecin-13,2'-[2H]-pyran]-7-yl 2,6-dideoxy-4-O-(2,6-dideoxy-3-O-methyl-α-L-arabinohexopyranosyl)-3-O-methylL-arabinohexopyranoside.

25-cyclohexyl-5-O-demethyl-25-de(1-methylpropyl)avermectin A_{1a}. Endectocide, used to treat Sheep scabies (veterinary). mp= 116-119°. *Pfizer Intl.*

8078 Lime Sulfurated Solution

5516

Calcium oxysulfide.

Vleminckx's Solution; Vleminckx's Lotion; calcium oxysulfide solution. Ectoparasiticide.

8079 Lindane

58-89-9 5526 200-401-2

$C_6H_6Cl_6$

σ-1,2,3,4,5,6-Hexachlorocyclohexane.

Scabene; Aparasin; Aphtiria; Esoderm; Gammalin; Gamene; Gammexane; Gexane; Jacutin; Kwell; Lindafor; Lindatox; Lorexane; Quellada; Streunex; Tri-6; Viton. Pediculicide; scabicide; ectoparasiticide. mp = 112.5°; soluble in Me_2CO (43.5 g/199 ml), C_6H_6 (28.9 g/100 ml), $CHCl_3$ (24.0 g/100 ml), Et_2O (20.8 g/100 ml), EtOH (6.4 g/100 ml); insoluble in H_2O; LD_{50} (mrat orl) = 88 mg/kg, (frat orl) = 91 mg/kg. *ICI.*

8080 Lufenuron

103055-07-8 5624

$C_{17}H_8Cl_2F_8N_2O_3$

1-[2,5-Dichloro-4-(1,1,2,3,3,3-hexafluoropropoxy)-phenyl]-3-(2,6-difluorobenzoyl)urea.

fluphenacur; CGA-184699; Program. Used in veterinary medicine for control of fleas. Ectoparaciticide. mp = 174°; insoluble in H_2O; soluble in organic solvents; LD_{50} (rat orl) > 2000 mg/kg; LC_{50} (rat ihl) > 2350 mg/m^3). *Ciba-Geigy Corp.*

8081 Malathion

121-75-5 5740 204-497-7

$C_{10}H_{19}O_6PS_2$

Diethyl mercaptosuccinate S-ester with O,O-dimethyl phosphorodithioate.

ENT-17034; Cythion; Derbac-M; Malamar 50; Malaspray; Organoderm; Prioderm; Suleo-M. Pediculicide; ectoparasiticide. mp = 2.9°; $bp_{0.7}$ = 156-157°; slightly soluble in H_2O (0.0145 g/100 ml), soluble in most organic solvents; LD_{50} (frat orl) = 1000 mg/kg, (mrat orl) = 1375 mg/kg. *Purdue Pharma L.P.*

8082 Mercuric Oleate
1191-80-6 5934 214-741-4
$C_{36}H_{66}HgO_4$
9-Octadecenoic acid mercury(2+) salt.
oleate of mercury. Formerly used as an ectoparasiticide. Insoluble in H_2O; slightly soluble in EtOH, Et_2O.

8083 Mesulfen
135-58-0 5977 205-202-4
$C_{14}H_{12}S_2$
2,7-Dimethylthianthrene.
mesulphen; Mitigal; Odylen; Sudermo; Peligal; Neosulfine. Ectoparasiticide; antipruritic; scabicide. mp = 123°; bp_3 = 184°, bp_{14} = 228-231°; insoluble in H_2O; freely soluble in Me_2CO, Et_2O, $CHCl_3$, petroleum ether; moderately soluble in EtOH, EtOAc.

8084 Nifluridide
61444-62-0
$C_{10}H_6F_7N_3O_3$
6'-Amino-α,α,α,2,2,3,3-heptafluoro-5'-nitro-m-propionotoluidide.
compound 109168. Ectoparasiticide. *Eli Lilly & Co.*

8085 Permethrin
52645-53-1 7321 258-067-9
$C_{21}H_{20}Cl_2O_3$
m-Phenoxybenzyl (±)-3-(2,2-dichlorovinyl)-2,2-dimethyl-cyclopropanecarboxylate.
Elimite; Nix; FMC-33297; NIA-33297; NRDC-143; PP-557; SBP-1513; S-3151; Ambush; Corsair; Dragnet; Ectiban; Eksmin; Pulvex; Pounce; Pynosect; Ridect Pour-On. Ectoparasiticide. mp → 35°; $bp_{0.05}$ = 220°; d^{20} = 1.190-1.272; almost insoluble in H_2O, soluble in all organic solvents except ethylene glycol; LD_{50} (frat orl) = 3801 mg/kg, (8 day old rat orl) = 340.5 mg/kg, (mrat orl) = 1500.0 mg/kg; toxic to bees and fish. *Allergan Inc.; Glaxo Wellcome Inc.; Sumitomo Pharm. Co. Ltd.*

8086 Quintiofos
1776-83-6 217-208-4
$C_{17}H_{16}NO_2PS$
O-Ethyl O-(8-quinolyl)phenylphosphonothioate.
Bayer 9037. Ixodicide. *Bayer AG.*

8087 Sulfiram
95-05-6 9123 202-387-3
$C_{10}H_{20}N_2S_3$
Bis(diethylthiocarbamyl) sulfide.
monosulfiram; TTMS; Kutkasin; Tetmosol. Ectoparasiticide. mp = 32°.

8088 Sulfur, Pharmaceutical
7704-34-9 9150 231-722-6
S
Sulfur, pharmaceutical.
Liquamat; Sastid; Sulfur Soap; component of: Acnomel, Bensulfoid, Fostril, Pernox, Resamid, Salicylic Acid and Sulfur Soap, Sebulex, Sulforcin, Sulfoxyl, Transact. Scabicide; ectoparasiticide. Soluble in CS_2, lanolin, olive oil. *Dermik Labs. Inc.; ECR Pharm.; Galderma Labs Inc.; Menley & James Labs Inc.; Stiefel Labs Inc.; Westwood-Squibb Pharm. Inc.*

8089 Temephos
3383-96-8 9286 222-191-1
$C_{16}H_{20}O_6P_2S_3$
O,O'-(Thiodi-p-phenylene O,O,O',O'-tetramethyl bis(phosphorothioate).
ENT-27165; AC-52160; Abate; Biothon. Ectoparasiticide. mp = 30.0-30.5°; soluble in CH_3CN, CCl_4, Et_2O, dichloroethane, C_7H_8; insoluble in H_2O, C_6H_{14}; LD_{50} (mrat orl) = 8600 mg/kg, (frat orl) = 13000 mg/kg. *Am. Cyanamid.*

Emetics

8090 Apocodeine
641-36-1 781
$C_{18}H_{19}NO_2$
5,6,6a,7-Tetrahydro10-methoxy-6-methyl-4H-dibenzo[de,g]quinolin-11-ol.
Emetic. mp = 124°; $[\alpha]_D^{24}$ = -97° (c = 0.45); slightly soluble in H_2O; soluble in EtOH, Et_2O, dilute acids.

8091 Apocodeine Hydrochloride
6377-14-6 781 228-947-7
$C_{18}H_{20}ClNO_2$
5,6,6a,7-Tetrahydro10-methoxy-6-methyl-4H-dibenzo[de,g]quinolin-11-ol hydrochloride.
Emetic. mp = 260-263° (dec); $[\alpha]_D^{22}$ = -43° (c = 0.51); very soluble in H_2O, soluble in EtOH.

8092 Apomorphine
58-00-4 787 200-360-0
$C_{17}H_{17}NO_2$
5,6,6a,7-Tetrahydro-6-methyl-4H-dibenzo[de,g]quinolin-10,11-diol.
Emetic. mp = 195°; soluble in EtOH, Me_2CO, $CHCl_3$; slightly soluble in H_2O, C_6H_6, Et_2O, petroleum ether; λ_m = 336, 399 nm; LD_{50} (mus ip) = 160 mg/kg.

8093 Apomorphine Hydrochloride
41372-20-7 787
$C_{17}H_{18}ClNO_2$
5,6,6a,7-Tetrahydro-6-methyl-4H-dibenzo[de,g]quinoline-10,11-diol hydrochloride.
Emetic. $[\alpha]_D^{25}$ = -48° (c = 1.2); soluble in H_2O (2 g/100 ml at 20°, 5.88 g/100 ml at 80°, EtOH (2 g/100 ml); slightly soluble in $CHCl_3$, Et_2O; LD_{50} (mus ip) = 44 mg/kg.

8094 Cephaeline
483-17-0 2020 207-591-6
$C_{28}H_{38}N_2O_4$
7',10,11-Trimethoxyemetan-6'-ol.
dihydropsychotrine; desmethylemetine. Antiamebic; emetic. Alkaloid of ipecac. mp= 115-116°; $[\alpha]_D^{20}$ = -43.4° (c = 2 $CHCl_3$); insoluble in H_2O; soluble in MeOH, EtOH, Me_2CO, $CHCl_3$; less soluble in Et_2O, petroleum ether; [dihydrochloride heptahydrate]: mp = 270°; $[\alpha]_D^{20}$ = +25.0° (c = 2); soluble in H_2O; moderately soluble in Me_2CO, $CHCl_3$; nearly insoluble in C_6H_6; [Dihydrobromide heptahydrate]: mp = 293°; soluble in H_2O; moderately soluble in alcohol; Me_2CO; nearly insoluble in C_6H_6.

8095 Ipecac
8012-96-2 5086 232-385-8
Ipsatol; component of: Dasin. Emetic. Dried rhizome and roots of *Caphaelis ipecauanha*. Contains 2-2.5% alkaloids, including emetine, cephaeline; emetamine; psychotrine; methyl psychotrine; protoemetine and resin. *Key Pharm.; SmithKline Beecham Pharm.*

8096 Sodium Chloride
7647-14-5 8742 231-598-3
NaCl
Hydrochloric acid sodium salt.
Adsorbanac; Ayr; Ringer's Injection; common salt; table salt; rock salt; component of: Arm-A-Vial, Colyte. Electrolyte replenisher; emetic; topical anti-inflammatory. d = 2.17; mp = 804°; soluble in H_2O (35.7 g/100 ml at 25°, 38.5 g/100 ml at 100°), glycerol (10 g/100 ml); slightly soluble in EtOH; LD_{50} (rat orl) = 3750 ± 430 mg/kg. *Alcon Labs; Ascher B.F. & Co.; Astra USA Inc.; Elkins-Sinn; Schwarz Pharma Kremers Urban Co.; Wyeth-Ayerst Labs.*

8097 Zinc Acetate
557-34-6 10256 209-170-2
$C_4H_6O_4Zn.2H_2O$
Acetic acid zinc salt dihydrate.
Emetic; astringent. 120 153 mp = 237°; d = 1.735; soluble in H_2O (43.4 g/100 at 25°, 62.5 g/100 ml at 100°), EtOH (3.3 g/100 ml at 25°, 100 g/100 ml at 76°); LD_{50} (rat orl) = 2460 mg/kg.

8098 Zinc Acetate, Basic
82279-57-0
$C_{12}H_{18}O_{13}Zn_4$
Hexakis(μ-acetato)-μ4-oxotetrazinc.
Used for the treatment and prophylaxis of hepatic copper toxicosis; for maintenance therapy in patients initially treated with a chelating agent. Formerly used as an emetic.

Enkephalinase Inhibitors

8099 Acetorphan
81110-73-8 72
$C_{21}H_{23}NO_4S$
(±)-N-[2-[(Acetylthio)methyl]-1-oxo-3-phenylpropyl]-glycine phenylmethyl ester.
Tiorfan; ecadotril [as (S)-form]; sinorphan [as (S)-form]. The racemate is an antidiarrheal, the (S)-form is an antihypertensive. Antisecretory enkephalinase inhibitor. mp = 89°; [(S)-form]: mp = 71°; $[\alpha]_D^{25}$ = -24.1° (c = 1.2 MeOH); LD_{50} (mus iv) > 100 mg/kg. *Bioproject.*

8100 (S)-Acetorphan
72
$C_{21}H_{28}3NO_4S$
(S)-N-[(R,S)-3-Acetylthio-2-benzopropanoyl]glycine benzyl ester.
ecadotril; sinorphan. The racemate is an antidiarrheal, the (S)-form is an antihypertensive. Antisecretory enkephalinase inhibitor. mp = 71°; $[\alpha]_D^{25}$ = -24.1° (c = 1.2 MeOH); LD_{50} (mus iv) > 100 mg/kg. *Bioproject.*

Enzymes

8101 α-Amylase, Bacterial
9000-85-5 640 232-560-9
Digestive aid. Generally derived from bacteria such as *Bacillus subtilis*. The amylases are enzymes which catalyze the hydrolysis of α-1→4 glucosidic linkages in polysaccharides.

8102 α-Amylase, Porcine
9000-90-2 640 232-565-6
Buclamase; Maxilase. Digestive aid. Enzyme derived from swine pancrease. Molecular weight ≅ 45,000.

8103 α-Amylase, Sweet potato
9000-91-3 640 232-566-1
Digestive aid. Enzyme from sweet potato, molecular weight ≅ 152,000.

8104 α-Chymostrypsin
9004-07-3 2320 232-671-2
Avazyme; Catarase; Enzeon; Zolyse; component of: Orenzyme. Proteolytic enzyme. *Alcon Labs; CIBA Vision AG; Marion Merrell Dow Inc.; Sterling Winthrop Inc.; Wallace Labs.*

8105 Amylase
9000-92-4 640 232-567-7
Digestive aid. Usually refers to α-amylase, an enzyme derived from bacteria.

8106 Cellulase
9012-54-8 232-734-4
component of: Kutrase, Ku-zyme. A preparation of cellulose-cleaving enxymes from *Aspergillus niger*. *Schwarz Pharma Kremers Urban Co.*

8107 Chymopapain
9001-09-6 2319 232-580-8
Chymodiactin; BAX-1526; NSC-107079. Proteolytic enzyme isolated from papaya latex; distihnct from papain. Molecular weight → 27000. *Knoll Pharm. Co.*

8108 Collagenase
9001-12-1 2544 232-582-9
Clostridiopeptidase A; Iruxol; Santyl. A rare proteolytic enzyme capable of digesting undenatured collagen. *Worthington Biochemical.*

8109 Lipase
9001-62-1 5536 232-619-9
Triacylglycerol lipase.
Digestive aid. An esterase which hydrolyzes fats, producing fatty acids and glycerol.

8110 Lysozyme
9001-63-2 5671 232-620-4
N-Acetylmuramide glycanohydrolase.
Muramidase; N-acetylmuramyl hydrolase; globulin G_2. A mucolytic enzyme with antibiotic properties. Derivative of lysozyme.

8111 Lysozyme Hydrochloride
9066-59-5 5671 232-954-0
N-Acetylmuramide glycanohydrolase hydrochloride.
Acdeam; Antalzyme; Immunozima; Lanzyme; Leftose;

Likinozym; Lisozima; Murazyme; Neutase; Neuzyme; Toyolysom-DS. A mucolytic enzyme with antibiotic properties. Derivative of lysozyme.

8112 Pancreatin

8049-47-6 7137 232-468-9

Diastase vera; Creon; Pancrease; Pancrex-Vet; Pankrotanon; Panzytrat; Zypanar. Digestive aid. Used in veterinary medicine in treatment of pancreatic enzyme deficiency. Derived from pancreas of hog or ox. Contains the enzymes amylopsin, steapsin, and trypsin. Converts strach into soluble carbohydrates. Insoluble in EtOH; partly soluble in H_2O; highest activity in neutral or slightly alkaline media.

8113 Pancrelipase

53608-75-6 7138 258-659-7

Accelerase; Cotazym; Ilozyme; Ku-Zyme HP; Pancrease; Viokase. Digestive aid. Enzyme concentrate containing primarily lipase, and also amylase and proteases.

8114 Papain

9001-73-4 7148 232-627-2

vegetable pepsin; Arbuz; Caroid; Nematolyt; Papayotin; Summetrin; Tromasin; Velardon; Vermizym; component of: Panafil. Proteolytic enzyme. Digestive aid. A proteolytic enzyme isolated from the fruit and leaves of *Carica papaya.* Used as a digestive aid, also as a debridant, an anthelmintic targeting Nematodes and as an agent which can prevent adhesions. λ_m = 278 nm ($A^{1\%}_{1\,cm}$ 25.0); insoluble in most organic solvents. *Rystan Co. Inc.; Sterling Winthrop Inc.*

8115 Penicillinase

9001-74-5 7219 232-628-8

β-lactamase.

Neutrapen. Enzyme obtained by fermentation from cultures of *Bacillus cereus.* Destroys penicillins and cephalosporins by catalyzing the hydrolysis of the amide bond in the β-lactam ring. *Schenley.*

8116 Pepsin

9001-75-6 7289 232-629-3

Puerzym. Digestive aid. Has been used to treat deficiency of pepsin secretion. The principal digestive enzyme in gastric juice; controls the degradation of proteins to proteoses and peptones hydrolyzing only peptide bonds. $[\alpha]_D^{26}$= -64.5° (pH 4.6); freely soluble in H_2O; insoluble in EtOH, $CHCl_3$, Et_2O.

8117 Rennin

9001-98-3 8303 232-645-0

Chymosin; rennase; lab; abomasal enzyme. Digestive aid. The milk-clotting enzyme from the stomach of the calf. Secreted as prorennin which is converted to rennin in acid. Rennet, a dried extract containing rennin is used in themanufacture of cheese and rennet casein.

8118 Sutilains

12211-28-8 235-390-3

Travase; BAX-1515. Proteolytic enzymes derived from *Bacillus subtilis. Knoll Pharm. Co.*

8119 Trypsin

9002-07-7 9926 232-650-8

Parenzyme; Parenzymol; Typtar; Trypure. Proteolytic enzyme crystallized from an extract of the pancreas gland of the ox. *Marion Merrell Dow Inc.; National Drug Co.*

Enzyme Cofactors

8120 Acetiamine

299-89-8 51

$C_{16}H_{22}N_4O_4S$

N-[(4-Amino-2-methyl-5-pyrimidinyl)methyl]-N-(4-hydroxy-2-mercapto-1-methyl-1-butenyl)formamide O,S-diacetate.

D.A.T.; vitamin B_1; Thianeuron. Vitamin (enzyme cofactor). mp = 122-123°, 123-124°; soluble in H_2O, MeOH, EtOH.

8121 Acetiamine Hydrochloride

28008-04-0 51 248-774-0

$C_{16}H_{23}ClN_4O_4S$

N-[(4-Amino-2-methyl-5-pyrimidinyl)methyl]-N-(4-hydroxy-2-mercapto-1-methyl-1-butenyl)formamide O,S-diacetate hydrochloride.

Neviton Comprimés. Vitamin (enzyme cofactor).

8122 Benfotiamine

22457-89-2 1068 245-013-4

$C_{19}H_{23}N_4O_6PS$

N-[(4-Amino-2-methyl-5-pyrimidinyl)methyl]-N-(4-hydroxy-2-mercapto-1-methyl-1-butenyl)formamide S-benzoate O-phosphate.

8088 C.B.; S-benzoylthiamine monophosphate; BTMP; Biotamin; Vitanevril. Vitamin (enzyme cofactor). mp = 165° (dec).

8123 Bisbentiamine

2667-89-2 1286 220-206-6

$C_{38}H_{42}N_8O_6S_2$

N,N'-[Dithiobis[2-(2-benzoyloxyethyl)-1-methylvinylene]]bis[N-[(4-amino-2-methyl-5-purimidinyl)methyl]formamide].

Béprocin; Beston. Vitamin (enzyme cofactor). Vitamin B_1 source. mp = 146-147°. *Tanabe Seiyaku Co. Ltd.*

8124 Calcium Pantothenate

137-08-6 7147 205-278-9

$C_{18}H_{32}CaN_2O_{10}$

(R)-N-(2,4-Dihydroxy-3,3-dimethyl-1-butyl)-β-alanine calcium salt (2:1).

Calpan; Pantholin; Calpanate; component of: Stuartinic. Vitamin (enzyme cofactor). mp = 195-196° (dec); $[\alpha]_D^{20}$ = 28.2° (c = 5); soluble in H_2O (35.7 g/100 ml), glycerol; slightly soluble in EtOH, Me_2CO. *BASF Corp.; Eli Lilly & Co.; Johnson & Johnson-Merck Consumer Pharm. ; Pharmacia & Upjohn.*

8125 Calcium Pantothenate, Racemic

6381-63-1 7147 6381-63-1

$C_{18}H_{32}CaN_2O_{10}$

(±)-N-(2,4-Dihydroxy-3,3-dimethyl-1-butyl)-β-alanine calcium salt (2:1) .

(±)-calcium pantothenate. Vitamin (enzyme cofactor). *BASF Corp.; Eli Lilly & Co.; Johnson & Johnson-Merck Consumer Pharm. ; Pharmacia & Upjohn.*

8126 Cetotiamine
137-76-8 2064
$C_{18}H_{26}N_4O_6S$
S-Ester of O-ethyl thiocarbonate with N-[(4-amino-2-methyl-5-pyrimidinyl)methyl]-N-(4-hydroxy-2-mercapto-1-methyl-1-butenyl)formamide ethyl carbonate.
O,S-dicarbethoxythiamine; DCET. Vitamin (enzyme cofactor). Vitamin B_1 source. mp = 113.5-114.5°. *Shionogi & Co. Ltd.*

8127 Cyclotiamine
6092-18-8 2824
$C_{13}H_{16}N_4O_3S$
N-[1-(2-Oxo-1,3-oxathian-4-ylidene)ethyl]-N-(4-amino-2-methyl-5-pyrimidinyl)methyl]formamide.
cyclothiamine; carbothiamine; cyclotiamine; CCT; Cometamine; Commetamin. Vitamin (enzyme cofactor). mp = 175.5° (dec); LD_{50} (mus orl) = 13390 mg/kg; [hydrochloride ($C_{13}H_{17}ClN_4O_3S$)]: mp = 179-180° (dec). *Yamanouchi U.S.A. Inc.*

8128 Dexpanthenol
81-13-0 2988 201-327-3
$C_9H_{19}NO_4$
D-(+)-2,4-Dihydroxy-N-(3-hydroxypropyl)-3,3-dimethylbutyramide.
D-Panthenol 50; Ilopan; Motilyn; pantothenol; pantothenyl alcohol; Alcopan-250; Intrapan; Pantenyl; Panthoderm; Bepanthen; Cozyme; Urupan; component of: Ilopan-Choline. Cholinergic agent. The dl-form is used as a vitamin. $bp_{0.02}$ = 118-120°; d_{20}^{20} = 1.2; $[\alpha]_D^{20}$= 29.5° (c = 5); freely soluble in H_2O, EtOH, MeOH; slightly soluble in Et_2O. *Abbott Labs Inc.; BASF Corp.; Hoffmann-LaRoche Inc.; Savage Labs.*

8129 Fursultiamine
804-30-8 4333 212-357-1
$C_{17}H_{26}N_4O_3S_2$
N-[(4-Amino-2-methyl-5-pyrimidinyl)methyl]-N-[4-hydroxy-1-methyl-2-[(tetrahydrofurfuryl)dithio]-1-butenyl]formamide.
TTFD; Alinamin F; Diteftin; Judolor. Vitamin (enzyme cofactor). mp = 132° (dec); d = 1.29; sparingly soluble in H_2O; soluble in organic solvents, mineral acids; LD_{50} (rat orl) = 2200 mg/kg, (rat ip) = 540 mg/kg. *Takeda Chem. Ind. Ltd.*

8130 Methylol Riboflavin
8002-52-6 6179
Hyflavin. Vitamin (enzyme cofactor). Mixture of methylol (1-3 methylol groups) derivatives of riboflavin. Soluble in H_2O, insoluble in organic solvents; dextrorotatory. *Endo Pharm. Inc.*

8131 Niacin
59-67-6 6612 200-441-0
$C_6H_5NO_2$
3-Pyridinecarboxylic acid.
Nicotinic acid; Niac; Nicamin; Nicobid; Nicolar; Wampocap; P.P. Factor; Akotin; Daskil; Niacor; Nicacid; Nicangin; Niconacid; NicoSpan. Vitamin (enzyme cofactor). At high doses, decreases hepatic secretion of very-high-density lipoproteins as a result of reduced triglyceride synthesis. Used in treatment of hypercholesterolemias and hypertriglyceridemias. mp = 236.6°; λ_m = 263 nm; soluble in H_2O (1.67 g/100 ml); freely soluble in H_2O at 100°, EtOH at 76°; insoluble in Et_2O; LD_{50} (raty sc) = 5000 mg/kg. *Abbott Labs Inc.; Apothecon; Forest Pharm. Inc.; Marion Merrell Dow Inc.; Rhône-Poulenc Rorer Pharm. Inc.; Wallace Labs.*

8132 Niacinamide
98-92-0 6574 202-713-4
$C_6H_6N_2O$
3-Pyridinecarboxamide.
Nicotinamide; Aminicotin; Benicot; Dipegyl; Nicamindon, Nicobion; Nictoamide; Nicotilamide; Pelmin; component of: Medriatric. Vitamin (enzyme cofactor). Precursor of the coenzyme NAD and NADP. mp = 128-131°; $bp_{0.0005}$ = 150-160°; λ_m 261 nm ($A_{1\ cm}^{1\%}$ 451); soluble in H_2O (100 g/100 ml), EtOH (66.6 g/100 ml), glycerol (10 g/100 ml); LD_{50} (rat sc) = 1680 mg/kg. *Wyeth-Ayerst Labs.*

8133 Nicotinamide Ascorbate
1987-71-9 6610
L-Ascorbic acid mixture with 3-pyridinecarboxamide.
merpress; nicoscorbine; Nicastubin. Vitamin (enzyme cofactor). mp = 141-145°; $[\alpha]_D^{20}$ = 27.5° (c = 8 H_2O); soluble in H_2O (40 g/100 ml), EtOH (2.4 g/100 ml), MeOH (10 g/100 ml); sparingly soluble in Me_2CO; insoluble in C_6H_6, Et_2O. *Gelatin Products.*

8134 Octotiamine
137-86-0 6857
$C_{23}H_{36}N_4O_5S_3$
8-[[2-[N-[(4-Amino-2-methyl-5-pyrimidinyl)methyl]-formamido]-1-(2-hydroxyethyl)propenyl]dithio]-6-mercaptooctanoic acid methyl ester acetate.
Gerostop; Neuvitan; TATD. Vitamin (enzyme cofactor). mp = 106-109°; λ_m = 234. 277 nm (ε 16200, 5820); [hydrochloride ($C_{23}H_{37}ClN_4O_5S_3$)]: mp = 134.5-135°; λ_m = 233 nm (ε 23000). *Fujisawa Pharm. USA Inc.*

8135 Pantothenic Acid
79-83-4 7147 201-229-0
$C_9H_{17}NO_5$
N-(2,4-Dihydroxy-3,3-dimethyl-1-butyl)-β-alanine.
chick antidermatitis factor; Vitamin B_5. Vitamin (enzyme cofactor). $[\alpha]_D^{25}$ = 37.5°; freely soluble in H_2O, EtOAc, dioxane, AcOH; moderately soluble in Et_2O, amyl alcohol; insoluble in C_6H_6, $CHCl_3$.

8136 Pantothenic Acid Sodium Salt
867-81-2 7147 212-768-6
$C_9H_{16}NNaO_5$
N-(2,4-Dihydroxy-3,3-dimethyl-1-butyl)-β-alanine sodium salt.
Vitamin (enzyme cofactor). $[\alpha]_D^{25}$= 27.1° (c = 2).

8137 Prosultiamine
59-58-5 8068 200-436-3
$C_{15}H_{24}N_4O_2S_2$
N-[(4-Amino-2-methyl-5-pyrimidinyl)methyl]-N-(4-hydroxy-1-methyl-2-(propyldithio)-1-butenyl]formamide.
dithiopropylthiamine; DTPT; TPD; Alinamin; Binova. Vitamin (enzyme cofactor). mp = 128-129° (dec); soluble

in organic solvents, sparingly soluble in H_2O. *Takeda Chem. Ind. Ltd.*

8138 Prosultiamine Hydrochloride

973-99-9 8068 213-547-7

$C_{15}H_{25}ClN_4O_2S_2$

N-[(4-Amino-2-methyl-5-pyrimidinyl)methyl]-N-(4-hydroxy-1-methyl-2-(propyldithio)-1-butenyl]formamide hydrochloride.

Vitamin (enzyme cofactor). mp = 160-161° (dec).

8139 Pyridoxal 5-Phosphate

54-47-7 8163 200-208-3

$C_8H_{10}NO_6P.H_2O$

3-Hydroxy-2-methyl-5-[(phosphonoxy)methyl]-4-pyridinecarboxaldehyde hydrate.

Pyromijin; Sechvitan; Vitazechs. Vitamin (enzyme cofactor). Soluble in H_2O; λ_m = 390 nm (E_m 3.7 pH > 7), 295 nm (E_m 5.1 pH < 7). *E. Merck.*

8140 Pyridoxine Hydrochloride

58-56-0 8166 200-386-2

$C_8H_{12}ClNO_3$

5-Hydroxy-6-methyl-3,4-pyridinedimethanol hydrochloride.

Beesix; Hexa-Betalin; Hexavibex; Bonasanit; Hexabione hydrochloride; Pyridipca; Pyridox; Bécilan; Benadon; Hexermin; component of: Bendectin, Spondylonal. Vitamin, vitamin source. Enzyme cofactor. May improve hematopoiesis in patients with sideroblastic anemias. Bendectin is useful in amelioration of morning sickness. mp = 205-212° (dec); λ_m = 290 nm (ε 8400 0.1N HCl), 253, 325 nm (ε 3700, 7100, pH 7); soluble in H_2O (22.2 g/100 ml), EtOH (1.1 g/100 ml), propylene glycol; sparingly soluble in Me_2CO; insoluble in Et_2O, $CHCl_3$. *BASF Corp.; Eli Lilly & Co.; Forest Pharm. Inc.; General Aniline; Lederle Labs.; Marion Merrell Dow Inc.; Merck & Co.Inc.; Parke-Davis.*

8141 Riboflavin

83-88-5 8367 201-507-1

$C_{17}H_{20}N_4O_6$

7,8-Dimethyl-10-(D-ribo-2,3,4,5-tetrahydroxypentyl)isoalloxazine.

Vitamin B_2; lactoflavine; Vitamin G; Beflavin; Flavaxin. Vitamin (enzyme cofactor). mp = 278-282° (dec); has three crystal-forms with different aqueous solubilities; $[\alpha]_D^{25}$ = -112° to -122° (c = 0.5 0.02N NaOH/EtOH); λ_m 2290225, 266, 371, 444, 475 nm; soluble in H_2O (0.033 g/100 ml - 0.006 g/100 ml), EtOH (0.0045 g/100 ml); slightly soluble in cyclohexanol, amyl acetate, benzyl alcohol, phenol; LD_{50} (rat orl) > 10000 mg/kg, (rat sc) = 5000 mg/kg, (rat ip) = 560 mg/kg. *BASF Corp.; Sterling Winthrop Inc.*

8142 Riboflavin 2,3,4,5-Tetrabutyrate

752-56-7 8367 212-034-5

$C_{33}H_{44}N_4O_{10}$

7,8-Dimethyl-10-(D-ribo-2,3,4,5-tetrahydroxypentyl)isoalloxazine 2,3,4,5-tetrabutyrate.

Bituvitan; Eyekas; Hibon; Lacflavin; Ribolact; Wakaflavin L; Viras. Vitamin (enzyme cofactor). *BASF Corp.; Sterling Winthrop Inc.*

8143 Riboflavin Monophosphate

146-17-8 8368 205-664-7

$C_{17}H_{21}N_4O_9P$

7,8-Dimethyl-10-(D-ribo-2,3,4,5-tetrahydroxypentyl)-isoalloxazine 5'-(dihydrogen phosphate).

flavin mononucleotide; FMN; vitamin B_2 phosphate. Vitamin, vitamin source. Enzyme cofactor. May be helpful in management of patients with hypoproliferative anemia. *Hoffmann-LaRoche Inc.; Takeda Chem. Ind. Ltd.*

8144 Riboflavin Monophosphate Monosodium Salt

130-40-5 8368 204-988-6

$C_{17}H_{20}N_4NaO_9P$

7,8-Dimethyl-10-(D-ribo-2,3,4,5-tetrahydroxypentyl)-isoalloxazine 5'-(dihydrogen phosphate) monosodium salt.

Hyryl; Ribo. Vitamin, vitamin source. Enzyme cofactor. May be helpful in management of patients with hypoproliferative anemia. [dihydrate]: soluble in H_2O (11.2 g/100 ml). *Hoffmann-LaRoche Inc.; Takeda Chem. Ind. Ltd.*

8145 Sapropterin

62989-33-7 8515

$C_9H_{15}N_5O_3$

(-)-(6R)-2-Amino-6-[(1R,2S)-1,2-dihydroxypropyl]-5,6,7,8-tetrahydro-4(3H)-pteridinone.

dapropterin; R-THBP; 6-BrBH$_4$. Vitamin (enzyme cofactor). Used in treatment of hyperphenylalaninemia. λ_m 265 nm (ε 14000 0.1N HCl). *Suntory Ltd.*

8146 Sapropterin Dihydrochloride

69056-38-8 8515

$C_9H_{13}Cl_2N_5O_3$

(-)-(6R)-2-Amino-6-[(1R,2S)-1,2-dihydroxypropyl]-5,6,7,8-tetrahydro-4(3H)-pteridinone dihydrochloride.

SUN-0588; Biopten. Vitamin (enzyme cofactor). Used in treatment of hyperphenylalaninemia. mp = 245-246° (dec); $[\alpha]_D^{25}$ = -6.81° (c = 0.665 0.1M HCl); λ_m 264 nm (ε 16770 2M HCl). *Suntory Ltd.*

8147 Thiamine

59-43-8 9430 200-425-3

$C_{12}H_{17}N_4OS$

3-[(4-Amino-2-methyl-5-pyrimidinyl)methyl]-5-(2-hydroxyethyl)-4-methylthiazolium chloride monohydrochloride.

vitamin B_1, aneurin; thiaminium chloride. Vitamin (enzyme cofactor). Essential nutrient for carbohydrate metabolism and nerve function.

8148 Thiamine Diphosphate

154-87-0 9431 205-836-1

$C_{12}H_{19}ClN_4O_7P_2S$

3-[(4-Amino-2-methyl-5-pyrimidinyl)methyl]-5-(2-hydroxyethyl)-4-methylthiazolium chloride P,P'-dioxide.

Vitamin (enzyme cofactor). Coenzyme required for decarboxylation of α-keto acids. mp = 238-240° (dec), 240-244° (dec); λ_m 242 nm; soluble in H_2O. *E. Merck.*

8149 Thiamine Disulfide
67-16-3 9432 200-644-4
$C_{24}H_{34}N_8O_4S_2$
N,N'-[Dithiobis[2-(2-hydroxyethyl)-1-methyl-2,1-ethenediyl]]bis[N-[(4-amino-2-methyl-5-pyrimidinyl)methyl]formamide].
aneurin disulfide; Aktivin; Neolamin. Vitamin (enzyme cofactor). mp = 177°; sparingly soluble in C_6H_6, Me_2CO, Et_2O, EtOH. *Hoffmann-LaRoche Inc.*

8150 Thiamine Hydrochloride
67-03-8 9430 200-641-8
$C_{12}H_{17}ClN_4OS$
3-[(4-Amino-2-methyl-5-pyrimidinyl)methyl]-5-(2-hydroxyethyl)-4-methylthiazolium chloride monohydrochloride.
Betalin S; Betaxin; Biamine; Benerva; Betabion; Bewon; Metabolin; Vitaneurin; component of: Spondylonal, Troph-Iron, Trophite, Trophite+Iron. Vitamin, vitamin source. Enzyme cofactor. mp = 248° (dec); soluble in H_2O (100 g/100 ml), glycerol (5.5 g/100 ml), 95% EtOH (1 g/100 ml), EtOH (0.32 g/100 ml); more soluble in MeOH; soluble in propylene glycol; insoluble in Et_2O, C_6H_6, $CHCl_3$, C_6H_{14}; LD_{50} (mus iv) = 89.2 mg/kg, (mus orl) = 8224 mg/kg. *BASF Corp.; Eli Lilly & Co.; Elkins-Sinn; Lederle Labs.; Menley & James Labs Inc.; Sterling Winthrop Inc.; Wyeth-Ayerst Labs.*

8151 Thiamine Mononitrate
532-43-4 9430 208-537-4
$C_{12}H_{17}N_5O_4S$
3-[(4-Amino-2-methyl-5-pyrimidinyl)methyl]-5-(2-hydroxyethyl)-4-methylthiazolium chloride nitrate (salt).
component of: Stuartinic. Vitamin (enzyme cofactor). Used to enrich flours and animal feeds. mp = 196-200° (dec); soluble in H_2O (2.7 g/100 ml at 25°, 30 g/100 ml at 100°). *Am. Cyanamid; BASF Corp.; Johnson & Johnson-Merck Consumer Pharm.*

8152 Vintiamol
26242-33-1 10129
$C_{21}H_{24}N_4O_3S$
N-[(4-Amino-2-methyl-5-pyrimidinyl)methyl]-N-[2-[(2-benzoylvinyl)thio]-4-hydroxy-1-methyl-1-butenyl]formamide.
Vitamin (enzyme cofactor). Two isomers, mp = 202-204° and mp = 100-103°.

Estrogens

8153 Benzestrol
85-95-0 1102
$C_{20}H_{26}O_2$
4,4'-(1,2-Diethyl-3-methyltrimethylene)-diphenol.
Octofollin; Ocestrol; Chemestrogèn. Estrogen mp = 162-166°; freely soluble in Me_2CO, Et_2O, EtOH, MeOH; slightly soluble in Et_2O, C_6H_6, $CHCl_3$, petroleum ether; insoluble in H_2O; [dibenzoate]: mp = 118-120°; [dimethyl ether]: mp = 56°.

8154 Broparoestrol
479-68-5 1471 207-537-1
$C_{22}H_{19}Br$
1-(2-Bromo-1,2-diphenylethenyl)-4-ethylbenzene.
B.D.P.E.; LN-107; Acnestrol; Longestrol. Estrogen. Commercial product is a mixture of cis- and trans-isomers. [cis isomer]: mp = 112-113.5°; [trans isomer]: mp = 111.5-112°; soluble in Et_2O, C_6H_6, $CHCl_3$; less soluble in EtOH.

8155 Chlorotrianisene
569-57-3 2225 209-318-6
$C_{23}H_{21}ClO_3$
1,1',1(1-Chloro-1-ethenyl-2-ylidene)tris[4-methoxybenzene].
tris-(p-methoxyphenyl)ethylene; Hormonisene; Merbentul; Tace. Estrogen. mp = 114-116°; λ_m = 310 nm (E1%;$Ks_{1\ cm}$ = 423 $CHCl_3$); insoluble in H_2O; soluble in EtOH (0.28 g/100 ml), Et_2O (3.6 g/100 ml), AcOH, Me_2CO, $CHCl_3$, CCl_4, C_6H_6, vegetable oils. *Marion Merrell Dow Inc.*

8156 Cloxestradiol
54063-33-1
$C_{20}H_{25}Cl_3O_3$
17β-(2,2,2-Trichloro-1-hydroxyethoxy)estra-1,3,5(10)-trien-3-ol.
Estrogen.

8157 Colpormon
1247-71-8 2554 214-997-7
$C_{22}H_{26}O_5$
16α-Hydroxyestrone diacetate.
Colpogynon. Estrogen. mp = 179-180°; $[\alpha]_D^{28}$ = 122° ($CHCl_3$). *Lab. Albert Rolland.*

8158 Dienestrol
84-17-3 3153 201-519-7
$C_{18}H_{18}O_2$
4,4'-(1,2-Diethylidene-1,2-ethanediyl)bisphenol.
dienoestrol; estrodienol; Cycladiene; Dienol; Dinovex; DV; Estroral; Gynefollin; Hormofemin; Oestrasid; Oestrodiene; Oestroral; Restrol; Retalon; Synestrol. Estrogen. mp = 227-228°, 231-234°; freely soluble in EtOH, MeOH, Et_2O, Me_2CO, propylene glycol, soluble in $CHCl_3$; insoluble in H_2O. *Boots Pharmaceuticals Inc.; Hoffmann-LaRoche Inc.*

8159 Dienestrol Diacetate
84-19-5 3153 201-520-2
$C_{22}H_{22}O_4$
4,4'-(1,2-Diethylidene-1,2-ethanediyl)bisphenol diacetate.
Lipamone; Retalon-Oral. Estrogen. mp = 119-120°. *Boots Pharmaceuticals Inc.; Hoffmann-LaRoche Inc.*

8160 Diethylstilbestrol
56-53-1 3177 200-278-5
$C_{18}H_{20}O_2$
(E)-4,4'-(1,2-Diethyl-1,2-ethenediyl)bisphenol.
DES; Antigestil; Bufon; Cyren A; Domestrol; Estrobene; Estrosyn; Oestromensyl; Oestromon; Palestrol; Serral; Sexocretin; Sibol; Stilbetin; Stilboefral; ; Stilboestroform;

Stilkap; Synestrin (tablets); Synthoestrin; Vagestrol. Estrogen. mp = 169-172°; λ_m = 259 nm ($E^{1\%}_{1\,cm}$ 654 0.1N NaOH); insoluble in H_2O; soluble in EtOH, Et_2O, $CHCl_3$, fatty oils. *Ayerst; Bayer AG; Bristol-Myers Squibb Co; Cooper Vision Inc.; E. Merck; Mallinckrodt Inc.; Norwich.*

8161 Diethylstilbestrol Diphosphate

522-40-7 3177 208-328-8

$C_{18}H_{22}O_8P_2$

(E)-4,4'-(1,2-Diethyl-1,2-ethenediyl)bisphenol diphosphate.

Stilphostrol. Estrogen. mp = 204-206°; sparingly soluble in cold H_2O. *Asta-Werke AG; Miles Inc.*

8162 Dimestrol

130-79-0 3257 204-994-9

$C_{20}H_{24}O_2$

(E)-1,1'-(1,2-Diethyl-1,2-ethenediyl)bis[4-methoxy-benzene].

Depot-Oestromenine; Depot-Oestromon; Synthila. Estrogen. mp = 124°; insoluble in H_2O; soluble in EtOH, Me_2CO, Et_2O.

8163 Equilenin

517-09-9 3675 208-230-5

$C_{18}H_{18}O_2$

3-Hydroxyestra-1,3,5,7,9-pentaen-17-one.

Estrogen. mp = 258-259°; $[\alpha]_D^{16}$ = 87° dioxane); λ_m = 231, 270, 282, 292, 325, 340 nm; soluble in EtOH (.63 g/100 ml at 20°, 2.5 g/100 at 76°).

8164 Equilin

474-86-2 3676 207-488-6

$C_{18}H_{20}O_2$

3-Hydroxyestra-1,3,5(10),7-tetraen-17-one.

Estrogen. mp = 238-240°; $[\alpha]_D^{25}$ = 308° (c = 2 dioxane); λ_m = 283-285 nm; soluble in EtOH, dioxane, Me_2CO, EtOAc; sparingly soluble in H_2O; [benzoate]: mp = 196-197°h; [methyl ether]: mp = 161-162°. *Schering-Plough HealthCare Products.*

8165 Estradiol

50-28-2 3746 200-023-8

$C_{18}H_{24}O_2$

(17β)-Estra-1,3,5(10)-triene-3,17-diol.

dihydrofollicular hormone; dihydrofolliculin; Dihydroxyestrin; dihydrotheelin; Dimenformon; Diogyn; Estrace; Estraderm; Estroclim; Evorel; Gynoestryl; Macrodiol; Menorest; Oestrogel; Ovocyclin; Ovocylin; Profoliol B; Progynon; Systen; Vagifem; Zumenon; Aquadiol; Climara; Diogyn; Diogynets; Estrace; Estraderm TTS; Estring Vaginal Ring; Progynon; Vivelle; NSC-9895; NSC-20293. Estrogen. mp = 173-179°; $[\alpha]_D^{25}$ = 76° to 83° (dioxane); λ_m = 225, 280 nm; insoluble in H_2O; soluble in EtOH, Me_2CO, dioxane. *Apothecon; Berlex Labs Inc.; Ciba-Geigy Corp.; Marion Merrell Dow Inc.; Mead Johnson Labs.; Pfizer Inc.; Pharmacia & Upjohn; Schering-Plough HealthCare Products.*

8166 Estradiol Benzoate

50-50-0 3746 200-043-7

$C_{25}H_{28}O_3$

(17β)-Estra-1,3,5(10)-triene-3,17-diol 3-benzoate.

Agofollin; Benzo-Gynoestryl; Benztrone; Pelanin benzoate; Progynon B; Ovahormon Benzoate; NSC-9566. Estrogen. mp = 191-196°; $[\alpha]_D^{25}$ = 58° to 63° (dioxane); λ_m = 225, 280 nm (c = 2, dioxane); soluble in EtOH, Me_2CO, dioxane; slightly soluble in Et_2O. *Schering-Plough HealthCare Products.*

8167 Estradiol Cypionate

313-06-4 3746 206-237-8

$C_{26}H_{36}O_3$

(17β)-Estra-1,3,5(10)-triene-3,17-diol 17β-cyclopentanepropanoate.

ECP; Depgynogen; Depogen; Estrofem; component of: Depo-Testadiol. Estrogen. mp = 151-152°; $[\alpha]_D^{25}$ = 45° ($CHCl_3$); soluble in EtOH, MeOH, C_67H_6, $CHCl_3$, vegetable oils. *Pharmacia & Upjohn.*

8168 Estradiol Dipropionate

113-38-2 3746 204-026-5

$C_{24}H_{32}O_4$

(17β)-Estra-1,3,5(10)-triene-3,17-diol 3,17-dipropionate.

Ovahormon Depot. Estrogen. mp = 104-105°. *Ciba-Geigy Corp.*

8169 Estradiol Enanthate

4956-37-0 3746 225-599-8

$C_{25}H_{36}O_3$

(17β)-Estra-1,3,5(10)-triene-3,17-diol 17-heptanoate.

SQ-16150. Estrogen. mp = 94-96°. *Bristol-Myers Squibb Co.*

8170 Estradiol Undecylate

3571-53-7 3746 222-677-3

$C_{29}H_{44}O_3$

(17β)-Estra-1,3,5(10)-triene-3,17-diol 17-undecanoate.

SQ-9993; Delestrec. Estrogen. *Bristol-Myers Squibb Co.*

8171 Estradiol Valerate

979-32-8 3746 213-559-2

$C_{23}H_{32}O_3$

(17β)-Estra-1,3,5(10)-triene-3,17-diol 17-valerate.

Climaval; Cyclacur; Delestrogen; Gynogen LA; Pelanin Depot; Progynon Depot; Progynova; Primofol; Valergen; Deladumone; Delestrogen; Gynogen L.A. 40; NSC-17590; component of: Deluteval 2X, Ditate. Estrogen. mp = 144-145°. *Bristol-Myers Squibb Co; Ciba-Geigy Corp.; Forest Pharm. Inc.; Mead Johnson Labs.; Savage Labs.*

8172 Estrapronicate

4140-20-9

$C_{27}H_{31}NO_4$

Estradiol 17-nicotinate 3-propionate.

Estrogen.

8173 Estrazinol Hydrobromide

15179-97-2

$C_{20}H_{26}BrNO_2$

(±)-3-Methoxy-8-aza-19-nor-17α-pregna-1,3,5(10)-trien-20-yn-17-ol hydrobromide.

W-4454A. Estrogen. *Parke-Davis.*

8174 Estriol
50-27-1 3750 200-022-2
$C_{18}H_{24}O_3$
Estra-1,3,5(10)-triene-3,16α,17β-triol.
Theelol; follicular hormone hydrate; oestriol; trihydroxyestrin; Aacifemine; Colpogyn; Destriol; Gynasan; Hormomed; Klimax E; Klimoral; Oekolp; Ortho-Gynest; Ovesterin; Ovestin; Ovo-Vinces; Tridestrin; Triovex. Estrogen. Used in estrogenic hormone therapy. mp = 282°; d = 1.27; $[\alpha]_D^{25}$ = 58° ± 5° (c = 4 dioxane); λ_m = 280 nm; insoluble in H_2O; soluble in EtOH, dioxane, $CHCl_3$, Et_2O, vegetable oils, C_5H_5N. *Parke-Davis.*

8175 Estriol Succinate
514-68-1 3750 208-185-1
$C_{26}H_{30}Na_2O_9$
Estra-1,3,5(10)-triene-3,16α,17β-triol 16,17-bis(sodium hemisuccinate).
Orgastyptin; Stiptanon; Synapause. Estrogen. Used in estrogenic hormone therapy. *Parke-Davis.*

8176 Estrofurate
10322-73-3
$C_{24}H_{26}O_4$
21,23-Epoxy-19,24-dinor-17α-chola-1,3,5(10)-7,20,22-hexaene-3,17-diol 3-acetate.
AY-11483. Estrogen. *Wyeth-Ayerst Labs.*

8177 Estrogens, Conjugated
Premarin; component of: PMB-2000, PMB-4000. Estrogen. *Wyeth-Ayerst Labs.*

8178 Estrogens, Esterified
Amnestrogen; Estratab; Menest; component of: Estratest, Menrium. Estrogen. *Bristol-Myers Squibb Co; Hoffmann-LaRoche Inc.; SmithKline Beecham Pharm.; Solvay Pharm. Inc.*

8179 Estrone
53-16-7 3751 200-164-5
$C_{18}H_{22}O_2$
3-Hydroxyestra-1,3,5(10)-17-one.
Theelin; oestrone; folliculin; follicular hormone; tokokin; thelykinin; ketohydroxyestrin; Hiestrone; Menformon; Glandubolin; Cristallovar; Destrone; Endofolliculina; Estrol; Fermidyn; Folikrin; Kolpon; Crinovaryl; Folisan; Disynformon; Hormovarine; Oestroperos; Wynestron; Thelestrin; Kestrone; Estrusol; Estrugenone; Femestrone Inj.; Folipex; Follestrine; Follidrin (tablets); Follicunodis; Hormofollin; Oestrin; Ovifollin; Perlatan; Ketodestrin. Estrogen. An aromatic steroid. A metabolite of 17β-estradiol which has a considerably lower biological activity. mp = 254.5-256°; $[\alpha]_D^{22}$ = 152° (c = 0.995 $CHCl_3$); λ_m = 282, 296 nm (ε 2300, 2130 dioxane), 300, 450 nm (conc. H_2SO_4), 239, 293 nm (0.1M NaOH); soluble in H_2O (0.003 g/100 ml), EtOH (0.4 g/100 ml at 15°, 2 g/100 ml at 76°), Me_2CO (2 g/100 ml at 15°), $CHCl_3$ (0.0.91 g/100 ml at 15°), C_6H_6 (0.69 g/100 ml at 80°), dioxane, C_5H_5N, vegetable oils, slightly soluble in Et_2O. *Parke-Davis; Savage Labs; Wyeth-Ayerst Labs.*

8180 Estrone Acetate
901-93-9 3751
$C_{20}H_{24}O_3$
3-Hydroxyestra-1,3,5(10)-17-one acetate.
Hogival. Estrogen. mp = 125-127°. *Parke-Davis.*

8181 Estrone Methyl Ether
1091-94-7 3751
$C_{19}H_{24}O_2$
3-Hydroxyestra-1,3,5(10)-17-one methyl ether.
Estrogen. mp = 164-165°; [dl-form]: mp = 143.2-144.2°. *Parke-Davis.*

8182 Estrone Propionate
975-64-4 3751
$C_{21}H_{26}O_3$
3-Hydroxyestra-1,3,5(10)-17-one propionate.
Estrogen. mp = 134-135°. *Parke-Davis.*

8183 Estrone Sulfate Piperazine Salt
7280-37-7 3751 230-696-3
$C_{22}H_{32}N_2O_5S$
3-Hydroxyestra-1,3,5(10)-17-one sulfate piperazine (salt).
estropipate; piperazine estrone sulfate; Harmogen; Ogen; Sulestrex Piperazine. Estrogen. mp = 190°, 245°; $[\alpha]_D^{25}$ = 87.8° (c = 1 0.4% NaOH); λ_m = 275, 268 nm (ε = 838, 851 0.4% NaOH). *Parke-Davis.*

8184 Ethinyl Estradiol
57-63-6 3780 200-342-2
$C_{20}H_{24}O_2$
19-Nor-17α-pregna-1,3,5(10)-trien-20-yn-3,17-diol.
Diogyn E; Estinyl; NSC-10973; component of: Brevicon, Demulen, Desogen, Estopherol, Estostep, Levlen, Loestrin, Lo/Ovral, ModiCon, Nordette, Norethrin 1/35E, Norlestrin, Ortho-Cyclen, Ortho-Novum, Ovcon, Ovral, Tri-Levlen, Triphasil. Estrogen. In combination with a progestogen (Desogestrel, Gestodene; Lynestrenol or Norgestimate), used as an oral contraceptive[71138-35-7], [109852-02-0], [8064-76-4], [79871-54-8], respectively. mp = 141-146°; $[\alpha]_D^{25}$ = 0 ± 1° (dioxane); λ_m = 281 nm (ε 2040 ± 60 EtOH); insoluble in H_2O; soluble in EtOH (16.6 g/100 ml), Et_2O (25 g/100 ml), Me_2CO (20 g/100 ml), dioxane (25 g/100 ml), $CHCl_3$ (5 g/100 ml); LD_{50} (rat orl) = 2952 mg/kg, (mus orl) = 1737 mg/kg. *Astra Chem. Ltd.; Berlex Labs Inc.; Eprova AG; Marion Merrell Dow Inc.; Mead Johnson Labs.; Organon Inc.; Parke-Davis; Pfizer Intl.; Roberts Pharm. Corp.; Schering-Plough HealthCare Products; Searle G.D. & Co.; Syntex Intl. Ltd.; Wyeth-Ayerst Labs.*

8185 Fenestrel
7698-97-7
$C_{16}H_{20}O_2$
5-Ethyl-6-methyl-4-phenyl-3-cyclohexene-1-carboxylic acid.
Estrogen. *Ortho Pharm. Corp.*

8186 Fosfestrol
522-40-7 4279 208-328-8
$C_{18}H_{22}O_8P_2$
α,α'-Diethyl-(E)-4,4'-stilbenediol bis(dihydrogen phosphate).
DESDP; diethylstilbesterol bisphosphate; diethylstilbestryl

bisphosphate; stilbestrol diphosphate; Stilphostrol. Estrogen; antineoplastic (hormonal). mp = 204-206° (dec); poorly soluble in H_2O. *Asta-Werke AG; Bayer AG; Bayer Corp. Pharm. Div.; Miles Inc.*

8187 Fosfestrol Disodium Salt
5965-09-3 4279 227-746-1

$C_{18}H_{20}Na_2O_8P_2$
α,α'-Diethyl-(E)-4,4'-stilbenediol bis(dihydrogen phosphate) disodium salt.
Estrogen; antineoplastic (hormonal). mp = 230°; soluble in H_20. *Asta-Werke AG; Bayer AG; Bayer Corp. Pharm. Div.; Miles Inc.*

8188 Fosfestrol Tetrasodium Salt
4719-75-9 4279 225-209-6

$C_{18}H_{18}Na_{28}P_2$
α,α'-Diethyl-(E)-4,4'-stilbenediol bis(dihydrogen phosphate) tetrasodium salt.
st52-asta; ST-51; Cytonal; Honvan; Honvol; Stilbostatin. Estrogen; antineoplastic (hormonal). *Asta-Werke AG; Bayer AG; Bayer Corp. Pharm. Div.; Miles Inc.*

8189 Hexestrol
84-16-2 4738 201-518-1

$C_{18}H_{22}O_2$
4,4'-(1,2-Diethylethylene)diphenol.
NSC-9894; hexoestrol; Synthovo; Cycloestrol; Hexanoestrol; Hormoestrol; Syntrogene. Estrogen. Also used as a hormonal antineoplastic. mp = 185-188°; soluble in Et_2O, Me_2CO, EtOH, MeOH; slightly soluble in C_6H_6, $CHCl_3$, insoluble in H_2O; [diacetate (Retalon-Lingual)]: mp = 137-139°; [dipropionate (Retalon Oleosum)]: mp = 127-128°. *Hoffmann-LaRoche Inc.*

8190 Hexestrol Diphosphate
14188-82-0 4738

$C_{18}H_{24}O_8P_2$
4,4'-(1,2-Diethylethylene)diphenol diphosphate.
Cytostatin; hexestrol 4,4'-diphosphoric acid. Estrogen. Hormonal antineoplastic. *Hoffmann-LaRoche Inc.*

8191 Hydromadinone
16469-74-2

$C_{21}H_{29}ClO_3$
6α-Chloro-17-hydroxyprogesterone.
Estrogen.

8192 Mestranol
72-33-3 5976 200-777-8

$C_{21}H_{26}O_2$
3-Methoxy-19-nor-17α-pregna-1,3,5(10)-trien-20-yn-17-ol.
17α-ethynylestradiol 3-methyl ether; EE_3ME; 33355; Menophase; Norquen; Ovastol; component of: Enovid, Norethrin 1/50M, Norinyl, Norquen, Ortho-Novum, Ovulen. Estrogen. Used in combination with a progestogen as an oral contraceptive. May be a carcinogen. mp = 150-151°; λ_m = 279, 287.5 nm ($E_{1\,cm}^{1\%}$ 82, 14.4 MeOH); insoluble in H_2O; poorly soluble in MeOH; soluble in EtOH, $CHCl_3$, Et_2O, Me_2CO, dioxane. *Ortho Pharm. Corp.; Roberts Pharm. Corp.; Searle G.D. & Co.; Syntex Intl. Ltd.*

8193 Methallenestrol
517-18-0 6012 208-232-6

$C_{18}H_{22}O_3$
3-(6-Methoxy-2-naphthyl)-2,2-dimethylpentanoic acid.
Vallestril. Estrogen. mp = 139-140°; soluble in Et_2O, vegetable oils.

8194 Methestrol
130-73-4 6045

$C_{20}H_{26}O_2$
4,4'-(1,2-Diethylethylene)di-o-cresol.
dimethylhexestrol; σ-promethestrol; promethestrol. Estrogen. mp = 145°; [dipropionate (meprane diproipionate)]: mp = 115°; soluble in Et_2O, EtOAc, C_6H_6; slightly soluble in EtOH; insoluble in H_2O.

8195 Moxestrol
34816-55-2 6372

$C_{21}H_{26}O_3$
11β-Methoxy-19-nor-17α-pregna-1,3,5(10)-trien-20-yne-3,17-diol.
R-2858; Surestryl. Estrogen. mp = 280°; $[\alpha]_D^{20}$ = 29° (c = 0.6 EtOH); λ_m = 280 nm ($E_{1\,cm}^{1\%}$ = 58.4 EtOH). *Roussel-UCLAF.*

8196 Mytatrienediol
5108-94-1 6424

$C_{20}H_{28}O_3$
3-Methoxy-16-methyl-1,3,5(10)-estratriene-16β,17β-diol.
SC-6924; Anvene; Manvene. Estrogen. Also has antilipemic and hypocalciuric properties. mp = 179-181°; $[\alpha]_D^{20}$ = 71° (dioxane). *Searle G.D. & Co.*

8197 Nylestriol
39791-20-3

$C_{25}H_{32}O_3$
17α-Ethynylestra-1,3,5(10)-triene-3,16α,17β-triol 3-cyclopentyl ether.
49825. Estrogen. *Eli Lilly & Co.*

8198 Pregnenolone Succinate
4598-67-8 225-001-5

$C_{25}H_{36}O_5$
3β-Hydroxypregn-5-en-20-one hydrogen succinate.
Formula 405. Non-hormonal sterol derivative. A neurosteroid. Forms a progesterone analog on dehydrogenation. *Doak Pharmacal Co. Inc.*

8199 Promestriene
39219-28-8 254-361-6

$C_{22}H_{32}O_2$
17β-Methoxy-3-propoxyestra-1,3,5(10)-triene.
An estrogen studied for topical treatment of hyperseborrhea.

8200 Quinestradiol
1169-79-5 8238 214-623-2

$C_{23}H_{32}O_3$
3-(Cyclopentyloxy)estra-1,3,5(10)-triene-16α,17β-diol.
quinestradol; estriol 3-cyclopentyl ether; Colpovis; Pentovis. Estrogen. mp = 98-100°. *Vismara.*

8201 Quinestrol

152-43-2 8239 205-803-1

$C_{25}H_{32}O_2$

3-(Cyclopentyloxy)19-norpregna-1,3,5(10)-triene-20-yn-17β-ol.

W-3566; Estrovis. Estrogen. mp = 107-108°; $[\alpha]_D^{25}$ = 5° (c = 0.5 dioxane). *Vismara.*

Expectorants

8202 Acetylcysteine

616-91-1 89 210-498-3

$C_5H_9NO_3S$

L-α-Acetamido-β-mercaptopropionic acid.

N-acetyl-L-cysteine; 5052; NSC-111180; Airbron; Broncholysin; Brunac; Fabrol; Fluatox; Fluimucil; Fluimucetin; Fluprowit; Mucocedyl; Mucolator; Mucolyticum; Mucomyst; Muco Sanigen; Mucosolvin; Mucret; NAC; Neo-Fluimucil; Parvolex; Respaire; Tixair. Mucolytic; corneal vulnerary; antidote to acetaminophen poisoning. mp = 109-110°; LD_{50} (rat orl) = 5050 mg/kg. *Bristol Laboratories; DuPont-Merck Pharm.; Mead Johnson Labs.; Mead Johnson Pharmaceuticals.*

8203 Ambroxol

18683-91-5 401 242-500-3

$C_{13}H_{18}Br_2N_2O$

trans-4-[(2-Amino-3,5-dibromobenzyl)amino]-cyclohexanol.

NA-872. Expectorant. Metabolite of bromhexine. *Thomae GmbH Dr. Karl.*

8204 Ambroxol Hydrochloride

23828-92-4 401

$C_{13}H_{19}Br_2ClN_2O$

trans-4-[(2-Amino-3,5-dibromobenzyl)amino]-cyclohexanol hydrochloride.

Abramen; Ambril; Bronchopront; Duramucal; Fluibron; Fluixol; Frenopect; Lindoxyl; Motosol; Muco-Burg; Mucofar; Mucasan; Mucosolvan; Mucoclear; Mucovent; Pect; Solvolan; Stas-Hustenlo+u7ser; Surbronc; Surfactal. Expectorant. mp = 233-234.5° (dec); LD_{50} (mus ip) = 268 mg/kg, (rat ip) = 380 mg/kg, (mus orl) = 2720 mg/kg, (rat orl) = 13400 mg/kg. *C.M. Ind.; Thomae GmbH Dr. Karl.*

8205 Ammonium Carbonate

10361-29-2 534 233-786-0

34% NH_3, 45% CO_2

Carbonic acid ammonium salt.

Hartshorn. Expectorant. Mixture of ammonium bicarbonate and ammonium carbamate. Also used as a pharmaceutic aid (ammonia source), in baking powder and smelling salts, as an analytical chemical reagent, and in the processing of textiles, rubber, and dyes. Dec on exposure to air; volatilizes 60°; incompatible with acids, acidic salts, iron salts, zinc salts, alkaloids, alum, calomel, tartar emetic.

8206 Bromhexine

3572-43-8 1412 210-280-8

$C_{14}H_{20}Br_2N_2$

2-Amino-3,5-dibromo-N-cyclohexyl-N-methyl benzenemethanamine.

Expectorant; mucolytic. Dec 237°; soluble in H_2O (0.4 g/100 ml) or 10% EtOH (0.4 g/100 ml); LD_{50} (rbt orl) > 10000 mg/kg.

8207 Bromhexine Hydrochloride

611-75-6 1412 210-280-8

$C_{14}H_{21}Br_2ClN_2$

2-Amino-3,5-dibromo-N-cyclohexyl-N-methylbenzene-methanamine monohydrochloride.

NA-274; Auxit; Bisolvon; Ophtosol; Quentan. Expectorant; mucolytic.

8208 Brovanexine

54340-61-3

$C_{24}H_{28}Br_2N_2O_4$

2',4'-Dibromo-α-(cyclohexylmethylamino)-o-vanillo-toluidide acetate (ester).

An expectorant.

8209 Calcium Iodide

10102-68-8 1720 233-276-8

CaI_2

Calcium iodide.

Expectorant. mp = 740°; bp = 1100°; soluble in H_2O, EtOH, Me_2CO; insoluble in Et_2O, dioxane.

8210 Carbocysteine

638-23-3 1850 211-327-5

$C_5H_9NO_4S$

S-(Carboxymethyl)-L-cysteine.

3-[(carboxymethyl)thio]alanine; carboxymethylcysteine; AHR-3053; LJ-206; Carbocit; Fluifort; Lisil; Lisomucil; Loviscol; Muciclar; Mucocis; Mucodyne; Mucofan; Mucolase; Mucolex; Mucopront; Mucotab; Mukinyl; Pectox; Pulmoclase; Reomucil; Rhinathiol; Siroxyl; Thiodril; Transbronchin. Mucolytic; expectorant. [L-form]: mp = 204-207°; $[\alpha]_D^{24}$ = 0.5° (1N HCl). *Lederle Labs.*

8211 Cistinexine

86042-50-4

$C_{50}H_{60}Br_4N_6O_6S_2$

Dibenzyl [dithiobis[(R)-1-[4,6-dibromo-α-(cyclohexyl-methylamino)-o-tolyl]carbamoyl]ethylene]dicarbamate.

Expectorant .

8212 Guaiacol

90-05-1 4575 201-964-7

$C_7H_8O_2$

2-Methoxyphenol.

methylcatechol; o-hydroxyanisole; Anastil. Expectorant. Isolated from guaiac resin. bp = 204-206°; bp_4 = 53-55°; soluble in H_2O (1.54 g/100 ml), glycerol (100 g/100 ml), NaOH solution; miscible with EtOH, $CHCl_3$, Et_2O, oils; slightly soluble in petroleum ether; LD_{50} (rat orl) = 725 mg/kg. *Hoechst AG.*

8213 Guaiacol Benzoate

531-37-3 4576 208-507-0

$C_{14}H_{12}O_3$

2-Methoxyphenyl benzoate.

benzoguaiacol; Benzosol. Expectorant. mp = 57-58°; slightly soluble in H_2O; soluble in hot EtOH, $CHCl_3$, Et_2O. *Bayer AG.*

8214 Guaiacol Carbonate
553-17-3 4577 209-034-2
$C_{15}H_{14}O_5$
Carbonic acid bis(2-methyoxyphenyl) ester.
diguaiacyl carbonate; guaiacol carbonic acid neutral ester; carbonic acid guiacol ether; Duotal. Expectorant. mp = 88.1°; nearly insoluble in H_2O; soluble in EtOH (1.6 g/100 ml), $CHCl_3$ (100 g/100 ml), Et_2O (5.6 g/100 ml); slightly soluble in liquid fatty acids.

8215 Guaiacol Phenylacetate
4112-89-4 4575 223-898-8
$C_{15}H_{14}O_3$
2-Methoxyphenol phenylacetate.
Gujaphenyl; Gunyl. Expectorant. *Hoechst AG.*

8216 Guaifenesin
93-14-1 4582 202-222-5
$C_{10}H_{14}O_4$
3-(2-Methoxyphenoxy)-1,2-propanediol.
guaiacyl glyceryl ether; glyceryl guaiacyl ether; glycerol guaiacolate; α-glyceryl guaiacol ether; o-methoxyphenyl glyceryl ether; guaiacol glyceryl ether; guaiphenesin; guaiacuran; MY-301; XL-90; 2-G; Actifed-C; amonidren; Calmipan; Colrex Expectorant; Equicol; Glycodex; Guaiamar; Guayanesin; Miocaina; Myocaine; Myoscain; Oresol; Oreson; Relaxil G; Reorganin; Respenyl; Resyl; Robitussin; Sirotol; Tenntus; Tulyn; component of: Brexin EX, of Brondecon, Congestac, Contac Cough Formula, Contac Cough & Sore Throat Formula, Coricidin, Dilor G, Dimacol, ENTEX, Fedahist, Histalet X Tablets, Hycotuss Tablets, Isoclor Expectorant C, Kwelcof, Lufyllin-GG, Naldecon-CX, Naldecon-DX, Naldecon-EX, Neothylline-GG, Nucofed Expectorant, PV Tussin Tablet, Quibron, Quibron Plus, Ru-Tuss DE Tablets, Ru-Tuss Expectorant, Slo-Phyllin GG, Sudafed Cough Syrup, Tedral Expectorant, Theolair Plus, Triaminic, Tussar-2, Tussar SF. Expectorant. Centrally acting muscle relaxant with expectorant properties. mp = 78.5-79°; bp_{19} = 215°; soluble in H_2O (5 g/100 ml at 25°), EtOH, $CHCl_3$, glycerol, propylene glycol, DMF; moderately soluble in C_6H_6; nearly insoluble in petroleum ether. *3M Pharm.; Ascher B.F. & Co.; Boots Pharmaceuticals Inc.; Bristol Laboratories; Bristol-Myers Squibb Co; Burroughs Wellcome Inc.; Degussa-Hüls Corp.; Forest Pharm. Inc.; Lemmon Co.; Marion Merrell Dow Inc.; Menley & James Labs Inc.; Norwich; Parke-Davis; Rhône-Poulenc; Robins, A. H. Co.; Sandoz Pharm. Corp.; Savage Labs; Schering AG; Schwarz Pharma Kremers Urban Co.; SmithKline Beecham Pharm.; Solvay Pharm. Inc.; Wallace Labs; Xttrium Labs. Inc.*; 144;166.

8217 Guaithylline
5634-38-8 4582 227-078-0
$C_{17}H_22N_4O_6$
3,7-Dihydro-1,3-dimethyl-1H-Purine-2,6-dione with 3-(2-methoxyphenoxy)-1,2-propanediol (1:1).
guaifylline; Eclabron. Bronchodilator; expectorant. Xanthine derivative. *See* guaifenesin. *Degussa-Hüls Corp.; Forest Pharm. Inc.; Key Pharm.; Robins, A. H. Co.; Xttrium Labs. Inc.*

8218 Potassium Guaicolsulfonate
1321-14-8 7798 215-314-5
$C_7H_7KO_5S$
Potassium 3-hydroxy-4-methoxybenzenesulfonate.
sulfoguaiacol; Tiocol; Orthocoll. Expectorant. Mixture of potassium salts of 4- and 5-guaiacolsulfonic acid. Soluble in H_2O (13.3g/100 ml), insoluble in Et_2O, nearly insoluble in EtOH.

8219 Potassium Iodide
7681-11-0 7809 231-659-4
IK
Jodid; Thyro-Block; Thyrojod; component of: Mudrane Tablets, Mudrane-2 Tablets, quadrinal. Antifungal; expectorant; iodine supplement. Used in veterinary medicine (orally) to treat goiter, actinobacillosis, actinomycosis, idodine deficiency, lead or mercury poisoning. d = 3.12; mp = 680°; soluble in H_2O (142.8 g/100 ml at 20°, 200 g/100 ml at 100°), EtOH (4.5 g/100 ml at 20°, 12.5 g/100 ml at 76°), absolute EtOH (1.96 g/100 ml), MeOH (12.5 g/100 ml), Me_2CO (1.3 g/100 ml), glycerol (50 g/100 ml), ethylene glycol (40 g/100 ml); LD_{50} (rat iv) = 285 mg/kg. *ECR Pharm.; Knoll Pharm. Co.; Wallace Labs.*

8220 Quillaic Acid
631-01-6 8221 211-149-8
$C_{30}H_{46}O_5$
(3β,4α,16α)-3,16-Dihydroxy-23-oxoolean-12-en-28-oic acid.
quillaja sapogenin. mp = 292-293°; $[\alpha]_D^{20}$ = +56.1° (c = 2.9 in C_5H_5N); soluble in EtOH, Et_2O, Me_2CO, EtOAc, AcOH.

8221 Sodium Citrate
68-04-2 8746 200-675-3
$C_6H_5Na_3O_7$
Trisodium citrate.
Citrosodine; Citnatin; Cystemme; Urisal. Systemic alkalizer; diuretic; expectorant; sudorific. Used in veterinary medicine as an anticoagulant in blood collection. [dihydrate]: soluble in H_2O; insoluble in EtOH; pH 8; LD_{50} (rat ip) = 1548 mg/kg; [pentahydrate]: not as stable as the dihydrate.

8222 Sodium Iodide
7681-82-5 8777 231-679-3
INa
Ioduril; Anayodin. Iodine supplement; expectorant. d = 3.67; mp = 561°; soluble in H_2O (200 g/100 ml), EtOH (50 g/100 ml), glycerol (100 g/100 ml); MLD (rat iv) = 1300 mg/kg.

8223 Terpin
80-53-5 9314 201-288-2
$C_{10}H_{20}O_2$
p-Menthane-1,8-diol.
dipenteneglycol. Expectorant (*cis*-form hydrate).

8224 Terpin Hydrate
1256-01-6 9314
$C_{10}H_{22}O_3$
p-Menthane-1,8-diol monohydrate.
terpinol. Expectorant.

8225 Tiopronin
1953-02-2 9597 217-778-4
N-(2-mercapto-1-oxopropyl)glycine.

Acadione; BRN 1859822; Capen; Captimer; CCRIS 1935; Epatiol; Meprin (detoxicant); Mercaptopropionylglycine; (2-Mercaptopropionyl)glycine; Mucolysin; N-(2-Mercapto-1-oxopropyl)glycine; N-(2-Mercaptopropionyl)-glycine; Sutilan; Thiola; Thiolpropionamidoacetic acid; Thiopronin; Thiopronine; Thiosol; Tioglis; Tiopronin; Tiopronine; Tiopronino; Tioproninum; Vincol. Mucolytic; expectorant. Used as an antidote in heavy metal poisoning. mp = 95-97°; LD_{50} (mus iv) = 2100 mg/kg. *Boehringer Ingelheim Ltd.*

Fibrinogen Receptor Antagonists

8226 Fradafiban

148396-36-5

$C_{20}H_{21}N_3O_4$

(3S,5S)-5-[[(4'-Amidino-4-biphenylyl)oxy]methyl]-2-oxo-3-pyrrolidoneacetic acid.

Fibrinogen receptor antagonist.

8227 Lamifiban

144412-49-7 5362

$C_{24}H_{28}N_4O_6$

[[1-[N-(p-Amidinobenzoyl)-L-tyrosyl]-4-piperidyl]oxy]-acetic acid.

Ro-44-9883/000. Antithrombotic. Specific nonpeptide platelet fibrinogen receptor (GPIIb/IIIa) antagonist. mp > 200° (dec); $[\alpha]_D^{20}$ = 29.8° (c = 0.86 1N HCl). *Hoffmann-LaRoche Inc.*

8228 Lamifiban Trifluoracetate Salt

144412-50-0 5362

$C_{26}H_{29}F_3N_4O_8$

[[1-[N-(p-Amidinobenzoyl)-L-tyrosyl]-4-piperidyl]oxy]-acetic acid trifluoroacetate salt.

Antithrombotic. Specific nonpeptide platelet fibrinogen receptor (GPIIb/IIIa) antagonist. mp = 125-130° (dec); LD_{50} (mus iv) = 250 mg/kg. *Hoffmann-LaRoche Inc.*

8229 Tirofiban

144494-65-5 9605

$C_{22}H_{36}N_2O_5S$

N-(Butylsulfonyl)-4-[4-(4-piperidyl)butoxy]-L-phenylalanine .

Fibrinogen receptor antagonist. Used as an antithrombotic and in the treatment of unstable angina. mp = 223-225°. *Merck & Co.Inc.*

8230 Tirofiban Hydrochloride

142373-60-2 9605

$C_{22}H_{37}ClN_2O_5S$

N-(Butylsulfonyl)-4-[4-(4-piperidyl)butoxy]-L-phenylalanine hydrochloride.

Fibrinogen receptor antagonist. Used as an antithrombotic and in the treatment of unstable angina. *Merck & Co.Inc.*

8231 Tirofiban Hydrochloride Monohydrate

150915-40-5 9605

$C_{22}H_{37}ClN_2O_5S.H_2O$

N-(Butylsulfonyl)-4-[4-(4-piperidyl)butoxy]-L-phenylalanine hydrochloride monohydrate.

L-700462; MK-383; Aggrastat. Fibrinogen receptor antagonist. Used as an antithrombotic and in the treatment of unstable angina. mp = 131-132°; $[\alpha]_D^{25}$= -14.4° (c = 0.92 MeOH). *Merck & Co.Inc.*

Ganglionic Blockers

8232 Azamethonium Bromide

306-53-6 929 206-186-1

$C_{13}H_{33}Br_2N_3$

[(Methylimino)diethylene]bis(ethyldimethylammonium bromide).

pentamethazine dibromide; Präparat 9295; Ciba 9295; Pendoimid; Azameton; Azamethone; Ganlion; Pentaméthazène. Antihypertensive agent. Ganglion blocking agent. mp = 212-215°; freely soluble in H_2O; LD_{50} (mus orl) = 2500 mg/kg, (mus iv) = 60 mg/kg, (rbt orl) = 3000 mg/kg, (rbt sc) = 160 mg/kg, (rbt iv) = 75 mg/kg. *Ciba-Geigy Corp.*

8233 Chlorisondamine Chloride

69-27-2 2151

$C_{14}H_{20}Cl_6N_2$

4,5,6,7-Tetrachloro-2-(2-dimethylaminoethyl)-2-methylisoindolinium chloride methochloride.

chlorisondamine dimethochloride; Su-3088; Ecolid; Ecolid chloride. Antihypertensive. Ganglion blocking agent. mp = 258-265°; soluble in H_2O, EtOH, MeOH. *Ciba-Geigy Corp.*

8234 Guanadrel

40580-59-4 4587

$C_{10}H_{19}N_3O_2$

(1,4-Dioxaspiro[4.5]dec-2-ylmethyl)guanidine.

Orally active postganglionic sympathetic inhibitor. Used as an antihypertensive agent. *Cutter Labs; Fisons plc.*

8235 Guanadrel Sulfate

22195-34-2 4587

$C_{20}H_{40}N_6O_8S$

(1,4-Dioxaspiro[4.5]dec-2-ylmethyl)guanidine sulfate.

Hylorel; CL-1388R; U-28288D; Anarel. Orally active postganglionic sympathetic inhibitor. Used as an antihypertensive agent. mp = 213.5-215°. *Cutter Labs; Fisons plc.*

8236 Hexamethonium

60-26-4 4724

$[C_{12}H_{30}N_2]^+$

N,N,N,N',N',N'-Hexamethyl-1,6-hexanediaminium.

hexathonide; hexamethone; Hexathide [as iodide]; Vegolysen-T [as tartrate]. Ganglion blocker used as an antihypertensive agent. *May & Baker Ltd.*

8237 Hexamethonium Bromide

55-97-0 4724 200-249-7

$C_{12}H_{30}Br_2N_2$

N,N,N,N',N',N'-Hexamethyl-1,6-hexanediaminium bromide.

Bistrium bromide; Esametina; Gangliostat; Hexameton bromide; Hexanium bromide; Simpatoblock; Vegolysen; Vegolysin. Ganglion blocker used as an antihypertensive agent. mp = 274-276°; soluble in H_2O, EtOH; insoluble in Me_2CO, $CHCl_3$, Et_2O. *May & Baker Ltd.*

8238 Hexamethonium Chloride
60-25-3 4724 200-465-1
$C_{12}H_{30}Cl_2N_2$
N,N,N,N',N',N'-Hexamethyl-1,6-hexanediaminium chloride.
Bistrium chloride; Chloor-Hexaviet; Esomid chloride; Hestrium chloride; Hexameton chloride; Hexone chloride; Hiohex chloride; Methium chloride; Meton. Ganglion blocker used as an antihypertensive agent. mp = 289-292° (dec); soluble in H_2O, EtOH; insoluble in $CHCl_3$, Et_2O. *May & Baker Ltd.*

8239 Mecamylamine
60-40-2 5814 200-476-1
$C_{11}H_{21}N$
N,2,3,3-Tetramethyl-2-norbornamine.
mecamine. Ganglion blocking agent with antihypertensive properties. Oily liquid; $bp_{4.0}$ = 72°; slightly soluble in H_2O. *Merck & Co.Inc.*

8240 Mecamylamine Hydrochloride
826-39-1 5814 212-555-8
$C_{11}H_{22}ClN$
N,2,3,3-Tetramethyl-2-norbornamine hydrochloride.
Inversine; Mevasine. Ganglion blocking agent with antihypertensive properties. mp - 245.5-246.5°; soluble in H_2O (21.2 g/100 ml), EtOH (8.2 g/100 ml), glycerol (10.4 g/100 ml), iPrOH (2.1 g/100 ml). *Merck & Co.Inc.*

8241 Nicotine
54-11-5 6611 200-193-3
$C_{10}H_{14}N_2$
(S)-3-(1-Methyl-2-pyrrolidinyl)pyridine.
Habitrol; Nicabate; Nicoderm; Nicolan; Nicopatch; Nicotell TTS; Nicotinell; Tabazur. Smoking cessation adjunct. Ganglionic stimulant which also stimulates the CNS. Also used as an insecticide and fumigant. Severe nicotine poisoning can cause nausea, salivation, abdominal pain, vomiting, diarrhea, cold sweat, dizziness, mental confusion, weakness, drop in blood pressure and pulse, convulsions, paralysis of respiratory muscles. From dried leaves of *Nicotiana tabacum and N. Rustica*. Oily liquid; bp_{745} = 247° (partial dec), bp_{17} = 123-125°; n_D^{20} = 1.5282; d_D^{20} = 1.0097; $[\alpha]_D^{20}$ = -169°; miscible with H_2O; soluble in alcohol, $CHCl_3$, Et_2O, petroleum ether, kerosene, oils.

8242 Pempidine
79-55-0 7207 201-211-2
$C_{10}H_{21}N$
1,2,2,6,6-Pentamethylpiperidine.
Ganglion blocking agent with antihypertensive properties. bp = 147°; [p-toluenesulfonate]: mp = 162-163°. *May & Baker Ltd.*

8243 Pempidine Tartrate
546-48-5 7207 208-902-8
$C_{14}H_{27}NO_6$
1,2,2,6,6-Pentamethylpiperidine tartrate.
M&B-4486; Pempidil; Pempiten; Perolysen; Tenormal; Tensinol; Tensoral. Ganglion blocking agent with antihypertensive properties. mp = 160°; soluble in EtOH, moderately soluble in H_2O. *May & Baker Ltd.*

8244 Pentacynium Bis(methyl sulfate)
3810-83-1 7243
$C_{29}H_{45}N_3O_9S_2$
4-[2-[(5-Cyano-5,5-diphenylpentyl)methylamino]ethyl]-4-methylmorpholinium bis(methylsulfate).
pentacyone mesylate; Presidal. Ganglion blocking agent with antihypertensive properties. mp = 173-175°; soluble in H_2O. *Glaxo Wellcome Inc.*

8245 Pentamethonium Bromide
541-20-8 7253 208-771-7
$C_{11}H_{28}Br_2N_2$
N,N,N,N',N',N'-Hexamethyl-1,5-pentanediaminium bromide.
C-5; Penthonium; Lytensium. Ganglion blocking agent with antihypertensive properties.

8246 Pentolinium Tartrate
52-62-0 7274 200-146-7
$C_{23}H_{42}N_2O_{12}$
1,1'-(1,5-Pentanediyl)bis[1-methyl-pyrrolidinium] salt with [R-(R*, R*)]-2,3-dihydroxybutanedioic acid.
pentapyrrolidinium bitartrate; M&B-2050A; Ansolysen tartrate; Ansolysen bitartrate; Pentilium. Ganglion blocking agent with antihypertensive properties. mp = 203° (dec); freely soluble in H_2O (250 g/100 ml); soluble in EtOH (0.12 g/100 ml); insoluble in Et_2O, $CHCl_3$. *May & Baker Ltd.*

8247 Phenactropinium Bromide
3784-89-2 7346
$C_{24}H_{28}ClNO_4$
3-[(Hydroxyphenylacetyl)oxy]-8-methyl-8-(2-oxo-2-phenylethyl)-8-azoniabicyclo[3.2.1]octane chloride.
N-phenacylhomatropinium chloride; Trophenium. Ganglion blocking agent with antihypertensive properties. mp = 195-197°. *Smith T&H.*

8248 Tetramethylammonium
51-92-3
$C_4H_{12}N^+$
Tetramethylammonium.
TMA. Ganglion blocker.

8249 Trimethaphan Camsylate
68-91-7 9837 200-696-8
$C_{32}H_{40}N_2O_5S_2$
(+)-1,3-Dibenzyldecahydro-2-oxoimidazo[4,5-c]thieno[1,2-a]thiolium 2-oxo-10-bornanesulfonate (1:1).
Arfonad; Nu-2222. Ganglion blocking agent with antihypertensive properties. mp = 245° (dec); $[\alpha]_D^{20}$ = 22.0° (c = 4 H_2O); soluble in H_2O (20 g/100 ml), EtOH (50 g/100ml); slightly soluble in Me_2CO, Et_2O. *Hoffmann-LaRoche Inc.*

8250 Trimethidinium Methosulfate
14149-43-0 9838 237-994-2
$C_{19}H_{42}N_2O_8S_2$
1,3,8,8-Tetramethyl-3-[3-(trimethylammonio)propyl]-3-azoniabicyclo[3.2.1]octane bis(methyl sulfate).
Ganglion blocking agent with antihypertensive properties. mp = 192-193°. *Boehringer Ingelheim GmbH.*

Gastric Proton Pump Inhibitors

8251 Lansoprazole
103577-45-3 5373
$C_{16}H_{14}F_3N_3O_2S$
2-[[[3-Methyl-4-(2,2,2-trifluoroethoxy)-2-pyridyl]methyl]sulfinyl]benzimidazole.
Prevacid; AG-1749; A-65006; Agopton; Lanzor; Ogast; Takepron; Zoton. Gastric proton pump inhibitor. Used as an antiulcerative. mp = 178-182°. *Takeda Chem. Ind. Ltd.; TAP Pharm. Inc.*

8252 Omeprazole
73590-58-6 6977
$C_{17}H_{19}N_3O_3S$
5-Methoxy-2-[[(4-methoxy-3,5-dimethyl-2-pyridyl)methyl]sulfinyl]benzimidazole.
Prilosec; Antra; Gastroloc; Losec; Mepral; Mopral; Omepral; Omeprazen; Parizac; Pepticum. Antiulcerative. Gastric antisecretory agent (proton pump inhibitor). Used in treatment of Zollinger-Ellison syndrome. mp = 156°; LD_{50} (mus iv) = 80 mg/kg, (mus orl) > 4000 mg/kg, (rat iv) > 50 mg/kg, (rat orl) > 4000 mg/kg. *Astra Hassle AB; Astra Sweden; Merck & Co.Inc.*

8253 Omeprazole Sodium
95510-70-6 6977
$C_{17}H_{18}N_3NaO_3S$
5-Methoxy-2-[[(4-methoxy-3,5-dimethyl-2-pyridyl)methyl]sulfinyl]benzimidazole sodium salt.
Losec Sodium; H 168/68. Antiulcerative. Gastric antisecretory agent (proton pump inhibitor). *Astra Hassle AB; Astra Sweden; Merck & Co.Inc.*

8254 Pantoprazole
102625-70-7 7146
$C_{16}H_{15}F_2N_3O_4S$
5-(Difluoromethoxy)-2-[[(3,4-dimethoxy-2-pyridyl)methyl]sulfinyl]benzimidazole.
SK&F-96022; BY-1023. Antiulcerative. Gastric proton pump inhibitor. mp = 139-140° (dec); [sodium salt]: mp >130° (dec); λ_m = 289 (ε 16400). *Byk Gulden Lomberg GmbH; SmithKline Beecham Pharm.*

8255 Pantoprazole Sodium Salt
138786-67-1 7146
$C_{16}H_{14}F_2N_3NaO_34S$
5-(Difluoromethoxy)-2-[[(3,4-dimethoxy-2-pyridyl)methyl]sulfinyl]benzimidazole sodium salt.
Gastric proton pump inhibitor. Used as an antiulcerative. mp > 130° (dec); λ_m 289 nm (ε 16400 MeOH). *Byk Gulden Lomberg GmbH; SmithKline Beecham Pharm.*

8256 Rabeprazole Sodium
117976-90-6 8272
$C_{18}H_{20}N_3NaO_3S$
2-[[[4-(3-Methoxypropoxy)-3-methyl-2-pyridinyl]methyl]sulfinyl]benzimidazole sodium salt.
LY-307640 sodium; E-3810. Antiulcerative. Gastric proton pump inhibitor. mp = 140-141° (dec). *Eisai Co. Ltd.; Eli Lilly & Co.*

8257 Rabeprazole Sodium
117976-90-6 8272
$C_{18}H_{20}N_3NaO_3S$
2-[[[4-(3-Methoxypropoxy)-3-methyl-2-pyridinyl]-methyl]sulfinyl]benzimidazole sodium salt.
LY-307640 sodium; E-3810. Antiulcerative. Gastric proton pump inhibitor. mp = 140-141° (dec). *Eisai Co. Ltd.; Eli Lilly & Co.*

8258 Timoprazole
57237-97-5
$C_{13}H_{11}N_3OS$
2-[(2-Pyridylmethyl)sulfinyl]benzimidazole.
H+,K(+)-ATPase ion pump and gastric secretion inhibitor.

Gastric Secretion Inhibitors

8259 Arbaprostil
55028-70-1 808
$C_{21}H_{34}O_5$
(E,Z)-(1R,2R,3R)-7-[3-Hydroxy-2-[(3R)-(3-hydroxy-3-methyl-1-octenyl)]-5-oxocyclopentyl]-5-heptenoic acid.
15-methylprostaglandin E_2; U-42842; Arbacet. Gastric antisecretory agent. Antiulcerative. Prostaglandin. [(15S)-methyl ester]: $[\alpha]_D$ = -79° (c = 1.3 $CHCl_3$); λ_m = 278 nm (ε 25250 EtOH); [(15R)-methyl ester ($C_{22}H_{36}O_5$)]: $[\alpha]_D$ = -74° (c = 1.0 $CHCl_3$); λ_m = 278 nm (ε 25200 EtOH). *Pharmacia & Upjohn.*

8260 Casokefamide
98815-38-4
$C_{33}H_{40}N_6O_7$
L-Tyrosyl-D-alanyl-L-phenylalanyl-D-alanyl-L-tyrosinamide.
Gastric secretion inhibitor. Increases the resistance to gastric proteases.

8261 Deprostil
33813-84-2
$C_{21}H_{38}O_4$
(1R,2S)-2-(3-Hydroxy-3-methyloctyl)-5-oxocyclopentaneheptanoic acid.
AY-22469. Gastric antisecretory agent.

8262 Enterogastrone
9007-67-4 3632 232-703-5
Anthelone E.
enteroanthelone; Duosan; Ileogastrone. Gastric antisecretory agent. Freely soluble in H_2O. *Res. Corp.*

8263 Fenoctimin
69365-65-7
$C_{27}H_{38}N_2$
4-(Diphenylmethyl)-1-(N-octylformimidoyl)piperidine .
Gastric antisecretory agent.

8264 Fenoctimin Sulfate Hemihydrate
69365-66-8
$C_{27}H_{40}N_2O_4S.0.5H_2O$
4-(Diphenylmethyl)-1-(N-octylformimidoyl)piperidine sulfate (1:1) hemihydrate.
McN-4097-12-98. Gastric antisecretory agent.

8265 Fenoctimine Sulfate [anhydrous]
69365-67-9
$C_{27}H_{40}N_2O_4S$
4-(Diphenylmethyl)-1-(N-octylformimidoyl)piperidine sulfate (1:1).
Gastric antisecretory agent.

8266 Octreotide
83150-76-9 6859
$C_{49}H_{66}N_{10}O_{10}S_2$
D-Phenylalanyl-L-cysteinyl-L-phenylalanyl-D-tryptophyl-L-lysyl-L-threonyl-N-[(1R,2R)-2-hydroxy-1-(hydroxymethyl)propyl]-L-cysteinamide cyclic (2→7)-disulfide.
Sandostatin; SMS-201-995; Longastatin. Gastric antisecretory agent. Used for treatment of agromegaly. $[\alpha]_D^{20}$ = -42° (c = 0.5 95% AcOH). *Sandoz Pharm. Corp.*

8267 Octreotide Acetate
79517-01-4 6859
$C_{49}H_{66}N_{10}O_{10}S_2.xC_2H_4O_2$
D-Phenylalanyl-L-cysteinyl-L-phenylalanyl-D-tryptophyl-L-lysyl-L-threonyl-N-[(1R,2R)-2-hydroxy-1-(hydroxymethyl)propyl]-L-cysteinamide cyclic (2→7)-disulfide acetate (salt).
Sandostatin; SMS-201-995 ac. Gastric antisecretory agent. Used for treatment of agromegaly. *Sandoz Pharm. Corp.*

8268 Omeprazole
73590-58-6 6977
$C_{17}H_{19}N_3O_3S$
5-Methoxy-2-[[(4-methoxy-3,5-dimethyl-2-pyridyl)methyl]sulfinyl]benzimidazole.
Prilosec; Antra; Gastroloc; Losec; Mepral; Mopral; Omepral; Omeprazen; Parizac; Pepticum. Antiulcerative. Gastric antisecretory agent (proton pump inhibitor). Used in treatment of Zollinger-Ellison syndrome. mp = 156°; LD_{50} (mus iv) = 80 mg/kg, (mus orl) > 4000 mg/kg, (rat iv) > 50 mg/kg, (rat orl) > 4000 mg/kg. *Astra Hassle AB; Astra Sweden; Merck & Co.Inc.*

8269 Omeprazole Sodium
95510-70-6 6977
$C_{17}H_{18}N_3NaO_3S$
5-Methoxy-2-[[(4-methoxy-3,5-dimethyl-2-pyridyl)methyl]sulfinyl]benzimidazole sodium salt.
Losec Sodium; H 168/68. Antiulcerative. Gastric antisecretory agent (proton pump inhibitor). *Astra Hassle AB; Astra Sweden; Merck & Co.Inc.*

8270 Rabeprazole
117976-89-3 8272
$C_{18}H_{21}N_3O_3S$
2-[[[4-(3-Methoxypropoxy)-3-methyl-2-pyridyl]methyl]sulfinyl]benzimidazole.
LY307640 . Antisecretory agent, anti-ulcerative, gastric acid pump inhibitor. mp = 99-100° (dec). *Eisai Corp. of North Am.; Eli Lilly & Co.; Janssen Chimica.*

8271 Rabeprazole Sodium
117976-90-6 8272
$C_{18}H_{20}N_3NaO_3S$
2-[[[4-(3-Methoxypropoxy)-3-methyl-2-pyridyl]methyl]sulfinyl]benzimidazole sodium salt.
Acephex; LY307640 sodium; E-3810. Antisecretory agent, anti-ulcerative, gastric acid pump inhibitor. mp = 140-141° (dec). *Eisai Corp. of North Am.; Eli Lilly & Co.; Janssen Chimica.*

8272 Rioprostil
77287-05-9 8395
$C_{21}H_{38}O_4$
(2R,3R,4R)-4-Hydroxy-2-(7-hydroxyheptyl)-3-[(E)-(4RS)-(4-hydroxy-4-methyl-1-octenyl)]-cyclopentanone.
TR-4698; ORF-15927; RWJ-15927; Bay o 6893; Rostil. Antiulcerative. Prostaglandin. Gastric antisecretory agent. $[\alpha]_D$ = -58.6° (c = 1 $CHCl_3$). *Bayer Corp. Pharm. Div.; Miles Inc.*

8273 Saviprazole
121617-11-6
$C_{15}H_{10}F_7N_3O_2S_2$
2-[[[4-(2,2,3,3,4,4,4-Heptafluorobutoxy)-2-pyridyl]methyl]sulfinyl]-1H-thieno[3,4-d]-imidazole.
HOE-731. Gastric H+/K(+)-ATPase inhibitor. A substituted thienoimidazole.

8274 Somatostatin
38916-34-6 8863 254-186-5
$C_{76}H_{104}N_{18}O_{19}S_2$
Ala-Gly-Cys-Lys-Asn-Phe-Phe-Trp-Lys-Thr-Phe-Thr-Ser-Cys (Cys-Cys disulfide).
GH-RIF; growth hormone-release inhibiting factor; somatotropin release inhibiting factor; SRIF; SRIF-14. Gastric antisecretory agent. Used for treatment of hemorrhaging of gastro-duodenal ulcers. Growth hormone inhibitor. Studied for use as an antidiabetic. A cyclic tetradecapeptide which inhibits the release of growth hormone, insulin and glucagon. *Genentech Inc.*

8275 Somatostatin Acetate
8863
$C_{76}H_{104}N_{18}O_{19}S_2 \cdot C_2H_4O_2$
Ala-Gly-Cys-Lys-Asn-Phe-Phe-Trp-Lys-Thr-Phe-Thr-Ser-Cys (Cys-Cys disulfide) acetate.
SRIF-A; Aminopan; Modustatina; Somatofalk; Stilamin. Gastric antisecretory agent. Used for treatment of hemorrhaging of gastro-duodenal ulcers. Growth hormone inhibitor. Studied for use as an antidiabetic. A cyclic tetradecapeptide which inhibits the release of growth hormone, insulin and glucagon. *Genentech Inc.*

8276 Telenzepine
80880-90-6 9270
$C_{19}H_{22}N_4O_2S$
4,9-Dihydro-3-methyl-4-[(4-methyl-1-piperazinyl)acetyl]-10H-thieno[3,4-b][1,5]benzodiazepin-10-one.
Antiulcerative; anticholinergic. Selective muscarinic M_1-receptor antagonist. Gastric acid inhibitor. mp = 263-264°. *Byk Gulden Lomberg GmbH.*

8277 Triletide
62087-96-1 263-401-1
$C_{27}H_{31}N_5O_5$
N-[N-(N-Acetyl-3-phenyl-L-alanyl)-3-phenyl-L-alanyl]-L-histidine methyl ester.
ZAMI-420. A tripeptide thromboxane synthesis inhibitor. Gastric secretion inhibitor.

8278 Trimoprostil
69900-72-7 9853
$C_{23}H_{38}O_4$
(Z)-7-[(1R,2R,3R)-2-[(E)-(3R)-3-Hydroxy-4,4-dimethyl-1-octenyl]-3-methyl-5-oxocyclopentyl]-5-heptenoic acid.
Ro-21-6937/000; TM-PGE$_2$; Ulstar. Antiulcerative. Synthetic prostaglandin E_2 analog with gastric antisecretory activity. $[\alpha]_D$ = -51.54° (c = 1 $CHCl_3$); LD_{50} mus orl) = 41 mg/kg, (mis ip) = 70 mg/kg, (mus sc) = 68 mg/kg, (rat orl) = 23 mg/kg, (rat ip) = 21 mg/kg, (rat sc) = 29 mg/kg. *Hoffmann-LaRoche Inc.*

8279 Urogastrone
9010-53-1
An inhibitory factor of gastric secretion obtained from human urine. Gastric secretion inhibitor.

8280 Zolenzepine
78208-13-6
$C_{19}H_{24}N_6O_2$
4,9-Dihydro-1,3-dimethyl-4-[(4-methyl-1-piperazinyl)-acetyl]pyrazolo[4,3-b][1,5]benzodiazepin-10H)-one.
Antimuscarinic. Gastric secretion inhibitor.

Gastroprokinetics

8281 Cinitapride
66564-14-5 2352
$C_{21}H_{30}N_4O_4$
4-Amino-N-[1-(3-cyclohexen-1-ylmethyl)-4-piperidyl]-2-ethoxy-5-nitrobenzamide.
Gastrointestinal prokinetic agent. *Anphar; Walton Pharm.*

8282 Cinitapride Tartrate
96623-56-2 2352
$C_{25}H_{36}N_4O_{10}$
4-Amino-N-[1-(3-cyclohexen-1-ylmethyl)-4-piperidyl]-2-ethoxy-5-nitrobenzamide tartrate.
LAS-17177; Cidine. Gastrointestinal prokinetic agent. *Anphar; Walton Pharm.*

8283 Cisapride
81098-60-4 2377 279-689-7
$C_{23}H_{29}ClFN_3O_4$
cis-4-Amino-5-chloro-N-[1-[3-(p-fluorophenoxy)-propyl]-3-methoxy-4-piperidyl]-o-anisamide.
R-51619; Propulsid; Acenalin; Alimix; Cipril; Prepulsid; Propulsin; Risamal. Peristaltic stimulant. Used as a gastrointestinal prokinetic agent. mp = 109.8°. *Janssen Pharm. Inc.*

8284 Ecabapide
104775-36-2
$C_{20}H_{25}N_3O_4$
m-[[[(3,4-Dimethoxyphenethyl)carbamoyl]methyl]-amino]-N-methylbenzamide.
A gastroprokinetic drug.

8285 Fedotozine
123618-00-8 3987
$C_{22}H_{31}NO_4$
(+)-(R)-α-Ethyl-N,N-dimethyl-α-[[(3,4,5-trimethoxy-benzyl)oxy]methyl]benzylamine.
Peripheral kappa opioid receptor agonist. Used as a gastrointestinal prokinetic agent. $[\alpha]_D$ = 16.5° (c = 6 EtOH). *Jouveinal.*

8286 Fedotozine D-(-)-Tartrate
133267-27-3 3987
$C_{26}H_{37}NO_{12}0$
(+)-(R)-α-Ethyl-N,N-dimethyl-α-[[(3,4,5-trimethoxy-benzyl)oxy]methyl]benzylamine D-(-)-tartrate.
JO-1196. Used as a gastrointestinal prokinetic agent. mp = 147°; $[\alpha]_D^{25}$ = 14.5° (c = 5 HCl). *Jouveinal.*

8287 Itopride
122898-67-3
$C_{20}H_{26}N_2O_4$
N-[p-[2-(Dimethylamino)ethoxy]benzyl]veratramide.
Gastroprokinetic.

8288 Loxiglumide
107097-80-3 5618
$C_{21}H_{30}Cl_2N_2O_5$
(±)-4-(3,4-Dichlorobenzamido)-N-(3-methoxypropyl)-N-pentylglutaramic acid.
CR-1505. Cholecystokinin A antagonist. Used as a gastrointestinal prokinetic agent. mp = 113-115°; slightly soluble in H_2O (0.01 g/100 ml). *Rotta Pharm.*

8289 Mosapride
112885-41-3
$C_{21}H_{25}ClFN_3O_3$
(±)-4-Amino-5-chloro-2-ethoxy-N-[[4-(4-fluorobenzyl)-2-morpholinyl]methyl] benzamide.
AS-4370. Sulpiride derivative. Enhances gastrointestinal motility by stimulating the 5-hydroxytryptamine 4 (5-HT4) receptor. Gastroprokinetic.

8290 Renzapride
88721-77-1
$C_{16}H_{22}ClN_3O_2$
[(±)-endo]-4-Amino-N-(1-azabicyclo[3.3.1]non-4-yl)-5-chloroanisamide.
BRL-20627. Serotonin 5-HT4 receptor agonist. Gastroprokinetic.

Gonad Stimulating Principals

8291 Buserelin
57982-77-1 1527 261-061-9
$C_{60}H_{86}N_{16}O_{13}$
6-[O-(1,1-Dimethylethyl)-D-serine]-9-(N-ethyl-L-prolinamide)-10-deglycinamide luteinizing hormone-releasing factor (pig).
5-oxoPro-His-Trp-Ser-Tyr-D-Ser(t-Bu)-Leu-Arg-ProNH-CH_2CH_3. Hormonal antineoplastic; gonad-stimulating principle. Nonapeptide agonist of LH-RH. $[\alpha]_D^{20}$ = -40.4° (c = 1 dimethylacetamide). *Hoechst AG.*

8292 Buserelin Acetate
68630-75-1 1527
$C_{62}H_{88}N_{16}O_{14}$
6-[O-(1,1-Dimethylethyl)-D-serine]-9-(N-ethyl-L-prolinamide)-10-deglycinamide luteinizing hormone-releasing factor (pig) acetate.

HOE-766; Receptal; Suprecur; Suprefact; Suprafact. Hormonal antineoplastic; gonad-stimulating principle. Nonapeptide agonist of LH-RH. *Hoechst AG.*

8293 Cetrorelix

120287-85-6

$C_{70}H_{92}ClN_{17}O_{14}$

N-Acetyl-3-(2-naphthyl)-D-alanyl-p-chloro-D-phenylalanyl-3-(3-pyridyl)-D-alanyl-L-seryl-L-tyrosyl-N^5-carbamoyl-D-ornithyl-L-leucyl-L-arginyl-L-prolyl-D-alaninamide.

A gonadotrophin-releasing hormone antagonist.

8294 Chorionic Gonadotropin

2273

Choriogonadotropin; CG; Human Chorionic hormone; HCG; Ambinon; Antuitrin S; A.P.L.; Choragon; Choriogonin; Chorex; Choron; Coriantin; Coriovis; Corulon; Endocorion; Follutein; Glukor; Gonadotraphon L.H.; Gonic; Libigen; Luteogonin B; Physex; Predalon; Pregnesin; Pregnyl; Primogonyl; Profasi; Profasi HP; Progon. Gonad stimulating principle. Glycoprotein hormone synthesized by chorionic tissue of the placenta. Freely soluble in H_2O, glycerol, glycols; insoluble in EtOH, Me_2CO, Et_2O; molecular weight → 39,500; has two subunits, α-subunit with 92 residues and β-subunit with 145 residues.

8295 Clomiphene

911-45-5 2446 213-008-6

$C_{26}H_{28}ClNO$

2-[p-(2-Chloro-1,2-diphenylvinyl)phenoxy]triethylamine.

clomifene; chloramiphene; MRL-41. Antiestrogen. Gonad stimulating principal. Synthetic estrogen agonist-antagonist. *Lemmon Co.; Merrell Pharm. Inc.; Serono Labs Inc.*

8296 Clomiphene Citrate

50-41-9 2446 200-035-3

$C_{32}H_{36}ClNO_8$

2-[4-(2-Chloro-1,2-diphenylethenyl)phenoxy]-N,N-diethylethanamine citrate.

clomifene citrate; MRL-41; MER-41; NSC-35770; Clomid; Clomphid; Clomivid; Clostilbegyt; Dyneric; Ikaclomine; Pergotime; Serophene; zuclomiphene [cis-form]; enclomiphene [trans-form]. Gonad stimulating principal. Antiestrogen. mp = 116.5-118°; slightly soluble in H_2O, $CHCl_3$, EtOH; freely soluble in MeOH; insoluble in Et_2O. *Lemmon Co.; Merrell Pharm. Inc.; Serono Labs Inc.*

8297 Cyclofenil

2624-43-3 2789 220-089-1

$C_{23}H_{24}O_4$

4-[[4-(Acetyloxy)phenyl]cyclohexylidenemethyl]phenol acetate.

H-3452; ICI-48213; Fertodur; Neoclym; Ondonid; Ondogyne; Rehibin; Sanocrisin; Sexadieno; Sexovid. Gonad stimulating principle. mp = 135-136°; λ_m 247 nm (ε 17000 EtOH); [free diol (F-6060)]: mp = 235-236°. *ICI.*

8298 Epimestrol

7004-98-0 3654 230-278-0

$C_{19}H_{26}O_3$

3-Methoxyestra-1,3,5(10)-triene-16α,17α-diol.

Org-817; NSC-55975; Stimovul. Anterior pituitary activator. Gonad stimulating principle. mp = 158-160°; $[\alpha]_D^{20}$ = 48° ($CHCl_3$). *Organon Inc.*

8299 FSH

9002-68-0 4299 232-662-3

Follicle-stimulating hormone; urofollitrophin; Fertinorm; Follitropin; Luteoantine; Metrodin. Gonad stimulating principle. Glycoprotein gonadotropic hormone found in the pituitary tissue of mammals. Directly regulates the metabolic activity of the granulosa cells of the ovaries and the Sertoli cells of the testis. Solid, soluble in H_2O.

8300 Ganirelix

124904-93-4

$C_{80}H_{113}ClN_{18}O_{13}$

N-Acetyl-3-(2-naphthyl)-D-alanyl-p-chloro-D-phenylalanyl-3-(3-pyridyl)-D-alanyl-L-seryl-L-tyrosyl-N^6-(N,N'-diethylamidino)-D-lysyl-L-leucyl-N^6-(N,N'-deithylamidino)-L-lysyl-L-propyl-D-alaninamide.

Gonad stimulating principle. *Syntex Intl. Ltd.*

8301 Ganirelix Acetate

129311-55-3

$C_{84}H_{121}ClN_{18}O_{17}$

N-Acetyl-3-(2-naphthyl)-D-alanyl-p-chloro-D-phenylalanyl-3-(3-pyridyl)-D-alanyl-L-seryl-L-tyrosyl-N^6-(N,N'-diethylamidino)-D-lysyl-L-leucyl-N^6-(N,N'-diethylamidino)-L-lysyl-L-propyl-D-alaninamide diacetate (salt).

RS-26306. Gonad stimulating principle. *Syntex Intl. Ltd.*

8302 Gonadorelin

33515-09-2 251-553-1

$C_{55}H_{75}N_{17}O_{13}$

Gonad stimulating principle. *Hoechst AG.*

8303 Gonadorelin Acetate

52699-48-6

$C_{55}H_{75}N_{17}O_{13}.xC_2H_4O_2.yH_2O$

Cystorelin; Hypocrine; Lutrelef; Lutrepulse. Gonad stimulating principle. *Hoechst AG.*

8304 Histrelin

76712-82-8 4760

$C_{66}H_{86}N_{18}O_{12}$

6-[1-(Phenylmethyl)-D-histidine]-9(n-ethyl-L-prolinamide)-10-deglycineamide luteinizing hormone-releasing factor (pig).

5-oxo-L-prolyl-L-histidyl-L-tryptophyl-L-seryl-L-tyrosyl-N-benzyl-D-histidyl-L-leucyl-L-arginyl-N-ethyl-L-prolinamide; [(im-Bzl)-D-His^6,Pro^9-Net]-gonadotropin-releasing hormone; ORF-17070; RWJ-17070; Supprelin. Used in treatment of precocious puberty. Gonadotropin-releasing hormone. Nonapeptide synthetic LH-RH agonist. $[\alpha]_D^{20}$ = -33.9° (c = 1 AcOH). *Ortho Pharm. Corp.; Roberts Pharm. Corp.*

8305 LH

9002-67-9 5499 232-661-8

Luteinizing Hormone.

ICSH. A glycoprotein gonadotropic hormoine found in the anterior lobe of the pituitary gland. Gonad stimulating principle. White powder, soluble in H_2O.

8306 LH-RH
9034-40-6 5500 232-895-0
$C_{55}H_{75}N_{17}O_{13}$
Luteinizing Hormone-Releasing Factor.
LH-RF; luteinizing hormone-releasing hormone; LRF; LRH; gonadorelin; LH-RH/FSH-RH; Fertagyl; Fertiral; Kryptocur; Relefact LH-RH; gonadoliberin; luliberin; 5-oxoPro-His-Trp-Ser-Tyr-Gly-Leu-Arg-Pro-GlyNH_2. Gonad stimulating principle. $[\alpha]_D^{25}$ = -50° (1% AcOH); destroyed by chymotrypsin, papain, subtilisin and thermolysin. *Hoechst AG.*

8307 LH-RH Hydrochloride
51952-41-1 5500
$C_{55}H_{75}N_{17}O_{13}.xC_2H_4O_2.yH_2O$
Gonadorelin acetate.
Gonad stimulating principle.

8308 Zuclomiphene
15690-55-8 2446
$C_{26}H_{28}ClNO$
(Z)-2-[p-(2-Chloro-1,2-diphenylvinyl)phenoxy]-triethylamine.
Isomer A; RMI-16312; 224. cis-Form of clomiphene. Gonad-stimulating principle.

Growth Hormone Releasing Factors

8309 Cetermin
157238-32-9
$C_{1132}H_{1716}N_{298}O_{330}S_{20}$
Transforming growth factor β2 (human).

8310 Pralmorelin Dihydrochloride
158827-34-0
$C_{45}H_{57}Cl_2N_9O_6$
D-alanyl-3-(2-naphthyl)-D-alanyl-L-alanyl-L-tryptophyl-D-phenylalanyl-L-lysinamide dihydrochloride.
WAY-GPA-748. Growth hormone-releasing factor. *Kakenyaku Kako.*

8311 Sermorelin
86168-78-7 8605
$C_{149}H_{246}N_{44}O_{42}S$
human growth hormone-releasing factor(1-29)amide.
human pancreatic somatoliberin(1-29)amide; GRF(1-29)NH_2; hpGRF(1-29)NH_2; SM-8144; Geref; Groliberin; Tyr-Ala-Asp-Ala-Ile-Phe-Thr-Asn-Ser-Tyr-Arg-Lys-Val-Leu-Gly-Gln-Leu-Ser-Ala-Arg-Lys-Leu-Leu-Gln-Asp-Ile-Met-Ser-Arg-NH_2. Growth hormone-releasing factor. $[\alpha]_D^{20}$ = -63.1° (c = 1 30% AcOH). *Serono Labs Inc.*

8312 Somatorelin
83930-13-6 8861
$C_{215}H_{358}N_{72}O_{66}S$
Growth hormone-releasing factor (human).

Growth Stimulants

8313 Actaplanin
37305-75-2
Kamoran; A-4696. Veterinary growth stimulant. Glycopeptide antibiotics derived from *Actinoplanes* strain ATCC 23342. *Eli Lilly & Co.*

8314 Alexomycin
165101-50-8
U-82127. Veterinary growth stimulant. Mixture of cyclic sulfur-containing peptides obtained from *Streptomyces arginensis. Pharmacia & Upjohn.*

8315 Efrotomycin
56592-32-6 3567
$C_{59}H_{88}N_2O_{20}$
31-O-[6-Deoxy-4-O-(6-deoxy-2,4-di-O-methyl-α-L-mannopyranosyl)-3-O-methyl-β-D-allopyranosyl]-1-methylmocimycin.
Producil; MK-621; FR-02A. Veterinary growth stimulant. Yellow solid; λ_m 232, 327 nm ($E^{1\%}_{1\ cm}$ 464 216 pH 7); LD_{50} (mus orl) > 4000 mg/kg, (mus sc) > 2000 mg/kg. *Merck & Co.Inc.*

8316 Laidlomycin Propionate Potassium
84799-02-0 5361
$C_{40}H_{65}KO_{13}$
16-Deethyl-3-O-demethyl-16-methyl-3-O-(1-oxopropyl)-monensin 26-propanoate monopotassium salt.
Cattlyst; RS-11988. Veterinary growth stimulant. mp = 190-192°. *Syntex Intl. Ltd.*

8317 Mecasermin
68562-41-4
Insulin-like growth factor I.
IGF-I. An endogenous growth hormone.

8318 Narasin
55134-13-9 6506
$C_{43}H_{72}O_{11}$
α-Ethyl-6-[5-[2-(5-ethyltetrahydro-5-hydroxy-6-methyl-2H-pyran-2-yl)-15-hydroxy-2,10,12-trimethyl-1,6,8-trioxadispiro[4.1.5.3]pentadec-13-en-9-yl]-2-hydroxy-1,3-dimethyl-4-oxoheptyl]tetrahydro-3,5-dimethyl-2H-pyran-2-acetic acid.
Natacyn; CL-12625; Antibiotic A-5283; Monteban; C-7819B. Veterinary growth stimulant. mp = 98-100°, 198-200°; λ_m 285 nm (ε 58 EtOH); $[\alpha]_D^{25}$ = -54° (c = 0.2 MeOH); insoluble in H_2O, soluble in organic solvents; LD_{50} (mus ip) = 7.15 mg/kg. *Alcon Labs; Eli Lilly & Co.*

8319 Nosiheptide
56377-79-8 6816 260-138-4
$C_{51}H_{43}N_{13}O_{12}S_6$
N-[1-(Aminocarbonyl)ethenyl]-2-[14-ethylidene-9,10,11,12,13,14,19,20,21,22,23,24,26,33,35,36-hexadecahydro-3,23-dihydroxy-11-(1-hydroxyethyl)-31-methyl-9,12,19,24,33,43-hexaoxo-30,32-imino-8,5:18,15:40,37-trinitrilo-21,36-([2,4]-endo-thiazolomethanimino)-5H,15H,37H-pyrido[3,2-w][2,11,21,27,31,7,14,17]benzoxatetrathiatriazacyclohexatriacontin-2-yl]-4-thiazolecarboxamide.
RP-9671; Multhiomycin; Primofax. Veterinary growth stimulant. mp = 310-320° (dec); $[\alpha]_D^{20}$ = 38° (c = 1 C_5H_5N); λ_m 242, 322 nm ($E^{1\%}_{1cm}$ 525, 229 H_2O/DMF); soluble in $CHCl_3$, dioxane; C_5H_5N, DMF, DMSO; slightly soluble in MeOH, EtOH, EtOAc, C_6H_6; insoluble in H_2O, petroleum ether. *Rhône-Poulenc Rorer Pharm. Inc.*

8320 Plauracin
62107-94-2
$C_{71}H_{88}N_{10}O_{18}$
Antibiotic complex produced by *Actinoplanes auranticolor* (ATCC 31011).
CP-38754. Veterinary growth stimulant. *Pfizer Intl.*

8321 Ractopamine
97825-25-7 8275
$C_{18}H_{23}NO_3$
(±)-all-rac-p-Hydroxy-α-[[[3-(p-hydroxyphenyl)-1-methylpropyl]amino]methyl]benzyl alcohol.
Veterinary growth stimulant. *Eli Lilly & Co.*

8322 Ractopamine Hydrochloride
90274-24-1 8275
$C_{18}H_{24}ClNO_3$
(±)-all-rac-p-Hydroxy-α-[[[3-(p-hydroxyphenyl)-1-methylpropyl]amino]methyl]benzyl alcohol hydrochloride.
LY-O31537; EL-737; Paylean. Veterinary growth stimulant. [mixture containing 51% RR,SS- and 49% RS,SR-forms]: mp 124-129°; [RR-form hydrochloride]: mp = 176-176.5°; $[\alpha]_D$ = -22.7°, $[\alpha]_{365}$ = -71.2° (c = 0.37 MeOH). *Eli Lilly & Co.*

8323 Somatotropin
9002-72-6 8864 232-666-5
Adenohypophyseal growth hormone.
GH; hypophyseal growth hormone; anterior pituitary growth hormone; phyone; pituitary growth hormone; somatotropic hormone; SH. Growth stimulant. Single-chain polypeptide of 191 amino acids; $[\alpha]_D^{20}$= -38.7° (c = 0.1M acetic acid). *Pharmacia & Upjohn.*

8324 Somatotropin, Recombinant
9002-72-6 8864 232-666-5
Adenohypophyseal growth hormone (recombinant).
Serostim. Growth stimulant. *Serono Labs Inc.*

8325 Somavubove
126752-39-4
$C_{976}H_{1533}N_{265}O_{286}S_8$
127-L-Leucine growth hormone (ox).
recombinantly derived bovine somatotropin. Recombinant bovine growth hormone. Galactopoietic agent (veterinary). *Upjohn Ltd.*

8326 Sometribove
102744-97-8 8864
$C_{976}H_{1533}N_{268}O_{291}S_9$
Bovine somatotropin, produced by recombinant technology. Veterinary growth stimulant. *Monsanto Co.*

8327 Sometripor
102733-72-2
$C_{979}H_{1527}N_{265}O_{287}S_8$
Porcine somatotropin, produced by recombinant technology. Veterinary growth stimulant. *Monsanto Co.*

8328 Somfasepor
129566-95-6
$C_{938}H_{1465}N_{257}O_{278}S_5$
8-190-Growth hormone (pig).
8-190-Somatotropin (pig clone pPGH-1); Grolene; Leanstar; P-3232; P-3895. Veterinary growth stimulant. *Pitman-Moore Inc.*

8329 Sulbenox
58095-31-1 9060
$C_9H_{10}N_2O_2S$
(4,5,6,7-Tetrahydro-7-oxobenzo[b]thien-4-yl)urea.
CL-206576; Vigazoo. Veterinary growth stimulant. mp = 245-246°; LD_{50} (rat orl) > 5000 mg/kg. *Am. Cyanamid.*

8330 Temodox
34499-96-2
$C_{12}H_{12}N_2O_5$
2-Hydroxyethyl 3-methyl-2-quinoxalinecarboxylate 1,4-dioxide.
CP-22341. Veterinary growth stimulant. *Pfizer Inc.*

8331 Trafermin
131094-16-1
$C_{764}H_{1201}N_{217}O_{219}S_6$
2-155-Basic fibroblast growth factor (human clone λ-KB7/λHFL1 precursor reduced); recombinant human basic fibroblast growth factor. Growth stimulant.

Hematinics

8332 Ammonium Ferric Citrate
1185-57-5 547
Ferric ammonium citrate.
iron ammonium citrate; ammonium ferric citrate; ferric ammonium citrate; 2-hydroxy-1,2,3-propanetricarboxylic acid, ammonium iron (3+) salt; FAC; ammonium iron (III) citrate; Soluble Ferric Citrate; prothoate+; iron (III) ammonium citrate; ammonium iron(III) citrate, brown. Hematinic. Very soluble in H_2O, insoluble in EtOH. *Mallinckrodt Inc.*

8333 Aquacobalamin
13422-52-1 4854 236-534-8
$C_{62}H_{90}CoN_{13}O_{15}P.OH$
Cobinamide hydroxide monohydrate dihydrogen phosphate (ester) inner salt 3'-ester with 5,6-dimethyl-1-α-D-ribofuranoylbenzimidazole.
aquocobalamin; vitamin B_{12b}; vitamin B_{12d}. Vitamin, vitamin source. May be used in management of patients with megaloblastic anemia (early sign of vitamin B_{12} deficiency). λ_m 274, 317, 351, 499, 527 nm (ε 20600, 6100, 26500, 8100,8500 H_2O). *Merck & Co.Inc.*

8334 Calcium Ferrous Citrate
53684-61-0 1708
$C_{12}H_{10}Ca_2FeO_{14}$
Ferrous calcium citrate.
Ferrocal; Rarical. Hematinic. Tetrahydrate, tasteless. *Ortho Pharm. Corp.*

8335 Cobaltous Chloride
7646-79-9 2498
Cl_2Co
Cobalt dichloride.
Hematinic. Hexahydrate; mp = 87°; d^{20} = 1.924; MLD (rbt sc) = 200 mg/kg.

8336 Dextran Iron Complex
9004-66-4 2991
A complex of trivalent iron and dextran.
Fenate; Imferon; Ironorm. Hematinic. Used in veterinary medicine and as an antianemic factor. LD_{50} (mus iv) = 2240 mg/kg. *Fermenta Animal Health Co.; Fisons Pharm. Div.*

8337 Epoetin-α
113427-24-0 3729
$C_{809}H_{1301}N_{229}O_{240}S_5$
1-165 Erythropoietin (human clone λHEPOFL13 protein moiety), glycoform α.
Epogen; Procrit; Epoade; Eprex; Erypo; Espo; A 165-mer glycoprotein. Antianemic; hematinic. *Amgen Inc.; Ortho Biotech Inc.*

8338 Epoetin-β
122312-54-3 3729
$C_{809}H_{1301}N_{229}O_{240}S_5$
1-165 Erythropoietin (human clone λHEPOFL13 protein moiety), glycoform β.
Marogen; EPOCH; BM-06.019; Epogin; Recormon; A 165-mer glycoprotein. antianemic; hematinic. *Chugai Pharm. Co. Ltd.*

8339 Ferric Albuminate
8001-11-4 4058
Albumized iron.
Combination of egg albumin and iron with 17-19% Fe. Hematinic. Freely soluble in H_2O, insoluble in EtOH.

8340 Ferric and Ammonium Acetate Solution
8006-27-7 4059
Basham's mixture. Contains 0.16-0.20% Fe and 3.5 % ammoium acetate. Hematinic.

8341 Ferric Citrate
2338-05-8 4063
Combination of iron and citric acid; indefinite composition.
Hematinic. Poorly soluble in cold H_2O, more soluble in hot H_2O, insoluble in EtOH.

8342 Ferric Fructose
12286-76-9 4295
$(C_6H_{10}FeO_7)_nK_{n/2}$ (n = 2- 100)
D-Fructose iron(3+)-containing complex potassium salt (2:1).
CB-302. Hematinic.

8343 Ferric Oxide, Saccharated
8047-67-4 4073
saccharated iron; iron sugar; Colliron I.V.; Feojectin; Ferrivenin; Ferum Hausmann; Iviron; Neo-Ferrum; Proferrin; Sucrofer. Solution containing about 2% Fe, suitable for iv injection. Hematinic.

8344 Ferric Pyrophosphate
10058-44-3 4075
$Fe_4O_{21}P_6$
Iron(3+) pyrophosphate.
Hematinic. Insoluble in H_2O. *Am. Oil Co.*

8345 Ferric Sodium Edetate
15708-41-5 4076
$C_{10}H_{12}FeN_2NaO_8$
Sodium [(ethylenedinitrilo)tetraacetato]ferrate(1-).
Ferrostrane; Ferrostrene; Sybron. Hematinic.

8346 Ferriclate Calcium Sodium
34150-62-4 4070
$C_{12}H_{44}CaFe_6Na_4O_{36}$
Monocalcium tetrasodium bis-[pentaaqua[D-gluconato(4-)]-tetra-μ-hydroxydioxotriferrate-(3-)].
Kelfer. Hematinic. *Lab. Mauricio Villela S.A.*

8347 Ferritin
9007-73-2 4083
Epadora; Ferrofolin; Ferrol; Ferrosprint; Ferrostar; Sanifer; Sideros; Unifer. An iron-storage protein found in many biological systems. Hematinic. Soluble in H_2O.

8348 Ferrocholinate
1336-80-7 4085
$C_{11}H_{20}FeNO_9$
2-Hydroxy-N,N,N-trimethylethanaminium (OC-6-44)-triaqua[2-hydroxy-1,2,3-propane-tricarboxylato(4-)]ferrate(1-).
Chelafer; Chel-Iron; Ferrolip. Chelate of ferric hydroxide and choline dihydrogen citrate, used as a hematinic. Freely soluble in H_2O. *Flint-Eaton.*

8349 Ferroglycine Sulfate
17169-60-7 4086
Plesmet; Kelferon; Ferronord; Ferro sanol; Glyferro; Pleniron. Hematinic. *Schwarz Arztnelmittelfabrik.*

8350 Ferrous Carbonate Mass
8030-35-1 4089
Blaud's Mass; Vallet's Mass; Fecarb. Contains 36-41% Fe, the remainder is honey and sugar. Hematinic. Insoluble in H_2O.

8351 Ferrous Carbonate Saccharated
8001-10-3 4090
Freshly precipitated $FeCO_3$, mixed with sugar, contains at least 15% Fe. Hematinic. Partially soluble in H_2O, soluble in mineral acids.

8352 Ferrous Citrate
23383-11-1 4092 245-625-1
Prepared from iron powder and citric acid. Hematinic. Monohydrate or decahydrate; insoluble in H_2O, Me_2Co. *Ortho Pharm. Corp.*

8353 Ferrous Fumarate
141-01-5 4094 205-447-7
$C_4H_2FeO_4$
Iron(2+) fumarate.
Fepstat; Toleron; Cpiron; Erco-Fer; Ferrofume; Ferronat; Ferrone; Ferrotemp; Ferrum; Fresamal; Firon; Fumafer; Fumar F; Fumiron; Galfer; Heferol; Ircon; Meterfer; One-Iron; Tolferain; Tolifer; component of: Chromagen, Ferancee, Stuartinic, Tolfrinic. Hematinic. mp > 280°; d^{25} = 2.435; soluble in H_2O (0.14 g/100 ml), EtOH (< 0.01 g/100 ml); LD_{50} (mus ip) = 480 mg/kg. *Ascher B.F. & Co.; Forest Pharm. Inc.; Johnson & Johnson-Merck*

Consumer Pharm. ; Mallinckrodt Inc.; Marion Merrell Dow Inc.; Parke-Davis; Savage Labs.

8354 Ferrous Gluconate
299-29-6 4095 206-076-3
$C_{12}H_{22}FeO_{14}$
Iron(2+) gluconate (1:2).
Fergon; Ferlucon; Ferronicum; Iromon; Irox; Nionate.
Hematinic. Soluble in H_2O, insoluble in EtOH; LD_{50} (mus iv) = 114 mg/kg, (mus orl) = 3700 mg/kg. *Sterling Health U.S.A.*

8355 Ferrous Lactate
5905-52-2 4098 227-608-0
$C_6H_{10}FeO_6$
Iron(2+) lactate.
Ferro-Drops. Hematinic. Soluble in H_2O, insoluble in EtOH; LD (rbt sc) = 578 mg/kg, (rbt iv) = 287 mg/kg. *Parke-Davis.*

8356 Ferrous Succinate
10030-90-7 4104 233-082-3
$C_4H_4FeO_4$
Iron(2+) succinate.
Cerevon; Ferromyn. Hematinic. Tetrahydrate or dihydrate; sparingly soluble in H_2O.

8357 Ferrous Sulfate Heptahydrate
7782-63-0 4105 231-753-5
$FeSO_4.7H_2O$
Iron(2+) sulfate (1:1) heptahydrate.
Feosol; Fero-Gradumet; Mol-Iron; Natabec; Slow-Fe; copperas green; green vitriol; iron vitriol; Feospan; Fesotyme; Fer-in-Sol; Haemofort; Ironate; Presfersul; Sulferrous; component of: Plastulen-N. antianemic; hematinic. d = 1.897; soluble in H_2O, insoluble in EtOH; LD_{50} (mus iv) = 65 mg/kg, (mus orl) = 1520 mg/kg. *Abbott Labs Inc.; Ciba-Geigy Corp.; Lederle Labs.; Parke-Davis; Schering-Plough Pharm.; SmithKline Beecham Pharm.*

8358 Ferrous Sulfate, Dried
13463-43-9 4105 231-753-5
$FeSO_4.xH_2O$
Iron(2+) sulfate (1:1) hydrate.
Feromax; Feroritard; Ferro-Gradumet; Fespan; Tetucur; component of: Cytoferin, Mediatric. antianemic; hematinic. Soluble in H_2O. *Wyeth-Ayerst Labs.*

8359 Folic Acid
59-30-3 4253 200-419-0
$C_{19}H_{19}N_7O_6$
N-[p-[[(2-Amino-4-hydroxy-6-pteridinyl)methyl]-amino]benzoyl]-L-glutamic acid.
Folicet; Vitamin M; Cytofol; Folacin; Foldine; Foliamin; Folipac; Folettes; Folsan; Folvite; Incafolic; Millafol; component of: Mission Prenatal, Plastulen-N. Vitamin, vitamin source. Hematopoietic. May be used in management of patients with megaloblastic anemia or with folate deficiency caused by drugs that inhibit dihydrofolate reductase or that interfere with the absorption and storage of folate. mp > 250°; $[\alpha]_D^{25}$ = 23° (c = 0.5 in 0.1N NaOH); λ_m = 256, 283 368 nm (log ε 4.43, 4.40, 3.96 pH 13); slightly soluble on H_2O (0.00016 g/100 ml at 20°, 1 g/100 ml at 100°), MeOH; less soluble in EtOH, BuOH; insoluble in $CHCl_3$, Me_2CO, Et_2O, C_6H_6. *Bristol-Myers Squibb Pharm. R&D; Lederle Labs.; Mission Pharmacal Co.*

8360 Glusoferron
56959-18-3
D-Gluconic acid polymer with D-glucitol iron(3+) salt.
Ferastral. Hematinic.

8361 Iron Sorbitex
1338-16-5 5112
A sterile, colloidal solution of a complex of trivalent iron, sorbitol and citric acid, stabilized with dextrin and sorbitol.
Astra 1572. *Astra USA Inc.*

8362 Liver Extract
5575
An extract from mammalian liver.
component of: Intraheptol, Pernaemon, Desiver, Anahaemin, Campolon, Cromaton, Curethyl, Examen, Ficalon, Hepalon, Hepatopron, Hormantoxone, Hoban, Neo-Heptatex, Pernaemyl, Pernexin, Perniciosan, Plexan, Reticulogen, Ripason, Sykoton, Tenelon, Hepol.

8363 Peptonized Iron
7292
A compound of iron oxide and peptone, mixed with sodium citrate to improve water solubility.
Saferon. Soluble in H_2O, insoluble in EtOH.

8364 Polyferose
9009-29-4 7731
Iron complex with polymer of β-D-fructosanosyl -α-D-glucopyranoside.
Jefron. A chelate complex of iron and a polymerized derivative of sucrose. *Marion Merrell Dow Inc.*

8365 Pyridoxine Hydrochloride
58-56-0 8166 200-386-2
$C_8H_{12}ClNO_3$
5-Hydroxy-6-methyl-3,4-pyridinedimethanol hydrochloride.
Beesix; Hexa-Betalin; Hexavibex; Bonasanit; Hexabione hydrochloride; Pyridipca; Pyridox; Bécilan; Benadon; Hexermin; component of: Bendectin, Spondylonal. Vitamin, vitamin source. Enzyme cofactor. May improve hematopoiesis in patients with sideroblastic anemias. mp = 205-212° (dec); λ_m = 290 nm (ε 8400 0.1N HCl), 253, 325 nm (ε 3700, 7100, pH 7); soluble in H_2O (22.2 g/100 ml), EtOH (1.1 g/100 ml), propylene glycol; sparingly soluble in Me_2CO; insoluble in Et_2O, $CHCl_3$. *BASF Corp.; Eli Lilly & Co.; Forest Pharm. Inc.; General Aniline; Lederle Labs.; Marion Merrell Dow Inc.; Merck & Co.Inc.; Parke-Davis.*

8366 Riboflavin Monophosphate
146-17-8 8368 205-664-7
$C_{17}H_{21}N_4O_9P$
7,8-Dimethyl-10-(D-ribo-2,3,4,5-tetrahydroxypentyl)-isoalloxazine 5'-(dihydrogen phosphate).
flavin mononucleotide; FMN; vitamin B_2 phosphate. Vitamin, vitamin source. Enzyme cofactor. May be helpful in management of patients with hypoproliferative anemia. *Hoffmann-LaRoche Inc.; Takeda Chem. Ind. Ltd.*

8367 Riboflavin Monophosphate Monosodium Salt
130-40-5 8368 204-988-6
$C_{17}H_{20}N_4NaO_9P$
7,8-Dimethyl-10-(D-ribo-2,3,4,5-tetrahydroxypentyl)isoalloxazine 5'-(dihydrogen phosphate) monosodium salt.
Hyryl; Ribo. Vitamin, vitamin source. Enzyme cofactor. May help in management of patients with hypoproliferative anemia. [dihydrate]: soluble in H_2O (11.2 g/100 ml). *Hoffmann-LaRoche Inc.; Takeda Chem. Ind. Ltd.*

Hematopoietics

8368 Cilmostim
148637-05-2
Macrophage colony-stimulating factor.
Macstim. Hematopoietic. *Genetics Institute Inc.*

8369 Erythropoietin
11096-26-7 3729 234-317-2
Erythropoiesis stimulating factor.
hempoietine; ESF; Ep; Epo. Hematopoietic. A glycoprotein hormone which stimulates red blood cell-formation. Has two components, Epo-α and Epo-β, both produced by recombinant technology. Epo-α (Epoetin alfa; Epoade; Epogen; Eprex; Erypo; Espo; Procrit), has 165 residues, molecular weight 30,400. Epo-β (Epoetin beta; Epogin; Marogen; Recormon), has 165 residues, molecular weight 30,000. They differ only in their carbohydrate moieties.

8370 Granulocyte Colony-Stimulating Factor
143011-72-7 4558
Hematopoietic growth factor.
CSF-β; G-CSF; GM-DF; MGI-2; pluripoietin. Hematopoietic stimulant; antineutropenic. Stimulates development of neutrophils. Enhances the functional activities of the mature end-cell. A glycoprotein.

8371 Granulocyte-Macrophage Colony Stimulating Factor
83869-56-1 4559
Colony-stimulating factor 2.
CSF-2; CSFα; GM-CSF; NIF-T. Hematopoietic stimulant; antineutropenic. Promotes the proliferation and development of early erythroid, megakayocytic and eosinophilic progenitor cells. Inhibits neutrophil migration. Enhances activities of mature end-cells.

8372 Interleukin-3
5021
Multipotent colony-stimulating factor.
IL-3; multi-CSF. Hematopoietic. Growth factor that promotes proliferation of multipotent hematopoietic stem cells and progenitor cells of megakariocyte, granulocyte-macrophage, erythroid, eosinophil, basophil, and mast-cell lineages.

8373 Macrophage Colony-Stimulating Factor
81627-83-0 5678
Colony-stimulating factor 1.
M-CSF; CSF-1. Immunomodulator; hematopoietic. Heavily glycosylated homodimer isolated from human urine. Growth factor that stimulates development of progenitor cells to monocytes or macrophages. Potentiates phagocytic activity and moncyte-mediated tumor cell cytotoxicity.

8374 Mirimostim
121547-04-4 5678
$C_{1058}H_{1651}N_{277}O_{341}S_{14}$
1-214-Colony-stimulating factor 1 (human clone p3ACSF-69 protein moiety reduced) homodimer.
Costilate; Leukoprol. Immunomodulator; hematopoietic. Non-glycosylated protein.

8375 Sodium Phenylbutyrate
1716-12-7
$C_{10}H_{11}NaO_2$
Sodium 4-phenylbutyrate.
PBA; Buphenyl; TriButyrate. A short-chain aromatic fatty acid that inhibits cell proliferation and induces apoptosis. Also used in treatment of ornithine transcarbamylase deficiency as a vehicle for waste nitrogen excretion in patients with inborn errors of urea synthesis. Antineoplastic, hematopoietic and antihyperammonemic. Used to reduce levels of ammonia in the blood.

Hemolytics

8376 Phenylhydrazine
100-63-0 7447 202-873-5
$C_6H_8N_2$
Hydrazinobenzene.
Hemolytic agent. mp = 19.5°; bp = 243.5°, bp_{100} = 173.5°, bp_{40} = 148.2°, bp_{20} = 131.5°, bp_{10} = 115.8°, $bp_{1.0}$ = 71.8°; soluble in EtOH, Et_2O, $CHCl_3$, C_6H_6; sparingly soluble in H_2O, petroleum ether. *Hoechst AG.*

8377 Phenylhydrazine Hydrochloride
59-88-1 7447
$C_6H_9ClN_2$
Hydrazinobenzene hydrochloride.
Hemolytic agent. mp = 243-246°; freely soluble in H_2O, soluble in EtOH, insoluble in Et_2O. *Hoechst AG.*

Hemostatics

8378 Adrenalone
99-45-6 170 202-756-9
$C_9H_{11}NO_3$
3',4'-Dihydroxy-2-(methylamino)acetophenone.
Kephrine; Stryphnone; Stypnone. Adrenergic (ophthalmic). Hemostatic. An α-adrenergic agonist. mp = 235-236° (dec); sparingly soluble in H_2O, EtOH, Et_2O. *Bayer Corp. Pharm. Div.; Hoechst AG.*

8379 Adrenalone Hydrochloride
62-13-5 170 200-525-7
$C_9H_{12}ClNO_3$
3',4'-Dihydroxy-2-(methylamino)acetophenone hydrochloride.
Styphnonasal. Adrenergic (ophthalmic); hemostatic. An α-adrenergic agonist. mp = 243°; soluble in H_2O, EtOH; insoluble in Et_2O. *Bayer Corp. Pharm. Div.; Hoechst AG.*

8380 Adrenochrome
54-06-8 171 200-192-8
$C_9H_9NO_3$
2,3-Dihydro-3-hydroxy-1-methyl-1H-indole-5,6-dione.
Hemostatic. mp = 115-120° (dec); λ_m 220, 300, 485 nm (log ε 4.33, 4.01, 3.64 H_2O); freely soluble in H_2O, fairly soluble in EtOH, insoluble in C_6H_6, Et_2O.

8381 Adrenochrome Monosemicarbazone
69-81-8 171 200-717-0
$C_{10}H_{12}N_4O_3$
2,3-Dihydro-3-hydroxy-1-methyl-1H-indole-5,6-dione-5-semicarbazone.
carbazochrome; Adrenoxyl; Cromadrenal; Cromosil. Hemostatic. *Byk Gulden Lomberg GmbH.*

8382 Adrenochrome Oxime Sesquihydrate
6055-73-8 171
$C_9H_{10}N_2O_3.1.5H_2O$
2,3-Dihydro-3-hydroxy-1-methyl-1H-indole-5,6-dione-5-oxime sesquihydrate.
Hemostatic. mp = 278°; more stable than adrenochrome.

8383 Adrenochrome Thiosemicarbazone
113185-69-6 171
$C_{10}H_{12}N_4O_2S$
2,3-Dihydro-3-hydroxy-1-methyl-1H-indole-5,6-dione 5-thiosemicarbazone.
Hemostatic. mp = 215-220°. *International Hormones.*

8384 Algin
9005-38-3 240
Sodium polymannuronate.
sodium alginate; Alto; Alman; Alloid; Allose; Kelgin; Protanal. A gelling polysaccharide derived from giant brown seaweed. Hemostatic. Soluble in H_2O, insoluble in organic solvents.

8385 Alginic Acid
9005-32-7 241 232-680-1
Polymannuronic acid.
Norgine. Very slightly soluble in H_2O.

8386 Alginic Acid Calcium Salt
9005-35-0 241
Calcium polymannuronate.
Sorbsan. Hemostatic. Forms gelatinous precipitate in H_2O.

8387 Alginic Acid Potassium Salt
9005-36-1 241
Potassium polymannuronate.
Stercofuge.

8388 ε-Aminocaproic Acid
60-32-2 451 200-469-3
$C_6H_{13}NO_2$
6-Aminohexanoic acid.
Amicar; CL-10304; CY-116; EACA; 177 J.D.;NSC-26154; Ipsilon; Hemocaprol; Caprocid; Capramol; Afibrin; Epsikapron; Hepin. Hemostatic. Antifbrinolytic. mp = 204-206°; freely soluble in H_2O, sparingly soluble in MeOH, insoluble in EtOH; LD_{50} (rat ip) → 7000 mg/kg, (rat iv) → 3300 mg/kg; [hydrobromide]: mp = 105°; [hydrochloride]: mp = 128-129°. *Elkins-Sinn; Immunex Corp.*

8389 Aminochromes
456
A family of 2,3-dihydroindole-5,6-quinones. Hemostatic. Obtained by oxidative cyclization of catecholamines.

8390 Batroxobin
9039-61-6 1038 232-918-4
Bothrops atrox serine proteinase.
bothrops venom proteinase; Botropase; Defibrase. A thrombin-like enzyme obtained from the venom of *Bothrops atrox*, a South American pit viper. Hemostatic at low doses, anticoagulant at high doses. Soluble in saline; nearly insoluble in H_20. *Pentapharm.*

8391 Carbazochrome Salicylate
13051-01-9 1832 235-927-1
$C_{18}H_{17}N_4NaO_5$
3-Hydroxy-1-methyl-5,6-indolinedione semicarbazone compound with sodium salicylate.
Amicar; CL-10304; CY-116; EACA; 177 J.D.; NSC-26154; Ipsilon; Hemocaprol; Caprocid; Capramol; Afibrin; Epsikapron; Hepin; Adenogen; Adrenosem; Adrestat-F; Statimo. Hemostatic. mp = 196-197.5° (dec); insoluble in Et_2O, $CHCl_3$, soluble in H_2O (0.061 g/100 ml). *Tanabe Seiyaku Co. Ltd.*

8392 Carbazochrome Sodium Sulfonate
51460-26-5 1833 257-217-0
$C_{10}H_{11}N_4NaO_5S$
Sodium 5,6-dihydro-1-methyl-5,6-dioxo-3-indoline-5-semicarbazone sulfonate.
AC-17; Adenaron; Adona; Carbazon; Donaseven; Emex; Odanon; Tazin. Hemostatic. mp = 227-228° (dec); soluble in H_2O; [free acid ($C_{10}H_{12}N_4O_5S$)]: mp = 195°. *Tanabe Seiyaku Co. Ltd.*

8393 Cephalins
2022
Phosphatidylethanolamine.
kephalins. Hemostatic. Also used as a clinical reagent in liver function testing. Glycerol with fatty acid esters at C_2 and C_2 and an ethanolamine phosphate at C_1. Insoluble in H_2O, Me_2CO, freely in $CHCl_3$, Et_2O, partially soluble in EtOH. *U.S. Government.*

8394 Cotarnine
82-54-2 2618 201-429-8
$C_{12}H_{15}NO_4$
5,6,7,8-Tetrahydro-4-methoxy-6-methyl-1,3-dioxolo[4,5-g]isoquinolin-5-ol.
Hemostatic. dec 132-133°; slightly soluble in H_2O, soluble in organic solvents.

8395 Cotarnine Chloride
10018-19-6 2618 233-012-1
$C_{12}H_{14}ClNO_3$
7,8-Dihydro-4-methoxy-6-methyl-1,3-dioxolo[4,5-g]-isoquinolinium chloride.
cotarninium chloride; Stypticin. Hemostatic. Soluble in H_2O (100 g/100 ml), EtOH (25 g/100 ml).

8396 Cotarnine Hydrochloride
36647-02-6 2618
$C_{12}H_{16}ClNO_4$
5,6,7,8-Tetrahydro-4-methoxy-6-methyl-1,3-dioxolo-[4,5-g]isoquinolin-5-ol hydrochloride.
Secalysat. Hemostatic.

8397 Cotarnine Phthalate
6190-36-9 2618 228-234-0
$C_{32}H_{32}N_2O_{10}$
5,6,7,8-Tetrahydro-4-methoxy-6-methyl-1,3-dioxolo-[4,5-g]isoquinolin-5-ol phthalate.
Styptol. Hemostatic.

8398 Deferiprone
30652-11-0
$C_7H_9NO_2$
3-Hydroxy-1,2-dimethyl-4(1H)-pyridone.
L1. Chelating agent. Used to treat thalassemia.

8399 Ellagic Acid
476-66-4 3588 207-508-3
$C_{14}H_6O_8$
2,3,7,8-Tetrahydroxy[1]benzopyrano[5,4,3-cde]-[1]benzopyran-5,10-dione.
benzoaric acid. Hemostatic. mp > 360°; λ_m 366, 255 nm (log ε 3.93, 4.60, EtOH); soluble in C_5H_5N, slightly soluble in H_2O, EtOH, insoluble in Et_2O; [tetracetate ($C_{22}H_{14}O_{12}$)]: mp = 340°.

8400 Ethamsylate
2624-44-4 3766 220-090-7
$C_{10}H_{17}NO_5S$
2,5-Dihydroxybenzenesulfonic acid compound with diethylamine (1:1).
MD-141; E-141; cyclonamine; etamsylate; Aglumin; Altodor; Biosinon; Dicynene; Dicynone. Hemostatic. mp = 125°; LD_{50} (mus iv) = 800 mg/kg, (rat iv) = 1350 mg/kg. *Esteve Group; Labs. O.M.*

8401 Factor IX
9001-28-9 3964 232-594-4
Blood coagulation factor IX.
Christmas factor; PTC; plasma thromboplastin component; autoprothrombin II. A glycoprotein that participates in the middle phases of blood coagulation. Hemostatic.

8402 Factor VIII
9001-27-8 3963 232-593-9
Blood coagulation factor VIII.
antihemophilic globulin; AHG; AHF; Factorate; Hemofil; Humafac; Koate-HP; Monoclate-P; Nordiocto; Profilate. A preparation of human blood containing coagulating factor VIII. Forms thromboplastin by reaction with Factor X. Hemostatic.

8403 Factor XIII
9013-56-3 3968
Blood coagulation factor XIII.
fibrin-stabilizing factor; FSF; fibrinase; Laki-Lorand factor; LLF; Fibrogamin. Enzyme precursor. When activated by thrombin/Ca^{2+}, converts soluble fibrin gel to a tough insoluble clot. Hemostatic. Antihemorrhagic. α_m = 280 nm ($E^{1\%}$ 13.8); does not dialyse out of plasma.

8404 Fibrinogen
9001-32-5 4116 232-598-6
Factor 1; Parenogen. A plasma glycoprotein essential to the clotting of blood. Hemostatic. Coagulant (cloting factor). Molecular weight ≅ 400,000; sparingly soluble in H_20; aqueous solutions are viscous. *Marion Merrell Dow Inc.*

8405 1,2-Naphthoquinone
524-42-5 6481 208-360-2
$C_{10}H_6O_2$
1,2-Naphthalenedione.
β-naphthoquinone. The 2-semicarbazone is used as a hemostatic. Hemostatic. mp = 145-147° (dec); λ_m 250, 340, 405 nm (log ε 4.35, 3.40, 3.40 EtOH); soluble in EtOH, C_6H_6, Et_2O; insoluble in H_2O. *Res. Corp.*

8406 1,2-Naphthoquinone-2-semicarbazone
31853-38-0 6481
$C_{11}H_9N_3O_2$
1,2-Naphthalenedione-2-semicarbazone.
naftazone; Haemostop Injection; Mediaven; Karbinon. Hemostatic. *Res. Corp.*

8407 1-Napthylamine-4-Sulfonic Acid
84-86-6 6490 201-567-9
$C_{10}H_9NO_3S$
4-Amino-1-naphthalenesulfonic acid.
The sodium salt is used as a hemostatic. [sesquihydrate]: d_4^{25}= 1.673; soluble in H_2O (0.029 g/100 ml at 10°, 0.031 g/100 ml at 20°, 0.059 at 50°, 0.23 g/100 ml at 100°); sparingly soluble in EtOH, ET_2O; insoluble in AcOH; [sodium salt tetrahydrate ($C_{10}H_8NNaO_3S.4H_20$; naphthionine, 101-E)]: freely soluble in H_2O, EtOH; insoluble in Et_2O.

8408 Oxamarin
15301-80-1 7047
$C_{22}H_{34}N_2O_4$
6,7-Bis[2-(diethylamino)ethoxy]-4-methylcoumarin.
Hemostatic. $bp_{0.5}$ = 195°. *Maggioni Farmaceutici S.p.A.*

8409 Oxamarin Dihydrochloride
6830-17-7 7047
$C_{22}H_{36}Cl_2N_2O_3$
6,7-Bis[2-(diethylamino)ethoxy]-4-methylcoumarin dihydrochloride.
Idro-P_3; M.G. 652. Hemostatic. mp = 224-226°, 234-246°. *Maggioni Farmaceutici S.p.A.*

8410 Oxidized Cellulose
7073
Collodion (4 g pyroxylin in 100 ml of Et_2O/EtOH (3:1), mixed with 2% camphor and 3% castor oil and about 18% tannic acid.
Absorbable cellulose; cellulosic acid; polyanhydroglucuronic acid; Oxycel; Hemo-Pak. Hemostatic. Prepared by oxidation of cellulose with N_2O_4.

8411 Styptic Collodion
2547
Mixture of flexible collodion with 18% tannic acid w/w. Hemostatic.

8412 Sulmarin
29334-07-4 9158 249-567-8
$C_{10}H_8O_{10}S_2$
6,7-Dihydroxy-4-methylcoumarin bis(hydrogen sulfate).
Idro P_2; M.G. 143. Hemostatic. [disodium salt trihydrate ($C_{10}H_6Na_2O_{10}S_2.3H_2O$)]: mp = 252-253° (dec); λ_m = 304 nm (pH 11.85). *Maggioni Farmaceutici S.p.A.*

8413 Thrombin
9002-04-4 9525 232-648-7
Blood Coagulation Factor IIa.
Thrombostat; fibrinogenase; [Standardized preparation of bovine thrombin]: topical thrombin; Thrombinar; Thrombogen; Thrombostat. Hemostatic (local). Key enzyme in the coagulation cascade. Converts fibrinogen to fibrin; activates factor XIII, which cross-links and stabilizes the fibrin polymer. *Parke-Davis.*

8414 Thromboplastin
9035-58-9 9527 232-903-2
Blood Coagulation Factor III.
cytozyme; thrombokinase; tissue factor; zymoplastic substance; trombostop; Tachostyptan. A membrane glycoprotein which, in the presence of Ca^{2+}, initiates coagulation by augmenting the proteolytic attack of Factor VII on Factors IX and X. Hemostatic.

8415 Tolonium Chloride
92-31-9 9658 202-146-2
$C_{15}H_{16}ClN_3S$
3-Amino-7-dimethylamino-2-methylphenazathionium.
Blutene; C.I. Basic Blue 17; C.I. 52040; Klot; Tolazul. Hemostatic. Soluble in H_2O (3.82 g/100 ml), EtOH (0.57 g/100 ml); λ_m = 640.4 nm (H_2O); LD_{50} (mus iv) = 27.56 mg/kg, (rat iv) = 28.93 mg/kg, (rbt iv) = 13.44 mg/kg. *Abbott Labs Inc.*

8416 Tranexamic Acid
1197-18-8 9704 214-818-2
$C_8H_{15}NO_2$
trans-4-(Aminomethyl)cyclohexanecarboxylic acid.
Cyklokapron; Trans-AMCHA; CL-65336. Hemostatic. Antifibrinolytic. mp = 386-392° (dec); soluble in H_2O (16.6 g/100 ml); slightly soluble in Et_2O, EtOH; insoluble in other organic solvents; LD_{50} (mus iv) = 1500 mg/kg, (rat iv) = 1200 mg/kg. *Daiichi Pharm. Corp.; Mitsubishi Chem. Corp.; Pharmacia & Upjohn.*

8417 cis-Tranexamic Acid
1197-17-7 9704
$C_8H_{15}NO_2$
cis-4-(Aminomethyl)cyclohexanecarboxylic acid.
Cis-AMCHA. Hemostatic. Antifibrinolytic. mp = 236-238° (dec). *Pharmacia & Upjohn.*

8418 Vasopressin, Arginine form
113-79-1 10073 204-035-4
$C_{46}H_{65}N_{15}O_{12}S_2$
8-L-Arginine-vasopressin.
arginine vasopressin; argipressin; rindervasopressin; antidiuretic hormone; β-hypophamine; Leiormone; Tonephin; Vasophysin. Antidiuretic; vasopressor hormone. Hemostatic. *Parke-Davis.*

8419 Vasopressin, Lysine form
50-57-7 5661 200-050-5
$C_{46}H_{65}N_{13}O_{12}S_2$
8-L-Lysine-vasopressin.
Diapid; Lypressin; Phe^3-Lys^8-oxytocin; Pitressin; Postaction; Schweine-Vasopressin; Syntopressin. Antidiuretic; vasopressor hormone. Hemostatic. See *Lypressin. Parke-Davis.*

Hepatoprotectants

8420 S-Adenosylmethionine
29908-03-0 155 249-946-8
$C_{15}H_{22}N_6O_5S$
(3S)-5'-[(3-Amino-3-carboxypropyl)methylsulfonio]-5'-deoxyadenosine inner salt.
active methionine; ademetionine; AdoMet; SAMe; Donamet; S. Amet; Gumbaral [as disulfate ditosylate]; Samyr [as disulfate ditosylate]. Anti-inflammatory; hepatoprotectant. Endogenous methyl donor involved in ezymatic transmethylation reactions. Used in treatment of chronic liver disease. [chloride ($C_{15}H_{23}ClN_6O_5S$)]: λ_m 260 nm (H_2O); $[\alpha]_D^{25}$ = 32°, (c = 3.3 H_2O); [disulfate ditosylate (Gumbaral, Samyr, $C_{29}H_{42}N_6O_{19}S_5$)]: LD_{50} (mus iv) = 560 mg/kg, (mus ip) = 2500 mg/kg, (mus orl) > 6000 mg/kg. *Ajinomoto Co. Inc.; Merck & Co.Inc.; Yamaza Shoyu.*

8421 Ataprost
83997-19-7
$C_{21}H_{32}O_4$
(+)-(2E,3aS,4R,5R,6aS)-4-[(1E,3S)-3-Cyclopentyl-3-hydroxypropenyl]-3,3a,4,5,6,6a-hexahydro-5-hydroxy-$\Delta^{2(1H),\delta}$-pentalenevaleric acid.
Prostaglandin with hepatoprotectant properties.

8422 Betaine
107-43-7 1225 203-490-6
$C_5H_{11}NO_2$
1-Carboxy-N,N,N-trimethylmethanaminium inner salt.
glycine betaine; glycocoll betaine; lycine; oxyneurine. Hepatoprotectant. Dec ≅ 310°; soluble in H_2O (160 g/100 ml), MeOH (55 g/100 ml), EtOH (8.7 g/100 ml); sparingly soluble in Et_2O. *Sterling Winthrop Inc.*

8423 Betaine Hydrochloride
590-46-5 1225 209-683-1
$C_5H_{12}ClNO_2$
1-Carboxy-N,N,N-trimethylmethanaminium chloride.
pluchine; Acidol. Hepatoprotectant. mp = 227-228° (dec), 232° (dec); soluble in H_2O (64.7 g/100 ml), EtOH (5.0 g/100 ml); insoluble in $CHCl_3$, Et_2O. *Sterling Winthrop Inc.*

8424 Betaine Sodium Aspartate
93227-64-6 1225
$C_9H_{17}N_2NaO_6$
1-Carboxy-N,N,N-trimethylmethanaminium sodium aspartate.

Somatyl. Hepatoprotectant. mp = 160-170° (dec);soluble in H_2O, insoluble in Me_2CO. *Sterling Winthrop Inc.*

8425 Catechin
154-23-4 1950 205-825-1
$C_{15}H_{14}O_6$
(2R-trans)-2-(3,4-Dihydroxyphenyl)-3,4-dihydro-2H-1-benzopyran-3,5,7-triol.
catechol; 3,3',4',5,7-flavanpentol; catechinic acid; catechuic acid; dexcyanidanol; cyanidol; Catergen. Antidiarrheal. Hepatoprotectant. mp = 93-96°, 175-177°; $[\alpha]_D^{18}$ = 16 to 18.4°.

8426 dl-Catechin
7295-85-4 1950 230-731-2
$C_{15}H_{14}O_6$
(±)-(2R-trans)-2-(3,4-Dihydroxyphenyl)-3,4-dihydro-2H-1-benzopyran-3,5,7-triol.
3,3',4',5,7-flavanpentol; dl-catechol; catechinic acid; catechuic acid; cyanidol; Catergen. Antidiarrheal. Hepatoprotectant. mp = 212-216°; slightly soluble in cold H_2O, Et_2O; soluble in hot H_2O, EtOH, Me_2CO, AcOH; insoluble in C_6H_6, $CHCl_3$, petroleum ether.

8427 l-Catechin
18829-70-4 1950 242-611-7
$C_{15}H_{14}O_6$
l-(2R-trans)-2-(3,4-Dihydroxyphenyl)-3,4-dihydro-2H-1-benzopyran-3,5,7-triol.
catechol; l-catechol; 3,3',4',5,7-flavanpentol; catechinic acid; catechuic acid; (-)-cyanidanol-3; cyanidol; Catergen. Antidiarrheal. Hepatoprotectant. mp = 93-96°, 175-177°; $[\alpha]_D$ = -16.8°.

8428 Cinametic acid
35703-32-3 2341
$C_{12}H_{14}O_5$
4-(2-Hydroxyethoxy)-3-methoxycinnamic acid.
Transoddi. Hepatobiliary regulator. Used in preoperative management of peptic ulcer patients. *Anphar.*

8429 Citiolone
1195-16-0 2381 214-793-8
$C_6H_9NO_2S$
2-Acetamido-4-mercaptobutyric acid σ-thiolactone.
AHCTL; BO-714; Citiolase; Thioxidrene. Hepatoprotectant. mp = 111.5-112.5°; λ_m 238 nm (ε 4400); LD_{50} (mus iv) = 1200 mg/kg, (rat ip) = 1950 mg/kg. *Degussa-Hüls Corp.; Sumitomo Pharm. Co. Ltd.*

8430 Cynarine
1884-24-8
$C_{25}H_{24}O_{12}$
3,4-Dihydroxycinnamic acid 1-carboxy-4,5-dihydroxy-1,3-cyclohexylene ester.
Hepatoprotectant.

8431 Malotilate
59937-28-9 5751 261-987-3
$C_{12}H_{16}O_4S_2$
1,3-Dithiol-2-ylidenepropanedioic acid bis(1-methylethyl) ester.
NKK-105; Hepation; Kantec. Hepatoprotectant. mp = 60.5°; soluble in C_6H_6, cyclohexane, C_6H_{14}, Et_2O. *Nihon Nohyaku Co. Ltd.*

8432 Methionine
63-68-3 6053 200-562-9
$C_5H_{11}NO_2S$
L-2-amino-4-(methylthio)butyric acid.
L-methionine; Met; M; Acimethin. Hepatoprotectant; antidote for acetaminophen poisoning. Urinary acidifier. An essential amino acid. mp = 280-282° (dec); $[\alpha]_D^{25}$= -8.11° (c = 0.8), $[\alpha]_D^{20}$= 23.40° (c = 5.0 3N HCl).

8433 D-Methionine
348-67-4 6053 206-483-6
$C_5H_{11}NO_2S$
D-2-Amino-4-(methylthio)butyric acid.
D-methionine. Hepatoprotectant; antidote for acetaminophen poisoning. Urinary acidifier. $[\alpha]_d^{25}$= 8.12° (c = 0.8), $[\alpha]_D^{25}$= -21.18° (c = 0.8 0.2N HCl).

8434 DL-Methionine
59-51-8 6053 200-432-1
$C_5H_{11}NO_2S$
DL-2-Amino-4-(methylthio)butyric acid.
racemethionine; DL-methionine; Amurex; Banthionine; Dyprin; Lobamine; Metione; Pedameth; Urimeth. Hepatoprotectant; antidote for acetaminophen poisoning. Urinary acidifier. mp = 281° (dec); d = 1.340; soluble in H_2O (1.82 g/100 ml at 0°, 3.38 g/100 ml at 25°, 6.07 g/100 ml at 60°, 10.52 g/100 ml at 75°, 17.60 g/100 ml at 100°); slightly soluble in EtOH; insoluble in Et_2O.

8435 Orazamide
2574-78-9 6993 219-923-7
$C_9H_{10}N_6O_5$
1,2,3,6-Tetrahydro-2,6-dioxo-4-pyrimidinecarboxylic acid with 5-amino-1H-imidazole-4-carboxamide (1:1).
5-aminoimidazole-4-carboxamide orotate; AICA orotate; Aicamin; Aicorat. Hepatoprotectant. mp = 284-285° (dec); LD_{50} (mus ip) = 600 mg/kg, (mus orl) > 4000 mg/kg. *Fujisawa Pharm. USA Inc.*

8436 Orazamide Dihydrate
60104-30-5 6993
$C_9H_{14}N_6O_7$
5-Aminoimidazole-4-carboxamide orotate.
AICA orotate; Aicamin; Aicorat; 170. Used in treatment of chronic hepatitis. Hepatoprotectant. LD_{50} (mus ip) = 0.6 g/kg, (mus orl) = 4.0 g/kg; [dihydrate]: dec 284-285°.

8437 Phosphorylcholine
107-73-3 7519 203-516-6
$C_5H_{15}ClNO_4P$
N,N,N-Trimethyl-2-(phosphonooxy)ethanaminium chloride.
choline chloride dihydrogen phosphate. Hepatoprotectant.

8438 Phosphorylcholine Barium Salt
6484-71-5 7519
$C_5H_{13}BaClNO_4P$
N,N,N-Trimethyl-2-(phosphonooxy)ethanaminium chloride barium salt.

Hepatoprotectant. Freely soluble in H_2O; insoluble in EtOH, Et_2O.

8439 Phosphorylcholine Calcium Salt
4826-71-5 7519 225-403-0
$C_5H_{13}CaClNO_4P$
N,N,N-Trimethyl-2-(phosphonooxy)ethanaminium chloride calcium salt.
Colifos; Epafosforil; Fosfocolina; Isocolin. Hepatoprotectant.

8440 Phosphorylcholine Magnesium Salt
17032-39-2 7519
$C_5H_{13}ClMgNO_4P$
N,N,N-Trimethyl-2-(phosphonooxy)ethanaminium chloride magnesium salt.
Heparexine. Hepatoprotectant.

8441 Protoporphyrin IX
553-12-8 8084 209-033-7
$C_{34}H_{34}N_4O_4$
7,12-Diethenyl-3,8,13,17-tetramethyl-21H,23H-porphine-2,18-dipropanoic acid.
ooporphyrin; Kammerer's porphyrin. Hepatoprotectant. The natural precursor to blood and plant pigments. Chelates with ferrous iron to form heme and with ferric iron to form hematin. Yellow crystals; λ_m 602.4, 582.2, 557.2 nm (25% HCl); freely soluble in $CHCl_3$, AcOH, EtOH/HCl, Et_2O/AcOH.

8442 Protoporphyrin IX Dimethyl Ester
5522-66-7 8084 226-870-3
$C_{36}H_{38}N_4O_4$
Dimethyl 7,12-diethenyl-3,8,13,17-tetramethyl-21H,23H-porphine-2,18-dipropanoate.
Hepatoprotectant. mp = 228-230°; λ_m 601, 556, 409 nm (25% HCl); soluble in $CHCl_3$, slightly soluble in MeOH, insoluble in $NaHCO_3$ solution.

8443 Protoporphyrin IX Disodium Salt
50865-01-5 8084 256-815-9
$C_{34}H_{32}N_4Na_2O_4$
Sodium 7,12-diethenyl-3,8,13,17-tetramethyl-21H,23H-porphine-2,18-dipropanoate.
Depocolin-S. Hepatoprotectant.

8444 Silibinin
22888-70-6 8680 245-302-5
$C_{25}H_{22}O_{10}$
[2R-[2α,3β,6(2R*,3R*)]]-2-[2,3-Dihydro-3-(4-hydroxy-3-methoxyphenyl)-2-(hydroxymethyl)-1,4-benzodioxin-6-yl]-2,3-dihydro-3,5,7-trihydroxy-4H-1-benzopyran-4-one.
silybin; silybum substance E_6; silymarin I. Flavonolignan from the milk thistle *Silybum marianum. Inhibits lipid peroxidation. Stimulates cellular metabolic activity. See* silymarin. Hepatoprotectant. mp = 158°; mp [monohydrate] = 167°; dec 180°; $[\alpha]_D^{20}$ = +11° (c = 0.25 in Me_2CO, EtOH); λ_m = 288 nm (log ε 4.33 in MeOH); soluble in Me_2CO, EtOAc, MeOH, EtOH; sparingly soluble in $CHCl_3$; nearly insoluble in H_2O.

8445 Silicristin
33889-69-9 251-720-9
$C_{25}H_{22}O_{10}$
2-[2,3-Dihydro-7-hydroxy-2-(4-hydroxy-3-methoxyphenyl)-3-(hydroxymethyl)-5-benzofuranyl]-3,5,7-trihydroxy-4-chromanone.
silycristin; silichristin. Flavonolignan from the milk thistle *Silybum marianum. Inhibits lipid peroxidation. Stimulates cellular metabolic activity. See* silymarin. Hepatoprotectant.

8446 Silidianin
29782-68-1 249-848-5
$C_{25}H_{22}O_{10}$
(+)-2,3α,3aα,7a-Tetrahydro-7aα-hydroxy-8-(4-hydroxy-3-methoxyphenyl)-4-(3α,5,7-trihydroxy-4-oxo-2β-chromanyl)-3,6-methanobenzofuran-7(6αH)-one.
silydianin. Flavonolignan from the milk thistle *Silybum marianum.* Inhibits lipid peroxidation. Stimulates cellular metabolic activity. See silymarin. Hepatoprotectant.

8447 Silymarin Group
65666-07-1 8680
$C_{25}H_{22}O_{10}$
Silybin.
silymarin I; Apihepar; Laragon; Legalon; Pluropon; Silarin; Silepar; Silirex; Silliver; Silmar. Antihepatotoxic principle isolated from the seeds of the milk thistle *Ailybum marianum* (L.) Gaertn. A mixtures of three flavonolignans, silybin; silidianin and silicristin. Hepatoprotectant. mp = 158°; dec 180°; $[\alpha]_D^{20}$ = 11° (c = 0.25 Me_2CO/EtOH); λ_m 288 nm (log ε 4.33 MeOH); insoluble in H_2O; soluble in EtOAc, Me_2CO, MeOH, EtOH; sparingly soluble in $CHCl_3$.

8448 Thioctic Acid
62-46-4 9462 200-534-6
$C_8H_{14}O_2S_2$
1,2-Dithiolane-3-pentanoic acid.
pyruvate oxidation factor, α-lipoic acid; Biletan; Thioctacid; Thioctan; Tioctan. Hepatoprotectant. mp = 60-61°; bp = 160-165°; insoluble in H_2O; soluble in oils, fats; forms an H_2O-soluble sodium salt. *Res. Corp.; Yamanouchi U.S.A. Inc.*

8449 Thioctic Acid Sodium Salt
2319-84-8 9462
$C_8H_{13}NaO_2S_2$
Sodium 1,2-dithiolane-3-pentanoate.
Hepatoprotectant. Soluble in H_2O. *Res. Corp.; Yamanouchi U.S.A. Inc.*

8450 d-Thioctic Acid
1200-22-2 9462
$C_8H_{14}O_2S_2$
d-1,2-Dithiolane-3-pentanoic acid.
pyruvate oxidation factor, α-lipoic acid; Biletan; Thioctacid; Thioctan; Tioctan. Hepatoprotectant. mp = 46-48°; $[\alpha]_D^{23}$ = 104° (c = 0.88 C_6H_6); λ_m 333 nm (ε 150 MeOH); insoluble in H_2O; soluble in oils, fats. *Res. Corp.; Yamanouchi U.S.A. Inc.*

8451 dl-Thioctic Acid
1077-28-7 9462 214-071-2
$C_8H_{14}O_2S_2$
dl-1,2-Dithiolane-3-pentanoic acid.
pyruvate oxidation factor, α-lipoic acid; Biletan; Thioctacid; Thioctan; Tioctan. Hepatoprotectant. mp = 60-61°; bp = 160-165°; insoluble in H_2O; soluble in oils, fats; forms an H_2O-soluble sodium salt. *Res. Corp.; Yamanouchi U.S.A. Inc.*

8452 l-Thioctic Acid
1077-27-6 9462
$C_8H_{14}O_2S_2$
l-1,2-Dithiolane-3-pentanoic acid.
pyruvate oxidation factor, α-lipoic acid; Biletan; Thioctacid; Thioctan; Tioctan. Hepatoprotectant. mp = 45-47.5°; $[\alpha]_D^{23}$ = -113° (c = 1.88 C_6H_6); λ_m = 330 nm (ε 140 MeOH). *Res. Corp.; Yamanouchi U.S.A. Inc.*

8453 Timonacic
444-27-9 9586 207-146-6
$C_4H_7NO_2S$
4-Thiazolidinecarboxylic acid.
ATC; norgamen; thioproline; NSC-25855; Detoxepa; Hepalidine; Heparegen; Tiazolidin. Hepatoprotectant. [dl-form]: mp = 195°; LD_{50} (mus orl) = 400 mg/kg; [l-form]: mp = 196-197° (dec); insoluble in EtOH; readily soluble in hot H_2O, acid, alkali. *Sogespar S.A.*

8454 Tiopronin
1953-02-2 9597 217-778-4
$C_5H_9NO_3S$
N-(2-Mercaptopropionyl)glycine.
N-(2-mercaptopropionyl)glycine; α-mercaptopropionylglycine; Acadione; Capen; Epatiol; Mucolysin; Thiola; Thiosol. A mucolytic; hepatoprotectant. Also used as an antidote for heavy metal poisoning. mp = 95-97°; LD_{50} (mus iv) = 2100 mg/kg. *C.M. Ind.; Santen Pharm. Co. Ltd.*

Histamine H_2 Receptor Antagonists

8455 Cimetidine
51481-61-9 2337 257-232-2
$C_{10}H_{16}N_6S$
2-Cyano-1-methyl-3-[2-[[(5-methylimidazol-4-yl)methyl]thio]ethyl]guanidine.
SKF-92334; Acibilin; Acinil; Cimal; Cimetag; Cimetum; Edalene; Dyspamet; Eureceptor; Gastromet; Peptol; Tagamet; Tametin; Tratul; Ulcedin; Ulcedine; Ulcerfen; Ulcimet; Ulcofalk; Ulcomedina; Ulcomet; Ulhys. Antiulcerative. Histamine H_2-receptor antagonist. mp = 141-143°; soluble in H_2O (11.4 g/l 37°); LD_{50} (mus orl) = 2600 mg/kg, (mus iv) = 150 mg/kg, (mus ip) = 470 mg/kg, (rat orl) = 5000 mg/kg, (rat iv) = 106 mg/kg, (rat ip) = 650 mg/kg. *Apothecon; Lemmon Co.; SmithKline Beecham Pharm.*

8456 Cimetidine Hydrochloride
70059-30-2 2337 274-297-2
$C_{10}H_{17}ClN_6S$
2-Cyano-1-methyl-3-[2-[[(5-methylimidazol-4-yl)methyl]thio]ethyl]guanidine monohydrochloride.
Tagamet Injection; Tagamet Liquid; Aciloc; Biomag; Brumetidina; Notul. Antiulcerative. Histamine H_2-receptor antagonist. *Apothecon; Lemmon Co.; SmithKline Beecham Pharm.*

8457 Donetidine
99248-32-5
$C_{20}H_{25}N_5O_3S$
5-[(1,2-Dihydro-2-oxo-4-pyridyl)methyl]-2-[[2-[[5-[(dimethylamino)methyl]furfuryl]thio]ethyl]amino]-4(1H)-pyrimidinone.
SK&F-93574. Histamine H_2-receptor Antagonist. *SmithKline Beecham Pharm.*

8458 Etintidine
69539-53-3
$C_{12}H_{16}N_6S$
2-Cyano-1-[2-[[(5-methylimidazol-4-yl)methyl]thio]ethyl]-3-(2-propynyl)guanidine.
Histamine H_2-receptor Antagonist. *Bristol-Myers Squibb Pharm. R&D.*

8459 Etintidine Hydrochloride
71807-56-2
$C_{12}H_{17}ClN_6S$
2-Cyano-1-[2-[[(5-methylimidazol-4-yl)methyl]thio]ethyl]-3-(2-propynyl)guanidine monohydrochloride.
BL-5641A. Histamine H_2-receptor Antagonist. *Bristol-Myers Squibb Pharm. R&D.*

8460 Famotidine
76824-35-6 3972
$C_8H_{15}N_7O_2S_3$
[1-Amino-3-[[[2-[(diaminomethylene)amino]-4-thiazolyl]metyhl]thio]propylidene]sulfamide.
MK-208; YM-11170; Amfamox; Dispromil; Famodil; Famodine; Famosan; Famoxal; Fanosin; Fibonel; Ganor; Gaster; Gastridin; Gastropen; Ifada; Lecedil; Motiax; Muclox; Nulcerin; Pepcid®; Pepcid AC; Pepcidina; Pepcidine; Pepdine; Pepdul; Peptan; Ulcetrax; Ulfamid; Ulfinol. Antiulcerative. Histamine H_2 receptor antagonist. Used for short term treatment of active duodenal ulcers. mp = 163-164°; soluble in DMF (80 g/100 ml), AcOH (50 g/100 ml), MeOH (0.3 g/100 ml), H_2O (0.1 g/100 ml); insoluble in EtOH, EtOAc, $CHCl_3$; LD_{50} (mus iv) = 244.4 mg/kg. *Johnson & Johnson-Merck Consumer Pharm.; Merck & Co.Inc.*

8461 Metiamide
34839-70-8
$C_9H_{16}N_4S_2$
1-Methyl-3-[2-[[(5-methylimidazol-4-yl)methyl]thio]ethyl]-2-thiourea.
SK&F-92058. Antiulcerative. Histamine H_2-receptor antagonist. *SmithKline Beecham Pharm.*

8462 Oxmetidine
72830-39-8
$C_{19}H_{21}N_5O_3S$
2-[[2-[[(5-Methylimidazol-4-yl)methyl]thio]ethyl]amino]-5-piperonyl-4-(1H)-pyrimidinone.
Antiulcerative. Histamine H_2-receptor antagonist. *SmithKline Beecham Pharm.*

8463 Oxmetidine Hydrochloride
63204-23-9
$C_{19}H_{23}Cl_2N_5O_3S$
2-[[2-[[(5-Methylimidazol-4-yl)methyl]thio]ethyl]amino]-5-piperonyl-4-(1H)-pyrimidinone dihydrochloride.
SK&F-92994-A_2. Antiulcerative. Histamine H_2-receptor antagonist. *SmithKline Beecham Pharm.*

8464 Oxmetidine Mesylate
84455-52-7
$C_{21}H_{29}N_5O_9S_3$
2-[[2-[[(5-Methylimidazol-4-yl)methyl]thio]ethyl]amino]-5-piperonyl-4-(1H)-pyrimidinone dimethanesulfonate.
SK&F-92994-J_2. Antiulcerative. Histamine H_2-receptor antagonist. *SmithKline Beecham Pharm.*

8465 Ranitidine
66357-35-5 8286 266-332-5
$C_{13}H_{22}N_4O_3S$
N-[2-[[5-[(Dimethylamino)methyl]furfuryl]thio]ethyl]-N'-methyl-2-nitro-1,1-ethenediamine.
Antiulcerative. Histamine H_2 receptor antagonist; inhibits gastric secretion. mp = 69-70°. *Glaxo Labs.*

8466 Ranitidine Bismuth Citrate
128345-62-0 8286
$C_{19}H_{27}BiN_4O_{10}S$
N-[2-[[5-[(Dimethylamino)methyl]furfuryl]thio]ethyl]-N'-methyl-2-nitro-1,1-ethenediamine compound with bismuth (3+) citrate (1:1).
Ranitidine bismutrex; GR-122311X; Pylorid. Histamine H_2-receptor Antagonist. Used as an antiulcerative. *Glaxo Wellcome plc.*

8467 Ranitidine Hydrochloride
66357-59-3 8286 266-333-0
$C_{13}H_{23}ClN_4O_3S$
N-[2-[[5-[(Dimethylamino)methyl]furfuryl]thio]ethyl]-N'-methyl-2-nitro-1,1-ethenediamine hydrochloride.
AH-19065; Azantac; Melfax; Noctone; Raniben; Ranidil; Raniplex; Sostril; Taural; Terposen; Trigger; Ulcex; Ultidine; Zantac; Zantic. Antiulcerative. Histamine H_2 receptor antagonist; inhibits gastric secretion. mp = 133-134°; soluble in H_2O, AcOH; less soluble in EtOH, MeOH; insoluble in organic solvents. *Glaxo Labs.*

8468 Sopromidine
79313-75-0
$C_{14}H_{23}N_7S$
(-)-1-[(R)-2-Imidazol-4-yl-1-methylethyl]-3-[2-[[5-methylimidazol-4-yl)methyl]thio]ethyl]guanidine.
Histamine H2 receptor agonist.

8469 Sufotidine
80343-63-1
$C_{20}H_{31}N_5O_3S$
1-[m-[3-[[1-Methyl-3-[(methylsulfonyl)methyl]-1H-1,2,4-triazol-5-yl]amino]propoxy]benzyl]-piperidine.
AH-25352X. Antiulcerative. Histamine H_2-receptor antagonist. *Glaxo Wellcome plc.*

8470 Tiotidine
69014-14-8
$C_{10}H_{16}N_8S_2$
2-Cyano-1-[2-[[[2-[(diaminomethylene)amino]-4-thiazolyl]methyl]thio]ethyl]-3-methyl-guanidine.
ICI-125211. Antiulcerative. Histamine H_2-receptor antagonist. *ICI.*

8471 Zaltidine Hydrochloride
90274-23-0
$C_8H_{12}Cl_2N_6S$
[4-(2-Methylimidazol-5-yl)-2-thiazolyl]guanidine dihydrochloride.
CP-57361-01. Antiulcerative. Histamine H_2-receptor antagonist. *Pfizer Inc.*

Immunomodulators

8472 Acemannan
110042-95-0 26
Polymanoacetate.
Carraklenz Wound & Skin Cleaner; Carrisyn; Snow & Sun Sports Gel; component of: Carraklenz Incontinence Skin Care Kit, Moisture Barrier Cream with Zinc, Moisture Guard, Skin Balm, Snow & Sun Sunburn Spray. Antiviral agent with immunomodulatory activity. From mucilage of *Aloe barbadensis* (aloe vera). A polydispersed, acetylated, linear mannan. MW: 1-2 x 10^6 daltons. *Carrington Labs Inc.*

8473 Amiprilose
56824-20-5 506
$C_{14}H_{27}NO_6$
3-O-[3-(Dimethylamino)propyl]-1,2-O-isopropylidene-α-D-glucofuranose.
SM-1213. Immunomodulator. Colorless viscous oil. *Greenwich Pharm. Inc.*

8474 Amiprilose Hydrochloride
60414-06-4 506
$C_{14}H_{28}ClNO_6$
3-O-[3-(Dimethylamino)propyl]-1,2-O-isopropylidene-α-D-glucofuranose hydrochloride.
Therafectin. Immunomodulator. mp = 181-183°; soluble in H_2O, MeOH, hot EtOH. *Greenwich Pharm. Inc.*

8475 Atiprimod Dimaleate
183063-72-1
$C_{30}H_{52}N_2O_8$
2-[3-(Diethylamino)propyl]-8,8-dipropyl-2-azaspiro[4,5]decane maleate (1:2).
SKF-106615-A12. Antirheumatic agent with anti-inflammatory properties. *SmithKline Beecham Pharm.*

8476 Azimexon
64118-86-1 264-679-7
$C_{19}H_{14}N_4O$
1-[(1-(2-Cyano-1-aziridinyl)-1-methylethyl]-2-aziridinecarboxamide.
Immunomodulator. Used in cancer treatment.

8477 Bucillamine
65002-17-7 1481
$C_7H_{13}NO_3S_2$
N-(2-Mercapto-2-methylpropionyl)-L-cysteine.
tiobutarit; thiobutarit; DE-019; SA-96; Rimatil. Immunomodulator; antirheumatic. mp = 139-140°; $[\alpha]_D^{25}$ = 32.3° (c = 1.0 EtOH); LD_{50} (mus ip) = 2285 mg/kg, (mus iv) = 989.6 mg/kg. *Santen Pharm. Co. Ltd.*

8478 Celmoleukin
94218-72-1
$C_{693}H_{1118}N_{178}O_{203}S_7$
Interleukin 2 (human clone pTIL2-21a), protein moiety.

8479 Copovithane
68045-74-9
Copolymer of 2-methylenetrimethylene-bis(methylcarbamate) and 1-vinyl-2-pyrrolidone (≅ 1:4).
BAY i 7433. Immunomodulator.

8480 Creosote Carbonate
8001-59-0
Shows antiviral activity.

8481 Dimepranol
53657-16-2
$C_5H_{13}NO$
(±)-1-(Dimethylamino)-2-propanol.
Immunomodulator. *Newport.*

8482 Dimepranol Acedoben
61990-51-0 263-361-5
$C_{14}H_{22}N_2O_4$
(±)-1-(Dimethylamino)-2-propanol p-acetamidobenzoate (salt).
Immunomodulator. *Newport.*

8483 Ditiocarb Sodium
148-18-5 3443 205-710-6
$C_5H_{10}NNaS_2.3H_2O$
Sodium diethyldithiocarbamate.
dithiocarb; DTC; DDC; DEDC; DDTC; DeDTC; Imuthiol. Immunomodulator, chelating agent and antidote, especially for nickel and cadmium poisoning. Used to counter platinum toxicity is pateints tretaed with cis-Platin or to treat Wilson's disease. [trihydrate]: mp = 90-92°, 94-102°; freely soluble in H_2O, EtOH, MeOH, Me_2CO, insoluble in Et_2O, C_6H_6; λ_m = 257, 290 nm (ε 1200, 13000 H_2O); LD_{50} (rat orl) = 2830 mg/kg, (mus orl) = 1870 mg/kg, (mus iv) > 1000 mg/kg.

8484 Etanercept
185243-69-0
Recombinant human TNF.
TNFR:Fc; Enbrel. Recombinant human tumor necrosis factor receptor p75 Fc fusion protein. An immunomodulator, used in treatment of arthritis. A dimeric fusion protein consisting of the extracellular ligad-binding portion of the human 75 kilodalton (p75) tumor necrosis factor receptor linked to the Fc portion of human IgG1. *Immunex Corp.*

8485 Glatiramer Acetate
147245-92-9
$(C_5H_9NO_4.C_3H_7NO_2.C_6H_{14}N_2O_2.C_9H_{11}NO_3)_x.xC_2H_4O_2$
L-Glutamic acid polymer with L-alanine, L-lysine, L-tyrosine, acetate (salt).
Copaxone; COP-1. Immunomodulator. *Lemmon Co.*

8486 Imiquimod
99011-02-6 4957
$C_{14}H_{16}N_4$
4-Amino-1-isobutyl-1H-imidazo[4,5-c]quinoline.
R-837; S-26308. Immunomodulator. mp = 292-294°. *3M Pharm.*

8487 Inosine Pranobex
36703-88-5 5006 253-162-1
$C_{52}H_{78}N_{10}O_{17}$
Inosine-2-hydroxypropyldimethylammonium 4-acetamidobenzoate (1:3).
methisoprinol; NP-113; NPT-10381; Aviral; Delimmun; Imunoviral; Inosiplex; Isoprinosin; Isoprinosina; Isoprinosine; Isoviral; Methisoprinol; Modimmunal; Pranosina; Pranosine; Viruxan. Antiviral agent; immunomodulator. Soluble in H_2O; LD_{50} (mus,rat orl,ip) > 4000 mg/kg. *Newport.*

8488 Interferon α
74899-72-2 5016 232-710-3
Alpha interferon.
alfa-interferon; alferon; OFN-α; LeIF; leukocyte interferon; lymphoblastoid interferon; Berfor [α-2C]; Alpha 2 [α-2C]; NSC-339140 [α-n1]; Suniferon [α-n1]; Wellferon [α-n1]; Alferon [α-n3]. Antineoplastic; antiviral; immunomodulator. In combination with nevirapine, has been used to treat hepatitis C successfully. Family of homologous proteins that inhibit cell proliferation and viral replication; modulates immune response. Produced by human lymphoblastoid cells induced with *Sendai* virus. MW = 18-20 kD.

8489 Interferon α-2a
76543-88-9 5016
$C_{860}H_{1353}N_{227}O_{255}S_9$
Interferon alfa-2a.
interferon αA; IFN-αA; Ro-22-8181; Canferon; Roferon-A; Ro-22-8181; roferon A. Antineoplastic; antiviral; immunomodulator. Alpha interferon, natural (injectable form); used for the treatment of genital warts. *Hoffmann-LaRoche Inc.; Interferon Sciences Inc.*

8490 Interferon α-2b
99210-65-8 5016
$C_{860}H_{1353}N_{229}O_{255}S_9$
Interferon alfa-2b.
INF-α_2; Sch-30500; YM-14090; Cibian; Introna; Intron A; Viraferon. Antineoplastic; antiviral; immunomodulator. *Schering-Plough HealthCare Products.*

8491 Interferon β
74899-71-1 5017 232-710-3
Beta interferon.
fibroblast interferon; FIF; IFN-β; Feron; Fiblaferon; Frone; Nafron. Antineoplastic; antiviral; immunomodulator.

8492 Interferon β-1a
145258-61-3 5017
$C_{908}H_{1406}N_{246}O_{252}S_7$
Interferon beta-1 (human fibroblast protein moiety).
Neoferon. Antineoplastic; antiviral; immunomodulator. A glycosylate polypeptide of 166 amino acid residues. From cultured Chinese Hamster ovary containing the engineered gene for human interferon beta. *Biogen, Inc.*

8493 Interferon β-1b
145155-23-3 5017
$C_{903}H_{1399}N_{245}O_{252}S_5$
17-L-Serine-2-166 interferon β1 (human fibroblast reduced).
Betaseron. Antineoplastic; antiviral; immunomodulator. A non-glycosylated polypeptide of 165 amino acid residues. Produced by *E. Coli. Berlex Labs Inc.*

8494 Interferon γ
9008-11-1 5018 232-710-3
Gamma interferon.
IFN-γ; immune IFN; ImIFN; type II interferon; S-6810 [as cys-tyr-cyc-interferon γ]; Actimmune [as cys-tyr-cyc-interferon γ]; Gammaferon [as cys-tyr-cyc-interferon γ]; Immuneron [as cys-tyr-cyc-interferon γ]; Polyferon [as cys-tyr-cyc-interferon γ]. Antiviral; antineoplastic; immunomodulator. Lymphokine produced by T-cells, structurally unrelated to interferon-α or -β. Molecular weight 16,000-25,000 daltons.

8495 Interferon γ-1a
98059-18-8 5018
$C_{761}H_{1206}N_{214}O_{225}S_6$
Interferon gamma-a.
Antineoplastic; antiviral; immunomodulator. Polypeptide of 146 amino acid residues. Produced in *E coli* K-12 (C600) by expression interferon gamma cDNA derived from human splenic lymphosyte mRNA. *Berlex Labs Inc.*

8496 Interferon γ-1b
98059-61-1 5018
$C_{734}H_{1166}N_{204}O_{204}S_5$
N^2-L-Methionyl-1-139-interferon-γ (human lymphocyte protein moiety reduced).
Actimmune; Immukin. Antineoplastic; antiviral; immunomodulator. *Genentech Inc.*

8497 Interleukin-2
85898-30-2 5020
$C_{690}H_{1115}N_{177}O_{203}S_6$
2-133-Interleukin 2 (human reduced), 125 L-serine.
Aldesleukin; Proleukin; TCGF; Costimulator; IL-2. Immunomodulator. *Cetus Corp.*

8498 Lentinan
37339-90-5 5462
$(C_6H_{10}O_5)_n$
LC-33. Immunomodulator. Neutral poysaccharides from the edible mushroom, *Lentinus edodes*. Dec 250°; insoluble in cold H_2O; slightly soluble in hot H_2O, DMSO; insoluble in organic solvents; $[\alpha]_D^{20}$ = 13.5 - 14.5° (2% NaOH), 19.5 - 21.5° (10% NaOH). *Ajinomoto Co. Inc.*

8499 Levamisole
14769-73-4 5486 238-836-5
$C_{11}H_{12}N_2S$
(-)-2,3,5,6-Tetrahydro-6-phenylimidazol[2,1-b]thiazole.
Levovermax; Totalon; tetramisole [as DL-form]; tetramizole [as DL-form]; dexamisole [as D(+)-form]. Immunomodulator with anthelmintic activity (targets nematodes). mp = 60-61.5°; $[\alpha]_D^{25}$ = -85.1° (c = 10 $CHCl_3$); [DL-form]: mp = 87-89°; [D-(+)-form]: mp = 60-61.5°; $[\alpha]_D^{25}$ = 85.1° (c = 10 $CHCl_3$). *Am. Cyanamid; Bayer Corp. Pharm. Div.; ICI ; Janssen Pharm. Inc.; McNeil Pharm.*

8500 Levamisole Hydrochloride
16595-80-5 5486 240-654-6
$C_{11}H_{13}Cl2N_2S$
(-)-2,3,5,6-Tetrahydro-6-phenylimidazol[2,1-b]thiazole hydrochloride.
R-12654; Ascaridil; Decaris; Ergamisol; Ergamisole; Levacide; Levadin; Levasole; Meglum; Nemicide; Nilverm; Ripercol; Solaskil; Spartakon; Tramisol; Bayer 9051 [as DL-form]; McN-JR-8299 [as DL-form]; R-8299 [as DL-form]; R-12563 [as D-(+)-form]. Immunomodulator with anthelmintic activity (targets nematodes). mp = 227-229°; $[\alpha]_D^{20}$ = -124° ± 2° (c = 0.9 H_2O); soluble in H_2O; [DL-form]: mp = 264-265°; soluble in H_2O (21 g/100 ml), MeOH, propylene glycol; sparingly soluble in EtOH, $CHCl_3$, C_6H_{14}, Me_2CO; LD_{50} (mus iv) = 22 mg/kg, (mus sc) = 84 mg/kg, (mus orl) = 210 mg/kg, (rat iv) = 24 mg/kg, (rat sc) = 130 mg/kg, (rat orl) = 480 mg/kg; [D-(+)-form]: mp = 227-227.5°; $[\alpha]_D^{20}$ = 125° (c = 0.7 H_2O). *Am. Cyanamid; Bayer Corp. Pharm. Div.; ICI ; Janssen Pharm. Inc.; McNeil Pharm.*

8501 Lisofylline
100324-81-0
$C_{13}H_{20}N_4O_3$
1-[(R)-5-Hydroxyhexyl]theobromine.
ProTec; CT-1501R. Immunomodulator. *Cell Therapeutics Inc.*

8502 Macrophage Colony-Stimulating Factor
81627-83-0 5678
Colony-stimulating factor 1.
M-CSF; CSF-1. Immunomodulator; hematopoietic. Glycosylated homodimer isolated from human urine. Growth factor that stimulates development of progenitor cells to monocytes or macrophages. Potentiates phagocytic activity and moncyte-mediated tumor cell cytotoxicity.

8503 Murabutide
74817-61-1
$C_{23}H_{40}N_4O_{11}$
2-Acetamido-3-O-[(R)-1-[[(S)-1-[[(R)-3-carbamoyl-1-carboxypropyl]carbamoyl]ethyl]carbamoyl]ethyl]-2-deoxy-D-glucopyranose butyl ester.
A muramyl dipeptide analog that stimulates antibody activity; used as an immunoadjuvant for synthetic vaccines. Immunomodulator.

8504 Mycophenolate Mofetil
115007-34-6
$C_{23}H_{31}NO_7$
2-Morpholinoethyl (E)-6-(4-hydroxy-6-methoxy-7-methyl-3-oxo-5-phthalanyl)-4-methyl-4-hexenoate.
RS-61443. Immunomodulator. *Syntex Intl. Ltd.*

8505 Pidotimod
121808-62-6 7574
$C_9H_{12}N_2O_4S$
(R)-3-[(S)-5-Oxoprolyl]-4-thiazolidinecarboxylic acid.
Axil; Onaka; Pigitil; Polimod. Immunomodulator. mp = 194-198° (dec), 192-194° (dec); $[\alpha]_D^{25}$ = -150° (c = 2 5N HCl); soluble in H_2O (3.78 g/100 ml), MeOH (1.38 g/100 ml), EtOH (0.44 g/100 ml), DMF (7.24 g/100 ml); insoluble in $CHCl_3$, C_6H_{14}; LD_{50} (mus,rat iv) > 4000 mg/kg, (mus,rat im) > 4000 mg/kg, (mus,rat ip) > 8000 mg/kg, (mus,rat orl) > 8000 mg/kg.

8506 Pimelautide
78512-63-7
$C_{29}H_{52}N_6O_9$
threo-6-Carbamoyl-N^2-[N-(N-lauroyl-L-alanyl)-D-γ-glutamyl]-N^6-glycyl-DL-lysine.
RP-44102. Murein-derived lauroyl-peptide with immunistimulating activity. Immunomodulator.

8507 Platonin
3571-88-8 7690 222-681-5
$C_{38}H_{61}I_2N_3S_3$
2,2'-[3-[(3-Heptyl-4-methyl-2(3H)-thiazolylidene)-ethylidene]-1-propene-1,3-diyl]bis[3-heptyl-4-methylthiazolium]diiodide.
platonin J; NK-19; Kankohso 101; Photosensitizer 101. Immunomodulator. mp = 204°; soluble in H_2O, EtOH; LD_{50} (mmus ip) = 46.9 mg/kg, (mmus orl) = 1539 mg/kg, (fmus ip) = 50.5 mg/kg, (fmus orl) = 1571 mg/kg.

8508 Prezatide Copper Acetate
130120-57-9
$C_{32}H_{54}CuN_{12}O_{12}$
Hydrogen [N^2-(N-glycyl-L-histidyl)-L-lysinato] [N^2-(N-glycyl-L-histidyl)-L-lysinato(2-)]cuprate(1-) diacetate.
lamin; PC1020 acetate. Immunomodulator. *Procyte Corp.*

8509 Procodazole
23249-97-0 7943
$C_{10}H_{10}N_2O_2$
2-Benzimidazolepropionic acid.
propazol; AL-1241; Estimulocel. Immunomodulator. mp = 228° (dec); soluble in EtOH, warm H_2O; insoluble in Et_2O, C_6H_6. *Lafarquim.*

8510 Propagermanium
12758-40-6 7979 235-800-0
$(C_3H_5GeO_{3.5})_n$
3-(Trihydroxygermanyl)propionic acid polymer.
proxigermanium; repagergermanium; GE-132; SK-818; Serocion. Immunomodulator. mp = 270° (dec); soluble in H_2O (1.1 g/100 ml); poorly soluble in organic solvents; LD_{50} (mus ip) = 2800 mg/kg.

8511 Romurtide
78113-36-7 8412
$C_{43}H_{78}N_6O_{13}$
2-Acetamido-3-O-[(R)-1-[[(S)-1-[[(R)-1-carbamoyl-3-[(S)-1-carboxy-5-stearamidopentyl]carbamoyl]propyl]-carbamoyl]ethyl]carbamoyl]ethyl]-2-deoxy-D-glucopyranose.
MDP-Lys (L18); muroctasin; DJ-7041; Nopia. Immunomodulator. mp = 176-178°; $[\alpha]_D^{20}$ = 24.9° (c = 1 MeOH); soluble in DMF (1.38 g/100 ml), MeOH (0.53 g/100 ml), EtOH (0.30 g/100 ml); insoluble in Me_2CO, CH_3CN, $CHCl_3$, H_2O; LD_{50} (mmus sc) = 436 mg/kg, (fmus sc) = 625 mg/kg, (mrat sc) = 761 761 mg/kg, (frat sc) = 801 mg/kg, (dog sc) > 200 mg/kg. *Daiichi Pharm. Corp.*

8512 Roquinimex
84088-42-6
$C_{18}H_{16}N_2O_3$
1,2-Dihydro-4-hydroxy-N,1-dimethyl-2-oxo-3-quinolinecarboxanilide.
LS-2616; FCF-89. Immunomodulator.

8513 Susalimod
149556-49-0
$C_{21}H_{16}N_2O_5S$
5-[[p-[(3-Methyl-2-pyridyl)sulfamoyl]phenyl]-ethynyl]salicylic acid.
Sulfasalazine analog. Immunomodulator.

8514 Temarotene
75078-91-0
$C_{23}H_{28}$
1,2,3,4-Tetrahydro-1,1,4,4-tetramethyl-6-[(E)-α-methylstyryl]naphthalene.
Ro-15-0778. A synthetic retinoid chemopreventive and chemotherapeutic agent. Immunomodulator; antiacne.

8515 Thymocartin
85466-18-8
$C_{21}H_{40}N_8O_7$
N-[N-(N^2-L-Arginyl-L-lysyl)-L-α-aspartyl]-L-valine.
RGH-0206; TP-4. Tetrapeptide. The 32-35 fragment of the naturally occurring thymic factor (thymopoietin). Immunostimulant.

8516 Thymomodulin
90803-92-2 9543
A cell-free thymic hormone preparation from calf thymus.
Leucotrofina. Immunomodulator.

8517 Thymopentin
69558-55-0 9544
$C_{30}H_{49}N_9O_9$
N-[N-[N-[N^2-L-Arginyl-L-lysyl)-L-α-aspartyl]-L-valyl-L-tyrosine.
H_2N-Arg-Lys-Asp-Val-Tyr-COOH; ORF-15244; TP-5. Immunomodulator. *Ortho Pharm. Corp.*

8518 Thymotrinan
85465-82-3
$C_{16}H_{31}N_7O_6$
N-(N^2-L-Arginyl-L-lysyl)-L-aspartic acid.
RGH-0205; TP-3. Tripeptide. Synthetic thymopoietin derivate. Immunostimulant.

8519 Tiprotimod
105523-37-3
$C_{10}H_{13}NO_4S_2$
2-[(3-Carboxypropyl)thio]-4-methyl-5-thiazoleacetic acid.
HBW-538. A synthetic thiazole derivative. Immunomodulator.

8520 Tucaresol
84290-27-7
$C_{15}H_{12}O_5$
α-(2-Formyl-3-hydroxyphenoxy)-p-toluic acid.
A Schiff base-forming molecule that mimics a natural carbonyl donor and enhances immune response. Immunomodulator.

8521 Ubenimex
58970-76-6 9973 261-529-2
$C_{16}H_{24}N_2O_4$
(-)-N-[(2S,3R)-3-Amino-2-hydroxy-4-phenylbutyryl]-L-leucine.
Bestatin; NK-421; NSC-265489. Immunomodulator. Antineoplastic agent. A dipeptide antitumor antibiotic obtained from *Streptomyces olivoreticuli*. mp = 233-236°; $[\alpha]_D^{20}$ = -15.5° (c = 1.0 1N HcL); λ_m 241.5, 248,253,258,264.5, 268 nm ($E^{1\%}_{1\,cm}$ 3.8, 4.0, 5.0, 6.0, 4.6, 2.7); soluble in AcOH, DMSO, MeOH; less soluble in H_2O; insoluble in EtOAc, C_6H_6, C_6H_1. *Microbiochem. Res. Found.*

Immunosuppressants

8522 Asobamast
104777-03-9
$C_{13}H_{15}N_3O_5S$
2-Ethoxyethyl [4-(3-methyl-5-isoxazolyl)-2-thiazolyl]-oxamate.
An antianaphylactic agent.

8523 Azastene
13074-00-5
$C_{23}H_{33}NO_2$
4,4,17-Trimethylandrosta-2,5-dieno[2,3-d]isoxazol-17β-ol.
Luteolytic. Inhibits steroid biogenesis and is immunosuppressive.

8524 Azathioprine
446-86-6 935 207-175-4
$C_9H_7N_7O_2S$
6-[(1-Methyl-4-nitroimidazol-5-yl)thio]purine.
BW-57-322; NSC-39084; Azamune; Azanin; Azoran; Imuran; Imurel. Immunosuppressant and antirheumatic agent. Used in combination with cyclophosphamide and hydroxychloroquine in treatment of rheumatoid arthritis. Listed as a known carcinogen. Dec 243-244°; λ_m 276 nm (ε 18200 MeOH), 280 nm (ε 17300 0.1N HCl), 285 nm (ε 15500 0.1N NaOH); slightly soluble in H_2O, $CHCl_3$, EtOH. *Glaxo Wellcome Inc.*

8525 Azathioprine Sodium
55774-33-9 935
$C_9H_6N_7NaO_2S$
6-[(1-Methyl-4-nitroimidazol-5-yl)thio]purine sodium salt.
Imuran Injection; Imurek. Immunosuppressant and antirheumatic agent. Has been used in treatment of rheumatoid arthritis. Soluble in H_2O. *Glaxo Wellcome Inc.*

8526 Brequinar
96187-53-0 1394
$C_{23}H_{15}F_2NO_2$
6-Fluoro-2-(2'-fluoro-[1,1'-biphenyl]-4-yl)-3-methyl-4-quinolinecarboxylic acid.
Biphenquinate; BPQ. Antineoplastic agent; immunosuppressant. Dihydroorotate dehydrogenase inhibitor. mp = 315-317°; soluble in H_2O, DMF. *E. I. DuPont de Nemours Inc.*

8527 Brequinar Sodium
96201-88-6 1394
$C_{23}H_{14}F_2NNaO_2$
6-Fluoro-2-(2'-fluoro-[1,1'-biphenyl]-4-yl)-3-methyl-4-quinolinecarboxylic acid sodium salt.
Dup-785; NSC-368390. Antineoplastic agent; immunosuppressant. Dihydroorotate dehydrogenase inhibitor. mp > 360°; soluble in H_2O. *E. I. DuPont de Nemours Inc.*

8528 Cedelizumab
156586-90-2
Immunoglobulin G 4 (human-mouse monoclonal OKTcdr4A complementary determining region-grafted σ-chain anti-human CD4 antigen) disulfide with human-mouse monoclonal OKTcdr4a complementary determining region-grafted kappa chain, dimer. RWJ-49004. Monoclonal antibody, designed as an immunosuppressant. *Johnson R. W. Pharm. Res. Institute.*

8529 Ciamexon
75985-31-8
$C_{11}H_{13}N_3O$
(±)-1-[(2-Methoxy-6-methyl-3-pyridyl)methyl]-2-aziridinecarbonitrile.
Immunosuppressant.

8530 Cyclophosphamide
50-18-0 2816 200-015-4
$C_7H_{15}Cl_2N_2O_2P$
(Bis(chloro-2-ethyl)amino)-2-tetrahydro-3,4,5,6-oxazaphosphorine-1,3,2-oxide-2 hydrate.
Asta-B-518; Clafen; Claphene; Cyclophosphamid; Cyclophosphamide; Cyclophosphamidum; Cyclophosphan; Cyclophosphane; Cyclostin; Cytophosphan; Cytoxan; NSC-26271; CB-4564; CP; CPA; CTX; CY; Endoxan; Endoxan R; Endoxan-Asta; Endoxana; Endoxanal; Endoxane; Enduxan; Genoxal; Hexadrin; Mitoxan; Neosar; NCI-C04900; Procytox; Semdoxan; Sendoxan; Senduxan; SK-20501; Zyklophosphamid. Immunosuppressant; antineoplastic agent. Listed as a known carcinogen. mp = 41-45°; soluble in H_2O (40 g/l), less soluble in oganic solvents; LD_{50} (rat orl) = 94 mg/kg. *Bristol-Myers Oncology; Pharmacia & Upjohn.*

8531 Cyclophosphamide Monohydrate
6055-19-2 2816
$C_7H_{15}Cl_2N_2O_2P.H_2O$
N,N-Bis(2-chloroethyl)tetrahydro-2H-1,3,2-oxazaphosphorin-2-amine 2-oxide monohydrate.
Cycloblastin; Cyclostin; Endoxan; Procytox; Sendoxan; Cytoxan; NSC-26271. Immunosuppressant; antineoplastic agent. mp = 41-45°; soluble in H_2O (40 g/l), less soluble in oganic solvents; LD_{50} (rat orl) = 94 mg/kg. *Degussa-Hüls Corp.*

8532 Cyclosporin A
59865-13-3 2821
$C_{62}H71_{11}N_{11}O_{12}$
Cyclosporine.
ciclosporin; 27-400; Sandimmun; Sandimmune; Neoral. Immunosuppressant. mp = 148-151°; $[\alpha]_D^{20}$ = -244° (c = 0.6 $CHCl_3$), -189° (c = 0.5 MeOH), soluble in MeOH, EtOH, Me_2CO, Et_2O, $CHCl_3$; slightly soluble in H_2O and saturated hydrocarbons; LD_{50} (mus iv) = 107 mg/kg, (mus orl) = 2329 mg/kg, (rat iv) = 25 mg/kg, (rat orl) = 1480 mg/kg, (rbt iv) > 10 mg/kg, (rbt orl) > 1000 mg/kg. *Sandoz Pharm. Corp.*

8533 Cyclosporin B
63775-95-1 2821
$C_{61}H71_{09}N_{11}O_{12}$
Ala^2-Cyclosporine.
Related to the immunosuppressant cyclosporin A. mp = 149-152°; $[\alpha]_D^{20}$ = -238° (c = 0.62 $CHCl_3$), -168° (c = 0.56 MeOH). *Sandoz Pharm. Corp.*

8534 Cyclosporin C
59787-61-0 2821
$C_{62}H71_{11}N_{11}O_{13}$
Thr^2-Cyclosporine.
Related to the immunosuppressant cyclosporin A. mp = 152-155°; $[\alpha]_D^{20}$ = -255° (c = 0.5 $CHCl_3$), -182° (c = 0.5 MeOH); solubility similar to that of cyclosporin A. *Sandoz Pharm. Corp.*

8535 Cyclosporin D
63775-96-2 2821
$C_{63}H71_{13}N_{11}O_{12}$
Val^2-Cyclosporine.
Related to the immunosuppressant cyclosporin A. mp = 148-151°; $[\alpha]_D^{20}$ = -245° (c = 0.52 $CHCl_3$), -211° (c = 0.51 MeOH). *Sandoz Pharm. Corp.*

8536 Cyclosporin G
74436-00-3 2821
$C_{62}H71_{11}N_{11}O_{12}$
Nva^2-Cyclosporine.
7-L-norvaline cyclosporin A. Related to the immunosuppressant cyclosporin A. mp = 196-197°; $[\alpha]_D^{20}$ = -245° (c = 1.0 $CHCl_3$), -191° (c = 1.04 MeOH). *Sandoz Pharm. Corp.*

8537 Daltroban
79094-20-5 2871
$C_{16}H_{16}ClNO_4S$
[p-[2-(p-Chlorobenzenesulfonamido)ethyl]phenyl]-acetic acid.
SKF-96148; BM-13505. Antithrombotic; immunosuppressant. Thromboxane synthetase inhibitor. *Boehringer Mannheim GmbH; SmithKline Beecham Pharm.*

8538 Gusperimus
104317-84-2 4610
$C_{17}H_{37}N_7O_3$
(±)-N-[[[4-[(3-Aminopropyl)amino]butyl]carbamoyl]hydroxymethyl]-7-guanidinoheptanamide.
deoxyspergualin; (±)-15-deoxyspergualin. Immunosuppressant. *Bristol-Myers Oncology.*

8539 Gusperimus Trihydrochloride
85468-01-5 4610
$C_{17}H_{40}Cl_3N_7O_3$
(±)-N-[[[4-[(3-Aminopropyl)amino]butyl]carbamoyl]-hydroxymethyl]-7-guanidinoheptanamide trihydrochloride.
BMS-181173; BMY-42215-1; NKT-01; NSC-356894; Spanidin. Immunosuppressant. [Sesquihdrate]: Soluble in H_2O; LD_{50} (mus ip) = 25-50 mg/kg. *Bristol-Myers Oncology.*

8540 (-)-Gusperimus Trihydrochloride
84937-45-1 4610
$C_{17}H_{40}Cl_3N_7O_3$
(-)-N-[[[4-[(3-Aminopropyl)amino]butyl]carbamoyl]-hydroxymethyl]-7-guanidinoheptanamide trihydrochloride.
BMS-181173; BMY-42215-1; NKT-01; NSC-356894. Immunosuppressant. [Dihydrate]: $[\alpha]_D^{25}$ = -7.3° (c = 1 in H_2O); LD_{50} (mus ip,iv) = 35-40 mg/kg. *Bristol-Myers Oncology.*

8541 Leflunomide
75706-12-6
$C_{12}H_9F_3N_2O_2$
α,α,α-Trifluoro-5-methyl-4-isoxazolecarboxy-p-toluidide.
Cyclo-oxygenase-2 inhibitor. Used in the treatment of rheumatoid arthritis. Immunosuppressant.

8542 Lenercept
156679-34-4
$C_{1993}H_{3112}N_{562}O_{624}S_{34}$
1-182 Tumor necrosis factor receptor (human reduced), (182-104')-protein with 104-330-immunoglobulin G 1 (human clone pTJ5 Cσ 1 reduced).
Tumor necrosis factor receptor-immunoglobulin fusion protein. Immunosuppressant.

8543 Mafosfamide
88859-04-5
$C_9H_{19}Cl_2N_2O_5PS_2$
(±)-2-[[2-[Bis(2-chloroethyl)amino]tetrahydro-2H-1,3,2-oxazaphosphorin-4-yl]thio]ethanesulfonic acid P-cis-oxide.
Antineoplastic; immunosuppressant. A cyclophosphamide.

8544 6-Mercaptopurine
6112-76-1 5919
$C_5H_4N_4S$
1,7-Dihydro-6H-purine-6-thione monohydrate.
6-purinethiol hydrate; purine-6-thiol monohydrate; NSC-755. Antineoplastic; immunosuppressant. Dec 313-314°; λ_m = 230, 312 nm (ε 14000, 19600 0.1N NaOH); insoluble in H_2O, organic solvents; slightly soluble in EtOH; LD_{50} (mus ip) = 157 mg/kg.

8545 6-Mercaptopurine [anhydrous]
50-44-2 5919 200-037-4
$C_5H_4N_4S$
1,7-Dihydro-6H-purine-6-thione.
purine-6-thiol; thiohypoxanthine; 3H-purine-6-thiol;

6MP; 6-purinethiol; 6-thioxopurine; 7-mercapto-1,3,4,6-tetrazaindene; U-4748; Ismipur; Leukeran; Leukerin; Leupurin; Mercaleukim; Mercaleukin; Mercaptopurine; Mercapurin; Mern; Puri-Nethol; Purimethol; Purinethiol. Antineoplastic; immunosuppressant.

8546 Mizoribine

50924-49-7 6306

$C_9H_{13}N_3O_6$

5-Hydroxy-1-β-D-ribofuranosylimidazole-4-carboxamide. HE-69; Bredinin. Immunosuppressive with antiarthritis activity. mp > 200° (dec); $[\alpha]_D^{27}$ = -35° (c = 0.8 H_2O); λ_m = 245, 279 nm (E 250, 580 H_2O); soluble in H_2O; slightly soluble in MeOH, EtOH; insoluble in other organic solvents; LD_{50} (mus iv) > 1500 mg/kg, (mus ip) > 2400 mg/kg. *Sumitomo Pharm. Co. Ltd.; Toyo Jozo.*

8547 Mycophenolate Mofetil

128794-94-5

$C_{23}H_{31}NO_7$

(E)-6-(1,3-dihydro-4-hydroxy-6-methoxy-7-methyl-3-oxo-5-isobenzofuranyl)-4-methyl-4-hexenoic acid 2-(4-morpholinyl)ethyl ester,.

Cellcept (tablet, capsule, suspension); Mycophenolate mofetil; Mycophenolic acid morpholinoethyl ester; RS-61443; RS 61443 . Immunosuppressant. For the prevention of acute rejection in pediatric renal transplant patients. *Roche Labs.*

8548 Rapamycin

53123-88-9 8288

$C_{51}H_{79}NO_{13}$

(3S,6R,7E,9R,10R,12R,14S,15E,17E,19E,21S,23S,26R,-27R,34aS)-9,10,12,13,14,21,22,23,24,25,26,27,32,33,-34,34a-Hexadecahydro-9,27-dihydroxy-3-[(1R)-2[(1S,3R,4R)-4-hydroxy-3-methoxycyclohexyl]-1-methylethyl]-10,21-dimethoxy-6,8,12,14,20,26-hexamethyl-23,27-epoxy-3H-pyrido[2,1-c][1,4]oxaazacyclohentriacontine-1,5,11,28,29(4H,6H,31H)pentone.

Sirolimus; Rapamune; AY-22989; WY-090217; RAPA; RPM; NSC-226080. Immunosuppressive with antiarthritis activity. mp = 183-185°; λ_m 267, 277 288 nm ($E^{1\%}_{1\ cm}$ 417, 541, 416 95% EtOH); $[\alpha]_D^{25}$ = -58.2° (MeOH); soluble in Et_2O, $CHCl_3$, Me_2CO, MeOH, DMF; sparingly soluble in C_6H_{14}, petroleum ether; insoluble in H_2O; LD_{50} (mus ip) ≅ 600 mg/kg, (mus orl) > 2500 mg/kg. *Wyeth-Ayerst Labs.*

8549 Sufosfamide

37753-10-9

$C_8H_{18}ClN_2O_5PS$

2-[[3-(2-Chloroethyl)tetrahydro-2H-1,3,2-oxazaphosphorin-2-yl]amino]ethanol methanesulfonate (ester).

An oxazaphosphorine derivative. Antineoplastic; immunosuppressant.

8550 Tacrolimus

104987-11-3 9200

$C_{44}H_{69}NO_{12}.H_2O$

(-)-(3S,4R,5S,8R,9E12S,14S,15R,16R,18R,19R26aS)-8-allyl-5,6,8,11,12,13,14,15,16,17,18,19,24,25,26,26a-Hexadecahydro-5,19-dihydroxy-3-[(E)-2-[(1R,3R,4R)-4-hydroxy-3-methoxycyclohexyl]-1-methylvinyl]-14,16-dimethoxy-4,10,12,18-tetramethyl-15,19-epoxy-3H-pyrido[2,1-c][1,4]oxaazacyclotricosine-1,7,20,21(4H,23H)-tetrone)monohydrate.

Prograf; Protopic; FK-506; FR-900506. Immunosuppressant. Used to treat eczema. [monohydrate]: mp = 127-129°; $[\alpha]_D^{23}$ = -84.4° (c = 1.02 $CHCl_3$); soluble in MeOH, EtOH, Me_2CO, EtOAc, $CHCl_3$; Et_2O; sparingly soluble in C_6H_{14}, petroleum ether; insoluble in H_2O; LD_{50} (mus ip) > 200 mg/kg, (mrat iv) = 57 mg/kg, (mrat orl) = 134 mg/kg, (frat iv) = 23.6 mg/kg, (frat orl) = 194 mg/kg. *Fujisawa Pharm. USA Inc.*

8551 Tresperimus

160677-67-8

$C_{17}H_{37}N_7O_3$

[4-[(3-Aminopropyl)amino]butyl]carbamic acid ester with N-(6-guanidinohexyl)glycolamide.

LF-08-0299. Analog of gusperimus, an immunosuppressant.

Insulin Sensitizers

8552 Pioglitazone

111025-46-8 7605

$C_{19}H_{20}N_2O_3S$

(±)-5-[p-[2-[(5-Ethyl-2-pyridyl)ethoxy]benzyl]-2,4-thiazolidinedione.

AD-4833. Insulin sensitizer. Antidiabetic agent. mp = 183-184°. *Takeda Chem. Ind. Ltd.*

8553 Pioglitazone Hydrochloride

112529-15-4 7605

$C_{19}H_{21}ClN_2O_3S$

(±)-5-[p-[2-[(5-Ethyl-2-pyridyl)ethoxy]benzyl]-2,4-thiazolidinedione monohydrochloride.

U-72107A. Insulin sensitizer. Antidiabetic agent. mp = 193-194°. *Takeda Chem. Ind. Ltd.*

8554 Troglitazone

97322-87-7 9898

$C_{24}H_{27}NO_5S$

(±)-all rac-5-[p-[(6-Hydroxy-2,5,7,8-tetramethyl-2-chromanyl)methoxy]benzyl]-2,4-thiazolidinedione.

CS-045; CI-991; GR-92132X; romglizone. Insulin sensitizer. Antidiabetic agent. mp = 184-186°. *Pfizer Inc.; Sankyo Co. Ltd.*

Ion Exchangers

8555 Carbacrylic Resins

1824

Cross-linked polyacrylic polycarboxylic ion exchange resins.

8556 Cholestyramine Resin

11041-12-6 2257

Cholestyramin.

colestyramine; Cholybar; Duolite AP-143 Resin; Questran; Questran Light; Dowex 1-X2-Cl; MK-135; Cuemid; Quantalan. Ion exchange resin, binds bile acids and is used as a hypolipidemic agent. Bile acid sequestrant. Ion exchange resin; insoluble in H_2O, organic solvents. *Bristol Laboratories; Parke-Davis; Rohm & Haas Co.*

8557 Colestipol
26658-42-4 2538
N-(2-Aminoethyl)-N'-[(2-aminoethyl]amino]ethyl]-1,2-ethanediamine polymer with (chloromethyl)oxirane.
Hypolipidemic agent. Bile acid sequestrant. *Pharmacia & Upjohn.*

8558 Colestipol Hydrochloride
37296-80-3 2538
N-(2-Aminoethyl)-N'-[(2-aminoethyl]amino]ethyl]-1,2-ethanediamine polymer with (chloromethyl)oxirane hydrochloride salt.
U-26597A; Cholestabyl; Lestid; Colestid; Colestid. Hypolipidemic agent. Bile acid sequestrant. A basic anion-exchange resin, highly cross-linked and insoluble. LD_{50} (rat orl) > 1000 mg/kg, (rat ip) > 4000 mg/kg. *Pharmacia & Upjohn.*

8559 Polidexide
9064-92-0 7716
Dextran 2-(diethylamino)ethyl 2-[[2-(diethylamino)ethyl]-diethylammonio] ethyl ether sulfate epichlorohydrin cross-linked.
sephadex 2-(diethylamino)ethyl ether; PDX chloride; DEAE Sephadex; Secholex. Hypolipidemic agent. An anion exchange resin containing quaternary ammonium groups which reduces serum cholesterol by binding bile acids in the intestine, facilitating their excretion. *Pharmacia.*

8560 Resodec
9012-13-9 8321
A polycarboxylic cation exchange resin.
Intended to remove sodium from the contents of the intestinal tract. An inert, nonabsorbable powder.

8561 Sodium Polystyrene Sulfonate
9003-59-2 8815
Sulfonated divinylbenzene copolymer with styrene sodium salt.
Amberlite IRP-69 Resin; Kayexalate. Can remove potassium ions selectively. *Rohm & Haas Co.; Sterling Winthrop Inc.*

Keratolytics

8562 Alcloxa
1317-25-5 215-262-3
$C_4H_9Al_2ClN_4O_7$
Chlorotetrahydroxy[(2-hydroxy-5-oxo-2-imidazolin-4-yl)-ureato]dialuminum.
aluminum chlorhydroxy allantoinate; ALCA; RC-173; 198. Astringent; keratolytic. *ICI Americas Inc.*

8563 Aldioxa
5579-81-7 224 226-964-4
$C_4H_7AlN_4O_5$
Dihydroxy[(2-hydroxy-5-oxo-2-imidazolin-4-yl)ureato]aluminum.
aluminum dihydroxy allantoinate; ALDA; RC-172; Alanetorin; Alusa; Arlanto; Ascomp; Chlokale; Isalon; Nische; Peptilate. Astringent; keratolytic. Antiulcerative. mp = 230°, insoluble in polar and non-polar solvents. *ICI Americas Inc.*

8564 Benzoyl Peroxide
94-36-0 1149 202-327-6
$C_{14}H_{10}O_4$
Dibenzoyl peroxide.
Acne-Aid Cream; Benoxyl; Benzac; Benzac W; Benzagel; Benzamycin; Brevoxyl; Clear By Design; Desquam-X; Dry and Clear; Epi-Clear; Fostex BPO Bar, Gel and Wash; Loroxide; Panoxyl; Persa-Gel; pHisoAc BP; Sulfoxyl; Vanoxide; benzoyl superoxide; Acetoxyl; Acnegel; Akneroxide L; Benoxyl; Benzagel 10; Benzaknen; Debroxide; Desanden; Lucidol; Nericur; Oxy-5; Oxy-L; PanOxyl; Peroxyderm; Preoxydex; Persadox; Sanoxit; Theraderm; Xerac BP 5; Xerac BP 10. Used to treat acne. Keratolytic. mp = 103-106°; sparingly soluble in H_2O, EtOH; soluble in C_6H_6, $CHCl_3$, Et_2O, CS_2 (2.5 g/100 ml), olive oil (2 g/100 ml). *Bristol-Myers Squibb Co; Dermik Labs. Inc.; Galderma Labs Inc.; Hyland Div. Baxter ; Ortho Pharm. Corp.; SmithKline Beecham Pharm.; Sterling Winthrop Inc.; Taiho; Westwood-Squibb Pharm. Inc.*

8565 Dibenzothiophene
132-65-0 205-072-9
$C_{12}H_8S$
Dibenzothiophene.
Keratolytic.

8566 Dichloroacetic Acid
79-43-6 3100 201-207-0
$C_2H_2Cl_2O_2$
Dichloroethanoic acid.
bichloracetic acid. DCA. Keratolytic; astringent. Liquid; bp = 193-194°; d_4^{20} = 1.563; two crystalline forms, mp = 9.7° and -4°; soluble in H_2O, EtOH, Et_2O; LD_{50} (rat orl) = 2820 mg/kg; [ethyl ester ($C_4H_6Cl_2O_2$)]: bp = 158.3-158.7°; d_4^{20} = 1.282; slightly soluble in H_2O; miscible with EtOH; Et_2O.

8567 Etarotene
87719-32-2
$C_{25}H_{32}O_2S$
6-[(E)-p-(Ethylsulfonyl)-α-methylstyryl]-1,2,3,4-tetrahydronaphthalene.
Ro-15-1570/000; 193. Keratolytic. *Hoffmann-LaRoche Inc.*

8568 Motretinide
56281-36-8 6367 260-094-6
$C_{23}H_{31}NO_2$
all-trans-N-Ethyl-9-(4-methoxy-2,3,6-trimethyl-phenyl)-3,7-dimethyl-2,4,6,8-nonatetra-enamide.
Tasmaderm; Ro-11-1430. Keratolytic. Used to treat acne. mp = 179-180°. *Hoffmann-LaRoche Inc.*

8569 Picotrin Diolamine
64063-83-8
$C_{29}H_{40}N_2O_4$
5-Tritylpicolinic acid compound with 2,2'-imidodiethanol.
Sch-19741; 309. Keratolytic. *Schering-Plough HealthCare Products.*

8570 Resorcinol

108-46-3 8323 203-585-2

$C_6H_6O_2$

1,3-Benzenediol.

Resorcin; m-dihydroxybenzene; component of: Acnomel, Rezamid, Sulforcin. Keratolytic; to treat acne. mp = 109-111°; bp = 280°; d = 1.272; soluble in H_2O (111.1 g/100 ml at 20°, 500 g/100 ml at 80°), EtOH (111.1 g/100 ml); freely soluble in Et_2O, glycerol; slightly soluble in $CHCl_3$. *Dermik Labs. Inc.; Galderma Labs Inc.; Menley & James Labs Inc.*

8571 Resorcinol Monoacetate

102-29-4 8323 203-022-0

$C_8H_8O_3$

1,3-Benzenediol monoacetate.

Euresol; 226. Keratolytic; antiseborrheic. bp ≅ 283° (dec); insoluble in H_2O; soluble in EtOH, C_6H_6, Me_2CO, $CHCl_3$. *Knoll Pharm. Co.*

8572 Retinoic Acid

302-79-4 8333 206-129-0

$C_{20}H_{28}O_2$

[all-E]-3,7-Dimethyl-9-(2,6,6-trimethyl-1-cyclohexen-1-yl)-2,4,6,8-nonatrienoic acid.

all-trans-Retinoic acid; vitamin A acid; tretinoin; NSC-122578; Aberel; Airol; Aknoten; Cordes Vas; Dermairol; Epi-Aberel; Eudyna; Retin-A; Vesanoid; Vesnaroid. Antineoplastic agent. Also a keratolytic, used to treat acne. mp = 180-182°; λ_m = 351 nm (ε 45000 MeOH); LD_{50} (10 day) (mus ip) = 790 mg/kg, (mus orl) = 2200 mg/kg, (rat ip) = 790 mg/kg, (rat orl) = 2000 mg/kg. *Degussa Ltd.; Hoffmann-LaRoche Inc.*

8573 13-cis-Retinoic Acid

4759-48-2 8333 225-296-0

$C_{20}H_{28}O_2$

(13-cis)-3,7-Dimethyl-9-(2,6,6-trimethyl-1-cyclohexen-1-yl)-2,4,6,8-nonatrienoic acid.

13-RA; Ro-4-3780; retinoic acid, 9Z form; 13-cis-Vitamin A acid; 13-cis-retinoic acid; cis-retinoic acid; neovitamin a acid; isotretinoin; Ro-4-3780; Accutane; Isotrex; Roaccutane; Tasmar; Isotretinoin; Accure; IsotrexGel; Roaccutane; 193. Keratolytic. Used to treat acne. mp = 174-175°; λ_m = 354 nm (ε 39800); LD_{50} (20 day) (mus ip) = 904 mg/kg, (mus orl) = 3389 mg/kg, (rat ip) = 901 mg/kg, (rat orl) > 4000 mg/kg. *Hoffmann-LaRoche Inc.*

8574 Salicylic Acid

69-72-7 8484 200-712-3

$C_7H_6O_3$

2-Hydroxybenzoic acid.

Advanced Pain Relief Callus Removers; Advanced Pain Relief Corn Removers; Clear away Wart Remover; Compound W; Dr. Scholl's Callus Removers; Dr. Scholl's Corn Removers; Dr. Scholl's Wart Remover Kit; Duofil Wart Remover; Duoplant; Freezone; Ionil; Ionil Plus; Salicylic Acid Soap; Saligel; Stri-Dex; component of: Domerine, Fostex Medicated Bar and Cream, Keralyt, Pernox, Salicylic Acid & Sulfur Soap, Sebucare, Sebulex, Solarcaine First Aid Spray; 51; 171; 311; 334; 337; 380; k537; k557. Keratolytic. *Bayer AG; Galderma Labs Inc.; Nippon Shinyaku Japan; Schering-Plough Pharm.; Sterling Health U.S.A.; Stiefel Labs Inc.; Westwood-Squibb Pharm. Inc.; Whitehall-Robins.*

8575 Sumarotene

105687-93-2

$C_{24}H_{30}O_2S$

1,2,3,4-Tetrahydro-1,1,4,4-tetramethyl-6-[(E)-α-methyl-p-(methylsulfonyl)styryl]naphthalene.

Ro-14-9706/000; 193. Keratolytic. *Hoffmann-LaRoche Inc.*

8576 Tazarotene

118292-40-3 9249

$C_{21}H_{21}NO_2S$

Ethyl 6-[(4,4dimethylthiochroman-6-yl)ethynyl]-nicotinate.

Zorac; AGN-190168. Keratolytic. Used to treat acne and psoriasis. White solid. *Allergan Inc.*

8577 Tetroquinone

319-89-1 9385 206-275-5

$C_6H_4O_6$

2,3,5,6-Tetrahydroxy-2,5-cyclohexadien-1,4-dione.

Tetrahydroxy-p-benzoquinone; HPEK-1; THQ; NSC-112931; Kelox. Systemic keratolytic. Used to treat acne. Slightly soluble in cold H_2O, freely soluble in hot H_2O or EtOH, poorly soluble in Et_2O; behaves like a dibasic acid.

Lactation Stimulating Hormones

8578 Prolactin

9002-62-4 7961

Pituitary lactogenic hormone.

adenohypophysial luteotropin; Anterior pituitary luteotropin; lactogen; Galactin; mammotropin; luteotropic hormone; luteotropin; LTH; Ferolactan. Polypeptide hormone of molecular weight 23,000 responsible for induction of lactation in mammlas at parturition. $[\alpha]_D^{25}$ = -40.5 (c = 1 in buffer at pH 7); poorly soluble in H_2O (0.01 g/100 ml); forms a water-soluble hydrochloride salt.

8579 Somavubove

126752-39-4

$C_{976}H_{1533}N_{265}O_{286}S_8$

127-L-Leucine growth hormone (ox).

recombinantly derived bovine somatotropin. Recombinant bovine growth hormone. Galactopoietic agent (veterinary). *Upjohn Ltd.*

Laxatives and Cathartics

8580 Agar

9002-18-0 182 232-658-1

Polysaccharide complex extracted from the agarocytes of algae of the *Rhodophyceae*.

Agar-agar; gelose; Japan Agar; Bengal Isinglass; Ceylon Isinglass; Chinese Isinglass; Japan Isinglass; Layor Carang. Suspending agent, used as a pharmaceutical aid. Laxative-cathartic. Powder; insoluble in cold H_2O, EtOH; slowly soluble in hot H_2O; a 1% solution forms a stiff jelly on cooling.

8581 Aloe-Emodin
481-72-1 313 207-571-7
$C_{15}H_{10}O_5$
1,8-Dihydroxy-3-(hydroxymethyl)-9,10-anthracenedione.
3-hydroxymethylchrysazine; rhabarberone; Found as the glycoside in rhubarb, senna and various species of aloe. Laxative-cathartic. mp = 223-224°; freely soluble in hot EtOH, Et_2O, C_6H_6, NH_4OH, H_2SO_4; [trimethyl ether ($C_{18}H_{16}O_5$)]: mp = 163°; [triacetate ($C_{21}H_{16}O_8$)]: mp = 175-177°.

8582 Aloin
1415-73-2 314 215-808-0
$C_{20}H_{14}O_4$
10-β-D-Glucopyranosyl-1,8-dihydroxy-3-hydroxymethyl-9(10H)-anthrone.
barbaloin. Laxative-cathartic. mp = 148-149°; at 18°, soluble in C_5H_5N (57 g/100 ml), AcOH (7.3 g/100 ml), MeOH (5.4 g/100 ml), Me_2CO (3.2 g/100 ml), MeOAc (2.8 g/100 ml), EtOH (1.9 g/100 ml), H_2O (1.8 g/100 ml), EtOAc (0.78 g/100 ml), iPrOH (0.27 g/100 ml); very slightly soluble in iBuOH, $CHCl_3$, CS_2, Et_2O.

8583 Bisacodyl
603-50-9 1282 210-044-4
$C_{22}H_{19}NO_4$
4,4'-(2-Pyridylmethylene)diphenol diacetate.
Correctol Tablets; Correctol Caplets; Dulcolax; Feen-a-Mint Tablets; SK-Bisacodyl; Theralax; Bicol; Broxalax; Contralax; DAMP; Dulcolan; Durolax; Endokolat; Eulaxan; Godalax; Laxadin; Laxanin N; Laxorex; Nigalax; Perilax; Pyrilax; Stadalax; Telemin; Ulcol. Laxative-cathartic. mp = 138°; insoluble in H_2O, alkaline solutions; soluble in acids, EtOH, Me_2CO, propylene glycol; LD_{50} (rat orl) > 3000 mg/kg. *Boehringer Ingelheim Pharm. Inc.; Schering-Plough Pharm.; SmithKline Beecham Pharm.; Thomae GmbH Dr. Karl.*

8584 Bisacodyl Tannex
1336-29-4 1282
4,4'-(2-Pyridylmethylene)diphenol diacetate (ester) complex with tannic acid.
Bisacodyl Complex with Tannin; Water-soluble complex of Bisacodyl and tannic acid; Clysodrast. Laxative-cathartic. *Rhône-Poulenc Rorer Pharm. Inc.*

8585 Bisoxatin
17692-24-9 1337
$C_{22}H_{17}NO_5$
2,2-Bis(p-hydroxyphenyl)-2H-1,4-benzoxazin-3(4H)-one.
Laxative-cathartic. *Wyeth-Ayerst Labs.*

8586 Bisoxatin Acetate
14008-48-1 1337 237-820-5
$C_{24}H_{19}NO_6$
2,2-Bis(p-hydroxyphenyl)-2H-1,4-benzoxazin-3(4H)-one acetate.
Wy-8138; Laxonalin; Maratan; Talsis; Tasis. Laxative-cathartic. mp = 190°; LD_{50} (rat orl) = 8000 mg/kg, (mus orl) > 10000 mg/kg. *Thomae GmbH Dr. Karl ; Wyeth-Ayerst Labs.*

8587 Calcium Polycarbophil
9003-97-8 1744
Calcium salt of a loosely cross-linked, hydrophilic resin of the polycarboxylic type. Mitrolan; Noveon CA-1; Noveon CA-2; Sorboquel; WL 140; Carbofil; Quival. Laxative-cathartic. *Goodrich B.F. Co.; Robins, A. H. Co.; Schering-Plough HealthCare Products.*

8588 Calomelol
8011-82-3 1768
Colloidal calomel.
Soluble calomel; A mixture of 80% HgCl and 20% proteins. Laxative-cathartic. Slightly soluble in H_2O; insoluble in EtOH, Et_2O.

8589 Carlsbad Salt, Artificial
8007-49-6 1890
A mixture of 1 part K_2SO_4, 9 parts NaCl, 18 parts $NaHCO_3$ and 22 parts anhydrous Na_2SO_4. Laxative-cathartic.

8590 Casanthranol
8024-48-4 1929
A purified mixture of anthranol glycosides derived from *Cascara sagrada*. Cantralax; Peristim; component of: D-S-S Plus, Peri-Colace, Casakol. Laxative-cathartic. *Hoffmann-LaRoche Inc.; Parke-Davis; Roberts Pharm. Corp.*

8591 Castor Oil
8001-79-4 1946 232-293-8
Triglyceride of fatty acids (ricinoleic, oleic, linoleic, palmitic, stearic and dihydroxtstearic acids). Ricinus oil; oil of Palma Christi; tangantangan oil; Neoloid. Laxative-cathartic. $d_{15.5}^{15.5}$ = 0.961 - 0.963; miscible with EtOH, MeOH, Et_2O, $CHCl_3$, AcOH.

8592 Cellulose Ethyl Hydroxyethyl Ether
9004-58-4 2014
Ethyl 2-hydroxyethyl ether cellulose.
ethyl hydroxyethyl cellulose; Etulos. Laxative-cathartic.

8593 Colocynthin
1398-78-3 2551 215-745-9
$C_{38}H_{54}O_{13}$
25-(Acetyloxy)-2-(β-D-glucopyranolsyloxy)-16,20-dihydroxy-9-methyl-19-norlanosta-1,5,23-triene-3,11,22-trione.
2-O-β-D-glucopyranosylcucurbitacin E. Laxative-cathartic. mp = 158-160°; $[\alpha]_D$ = 50° (c = 0.4 EtOH); λ_m 234-236 nm (log ε 4.11); soluble in EtOH, Me_2CO, $CHCl_3$; slightly soluble in Et_2O, H_2O.

8594 Danthron
117-10-2 2878 204-173-5
$C_{14}H_8O_84$
1,8-Dihydroxyanthraquinone.
chrysazin; dantron; Altan; Antrapurol; Diaquone; Dorbane; Istizin; Modane. Laxative-cathartic. mp = 193-197°; λ_m 430 250 nm (log ε 4.35, 4.60); almost insoluble in H_2O (0.000156 g/100 ml at 25°); soluble in EtOH 0.05 g/100 ml), Et_2O (0.2 g/100 ml); soluble in $CHCl_3$; soluble in AcOH (10 g/100 ml at 75°; slightly soluble in alkaline

solutions (0.8 g/100 ml of 0.5N NaOH); LD_{50} (mus ip) = 500 mg/kg.

8595 Docusate Calcium

128-49-4 3459 204-889-8

$C_{40}H_{74}CaO_{14}S_2$

1,4-Bis(2-ethylhexyl)sulfosuccinate calcium salt.

Surfak; component of: Doxidan. Laxative-cathartic. Soluble in mineral and vegetable oils, polyethylene glycol; insoluble in glycerol. *Hoechst Roussel Pharm. Inc.*

8596 Docusate Potassium

7491-09-0 3460 231-308-5

$C_{20}H_{37}KO_7S$

1,4-Bis(2-ethylhexyl)sulfosuccinate potassium salt.

Rectalad Enema. Laxative-cathartic. *Wallace Labs.*

8597 Docusate Sodium

577-11-7 3460 209-406-4

$C_{20}H_{37}NaO_7S$

1,4-Bis(2-ethylhexyl)sulfosuccinate sodium salt.

Colace; Correctol Stool Softener Laxative; Dialose; Doxinate; D.S.S.; Modane Soft; Molofac; Aerosol OT; Comfolax; Coprola; Dioctylal; Dioctyl; Diotilan; Disonate; Disposaject; Doxol; Dulcivac; Jamylène; Molatoc; Molcer; Nevax; Regutol; Soliwax; Velmol; Waxsol; Yal; component of: Correctol Caplets, Correctol Tablets, Dialose Plus, Dorbantil, Doxan, D-S-S Plus, Feen-a-Mint Pills, Geriplex, Laxcaps, Modane Plus, Peri-Colace, Phillips Gelcaps, Senokap DSS, Senokot S, Unilax. Laxative-cathartic. Soluble in H_2O (1.5 g/100 ml at 25°, 2.3 g/100 ml at 40°, 3.0 g/100 ml at 50° 5.5 g/100 ml at 70°), CCl_4, petroleum ether, naphtha, xylene, dibutyl phthalate, liquid petrolatum; Me_2CO, EtOH, vegetable oils; very soluble in H_2O + EtOH or H_2O-miscible solvents. *3M Pharm.; Ascher B.F. & Co.; Bristol-Myers Squibb Pharm. R&D; Hoechst Roussel Pharm. Inc.; Johnson & Johnson-Merck Consumer Pharm. ; Mallinckrodt Inc.; Parke-Davis; Purdue Pharma L.P.; Roberts Pharm. Corp.; Savage Labs; Schering-Plough Pharm.; Sterling Health U.S.A.*

8598 Emodin

518-82-1 3602 208-258-8

$C_{15}H_{10}O_5$

1,3,8-Trihydroxy-6-methyl-9,10-anthracenedione.

frangula emodin; rheum emodin; archin; frangulic acid; Occurs as the rhamnoside in rhubarb. Laxative-cathartic. mp = 256-257°; λ_m 222, 252, 265, 289, 437 nm (log ε 4.55, 4.26, 4.27, 4.34, 4.10 EtOH); insoluble in H_2O; soluble in EtOH and alkline solutions; soluble at 25° in Et_2O (0.14 g/100 ml), $CHCl_3$ (0.071 g/100 ml), CCl_4 (0.01 g/100 ml), CS_2 (0.009 g/100 ml)C_6H_6 (0.041 g/100 ml); [3-methyl ether ($C_{16}H_{12}O_5$; rheochrysidin; physcione; parietin)]: mp = 207°.

8599 Frangulin

60529-33-1 4288

Franguloside.

avornin; Cascarin; Consists mainly of two glycoisdes, Frangulin A and Frangulin B. Laxative-cathartic.

8600 Frangulin A

521-62-0 4288 208-316-2

$C_{21}H_{20}O_9$

1,3,8-Trihydroxy-6-methylanthraquinone-1-rhamnoside.

emodin-l-rhamnoside; rhamnoxanthin. Laxative-cathartic. mp = 228°; λ_m 225, 264, 282, 300, 430 nm (log ε 4.52, 4.28, 4.15, 3.97, 4.05).

8601 Frangulin B

14101-04-3 4288 237-953-9

$C_{20}H_{18}O_9$

6-O-(D-Apiofuranosyl)-1,6,8-trihydroxy-3-methyl-anthraquinone.

Laxative-cathartic. mp = 196°.

8602 Glucofrangulin

52731-38-1 4460

$C_{27}H_{30}O_{14}$

Erbalax N; Irgalax; Anthraquinone glycoside from bark of the alder buckthorn. Consists of two isomers, glucofrangulin A and B, which differ in the linkage between the sugar and the 3 position of the aglycone. Laxative-cathartic. [glucofrangulin A ocatacetate ($C_{43}H_{46}O_{22}$)]: mp = 228-230°; $[\alpha]_D^{20}$ = -124° (c = 1.16 Me_2CO); λ_m 212, 264, 360 nm (log ε 4.57, 4.56, 4.26).

8603 Lactitol

585-86-4 5352 209-566-5

$C_{12}H_{24}O_{11}$

4-O-β-D-Galactopyranosyl-D-glucitol.

lactit; lactit M; lactite; lactiobiosit; lactosit; lactositol. Sweetener. Laxative-cathartic. mp = 146°; $[\alpha]_D^{23}$ = 14° (c = 4 H_2O); soluble in H_2O, DMSO, DMF; slightly soluble in EtOH, Et_2O.

8604 Lactitol Dihydrate

81025-03-8 5352

$C_{12}H_{24}O_{11}.2H_2O$

4-O-β-D-Galactopyranosyl-D-glucitol dihydrate.

Lacty. Sweetener. Laxative-cathartic. mp = 75°; $[\alpha]_D^{25}$ = 13.5° - 15.0°; soluble in H_2O (140 g/100 ml at 25°).

8605 Lactitol Monohydrate

81025-04-9 5352

$C_{12}H_{24}O_{11}.H_2O$

4-O-β-D-Galactopyranosyl-D-glucitol monohydrate.

Importal; Portolac. Sweetener. Laxative-cathartic. mp = 94-97°, 120°; $[\alpha]_D^{22}$ = 12.3°; soluble at 25° in H_2O (206 g/100 ml), EtOH (0.75 g/100 ml), Et_2O (0.4 g/100 ml), DMSO (233 g/100 ml), DMF (39 g/100 ml), 50° H_2O (512 g/100 ml), EtOH (0.88 g/100 ml), 75° H_2O (917 g/100 ml).

8606 Lactulose

4618-18-2 5360 225-027-7

$C_{12}H_{22}O_{11}$

4-O-β-D-Galactopyranosyl-D-fructofuranose.

Bifiteral; Cephulac; Duphalac; Generlac; Lactuflor; Laevilac; Normase. Laxative-cathartic. mp = 169°; soluble in H_2O (76.4 g/100 ml at 30°, 81.0 g/100 ml at 60°, >86 g/100 ml at 90°).

8607 Magnesium Carbonate Hydroxide
39409-82-0 5696
$(MgCO_3)_4.Mg(OH)_2.5H_2O$
Carbonic acid magnesium salt (1:1) mixture with magnesium hydroxide hydrate.
Marinco C; component of: Bufferin Arthritis Styrength, Bufferin Extra Strength, Bufferin Regular, Maalox HRF. Antacid. Laxative-cathartic. Slightly soluble in H_2O (0.03 g/100 ml), insoluble in EtOH. *Bristol-Myers Squibb Pharm. R&D; Rhône-Poulenc Rorer Pharm. Inc.*

8608 Magnesium Chloride [anhydrous]
7786-30-3 5698 232-094-6
Cl_2Mg
Hydrochloric acid magnesium salt (2:1).
Magnogene. Electrolyte replenisher. Laxative-cathartic. mp = 712°; d = 2.41, 2.325; soluble in H_2O (exothermic).

8609 Magnesium Chloride Hexahydrate
7791-18-6 5698
$Cl_2Mg.6H_2O$
Hydrochloric acid magnesium salt (2:1) hexahydrate.
Electrolyte replenisher. Laxative-cathartic. Dec 118°; d = 1.56; soluble in H_2O 166.6 g/100 ml at 20°, 333.2 g/100 ml at 100°), EtOH (50 g/100 ml); LD_{50} (rat orl) = 8100 mg/kg.

8610 Magnesium Citrate
3344-18-1 5699 222-093-9
$C_{12}H_{10}Mg_3O_{14}$
Hydroxy-1,2,3-propanetricarboxylic acid magnesium salt (2:3).
Citramag. Laxative-cathartic. [tetradecahydrate]: soluble in H_2O.

8611 Magnesium Hydroxide
1309-42-8 5706 215-170-3
H_2MgO_2
Magnesium hydrate.
Mint-O-Mag; Marinco H; Betalgil; Lederscon; Mucaine; Muthesa; Oxaine M; Phillips Magnesia Tablets; Phillips Milk of Magnesia Liquid; Simeco; Tepilta; component of: Arthritis Pain Formula Maximum Strength, Ascriptin, Calcitrel, Camalox, Di-Gel Liquid, Di-Gel Tablets, Gelusil, Haley's MO, Kestyomatin, Kudrox, Maalox, Maalox Plus, Mylanta Gelcaps, Mylanta Liquid, Mylanta Tablets, Simeco Suspension, Wingel. Antacid; cathartic; laxative (veterinary). Insoluble in H_2O, soluble in dilute acids; absorbs CO_2 in presence of H_2O. *Bristol-Myers Squibb Pharm. R&D; Johnson & Johnson-Merck Consumer Pharm. ; Lederle Labs.; Parke-Davis; Rhône-Poulenc Rorer Pharm. Inc.; Schering-Plough Pharm.; Schwarz Pharma Kremers Urban Co.; Sterling Health U.S.A.; Sterling Winthrop Inc.; Whitehall-Robins.*

8612 Magnesium Lactate
18917-93-6 5708 242-671-4
$C_6H_{10}MgO_6$
2-Hydroxypropanoic acid magnesium salt.
Laxative-cathartic. Soluble in H_2O (4 g/100 ml at 25°, 28.6 g/100 ml at 100°), slightly soluble in EtOH.

8613 Magnesium Phosphate, Dibasic
7757-86-0 5718 231-823-5
$HMgO_4P$
Magnesium hydrogen phosphate.
secondary magnesium phosphate. Laxative-cathartic. White crystalline powder, d = 2.13; slightly soluble in H_2O, soluble in dilute acids.

8614 Magnesium Sulfate [anhydrous]
7487-88-9 5731 231-298-2
$MgSO_4$
Sulfuric acid magnesium salt (1:1).
Kieserite [as monohydrate]. Magnesium replenisher. The heptahydrate is an anticonvulsant and cathartic. Also effective in termination of refractory ventricular tachyarrhythmias. Loss of deep tendon reflex is a sign of overdose. *Astra USA Inc.; Dow Chem. U.S.A.; McGaw Inc.*

8615 Magnesium Sulfate Heptahydrate
10034-99-8 5731
$MgSO_4.7H_2O$
Sulfuric acid magnesium salt (1:1) heptahydrate.
bitter salts; Epsom salts. Magnesium replenisher. The heptahydrate is an anticonvulsant and cathartic. Also effective in termination of refractory ventricular tachyarrhythmias. Loss of deep tendon reflex is a sign of overdose. d = 1.67; soluble in H_2O (71 g/100 ml at 20°, 91 g/100 ml at 40°), slightly soluble in EtOH. *Astra USA Inc.; Dow Chem. U.S.A.; McGaw Inc.*

8616 Magnesium Sulfate Trihydrate
15320-30-6 5731
$MgO_4S.3H_2O$
Sulfuric acid magnesium salt (1:1) trihydrate.
Laxative-cathartic. Colorless crystalline material. *Astra USA Inc.; Dow Chem. U.S.A.; McGaw Inc.*

8617 Mercurous Chloride
10112-91-1 5951 233-307-5
Hg_2Cl_2
Dimercury dichloride.
calomel; mild mercury chloride; mercury monochloride; mercury subchloride; precipité blanc; Calogreen. Antiseptic; cathartic; diuretic; antisyphilitic. Used as a cathartic, local antiseptic and desiccant in veterinary medicine. Mercurial compound. CAUTION: Excessive doses can cause mercury poisoning. Sublimes at 400-500°; d = 7.15; practically insoluble in H_2O; insoluble in alcohol, Et_2O; incompatible with bromides, iodides, alkali chlorides, sulfates, sulfites, carbonates, hydroxides, lime H_2O, acacia, ammonia, golden antimony sulfide, cocain.

8618 Mercury Mass
7439-97-6 5958
blue pill; blue mass; A mixture of honey; licorice; althea, glycerol, some mercury oleate and 32-34% metallic mercury. Laxative-cathartic.

8619 Oxyphenisatin Acetate
115-33-3 7108 204-083-6
$C_{24}H_{19}NO_5$
3,3-Bis(p-hydroxyphenyl)-2-indolinone diacetate (ester).
Isocrin; Lavema; NSC-59687; diphesatin; Isacen; Isaphen; Laxo-Isatin; Promassolax; Sanapert; Bydolax; Cirotyl; Lis-

agal; Contax; Prulet; Purgaceen; Bisatin. Laxative-cathartic. mp = 242°; insoluble in H_2O, Et_2O, dilute HCl; slightly soluble in EtOH. *Parke-Davis; Sterling Winthrop Inc.*

8620 Petrolatum, Liquid
7327
A mixture of liquid hydrocarbons from petroleum.
liquid paraffin; mineral oil; white mineral oil; paraffin oil; Clearteck; Drakeol; Hevyteck; Kremol; Kaydol; Alboline; Nujol; Paroleine; Saxol; Adepsine oil; Glymol. Laxative-cathartic. d (light oil) = 0.83 - 0.86; d (heavy oil) = 0.875-0.905; insoluble in H_2O, EtOH; soluble in C_6H_6, $CHCl_3$, Et_2O, CS_2, petroleum ether, oils.

8621 Phenolphthalein
81-90-3 7392 201-384-4
$C_{20}H_{14}O_4$
3,3-Bis(p-hydroxyphenyl)phthalide.
Evac-Q-Tabs; Ex-Lax; Modane; Prulet; Chocolax; Darmol; component of Agoral, Alophen, Doxan, Doxidan, Feen-A-Mint Laxative Mints, Laxcaps, Modane Plus, Phillips Gelcaps. Laxative-cathartic. mp = 258-262°; d = 1.299; soluble in EtOH (8.3 g/100 ml), Et_2O (1 g/100 ml); slightly soluble in $CHCl_3$; insoluble in H_2O; λ_m 205, 229, 276 nm (ε 27261, 14692, 2006 MeOH). *Hoechst Roussel Pharm. Inc.; Mission Pharmacal Co.; Parke-Davis; Sandoz Pharm. Corp.; Savage Labs; Schering-Plough Pharm.; Sterling Health U.S.A.*

8622 Phenolphthalol
81-92-5 7395 201-386-5
$C_{20}H_{18}O_3$
2-[Bis(4-hydroxyphenyl)methyl]benzenemethanol.
Egmol; Regolax. Laxative-cathartic. mp = 201-202°; [monoacetate]: mp = 171-174°; [triacetate ($C_{26}H_{24}O_6$)]: mp = 104-106°.

8623 Phenoltetrachlorophthalein
639-44-1 7398 211-354-2
$C_{20}H_{10}Cl_4O_4$
4,5,6,7-Tetrachloro-3,3-bis(4-hydroxyphenyl)-1(3H)-isobenzofuranone.
Laxative-cathartic. Dec >300°; insoluble in H_2O, $CHCl_3$, C_6H_6; soluble in EtOH, Et_2O, Me_2CO, AcOH.

8624 Phenoltetrachlorophthalein Disodium Salt
128-71-2 7398
$C_{20}H_8Na_2Cl_4O_4$
4,5,6,7-Tetrachloro-3,3-bis(4-hydroxyphenyl)-1(3H)-isobenzofuranone disodium salt.
Chlor-Tetragnost. Used as a diagnostic aid for hepatic function. Laxative-cathartic. Freely soluble in H_2O.

8625 Picosulfate Sodium
10040-45-6 7559 233-120-9
$C_{18}H_{13}NNa_2O_8S_2$
4,4'-(2-Pyridylmethylene)diphenolbis(hydrogen sulfate) disodium salt.
LA-391; DA-1773; Guttalax-Fher; Laxoberal; Laxoberon; Neopax; Pico-Salax. Laxative-cathartic. mp = 272-275°; λ_m 218 262 nm (ε 20450, 6075 H_2O); readily soluble in H_2O, slightly soluble in EtOH, insoluble in most organic solvents. *Ist. De Angeli.*

8626 Poloxamers
106392-12-5 7722
$HO(C_2H_4O)_a(C_2H_4O)_b(C_2H_4O)_aH$
Block copolymer of oxirane with methyloxirane.
Lutrol F; Pluracare; Pluronic. Base for ointments or suppositories, surfactant, tablet binder, emulsifying agent or tablet coating agent. Laxative-cathartic. Soluble in H_2O, (more at lower temperatures), aromatic solvents, chlorinated solvents, Me_2CO, EtOH, propylene glycol, hexylene glycol, butyl cellosolve, butyl carbitol, MEK, cyclohexanone; insoluble in ethylene glycol, kerosene, mineral oil. *BASF Corp.*

8627 Potassium Bisulfate
7646-93-7 7774 231-594-1
HKO_4S
Potassium hydrogen sulfate.
potassium acid sulfate. Laxative-cathartic. d = 2.24; mp = 197°; soluble in H_2O (55.5 g/100 ml at 25°, 117.6 g/100 ml at 100°).

8628 Potassium Bitartrate
868-14-4 7776 212-769-1
$C_4H_5KO_6$
[R-(R*,R*)]-2,3-Dihydroxybutanedioic acid monopotassium salt.
argol; potassium hydrogen tartrate; cream of tartar; cremor tartari; faecula; faecla. Diuretic; laxative; cathartic. A crystalline crust deposited on the sides of the vat in which grape juice has been fermented; it contains 40-70% tartaric acid, principally as potassium hydrogen tartrate. Used as a laxative, cathartic and diuretic. Soluble in H_2O (0.62 g/100 ml at 25°, 6.25 g/100 ml at 100°), EtOH (0.011 g/100 ml), dilute mineral acids, alkaline solutions.

8629 Potassium Phosphate, Dibasic
7758-11-4 7828 231-834-5
HK_2O_4P
Dipotassium hydrogen phosphate.
dipotassium phosphate; dikalium phosphate; DKP; Isolyte. Cathartic; calcium regulator. Also used as a buffering agent. Soluble in H_2O (150 g/100 ml at 25°); slightly soluble in EtOH. *McGaw Inc.*

8630 Potassium Sodium Tartrate
304-59-6 7840 206-156-8
$C_4H_4KNaO_6.4H_2O$
[R-(R*,R*)]-2,3-Dihydroxybutanedioic acid monopotassium monosodium salt tetrahydrate.
Rochelle salt; Seignette salt. Laxative-cathartic. d = 1.79; mp = 78-80°; soluble in H_2O (111 g/100 ml), insoluble in EtOH.

8631 Potassium Sulfate
7778-80-5 7845 231-915-5
K_2O_4S
Sulfuric acid potassium salt.
Sal polychrestum; arcanum duplicatum; tartarus vitriolatus. Laxative-cathartic. d = 2.66; mp = 1067°; soluble in H_2O (12 g/100 ml at 25°, 25 g/100 ml at 100°), glycerol (1.3 g/100 ml); insoluble in EtOH.

8632 Potassium Sulfite
10117-38-1 7847 233-321-1
K_2O_3S
Laxative-cathartic. Soluble in H_2O (28.5 g/100 ml), slightly soluble in EtOH.

8633 Potassium Tartrate
921-53-9 7849 213-067-8
$C_4H_4K_2O_6$
soluble tartar. Laxative-cathartic. d = 1.98; soluble in H_2O (200 g/100 ml), slightly soluble in EtOH.

8634 Psyllium Husk
7674
Plantago seed; Plantain seed; Flea seed; Hydrocil Instant; Modane Bulk; Mylanta Fiber; Syllact; Metamucil; Seratan; Fibrolax; Fibrogel; Isogel; Regulan; Effer-Syllium; Fiberall; component of: Perdiem, Perdiem Fiber. Laxative-cathartic. The seed from *Plantago ovata Forsk.*, *Plantaginaceae* (Indian Plantago). *Carter-Wallace; Johnson & Johnson-Merck Consumer Pharm.; Rhône-Poulenc Rorer Pharm. Inc.; Savage Labs; Solvay Pharm. Inc.*

8635 Seidlitz Mixture
8014-63-9 8567
A mixture of 3 parts Rochelle salt and 1 part sodium bicarbonate; a Seidlitz powder consists of 10g of the mixture with 2.17g tartaric acid. Laxative-cathartic.

8636 Senna
8599
Dried leaves from various plants such as *Cassia senna*. Senokot; Perdiem. Laxative-cathartic. *Purdue Pharma L.P.; Rhône-Poulenc Rorer Pharm. Inc.*

8637 Sennoside A
81-27-6 8600 201-339-9
$C_{42}H_{38}O_{20}$
(R*,R*)-5,5'-Bis(β-D-glucopyranosyloxy)-9,9',10,10'-tetrahydro-4,4'-dihydroxy-10,10'-dioxo[9,9'-bi-anthracene]-2,2'-dicarboxylic acid.
Gentle Nature; Glysennid. Laxative-cathartic. Dec 200-240°; $[\alpha]_D^{20}$ = -164° (c = 0.1 in 60% Me_2CO), -147° (c = 0.1 in 70% Me_2CO), -24° (c = 0.2 in 70% dioxane); insoluble in H_2O, C_6H_6, Et_2O, $CHCl_3$; sparingly soluble in MeOH, carbitol, Me_2CO, dioxane; isomerized slowly to Sennoside B in $NaHCO_3$ solution. *Sandoz Pharm. Corp.*

8638 Sennoside B
128-57-4 8600 204-895-0
$C_{42}H_{38}O_{20}$
(R*,S*)-5,5'-Bis(β-D-glucopyranosyloxy)-9,9',10,10'-tetrahydro-4,4'-dihydroxy-10,10'-dioxo[9,9'-bi-anthracene]-2,2'-dicarboxylic acid.
Laxative-cathartic. Dec 180-186°; $[\alpha]_D^{20}$ = -100° (c = 0.2 in 70% Me_2CO), -67° (c = 0.4 in 70% dioxane); λ_m 270 308 355 nm (0.5% $NaHCO_3$); generally more soluble than Sennoside A.

8639 Sodium Phosphate, Dibasic [anhydrous]
7558-79-4 8805
Na_2HPO_4
Disodium hydrogen phosphate.
exsiccated sodium phosphate; component of: Metro I.V. Laxative-cathartic. Soluble in H_2O (12.5 g/100 ml at 25°, much more in hot H_2O), insoluble in EtOH. *McGaw Inc.*

8640 Sodium Phosphate, Dibasic, Dihydrate
10028-24-7 8805
$Na_2HPO_4.2H_2O$
Disodium hydrogen phosphate dihydrate.
Sorenson's phosphate; Sorenson's sodium phosphate; component of: Metro I.V. Laxative-cathartic. *McGaw Inc.*

8641 Sodium Phosphate, Dibasic, Dodecahydrate
10039-32-4 8805
$Na_2HPO_4.12H_2O$
Disodium hydrogen phosphate dodecahydrate.
component of: Metro I.V. Laxative-cathartic. mp = 34-35°; readily loses 5 moles of H_2O to give the heptahydrate; d = 1.5; soluble in H_2O (33 g/100 ml), insoluble in EtOH. *McGaw Inc.*

8642 Sodium Phosphate, Dibasic, Heptahydrate
7782-85-6 8805 231-448-7
$Na_2HPO_4.7H_2O$
Disodium hydrogen phosphate heptahydrate.
component of: Metro I.V. Laxative-cathartic. d = 1.7; soluble in H_2O (25 g/100 ml), insoluble in EtOH; LD_{50} (rat orl) = 12930 mg/kg. *McGaw Inc.*

8643 Sodium Succinate
150-90-3 8828 205-778-7
$C_4H_4Na_2O_4$
Succinic acid sodium salt.
disodium succinate; Soduxin. Respratory stimulant; analeptic; urinary alkalizer; diuretic. Laxative-cathartic. Soluble in H_2O (20 g/100 ml), insoluble in EtOH; LD_{50} (mus iv) = 4500 mg/kg.

8644 Sodium Sulfate [anhydrous]
7757-82-6 8829 231-820-9
Na_2SO_4
Sulfuric acid disodium salt.
mirabilite; thenardite; salt cake; component of: Colyte. Calcium regulator . Laxative-cathartic. mp = 800°; d = 2.7; soluble in H_2O (50 g/100 ml at 33°; solubility decreases with temperature to 41.6 g/100 ml at 100°); insoluble in EtOH. *Schwarz Pharma Kremers Urban Co.*

8645 Sodium Sulfate Decahydrate
7727-73-3 8829
$Na_2SO_4.10H_2O$
Sulfuric acid disodium salt decahydrate.
Glauber's salt; component of: Colyte. Calcium regulator. Laxative-cathartic. mp = 32.4°; d = 1.46; loses all H_2O at 100°; soluble in H_2O (66 g/100 ml at 25°, 30 g/100 ml at 15°); soluble in glycerol; insoluble in EtOH. *Schwarz Pharma Kremers Urban Co.*

8646 Sodium Tartrate
868-18-8 8833 212-773-3
$C_4H_4Na_2O_6$
Tartaric acid disodium salt.
Laxative-cathartic. Loses H_2O at 120°; d = 1.82; soluble

in H_2O (33 g/100 ml at 25°, 66 g/100 ml at 100°), insoluble in EtOH.

8647 Sorbitol
50-70-4 8873 200-061-5
$C_6H_{14}O_6$
D-Glucitol.
Sorbo; L-gulitol; Cystosol; Resulax; Sorbilax; Sorbitur; Sorbostyl; Sorbilande; component of: Probilagol. Flavor, tablet excipient. Laxative-cathartic. mp = 110-112°; $[\alpha]_D^{20}$= -2° (H_2O); freely soluble in H_2O (83 g/100 ml); soluble in hot EtOH; less in cold EtOH; soluble in MeOH, iPrOH, BuOH, cyclohexanol, phenol, Me_2CO, AcOH, DMF, C_5H_5N, acetamide; insoluble in other organic solvents. *ICI Americas Inc.; Pfanstiehl Labs Inc.; Purdue Pharma L.P.*

8648 Sulisatin
54935-03-4 9156
$C_{21}H_{17}NO_9S_2$
3,3-Bis(p-hydroxyphenyl)-7-methyl-2-indolinone bis(hydrogen sulfate) (ester).
Stimulates the motility of the large intestine. Laxative-cathartic.

8649 Sulisatin Disodium Salt
54935-04-5 9156 259-399-7
$C_{21}H_{15}NNa_2O_9S_2$
3,3-Bis(p-hydroxyphenyl)-7-methyl-2-indolinone bis(hydrogen sulfate) (ester) disodium salt.
DAN-603; Laxitex. Stimulates the motility of the large intestine. Laxative-cathartic. mp > 360°.

8650 Yellow Phenolphthalein
8053-05-2 10230
A by-product in the manufacture of phenolphthalein; more effective laxative than phenolphthalein.
Feen-A-Mint Gum; component of: Correctol Caplets, Correctol Tablets, Dialose Plus, Feen-A-Mint Pills, Unilax. Laxative-cathartic. d_4^{20}= 1.290 - 1.296; mp = 255-260°; soluble in EtOH (8.3 g/100 ml), Et_2O (1 g/100 ml). *Ascher B.F. & Co.; Johnson & Johnson-Merck Consumer Pharm.; Schering-Plough Pharm.*

Leukotriene Antagonists

8651 Ablukast
96566-25-5
$C_{28}H_{34}O_8$
(±)-6-Acetyl-7-[[5-(4-acetyl-3-hydroxy-2-propylphenoxy)pentyl]oxy]-2-chromancarboxylic acid.
Ro-23-3544/000. Leukotriene antagonist. Used as an antiasthmatic. *Hoffmann-LaRoche Inc.*

8652 Ablukast Sodium
96565-55-8
$C_{28}H_{33}NaO_8$
Sodium (±)-6-acetyl-7-[[5-(4-acetyl-3-hydroxy-2-propylphenoxy)pentyl]oxy]-2-chroman-carboxylate.
Ro-23-3544/001; Ulpax. Leukotriene antagonist. Used as an antiasthmatic. *Hoffmann-LaRoche Inc.*

8653 Cinalukast
128312-51-6
$C_{23}H_{28}N_2O3S$
3'-[(E)-2-(4-Cyclobutyl-2-thiazolyl)vinyl]-2,2-diethylsuccinanilic acid.
Ro-24-5913. Leukotriene antagonist. Used as an antiasthmatic. *Hoffmann-LaRoche Inc.*

8654 Ibudilast
50847-11-5 4923
$C_{14}H_{18}N_2O$
1-(2-Isopropylpyrazolo[1,5-a]pyridin-3-yl)-2-methyl-1-propanone.
KC-404; Ketas. Antiallergic; antiasthmatic; vasodilator (cerebral). Phosphodiesterase inhibitor. Leukotriene D_4 antagonist. mp = 53.5-54°; slightly soluble in H_2O, freely soluble in organic solvents; LD_{50} (mus iv) = 260 mg/kg. *Kyorin Pharm. Co. Ltd.*

8655 Linetastine
159776-68-8
$C_{35}H_{40}N_2O_6$
1-[5'-(3-Methoxy-4-ethoxycarbonyloxyphenyl)-2',4'-pentadienoyl aminoethyl]-4-diphenylmethoxy-piperidine.
TMK-688. Antihistaminic; leukotriene antagonist. Antihistaminic; leukotriene antagonist.

8656 Montelukast
158966-92-8 6340
$C_{35}H_{36}ClNO_3S$
1-[[[(R)-m-[(E)-2-(7-Chloro-2-quinolyl)vinyl]-α-[o-(1-hydroxy-1-methylethyl)phenethyl]benzyl]thio]-methyl]cyclopropane acetic acid.
Selective leukotriene D_4 receptor antagonist. Used as an antiasthmatic. *Merck & Co.Inc.*

8657 Montelukast Sodium
151767-02-1 6340
$C_{35}H_{35}ClNNaO_3S$
Sodium 1-[[[(R)-m-[(E)-2-(7-chloro-2-quinolyl)vinyl]-α-[o-(1-hydroxy-1-methylethyl)phenethyl]-benzyl]thio]-methyl]cyclopropane-acetate.
Singulair; MK-476. Selective leukotriene D_4 receptor antagonist. Used as an antiasthmatic. *Merck & Co.Inc.*

8658 Ontazolast
147432-77-7
$C_{21}H_{25}N_3O$
2-[[(S)-2-Cyclohexyl-1-(2-pyridyl)ethyl]amino]-5-methylbenzoxazole.
BIRM-270. Leukotriene antagonist. Used as an anti-asthmatic. *Boehringer Ingelheim Pharm. Inc.*

8659 Pobilukast
107023-41-6
$C_{26}H_{44}O_5S$
(2S,3R)-3-[(2-Carboxyethyl)thio]-3-[o-(8-phenyloctyl)-phenyl]lactic acid.
Leukotriene antagonist. Used as antiasthmatic. *SmithKline Beecham Pharm.*

8660 Pobilukast Edamine
137232-03-2
$C_{28}H_{42}N_2O_5S.H_2O$
(2S,3R)-3-[(2-Carboxyethyl)thio]-3-[o-(8-phenyloctyl)-phenyl]lactic acid compound with ethylene diamine (1:1) monohydrate.
SK&F-104353-Q. Leukotriene antagonist. Used as an antiasthmatic. *SmithKline Beecham Pharm.*

8661 Pranlukast
103177-37-3 7889
$C_{27}H_{23}N_5O_4$
N-[4-Oxo-2-(1H-tetrazol-5-yl)-4H-1-benzopyran-8-yl]-p-(4-phenylbutoxy)benzamide.
ONO-1078; ONO-RS-411; SB-205312; Onon. Leukotriene antagonist. Used as an antiasthmatic. mp = 244-245°; LD_{50} (mmus iv) > 1000 mg/kg.

8662 Ritolukast
111974-60-8
$C_{17}H_{13}F_3N_2O_3S$
1,1,1-Trifluoro-α-2-quinolylmethanesulfon-m-anisidide.
WY-48252. Leukotriene antagonist. Used as an antiasthmatic. *Wyeth-Ayerst Labs.*

8663 Sulukast
98116-53-1
$C_{25}H_{36}N_4O_3S$
3-[[(1R,2E,4Z)-1-[(αS)-α-Hydroxy-m-1H-tetrazol-5-ylbenzyl]-2,4-tetradecadienyl]thio]-propionic acid.
LY-170680. Leukotriene antagonist. Used as an antiasthmatic. *Eli Lilly & Co.*

8664 Tomelukast
88107-10-2
$C_{16}H_{22}N_4O_3$
2'-Hydroxy-3'-propyl-4'-[4-(1H-tetrazol-5-yl)butoxy]-acetophenone.
LY-171883. Leukotriene antagonist. Used as an antiasthmatic. *Eli Lilly & Co.*

8665 Verlukast
120443-16-5
$C_{26}H_{27}ClN_2O_3S_2$
3-[[(αR)-m-[(E)-2-(7-Chloro-2-quinolyl)vinyl]-α-[[2-(dimethylcarbamoyl)ethyl]thio]benzyl]thio]-propionic acid.
MK-679; L-668019. Leukotriene antagonist. Used as an antiasthmatic. *Merck & Co.Inc.*

8666 Zafirlukast
107753-78-6 10241
$C_{31}H_{33}N_3O_6S$
Cyclopentyl 3-[2-methoxy-4-[(o-tolylsulfonyl)-carbamoyl]benzyl]-1-methylindole-5-carbamate.
N-[4-[5-(cyclopentyloxycarbonyl)amino-1-methylindol-3-ylmethyl]-3-methoxybenzoyl]-2-methylbenzenesulfonate;
Accolate; ICI-204219. Leukotriene D_4 antagonist. Used as an antiasthmatic. mp = 138-140°. *ICI; Zeneca Pharm.*

LH-RH Agonists

8667 Buserelin
57982-77-1 1527 261-061-9
$C_{60}H_{86}N_{16}O_{13}$
6-[O-(1,1-Dimethylethyl)-D-serine]-9-(N-ethyl-L-prolinamide)-10-deglycinamideluteinizing hormone-releasing factor (pig).
5-oxoPro-His-Trp-Ser-Tyr-D-Ser(t-Bu)-Leu-Arg-$ProNHCH_2CH_3$. Hormonal antineoplastic; gonad-stimulating principle. Nonapeptide agonist of LH-RH. $[\alpha]_D^{20}$ = -40.4° (c = 1 dimethylacetamide). *Hoechst AG.*

8668 Buserelin Acetate
68630-75-1 1527
$C_{62}H_{88}N_{16}O_{14}$
6-[O-(1,1-Dimethylethyl)-D-serine]-9-(N-ethyl-L-prolinamide)-10-deglycinamideluteinizing hormone-releasing factor (pig) acetate.
HOE-766; Receptal; Suprecur; Suprefact; Suprafact. Hormonal antineoplastic; gonad-stimulating principle. Nonapeptide agonist of LH-RH. *Hoechst AG.*

8669 Deslorelin
57773-65-6 2968
$C_{64}H_{83}N_{17}O_{12}$
6-D-Tryptophan-9-(N-ethyl-prolinamide)-10-deglycinamideluteinizing hormone-releasing factor (pig).
5-oxoPro-His-Trp-Ser-Tyr-D-Trp-Leu-Arg-Pro-$NHCH_2$-CH_3; D-Trp^6,des-Gly^{10}-LH-RH ethylamide; Somagorad. LH-RH agonist. Used in treatment of precocious puberty. $[\alpha]_D^{24}$ = -61° (c = 0.37 0.1M AcOH).

8670 Detirelix
89662-30-6
$C_{78}H_{105}ClN_{18}O_{13}$
N-Acetyl-3-(2-naphthalenyl)-D-alanyl-4-chloro-D-phenylalanyl-D-tryptophyl-L-seryl-L-tyrosyl-N^6-[(ethylamino)(ethylimino)methyl]-D-lysyl-L-luecyl-L-arginyl-L-prolyl-D-alaninamide.
LH-RH agonist. *Syntex Intl. Ltd.*

8671 Detirelix Acetate
102583-46-0
$C_{82}H_{113}Cl\ N_{18}O_{17}$
N-Acetyl-3-(2-naphthalenyl)-D-alanyl-4-chloro-D-phenylalanyl-D-tryptophyl-L-seryl-L-tyrosyl-N^6-[(ethylamino)(ethylimino)methyl]-D-lysyl-L-luecyl-L-arginyl-L-prolyl-D-alaninamide diacetate (salt).
RS-68439. LH-RH agonist. *Syntex Intl. Ltd.*

8672 Goserelin
65807-02-5 4547
$C_{59}H_{84}N_{18}O_{14}$
6-[O-(1,1-Dimethylethyl)-D-serine]-10-deglycineanide luteinizing hormone-releasing factor (pig)-2-(aminocarbonyl)hydrazide.
D-Ser-$(Bu^t)^6Azgly^{10}$-gonadorelin; D-Ser-$(Bu^t)^6Azgly^{10}$-luliberin; ICI-118630; Zoladex; 5-oxoPro-His-Trp-Ser-Tyr-D-Ser(t-Bu)-Lue-Arg-Pro-$NHNHCONH_2$. LH-RH agonist. Hormonal antineoplastic used in treatment of prostatic carcinoma. *ICI.*

8673 Histrelin

76712-82-8 4760

$C_{66}H_{86}N_{18}O_{12}$

6-[1-(Phenylmethyl)-D-histidine]-9(n-ethyl-L-prolinamide)-10-deglycineamide luteinizing hormone-releasing factor (pig).

5-oxo-L-prolyl-L-histidyl-L-tryptophyl-L-seryl-L-tyrosyl-N-benzyl-D-histidyl-L-leucyl-L-arginyl-N-ethyl-L-prolinamide; [(im-Bzl)-D-His^6,Pro^9-Net]-gonadotropin-releasing hormone; ORF-17070; RWJ-17070; Supprelin. Used in treatment of precocious puberty. Gonadotropin-releasing hormone. Nonapeptide synthetic LH-RH agonist. $[\alpha]_D^{20}$ = -33.9° (c = 1 AcOH). *Ortho Pharm. Corp.; Roberts Pharm. Corp.*

8674 Leuprolide

53714-56-0 5484

$C_{59}H_{84}N_{16}O_{12}$

6-D-Leucine-9-(N-ethylprolinamide)-10-deglycineamide luteinizing hormone-releasing factor (pig).

leuprorelin; (D-leu^6)-des-Gly^{10}-LH-RH-ethylamide. LH-RH agonist. Hormonal antineoplastic used in treatment of prostatic carcinoma. $[\alpha]_D^{25}$ = -31.7° c = 1 1% AcOH). *Abbott Labs Inc.; Takeda Chem. Ind. Ltd.*

8675 Leuprolide Acetate

74381-53-6 5484

$C_{61}H_{88}N_{16}O_{14}$

6-D-Leucine-9-(N-ethylprolinamide)-10-deglycineamide luteinizing hormone-releasing factor (pig) monoacetate.

leuprolide acetate; Abbott 43818; A-43818; TAP-144; Carcinil; Enantone; Leuplin; Lucrin; Lupron; Prostap. LH-RH agonist.

8676 Lutropin Alfa

152923-57-4

Luteinizing hormone (human α-subunit reduced complex human β-subunit reduced), glycoform α.

Luteinizing hormone; LH; interstitial cell stimulating hormone. A pituitary hormone that in the female stimulates development of corpora lutea and contributes to progesterone secretion and in the male stimulates development of interstitial tissue in the testis and secretion of testosterone. Luteolytic.

8677 Nafareline

76932-56-4 6437

$C_{66}H_{83}N_{17}O_{13}$

6-[3-(2-Naphthalenyl)-D-alanine] luteinizing hormone-releasing factor (pig).

5-oxo-L-prolyl-L-histidyl-L-tryptophyl-L-seryl-L-tyrosyl-3-(2-naphthyl)-D-alanyl-L-leucyl-L-arginyl-L-prolylglycinamide; [6-[3-(2-naphthyl)-D-alanine] LHRH; NAG; D-nal(2)^6-LHRH. LH-RH agonist. Used in treatment of endometriosis. *Syntex Intl. Ltd.*

8678 Nafareline Acetate Hydrate

86220-42-0 6437

$C_{66}H_{83}N_{17}O_{13}.xC_2H_4O_2.yH_2O$

6-[3-(2-Naphthalenyl)-D-alanine]luteinizing hormone-releasing factor (pig) acetate hydrate.

RS-94991-298; Nacenyl; Nasanyl; Synarel; Synrelina. LH-RH agonist. *Syntex Intl. Ltd.*

8679 Teverelix

144743-92-0

$C_{74}H_{100}ClN_{15}O_{14}$

N-Acetyl-3-(2-naphthyl)-D-alanyl-p-chloro-L-phenylalanyl-3-(3-pyridyl)-D-alanyl-L-seryl-L-tyrosyl-N^6-carbamoyl-D-lysyl-L-leucyl-N^6-isopropyl-L-lysyl-L-prolyl-D-alaninamide.

Antarelix. Potent LH-RH antagonist.

8680 Triptorelin

57773-63-4 9878

$C_{64}H_{82}N_{18}O_{13}$

6-D-Tryptophan luteinizing hormone-releasing factor (pig).

6-D-tryptophan-LH-RH; D-trp^6-LHRH; D-Trp^{6}LRH; D-trp^6-gonaorelin; Ay-25650; Wy-42462; Wy-42422. LH-RH agonist. Hormonal antineoplastic used in treatment of prostatic carcinoma. $[\alpha]_D^{23}$ = -58.8° (c = 0.33 AcOH). *Wyeth-Ayerst Labs.*

8681 Triptorelin Acetate

140194-24-7 9878

$C_{66}H_{86}N_{18}O_{15}$

6-D-Tryptophanluteinizing hormone-releasing factor (pig) acetate.

Decapeptyl. LH-RH agonist. Hormonal antineoplastic used in treatment of prostatic carcinoma.

Lipotropics

8682 DL-N-Acetylmethionine

65-82-7 97 200-617-7

$C_7H_{13}NO_3S$

N-Acetylmethionine.

Methionamine. The DL-form is a lipotropic agent. mp = 114-115°; [D-(+)-form)]: mp = 104-105°; $[\alpha]_D^{25}$ = 20.3 (c = 4 H_2O); [L-(-)-form]: mp = 104 °; $[\alpha]_D^{25}$ = -20.3°.

8683 Choline Chloride

67-48-1 2261 200-655-4

$C_5H_{14}ClNO$

2-Hydroxy-N,N,N-trimethylethanaminium chloride.

Biocolina; Hepacholine; Lipotril. A lipotropic agent. Used in veterinary medicine as a nutritional factor and a dietary source of choline. Very soluble in H_2O, EtOH; LD_{50} (rat ip) = 400 mg/kg, (rat orl) = 6640 mg/kg.

8684 Choline Dehydrocholate

4201-78-9 2262 224-106-3

$C_{29}H_{48}NO_6$

Dehydrocholic acid salt of choline.

Biscolan. A lipotropic agent. mp = 196-198°.

8685 Choline Dihydrogen Citrate

77-91-8 2263 201-068-6

$C_{11}H_{21}NO_8$

(2-Hydroxyethyl)trimethylammonium citrate.

Chothyn; Cirrocolina; Citracholine. A lipotropic agent. Used in veterinary medicine as a nutritional factor and a dietary source of choline. mp = 105-107.5°; freely soluble in H_2O; slightly soluble in EtOH; insoluble in C_6H_6, $CHCl_3$, Et_2O.

8686 Inositol
87-89-8 5008 201-781-2
$C_6H_{12}O_6$
Hexahydroxycyclohexane.
meso-inositol; I-inositol; cyclohexanehexol; cyclohexitol; meat sugar; inosite; mesoinosite; phaseomannite; dambose; nucite; bios I; rat antispectacled eye factor; mouse antialopecia factor. A lipotropic agent. Vitamin, vitamin source. mp = 225-227°; d = 1.752; soluble in H_2O (14 g/100 ml at 25°, 28 g/100 ml at 60°, slightly soluble in EtOH, insoluble on other organic solvents.

8687 Lecithin
8002-43-5 5452 232-307-2
Phosphatidylcholine.
Lecithol; Vitellin; Kelecin; Granulestin. A lipotropic agent. Insoluble in H_2O; soluble in EtOH, $CHCl_3$, Et_2O, petroleum ether; sparingly soluble in C_6H_6; insoluble in Me_2CO; d_4^{24}= 1.0305.

8688 Methionine
63-68-3 6053 200-562-9
$C_5H_{11}NO_2S$
L-2-amino-4-(methylthio)butyric acid.
L-methionine; Met; M; Acimethin. Hepatoprotectant; antidote for acetaminophen poisoning. Urinary acidifier. An essential amino acid. mp = 280-282° (dec); $[\alpha]_D^{25}$= -8.11° (c = 0.8), $[\alpha]_D^{20}$= 23.40° (c = 5.0 3N HCl).

8689 D-Methionine
348-67-4 6053 206-483-6
$C_5H_{11}NO_2S$
D-2-Amino-4-(methylthio)butyric acid.
D-methionine. Hepatoprotectant; antidote for acetaminophen poisoning. Urinary acidifier. $[\alpha]_d^{25}$= 8.12° (c = 0.8), $[\alpha]_D^{25}$= -21.18° (c = 0.8 0.2N HCl).

8690 DL-Methionine
59-51-8 6053 200-432-1
$C_5H_{11}NO_2S$
DL-2-Amino-4-(methylthio)butyric acid.
racemethionine; DL-methionine; Amurex; Banthionine; Dyprin; Lobamine; Metione; Pedameth; Urimeth. Hepatoprotectant; antidote for acetaminophen poisoning. Urinary acidifier. mp = 281° (dec); d = 1.340; soluble in H_2O (1.82 g/100 ml at 0°, 3.38 g/100 ml at 25°, 6.07 g/100 ml at 60°, 10.52 g/100 ml at 75°, 17.60 g/100 ml at 100°); slightly soluble in EtOH; insoluble in Et_2O.

Lupus Erythematosus Suppressants

8691 Bismuth Sodium Trigylcollamate
5798-43-6 1322
$C_{24}H_{28}BiN_4Na_7O_{25}$
Nitrilotriacetic acid bismuth complex sodium salt.
Bistrimate. Lupus erythematosus suppressant. Very soluble in H_2O; insoluble in Me_2CO, Et_2O, C_6H_6.

8692 Bismuth Subsalicylate
14882-18-9 1327 238-953-1
$C_7H_5BiO_4$
2-Hydroxybenzoic acid bismuth (3+) salt.
basic bismuth salicylate; oxo(salicylato)bismuth; Bismogenol Tosse Inj.; Stabisol. Antidiarrheal; antacid; antiulcerative. Also used as a lupus erythematosus suppressant. Insoluble in H_2O, EtOH. *Mobay.*

8693 Chloroquine
54-05-7 2215 200-191-2
$C_{18}H_{26}ClN_3$
7-Chloro-4-[[4-(diethylamino)-1-methylbutyl]amino]-quinoline.
Aralen; SN-7618; RP-3377; Artrichin; Bemaphate; Capquin; Nivaquine B; Resoquine; Reumachlor; Sanoquin; nivaquine [as sulfate]. Antiamebic; antimalarial; antirheumatic. Also used as a lupus erythematosus suppressant. mp = 87°. *Sterling Winthrop Inc.*

8694 Chloroquine Diphosphate
50-63-5 2215 200-055-2
$C_{18}H_{32}ClN_3P_2O_8$
7-Chloro-4-[[4-(diethylamino)-1-methylbutyl]amino]-quinoline diphosphate.
Aralen phosphate; Arechin; Avloclor; Imagon; Malaquin; Resochin; Tresochin. Antiamebic; antimalarial; antirheumatic. Also has activity as a lupus erythematosus suppressant. Also used as a lupus erythematosus suppressant. mp = 193-195°, 215-218°; soluble in H_2O, insoluble in organic solvents; pH (1% solution) ≅ 4.5. *Sterling Winthrop Inc.*

8695 Hydroxychloroquine
118-42-3 4863 204-249-8
$C_{18}H_{26}ClN_3O$
2-[[4-[(7-Chloro-4-quinolyl)amino]pentyl]ethylamino]-ethanol.
oxychloroquine; oxichloroquine. Antimalarial and lupus erythematosus suppressant. Antirheumatic agent. Also an antiamebic, antimalarial. Used in combination with cyclophosphamide and azathioprine in treatment of rheumatoid arthritis. mp = 89-91°. *Apothecon; Sterling Winthrop Inc.*

8696 Hydroxychloroquine Sulfate
747-36-4 4863 212-019-3
$C_{18}H_{28}ClN_3O_5S$
2-[[4-[(7-Chloro-4-quinolyl)amino]pentyl]ethylamino]-ethanol sulfate (1:1) (salt).
Plaquenil Sulfate; Ercoquin; Quensyl. Antimalarial and lupus erythematosus suppressant. mp = 198°, 240°; freely soluble in H_2O; insoluble in EtOH, $CHCl_3$, Et_2O; pH (aqueous solutions) ≅ 4.5. *Apothecon; Sterling Winthrop Inc.*

Mineralocorticoids

8697 Aldosterone
52-39-1 226 200-139-9
$C_{21}H_{28}O_5$
11β,21-Dihydroxy-3,20-dioxo-pregn-4-en-18-al.
Aldocorten; Aldocortin; Electrocortin; Aldocortene. Mineralocorticoid. mp = 108-112°; $[\alpha]_D^{23}$= +152° (c = 2 in Me_2CO); λ_m = 240 nm (log ε 420).

8698 Amcinonid
51022-69-6 407 256-915-2
$C_{28}H_{35}FO_7$
(11β,16α)-21-(Acetyloxy)-16,17-[cyclopentylidene-bis(oxy)]-9-fluoro-11-hydroxypregna-1,4-diene-3,20-dione.
CL-34699; Amciderm; Cyclocort; Penticort. Glucocorticoid. Used to treat eczema. A high-potency topical and systemic corticosteroid. *Am. Cyanamid; Lederle Labs.*

8699 Betamethasone
378-44-9 1226 206-825-4
$C_{22}H_{29}FO_5$
(11β,16β)-9-Fluoro-11,17,21-trihydroxy-16-methyl-pregna-1,4-diene-3,20-dione.
betadexamethasone; flubenisolone; Sch-4831; NSC-39470; beta-Corlan; Becort; Betasolon; Betnelan; Celestene; Celestone; Dermabet; Diprolene; Visubeta. Glucocorticoid. mp = 231-234° (dec); $[\alpha]_D$ = +108° (in Me_2CO); λ_m = 238 nm (ε 15200 in EtOH). *Merck & Co.Inc.; Roussel-UCLAF.*

8700 Betamethasone Acetate
987-24-6 1226 213-578-6
$C_{24}H_{31}FO_6$
(11β,16β)-9-Fluoro-11,17,21-trihydroxy-16-methyl-pregna-1,4-diene-3,20-dione 21-acetate.
Betafluorene, Celestovet; component of: Betavet Soluspan, Celestone Soluspan. Glucocorticoid. mp = 205-208° [also reported as 196-201°]; $[\alpha]_D$ = +140° (in $CHCl_3$); λ_m = 238 nm (ε 14800). *Merck & Co.Inc.; Roussel-UCLAF; Schering AG.*

8701 Betamethasone Acibutate
5534-05-4 226-885-5
$C_{28}H_{37}FO_7$
(11β,16β)-9-Fluoro-11,17,21-trihydroxy-16-methyl-pregna-1,4-diene-3,20-dione 21-acetate 17-isobutyrate.
Glucocorticoid. *Merck & Co.Inc.; Roussel-UCLAF.*

8702 Betamethasone Adamantoate
1226
$C_{33}H_{43}FO_6$
(11β,16β)-9-Fluoro-11,17,21-trihydroxy-16-methyl-pregna-1,4-diene-3,20-dione 21-adamantoate.
Betsovet. Glucocorticoid. *Merck & Co.Inc.; Roussel-UCLAF.*

8703 Betamethasone Benzoate
22298-29-9 1226 244-897-9
$C_{29}H_{33}FO_6$
(11β,16β)-9-Fluoro-11,17,21-trihydroxy-16-methyl-pregna-1,4-diene-3,20-dione 17-benzoate.
W-5975; Bebate; Beben; Benisone; Euvaderm; Flurobate; Parbetan; Uticort. Glucocorticoid. mp = 225-228°; $[\alpha]_D^{24}$ = +63.5° (in dioxane). *Merck & Co.Inc.; Parke-Davis; Roussel-UCLAF.*

8704 Betamethasone Dipropionate
5593-20-4 1226 227-005-2
$C_{28}H_{37}FO_7$
(11β,16β)-9-Fluoro-11,17,21-trihydroxy-16-methylpregna-1,4-diene-3,20-dione 17,21-dipropionate.
betamethasone dipropionate; Sch-11460; Diprolene; Diproderm; Diprophos; Diprosis; Diprosone; Maxivate; Psorion; Rinderon-DP; component of: Alphatrex, Betasone, Lotrisone. Glucocorticoid. mp = 170-179° (dec); $[\alpha]_D^{26}$ = +65.7° (in dioxane); λ_m = 238 nm (ε 15700 in MeOH). *ICN Pharm. Inc.; Lemmon Co.; Merck & Co.Inc.; Roussel-UCLAF; Savage Labs; Schering AG.*

8705 Betamethasone Sodium Phosphate
151-73-5 1226 205-797-0
$C_{22}H_{28}FNa_2O_8P$
(11β,16β)-9-Fluoro-11,17,21-trihydroxy-16-methyl-pregna-1,4-diene-3,20-dione 21-phosphate disodium salt.
betamethasone 21-(dihydrogen phospate) disodium salt; Bentelan; Betnesol; Celestan; Durabetason; Vista-Methasone; component of: Betasone, Betavet Soluspan, Celestone Soluspan. Glucocorticoid. *Merck & Co.Inc.; Roussel-UCLAF; Schering AG.*

8706 Betamethasone Valerate
2152-44-5 1226 218-439-3
$C_{27}H_{37}FO_6$
(11β,16β)-9-Fluoro-11,17,21-trihydroxy-16-methyl-pregna-1,4-diene-3,20-dione 17-valerate.
Bedermin; Beta-Val; Betnesol-V; Betneval; Betnovate; Bextasol; Celestan-V; Celestoderm-V; Dermosol; Dermovaleas; Ecoval 70; Hormezon; Tokuderm; Valisone; component of: Betatrex, Gentocin, Topagen. Glucocorticoid. mp = 183-184°; $[\alpha]_D = _7$° (in dioxane); λ_m = 239 nm (ε 15920 in dioxane). *Merck & Co.Inc.; Roussel-UCLAF; Savage Labs; Schering AG.*

8707 Chloroprednisone
52080-57-6 2209 257-644-2
$C_{21}H_{25}ClO_5$
(6α)-Chloro-17,21-dihydroxypregna-1,4-diene-3,11,20-trione.
Glucocorticoid (topical). *Syntex Labs. Inc.*

8708 Chloroprednisone Acetate
14066-79-6 2209 237-919-3
$C_{23}H_{27}ClO_6$
(6α)-Chloro-17,21-dihydroxypregna-1,4-diene-3,11,20-trione 21-acetate.
Topilan. Glucocorticoid (topical). mp = 217-219°; $[\alpha]_D$= +144° (in $CHCl_3$); λ_m = 237 nm (log ε 4.19 in EtOH). *Syntex Labs. Inc.*

8709 Cicortonide
19705-61-4
$C_{29}H_{37}ClFNO_7$
3-(2-Chloroethoxy)-9-fluoro-11β,16α,17,21-tetrahydroxy-20-oxopregna-3,5-diene-6-carbonitrile cyclic 16,17-acetal with acetone.
Glucocorticoid. *Syntex Labs. Inc.*

8710 Ciprocinonide
58524-83-7 261-307-5
$C_{28}H_{34}F_2O_7$
(6α,11β,16α)-21-[(Cyclopropylcarbonyl)oxy]-6,9-difluoro-11-hydroxy-16,17-[(1-methylethylidene)-bis(oxy)]pregna-1,4-diene-3,20-dione.
RS-2386. Adrenocortical steroid. *Syntex Labs. Inc.*

8711 Clobetasol
25122-41-2 2423 246-633-8
$C_{22}H_{28}ClFO_4$
(11β,16β)-21-Chloro-9-fluoro-11,17-dihydroxy-16-methylpregna-1,4-diene-3,20-dione.
Anti-inflammatory. Glucocorticoid. *Glaxo Labs.*

8712 Clobetasol Propionate
25122-46-7 2423 246-634-3
$C_{25}H_{32}ClFO_5$
(11β,16β)-21-Chloro-9-fluoro-11,17-dihydroxy-16-methylpregna-1,4-diene-3,20-dione 17-propionate.
CCl-4725; GR-2/925; Clobesol; Dermoval; Dermovate; Dermoxin; Dermoxinale; Temovate. Anti-inflammatory. Glucocorticoid. mp = 195.5-197°; $[\alpha]_D$ = +103.8° (c = 1.04 in dioxane); λ_m = 237 nm (ε 15000 EtOH). *Glaxo Labs.*

8713 Clobetasone
54063-32-0 2424 258-953-5
$C_{22}H_{26}ClFO_4$
(16β)-21-Chloro-9-fluoro-17-hydroxy-16-methylpregna-1,4-diene-3,11,20-trione.
21-chloro-11-dehydrobetamethasone. Anti-inflammatory. Glucocorticoid. *Glaxo Labs.*

8714 Clobetasone Butyrate
25122-57-0 2424 246-635-9
$C_{26}H_{32}ClFO_5$
(16β)-21-Chloro-9-fluoro-17-hydroxy-16-methylpregna-1,4-diene-3,11,20-trione 17-butyrate.
CCl-5537; GR-2/1214; Emovate; Eumovate; Molivate. Anti-inflammatory. Glucocorticoid. mp = 90-100°. *Glaxo Labs.*

8715 Clocortolone
4828-27-7 2430 225-406-7
$C_{27}H_{36}ClFO_5$
9-Chloro-6α-fluoro-11β,21-dihydroxy-16α-methylpregna-1,4-diene-3,20-dione.
Glucocorticoid. mp = 254° (dec). *Schering AG.*

8716 Clocortolone Acetate
4258-85-9 2430
$C_{22}H_{28}ClFO_4$
(6α,11β,16α)-9-Chloro-6-fluoro-11,21-dihydroxy-16-methylpregna-1,4-diene-3,20-dione.
SH-818. Glucocorticoid. mp = 252° (dec). *Schering AG.*

8717 Clocortolone Pivalate
34097-16-0 2430 251-826-5
$C_{24}H_{30}ClFO_5$
(6α,11β,16α)-9-Chloro-6-fluoro-11,21-dihydroxy-16-methylpregna-1,4-diene-3,20-dione 21-acetate.
CL-68; SH-863; Cloderm; Purantix. Glucocorticoid. *Schering AG.*

8718 Clomegestone Acetate
424-89-5
$C_{24}H_{31}ClO_4$
(16α)-17-(Acetyloxy)-6-chloro-16-methylpregna-1,4-diene-3,20-dione.
SH-741. Progestin. *Schering AG.*

8719 Cloprednol
5251-34-3 2460 226-052-6
$C_{21}H_{25}ClO_5$
(11β)-6-Chloro-11,17,21-trihydroxypregna-1,4,6-triene-3,20-dione.
RS-4691; Cloradryn; Novacort; Synestan. Glucocorticoid. *Syntex Labs. Inc.*

8720 Corticosterone
50-22-6 2601 200-019-6
$C_{21}H_{30}O_4$
(11β)-11,21-Dihydroxypregn-4-ene-3,20-dione.
Kendall's compound B; Reichstein's substance H; compound B. Glucocorticoid. mp = 180-182°; $[\alpha]_D^{15}$ = +223° (c = 1.1 in alcohol); λ_m = 240 nm; insoluble in H_2O; soluble in common organic solvents. *Organon Inc.; Res. Corp.; Schering AG; Searle G.D. & Co.; Syntex Labs. Inc.*

8721 Cortisone
53-06-5 2602 200-162-4
$C_{21}H_{28}O_5$
17,21-Dihydroxypregn-4-ene-3,11,20-trione.
Kendall's compound E; Wintersteiner's compound F; Reichstein's substance Fa; KE; Adreson; Corlin; Cortadren; Cortogen; Cortone; Incortin; Scheroson. Glucocorticoid. mp = 220-224°; $[\alpha]_D^{25}$ = +209° (c = 1.2 in 95% EtOH); λ_m = 237 nm (ε = 14000); fairly soluble in polar organic solvents.

8722 Cortisone Acetate
50-04-4 2602 200-006-5
$C_{23}H_{30}O_6$
17,21-Dihydroxypregn-4-ene-3,11,20-trione 21-acetate.
Cortelan; Cortistab; Cortisyl Artriona. Glucocorticoid. mp = 235-238°; $[\alpha]_D^{25}$ = +164° (c = 0.5 in Me_2CO); λ_m = 238 nm (ε = 15800); soluble in H_2O (2.2 mg/100 ml); more soluble in organic solvents. *Merck & Co.Inc.; Schering AG; Upjohn Ltd.*

8723 Cortivazol
1110-40-3 2603 214-175-8
$C_{32}H_{38}N_2O_5$
(11β,16α)-21-(Acetyloxy)-11,17-dihydroxy-6,16-dimethyl-2'-phenyl-2'H-pregna-2,4,6-trieno[3,2-c]pyrazol-20-one.
NSC-80998; MK-650; H-3625; Altim; Diaster; Dilaster. Glucocorticoid. mp = 160-165°; $[\alpha]_D^{23}$ = +14° (in $CHCl_3$); λ_m = 238, 315 nm (ε = 15700, 19000). *Merck & Co.Inc.*

8724 Deflazacort
14484-47-0 2916 238-483-7
$C_{25}H_{31}NO_6$
(11β,16β)-21-(Acetyloxy)-11-hydroxy-2'-methyl-5'H-pregna-1,4-dieno[16,17-doxazole-3,20-dione.
oxazacort; azacort; DL-458-IT; L-5458; MDL-458; Calcort; Deflan; Dezacor; Flantadin; Lantadin. Anti-inflammatory; glucocorticoid. Systemic corticosteroid; oxazoline derivative of prednisolone. mp = 255-256.5°; $[\alpha]_D$ = +62.3° (c = 0.5 $CHCl_3$); λ_m = 241-242 nm ($E^{1\%}_{1cm}$ 352.5 MeOH); LD_{50} (mus orl) = 5200 mg/kg. *Gruppo Lepetit S.p.A.; Marion Merrell Dow Inc.*

8725 Deoxycorticosterone

64-85-7 2947 200-596-4

$C_{21}H_{30}O_3$

21-Hydroxypregn-4-ene-3,20-dione.

desoxycorticosterone; 21-hydroxyprogesterone; 11-deoxycorticosterone; cortexone; desoxycortone; Kendall's desoxy compound B; Reichstein's substance Q. Mineralocorticoid. Occurs in the adrenal cortex. mp = 141-142°; $[\alpha]_D^{22}$= +178° (in alcohol); λ_m = 240 nm; soluble in alcohol, Me_2CO. *Ciba-Geigy Corp.; Hoechst AG; Organon Inc.; Schering AG.*

8726 Deoxycorticosterone Acetate

56-47-3 2948 200-275-9

$C_{21}H_{30}O_3$

21-Acetyloxypregn-4-ene-3,20-dione.

desoxycorticosterone acetate; cortexone acetate; deoxycortone acetate; desoxycortone acetate; DCA; Cortate; Cortiron; Decosteron; Doca; Dorcostrin; Percorten; Syncortyl. Mineralocorticoid. mp = 154-160°; $[\alpha]_D^{20}$= +168-176° (in dioxane); nearly insoluble in H_2O; more soluble in organic solvents. *Ciba-Geigy Corp.; Hoechst AG; Organon Inc.; Schering AG.*

8727 Deoxycorticosterone Pivalate

808-48-0 212-366-0

$C_{26}H_{38}O_4$

21-(2,2-Dimethyl-1-oxopropoxy)pregn-4-ene-3,20-dione.

desoxycorticosterone pivalate; desoxycortone pivalate; Percorten Pivalate. Mineralocorticoid. *Ciba-Geigy Corp.; Hoechst AG; Organon Inc.; Schering AG.*

8728 Descinolone Acetonide

2135-14-0 218-368-8

$C_{24}H_{31}FO_5$

(11β,16α)-9-Fluoro-11-hydroxy-16,17-[(1-methylethylidene)bis(oxy)]pregna-1,4-diene-3,20-dione.

CL-27071; NSC-44827. Glucocorticoid. *Lederle Labs.*

8729 Desoximetasone

382-67-2 2975 206-845-3

$C_{22}H_{29}FO_4$

(11β,16α)-9-Fluoro-11,21-dihydroxy-16-methylpregna-1,4-diene-3,20-dione.

desoxymethasone; A-41-304; R-2113; HOE-304; Esperson; Stidex; Topicorte; Topisolon. Anti-inflammatory; glucocorticoid. mp = 217°; $[\alpha]_D$ = +109° (in $CHCl_3$); λ_m = 238 nm (ε 15750); soluble in alcohol, Me_2CO, $CHCl_3$, hot EtOAc; slightly soluble in Et_2O, C_6H_6; insoluble in H_2O, dilute aqueous acids, alkalies. *Hoechst Roussel Pharm. Inc.; Roussel-UCLAF; Schering AG.*

8730 Dexamethasone

50-02-2 2986 200-003-9

$C_{22}H_{29}FO_5$

9-Fluoro-11β,17,21-trihydroxy-16α-methylpregna-1,4-diene-3,20-dione.

hexadecadrol; Aeroseb-Dex; Corson; Cortisumman; Decacort; Decaderm; Decadron; Decalix; Decasone; Dekacort; Deltafluorene; Deronil; Deseronil; Dexacortal; Dexacortin; Dexafarma; Dexa-Mamallet; Dexameth; Dexamonozon; Dexapos; Dexa-sine; Dexasone; Dexinoral; Dinormon; Fluormone; Isopto-Dex; Lokalison F; Loverine; Luxazone; Maxidex; Millicorten; Pet-Derm III; component of: Azimycin, Azium, Deronil, Dexacidin, Fulvidex, Maxitrol, Naquasone, Tobradex, Tresaderm. Anti-inflammatory; glucocorticoid; diagnostic aid (Cushing's Syndrome, depression). mp = 262-264°, 268-271°; $[\alpha]_D^{25}$ = ,.5° (in dioxane); somewhat soluble in H_2O (0.1 mg/ml 25°); soluble in Me_2CO, EtOH, $CHCl_3$. *Alcon Labs; Herbert; Iolab; Merck & Co.Inc.; Organon Inc.; Schering-Plough HealthCare Products; Solvay Pharm. Inc.; Upjohn Ltd.*

8731 Dexamethasone 21-Acetate

1177-87-3 2986 214-646-8

$C_{24}H_{31}FO_6$

9-Fluoro-11β,17,21-trihydroxy-16α-methylpregna-1,4-diene-3,20-dione 21-acetate.

Dalalone DP; Dalalone LA; Decadronal; Decadron-LA; Dectancyl; Dexacortisyl. Anti-inflammatory; glucocorticoid; diagnostic aid (Cushing's Syndrome, depression). mp = 215-221°, 229-231°, 238-240°; $[\alpha]_D^{25}$ = +73° (in $CHCl_3$); λ_m = 239 nm (ε 14900). *Forest Pharm. Inc.; Merck & Co.Inc.*

8732 Dexamethasone 21-Acetate Monohydrate

55821-90-3 2986

$C_{24}H_{33}FO_7$

9-Fluoro-11β,17,21-trihydroxy-16α-methylpregna-1,4-diene-3,20-dione 21-acetate monohydrate.

Dalalone DP; Dalalone LA; Decadron-LA. Anti-inflammatory; glucocorticoid; diagnostic aid (Cushing's Syndrome, depression). *Forest Pharm. Inc.; Merck & Co.Inc.*

8733 Dexamethasone 21-Phosphate Disodium Salt

2392-39-4 2986 219-243-0

$C_{22}H_{28}FNa_2O_8P$

9-Fluoro-11β,17,21-trihydroxy-16α-methylpregna-1,4-diene-3,20-dione 21-phosphate disodium salt.

dexamethasone sodium; dexamethasone 21-phosphate disodium salt; dexamethasone 21-(dihydrogen phosphate) disodium salt; Ak-Dex; Baldex; Colvasone; Dalalone; Decadron; Dexabene; Dezone; Fortecortin; Hexadrol; Maxidex Ointment; Oradexon; Orgadrone; Solu-Deca; component of: NeoDecadron. Anti-inflammatory; glucocorticoid (antiasthmatic); diagnostic aid (Cushing's Syndrome, depression). Used in diagnosis of Cushing's syndrome and depression. mp = 233-235°; $[\alpha]_D$ = 257° (H_2O); $[\alpha]_D^{25}$ = 74° ± 4° (c = 1); λ_m = 238-239 nm (ε 14000); soluble in H_2O. *Alcon Labs; Elkins-Sinn; Forest Pharm. Inc.; Merck & Co.Inc.; Organon Inc.*

8734 Dexamethasone Acefurate

83880-70-0 2986

$C_{29}H_{33}FO_8$

9-Fluoro-11β,17,21-trihydroxy-16α-methylpregna-1,4-diene-3,20-dione 17-(2-furoate).

Sch-31353. Anti-inflammatory; glucocorticoid; diagnostic aid (Cushing's Syndrome, depression). *Schering-Plough HealthCare Products.*

8735 Dexamethasone Diethylaminoacetate

2986

$C_{28}H_{41}FNO_6$

9-Fluoro-11β,17,21-trihydroxy-16α-methylpregna-1,4-diene-3,20-dione 21-diethylaminoacetate.

Solu-Forte-Cortin. Anti-inflammatory; glucocorticoid; diagnostic aid (Cushing's Syndrome, depression). *Merck & Co.Inc.*

8736 Dexamethasone Dimethylbutyrate

2986

$C_{28}H_{39}FO_6$

9-Fluoro-11β,17,21-trihydroxy-16α-methylpregna-1,4-diene-3,20-dione 21-(3,3-dimethylbutyrate).

dexamethasone tert-butylacetate; Decadron TBA. Anti-inflammatory; glucocorticoid; diagnostic aid (Cushing's Syndrome, depression). *Merck & Co.Inc.*

8737 Dexamethasone Dipropionate

55541-30-5 2986 259-699-8

$C_{28}H_{37}FO_7$

9-Fluoro-11β,17,21-trihydroxy-16α-methylpregna-1,4-diene-3,20-dione 17,21-dipropionate.

ST12; THS-101; Methaderm. Anti-inflammatory; glucocorticoid; diagnostic aid (Cushing's Syndrome, depression). *Innothera.*

8738 Dexamethasone Isonicotinate

2265-64-7 2986 218-866-5

$C_{28}H_{32}FNO_6$

9-Fluoro-11β,17,21-trihydroxy-16α-methylpregna-1,4-diene-3,20-dione 21-(4-pyridinecarboxylate).

dexamethasone 21-isonicotinate; Auxiloson; Ausixone; Voren. Anti-inflammatory; glucocorticoid; diagnostic aid (Cushing's Syndrome, depression). mp = 250-252°; $[\alpha]_D^{27}$ = +183.5° (in dioxane). *Merck & Co.Inc.*

8739 Dexamethasone Palmitate

14899-36-6 2986

$C_{38}H_{37}FO_7$

9-Fluoro-11β,17,21-trihydroxy-16α-methylpregna-1,4-diene-3,20-dione 21-palmitate.

Limethasone. Anti-inflammatory; glucocorticoid; diagnostic aid (Cushing's Syndrome, depression). *Merck & Co.Inc.*

8740 Diflorasone

2557-49-5 3186 219-875-7

$C_{22}H_{28}F_2O_5$

(6α,11β,16β)-6,9-Difluoro-11,17,21-trihydroxy-16-methylpregna-1,4-diene-3,20-dione.

6α,9α-difluoro-16β-methylprednisolone. Anti-inflammatory (topical); glucocorticoid. The 16β-analog of flumethasone, a glucocorticoid. *Merck & Co.Inc.; Pfizer Inc.; Upjohn Ltd.*

8741 Diflorasone Diacetate

33564-31-7 3186 251-575-1

$C_{26}H_{32}F_2O_7$

(6α,11β,16β)-17,21-Bis(acetyloxy)-6,9-difluoro-11-hydroxy-16-methyl-pregna-1,4-diene-3,20-dione.

6α,9α-difluoro-16β-methylprednisolone acetate; U-24865; Dermaflor; Diacort; Difulal; Florone; Maxiflor; Psorcon; Soriflor. Anti-inflammatory (topical); glucocorticoid. mp = 221-223 (dec); $[\alpha]_D$ = +61° (in $CHCl_3$); λ_m = 238 nm (ε 17250 in alcohol). *ABIC; Dermik Labs. Inc.; Merck & Co.Inc.; Pfizer Inc.; Upjohn Ltd.*

8742 Diflucortolone

2607-06-9 3189

$C_{22}H_{28}F_2O_4$

(6α,11β,16α)-6,9-Difluoro-11,21-dihydroxy-16-methylpregna-1,4-diene-3,20-dione.

Anti-inflammatory. Glucocorticoid. The 9α-fluoro derivative of fluocortolone. mp = 240-244°, 248-249°; $[\alpha]_D^{22}$ = +111° (in MeOH); λ_m = 237 nm (ε 16600). *Schering-Plough HealthCare Products.*

8743 Diflucortolone Pivalate

15845-96-2 3189

$C_{27}H_{36}F_2O_5$

(6α,11β,16α)-6,9-Difluoro-11,21-dihydroxy-16-methylpregna-1,4-diene-3,20-dione 21 pivalate.

SH-968; Neribas; Neriforte; Nerisona; Nerisone; Temetex; Texmeten. Anti-inflammatory. Glucocorticoid. mp = 195-195.5°; $[\alpha]_D^{22}$ = +100.8° (in dioxane); LD_{50} (mus orl) > 4000 mg/kg, (rat orl) = 3100 mg/kg, (mus sc) = 180 mg/kg, (rat sc) = 13 mg/kg, (mus ip) = 450 mg/kg, (rat ip) = 98 mg/kg. *Schering-Plough HealthCare Products.*

8744 Flucloronide

3693-39-8 4157 223-010-9

$C_{24}H_{29}Cl_2FO_5$

(6α,11β,16α)-9,11-Dichloro-6-fluoro-21-hydroxy-16,17-[(1-methylethylidene)bis(oxy)]pregna-1,4-diene-3,20-dione.

fluclorolone acetonide; RS-2252; Topilar. Glucocorticoid. *Syntex Labs. Inc.*

8745 Fludrocortisone

127-31-1 4166 204-833-2

$C_{21}H_{29}FO_5$

(11β)-9-Fluoro-11,17,21-trihydroxypregn-4-ene-3,20-dione.

9α-fluorcortisol; fluodrocortisone; fluohydrisone; fluohydrocortisone; Astonin H. Mineralocorticoid. mp = 260-262°; $[\alpha]_D^{23}$ = +139° (c = 0.55 95% EtOH); λ_m = 239 nm (ε 17600 MeOH); soluble in H_2O (0.14 mg/ml). *Olin Mathieson.*

8746 Fludrocortisone Acetate

514-36-3 4166 208-180-4

$C_{23}H_{31}FO_6$

(11β)-9-Fluoro-11,17,21-trihydroxypregn-4-ene-3,20-dione 21-acetate.

fludrocortisone 21-acetate; Alflorone; F-Cortef; Florinef. Mineralocorticoid. mp = 233-234°; $[\alpha]_D^{23}$= +123° (c = 0.64 $CHCl_3$); λ_m = 238 nm (ε 16800 MeOH); very slightly soluble in H_2O (0.04 mg/ml); more soluble in Me_2CO (56 mg/ml), $CHCl_3$ (20 mg/ml), Et_2O (4 mg/ml). *Bristol-Myers Squibb Co; Olin Mathieson; Upjohn Ltd.*

8747 Flumethasone

2135-17-3 4173 218-370-9

$C_{22}H_{28}F_2O_5$

(6α,11β,16α)-6,9-Difluoro-11,17,21-trihydroxy-16-methylpregna-1,4-diene-3,20-dione.

flumetasone; 6α-fluorodexamethazone; U-10974; NSC-5402; Aniprome; Cortexilar; Flucort; Methagon. Glucocorticoid; anti-inflammatory. *Upjohn Ltd.*

8748 Flumethasone Acetate
4173
$C_{24}H_{30}F_2O_6$
(6α,11β,16α)-6,9-Difluoro-11,17,21-trihydroxy-16-methylpregna-1,4-diene-3,20-dione 21-acetate.
flumetasone 1-acetate. Glucocorticoid; anti-inflammatory. mp = 260-264°; $[\alpha]_D$ = 91° (in EtOH); λ_m = 237 nm (log ε 4.16). *Upjohn Ltd.*

8749 Flumethasone Pivalate
2002-29-1 4173 218-370-9
$C_{27}H_{36}F_2O_6$
(6α,11β,16α)-6,9-Difluoro-11,17,21-trihydroxy-16-methylpregna-1,4-diene-3,20-dione 21-pivalate.
flumetasone 21-pivalate; NSC-107680; Locacorten (obsolete); Locorten; Lorinden; Losalen. Glucocorticoid; anti-inflammatory. *Upjohn Ltd.*

8750 Flumoxonide
60135-22-0 262-074-2
$C_{26}H_{34}F_2O_7$
(6α,11β,16α)-6,9-Difluoro-11-hydroxy-21,21-dimethoxy-16,17-[(1-methylethylidene)bis(oxy)]pregna-1,4-diene-3,20-dione.
RS-40584. Adrenocortical steroid. *Syntex Labs. Inc.*

8751 Flunisolide
3385-03-3 4180 222-193-2
$C_{24}H_{31}FO_6$
(6α,11β,16α)-6-Fluoro-11,21-dihydroxy-16,17-[(1-methylethylidene)bis(oxy)]pregna-1,4-diene-3,20-dione.
RS-3999; Aerobid; Bronalide; Lunis; Nasalide; Rhinalar; Synaclyn; Syntaris. Glucocoriticoid; antiasthmatic. Synthetic fluorinated corticosteroid related to prednisolone. *Syntex Labs. Inc.*; 21;169.

8752 Flunisolide 21-Acetate
4533-89-5 4180 224-871-3
$C_{26}H_{33}FO_7$
(6α,11β,16α)-6-Fluoro-11,21-dihydroxy-16,17-[(1-methylethylidene)bis(oxy)]pregna-1,4-diene-3,20-dione 21-acetate.
RS-1320. Glucocoriticoid; antiasthmatic; anti-inflammatory. *Syntex Labs. Inc.*

8753 Flunisolide Hemihydrate
77326-96-6
$C_{24}H_{31}FO_6 \cdot 1/2H_2O$
(6α,11β,16α)-6-Fluoro-11,21-dihydroxy-16,17-[(1-methylethylidene)bis(oxy)]pregna-1,4-diene-3,20-dione hemihydrate.
Glucocoriticoid; antiasthmatic. *Syntex Labs. Inc.*

8754 Fluocinolone Acetonide
67-73-2 4185 200-668-5
$C_{24}H_{30}72O_6$
(6α,11β,16α)-6,9-Difluoro-11,21-dihydroxy-16,17-[(1-methylethylidene)bis(oxy)]pregna-1,4-diene-3,20-dione.
NSC-92339; Coriphate; Cortiplastol; Dermalar; Fluonid; Fluovitef; Fluvean; Fluzon; Jellin; Localyn; Synalar; Synamol; Synandone; Synemol; Synotic; Synsac; component of: N-Synalar. Glucocoriticoid; anti-inflammatory. mp = 265-266°; $[\alpha]_D$ = +95° (in $CHCl_3$); λ_m = 238 nm (log ε 4.21). *Lemmon Co.; Syntex Labs. Inc.*

8755 Fluocortolone
152-97-6 4188 205-811-5
$C_{22}H_{29}FO_4$
(6α,11β,16α)-6-Fluoro-11,21-dihydroxy-16-methylpregna-1,4-diene-3,20-dione.
SH-742; Ultralan oral. Glucocorticoid. mp = 188-190.5°; $[\alpha]_D^{20}$ = +100° (in dioxane); λ_m = 242 nm (ε 16300); somewhat soluble in H_2O (295 mg/l at 37°), EtOH (120 mg/l at 20°), C_7H_8 (440 mg/l at 20°). *Schering AG.*

8756 Fluocortolone Caproate
303-40-2 4188 206-140-0
$C_{28}H_{39}FO_5$
(6α,11β,16α)-6-Fluoro-11,21-dihydroxy-16-methylpregna-1,4-diene-3,20-dione 21-hexanoate.
fluocortolone 21-hexanoate; SH-770; Ficoid; Ultralanum. Glucocorticoid. mp = 242-245°; $[\alpha]_D^{20}$ = +98.5° (in dioxane); λ_m = 242 nm (ε 16200); slightly soluble in H_2O (7.8 mg/l at 37°); more soluble in EtOH (450 mg/l at 20°), C_7H_8 (440 mg/l at 20°). *Schering AG.*

8757 Fluocortolone Pivalate
4188
$C_{22}H_{37}FO_5$
(6α,11β,16α)-6-Fluoro-11,21-dihydroxy-16-methylpregna-1,4-diene-3,20-dione 21-pivalate.
fluocortolone trimethylacetate. Glucocorticoid. mp = 187°; nearly insoluble in H_2O; soluble in $CHCl_3$, MeOH; slightly soluble in Et_2O. *Schering AG.*

8758 Fluorometholone
426-13-1 4213 207-041-5
$C_{22}H_{29}FO_4$
(6α,11β)-9-Fluoro-11,17-dihydroxy-6-methylpregna-1,4-diene-3,20-dione.
fluormetholon; Cortilet; Delmeson; Efflumidex; Fluaton; Flumetholon; Fluor-Op; FML; FML Forte; FML Liquifilm; FML S.O.P.; Loticort; Oxylone; Ursnon; component of: FML-S Liquifilm, Neo-Oxylone. Glucocorticoid; anti-inflammatory. mp = 292-303°. *Allergan Inc.; Iolab; Upjohn Ltd.*

8759 Fluorometholone Acetate
3801-06-7 4213 223-270-3
$C_{24}H_{31}FO_5$
(6α,11β)-9-Fluoro-11,17-dihydroxy-6-methylpregna-1,4-diene-3,20-dione 17-acetate.
U-17323; Flarex; component of: Tobrasone. Glucocorticoid; anti-inflammatory. mp = 230-232°; $[\alpha]_D$ = +28° (in $CHCl_3$). *Am. Cyanamid; Upjohn Ltd.*

8760 Fluperolone Acetate
2119-75-7 4225 218-327-4
$C_{24}H_{31}FO_6$
[11β,17α,17(S)]-17-[2-(Acetyloxy)-1-oxopropyl]-9-fluoro-11,17-dihydroxyandrosta-1,4-dien-3-one.
P-1742; ALAcortril; Methral. Glucocoriticoid; anti-inflammatory. mp = 251-253°; $[\alpha]_D$ = 87°; λ_m = 239 nm (ε 15350). *Pfizer Inc.*

8761 Fluprednisolone
53-34-9 4229 200-170-8
$C_{21}H_{27}FO_5$
(6α,11β)-6-Fluoro-11,17,21-trihydroxypregna-1,4-diene-3,20-dione.
U-7800; NSC-46439; Alphadrol; Etadrol. Glucocoriticoid; anti-inflammatory. mp = 208-213°; $[\alpha]_D$ = +92°. *Bayer AG; Syntex Labs. Inc.; Upjohn Ltd.*

8762 Fluprednisolone Valerate
23257-44-5 245-535-2
$C_{26}H_{35}FO_6$
(6α,11β)-6-Fluoro-11,21-dihydroxy-17-[(1-oxopentyl)oxy]pregna-1,4-diene-3,20-dione.
Glucocoriticoid. *Miles Inc.*

8763 Flurandrenolide
1524-88-5 4232 216-196-8
$C_{24}H_{33}F_2O_6$
(6α,11β,16α)-6-Fluoro-11,21-dihydroxy-16,17-[(1-methylethylidene)bis(oxy)]pregn-4-ene-3,20-dione.
fluorandrenolone; flurandrenolone; flurandrenolone acetate (obsolete); fludroxycortide; Cordran; Drenison; Drocort; Haelan; Sermaka. Glucocoriticoid; anti-inflammatory. mp = 247-255°; $[\alpha]_D^{25}$ = +140-150° (in $CHCl_3$); λ_m = 236 nm (log ε 4.17). *Eli Lilly & Co.; Syntex Labs. Inc.*

8764 Formocortal
2825-60-7 4270 220-584-2
$C_{29}H_{38}ClFO_8$
(11β,16α)-21-(Acetyloxy)-3-(2-chloroethoxy)-9-fluoro-11-hydroxy-16,17-[(1-methylethylidene)bis(oxy)]-20-oxopregna-3,5-diene-6-carboxaldehyde.
fluoformylon; FI-6341; Cortocin-F; Cutisterol; Deflamene; Fluderma. Glucocoriticoid. mp = 180-182°; $[\alpha]_D^{20}$ = +26° (in $CHCl_3$); λ_m = 216, 324 nm (ε 12100, 12100 EtOH); LD_{50} (mus orl) >2000 mg/kg, (mus ip) = 537 mg/kg. *Farmitalia Carlo Erba Ltd.*

8765 Hydrocortamate
76-47-1 4827 200-963-9
$C_{27}H_{41}NO_6$
N,N-Diethylglycine (11β)-11,17-dihydroxy-3,20-dioxo-pregn-4-en-21-yl ester.
cortisol 21-ester with N,N-diethylglycine; hydrocortisone 21-diethylaminoacetate; Ulcort. Glucocoriticoid. mp = 162-163°. *Pfizer Inc.; Schering AG.*

8766 Hydrocortamate Hydrochloride
125-03-1 4827 204-723-4
$C_{27}H_{42}ClNO_6$
N,N-Diethylglycine (11β)-11,17-dihydroxy-3,20-dioxo-pregn-4-en-21-yl ester hydrochloride.
ethamicort; Magnacort. Glucocoriticoid. Dec 222°. *Pfizer Inc.; Schering AG.*

8767 Hydrocortisone
50-23-7 4828 200-020-1
$C_{21}H_{30}O_5$
(11β)-11,17,21-Trihydroxypregna-1,4-diene-3,20-dione.
cortisol; 17-hydroxycorticosterone; anti-inflammatory hormone; Kendall's compound F; Reichstein's substance M; Aeroseb-HC; Ala-Cort; Anflam; Cetacort; Cort-Dome; Cortef; Cortenema; Cortril; Dermacort; Dermocortal; Dermolate; Dioderm; Efcortelan; Evacort; Ficortril; Hydracort; Hydro-Adreson; Hydrocort; Hydrocortisyl; Hydrocortone; Hytone; Lacticare-HC; Medicort; Mildison; Nutracort; Penecort; Proctocort; Scheroson F; Synacort; Texacort; Timocort; Zenoxone; Acticort; Alphaderm; CaldeCort Spray; Eldecort; component of: Cor-Tar-Quin, Cort-Quin Cortisporin, Drotic, Neo-Cort-Dome, Nystaform-HC, Otalgine, Otic-Neo-Cort-Dome, Otobiotic, Otocort, Pediotic Suspension, Racet, Vioform-Hydrocortisone, VoSol HC, Vytone. Glucocoriticoid. Principal glucocorticoid hormone produced by the adrenal cortex. mp = 217-220°; $[\alpha]_D^{22}$ = +167° (in absolute EtOH); λ_m = 242 nm ($E^{1\%}_{1cm}$ 445); slightly soluble in H_2O (0.28 mg/ml); more soluble in EtOH, MeOH, Me_2OH, $CHCl_3$, propylene glycol, Et_2O; soluble in concentrated sulfuric acid. *Dermik Labs. Inc.; Fisons plc; Herbert; ICN Pharm. Inc.; Key Pharm.; Lemmon Co.; Marion Merrell Dow Inc.; Merck & Co.Inc.; Miles Inc.; Pfizer Inc.; Schering AG; Stiefel Labs Inc.; Upjohn Ltd.; Whitehall Labs. Inc.*

8768 Hydrocortisone Aceponate
74050-20-7
$C_{26}H_{36}O_7$
(11β)-11,17,21-Trihydroxypregna-1,4-diene-3,20-dione 21-acetate 17-propionate.
Glucocoriticoid.

8769 Hydrocortisone Acetate
50-03-3 4828 200-004-4
$C_{23}H_{32}O_6$
(11β)-21-(Acetoxy)-11,17,21-trihydroxypregna-1,4-diene-3,20-dione.
Anusol-HC; CaldeCort; Colifoam; Colofoam; Cortaid; Cordes; Cortef; Cortifoam; Cortril Acetate-AS; Efcolin; Hc45; Hydrin-2; Hydrocal; Hydrocortistab; Hydrocortone Acetate; Lanacort; Lenirit; Sigmacort; Sintotrat; Velopural; component of: Chloromycetin Hydrocortisone Ophthalmic, Clear-Aid, Coly-MycinS Otic, Cirticaine Cream, Cortisporin Cream, Epifoam, Lidaform-HC, Lidamantle-HC, Mantadil, Neo-Cortef, Ophthocort, ProctoFoam, Protef. Glucocorticoid. Dec 223°; d_4^{20} = 1.289; $[\alpha]_D^{25}$ = +166° (c = 0.4 in dioxane); $[\alpha]_D^{25}$ = +150.7° (c = 0.4 in Me_2OH); λ_m = 242 nm ($E^{1\%}_{1cm}$ 390 in MeOH); slightly soluble in H_2O (1 mg/100 ml); more soluble in EtOH, MeOH, Me_2CO, $CHCl_3$, Et_2O; very soluble in DMF, dioxane. *Broemmel Pharm.; Fisons plc; Merck & Co.Inc.; Parke-Davis; Upjohn Ltd.*

8770 Hydrocortisone Buteprate
72590-77-3 276-726-9
$C_{28}H_{40}O_7$
(11β)-11-Hydroxy-17-(1-oxobutoxy)-21-(1-oxopropoxy)pregna-1,4-diene-3,20-dione.
TS-408; Pandel. Glucocorticoid. *Taisho.*

8771 Hydrocortisone Butyrate
13609-67-1 4828 237-093-4
$C_{28}H_{40}O_7$
(11β)-11,21-Dihydroxy-17-(1-oxobutoxy)pregna-1,4-diene-3,20-dione.
hydrocortisone 17-butyrate; Alfason; Laticort; Locoid; Plancol. Glucocoriticoid.

8772 Hydrocortisone Cypionate
508-99-6 208-091-0
$C_{29}H_{42}O_6$
(11β)-21-(3-Cyclopentyl-1-oxopropoxy)-11,17-dihydroxypregna-1,4-diene-3,20-dione.
hydrocortisone cyclopentylpropionate; Cortef Oral Suspension. Glucocoriticoid. *Upjohn Ltd.*

8773 Hydrocortisone Dihydrogen Phosphate
3863-59-0 223-382-2
(11β)-11,17-Dihydroxy-21-(phosphonooxy)pregna-1,4-diene-3,20-dione.
hydrocortisone 21-(dihydrogen phosphate). Glucocorticoid.

8774 Hydrocortisone Hemisuccinate
83784-20-7
$C_{25}H_{36}O_9$
(11β)-21-(3-Carboxy-1-oxopropoxy)-11,17-dihydroxypregna-1,4-diene-3,20-dione monohydrate.
Adrenocortical steroid.

8775 Hydrocortisone Hemisuccinate [anhydrous]
2203-97-6 218-612-3
$C_{25}H_{34}O_8$
(11β)-21-(3-Carboxy-1-oxopropoxy)-11,17-dihydroxypregna-1,4-diene-3,20-dione.
Adrenocortical steroid.

8776 Hydrocortisone Sodium Phosphate
6000-74-4 4828 227-843-9
$C_{21}H_{29}Na_2O_8P$
(11β)-11,17-Dihydroxy-21-(phosphonooxy)pregna-1,4-diene-3,20-dione disodium salt.
hydrocortisone sodium phosphate; hydrocortisone phosphate; hydrocortisone 21-disodium salt; Cleiton; Efcortesol; Hydrocortone Phosphate. Glucocorticoid. $[\alpha]_D^{25}$ = +120° (in H_2O); λ_m = 242 nm ($A_{1cm}^{1\%}$ 298-341 MeOH); soluble in H_2O (>500 mg/ml); pH (1% aqueous solution) 7.5-8.5. *Merck & Co.Inc.*

8777 Hydrocortisone Sodium Succinate
125-04-2 4828 204-725-5
$C_{25}H_{33}NaO_8$
(11β)-21-(3-Carboxy-1-oxopropoxy)-11,17-dihydroxypregna-1,4-diene-3,20-dione.
hydrocortisone 21-(sodium succinate); hydrocortisone hemisuccinate sodium salt; A-hydroCort; Buccalsone; Corlan; Efcortelan Soluble; Saxizon; Solu-Cortef; Solu-Glyc. Glucocorticoid. mp = 169-172°; soluble in H_2O (500 mg/ml), MeOH, EtOH; less soluble in $CHCl_3$. *Abbott Labs Inc.; Upjohn Ltd.*

8778 Hydrocortisone Valerate
57524-89-7 4828 260-786-8
$C_{26}H_{38}O_6$
(11β)-11,21-Dihydroxy-17-[(1-oxopentyl)oxy]pregna-1,4-diene-3,20-dione.
hydrocortisone 17-valerate; Westcort. Glucocorticoid. *Westwood-Squibb Pharm. Inc.*

8779 Medrysone
2668-66-8 5840 220-208-7
$C_{22}H_{32}O_3$
(6α,11β)-11-Hydroxy-6-methylpregn-4-ene-3,20-dione.
hydroxymesterone; U-8471; HMS; Medrocort; Ophtocortin; Spectamedryn. Glucocorticoid. mp = 155-158°; $[\alpha]_D$ = 189° (in $CHCl_3$). *Allergan Inc.; Upjohn Ltd.*

8780 Meprednisone
1247-42-3 5907 214-996-1
$C_{22}H_{28}O_5$
(16β)-17,21-Dihydroxy-16-methylpregna-1,4-diene-3,11,20-trione.
16β-methylprednisone; NSC-527579; Sch-4358; Betapred; Betapar; Deltacortene; Beta; Betalone; Deltisona B. Glucocorticoid. mp = 200-205°; $[\alpha]_D$ = +200° (dioxane); λ_m = 239 nm ($E_{1cm}^{1\%}$ 416 MeOH). *Parke-Davis; Schering-Plough HealthCare Products.*

8781 Methylprednisolone
83-43-2 6189 201-476-4
$C_{22}H_{30}O_5$
(6α,11β)-11,17,21-Trihydroxy-6-methylpregna-1,4-diene-3,20-dione.
NSC-19987; Medrate; Medrol; Medrone; Metastab; Metrisone; Promacortine; Suprametil; Urbason. Glucocorticoid. mp = 228-237°; $[\alpha]_D^{20}$ = 83° (dioxane); λ_m = 243 nm (α_M 14875 95% EtOH). *Schering-Plough HealthCare Products; Upjohn Ltd.*

8782 Methylprednisolone Aceponate
86401-95-8
$C_{27}H_{36}O_7$
(6α,11β)-11,17,21-Trihydroxy-6-methylpregna-1,4-diene-3,20-dione 21-acetate 17-propionate.
Advantan. Glucocorticoid.

8783 Methylprednisolone Acetate
53-36-1 6189 200-171-3
$C_{24}H_{32}O_6$
(6α,11β)-21-(Acetyloxy)-11,17-dihydroxy-6-methylpregna-1,4-diene-3,20-dione.
methylprednisolone 21-acetate; depMedalone 40; depMedalone 80; Depo-Medrate; Depo-Medrol; Depo-Medrone; Mepred; Vetacortyl; component of: NeoMedrol. Glucocorticoid. mp = 205-208°; $[\alpha]_D^{20}$ = +101° (dioxane); λ_m = 243 nm (α_M 14825 95% EtOH); nearly insoluble in H_2O. *Forest Pharm. Inc.; Upjohn Ltd.*

8784 Methylprednisolone Dihydrogen Phosphate
22252-38-6 244-869-6
(6α,11β)-11,17-Dihydroxy-6-methyl-21-(phosphonooxy)-pregna-1,4-diene-3,20-dione.
Glucocorticoid.

8785 Methylprednisolone Hemisuccinate
2921-57-5 220-863-9
$C_{26}H_{34}O_8$
(6α,11β)-21-(3-Carboxy-1-oxopropoxy)-11,17-dihydroxy-6-methylpregna-1,4-diene-3,20-dione.
methylprednisolone 21-(hydrogen succinate).
Adrenocortical steroid.

8786 Methylprednisolone Sodium Phosphate
5015-36-1 6189 225-694-4
$C_{22}H_{29}Na_2O_8P$
(6α,11β)-11,17-Dihydroxy-6-methyl-21-(phosphonooxy)pregna-1,4-diene-3,20-dione disodium salt.
U-12019E; Medrol Stabisol. Glucocorticoid. *Upjohn Ltd.*

8787 Methylprednisolone Sodium Succinate
2375-03-3 6189 219-156-8
$C_{26}H_{33}NaO_8$
(6α,11β)-21-(3-Carboxy-1-oxopropoxy)-11,17-dihydroxy-6-methylpregna-1,4-diene-3,20-dione.
Urbason-Solubile; Solu-Medrol. Glucocorticoid. Soluble in H_2O. *Abbott Labs Inc.; Elkins-Sinn; Upjohn Ltd.*

8788 Methylprednisolone Suleptanate
90350-40-6
$C_{33}H_{48}NNaO_{10}S$
(6α,11β)-11,17-Dihydroxy-6-methyl-21-[[8-[methyl(2-sulfoethyl)amino]-1,8-dioxooctyl]oxy]pregna-1,4-diene-3,20-dione monosodium salt.
Medrosol. Anti-inflammatory; glucocorticoid. *Upjohn Ltd.*

8789 Naflocort
80738-47-2
$C_{29}H_{35}FO_5$
(6β,11β)-9-Fluoro-1',4'-dihydro-11,21-dihydroxy-2'H-naptho[2',3':16,17]pregna-1,4-diene-3,20-dione monohydrate.
SQ-26490. Adrenocortical steroid (topical). *Bristol-Myers Squibb Co.*

8790 Naflocort [anhydrous]
59497-39-1
$C_{29}H_{33}FO_4$
(6β,11β)-9-Fluoro-1',4'-dihydro-11,21-dihydroxy-2'H-naptho[2',3':16,17]pregna-1,4-diene-3,20-dione.
Adrenocortical steroid (topical). *Bristol-Myers Squibb Co.*

8791 Nivazol
24358-76-7 246-202-4
$C_{28}H_{31}FN_2O$
(17α)-2'-(4-Fluorophenyl)-2'H-pregna-2,4-dien-20-yno[3,2-c]pyrazol-17-ol.
nivacortol; Win-27914. Glucocorticoid. *Sterling Winthrop Inc.*

8792 Paramethasone
53-33-8 7162 200-169-2
$C_{22}H_{29}FO_5$
(6α,11β,16α)-6-Fluoro-11,17,21-trihydroxy-16-methylpregna-1,4-diene-3,20-dione.
16α-methylprednisolone. Glucocorticoid.

8793 Paramethasone Acetate
1597-82-6 7162 216-486-4
$C_{24}H_{31}FO_6$
(6α,11β,16α)-21-(Acetyloxy)-6-fluoro-11,17,21-trihydroxy-16-methylpregna-1,4-diene-3,20-dione.
paramethasone 21-acetate; Cortidene; Dilar; Dillar; Haldrate; Haldrone; Metilar; Monocortin; Paramezone; Syntecort; Stemex. Glucocorticoid. mp = 228-241° (dec); $[\alpha]_D$ = +85°; λ_m = 243 nm (log ε 4.16 MeOH); soluble in Me_2OH, EtOH; slightly soluble in H_2O; LD_{50} (rat ip) = 392 mg/kg. *Eli Lilly & Co.; Syntex Labs. Inc.*

8794 Prednicarbate
73771-04-7 7899 277-590-3
$C_{27}H_{36}O_8$
(11β)-17-[(Ethoxycarbonyl)oxy]-11-hydroxy-21-(1-oxopropoxy)pregna-1,4-diene-3,20-dione.
prednisolone; 17-ethylcarbonate 21-propionate; HOE-777; S-77-0777; Dermatop; EsCort; Prednitop; Regenit. Glucocorticoid; anti-inflammatory (topical). mp = 110-112°; $[\alpha]_D^{20}$ = +63° (c = 0.1 EtOH); λ_m = 241 nm (ε 15000). *Hoechst Roussel Pharm. Inc.*

8795 Prednisolamate
5626-34-6 7902
$C_{27}H_{30}O_5$
(11β)-11,17,21-Trihydroxy-6-methylpregna-1,4-diene-3,20-dione 21-n,n-diethylglycine ester.
prednisolone 21-diethylaminoacetate; 21-N,N-diethylglycinate. Glucocorticoid. mp = 175-177°. *Pfizer Inc.*

8796 Prednisolamate Hydrochloride
17140-01-1 7902
$C_{27}H_{30}O_5$
(11β)-11,17,21-Trihydroxy-6-methylpregna-1,4-diene-3,20-dione 21-n,n-diethylglycine ester.
prednisolone 21-diethylaminoacetate hydrochloride; Deltacortril DA. Glucocorticoid. mp = 237.4-239.8°; $[\alpha]_D$ = +120.7° (in H_2O). *Pfizer Inc.*

8797 Prednisolone
50-24-8 7901 200-021-7
$C_{21}H_{28}O_5$
(11β)-11,17,21-Trihydroxypregna-1,4-diene-3,20-dione.
metacortandralone; delta F; δ^1-dehydrocortisol; δ^1-hydrocortisone; δ^1-dehydrohydrocortisone; hydroretrocortine; NSC-9120; Codelcortone; Cortalone; Decaprednil; Decortin H; Delta-Cortef; Deltacortril Enteric; Deltastab; Deltasolone; Flamasone; Hydeltra; Hydrodeltalone; Klismacort; Meticortelone; Meti-Derm; Paracortol; Precortancyl; Precortilon; Precortisyl; Prednelan; Prednicen; Predni-Dome; Predniretard; Predonine; Solone; Sterolone; component of: K-Predne-Dome. Glucocorticoid. Synthetic corticosteroid; metabolically interconvertible with prednisone. mp = 240-241° (dec); $[\alpha]_D^{25}$ = +102° (dioxane); λ_m = 242 nm (ε 15,000 MeOH); slightly soluble in H_2O; soluble in alcohol, $CHCl_3$, Me_2OH, MeOH, dioxane. *Miles Inc.; Parke-Davis; Schering-Plough HealthCare Products; Upjohn Ltd.*

8798 Prednisolone 21-tert-Butylacetate
7681-14-3 7901 231-661-5
$C_{27}H_{38}O_6$
(11β)-21-(3,3-Dimethyl-1-oxobutoxy)-11,17-dihydroxypregna-1,4-diene-3,20-dione.
prednisolone tebutate; Codelcortone-T.B.A.; Hydeltra-T.B.A.; Predalone-T.B.A. Glucocorticoid. mp = 266-273°. *Forest Pharm. Inc.; Merck & Co.Inc.*

8799 Prednisolone 21-Trimethylacetate
1107-99-9 7901 214-172-1
$C_{26}H_{36}O_6$
(11β)-21-(2,2-Dimethyl-1-oxopropoxy)-11,17-dihydroxypregna-1,4-diene-3,20-dione.
prednisolone pivalate; Ultracortenol. Glucocorticoid. mp = 233-236°; $[\alpha]_D^{26}$ = +103° (c = 1.208 $CHCl_3$); λ_m = 244 nm (ε 14700).

8800 Prednisolone Acetate
52-21-1 7901 200-134-1
$C_{23}H_{30}O_6$
(11β)-21-(Acetyloxy)-11,17-dihydroxypregna-1,4-diene-3,20-dione.
Ak-Tate; Econopred; Hostacortin H; Inflanefran; Meticotelone Acetate; Predalone 50; Pred Forte; Pred Mild; Scherisolon; Sterane; component of: Blephamide Liquifilm, Blephamide SOP, Cetapred Ointment, Isopto Cetapred, Metimyd, Mydapred, Neo-Delta-Cortef, Poly-Pred, Pred-G Liquifilm, Pred-G SOP, Vasocidin Ointment. Glucocortin. Dec 237-239°; $[\alpha]_D^{25}$ = +116° (in dioxane). *Alcon Labs; Allergan Inc.; Iolab; Miles Inc.; Upjohn Ltd.*

8801 Prednisolone Hemisuccinate
2920-86-7 7901 220-861-8
$C_{25}H_{32}O_8$
(11β)-21-(3-Carboxy-1-oxopropoxy)-11,17-dihydroxy-pregan-1,4-diene-3,20-dione.
prednisolone 21-(hydrogen succinate); Fiasone (amp). Glucocorticoid.

8802 Prednisolone Metasulfobenzoate Sodium
630-67-1 7901 211-141-4
$C_{28}H_{31}NaO_9S$
(11β)-11,17-Dihydroxy-21-[(3-sulfobenzoyl)oxy]pregna-1,4-diene-3,20-dione monosodium salt.
prednisolone sodium metasulfobenzoate; prednisolone 21-(3-sodium-sulphobenzoate); Cortico-Sol; Predenema; Predfoam; Solupred. Glucocorticoid. mp = 293-295° (dec); $[\alpha]_D^{20}$ = 170° (in H_2O). *Allergan Inc.; O'Neal Jones & Feldman Pharm.*

8803 Prednisolone Sodium Phosphate
125-02-0 7903 204-722-9
$C_{21}H_{27}Na_2O_8P$
(11β)-11,17-Dihydroxy-21-(phosphonooxy)pregna-1,4-diene-3,20-dione disodium salt.
21-prednisolonephosphoric acid disodium salt; prednisolone phosphate disodium; disodium prednisolone 21-phosphate; Ak-pred; Codelsol; Hefasolon; Hydeltrasol; Inflamase; Metreton; Pediapred; Prednesol; Predsol; Solucort; Solu-Predalone; component of: Optimyd, Vasocidin Solution. Glucocorticoid. $[\alpha]_D^{25}$ = +102.5°; λ_m = 243 nm ($A_{1cm}^{1\%}$ 308 MeOH); soluble in H_2O, MeOH, EtOH; pH (1% aqueous solution) = 7.5-8.5. *Fisons plc; Iolab; Merck & Co.Inc.; Schering-Plough HealthCare Products.*

8804 Prednisolone Sodium Succinate
1715-33-9 7901 216-995-1
$C_{25}H_{31}NaO_8$
(11β)-21-(3-Carboxy-1-oxopropoxy)-11,17-dihydroxypregna-1,4-diene-3,20-dione monosodium salt.
prednisolone 21-succinate sodium salt; Di-Adreson-F; Meticotelone Soluble; Solu-Decortin-H. Glucocorticoid. *Schering-Plough HealthCare Products; Upjohn Ltd.*

8805 Prednisolone Steaglate
5060-55-9 7901 225-763-9
$C_{41}H_{64}O_8$
(11β)-11,17-Dihydroxy-21-[[[(1-oxooctadecyl)oxy]-acetyl]oxy]pregna-1,4-diene-3,20-dione.
prednisolone 21-stearoylglycolate; Sintisone. Glucocorticoid. mp = 105-107°; $[\alpha]_D^{20}$ = +57-63° (dioxane); λ_m = 244 nm ($E_{1cm}^{1\%}$ 212±10 MeOH).

8806 Prednisone
53-03-2 7904 200-160-3
$C_{21}H_{26}O_5$
17,21-Dihydroxypregna-1,4-diene-3,11,20-trione.
δ^1-dehydrocortisone; δ^1-cortisone; deltacortisone; delta E; metacortandracin; retrocortine; NSC-10023; Ancortone; Colisone; Cortancyl; Dacortin; Decortancyl; Decortin; Delcortin; Deltacortone; Deltasone; Deltison; Di-Adreson; Encorton; Meticorten; Nurison; Orasone; Paracort; Prednilonga; Pronison; Rectodelt; Sone; Ultracorten. Glucocorticoid. Dec 233-235°; $[\alpha]_D^{25}$ = +172° (in dioxane); λ_m = 238 nm (ε 15500); slightly soluble in H_2O; soluble in alcohol (0.67 g/100 ml), $CHCl_3$ (0.2 g/100 ml); more soluble in organic solvents. *Miles Inc.; Parke-Davis; Schering-Plough HealthCare Products; Solvay Pharm. Inc.; Upjohn Ltd.*

8807 Prednisone 21-Acetate
125-10-0 7904 204-726-0
$C_{23}H_{28}O_6$
21-(Acetyloxy)-17-hydroxypregna-1,4-diene-3,11,20-trione.
Delta-Cortelan; Hostacortin. Glucocorticoid. Dec 226-232°; $[\alpha]_D^{25}$ = +186° (dioxane); λ_m = 238 nm (ε 16100 EtOH). *Schering-Plough HealthCare Products.*

8808 Prednival
15180-00-4 7905 239-228-2
$C_{26}H_{36}O_6$
(11β)-11,21-Dihydroxy-17-[(1-oxopentyl)oxy]pregna-1,4-diene-3,20-dione.
prednisolone 17-valerate; W-4869. Glucocorticoid. mp = 210-213°; $[\alpha]_D$ = +3.5° (dioxane). *Parke-Davis; Vismara.*

8809 Prednival 21-Acetate
72064-79-0 7905 276-312-8
$C_{28}H_{38}O_7$
(11β)-21-(Acetyloxy)-11-hydroxy-17-[(1-oxopentyl)oxy]-pregna-1,4-diene-3,20-dione.
Acepreval. Glucocorticoid. *Parke-Davis.*

8810 Prednylidene
599-33-7 7906 209-964-9
$C_{22}H_{28}O_5$
(11β)-11,17,21-Trihydroxy-16-methylenepregna-1,4-diene-3,20-dione.
16-methylprednisolone; Dacortilen; Decortilen; Sterocort. Glucocorticoid. mp = 233-235°; $[\alpha]_D^{23}$ = +31° (dioxane); λ_m = 242 nm (ε 15900). *Merck & Co.Inc.*

8811 Prednylidene 21-Diethylaminoacetate Hydrochloride
22887-42-9 7906 245-299-0
$C_{28}H_{40}ClNO_6$
(11β)-11,17-Dihydroxy-16-methylene-3,20-dioxopregna-1,4-dien-21-yl N,N-diethylaminoacetate hydrochloride.
Decortilen soluble. Glucocorticoid. mp = 245-246°; $[\alpha]_D$ = +45° (H_2O); λ_m = 246-247 nm ($E^{1\%}_{1cm}$ 300). *Merck & Co.Inc.*

8812 Procinonide
58497-00-0 261-289-9
$C_{27}H_{34}F_2O_7$
(6α,11β,16α)-6,9-Difluoro-11-hydroxy-16,17-[(1-methylethylidene)bis(oxy)]-21-(1-oxopropoxy)pregna-1,4-diene-3,20-dione.
RS-2352. Adrenocortical steroid. *Syntex Labs. Inc.*

8813 Ticabesone
74131-77-4
$C_{22}H_{28}F_2O_4S$
(6α,11β,16α)-6,9-Difluoro-11-hydroxy-16-methyl-3-oxo-androsta-1,4-diene-17-carbothioic acid S-methyl ester.
Glucocorticoid. *Syntex Labs. Inc.*

8814 Ticabesone Propionate
73205-13-7
$C_{25}H_{32}F_2O_5S$
(6α,11β,16α)-6,9-Difluoro-11-hydroxy-16-methyl-3-oxo-17-(oxopropoxy)androsta-1,4-diene-17-carbothioic acid S-methyl ester.
RS-35909-00-00-0. Glucocorticoid. *Syntex Labs. Inc.*

8815 Timobesone
87116-72-1
$C_{22}H_{29}FO_3S$
(11β,16β,17α)-17-(Acetoxy)-9-fluoro-11-hydroxy-16-methyl-3-oxoandrosta-1,4-diene-17-carbothioic acid S-methyl ester.
Adrenocortical steroid. *Syntex Labs. Inc.*

8816 Timobesone Acetate
79578-14-6
$C_{24}H_{31}FO_4S$
(11β,16β,17α)-9-Fluoro-11-hydroxy-16-methyl-3-oxo-androsta-1,4-diene-17-carbothioic acid S-methyl ester.
RS-85446-007. Adrenocortical steroid. *Syntex Labs. Inc.*

8817 Tipredane
85197-77-9
$C_{22}H_{31}FO_2S_2$
(11β,17α)-17-(Ethylthio)-9-fluoro-11-hydroxy-17-(methylthio)androsta-1,4-diene-3-one.
SQ-27239. Adrenocortical steroid (topical). *Bristol-Myers Squibb Co.*

8818 Tralonide
21365-49-1
$C_{24}H_{28}Cl_2F_2O_4$
(6α,11β,16α)-9,11-Dichloro-6,21-difluoro-16,17-[(1-methylethylidene)bis(oxy)]pregna-1,4-diene-3,20-dione.
Glucocorticoid. *Eli Lilly & Co.; Syntex Labs. Inc.*

8819 Triamcinolone
124-94-7 9727 204-718-7
$C_{21}H_{27}FO_6$
(11β,16α)-9-Fluoro-11,16,17,21-tetrahydroxypregna-1,4-diene-3,20-dione.
CL-19823; Aristocort; Cinolone; Kenacort; Ledercort (tablets); Omicilon; Tricortale; Volon. Glucocorticoid. mp = 269-271°; $[\alpha]^{25}_D$ = +75° (Me_2OH); λ_m = 238 nm (ε 15800). *Am. Cyanamid; Bristol-Myers Squibb Co; Fujisawa Pharm. USA Inc.*

8820 Triamcinolone Acetonide
76-25-5 9728 200-948-7
$C_{24}H_{31}FO_6$
(11β,16α)-9-Fluoro-11,21-dihydroxy-16,17-[(1-methylethylidene)bis(oxy)]pregna-1,4-diene-3,20-dione.
9α-fluoro-16α-hydroxyprednisolone; Adcortyl; Azmacort; Delphicort; Extracort; Ftorocort; Kenacort-A; Kenalog; Ledercort Cream; Nasacort; Respicort; Rineton; Solodelf; TAC-3; TAC-40; Tramacin; Triacet; Triam; Triamonide 40; Tricinolon; Trymex; Vetalog; Volon A; Volonimat; component of: Mycolog II, Myco-Triacet II, Mytrex, Panolog. Glucocorticoid; antiasthmatic (inhalant); antiallergic (nasal). mp = 292-294°; $[\alpha]^{23}_D$ = +109° (c = 0.75 $CHCl_3$); λ_m = 238nm (ε 14600 EtOH); sparingly soluble in MeOH, Me_2OH, EtOAc. 3261; *Am. Cyanamid; Bristol-Myers Squibb Co; Forest Pharm. Inc.; Herbert; Johnson & Johnson Med. Inc.; Lemmon Co.; Olin Mathieson; Rhône-Poulenc Rorer Pharm. Inc.; Savage Labs.*

8821 Triamcinolone Acetonide 21-Hemisuccinate
9728
$C_{28}H_{35}FO_9$
(11β,16α)-9-Fluoro-11-hydroxy-16,17-[(1-methylethylidene)bis(oxy)]pregna-1,4-diene-3,20-dione 21-hemisuccinate.
Solutedarol. Glucocorticoid; antiasthmatic (inhalant); antiallergic (nasal). *Am. Cyanamid; Olin Mathieson.*

8822 Triamcinolone Acetonide Sodium Phosphate
1997-15-5 9728 217-878-8
$C_{24}H_{30}FNa_2O_9P$
(11β,16α)-9-Fluoro-11-hydroxy-16,17-[1-methylethylidinebis(oxy)]-21-(phosphonooxy)pregna-1,4-diene3,20-dione disodium salt.
CL-61965; CL-106359; Aristosol. Glucocorticoid; antiasthmatic (inhalant); antiallergic (nasal). *Am. Cyanamid; Olin Mathieson.*

8823 Triamcinolone Benetonide
31002-79-6 9729 250-427-3
$C_{35}H_{42}FNO_8$
(11β,16α)-21-[3-(Benzoylamino)-2-methyl-1-oxopropoxy]-9-fluoro-11-hydroxy-16,17-[(1-methylethylidene)bis(oxy)]pregna-1,4-diene-3,20-dione.
triamcinolone acetonide β-benzoylaminoisobutyrate; TBI; Tibicorten. Glucocorticoid; anti-inflammatory (topical). mp = 203-207°; $[\alpha]^{20}_D$ = +96° (c = 1EtOH); soluble in MeOH, EtOH, Me_2OH, dioxane, C_5H_5N, DMF, $CHCl_3$; insoluble in H_2O. *Sigma-Tau Pharm. Inc.*

8824 Triamcinolone Diacetate
67-78-7 9727 200-669-0
$C_{25}H_{31}FO_6$
(11β,16α)-16,21-Bis(acetyloxy)-9-fluoro-11,17-dihydroxypregna-1,4-diene-3,20-dione.
triamcinolone 16,21-diacetate; Aristocort Forte Parenteral; Aristocort Syrup; Cenacort; CINO-40; Kenacort Diacetate Syrup; TAC-D; Tracilon; Triamolone 40. Glucocorticoid. mp = 186-188°; $[\alpha]_D^{25}$ = +22° (in $CHCl_3$); λ_m = 239 nm (ε 15200). *Bristol-Myers Squibb Co; Forest Pharm. Inc.; Fujisawa Pharm. USA Inc.; Herbert.*

8825 Triamcinolone Furetonide
4989-94-0
$C_{33}H_{35}FO_8$
(11β,16α)-9-Fluoro-11,16,17,21-tetrahydroxypregna-1,4-diene-3,20-dione with acetone 21-(2-benzofurancarboxylate).
Glucocorticoid.

8826 Triamcinolone Hexacetonide
5611-51-8 9730 227-031-4
$C_{30}H_{41}FO_7$
(11β,16α)-21-(3,3-Dimethyl-1-oxobutoxy)-9-fluoro-11-hydroxy-16,17-[(1-methylethylidene)bis(oxy)]pregna-1,4-diene-3,20-dione.
triamcinolone acetonide tert-butyl acetate; TATBA; CL-34433; Aristospan; Hexatrione; Lederlon; Lederspan. Anti-inflammatory. The hexacetonide ester of the potent glucocorticoid triamcinolone. mp = 295-296°; $[\alpha]_D^{25}$ = +90° (c = 1.13% $CHCl_3$); λ_m = 238nm (ε 15500 EtOH); somewhat soluble in $CHCl_3$, dimethylacetamide; sparingly soluble EtOAc, MeOH, diethyl carbonate, glycerin, propylene glycol; nearly insoluble in H_2O, absolute alcohol. *Am. Cyanamid; Herbert.*

Miotics

8827 Acetylcholine Bromide
66-23-9 87 200-622-4
$C_7H_{16}BrNO_2$
2-Acetyloxy-N,N,N-trimethylethanaminium bromide.
Pragmoline; Tonocholin B. Cholinergic agent. Also has miotic properties. Very soluble in cold H_2O, decomposed by hot H_2O, soluble in EtOH, insoluble in Et_2O.

8828 Acetylcholine Chloride
60-31-1 88 200-468-8
$C_7H_{16}ClNO_2$
2-Acetyloxy-N,N,N-trimethylethanaminium chloride.
Acecoline; Arterocoline; Miochol; Ovisot. Cholinergic; cardiac depressant; miotic; vasodilator (peripheral). mp = 149-152°; very soluble in cold H_2O, decomposed by hot H_2O, soluble in EtOH, insoluble in Et_2O; LD_{50} (rat sc) = 250 mg/kg. *Iolab.*

8829 Carbachol
51-83-2 1823 200-127-3
$C_6H_{15}ClN_2O_2$
2-[(Aminocarbonyl)oxy]-N,N,N-trimethylethanaminium chloride.
choline chloride carbamate; Carbastat Intraocular; Isopto Carbachol; Miostat; Carcholin; Moryl; Doryl; Coletyl; Lentin. Cholinergic agent. Also has miotic properties. mp = 200-203°, 204-205°, 210-212°; soluble in H_2O (100 g/100 ml), MeOH (10 g/100 ml), insoluble in $CHCl_3$, Et_2O; LD_{50} (mus orl) = 15 mg/kg, (mus iv) = 0.3 mg/kg. *Alcon Labs; CIBA Vision AG.*

8830 Neostigmine
59-99-4 6553
$[C_{12}H_{19}N_2O_2]^+$
3-[[(Dimethylamino)carbonyl]oxy]-N,N,N-trimethylbenzenaminium.
synstigmin; proserine. Cholinergic; miotic; antidote for curare poisoning. *Apothecon; Elkins-Sinn; Hoffmann-LaRoche Inc.; ICN Pharm. Inc.*

8831 Neostigmine Bromide
114-80-7 6553 204-054-8
$C_{12}H_{19}BrN_2O_2$
3-[[(Dimethylamino)carbonyl]oxy]-N,N,N-trimethylbenzenaminium bromide.
Juvastigmin (tabl.); Neoesserin; Neostigmin (tabl.); Normastigmin (tabl.); Prostigmin. Cholinergic agent. Cholinesterase inhibitor. Also a miotic and used as an antidote for curare poisoning. mp = 167° (dec); soluble in H_2O (100 g/100 ml), EtOH. *Apothecon; Elkins-Sinn; Hoffmann-LaRoche Inc.; ICN Pharm. Inc.*

8832 Neostigmine Methylsulfate
51-60-5 6553 200-109-5
$C_{13}H_{22}N_2O_6S$
3-[[(Dimethylamino)carbonyl]oxy]-N,N,N-trimethylbenzenaminium methyl sulfate.
Intrastigmina; Juvastigmin (amp.); Metastigmin; Neostigmin (inj.); Normastigmin (amp.); Prostigmin (amp.); Stiglyn. Cholinergic; miotic; antidote for curare poisoning. Cholinesterase inhibitor. mp = 142-145°; soluble in H_2O (10 g/100 ml), less soluble in EtOH; LD_{50} (mus iv) = 0.16 mg/kg, (mus sc) = 0.42 mg/kg, (mus orl) = 7.5 mg/kg. *Apothecon; Elkins-Sinn; Hoffmann-LaRoche Inc.; ICN Pharm. Inc.*

8833 Physostigmine
57-47-6 7540 200-332-8
$C_{15}H_{21}N_3O_2$
(3aS-cis)-1,2,3,3a,8,8a-Hexahydro-1,3a,8-trimethylpyrrolo[2,3-b]indol-5-ol methylcarbamate (ester).
Cogmine; Eserine. Cholinergic agent (cholinesterase inhibitor). Also a miotic. Obtained from Calabar beans (*Physostigma venenosum*). mp = 105-106°; $[\alpha]_D^{17}$ = -76° (c = 1.3 $CHCl_3$), $[\alpha]_D^{25}$ = -120° (C_6H_6); soluble in EtOH, C_6H_6, $CHCl_3$; slightly soluble in H_2O; pKa_1 = 6.12, pka_2 = 12.24; may oxidize to form eserine blue ($C_{26}H_{31}N_5O_2$); LD_{50} (mus orl) = 4.5 mg/kg.

8834 Physostigmine Salicylate
57-64-7 7540 200-343-8
$C_{22}H_{27}N_3O_5$
(3aS-cis)-1,2,3,3a,8,8a-Hexahydro-1,3a,8-trimethylpyrrolo[2,3-b]indol-5-ol methylcarbamate (ester) salicylate.
Antilirium; Isopto Eserine. Cholinergic agent (cholinesterase inhibitor). Also a miotic. mp = 185-187°;

λ_m = 239, 252, 303 nm (log ε 4.09, 4.04, 3.78 MeOH); soluble in H_2O (1.33 g/100 ml at 25°, 6.25 g/100 ml at 80°), EtOH (6.25 g/100 ml at 25°, 20 g/100 ml at 76°), $CHCl_3$ (16.6 g/100 ml), Et_2O (0.4 g/100 ml); LD_{50} (mus ip) = 0.64 mg/kg. *Alcon Labs; Forest Pharm. Inc.*

8835 Physostigmine Sulfate
64-47-1 7540 200-585-4
$C_{30}H_{44}N_6O_8S$
(3aS-cis)-1,2,3,3a,8,8a-Hexahydro-1,3a,8-trimethyl-pyrrolo[2,3-b]indol-5-ol methylcarbamate (ester) sulfate.
Eserine sulfate. Cholinergic agent (cholinesterase inhibitor). Also a miotic. mp = 140°; soluble in EtOH (250 g/100 ml), H_2O (25 g/100 ml), Et_2O (0.83 g/100 ml). *CIBA Vision AG.*

8836 Pilocarpine
92-13-7 7578 202-128-4
$C_{11}H_{16}N_2O_2$
3-Ethyldihydro-4-[(1-methyl-1H-imidazol-5-yl)methyl]-2(3H)-furanone.
Ocusert Pilo. Antiglaucoma agent; cholinergic (ophthalmic). Also a miotic. mp = 34°; bp_5 = 260°; $[\alpha]_D^{18}$ = 106° (c = 2); soluble in H_2O, EtOH, $CHCl_3$; insoluble in Et_2O, C_6H_6, petroleum ether. *Alcon Labs; CIBA Vision AG; Wyeth Labs.*

8837 Pilocarpine Hydrochloride
54-71-7 7578 200-212-5
$C_{11}H_{17}ClN_2O_2$
3-Ethyldihydro-4-[(1-methyl-1H-imidazol-5-yl)methyl]-2(3H)-furanone hydrochloride.
Adsorbocarpine; Almocarpine; Isopto Carpine; component of: E-Pilo, Pilocar, Pilopine HS Gel. Antiglaucoma agent; cholinergic (ophthalmic). Also a miotic. mp = 204-205°; $[\alpha]_D^{18}$ = 91° (c = 2); freely soluble in H_2O, EtOH; insoluble in Et_2O, $CHCl_3$. *Alcon Labs; CIBA Vision AG; Wyeth Labs.*

Monoamine Oxidase Inhibitors

8838 Almoxatone
84145-89-1
$C_{18}H_{19}ClN_2O_3$
(+)-(R)-3-[p-[(m-Chlorobenzyl)oxy]phenyl]-5-[(methyl-amino)methyl]-2-oxazolidinone.
A short-acting MAO inhibitor.

8839 Amiflamine
77518-07-1
$C_{12}H_{20}N_2$
S-(+)-4-Dimethylamino-α,2-dimethylphenethylamine.
A monoamine oxidase inhibitor.

8840 Bazinaprine
94011-82-2
$C_{17}H_{19}N_5O$
3-(2-Morpholinoethylamino)-4-cyano-6-phenylpyridazine.
SR-95191. A type A monoamine oxidase inhibitor.

8841 Debrisoquin
1131-64-2 2901 214-470-1
$C_{10}H_{13}N_3$
3,4-Dihydro-2(1H)-isoquinolinecarboxamidine.
isocaramidine. Antihypertensive. Monoamine oxidase inhibitor. *Hoffmann-LaRoche Inc.*

8842 Debrisoquin Sulfate
581-88-4 2901 209-472-4
$C_{20}H_{28}N_6O_4S$
3,4-Dihydro-2(1H)-isoquinolinecarboxamidine sulfate (2:1).
Declinax; Ro-5-3307/1. Antihypertensive. Monoamine oxidase inhibitor. mp = 278-280°, 284-285°, 266-268°; freely soluble in H_2O; LD_{50} (neonate rat orl) = 88 ± 18 mg/kg, (rat orl) = 1580 ± 163 mg/kg. *Hoffmann-LaRoche Inc.*

8843 Iproclozide
3544-35-2 5092 222-589-5
$C_{11}H_{15}ClN_2O_2$
(p-Chlorophenoxy)acetic acid 2-isopropyl-hydrazide.
PC-603; Sursum. Monoamine oxidase inhibitor. Antidepressant. mp = 93-94°.

8844 Iproniazid
54-92-2 5094 200-218-8
$C_9H_{13}N_3O$
Isonicotinic acid 2-isopropylhydrazide.
Antidepressant. A monoamine oxidase inhibitor. mp = 112.5-113.5°; soluble in H_2O, EtOH; pH (aqueous solution) = 6.7; [Dihydrochloride]: Dec 227-228°; soluble in H_2O.

8845 Iproniazid Phosphate
305-33-9 5094 206-164-1
$C_9H_{16}N_3PO_5$
Isonicotinic acid 2-isopropylhydrazide phosphate.
Marsilid. Antidepressant. A monoamine oxidase inhibitor. mp = 175-184°; λ_m = 265 nm ($A_{1\,cm}^{1\%}$ 166 MeOH), 264 nm ($A_{1\,cm}^{1\%}$ 176 H_2O), 267 nm ($A_{1\,cm}^{1\%}$ 179 0.1N HCl), 244, 272, 308 nm ($A_{1\,cm}^{1\%}$ 113, 121, 135 0.1N KOH); soluble in H_2O (18.8 g/100 ml), MeOH (2.1 g/100 ml), EtOH (0.9 g/100 ml), $CHCl_3$ (0.06 g/100 ml); insoluble in Et_2O, C_6H_{14}.

8846 Isocarboxazid
59-63-2 5172 200-438-4
$C_{12}H_{13}N_3O_2$
5-Methyl-3-isoxazolecarboxylic acid 2-benzyl-hydrazide.
Ro-5-0831; Marplan. Antidepressant. A monoamine oxidase inhibitor. mp = 105-106°; barely soluble in H_2O (0.05%); slightly more soluble in 95% alcohol (1-2%), glycerol, propylene glycol. *Hoffmann-LaRoche Inc.*

8847 Lazabemide
103878-84-8 5407
$C_8H_{10}ClN_3O$
N-(2-Aminoethyl)-5-chloropicolinamide.
Ro-19-6327/000. Monoamine oxidase inhibitor. Used as an antiparkinsonian. *Hoffmann-LaRoche Inc.*

8848 Lazabemide Hydrochloride
103878-83-7 5407
$C_8H_{11}Cl_2N_3O$
N-(2-Aminoethyl)-5-chloropicolinamide hydrochloride.
Ro-19-6327. Monoamine oxidase inhibitor. Used as an antiparkinsonian. mp = 193-195°; LD_{50} (mus orl) = 1000-2000 mg/kg. *Hoffmann-LaRoche Inc.*

8849 Moclobemide
71320-77-9 6309
$C_{13}H_{17}ClN_2O_2$
p-Chloro-N-(2-morpholinoethyl)benzamide.
Ro-11-1163/000; Aurorix; Manerix; Moclaime. Antidepressant. A reversible monoamine oxidase A inhibitor. mp = 137°; LD_{50} (rat orl) = 707 mg/kg; [monohydrochloride]: mp = 208°. *C.M. Ind.; Hoffmann-LaRoche Inc.*

8850 Mofegiline
119386-96-8 6313
$C_{11}H_{13}F_2N$
(E)-2-(Fluoromethylene)-4-(p-fluorophenyl)butylamine.
Antiparkinsonian. Selective and irreversible monoamine oxidase B inhibitor. *Marion Merrell Dow Inc.*

8851 Mofegiline Hydrochloride
120635-25-8 6313
$C_{11}H_{14}ClF_2N$
(E)-2-(Fluoromethylene)-4-(p-fluorophenyl)butylamine hydrochloride.
MDL-72974A. Antiparkinsonian. Selective and irreversible monoamine oxidase B inhibitor. mp = 131°. *Marion Merrell Dow Inc.*

8852 Nialamide
51-12-7 6575 200-079-3
$C_{16}H_{18}N_4O_2$
Isonicotinic acid 2-[(2-benzylcarbamoyl)ethyl]hydrazide.
NF-XIII; Espril; Niamid; Niamidal; Niaquitil; Nuredal; Nyezin. Antidepressant. A monoamine oxidase inhibitor. mp = 152-153°; slightly soluble in H_2O, more soluble in acids; LD_{50} (mus orl) = 590 mg/kg, (mus ip) = 435 mg/kg, (mus iv) = 120 mg/kg. *Pfizer Inc.*

8853 Octamoxin
4684-87-1 6845
$C_8H_{20}N_2$
(1-Methylheptyl)hydrazine.
2-hydrazinooctane; octomoxine; Ximaol. Antidepressant. A monoamine oxidase inhibitor. *Octel Chem. Ltd.*

8854 Octamoxin Sulfate
3845-07-6 6845
$C_8H_{22}N_2O_4S$
(1-Methylheptyl)hydrazine sulfate.
Nimaol. Monoamine oxidase inhibitor. Antidepressant. mp = 78-80°. *Octel Chem. Ltd.*

8855 Pargyline
555-57-7 7172 209-101-6
$C_{11}H_{13}N$
N-Methyl-N-2-propynylbenzylamine.
MO-911; A-19120; Eudatin; Supirdyl. Monoamine oxidase inhibitor. Antihypertensive bp_{11} = 96-97°. *Abbott Labs Inc.*

8856 Pargyline Hydrochloride
306-07-0 7172 206-175-1
$C_{11}H_{14}ClN$
N-Methyl-N-2-propynylbenzylamine hydrochloride.
Eutonyl; A-19120; MO-911; NSC-43798; component of: Eutron. Monoamine oxidase inhibitor with antihypertensive properties. mp = 154-155°; readily soluble in H_2O. *Abbott Labs Inc.*

8857 Phenelzine
51-71-8 7366 200-117-9
$C_8H_{12}N_2$
β-Phenylethylhydrazine.
phenethylhydrazine; phenalzine. Antidepressant. A monoamine oxidase inhibitor. $bp_{0.1}$ = 74°; n_D^{20} = 1.5494; [hydrochloride]: mp = 174°. *Lakeside BioTechnology.*

8858 Phenelzine Dihydrogen Sulfate
156-51-4 7366 200-117-9
$C_8H_{14}N_2O_4S$
β-Phenylethylhydrazine sulfate (1:1).
phenelzine acid sulfate; W-1544a; Nardelzine; Nardil; Stinerval. Antidepressant. A monoamine oxidase inhibitor. Soluble in H_2O; LD_{50} (mus orl) = 156 mg/kg. *Lakeside BioTechnology; Parke-Davis.*

8859 Phenoxypropazine
3818-37-9 7411
$C_9H_{14}N_2O$
(1-Methyl-2-phenoxyethyl)hydrazine.
fenoxypropazine. Antidepressant. A monoamine oxidase inhibitor. $bp_{0.2}$ = 98-102°.

8860 Phenoxypropazine Maleate
3941-06-8 7411
$C_{13}H_{18}N_2O_5$
(1-Methyl-2-phenoxyethyl)hydrazine maleate (1:1).
Drazine. Antidepressant. A monoamine oxidase inhibitor. mp = 107-108°; very soluble in H_2O, less soluble in EtOH, iPrOH.

8861 Pirlindole
60762-57-4
$C_{15}H_{18}N_2$
2,3,3a,4,5,6-Hexahydro-8-methyl-1H-pyrazino[3,2,1-jk]carbazole.
pyrazidole. A selective and reversible monoamine oxidase A inhibitor. Antidepressant.

8862 Pivalylbenzhydrazine
306-19-4 7668 206-180-9
$C_{12}H_{18}N_2O$
2,2-Dimethylpropanoic acid.
pivhydrazine; Ro-4-1634; Tersavid. Monoamine oxidase inhibitor. mp = 68-69°. *Hoffmann-LaRoche Inc.*

8863 Prodipine
31314-38-2 7949
$C_{20}H_{25}N$
1-(1-Methylethyl)-4,4-diphenylpiperidine.
CNS stimulating agent with antidepressant properties.

Used as an antiparkinsonian. Monoamine oxidase inhibitor. $bp_{0.01}$ = 117-125°. *Byk Gulden Lomberg GmbH.*

8864 Prodipine Hydrochloride
31314-39-3 7949
$C_{20}H_{26}ClN$
1-(1-Methylethyl)-4,4-diphenylpiperidine hydrochloride.
Anthen. CNS stimulating agent with antidepressant properties. Used as an antiparkinsonian. Monoamine oxidase inhibitor. mp = 267°. *Byk Gulden Lomberg GmbH.*

8865 Safrazine
33419-68-0
$C_{11}H_{16}N_2O_2$
β-Piperonylisopropyl hydrazine.
Monoamine oxidase inhibitor.

8866 Safrazine Hydrochloride
7296-30-2
$C_{11}H_{17}ClN_2O_2$
β-Piperonylisopropyl hydrazine hydrochloride.
Monoamine oxidase inhibitor.

8867 Selegiline
14611-51-9 8569
$C_{13}H_{17}N$
(-)-(R)-N-α-Dimethyl-N-2-propynylphenethylamine.
(-)-deprenil; L-deprenyl; deprenyl [as (±)-form]; phenylisopropyl-N-methylpropinylamine [as (±)-form]. Antidyskinetic; antiparkinsonian. Monoamine oxidase B inhibitor. Related to pargyline. $bp_{0.8}$ = 92-93°; $[\alpha]_D^{20}$ = -11.2°; [(dl-form)]: bp_5 = 103-110°; n_D^{20} = 1.5224. *Chinoin; Somerset Pharm. Inc.*

8868 Selegiline Hydrochloride
14611-52-0 8569
$C_{13}H_{18}ClN$
(-)-(R)-N-α-Dimethyl-N-2-propynylphenethylamine hydrochloride.
deprenyl; Eldepryl; Déprényl; Eldéprine; Eldepryl; Jumex; Movergan; Plurimen; E-250 [as (±)-form hydrochloride]. Antidyskinetic; antiparkinsonian. Monoamine oxidase B inhibitor. Related to pargyline. mp = 141-142°; $[\alpha]_D^{25}$ = -10.8° (c = 6.48 H_2O); LD_{50} (rat iv) = 81 mg/kg, (rat sc) = 280 mg/kg; [(±)-form hydrochloride]: LD_{50} (rat iv) = 63 mg/kg, (rat sc) = 126 mg/kg, (rat orl) = 385 mg/kg. *Chinoin; Somerset Pharm. Inc.*

8869 Toloxatone
29218-27-7 9659 249-522-2
$C_{11}H_{13}NO_3$
5-(Hydroxymethyl)-3-m-tolyl-2-oxazolidinone.
MD-69276; Humoryl; Perenum. Antidepressant. A reversible monoamine oxidase A inhibitor. mp = 76°; LD_{50} (mus orl) = 1850 mg/kg, 1500 mg/kg. *Delalande; Delandale Labs. Ltd.*

8870 Tranylcypromine
155-09-9 9708 205-841-9
$C_9H_{11}N$
(±)-trans-2-Phenylcyclopropanamine.
SKF-trans-385. Antidepressant. A monoamine oxidase inhibitor. Liquid; $bp_{1.5-1.6}$ = 79-80°; [hydrochloride]: mp = 164-166°. *SmithKline Beecham Pharm.*

8871 Tranylcypromine Sulfate
13492-01-8 9708 236-807-1
$C_{18}H_{24}N_2O_4S$
(±)-trans-2-Phenylcyclopropylamine sulfate (2:1).
Parnate; Tylciprine. Monoamine oxidase inhibitor. Antidepressant. Soluble in H_2O; less soluble in EtOH, Et_2O; insoluble in $CHCl_3$. *SmithKline Beecham Pharm.*

Mucolytics

8872 Acetylcysteine
616-91-1 89 210-498-3
$C_5H_9NO_3S$
L-α-Acetamido-β-mercaptopropionic acid.
N-acetyl-L-cysteine; 5052; NSC-111180; Airbron; Broncholysin; Brunac; Fabrol; Fluatox; Fluimucil; Fluimucetin; Fluprowit; Mucocedyl; Mucolator; Mucolyticum; Mucomyst; Muco Sanigen; Mucosolvin; Mucret; NAC; Neo-Fluimucil; Parvolex; Respaire; Tixair. Mucolytic; corneal vulnerary; antidote to acetaminophen poisoning. mp = 109-110°; LD_{50} (rat orl) = 5050 mg/kg. *Bristol Laboratories; DuPont-Merck Pharm.; Mead Johnson Labs.; Mead Johnson Pharmaceuticals.*

8873 Bromhexine
3572-43-8 1412 222-684-1
$C_{14}H_{20}Br_2N_2$
2-Amino-3,5-dibromo-N-cyclohexyl-N-methylbenzene-methanamine.
An expectorant; mucolytic. mp = 237.5-238° (dec); soluble in H_2O (0.4 g/100 ml); LD_{50} (rbt orl) >10000 mg/kg.

8874 Bromhexine Hydrochloride
611-75-6 1412 210-280-8
$C_{14}H_{21}Br_2ClN_2$
2-Amino-3,5-dibromo-N-cyclohexyl-N-methylbenzene-methanamine monohydrochloride.
NA-274; Auxit; Bisolvon; Ophtosol; Quentan. Expectorant; mucolytic.

8875 Carbocysteine
638-23-3 1850 211-327-5
$C_5H_9NO_4S$
S-(Carboxymethyl)-L-cysteine.
3-[(carboxymethyl)thio]alanine; carboxymethylcysteine; AHR-3053; LJ-206; Carbocit; Fluifort; Lisil; Lisomucil; Loviscol; Muciclar; Mucocis; Mucodyne; Mucofan; Mucolase; Mucolex; Mucopront; Mucotab; Mukinyl; Pectox; Pulmoclase; Reomucil; Rhinathiol; Siroxyl; Thiodril; Transbronchin. Mucolytic; expectorant. [L-form]: mp = 204-207°; $[\alpha]_D^{24}$ = 0.5° (1N HCl). *Lederle Labs.*

8876 Dembrexine
83200-09-3
$C_{13}H_{17}Br_2NO_2$
trans-4-[(3,5-Dibromosalicyl)amino]cyclohexanol.
Mucolytic.

8877 Dimesna
16208-51-8
$C_4H_8Na_2O_6S_4$
Disodium 2,2'-dithiodiethanesulfonate.
Mucolytic, detoxifying agent.

8878 Domiodol
61869-07-6 3473
$C_5H_9IO_3$
2-(Iodomethyl)-1,3-dioxolane-4-methanol.
M.G. 13608; Mucolitico. A mucolytic. LD_{50} (mmus ip) = 79 mg/kg, (mmus orl) = 140 mg/kg, (fmus ip) = 89 mg/kg, (fmus orl) = 145 mg/kg; [cis form]: $bp_{0.2}$ = 106-108°; LD_{50} (mmus orl) = 135 mg/kg; [trans form]: $bp_{0.2}$ = 114-116°; LD_{50} (mmus orl) = 150 mg/kg. *Maggioni Farmaceutici S.p.A.*

8879 Erdosteine
84611-23-4 3684
$C_8H_{11}NO_4S_2$
(±)-[[[(Tetrahydro-2-oxo-3-thienyl)carbamoyl]-methyl]thio]acetic acid.
dithiosteine; RV-144; Secresolv. A mucolytic. mp = 156-158°; LD_{50} (mus orl) > 10000 mg/kg, (rat orl) > 3500 mg/kg. *Refarmed.*

8880 Guaimesal
81674-79-5
$C_{16}H_{14}O_5$
(±)-2-(o-Methoxyphenoxy)-2-methyl-1,3-benzodioxan-4-one.
Mucolytic.

8881 Letosteine
53943-88-7 5473 258-879-3
$C_{10}H_{17}NO_4S_2$
2-[2-[(Carboxymethyl)thio]ethyl]-4-thiazolidinecarboxylic acid 2-ethyl ester.
Visoctiol. A mucolytic. mp = 142°. *Ferlux-Chemie.*

8882 Lysozyme
9001-63-2 5671 232-620-4
N-Acetylmuramide glycanohydrolase.
Muramidase; N-acetylmuramyl hydrolase; Globulin G_2. Enzyme of molecular weight 14,400 ± 100. Found in tears, nasal mucus, milk, blood serum and saliva. A mucolytic.

8883 Lysozyme Hydrochloride
9066-59-5 5671 232-954-0
N-Acetylmuramide glycanohydrolase hydrochloride.
Acdeam; Antalzyme; Immunozima; Lanzyme; Leftose; Likinozym; Lisozima; Murazyme; Neutase; Neuzyme; Toyolysom-DS. A mucolytic enzyme with antibiotic properties. Derivative of lysozyme.

8884 Mecysteine Hydrochloride
18598-63-5 5828 242-435-0
$C_4H_{10}ClNO_2S$
Methyl α-amino-β-mercaptopropionate hydrochloride.
LJ-48; Acdrile; Visclair. A mucolytic. mp = 140-141°; $[\alpha]_D^{20}$ = -2.9° (MeOH).

8885 Mesna
19767-45-4 5969 243-285-9
$C_2H_5NaO_3S_2$
Sodium 2-mercaptoethanesulfonate.
D-7093; Mesnex; UCB-3983; Mistabron; Mistabronco; Mucofluid; Uromitexan. A mucolytic. Also used as a detoxifying agent, particularly in connection with alkylating agents used in cancer chemotherapy. Freely soluble in H_2O, less soluble in organic solvents; LD_{50} (mmus iv) = 1887 mg/kg, (mmus ip) = 2005 mg/kg, (mmus orl) = 6102 mg/kg, (fmus iv) = 2048 mg/kg, (fmus ip) = 2098 mg/kg, (fmus orl) > 7200 mg/kg, (mrat iv) = 2098 mg/kg, (mrat ip) = 1529, (mrat orl) = 4440 mg/kg, (frat iv) = 1683 mg/kg, (frat ip) = 1251 mg/kg, (frat orl) = 4679 mg/kg. *Asta-Werke AG; Bristol-Myers Oncology; UCB Pharma.*

8886 Nesosteine
84233-61-4
$C_{11}H_{11}NO_3S$
o-(3-Thiazolidinylcarbonyl)benzoic acid.
CO-1177. Reduces mucus viscosity and induces improvement in mucociliary clearance. Mucolytic.

8887 Sobrerol
498-71-5 8707 207-868-1
$C_{10}H_{18}O_2$
5-Hydroxy-α,α,4-trimethyl-3-cyclohexene-1-methanol.
6,8-carvomenthenediol; l-p-menthene-6,8-diol; p-menth-6-ene-2,8-diol; pinol hydrate. A mucolytic. The trans form is the naturally occuring form. Formed by the auto-oxidation of α-pinene. [dl-trans-form]: mp = 130-131.5°; bp = 270-271°; soluble in H_2O (3.3 g/100 ml at 15°); LD_{50} (mus,rat iv) = 580 mg/kg; [d- or l-trans-form]: mp = 150°; $[\alpha]_{DS}^{20}$ = (±) 150° resp. (c = 10 EtOH). *Camillo-Corvi; Glidden Co.*

8888 Stepronin
72324-18-6 8963 276-587-4
$C_{10}H_{11}NO_4S_2$
N-(2-Mercaptopropionyl)glycine 2-thiophene-carboxylate (ester).
TTPG; prostenoglycine; Tiase. A mucolytic. mp = 168-170°; λ_m = 292 nm; LD_{50} (mus orl) > 2500 mg/kg, (mus iv) > 1250 mg/kg, (rat orl) > 2500 mg/kg, (rat im) = 1801 mg/kg. *Mediolanum Farmaceutici S.p.A.*

8889 Stepronin Lysine Salt
113790-28-6 8963
$C_{16}H_{25}N_3O_6S_2$
N-(2-Mercaptopropionyl)glycine 2-thiophene carboxylate (ester) lysine (salt).
Masor; Mucodil. A mucolytic. *Mediolanum Farmaceutici S.p.A.*

8890 Stepronin Sodium Salt
78126-10-0 8963
$C_{10}H_{10}NNaO_4S_2$
N-(2-Mercaptopropionyl)glycine 2-thiophene-carboxylate (ester) sodium salt.
Broncoplus; Tioten. A mucolytic. *Mediolanum Farmaceutici S.p.A.*

8891 Tasuldine
88579-39-9
$C_{10}H_9N_3S$
2-[(3-Pyridinylmethyl)thio]pyrimidine.
HE-10004. Bronchosecretolytic agent; systemic-mucolytic.

8892 Tiopronin
1953-02-2 9597 217-778-4
$C_{15}H_{17}NS_2$
N-(2-mercapto-1-oxopropyl)glycine.
Acadione; BRN 1859822; Capen; Captimer; CCRIS 1935; Epatiol; Meprin (detoxicant); Mercaptopropionylglycine; (2-Mercaptopropionyl)glycine; Mucolysin; N-(2-Mercapto-1-oxopropyl)glycine; N-(2-Mercaptopropionyl)-glycine; Sutilan; Thiola; Thiolpropionamidoacetic acid; Thiopronin; Thiopronine; Thiosol; Tioglis; Tiopronin; Tiopronine; Tiopronino; Tioproninum; Vincol. Mucolytic; expectorant. Used as an antidote in heavy metal poisoning. mp = 95-97°; LD_{50} (mus iv) = 2100 mg/kg. *Boehringer Ingelheim Ltd.*

8893 Tyloxapol
25301-02-4 9964
p-(1,1,3,3-Tetramethylbutyl)phenol polymer with ethylene oxide and formaldehyde.
oxyethylated tertiary octylphenol-formaldehyde polymer; tyloxypal; Alevaire; Superinone; Triton A-20; Triton WR-1339; component of: Enuclene, Exosurf. A nonionic detergent with surface-tension-reducing properties, used in pulmonary surfactants. Used as a mucolytic. Freely soluble in H_2O; soluble in C_6H_6, C_7H_8, $CHCl_3$, CCl_4, CS_2, AcOH. *Alcon Labs; Glaxo Wellcome Inc.; Rohm & Haas Co.; Sterling Winthrop Inc.*

Muscle Relaxants

8894 Adiphenine
64-95-9 160 200-599-0
$C_{20}H_{25}NO_2$
2-(Diethylamino)ethyl diphenylacetate.
Trasentine; Diphacil; Difacil. Smooth muscle relaxant. Anticholinergic. *CIBA plc.*

8895 Adiphenine Hydrochloride
50-42-0 160 200-036-9
$C_{20}H_{26}ClNO_2$
2-(Diethylamino)ethyl diphenylacetate hydrochloride.
NSC-129224; Trasentine hydrochloride; Difacil hydrochloride; Diphacil hydrochloride; Patrovina; Spasmolytin. Anticholinergic. Smooth muscle relaxant. mp = 113-114°; freely soluble in H_2O; poorly soluble in EtOH, Et_2O. *Ciba-Geigy Corp.*

8896 Adiphenine Methobromide
6113-04-8 160 228-077-8
$C_{21}H_{28}BrNO_2$
2-(Diethylamino)ethyl diphenylacetate methobromide.
adiphenine methyl bromide; Lunal. Smooth muscle relaxant. Anticholinergic. LD_{50} (rbt iv) = 22.5-27.5 mg/kg. *CIBA plc.*

8897 Afloqualone
56287-74-2 181
$C_{16}H_{14}FN_3O$
6-Amino-2-(fluoromethyl)-3-o-tolyl-4-(3H)-quinazolinone.
HQ-495; Aroft; Arofuto. Skeletal muscle relaxant. mp = 195-196°; LD_{50} (mus ip) = 315.1 mg/kg. *Tanabe Seiyaku Co. Ltd.*

8898 Alcuronium
23214-96-2 222
$[C_{44}H_{50}N_4O_2]^{2+}$
4,4'-Didemethyl-4,4'-di-2-propenyl-toxiferine.
diallylnortoxiferine; diallyltoxiferine. Skeletal muscle relaxant. [di-iodide ($C_{44}H_{50}I_2N_4O_2$)]: λ_m = 291 nm (ε = 39900 MeOH). *Hoffmann-LaRoche Inc.*

8899 Alcuronium Dichloride
15180-03-7 222 239-229-8
$C_{44}H_{50}Cl_2N_4O_2$
N,N'-Diallylnortoxiferinium dichloride.
Ro-4-3816; Alloferin; Toxiferene. Skeletal muscle relaxant. $[\alpha]_D^{22}$ = -348° (MeOH); λ_m = 292 nm (ε 43000 MeOH). *Hoffmann-LaRoche Inc.*

8900 Ambenoxan
2455-84-7 219-531-6
$C_{14}H_{21}NO_4$
N-[2-(2-Methoxyethoxy)ethyl]-1,4-benzodiozan-2-methylamine.
Muscle relaxant.

8901 Ambuphylline
5634-34-4 404 227-077-5
$C_{11}H_{19}N_5O_3$
Theophylline compound with 2-amino-2-methyl-1-propanol (1:1).
2-amino-2-methyl-1-propanol compound with theophylline; bufylline (formerly); theophylline aminoisobutanol; Butaphyllamine; Buthoid; component of: Nethaphyl. Bronchodilator; diuretic; smooth muscle relaxant. Xanthine derivative. mp = 254-256°; soluble in H_2O (55 g/100 ml); LD_{50} (mus orl) = 600 mg/kg. *Marion Merrell Dow Inc.*

8902 Aminophylline
317-34-0 485 206-264-5
$C_{16}H_{24}N_{10}O_4$
3,7-Dihydro-1,3-dimethyl-1H-purine-2,6-dione compound with ethylenediamine (2:1).
theophylline compound with ethylenediamine (2:1); theophylline ethylenediamine; theophyllamine; Aminocardol; Aminodur; Aminophyllin; Cardiofilina; Cardiomin; Cardophylin; Carena; Diophyllin; Euphyllin CR; Genophyllin; Grifomin; Inophylline; Metaphyllin; Minaphol; Pecram; Peterphyllin; Phyllindon; Phyllocontin; Pylcardin; Rectalad Aminophylline; Somophyllin; Stenovasan; Tefamin; Theodrox; Theolamine; Theomin; Theophyldine; component of: Amesec, Mudrane GG Tablets, Mudrane GG-2 Tablets, Mudrane Tablets, Mudrane-2 Tablets. Smooth muscle relaxant, used as a bronchodilator. Xanthine derivative. CAUTION: Causes acute poisoning: restlessness,

anorexia, nausea, fever, vomiting, dehydration, tremors, delirium, coma; may cause cardiovascular and respiratory collapse, shock, cyanosis, death. [Dihydrate]: soluble in H_2O (20 g/100 ml); insoluble in EtOH, Et_2O; LD_{50} (mus orl) = 540 mg/kg. *ECR Pharm.; Elkins-Sinn; Fisons plc; Glaxo Labs.; Purdue Pharma L.P.; Searle G.D. & Co.; Wallace Labs.*

8903 Atracurium Besylate
64228-81-5 900 264-743-4
$C_{65}H_{82}N_2O_{18}S_2$
2-(2-Carboxyethyl)-1,2,3,4-tetrahydro-6,7-dimethoxy-2-methyl-1-veratrylisoquinolinium benzenesulfonate pentamethylene ester.
Tracrium; BW-33A; Wellcome 33-A-74. Skeletal muscle relaxant. mp = 85-90°. *Glaxo Wellcome Inc.*

8904 1R-cis, 1R'-cis-Atracurium Besylate
96946-42-8 900
$C_{65}H_{82}N_2O_{18}S_2$
(1R-cis,1R'-cis)-2-(2-Carboxyethyl)-1,2,3,4-tetrahydro-6,7-dimethoxy-2-methyl-1-veratrylisoquinolinium benzenesulfonate pentamethylene ester.
Tracrium; BW-33A; 51W89; Nimbex; cisatracurium besylate. Skeletal muscle relaxant. White solid. *Glaxo Wellcome Inc.*

8905 Atropine
51-55-8 907 200-104-8
$C_{17}H_{23}NO_3$
1αH,5αH-Tropan-3α-ol (±)-tropate (ester).
tropine tropate; dl-tropyl tropate. Anticholinergic. Parasympatholytic alkaloid from *Atropa belladona*. mp = 114-116°; soluble in H_2O (0.22 g/100 ml at 25°, 11.1 g/100 ml at 80°), EtOH (50 g/100 ml at 25°, 83.3 g/100 ml at 60°), glycerol (3.7 g/100 ml), Et_2O (4 g/100 ml), $CHCl_3$ (100 g/100 ml), C_6H_6; LD_{50} (rat orl) = 750 mg/kg.

8906 Azumolene
64748-79-4
$C_{13}H_9BrN_4O_3$
1-[[[5-(p-Bromophenyl)-2-oxazolyl]methylene]-amino]hydantoin.
Skeletal muscle relaxant.

8907 Azumolene Sodium
91524-18-4
$C_{13}H_8BrN_4O_3.2H_2O$
1-[[[5-(p-Bromophenyl)-2-oxazolyl]methylene]-amino]hydantoin sodium salt dihydrate.
EU-4093. Skeletal muscle relaxant.

8908 Baclofen
1134-47-0 967 214-486-9
$C_{10}H_{12}ClNO_2$
β-(Aminomethyl)-p-chlorohydrocinnamic acid.
Lioresal; Ba-34,647; Baclon. Skeletal muscle relaxant. Antispasmodic. mp = 206-208°, 189-191°; LD_{50} (mmus iv) = 45 mg/kg, (mmus sc) = 103 mg/kg, (mmus orl) = 200 mg/kg, (mrat iv) = 78 mg/kg, (mrat sc) = 115 mg/kg, (mrat orl) = 145 mg/kg; [hydrochloride]: mp = 179-181°. *C.M. Ind.; Ciba-Geigy Corp.*

8909 Benzoctamine
17243-39-9 1117
$C_{18}H_{19}N$
N-Methyl-9,10-ethanoanthracene-9(10H)-methylamine.
Skeletal muscle relaxant; sedative/hypnotic; anxiolytic. *Ciba-Geigy Corp.*

8910 Benzoctamine Hydrochloride
10085-81-1 1117 233-216-0
$C_{18}H_{20}ClN$
N-Methyl-9,10-ethanoanthracene-9(10H)-methylamine hydrochloride.
Ba-30803; Tacitin. Skeletal muscle relaxant; sedative/hypnotic; anxiolytic. mp = 320-322°; LD_{50} (rat orl) = 700 ± 170 mg/kg. *Ciba-Geigy Corp.*

8911 Benzoquinonium Chloride
311-09-1 1137
$C_{34}H_{50}Cl_2N_4O_2$
N,N'-[(3,6-Dioxo-1,4-cyclohexadiene-1,4-diyl)-bis(imino-3,1-propanediyl)]bis[N,N-diethylbenzenemethanaminium]dichloride.
Win-2747; Mytolon; Amilyt. Skeletal muscle relaxant. mp = 191-195°; freely soluble in H_2O. *Sterling Res. Labs.*

8912 C-Calebassine
7257-29-6 1762
$[C_{40}H_{48}N_4O_2]^{2+}$
C-Strychnotoxine.
C-Toxiferene II. Skeletal muscle relaxant. [dichloride ($C_{40}H_{48}Cl_2N_4O_2$)]: $[\alpha]_D^{20}$ = 72.1° (c = 0.67 H_2O); λ_m = 253, 302 nm (+log ε = 4.37, 3.77, H_2O); [dipicrate ($C_{52}H_{52}N_{10}O_{16}$)]: mp = 215°; insoluble in H_2O, EtOH, MeOH, dioxane, slightly soluble in Me_2CO.

8913 Candocuronium Iodide
54278-85-2
$C_{26}H_{46}I_2N_2$
17a,17a-Dimethyl-3β-(1-methylpyrrolidinio)-17a-azonia-D-homoandrost-5-ene diiodide.
chandonium iodide. A muscle relaxant.

8914 Carisoprodol
78-44-4 1889 201-118-7
$C_{12}H_{24}N_2O_4$
2-Methyl-2-propyl-1,3-propanediol carbamate isopropylcarbamate.
Rela; Soma; isopropyl meprobamate; carisoprodate; Apesan; Arusal; Caprodat; Carisoma; Domarax; Flexal; Flexartal; Miolisodal; Mioril; Relasom; Sanoma; Somadril; Somalgit. Skeletal muscle relaxant. mp = 92-93°; soluble in organic solvents, poorly soluble in H_2O (0.03 g/100 ml at 25°, 0.140 g/100 ml at 50°); LD_{50} (mus orl) = 2340 mg/kg, (mus ip) = 980 mg/kg, (rat orl) = 1320 mg/kg, (rat ip) = 450 mg/kg. *Schering-Plough HealthCare Products; Wallace Labs.*

8915 Chlormezanone
80-77-3 2155 201-307-4
$C_{11}H_{12}ClNO_3S$
2-(p-Chlorophenyl)tetrahydro-3-methyl-4H-1,3-thiazin-4-one 1,1-dioxide.
dichloromezanone; chlormethazanone; Alinam; Banabin-Sintyal; Fenarol; Lobak; Mio-Sed; Rexan; Rilansyl;

Rilaquil; Rilasol; Supotran; Suprotran; Tanafol; Trancote; Trancopal; Transanate. Anxiolytic; skeletal muscle relaxant. mp = 116.2-118.2°; soluble in H_2O (< 0.25 g/100 ml), EtOH (< 1 g/100 ml). *Sterling Winthrop Inc.*

8916 Chlorphenesin Carbamate
886-74-8 2231 212-954-7
$C_{10}H_{12}ClNO_4$
3-(p-Chlorophenoxy)-1,2-propanediol 1-carbamate.
Maolate; U-19646; Rinlaxer. Skeletal muscle relaxant. mp = 89-91°; insoluble in H_2O, C_6H_6, C_6H_{12}, soluble in EtOH, Me_2CO, EtOAc, dioxane; LD_{50} (rat orl) = 748 mg/kg, (mus iv) = 239 mg/kg. *Pharmacia & Upjohn.*

8917 Chlorproethazine
84-01-5 2236 201-510-8
$C_{19}H_{23}ClN_2S$
2-Chloro-10-(3-diethylaminopropyl)phenothiazine.
RP-4909. Muscle relaxant; antipsychotic. A phenothiazine. bp_1 = 225-240°. *Rhône-Poulenc Rorer Pharm. Inc.*

8918 Chlorproethazine Hydrochloride
4611-02-3 2236 225-018-8
$C_{19}H_{24}ClN_2S$
2-Chloro-10-(3-diethylaminopropyl)phenothiazine hydrochloride.
Neuriplege. Muscle relaxant; antipsychotic. A phenothiazine. mp = 178°; soluble in H_2O (1.67 g/100 ml), EtOH (0.33 g/100 ml), $CHCl_3$ (20 g/100 ml); insoluble in Me_2CO, C_6H_6, Et_2O. *Rhône-Poulenc Rorer Pharm. Inc.*

8919 Chlorzoxazone
95-25-0 2249 202-403-9
$C_7H_4ClNO_2$
5-Chloro-2-benzoxazolinone.
Paraflex; Parafon Forte DSC; Biomioran; Solaxin. Skeletal muscle relaxant. mp = 191-191.5°; sparingly soluble in H_2O; soluble in MeOH, EtOH, iPrOH; [suspension]: LD_{50} (mus orl) = 3650 mg/kg, (mus ip) = 380 mg/kg; [sodium salt]: LD_{50} (mus orl) = 440 mg/kg, (mus ip) = 183 mg/kg. *Lemmon Co.; McNeil Pharm.*

8920 Cinflumide
64379-93-7
$C_{12}H_{12}FNO$
(E)-N-Cyclopropyl-m-fluorocinnamamide.
BW-532U. Skeletal muscle relaxant. *Glaxo Wellcome Inc.*

8921 Cinnamedrine
90-86-8 2357 202-021-2
$C_{19}H_{23}NO$
α-[1-(Cinnamylmethylamino)ethyl]benzyl alcohol.
N-cinnamylephedrine. Smooth muscle relaxant; antispasmodic. Reduces epinephrine-induced autonomic activity. mp = 72-78°; [dl-hydrochloride]: mp = 180-185°; soluble in H_2O (1.25 g/100 ml at 20°), EtOH (10 g/100 ml at 20°), $CHCl_3$ (2 g/100 ml); sparingly soluble in Et_2O, C_6H_6. *Sterling Winthrop Inc.; Winthrop.*

8922 Clodanolene
14796-28-2
$C_{14}H_9Cl_2N_3O_3$
1-[[5-(3,4-Dichlorophenyl)furfurylidene]amino]hydantoin.
F-413. Skeletal muscle relaxant.

8923 Curare
8063-06-7 2741 232-511-1
Ourari; urari; woorari; wourara. Extracts from bark of various Strychnos and Chondodendron species. Used as an arrow poison. The main active ingredient is (+)-tubocurarine chloride. Skeletal muscle relaxant.

8924 Cyclarbamate
5779-54-4 2772 227-302-7
$C_{21}H_{24}N_2O_4$
1,1-Cyclopentanedimethanol dicarbanilate.
cyclopentaphene; C-1428; BSM-906M; Calmalone; Casmalon. Skeletal muscle relaxant; anxiolytic. mp = 151-152°; insoluble in H_2O; slightly soluble in EtOH, glycerol, propylene glycol. *Lab Cassenne Marion.*

8925 Cyclobenzaprine
303-53-7 2782 206-145-8
$C_{20}H_{21}N$
N,N-Dimethyl-5H-dibenzo[a,d]cycloheptene-$\Delta^{5,\sigma}$-propylamine.
MK-130; Ro-4-1577; RP-9715. Anxiolytic; skeletal muscle relaxant. bp_1 = 175-180°; λ_m = 224, 289 nm (log ε = 4.57, 4.02). *Apothecon; Hoffmann-LaRoche Inc.; Merck & Co.Inc.*

8926 Cyclobenzaprine Hydrochloride
6202-23-9 2782 228-264-4
$C_{20}H_{22}ClN$
N,N-Dimethyl-5H-dibenzo[a,d]cycloheptene-$\Delta^{5,\sigma}$-propylamine hydrochloride.
Flexeril; Flexiban. Anxiolytic; skeletal muscle relaxant. mp = 216-218°; soluble in H_2O (> 20 g/100ml); soluble in MeOH, EtOH; less soluble in iPrOH, $CHCl_3$, CH_2Cl_2; insoluble in hydrocarbons; λ_m = 226, 295 nm (ε 52300, 12000); LD_{50} (mus iv) = 35 mg/kg, (mus orl) = 250 mg/kg. *Apothecon; Hoffmann-LaRoche Inc.; Merck & Co.Inc.*

8927 Dacuronium Bromide
27115-86-2
$C_{33}H_{58}Br_2N_2O_3$
(3α,17β-Dihydroxy-5α-androstan-2β,16β-ylene)-bis(1-methylpiperidinium)dibromide 3-acetate.
Muscle relaxant.

8928 Dantrolene
7261-97-4 2879 230-684-8
$C_{14}H_{10}N_4O_5$
1-[[5-(p-Nitrophenyl)furfurylidene]amino]hydantoin.
F-368. Skeletal muscle relaxant. mp = 279-280°. *Norwich; Procter & Gamble Pharm. Inc.*

8929 Dantrolene Sodium Salt
24868-20-0 2879
$C_{14}H_9N_4NaO_5.3\ 1/2\ H_2O$
1-[[5-(p-Nitrophenyl)furfurylidene]amino]hydantoin sodium salt hydrate.

Dantrium [as hemiheptahydrate]; Dantamacrin; Dantrix [as tetrahydrate]. Skeletal muscle relaxant. Slightly soluble in H_2O. *Norwich; Procter & Gamble Pharm. Inc.*

8930 Decamethonium Bromide
541-22-0 2904 208-772-2
$C_{16}H_{38}Br_2N_2$
Decamethylenebis(trimethylammonium) dibromide.
C-10; Syncurine. Skeletal muscle relaxant. mp = 268-270° (dec); soluble in H_2O, EtOH; poorly soluble in $CHCl_3$; insoluble in Et_2O.

8931 Diazepam
439-14-5 3042 207-122-5
$C_{16}H_{13}ClN_2O$
7-Chloro-1,3-dihydro-1-methyl-5-phenyl-2H-1,4-benzodiazepin-2-one.
Alupram; Valium; Valrelease; LA-111; Ro- 5-2807; Wy-3467; NSC-77518; Apaurin; Atensine; Atilen; Bialzepam; Calmpose; Ceregulart; Dialar; Diazemuls; Dipam; Eridan; Eurosan; Evacalm; Faustan; Gewacalm; Horizon; Lamra; Lembrol; Levium; Mandrozep; Neurolytril; Noan; Novazam; Paceum; Pacitran; Paxate; Paxel; Pro-Pam; Q-Pam; Relanium; Sedapam; Seduxen; Servizepam; Setonil; Solis; Stesolid; Tranquase; Tranquo-Puren; Tranquo-Tablinen; Unisedil; Valaxona; Valiquid; Valium; Valium Injectable; Valrelease; Vival; Vivol. Anxiolytic; skeletal muscle relaxant; sedative/hypnotic. Also used as an intravenous anesthetic. Pharmaceutical preparation for the treatment of depression. A benzodiazepine. mp = 125-126°; soluble in DMF, $CHCl_3$, C_6H_6; Me_2CO, EtOH; slightly soluble in H_2O; LD_{50} (rat orl) = 710 mg/kg. *Berk Pharm. Ltd.; Hoffmann-LaRoche Inc.*

8932 Doxacurium Chloride
106819-53-8 3487
$C_{56}H_{78}Cl_2N_2O_{16}$
(1R,2S;1S,2R)-1,2,3,4-Tetrahydro-2-(3-hydroxypropyl)-6,7,8-trimethoxy-2-methyl-1-(3,4,5-trimethoxy-benzylisoquinolinium chloride succinate mixture with (±)-(1R*,2S*;1S*,2R*)-1,2,3,4-tetrahydro-2-(3-hydroxypropyl)-6,7,8-trimethoxy-2-methyl-1-(3,4,5-trimethoxybenzylisoquinolinium chloride succinate (2:1).
Nuromax; BW-A938U. Skeletal muscle relaxant. Neuromuscular blocking agent. Soluble in H_2O. *Glaxo Wellcome Inc.*

8933 Eperisone
64840-90-0 3642
$C_{17}H_{25}NO$
4'-Ethyl-2-methyl-3-piperidinopropio-phenone.
Skeletal muscle relaxant. *Eisai Co. Ltd.*

8934 Eperisone Hydrochloride
56839-43-1 3642
$C_{17}H_{26}ClNO$
4'-Ethyl-2-methyl-3-piperidinopropio-phenone hydrochloride.
EMPP; E-646; Mional; Myonal. Skeletal muscle relaxant. mp = 170-172°; LD_{50} (rat orl) = 1300-1850 mg/kg, (mus orl) = 1024 mg/kg. *Eisai Co. Ltd.*

8935 Fazadinium Bromide
49564-56-9 3981 256-378-4
$C_{28}H_{24}Br_2N_6$
1,1'-Azobis[3-methyl-2-phenyl-1H-imidazol[1,2-a]-pyridin-4-ium] dibromide.
AH-8165D; Fazadon. A short-acting curarimimetic. Used as a skeletal muscle relaxant. mp = 215-219°, 218-220°; λ_m = 285, 297 nm (log ε = 4.04, 4.34 H_2O); LD_{50} (mus iv) = 0.31 mg/kg. *Allen & Hanbury.*

8936 Fenalamide
4551-59-1 3995 224-917-2
$C_{19}H_{30}N_2O_3$
Ethyl N-[2-(diethylamino)ethyl]-2-ethyl-2-phenylmalonamate.
Spasmamide. Skeletal muscle relaxant. Used as an antispasmodic. bp_3 = 182-188°; insoluble in H_2O, soluble in EtOH, MeOH, EtOAc, C_6H_6, $CHCl_3$, Et_2O, mineral acids; [hydrobromide]: mp = 74-80° (dec); [hydrochloride]: mp = 71-74°. *Marion Merrell Dow Inc.; Schering-Plough HealthCare Products.*

8937 Fenyripol
3607-24-7
$C_{12}H_{13}N_3O$
α-[(2-Pyrimidinylamino)methyl]benzyl alcohol.
Skeletal muscle relaxant.

8938 Fenyripol Hydrochloride
2441-88-5
$C_{12}H_{14}ClN_3O$
α-[(2-Pyrimidinylamino)methyl]benzyl alcohol hydrochloride.
IN-836; NSC-43183. Skeletal muscle relaxant.

8939 Fetoxylate
54063-45-5
$C_{36}H_{36}N_2O_3$
2-Phenoxyethyl 1-(3-cyano-3,3-diphenylpropyl)-4-phenylisonipecotate.
Smooth muscle relaxant. *Abbott Laboratories Inc.; McNeil Pharm.*

8940 Fetoxylate Hydrochloride
23607-71-8
$C_{36}H_{37}ClN_2O_3$
2-Phenoxyethyl 1-(3-cyano-3,3-diphenylpropyl)-4-phenylisonipecotate hydrochloride.
McN-JR-13558-11; R-13558. Smooth muscle relaxant. *Abbott Laboratories Inc.; McNeil Pharm.*

8941 Flavoxate
15301-69-6 4135 239-337-5
$C_{24}H_{25}NO_4$
2-Piperidinoethyl 3-methyl-4-oxo-2-phenyl-4H-1-benzopyran-8-carboxylate.
2-piperidinoethyl 3-methylflavone-8-carboxylate. Anti-spasmodic; smooth muscle relaxant. A flavone derivative. Slightly soluble in H_2O (0.001 g/100 ml at 37°); soluble in EtOH, $CHCl_3$; LD_{50} (rat orl) = 1110 mg/kg, (rat iv) = 20.8 mg/kg. *Recordati Corp.; Seceph; SmithKline Beecham Pharm.*

8942 Flavoxate Hydrochloride
3717-88-2 4135 223-066-4
$C_{24}H_{26}ClNO_4$
2-Piperidinoethyl 3-methyl-4-oxo-2-phenyl-4H-1-benzopyran-8-carboxylate hydrochloride.
DW-61; NSC-114649; Rec-7-0040; Bladderon; Genurin; Patricin; Spasuret; Urispas. Antispasmodic; smooth muscle relaxant. A flavone derivative. mp = 232-234°; LD_{50} (rat iv) = 27.4 mg/kg. *Recordati Corp.; Seceph.*

8943 Flavoxate Succinate
28782-19-6 4135 249-217-4
$C_{28}H_{31}NO_8$
2-Piperidinoethyl 3-methyl-4-oxo-2-phenyl-4H-1-benzopyran-8-carboxylate succinate.
Antispasmodic; smooth muscle relaxant. A flavone derivative. Soluble in H_2O (33.7 g/100 ml at 37°). *Recordati Corp.; Seceph.*

8944 Flumetramide
7125-73-7 4176
$C_{11}H_{10}F_3NO_2$
6-(α,α,α-Trifluoro-p-tolyl)-3-morpholinone.
Duraflex; McN-1546. Skeletal muscle relaxant. mp = 115.5-116.5°. *McNeil Pharm.*

8945 Flurazepam
17617-23-1 4233 241-591-7
$C_{21}H_{23}ClFN_3O$
7-Chloro-1-[2-(diethylamino)ethyl]-5-(o-fluorophenyl)-1,3-dihydro-2H-1,4-benzodiazepin-2-one.
Felmane; Noctosom; Stauroderm. Anticonvulsant; sedative/hypnotic; muscle relaxant. A benzodiazepine. mp = 77-82°. *Hoffmann-LaRoche Inc.*

8946 Flurazepam
17617-23-1 4233 241-591-7
$C_{21}H_{23}ClFN_3O$
7-Chloro-1-[2-(diethylamino)ethyl]-5-(o-fluorophenyl)-1,3-dihydro-2H-1,4-benzodiazepin-2-one.
Felmane; Noctosom; Stauroderm. Sedative/hypnotic; muscle relaxant. mp = 77-82°. *Hoffmann-LaRoche Inc.*

8947 Flurazepam Hydrochloride
1172-18-5 4233 214-630-0
$C_{21}H_{25}Cl_3FN_3O$
7-Chloro-1-[2-(diethylamino)ethyl]-5-(o-fluorophenyl)-1,3-dihydro-2H-1,4-benzodiazepin-2-one dihydrochloride.
Dalmane; Ro-5-6901; NSC-78559; Benozil; Dalmadorm; Dalmate; Dormodor; Felson; Insumin; Lunipax; Somlan. An anticonvulsant with sedative/hypnotic properties. Muscle relaxant. A benzodiazepine. mp = 190-220°; LD_{50} (mus ip) = 290 mg/kg, (mus orl) = 870 mg/kg, (mus iv) = 84 mg/kg. *Hoffmann-LaRoche Inc.*

8948 Gallamine Triethiodide
65-29-2 4361 200-605-1
$C_{30}H_{60}I_3N_3O_3$
2,2',2-[1,2,3-Benzenetriyltris(oxy)]tris[N,N,N-triethylethanaminium] triiodide.
Flaxedil; benzcurine iodide; RP-3697; F-2559; Tricuran; Retensin; Relaxan. Skeletal muscle relaxant. mp = 152-153°; freely soluble in H_2O, EtOH, Me_2CO; sparingly soluble in Et_2O, C_6H_6, $CHCl_3$; incompatible with meperidine hydrochloride (a narcotic analgesic). *Davis & Geck Med. Dev. Div.; Rhône-Poulenc.*

8949 Hexacarbacholine Bromide
306-41-2 4713
$C_{18}H_{40}Br_2N_4O_4$
N,N,N,N',N',N'-Hexamethyl-4,13-dioxo-3,14-dioxa-5,12-diazahexadecane-1,16-diaminium dibromide.
BC-16; Imbretil. Skeletal muscle relaxant. mp = 174-176°. *Oesterreiche Stickstoffwerke.*

8950 Hexafluorenium Bromide
317-52-2 4721 206-265-0
$C_{36}H_{42}Br_2N_2$
N,N'-Di-9H-fluoren-9-yl-N,N,N',N'-tetramethyl-1,6-hexanediaminium dibromide.
Mylaxen. Succinylcholine synergist. Used as a skeletal muscle relaxant. mp = 188-189°. *Irwin Neissler.*

8951 Idrocilamide
6961-46-2 4935 230-155-1
$C_{11}H_{13}NO_2$
N-(2-Hydroxyethyl)cinnamamide.
LCB 29; Srilane; Brolitène. Skeletal muscle relaxant. mp = 100-102°; soluble in EtOH, slightly soluble in H_2O; LD_{50} (mus orl) > 2950 mg/kg, (rat orl) > 3000 mg/kg. *Lipha Pharm. Inc.*

8952 Inaperisone
99323-21-4 4963
$C_{16}H_{23}NO$
4'-Ethyl-2-methyl-3-(1-pyrrolidinyl)propiophenone.
Skeletal muscle relaxant. *Hokoriku.*

8953 (±)-Inaperisone
125676-39-3 4963
$C_{16}H_{23}NO$
(±)-4'-Ethyl-2-methyl-3-(1-pyrrolidinyl)propiophenone.
Skeletal muscle relaxant. *Hokoriku.*

8954 Inaperisone Hydrochloride
90878-85-6 4963
$C_{16}H_{24}ClNO$
4'-Ethyl-2-methyl-3-(1-pyrrolidinyl)propiophenone hydrochloride.
Skeletal muscle relaxant. *Hokoriku.*

8955 (±)-Inaperisone Hydrochloride
118230-97-0 4963
$C_{16}H_{24}ClNO$
(±)-4'-Ethyl-2-methyl-3-(1-pyrrolidinyl)propiophenone hydrochloride.
HSR-770; HY-770; Riran. Skeletal muscle relaxant. mp = 152-153°; soluble in H_2O, MeOH; LD_{50} (mus orl) = 425 mg/kg. *Hokoriku.*

8956 Isomylamine
28815-27-2
$C_{18}H_{35}NO_2$
2-(Diethylamino)ethyl 1-isopentylcyclohexanecarboxylate.
Smooth muscle relaxant.

8957 Isomylamine Hydrochloride
24357-98-0
$C_{18}H_{36}ClNO_2$
2-(Diethylamino)ethyl 1-isopentylcyclohexanecarboxylate hydrochloride.
NSC-78987. Smooth muscle relaxant.

8958 Lanperisone
116287-14-0
$C_{15}H_{18}F_3NO$
(-)-(R)-2-Methyl-3-(1-pyrrolidinyl)-4'-(trifluoromethyl)-propiophenone.
Centrally acting muscle relaxant.

8959 Laudexium Methyl Sulfate
3253-60-9 5392
$C_{54}H_{80}N_2O_{16}S_2$
2,2'-Decamethylenebis(1,2,3,4-tetrahydro)-6,7-dimethoxy-2-methyl-1-veratrylisoquinolinium methylsulfate.
Compd. 20; Laudissine; Laudolissin. Skeletal muscle relaxant. mp = 172-174°. *Allen & Hanbury.*

8960 Leptodactyline
13957-33-0 5467
$[C_{11}H_{18}NO]^+$
3-Hydroxy-N,N,N-trimethylbenzeneethanaminium.
Found in skin of amphibians of the genus *Leptodactyl.* Prepared as the chloride. A neuromuscular blocker. [chloride]: mp = 220°; soluble in H_2O, EtOH, HCl; insoluble in Et_2O, EtOAc.

8961 Levoxadrol
4792-18-1 3352
$C_{20}H_{23}NO_2$
(-)-2-(2,2-Diphenyl-1,3-dioxolan-4-yl)-piperidine.
Smooth muscle relaxant; local anesthetic. *Cutter Labs.*

8962 Levoxadrol Hydrochloride
23257-58-1 3352
$C_{20}H_{24}ClNO_2$
(-)-2-(2,2-Diphenyl-1,3-dioxolan-4-yl)piperidine hydrochloride.
CL-912C; NSC-526063; Levoxan. Smooth muscle relaxant; local anesthetic. mp = 248-254°; $[\alpha]_D^{20}$ = -34.5° (c = 2 MeOH); LD_{50} (mus orl) = 230 mg/kg. *Cutter Labs.*

8963 Lorbamate
24353-88-6
$C_{12}H_{22}N_2O_4$
2-(Hydroxymethyl)-2-methylpentyl cyclopropane-carbamate carbamate (ester).
Abbott 19957. Skeletal muscle relaxant. *Abbott Labs Inc.*

8964 Mebeverine
3625-06-7 5808 222-830-4
$C_{25}H_{35}NO_5$
4-[Ethyl(p-methoxy-α-methylphenethyl)amino]butyl veratrate.
Antispasmodic. Smooth muscle relaxant. A pyrazolone. *N.V. Philips.*

8965 Mebeverine Hydrochloride
2753-45-9 5808 220-400-0
$C_{25}H_{36}ClNO_5$
4-[Ethyl(p-methoxy-α-methylphenethyl)amino]butyl veratrate hydrochloride.
CSAG 144; Colofac; Duspatalin; Duspatal. Antispasmodic. Smooth muscle relaxant. A pyrazolone. mp = 125-127°, 129-131°. *N.V. Philips.*

8966 Mebezonium Iodide
7681-78-9 231-676-7
4,4'-Methylenebis(cyclohexyltrimethylammonium) diiodide.
Component of T-61, a veterinary euthanasic agent.

8967 Memantine
19982-08-2 5872
$C_{12}H_{21}N$
3,5-Dimethyl-1-adamantanamine.
DMAA; D-145. Skeletal muscle relaxant with antiparkinsonian properties. Derivative of adamantine. Oil; n_D^{25} = 1.4941. *Eli Lilly & Co.*

8968 Memantine Hydrochloride
41100-52-1 5872 255-219-6
$C_{12}H_{22}ClN$
3,5-Dimethyl-1-adamantanamine hydrochloride.
Akatinol. Skeletal muscle relaxant with antiparkinsonian properties. Derivative of adamantine. mp = 258°, 290-295°. *Eli Lilly & Co.*

8969 Mephenesin
59-47-2 5895 200-427-4
$C_{10}H_{14}O_3$
3-(o-Methylphenoxy)-1,2-propanediol.
Tolserol; BDH-312; Atensin; Avosyl; Avoxyl; Curythan; Daserol; Decontractyl; Dioloxol; Glyotol; Glykresin; Kinavosyl; Lissephen; Memphenesin; Mepherol; Mephesin; Mephson; Mervaldin; Myanesin; Myanol; Myodetensine; Myolysin; Myopan; Myoserol; Myoten; Oranixon; Prolax; Relaxar; Relaxil; Renarcol; Rhex; Sansdolor; Sinan; Spasmolyn; Stilalgin; Thoxidil; Tolansin; Tolcil; Tolhart; Tolosate; Toloxyn; Tolseron; Tolulexin; Tolulox; Tolyspaz; Walconesin. Skeletal muscle relaxant. mp = 70-71°; λ_m = 270 nm (E 0.395, H_2O); freely soluble in EtOH, $CHCl_3$, propylene glycol, soluble in H_2O (1.17 g/100 mlat 20°), Et_2O (9.1 g/100 ml at 20°); LD_{50} (mus ip) = 471 mg/kg, 515.6 mg/kg, (mus orl) = 990 mg/kg, 1918.6 mg/kg, (rat ip) = 283 mg/kg, (rat orl) = 945 mg/kg, (hmtr ip) = 322 mg/kg, (hmtr orl) = 821 mg/kg. *Bristol-Myers Squibb Co.*

8970 Mephenesin Carbamate
533-06-2 5895 208-553-1
$C_{11}H_{15}NO_4$
3-(o-Methylphenoxy)-1,2-propanediol carbamate.
Tolseram. Skeletal muscle relaxant. mp = 80-84°, 90-93°; λ_m = 271, 277 nm ($A_{1\,cm}^{1\%}$ 72.7, 64.1 EtOH); soluble in H_2O (0.3 g/100 ml), $CHCl_3$ (2 g/100 ml); LD_{50} (mus ip) = 604 mg/kg, (mus orl) = 2296 mg/kg. *Bristol-Myers Squibb Co.*

8971 Mephenoxalone
70-07-5 5896 200-723-3
$C_{11}H_{13}NO_4$
5-[(o-Methylphenoxy)methyl]-2-oxazolidinone.
metoxadone; methoxydon; methoxydone; methoxadone; AHR-233; OM-518; Control-Om; Dorsiflex; Dorsilon; Ekilan; Lenetran; Placidex; Riself; Tranpoise; Trepidone; Xerene. Tranquillizer. Skeletal muscle relaxant; anxiolytic. mp = 143-145°; insoluble in H_2O; LD_{50} (rat orl) = 3820 ± 17 mg/kg. *Marion Merrell Dow Inc.; Robins, A. H. Co.*

8972 Mesuprine
7541-30-2
$C_{19}H_{26}N_2O_5S$
2'-Hydroxy-5'-[1-hydroxy-2-[(p-methoxyphenethyl)-amino]propyl]methanesulfonanilide.
Vasodilator; smooth muscle relaxant.

8973 Mesuprine Hydrochloride
7660-71-1
$C_{19}H_{27}ClN_2O_5S$
2'-Hydroxy-5'-[1-hydroxy-2-[(p-methoxyphenethyl)-amino]propyl]methanesulfonanilide hydrochloride.
MJ-1987. Vasodilator; smooth muscle relaxant.

8974 Metaxalone
1665-48-1 5994 216-777-6
$C_{12}H_{15}NO_3$
5-[(3,5-Xylyloxy)methyl]-2-oxazolidinone.
AHR-438; Skelaxin. Skeletal muscle relaxant. mp = 121.5-123°. *Robins, A. H. Co.*

8975 Methixene
4969-02-2 6059 225-610-6
$C_{20}H_{23}NS$
1-Methyl-3-(thioxanthen-9-ylmethyl)piperidine.
Smooth muscle relaxant. Anticholinergic. $bp_{0.07}$ = 171-175°. *Wander Pharma.*

8976 Methixene Hydrochloride
1553-34-0 6059 216-300-1
$C_{20}H_{24}ClNS$
1-Methyl-3-(thioxanthen-9-ylmethyl)piperidine hydrochloride.
Smooth muscle relaxant. Anticholinergic.

8977 Methixene Hydrochloride Monohydrate
7081-40-5 6059
$C_{20}H_{24}ClNS.H_2O$
1-Methyl-3-(thioxanthen-9-ylmethyl)piperidine hydrochloride monohydrate.
SJ-1977; NSC-78194; Cholinfall; Methixart; Methyloxan; Tremaril; Tremarit; Tremonil; Tremoquil; Trest. Smooth muscle relaxant. Anticholinergic. mp = 215-217°; λ_m = 268 nm (ε 10250 dilute HCl); soluble in H_2O, EtOH, $CHCl_3$. *Wander Pharma.*

8978 Methocarbamol
532-03-6 6060 208-524-3
$C_{11}H_{15}NO_5$
3-(o-Methoxyphenoxy)-1,2-propanediol 1-carbamate.
AHR-85; Neuraxin; Miolaxene; Lumirelax; Etroflex; Delaxin; Robamol; Traumacut; Tresortil; Relestrid; Robaxin. Skeletal muscle relaxant. mp = 92-94°; soluble in H_2O (2.5 g/100 ml at 20°), EtOH, propylene glycol. *Robins, A. H. Co.*

8979 Metocurine Iodide
7601-55-0 6227 231-510-3
$C_{40}H_{48}I_2N_2O_6$
(+)-O,O'-Dimethylchondrocurarine diiodide.
d-tubocurrarine iodide dimethyl ether; tubocurarine dimethyl ether iodide; dimethyltubocurarine iodide; Metubine iodide. Skeletal muscle relaxant. Non-depolarizing, competitive neuromuscular blocking agent. mp = 257-267° (dec); $[\alpha]_D^{22}$ = 148° to 158° (c = 0.25); λ_M = 280 nm ($E_{1\,cm}^{1\%}$ 74); soluble in H_2O (0.3 g/100 ml), dil. HCl or NaOH; very slightly soluble in EtOH; insoluble in organic solvents. *Eli Lilly & Co.*

8980 Mivacurium Chloride
106861-44-3 6305
$C_{58}H_{80}Cl_2N_2O_{14}$
(R)-1,2,3,4-Tetrahydro-2-(3-hydroxypropyl)-6,7-dimethoxy-2-methyl-1-(3,4,5-trimethoxy-benzyl)isoquinolinium chloride.
Mivacron; BW-B1090U dichloride. Neuromuscular blocking agent. $[\alpha]_D^{20}$ = -62.7° (c = 1.9 H_2O). *Glaxo Wellcome Inc.*

8981 Nafomine
46263-35-8
$C_{12}H_{13}NO$
O-[(2-Methyl-1-naphthyl)methyl]]hydroxylamine.
Skeletal muscle relaxant. *Marion Merrell Dow Inc.*

8982 Nafomine Malate
23247-36-1
$C_{16}H_{19}NO_6$
O-[(2-Methyl-1-naphthyl)methyl]]hydroxylamine malate.
Skeletal muscle relaxant. *Marion Merrell Dow Inc.*

8983 Nelezaprine
69624-60-8
$C_{18}H_{21}ClN_2$
(E)-9-Chloro-11-[3-(dimethylamino)propylidene]-6,11-dihydro-5H-pyrrolo[2.1-b][3]benzazepine.
Skeletal muscle relaxant. *Merck & Co.Inc.*

8984 Nelezaprine Maleate
107407-62-5
$C_{22}H_{25}ClN_2O_4$
(E)-9-Chloro-11-[3-(dimethylamino)propylidene]-6,11-dihydro-5H-pyrrolo[2.1-b][3]benzazepine maleate (1:1).
L-637510. Skeletal muscle relaxant. *Merck & Co.Inc.*

8985 Nimetazepam
2011-67-8 6641 217-931-5
$C_{16}H_{13}N_3O_3$
1,3-Dihydro-1-methyl-7-nitro-5-phenyl-2H-1,4-benzodiazepin-2-one.
1-methylnitrazepam; nimetazam; S-1530; Elimin; Hypnon. Skeletal muscle relaxant with anticonvulsant properties. The desmethyl derivative of nitrazepam. Federally controlled substance (depressant). mp = 156.5-157.5°; λ_m = 259, 308 nm (ε = 15800, 9600 MeOH);

LD_{50} (mmus orl) = 910 mg/kg, (mmus ip) = 970 mg/kg, (mmus sc) = 1500 mg/kg, (fmus orl) = 750 mg/kg, (fmus ip) = 840 mg/kg, (fmus sc) = 1500 mg/kg, (mrat orl) = 1150 mg/kg, (mrat ip) = 970 mg/kg, (mrat sc) = 1000 mg/kg, (frat orl) = 970 mg/kg, (frat ip) = 980 mg/kg, (frat sc) = 1000 mg/kg. *Hoffmann-LaRoche Inc.; Sumitomo Pharm. Co. Ltd.*

8986 Orphenadrine

83-98-7 7007 201-509-2

$C_{18}H_{23}NO$

N,N-Dimethyl-2-[(o-methyl-α-phenylbenzyl)oxy]-ethylamine.

BS-5930; Biorphen; Brocasipal. Antihistaminic. Skeletal muscle relaxant. bp_{12} = 195°; [hydrochloride]: mp = 156-157°; soluble in H_2O, EtOH, $CHCl_3$; sparingly soluble in Me_2CO, C_6H_6; insoluble in Et_2O. *3M Pharm.; Forest Pharm. Inc.*

8987 Orphenadrine Citrate

4682-36-4 7007 225-137-5

$C_{24}H_{31}NO_8$

N,N-Dimethyl-2-[(o-methyl-α-phenylbenzyl)oxy]-ethylamine citrate (1:1).

Banflex; Disipal; Norflex; X-Otag; component of: Norgesic. Antihistaminic. Skeletal muscle relaxant with anticonvulsant properties. *3M Pharm.; Forest Pharm. Inc.*

8988 Pancuronium Bromide

15500-66-0 7139 239-532-5

$C_{35}H_{60}Br_2N_2O_4$

1,1'-(3α,17β-Dihydroxy-5α-androstan-2β,16β-ylene)-bis[1-methylpiperidinium] dibromide diacetate.

Pavulon; Org NA 97; NA 97; Mioblock; Poncuronium bromide. Skeletal muscle relaxant. Soluble in H_2O (100 g/100 ml), $CHCl_3$ (5 g/100 ml); LD_{50} (mus iv) = 0.047 mg/kg, (mus ip) = 0.152 mg/kg, (mus sc) = 0.167 mg/kg, (mus orl) = 21.9 mg/kg, (rat iv) = 0.153 mg/kg, (rbt iv) = 0.016 mg/kg. *Astra Sweden; Elkins-Sinn; Organon Inc.*

8989 Papaverine

58-74-2 7151 200-397-2

$C_{20}H_{21}NO_4$

1-[(3,4-Dimethoxyphenyl)methyl]-6,7-dimethoxy-isoquinoline.

6,7-dimethoxy-1-veratrylisoquinoline. Smooth muscle relaxant with cerebral vasodilator properties. Found in opium (0.8-1.0%). mp = 147°; d_4^{20} = 1.337; pK (25°) = 8.07; λ_m = 239, 278-280, 314,327 nm (log ε 4.83, 3.86, 3.60, 3.67 EtOH); nearly insoluble in H_2O; soluble in hot C_6H_6, glacial AcOH, Me_2CO; slightly soluble in $CHCl_3$, CCl_4, petroleum ether.

8990 Papaverine Hydrochloride

61-25-6 7151 200-502-1

$C_{20}H_{22}ClNO_4$

6,7-Dimethoxy-1-veratrylisoquinoline hydrochloride.

Artegodan; Cepaverin; Cerebid; Cerespan; Dynovas; Optenyl; Pameion; Panergon; Paptial T.R.; Pavabid; Pavacap; Pavacen; Pavadel; Pavagen; Pavakey; Pavased; Spasmo-Nit; Therapav; Vasal; Vasospan; component of: Copavin. Smooth muscle relaxant with cerebral vasodilator properties. mp = 220-225°; λ_m = 249-250, 280-282, 311 nm (log ε 4.69, 3.80, 3.82 EtOH); soluble in EtOH, EtOH; soluble in H_2O (2.25 g/100 ml); nearly insoluble in Et_2O; LD_{50} (mus iv) = 27.5 mg/kg, (mus sc) = 150 mg/kg, (rat sc) = 370 mg/kg. *Eli Lilly & Co.; Marion Merrell Dow Inc.*

8991 Papaverine Nitrite

132-40-1 7151

$C_{20}H_{22}N_2O_6$

6,7-Dimethoxy-1-veratrylisoquinoline nitrite.

Smooth muscle relaxant with cerebral vasodilator properties. Slightly soluble in H_2O, EtOH; freely soluble in $CHCl_3$, Me_2CO.

8992 Phenprobamate

673-31-4 7412 211-606-1

$C_{10}H_{13}NO_2$

3-Phenyl-1-propanol carbamate.

proformiphen; MH-532; Extacol; Palmita; Spantol; Ansepron; Gamaquil; Quamaquil. Anxiolytic; skeletal muscle relaxant. mp = 101-104°; soluble in EtOH, $CHCl_3$, propylene glycol, ethylenediamine, DMF; less soluble in Et_2O; insoluble in H_2O; LD_{50} (mus orl) = 840 mg/kg. *Siegfried AG.*

8993 Phenyramidol

553-69-5 7474 209-044-7

$C_{13}H_{14}N_2O$

α-[(2-Pyridylamino)methyl]benzyl alcohol.

fenyramidol. Analgesic; muscle relaxant (skeletal). mp = 82-85°; λ_m = 243, 303 nm (log ε = 4.24, 3.63 95% EtOH); [methiodide ($C_{14}H_{17}IN_2O$)]: mp = 164-166°.

8994 Phenyramidol Hydrochloride

326-43-2 7474 206-308-3

$C_{13}H_{15}ClN_2O$

α-[(2-Pyridylamino)methyl]benzyl alcohol hydrochloride.

IN-511; MJ-505; NSC-17777; Abbolexin; Anabloc; Analexin; Cabral; Miodar. Skeletal muscle relaxant; analgesic. mp = 140-142°; soluble in H_2O.

8995 Pipecurium Bromide

52212-02-9 7612 257-740-4

$C_{35}H_{62}Br_2N_4O_4$

4,4'-(3α,17β-Dihydroxy-5α-androst-2β,16β-ylene)bis[1,1-dimethylpiperazinium] dibromide diacetate (ester).

Arduan; RGH-1106. Skeletal muscle relaxant. Non-depolarizing curare-like agent. Derivative of androstane. mp = 262-264° (dec); $[\alpha]_D^{25}$ = 8.1° (c = 1 H_2O); LD_{50} (mus iv) = 29.7 mg/kg, (mus ip) = 70.6 mg/kg, (mus sc) = 60.5 mg/kg, (rat iv) = 172.6 mg/kg, (rat ip) = 449.6 mg/kg, (rat sc) = 455.8 mg/kg. *Gedeon Richter Chem. Works; Organon Inc.*

8996 Pipoxolan

23744-24-3 7637

$C_{22}H_{25}NO_3$

5,5-Diphenyl-2-(2-piperidinoethyl)-1,3-dioxolan-4-one.

Skeletal muscle relaxant used as an antispasmodic. *Rowa Ltd.*

8997 Pipoxolan Hydrochloride
18174-58-8 7637
$C_{22}H_{26}ClNO_3$
5,5-Diphenyl-2-(2-piperidinoethyl)-1,3-dioxolan-4-one hydrochloride.
BR-18; Rowapraxin. Skeletal muscle relaxant used as an antispasmodic. mp = 207-209°; soluble in H_2O; LD_{50} (rat orl) = 1500 mg/kg, (rat iv) = 60 mg/kg, (rat ip) = 130 mg/kg, (rat sc) > 300 mg/kg, (mus orl) = 700 mg/kg, (mus iv) = 35 mg/kg, (mus ip) = 130 mg/kg. *Rowa Ltd.*

8998 Procyclidine
77-37-2 7944 201-023-0
$C_{19}H_{29}NO$
α-Cyclohexyl-α-phenyl-1-pyrrolidine-propanol.
Skeletal muscle relaxant; anticholinergic; antiparkinsonian. mp = 85.5-86.5°;λ_m = 285.5 nm (ε 233 17% EtOH); [methosulfate ($C_{21}H_{35}NO_5S$)]: mp = 100°; soluble in H_2O, EtOH. *Eli Lilly & Co.; Glaxo Wellcome Inc.*

8999 Procyclidine Hydrochloride
1508-76-5 7944 216-141-8
$C_{19}H_{30}ClNO$
α-Cyclohexyl-α-phenyl-1-pyrrolidinepropanol hydrochloride.
Arpicolin; Kemadrin; Osnervan. Anticholinergic. Skeletal muscle relaxant. Antiparkinsonian. Dec 226-227°; soluble in H_2O (3 g/100 ml), organic solvents. *Eli Lilly & Co.; Glaxo Wellcome Inc.*

9000 Procyclidine Methochloride
30953-84-5 7944
$C_{19}H_{29}NO{\cdot}CH_3Cl$
1-(3-Cyclohexyl-3-hydroxy-3-phenylpropyl)-1-methylpyrrolidinium chloride.
tricyclamol chloride; Lergine; Tricoloid. Smooth muscle relaxant; anticholinergic agent. mp = 159-164°; soluble in H_2O, EtOH. *Eli Lilly & Co.; Glaxo Wellcome Inc.*

9001 Progabide
62666-20-0 7955 263-679-4
$C_{17}H_{16}ClFN_2O_2$
4-[[α-(p-Chlorophenyl)-5-fluorosalicylidene]-amino]butyramide.
halogabide; Gabren; Gabrene; SL-76002. Smooth muscle relaxant used as an anticonvulsant. An antagonist of σ-aminobutyric acid (GABA) possessing antiepileptic activity. mp = 133-135°, 142.5°; λ_m = 332, 250, 210 (ε 4200, 10800, 24000 MeOH); LD_{50} (mus ip) = 900 mg/kg. *Synthelabo Pharmacie.*

9002 Proxazole
5696-09-3 8090
$C_{17}H_{25}N_3O$
5-[2-(Diethylamino)ethyl]-3-(α-ethylbenzyl)-1,2,4-oxadiazole.
Aerbron; propoxaline. Antispasmodic. Smooth muscle relaxant used as an anticonvulsant. $bp_{0.2}$ = 132°; [nitrate]: mp = 127-128°. *Angelini Francesco.*

9003 Proxazole Citrate
132-35-4 8090 205-059-8
$C_{23}H_{33}N_3O_8$
5-[2-(Diethylamino)ethyl]-3-(α-ethylbenzyl)-1,2,4-oxadiazole citrate (1:1).
AF-634; Flou; Pirecin; Toness. Antispasmodic. Smooth muscle relaxant used as an anticonvulsant. LD_{50} (rat ip) = 39 mg/kg, (rat orl) = 60 mg/kg. *Angelini Pharm. Inc.*

9004 Quinetolate
5714-76-1 227-207-0
$C_{38}H_{57}N_5O_7$
6-(Diethylcarbamoyl)-3-cyclohexene-1-carboxylic acid compound with 4-[[2-(dimethylamino)ethyl]amino]-6-methoxyquinoline (2:1).
Smooth muscle relaxant.

9005 Quinine
130-95-0 8245 205-003-2
$C_{20}H_{24}N_2O_2$
(8S,9R)-6'-Methoxycinchonan-9-ol.
Skeletal muscle relaxant; antimalarial. Primary alkaloid of *Cinchona* spp. Optical isomer of quinidine. mp = 177° (dec); $[\alpha]_D^{15}$ = -169° (c = 2 97% EtOH), $[\alpha]_D^{17}$ = -117° (c = 1.5 $CHCl_3$), $[\alpha]_D^{15}$ = -285° (c = 0.4M 0.1N H_2SO_4); soluble in H_2O (0.05 g/100 ml at 25°, 0.13 g/100 ml at 100°), EtOH (125 g/100 ml), C_6H_6 (1.25 g/100 ml at 25°, 5.5 g/100 ml at 50°), $CHCl_3$ (83.3 g/100 ml), Et_2O (0.4 g/100 ml), glycerol (5 g/100 ml); insoluble in petroleum ether.

9006 Rapacuronium Bromide
156137-99-4
$C_{37}H_{61}BrN_2O_4$
1-Allyl-1-(3α,17β-dihydroxy-2β-piperidino-5α-androstan-16β-yl)piperidinium bromide 3-acetate 17-propionate.
Raplon; Org 9487. Neuromuscular blocking agent. Organon.

9007 Ritodrine
26652-09-5 8401 247-879-9
$C_{17}H_{21}NO_3$
erythro-p-Hydroxy-α-[1-[(p-hydroxyphenethyl)amino]ethyl]benzyl alcohol.
DU-21220. Smooth muscle relaxant; tocolytic agent. A β-adrenergic agonist. mp = 88-90°. *Astra Chem. Ltd.; Astra Sweden; Philips-Duphar B.V.; Teva Pharm. (USA).*

9008 Ritodrine Hydrochloride
23239-51-2 8401 245-514-8
$C_{17}H_{22}ClNO_3$
erythro-p-Hydroxy-α-[1-[(p-hydroxyphenethyl)-amino]ethyl]benzyl alcohol hydrochloride.
Pre-Par; Yutopar; DU-21220; Miolene; Prempar; Utemerin; Utopar. Smooth muscle relaxant. A β-adrenergic agonist and tocolytic. mp = 193-195° (dec); λ_m = 267.5 nm (ε 3310). *Astra Chem. Ltd.; Astra Sweden; Philips-Duphar B.V.; Teva Pharm. (USA).*

9009 Rocuronium
143558-00-3 8407
$[C_{32}H_{53}N_2O_4]^+$
1-Allyl-1-(3α,17β-dihydroxy-2β-morpholino-5α-androst-16β-yl)pyrrolidinium.

Skeletal muscle relaxant. Non-depolarizing neuromuscular blocking agent. *Akzo Chemie ; Organon Inc.*

9010 Rocuronium Bromide
119302-91-9 8407

$C_{32}H_{53}BrN_2O_4$

1-Allyl-1-(3α,17β-dihydroxy-2β-morpholino-5α-androst-16β-yl)pyrrolidinium bromide.

Zemuron; Esmeron; Org-9426. Skeletal muscle relaxant. Non-depolarizing neuromuscular blocking agent. mp = 161-169°; $[\alpha]_D^{20}$ = 18.7° (c = 1.03 $CHCl_3$). *Akzo Chemie ; Organon Inc.*

9011 Rolodine
1866-43-9

$C_{14}H_{14}N_4$

4-(Benzylamino)-2-methyl-7H-pyrrolo[2,3-d]-pyrimidine.

BW-58-271; NSC-106570. Skeletal muscle relaxant. *Glaxo Wellcome Inc.*

9012 Sildenafil
139755-83-2

$C_{22}H_{30}N_6O_4S$

1-[[3-(6,7-dihydro-1-methyl-7-oxo-3-propyl-1H-pyrazolo[4,3-d]pyrimidin-5-yl)-4-ethoxyphenyl]sulfonyl]-4-methylpiperazine.

A selective inhibitor of cyclic guanosine monophosphate (cGMP)-specific phosphodiesterase type 5 (PDE5), which allows smooth muscle relaxation and inflow of blood to the corpus cavernosum. An oral therapy for erectile dysfunction. *Pfizer Intl.*

9013 Sildenafil Citrate
171599-83-0

$C_{28}H_{38}N_6O_{11}S$

1-[[3-(6,7-dihydro-1-methyl-7-oxo-3-propyl-1H-pyrazolo[4,3-d]pyrimidin-5-yl)-4-ethoxyphenyl]sulfonyl]-4-methylpiperazine citrate.

Viagra; UK-92480. A selective inhibitor of cyclic guanosine monophosphate (cGMP)-specific phosphodiesterase type 5 (PDE5), which allows smooth muscle relaxation and inflow of blood to the corpus cavernosum. An oral therapy for erectile dysfunction. Soluble in H_20 (3.5 mg/ml). *Pfizer Intl.*

9014 Styramate
94-35-9 9027 202-326-0

$C_9H_{11}NO_3$

β-Hydroxyphenethyl carbamate.

Sinaxar; Linaxar. Skeletal muscle relaxant. mp = 111-112°; LD_{50} (mus orl) = 1240 mg/kg. *Armour Pharm. Co. Ltd.*

9015 Succinylcholine Bromide
55-94-7 9043 200-248-1

$C_{14}H_{30}Br_2N_2O_4$

Choline bromide succinate.

Suxamethonium bromide; IS-370; compd. 48/268; LT-1; M&B-2207; Brevidil M. Skeletal muscle relaxant. Depolarizing neuromuscular blocking agent. mp = 225°; freely soluble in H_2O.

9016 Succinylcholine Chloride
71-27-2 9044 200-747-4

$C_{14}H_{30}Cl_2N_2O_4$

Choline chloride succinate.

suxamethonium chloride; diacetyl-choline dichloride; choline succinate dichloride; Anectine; Querlicin; Sucostrin. Skeletal muscle relaxant. Depolarizing neuromuscular blocking agent. mp = 156-163°, 190°; soluble in H_2O (100 g/100 ml), EtOH (0.42 g/100 ml); sparingly soluble in C_6H_6, $CHCl_3$; insoluble in Et_2O; LD_{50} (mus iv) = 0.45 mg/kg. *Abbott Labs Inc.; Apothecon; Glaxo Wellcome Inc.; Organon Inc.*

9017 Succinylcholine Iodide
541-19-5 9045 208-770-1

$C_{14}H_{30}I_2N_2O_4$

Choline iodide succinate.

Suxamethonium iodide; bis(β-dimethyl-aminoethyl)succinate; Celocurine. Skeletal muscle relaxant; depolarizing neuromuscular blocker. mp = 243-245°; soluble in H_2O.

9018 Suxethonium Bromide
111-00-2 9184 203-824-0

$C_{16}H_{34}Cl_2N_2O_4$

Ethyl (2-hydroxyethyl)dimethylammonium bromide succinate.

IS-362; M-115; M&B-2210; Brevidil E. Neuromuscular blocking agent. Skeletal muscle relaxant. Used in endotracheal intubation. mp = 158°; freely soluble in H_2O. *May & Baker Ltd.*

9019 Suxethonium Chloride
54063-57-9 258-958-2

$C_{16}H_{34}Cl_2N_2O_4$

Ethyl (2-hydroxyethyl)dimethylammonium chloride succinate.

Neuromuscular blocker; used during endotrachial intubation.

9020 Tetrazepam
10379-14-3 9379 233-837-7

$C_{16}H_{17}ClN_2O$

7-Chloro-5-(cyclohexen-1-yl)-1,3-dihydro-1-methyl-2H-1,4-benzodiazepin-2-one.

CB-4261; Musaril; Myolastan. Skeletal muscle relaxant. mp = 144°; λ_m = 227 nm (ε 28500 EtOH); LD_{50} (mus ip) = 415 mg/kg, (mus orl) = 2000 mg/kg. *Clin-Byla France.*

9021 Theophylline Sodium Glycinate
8000-10-0 9421

$C_9H_{12}N_5NaO_4$

3,7-Dihydro-1,3-dimethyl-1H-purine-2,6-dione monosodium salt with glycine.

sodium theophyllin-7-ylglycinate; Biophylline; Englate; Glytheonate; Panophylline; Pemophyllin. Smooth muscle relaxant, used as a bronchodilator. Xanthine derivative. Dec 180°; d^{20} = 1.05; soluble in H_2O (18 g/100 ml); pH (saturated solution) = 8.7-9.1. *Marion Merrell Dow Inc.*

9022 Thiocolchicoside
602-41-5 9460 210-017-7

$C_{27}H_{33}NO_{10}S$

2-10-Di(demethoxy)-2-glucosyloxy-10-methylthiocolchicine.

2-demethoxy-2-glucosidoxythiocolchicine; Coltramyl; Coltrax; Miorel; Musco-Ril. Skeletal muscle relaxant. mp = 220° (dec); $[\alpha]_D$ = -609° (H_2O), -240° (EtOH).

9023 Thiphenamil
82-99-5 9509
$C_{20}H_{25}NOS$
S-[2-(Diethylamino)ethyl]diphenylthioacetate.
Smooth muscle relaxant used as an anticholinergic agent. *Sterling Res. Labs.*

9024 Thiphenamil Hydrochloride
548-68-5 9509 208-955-7
$C_{20}H_{26}ClNOS$
S-[2-(Diethylamino)ethyl]diphenylthioacetate hydrochloride.
Thiphen; Trocinate. Smooth muscle relaxant used as an anticholinergic agent. mp = 129-130°; soluble in H_2O. *Sterling Res. Labs.*

9025 Tizanidine
51322-75-9 9624
$C_9H_8ClN_5S$
5-Chloro-4-(2-imidazolin-2-ylamino)-2,1,3-benzothiadiazole.
DS-103-282; Sirdalud; Ternelin. Antispasmodic; centrally acting skeletal muscle relaxant. α_2 adrenoceptor agonist; centrally active myotonolytic. Has been investigated for antitremor activity. mp = 221-223°; LD_{50} (mus orl) = 235 mg/kg. *Athena Neurosciences Inc.; Wander Pharma.*

9026 Tizanidine Hydrochloride
64461-82-1 9624
$C_9H_9Cl_2N_5S$
5-Chloro-4-(2-imidazolin-2-ylamino)-2,1,3-benzothiadiazole monohydrochloride.
AN021; DS-103-282; Zanaflex. Antispasmodic; centrally acting skeletal muscle relaxant. α_2 adrenoceptor agonist; centrally active myotonolytic. Has been investigated for antitremor activity. *Athena Neurosciences Inc.; Wander Pharma.*

9027 Tolperisone
728-88-1 9660 211-976-4
$C_{16}H_{23}NO$
2,4'-Dimethyl-3-piperidinopropiophenone.
mydetone. Skeletal muscle relaxant. *Eisai Co. Ltd.*

9028 Tolperisone Hydrochloride
3644-61-9 9660 222-876-5
$C_{16}H_{24}ClNO$
2,4'-Dimethyl-3-piperidinopropiophenone hydrochloride.
N-553; Abbsa; Atmosgen; Arantoick; Besnoline; Isocalm; Kineorl; Menopatol; Metosomin; Minacalm; Muscalm; Mydocalm; Naismeritin; Tolisartine. Skeletal muscle relaxant. mp = 176-177°; LD_{50} (mus sc) = 620 mg/kg. *Eisai Co. Ltd.*

9029 Tubocurarine Chloride
57-94-3 9939 200-356-9
$C_{37}H_{42}Cl_2N_2O_6.5H_2O$
7',12'-Dihydroxy-6,6'-dimethoxy-2,2',2'-trimethyltubocuraranium chloride.
Jexin [d-form]; Tubadil; Delacurarine; Curarin-HAF; Tubarine; Introcostrin. Skeletal muscle relaxant. Non-depolarizing, competitive neuromuscular blocking agent. mp = 270° (dec); $[\alpha]_D^{22}$ = 190° (c = 0.5), $[\alpha]_D^{23}$ = 219° (c = 0.785 MeOH); λ_m = 280 nm ($E_{1\,cm}^{1\%}$ 118 H_2O); soluble in H_2O (5 g/100 ml at 25°, 2.5 g/100 ml, 1.3 g/100 ml), soluble in MeOH, insoluble in C_5H_5N, $CHCl_3$, C_6H_6, Me_2CO, Et_2O; LD_{50} (mus orl) = 33.2 mg/kg, (rat orl) = 27.8 mg/kg. *Apothecon.*

9030 Vecuronium Bromide
50700-72-6 10075 256-723-9
$C_{34}H_{57}BrN_2O_4$
1-(3α,17β-Dihydroxy-2β-piperidino-5α-androstan-16β,5α-yl)-1-methylpiperidinium bromide diacetate.
Norcuron; Org-NC45; NC45; Musculax. Neuromuscular blocking agent. Used as a skeletal muscle relaxant. mp = 227-229°; LD_{50} (mus iv) = 0.061 mg/kg. *Organon Inc.*

9031 Xilobam
50528-97-7
$C_{14}H_{19}N_3O$
1-(1-Methyl-2-pyrrolidinylidene)-3-(2,6-xylyl)urea.
McN-3113. Skeletal muscle relaxant. *McNeil Pharm.*

9032 Zoxazolamine
61-80-3 10328 200-519-4
$C_7H_5ClN_2O$
2-Amino-5-chlorobenzoxazole.
McN-485; Deflexol; Flexilon; Flexin; Zoxamin; Zoxine. Skeletal muscle relaxant; uricosuric agent. mp = 184-185°; λ_m = 244, 285 nm; slightly soluble in H_2O; soluble in EtOH, propylene glycol; LD_{50} (mus ip) = 376 mg/kg, (mus orl) = 678 mg/kg, (rat ip) = 102 mg/kg, (rat orl) = 730 mg/kg. *Dow Chem. U.S.A.; McNeil Pharm.*

Mydriatics

9033 Atropine Oxide
4438-22-6 908
$C_{17}H_{23}NO_4$
1αH,5αH-Tropan-3α-ol (±)-tropate (ester) 8-oxide.
atropine-N-oxide; atropine aminoxide; genatropine; aminoxytropine tropate. Anticholinergic. mp = 127-128°, dec 135°; soluble in EtOH, $CHCl_3$; insoluble in Et_2O.

9034 Atropine Oxide Hydrochloride
4574-60-1 908 224-959-1
$C_{17}H_{24}ClNO_4$
1αH,5αH-Tropan-3α-ol (±)-tropate (ester) 8-oxide hydrochloride.
Tropinox, Xtro. Anticholinergic. mp = 192-193°.

9035 Atropine Sulfate [anhydrous]
55-48-1 908 200-235-0
$C_{34}H_{48}N_2O_{10}S$
1αH,5αH-Tropan-3α-ol (±)-tropate (ester) sulfate (2:1) (salt).
Atropisol; Atropisol® Ophthalmic Solution. Anti-cholinergic; mydriatic. Used in pre-anesthetic medication. Veterinary use is as an antispasmodic and as an antidote for organophosphorus insecticide poisoning.

mp = 190-194°; soluble in H_2O (2.5 g/ml), less soluble in organic solvents; LD_{50} (rat orl) = 622 mg/kg.

9036 Atropine Sulfate Monohydrate
5908-99-6 908
$C_{34}H_{48}N_2O_{10}S.H_2O$
1αH,5αH-Tropan-3α-ol (±)-tropate (ester) sulfate (2:1) (salt) monohydrate.
Anticholinergic.

9037 Berefrine
105567-83-7
$C_{14}H_{21}NO_2$
m-[(2R,5R)-2-tert-Butyl-3-methyl-5-oxazolidinyl]phenol mixture with m-[(2S,5R)-2-tert-butyl-3-methyl-5-oxazolidinyl]phenol.
Mydriatic. *Angelini Pharm. Inc.*

9038 Cyclazocine
3572-80-3 2773 222-689-9
$C_{18}H_{25}NO$
3-(Cyclopropylmethyl)-1,2,3,4,5,6-hexahydro-6,11-dimethyl-2,6-methano-3-benzazocin-8-ol.
NSC-107429; Win-20740. Analgesic with mixed narcotic agonist-antagonist properties. mp = 210-204°. *Sterling Winthrop Inc.*

9039 Cyclopentolate
512-15-2 2815 208-136-4
$C_{17}H_{25}NO_3$
2-(Dimethylamino)ethyl 1-hydroxy-α-phenylcyclopentaneacetate.
Mydriatic. Anticholinergic (ophthalmic). *Alcon Labs.*

9040 Cyclopentolate Hydrochloride
5870-29-1 2815 227-521-8
$C_{17}H_{26}ClNO_3$
2-(Dimethylamino)ethyl 1-hydroxy-α-phenylcyclopentaneacetate hydrochloride.
Cyclogyl; Ak-Pentolate; Alnide; Mydplegic; Mydrilate; Zyklolat; component of: Cyclomydril. Anticholinergic (ophthalmic). Mydriatic. mp = 137-141°; freely soluble in H_2O, EtOH; insoluble in Et_2O. *Alcon Labs.*

9041 Cyprenorphine
4406-22-8 2838
$C_{26}H_{33}NO_4$
6,7,8,14-N-(Cyclopropylmethyl)tetrahydro-7α-(1-hydroxy-1-methylethyl)-6,14-endo-ethenonororipavine.
Analgesic; etorphine antagonist (veterinary). Narcotic antagonist closely related to diprenorphine. Federally controlled substance (opium derivative). mp = 234° (7α- or 7β-linkage unspecified). *Reckitt & Colman Inc.*

9042 Epinephrine
51-43-4 3656 200-098-7
$C_9H_{13}NO_3$
(R)-4-[1-Hydroxy-2-(methylamino)ethyl]-1,2-benzenediol.
l-methylaminoethanolcatechol; adrenalin; levorenen; Bronkaid Mist; Epifrin; Epiglaufrin; Eppy; Glaucon; Glauposine; Primatene Mist; Simplene; Sus-phrine; Suprarenaline; component of: Citanest Forte. Bronchodilator; cardiostimulant; mydriatic; antiglaucoma. Endogenous catecholamine with combined α- and β-agonist activity. Principal sympathomimetic hormone produced by the adrenal medulla. mp = 211-212°; dec 215°; $[\alpha]_D^{25}$ = -50.0° to -53.5° (in 0.6N HCl); slightly soluble in H_2O, EtOH; soluble in aqueous solutions of mineral acids; insoluble in aqueous solutions of ammonia and alkali carbonates; insoluble in Et_2O, Me_2CO, oils; LD_{50} (mus ip) = 4 mg/kg. *Alcon Labs; Allergan Inc.; Astra Sweden; Bristol-Myers Squibb Co; CIBA Vision AG; Elkins-Sinn; Evans Medical Ltd.; Parke-Davis; Sterling Health U.S.A.; Whitehall-Robins; Wyeth-Ayerst Labs.*

9043 Epinephrine d-Bitartrate
51-42-3 3656 200-097-1
$C_{13}H_{19}NO_9$
(R)-4-[1-Hydroxy-2-(methylamino)ethyl]-1,2-benzenediol [R-(R*,R*)]-2,3-dihydroxybutanedioate (1:1) (salt).
Asmatane Mist; Asthmahaler Epitrate; Bronitin Mist; Bronkaid Mist Suspension; Epitrate; Medihaler-Epi; Primatene Mist Suspension; Suprarenin; component of: Asthmahaler, E-Pilo. Bronchodilator; cardiostimulant; mydriatic; antiglaucoma. Adrenergic (opthalmic). mp = 147-154° (some decomposition); soluble in H_2O (30 g/100 ml); slightly soluble in EtOH. *3M Pharm.; CIBA Vision Corp.; Menley & James Labs Inc.; Sterling Health U.S.A.; Whitehall-Robins; Wyeth-Ayerst Labs.*

9044 Epinephrine Hydrochloride
55-31-2 3656
$C_9H_{14}ClNO_3$
(-)-3,4-Dihydroxy-α-[(methylamino)methyl]benzyl alcohol hydrochloride.
Adrenalin; Epifrin; Glaucon; Suprarenin. Bronchodilator; cardiostimulant; mydriatic; antiglaucoma. An α-adrenergic agonist. *Alcon Labs; Allergan Inc.; Astra Chem. Ltd.; Bristol-Myers Squibb Co; CIBA Vision AG; Elkins-Sinn; Forest Pharm. Inc.; Parke-Davis; Sterling Health U.S.A.; Whitehall Labs. Inc.; Wyeth-Ayerst Labs.*

9045 dl-Epinephrine
329-65-7 3656 206-347-6
$C_{20}H_{32}N_2O_6$
(±)-α-[1-(Methylamino)ethyl]benzenemethanol sulfate salt (2:1).
racepinefrine; racepinephrine. An α-adrenergic agonist. Used as a bronchodilator, cardiostimulant, mydriatic and antiglaucoma agent. Slightly soluble in H_2O, EtOH. *Alcon Labs; Allergan Inc.; Astra Chem. Ltd.; Bristol-Myers Squibb Co; CIBA Vision AG; Elkins-Sinn; Forest Pharm. Inc.; Parke-Davis; Sterling Health U.S.A.; Whitehall Labs. Inc.; Wyeth-Ayerst Labs.*

9046 dl-Epinephrine Hydrochloride
329-63-5 3656 206-346-0
$C_{20}H_{33}ClN_2O_6$
(±)-α-[1-(Methylamino)ethyl]benzenemethanol hydrochloride salt.
Asthmanefrin; Vaponefrin. An α-adrenergic agonist. Used as a bronchodilator, cardiostimulant, mydriatic and antiglaucoma agent. mp = 157°, soluble in H_2O; sparingly soluble in EtOH. *Alcon Labs; Allergan Inc.; Astra Chem. Ltd.; Bristol-Myers Squibb Co; CIBA Vision AG; Elkins-Sinn; Forest Pharm. Inc.; Parke-Davis; Sterling Health U.S.A.; Whitehall Labs. Inc.; Wyeth-Ayerst Labs.*

9047 Hydroxyamphetamine
1518-86-1 4855
$C_9H_{13}NO$
(±)-p-(2-Aminopropyl)phenol.
p-hydroxyphenylisopropylamine; α-methyltyramine; Paredrine; Paredrinex; Pulsoton. An α-adrenergic (ophthalmic) agonist. Used as a mydriatic agent. mp = 125-126°; soluble in H_2O, EtOH, $CHCl_3$, EtOAc; [iodide]: mp = 155°, soluble in H_2O, EtOH, Me_2CO; [hydrochloride]: mp = 171-172°; soluble in H_2O, EtOH; insoluble in Et_2O. *Allergan Inc.; SmithKline Beecham Pharm.*

9048 Hydroxyamphetamine Hydrobromide
306-21-8 4855 206-181-4
$C_9H_{14}BrNO$
(±)-p-(2-Aminopropyl)phenol hydrobromide.
Paredrine hydrobromide; component of: Paremyd. An α-adrenergic (ophthalmic) agonist. Used as a mydriatic agent. mp = 171-172°; soluble in H_2O, EtOH; insoluble in Et_2O. *Allergan Inc.; SmithKline Beecham Pharm.*

9049 Phenylephrine
59-42-7 7440 200-424-8
$C_9H_{13}NO_2$
(R)-3-Hydroxy-α-[(methylamino)methyl]benzenemethanol.
Mydriatic; decongestant. α-Adrenergic agonist. mp = 169-172°.

9050 Phenylephrine Hydrochloride
61-76-7 7440 200-517-3
$C_9H_{14}ClNO_2$
(R)-m-Hydroxy-α[(methylamino)methyl]benzyl alcohol hydrochloride.
metaoxedrin; Adrianol; Ak-Dilate; Ak-Nefrin; Alcon Efrin; Biomydrin; Isophrin; m-Sympatol; Mezaton; Mydfrin; Neophryn; Neo-Synephrine; Nostril; Pyracort D; Prefrin; component of: Afrin 4-Way Nasal Spray Regular and Menthol, Anaplex HD, Benzedrex Nasal Spray Regular, Cerose-DM, Cyclomydril, Dristan Cold Multi-Symptom, Dristan Nasal Spray, Entex Capsules and Liquid, Histalet Forte, Hycomine Compound, Isopto Frin, Naldecon, Novahistine, Op-Isophrin-Z, Phenergan VC, Prefrin Liquifilm, Preparation H Cream, Preparation H Ointment, PV Tussin Syrup, Relief, RU-Tuss, Tympagesic, Vasosulf, Zincfrin. Mydriatic; decongestant. α-Adrenergic agonist. mp = 140-145°; $[\alpha]_D^{25}$ = -46.2° to -47.2°; soluble in H_2O, EtOH; LD_{50} (rat ip) = 17 ± 1.1 mg/kg, (rat sc) = 33.0 ± 2 mg/kg. *Alcon Labs; Boehringer Ingelheim Ltd.; Boots Pharmaceuticals Inc.; Bristol-Myers Squibb Co; CIBA Vision AG; DuPont-Merck Pharm.; Fisons plc; Marion Merrell Dow Inc.; Norwich Eaton; Parke-Davis; Robins, A. H. Co.; Schering AG; Schering-Plough Pharm.; Solvay Animal Health Inc.; Sterling Winthrop Inc.; Whitehall Labs. Inc.; Wyeth-Ayerst Labs.*

9051 Yohimbine
146-48-5 10236 205-672-0
$C_{21}H_{26}N_2O_3$
17α-Hydroxyyohimban-16α-carboxylic acid.
quebrachine; corynine; aphrodine. Indole alkaloid with α_2-adrenergic blocking activity. Used as a mydriatic. mp = 234°, 235-237°; $[\alpha]_D^{20}$= 50.9 - 62.2° (EtOH), 108° (C_5H_5N); λ_m = 226, 280, 291 nm (log ε 4.56, 3.88, 3.80 MeOH); sparingly soluble in H_2O; soluble in EtOH, $CHCl_3$, hot C_6H_6; moderately soluble in Et_2O.

Narcotic Antagonists

9052 Cyprenorphine Hydrochloride
16550-22-4 2838
$C_{26}H_{34}ClNO_4$
6,7,8,14-N-(Cyclopropylmethyl)tetrahydro-7α-(1-hydroxy-1-methylethyl)-6,14-endo-ethenonororipavine hydrochloride.
M-285. Analgesic; etorphine antagonist (veterinary). Narcotic antagonist closely related to diprenorphine. Federally controlled substance (opium derivative). mp = 248°. *Reckitt & Colman Inc.*

9053 Diprenorphine
14357-78-9 3403 238-325-7
$C_{26}H_{35}NO_4$
21-Cyclopropyl-6,7,8,14-tetrahydro-7α-(1-hydroxy-1-methylethyl)-6,14-endoethanooripavine.
M.5050; M.5050 Injection; RX-5050M. Narcotic antagonist, related to cyprenorphine. mp = 185°; [hydrochloride]: LD_{50} (mus sc) = 316 mg/kg. *Lemmon Co.*

9054 Fenmetozole
41473-09-0
$C_{10}H_{10}Cl_2N_2O$
2-[(3,4-Dichlorophenoxy)methyl]-2-imidazoline.
Narcotic antagonist. Used as an antidepressant. *Marion Merrell Dow Inc.*

9055 Fenmetozole Hydrochloride
23712-05-2
$C_{10}H_{11}Cl_3N_2O$
2-[(3,4-Dichlorophenoxy)methyl]-2-imidazoline monohydrochloride.
DH-524. Antidepressant; narcotic antagonist. *Marion Merrell Dow Inc.*

9056 Levallorphan
152-02-3 5485 205-799-1
$C_{19}H_{25}NO$
17-Allylmorphinan-3-ol.
Narcotic antagonist, related to cyprenorphine. mp = 180-182°; $[\alpha]_D^{20}$ = -88.9° (c = 3 MeOH). *Hoffmann-LaRoche Inc.*

9057 Levallorphan Tartrate
71-82-9 5485 200-767-3
$C_{23}H_{31}NO_7$
17-Allylmorphinan-3-ol tartrate (1:1).
Lorfan. Narcotic antagonist, related to cyprenorphine. mp = 176-177°; $[\alpha]_D^{16}$ = -39°; soluble in H_2O. *Hoffmann-LaRoche Inc.*

9058 Nadide
53-84-9 6429 200-184-4
$C_{21}H_{27}N_7O_{14}P_2$
3-Carbamoyl-1-β-D-ribofuranosylpyridinium hydroxide 5' ester with adenosine 5'-pyrophosphate.
DPN; CO-I; NAD; Codehydrogenase I; NSC-20272.

Narcotic antagonist. Alcohol antagonist. A biologically active form of nicotinic acid that acts as a coenzyme of the dehydrogenases. Occurs as α-NAD and β-NAD. The β-anomer is bioactive. Soluble in H_2O; $[\alpha]_D^{20}$ = -31.5° (c = 1.2 H_2O); λ_m = 260 nm (ε 18100).

9059 Nalmefene

55096-26-9 6447

$C_{21}H_{25}NO_3$

17-(Cyclopropylmethyl)-4,5α-epoxy-6-methylene-morphinan-3,14-diol.

JF-1; ORF-11676. Narcotic antagonist. mp = 188-190°. *Key Pharm.*

9060 Nalmexone

16676-26-9

$C_{21}H_{25}NO_4$

4,5α-Epoxy-3,14-dihydroxy-17-(3-methyl-2-butenyl)-morphinan-6-one.

Narcotic antagonist.

9061 Nalmexone Hydrochloride

16676-27-0

$C_{21}H_{26}ClNO_4$

4,5α-Epoxy-3,14-dihydroxy-17-(3-methyl-2-butenyl)-morphinan-6-one hydrochloride.

EN-1620A. Narcotic antagonist.

9062 Nalorphine

62-67-9 6448 200-546-1

$C_{19}H_{21}NO_3$

17-Allyl-7,8-didehydro-4,5α-epoxymorphinan-3,6α-diol.

Narcotic antagonist. mp = 208-209°; $[\alpha]_D^{25}$ = -155.3° (c = 3 MeOH); λ_m = 285 nm (acid), 298 nm (base); poorly soluble in H_2O, Et_2O; soluble in organic solvents. *Merck & Co.Inc.*

9063 Nalorphine Hydrochloride

57-29-4 6448 200-321-8

$C_{19}H_{22}ClNO_3$

17-Allyl-7,8-didehydro-4,5α-epoxymorphinan-3,6α-diol hydrochloride.

Nalline. Narcotic antagonist. mp = 260-263°; soluble in H_2O, EtOH; λ_m = 285 nm (H_2O); LD_{50} (rat sc) = 1460 mg/kg. *Merck & Co.Inc.*

9064 Naloxone

465-65-6 6449 207-365-7

$C_{19}H_{21}NO_4$

17-Allyl-4,5α-epoxy-3,14-dihydroxymorphinan-6-one.

Narcotic antagonist. mp = 184°, 177-178°; $[\alpha]_D^{20}$ = -194.5° (c = 0.93 $CHCl_3$); soluble in $CHCl_3$, insoluble in petroleum ether. *Abbott Labs Inc.; Astra USA Inc.; DuPont-Merck Pharm.; Wyeth-Ayerst Labs.*

9065 Naloxone Hydrochloride

357-08-4 6449 206-611-0

$C_{19}H_{22}ClNO_4$

17-Allyl-4,5α-epoxy-3,14-dihydroxymorphinan-6-one hydrochloride.

Narcan; EN-15304; Nalone; Narcanti. Narcotic antagonist. mp = 200-205°; soluble in H_2O, EtOH; insoluble in Et_2O. *Abbott Labs Inc.; Astra USA Inc.; DuPont-Merck Pharm.; Wyeth-Ayerst Labs.*

9066 Naltrexone

16590-41-3 6450 240-649-9

$C_{20}H_{23}NO_4$

17-(Cyclopropylmethyl)-4,5α-epoxy-3,14-dihydroxymorphinan-6-one.

UM-792; ReVia. Narcotic antagonist, related to cyprenorphine. Cogener of naloxone. Has opiate-blocking activity. mp = 168-170°; LD_{50} (mus sc) = 586 mg/kg. *DuPont-Merck Pharm.*

9067 Oxilorphan

42281-59-4 255-749-8

$C_{20}H_{27}NO_2$

(-)-17-(Cyclopropylmethyl)morphinan-3,14-diol.

levo-BC-2605. Narcotic antagonist.

9068 Quadazocine

71276-43-2

$C_{25}H_{37}NO_2$

(-)-(2R,6S,11S)-1-Cyclopentyl-5-(1,2,3,4,5,6-hexahydro-8-hydroxy-3,6,11-trimethyl-2,6-methano-3-benzazocin-11-yl)-3-pentanone.

Narcotic antagonist. *Sterling Winthrop Inc.*

9069 Quadazocine Mesylate

71276-44-3

$C_{26}H_{41}NO_5S$

(-)-(2R,6S,11S)-1-Cyclopentyl-5-(1,2,3,4,5,6-hexahydro-8-hydroxy-3,6,11-trimethyl-2,6-methano-3-benzazocin-11-yl)-3-pentanone methanesulfonate (salt).

Win-44441-3. Narcotic antagonist. *Sterling Winthrop Inc.*

Nootropics

9070 Aceglutamide

2490-97-3 25 219-647-7

$C_7H_{12}N_2O_4$

N^2-Acetyl-L-glutamine.

Acutil-S. Nootropic. mp = 197°; $[\alpha]_D^{20}$ = -12.5° (c = 2.9 H_2O). *Merck & Co.Inc.*

9071 Aceglutamide Aluminum

12607-92-0 25

$C_{35}H_{59}Al_3N_{10}O_{24}$

Pentakis (N^2-acetl-L-glutaminato)terahydroxytrialuminum.

Glumal; KW-110. Nootropic (free acid); antiulcerative (aluminum complex). mp = 221° (dec); soluble in H_2O; insoluble in MeOH, EtOH, Me_2CO; LD_{50} (mmus orl) = 14.3 g/kg, (mmus ip) = 5.0 g/kg, (mmus iv) = 0.46 g/kg, (mrat orl) > 14.5 g/kg, (mrat ip) = 4.2 g/kg, (mrat iv) = 0.40 g/kg. *Kyowa Hakko Kogyo Co. Ltd.*

9072 Acetylcarnitine

14992-62-2 85

$C_9H_{17}NO_4$

(3-Carboxy-2-hydroxypropyl)trimethylammonium hydroxide inner salt acetate.

vitamin B_T acetate. Nootropic. [(L)-form]: mp = 145° (dec); $[\alpha]_D^{20}$ = -19.52° (c = 6); very soluble in H_2O, EtOH;

insoluble in Et_2O; [(DL)-form hydrochloride]: mp = 187-188°.

9073 L-Acetylcarnitine Hydrochloride
5080-50-2 85
$C_9H_{18}ClNO_4$
(3-Carboxy-2-hydroxypropyl)trimethylammonium hydroxide inner salt acetate hydrochloride.
levacecarnine hydrochloride; ST-200; Alcar; Branigen; Nicetile; Normobren; Zibren. Nootropic. mp = 187° (dec), 181°; $[\alpha]_D^{20}$ =-28° (c = 2 H_2O); very soluble in H_2O, soluble in EtOH, insoluble in Et_2O.

9074 Adafenoxate
82168-26-1
$C_{20}H_{26}ClNO_3$
2-(1-Adamantylamino)ethyl (p-chlorophenoxy)acetate.
Nootropic.

9075 Aniracetam
72432-10-1 700
$C_{12}H_{13}NO_3$
1-p-Anisoyl-2-pyrrolidinone.
Ro-13-3057; Draganon; Sarpul. Nootropic. Cognition enhancer. Related to priacetam. mp = 121-122°; LD_{50} (mus orl) = 4500 mg/kg, (rat orl) = 5000 mg/kg. *Hoffmann-LaRoche Inc.*

9076 Besipirdine
119257-34-0 1223
$C_{16}H_{17}N_3$
1-(Propyl-4-pyridylamino)indole.
Nootropic. Cholinomimetic agent with noradrenergic properties. *Hoechst Roussel Pharm. Inc.*

9077 Besipirdine Hydrochloride
130953-69-4 1223
$C_{16}H_{18}ClN_3$
1-(Propyl-4-pyridylamino)indole hydrochloride.
HP-749. Nootropic. Cholinomimetic agent with noradrenergic properties. mp = 212-214°. *Hoechst Roussel Pharm. Inc.*

9078 Besipirdine Maleate
119257-40-8 1223
$C_{20}H_{21}N_3O_4$
1-(Propyl-4-pyridylamino)indole maleate.
Nootropic. Cholinomimetic agent with noradrenergic properties. mp = 115-116°. *Hoechst Roussel Pharm. Inc.*

9079 Bifemelane
90293-01-9 1255
$C_{18}H_{23}NO$
N-Methyl-4-[(α-phenyl-o-tolyl)oxy]butylamine.
Nootropic. Monoamine oxidase inhibitor. *Mitsubishi Chem. Corp.*

9080 Bifemelane Hydrochloride
62232-46-6 1255
$C_{18}H_{24}ClNO$
N-Methyl-4-[(α-phenyl-o-tolyl)oxy]butylamine hydrochloride.
E-0687; MCI-2016; Alnert; Celeport. Nootropic. Monoamine oxidase inhibitor. mp = 117-121°; LD_{50} (mus orl) = 1000 mg/kg, (mus ip) = 173 mg/kg, (rat orl) = 1080 mg/kg, (rat ip) = 130 mg/kg. *C.M. Ind.; Mitsubishi Chem. Corp.*

9081 Choline Alfoscerate
28319-77-9 2260 248-962-2
$C_8H_{20}NO_6P$
(R)-2,3-Dihydroxypropyl hydrogen phosphate choline hydroxide inner salt.
L-α-GPC; Brezal; Delecit; Gliatilin. Nootropic. Phospholipid. Precursor in choline synthesis; intermediate phos-phatidylcholine catabolism. mp = 142.5-143°; soluble in H_2O; $[\alpha]_D^{25}$= -2.7° (c = 2.7 H_2O pH 2.5), -2.8° (c = 2,6 H_2O pH 5.8).

9082 Citicoline
987-78-0 2380 213-580-7
$C_{14}H_{26}N_4O_{11}P_2$
Choline cytidine 5'-pyrophosphate ester.
CDP-choline; Audes; Cereb; Citifar; Colite; Corenalin; Cyscholin; Difosfocin; Emicholine-F; Ensign; Haocolin; Hornbest; Neucolis; Nicholin; Nicolin; Niticolin; Reagin; Recognan; Rexort; Sintoclar; Somazine; Suncholin. Vasodilator (cerebral). Naturally occurring nuceotide. Coenzyme in lecithin biosynthesis. $[\alpha]_D^{20}$ = +17.2° (in H_2O); $[\alpha]_D^{25}$= +19.0° (c = 0.86 in H_2O); λ_m = 280 nm (ε 12800 at pH 1); soluble in H_2O; nearly insoluble in most organic solvents; LD_{50} (mus iv) 4600 mg/kg; (mus orl) = 8000 mg/kg. *Abbott Laboratories Inc.*

9083 Dimiracetam
126100-97-8
$C_6H_8N_2O_2$
(±)-Dihydro-1H-pyrrolo[1,2-a]imidazole-2,5(3H,6H)-dione.
Cognition enhancer.

9084 Donepezil
120014-06-4
$C_{24}H_{29}NO_3$
(±)-2-[(1-Benzyl-4-piperidyl)methyl]-5,6-dimethoxy-1-indanone.
Nootropic. A reversible acetylcholine esterase inhibitor used to treat Alzheimer's disease. *Pfizer Inc.*

9085 Donepezil Hydrochloride
142057-77-0
$C_{24}H_{30}ClNO_3$
(±)-2-[(1-Benzyl-4-piperidyl)methyl]-5,6-dimethoxy-1-indanone hydrochloride.
E-2020; BNAG; Aricept. Nootropic. A reversible acetylcholine esterase inhibitor used to treat Alzheimer's disease. An adjunct in treatment of dimentia symptoms. A cognition adjuvant. Freely soluble in $CHCl_3$; soluble in H_2O, glacial acetic acid; slightly soluble in EtOH, acetonitrile; practically insoluble in EtOAc, n-hexane. *Pfizer Inc.*

9086 Dupracetam
59776-90-8
$C_{12}H_{18}N_4O_4$
1,2-Bis[(2-oxo-1-pyrrolidinyl)acetyl]hydrazine.
Nootropic.

9087 E-2020
120011-70-3 3533
$C_{24}H_{30}ClNO_3$
2,3-Dihydro-5,6-dimethoxy-2-[[1-(phenylmethyl)-4-piperidinyl]methyl]-1H-inden-1-one hydrochloride.
donepezil hydrochloride; Aricept. Nootropic. Cognition enhancer. Acetylcholinesterase inhibitor. mp = 211-212°. *Eisai Co. Ltd.*

9088 Exifone
52479-85-3 3958 257-945-9
$C_{13}H_{10}O_7$
2,3,3',4,4',5'-Hexahydroxybenzophenone.
Adlone. Nootropic. mp = 270°, 272-273°; LD_{50} (rat ip) = 355 mg/kg, (rat orl) = 1425 mg/kg. *BASF Corp.*

9089 Fipexide
34161-24-5 4126 251-857-4
$C_{20}H_{21}ClN_2O_4$
1-[(p-Chlorophenoxy)acetyl]-4-piperonylpiperazine.
Nootropic. *Lab. Bouchara.*

9090 Fipexide Hydrochloride
34161-23-4 4126 251-856-9
$C_{20}H_{22}Cl_2N_2O_4$
1-[(p-Chlorophenoxy)acetyl]-4-piperonylpiperazine hydrochloride.
BP-662; Attentil; Vigilor. Nootropic. mp = 230-232°; LD_{50} (mus orl) = 4150 mg/kg, (mus ip) = 499 mg/kg, (rat orl) = 4482, 7000 mg/kg, (rat ip) = 537, 450 mg/kg. *Lab. Bouchara.*

9091 Idebenone
58186-27-9 4932
$C_{19}H_{30}O_5$
2-(10-Hydroxydecyl)-5,6-dimethoxy-3-methyl-p-benzoquinone.
CV-2619; Avan; Daruma; Mnesis. Nootropic. Protective against cerebral ischemia. Ubiquinone derivative. mp = 46-50°, 52-53°; insoluble in H_2O, soluble in organic solvents.

9092 Indeloxazine Hydrochloride
65043-22-3 4972
$C_{14}H_{18}ClNO_2$
(±)-2-[(Inden-7-yloxy)methyl]morpholine hydrochloride.
CI-874; YM-08054-1; Elen; Noin. Antidepressant; nootropic. Serotonin uptake inhibitor. mp = 155-156°, 169-170°; LD_{50} (mus iv) = 47 mg/kg; [(+)-form]: mp = 112-113°; $[\alpha]_D^{21}$ = = 4.9° (c = 5 MeOH); [(-)-form]: mp = 142-142.5°; $[\alpha]_D^{20}$ = -4.9° (c = 5 MeOH). *Parke-Davis; Yamanouchi U.S.A. Inc.*

9093 Ipidacrine
62732-44-9
$C_{12}H_{16}N_2$
9-Amino-2,3,5,6,7,8-hexahydro-1H-cyclopent[b]quinoline.
NIK-247. Nootropic. Antiparkinsonian. Improves scopolamine-induced amnesia. Under study for use in treatment of Alzheimer's disease.

9094 Meclofenoxate
51-68-3 5820 200-116-3
$C_{12}H_{16}ClNO_3$
2-(Dimethylamino)ethyl(p-chlorophenoxy)acetate.
centrophenoxine; meclofenoxane; acephen; ANP-235; Analux; Cetrexin; Proseryl. CNS stimulant. Also used as a plant growth regulator. *CNRS.*

9095 Meclofenoxate Hydrochloride
3685-84-5 5820 222-975-3
$C_{12}H_{17}Cl_2NO_3$
2-(Dimethylamino)ethyl(p-chlorophenoxy)acetate hydrochloride.
Cellative; Clocete; Lucidril; Methocynal; Proserout; Brenal; Marucotol; Helfergin. CNS stimulant. Also used as a plant growth regulator. mp = 135-139°; soluble in H_2O, iPrOH, Me_2CO; insoluble in C_6H_6, Et_2O, $CHCl_3$; LD_{50} (mus iv) = 330 mg/kg, (mus ip) = 845 mg/kg, (mus orl) = 1750 mg/kg. *CNRS.*

9096 Nebracetam
97205-34-0 6520
$C_{12}H_{16}N_2O$
4-(Aminomethyl)-1-benzyl-2-pyrrolidinone.
Nootropic. Cognition enhancer. *Boehringer Ingelheim Pharm. Inc.*

9097 dl-Nebracetam
116041-13-5 6520
$C_{12}H_{16}N_2O$
(±)-4-(Aminomethyl)-1-benzyl-2-pyrrolidinone.
WEB-1881. Nootropic. Cognition enhancer. *Boehringer Ingelheim Pharm. Inc.*

9098 Nebracetam Fumarate
97205-35-1 6520
$C_{28}H_{36}N_2O_6$
4-(Aminomethyl)-1-benzyl-2-pyrrolidinone fumarate.
WEB-1881FU. Nootropic. Cognition enhancer. mp = 192-194°; LD_{50} (mrat iv) = 257 mg/kg, (mrat orl) = 2884 mg/kg, (frat iv) = 257 mg/kg, (frat orl) = 3622 mg/kg. *Boehringer Ingelheim Pharm. Inc.*

9099 Nefiracetam
77191-36-7 6528
$C_{14}H_{18}N_2O_2$
2-Oxo-1-pyrrolidineaceto-2',6'-xylidide.
DMMPA; DM-9384; DZL-221; Translon. Nootropic. Cognition enhancer. mp = 153°; LD_{50} (mus iv) = 421 mg/kg, (mus orl) = 1766 mg/kg. *Natterman.*

9100 Nizofenone
54533-85-6 6759
$C_{21}H_{21}ClN_4O_3$
2'-Chloro-2-[2-[(diethylamino)methyl]imidazol-1-yl]-5-nitrobenzophenone.
Nootropic. mp = 75-76°. *Yoshitomi.*

9101 Nizofenone Fumarate
54533-86-7 6759
$C_{25}H_{25}ClN_4O_7$
2'-Chloro-2-[2-[(diethylamino)methyl]imidazol-1-yl]-5-nitrobenzophenone fumarate.
midafenone; Y-9179; Ekonal. Nootropic. mp = 157-158°;

LD$_{50}$ (mmus orl) = 495 mg/kg, (mmus iv) = 62 mg/kg, (mmus sc) = 270 mg/kg, (fmus orl) = 504 mg/kg, (fmus iv) = 70 mg/kg, (fmus sc) = 278 mg/kg, (mrat orl) = 1711 mg/kg, (mrat iv) = 63 mg/kg, (mrat sc) = 1830 mg/kg, (frat orl) = 1580 mg/kg, (frat iv) = 65 mg/kg, (frat sc) = 1629 mg/kg. *Yoshitomi.*

9102 Nonivamide
2444-46-4 219-484-1
$C_{17}H_{27}NO_3$
N-Vanillylnonamide.
nonylic acid vanillylamide; nonanoyl vanillylamide. Topical analgesic. Antinociceptive. Capsaicin derivative. Acts by depleting stores of substance P from sensory neurons.

9103 Oxiracetam
62613-82-5 7076
$C_6H_{10}N_2O_3$
4-Hydroxy-2-oxo-1-pyrrolidineacetamide.
CGP-21690E; CT-848; ISF-2522; Neuractiv; Neuromet. Nootropic. mp = 165-168°; [(R)-form]: mp = 135-136°; $[\alpha]_D$ = 36.2° (c = 1.00 H_2O); [(S)-form]: mp = 135-136°; $[\alpha]_D$ = -36.0 (c = 1.00 H_2O). *I.S.F.*

9104 Piracetam
7491-74-9 7641 231-312-7
$C_6H_{10}N_2O_2$
2-Oxo-1-pyrrolidineacetamide.
CI-781; UCB-6215; Avigilen; Axonyl; Cerebroforte; Encetrop; Gabacet; Genogris; Geram; Nootron; Nootrop; Nootropil; Nootropyl; Norzetam; Normabrain; Pirroxil. Nootropic. mp = 151.5-152.5°. *UCB Pharma.*

9105 Piridoxilate
24340-35-0
$C_{20}H_{26}N_2O_{12}$
[[5-Hydroxy-4-(hydroxymethyl)-6-methyl-3-pyridyl]-methoxy]glycolic acid compound with [[4,5-bis(hydroxymethyl)-2-methyl-3-pyridyl]oxy]-glycolic acid (1:1).
Putative physiological regulator of cell respiration. Protective against cardiac and cerebral hypoxia. Long-term treatment may lead to calcium oxalate nephrolithiasis. Antianoxic; nootropic.

9106 Pramiracetam
68497-61-1 7886
$C_{14}H_{27}N_3O_2$
N-[2-(Diisopropylamino)ethyl]-2-oxo-1-pyrrolidine-acetamide.
U-98528E; SUD919CL2Y; amacetam. Nootropic. $bp_{0.15}$ = 162-164°; [monohydrate]: mp = 47-48°. *Boehringer Ingelheim Pharm. Inc.; Parke-Davis.*

9107 Pramiracetam Sulfate
72869-16-0 7886
$C_{14}H_{29}N_3O_6S$
N-[2-(Diisopropylamino)ethyl]-2-oxo-1-pyrrolidine-acetamide sulfate.
CI-879. Nootropic. LD$_{50}$ (mmus orl) = 5434 mg/kg, (fmus orl) = 4355 mg/kg. *Boehringer Ingelheim Pharm. Inc.; Parke-Davis.*

9108 Propentofylline
55242-55-2 7997
$C_{15}H_{22}N_4O_3$
3-Methyl-1-(5-oxohexyl)-7-propylxanthine.
HWA-285; Albert-285; HOE-285; Hextol; Karsivan. Nootropic. Peripheral vasodilator (veterinary). Inhibits cyclic-AMP phos-phodiesterase. mp = 69-70°; soluble in H_2O (3.2 g/100 ml at 25°), EtOH (> 10 g/100ml), DMSO (> 10 g/100 ml); LD$_{50}$ (mmus orl) = 900 mg/kg, (mmus iv) = 168 mg/kg, (mmus ip) = 375 mg/kg, (mmus sc) = 450 mg/kg, (fmus orl) = 780 mg/kg, (fmus iv) = 170 mg/kg, (fmus ip) = 346 mg/kg, (fmus sc) = 508 mg/lg), (mrat orl) = 1150 mg/kg, (mrat iv) = 180 mg/klg, (mrat ip) = 199 mg/kg, (mrat sc) = 400 mg/kg, (frat orl) = 940 mg/kg, (frat iv) = 195 mg/kg, (frat ip) = 196 mg/kg, (frat sc) = 338 mg/kg. *Hoechst AG.*

9109 Pyritinol
1098-97-1 8180 214-150-1
$C_{16}H_{20}N_2O_4S_2$
3,3'-(Dithiodimethylene)bis(5-hydroxy-6-methyl-4-pyridinemethanol).
pyroxidine-5-disulfide; pyrithioxin; Bonifen; Epocan. Nootropic. mp = 218-220°. *E. Merck.*

9110 Pyritinol Hydrochloride
10049-83-9 8180 233-178-5
$C_{16}H_{24}Cl_2N_2O_5S_2$
3,3'-(Dithiodimethylene)bis(5-hydroxy-6-methyl-4-pyridinemethanol) dihydrochloride monohydrate.
Biocefalin; Enbol; Encephabol; Enerbol; Life. Nootropic. mp = 184°. *E. Merck.*

9111 Rolziracetam
18356-28-0
$C_7H_9NO_2$
Dihydro-1H-pyrrolizine-3,5(2H,6H)-dione.
CI-911. Contributes to reversal of amnesia; protective against barbiturate intoxication; may benefit patients with moderate dementia. Nootropic; cognition adjuvant.

9112 Sabcomeline
159912-53-5
$C_{10}H_{15}N_3O$
R-(Z)-α-(Methoxyimino)-1-azabicyclo[2.2.2]octane-3-acetonitrile.
SB-202026. Partial muscarinic receptor agonist. Cognitive adjuvant; nootropic.

9113 Sabeluzole
104383-17-7 8462
$C_{22}H_{26}FN_3O_2S$
(±)-4-(2-Benzothiazolylmethylamino)-α-[(p-fluorophenoxy)methyl]-1-piperidineethanol.
R-58735; Reminyl. Anticonvulsant. Nootropic. mp = 101.7°. *Janssen Pharm. Inc.*

9114 Sibopirdine
139781-09-2
$C_{23}H_{18}N_4.H_2O$
5,5-Bis(4-pyridylmethyl)-5H-cyclopenta[2,1-b:3,4-b']dipyridine monohydrate.
DuP-921. Nootropic. *DuPont-Merck Pharm.*

9115 Tacrine

321-64-2 9199 206-291-2

$C_{13}H_{14}N_2$

9-Amino-1,2,3,4-tetrahydroacridine.

Centrally active anticholinesterase. Nootropic. Respiratory stimulant. Cognition adjuvant and antidote for curare poisoning. mp = 183-184°. *Parke-Davis.*

9116 Tacrine Hydrochloride

1684-40-8 9199 216-867-5

$C_{13}H_{15}ClN_2$

9-Amino-1,2,3,4-tetrahydroacridine monohydrochloride.

Cognex; CI-970; THA. Nootropic. Respiratory stimulant. Centrally active anticholinesterase. Cognition adjuvant and antidote for curare poisoning. mp = 283-284°; soluble in H_2O. *Parke-Davis.*

9117 Tamitinol

59429-50-4

$C_{11}H_{18}N_2OS$

4-[(Ethylamino)methyl]-2-methyl-5-[(methylthio)methyl]-3-pyridinol.

Nootropic.

9118 Teniloxazine

62473-79-4

$C_{16}H_{19}NO_2S$

(±)-2-[[(α-2-Thienyl-o-tolyl)oxy]methyl]morpholine.

Y-8894 [as maleate]. Antidepressant; nootropic; antianoxic.

9119 Tenilsetam

86696-86-8

$C_8H_{10}N_2OS$

(±)-3-(2-Thienyl)-2-piperazinone.

A cognition-enhancing drug used for treatment of Alzheimer's disease. Advanced glycosylation endproducts inhibitor. Prevents protein crosslinking of amyloid plaques. Antidementia; nootropic; antihypoxidotic.

9120 Tepirindole

72808-81-2

$C_{16}H_{19}ClN_2$

5-Chloro-3-(1,2,3,6-tetrahydro-1-propyl-4-pyridyl)-indole.

HR-592. A derivative of indole. Nootropic; tranquilizer.

9121 Velnacrine

104675-29-8 10077

$C_{13}H_{14}N_2O$

(±)-9-Amino-1,2,3,4-tetrahydro-1-acridinol.

Nootropic. Cognition-enhancing acetylcholinesterase inhibitor. mp = 245°. *Hoechst Roussel Pharm. Inc.*

9122 Velnacrine Maleate

118909-22-1 10077

$C_{17}H_{18}N_2O_5$

(±)-9-Amino-1,2,3,4-tetrahydro-1-acridinol maleate.

Mentane; HP-029; P83 6029A. Nootropic. Cognition-enhancing acetylcholinesterase inhibitor. mp = 171-173°; LD_{50} (mus orl) = 136 mg/kg, (mus ip) = 162 mg/kg. *Hoechst Roussel Pharm. Inc.*

9123 Vinconate

70704-03-9 10123 274-789-7

$C_{18}H_{20}N_2O_2$

(±)-Methyl-3-ethyl2,3,3a,4-tetrahydro-1H-indolo-[3,2,1-de][1,5]naphthyridine-6-carboxylate.

chanodesethylapovincamine. Nootropic. An analog of vincamine. A synthetic hexahydrocanthane alkaloid. mp = 120-121°. *Omnium Chim.*

9124 Vinconate Hydrochloride

119600-43-0 10123

$C_{18}H_{21}ClN_2O_2$

(±)-Methyl-3-ethyl2,3,3a,4-tetrahydro-1H-indolo-[3,2,1-de][1,5]naphthyridine-6-carboxylate hydrochloride.

OC-340; OM-853. Nootropic. A synthetic analog of vincamine hydrochloride. mp = 194-195°; LD_{50} (mmus orl) = 947 mg/kg, (mmus iv) = 127 mg/kg, (mmus sc) = 4073 mg/kg, (fmus orl) = 699 mg/kg, (fmus iv) = 131 mg/kg, (fmus sc) = 3446 mg/kg, (mrat orl) = 2582 mg/kg, (mrat iv) = 112 mg/kg, (mrat sc) > 6000 mg/kg, (frat orl) = 2348 mg/kg, (frat iv) = 117 mg/kg, (frat sc) > 6000 mg/kg. *Omnium Chim.*

9125 Vindeburnol

74709-54-9

$C_{17}H_{20}N_2O$

(±)-20,21-Dinor-16α-eburnamenine.

RU-24722. An eburnamenine derivative with hepatotoxic properties. A putative phasic activator of catecholaminergic systems. Protective against cerebral anoxia and ischemia. Nootropic.

9126 Xanomeline

131986-45-3 10190

$C_{14}H_{23}N_3OS$

3-[4-(Hexyloxy)-1,2,5-thiadiazol-3-yl]-1,2,5,6-tetrahydro-1-methylpyridine.

LY-246708. Cholinergic; nootropic. Selective muscarinic M_2-receptor agonist. *Eli Lilly & Co.; Novo Nordisk Pharm. Inc.*

9127 Xanomeline Oxalate

141064-23-5 10190

$C_{16}H_{25}N_3O_5S$

3-[4-(Hexyloxy)-1,2,5-thiadiazol-3-yl]-1,2,5,6-tetrahydro-1-methylpyridine oxalate.

Cholinergic; nootropic. Selective muscarinic M_2-receptor agonist. mp = 148°. *Eli Lilly & Co.; Novo Nordisk Pharm. Inc.*

9128 Xanomeline Tartrate

152854-19-8 10190

$C_{18}H_{29}N_3O_7S$

3-[4-(Hexyloxy)-1,2,5-thiadiazol-3-yl]-1,2,5,6-tetrahydro-1-methylpyridine tartrate.

xanomeline (+)-L-hydrogen tartrate; LY-246708 tartrate; NNC-11-0232; Lomeron; Memcor. Cholinergic; nootropic. Selective muscarinic M_2-receptor agonist. mp = 95.5°. *Eli Lilly & Co.; Novo Nordisk Pharm. Inc.*

Oxytocics

9129 Carboprost
35700-23-3 1871

$C_{21}H_{36}O_5$

(E,Z)-(1R,2R,3R,5S)-7-[3,5-Dihydroxy-2-[(3S)-(3-hydroxy-3-methyl-1-octenyl)]cyclopentyl]-5-heptenoic acid.

U-32921; (15S)-15-methylprostaglandin $F_{2\alpha}$. Oxytocic. Prostaglandin. *Pharmacia & Upjohn.*

9130 Carboprost Methyl
35700-21-1 1871

$C_{22}H_{38}O_5$

Methyl (E,Z)-(1R,2R,3R,5S)-7-[3,5-dihydroxy-2-[(3S)-(3-hydroxy-3-methyl-1-octenyl)]cyclopentyl]-5-heptenoate.

U-36384; carboprost methyl ester. Oxytocic. Prostaglandin. mp = 55-56°; $[\alpha]_D$ = 24° (c = 0.81 EtOH). *Pharmacia & Upjohn.*

9131 Carboprost Tromethamine Salt
58551-69-2 1871

$C_{25}H_{47}NO_8$

(E,Z)-(1R,2R,3R,5S)-7-[3,5-Dihydroxy-2-[(3S)-(3-hydroxy-3-methyl-1-octenyl)]cyclopentyl]-5-heptenoic acid compound with 2-amino-2-(hydroxymethyl)-1,3-propanediol.

carboprost trometamol; U-32921E;U-36384; Hemabate; Prostin/15M. Prostaglandin. Oxytocic. *Pharmacia & Upjohn.*

9132 Cargutocin
33605-67-3 1887

$C_{42}H_{65}N_{11}O_{12}$

1-Butyric acid-6-(L-2-aminobutyric acid)-7-glycineoxytocin.

Y-5350; Statocin. Oxytocic. $[\alpha]_D^{25}$= -44.0° (c = 0.55 H_2O). *Yoshitomi.*

9133 Deaminooxytocin
113-78-0 2899 204-034-9

$C_{43}H_{65}N_{11}O_{12}S_2$

1-(3-Mercaptopropanoic acid)oxytocin.

demoxytocin; desaminooxytocin; ODA-914; Sandopart. Oxytocic. [(L)-isomer]: mp = 179°, 182-183°; $[\alpha]_D^{20}$ = -88.3°, $[\alpha]_D^{21}$ = -107°, $[\alpha]_D^{25}$ = -95.1 (all c = 0.5 1N AcOH); [(D)-isomer]: $\alpha]_D^{20}$ = 104° (c = 0.5 in 1N AcOH).

9134 Ergonovine
60-79-7 3694 200-485-0

$C_{19}H_{23}N_3O_2$

9,10-Didehydro-N-[(S)-2-hydroxy-1-methylethyl]-6-methylergoline-8β-carboxamide.

Ergotrate; ergometrine; Ergobasin; Ergotocine; Ergostetrine; Ergotrate; Ergoklinine; Syntometrine. Oxytocic. mp = 162°; $[\alpha]_D^{20}$ = 90° (H_2O); freely soluble in MeOH, EtOH, EtOAc, Me_2CO; soluble in H_2O; slightly soluble in $CHCl_3$. *Eli Lilly & Co.*

9135 Ergonovine Maleate
129-51-1 3694 204-953-5

$C_{23}H_{27}N_3O_6$

9,10-Didehydro-N-[(S)-2-hydroxy-1-methylethyl]-6-methylergoline-8β-carboxamide maleate (1:1) (salt).

Ergotrate Maleate; Cornocentin; Ermetrine. Oxytocic. mp = 167° (dec); $[\alpha]_D^{25}$ = 48° to 57°; soluble in H_2O (2.8 g/100 ml), EtOH (0.83 g/100 ml); LD_{50} (mus iv) = 8.26 mg/kg. *Eli Lilly & Co.*

9136 Gemeprost
64318-79-2 4393 264-775-9

$C_{23}H_{38}O_5$

Methyl (E)-7-[(1R,2R,3R)-3-hydroxy-2-[(E)-(3R)-3-hydroxy-4,4-dimethyl-1-octenyl]-5-oxocyclopentyl]-2-heptenoate.

16,16-dimethyl-trans-Δ^2-PGE_1 methyl ester; ONO-802; SC-37681; Cergem; Cervagem; Cervageme; Preglandin. Abortifacient; oxytocic; prostaglandin. Analog of prostaglandin E_1. Used in termination of first trimester pregnancy. *Ono Pharm.; Searle G.D. & Co.*

9137 Meteneprost
61263-35-2

$C_{23}H_{38}O_4$

(Z)-7-[(1R,2R,3R)-3-Hydroxy-2-[(E)-(3R)-3-hydroxy-4,4-dimethyl-1-octenyl]-5-methylenecyclopentyl]-5-heptenoic acid.

U-46785. Oxytocic. Prostaglandin. *Pharmacia & Upjohn.*

9138 Methylergonovine
113-42-8 6147 204-027-0

$C_{20}H_{25}N_3O_2$

9,10-Didehydro-N-[(S)-1-(hydroxymethyl)propyl]-6-methylergoline-8β-carboxamide.

methylergometrine; methylergobasine. Oxytocic. mp = 172° (dec); $[\alpha]_D^{20}$ = -45° (c = 0.5 C_5H_5N); freely soluble in EtOH, Me_2CO; sparingly soluble in H_2O. *Sandoz Pharm. Corp.*

9139 Methylergonovine Maleate
57432-61-8 6147 260-734-4

$C_{24}H_{29}N_3O_6$

9,10-Didehydro-N-[(S)-1-(hydroxymethyl)propyl]-6-methylergoline-8β-carboxamide maleate (1:1) (salt).

Methergine; Methergin; Basofortina; Metenarin; Methylergobrevin; Ryegonovin; Spametrin-F. Oxytocic. Slightly soluble in H_2O, EtOH; sparingly soluble in $CHCl_3$, Et_2O. *Sandoz Pharm. Corp.*

9140 Methylergonovine Tartrate
6209-37-6 6147

$C_{44}H_{54}N_6O_{10}$

9,10-Didehydro-N-[(S)-1-(hydroxymethyl)propyl]-6-methylergoline-8β-carboxamide tartrate (2:1) (salt).

Oxytocic. Freely soluble in H_2O, EtOH; slightly soluble in Et_2O, $CHCl_3$. *Sandoz Pharm. Corp.*

9141 Nacartocin
77727-10-7

$C_{46}H_{71}N_{11}O_{11}S$

1-(3-Mercaptopropionic acid)-2-[2-(p-ethylphenyl)-L-alanine]-6-(L-2-aminobutyric acid)oxytocin.

Synthetic oxytocin analog with natriuretic activity. Oxytotic.

9142 Oxytocin
50-56-6 7114 200-048-4
$C_{44}H_{66}N_{12}O_{12}S_2$
Oxytocin.
Pitocin; Syntocinon; Uteracon; Alpha-hypophamine; ocytocin; Intertocine-S; Perlacton; Orasthin; Oxystin; Partocon; Synpitan; H-Cys-Tyr-Ile-Glu-Asp-Cys-Pro-Leu-Gly-NH_2 cyclic (1→6)-disulfide. Oxytocic. $[\alpha]_D^{22}$ = -26.2° (c = 0.53); soluble in H_2O, 1-BuOH, 2-BuOH. *Fermenta Animal Health Co.; Hoechst Roussel Pharm. Inc.; Parke-Davis; Sandoz Pharm. Corp.; Wyeth-Ayerst Labs.*

9143 Oxytocin Citrate
7563-62-4 7114
$C_{50}H_{74}N_{12}O_{19}S_2$
Oxytocin citrate (salt).
Pitocin-Buccal. Oxytocic.

9144 Pituitary, Posterior
7666
Pituitary extract (posterior).
Pituamin; Di-Sipidin; Pituitrin. Oxytocic; antidiuretic. Desiccated hypophysis. From the posterior lobe of pituitary body of domesticated animals. Contains oxytocin, vasopressin. Partially soluble in H_2O. *Parke-Davis.*

9145 Prostaglandin E_2
363-24-6 8064 206-656-6
$C_{20}H_{32}O_5$
(5Z,11α,13E,15S)-11,15-Dihydroxy-9-oxoprosta-5,13-dien-1-oic acid.
dinoprostone; PGE_2; U-12062; Cerviprost; Dinoprost; Enzaprost F; Glandin; Minprositin E_2; Prepidil; Propess; Prostarmon-E; Prostin E_2. Abortifacient; oxytocic. Prostaglandin. mp = 66-68°; $[\alpha]_D^{26}$ = -61° (c = 1 THF). *Pharmacia & Upjohn.*

9146 Prostaglandin $F_{2\alpha}$
551-11-1 8065
$C_{20}H_{34}O_5$
(E,Z)-(1R,2R,3R,5S)-7-[3,5-Dihydroxy-2-[(3S)-(3-hydroxy-1-octenyl)]cyclopentyl]-5-heptenoic acid.
dinoprost; $PGF_{2\alpha}$; U-14583; Enzaprost F; Glandin; Prostarmon F. Abortifacient; oxytocic. Prostaglandin. mp = 25-35°; $[\alpha]_D^{25}$ = 23.5° (c = 1 THF); freely soluble in MeOH, EtOH, EtOAc, $CHCl_3$; slightly soluble in H_2O; LD_{50} (rbt iv) = 2.5-5.0 mg/kg, (rbt im) = 2.5-5.0 mg/kg. *Pharmacia & Upjohn.*

9147 Prostaglandin F_{2a} Tromethamine Salt
38562-01-5 8065 254-002-3
$C_{24}H_{45}NO_8$
(E,Z)-(1R,2R,3R,5S)-7-[3,5-Dihydroxy-2-[(3S)-(3-hydroxy-1-octenyl)]cyclopentyl]-5-heptenoic acid compound with 2-amino-2-(hydroxymethyl)-1,3-propanediol (1:1).
$PGF_{2\alpha}$ THAM; U-14583E; Lutalyse; Prostin F_2 Alpha. Abortifacient; oxytocic. Prostaglandin. mp = 100-101°; freely soluble in H_2O (> 20 g/100 ml). *Pharmacia & Upjohn.*

9148 Quipazine Maleate
5786-68-5 227-314-2
$C_{17}H_{19}N_3O_4$
2-(1-Piperazinyl)quinoline maleate (1:1).
MA-1291. Antidepressant; oxytoxic. *Bayer Corp. Pharm. Div.; Miles Inc.*

9149 Sparteine
90-39-1 8887 201-988-8
$C_{15}H_{26}N_2$
Dodecahydro [7S-(7α,7aα,14α,14aβ)]-7,14-methano-2H,6H-dipyrido[1,2-a:1',2'-e][1,5]diazocine.
l-sparteine; lupinidine. Oxytocic. bp_8 = 173°; $[\alpha]_D^{21}$ = -16.4° (c = 10 EtOH); d_4^{20} = 1.020; soluble in H_2O (0.31 g/100 ml); freely soluble in EtOH, $CHCl_3$, Et_2O.

9150 Sparteine Sulfate Pentahydrate
6160-12-9 8887
$C_{15}H_{28}N_2O_4S.5H_2O$
Dodecahydro [7S-(7α,7aα,14α,14aβ)]-7,14-methano-2H,6H-dipyrido[1,2-a:1',2'-e][1,5]diazocine sulfate (1:1) pentahydrate.
Depasan; Tocosamine. Oxytocic. dec 136°; soluble in H_2O (9.1 g/100 ml), EtOH (3.3 g/100 ml); insoluble in $CHCl_3$, Et_2O.

Pigmentation Agents

9151 Methoxsalen
298-81-7 6068 206-066-9
$C_{12}H_8O_4$
9-Methoxy-7H-furo[3,2-g][1]benzopyran-7-one.
Methoxa-Dome; 8-MOP; 8-MP; Oxsoralen Lotion; Oxsoralen Ultra; Uvadex; ammodin; xanthotoxin; 8-methoxypsoralen; Meladinine; Meloxine; Oxsoralen. Pigmentation agent. Used in treatment of psoriasis and mycosis fungoides. mp = 148°; λ_m = 219, 249, 300 nm (log ε 4.32, 435, 4.06); insoluble in cold H_2O; sparingly soluble in boiling H_2O; soluble in Me_2CO, AcOH, propylene glycol, C_6H_6, boiling EtOH; freely soluble in $CHCl_3$; LD_{50} (rat ip) = 470 ± 30 mg/kg. *Bayer Corp. Pharm. Div.; Hoffmann-LaRoche Inc.; ICN Pharm. Inc.*

9152 Trioxsalen
3902-71-4 9864 223-459-0
$C_{14}H_{12}O_3$
2,5,9-Trimethyl-7H-furo[3,2-g][1]benzopyran-7-one.
Trisoralen; NSC-71047. Pigmentation agent. mp = 234.5-235°; λ_m 250 295 335 nm (log ε 4.35, 3.99, 3.80 MeOH); slightly soluble in EtOH, $CHCl_3$; soluble in CH_2Cl_2; insoluble in H_2O. *ICN Pharm. Inc.*

Plasma Volume Expanders

9153 Dextran
9004-54-0 2989 232-677-5
Dextraven; Gentran; Hemodex; Intradex; Macrose; Onkotin; Plavolex; Polyglucin; Promit. Plasma volume extender. Polysaccharide composed primarily of D-glucose units linked α-D-(1→6); produced by *Leucono-*

stoc mesenteroides, L. *dextranicum (Lactobacteriaceae). Baxter Healthcare Systems; CIBA Vision AG; McGaw Inc.*

9154 Dextran 40
9004-54-0 2989 232-677-5
LMD; LMWD; LVD, Eudextran; Gentran 40; Rheomacrodex; Rheotran. Plasma volume extender; blood flow adjuvant. Polysaccharide produced by the action of *Leuconostoc mesenteroides* on sucrose. Average molecular weight 40,000. *Baxter Healthcare Systems; McGaw Inc.*

9155 Dextran 70
2989
Hyskon; Macrodex; Aquasite; component of: Biontears, Estivin II, Tears Naturale, Tears Naturale II, Tears Naturale Free. Plasma volume extender. Polysaccharide produced by the action of *Leuconostoc mesenteroides* on sucrose. Average molecular weight 70,000. *Alcon Labs; CIBA Vision AG; McGaw Inc.*

9156 Dextran 75
2989
Gentran 75. Plasma volume extender. Polysaccharide produced by *Leuconostoc mesenteroides.* Average molecular weight 75,000. *Baxter Healthcare Systems.*

9157 Hetastarch
9004-62-0 4707
Starch 2-hydroxyethyl ether.
Hydroxyethyl starch; HES; 6-H.E.S.; Hespan; Hespander; Hestar; Hestat; Hestsol; Plasmasteril; Volex. An amylopectin derivative used as a cryoprotective agent for erythrocytes and as a plasma volume extender.

9158 Plasma Protein Fraction
Plasma protein fraction.
plasma protein fraction, human (formerly); Plasmanate; Plasma-Plex; Plasmatein; Protenate. Plasma volume supporter. *Abbott Labs Inc.; Alpha 1 Biomedicals Inc.; Cutter Labs; Hyland Div. Baxter ; Rhône-Poulenc Rorer Pharm. Inc.*

9159 Polygeline
9015-56-9
Haemaccel. Gelatin-based plasma volume expander. Polymer of urea and polypeptides derived from denatured gelatin.

9160 Serum Albumin
9048-46-8 8613 232-936-2
A plasma protein.
Human serum albumin; HAS; Albuminate; Albuminar; Albumisol; Albumispan; Buminate; Pro-Bumin; Proserum. Plasma volume extender. Also used in the treatment of hypoproteinemia.

Potassium Channel Activators

9161 Aprikalim
92569-65-8
$C_{12}H_{16}N_2OS_2$
(-)-(1R,2R)-Tetrahydro-N-methyl-2-(3-pyridyl)thio-2H-thiopyran-2-carboxamide 1-oxide.
A direct activator of ATP-sensitive potassium channels.

9162 Emakalim
129729-66-4
$C_{17}H_{16}N_2O_3$
(-)-(3S,4R)-3-Hydroxy-2,2-dimethyl-4-(2-oxo-1(2H)-pyridyl)-6-chromancarbonitrile.
Potassium channel activator.

9163 Levcromakalim
94535-50-9 5487
$C_{16}H_{18}N_2O_3$
(3S,4R)-3-Hydroxy-2,2-dimethyl-4-(2-oxo-1-pyrrolidinyl)-6-chromancarbonitrile.
BRL-38227; levkromakalim; lemakalim; (-)-cromakalim; BRL-34915 [(±)-form]; cromakalim [(±)-form]. Antihypertensive; antiasthmatic. Potassium channel activator. mp = 242-244°; $[\alpha]_D^{26}$ = -52.2° (c = 1 $CHCl_3$); [(+)-form]: mp = 243-245°; $[\alpha]_D^{26}$ = 53.5° (c = 1 $CHCl_3$); [(±)-form]: mp = 230-231°. *SmithKline Beecham Pharm.*

9164 Mazokalim
164178-54-5
$C_{23}H_{28}N_6O_6$
Ethyl 5-[(3S,4R)-4-[(1,6-dihydro-6-oxo-3-pyridazinyl)oxy]-3-hydroxy-2,2,3-trimethyl-6-chromanyl]-1H-tetrazole-1-butyrate.
Potassium channel activator.

9165 Nicorandil
65141-46-0 6608 265-514-1
$C_8H_9N_3O_4$
N-(2-Hydroxyethyl)nicotinamide nitrate (ester).
SG-75; Ikorel; Perisalol; Sigmart. Coronary vasodilator used as an antianginal agent. Nicotinamide derivative with dual mechanism as both nitrovasodilator and potassium channel activator. mp = 92-93°; LD_{50} (rat orl) = 1200-1300 mg/kg, (rat iv) = 800-1000 mg/kg. *Chugai Pharm. Co. Ltd.; Upjohn Ltd.*

9166 Pinacidil
60560-33-0 7592 262-294-9
$C_{13}H_{19}N_5.H_2O$
(±)-2-Cyano-1-(4-pyridyl)-3-(1,2,2-trimethylpropyl)-guanidine monohydrate.
Pindac; P-1134. Potassium channel activator. Used as an antihypertensive. mp = 164-165°; LD_{50} (mus orl) = 600 mg/kg, (rat orl) = 570 mg/kg. *Eli Lilly & Co.*

9167 Rilmakalim
132014-21-2
$C_{21}H_{23}NO_5S$
(+)-1[(3S,4R)-3-Hydroxy-2,2-dimethyl-6-(phenylsulfonyl)-4-chrmanyl]-2-pyrrolidinone.
HOE-234 [as hemihydrate]. Activator of ATP-sensitive potassium channels.

Progestogens

9168 Algestone Acetophenide
24356-94-3 239 246-195-8
$C_{29}H_{36}O_4$
(R)-16α,17-Dihydroxypregn-4-ene-3,20-dione cyclic acetal with acetophenone.
Neolutin Depositum; P-DHP; SQ-15,101; Deladroxone;

Droxone. Progestogen. Used to treat acne. mp = 150-151°; $[\alpha]_D^{23}$ = 51° ($CHCl_3$). *Bristol-Myers Squibb Co.*

9169 Allylestrenol
432-60-0 299
$C_{21}H_{32}O$
17α-Allyl-17β-hydroxy-19-nor-4-androstene.
Gestanin; Gestanol; Orageston; Gestanyn; Gestanon; Turinal. Progestogen. mp = 769.5-80°; insoluble in H_2O; soluble in EtOH, Et_2O, Me_2CO, $CHCl_3$. *Organon Inc.*

9170 Altrenogest
850-52-2 327
$C_{21}H_{26}O_2$
17α-Allyl-17β-hydroxyestra-4,9,11-trien-3-one.
A-35957; RU-2267; Regumate. Progestogen. mp = 120°; $[\alpha]_D^{20}$ = -72° (c = 0.5 EtOH). *Roussel-UCLAF.*

9171 Amadinone
30781-27-2
$C_{20}H_{25}ClO_3$
6-Chloro-17α-hydroxy-19-norpregna-4,6-diene-3,20-dione.
Progestogen. *Syntex Labs. Inc.*

9172 Amadinone Acetate
22304-40-3
$C_{22}H_{27}ClO_4$
6-Chloro-17α-hydroxy-19-norpregna-4,6-diene-3,20-dione acetate.
RS-2208. Progestogen. *Syntex Labs. Inc.*

9173 Anagestone
2740-52-5 664
$C_{22}H_{34}O_2$
17α-Hydroxy-6α-methylpregn-4-ene-20-one.
Progestogen. mp = 190-193°; $[\alpha]_D^{20}$ = 51° ($CHGCl_3$). *Ortho Pharm. Corp.*

9174 Anagestone Acetate
3137-73-3 664
$C_{24}H_{36}O_3$
17α-Hydroxy-6α-methylpregn-4-ene-20-one acetate.
Anatropin. Progestogen. mp = 173-175°; $[\alpha]_D^{20}$ = 24° ($CHCl_3$). *Ortho Pharm. Corp.*

9175 Chlormadinone Acetate
302-22-7 2152
$C_{23}H_{29}ClO_4$
6-Chloro-17α-hydroxypregna-4,6-diene-3,20-dione acetate.
NSC-92338; Chronosyn; Cyclonorm; Fertiletten; Gestafortin; Lormin; Luteran; Lutoral; Natrol; Normenon; Menstridyl; Prostal; Traslan; Verton; component of: Amenyl, Lutestral, Menova, C-Quens, Gestamestrol, Sequens. An orally active progestogen with antiandrogenic activity. Used as a component of oral contraceptives. Used as a hormonal antineoplastic and, in veterinary medicine, as an estrus regulator. mp = 212-214°; $[\alpha]_D$ = 6° (c = 1 $CHCl_3$); λ_m = 283.5, 286 nm (ε = 23400, 22100). *Syntex Labs. Inc.*

9176 Cingestol
16915-71-2
$C_{20}H_{28}O$
19-Nor-17α-pregn-5-en-20-yn-17β-ol.
Progestogen.

9177 Clogestone
20047-75-0
$C_{23}H_{31}ClO_4$
6-Chloro-3β,17-dihydroxypregna-4,6-dien-20-one.
Progestogen. *Wyeth-Ayerst Labs.*

9178 Clogestone Acetate
3044-32-4
$C_{25}H_{33}ClO_5$
6-Chloro-3β,17-dihydroxypregna-4,6-dien-20-one diacetate.
AY-11440. Progestogen. *Wyeth-Ayerst Labs.*

9179 Clomegestone
5367-84-0
$C_{22}H_{29}ClO_3$
6-Chloro-17α-hydroxy-16α-methylpregna-4,6-diene-3,20-dione.
Progestogen. *Schering AG.*

9180 Clomegestone Acetate
424-89-5
$C_{24}H_{31}ClO_4$
6-Chloro-17α-hydroxy-16α-methylpregna-4,6-diene-3,20-dione acetate.
SH-741. Progestogen. *Schering AG.*

9181 Delmadinone
15262-77-8 2930 239-306-6
$C_{21}H_{25}ClO_3$
6-Chloro-17-hydroxypregna-1,4,6-triene-3,20-dione.
1,6-bisdehydro-6-chloro-17α-acetoxyprogesterone; Δ^1-chlormadinone. Progestogen with antiestrogenic and antiandrogenic activity. *Pharmacia & Upjohn.*

9182 Delmadinone Acetate
13698-49-2 2930 237-219-8
$C_{23}H_{27}ClO_4$
6-Chloro-17-hydroxypregna-1,4,6-triene-3,20-dione acetate.
RS-1301; Δ^1-chlormadinone acetate; Delminal; Estrex; Tardastrex; Tarden; Zenadrex. Progestogen with antiestrogenic and antiandrogenic activity. mp = 168-170°; $[\alpha]_D$ = -83° ($CHCl_3$); λ_m = 229, 258, 297 nm (log ε 4.00, 4.00, 4.03 EtOH). *Pharmacia & Upjohn.*

9183 Demegestone
10116-22-0 2939
$C_{21}H_{28}O_2$
17α-Methyl-19-nor-$\Delta^{4,9}$-pregnadiene-3,20-dione.
R-2453; Lutionex. Progestogen. mp = 106°; $[\alpha]_D$ = -275° (c = 0.5 EtOH); λ_m = 214, 302 nm (ε 6350, 21000 EtOH). *Roussel-UCLAF.*

9184 Desogestrel
54024-22-5 2971

$C_{22}H_{30}O$

13-Ethyl-11-methylene-18,19-dinor-17α-pregn-4-en-20-yn-17-ol.

Org-2969; Cyclosa; Dicromil; Marvelon 150/320; Mercilon; Ortho-Cept; Oviol; Varnoline; component of: Desogen. Progestogen with low androgenic potency. In combination with an estrogen (ethinylestradiol), used as an oral contraceptive [71138-35-7]. mp = 109-110°; $[\alpha]_D^{20}$= 55° ($CHCl_3$). *Organon Inc.*

9185 Dienogest
65928-58-7

$C_{20}H_{25}NO_2$

17-Hydroxy-3-oxo-19-nor-17α-pregna-4,9-diene-21-nitrile.

Dienol; DV; Estrodienol; Cycladiene; Dinovex; Estroral; Gynefollin; hormofemin; Oestrasid; Oestrodiene; Oestroral; Restrol; Retalon; Synestrol. Progestogen.

9186 Dimethisterone
79-64-1 3268

$C_{23}H_372O_2$

6α,21-Dimethyl-17β-hydroxy-17α-pregn-4-en-20-yn-3-one.

Secrosteron; component of: Oracon, Ovin, Tova. Progestogen. Formerly used with estrogens in oral contraceptives. mp = 102°; $[\alpha]_D^{20}$= 10° (c = 1 $CHCl_3$); λ_m = 240 nm ($E_{1\ cm}^{1\%}$ 450 iPrOH); insoluble in H_2O; soluble in EtOH; less soluble in Me_2CO, $CHCl_3$. *BDH Laboratory Supplies.*

9187 Drospirenone
67392-87-4 3510

$C_{24}H_{30}O_3$

6β,7β,15β,16β-Dimethylene-3-oxo-4-androstene-[17-β-1')-spiro-5']perhydrofuran-2'-one.

ZK-30595; dihydrospirorenone. Progestogen with antimineralcorticpoid and antiandrogenic activity. mp = 201.3°; $[\alpha]_D^{22}$= -8182° (c = 0.5 $CHCl_3$); λ_m = 265 nm (ε 19000 MeOH). *Schering AG.*

9188 Dydrogesterone
152-62-5 3524

$C_{21}H_{28}O_2$

10α-Pregna-4,6-diene-3,20-dione.

Dufaston; Duphaston; Gestatron; Gynorest; Prodel; Retrone; NSC-92336. Progestogen. mp = 169-170°; $[\alpha]_D^{25}$= -484.5 ($CHCl_3$); λ_m = 286.5 (ε 26400).

9189 Ethisterone
434-03-7 3787

$C_{21}H_{28}O_2$

17α-Ethynyl-17β-hydroxyandrost-4-en-3-one.

Ora-Lutin; Pranone; Syngestrotabs; Trosinone; NSC-9565; Pregneninolone; Lutocyclin; Lutocyclol; Progestoral. Progestogen. mp = 269-275°; $[\alpha]_D^{23}$ = 23.8° (dioxane), $[\alpha]_D^{25}$= -32.0° (C_5H_5N); λ_m = 241 nm ($E_{1\ cm}^{1\%}$ 513 MeOH); insoluble in H_2O; soluble in EtOH, Me_2CO, Et_2O, $CHCl_3$, vegetable oils. *Abbott Labs Inc.; Parke-Davis; Pfizer Inc.; Schering-Plough HealthCare Products.*

9190 Ethynerone
3214-93-4

$C_{20}H_{23}ClO_2$

21-Chloro-17-hydroxy-19-nor-17α-pregna-4,9-dien-20-yn-3-one.

Progestogen.

9191 Ethynodiol
1231-93-2 3905 214-971-5

$C_{22}H_{30}O_3$

19-Nor-17α-pregn-4-en-20-yne-3β,17-diol.

ED. Progestogen. *Searle G.D. & Co.*

9192 Ethynodiol Diacetate
297-76-7 3905 206-044-9

$C_{24}H_{32}O_4$

19-Nor-17α-pregn-4-en-20-yne-3β,17-diol diacetate.

SC-11800; Femulen; Luteonorm; Luto-Metrodiol; Metrodiol; component of: Demulen, Metrulen, Ovulen, Luteolas, Ovaras, Conova, Miniluteolas. Progestogen. Used in combination with an estrogen (mestranol, ethinyl estradiol) as an oral contraceptive. mp - 126-127°; $[\alpha]_D$ = -72.5° ($CHCl_3$). *Searle G.D. & Co.*

9193 Etonogestrel
54048-10-1

$C_{22}H_{28}O_2$

13-Ethyl-17-hydroxy-11-methylene-18,19-dinor-17α-pregn-4-en-20-yn-3-one.

Org-3236. Progestogen. *Organon Inc.*

9194 Flurogestone Acetate
2529-45-5 4235

$C_{23}H_{31}FO_5$

9-Fluoro-11β,17-dihydroxypregn-4-ene-3,20-dione 17-acetate.

SC-9880; NSC-65411; Chronogest; Cronolone; Synchronate; component of: Syncro-Mate. Progestogen. Used in veterinary medicine for estrus regulation. mp = 266-269°; $[\alpha]_D$ = 77.6° ($CHCl_3$); λ_m = 238 nm (ε 17500 MeOH). *Searle G.D. & Co.*

9195 Gestaclone
19291-69-1

$C_{23}H_{27}ClO_2$

6-Chloro-1α,2α:16α,17-bismethylene-4,6-pregnadiene-3,20-dione.

SH-1040. Progestogen. *Schering AG.*

9196 Gestodene
60282-87-3 4421 262-145-8

$C_{21}H_{26}O_2$

13-Ethyl-17-hydroxy-18,19-dinor-17α-pregna-4,15-dien-20-yn-3-one.

SH B 331; component of: Femodene, Femovan, Ginoden, Gynera, Milvane, Minulet, Monodie, Phaeva, Triminulet. Progestogen. Used in combination with ethinyl estradiol [109852-02-0] as an oral contraceptive. mp = 197.9°. *Schering AG.*

9197 Gestonorone Caproate
1253-28-7 4422 215-010-2

$C_{26}H_{38}O_4$

17-Hydroxy-19-norpregn-4-ene-3,20-dione hexanoate.

SH-582; NSC-84054; gestronol caproate; Depostat. Progestogen. Used in treatment of benign prostatic hypertrophy. mp = 123-124°; $[\alpha]_D$ = 13° ($CHCl_3$); λ_m = 239 nm (ε 17540). *Schering AG.*

9198 Gestrinone
40542-65-2

$C_{21}H_{24}O_2$

13-Ethyl-17-hydroxy-18,19-dinor-17α-pregn-4,9,11-trien-20-yn-3-one.

R-2323; RU-2323; A-46745. Progestogen. *Hoechst Roussel Pharm. Inc.*

9199 Haloprogesterone
3538-57-6

$C_{21}H_{28}BrFO_2$

17-Bromo-6α-fluoropregn-4-ene-3,20-dione.

17α-bromo-6α-fluoroprogesterone. Progestogen.

9200 17-Hydroxy-16-methylene-Δ^6-progesterone
10087-54-4 4878

$C_{32}H_{28}O_3$

17-Hydroxy-16-methylene-Δ^6-progesterone.

Progestogen. mp = 196.5-197°; $[\alpha]_D^{20}$ = -72.5°; λ_m = 283 nm (log ε 4.51). *E. Merck.*

9201 17-Hydroxy-16-methylene-Δ^6-progesterone Acetate
805-84-5 4878

$C_{34}H_{30}O_4$

17-Hydroxy-16-methylene-Δ^6-progesterone acetate.

Superlutin. Progestogen. mp = 233-234°; $[\alpha]_D^{20}$ = -132°; λ_m = 283 nm (lpg ε 4.50). *E. Merck.*

9202 17α-Hydroxyprogesterone
68-96-2 4886

$C_{21}H_{30}O_3$

17-Hydroxypregn-4-ene-3,20-dione.

Gestageno; Prodox. Progestogen. Used in veterinary medicine for estrus regulation. mp = 222-223°; $[\alpha]_D^{17}$ = 105.6° (c = 1.0417 $CHCl_3$); [acetate]: mp = 239-240°; λ_m = 240 nm (log ε 4.33). *Merck & Co.Inc.; Pharmacia & Upjohn; Syntex Labs. Inc.*

9203 17α-Hydroxyprogesterone Caproate
630-56-8 4886

$C_{27}H_{40}O_4$

17-Hydroxypregn-4-ene-3,20-dione caproate.

Delalutin; Hyproval P.A.; Lewntogest; Pharlon; Proge; Proluton Depot; Teralutil. Progestogen. Used in veterinary medicine for estrus regulation. mp = 119-121°; $[\alpha]_D^{20}$ = 61° (c = 1 $CHCl_3$). *Schering AG.*

9204 Lilopristone
97747-88-1

$C_{22}H_{36}O_5$

11β-[p-(Dimethylamino)phenyl]-17β-hydroxy-17-[(Z)-3-hydroxypropenyl]estra-4,9-diene-3-one.

13β-configured (type II) progestin antagonist. Anti-progestin; antigestagin; abortifacient.

9205 Lynestrenol
52-76-6 5659 200-151-4

$C_{20}H_{28}O$

17α-Ethinyl-17β-hydroxyestr-4-ene.

NSC-37725; Exluton; Exlutona; Exlutena; Orgametril; Orgametil; component of: Anacyclin; Fysionorm; Minilyn; Noracyclin; Ovanon; Ovoresta; Yermonil; Lyndiol. Progestogen. Used in combination with estrogens (ethinyl estradiol, mestranol), used as an oral contraceptive [109852-02-0], [8015-14-3] respectively. mp = 158-160°; $[\alpha]_D$ = -13° ($CHCl_3$).

9206 Medrogestone
977-79-7 5836

$C_{23}H_{32}O_2$

6,17-Dimethylpregna-4,6-diene-3,20-dione.

AY-62022; NSC-123018; Colpro; Colprone; Prothil. Progestogen. mp = 144-146°; $[\alpha]_D^{23}$ = 79° (c = 1 $CHCl_3$); λ_m = 288 nm (ε 25000). *Wyeth-Ayerst Labs.*

9207 Medroxyprogesterone
520-85-4 5838 208-298-6

$C_{22}H_{32}O_3$

17α-Hydroxy-6α-methylprogesterone.

Progestogen. Orally active progestogen used with estrogens (eg. ethinyl estradiol) in oral contraceptives. Used in veterinary medicine for estrus regulation. mp = 220-223.5°; $[\alpha]_D^{25}$ = 75° ($CHCl_3$); λ_m = 241 nm (ε 16000 EtOH). *Farmitalia Carlo Erba Ltd.; Pharmacia & Upjohn.*

9208 Medroxyprogesterone Acetate
71-58-9 5838

$C_{24}H_{34}O_4$

17α-Hydroxy-6α-methylprogesterone 17-acetate.

MAP; Amen; Clinovir; Curretab; Cycrin; Depo-Clinovir; Depo-Provera; Farlutal; Gestapuran; G-Farlutal; Hysron; Lutoral; Nadigest; Nidaxin; Oragest; Perlutex; Prodasone; Provera; Sodelut G; Veramix; component of: Provest. Orally active progestogen once used with estrogens in oral contraceptives. Used in veterinary medicine for estrus regulation. Progestogen. mp = 207-209°; $[\alpha]_D$ = 61° ($CHCl_3$); λ_m = 240 nm (ε 15900 EtOH). *Farmitalia Carlo Erba Ltd.; Pharmacia & Upjohn; Solvay Pharm. Inc.; Wyeth-Ayerst Labs.*

9209 Megestrol Acetate
595-33-5 5849

$C_{24}H_{32}O_4$

17-Hydroxy-6-methylpregna-4,6-diene-3,20-dione acetate.

Maygace; Megace; Megestat; Megestil; Nia; Niagestin; Ovaban; component of: Co-Ervonum, Kombiquens, Noval, Nuvacon, Planovin, Triu-Ervonum, Volidan, Weradys, Delpregnin. Progestogen. Used for palliative treatment of breast cancer. mp = 214-216°; $[\alpha]_D^{24}$ = 5° ($CHCl_3$); λ_m = 287 nm (log ε 4.40); soluble in H_2O (2 σg/ml), plasma (24 σg/ml). *BDH Laboratory Supplies; Searle G.D. & Co.*

9210 Melengestrol
5633-18-1 5859

$C_{23}H_{30}O_3$

17α-Hydroxy-6-methyl-16-methylenepregna-4,6-diene-3,20-dione.

Progestogen; antineoplastic agent. *BDH Laboratory Supplies.*

9211 Melengestrol Acetate
2919-66-6 5859
$C_{25}H_{32}O_4$
17α-Hydroxy-6-methyl-16-methylenepregna-4,6-diene-3,20-dione acetate.
MGA. Progestogen; antineoplastic agent. mp = 224-226°; $[\alpha]_D^{23}$ = -127° (c = 0.31 $CHCl_3$); λ_m = 287 nm (log ε 4.35 EtOH). *BDH Laboratory Supplies.*

9212 Methynodiol
23163-42-0
$C_{21}H_{34}O_2$
11β-Methyl-19-nor-17α-pregn-4-ene-20-yne 3β,17-diol.
Progestogen. *Searle G.D. & Co.*

9213 Methynodiol Diacetate
23163-51-1
$C_{25}H_{34}O_4$
11β-Methyl-19-nor-17α-pregn-4-ene-20-yne 3β,17-diol diacetate.
SC-19198. Progestogen. *Searle G.D. & Co.*

9214 Nomegestrol
58691-88-6
$C_{21}H_{28}O_3$
17-Hydroxy-6-methyl-19-norpregna-4,6-diene-3,20-dione.
TX-066 [as acetate]; Lutenyl [as acetate]; Uniplant [as acetate]; Thermex; Monaco. A synthetic progestin with a high affinity for the progesterone receptor used, in the acetate form, as a subdermal contraceptive. Nonandrogenic progestogen.

9215 Norethindrone
68-22-4 6790 200-681-6
$C_{20}H_{26}O_2$
17-Hydroxy-19-nor-17α-pregn-4-en-20-yn-3-one.
anhydrohydroxynorprogesterone; 19-norethisterone; norpregneninolone; NSC-9564; Conludag; Menzol; Micronor; Micronovum; Mini-Pe; mini-pill; Norcolut; Noriday; Norluten; Norlutin; Nor-QD; Primolut N; Utovlan; component of: Binovum, Brevicon, Brevinor, Conceplan, Modicon, Neocon 1/35, Norimin, Norinyl 1+35, Norquentiel, Ortho-Novum 1/35, Ortho-Novum 7/7/7, Ovcon, Ovysmen, Synphase, Tri-Norinyl, Trinovum, Norinyl-1, Ortho-Novin 1/50, Ortho-Novum 1/50. Progestogen. Used in combination with estrogens as an oral contraceptive. mp = 203-204°; $[\alpha]_D^{20}$ = -31.7° ($CHCl_3$); λ_m = 240 nm (log ε 4.24). *Syntex Labs. Inc.*

9216 Norethindrone Acetate
51-98-9 6790 200-132-0
$C_{22}H_{28}O_3$
17β-Hydroxy-19-norpregn-4-en-20-yn-3-one acetate.
Aygestin; Milligynon; Norlutate; Primolut-Nor; component of: Anovlar, Estrostep, Etalontin, Gynovlar, Loestrin, Minovlar, Norlestrin, Primosiston. Progestogen. Used in combination with estrogens as an oral contraceptive. mp = 161-162°; λ_m = 240 nm (ε 18690). *Bristol-Myers Squibb Pharm. R&D; Parke-Davis; Schering AG; Syntex Labs. Inc.; Wyeth-Ayerst Labs.*

9217 Norethynodrel
68-23-5 6791 200-682-1
$C_{20}H_{26}O_2$
17-Hydroxy-19-nor-17α-pregn-5(10)-en-20-yn-3-one.
SC-4642; NSC-15432; component of: Conovid E, Enavid, Enovid. Progestogen. Used in combination with estrogens as an oral contraceptive. mp = 169-170°; $[\alpha]_D$ = 108° (c = 1 $CHCl_3$). *Searle G.D. & Co.*

9218 Norgesterone
13563-60-5 6795
$C_{20}H_{28}O_2$
17α-Hydroxy-19-norpregna-5(10),20-diene-3-one.
norvinodrel; vinylestrenolone; component of: Vestalin. Progestogen. Used in combination with estrogens as an oral contraceptive. mp = 142-143°; $[\alpha]_D$ = 161° ($CHCl_3$). *Richter.*

9219 Norgestimate
35189-28-7 6796
$C_{23}H_{31}NO_3$
17α-(Acetyloxy)-13-ethyl-18,19-dinorpregn-4-en-20-yn-3-one 3-oxime.
D-138; ORF-10131; RWJ 10131; component of: Ortho-Cyclen, Cilest, Ortho-Tri-Cyclen, Ortrel, TriCilest. Progestogen. Used in combination with estrogens as an oral contraceptive. mp = 214-218°; $[\alpha]_D^{25}$ = 110°. *Ortho Pharm. Corp.*

9220 Norgestomet
25092-41-5
$C_{23}H_{32}O_4$
17-Hydroxy-11β-methyl-19-norpregn-4-ene-3,20-dione acetate.
SC-21009. Progestogen. *Searle G.D. & Co.*

9221 Norgestrel
6533-00-2 6797 229-433-5
$C_{21}H_{28}O_2$
(±)-13-Ethyl-17α-hydroxy-18,19-dinorpregn-4-en-20-yn-3-one.
WY-3707; Neogest; Ovrette; component of: Lo/Ovral, Ovral, Stediril. Progestogen. Used in combination with an estrogen (ethinyl estradiol) as an oral contraceptive. The levorotatory isomer is biologically active. mp = 205-207°; λ_m = 241 nm (ε 16700 EtOH). *Wyeth-Ayerst Labs.*

9222 (-)-Norgestrel
797-63-7 6797
$C_{21}H_{28}O_2$
(-)-13-Ethyl-17α-hydroxy-18,19-dinorpregn-4-en-20-yn-3-one.
levonorgestrel; D-norgestrel; dexnorgestrel; Wy-5104; Microlut; Microval; Norgeston; Norplant; component of: Levlen, Logynon, Microgynon, Nordette, Ovran, Ovranette, Tetragynon, Tri-Levlen, Trinordiol, Triphasil. Progestogen. Used in implants as an oral contraceptive. The levorotatory isomer is biologically active. mp = 235-237°; $[\alpha]_D^{20}$ = -32.4° (c = 0.496 $CHCl_3$); λ_m = 241 nm (ε 16770 MeOH). *Berlex Labs Inc.; Wyeth-Ayerst Labs.*

9223 (+)-Norgestrel

797-64-8 6797

$C_{21}H_{28}O_2$

(8α,9β,10α,13α,14β)-13-Ethyl-17-hydroxy-18,19-norpregn-4-en-20-yn-3-one.

dextronorgestrel. Progestogen. The levorotatory isomer is biologically active. mp = 238-242°; $[\alpha]_D^{25}$ = 40.7° ($CHCl_3$). *Wyeth-Ayerst Labs.*

9224 Norgestrienone

848-21-5 6798

$C_{20}H_{22}O_2$

17α-Hydroxy-19-norpregna-4,9,11-trien-20-yn-3-one.

Ogyline. Progestogen. mp = 169°; $[\alpha]_D^{2\cdot}$ = 63° (c = 0.5 EtOH); λ_m = 342, 238 nm (ε 29100, 5920); soluble in alcohols, Et_2O, Me_2CO, C_6H_6, $CHCl_3$; insoluble in H_2O. *Roussel-UCLAF.*

9225 Norvinisterone

6795-60-4 6814

$C_{20}H_{28}O_2$

17α-Hydroxy-19-norpregna-4,20-dien-3-one.

Nor-Progestelea. Progestogen. mp = 169-171°; $[\alpha]_D$ = 36°. *Searle G.D. & Co.*

9226 Onapristone

96346-61-1

$C_{29}H_{39}NO_3$

11β-[p-(Dimethylamino)phenyl]-17α-hydroxy-17-[3-hydroxypropyl]-13α-estra-4,9-diene-3-one.

Progesterone receptor antagonist. Antiprogestin.

9227 Oxogestone

3643-00-3

$C_{20}H_{30}O_2$

20β-Hydroxy-19-norpregna-4-en-3-one.

Progestogen. *Organon Inc.*

9228 Oxogestone Phenpropionate

16915-80-3

$C_{29}H_{38}O_3$

20β-Hydroxy-19-norpregna-4-en-3-one hydrocinnamate.

Progestogen. *Organon Inc.*

9229 Pentagestrone

7001-56-1 7251

$C_{26}H_{38}O_3$

3-(Cyclopentyloxy)-17-hydroxypregna-3,5-dien-20-one.

Progestogen. mp = 184.5-186.5°; $[\alpha]_D$ = -115°. *Vismara.*

9230 Pentagestrone Acetate

1178-60-5 7251

$C_{28}H_{40}O_4$

3-(Cyclopentyloxy)-17-hydroxypregna-3,5-dien-20-one acetate.

Gestovis. Progestogen. mp = 137-138°, 157-158°; $[\alpha]_D$ = -147° (dioxane). *Vismara.*

9231 Pregnenolone Succinate

4598-67-8 225-001-5

$C_{25}H_{36}O_5$

3β-Hydroxypregn-5-en-20-one hydrogen succinate.

Formula 405. Non-hormonal sterol derivative. A neurosteroid. Forms a progesterone analog on dehydrogenation. *Doak Pharmacal Co. Inc.*

9232 Progesterone

57-83-0 7956

$C_{21}H_{30}O_2$

Pregn-4-en-3,20-dione.

corpus luteum hormone; luteohormone; Corlutina; Corluvite; Cyclogest; Gestiron; Gestone; Lipo-Lutin; Lutocuclin M; Lutogyl; Lutromone; Progestasert; Progestin; Progestogel; Progestol; Progeston; Prolidon; Proluton; Syngesterone; Utrogestan. Progestogen. Active principle of the corpus luteum. When adminstered between days 5-25 of menstrual cycle, exhibits antiovulatory activity. [α form]: mp = 127-131°; d^{23} = 1.166; [β form]: mp = 121°; d^{20} = 1.171; $[\alpha]_D^{20}$ = 172° to 182° (c = 2 dioxane); λ_m = 240 nm; soluble in EtOH, Me_2CO, dioxane.

9233 Proligestone

23873-85-0 245-922-6

$C_{24}H_{34}O_4$

14,17-Dihydroxypregn-4-ene-3,20-dione cyclic acetal with propionaldehyde.

A progestin.

9234 Promegestone

34184-77-5 7969

$C_{22}H_{30}O_2$

17β-Methyl-17-(1-oxopropyl)estra-4,9-dien-3-one.

R-5020; RU-5020; Surgestone. Synthetic progestogen. Has high affinity for the progesterone receptor and no androgenic activity. Used as a radioligand for the progestogen receptor. mp = 512°; soluble in Me_2CO, C_6H_6; insoluble in H_2O; $[\alpha]_D^{20}$ = -262° (c = 0.5 EtOH); λ_m = 215, 305 nm ($E^{1\%}_{1\,cm}$ 202, 648 EtOH). *Roussel-UCLAF.*

9235 Quingestanol

10592-65-1

$C_{25}H_{34}O_2$

3-(Cyclopentyloxy)-19-nor-17α-pregna-3,5-dien-20-yn-17-ol.

Progestogen. *Parke-Davis.*

9236 Quingestanol Acetate

300-39-3

$C_{27}H_{36}O_3$

3-(Cyclopentyloxy)-19-nor-17α-pregna-3,5-dien-20-yn-17-ol acetate.

W-4540. Progestogen. *Parke-Davis.*

9237 Quingestrone

67-95-8

$C_{26}H_{38}O_2$

3-(Cyclopentyloxy)pregna-3,5-dien-20-one.

W-3399. Progestogen. *Parke-Davis.*

9238 Tigestol

896-71-9

$C_{20}H_{28}O$

19-Nor-17α-pregn-5(10)-en-20-yn-17-ol.

Progestogen.

9239 Trengestone
5192-84-7 9717
$C_{21}H_{25}ClO_2$
(9β,10α)-6-Chloropregna-1,4,6-triene-3,20-dione.
Ro-4-8347; Retroid. Progestogen. A retrosteroid. mp = 208-209°; λ_m = 229, 253, 302 nm (ε 11500, 10520, 10650). *Hoffmann-LaRoche Inc.*

9240 Trimegestone
74513-62-5
$C_{22}H_{30}O_3$
(S)-Lactoyl-17α-methylestra-4,9-dien-3-one.
RU-27987. Progestogen. *Roussel-UCLAF.*

Prolactin Inhibitors

9241 Bromocriptine
25614-03-3 1437 247-128-5
$C_{32}H_{40}BrN_5O_5$
2-Bromo-12'-hydroxy-2'-(1-methylethyl)-5'-(2-methylpropyl)-5'α-ergotaman-3',6',18-trione.
CB-154. Prolactin inhibitor. Antiparkinsonian. mp = 215-218° (dec); $[\alpha]_D^{20}$ = -195° (c = 1 CH_2Cl_2); LD_{50} (rbt orl) > 1000 mg/kg, (rbt iv) = 12 mg/kg. *Sandoz Pharm. Corp.*

9242 Bromocriptine Mesylate
22260-51-1 1437 244-881-1
$C_{33}H_{44}BrN_5O_8S$
2-Bromo-12'-hydroxy-2'-(1-methylethyl)-5'-(2-methylpropyl)-5'α-ergotaman-3',6',18-trione monomethanesulfonate (salt).
bromocriptine methanesulfonate; CB-154 mesylate; Parlodel; Bagren; Pravidel. Prolactin inhibitor used to treat Parkinson's disease. Dopamine receptor agonist. mp = 192-196° (dec); $[\alpha]_D^{20}$ = 95° (c = 1 MeOH/CH_2Cl_2); soluble in MeOH (91 g/100 ml), EtOH (2.3 g/100 ml), H_2O (0.08 g/100 ml), $CHCl_3$ (0.045 g/100ml)C_6H_6 (< 0.01 g/100 ml). *Sandoz Pharm. Corp.*

9243 Cabergoline
81409-90-7 1637
$C_{26}H_{37}N_5O_2$
1-[(6-Allylergolin-8β-yl)carbonyl]-1-[3-(dimethylamino)-propyl]-3-ethylurea.
FCE 21336; Dostinex. Antidyskinetic; antihyperprolactinemic. Dopamine receptor agonist; prolactin inhibitor. mp = 102-104°; LD_{50} (mus orl) > 400 mg/kg; [diphosphate]: mp = 153-155°. *Farmitalia Carlo Erba Ltd.*

9244 Lergotrile
36945-03-6
$C_{17}H_{18}ClN_3$
2-Chloro-6-methylergoline-8β-acetonitrile.
Prolactin inhibitor. *Eli Lilly & Co.*

9245 Lergotrile Mesylate
51473-23-5 257-225-4
$C_{18}H_{22}ClN_3O_3S$
2-Chloro-6-methylergoline-8β-acetonitrile monomethanesulfonate.
Prolactin inhibitor. *Eli Lilly & Co.*

9246 Lisuride
18016-80-3 5541 241-925-1
$C_{20}H_{26}N_4O$
3-(9,10-Didehydro-6-methylergolin-8α-yl)-1,1-diethylurea.
methylergol carbamide; lysuride. Antimigraine; prolactin inhibitor; antiparkinsonian. Dopamine D_2 receptor agonist. mp = 186°; $[\alpha]_D^{20}$ = 313° (c = 0.60 C_5H_5N).

9247 Lisuride Maleate
19875-60-6 5541 243-387-3
$C_{24}H_{30}N_4O_5$
3-(9,10-Didehydro-6-methylergolin-8α-yl)-1,1-diethylurea maleate.
Apodel; Cuvalit; Dopergin; Eunal; Lysenyl; Revanil. Antimigraine; prolactin inhibitor; antiparkinsonian. Dopamine D_2 receptor agonist mp = 200° (dec); $[\alpha]_D^{20}$ = 288° (c = 0.5 MeOH); λ_m = 313 nm (MeOH); LD_{50} (mus iv) = 14.4 mg/kg.

9248 Metergoline
17692-51-2 5999 241-686-3
$C_{25}H_{29}N_3O_2$
(+)-N-(Carboxy)-1-methyl-9,10-dihydrolysergamine benzyl ester.
methergoline; FI-6337; MCE; Liserdol; Contralac. Seratonin 5-hydroxytryptamine receptor antagonist; prolactin inhibitor. mp = 146-149°; $[\alpha]_D^{28}$ = -7±2°; λ_m = 291 nm ($E_{1cm}^{1\%}$ 165; soluble in C_5H_5N, alcohol, Me_2OH $CHCl_3$; nearly insoluble in C_6H_6, H_2O, Et_2O; LD_{50} (mus orl) = 430 mg/kg. *Farmitalia Carlo Erba Ltd.*

9249 Quinagolide
87056-78-8 8226
$C_{20}H_{33}N_3O_3S$
(±)-N,N-Diethyl-N'-[(3R*,4aR*,10aS*)-1,2,3,4,4a,5,10,10a-octahydro-6-hydroxy-1-propylbenzo[g]quinolin-3-yl]sulfamide.
Prolactin inhibitor. Dopamine D_2 receptor agonist. mp = 122.5-124°. *Sandoz Pharm. Corp.*

9250 Quinagolide Hydrochloride
94424-50-7 8226
$C_{20}H_{34}ClN_3O_3S$
(±)-N,N-Diethyl-N'-[(3R*,4aR*,10aS*)-1,2,3,4,4a,5,10,10a-octahydro-6-hydroxy-1-propylbenzo[g]quinolin-3-yl]sulfamide hydrochloride.
CV-205-502; SDZ-205-502; Norprolac. Prolactin inhibitor. Dopamine D_2 receptor agonist. mp = 234-236°. *Sandoz Pharm. Corp.*

9251 Terguride
37686-84-3 9308 253-624-2
$C_{20}H_{28}N_4O$
1,1-Diethyl-3-(6-methylergolin-8α-yl)urea.
TDHL. Prolactin inhibitor. Antiparkinsonian. Ergot derivative with dopamine agonist and antagonist activity. mp = 203-204° (dec), 205-207° (dec); $[\alpha]_D^{20}$ = 30° (c = 1 C_5H_5N); 29° (c = 0.2 C_5H_5N); λ_m = 292, 281, 224 nm (log ε 3.72, 3.81, 4.42 C_5H_5N); insoluble in H_2O. *SPOFA.*

9252 Terguride Hydrogen Maleate
37686-85-4 9308 253-625-8
$C_{24}H_{32}N_4O_5$
1,1-Diethyl-3-(6-methylergolin-8α-yl)urea maleate.
SH-406; VUFB-6638; ZK-31224; Dironyl; Mysalfon. Prolactin inhibitor. Antiparkinsonian. Ergot derivative with dopamine agonist and antagonist activity. mp = 190-191°; [monohydrate]: mp = 150-153°; $[\alpha]_D^{20}$ = -15.0° (c = 0.1 H_2O); soluble in H_2O (1.26 g/100 ml). *SPOFA.*

Prostaglandins

9253 Arbaprostil
55028-70-1 808
$C_{21}H_{34}O_5$
(E,Z)-(1R,2R,3R)-7-[3-Hydroxy-2-[(3R)-(3-hydroxy-3-methyl-1-octenyl)]-5-oxocyclopentyl]-5-heptenoic acid.
15-methylprostaglandin E_2; U-42842; Arbacet. Gastric antisecretory agent. Antiulcerative. Prostaglandin. [(15S)-methyl ester ($C_{22}H_{36}O_5$)]: $[\alpha]_D$ = -79° (c = 1.3 $CHCl_3$); λ_m = 278 nm (ε 25250 EtOH); [(15R)-methyl ester ($C_{22}H_{36}O_5$)]: $[\alpha]_D$ = -74° (c = 1.0 $CHCl_3$); λ_m = 278 nm (ε 25200 EtOH). *Pharmacia & Upjohn.*

9254 Ataprost
83997-19-7
$C_{21}H_{32}O_4$
(+)-(2E,3aS,4R,5R,6aS)-4-[(1E,3S)-3-Cyclopentyl-3-hydroxypropenyl]-3,3a,4,5,6,6a-hexahydro-5-hydroxy-$\Delta^{2(1H),\delta}$-pentalenevaleric acid.
Prostaglandin with hepatoprotectant properties.

9255 Beraprost
88430-50-6 1190
$C_{24}H_{30}O_5$
(±)-(1R,2R,3aS,8bS)-2,3,3a,8b-Tetrahydro-2-hydroxy-1-[(e)-(3S,4RS)-3-hydroxy-4-methyl-1-octen-6-ynyl]-1H-cyclopenta[b]benzofuran-5-butyric acid.
ML-1229; MDL-201229. Prostaglandin. Platelet aggregation inhibitor. *Marion Merrell Dow Inc.*

9256 Beraprost Sodium Salt
88475-69-8 1190
$C_{24}H_{29}NaO_5$
Sodium (±)-(1R,2R,3aS,8bS)-2,3,3a,8b-tetrahydro-2-hydroxy-1-[(e)-(3S,4RS)-3-hydroxy-4-methyl-1-octen-6-ynyl]-1H-cyclopenta[b]benzofuran-5-butyrate.
TRK-100; ML-1129; MDL-201129; Dorner; Procyclin. Prostaglandin. Platelet aggregation inhibitor. *Marion Merrell Dow Inc.*

9257 Carboprost
35700-23-3 1871
$C_{21}H_{36}O_5$
(E,Z)-(1R,2R,3R,5S)-7-[3,5-Dihydroxy-2-[(3S)-(3-hydroxy-3-methyl-1-octenyl)]cyclopentyl]-5-heptenoic acid.
U-32921; (15S)-15-methylprostaglandin $F_{2\alpha}$. Prostaglandin. Oxytocic. *Pharmacia & Upjohn.*

9258 Carboprost Methyl Ester
35700-21-1 1871
$C_{22}H_{38}O_5$
Methyl (E,Z)-(1R,2R,3R,5S)-7-[3,5-dihydroxy-2-[(3S)-3-hydroxy-3-methyl-1-octenyl)]cyclopentyl]-5-heptenoate.
U-36384; carboprost methyl. Prostaglandin. Oxytocic. mp = 55-56°; $[\alpha]_D$ = 24° (c = 0.81 EtOH). *Pharmacia & Upjohn.*

9259 Carboprost Tromethamine Salt
58551-69-2 1871
$C_{25}H_{47}NO_8$
(E,Z)-(1R,2R,3R,5S)-7-[3,5-Dihydroxy-2-[(3S)-(3-hydroxy-3-methyl-1-octenyl)]cyclopentyl]-5-heptenoic acid with 2-amino-2-(hydroxymethyl)-1,3-propanediol.
carboprost trometamol; U-32921E;U-36384; Hemabate; Prostin/15M. Prostaglandin. Oxytocic. *Pharmacia & Upjohn.*

9260 Cloprostenol
40665-92-7 2461 255-028-8
$C_{22}H_{29}ClO_6$
(±)-(Z)-7-[(1R*,2R*,3R*,5S*)2-[(E)-(3R*)-4-(m-Chlorophenoxy)-3-hydroxy-1-butenyl]-3,5-dihydroxycyclopentyl]-5-heptenoic acid.
Prostaglandin. Used in treatment of infertility in farm animals and in synchronization of estrus. *Bayer Corp., Ag. Div., Animal Health; ICI.*

9261 Cloprostenol Sodium
55028-72-3 2461 259-439-3
$C_{22}H_{28}ClNaO_6$
Sodium (±)-(Z)-7-[(1R*,2R*,3R*,5S*)2-[(E)-(3R*)-4-(m-chlorophenoxy)-3-hydroxy-1-butenyl]-3,5-dihydroxycyclopentyl]-5-heptenoate.
Estrumate; ICI-80996; Planate. Prostaglandin. Used to treat infertility in farm animals and in synchronization of estrus. *Bayer Corp., Ag. Div., Animal Health; ICI.*

9262 Dinoprost Tromethamine
38562-01-5 8065 254-002-3
$C_{24}H_{45}NO_8$
(E,Z)-(1R,2R,3R,5S)-7-[3,5-Dihydroxy-2-[(3S)-(3-hydroxy-1-octenyl)]cyclopentyl]-5-heptenoic acid compound with 2-amino-2-(hydroxyethyl)-1,3-propanediol.
Prostaglandin $F_{2\alpha}$ tromethamine; $PGF_{2\alpha}$ THAM; U-14583E. Prostaglandin. Oxytocic. mp = 25-35°; $[\alpha]_D^{25}$ = 23.5° (c = 1 THF); freely soluble in MeOH, EtOH, EtOAc, $CHCl_3$, slightly soluble in H_2O; LD_{50} (rbt iv, im) = 2.5-5.0 mg/kg; [tromethamine salt ($C_{24}H_{45}NO_8$)]: mp = 100-101°, soluble in H_2O (> 200 mg/ml). *Pharmacia & Upjohn.*

9263 Enprostil
73121-56-9 3629
$C_{23}H_{28}O_6$
Methyl 7-[(1R*,2R*,3R*)-3-hydroxy-2-[(E)-(3R*)-3-hydroxy-4-phenoxy-1-butenyl]-5-oxocyclopentyl]-4,5-heptadienoate.
RS-84135; Camleed; Fundyl; Gardrin; Gardrine; Syngard. Antiulcerative. Gastric antisecretory. Prostaglandin. mp < 46°; slightly soluble in H_2O; soluble in EtOH, propylene glycol, propylene carbonate; λ_m = 220, 265, 271, 277 nm (log ε 4.01, 3.14, 3.24, 3.16 MeOH). *Syntex Labs. Inc.*

9264 Fluprostenol
40666-16-8 4231 255-029-3
$C_{23}H_{29}F_3O_6$
(±)-(Z)-7-[(1R*,2R*,3R*,5S*)-3,5-Dihydroxy-2-[(E)-(3R*)-3-hydroxy-4-[(α,α,α-trifluoro-m-tolyl)oxy]-1-butenyl]cyclopentyl]-5-heptenoic acid.
Equimate; ICI-81008. Prostaglandin. Used to treat infertility in mares. *Bayer Corp., Ag. Div., Animal Health; ICI.*

9265 Fluprostenol Sodium
55028-71-2 4231 259-438-8
$C_{23}H_{28}F_3NaO_6$
Sodium (±)-(Z)-7-[(1R*,2R*,3R*,5S*)-3,5-dihydroxy-2-[(E)-(3R*)-3-hydroxy-4-[(α,α,α-trifluoro-m-tolyl)oxy]-1-butenyl]cyclopentyl]-5-heptenoate.
ICI-81008. Prostaglandin. Used in treatment of infertility in mares. *Bayer Corp., Ag. Div., Animal Health; ICI.*

9266 Gemeprost
64318-79-2 4393 264-775-9
$C_{23}H_{38}O_5$
Methyl (E)-7-[(1R,2R,3R)-3-hydroxy-2-[(E)-(3R)-3-hydroxy-4,4-dimethyl-1-octenyl]-5-oxocyclopentyl]-2-heptenoate.
16,16-dimethyl-trans-Δ^2-PGE_1 methyl ester; ONO-802; SC-37681; Cergem; Cervagem; Cervageme; Preglandin. Abortifacient; oxytocic; prostaglandin. Analog of prostaglandin E_1. Used in termination of first trimester pregnancy. *Ono Pharm.; Searle G.D. & Co.*

9267 Latanoprost
130209-82-4 5387
$C_{26}H_{40}O_5$
Isopropyl (Z)-7-[(1R,2R,3R,5S)-3,5-dihydroxy-2-[(3R)-3-hydroxy-5-phenylpentyl]-cyclopentyl]-6-heptenoate.
Xalatan; PHXA41; XA41. Prostaglandin. Antiglaucoma agent. $[\alpha]_D^{20}$ = 31.57° (c = 0.91 CH_3CN). *Kabi Pharmacia Diagnostics; Pharmacia & Upjohn.*

9268 Limaprost
74397-12-9 5514
$C_{22}H_{36}O_5$
(E)-7-[(1R,2R,3R)-3-Hydroxy-2-[(E)-(3S,5S)-3-hydroxy-5-methyl-1-nonenyl]-5-oxocyclopentyl]-2-heptenoic acid.
Prostaglandin. Antianginal. mp = 97-100°. *Ono Pharm.; Warner-Lambert.*

9269 Luprostiol
67110-79-6 5368
$C_{21}H_{29}ClO_6S$
(±)-(Z)-7-[(1R*,2S*,3S*,5R*)-2-[[(2R*)-3-(m-Chlorophenoxy)-2-hydroxypropyl]thio]-3,5-dihydroxycyclopentyl]-5-heptenoic acid.
prostianol;EMD-34946; Pronilin; Prosolvin; Reprodin. Analog of prostaglandin $F_{2\alpha}$ used in veterinary medicine for induction of luteolysis. Synthetic prostaglandin; luteolytic (veterinary).

9270 Meteneprost
61263-35-2
$C_{23}H_{38}O_4$
(Z)-7-[(1R,2R,3R)-3-Hydroxy-2-[(E)-(3R)-3-hydroxy-4,4-dimethyl-1-octenyl]-5-methylenecyclopentyl]-5-heptenoic acid.
U-46785. Oxytocic. Prostaglandin. *Pharmacia & Upjohn.*

9271 Misoprostol
59122-46-2 6297
$C_{22}H_{38}O_5$
(±)-Methyl (1R,2R,3R)-3-hydroxy-2-[(E)-(4RS)-4-hydroxy-4-methyl-1-octenyl]-5-oxocyclopentaneheptanoate.
Cytotec; SC-29333. Antiulcerative. Cytoprotective prostaglandin PGE_1 analog. Soluble in H_2O; LD_{50} (rat ip) = 40-62 mg/kg, (rat orl) = 81-100 mg/kg, (mus ip) = 70-160 mg/kg, (mus orl) = 27-138 mg/kg. *Searle G.D. & Co.*

9272 Naxaprostene
87269-59-8
$C_{25}H_{33}O_4$
α-[(2E,3aS,4R,6aS)-4-[(1E,3S)-3-Cyclohexyl-3-hydroxypropenyl]hexahydry-5-hydroxy-2(1H)-pentalenylidene]-m-toluic acid.
Prostacyclin analog with antithrombotic and antihypertensive activities. Prostaglandin.

9273 Nileprost
71097-83-1
$C_{22}H_{32}NO_5$
(E)-(3aR,4R,5R,6aS)-δ-Cyano-3,3a,4,5,6,6a-hexahydro-5-hydroxy-4-[(E)-(3S,4RS)-3-hydroxy-4-methyl-1-octenyl]-2H-cyclopenta[b]furan-$\Delta^{2,\delta}$-valeric acid.
A 5-cyano-16-methyl-prostacyclin. Prostaglandin antagonist.

9274 Ornoprostil
70667-26-4 7003
$C_{23}H_{38}O_6$
Methyl (-)-(1R,2R,3R)-3-hydroxy-2-[(e)-(3S,5S)-3-hydroxy-5-methyl-1-nonenyl]-ε,5dioxocyclopentaneheptenoate.
ronoprost; ONO-1308; OU-1308; Alloca; Ronok. Prostaglandin. Antiulcerative. *Ono Pharm.*

9275 Prostacyclin
35121-78-9 8061
$C_{20}H_{32}O_5$
(Z)-(3aR,4R,5R,6aS)-Hexahydro-5-hydroxy-4-[(E)-(3S)-3-hydroxy-1-octenyl]-2H-cyclopenta[b]furan-$\Delta^{2,\delta}$-valeric acid.
Epoprostenol; PGI_2; Prostaglandin X; PGX; U-53217. Prostaglandin. Platelet aggregation inhibitor. Soluble, unstable in H_2O; loses antiaggregation activity within 30 seconds at 100° or 10 minutes at 37°. *Pharmacia & Upjohn.*

9276 Prostacyclin Sodium Salt
61849-14-7 8061 263-273-7
$C_{20}H_{31}NaO_5$
Sodium (Z)-(3aR,4R,5R,6aS)-hexahydro-5-hydroxy-4-[(E)-(3S)-3-hydroxy-1-octenyl]-2H-cyclopenta[b]furan-$\Delta^{2,\delta}$-valerate.
Flolan. Prostaglandin. Platelet aggregation inhibitor. Stable for 2 hours at -30° if kept dry. *Glaxo Wellcome Inc.*

9277 Prostaglandin E_1
745-65-3 8063 212-017-2
$C_{20}H_{34}O_5$
(1R,2R,3R)-3-Hydroxy-2-[(E)-(3S)-3-hydroxy-1-octenyl]-5-oxocyclopentaneheptanoic acid.
Alprostadil; Caverject; Prostin VR Pediatric; PGE_2; Palux;

Prostandin; Liple; Minprog; Prostivas. Prostaglandin. Peripheral vasodilator. mp = 115-116°; $[\alpha]_{578}$ = -61.6° (c = 0.56 THF). *Pharmacia & Upjohn.*

9278 Prostaglandin E_2
363-24-6 8064 206-656-6
$C_{20}H_{32}O_5$
(5Z,11α,13E,15S)-11,15-Dihydroxy-9-oxoprosta-5,13-dien-1-oic acid.
dinoprostone; PGE_2; U-12062; Cerviprost; Dinoprost; Enzaprost F; Glandin; Minprositin E_2; Prepidil; Propess; Prostarmon-E; Prostin E_2. Abortifacient; oxytocic. Prostaglandin. mp = 66-68°; $[\alpha]_D^{26}$ = -61° (c = 1 THF). *Pharmacia & Upjohn.*

9279 Prostaglandin $F_{2\alpha}$
551-11-1 8065
$C_{20}H_{34}O_5$
(E,Z)-(1R,2R,3R,5S)-7-[3,5-Dihydroxy-2-[(3S)-(3-hydroxy-1-octenyl)]cyclopentyl]-5-heptenoic acid.
dinoprost; $PGF_{2\alpha}$; U-14583; Enzaprost F; Glandin; Prostarmon F. Abortifacient; oxytocic. Prostaglandin. mp = 25-35°; $[\alpha]_D^{25}$ = 23.5° (c = 1 THF); freely soluble in MeOH, EtOH, EtOAc, $CHCl_3$; slightly soluble in H_2O; LD_{50} (rbt iv) = 2.5-5.0 mg/kg, (rbt im) = 2.5-5.0 mg/kg. *Pharmacia & Upjohn.*

9280 Prostalene
54120-61-5 8066 258-984-4
$C_{22}H_{36}O_5$
(±)-Methyl 7-[(1R*,2R*,3R*,5S*)-3,5-dihydroxy-2-[((E)-3-hydroxy-3-methyl-1-octenyl]cyclopentyl]-4,5-heptadienoate.
Synchrocept; RS-9390. Prostaglandin. Luteolytic. *Syntex Intl. Ltd.*

9281 Rioprostil
77287-05-9 8395
$C_{21}H_{38}O_4$
(2R,3R,4R)-4-Hydroxy-2-(7-hydroxyheptyl)-3-[(E)-(4RS)-(4-hydroxy-4-methyl-1-octenyl)]cyclopentanone.
TR-4698; ORF-15927; RWJ-15927; Bay o 6893; Rostil. Antiulcerative. Prostaglandin. Gastric antisecretory agent. $[\alpha]_D$ = -58.6° (c = 1 $CHCl_3$). *Bayer Corp. Pharm. Div.; Miles Inc.*

9282 Rosaprostol
56695-65-9 8419 260-341-8
$C_{18}H_{34}O_3$
(1RS,2SR,5RS)-2-Hexyl-5-hydrocyclopentaneheptanoic acid.
2-hexyl-5-hydroxycyclopentaneheptanoic acid; C-83; IBI-C83; Rosal. Gastric antisecretory; cytoprotectant. Antiulcerative. Prostaglandin. Oil; [sodium salt ($C_{18}H_{33}N_O3$)]: LD_{50} (mus orl) = 3000 mg/kg, (rat orl) > 5000 mg/kg. *Ist. Biochim.*

9283 Sulprostone
60325-46-4 9165 262-173-0
$C_{23}H_{31}NO_7S$
(Z)-7-[(1R,2R,3R)-3-Hydroxy-2-[(E)-(3R)-3-hydroxy-4-phenoxy-1-butenyl)]-5-oxocyclopentyl]-N-(methylsulfonyl)-5-heptenamide.
CP-34089; SHB-286; ZK-57671; Nalador. Prostaglandin. Abortifacient. Analog of prostaglandin E_2. Uterine stimulant. Colorless oil. *Pfizer Intl.; Schering AG.*

9284 Tilsuprost
80225-28-1
$C_{20}H_{33}NO_4S$
Methyl (±)-4-[[(3aR*,4R*,5R*,6aS*)-3,3a,4,5,6,6a-hexahydro-5-hydroxy-4-[(E)-(3S*)-3-hydroxy-1-octenyl]cyclopenta[b]pyrrol-2-yl]thio]-butyrate.
Prostacyclin analog.

9285 Trimoprostil
69900-72-7 9853
$C_{23}H_{38}O_4$
(Z)-7-[(1R,2R,3R)-2-[(E)-(3R)-3-Hydroxy-4,4-dimethyl-1-octenyl]-3-methyl-5-oxocyclopentyl]-5-heptenoic acid.
Ro-21-6937/000; TM-PGE_2; Ulstar. Antiulcerative. Synthetic prostaglandin E_2 analog with gastric antisecretory activity. $[\alpha]_D$ = -51.54° (c = 1 $CHCl_3$); LD_{50} mus orl) = 41 mg/kg, (mis ip) = 70 mg/kg, (mus sc) = 68 mg/kg, (rat orl) = 23 mg/kg, (rat ip) = 21 mg/kg, (rat sc) = 29 mg/kg. *Hoffmann-LaRoche Inc.*

9286 Unoprostone
120373-36-6 9984
$C_{22}H_{38}O_5$
(+)-(Z)-7-[(1R,2R,3R,5S)-3,5-Dihydroxy-2-(3-oxodecyl)cyclopentyl]-5-heptenoic acid.
13,14-dihydro-15-keto-20-ethyl-$PGF_{2\alpha}$; Prostaglandin. Antiglaucoma agent; used in treatent of intraocular hypertension. *Ueno Fine Chem. Industry Ltd.*

9287 Unoprostone Isopropyl Ester
120373-24-2 9984
$C_{25}H_{44}O_5$
Isopropyl (+)-(Z)-7-[(1R,2R,3R,5S)-3,5-dihydroxy-2-(3-oxodecyl)cyclopentyl]-5-heptenoate.
UF-021; Rescula. Prostaglandin. Antiglaucoma agent; used in treatent of intraocular hypertension. *Ueno Fine Chem. Industry Ltd.*

Protease Inhibitors

9288 Aprotinin
9087-70-1 796 232-994-9
$C_{284}H_{432}N_{84}O_{79}S_7$
Pancreatic basic trypsin inhibitor.
pancreatic trypsin inhibitor; Bayer A 128; Kiker 52G; RP-9921; Antagosan; Antikrein; Fosten; Iniprol; Kir Richter; Onquinin; Repulson; Trasuylol; Trazinin; Zymofren. Protease Inhibitor. An inhibitor of kallikrein which also inhibits chymotrypsin, trypsin, plasmin and other intracellular proteases. Prepared from bovine lung. Used mainly as a proteolytic inhibitor in radioimmunoassays of polypeptide hormones. λ_m (pH 5.9) = 280 nm; stable in neutral or acid at high temperature; partially denatured on treatment with 8M urea; LD_{50} (mus iv) = 2.5 x 10^6 kallikrein inhibitor units/kg. *Bayer AG; Miles Inc.; Riker Labs.*

9289 Camostat
59721-28-7 1775
$C_{20}H_{22}N_4O_5$
N,N-Dimethylcarbamoylmethyl-p-(p-guanidinobenzoyloxy)phenylacetate.
Protease Inhibitor. Orally active, non-peptide proteolytic enzyme inhibitor. Has anti-trypsin and anti-plasmin activities. Related structurally to gabexate. *Ono Pharm.*

9290 Camostat Monomethanesulfonate
59721-29-8 1775
$C_{20}H_{22}N_4O_5.CH_3SO_3H$
N,N-Dimethylcarbamoylmethyl-p-(p-guanidinobenzoyloxy)phenylacetate monomethanesulfonate.
camostat mesylate; FOY-305; Foipan. Protease Inhibitor. Orally active, non-peptide proteolytic enzyme inhibitor. Has anti-trypsin and anti-plasmin activities. mp = 150-155°; soluble in H_20. *Ono Pharm.*

9291 Gabexate
39492-01-8 4344
$C_{16}H_{23}N_3O_4$
p-Hydroxybenzoic acid ethyl ester 6-guanidinohexanoate.
p-carbethoxyphenyl ε-guanidineocaproate. Protease Inhibitor. Orally active, non-peptide proteolytic enzyme inhibitor. Also inhibits the hydrolytic effects of thrombin, plasmin, kallikrein, and trypsin (but not chymotrypsin). *Ono Pharm.*

9292 Gabexate Monomethanesulfonate
56974-61-9 4344
$C_{16}H_{23}N_3O_4.CH_3SO_3H$
p-Hydroxybenzoic acid ethyl ester 6-guanidinohexanoate.
gabexate mesylate; FOY; Megacert. Protease Inhibitor. Orally active, non-peptide proteolytic enzyme inhibitor. Also inhibits the hydrolytic effects of thrombin, plasmin, kallikrein, and trypsin (but not chymotrypsin). Soluble in H_2O, EtOH, $CHCl_3$; slightly soluble in Me_2CO; nearly insoluble in Et_2O; LD_{50} (mus orl) = 8000 mg/kg, (mus sc) = 4700 mg/kg, (mus iv) = 25 mg/kg. *Ono Pharm.*

9293 Nafamostat
81525-10-2 6436
$C_{19}H_{17}N_5O_2$
6-Amidino-2-naphthyl-4-guanidinobenzoate.
nafamstat. Protease Inhibitor. Non-peptide enzyme inhibitor with inhibitory effects on trypsin, thrombin, kallikrein, plasmin, and complement-mediated hemolysis. *Torii Pharm. Co. Ltd.*

9294 Nafamostat Dimethanesulfonate
82956-11-4 6436
$C_{19}H_{17}N_5O_2.2CH_3SO_3H$
6-Amidino-2-naphthyl-4-guanidinobenzoate dimethanesulfonate.
nafamastat mesylate; FUT-175; Futhan. Used as an anticoagulant. Protease Inhibitor. Non-peptide enzyme inhibitor with inhibitory effects on trypsin, thrombin, kallikrein, plasmin, and complement-mediated hemolysis. mp = 217-220°, 260° (dec). *Torii Pharm. Co. Ltd.*

9295 Urinastatin
80449-31-6 10021
Bikunin trypsin inhibitor.
mingin; urinary trypsin inhibitor; UTI; ulinastatin; Miraclid. Protease Inhibitor. Acid-stable glycoprotein. Single polypeptide chain of 147 amino acid residues.

Pulmonary Surfactants

9296 Beractant
108778-82-1 1189 200-567-6
A modified bovine lung solvent extract.
Survanta; A-60386-X; Surfactant TA; A-60386X; Surfacten. Pulmonary surfactant. Contains primarily phospholipids, modified by the addition of dipalmitoylphosphatidylcholine, palmitic acid and tripalmitin. *Ross Products.*

9297 Calfactant
183325-78-2
Infasurf. Pulmonary surfactant, used in treatment of neonatal respiratory distress syndrome. An unmodifed calf lung lavage extract containing mostly phospholipds and surfactant-specific proteins. *Forest Pharm. Inc.*

9298 Colfosceril Palmitate
63-89-8 2540
$C_{40}H_{80}NO_8P$
Choline hydroxide dihydrogen phosphate inner salt ester with L-1,2-dipalmitin.
129Y83; component of: Exosurf. Pulmonary surfactant. Also used as a diagnostic aid in connection with fetal lung maturity. mp = 234-235°; [α]23:kls$_D$ = 6.6° (c = 4.2 in 1:1 $CHCl_3$/MeOH); readily soluble in $CHCl_3$, hot dioxane, hot diisobutyl ketone; soluble in EtOH (1.5 g/100 ml), Et_2O (0.02 g/100 ml), Me_2CO (0.02 g/100 ml), C_5H_5N (1.1 g/100 ml), AcOH (4.0 g/100 ml), MeOH (1.4 g/100 ml). *Glaxo Wellcome Inc.*

9299 Exosurf®
99732-49-7 3960
(R)-4-Hydroxy-N,N,N-trimethyl-10-oxo-7-[(1-oxo-hexadecl)oxy]-3,5,9-trioxa-4-phosphapentacosn-1-aminium hydroxide inner salt, 4-oxide mixture with formaldehyde polymer with oxirane and 4-(1,1,3,3)-tetramethylbutyl)phenol and 1-hexadecanol.
Surfexo. Pulmonary surfactant. A combination of colfosceril palmitate and tyloxapol. *Glaxo Wellcome Inc.*

9300 Infasurf
5002
A $CHCl_3$/MeOH extract of calf lung lavage fluid.
CLSE; CLL. Pulmonary surfactant. Also used in the treatment of respiratory distress syndrome. Consists of 35 mg/ml phospholipid, 55-70% of which is dipalmitoylphosphatidylcholine. Also contains low molecular weight surfactant proteins, fatty acids and neutral lipids.

9301 Neltenexine
99453-84-6
$C_{18}H_{20}Br_2N_2O_2S$
4',6'-Dibromo-α-[(trans-4-hydroxycyclohexyl)amino]-2-thiophene-carboxy-o-toluidide.

A compound with surfactant production activity which may be useful in pulmonary emphysema.

9302 Poractant Alfa

129069-19-8 7754

An extract of porcine lung.

PLS; Curosurf. Pulmonary surfactant. Also used in the treatment of respiratory distress syndrome. Containing > 90% lipids, about 1% of the hydrophobic proteins SP-B and SP-C. Contains 9.9% sphingomyelin, which distinguishes it from other pulmonary surfactants.

9303 Pumactant

8126

ALEC; artificial lung expanding compound. Pulmonary surfactant. A mixture of 1,2-dipalmitoyl-sn-glycero(3)-phosphocholine (70%) and 2-oleoyl-1-palmitoyl-cn-glycero(3)phospho(1)-sn-glycerol. Crystalline powder.

9304 Tyloxapol

25301-02-4 9964

p-(1,1,3,3-Tetramethylbutyl)phenol polymer with ethylene oxide and formaldehyde.

oxyethylated tertiary octylphenol formaldehyde polymer; tyloxypal; Alevaire; Superinone; Triton A-20; Triton WR-1339; component of: Enuclene, Exosurf. A nonionic detergent with surface-tension-reducing properties, used in pulmonary surfactants. Used as a mucolytic. Freely soluble in H_2O; soluble in C_6H_6; C_7H_8, $CHCl_3$, CCl_4, CS_2, AcOH. *Alcon Labs; Glaxo Wellcome Inc.; Rohm & Haas Co.; Sterling Winthrop Inc.*

Replenishers

9305 Betasine

3734-24-5 1228

$C_9H_9I_2NO_3$

β-Amino-4-hydroxy-3,5-diiodobenzenepropanoic acid.

betasinum; betazine. Iodine replenisher. mp = 178-179° (dec); sparingly soluble in H_2O, insoluble in organic solvents; [ammonium salt]: mp = 151-152°; slightly soluble in cold H_2O.

9306 Calcium Carbonate

471-34-1 1697 207-439-9

$CaCO_3$

Carbonic acid calcium salt (1:1).

Cal-Sup; Children's Mylanta Upset Stomach Relief; Chooz; Mylanta Soothing Lozenges; Calcit; Calcichew; Calcidia; Citrical; component of: Bufferin Arthritis Strength, Bufferin Extra Strength, Bufferin Regular, Calcitrel, Camalox, Di-Gel Tablets, Mylanta Gelcaps, Mylanta Tablets, Titralac, Tylenol Headache Plus. Calcium replenisher; antacid. mp = 825°, 1339° (102.5 atm); $d_{25.2}$ = 2.711; insoluble in H_2O. *3M Pharm.; Bristol-Myers Squibb HIV Products; Johnson & Johnson-Merck Consumer Pharm. ; McNeil Consumer Products Co.; Rhône-Poulenc Rorer Pharm. Inc.; Schering-Plough Pharm.; Sterling Winthrop Inc.*

9307 Calcium Chloride

10043-52-4 1699 233-140-8

$CaCl_2$

Hydrochloric acid calcium salt (2:1).

CalPlus; Intergravin-orales. Calcium replenisher. mp = 772°; bp > 1600°; d_4^{15} = 2.152; freely soluble in H_2O (exothermic), EtOH; LD_{50} (mus iv) = 42.2 mg/kg. *Astra Chem. Ltd.; Mallinckrodt Inc.*

9308 Calcium Glubionate

12569-38-9

$C_{18}H_{32}CaO_{19}.H_2O$

(D-Gluconato)(lactobionato)calcium monohydrate.

Neo-Calglucon. Calcium replenisher. *Sandoz Pharm. Corp.*

9309 Calcium Gluceptate

29039-00-7 4463 249-383-8

$C_{14}H_{26}CaO_{16}$

Calcium glucoheptonate (1:2).

Calcium replenisher. *Pfanstiehl Labs Inc.*

9310 Calcium Gluconate

299-28-5 1712 206-075-8

$C_{12}H_{22}CaO_{14}$

D-Gluconic acid calcium salt (2:1).

Calciofon; Calglucon; Ebucin; Glucal; Glucobiogen. Calcium replenisher. $[\alpha]_D^{20}$ = 6°; soluble in H_2O (3.3 g/100 ml at 20°, 20 g/100 ml at 100°), insoluble in organic solvents. *Astra Chem. Ltd.; Mission Pharmacal Co.; Parke-Davis; Sandoz Pharm. Corp.*

9311 Calcium Glycerophosphate

27214-00-2 1713 248-328-5

$C_3H_7CaO_6P$

1,2,3-Propanetriol mono(dihydrogen phosphate) calcium salt (1:1).

Neurosin; component of Claphosan. Calcium replenisher. Dec > 170°; soluble in H_2O (2 g/100 ml), almost insoluble in EtOH, boiling H_2O. *Glenwood Inc.*

9312 Calcium Hypophosphite

7789-79-9 1718 232-190-8

$CaH_4O_4P_2$

Calcium replenisher. Soluble in H_2O, slightly soluble in glycerol, insoluble in EtOH.

9313 Calcium Iodostearate

1301-16-2 1722

$C_{36}H_{68}CaI_2O_4$

2-Iodooctadecanoic acid calcium salt (2:1).

stearodine. Iodine source. Insoluble in H_2O, EtOH; soluble in C_6H_6, $CHCl_3$, Et_2O.

9314 Calcium Lactate

814-80-2 1723 212-406-7

$C_6H_{10}CaO_6$

2-Hydroxypropanoic acid calcium salt (2:1).

Prequist Powder; component of: Calcet. Calcium replenisher. Soluble in H_2O, insoluble in EtOH. *Mission Pharmacal Co.; Parke-Davis.*

9315 Calcium Levulinate

591-64-0 1724 209-725-9

$C_{10}H_{14}CaO_6.2H_2O$

4-Oxopentanoic acid calcium salt (2:1).

Calcium replenisher. mp = 125°; very soluble in H_2O.

9316 Calcium Phosphate, Dibasic
7757-93-9 1739 231-826-1
$CaHPO_4$
Calcium phosphate (1:1).
CalStar; D.C.P. Calcium replenisher. d = 2.31; insoluble in H_2O, EtOH; soluble in dilute HCl, HNO_3; slightly soluble in dilute AcOH. *FMC Corp. Pharm. Div.; Parke-Davis.*

9317 Calcium Phosphate, Tribasic
7758-87-4 1741 231-840-8
$Ca_5(OH)(PO_4)_2$
Calcium hydroxide phosphate.
Hydroxyapatite; Hydroxylapatite. Calcium replenisher. d = 3.14; mp = 1670°, insoluble in H_2O, EtOH, AcOH; soluble in dilute HCl, HNO_3.

9318 Dextrose
77029-61-9 4467
$C_6H_{12}O_6.H_2O$
D-Glucose monohydrate.
Cartose; D-glucopyranose. Fluid; nutrient replenisher. [α-form, anhydrous]: mp = 146°; $[\alpha]_D$ = 112.2° → 52.7° (c = 10, H_2O); soluble in H_2O (90.9 g/100 ml at 25°, 125 g/100 ml at 30°, 243.9 g/100 ml at 50°, 357.1 g/100 ml at 70°, 555.5 g/100 ml at 90°), MeOH (0.83 g/100 ml at 20°); sparingly soluble in EtOH, Et_2O, Me_2CO; soluble in AcOH, C_5H_5N; LD_{50} (rbt iv) = 35000 mg/kg. *Astra Chem. Ltd.; Bristol-Myers Squibb Co; McGaw Inc.; Pfanstiehl Labs Inc.; Sterling Winthrop Inc.*

9319 Durapatite
1306-06-5 3519 215-145-7
$Ca_5HO_{13}P_3$
Calcium phosphate hydroxide.
Hydroxyapatite; Alveograf; Periograf; Win-40350; Ossopan. Calcium replenisher. Dec >100°; insoluble in H_2O. *Sterling Winthrop Inc.*

9320 l-Folinic Acid Calcium Salt
80433-71-2 4254
$C_{20}H_{21}CaN_7O_7$
Calcium N-p-[[[(6RS)-2-amino-5-formyl-5,6,7,8-tetrahydro-4-hydroxy-6-pteridinyl]methyl]amino]-benzoyl]-L-glutamate (1:1).
calcium (6S) folinate; calcium levofolinate; levoleucovorin calcium; CL-307782; Elvorine. Antianemic (folate deficiency); antidote to folic acid antagonists. $[\alpha]_D^{21}$ = -15.1° (c = 1.82). *Elkins-Sinn; Glaxo Wellcome plc; Immunex Corp.*

9321 Levocarnitine
541-15-1 1898 208-768-0
$C_7H_{15}NO_3$
L-(3-Carboxy-2-hydroxypropyl)trimethylammonium hydroxide inner salt.
3-Hydroxy-4-trimethylammoniobutanoate; γ-Trimethyl-β-hydroxybutyrobetaine; levocarnitine; vitamin B_7; Cardiogen; Carnitene; Carnicor; Carnitor; Carnum; Carrier; Miocor; Miotonal; Vitacarn. Hypolipidemic agent. Carnitine replenisher. Dec 197-198°; $[\alpha]_D^{30}$ = -23.9° (c = 0.86 H_2O); soluble in H_2O, hot EtOH; insoluble in Me_2CO, Et_2O, C_6H_6. *Sigma-Tau Pharm. Inc.*

9322 Magnesium Chloride [anhydrous]
7786-30-3 5698 232-094-6
Cl_2Mg
Hydrochloric acid magnesium salt (2:1).
Magnogene. Electrolyte replenisher. Laxative-cathartic. mp = 712°; d = 2.41, 2.325; soluble in H_2O (exothermic).

9323 Magnesium Chloride Hexahydrate
7791-18-6 5698
$Cl_2Mg.6H_2O$
Hydrochloric acid magnesium salt (2:1) hexahydrate.
Electrolyte replenisher. Laxative-cathartic. Dec 118°; d = 1.56; soluble in H_2O 166.6 g/100 ml at 20°, 333.2 g/100 ml at 100°), EtOH (50 g/100 ml); LD_{50} (rat orl) = 8100 mg/kg.

9324 Magnesium Gluconate
3632-91-5 4464 222-848-2
$C_{12}H_{22}MgO_{14}$
Gluconic acid magnesium salt (2:1).
Magnesium replenisher. *Forest Pharm. Inc.*

9325 Magnesium Gluconate Dihydrate
59625-89-7 4464
$C_{12}H_{22}MgO_{14}.2H_2O$
Gluconic acid magnesium salt (2:1) dihydrate.
Almora; Ulta-Mg. Magnesium replenisher. *Forest Pharm. Inc.*

9326 Magnesium Sulfate [anhydrous]
7487-88-9 5731 231-298-2
$MgSO_4$
Sulfuric acid magnesium salt (1:1).
Kieserite [as monohydrate]. Magnesium replenisher. The heptahydrate is an anticonvulsant and cathartic. Also effective in termination of refractory ventricular tachyarrhythmias. Loss of deep tendon reflex is a sign of overdose. *Astra USA Inc.; Dow Chem. U.S.A.; McGaw Inc.*

9327 Magnesium Sulfate Heptahydrate
10034-99-8 5731
$MgSO_4.7H_2O$
Sulfuric acid magnesium salt (1:1) heptahydrate.
bitter salts; Epsom salts. Magnesium replenisher. The heptahydrate is an anticonvulsant and cathartic. Also effective in termination of refractory ventricular tachyarrhythmias. Loss of deep tendon reflex is a sign of overdose. d = 1.67; soluble in H_2O (71 g/100 ml at 20°, 91 g/100 ml at 40°), slightly soluble in EtOH. *Astra USA Inc.; Dow Chem. U.S.A.; McGaw Inc.*

9328 Manganese Chloride
13446-34-9 5768
$Cl_2Mn\cdot 4H_2O$
Manganese chloride tetrahydrate.
Supplement (trace mineral). [anhydrous]: LD_{50} (mus sc) = 180-250 mg/kg, (dog iv) = 201.6 mg/kg.

9329 Manganese Sulfate
10034-96-5 5783
$MnSO_4.H_2O$
Manganese sulfate (1:1) monohydrate.
Supplement (trace mineral).

9330 Methenamine Tetraiodine
12001-65-9 6043

$C_6H_{12}I_4N_4$
Hexamethylenetetramine tetraiodide.
Iodoformine; Mirion; Siomine. Antiseptic; Iodine source. Prepared from potassium mercuric iodide and methenamine. Reddish powder; deflagrates at 138°; nearly insoluble in H_2O; slightly soluble in alcohol, $CHCl_3$, Et_2O, carbon disulfide; soluble in aqueous solutions of sodium or potassium iodides, sodium thiosulfate, dilute HCl; decomposition likely in aqueous .

9331 Periodyl
53586-99-5 7313

$C_{18}H_{32}I_2O_3$
12-Hydroxy-9,10-diiodo-9-octadecenoic acid.
diiodoricinstearolic acid; ricinstearolic acid diiodide; Diiodyl; Joristen. Iodine replenisher. mp = 62°; insoluble in H_2O, acids, soluble in alkalies, EtOH, Et_2O, $CHCl_3$; slightly soluble in C_6H_6. *Riedel de Haen (Chinosolfabrik).*

9332 Potassium Acetate
127-08-2 7764 204-822-2

$C_2H_3KO_2$
Acetic acid potassium salt (1:1).
Electrolyte replenisher. d^{25} = 1.57; mp = 292°; soluble in H_2O (200 g/100 ml at 20°, 500 g/100 ml at 100°), EtOH (34.5 g/100 ml); LD_{50} (rat orl) = 3250 mg/kg.

9333 Potassium Bicarbonate
298-14-6 7770 206-059-0

$KHCO_3$
Carbonic acid monopotassium salt.
Kafylox; K-Lyte; component of: K-Lyte, K-Lyte/Cl, K-Lyte DS. Electrolyte replenisher. Soluble in H_2O (35.7 g/100 ml at 20°, 50 g/100 ml at 50°), insoluble in EtOH. *Bristol Laboratories.*

9334 Potassium Chloride
7447-40-7 7783 231-211-8

KCl
Chloropotassuril; Diffu-K; Emplets Potassium Chloride; Enseal; Kaleorid; Kalitabs; Kalium-Duriles; Kaochlor; Kaon-Cl; Kaon-Cl 10; Kaskay; Kato; Kayback; Kay-Cee-L; Kay-Ciel; Kaysay; K-Contin; K-Dur; K-Lease; K-Lor; ; Klor-Con; Klorvess; Klotrix; K-Norm; K-Tab; Lento-Kalium; Micro-K; Nu-K; Peter-Kal; PfiKlor; Rekawan; Repone-K; Slow-K; Span-K; Ten-K; component of: Colyte, Infalyte, K-Lyte/Cl, Kolyum, K-Predne-Dom. Electrolyte replenisher. Used in veterinary medicine as a potassium supplement. Large doses may cause gastrointestinal irritation, weakness, circulatory disturbances. d = 1.98; mp = 773°; pH 7; soluble in H_2O (35.7 g/100 ml at 20°, 55.5 g/100 ml at 100°), glycerol (7.1 g/100 ml), EtOH (0.4 g/100 ml). *Abbott Labs Inc.; Apothecon; Bayer Corp.; Bristol Laboratories; Ciba-Geigy Corp.; Fisons Pharm. Div.; Forest Pharm. Inc.; ICN Pharm. Inc.; Key Pharm.; McGaw Inc.; Parke-Davis; Robins, A. H. Co.; Sandoz Pharm. Corp.; Savage Labs; Schwarz Pharma Kremers Urban Co.*

9335 Potassium Gluconate
299-27-4 7796 206-074-2

$C_6H_{11}KO_7$
D-Gluconic acid monopotassium salt.
Kaon; Gluconsan K; Kalimozan; Katrin; Potasoral; Potassuril; K-IAO; Tumil-K; component of: Kolyum, Twin-K. Electrolyte replenisher. Dec 180°; freely soluble in H_2O; insoluble in EtOH, Et_2O, C_6H_6, $CHCl_3$. *Fisons plc; Knoll Pharm. Co.; Savage Labs.*

9336 Potassium Iodide
7681-11-0 7809 231-659-4

IK
Jodid; Thyro-Block; Thyrojod; component of: Mudrane Tablets, Mudrane-2 Tablets, quadrinal. Antifungal; expectorant; iodine supplement. Used in veterinary medicine (orally) to treat goiter, actinobacillosis, actinomycosis, idodine deficiency, lead or mercury poisoning. d = 3.12; mp = 680°; soluble in H_2O (142.8 g/100 ml at 20°, 200 g/100 ml at 100°), EtOH (4.5 g/100 ml at 20°, 12.5 g/100 ml at 76°), absolute EtOH (1.96 g/100 ml), MeOH (12.5 g/100 ml), Me_2CO (1.3 g/100 ml), glycerol (50 g/100 ml), ethylene glycol (40 g/100 ml); LD_{50} (rat iv) = 285 mg/kg. *ECR Pharm.; Knoll Pharm. Co.; Wallace Labs.*

9337 Potassium/Magnesium Aspartate
14842-81-0

($C_4H_6KNO_4 \cdot 1/2H_2O$) with ($C_8H_{12}MgN_2O_8 \cdot 4H_2O$)
Wy-2837; Wy-2838. Nutrient. A mixture of potassium aspartate and magnesium aspartate. *Wyeth Labs.*

9338 Prolonium Iodide
123-47-7 7965 204-630-9

$C_9H_{24}I_2N_2O$
(2-Hydroxytrimethylene)bis(trimethylammonium iodide).
Entodon; Endojodin. Iodine replenisher. Dec 275°; freely soluble in H_2O; soluble in EtOH; insoluble in Et_2O, Me_2CO. *Sterling Winthrop Inc.*

9339 Protein Hydrolysates
8071

Sterile solution of amino acids and short peptides from hydrolysis of proteins.
Amigen; Aminokrovin; Aminosol; Bioplex; Dekamin; Parenamine; Parnetamin; Protigényl; Travamin. Fluid and nutrient replenisher.

9340 Rubidium Iodide
7790-29-6 8443 232-198-1

IRb
Hydriodic acid rubidium salt.
Iodine source. d = 3.55; mp = 642°, bp = 1300°; soluble in H_2O (151.5 g/100 ml), EtOH.

9341 Sodium Chloride
7647-14-5 8742 231-598-3

NaCl
Hydrochloric acid sodium salt.
Adsorbanac; Ayr; Ringer's Injection; common salt; table salt; rock salt; component of: Arm-A-Vial, Colyte. Electrolyte replenisher; emetic; topical anti-inflammatory. d = 2.17; mp = 804°; soluble in H_2O (35.7 g/100 ml at 25°, 38.5 g/100 ml at 100°), glycerol (10 g/100 ml); slightly soluble in EtOH; LD_{50} (rat orl) = 3750 ± 430 mg/kg. *Alcon Labs; Ascher B.F. & Co.; Astra USA Inc.; Elkins-Sinn; Schwarz Pharma Kremers Urban Co.; Wyeth-Ayerst Labs.*

9342 Sodium Gluconate
527-07-1 8766 208-407-7
$C_6H_{11}NaO_7$
D-Gluconic acid monosodium salt.
gluconic acid sodium salt. Electrolyte replenisher. Soluble in H_2O (59 g/100 ml at 25°), sparingly soluble in EtOH, insoluble in Et_2O. *Pfanstiehl Labs Inc.*

9343 Sodium Glycerophosphate
55073-41-1
$C_3H_7Na_2O_6P$
1,2,3-Propanetriol mono(dihydrogen phosphate) disodium salt.
Tonic.

9344 Sodium Iodide
7681-82-5 8777 231-679-3
INa
Ioduril; Anayodin. Iodine supplement; expectorant. d = 3.67; mp = 561°; soluble in H_2O (200 g/100 ml), EtOH (50 g/100 ml), glycerol (100 g/100 ml); MLD (rat iv) = 1300 mg/kg.

9345 Sodium Lactate
72-17-3 8781 200-772-0
$C_3H_5NaO_3$
2-Hydroxypropanoic acid monosodium salt.
Lacolin. Electrolyte replenisher. Miscible with H_2O, EtOH. *McGaw Inc.; Pfanstiehl Labs Inc.*

9346 Strontium Iodide
10476-86-5 9005 233-972-1
I_2Sr
Hydriodic acid strontium salt (2:1).
Iodine replenisher. mp = 402°; d = 4.42; soluble in H_2O (500 g/100 ml), EtOH; LD_{50} (rat ip) = 800 mg/kg.

9347 Zinc Sulfate
7733-02-0 10293 231-793-3
$ZnSO_4$
Sulfuric acid zinc salt (1:1).
Verazinc; white vitriol; zinc vitriol; Kreatol; Optraex; Solvezink; Solvazinc; Zincaps; Zincate; Zincomed; Z-Span; component of: VasoClear A, Zincfrin. Electrolyte replenisher. Zinc supplement. Astringent. Soluble in H_2O, insoluble in EtOH. *Alcon Labs; CIBA Vision AG; Forest Pharm. Inc.*

9348 Zinc Sulfate Heptahydrate
7446-20-0 10293
$ZnSO_4.7H_2O$
Sulfuric acid zinc salt (1:1) heptahydrate.
Op-Thal-Zin; Redeema. Electrolyte replenisher. Zinc supplement. Astringent. d = 1.97; mp = 100°; soluble in H_2O (166.6 g/100 ml), glycerol (40 g/100 ml); insoluble in EtOH. *Alcon Labs; CIBA Vision AG; Forest Pharm. Inc.*

Respiratory Stimulants

9349 Almitrine
27469-53-0 309 248-475-5
$C_{26}H_{29}F_2N_7$
2,4-Bis(allylamino)-6-[4-[bis(p-fluorophenyl)methyl]-1-piperazinyl]-s-triazine.
S-2620. Respiratory stimulant. A potent stimulant of the carotid bodies that produces a long-lasting enhancement of alveolar ventilation. mp = 181°. *Sci. Union et Cie France.*

9350 Almitrine Dimethanesulfonate
29608-49-9 309 249-726-1
$C_{28}H_{37}F_2N_7O_6S_2$
2,4-Bis(allylamino)-6-[4-[bis(p-fluorophenyl)methyl]-1-piperazinyl]-s-triazine dimethanesulfonate.
Vectarion. Respiratory stimulant. mp = 243° (dec); λ_m 227, 246 nm (log ε 4.52, 4.53 EtOH); LD_{50} (mus iv) = 210 mg/kg, (mus ip) = 390 mg/kg, (mus orl) > 2000 mg/kg. *Sci. Union et Cie France.*

9351 Bemegride
64-65-3 1054 200-588-0
$C_8H_{13}NO_2$
3-Ethyl-3-methylglutarimide.
NP-13; Megimide; Mikedimide; Eukraton; Malysol. Respiratory stimulant. CNS stimulant; antidote, barbiturate poisoning. mp = 127°; soluble in H_2O, Me_2CO; LD_{50} (mus iv) = 18.8 mg/kg, (rat iv) = 17.0 mg/kg. *Abbott Labs Inc.*

9352 Carbon Dioxide
124-38-9 1857 204-696-9
CO_2
Carbonic anhydride.
Respiratory stimulant. d (gas) = 1.976 g/l at 0°; soluble in H_2O (171 ml gas/100 ml at 0°, 88 ml/100 ml at 20°, 36 ml/100 ml at 60°).

9353 Cropropamide
633-47-6 2659 211-193-8
$C_{13}H_{24}N_2O_2$
N-[1-(Dimethylcarbonylpropyl)-N-propylcrotonamide.
component of: prethcamide, Micoren, Respirot. Analgesic. $bp_{0.25}$ = 128-130°; soluble in H_2O, Et_2O. *Ciba-Geigy Corp.*

9354 Crotetamide
6168-76-9 2662 228-208-9
$C_{12}H_{22}N_2O_2$
N-[1-(Dimethylcarbamoyl)propyl]-N-ethylcrotonamide.
crotethamid; crotethamide; component of: prethcamide, Micoren, Respirot. Analgesic. $bp_{0.03}$ = 132-134°; soluble in H_2O, Et_2O. *Ciba-Geigy Corp.*

9355 Dimefline
1165-48-6 3249 214-616-4
$C_{20}H_{21}NO_3$
8-[(Dimethylamino)methyl]-7-methoxy-3-methylflavone.
Respiratory stimulant. Analeptic. *Recordati Corp.; Wallace Labs.*

9356 Dimefline Hydrochloride
2740-04-7 3249 220-366-7
$C_{20}H_{22}ClNO_3$
8-[(Dimethylamino)methyl]-7-methoxy-3-methylflavone hydrochloride.
Remeflin; DW-62; Rec-7/0267; NSC-114650. Respiratory stimulant. Analeptic. Dec 213-214°. *Recordati Corp.; Wallace Labs.*

9357 Dimorpholamine
119-48-2 3317 204-328-7
$C_{20}H_{38}N_4O_4$
N,N'-1,2-Ethanediylbis[n-butyl-4-morpholinecarboxamide].
TH-1064; Amipan T; Théraleptique; Theraptique. CNS stimulant (analeptic). Respiratory stimulant. mp = 41-42°; $bp_{0.4}$ = 229°; soluble in H_2O (50 g/100 ml); LD_{50} (mus iv) = 54 mg/kg, (mus ip) = 80 mg/kg, (mus sc) = 104 mg/kg, (mus orl) = 380 mg/kg. *ICI.*

9358 Diniprofylline
17692-30-7
$C_{22}H_{20}N_6O_6$
7-(2,3-Dihydroxypropyl)theophylline bis(nicotinate ester). Respiratory stimulant.

9359 Doxapram
309-29-5 3488 206-216-3
$C_{24}H_{30}N_2O_2$
1-Ethyl-4-(2-morpholinoethyl)-3,3-diphenyl-2-pyrrolidinone.
Respiratory stimulant. CNS stimulant. *Robins, A. H. Co.*

9360 Doxapram Hydrochloride Monohydrate
7081-53-0 3488
$C_{24}H_{31}ClN_2O_2,H_2O$
1-Ethyl-4-(2-morpholinoethyl)-3,3-diphenyl-2-pyrrolidinone monohydrochloride monohydrate.
Dopram; AHR-619; Doxapril; Stimulexin. Respiratory stimulant. CNS stimulant. mp = 217-219°; soluble in H_2O; slightly soluble in EtOH, $CHCl_3$; LD_{50} (rat orl) = 261 mg/kg. *Robins, A. H. Co.*

9361 Ethamivan
304-84-7 3765 206-157-3
$C_{12}H_{17}NO_3$
N,N-Diethylvanillamide.
vanillic acid diethylamide; vanillic diethylamide; NSC-406087; Cardiovanil; Emivan; Vandid. CNS and respiratory stimulant. mp = 95-95.5°; LD_{50} (rat ip) = 28 mg/kg. *3M Pharm.; Oesterreiche Stickstoffwerke.*

9362 Fominoben
18053-31-1 4258 241-964-4
$C_{21}H_{24}ClN_3O_3$
3'-Chloro-α-[methyl[(morpholinocarbonyl)methyl]amino]-o-benzotoluidide.
Antitussive; respiratory stimulant. mp = 122.5-123°. *Boehringer Ingelheim Ltd.; Thomae GmbH Dr. Karl.*

9363 Fominoben Hydrochloride
24600-36-0 4258 246-344-7
$C_{21}H_{25}Cl_2N_3O_3$
3'-Chloro-α-[methyl[(morpholinocarbonyl)methyl]amino]-o-benzotoluidide hydrochloride.
PB-89; Finaten; Noleptan; Oleptan; Terion; Tussirama. Antitussive; respiratory stimulant. mp = 206-208° (dec); soluble in H_2O (0.1 mg/100 ml); LD_{50} (mus ip) = 630 mg/kg, (rat ip) = 1201 mg/kg, (mus orl) = 2200 mg/kg, (rat orl) = 1250 mg/kg. *Boehringer Ingelheim Ltd.; Thomae GmbH Dr. Karl.*

9364 Lobeline
90-69-7 5580 202-012-3
$C_{22}H_{27}NO_2$
2-[6-(β-Hydroxyphenethyl)-1-methyl-2-piperidyl]acetophenone.
inflatine; α-lobeline. Respiratory stimulant. Ganglionic stimulant. Found in the leaves and seeds of *Lobelia infata L. Lobeliaceae* (Indian tobacco). mp = 130-131°; $[\alpha]_D^{15}$ = -43° (EtOH); very slightly soluble in H_2O, petroleum ether; soluble in hot EtOH, $CHCl_3$, C_6H_6, Et_2O.

9365 (±)-Lobeline
134-65-6 5580
$C_{22}H_{27}NO_2$
(±)-2-[6-(β-Hydroxyphenethyl)-1-methyl-2-piperidyl]acetophenone.
lobelidine. Respiratory stimulant. mp = 110°.

9366 Lobeline Hydrochloride
134-63-4 5580 205-150-2
$C_{22}H_{28}ClNO_2$
2-[6-(β-Hydroxyphenethyl)-1-methyl-2-piperidyl]acetophenone hydrochloride.
Zoolobelin; Lobron. Respiratory stimulant. mp = 178-180°; $[\alpha]_D^{20}$ = -43° (c = 2); λ_m 245 280 nm (log ε 4.08, 3.05 MeOH); soluble in H_2O (2.5 g/100 ml), EtOH (8.3 g/100 ml); very soluble in $CHCl_3$; slightly soluble in Et_2O.

9367 Lobeline Sulfate
134-64-5 5580 205-151-8
$C_{44}H_{56}N_2O_8S$
2-[6-(β-Hydroxyphenethyl)-1-methyl-2-piperidyl]acetophenone sulfate.
Bantron, Unilobin; Lobeton; Lobidan; Toban. Respiratory stimulant. $[\alpha]_D^{20}$ = -25° (c = 2); soluble in H_2O (3.3 g/100 ml), slightly soluble in EtOH.

9368 Mepixanox
17854-59-0 5906 241-810-6
$C_{20}H_{21}NO_3$
3-Methoxy-4-(piperidinomethyl)xanthen-9-one.
mepixanthone; Pimexone. Respiratory stimulant. mp = 159-160°; LD_{50} (mus ip) = 70.73 mg/kg; [hydrochloride ($C_{20}H_{22}ClNO_3$)]: mp > 200° (dec). *Mondi.*

9369 Nikethamide
59-26-7 6635 200-418-5
$C_{10}H_{14}N_2O$
N,N-Diethyl-3-pyridinecarboxamide.
N,N-diethylnicotinamide; Anacardone; Astrocar; Carbamidal; Cardamine; Cardiamid; Cardimon; Coracon; Coractiv N; Coramine; Cordiamin; Corediol; Cormed; Cormid; Corvitol; Corvotone; Dynacoryl; Ecoran; Inicardio; Niamine; Nicamide; Nicor; Nicorine; Nikardin; Pyricardyl; Salvacard; Stimulin; Ventramine. CNS and respiratory stimulant. mp = 24-26°; bp = 296-300°, bp_{10} = 158-159°, bp_3 = 128-129°, $bp_{0.4}$ = 115°; d_4^{25} = 1.058-1.066; miscible with H_2O, Et_2O, $CHCl_3$, Me_2CO, EtOH; LD_{50} (rat ip) = 272 mg/kg. *Abbott Labs Inc.; Ciba-Geigy Corp.*

9370 Picrotoxin
124-87-8 7570 204-716-6

$C_{30}H_{34}O_{13}$
Cocculin.
Fish berry; Cocculine. CNS and respiratory stimulant. Isolated from seed of *Animirta cocculus*. Used as an antidote for barbiturate poisoning and as a fish poison. CAUTION: Extremely poisonous. mp = 203°; $[\alpha]_D^{16}$ = -29.3° (c = 4 EtOH); soluble in H_2O (0.285 g/100 ml at 25°, 20 g/100 ml at 100°), EtOH (7.4 g/100 ml at 25°, 33 g/100 ml at 778°); sparingly soluble in Et_2O, $CHCl_3$; LD_{50} (mus ip) = 7.2 mg/kg; highly toxic to fish. *Indofine Chem. Co.; Pfalz & Bauer; TCI America.*

9371 Pimeclone
534-84-9 7583

$C_{12}H_{21}NO$
2-(Piperidinomethyl)cyclohexanone.
NA-66; Nu-582. Respiratory stimulant. bp_{14} = 118-120°, bp_6 = 124°.

9372 Pimeclone Hydrochloride
6966-09-2 7583

$C_{12}H_{22}ClNO$
2-(Piperidinomethyl)cyclohexanone hydrochloride.
Karion; Spiractin. Respiratory stimulant. mp = 161-165°.

9373 Prethcamide
8015-51-8 2659, 2662

N-[1-(Dimethylcarbamoyl)propyl]-N-propylcrotonamide mixture with N-[1-(dimethylcarbamoyl)propyl]-N-ethylcrotonamide.
Micoren, Respirot. Respiratory stimulant; analgesic. Soluble in H_2O. *Ciba-Geigy Corp.*

9374 Pyridofylline
53403-97-7 8159 258-521-6

$C_{17}H_{23}N_5O_9S$
Pyridoxol salt of 7-(2-hydroxyethyl)theophylline hydrogen sulfate ester.
Atherophylline. Respiratory stimulant. Vasodilator (coronary). mp = 144-146°; LD_{50} (mus iv) = 1000 mg/kg, (mus orl) = 1600 mg/kg.

9375 Sodium Succinate
150-90-3 8828 205-778-7

$C_4H_4Na_2O_4$
Succinic acid sodium salt.
disodium succinate; Soduxin. Respratory stimulant; analeptic; urinary alkalizer; diuretic. Laxative-cathartic. Soluble in H_2O (20 g/100 ml), insoluble in EtOH; LD_{50} (mus iv) = 4500 mg/kg.

9376 Tacrine
321-64-2 9199 206-291-2

$C_{13}H_{14}N_2$
5-Amino-1,2,3,4-tetrahydroacridine.
9-Amino-1,2,3,4-tetrahydroacridine. Centrally active anticholinesterase. Nootropic. Respiratory stimulant. Cognition adjuvant and antidote for curare poisoning. mp = 183-184°. *Parke-Davis.*

9377 Tacrine Hydrochloride
1684-40-8 9199 216-867-5

$C_{13}H_{15}ClN_2$
5-Amino-1,2,3,4-tetrahydroacridine monohydrochloride.
Cognex; CI-970; THA. Nootropic. Respiratory stimulant. Centrally active anticholinesterase. Cognition adjuvant and antidote for curare poisoning. mp = 283-284°; soluble in H_2O. *Parke-Davis.*

Sclerosing Agents

9378 Ethanolamine
141-43-5 3772 205-483-3

C_2H_7NO
2-Aminoethanol.
ethylolamine; colamine; β-hydroxyethylamine; 2-hydroxyethylamine; monoethanolamine; β-aminoethyl alcohol. Sclerosing agent. *I.G. Farben.*

9379 Ethanolamine Oleate
2272-11-9 3772 218-878-0

$C_{20}H_{41}NO_3$
9-Octadienoic acid (Z) compound with 2-aminoethanol.
Ethamolin; Antivariz; Esclerosina. Sclerosing agent. *Glaxo Labs.; Schwarz Pharma Kremers Urban Co.*

9380 Ethylamine
75-04-7 3808 200-834-7

C_2H_7N
Aminoethane.
ethanamine; monoethylamine. Sclerosing agent. d_{15}^{15} = 0.689; mp = -80°; bp = 16.6°; miscible with H_2O, EtOH, Et_2O; LD_{50} (rat orl) = 400 mg/kg.

9381 Ethylamine Hydriodide
506-58-1 3808

C_2H_8IN
Aminoethane hydriodide.
Sclerosing agent. mp = 188°; d = 2.10; freely soluble in H_2O, EtOH; insoluble in Et_2O, $CHCl_3$.

9382 Ethylamine Hydrochloride
557-66-4 3808 209-182-8

C_2H_8ClN
Aminoethane hydrochloride.
Sclerosing agent. mp = 110°; d = 1.22; soluble in H_2O (250 g/100 ml); freely soluble in EtOH; slightly soluble in $CHCl_3$, Et_2O; insoluble in Et_2O.

9383 Ethylamine Oleate
39664-27-2 3808

$C_{20}H_{41}NO_2$
Aminoethane 9-octadecenoate (Z).
Etalate. Sclerosing agent.

9384 2-Hexyldecanoic Acid
25354-97-6 4747 246-885-9

$C_{16}H_{32}O_2$
Sclerosing agent. $bp_{0.02}$ = 140-150°.

9385 2-Hexyldecanoic Acid Sodium Salt
536-37-8 4747 208-629-4
$C_{16}H_{31}NaO_2$
Devaricin. Sclerosing agent. Soluble in H_2O.

9386 Laureth-9
9002-92-0 7717 500-002-6
$C_{30}H_{62}O_{10}$
Polyethyleneglycol monododecyl ether.
nonaethylene glycol monododecyl ether; polyoxyethylene lauryl ether; Aethoxysklerol; Atlas G-4829; Hetoxol L-9; Lipal 9-LA; Polidocanol; Thesit. Sclerosing agent. Spermaticide. Soluble in H_2O, EtOH, C_7H_8; miscible with hot mineral, natural and synthetic oils, fats and fatty alcohols; LD_{50} (mus orl) = 1170 mg/kg, (mus iv) = 125 mg/kg.

9387 Sodium Ricinoleate
5323-95-5 8378 226-191-2
$C_{18}H_{33}NaO_3$
Sodium [R-(Z)]-12-hydroxy-9-octadecenoate.
Soricin; Colidosan. Sclerosing agent. Soluble in H_2O, EtOH.

9388 Sodium Tetradecyl Sulfate
139-88-8 8838 205-380-3
$C_{14}H_{29}NaSO_4$
Sodium 7-ethyl-2-methyl-4-undecanol hydrogen sulfate.
Sotradecol; Tergitol 4; Trombavar; Trombovar. Wetting agent. Sclerosing agent. LD_{50} (rat orl) = 4950 mg/kg.

9389 Tribenoside
10310-32-4 9738 233-687-2
$C_{29}H_{34}O_6$
Ethyl 3,5,6-tri-O-benzyl-D-glucofuranoside.
21-401 Ba; Alven; Flebosan; Glyvenol; Hemocuron; Venex. Sclerosing agent. $bp_{1.2}$ = 270-280°; $[\alpha]_D^{26}$ = 8° ($CHCl_3$). *Bayer Corp. Pharm. Div.*

Sedatives and Hypnotics

9390 Acecarbromal
77-66-7 18 201-047-1
$C_9H_{15}BrN_2O_3$
1-Acetyl-3-(α-bromo-α-ethylbutyryl)urea.
Abasin; Carbased; Sedamyl; Acetyl Adalin. Sedative/hypnotic. mp = 109°; slightly soluble in H_2O; freely soluble in EtOAC, EtOH. *3M Pharm.*

9391 Acetal
105-57-7 36 203-310-6
$C_6H_{14}O_2$
1,1-Diethoxyethane.
diethyl acetal; acetaldehyde diethyl acetal; ethylidene diethyl ether. Sedative/hypnotic. d_4^{20} = 0.8254; bp = 102.7°, bp_{200} = 66.3°, bp_{60} = 39.8°, bp_{40} = 31.9°, bp_{20} = 19.6°, bp_{10} = 8.0°, bp_5 = -2.3°, bp_1 = -23.0°; soluble in H_2O (5 g/100 ml); soluble in EtOH, Et_2O, C_7H_{16}, EtOAc; LD_{50} (rat orl) = 4570 mg/kg.

9392 Acetophenone
98-86-2 71 202-708-7
C_8H_8O
1-Phenylethanone.
acetylbenzene; hypnone. Sedative/hypnotic. mp = 20.5°; bp = 202°; d_{15}^{15} = 1.033; slightly soluble in H_2O; freely soluble in EtOH, $CHCl_3$, Et_2O, glycerol; LD_{50} (rat orl) = 900 mg/kg.

9393 Acevaltrate
25161-41-5 246-685-1
$C_{24}H_{32}O_{10}$
1,7a-Dihydro-1,6-dihydroxyspiro[cyclopenta[c]pyran-7(6H),2'-oxirane]-4-methanol-4-acetate 1(or 6)-isovalerate 6(or 1)-(3-hydroxy-3-methylbutyrate acetate).
Sedative.

9394 Adinazolam
37115-32-5 159
$C_{19}H_{18}ClN_5$
8-Chloro-1-[(dimethylamino)methyl]-6-phenyl-4H-s-triazolo[4,3-a][1,4]benzodiazepine.
U-41123. Antidepressant with anxiolytic properties. Sedative. A triazolobenzodiazepine. Dimethylamino derivative of alprazolam. mp = 171-172.5°. *Ciba-Geigy Corp.; Pratt Pharm.; Upjohn Ltd.*

9395 Adinazolam Mesylate
57938-82-6 159
$C_{20}H_{22}ClN_5O_3S$
8-Chloro-1-[(dimethylamino)methyl]-6-phenyl-4H-s-triazolo[4,3-a][1,4]benzodiazepine monomethanesulfonate.
U-41123F; adinazolam monomethanesulfonate; Deracyn. Antidepressant with anxiolytic properties. Sedative/hypnotic. Tricyclic. mp = 230-244°. *Ciba-Geigy Corp.; Pharmacia & Upjohn; Upjohn Ltd.*

9396 Aldol
107-89-1 225 203-530-2
$C_4H_8O_2$
3-Hydroxybutanal.
acetaldol. Sedative/hypnotic. d^{16} = 1.109; bp_{20} = 83°; dec 85°; soluble in H_2O, EtOH, Et_2O; LD_{50} (rat orl) = 2180 mg/kg.

9397 Allobarbital
52-43-7 261 200-140-4
$C_{10}H_{12}N_2O_3$
5,5-Di-2-propenyl-2,4,6(1H,3H,5H)-pyrimidinetrione.
NSC-9324; allobarbitone; Malilum; Diadol; Dial. Sedative/hypnotic. mp = 171-173°; soluble in H_20 (0.33 g/100 ml at 20°, 2 g/100 ml at 100°), EtOH (5 g/100 ml at 20°), Et_2O (5 g/100 ml), Me_2CO, EtOAc; insoluble in hydrocarbons; LD_{50} (rat ip) = 127.3 mg/kg.

9398 Alonimid
2897-83-8
$C_{14}H_{13}NO_3$
2,3-Dihydrospiro[naphthalene-1(4H),3'-piperidine]-2',4,6'-trione.
Sedative/hypnotic. *Marion Merrell Dow Inc.*

9399 Alprazolam
28981-97-7 320 249-349-2
$C_{17}H_{13}ClN_4$
8-Chloro-1-methyl-6-phenyl-4H-s-triazolo[4,3-a][1,4]benzodiazepine.
D-65MT; U-31889; Alplax; Tafil; Tranquinal; Tranquinal; Trankimazin; Xanax; Xanor. Anxiolytic. Sedative/hypnotic. A benzodiazepine. mp = 228-228.5°; λ_m = 222 nm (ε 40250 EtOH); soluble in EtOH, insoluble in H_2O; LD_{50} (mus orl) = 1020 mg/kg, (mus ip) = 540 mg/kg, (rat orl) > 2000 mg/kg, (rat ip) = 610 mg/kg. *Pharmacia & Upjohn.*

9400 Ammonium Bromide
12124-97-9 531 235-183-8
BrH_4N
Ammonium bromide.
Sedative/hypnotic. White hygroscopic crystals; sublimes at high temperature; d^{25} = 2.429; freely soluble in H_2O, EtOH, MeOH, Me_2CO; slightly soluble in Et_2O; insoluble in EtOAc.

9401 Ammonium Valerate
42739-38-8 604 255-923-3
$C_5H_{13}NO_2$
Pentanoic acid ammonium salt.
ammonium valerianate; valeric acid ammonium salt. Formerly used as a sedative/hypnotic. mp = 108°; very soluble in H_2O, EtOH, soluble in Et_2O.

9402 Amobarbital
57-43-2 607 200-330-7
$C_{11}H_{18}N_2O_3$
5-Ethyl-5-(3-methylbutyl)-2,4,6(1H,3H,5H)-pyrimidinetrione.
Amytal; pentymal; barbamil; amylobarbitone; Somnal; Dormytal; Isomytal; Eunoctal; Amal; Mylodorm; Sednotic; Amasust; Stadadorm; component of: Dexamyl. Sedative/hypnotic. mp = 156-158°; soluble in H_2O (0.077 g/100 ml), EtOH (20 g/100 ml), $CHCl_3$ (5.88 g/100 ml), Et_2O (16.67 g/100 ml); freely soluble in C_6H_6; insoluble in petroleum ether; LD_{50} (mus sc) = 212 mg/kg. *Eli Lilly & Co.; SmithKline Beecham Pharm.*

9403 Amobarbital Sodium
64-43-7 607 200-584-9
$C_{11}H_{17}N_2NaO_3$
5-Ethyl-5-(3-methylbutyl)-2,4,6(1H,3H,5H)-pyrimidinetrione sodium salt.
Amytal Sodium; Talamo; Amsebarb; Barbamyl; Dorminal; Inmetal; component of: tuinal. Sedative/hypnotic. Very soluble in H_2O, soluble in EtOH (100 g/100 ml), insoluble in Et_2O. *Eli Lilly & Co.; Marion Merrell Dow Inc.*

9404 Amphenidone
134-37-2 621
$C_{11}H_{10}N_2O$
1-(m-Aminophenyl)-2(1H)-pyridone.
Dornwal. Sedative/hypnotic. mp = 182.5-184.5°; LD_{50} (mus orl) = 1300 mg/kg, (rat orl) = 3200 mg/kg. *Wallace & Tiernan Inc.*

9405 Aprobarbital
77-02-1 794 200-997-4
$C_{10}H_{14}N_2O_3$
5-Allyl-5-isopropyl-2,4,6(1H,3H,5H)-pyrimidinetrione.
Alurate; Numal; Allypropymal; Aprozal. Sedative/hypnotic. mp = 140-141.5°; insoluble in H_2O, petroleum ether; soluble in EtOH, $CHCl_3$, Et_2O, Me_2CO, C_6H_6, AcOH; LD_{50} (mus ip) = 200 mg/kg. *Hoffmann-LaRoche Inc.*

9406 Aprobarbital Sodium
125-88-2 794 204-760-6
$C_{10}H_{13}N_2NaO_3$
5-Allyl-5-isopropyl-2,4,6(1H,3H,5H)-pyrimidinetrione sodium salt.
aprobarbital sodium; Alurate Sodium; Somnipron. Sedative/hypnotic. Soluble in H_2O, slightly soluble in EtOH, insoluble in Et_2O. *Hoffmann-LaRoche Inc.*

9407 Apronalide
528-92-7 795 208-443-3
$C_9H_{16}N_2O_2$
N-(Aminocarbonyl)-2-(1-methylethyl)-4-pentenamide.
Apronal; Isodormid; Sedormid. Sedative/hypnotic. mp = 194°; soluble in H_2O (0.03 g/100 ml at 20°, 0.48 g/100 ml at 100°), EtOH (10 g/100 ml), Et_2O (1.33 g/100 ml).

9408 Barbital
57-44-3 989 200-331-2
$C_8H_{12}N_2O_3$
5,5-Diethyl-2,4,6(1H,3H,5H)-pyrimidinetrione.
5,5-diethylbarbiturate; diethylmalonyl-urea; Barbitone; Malonal; Veroletten; Sédeval; Dormonal; Hypnogène; Deba; Vespéral; Uronal. Sedative/hypnotic. mp = 188-192°; soluble in H_2O (0.77 g/100 ml at 20°, 7.7 g/100 ml at 100°), EtOH (7.14 g/100 ml), $CHCl_3$ (1.66 g/100 ml), Et_2O (2.86 g/100 ml), Me_2CO, EtOAc, petroleum ether, AcOH, C_5H_5N.

9409 Barbital Sodium
144-02-5 989 205-613-9
$C_8H_{11}N_2NaO_3$
5,5-Diethyl-2,4,6(1H,3H,5H)-pyrimidinetrione sodium salt.
sodium 5,5-diethylbarbiturate; barbital sodium; barbitone sodium; soluble barbital; Veronal sodium; Medina; Embinal. Sedative/hypnotic. Soluble in H_2O (20 g/100 ml at 20°, 40 g/100 ml at 100°), EtOH (0.25 g/100 ml).

9410 Barbituric Acid
67-52-7 990 200-658-0
$C_4H_4N_2O_3$
2,4,6(1H,3H,5H)-Pyrimidinetrione.
malonylurea. Sedative/hypnotic. mp ≅ 248°; poorly soluble in cold H_2O, freely soluble in hot H_2O; LD_{50} (rat orl) > 5000 mg/kg.

9411 Bentazepam
29462-18-8
$C_{17}H_{16}N_2OS$
1,2,3,6,7,8,9-Hexahydro-5-phenyl-2H-[1]-benzothieno[2,3-e]-1,4-diazepin-2-one.
CI-718. Sedative/hypnotic. *Parke-Davis.*

9412 Benzoctamine
17243-39-9 1117
$C_{18}H_{19}N$
N-Methyl-9,10-ethanoanthracene-9(10H)-methylamine.
Skeletal muscle relaxant; sedative/hypnotic; anxiolytic. *Ciba-Geigy Corp.*

9413 Benzoctamine Hydrochloride
10085-81-1 1117 233-216-0
$C_{18}H_{20}ClN$
N-Methyl-9,10-ethanoanthracene-9(10H)-methylamine hydrochloride.
Ba-30803; Tacitin. Skeletal muscle relaxant; sedative/hypnotic; anxiolytic. mp = 320-322°; LD_{50} (rat orl) = 700 ± 170 mg/kg. *Ciba-Geigy Corp.*

9414 d-Bornyl-α-Bromoisovalerate
52964-40-6 1368
$C_{15}H_{25}BrO_2$
1,7,7-Trimethylbicyclo[2.2.1]hept-2-yl (1R, endo)-2-bromo-3-methylbutanoate.
Brovalol; Eubornyl; Valisan. Sedative/hypnotic. bp = 163°; d = 1.18; soluble in EtOH, Et_2O, $CHCl_3$; insoluble in H_2O.

9415 d-Bornyl Isovalerate
53022-14-3 1370
$C_{15}H_{26}O_2$
1,7,7-Trimethylbicyclo[2.2.1]hept-2-yl (1R, endo)-3-methylbutanoate.
borneol isovalerate; Bornyval. Sedative/hypnotic. bp = 255-260°; $[\alpha]_D^{20}$ = 27-28°; d = 0.955; soluble in EtOH, Et_2O; insoluble in H_2O.

9416 Brallobarbital
561-86-4 1387 209-225-0
$C_{10}H_{11}BrN_2O_3$
5-Allyl-5-(2-bromoallyl)-2,4,6(1H,3H,5H)-pyrimidinetrione.
Vesperone. Sedative/hypnotic. mp = 168-169°.

9417 Bromoform
75-25-2 1441 200-854-6
$CHBr_3$
Tribromomethane.
Antitussive; sedative. Federally controlled substance. CAUTION: Dangerous substance; irratates skin, eyes, respiratory system; depresses central nercous system; causes liver damage. May be habit forming. Liquid; bp = 149-150°; mp = 7.5°; d_D^{15} = 2.9035; n_D^{15} = 1.6005; soluble in H_2O (0.125 g/100 ml); miscible with EtOH, C_6H_6, $CHCl_3$, Et_2O, Me_2CO, oils; incompatible with caustic alkalies; protect from light; LD_{50} (mus sc) = 1814 mg/kg.

9418 Bromoisovalum
496-67-3 1418 207-825-7
$C_6H_{11}BrN_2O_2$
N-(Aminocarbonyl)-2-bromo-3-methylbutanamide.
bromvaletone; B.V.U.; Bromural; Bromisoval; Uvaleral; Bromuvan; Somnurol; Brovalurea; Dormigene; Isobromyl; Alluval; Pivadorm. Sedative/hypnotic. mp = 147-149°; freely soluble in hot H_2O; readily soluble in EtOH, Et_2O; LD_{50} (mmus ip) = 724.9 mg/kg.

9419 Brotizolam
57801-81-7 1472 260-964-5
$C_{15}H_{10}BrClN_4S$
2-Bromo-4-(o-chlorophenyl)-9-methyl-6H-thieno[3,2-f]-s-triazolo[4,3-a][1,4]diazepine.
Lendorm; We-941; Lendormin; Mederantil; Sintonal. Sedative/hypnotic. mp = 212-214°; LD_{50} (mus orl) > 10000 mg/kg, (mus ip) = 920 mg/kg, (rat orl) > 10000 mg/kg, (rat ip) = 1000 mg/kg. *Boehringer Ingelheim Pharm. Inc.*

9420 Butabarbital
125-40-6 1530 204-738-6
$C_{10}H_{16}N_2O_3$
5-Ethyl-5-(1-methylpropyl)-2,4,6(1H,3H,5H)-pyrimidinetrione.
component of: Pyridium Plus. Sedative/hypnotic. *Parke-Davis.*

9421 Butabarbital Sodium
143-81-7 1530 205-611-8
$C_{10}H_{15}N_2NaO_3$
5-Ethyl-5-(1-methylpropyl)-2,4,6(1H,3H,5H)-pyrimidinetrione sodium salt.
Buticaps; Butisol Sodium; secbutobarbitone sodium; Astrudion; Bubarbital sodium; Butabarbitone sodium; Busodium; Busotran; Butabon; Butabar; Butak; Buta-Kay; Butrate; Butte; Buticaps; Butalan; Butanotic; Butex; Butatran; Butazem; Carrbutabarb; Butased; Butisol sodium; Loubarb; Nervan; Prelital; Sarisol; component of: Butibel. Sedative/hypnotic. mp = 165-168°; soluble in H_2O (50 g/100 ml), EtOH (14.3 g/100 ml); insoluble in Et_2O, C_6H_6. *Wallace Labs.*

9422 Butalbital
77-26-9 1536 201-017-8
$C_{11}H_{16}N_2O_3$
5-Allyl-5-isobutyl-2,4,6(1H,3H,5H)-pyrimidinetrione.
Sandoptal; alisobumal; allylbarbital; itobarbital; tetrallobarbital; component of: Axocet, Axotal, Fiorinal. Sedative/hypnotic. mp = 138-139°; insoluble in H_2O, petroleum ether; soluble in EtOH, $CHCl_3$, Et_2O, Me_2CO, AcOH. *Sandoz Pharm. Corp.; Savage Labs.*

9423 Butallylonal
1142-70-7 1537 214-537-5
$C_{11}H_{15}BrN_2O_3$
5-(2-Bromo-2-propenyl)-5-(1-methylpropyl)-2,4,6(1H,3H,5H)-pyrimidinetrione.
sonbutal. Sedative/hypnotic. mp = 132-133°; insoluble in H_2O; freely soluble in EtOH, Et_2O.

9424 Butallylonal Sodium
3486-86-0 1537
$C_{11}H_{14}BrN_2NaO_3$
5-(2-Bromo-2-propenyl)-5-(1-methylpropyl)-2,4,6(1H,3H,5H)-pyrimidinetrione sodium salt.
Sedative/hypnotic. Soluble in H_2O, EtOH; slightly soluble in Et_2O, $CHCl_3$.

9425 Butethal
77-28-1 1550 201-019-9
$C_{10}H_{16}N_2O_3$
5-Butyl-5-ethyl-2,4,6(1H,3H,5H)-pyrimidinetrione.
butobarbitone; Soneryl; Neonal; Butobarbital; Etoval.

Sedative/hypnotic. mp = 124-127°; soluble in EtOH (20 g/100 ml), Et_2O (10 g/100 ml); insoluble in H_2O, petroleum ether; LD_{50} (mus ip) = 319.6 mg/kg.

9426 Butoctamide
32838-26-9 1562
$C_{12}H_{25}NO_2$
N-(2-Ethylhexyl)-3-hydroxybutyramide.
hexobutyramide. Sedative/hypnotic. $bp_{0.30}$ = 149-150°. *Lion Dentrifice.*

9427 Butoctamide Hydrogen Succinate
32838-28-1 1562
$C_{16}H_{29}NO_5$
Butanedioic acid mono-[3-[(2-ethylhexyl)amino]-1-methyl-3-oxopropyl] ester.
BAHS; M-2H; Listomin S. Sedative/hypnotic. mp = 46.5°. *Lion Dentrifice.*

9428 Butoctamide Hydrogen Succinate Calcium Salt
32266-82-3 1562
$C_{32}H_{58}CaN_2O_{10}$
Butanedioic acid mono-[3-[(2-ethylhexyl)amino]-1-methyl-3-oxopropyl] ester calcium salt.
Sedative/hypnotic. λ_m = 192 nm ($E^{1\%}_{1\ cm}$ 290); LD_{50} (mus ip) = 476 mg/kg, (mus orl) = 2000 mg/kg. *Lion Dentrifice.*

9429 Calcium Bromide
7789-41-5 1693 232-164-6
Br_2Ca
Calcium dibromide.
Anticonvulsant; sedative. Also used to treat hypocalcemia in veterinary medicine. mp = 730°; d_4^{25} = 3.353; soluble in H_2O, MeOH, EtOH, Me_2CO; insoluble in dioxane, $CHCl_3$, Et_2O.

9430 Calcium Bromolactobionate
33659-28-8 1694 251-616-3
$C_{24}H_{42}Br_2Ca_2O_{24}$
D-Gluconic acid-4-O-β-D-galactopyranosyl calcium salt (2:1) with calcium bromide (1:1).
Brocalcin; Calabron; Calcibromin; Calcibronat; Calciobrom. Sedative/hypnotic. Freely soluble in H_2O.

9431 Calcium 2-Ethylbutanoate
136-91-4 1707 205-266-3
$C_{12}H_{22}CaO_4$
2-Ethylbutanoic acid calcium salt.
Ethanion. Sedative/hypnotic. Soluble in H_2O. *Merck & Co.Inc.*

9432 Capuride
5579-13-5 1818
$C_9H_{18}N_2O_2$
(2-Ethyl-3-methylvaleryl)urea.
Pacinox; McN-X-94; NSC-27178. Sedative/hypnotic. mp = 172°. *McNeil Pharm.*

9433 Carbocloral
541-79-7 1849
$C_5H_8Cl_3NO_3$
Ethyl (2,2,2-trichloro-1-hydroxyethyl)carbamate.
Prodorm; CI-336; CN-16146; HY-185; NSC-33077; Ural; Uraline; Uralium. Sedative/hypnotic. mp ≅ 103°; freely soluble in EtOH, Et_2O; insoluble in H_2O. *Parke-Davis.*

9434 Carbromal
77-65-6 1879 201-046-6
$C_7H_{13}BrN_2O_2$
2-Bromo-2-ethylbutyrylurea.
Adalin; Tildin; Planadalin; Diacid; Adisomnol; Bromadal; Uradal; Nyctal; component of: Carbrital. Sedative/hypnotic. mp = 116-119°; soluble in H_2O (0.03 g/100 ml), EtOH (5.5 g/100 ml), $CHCl_3$ (33.3 g/100 ml), Et_2O (7.1 g/100 ml). *Parke-Davis.*

9435 Carbubarb
960-05-4 1880
$C_{11}H_{17}N_3O_5$
5-Butyl-5-(2-carbamoyloxyethyl)-2,4,6(1H,3H,5H)-pyrimidinetrione.
Nogexan. Sedative/hypnotic. mp = 192-194°; LD_{50} (mus sc) = 1400 mg/kg.

9436 Carfimate
3567-38-2 1886
$C_{10}H_9NO_2$
1-Phenyl-2-propynyl carbamate.
phenylethynylcarbinol carbamate; Equilium; Nirvotin. Sedative/hypnotic. mp = 86-87°. *Farmitalia Carlo Erba Ltd.*

9437 α-Chloralose
15879-93-3 2114 240-016-7
$C_8H_{11}Cl_3O_6$
(R)-1,2-O-(2,2,2-Trichloroethylidene)-α-D-glucofuranose.
glucochloral; anhydroglucochloral; chloralosane; Alphakil; Dorcalm; Somio. Sedative/hypnotic. Has been used as a rodenticide. mp = 187°; $[\alpha]^2_D$ = 19° (c = 5 EtOH); soluble in H_2O (0.44 g/100 ml ar 15°, 0.83 g/100 ml at 37°), EtOH (6.67 g/100 ml at 25°), soluble in Et_2O, AcOH; slightly soluble in $CHCl_3$; insoluble in petroleum ether; LD_{50} (mus orl) = 400 mg/kg.

9438 Chloral Betaine
2218-68-0 2111 218-722-1
$C_7H_{14}Cl_3NO_4$
1-Carboxy-N,N,N-trimethylmethanaminium hydroxide inner salt compound with 2,2,2-trichloro-1,1-ethanediol (1:1).
Beta-Chlor; Somilan; 5107. Sedative/hypnotic. mp = 122.5-124.5°. *BDH Laboratory Supplies.*

9439 Chloral Formamide
515-82-2 2112 208-210-6
$C_3H_4Cl_3NO_2$
N-(2,2,2-Trichloro-1-hydroxyethyl)-formamide.
chloralamide; chloramide. Sedative/hypnotic. mp = 115-116°, 124-126°; readily soluble in EtOH, Et_2O, Me_2CO, EtOAc, glycerol; soluble in H_2O. *Boehringer Ingelheim GmbH.*

9440 Chloral Hydrate
302-17-0 2113 206-117-5
$C_2H_3Cl_3O_2$
2,2,2-Trichloro-1,1-ethanediol.
Noctec; SK Chloral Hydrate; Escre; Nycton; Somnos; Chloraldurat. Sedative/hypnotic. A hydrated aldehyde. mp = 57°; d = 1.91; bp = 98°; freely soluble in H_2O (240 g/100 ml at 0°, 830 g/100 ml at 25°, 1430 g at 40°), EtOH (76.9 g/100 ml), $CHCl_3$ (50 g/100 ml), Et_2O (66.6 g/100 ml), glycerol (200 g/100 ml), CS_2 (1.47 g/100 ml); sparingly soluble in CCl_4, C_6H_6, C_7H_8; LD_{50} (rat orl) = 479 mg/kg. *Bristol-Myers Squibb Pharm. R&D; SmithKline Beecham Pharm.*

9441 Chlordiazepoxide
58-25-3 2132 200-371-0
$C_{16}H_{14}ClN_3O$
7-Chloro-2-(methylamino)-5-phenyl-3H-1,4-benzodiazepin 4-oxide.
Abboxide; Librelease; Libritabs; Limbitrol; Menrium; methaminodiazepoxide; Clopoxide; Helogaphen; Multum; Risolid; Silibrin; Tropium. Anxiolytic. Sedative/hypnotic. A benzodiazepine. mp = 236-236.5°. *Abbott Labs Inc.; Hoffmann-LaRoche Inc.; Roche Puerto Rico.*

9442 Chlordiazepoxide Hydrochloride
438-41-5 2132 207-117-8
$C_{16}H_{15}Cl_2N_3O$
7-Chloro-2-(methylamino)-5-phenyl-3H-1,4-benzodiazepin 4-oxide hydrochloride.
A-Poxide; Librium; SK-Lygen; Librax; Ro-5-0690; NSC-115748; Ansiacal; Balance; Benzodiapin; Cebrum; Corax; Disarim; Elenium; Equibral; Labican; Lentotran; Librium; O.C.M.; Psichial; Psicoterina; Reliberan; Seren Vita; Viansin. Anxiolytic. Sedative/hypnotic. A benzodiazepine. mp = 213°; soluble in H_2O (0.15 g/100 ml). *Abbott Labs Inc.; Hoffmann-LaRoche Inc.; Parke-Davis; Roche Puerto Rico; SmithKline Beecham Pharm.*

9443 Chlorethate
5634-37-7
$C_5H_4Cl_6O_3$
Bis(2,2,2-trichloroethyl)carbonate.
SK&F-12866. Sedative/hypnotic. *SmithKline Beecham Pharm.*

9444 Chlorhexadol
3563-58-4 2139 222-634-9
$C_8H_{15}Cl_3O_3$
Methyl-4-(2,2,2-trichloro-1-hydroxyethoxy)pentan-2-ol.
Lora; Chloralodol; Mecoral; Medodorm. Sedative/hypnotic. Federally controlled substance (depressant). mp = 102-104°; readily soluble in EtOH, $CHCl_3$; soluble in Et_2O; slightly soluble in CCl_4. *Wallace Labs.*

9445 Cinolazepam
75696-02-5 2368
$C_{18}H_{13}ClFN_3O_2$
7-Chloro-5-(o-fluorophenyl)-2,3-dihydro-3-hydroxy-2-oxo-1H-1,4-benzodiazepine-1-propionitrile.
OX-373; Gerodorm. Sedative/hypnotic. mp = 190-193°. *Gerot Pharmazeutika.*

9446 Clomethiazole
533-45-9 2444 208-565-7
C_6H_8ClNS
5-(2-Chloroethyl)-4-methylthiazole.
S.C.T.Z.; chlorethiazol; chlormethiazole. A sedative/hypnotic; anticonvulsant. Oily liquid, d_4^{25} = 1.233; bp_7 = 92°; [hydrochloride]: mp = 130°; soluble in H_2O, EtOH; [methanedilsulfonate]: mp = 120°. *Hoffmann-LaRoche Inc.*

9447 Clomethiazole Ethanedisulfonate
1867-58-9 2444 217-483-0
$C_{14}H_{22}Cl_2N_2O_6S_4$
5-(2-Chloroethyl)-4-methylthiazole ethanedisulfonate.
SCTZ; Distraneurin; Hemineurin; Heminevrin. Anticonvulsant. Sedative/hypnotic. mp = 124°. *Hoffmann-LaRoche Inc.*

9448 Clomethiazole Hydrochloride
6001-74-7 2444
$C_6H_9Cl_2NS$
5-(2-Chloroethyl)-4-methylthiazole hydrochloride.
Anticonvulsant. Sedative/hypnotic. mp = 130°; soluble in H_2O, EtOH.

9449 Cloperidone
4052-13-5
$C_{21}H_{23}ClN_4O_2$
3-[3-[4-(m-Chlorophenyl)-1-piperazinyl]propyl]-2,4(1H,3H)-quinazolinedione.
Sedative/hypnotic.

9450 Cloperidone Hydrochloride
525-26-8
$C_{21}H_{24}Cl_2N_4O_2$
3-[3-[4-(m-Chlorophenyl)-1-piperazinyl]propyl]-2,4(1H,3H)-quinazolinedione hydrochloride.
MA-1337. Sedative/hypnotic. *Bayer Corp. Pharm. Div.*

9451 Cyclobarbital
52-31-3 2780 200-138-3
$C_{12}H_{16}N_2O_3$
5-(1-Cyclohexen-1-yl)-5-ethyl-2,4,6(1H,3H,5H)-pyrimidinetrione.
Phanodorn; cyclobarbitone; hexemal; tetrahydrophenobarbital; Cavonyl; Cyclodorm; Cyklodorm; Fanodormo; Irifan; Namuron; Palinum; Phanodorn; Philodorm; Prälumin; Pro-Sonil; Sonoform. Sedative/hypnotic. mp = 171-174°; slightly soluble in cold H_2O, appreciably soluble in hot H_2O, soluble in EtOH (20 g/100 ml), Et_2O (5 g/100 ml); LD_{50} (mus ip) = 350 mg/kg, (rat ip) = 290 mg/kg. *Bayer Corp. Pharm. Div.; Sterling Winthrop Inc.*

9452 Cyclobarbital Calcium
143-76-0 2780 205-610-2
$C_{24}H_{30}CaN_4O_6$
5-(1-Cyclohexen-1-yl)-5-ethyl-2,4,6(1H,3H,5H)-pyrimidinetrione calcium salt.
Phanodorn Calcium; hexemal calcium; Itridal; Kollerdormfix; Pronox. Sedative/hypnotic. Soluble in H_2O (1.43 g/100 ml), EtOH (0.2 g/100 ml), insoluble in Et_2O, $CHCl_3$. *Bayer Corp. Pharm. Div.; Sterling Winthrop Inc.*

9453 Cyclopentobarbital
76-68-6 2814 200-979-6
$C_{12}H_{14}N_2O_3$
5-(2-Cyclopenten-1-yl)-5-(2-propenyl)-2,4,6(1H,3H,5H)-pyrimidinetrione.
Cyclopal. Sedative/hypnotic. mp = 139-140°; slightly soluble in cold H_2O; moderately soluble in hot H_2O; freely soluble in EtOH, organic solvents.

9454 Cyclopentobarbital Sodium
302-34-1 2814
$C_{12}H_{13}N_2NaO_3$
5-(2-Cyclopenten-1-yl)-5-(2-propenyl)-2,4,6(1H,3H,5H)-pyrimidinetrione sodium .
Cyclopal sodium. Sedative/hypnotic. Very soluble in H_2O.

9455 Cyprazepam
15687-07-7
$C_{19}H_{18}ClN_3O$
7-Chloro-2-[(cyclopropylmethyl)amino]-5-phenyl-3H-1,4-benzodiazepin 4-oxide.
W-3623. Sedative/hypnotic.

9456 Cypripedium
2839
Dried rhizome and roots of *Calypso bulbosa, Cypripedium pubescens or Cypripedium parviflorum*.
Lady's Slipper; American Valerian; nerve root; yellow moccasin flower; Noah's ark. Sedative/hypnotic.

9457 Detomidine
76631-46-4 2981
$C_{12}H_{14}N_2$
4-(2,3-Dimethylbenzyl)imidazole.
MPV-253 All. An α_2-adrenergic agonist with sedative and analgesic properties. mp = 114-116°; LD_{50} (mus iv) = 35 mg/kg; [hydrochloride (Domosedan)]: mp = 160°. *Farmos Group Ltd.*

9458 Detomidine Hydrochloride
90038-01-0 2981
$C_{12}H_{15}ClN_2$
4-(2,3-Dimethylbenzyl)imidazole hydrochloride.
MPV-253 All; Domosedan. An α_2 adrenoceptor agonist with sedative properties. mp = 160°. *Farmos Group Ltd.*

9459 Dexclamol
52340-25-7
$C_{24}H_{29}NO$
(+)-2,3,4,4aβ,8,9,13bα,14-Octahydro-3α-isopropyl-1H-benzo[6,7]cyclohepta[1,2,3-de]pyrido[2,1-a]isoquinolin-3-ol.
Sedative/hypnotic.

9460 Dexclamol Hydrochloride
52389-27-2
$C_{24}H_{30}ClNO$
(+)-2,3,4,4aβ,8,9,13bα,14-Octahydro-3α-isopropyl-1H-benzo[6,7]cyclohepta[1,2,3-de]pyrido[2,1-a]isoquinolin-3-ol hydrochloride.
AY-24169. Sedative/hypnotic.

9461 Diazepam
439-14-5 3042 207-122-5
$C_{16}H_{13}ClN_2O$
7-Chloro-1,3-dihydro-1-methyl-5-phenyl-2H-1,4-benzodiazepin-2-one.
Alupram; Valium; Valrelease; LA-111; Ro- 5-2807; Wy-3467; NSC-77518; Apaurin; Atensine; Atilen; Bialzepam; Calmpose; Ceregulart; Dialar; Diazemuls; Dipam; Eridan; Eurosan; Evacalm; Faustan; Gewacalm; Horizon; Lamra; Lembrol; Levium; Mandrozep; Neurolytril; Noan; Novazam; Paceum; Pacitran; Paxate; Paxel; Pro-Pam; Q-Pam; Relanium; Sedapam; Seduxen; Servizepam; Setonil; Solis; Stesolid; Tranquase; Tranquo-Puren; Tranquo-Tablinen; Unisedil; Valaxona; Valiquid; Valium; Valium Injectable; Valrelease; Vival; Vivol. Anxiolytic; skeletal muscle relaxant; sedative/hypnotic. Also used as an intravenous anesthetic. Pharmaceutical preparation for the treatment of depression. A benzodiazepine. mp = 125-126°; soluble in DMF, $CHCl_3$, C_6H_6; Me_2CO, EtOH; slightly soluble in H_2O; LD_{50} (rat orl) = 710 mg/kg. *Berk Pharm. Ltd.; Hoffmann-LaRoche Inc.*

9462 Dichloralphenazone
480-30-8 3097 207-546-0
$C_{15}H_{18}Cl_6N_2O_5$
1,2-Dihydro-1,5-dimethyl-2-phenyl-3H-pyrazol-3-one compound with 2,2,2-trichloro-1,1-ethanediol.
Dichloralantipyrine; Bihypnal; Bonadorm; Dormwell; Sedor; Sominat; Welldorm. Compound and chloral and antipyrine. Sedative/hypnotic. mp = 68°.

9463 Didrovaltrate
18296-45-2
$C_{22}H_{32}O_8$
1,4a,5,7a-Tetrahydro-16,-dihydroxyspiro[cyclopenta[c]pyran-7(6H)-,2'-oxirane]-4-methanol 6-acetate 1,4-diisovalerate.
Sedative; antispasmodic.

9464 Diethylbromoacetamide
511-70-6 3164 208-132-2
$C_6H_{12}BrNO$
2-Bromo-2-ethylbutanamide.
carbromide. Sedative/hypnotic. mp = 67°; soluble in H_2O (0.87 g/100 ml); freely soluble in hot H_2O; soluble in EtOH, Et_2O, C_6H_6, oils. *Kalle BV.*

9465 Difebarbamate
15687-09-9
$C_{28}H_{42}N_4O_9$
1,3-Bis(3-butoxy-2-hydroxypropyl)-5-ethyl-5-phenylbarbituric acid dicarbamate ester.
component of: Atrium. Sedative.

9466 Doxefazepam
40762-15-0 3490
$C_{17}H_{14}ClFN_2O_3$
7-Chloro-5-(o-fluorophenyl)-1,3-dihydro-3-hydroxy-1-(2-hydroxyethyl)-2H-1,4-benzodiazepine-2-one.
SAS-643; Doxans. Sedative/hypnotic. mp = 138-140°; LD_{50} (rat orl) = 2550 mg/kg, (rat ip) = 586 mg/kg, (mus orl) > 1500 mg/kg, (mus ip) = 774 mg/kg. *Schiapparelli.*

9467 Doxylamine
469-21-6 3497 207-414-2

$C_{17}H_{22}N_2O$

2-[α-[2-(Dimethylamino)ethoxy]-α-methylbenzyl]-pyridine.

Antihistaminic. Sedative/hypnotic. $bp_{0.5}$ = 137-141°; soluble in acids. *Marion Merrell Dow Inc.; Pfizer Intl.; Whitehall Labs. Inc.*

9468 Doxylamine Succinate
562-10-7 3497 209-228-7

$C_{21}H_{28}N_2O_5$

2-[α-[2-(Dimethylamino)ethoxy]-α-methylbenzyl]pyridine succinate.

Decapryn Succinate; Unisom; Mereprine; Alsodorm; Gittalun; Hoggar N; Sedaplus; component of: Robitussin Night Time Cold Formula. Sedative/hypnotic; antihistaminic. mp = 100-104°; soluble in H_2O (100 g/100 ml), EtOH (50 g/100 ml), $CHCl_3$ (50 g/100 ml); slightly soluble in C_6H_6, Et_2O; LD_{50} (mus orl) = 470 mg/kg, (mus iv) = 62 mg/kg, (mus sc) = 460 mg/kg, (rbt orl) = 250 mg/kg, (rbt iv) = 49 mg/kg, (mrat sc) = 440 mg/kg, (frat sc) = 445 mg/kg. *Marion Merrell Dow Inc.; Pfizer Intl.; Whitehall Labs. Inc.*

9469 Ectylurea
95-04-5 3552 202-386-8

$C_7H_{12}N_2O_2$

cis-(2-Ethylcrotonyl)urea.

Astyn; Ektyl; Levanil; Nostal; Nostyn; Cronil; Neuroprocin; Pacetyn. Sedative/hypnotic. mp = 158°, 191-193°, 198°; slightly soluble in Et_2O, hot EtOH, other organic solvents; LD_{50} (rat orl) = 2500 mg/kg, (rat ip) = 900 mg/kg. *Miles Inc.*

9470 Enallylpropymal
1861-21-8 3607 217-463-1

$C_{11}H_{16}N_2O_3$

1-Methyl-5-(1-methylethyl)-5-(2-propenyl)-2,4,6(1H,3H,5H)-pyrimidinetrione.

Sedative/hypnotic. mp = 56-57°, bp_{12} = 176-178°; soluble in organic solvents. *Hoffmann-LaRoche Inc.*

9471 Enallylpropymal Sodium Salt
59005-68-4 3607

$C_{11}H_{15}N_2NaO_3$

1-Methyl-5-(1-methylethyl)-5-(2-propenyl)-2,4,6(1H,3H,5H)-pyrimidinetrione sodium.

Narconumal. Sedative/hypnotic. Soluble in H_2O. *Hoffmann-LaRoche Inc.*

9472 Estazolam
29975-16-4 3744 249-982-4

$C_{16}H_{11}ClN_4$

8-Chloro-6-phenyl-4H-s-triazolo[4,3-a][1,4]benzodiazepine.

ProSom; Abbott-47631; D-40TA; Cannoc; Esilgan; Eurodin; Julodin; Nemurel; Nuctalon; ProSom; Somnatrol. Sedative/hypnotic. mp = 228-229°; LD_{50} (mmus orl) = 740 mg/kg, (rat orl) = 3200 mg/kg, (rbt orl) = 300 mg/kg. *Abbott Labs Inc.; Pharmacia & Upjohn; Takeda Chem. Ind. Ltd.; Toyama Chem. Co. Ltd.*

9473 Etaqualone
7432-25-9 3757

$C_{17}H_{16}N_2O$

3-(o-Ethylphenyl)-2-methyl-4(3H)-quinazolinone.

Aolan. Sedative/hypnotic. mp = 81°; [hydrochloride (C_1=77$H_{17}ClN_2O$)]: mp = 247°. *Beiersdorf AG.*

9474 Ethchlorvynol
113-18-8 3774

C_7H_9ClO

1-Chloro-3-ethyl-1-penten-4-yn-3-ol.

Placidyl; Arvynol; Serenesil; Roeridorm; Normoson. Sedative/hypnotic. bp = 173-174°, 181°, $bp_{0.1}$ = 28.5-30°; d_4^{25}= 1.065 - 1.070; insoluble in H_2O, soluble in most organic solvents; LD_{50} (mus orl) = 290 mg/kg. *Abbott Labs Inc.; Pfizer Intl.*

9475 Ethinamate
126-52-3 3779 204-789-4

$C_9H_{13}NO_2$

1-Ethynylcyclohexanol carbamate.

Valmid; Valmidate; Valamin. Sedative/hypnotic. mp = 96-98°; bp_3 = 118-122°; not optically active; soluble in H_2O (0.25 g/100 ml), EtOH (35 g/100 ml), C_6H_{16} (2 g/100 ml). *Eli Lilly & Co.; Schering AG.*

9476 Etodroxizine
17692-34-1 3921

$C_{23}H_{31}ClN_2O_3$

2-[2-[2-[4-(p-Chloro-α-phenylbenzyl)-1-piperazinyl]-ethoxy]ethoxy] ethanol.

Sedative/hypnotic. $bp_{0.01}$ = 250°.

9477 Etodroxizine Dimaleate
56335-21-8 3921

$C_{31}H_{39}ClN_2O_{11}$

2-[2-[2-[4-(p-Chloro-α-phenylbenzyl)-1-piperazinyl]-ethoxy]ethoxy] ethanol dimaleate.

Indunox; Drimyl. Sedative/hypnotic. LD_{50} (rat orl) = 920 mg/kg.

9478 Etomidate
33125-97-2 3927 251-385-9

$C_{14}H_{16}N_2O_2$

(+)-Ethyl-1-(α-methylbenzyl)imidazole-5-carboxylate.

Amidate; R-16659; Hypnomidate. Sedative/hypnotic. Carboxilated derivative of imidazole. Also used as an intravenous anesethetic and as a supplement to maintain anesthesia. mp = 67°; $[\alpha]_D^{20}$ = 66° (c = 1 EtOH); λ_m = 240 nm (ε 12200 iPrOH); soluble in H_2O (0.0045 mg/100 ml), $CHCl_3$, MeOH, EtOH, Me_2CO, propylene glycol; LD_{50} (mus iv) = 29.5 mg/kg, (rat iv) = 14.8 - 24.3 mg/kg. *Abbott Laboratories Inc.; Abbott Labs Inc.*

9479 Fenadiazole
1008-65-7 3994

$C_8H_6N_2O_2$

o-1,3,4-Oxodiazol-2-ylphenol.

JL-512; Hypnazol. Sedative/hypnotic. mp = 111-112°; LD_{50} (mus ip) = 940 mg/kg. *Lab. Jacques Logeais.*

9480 Fenobam
63540-28-3
$C_{11}H_{11}ClN_4O_2.H_2O$
1-(m-Chlorophenyl)-3-(1-methyl-4-oxo-2-imidazolin-2-yl)urea monohydrate.
McN-3377-98. Sedative/hypnotic. *McNeil Pharm.*

9481 Flunitrazepam
1622-62-4 4181 216-597-8
$C_{16}H_{12}FN_3O_3$
5-(o-Fluorophenyl)-1,3-dihydro-1-methyl-7-nitro-2H-1,4-benzodiazepin-2-one.
Rohypnol; 5-4200; Narcozep; Roipnol. Sedative/hypnotic. mp = 166-167°, 170-172°. *Hoffmann-LaRoche Inc.*

9482 Flurazepam
17617-23-1 4233 241-591-7
$C_{21}H_{23}ClFN_3O$
7-Chloro-1-[2-(diethylamino)ethyl]-5-(o-fluorophenyl)-1,3-dihydro-2H-1,4-benzodiazepin-2-one.
Felmane; Noctosom; Stauroderm. Anticonvulsant; sedative/hypnotic; muscle relaxant. A benzodiazepine. mp = 77-82°. *Hoffmann-LaRoche Inc.*

9483 Flurazepam Hydrochloride
1172-18-5 4233 214-630-0
$C_{21}H_{25}Cl_3FN_3O$
7-Chloro-1-[2-(diethylamino)ethyl]-5-(o-fluorophenyl)-1,3-dihydro-2H-1,4-benzodiazepin-2-one dihydrochloride.
Dalmane; Ro-5-6901; NSC-78559; Benozil; Dalmadorm; Dalmate; Dormodor; Felson; Insumin; Lunipax; Somlan. An anticonvulsant with sedative/hypnotic properties. Muscle relaxant. A benzodiazepine. mp = 190-220°; LD_{50} (mus ip) = 290 mg/kg, (mus orl) = 870 mg/kg, (mus iv) = 84 mg/kg. *Hoffmann-LaRoche Inc.*

9484 Fosazepam
35322-07-7
$C_{18}H_{18}ClN_2O_2P$
7-Chloro-1-[(dimethylphosphinyl)methyl]-1,3-dihydro-5-phenyl-2H-1,4-benzodiazepin-2-one.
HR-930. Sedative/hypnotic. A benzodiazepine. *Hoechst Roussel Pharm. Inc.*

9485 5-Furfuryl-5-isopropylbarbituric Acid
1146-21-0 4326 214-549-0
$C_{12}H_{14}N_2O_4$
5-(2-Furanylmethyl)-5-(1-methylethyl)-2,4,6(1H,3H,5H)-pyrimidinetrione.
Dormovit. Sedative/hypnotic. mp = 168-170°; insoluble in cold H_2O; soluble in hot H_2O, EtOH, Et_2O, $CHCl_3$. *Wiernik AG.*

9486 Glutethimide
77-21-4 4485 201-012-0
$C_{13}H_{15}NO_2$
2-Ethyl-2-phenylglutarimide.
Elrodorm; Doriden; Doriden-Sed. Sedative/hypnotic. [dl-form]: mp = 84°; λ_m 251, 257, 263 nm ((MeOH); freely soluble in EtOAc, Me_2CO, Et_2O, $CHCl_3$; soluble in EtOH, MeOH; insoluble in H_2O; [d-form]: mp = 102.5-103°; $[\alpha]_D^{20}$ = 176° (MeOH); [l-form]: mp = 102-103°; $[\alpha]_D^{20}$= -181° (MeOH). *Ciba-Geigy Corp.*

9487 Halazepam
23092-17-3 4619 245-425-4
$C_{17}H_{12}ClF_3N_2O$
7-Chloro-1,3-dihydro-5-phenyl-1-(2,2,2-trifluoroethyl)-2H-1,4-benzodiazepin-2-one.
Paxipam; Sch-12041. Anxiolytic with sedative and hypnotic properties. A benzodiazepine. mp = 164-166°; LD_{50} (mus orl) > 4000 mg/kg. *Schering-Plough HealthCare Products.*

9488 Haloxazolam
59128-97-1 4635
$C_{17}H_{14}BrFN_2O_2$
10-Bromo-11b-(o-fluorophenyl)-2,3,7,11b-tetra-hydrooxazolo[3,2-d][1,4]benzodiazepin-6(5H)-one.
CS-430; Somelin. Sedative/hypnotic. mp = 185°; sparingly soluble in H_2O, LD_{50} (mus orl) = 1850 mg/kg.

9489 Heptabarbital
509-86-4 4689 208-107-6
$C_{13}H_{18}N_2O_3$
5-(Cyclohepten-1-yl)-5-ethylbarbituric acid.
heptabarb; heptabarbitone; Heptadorm; Medomin. Sedative/hypnotic. Federally controlled substance (depressant). CAUTION: may be habit forming. mp = 174°; λ_m = 218.5, 254 nm (0.2N NaOH); sparingly soluble in H_2O; soluble in EtOH (4.0 g/100 ml), Me_2CO (5.7 g/100 ml), $CHCl_3$ (1.4 g/100 ml). *Ciba-Geigy Corp.*

9490 Hexapropymate
358-52-1 4733 206-618-9
$C_{10}H_{15}NO_2$
1-(2-Propynyl)cyclohexanol carbamate.
hexopropynate; L-2103; Modirax; Lunamin; Merinax. Sedative/hypnotic. mp = 99°; soluble in EtOH, glycerol, propylene glycol. *Soc. Belge des Labs. Labaz.*

9491 Hexethal Sodium
144-00-3 4740
$C_{12}H_{19}N_2NaO_3$
5-Ethyl-5-hexyl-2,4,6(1H,3H,5H)-pyrimidinetrione sodium salt.
Ortal Sodium; Heberal. Sedative/hypnotic. mp = 126°; soluble in H_2O, EtOH; insoluble in Et_2O, C_6H_6; MLD (rat ip) = 240-250 mg/kg.

9492 Hexobarbital
56-29-1 4742 200-264-9
$C_{12}H_{16}N_2O_3$
5-(1-Cyclohexen-1-yl)-1,5-dimethyl-2,4,6(1H,3H,5H)-pyrimidinetrione.
methylhexabarbital; methexenyl; enhexymal; hexobarbitone; Citodon; Citopan; Cyclonal; Dorico; Evipal; Evipan; Hexanastab Oral; Noctivane; Sombucaps; Sombulex; Somnalert. Sedative and hypnotic. Intravenous anesthetic. mp = 145-147°; poorly soluble in H_2O (0.033 g/100 ml); soluble in EtOH, MeOH, Et_2O, $CHCl_3$, Me_2CO, C_6H_6. *3M Pharm.; Sterling Winthrop Inc.*

9493 Hexobarbital Sodium
50-09-9 4742 200-009-1
$C_{12}H_{15}N_2NaO_3$
5-(1-Cyclohexen-1-yl)-1,5-dimethyl-2,4,6(1H,3H,5H)-pyrimidinetrione sodium salt.
hexobarbital soluble; Cyclonal Sodium; Dorico Soluble; Evipal Sodium; Evipan Sodium; Hexanastab; Hexenal; Methexenyl Sodium; Noctivane Sodium; Narcosan Soluble; Privenal. Sedative/hypnotic. Intravenous anesthetic. Federally controlled substance. CAUTION: may be habit forming. Soluble in H_2O, EtOH, MeOH, Me_2CO; insoluble in $CHCl_3$, Et_2O, C_6H_6. *3M Pharm.; Sterling Winthrop Inc.*

9494 Homofenazine
3833-99-6 4774
$C_{23}H_{28}F_3N_3OS$
Hexahydro-4-[3-[2-(trifluoromethyl)phenothiazin-10-yl]propyl]-1H-1,4-diazepin-1-ethanol.
D-775; HFZ. Sedative/hypnotic. bp_1 = 230-240°. *Degussa-Hüls Corp.*

9495 Homofenazine Difumarate
3093-71-8 4774
$C_{31}H_{36}F_3N_3O_9S$
Hexahydro-4-[3-[2-(trifluoromethyl)phenothiazin-10-yl]propyl]-1H-1,4-diazepin-1-ethanol difumarate.
Sedative/hypnotic. mp = 148°. *Degussa-Hüls Corp.*

9496 Homofenazine Dihydrochloride
1256-01-5 4774 215-017-0
$C_{23}H_{30}Cl_2F_3N_3OS$
Hexahydro-4-[3-[2-(trifluoromethyl)phenothiazin-10-yl]propyl]-1H-1,4-diazepin-1-ethanol dihydrochloride.
Pasaden. Sedative/hypnotic. *Degussa-Hüls Corp.*

9497 Hydrobromic Acid
10035-10-6 4819 233-113-0
HBr
Hydrogen bromide.
Sedative/hypnotic. Aqueous solution, marketed in various concentrations: 50% HBr (d = 1.517), 40% HBr (d = 1.38), 34% HBr (d = 1.31), 1-% HBr (d = 1,08); solution contains up to 68.85% HBr in H_2O; forms azeotropes with H_2O, bp_{100} = 74.12° (49.80%), bp_{400} = 107.00° (48.47%), bp_{700} = 122.0° (47.74%), bp_{800} = 125.79° (47.56%).

9498 Isovaleryl Diethylamide
533-32-4 5253 208-562-0
$C_9H_{19}NO$
N,N-Diethyl-3-methylbutyramide.
Valy; Xalyl. Sedative/hypnotic. bp = 210-212°, bp_{14} = 93-95°, bp_5 = 77=78°; d_{20} = 0.8764; soluble in H_2O (4 g/100 ml), EtOH.

9499 Lithium Bromide
7550-35-8 5551 231-439-8
BrLi
Hydrobromic acid lithium salt.
Sedative/hypnotic. mp = 547°; soluble in H_2O (166 g/100 ml at 20°, 250 g/100 ml at 100°); freely soluble in EtOH, ethylene glycol; soluble in Et_2O, amyl alcohol.

9500 Lopirazepam
42863-81-0 255-974-1
$C_{14}H_9Cl_2N_3O_2$
7-Chloro-5-(o-chlorophenyl)-1,3-dihydro-3-hydroxy-2H-pyrido[3,2-e]-1,4-diazepin-2-one.
D-12524. Sedative; anxiolytic. A benzodiazepine.

9501 Loprazolam
61197-73-7 5604
$C_{23}H_{21}ClN_6O_3$
(Z)-6-(o-Chlorophenyl)-2,4-dihydro-2-[(4-methyl-1-piperazinyl)methylene]-8-nitro-1H-imidazo[1,2-a][1,4]benzodiazepin-1-one.
Sedative/hypnotic. A benzodiazepine. mp = 214-215°; LD_{50} (mus orl) > 1000 mg/kg. *Roussel-UCLAF.*

9502 Loprazolam Methanesulfonate
70111-54-5 5604
$C_{24}H_{25}ClN_6O_6S$
(Z)-6-(o-Chlorophenyl)-2,4-dihydro-2-[(4-methyl-1-piperazinyl)methylene]-8-nitro-1H-imidazo[1,2-a]-[1,4]benzodiazepin-1-one methanesulfonate.
loprazolam mesylate; HR-158; RU-31158; Dormonoct; Havlane; Somnovit; Sonin. Sedative/hypnotic. A benzodiazepine. mp = 205=210°. *Roussel-UCLAF.*

9503 Lormetazepam
848-75-9 5611 212-700-5
$C_{16}H_{12}Cl_2N_2O_2$
7-Chloro-5-(o-chlorophenyl)-1,3-dihydro-3-hydroxy-1-methyl-2H-1,4-benzodiazepin-2-one.
WY-4082; Ergocalm; Loramet; Noctamid. Sedative/hypnotic. A benzodiazepine. mp = 205-207°. *Am. Home Products.*

9504 Magnesium Bromide
7789-48-2 5695 232-170-9
Br_2Mg
Magnesium dibromide.
Anticonvulsant; sedative. Also used in organic synthesis. mp = 165° (dec); soluble in H_2O (333 g/100 ml), soluble in EtOH.

9505 Mecloqualone
340-57-8 5821 206-432-8
$C_{15}H_{11}ClN_2O$
3-(o-Chlorophenyl)-2-methyl-4(3H)-quinazolinone.
W-4744; NSC-142005; Nubarene. Sedative/hypnotic. mp = 126-128°; [hydrochloride ($C_{15}H_{12}Cl_2N_2O$)]: mp = 239-241°.

9506 Mecloxamine
5668-06-4 5823
$C_{19}H_{24}ClNO$
2-[(p-Chloro-α-methyl-α-phenylbenzyl)oxy]-N,N-dimethylpropylamine.
Anticholinergic; sedative/hypnotic. $bp_{0.6}$ = 154-160°. *Astra USA Inc.*

9507 Mecloxamine Citrate
5823
$C_{25}H_{32}ClNO_8$
2-[(p-Chloro-α-methyl-α-phenylbenzyl)oxy]-N,N-dimethylpropylamine citrate.

component of: Melidorm. Sedative/hypnotic properties. Anticholinergic. mp = 120-124°. *Astra USA Inc.*

9508 Menthyl Valerate

89-47-4 5887 201-910-2

$C_{17}H_{24}O_3$

(1α,2β,5α)-3-Methylbutanoic acid 5-methyl-2-(1-methylethyl)cyclohexyl ester.

Validol. Sedative/hypnotic. d = 0.906-0.908; insoluble in H_2O; freely soluble in EtOH, $CHCl_3$, Et_2O, oils.

9509 Meparfynol

77-75-8 5890 201-055-5

$C_6H_{10}O$

3-Methyl-1-pentyn-3-ol.

methylparafynol; methylpentynol; Allotropal; Anti-stress; Apridol; Atemorin; Atempol; Dalgol; Dorison; Dormalest; Dormidin; Dormigen; Dormiphen; Dormison; Dormosan; Formison; Hesofen; Hexofen; Imnudorm; Oblivon; Pentadorm; Perlopal; Riposon; Seral; Somnesin. Sedative/hypnotic. d_4^{20} = 0.8688; d_{20}^{20} = 0.8721; bp = 121-122°; bp_{37} = 50°, $bp_{6.5}$ = 20°; mp = -30.6°; soluble in H_2O (12.8 g/100 ml), Et_2O, Me_2CO, CCl_4, C_6H_6, EtOAc, MEK, most other organic solvents; LD_{50} (mus,rat, gpg orl) = 600-900 mg/kg. *Bayer Corp. Pharm. Div.; Schering-Plough HealthCare Products.*

9510 Meparfynol Carbamate

302-66-9 5890 206-125-9

$C_7H_{11}O_2$

3-Methyl-1-pentyn-3-ol carbamate.

N-Oblivon; Oblivon C; Trusono. Sedative/hypnotic. mp = 55.8-57°, 53.5-55°; bp_{16} = 120-121°, $bp_{0.01}$ = 95°; soluble in H_2O (1.6 g/100 ml); LD_{50} (mus sc 4 hr) = 560 mg/kg. *Schering-Plough HealthCare Products.*

9511 Mephobarbital

115-38-8 5899 204-085-7

$C_{13}H_{14}N_2O_3$

5-Ethyl-1-methyl-5-phenyl-2,4,6(1H,3H,5H)-pyrimidinetrione.

Mebaral; Menta-Bal; Phemiton; Prominal; Isonal; component of: Mebroin. An anticonvulsant; sedative; hypnotic. The N-methyl analog of phenobarbital. mp = 176°; slightly soluble in cold H_2O; freely soluble in hot H_2O, EtOH. *Marion Merrell Dow Inc.; Sterling Winthrop Inc.*

9512 Meprobamate

57-53-4 5908 200-337-5

$C_9H_{18}N_2O_4$

2-Methyl-2-propyl-1,3-propanediol dicarbamate.

Equanil; Meprospan; Meprotabs; Miltown; Tamate; Appetrol; Deprol; Equagesic; Micrainin; Milpath; Milprem; Miltrate; PMB-200; PMB-400; procalmadiol; procalmidol; Amosene; Andaxin; Aneural; Artolon; Atraxin; Ayeramate; Bamo 400; Biobamat; Calmiren; Cirpon; Cyrpon; Ecuanil; Fas-Cile 200; Gadexyl; Holbamate; Kesso-Bamate; Klort; Mar-Bate; Mepavlon; Meposed; Meprin; Meprindon; Meprobam; Mepr; Meprotabs; Meproten; Meprotil; Meptran; Mesmar; Miltaun; Morbam; My-trans; Nervonus; Oasil; Panediol; Perequil; Pertranquil; Placidon; Probamyl; Promate; Quaname; Quanil; Reostral; Restenil; Restran; Sowell; Trankvilan; Tranlisant; Tranquilan; Tranquiline; Urbilat. Anxiolytic with sedative and hypnotic properties. A propanediol carbamate. mp = 104-106°; soluble in H_2O (0.34 g/100 ml), more soluble in organic solvents; stable in dilute acid or alkali; LD_{50} (mus ip) = 800 mg/kg. *Lederle Labs.; Marion Merrell Dow Inc.; Wallace Labs; Wyeth-Ayerst Labs.*

9513 Methaqualone

72-44-6 6028 200-780-4

$C_{16}H_{14}N_2O$

2-Methyl-3-o-tolyl-4(3H)-quinazolinone.

CI-705; CN-38703; QZ-2; RIC-272; R-148; TR-495; metolquizolone; MAOA; MTQ; ortonal; Rorer 148; Cateudyl; Citexal; Dormigoa; Dormogen; Dormutil; Dorsedin; Fadormir; Holodorm; Hyminal; Hypcol; Hyptor; Ipnofil; Melsomin; Mequin; Mollinox; Motolon; Nobedorm; Noctilene; Normi-Nox; Omnyl; Optinoxan; Parminal; Parest; Quaalude; Roulone; Rouqualone; Sindesvel; Somnafac; Sonal; Somberol; Somnomed; Soverin; Torinal; Tuazol; Tuazolone. Sedative/hypnotic. mp = 120°, 114-116°; λ_m = 225, 263, 304, 316 nm (EtOH), 234, 269 nm (0.1N HCl); soluble in EtOH, Et_2O, $CHCl_3$;insoluble in H_2O; LD_{50} (rat orl) = 255 mg/kg. *Parke-Davis; Wallace Labs.*

9514 Methaqualone Hydrochloride

340-56-7 6028 206-431-2

$C_{16}H_{15}ClN_2O$

2-Methyl-3-o-tolyl-4(3H)-quinazolinone hydrochloride.

Optimil; Parest; Melsedin; Mequelon; Metasdorm; Methased; Optimil; Paxidorm; Revonal; Riporest; Sedaquin; Sleepinal; Somnium; Toquilone; Toraflon. Sedative/hypnotic. mp = 255-265°; soluble in Et_2O, EtOH; insoluble in H_2O. *Parke-Davis; Wallace Labs.*

9515 Methitural

467-43-6 6058

$C_{12}H_{20}N_2O_2S_2$

5-(1-Methylbutyl)-5-[2-(methylthio)ethyl]-2-thiobarbituric acid.

Sedative/hypnotic. *E. Merck.*

9516 Methitural Sodium Salt

730-68-7 6058 211-985-3

$C_{12}H_{19}N_2NaO_2S_2$

Sodium 5-(1-methylbutyl)-5-[2-(methylthio)ethyl]-2-thiobarbiturate.

methioturiate; AM-109; Sch-3132; Neraval; Thiogenal. Sedative/hypnotic. Freely soluble in H_2O. *E. Merck.*

9517 Methyprylon

125-64-4 6216 204-745-4

$C_{10}H_{17}NO_2$

3,3-Diethyl-2-methyl-2,4-piperidinedione.

Noludar; Noctan; Dimerin. Sedative/hypnotic. mp = 74-77°; λ_m = 295 nm ($A_{1\,cm}^{1\%}$ 2.0 EtOH); soluble in H_2O, EtOH, C_6H_6, $CHCl_3$. *Hoffmann-LaRoche Inc.*

9518 4-Methyl-5-thiazoleethanol

137-00-8 6208 205-272-6

C_6H_9NOS

5-Hydroxyethyl-4-methylthiazole.

Sedative/hypnotic. d_4^{24}= 1.196; bp_7 = 135°, bp_3 = 123-124°; bp_1 = 103°; soluble in EtOH, Et_2O, C_6H_6, $CHCl_3$, H_2O; [hydrochloride ($C_6H_{10}ClNOS$)]: soluble in H_2O, EtOH. *E. I. DuPont de Nemours Inc.; Merck & Co.Inc.*

9519 Midaflur
23757-42-8
$C_7H_3F_{12}N_3$
4-Amino-2,2,5,5-tetrakis(trifluoromethyl)-3-imidazoline.
EXP-338. Sedative/hypnotic. *DuPont-Merck Pharm.*

9520 Narcobarbital
125-55-3 6509
$C_{11}H_{15}BrN_2O_3$
5-(2-Bromo-2-propenyl)-1-methyl-5-(1-methylethyl)-2,4,6(1H,3H,5H)-pyrimidinetrione.
enibomal. Sedative/hypnotic; anticonvulsant. Federally controlled substance (depressant). mp = 115°; sparingly soluble in H_2O; soluble in EtOH, MeOH, C_5H_5N. *Riedel de Haen (Chinosolfabrik).*

9521 Narcobarbital Sodium Salt
3329-16-6 6509 222-050-4
$C_{11}H_{14}BrN_2NaO_3$
5-(2-Bromo-2-propenyl)-1-methyl-5-(1-methylethyl)-2,4,6-(1H,3H,5H)-pyrimidinetrione sodium salt.
Eunarcon; Narcotal. Sedative/hypnotic. Anticonvulsant. Federally controlled substance (depressant). Soluble in H_2O. *Riedel de Haen (Chinosolfabrik).*

9522 Nealbarbital
561-83-1 6516 209-224-5
$C_{12}H_{18}N_2O_3$
5-(2,2-Dimethylpropyl)-5-(2-propenyl)-2,4,6(1H,3H,5H)-pyrimidinetrione.
neallymal; Nevental; nealbarbitone; Censedal. Sedative/hypnotic. Federally controlled substance (depressant). mp = 155-157°; insoluble in H_2O, petroleum ether; moderately soluble in $CHCl_3$; freely soluble in EtOH, Et_2O, Me_2CO. *Pharmacia & Upjohn.*

9523 Niaprazine
27367-90-4 6576 248-431-5
$C_{20}H_{25}FN_4O$
N-[3-[4-(p-Fluorophenyl)-1-piperazinyl]-1-methylpropyl]nicotinamide.
1709 CERM; Nopron. Sedative/hypnotic. mp = 131°; LD_{50} (mus orl) = 890 mg/kg, (mus iv) = 145°. *C.E.R.M.*

9524 Nisobamate
25269-04-9
$C_{13}H_{26}N_2O_4$
2-(Hydroxymethyl)-2,3-dimethylpentyl isopropylcarbamate carbamate (ester).
W-1015. Anxiolytic with sedative/hypnotic properties. *Wallace Labs.*

9525 Nitrazepam
146-22-5 6667 205-665-2
$C_{15}H_{11}N_3O_3$
1,3-Dihydro-7-nitro-5-phenyl-2H-1,4benzodiazepin-2-one.
Mogadon; Ro-4-5360; Ro-5-3059; NSC-58775; Benzalin; Calsmin; Eatan; Eunoctin; Imeson; Insomin; Ipersed; Mogadan; Nelbon; Neuchlonic; Nitrados; Nitrenpax; Noctesed; Pelson; Radedorm; Remnos; Somnased; Somnibel; Somnite; Sonebon; Surem; Unisomnia. An anticonvulsant and sedative/hypnotic. A benzodiazepine. mp = 224-226°; λ_m = 277.5 nm ($e_{1\ cm}^{1\%}$ = 1500 0.1N H_2SO_4); soluble in EtOH, Me_2CO, EtOAc, $CHCl_3$; insoluble in H_2O, Et_2O, C_6H_6, C_6H_{14}; LD_{50} (rat orl) = 825 ± 80 mg/kg. *Hoffmann-LaRoche Inc.*

9526 Noreximide
6319-06-8
$C_9H_9NO_2$
cis-Exo-5-Norbornene-2,3-dicarboximide.
Sedative.

9527 Novonal
512-48-1 6820 208-143-2
$C_9H_{17}NO$
2,2-Diethyl-4-pentenamide.
Epinoval. Sedative/hypnotic. mp = 75-76°; soluble in H_2O (8.3 g/100 ml), EtOH, Et_2O.

9528 Opium
6986
gum opium; crude opium. Sedative/hypnotic. Air-dried, milky exudate from incised unripe capsules of *Papaver somniferum*. Contains about 20 alkaloids, the most important being morphine (10-16%). Others are noscapine (4-8%), codeine (0.8-2.5%), papaverine (0.5-2.5%) and thebaine (0.5-2%).

9529 Paraldehyde
123-63-7 7160 204-639-8
$C_6H_{12}O_3$
2,4,6-Trimethyl-s-trioxane.
Paral. Sedative/hypnotic. mp = 12°; bp ≅ 124°; d_{25}^{25} ≅ 0.994; soluble in H_2O (12.5 g/100 ml at 25°, 5.8 g/100 ml at 100°); soluble in EtOH, $CHCl_3$, Et_2O, oils; LD_{50} (rat orl) = 1650 mg/kg. *Forest Pharm. Inc.; Phillips 66.*

9530 Pentaerythritol Chloral
78-12-6 7246
$C_{13}H_{16}Cl_{12}O_8$
1,1'-[2,2-Bis[(2,2,2-trichloro-1-hydroxyethoxy)methyl]-1,3-propanediylbis(oxy)]bis[2,2,2-trichloroethanol].
Petrichloral; Perichlor. Sedative/hypnotic. mp = 52-54°; soluble in H_2O, EtOAc, formamide, EtOH, $CHCl_3$, C_6H_6 (all > 30 g/100 ml), tetrachlorethylene, iPrOH (5 g/100 ml). *Am. Home Products.*

9531 Pentobarbital
76-74-4 7272 200-983-8
$C_{11}H_{18}N_2O_3$
5-Ethyl-5-(1-methylbutyl)-2,4,6(1H,3H,5H)-pyrimidinetrione.
Neodorm; pentobarbitone; mebubarbital. Sedative/hypnotic. mp = 129-130°. *Abbott Labs Inc.*

9532 Pentobarbital Calcium Salt
7563-42-0 7272 231-460-2
$C_{22}H_{34}CaN_4O_6$
5-Ethyl-5-(1-methylbutyl)-2,4,6(1H,3H,5H)-pyrimidinetrione sodium salt.

pentobarbital calcium; Repocal. Sedative/hypnotic. *Abbott Labs Inc.*

9533 Pentobarbital Sodium
57-33-0 7272 200-323-9
$C_{11}H_{17}N_2NaO_3$
5-Ethyl-5-(1-methylbutyl)-2,4,6(1H,3H,5H)-pyrimidinetrione calcium salt.
Nembutal; pentobarbital sodium; soluble pentobarbital; Carbrital; Narcoren; Pentone; Praecicalm; Somnopentyl; Sopental; component of: Beuthanasia-D, Carbrital, Sedalixir. Sedative/hypnotic. mp ≅ 127° (dec); freely soluble in H_2O, EtOH; insoluble in C_6H_6, Et_2O; LD_{50} (rat orl) = 118 mg/kg. *Abbott Labs Inc.; Marion Merrell Dow Inc.; Parke-Davis; Schering-Plough Animal Health; Wyeth-Ayerst Labs.*

9534 tert-Pentyl Alcohol
75-85-4 7282 200-908-9
$C_5H_{12}O$
2-Methyl-2-butanol.
dimethyl ethyl carbinol; amylene hydrate; t-amyl alcohol. Sedative/hypnotic. bp_{765} = 102.5°; mp = -9.0°; d^{20} = 0.8084; soluble in H_2O (12.5 g/100 ml), EtOH, Et_2O, C_6H_6, $CHGCl_3$, glycerol, oils; LD_{50} (rat orl) = 1000mg /kg.

9535 Perlapine
1977-11-3 7320
$C_{19}H_{21}N_3$
6-(4-Methyl-1-piperazinyl)morphanthridine.
AW-142333; HF-2333; MP-11; Hypnodin. Sedative/hypnotic. mp = 136-138°; LD_{50} (mus orl) = 415 mg/kg. *Sandoz Pharm. Corp.; Wander Pharma.*

9536 Phenallymal
115-43-5 7349 204-089-9
$C_{13}H_{12}N_2O_3$
5-Phenyl-5-(2-propenyl)-2,4,6(1H,3H,5H)-pyrimidinetrione.
phenallymalum; alphenal; prophenal; Alphenate; Allofenyl; Fenallymal. Sedative/hypnotic. mp = 156-157.5°; readily soluble in EtOH, $CHCl_3$; soluble in H_2O (0.17 g/100 ml), Et_2O (10 g/100 ml), C_6H_6 (0.2 g/100 ml), CCl_4 (0.025 g/100 ml), petroleum ether (0.0057 g/100 ml).

9537 Phenobarbital
50-06-6 7386 200-007-0
$C_{12}H_{12}N_2O_3$
5-Ethyl-5-phenyl-2,4,6(1H,3H,5H)-pyrimidinetrione.
5-ethyl-5-phenylbarbituric acid; phenobarbitone; Agrypnal; Barbiphenyl; Barbipil; Eskabarb; Gardenal; Luminal; Phenobal; Solfoton; Talpheno; component of: Antrocol, Barbidonna, Bronkotabs, Chardonna-2, Donnatal, Donnazyme, Hydantal, Kinesed, Levsin PB Drops and Tablets, Quadrinal, Tedral. An anticonvulsant, sedative and hypnotic. A long-acting barbiturate. mp = 174-178°; λ_m = 240 nm ($A^{1\%}_{1\ cm}$ 431 pH10), 256 nm ($A^{1\%}_{1\ cm}$ 314 0.1N NaOH); soluble in H_2O (0.1 g/100 ml), EtOH (12.5 g/100 ml), $CHCl_3$ (2.5 g/100 ml), Et_2O (7.7 g/100 ml), C_6H_6 (0.14 g/100 ml); LD_{50} (rat . *ECR Pharm.; Knoll Pharm. Co.; Marion Merrell Dow Inc.; Parke-Davis; Robins, A. H. Co.; Sandoz Pharm. Corp.; Schwarz Pharma Kremers Urban Co.; SmithKline Beecham Pharm.; Sterling Winthrop Inc.; Wallace Labs; Zeneca Pharm.*

9538 Phenobarbital Sodium
57-30-7 7386 200-322-3
$C_{12}H_{11}N_2NaO_3$
5-Ethyl-5-phenyl-2,4,6(1H,3H,5H)-pyrimidinetrione sodium salt.
5-ethyl-5-phenylbarbituric acid sodium salt; Luminal Sodium; sol phenobarbital; sol phenobarbitone. Anticonvulsant sedative/hypnotic. A long-acting barbiturate. Soluble in H_2O (100 g/100 ml), EtOH (10 g/100 ml); insoluble in Et_2O, $CHCl_3$; LD_{50} (rat orl) = 660 mg/kg. *Sterling Winthrop Inc.; Wyeth-Ayerst Labs.*

9539 Phenylmethylbarbituric Acid
76-94-8 7457 200-994-8
$C_{11}H_{10}N_2O_3$
5-Methyl-5-phenyl-2,4,6(1H,3H,5H)-pyrimidinetrione.
Rutonal. An anticonvulsant; sedative/hypnotic. mp = 226°; insoluble in H_2O; soluble in EtOH, Et_2O; has water-soluble sodium salt. *Bayer Corp.*

9540 Piperidione
77-03-2 7622 200-999-5
$C_9H_{15}NO_2$
3,3-Diethyl-2,4-piperidinedione.
dihyprylone; Sedulon; Tusseval. Antitussive; sedative/hypnotic. mp = 102-107°; soluble in H_2O, EtOH, $CHCl_3$.

9541 Potassium Bromide
7758-02-3 7780 231-830-3
BrK
Anticonvulsant; sedative. CAUTION: Large doses may cause CNS suppression. mp = 730°; d = 2.75; soluble in H_2O (66.6 g/100 ml at 25°, 100 g/100 ml at 100°), EtOH (0.4 g/100 ml), glycerol (21.7 g/100 ml).

9542 Prazepam
2955-38-6 7895 220-975-8
$C_{19}H_{17}ClN_2O$
7-Chloro-1-(cyclopropylmethyl)-1,3-dihydro-5-phenyl-2H-1,4-benzodiazepin-2-one.
W-4020; Centrax; Demetrin; Lysanxia; Prazene; Sedapran; Settima; Trepidan; Verstran. Anxiolytic with sedative/hypnotic properties. mp = 145-146°. *Parke-Davis.*

9543 Propallylonal
545-93-7 7980 208-896-7
$C_{10}H_{13}BrN_2O_3$
5-(2-Bromo-2-propenyl)-5-(1-methylethyl)-2,4,6(1H,3H,5H)-pyrimidinetrione.
Noctal. Sedative/hypnotic. mp = 177-179°; slightly soluble in H_2O; soluble in Et_2O, $CHCl_3$, C_6H_6; freely soluble in EtOH, Me_2CO, AcOH; MLD (rbt orl) = 300-350 mg/kg.

9544 Propiomazine
362-29-8 8007 206-646-1
$C_{20}H_{24}N_2OS$
1-[10-[2-(Dimethylamino)propyl]-10H-phenothiazin-2-yl]-1-propanone.

Wy-1359; propionylpromethazine; Indorm [as maleate]; Propavan [as maleate]; CB-1678 [as maleate]. Sedative/hypnotic. $bp_{0.5}$ = 235-245°; [maleate]: mp = 160-161°. *Wyeth-Ayerst Labs.*

9545 Propiomazine Hydrochloride
1240-15-9 8007

$C_{20}H_{25}ClN_2OS$

1-[10-[2-(Dimethylamino)propyl]-10H-phenothiazin-2-yl]-1-propanone hydrochloride.

Largon. Sedative/hypnotic. Very soluble in H_2O, EtOH; insoluble in C_6H_6. *Wyeth-Ayerst Labs.*

9546 Proxibarbal
2537-29-3 8091 219-803-4

$C_{10}H_{14}N_2O_4$

5-Allyl-5-(2-hydroxypropyl)-2,4,6(1H,3H,5H)-pyrimidinetrione.

proxibarbital; HH-184; Axeen; Centralgol; Centralgyl; Ipronal. Sedative/hypnotic. Federally controlled substance (depressant). mp = 157-158°, 166.5-168.5°; moderately soluble in H_2O. *Hommel GmbH.*

9547 Pyrithyldione
77-04-3 8179 201-000-5

$C_9H_{13}NO_2$

3,3-Diethyl-2,4(1H,3H)-pyridinedione.

Presidon; Persedon; Tetridin; Benedorm. Sedative/hypnotic. Three crystal modifications, mp = 92-93°, 97-98°, 81-86°; bp_{14} = 187-189°; moderately soluble in H_2O, freely soluble in polar organic solvents. *Hoffmann-LaRoche Inc.*

9548 Quazepam
36735-22-5 8211 253-179-4

$C_{17}H_{11}ClF_4N_2S$

7-Chloro-5-(o-fluorophenyl)-1,3-dihydro-1-(2,2,2-trifluoroethyl)-2H-1,4-benzodiazepine-2-thione.

Dormalin; Sch-16134; Doral; Oniria; Prosedar; Quazium; Selepam. Sedative/hypnotic. Federally controlled substance (depressant). mp = 137.5-139°; LD_{50} (mus iv) > 1370 mg/kg, (mus orl) > 1500 mg/kg, (mmus ip) = 845 mg/kg, (fmus ip) = 921 mg/kg, (mrat ip) = 3072 mg/kg, (frat ip) = 2749 mg/kg. *Schering-Plough HealthCare Products.*

9549 Reclazepam
76053-16-2

$C_{18}H_{13}Cl_2N_3O_2$

2-[7-Chloro-5-(o-chlorophenyl)-2,3-dihydro-1H-1,4-benzodiazepin-1-yl]-2-oxazolin-4-one.

SC-33963. Sedative/hypnotic. A benzodiazepine. *Searle G.D. & Co.*

9550 Reposal
3625-25-0 8306

$C_{14}H_{18}N_2O_3$

5-Bicyclo[3.2.1]oct-2-en-3-yl-5-ethyl-2,4,6(1H,3H,5H)-pyrimidinetrione.

WT-161; Reposamal. Sedative/hypnotic. mp = 213°; LD_{50} (mus ip) = 175 mg/kg. *Calanda Stiftung.*

9551 Rilmazafone
99593-25-6 8387

$C_{21}H_{20}Cl_2N_6O_3$

5-[(2-Aminoacetamido)methyl]-1-[4-chloro-2-(o-chlorobenzoyl)phenyl]-N,N-dimethyl-1H-1,2,4-triazole-3-carboxamide.

450191-S [as hydrochloride monohydrate]; Rhythmy [as hydrochloride monohydrate]. Sedative/hypnotic. [hydrochloride dihydrate ($C_{21}H_{21}Cl_3N_6O_3.2H_2O$; 450191-S, Rhythmy)]: mp = 107°; LD_{50} (mus orl) > 1500 mg/kg. *Shionogi & Co. Ltd.*

9552 Roletamide
10078-46-3

$C_{16}H_{19}NO_4$

3',4',5'-Trimethoxy-3-(3-pyrrolin-1-yl)acrylophenone.

CL-59112. Sedative/hypnotic.

9553 Romifidine
65896-16-4

$C_9H_9BrFN_3$

2-(2-Bromo-6-fluoroanilino)-2-imidazoline.

STH-2130. alpha Adrenergic agonist; sedative (veterinary). *Boehringer Ingelheim Pharm. Inc.*

9554 Romifidine Hydrochloride
65896-14-2

$C_9H_{10}BrClFN_3$

2-(2-Bromo-6-fluoroanilino)-2-imidazoline hydrochloride.

STH-2130-Cl; Sedivet. alpha Adrenergic agonist; sedative (veterinary). *Boehringer Ingelheim Pharm. Inc.*

9555 Secobarbital
76-73-3 8563 200-982-2

$C_{12}H_{18}N_2O_3$

5-Allyl-5-(1-methylbutyl)-2,4,6(1H,3H,5H)-pyrimidinetrione.

Seconal; component of: Hyptran. Short acting sedative/hypnotic. mp = 100°. *Eli Lilly & Co.; Wallace Labs.*

9556 Secobarbital Sodium
309-43-3 8563 206-218-4

$C_{12}H_{17}N_2NaO_3$

5-Allyl-5-(1-methylbutyl)-2,4,6(1H,3H,5H)-pyrimidinetrione sodium salt.

meballymal sodium; quinalbarbitone sodium; Barbosec; Immenoctal; Pramil; Quinalspan; Sedutain; Seconal Sodium; seotalnatrium. Short acting sedative/hypnotic. Very soluble in H_2O, soluble in EtOH, insoluble in Et_2O; LD_{50} (rat orl) = 125 mg/kg. *Eli Lilly & Co.; Wyeth-Ayerst Labs.*

9557 Sodium Bromide
7647-15-6 8737 231-599-9

BrNa

Sedoneural. Sedative/hypnotic; anticonvulsant. Used in veterinary medicine as a seditive and to control convulsions. mp = 755°; d = 3.21; soluble in H_2O (91 g/100 ml), EtOH (6.25 g/100 ml), MeOH (16.6 g.100 ml); cyrstallizes as a dihydrate; LD_{50} (rat orl) = 3500 mg/kg.

9558 Sulfonethylmethane
76-20-0 9131 200-942-4
$C_8H_{18}O_4S_2$
2,2-Bis(ethylsulfonyl)butane.
methylsulfonal; Trional. Sedative/hypnotic. mp = 74-76°; soluble in H_2O (0.5 g/100 ml at 20°, 3.3 g/100 ml at 100°), EtOH (12.5 g/100 ml), Et_2O.

9559 Sulfonmethane
115-24-2 9133 204-074-7
$C_7H_{16}O_4S_2$
2,2-Bis(ethylsulfonyl)propane.
sulfonal; propane-diethylsulfone. Sedative/hypnotic. mp = 124-126°; bp = 300°; soluble in H_2O (0.27 g/100 ml at 20°, 6.25 g/100 ml at 100°), EtOH (1.67 g/100 ml at 20°, 33.3 g/100 ml at 76°), Et_2O (1.56 g/100 ml), $CHCl_3$ (9.1 g/100 ml), C_6H_6; insoluble in glycerol.

9560 Supidimide
49785-74-2 256-490-3
$C_{12}H_{12}N_2O_4S$
2-(2-Oxo-3-piperidyl)-1,2-benzisothiazoline-3-one-1,1-dioxide.
Sedative/hypnotic. Derivative of thalidimide.

9561 Suproclone
77590-92-2
$C_{22}H_{22}ClN_5O_4S_2$
4-Propionyl-1-piperazinecarboxylic acid, ester with (±)-6-(7-chloro-1,8-naphthyridin-2-yl)-2,3,6,7-tetrahydro-7-hydroxy-5H-p-dithiino[2,3-c]pyrrol-5-one.
Sedative/hypnotic.

9562 Taglutimide
14166-26-8
$C_{14}H_{16}N_2O_4$
cis-Endo-N-(2,6-Dioxo-3-piperidyl)-2,3-norborandedicarboximide.
K-2004. Sedative/hypnotic. A glutarimide derivative.

9563 Talbutal
115-44-6 9206 204-090-4
$C_{11}H_{16}N_2O_3$
5-Allyl-5-sec-butyl-2,4,6(1H,3H,5H)-pyrimidinetrione.
Lotusate. Sedative/hypnotic. Federally controlled substance (depressant). mp = 108-110°; insoluble in H_2O, petroleum ether; soluble in EtOH, $CHCl_3$, Et_2O, Me_2CO, AcOH; LD_{50} (rat orl) = 57.5 mg/kg. *Sterling Winthrop Inc.*

9564 Temazepam
846-50-4 9285 212-688-1
$C_{16}H_{13}ClN_2O_2$
7-Chloro-1,3-dihydro-3-hydroxy-1-methyl-5-phenyl-2H-1,4-benzodiazepin-2-one.
oxydiazepam; Wy-3917; ER-115; K-3917; Ro-5-5354; Euhypnos; Euipnos; Gelthix; Levanxene; Levanxol; Normison; Perdorm; Planum; Remestan; Restoril. Sedative/hypnotic; anxiolytic. A benzodiazepine. mp = 119-121°. *Hoffmann-LaRoche Inc.; Sandoz Pharm. Corp.*

9565 Tepirindole
72808-81-2
$C_{16}H_{19}ClN_2$
5-Chloro-3-(1,2,3,6-tetrahydro-1-propyl-4-pyridyl)indole.
HR-592. Nootropic; tranquilizer. A derivative of indole.

9566 Tetrabarbital
76-23-3 9324 200-946-6
$C_{12}H_{20}N_2O_3$
5-Ethyl-5-(1-ethylbutyl)-2,4,6(1H,3H,5H)-pyrimidinetrione.
JL-991; Butysal; Butysedal. Sedative/hypnotic. mp = 122°.

9567 Thalidomide
50-35-1 9390 200-031-1
$C_{13}H_{10}N_2O_4$
N-(2,6-Dioxo-3-piperidyl)phthalimide.
Kevadon; K-17; NSC-66847; Distaval; Softenon; Sedalis; Talimol; Pantosediv; Neurosedyn; Contergan. Formerly used as a sedative/hypnotic but discontinued because of its marked teratogenicity. Has recently been shown to be effective against bone cancer. mp = 269-271°; λ_m 220, 300 nm (pH 7); poorly soluble in H_2O (0.005 g/100 ml); sparingly soluble in H_2O, MeOH, EtOH, EtOAc, AcOH; very soluble in C_5H_5N, DMF, dioxane; insoluble in Et_2O, $CHCl_3$, C_6H_6. *Grunenthal; Marion Merrell Dow Inc.*

9568 Thioridazine
50-52-2 9497 200-044-2
$C_{21}H_{26}N_2S_2$
10-[2-(1-Methyl-2-piperidyl)ethyl]-2-(methylthio)-phenothiazine.
Mellaril-S. Antipsychotic; sedative. A dopamine receptor blocker; parent compound of sulforidazine and mesoridazine. A phenothiazine derivative. mp = 72-74°; $bp_{0.02}$ - 230°; λ_m 263, 314 nm (ε 38172, 4595 EtOH), 230, 263 nm (ε 20939, 45954 0.1N HCl), 313 nm (ε 5226 0.1N NaOH); insoluble in H_2O, soluble in EtOH 16.6 g/100 ml), $CHCl_3$ 123 g/100 ml), Et_2O (33.3 g/100 ml); L. *Sandoz Pharm. Corp.*

9569 Thioridazine Hydrochloride
130-61-0 9497 204-992-8
$C_{21}H_{27}ClN_2S_2$
10-[2-(1-Methyl-2-piperidyl)ethyl]-2-(methylthio)-phenothiazine monohydrochloride.
TP-21; Aldazine; Mallorol; Mellaril; Melleretten; Melleril; Novoridazine; Orsanil; Ridazin; Stalleril. Antipsychotic, used also as a sedative/hypnotic. A phenothiazine dopamine receptor blocker. mp = 158-160°; λ_m = 262, 310 nm (ε 41842, 3215 H_2O), 264, 310 nm (ε 41598, 3256 EtOH), 264 305 nm (ε 42371 5495 0.1N HCl), 263 nm (ε 18392 0.1N NaOH); soluble in H_2O (11.1 g/100 ml), EtOH (10 g/100 ml), MeOH, $CHCl_3$ (20 g/100 ml). *Sandoz Pharm. Corp.*

9570 Tracazolate
41094-88-6 255-214-9
$C_{16}H_{24}N_4O_2$
Ethyl 4-(butylamino)-1-ethyl-6-methyl-1H-pyrazolo[3,4-b]pyridine-5-carboxylate.
ICI-136753. Sedative/hypnotic. *ICI.*

9571 Trepipam Maleate
39624-66-3 254-546-1
$C_{23}H_{27}NO_6$
(+)-2,3,4,5-Tetrahydro-7,8-dimethoxy-3-methyl-1-phenyl-1H-3-benzazepine maleate (1:1).
Sch-12679; Trimopam Maleate. Sedative/hypnotic. *Schering-Plough HealthCare Products.*

9572 Triazolam
28911-01-5 9734 249-307-3
$C_{17}H_{12}Cl_2N_4$
8-Chloro-6-(o-chlorophenyl)-1-methyl-4H-s-triazolo-[4,3-a][1,4]benzodiazepine.
Halcion; U-33030; Novodorm; Songar. Sedative/hypnotic. A benzodiazepine. mp = 233-235°; LD_{50} (mus orl) > 100 mg/kg, (rat orl) > 5000 mg/kg. *Pharmacia & Upjohn.*

9573 Tricetamide
363-20-2
$C_{16}H_{24}N_2O_5$
N-[(Diethylcarbamoyl)methyl]-3,4,5-trimethoxy-benzamide.
R-548. Sedative/hypnotic. *3M Pharm.*

9574 2,2,2-Trichloroethanol
115-20-8 9768 204-071-0
$C_2H_3Cl_3O$
Trichloroethyl alcohol.
Sedative/hypnotic. mp = 18°; bp = 151-153°; d_{20}^{20} = 1.55; soluble in H_2O (8.3 g/100 ml), EtOH, Et_2O; LD_{50} (rat orl) = 600 mg/kg. *Callery Chem. Co.*

9575 Trichlorourethan
107-69-7 9778 203-511-9
$C_3H_4Cl_3NO_2$
2,2,2-Trichloroethanol carbamate.
carbamic acid trichloroethyl ester. Sedative/hypnotic. mp = 64-65°; soluble in H_2O (1 g/100 ml), very soluble in $CHCl_3$, EtOH, Et_2O; soluble in C_6H_6; sparingly soluble in petroleum ether.

9576 Triclofos
306-52-5 9788 206-185-6
$C_2H_4Cl_3O_4P$
2,2,2-Trichlorethanol dihydrogen phosphate.
trichloroethyl phosphate. Sedative/hypnotic. *Marion Merrell Dow Inc.; Schering-Plough HealthCare Products.*

9577 Triclofos Sodium
7246-20-0 9788 230-652-3
$C_2H_3Cl_3NaO_4P$
2,2,2-Trichlorethanol dihydrogen phosphate monosodium salt.
Triclos; Sch-10159; Trichloryl; Tricloryl. Sedative/hypnotic. Soluble in H_2O, LD_{50} (mus orl) = 1400 mg/kg. *Marion Merrell Dow Inc.; Schering-Plough HealthCare Products.*

9578 Trifluoperazine
117-89-5 9811 204-219-4
$C_{21}H_{24}F_3N_3S$
10-[3-(4-Methyl-1-piperazinyl))propyl]-2-(trifluoromethyl)phenothiazine.
Antipsychotic, used also as a sedative/hypnotic. A phenothiazine. $bp_{0.6}$ = 202-210°; λ_m 258, 307.5 (log ε 4.50 3.50 EtOH); LD_{50} (rat orl) = 542.7 mg/kg, (mus orl) = 424.0 mg/kg. *Apothecon; SmithKline Beecham Pharm.*

9579 Trifluoperazine Dihydrochloride
440-17-5 9811 207-123-0
$C_{21}H_{26}Cl_2F_3N_3S$
10-[3-(4-Methyl-1-piperazinyl)propyl]-2-(trifluoromethyl)phenothiazine dihydrochloride.
triftazin; triphthasine; Eskazinyl; Eskazine; Jatroneural; Modalina; Stelazine; Terfluzine. Antipsychotic, used also as a sedative/hypnotic. A phenothiazine. mp = 242-243°; freely soluble in H_2O, insoluble in Et_2O, C_6H_6; hygroscopic. *Labaz S.A.; SmithKline Beecham Pharm.*

9580 Trimetozine
635-41-6 9850 211-236-0
$C_{14}H_{19}NO_5$
4-(3,4,5-Trimethoxybenzoyl)morpholine.
Abbott-22370; PS-2383; NSC-62939; V-7; Opalène; Trioxazine. Anxiolytic; sedative/hypnotic. mp = 120-122°; slightly soluble in H_2O, EtOH. *Abbott Labs Inc.*

9581 Uldazepam
28546-58-9
$C_{18}H_{15}Cl_2N_3O$
2-[(Allyloxy)amino]-7-chloro-5-(o-chlorophenyl)-3H-1,4-benzodiazepine.
U-31920. Anxiolytic with sedative/hypnotic properties.

9582 Valofane
3258-51-3 221-857-9
$C_{10}H_{14}N_2O_4$
(3-Allyltetrahydro-5-methyl-2-oxo-3-furoyl)urea.
valofan. An atypical barbiturate. A tautomer (and prodrug) of proxibarbal.

9583 Valtrate
18296-44-1 242-174-2
$C_{22}H_{30}O_8$
1,7a-Dihydro-1,6-dihydroxyspiro[cyclopenta[c]pyran-7-(6H),2'-oxirane]-4-methanol 4-acetate 1,6-diisovalerate.
Sedative. One of the valepotriates (present in various Valeriana species), a group of chemically unstable iridoid triesters possessing sedative activity.

9584 Vinbarbital Sodium
125-44-0 10118
$C_{11}H_{15}N_2NaO_3$
5-Ethyl-5-(1-methyl-1-butenyl)-2,4,6(1H,3H,5H)-pyrimidinetrione.
Delvinal Sodium. Sedative/hypnotic. Slightly soluble in Et_2O, $CHCl_3$; LD_{50} (rat orl) = 130 mg/kg; [free acid ($C_{11}H_{15}N_2O_3$)]: mp = 161-163°. *Merck & Co.Inc.*

9585 Vinylbital
2430-49-1 10131 219-395-8
$C_{11}H_{16}N_2O_3$
5-(1-Methylbutyl)-5-vinyl-2,4,6(1H,3H,5H)-pyrimidinetrione.
JD-96; butyvinal; Speda; Optanox. Sedative/hypnotic. Federally controlled substance (depressant). mp = 90-91.5°.

9586 Zaleplon
151319-34-5
$C_{17}H_{15}N_5O$
3'-(3-Cyanopyrazolo[1,5-α]pyrimidin-7-yl)-N-ethylacetanilide.
CL-284846. Sedative/hypnotic.

9587 Zolazepam
31352-82-6
$C_{15}H_{15}FN_4O$
4-(o-Fluorophenyl)-6,8-dihydro-1,3,8-trimethylpyrazolo[3,4-e][1,4]diazepin-7(1H)-one.
Sedative/hypnotic. A benzodiazepine. *Parke-Davis.*

9588 Zolazepam Hydrochloride
33754-49-3 251-668-7
$C_{15}H_{16}ClFN_4O$
4-(o-Fluorophenyl)-6,8-dihydro-1,3,8-trimethylpyrazolo[3,4-e][1,4]diazepin-7(1H)-one hydrochloride.
CI-716; component of: Telazol. Sedative/hypnotic. A benzodiazepine. *Parke-Davis.*

9589 Zolpidem
82626-48-0 10321
$C_{19}H_{21}N_3O$
N,N,6-Trimethyl-2-p-tolylimidazo[1,2-a]pyridine-3-acetamide.
SL-800750. Sedative/hypnotic. mp = 196°. *Synthelabo Pharmacie.*

9590 Zolpidem Tartrate
99294-93-6 10321
$C_{42}H_{48}N_6O_8$
N,N,6-Trimethyl-2-p-tolylimidazo[1,2-a]pyridine-3-acetamide L-(+)-tartrate (2:1).
SL-800750-23N; Ambien; Ivadal; Niotal; Stilnoct; Stilnox. Sedative/hypnotic. Soluble in H_2O (2.3 g/100 ml). *Synthelabo Pharmacie.*

9591 Zopiclone
43200-80-2 10324 256-138-9
$C_{17}H_{17}ClN_6O_3$
4-Methyl-1-piperazinecarboxylic acid ester with 6-(5-chloro-2-pyridyl)-6,7-dihydro-7-hydroxy-5H-pyrrolo[3,4-b]pyrazin-5-one.
RP-27267; Amoban; Amovane; Imovance; Imovane; Sopivan; Zimovane. Sedative/hypnotic. mp = 178°. *Rhône-Poulenc Rorer Pharm. Inc.*

Serenics

9592 Eltoprazine
98224-03-4 3592
$C_{12}H_{16}N_2O_2$
1-(1,4-Benzodioxan-5-yl)piperazine.
Serenic. Serotonin receptor agonist. *Philips-Duphar B.V.*

9593 Eltoprazine Hydrochloride
98206-09-8 3592
$C_{12}H_{17}ClN_2O_2$
1-(1,4-Benzodioxan-5-yl)piperazine hydrochloride.
DU-28853. Serenic. Serotonin receptor agonist. mp = 256-258°; soluble in H_2O (19 g/100 ml). *Philips-Duphar B.V.*

9594 Fluprazine
76716-60-4
$C_{14}H_{19}F_3N_4O$
[2-[4-(α,α,α-Trifluoro-m-tolyl)-1-piperazinyl]ethyl]urea.
Serenic.

Serotonin Receptor Agonists

9595 Buspirone
36505-84-7 1528 253-072-2
$C_{21}H_{31}N_5O_2$
N-[4-[4-(2-Pyrimidinyl)-1-piperazinyl]butyl]-1,1-cyclopentanediacetamide hydrochloride.
Serotonin receptor agonist. Non benzodiazepine anxiolytic, used as a minor tranquillizer. An azapirone. *Mead Johnson Pharmaceuticals.*

9596 Buspirone Hydrochloride
33386-08-2 1528 251-489-4
$C_{21}H_{32}ClN_5O_2$
N-[4-[4-(2-Pyrimidinyl)-1-piperazinyl]butyl]-1,1-cyclopentanediacetamide hydrochloride.
Ansial; Ansiced; Axoren; Bespar; Buspar; Buspimem; Buspinol; Buspisal; Censpar; Lucelan; Narol; Travin. Non-benzodiazepine anxiolytic, used as a minor tranquillizer. An azapirone. Serotonin receptor agonist. mp = 201.5-202.5°; LD_{50} (rat ipr) = 136 mg/kg. *Mead Johnson Pharmaceuticals.*

9597 Dihydroergotamine
511-12-6 3217 208-123-3
$C_{33}H_{37}N_5O_5$
9,10-Dihydro-12'-hydroxy-2'-methyl-5'-(phenylmethyl)-ergotoman-3',6',18-trione.
Divegal [as tartrate]. Antiadrenergic; antimigraine. α-Adrenergic blocker with venoconstrictor activity. Also binds to serotonin $5HT_1$-receptors. mp = 239°; $[\alpha]_D^{20}$= -64°, $[\alpha]_{546}^{20}$ = -79° (c = 0.5 C_5H_5N); insoluble in H_2O; sparingly soluble in EtOH, MeOH, $CHCl_3$, C_6H_6; Dec 210-215°. *Sandoz Pharm. Corp.*

9598 Dihydroergotamine Methanesulfonate
6190-39-2 3217 228-235-6
$C_{34}H_{41}N_5O_8S$
9,10-Dihydro-12'-hydroxy-2'-methyl-5'-(phenylmethyl)-ergotoman-3',6',18-trione monomethanesulfonate.
DHE-45; Agit; Angionorm; Dergotamine; DET MS; D.H.E. 45; Diergo; Dihydergot; Dirgotarl; Endophleban; Ergomimet; Ergont; Ergotonin; Ikaran; Migranal; Morena; Orstanorm; Séglor; Tonopres; Verladyn. Antiadrenergic; antimigraine; α-adrenergic blocker; venoconstrictor. Also binds to serotonin $5HT_1$-receptors. mp = 230-235°; moderatley soluble in H_2O. *Sandoz Pharm. Corp.*

9599 Eltoprazine
98224-03-4 3592
$C_{12}H_{16}N_2O_2$
1-(1,4-Benzodioxan-5-yl)piperazine.
Serenic. Serotonin receptor agonist. *Philips-Duphar B.V.*

9600 Eltoprazine Hydrochloride
98206-09-8 3592
$C_{12}H_{17}ClN_2O_2$
1-(1,4-Benzodioxan-5-yl)piperazine hydrochloride.
DU-28853. Serenic. Serotonin receptor agonist. mp = 256-258°; soluble in H_2O (19 g/100 ml). *Philips-Duphar B.V.*

9601 Ergotamine
113-15-5 3703 204-023-9
$C_{33}H_{35}N_5O_5$
(5'α)-12'-Hydroxy-2'-methyl-5'-(phenylmethyl)-ergotoman-3',6',18-trione.
Ergoton-A [as succinate]. Serotonin receptor agonist. Analgesic, specific to migraine. An alkaloid from ergot, a dark colored fungus, which attacks damp rye and other grasses, and which, when contained in flour, causes ergotism. Used in medicine as a vasoconstrictor. Dec 212-214°; $[\alpha]_D^{20}$ = -160° ($CHCl_3$); insoluble in H_2O, petroleum ether; soluble in MeOH (1.4 g/100 ml), EtOH (0.33 g/100 ml), Me_2CO (0.67 g/100 ml); freely soluble in $CHCl_3$, C_5H_5N, AcOH; soluble in EtOAc; slightly soluble in. *3M Pharm.; Fisons plc; Organon Inc.; Parke-Davis; Sandoz Pharm. Corp.*

9602 Ergotamine Tartrate
379-79-3 3703 206-835-9
$C_{70}H_{76}N_{10}O_{16}$
12'-Hydroxy-2'-methyl-5'α-(phenylmethyl)ergotaman-3',6',18-trione [R-(R*,R*)]-2,3-dihydroxy-butanedioate (2:1) salt.
Ergate; Ergomar; Ergostat; Ergotartrat; Exmigra; Femergin; Gynergin; Lingraine; Lingran; Medihaler Ergotamine; component of: Cafergot, Wigraine. Analgesic, specific to migraine. Antimigraine. Has also been used as an oxytocic in veterinary medicine. Serotonin receptor agonist. mp = 203° (dec); $[\alpha]_D^{25}$ = -125 to -155° (c = 0.4 $CHCl_3$); soluble in H_2O (2 mg/ml), EtOH; LD_{50} (rat iv) = 80 mg/kg. *3M Pharm.; Fisons Pharm. Div.; Organon Inc.; Parke-Davis; Sandoz Pharm. Corp.*

9603 Flesinoxan
98206-10-1 4138
$C_{22}H_{26}FN_3O_4$
(+)-(S)-p-Fluoro-N-[2-[4-[2-(hydroxymethyl)-1,4-benzodioxan-5-yl]-1-piperazinyl]ethyl]-benzamide.
Anxiolytic. Serotonin $5\text{-}HT_{1A}$ receptor agonist. A phenylpiperazine derivative. *Solvay Duphar Labs Ltd.*

9604 Flesinoxan Hydrochloride
98205-89-1 4138
$C_{22}H_{27}ClFN_3O_4$
(+)-(S)-p-Fluoro-N-[2-[4-[2-(hydroxymethyl)-1,4-benzodioxan-5-yl]-1-piperazinyl]ethyl]benzamide hydrochloride.
DU-29373. Anxiolytic. Serotonin $5\text{-}HT_{1A}$ receptor agonist. A phenylpiperazine derivative. mp = 184.5-185.5°, 183-184°; $[\alpha]_D$ = 25° (c = 1 MeOH); freely soluble in H_2O, sparingly soluble in EtOH. *Solvay Duphar Labs Ltd.*

9605 Ipsapirone
95847-70-4 5096
$C_{19}H_{23}N_5O_3S$
2-[4-[4-(2-Pyrimidinyl)-1-piperazinyl]butyl]-1,2-benzoisothiazolin-3-one 1,1-dioxide.
Isapirone. A serotonin (5-hydroxytryptamine, $5\text{-}HT_1$) receptor antagonist. Non-benzodiazepine anxiolytic. mp = 137-138°. *Bayer AG.*

9606 Ipsapirone Hydrochloride
92589-98-5 5096
$C_{19}H_{24}ClN_5O_3S$
2-[4-[4-(2-Pyrimidinyl)-1-piperazinyl]butyl]-1,2-benzoisothiazolin-3-one 1,1-dioxide monohydrochloride.
Bay q 7821; TVX Q 7821. Non-benzodiazepine anxiolytic. A serotonin (5-hydroxytryptamine, $5\text{-}HT_1$) receptor agonist. mp = 221-222°. *Bayer AG.*

9607 Lesopitron
132449-46-8 5470
$C_{15}H_{21}ClN_6$
2-[4-[4-(4-Chloropyrazol-1-yl)butyl]-1-piperazinyl]pyrimidine.
Anxiolytic. Selective serotonin $5HT_{1A}$ receptor agonist. *Esteve Group.*

9608 Lesopitron Dihydrochloride
132449-89-9 5470
$C_{15}H_{23}Cl_3N_6$
2-[4-[4-(4-Chloropyrazol-1-yl)butyl]-1-piperazinyl]pyrimidine dihydrochloride.
E-4424. Anxiolytic. Selective serotonin $5HT_{1A}$ receptor agonist. mp = 194-197.5°. *Esteve Group.*

9609 Methysergide
361-37-5 6217 206-644-0
$C_{21}H_{27}N_3O_2$
(+)-9,10-Didehydro-N-[1-(hydroxymethyl)propyl]-1-methyl-D-lysergamide.
UML-491. Serotonin receptor agonist. Vasoconstrictor, specific to migraine. Antimigraine. mp = 194-196°; $[\alpha]_D^{20}$ = -45° (c = 0.5 C_5H_5N); [dimaleate]: mp = 165°; soluble in MeOH, less soluble in H_2O, insoluble in EtOH. *Sandoz Pharm. Corp.*

9610 Methysergide Hydrogen Maleate
129-49-7 6217 204-950-9
$C_{25}H_{31}N_3O_6$
(+)-9,10-Didehydro-N-[1-(hydroxymethyl)propyl]-1-methyl-D-lysergamide hydrogen maleate.
Sansert; Deseril; Désernil. Serotonin receptor agonist. Vasoconstrictor, specific to migraine. Antimigraine. *Sandoz Pharm. Corp.*

9611 Sumatriptan
103628-46-2 9172
$C_{14}H_{21}N_3O_2S$
3-[2-(Dimethylamino)ethyl]-N-methylindole-5-methanesulfonamide.
GR-43175. Serotonin $5HT_1$ receptor agonist. Antimigraine. mp = 169-171°. *Glaxo Wellcome Inc.*

9612 Sumatriptan Succinate
103628-48-4 9172
$C_{18}H_{27}N_3O_6S$
3-[2-(Dimethylamino)ethyl]-N-methylindole-5-methanesulfonamide succinate (1:1).
Imigran; Imitrex; GR-43175C. Serotonin $5HT_1$ receptor agonist. Antimigraine. mp = 165-166°. *Glaxo Wellcome Inc.*

9613 Tandospirone
87760-53-0 9219
$C_{21}H_{29}N_5O_2$
(3aα,4β,7β,7aα)-Hexahydro-2-[4-[4-(2-pyrimidinyl)-1-piperazinyl]butyl]-4,7-methano-1H-isoindole-1,3(2H)-dione.
Anxiolytic; antidepressant. A serotonin receptor agonist. mp = 112-113.5°. *Pfizer Inc.; Sumitomo Pharm. Co. Ltd.*

9614 Tandospirone Citrate
112457-95-1 9219
$C_{27}H_{37}N_5O_9$
(3aα,4β,7β,7aα)-Hexahydro-2-[4-[4-(2-pyrimidinyl)-1-piperazinyl]butyl]-4,7-methano-1H-isoindole-1,3(2H)-dione 2-hydroxy-1,2,3-propanetricarboxylate (1:1).
SM-3997. Serotonin receptor agonist. Anxiolytic; antidepressant. mp = 169.5-170°; [hydrochloride]: mp = 227-229°. *Pfizer Inc.; Sumitomo Pharm. Co. Ltd.*

Serotonin Receptor Antagonists

9615 Azasetron
123040-69-7 933
$C_{17}H_{20}ClN_3O_3$
(±)-6-Chloro-3,4-dihydro-4-methyl-3-oxo-N-3-quinuclidinyl-2H-1,4-benzoxazine-8-carboxamide.
nazasetron. Antiemetic. Specific serotonin $5HT_3$ receptor antagonist. *Yoshitomi.*

9616 Azasetron Hydrochloride
141922-90-9 933
$C_{17}H_{21}Cl_2N_3O_3$
(±)-6-Chloro-3,4-dihydro-4-methyl-3-oxo-N-3-quinuclidinyl-2H-1,4-benzoxazine-8-carboxamide hydrochloride.
Y-25130; Serotone. Antiemetic. Specific serotonin $5HT_3$ receptor antagonist. mp= 281° (dec), 305° (dec); LD_{50} (mrat iv) = 135 mg/kg, (frat iv) = 132 mg/kg. *Yoshitomi.*

9617 Dolasetron
115956-12-2 3471
$C_{19}H_{20}N_2O_3$
Indole-3-carboxylic acid ester with (8r)-hexahydro-8-hydroxy-2,6-methano-2H-quinolizin-3(4H)-one.
MDL-73147. Antiemetic; antimigraine. Serotonin receptor antagonist. *Hoechst Marion Roussel Inc.*

9618 Dolasetron Mesylate
115956-13-3 3471
$C_{20}H_{24}N_2O_6S$
Indole-3-carboxylic acid ester with (8r)-hexahydro-8-hydroxy-2,6-methano-2H-quinolizin-3(4H)-one monomethanesulfonate.
Anxemet; MDL-73147EF. Antiemetic; antimigraine. Serotonin receptor antagonist. mp = 278°. *Hoechst Marion Roussel Inc.*

9619 Granisetron
109889-09-0 4557
$C_{18}H_{24}N_4O$
1-Methyl-N-(9-methyl-endo-9-azabicyclo[3.3.1]non-3-yl)-1H-imidazole-3-carboxamide.
BRL-43694. Antiemetic. Specific serotonin $5HT_3$ receptor antagonist. Used in treatment of nausea in cancer patients. *SmithKline Beecham Pharm.*

9620 Granisetron Hydrochloride
107007-99-8 4557
$C_{18}H_{25}ClN_4O$
1-Methyl-N-(9-methyl-endo-9-azabicyclo[3.3.1]non-3-yl)-1H-imidazole-3-carboxamide hydrochloride.
BRL-43694A; Kytril. Antiemetic. Specific serotonin $5HT_3$ receptor antagonist. Used to treat nausea in cancer patients. mp = 290-292°. *SmithKline Beecham Pharm.*

9621 Ketanserin
74050-98-9 5307 277-680-2
$C_{22}H_{22}FN_3O_3$
3-[2-[4-(p-Fluorobenzoyl)piperidino]ethyl]-2,4-(1H,3H)-quinazolinedione.
R-41468. Specific serotonin $5HT_2$-receptor antagonist used as an antihypertensive agent. mp = 227-235°; soluble in H_2O (0.001 g/100 ml), EtOH (0.038 g/100 ml), DMF (2.34 g/100 ml). *Janssen Pharm. Inc.*

9622 Ketanserin Tartrate
83846-83-7 5307 281-062-8
$C_{26}H_{28}FN_3O_9$
3-[2-[4-(p-Fluorobenzoyl)piperidino]ethyl]-2,4-(1H,3H)-quinazolinedione tartrate.
R-49945; Ket; Perketan; Serepress; Sufrexal. Specific serotonin $5HT_2$-receptor antagonist used as an antihypertensive agent. *Janssen Pharm. Inc.*

9623 Mesulergine
64795-35-3
$C_{18}H_{26}N_4O_2S$
N'-(1,6-Dimethylergolin-8α-yl)-N,N-dimethylsulfamide.
Serotonin 5-hydroxytryptamine (5-HT) receptor antagonist.

9624 Metergoline
17692-51-2 5999 241-686-3
$C_{25}H_{29}N_3O_2$
(+)-N-(Carboxy)-1-methyl-9,10-dihydrolysergamine benzyl ester.
methergoline; FI-6337; MCE; Liserdol; Contralac. Seratonin 5-hydroxytryptamine receptor antagonist; prolactin inhibitor. mp = 146-149°; $[\alpha]_D^{28}$ = -7±2°; λ_m = 291 nm ($E_{1cm}^{1\%}$ 165; soluble in C_5H_5N, alcohol, Me_2OH $CHCl_3$; nearly insoluble in C_6H_6, H_2O, Et_2O; LD_{50} (mus orl) = 430 mg/kg. *Farmitalia Carlo Erba Ltd.*

9625 Methysergide
361-37-5 6217 206-644-0
$C_{21}H_{27}N_3O_2$
(+)-9,10-Didehydro-N-[1-(hydroxymethyl)propyl]-1-methyl-D-lysergamide.
UML-491. Serotonin receptor antagonist. Vasoconstrictor, specific to migraine. Antimigraine. mp = 194-196°; $[\alpha]_D^{20}$=

-45° (c = 0.5 C_5H_5N); [dimaleate]: mp = 165°; soluble in MeOH, less soluble in H_2O, insoluble in EtOH. *Sandoz Pharm. Corp.*

9626 Methysergide Hydrogen Maleate
129-49-7 6217 204-950-9
$C_{25}H_{31}N_3O_6$
(+)-9,10-Didehydro-N-[1-(hydroxymethyl)propyl]-1-methyl-D-lysergamide hydrogen maleate.
Sansert; Deseril; Désernil. Serotonin receptor antagonist. Vasoconstrictor, specific to migraine. Antimigraine. *Sandoz Pharm. Corp.*

9627 Naftopidil
57149-07-2 6443
$C_{24}H_{28}N_2O_3$
4-(o-Methoxyphenyl)-α-[(1-naphthyloxy)methyl]-1-piperazineethanol.
KT-611; Avishot; Flivas. An α_1 adrenergic blocker and serotonin $5HT_{1A}$ receptor agonist. Used as an antihypertensive agent. Also used in treatment of BPH. mp = 125-126°, 125-129°; insoluble in H_2O; LD_{50} (mus orl) = 1300 mg/kg, (rat orl) = 6400 mg/kg. *Boehringer Ingelheim Ltd.; Boehringer Mannheim GmbH; C.M. Ind.*

9628 Naftopidil Dihydrochloride
57149-08-3 6443
$C_{24}H_{30}Cl_2N_2O_3$
4-(o-Methoxyphenyl)-α-[(1-naphthyloxy)methyl]-1-piperazineethanol dihydrochloride.
An α_1 adrenergic blocker and serotonin $5HT_{1A}$ receptor agonist. Used as an antihypertensive agent. Also used in treatment of BPH. mp = 212-213°. *Boehringer Mannheim GmbH.*

9629 Nefazodone
83366-66-9 6527
$C_{25}H_{32}ClN_5O_2$
2-[3-[4-(m-Chlorophenyl)-1-piperazinyl]propyl]-3-ethyl-4-(2-phenoxyethyl)-Δ^2-1,2,4-triazolin-5-one.
Antidepressant. Selective serotonin $5\text{-}HT_2$ receptor antagonist. mp = 83-84°. *Bristol-Myers Squibb Pharm. R&D.*

9630 Nefazodone Hydrochloride
82752-99-6 6527
$C_{25}H_{33}Cl_2N_5O_2$
2-[3-[4-(3-Chlorophenyl)-1-piperazinyl]propyl]-5-ethyl-2,4-dihydro-4-(2-phenoxyethyl)-3H-1,2,4-triazol-3-one monohydrochloride.
BMY-13754; MJ-13754-1; Dutonin; Serazone. Antidepressant. Selective serotonin $5\text{-}HT_2$ receptor antagonist. mp (crystals from iPrOH) = 186-87°, mp (crystals from EtOH) = 175-177°. *Bristol-Myers Squibb Pharm. R&D; Mead Johnson Pharmaceuticals.*

9631 Nexopamil
136033-49-3
$C_{24}H_{40}N_2O_3$
(2S)-5-(Hexylmethylamino)-2-isopropyl-2-(3,4,5-trimethoxyphenyl)valeronitrile.
LU-49938. Serotonin 5-HT2 receptor and calcium channel antagonist. A verapamil derivative.

9632 Ondansetron
116002-70-1 6979
$C_{18}H_{19}N_3O$
(±)-2,3-Dihydro-9-methyl-3-[(2-methylimidazol-1-yl)-methyl]carbazol-4(1H)-one.
Anxiolytic; antiemetic; antischizophrenic. Specific serotonin receptor ($5HT_3$) antagonist. mp = 231-232°; [3S-form]: $[\alpha]_D^{25}$ = -14° (c = 0.19, MeOH); [3R-form]: $[\alpha]_D^{24}$= +16° (c = 0.34, MeOH). *Glaxo Wellcome Inc.*

9633 Ondansetron Hydrochloride
103639-04-9 6979
$C_{18}H_{20}ClN_3O.2H_2O$
(±)-2,3-Dihydro-9-methyl-3-[(2-methylimidazol-1-yl)-methyl]carbazol-4(1H)-one monohydrochloride dihydrate.
Zofran; GR-38032F. Serotonin receptor antagonist. Anxiolytic; antiemetic; antischizophrenic. mp = 178.5-179.5°. *Glaxo Wellcome Inc.*

9634 Oxetorone
26020-55-3 7068 247-411-3
$C_{21}H_{21}NO_2$
3-Benzofuro[3,2-c][1]benzoxepin-6(12H)-ylidene-N,N-dimethyl-1-propanamine.
Serotonin and histamine antagonist with antimigraine activity. *Labaz S.A.*

9635 Oxetorone Fumarate
34522-46-8 7068
$C_{25}H_{25}NO_6$
3-Benzofuro[3,2-c][1]benzoxepin-6(12H)-ylidene-N,N-dimethyl-1-propanamine fumarate.
L-6257; Nocertone; Oxedix. Serotonin, histamine antagonist with antimigraine activity. mp = 160°. *Labaz S.A.*

9636 Ramosetron
132036-88-5
$C_{17}H_{17}N_3O$
(-)-(R)-Methylindol-3-yl-4,5,6,7-tetrahydro-5-benzimidazolyl ketone.
YM-060. Potent serotonin 5-HT3-receptor antagonist.

9637 Ritanserin
87051-43-2 8399
$C_{27}H_{25}F_2N_3OS$
6-[2-[4-[Bis(p-fluorophenyl)methylene]piperidino]ethyl]-7-methyl-5H-thiazolo[3,2-a]pyrimidin-5-one.
R-55667; Tiserton. Anxiolytic; antidepressant. A selective serotonin receptor antagonist. mp = 145.5°; LD_{50} (mmus iv) = 28.2 mg/kg, (mmus orl) = 626 mg/kg, (fmus iv) = 28.2 mg/kg, (fmus orl) = 993 mg/kg, (mrat iv) = 20.0 mg/kg, (mrat orl) = 856 mg/kg, (frat iv) = 22.2 mg/kg, (frat orl) = 515 mg/kg, (mdog iv) = 24.1 mg/kg, (mdog orl) = 1280 mg/kg, (fdog iv) = 33.2 mg/kg, (fdog orl) = 640-1280 mg/kg. *Janssen Pharm. Inc.*

9638 Ritanserin L-Tartrate
93076-39-2 8399
$C_{31}H_{31}F_2N_3O_7S$
6-[2-[4-[Bis(p-fluorophenyl)methylene]-piperidino]ethyl]-7-methyl-5H-thiazolo-[3,2-a]pyrimidin-5-one L-tartrate.
Serotonin receptor antagonist. Anxiolytic, antidepressant. mp = 198.7°. *Janssen Pharm. Inc.*

9639 Sarpogrelate
125926-17-2
$C_{24}H_{31}NO_6$
(±)-2-(Dimethylamino)-1-[[o-(m-methoxyphenethyl)-phenoxy]ethyl hydrogen succinate.
MCI-9042 [as hydrochloride]; Anplag [as hydrochloride]. A serotonin 2A (5-HT2A) receptor antagonist. Antiplatelet agent. Antithromobotic.

9640 Seganserin
87729-89-3
$C_{29}H_{27}F_2N_3O$
3-[2-[4-[Bis(p-fluorophenyl)methylene]piperidino]ethyl]-2-methyl-4H-pyrido[1,2-a]pyrimidin-4-one.
R-56413. A serotonin 5-HT2 receptor antagonist.

9641 Tipindole
7489-66-9
$C_{16}H_{20}N_2O_2S$
2-(Dimethylamino)ethyl 1,3,4,5-tetrahydrothio-pyrano[4,3-b]indole-8-carboxylate.
Serotonin receptor antagonist.

9642 Tropisetron
89565-68-4 9914
$C_{17}H_{20}N_2O_2$
1αH,5αH-Tropan-3α-yl indole-3-carboxylate.
ICS-205-930. Antiemetic. Serotonin receptor antagonist. mp = 201-202°. *Sandoz Pharm. Corp.*

9643 Tropisetron Hydrochloride
105826-92-4 9914
$C_{17}H_{21}ClN_2O_2$
1αH,5αH-Tropan-3α-yl indole-3-carboxylate hydrochloride.
Navoban; Novaban. Serotonin receptor antagonist. Antiemetic. mp = 283-285°. *Sandoz Pharm. Corp.*

Serotonin Uptake Inhibitors

9644 Femoxetine
59859-58-4 3993
$C_{20}H_{25}NO_2$
(+)-trans-3-[(p-Methoxyphenoxy)methyl]-1-methyl-4-phenylpiperidine.
Antidepressant. Serotonin uptake inhibitor. *Fabre Pierre; Ferrosan A/S.*

9645 Femoxetine Hydrochloride
56222-04-9 3993
$C_{20}H_{26}ClNO_2$
(+)-trans-3-[(p-Methoxyphenoxy)methyl]-1-methyl-4-phenylpiperidine hydrochloride.
FG-4963; Malexil. Antidepressant. Serotonin uptake inhibitor. LD_{50} (fmus iv) = 48 mg/kg, (fmus sc) = 941 mg/kg, (fmus orl) = 1408 mg/kg, (mmus iv) = 45 mg/kg, (mmus sc) = 723 mg/kg, (mmus orl) = 1687 mg/kg. *Fabre Pierre; Ferrosan A/S.*

9646 Fluoxetine
54910-89-3 4222
$C_{17}H_{18}F_3NO$
(±)-N-Methyl-3-phenyl-3-[(α,α,α-trifluoro-p-tolyl)oxy]propylamine.
Serotonin uptake inhibitor. Antidepressant. *C.M. Ind.; Eli Lilly & Co.*

9647 Fluoxetine Hydrochloride
59333-67-4 4222 260-101-2
$C_{17}H_{19}ClF_3NO$
(±)-N-Methyl-3-phenyl-3-[(α,α,α-trifluoro-p-tolyl)oxy]propylamine hydrochloride.
LY-110140; Adofen; Fluctin; Fluoxeren; Fontex; Foxetin; Prozac; Reneuron. Antidepressant. Serotonin uptake inhibitor. mp = 158-159°; slightly soluble in H_2O (1-2 mg/ml), ethyl acetate, toluene, $CHCl_3$, C_6H_{14} (0.5-0.77 mg/ml); soluble in MeOH, ETOH (>100 mg/mg), acetonitrile (33-100 mg/ml), C_6H_6; λ_m = 227, 264, 268, 275 nm ($E^{1\%}_{1\,cm}$ 372, 29,29,22 MeOH); LD_{50} (rat orl) = 452 mg/kg. *C.M. Ind.; Eli Lilly & Co.*

9648 Fluvoxamine
54739-18-3 4251
$C_{15}H_{21}F_3N_2O_2$
5-Methoxy-4'-(trifluoromethyl)valerophenone (E)-*O*-(2-aminoethyl)oxime.
Antidepressant; anxiolytic; antiobsessional agent. Serotonin uptake inhibitor. *Philips-Duphar B.V.; Solvay Duphar Labs Ltd.*

9649 Fluvoxamine Maleate
61718-82-9 4251
$C_{19}H_{25}F_3N_2O_6$
5-Methoxy-4'-(trifluoromethyl)valerophenone (E)-*O*-(2-aminoethyl)oxime maleate.
DU-23000; MK-264; Dumirox; Faverin; Fevarin; Floxyfral; Luvox; Maveral. Antidepressant; anxiolytic; antiobsessional agent. Serotonin uptake inhibitor. mp = 120-121.5°. *Philips-Duphar B.V.; Solvay Pharm. Inc.*

9650 Indalpine
63758-79-2 4965 264-445-4
$C_{15}H_{20}N_2$
3-[2-(4-Piperidyl)ethyl]indole.
LM-5008; Upstene. Antidepressant. A selective serotonin uptake inhibitor. [monohydrochloride ($C_{15}H_{21}ClN_2$)]: mp = 167°. *Mar-Pha Soc. Etud. Exploit. Marques.*

9651 Indeloxazine Hydrochloride
65043-22-3 4972
$C_{14}H_{18}ClNO_2$
(±)-2-[(Inden-7-yloxy)methyl]morpholine hydrochloride.
CI-874; YM-08054-1; Elen; Noin. Antidepressant; nootropic. Inhibits synaptosomal uptake of serotonin and noradrenaline. mp = 155-156°, 169-170°; LD_{50} (mus iv) = 47 mg/kg; [(+)-form]: mp = 112-113°; $[\alpha]_D^{21}$ = = 4.9° (c = 5 MeOH); [(-)-form]: mp = 142-142.5°; $[\alpha]_D^{20}$ = -4.9° (c = 5 MeOH). *Parke-Davis; Yamanouchi U.S.A. Inc.*

9652 Milnacipran
92623-85-3 6281
$C_{15}H_{22}N_2O$
cis-(±)-2-(Aminomethyl)-N,N-diethyl-1-phenyl-cyclopropanecarboxamide.
midalcipran. Antidepressant. Serotonin and nor-epinephrine uptake inhibitor. *Fabre Pierre.*

9653 Milnacipran Hydrochloride
101152-94-7 6281
$C_{15}H_{23}ClN_2O$
cis-(±)-2-(Aminomethyl)-N,N-diethyl-1-phenylcyclopropanecarboxamide hydrochloride.
F-2207. Antidepressant. Serotonin and norepinephrine uptake inhibitor. mp = 179-181°; LD_{50} (mus orl) = 237°. *Fabre Pierre.*

9654 Paroxetine
61869-08-7 7175
$C_{19}H_{20}FNO_3$
(-)-(3S,4R)-4-[(p-Fluorophenyl)-3-[(3,4-methylenedioxy)-phenoxy]methyl]piperidine.
BRL-29060; FG-7051; Aropax; Paxil; Seroxat. Antidepressant. Serotonin uptake inhibitor. [hydrochloride ($C_{19}H_{21}ClFNO_3$)]: mp = 118°; [hydrochloride hemihydrate ($C_{19}H_{21}ClFNO_3.0.5H_2O$)]: mp = mp = 129-131°; [maleate]: mp = 136-138°; $[\alpha]_D$ = -87° (c = 5 EtOH); LD_{50} (mus sc) = 845 mg/kg, (mus orl) = 500 mg/k. *Beecham Res. Labs. UK; Ferrosan A/S; SmithKline Beecham Pharm.*

9655 Sertraline
79617-96-2 8612
$C_{17}H_{17}Cl_2N$
(1S,4S)-4-(3,4-Dichlorophenyl)-1,2,3,4-tetrahydro-N-methyl-1-naphthalenamine.
Antidepressant. Selective serotonin uptake inhibitor. *Pfizer Inc.; Roerig Div. Pfizer Pharm.*

9656 Sertraline Hydrochloride
79559-97-0 8612
$C_{17}H_{18}Cl_3N$
(1S,4S)-4-(3,4-Dichlorophenyl)-1,2,3,4-tetrahydro-N-methyl-1-naphthalenamine hydrochloride.
Cp-51974-1; Lustral; Zoloft. Antidepressant. Serotonin uptake inhibitor. mp = 243-245°; $[\alpha]_D^{23}$ = 38° (c = 2 MeOH). *Pfizer Inc.; Roerig Div. Pfizer Pharm.*

9657 Sibutramine
106650-56-0 8629
$C_{17}H_{26}ClN$
(±)-1-(p-Chlorophenyl)-α-isobutyl-N,N-dimethylcyclobutanemethylamine.
Monoamine, serotonin uptake inhibitor. Anorexic and antidepressant. *Boots Pharmaceuticals Inc.*

9658 Sibutramine Hydrochloride [anhydrous]
84485-00-7 8629
$C_{17}H_{27}Cl_2N$
(±)-1-(p-Chlorophenyl)-α-isobutyl-N,N-dimethylcyclobutanemethylamine hydrochloride.
Anorexic; antidepressant. Monoamine reuptake inhibitor. *Boots Co.; Boots Pharmaceuticals Inc.*

9659 Sibutramine Hydrochloride Monohydrate
125494-59-9 8629
$C_{17}H_{27}Cl_2N.H_2O$
(±)-1-(p-Chlorophenyl)-α-isobutyl-N,N-dimethylcyclobutanemethylamine hydrochloride monohydrate.
BTS-54524; Reductil; Meridia. Anorexic; antidepressant. Monoamine reuptake inhibitor. mp = 193-195.5°. *Boots Co.; Boots Pharmaceuticals Inc.*

9660 Viqualine
72714-74-0
$C_{20}H_{26}N_2O$
6-Methoxy-4-[3-[(3R,4R)-3-vinyl-4-piperidyl]propyl]-quinoline.
PK-5078. Used in treatment of alcoholism. Serotonin releaser and uptake inhibitor; antidepressant.

9661 Zimeldine
56775-88-3 10254
$C_{16}H_{17}BrN_2$
(Z)-3-[1-(p-Bromophenyl)-3-(dimethylamino)propenyl]-pyridine.
H-102/09; Zimelidine. Antidepressant that inhibits membranal 5-hydroxytryptamine uptake. *Astra Hassle AB.*

9662 Zimeldine Hydrochloride
61129-30-4 10254 262-279-7
$C_{16}H_{21}BrCl_2N_2O$
(Z)-3-[1-(p-Bromophenyl)-3-(dimethylamino)propenyl]-pyridine dihydrochloride monohydrate.
Normud; Zelmid; Zimelidine Hydrochloride. Antidepressant. Inhibits membranal 5-hydroxytryptamine (serotonin) uptake. mp = 193°. *Astra Hassle AB; Astra Sweden; Merck & Co.Inc.*

9663 Zimeldine Hydrochloride [anhydrous]
60525-15-7 10254
$C_{16}H_{19}BrCl_2N_2$
(Z)-3-[1-(p-Bromophenyl)-3-(dimethylamino)propenyl]-pyridine dihydrochloride.
Antidepressant. Inhibits membranal 5-hydroxytryptamine (serotonin) uptake. *Astra Hassle AB.*

Spermaticides

9664 Chlorindanol
145-94-8 2144
C_9H_9ClO
7-Chloro-2,3-dihydro-1H-inden-4-ol.
NSC-158565. Spermaticide. mp = 91-93°. *Esta Med. Labs.*

9665 Laureth-10S
9014-89-5 7717
$C_{32}H_{66}O_{10}S$
Polyethylene glycol mono[2-(dodecylthio)ethyl] ether.
polyoxyethylene lauryl ether. Spermaticide. Soluble in H_2O, EtOH, C_7H_8; miscible with hot mineral, natural and synthetic oils, fats and fatty alcohols; LD_{50} (mus orl) = 1170 mg/kg, (mus iv) = 125 mg/kg.

9666 Laureth-9
9002-92-0 7717 500-002-6
$C_{30}H_{62}O_{10}$
Polyethyleneglycol monododecyl ether.
nonaethylene glycol monododecyl ether; polyoxyethylene lauryl ether; Aethoxysklerol; Atlas G-4829; Hetoxol L-9; Lipal 9-LA; Polidocanol; Thesit. Sclerosing agent. Spermaticide. Soluble in H_2O, EtOH, C_7H_8; miscible with hot mineral, natural and synthetic oils, fats and fatty alcohols; LD_{50} (mus orl) = 1170 mg/kg, (mus iv) = 125 mg/kg.

9667 Menfegol
57821-32-6
$(C_2H_4O)_nC_{16}H_{24}O$
α-[p-(p-Menthyl)phenyl]-ω-hydroxypoly(oxyethylene).
Neo Sampoon. Foaming vaginal contraceptive tablet. Spermaticide.

9668 Nonoxynol-9
26027-38-3 6772
$C_{15}H_{24}O(C_2H_4O)_n$ (n≅ 9)
Polyethylene glycol mono(p-nonylphenyl) ether.
Conceptrol; Emko; Gynol II; Intercept; Semicid; Today Sponge; C-Film; Conco NI-90; Dowfax 9N9; Encare; Igepal CO-630; Intercept; Neutronyx 611; Semicid; Staycept; Tergitol TP-9; component of: Delfen, Gentersal, Ortho-Crème, Conceptrol. Spermaticide. Average molecular weight = 617; d_4^{25} = 1.06; soluble in H_2O, EtOH, ethylene glycol, ethylene chloride, xylene, corn oil; insoluble in kerosene, mineral oil. *Ortho Pharm. Corp.; Schering-Plough Pharm.; Whitehall-Robins.*

9669 Polidocanol
3055-99-0 7717 221-284-4
$C_{12}H_{35}(OCH_2CH_2)_nOH$ (average polymer, n = 9)
α-Dodecyl-ω-hydroxypoly(oxy-1,2-ethanediyl).
polyethylene glycol (9) monodecyl ether; dodecyl alcohol polyoxyethylene ether; hydroxypolyethoxydodecane; laureth 9; polyoxyethylene lauryl ether; polyethylene glycol monododecyl ether; Aethoxysklerol; Aetoxisclerol; Atlas G-4829; Hetoxol L-9; Lipal 9LA; Thesit. Sclerosing agent; antipruritic; anesthetic (topical). Also used as a solvent, non-ionic emulsifier, pharmaceutic aid (surfactant), spermatacide. Soluble in H_2O, EtOH, C_7H_8; miscible with hot mineral, natural and synthetic oils; miscible with fats and fatty alcohols; LD_{50} (mus orl) = 1170 mg/kg, (mus iv) 125 mg/kg. *Kreussler Chemische-Fabrik.*

9670 Tolnidamine
50454-68-7
$C_{16}H_{13}ClN_2O_2$
1-(4-Chloro-2-methylbenzyl)-1H-indazole-3-carboxylic acid.
Antispermatogenic.

Substance Abuse Deterrents

9671 Acamprosate 6473
77337-76-9 278-667-4
$C_5H_{11}NO_4S$
3-Acetamido-1-propanesulfonic acid.
Used to treat acoholism. *Lab. Meram.*

9672 Acamprosate Calcium
77337-73-6 14
$C_{10}H_{20}CaN_2O_8S_2$
3-(Acetylamino)-1-propanesulfonic acid calcium salt (2:1).
Ca-AOTA; Aotal; calcium bisacetyl homotaurine. A GABA agonist. Used to treat acoholism. mp = 270°; λ_m 192 nm (ε 7360 H_2O); LD_{50} (mmus ip) = 1870 mg/kg. *Lab. Meram.*

9673 Bupropion
34911-55-2 1523
$C_{13}H_{18}ClNO$
(±)-2-(tert-Butylamino)-3'-chloropropiophenone.
amfebutamon; amfebutamone. Antidepressant with action similar to that of the tricyclic antidepressants. Pale yellow oil; bp = 52°; soluble in MeOH, EtOH, Me_2CO, Et_2O, C_6H_6. *Burroughs Wellcome Inc.*

9674 Bupropion Hydrochloride
31677-93-7 1523 250-759-9
$C_{13}H_{19}Cl_2NO$
(±)-2-(tert-Butylamino)-3'-chloropropiophenone hydrochloride.
Wellbatrin; Wellbutrin; Zyban. Antidepressant. Also used in smoking cessation therapy. mp = 233-234°; solubility (in H_2O) = 32 mg/ml, (in alcohol) = 193 mg/ml, (in 0.1 N HCl) = 333 mg/ml; LD_{50} (mus ip) = 230 mg/kg, (rat ip) = 210 mg/kg, (mus orl) = 575 mg/kg, (rat orl) = 600 mg/kg. *Burroughs Wellcome Inc.*

9675 Calcium Cyanamide, Citrated
8013-88-5 1703
Citrated calcium carbimide.
CCC; carbimide; Colme; Dipsan; Abstem; Temposil. Aldehyde dehydrogenase inhibitor. Alcohol deterrent. *Am. Cyanamid.*

9676 1,1-Dimethyl-4-phenylpiperazinium iodide
54-77-3 200-213-0
$C_{12}H_{19}IN^{+2}$
1,1-Dimethyl-4-phenylpiperazinium iodide.
DMPP-iodide. Ganglionic stimulant.

9677 Disulfiram
97-77-8 3428
$C_{10}H_{20}N_2S_4$
Tetraethylthioperoxydicarbonic diamide.
tetraethylthiuram disulfide; TTD; Cronetal; Abstensil; Stopetyl; Contralin; Antadix; Antietanol; Exhoran; Ethyl Thiurad; Antabuse; Etabus; Abstinyl; Thiuranide; Esperal; Tetradine; Noxal; Tetraetil. Vulcanizing agent also used as an alcohol deterrent. mp = 70°; d = 1.30; poorly soluble in H_2O (0.02 g/100 ml); soluble in EtOH (3.82 g/100 ml), Et_2O (7.14 g/100 ml), Me_2CO, C_6H_6, $CHCl_3$, CS_2; LD_{50} (rat orl) = 8600 mg/kg. *Naugatuck; Wyeth Labs.*

9678 Nicotine
54-11-5 6611 200-193-3
$C_{10}H_{14}N_2$
(S)-3-(1-Methyl-2-pyrrolidinyl)pyridine.
Habitrol; Nicabate; Nicoderm; Nicolan; Nicopatch; Nicotell TTS; Nicotinell; Tabazur. Smoking cessation adjunct. Ganglionic stimulant which also stimulates the CNS. Also used as an insecticide and fumigant. Severe nicotine poisoning can cause nausea, abdominal pain, vomiting, diarrhea, cold sweat, dizziness, mental confusion, weakness, drop in blood pressure and pulse, convulsions, paralysis of respiratory muscles. From dried leaves of *Nicotiana tabacum and N. Rustica.* Oily liquid; bp_{745} = 247° (partial dec), bp_{17} = 123-125°; n_D^{20} = 1.5282; d_D^{20} = 1.0097; $[\alpha]_D^{20}$ = -169°; miscible with H_2O; soluble in alcohol, $CHCl_3$, Et_2O, petroleum ether, kerosene, oils.

9679 Nicotine Polacrilex
96055-45-7 6611
$[(C_4H_6O_2)_x(C_{10}H_{10})_y](C_{10}H_{14}N_2)$
Methacrylic acid polymer with divinylbenzene complex with nicotine.
Nicorette. Smoking cessation adjunct. *Marion Merrell Dow Inc.*

9680 Nitrefazole
21721-92-6 6668
$C_{10}H_8N_4O_4$
2-Methyl-4-nitro-1-(4-nitrophenyl)-1H-imidazole.
EMD-15700; Altimol. Alcohol dehydrogernase inhibitor. Used as an alcohol deterrent. mp = 185-187°. *E. Merck.*

Thrombolytics

9681 Anistreplase
81669-57-0 712
BRL-26921; APSAC; Eminase. Fibrinolytic. Thrombolytic enzyme. p-Anisolyated derivative of the primary (human) lys-plasminogen streptokinase complex (1:1). Streptokinase in a noncovalent 1:1 complex with plasminogen. Complex of streptokinase and plasminogen. The catalytic site is blocked by anisolyation while the fibrin-binding site is unaffected. *SmithKline Beecham Pharm.*

9682 Lanoteplase
171870-23-8
$C_{2184}H_{3323}N_{633}O_{666}S_{29}$
N-[N^2-(N-Glycyl-L-alanyl)-L-arginyl]-117-L-glutamine-245-L-methionine-(1-5)-(87-527)-plasminogen activator (human tissue type protein moiety).
BMS-200980; SUN-9216. Fibrinolytic. Thrombolytic enzyme. A tissue plasminogen activator protein derived from human t-PA by deletion of the fibronectin-like and EGF-like domains and mutation of Asn-117 to GFln-117. *Bristol-Myers Squibb Pharm. R&D.*

9683 Nasaruplase
99821-44-0
$C_{2031}H_{3121}N_{585}O_{601}S_{31}$
Pro-urokinase (enzyme-activating) (human clone pA3/pD2/pF1 protein moiety), glycosylated.
A plasminogen pro-activator. Thrombolytic.

9684 Octimibate
89838-96-0
$C_{29}H_{30}N_2O_3$
8-[(1,4,5-Triphenylimidazol-2-yl)oxy]octanoic acid.
BMY-22389. Nonprostanoid prostacyclin antagonist that inhibits platelet aggregation. Thrombolytic.

9685 Pamicogrel
101001-34-7
$C_{25}H_{24}N_2O_4S$
Ethyl 2-[4,5-bis(p-methoxyphenyl)-2-thiazolyl]pyrrole-1-acetate.
Platelet antiaggregant.

9686 Plasmin
9001-90-5 7678 232-640-3
Human fibrinolysin.
Actase; serum tryptase; Fibrinolysin; component of: Elase, Elase-Chloromycetin. Fibrinolytic. Thrombolytic enzyme. Enzyme obtained from human plasma by conversion of profibrinolysin to fibrinolysin with streptokinase. *Cutter Labs; Fujisawa Pharm. USA Inc.; Ortho Pharm. Corp.; Parke-Davis.*

9687 Pro-Urokinase
82657-92-9 8089
Prourokinase (enzyme-activating).
scu-PA; pro-UK; pro u-PA; PUK; Sandolase; Thombolyse; Tomieze. Fibrinolytic. Thrombolytic enzyme. A single-chain proenzyme form of urokinase. *Genentech Inc.*

9688 Reteplase
133652-38-7
$C_{1736}H_{2653}N_{499}O_{522}S_{22}$
173-L-Serine-174-L-tyrosine-175-L-glutamine-173-527-plasminogen activator (mutant of human tissue-type).
BM-06022. Derived from human tissue plasminogen activator. Thrombolytic. Used in treatment of myocardial infarction. *Boehringer Mannheim GmbH.*

9689 Saruplase
99149-95-8
$C_{2031}H_{3121}N_{585}O_{601}S_{31}$
Prourokinase (enzyme-activating) (human clone pUK4/pUK18).
Sandolase; recombinant single-chain urokinase-type plasminogen activator; rscu-PA; pro-urokinase. Thromolytic. Plasminogen activator. Has a high binding affinity for fibrin.

9690 Silteplase
131081-40-8
$C_{2580}H_{3948}N_{752}O_{784}S_{40}$
N-[N^2-(N-glycyl-L-alanyl)-L-argilnyl]plasminogen activator, glycoform.
Reduced human tissue plasminogen activator. Non-glycosylated protein. Thrombolytic.

9691 Streptokinase
9002-01-1 8981 232-647-1
Streptococcal fibrinolysin.
plasminokinase; Kabikinase; Streptase. Fibrinolytic; thrombolytic. Thrombolytic coenzyme obtained from cultures of various strains of *Streptococcus haemolyticus.* Plasminogen activator. Activates plasminogen to produce plasmin which dissolves fibrin. *Am. Cyanamid; Astra Chem. Ltd.; Pharmacia & Upjohn.*

9692 Tissue Plasminogen Activator
105857-23-6 9608
$C_{2569}H_{389}74N_{746}O_{781}S_{40}$
Fibrinokinase.
Alteplase; Activase; recombinant human tissue-type plasminogen activator; rt-PA; TPA; A 527-mer serine protease. Fibrinolytic. Thrombolytic enzyme. *Genentech Inc.*

9693 Tulopafant
116289-53-3
$C_{25}H_{19}N_3O_2S$
(+)-3'-Benzoyl-3-(3-pyridyl)-1H,3H-pyrrolo[1,2-c]-thiazole-7-carboxanilide.
RP-59227. Platelet-activating factor antagonist. Thrombolytic.

9694 Urokinase
9039-53-6 10024 232-917-9
Abbokinase; Breokinase; Win-Kinase; Win-22005; Actosolv; Persolv; Purochin; Ukidan; Uronase. Fibrinolytic. Thrombolytic enzyme; a two-peptide chain serine protease that directly activates plasminogen. Isolated from human sources. *Abbott Labs Inc.; Sterling Winthrop Inc.*

Thromboxane Inhibitors

9695 Camonagrel
105920-77-2
$C_{15}H_{16}N_2O_3$
(±)-5-(2-Imidazol-1-ylethoxy)-1-indan-carboxylic acid.
A selective thromboxane synthase inhibitor.

9696 Daltroban
79094-20-5 2871
$C_{16}H_{16}ClNO_4S$
[p-[2-(p-Chlorobenzenesulfonamido)ethyl]phenyl]-acetic acid.
SKF-96148; BM-13505. Antithrombotic; immunosuppressant. Thromboxane synthetase inhibitor. *Boehringer Mannheim GmbH; SmithKline Beecham Pharm.*

9697 Isbogrel
89667-40-3 5120
$C_{18}H_{19}NO_2$
(E)-7-Phenyl-7-(3-pyridyl)-6-heptenoic acid.
CV-4151. Antithrombotic. Thromboxane synthetase inhibitor. mp = 114-115°. *Takeda Chem. Ind. Ltd.*

9698 Ozagrel
82571-53-7 7115
$C_{13}H_{12}N_2O_2$
(E)-p-(Imidazol-1-ylmethyl)cinnamic acid.
OKY-046. Antianginal; antithrombotic. Thromboxane synthetase inhibitor. mp = 223-224°. *Kissei; Ono Pharm.*

9699 Ozagrel Hydrochloride
78712-43-3 7115
$C_{13}H_{13}ClN_2O_2$
(E)-p-(Imidazol-1-ylmethyl)cinnamic acid hydrochloride.
Antianginal; antithrombotic. Thromboxane synthetase inhibitor. mp = 214-217°. *Kissei; Ono Pharm.*

9700 Ozagrel Sodium
7115
$C_{13}H_{11}N_2NaO_2$
Sodium (E)-p-(Imidazol-1-ylmethyl)cinnamic acid.
Cataclot; Xanbon. Antianginal; antithrombotic. Thromboxane synthetase inhibitor. LD_{50} (mmus iv) = 1940 mg/kg, (mmus orl) = 3800 mg/kg, (mmus sc) = 2450 mg/kg, (fmus iv) = 1580 mg/kg, (fmus orl) = 3600 mg/kg, (fmus sc) = 2100 mg/kg, (mrat iv) = 1150 mg/kg, (mrat orl) = 5900 mg/kg, (mrat sc) = 2300 mg/kg, (frat iv) = 1300 mg/kg, (frat orl) = 5700 mg/kg, (frat sc) = 2250 mg/kg. *Kissei; Ono Pharm.*

9701 Ridogrel
110140-89-1 8379
$C_{18}H_{17}F_3N_2O_3$
(E)-5-[[[α-3-Pyridyl-m-(trifluoromethyl)benzylidene]-amino]oxy]valeric acid.
R-68070. Thromboxane synthetase inhibitor. Used as an antithrombotic. mp = 70.3°. *Janssen Pharm. Inc.*

9702 Seratodrast
112665-43-7 8603
$C_{22}H_{26}O_4$
(±)-2,4,5-Trimethyl-3,6-dioxo-ζ-phenyl-1,4-cyclohexadiene-1-heptanoic acid.
AA-2414; A-73001; Abbott-73001; ABT-001. A thromboxane A_2-receptor antagonist. Used as an antiasthmatic. mp = 128-129°. *Abbott Labs Inc.*

9703 Vapiprost
85505-64-2
$C_{30}H_{39}NO_4$
(+)-(4Z)-7-[(1R,2R,3S,5S)-5-(4-Biphenylmethoxy)-3-hydroxy-2-piperidinocyclopentyl]-4-heptenoic acid.
GR-32191. Thromboxane synthetase inhibitor. *Glaxo Wellcome plc.*

9704 Vapiprost Hydrochloride
87248-13-3
$C_{30}H_{40}ClNO_4$
(+)-(4Z)-7-[(1R,2R,3S,5S)-5-(4-Biphenylmethoxy)-3-hydroxy-2-piperidinocyclopentyl]-4-heptenoic acid hydrochloride.
GR-32191B. Thromboxane synthetase inhibitor. *Glaxo Wellcome plc.*

Thyroid Hormones

9705 Levothyroxine Sodium
55-03-8 5497
$C_{15}H_{10}I_4NNaO_4$
O-(4-Hydroxy-3,5-diiodophenyl)-3,5-diiodo-L-tyrosine monosodium salt.
L-thyroxine sodium salt; Levothroid; Synthroid Sodium; Eltroxin; Euthyrox; Laevoxin; Letter; Levaxin; Levothyrox; Oroxine; Thyroxevan. Thyroid hormone. Used in treatment of hyperthyroidism. Sodium salt of the amino acid L-thyroxine. Obtained from thyroid gland of domesticated animals or synthesized. *Astra USA Inc.; Forest Pharm. Inc.; Knoll Pharm. Co.*

9706 Levothyroxine Sodium Pentahydrate
25416-65-3 5497
$C_{15}H_{10}I_4NNaO_4 \cdot 5H_2O$
O-(4-Hydroxy-3,5-diiodophenyl)-3,5-diiodo-L-tyrosine sodium salt pentahydrate.
L-thyroxine sodium salt pentahydrate. Thyroid hormone. Used in treatment of hyperthyroidism. Sodium salt of the amino acid L-thyroxine. Obtained from thyroid gland of domesticated animals or synthesized. d = 2.381; $[\alpha]_D^{20}$ = -4.4° (c= 3 70% EtOH); slightly soluble in H_2O (0.015 g/100 ml); more soluble in EtOH; very slightly soluble in $CHCl_3$, Et_2O. *Astra USA Inc.; Forest Pharm. Inc.; Knoll Pharm. Co.*

9707 Liothyronine
6893-02-3 5535 228-120-0
$C_{15}H_{12}I_3NO_4$
L-3-[4-(4-Hydroxy-3-iodophenoxy)-3,5-diiodophenyl]-alanine.
T-3. Thyroid hormone. Dec 236-237°; $[\alpha]_D^{29.5}$ = 21.5° (c = 4.75 in a mixture of 1 part 1N HCl and 2 parts EtOH); insoluble in H_2O, EtOH, propylene glycol; soluble in dilute alkalies. *SmithKline Beecham Pharm.*

9708 Liothyronine Sodium
55-06-1 5535 200-223-5
$C_{15}H_{11}I_3NNaO_4$
L-3-[4-(4-Hydroxy-3-iodophenoxy)-3,5-diiodophenyl]-alanine monosodium salt.
liothyroninide sodium salt; sodium L-triiodothyronine; Cytomel; Cytobin; Cytomine; Cyomel; Cynomel; Cytomel; Tertroxin; Triostat; Triothyrone. Thyroid hormone. *SmithKline Beecham Pharm.*

9709 Liotrix
8065-29-0
$[C_{15}H_{11}I_3NNaO_4].[C_{15}H_{10}I_4NNaO_4.xH_2O]$
A 1:4 mixture of liothyronine sodium and levothyroxine sodium.
Euthroid; Thyrolar. Thyroid hormone. *Forest Pharm. Inc.; Parke-Davis.*

9710 Nonathymulin
63958-90-7
$C_{33}H_{54}N_{12}O_{15}$
N^2-[N-[N-[N-[N^2-[N-[N^2-[N-(-5-Oxo-L-propyl)-L-alanyl]-L-lysyl]-L-seryl]-L-glutaminyl]glycyl]glycyl]-L-seryl]-L-asparagine.
facteur thymique serique. Synthetic serum thymic factor. Thyroid hormone.

9711 Thyroglobulin
9010-34-8 232-721-3
Substance extracted from hog thyroid glands.
Proloid; Thyractin; Thyroprotein. Thyroid hormone. *Parke-Davis; Sterling Winthrop Inc.*

9712 Thyroid
9551
Dried thyroid.
NSC-26492; Tiroidina; Thyradin; Thyrocrine; Tyroidina. Thyroid hormone. Dried and powdered thyroid gland of domesticated animals.

9713 Thyroidin
9552
Iodothyrin.
Thyroid hormone. An extract of the thyroid gland, diluted with (e.g.) milk sugar.

9714 Thyroxine
7488-70-2 9555
$C_{15}H_{11}I_4NO_4$
O-(4-Hydroxy-3,5-diiodophenyl)-3,5-diiodotyrosine.
3,5,3',5'-tetraiodothyronine. The L-form is a thyroid hormone; the D-form is antihyperlipoproteinemic. *Astra USA Inc.; Baxter Healthcare Systems; Forest Pharm. Inc.; Knoll Pharm. Co.*

9715 D-Thyroxine
51-49-0 9555 200-102-7
$C_{15}H_{11}I_4NO_4$
D-O-(4-Hydroxy-3,5-diiodophenyl)-3,5-diiodotyrosine.
dextrothyroxine; Debetrol. The L-form is a thyroid hormone; the D-form is antihyperlipoproteinemic. Optical isomer of the endogenous hormone L-thyroxine. Can produce modest lowering of plasma low-density lipoprotein. CAUTION: may cause serious cardiac toxicity. Dec 237°; $[\alpha]_{546}^{21}$ = 2.97° (c = 3.7 NaOH/EtOH). *Astra USA Inc.; Baxter Healthcare Systems; Forest Pharm. Inc.; Knoll Pharm. Co.*

9716 D-Thyroxine Sodium Salt
137-53-1 9555 205-301-2
$C_{15}H_{10}I_4NNaO_4$
D-O-(4-Hydroxy-3,5-diiodophenyl)-3,5-diiodotyrosine sodium salt.
dextrothyroxine sodium; Biotirmone; Choloxin; Detyroxin; Dethyrona; Dextroid; Dynothel; Eulipos. The L-form is a thyroid hormone, the D-form is antihyperlipoproteinemic. Thyroid hormone. *Astra USA Inc.; Baxter Healthcare Systems; Forest Pharm. Inc.; Knoll Pharm. Co.*

9717 L-Thyroxine
51-48-9 9555 200-101-1
$C_{15}H_{11}I_4NO_4$
L-O-(4-Hydroxy-3,5-diiodophenyl)-3,5-diodo-L-tyrosine.
Thyroid hormone. (The D-form is antihyperlipoproteinemic.) Dec 235-236°; $[\alpha]_{546}^{25}$ = -3.2° (90.66 g in 6.07 g of 0.5N NaOH and 13.03 g EtOH), $[\alpha]_D^{20}$ = -4.4° (c = 3 0.13N NaOH/70% EtOH). *Astra USA Inc.; Baxter Healthcare Systems; Forest Pharm. Inc.; Knoll Pharm. Co.*

9718 DL-Thyroxine
300-30-1 9555 206-088-9
$C_{15}H_{11}I_4NO_4$
(±)-O-(4-Hydroxy-3,5-diiodophenyl)-3,5-diiodotyrosine.
DL-thyroxine. The L-form is a thyroid hormone, the D-form is antihyperlipoproteinemic. Thyroid hormone. Dec 230-231°; insoluble in H_2O and most organic solvents. *Astra USA Inc.; Baxter Healthcare Systems; Forest Pharm. Inc.; Knoll Pharm. Co.*

Thyrotropic Hormones

9719 Montirelin
90243-66-6
$C_{17}H_{24}N_6O_4S$
N-[[(3R,6R)-6-Methyl-5-oxo-3-thiomorpholinyl]carbonyl]-L-histidyl-L-prolinamide.
NS-3 [as tetrahydrate]; CG-3703 [as tetrahydrate]. Thyrotropin-releasing hormone analog.

9720 Posatirelin
78664-73-0
$C_{17}H_{28}N_4O_4$
(2S)-N[(1S)-1-[[(2S)-2-Carbamoyl-1-pyrrolidinyl]-carbonyl]-3-methylbutyl]-6-oxopipercolamide.
RGH-2202; L-6-ketopiperidine-2-carbonyl-L-leucyl-L-proline amide. Thyrotropin releasing hormone (TRH) analog, a neuroactive peptide.

9721 Taltirelin
103300-74-9
$C_{17}H_{23}N_7O_5$
(-)-N-[(S)-Hexahydro-1-methyl-2,6-dioxo-4-pyrimidinyl-carbonyl]-L-histidyl-L-prolinamide tetrahydrate.
TA-0910 [as tetrahydrate]. A novel thyrotropin-releasing hormone analog.

9722 TRH
24305-27-9 9720 246-143-4
$C_{16}H_{22}N_6O_4$
5-Oxo-L-prolyl-L-histidyl-L-prolinamide.
Thyrotropin-releasing factor; thyrotropin-releasing hormone; Protirelin; Thypinone; Abbott 38579; Synthetic TRH; thyroliberin; Antepan; Stimu-TSH; Thyrefact. A hypothalamic neurohormone. Stimulates release and synthesis of thyroid-stimulating hormone (TSH). Thyrotropic hormone. Freely soluble in MeOH, soluble in $CHCl_3$, insoluble in C_5H_5N. *Abbott Labs Inc.*

9723 TRH Tartrate
54974-54-8 9720
$C_{16}H_{22}N_6O_4.xC_4H_6O_6$
5-Oxo-L-prolyl-L-histidyl-L-prolinamide tartrate.
Irtonin; Xantium. Thyrotropic hormone.

9724 TSH
9002-71-5 9931 232-664-4
Thyroid-stimulating hormone.
thyrotropin; thyrotropic hormone; thyreotrophic hormone; TTH; Dermathycin; Thytropar. Thyrotropic hormone. Diagnostic aid (thyroid function). A glycoprotein produced by anterior lobe of pituitary gland. Stimulates production of thyroxine, raising metabolic rate. Inactivated by heating, proteolysis, oxidizing agents.

Tocolytics

9725 Albuterol
18559-94-9 217 242-424-0
$C_{13}H_{21}NO_3$
2-(tert-Butylamino)-1-(4-hydroxy-3-hydroxymethyl-phenyl)ethanol.
salbutamol; Proventil Inhaler; Ventalin Inhaler. Bronchodilator; tocolytic. Ephedrine derivative. mp = 151°, 157-158°; soluble in most organic solvents. *Allen & Hanbury; Apothecon; Glaxo Wellcome Inc.; Key Pharm.*

9726 Albuterol Sulfate
51022-70-9 217 256-916-8
$C_{26}H_{44}N_2O_{10}S$
2-(tert-Butylamino)-1-(4-hydroxy-3-hydroxymethyl-phenyl)ethanol sulfate (2:1).
Sch-13949W Sulfate; Aerolin; Asmaven; Broncovaleas; Cetsim; Cobutolin; Ecovent; Loftan; Proventil; Salbumol; Salbutard; Salbutine; Salbuvent; Sultanol; Ventelin; Ventodiscks; Ventolin; Volma. Bronchodilator; tocolytic. Ephedrine derivative. *Allen & Hanbury; Apothecon; Glaxo Labs.; Key Pharm.; Lemmon Co.; Schering AG.*

9727 Fenoterol
13392-18-2 4022
$C_{17}H_{21}NO_4$
5-[1-Hydroxy-2-[[2-(4-hydroxyphenyl)-1-methylethyl]-amino]ethyl]-1,3-benzenediol.
TH-1165. Bronchodilator; tocolytic. β-Adrenergic agonist. Ephedrine derivative. *Boehringer Ingelheim Pharm. Inc.*

9728 Fenoterol Hydrobromide
1944-12-3 4022
$C_{17}H_{22}BrNO_4$
5-[1-Hydroxy-2-[[2-(4-hydroxyphenyl)-1-methylethyl]-amino]ethyl]-1,3-benzenediol hydrobromide.
TH-1165a; Airum; Berotec; Dosberotec; Partusisten. Bronchodilator; tocolytic. β-Adrenergic agonist. Ephedrine derivative. mp = 222-223°; LD_{50} (mus sc) = 1100 mg/kg, (mus orl) = 1990 mg/kg. *Boehringer Ingelheim Pharm. Inc.*

9729 Fenoterol Hydrochloride
1944-10-1 4022
$C_{17}H_{22}ClNO_4$
5-[1-Hydroxy-2-[[2-(4-hydroxyphenyl)-1-methylethyl]-amino]ethyl]-1,3-benzenediol hydrochloride.
Tocolytic. mp = 183°. *Boehringer Ingelheim Pharm. Inc.*

9730 Hexoprenaline
3215-70-1 4745
$C_{22}H_{32}N_2O_6$
α,α'-[Hexamethylene-bis(iminomethylene)]bis(3,4-dihydroxybenzyl alcohol).
BYK-1512. Bronchodilator; tocolytic. β-Adrenergic agonist. Ephedrine derivative. [hemihydrate]: mp = 162-165°. *Lentia; Oesterreiche Stickstoffwerke; OSSW; Savage Labs.*

9731 Hexoprenaline Dihydrochloride
4323-43-7 4745 224-354-2
$C_{22}H_{34}Cl_2N_2O_6$
α,α'-[Hexamethylene-bis(iminomethylene)]bis(3,4-dihydroxybenzyl alcohol) dihydrochloride.
ST-1512; Ipradol. Bronchodilator; tocolytic. β-Adrenergic agonist. Ephedrine derivative. mp = 197.5-198°. *Lentia; OSSW; Savage Labs.*

9732 Hexoprenaline Sulfate
32266-10-7 4745 250-974-8
$C_{22}H_{34}N_2O_{10}S$
α,α'-[Hexamethylene-bis(iminomethylene)]bis(3,4-dihydroxybenzyl alcohol) sulfate.
Bronalin; Delaprem; Etoscol; Gynipral; Ipradol; Leanol. Bronchodilator; tocolytic. β-Adrenergic agonist. Ephedrine derivative. mp = 222-228°. *Lentia; Oesterreiche Stickstoffwerke; OSSW; Savage Labs.*

9733 Ritodrine
26652-09-5 8401 247-879-9
$C_{17}H_{21}NO_3$
erythro-p-Hydroxy-α-[1-[(p-hydroxyphenethyl)amino]-ethyl]benzyl alcohol.
DU-21220. Smooth muscle relaxant; tocolytic agent. A β-adrenergic agonist. mp = 88-90°. *Astra Chem. Ltd.; Astra Sweden; Philips-Duphar B.V.; Teva Pharm. (USA).*

9734 Ritodrine Hydrochloride
23239-51-2 8401 245-514-8
$C_{17}H_{22}ClNO_3$
erythro-p-Hydroxy-α-[1-[(p-hydroxyphenethyl)amino]-ethyl]benzyl alcohol hydrochloride.
Pre-Par; Yutopar; DU-21220; Miolene; Prempar; Utemerin; Utopar. Smooth muscle relaxant. A β-

adrenergic agonist and tocolytic. mp = 193-195° (dec); λ_m = 267.5 nm (ε 3310). *Astra Chem. Ltd.; Astra Sweden; Philips-Duphar B.V.; Teva Pharm. (USA).*

9735 Terbutaline

23031-25-6 9302 245-385-8

$C_{12}H_{19}NO_3$

5-[2-[(1,1-Dimethylethyl)amino]-1-hydroxyethyl]-1,3-benzenediol.

Bronchodilator; tocolytic. β-Adrenergic agonist. Ephedrine derivative. mp = 119-120°. *Astra Draco AB; Ciba-Geigy Corp.; Merrell Pharm. Inc.*

9736 Terbutaline Sulfate

23031-32-5 9302 245-386-3

$C_{24}H_{40}N_2O_{10}S$

5-[2-[(1,1-Dimethylethyl)amino]-1-hydroxyethyl]-1,3-benzenediol sulfate (2:1) (salt).

KWD-2019; Brethaire; Brethine; Bricanyl; Butaliret; Monovent; Terbasmin; Terbul. Bronchodilator; tocolytic. β-Adrenergic agonist. Ephedrine derivative. mp = 246-248°; λ_m = 276 nm ($A^{1\%}_{1cm}$ 67.6 0.1 N HCl); pKa_1 = 8.8, pKa_2 = 10.1, pKa_3 = 11.2; soluble in H_2O (>2.0 g/100 ml), 0.1N HCl (>2.0 g/100 ml), 0.1N NaOH (>2.0 g/100 ml), EtOH (0.012 g/100 ml), 10% EtOH (>2.0 g/100 ml); MeOH (0.27 g.100 ml). *Astra Draco AB; Ciba-Geigy Corp.; Marion Merrell Dow Inc.; Merrell Pharm. Inc.*

Topical Protectants

9737 Allantoin

97-59-6 255 202-592-8

$C_4H_6N_4O_3$

(2,5-Dioxo-4-imidazolidinyl)urea.

glyoxyldiureide; cordianine; Psoralon; Septalan; component of: Skin-Balm. Vulnerary. Used to treat skin ulceration. [(dl)-form]: mp = 238°; soluble in H_2O (0.52 g/100 ml), EtOH (0.2 g/100 ml). *Carrington Labs Inc.; ICI Americas Inc.*

9738 Amifostine

20537-88-6 424

$C_5H_{15}N_2O_3PS$

2-[(3-Aminopropyl)amino]ethanethiol dihydrogen phosphate ester.

ethiofos; gammaphos; SAPEP; NSC-296961; WR-2721; YM-08310; Ethyol. Protectant. Used in cancer treatment as a radioprotector.

9739 Amifostine Monohydrate

63717-27-1 424

$C_5H_{15}N_2O_3PS.H2O$

2-[(3-Aminopropyl)amino]ethanethiol dihydrogen phosphate ester monohydrate.

Protectant. Used in cancer treatment as a radioprotector. mp = 160-161° (dec); LD_{50} (mus ip) = 700 mg/kg.

9740 Amifostine Trihydrate

112901-68-5 424

$C_5H_{15}N_2O_3PS.3H2O$

2-[(3-Aminopropyl)amino]ethanethiol dihydrogen phosphate ester trihydrate.

Protectant. Used in cancer treatment as a radioprotector. Soluble in H_2O (> 9 g/100 ml).

9741 Balsam Peru

976

Peruvian Balsam; Indian Balsam; China Oil; Black Balsam; Honduras Balsam; Surinam Balsam; Extract from *Toluifera pereirae* in forests near the Pacific coast of Central America. Consists largely of esters of cinnamic and benzoic acids. Used as a vulnerary and ectoparasiticide. d = 1.150 - 1.170; insoluble in H_2O, olive oil; soluble in EtOH, $CHCl_3$, AcOH; partially soluble in Et_2O, petroleum ether.

9742 Balsam Traumatic

978

Friar's Balsam; Turlington's Balsam; composed of benzoin (100 parts), storax (35), balsam Tolu (35), balsam Peru (16), aloe (8), myrrh (8), angelica (4) and EtOH (694). Topical protectant and expectorant.

9743 Benzoin

119-53-9 1124 204-331-3

$C_{14}H_{12}O_2$

2-Hydroxy-1,2-diphenylethanone.

bitter almond oil camphor. [(dl)-form]: mp = 137°; bp_{768} = 344°, bp_{12} = 194°; λ_m 247 nm (ε 14500 EtOH); soluble in H_2O (0.03 g/100 ml), C_5H_5N (20 g/100 ml), Me_2CO, boiling EtOH; slightly soluble in Et_2O; [methyl ether]: mp = 49°; [(l)-form]: mp = 132°; $[\alpha]_D^{12}$ =-118° (c = 1.2 Me_2CO); [(d)-form]: mp = 132°; $[\alpha]_D^{12}$= 120.5° (c = 1.2 Me_2CO).

9744 Bismuth Phosphate

10049-01-1 1316 233-161-2

BiO_4P

Phosphoric acid bismuth salt (1:1).

Bismugel. Antacid and protectant. d^{15} = 6.323; slightly soluble in H_2O, dilute acids; insoluble in EtOH, AcOH; soluble in HCl, HNO_3.

9745 Bismuth Subcarbonate

5892-10-4 1324 227-567-9

CBi_2O_5

basic bismuth carbonate; bismuth oxycarbonate. Topical protectant. Insoluble in H_2O, EtOH; soluble in mineral acids.

9746 Bismuth Subgallate

22650-86-8 1325

$C_7H_5BiO_6$

Basic bismuth gallate.

gallic acid bismuth basic salt; B.S.G.; Dermatol. Topical protectant; astringent; antacid. Topical protectant, astringent and antacid. Insoluble in H_2O, EtOH, $CHCl_3$, Et_2O; soluble in dilute alkaline solutions, hot mineral acids.

9747 Calamine

1675

Zinc oxide containing about 0.5% ferric oxide.

Eczederm. Topical protectant. Pink powder; insoluble in H_2O, soluble in mineral acids.

9748 Collodion
2547
A solution of pyroxylin (4 g) in 100 ml of $EtOH/Et_2O$ (1:3). Topical protectant. d_{25}^{25} = 0.765 - 0.775.

9749 Esculin
531-75-9 3739 208-517-5
$C_{15}H_{16}O_9$
6-(β-D-glucopyranosyloxy)-7-hydroxy-2H-1-benzopyran-2-one.
esculoside; bicolorin; enallachrome; polychrome; Escosyl. Skin protectant. mp = 204-206°; soluble in H_2O (0.17 g/100 ml at 20°, 7.7 g/100 ml at 100°) hot EtOH, MeOH, C_5H_5N, EtOAc, AcOH; $[\alpha]_D^{18}$ = -78.4° (c = 2.5 50% dioxane); [pentaacetate]: mp = 163-164°.

9750 Gum Benzoin
9000-05-9 4608 232-523-7
Balsamic resin from *Styrax benzoin*. resin benzoin; resin benjamin; gum benjamin. Topical protectant. Soluble in EtOH.

9751 Hydroxypropyl Cellulose
9004-64-2 4888
Cellulose 2-hydroxypropyl ether.
Klucel; Lacrisert. Topical protectant. Soluble in polar organic solvents. *Hercules Inc.*

9752 Iodochlorhydroxyquin
130-26-7 5052 204-984-4
C_9H_5ClINO
5-Chloro-7-iodo-8-quinolinol.
5-chloro-8-hydroxy-7-iodoquinoline; clioquinol; chloroiodoquin; iodochlorohydroxyquinoline; iodochloroxyquinoline; Amebil; Alchloquin; Amoenol; Bactol; Barquinol; Budoform; Chinoform; Clioquinol; Cliquinol; Cort-Quin; Eczecidin; Enteroquinol; Entero-Septol; Entero-Vioform; Enterozol; Entrokin;Hi-Eneterol; Iodoenterol; Nioform; Nystaform; Quinambicide; Quin-O-Crème; Rheaform Boluses; Rometin; Vioform; Vioformio; component of: Domeform-HC, Formtone-HC, Lidaform-HC, Nystaform, Nystaform-HC, Racet, Vioform-Hydrocortisone. Antiamebic; topical anti-infective. Also used as an intestinal anti-infective (veterinary). Quinoline. CAUTION: May be linked with the occurrence subacute myelo-optic neuropathy. Dec 178-179°; λ_m = 266 nm ($A_{1\,cm}^{1\,\%}$ 1120 in 0.1 N MeOHic NaOH), 269 nm ($A_{1\,cm}^{1\%}$ 1120 (MeOH/KOH), 255 nm ($A_{1\,cm}^{1\,\%}$ 1570 EtOH); soluble in boiling alcohol (1:43) or ethyl acetate (1:17), chloroform (1:128), cold acetic acid (1:170); slightly soluble in $CHCl_3$, AcOH; nearly insoluble in H_2O, cold alcohol, Et_2O; LD_{50} (cat orl) = 400 mg/kg. *Bayer AG; Ciba-Geigy Corp.; Dermik Labs. Inc.; Lemmon Co.; Marion Merrell Dow Inc.*

9753 Petrolatum, Hydrophilic
7326
Petroleum jelly; paraffin jelly; vasoliment; Stanolind; Vaseline; Saxoline; Cosmoline. Topical protectant. Purified mixture of semisolid hydrocarbons. mp = 38-54°; d_{25}^{60} = 0.820 - 0.865; insoluble in H_2O, glycerol, EtOH; soluble in C_6H_6, $CHCl_3$, Et_2O, petroleum ether, CS_2, oils.

9754 Petrolatum, White
8009-03-8 7327
Stanolene.
Topical protectant. Insoluble in H_2O, glycerol, EtOH; soluble in C_6H_6, $CHCl_3$, Et_2O, petroleum ether, CS_2, oils.

9755 Pyroxylin
9004-70-0 8195
Cellulose nitrate.
nitrocellulose; collodion cotton; soluble gun cotton; collodion wool; colloxylin; xyloidin; Parlodion. Topical protectant. Soluble in MeOH, Me_2CO, AcOH, amyl acetate.

9756 Shark Liver Oil
68990-63-6 8622 273-616-2
Oil expressed from the liver of sharks and related species. Contains squalene, pristane, vitamins A and D, fatty acid esters, glycerol ethers, triglycerides, cholesterol and fatty alcohols. Robecote. Topical protectant. d^{20} = 0.922.

9757 Storax
8023-62-9 8974
Balsam obtained from the trunk of *Liquidambar orientalis* (Americam Storax or Levant Storax). Styrax; sweet oriental gum. Topical protectant. Contains cinnamic acid esters of α- and β-storesin and cinnamate esters of phenol, benzyl alcohol and EtOH. Also traces of vanillin. Semiliquid; almost completely soluble in EtOH (100 g/100 ml); soluble in Et_2O, Me_2CO, CS_2.

9758 Titanium Dioxide
13463-67-7 9612 236-675-5
O_2Ti
Unitane; C.I. Pigment White 6; C.I. 77891. Topical protectant. White powder, widely used as a pigment; mp = 1855°; insoluble in H_2O or mineral acids.

9759 Zinc Oxide
1314-13-2 10279 215-222-5
OZn
flowers of zinc; philosopher's wool; zinc white; C.I. Pigment White 4; C.I. 77947. Topical protectant. Astringent. d = 5.67; d_4^{20} = 5.607; insoluble in H_2O, soluble in mineral acids, ammonia, ammonium carbonate or dilute AcOH.

Ultraviolet Screens

9760 Actinoquinol
15301-40-3 143 239-334-9
$C_{11}H_{11}NO_4S$
8-Ethoxy-5-quinolinesulfonic acid.
Ultraviolet screen. The sodium salt is the better uv screen. mp = 286-288° (dec); soluble in dilute $NaHCO_3$.

9761 Actinoquinol Sodium Salt
7246-07-3 143 230-651-8
$C_{11}H_{10}NNaO_4S$
Sodium 8-ethoxy-5-quinolinesulfonate.
sodium etoquinol; etoquinol sodium; Corodenin; Uviban. Ultraviolet screen. The sodium salt is the better uv screen.

9762 p-Aminobenzoic Acid
150-13-0 443 205-753-0
$C_7H_7NO_2$
4-Aminobenzoic acid.
para-aminobenzoic acid; vitamin B_x; bacterial vitamin H^1; chromotrichia factor; antichromotrichia factor; trichochromogenic factor; anticantic vitamin; PABA; Amben; Paraminol; Sunbrella; component of: Pabanol, PreSun. Ultra-violet screen; antirickettsial (formerly). A naturally occurring B complex factor. Found in baker's yeast (5-6 ppm) and brewer's yeast (10-100 ppm). mp = 187-187.5°; λ_m 266 nm ($E^{1\%}_{1\,cm}$ 1070 H_2O), 288 nm ($E^{1\%}_{1\,cm}$ 137 iPrOH); soluble in H_2O (0.59 g/100 mlat 25°, 1.1 g/100 ml at 100°), EtOH (12.5 g/100 ml), Et_2O (2 g/100 ml), EtOAc, AcOH; slightly soluble in C_6H_6; insoluble in petroleum ether; pKa = 4.65, 4.80; pH (0.5%) = 3.5; LD_{50} (mus orl) = 2850 mg/kg, (rat orl) > 6000 mg/kg, (rbt iv) = 2000 mg/kg, (rbt orl) = 1830 mg/kg. *Aktieselskabet Pharmacia; Dey Labs; DuPont-Merck Pharm.; Heyden Chem.; ICN Pharm. Inc.*

9763 p-Aminobenzoic Acid Diethylamine Salt
6018-84-4 443
$C_{11}H_{18}N_2O_2$
4-Aminobenzoic acid diethylamine salt.
Navanide. Ultraviolet screen. mp = 170-173°; very soluble in H_2O.

9764 Bornelone
2226-11-1 218-757-2
$C_{14}H_{20}O$
5-(3,3-Dimethyl-2-norbornylidene)-3-penten-2-one.
Ultraviolet screen.

9765 Bumetrizole
3896-11-5 223-445-4
$C_{17}H_{18}ClN_3O$
2-tert-Butyl-6-(5-chloro-2H-benzotriazol-2-yl)-p-cresol.
Ultraviolet screen.

9766 Butyl Methoxydibenzoylmethane
70365-09-1 1616
$C_{20}H_{22}O_3$
1-(p-tert-Butylphenyl)-3-(p-methoxyphenyl)-1,3-propanedione.
Avobenzone; Parsol 1789; Parsol A; component of: Photoplex. Ultraviolet screen. mp = 83.5°. *Allergan Inc.; Givaudan-Roure SA.*

9767 β-Carotene
7235-40-7 1902 230-636-6
$C_{40}H_{56}$
(all E)-1,1'-(3,7,12,16-Tetramethyl-1,3,5,7,9,11,13,15,17-octadecanonaene-1,18-diyl)bis[2,6,6-trimethylcyclohexene].
BetaVit; Lucaratin; Solatene; Carotaben; β,β-carotene; Provatene. Vitamin; ultraviolet screen. Vitamin A precursor. mp = 183°; λ_m 497, 466 nm; soluble in CS_2, C_6H_6, $CHCl_3$; moderatley soluble in Et_2O, petroleum ether, oils; soluble in C_6H_{14} (0.11 g/100 ml at 0°); poorly soluble in MeOH, EtOH; insoluble in H_2O. *BASF Corp.; Hoffmann-LaRoche Inc.*

9768 Cinoxate
104-28-9 2370 203-191-0
$C_{14}H_{18}O_4$
2-Ethoxyethyl p-methoxycinnamate.
Give-Tan; Sun-Dare. Ultraviolet screen. bp_2 = 184-187°; d^{25}_{25} = 1.102; soluble in glycerol (0.5 g/100 ml), propylene glycol (5 g/100 ml), alcohols, esters, vegetable oils; insoluble in H_2O (< 0.05 g/100 ml). *Givaudan-Roure SA.*

9769 4-(Dimethylamino)benzoic Acid
619-84-1 3282 210-615-8
$C_9H_{11}NO_2$
Ultraviolet screen. mp = 242.5=243.5°; soluble in EtOH, sparingly soluble in Et_2O, insoluble in AcOH.

9770 4-(Dimethylamino)benzoic Acid 2-Ethylhexyl Ester
21245-02-3 3282 244-289-3
$C_{17}H_{27}NO_2$
2-Ethylhexyl p-(dimethylamino)benzoate.
Arlatone UVB; Escalol 507; Padimate O; component of: Caraloe Sun & Snow Lip Balm, Photoplex, PreSun23, PreSun 46, Radiacare Lip Balm. Ultraviolet screen. *Allergan Herbert; Carrington Labs Inc.; ICI Americas Inc.; ISP Van Dyk Inc.; Westwood-Squibb Pharm. Inc.*

9771 4-(Dimethylamino)benzoic Acid 3-Methylbutyl Ester
21245-01-2 3282 244-288-8
$C_{14}H_{21}NO_2$
3-Methylbutyl p-(dimethylamino)benzoate.
Padimate A; Escalol 506. Ultraviolet screen.

9772 Dioxybenzone
131-53-3 3357 205-026-8
$C_{14}H_{12}O_4$
2,2'-Dihydroxy-4-methoxybenzophenone.
NSC-56769; benzophenone-8; Cyasorb UV 24; Spectra-Sorb UV 24; component of: Solaquin. Ultraviolet screen. mp = 68°; soluble in EtOH (21.8 g/100 ml), iPrOH (17 g/100 ml), propylene glycol (6.2 g/100 ml), ethylene glycol (3.0 g/100 ml), n-C_6H_{14} (1.5 g/100 ml). *Am. Cyanamid; ICN Pharm. Inc.*

9773 Drometrizole
2440-22-4 3503 219-470-5
$C_{13}H_{11}N_3O$
2-(2H-Benzotriazol-2-yl)-p-cresol.
Tinuvin P. Ultraviolet screen. mp = 131-133°; bp_{10} = 225°; soluble in EtOAc, Me_2CO, dioctyl phthalate; caprolactam solutions. *Ciba-Geigy Corp.*

9774 Etocrylene
5232-99-5 226-029-0
$C_{18}H_{15}NO_2$
Ethyl 2-cyano-3,3-diphenylacrylate.
Ultraviolet screen.

9775 Homosalate
118-56-9 4776 204-260-8
$C_{16}H_{22}O_3$
3,3,5-Trimethylcyclohexyl salicylate.

homomenthyl salicylate; Heliophan. Ultraviolet screen. bp_4 = 161-165°; d_{25}^{25} = 1.045. *Greeff R.W. & Co.*

9776 Lawsone
83-72-7 5406 201-496-3
$C_{10}H_6O_3$
2-Hydroxy-1,4-naphthalenedione.
Ultraviolet screen. From leaves of *Lawsonia inermis* L. dec 195-196°.

9777 Mexenone
1641-17-4 6255 216-688-2
$C_{15}H_{14}O_3$
2-Hydroxy-4-methoxy-4'-methylbenzophenone.
benzophenone 10; Uvistat. Ultraviolet screen. *Am. Cyanamid.*

9778 Octabenzone
1843-05-6 6838 217-421-2
$C_{21}H_{26}O_3$
2-Hydroxy-4-(octyloxy)benzophenone.
benzophenone 12; Specta-Sorb UV 531. Ultraviolet screen. mp = 45-46°. *E. I. DuPont de Nemours Inc.*

9779 Octocrylene
6197-30-4 228-250-8
$C_{24}H_{27}NO_2$
2-Ethylhexyl 2-cyano-3,3-diphenylacrylate.
Ultraviolet screen.

9780 Octrizole
3147-75-9 221-573-5
$C_{20}H_{25}N_3O$
2-(2H-Benzotriazol-2-yl)4-(1,1,3,3-tetramethylbutyl)-phenol.
Ultraviolet screen.

9781 Octyl Methoxycinnamate
5466-77-3 6864 226-775-7
$C_{18}H_{26}O_3$
3-(4-Methoxyphenyl)-2-propenoic acid 2-ethylhexyl ester.
Parsol MCX; Parsol MOX. Ultraviolet screen. $bp_{0.76}$ 185-195+s, $bp_{0.076}$ = 140-150°. *Givaudan-Roure SA.*

9782 Oxybenzone
131-57-7 7088 205-031-5
$C_{14}H_{12}O_3$
2-Hydroxy-4-methoxybenzophenone.
benzophenone 3; MOB; Cyasorb UV 9; Spectra-Sorb UV 9; Uvinul M40; NSC-7778; component of: Caraloe Snow & Sun Lip Balm, Durascreen, PreSun 15 Active, PreSun 15 Sensitive Skin, PreSun 23, PreSun 29, PreSun 30, PreSun 46,Radiacare Lip Balm, Solaquin. Ultraviolet screen. mp = 66°; soluble in most organic solvents; LD_{50} (rat orl) >12800 mg/kg. *BASF Corp.; Carrington Labs Inc.; ICN Pharm. Inc.; Schwarz Pharma Kremers Urban Co.; Westwood-Squibb Pharm. Inc.*

9783 Padimate A
14779-78-3 3282 238-849-6
$C_{14}H_{21}NO_2$
Pentyl p-(dimethylamino)benzoate.
Padimate A. Ultraviolet screen.

9784 Sulisobenzone
4065-45-6 9157 223-772-2
$C_{14}H_{12}O_6S$
5-Benzoyl-4-hydroxy-2-methoxybenzenesulfonic acid.
Sungard; NSC-60584; Spectra-Sorb UV 284; Uval; Uvinul MS-40. Ultraviolet screen. mp = 145°; soluble in MeOH (50 g/100 ml), EtOH (30 g/100 ml), H_2O (25 g/100 ml), EtOAc (1 g/100 ml). *Bayer Corp. Pharm. Div.; GAF.*

Uricosurics

9785 Benzbromarone
3562-84-3 1093 222-630-7
$C_{17}H_{12}Br_2O_3$
3,5-Dibromo-4-hydroxyphenyl-2-ethyl-3-benzofuranyl ketone.
MJ-10061; L-2214; Azubromaron; Besuric; Desuric; Max-Uric; Minuric; Narcaricin; Normurat; Uricovac; Urinorm. Uricosuric. mp = 151°. *Labaz S.A.*

9786 Ethebenecid
1213-06-5 3775 214-925-4
$C_{11}H_{15}NO_4S$
p-Diethylsulfamoylbenzoic acid.
Antidipsin; Longacid; Urelim. Uricosuric. Inhibits excretion of penicillin. mp = 192-194°.

9787 Halofenate
26718-25-2
$C_{19}H_{17}ClF_3NO_4$
(p-Chlorophenyl)[(α,α,α-trifluoro-m-tolyl)oxy]acetic acid ester with N-(2-hydroxyethyl)acetamide.
Lipivas. Uricosuric; antihyperlipoproteinemic.

9788 Irtemazole
115574-30-6
$C_{18}H_{16}N_4$
(±)-5-(α-Imidazol-1-ylbenzyl)-2-methylbenzimidazole.
R-60844. Uricosuric. *Janssen Pharm. Inc.*

9789 Orotic Acid
65-86-1 7004 200-619-8
$C_5H_4N_2O_4$
1,2,3,6-Tetrahydro-2,6-dioxo-4-pyrimidinecarboxylic acid.
animal galactose factor; Oropur; Orotyl. Uricosuric. mp = 345-346°. *Kyowa Hakko Kogyo Co. Ltd.; Rhône-Poulenc Rorer Pharm. Inc.*

9790 Orotic Acid Choline Salt
24381-49-5 7004 246-213-4
$C_{10}H_{18}N_3O_5$
1,2,3,6-Tetrahydro-2,6-dioxo-4-pyrimidinecarboxylic acid choline salt.
choline orotate; Cholergol. Uricosuric. *Kyowa Hakko Kogyo Co. Ltd.; Rhône-Poulenc Rorer Pharm. Inc.*

9791 Orotic Acid Ethyl Ester
1747-53-1 7004
$C_7H_8N_2O_4$
Ethyl 1,2,3,6-tetrahydro-2,6-dioxo-4-pyrimidinecarboxylate.
Uricosuric. mp = 188-189°. *Kyowa Hakko Kogyo Co. Ltd.; Rhône-Poulenc Rorer Pharm. Inc.*

9792 Orotic Acid Methyl Ester
6153-44-2 7004 228-171-9
$C_6H_6N_2O_4$
Methyl 1,2,3,6-tetrahydro-2,6-dioxo-4-pyrimidinecarboxylate.
Uricosuric. mp = 249°. *Kyowa Hakko Kogyo Co. Ltd.; Rhône-Poulenc Rorer Pharm. Inc.*

9793 Orotic Acid Monohydrate
50887-69-9 7004
$C_5H_4N_2O_4.H_2O$
1,2,3,6-Tetrahydro-2,6-dioxo-4-pyrimidine-carboxylic acid.
Lactinium; Oroturic. Uricosuric. mp = 334°; λ_m 282 nm; soluble in H_2O (0.17 g/100 ml). *Kyowa Hakko Kogyo Co. Ltd.; Rhône-Poulenc Rorer Pharm. Inc.*

9794 Oxycinchophen
485-89-2 7091 207-624-4
$C_{16}H_{11}NO_3$
3-Hydroxy-2-phenyl-4-quinoline-carboxylic acid.
3-hydroxy-2-phenylcinchoninic acid; 3-hydroxy-chinchophen; HCP; Fenidrone; Magnofenyl; Magnophenyl; Oxinofen; Reumalon. Antidiuretic; uricosuric. Dec 206-207°; sparingly soluble in H_2O, Et_2O; soluble in AcOH, alkalies, hot EtOH, C_6H_6. *Chemo Puro.*

9795 Probenecid
57-66-9 7934 200-344-3
$C_{13}H_{19}NO_4S$
p-(Dipropylsulfamoyl)-benzoic acid.
Benemid; Probecid; Proben; component of: Colbenemid, Polycillin-PRB. Uricosuric. Used to treat gout. mp = 194-196°; λ_m = 242.5 nm (0.1N NaOH); soluble in $CHCl_3$, insoluble in H_2O; LD_{50} (rat orl) = 1600 mg/kg. *Apothecon; Merck & Co.Inc.; Wyeth-Ayerst Labs.*

9796 Seclazone
29050-11-1
$C_{10}H_8ClNO_3$
7-Chloro-3,3a-dihydro-2-methyl-2H,9H-isoxazolo-[3,2-b][1,3]benzoxazin-9-one.
W-2352. Anti-inflammatory; uricosuric. *Wallace Labs.*

9797 Sulfinpyrazone
57-96-5 9121 200-357-4
$C_{23}H_{20}N_2O_3S$
1,2-Diphenyl-4-[2-(phenylsulfinyl)ethyl]-3,5-pyrazolidinedione.
G-28315; Anturan; Anturane; Anturano; Enturen. Uricosuric; antithrombotic. Used to treat gout. A nonsteroidal anti-inflammatory agent. Inhibits cyclooxygenase. Prolongs circulating platelet survival. mp = 136-137°; λ_m = 255 nm (1N NaOH); soluble in EtOAc, $CHCl_3$; slightly soluble in H_2O, EtOH, Et_2O, mineral oils; [d-form]: mp = 130-133°; $[\alpha]^2_D$ = 67.1° (c = 2.04 EtOH), $[\alpha]_D^{25}$ = 109.3 (c = 0.5 $CHCl_3$); [l-form]: mp = 130-133°; $[\alpha]_D^{23}$ = -64.2° (c = 2.14 EtOH), $[\alpha]_D^{26}$ = -104.5° (c = 0.5 $CHCl_3$). *Ciba-Geigy Corp.*

9798 Ticrynafen
40180-04-9 9570 254-826-3
$C_{13}H_8Cl_2O_4S$
[2,3-Dichloro-4-(2-thienylcarbonyl)phenoxy]acetic acid.
tienylic acid; tienilic acid; thienylic acid; ANP-3624; CE-3624; SKF-62698; Difluorex; Selacryn. Diuretic; uricosuric; antihypertensive. mp = 148-149°, 157°; LD_{50} (mus iv) = 225 mg/kg, (mus orl) = 1275 mg/kg. *SmithKline Beecham Pharm.*

9799 Zoxazolamine
61-80-3 10328 200-519-4
$C_7H_5ClN_2O$
2-Amino-5-chlorobenzoxazole.
McN-485; Deflexol; Flexilon; Flexin; Zoxamin; Zoxine. Skeletal muscle relaxant; uricosuric agent. mp = 184-185°; λ_m = 244, 285 nm; slightly soluble in H_2O; soluble in EtOH, propylene glycol; LD_{50} (mus ip) = 376 mg/kg, (mus orl) = 678 mg/kg, (rat ip) = 102 mg/kg, (rat orl) = 730 mg/kg. *Dow Chem. U.S.A.; McNeil Pharm.*

Vasodilators, Cerebral

9800 Alprostadil
745-65-3 8063 212-017-2
$C_{20}H_{34}O_5$
(1R,2R,3R)-3-Hydroxy-2-[(E)-(3S)-3-hydroxy-1-octenyl]-5-oxocyclopentaneheptanoic acid.
prostaglandin E_1; PGE_1; U-10136; Caverject; Liple; Minprog; Palux; Prostandin; Prostine VR; Prostin VR Pediatric; Prostivas. Vasodilator (peripheral). Primary prostaglandin isolated from purified biological extracts. mp = 115-116°; $[\alpha]_{578}$ = -61.6° (c = 0.56 THF). *Upjohn Ltd.*

9801 Amyl Nitrite
110-46-3 5137 203-770-8
$C_5H_{11}NO_2$
Isopentyl nitrite.
isoamyl nitrite; pentyl nitrite; Amyl Nitrite, Vaporole. Vasodilator; antianginal. Organic nitrate. Unstable, flammable liquid; decomposes on exposure to air, light; bp = 97-99° (volatilizes at lower temperatures); d_{25}^{25} = 0.875; n_D^{21} = 1.3781; slightly soluble in H_2O; miscible with alcohol, $CHCl_3$, Et_2O; incompatible with alcohol, antipyrine, caustic alkalies, alkaline carbonates, potassium iodide, bromides, ferrous salts. *Burroughs Wellcome Inc.*

9802 Bencyclane
2179-37-5 1060 218-547-0
$C_{19}H_{31}NO$
3-[(1-Benzylcycloheptyl)oxy]-N,N-dimethyl-propylamine.
benzcyclan. Vasodilator (peripheral, cerebral). bp_3 = 146-156°. *EGYT.*

9803 Bencyclane Fumarate
14286-84-1 1060 238-204-9
$C_{23}H_{35}NO_5$
3-[(1-Benzylcycloheptyl)oxy]-N,N-dimethyl propylamine fumarate.

EGYT-201; Angiociclan; Dantrium; Dilangio; Fludilat; Fluxema; Halidor; Vasorelax. Vasodilator (peripheral, cerebral). mp = 131-133°; soluble in H_2O (1 g/100 ml at 25°); readily soluble in EtOH; slightly soluble in Me_2CO; λ_m (pH 3.4-6.6) = 207 nm; LD_{50} (mus orl) = 445.6 mg/kg, (mus iv) = 49.9 mg/kg, (mus ip) = 132 mg/kg, (mus sc) = 203 mg/kg. *EGYT.*

9804 Ciclonicate

53449-58-4 2324 258-561-4

$C_{15}H_{21}NO_2$

trans-3,3,5-Trimethylcyclohexyl nicotinate.

cyclonicate; P-350; Bled. Vasodilator (cerebral, peripheral). Derivative of nicotinic acid. Liquid; $bp_{0.6}$ = 127-128°. *Takeda Chem. Ind. Ltd.*

9805 Cinnarizine

298-57-7 2365 206-064-8

$C_{26}H_{28}N_2$

1-Cinnamyl-4-(diphenylmethyl)piperazine.

cinnipirine; R-516; R-1575; 516-MD; Aplactan; Aplexal; Apotomin; Artate; Carecin; Cerebolan; Cerepar; Cinaperazine; Cinazyn; Cinnacet; Cinnageron; Corathiem; Denapol; Dimitron; Eglen; Folcodal; Giganten; Glanil; Hilactan; Ixterol; Katoseran; Labyrin; Midronal; Mitronal; Olamin; Processine; Sedatromin; Sepan; Siptazin; Spaderizine; Stugeron; Stutgeron; Stutgin; Toliman; component of: Emesazine. Antihistaminic; vasodilator (peripheral, cerebral). Calcium channel blocker with antiallergic and antivasoconstricting activity. [hydrochloride]: mp = 192° (dec); soluble in H_2O (20 g/l). *Abbott Laboratories Inc.*

9806 Citicoline

987-78-0 2380 213-580-7

$C_{14}H_{26}N_4O_{11}P_2$

Choline cytidine 5'-pyrophosphate ester.

CDP-choline; Audes; Cereb; Citifar; Colite; Corenalin; Cyscholin; Difosfocin; Emicholine-F; Ensign; Haocolin; Hornbest; Neucolis; Nicholin; Nicolin; Niticolin; Reagin; Recognan; Rexort; Sintoclar; Somazine; Suncholin. Vasodilator (cerebral). Naturally occurring nuceotide. Coenzyme in lecithin biosynthesis. $[\alpha]_D^{20}$ = +17.2° (in H_2O); $[\alpha]_D^{25}$ = +19.0° (c = 0.86 in H_2O); λ_m = 280 nm (ε 12800 at pH 1); soluble in H_2O; nearly insoluble in most organic solvents; LD_{50} (mus iv) 4600 mg/kg; (mus orl) = 8000 mg/kg. *Abbott Laboratories Inc.*

9807 Citicoline Sodium

33818-15-4 2380 251-689-1

$C_{14}H_{25}N_4NaO_{11}P_2$

Choline cytidine 5'-pyrophosphate ester sodium (salt).

Acticolin; Brassel; Cebroton; Cidifos; Flussorex; Gerolin; Logan; Neurotron; Sinkron. Vasodilator (cerebral). *Abbott Laboratories Inc.*

9808 Cyclandelate

456-59-7 2771 207-271-6

$C_{17}H_{24}O_3$

3,3,5-Trimethylcyclohexanol-α-phenyl-α-hydroxyacetate.

BS-572; Cyclergine; Cyclobral; Cyclolyt; Cyclomandol; Cyclospasmol; Natil; Novodil; Perebral; Spasmocyclon. Vasodilator (peripheral, cerebral). Calcium modulator. mp = 55.0-56.5°; bp_{14} = 192-194°; λ_m = 269, 258, 251 (ε 1575, 2020, 1630 in EtOH); soluble in $CHCl_3$, MeOH, CH_3CN, EtOAc, DMF, C_7H_8; insoluble in H_2O. *Brocades-Stheeman & Pharmacia; Wyeth-Ayerst Labs.*

9809 Diisopropylamine Dichloroacetate

660-27-5 3241 211-538-2

$C_8H_{17}Cl_2NO_2$

Dichloroacetic acid compound with N-(1-methylethyl)-2-propanamine (1:1).

diisopropylammonium dichloroacetate; diisopropylamine dichloroethanoate; DADA; DIPA-DCA; IS-401; Dapocel; Dedyl; DIEDI; Disotat; Kalodil; Oxypangam; Tensicor. Vasodilator (cerebral, peripheral); hypotensive. mp = 119-121°; soluble in H_2O (>50%); LD_{50} (mus orl) = 1700 mg/kg.

9810 Dotarizine

84625-59-2

$C_{29}H_{34}N_2O_2$

1-(Diphenylmethyl)-4-[3-(2-phenyl-1,3-dioxolan-2-yl)propyl]piperazine.

Has calcium channel blocking properties and inhibitory effects on 5-HT2A and 5-HT2C receptors. Cerebral and peripheral vasodilator.

9811 (-)-Eburnamonine

4880-88-0 3537 225-490-5

$C_{19}H_{22}N_2O$

(3α,16α)-Eburnamenin-14(15H)-one.

16-oxoeburnane; vincamone; CH-846; Cervoxan. Vasodilator (cerebral). Obtained by acid hydrolysis of vincamine. mp = 168-170°; 177-178°; $[\alpha]_D^{25}$ = -102° (in $CHCl_3$); [α]siD = -100° (c = 0.783 in $CHCl_3$); λ_m 205,240, 265, 290, 300 nm (log ε 4.28, 4.26, 3.90, 3.59, 3.57). *Gedeon Richter Chem. Works.*

9812 (-)-Eburnamonine Phosphate

94134-60-8 3537 302-830-1

$C_{19}H_{25}N_2O_5P$

(3α,16α)-Eburnamenin-14(15H)-one phosphate (1:1).

Eburnal. Vasodilator (cerebral). *Gedeon Richter Chem. Works.*

9813 Fasudil

103745-39-7 3980

$C_{14}H_{17}N_3O_2S$

Hexahydro-1-(5-isoquinolyl-1-sulfonyl)-1H-1,4-diazepine.

Vasodilator (cerebral). Intracellular calcium antagonist. *Asahi Chem. Industry.*

9814 Fasudil Hydrochloride

105629-07-7 3980

$C_{14}H_{18}ClN_3O_2S$

Hexahydro-1-(5-isoquinolyl-1-sulfonyl)-1H-1,4-diazepine monohydrochloride.

AT-877; HA-1077. Vasodilator (cerebral). Intracellular calcium antagonist. mp 220°; soluble in H_2O; LD_{50} (mus iv) = 67.5 mg/kg, (mus sc) = 124 mg/kg, (mus orl) = 273.9 mg/kg. *Asahi Chem. Industry.*

9815 Felodipine
86189-69-7 3991
$C_{18}H_{19}Cl_2NO_4$
Ethyl methyl 4-(2,3-dichlorophenyl)-1,4-dihydro-2,6-dimethyl-3,5-pyridinedicarboxylate.
Plendil; H 154/82; Agon; Feloday; Flodil; Hydac; Munobal; Prevex; Splendil. Antianginal; antihypertensive agent. Dihydropyridine calcium channel blocker sold as the racemate. mp= 145°. *Astra USA Inc.; Merck & Co.Inc.*

9816 Fenoxedil
54063-40-0 4026
$C_{28}H_{42}N_2O_5$
2-(p-Butoxyphenoxy)-N-(2,5-diethoxyphenyl)-N-[2-(diethylamino)ethyl]acetamide.
Vasodilator (cerebral).

9817 Fenoxedil Hydrochloride
27471-60-9 4026 248-478-1
$C_{28}H_{43}ClN_2O_5$
2-(p-Butoxyphenoxy)-N-(2,5-diethoxyphenyl)-N-[2-(diethylamino)ethyl]acetamide monohydrochloride.
Suplexedil. Vasodilator (cerebral). mp = 140°; LD_{50} (mus orl) = 750°, (mus iv) = 17 mg/kg.

9818 Flunarizine
52468-60-7 4179 257-937-5
$C_{26}H_{26}F_2N_2$
(E)-1-[Bis-(p-fluorophenyl)methyl]-4-cinnamylpiperazine.
Vasodilator (peripheral, cerebral). Calcium channel blocker. Fluoronated derivative of cinnarizine. Also binds to α-adrenoceptors. *Janssen Pharm. Inc.*

9819 Flunarizine Hydrochloride
30484-77-6 4179 250-216-6
$C_{26}H_{28}Cl_2F_2N_2$
(E)-1-[Bis-(p-fluorophenyl)methyl]-4-cinnamylpiperazine dihydrochloride.
R-14950; Dinaplex; Flugeral; Flunagen; Flunarl; Fluxarten; Gradient; Issium; Mondus; Sibelium. Vasodilator (peripheral, cerebral). Calcium channel blocker. mp = 251.5°. *Janssen Pharm. Inc.*

9820 Fostedil
75889-62-2
$C_{18}H_{20}NO_3PS$
Diethyl (p-2-benzothiazolylbenzyl)-phosphonate.
A-53986; KB-944. Vasodilator; calcium channel blocker. *Abbott Labs Inc.; Kanebo Pharm. Ltd.*

9821 Ibudilast
50847-11-5 4923
$C_{14}H_{18}N_2O$
1-(2-Isopropylpyrazolo[1,5-a]pyridin-3-yl)-2-methyl-1-propanone.
KC-404; Ketas. Antiallergic; antiasthmatic; vasodilator (cerebral). Phosphodiesterase inhibitor. Leukotriene D_4 antagonist. mp = 53.5-54°; slightly soluble in H_2O, freely soluble in organic solvents; LD_{50} (mus iv) = 260 mg/kg. *Kyorin Pharm. Co. Ltd.*

9822 Ifenprodil
23210-56-2 4936 245-491-4
$C_{21}H_{27}NO_2$
4-Benzyl-α-(p-hydroxyphenyl)-β-methyl-1-piperidineethanol.
RC-61-91. Vasodilator (peripheral, cerebral). mp = 114°. *Robert et Carriere.*

9823 Ifenprodil Tartrate
66157-43-5 4936
$C_{46}H_{40}N_2O_{10}$
4-Benzyl-α-(p-hydroxyphenyl)-β-methyl-1-piperidineethanol tartrate (2:1).
Cerocral; Dilvax; Vadilex. Vasodilator (cerebral, peripheral). mp = 178-180°; soluble in EtOH, H_2O; slightly soluble in Me_2CO $CHCl_3$; nearly insoluble in Et_2O; LD_{50} (mmus iv = 17 mg/kg, (mmus ip) = 120 mg/kg, (mmus orl) = 275 mg/kg. *Robert et Carriere.*

9824 Iproxamine
52403-19-7
$C_{18}H_{29}NO_4$
5-[2-(Dimethylamino)ethoxy]carvacryl isopropyl carbonate.
Vasodilator. *Parke-Davis.*

9825 Iproxamine Hydrochloride
51222-37-8
$C_{18}H_{30}ClNO_4$
5-[2-(Dimethylamino)ethoxy]carvacryl isopropyl carbonate hydrochloride.
Go-2782; W-42782. Vasodilator. *Parke-Davis.*

9826 Lomerizine
101477-55-8 5593
$C_{27}H_{30}F_2N_2O_3$
1-[Bis(4-fluorophenyl)methyl]-4-[(2,3,4-trimethoxyphenyl)methyl]methyl]piperazine.
Antimigraine; vasodilator (cerebral). Diphenylpiperazine calcium channel blocker; selective cerebral vasodilator. *Kanebo Pharm. Ltd.*

9827 Lomerizine Dihydrochloride
101477-54-7 5593
$C_{27}H_{32}Cl_2F_2N_2O_3$
1-[Bis(p-fluorophenyl)methyl]-4-(2,3,4-trimethoxybenzyl)piperazine dihydrochloride.
KB-2796. Antimigraine. Diphenylpiperazine calcium channel blocker. Selective cerebral vasodilator. mp = 214-218° (dec), 204-207° (dec); LD_{50} (mus orl) = 300 mg/kg. *Kanebo Pharm. Ltd.*

9828 Mefenidil
58261-91-9 221-452-7
$C_{12}H_{11}N_3$
5-Methyl-2-phenylimidazole-4-acetonitrile.
McN-2378. Vasodilator (cerebral).

9829 Mefenidil Fumarate
83153-38-2
$C_{16}H_{15}N_3O_4$
5-Methyl-2-phenylimidazole-4-acetonitrile fumarate (1:1).
McN-2378-46. Vasodilator (cerebral).

9830 Nafronyl
31329-57-4 6440 250-572-2
$C_{24}H_{33}NO_3$
2-(Diethylamino)ethyl ester tetrahydro-α-(1-naphthalenyl-methyl)-2-furanpropanic acid.
naftidrofuryl; Dubimax; Gevatran; Tridus. Vasodilator. $bp_{0.5}$ = 190°; d_4^{31} = 1.0465; n_D^{20} = 1.5513; LD_{50} (mus orl) = 365 mg/kg. *Lipha Pharm. Inc.*

9831 Nafronyl Oxalate
3200-06-4 6440 221-703-0
$C_{26}H_{35}NO_7$
2-(Diethylamino)ethyl ester tetrahydro-α-(1-naphthalenyl-methyl)-2-furanpropionate oxalate (1:1).
nafronyl acid oxalate; EU-1806; LS-121; Citoxid; Di-Actane; Dusodril; Praxilene. Vasodilator. mp = 110-111°; soluble in H_2O. *Lipha Pharm. Inc.*

9832 Nicametate
3099-52-3 6577 221-452-7
$C_{12}H_{18}N_2O_2$
2-(Diethylamino)ethyl nicotinate.
Eucast. Vasodilator (peripheral, cerebral). Liquid; bp_{10} = 155-157°; bp_2 = 120-125°.

9833 Nicametate Citrate Monohydrate
1641-74-3 6577
$C_{18}H_{28}N_2O_{10}$
2-(Diethylamino)ethyl nicotinate citrate monohydrate.
Euclidan; Nutrin; Soclidan. Vasodilator (peripheral, cerebral).

9834 Nicardipine
55985-32-5 6579 259-932-3
$C_{26}H_{29}N_3O_6$
2-(Benzylmethylamino) ethyl methyl 1,4-dihydro-2,6-dimethyl-4-(m-nitrophenyl)-3,5-pyridinedicarboxylate.
Antianginal; antihypertensive; vasodilator. Dihydropyridine calcium channel blocker. Has antihypertensive properties. *Syntex Labs. Inc.; Yamanouchi U.S.A. Inc.*

9835 Nicardipine Hydrochloride
54527-84-3 6579 259-198-4
$C_{26}H_{30}Cl9N_3O_6$
2-(Benzylmethylamino) ethyl methyl 1,4-dihydro-2,6-dimethyl-4-(m-nitrophenyl)-3,5-pyridine dicarboxylate monohydrochloride.
YC-93; RS-69216; RS-69216-XX-07-0; Barizin; Bionicard; Cardene; Dacarel; Lecibral; Lescodil; Loxen; Nerdipina; Nicant; Nicardal; Nicarpin; Nicapress; Nicodel; Nimicor; Perdipina; Perdipine; Ranvil; Ridene; Rycarden; Rydene; Vasodin; Vasonase. Antianginal; antihypertensive agent. Dihydropyridine calcium channel blocker. Has antihypertensive properties. [α form]: mp = 179-181°; [β form]: mp = 168-170°; LD_{50} (mrat orl) = 634 mg/kg), (mrat iv) = 18.1 mg/kg, (frat orl) = 557 mg/kg, (frat iv) = 25.0 mg/kg, (mmus orl) = 634 mg/kg, (mmus iv) = 20.7 mg/kg, (fmus orl) = 650 mg/kg, (fmus iv) = 19.9 mg. *Syntex Labs. Inc.; Yamanouchi U.S.A. Inc.*

9836 Nicergoline
27848-84-6 6580 248-694-6
$C_{24}H_{26}BrN_3O_3$
10-Methoxy-1,6-dimethylergoline-8β-methanol 5-bromonicotinate (ester).
nicotergoline; nimergoline, MNE; FI-6714; Cergodum; Circo-Maren; Dilasenil; Duracebrol; Ergotop; Ergobel; Memoq; Nicergolent; Sermion; Vasospan. Vasodilator (peripheral, cerebral). mp = 136-138°; LD_{50} (mmus orl) = 860, (mrat orl) = 2800, (mmus iv) = 46 mg/kg. *Farmitalia Carlo Erba Ltd.*

9837 Nimodipine
66085-59-4 6643 266-127-0
$C_{21}H_{26}N_2O_7$
Isopropyl 2-methoxyethyl 1,4-dihydro-2,6-dimethyl-4-(m-nitrophenyl)-3,5-pyridinedicarboxylate.
BAY e 9736; Admon; Nimotop; Periplum. Vasodilator (cerebral). Dihydropyridine calcium channel blocker mp = 125°; LD_{50} (mus orl) = 3562, (rat orl) = 6599, (mus iv) = 33 mg/kg, (rat iv) = 16 mg/kg; [(+)-form]: $[\alpha]_D^{20}$ = 9° (c = 0.439 in dioxane); [(-)-form]: $[\alpha]_D^{20}$ = -7.93° (c = 0.374 in dioxane). *Bayer AG; Miles Inc.*

9838 Oxfenicine
32462-30-9 251-061-7
$C_8H_9NO_3$
L-2-(p-Hydroxyphenyl)glycine.
UK-25842. Vasodilator. *Pfizer Inc.*

9839 Papaverine
58-74-2 7151 200-397-2
$C_{20}H_{21}NO_4$
1-[(3,4-Dimethoxyphenyl)methyl]-6,7-dimethoxyisoquinoline.
6,7-dimethoxy-1-veratrylisoquinoline. Smooth muscle relaxant with cerebral vasodilator properties. Found in opium (0.8-1.0%). mp = 147°; d_4^{20} = 1.337; pK (25°) = 8.07; λ_m = 239, 278-280, 314,327 nm (log ε 4.83, 3.86, 3.60, 3.67 EtOH); nearly insoluble in H_2O; soluble in hot C_6H_6, glacial AcOH, Me_2CO; slightly soluble in $CHCl_3$, CCl_4, petroleum ether.

9840 Papaverine Hydrochloride
61-25-6 7151 200-502-1
$C_{20}H_{22}ClNO_4$
6,7-Dimethoxy-1-veratrylisoquinoline hydrochloride.
Artegodan; Cepaverin; Cerebid; Cerespan; Dynovas; Optenyl; Pameion; Panergon; Paptial T.R.; Pavabid; Pavacap; Pavacen; Pavadel; Pavagen; Pavakey; Pavased; Spasmo-Nit; Therapav; Vasal; Vasospan; component of: Copavin. Smooth muscle relaxant with cerebral vasodilator properties. mp = 220-225°; λ_m = 249-250, 280-282, 311 nm (log ε 4.69, 3.80, 3.82 EtOH); soluble in EtOH, EtOH; soluble in H_2O (2.25 g/100 ml); nearly insoluble in Et_2O; LD_{50} (mus iv) = 27.5 mg/kg, (mus sc) = 150 mg/kg, (rat sc) = 370 mg/kg. *Eli Lilly & Co.; Marion Merrell Dow Inc.*

9841 Pentifylline
1028-33-7 7269 213-842-0
$C_{13}H_{20}N_4O_2$
1-Hexyl-3,7-dihydro-3,7-dimethyl-1H-purine-2,6-dione.

SK-7; component of: Cosaldon. Vasodilator (cerebral). Also used as a stabilizer of vitamin preparations. mp = 82-83°. *Chem. Werke Albert.*

9842 Pindolol
13523-86-9 7597 236-867-9
$C_{14}H_{20}N_2O_2$
1-(1H-Indol-4-yloxy)-3-[(1-methylethyl)amino]-2-propanol.
prinodolol; LB-46; Betapindol; Blockin L; Calvisken; Decreten; Durapindol; Glauco-Visken; Pectobloc; Pinbetol; Pynastin; Visken. Antianginal; antihypertensive; antiarrhythmic; antiglaucoma; vasodilator. β-Adrenergic blocker with partial agonist activity. mp = 171-173°. *Sandoz Pharm. Corp.*

9843 Tinofedrine
66788-41-8 9589 266-477-4
$C_{20}H_{21}NOS_2$
(+)-R-α-[(S)-1-(3,3-Di-3-thienylallyl)amino]ethyl]benzyl alcohol.
Homburg D8955. Vasodilator (cerebral). *Degussa Ltd.*

9844 Tinofedrine Hydrochloride
50776-39-1 9589
$C_{20}H_{22}ClNOS_2$
(+)-R-α-[(S)-1-(3,3-Di-3-thienylallyl)amino]ethyl]benzyl alcohol hydrochloride.
D-8955; Novocebrin. Vasodilator (cerebral). mp = 226-229° (dec); LD_{50} (mus orl) = 1890 mg/kg, (rat orl) = 6600 mg/kg, (mus iv) = 20.15 mg/kg, (rat iv) = 14.00 mg/kg. *Degussa Ltd.*

9845 Tipropidil
70895-45-3
$C_{20}H_{35}NO_2S$
1-[p-(Isopropylthio)phenoxy]-3-(octylamino)-2-propanol.
Vasodilator. *Bristol-Myers Squibb Co.*

9846 Tipropidil Hydrochloride
70895-39-5
$C_{20}H_{36}ClNO_2S$
1-[p-(Isopropylthio)phenoxy]-3-(octylamino)-2-propanol hydrochloride.
MJ-12880-1. Vasodilator. *Bristol-Myers Squibb Co.*

9847 Vincamine
1617-90-9 10120 216-576-3
$C_{21}H_{26}N_2O_3$
(3α,14β,16α)-14,15-Dihydro-14-hydroxyeburnamenine-14-carboxylic acid methyl ester.
Angiopac; Arteriovinca; Devincan; Equipur; Novicet; Ocu-vinc; Oxygeron; Perval; Pervincamine; Pervone; Sostenil; Tripervan; Vincadar; Vincafarm; Vincafolina; Vincafor; Vincagil; Vincalen; Vincamidol; Vincapront; Vinvasaunier; Vincimax; Vinodrel Retard; Vraap; NSC-91998. Antineoplastic agent. Vasodilator (cerebral). Alkaloid isolated from *vinca minor*. Other alkaloids from the vincamine fraction are vincine, vincaminine, and vincinine. Occurs naturally in d-form. mp = 232-233°; $[\alpha]_D^{23}$ = +41° (C_5H_5N); λ_m = 225, 278 nm (log ε 4.14, 3.61); LD_{50} (mus iv) = 75 mg/kg, (mus sc) > 1000 mg/kg, (mus orl) = 1000 mg/kg; [(±)-form]: mp = 228-229°.

9848 Vincamine Hydrochloride
10592-03-7 10120 234-197-1
$C_{21}H_{27}ClN_2O_3$
(3α,14β,16α)-14,15-Dihydro-14-hydroxyeburnamenine-14-carboxylic acid methyl ester hydrochloride.
Cetal; Esberidin. Vasodilator (cerebral).

9849 Vinpocetine
42971-09-5 10128 256-028-0
$C_{22}H_{26}N_2O_2$
Ethyl apovincamin-22-oate.
AY-27255; RGH-4405; Cavinton; Ceractin. Vasodilator (cerebral). mp = 147-153° (dec); $[\alpha]_D^{20}$ = +114° (c = 1 C_5H_5N); λ_m = 229, 275, 315 nm (log ε 4.45, 4.08, 3.85 96% EtOH); LD_{50} (mus orl) = 534 mg/kg, (mus ip) = 240 mg/kg, (mus iv) = 58.7 mg/kg. *Gedeon Richter Chem. Works.*

9850 Viquidil
84-55-9 10141 201-540-1
$C_{20}H_{24}N_2O_2$
1-(6-Methoxy-4-quinolyl)-3-(3-vinyl-4-piperidyl)-1-propanone.
chinicine; mequiverine; quinotoxine; quinotoxol; LM-192. Vasodilator; antiarrhythmic. Viscous oil; $[\alpha]_D$ = +43°; slightly soluble in H_2O; soluble in EtOH, $CHCl_3$, Et_2O. *Polaroid.*

9851 Viquidil Hydrochloride
52211-63-9 10141 257-739-9
$C_{20}H_{25}ClN_2O_2$
1-(6-Methoxy-4-quinolyl)-3-(3-vinyl-4-piperidyl)-1-propanone monohydrochloride.
Desclidium; Permiran. Vasodilator; antiarrhythmic. mp ≅ 184°; λ_m = 246, 355 nm (in $CHCl_3$); soluble in EtOH; sparingly soluble in H_2O; nearly insoluble in Me_2CO. *Polaroid.*

Vasodilators, Coronary

9852 Amotriphene
5585-64-8 615
$C_{26}H_{29}NO_3$
2,3,3-Tris(p-methoxyphenyl)-N,N-dimethylallylamine.
aminoxytryphine; Myordil. Vasodilator (coronary). *Sterling Winthrop Inc.*

9853 Azaclorzine
49864-70-2
$C_{22}H_{24}ClN_3OS$
2-Chloro-10-[3-(hexahydropyrrolo[1,2-a]pyrazin-2(1H)-yl)propionyl]phenothiazine.
nonachlazine. Vasodilator (coronary).

9854 Azaclorzine Hydrochloride
49780-10-1
$C_{22}H_{26}Cl_3N_3OS$
2-Chloro-10-[3-(hexahydropyrrolo[1,2-a]pyrazin-2(1H)-yl)propionyl]phenothiazine dihydrochloride.
AY-25329. Vasodilator (coronary).

9855 Bemarinone
92210-43-0
$C_{11}H_{12}N_2O_3$
5,6-Dimethyl-4-methyl-2(1H)-quinazolinone.
Cardiotonic (positive inotropic, vasodilator). A quinazoline. *Ortho Pharm. Corp.*

9856 Bemarinone Hydrochloride
101626-69-1
$C_{11}H_{13}ClN_2O_3$
5,6-Dimethyl-4-methyl-2(1H)-quinazolinone monohydrochloride.
ORF-16600; RWJ-16600. Cardiotonic (positive inotropic, vasodilator). A quinazoline. *Ortho Pharm. Corp.*

9857 Bendazol
621-72-7 1062 210-703-6
$C_{14}H_{12}N_2$
2-Benzylbenzimidazole.
bendazole. Vasodilator (coronary). mp = 187°; nearly insoluble in H_2O; soluble in glacial AcOH, EtOH, hot C_6H_6, propylene glycol.

9858 Bendazol Hydrochloride
1212-48-2 1062 214-921-2
$C_{14}H_{13}ClN_2$
2-Benzylbenzimidazole hydrochloride.
Dibasol; Dibasole; Tromasedan. Vasodilator (coronary). mp = 175°.

9859 Benfurodil Hemisuccinate
3447-95-8 1070 222-367-8
$C_{19}H_{18}O_7$
2-(1-Hydroxyethyl)-β-(hydroxymethyl)-3-methyl-5-benzofuranacrylic acid γ-lactone hydrogen succinate.
4091-CB; benzofurodil; Eucilat; Eudilat. Cardiotonic; vasodilator. mp = 144°; soluble in alkaline solutions; LD_{50} (mus orl) = 550 mg/kg. *Clin-Byla France.*

9860 Benziodarone
68-90-6 1113 200-695-2
$C_{17}H_{12}I_2O_3$
2-Ethylbenzofuranyl 4-hydroxy-3,5-diiodophenyl ketone.
L-2329; 2329 Labaz; Amplivix; Cardivix; Dilafurane; Dila-Vasal; Retrangor. Vasodilator (coronary). mp = 167°; slightly soluble in H_2O (0.2% at 25°, 1% at 45°); soluble in $CHCl_3$, Me_2CO. *Labaz S.A.; Soc. Belge Azote Prod. Chim. Marly.*

9861 Bepridil
64706-54-3 1188
$C_{24}H_{34}N_2O$
1-[2-(N-Benzylanilino)-1-(isobutoxymethyl)ethyl]-pyrrolidine.
Antianginal. Calcium channel blocker with antianginal and antiarrhythmic (class IV) properties. $bp_{0.1}$ = 184°; $bp_{0.5}$ = 192°; n_D^{20} = 1.5538. *McNeil Pharm.; Wallace Labs.*

9862 Bepridil Hydrochloride
74764-40-2 1188
$C_{24}H_{37}ClN_2O_3$
1-[2-(N-Benzylanilino)-1-(isobutoxymethyl)ethyl]-pyrrolidine monohydrochloride monohydrate.
CERM-1978; Angopril; Bepadin; Cordium; Vascor. Antianginal. Calcium channel blocker with antianginal and antiarrhythmic (class IV) properties. mp 91°; LD_{50} (mus orl) = 1955 mg/kg, (mus iv) = 23.5 mg/kg. *C.M. Ind.; McNeil Pharm.; Wallace Labs.*

9863 Chloracyzine
800-22-6 2108
$C_{19}H_{21}ClN_2OS$
2-Chloro-10-(3-diethylaminopropionyl)phenothiazine.
chloracizine; chloracysin; chlorocizin; khloratsizin; G-020. Vasodilator (coronary). *Rhône-Poulenc.*

9864 Chromonar
804-10-4 2296 212-356-6
$C_{20}H_{27}NO_5$
[[3-[2-(Diethylamino)ethyl]-4-methyl-2-oxo-2H-1-benzopyran-7-yl]oxy]acetic acid ethyl ester.
carbocromen. Vasodilator (coronary). Nearly insoluble in H_2O. *Cassella AG.*

9865 Chromonar Hydrochloride
655-35-6 2296 211-511-5
$C_{20}H_{28}ClNO_5$
[[3-[2-(Diethylamino)ethyl]-4-methyl-2-oxo-2H-1-benzopyran-7-yl]oxy]acetic acid ethyl ester hydrochloride.
AG-3; A-27053; NSC-110430; Antiangor; Cassella 4489; Intensain; Interkordin. Vasodilator (coronary). mp = 159-160°; soluble in H_2O, EtOH, CH_2Cl_2, $CHCl_3$; sparingly soluble in Me_2CO, MEK, C_6H_6, Et_2O; LD_{50} (mus orl) = 6300 mg/kg, (mus ip) = 528 mg/kg, (mus iv) = 35.5 mg/kg. *Abbott Labs Inc.; Cassella AG.*

9866 Clobenfurol
3611-72-1 2418 222-780-3
$C_{15}H_{11}ClO_2$
α-(4-Chlorophenyl)-2-benzofuranmethanol.
cloridarol; Menacor. Vasodilator (coronary). mp = 48-49°.

9867 Clonitrate
2612-33-1 2452
$C_3H_5ClN_2O_6$
3-Chloro-1,2-propanediol dinitrate.
dinitrochlorohydrin; Dylate. Vasodilator (coronary). Liquid; bp_{760} = 190-195° (some decomposition); bp_{15} = 121-123°; bp_{10} = 117.5°; d^9 = 1.5112; d^{15} = 1.5408; soluble in EtOH, Me_2CO, $CHCl_3$; nearly insoluble in H_2O, acids.

9868 Cloricromen
68206-94-0 2467
$C_{20}H_{26}ClNO_5$
Ethyl [[8-chloro-3-[2-(diethylamino)ethyl]-4-methyl-2-oxo-2H-1-benzopyran-7-yl]oxy]acetate.
AD_6; 8-chlorocarbochromen. Antithrombotic; vasodilator (coronary). Related to coumarin. mp = 147-148°. *Fidia Pharm.*

9869 Cloricromen Hydrochloride
74697-28-2 2467
$C_{20}H_{27}Cl_2NO_5$
Ethyl [[8-chloro-3-[2-(diethylamino)ethyl]-4-methyl-2-oxo-2H-1-benzopyran-7-yl]oxy]acetate monohydrochloride.

Cromocap; Proendotel. Antithrombotic; vasodilator (coronary). mp = 219-220°. *Fidia Pharm.*

9870 Dilazep
35898-87-4 3244

$C_{31}H_{44}N_2O_{10}$

Tetrahydro-1H-1,4-diazepine-1,4-(5H)-dipropanol 3,4,5-trimethoxybenzoate (diester).
Vasodilator (coronary). *Asta-Werke AG.*

9871 Dilazep Dihydrochloride
20153-98-4 3244 243-548-8

$C_{31}H_{46}Cl_2N_2O_{10}$

Tetrahydro-1H-1,4-diazepine-1,4-(5H)-dipropanol 3,4,5-trimethoxybenzoate (diester) dihydrochloride.
Asta C 4898; Comelian; Cormelian; Labitan. Vasodilator (coronary). [monohydrate]: mp = 194-198°; LD_{50} (mmus iv) = 26.6 mg/kg, (mmus ip) = 161 mg/kg, (mmus orl) = 3740 mg/kg. *Asta-Werke AG.*

9872 Diltiazem
42399-41-7 3247 255-796-4

$C_{22}H_{26}N_2O_4S$

(+)-5-[2-(Dimethylamino)ethyl]-cis-2,3-dihydro-3-hydroxy-2-(p-methoxyphenyl)-1,5-benzothiazepin-4(5H)-one acetate (ester).
Antianginal; antihypertensive; antiarrhythmic (class IV). Calcium channel blocker with coronary vasodilating properties. *Bristol Myers Squibb Pharm. Ltd.; Forest Pharm. Inc.; Hoechst Marion Roussel Inc.; Lemmon Co.; Rhône-Poulenc Rorer Pharm. Inc.; Shionogi & Co. Ltd.; Tanabe Seiyaku Co. Ltd.*

9873 d-cis-Diltiazem Hydrochloride
33286-22-5 3247 251-443-3

$C_{22}H_{27}ClN_2O_4S$

(+)-5-[2-(Dimethylamino)ethyl]-cis-2,3-dihydro-3-hydroxy-2-(p-methoxyphenyl)-1,5-benzothiazepin-4(5H)-one acetate (ester) monohydrochloride.
CRD-401; RG-83606; Adizem; Altiazem; Anginyl; Angizem; Britiazim; Bruzem; Calcicard; Cardizem; Citizem; Cormax; Deltazen; Diladel; Dilpral; Dilrene; Dilzem; Dilzene; Herbesser; Masdil; Tildiem. Antianginal; antihypertensive; class IV antiarrhythmic. Calcium channel blocker with coronary vasodilating properties. mp = 207.5-212°; $[\alpha]_D^{24}$ = +98.3 ± 1.4° (c = 1.002 in MeOH); soluble in H_2O, MeOH, $CHCl_3$; slightly soluble in absolute EtOH; practically insoluble in C_6H_6; LD_{50} (mmus iv) = 61 mg/kg, (mmus sc) = 260 mg/kg, (mmus orl) = 740 mg/kg, (fmus iv) = 58 mg/kg, (fmus sc) = 280 mg/kg, (fmus orl) = 640 mg/kg, (mrat iv) = 38 mg/kg, (mrat sc) = 520 mg/kg, (mrat orl) = 560 mg/kg, (frat iv) = 39 mg/kg, (frat sc) = 550 mg/kg, (frat orl) = 610 mg/kg. *Bristol Myers Squibb Pharm. Ltd.; Forest Pharm. Inc.; Hoechst Marion Roussel Inc.; Lemmon Co.; Rhône-Poulenc Rorer Pharm. Inc.; Tanabe Seiyaku Co. Ltd.*

9874 Dipyridamole
58-32-2 3410 200-374-7

$C_{24}H_{40}N_8O_4$

2,2',2'',2'''-[(4,8-Dipiperidinylpyrimido[5,4-d]pyrimidine-2,6-diyl)dinitrilo]tetraethanol.
NSC-515776; RA-8; Anginal; Cardoxil; Cleridium; Coridil; Coronarine; Curantyl; Dipyridan; Gulliostin; Natyl; Peridamol; Persantine; Piroan; Prandiol; Protangix. Vasodilator (coronary). Phosphodiesterase inhibitor. Decreases platelet aggregation to damaged endothelium. mp = 163°; slightly soluble in H_2O; soluble in MeOH, EtOH, $CHCl_3$; slightly soluble in Me_2CO, C_6H_6, EtOAc; LD_{50} (rat orl) = 8400 mg/kg, (rat iv) = 208 mg/kg. *Boehringer Ingelheim Ltd.; Thomae GmbH Dr. Karl.*

9875 Droprenilamine
57653-27-7 3506

$C_{24}H_{33}N$

N-(2-Cyclohexyl-1-methylethyl)-γ-phenylbenzene-propanamine.
droprenylamine. Vasodilator (coronary). Also has antiarrhythmic and antihypotensive activity. *Maggioni Farmaceutici S.p.A.*

9876 Droprenilamine Hydrochloride
59182-63-7 3506

$C_{24}H_{34}ClN$

N-(2-Cyclohexyl-1-methylethyl)-γ-phenylbenzene-propanamine monohydrochloride.
MG-8926; Valcor. Vasodilator (coronary). mp = 175-176°; LD_{50} (mus orl) = 2850 mg/kg, (mus ip) = 68 mg/kg. *Maggioni Farmaceutici S.p.A.*

9877 Efloxate
119-41-5 3565 204-321-9

$C_{19}H_{16}O_5$

Ethyl-[(4-oxo-2-phenyl-4H-1-benzopyran-7-yl)oxy]acetate.
oxyflavil; Re-1-0185; Recordil. Vasodilator (coronary). mp = 123-124°; soluble in common organic solvents; slightly soluble in H_2O; LD_{50} (rat ip) = 3200 mg/kg. *Recordati Corp.*

9878 Erythrityl Tetranitrate
7297-25-8 3716 230-734-9

$C_4H_6N_4O_{12}$

(R*,S*)-1,2,3,4-Butanetetroltetranitrate.
erythritol tetranitrate; erythrol tetranitrate; eritrityl tetranitrate; tetranitrol; tetranitrin; nitroerythrite; NSC-106566; Cardilate; Cardiloid. Coronary vasodilator used as an antianginal. Oral/sublingual/buccal tablets; for treatment of angina pectoris. An organic nitrate. CAUTION: Explosive. Sold in tablet form only; tablets are nonexplosive. mp = 61°; soluble in EtOH, Et_2O, glycerol; insoluble in H_2O; explodes on percussion. *Burroughs Wellcome Inc.; Glaxo Wellcome plc.*

9879 Etafenone
90-54-0 3753 202-002-9

$C_{21}H_{27}NO_2$

2'-[2-(Diethylamino)ethoxy]-3-phenyl]propylphenone.
LG-11457. Vasodilator (coronary). bp_{30} = 264-268°. *Guidotti.*

9880 Etafenone Hydrochloride
2192-21-4 3753

$C_{21}H_{28}ClNO_2$

2'-[2-(Diethylamino)ethoxy]-3-phenyl]propylphenone monohydrochloride.
heptaphenone; Asmedol; Baxacor; Corodilan; Dialicor; Pagano-Cor; Relicor. Vasodilator (coronary). mp = 129-130°; LD_{50} (rat orl) = 716 mg/kg, (rat iv) = 20.8 mg/kg. *Guidotti.*

9881 Felodipine
86189-69-7 3991
$C_{18}H_{19}Cl_2NO_4$
Ethyl methyl 4-(2,3-dichlorophenyl)-1,4-dihydro-2,6-dimethyl-3,5-pyridinedicarboxylate.
Plendil; H 154/82; Agon; Feloday; Flodil; Hydac; Munobal; Prevex; Splendil. Antianginal; antihypertensive agent. Dihydropyridine calcium channel blocker sold as the racemate. mp= 145°. *Astra USA Inc.; Merck & Co.Inc.*

9882 Fendiline
13042-18-7 4011 235-915-6
$C_{23}H_{25}N$
N-(3,3-Diphenylpropyl)-α-methylbenzylamine.
Vasodilator (coronary). Calcium blocking agent. bp_1 = 183-187°. *Chinoin.*

9883 Fendiline Hydrochloride
13636-18-5 4011 237-121-5
$C_{23}H_{25}N$
N-(3,3-Diphenylpropyl)-α-methylbenzylamine hydrochloride.
HK-137; Cordan; Fendilar; Sensit. Vasodilator (coronary). mp = 204-205°; slightly soluble in H_2O; soluble in MeOH, EtOH, $CHCl_3$; LD_{50} (mus iv) = 14.5 mg/kg, (mus orl) = 950 mg/kg. *Chinoin.*

9884 Floredil
53731-36-5 4144
$C_{16}H_{25}NO_4$
4-[2-(3,5-Diethoxyphenoxy)ethyl]morpholine.
Vasodilator (coronary). *Orsymonde.*

9885 Floredil Hydrochloride
30116-80-4 4144
$C_{16}H_{26}ClNO_4$
4-[2-(3,5-Diethoxyphenoxy)ethyl]morpholine monohydrochloride.
Carfonal. Vasodilator (coronary). *Orsymonde.*

9886 Ganglefene
299-61-6 4375
$C_{20}H_{33}NO_3$
3-Diethylamino-1,2-diemethylprophyl p-isobutoxybenzoate.
Vasodilator (coronary). LD_{50} (mus sc) = 530 mg/kg.

9887 Ganglefene Hydrochloride
4375
$C_{20}H_{34}ClNO_3$
3-Diethylamino-1,2-diemethylprophyl p-isobutoxybenzoate monohydrochloride.
p-isobutoxybenzoic acid α,β-dimetyl-γ-diethylaminopropyl ester; Gangleron. Vasodilator (coronary).

9888 Heart Muscle Extract
4651
Myocardone; Recosen; Rocosenin; Corhormon; Lysomiol; Hormocardiol; Herzolan; Cordiomon Injection. Vasodilator (coronary). Prepared from the hearts of calf embryos. *Upjohn Ltd.*

9889 Hexestrol Bis(β-diethylaminoethyl ester)
2691-45-4 4739 220-261-6
$C_{30}H_{48}N_2O_2$
2,2'-[(1,2-Diethyl-1,2-ethanediyl)bis(4,1-phenyleneoxy)]bis[N,N-diethylethanamine].
Vasodilator (coronary).

9890 Hexestrol Bis(β-diethylaminoethyl ester) Dihydrochloride
69-14-7 4739
$C_{30}H_{50}Cl_2N_2O_2$
2,2'-[(1,2-Diethyl-1,2-ethanediyl)bis(4,1-phenyleneoxy)]bis[N,N-diethylethanamine] Dihydrochloride.
Coragil; Coralgina. Vasodilator (coronary). mp = 226-227°; soluble in H_2O, MeOH, $CHCl_3$, hot EtOH.

9891 Hexobendine
54-03-5 4743 200-189-1
$C_{30}H_{44}N_2O_{10}$
3,4,5-Trimethoxybenzoic acid diester with 3,3'-[ethylenebis(methylimino)]di-1-propanol.
hexabendin. Vasodilator (coronary). mp = 75-77°. *OSSW.*

9892 Hexobendine Dihydrochloride
50-62-4 4743 200-054-7
$C_{30}H_{46}Cl_2N_2O_{10}$
3,4,5-Trimethoxybenzoic acid diester dihydrochloride with 3,3'-[ethylenebis(methylimino)]-di-1-propanol.
ST-7090; Andiamine; Reoxyl; Ustimon. Vasodilator (coronary). mp = 170-174°; λ_m = 267 mn; soluble in H_2O; less soluble in EtOH; nearly insoluble in Et_2O. *Marion Merrell Dow Inc.; OSSW.*

9893 Isosorbide
652-67-5 5244 211-492-3
$C_6H_{10}O_4$
1,4:3,6-Dianhydro-D-glucitol.
1,4:3,6-dianhydrosorbitol; AT-101; NSC-40725; Hydronol; Ismotic; Isobide. Osmotic diuretic. mp = 61-64°; $[\alpha]_D$ = +44°. *Alcon Labs.*

9894 Isosorbide Dinitrate
87-33-2 5245 201-740-9
$C_6H_8N_2O_8$
1,4:3,6-Dianhydro-D-glucitol dinitrate.
Astridine; Cardio 10; Cardis; Carvanil; Carvasin; Cedocard; Corovliss; Dignionitrat; Dilatrate; Diniket; Disorlon; Duranitrat; EureCor; Flindix; Frandol; Glentonin; IBD; Imtack; Isdin; Iso-Bid; Isocard; Isoket; Iso-Mack; Iso-Puren; Isorbid; Isordil; Isordil Tembids; Isostenase; Isotrate; Langoran; Laserdil; Maycor; Myorexon; Nitorol; Nitrol; Nitrosorbonl Nosim; Rifloc Retard; Rigedal; Risordan; Soni-Slo; Sorbangil; Sorbichew; Sorbid SA; Sorbitrate; Sorquad; Vascardin; Vasorbate; Vasotrate; SDM-25; SDM-40; component of: BiDil, Dilatrate-SR. Coronary vasodilator. An organic nitrate used to treat angina pectoris. mp = 70°; $[\alpha]_D^{20}$ = 135° (EtOH); soluble in H_2O (1.1 mg/ml), more soluble in organic solvents. *ICI ; Medco Res. Inc.; Reed & Carnrick; Schwarz Pharma Kremers Urban Co.; Tillots Pharma; Wyeth Labs; Wyeth-Ayerst Labs; Zeneca Pharm.*

9895 Isosorbide Mononitrate
16051-77-7 5245 240-197-2
$C_6H_9NO_6$
1,4:3,6-Dianhydro-D-glucitol-5-mononitrate.
isosorbide-5-mononitrate; Corangin; Elan; Elantan; Imdur; ISMO; Isomonat; Monicor; Monit; Mono-Cedocard; Monoclair; Monoket; Momo Mack; Monosorb; Olicard; Pentacard; BM-22.145; IS 5-MN; AHR-4698. Coronary vasodilator used as an antianginal agent. An organic nitrate used to treat angina pectoris. A metabolite of isosorbide dinitrate. mp = 88-91°. *Boehringer Mannheim GmbH; Key Pharm.; Schwarz Pharma Kremers Urban Co.; Wyeth-Ayerst Labs.*

9896 Itramin Tosylate
13445-63-1 5263
$C_9H_{14}N_2O_6S$
2-Aminoethanol nitrate (ester) p-toluenesulfonate.
itramine tosilate; Cardisan; Cardosan; Nilatil; Tostram. Vasodilator (coronary). mp = 132-133°. *Aktieselskabet Pharmacia.*

9897 Khellin
82-02-0 5321 201-392-8
$C_{14}H_{12}NO_5$
4,9-Dimethoxy-7-methyl-5H-furo[3,2-g][1]benzopyran-5-one.
visammin; Kellin; Kelamin; Kelicor; Gynokhellan; Kelicorin; Keloid; Norkel; Simeskellina; Vasokellina; Visnagalin; Visnagen; Methafrone; Eskel; Amicardien; Viscardan; Corafurone; Cardio-Khellin; Benecardin; Ammivisnagen; Hkelfren; Lynamine; Coronin; Ammicardine; Ammipuran; Ammivin. Vasodilator (coronary). mp = 154-155°; $bp_{0.05}$ = 180-200°; λ_m = 250, 338 nm ($E^{1\%}_{1cm}$ 1600, 200 EtOH); soluble (25°) in H_2O (2.5 g/ml), Me_2CO (300 g/ml), MeOH (260 g/ml), isopropanol (125 g/ml), Et_2O (50 g/ml); LD_{50} (rat orl) = 80 mg/k. *Hoechst AG; Key Pharm.*

9898 Lidoflazine
3416-26-0 5507 222-312-8
$C_{30}H_{35}F_2N_3O$
4-[4,4-Bis(p-fluorophenyl)butyl]-1-piperazineaceto-2',6'-xylidide.
McN-JR-7094; R-7904; Angex; Clinium; Klinium; Ordiflazine; Corflazine. Coronary vasodilator; antianginal agent. Used for treatment of angina pectoris. mp = 159-161°; soluble in $CHCl_3$, less soluble in other organic solvents, insoluble in H_2O. *Abbott Laboratories Inc.; Janssen Pharm. Inc.; McNeil Pharm.*

9899 Mannitol Hexanitrate
15825-70-4 5789 239-924-6
$C_6H_8N_6O_{18}$
Hexanitrate of D-mannitol.
mannitol nitrate; nitromannite; nitromannitol; Maxitate; Medemanol; Dilangil; Moloid; Mannitrin; Nitranitol; Manexin. Vasodilator (coronary). CAUTION: Explodes on percussion. Less stable than nitroglycerol at 75°. Nonexplosive at dilutions of 10% or less. mp = 106-108°; soluble in EtOH, Et_2O; insoluble in H_2O. *Madan.*

9900 Medibazine
53-31-6 5831 200-168-7
$C_{25}H_{26}N_2O_2$
1-(1,3-Benzodioxol-5-ylmethyl)-4-(diphenylmethyl)piperazine.
S-4105. Vasodilator (coronary); bronchodilator. N-substituted piperazine derivative.

9901 Medibazine Dihydrochloride
96588-03-3 5831
$C_{25}H_{28}Cl+2N_2O_2$
1-(1,3-Benzodioxol-5-ylmethyl)-4-(diphenylmethyl)-piperazine dihydrochloride.
Vialibran. Vasodilator (coronary); bronchodilator. N-substituted piperazine derivative. mp = 288°. *Sci. Union et Cie France.*

9902 Mioflazine
79467-23-5
$C_{29}H_{30}Cl_2F_2N_4O_2$
4-[4,4-Bis(p-fluorophenyl)butyl]-3-carbamoyl-2,6-dichloro-1-piperazineacetamide.
Vasodilator (coronary). *Abbott Laboratories Inc.*

9903 Mioflazine Hydrochloride
79467-24-6
$C_{29}H_{34}Cl_4F_2N_4O_3$
4-[4,4-Bis(p-fluorophenyl)butyl]-3-carbamoyl-2,6-dichloro-1-piperazineacetamide dihydrochloride monohydrate.
R-51469. Vasodilator (coronary). *Abbott Laboratories Inc.*

9904 Mixidine
27737-38-8
$C_{15}H_{22}N_2O_2$
2-[93,4-Dimethoxyphenethyl)imino]-1-methylpyrrolidine.
McN-1589. Vasodilator (coronary). *McNeil Consumer Products Co.*

9905 Molsidomine
25717-80-0 6316 247-207-4
$C_9H_{14}N_4O_4$
N-Carboxy-3-morpholinosynonimine ethyl ester.
morsydomine; SIN-10; CAS-276; Corvaton; Corvasal; Molsidolat; Morial; Motazomin. Coronary vasodilator; antianginal agent. Non-benzene aromatic, heterocyclic and mesoionic compound. mp = 140-141°; soluble in $CHCl_3$, dilute HCl, EtOH, EtOAc, MeOH; sparingly soluble in H_2O, Me_2CO, EtOH; slightly soluble in Et_2O, petroleum Et_2O; pK (100°) = 3.0 ± 0.1; λ_m = 326 nm; LD_{50} (mmus sc) = 780 mg/kg, (mmus iv) = 860 mg, (mmus ip) = 700 mg/kg, (mmus orl) = 830 mg/kg, (fmus sc) = 750 mg/kg, (fmus iv) = 800 mg/kg, (fmus ip) = 760 mg/kg, (fmus orl) = 840 mg/kg, (mrat sc) = 1380 mg/kg, (mrat iv) = 830 mg/kg, (mrat ip) = 1250 mg/kg, (mrat orl) = 1050 mg/kg, (frat sc) = 1350 mg/kg, (frat iv) = 760 mg/kg, (frat ip) = 1250 mg/kg, (frat orl) = 1200 mg/kg. *Hoechst Roussel Pharm. Inc.; Takeda Chem. Ind. Ltd.*

9906 Nicorandil
65141-46-0 6608 265-514-1
$C_8H_9N_3O_4$
N-(2-Hydroxyethyl)nicotinamide nitrate (ester).
SG-75; Ikorel; Perisalol; Sigmart. Coronary vasodilator used as an antianginal agent. Nicotinamide derivative

with dual mechanism as both nitrovasodilator and potassium channel activator. mp = 92-93°; LD_{50} (rat orl) = 1200-1300 mg/kg, (rat iv) = 800-1000 mg/kg. *Chugai Pharm. Co. Ltd.; Upjohn Ltd.*

9907 Nifedipine

21829-25-4 6617 244-598-3

$C_{17}H_{18}N_2O_6$

Dimethyl 1,4-dihydro-2,6-dimethyl-4-(o-nitrophenyl)-3,5-pyridinedicarboxylate.

Bay a 1040; Adalat; Adalate; Adapress; Aldipin; Alfadat; Anifed; Aprical; Bonacid; Camont; Chronadalate; Citilat; Coracten; Cordicant; Cordilan; Corotrend; Duranifin; Ecodipi; Hexadilat; Introcar; Kordafen; Nifedicor; Nifedin; Nifelan; Nifelat; Nifensar XL; Orix; Oxcord; Pidilat; Procardia; Sepamit; Tibricol; Zenusin. Coronary vasodilator used as an antianginal agent. Dihydropyridine calcium channel blocker. mp = 172-174°; λ_m = 340, 235 nm (ε 5010, 21590 MeOH), 338, 238 nm (ε 5740, 20600 0.1N HCl), 340, 238 nm (5740, 20510 0.1N NaOH); soluble in Me_2CO (25.0 g/100 ml), CH_2Cl_2 (16 g/100ml), $CHCl_3$ (14 g/100 ml), EtOAc (5 g/100 ml), MeOH (2.6 g/100 ml), EtOH (1.7 g/100 ml); LD_{50} (mus orl) = 494 mg/kg, (mus iv) = 4.2 mg/kg, (rat orl) = 1022 mg/kg, (rat iv) = 15.5 mg/kg. *Bayer AG; Miles Inc.; Pfizer Inc.; Pratt Pharm.*

9908 Nisoldipine

63675-72-9 6658 264-407-7

$C_{20}H_{24}N_2O_6$

1,4-Dihydro-2,6-dimethyl-4-(2-nitrophenyl)-3,5-pyridinedicarboxylic acid methyl 2-methylpropyl ester.

Bay k 5552; Baymycard; Nisocor; Norvasc; Sular; Syscor; Zadipina. Antianginal; antihypertensive agent. Calcium channel blocker. Dihydropyridine calcium channel blocker. mp = 151-152°. *Bayer AG; Miles Inc.; Zeneca Pharm.*

9909 Nitroglycerin

55-63-0 6704 200-240-8

$C_3H_5N_3O_9$

1,2,3-Propanetriol trinitrate.

glyceryl trinitrate; glycerol nitric acid triester; nitroglycerol; trinitroglycerol; glonoin; trinitrin; blasting gelatin; blasting oil; SNG; SDM-7; SDM-17; SDM-20; SDM-26; SDM-27; SDM-36; SDM-47; SDM-71; SDM-74; SDM-75; SDM-77; SDM-78; SDM-79; Adesitrin; Angibid; Angiolingual; Anginine; Angorin; Aquo-Trinitrosan; Cardamist; Cordipatch; Coro-Nitro; Corditrine; Deponit; Diafusor; Discotrine; Gilucor; GTN; Klavikordal; Lenitral; Lentonitrina; Millisrol; Minitran; Myoglycerin; Niong; Nitrobid; Nitrodisc; Nitro-Dur; Nitrofortin; Nitrogard; Nitro-Gesani; Nitroglin; Nitroglyn; Nitrolan; Nitrolande; Nitrolar; Nitrolent; Nitrolingual; Nitrol Ointment; Nitro Mack; Nitromex; Nitronal; Nitrong; Nitro-PRN; Nitrorectal; Nitroretard; Nitrosigma; Nitrospan; Nitrostat; Nitrozell-retard; Nysconitrine; Percutol; Perlinganit; Reminitrol; Susadrin; Suscard; Sustac; Sutonit; Transderm-Nitro; Transderm-Nitro TTS; Transiderm-Nitro; Tridil; Trinalgon; Trinitrosan; Vasoglyn; component of: SDM No. 27, No. 37. Antianginal. A coronary vasodilator used in the treatment of angina pectoris. Also used in manufacture of dynamite. CAUTION: Accute poisoning can cause nausea, vomiting abdominal cramps, headache, mental confusion, delirium, bradypnea, bradycardia, paralysis, convulsions, methemoglobinemia, cyanosis, circulatory collapse, death. Chronic poisoning can cause severe headaches, hallucinations, skin rashes. Alcohol aggravates symptoms. Toxic effects may occur by ingestion, inhalation, absorption. [labile form]:mp = 2.8°; [stable form]: mp = 13.5°; begins to decompose at ≅ 50°; d_{15}^{15} = 1.5991; n_D^{15} = 1.474; heat of combustion = 1580 cal/g; slightly soluble in H_2O (0.125 g/100 ml), EtOH (0.5 g/100 ml); more soluble in MeOH (2.25 g/100 ml), CS_2 (15 g/100 ml); miscible with Et_2O, Me_2CO, glacial AcOH, EtOAc, C_6H_6, nitrobenzene, C_5H_5N, $CHCl_3$, ethylene bromide, dichloroethylene; sparingly soluble in petroleum Et_2O, liquid petrolatum, glycerol. *3M Pharm.; Ciba-Geigy Corp.; Hoechst Marion Roussel Inc.; ICI Americas Inc.; Key Pharm.; KV Pharm.; Marion Merrell Dow Inc.; Parke-Davis; Rhône-Poulenc Rorer Pharm. Inc.; Schwarz Pharma Kremers Urban Co.; Searle G.D. & Co.; U.S. Ethicals Inc.; Zeneca Pharm.*

9910 Oxprenolol

6452-71-7 7086 229-257-9

$C_{15}H_{23}NO_3$

1-(o-Allyloxyphenoxy)-3-isopropylamino-2-propanol.

Antianginal; antihypertensive; class IV antiarrhythmic. A β-adrenergic blocker. mp = 78-80°. *Bayer AG; Ciba-Geigy Corp.*

9911 Oxprenolol Hydrochloride

6452-73-9 7086 229-260-5

$C_{15}H_{24}ClNO_3$

1-(o-Allyloxyphenoxy)-3-isopropylamino-2-propanol hydrochloride.

Ba-39089; Coretal; Laracor; Paritane; Slow-Pren; Trasicor; Trasacor. Antianginal; antihypertensive; class IV antiarrhythmic. A β-adrenergic blocker. mp = 107-109°. *Bayer AG; Ciba-Geigy Corp.*

9912 Pentaerythritol Tetranitrate

78-11-5 7249 201-084-3

$C_5H_8N_4O_{12}$

2,2-Bis(hydroxymethyl)-1,3-propanediol tetranitrate.

PETN; nitropentaerythritol; penthrit; niperyt; Angitet; Cardiacap; Dilcoran-80; Lentrat; Hasethrol; Metranil; Mycardol, Neo-Corovas; Nitropenta; Nitropenton; Pentral 80; Pentafilin; Pentafin; Pentitrate; Penthrit; Pentrite; Pentritol; Pentanitrine; Pentryate; Pergitral; Peritrate; Perityl; Prevangor; Quintrate; Subicard; Terpate; Vasodiatol; component of: Miltrate, SDM No. 23, SDM No. 35. Coronary vasodilator. Used as an antianginal. An organic nitrate. CAUTION: Explodes on percussion. More sensitive to shock than TNT. Dilution with an inert ingredient helps to prevent accidental explosions. mp = 140°; d_D^{20} = 1.773; soluble in Me_2CO; sparingly soluble in EtOH, Et_2O; nearly insoluble in H_2O. *ICI Americas Inc.; Parke-Davis; Rhône-Poulenc Rorer Pharm. Inc.; Wallace Labs.*

9913 Pentrinitrol

1607-17-6 7280 216-529-7

$C_5H_9N_3O_{10}$

Pentaerythritol trinitrate.

W-2197; Petrin. Vasodilator (coronary). Related to pentaerythritol tetranitrate. CAUTION: Explosive. Dilution with an inert ingredient helps to prevent accidental explosions. Viscous liquid; n_D^{20} = 1.4941; d_D^{20} = 1.554; slightly soluble in H_2O (0.705 g/100 ml); more soluble in C_6H_6 (21.40 g/100 ml); highly soluble in EtOH, Et_2O. *Parke-Davis; Warner-Lambert.*

9914 Perhexiline
6621-47-2 7305 229-569-5
$C_{19}H_{35}N$
2-(2,2-Dicyclohexylethyl)piperidine.
perhexilene. Vasodilator (coronary); diuretic. Calcium blocking agent. [hydrochloride]: mp= 243-245.5°. *Marion Merrell Dow Inc.*

9915 Perhexiline Maleate
6724-53-4 7305 229-775-5
$C_{23}H_{39}NO_4$
2-(2,2-Dicyclohexylethyl)piperidine maleate (1:1).
perhexilene maleate; Pexid. Vasodilator (coronary); diuretic. Calcium blocking agent. mp = 188.5-191°; LD_{50} (rat orl) > 7000 mg/kg, (mus orl) = 4370 mg/kg. *Marion Merrell Dow Inc.*

9916 Pimefylline
10001-43-1 7584
$C_{15}H_{18}N_6O_2$
7-[2-[(3-Pyridylmethyl)amino]ethyl]theophyilline.
pimephylline; ES-771. Vasodilator (coronary). mp = 111-112°; λ_m = 270 nm (in H_2O); soluble in H_2O, $CHCl_3$, Me_2CO, EtOH; LD_{50} (mus orl) = 1900 mg/kg, (mus iv) = 402 mg/kg. *Eprova AG.*

9917 Pimefylline Nicotinate
10058-07-8 7584 233-185-3
$C_{21}H_{23}N_7O_4$
7-[2-[(3-Pyridylmethyl)amino]ethyl]theophyilline nicotinate.
ES-902; Teonicon. Vasodilator (coronary). mp = 159-160°; λ_m = 267 nm (in H_2O); soluble in H_2O; slightly soluble in MeOH; nearly insoluble in Me_2CO, $CHCl_3$; LD_{50} (mus orl) = 2530 mg/kg, (mus iv) = 470 mg/kg. *Eprova AG.*

9918 Prenylamine
390-64-7 7919 206-869-4
$C_{24}H_{27}N$
N-(3,3-Diphenylpropyl)-α-methylphenethylamine.
B-436; Elecor. Vasodilator (coronary). mp = 36.5-37.5°. *Hoechst AG.*

9919 Prenylamine Lactate
69-43-2 7919 200-705-5
$C_{27}H_{33}NO_3$
N-(3,3-Diphenylpropyl)-α-methylphenethylamine lactate.
Angormin; Bismetin; Carditin-Same; Coredamin; Corontin; Crepasin; Daxauten; Hostaginan; Incoran; Irrorin; Lactamin; Plactamin; Reocorin; Roinin; Seccidin; Sedolaton; Segontin; Synadrin. Vasodilator (coronary). mp = 140-142°; λ_m = 260 nm ($E^{1\%}_{1cm}$ 170 in $CHCl_3$); sparingly soluble in H_2O (0.5%); soluble in organic solvents. *Hoechst AG.*

9920 Propatyl Nitrate
2921-92-8 7995 220-866-5
$C_6H_{11}N_3O_9$
2-Ethyl-2-(hydroxymethyl)-1,3-prpanediol trinitrate.
ettriol trinitrate; ETTN; Win-9317; Atrilon 5; Etrynit; Gina; Ginapect; Vassangor. Vasodilator (coronary). CAUTION: Explosive, but only slightly sensitive to shock. mp = 51-52°; soluble in Me_2CO, EtOH; nearly insoluble in H_2O; lowest explosive temperature is 220°; heat of combustion = 829.2 kcal/mol. *Sterling Winthrop Inc.*

9921 Pyridofylline
53403-97-7 8159 258-521-6
$C_{17}H_{23}N_5O_9S$
Pyridoxol salt of 7-(2-hydroxyethyl)theophylline hydrogen sulfate ester.
Atherophylline. Respiratory stimulant. Vasodilator (coronary). mp = 144-146°; LD_{50} (mus iv) = 1000 mg/kg, (mus orl) = 1600 mg/kg.

9922 Terodiline
15793-40-5 9311
$C_{20}H_{27}N$
N-tert-Butyl-1-methyl-3,3-diphenylpropylamine.
Coronary vasodilator used as an antianginal agent. Used also in treatment of urinary incontinence. Calcium antagonist with anticholinergic and vasodilatory activity. Liquid; $bp_{1.0}$ = 130-132°. *Marion Merrell Dow Inc.*

9923 Terodiline Hydrochloride
7082-21-5 9311
$C_{20}H_{28}ClN$
N-tert-Butyl-1-methyl-3,3-diphenylpropylamine hydrochloride.
Bicor; Mictrol; Micturin; Micturol. Coronary vasodilator used as an antianginal agent. Used also in treatment of urinary incontinence. Calcium antagonist with anticholinergic and vasodilatory activity. mp = 178-180°; soluble in EtOH, slightly soluble in Et_2O. *Marion Merrell Dow Inc.*

9924 Tolamolol
38103-61-6 253-783-8
$C_{19}H_{24}N_2O_4$
4-[2-[[2-Hydroxy-3-(2-methylphenoxy)propyl]amino]-ethoxy]benzamide.
Vasodilator (coronary); cardiac depressant (antiarrhythmic); antiadrenergic (β-receptor). *CIBA plc; Pfizer Inc.*

9925 Trapidil
15421-84-8 9709 239-434-2
$C_{10}H_{15}N_5$
7-(Diethylamino)-5-methyl-2-triazolo[1,5-a]pyrimidine.
trapymin; AR-12008; Avantrin; Rocornal. Vasodilator (coronary). A triazolopyrimidine platelet-derived growth factor antagonist. mp = 98-99.4°, 102-104°; λ_m = 222, 270, 307 nm (log ε 4.28, 3.83, 4.28 MeOH); pK_s = 2.79; soluble in H_2O, 1N sulfuric acid, 10% ammonium hydroxide, MeOH, iPrOH, n-BuOH, $CHCl_3$, C_6H_6, Et_2O; nearly insoluble in C_6H_{14}, C_7H_{16}; very stable except under extremely alkaline conditions; LD_{50} (mus iv) = 115 mg/kg, (mus orl) = 380 mg/kg, (mus ip) = 155 mg/kg, (mus sc) = 132 mg/kg.

9926 Tricromyl
85-90-5 9791 201-641-0
$C_{10}H_8O_2$
3-Methyl-4H-1-benzoxopyran-4-one.

methylchromone; Cromonalgina. Antispasmodic; coronary vasodilator. Originated from an ancient Egyptian drug now termed *bezr el khelda*. mp = 68°; λ_m = 304 nm (in EtOH). *Lab. Franc. Chimiother.*

9927 Trimetazidine

5011-34-7 9835 225-690-2

$C_{14}H_{22}N_2O_3$

1-(2,3,4-Trimethoxybenzyl)piperazine.

40045. Antianginal; vasodilator (coronary). $bp_{2.0}$ = 200-205°. *Sci. Union et Cie France.*

9928 Trimetazidine Dihydrochloride

13171-25-0 9835 236-117-0

$C_{14}H_{24}Cl_2N_2O_3$

1-(2,3,4-Trimethoxybenzyl)piperazine dihydrochloride.

Kyurinett; Vastarel F; Yoshimilon. Antianginal; vasodilator (coronary). mp = 225-228°; LD_{50} (mmus iv) = 91 mg/kg, (mmus ip) = 264 mg/kg, (mmus orl) = 608 mg/kg, (fmus iv) = 107 mg/kg, (fmus ip) = 245 mg/kg, (fmus orl) = 608 mg/kg, (mrat iv) = 124 mg/kg, (mrat ip) = 327 mg/kg, (mrat orl) = 1147 mg/kg; (frat iv) = 124 mg/kg, (frat ip) = 288 mg/kg, (frat orl) = 987 mg/kg. *Sci. Union et Cie France.*

9929 Trolnitrate Phosphate

588-42-1 9900 209-617-1

$C_6H_{18}N_4O_{17}P_2$

2,2',2''-Nitrilotrisethanol trinitrate (ester) phosphate (1:2) salt.

triethanolamine trinitrate biphosphate; trinitrotriethanolamine diphosphate; Angitrit; Bentonyl; Duronitrin; Metamed; Nitretamin; Nitroduran; Ortin; Praentiron; Vasomed. Antianginal. mp = 107-109°. *Bristol-Myers Squibb Co; Schering-Plough HealthCare Products.*

9930 Visnadine

477-32-7 10147 207-515-1

$C_{21}H_{24}O_7$

3,4,5-Trihydroxy-2,2-dimethyl-6-chromanacrylic acid δ-lactone 4-acetate 3-(2-methylbutyrate).

Cardine; Carduben; Vibeline; Visnamine. Vasodilator (coronary). mp = 85-88°; $[\alpha]_D^{20}$ = +9.2° (EtOH); $[\alpha]_D^{30}$ = +42.5° (c = 2 in dioxane); slightly soluble in H_2O; very soluble in EtOH, MeOH, $CHCl_3$, Me_2CO, Et_2O, C_6H_6, DMF; LD_{50} (mus orl) = 2240 mg/kg, (mus sc) >370 mg/kg. *Penick.*

Vasodilators, Peripheral

9931 Acetylcholine Chloride

60-31-1 88 200-468-8

$C_7H_{16}ClNO_2$

2-Acetyloxy-N,N,N-trimethylethanaminium chloride.

Acecoline; Arterocoline; Miochol; Ovisot. Cholinergic; cardiac depressant; miotic; vasodilator (peripheral). mp = 149-152°; very soluble in cold H_2O, decomposed by hot H_2O, soluble in EtOH, insoluble in Et_2O; LD_{50} (rat sc) = 250 mg/kg. *Iolab.*

9932 Alprostadil

745-65-3 8063 212-017-2

$C_{20}H_{34}O_5$

(1R,2R,3R)-3-Hydroxy-2-[(E)-(3S)-3-hydroxy-1-octenyl]-5-oxocyclopentaneheptanoic acid.

prostaglandin E_1; PGE_1; U-10136; Caverject; Liple; Minprog; Palux; Prostandin; Prostine VR; Prostin VR Pediatric; Prostivas. Vasodilator (peripheral). Primary prostaglandin isolated from purified biological extracts. mp = 115-116°; $[\alpha]_{578}$ = -61.6° (c = 0.56 in THF). *Upjohn Ltd.*

9933 Aluminum Nicotinate

1976-28-9 364 217-832-7

$C_{18}H_{12}AlN_3O_6$

3-Pyridinecarboxylic acid aluminum salt.

nicotinic acid aluminum salt; Alunitine Nicolex. Vasodilator (peripheral); antihyperlipoproteinemic. *Walker Labs.*

9934 Amyl Nitrite

110-46-3 5137 203-770-8

$C_5H_{11}NO_2$

Isopentyl nitrite.

isoamyl nitrite; pentyl nitrite; Amyl Nitrite, Vaporole. Vasodilator; antianginal. Organic nitrate. Unstable, flammable liquid; decomposes on exposure to air, light; bp = 97-99° (volatilizes at lower temperatures); d_{25}^{25} = 0.875; n_D^{21} = 1.3781; slightly soluble in H_2O; miscible with EtOH, $CHCl_3$, Et_2O; incompatible with alcohol, antipyrine, caustic alkalies, alkaline carbonates, potassium iodide, bromides, ferrous salts. *Burroughs Wellcome Inc.*

9935 Bamethan

3703-79-5 981 223-043-9

$C_{12}H_{19}NO_2$

α-[(Butylamino)methyl]-p-hydroxybenzyl alcohol.

Butyl-Nor-Sympatol. Vasodilator (peripheral). mp = 123.5-125°.

9936 Bamethan Sulfate

5716-20-1 981 227-214-9

$C_{24}H_{40}N_2O_8S$

α-[(Butylamino)methyl]-p-hydroxybenzyl alcohol sulfate (2:1) (salt).

Bupatol; Butedrin; Garmian; Rotesar; Vasculat; Vasculit; Vascunicol. Vasodilator (peripheral).

9937 Bencyclane

2179-37-5 1060 218-547-0

$C_{19}H_{31}NO$

3-[(1-Benzylcycloheptyl)oxy]-N,N-dimethylpropylamine.

benzcyclan. Vasodilator (peripheral, cerebral). bp_3 = 146-156°. *EGYT.*

9938 Bencyclane Fumarate

14286-84-1 1060 238-204-9

$C_{23}H_{35}NO_5$

3-[(1-Benzylcycloheptyl)oxy]-N,N-dimethylpropylamine fumarate.

EGYT-201; Angiociclan; Dantrium; Dilangio; Fludilat;

Fluxema; Halidor; Vasorelax. Vasodilator (peripheral, cerebral). mp = 131-133°; soluble in H_2O (1 g/100 ml at 25°); readily soluble in EtOH; slightly soluble in Me_2CO; λ_m (pH 3.4-6.6) = 207 nm; LD_{50} (mus orl) = 445.6 mg/kg, (mus iv) = 49.9 mg/kg, (mus ip) = 132 mg/kg, (mus sc) = 203 mg/kg. *EGYT.*

9939 Betahistine
5638-76-6 1224 227-086-4
$C_8H_{12}N_2$
2-[2-(Methylamino)ethyl]pyridine.
Vasodilator (peripheral). Liquid; bp_{30} = 113-114°; soluble in H_2O, EtOH, Et_2O, $CHCl_3$.

9940 Betahistine Hydrochloride
5579-84-0 1224 226-966-5
$C_8H_{14}Cl_2N_2$
2-[2-(Methylamino)ethyl]pyridine dihydrochloride.
Betaserc; Serc; Vasomotal. Vasodilator (peripheral). mp = 148-149°. *Unimed Pharm. Inc.*

9941 Betahistine Maleate
1224
$C_{12}H_{16}N_2O_4$
2-[2-(Methylamino)ethyl]pyridine maleate (1:1).
Suzutolon. Vasodilator (peripheral).

9942 Biclodil
85125-49-1
$C_8H_8Cl_2N_4O$
[(2,6-Dichlorophenyl)amidino]urea.
Antihypertensive (vasodilator).

9943 Biclodil Hydrochloride
75564-40-8
$C_8H_9Cl_3N_4O$
[(2,6-Dichlorophenyl)amidino]urea monohydrochloride.
WHR-1051B. Antihypertensive (vasodilator).

9944 Bradykinin
58-82-2 1386 200-398-8
$C_{50}H_{73}N_{15}O_{11}$
Kallidin I.
callidin I; kallidin-9; BRS-640. Vasodilator. A tissue hormone; a member of the plasma kinin family, a group of hypotensive peptides. Acts on smooth muscle, dilates peripheral vessels, increases capillary permeability. A potent pain-producing agent. Amorphous precipitate; $[\alpha]_D^{25}$ = -76.5° (c = 1.37 in 1N AcOH); soluble in glacial AcOH, 10% trichloroacetic acid, 70% EtOH; less soluble in 90% EtOH, cold MeOH; nearly insoluble in nonpolor organic solvents. *Squibb E.R. & Sons.*

9945 Brovincamine
57475-17-9 1473
$C_{21}H_{25}BrN_2O_3$
(3α,14β,16α)-11-Bromo-14,15-dihydro-14-hydroxy-eburamenine-14-carboxylic acid methyl ester.
11-brovincamine. Vasodilator (peripheral). A vincamine derivative. mp = 214° (dec); $[\alpha]_D^{20}$ = +8.7° (1% in $CHCl_3$). *Sandoz Pharm. Corp.*

9946 Brovincamine Hydrogen Fumarate
84964-12-5 1473
$C_{25}H_{29}BrN_2O_7$
(3α,14β,16α)-11-Bromo-14,15-dihydro-14-hydroxy-eburamenine-14-carboxylic acid methyl ester hydrogen fumarate.
BV-26-723; Sabromin; Zabromin. Vasodilator (peripheral). A vincamine derivative. mp = 144°; $[\alpha]_D^{20}$ = +4.7° (0.388% in H_2O). *Sandoz Pharm. Corp.*

9947 Budralazine
36798-79-5 1492
$C_{14}H_{16}N_4$
4-Methyl-3-penten-2-one (1-phthalazinyl)hydrazone.
DJ-1461; Buterazine. An α-adrenergic agonist. Used as an antihypertensive agent. Direct-acting vasodilator with central sympathoinhibitory activity. A derivative of hydralazine. mp = 132-133°; λ_m = 208, 240, 289, 357 nm (ε = 27000, 89000, 20000, 15000 MeOH); LD_{50} (mus orl) = 1820 mg/kg, (mus ip) = 4020 mg/kg, (rat orl) = 620 mg/kg, (rat ip) = 3570 mg/kg. *Daiichi Seiyaku.*

9948 Bufeniode
22103-14-6 1495 244-781-8
$C_{18}H_{23}I_2NO_2$
4-Hydroxy-3,5-diiodo-α-[1[(1-methyl-3-phenylpropyl)-amino]ethyl] benzyl alcohol.
HF-241; diiodobuphenine; Diastal; Proclival. Peripheral vasodilator, used as an antihypertensive agent. mp (slow heating) = 185° (some decomposition); mp (fast heating) = 212°; LD_{50} (mus ip) > 600 mg/kg, (mus orl) > 2000 mg/kg. *Lab. Houde.*

9949 Buflomedil
55837-25-7 1498 259-851-3
$C_{17}H_{25}NO_4$
2',4',6'-Trimethoxy-4-(1-pyrrolidinyl)butyrophenone.
Vasodilator (peripheral). A competitive, nonselective inhibitor of α-adrenergic receptors and a weak calcium antagonist. *Orsymonde.*

9950 Buflomedil Hydrochloride
35543-24-9 1498 252-611-9
$C_{17}H_{26}ClNO_4$
2',4',6'-Trimethoxy-4-(1-pyrrolidinyl)butyrophenone monohydrochloride.
LL-1656; Bufedil; Buflan; Buflocit; Buflonat; Fonzylane; Irrodan; Lofton; Loftyl; Provas. Vasodilator (peripheral). A competitive, nonselective inhibitor of α-adrenergic receptors and a weak calcium antagonist. mp = 192-193°; LD_{50} (mus iv) ≅ 80 mg/kg. *C.M. Ind.; Orsymonde.*

9951 Butalamine
22131-35-7 1535 244-794-9
$C_{18}H_{28}N_4O$
N,N-Dibutyl-N'-(3-phenyl-1,2,4-oxadiazol-5-yl)-1,2-ethanediamine.
Vasodilator (peripheral).

9952 Butalamine Hydrochloride
56974-46-0 1535
$C_{18}H_{29}ClN_4O$
N,N-Dibutyl-N'-(3-phenyl-1,2,4-oxadiazol-5-yl)-1,2-ethanediamine hydrochloride.

LA-1221; Adrevil; Adrevil forte; Hemotrope; Surem; Surheme. Vasodilator (peripheral). mp = 145°; LD_{50} (mus iv) = 43 mg/kg, (mus sc) = 2500 mg/kg, (mus orl) = 625 mg/kg, (rat sc) > 4000 mg/kg, (rat orl) = 1600 mg/kg.

9953 Buterizine
68741-18-4
$C_{31}H_{38}N_4$
2-Butyl-5-[[4-(diphenylmethyl)-1-piperazinyl]methyl]-1-ethylbenzimidazole.
R-38198. Vasodilator (peripheral). *Abbott Labs Inc.*

9954 Cetiedil
14176-10-4 2062 238-028-2
$C_{20}H_{31}NO_2S$
2-(Hexahydro-1H-azepin-1-yl)ethyl α-cyclohexyl-3-thiopheneacetate.
Vasodilator (peripheral). Inhibits the vesicular loading of acetylcholine. *Innothera.*

9955 Cetiedil Citrate
16286-69-4 2062 240-381-2
$C_{26}H_{39}NO_9S$
2-(Hexahydro-1H-azepin-1-yl)ethyl α-cyclohexyl-3-thiopheneacetate citrate (1:1).
Celsis; Stratene; Vasocet. Vasodilator (peripheral). Inhibits the vesicular loading of acetylcholine. mp = 115°. *Innothera; McNeil Pharm.*

9956 Ciclonicate
53449-58-4 2324 258-561-4
$C_{15}H_{21}NO_2$
trans-3,3,5-Trimethylcyclohexyl nicotinate.
cyclonicate; P-350; Bled. Vasodilator (cerebral, peripheral). Derivative of nicotinic acid. Liquid; $bp_{0.6}$ = 127-128°. *Takeda Chem. Ind. Ltd.*

9957 Cinepazide
23887-46-9 2350 245-928-9
$C_{22}H_{31}N_3O_5$
1-[2-Oxo-2-(1-pyrrolidinyl)ethyl]-4-[1-oxo-3-(3,4,5-trimethoxyphenyl)-2-propenyl]piperazine.
Vasodilator (peripheral). *Delalande.*

9958 Cinepazide Maleate
26328-04-1 2350 247-613-1
$C_{26}H_{35}N_3O_9$
1-[2-Oxo-2-(1-pyrrolidinyl)ethyl]-4-[1-oxo-3-(3,4,5-trimethoxyphenyl)-2-propenyl]piperazine maleate (1:1).
MD-67350; Brendil; Vasodistal. Vasodilator (peripheral). mp = 135°; LD_{50} (mus orl) = 1000 mg/kg, (mus iv) = 617 mg/kg. *Delalande.*

9959 Cinnarizine
298-57-7 2365 206-064-8
$C_{26}H_{28}N_2$
1-Cinnamyl-4-(diphenylmethyl)piperazine.
cinnipirine; R-516; R-1575; 516-MD; Aplactan; Aplexal; Apotomin; Artate; Carecin; Cerebolan; Cerepar; Cinaperazine; Cinazyn; Cinnacet; Cinnageron; Corathiem; Denapol; Dimitron; Eglen; Folcodal; Giganten; Glanil; Hilactan; Ixterol; Katoseran; Labyrin; Midronal; Mitronal; Olamin; Processine; Sedatromin; Sepan; Siptazin; Spaderizine; Stugeron; Stutgeron; Stutgin; Toliman; component of: Emesazine. Antihistaminic; vasodilator (peripheral, cerebral). Calcium channel blocker with antiallergic and antivasoconstricting activity. [hydrochloride]: mp = 192° (dec); soluble in H_2O (20 g/l). *Abbott Laboratories Inc.*

9960 Diisopropylamine Dichloroacetate
660-27-5 3241 211-538-2
$C_8H_{17}Cl_2NO_2$
Dichloroacetic acid compound with N-(1-methylethyl)-2-propanamine (1:1).
diisopropylammonium dichloroacetate; diisopropylamine dichloroethanoate; DADA; DIPA-DCA; IS-401; Dapocel; Dedyl; DIEDI; Disotat; Kalodil; Oxypangam; Tensicor. Vasodilator (cerebral, peripheral); hypotensive. mp = 119-121°; soluble in H_2O (>50%); LD_{50} (mus orl) = 1700 mg/kg.

9961 Dimoxyline
147-27-3 3318
$C_{22}H_{25}NO_4$
1-(4-Ethoxy-3-methoxybenzyl)-6,7-dimethoxy-3-methylisoquinoline.
dioxyline; Paveril. Vasodilator (peripheral). mp = 124-125°; [hydrochloride]: dec 196-208°; [phosphate]: dec 197-199°; more soluble than the hydrochloride; LD_{50} (mus iv) = 112.7 mg/kg. *Eli Lilly & Co.*

9962 Eledoisin
69-25-0 3579
$C_{54}H_{85}N_{13}O_{15}S$
5-Oxo-L-propyl-L-seryl-L-lysyl-L-aspartyl-L-alanyl-L-phenylalanyl-L-isoeucylglycyl-L-leucyl-L-methioninamide.
ELD-950. Hypotensive; stimulator of lacrimal secretion; vasodilator. A hendecapeptide from the posterior salivary glands of eledone spp (a small octopus). Possesses physiological action similar to other tachykinins: stimulates extravascular smooth muscle, acts as a potent vasodilator and hypotensive agent. Sesquihydrate powder dec at 230°; $[\alpha]_D^{22}$ = -44° (c = 1 in 95% AcOH); slowly loses activity when incubated in blood. *Farmitalia Carlo Erba Ltd.*

9963 Endralazine
39715-02-1 3617
$C_{14}H_{15}N_5O$
6-Benzoyl-5,6,7,8-tetrahydropyrido[4,3-c]pyridazin-3(2H)-one hydrazone.
Antihypertensive agent with peripheral vasodilating properties. mp = 220-223° (dec). *Sandoz Pharm. Corp.*

9964 Endralazine Monomethanesulfonate
65322-72-7 3617
$C_{15}H_{19}N_5O_4S$
6-Benzoyl-5,6,7,8-tetrahydropyrido[4,3-c]pyridazin-3(2H)-one hydrazone monomethanesulfonate.
Migranal; endralazine mesylate; BQ-22-708; Miretilan. Antihypertensive agent with peripheral vasodilating properties. mp = 185-188° (dec). *Sandoz Pharm. Corp.*

9965 Erythrityl Tetranitrate
7297-25-8 3716 230-734-9
$C_4H_6N_4O_{12}$
(R*,S*)-1,2,3,4-Butanetetroltetranitrate.
erythritol tetranitrate; erythrol tetranitrate; eritrityl

tetranitrate; tetranitrol; tetranitrin; nitroerythrite; NSC-106566; Cardilate; Cardiloid. Coronary vasodilator used as an antianginal. Oral/sublingual/buccal tablets; for treatment of angina pectoris. An organic nitrate. CAUTION: Explosive. Sold in tablet form only; tablets are nonexplosive. mp = 61°; soluble in EtOH, Et_2O, glycerol; insoluble in H_2O; explodes on percussion. *Burroughs Wellcome Inc.; Glaxo Wellcome plc.*

9966 Felodipine
86189-69-7 3991

$C_{18}H_{19}Cl_2NO_4$

Ethyl methyl 4-(2,3-dichlorophenyl)-1,4-dihydro-2,6-dimethyl-3,5-pyridinedicarboxylate.

Plendil; H 154/82; Agon; Feloday; Flodil; Hydac; Munobal; Prevex; Splendil. Antianginal; antihypertensive agent. Dihydropyridine calcium channel blocker sold as the racemate. mp= 145°. *Astra USA Inc.; Merck & Co.Inc.*

9967 Fenoxedil
54063-40-0 4026

$C_{28}H_{42}N_2O_5$

2-(p-Butoxyphenoxy)-N-(2,5-diethoxyphenyl)-N-[2-(diethylamino)ethyl]acetamide.

Vasodilator (cerebral).

9968 Fenoxedil Hydrochloride
27471-60-9 4026 248-478-1

$C_{28}H_{43}ClN_2O_5$

2-(p-Butoxyphenoxy)-N-(2,5-diethoxyphenyl)-N-[2-(diethylamino)ethyl]acetamide monohydrochloride.

Suplexedil. Vasodilator (cerebral). mp = 140°; LD_{50} (mus orl) = 750°, (mus iv) = 17 mg/kg.

9969 Flunarizine
52468-60-7 4179 257-937-5

$C_{26}H_{26}F_2N_2$

(E)-1-[Bis-(p-fluorophenyl)methyl]-4-cinnamylpiperazine.

Vasodilator (peripheral, cerebral). Calcium channel blocker. Fluoronated derivative of cinnarizine. Also binds to α-adrenoceptors. *Janssen Pharm. Inc.*

9970 Flunarizine Hydrochloride
30484-77-6 4179 250-216-6

$C_{26}H_{28}Cl_2F_2N_2$

(E)-1-[Bis-(p-fluorophenyl)methyl]-4-cinnamylpiperazine dihydrochloride.

R-14950; Dinaplex; Flugeral; Flunagen; Flunarl; Fluxarten; Gradient; Issium; Mondus; Sibelium. Vasodilator (peripheral, cerebral). Calcium channel blocker. mp = 251.5°. *Janssen Pharm. Inc.*

9971 Fostedil
75889-62-2

$C_{18}H_{20}NO_3PS$

Diethyl (p-2-benzothiazolylbenzyl)phosphonate.

A-53986; KB-944. Vasodilator (peripheral). Calcium channel blocker. *Abbott Labs Inc.; Kanebo Pharm. Ltd.*

9972 Hepronicate
7237-81-2 4688

$C_{28}H_{31}N_3O_6$

2-Hexyl-2(hydroxymethyl)-1,3-propanediol trinicotinate.

Megrin. Vasodilator (peripheral). mp = 94-96°. *Yoshitomi.*

9973 Histamine Phosphate
51-74-1 4756 200-118-4

$C_5H_{15}N_3O_8P_2$

1H-Imidazole-4-ethanamine phosphate (1:2).

Histapon. Gastric secretion stimulant; antiallergic (hyposensitization therapy); vasodilator. Potent vasodilator found in normal tissues and blood; stimulates secretion of pepsin and acid by stomach. mp = 140°; soluble in H_2O.

9974 Ifenprodil
23210-56-2 4936 245-491-4

$C_{21}H_{27}NO_2$

4-Benzyl-α-(p-hydroxyphenyl)-β-methyl-1-piperidineethanol.

RC-61-91. Vasodilator (cerebral, peripheral). mp = 114°. *Robert et Carriere.*

9975 Ifenprodil Tartrate
66157-43-5 4936

$C_{46}H_{40}N_2O_{10}$

4-Benzyl-α-(p-hydroxyphenyl)-β-methyl-1-piperidineethanol tartrate (2:1).

Cerocral; Dilvax; Vadilex. Vasodilator (cerebral, peripheral). mp = 178-180°; soluble in EtOH, H_2O; slightly soluble in Me_2CO $CHCl_3$; nearly insoluble in Et_2O; LD_{50} (mmus iv = 17 mg/kg, (mmus ip) = 120 mg/kg, (mmus orl) = 275 mg/kg. *Robert et Carriere.*

9976 Iloprost
73873-87-7 4940

$C_{22}H_{32}O_4$

5-[Hexahydro-5-hydroxy-4-(3-hydroxy-4-methyl-1-octen-6-ynyl)-2(1H)-pentalenylidene]pentanoic acid.

ciloprost; ZK-36374. Antithrombotic; vasodilator (peripheral). Prostacyclin analog; 1:1 mixture of 16α- and 16β-methyl diastereomers. Colorless oil. *Schering AG.*

9977 Iloprost Tromethamine
4940

$C_{26}H_{43}NO_7$

5-[Hexahydro-5-hydroxy-4-(3-hydroxy-4-methyl-1-octen-6-ynyl)-2(1H)-pentalenylidene]pentanoic acid tromethamine.

Endoprost; Ilomedin. Antithrombotic; vasodilator (peripheral). *Schering AG.*

9978 Inositol Niacinate
6556-11-2 5009 229-485-9

$C_{42}H_{30}N_6O_{12}$

myo-Inositol hexa-3-pyridinecarboxylate.

inositol nicotinate; hexanicotinoyl inositol; inositol hexanicotinate; meso-inositol hexanicotinate; Dilcit; Dilexpal; Esantene; Hämovannid; Hexanicit; Hexopal; Linodil; Mesonex; Mesotal; Palohex. Vasodilator (peripheral). mp = 254.3-254.9°; nearly insoluble in H_2O; soluble in dilute acids.

9979 Iproxamine
52403-19-7

$C_{18}H_{29}NO_4$

5-[2-(Dimethylamino)ethoxy]carvacryl isopropyl carbonate.

Vasodilator. *Parke-Davis.*

9980 Iproxamine Hydrochloride
51222-37-8
$C_{18}H_{30}ClNO_4$
5-[2-(Dimethylamino)ethoxy]carvacryl isopropyl carbonate hydrochloride.
Go-2782; W-42782. Vasodilator. *Parke-Davis.*

9981 Isoxsuprine
395-28-8 5259 206-898-2
$C_{18}H_{23}NO_3$
4-Hydroxy-α-[1-[(1-methyl-2-phenoxyethyl)-amino]ethyl]benzyl alcohol.
Vasodilator (peripheral). mp = 102.5-103.5°. *N. Am. Philips.*

9982 Isoxsuprine Hydrochloride
579-56-6 5259 209-443-6
$C_{18}H_{24}ClNO_3$
4-Hydroxy-α-[1-[(1-methyl-2-phenoxyethyl)-amino]ethyl]benzyl alcohol hydrochloride.
Dilavase; Divadilan; Duviculine; Isolait; Navilox; Suprilent; Vadosilan; Vasodilan; Vasoplex; Vasotran. Vasodilator (peripheral). mp = 203-204°; slightly soluble in H_2O (2% at 25°); soluble in EtOH; LD_{50} (rat orl) = 1750 mg/kg, (rat ip) = 164 mg/kg. *Lemmon Co.; Mead Johnson Labs.; N. Am. Philips.*

9983 Isoxsuprine Resinate
5259
Defencin. Vasodilator (peripheral). *N. Am. Philips.*

9984 Kallidin
342-10-9 5290 206-438-0
$C_{56}H_{85}N_{17}O_{12}$
N^2-L-Lysylbradykinin.
kallidin-10; kallidin-II. Vasodilator. Hypotensive and smooth muscle-stimulating principle. Decapeptide with structure homologous to bradykinin. Amorphous precipitate; $[\alpha]_D^{25}$ = -57° (c = 1 in 1N AcOH).

9985 Kallikrein
9001-01-8 5291 232-574-5
Kallidinogenase.
Callicrein; Padreatin; Padukrein; Glumorin; Depot-Glumorin; Circuletin; Kalirechin; Onokrein P; Padutin; Prokrein; Promotin. Vasodilator. Hypotensive enzyme that releases kinins from plasma proteins. *Bayer AG.*

9986 Mesuprine
7541-30-2
$C_{19}H_{26}N_2O_5S$
2'-Hydroxy-5'-[1-hydroxy-2-[(p-methoxy-phenethyl)amino]propyl]methanesulfonanilide.
Vasodilator; smooth muscle relaxant.

9987 Mesuprine Hydrochloride
7660-71-1
$C_{19}H_{27}ClN_2O_5S$
2'-Hydroxy-5'-[1-hydroxy-2-[(p-methoxy-phenethyl)amino]propyl]methanesulfonanilide hydrochloride.
MJ-1987. Vasodilator; smooth muscle relaxant.

9988 Moxisylyte
54-32-0 6374 200-204-1
$C_{16}H_{25}NO_3$
[2-(4-Acetoxy-2-isopropyl-5-methylphenoxy)-ethyl]dimethylamine.
thymoxamine. Vasodilator (peripheral). A selective α_{1A}-adrenergic blocker. *C.M. Ind.*

9989 Moxisylyte Hydrochloride
964-52-3 6374 213-519-4
$C_{16}H_{26}ClNO_3$
[2-(4-Acetoxy-2-isopropyl-5-methylphenoxy)-ethyl]dimethylamine monohydrochloride.
Arlitine; Moxyl; Opilon; Uroalpha; Vasoklin. Vasodilator (peripheral). A selective α_{1A}-adrenergic blocker. mp = 208-210°; LD_{50} (mus orl) ≅ 265, (rat orl) ≅ 740 mg/kg, (mus sc) ≅ 200 mg/kg, (rat sc) ≅ 190 mg/kg. *C.M. Ind.*

9990 Nafronyl
31329-57-4 6440 250-572-2
$C_{24}H_{33}NO_3$
2-(Diethylamino)ethyl ester tetrahydro-α-(1-naphthalenyl-methyl)-2-furanpropanic acid.
naftidrofuryl; Dubimax; Gevatran; Tridus. Vasodilator. $bp_{0.5}$ = 190°; d_4^{31} = 1.0465; n_D^{20} = 1.5513; LD_{50} (mus orl) = 365 mg/kg. *Lipha Pharm. Inc.*

9991 Nafronyl Oxalate
3200-06-4 6440 221-703-0
$C_{26}H_{35}NO_7$
2-(Diethylamino)ethyl ester tetrahydro-α-(1-naphthalenyl-methyl)-2-furanpropionate oxalate (1:1).
nafronyl acid oxalate; EU-1806; LS-121; Citoxid; Di-Actane; Dusodril; Praxilene. Vasodilator. mp = 110-111°; soluble in H_2O. *Lipha Pharm. Inc.*

9992 Nicametate
3099-52-3 6577 221-452-7
$C_{12}H_{18}N_2O_2$
2-(Diethylamino)ethyl nicotinate.
Eucast. Vasodilator (peripheral, cerebral). Liquid; bp_{10} = 155-157°; bp_2 = 120-125°.

9993 Nicametate Citrate Monohydrate
1641-74-3 6577
$C_{18}H_{28}N_2O_{10}$
2-(Diethylamino)ethyl nicotinate citrate monohydrate.
Euclidan; Nutrin; Soclidan. Vasodilator (peripheral, cerebral).

9994 Nicardipine
55985-32-5 6579 259-932-3
$C_{26}H_{29}N_3O_6$
2-(Benzylmethylamino) ethyl methyl 1,4-dihydro-2,6-dimethyl-4-(m-nitrophenyl)-3,5-pyridinedicarboxylate.
Antianginal; antihypertensive; vasodilator. Dihydropyridine calcium channel blocker. Has antihypertensive properties. *Syntex Labs. Inc.; Yamanouchi U.S.A. Inc.*

9995 Nicardipine Hydrochloride
54527-84-3 6579 259-198-4
$C_{26}H_{30}Cl9N_3O_6$
2-(Benzylmethylamino) ethyl methyl 1,4-dihydro-2,6-dimethyl-4-(m-nitrophenyl)-3,5-pyridinedicarboxylate monohydrochloride.
YC-93; RS-69216; RS-69216-XX-07-0; Barizin; Bionicard; Cardene; Dacarel; Lecibral; Lescodil; Loxen; Nerdipina; Nicant; Nicardal; Nicarpin; Nicapress; Nicodel; Nimicor; Perdipina; Perdipine; Ranvil; Ridene; Rycarden; Rydene; Vasodin; Vasonase. Antianginal; antihypertensive agent. Dihydropyridine calcium channel blocker. Has antihypertensive properties. [α form]: mp = 179-181°; [β form]: mp = 168-170°; LD_{50} (mrat orl) = 634 mg/kg), (mrat iv) = 18.1 mg/kg, (frat orl) = 557 mg/kg, (frat iv) = 25.0 mg/kg, (mmus orl) = 634 mg/kg, (mmus iv) = 20.7 mg/kg, (fmus orl) = 650 mg/kg, (fmus iv) = 19.9 mg. *Syntex Labs. Inc.; Yamanouchi U.S.A. Inc.*

9996 Nicergoline
27848-84-6 6580 248-694-6
$C_{24}H_{26}BrN_3O_3$
10-Methoxy-1,6-dimethylergoline-8β-methanol 5-bromonicotinate (ester).
nicotergoline; nimergoline, MNE; FI-6714; Cergodum; Circo-Maren; Dilasenil; Duracebrol; Ergotop; Ergobel; Memoq; Nicergolent; Sermion; Vasospan. Vasodilator (peripheral, cerebral). mp = 136-138°; LD_{50} (mmus orl) = 860, (mrat orl) = 2800, (mmus iv) = 46 mg/kg. *Farmitalia Carlo Erba Ltd.*

9997 Nicofuranose
15351-13-0 6605 239-385-7
$C_{30}H_{24}N_4O_{10}$
D-Fructofuranose 1,3,4,6-tetranicotinate.
Vasperdil; Bradilan. Vasodilator (peripheral). mp = 140-142°; $[\alpha]_D^{18}$= -8.5° (in $CHCl_3$). *Eprova AG.*

9998 Nicotinyl Alcohol
100-55-0 6614 202-864-6
C_6H_7NO
3-Pyridinemethanol.
nicotinic alcohol; Nu-2121; Roniacol; Ronicol. Vasodilator (peripheral). May bind to adrenoceptors. Liquid; bp_{28} = 154°; bp_{16} = 144-145°; bp_{12} = 114°; bp0.1:ls = 110°; soluble in H_2O, Et_2O; sparingly soluble in petroleum ether. *Hoffmann-LaRoche Inc.*

9999 Nicotinyl Tartrate
6164-87-0 6614 228-199-1
$C_{10}H_{13}NO_7$
3-Pyridinemethanol D-tartrate.
Roniacol Tartrate; Radecol; Niltuvin. Vasodilator (peripheral). mp = 147-148°; soluble in H_2O, EtOH. *Hoffmann-LaRoche Inc.*

10000 Nylidrin
447-41-6 6830 207-182-2
$C_{19}H_{25}NO_2$
p-Hydroxy-α-[1-[(methyl-3-phenylpropyl)amino]ethyl]benzyl alcohol.
phenyl-sec-butyl norsuprifen. Vasodilator (peripheral). mp = 111-112°. *Troponwerke Dinklage.*

10001 Nylidrin Hydrochloride
849-55-8 6830 212-701-0
$C_{19}H_{26}ClNO_2$
p-Hydroxy-α-[1-[(methyl-3-phenylpropyl)amino]ethyl]benzyl alcohol hydrochloride.
SKF-1700-A; Arlidin; Bufedon; Buphedrin; Dilatal; Dilatol; Dilatropon; Dilydrin; Opino; Penitardon; Perdilatal; Rudilin; Rydrin; Tocodilydrin; Tocodrin. Vasodilator (peripheral). Sparingly soluble in H_2O; slightly soluble in EtOH; nearly insoluble in Et_2O, $CHCl_3$, C_6H_6. *Troponwerke Dinklage.*

10002 Pentifylline
1028-33-7 7269 213-842-0
$C_{13}H_{20}N_4O_2$
1-Hexyl-3,7-dihydro-3,7-dimethyl-1H-purine-2,6-dione.
SK-7; component of: Cosaldon. Vasodilator (cerebral). Also used as a stabilizer of vitamin preparations. mp = 82-83°. *Chem. Werke Albert.*

10003 Pentoxifylline
6493-05-6 7278 229-374-5
$C_{13}H_{18}N_4O_3$
3,7-Dihydro-3,7-dimethyl-1-(5-oxohexyl)-1H-purine-2,6-dione.
oxpentifylline; vazofirin; BL-191; Azupentat; Durapental; Rentylin; Torental; Trental. Vasodilator. mp = 105°; λ_m = 273, 208 nm ($E_{1\,cm}^{1\%}$ 365, 935); soluble in H_2O (0.0077 g/100 l), more soluble in organic solvents; LD_{50} (mus orl) = 1385 mg/kg. *Chem. Werke Albert.*

10004 Pindolol
13523-86-9 7597 236-867-9
$C_{14}H_{20}N_2O_2$
1-(1H-Indol-4-yloxy)-3-[(1-methylethyl)amino]-2-propanol.
prinodolol; LB-46; Betapindol; Blockin L; Calvisken; Decreten; Durapindol; Glauco-Visken; Pectobloc; Pinbetol; Pynastin; Visken. Antianginal; antihypertensive; antiarrhythmic; antiglaucoma; vasodilator. β-Adrenergic blocker with partial agonist activity. mp = 171-173°. *Sandoz Pharm. Corp.*

10005 Piribedil
3605-01-4 7648 222-764-6
$C_{16}H_{18}N_4O_2$
2-[4-(1,3-Benzodioxol-5-ylmethyl)-1-piperazinyl]pyrimidine.
EU-4200; ET-495; Trivastal. Vasodilator (peripheral). Central dopanimergic agonist. mp = 98°; LD_{50} (mus ip) = 690.3 mg/kg. *Sci. Union et Cie France.*

10006 Pirsidomine
132722-74-8
$C_{17}H_{22}N_4O_3$
N-p-Anisoyl-3-(cis-2,6-dimethylpiperidino)sydnone imine.
CAS 936. Vasodilator. *Hoechst Roussel Pharm. Inc.*

10007 Prazosin
19216-56-9 7897 242-885-8
$C_{19}H_{21}N_5O_4$
1-(4-Amino-6,7-dimethoxy-2-quinazolinyl)-4-(2-furoyl)piperazine.

furazosin. An α_1-adrenergic blocking agent used as an antihypertensive and also in treatment of BPH. mp = 278-280°. *Brocades-Stheeman & Pharmacia; Pfizer Intl.*

10008 Prazosin Hydrochloride

19237-84-4 7897 242-903-4

$C_{19}H_{22}ClN_5O_4$

1-(4-Amino-6,7-dimethoxy-2-quinazolinyl)-4-(2-furoyl)piperazine hydrochloride.

Minipress; CP-12299-1; Alpress LP; Duramipress; Eurex; Hypovase; Peripress; Sinetens. An α_1-adrenergic blocking agent used as an antihypertensive agent and also in treatment of BPH. Soluble in Me_2CO (0.72 mg/100 ml), MeOH (640 mg/100 ml), EtOH (84 mg/100 ml), DMF (130 mg/100 ml), dimethylacetamide (120 mg/100 ml), H_2O (140 mg/100 ml at pH 3.5), $CHCl_3$ (4.1 mg/100 ml); λ_m = 246, 329 nm (a_M = 137 ± 3, 27.6 ± 0.3 MeOH/1% HCl). *Brocades-Stheeman & Pharmacia; Pfizer Inc.*

10009 Sodium Nitrite

7632-00-0 8793 231-555-9

$NaNO_2$

Nitrous acid sodium salt.

erinitrit. Antidote to cyanide poisoning; vasodilator. Also used as a reagent in manufacture of inorganic and organic compounds, as well as in textile dyeing and printing, photography, meat curing and preserving. mp = 217°; dec above 320°; d = 2.17; soluble in H_2O; slightly soluble in EtOH; aqueous solution is alkaline; oxidizes to nitrate in air; LD_{50} (rat orl) = 180 mg/kg.

10010 Sodium Nitroprusside

14402-89-2 8794 238-373-9

$C_5FeN_6Na_2O$

Disodium pentacyanonitrosyl ferrate(2-) dihydrate.

sodium nitroferricyanide; sodium nitroprussiate; Nipruss; Nipride [dihydrate]; Nitropress [dihydrate]. Antihypertensive. A potent, directly acting vasodilator used intravenously for treatment of hypertensive emergencies. Also used as a reagent in the detection of many organic compounds and alkali sulfides. *Abbott Labs Inc.; Elkins-Sinn; Hoffmann-LaRoche Inc.*

10011 Suloctidil

54063-56-8 9160 258-957-7

$C_{20}H_{35}NOS$

erythro-p-(Isopropylthio)α-[1-octylamino)ethyl]-benzyl alcohol.

MJF-12637; CP-556S; Bemperil; Cerebro; Circleton; Dulasi; Dulcotil; Euvasal; Fluversin; Fluvisco; Hemoantin; Iangene; Loctidon; Locton; Octamet; Polivasal; Sudil; Sulocton; Sulodene. Vasodilator (peripheral). mp = 62-63°; LD_{50} (mus orl) = 3700 mg/kg. *Continental Pharma Inc.*

10012 Tipropidil

70895-45-3

$C_{20}H_{35}NO_2S$

1-[p-(Isopropylthio)phenoxy]-3-(octylamino)-2-propanol.

Vasodilator. *Bristol-Myers Squibb Co.*

10013 Tipropidil Hydrochloride

70895-39-5

$C_{20}H_{36}ClNO_2S$

1-[p-(Isopropylthio)phenoxy]-3-(octylamino)-2-propanol hydrochloride.

MJ-12880-1. Vasodilator. *Bristol-Myers Squibb Co.*

10014 Tolazoline

59-98-3 9645 200-448-9

$C_{10}H_{12}N_2$

2-Benzyl-2-imidazoline.

phenylmethylimidazoline. Vasodilator (peripheral). An imidazoline α-adrenergic blocker. *CIBA plc.*

10015 Tolazoline Hydrochloride

59-97-2 9645 200-447-3

$C_{10}H_{13}ClN_2$

2-Benzyl-2-imidazoline hydrochloride.

priscoline hydrochloride; Lambral; Priscol; Priscoline; Vaso-Dilatan. Vasodilator (peripheral). An imidazoline α-adrenergic blocker. mp = 174°; soluble in H_2O, EtOH, $CHCl_3$; slightly soluble in Et_2O, EtOAc; pH (2.5%) = 4.9-5.3. *CIBA plc.*

10016 Viagra

171599-83-0

$C_{28}H_{38}N_6O_{11}S$

1-[[3-(6,7-dihydro-1-methyl-7-oxo-3-propyl-1H-pyrazolo[4,3-d]pyrimidin-5-yl)-4-ethoxyphenyl]sulfonyl]-4-methylpiperazine citrate.

Sildenafil; 1-((3-(4,7-dihydro-1-methyl-7-oxo-3-propyl-1H-pyrazolo(4,3-d)pyrimidin-5-yl)-4-ethoxyphenyl). A selective inhibitor of cyclic guanosine monophosphate (cGMP)-specific phosphodiesterase type 5 (PDE5). Indicated for the treatment of erectile dysfunction. Studies that established benefit demonstrated improvements in success rates for sexual intercourse compared with placebo. *Pfizer Intl.*

10017 Viprostol

73647-73-1

$C_{23}H_{36}O_5$

(±)-Methyl (Z)-7-[(1R,2R,3R)-2-[(E)-(4RS)-4-butyl-4-hydroxy-1,5-hexadienyl]-3-hydroxy-5-oxo-cyclopentyl]-5-heptenoate.

CL-115347; sulfonyl)-4-methyl-piperazine. Hypotensive; vasodilator. *Lederle Labs.*

10018 Xanthinol Niacinate

437-74-1 10194 207-115-7

$C_{19}H_{26}N_6O_8$

7-[2-Hydroxy-3-[(2-hydroxyethyl)-methylamino]propyl]theophylline nicotinate.

xanthinol nicotinate; SK-331-A; Angiomin; Complamin; Sadamin; Xavin. Vasodilator (peripheral). mp = 180°; soluble in H_2O. *Roerig Div. Pfizer Pharm.*

10019 Zolertine

4004-94-8

$C_{13}H_{18}N_6$

1-Phenyl-4-[2-(1H-tetrazol-5-yl)ethyl]piperazine.

Antiadrenergic; vasodilator.

10020 Zolertine Hydrochloride
7241-94-3
$C_{13}H_{19}ClN_6$
1-Phenyl-4-[2-(1H-tetrazol-5-yl)ethyl]piperazine monohydrochloride.
MA-1277. Antiadrenergic; vasodilator. *Miles Inc.*

Vasoprotectants

10021 Azetirelin
95729-65-0
$C_{15}H_{20}N_6O_4$
(-)-N-[[(2S)-4-Oxo-2-azetidinyl]carbonyl]-L-histidyl-L-prolinamide.
Thyrotropin-releasing hormone.

10022 Benzarone
1477-19-6 1091 216-026-2
$C_{17}H_{14}O_3$
2-Ethyl-3-benzofuranyl p-hydroxy-phenylketone.
L-2197. Vasoprotectant. Capillary protectant. mp = 170°; soluble in AcOH (0.52 g/100 ml), C_6H_6 (1.61 g/100 ml), chlorobenzene (2.05 g/100 ml); soluble in H_2SO_4; LD_{50} (rbt der) > 3000 mg/kg, (rat ip) = 1500 mg/kg, (mus ip) 290 mg/kg.

10023 Bimoclomol
130493-03-7
$C_{14}H_{20}ClN_3O_2$
(N-[2-Hydroxy-3-(1-piperidinyl) propoxy]-3-pyridine-carboximidoyl-chloride).
Vasoprotectant. Also ameliorates peripheral neuropathy.

10024 Bioflavonoids
1269
Vitamin P complex; citrul flavonoid compounds; Arliflav; Pecitrol Veinogène; CVP. Vasoprotectant. Capillary protectant. *U.S. Vitamin.*

10025 Chromocarb
4940-39-0 2294 225-583-0
$C_{10}H_6O_4$
4-Oxo-4H-1-benzopyran-2-carboxylic acid.
Angiophtal [as diethylamine salt]; Campel; Fludarene. Vasoprotectant. The dimethylamine salt is used as a capillary protectant. mp = 250-251°; 255-256°; λ_m 230 305 nm (ε 20220 8075); soluble in EtOH, NH_3; sparingly soluble in H_2O; [diethylamine salt ($C_{14}H_{17}NO_4$)]: mp = 138°; soluble in H_2O; LD_{50} (mus iv) = 800 mg/kg, (mus orl) > 5000 mg /kg.

10026 Clobenoside
29899-95-4 2419 249-940-5
$C_{25}H_{32}Cl_2O_6$
Ethyl 5,6-bis-O-(p-chlorobenzyl)-3-O-propyl-d-glucofuranoside.
43853; ZY-15028; Arvigol; Floganol. Vasoprotectant. $[\alpha]_D^{20}$ = -17° (c = 1 $CHCl_3$); poorly soluble in H_2O. *Ciba-Geigy Corp.*

10027 Diosmin
520-27-4 3350 208-289-7
$C_{28}H_{32}O_{15}$
7-[[6-O-(6-Deoxy-α-L-mannopyranosyl)-β-D-glucopyranosyl]oxy]-5-hydroxy-2-(3-hydroxy-4-methoxyphenyl)-4H-1-benzopyran-4-one.
barosmin; buchu resin; Diosmil; Diosven; Diovenor; Flebosmil; Flebosten; Hemerven; Insuven; Litosmil; Tovene; Varinon; Ven-Detrex; Venosmine. Vasoprotectant. Capillary protectant. A naturally occurring flavonic glycoside isolated from plant sources. [monohydrate]: mp = 275-277° (dec), 283° (dec); λ_m 255 268 345 nm (log ε 4.28, 4.25, 4.30 EtOH); insoluble in H_2O, EtOH.

10028 Dobesilate Calcium
20123-80-2 3455 243-531-5
$C_{12}H_{10}CaO_{10}S_2$
2,5-Dihydroxybenzenesulfonic acid calcium salt.
hydroquinone calcium sulfate; Dexium; Doxium. Vasoprotectant. Vasotropic. mp > 300° (dec); very soluble in H_2O, EtOH; insoluble in Et_2O, C_6H_6, $CHCl_3$; LD_{50} (mus) = 700 mg/kg. *Labs. O.M.*

10029 Escin
6805-41-0 3737 229-880-6
Aescin; Aescusan; Reparil. Vasoprotectant. Mixture of saponins from the seed of the horse chestnut tree. Used in treatment of peripheral vascular disorders.

10030 α-Escin
66795-86-6 3737 266-482-1
Vasoprotectant. Used in treatment of peripheral vascular disorders. mp = 225-227°; $[\alpha]_D^{25}$ = -13.5° (c = 5 MeOH); very soluble in H_2O; LD_{50} (mus iv) = 3.2 mg/kg, (mus orl) = 320 mg/kg, (rat iv) = 5.4 mg/kg, (rat orl) = 720 mg/kg, (gpg iv) = 15.2 mg/kg, (gpg orl) =475 mg/kg; [sodium salt]: mp = 251.

10031 β-Escin
11072-93-8 3737
Flogencyl. Vasoprotectant. Used in treatment of peripheral vascular disorders. β-Escin is the natural form. mp = 222-223°; $[\alpha]_D^{27}$ = -23.7° (c = 5 MeOH); insoluble in H_2O; LD_{50} (mus iv) = 1.4 mg/kg, (mus orl) = 134 mg/kg, (rat iv) = 2.0 mg/kg, (rat orl) = 400 mg/lg, (gpg iv) = 7.2 mg/kg, (gpg orl) = 475 mg/kg.

10032 Folescutol
15687-22-6 4252 239-783-0
$C_{14}H_{15}NO_5$
6,7-Dihydroxy-4-(morpholinomethyl)coumarin.
Pholescutol. Vasoprotectant. Capillary protectant. mp = 232°. *Lab. Dausse.*

10033 Folescutol Hydrochloride
36002-19-4 4252 252-831-5
$C_{14}H_{16}ClNO_5$
6,7-Dihydroxy-4-(morpholinomethyl)coumarin hydrochloride.
LD-2988; Covalan. Vasoprotectant. Capillary protectant. mp = 259-261°. *Lab. Dausse.*

10034 Leucocyanidin
480-17-1 5477
$C_{15}H_{14}O_7$
3,3',4,4',5,7-Flavanhexol.
3,3',4,4',5,7-hexahydroxyflavane; leucocianidol; Flavan; Hamaméliode P; Résivit. Vasoprotectant. Capillary protectant. [monohydrate]: mp > 355°; λ_m 285 nm (EtOH); soluble in H_2O, EtOH, Me_2CO; insoluble in Et_2O, $CHCl_3$, petroleum ether.

10035 Metescufylline
15518-82-8 6000 239-550-3
$C_{25}H_{31}N_5O_8$
7-[2-(Dimethylamino)ethyl]theophylline [(7-hydroxy-4-methyl-2-oxo-2H-1-benzopyran-6-yl)oxy]acetate.
methesacufylline; Veinartan; Venarterin. Vasoprotectant. Capillary protectant. mp = 124°; slightly soluble in cold H_2O, freely soluble in hot H_2O, insoluble in Me_2CO; LD_{50} (mus iv) = 260 mg/kg. *Lab. Dausse.*

10036 Naftazone
15687-37-3 239-785-1
$C_{11}H_9N_3O_2$
1,2-Naphthoquinone 2-semicarbazone.
Etioven; Mediaven. Vasoprotectant. Insoluble in H_2O.

10037 Quercetin Dihydrate
6151-25-3 8216 204-187-1
$C_{15}H_{10}O_7$
2-(3,4-Dihydroxyphenyl)-3,5,7-trihydroxy-4H-1-benzopyran-4-one.
meletin; sophoretin; cyanidenolon 1522; Natural flavanoid isolated from a variety of plant sources such as ragweed pollen or clover blossom. Vasoprotectant. Capillary protectant. Dec 314°; λ_m 258, 375 nm (log ε 2.75, 2.75 EtOH); soluble in absolute EtOH (0.34 g/100 ml at 20°, 4.35 g/100 ml at 76°), soluble in AcOH, insoluble in H_2O; LD_{50} (mus orl) = 160 mg/kg.

10038 Quercetin Pentabenzyl Ether
13157-90-9 8216
$C_{50}H_{40}O_7$
2-(3,4-Dihydroxyphenyl)-3,5,7-trihydroxy-4H-1-benzopyran-4-one pentabenzyl ether.
penta-O-benzylquercetin; Parietrope. Vasoprotectant. Capillary protectant. mp = 123-125°; λ_m 249, 343 nm (log ε 4.43, 4.14 $CHCl_3$).

10039 Rutin
153-18-4 8456 205-814-1
$C_{27}H_{30}O_{16}.3H_2O$
3-[[6-O-(6-Deoxy-α-L-mannopyranosyl)-β-D-glucopyranosyl]oxy]-2-(3,4-dihydroxyphenyl)-5,7-dihydroxy-4H-1-benzopyran-4-one.
rutoside; melin; phytomelin; eldrin; ilixathin; sophorin; globularicitrin; paliuroside; osyritrin; myrticolorin; violaquercetrin; Birutan; component of: Veliten; Found in many plants. Vasoprotectant. Capillary protectant. Dec 214-215°; $[\alpha]_D^{23}$ = 13.82° (EtOH), -39.43° (C_5H_5N); soluble in H_2O (0.01 g/100 ml at 20°, 0.5 g/100 ml at 100°), MeOH (14.3 g/100 ml at 65°), C_5H_5N, formamide and alkaline solutions; slightly soluble in EtOH, Me_2CO, EtOAc; insoluble in $CHCl_3$, CS_2, Et_2O, C_6H_6, petroleum ether; LD_{50} (mus iv) = 950 mg/kg. *Lederle Labs.*

10040 Siagoside
100345-64-0
$C_{73}H_{129}N_3O_{30}$
N-(II^3-N-Acetylneuraminosylgangliotetraosyl)ceramid intramolecular ester.
Ganglioside. May attenuate some morphological and functional deficits related to striatal damage after acute cerebral ischemia.

10041 Troxerutin
7085-55-4 9920 230-389-4
$C_{33}H_{42}O_{19}$
3',4',7-Tris(hydroxyethyl)rutin.
Z 6000; THR; trioxyethylrutin; Posorutin; Ruven; Vastribil; Veinamitol; Veniten. Vasoprotectant. Used in treatment of veinous disorders. mp = 181°; soluble in H_2O, glycerol, propylene glycol; insoluble in MeOH, EtOH, Et_2O, C_6H_6, $CHCl_3$.

Vitamins, Vitamin Sources

10042 Acetatocobalamin
22465-48-1 4854 245-019-7
$C_{64}H_{91}CoN_{13}O_{16}P$
Cobinamide acetate (salt) dihydrogen phosphate (ester) inner salt 3'-ester with 5,6-dimethyl-1-α-D-ribofuranoylbenzimidazole.
Depo-gamma; Docelan; Novidroxin. Vitamin, vitamin source. *Merck & Co.Inc.*

10043 N-Acetyl Analog
523-68-2 10162 208-348-7
$C_{13}H_{13}NO_2$
4-Acetamido-2-methyl-1-naphthol.
K-Vitrat. Vitamin source. Prothrombogenic. mp = 208-210°.

10044 Adenine
73-24-5 150 200-796-1
$C_5H_5N_5$
1,6-Dihydro-6-iminopurine.
6-aminopurine; Leuco-4; vitamin B_4. Vitamin, vitamin source. Dec 360-365°; λ_m 207, 260.5 nm (23200, 13400 pH 7.0); soluble in H_2O (0.05 g/100 ml at 20°, 2.5 g/100 ml at 100°), slightly soluble in EtOH; insoluble in Et_2O, $CHCl_3$; LD_{50} (rat orl) = 745 mg/kg.

10045 Aquacobalamin
13422-52-1 4854 236-534-8
$C_{62}H_{90}CoN_{13}O_{15}P.OH$
Cobinamide hydroxide monohydrate dihydrogen phosphate (ester) inner salt 3'-ester with 5,6-dimethyl-1-α-D-ribofuranoylbenzimidazole.
aquocobalamin; vitamin B_{12b}; vitamin B_{12d}. Vitamin, vitamin source. May be used in management of patients with megaloblastic anemia (early sign of vitamin B_{12} deficiency). λ_m 274, 317, 351, 499, 527 nm (ε 20600, 6100, 26500, 8100,8500 H_2O). *Merck & Co.Inc.*

10046 Ascorbic Acid

50-81-7 867 200-066-2

$C_6H_8O_6$

L-Ascorbic acid.

Ascorbicap; Cebione; Cecon; Cenolate; Cetane; Cetane-Caps TC; Cevalin; Cevex; Adenex; Allercorb; Ascorin; Ascorteal; Ascorvit; Cantan; Cantaxin; Catavin C; Cebicure; Cebion; Cecon; Cegiolan; Celaskon; Celin; Cenetone; Cereon; Cergona; Cescorbat; Cetamid; Cetebe; Cetemican; Cevalin; Cevatine; Cevex; Cevimin; Ce-Vi-Sol; Cevitan; Cevitex; Cewin; Ciamin; Cipca; Concemin; C-Vimin; Davitamon C; component of: Chromagen, Freancee, Mediatric, Stuartinic, Tolfrinic, Veliten. Vitamin, vitamin source. mp = 190-192°; d =1.65; $[\alpha]_D^{25}$ = 20.5 - 21.5° (c = 1), $[\alpha]_D^{23}$ =48° (c = 1 MeOH)λ_m = 245 nm (acid solution), 265 nm (neutral solution); soluble in H_2O (33 g/100 ml, 40 g/100 ml at 45°, 80 g/100 ml at 100°), EtOH (3.3 g/100 ml), absolute EtOH (2 g/100 ml); glycerol (1 g/100 ml), propylene glycol (5 g/100 ml), insoluble in Et_2O, $CHCl_3$, C_6H_6, petroleum ether. *Abbott Labs Inc.; Ascher B.F. & Co.; BASF Corp.; Eli Lilly & Co.; Forest Pharm. Inc.; ICN Pharm. Inc.; Johnson & Johnson-Merck Consumer Pharm. ; Lederle Labs.; Marion Merrell Dow Inc.; Savage Labs; Wyeth-Ayerst Labs.*

10047 Biotin

58-85-5 1272 200-399-3

$C_{10}H_{16}N_2O_3S$

(3aS,4S,6aR)-Hexahydro-2-oxo-1H-thieno[3,4-d]-imidazole-4-valeric acid.

cis-tetrahydro-2-oxothieno[3,4-d]imidazoline-4-valeric acid. Bioepiderm; vitamin H; coenzyme R; bios II; Biodermatin. Vitamin, vitamin source. A growth factor present in all living cells. Plays a role in carboxylation reactions. mp = 232-233°; $[\alpha]_D^{21}$ = 91° (c = 1 in 0.1N NaOH); soluble in H_2O (0.022 g/100 ml at 25°), EtOH 0.08 g/100 ml at 25°). *Sterling Winthrop Inc.*

10048 Calcipotriene

112965-21-6 1679

$C_{27}H_{40}O_3$

(5Z,7E,22E,24S)-24-Cyclopropyl-9,10-secochola-5,7,10(19),22-tetraene-1α,3β,24-triol.

MC-903; calciptriol; Daivonex; Dovonex; component of: Dovex. Vitamin, vitamin source. Antipsoriatic. mp = 166-168°; λ_m 264 nm (ε 17200 96% EtOH). *LeoAB; Westwood-Squibb Pharm. Inc.*

10049 Calcitriol

32222-06-3 1681 250-963-8

$C_{27}H_{44}O_3$

(5Z,7E)-9,10-Secocholesta-5,7,10(19)-triene-1α,3β,25-triol.

1α,25-dihydroxycholecalciferol; 1α,25-dihydroxyvitamin D_3; Calcijex; Rocaltrol; Toptriol; Ro-21-5535. Calcium regulator; vitamin (antirachitic). Biologically active form of vitamin D_3, involved in intestinal calcium transport, bone calcium resorption, formed by hydroxylation of vitamin D_3 in liver and kidney. mp = 111-115°; λ_m = 264 nm (ε 19000 EtOH); $[\alpha]_D^{25}$ = 48° (MeOH); slightly soluble in MeOH, EtOH, EtOAc, THF; air, light sensitive. *Abbott Labs Inc.; Hoffmann-LaRoche Inc.*

10050 Calcium Ascorbate

5743-27-1 1688 227-261-5

$C_{12}H_{14}CaO_{12}.2H_2O$

L-Ascorbic acid calcium salt dihydrate.

Vitamin, vitamin source. Nutritional supplement. $[\alpha]_D^{20}$ = 95.6° (c = 2.4); freely soluble in H_2O; insoluble in MeOH, EtOH.

10051 Calcium Pantothenate

137-08-6 7147 205-278-9

$C_{18}H_{32}CaN_2O_{10}$

(R)-N-(2,4-Dihydroxy-3,3-dimethyl-1-butyl)-β-alanine calcium salt (2:1).

Calpan; Pantholin; Calpanate; component of: Stuartinic. Vitamin (enzyme cofactor). mp = 195-196° (dec); $[\alpha]_D^{20}$ = 28.2° (c = 5); soluble in H_2O (35.7 g/100 ml), glycerol; slightly soluble in EtOH, Me_2CO. *BASF Corp.; Eli Lilly & Co.; Johnson & Johnson-Merck Consumer Pharm.; Pharmacia & Upjohn.*

10052 Calcium Pantothenate, Racemic

6381-63-1 7147 6381-63-1

$C_{18}H_{32}CaN_2O_{10}$

(±)-N-(2,4-Dihydroxy-3,3-dimethyl-1-butyl)-β-alanine calcium salt (2:1) .

(±)-calcium pantothenate. Vitamin (enzyme cofactor). *BASF Corp.; Eli Lilly & Co.; Johnson & Johnson-Merck Consumer Pharm. ; Pharmacia & Upjohn.*

10053 α-Carotene

7488-99-5 1901

$C_{40}H_{56}$

Provitamin A. Vitamin A precursor. mp = 187.5°; $[\alpha]_{643}^{18}$ = 385° (c = 0.08 C_6H_6); λ_m = 485, 454 nm; freely soluble in CS_2, $CHCl_3$; soluble in C_6H_6, Et_2O; slightly soluble in EtOH, petroleum ether; insoluble in H_2O.

10054 β-Carotene

7235-40-7 1902 230-636-6

$C_{40}H_{56}$

(all E)-1,1'-(3,7,12,16-Tetramethyl-1,3,5,7,9,11,13,15,17-octadecanonaene-1,18-diyl)bis[2,6,6-trimethylcyclohexene].

BetaVit; Lucaratin; Solatene; Carotaben; β,β-carotene; Provatene. Vitamin; ultraviolet screen. Vitamin A precursor. mp = 183°; λ_m 497, 466 nm; soluble in CS_2, C_6H_6, $CHCl_3$; moderatley soluble in Et_2O, petroleum ether, oils; soluble in C_6H_{14} (0.11 g/100 ml at 0°); poorly soluble in MeOH, EtOH; insoluble in H_2O. *BASF Corp.; Hoffmann-LaRoche Inc.*

10055 δ-Carotene

472-92-4 1903

$C_{40}H_{56}$

Vitamin source. mp = 140.5°; $[\alpha]_D^{25}$ = 352° ± 16° (C_6H_{14}).

10056 Cholecalciferol

67-97-0 10157 200-673-2

$C_{27}H_{44}O$

9,10-Secocholesta-5,7,10(19)-trien-3-ol.

Vitamin D_3; 7-dehydrocholesterol. Vitamin. Antirachitic.

mp = 84-85°; $[\alpha]_D^{20}$ = 84.8° (c = 1.6 Me_2CO), 51.9° (c = 1.6 $CHCl_3$); λ_m = 264.5 nm ($E_{1\ cm}^{1\%}$ 450-490 EtOH or C_6H_{14}); soluble in most organic solvents, insoluble in H_2O.

10057 Chromic Chloride

10060-12-5

$CrCl_3.6H_2O$

Chromium(3+) chloride hexahydrate.
Trace mineral supplement.

10058 Cobamamide

13870-90-1 2513 237-627-6

$C_{72}H_{11}CoN_{18}O_{17}P$

Inner salt of the Co-(5'-deoxyadenosine-5') derivative of the 3'ester of cobinamide phosphate with 5,6-dimethyl-1-α-D-ribofuranosylbenzimidazole.

LM176; coenzyme B+si12; adenosyl B_{12}; DBC; dibencozide; dibenzcozamide; dimebenzcozamide; Actimide; Anabasi; Betarin; Calomide; Cobalion; Cobatlamin S; Cobanzyme; Cobazymase; Coenzile; Dolonevran; Enzicoba; Héraclène; Hi-Fresmin; Hycobal; Indusil; Represil; Sabalamin; Xobaline. Vitamin, vitamin source. Hematopoietic. λ_m 260, 375, 522 nm (H_2O); soluble in H_2O (2.59 g/100 ml); insoluble in Me_2CO, Et_2O, dichloroethylene, dioxane.

10059 Cyanocobalamin

68-19-9 10152 200-680-0

$C_{63}H_{88}CoN_{14}O_{14}P$

Vitamin B_{12}.

Betalin 12 Crystalline; Cyomin; Cyredin; Dicopac Kit; Docibin; Redisol; Rubramin PC; Sytobex; Vibalt; Vitapsrint B12; Anacobin; Antipernicin; Bedodeka; Bedoz; Behepan; Berubi; Berubigen; Betalin-12; Betolvex; Cobalin; Crystamine; Cykobemin; Cytacon; Cytamen, Cytobion; Docémine; Docibin; Docigram; Docivit; Dodex; Fresmin; Macrabin; Millevit; Redisol; Rubesol; Rubramin PC; Sytobex; Vibalt; Vitarubin; component of: Chromagen, Geriplex, Spondylonal, Tolfrinic, Troph-Iron, Trophite, Trophite+Iron. Vitamin, vitamin source. Hematopoietic. mp > 300°; $[\alpha]_{656}^{23}$ = -59° ± 9° (dilute aqueous solution); λ_m = 278, 361, 550 nm ($A_{1\ cm}^{1\%}$ 115, 204, 64 H_2O). *Amersham Corp.; Ascher B.F. & Co.; Eli Lilly & Co.; Elkins-Sinn; Forest Pharm. Inc.; Lederle Labs.; Marion Merrell Dow Inc.; Menley & James Labs Inc.; Merck & Co.Inc.; Parke-Davis; Roerig Div. Pfizer Pharm.; Savage Labs; Wyeth-Ayerst Labs.*

10060 Ergocalciferol

50-14-6 10156 200-014-9

$C_{28}H_{44}O$

(3β,5Z,7E,22E)-9,10-Secoerga-5,7,10(19),22-tetraen-3-ol.

Vitamin D_2; Calciferol; Deltalin; Drisdol; Oleovitamin D; Condol; Decaps; Dee-Ron; De-Rat Concentrate; Deratol; Detalup; Diactol; Divit Urto; Drisdol; D-Tracetten; Ergorone; Ertron; Fortodyl; Hi-Deratol; Infron; Metadee; Mina D_2; Mulsiferol; Mykostin; Osteil; Radiostol; Radsterin; Shock-Ferol; Sterogyl; Uvesterol-D; Vio-D; component of: Haliver. Vitamin, vitamin source. Antirachitic. Also used as a rodenticide. mp = 115-118°; $[\alpha]_D^{25}$ = 82.6° (c = 3 Me_2CO), $[\alpha]_D^{20}$ = 102.5° (EtOH), $[\alpha]_D^{20}$ = 52° ($CHCl_3$); λ_m = 264,5 nm ($E_{1\ cm}^{1\%}$ 4587.9 ± 7.5 C_6H_{14}); soluble in most organic solvents, insoluble in H_2O; soluble in EtOH (0.15 g/100 ml at 20°, 2.2 g/100 ml at 86°, Et_2O (1.43 g/100 ml at 20°, 2.56 g/100 ml at 35°), $CHCl_3$ (3.22 g/100 ml); [acetate ($C_{30}H_{46}O_2$)]: mp = 179°; [benzoate ($C_{35}H_{48}O_2$)]: mp = 169-171°; $[\alpha]_D^{23}$ = -71° (c = 1.1 $CHCl_3$), $[\alpha]_{546}$ = -88° ($CHCl_3$). *Eli Lilly & Co.; Parke-Davis; Schwarz Pharma Kremers Urban Co.*

10061 Ergosterol

57-87-4 3701 200-352-7

$C_{28}H_{44}O$

(3β,22E)-Ergosta-5,7,22-trien-3-ol.

ergosterin; Provitamin D. Vitamin D source. mp = 166-183°; $bp_{0.01}$ = 250°; $[\alpha]_D^{20}$ = -135°; (c = 1.2 $CHCl_3$), $[\alpha]_{546}^{20}$ = -171° ($CHCl_3$); λ_m = 262, 271, 282, 293 nm; insoluble in H_2O; soluble in EtOH (0.15 g/100 ml at 20°, 2.2 g/100 ml at 86°, Et_2O (1.43 g/100 ml at 20°, 2.56 g/100 ml at 35°), $CHCl_3$ (3.22 g/100 ml); [acetate ($C_{30}H_{46}O_2$)]: mp = 179°; [benzoate ($C_{35}H_{48}O_2$)]: mp = 169-171°; $[\alpha]_D^{23}$ = -71° (c = 1.1 $CHCl_3$), $[\alpha]_{546}$ = -88° ($CHCl_3$).

10062 Folic Acid

59-30-3 4253 200-419-0

$C_{19}H_{19}N_7O_6$

N-[p-[[(2-Amino-4-hydroxy-6-pteridinyl)-methyl]amino]benzoyl]-L-glutamic acid.

Folicet; Vitamin M; Cytofol; Folacin; Foldine; Foliamin; Folipac; Folettes; Folsan; Folvite; Incafolic; Millafol; component of: Mission Prenatal, Plastulen-N. Vitamin, vitamin source. Hematopoietic. May be used in management of patients with megaloblastic anemia or with folate deficiency caused by drugs that inhibit dihydrofolate reductase or that interfere with the absorption and storage of folate. mp > 250°; $[\alpha]_D^{25}$ = 23° (c = 0.5 in 0.1N NaOH); λ_m = 256, 283 368 nm (log ε 4.43, 4.40, 3.96 pH 13); slightly soluble on H_2O (0.00016 g/100 ml at 20°, 1 g/100 ml at 100°), MeOH; less soluble in EtOH, BuOH; insoluble in $CHCl_3$, Me_2CO, Et_2O, C_6H_6. *Bristol-Myers Squibb Pharm. R&D; Lederle Labs.; Mission Pharmacal Co.*

10063 Hydroxocobalamin

13422-51-0 4854 236-533-2

$C_{62}H_{89}CoN_{13}O_{15}P$

Cobinamide dihydroxide dihydrogen phosphate (ester) mono(inner salt) 3'-ester with 5,6-dimethyl-1-α-D-ribofuranosylbenzimidazole.

AlphaRedisol; Codroxomin; Ducobee-Hy; Neo-Betalin 12; Redisol-H; Sytobex-H; Axlon; Cobalin-H; Codroxomin; Droxomin; Duradoce; Duralta-12; Neo-Cytamen. Vitamin, vitamin source. mp > 300°; λ_m 279, 325, 359, 516, 537 nm (ε 19000, 11400, 20600, 8900, 9500 H_2O); moderatley soluble in H_2O, MeOH, EtOH; insoluble in Me_2CO, Et_2O, petroleum ether, C_6H_6, halogenated hydrocarbons. *Eli Lilly & Co.; Forest Pharm. Inc.; Merck & Co.Inc.; Parke-Davis; Sterling Winthrop Inc.*

10064 1α-Hydroxycholecalciferol

41294-56-8 4864 255-297-1

$C_{27}H_{44}O_2$

(1α,3β,5Z,7E)-9,10-Secocholesta-5,7,10(19)-triene-1,3-diol.

1α-OH-CC; alfacalcidol; Alfarol; Alpha D_3; EinsAlpha; Etalpha; One-Alpha, Vetalpha. Vitamin D source. mp = 134-136°, 138-139.5°; $[\alpha]_D^{25}$ = 28° (Et_2O); λ_m 264 nm (ε 18000 Et_2O).

10065 Inositol
87-89-8 5008 201-781-2
$C_6H_{12}O_6$
Hexahydroxycyclohexane.
meso-inositol; l-inositol; cyclohexanehexol; cyclohexitol; meat sugar; inosite; mesoinosite; phaseomannite; dambose; nucite; bios I; rat antispectacled eye factor; mouse antialopecia factor. Vitamin, vitamin source. A lipotropic agent. mp = 225-227°; d = 1.752; soluble in H_2O (14 g/100 ml at 25°, 28 g/100 ml at 60°, slightly soluble in EtOH, insoluble on other organic solvents.

10066 Manganese Chloride
13446-34-9 5768
$Cl_2Mn{\cdot}4H_2O$
Manganese chloride tetrahydrate.
Supplement (trace mineral). [anhydrous]: LD_{50} (mus sc) = 180-250 mg/kg, (dog iv) = 201.6 mg/kg.

10067 Manganese Sulfate
10034-96-5 5783
$H_8Cl_2MnO_4$
Manganese sulfate (1:1) monohydrate.
Supplement (trace mineral).

10068 Maxacalcitol
103909-75-7
$C_{26}H_{42}O_4$
1α 25-Dihydroxy-22-oxacalcitriol.
Vitamin D analog with antiproliferative properties. Antipsoriatic.

10069 Mecobalamin
13422-55-4 6125 236-535-3
$C_{63}H_{91}CoN_{13}O_{14}P$
Cobinamide Co-methyl derivative hydroxide dihydrogen phosphate (ester) inner salt 3'-ester with 5,6-dimethyl-1-α-D-ribofuranosylbenzimidazole.
methyl vitamin B_{12}; MeCbl; Algobaz; Hitocobamin-M; Methycobal; Methylcobaz. Vitamin, vitamin source. Hematopoietic. λ_m 522, 342, 266 nm (ε 9357, 14416, 19897 pH 7), 462, 304, 264 nm (ε 9599, 22855, 24737 0.1N HCl).

10070 Menadiol
481-85-6 5873 207-573-8
$C_{11}H_{10}O_2$
2-Methyl-1,4-naphthalenediol.
dihydrovitamin K_3. Vitamin, vitamin source. Prothrombogenic. mp = 168-170°, 181°; slightly soluble in C_6H_6, $CHCl_3$; readily soluble in Me_2CO, EtOH. *Eli Lilly & Co.; Hoffmann-LaRoche Inc.*

10071 Menadiol Diacetate
573-20-6 5873 209-352-1
$C_{15}H_{14}O_4$
2-Methyl-1,4-naphthalenediol diacetate.
vitamin K_4; Kapilin; Prokayvit Oral; Vitavel K. Vitamin, vitamin source. Prothrombogenic. mp = 112-114°; insoluble in H_2O; slightly soluble in EtOH at 20°; soluble in EtOH (30 g/100 ml at 86°); soluble in AcOH. *Eli Lilly & Co.; Hoffmann-LaRoche Inc.*

10072 Menadiol Dibutyrate
53370-40-8 5873
$C_{19}H_{22}O_4$
2-Methyl-1,4-naphthalenediol dibutyrate.
Karanum. Vitamin, vitamin source. Prothrombogenic. mp = 53°; insoluble in H_2O; soluble in EtOH, C_6H_6, oils, fats. *E. Merck; Eli Lilly & Co.; Hoffmann-LaRoche Inc.*

10073 Menadiol Sodium Diphosphate
131-13-5 5873 205-012-1
$C_{11}H_8Na_4O_8P_2.6H_2O$
2-Methyl-1,4-naphthalenediol bis(dihydrogen phosphate) tetrasodium salt hexahydrate.
Kappadione; Synkayvite; Synkavit. Vitamin, vitamin source. Prothrombogenic. Very soluble in H_2O; insoluble in MeOH, EtOH, Et_2O, Me_2CO. *Eli Lilly & Co.; Hoffmann-LaRoche Inc.*

10074 Menadione
58-27-5 5874 200-372-6
$C_{11}H_8O_2$
2-Methyl-1,4-naphthoquinone.
Kappaxin; Kayquinone; Thyloquinone; Vitamin $K_{2(0)}$; Vitamin K_3; Kaynone; Kayklot; Klottone; Kolklot. Vitamin, vitamin source. Prothrombogenic. mp = 105-107°; insoluble in H_2O; soluble in EtOH (1.66 g/100 ml), C_6H_6 (10 g/100 ml), vegetable oils (2 g/100 ml); moderately soluble in $CHCl_3$, CCl_4; LD_{50} (mus orl) = 500 mg/kg. *Abbott Labs Inc.; Bristol-Myers Squibb Pharm. R&D; Sterling Winthrop Inc.*

10075 Menadione Dimethylpyrimidinol Bisulfite
14451-99-1 5874 238-435-5
$C_{17}H_{18}N_2O_6S$
2-Methyl-1,4-naphthoquinone dimethylpyrimidinol bisulfite.
Hetrazeen. Vitamin, vitamin source. Prothrombogenic. mp = 215-217°; soluble in H_2O (1 g/100 ml); slightly soluble in EtOH; insoluble in Et_2O, C_6H_6.

10076 Menadione Sodium Bisulfite
130-37-0 5874 204-987-0
$C_{11}H_9NaO_5S$
Sodium 1,2,3,4-tetrahydro-2-methyl-1,4-dioxo-2-naphthalenesulfonate.
Vitamin, vitamin source. Prothrombogenic. *Abbott Labs Inc.*

10077 Menadione Sodium Bisulfite Trihydrate
6147-37-1 5874
$C_{11}H_9NaO_5S.3H_2O$
Sodium 1,2,3,4-tetrahydro-2-methyl-1,4-dioxo-2-naphthalenesulfonate trihydrate.
Hykinone; Klotogen. Vitamin, vitamin source. Prothrombogenic. Soluble in H_2O (50 g/100 ml); slightly soluble in EtOH; insoluble in Et_2O, C_6H_6. *Abbott Labs Inc.*

10078 Menadoxime
573-01-3 5875 209-345-3
$C_{13}H_{14}N_2O_4$
Ammonium salt of 2-methylnaphthoquinone 4-oxime O-carboxymethyl ether.
Kapilon injectable. Vitamin, vitamin source. Prothrombogenic. Soluble in H_2O; [free acid ($C_{13}H_{11}NO_4$)]: mp = 162-163°; forms water-soluble salts. *Glaxo Wellcome plc.*

10079 Menaquinone 4
863-61-6 5876
$C_{31}H_{40}O_2$
(E,E,E)-2-Methyl-3-(3,7,11,15-tetramethyl-2,6,10,14-hexadecatetraenyl)-1,4-naphthalenedione.
menatetrenone; MK 4; Vitamin $K_{2(20)}$; Kaytwo. Vitamin, vitamin source. mp = 35°; λ_m = 248 nm ($E^{1\%}_{1\ cm}$ 439).

10080 Menaquinone 6
84-81-1 5876
$C_{41}H_{56}O_2$
(all-E)-2-(3,7,11,15,19,23,27-Heptamethyl-2,6,10,14,18,22,26-octacosaheptaenyl)-3-methyl-1,4-naphthalenedione.
MK 6; Vitamin $K_{2(30)}$. Vitamin, vitamin source. mp = 50°; λ_m 243, 248, 261, 270, 325-328 nm ($E^{1\%}_{1cm}$ 304, 320, 290, 292, 53 petroleum ether).

10081 Menaquinone 7
2124-57-4 5876
$C_{46}H_{64}O_2$
(all-E)-2-(3,7,11,15,19,23,27-Heptamethyl-2,6,10,14,18,22,26-octacosaheptaenyl)-3-methyl-1,4-naphthalenedione.
Vitamin $K_{2(35)}$. Vitamin source. mp = 54°; $bp_{0.0002}$ = 200° (dec); λ_m 243, 248, 261, 270, 325-328 nm ($E^{1\%}_{1\ cm}$ 278, 195, 266, 267, 48); slightly less soluble than vitamin K_1 in the same solvents.

10082 Menaquinones
5876
2-Methyl-3-all-trans-polyprenol 1,4-naphthoquinones.
Vitamin K_2; A group of prenylated naphthoquinones having the same physiological properties as Vitamin K and containing specific menaquinones, such as menaquinone 4, menaquinone 6 and menaquinone 7.

10083 Niacin
59-67-6 6612 200-441-0
$C_6H_5NO_2$
3-Pyridinecarboxylic acid.
Nicotinic acid; Niac; Nicamin; Nicobid; Nicolar; Wampocap; P.P. Factor; Akotin; Daskil; Niacor; Nicacid; Nicangin; Niconacid; NicoSpan. Vitamin (enzyme cofactor). At high doses, decreases hepatic secretion of very-high-density lipoproteins as a result of reduced triglyceride synthesis. Used in treatment of hypercholesterolemias and hypertriglyceridemias. mp = 236.6°; λ_m = 263 nm; soluble in H_2O (1.67 g/100 ml); freely soluble in H_2O at 100°, EtOH at 76°; insoluble in Et_2O; LD_{50} (raty sc) = 5000 mg/kg. *Abbott Labs Inc.; Apothecon; Forest Pharm. Inc.; Marion Merrell Dow Inc.; Rhône-Poulenc Rorer Pharm. Inc.; Wallace Labs.*

10084 Niacinamide
98-92-0 6574 202-713-4
$C_6H_6N_2O$
3-Pyridinecarboxamide.
Nicotinamide; Aminicotin; Benicot; Dipegyl; Nicamindon, Nicobion; Nictoamide; Nicotilamide; Pelmin; component of: Medriatric. Vitamin (enzyme cofactor). Precursor of the coenzymes NAD and NADP. mp = 128-131°; $bp_{0.0005}$ = 150-160°; λ_m 261 nm ($A^{1\%}_{1\ cm}$ 451); soluble in H_2O (100 g/100 ml), EtOH (66.6 g/100 ml), glycerol (10 g/100 ml); LD_{50} (rat sc) = 1680 mg/kg. *Wyeth-Ayerst Labs.*

10085 Nicotinamide Ascorbate
1987-71-9 6610
L-Ascorbic acid mixture with 3-pyridinecarboxamide.
merpress; nicoscorbine; Nicastubin. Vitamin, vitamin source. Enzyme cofactor. mp = 141-145°; $[\alpha]_D^{20}$ = 27.5° (c = 8 H_2O); soluble in H_2O (40 g/100 ml), EtOH (2.4 g/100 ml), MeOH (10 g/100 ml); sparingly soluble in Me_2CO; insoluble in C_6H_6, Et_2O. *Gelatin Products.*

10086 Nitrocobalamin
20623-13-6 4854 243-924-1
$C_{62}H_{88}CoN_{14}O_{16}P$
Cobinamide nitrite (salt) dihydrogen phosphate (ester) inner salt 3'-ester with 5,6-dimethyl-1-α-D-ribofuranoylbenzimidazole.
nitritocobalamin; vitamin B_{12c}. Vitamin, vitamin source. λ_m 352 527.5 nm ($E^{1\%}_{1cm}$ 153.2, 59.5 H_2O), 357, 535 nm ($E^{1\%}_{1cm}$ 139 62.5 0.1N NaOH). *Merck & Co.Inc.*

10087 Panthenol
16485-10-2 240-540-6
$C_9H_{19}NO_4$
(±)-2,4-Dihydroxy-N-(3-hydroxypropyl)-3,3-dimethylbutyramide.
(±)-pantothenyl alcohol; component of: Zentinic. Vitamin, vitamin source. *Eli Lilly & Co.*

10088 Phylloquinone
84-80-0 7536 201-564-2
$C_{31}H_{46}O_2$
[R-[R*,R*-(E)]]-2-Methyl-3-(3,7,11,15-tetramethyl-2-hexadecenyl)-1,4-naphthalenedione.
Aquamephyton; Konakion; Mephyton; Mono-kay; Phytonadione; vitamin K_1; 3-phytylmenadione; Veda-K_1; Veta-K_1. Vitamin, vitamin source. Prothrombogenic. $[\alpha]_D^{25}$ = -0.28° (dioxane); λ_m = 242, 248, 260, 269, 325 nm ($E^{1\%}_{1\ cm}$ 396, 4319, 383, 387, 68 petroleum ether); insoluble in H_2O; sparingly soluble in MeOH; soluble in EtOH, Me_2CO, C_6H_6, petroleum ether, C_6H_{14}, dioxane, $CHCl_3$, Et_2O, fats and oils; [dihydro form (phytonadiol; dihydrovitamin K_1)]: insoluble in H_2O, sparingly soluble in petroleum ether, freely soluble in Et_2O; [dihydro form sodium diphosphate ($C_{31}H_{48}Na_2O_8P_2$; phytonadiol sodium diphosphate; Kayhydrin)]: mp = 138°; soluble in H_2O, MeOH. *Abbott Labs Inc.; Hoffmann-LaRoche Inc.; Merck & Co.Inc.*

10089 Phylloquinone 2,3-Epoxide

25486-55-9 7536 247-022-9

$C_{31}H_{46}O_3$

[R-[R*,R*-(E)]]-2-Methyl-3-(3,7,11,15-tetramethyl-2-hexadecenyl)-1,4-naphthalenedione 2,3-epoxide.

vitamin K_1 epoxide; vitamin K_1 oxide. Vitamin, vitamin source. Prothrombogenic. Colorless oil; λ_m 259, 305 nm (log E_M 3.79, 3.31 95% EtOH). *Abbott Labs Inc.; Hoffmann-LaRoche Inc.; Merck & Co.Inc.*

10090 Pyridoxine Hydrochloride

58-56-0 8166 200-386-2

$C_8H_{12}ClNO_3$

5-Hydroxy-6-methyl-3,4-pyridinedimethanol hydrochloride.

Beesix; Hexa-Betalin; Hexavibex; Bonasanit; Hexabione hydrochloride; Pyridipca; Pyridox; Bécilan; Benadon; Hexermin; component of: Bendectin, Spondylonal. Vitamin, vitamin source. Enzyme cofactor. May improve hematopoiesis in patients with sideroblastic anemias. mp = 205-212° (dec); λ_m = 290 nm (ε 8400 0.1N HCl), 253, 325 nm (ε 3700, 7100, pH 7); soluble in H_2O (22.2 g/100 ml), EtOH (1.1 g/100 ml), propylene glycol; sparingly soluble in Me_2CO; insoluble in Et_2O, $CHCl_3$. *BASF Corp.; Eli Lilly & Co.; Forest Pharm. Inc.; General Aniline; Lederle Labs.; Marion Merrell Dow Inc.; Merck & Co.Inc.; Parke-Davis.*

10091 Raxofelast

128232-14-4

$C_{15}H_{18}O_5$

2-(2,3-Dihydro-5-acetoxy-4,6,7-tribenzofuranyl)-acetic acid.

IRFI-016. A hydrophilic vitamin E analog. An antioxidant which inhibits hepatic lipid peroxidation and the formation of gastric lesions. Protects against ischemic damage.

10092 Riboflavin Monophosphate

146-17-8 8368 205-664-7

$C_{17}H_{21}N_4O_9P$

7,8-Dimethyl-10-(D-ribo-2,3,4,5-tetrahydroxy-pentyl)isoalloxazine 5'-(dihydrogen phosphate).

flavin mononucleotide; FMN; vitamin B_2 phosphate. Vitamin, vitamin source. Enzyme cofactor. May be helpful in management of patients with hypoproliferative anemia. *Hoffmann-LaRoche Inc.; Takeda Chem. Ind. Ltd.*

10093 Riboflavin Monophosphate Monosodium Salt

130-40-5 8368 204-988-6

$C_{17}H_{20}N_4NaO_9P$

7,8-Dimethyl-10-(D-ribo-2,3,4,5-tetrahydroxy-pentyl)isoalloxazine 5'-(dihydrogen phosphate) monosodium salt.

Hyryl; Ribo. Vitamin, vitamin source. Enzyme cofactor. May be helpful in management of patients with hypoproliferative anemia. [dihydrate]: soluble in H_2O (11.2 g/100 ml). *Hoffmann-LaRoche Inc.; Takeda Chem. Ind. Ltd.*

10094 Sodium Ascorbate

134-03-2 8723 205-126-1

$C_6H_7NaO_6$

Monosodium L-ascorbate.

Ascorbin; Cevalin; Sodascorbate; Natrascorb; Cenolate; Ascorbicin; Cebitate. Vitamin, vitamin source. Antiscorbutic. Dec 218°; $[\alpha]_D^{20}$ = 104.4°; freely soluble in H_2O at 25° (62 g/100 ml H_2O, 78 g/100 ml at 75°). *BASF Corp.; Eli Lilly & Co.; Marion Merrell Dow Inc.*

10095 Sodium Folate

6484-89-5 8763 229-348-3

$C_{19}H_{18}N_7NaO_6$

Folic acid sodium salt.

sodium Folvite. Vitamin, vitamin source. Sold as a sterile aqueous solution with pH between 8.5 and 11.0.

10096 Sulfiotcobalamine

15671-27-9 4854 239-747-4

$C_{62}H_{89}CoN_{13}O_{17}PS$

Cobinamide sulfite (1:1) (salt) dihydrogen phosphate (ester) inner salt 3'-ester with 5,6-dimethyl-1-α-D-ribofuranoylbenzimidazole.

cobalaminsulfonic acid. Vitamin, vitamin source. λ_m 275, 365, 418, 516 nm ($E_{1\,cm}^{1\%}$ 328, 130, 49, 61). *Abbott Labs Inc.; Merck & Co.Inc.*

10097 Thiamine Hydrochloride

67-03-8 9430 200-641-8

$C_{12}H_{17}ClN_4OS$

3-[(4-Amino-2-methyl-5-pyrimidinyl)methyl]-5-(2-hydroxyethyl)-4-methylthiazolium chloride monohydrochloride.

Betalin S; Betaxin; Biamine; Benerva; Betabion; Bewon; Metabolin; Vitaneurin; component of: Spondylonal, Troph-Iron, Trophite, Trophite+Iron. Vitamin, vitamin source. Enzyme cofactor. mp = 248° (dec); soluble in H_2O (100 g/100 ml), glycerol (5.5 g/100 ml), 95% EtOH (1 g/100 ml), EtOH (0.32 g/100 ml); more soluble in MeOH; soluble in propylene glycol; insoluble in Et_2O, C_6H_6, $CHCl_3$, C_6H_{14}; LD_{50} (mus iv) = 89.2 mg/kg, (mus orl) = 8224 mg/kg. *BASF Corp.; Eli Lilly & Co.; Elkins-Sinn; Lederle Labs.; Menley & James Labs Inc.; Sterling Winthrop Inc.; Wyeth-Ayerst Labs.*

10098 Thiamine Mononitrate

532-43-4 9430 208-537-4

$C_{12}H_{17}N_5O_4S$

3-[(4-Amino-2-methyl-5-pyrimidinyl)methyl]-5-(2-hydroxyethyl)-4-methylthiazolium chloride nitrate (salt).

component of: Stuartinic. Vitamin (enzyme cofactor). Used to enrich flours and animal feeds. mp = 196-200° (dec); soluble in H_2O (2.7 g/100 ml at 25°, 30 g/100 ml at 100°). *Am. Cyanamid; BASF Corp.; Johnson & Johnson-Merck Consumer Pharm.*

10099 β-Tocopherol

148-03-8 9632 205-708-5

$C_{28}H_{48}O_2$

3,4-Dihydro-2,5,8-trimethyl-2-(4,8,12-trimethyltridecyl)-2H-1-benzopyran-6-ol.

5,8-dimethyltocol; cumotocopherol; neotocopherol;

p-xylotocopherol. Vitamin, vitamin source. $bp_{0.1}$ = 200-210°; $[\alpha]_D^{20}$ = 6.37°; λ_m 297 nm ($E_{1\,cm}^{1\%}$ 86.4); insoluble in H_2O; freely soluble in oils, fats, Me_2CO, EtOH, $CHCl_3$, Et_2O; [allophanate ($C_{30}H_{50}N_2O_4$)]: mp = 138-139°; $[\alpha]_D^{18}$ = 6.7° ($CHCl_3$); [acetate ($C_{30}H_{50}O_3$)]: $bp_{0.3}$ = 215-220°; insoluble in H_2O; freely soluble in Me_2CO, $CHCl_3$, Et_2O; less soluble in EtOH.

10100 σ-Tocopherol

7616-22-0 9633 231-523-4

$C_{28}H_{48}O_2$

3,4-Dihydro-2,7,8-trimethyl-2-(4,8,12-trimethyltridecyl)-2H-1-benzopyran-6-ol.

7,8-dimethyltocol; o-xylotocopherol. Vitamin, vitamin source. $bp_{0.1}$ = 200-210°; $[\alpha]_D^{20}$ = -2.4° (EtOH); λ_m 298 nm ($E_{1\,cm}^{1\%}$ 92.8); insoluble in H_2O; freely soluble in oils, fats, Me_2CO, EtOH, $CHCl_3$, Et_2O; [allophante ($C_{30}H_{50}N_2O_4$)]: mp = 136-138°; $[\alpha]_D^{18}$ = +3.4° (chloroform).

10101 δ-Tocopherol

119-13-1 9634 204-299-0

$C_{27}H_{46}O_2$

3,4-Dihydro-2,8-dimethyl-2-(4,8,12-trimethyltridecyl)-2H-1-benzopyran-6-ol.

8-methyltocol. Vitamin source. $[\alpha]_{546}^{25}$ = 3.4° (c = 15.5 EtOH); $[\alpha]_{546}^{25}$ = 1.1° (c = 10.9 C_6H_6); λ_m 298 nm ($E_{1\,cm}^{1\%}$ 91.2).

10102 Tocophersolan

30999-06-5

$C_{33}H_{54}O_5(C_2H_4O)_n$

Mono-[2,5,7,8-tetramethyl-2-(4,8,12,trimethyltridecyl)-6-chromanyl] succinate polyethylene glycol monoester.

Eastman Vitamin ETPGS; TPGS. Vitamin, vitamin source. *Eastman Chem. Co.*

10103 Vitamin A

68-26-8 10150 200-683-7

$C_{20}H_{30}O$

3,7-Dimethyl-9-(2,6,6,-trimethyl-1-cyclohexen-1-yl)-2,4,6,8-nonatetraen-1-ol.

Alphalin; Anatola; Aquasol A; Aquasol A Parenteral; A-Sol; A-Vitan; Vi-Dom-A; Retinol; Acon; Afaxin; Agiolan; Aoral; Apexol; Apostavit; Atav; Avibon; Avita; Avitol; Axerol; Dohyfral A; Epiteliol; Nio-A-Let; Prepalin; Testavol; Vaflol; Vi-Alpha; Vitpex; Vogan; Vogan-Neu. Vitamin, vitamin source. Antixerophthalmic. mp = 62-64°; $bp_{0.005}$ = 120-125°; λ_m 324-325 nm ($E_{1\,cm}^{1\%}$ 1835; insoluble in H_2O, glycerol; soluble in EtOH, MeOH, $CHCl_3$, fats and oils; LD_{50} (10 day) (mus ip) = 1510 mg/kg, (mus orl) = 2570 mg/kg. *Abbott Laboratories Inc.; Astra Chem. Ltd.; BASF Corp.; Eli Lilly & Co.; Parke-Davis; Purdue Pharma L.P.*

10104 Vitamin A Acetate

127-47-9 10150 204-844-2

$C_{22}H_{32}O_2$

3,7-Dimethyl-9-(2,6,6,-trimethyl-1-cyclohexen-1-yl)-2,4,6,8-nonatetraen-1-ol acetate.

Vitamin, vitamin source. Antixerophthalmic. mp = 57-58°; λ_m 326 nm ($E_{1cm}^{1\%}$ 1550 EtOH); LD_{50} (10 day) (mus orl) = 4100 mg/kg.

10105 Vitamin A Palmitate

79-81-2 10150 201-228-5

$C_{36}H_{60}O_2$

3,7-Dimethyl-9-(2,6,6,-trimethyl-1-cyclohexen-1-yl)-2,4,6,8-nonatetraen-1-ol palmitate.

Arovit; Optovit-A; The ester found in fish liver oils. Vitamin, vitamin source. Antixerophthalmic. mp = 28-29°; λ_m 325-328 nm ($E_{1\,cm}^{1\%}$ 975 EtOH); LD_{50} (10 day) (mus orl) = 6060 mg/kg, (rat orl) = 7910 mg/kg; biopotency 3.33×10^6 I.U./g.

10106 Vitamin E

59-02-9 10159 200-412-2

$C_{29}H_{50}O_2$

[2R-2R*(4R*,8R*)]-3,4-Dihydro-2,5,7,8-tetramethyl-2-(4,8,12-trimethyltridecyl)-2H-1-benzopyran-6-ol.

(+)-α-tocopherol; α-tocopherol; antisterility vitamin. Vitamin, vitamin source. $[\alpha]_{5461}^{25}$ = -3.0° (C_6H_6), $[\alpha]_{5461}^{25}$ = 0.32° (EtOH).

10107 d-Vitamin E Acetate

58-95-7 10160 200-405-4

$C_{31}H_{52}O_3$

[2R-[2R*(4R8,8R*)]]-3,4-Dihydro-2,5,7,8-tetramethyl-2-(4,8,12-trimethyltridecyl)-2H-1-benzopyran-6-ol acetate.

E-Vicotrat; Spondyvit; Tocopherex. Vitamin, vitamin source. mp = 28°; $[\alpha]_D^{25}$ = 0.25° (c = 10 $CHCl_3$), 3.2° (EtOH); [l-form]: mp = 23°; $[\alpha]_D^{25}$ = -2.0° (EtOH). *Hoffmann-LaRoche Inc.*

10108 dl-Vitamin E Acetate

52225-20-4 10160 257-757-7

$C_{31}H_{52}O_3$

(±)-[2R*(4R*,8R*)]-3,4-Dihydro-2,5,7,8-tetramethyl-2-(4,8,12-trimethyltridecyl)-H-1-benzopyran-6-ol acetate.

Detulin; Ephynal; Eprolin; Epsilan-M; Eusovit; E-Vimin; Evion; Juvela; OptoVit-E; Toco500; Vitagutt; α-tocopherol acetate; α-tocopheryl acetate. Vitamin, vitamin source. mp = -27.5°; $d_4^{21.3}$ = 0.9533; $bp_{0.01}$ = 184°, $bp_{0.025}$ = 194°, $bp_{0.3}$ = 224°; insoluble in H_2O; freely soluble in Me_2CO, $CHCl_3$, Et_2O; less soluble in EtOH. *Hoffmann-LaRoche Inc.*

10109 Vitamin E Nicotinate

43119-47-7 10159 256-101-7

$C_{35}H_{53}NO_3$

[2R-2R*(4R*,8R*)]-3,4-Dihydro-2,5,7,8-tetramethyl-2-(4,8,12-trimethyltridecyl)-2H-1-benzopyran-6-ol nicotinate.

Hijuven; Juvela Nicotinate; Renascin. Vitamin, vitamin source.

10110 Vitamin E Succinate

4345-03-3 10159 224-403-8

$C_{33}H_{56}O_6$

[2R-2R*(4R*,8R*)]-3,4-Dihydro-2,5,7,8-tetramethyl-2-(4,8,12-trimethyltridecyl)-2H-1-benzopyran-6-ol succinate.

Vitamin E acid succinate. Vitamin, vitamin source. mp = 76-77°; λ_m 286 nm ($E_{1\,cm}^{1\%}$ 38.5 EtOH).

10111 Vitamin K_5
83-70-5 10162
$C_{11}H_{11}NO$
4-Amino-2-methyl-1-naphthalenol.
Synkamin. Vitamin, vitamin source. Prothrombogenic.

10112 Vitamin K_5 Hydrochloride
130-24-5 10162 204-983-9
$C_{11}H_{12}ClNO$
4-Amino-2-methyl-1-naphthalenol hydrochloride.
Kayvisyn. Vitamin, vitamin source. Prothrombogenic. dec 280-282°; ; freely soluble in H_2O, soluble in EtOH, insoluble in Et_2O.

10113 Vitamin U
1115-84-0 10165 214-231-1
$C_6H_{14}ClNO_2S$
(S)-(3-Amino-3-carboxypropyl)dimethylsulfonium chloride.
methylmethioninesulfonium chloride; MMSC; Cabagin-U; Epadyn-U; Vitas-U. Vitamin, vitamin source. An antiulcer vitamin found in leaves of cabbage and other green vegetables. mp = 134° (dec); LD_{50} (mus iv) = 2760 mg/kg; [bromide analog ($C_6H_{14}BrNO_2S$, methylmethioninesulfonium bromide; Ardésyl)]: mp =139° (dec).

10114 Zinc Gluconate
4468-02-4 224-736-9
$C_{12}H_{22}O_{14}Zn$
Zinc D-gluconate (1:2).
Used in cough lozenges to putatively to reduce symptoms of the common cold. Supplement (trace mineral).

Vulnerary Agents

10115 Acetylcysteine
616-91-1 89 210-498-3
$C_5H_9NO_3S$
L-α-Acetamido-β-mercaptopropionic acid.
N-acetyl-L-cysteine; 5052; NSC-111180; Airbron; Broncholysin; Brunac; Fabrol; Fluatox; Fluimucil; Fluimucetin; Fluprowit; Mucocedyl; Mucolator; Mucolyticum; Mucomyst; Muco Sanigen; Mucosolvin; Mucret; NAC; Neo-Fluimucil; Parvolex; Respaire; Tixair. Mucolytic; corneal vulnerary; antidote to acetaminophen poisoning. mp = 109-110°; LD_{50} (rat orl) = 5050 mg/kg. *Bristol Laboratories; DuPont-Merck Pharm.; Mead Johnson Labs.; Mead Johnson Pharmaceuticals.*

10116 Allantoin
97-59-6 255 202-592-8
$C_4H_6N_4O_3$
(2,5-Dioxo-4-imidazolidinyl)urea.
glyoxyldiureide; cordianine; Psoralon; Septalan; component of: Skin-Balm. Vulnerary. Used to treat skin ulceration. [(dl)-form]: mp = 238°; soluble in H_2O (0.52 g/100 ml), EtOH (0.2 g/100 ml). *Carrington Labs Inc.; ICI Americas Inc.*

10117 Asiaticoside
16830-15-2 869 240-851-7
$C_{48}H_{78}O_{19}$
(2α,3β,4α)-2,3,23-Trihydroxyurs-12-en-28-oic acid O-6-deoxy-α-L-mannopyranosyl-(1→4)-O-β-D-glucopyranosyl-(1→6)-O-β-D-glucopyranosyl ester.
Centelase Dermatologico; Madecassol. Vulnerary. mp = 230-233°; insoluble in H_2O; soluble in EtOH, C_5H_5N; $[\alpha]_D^{20}$ = -14° (EtOH).

10118 Cadexomer Iodine
94820-09-4 1646
Product of reaction of dextrin with epichlorohydrin coupled with ino-exchange groups and with iodine.
Iodosorb. Antiseptic; antiulcerative. Vulnerary; used to treat venous ulcers. *Perstorp AB.*

10119 Chitin
1398-61-4 2105 215-744-3
$(C_8H_{13}NO_5)_n$
Biopolymer consisting predominantly of N-acetyl-D-glucosamine chains.
Vulnerary. A major component of exoskeletons. Amorphous solid; insoluble in all solvents; soluble in concentrated HCL, H_2SO_4.

10120 Dextranomer
56087-11-7 2992
Dextran 2,3-dihydroxypropyl 2-hydroxy-1,3-propanediyl ether.
Debrisan; Debrisorb. Vulnerary. Insoluble in all solvents.

10121 Enbucrilate
6606-65-1
$C_8H_{11}NO_2$
Butyl 2-cyanoacrylate.
A tissue adhesive used in surgery.

10122 Oxaceprol
33996-33-7 7035 251-780-6
$C_7H_{11}NO_4$
trans-1-Acetyl-4-hydroxy-L-proline.
CO-61; AHP-200; Jonctum. Anti-inflammatory; vulnerary. Derivative of hydroproline. mp = 133-134°, 126-128°; $[\alpha]_D^{20}$ = -116.5° (c = 3.2); $[\alpha]_D^{18}$ = -119.5° (c = 3.75); soluble in alcohol, H_2O, MeOH; insoluble in $CHCl_3$. *Richardson-Merrell.*

10123 Tocoretinate
40516-48-1 9639
$C_{49}H_{76}O_3$
(±)-(2R*)-2,5,7,8-Tetramethyl-2-[(4R*,8R*)-4,8,12-trimethyltridecyl]-6-chromanyl retinoate.
Tretinoin tocoferil; L-300; N-021; Olcenon. Vulnerary. Yellow oil; λ_m 365 nm ($E^{1\%}_{1\,cm}$ 642 EtOH); LD_{50} (mus iv) > 1000 mg/kg, (mus orl) > 2000 mg/kg. *Nisshin Denka K.K.*

PART II

INDEXES

CAS REGISTRY NUMBER INDEX

CAS Registry Number Index

CAS Registry Number Index

CAS Registry Number Index

CAS Registry Number Index

CAS Registry Number Index

CAS Registry Number Index

CAS Registry Number Index

CAS Registry Number Index

CAS Registry Number Index

CAS Registry Number Index

CAS Registry Number Index

CAS Registry Number Index

CAS Registry Number Index

CAS Registry Number Index

CAS Registry Number Index

CAS Registry Number Index

CAS Registry Number Index

CAS Registry Number Index

CAS Registry Number Index

CAS Registry Number Index

CAS Registry Number Index

CAS Registry Number Index

CAS Registry Number Index

CAS Registry Number Index

CAS Registry Number Index

CAS Registry Number Index

CAS Registry Number Index

CAS Registry Number Index

CAS Registry Number Index

CAS Registry Number Index

CAS Registry Number Index

CAS Registry Number Index

CAS Registry Number Index

CAS Registry Number Index

CAS Registry Number Index

CAS Registry Number Index

CAS Registry Number Index

CAS Registry Number Index

CAS Registry Number Index

CAS Registry Number Index

EINECS NUMBER INDEX

EINECS No.	Name	Record Number
200-003-9	Dexamethasone	1778, 4855, 8730
200-004-4	Hydrocortisone Acetate	8769
200-006-5	Cortisone Acetate	8722
200-007-0	Phenobarbital	3014, 9537
200-008-6	Mitomycin	5252
200-009-1	Hexobarbital Sodium	887, 9493
200-010-7	Oxyphenonium Bromide	2822
200-011-2	Metharbital	2994
200-012-8	Mephenytoin	2992
200-013-3	Meperidine Hydrochloride	394
200-014-9	Ergocalciferol	10060
200-015-4	Cyclophosphamide	5109, 8530
200-017-5	Hydroxyphenamate	6754
200-019-6	Corticosterone	8720
200-020-1	Hydrocortisone	8767
200-021-7	Prednisolone	8797
200-022-2	Estriol	8174
200-023-8	Estradiol	8165
200-024-3	DDT	8075
200-030-6	Propantheline Bromide	2850
200-031-1	Thalidomide	5349, 9567
200-032-7	Cocaine	941
200-033-2	Lysergide	5797
200-034-8	Protheobromine	7964
200-035-3	Clomiphene Citrate	3627, 8296
200-036-9	Adiphenine Hydrochloride	2677, 8895
200-037-4	6-Mercaptopurine [anhydrous]	5233, 8545
200-040-0	Desipramine	3124
200-041-6	Amitriptyline	3061
200-042-1	Imipramine	3183
200-043-7	Estradiol Benzoate	8166
200-044-2	Thioridazine	5882, 9568
200-045-8	Chlorpromazine	3551, 5730
200-046-3	Quinine Sulfate Dihydrate	4966
200-046-3	Quinidine Sulfate [anhydrous]	1682, 1683
200-047-9	Reserpine	4426
200-048-4	Oxytocin	9142
200-050-5	Lypressin	3467, 3475, 8419
200-051-0	Phendimetrazine Bitartrate	1111
200-052-6	Cephaloridine	2048
200-053-1	Phentolamine	231, 4397, 5534
200-054-7	Hexobendine Dihydrochloride	9892
200-055-2	Chloroquine Diphosphate	1332, 1733 4929, 8694
200-056-8	Niclosamide	1209
200-061-5	Sorbitol	8647
200-064-1	Acetylsalicylic Acid	482, 511, 4563 5919, 6300
200-066-2	Ascorbic Acid	10046
200-074-6	l-Ephedrine Hydrochloride	108, 6924, 7169
200-077-2	Procaine Hydrochloride	1016
200-078-8	Procainamide	1667
200-079-3	Nialamide	3243, 8852
200-080-9	Pralidoxime Chloride	7667
200-085-6	Fluorouracil	5171
200-086-1	Tiratricol	4539
200-089-8	Isoproterenol Hydrochloride	6951, 7211
200-090-3	Scopolamine Hydrobromide	2857
200-090-3	Scopolamine	2856, 3608
200-096-6	Norepinephrine	152, 4517

EINECS Number Index

EINECS Number Index

EINECS Number Index

212-752-9	Cupric Citrate	6867
212-755-5	Potassium Citrate	1151
212-768-6	Pantothenic Acid Sodium Salt	8136
212-769-1	Potassium Bitartrate	7961, 8628
212-773-3	Sodium Tartrate	8646
212-925-9	Clofibric Acid	4061
212-948-4	Diphencyprone	1320
212-954-7	Chlorphenesin Carbamate	8916
212-973-0	Nortriptyline Hydrochloride	3250
212-978-8	Dothiepin Hydrochloride	3138
212-992-4	Captodiamine Hydrochloride	6708
213-000-2	Opipramol Hydrochloride	3258, 5827
213-006-5	Tiocarlide	2652
213-008-6	Clomiphene	3626, 8295
213-009-1	Etonitazene	368
213-013-3	Nandrolone Cyclohexane-propionate	276, 827
213-014-9	Noscapine Hydrochloride	6443
213-017-5	Methacycline	2586
213-020-1	Diphenoxylate	3449
213-045-8	Lithium Citrate	4975
213-067-8	Potassium Tartrate	8633
213-346-4	Ethenzamide	587
213-425-3	Thiobutabarbital Sodium Salt	906
213-460-4	Morphazinamide	2635
213-475-6	Meprylcaine Hydrochloride	985
213-484-5	Fluindione	2918
213-493-4	Oxolamine	4738, 6446
213-496-0	Sotalol Hydrochloride	1484, 1694 4436, 7092
213-515-2	Demoxepam	6726
213-519-4	Moxisylyte Hydrochloride	9989
213-522-0	Nifuroxazide	2262
213-523-6	Ethylestrenol	265
213-529-9	Pridinol Hydrochloride	2844, 5507
213-530-4	Acetohexamide	3351
213-534-6	Teslac	5348
213-540-9	Diphenidol	3562
213-547-7	Prosultiamine Hydrochloride	8138
213-554-5	Canrenone	76, 7895
213-559-2	Estradiol Valerate	8171
213-566-0	Clopenthixol	5740
213-567-6	Prenoxdiazine Hydrochloride	6458
213-568-1	Chloramphenicol Mono-succinate Sodium Salt	1901, 5993
213-571-8	Penamecillin	2112
213-573-9	Ethaverine Hydrochloride	6206
213-574-4	Nafcillin Sodium	2103
213-577-0	Flumedroxone Acetate	5000
213-578-6	Betamethasone Acetate	8700
213-580-7	Citicoline	9082, 9806
213-588-0	Fentanyl Citrate	372
213-592-2	Lymecycline	2583
213-812-7	Pyrrolnitrin	3771
213-842-0	Pentifylline	9841, 10002
213-859-3	Sulfadimethoxine Sodium Salt	2496
213-872-4	Pirenoxine	3848
213-876-6	Testosterone Acetate	846
213-885-5	Benzilonium Bromide	2706
213-887-6	Moperone	5814
213-900-5	Butaperazine Dimaleate	5723
213-907-3	Colistin	2349
213-970-7	Ethambutol Dihydrochloride	2623
214-071-2	dl-Thioctic Acid	8451
214-105-6	Tramazoline	194, 7839
214-109-8	Bucetin	526
214-112-4	Meticrane	4341, 7942
214-123-4	Nordazepam	6789
214-126-0	Nifurtoinol	2271
214-133-9	Clostebol	263
214-134-4	Ethopropazine Hydrochloride	2766, 5483
214-140-7	Methadone Hydrochloride	399
214-149-6	Triflupromazine Hydrochloride	5901
214-150-1	Pyritinol	9109
214-152-2	Prasterone Sodium Sulfate	841
214-164-8	Meclizine Hydrochloride	3581
214-172-1	Prednisolone 21-Trimethylacetate	8799
214-175-8	Cortivazol	8723
214-176-3	Pipacycline	2595
214-178-4	Acetyldigitoxin	7428
214-180-5	Bietaserpine Bitartrate	4171
214-203-9	L-Penicillamine	1761, 7555
214-230-6	Metformin Hydrochloride	3413
214-231-1	Vitamin U	10113
214-328-9	Pyrithione	3769
214-458-6	Cyclobutyrol Sodium Salt	7575
214-470-1	Debrisoquin	4228, 8841
214-486-9	Baclofen	8908
214-537-5	Butallylonal	9423
214-538-0	Anthralin	5671
214-549-0	5-Furfuryl-5-isopropyl-barbituric Acid	9485
214-553-2	Clorindione	2908
214-556-9	Pyrovalerone Hydrochloride	7761
214-587-8	Phenglutarimide	2831
214-588-3	Tolanase	3433
214-600-7	Sulfatolamide	2549
214-607-5	Moxaverine Hydrochloride	6231
214-616-4	Dimefline	9355
214-623-2	Quinestradiol	8200
214-630-0	Flurazepam Hydrochloride	2974, 8947, 9483
214-636-3	Oxacillin Sodium	2106
214-643-1	Pseudococaine Tartrate	1028
214-646-8	Dexamethasone 21-Acetate	1779, 4856, 8731
214-648-9	Thiethylperazine Dimaleate	3614
214-723-6	Buformin Hydrochloride	3357
214-741-4	Mercuric Oleate	8082
214-793-8	Citiolone	8429
214-818-2	Tranexamic Acid	8416
214-821-9	Phentermine Hydrochloride	1115, 7754
214-917-0	Prolintane Hydrochloride	3283, 7759
214-921-2	Bendazol Hydrochloride	9858
214-925-4	Ethebenecid	9786
214-931-7	Clobutinol Hydrochloride	6400
214-936-4	Xylometazoline Hydrochloride	202, 7845
214-947-4	Zolimidine	6560, 7807
214-956-3	Protriptyline Hydrochloride	3287
214-957-9	Isothipendyl	3959
214-958-4	Prothipendyl Hydrochloride	5858
214-963-1	Mefexamide	7741
214-966-8	Doxepin Hydrochloride	3140
214-967-3	Methdilazine Hydrochloride	5662
214-971-5	Ethynodiol	7770, 9191
214-976-2	Biperiden Hydrochloride	2713, 5462
214-993-5	Propicillin Potassium	2152
214-996-1	Meprednisone	8780
214-997-7	Colpormon	8157
215-010-2	Gestonorone Caproate	5550, 9197
215-011-8	Oxabolone 17-Cyclopentanepropionate	285
215-017-0	Homofenazine Dihydrochloride	9496
215-019-1	Prochlorperazine Edisylate	3602, 5852
215-031-7	Paromomycin Sulfate	1355, 1887

EINECS Number Index

EINECS Number Index

EINECS Number Index

EINECS Number Index

EINECS Number Index

EINECS Number Index

EINECS No.	Name	Entry No.
251-501-8	Pyrantel Tartrate	1226
251-502-3	Metaraminol Bitartrate	135, 4511
251-553-1	Gonadorelin	8302
251-574-6	Cefoxitin Sodium	2002
251-575-1	Diflorasone Diacetate	4866, 8741
251-578-8	Tertatolol Hydrochloride	4453, 7099
251-591-9	Pirisudanol	3274
251-616-3	Calcium Bromolactobionate	9430
251-627-3	Clotiazepam	6720
251-668-7	Zolazepam Hydrochloride	9588
251-688-6	Pivampicillin	2146
251-689-1	Citicoline Sodium	9807
251-720-9	Silicristin	8445
251-780-6	Oxaceprol	4733, 10122
251-801-9	Auranofin	1723
251-826-5	Clocortolone Pivalate	8717
251-831-2	Acecainide Hydrochloride	1507
251-845-9	Trimebutine Maleate	2888, 6269
251-856-9	Fipexide Hydrochloride	9090
251-857-4	Fipexide 9089	
251-867-9	Propafenone Hydrochloride	1672
251-980-3	Acebutolol Hydrochloride	1374, 1505 4129, 6996
252-030-0	Cefamandole	1950
252-064-6	Dibekacin	1853
252-070-9	Antimony Sodium Tartrate	1164
252-075-6	Fenalcomine Hydrochloride	968, 7476
252-082-4	Loperamide Hydrochloride	3455
252-084-5	Isoxicam	4686
252-099-7	Ketotifen	1789
252-100-0	Ketotifen Fumarate	1790
252-126-2	Fenclofenac	4637
252-130-4	Urapidil	4477
252-191-7	Zipeprol	6472
252-192-2	Zipeprol Dihydrochloride	6473
252-213-5	Ticarcillin	2180
252-255-4	Carbuterol Hydrochloride	6905, 7148
252-257-5	Carbuterol	6904, 7147
252-362-6	Endrysone	4871
252-369-4	Bufetolol Hydrochloride	1398, 1551, 7023
252-578-0	Midecamycin A1	2228
252-607-7	Fludalanine	2294
252-611-9	Buflomedil Hydrochloride	9950
252-615-0	Enilconazole	3700
252-638-6	Viloxazine Hydrochloride	3343
252-641-2	Cefoxitin	2001
252-645-4	Tritiozine	6558
252-687-3	Tolmetin Sodium [anhydrous]	4817
252-732-7	Trimazosin	245, 4471
252-742-1	Rosaramicin	2238
252-802-7	Tiaramide Hydrochloride	1820, 4808
252-825-2	Nadoxolol Hydrochloride	1642, 7068
252-831-5	Folescutol Hydrochloride	10033
252-866-6	Camazepam	6705
252-895-4	Halofantrine Hydrochloride	4941
252-958-6	Dimemorfan Phosphate	6413
252-963-3	Dimemorfan	6412
252-974-3	Piroxicam	729, 4759
252-979-0	Fenbufen	4636
253-014-6	Norpseudoephedrine	1104
253-070-1	Nictindole	6342
253-072-2	Buspirone	6703, 9595
253-143-8	Etidocaine	961
253-144-3	Etidocaine Hydrochloride	962
253-162-1	Inosine Pranobex	6627, 8487
253-179-4	Quazepam	9548
253-258-3	Labetalol	222, 4319, 7053
253-287-1	Cicloprofen	4598
253-297-6	Fosfonet Disodium [anhydrous]	6617
253-347-7	Azlocillin Sodium	1928
253-348-2	Azlocillin	1927
253-366-0	Clenbuterol	6906, 7150
253-460-1	Apramycin	1842
253-483-7	Metoprolol	1445, 1628 4345, 7060
253-536-4	Pipotiazine Palmitate	5846
253-538-5	Amikacin	1840
253-539-0	Acebutolol	1373, 1504 4128, 6995
253-552-1	Fenoverine	6211
253-580-4	Bacampicillin Hydrochloride	1931
253-624-2	Terguride	5523, 9251
253-625-8	Terguride Hydrogen Maleate	5524, 9252
253-632-6	Clofoctol	2287
253-694-4	Azanator	7136
253-751-3	Pirbuterol Hydrochloride	6968, 7240
253-774-9	Carmantadine	5469
253-783-8	Tolamolol	1489, 1710, 9924
253-819-2	Sulindac	4791
253-854-3	Trientine Dihydrochloride	7560
253-874-2	Minoxidil	1322, 4353
253-906-5	Penbutolol Sulfate	1471, 1650 4386, 7079
254-002-3	Prostaglandin F2a Tromethamine Salt	7, 9147, 9262
254-136-2	Indoramin Hydrochloride	221, 4313
254-137-8	Cephradine	2055
254-186-5	Somatostatin	6545, 8274
254-220-9	Emorfazone	578, 4623
254-309-2	Trimebutine	2887, 6268
254-361-6	Promestriene	8199
254-464-6	Betasizofiran	6582
254-513-1	Nitrendipine	4376, 7364
254-546-1	Trepipam Maleate	9571
254-604-6	Alminoprofen	484, 4549
254-620-3	Tifemoxone	3324
254-648-6	Amikacin Sulfate	1841
254-659-6	Pipotiazine	5845
254-697-3	Dopamantine	5477
254-728-0	Nifurzide	2273
254-758-4	Rosoxacin	2451
254-826-3	Ticrynafen	7977, 9798
254-834-7	Tioxaprofen	4814, 6370
254-905-2	Pamidronic Acid	5426, 5445, 7115
254-945-0	Indanazoline Hydrochloride	126, 7819
255-028-8	Cloprostenol	9260
255-029-3	Fluprostenol	9264
255-035-6	Piprofurol	7370
255-096-9	Suprofen	772, 4792
255-140-7	Josamycin Propionate	2223
255-201-8	Enprofylline	7164
255-214-9	Tracazolate	9570
255-219-6	Memantine Hydrochloride	8968
255-297-1	1α-Hydroxycholecalciferol	10064
255-464-9	Ciclopirox Olamine	3680
255-505-0	Tocainide	1706
255-528-6	Sulbenicillin	2164
255-543-8	Fluocortin Butyl	4880
255-555-3	Lonaprofen	645, 4692
255-567-9	Bezafibrate	4046
255-663-0	Carnidazole	5578
255-706-3	Nadolol	1451, 1640 4365, 7066
255-749-8	Oxilorphan	9067

NAME AND SYNONYM INDEX

Name and Synonym Index

Name and Synonym Index

Name and Synonym Index

Name and Synonym Index

Name and Synonym Index

Name and Synonym Index

Name and Synonym Index

Name and Synonym Index

Name and Synonym Index

Name and Synonym Index

Name and Synonym Index

Name and Synonym Index

PART III

MANUFACTURER AND SUPPLIER DIRECTORY

MANUFACTURERS AND SUPPLIERS

3M Company
3M Center
St Paul, MN 55144
USA
Tel: +1 (612) 733-1110

3M Health Care
3M Center
St Paul, MN 55144
USA
Tel: +1 (612) 733-1110

3M Health Care Ltd
1 Morley Street
Loughborough,
Leics LE11 1EP
England
Tel: +44 (01509) 611611

3M Pharmaceuticals
3M Center 2751
St Paul, MN 55144-1000
USA
Tel: +1 (612) 733-0266
Fax: +1 (612) 737-2759

Abbott Laboratories
100 Abbott Park Rd
Abbott Park, IL 60064
USA
Tel: +1 (847) 937-6100
Fax: +1 (847) 937-1511

Abbott Laboratories Ltd
Abbott House
Moorbridge Rd
Maidenhead, Berks SL6 8JG
England
Tel: +44 (01628) 773355

ABIC
Address Unknown

Adria Labs
Direct Inquiries to Pharmacia & Upjohn

Advanced Magnetics Inc
Corporate Headquarters
61 Mooney St
Cambridge, MA 02138
USA
Tel: +1 (617) 497-2070
Fax: +1 (617) 547-2445

Agouron Pharmaceuticals Inc
10350 North Torrey Pine Rd
La Jolla, CA 92037
USA
Tel: +1 (858) 622-3000

Ajinomoto Co Inc
1-15-1, Kyobashi
Chuo-ku Tokyo 104
Japan
Tel: +81 (3) 5250-8111

Ajinomoto-Takara Corp
2-17-11, Kyobashi
Chuo-ku Tokyo 104
Japan
Tel: +81 (3) 3563-7589
Fax: +81 (3) 3535-3689

Aktieselskabet Pharmacia
Direct Inquiries to Pharmacia & Upjohn

Akzo Chemie
Stationsplein 4
PO Box 247
NL-3800 Le Amersfort
The Netherlands

Akzo Nobel
Terhulpsesteenweg 166
Chee de la Hulpe 166
Brussels
Belgium
Tel: +32 (2) 663 5533

Albemarle Asano Corp
16th Floor
Fukoku Seimei Bldg
2-2, Uchisaiwaicho, 2-Chome
Chiyoda-ku, Tokyo 100
Japan
Tel: +81 (3) 5251-0791
Fax: +81 (3)3500-5623

Albemarle Asia Pacific Corp
111 Somerset Road #13-03
Singapore 238164
Singapore
Tel: +65 732-6286
Fax: +65 737-4155

Albemarle Corp
451 Florida St
Baton Rouge, LA
70801-1785
USA
Tel: +1 (225) 388-7402
Fax: +1 (225) 388-7848

Albemarle SA
Parc Scientifique Einstein
Rue du Bosquet 9
B-1348 Louvain La Neuve Sud
Belgium
Tel: +32 (10) 48-1711
Fax: +32 (10) 48-1717

Albright & Wilson Americas Inc
4851 Lake Brook Dr
PO Box 4439
Glen Allen, VA 23060
USA
Tel: +1 (804) 968-6300
Fax: +1 (804) 968-6385

Albright & Wilson Ltd
PO Box 3
210-222 Hagley Rd
West Oldbury
W Midlands B68 ONN
England
Tel: +44 (0121) 429 4942
Fax: +44 (0121) 420 5151

Alcon Japan Ltd
Koraku Kokusai Bldg
1-5-3, Koraku, Bunkyo-ku
Tokyo 112
Japan
Tel: +81 (3) 3812-7881
Fax: +81 (3)3812-0188

Alcon Laboratories
PO Box 6600
6201 South Freeway
Fort Worth, TX 76115
USA
Tel: +1 (817) 293 0450

Alfa Wassermann SpA
Viale Sarca 223
20173 Milano
Italy
Tel: +39 (02) 64222-310

Allchem Industries
6010 NW First Place
Gainesville, FL 32607
USA
Tel: +1 (352) 378-9696
Fax: +1 (352) 338-0400

Allen & Hanbury
Direct Inquiries to Glaxo Wellcome

Allergan Herbert
2525 DuPont Dr
Irvine, CA 92713
USA
Tel: +1 (714) 246-4500
Fax: +1 (714) 246-6987

Allergan Inc
2525 Dupont Dr
PO Box 19534
Irvine, CA 92623-9534
USA
Tel: +1 (714) 246-4500
Fax: +1 (714) 246-6987

Alliance Pharm Corp
3040 Science Pk Dr
San Diego, CA 92121
USA
Tel: +1 (858) 410-5200
Fax: +1 (858) 410-5201

Alpha 1 Biomedicals Inc
Two Democracy Center
6903 Rockledge Dr
Bethesda, MD 20817-1129
USA
Tel: +1 (301) 564-4400
Fax: +1 (301) 564-4424

Altana Inc
60 Baylis Rd
Melville, NY 11747
USA
Tel: +1 (516) 454-7677
Fax: +1 (516) 454-0732

American Cyanamid
5 Garret Mountain Plaza
West Patterson, NJ 07470
USA
Tel: +1 (973) 357-3100

American Home Products
Five Giralda Farms
Madison, NJ 07940
USA
Tel: +1 (973) 660-5000
Fax: +1 (973) 660-5771

American Hospital Supply
20 Wiggins Ave
Bedford, MA 01730
USA
Tel: +1 (781) 275-1100

Amersham Corp
2636 South Clearbrook Dr
Arlington Heights, IL 60005
USA
Tel: +1 (847) 593-6300
Fax: +1 (847) 593-8075

Amersham International plc
Amersham Place
Little Chalfont
Amersham
Bucks HP7 9NA
England
Tel: +44 (01494) 544000

Amgen Inc
Amgen Center
Thousand Oaks, CA
91320-1799
USA
Tel: +1 (805) 447-1000
Fax: +1 (805) 447-1010

Amylin Pharmaceuticals Inc
9373 Town Center Dr
San Diego, CA 92121
USA
Tel: +1 (858) 552-2200
Fax: +1 (858) 552-2212

Anaquest
Address Unknown

Angelini Francesco
Address Unknown

Angelini Group, Italy
Viale Amelia 70
00181 Rome
Italy
Tel: +39 (06) 78053-1
Fax: +39 (06) 78053-291

Angelini Pharmaceuticals Inc
70 Grande Ave
River Edge, NJ 07661
USA
Tel: +1 (201) 489-4100

Anphar
Address Unknown

Anphar-Rolland
BP 203
91007 Evry Cedex
France
Tel: +33 (1) 64 97 20 30
Fax: +33 (1) 64 97 05 84

Antibiotice SA
1 Valea Lupului Street
Lasi 6600
Romania
Tel: +40 (32) 211010
Fax: +40 (32) 211020

Apothecon
Direct Inquiries to
Bristol-Myers Squibb Co

Apothekernes
Direct Inquiries to ASTRA USA Inc

Arizona
1001 E Business 98
Panama City, FL 32401
USA
Tel: +1 (850) 785-6700
Fax: +1 (850) 785-2203

Armour Pharmaceuticals Co Ltd
St Leonards Road
Eastbourne
East Sussex BN21 3YG
England
Tel: +44 (01323) 410200

Asahi Chem Industry
Lyoner Str 44-48
D-60528 Frankfurt
Germany

Ascher, BF & Co
15501 W 109th St
PO Box 717
Shawnee Mission, KS 66201
USA
Tel: +1 (913) 888-1880

Asta Chemische Fabrik
Direct Inquiries to ASTA Medica

Asta Medica AB
Kemistvagen 17
SE-18379 Taby
Sweden

Asta Medica AG
Weissmullerstr 45
D-60314 Frankfurt am Main
Germany
Tel: +49 69 400101
Fax: +49 69 40012740

ASTA Medica Inc
Continental Plaza, Tower 1
401 Hackensack Ave
Hackensack, NJ 07601
USA
Tel: +1 (201) 525-2680
Fax: +1 (201) 488-8595

ASTA Medica Ltd
168 Cowley Road
Cambridge CB4 0DL
England
Tel: +44 (01223) 423434
Fax: +44 (01223) 420943

Asta-Werke AG
Direct Inquiries to Asta Medica

Astra Chemicals Ltd
Direct Inquiries to AstraZeneca

Astra Draco AB
BO Box 34
Lund SE-221 00
Sweden
Tel: +46 (46) 336000

Astra Hässle AB
Karragatan 5
Molndal SE 431 83
Sweden
Tel: +46 (31) 7761000

Astra Pharmaceuticals Ltd
Home Park Estate
King's Langley,
Herts WD4 8DH
England
Tel: +44 (01923) 266191
Fax: +44 (01923) 260431

Astra USA Inc
Direct Inquiries to Astra Zeneca

AstraZeneca
1800 Concord Pike
PO Box 15437
Wilmington, DE 19850
USA
Tel: +1 (302) 886-3000
Fax: +1 (302) 886-2972

Athena Neurosciences Inc
800 Gateway Blvd
S. San Francisco, CA 94080
USA
Tel: +1 (650) 877-0900
Fax: +1 (650) 877-8370

Atrix Laboratories
2579 Midpoint Dr
Fort Collins, CO 80525-4417
USA
Tel: +1 (970) 482-5868
Fax: +1 (970) 482-9735

Ayerst
Direct Inquiries to
Wyeth-Ayerst Laboratories

Ayrton Saunders plc
34 Hanover Street
Liverpool
Merseyside
England

Bacillofabrik Dr Bode & Co
Address Unknown

BASF Corp
3000 Continental Dr
Mt Olive, NJ 07828
USA
Tel: +1 (973) 426-2800
Fax: +1 (973) 426-2810

Basic Inc
Address Unknown

Battle Hayward & Bower Ltd
Crofton Drive
Allenby Rd Industrial Estate
Lincoln
Lincs LN3 4NP
England
Tel: +44 (01522) 529206

Bausch & Lomb Pharmaceuticals Inc
One Bausch & Lomb Place
Rochester, NY 14604
USA
Tel: +1 (716) 338-6000

Bausch & Lomb Vision Care Division
1400 N Goodman St
Tampa, FL 33637
USA
Tel: +1 (813) 975-7700

Baxter Healthcare Corp Hyland Div
One Baxter Parkway
Deerfield, IL 60015
USA
Tel: +1 (847) 948-4731

Baxter Healthcare Systems
One Baxter Parkway
Deerfield, IL 60015
USA
Tel: +1 (847) 948-4731

Bayer AG
Werk Leverkusen
D-51368 Leverkusen
Germany
Tel: +49 214 301
Fax: +49 214 306 6328

Bayer Animal Health
12707 Shawnee Mission Pk
PO Box 390
Shawnee Mission, KS 66201
USA
Tel: +1 (913) 631-4800

Bayer Corp
Pharmaceutical Div
400 Morgan Lane
West Haven, CT 06516
USA
Tel: +1 (203) 937-2000

BDH Laboratory Supplies
Broom Road
Parkstone
Poole
Dorset BH15 1TD
England
Tel: +44 (01202) 660444
Fax: +44 (01202) 666856

Becton Dickinson Microbiology Systems
1 Becton Dr
Franklin Lakes, NJ 07417
USA
Tel: +1 (201) 847-6800

Beecham Group plc
Four New Horizons Court
Harlequin Ave
Brentford
Middx TW8 9EP England
Tel: +44 (020) 8975 2000

Beecham Research Labs,
Direct Inquiries Beecham Group plc

Beiersdorf AG
Aliothstr 40
CH-4142 Münchenstein 2
Switzerland
Tel: +41 (61) 415-6111
Fax: +41 (61) 415-6332

Beiersdorf AG
Unnastr 48
D020245 Hamburg
Germany
Tel: +49 40 49090
Fax: +49 40 49093434

Beiersdorf Inc
Wilton Corporate Center
187 Danbury Rd
Wilton, CT 06897
USA
Tel: +1 (203) 563-5800
Fax: +1 (203) 563-5895

Beiersdorf NV
Boulevard Industriel 30
B-1070 Bruxelles
Belgium
Tel: +32 (2) 526-5211
Fax: +32 (2) 526-5219

Beiersdorf Ltd
Yeomans Drive, Blakelands
Milton Keynes
Bucks MK14 5LS
England
Tel: +44 (01908) 211333
Fax: +44 (01908) 211555

Benz Research and Dev Corp
6447 Parkland Dr
PO Box 1839
Sarasota, FL 34230-1839
USA
Tel: +1 (941) 758-8256

Berk Pharmaceuticals Ltd
Brampton Road
Eastbourne
East Sussex BN22 9AG
England
Tel: +44 (01323) 501111

Berlex Laboratories Inc
300 Fairfield Rd
Wayne, NJ 07470-7358
USA
Tel: +1 (973) 694-4100

Bilhuber
Address Unknown

BioCryst Pharmaceuticals Inc
2190 Parkway Lake Dr
Birmingham, AL 35244
USA
Tel: +1 (205) 444-4600

BioDevelopment Corp
8180 Greensboro Dr #1000
McLean, VA 22102
USA

Biofarma A/S
Naverland 22
DK-2600 Glostrup
Denmark
Tel: +45 4 327-0313

Biona A/S
DK-2860 Soeborg
Denmark
Tel: +45 3 969-2400
Fax: +45 3 969-2199

Bioproject
30, rue des Francs-Bourgeois
75003 Paris
France
Tel: +33 (4) 42 71 71 16
Fax: +33 (4) 42 71 39 56

Biorex
PO Box 348
8201Vesprem-Szabadsapuszta
Hungary
Tel: +36 88-421-629
Fax: +36 88-429-237

Biorex Laboratories Ltd
2 Crossfield Chambers
Gladbeck Way
Enfield,
Middx EN2 7HT
England
Tel: +44 (020) 8366 9301

Boehringer Ingelheim Ltd
Ellesfield Avenue
Bracknell
Berks RG12 8YS
England
Tel: +44 (01344) 424600

Boehringer Ingelheim Pharmaceuticals Inc
900 Ridgebury Rd
Ridgefield Park, CT
06877-0103
USA
Tel: +1 (203) 798-9988

Boehringer Ingelheim GmbH
Binger Str 173
D-55216 Ingelheim am Rhein
Germany
Tel: +49 61 3277 5063
Fax: +49 61 3277 4225

Boehringer Mannheim GmbH
Simpson Parkway
Kirton Campus
Livingston
West Lothian EH54 7BH
England
Tel: +44 (01589) 412512

Boots Company, The
1 Thane Road West
Nottingham
Oxon NG2 3AA
England
Tel: +44 (01602) 506111

Bottu
20, avenue Raymond Aron
92165 Antony Cedex
France
Tel: +33 140 91 61 23

Bracco Diagnostics Inc
107 College Road E
Princeton, NJ 08540
USA

Bristol-Myers Nutritional Group
725 E Main
Zeeland, MI 49464-0136
USA
Tel: +1 (616) 748-7100

Bristol-Myers Squibb Co
PO Box 4000
Princeton, NJ 08540
USA
Tel: +1 (609) 921-4000

Bristol-Myers Squibb Europe
Le Grande Arche Nord
Paris La Défense Cedex
92044 Paris
France
Tel: +33 (1) 4090 6000
Fax: +33 (1) 4090 6100

Bristol-Myers Squibb HIV Products
345 Park Ave
New York, NY 10154-0000
USA
Tel: +1 (212) 546-2856

Bristol-Myers Squibb Pharmaceutical Res and Dev
1 Squibb Drive
New Brunswick, NJ 08901
USA
Tel: +1 (201) 519-2000

Bristol Myers Squibb Pharmaceuticals Ltd
Bristol Myers Squibb House
141-149 Staines Rd
Hounslow, Middx TW3 3JA
England
Tel: +44 (020) 8572 7422

British Biotechnology Ltd
Watlington Rd
Oxford OX4 5LY
England
Tel: +44 (01865) 748747
Fax: +44 (01865) 781047

British Drug Houses
Direct Inquires to Merck

Brocades Ltd
Brocades House,
Pyrford Road
West Byfleet, Weybridge,
Surrey KT14 6RA
England
Tel: +44 (01932) 342291

Brocades-Stheeman & Pharmacia
Direct Inquiries to Pharmacia & Upjohn

Broemmel Pharmaceuticals
3M Pharmaceuticals
3M Center, 275-3W01
St Paul, MN 55133-3275
USA

Buckeye Technologies
1001 Tillman St
PO Box 8407
Memphis, TN 38108
USA
Tel: +1 (901) 320-8100

Burroughs Wellcome
Direct Inquiries to
GlaxoWellcome

Byk Gulden Lomberg GmbH
Byk-Gulden-Str 2
Postfach 100310
7750 Konstanz
Germany
Tel: +49 7531 84 0
Fax: +49 7531 84 2474

C H Boehringer Sohn
Direct Inquiries to Boehringer
Ingelheim

CERM
Address Unknown

CM Industries
Erregierre Industria Chimica
SpA
Via Francesco Baracca, 57
24060 San Paolo D'Argon
(BG)
Italy
Tel: +39 (03) 595022

Cadus Pharmaceutical Corp
777 Old Saw Mill River Rd
Tarrytown, NY 10591-6705
USA
Tel: +1 (914) 345-3344
Fax: +1 (914) 345-3565

Calanda Stiftung
Address Unknown

California Research Co
Address Unknown

Callery Chemical
1420 Mars-Evans City Rd
Evans City, PA 16033
USA
Tel: +1 (412) 967-4141
Fax: +1 (412) 967-4140

Cambridge NeuroScience Inc
One Kendall Square
Bldg 700
Cambridge, MA 02139
USA
Tel: +1 (617) 225-0600
Fax: +1 (617) 225-2741

Camillo-Corvi
Address Unknown

Carbide & Carbon Chem
Address Unknown

Carlo Erba Reagenti
Strada Rivoltana KM 6/7
20090 Rodano (Mi)
Italy
Tel: +39 (02) 9523 1
Fax: +39 (02) 95235904

Carrington Laboratories Inc
2001 Walnut Hill Lane
Irving, TX 75038
USA
Tel: +1 (800) 527-5216
Fax: +1 (972) 518-1020

Carter-Wallace
PO Box 1001
Cranbury, NJ 08512
USA
Tel: +1 (609) 655-6000

Cassella AG
Hanauer Landstrasse 526
D-60386 Frankfurt
Germany
Tel: +49 (69) 4109 01
Fax: +49 (69) 4109 2650

CBD Corp
Address Unknown

Cell Therapeutics Inc
201 Elliott Ave West, Ste 400
Seattle, WA 98119-4230
USA
Tel: +1 (206) 282-7100
Fax: +1 (206) 284-6206

Centeon LLC
1020 First Ave
King of Prussia, PA 19406
USA
Tel: +1 (610) 878-4000
Fax: +1 (610) 878-4009

Centocor Inc
200 Great Valley Parkeway
Malvern, PA 19355
USA
Tel: +1 (610) 651-6000
Fax: +1 (610) 889-4701

Centre d'Études l'Ind Pharm
Address Unknown

Cetus Corp
4560 Horton St
Emeryville, CA 94608-2997
USA
Tel: +1 (510) 653-5948

Chantal Pharmaceutical Corp
12121 Wilshire Blvd 1120
Los Angeles, CA 90025-1123
USA
Tel: +1 (310) 207-1950
Fax: +1 (310) 826-4214

Chantereau
Address Unknown

Chem Werke Albert
Address Unknown

Chem-Pharm Fabrik
Bahnhofstr 33-35 + 40
73033 Goeppingen
Germany
Tel: +49 7161 676-0
Fax: +49 7161 676-298

Chemex Pharmaceuticals
660 White Plains Rd
Ste 400
Tarrytown, NY 10591
USA
Tel: +1 (914) 332-8633

Chemiewerk Homburg
Address Unknown

Chemo Puro
Address Unknown

Chemoterapico
Address Unknown

Chimie et Atomistique
Address Unknown

Chinoin
1325 Budapest, Pf 110
H-1045 Budapest
Hungary
Tel: +36 (1) 169-0900
Fax: +36 (1) 169-0293

Chiron Corp
4560 Horton St
Emerville, CA 94608-2916
USA
Tel: +1 (510) 655-8730
Fax: +1 (510) 655-9910

Christiaens SA
Address Unknown

Chugai Pharmaceutical Co Ltd
Mulliner House, Flanders Rd
Turnham Green
London, W4 1NN
England
Tel: +44 (020) 8987-5600

CIBA plc
Direct Inquiries to Novartis

CIBA Vision AG
Grenzstr 10
CH-8180 Buelach
Switzerland
Tel: +41 (084) 880-8488
Fax: +41 (084) 880-8489

CIBA Vision Corp
11460 Johns Creek Parkway
Duluth, GA 30097-1556
USA

CIBA Vision Ltd
Park West
Royal London Park
Flanders Rd, Hedge End
Southampton,
Hants SO30 2LG
England
Tel: +44 (01489) 785580
Fax: +44 (01489) 786802

CIBA Vision Optics NL
4 Prinsenkade
NL-4811VB Breda
The Netherlands
Tel: +31 76-5245600
Fax: +31 76-5245620

Ciba-Geigy Corp
Direct Inquiries to Novartis

Cilag-Chemie Ltd
Saunderton
High Wycombe,
Bucks HP14 4HJ
England
Tel: +44 (01494) 563541

CIS-US Inc
10 DeAngelo Dr
Bedford, MA 01730
USA
Tel: +1 (781) 275-7120
Fax: +1 (781) 275-2634

CK Witco (Europe) SA
7, rue du Pre-Bouvier
CH-1217 Meyrin
Switzerland
Tel: +41 (22) 989-2392

CK Witco Asia Pacific Pte Ltd
12 Science Park Dr
118225 Singapore
Singapore
Tel: +65 770-5146

CK Witco Canada Ltd
565 Coronation Dr
West Hill, ON M1W 2K3
Canada
Tel: +1 (416) 284-6077

CK Witco Chemical Corp
One American Lane
Greenwich, CT
USA
Tel: +1 (203) 552-2747
Fax: +1 (203) 552-2882

CK Witco Chemical Ltd
Direct Inquires to
CK Witco (Europe) SA

Clin-Byk France
593, route de Boissise
77350 Le Mee-Sur-Seine
France
Tel: +33 (1) 64 41 22 22
Fax: +33 (1) 64 41 22 00

Clin-Midy
9, rue du President Allende
94256 Gentilly Cedex
France
Tel: +33 (1) 40 73 40 73
Fax: +33 (1) 40 73 93 00

CNRS
16, rue Pierre et Marie Curie
75005 Paris
France
Tel: +33 (1) 42 34 94 00
Fax: +33 (1) 43 26 87 23

Colgate-Palmolive
One Colgate Way
Canton, MA 02021
USA
Tel: +1 (908) 878-7500

Consiglio Nazionale delle Ricerche
Via Tiburtina, 770
I-00159 Rome
Italy
Tel: +39 (06) 49932538
Fax: +39 (06) 49932440

Continental Pharma Inc
Address Unknown

Cook Imaging Corp
927 S Curry Pike B
Bloomington, IN 47403
USA
Tel: +1 (812) 333-0887
Fax: +1 (812) 332-3079

Cook-Waite Labs Inc
Direct Inquires to Eastman Kodak Co

Cooper Companies Inc, The
10 Faraday
Irvine, CA 92618-1850
USA
Tel: +1 (949) 597-4700
Fax: +1 (949) 597-0662

Cooper Vision Inc
200 Willow Brook Office Park
Fairport, NY 14450
USA

Corbiere
Address Unknown

Cortech Inc
376 Main St
PO Box 74
Bedminster, NJ 07921
USA
Tel: +1 (908) 234-1881

Council of Scientific and Industrial Research, New Delhi
Address Unknown

Crinos
Piazza XX Settembre, 2
22079 Villa Guardia (CO)
Italy
Tel: +39 (031) 385111
Fax: +39 (031) 481784

Crookes Healthcare Ltd
1 Thane Road West
Nottingham
NG2 3AA
England
Tel: +44 (01602) 506111

Cutter Laboratories
Direct Inquiries to Bayer Corp

Cypros Pharmaceutical Corp
2714 Loker Ave West
Carlsbad, CA 92008
USA
Tel: +1 (760) 929-9500
Fax: +1 (760) 929-8038

Cytogen Corp
600 College Rd
E Princeton, NJ 08540
USA
Tel: +1 (609) 987-8270
Fax: +1 (609) 951-9298

Daiichi Pharmaceutical Co Ltd
3-14-10, Nihonbashi
Chuo-ku, Tokyo 103
Japan
Tel: +81 (3) 3272-0611
Fax: +81 (3) 3272-8427

Daiichi Pharmaceutical Corp
11 Philips Parkway
Montvale, NJ 07645
USA
Tel: +1 (201) 573-7000

Daiichi Seiyaku
3-14-10, Nihonbashi
Chuo-ku, Tokyo 103
Japan
Tel: +81 (3) 3272-0611
Fax: +81 (3) 3272-8427

Dainippon Pharmaceutical
2-6-8, Dosho-machi
Chuo-ku, Osaka 541
Japan
Tel: +81 (6) 6203-5321
Fax: +81 (6) 6203-6581

Dautreville & Lebas
Address Unknown

Davis & Geck Medical Device Div
Direct Inquiries to
Wyeth-Ayerst Laboratories

DDSA Pharmaceuticals Ltd
Address Unknown

DeAngeli
Address Unknown

Degussa Ltd
Direct Inquires to
Degussa-Huls AG

Degussa-Huls AG
65 Challenger Rd
Ridgefield Park, NJ 07660
USA
Tel: +1 (201) 641-6100
Fax: +1 (201) 807-3183

Degussa-Huls AG
Headquarters
Weissfrauenstrasse 9
D-60311 Frankfurt am Main
Germany
Tel: +49 (69) 218-3618
Fax: +49 (69) 218-3849

Delagrange
1, avenue Pierre Brossolette
91380 Chilly Mazarin
France
Tel: +33 (1) 69 79 77 77
Fax: +33 (1) 69 79 75 75

Delandale Labs Ltd
16, rue Henri Regnault
La Defense 6
92400 Courbevoie
France
Tel: +33 (1) 45 37 55 55
Fax: +33 (1) 49 00 02 93

Dermik Labs Inc
Direct Inquires to
Rhône-Poulenc Rorer

Deutsche Hydrierwerke
Address Unknown

Dey Laboratories
2751 Napa Valley Corp Dr
Napa, CA 92558
USA
Tel: +1 (707) 224-3200
Fax: +1 (707) 224-3235

Dickinson, E E, Co
2 Enterprise Dr
Shelton, CT 06484-4666
USA
Tel: +1 (860) 388 3952

Diosynth BV
Vlijtseweg 130
PO Box 407
NL-7300 AK Apeldoorn
The Netherlands
Tel: +31 (55) 5286144
Fax: +31 (55) 5218808

Diosynth France SA
92821 Puteaux Cedex
France
Tel: +33 (1) 55 23 51 75

Dista Products Ltd
PO Box 25768
Alexandria, VA 22313
USA
Tel: +1 (800) 545-5979

Doak Pharmacal Co Inc
67 Sylvester St
Westbury, NY 11590-4910
USA
Tel: +1 (516) 333-7222

Dome/Hollister-Stier
Direct Inquiries to Bayer plc

Donau Pharm
Address Unknown

Dott Inverni & Della Beffa
Address Unknown

Dow Chemical USA
1803 Bldg
Midland, MI 48674
USA
Tel: +1 (517) 832-1000

Dumex Canada
104 Shorting Road
Toronto, ON M1S 3S4
Canada
Tel: +1 (416) 299-4003
Fax: +1 (416) 299-4912

Dumex USA
2250 Military Rd
Tonawanda, NY 14150
USA
Tel: +1 (800) 463-0106
Fax: +1 (716) 842-0707

DuPont Pharmaceutical Co
Experimental Sta 400/2413
PO Box 80400
Wilmington, DE 19880-0400
USA
Tel: +1 (302) 992-5000

DuPont Pharmaceuticals Ltd
Wedgwood Way
Stevenage
Herts SG1 4QN
England
Tel: +44 (01438) 842500

DuPont-Merck Pharmaceuticals
Direct Inquiries to DuPont Pharmaceuticals

DuPont-Merck, Radiopharmaceutical Div
Direct Inquiries to DuPont Pharmaceuticals

Dura Pharmaceuticals Inc
7475 Lusk Blvd
San Diego, CA 92121
USA
Tel: +1 (619) 457-2553

Dynamit Nobel AG
Kaiserstr 1
Postfach 12 61
53839 Troisdorf
Germany
Tel: +49 (22) 41 89-0
Fax: +49 (22) 41 89-15 40

E Fougera & Co
60 Baylis Road
Melville, NY 11747
USA
Tel: +1 (516) 454-6996
Fax: +1 (516) 756-7017

E Geistlich Sohne
CH-6110 Wolhusen
Switzerland
Tel: + 41 710333

E I Du Pont de Nemours Inc
1007 Market Street
Wilmington, DE 19898
USA
Tel: +1 (302) 774-7573

E Merck
Frankfurter Str 250
D-64293 Darmstadt
Germany
Tel: +49 61 51 72 0
Fax: +49 61 51 72 2000

ERASME
Address Unknown

Eastman Chemical Co
Fine Chemicals
PO Box 431
Kingsport, TN 37662
USA
Tel: +1 (423) 229-8124
Fax: +1 (423) 229-8133

Eastman Kodak
2/15/KO- Mailstop: 00539
343 State St
Rochester, NY 14650
USA
Tel: +1 (716) 724-4513
Fax: +1 (716) 724-0964

Eaton Labs
Address Unknown

ECR Pharmaceuticals
3981 Deep Rock Rd
PO Box 71600
Richmond, VA 23233-0141
USA
Tel: +1 (804) 527-1950

EGYT
Address Unknown

Eisai Co Ltd
4-6-10, Koishikawa
Bunkyo-ku, Tokyo 112-88
Japan
Tel: +81 (3) 3817-3700
Fax: +81 (3) 3811-3305

Eisai Corp of North Am
300 Frank W Burr Blvd
Teaneck, NJ 07666
USA
Tel: +1 (201) 692-9160

Eisai Merrimack Valley Laboratories Inc
100 Federal Street
Andover, MA 01810-0103
USA
Tel: +1 (978) 989-9911

Elan Pharmaceutical Research Corp
Lincoln House
Lincoln Place
Dublin 2
Ireland
Tel: +353 1 709-4000
Fax: +353 1 671-0920

Eli Lilly & Co
Lilly Corporate Center
Indianapolis, IN 46285
USA
Tel: +1 (317) 276-2000

Eli Lilly (Suisse) SA
PP Box 580
CH -1214 Venier/Geneva
Switzerland
Tel: +41 22-30-60-401

Eli Lilly Asia Pacific Pte Ltd
583 Orchard Road
#12-01/04
Forum
Singapore 238884
Tel: +65 732-2066

Eli Lilly Asia Inc
Room 408, Man Po International Center
660 Xin Hua Rd
Shanghai 200052
PR China
Tel: +86 21-6282-6008

Eli Lilly GmbH
Barichgasse 40-42
A-1030 Vienna
Austria
Tel: +43 (1) 711-780

Eli Lilly Group Ltd
Kingsclere Road
Basingstoke
Hants RG1 2XA
England
Tel: +44 (01256) 473241

Eli Lilly International Corporation
Lilly House
13 Hanover Square
London W1R OPA
England
Tel: +44 (020) 7409 4839

Eli Lilly Japan KK
Sannomiya Plaza Bldg
7-1-5, Isogami-dori
Chuo-ku, Kobe 651
Japan
Tel: +81 (8178) 242-9000

Elizabeth Arden
Direct Inquires to Eli Lilly

Elkins-Sinn
2 Esterbrook Lane
Cherry Hill, NJ 08002-4009
USA
Tel: +1 (610) 688-4400

EM Industries Inc
Direct Inquiries to Merck
Hawthorne, NY 10532
USA
Tel: +1 (914) 592-4660
Fax: +1 (914) 592-9469

Endo Pharmaceuticals Inc
220 Lake Dr
Newark, DE 19702
USA
Tel: +1 (800) 462-3636
Fax: +1 (877) 329-3636

Enzon Inc
40 Kingsbridge Rd
Piscataway, NJ 08854
USA
Tel: +1 (732) 980-4500
Fax: +1 (732) 980-5911

Enzypharm BV
Industrieweg 17
NL-3762 EG Soest
The Netherlands
Tel: +31 (35) 6030051
Fax: +31 (35) 6029962

Epoch Pharmaceuticals Inc
1725 220th St SE, Ste 104
Bothell, WA 98021
USA
Tel: +1 (425) 485-8566

Eprova AG
Im Laternenacker 5
CH -8200 Schaffhausen
Switzerland
Tel: +41 (52) 630 7272
Fax: +41 (52) 630-7255

Esta Med Labs
Address Unknown

Esteve Group
Av Mare de Deu de
Montserrat, 221
8041 Barcelona
Spain
Tel: +34 93 446-6053
Fax: +34 93 433-0072

Esteve Group
Av Mare de Deu de
Montserrat, 12
8024 Barecelona
Spain
Tel: +34 93 284-6000
Fax: +34 93 284-6850

Ethicon Inc
Route 22
Somerville, NJ 08876
USA
Tel: +1 (908) 218-0707

Ethyl Corp
330 South Fourth St
PO Box 2189
Richmond, VA 23218
USA
Tel: +1 (804) 788-5000
Fax: +1 (804) 788-5688

Evans Medical Ltd
Evans House
Regent Park, Kingston Rd
Leatherhead
Surrey KT22 7PQ
England
Tel: +44 (01372) 364000

F Hoffmann-LaRoche Ltd
CH-4070 Basel
Switzerland
Tel: +41 (61) 688 88 88
Fax: +41 (61) 688 27 75

Farbenfabriken Bayer AG
Address Unknown

Farmitalia Carlo Erba Ltd
Italia House
23 Grosvenor Rd
St Albans
Herts AL1 3AW
England
Tel: +44 (01727) 40041

Farmitalia, Societa Farmaceutici
Address Unknown

Farmos Group Ltd
PO Box 425
FIN-20101 Turku
Finland
Tel: +358 21 66 22 11

Ferlux-Chemie
24, Avenue d'Aubiere
63804 Cournon d'Auvergne
France
Tel: +33 (4) 73 84 21 84
Fax: +33 (4) 73 84 21 80

Fermenta Animal Health Co
15th & Oak Street
PO Box 338
Elwood, KS 66024
USA

Ferrer
Address Unknown

Ferring Pharmaceuticals Inc
120 White Plains Rd
Tarrytown, NY 10591
USA
Tel: +1 (888) 337-7464

Ferrosan A/S
Corporate Headquarters
Sydmarken 5
DK-2860 Soeborg
Denmark
Tel: +45 3 969-2111
Fax: +45 3 969-6518

Ferrosan AB
Grynbodgatan 14
SE-21 33 Malmo
Sweden
Tel: +46 (40) 6607070
Fax: +46 (40) 6607089

Ferrosan AB
Kutojantie 11 (Vanvarsvagen)
FIN-02630 Espoo
Finland
Tel: +358 9 525 9050
Fax: +358 9 520 236

Ferrosan Ltd
69 Monmouth Street
London WC2H 9DG
England
Tel: +44 (020) 7240-2122
Fax: +44 (020) 7240-2188

Ferrosan Norge AS
Grini Naeringspark 1
1361 Osteras
Norway
Tel: +47 (6) 714-9505
Fax: +47 (6) 714-9530

Fidia Pharmaceuticals
Address Unknown

Fisons Pharmaceuticals Div
Rhône Poulenc Rorer
Mailstop 4C29, Box 5094
Collegeville, PA 19426-0998
USA
Tel: +1 (610) 454-8110

Fisons plc
Fison House
Princes St
Ipswich
Suffolk IP1 1QH
England
Tel: +44 (01473) 232525

Flint-Eaton
Address Unknown

FMC Corp, Pharm Div
1735 Market St
Philadelphia, PA 19103
USA
Tel: +1 (215) 299-6534
Fax: +1 (215) 299-6821

Forest Pharmaceuticals Inc
13600 Shoreline Dr
St Louis, MO 63045
USA
Tel: +1 (800) 678-1605
Fax: +1 (314) 493-7450

Fujirebio Inc
2-7-1, Nishi-shinjuku
Shinjuku-ku
Tokyo 163-07
Japan
Tel: +81 (3) 3348-0691
Fax: +81 (3) 3342-6220

Fujisawa Pharmaceuticals Co Ltd
3-4-7, Doso-machi
Chuo-ku, Osaka 541
Japan
Tel: +81 (6) 6202-1141
Fax: +81 (6) 6222-4988

Fujisawa Pharmaceuticals USA Inc
3 Parkway North Center
Deerfield, IL 60015
USA
Tel: +1 (708) 317-0600

GAF
Direct Inquiries to Intl Specialty Products Inc

Galderma Canada Inc
7300 Warden Ave, Ste 210
Markham, ON L3R 9Z6
Canada
Tel: +1 (905) 944-0717
Fax: +1 (905) 944-0790

Galderma Laboratories Inc
3000 Alta Mesa Blvd
Ste 300
Fort Worth, TX 76133
USA
Tel: +1 (817) 263-2600
Fax: +1 (817) 263-2609

Gea A/S
Holger Danskes Vej 89
DK-2860 Frederiksberg
Denmark
Tel: +45 38 34 42 42
Fax: +45 38 34 11 23

Gedeon Richter Chem Works
Gyomroi ût 19-21
H-1103 Budapest
Hungary
Tel: +36 (1) 261 2199

Gelatin Products
Address Unknown

GenDerm
Medicis Pharmaceutical Corp
4343 E Camelback Rd
Phoenix, AZ 85018
USA
Tel: +1 (602) 808-8800
Fax: +1 (602) 808-0822

Genentech Inc
1 DNA Way
So San Francisco, CA 94080
USA
Tel: +1 (650) 225-1000
Fax: +1 (650) 225-6000

General Aniline
Address Unknown

Genetics Institute Inc
35 Cambridge Park Dr
Cambridge, MA 02140-2325
USA
Tel: +1 (617) 876-1170

Genta Inc
99 Hayden Ave, Ste 200
Lexington, MA
USA
Tel: +1 (781) 860-5150

Genzyme Corp
One Kendal Square
Cambridge, MA 02139
USA
Tel: +1 (617) 252-7500
Fax: +1 (617) 252-7600

Genzyme Ltd
37 Hollands Road
Haverhill
Suffolk CB9 8PU
England
Tel: +44 (01440) 703522

Gerda
6, rue Childebert
69002 Lyon
France
Tel: +33 (4) 72 77 69 19
Fax: +33 (4) 72 77 69 13

Gerot Pharmazeutika
Arnethgasse 3
A-1160 Vienna
Austria
Tel: +43 (1) 485 3505
Fax: +43 (1) 485 8932

Gilead Sciences Inc
333 Lakeside Dr
Foster City, CA 94404
USA
Tel: +1 (650) 574-3000
Fax: +1 (650) 578-9264

Gist-Brocades International
PO 241068
8270 Red Oak Blvd, Ste 401
Charlotte, NC 28217
USA
Tel: +1 (704) 527-9000
Fax: +1 (704) 527-8844

Giuliani SpA
Via Palagi
2-20129 Milano
Italy
Tel: +39 (02) 20541
Fax: +39 (02) 29401341

Givaudan-Roure SA
55, rue de la Voie des Bancs
95100 Argenteuil
France
Tel: +33 (139) 98 15 15
Fax: +33 (139) 82 00 15

Glaxo Labs
Direct Inquiries to Glaxo Wellcome

Glaxo Wellcome Inc
Five Moore Dr
PO Box 13398
Res Triangle Pk, NC 27709
USA
Tel: +1 (919) 248-2100
Fax: +1 (919) 248-7699

Glaxo Wellcome plc
Glaxo Wellcome House
Berkley Ave
Greenford
Middx UB6 0NN
England
Tel: +44 (0171) 4934060

Glenwood Inc
83 N Summit St
Tenafly, NJ 07670-0051
USA
Tel: +1 (201) 569-0050

Glidden Co
1900 Josey Lane
Carrolton, TX 75007
USA
Tel: +1 (214) 417-7400

Goodrich, BF, Co
Specialty Chemicals
9911 Brecksville Rd
Cleveland, OH 44141
USA
Tel: +1 (216) 447-6220
Fax: +1 (216) 447-6760

Goodrich, BF, Co, Europe
Specialty Chemicals
Rue de Verdun/straat 742
B-1130 Brussels
Belgium
Tel: +32 (2) 247-1911
Fax: +32 (2) 247-1990

Grace, WR & Co
Dewey & Almy Chemical Div
5225 Phillip Lee Dr
Altanta, GA 30336
USA
Tel: +1 (404) 691-8646

Greeff, RW & Co, LLC
777 West Putnam Ave
Greenwich, CT 06830
USA
Tel: +1 (203) 532-2900
Fax: +1 (203) 532-2980

Greenwich Pharmaceuticals Inc
501 Office Center Drive
Ft Washington, PA 19034
USA

Grünenthal
Postfach 50 04 414
D-52088 Aachen
Germany
Fax: +49 0241 569-0

Grupo Farmaceutico Almirall SA
Maximo Aguirre 14
480940 Leioa
Spain
Tel: +34 94 4639000
Fax: +34 94 4646110

Gruppo Lepetit SpA
Via Murat 23
I-20159 Milano
Italy
Tel: +39 (2) 27 77 1

Guardian Laboratories
230 Marcus Blvd
PO Box 18050
Hauppauge, NY 11788
USA

Guilford Pharmaceuticals Inc
6611 Tributary St
Baltimore, MD 21224
USA
Tel: +1 (410) 631-6302
Fax: +1 (410) 631-6338

Hamari Chemicals Ltd
1-4-29, Shibajima
Higashiyodogawa-ku
Osaka 533
Japan
Tel: +81 (6) 6322-0191

Helopharm
Address Unknown

Herbert
Direct Inquiries to DuPont Pharmaceuticals

Hercules Inc
1313 North Market St
Wilmington, DE 19894
USA
Tel: +1 (302) 594-5000
Fax: +1 (302) 594-5400

Hermes (GB) Ltd
7-9 Colville Road
London W3 8BL
England
Tel: +44 (020) 7259 5191

Heumann Pharma GmbH
Heideloffstr 18-28
90478 Neurnberg
Germany
Tel: +49 911 430 20
Fax: +49 911 430 24 15

Hexachemie
Address Unknown

Hexcel
Two Stamford Plaza
281 Tresser Blvd
Stamford, CT 06901
USA
Tel: +1 (203) 969-0666
Fax: +1 (203) 358-3977

Heyden Chemical
Address Unknown

Hindustan Antibiotics Ltd
Pune, Maharashtra
India

Hisamitsu Pharmaceutical Co Ltd
408 Tashirio Daikan-machi
Tosu-shi, Saga 841
Japan
Tel: +81 (942) 83 2101
Fax: +81 (942) 83 6119

Hoechst AG
D-65926 Frankfurt am Main
Germany
Tel: +49 69 305-2318
Fax: +49 69 305-83376

Hoechst AG (USA)
3 Park Ave
New York, NY 10016
USA
Tel: +1 (212) 251-8088
Fax: +1 (212) 251-8011

Hoechst Marion Roussel Inc
10236 Marion Park Dr
Kansas City, MO 64137-1405
USA
Tel: +1 (816) 966-4000
Fax: +1 (816) 966-3270

Hoechst Roussel Pharmaceuticals Inc
2110 East Galbraith
Cincinnati, OH 45215
USA
Tel: +1 (513) 948-9111

Hoechst Ltd
Hoechst House
Salisbury Rd
Hounslow
Middx TW4 6JH
England
Tel: +44 (020) 8570 7712

Hoffmann-LaRoche Inc
340 Kingsland St
Nutley, NJ 07110
USA
Tel: +1 (973) 235-5000

Hoffmann-LaRoche Ltd
CH-4070 Basel
Switzerland
Tel: +41 61 688 1111
Fax: +41 61 691 9391

Hokoriku
Address Unknown

Holding Ceresia
Address Unknown

Hommel GmbH
Postfach 1662
59336 Ludinghausen
Germany
Tel: +49 2591 23050
Fax: +49 02591 4413

Hooker Chemical
Direct Inquires to
Occidental Chemical Corp

Hovione
Sete Casas
2674-506 Loures
Portugal
Tel: +351 21 982 9000
Fax: +351 21 982 9388

Hybridon, Inc
155 Fortune Blvd
Milford, MA 01757
USA
Tel: +1 (508) 482-7500
Fax: +1 (508) 482-7510

Hynson, Westcott & Dunning
Charles and Chase Sts
Baltimore, MD 21201
USA

IG Farben
Address Unknown

ISF
Address Unknown

Ibis Therapeutics
2292 Faraday Ave
Carlsbad, CA 92008
USA
Tel: +1 (760) 603-2700

ICI Americas Inc
Concord Plaza
3411 Silverside Rd
Wilmington, DE 19850
USA
Tel: +1 (302) 887-3000

ICI Chemicals and Polymers Ltd
1900 Josey Lane
Carrolton, TX 75007
USA
Tel: +1 (214) 417-7400

ICN Pharmaceuticals Inc
ICN Plaza
3300 Hyland Ave
Costa Mesa, CA 92626
USA
Tel: +1 (714) 545-0100
Fax: +1 (714) 556-0131

IDEC Pharmaceuticals Corp
11011 Torreyana Rd
San Diego, CA 92121
USA
Tel: +1 (619) 550-8500
Fax: +1 (618) 550-8750

Illumina
15817 Bernardo Center Dr
Ste 102
San Diego, CA 92127-2322
USA
Tel: +1 (619) 672-0419
Fax: +1 (619) 672-2325

Ilon Labs
Address Unknown

Immunetech Pharmaceuticals
Direct Inquiries to Dura
Pharmaceuticals

Immunex Corp
51 University St
Seattle, WA 98101
USA
Tel: +1 (206) 587-0430
Fax: +1 (206) 587-0606

Immunomedics Inc
300 American Rd
Morris Plains, NJ 07950
USA
Tel: +1 (973) 605-8200
Fax: +1 (973) 605-8282

Imutec Pharma Inc
Direct Inquiries to Lorus Therapeutics Inc

INDOFINE Chemical Co
PO Box 473
Somerville, NJ 08876
USA
Tel: +1 (908) 359-6778
Fax: +1 (908) 359-1179

Inex Pharmaceuticals Corp
1779 West 75th Avenue
V6P 6P2 Vancouver, BC
Canada
Tel: +1 (604) 264-9959

Innothera
7-9, avenue Francois-Vincent Raspail BP 12
94111 Arcueil Cedex
France
Tel: +33 (1) 46 15 18 00
Fax: +33 (1) 46 63 43 60

Inst Chemioter
Address Unknown

Inst Gentili SpA
Address Unknown

Inst Invest Desarr
Address Unknown

Inst Phys & Chem Res
Address Unknown

Interco Fribourg
Address Unknown

Interferon Sciences Inc
783 Jersey Ave
New Brunswick, NJ
08901-3660
USA
Tel: +1 (732) 249-3250
Fax: +1 (732) 249-6895

International Specialty Products Inc (ISP)
1361 Alps Rd
Wayne, NJ 07470
USA
Tel: +1 (201) 628-4000
Fax +1 (201) 628-4117

Interneuron Pharmaceuticals Inc
1 Ledgemont Center
99 Hayden Ave, Ste 340
Lexington, MA 02173
USA
Tel: +1 (617) 861-8444
Fax: +1 (617) 861-3830

Investigacion Tecnica y Aplicada
Address Unknown

Iolab
2, Central Parc-Avenue Sully Prudhomme
92298 Chatenay Malabry Cedex
France
Tel: +33 (1) 43 50 80 80
Fax: +33 (1) 43 50 96

Irwin, Neissler
Address Unknown

Isis Pharmaceuticals Inc
2292 Faraday Ave
Carlsbad, CA 92008
USA
Tel: +1 (619) 931-9200
Fax: +1 (619) 931-9639

ISP Van Dyk Inc
Address Unknown

Ist Biochim
Address Unknown

Ist De Angeli
Address Unknown

Italfarmaco SpA
Via dei Lavoratori, 54
20092 Cinisello Balsamo (MI)
Italy
Tel: +39 (02) 64432301
Fax: +39 (02) 64432305

Janssen Pharmaceutical Inc
1125 Trenton-Harbourton Rd
PO Box 200
Titusville, NJ 08560
USA
Tel: +1 (609) 730-2000

Janssen Pharmaceutical Ltd
Grove
Wantage
Oxon OX12 0DQ
England
Tel: +44 (01235) 777333

Johnson & Johnson Medical Inc
One Johnson & Johnson Plaza
New Brunswick, NJ 08933
USA
Tel: +1 (732) 524-0400

Johnson & Johnson-Merck Consumer Pharmaceuticals
Camp Hill Rd
Fort Washington, PA 19034
USA

Jouveinal
1, rue des Moissons - BP 100
94265 Fresnes Cedex
France
Tel: +33 (1) 40 96 74 00
Fax: +33 (1) 46 68 16 44

Julian
Address Unknown

Juvantia Pharma Ltd
Tykistokatu 6A
FIN-20520 Turku
Finland
Tel: +358 2 333 7684
Fax: +358 2 333 7680

Kabi Pharmacia Diagnostics
800 Centiennial Ave
Piscataway, NJ 08540
USA

KabiVitrum AB
Direct Inquiries to Pharmacia & Upjohn

Kaken Pharmaceutical Co Ltd
1 Hinode
Urayasu-shi, Chiba 279
Japan
Tel: +81 (473) 90-6140
Fax: +81 (473) 90-6161

Kakenyaku Kako
Address Unknown

Kali-Chemie
Hans-Bockler-Allee 20
D-30173 Hannover
Germany
Tel: +49 511 8571
Fax: +49 511 282126

Kalle BV
Wetering 20
NL-6002 SM Weert
The Netherlands
Tel: +31 (495) 45 84 58
Fax: +31 (495) 45 87 44

Kanebo Cosmetics Ltd
Bone Lane
Newbury
Berks RG14 5TD
England
Tel: +44 (01635) 46362

Kanebo Pharmaceuticals Ltd
1-3-12, Motoakasaka
Minato-ku, Tokyo 107
Japan
Tel: +81 (3) 5411-3530
Fax: +81 (3) 5411-3568

Kefalas A/S
Address Unknown

Kendall McGaw Inc
2525 McGaw Ave
Irvine, CA 92614
USA
Tel: +1 (949) 660-2000

Key Pharmaceuticals
Direct Inquiries to
Schering-Plough

Keystone Chemurgic
Address Unknown

Kissei
Address Unknown

Klinge Pharma GmbH
Berg-am-Laim Str 129
81673 Munich
Germany
Tel: +49 69 4544-01
Fax: +49 69 4544-1329

Knoll Ltd
Fleming House
71 King St
Maidenhead
Berks SL6 1DU
England
Tel: +44 (01628) 776360

Knoll Pharmaceutical Co
3000 Continental Dr, North
Mt Olive, NJ 07828-1234
USA
Tel: +1 (800) 524-2474

Kobayashi Pharmaceutical Co Ltd
2-7-16, Shoji-higashi
Ikuno-ku, Osaka 544
Japan
Tel: +81 (6) 6754-9522

Kowa Chemical Industries Co Ltd
6-1-1, Heiwajima
Ohta-ku, Tokyo 143
Japan
Tel: +81 (3) 3767-3561
Fax: +81 (3) 3767-3917

Kreussler, Chemische-Fabrik
Rheingaustr 87-93
D-65203 Wiesbaden
Germany
Tel: +49 611 92710
Fax: +49 611 9271-111

KV Pharmaceutical
2503 S Hanley Rd
Saint Louis, MO 63144-2555
USA
Tel: +1 (314) 645-6600

Kyorin Pharmaceutical Co Ltd
2-5, Kanda Surugadai
Chiyoda-ku, Tokyo 101
Japan
Tel: +81 (3) 3293-3411
Fax: +81 (3) 3293-6588

Kyowa Hakko Kogyo Co Ltd
Ohtemachi Bldg
1-6-1 Ohte-machi
Chiyoda-ku, Tokyo 100
Japan
Tel: +81 (3) 3282-0007
Fax: +81 (3) 3284-1968

L Merckle GmbH
Graf-Arco-Str 3
89079 Ulm (Donau)
Germany
Tel: +49 731 402-01
Fax: +49 731 402-7832

Lab Albert Rolland
France Evry - Tour Lorraine
BP 203
91007 Evry Cedex
France
Tel: +33 (1) 64 97 20 30
Fax: +33 (1) 64 97 05 84

Lab Bouchara
66, rue Marjolin
92300 Levallois Perret
France
Tel: +33 (1) 45 19 10 00
Fax: +33 (1) 45 46 82 95

Lab Cassenne Marion
Tour Roussel-Hoechst
1, terrasse Bellini
92910 Paris La Defense Cedex
France
Tel: +33 (1) 40 81 55 00
Fax: +33 (1) 40 81 40 82

Lab Dausse
Address Unknown

Lab Franc Chimiother
Address Unknown

Lab Houdé
Tour Roussel-Hoechst
1, terrasse Bellini
92910 Paris La Defense
Cedex
France
Tel: +33 (1) 40 81 42 00
Fax: +33 (1) 40 81 51 43

Lab Jacques Logeais
71, avenue du General de
Gaulle
92137 Issy Les Moulineaux
Cedex
France
Tel: +33 (1) 46 45 21 99

Lab Laborec
Address Unknown

Lab Lafon, France
20, rue Charles Martigny BP22
94701 Maisons Alfort
France
Tel: +33 (1) 49 81 81 00
Fax: +33 (1) 48 98 13 72

Lab Mauricio Villela SA
Address Unknown

Lab Meram
Avenue de la Liberation
77020 Melun Cedex
France
Tel: +33 (1) 64 87 20 50
Fax: +33 (1) 64 87 20 78

Lab Prod Biol Braglia
Address Unknown

Lab ProTer
Address Unknown

Labaz (Labs)
1, rue de la Viegre
33003 Bordeaux Cedex
France
Tel: +33 (56) 90 91 93

Labaz SA
9, rue du President Allende
94258 Gentilly Cedex
France
Tel: +33 (1) 40 73 63 00
Fax: +33 (1) 40 73 48 57

Laboratoire UPSA
128, rue Danton BP 325
92506 Rueil Malmaison
Cedex
France
Tel: +33 (1) 47 16 87 72
Fax: +33 (1) 47 16 87 78

Laboratoires Biocodex
19, rue Barbes
92126 Montrouge Cedex
France
Tel: +33 (1) 46 56 67 89
Fax: +33 (1) 40 92 17 61

Laboratorio Bago, SA
Address Unknown

Labs Fher SA
Address Unknown

Labs Franca Inc
Address Unknown

Labs OM
Address Unknown

Labs Sapos
Address Unknown

Lakeside BioTechnology
Address Unknown

Langley Smith Ltd
Address Unknown

Lark, SpA
Address Unknown

Laroche-Navarron
Address Unknown

Lederle Labs
Direct Inquiries to
Wyeth-Ayerst

Lee Laboratories
1475 Athens Highway
Grayson, GA 30221
USA
Tel: +1 (770) 972-4450
Fax: +1 (770) 979-9570

Lemmon Co
Direct Inquiries to Teva
Pharmaceuticals

Lentia
Address Unknown

Leo AB
55 Industriparken
Ballerup
DK-2750 Copenhagen
Denmark
Tel: +45 44 923 800
Fax: +45 44 943 040

Lever Brothers
Direct Inquiries to Unilever

Licencia Budapest
Address Unknown

Lion Dentrifice
Address Unknown

Lipha Pharmaceuticals Inc
1114 Ave of the Americas
41st Floor
New York, NY 10036
USA
Tel: +1 (212) 398-4602
Fax: +1 (212) 398-5021

Lipha Pharmaceuticals Ltd
Harrier House
High St, Yiewsley
West Drayton
Middx UB7 7QG
England
Tel: +44 (01895) 452200
Fax: +44 (01895) 420605

Lloyd, Hamol Ltd
Direct Inquiries to Reckitt &
Colman

Lombart Lenses Ltd Inc
1215 Boissevain Ave
PO Box 1693
Norfolk, VA 23501
USA
Tel: +1 (757) 625-7866

Lorus Therapeutics Inc
7100 Woodbine Ave
Ste 215
Markham ON L3R 5J2
Canada
Tel: +1 (905) 305-1100
Fax: +1 (905) 305-1584

Lovens Komiske Fabrik AS
Ramstadsletta 15
1322 Hovik
Norway
Tel: +47 (67) 12 30 03
Fax: +47 (67) 12 30 33

Lundbeck
37, ave Pierre 1er de Serbie
75008 Paris
France
Tel: +33 (1) 53 67 42 00

Lundbeck GmbH & Co
Amsinckstrβe 59
20097 Hamnburg
Germany
Tel: +49 40 236 49 0
Fax: +49 40 236 49 255

Lusofarmico
Address Unknown

Madan
Address Unknown

Maggioni Farmaceutici SpA
Address Unknown

Mallinckrodt Inc
7733 Forsyth Blvd
St Louis, MO 63105-1820
USA
Tel: +1 (314) 654-2000
Fax: +1 (314) 654-6510

Maltbie Chem
Address Unknown

Marion Merrell Dow Inc
Direct Inquires to Hoechst Marion Roussel Inc

Mar-Pha Soc Etud Exploit Marques
Address Unknown

Martin Dennis
Address Unknown

Maro Seiyaku
Address Unknown

Matieres Colorantes
255, rue de Paris
93100 Montreuil
France
Tel: +33 (1) 42 87 29 45
Fax: +33 (1) 42 87 10 39

Mauvernay
Address Unknown

May & Baker Ltd
Address Unknown

McNeil Consumer Products Co
7050 Camp Hill Rd
Fort Washington, PA 19034
USA
Tel: +1 (215) 233 7000

McNeil Pharmaceutical
McKean and Welsh Rds
PO Box 13886
Spring House, PA 19477
USA

Mead Johnson Labs
Direct Inquiries to Bristol-Myers Squibb Co

Mead Johnson Nutritionals
Direct Inquiries to Bristol-Myers Squibb Co

Medco Research Inc
85 T Alexander Dr
PO Box 13886
Res Triangle Pk, NC 27709
USA
Tel: +1 (919) 549-8117
Fax: +1 (919) 549-7515

Medical Market Specialties Inc
Address Unknown

Medicis Pharmaceutical Corp
4343 E Camelback Rd
Phoenix, AZ 85018
USA
Tel: +1 (602) 808-8800
Fax: +1 (602) 808-0822

Mediolanum Farmaceutici SpA
Via SG Cottolengo, 15
20143 Milan
Italy
Tel: +39 (02) 8912-2232
Fax: + 39 (02) 8913-2375

Medi-Physics Inc
2320 W Peoria Ave
Ste B-140-A
Phoenix, AZ 85029
USA
Tel: +1 (602) 371-8021

Medi-Physics Inc
1341 Gene Autry Way
Anaheim, CA 92805
USA
Tel: +1 (714) 634-9633

Meiji Milk Products Co Ltd
2-3-6, Kyobashi
Chuo-ku, Tokyo 104
Japan
Tel: +81 (3) 3281-6118
Fax: +81 (3) 3281-4717

Meiji Seika Kaisha Ltd
2-4-16, Kyobashi
Chuo-ku, Tokyo 104
Japan
Tel: +81 (3) 3272-6511
Fax: +81 (3) 3271-5792

Menley & James Laboratories Inc
100 Tournament Dr
Horsham, PA 19044
USA
Tel: +1 (215) 441-6500
Fax: +1 (215) 441-6576

Merck & Co Inc
One Merck Dr
PO Box 100
Whitehouse Sta, NJ 08889
USA
Tel: +1 (908) 423-1000
Fax: +1 (908) 594-4662

Merck KGaA
Frankfurter Str 250
D-64293 Darmstadt
Germany
Tel: +49 61 51-72-0
Fax: +49 61 51-72-2000

Merck Ltd
Merck House
Poole
Dorset BH15 1TD
England
Tel: +44 (01202) 669700

Merck Pharmaceuticals Ltd
Harrier House
High St
West Drayton
Middx UB7 7QG
England
Tel: +44 (01895) 452200
Fax: +44 (01895) 420605

Merck Sharpe & Dohme Research Labs
Hillsborough Rd
Three Bridges, NJ 08887
USA
Tel: +1 (908) 369-4900

Merrell Dow Pharmaceuticals Inc
PO Box 9627
Kansas City, MO 64134
USA

Merrell Pharmaceuticals
Address Unknown

Microbiochem Res Found
Address Unknown

Miles Inc
One Mellon Center
500 Grant St
Pittsburgh, PA 15219-2502
USA
Tel: +1 (412) 394-5500
Fax: +1 (412) 394-5579

Mission Pharmacal Co
1325 East Durango Blvd
San Antonio, TX 78210-1771
USA
Tel: +1 (210) 553-7118

Mitsubishi Chemical Corp
Mitsubishi Bldg
5-2 Marunouchi 2-chome
Chiyoda-ku, Tokyo 100
Japan
Tel: +81 (3) 3283-6254
Fax: +81 (3) 3283-6287

Mitsubishi Kasei
Address Unknown

Mitsui Pharmaceuticals Inc
3-12-2, Nihonbashi
Chuo-ku, Tokyo 103
Japan
Tel: +81 (3) 3274-4711
Fax: +81 (3) 3281-4670

Mitsui Toatsu
Address Unknown

Mizzy
Address Unknown

Mobay
Direct Inquiries to Monsanto

Mondi
Address Unknown

Monsanto Co
800 North Lindbergh Blvd
St Louis, MO 63167
USA
Tel: +1 (314) 694-1000

Mundipharma AG
Mundipharma Str 6
65549 Limburg (Lahn)
Germany

Muro Pharmaceuticals Inc
890 East St
Tewksbury, MA 01876-1496
USA
Tel: +1 (978) 851-5981
Fax: +1 (978) 851-7346

N Am Philips
Address Unknown

NV Nederlandsche Comb Chem Ind
Address Unknown

NV Amsterdamsche Chininefabriek
Address Unknown

NV Philips
Address Unknown

National Cancer Institute
Building, 31, Room 10A03 31
Center Drive
MSC 2580
Bethesda, MD 20892-2580
USA
Tel: +1 (301) 435-3848

National Drug Co
Address Unknown

National Foundation for Cancer Research
Address Unknown

National Research Dev Corp
Address Unknown

Natterman
Address Unknown

Naugatuck
Address Unknown

Newport
Address Unknown

Nicholas Labs Ltd
Address Unknown

Nihon Nohyaku Co Ltd
Eitaro Bldg
1-2-5 Nihonbashi
Chuo-ku, Tokyo 103
Japan
Tel: +81 (3) 3278-0461
Fax: +81 (3) 3281-5462

Nippon Chemiphar
2-2-3, Iwamoto-cho
Chiyoda-ku, Tokyo 101
Japan
Tel: +81 (3) 3863-1211
Fax: +81 (3) 3864-5940

Nippon Kayaku Co Ltd
Tokyo Fujimi Bldg
1-11-2 Fujimi
Chiyoda-ku, Tokyo 102
Japan
Tel: +81 (3) 3237-5111
Fax: +81 (3) 3237-5091

Nippon Shinyaku, Japan
Hachijo Sagaru, Nishiohji
Minami-ku, Kyoto 601
Japan
Tel: +81 (75) 321-9105
Fax: +81 (75) 321-0400

Nissan Kenzai Co Ltd
C/O Nissan Chemical Industries, Toyama Factory
635, Sakakura, Fuchu-machi
Nei-gun, Toyama 939-27
Japan
Tel: +81 (764) 65-6300
Fax: +81 (764) 65-6303

Nisshin Denka KK
2-2-1, Ohama
Sakata-shi, Yamagata 998
Japan
Tel: +81 (0234) 33-2121

Nisshin Kasei Co Ltd
11-5, Senju Kawara-machi
Adachi-ku, Tokyo 120
Japan
Tel: +81 (3) 3888-1181
Fax: +81 (3) 3870-2121

Nopco
Address Unknown

Nordmark
Address Unknown

Norton, HN
Gemini House
Flex Meadows
Harlow
Essex CM19 5TJ
England
Tel: +44 (01279) 426666

Norwich
Direct Inquiries to Procter & Gamble

Norwich Eaton
Direct Inquiries to Procter & Gamble

Novartis Pharmaceuticals, Corp
59 Route 10
East Hanover, NJ 07936-1011
USA
Tel: +1 (908) 503-7500

Novo Nordisk Biotech Inc
1445 Drew Ave
Davis, CA 95616
USA

Novo Nordisk Pharmaceuticals Inc
100 Overlook Center #2
Princeton, NJ 08540-7814
USA
Tel: +1 (609) 987-5800

Novocol Chem
Address Unknown

Novopharm Biotech Inc
147 Hamelin Street
Winnipeg, MB R3T 3Z1
Canada
Tel: +1 (204) 478-1023
Fax: +1 (204) 452-7721

Occidental Chemical Corp
Occidental Tower
5005 LBJ Freeway
Dallas, TX 75244
USA
Tel: +1 (972) 404 3800

Oclassen Pharmaceuticals Inc
100 Pelican Way
San Rafael, CA 94901
USA
Tel: +1 (415) 258-4500
Fax: +1 (415) 258-4550

Octel Chemicals Ltd
PO Box 17, Oil Sites Road
Ellesmere Port
South Wirral L65 4HF
England
Tel: +44 (0151) 3553611

Oesterreiche Stickstoffwerke
Address Unknown

Ohio State University
Address Unknown

Olin Mathieson
Address Unknown

Olin Research Ctr
350 Knotter Dr
PO Box 586
Cheshire, CT 06410
USA
Tel: +1 (203) 271-4316
Fax: +1 (203) 271-4060

Omnium Chim
Address Unknown

O'Neal, Jones & Feldman Pharmaceuticals
Address Unknown

Ono Pharmaceutical
2-1-5, Dosho-machi
Chuo-ku, Osaka 541
Japan
Tel: +81 (6) 6222-5551
Fax: +81 (6) 6222-5706

Optacryl Inc
2890 S Tejon St
Englewood, CO 80110-0120
USA
Tel: +1 (303) 789-0933

Optech Inc
6341 Troy Circle
Englewood, CO 80111-0641
USA
Tel: +1 (303) 708-1390

Orgamol, SA
Address Unknown

Organon Inc
375 Mount Pleasant Ave
West Orange, NJ 07052
USA
Tel: +1 (201) 325-4500

Organon Laboratories Ltd
Science Park
Milton Rd
Cambridge CB4 4FL
England
Tel: +44 (01223) 423445

Orion Pharma
Orionintie 1
PO Box 65
FIN-02101 Espoo
Finland
Tel: +358 9 4291
Fax: +358 9 4293815

Orsymonde
Address Unknown

Ortho Biotech Inc
PO Box 670
700 US Highway 202 South
Raritan, NJ 08869-0670
USA
Tel: +1 (908) 704-5000

Ortho Diagnostic Systems Inc
US Route 202
Raritan, NJ 08869
USA
Tel: +1 (908) 218-8000

Ortho Pharmaceutical Corp
Route 202 South
Raritan, NJ 08869
USA
Tel: +1 (908) 704-1500
Fax: +1 (908) 526-4997

OSI Pharmaceuticals
106 Charles Lindbergh Blvd
Uniondale, NY 11553-3649
USA
Tel: +1 (516) 222-0023
Fax: +1 (516) 222-0114

OSSW
Address Unknown

Otsuka America Pharmaceutical
2440 Research Blvd Ste 500
Rockville, MD 20850
USA
Tel: +1 (301) 990-0030

Otsuka Pharmaceuticals Co Ltd
2-9, Kanda Tsukasa-cho
Chiyoda-ku
Tokyo 101-8535
Japan
Tel: +81 (3) 3292-0021

OXIS International Inc
6040 North Cutter Circle
Ste 317
Portland, OR 97217
USA
Tel: +1 (503) 283-3911
Fax: +1 (503) 283-4058

Paines & Byme Ltd
Address Unknown

Paragon Vision Sciences
947 Elm Avenue
Mesa, AZ 85204
USA
Tel: +1 (480) 892 7602

Parke Davis & Co Ltd
Lambert Court
Chestnut Ave
Eastleigh Hamps SO5 3ZQ
England
Tel: +44 (01703) 620500

Parke-Davis
2800 Plymouth Rd
Ann Arbor, MI 48105
USA
Tel: +1 (734) 622-7000
Fax: +1 (734) 622-5229

Patchem, AG
Address Unknown

PCAS
Address Unknown

Penederm Inc
320 Lakeside Dr, Ste A
FosterCity, CA 94404
USA
Tel: +1 (415) 358-0100
Fax: +1 (415) 358-0101

Penick
Address Unknown

Penta Mfg
PO Box 1448
Fairfield, NJ 07007
USA
Tel: +1 (201) 740-2300
Fax: +1 (201) 740-1839

Pentapharm
Engelgasse 109
CH-4002 Basel
Switzerland
Tel: +41 (61) 706-9848
Fax: +41 (61) 319-9619

PerImmune Inc
1330 Piccard Dr
Rockville, MD 20850-4396
USA
Tel: +1 (301) 258-5200

Permeable Technologies Inc
712 Ginesi Dr
Morganville, NJ 07751
USA

Person & Covey Inc
616 Allen Ave
Glendale, CA 91201-0201
USA
Tel: +1 (818) 240-1030

Perstorp AB
SE-28 4 80 Perstorp
Sweden
Tel: +46 (0) 435 3800
Fax: +46 (0) 435 3810

Pfalz & Bauer
172 E Aurora St
Waterbury, CT 06708
USA
Tel: +1 (203) 574-0075
Fax: +1 (203) 574-3181

Pfanstiehl Laboratories Inc
1219 Glen Rock Ave
Waukegan, IL 60085
USA
Tel: +1 (847) 623-0370
Fax: +1 (847) 623-9173

Pfizer Group Ltd
PO Box 2
Ramsgate Rd
Sandwich
Kent CT13 9NJ
England
Tel: +44 (01304) 616161

Pfizer Inc
Central Research
Eastern Point Rd
Groton, CT 06340
USA
Tel: +1 (860) 441-4100

Pfizer International
235 E 42nd St
New York, NY 10017-5755
USA

Pfizer Pharmaceuticals Roerig Div,
235 E 42nd St
New York, NY 10017-2399
USA

Pfleger (Dr R Pfleger)
96045 Bamberg
Germany
Tel: +49 951 60430
Fax: +49 951 604329

Pharm Res Products
Address Unknown

Pharmachemie
Swensweg 5
PO Box 552
2003 RN Haarlem
The Netherlands
Tel: +31 23 524 77 90
Fax: +31 23 514 77 74

Pharmacia
Direct Inquiries to Pharmacia & Upjohn

Pharmacia & Upjohn
95 Corporate Dr
Bridgewater, NJ 08807-1265
USA
Tel: +1 (908) 306-4400
Fax: +1 (908) 306-4433

Pharmacia & Upjohn AB
Lindhagensgatan 133
SE-112 87 Stockholm
Sweden
Tel: +46 (08) 695 8000
Fax: +46 (08) 618 8607

Pharmacia & Upjohn Inc
301 Henrietta St
Kalamazoo, MI 49001
USA
Tel: +1 (616) 323-4000
Fax: +1 (616) 323-4077

Pharmacia Hepar Inc
150 Industrial Dr
Franklin, OH 45005
USA
Tel: +1 (513) 746-3603

Pharmos Corp
Two Innovation Dr
Alachua, FL 32615
USA
Tel: +1 (904) 462-1210
Fax: +1 (904) 762-5401

Philips-Duphar BV
Address Unknown

Phillips
Specialty Chemicals
874 Adams Bldg
Bartlesville, OK 74004
USA
Tel: +1 (918) 661-9092
Fax: +1 (918) 661-8379

Pierre Fabre
5, ave Napoleon III - BP 497
74164 St Julien en Genevois
Cedex
France
Tel: +33 (4) 50 35 35 55
Fax: +33 (4) 50 35 35 90

Pierre Fabre
45, place Abel-Gance
92654 Boulogne Cedex
France
Tel: +33 (1) 49 10 80 00
Fax: +33 (5) 61 39 15 98

Pierrel SpA
Address Unknown

Pilkington Barnes Hind
810 Kifer Rd
Sunnyvale, CA 94086
USA
Tel: +1 (858) 614-7600

Pineapple Research Inst
Address Unknown

Pitman Moore Europe Ltd
Breakspear Road South
Harefield
Uxbridge
Middx UB9 6LS
England
Tel: +44 (01895) 626000

Pitman-Moore Inc
1201 Douglas Ave
Kansas City, KS 66103-0140
USA
Tel: +1 (913) 321-1070

Polaroid
Address Unknown

Polfa
Address Unknown

Polichimica SpA
Address Unknown

Poythress
Address Unknown

Pratt Pharmaceuticals
Pfizer Inc
235 E 42nd St
New York, NY 10017-5755
USA

Procter & Gamble Pharmaceuticals Inc
11810 East Miami River Rd
Ross, OH 45061
USA
Tel: +1 (513) 983-1100

ProCyte Corp
12040 115th Ave NE
Ste 210
Kirkland, WA 98034-6900
USA
Tel: +1 (206) 820-4548
Fax: +1 (206) 820-4111

Promonta
Direct Inquiries to Lundbeck GmbH

Provesan SA
Address Unknown

Purdue Pharma LP
100 Connecticut Ave
Norwalk, CT 06856
USA
Tel: +1 (203) 853-0123
Fax: +1 (203) 838-1576

Quimicobiol
Address Unknown

Quinoderm Ltd
Address Unknown

RW Johnson Pharmaceutical Research Institute
Route 202 South
PO Box 300
Raritan, NJ 08869-0602
USA
Tel: +1 (908) 704-4000

Raschig GmbH
Ludwigshafen
Germany

Ravensberg
Address Unknown

Ravizza
Address Unknown

Recherche et Ind Therap
Address Unknown

Reckitt & Colman Europe
One Burlington Lane
London W4 2RW
England
Tel: +44 (0181) 994-6464
Fax: +44 (0181) 944-8940

Reckitt & Colman Inc
1655 Valley Rd
Wayne, NJ 07470
USA
Tel: +1 (020) 8633 3600
Fax: +1 (020) 8633 3633

Recordati Corp
110 Commerce Dr
Allendale, NJ 07401
USA
Tel: +1 (212) 236-3669
Fax: +1 (212) 236-9404

Recordati Industria Chimica E Pharmaceutica SpA
Via M Civitali, 1
1-20148 Milano
Italy
Tel: +39 (02) 487 87536
Fax: +39 (02) 487 05223

Reed & Carnrick
65 Horse Hill Rd
Cedar Knolls, NJ 07927
USA
Tel: +1 (973) 267-2670

Refarmed
Address Unknown

Res Inst Pharm Chem
Address Unknown

Research Corp
Address Unknown

Resfar SRL
Address Unknown

Rexall Sundown, Inc
6111 Broken Sound Parkway
Boca Raton, FL 33487
USA
Tel: +1 (561) 241-9400
Fax: +1 (561) 995-0197

Rhinepreussen AG
Address Unknown

Rhône-Poulenc
Direct Inquiries to
Rhône-Poulenc Rorer

Rhône-Poulenc Rorer
20, avenue Raymond Aron
92165 Antony Cedex
France
Tel: +33 (1) 55 71 71 71

Rhône-Poulenc Rorer Holdings Ltd
St Leonards House
52 St Leonard Rd
Eastbourne
East Sussex BN21 3YG
England
Tel: +44 (01323) 721422

Rhône-Poulenc Rorer Pharmaceuticals Inc
PO Box 1200
Collegeville, PA 19426-0107
USA

Richardson-Merrell
Direct Inquiries to Hoechst
Marion Roussel

Richardson-Vicks Inc
Direct Inquiries to Hoechst
Marion Roussel

Riedel de Haen (Chinosolfabrik)
Wunstorfer Str 40
30926 Seeize
Germany
Tel: +49 5137 999258
Fax: +49 5137 999674

Riker Labs
Direct Inquiries to 3M
Pharmaceuticals

Robert et Carriere
Address Unknown

Roberts Pharmaceutical Corp
4 Industrial Way West
Eatontown, NJ 07724
USA
Tel: +1 (732) 676-1200
Fax: +1 (732) 676-1300

Roche Laboratories
340 Kingsland St
Nutley, NJ 07110-1199
USA
Tel: +1 (973) 235-5000

Roche Products Ltd
40 Broadwater Road
Welwyn Garden City
Herts AL7 3AY
England
Tel: +44 (01707) 328128

Roche Puerto Rico
Direct Inquires to ICN
Pharmaceuticals

Rohm and Haas Co
100 Independence Mall W
Philadelphia, PA 19106-2399
USA
Tel: +1 (215) 785-8000

Rorer
Direct Inquiries to
Rhône-Poulenc Rorer

Ross Products
US Highway 29 North
PO Drawer 479
Altavista, VA 24517
USA
Tel: +1 (804) 369-3100

Roswell Park Memorial Inst
Buffalo, NY 14203
USA
Tel: +1 (716) 845-2300

Rotta Pharm
6, rue Casimir-Delavigne
75006 Paris
France
Tel: +33 (1) 44 07 12 44

Roussel Laboratories Ltd
Broadwater Park
North Orbital Rd, Denham
Uxbridge
Middx UB9 5HP
England
Tel: +44 (01895) 834343

Roussel-UCLAF
Direct Inquiries to Hoechst
Marion Roussel

Rowa Ltd
Newtown
Bantry, Cork
Ireland
Tel: +353 (027) 50077

Rowa-Wagner
Frankenforster Str 77
51427 Bergisch Gladbach
Germany
Tel: +49 2204 61081
Fax: +49 2204 61084

RW Johnson Pharmaceutical Research Institute, The
920 Route 202
PO Box 300
Raritan, NJ 08869-0602
USA
Tel: +1 (908) 704-4000

Rybar Labs Ltd
Address Unknown

Rystan Co Inc
PO Box 214
Little Falls, NJ 07420-0214
USA
Tel: +1 (973) 256-3737

SIFA
Address Unknown

Salix Pharmaceuticals Inc
3600 W Bayshore Rd
Ste 205
Palo Alto, CA 94303
USA
Tel: +1 (650) 856-1550

San NopCo Ltd
1-5-9, Nihonbashi Hon-cho
Chuo-ku, Tokyo 103
Japan
Tel: +81 (3) 3279-3030
Fax: +81 (3) 3246-0550

Sandoz Pharmaceuticals Corp
Direct Inquires to Novartis
Pharmaceuticals

Sankyo Co Ltd
3-5-1, Nihonbashi Hon-cho
Chuo-ku, Tokyo 103
Japan
Tel: +81 (3) 5255-7111
Fax: +81 (3) 5255-7035

Sanofi Winthrop
301 Oxford Valley Rd
Morrisville, PA 19067-7706
USA
Tel: +1 (215) 321-7560

Sanofi Winthrop France
9, rue du President Allende
94258 Gentilly Cedex
France
Tel: +33 (1) 41 24 60 00
Fax: +33 (1) 41 24 63 00

Santen Pharmaceutical Co Ltd
3-9-19, Shimoshinjo
Higashiyodogawa-ku
Osaka 533
Japan
Tel: +81 (6) 6321-7045
Fax: +81 (6) 6325-8209

Savage Laboratories
60 Baylis Rd
Melville, NY 11747
USA
Tel: +1 (516) 454-7677
Fax: +1 (516) 454-0732

Schein Pharmaceutical Inc
620 N 51st Ave
Phoenix, AZ 85043-4705
USA
Tel: +1 (602) 278-1400
Fax: +1 (602) 447-3385

Schenley
Address Unknown

Schering AG
Muellerstr 170-178
D-13342 Berlin
Germany
Tel: +49 30 4681 111
Fax: +49 30 4681 5305

Schering Health Care Ltd
The Brow, Burgess Hill
West Sussex RH15 9BS
England
Tel: +44 (01444) 232323

Schering-Plough HealthCare Products
110 Allen Road
Liberty Corner, NJ 07938
USA
Tel: +1 (908) 604-1640

Schering Plough Ltd
Chiswick Avenue, Field Road
Industrial Estate Mildenhall
Bury St Edmunds
Suffolk IP28 7AX
England
Tel: +44 (01638) 716321

Schering-Plough Pharmaceuticals
2015 Galloping Hill Rd
Kenilworth, NJ 07033-0530
USA
Tel: +1 (908) 298-4000

Schevico
Address Unknown

Schiapparelli
Direct Inquiries to Alfa
Wassermann

Schwartz's Essencefabriken
Address Unknown

Schwarz Arztnelmittelfabrik
Address Unknown

Schwarz Pharma Kremers Urban Co
6140 Est Executive Dr
Mequon, WI 53092
USA

Schwarz Pharma Ltd
Schwarz House
East St
Chesham
Bucks HP5 1DG England
Tel: +44 (01494) 772071

Sci Union et Cie, France
Address Unknown

SciClone Pharmaceuticals Inc
901 Mariners Island Blvd
San Mateo, CA 94404-1593
USA
Tel: +1 (415) 358-3456
Fax: +1 (415) 358-3469

Scios Nova Inc
820 W Maude Ave
Sunnyvale, CA 94086
USA
Tel: +1 (408) 481-9177
Fax: +1 (408) 481-9188

Scotia Pharmaceuticals Ltd
Address Unknown

SCS Pharmaceuticals
Address Unknown

Searle Ltd
PO Box 53
Lane End Rd
High Wycombe
Bucks HP12 4HL
England
Tel: +44 (01494) 521124
Fax: +44 (01494) 447872

Searle, GD & Co
5200 Old Orchard Rd
Skokie, IL 60077
USA
Tel: +1 (847) 982-7000
Fax: +1 (847) 470-1480

Seceph
Address Unknown

Selvi
Address Unknown

Serono Laboratories Inc
100 Longwater Circle
Norwell, MA 02061-0163
USA
Tel: +1 (781) 982-9000

Serono Laboratories Ltd
99 Bridge Road East
Welwyn Garden City
Herts AL7 1BG
England
Tel: +44 (01707) 331972

Shell
One Shell Plaza
Houston, TX 77252-2463
USA
Tel: +1 (713) 241-6161
Fax: +1 (713) 241-4043

Shionogi & Co Ltd
3-1-8, Dosho-machi
Chuo-ku, Osaka 541
Japan
Tel: +81 (6) 6202-2161
Fax: +81 (6) 6229-9596

Siegfried AG
Address Unknown

Sigma-Tau Pharmaceuticals Inc
800 South Frederick Ave
Ste 300
Gaithersburg, MD 20877
USA
Tel: +1 (301) 948-1041
Fax: +1 (301) 948-3194

Sigma-Tau SpA
Industrie famaceutiche riunite
Viale Shakespeare, 47
00144 Rome
Italy
Tel: +39 (6) 592-6443

Simes SpA
Address Unknown

Smith, T&H
Address Unknown

SmithKline Beecham Animal Health
Direct Inquiries to Pfizer Inc

SmithKline Beecham Pharmaceuticals
One Franklin Place
Philadelphia, PA 19102
USA
Tel: +1 (215) 751-3415
Fax: +1 (215) 751-7655

Snow Brand Milk Products Co Ltd
44 Montgomery St
San Francisco, CA 94104
USA
Tel: +1 (415) 677-0914

Soc Belge Azote Prod Chim Marly
Address Unknown

Soc Belge des Labs Labaz
Address Unknown

Soc Chim des Usines du Rhône
Address Unknown

Soc Chim Org Biol
Address Unknown

Soc Etudes Sci Ind L'Île de France
Address Unknown

Soc Farmaceutici Italia
Address Unknown

Soc Franc Recherches Biochim
Address Unknown

Soc Ind Fabric Antiboit
Address Unknown

Soc Italo-Brit L Manetti
Address Unknown

Soc Italo-Brit L Manetti-H Roberts
Address Unknown

Societa Prodiotti Antibiotici, Italy
Address Unknown

Societe Belge de l'azote
Address Unknown

Societe Berri-Balzac
Address Unknown

Sogeras
Address Unknown

Sola/Barnes-Hind
Direct Inquiries to Allergan Inc

Solvay America Inc
3333 Richmond Ave
Houston, TX 77098-3009
USA
Tel: +1 (713) 525-6000
Fax: +1 (713) 525-7887

Solvay Animal Health Inc
1201 Northland Dr
Mendota Heights, MN 55120
USA
Tel: +1 (651) 681-3880
Fax: +1 (651) 681-9425

Solvay Deutschland GmbH
Hans-Bockler-Allee 20
D-30173 Hannover
Germany
Tel: +49 511-85-70
Fax: +49 511-28-21-26

Solvay Duphar Laboratories Ltd
Duphar House, Gaters Hill
West End, Southampton,
Hamps SO3 3JD
England

Solvay Pharmaceuticals SA
33, rue du Prince Albert
B-1050 Brussels
Belgium
Tel: +32 (2) 509 6111
Fax: +32 (2) 509 6304

Solvay Pharmaceuticals Inc
901 Sawyer Rd
Marietta, GA 30062
USA
Tel: +1 (770) 578-9000

Solvay Holding Co Ltd
Grovelands Business Centre
Boundary Way
GB Hemel Hempstead
Herts HP2 7TE
England
Tel: +44 (01442) 236555
Fax: +44 (01442) 238770

Somerset Pharmaceuticals Inc
5215 W Laurel St
Tampa, FL 33607-0172
USA
Tel: +1 (813) 288-0040

Sonus Pharmaceuticals Inc
22026 20th Ave SE
Bothell, WA 98021-4405
USA
Tel: +1 (206) 487-9500

SPA
Address Unknown

Sphinx Pharmaceutical Corp
20 T W Alexander Dr
Res Triangle PK, NC 27709
USA
Tel: +1 (919) 314-4000
Fax: +1 (919) 314-4350

SPOFA
Husinecka lla
130 00 Praha 3
Czech Republic
Tel: +42 (2) 6278502
Fax: +42 (2) 6278320

Spojene
Direct Inquires to SPOFA

Squibb, ER & Sons
Direct Inquiries to
Bristol-Myers Squibb Co

Standard Oil Co, Indiana
Division of AMOCO Oil
Hc 331 Box S
Bremen, IN 46506
USA
Tel: +1 (219) 546-4342

Stauffer Chemical Co
Address Unknown

Stem Corporation
Woodrolfe Road
Tollesbury
Essex CM9 8SJ
England
Tel: +44 (01621) 868685
Fax: +44 (01621) 868445

Sterling Health USA
Direct Inquiries to Sanofi
Winthrop

Sterling Research Labs
Direct Inquiries to Sanofi
Winthrop

Sterling Winthrop Inc
Direct Inquiries to Sanofi
Winthrop

Stiefel France
ZI du Petit Nantere
15, rue des Grands Pres
92007 Nanterre Cedex
France
Tel: +33 (1) 46 49 80 50
Fax: +33 (1) 47 82 99 72

Stiefel Laboratories Inc
255 Alhambra Circle
Coral Gables, FL 33134
USA
Tel: +1 (305) 443-3800
Fax: +1 (305) 443-3467

Stokely-Van Camp
Oakland, CA 94601
USA
Tel: +1 (510) 261-3672

Stuart
Direct Inquiries to AstraZeneca

Sumitomo Pharmaceuticals Co Ltd
2-2-8, Dosho-machi
Chuo-ku, Osaka 541
Japan
Tel: +81 (6) 6229-5775
Fax: +81 (6) 6233-2399

Sun Pharmaceuticals Corp
1345 Pine Ave
Orlando, FL 32824-7942
USA
Tel: +1 (407) 859-3162

SunPharm Corp
4651 Salisbury Rd Ste 205
Jacksonville, FL 32256
USA
Tel: +1 (904) 296-3320

Suntory Ltd
2-1-40, Dojimahama
Kita-ku, Osaka 530
Japan
Tel: +81 (6) 6346-1131
Fax: +81 (6) 6345-1169

Synaptic Pharmaceutical Corp
215 College Rd
Paramus, NJ 07652
USA
Tel: +1 (201) 261-1331
Fax: +1 (201) 261-0623

Synergen Inc
1885 33rd St
Boulder, CO 80301-2505
USA
Tel: +1 (303) 938-6200
Fax: +1 (303) 441-5535

Syntex International Ltd
Direct Inquiries to Hoffman
LaRoche

Syntex Labs Inc
Boulder, CO
USA

Syntex Pharmaceuticalsl Ltd
Syntex House
St Ives Rd
Maidenhead
Berks SL6 1RD
England
Tel: +44 (01628) 33191

Synthelabo Pharmacie
Lindberghstr 1
82178 Puchheim
Germany
Tel: +49 89 89017-0
Fax: +49 89 89017-299

Taiho
1-27, Kanda Nishiki-cho
Chiyoda-ku, Tokyo 101
Japan
Tel: +81 (3) 3294-4527
Fax: +81 (3) 3233-4318

Taisho
3-24-1, Takata
Toshima-ku, Tokyo 171
Japan
Tel: +81 (3) 3985-1111
Fax: +81 (3) 3982-9701

Takeda Chemical Industries Ltd
4-1-1, Dosho-machi
Chuo-ku, Osaka 541
Japan
Tel: +81 (6) 6204-2111
Fax: +81 (6) 6204-2880

Tanabe Research Laboratories, USA Inc
4540 Towne Centre Ct
San Diego, CA 92121
USA
Tel: +1 (619) 558-9211

Tanabe Seiyaku
Address Unknown

TAP Pharmaceuticals Inc
Bannockburn Lake Office
Plaza
2355 Waukegan Rd
Deerfield, IL 60015
USA
Tel: +1 (847) 236-2270

TCI America
9211 North Harborgate St
Portland, OR 97203
USA
Tel: +1 (800) 423-8616
Fax: +1 (503) 283-1987

TechAmerica
Address Unknown

Teijin Ltd
Teijin Bldg
1-6-7, Minami-honmachi
Chuo-ku, Osaka 541
Japan
Tel: +81 (6) 6268-2132
Fax: +81 (6) 6266-1481

Teikoku Hormone Mfg Co Ltd
2-5-1, Akasaka
Minato-ku, Tokyo 107
Japan
Tel: +81 (3) 3583-8361
Fax: +81 (3) 3583-3328

Telios Pharmaceuticals Inc
4757 Nexus Centre Dr
San Diego, CA 92121
USA
Tel: +1 (619) 622-2600

Teva Pharmaceuticals (USA)
650 Cathill Rd
PO Box 904
Sellersville, PA 18960
USA
Tel: +1 (215) 256-8400
Fax: +1 (215) 721-9669

Theraplix
Rhône-Poulenc Rorer
46-52, rue Albert
75640 Paris Cedex 13
France
Tel: +33 (1) 40 77 30 00
Fax: +33 (1) 40 77 322 20

Thomae GmbH, Dr Karl
Birkendorfer Str 65
88937 Biberach
Germany
Tel: +49 07351/54-0
Fax: +49 07351/54-4600

Tillots Pharma
Hauptstr 27
CH-4417 Ziefen
Switzerland

Torii Pharmaceutical Co Ltd
3-4-1, Nihonbashi Hon-cho
Chuo-ku, Tokyo 103
Japan
Tel: +81 (3) 3231-6811
Fax: +81 (3) 5203-7333

Toyama Chemical Co Ltd
3-2-5, Nishi-shinj u
Shinj u-ku, Tokyo 160
Japan
Tel: +81 (3) 5381-3889
Fax: +81 (3) 3348-6460

Toyo Jozo
Direct Inquiries to Asahi
Chemical

Toyo Koatsu Co Ltd
Hiroshima
Japan

Toyo Pharmachemicals Co Ltd
Tokyo Bldg
2-7-3, Marunouchi
Chiyoda-ku, Tokyo 100
Japan
Tel: +81 (3) 3211-8621
Fax: +81 (3) 3211-8625

Trega Biosciences Inc
3550 General Atomics Ct
San Diego, CA 92121
USA
Tel: +1 (619) 455-3814
Fax: +1 (619) 455-2544

Triple Crown America Inc
13 N 7th St
Perkasie, PA 18944
USA
Tel: +1 (215) 453-2500
Fax: +1 (215) 453-2508

Troponwerke Dinklage
Address Unknown

US Bioscience Corp
One Tower Bridge
100 Front St
W Conshohocken, PA 19428
USA
Tel: +1 (610) 832-0570
Fax: +1 (610) 832-4500

US Ethicals Inc
Address Unknown

US Vitamin
Address Unknown

UCB Pharma
Allee de la Recherche 60
Brussels
Belgium
Tel: +32 (2) 559 9999
Fax: +32 (2) 559 9900

UCB Pharma
21, rue de Neuilly
92003 Nanterre Cedex
France
Tel: +33 (1) 47 29 44 35
Fax: +33 (1) 47 25 47 20

UCB Pharma oy Finland
Maistraatinporti 2
FIN-0020 Helsinka
Finland

UCB Research Inc
840 Memorial Dr
Cambridge, MA 02139
USA
Tel: + 1 (617) 547-8481

Ucyclyd Pharma Inc
Direct Inquiries to Medicis
Pharmaceutical Corp

Ueno Fine Chemicals Industry Ltd
2-4-8, Koraibashi
Chuo-ku, Osaka 541
Japan
Tel: +81 (6) 6203-0761
Fax: +81 (6) 6222-2413

Ueno Kagaku Kogyo KK
3-3-2, Shodai Tajika
Hirakata-shi, Osaka 573
Japan
Tel: +81 (7) 20 56-2281

Ugine Kuhlmann
Direct Inquires to Rhône
Poulenc

Unicler
Address Unknown

Unilab Corp
401 Hackensack Ave
Hackensack, NJ 07601-6411
USA
Tel: +1 (201) 525-1000

Unilever International
Greyfriars
Lewins Mead
Bristol Avon BS1 2JJ
England
Tel: +44 (01272) 276276

Unimed Pharmaceuticals Inc
2150 East Lake Cook Rd
Ste 210
Buffalo Grove, IL 60089-1862
USA
Tel: +1 (847) 541-2525
Fax: +1 (847) 541-2569

Union Carbide Corp
Address Unknown
Danbury, CT
USA
Tel: +1 (203) 794-7024

United Catalysts Inc
PO Box 32370
Louisville, KY 40232
USA
Tel: +1 (502) 634-7200
Fax: +1 (502) 637-3132

Upjohn Ltd
Direct Inquiries to Pharmacia
& Upjohn

Uriach
Address Unknown

Usines de Melle
Direct Inquiries to Rhône
Poulenc

Valeas
via Vallisneri, 10
20133 Milano
Italy

Vanderbilt, RT Co Inc
30 Winfield
Enfield, CT 06082
USA
Tel: +1 (203) 853-1400

VEB Arzneimittelwerk
Address Unknown

VEB Farbenfabrik Wolfen
Address Unknown

Vismara
Address Unknown

Vistakon Inc
4500 Salisbury Rd
Ste 300
Jackson, FL 32216
USA
Tel: +1 (904) 443-1000

Wakamoto Pharmaceutical Co Ltd
1-5-3, Nihonbahi Muro-machi
Chuo-ku, Tokyo 103
Japan
Tel: +81 (3) 3279-0371
Fax: +81 (3) 3279-0393

Walker Labs
Address Unknown

Wallace & Tiernan Inc
P O Box 178
Newark, NJ 07101-9976
USA
Tel: +1 (973) 759-8000
Fax: +1 (973) 751-6589

Wallace & Tiernan Ltd
Priory Works
Tonbridge, Kent TN11 0QL
England
Tel: +44 (01732) 771777
Fax: +44 (01732) 77190

Wallace Laboratories
10200 E Girard Ave
Denver, CO 80231-0550
USA
Tel: +1 (303) 745-4676

Walter Reed Army Institute of Research
16th Street NW
Washington, DC 20307
USA

Walton Pharmaceuticals
PO Box 76
East Horsley
Surrey KT24 5YW
England
Tel: +44 (01483) 280001

Wander Pharma
Deutschherrnstr 15
90429 Nuernberg
Germany
Tel: +49 911 2730
Fax: +49 911 273653

Ward Blenkinsop
Address Unknown

Warner Lambert
201 Tabor Rd
Morris Plains, NJ 07950
USA
Tel: +1 (973) 385-2000

Wellcome Foundation Ltd, The
PO Box 129
Unicorn House
160 Euston Rd
London, NW1 2BP
England
Tel: +44 (020) 7387 4477

Wellcome plc
Unicorn House
160 Euston Rd
London, NW1 2BP
England
Tel: +44 (020) 7387 4477

Wesley-Jessen
333 East Howard Ave
Des Plaines, IL 60018
USA
Tel: +1 (847) 294-3000
Fax: +1 (847) 294-3434

Westwood-Squibb Pharmaceuticals Inc
100 Forest Ave
Buffalo, NY 14213
USA
Tel: +1 (716) 887-3400

Whitefin Holding
Address Unknown

Whitehall
111, rue des Chateau des Rentiers
75013 Paris
France
Tel: +33 (1) 44 06 43 21
Fax: +33 (1) 44 06 43 69

Whitehall Laboratories Ltd
Huntercombe Lane South
Taplow
Maidenhead,
Berks SL6 0PH
England
Tel: +44 (01628) 669011

Whitehall-Robins
PO Box 8299
Philadelphia, PA 19101
USA
Tel: +1 (973) 660-6805

Wiernik AG
Address Unknown

Windsor Healthcare Ltd
Ellesfield Avenue
Bracknell
Berks RG12 8YS
England
Tel: +44 (01344) 484448

Winthrop
Direct Inquiries to Sanofi Winthrop

Winthrop-Stearns
Direct Inquiries to Sanofi Winthrop

Wisconsin Alumni Research Foundation
Address Unknown

Worthington Biochemical
Address Unknown

Wyeth Laboratories
Direct Inquires to
Wyeth-Ayerst Laboratories

Wyeth-Ayerst Laboratories
PO Box 8299
Philadelphia, PA 19101
USA
Tel: +1 (610) 971-4980

Xenon Vision
Address Unknown

Xoma Corp
2910 Seventh St
Berkeley, CA 94710
USA
Tel: +1 (310) 829-7681

Xttrium Labs Inc
415 West Pershing Rd
Chicago, IL 60609
USA
Tel: +1 (773) 268-5800
Fax: +1 (773) 924-6002

Yamanouchi Europe BV
PO Box 108
NL-2350 A C Leiderdrop
The Netherlands
Tel: +31 7154 55745
Fax: +31 7154 800

Yamanouchi Pharma
10, pl de la Coupole - BP 105
94223 Charenton Le Pont
Cedex
France
Tel: +33 (1) 46 76 64 00
Fax: +33 (1) 46 76 64 99

Yamanouchi USA Inc
4747 Willow Rd
Pleasanton, CA 94588
USA
Tel: +1 (925) 924-2000

Yoshitomi
2-6-9, Hirano-machi
Chuo-ku, Osaka 541
Japan
Tel: +81 (6) 6201-2646
Fax: +81 (6) 6232-0910

Zambeletti
Address Unknown

Zambon France
46/48, avenue du General Leclerc
92100 Boulogne-Billancourt
France
Tel:+33 (1) 46 99 15 60

Zambon Group
Via Lillo del Duca, 10
Bresso
20091 Milano
Italy
Tel: +39 (02) 665241
Fax: +39 (02) 66501492

Zeeland Chemicals
215 N Centennial St
Zeeland, MI 49464
USA
Tel: +1 (616) 772-2193
Fax: +1 (616) 772-6554

Zeneca Pharmaceuticals
Alderley Park
Macclesfield
Cheshire SK10 4TF
England
Tel: +44 (01625) 582828

Zeneca Pharmaceuticals
Kings Court
Water Lane
Wilmslow
Cheshire SK9 5AZ
England
Tel: +44 (01625) 712712